MICROBIOLOGY

ABRAHAM I. BRAUDE, M.D., Ph.D.

Professor of Medicine and Pathology
University of California, San Diego

Associate Editors

CHARLES E. DAVIS, M.D.

Professor of Pathology
University of California, San Diego

JOSHUA FIERER, M.D.

Associate Professor of Medicine
University of California, San Diego

1982

W. B. SAUNDERS COMPANY Philadelphia
London
Toronto
Mexico City
Sydney
Tokyo

W. B. Saunders Company: West Washington Square
Philadelphia, PA 19105

1 St. Anne's Road
Eastbourne, East Sussex BN21 3UN, England

1 Goldthorne Avenue
Toronto, Ontario M8Z 5T9, Canada

Cedro 512
Mexico 4, D.F. Mexico

9 Waltham Street
Artarmon, N.S.W. 2064, Australia

Ichibancho, Central Bldg., 22-1
Chiyoda-ku, Tokyo 102, Japan

Microbiology ISBN 0-7216-1920-7

Library of Congress catalog card number 81-40588.

Last digit is the print number: 9 8 7 6 5 4 3 2 1

CONTRIBUTORS

YOUSEF AL-DOORY, Ph.D.
Associate Professor of Pathology, George Washington University School of Medicine; Chief, Sections of Mycology and Serology, Division of Laboratory Medicine, George Washington University Hospital, Washington, D.C.
DEMATIACEAE: AGENTS OF CHROMOMYCOSIS

AARON D. ALEXANDER, Ph.D.
Professor of Microbiology, Department of Microbiology, Chicago College of Osteopathic Medicine, Chicago, Illinois.
LEPTOSPIRA

A. O. ANYA, Ph.D.
Professor of Zoology and Director, School of Postgraduate Studies, University of Nigeria, Nsukka, Nigeria.
NEMATHELMINTHES

HOWARD ROBERT ATTEBERY, D.D.S.
Research Microbiologist, Veterans Administration Medical Center, San Diego, California.
FUSOBACTERIA; GRAM-POSITIVE COCCI: PEPTOCOCCUS, PEPTOSTREPTOCOCCUS, STREPTOCOCCUS (ANAEROBIC), AND SARCINIA; VEILLONELLA

HERMAN BAER, M.D.
Associate Professor of Pathology and Medical Microbiology; Director, Clinical Microbiology Laboratories, Shands Teaching Hospital and Clinics, Miller Health Center, University of Florida, College of Medicine, Gainesville, Florida.
CLASSIFICATION OF BACTERIA; MORAXELLA AND ACINETOBACTER

LANE BARKSDALE, M.D.
Professor of Microbiology, New York University School of Medicine and Medical Center, New York, New York.
DIPHTHERIA BACILLI AND OTHER CORYNEBACTERIA

WILLIAM R. BEISEL, A.B., M.D., F.A.C.P.
Deputy for Science, US Army Medical Research Institute of Infectious Diseases, Fort Dietrick, Maryland.
METABOLIC EFFECTS OF INFECTION

ABRAHAM I. BRAUDE, M.D., Ph.D.
Professor of Medicine and Pathology, University of California, San Diego, California.
THE ASPERGILLI; BACTERIAL ENDOTOXINS; CANDIDA TORULOPSIS; DESCRIPTION OF ANTIMICROBIAL DRUGS; MECHANISMS OF ACQUIRED RESISTANCE TO INFECTION; MECHANISMS OF ACTION OF ANTIMICROBIALS; MECHANISMS OF IMMUNOLOGIC INJURY IN INFECTIOUS DISEASES; MECHANISMS OF NATURAL RESISTANCE TO INFECTION; MISCELLANEOUS FUNGI: THE AGENTS OF MYCETOMA AND RHINOSPORIDIUM; POXVIRUSES; RESISTANCE TO ANTIMICROBIAL DRUGS; SPOROTRICHOSIS; THE ZYGOMYCETES

JOHN J. S. BURTON, Ph.D.
University of California ICMR, Institute for Medical Research, Kuala Lumpur, Malaysia.
ARTHROPODS OF MEDICAL IMPORTANCE

BARUN DEB CHATTERJEE, M.B.B.S., Ph.D.
Professor and Head, Department of Bacteriology and Serology, School of Tropical Medicine, Calcutta, India.
VIBRIOS

T. H. CHEN, M.D.
Research Microbiologist, Department of Biomedical and Environmental Health Sciences, School of Public Health, University of California, Berkeley, California.
YERSINIA, PASTEURELLA, AND FRANCISELLA

RICHARD W. COMPANS, Ph.D.
Professor of Microbiology, University of Alabama Medical Center, Birmingham, Alabama.
MORPHOLOGY AND STRUCTURE OF VIRUSES

EUGENE H. COTA-ROBLES, Ph.D.
Professor of Biology, University of California, Santa Cruz, California.
THE STRUCTURE OF THE BACTERIAL CELL

MANUEL CUADRA, M.D.
Professor (Retired) of Infectious, Parasitic, and Tropical Diseases, Universidad Nacional Mayor de San Marcos, Facultad de Medicina, Lima, Peru; Staff Member, Robert Koch Institut, Associated with Virchow Krankenhaus, Berlin, West Germany.
BARTONELLA BACILLIFORMIS

CHARLES E. DAVIS, M.D.
Professor of Pathology, University of California, San Diego, School of Medicine; Associate Director of Microbiology, University Hospital, San Diego, California.
CLASSIFICATION AND IDENTIFICATION OF BACTERIA; CRYPTOCOCCUS; ERYSIPELOTHRIX RHUSOPATHIAE

DENNIS M. DIXON, Ph.D.
Assistant Professor, Department of Biology, Loyola College of Maryland, Baltimore, Maryland.
BLASTOMYCES AND PARACOCCIDIOIDES

ALEXANDER L. DOHANY, Ph.D., MAJ, MSC
Chief, Department of Acarology, US Army Medical Research Unit, Kuala Lumpur, Department of State, Washington, D.C.
ARTHROPODS OF MEDICAL IMPORTANCE

JEAN M. DOLBY, Ph.D.
Member of the Senior Scientific Staff of the Medical Research Council, Clinical Research Centre, Harrow, Middlesex, England.
BORDETELLA

SANFORD S. ELBERG, Ph.D.
Professor of Microbiology and Immunology, Department of Biomedical and Environmental Health Sciences, School of Public Health, University of California, Berkeley, California.
YERSINIA, PASTEURELLA, AND FRANCISELLA

MAKOTO ENOMOTO, M.D.
Lecturer of Pathology, St. Marianna University, School of Medicine; Consulting Pathologist, Sagamihara Kyodo Hospital, Sagamihara, Japan.
FUNGAL TOXINS

VLADIMIR FARKAŠ, Ph.D.
Researcher, Institute of Chemistry, Department of Biochemistry of Saccharides, Slovak Academy of Sciences, Bratislava, Czechoslovakia.
MORPHOLOGY AND STRUCTURE OF FUNGI

JOSHUA FIERER, M.D.
Associate Professor of Medicine, University of California, San Diego, School of Medicine; Chief, Microbiology Laboratory, Veterans Administration Medical Center; Attending Physician, University Hospital, San Diego, California.
MONILIFORMIS

ROBERT G. GARRISON, Ph.D.
Associate Professor of Microbiology, Department of Microbiology, University of Kansas Medical Center; Research Microbiologist, Veterans Administration Medical Center, Kansas City, Missouri.
SPOROTHRIX SCHENKII

ISAAC GINSBURG, M.Sc., Ph.D.
Professor of Microbiology, The Hebrew University Hadassah, Faculty of Dental Medicine Founded by the Alpha Omega Fraternity; Chairman, Department of Oral Biology, Faculty of Dental Medicine, Hebrew University; Member, Institute for Microbiology, Faculty of Medicine, Hebrew University Hadassah Medical Center, Jerusalem, Israel.
STREPTOCOCCUS

RUTH E. GORDON, Ph.D.
Professor of Microbiology, Waksman Institute of Microbiology, Rutgers, The State University of New Jersey, Piscataway, New Jersey.
NOCARDIA AND STREPTOMYCES

NANCY K. HALL, Ph.D.
Assistant Professor of Pathology, Health Science Center, University of Oklahoma, Oklahoma City, Oklahoma.
HISTOPLASMA CAPSULATUM

HERBERT S. HEINEMAN, M.D.
Clinical Professor of Medicine, Jefferson Medical College of Thomas Jefferson University; Attending Physician and Consultant in Infectious Diseases, Mercy Catholic Medical Center; Director, Public Health Laboratory, Department of Public Health, Philadelphia, Pennsylvania.
SHOCK IN INFECTIOUS DISEASES

SHALOM Z. HIRSCHMAN, M.D.
Professor of Medicine and Director, Division of Infectious Diseases, The Mount Sinai School of Medicine of The University of New York; At-

tending Physician, The Mount Sinai Hospital, New York, New York.
HEPATITIS VIRUSES

MONTO HO, M.D.
Professor of Medicine and Microbiology, University of Pittsburgh; Chief, Division of Infectious Diseases, Presbyterian-University Hospital, Pittsburgh, Pennsylvania.
INTERFERON AND INTERFERENCE

BETTY C. HOBBS, D.Sc., Ph.D., Dip. Bact., F.R.C.Path., F.R.S.H.
Director (Retired), Food Hygiene Laboratory, Central Public Health Laboratory, London, England; Consultant Microbiologist, Christian Medical and Brown Memorial Hospital, Punjab, India.
THE CLOSTRIDIA

PATRICIA A. HOFFEE, Ph.D.
Professor of Microbiology, University of Pittsburgh, School of Medicine, Pittsburgh, Pennsylvania.
BACTERIAL GENETICS

TOR HOFSTAD, M.D.
Associate Professor of Microbiology, Faculty of Medicine, University of Bergen; Chief Physician, Department of Microbiology, Haukeland Hospital, Bergen, Norway.
BACTEROIDES

MARIAN C. HORZINEK, D.V.M., Ph.D.
Professor of Virology and Head, Department of Virology, Veterinary Faculty, State University, Utrecht, The Netherlands.
TOGAVIRUSES

GERALD T. KEUSCH, M.D.
Professor of Medicine, Tufts University School of Medicine; Attending Physician, New England Medical Center Hospitals, Boston, Massachusetts.
MALNUTRITION AND INFECTION

MOGENS KILIAN, D.D.S., Ph.D.
Associate Professor of Microbiology, Royal Dental College, Aarhus, Denmark.
HAEMOPHILUS

HOWARD W. LARSH, Ph.D.
Research Professor of Microbiology, University of Oklahoma, Norman, Oklahoma; Director of Research and Laboratories, Missouri State Chest Hospital, Mt. Vernon, Missouri.
HISTOPLASMA CAPSULATUM

D. L. LEE, B.Sc., Ph.D.
Professor of Agricultural Zoology and Head of Department of Pure and Applied Zoology, University of Leeds, Leeds, England.
CLASSIFICATION AND ANATOMY OF PARASITES

FRITZ LEHMANN-GRUBE, M.D.
Professor of Virology, University of Hamburg, Hamburg, Germany.
ARENAVIRUSES

STANLEY M. LEMON, M.D.
Assistant Professor of Medicine, Department of Virus Diseases, Walter Reed Army Institute of Research, Walter Reed Army Medical Center; Staff Physician, Walter Reed General Hospital; Staff Physician, Walter Reed Army Medical Center, Washington, D.C.
THE HERPESVIRUSES

G. PHILIP MANIRE, Ph.D.
Kenan Professor of Bacteriology and Immunology, Vice Chancellor and Dean of Graduate School, University of North Carolina, Chapel Hill, North Carolina.
THE CHLAMYDIAE

FRANCOIS MARIAT, Dr. es Sciences (Paris)
Professor, Institut Pasteur, Paris, France.
SPOROTHRIX SCHENCKII

ZELL A. McGEE, M.D.
Professor of Medicine, Vanderbilt University, School of Medicine; Director, George Hunt Laboratory; Chief, Division of Infectious Diseases, Department of Medicine, Vanderbilt University Hospital, Vanderbilt, Tennessee.
CELL WALL–DEFECTIVE BACTERIA; MYCOPLASMAS

JOSEPH L. MELNICK, Ph.D.
Distinguished Service Professor and Chairman, Department of Virology & Epidemiology, Baylor College of Medicine, Houston, Texas.
CLASSIFICATION OF VIRUSES

JOSEPH H. MILLER, M.S., Ph.D.
Professor of Medical Parasitology, Louisiana State University Medical Center; Visiting Scientist, The Charity Hospital in New Orleans, New Orleans, Louisiana.
THE PROTOZOA

SAROJ K. MISHRA, Ph.D.
Chief, Serodiagnostic Division, Robert Koch Institute, Berlin, Germany.
NOCARDIA AND STREPTOMYCES

SUSUMU MITSUHASHI, Ph.D. Dr. Med. Sci.
Professor, Department of Microbiology, and Director, Laboratory of Bacterial Resistance,

Gunma University, School of Medicine, Maebishi City, Japan.
RESISTANCE TO ANTIMICROBIAL DRUGS

STEPHEN A. MORSE, Ph.D.
Associate Professor, Department of Microbiology and Immunology, University of Oregon Health Sciences Center, Portland, Oregon.
NEISSERIA

STEPHEN I. MORSE, M.D. (Deceased)
Professor and Chairman, Microbiology and Immunology, State University of New York, Downstate Medical Center, Brooklyn, New York.
STAPHYLOCOCCI

DANIEL M. MUSHER, M.D.
Professor of Medicine, Professor of Microbiology and Immunology, Baylor College of Medicine; Chief, Infectious Disease Section, Veterans Administration Hospital, Houston, Texas.
SPIROCHETES: TREPONEMA AND BORRELIA

HAROLD C. NEU
Professor of Medicine and Pharmacology, Columbia University, College of Physicians and Surgeons; Chief of Infectious Diseases Columbia-Presbyterian Medical Center, New York, New York.
PHARMACOLOGY AND TOXICOLOGY OF ANTIMICROBIAL AGENTS

FRITS ØRSKOV, M.D.
Collaborative Centre for Reference and Research on Escherichia (WHO), Statens Seruminstitut, Copenhagen, Denmark.
ENTEROBACTERIACEAE

IDA ØRSKOV, M.D.
Collaborative Centre for Reference and Research on Escherichia (WHO), Statens Seruminstitut, Copenhagen, Denmark.
ENTEROBACTERIACEAE

MICHAEL N. OXMAN, M.D.
Professor of Medicine and Pathology, University of California, San Diego, School of Medicine; Chief, Infectious Diseases and Clinical Virology Sections, Veterans Administration, Administration Medical Center, San Diego, California.
PAPOVAVIRUSES

JOSEPH S. PAGANO, M.D.
Director, Cancer Research Center, and Professor of Medicine and of Bacteriology and Immunology, University of North Carolina at Chapel Hill School of Medicine; Attending Physician in Medicine and Infectious Disease, North Carolina Memorial Hospital, Chapel Hill, North Carolina.
THE HERPESVIRUSES

ZBIGNIEW S. PAWLOWSKI, M.D., D.T.M.&H.
Professor of Medical Parasitology, Medical Faculty, Academy of Medicine, Poznan, Poland; Chief, Clinic of Parasitic and Tropical Diseases, Academy of Medicine, Poznan, Poland.
PLATYHELMINTHES

LENNART PHILIPSON, M.D., Dr. Med. Sci.
Professor of Microbiology, Uppsala University, Uppsala, Sweden.
ADENOVIRUS

M. J. PICKETT, Ph.D.
Professor of Microbiology, University of California, Los Angeles, California.
GENUS PSEUDOMONAS

LEO PINE, B.S., M.S., Ph.D.
Chief, Products Development Branch, Biological Products Division, Center for Disease Control, Atlanta, Georgia.
ACTINOMYCES AND MICROAEROPHILIC ACTINOMYCETES

BOSKO POSTIC, M.D.
Professor of Medicine, University of South Carolina, School of Medicine; Chief, Medical Service, William Jennings Bryan Dorn Veterans Hospital, Columbia, South Carolina.
RHABDOVIRUS

C. R. PRINGLE, B.Sc., Ph.D.
Honorary Lecturer, University of Glasgow; Member of the Medical Research Council, MRC Virology Unit, Glasgow, Scotland.
THE GENETICS OF VIRUSES

T. RAMAKRISHNAN, M.Sc., Ph.D.
Professor, Microbiology and Cell Biology Laboratory, Indian Institute of Science, Bangalore, India.
BACTERIAL PHYSIOLOGY

G. RAMANANDA RAO, M. Pharm., Ph.D.
Assistant Professor, Microbiology and Cell Biology Laboratory, Indian Institute of Science, Bangalore, India.
BACTERIAL PHYSIOLOGY

GARRISON RAPMUND, M.D.
Brigadier General, MC, US Army; Assistant Surgeon General (Research and Development), Department of the Army, Washington, D.C.
RICKETTSIA

DOUGLAS D. RICHMAN, M.D.
Assistant Professor of Pathology and Medicine, University of California, San Diego, School of Medicine; Attending Physician, Veterans Administration Medical Center, San Diego, California.
ORTHOMYXOVIRUSES AND PARAMYXOVIRUSES

DAVID L. RINGO, Ph.D.
Assistant Research Biologist, University of California, Santa Cruz, California.
THE STRUCTURE OF THE BACTERIAL CELL

HOWARD J. SAZ, Ph.D.
Professor of Biology, University of Notre Dame, Notre Dame, Indiana.
BIOCHEMISTRY OF PARASITES: HELMINTHS

H. P. R. SEELIGER, M.D.
Full Professor of Hygiene and Microbiology and Director of Institute of Hygiene, Bayerische Julius-Maximilians-Universität, Wuerzburg, Federal Republic of Germany.
LISTERIA MONOCYTOGENES

SMITH SHADOMY, Ph.D.
Professor of Medicine and Microbiology, Medical College of Virginia, Virginia Commonwealth University, Richmond, Virginia.
BLASTOMYCES AND PARACOCCIDIOIDES

RUDOLF SIEGERT, M.D.
Professor of Medical Microbiology, Institute of Hygiene, Philipps-Universität, Marburg, Federal Republic of Germany.
MARBURG VIRUS AND EBOLA VIRUS

SAMUEL C. SILVERSTEIN, M.D.
Associate Professor and Physician, Laboratory of Cellular Physiology and Immunology, Rockefeller University; Physician, Rockefeller University Hospital, New York, New York.
VIRAL REPLICATION

IRVING J. SLOTNICK, Ph.D.
Chief Microbiologist, Cedars Sinai Medical Center, Los Angeles, California.
ACTINOBACILLUS AND CARDIOBACTERIUM

DONALD W. SMITH, Ph.D.
Professor of Medical Microbiology, University of Wisconsin, Madison, Wisconsin.
MYCOBACTERIA

STEPHEN A. SPECTOR, M.D.
Assistant Professor of Pediatrics, University of California, San Diego, School of Medicine; Assistant Professor of Pediatrics, Division of Infectious Diseases, University of California, San Diego, University Hospital, San Diego, California.
IMMUNOPROPHYLAXIS AND IMMUNOTHERAPY

WESLEY W. SPINK, A.B., M.D.
Emeritus Regents' Professor of Medicine and Comparative Medicine, University of Minnesota Medical School; Staff Member, University of Minnesota Hospitals, Minneapolis, Minnesota.
BRUCELLA

NEVILLE F. STANLEY, B.Sc., D.Sc., F.R.A.C.P.(Hon.), M.A.S.M.
Professor and Chairman, Department of Microbiology, University of Western Australia; Director, Clinical Microbiology Services, Queen Elizabeth II Medical Centre, Perth, Western Australia.
REOVIRIDAE PATHOGENIC FOR MAN

WILLARD ALLEN TABER
Professor, Department of Biology, Texas A & M University, College of Sciences, College Station, Texas.
CLASSIFICATION OF FUNGI

DAVID TAYLOR-ROBINSON, M.D.
Division of Communicable Diseases, M.R.C. Clinical Research Centre; Hon. Consultant Microbiologist, Northwick Park Hospital, Harrow, Middlesex, England.
MYCOPLASMAS

D. A. J. TYRRELL, C.B.E., M.D., D.Sc., F.R.C.P., F.R.C.Path., F.R.S.
Head, Division of Communicable Diseases, Clinical Research Centre; Consultant Physician, Northwick Park Hospital, Harrow, Middlesex, London, England.
CORONAVIRUS

WILLIAM EDWARD VAN HEYNINGEN, Ph.D., Sc.D., D.Sc.
Emeritus Reader in Bacterial Chemistry, Sir William Dunn School of Pathology, University of Oxford, Oxford, England.
BACTERIAL EXOTOXINS

HENRY A. WALCH, Ph.D.
Professor of Microbiology, San Diego State University, San Diego, California.
COCCIDIOIDES IMMITIS

TADEUSZ J. WIKTOR, D.V.M.
Member, The Wistar Institute, Philadelphia, Pennsylvania.
RHABDOVIRUS

ROBERT P. WILLIAMS
Professor of Microbiology and Immunology, Baylor College of Medicine, Houston, Texas.
BACILLUS ANTHRACIS AND OTHER AEROBIC SPORE–FORMING BACILLI

SHELDON M. WOLFF, M.D.
Endicott Professor and Chairman, Department of Medicine, Tufts University School of Medicine; Physician-in-Chief, New England Medical Center Hospital, Boston Massachusetts.
FEVER

NATHANIEL A. YOUNG, M.D. (Deceased)
Head, Viral Oncology and Molecular Pathology Sections, Laboratory of Pathology, National Institutes of Health; Senior Attending Physician in Infectious Diseases, Clinical Center, National Institutes of Health, Bethesda, Maryland.
PICORNAVIRUSES

JULIUS S. YOUNGNER, Sc.D.
Professor and Chairman, Department of Microbiology, University of Pittsburgh, School of Medicine, Pittsburgh, Pennsylvania.
PERSISTENT VIRAL INFECTIONS

PREFACE

Soon after we set out to prepare a comprehensive textbook of microbiology we realized that the subject matter was too vast for a few authors to handle. We decided instead to call upon leading experts throughout the world to write individual chapters on subjects that belong to their areas of acknowledged authority. This international approach allowed us to go beyond the limitations imposed by the usual institutional, regional, or even national boundaries that most textbooks must observe in their authorship and to produce the first multi-authored international textbook of microbiology. Seventy-one authors have contributed 88 chapters to this new systematic textbook.

In addition to striving for authenticity, we have tried to give students a cohesive account of all aspects of microbiology. This was done by presenting the information in three sections that would give a logical progression from general principles to specific information. Because the contributors are both scientists and teachers, they have presented their material in a way that lets even the beginning student understand this complicated subject. Students concerned primarily with the basic aspects of microbiology will find complete presentations of microbial genetics, biochemistry, morphogenesis, physiology, latency, and taxonomy. These fundamental topics are covered not only with respect to bacteria and viruses but to mycology and parasitology as well. Students with a special interest in fungi or parasites should find this textbook valuable because it satisfies their need for information in their own fields and also allows them to refer to closely related information dealing with the other microbial agents. It should be noted that the parasitology in this book is not restricted to invasive agents but presents extensive coverage of ectoparasites and other arthropods important to entomologists.

The chapters on specific microbial agents will meet the needs of the special areas of microbiology. For the medical technologist there are complete diagnostic data on every organism of medical importance; for the experimental pathologist and immunologist there is an extensive analysis of the pathogenic properties and immune reactions for each pathogenic agent; for the pharmacist and pharmacologist there are the essentials of antimicrobial chemotherapy; and for the epidemiologist there are sections in each chapter dealing with the prevalence, mode of spread, and protection against each specific microbial agent.

This textbook plus a larger section on specific infectious diseases is also published under the title of *Medical Microbiology and Infectious Diseases* by the same publisher.

For all this we thank our many distinguished authors who gave generously of their time, expert knowledge, and teaching skills. We hope this book will justify their valuable efforts by helping students learn microbiology throughout the world.

CONTENTS

I. MICROBIOLOGY

A. General Microbiology

Bacteriology

Mycology

I Microbiology

A. GENERAL MICROBIOLOGY

Bacteriology

THE STRUCTURE OF THE BACTERIAL CELL 1

Eugene H. Cota-Robles, Ph.D.
and David L. Ringo, Ph.D.

The structure of bacteria can be best understood by contrasting their cellular organization to that of the higher biologic forms — animals, plants, protozoa, algae, and fungi. In these life forms the cells are divided into internal compartments by membrane systems. These compartments (cellular *organelles)* have specific activities for the maintenance of the life functions of the cell. The nuclear compartment *(nucleus)* contains the cell's hereditary material, *deoxyribonucleic acid* (DNA). The nucleus is separated by a nuclear membrane from the rest of the cell, the *cytoplasm.* The DNA of higher cells is organized in two or more chromosomes.

Bacterial cells are not compartmentalized and therefore are considered to lack organelles. The hereditary material of bacteria consists of a single chromosome. This bacterial *nucleoid* is a single, tightly bundled, long strand of DNA that lies within the cytoplasm not surrounded by a nuclear membrane. For this reason bacteria are referred to as *prokaryotic,* that is, having a primitive nuclear structure; and higher cells are designated as *eukaryotic,* literally, having a true nucleus. All bacteria (as well as the blue-green algae) possess the prokaryotic form of organization.

In a broad sense, all types of cells function in the same manner. The information stored in the DNA is transcribed as *messenger ribonucleic acid* (mRNA), which moves from the nucleus into the cytoplasm. Here the messenger RNA attaches to ribosomes, where the genetic information is translated into specific proteins by the polymerization of amino acids. The synthesis of proteins is one of the most basic of cell processes, since virtually every metabolic activity of the cell is mediated by enzymes, the proteins that catalyze specific chemical reactions.

The basic function of a bacterial cell is to assimilate chemicals from its environment in order to grow and divide. Some bacteria perform these functions by using the simplest inorganic chemicals (carbon dioxide and minerals), and are termed *autotrophs.* Certain members of this group, photosynthetic bacteria and blue-green algae, rely on light as a source of energy and thus resemble plants in their photoautotrophic metabolism. Another and larger group of bacteria, the *heterotrophs,* use simple organic molecules from their environment as a source of energy and as building blocks for cellular material. Many of this latter group have adapted to growth within the animal body. Some are harmless symbionts, such as *Escherichia coli,* which normally inhabits the mammalian gut. Others are harmful and may cause human disease. These bacteria are termed *pathogens,* and their infection of body tissues results in a characteristic illness. Many of the features of bacterial cell structure are related, in one way or another, to their role as disease organisms and to the ways in which these diseases are treated and controlled.

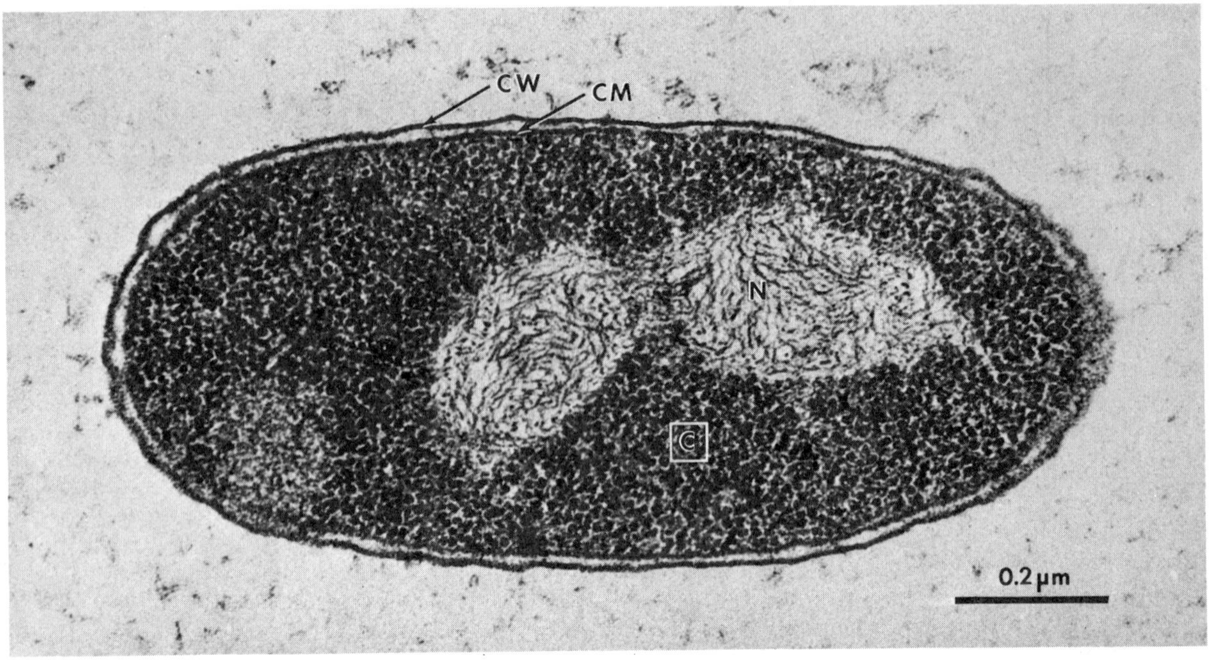

FIGURE 1. *This electron micrograph outlines the organization that is common to all bacterial cells. The nucleoid (N) occupies the central area of the cell; the cytoplasm (C) contains many ribosomes (individual dark granules); a cytoplasmic membrane (CM) and cell wall (CW) surround the cell. This bacterium,* Alteromonas espejiana, *is a marine pseudomonad that possesses a very simple wall structure.*

The generalized structure of a typical bacterial cell is shown in Figure 1. The prokaryotic cell consists of a cell wall, a cytoplasmic membrane surrounding a cytoplasm packed with ribosomes, and a more or less central nucleoid. The figure is an example of the image produced by the electron microscope of a very thin section cut through a chemically preserved (fixed) bacterium. It is possible to determine the overall shape and size of such a bacterial cell with the light microscope, but details that are smaller than about 0.2 μm cannot be resolved (Table 1). Thus, the internal structure of the bacterium can be deduced only through the use of the higher resolving power of the electron microscope, and all such detailed knowledge has been obtained only within the past 25 years.

TABLE 1. Units Used to Describe Bacteria and their Structures, and the Useful Limits of Resolution of the Human Eye, the Light Microscope, and the Electron Microscope

Units:
1 m × 10^{-3} = 1 mm (millimeter)
1 mm × 10^{-3} = 1 μm (micrometer or micron)
1 μm × 10^{-3} = 1 nm (nanometer)
Limits of Resolving Power:
Unaided human eye: 0.2 mm
Compound light microscope: 0.2 μm
Transmission electron microscope: 2 nm to 0.2 nm

NUCLEOID

Embedded in the ribosome-rich cytoplasm is the bacterial genetic apparatus, the nucleoid. No boundary separates it from the rest of the cytoplasm; the nucleoid's physical segregation is maintained by the very nature of the DNA that is its make up. The DNA consists of one very long and narrow molecule — a double helix — which if stretched to its full length would measure more than 1.0 mm long and only 0.000002 mm (2 nm) wide. Genetic analysis has shown that most of the bacterial genes are linked to one another in an orderly fashion, and, moreover, that the linkage is continuous and circular. We therefore describe the bacterial nucleoid as a single, circular chromosome. Physical studies also demonstrate that the DNA strand of the chromosome is a circle that is tightly coiled into a bundle to produce the nucleoid image shown with the electron microscope. The chromosome replicates in the growing cell in preparation for cell division. Bacterial cells divide by binary fission into two daughter cells, each of which retains a copy of the chromosome. A small percentage of the cell's genetic information is present as much smaller DNA molecules, *plasmids,* which may be present in many copies and replicate independently. Bacterial plasmids often carry the genes involved in resistance to chemotherapeutic drugs. Plasmids cannot be observed within cells by the electron microscope because they are masked by the dense cytoplasm.

CYTOPLASM

Although certain autotrophic bacteria contain internal membranes associated with photosynthesis and other special metabolic functions, the cytoplasm of most heterotrophic bacteria contains only ribosomes and storage granules. Ribosomes are the site of protein synthesis, and their large numbers reflect the importance of this synthetic role in the bacterial cell. Many enzymes are present in the cytoplasm, whereas others are bound to the cell membrane. It is impossible, however, to recognize individual protein molecules with the electron microscope, except when special preparative techniques are used.

Each ribosome is a complex structure about 20 nm in diameter, and is an aggregate of several RNA molecules and many protein molecules. The ribosome consists of two major subunits, one somewhat larger than the other. Although the ribosomes of all types of cells function in a similar manner in protein synthesis, it is possible to distinguish the ribosomes of prokaryotic cells from those of eukaryotic cells by their different sedimentation velocity in the ultracentrifuge. Bacterial ribosomes have a sedimentation value of 70S (Svedberg units), whereas eukaryotic ribosomes are slightly larger and heavier with a sedimentation value of 80S. The antibiotics tetracycline, streptomycin, and chloramphenicol act by interfering with one stage or other of protein synthesis by bacterial (70S) ribosomes.

A second class of cytoplasmic particles, which are frequently but not always found, are storage granules (Fig. 2). The presence and amount of these particles vary with the type of bacterium and its level of metabolic activity. Storage granules represent a mechanism for temporary storage of excess metabolites. Three types of such granules are common: glycogen, a glucose polymer that serves as a carbohydrate reserve; poly-β-hydroxybutyric acid, a lipid reserve; and volutin, a polyphosphate compound. These granules are often large enough to be seen with the light microscope. Volutin granules are also known as metachromatic granules because of their staining behavior; that is, they appear red when stained with methylene blue dye.

CELL MEMBRANE

The cytoplasmic membrane, a delicate structure that is only 6 nm thick, represents the true

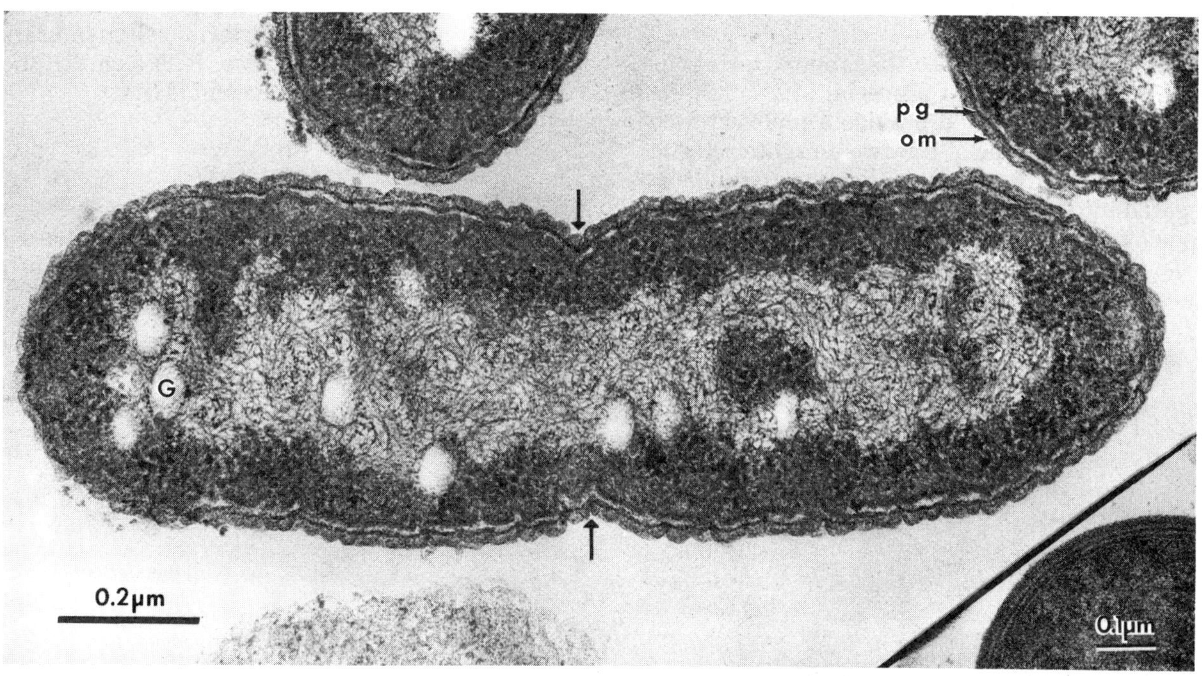

FIGURE 2. *A typical rod-shaped gram-negative cell,* Chromobacterium violaceum, *is seen here in longitudinal and cross-section. It contains several storage granules (G), probably poly-β-hydroxybutyric acid. This cell is in the process of division, as indicated by the inward constriction of the cell wall around the middle of the cell (arrows). The outer membrane and peptidoglycan layers of the cell wall may be seen. The insert at the lower right shows the wall of a gram-positive cell,* Bacillus sphaericus, *for comparison. (Insert courtesy of Drs. Stanley Holt and Donald Tipper, University of Massachusetts.)*

barrier between the inside and the outside of the cell. In addition to a membrane, the bacterial cell requires the presence of a cell wall to permit it to withstand the cell's osmotic pressure and other forces encountered in nature. Removal of the cell wall normally results in the rupture of the cell membrane, causing *lysis,* the dissolution and death of the cell. In an osmotically protected environment the cell may remain intact and alive even after the cell wall is removed. A cell having no wall is called a *protoplast,* if the wall is completely removed, or a *spheroplast,* if some wall fragments remain.

The selective permeability of the cell membrane determines which molecules from the cell's environment can enter the cell and which are excluded. Some molecules diffuse passively into the cell; others require active transport across the membrane by specific enzymes. Enzymes involved in cell wall synthesis and synthesis of extracellular products are lodged in the cytoplasmic membrane. The antibiotic properties of polymyxin-B result from its interference with normal membrane permeability.

The electron transport system, the primary energy-generating system of the cell, is also located on the cell membrane. Thus, the membrane is the site where most chemical energy available through metabolism is converted into a form of energy (ATP) that can be used in the biosynthesis of cell materials or in other energy-requiring cellular reactions.

There is some evidence that the bacterial chromosome is attached to a special site on the cell membrane. This would provide a physical mechanism by which one of the two daughter chromosomes produced by DNA replication could be segregated into each of the daughter cells during cell division.

Generally, the bacterial membrane is found solely at the cell periphery. However, some bacterial cells exhibit complex infoldings or invaginations of the membrane into the cytoplasmic space (Fig. 3). These invaginations have been named *mesosomes* to distinguish them from the membrane proper and from the cytoplasm. This distinction may not be necessary, since mesosomes are not really separate bodies but rather extensions of the cell membrane. A number of efforts have been made to ascertain whether mesosomal invaginations have unique or specific functions. It has been suggested that they serve to increase the surface area of the cell membrane or that they represent specialized areas of the membrane; they are often seen in association with cell division or spore formation. However, to date no specialized functions have been conclusively demonstrated to be associated with the mesosomal membrane. Mesosomes have been observed only with the electron microscope, and some evidence suggests that the structures may be artifacts of the preparation of the cells for electron microscopy.

The appearance and chemical composition of the membrane of prokaryotic cells are quite similar to those of the cell membrane of eukaryotic cells. One notable chemical difference is that most bacterial membranes do not contain sterols (e.g., cholesterol) as do the membranes of higher life forms. In the bacterial cell, the cytoplasmic membrane performs many functions that are localized in the specialized organelles of higher cells. Thus, although the bacterial cell membrane may be simple in appearance, it is a multifunctional structure of extreme complexity.

FLAGELLA

A number of bacterial cell types demonstrate a striking rapid motility that is the product of a specialized cell structure, the bacterial *flagellum.*

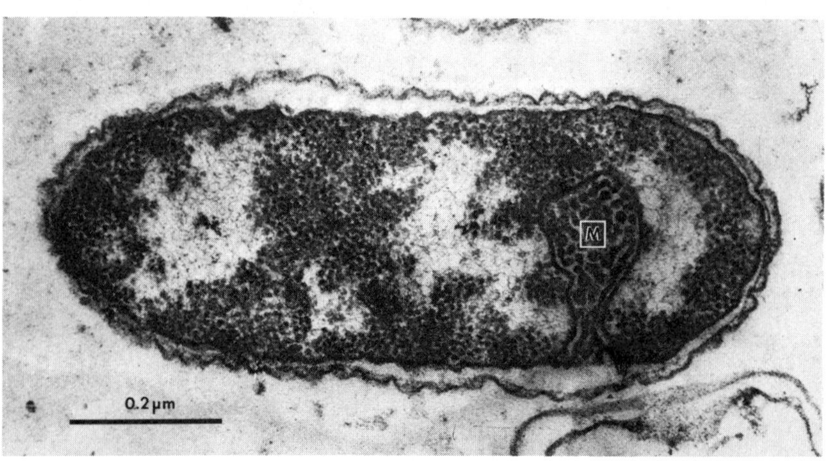

FIGURE 3. *A* Chromobacterium violaceum *cell, showing a single large mesosome (M), a complex invagination of the cytoplasmic membrane.*

Bacterial flagella are completely different in their structure and mechanism of function from the flagella and cilia of protozoa and other eukaryotic cells. In the higher cells a planar wave motion is generated within a membrane-bound flagellum by the sliding action of multiple protein filaments, the flagellar microtubules. In bacteria, a single proteinaceous filament, about 15 nm in diameter, projects from the cell. At its base a complex structure, which is highly integrated with the cell wall and membrane, produces a rotary motion in the flagellar filament. The cell is driven forward by the propeller-like motion of the flagellum. The mechanism of operation of this fascinating, chemically driven rotary motor is currently under investigation.

Motility is also linked to membrane-bound receptors that identify the presence of specific molecules in the cell's environment. This phenomenon of *chemotaxis* allows bacteria to move toward higher concentrations of nutrients and away from harmful substances.

Bacterial flagella may be arranged in a diverse manner in different cell types. Some bacteria possess many flagella distributed over the entire surface of the cell (*peritrichous* flagella), whereas others possess a single flagellum or a bundle of flagella located at one pole of the cell (*polar* flagella). Special flagellar stains are used to enlarge the dimensions of the flagella artificially so that they may be observed with the light microscope. Motility versus nonmotility and the arrangement of flagella are useful diagnostic characteristics for identifying bacteria.

FIMBRIAE OR PILI

Electron microscopic examination of intact gram-negative cells frequently reveals the presence of long, slender projections extending from the surface of the cells. These projections, which originate at the cell membrane and extend through the cell wall, are of two types: the first are called *fimbriae,* or *common pili,* and the second are called *sex pili.*

The most common of these projections are the fimbriae, or common pili. Fimbriae are frequently found in large numbers (as many as several hundred per cell) projecting from the entire surface of freshly isolated gram-negative bacteria. They are made of a single protein, whose molecules are arranged in a helix to form a long filament. The function of fimbriae has not been completely clarified, but there is mounting evidence that these slender projections play a role in facilitating the adherence of bacterial cells to other surfaces. Fimbriae are rather stable and are not easily removed from the surface of the bacterial cell. The length and width of a common pilus vary greatly among different bacterial types but generally fall within the range of 3 to 15 nm in width and from 0.2 μm to a few micrometers in length.

Sex pili are less common than fimbriae; only a few gram-negative species are known to possess them. They act as specific bridges between bacterial cells during *conjugation,* a process in which unidirectional transfer of genetic information takes place between cells. The sex pilus has the form of a hollow tube, and it is believed that bacterial DNA migrates through the narrow central space of the pilus during conjugation. Sex pili are long (1 μm or more), relatively thick (about 25 nm), and very fragile. They can be broken by simple physical treatment. Very few (no more than ten) sex pili are found on bacteria that possess these fragile appendages.

CELL WALL

The wall of the bacterium confers protective rigidity to the cell. The chemical composition and organization of the prokaryotic cell wall is a unique structure that is not found in eukaryotic cells. Its basic component is a mixed polymer called *peptidoglycan.* The peptidoglycan is made of two components: Linear polymers, called glycan strands, are made of two substituted hexose sugars — *N*-acetylglucosamine and *N*-acetylmuramic acid. These sugars are present in equal amounts and occupy alternate positions along the chains, which may contain dozens of sugar molecules. The second component is short peptides — identical chains of four amino acids — that attach to the glycan strand (but only to the *N*-acetylmuramic acid residues), thus forming side branches to the glycan strands. Some of these tetrapeptides are, in turn, linked to one another by other short peptides, forming cross-bridges between the glycan strands (Fig. 4). Peptidoglycan forms a single, bag-shaped macromolecule that surrounds the cytoplasmic membrane — conferring cell shape, countering the osmotic force exerted by the cell, and forming a physical barrier against the outside environment.

Besides the occurrence in peptidoglycan of the two unusual *N*-acetyl sugars, the peptide chains contain some amino acids that are found nowhere else except in bacterial cell walls. Among these are *meso*-diaminopimelic acid and the D-isomers of glutamic acid and alanine. (The amino acids found in all proteins are the L-isomers.)

Peptidoglycan is a component in the walls of almost all bacterial cells. Two notable exceptions are the cells of *Mycoplasma* and *Halobacteria.* *Mycoplasma* are very small bacteria that actually have no cell wall, being surrounded by a cytoplas-

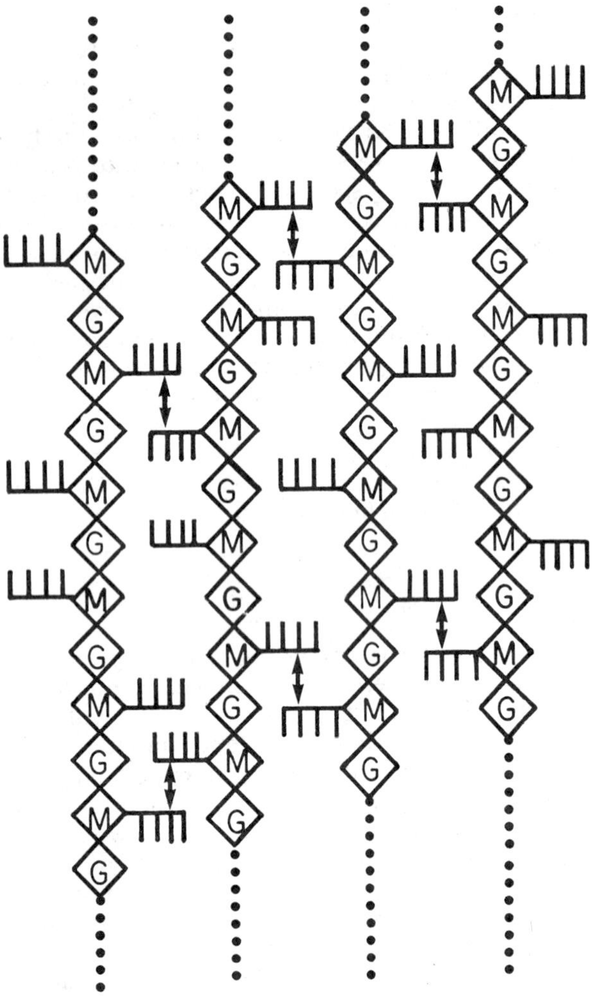

FIGURE 4. *This diagram schematically represents the structure of peptidoglycan. The glycan strands are composed of alternating units of N-acetylglucosamine (G) and N-acetylmuramic acid (M). Tetrapeptide side chains connect to the muramic acid residues. Some of the tetrapeptides are joined by other peptides (short arrows) to form cross-bridges between the glycan strands.*

mic membrane only. Halobacteria, which live in high-salt environments, have a wall that lacks peptidoglycan. The composition of peptidoglycan is not absolutely uniform among all types of bacteria; however, the variation is not in the carbohydrate backbone of the glycan strand but in the type of amino acids in the tetrapeptide and in the types and number of amino acids that cross-link the tetrapeptides to one another.

Bacteria can be divided into two major groups on the basis of their cell wall structure. The first, *gram-negative* organisms, have a thin peptidoglycan layer adjacent to the cytoplasmic membrane, plus other major cell wall components exterior to the peptidoglycan. Examples of gram-negative bacteria are *Escherichia, Salmonella, Pseudomonas, Treponema,* and *Chlamydia.* The second, *gram-positive* organisms, have a much thicker peptidoglycan structure, consisting of many layers (although they are not visible as such with the electron microscope) of peptidoglycan, making a total cell wall thickness of 10 to 100 nm (Figs. 2 and 5). Peptidoglycan represents the major component of the gram-positive cell wall, although minor components known as teichoic acids are also present. Examples of gram-positive bacteria include *Bacillus, Streptococcus, Staphylococcus,* and *Clostridium.*

The original basis of division of bacteria into gram-negative and gram-positive forms was the staining procedure devised almost 100 years ago by the Danish physician Christian Gram. This stain, which is still in common clinical use for diagnostic purposes, involves four steps: Cells are 1) stained with crystal violet; 2) treated with iodine to form a crystal violet–iodine complex; 3) washed with an organic solvent such as alcohol; and 4) stained again with safranin. In gram-positive cells the purple dye complex is retained, whereas in gram-negative cells the dye complex is removed by the solvent. Thus gram-positive cells appear purple under the microscope; gram-negative cells are counterstained by the safranin and appear red. Although the exact mechanism of the staining reaction is not known, it appears that the staining occurs within the cell, not on the cell wall, and that the thicker peptidoglycan wall of the gram-positive cell protects the dye-iodine complex to a greater extent from the action of the solvent.

Since the cell wall of gram-positive bacteria consists of a layer of peptidoglycan that may be many times thicker than the peptidoglycan of the gram-negative cell wall, gram-positive cells tend to be sturdier and less susceptible to breakage by physical forces. In contrast, the presence in the gram-negative cell of other cell wall components in addition to peptidoglycan gives those bacteria certain advantages.

The gram-negative cell wall contains significant amounts of lipoprotein and lipopolysaccharide in association with the thin peptidoglycan layer. This layer of lipid material lies outside the peptidoglycan and is frequently called the *outer membrane.* Although it appears as a membrane in electron micrographs and has the property of excluding some large molecules, it appears to contain no enzymes, and its function and organization are much less complex than those of the cytoplasmic membrane.

An example of an advantage that gram-negative cells possess is a lower susceptibility to cell lysis by the naturally occurring hydrolytic enzyme *lysozyme.* Lysozyme is widespread in nature and specifically attacks peptidoglycan by cleaving the bonds of the glycan strand between *N*-acetylmuramic acid and *N*-acetylglucosamine.

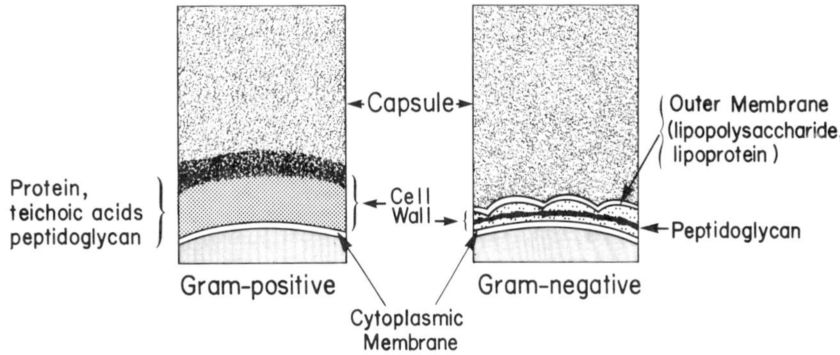

FIGURE 5. *These diagrams compare the structures of the gram-positive and the gram-negative bacterial cell walls, indicating the relative positions of the cytoplasmic membrane and the bacterial capsule.*

In the absence of the protective peptidoglycan layer, the bacterial cell swells excessively owing to osmotic forces. Eventually the cytoplasmic membrane ruptures from the internal pressure. The walls of gram-positive cells are generally quite sensitive to degradation by lysozyme. Gram-negative bacteria are less susceptible to such attack, since the outer membrane excludes lysozyme and prevents it from binding to its site of action on the cell wall's peptidoglycan.

The uniqueness of bacterial peptidoglycan has proved to be particularly useful in the treatment of many bacterial diseases. The biosynthesis of peptidoglycan can be disturbed by a number of antibiotics, including penicillin. The sensitive target of penicillin action is one of the final chemical reactions in peptidoglycan synthesis — the one that effects the linking of the peptide side chains. Thus, the drug prevents cross-bridges from being formed between the glycan strands. Without this bonding the integrity of the peptidoglycan is weakened, producing a fragile wall that cannot protect the cell adequately.

Since penicillin interferes with peptidoglycan synthesis, and since such synthesis does not occur in eukaryotic cells, cells of the animal body are resistant to the primary action of penicillin. Penicillin affects sensitive bacteria only when the cell wall is being synthesized, since it blocks synthesis rather than attacking existing peptidoglycan, as does lysozyme. Thus, if a penicillin-sensitive cell is not growing at the time the antibiotic is used, the cell wall will not be affected. Gram-negative bacteria are significantly less susceptible to the antibiotic action of penicillin than are gram-positive bacteria. This difference, as with lysozyme, is primarily the result of the protective effect of the outer membrane of the gram-negative cell wall, which makes it difficult for the penicillin molecule to reach its site of inhibition. However, forms of penicillin have been synthesized that can penetrate this barrier and attack gram-negative organisms.

CAPSULE

A few bacterial types, including a number of disease-producing bacteria, manufacture an inert capsule that surrounds the entire cell. Capsular material is easily visualized with the light microscope by negative stains such as India ink. The role of the capsule in the life of bacteria has yet to be established; however, for some bacteria the capsule serves as a protective structure. Certain encapsulated cells are resistant to ingestion by animal phagocytes; thus, these bacteria cannot be destroyed by the phagocytic line of defense of animal blood.

Capsules have a defined chemical composition that is characteristic of the cell type that produces it. In most bacteria the capsule is composed of polysaccharides. A few bacteria produce capsules composed of polypeptides containing D amino acids. The structure of the bacterial capsule is not a highly ordered one as is the structure of the cell wall, thus a general form of capsule structure is not recognized.

VARIATIONS IN BACTERIAL FORM

Within the large group of prokaryotic organisms, there is a great range in the size and shape of individual cells (Fig. 6). Some are at the limit of visibility of the light microscope; others are as large as some eukaryotic cells. Common cell shapes include spheres (cocci), rods (bacilli), comma shapes (vibrio), and spirals (spirillum, spirochete). Cell size and shape are valuable diagnostic characteristics and are readily seen with the light microscope.

Another diagnostic characteristic of bacteria is the way in which cells do or do not remain associated during growth. They may exist as solitary individuals, or they may remain attached in random clusters, in linear chains, or in ordered sheets or packets.

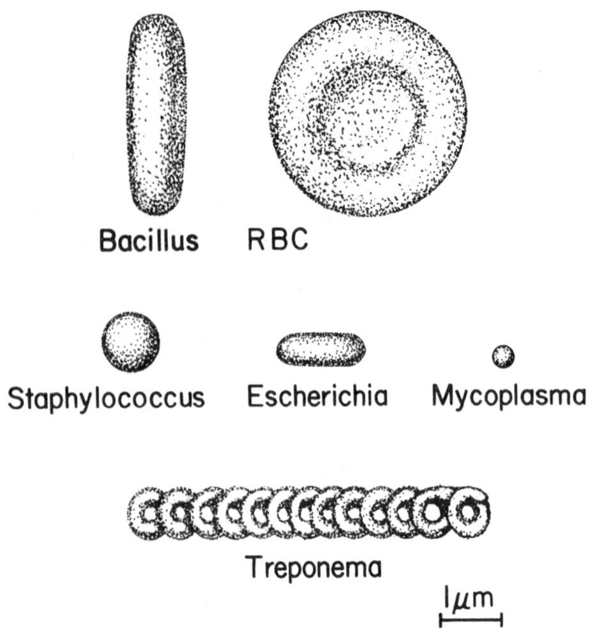

FIGURE 6. *Schematic drawings of common bacteria of different sizes and shapes are illustrated here with a human red blood cell to show their relative size.*

Although all the cells of an individual animal originate from a single fertilized egg cell, they show many forms in the adult, meaning that they have *differentiated* to perform certain specialized functions within the organism. When a single bacterial cell divides many times to form a colony of cells, the cells in that colony may be associated in characteristic spatial arrangements, as described previously, but all individuals are usually identical in form. Differentiation of cells as found in higher organisms does not occur in bacteria; but something akin to it occurs when special cells are formed with certain reproductive functions. The myxobacteria and the actinomycetes are examples of bacteria that have elaborate growth habits and produce fruiting bodies that bear spores.

Of more practical importance are the formation of *endospores* by certain gram-positive organisms. An endospore results when part of a cell, including the nucleoid and some cytoplasm, is walled off from the rest of the mother cell. A complex multilayered wall structure then forms around it (Fig. 7). The total process is called *sporulation.* The resulting endospore demonstrates no metabolic activity and is highly resistant to radiation, chemicals, dessication, and heat. This dormant stage represents a powerful mechanism for both dispersal and cell survival under adverse conditions. Since spores are relatively large refractile bodies, they may be seen with the light microscope. Their presence or absence and the position of the spore within the mother cell (median or terminal) are another set of useful diagnostic characteristics.

Under favorable circumstances the endospore is able to germinate rapidly. Metabolic activity resumes, and the spore swells and breaks open to release a cell identical to the one that originally produced the endospore. Such a cell, although again sensitive to heat and other adverse environmental conditions, can resume normal growth.

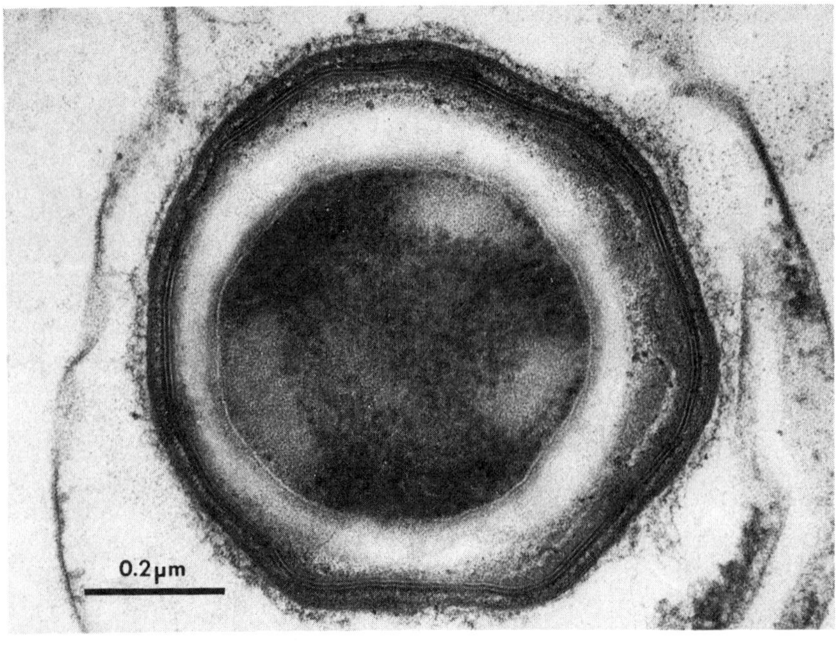

FIGURE 7. *This electron micrograph of a bacterial endospore of* Bacillus sphaericus *illustrates the complex layer structure of the spore wall. The spore is surrounded by remnants of the mother cell. (Courtesy of Drs. Stanley Holt and Donald Tipper, University of Massachusetts.)*

The organization of endospores differs substantially from that of vegetative (growing) cells. New structures are produced during sporulation as a result of the expression of specific bacterial genes. Among the new structures formed is a spore coat composed of a thin yet highly organized layer of extremely hydrophobic proteins. Also produced is the spore cortex, which lies between the spore coat and the significantly reduced cell protoplast. The cortex may be thick and is composed of a unique peptidoglycan; as such it may be analogous to the cell wall of the vegetative cell. As the spore is forming, large amounts of calcium are taken up, and corresponding quantities of dipicolinic acid are synthesized. This chemical combination appears to be responsible for the spore's heat resistance.

The production of endospores is limited to a small number of bacterial types, primarily *Bacillus, Clostridium,* and *Sporosarcina*. However, because of the extreme resistance of endospores to inactivation by physical and chemical means, endospore-forming bacteria are of major importance to human society. Most nonspore-forming bacteria are readily destroyed by boiling, but much higher temperatures are required to kill endospores (for example, autoclaving at 121° C for 20 minutes). Thus, sterilization procedures must acknowledge the high resistance of bacterial endospores even though the actual number of endospores in a given environment may be rather small.

SUMMARY

Bacteria are enormously diverse in their form, habitat, and metabolic patterns. The outline of bacterial cell structure presented here represents a series of rather broad generalizations. They are based, of necessity, on those organisms that have been the subject of the most thorough investigation because of their importance in medicine, industry, or laboratory research. Consequently, gaps in our knowledge and exceptions to our generalizations will occur, particularly among the less common and more poorly studied groups of bacteria.

We can draw an overall conclusion, however: Although bacteria are generally smaller and appear simpler in their organization than the cells of animals and plants, this simplicity is only relative. Bacteria perform a variety of cellular functions common to all living systems and other functions that are peculiar to the prokaryotic kingdom; their biochemical and physiologic machinery is quite complex.

CLASSIFICATION AND IDENTIFICATION OF BACTERIA 2

Herman Baer, M.D.
and Charles E. Davis, M.D.

Classification is the orderly arrangement of sets of organisms into a system; nomenclature deals with the naming or labeling of these organisms; and the recognition and allocation of an unknown organism is called identification. Classification, nomenclature, and identification are the three interdependent essentials of taxonomy.

BACTERIAL NOMENCLATURE

The naming of bacteria is subject to the rules and recommendations of the Bacteriological Code; the need to comply with the Code has been responsible for many changes in bacterial nomenclature. As with the nomenclature of higher organisms, bacterial nomenclature is binomial: The designation of the genus is given first, with the first letter capitalized; the species designation, which is given a lower case first letter, follows the designation of the genus.

The shifting of an organism into a different genus will automatically change its generic name. Since there are no rules for classification, the rearrangement of organisms is more or less left to the whim of the taxonomist. Frequent changes of the generic designations of bacteria are the unfortunate consequence.

BACTERIAL CLASSIFICATION

The classification of higher organisms is designed to reflect natural relatedness and phylogenetic relationships. The classification is hierarchic, proceeding from the large taxonomic groups to the species. In bacteria, studies of phylogenetic relationships are only beginning (see further on). In general, the hierarchic approach has met with limited success in bacterial classification, and its use has been sharply reduced in the latest edition of *Bergey's Manual of Determinative Bacteriology* (1974), the standard work on bacterial classification. The species is the fundamental unit of bacterial classification, and taxonomic ranks higher than the genus have limited importance, with the exception of the family Enterobacteriaceae (see Chapter 31).

In the conventional approach to bacterial classification, exemplified in *Bergey's Manual,* the genera of medically important bacteria are grouped together according to certain easily observable and relatively constant properties.

Staining Reactions

The *gram reaction* is by far the most important of these properties. The Gram stain, which was devised by Christian Gram in 1884, divides bacteria into two fundamentally different categories, gram-positive and gram-negative. Infected material obtained directly from patients or from small portions of bacterial colonies or liquid cultures are allowed to dry in air and are heat fixed on glass slides before treatment with Gram's reagents. All conventional bacteria are stained purple by crystal violet, the primary dye of the Gram stain. After treatment with Gram's iodine (3 per cent I_2-KI at pH 8.0), which complexes with the crystal violet, only gram-positive bacteria resist decolorization with alcohol, ether, or acetone. Because they have retained the crystal violet, gram-positive bacteria do not color with safranin, the red counterstain used in the last step of the Gram stain. Gram-negative bacteria stain red from the safranin because they lose the crystal violet when treated with decolorizing agents. The gram-positive cell wall itself does not stain. Instead, it presents a permeability barrier to elution of the crystal violet–iodine complex from the interior of the cell by the decolorizing agent. Gram-negative walls contain lipopolysaccharide and thinner, less complex peptidoglycan (see Chapter 1) that do not resist penetration of the cell by the decolorizers.

Another staining reaction used in the classification of bacteria is the *Ziehl-Neelson (acid-fast) stain.* Mycobacteria (Chapter 40) are difficult to stain by ordinary techniques because of their high content of waxes and lipids. They are stained red, however, when they are flooded for five to seven minutes with carbol fuchsin that is kept steaming hot by placing the glass slide over a flame or on a hot plate during the staining. All bacteria are stained by this procedure, but only mycobacteria retain the carbol fuchsin when treated with 3 per cent HCl in alcohol (acid-fast). After decolorization, other bacteria are stained by a cold counterstain such as methylene blue to contrast with the red mycobacteria. Nocardia (Chapter 41) do not stain with the conventional Ziehl-Neelsen stain but remain red (acid-fast) if they are decolorized either by 1 per cent sulfuric acid or brief exposure to 3 per cent HCl. The "firmly bound" lipids of mycobacteria, which are resistant to extraction by organic solvents, are responsible for the acid-fastness of mycobacteria (see Chapter 40).

MORPHOLOGY

Microscopically, bacterial cells can be identified as either spherical (cocci) or rod-shaped. Cocci may be arranged singly, or in pairs, clusters, or chains. Rod-shaped bacteria may be straight and regular or they may be club-shaped, curved, or spiral-shaped, or they may have pointed ends. Microscopic examination reveals the presence of spores. With the help of special staining procedures, the presence, number, and arrangement of flagella are noted. Flagella may be distributed over the entire surface of the cell (peritrichous flagellation), or they may be limited to the poles (polar flagellation).

Metabolic Reactions

The reaction to atmospheric oxygen separates bacteria into aerobic, facultative (capable of either aerobic or anaerobic growth) and strictly anaerobic organisms. The study of nutritional requirements, or the ability to grow on certain kinds of culture media, provides useful information for classification. The mode of attack on carbohydrates, usually glucose, is of taxonomic significance: Glucose is degraded anaerobically in the absence of atmospheric oxygen by the so-called fermentative organisms. Oxidative organisms require atmospheric oxygen for the degradation of glucose. Other organisms fail to metabolize glucose at all. This fermentative (F) versus oxidative (O) metabolism of carbohydrates is determined in the so-called OF test in which part of the media is exposed to air and part is anaerobic. Other characteristics thought to be of fundamental taxonomic significance include the enzyme catalase, which decomposes H_2O_2, and oxidase, which transfers hydrogen directly from its substrate to oxygen.

The properties just described and others have

been used successfully by bacteriologists for many years to classify bacteria and have been accepted as fundamental and important; that is, they were weighted heavily in the classification. With the help of these properties, medically important bacteria can often be divided to the generic level, or placed into a group of genera. This is illustrated in the following list in an oversimplified fashion for the classification of some medically important gram-positive cocci:

Facultatively anaerobic, arranged in clusters, catalase-positive:
  fermentative: *Staphylococcus*
  oxidative: *Micrococcus*
Facultatively anaerobic, arranged in pairs or chains, catalase-negative: *Streptococcus*
Anaerobic, arranged in clusters, catalase-variable: *Peptococcus*
Arranged in pairs or chains, catalase-negative: *Peptostreptococcus*

The definition of the various species within a genus is based on a set of physiologic and biochemical characteristics, including the degradation of carbohydrates, amino acids, and a variety of other substrates. Certain characteristics are quite constant and highly discriminating — they separate otherwise very similar species. For example, among all species of staphylococci, the production of coagulase is unique to *Staphylococcus aureus* and has become part of the definition of the species.

Coagulase is an enzyme that causes clotting of citrated or oxalated plasma by reacting with the coagulase-reacting factor (prothrombin) to activate thrombin. Coagulase-positive staphylococci usually also produce clumping factor ("bound coagulase") that causes macroscopic agglutination of the staphylococci when they are mixed with plasma on a glass slide. More commonly, bacterial species are defined by means of a whole pattern of biochemical properties.

Many bacterial species have been subdivided further by a variety of techniques, including antigen analysis, biochemical reactions, and susceptibility to bacteriophages — viruses that penetrate and lyse bacteria. The resulting serotypes, biotypes, and phage types have considerable clinical and epidemiologic significance.

The fact that the conventional scheme of bacterial classification is artificial and based on arbitrarily chosen criteria has not detracted from its great practical usefulness. Although it has never been possible to define the concept of the bacterial species satisfactorily, the operational value of the concept of bacterial species to medicine and epidemiology has been immense.

In recent years, new approaches to bacterial classification have evolved in an effort to supplant the conventional taxonomy by a more rational system. These include the Adansonian method of classification, or numerical classification, and the study of DNA-relatedness.

Adansonian and Numerical Classification

The numerical classification was developed by Adanson and applied by him successfully to the taxonomy of groups of higher organisms. Instead of studying a relatively small number of properties considered to be important, a very large number of characteristics is examined, each of which is assigned equal weight. Taxonomic groups are established on the basis of the overall similarity of organisms. Similarity between two organisms can be expressed as the similarity coefficient, which is defined as the percentage of shared properties among all the properties tested. (Only those tests that give a positive result with at least one of the organisms are used for determination of the similarity coefficient.) If two organisms show a similarity coefficient of about 90, they are likely to belong to the same species.

The massive amounts of data generated by the study of a large number of properties of a great many strains is best handled by a computer. The greatest value of this so-called numerical taxonomy or numerical classification system lies in the re-evaluation of the conventional classification of groups of bacteria. Numerical classification has confirmed the appropriateness of the classification of many organisms and has suggested the reclassification of others.

Classification by DNA-relatedness

The emerging field of genetic or molecular taxonomy is based on the analysis of bacterial DNA (Brenner and Falkow, 1971). In contrast to conventional taxonomy, this method can be expected to discover phylogenetic relationships among organisms and therefore provide a truly natural classification.

The degree to which organisms are related can be evaluated by the number of genes they have in common. Organisms of the same species would be expected to share most of their genes. As a consequence, the nucleotide composition and sequence of their DNA should be very similar. Techniques for determining the nucleotide sequence of nucleic acids are available, and the complete base sequence of several small viruses is known. It seems likely that this will have a major impact on the recognition of phylogenetic relationships among viruses. However, it is not practical to determine the base sequence of the much larger bacterial DNA at the present time. Other criteria to determine the DNA-relatedness of two bacteria include the comparison of the nucleotide base composition of their DNA, the degree to which their DNA will hybridize, and the ability of their DNA to undergo recombination.

The nucleotide base composition of DNA varies widely among different groups of bacteria and is characteristic for bacteria of the same genus. Since adenine (A) pairs with thymine (T), and guanine (G) pairs with cytosine (C), the percentage of each nucleotide in double-stranded DNA can be expressed by the formula AT + GC = 100 per cent. It is customary to express the base composition of DNA in terms of its GC content, which can vary from 22 per cent to 74 per cent for bacterial DNA. Bacteria with a different GC content have few base sequences in common and therefore have different genes and cannot be closely related. Conversely, organisms with a similar GC content may or may not be closely related, since a similarity in GC content does not necessarily mean that the base sequences are similar. For example, *E. coli,* a gram-negative bacillus of the family Enterobacteriaceae, and certain species of *Bacillus,* spore-forming gram-positive bacilli, have a GC content of 50 per cent, but are clearly unrelated. Determination of the nucleotide base composition is a powerful tool in reassessing the conventional classification of bacteria, and it is widely used for that purpose. For example, the genus *Micrococcus* appeared reasonably homogeneous by the usual bacteriologic criteria; however, some micrococci were found to have a GC content of 30 to 37 per cent, which corresponds to the GC content of staphylococci, instead of 66 to 75 per cent, which is typical of the other micrococci. Hence, these micrococci have recently been reclassified as staphylococci.

Hybridization is used to determine the degree to which base sequences of the DNA of two organisms are shared, or "homologous." The procedure for the hybridization experiment is performed as follows: The DNA of one organism is made radioactive biosynthetically, extracted from the bacteria, sheared into small fragments, and mixed with a large excess of nonradioactive DNA from the other organism. The mixture is heated to achieve separation of the DNA into single strands. After cooling, the complementary strands of DNA reanneal into double-stranded DNA. If the two organisms are related, they have many base sequences in common, and radioactive fragments will reanneal with nonradioactive complementary DNA. The percentage of radioactive DNA bound to nonradioactive DNA can be used to estimate the degree of complementarity and therefore the genetic relatedness between the organisms. Hybridization has been used extensively to study DNA homology between members of the family Enterobacteriaceae and has great potential for supplying new information on the taxonomy of other groups of organisms.

Under certain conditions, DNA can be transferred from one organism to another by the mechanisms of transformation, transduction, and conjugation (see Chapter 4). The transferred (donor) DNA can recombine with the resident (host) DNA to produce a genetic change that may manifest itself as a newly acquired property — for example, antibiotic resistance. It is assumed that two organisms are closely related if their DNA can recombine (Jones and Sneath, 1970). Recombination generally involves small pieces of DNA that are equivalent to only a few genes, and many factors other than gene similarity influence recombination rates. It must also be kept in mind that transfer of extrachromosomal genetic material (plasmids) can occur between organisms that are taxonomically unrelated.

Transformation and recombination have been used successfully for taxonomic studies of *Neisseria, Moraxella,* and related organisms, but recombination is less universally applicable than other methods of determining DNA-relatedness.

BACTERIAL IDENTIFICATION

Bacterial classification is a basic science; in contrast, bacterial identification is an applied discipline of microbiology with great practical significance. The identification of a bacterium suspected of being the cause of an infection is one of the major tools for the objective diagnosis of infectious diseases. The approach to the identification of bacterial isolates in the diagnostic laboratory is quite different from that taken by the bacterial taxonomist in the classification of a new or unknown organism. The taxonomist will study as many characteristics as possible; time and cost are of little concern. In contrast, in order to be clinically useful and relevant, the identification of a clinical isolate must be provided as quickly as possible. Economics and practicality dictate the use of only a minimal number of diagnostic tests (Bartlett, 1974). Thus, by necessity, identification in the clinical laboratory will always represent a compromise between accuracy on the one hand, and speed and economy on the other. Just how this compromise is made will determine the quality of a laboratory.

The practical limitation on the number of properties that can be used for identification of bacteria makes the selection of the diagnostic tests very important. The same properties of bacteria that are used in the conventional classification are employed in the identification of clinical isolates. For this reason, identification is sometimes referred to as "classification in reverse."

A hierarchic set of tables permits the identification process to proceed in a rational, stepwise fashion. A logical, practical example of a hierarchic set of diagnostic tables for medically important bacteria is presented by Cowan and Steel (1974). In the first stage of identification, the unknown organism is assigned either to a single genus or to a group of genera with the help of a small number of characteristics. The groups of organisms established by the first set of tests are differentiated to the species level by one or two additional sets of biochemical or serologic tests. The obvious advantage of this stepwise approach to identification is that it requires fewer tests than the shotgun approach, in which a very large number of tests is performed initially.

A diagnostic key is indispensible for the objective identification of clinical isolates in the modern laboratory. Nevertheless, a key is useless without consideration of the time-honored methods of the Gram stain, colonial morphology, and growth characteristics.

Examples of diagnostic keys or flow sheets based on these stable characteristics are given in Tables 1 through 7. The first-stage tables (Tables 1 and 2) classify medically important bacteria according to their gram reactions and microscopic morphology. This type of classification system reflects the thinking of the experienced microbiologist when he analyzes Gram stains of primary material from patients and is often the only guide to treatment during the first 24 hours. The second-stage tables (Table 3 through 7) illustrate the identification of medically important bacteria according to growth characteristics, ability to grow in the presence of oxygen, colonial morphology, pigment production, motility, and the presence of certain stable enzymes such as catalase, oxidase, or coagulase. These simple observations permit the rapid determination of the genera of many bacteria and both the genera and species of a suprisingly large number of bacteria. More important, attention to these major, easily established properties prevents naive errors. Laboratories that use the Gram stain and pure single colony bacteriology as the cornerstones of bacterial identification will not misidentify a gram-negative bacillus as a gram-positive bacillus or a coccus just because the biochemical reactions are similar.

Third-stage tables that involve second or third subcultures into biochemical media are not included in this chapter. When a subculture into one or two simple, defined media permits identification of a bacterium to the species level, this step is included in the second-stage tables. The appendix to this chapter provides a glossary explaining some of these reactions. Additional information about the definitive identification of medically important species is provided in Part One of this book.

Gram-positive bacteria (Table 1) are easily divided into cocci and bacilli on the basis of microscopic morphology. Careful consideration of cell shape as well as the usual search for chains or tetrads virtually always separates streptococci from staphylococci. *Clostridium* and *Bacillus* cannot be differentiated from one another by Gram stain and microscopic morphology, but are readily differentiated from the other Gram-positive bacilli. The typical club-shaped, palisading *Corynebacterium* and the extensively branched, beaded, filamentous *Nocardia* or *Actinomyces israelii* are also distinctive, but the remainder of the gram-positive rods are more difficult to separate microscopically. *Propionobacterium, Listeria,* and *Erysipelothrix* are usually indistinguishable from diptheroids (corynebacteria). In contrast, *Propionobacterium* and *Erysipelothrix* may also be filamentous and indistinguishable from other filamentous bacteria with rudimentary branching. Although extensive branching is a uniform characteristic of *A. israelii* and *Nocardia,* the other *Actinomyces* species and *Arachnia* may exhibit only rudimentary branching. *Nocardia* may be separated from *Actinomyces* and *Streptomyces* by its tendency to retain carbol fuchsin in the acid-fast stain. The rapid-growing, atypical mycobacteria, which take the gram stain well, may be distinguished from other club-shaped bacteria by the Ziehl-Neelsen stain.

The gram-negative bacteria shown in Table 2 are more difficult to separate by microscopic morphology, but can readily be separated into cocci, the pleomorphic group, and the uniformly-shaped, larger Enterobacteriaceae-Pseudomonas group. In addition, the following bacteria can frequently be identified accurately by microscopy: *Neisseria* (kidney-shaped diplococci with flattened apposing edges), the *Acinetobacter-Moraxella* group (a mixture of *Neisseria*-like and bacillary forms), *Brucella* (tiny, faintly-staining bacilli), *Bacteroides* (small, uniformly encapsulated bacilli with occasional filamentous forms), *Haemophilus* (similar to *Bacteroides* but more pleomorphic with a larger population of filamentous forms), *Fusobacterium nucleatum* (severely pointed with a rigid, crystalline appearance), *Fusobacterium necrophorum* (spheroplastic), and *Vibrio* (when markedly comma-shaped). Generally, the Enterobacteriaceae cannot be separated microscopically from one another or from the other bacteria listed in the group with uniform shape. Although most of the coccobacillary, pleomorphic bacteria may be confused with one another, the source of the clinical specimen aids the microbiologist. A spinal fluid is more likely to contain *Haemophilus influenzae;* a surgical wound, *Bacteroides;* and a cat or dog bite, *Pasteurella multocida.*

TABLE 1. Classification of Medically Important Bacteria by Staining Reactions and Microscopic Morphology*
First-stage Table

GRAM-POSITIVE

- Cocci
  - Round, in clusters & tetrads
    - *Staphylococcus*
    - *Micrococcus*
    - *Peptococcus*
  - Oval Shape, chains
    - *Streptococcus*
    - *Peptostreptococcus*
- Bacilli
  - Club-shaped palisading
    - *Corynebacterium*
    - *Listeria*
    - *Erysipelothrix*
    - *Mycobacterium* (rapid-growers)
    - *Propionobacterium* *(Eubacterium)*
  - Acid-fast
    - *Mycobacterium*
  - Filamentous
    - Branching rudimentary or absent
      - *Erysipelothrix*
      - *Lactobacillus*
      - *Eubacterium*
      - *(Actinomyces)*
      - *(Propionobacterium)*
    - Extensive branching
      - *Actinomyces*
      - *Arachnia*
      - *Nocardia*
      - *Streptomyces*
    - Partially acid-fast
      - *Nocardia*
  - Spore-bearing large, uniform
    - *Bacillus*
    - *Clostridium*

*Bacteria are listed in parenthesis when another shape or arrangement occurs more commonly.

TABLE 2. Classification of Medically Important Bacteria by Gram Stain and Microscopic Morphology
First-stage Table

GRAM-NEGATIVE

- Cocci
  - *Neisseria*
  - *Veillonella*
- Bacilli
  - Medium-sized coccobacillary
    - *Acinetobacter*
    - *Moraxella*
  - Tiny-coccobacillary
    - *Brucella*
    - *Bordetella*
    - *Bacteroides*
  - Pleomorphic, coccobacillary and filamentous
    - *Haemophilus*
    - *Bacteroides*
    - *Pasteurella*
    - *Francisella*
    - *Actinobacillus*
    - *Eikenella*
    - *Cardiobacterium*
    - *Flavobacterium*
  - Exaggerated pointed ends
    - *Fusobacterium nucleatum*
  - Coiled and spheroplastic
    - *Streptobacillus*
    - *Fusobacterium*
  - Curved, comma-shaped
    - *Vibrio*
    - *Campylobacter*
  - Shape uniformly bacillary
    - Enterobacteriaceae
    - *Pseudomonas*
    - *Aeromonas*
    - *Alcaligenes*
    - *Chromobacterium*

The second-stage table for identification of gram-positive cocci best illustrates the usefulness of microscopic and colonial morphology, selective media, and simple techniques for detection of enzymes. As shown in Table 3, virtually all gram-positive cocci can be identified from properly chosen primary or secondary cultures. Tables 4, 5, 6, and 7 — the second-stage tables for gram-positive bacilli, anaerobic gram-negative bacilli, and facultative gram-negative bacilli, respectively, demonstrate the other end of the spectrum. Identification of many of these bacteria require complete diagnostic tables listing fermentation and decarboxylation patterns, end products of glucose metabolism, nitrate reductase activity, urease activity, end products of tryptophane metabolism, and many other properties. Nevertheless, there are a large number of distinctive organisms among these groups that can be identified from primary or secondary cultures. Furthermore, this approach permits the intelligent selection of further tests for definitive identification. Bacteria that can be definitively identified from primary culture plates are indicated in Tables 4, 5, 6, and 7 either by giving the genus and species names or by listing a group of bacteria under the heading of Unique Characteristics.

The taxonomy of the anaerobic nonspore–forming gram-positive bacilli (Table 4) is currently in a most confused state. Many organisms such as the ramibacteria, the catenabacteria, and the bifidobacteria have been listed as separate genera, as *Actinomyces,* as *Lactobacillus,* and as *Eubacterium.* The eubacteria are given generic status with characteristics intermediate between the propionobacteria and the *Actinomyces.* Catenabacteria and ramibacteria have found a temporary home among the lactobacilli, and *Bifidobacterium* is regarded as a separate genus among the Actinomycetaceae family.

Taxonomists have recently divided the fusobacteria and the *Bacteroides* (Table 5) on the basis of end products of glucose metabolism. Fusobacteria produce predominantly butyric acid from glucose metabolism, whereas *Bacteroides* produce a mixed acid pattern. This rational approach to tax-

TABLE 3. Identification of Gram-Positive Cocci
Second-stage Table

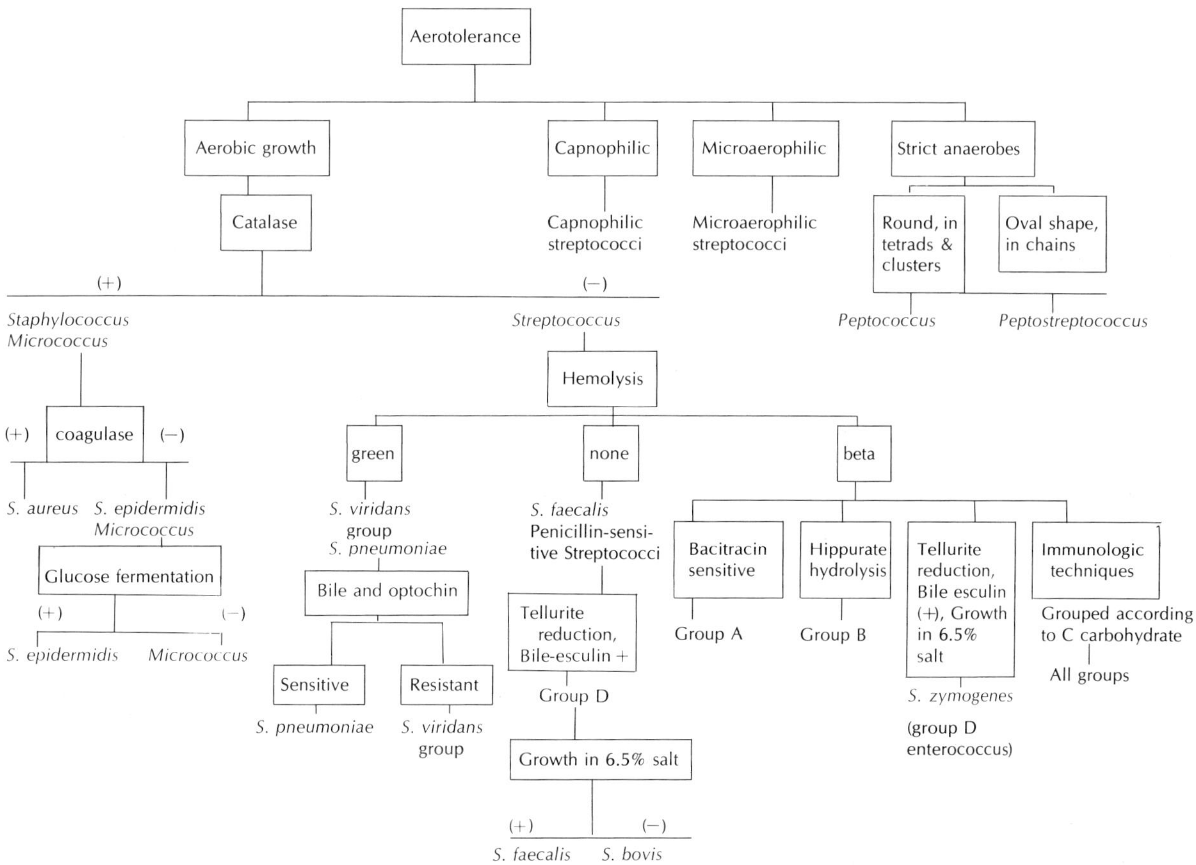

TABLE 4. Identification of Gram-positive Bacilli
Second-stage Table

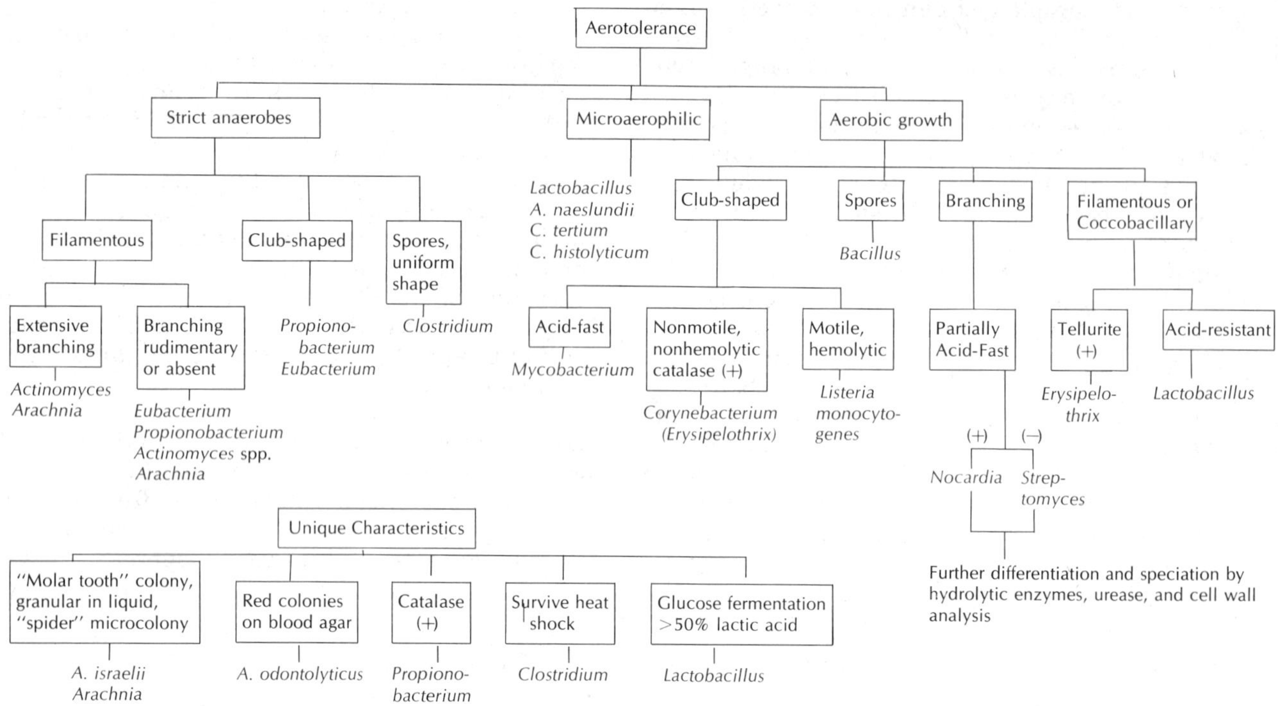

Further differentiation of anaerobic gram-positive bacilli by biochemical reactions and GLC of metabolic products

onomy with a precedent among the Enterobacteriaceae (butylene glycol versus mixed acid fermentation patterns — see Chapter 31) has not greatly disturbed the older, morphologic method of characterizing fusobacteria as the anaerobic gram-negative bacilli with pointed ends. The pleomorphic, spheroplastic bacterium that was best known to microbiologists as *Bacteroides funduliformis* has now been reclassified as *F. necrophorum* because it produces primarily butyric acid from glucose fermentation. The tendency to split *Bacteroides* into multiple species on the basis of minor differences in fermentation patterns and end products of glucose metabolism has resulted in confusion, however. For example, the bile-resistant (or enhanced) *Bacteroides* that were considered to be one species ten years ago *(B. fragilis)* were first divided into five subspecies of *B. fragilis,* and then each subspecies was granted full species status about three years ago. Despite this confusion, the majority of *Bacteroides* that are clearly capable of causing disease can be assigned to a definable species.

Tables 6 and 7 are the most extensive and complicated because there are more recognized species of facultative and aerobic gram-negative bacilli than there are in the other groups. Nevertheless, this approach to identification works well. The fastidious bacteria listed in Table 6 generally produce smaller, more delicate colonies on blood agar than the Enterobacteriaceae, *Pseudomonas,* and vibrios (Table 7). Furthermore, they are inhibited or fail to grow on selective media such as eosin–methylene blue (EMB) or MacConkey agar. Although most of these bacteria cannot be differentiated from one another by gram stain, the group can be distinguished from the hardier members of the Gram-negative bacilli. Determination of the oxidase reaction, catalase production, enhancement by CO_2, and nutritional requirements permit the identification of most of these bacteria to the generic level and some of them to the level of species. The nutritionally versatile, hardier bacteria in Table 7 can be divided into two major groups by the oxidase reaction. Enterobacteriaceae are oxidase-negative, glucose-fermenting, nitrate-reducing bacteria that have peritrichous flagella when motile. The *Yersinia* species *(pestis, pseudotuberculosis,* and *enterocolitica)* and certain *Erwinia* species fit this definition and are now classified with the Enterobacteriaceae. The vibrios and *Campy-*

lobacter (formerly *V. fetus)* can sometimes be differentiated from the other oxidase-positive, hardy gram-negative bacilli by Gram stain. Otherwise, these nutritionally versatile bacteria are differentiated from one another as indicated in Table 7 by pigment production, oxidation and fermentation patterns, and other biochemical reactions.

Bacterial Identification and Clinical Relevance

The function of the diagnostic bacteriology laboratory is to provide clinically useful information at a reasonable cost in the shortest length of time (Bartlett, 1974). It is clearly not the function of the laboratory to do sophisticated bacterial identification as an end in itself. For this reason, a question that each laboratory must constantly ask itself is how far to go with the identification of bacterial isolates obtained from clinical specimens. Ideally, each organism could be identified fully to reflect the current status of bacterial classification, but this is not always feasible. The amount of work and the cost involved could be prohibitive. Furthermore, this approach would generate an enormous amount of clinically irrelevant data. Reporting the identity of all organisms regardless of potential pathogenic significance from a site normally inhabited by indigenous flora will confuse and frustrate most physicians. It is customary in most diagnostic laboratories to establish the identity of all organisms isolated from a specimen of a normally sterile anatomic site, but to identify and report only known pathogens from sites normally inhabited by indigenous flora. This approach is universally accepted for certain specimens (e.g., feces) but remains controversial for other specimens (e.g., throat and lower respiratory tract). This decision becomes especially important for specimens from immunocompromised patients who may be infected with opportunistic members of the normal flora.

TABLE 5. Identification of Strictly Anaerobic Gram-Negative Bacteria
Second-stage Table

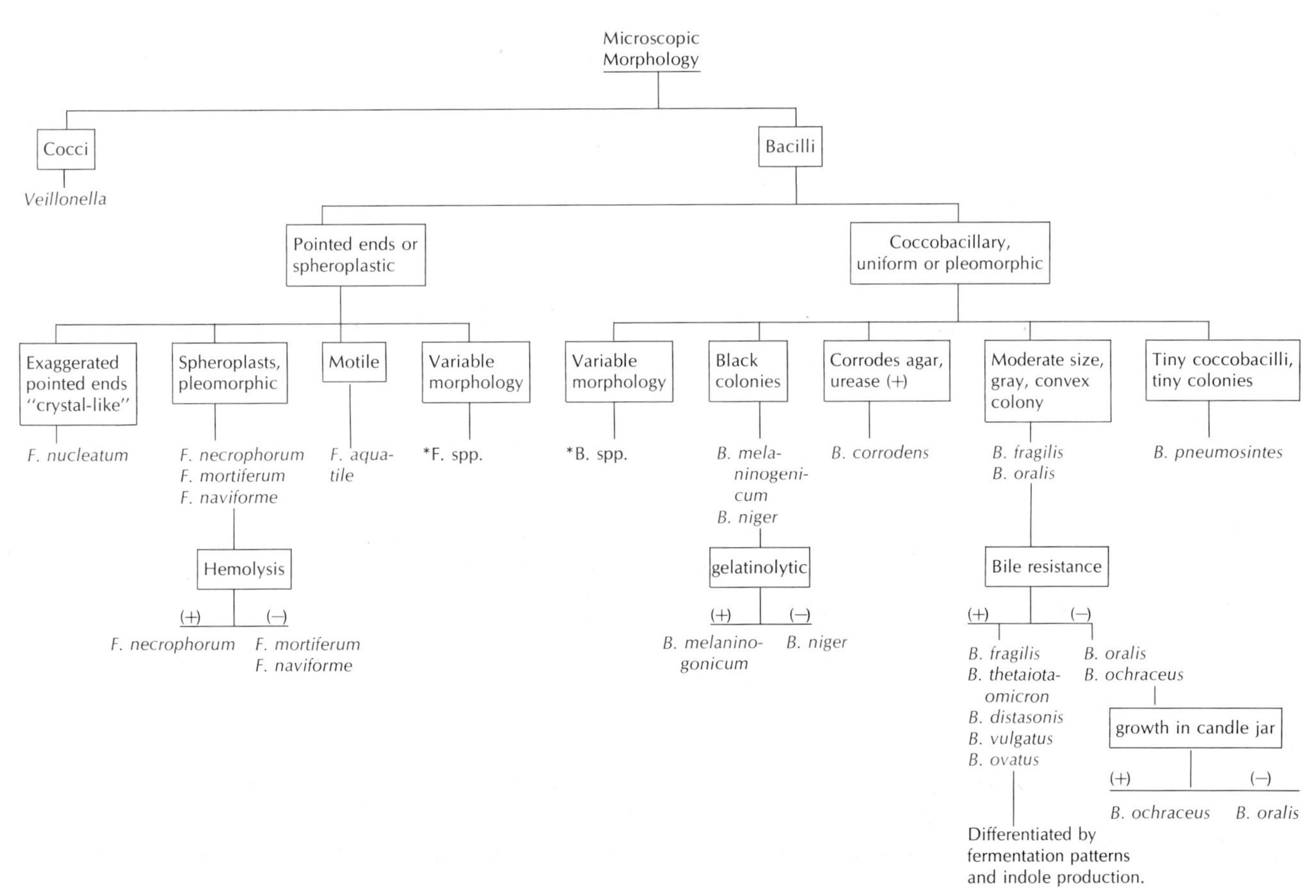

*F. species and B. species are differentiated by biochemical profiles and gas-liquid chromatography of metabolic products of spent cultures. Fusobacteria produce predominantly butyric acid from glucose fermentation. Bacteroides produce a mixed pattern of succinic, acetic, formic, and others.

TABLE 6. Identification of Facultative and Aerobic Gram-negative Bacteria Second-stage Table for Fastidious Genera*

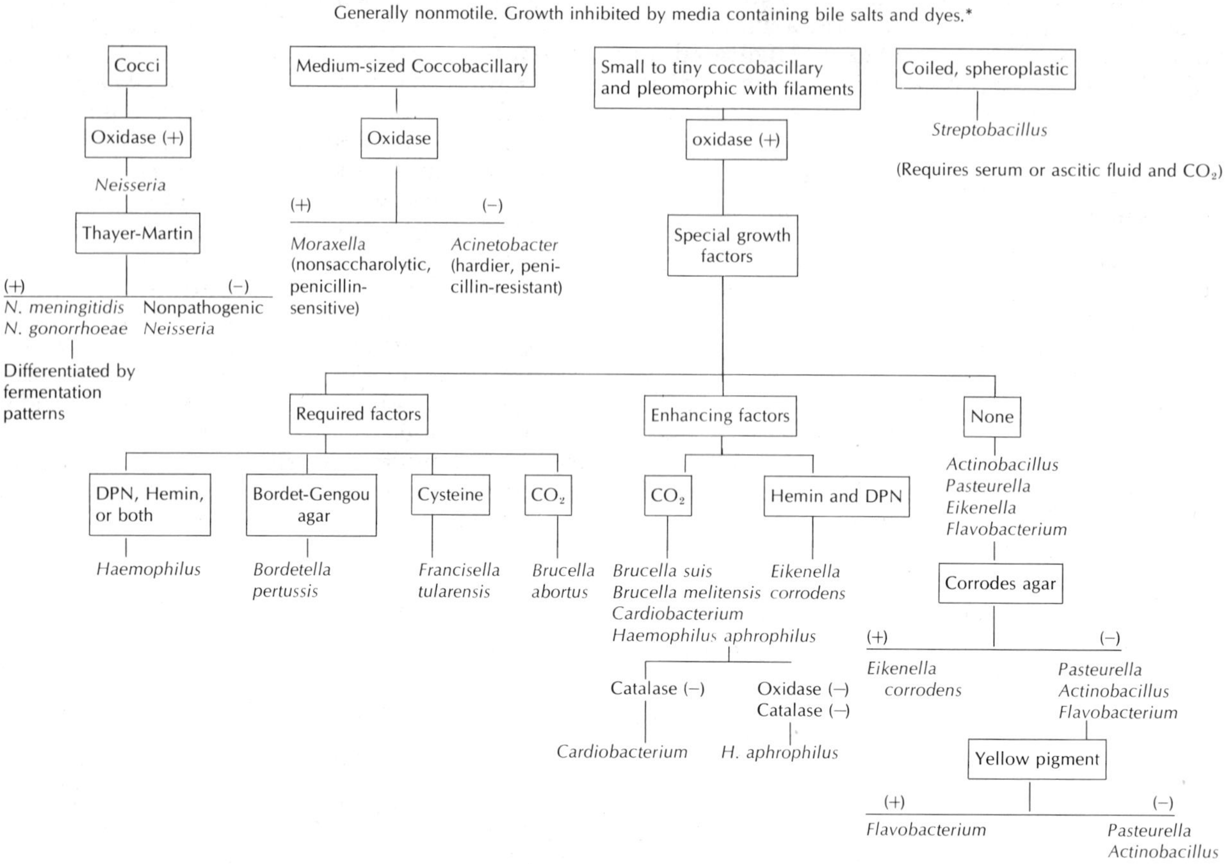

*See Table 7 for nutritionally versatile genera.

References

Bartlett, R. C.: Medical Microbiology. Quality, Cost and Clinical Relevance. New York, John Wiley & Sons, 1974.

Brenner, D. J., and Falkow, S.: Molecular relationships among members of the enterobacteriaceae. In Caspari, E. W. (ed.): Advances in Genetics, Vol. 16. New York, Academic Press, 1971, pp. 81–118.

Buchanan, R. E., and Gibbons, N. E. (eds.): Bergey's Manual of Determinative Bacteriology. 8th ed. Baltimore, Williams & Wilkins Company, 1974.

Cowan, S. T., and Steel, K. J.: Manual for the Identification of Medical Bacteria. 2nd ed. Cambridge, Cambridge University Press, 1974.

Jones, D., and Sneath, P. H. A.: Genetic transfer and bacterial taxonomy. Bacteriol Rev. 34:40–81, 1970.

Lennette, E. H., Spaulding, E. H., and Truant, J. P. (eds.): Manual of Clinical Microbiology. 2nd ed. Washington, D.C., American Society for Microbiology, 1974.

APPENDIX GLOSSARY FOR TERMS AND REACTIONS USED IN TABLES 1–7

Acid-fast: The property of mycobacteria to resist decolorization by 3 per cent HCl in 95 per cent alcohol after staining with hot carbol fuchsin (Ziehl-Neelsen stain). *Nocardia,* but not *Actinomyces species,* resist decolorization with 1 per cent H_2SO_4 in alcohol or with very brief exposures to 3 per cent HCl after treatment with carbol fuchsin. Since they are decolorized by the conventional Ziehl-Neelsen stain, *Nocardia* are referred to as *partially acid-fast.* "Firmly-bound" lipids, which are not extractable by organic solvents, are responsible for the acid-fastness of mycobacteria.

Bacitracin sensitivity: Group A β-hemolytic streptococci are differentiated from other β-hemolytic streptococci by inhibition of growth around a paper disc containing 0.04 units of bacitracin. Any zone of inhibition is considered a positive test. Many green streptococci (α-hemolytic streptococci) are sensitive to this concentration, so hemolysis must be carefully evaluated. Other β-hemolytic streptococci are inhibited only rarely by the bacitracin disc.

Bile-"positive": Capable of growth in bile. Group

D streptococci grow on agar containing 40 per cent oxgall bile. *Streptococcus faecalis* and *S. zymogenes* (the penicillin-resistant enterococci) can be differentiated from other group D streptococci (such as *S. bovis)* because the enterococci will also grow in broth containing 6.5 per cent NaCl. Growth of the intestinal *Bacteroides,* especially *B. fragilis,* is enhanced in the presence of 20 per cent oxgall bile. The growth of other *Bacteroides* is inhibited.

Bile soluble: The addition of an equal volume of 10 per cent oxgall bile or deoxycholate to turbid broth cultures or saline suspensions of *S. pneumoniae* (the pneumococcus), but not to *S. viridans* or other α-hemolytic streptococci, clears the cultures within 30 minutes because the pneumococci are lysed. Bile and optochin (see further on) activate an autolytic enzyme of pneumococci, L-alanine-muramyl amidase, which acts on the peptidoglycan and actually causes the lysis.

Bile-esculin: A medium that differentiates group D streptococci from all other streptococci because only group D grows on bile *and* hydrolyzes esculin. It does not differentiate the enterococci from the other group D streptococci. (See *Bile-"positive"*).

Bordet-Gengou agar: An enriched medium containing 20 per cent sheep blood for isolation and identification of *Bordetella pertussis,* the cause of whooping cough. Penicillin (0.25 μg/ml) is usually added to a portion of the media, and the patient's secretions are inoculated onto Bordet-Genjou agar, both with and without penicillin, to inhibit the normal bacterial flora. Plates are frequently inoculated by permitting patients to cough directly onto the medium.

C carbohydrate (group-specific C antigens): An antigenic layer of the cell wall of streptococci that lies just outside the peptidoglycan layer. β-hemolytic streptococci may be separated into groups A through O by precipitin reactions of acid extracts of the bacteria against group-specific antisera prepared in rabbits (Lancefield grouping scheme).

Capnophilic: Bacteria are called capnophilic when their growth either requires or is enhanced by CO_2.

Catalase: An enzyme that decomposes hydrogen peroxide to H_2O and free oxygen. In the laboratory, catalase is usually tested for by adding 3 per cent H_2O_2 directly to colonies on solid agar or by transferring a colony directly to a drop of 30 per cent H_2O_2. Immediate bubbling consti-

TABLE 7. Identification of Facultative and Aerobic Gram-negative Bacteria Second-stage Table for Nutritionally Versatile Genera*

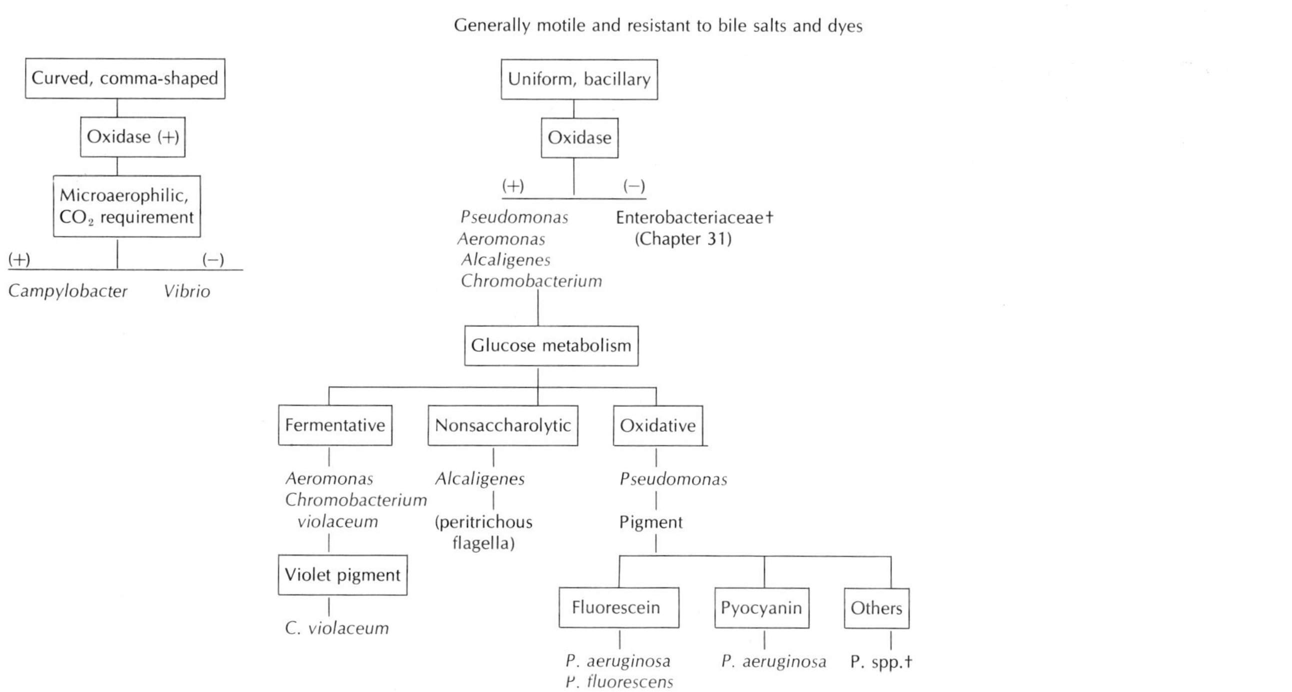

*See Table 6 for fastidious genera.

†Most organisms identified to species by fermentation patterns, oxidation patterns, production of nitrate reductase, urease, and occasionally by characteristic pigment.

tutes a positive test. Catalase should not be tested for on blood agar plates because red blood cells contain catalase.

Coagulase: An enzyme synthesized only by *Staphylococcus aureus* that clots plasma by reacting with the coagulase-reacting factor (prothrombin) to activate thrombin. Coagulase-positive staphylococci usually also produce clumping factor ("bound coagulase"), which causes macroscopic agglutination of the staphylococci when they are mixed with plasma on a glass slide.

Cysteine: An amino acid derived from cystine that enhances the growth of several bacteria. *Francisella tularensis* will grow only on media supplemented with either cysteine or cystine.

Facultative (facultative anaerobe): As used in bacteriology, facultative refers to bacteria that will grow either anerobically or in the presence of atmospheric concentrations of oxygen.

Fluorescein: A greenish yellow, water-diffusible pigment produced by *Pseudomonas aeruginosa* and *P. fluorescens.*

Gelatinolytic: Property of liquefying boiled gelatin; used in bacteriology as an identifying characteristic of some proteolytic bacteria.

Gram stain: Stain devised by Christian Gram in 1884 that divides bacteria into two fundamentally different categories, gram-positive and gram-negative. All bacteria are stained purple by crystal violet, the primary dye of the Gram stain. After treatment with 3 per cent I_2-KI, only gram-positive bacteria resist decolorization with alcohol, ether, or acetone. Gram-negative bacteria, which would otherwise be colorless, are stained by treatment with a red counterstain, usually safranin. The gram-positive cell wall acts as a permeability barrier to elution of the crystal violet–iodine complex by the decolorizing agent. Gram-negative walls, which are lipid-rich and glycopeptide-poor, do not prevent access of the decolorizing agents to the interior of the cell.

Hippurate hydrolysis: Cultures of group B streptococci, but not other streptococci, break down sodium hippurate to benzoic acid and glycine. A positive test is detected by the addition of ninhydrin, which is reduced by glycine to a purple color. Occasional strains of enterococci will hydrolyze hippurate but are easily differentiated from group B streptococci by other tests (see *Bile sculin).*

Microaerophilic: Generally used to designate organisms that grow better anaerobically than aerobically, such as many strains of streptococci and lactobacilli. Some bacteria, such as *Campylobacter* (vibrio) *fetus* actually grow better at reduced concentrations of oxygen than they do either in atmospheric concentrations or in the absence of oxygen.

Optochin: Ethylhydrocupreine hydrochloride. The growth of *S. pneumoniae,* but not other α-hemolytic streptococci, are inhibited around a paper disc containing a 1:4000 concentration of optochin. The size of the zone of inhibition is larger when the pneumococcus is incubated in an environment without added CO_2. Zones of less than 12 to 15 mm should be confirmed by bile solubility (see earlier definition).

Oxidase test: When colonies of bacteria containing cytochrome C are flooded with methylphenylenediamine compounds, they turn black within seconds because of the production of a colored compound, indophenol oxide. *Neisseria* and *Pseudomonas species* are examples of oxidase-positive bacteria. Many bacteria with functional electron transport systems contain cytochromes other than cytochrome C and are oxidase-negative in spite of their ability to use oxidative phosphorylation as a major energy source.

Pyocyanin: A bluish green, diffusible phenazine pigment produced only by *Pseudomonas aeruginosa.*

Spheroplasts (spheroplastic): Bacteria that assume bizarre, swollen shapes because of a lack of integrity of the peptidoglycan. This rigid layer of the cell wall encases the cell contents of bacteria, which are always under higher internal osmotic pressure than body fluids or conventional bacteriologic media. Spheroplasts usually form because of damage to the peptidoglycan by cell wall–active antibiotics such as penicillin, but occur spontaneously in *Streptobacillus moniliformis* and *Fusobacterium necrophorum.*

Tellurite (sodium tellurite): Diphtheria bacilli, group D streptococci, *Erysipelothrix,* and a few other bacteria can reduce tellurite and form black colonies in media containing this compound.

Thayer-Martin media: Chocolate agar (laked blood agar or blood agar with enrichments) with the antibiotics colistin, vancomycin, and nystatin added to inhibit the normal bacterial flora of the mouth and vagina. Pathogenic *Neisseria* (the gonococcus and meningococcus), but not nonpathogenic *Neisseria,* grow on Thayer-Martin agar.

Ziehl-Neelsen stain: See *Acid-fast.*

BACTERIAL PHYSIOLOGY 3

T. Ramakrishnan, M.Sc., Ph.D.
and G. Ramananda Rao, M. Pharm., Ph.D.

The origin of all free energy reaching this planet is the nuclear fusion reactions taking place in the sun. With the exception of chemoautotrophic bacteria, all living organisms draw energy from this source — either directly or indirectly. As in all other living organisms the requirement of energy for microorganisms is fundamental and essential. This need is explained by the fact that several processes involved in the biosynthesis of proteins, nucleic acids, and polysaccharides require the input of energy (endergonic), that is, having a positive change in free energy ($\Delta F'$ value). These thermodynamically unfavorable reactions are made possible under conditions in living organisms by coupling them with energy-yielding, or exergonic, reactions (negative change in $\Delta F'$ value).

The release of energy in exergonic reactions and the manner in which energy is trapped, stored, and utilized for endergonic reactions, are some of the fascinating aspects of bacterial metabolism. The central role of adenosine triphosphate (ATP) in energy exchanges in biologic systems was postulated by Fritz Lipmann and Hermann Kalckar in 1941 (Lipmann, 1941). It consists of three molecular species, adenine, D-ribose, and a triphosphate unit (Fig. 1).

The α- and β-phosphates and the β- and γ-phosphates are connected by acid anhydride linkages called energy-rich bonds and are symbolized by a wriggle bond (~). The electrostatic repulsion between the negatively charged phosphate groups, which results from their close proximity, is reduced when ATP is hydrolyzed. The high potential energy of ATP is due to the fact that adenosine diphosphate (ADP) and P_i enjoy greater resonance stabilization than ATP. The presence of this energy in ATP makes it the most functionally important compound in biologic systems.

In addition to ATP, a variety of other energy-rich compounds occur in biologic systems. They include 1) derivatives of phosphoric acid such as triphosphates of guanosine (GTP), cytidine (CTP), and uridine (UTP); and 2) derivatives of carboxylic acids, which include acyl thioesters such as acetyl coenzyme A. In fact, some of these compounds, such as phosphoenolpyruvate and creatine phosphate, have a higher potential energy than ATP.

ATP is formed by a number of reactions that are all of the general type ADP + X ~ P → ATP + X, where X ~ P is a high-energy intermediate. ATP is formed by two mechanisms in bacteria — substrate level phosphorylation and oxidative phosphorylation.

Substrate Level Phosphorylation

In this mechanism the energy required for the conversion of ADP to ATP is supplied by high-energy metabolic intermediates such as phosphoenolpyruvate, succinyl coenzyme A, or creatine phosphate — a compound regarded as a store of high-energy phosphate.

$$\text{Phosphoenolpyruvate} + \text{ADP} \longrightarrow \text{Pyruvate} + \text{ATP}$$

$$\text{Succinyl coenzyme A} + \text{ADP} \xrightarrow{\text{GDP}} \text{Succinate} + \text{ATP} + \text{Coenzyme A}$$

$$\text{Creatine phosphate} + \text{ADP} \longrightarrow \text{Creatine} + \text{ATP}$$

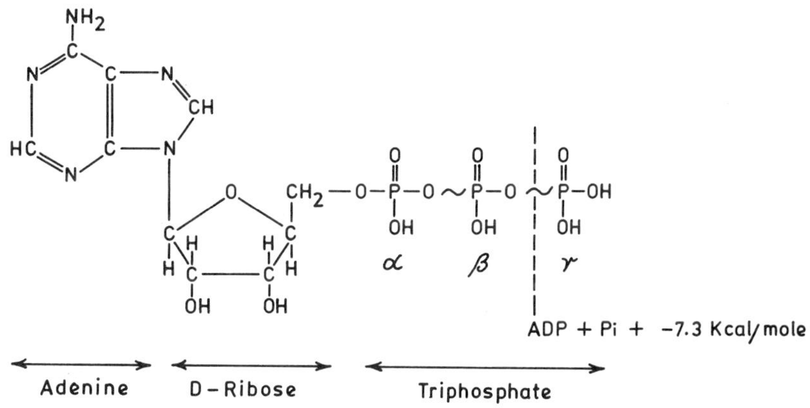

FIGURE 1. *Structure of ATP and energy release.*

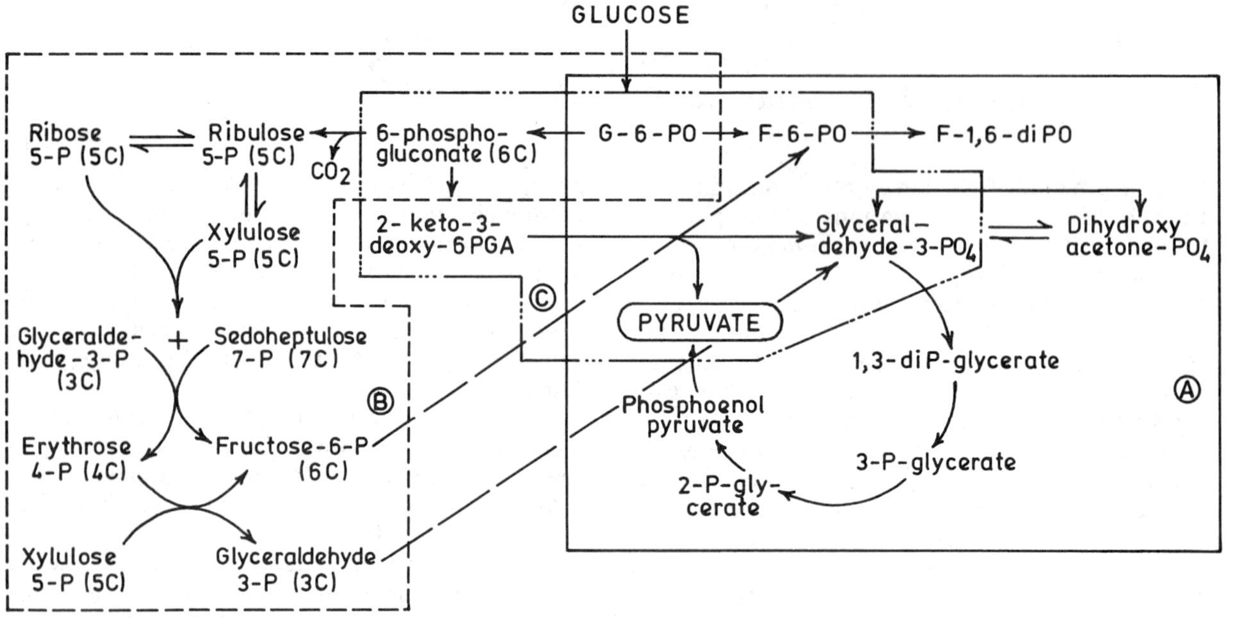

FIGURE 2. *Pathways involved in the metabolism of glucose to pyruvate A, Embden Myerhoff pathway (EMP); B, pentose-phosphate pathway; and C, Entner-Doudoroff pathway. In EMP, one molecule of fructose-1,6 diphosphate is split into two three-carbon compounds, glyceraldehyde 3-phosphate. Two molecules of ATP are needed to start the pathway, but four molecules of ATP are formed by the fermentation of glucose to pyruvic acid.*

The energy is released from the high-energy phosphate bonds in the substrate by means of substrate oxidation. This can be illustrated by reactions occurring in the glycolytic cycle (Fig. 2). The oxidation of glyceraldehyde 3-phosphate to 1,3 diphosphoglyceric acid by nicotinamide-adenine dinucleotide (NAD) produces a high-energy phosphate bond through the carboxy-linkage (as indicated by ↓) in the following manner:

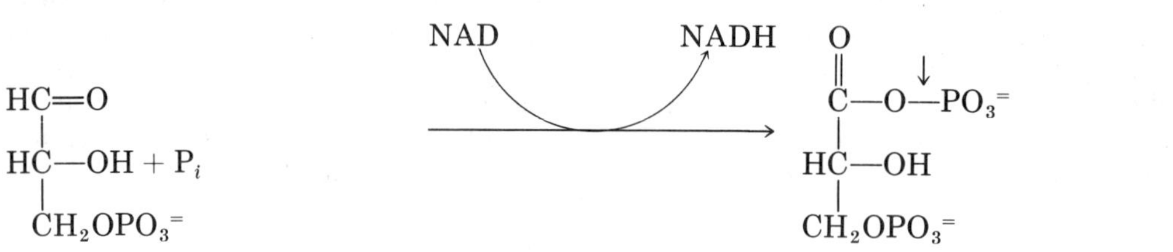

Glyceraldehyde-3-phosphate 1,3-Diphosphoglyceric acid

This high-energy phosphate bond is then transferred to ADP to form a molecule of ATP, as follows:

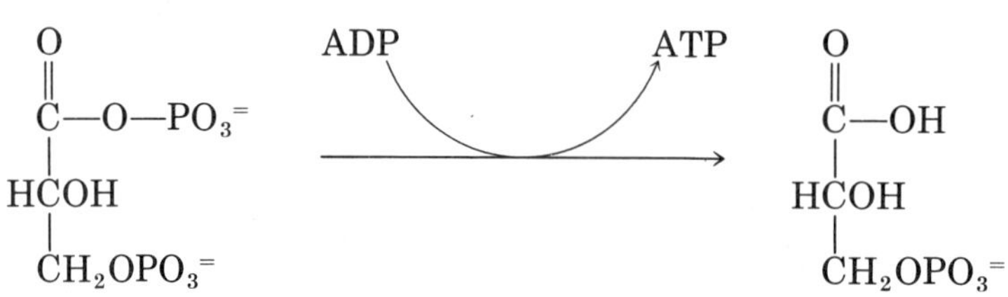

It should be noted that formation of the high-energy phosphate bond in 1,3-diphosphoglyceric acid is accompanied by the release of electrons to NAD to form NADH (reduced NAD).

Oxidative Phosphorylation

This phosphorylation is associated with electron transport. It is a coupled process wherein reduced substrates are oxidized with concomitant phosphorylation of ADP to ATP. During this process hydrogen atoms or electrons pass through oxidation-reduction reactions mediated by closely linked respiratory enzymes, especially flavoproteins and cytochromes, until the electrons reach a final acceptor, usually oxygen (see section on Aerobic Respiration). The energy present in reduced coenzymes is not released all at once but in a stepwise manner. This liberated energy is trapped in an unidentified form, $X \sim P$, which in turn, is utilized in the phosphorylation of ADP to ATP.

How Do Bacteria Derive Energy?

Bacteria are classified into autotrophs and heterotrophs, depending on the mode in which they derive energy. Autotrophs (Greek *autos* = self, *trophe* = nutrition) such as *Nitrosomonas, Nitrobacter,* and *Thiobacillus* obtain energy by oxidizing NH^+_4, HNO_2, and S, respectively, and by utilizing molecular oxygen to do so (see section on Autotrophic Nutrition). In heterotrophs (Greek *heteros* = another), preformed organic compounds such as polysaccharides, fats, and proteins are oxidized either aerobically (respiration) or anaerobically (fermentation), with release of energy. In respiration, molecular oxygen serves as the ultimate hydrogen acceptor, whereas in fermentation (which was defined by Pasteur as life without air) organic compounds serve as both electron donors and acceptors.

The brilliant contributions of Embden, Meyerhof, and Parnas led to the elucidation of the major pathway used by bacteria for the fermentation of glucose (Meyerhof, 1942). This pathway is termed the Embden-Meyerhof-Parnas (EMP) pathway, or glycolysis, and describes the conversion of glucose to pyruvic acid (Fig. 2*A*). This series of reactions also occurs in mammalian cells. Pyruvic acid is further metabolized to alcohol in yeast and to lactic acid in muscle. In the vast majority of bacteria, glucose is metabolized through the EMP pathway to pyruvate (Fig. 2*A*).

In obligate aerobic bacteria, which do not possess the EMP pathway, glucose is metabolized to pyruvate through the pentose phosphate pathway (hexose monophosphate shunt) (Fig. 2*B*). The

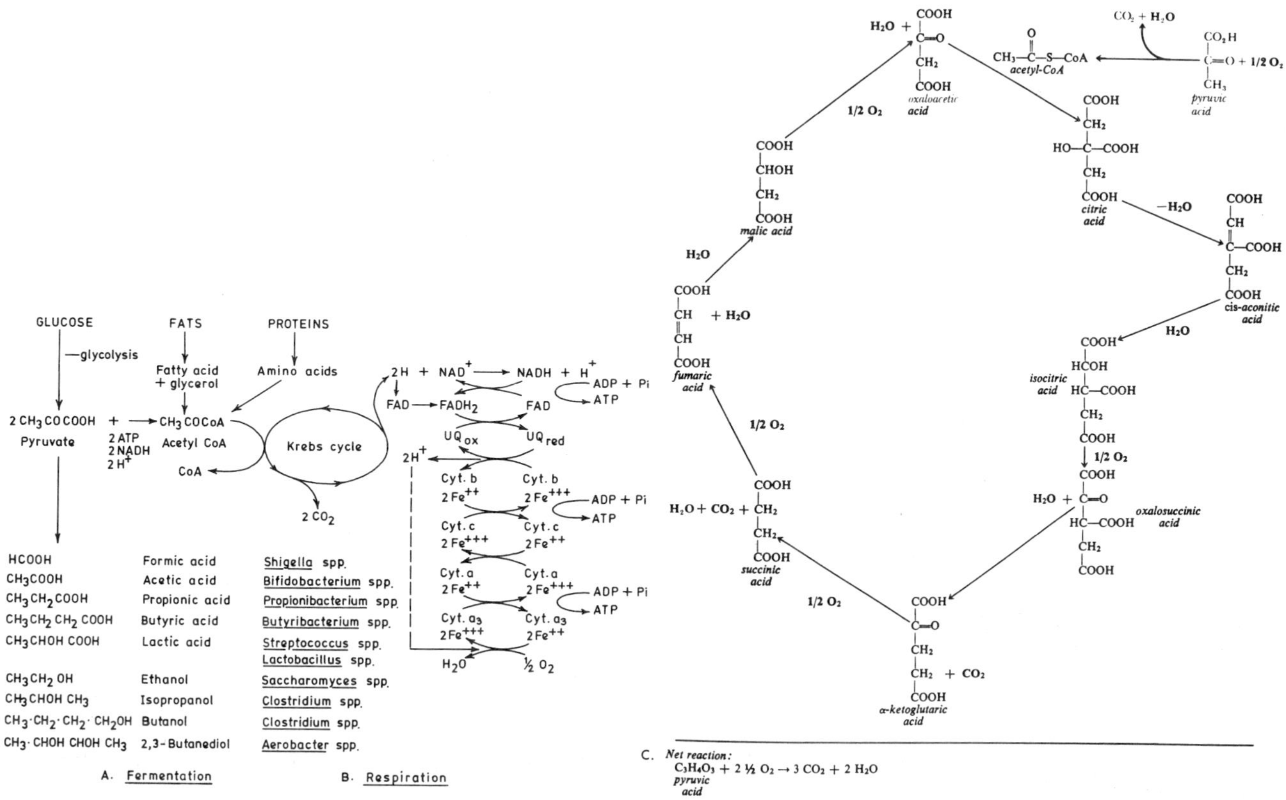

FIGURE 3. *A, Key role of pyruvate in the formation of fermentatation products and B, its oxidation through respiration, yielding ATP.*

reaction sequence of this pathway provides the ribose and erythrose-4-phosphate essential for the biosynthesis of nucleic acids and aromatic amino acids, respectively (Horecker, 1962). Another intermediate of this pathway — namely, ribulose-5-phosphate — is converted to ribulose-1,5-diP, the acceptor of CO_2 in autotrophic fixation of CO_2. The net yield of ATP is half that characteristic of the EMP pathway.

A facultative aerobic organism such as *Escherichia coli* has the ability to degrade glucose by either the EMP pathway or the pentose phosphate pathway. Whereas the EMP pathway gives more energy, the pentose phosphate pathway provides more reducing power in the form of NADPH (reduced nicotinamide adenine dinucleotide phosphate). However, a balance between these two pathways is controlled by phosphofructokinase, which phosphorylates fructose-6-P to fructose-1,6-diP. When the intracellular level of ATP increases, phosphofructokinase is strongly inhibited. As a result, phosphorylation of fructose-6-P is restricted, and the carbon flows via the pentose phosphate pathway. Increased levels of ATP similarly inhibit carbon flow through the EMP pathway under aerobic conditions and is termed the *Pasteur effect.*

In many species of aerobic bacteria glucose is metabolized by the Entner-Duodoroff pathway (Fig. 2*C*). In this pathway, 6-phosphogluconate of the pentose phosphate pathway is converted to 2-keto-3-deoxy-6-phosphogluconate (instead of ribulose 5-P as in the pentose phosphate pathway). The enzyme aldolase cleaves this into pyruvate and glyceraldehyde-3-P, intermediates in glycolysis. The latter is metabolized via the EMP pathway.

Pyruvate occupies a key position in carbohydrate fermentation. As a result of the conversion of glucose to pyruvate, NAD is reduced. Additional fermentations yield different products from pyruvate (Fig. 3*A*) and cause reoxidation of NADH, so that the oxidation-reduction balance of the cell is restored. In medical microbiology, one of the more important fermentations of pyruvate is the mixed acid fermentation by Enterobacteriaceae, such as *Escherichia coli* and *Shigella* spp. The reaction produces formic, lactic, and succinic acids, as well as a number of other products. The production of formate and acetate involves decarboxylation of pyruvate by the following series of reactions:

$$(1)\quad \underset{\text{Pyruvate}}{CH_3COCO_2H} + \underset{\text{Coenzyme A}}{CoASH} \rightarrow \underset{\text{Acetyl Coenzyme A}}{H_3\overset{\overset{O}{\|}}{C}\,SCOA} + \underset{\text{Formic acid}}{HCO_2H}$$

$$(2)\quad \underset{\text{Acetyl COA}}{H_3CO\,SCOA} + H_3PO_4 \rightarrow \underset{\text{Acetyl Phosphate}}{CH_3-\overset{\overset{O}{\|}}{C}-O-\overset{\overset{O}{\|}}{\underset{\underset{OH}{|}}{P}}-OH}$$

$$(3)\quad \text{Acetyl-phosphate} + ADP \rightarrow \underset{\text{Acetic acid}}{CH_3-\overset{\overset{O}{\|}}{C}-OH} + ATP$$

E. coli and *Salmonella spp.* have an enzyme system (formic hydrogenylase) that produces molecular hydrogen and CO_2 from formate: $HCO_2 \rightarrow H_2 + CO_2$. The production of large amounts of hydrogen ion in the mixed acid fermentation is important in the laboratory identification of *E. coli.* Similarly, the ability of *E. coli* and most *Salmonella* spp. to produce H_2 and CO_2 from formate distinguish them from *S. typhi* and *Shigella* spp., which cannot break down formate to these gases.

Aerobic Respiration

The link between glycolysis and the tricarboxylic acid (TCA) cycle is the oxidative decarboxylation of pyruvate to form acetyl coenzyme A. The acetyl unit of acetyl coenzyme A is completely oxidized to CO_2 through a cyclic reaction called the Krebs cycle (Fig. 3*B*). During the oxidative process electrons flow from reduced NAD and FAD (Flavin adenine dinucleotide) to molecular oxygen through a chain of electron carriers with coupled synthesis of ATP at two or three sites, depending upon the substrate oxidized.

Glucose oxidation results in the production of 38 moles of ATP. These are derived from the net production of 2 ATP moles during the conversion of 1 mole of glucose to 2 moles of pyruvate and 36 moles of ATP during the conversion of 2 moles of pyruvate to 6 moles each of CO_2 and H_2O. The free energy change of glucose oxidation is 686,000 cal./mole. If 7000 calories are required in the synthesis of 1 mole of ATP, the efficiency of energy trapped is $\frac{7000 \times 38}{686{,}000} \times 100 = 39$ per cent. Since free energy of ATP synthesis may approach 12,000 cal./mole, depending on the concentrations of ADP, ATP, and phosphate at the actual site on the enzyme catalyzing the reactions, the efficiency could be as high as 60 per cent.

The Glyoxylate Cycle

In addition to its role in terminal respiration, the TCA cycle is also important in providing the compounds required for biosyntheses. One of these is α-oxoglutarate, which is the precursor of glutamic acid, a key compound in protein synthesis. During biosynthesis α-oxoglutarate and other important intermediates, such as 4-carbon dicarboxylic acids, are continually removed from the TCA cycle and must be resynthesized from acetate. For this resynthesis of TCA intermediates, bacteria use a modified TCA cycle in which the reaction sequence between isocitric acid and malic acid (Fig. 3*C*) is changed through the glyoxalate bypass, which short-circuits the reactions that lead to the evolution of carbon dioxide. The bypass excludes the following reactions:

$$\text{Isocitrate} \rightarrow \alpha\text{-oxoglutarate} \rightarrow \text{succinate} + CO_2$$

This bypass consists of two enzymatic reactions:

$$\text{1) Isocitrate} \xrightleftharpoons{\text{isocitrate lyase}} \text{succinate} + \text{glyoxylate}$$

$$\text{2) Acetyl COA} + \text{glyoxalate} \xrightarrow{\text{malate synthase}} \text{malate} + \text{COA-SH}$$

Thus, acetate is not used as a source of energy, but rather as a precursor of malic COOH and other 4-carbon dicarboxylic acids.

CH_2
CHOH
|
COOH

BIOSYNTHESIS

Microorganisms offer several advantages for biosynthetic studies, including rapid growth, the ability to synthesize all cell components, and the homogeneity of the cell population. Beadle and Tatum (1941) isolated mutants having a block in a metabolic pathway resulting from a lack of or defective activity of key enzymes. Such mutants (auxotrophic mutants) require for their growth the product of the blocked reaction, which the wild-type strain (prototroph) can synthesize de novo. The discovery that auxotrophic mutants accumulate previously undetectable compounds that were intermediates in the biosynthetic pathways for the formation of various macromolecules led to the elucidation of many of these pathways. These macromolecules that bacteria synthesize are classified as 1) repeating units (capsules and cell walls), 2) information molecules such as deoxyribonucleic acid (DNA) and ribonucleic acid (RNA), 3) proteins, and 4) structurally and functionally important lipids and polysaccharides.

Repeating Polymers

The synthesis of peptidoglycan and lipopolysaccharide are similar in that the structural subunits of each are assembled on a lipid carrier in the cell membrane and are then transferred to growing ends of the polymer in the cell wall. The structural units, biosynthetic precursors, and polymerization sites of these polymers, as well as the capsular polysaccharides, are shown in Table 1. The biosynthesis of the lipopolysaccharide is discussed in Chapter 6.

The peptidoglycan is the most important of these polymers because it is an indispensable structural component of all bacteria. It is necessary for the structural integrity of the cell wall, it provides the characteristic shape of the bacterial cell, and it acts as a rigid corset that prevents the rupture of the cell. The internal osmotic pressure reaches 5 to 20 atmospheres in most bacteria as a result of active transport of solutes and would cause the cell to burst in water and other hypotonic environments if it were not for the strength

TABLE 1. Capsular and Cell Wall Polysaccharides in Bacteria

1 POLYSACCHARIDES	2 REPEATING UNIT	3 LINKAGE	4 PRECURSORS	5 SITE OF POLYMERIZATION	6 ORGANISMS
Capsular					
Dextran	Glucose	α-1,6	Sugar-NDP (nucleoside diphosphates such as ADP, UDP, GDP, CDP)	Outer surface of the cell	*Leuconostoc*
Levan	Fructose	a-1,6		Syntheisized within the cell but excreted outside	*Acetobacter xylinum*
Cellulose	Glucose	β-1,4			
Glycogen	Glucose	α-1,4			*Clostridium* spp.
Poly-L-glutamic acid	Glutamic acid	γ-peptide bond	Glutamate, ATP	Cytoplasm	*Bacillus anthracis*
Pneumococcus type III polysaccharide	Glucuronic acid, glucose	β-1,4	UDP-glucuronic acid, UDP-glucose		*Streptococcus pneumoniae*
Cell wall					
Glycopeptide	GlcNAc-MurNAc-pentapeptide	β-1,4 (between sugars). Tetrapeptide side chain is attached to the -COOH of the lactic group of MurNAc	UDP-MurNAc UDP-GlcNAc Glycyl tRNA	Cell membrane murein subunit is added to the growing chains in the cell wall through the participation of bactoprenol-P-P	Gram-positive bacteria such as *Staphylococcus aureus, Micrococcus lysodeikticus* (high content)
Lipopolysaccharide	*Core:* (KDO-heptose)-glu-gal-glu-GlcNAc *Specific side chains:* (-rha-man-gal-)$_n$ abequose	Side chain sugars are attached with their reducing groups toward the core (1→ 4, 1 → 6 and so forth)	*Core:* UDP-(glu-gal-GlcNAc) *Side chain:* GDP-(gal-man)-TDP-rha	Repeating side chain subunits are constructed by attachment to bactoprenol on membrane and then transferred to the "open" end of the growing polymer of the wall	Gram-negative true bacteria. Best studied in *Salmonella* spp.
Cell wall teichoic acids	Glycerol or ribitol substituted at OH groups with sugars or amino acids	Phosphodiester	Glycerol or ribitol, D-alanine, glucose succinate, oligosaccharides	Covalently linked to peptidoglycan at C6-hydroxyl of *N*-acetylmuramic acid	Gram-positive bacteria such as staphylococci, streptococci, *Lactobacillus*, and *Bacillus*
Lipoteichoic acids	Glycerol substituents	Phosphodiester	Fatty acid substituted glycerophosphate, & glycerol teichoic acid	Membrane	All gram-positive bacteria

of the peptidoglycan. As the name implies, the peptidoglycan is composed of sugar and peptides. The basic unit of the sugar is a disaccharide containing *N*-acetylglucosamine and muramic acid:

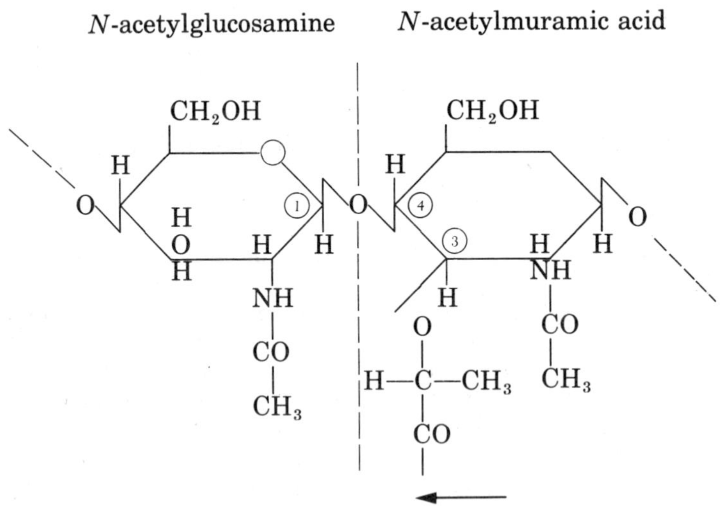

Muramic acid is a derivative of *N*-acetylglucosamine, to which lactic acid is incorporated at the 3-position. The two sugars in the disaccharide are linked at the β-1,4-position. The peptide is attached at the arrow through an amide linkage between L-alanine and the carboxyl group of muramic acid. The peptidoglycan is thus made up of polysaccharide chains linked together by peptides. The polysaccharide chains are composed of repeating units of the muramic acid *N*-acetylglucosamine disaccharide. In *E. coli* the cross-linking peptide strands consist of L-alanine, D-glutamic acid, α-ϵ = diaminopimelic acid (DAP), and finally D-alanine to form a tetrapeptide (Fig. 4). The free carboxyl groups in the D-alanine residues are also attached to the NH_2 group of the diaminopimelic acid in adjacent tetrapeptides to produce a cross-linked structure (Fig. 4). In staph-

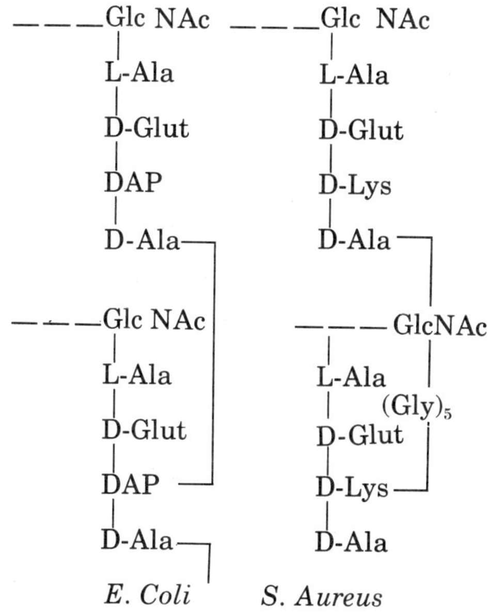

FIGURE 4. *Comparison of peptidoglycans in* S. aureus *and* E. coli. *Both bacteria contain linear sugar polymers (GlcNac + MurNAc) attached to pentapeptides through an amide linkage between N-acetylmuramic acid and L-alanine. In* E. coli *these sugar peptides are cross-linked by a peptide bond between the terminal D-alanine and diaminopimelic acid. In staphylococci, the cross-linkage between strands is achieved through a pentaglycine bridge that connects the terminal D-alanine of one peptide to the preterminal L-lysine of the neighboring peptide.*

ylococci a second chain composed of five glycine molecules — (Gly) 5 — is used to connect neighboring peptides.

Synthesis of the peptidoglycan begins in the cytoplasm with formation of *N*-acetyl glucosamine-6-phosphate (Glc Nac-6-P) from fructose-6-phosphate, glutamine, and acetyl COA.

Glc Nac-6-P is attached to a nucleoside diphosphate (as generally occurs when sugars are polymerized) to form uridine diphospho-*N*-acetylglucosamine (UDP Glc Nac). UDP-*N*-acetyl muramic acid (UDP-Mur Nac) is then produced by the following reaction involving phosphoenolpyruvate and 2H:

$$\text{UDP Glc Nac} + \underset{\text{phosphoenolpyruvate}}{CH_2 = C\text{-}CO_2H} \xrightarrow{2(H)} \text{UDP-Mur Nac}$$

The cross-linking pentapeptide is then attached, first by addition of L-alanine to the lactyl carboxyl group of UDP-Mur Nac, followed by serial addition of the remaining four amino acids. The formation of each peptide bond requires ATP and a specific enzyme.

The second step in peptidoglycan synthesis is the transfer of UDP Glc Nac and the muramyl pentapeptide to the carrier lipid, bactoprenol — a C_{55} polyisoprenoid alcohol with the formula $H_2O_3P-O(CH_2=CH\text{-}CH_2)_{11}H$. This carrier lipid is the site of disaccharide formation, which results from the addition of acetylglucosamine (from UDP Glc Nac) to the C_4 hydroxyl group of muramic acid. At this stage the β-1,4-linked disaccharide forms the following complex with the pentapeptide and carrier lipid:

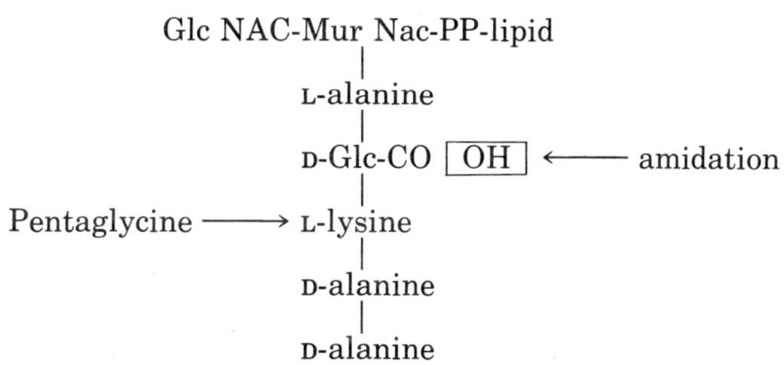

Note that the muramyl pentapeptide is connected to the lipid by a pyrophosphate bridge (PP). Further changes occur, depending on the species. In *S. aureus,* for example, the pentaglycine bridge is attached to L-lysine by sequential addition of glycine from glycyl-tRNA. There is also amidation of the γ-carboxyl group of D-glutamic acid.

After the disaccharide-pentapeptide unit is constructed, it is transferred from the carrier lipid to the peptidoglycan in the cell wall, where the peptidoglycan is elongated and cross-linked by transpeptidation between peptide bridges. This is achieved in *S. aureus* when D-alanine is split from D-alanyl-D-alanine, so that a peptide bond can form between the carboxyl group of the residual D-alanine and the terminal amino group of a neighboring pentaglycine chain (see Figure 4). This reaction completes the synthesis of the cell wall peptidoglycan.

In addition to the differences in structure of the peptidoglycan of *E. coli* and *S. aureus*, there are differences in other cell wall components. Characteristic lipopolysaccharides are found in *E. coli* and all other gram-negative bacteria, but never in gram-positive bacteria. In contrast, teichoic acids are found only in gram-positive bacteria. Teichoic acids are cell wall polymers composed of long chains of either glycerol or ribitol. These chains are linked (to each other) by phosphodiester bonds and contain substitutions of both amino acids and monosaccharides. In some bacteria they are connected to the cell wall through muramic acid-6-phosphate. Glycerol teichoic acids have the general formula

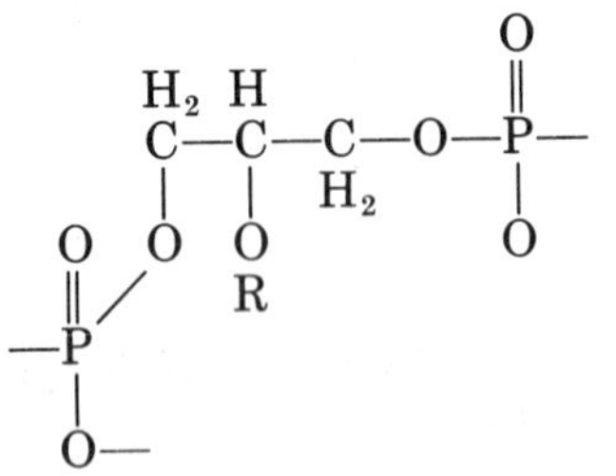

The R group is often D-alanine. The formula for ribitol teichoic acid is generally

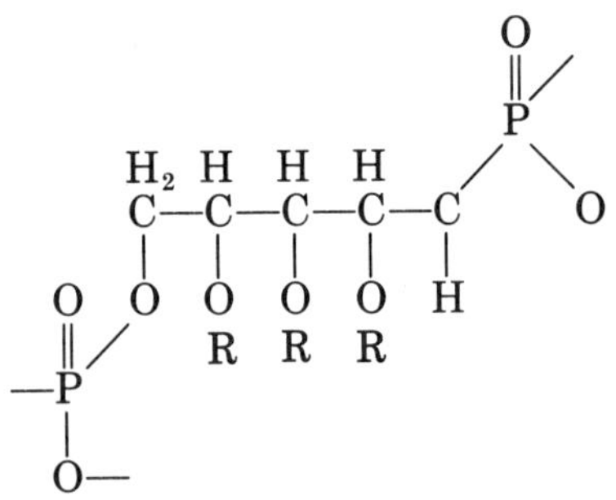

The R group substitution may also be D-alanine but is often *N*-acetylglucosamine, glucose, oligosaccharides, or succinate. Both types of teichoic acid are bound to the peptidoglycan and may be important surface antigens in streptococci, staphylococci, and other gram-positive bacteria. In addition, glycerol teichoic acid is bound to a glycolipid in the cytoplasmic membrane of all gram-positive bacteria and is known as lipoteichoic acid. In spite of their widespread occurrence, the physiologic role of lipoteichoic acids are not well understood. They may participate in the synthesis of peptidoglycan-associated teichoic acids, in binding Mg^{++} ions at the cell surface, and in regulating cell division. In the pneumococci, the lipoteichoic acid, which contains choline, inhibits autolysis of the cell wall and prevents pneumococcal cells from separating, so that they form chains. This is brought about by inhibition of the autolytic enzyme mucopeptide amidohydrolase (Höltje and Tomasz, 1975).

Information Molecules

DNA, RNA, and proteins are polymers composed of subunits (monomers). The subunits are linked to one another in a linear sequence by a phosphodiester bond (DNA and RNA) or a peptide bond (protein). The biosynthetic relationship between these macromolecules is shown in Figure 5. Three types of reactions are involved in the genetic flow from DNA→RNA→protein: replication, transcription, and translation.

In replication, free deoxyribonucleotides are assembled linearly to form an identical replica of the original DNA structure for hereditary transmission. The basis for the exact replication of a DNA strand is the base complementarity between adenine and thymine and between guanine and cytosine as proposed by the Watson and Crick model. DNA polymerase in extracts of *E. coli* has been found to catalyze the sequential addition of deoxyribonucleotides to the free 3′-OH ends using the opposite intact strand as template (Fig. 6*A*). There are three different DNA polymerases in *E. coli,* designated I, II, and III.*

DNA-dependent RNA polymerase brings about polymerization of ribonucleotide triphosphates to form a polynucleotide strand having complemen-

*The function of different polymerases is not clear, except with respect to DNA repair (repair of the patch) by polymerases and ligase. Polymerases II and III can repair more extensive genetic damage than polymerase I. Polymerases II and III restore wide areas of DNA injury (1000 to 3000 nucleotides), a process known as "long patch" repair. Polymerase I carries out only "short patch" repair limited to 10 to 30 nucleotides of an injured DNA molecule. Such injury can occur from ultraviolet irradiation that causes pyrimidine dimer formation (e.g., thymine-thymine, thymine-cytosine, or cytosine-cytosine dimers). These dimers form through the linkage of 5, 6 unsaturated bonds of adjacent pyrimidines to form a cyclobutane ring, which distorts the DNA helix and causes replication errors. Repair in the light involves separation of the dimers by photoligase in the presence of light (photoreactivation). Repair in the dark is more complex and requires excision of the dimer by endonucleases and re-establishment of the continuity of the DNA molecule.

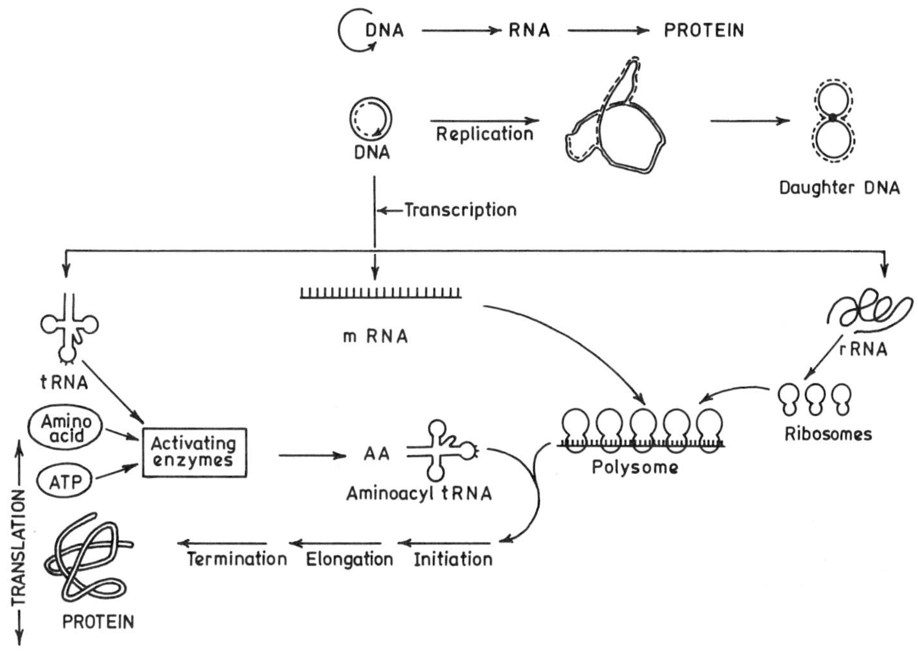

FIGURE 5. *Biosynthetic relationship among DNA, RNA, and proteins.*

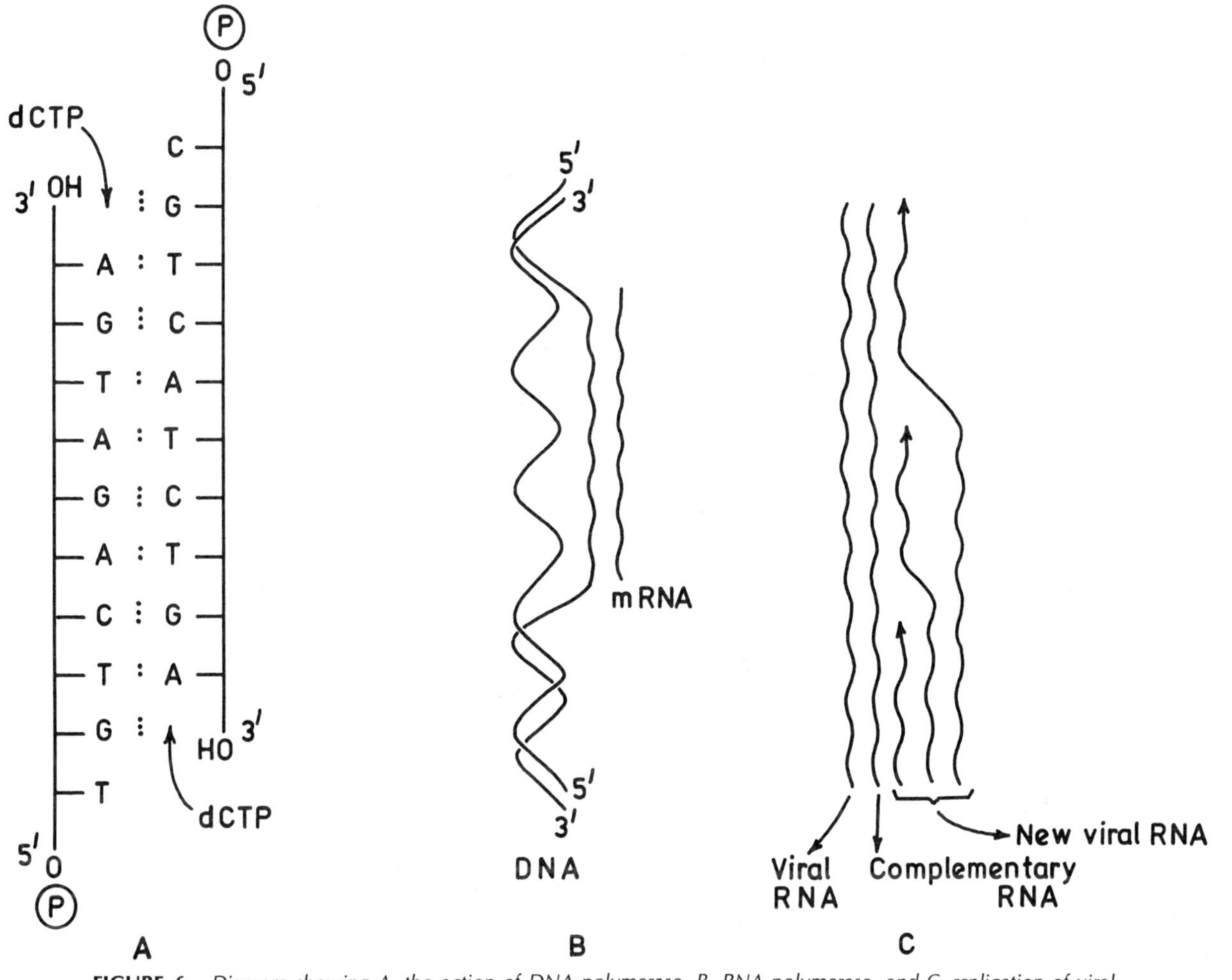

FIGURE 6. *Diagram showing A, the action of DNA polymerase, B, RNA polymerase, and C, replication of viral RNA.*

tarity with the DNA strand that served as a template. This process is called transcription (Fig. 6B). In viruses in which the genotype is RNA, its replication involves the synthesis of a complementary strand, which then serves as a template for production of new viral RNA (Fig. 6C). In RNA synthesis, adenine pairs with uracil and guanine pairs with cytosine. As shown in Figure 5, three classes of RNAs are formed in the cells by transcription: ribosomal RNA (rRNA), transfer RNA (tRNA), and messenger RNA (mRNA).

Proteins

The phenotype of microorganisms is determined and manifested both directly and indirectly by its structural and enzymic proteins. The mRNA specifies the sequence of amino acids in a protein. Ribosomes provide nonspecific surfaces on which the charged (aminoacyl) tRNA molecules bind, transfer their amino acids to the nascent polypeptide, and are released as uncharged tRNAs (Fig. 5). The sequence of three nucleotide bases (triplet code) in mRNA specifies which amino acid is to be added onto the growing chain. Since there are 64 possible codons from A, U, G, and C, some of the 20 amino acids are specified by multiple codons.

The sequence of molecular events that occur in the lengthening of a polypeptide chain on the surface of 70S ribosome are summarized in Chapter 20.

The amino acids required for protein synthesis are either provided in the medium or synthesized from precursors. All open chain amino acids are synthesized from four precursors: oxalacetate, pyruvate, α-oxoglutarate, and 3-phosphoglycerate. Among the aromatic amino acids, histidine is derived from pentose, phosphate, and others (for example, tyrosine and tryptophane) from a condensation of D-erythrose-4-phosphate and phosphoenolpyruvate. When amino acids are present in the growth medium, precursor synthesis is prevented through end-product inhibition (see section on Regulation of Growth at end of chapter).

CHEMICAL COMPOSITION AND NUTRITION

Chemical Composition of Bacteria

The main chemical elements in the bacterial cell are nitrogen, carbon, oxygen, and hydrogen. The percentage (dry matter) of nitrogen is 8 to 15 and the percentage of carbon is 45 to 55.

From the various elements and their compounds, bacteria synthesize nucleic acids, proteins, carbohydrates, lipids, glycoproteins, lipoproteins, nucleoproteins, enzymes, and vitamins.

Water. The water content in the cytoplasm of most species of bacteria varies from 75 per cent *(E. coli)* to 85 per cent *(Corynebacterium diphtheria, Mycobacterium tuberculosis, Vibrio cholerae).* Water is the main component of the cell and is found free or bound with other component substances.

Mineral Substances. Inorganic substances (phosphorus, sulphur, sodium, magnesium, potassium, calcium, iron, silicon, chlorine) and trace elements (molybdenum, cobalt, boron, manganese, zinc, copper) are also found in the bacterial cell. The total amount of mineral substances in bacteria grown on standard nutrient media varies from 2 to 14 per cent of the microbial mass.

Dry Matter. The organic part of the dry matter of bacteria consists of proteins, nucleic acids, carbohydrates, lipids, and other compounds.

Proteins. More than 50 to 80 per cent of the dry matter of the bacterial cell is made up of proteins found in the cytoplasm, nucleoid, cytoplasmic membrane, and other cell structures (Fig. 7).

In nucleoproteins, the prosthetic group is made up of nucleic acids. Similarly, in lipoproteins the prosthetic group is made up of either fats (lipids) or fat-like substances (lipoids). Lipoproteins are found within the cell as semisolid inclusions. The lipoproteins of the cytoplasmic membrane regulate the substances entering the bacterial cell.

Enzymes are proteins with active groups that catalyze biochemical reactions. The protein part of the enzyme is known as the apoenzyme and the active (prosthetic) group catalyzes the specific chemical reaction. In some cases the active, or prosthetic, groups are not bound firmly to the protein (apoenzyme) and are easily separated from it, whereas others can bind themselves to different proteins. These freely existing nonprotein catalysts involved in biochemical transformations are known as coenzymes. Another group of enzymes contain hemin compounds as the active group. Enzymes concerned with oxidation belong to this group. Enzymes act as oxidoreductases, transferases, hydrolases, lyases, isomerases, and ligases. Their functions in bacteria are summarized in Table 2.

Nucleic Acids. The amount of nucleic acids in the bacterial cell depends on the bacterial species and the nutrient medium, and it varies within 10 to 30 per cent of the dry matter. Ribonucleic acid (RNA) takes part in the synthesis of proteins, and deoxyribonucleic acid (DNA) determines hereditary properties. DNA is composed of adenine (A), guanine (G), cytosine (C), thymine (T), phosphoric acid, and ribose. Thus the difference between these two nucleic acids is that DNA contains the nitrogenous base, thymine, and deoxyribose, whereas RNA contains uracil and ribose.

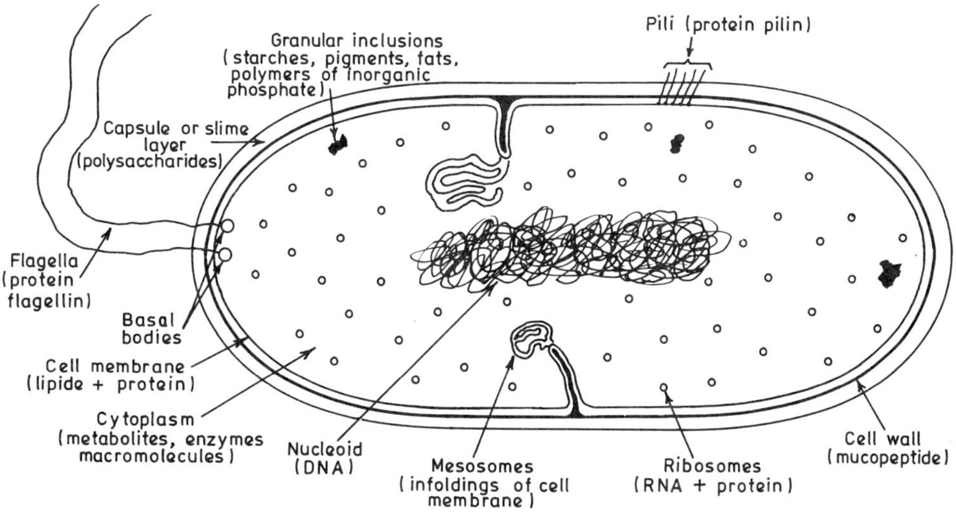

FIGURE 7. *Outline of generalized ultrastructure of a bacterial cell to indicate structural proteins (see Chapter 1).*

The structure of DNA was elucidated by Watson and Crick, who established that DNA is composed of two polynucleotide chains spirally wound and held together by hydrogen bonds between guanine and cytosine on one hand and adenine and thymine on the other. For every species of bacteria there is a definite ratio of paired bases $\frac{\text{guanine} + \text{cytosine}}{\text{adenine} + \text{thymine}}$. The guanine + cytosine (GC) content of bacteria varies from 28 per cent (some lactobacilli species) to 73 per cent (certain mycobacteria).

Carbohydrates. Carbohydrates and polyatomic alcohols compose 12 to 18 per cent of the dry matter in the bacterial cell. Most of the carbohydrate is a polysaccharide complex, sometimes bound to proteins and lipids, that is found in the cell wall and slime layer. The cytoplasm of many bacteria has a comparatively large amount of inclusions, chemically resembling glycogen or starch.

The polysaccharides of the capsules of types II, III, and VIII pneumococcus belong to the group of compounds that do not contain nitrogen. They are polymers of aldobionic acids, and during complete hydrolysis break down into glucose and glucuronic acid. The polysaccharides of other microorganisms include dextrans, levans (fructosans), and cellulose (Table 1).

Some bacteria have hexosamines that on hydrolysis break down into monosaccharides, aminosaccharides, and amino acids (types I, IV, and XIV pneumococcus, *C. diphtheriae, M. tuberculosis*). Acid hydrolysis of polysaccharides releases galactose, glucose, fructose, and other monosaccharides.

The type specificity in *Salmonella* spp depends on the polysaccharide side chains of their lipopolysaccharides. This is of great significance in laboratory identification.

Lipids. In those bacteria that do not store fat in the form of inclusions, lipids constitute 10 per cent of the dry matter (in *C. diphtheriae* it is only 5 per cent). In those bacteria that store fats as

TABLE 2. Enzymes Found in Bacteria

TYPE	REACTION CATALYZED	EXAMPLES
Oxidoreductase	Oxidation and reduction	Dehydrogenase, oxidase, peroxidase
Transferase	Transfer of a group containing C, N, P, or S from one substrate to another	Transaminase, transferase, transmethylase, transketolase
Hydrolase	Hydrolytic cleavage	Esterase, amidase, peptidase, phosphatase
Lyase	Nonhydrolytic removal of chemical group, usually from a double bond	Decarboxylases, deaminase, aldolase
Isomerase	Intramolecular rearrangements as in the interconversion of one isomer into the other	Isomerase, racemase, epimerase, mutase
Ligase	Joining together of two different molecules, or of two ends in the same molecule. The reaction catalyzed by these enzymes involves the cleavage of a high-energy phosphate bond in ATP or another energy donor.	Synthetases

special inclusions, the amount of lipids reaches 40 per cent. Bacterial lipids are made up of free fatty acids (26 to 28 per cent), neutral fats, waxes, and phospholipids.

The lipids of typhoid bacteria are almost exclusively free fatty acids (palmitic, stearic, oleic, lauric, myristic, tetracosanoic, butyric, caproic, and others). Tuberculostearic, ketostearic, palmitostearic, phthienic, and phthioic acids have been extracted from tubercle bacilli, and diphtheric acid have been extracted from diphtheria bacilli.

Tubercle bacilli contain bound lipids (12 to 15 per cent) composed of mycolic acid linked to carbohydrates. They contain a large amount of wax (wax D) that on saponification releases up to 84 per cent of high molecular weight fatty acids. These fatty acids possess a characteristically high resistance to the action of minerals, alcohols, and alkalies. A modicum of other specific components have been recognized in *M. tuberculosis:* phthiocerol dimycocerosate, and cardiolipin, which along with wax D may be important in pathogenicity.

The chemical composition of the microbial cells depends on the composition of the medium and the environmental conditions. In the case of pathogens isolated from their hosts, the extra-and intracellular milieu may determine the bacterial composition to a large extent.

Nutrition

Transport of Nutrients. The phospholipid bilayer, which constitutes the bacterial cell membrane, is a barrier to the passage of most nutrients into the cytoplasm because nutrients are largely water soluble (polar). There are, however, proteins embedded in the membrane, with an average molecular weight of 30,000, that can actively transport into the cell a variety of molecules, including amino acids, lactose, glucose, galactose, and sulfate. This active transport allows the nutrient to reach a concentration within the cell that is thousands of times greater than on its exterior. In order to meet the work requirement for such transport against a concentration gradient, energy must be supplied by metabolic processes within the cell.

Another set of proteins, known as permeases, also facilitate transfer of nutrients into the cytoplasm, but this process requires no energy because the substrate moves from a higher concentration outside the cell to a lower concentration inside. The process is sometimes called "facilitated diffusion" and differs from simple diffusion in that the permease discriminates among substrates (Cohen and Monod, 1957). In other words, a specific permease selectively binds certain substrates, but not others, on the outside of the cell and catalyzes their movement across the cell membrane into the cell. Facilitated diffusion is much less important for transport in bacteria than is active transport. Transport of glycerol into *E. coli* is an example of facilitated diffusion.

Specific permeases are also responsible for a third type of transport known as "group translocation." In contrast to facilitated diffusion, group translocation involves a chemical change in the substrate, and the reaction requires metabolic energy. Group translocation is typified by the phosphotransferase system that can transfer fructose, glucose, mannose, mannitol, and other sugars against a concentration gradient. The system is composed of two enzymes and a small heat stable carrier protein, HPr. Energy is provided by the energy-rich phosphate bond of phosphoenolpyruvate. The first enzyme, enzyme I, catalyzes the transfer of phosphate (P) to the carrier protein HPr in the following reaction:

$$\text{P-enolpyruvate} + \text{HPr} \rightarrow \text{pyruvate} + \text{P-HPr}$$

The activated carrier protein, P-HPr, then reacts with free hexose so as to carry hexose across the membrane into the cell as hexose-6-phosphate. A second enzyme, enzyme II, catalyzes phosphorylation of the sugar as follows:

$$\text{P-HPr} + \text{hexose} \rightarrow \text{hexose-6-P} + \text{HPr}$$

There is a specific enzyme II for each hexose.

Species of bacteria vary in the method of transport used to carry nutrients into a cell. In addition, simple diffusion allows passage into the cell of substances present in high concentration outside the cell. Water is the most important substance that passes across the cell membrane by passive diffusion.

Types of Nutrition. *Autotrophic,* chemosynthetic, and photosynthetic bacteria are able to produce organic substances from inorganic compounds. They do not require organic carbon compounds and synthesize the component parts of their cell by absorbing carbon dioxide, in addition to water and simple nitrogen compounds (ammonia and its salts, nitrate, and others). Nitrifying bacteria and many sulphur bacteria belong to the autotrophic microbes. They synthesize complex substances at the expense of the energy they receive from the oxidation of ammonia to nitrites (*Nitrosomonas*) and nitrates (*Nitrobacter*) and the oxidation of sulphur, sulphides, and thiosulphates to sulphuric acid and its salts (*Thiobacillus thiooxidans*).

Some species of microorganisms — anaerobic purple and green sulphur bacteria (*Thiorhodaceae, Chlorobacteriaceae*) — contain chlorophyll and use radiant energy for photosynthesis.

The autotrophic bacteria use carbon dioxide as the sole source of carbon and are unable to absorb more complex carbon compounds. For this reason, such organisms cannot be pathogenic for humans and animals.

Heterotrophic bacteria require organic carbon (carbohydrates and keto, amino, and fatty acids), inorganic substances, trace elements, and vitamins. Heterotrophic bacteria can be subdivided into 1) saprophytes and 2) parasites.

1) *Saprophytes* (Greek *sapros* = decaying, phyton = plant) live at the expense of organic substances found in the surrounding environment. These include most species of bacteria inhabiting our planet.

2) *Parasites* make up a comparatively small number of species of bacteria that, in the process of evolution, have adapted themselves to a parasitic mode of life. However, this division of heterotrophic bacteria into saphrophytes and parasites is not absolute. Certain species of bacteria pathogenic to man can exist in the environment as saprophytes, and, conversely, some saprophytes under unfavorable conditions can cause disease in humans and animals.

Nutrients Essential for Growth. The majority of bacteria develop only on complex media containing peptone (a product of enzymatic breakdown of meat and other protein substances), meat extract, and products of similar biologic origin, which contain all the nutrients in the form of high molecular weight compounds essential for growth.

NITROGEN. According to the character of nitrogen nutrition, bacteria have been subdivided into a number of groups including those that function by 1) fixing atmospheric nitrogen; 2) absorbing mineral forms of nitrogen (ammonium sulphate); 3) assimilating ammonium salts, nitrates, or nitrates in the presence of amino acids or purines; 4) growing in the presence of individual amino acids or their mixtures; and 5) growing in protein nutrient media.

CARBON. The sources of carbon for bacteria may be different carbohydrates, polyhydric alcohols, or organic acids. According to their ability to synthesize complex compounds, bacteria may be divided into four groups. 1) Those that obtain carbon from carbon dioxide and nitrogen from inorganic compounds. These organisms use radiant energy (light). Autotrophs that are capable of chemosynthesis obtain energy by the simple process of oxidation of inorganic compounds (nitrifying bacteria, sulphur bacteria, some iron bacteria). 2) Those that derive carbon and obtain energy from organic carbon compounds and nitrogen from its inorganic compounds (the majority of saprophytes). 3) Those that obtain carbon and energy from organic carbon compounds and nitrogen from amino acids (majority of commensals). 4) Those that absorb carbon and obtain energy from organic compounds, obtain nitrogen from a complex of many amino acids, and require one or more vitamins (pathogenic bacteria).

VITAMINS. Besides peptones, carbohydrates, fatty acids, and inorganic elements, bacteria require special substances — vitamins or growth factors that function as coenzymes. As coenzymes, vitamins act as acceptors or donors of chemical groups or H atoms, which are transferred to or from the substrate by various enzymes. These are summarized in Table 3.

TABLE 3. Vitamins and Their Coenzyme Function in Bacteria

VITAMIN	COENZYME FORM	TRANSFER REACTION IN WHICH COENZYME COLLABORATES
Thiamine	Thiamine pyrophosphate (TPP)	Oxidative decarboxylation of α-keto acids
Nicotinamide	Nicotinamide-adenine dinucleotide (NAD)	Hydrogen transfer in fermentation and respiration
	Nicotinamide-adenine dinucleotide phosphate (NADP)	NAD reversibly transfers electrons by coupling to different enzymes
Riboflavin	Flavin mononucleotide (FMN) Flavin adenine dinucleotide (FAD)	Hydrogen transfer in respiration. FAD is an electron acceptor for the oxidizing enzymes known as flavoproteins. FAD similar in action to NAD
Pyridoxine	Pyridoxal phosphate (PALP)	Catalyzes completely different reactions varying from amino-transfer and decarboxylation to racemization. It is the coenzyme of amino acid metabolism and the active group for amino transferases, decarboxylases, lyases, and synthetases. In all cases it acts by combination of its aldehyde group with the amino group of the substrate
Pantothenic acid	Coenzyme A (CoA) adenosine 3′5′ diphosphate + panthotheine phosphate	Transfer of acyl groups—All acyltransferases transfer acyl groups to or from CoA
Folic acid	Tetrahydrofolate (THFA)	Acts as a carrier of methyl, hydroxymethyl, formyl, or formimino groups
Biotin	Biotin	Carboxyl transfer; energy is derived from ATP
Cobalamine	B_{12} coenzyme	Methyl transfer; isomerase reactions

Some bacteria do not require a supplement of vitamins to the nutrient medium because they can synthesize these compounds. Others grow poorly on vitamin-free media, but their growth is enhanced upon the addition of vitamins. Bacteria such as pneumococcus and hemolytic streptococcus cannot be cultivated without vitamins. *Hemophilus influenzae* requires supplements of hemin and NAD for growth.

The amount of vitamins in the nutrient medium is expressed in micrograms, and they are required in concentrations varying from 0.001 to 10 mg/L. The concentrations of vitamins in bacterial cells (parts per million of dry weight) vary with different species but have the following ranges: nicotinic acid 210 to 250, riboflavin 44 to 67, thiamine 9 to 26, pyridoxine 6 or 7, pantothenic acid 90 to 140, and folic acid 3 to 15.

Intestinal microflora supply humans and animals directly with vitamins. Many bacteria participate in the vitamin metabolism of plants, in enriching food products with vitamins, and in producing vitamins.

INORGANIC SUBSTANCES. Potassium exerts a catalytic action and activates enzyme systems. Calcium participates in nitrification, in nitrogen fixation by soil microorganisms (*Azotobacter*), and in the production of gelatinase. Iron is found in the respiratory enzymes and functions as a catalyst in oxidation processes. Trace elements are incorporated into the structure of the active groups of some enzymes. Sulfur is a constituent of cysteine and methionine and of the coenzymes CoA and cocarboxylase. Phosphorus is a constituent of nucleic acids, phospholipids, ATP, and NADP. Magnesium is a cofactor for enzymes, and serves to bind enzymes to substrates. Cobalt is a constituent of vitamin B_{12}. Manganese, zinc, and copper are also essential for the activity of certain enzymes.

BACTERIAL GROWTH

The growth of bacteria represents the increase in mass of bacterial cytoplasm as a result of the synthesis of cellular material.

Bacteria reproduce by simple transverse division, vegetative reproduction that occurs in different planes and produces many different cellular arrangements (clusters, chains, pairs). The transverse division of bacteria is not only a process of cell division of one mother cell into two equal daughter cells, but also represents a continuous separation of daughter cells from the mother cell so that the former in their turn become mother cells.

The rate of cell division differs among bacteria. It depends on the species of microorganism, age of culture, nutrient medium, temperature, concentration of carbon dioxide, and other factors.

The length of the generation of *E. coli, Clostridium perfringens,* and *Streptococcus faecalis* is 20 minutes in nutrient broth medium, whereas for the cells of a mammalian tissue culture it is 24 hours. Thus, bacteria reproduce about 100 times faster than a tissue in culture. The increase in the number of cells can be expressed in the following way:

1-2-4-8-16-32	N	(number of cells)
0-1-2-3-4-5	n	(number of generations)

The total amount of bacteria (N) after n generations will be equal to 2^n per cell of seeded material. If we take the original amount of bacteria inoculated into the nutrient medium as a single individual, and the time for one division as 30 minutes, theoretically, the total amount of bacteria produced per 24 hours would be $N = 2^{48}$. Given a division every 20 minutes, in 36 hours the microbial mass will be equal to 400 tons. Thermophilic microbes divide even more rapidly.

However, in natural as well as in artificial conditions the reproduction of bacteria is on a considerably smaller scale. It is limited by a number of internal and external factors described in the section on regulation of growth. Figure 8 illustrates schematically the rate of growth expressed as the number of cells per milliliter of the medium.

There are six principal phases of reproduction that are designated in Figure 8 by Roman numerals:

I. An initial *lag phase* represents a period of physiologic adjustment, during which the cells may need to synthesize new enzymes and reestablish minimal intracellular concentrations of substrates, enzymes, and inorganic ions.

II. A phase of *increasing growth* rate reflects the randomness of adjustment of individuals in the population and the increase in the rate of growth of each individual as the rates of its separate metabolic processes become maximum.

III. A phase of *logarithmic growth* is characterized by a maximal division rate and decrease in cell size. The "growth rate" of a culture is usually expressed in terms of cell doubling time, which is equated with generation time. This rate is influenced by temperature, the nature of the carbon source, the concentration of an essential nutrient if very low, the variety of nutrients available, and, for facultative anaerobes, the O_2 tension.

IV. A phase of *declining growth* rate represents the cessation of growth as a consequence of the exhaustion of the various nutrients in the medium.

V. A maximal *stationary phase* occurs when the number of newly produced bacteria is almost

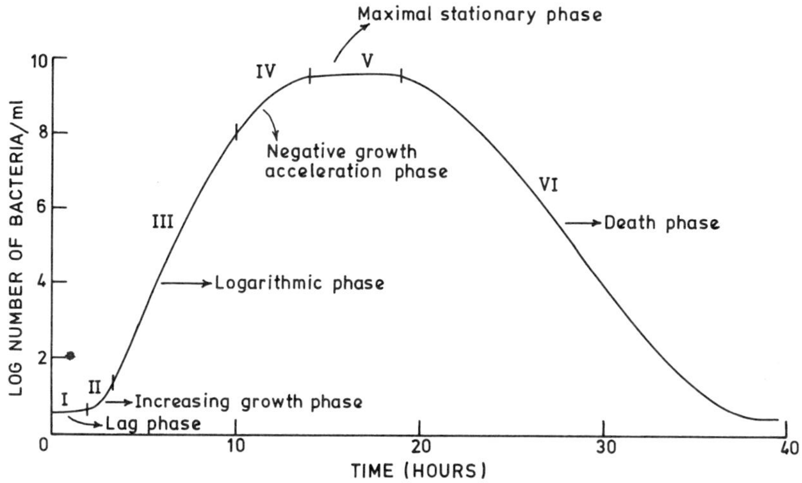

FIGURE 8. *Bacterial growth curve.*

equal to the number of organisms that die because of the lack of essential nutrients.

VI. The *death phase* is a consequence of the loss of selective permeability and of lysis — disintegration of cells.

The length of these phases is arbitrary, for it can vary depending on the species of bacteria and the conditions of cultivation. Thus, *E. coli* divide every 20 minutes, *Salmonella typhosa* divide every 25 minutes, *Streptococcus pyogenes* divide every 30 minutes, and *M. tuberculosis* divide every 18 hours under optimal growth conditions.

Regulation of Growth

The biosynthetic pathways involved in growth are controlled by *feedback regulation*. The enzyme that mediates the first reaction in the pathway is inhibited by the end-product or end-products of the pathway, a process designated *end-product inhibition*. This process prevents an accumulation of the end-product or its intermediates. It assures the balanced operation of both catabolic and biosynthetic pathways.

The mechanism of end-product inhibition is not known, but it is thought that the binding of the end-product distorts the enzyme and creates a conformational change in the site for substrate attachment, so that the enzyme no longer binds properly to its substrate. An important example of this phenomenon is the inhibition of pyrimidine biosynthesis by cytidine triphosphate (CTP). The synthesis of cytidine begins when the enzyme aspartic transcarbamylase (ATCase) catalyzes the condensation of carbamyl phosphate and aspartic acid with the production of carbamyl aspartic acid as follows:

$$\text{carbamyl phosphate} + \text{aspartic acid} \xrightarrow{\text{ATCase}} \text{carbamyl aspartic acid}$$

The end-product of this reaction is CTP, which inhibits aspartic transcarbamylase and prevents overproduction of the pyrimidine. This control of enzyme function is called *allosteric*, a term that refers to the fact that the enzyme inhibitor has a different shape, or structure, than the substrate, whose attachment to the enzyme is prevented (Monod et al., 1963). Another device for controlling metabolic pathways is known as *catabolite repression*. This phenomenon, described in Chapter 4, represses synthesis of the enzyme.

References

Beadle, G.W., and Tatum, E.L.: Genetic control of biochemical reactions in Neurospora. Proc. Natl Acad Sci USA 27:499, 1941.

Cohen, G.N., and Monod, J.: Bacterial permeases. Bacteriol Rev 21:169, 1957.

Höltje, J.-V., and Tomasz, A.: Lipoteichoic acid: a specific inhibitor of autolysin activity in *Pneumococcus*. Proc Natl Acad Sci USA, 72:1690, 1975.

Horecker, B.L.: Pentose metabolism in bacteria. New York, John Wiley & Sons, 1962.

Lipmann, F.: Metabolic generation and utilization of phosphate bond energy. Adv Enzymol 1:99, 1941.

Meyerhof, O.: Intermediate carbohydrate metabolism. *In* A Symposium on Respiratory Enzymes. Vol. 3. Madison, University of Wisconsin Press, 1942.

Monod, J., Changeux, J., and Jacob, F.: Allosteric proteins and cellular control systems. J Mol Biol 6:306, 1963.

4 BACTERIAL GENETICS

Patricia A. Hoffee, Ph.D.

At first, bacteria were not considered to have a regular genetic apparatus such as that found in eukaryotic cells. This was mainly because of the lack of a well-defined nucleus and their small size. However, during the past 35 years extensive work has demonstrated that bacteria do have a form of genetic inheritance comparable to higher organisms, with the main exception that they contain a haploid genome in almost all instances. In fact, bacteria have been used as a model system to establish most of the basic concepts of molecular genetics that we have today (Hayes, 1968; Stanier et al., 1970; Watson, 1976).

MOLECULAR ASPECTS

The Bacterial Chromosome: Its Structure and Replication

In the intact bacterial cell there is a single, continuous molecule of deoxyribonucleic acid (DNA) that makes up the chromosome. This molecule has a molecular weight of about 3×10^9 and is a closed circle. The molecular structure of this chromosome was determined by the Nobel prize–winning work of Watson and Crick (Watson, 1976; Watson and Crick, 1953). It is composed of a double helix made of two complementary strands of polynucleotides that contain purine and pyrimidine bases arranged along a backbone of alternating deoxyribose and phosphate groups (Fig. 1). The two strands are held together by hydrogen bonds that occur between a purine and a pyrimidine. This bonding is formed specifically between the pyrimidine thymine and the purine adenine (A-T base pair), or between the pyrimidine cytosine and the purine guanine (G-C base pair) (Fig. 2). Thus, the sequence of bases on one strand always has a complementary sequence of bases on the opposite strand; that is, the sequence ATTATCCG will have a complementary sequence on the opposite strand of TAATAGGC. The presence of the complementary strands allows for faithful replication of the chromosome during division. The two strands separate at a particular origin of replication, each strand acting as its own template, and a new complementary strand is made by the enzymatic polymerization of the deoxyribonucleotide subunits — dATP, dTTP, dCTP, and dGTP. Because of the specificity of the bonding, the new strands will be the exact complement of the template strand, and the genetic information will be faithfully transmitted to the daughter cells. This mode of replication is referred to as *semiconservative* and is illustrated in Figure 3. By this mechanism the daughter cells in the first generation will have one strand of the double-stranded DNA molecule that originated from the

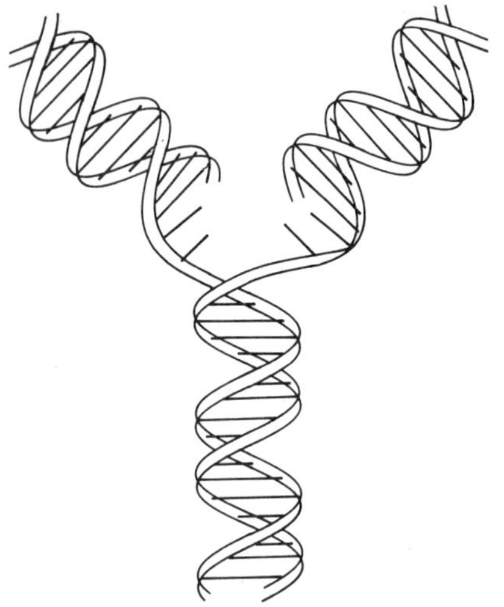

FIGURE 1. *The Watson-Crick model of DNA structure and replication. Two complementary strands are held together by the hydrogen bonding of adenine with thymine, and guanine with cytosine. The strands unwind during replication, and each strand acts as a template for the synthesis of the new daughter strands.*

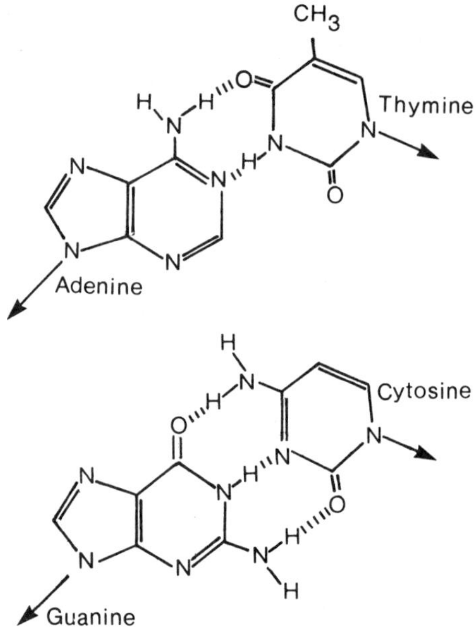

FIGURE 2. *Hydrogen bonding forms base pairs in the DNA. A-T pairs form two hydrogen bonds, whereas G-C pairs form three hydrogen bonds. The arrow indicates the position that binds to deoxyribose.*

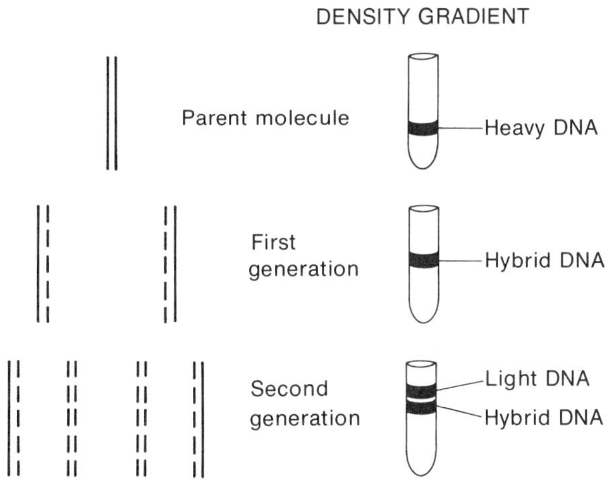

FIGURE 3. *Semiconservative replication of DNA in* E. coli *as demonstrated by the experiment of Meselson and Stahl (1958). The DNA of the parents is labeled with* ^{15}N*, so that it appears heavy in a density gradient. After one generation in* ^{14}N *medium, the DNA from the first generation progeny is hybrid in density, since it has one strand of heavy DNA and one strand of light DNA. After the second generation, DNA or light density begins to appear along with the DNA of hybrid density. This pattern of density bands is consistent with semiconservative replication.*

parent cell and one new strand. In the second generation, two of the daughter cells will have no strand from the original parent cell. The mechanism of semiconservative replication was demonstrated by the experiment of Meselson and Stahl (1958) in which the parent cell DNA was labeled with heavy nitrogen and the pattern of distribution of the heavy label was then followed through several generations (Fig. 3).

In order for each daughter cell to receive a copy of the chromosome at cell division some control must be maintained over the replication of the chromosome. It is hypothesized that the bacterial chromosome has some attachment point on the cell membrane so that replication can be synchronized with cell division. A model proposed by Jacob and Brenner (1963), referred to as the *replicon model,* proposes that there is a specific point on the chromosome that is activated by an "initiator" protein. Replication begins at this point on the chromosome and then proceeds in a bidirectional manner until the whole chromosome has been replicated. Numerous proteins are involved in the replication process, and a number of them have been identified by the isolation of mutants of *Escherichia coli,* which are temperature sensitive for DNA replication (a temperature-sensitive mutant is one that can function normally at a permissive, or lower, temperature and is unable to do so at a nonpermissive, or higher, temperature; see section on Mutation). At least five specific genes have already been identified as being involved in the process of replication, and several other ones have been postulated but have yet to be identified.

The Chromosome as a Functioning Unit

All the genetic potential of the bacterial cell is contained in the base sequence of the DNA molecule. (However, many bacteria contain extrachromosomal elements also of DNA that can confer additional genetic potential on a cell; see section on Gene Transfer in Bacteria). In order for this information to be expressed in the cellular behavior of the cell a number of complex processes must occur. The first process involves the transcription of the DNA base sequence into a ribonucleic acid molecule that will serve as a message — hence, the name messenger ribonucleic acid (mRNA). Segments of the DNA molecules are transcribed into mRNA by the action of an enzyme called RNA polymerase. This enzyme can polymerize ribonucleotides (ATP, GTP, CTP, and UTP) into a complementary strand of RNA, so that the presence of A in the DNA will result in U in the RNA, G in the DNA will give C in the RNA, T in the DNA will result in A in the RNA, and C in the DNA will give G in the RNA. Each discrete segment of DNA will specify an RNA message that will eventually be translated into a protein molecule. Such a segment or sequence of bases in the DNA is referred to as a *gene.* The mRNA molecule is translated into protein by the protein-synthesizing machinery of the cell, which is made up of activated amino acids, transfer RNA (tRNA), and ribosomes. The mRNA and the tRNA, which has a specific amino acid attached, come together on the surface of the ribosome. The tRNA molecule has a triplet of bases on one end that is complementary to a triplet of bases on the mRNA. Each tRNA molecule finds its complementary triplet on the mRNA, and the amino acid that it carries is put into a peptide linkage with the amino acid of the preceding tRNA molecule. As the ribosome moves along the mRNA, the peptide grows by the addition of each amino acid in a sequential manner until the complete mRNA has been translated into a sequence of amino acids. The sequence of bases in the DNA represents a code that was broken some 15 years ago. Each three bases in the DNA (a *codon*), and thus in the mRNA, specifies a particular amino acid. So whenever this codon appears in the DNA, a particular amino acid will be found in a particular position in the protein coded for by that segment of DNA. This genetic code is presented in Figure 4. Any possible sequence of three bases specifies an amino acid, thus AAU will specify asparagine, whereas AAG will specify lysine. The exceptions are the three codons UAG (amber), UAA (ochre), and UGA (opal). These codons do not specify an amino acid and are involved in the termination of

First	Second: U	C	A	G	Third
U	phe	ser	tyr	cys	U
	phe	ser	tyr	cys	C
	leu	ser	(ochre)	(opal)	A
	leu	ser	(amber)	trp	G
C	leu	pro	his	arg	U
	leu	pro	his	arg	C
	leu	pro	gln	arg	A
	leu	pro	gln	arg	G
A	ile	thr	asn	ser	U
	ile	thr	asn	ser	C
	ile	thr	lys	arg	A
	met, fmet	thr	lys	arg	G
G	val	ala	asp	gly	U
	val	ala	asp	gly	C
	val	ala	glu	gly	A
	val	ala	glu	gly	G

FIGURE 4. *The genetic code. The combinations of the four bases gives 64 possible triple codons, which are listed with their amino acid assignments. Three codons, UAA (ochre), UAG (amber), and UGA (opal), do not code for any amino acid and are the nonsense codons that are used for termination signals.*

peptide synthesis. They are called *nonsense codons.* In order for the proper DNA sequences to be read at the appropriate times and in the right way, various mechanisms of control of transcription have evolved in bacterial cells. Some of these mechanisms will be discussed later in this chapter.

Although the detailed work on the genetic code and its translation and transcription led to a universal hypothesis of colinearity of the gene and its protein, recent work with viruses and mammalian cells suggests that other mechanisms of gene structure exist. It is now known that in some mammalian viruses such as SV-40 there exist overlapping genes in which codons are read in different phases, so that the same sequence of bases in the DNA may code for more than one protein. Another mechanism of interest is the recent finding of genes in pieces, or gene splicing, in which discontinuous sequences of bases in the DNA can code for a single polypeptide, such as in antibody synthesis. Thus, in addition to the "one-gene, one-polypeptide" hypothesis there is now a "one-gene, two-polypeptide" hypothesis as well as a "three-gene, one-polypeptide" hypothesis.

MUTATION

Molecular Basis of Mutation

A mutation is defined as a change in the base sequence of the DNA. In many cases such a base change will result in an altered amino acid sequence of a protein that will, in turn, alter the normal functioning of that protein. In addition, the base change will be propagated when the DNA is replicated. Thus, a mutation is characterized by its effect on cell growth or metabolism and by its stability in the progeny cells. The base change can occur in the DNA by one of two mechanisms: 1) substitution of one base pair by a different base pair, or 2) an addition or deletion of a base pair or a segment of DNA during breakage of the sugar-phosphate backbone of the DNA. Single base pair changes are referred to as point mutations. Such mutations can occur spontaneously during replication or repair of the DNA or can be increased in rate of appearance by chemicals or physical agents that can interact with the DNA molecule. Chemicals that can enhance the rate of mutation are called mutagens. The first group of mutagens includes base analogues, which are incorporated into the DNA in place of the natural base. These analogues have an increased tendency to pair with the wrong base during replication, which results in the replacement of one base pair by a different base pair. Such base analogues include compounds such as bromouracil, an analogue of thymine, and 2-aminopurine, an analogue of adenine. A second group is composed of compounds that chemically alter bases in the DNA and thus alter their pairing ability. These compounds include chemicals such as nitrous acid and alkylating reagents, of which the most powerful is *N*-methyl-*N*-nitroso-*N′*-nitroguanidine (nitrosoguanidine). The third group consists of still other chemicals, such as the acridine dyes, that cause mutation by intercalating between stacked base pairs in the DNA, which results in the insertion or deletion of base pairs during replication.

Substitution of one base pair by another may lead to a change in a codon that results in the replacement of the original amino acid in the protein by a new amino acid. For example, if there is a change from a A-T pair to a G-C pair, one might have a change in the anticodon (the name given to the codon in the mRNA) from UCG to CCG and thus a change in the amino acid sequence in which serine is replaced by proline. This type of change, in which one amino acid replaces another in the protein, is referred to as a *missense mutation.* If the base change results in one of the anticodons — UAA, UGA, or UAG — being produced, no amino acid is placed in the polypeptide chain, and termination of peptide synthesis occurs. This type of change, in which no amino acid is designated by the anticodon, is termed a *nonsense mutation.*

Whenever additions or deletions of single base pairs occur, such as with acridine dyes, there is a change in the reading frame of the mRNA so that all codons from the point of insertion or deletion are misread. For example, if the base sequence *UGG-UGG-UGG-UGG* occurs in the mRNA it will be translated in the protein as the amino acid sequence trp-trp-trp-trp. When a base addition occurs, such as the insertion of C, the base sequence becomes UGG-CUG-GUG-GUG, and each triplet is then misread from the point of insertion, giving the amino acid sequence trp-leu-val-val. Such mutants are termed *frameshift mutants* and are found when additions or deletions of one or two bases occur.

Another type of mutation is found when a large segment of DNA is either deleted or inserted. Mutations caused by large deletions are referred to as *deletion mutants* and those caused by large additions are termed *insertion mutants*. Insertion mutants are a relatively new concept, although such mutations have been described in the literature for some time. They are caused by the insertions of specific base sequences of about 800 to 1400 base pairs that are called *insertion sequences* (IS). These IS elements have now been found in the chromosome of *E. coli,* in various bacterial viruses, and in plasmids, and they act as a mechanism for joining pieces of DNA that occur in bacterial cells.

Just as mutations occur in the forward direction, as described previously, similar chemical changes occur that can result in getting back the original or a pseudooriginal phenotype. These changes, referred to as *reversions,* will be dependent in type on the kind of forward mutation that initially occurred. For example, in the case of base substitution, a reversion can simply be the change of the same base pair back to the one that was originally present. If the triplet UCG mutated to CCG, it could revert back to UCG by a change in the same base pair. Such revertants are termed *true revertants* and are genetically identical to the original parent. In addition, there are revertants that result not from changing the mutated initial base pair but from changing a base pair at another site in the DNA. Such a revertant, in which a second mutation compensates for the original mutation, is referred to as a *suppressor mutation*. They can occur within the same gene as the original mutation (*intragenic*) or they can occur in a gene outside the gene in which the initial mutation occurred *(extragenic)*. Intragenic suppression can correct both point mutations and frameshift mutations. Extragenic suppression is usually the result of a mutation occurring in a gene that codes for a product involved in translation, such as a tRNA molecule. The secondary mutation corrects the original defect by recognizing this defect as normal for translation. An important class of extragenic suppressors are those that correct a nonsense mutation. Some of these suppressors have been shown to be the result of a mutant tRNA that now recognizes the nonsense codon as one of the sense codons, and allows an amino acid to be inserted into the peptide chain and its synthesis to continue. Deletion mutants are not capable of reversion, but insertion mutants can revert to the original genotype of the parent cell by a simple loss of the inserted sequence.

The Phenotypic Expression of Mutation

The preceding discussion gives one an understanding of what mutation is at the molecular level. However, such mutations, or changes in the base sequence of DNA, can be recognized only if they have an effect on the phenotype of the bacterial cell or virus that is being studied. Early in the study of bacteriology the apparent rapidity with which mutations occurred in bacteria and the selection of such mutants by the environment

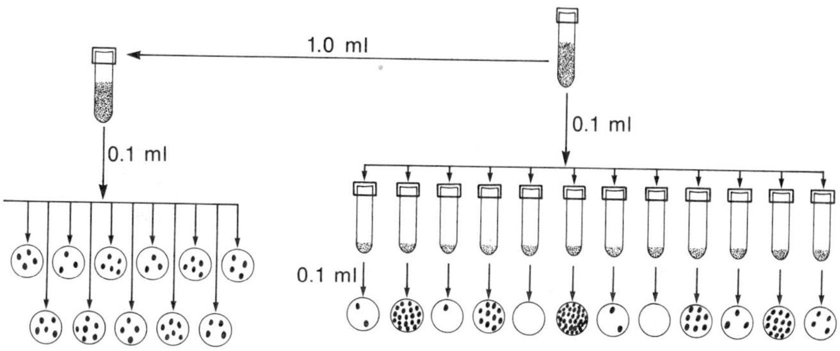

FIGURE 5. *The fluctuation test of Luria and Delbrück (1943). A series of individual culture tubes containing 1 ml of nutrient medium are inoculated at low density with equal numbers of wild-type bacteria. After incubation overnight, a sample of each tube is plated on nutrient agar containing a selective agent such as streptomycin. Resistant cells will appear as colonies on the plates. A large fluctuation is seen in the numbers of resistant cells from different tubes, supporting the theory of spontaneous mutation. The control tube on the left is sampled multiple times to determine the fluctuation caused by random error.*

made it difficult to accept that bacteria had a genetic apparatus like that found in eukaryotic systems. It was as if the bacterial cell was altered as a direct response to environmental changes. The acceptance of bacterial cells as having a genetic makeup similar to other living organisms came with the classic experiment of Luria and Delbrück in 1943. This experiment, called the *fluctuation test,* is diagrammed in Figure 5. To understand the fluctuation test one must remember that a culture of bacteria grows exponentially, with each cell giving rise to two daughter cells until the culture reaches stationary phase. If, during growth, mutations arise spontaneously in bacteria, the point of time during the growth cycle at which the mutation occurs will determine how many cells will carry that mutation at the end of the growth cycle. For example, if we start with a dilute culture of *E. coli* that is sensitive to the presence of streptomycin (str^s will not grow in the presence of streptomycin), and if the theory of spontaneous mutation is true, there is the probability a mutation that will make a cell resistant to the action of streptomycin (str^r) will occur at any point in the growth cycle. If this mutation occurs early during the growth of the culture, at the end of the growth phase there will be a large number of str^r cells that will have been derived from the original mutant cell. If the mutation occurs late during the growth of the culture, at the end of the growth phase there will be only a few cells that are str^r. The number of str^r cells present in the population at the end of the growth phase is determined by plating out an aliquot of the cells on a nutrient agar plate containing streptomycin. The str^s cells will be killed, and only the str^r cells will grow into colonies. If mutation in bacteria does not occur spontaneously, but instead is directed by the environment, the number of cells that become str^r at the end of the growth phase will be independent of time. The fluctuation test sets up a large number of independent cultures, each started from a few str^s cells, to test this hypothesis. The cultures are allowed to grow for a definitive amount of time in the absence of streptomycin, and a sample from each culture is then plated on a nutrient agar plate containing streptomycin. If mutation is spontaneous, the number of resistant cells present in the independent tubes at the time of plating will show a large fluctuation, the number present being dependent on the time at which the mutation occurred. If mutation is not a random event, all the cultures will have the same number of resistant colonies, since exposure to streptomycin occurs at the same time, and all cells have the same probability of becoming resistant. In all instances that were tested, a large fluctuation in the number of mutant cells occurred, providing evidence for the theory of spontaneous mutation in bacteria. This type of analysis is important today as one of the best methods for providing evidence for the occurrence of mutation, particularly in the newly developing field of somatic cell genetics.

Although the results of the fluctuation test convinced most workers that bacteria undergo spontaneous mutation, there were a few workers who objected to the statistical nature of the data. These doubters were finally convinced by the development of the replica plating or indirect selection technique of Lederberg and Lederberg in 1952. The technique of replica plating is diagrammed in Figure 6. Although it was initially used to provide evidence that one could select a mutant cell without ever having that cell come into contact with the selecting agent, replica plating is now a valuable technique for the scoring of numerous genetic markers in a large number of colonies with a minimum amount of work. The technique involves using a square of velveteen, which has a raised surface, and placing it over the

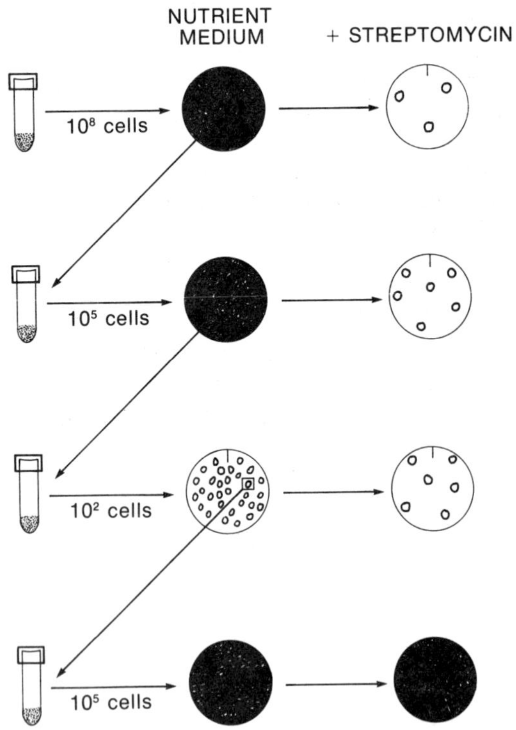

FIGURE 6. *Replica plating or indirect selection of Lederberg and Lederberg (1952). Streptomycin-sensitive cells, 10^8, are spread on a nutrient medium and the plate is incubated to allow growth that will be confluent. The cells on the plate are then replicated using a sterile velveteen cloth to a sterile plate containing streptomycin. A few colonies develop after overnight incubation. The location on the original plate of resistant cells is determined by comparing the plates. The cells on this area are removed and transferred to fresh nutrient broth and allowed to grow. The procedure is repeated until an isolated clone is picked, the progeny of which are all streptomycin resistant.*

flattened surface of a wood block with a diameter slightly less than the size of a petri dish. A plate containing colonies of bacterial cells is pressed against the velveteen, and the cells are transferred to the cloth. Sterile plates containing various selective media are then inverted onto the fabric and replicas of the initial plate are transferred to the selective media. By sequential plating, picking, and growing of the cells, as shown in Figure 6, one can eventually isolate a colony of bacteria, the progeny of which are all resistant to the selective agent, even though they never came into contact with the agent. Thus, bacteria are constantly undergoing mutation, and any change in the environment can select a cell that has a growth advantage over the parent cells.

Mutation Rate. The rate at which mutation occurs is expressed as a probability of the event occurring per cell generation, or each time one cell divides to form two new cells. To determine the mutation rate it is necessary to know within a specified time the increase that has occurred in the number of cells and in the number of mutations. The mutation rate (a) can be expressed as $a = \frac{M_t - M_0}{N_t - N_0}$, where M_0 is the number of mutations present at time zero and M_t is the number present at time t, and N_0 is the number of cells present at time zero and N_t is the number of cells present at time t. Another method for the determination of mutation rate is based on the use of the fluctuation test. If there are tubes found in the fluctuation test in which no mutations are present at the time of plating, the mutation rate can be determined from the Poisson distribution, where $\ln P_0 = -m$. Although it is difficult to get an absolute value for the mutation rate because of various problems in lag of expression of the phenotype, in general, values range from 10^{-6} to 10^{-9}; that is, there is a probability that a given type of mutant will be present to the extent of 1 mutant cell per 10^6 to 10^9 cell divisions. Even though mutant cells may be present at a low frequency, if the environment of such a culture is altered to give these cells a growth advantage, eventually this cell type will take over the population. For example, if one cell that is resistant to streptomycin is present in a population of 10^8 streptomycin-sensitive cells, this one cell will survive if the culture is exposed to the antibiotic. That one mutant cell will eventually develop into a population of streptomycin-resistant organisms.

Selection Procedures. In the study of microbial genetics, selection of specific mutant phenotypes depends on the ability to alter the environment of bacterial cells. It is from the study of mutants with altered growth properties and altered control mechanisms that one can begin to understand the normal and infectious properties of microorganisms. Thus, the basis for the study of microbial genetics is the ability to select from a large population a few cells that have the genetic characteristics one wishes to study. To this end various standard selection procedures have been developed to enable one to isolate particular types of mutant cells. Most of these studies have been done with *E. coli* or *Salmonella typhimurium*, because of the ability of these gram-negative enteric bacteria to grow on a minimal medium consisting of only salts, trace metals, and a carbon and nitrogen source. Provided with one of a variety of carbohydrates these bacteria have all the necessary enzymes and proteins to make all their necessary growth requirements, including amino acids, purines, pyrimidines, vitamins, and other metabolic intermediates. The organisms (as isolated from nature) with these genetic capabilities are referred to as *wild type* cells. Cells isolated from the wild type that differ in one or more genes are designated as *mutants*. Mutant strains with additional growth requirements are termed *auxotrophs*. The nomenclature used to designate these strains is given in Table 1. The compound usually is listed by the first three letters of the name, hence *ara* for arabinose and *his* for histidine, followed by + or −. Resistance or sensitivity to a particular environment is designated as ^{s} or ^{r}. The following procedures have been developed to isolate various classes of mutants.

RESISTANCE. This is the easiest type of mutant to isolate. The procedure involves plating a large number of sensitive organisms in the presence of the selective agent. Surviving clones are purified and are resistant to that agent. This method can be used to isolate cells resistant to bacteriophages, antibiotics, chemicals, or physical agents such as ultraviolet light or x-rays.

AUXOTROPHS. This type of mutant is difficult to isolate because the parent cells grow as well as

TABLE 1. Notation of Genetic Markers

+: Ability to utilize a carbohydrate or synthesize an intermediate, usually wild-type

−: Lack of ability to utilize a carbohydrate or synthesize an intermediate, usually mutant

Examples:

- *ara*$^+$: Ability to utilize arabinose as a carbon and energy source
- *ara*$^-$: Inability to utilize arabinose as a carbon and energy source but will utilize other sugars
- *his*$^+$: Ability to synthesize histidine
- *his*$^-$: Inability to synthesize histine, requires histidine for growth, histidine auxotroph

s: Sensitivity to agent—chemical or antibiotic

r: Resistance to agent—chemical or antibiotic

Examples:

- *str*^{s}: Sensitive and will not grow in the presence of streptomycin
- *str*^{r}: Resistant and will grow in the presence or absence of streptomycin

the mutant under all conditions. The method for the isolation of auxotrophs makes use of the fact that the parent will grow under conditions in which the mutant will not. The antibiotic penicillin, which kills only growing cells, is used. Wild-type cells are placed in a minimal glucose medium containing penicillin and are incubated for 18 to 20 hours. During this time wild-type cells grow and are killed by the action of penicillin. Any auxotrophs present are not able to grow because of the lack of a specific growth factor and so are not killed by penicillin. The surviving cells are plated on a supplemented medium in the absence of penicillin. To select for a particular mutant among the survivors, the clones can be tested by replica plating on a large number of media containing different supplements; that is, a *his*$^-$ auxotroph will be selected for by looking for growth on a medium with histidine, but no growth will occur when histidine is omitted. Using the penicillin method of selecting auxotrophs, mutants in nearly all known biosynthetic pathways that are coded for by the bacterial chromosomes of *E. coli* and *S. typhimurium* are now available.

FERMENTATION-NEGATIVE MUTANTS. A fermentation-negative mutant is a strain that has lost the ability to use a particular carbohydrate as a carbon and energy source but can still use other carbohydrates for growth. These mutants generally are deficient in the specific enzymes necessary for the catabolism of the specific carbohydrate. For example, mutants unable to use the carbohydrate lactose are deficient either in a protein necessary for the transport of the sugar into the cell or in the enzyme β-galactosidase, which catalyzes the breakdown of lactose into galactose and glucose. Such mutants can be isolated by combining the penicillin selection procedure with the identification of the specific mutant on a differential medium such as eosin-methylene blue (EMB) lactose medium or McConkey's lactose medium, in which the lactose-fermenting colonies give a color reaction (due to acid production) and the lactose-negative cells (*lac*$^-$) show a lack of color. For the penicillin selection, the sugar lactose is used in place of glucose in the minimal medium with penicillin. Lactose-negative cells that are not growing survive, whereas the growing lactose-positive cells are killed.

CONDITIONAL LETHAL MUTANTS. Conditional lethal mutants have a mutation that is expressed under one set of environmental conditions but not under others. The environmental conditions in which the mutation is expressed do not permit growth and are referred to as *nonpermissive conditions*. Similarly, when the mutation is not expressed, conditions permit growth and are called *permissive conditions*. A major class of such mutants are temperature-sensitive mutants (*ts*). Such *ts* mutants will grow at a lower temperature (25° C) but will not survive at a higher temperature (42° C), whereas the wild-type cells are capable of growing at both temperatures. These mutants usually have a missense mutation that results in the production of a protein that will function normally at a low temperature but not at the higher temperature. A second class of conditional lethal mutants result from a nonsense mutation. Such mutants can grow only in the presence of a secondary mutation or in a suppressor strain that allows for correction of the mutational defect at the translational level. Conditional lethal mutants are extremely valuable, since they allow the isolation, where no other selection procedure exists, of mutants defective in essential functions. A *ts* mutant can be isolated by plating cells at the permissive temperature and then, by replica plating, screening clones that will not grow at the nonpermissive temperature. The defective function must then be determined by biochemical analysis of the *ts* mutants.

GENE TRANSFER IN BACTERIA

Bacteria can exchange genetic material; however, the mechanisms of exchange differ strikingly from genetic exchange in eukaryotes. The three methods of gene transfer that occur in bacteria — *transformation, conjugation,* and *transduction* — have several features in common: 1) Only fragments of the chromosome of one parent (called the donor cell) are usually transferred to a second cell (called the recipient cell). 2) Fragments of the donor chromosome are assumed to pair with the homologous regions of the recipient chromosome. 3) After pairing, the genetic material of the donor usually replaces its allelic material on the recipient chromosome to yield haploid recombinants. This is referred to as *replacement integration*. 4) In some special instances, certain genetic elements (plasmids or bacteriophages) can recombine in total with the bacterial chromosome to form a composite molecule containing all the genetic material from both DNA molecules. This type of recombination is referred to as *additive recombination*.

Transformation

Transformation is gene transfer resulting from the uptake by a recipient cell of naked DNA released by a donor cell. For example, DNA released or purified in the laboratory from a donor strain that is *str*^{r} is mixed with cells that are *str*^{s}. An aliquot of the mixture is plated on a nutrient medium containing streptomycin, and approximately 1 in 100 cells plated are capable of growth.

Thus, resistance to streptomycin has appeared with a frequency of 10^{-2}. These *str*^{r} cells do not arise from mutations, since mutation to *str*^{r} occurs with a frequency of only 1 in 10^{8} cells plated, but rather must result from the transfer of the *str*^{r} DNA to the recipients. This type of gene transfer was originally described by Griffith in 1928 while working with the pneumococcus. This organism is virulent to mice only if the strains produce a polysaccharide capsule and hence have a smooth appearance during colony growth. If the capsule is absent, as in rough strains, virulence is also absent. Griffith showed that injection into a mouse of heat-killed smooth strains of pneumococci plus live rough strains killed the mouse and that live smooth pneumococci appeared in the dead mouse. He did not understand the mechanisms involved but suggested the name of "transforming factor" for the agent responsible for this phenomenon. In 1944, Avery, Macleod, and McCarty showed that the factor involved was DNA.

Transformation takes place in a large number of gram-negative and gram-positive bacteria, including *Pneumococcus, Hemophilus, Bacillus, Neisseria, Streptococcus, Xanthomonas,* and *Rhizobium.* Studies on transformation have been carried out in the laboratory, in most cases with *Pneumococcus, Hemophilus,* or *Bacillus.* More recently it has been found that *E. coli* also can undergo transformation if high concentrations of calcium ion are present.

In the interaction of transforming DNA with recipient cells, three determinants appear to be of major importance:

The size of the DNA. DNA that can transform cells has a minimum molecular weight of about 5×10^5 but may be as large as 10^8. Double-stranded DNA is many times more efficient for transformation than are single-stranded DNA structures.

The concentration of the DNA. Transformation is a function of DNA concentration. At a concentration of DNA below 100 mμg/ml, the number of transformants is proportional to the DNA concentration. This indicates that each transformant arises from the interaction of a recipient with a single molecule of transforming DNA. DNA that is foreign to the bacterial recipient DNA can also be taken up if it is double stranded. Thus, *Pneumococcus* can take up calf thymus DNA. Such foreign DNA, however, does not integrate into the recipient DNA and thus cannot transform.

The physiologic state of the recipient cell. In order for a recipient cell to take up DNA it must be in a physiologic state called *competence.* The period in the growth cycle when competence appears can vary with each bacterial species but generally appears near the end of the growth phase, just before the stationary phase (Fig. 7).

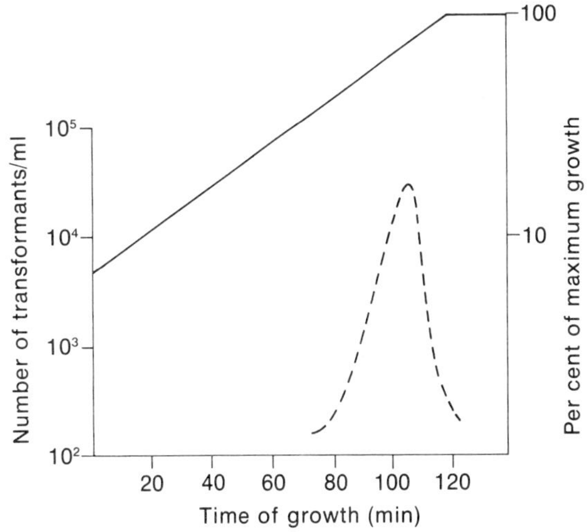

FIGURE 7. *The appearance of competence during cell growth. During cell growth (——) samples are removed from the culture and tested for their ability to be transformed (---). Competence peaks in late log phase.*

This is usually a transient state, but cells can be maintained in the competent state by freezing at −40° C in 10 to 15 per cent glycerol. Competence is genetically controlled, and a protein can be extracted from competent cells that can make noncompetent cells competent. The mechanism of action of this protein, as well as the biochemical basis for competence, are unknown.

In the process of transformation, double-stranded DNA binds to the surface of the recipient cells. This DNA is cut by a membrane-bound endonuclease, and entry of the DNA is initiated. The uptake of the DNA occurs with digestion of one of the two complementary strands and the entrance into the cell of the other strand of DNA. This single-stranded piece of DNA is then integrated into the recipient cell at the homologous region of DNA replacing the recipient allele. The initial recombinant formed has a short region of the DNA that is part donor DNA and part recipient DNA. However, during the first replication cycle each strand of DNA is replicated faithfully and a new haploid cell emerges that now carries some new genetic information derived from the donor cell.

Conjugation

Conjugation is gene transfer that occurs between sexually differentiated bacteria. It requires cell-to-cell contact between two viable cells, a donor, or male cell, and a recipient, or female cell. Conjugation was discovered in 1946 by Lederberg and Tatum and was shown to differ from transformation. The gene transfer they described was resistant to the action of DNAse and needed physical contact between two cells.

In conjugation, the donor, or male strain, always transfers its genetic material to the recipient, or female strain. Genetic material does not go from the female to the male cell.

The Sex Factor. Male cells differ from female cells by possessing a factor called the fertility factor, or *F factor*. The F factor, a typical plasmid, is a small, circular DNA molecule, about 2 per cent of the size of the bacterial chromosome. It can exist within bacterial cells and replicate independently of the host chromosome. Male cells carrying the F factor can be recognized by the presence on the cell surface of a protein appendage referred to as an *F pilus* (see Chap. 1). The F pilus serves as a receptor site for the binding of various male-specific bacteriophages that can contain either DNA or RNA. The genetic information for the synthesis of the F pilus is contained in a series of genes located in the F factor. When the F factor is in the autonomous state, that is, not associated with the chromosome, the host cells are referred to as *F^+ cells.* F^+ cells are capable of transferring the F factor to recipient, or *F^- cells,* upon cell-to-cell contact. Within a short time as many as 70 per cent of the F^- cells can become F^+ cells. However, the F^+ cells remain F^+, suggesting that the factor must be replicated sometime during or before the transfer to the F^- cells. In a cross between an F^+ and an F^- cell one can detect a few recombinants for some bacterial markers with a frequency of about 10^{-5}.

Hfr Strains. In 1950, Cavelli isolated a male strain of bacteria that was unusual in that it had a 1000-fold increase in the frequency with which recombinants appeared, and it did not transfer the F factor to the recipients except in very rare instances. Such strains, and there are many now that have been isolated, are called *Hfr strains,* or high frequency of recombination strains. These strains arise in F^+ populations and are the result of the association of the F factor with the bacterial chromosome (see later discussion).

With the finding of Hfr strains that showed such high recombination values, the kinetics of the conjugation process could be investigated. Such studies were initiated in the late 1950s by Jacob and Wollman with the development of interrupted mating experiments. An interrupted mating experiment is performed by mixing together male and female cells in a ratio of 1 Hfr to 20 female cells. The female cells contain a series of genetic defects, for example: *thr*$^-$, auxotrophic for threonine; *leu*$^-$, auxotrophic for leucine; *lac*$^-$, unable to ferment lactose; *gal*$^-$, unable to ferment galactose; *azi*^{s}, sensitive to azide; $T_1{}^s$, sensitive to the phage T1; and *str*^{r}, resistant to streptomycin. In contrast, the male strain carries the opposite markers, that is, sensitive to streptomycin, resistant to azide and T1; prototrophic for threonine and leucine, and ferments both lactose and galactose. The two cultures are mixed at high density for five minutes, and are then diluted to stop additional pair formation. At this point samples are removed at timed intervals. The samples are agitated violently to break up the mating pairs, then aliquots are plated on various selective media to determine what genetic information has been transferred from the Hfr strain to the recipient strain. The media are made so that neither parent strain is able to grow. For example, including streptomycin will eliminate growth of the Hfr; the absence of threonine or leucine will prevent growth of F^- cells unless they have received the respective genes from the donor. Data from a typical mating experiment are shown in Figure 8. Several things are characteristic of these experiments: 1) The curves do not go through the origin. Thus, if the mating is interrupted at zero time no recombinants are formed. However, the longer the mating time, the more genetic material is transferred, so that at 30 minutes all the tested markers have appeared in the recipient — some to a greater and some to a lesser extent. 2) The time that a curve begins to rise represents the time of entry of the marker into the recipient. Thus, 18 minutes is the time that lactose enters the recipient. 3) The timed order of transfer reflects the order of genes on the chromosome and is characteristic of a particular Hfr strain. In the example in Figure 8, the first marker to enter the recipient is the azide gene, followed by T1, *lac*,

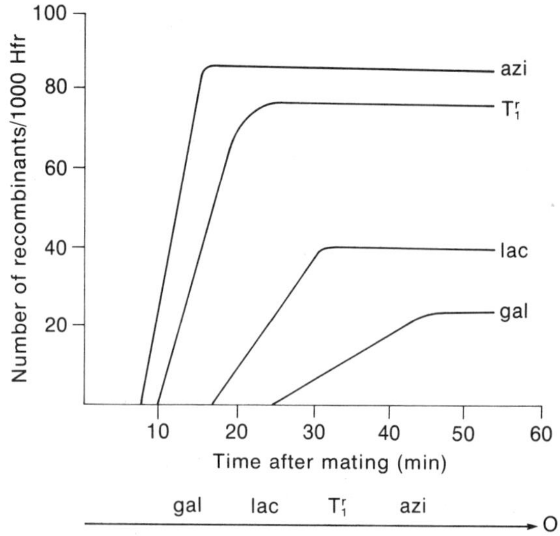

FIGURE 8. *Interrupted mating between Hfr and F^- cells. An Hfr strain is mixed with a genetically marked F^- strain (see text). At the specified time, samples are taken, the mating is stopped by vigorous agitation, and aliquots are plated on various selective media. The data are summarized in the bottom arrow and give the order of transfer of the markers from the Hfr to the F^- cell.*

and *gal*, in that order. Since the time of entrance is dependent on their location on the chromosome, one can construct a chromosome map by timing the genes entering the F^- cell. Such mapping is usually only accurate within one minute.

Different Hfr strains vary in the time that certain markers enter the recipient and in their point of origin (defined as the part of the chromosome that is first to enter the recipient), but the order of genetic markers with respect to each other does not change. Let us assume that we have three different Hfr strains, and the genetic markers are represented as A, B, C, D, and E, with O representing the point of origin. Interrupted mating experiments between each Hfr and the same recipient give the following pattern of gene transfer. Hfr-1 transfers the markers as — O-D-C-B-A-E, Hfr-2 transfers the markers as — O-A-B-C-D-E, and Hfr-3 transfers the markers as — O-B-C-D-E-A. All the data, however, are consistent if we assume that the chromosome is a circular structure that can be broken at any point. The point of rupture becomes the point of origin of transfer, but the direction of transfer can be either clockwise or counterclockwise, as shown in Figure 9. Indeed, this is exactly what happens. The F factor, which exists in the cytoplasm, will sometimes associate itself with the bacterial chromosome. It does so as illustrated in Figure 10. The circular DNA structure that is the F factor will associate with specific areas on the chromosome. The two circular structures will break and recombine with one another, allowing complete integration of the F factor DNA into the chromosomal DNA. The point at which the F factor integrates will determine the point of origin of a particular Hfr strain as well as the direction of marker transfer. During the transfer, part of the F factor will be the last marker to enter the F^- cell. The entire chromosome is not usually transferred during the mating, since it is ruptured before the end is transferred explaining why the F^- recipient in a cross with an Hfr strain does not become male.

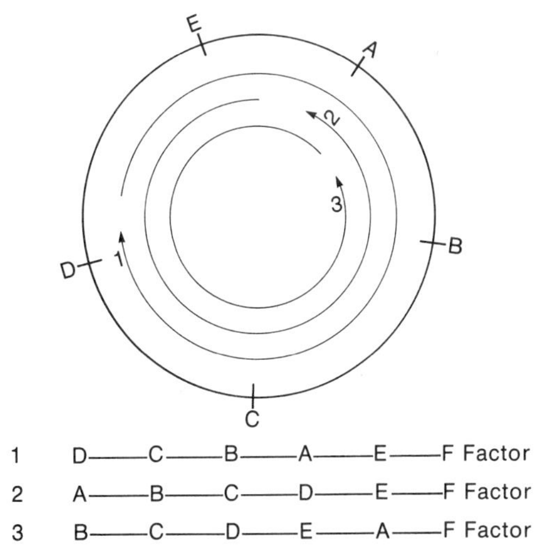

FIGURE 9. *Genetic evidence that the bacterial chromosome is a circular structure. Data from interrupted mating experiments with Hfr 1, Hfr 2, or Hfr 3 and the same F^- strain give the order of transfer of markers as shown, with the F factor the last marker to enter. These data are consistent with a circular chromosome, which can be broken at various points.*

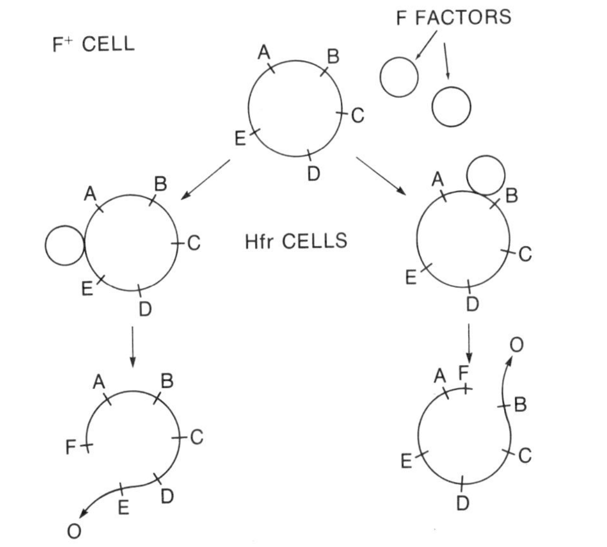

FIGURE 10. *Association of the F factor with the bacterial chromosome. The F factor can integrate into the chromosome at various sites that have an area of base pair homology. The site of intergration will determine the point of origin for the Hfr during transfer of the DNA to an F^- cell.*

F-PRIME FACTORS. Hfr strains are not always stable, and sometimes the F factor will be lost from the chromosome, once again giving an F^+ cell. In most cases this reversal results in the original F factor as well as the complete bacterial chromosome. Occasionally, however, the F factor will bring with it part of the bacterial chromosome and is then called an F-prime (F′) factor. These F′-factors are designated by the bacterial genes they carry. For example, if an F factor carries the genes for lactose fermentation it is referred to as an F′-*lac* plasmid. These factors can transfer the attached bacterial genes with high frequency to recipient cells and have been used extensively to do complementation analyses in bacteria cells.

KINETICS OF TRANSFER. When a culture of Hfr cells is mixed with an excess of F^- cells, the Hfr cell attaches to the F^- cell by way of the F-pilus. The tip of the pilus attaches to the gram-negative wall of the recipient, and the cells are brought into direct contact — presumably by retraction of the F pilus. When pair formation occurs there is some form of signal that initiates replication of the chromosome at the site of the F factor integration. As the chromosome is replicated in the Hfr strain, one parental strand of DNA passes into the recipient cell, and the other remains in the donor cell (see Figure 11). In both the donor and the recipient cells complementary strands are synthesized during the transfer. Rarely is the entire chromosome transferred, although plasmids not associated with the chromosome are transferred in total by the same mechanisms. The piece of chromosome that is transferred to the recipient is presumed to align itself with the homologous region on the recipient chromosome, followed by replacement of the recipient allele by the donor allele. The progeny of the initial recombinant event are thus haploid. In the case in which F′ factors are transferred to the recipient cells, the progeny of the recombination event become partially diploid for the genes carried on the F′ factor. This is true even if the F′ factor integrates into the chromosome, because it does so by additive recombination.

Plasmids

The F factor discussed previously is a prototype of other small extrachromosomal genetic elements that exist in bacterial cells. These elements, termed *plasmids,* replicate autonomously and often confer new genetic properties on the bacterial host cell. The plasmids can be divided into two major classes: *transmissible plasmids,* which have the ability to initiate their own transfer by cell-to-cell contact; and *nontransmissible plasmids,* which lack the ability to promote their own transfer by cell-to-cell contact. Nontransmissible plasmids can be transferred from one bacterial cell to another by transduction, or transformation, or mobilization by a self-transmissible plasmid. Among the best characterized transmissible plasmids are F factors, antibiotic resistance factors (R factors), and some of the Colicinogenic factors (Col factors).

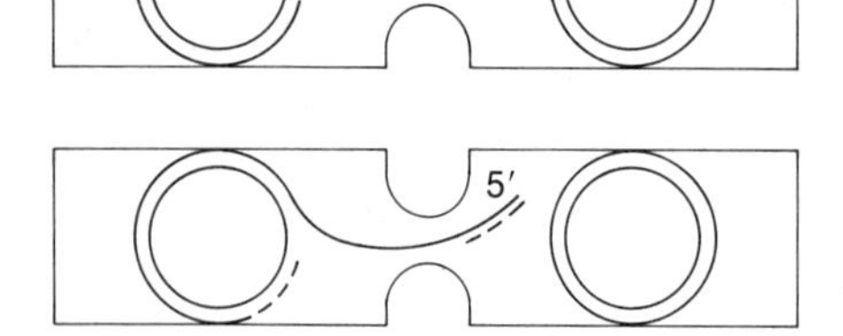

FIGURE 11. *Transfer of DNA from an Hfr cell to F⁻ cell. The cells become connected by way of the F pilus, which can act as a grappling hook to bring the two cells in contact. The chromosome in the donor breaks and begins replication with the transfer of one of the original donor DNA strands into the F⁻ cell. The DNA enters the F⁻ cells with a leading 5′ end and is replicated inside the F⁻ cell. Simultaneous replication of the donor DNA occurs in the Hfr cell.*

Transmissible Plasmids

F FACTORS. The sex factor is best characterized for its mediation of conjugation and is a circular double-stranded DNA molecule with a molecular weight of 60×10^6. The F factor carries a number of genes, 13 of which are involved in the transmission of the plasmid (*tra* genes). The plasmid controls its own replication in the autonomous state allowing only one or two copies of the plasmid per cell. Transmission of the F factor is dependent on the presence of the F pilus. The pilus is expressed in all cells carrying the F factor and can be identified by the sensitivity of such cells to male-specific viruses that use the pilus as a receptor site. F factors cannot coexist in the same cell with some types of plasmids. Two plasmids that cannot coexist are said to be in the same incompatibility group. In addition, autonomous F factors cannot coexist in a cell that has an integrated F factor (Hfr strain). All transmissible plasmids studied to date appear to have analogous *tra* systems for the transfer of the plasmid from cell to cell.

R FACTORS. The R factors are plasmids that carry genes determining resistance to antibiotics. Some R plasmids consist of two components, the RTF, or resistance transfer factor (analogous to the F factor), and the r determinants. The r determinant portion of the R factor carries one or more genes that code for proteins that abolish the effectiveness of various antibiotics. The RTF component is similar in size to the F factor, but the r determinant component varies in size, depending on the number of genes carried. In general, the composite R factors have a molecular weight of about 60 to 70×10^6. The two components of the R factor are connected by insertion sequences (see further on) on both sides of the r determinant to give a single, closed double-stranded circular DNA molecule. Transfer of the R factor from cell to cell is dependent on a pilus structure. Unlike the F^+ cells, however, cells carrying the R factor do not usually express the pilus on the surface of the cell. Only newly infected cells express the pilus, allowing for a rapid transfer of the R factor from R^+ to R^- cells, followed by a repression of the pilus formation after several generations. The structure of the pilus of the R factor differs from that of the F pilus. One group of R factors, however, has an F-like pilus (that is, sensitive to

F-specific viruses), whereas other R factors code for other types of pili (that is, I-like pili). R factors can be grouped by incompatibility or compatibility with each other and with the F factor. (See Chapter 21 for further discussion of R factors.)

Other transmissible plasmids in *E. coli* have been associated with toxins that produce diarrhea in animals and in man. Certain *E. coli* produce two enterotoxins, a low molecular weight heat-stable toxin (ST) and a heat-labile toxin (LT), that have many properties similar to cholera toxin. The *E. coli* ST and LT are both coded for by plasmid genes and are often associated with the presence of drug-resistant factors and, in some porcine strains, the production of alpha hemolysin and the K88 antigen responsible for attachment to the bowel mucosa. It appears that LT and ST may be coded for by the same plasmid.

Col Factors. Colicinogenic factors are plasmids that code for bactericidal substances called *colicins* or *bacteriocins*. These substances are proteins that kill closely related strains of bacteria that do not carry the same plasmid. As with the F factor and the R factor, the Col factors are closed, double-stranded circular DNA molecules. Col factors that are transmissible and classified as group II have a molecular weight of 60 to 110 $\times$ 10^6. They carry a *tra* system analogous to the F factor, allowing transfer by cell-to-cell contact. The Col factors are designated by capital letters. Group II contains the factors B, I, and V. These factors show controlled replication and have one or two copies per cell. Like the F factor they code for a pilus structure that is involved in their transmission from cell to cell. Col factors can also transfer chromosomal genes and form strains analogous to Hfr strains. The colicins produced act by binding to receptors in the outer membrane of sensitive bacterial cells. After some rearrangement, they come into contact with the cytoplasmic membrane and exert their effect. The different colicins act in killing sensitive cells by a variety of mechanisms that include degradation of DNA and RNA, interference with the energy system of the cells, and inactivation of ribosome structure with inhibition of protein synthesis. It appears that only a few cells in a Col^+ population actually produce the colicin, whereas most of the cells are in a repressed state similar to that seen in lysogenic cells. Like lysogenic cells, the repression is destroyed by exposing the cells to agents such as mitomycin C. Although most of the studies on colicins have been in gram-negative organisms, recent work has focused on similar substances produced by gram-positive organisms. For example, in the group D streptococcus there has been described a streptocin that appears to be coded for by genes on a self-transmissible plasmid.

Nontransmissible Plasmids

Col Factors. The nontransmissible Col factors include the factors E_1, E_2, E_3, K, and D. These plasmids are much smaller than the transmissible plasmids and have molecular weights from 3 to 6 $\times$ 10^6. This group of Col factors also differs from group II by having between 5 and 25 copies of the factor produced per chromosome-equivalent under a variety of growth conditions, suggesting a relaxed control of replication.

Staphylococcal Plasmids. The best known plasmid occurring in *Staphylococcus* carries a gene for the determination of penicillinase, rendering the bacterial cell resistant to the action of penicillin. Other plasmids also occur in this organism, giving resistance to other antimicrobial agents. Self-transmissable plasmids have not been described in the staphylococcus, but transfer of plasmids by transduction has been shown to occur both in vitro and in vivo.

Insertion Sequences and Recombination of Plasmids

Plasmids have the ability to recombine with each other as well as with the bacterial host chromosome. This ability is mainly due to the presence in the plasmids and in the bacterial chromosome of insertion sequences (IS). Insertion sequences are segments of DNA that have a specific base sequence and range in length from 800 to 1400 base pairs. They have been designated as IS1, IS2, IS3, IS4, and so on. The insertion sequences IS1, IS2, and IS3 are found at numerous points on the *E. coli* chromosome. Identical IS elements are found in various R factors, lambda phages, and F factors. The presence of these common base sequences allows for these various genomes to recombine and form new combinations of genes. Insertion sequences belong to a group of elements known as transposable elements, defined as segments of DNA that, as discrete genetic and physical entities, can move from one position in a genome to another position in the same or different genome. The IS elements are elements containing no known genes unrelated to the insertion function. More complex elements are the *Tn elements,* which contain IS segments usually as an inverted repeat on either end of the DNA fragment and additional genes unrelated to the insertion function. Many of the Tn elements are translocatable drug-resistance elements such as Tn10 (a tetracycline-resistant element) that can translocate from its position on an R factor to the bacterial chromosome and then to a bacteriophage. Other drug-resistance Tn elements have been described, such as Tn5 (kanamycin resistance) and Tn9 (chloramphenicol resistance), and are often referred to as *transposons*. More complex transposable elements are the plasmids or viruses that can contain both IS segments and Tn elements in addition to other genes needed for replication and transmission, that is, R factors. Thus, insertion sequence elements can act as sites for joining various DNA segments and allowing for new arrangements of genes in bacteria. This process results in optimal growth of the bacteria under a variety of changing environmental conditions and helps ensure survival.

Transduction

Transduction is the transfer of genetic information from a donor cell to a recipient cell by way of a virus vector. For the most part transduction is carried out by temperate DNA-containing bacteriophages. A temperature phage is a virus that can either lyse or lysogenize a host cell upon infection. In the lysogenic response, the genes of the virus that are responsible for initiating the lytic response are repressed, and the viral DNA is integrated into the host DNA by a mechanism similar to the integration of the F factor. In the integrated, or *prophage,* state the virus genes replicate with the host cell genes and all daughter cells will have a viral gene integrated into their chromosome. Bacterial cells that have a viral genome integrated into their chromosome are called *lysogenic cells.* They differ from nonlysogenic cells in two respects: 1) they are immune to superinfection by the same type virus, and 2) they usually can be induced to produce virus particles without infection from the outside. A repressor gene product produced by the virus maintains the repressed and integrated state by preventing transcription of the lytic genes of the virus. Induction of a lysogenic cell to enter the lytic cycle occurs when this repressor protein is destroyed or prevented from functioning.

There are three major types of transduction: 1) *generalized transduction,* in which any genetic marker can be transduced, and the donor DNA is integrated by replacement of the recipient allele; 2) *specialized* or *restricted transduction,* in which only markers that are adjacent to the site where the viral DNA is integrated into the host DNA are transduced, and integration is usually by additive recombination; 3) *abortive transduction,* in which any marker can be transduced, but the DNA is not integrated into the recipient chromosome.

Generalized Transduction. To carry out generalized transduction, a bacterial culture is infected with a temperate virus under conditions of low multiplicity of infection (moi) of about .01 to 0.1 phage particles per cell. The culture is then allowed to grow through several cycles of phage development until most of the cells have been infected and lysed. During the maturation of the viral particles an occasional mistake is made and instead of packaging viral DNA, host DNA of the same size is put into the viral protein coat. These particles, which are now the *transducing particles,* are characterized by the fact that they contain only host DNA and no phage DNA. When an appropriately marked recipient strain is then mixed with this lysate at an moi of 5 phage particles to each cell, some cells will be infected by both a normal particle and a transducing, or defective, particle. By plating aliquots of the infected culture on appropriate selective media, one selects for transductants that now have donor DNA incorporated into the recipient DNA. Such generalized transductants can be selected for any genetic marker and arise at a frequency of about 10^{-5} or one transductant for every 10^5 phage particles added. The transductants are characterized by being stable once they are formed, which is a result of the fact that the donor DNA has replaced the recipient DNA during the process. The size of the DNA that can be transduced is dependent on the size of the virus but is generally about 1 per cent of the size of the bacterial chromosome. Only closely linked genetic markers can be cotransduced on the same DNA particle, which enables transduction to be used for the genetic mapping of closely linked genes.

Abortive Transduction. During the process of generalized transduction, as described previously, there are instances when the donor DNA that is injected into the recipient cells does not integrate into the recipient DNA. These cells are referred to as *abortive transductants.* The piece of donor DNA exists in the recipient cell and is capable of functioning normally; however, it is not capable of being replicated. The result is that only one cell at any time in the population derived from the initially infected cell carries the DNA piece. The other daughter cells will receive, upon cell division, products coded for by the DNA and thus will be capable of carrying out several cell divisions until the protein product has become diluted out. As a result only minute colonies are formed by abortive transduction, although they may appear at 100 times the frequency of stable generalized transductants. Abortive transductants are seen only if two negative phenotypes happen to be defective in different genes, and thus, it is a useful method for carrying out complementation analyses.

Specialized Transduction. Specialized transduction is limited to genes that are adjacent to the site of prophage integration. This is clearly due to the mechanism by which specialized transducing particles are formed. Such particles can only be generated by starting with a bacteria culture that has been made lysogenic. Such a culture of cells when treated with short doses of ultraviolet light will lose the active viral repressor protein and begin to enter the lytic cycle. During this process the viral DNA is detached from the host DNA by a reversal of the mechanism by which it had been integrated. In the majority of cases this detachment is performed without any mistake, and a normal virus genome is released. However, in 1 in 10^6 cells this excision will result in a mistake, and a particle will be released that now carries part of the viral DNA and part of the bacterial DNA (Fig. 12). In the case of lambda phage, the host DNA carried will be either the genes for galactose fermentation or the genes for biotin synthesis. These two loci occur on either side of the integrated virus. These particles are referred to as λ*dgal* or λ*dbio,* which stands for lambda-defective galactose or biotin. The particles are termed defective because they are missing part of the viral genes and cannot carry out a normal productive infection of a bacterial cell without the presence of a normal viral particle. They will, however, inject their hybrid DNA into a recipient cell. The DNA will integrate at the homologous site on the recipient DNA by adding to the chromosome rather than by replacing the recipient allele. The transductants produced are therefore partially diploid for this region of the bacterial chromosome and are termed *heterogenotes*. In addition to carrying the defective particle, most of these transductants will also carry a normal phage genome and thus can be considered to be doubly lysogenic. The presence of the donor DNA is detected by infecting a gal^- recipient with the lysate produced by ultraviolet irradiation of the lysogenic cells and then plating samples onto plates containing EMB-galactose medium. Cells that have received the galactose genes will give dark clones with a green sheen on this medium. If these galactose-positive clones are picked and restreaked, it is found that they are unstable. That is, in 1 in 10^3 cell divisions, segregation of a galactose-negative clone will result. This is due to the loss of the λ*dgal* particle in some cells at division. The initial number of galactose-positive transductants found is about 1 in 10^5 galactose-negative cells infected. If, however, one starts with a heterogenote clone and treats it with ultraviolet light to produce the lysate, a lysate is found that can now transduce the *gal* genes at a very high frequency. About 1 in 10 gal^- cells infected can become galactose-positive. These lysates are referred to as *HFT*, or *high frequency transducing,* lysates. They occur because the lysate from the heterozygote has many particles carrying the *gal* genes, approximately one half of the viral particles produced. This is a result of the presence in the heterogenote of a normal virus particle that allows for replication of the defective particle in each cell in the culture.

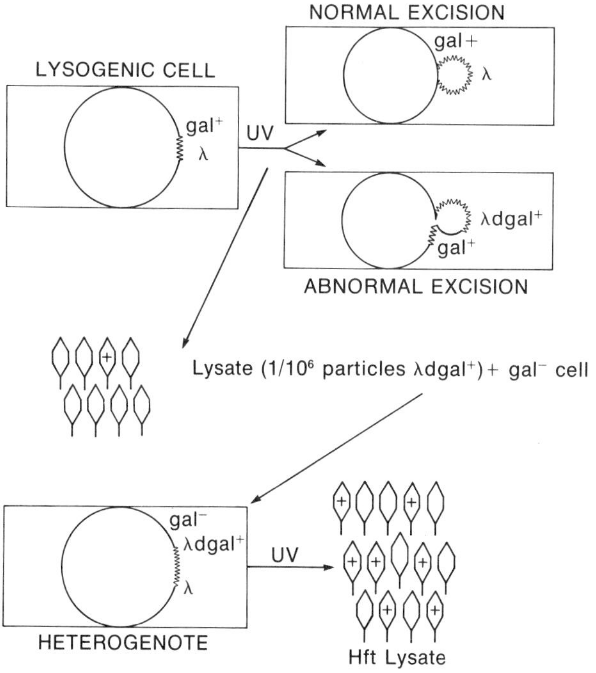

FIGURE 12. *Specialized transduction of the galactose genes in* E. coli *by lambda phage. A lysogenic cell is treated with UV light to cause cells to undergo the lytic cycle. In nearly all cells, excision of the virus is normal, and normal virus particles result. In one in 10^6 cells, abnormal excision takes place giving a defective particle λdgal⁺. The presence of λdgal⁺ particles in the lysate can be shown by mixing the lysate with a gal⁻ strain and selecting for gal⁺ transductants. Many of these transductants will be heterogenotes and will carry both λdgal⁺ and normal λ. Ultraviolet irradiation of these cells yields a lysate that is half normal particle and half λdgal⁺ particles or an HFT lysate. These result because the normal λ particles will provide the functions missing in λdgal⁺ and allow its replication in each cell.*

REGULATION OF GENE EXPRESSION

Bacteria have evolved rather intricate and elegant mechanisms to control the phenotypic expression of their genotype. In catabolic pathways, the enzymes needed for the breakdown of a substrate are expressed only if the substrate is available in the environment. In anabolic pathways, if the end-product is available in the environment the enzymes needed for the synthesis of that product are not expressed. When mutants became available that had lost their ability to control the synthesis of the enzymes involved in these pathways, an understanding of what was occurring at the molecular level began. The basic mechanism of control was elaborated by the work of Jacob and Monod in 1961 and is referred to as the *operon theory*. In this theory they describe two major classes of genes: 1) *structural genes* — genes that specify the structure of a specific protein such as an enzyme, and 2) *regulatory genes, or sites* — genes, or sites, on the chromosome that control the rate of synthesis of the structural genes. The operon is defined as a group of contiguous structural genes showing coordinate expression, together with their closely linked controlling sites. The best understood operon, that concerned with the catabolism of lactose, is diagrammed in Figure 13. This type of control is referred to as *negative control,* since the presence of a repressor protein, which is bound to a site on the DNA called the operator site, prevents the synthesis of the mRNA from the structural genes. In the presence of the substrate or inducer, the structure of the repressor product is altered so that it no longer binds to the DNA, and mRNA production begins. The site at which mRNA synthesis is initiated by RNA polymerase is called the promoter site. This is also the site for the action of another regulator molecule, the cAMP activator protein – cAMP complex.

Mutations that occur in the structural genes usually result in the loss of enzyme activity caused by the alteration of the protein structure. Mutations that occur in the regulatory gene or controlling sites result either in strains that cannot be induced to produce enzymes *(noninducible)* or in strains that always produce the enzymes, even in the absence of the inducer *(constitutive).* To determine whether such mutations have occurred in the regulator gene or in the operator site, partial diploid studies are carried out using *F′lac* factors that carry different forms of the mutated genes (Fig. 14). In a diploid study, constitutivity is recessive to the wild-type inducible state if the regulator has been altered but is dominant

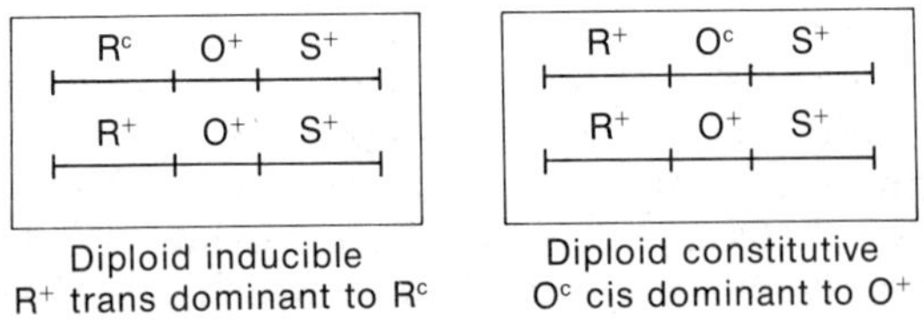

FIGURE 14. *Differentiation of constitutive mutants as mutations in the regulator gene (R^c) or the operator site (O^c) in an operon under negative control. Partial diploids are constructed with F′-lac plasmids and then assayed for S enzyme activity after growth in the presence and absence of the inducer. Mutations in the R gene are constitutive because of the lack of repressor, and these are recessive and inducible in the presence of a wild-type repressor. Mutations in the O site are constitutive because they do not bind repressor and thus are dominant and still constitutive in the presence of the wild-type repressor.*

to the wild-type inducible state if the alteration has occurred in the operator site. If the regulator has been altered to give a noninducible state, this is dominant to both inducibility and constitutivity (Fig. 15).

A more complex but similar control pattern is seen in the pathway for the catabolism of arabinose. This operon is under *positive control.* Positive control differs from negative control in that it requires the presence of an activator protein to initiate mRNA transcription by RNA polymerase. Diploid studies in a positive control system give a different pattern than that seen in a negative control system. In this case alteration of the regulator to give the noninducible state is reces-

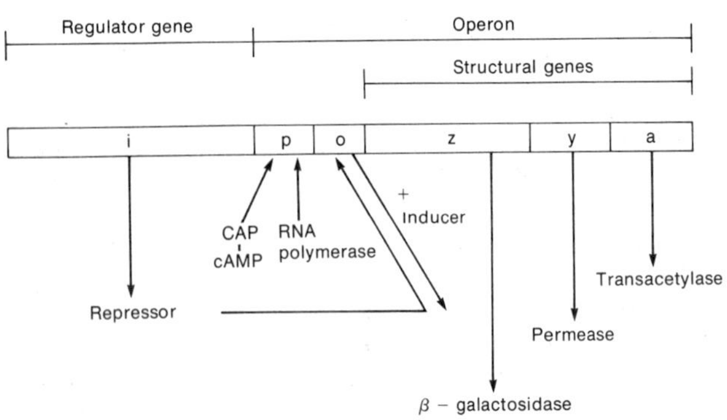

FIGURE 13. *The operon model for regulation of lactose fermentation. The repressor, the product of the regulator gene, prevents the synthesis of mRNA for the structural genes, z, y, and a, by binding to the operator site O. Addition of the inducer prevents the repressor from binding at O, and mRNA synthesis is initiated at the promoter, or p, site by RNA polymerase. The promotor site also has a binding site for the CAP-cAMP complex.*

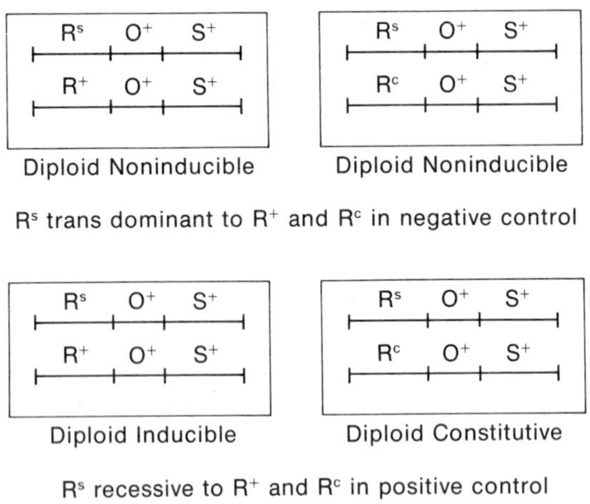

FIGURE 15. *Differentiation of negative and positive control by partial diploid studies. R^s, or noninducible, strains in a negative control system result from the production of a mutated repressor that no longer binds the inducer and thus remains attached to the operative site. This R^s repressor will also bind to the operative site on the opposite DNA strand and thus is dominant both to wild type and constitutivity. The R^s, or noninducible, state in positive control systems results from lack of production of an activator protein. Thus, the absence of the product makes the R^s state recessive to either the presence of the wild-type activator or an altered R^c constitutive activator.*

sive to the wild-type inducible state. Likewise, constitutivity is dominant to noninducibility (Fig. 15). These results are consistent with a requirement for the presence of an activator protein to get expression of the operon.

Catabolite repression is an additional control mechanism that has evolved in bacterial cells and is superimposed on the operator-repressor interaction seen in both negative and positive control. It has been known for many years that bacterial cells growing in the presence of glucose will not use a carbon source that requires inducible enzyme synthesis until all of the glucose has been used. It is now known that in the presence of glucose, the level of cAMP in bacterial cells drops from 10^{-4} M to 10^{-7} M. cAMP is essential for the initiation of mRNA transcription. This compound binds to a protein called CAP or *catabolite gene activator product,* and this complex interacts with the DNA at the promotor site to stimulate transcription. This is a general control mechanism added to the very specific controls seen within each catabolic pathway.

References

Avery, O. T., Macleod, C. M., and McCarty, M.: Studies on the chemical nature of the substance inducing transformation of pneumococcal types. J Exp Med 79:137, 1944.

Griffith, F.: The significance of pneumococcal types. J Hyg 27:113, 1928.

Hayes, W.: The Genetics of Bacteria and their Viruses. 2nd ed. New York, John Wiley & Sons, 1968.

Jacob, F., and Brenner, S.: Sur la régulation de la synthése du DNA chez les bacteries: l'hypothèse du replicon. C R Acad Sci (Paris), 256:298, 1963.

Jacob, F., and Monod, J.: Genetic regulatory mechanisms in the synthesis of proteins. J Mol Biol 3:318, 1961.

Jacob, F., and Wollman, F.: Sur les processus de conjugaison et de recombination genetique chez *E. coli.* Ann Inst Pasteur 91:486, 1956.

Lederberg, J., and Lederberg, E. M.: Replica plating and indirect selection of bacterial mutants. J Bacteriol 63:399, 1952.

Lederberg, J., and Tatum, E.L.: Gene recombination in *E. coli.* Nature 158:582, 1946.

Luria, S.E., and Delbrück, M.: Mutations of bacteria from virus sensitivity to virus resistance. Genetics 28:491, 1943.

Meselson, M., and Stahl, F.: The replication of DNA in *Escherichia coli.* Proc Natl Acad Sci USA 44:671, 1958.

Stanier, R., Doudoroff, M., and Adelberg, E. A.: The Microbial World. 3rd ed. Englewood Cliffs, N.J., Prentice Hall, 1970.

Watson, J. D.: The Molecular Biology of the Gene. 3rd ed. New York, Benjamin Company, Inc., 1976.

Watson, J.D., and Crick, F.H.C.: Genetic implications of the structure of DNA. Nature 171:964, 1953.

BACTERIAL EXOTOXINS 5

William Edward van Heyningen, Ph.D., Sc.D.

There are not more than three or four infectious diseases for which we have a reasonably clear understanding of the means by which the infecting organism brings about its harmful effects. The ability of the organism to grow in the body is not the same thing as its capacity to produce disease, for there is no intrinsic reason why the mere presence of a very small weight of infecting organisms in the body of the patient — from a few micrograms in tetanus to 100 mg at most in anthrax — should be harmful. The idea that pathogenic organisms produce poisons is an old and obvious one, and it is indeed a fact that many pathogenic bacteria do produce antigenic poisons,

TABLE 1. Differences Between Bacterial Exotoxins and Endotoxins

	EXOTOXINS	ENDOTOXINS
Parent organisms	gram-positive and gram-negative	gram-negative
Within or without parent organism	within and without	within
Chemical nature	simple protein	protein-lipid-polysaccharide
Stability to heating (100°)	labile	stable
Detoxification by formaldehyde	detoxified	not detoxified
Neutralization by homologous antibody	complete	partial
Biologic activity	individual to toxin	same for all toxins
Toxicity compared with strychnine as 1	100 to 1,000,000	0.1

usually known as toxins. There are two types of toxins — the exotoxins, which are discussed in this chapter, and the endotoxins, which are discussed in Chapter 6. The differences between these two types of toxin are shown in Table 1. The harmful effects of endotoxins are perhaps due more to their immunologic properties than to their comparatively low toxicity, but their pathologic importance must not be underestimated.

Table 2 lists the better known bacterial exotoxins. The fact that a pathogenic organism produces a particular exotoxin does not necessarily mean that this toxin plays an important part in producing the harmful effects of the infectious disease.

TABLE 2. Exotoxins of Pathogenic Bacteria

BACTERIUM	DISEASE CAUSED IN MAN	TOXINS
Bacillus anthracis	anthrax	complex lethal edema-producing toxin
Bordetella pertussis	whooping cough	lethal, dermonecrotizing toxin
Clostridium botulinum	botulism	**6 type-specific lethal neurotoxins**
Cl. oedematiens	gas gangrene	1. alpha, lethal, dermonecrotizing 2. beta, lethal, dermonecrotizing, hemolytic 3. gamma, lethal, dermonecrotizing, hemolytic 4. delta, hemolytic 5. epsilon, lethal, hemolytic 6. zeta, hemolytic
Cl. perfringens	gas gangrene and enteritis necroticans	1. **alpha, lethal, dermonecrotizing, hemolytic** 2. beta, lethal 3. gamma, lethal 4. delta, lethal 5. epsilon, lethal, dermonecrotizing 6. eta, lethal (?) 7. iota, lethal, dermonecrotizing 8. theta, lethal, cardiotoxic, hemolytic 9. kappa, lethal, proteolytic 10. **enterotoxin**
Cl. septicum	gas gangrene	alpha, lethal, hemolytic
Cl. sordellii	gas gangrene	1. edema-producing toxin 2. hemorrhagic toxin
Cl. tetani	tetanus	1. **tetanospasmin, lethal, neurotoxic** 2. neurotoxin, nonspasmogenic 3. tetanolysin, lethal, cardiotoxic, hemolytic
corynebacterium diphtheriae	diphtheria	**diphtheria toxin, lethal, dermonecrotizing**
Escherichia coli	diarrhea	1. **heat-labile enterotoxin** (2. heat-stable enterotoxin)
Pseudomonas aruginosa	pyogenic infections	**exotoxin A**
Staphylococcus aureus	pyogenic infections, enterotoxaemia	1. alpha, lethal, dermonecrotizing, hemolytic 2. beta, lethal, hemolytic 3. gamma, lethal, hemolytic 4. delta, hemolytic 5. **exfoliating toxin** 6. **enterotoxin**
Streptococcus pyogenes	pyogenic infections, scarlet fever, rheumatic fever	1. Dick toxin, erythrogenic, nonlethal 2. streptolysin O, lethal, hemolytic, cardiotoxic 3. streptolysin S, lethal, hemolytic
Vibrio cholerae *Vibrio El Tor*	cholera	**cholera toxin, lethal, enterotoxic**
Salmonella typhimurium	enteritis	**enterotoxin?**
Shigella shiga	dysentery	**enterotoxin**
Yersinia pestis	plague	murine toxin

TABLE 3. Criteria for Judging Whether Toxins are Responsible for the Harmful Effects of an Infectious Disease

1. The organism produces a toxin.
2. Virulent strains produce the toxin; avirulent do not.
3. The organism produces disease without multiplying profusely or spreading extensively.
4. The blood and lymph are sterile.
5. Organs and tissue at a distance from the site of infection are affected.
6. Introduction of sterile cell-free toxin into animals produces symptoms mimicking the disease.
7. The clinician sees toxic effects in the patient: peripheral vascular collapse; direct action on the heart muscle; central or peripheral nervous system affected (see Figure 1).
8. The disease can be prevented by immunization against the toxin.

In most cases there is no good reason for concluding whether it does or does not. The criteria that must be considered in coming to the conclusion that it does are listed in Table 3. These criteria need not all be fulfilled, and the toxins that fulfill at least some of them, or are relevant to those that do, are shown in bold type in Table 2. These are the toxins that we will discuss in this chapter, and in doing so we will be obliged to dismiss a number of well-known toxins, such as the staphylococcal and streptococcal hemolysins, that are of great scientific interest but as yet of no proven medical relevance.

Further reading (the references in this chapter should be consulted as much for their lists of references as for their texts): van Heyningen, 1970.

NEUROTOXINS

Tetanus Toxin (Tetanospasmin)

The harmful effects of tetanus are entirely due to this toxin, which fulfills all of the criteria in Table 3. Tetanus toxin is one of the most poisonous toxins known — purified preparations of the toxin may contain 100 million lethal doses (mouse) per milligram (i.e., 1 g kills two million tons of living matter). It is also one of the most dangerous toxins known. Reliable statistics on tetanus do not exist, but it is possible that the toxin kills perhaps two million people every year, mainly newborn babies (*tetanus neonatorum,* or umbilical tetanus), in the underdeveloped countries of Asia, Africa, and South America.

The toxin acts only on the nervous system, without producing any morphologic changes. The disease is characterized by a spastic paralysis, which nearly always involves the jaw muscles (lockjaw) and the muscles of deglutition, and eventually leads to generalized convulsions (Fig. 1). The spasticity is due to the action of the toxin in blocking pre- and postsynaptic inhibition in the central nervous system, particularly in the spinal cord and cerebellum, which it reaches by ascending in the nerves by retrograde axonal transport, after uptake from the anaerobically replicating organisms near the wound of entry (which is often so small as to be undetectable *post mortem*). Apart from eradication of the organisms with penicillin to prevent formation of more toxin, treatment of tetanus is aimed at controlling convulsions by curarization. However, the toxin can also block neuromuscular transmission (like botulinum toxin; see below), and it also, perhaps indirectly, causes overactivity in the sympathetic nervous system. The latter effects involve drastic swings in blood pressure and heart rate, and a high mortality rate is associated with the

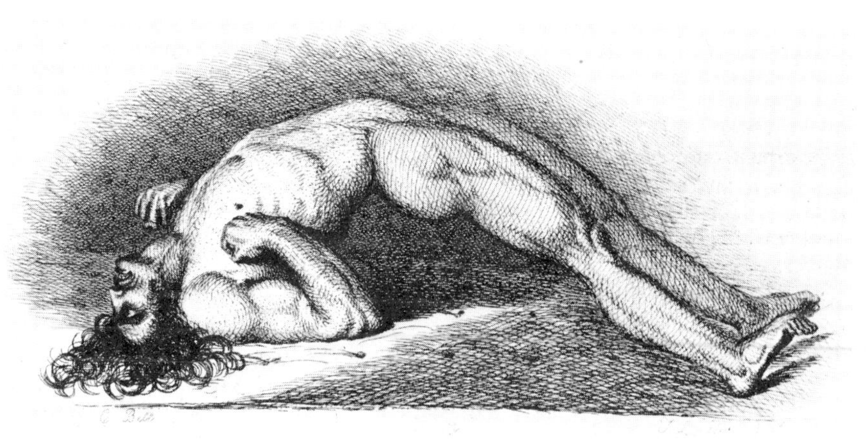

FIGURE 1. *Sir Charles Bell's drawing (1832) of a British soldier suffering from tetanus (lockjaw) as a result of wounds received at the Battle of Corunna. Note the spastic paralysis resulting from opposing muscles pulling against each other, owing to tetanus toxin blocking synaptic inhibition. This man will certainly die.*

appearance of these symptoms, unless they are controlled.

The toxin blocks both evoked and spontaneous transmitter release at the central and peripheral synapses on which it acts. The mechanism of action is not yet understood, and neither are the factors that determine its specificity with respect to particular synapses. Its effects are long lasting, and it appears that, at least at peripheral synapses, restoration of function requires sprouting of new nerve terminals.

What is known about the structure of the toxin is shown in Figure 2. The toxin is a simple protein, apparently consisting of two components, a "heavy" and a "light" polypeptide chain bound together by a peptide bond that is easily "nicked" by proteolytic enzymes in the culture filtrate, and a disulfide bond that is easily cleaved by reduction. The two chains are further held together by noncovalent forces that can be broken with urea or sodium dodecyl sulfate. These chains are inert and antigenically distinct when separated, but can be recombined to form the fully toxic protein.

The mode of action of the toxin at the molecular level has not yet been determined, but it has been shown that the well-known specific binding of tetanus toxin to nervous tissue (Wassermann-Takaki phenomenon, 1898) is due to the binding of the toxin to certain gangliosides. The gangliosides are components of the cell membranes of most tissues, and are particularly concentrated in nervous tissue. Since they consist of a water-insoluble ceramide moiety and a water-soluble oligosaccharide moiety (Fig. 3), they are amphipathic — i.e., they are both water soluble and fat soluble, and are therefore well suited to play a role in cell membranes. Tetanus toxin shows a particular affinity for the sialidase-sensitive ganglioside GGnSSLC, which contains two sialosyl residues attached to the internal galactosyl residue, and can bind with it in a ratio of 3 to 4

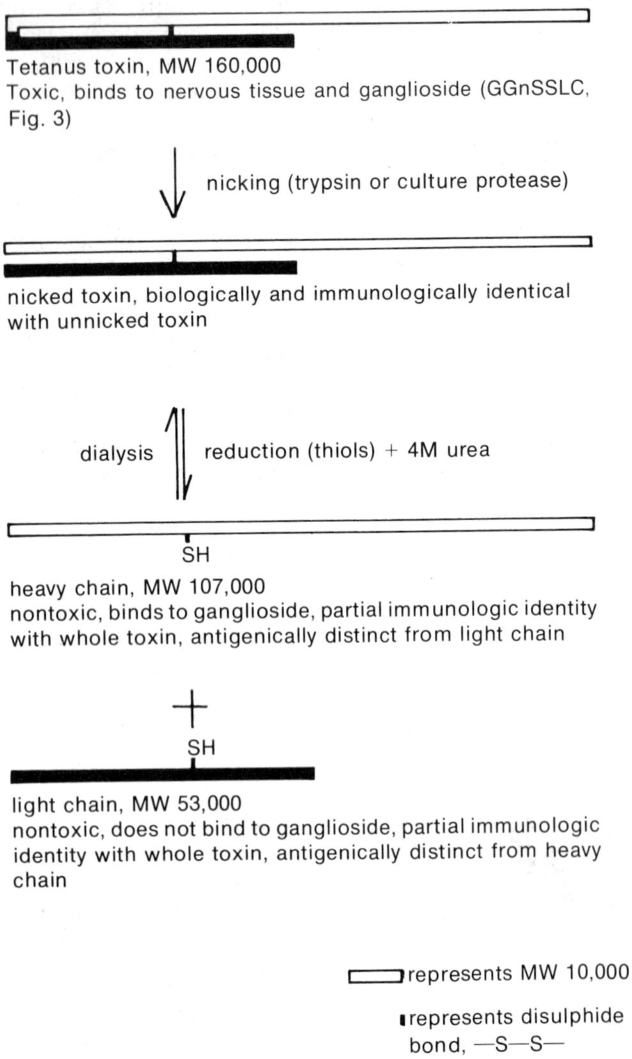

FIGURE 2. *Diagrammatic representation of the structure of tetanus toxin. As in Figures 4, 5, and 6, the black and white bands represent polypeptide chains. In each figure the lengths of these bands are proportional to their relative molecular weights, but note the different scales. Folding of chains by noncovalent bonds is not depicted and the only folding shown is that known to be due to disulfide bonds (—S —S—) from one part of a chain to another. These disulfide bonds can be broken by reduction with thiols (to convert —S—S to —SH=SH—), and restored by dialysing away the thiols. The fact that a protein denaturing agent (urea) is also necessary to separate the light and heavy chains of tetanus toxin suggests that they are held together by noncovalent bonds in addition to the covalent disulfide and peptide bond (see cholera toxin, Fig. 6).*

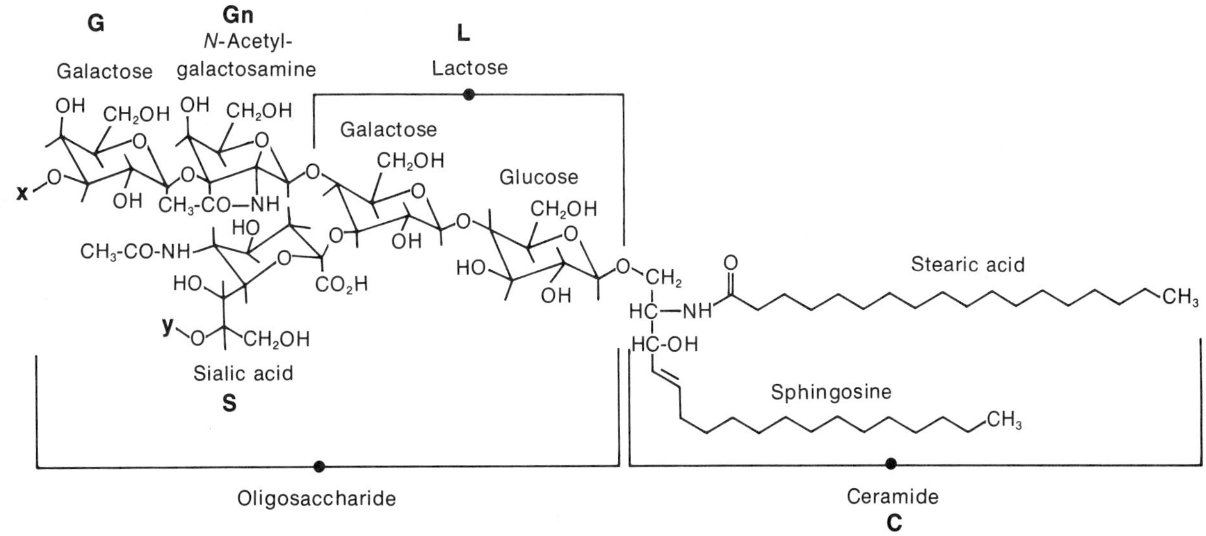

FIGURE 3. *Structures of some of the best-known gangliosides. The bond attaching the sialosyl (S) residue to the galactose moiety of the lactose (L) residue is sialidase- (that is, neuraminidase) insensitive. When sialosyl residues substitute for X or Y they are attached by sialidase-sensitive bonds.*
GGn = H; y = H: Monosialosylganglioside SLC (GM3)
G = H; y = H: Monosialosylganglioside GnSLC (GM2)
x = H; y = H: Monosialosylganglioside GGnSLC (GM1)
x = S; y = H: Disialosylganglioside SGGnSLC (GD1a)
x = H; y = S: disialosylganglioside GGnSSCL (GD1b)
x = S; y = S: trisialosylganglioside SGGnSSLC (GT1)
(The shorthand names in brackets are more widely used but are not self-explanatory.)

molecules ganglioside to 1 molecule toxin (Table 4). It is the only "heavy" component of the tetanus toxin molecule that combines with ganglioside.

Other bacterial toxins, notably cholera toxin, and certain other biologically active proteins, also bind to gangliosides (Table 4). The significance of this will be discussed below.

Further reading: Bizzini, 1977; Helting et al., 1977; S. van Heyningen, 1976; van Heyningen and Mellanby, 1971; Mellanby and Pope, 1976; and van Heyningen, 1963.

Botulinum Toxins

Botulism generally assumes the form of food poisoning resulting from the ingestion of toxins preformed by *Clostridium botulinum* growing on food. There are six serologic types, A to F, of the organism, each producing an immunologically type-specific toxin. These toxins have different molecular weights, but they all have apparently the same neurotoxic activity. Only Types A, B, E, and F are known to affect man. Figure 4 shows

TABLE 4. Biologically Active Proteins Binding to Ganglioside Receptors

PROTEIN	RECEPTOR	TARGET
Tetanus toxin[1]	GGnSSLC>SGGnSSLC>SGGnSLC>GGnSLC	synaptic transmission
Botulinum toxin[2]	not identified—sialidase labile?	synaptic transmission
Cholera enterotoxin[3]	GGnSLC only	adenylate cyclase
E. coli enterotoxin[4]	GGnSLC only	adenylate cyclase
Staphylococcus alpha-toxin[5]	SGGlnLC[6]	cell membrane?
Vibrio parahaemolyticus toxin[7]	SGGnSSLC	?
Thyrotropin hormone[8]	GGnSSLC>SGGnSSLC>GGnSLC>GnSLC=SLC>SGGnSLC	adenylate cyclase
Human chorionic gonadotropin[9]	SGGnSSLC>SGGnSLC>GGnSSLC>GnSLC>GGnSLC	adenylate cyclase
Interferon[10]	GnSLC	cell reaction to virus
Sendai virus[11]	not identified—sialidase labile?	?

1. van Heyningen (1963); 2. at pH 5, Mellanby and Pope (1976); 3. van Heyningen (1973); 4. Pierce (1973); 5. Kato and Naiki (1976); 6. Gln = N acetyl-glucosamine (this ganglioside so far found only in red blood cells); 7. Takeda et al. (1976); 8. Mullin et al. (1976); 9. Lee et al. (1976); 10. Besancon et al. (1976); 11. Haywood (1974).

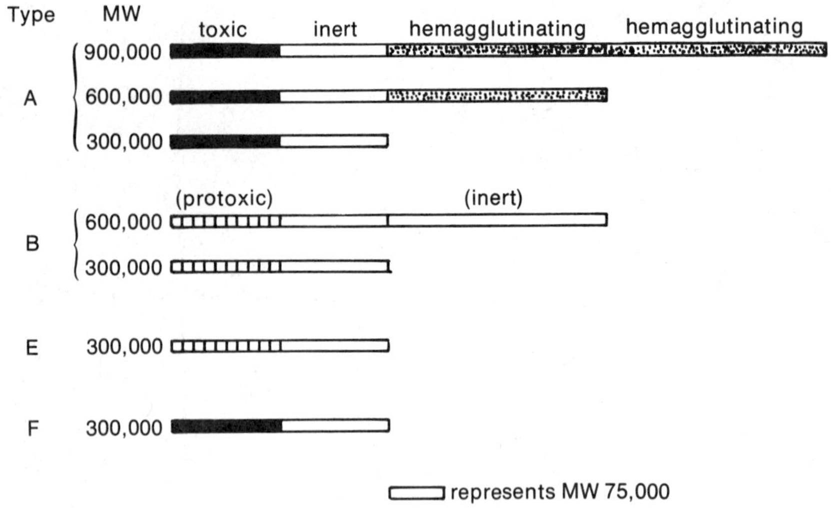

FIGURE 4. *Diagrammatic representation of the composition of botulinum toxins (see legend to Fig. 2) as they appear to exist naturally. Under certain conditions of pH and ionic strength smaller toxic components (MW about 150,000) may be found in laboratory preparations. The toxins from* Cl. botulinum *types B and E appear as nontoxic protoxins that are converted to active toxins by proteolytic enzymes without appreciable change in molecular weight.*

that these toxins appear to be aggregates of two or three kinds of protein components — toxic, inert, and hemagglutinating. Only the Type A toxin contains the hemagglutinating component, and in the Types B and E toxins the toxic components are produced by the organism as inert protoxins that are converted by proteolytic enzymes into active toxins with no detectable change in molecular weight.

Botulinum toxin is as toxic as tetanus toxin; like tetanus toxin, it acts only on nervous tissue, and does not directly cause any detectable morphologic change in this tissue. It apparently has no central action, and seems to act only peripherally at the neuromuscular junctions, where (like tetanus toxin) it blocks both evoked and spontaneous release of acetylcholine from cholinergic motor nerve endings.

Nothing is yet known of the action of the toxin at the molecular level. The toxic component of the toxin (Fig. 4) has a molecular weight of about 150,000 (similar to that of tetanus toxin). It can be separated from the inert and hemagglutinating components without loss or change in character of the toxicity. Nothing is yet known of the structure of this toxic component.

It used to be thought that botulinum toxin does not bind to nervous tissue, or to gangliosides, but recent work has demonstrated that it is bound to synaptic membranes and to gangliosides under different physical conditions (pH, salt concentration) from those suitable for the binding of tetanus toxin. Perhaps these differences may be related to the different conditions under which these toxins block synaptic transmission in the laboratory. Thus, the in vitro effect of botulinum toxin on a phrenic nerve diaphragm preparation can be seen within 30 minutes of adding the toxin to the bathing medium, whereas the similar effect of tetanus toxin can be reproduced only by injecting the toxin into the living animal and studying a nerve-muscle preparation isolated some hours later.

Recently it has been shown that the cause of the Sudden Infant Death Syndrome (crib death) can sometimes be due to botulism resulting from ingestion of *Cl. botulinum* spores, their survival of the passage through the infant stomach, followed by their vegetation and consequent toxin production in the gut. This means that, contrary to previously held opinion, botulism can be an infectious disease (in infants) and an exotoxinosis.

Further reading: Arnon *et al,* 1978; and Boroff and DasGupta, 1971.

TOXINS BLOCKING PROTEIN SYNTHESIS

Diphtheria Toxin

Diphtheria toxin is produced only by lysogenic strains of *Corynebacterium diphtheriae* infected with a bacteriophage carrying the *tox* gene, which carries the structural information for the synthesis of the toxin. The expression of this information is regulated, at least in part, by the bacterial host. The toxin must be considered responsible for the harmful effects of diphtheria since it fulfills all the criteria listed in Table 3. Unlike tetanus and botulinum toxins, diphtheria toxin appears to act on all tissues of susceptible animals. It has a necrotizing effect on cells, as may be seen when it is injected intracutaneously, as in the Schick test. In the diphtheria patient or the experimentally injected animal, the effect of the toxin is best seen in the heart muscles and the adrenal glands.

The mode of action of the toxin at the molecular level is well understood. It acts by blocking protein synthesis in susceptible cells by inhibiting polypeptide synthesis. The toxin inhibits the polypeptide synthesis by catalyzing the following reaction:

$$\underset{\text{active}}{NAD^+ + EF\text{-}2} \rightleftharpoons \underset{\text{inactive}}{ADPR\text{—}EF\text{-}2} + \text{nicotinamide} + H^+$$

EF-2 is the Elongation Factor-2, which is required for translocating polypeptide-transfer RNA from the acceptor site to the donor site on the eukaryotic ribosome. It is inactivated by being coupled with the adenosine diphosphate ribose (ADPR) resulting from the cleavage of nicotinamide adenine diphosphate (NAD).

The structure of diphtheria toxin is shown in Figure 5. It consists of two peptide chains, A and B, linked together by easily cleavable peptide and disulfide bonds. Only the A (active) part of the diphtheria toxin molecule is responsible for the reaction above, resulting in the inactivation of Elongation Factor-2, and this active part must therefore be considered as an enzyme. The other part (B, binding) of the toxin molecule is responsible for binding the toxin to the membranes of susceptible cells and facilitating the entry of A into the cytoplasm of the cell, where it catalyzes the reaction above. The whole toxin AB is toxic to susceptible animals or whole cells, whereas part A is not toxic to animals or whole cells, but will catalyze the reaction in cell extracts. On the other hand, the whole toxin AB is ineffective in cell extracts. Cells from species of animals that are relatively insusceptible to diphtheria toxin (e.g., mice) do not contain the receptor for the B-component and therefore do not bind toxin, but protein synthesis in extracts from these cells is blocked by A. This important concept of a bipartite (AB) toxin was first revealed with diphtheria toxin, and has led to similar concepts for other toxins and biologically active proteins (see below).

Although the target of diphtheria toxin — that is, the substrate for the enzymic A component — has been identified (NAD + EF-2), the receptor for the toxin — that is, the substance in the susceptible cell membrane to which the B component binds — has not yet been identified. It may be very difficult to do so, since the affinity constant of diphtheria toxin for susceptible cells is low

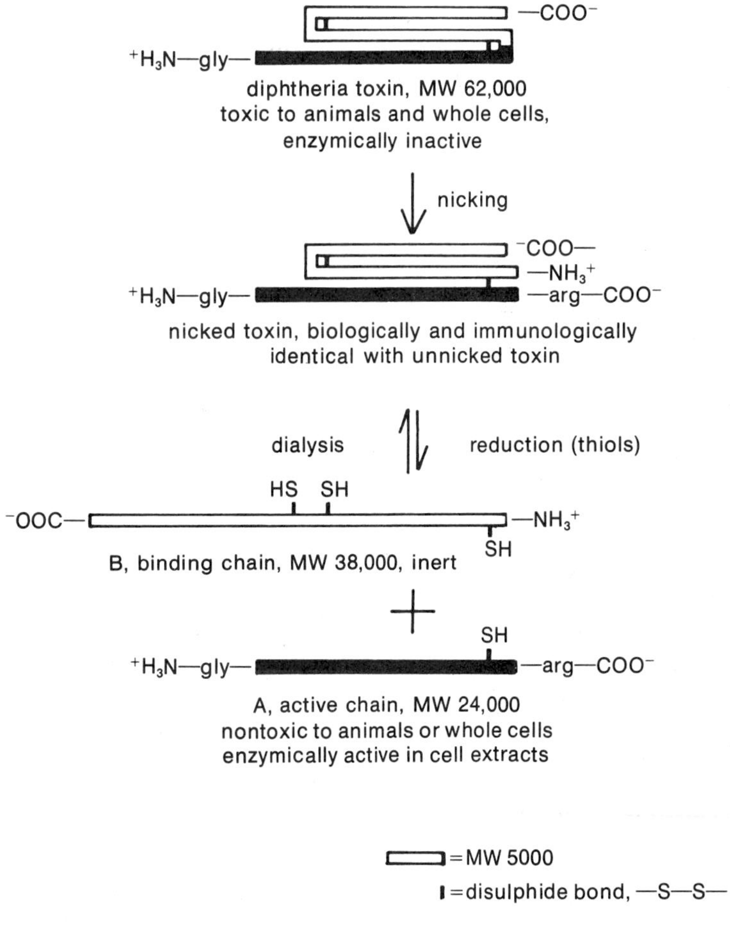

FIGURE 5. *Diagrammatic representation of the structure of diphtheria toxin (see legend to Fig. 2).*

compared with that of other toxins and since very few molecules of toxin are bound per cell.

Further reading: Boquet and Pappenheimer, 1976; Collier, 1975; Collier, 1977; Murphy, 1976; and Pappenheimer and Gill, 1973.

Pseudomonas Aeruginosa Exotoxin A

Recently a lethal, necrotizing toxin (Exotoxin A) has been shown to be produced by most clinical isolates of *Pseudomonas aeruginosa,* which is a major source of hospital infection. The organism is an "opportunistic" pathogen that commonly infects patients whose defense mechanism is impaired because of genetic immunodeficiency, or because they are being treated with immunosuppressive drugs, or because of extensive burns. Exotoxin A is mentioned here because it is similar to diphtheria toxin insofar as it catalyzes the reaction of NAD with EF-2 to form ADPR — EF-2 and thereby blocks protein synthesis in animals, cells, and cell extracts. But there are differences. Unlike whole diphtheria toxin, the whole exotoxin A (MW about 66,000) is active in cell extracts as well as on whole cells; and although a smaller product of natural proteolysis (MW about 33,000) appears to be more active than the whole toxin in cell extracts, it is also active on whole cells. The toxins also differ in their relative toxicity to different species of animals and cell lines, which suggests that they bind to different receptors. There is no immunologic relationship between the two toxins.

Further reading: Collier, 1975.

ENTEROTOXINS

An enterotoxin is a toxin produced in the intestinal tract by an infecting organism. This toxin causes diarrhea or vomiting. This effect generally results from the direct action of the toxin on the intestinal wall (e.g., cholera), but may also follow the absorption of the toxin into the bloodstream and its action elsewhere, as is thought to be the case with the staphylococcal enterotoxins. It is now recognized that acute gastroenteritis is a global medical problem, particularly severe in terms of morbidity, mortality, and economic impact in the less developed countries of the world. It is also being recognized that many of these enteropathies are enterotoxinoses.

Further reading: Banwell and Sherr, 1973; and Craig, 1972.

Cholera Toxin

Cholera toxin, diphtheria toxin, and tetanus toxin are the three toxins that are quite certainly responsible for the harmful effects of the diseases caused by their parent organisms. Cholera is a disease of acute diarrhea, which may be so severe that a patient may lose up to 30 liters of watery stools in a day. Death follows the consequent dehydration. The toxic nature of cholera was recognized by John Snow in 1849, well before the germ theory of disease was established, and then reaffirmed 35 years later by Robert Koch who postulated a cholera toxin; but the existence of such a toxin was not proved until a further 75 years later, when S. N. De showed that cell-free culture filtrates of *Vibrio cholerae* promoted fluid accumulation when introduced directly into the gut. Since then the toxin has been isolated and subjected to a great deal of study all over the world. The toxin gene resides in the chromosomes of both biotypes, *V. cholerae* and *V. El Tor,* and both serotypes, Ogawa and Inaba. Cholera toxin causes diarrhea by stimulating a net output of chloride ions by the brush border cells of the small intestine. The stimulation of chloride secretion follows an increased output of adenosine-3′:5′-cyclic monophosphate (cyclic AMP), which results from stimulation of the enzyme adenylate cyclase by the toxin. The toxin acts on all eukaryotic cells containing this enzyme (and containing a particular ganglioside in the membrane; see below), besides those of the intestinal mucosa, and brings about a great diversity of effects, including increase in capillary permeability when injected into the skin, morphologic changes in Chinese hamster ovary (CHO) cells and in adrenal tumor cells (together with increased steroidogenesis), increased glycogenolysis in liver cells, and lipolysis in fat cells. The stimulation of adenylate cyclase by cholera toxin differs from that caused by hormones such as epinephrine and thyrotropin in that it involves a permanent change in the enzyme. For this reason recovery from cholera depends on the replacement of the affected cells of the intestinal wall, which takes place within hours.

Like diphtheria toxin, cholera toxin is a complex of two parts, A and B, but unlike diphtheria toxin these parts are held together by noncovalent bonds (hydrogen bonds, ionic bonds, hydrophobic interactions, and van der Waals forces, Fig. 6). Under fairly mild conditions the bonds linking A and B are broken, yielding one molecule of A and an aggregate of five molecules of subunits of B. This aggregate of subunits of B may occur along with the intact toxin during purification, and is known as "choleragenoid"; it is biologically inert, and immunologically almost (but not quite) identical with the whole toxin. Part A comprises two parts, held together by a peptide bond and a sulfhydryl bond. The peptide bond is usually nicked by natural proteolysis, and

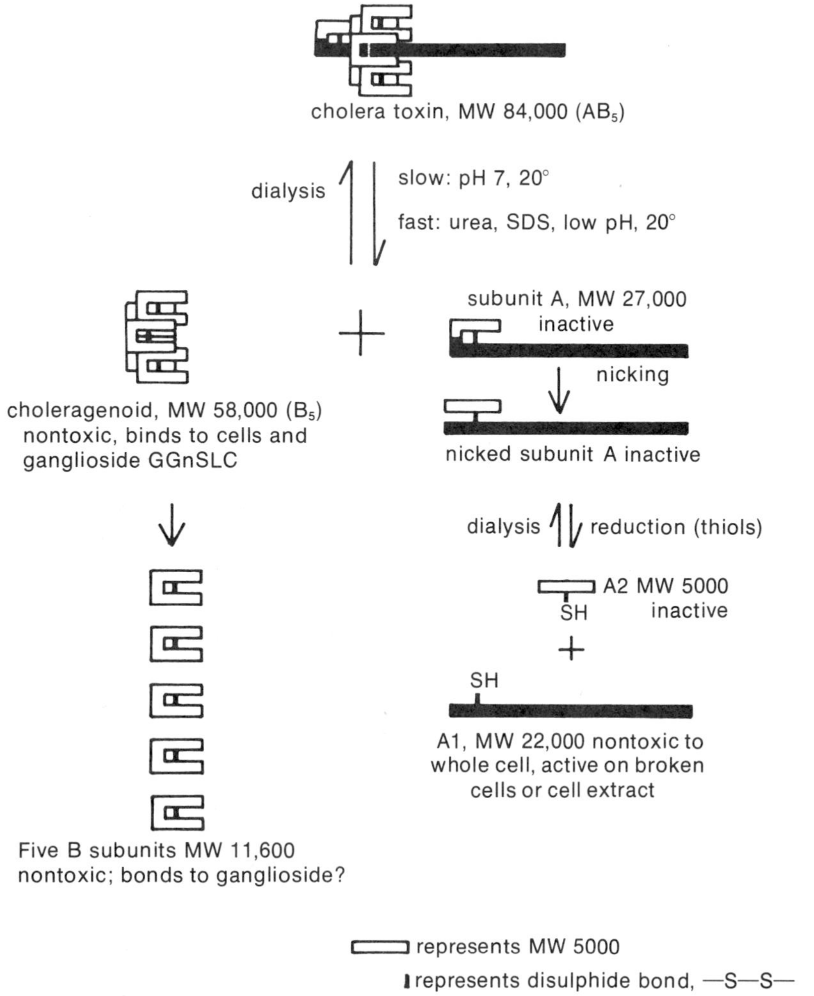

FIGURE 6. *Diagrammatic representation of structure of cholera toxin (see legend to Fig. 2). The A component is held to the B (choleragenoid) component by noncovalent bonds (hydrogen bonds, van der Waals forces, etc.) as are the subunits of the B component. The A2 and A1 fragments of the A component are held together by a nickable peptide bond and a reducible disulfide bond.*

on reduction with a thiol the sulfhydryl bond is also broken to yield a smaller fragment A2 and a larger fragment A1. The biologic activity of cholera toxin resides in the fragment A1. The function of the "choleragenoid" aggregate, like that of the B component of diphtheria toxin, is to bind the toxin to the cell membrane and thus to facilitate the entry of the active component A1 into the cell.

The cell membrane receptor that recognizes and binds the choleragenoid part of cholera toxin has been identified as the sialidase- (or neuraminidase) stable ganglioside GGnSLC, generally known as GM1 (Table 4). Cells that do not contain this ganglioside are not susceptible to the action of the toxin, but if the ganglioside is simply added to the cells it becomes incorporated into their membranes and they become susceptible to this toxin.

The A1 fragment, like diphtheria toxin fragment A, is inactive on the whole cell but active on broken or lysed cells and on extract of the cells; it appears also to be an enzyme catalyzing the cleavage of NAD into ADPR and nicotinamide. The ADPR is bound not to EF2, but probably to some arginine-containing protein, possibly adenylate cyclase itself, and as a result of this the adenylate cyclase is permanently stimulated. The strong evidence that the action of the A1 component of cholera toxin is enzymic refutes the suggestion (Bennett and Cuatrecasas, 1977) that it stimulates adenylate cyclase by combining with it directly.

An important concept with wide implications is revealed by diphtheria and cholera toxins. This concept holds that biologically active proteins can have two distinct and separable parts, one part binding to the cell surface receptor and the other acting on a target within the cell. It is likely that the *Escherichia coli* enterotoxin is similarly constituted (see below); and it may be significant that tetanus toxin appears to consist of two separable components that are inactive singly but active again when rejoined, and that one of these components is responsible for the binding of tetanus toxin to the ganglioside that appears to be the receptor for the toxin in nervous tissue. The concept of a bipartite active protein is not restricted to bacterial toxins. The toxic proteins abrin and ricin from the seeds of two unrelated plant families are similarly composed. The A components both block protein synthesis by interfering with polypeptide chain elongation (as does diphtheria toxin), in this case by inactivating the 60S ribosomal subunits. The cell membrane receptors for the B chains of both toxins are similar and appear to contain lactose residues. Fully active hybrids of ricin A/abrin B and abrin A/ricin B can be constituted. It is not only these toxic bacterial and plant proteins that are so constituted — certain glycoprotein tropic hormones (including thyrotropin and gonadotropin) are also known to consist of alpha and beta units. The alpha units stimulate adenylate cyclase (but not permanently, the way cholera toxin does) and have some amino acid sequence homologies with the A component of cholera toxin; the beta units are responsible for the specific binding of the hormones to their susceptible tissues and have amino acid sequence homologies with the B component of cholera toxin. Moreover, these hormones bind to certain gangliosides and show some specificity in this respect (Table 4).

Further reading: Bennett and Cuatrecasas, 1977; Craig et al., 1976; Finkelstein, 1973; Finkelstein, 1976; Fishman and Brady, 1976; Flores and Sharp, 1975; Moss et al., 1976; Olsnes and Pihl, 1977; and van Heyningen, 1974.

Escherichia Coli Enterotoxin

Enteropathogenicity and the ability to produce both a heat-stable small molecular and a heat-labile protein enterotoxin is conferred on *E. coli* by an enterotoxin plasmid (Ent$^+$). It is likely that these strains of *E. coli* contribute very heavily to enterotoxic enteropathies throughout the world. We will be concerned only with the heat-labile toxin that bears many resemblances to cholera toxin: (1) it stimulates adenylate cyclase, with the same consequences; (2) it cross-reacts immunologically with cholera toxin and choleragenoid, i.e., antisera to cholera toxin and choleragenoid neutralize both toxins, antisera to coli toxin neutralizes cholera toxin poorly; (3) both toxins bind to the ganglioside GGnSLC (but see below). There are, however, some differences. Coli toxin preparations vary considerably in molecular weight and have not yet been properly purified. This may be due to association of the toxin with other materials in the culture filtrate. It is also several orders of magnitude less toxic than cholera toxin, even taking into account the impurity of the preparations tested. Coli toxin appears to be synthesized as a protoxin, which is converted to the active form by proteolytic enzymes (thus it may be fully active in the gut, which contains trypsin-like enzymes, and inactive in the skin). Partially purified coli toxin preparations from several laboratories show faint bands in SDS gel corresponding to molecular weights of 23,000 to 30,000, and when *E. coli* is treated with the antibiotic polymyxin a protein of MW 23,000 is obtained that is very similar to cholera toxin A1 active component, but has some activity on whole cells.

As to the cell receptor for coli toxin, some workers have found that the ganglioside GGnSLC binds coli toxin as well as it binds cholera toxin; others find that coli toxin is far less readily bound. The results so far obtained seem to suggest that although coli and cholera toxins may have the same active component, and although the binding components may bind to the same receptor, they differ in their binding capacity. This might be connected with the variations found in the structure of coli toxin, and the possibility that the toxin might be associated with other components of culture filtrates. These other components may block binding capacity to varying degrees, and thus affect toxicity.

Further reading: Gill et al., 1976; and Craig et al., 1976.

Other Enterotoxins

Salmonella enteritidis and *S. typhimurium* have been reported to produce enterotoxins possessing similarities to the *E. coli* enterotoxin, and *S. typhimurium* has been reported to produce a heat-labile factor inducing capillary permeability in rabbit skin, like *V. cholerae* and *E. coli* enterotoxins. These observations have not yet been confirmed, but they might prove to be important in the future in considerations of the intestinal effects of salmonella infections.

Shigella dysenteriae has been known for a long time to produce an exotoxin that appears to be as potent as botulinum toxin in producing flaccid paralysis in rabbits. The toxin is in fact not a neurotoxin but causes hemorrhages in the spinal cord of the rabbit, which leads to edema, and thus to pressure on nerves. Recently it has been shown that this toxin is a cytotoxic enterotoxin capable of causing fluid accumulation in ligated ileal loops of the rabbit. The role of this toxin in the pathogenesis of shigellosis has not yet been determined. There has been some evidence that toxin-producing strains of *S. dysenteriae* are no more pathogenic than invasive atoxic strains, but it has been proposed that both toxinogenesis and invasiveness play a role in the disease.

Food poisoning due to contamination by *Bacillus cereus, Clostridium perfringens, Staphylococ-*

cus aureus, and *Vibrio parahaemolyticus* is now recognized as a health hazard. *B. cereus* culture filtrate induces fluid accumulation in ligated rabbit ileal loops and an increase in capillary permeability in the skin. Apparently an enterotoxin is produced, but as yet little is known about its nature or its medical significance. Sporulating cultures of *Cl. perfringens* cause fluid accumulation in the gut, possibly owing to decreased absorption rather than increased secretion. A diarrheagenic and emetic toxin, different from the other known toxins of *Cl. perfringens,* has been isolated, with a molecular weight of 35,000. Little is known about the mode of action of the toxin, but it is likely that it is an important factor in *Cl. perfringens* food poisoning. The staphylococcus produces five enterotoxins, A to E, that can be differentiated serologically but have similar physicochemical properties (MW 26,000 to 30,000) and gastrointestinal effects, causing diarrhea and vomiting. The emetic effect is neurologically mediated. The site of the emetic action of these toxins has been shown to be in the abdominal viscera, and the sensory emetic stimulus reaches the vomiting center via the vagus and sympathetic nerves. The diarrheagenic mechanism has not yet been elucidated. The toxins seem most active on the middle segment of the small intestine where they enhance secretory activity without blocking absorption of sodium or water.

Vibrio parahaemolyticus is a major cause of seafood-borne gastroenteritis in Japan and elsewhere. The organism produces a heat-stable protein of MW 45,000 that is hemolytic, cardiotoxic, and capable of causing fluid accumulation in ligated rabbit ileal loops, causing degenerative changes at the same time (unlike cholera toxin). This toxin is blocked by a sialidase-sensitive ganglioside (Table 4).

Further reading: Finkelstein, 1976; Takeda et al., 1976; and van Heyningen, 1971.

OTHER TOXINS

Clostridium Perfringens Alpha Toxin

This toxin is discussed in order to show that a fatal disease resulting from an infection by an organism producing a lethal exotoxin need not necessarily be an exotoxinosis. *Clostridium perfringens* is the main causative organism of gas gangrene, a disease resulting from the infection of deep penetrating wounds, and characterized by tissue destruction and toxemia and shock. Several toxins are produced, including a proteolytic collagenase (kappa), a cardiotoxic oxygen-labile hemolysin (theta), and a tissue-necrotizing oxygen-stable hemolysin (alpha). The alpha toxin is an enzyme, a phospholipase C, which catalyzes the cleavage of phosphyryl choline, phosphoryl ethanolamine, and other compounds from phospholipids. It has been shown that gas gangrene can be prevented in experimentally infected animals by prophylactic active or passive immunization against the alpha toxin, but not against the other toxins. In view of this, and of the undoubted fact that the injection of a lethal dose of alpha toxin can produce symptoms of shock, it is not unreasonable to suggest that *Cl. perfringens* gas gangrene is an exotoxinosis, like tetanus, diphtheria, and cholera, which were discussed earlier.

However, the situation in gas gangrene is much more complicated. Although alpha antitoxin gives good protection when infection is initiated in slightly damaged muscle, it gives little or no protection when infection is initiated in severely damaged muscle, or in slightly damaged muscle when the bacteriostatic action of serum proteins (transferrin) is lowered by pretreating the experimental animals with iron compounds. It would seem that the protective role of alpha antitoxin is not so much due to its neutralizing the lethality of the toxin, but rather to its neutralizing the ability of the toxin to cause tissue necrosis and so pave the way for further growth of the organism. In effect, the role of antitoxin is bacteriostatic. But if other factors prevail that favor bacterial growth, such as severe tissue damage, or if the bacteriostatic action of serum proteins is lowered, then massive growth of the organism results in a number of effects, such as removal of oxygen from the environment; fall in redox potential; marked changes in vascular permeability; loss of plasma protein; and severe hemoconcentration. It may be that these conditions are responsible for the fatal shock of gas gangrene.

Further reading: Bullen, 1970.

Staphylococcal Exfoliative Toxin

Some phage group II strains of staphylococci are responsible for an exfoliative dermatitis (staphylococcal scalded skin syndrome, SSSS) in very young infants, and more rarely in older children and adults. The epidermis can be displaced at the slightest touch, like the skin of a ripe peach, and more than half the body may be denuded to give an appearance resembling severely scalded skin. It is distressing to endure, and to see, but it passes quickly and is rarely fatal, since the denuded areas dry up and the skin is rapidly replaced. The effect can be reproduced experimentally in newborn mice by infecting the skin with group II staphylococci or by injecting sterile culture filtrates. In children and adults exfoliation may also result from sensitivity to certain drugs, including sulfa compounds, phenylbutazone and barbiturates, but in these latter cases the splitting of the skin occurs at the dermoepidermal junction, whereas SSSS is characterized by intraepidermal splitting.

The exfoliative effect of group II staphylococcal skin infection is a clear case of a harmful effect of an infection being due to an exotoxin. This toxin (exfoliatin) has been freed of other active staphylococcal products and purified. It is a protein with reported molecular weights ranging from 24,000 to 33,000. Reports on its heat stability also vary, and it may be that there is more than one form of the toxin. Nothing is as yet known about the mode of action of the toxin at the molecular level.

Further reading: Taylor, 1976.

IMMUNIZATION AGAINST EXOTOXINOSES

Since some or all of the harmful effects of some infections are entirely due to exotoxins, and since these exotoxins are antigenic, the question arises whether protection against the diseases can be obtained by passive or active immunization against the relevant exotoxins. Active immunization obviously has only prophylactic possibilities because the production of sufficiently high levels of antibodies can be attained only some weeks after injection of the antigen. Therapeutic passive immunization is of limited value because exotoxins are rapidly fixed to their susceptible cells, and once fixed they cannot be neutralized by antitoxin. In fact, the bound toxin may not remain on the surface of the cell, where it can be reached by antitoxin, for more than a few minutes. In the case of diphtheria and cholera toxins we know that the active components, A, of the toxins are rapidly "eclipsed," and pass into the interior of the cell where they are unavailable to antitoxin. In the case of cholera toxin we know that the cell-binding component, B, which also has most of the antitoxin binding capacity, remains on the surface of the cell while the A component acts inside; in the case of diphtheria toxin it is not yet known what happens to the B component; in either case the cell-damaging A component is sheltered from antitoxin.

Although antitoxin probably has little therapeutic value, there may be circumstances in which it might have some prophylactic value if it is administered soon enough, especially in suspected tetanus and diphtheria, before all the toxin has reached its susceptible tissue.

Active immunization can be achieved in advance of infection by injecting toxoid, that is toxin that has been rendered nontoxic without affecting its antigenicity, by treatment with agents such as formaldehyde or glutaraldehyde. Obviously it is feasible to consider actively immunizing only subjects that are at risk, such as children against diphtheria and tetanus, and soldiers against tetanus and gas gangrene. Pregnant women should also be immunized against tetanus to protect their babies from umbilical tetanus. It is not feasible to immunize infants against SSSS, for example, because the disease is very rare. It may be possible to immunize soldiers against gas gangrene toxin, but, as we have seen, it is likely that only those who receive wounds that do not bring about much tissue damage will be protected. This might still be worthwhile.

It might be instructive to compare the problems of active immunization against the three frank exotoxinoses we have already discussed: diphtheria, tetanus, and cholera.

People who have had diphtheria are probably immune to it, or at least to its worst effects, for the rest of their lives; they have been exposed to enough toxin to ensure this. The same state of affairs can be achieved by active immunization with purified toxoid. In developed countries where active immunization with diphtheria toxoid is carried out extensively, diphtheria has more or less disappeared. In countries where active immunization is not practiced, it is rife, and only survivors of the disease, or people who have had frequent contact with the organism without getting diseased, are immune.

People who have had tetanus and survived are no more immune to the disease than people who have not had it. The reason for this is that a sublethal (or lethal) dose of tetanus toxin is such a small amount of antigen (a few picograms or nanograms) that it is quite inadequate to elicit immunity. In Brazil and in India there is some evidence that some members of the indigenous populations may have protective levels of antitoxin in their sera, probably as a result of frequent experience of sublethal amounts of toxin. Active immunity to the disease is easily attained by immunization with the purified toxoid. In the early phase of World War I, when immunization was not practiced, the incidence of tetanus in the British Army was nearly 7 in every 1000 wounded; in World War II only 35 British soldiers died of tetanus and there are good reasons to doubt whether half of them had in fact been immunized.

How effective is antitoxic active immunization against cholera? The answer is still being sought. Although cholera is a frank exotoxinosis, it is otherwise a very different disease from either diphtheria or tetanus. For one thing, the toxin acts enterally not parenterally as in the other two diseases, and for another, the disease is very quick, with the time interval between onset and death being measured in hours rather than days.

The enteral nature of the disease suggests that for the actively produced antitoxin to be effective it must be present in the small intestine, secreted from the wall of the intestine in the form of IgA antibodies. Research is at present being conducted on the practical means of bringing about such enteral active immunity to the toxin. The rapid course of the disease means that there is not enough time for the anamnestic response of antibody production to secondary antigenic stimulus to play an important part. Protection therefore must probably depend on there being sufficient antibody present enterally at the time of infection.

If effective antitoxic immunity to cholera toxin can ever be achieved, it will have the very important added advantage that such immunity will also be effective against many *E. coli* diarrheas, since cholera antitoxin neutralizes *E. coli* toxin.

References

Arnon, S. S., Midura, T. F., Damus, D., Wood, R. M., and Chin, J.: Intestinal infection and toxin production by *Clostridium botulinum* as one cause of sudden infant death syndrome. Lancet 1:1273, 1978.

Banwell, J. G., and Sherr, H.: The effect of bacterial enterotoxins on the gastrointestinal tract. Gastroenterology 65:467, 1973.

Bennett, F., and Cuatrecasas, P.: Cholera toxin: Membrane gangliosides and activation of adenylate cyclase. In Cuatrecasas, P. (ed.): The Specificity and Action of Animal, Bacterial and Plant Toxins. London, Chapman and Hall, 1977, p. 1.

Bizzini, B.: Tetanus toxin structure as a basis for elucidating its immunological and neuropharmacological activities. In Cuatrecasas, P. (ed.): The Specificity and Action of Animal, Bacterial and Plant Toxins. London, Chapman and Hall, 1977, p. 175.

Boroff, D. A., and DasGupta, B. R.: Botulinum toxin. In Kadis, S., Montie, T. C., and Ajl, S. J. (eds.): Microbial Toxins Volume IIA. New York and London, Academic Press, 1971, p. 1.

Boquet, P., and Pappenheimer, Jr., A. M.: Interaction of diphtheria toxin with mammalian cell membranes. J Biol Chem 251:5770, 1976.

Bullen, J. J.: Role of toxins in host-parasite relationships. In Ajl, S. J., Kadis, S., and Montie, T. C. (eds.): Microbial Toxins I. New York and London, Academic Press, 1970, p. 233.

Collier, R. J.: Diphtheria toxin: Mode of action and structure. Bacter Rev 39:54, 1975.

Collier, R. J.: Inhibition of protein synthesis by exotoxins from *Corynebacterium diphtheriae* and *Pseudomonas aeruginosa*. In Cuatrecasas, P. (ed.): The Specificity and Action of Animal, Bacterial and Plant Toxins. London, Chapman and Hall, 1977, p. 67.

Craig, J. P.: The enterotoxic enteropathies. In Symposia of the Society for General Microbiology XXII, Microbial Pathogenicity in Man and Animals. Cambridge, The University Press, 1972, p. 129.

Craig, J. P., Benenson, A. S., Hardegree, M. C., Pierce, N. F., and Richardson, S. H. (eds.): The Structure and Functions of Enterotoxins. J Infect Dis 133: March Supplement, 1976.

Finkelstein, R. A.: Cholera. CRC Critical Reviews in Microbiology 2:553, 1973.

Finkelstein, R. A.: Progress in the study of cholera and related enterotoxins. In Bernheimer, A. W. (ed.): Mechanisms in Bacterial Toxinology. New York, London, Sydney, Toronto, John Wiley and Sons, 1976, p. 53.

Fishman, P. H., and Brady, R. O.: Biosynthesis and function of gangliosides. Science 194:906, 1976.

Flores, J., and Sharp, W. G.: Effects of cholera toxin on adenylate cyclase. Studies with guanylimidodiphosphate. J Clin Invest 56:1345, 1975.

Gill, D. M., Evans, Jr., D. J., and Evans, D. G.: Mechanism of activation of adenylate cyclase in vitro by polymyxin-released heat-labile enterotoxin of *Escherichia coli*. J Infect Dis 33:S103, 1976.

Helting, T. B., Swizler, O., and Wiegandt, H.: Structure of Tetanus toxin. II Toxin binding to ganglioside. J Biol Chem 252:194, 1977.

Mellanby, J., and Pope, D.: The relationship between the action of tetanus toxin and its binding to membranes and gangliosides. In Porcellati, G., Ceccarelli, B., and Tettamanti, G. (eds.): Ganglioside Function. New York and London, Plenum Press, 1976, p. 215.

Moss, J., Manganiello, V. C., and Vaughan, M.: Hydrolysis of nicotinamide adenine dinucleotide by choleragen and its A protomer: Possible role in the activation of adenylate cyclase. Proc Nat Acad Sci U.S.A. 73:4424, 1976.

Murphy, J. R.: Structure activity relationships of diphtheria toxin. In Bernheimer, A. W. (ed.): Mechanisms in Bacterial Toxinology. New York, London, Sydney, Toronto, John Wiley and Sons, 1976, p. 31.

Olsnes, S., and Pihl, A.: Abrin, ricin, and their associated agglutinins. In Cuatrecasas, P. (ed.): The Specificity and Action of Animal, Bacterial and Plant Toxins. London, Chapman and Hall, 1977, p. 129.

Pappenheimer, Jr., A. M., and Gill, D. M.: Diphtheria. Science 182:353, 1973.

Takeda, Y., Takeda, T., Honda, T., and Miwatani, T.: Inactivation of the biological activities of the thermostable direct haemolysin of Vibrio parahaemolyticus by ganglioside GT1. Infect and Immun 14:1, 1976.

Taylor, A. G.: Toxins and the genesis of specific lesions: enterotoxin and exfoliatin. In Bernheimer, A. W. (ed.): Mechanisms in Bacterial Toxinology. New York, London, Sydney, Toronto, John Wiley and Sons, 1976, p. 195.

van Heyningen, S.: Binding of ganglioside by the chains of tetanus toxin. FEBS Letters 68:5, 1976.

van Heyningen, W. E.: The fixation of tetanus toxin, strychnine, serotonin and other substances by ganglioside. J Gen Microbiol 31:375, 1963.

van Heyningen, W. E.: General characteristics. In Ajl, S. J., Kadis, S., and Montie, T. C. (eds.): Microbial Toxins Volume I. New York and London, Academic Press, 1970, p. 1.

van Heyningen, W. E.: The exotoxin of *Shigella dysenteriae*. In Kadis, S., Montie, T. C., and Ajl, S. J. (eds.): Microbial Toxins Volume IIA. New York and London, Academic Press, 1971, p. 255.

van Heyningen, W. E.: Gangliosides as membrane receptors for tetanus toxin, cholera toxin and serotonin. Nature 249:415, 1974.

van Heyningen, W. E., and Mellanby, J.: Tetanus toxin. In Kadis, S., Montie, T. C., and Ajl, S. J. (eds.): Microbial Toxins Volume IIA. New York and London, Academic Press, 1971, p. 69.

BACTERIAL ENDOTOXINS 6

Abraham I. Braude, M.D.

The lipopolysaccharides (LPS) in the cell wall of gram-negative bacteria can cause hypotension, shock, fever, intravascular coagulation, and death. Because of this toxicity, and because they are incorporated within the bacterial cell wall, they are called endotoxins. Exotoxins, by contrast, are not structural components of the cell, and are released from the bacteria, so that their

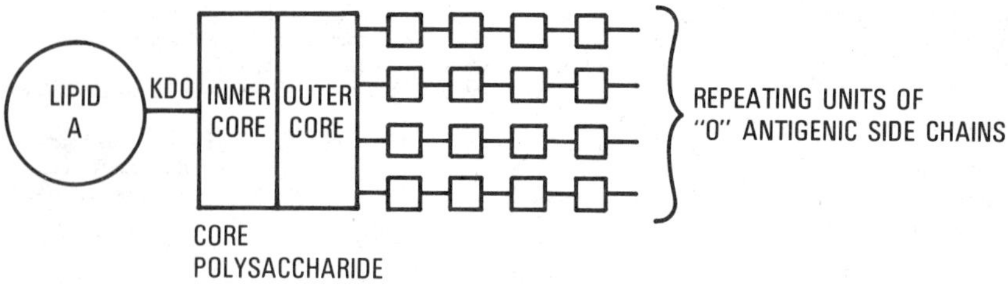

FIGURE 1. *Diagram of lipopolysaccharide showing three main regions: lipid A, core, and "O" antigens. Note key position of KDO as a link between lipid A and polysaccharides. KDO is attached via a ketoacidic linkage to glucosamine residues of lipid A.*

biologic activity is far greater in the culture filtrates than in the parent cell suspension. Although endotoxin may also escape into surrounding fluids, the whole cell of gram-negative bacteria retains the major portion of the endotoxic activity.

CHEMICAL STRUCTURE OF ENDOTOXINS

The toxic LPS or endotoxins of smooth bacteria are macromolecules composed of three main regions: Lipid A, core polysaccharide, and "O" antigens (Fig. 1).

Each of these regions has a unique composition. The "O" antigens are made up of a series of repeating oligosaccharide units, each composed of three or four different hexoses. There may be as many as 10 such units in the smoothest gram-negative bacteria. Although the lipid A and core polysaccharide of most gram-negative bacteria share a similar, if not identical, structure, the "O" antigens vary with each species and serologic type of organism. In rough bacteria, the "O" antigens are lost through a mutation that deprives the bacteria of enzymes required to synthesize the "O" antigen or attach them to the core. These rough bacteria thus lose hydrophilic surface properties provided by the abundant external sugars and no longer form smooth suspensions in liquid cultures (Fig. 2) or smooth colonies on solid media.

Lipid A

The key to isolation of lipid A is KDO, a unique sugar with the formula 2-keto-3-deoxyoctonate (Heath and Ghalambor, 1963). A trisaccharide of KDO links lipid A to the core region, and mild acid hydrolysis will cleave the ketosidic linkage

FIGURE 2. *Comparison of smooth parent culture of E. coli 0113 with rough mutant. The rough mutant is galactose deficient and cannot attach the "O" antigenic sugars to the core. The absence of "O" antigen deprives the LPS of abundant hydrophilic surface sugars and exposes the central hydrophobic lipids that are insoluble in water. (Rough mutant is on the right.)*

FIGURE 3. *Diagram of proposed structure of lipid A from S. minnesota. Note the following key structural features as indicated by arrows: 1) amide-linked β hydroxymyristic acid, 2) β-1,6 linkage of glucosamine disaccharide, 3) ester-linked long-chain fatty acids (R) at hydroxyl groups in positions 3,4, and 6 — these fatty acids are either lauric, palmitic, and 3-myristoxymystic, 4) phosphate groups at 1 and 4′ may serve as bridges connecting the dissacharide subunits, and 5) KDO trisaccharide.*

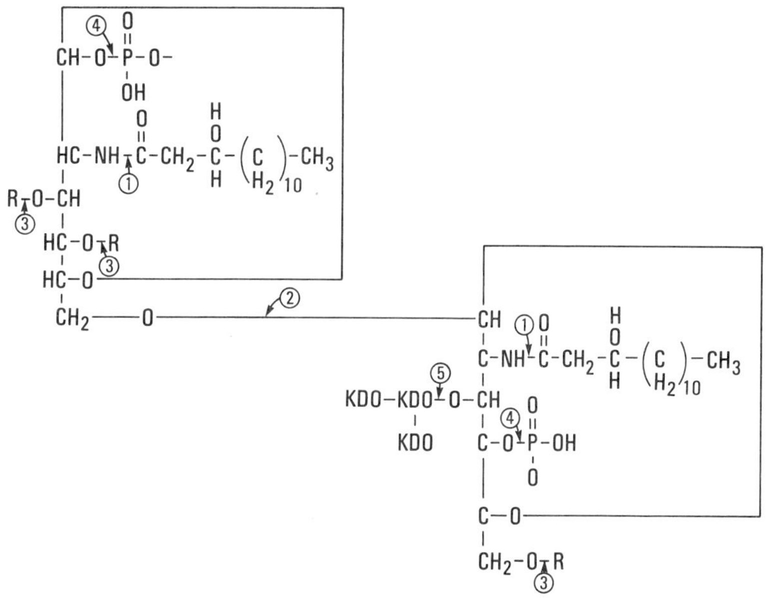

of KDO to lipid A, yielding water-insoluble preparations of lipid A with a molecular weight of approximately 2000. Another approach to the isolation of lipid A is through the use of mutants that synthesize a defective polysaccharide containing only lipid A and KDO. Mutants without lipid A have never been isolated, and therefore it appears indispensable for survival of bacteria.

Lipid A in most gram-negative bacilli of medical importance is composed of glucosamine-4-phosphate, long-chain fatty acids, and ethanolamine. [In certain saprophytes such as *Rhodopseudomonas,* glucosamine is absent from the lipid A (Hase and Rietschel, 1976)]. Thus, lipid A is an unusual phospholipid that uses D-glucosamine instead of glycerol as a skeleton. Disaccharides of glucosamine have β1,6 linkages in *Salmonella, Shigella, Escherichia coli, Yersinia, Fusobacterium, Pseudomonas,* and *Proteus.* The disaccharide subunits appear to be connected to each other through pyrophosphate or phosphodiester groups. The fatty acids are attached to the glucosamine subunit by ester or amide linkages (Fig. 3). In lipid A of *Salmonella minnesota,* for example, the hydroxyl groups in positions 6′, 3, and 4 are esterified by lauric, palmitic, and 3-myristoxymyristic acid. The amino groups, on the other hand, are acylated with β-hydroxymyristic acid (3-hydroxytetradecanoic acid), a 14-carbon compound characteristically present in enterobacterial lipid A. Other β-hydroxy-fatty acids containing 10 to 17 carbon atoms are characteristic of different bacteria. In *Pseudomonas,* for example, the lipid A contains β-hydroxymyristic as in the enteric bacteria. These long-chain fatty acids give lipid A its lipoidal characteristics.

Core Polysaccharide

This region contains not only KDO, the link to lipid A, but also heptose, phosphate, ethanolamine, and three hexoses. The hexoses, which consist of galactose, glucose and *N*-acetyl glucosamine, are designated the outer core, whereas the remainder is called the inner core. Much of the information on core sugars was obtained by examination of *Salmonella* mutants that were blocked in different steps in the biosynthesis of the core. These blocks resulted from defective activity of enzymes involved in transfer (transferase) or synthesis (synthetase; epimerase) of sugars. The sugars are transferred as sugar-nucleotides, usually as derivatives of uridine diphosphate (UDP) or guanosine diphosphate (GDP). The enzymes are associated with the cytoplasmic membrane where they catalyze the stepwise addition of the sugar-nucleotides to the nonreducing terminus of the growing core polymer. The inner core is constructed in the membrane by stepwise addition of KDO, heptose phosphate, and ethanolamine, followed by the hexose units of the outer core (Nikaido, 1973).

In rough mutants containing only KDO and heptose in the core, the LPS is designated chemotype Rd (Lüderitz et al., 1966). Such mutants lack the enzyme UDP-glucose synthetase so that synthesis of the core is not completed. In mutants with LPS designated Rc, the enzyme UDP galactose-4-epimerase is deficient so that glucose, but not galactose, is synthesized and incorporated into the core. The molecular weight of the Rc mutant is about 10,000. In chemotype Rb, the LPS contains all basal sugars except *N*-

acetylglucosamine, but the deficient enzyme has not been identified. In chemotype Ra, the core LPS is completed but the "O" side chains cannot be attached. The stepwise buildup of the core can be appreciated from its sugar sequences in the following diagrams showing each of the chemotypes of *Salmonella* LPS:

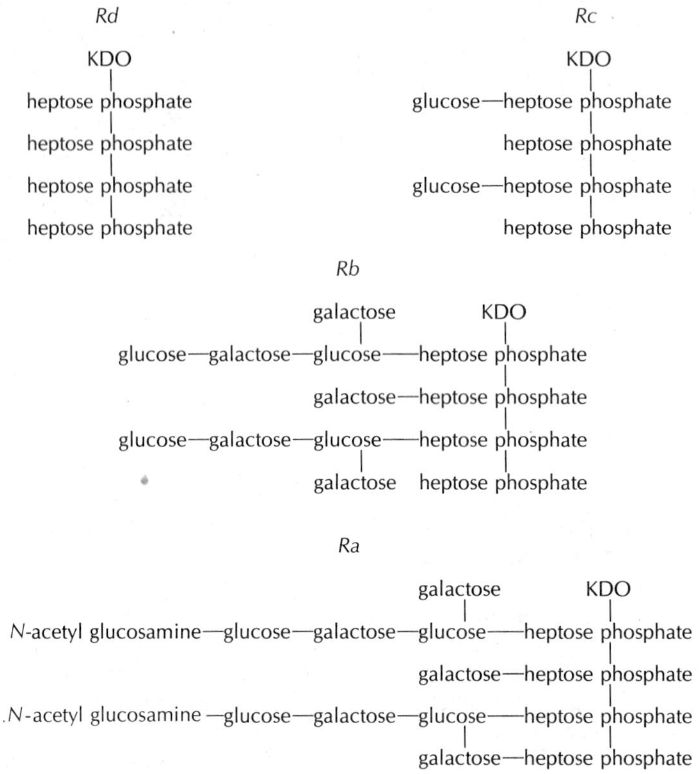

"O" Antigens

The outermost region of the LPS molecule consists of repeating oligosaccharide units containing usually three to five sugars each. Among these are certain unusual sugars that are rare except in gram-negative bacteria. Such rare sugars as 6-deoxyhexoses and 3,6-dideoxyhexoses are among the antigenic determinants that help classify the LPS and bacteria into serotypes through the use of specific antisera. Fucose and rhamnose are 6-deoxyhexoses commonly present in "O" side chains. Paratose, abequose, colitose, and tyvelose are important α-3,6-dideoxyhexoses (Table 1). The synthesis of "O" side chains apparently begins on the inner surface of the cytoplasmic membrane where the oligosaccharide units are constructed (Nikaido, 1973). In *S. typhimurium,* for example, the unit is a tetrasaccharide built up by sequential addition of galactose, rhamnose, mannose, and the 3,6-dideoxyhexose abequose. The sugars are transported as nucleotides, and the tetrasaccharide is synthesized on a carrier lipid. The carrier lipid apparently transfers the tetrasaccharide across the cytoplasmic membrane where the units are polymerized into "O" side chains before attachment to the core region. The carrier lipid is a phosphorylated C55 polyisoprenoid alcohol (undecaprenol) with the following formula:

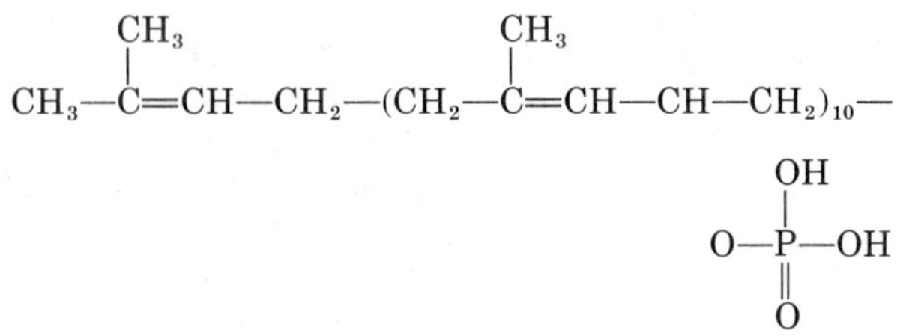

This carrier lipid is probably identical to that which functions in the biosynthesis of cell wall peptidoglycan.

The chemical structure of many O-specific side chains has been determined by biosynthetic studies. The composition of the repeating oligosaccharide units in these side chains is given for certain pathogenic gram-negative bacilli in Table

TABLE 1. Composition of Repeating Oligosaccharide Units in Certain Pathogenic Gram-Negative Bacteria

BACTERIA	STRUCTURE	IMMUNE DOMINANT SUGAR	GROUP OR SEROGROUP
Salmonella typhosa	α-Tyv (on Man), α-Glc (on Gal) 2—(α)—Man—1,4—Rha—1,3 α-Gal—1	α-Tyvelose	D
Salmonella typhimurium	α-Abe (on Man), α-Glc$_{1,4}$ (on Gal) 2—d—Man—1,4—Rha → α-Gal—1	α-Abequose	B
Salmonella paratyphi	α-Par$_{1,3}$ (on Man), α-Glc$_{1,4}$ (on Gal) 2—α-Man—1,4—Rha—1,3—α-Gal—1	Paratose	A
Salmonella thompson	Glc (on third Man) Man—1,2—Man, 1,2—Man 1,2—Man—1	Glucose	C_1
Salmonella anatum	6—B Man—1,4—Rha—1,3 α-Gal—1	Mannose	E_1
Shigella flexneri	Ac 6—Glc Nac—1,2—Rha—1,4—Rha—1	Rhamnose	2a

Glc = glucose; Gal = galactose; Rha = rhamnose; Man = mannose; Par = paratose; Glc Nac = *N*-acetyl glucosamine; Ab = abequose; Tyv = tyvelose; Ac = acetyl

1. It should be noted that for each repeating unit there is one or more immunodominant sugar that is most responsible for the serologic specificity of antiserum against each "O" antigen (Lüderitz et al., 1966). It is the sugar with the highest affinity for the reactive site of the "O" antibody, as determined by its ability to inhibit serologic reactions. From Table 1 it can be seen that the immunodominant sugar may be a terminal nonreducing sugar in the side chain (α-tyvelose in *S. typhosa*) or main chain (rhamnose in *S. flexneri*). It may also be a nonterminal sugar in the main chain.

BIOLOGIC ACTIVITY OF ENDOTOXINS

Injections of small doses of endotoxins into experimental animals bring about dramatic changes in blood pressure, clotting, body temperature, circulating blood cells, metabolism, humoral immunity, cellular immunity, and resistance to infection. A large dose is lethal. These changes can be elicited with LPS from gram-negative bacteria or the intact organisms. There are many ways to isolate LPS but the most widely used is the phenol-water method of Westphal et al. (1952). The organisms are extracted with 45 per cent aqueous phenol at 65 to 68° C. After cooling, the LPS is recovered in the aqueous phase, and must then be separated by ultracentrifugation from RNA, which also partitions into the aqueous layer. In rough organisms, the LPS is hydrophobic and is found mainly in the phenol phase; recovery from the aqueous phase is poor. Rough LPS is best extracted by a solvent containing phenol, petroleum ether, and chloroform and recovered by ultracentrifugation (Galanos et al., 1969).

General Response to LPS

After intravenous injection of LPS the physiologic and biochemical changes depend on its dose and the animal species. In most animals (rabbits, dogs, monkeys, swine) injection of LPS in doses ranging from 0.1 to 100 μg will cause rapid onset of fever, neutropenia, and hypotension. The onset of neutropenia occurs almost immediately after injection of LPS, but fever and hypotension follow quickly, usually within 30 minutes. With large doses of LPS the hypotension may become irreversible and cause fatal shock. The pattern of fever, neutropenia, and hypotension has also been seen in patients given LPS or intravenous typhoid vaccine for fever therapy.

Fever

The most detailed studies of fever after injection of LPS have been carried out in rabbits. After intravenous injection of 0.1 μg LPS from most smooth bacteria the temperature begins to rise after a latent period of 10 to 20 minutes and reaches a peak at about 70 minutes. After the first peak the temperature declines slightly but starts to rise again at 2 hours and reaches a second higher peak at about 3 hours. The second peak alone is dose dependent; it is absent with minimal pyogenic doses of LPS and gradually increases with rising doses until it reaches a ceiling that cannot be exceeded. According to current concepts the fever occurring after intravenous injection of endotoxin is mediated by a protein released from neutrophils or monocytes and designated endogenous pyrogen (EP) (Atkins and Bodel, 1972). No latent period precedes the rise in fever after injection of EP because it stimulates the fever centers in the hypothalamus directly without the intermediate reaction between pyrogen and leukocyte that occurs with LPS. It is possible, however, that endotoxin itself can stimulate the fever centers directly because the amount of LPS needed to cause fever after injection into the cerebrospinal fluid is only one-one thousandth that required by intravenous injection, and the latent period is less after intraspinal injection (Bennett et al., 1957). These findings would suggest that fever production requires EP in bacteremia but not in meningitis (e.g., meningococcemia vs meningococcal meningitis).

One of the classic features of endotoxin fever is tolerance (Fig. 4). The tolerance phenomenon occurs after repeated injections of the same dose of LPS and is characterized by a diminution in the height and duration of fever with successive injections. In rabbits this causes mainly a decline in the second peak. Two forms of tolerance are seen: (1) *early* tolerance, which is seen within 12 hours of the first injection; and (2) *late* tolerance, which appears gradually over several days. Early tolerance has been attributed to depletion of EP, or to a refractory response of the heat centers, which protects against sustained hypothermia. Late tolerance results from an immune reaction in which circulating antibody neutralizes LPS. Late tolerance can be transferred passively with serum, and causes accelerated disappearance of LPS into the reticuloendothelial system (Carey et al., 1958).

These febrile reactions are seen in nearly all mammals, including humans. In fact, man is far more sensitive to the pyrogenic action of LPS than any other animal, requiring 10 times less LPS than rabbits for the same febrile response (Wolff, 1973). Man also exhibits a longer latent period between injection of LPS and onset of fever and the fever curves are always monophasic. In tolerant human subjects the monophasic fever disappears completely, whereas tolerant rabbits maintain a monophasic response.

Rats and mice are exceptions to the rule that LPS injection causes fever in mammals. These two rodents respond to injection of LPS by a depression in body temperature.

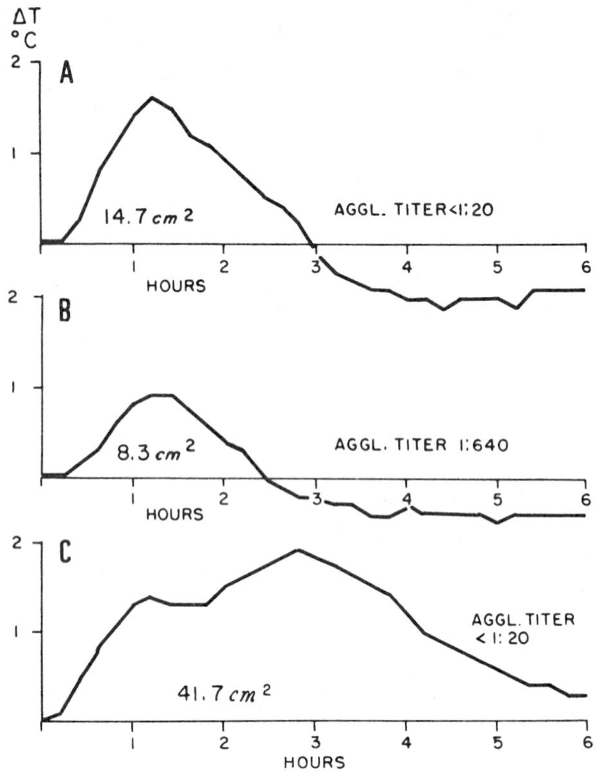

FIGURE 4. *A comparison in rabbits of A, early tolerance, B, late tolerance, and C, no tolerance to the pyrogenic activity of lipopolysaccharides. Note biphasic fever peaks in nontolerant rabbits and disappearance of second peak after tolerance develops. Curve "A" is produced in rabbits challenged 24 hours after one I.V. injection and curve "B" after the last of seven daily I.V. injections of 1.0 μg of endotoxin. Curve "C" was obtained in control rabbits with no prior injections (Milner, 1973.)*

Hypotension

The blood pressure may fall in animals given the small doses of endotoxin required to produce fever, but sustained hypotension and shock usually require larger doses of LPS. For example, 100 μg *Escherichia coli* endotoxin will induce severe hemodynamic disturbances in 75 per cent of rabbits after intravenous injection.

Severe hypotension occurs about 30 minutes after injection of LPS and is the terminal manifestation of a series of preceding hypodynamic disturbances. The first of these in rabbits is a rise in pulmonary artery pressure due to an increase in acute resistance in the pulmonary veins. This obstruction to flow causes a drop in cardiac output with a secondary fall in arterial pressure as seen in Figure 5 (Neuhof, 1975). In dogs the first reaction is constriction of hepatic and small mesenteric veins so that blood is pooled in the mesenteric circulation. The resultant decrease in ve-

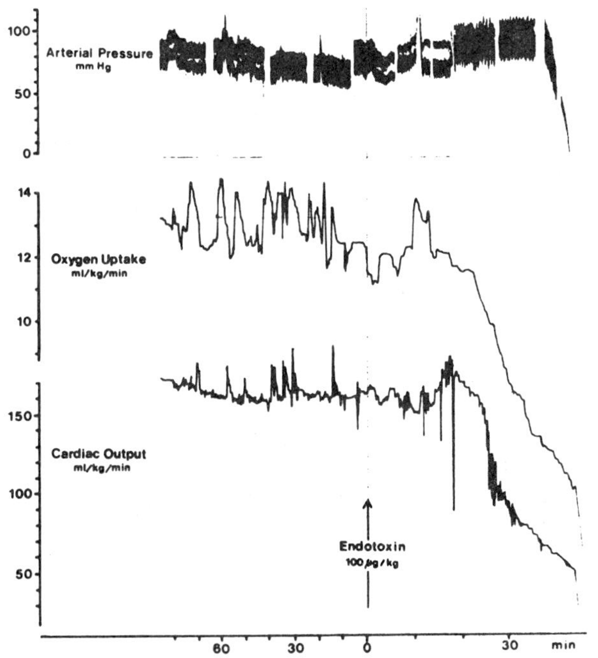

FIGURE 5. *Early phase of endotoxin shock in the rabbit. Fall in cardiac output precedes drop in blood pressure (Neuhof, 1975).*

nous return to the heart lowers the cardiac output but prevents pulmonary hypertension despite increased resistance in the pulmonary vascular system. The fall in cardiac output in all animals undergoing endotoxic shock results in reduced oxygen supply to the tissues. The low arterial pressure causes, in turn, reduced coronary perfusion.

In contrast to shock from fluid loss by hemorrhage, diarrhea, sweating, or vomiting, the hypotension in endotoxic shock cannot be corrected by administration of intravenous saline or blood. Fluid administration is likewise ineffective in overcoming shock in patients with heavy bacteremia due to gram-negative organisms.

Intravascular Coagulation

The local and generalized Shwartzman reactions are the classic manifestations of abnormal clotting from endotoxin (Table 2). To produce either phenomenon animals are given two injections of endotoxin 12 to 18 hours apart. In the local Shwartzman reaction, the first dose is injected intradermally and the second intravenously. An area of hemorrhagic necrosis occurs at the site of intradermal injection within a few hours after the second injection. In the generalized Shwartzman reaction both injections of LPS are given intravenously and the animals develop bilateral cortical necrosis of the kidneys within a few hours after the second dose.

Both reactions are the result of intravascular coagulation and the selective occlusion of small vessels by fibrin. The mechanism proposed for the local reaction involves two steps: first, the intravenous dose initiates intravascular coagulation with production of fibrin polymers; second, these polymers are trapped in the skin vessels that had been injured by the inflammatory reaction around the intradermal dose of LPS. There is no renal cortical necrosis because most of the circulating fibrin is cleared by the reticuloendothelial system (RE). In the generalized Shwartzman reaction, the first intravenous dose of LPS initiates coagulation and the second blocks RE removal of fibrin. The excess circulating fibrin is then filtered by the glomerular capillaries, which become occluded so that the renal cortex is infarcted. Small vessels in other organs may also filter fibrin. Both local and generalized reactions can be prevented by anticoagulants such as heparin.

The mechanism by which endotoxin initiates coagulation is not clear but there is good evidence that both intrinsic (plasma) and extrinsic (tissue) factors are involved (Morrison and Ulevitch, 1978). Clotting via the *intrinsic* pathway appears to be initiated when endotoxin activates Hageman factor (XIII) through a reaction involving the negatively charged phosphate residues of

TABLE 2. Comparison of Local and Generalized Shwartzman Reactions

TYPE OF REACTION	FIRST DOSE	SECOND DOSE	INTERVAL BETWEEN DOSES (Hours)	DOSE OF ENDOTOXIN (mg), e.g., *E. coli* 0111 LPS First	Second	MANIFESTATIONS OF REACTION	CLINICAL EXAMPLE
Local Shwartzman	Intradermal	Intravenous	21–24	0.25	0.03	Hemorrhagic necrosis of skin	Meningococcal purpura
Generalized Shwartzman	Intravenous	Intravenous	21–24	0.125	0.03	Bilateral renal cortical necrosis	Bilateral renal cortical necrosis in *E. coli* septicemia of pregnancy

lipid A. Hageman activation requires that a complex be formed between endotoxin and Hageman factor, probably by binding it to lipid A. Activation of Hageman factor eventuates in clotting because it initiates steps leading to the conversion of prothrombin to thrombin. The *extrinsic* pathway has been implicated because blood leukocytes are needed for both the local and generalized Shwartzman reactions, which are prevented by nitrogen mustard and other agents that cause leukopenia. Among the leukocytes, monocytes are currently regarded as the important circulating mediator cell for releasing the tissue factor (TF) that initiates clotting through the extrinsic pathway. TF is a lipoprotein in the plasma membrane of various cells, which complexes with factor VII and calcium ions to activate factor X. Activated factor X then converts prothrombin to thrombin.

The intravascular coagulation that leads to the generalized Shwartzman reaction depletes clotting factors such as fibrinogen, platelets, and prothrombin. During intravascular coagulation, fibrinolysis is activated so that fibrin-degradation fragments (or fibrin split-products) accumulate in the blood. These fragments become anticoagulants since they inhibit both proteolysis of fibrinogen by thrombin and the polymerization of fibrin monomer to form a clot. In addition, the split-products inhibit platelet aggregation. This combined effect of split-products on clotting and platelet function can cause serious bleeding in patients whose clotting factors are consumed by disseminated intravascular coagulation in meningococcemia and other severe infections caused by gram-negative bacteria (Davis and Arnold, 1974).

In Vitro Coagulation

The addition of endotoxin to whole blood markedly shortens clotting time by activating Hageman factor and by stimulating leukocytes to release TF.

Endotoxin also clots the lysate prepared from the amebocytes of the horseshoe crab, *Limulus polyphemus* (Levin and Bang, 1964). This reaction has become important because clotting occurs in the presence of extremely small concentrations of endotoxin. The limulus lysate contains a clottable protein that forms a gel when exposed to as little as 0.0005 μg/ml of LPS. Endotoxin activates a high molecular weight (84,000) enzyme, which then reacts with a clottable protein to produce a gel. This reaction has been used to detect endotoxin in drugs, fluids, and other materials prepared for injection into patients. Limulus lysate has also been used to demonstrate endotoxin in the blood and spinal fluid of patients with gram-negative bacterial infections.

Disturbances in Circulating Blood Cells

Severe neutropenia occurs within minutes after injection of LPS into rabbits and other animals and persists at levels less than 10 per cent of normal for 4 hours (Fig. 6). The neutropenia is followed by leukocytosis, with normoblasts and many other immature cells appearing in the circulation. The neutropenia results from sequestration of neutrophils in capillaries of the lung and other organs; and the leukocytosis results from the release of granulocytes from the bone marrow. Higher doses of LPS are required for neutropenia than for leukocytosis, probably because the numbers released from the marrow mask the mild sequestration of neutrophils produced by small doses. Because human beings cannot tolerate the relatively large doses given to rabbits and other animals, neutropenia has not always been observed in human studies but leukocytosis is constant and reproducible in them (Wolff, 1973).

The number of circulating platelets also falls rapidly after intravenous injection of LPS, and they are found in leukocyte-platelet thrombi or platelet-aggregates in small vessels of the lung and liver. As the platelets fall, there is a simultaneous release from them of platelet constituents such as serotonin and platelet factor 8. This rapid drop in platelets requires complement. From in vitro experiments it appears that endotoxin also causes platelet lysis through a reaction that requires the alternative pathway of complement. Primates, including man, appear to be much less susceptible to thrombocytopenia after endotoxin injection than are lower mammals, and primate platelets do not seem to react to endotoxin in vitro. In contrast to rabbit platelets, which undergo massive aggregation in platelet-rich plasma, human platelets do not aggregate when exposed to endotoxin (Morrison and Ulevitch, 1978).

Lymphocytes fall less precipitously than granulocytes and platelets and return more slowly to the circulation. They are also unaffected by tolerance. Daily injections of the same dose of endotoxin elicit a shorter period of neutrophil response. These tolerant animals develop only a brief period of neutropenia and a brisk rise in circulating neutrophils that begins between 1 and 2 hours after injection of endotoxin (Fig. 6). Lymphocytes, on the other hand, maintain their slow decline in tolerant animals.

Immune Reactions: Cellular

Lymphocytes. Endotoxin stimulates proliferation of bone marrow-derived (B) lymphocytes (Peavy et al., 1973). This mitogenic effect of B lymphocytes results from the action of lipid A. It is demonstrated morphologically by transforma-

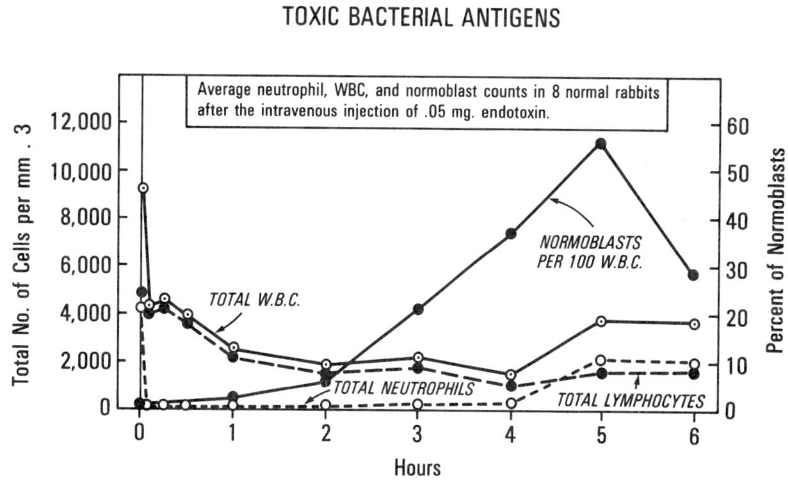

A

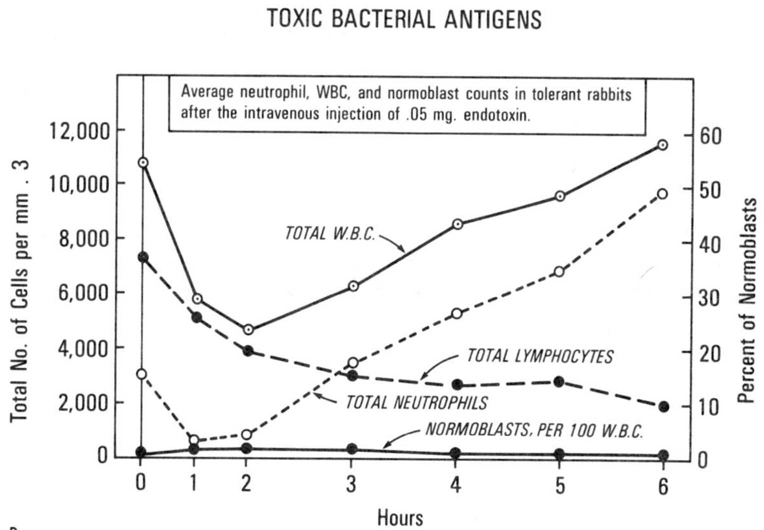

B

FIGURE 6. *Comparison of hematologic responses to intravenous lipopolysaccharide in A, normal (nontolerant) and B, tolerant rabbits.*

tion of B lymphocytes to blast cells and biochemically by increased synthesis of RNA, DNA, and protein. The phenomenon is seen in vitro with mouse lymphocytes, but not human or rabbit lymphocytes. It appears that stimulation by LPS of B cell mitosis results in division and differentiation into cells that secrete at a high rate the specific antibodies that are genetically determined for each cell. If such a reaction occurs in vivo, it could explain the increased nonspecific resistance to infection that occurs in mice shortly after an injection of endotoxin.

This idea is supported by the fact that mice (C3H/H$_e$J) whose B lymphocytes do not proliferate when exposed to LPS also fail to develop increased resistance to infection when injected with LPS. The defective response of B lymphocytes to LPS is genetic and is accompanied by increased resistance of the mice to LPS toxicity (Sulzer and Goodman, 1977).

Macrophages. LPS activates macrophages most likely by a direct effect of lipid A on the cell membrane and without the mediation of T lymphocytes (Rosenstreich et al., 1977). Activation of macrophages by LPS causes them to increase in size, become more adherent to glass, exhibit greater random migration, and carry out increased phagocytosis. There is increased phagocytosis of certain bacteria, complement- (C3b) coated particles, and IgG antibody-coated particles. Activated macrophages are metabolically more active, showing an increased consumption of glucose via the hexosemonophosphate shunt. They also increase their production of certain enzymes such as collagenase, lactic dehydrogenase (cytoplasmic), and acid phosphatase (lysosomal). Electron microscopy reveals hypertrophy of the Golgi region and an increase in the number and size of lysosomes. As little as 1 μg/ml of LPS will stimulate increased phagocytosis in vitro, and similar concentrations in vivo will cause activation of the reticuloendothelial system so that endotoxin and various colloidal particles are cleared more rapidly from the circulation. These properties of activated macrophages might also contribute to the nonspecific resistance to infection observed in animals after injection of endotoxin. The influence of endotoxin on macrophages in vivo can be seen in exudates; after intraperitoneal injection of endotoxin, macrophages that exhibit various properties typical of activation accumulate.

One effect of LPS on macrophages is increased release of certain lysosomal enzymes, such as acid phosphatase, into the surrounding medium (Allison et al., 1973). Proteolytic enzymes, such as collagenase and plasminogen activator, are also released. Since plasmin (activated plasminogen) would cause fibrinolysis and collagenase could disrupt connective tissue, it is easy to imagine how macrophages might mediate certain toxic effects of LPS. Prostaglandins, a group of biologically active aliphatic acids, have also been implicated as mediators of LPS toxicity upon release from macrophages. Prostaglandins E and F are synthesized and released in large amounts from mouse macrophages after exposure to LPS. Since prostaglandins cause increased vascular permeability, vasodilatation, vasoconstriction (especially by prostaglandin F_2 in the pulmonary circulation), decreased cardiac output, and contraction of smooth muscle (uterus, bowel, bronchi), they could account for the cardiovascular effects, abortion, and diarrhea produced by endotoxin. This idea gains support from the finding that in C3H/H_eJ mice, which are genetically resistant to LPS toxicity, the macrophages do not secrete increased levels of prostaglandins upon exposure to endotoxin. Moreover, levels of prostaglandins are elevated in susceptible animals after injection of endotoxin, and indomethacin, which inhibits prostaglandin synthesis, also prevents certain toxic effects of endotoxin.

Endotoxin is taken up by macrophages by pinocytosis. The endotoxin is first adsorbed onto the cell membrane and then enters the cell through membrane invaginations. Peritoneal macrophages of guinea pigs are reported not to detoxify ingested endotoxin in vitro, but information on macrophages of other species in vitro and in vivo is too limited for general conclusions on cellular detoxification.

Large doses of endotoxin (e.g., 50 μg/ml *Escherichia coli* LPS) will kill macrophages in vitro.

Immune Reactions: Humoral

Antibody. Injection of endotoxin or infection with gram-negative bacteria stimulates circulating antibody to LPS. Smooth endotoxins stimulate primarily antibody to the "O" antigenic side chains and endotoxin from rough bacteria, antibody to the core. Antibody to lipid A has also been described after injection of bacterial cells of *Salmonella minnesota* R595, which have the glycolipid KDO-lipid A on their surface. Antibody to each of these antigens ("O", core, lipid A) can be demonstrated by adsorbing LPS from smooth or rough organisms to the surface of red cells and then testing the antisera for their ability to cause hemolysis (in the presence of complement) or hemagglutination. Antibody to these antigens can also be demonstrated by precipitation reactions, precipitation-inhibition with individual sugars, enzyme-linked immunoabsorption, and bactericidal reactions in the presence of complement. Antibodies against LPS are both IgM and IgG. When lipid A is removed from the "O" polysaccharides by mild acid treatment, the "O" antigenic unit can no longer stimulate the formation of antibodies unless it is coupled to protein. The uncoupled, lipid-free O antigen is thus a nonantigenic hapten and can be identified only by in vitro reactions with specific antibody.

Antibody to "O" or "core" antigens can prevent toxicity of LPS, including the local Shwartzman and generalized Shwartzman reactions and death. The protection by "O" antibody is specific and limited to homologous LPS, whereas that from core antibody provides broad protection against LPS from a wide range of bacteria with unrelated "O" antigens (Braude et al., 1977; Ziegler et al., 1973).

The ability of LPS to stimulate antibody is independent of T lymphocytes. In this respect, LPS differs from most protein and cellular antigens that require a cooperation between T and B lymphocytes, in which the T lymphocytes activate the B lymphocytes to synthesize immunoglobulin.

Complement. Endotoxins can activate the complement system so that the various components are consumed and their activity disappears from serum (Mergenhagen et al., 1973). This activation can occur through either the classic or the alternative pathways, depending on the preparation of LPS. Activation of the classic pathway begins with activating of C1, which in turn catalyzes the assembly of C4,2 into the enzyme C3 convertase. C3 convertase activates C3 and the remaining components of complement. The first step in the reaction occurs in the presence of specific antibody against LPS, i.e., either IgG or IgM. After LPS combines with the Fab portion of the antibody, its Fc portion binds with C1 through its subunit C1q, and C1 becomes activated. There is also evidence that LPS from rough strains, as well as lipid A, can bind C1 directly and thereby activate the classic pathway without participation of antibody to LPS (Morrison and Ulevitch, 1978).

When the alternative pathway is activated by

LPS, the first two steps of the classic pathway are omitted. Instead, LPS causes the enzymatic activation of a protein that can cleave C3. The activated protein, designated C3 activator, splits C3 into C3A and C3B. The C3B activates the remaining complement components.

Activation of complement by either pathway can kill gram-negative (but not gram-positive) bacteria. Complement activated by the classic pathway kills bacteria more rapidly, and rough bacteria are generally more susceptible than smooth bacteria to this bactericidal action. It has been suggested that this difference in susceptibility to killing by the complement system is due to the greater ability of LPS from rough bacteria to activate the more efficient classic pathway.

Complement can also be activated by endotoxin in vivo (Spink et al., 1964; Gilbert and Braude, 1962). Endotoxins obtained by trichloroacetic extraction (Boivin preparation) are more effective in lowering complement levels in vivo. The initial hypotensive response to endotoxin is dependent on complement activation, and in patients with shock due to gram-negative bacteremia the levels of C3 are reduced (McCabe, 1973). These findings suggest that bacteremic shock may be mediated to some extent through a reaction between endotoxin and complement.

Adjuvant Effect. When endotoxin is injected intravenously along with a protein antigen, the subsequent antibody response to the protein antigen begins earlier, increases faster, and reaches a higher level than in controls given no endotoxin (Johnson, 1964). This enhancement of antibody response by endotoxin in known as its adjuvant effect, and explains the well known clinical observation that incorporation of typhoid vaccine with diphtheria or tetanus toxoid produces more antitoxin. Endotoxin does not need to be given by the same route as the protein antigen, but for maximal effect it should be given at the same time or within 6 hours of the antigen. Repeated injections of endotoxin, leading to tolerance to LPS toxicity, also abolish the adjuvant effect. The mechanism of the adjuvant effect of LPS is unknown. Since LPS can be detoxified by succinylation without losing its adjuvant activity, LPS toxicity does not seem essential for the adjuvant effect.

Stimulation of Interferon Activity

When endotoxin is injected intravenously into experimental animals peak levels of interferon in the serum are reached within 2 to 3 hours (Youngner and Stinebring, 1966). Earlier data suggested that endotoxin caused the release of preformed interferon, but more recent evidence indicates that both endotoxin and viruses stimulate synthesis of interferon.

Other Effects

Abortion. Injections of sublethal amounts of endotoxin can interrupt pregnancy in mice (Zahl and Bjerknes, 1943). The abortion is accompanied by placental hemorrhage and is apparently related to serotonin release. Endotoxins obtained from smooth or rough gram-negative bacteria, as well as lipid A, are abortifacient.

Tumor Necrosis. Endotoxin causes hemorrhagic necrosis of tumors in guinea pigs, rats, mice, and man. Most studies have been done with transplantable sarcomas of mice. The tumor becomes hemorrhagic 4 hours after the intravenous injection of crude or purified endotoxins and may undergo extensive necrosis and complete regression. The appearance of the necrotic tumor has suggested a similarity to the local Shwartzman reaction, but no thrombosis is observed in the hemorrhagic tumor.

Metabolic Effects. Intravenous injection of LPS causes changes in carbohydrates, lipids, iron, and sensitivity to epinephrine (Fiser et al., 1974). The level of blood glucose rises to a maximum within 2 hours after injection of LPS and then declines to severe hypoglycemic levels in shocked animals. This reaction is accompanied by rapid depletion of total body carbohydrate and impaired synthesis of glucose and glycogen (Berry, 1975). Administration of glucose or pyruvate fails to restore gluconeogenesis or glycogen synthesis to normal.

Hyperglyceridemia has been found in rabbits and rhesus monkeys after intravenous endotoxin (Lequire et al., 1959). It has also been seen in patients during septicemia due to gram-negative bacteria, but not gram-positive bacteria. It appears that endotoxin elevates serum triglyceride concentrations by interfering with the activation of lipid-clearing enzymes and the clearance of lipids by the reticuloendothelial system.

Hypoferremia occurs in mice and rats after injection of small doses (0.1 μg) of LPS and can be used as a bioassay for endotoxin.

The increased susceptibility to epinephrine is manifested by its ability to produce skin hemorrhages after intradermal injection in rabbits given intravenous LPS simultaneously, or after intradermal infections of mixtures of LPS and epinephrine (Thomas, 1956). As little as 5 μg of intradermal epinephrine and 1 μg of intravenous LPS can produce hemorrhagic necrosis resembling the local Shwartzman reaction. In contrast to the Shwartzman reaction, the epinephrine lesion is not blocked by heparin or nitrogen mustard, but is prevented by cortisone, dibenzaline, and chlorpromazine, which do not prevent the Shwartzman. This reaction to epinephrine may reflect either an increased sensitivity of the arterioles to adrenergic stimulation or an increased release of catecholamines at the peripheral nerve endings resulting from endotoxin activation of the sympathetic nervous system. The combined effects of local catecholamines and injected epinephrine might cause sufficient vasoconstriction to infarct the skin, whereas neither could do so alone.

ALTERATION OF LPS TOXICITY

Reduced Toxicity

The toxicity of LPS can be reduced in vivo with antiserum or by inducing tolerance, as described earlier. It can also be reduced by the administration of cortisone (Chedid and Boyer, 1955) or polymyxin B sulfate. Cortisone prevents death when given before lethal doses of endotoxin are injected. When polymyxin B is injected before or together with endotoxin it can neutralize the Shwartzman reactions and prevent death of mice from endotoxin (Rifkind and Hill, 1967). Polymyxin B acts by disrupting the structure of LPS.

Although relatively resistant to heat, the toxicity of endotoxin can be abolished in vitro by dry heat at 170° C for 3 hours. It is also detoxified by acid hydrolysis at elevated temperatures (0.1 N acid), by alkaline hydrolysis, and by oxidation with periodate, H_2O_2, or permanganate. It can be reversibly inactivated by a heat-labile factor in serum that appears to depolymerize LPS. Sodium dodecyl sulfate and sodium deoxycholate are also reputed to inactivate LPS by depolymerizing it; after removal of these agents the LPS reaggregates and toxicity is restored.

Potentiation of Toxicity

A variety of procedures can potentiate LPS toxicity in experimental animals. One mechanism is reticuloendothelial blockage by thorium dioxide, which enhances the ability of endotoxin to kill animals and eliminates the increased resistance to the Shwartzman reaction and fever (tolerance) that follows repeated injections of endotoxins (Beeson, 1947). Colloidal iron saccharate, trypan blue, and lead acetate all enhance LPS toxicity when injected. A high molecular weight polygalactose, known as carrageenan, can increase the lethality of endotoxin by 100- to more than 3000-fold (Becker and Rudbach, 1978). Carrageenan is believed to function as a macrophage toxin by destabilizing the lysosomal membrane. All these agents probably enhance the toxicity of LPS by impairing macrophage function and blocking the reticuloendothelial system.

Other techniques for potentiating LPS toxicity are adrenalectomy (Parant and Chedid, 1971), mycobacterial infection, and high environmental temperatures. The mechanisms involved in lowering resistance to endotoxin by these procedures are unknown. Hypersensitivity to endotoxin, induced by repeated small intraperitoneal injections, increases the lethality of endotoxin for mice. These hypersensitivity deaths resemble anaphylaxis, occurring within 2 hours after challenge with endotoxin (Braude, 1975).

References

Allison, A., Davies, P., and Page, R.: Effect of endotoxin on macrophages and other lymphoreticular cells. J Infect Dis 128(S):204, 1973.

Atkins, E., and Bodel, P.: Fever. N Engl J Med 286:27, 1972.

Becker, L., and Rudbach, J. A.: Potentiation of endotoxicity by carrageenan. Infect Immun 19:1099, 1978.

Beeson, P.: Effect of reticuloendothelial blockage on immunity to the Shwartzman phenomenon. Proc Soc Exp Biol Med 64:146, 1947.

Bennett, I. L., Jr., Petersdorf, R. G., and Keene, W. R.: Pathogenesis of fever: evidence for direct cerebral action of bacterial endotoxins. Trans Assoc Am Physicians 70:64, 1957.

Berry, L. J.: Metabolic effects of endotoxin. In Schlesinger, D. (ed.): Microbiology 1975. Washington D.C., American Society for Microbiology, 1975, p. 315.

Braude, A.: Opposing effects of immunity to endotoxin: hypersensitivity versus protection. In Urbaschek, B., Urbaschek, R., and Neter, E. (eds.): Gram-Negative Bacterial Infections. New York, Springer-Verlag, 1975, p. 69.

Braude, A., Ziegler, E., Douglas, H., and McCutchan, J.: Antibody to cell wall glycolipid of Gram-negative bacteria: induction of immunity to bacteremia and endotoxemia. J Infect Dis 136(S):167, 1977.

Carey, F., Zalesky, M., and Braude, A.: Studies with radioactive endotoxin. III. The effect of tolerance on the distribution of radioactivity after intravenous injection of *Escherichia coli* endotoxin labeled with Cr^{51}. J Clin Invest 37:441, 1958.

Chedid, L., and Boyer, F.: Étude comparative du pouvoir antitoxique de la cortisone et de la chlorpromazine. Ann Inst Pasteur 88:336, 1955.

Davis, C., and Arnold, K.: Role of meningococcal endotoxin in meningococcal purpura. J Exp Med 140:159, 1974.

Fiser, R., Denniston, J., and Beisel, W.: Endotoxemia in the rhesus monkey; alterations in host lipid and carbohydrate metabolism. Pediatr Res 8:13, 1974.

Galanos, C., Lüderitz, O., and Westphal, O.: A new method for the extraction of R lipopolysaccharides. Eur J Biochem 9:245, 1969.

Gilbert, V., and Braude, A.: Reduction in serum complement in rabbits after injection of endotoxin. J Exp Med 116:477, 1962.

Hase, S., and Rietschel, E. T.: Isolation and analysis of the lipid A backbone, lipid A structure of lipopolysaccharides from various bacterial groups. Eur J Biochem 63:101, 1976.

Heath, E. C., and Ghalambor, M. A.: 2-keto-3-deoxyoctonate, a constituent of cell wall lipopolysaccharide preparations obtained from *E. coli*. Biochem Biophys Res Commun 10:340, 1963.

Johnson, A.: Adjuvant action of bacterial endotoxin of the primary antibody response. In Landy, M., and Braun, W. (eds.): Bacterial Endotoxins. New Brunswick, N. J., Rutgers University Press, 1964, p. 252.

Lequire, V., Hutcherson, J., Hamilton, R., and Gray, M.: The effects of bacterial endotoxin on lipid metabolism. I. The responses of the serum lipids in rabbits to single and multiple injections of Shear's polysaccharide. J Exp Med 110:293, 1959.

Levin, J., and Bang, F.: The role of endotoxin in the extracellular coagulation of limulus blood. Bull Johns Hopkins Hosp 115:265, 1964.

Lüderitz, O., Staub, A. M., and Westphal, O.: Immunochemistry of O and R antigens of Salmonella and related Enterobacteriaceae. Bacteriol Rev 30:192, 1966.

McCabe, W.: Serum complement levels in bacteremia due to gram negative organisms. N Engl J Med 288:21, 1973.

Mergenhagen, S., Snyderman, R., and Phillips, K.: Activation of complement by endotoxin. J Infect Dis 128(S):78, 1973.

Milner, K.: Patterns of tolerance to endotoxin. J Infect Dis 128(S):229, 1973.
Morrison, D., and Ulevitch, R.: The effects of bacterial endotoxins on host mediation systems: A review. Am J Path 93:527, 1978.
Neuhof, H.: Changes in hemodynamics and gas metabolism after endotoxin injection. In Urbaschek, B., Urbaschek, R., and Neter, E. (eds.): Gram-Negative Bacterial Infections. New York, Springer-Verlag, 1975, p. 259.
Nikaido, H.: Biosynthesis and assembly of lipopolysaccharide and the outer membrane layer of Gram-negative cell wall. In Leive, L. (ed.): Bacterial Membranes and Walls. New York, Marcel Dekker, Inc., 1973, p. 131.
Parant, M., and Chedid, L.: Sensibilization de la souris aux endotoxines par la surrénalectomie ou par l'administration d'actinomycine D. C R Acad Sci (D) (Paris) 272:1308, 1971.
Peavy, D., Shands, J., Adler, W., and Smith, R.: Selective effects of bacterial endotoxins on various subpopulations of lymphoreticular cells. J Infect Dis 128(S):83, 1973.
Rifkind, D., and Hill, R.: Neutralization of the Shwartzman reactions by polymyxin B. J Immunol 99:564, 1967.
Rosenstreich, D., Glode, L., Wahl, L., Sandberg, A., and Mergenhagen, S.: Analysis of the cellular defects of endotoxin-unresponsive mice. In Schlesinger, D. (ed.): Microbiology 1977. Washington, D.C., American Society for Microbiology, 1977, p. 314.
Spink, W., Davis, R., Potter, R., and Chartrand, S.: The initial stage of canine endotoxin shock as an expression of anaphylactic shock: studies on complement titers and plasma histamine concentrations. J Clin Invest 43:696, 1964.
Sulzer, B., and Goodman, G.: Characteristics of Endotoxin. Resistant Low-Responder Mice. In Schlesinger, D. (ed.): Microbiology 1977. Washington D.C., American Society for Microbiology, 1977, p. 304.
Thomas, L.: The role of epinephrine in the reactions produced by the endotoxins of Gram-negative bacteria. I. Hemorrhagic necrosis produced by epinephrine in the skin of endotoxin-treated rabbits. J Exp Med 104:865, 1956.
Westphal, O., Lüderitz, O., and Bister, F.: Uber die extraktion von Bakterien mit Phenol/Wasser. Z Naturforsch 7b:148, 1952.
Wolff, S. M.: Biologic effects of bacterial endotoxins in man. J Infect Dis 128(S):251, 1973.
Youngner, J. S., and Stinebring, W. R.: Comparison of interferon production in mice by bacterial endotoxin and statolon. Virology 29:310, 1966.
Zahl, P., and Bjerknes, C.: Induction of decidua-placental hemorrhage in mice by the endotoxins of certain gram-negative bacteria. Proc Soc Exp Biol Med 54:329, 1943.
Ziegler, E., Douglas, H., and Braude, A.: Human antiserum for prevention of the local Shwartzman reaction and death from bacterial lipopolysaccharides. J Clin Invest 52:3236, 1973.

Virology

MORPHOLOGY AND STRUCTURE OF VIRUSES 7

Richard W. Compans, Ph.D.

Viruses were first distinguished from other microorganisms on the basis of size, as determined by filtration techniques. Most infectious virus particles, or *virions,* are smaller than any bacteria, and are below the limit of resolution of the light microscope. The smallest virions are 20 to 30 nanometers in diameter (1 nanometer, nm, = 10^{-9} meter) and are the simplest agents known to cause infectious diseases of man.

In addition to their size, virions may be distinguished from other microorganisms by the fact that they contain only a single type of nucleic acid, either DNA or RNA. RNA-containing viruses are unique in possessing this type of nucleic acid as their genetic material.

Virions lack metabolic activity and do not possess ribosomes or many of the enzymes necessary for macromolecular synthesis, although they may contain certain specialized enzymes, as discussed in Chapter 9. As a consequence, viruses depend on cellular processes for their replication and are obligate intracellular parasites. Their replication process involves intracellular dissolution of virions into their macromolecular components, followed by synthesis of viral proteins and nucleic acids, whose structures are specified by the viral genome. The individual viral macromolecules are then assembled into progeny virions in yields as high as thousands of particles per infected cell. This type of replication cycle differs markedly from division by binary fission, which is the process by which bacteria multiply.

METHODS FOR ANALYSIS OF VIRUS STRUCTURE

Filtration through membranes with graded pore sizes provided the earliest information on the size characteristics of different viruses. Similarly, ultracentrifugation studies enabled estimates of virus particle size before viruses could be visualized in the electron microscope. The ultracentrifuge is still essential for the concentration and purification of viruses, both of which are prerequisites for direct biochemical analyses. Differential centrifugation enables viruses to be concentrated and partially purified; alternatively, many viruses may be concentrated by precipitation with reagents such as ammonium sulfate or polyethylene glycol. Subsequently, most viruses may be purified by density gradient centrifugation in various media, the choice of which depends on the buoyant density as well as the stability of the virion. The most common are aqueous solutions of sucrose or potassium tartrate for the less dense and more fragile viruses, and cesium chloride solutions for the more stable viruses.

The structure of almost all animal viruses has now been examined directly by electron microscopy. If purified or concentrated virus suspensions are available, negative staining is usually the method of choice for specimen preparation. The procedure involves drying a suspension of virions in an aqueous solution of a heavy atom salt such as uranyl acetate or sodium phosphotungstate, which forms an electron-dense background. Virions, or other biologic structures, stand out as electron-lucent particles by this staining method, and it is possible to visualize the fine structure of the particles with high resolution (see Figs. 4 and 6). Virus-infected cells may also be examined in thin section by electron-microscopy after embedding in suitable polymers, and this approach has afforded much information on the intracellular aspects of virus replication.

The recent growth of *molecular virology* as an experimental science has yielded a wealth of information on the molecular structure and replication processes of viruses. Cell culture systems are available for propagation and assay of most animal viruses. Extensive use is made of radioisotopic labeling with specific precursors that are incorporated into viral macromolecules — for example, nucleosides into nucleic acids and amino acids into proteins. Viral macromolecules are analyzed by various procedures, of which density gradient centrifugation and gel electrophoresis are most popular. Since most virions contain only a few types of macromolecules, it has been possible to catalog the type and number of macromolecular components present and to study the events involved in their biosynthesis in infected cells.

GENERAL STRUCTURAL ORGANIZATION OF VIRUSES

The simplest virions consist solely of nucleic acid enclosed in a protein coat. Early electron microscopic observations indicated that most virions were either rodlike or roughly spherical particles, and chemical analyses of plant viruses demonstrated that they contained only a small percentage of nucleic acid — for example, 5 per cent for tobacco mosaic virus (TMV). Crick and Watson suggested that the viral coat consisted of multiple identical subunits, which would allow such a limited amount of nucleic acid to direct the synthesis of a large amount of coat protein. Subsequent biochemical analyses and x-ray diffraction studies of virus crystals have demonstrated such a subunit structure. Further, the protein subunits form a symmetric shell, termed the *capsid*. The capsid together with the nucleic acid is designated the *nucleocapsid*. Two types of capsid symmetry are observed in viruses: *helical* for the viruses that appeared rodlike, and *icosahedral* for the "spherical" viruses. Although the simplest virions consist solely of the naked nucleocapsid, many are more complex and possess an outer lipid-containing membrane, termed the *viral envelope*. Schematic diagrams of the structural components of the virion and their designations are shown in Figure 1.

The nucleic acid is contained within the nucleocapsid and is thus protected by the viral protein from degradation by nucleases. In addition to serving a protective function, the external proteins of the virion are important for specific interactions with receptors on host cells in the ini-

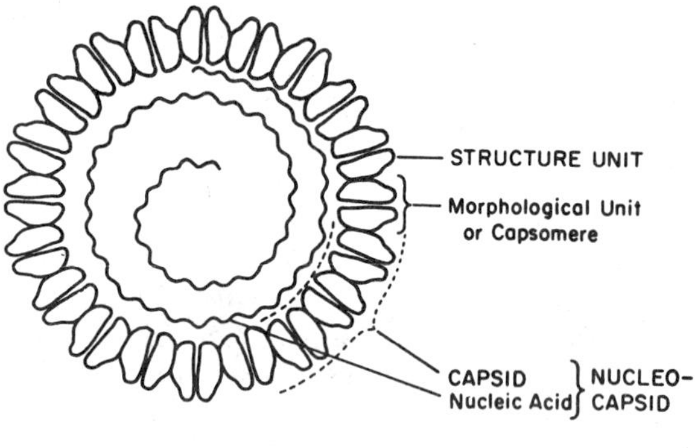

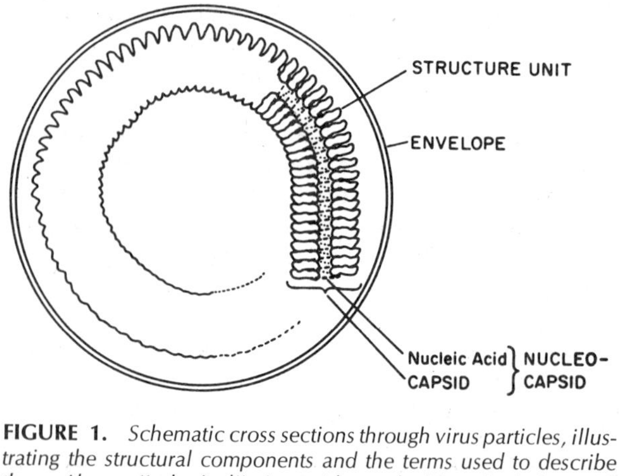

FIGURE 1. *Schematic cross sections through virus particles, illustrating the structural components and the terms used to describe them. Above: "spherical" virus with a naked icosahedral capsid. Below: enveloped virus with a helical nucleocapsid. (Modified from D. L. D. Caspar et al., Cold Spring Harbor Symp Quant Biol 27:49, 1962.)*

tial stages of the infection process and are the antigens that elicit an immune response following virus infection. Lipid-containing virions are much more heat labile than are virions consisting of naked nucleocapsids and are sensitive to lipid solvents or detergents.

Although there is no limit to the number of different sizes and shapes of viruses that could conceivably exist, only a few structural types have been observed. This finding forms the basis for a rational classification of animal viruses into families with common structural features, as discussed in Chapter 8. Classification is based on the size and structure of the viral nucleic acid, the size and symmetry of the capsid, and the presence or absence of an envelope. Fortunately, viruses that are classified together on the basis of structure also replicate by similar processes and frequently exhibit other similarities in their interactions with host cells. A total of 16 different families of animal viruses have been identified, five containing DNA genomes, and ten containing RNA genomes (Table 1). A more detailed description of the morphology and biologic properties of these agents is presented in Chapter 8.

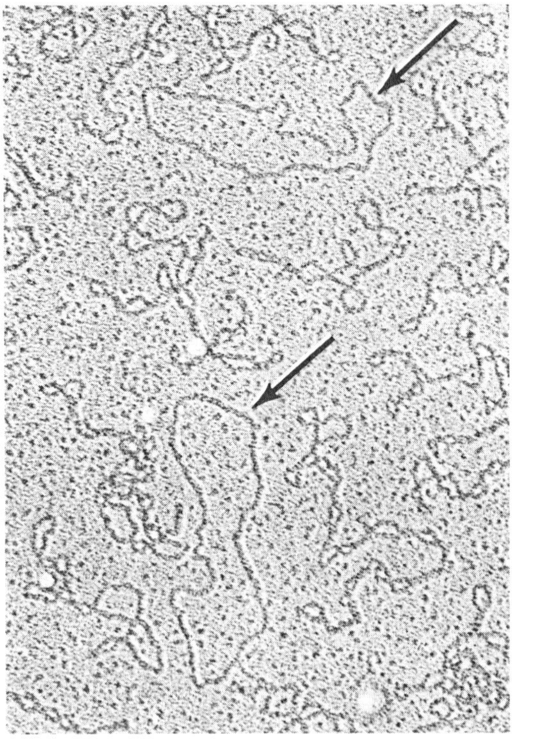

FIGURE 2. *Electron micrograph of the DNA genome of a papovavirus (simian virus 40). Most molecules are in a circular configuration containing several superhelical twists, although some molecules (arrows) are seen as open, untwisted circles. One strand of the DNA is broken in such molecules, which enables unwinding of the twists. Courtesy of Dr. Patricia Hale. Magnification: × 70,000.*

VIRAL NUCLEIC ACIDS

The molecular weight and configuration of the viral nucleic acid is an important characteristic of each virus family (Table 1). DNA viruses exhibit much greater variation in genome size than is observed with RNA-containing viruses. Most DNA-containing animal viruses contain double-stranded nucleic acid, whereas parvoviruses are exceptional in containing single-stranded DNA. Viral DNA may occur as linear molecules, or as covalently closed circular molecules (for example, in papovaviruses, Fig. 2).

The RNA viruses exhibit great variation in the configuration of their nucleic acids. The possibilities include one linear molecule of single-strand-

TABLE 1. Structural Properties of Animal Viruses

VIRUS FAMILY	MOLECULAR WEIGHT OF VIRAL NUCLEIC ACID $\times 10^{-6}$	CAPSID SYMMETRY	PRESENCE OF ENVELOPE	DIAMETER OF VIRION (nm)
DNA Viruses				
Parvoviridae	1.5–2	icosahedral	−	20
Papovaviridae	3–5	icosahedral	−	45–55
Adenoviridae	20–25	icosahedral	−	75
Herpesviridae	100	icosahedral	+	150
Poxviridae	160–200	complex	+	240 × 300
RNA Viruses				
Picornaviridae	2.6	icosahedral	−	25–30
Togaviridae	4	icosahedral	+	50–70
Arenaviridae	4–5	helical (?)	+	80–300
Bunyaviridae	4–5	helical (?)	+	80–100
Orthomyxoviridae	4–5	helical	+	90–120
Paramyxoviridae	6–7	helical	+	120–150
Rhabdoviridae	3.8	helical	+	70 × 170
Coronaviridae	5–6	helical	+	80–120
Retroviridae	6–7	icosahedral	+	100–120
Reoviridae	12–15	icosahedral	−	70–80

ed RNA (picorna-, toga-, paramyxo-, and rhabdoviruses), multiple segments of single-stranded RNA (arena-, bunya-, and orthomyxoviruses), multiple segments of double-stranded RNA (reovirus), and a complex of two identical copies of a single-stranded RNA (retroviruses). Thus far, no covalently closed circular molecules have been identified in RNA animal viruses. For several viruses with segmented RNA genomes, it has been shown that each segment functions as a gene that specifies a particular viral protein.

For some virus families, the nucleic acid alone is infectious. This provides the most convincing demonstration that the viral nucleic acid is the sole genetic information required for the formation of progeny virions. Infectious nucleic acids have been demonstrated only for those virus families in which the nucleic acid is present as a single molecule and in which no enzymatic activities essential for replication are present in the virion.

STRUCTURE OF ICOSAHEDRAL NUCLEOCAPSIDS

An icosahedron consists of 20 equilateral triangular faces, and has 12 vertices at which five triangular faces meet (Fig. 3). It exhibits axes of fivefold symmetry through each vertex, threefold symmetry at the center of each triangular face, and twofold symmetry at the midpoints of the edges between two faces. The simplest icosahedral viruses possess 60 identical asymmetric protein subunits (structure units), three of which are positioned on each of the triangular faces of the particle. The icosahedron is formed, and maintains its stability, by means of identical protein-protein bonds between adjacent subunits. Most icosahedral viruses, however, are larger and possess many more protein subunits. Frequently, such subunits are arranged in clusters, which form the *morphologic units,* or *capsomers,* that are the subunits visible by electron microscopy. The most common pattern of clustering results in formation of groups of five (pentamers) found at the vertices, and groups of six (hexamers) found on faces of the particle. The adenovirus capsid (Fig. 4) is an example of a complex icosahedral structure exhibiting this type of clustering.

The triangular facets of an icosahedron may be further subdivided into smaller equilateral triangles, and the term *triangulation number* is used to indicate the number of smaller units present on each face. Only certain triangulation numbers are possible, and they are given by the relationship $T = h^2 + hk + k^2$, where h and k are any integers. Thus, some of the possible triangulation numbers are 1, 3, 4, 7, 12, and so on. In terms of virus capsid structure, three structure units would be present on each of the smaller triangles. If these are clustered into pentamers and hexamers, the number of morphologic units in the capsid is equal to $10\ T + 2$. Thus, a capsid of triangulation number T will consist of 12 pentamers and $10\ (T-1)$ hexamers.

Although simpler viruses may have only one type of structural subunit, more complex particles contain several types of protein as capsid components. Each protein, however, is specified by a particular viral gene; thus, virions contain the same set of capsid proteins irrespective of the host

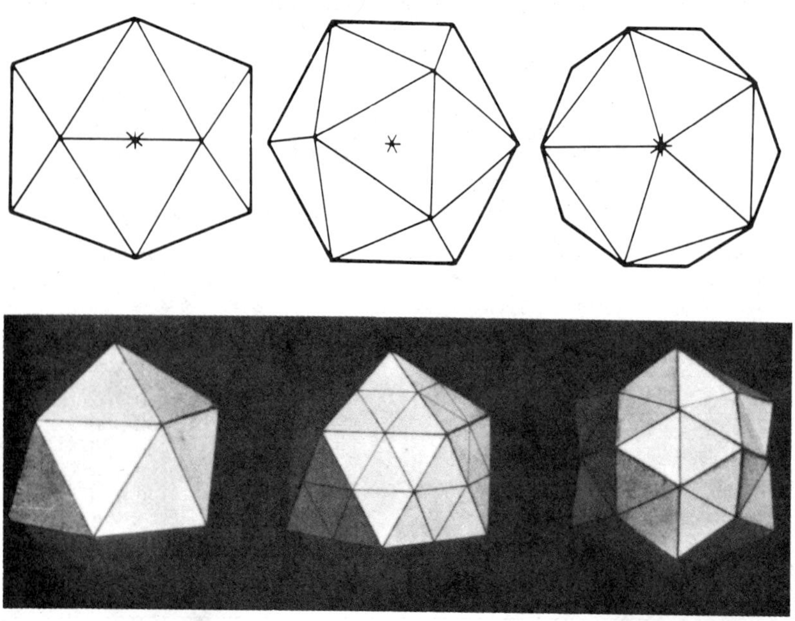

FIGURE 3. *Structural features of an icosahedron. Above, diagrams of an icosahedron viewed along different axes of symmetry (*). From left to right, the structure is viewed along a twofold, threefold, and fivefold axis of symmetry. Below, examples of the subdivision of an icosahedron into surfaces of different triangulation numbers. From left to right, T = 1; T = 4; T = 3. (The lower portion is taken from Caspar and Klug, Cold Spring Harbor Symp Quant Biol 27:1, 1962.)*

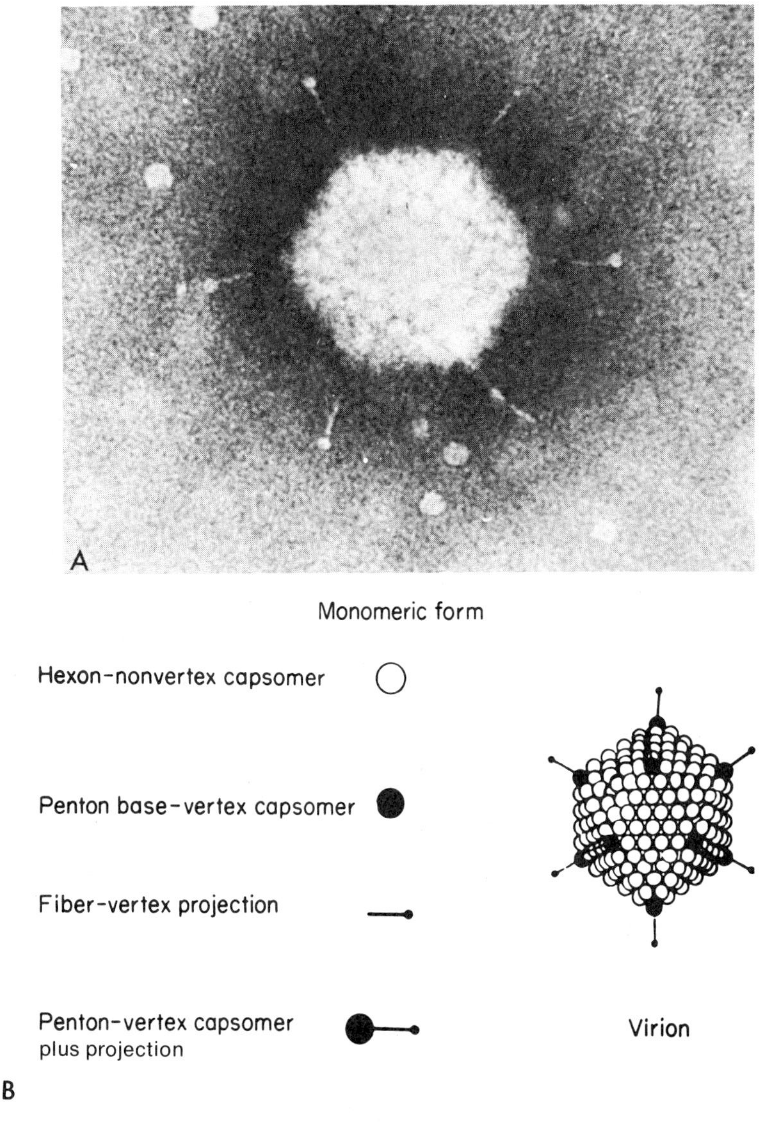

FIGURE 4. *A, adenovirus particle negatively stained with sodium silicotungstate, illustrating the icosahedral capsid structure with fibers projecting from each of the vertices. (From Valentine and Pereira: J Mol Biol 13:13, 1965; supplied courtesy of Dr. N. G. Wrigley). B, schematic diagram of the structure of an adenovirus particle, illustrating the different capsid proteins and their arrangement, as well as the terms used to describe the individual structural components. (From E. Norrby: J Gen Virol 5:221, 1969.)*

cell of origin. Viruses with large genomes can direct the synthesis of more different proteins than can viruses with more limited genetic information, and for DNA viruses there is a good correlation between genetic information content and complexity of the capsid. For adenoviruses, the *hexon* subunit found on the faces of the particle consists of a different polypeptide species from that contained in the *penton* (vertex) subunit, and another distinct protein, the *fiber*, radiates outward from each vertex of the particle. Some icosahedral particles contain internal core proteins associated with the nucleic acid that differ from those of the capsid, whereas in other virus types, the nucleic acid is directly associated with the capsid proteins. At least one icosahedral virus, reovirus, contains two distinct capsid layers. Treatment of reovirus particles with proteases removes the outer capsid layer and reveals an internal capsid that also appears to exhibit icosahedral symmetry.

For icosahedral viruses, the dimensions of each virion are uniform, as a result of identical protein-protein interactions that determine the structure of each particle. The process of assembly of an icosahedral virus is obviously designed to accommodate the viral genome in a capsid of suitable size; nucleic acids of a much larger size than a particular viral genome would probably not fit into its particular capsid. However, the presence of viral nucleic acid is not essential for the assembly of the capsid, and aberrant particles are produced by infected cells, including empty icosahedral capsids devoid of any nucleic acid. The size and organization of empty capsids is very similar to that of fully infectious virions; thus it is apparent that the capsid proteins possess the shapes and bonding characteristics that cause them to form a particular structure. Icosahedral capsids may also be produced that contain segments of cellular nucleic acid instead of the viral genome.

HELICAL NUCLEOCAPSIDS

A helical arrangement of protein subunits has been demonstrated by crystallographic analysis of plant viruses such as TMV. As shown in Fig. 5, the RNA itself is in a helical arrangement in the TMV particle, and lies in a groove at a radius of 4 nm from the cylindric axis. The particle itself has a central hollow core 4 nm in diameter and an overall particle diameter of 18 nm. Virus particles consist of a single type of structural protein arranged in the form of repeating units, with interactions between adjacent subunits as well as between the protein and the viral RNA. Helical nucleocapsids occur in many animal RNA viruses but have not been observed in DNA animal viruses. The helical ribonucleoproteins in animal viruses are flexible, whereas many plant viruses are rigid rods. The flexibility of such nucleocapsids requires that there be no firm bonds between protein subunits in adjacent turns of the helix. All of the animal viruses containing helical nucleocapsids also possess a lipid-containing envelope; examples of some enveloped viruses with helical nucleocapsids are shown in Figure 6. In such viruses, flexibility of the helix is important to enable folding of these nucleocapsids for enclosure within the viral membrane.

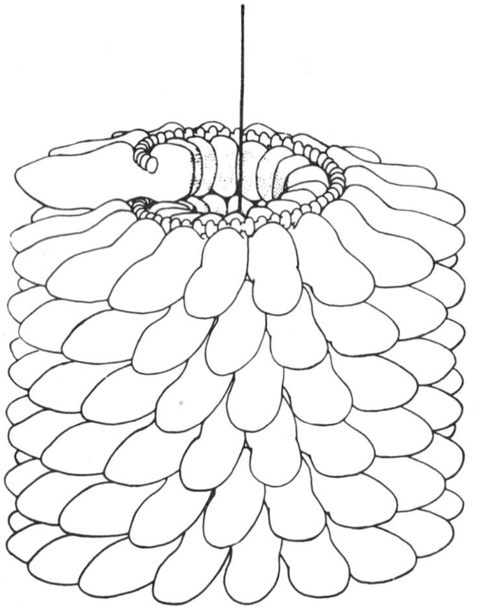

FIGURE 5. *Schematic diagram of the structure of tobacco mosaic virus based on x-ray diffraction studies. The RNA is coiled between the turns of the protein subunits; there are 49 nucleotides and about 16 1/3 protein subunits in each turn of the helix. The central hole is about 4 nm in diameter, and the overall diameter of the helix is 18 nm. (From Caspar and Klug: Cold Spring Harbor Symp Quant Biol 27:1, 1962.)*

The length of the helical nucleocapsid is determined by the length of the viral RNA molecule itself; for TMV, a particle length of 300 nm is observed. The TMV protein can aggregate by itself to form rods of the same diameter as the virion, but such particles are of random length, due to the absence of the RNA as a length-determining factor. It is also possible for TMV protein to form helices with RNA from other sources, or with synthetic polynucleotides, indicating that the protein does not show strict specificity for a particular nucleic acid sequence for the interactions that lead to the assembly of the helical virus.

VIRAL ENVELOPES

More than half of the established families of animal viruses contain lipids as major structural components, which are present in a limiting membrane, or *envelope*. The lipids are acquired during the maturation process, which usually occurs by budding at a cellular membrane. Thus the viral lipids are acquired from a membrane of the host cell and are similar to that membrane in composition; they are arranged in the form of a bilayer similar to that observed in other biologic membranes. The general arrangement of the other structural components of viral envelopes also closely resembles that observed in other biologic membranes in many respects. The viral envelope is asymmetric with respect to the distribution of carbohydrate components, as are cellular membranes: glycoproteins and glycolipids are found only on the external surface. The virion glycoproteins are virus-specific (that is, the polypeptide backbone is coded by the viral genome); however, the carbohydrate portion may vary with the host cell type and is synthesized by cellular enzymes. Viral glycoproteins form projections or spikes on the external surface of the virion. These glycoproteins are essential for viral infectivity and are responsible for adsorption of viruses to receptors on cell surfaces. They are also the antigens of importance in immunity to such viruses, in that antibody directed against the viral glycoproteins will neutralize virus infectivity. Host cell membrane proteins are effectively excluded from the viral membrane during the assembly of lipid-containing viruses.

Viral envelope glycoproteins appear to be amphipathic; that is, they have hydrophobic and hydrophilic domains. The isolated glycoproteins form rosette-like clusters (Fig. 6*B*) presumably because they aggregate by their hydrophobic ends. The hydrophobic end appears to be involved

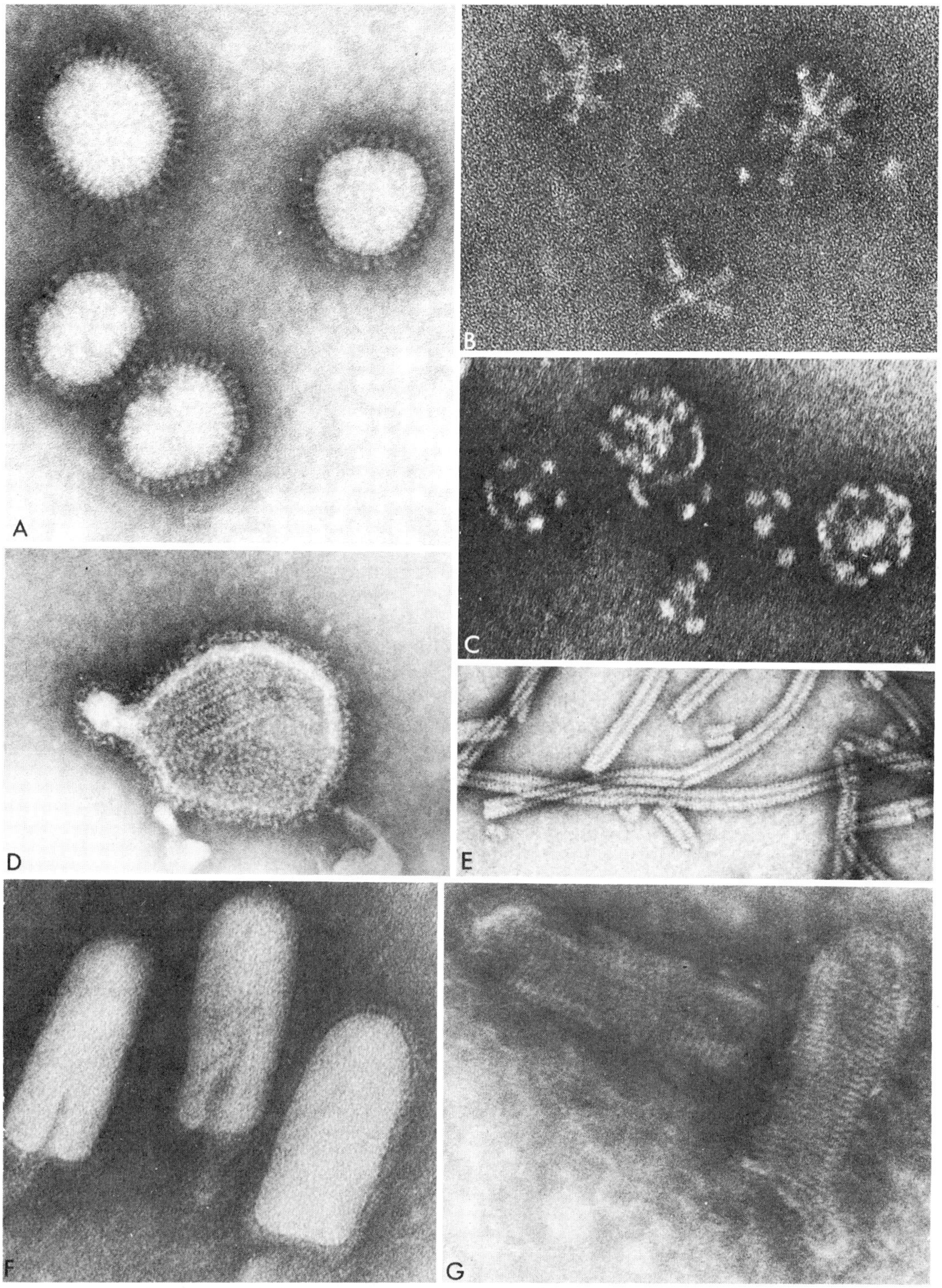

FIGURE 6. *Electron micrographs of some enveloped animal viruses and their structural components. A, Influenza virions, showing prominent surface spikes protruding from the viral membrane. The spikes consist of two types of glycoproteins, the hemagglutinin and neuraminidase. Isolated hemagglutinin spikes (B) consist of rodlike subunits that usually aggregate into rosette-like clusters because one end of the spike is hydrophobic; this end anchors the spike to the membrane. Isolated neuraminidase spikes (C) are mushroom-shaped, and also aggregate into characteristic clusters. D, Parainfluenza (SV5) particle penetrated by stain, revealing the internal helical nucleocapsid. E, Isolated SV5 nucleocapsids, some of which are broken into short fragments. The helical structure has a central hollow core, and the turns of the helix give the structure a herringbone-like appearance. F, Rhabdovirus (vesicular stomatitis) particles showing the bullet shape that is characteristic of this virus family. G, Partially disrupted vesicular stomatitis virions, in which the structure of the tightly coiled internal ribonucleoproteins can be observed. B and C are from Laver and Valentine, Virology 38:105, 1969 (Courtesy of Dr. W. G. Laver). D and E are from Compans and Nakamura, CRC Handbook Series in Clinical Laboratory Sciences, Section H, Volume 1, pp. 361–385, 1978. Magnifications: A, + 200,000; B and C, × 500,000; D, × 175,000; E, × 200,000; F and G, × 300,000.*

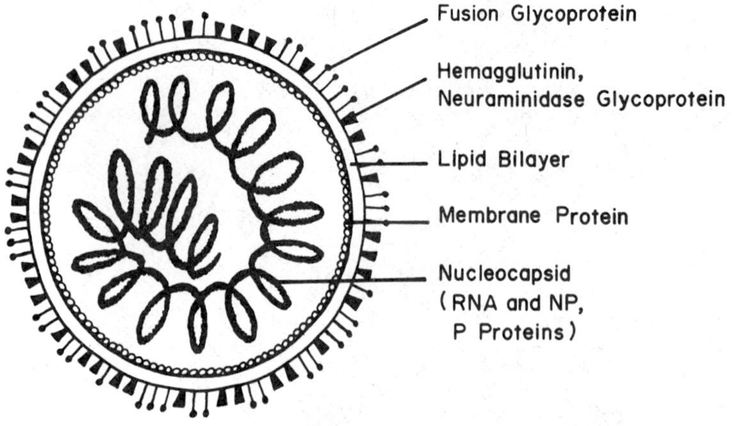

FIGURE 7. *Schematic diagram of the arrangement of structural proteins in a parainfluenza virion. Two types of glycoproteins form projections on the external surface of the lipid bilayer, and a membrane protein lines the inner surface of the bilayer. The helical nucleocapsid is coiled inside the envelope, and consists of the viral RNA and a nucleoprotein (NP) as well as other minor proteins (P) which may possess RNA polymerase activity. (From Compans and Nakamura, CRC Handbook Series in Clinical Laboratory Sciences, Section H, Volume 1, pp. 361–385, 1978).*

in binding of the glycoprotein to the viral membrane and may extend into or traverse the lipid bilayer to interact with proteins on the internal surface of the virion. Such interactions may be essential for the assembly process.

Virus-specific proteins are also components of the internal surface of the envelope. In the case of viruses with icosahedral nucleocapsids, the lipid membrane with its associated glycoproteins may be wrapped closely around the capsid itself, although intervening proteins may be present in some instances. For enveloped viruses with helical nucleocapsids, a particular type of protein is usually found in association with the inner surface of the envelope, and it is termed the M, or membrane, protein.

The morphology of some representative enveloped viruses is shown in Figure 6. The overall shapes and sizes of the virions are features characteristic of each virus family and are primarily determined by the arrangement of internal structural components. Some enveloped viruses are highly pleomorphic and may contain multiple nucleocapsids and genomes in a single

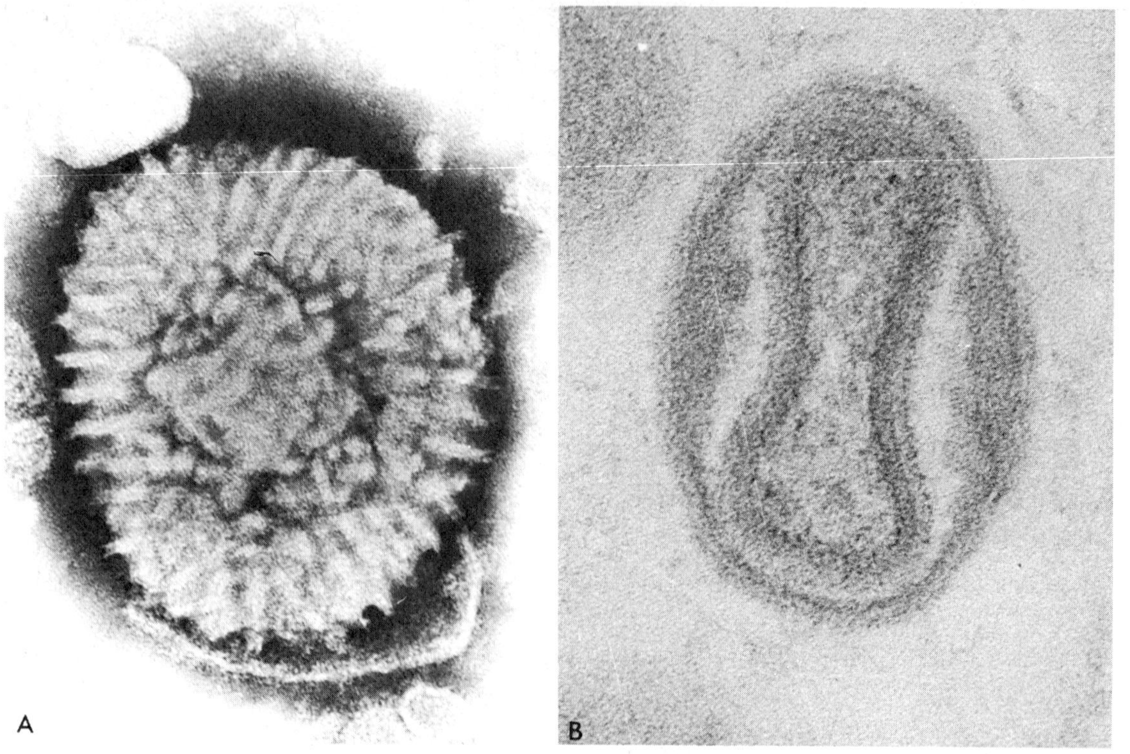

FIGURE 8. *Electron micrographs of vaccinia (poxvirus) virions. A, Negatively stained particle showing ridges or tubular elements covering the surface. B, Thin section of vaccinia virion showing a central biconcave core, two lateral bodies, and an outer membrane. A is from Dales, S.: J Cell Biol 18:51, 1963; B is from Pogo and Dales: Proc Natl Acad Sci U.S. 63:820, 1969. (Courtesy of Dr. Samuel Dales.) Magnification: A, × 228,000; B, × 220,000.*

envelope. All enveloped viruses possess glycoprotein spikes, but the shapes and lengths of these structures are also a distinctive property of each virus. The location of each structural protein has been elucidated in many enveloped viruses, as depicted for the parainfluenza virion in Figure 7.

OTHER TYPES OF VIRUS STRUCTURE

Poxviruses are the largest and most complex of any of the viruses infecting man, containing about 30 or more different types of protein in the virion. They do not contain a capsid with either helical or icosahedral symmetry (Fig. 8). The virion contains DNA encased in a core structure, with two structures termed lateral bodies adjacent to the core. A lipid layer is also present in the interior of the particle. The resulting particle is oblong, and the surface is covered with distinct ridges. Poxviruses are more stable than most other lipid-containing viruses.

Certain plant diseases are caused by infectious nucleic acid molecules that lack any protein coat, and the term *viroid* has been given to such infectious agents. The known plant viroids are small, covalently closed, circular RNA molecules. It is possible that viroids may also occur in certain diseases of animals or man in which no conventional virus has been identified.

References

General

Caspar, D. L. D., and Klug, A.: Physical principles in the construction of regular viruses. Cold Spring Harbor Symp Quant Biol 27: 1–24, 1962.

Crick, F. H. C., and Watson, J. D.: Structure of small viruses. Nature 177:473–475, 1956.

Methods

Brenner, S., and Horne, R. W.: A negative staining method for high resolution electron microscopy of viruses. Biochem Biophys Acta 34:103–110, 1959.

Brakke, M. K.: Density gradient centrifugation. In Maramorosch, R., and Koprowski, H. (eds.): Methods in Virology Vol. II. New York, Academic Press, pp. 93–117, 1967.

Elford, W. J.: The sizes of viruses and bacteriophages and methods for their determination. In Doerr, H., and Hallauer, C. (eds.): Handbuch der Virusforschung. Vienna, Springer.

Maizel, J. V.: Acrylamide gel electrophoresis of proteins and nucleic acids. In: Fundamental Techniques in Virology. Habel, K. and Salzman N. P. (eds.): New York, Academic Press, pp. 334–362, 1969.

Viral Nucleic Acids

Air, G. M.: DNA sequencing of viral genomes. In Fraenkel-Conrat. H., and Wagner, R. R. (eds.): Comprehensive Virology. New York, Plenum Press, Vol. 13, pp. 205–292, 1979.

Shatkin, A. J.: Animal RNA viruses: genome structure and function. Ann Rev Biochem 43:643–664, 1974.

Icosahedral Nucleocapsids

Finch, J. T., and Klug, A.: The structure of viruses of the papilloma-polyoma type. III. Structure of rabbit papilloma virus. J Mol Biol 13:1–12, 1965.

Valentine, R. C., and Pereira, H. G. Antigens and the structure of adenovirus. J Mol Biol 13:13–20, 1965.

Helical Nucleocapsids

Caspar, D. L. D.: Assembly and stability of the tobacco mosaic virus particle. Adv Prot Chem 18:37–121, 1963.

Compans, R. W., Mountcastle, W. E., and Choppin, P. W.: The sense of the helix of paramyxovirus nucleocapsids. J Mol Biol 65:167–169, 1971.

Finch, J. T., and Gibbs, A. J.: Observations on the structure of the nucleocapsids of some paramyxoviruses. J Gen Virol 6:141–150, 1970.

Viral Envelopes

Compans, R. W., and Klenk, H.-D.: Viral membranes. In Fraenkel-Conrat, H., and Wagner, R. R. (eds.): Comprehensive Virology. New York, Plenum Press, Vol. 13, pp. 293–407, 1979.

Harrison S. C., David, A., Jumblatt, J., and Darnell, J. E.: Lipid and protein organization in Sindbis virus. J Mol Biol 60:523–528, 1971.

Lenard, J., and Compans, R. W.: The membrane structure of lipid-containing viruses. Biochem Biophys Acta 344:51–94, 1974.

Other Types of Virus Structure

Dales, S.: The structure and replication of poxviruses as exemplified by vaccinia. In Dalton, A. J., and Haguenau, F. (eds.): Ultrastructure of Animal Viruses and Bacteriophages. New York, Academic Press, pp. 109–129, 1973.

Diener, T. O.: Viroids: The smallest known agent of infectious disease. Ann Rev Microbiol 28:23–40, 1974.

CLASSIFICATION OF VIRUSES 8

Joseph L. Melnick, Ph.D.

Until about 1950, so little was known about viruses other than their pathogenic effect in causing diseases that they were classified according to the diseases they caused rather than the properties of the virus particle. Now, we are approaching the end of an important phase of discovery and characterization of animal viruses. The knowledge thus gained concerning the viruses themselves has made it possible to establish broad groupings for these agents. It appears that most of the major groups of viruses of vertebrates — at least of man and the animals important to man — have been recognized and described. Many of these virus groupings, initially established on tentative and provisional bases, now appear to form real families and genera, in which the members are indeed related in fundamental ways. For example, the validity of the original grouping of the enteroviruses based on an enteric habitat and small size is being borne out by sophisticated techniques of molecular virology that compare the genetic makeup of different members of the group and their mode of replication.

The shift in emphasis — from sketching the broad outlines of the virus kingdom based on disease causation to filling in essential details about the viruses themselves — has been recognized by the change in the name of the International Committee on Nomenclature of Viruses (ICNV) to the International Committee on Taxonomy of Viruses (ICTV). The first report of the ICNV was published in 1971 (Wildy, 1971). Work of the Study Groups and Subcommittees of the ICTV is proceeding, and reports from these groups in their special areas of virology appear regularly in *Intervirology*, the journal of the Virology Section of the IAMS. Subsequent reports of the ICTV have been published (Fenner, 1976; Matthews, 1979).

Figures 1 and 2 serve as useful reference points for the following discussion of classification based upon properties of the virus particles. Comparison of these figures also shows how rapidly knowledge of virus composition and structure has advanced. Figure 1 is taken from a text published in 1967 (Davis et al., 1967); it remains fundamentally applicable in current virology. However, Figure 2, a diagram prepared about ten years

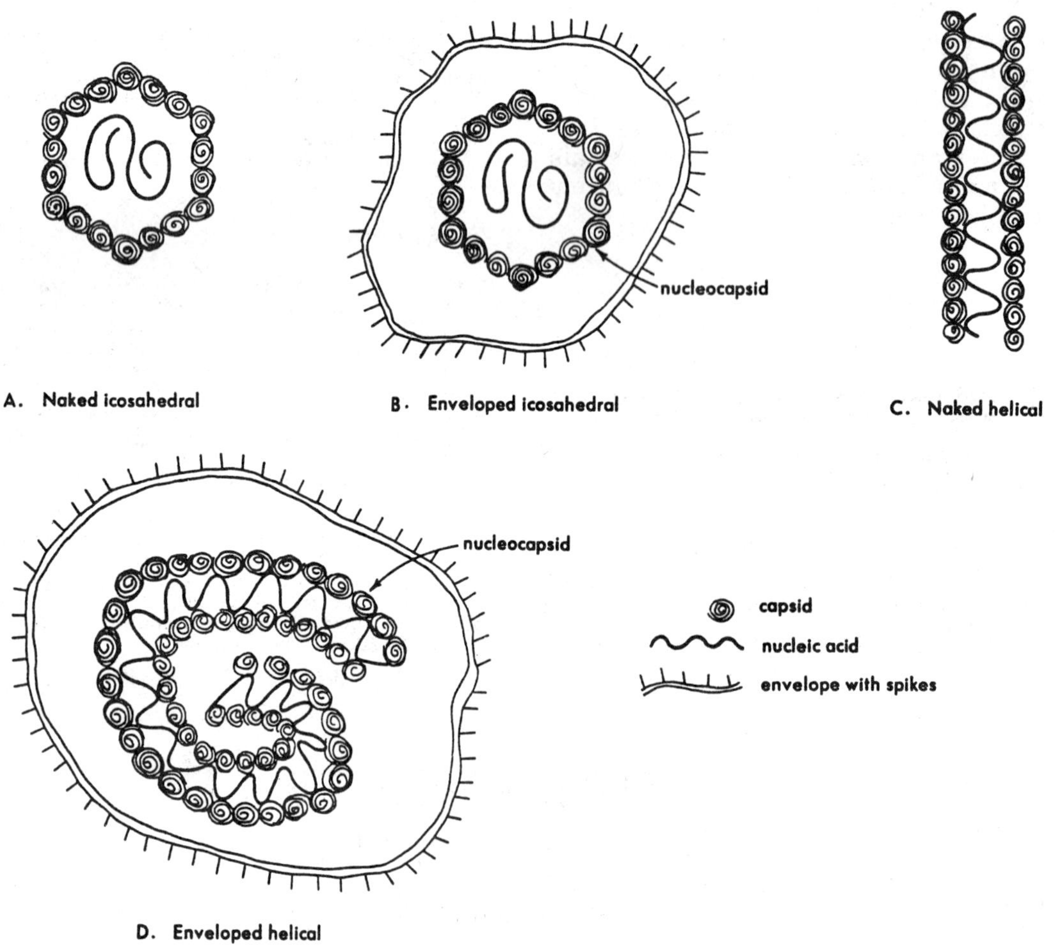

FIGURE 1. *Schematic diagram of simple forms of virions and their components. The naked icosahedral virions resemble small crystals; the naked helical virions resemble rods with a fine regular helical pattern in their surface. The enveloped icosahedral virions are made up of icosahedral nucleocapsids surrounded by the envelope; the enveloped helical virions are helical nucleocapsids bent to form a coarse, often irregular coil, within the envelope. (From Davis, B. D., et al.: Microbiology. New York, Hoeber Medical Division, 1967.)*

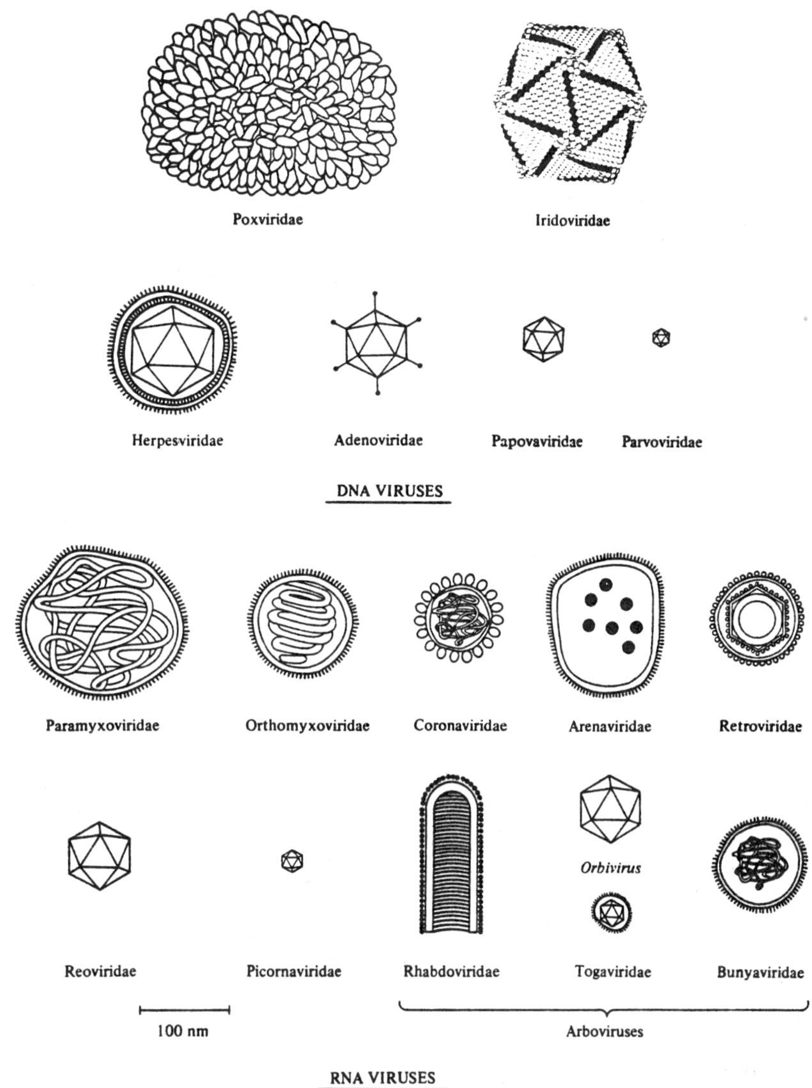

FIGURE 2. *Diagram illustrating the shapes and relative sizes of animal viruses of the major families (bar = 100 nm). From Fenner, F., and White, D. O.: Medical Virology. 2nd ed. New York, Academic Press, 1976.*

later (Fenner and White, 1976), not only draws upon additional information gained in the interim but also illustrates the wide variety of size and structure that is found among viruses of vertebrates.

Tables 1 to 5 are schematic diagrams showing separation of viruses of vertebrates into 16 families (Melnick, 1979). Table 1 describes viruses that have a DNA genome, cubic symmetry, and a naked nucleocapsid; Table 2 describes DNA-containing viruses with envelopes or complex coats. RNA-containing viruses are presented in three tables: Table 3, those with cubic capsid symmetry; Table 4, those with helical symmetry; and Table 5, those with capsid architecture either asymmetric or unknown. Commentaries follow on the viruses that have been definitely assigned to these groups. Also included are hepatitis viruses A and B, and some other agents whose classification is still tentative.

DNA VIRUSES

PARVOVIRIDAE
(Bachmann et al., 1975)

Originally named picodnaviruses to reflect their small size and DNA-containing genome (Mayor and Melnick, 1966), the family Parvoviridae now includes three named genera, *Parvovirus, Densovirus,* and *Adeno-associated-virus.* A typical member is adeno-associated satellite virus, several serotypes of which are indigenous

TABLE 1. DNA-Containing Viruses With Cubic Symmetry and Naked Nucleocapsid

Capsid symmetry	Virion: naked or enveloped	Site of capsid assembly	Reaction to ether	No. of capsomeres	Diam. of virion (nm)	Mol. wt. of NA in virion ($\times 10^6$)	Genus name	Family name
Cubic	Naked	Nucleus	Resistant	32	18–26	1.5–2.2	PARVOVIRUS	PARVOVIRIDAE
							ADENO-ASSOCIATED VIRUS	
							DENSOVIRUS	
				72	45–55	2.4–5.0	PAPILLOMAVIRUS (Papovavirus-A)	PAPOVAVIRIDAE
							POLYOMAVIRUS (Papovavirus-B)	
				252	70–90	20–30	MASTADENOVIRUS	ADENOVIRIDAE
							AVIADENOVIRUS	

to man — but as yet with no known disease association in human beings. Reading from the left-hand side of Table 1, these are DNA-containing viruses that have cubic symmetry and a naked (unenveloped) nucleocapsid; during replication, capsid assembly takes place in the nucleus of the host cell. For the DNA viruses whose capsid assembly takes place in the nucleus (not only parvoviruses, but also papovaviruses, adenoviruses, and herpesviruses), a phase of replication — that is, viral protein synthesis — occurs in the cytoplasm. Messenger RNA of these viruses is associated with polyribosomes.

Infectivity is resistant not only to ether and other lipid solvents but also to heat (56° C for 60 minutes). The capsid has 32 capsomeres, the diameter of the virus particle is 18 to 26 nm, and the molecular weight of the nucleic acid is 1.5 to 2.2×10^6. The genus *Parvovirus* includes autonomously replicating viruses of several animal species (cat, cow, dog, goose, mink, mouse, pig, rabbit, rat). Members of the *Densovirus* genus also replicate autonomously; they are viruses of insects, but also can produce cytopathic effects in cultures of certain vertebrate cells (L cells). In contrast, adeno-satellite viruses are defective; that is, they cannot multiply in the absence of a replicating adenovirus, which serves as a "helper virus" — herpesvirus can act as a partial helper. Satellite viruses occur in cow, chicken, dog, horse, man, and monkey hosts.

Parvoviridae are the only DNA-containing viruses of vertebrates whose DNA genome is single-stranded within the virion; all the others (see Tables 1 and 2) have double-stranded DNA. In the case of adeno-satellite viruses and densoviruses, separate virions contain single strands of positive or negative DNA; these strands are complementary and when isolated from the virion shells they come together to form a double strand. In contrast, for members of the genus *Parvovirus,* the DNA in the virion is a positive strand only. However, the single-stranded DNA molecule has a hairpin-like structure at both the 5′ and the 3′ ends. In members of this genus, about 1 per cent of the virions form double strands similar to the self-complementary strands of the other two genera. Members of the *Parvovirus* genus show marked preference for actively dividing cells, have been shown to be transmissible transplacentally, and are receiving attention for their special disease potential in fetuses and neonates (Kilham and Margolis, 1975). One member has been associated with acute viral gastroenteritis of man.

In the light of accumulating data, it is becoming clear that hepatitis virus type B has a number of important properties similar to those of representative members of the parvovirus family. The 42-nm "Dane" particle (found in the serum of individuals infected with hepatitis B) is now recognized as the virus of this disease. The morphology, nucleic acid type, and nucleic acid strandedness of the Dane particles place them in a class by themselves, unrelated to any other known viruses. However, the 25 nm central core, which is assembled in the nucleus of the infected hepatocyte and which can be readily released from Dane particles, shares biochemical and biophysical properties with several members of the family Parvoviridae.

PAPOVAVIRIDAE

(Melnick, 1962; Melnick et al., 1974a)

These relatively small, ether-resistant viruses contain double-stranded DNA in circular form. Many are unusually heat-stable, surviving temperatures that inactivate most viruses. The representatives that infect human beings are the papilloma or wart virus (of which at least two serotypes exist) and SV40-like viruses such as JC virus, which has been isolated from the brain tissue of patients with progressive multifocal leukoencephalopathy (PML), or BK virus, which has been isolated from the urine of immunosuppressed recipients of renal transplants. Other members include papilloma viruses of several vertebrate species, polyoma and K viruses of mice, and vacuolating viruses of monkeys (SV40) and of rabbits. These viruses have relatively slow growth cycles characterized by replication within the nucleus. Papovaviruses produce latent and chronic infections in their natural hosts. Many produce tumors, particularly in experimentally infected rodents, serving as model systems to study viral carcinogenesis.

There are two genera whose names are in dispute: *Papillomavirus* (or *Papovavirus-A*) and *Polyomavirus* (or *Papovavirus-B*). Serious concern has been expressed at the use of "polyomavirus" for the human viruses, since the group B papovaviruses of man have *not* been shown to be "polyoma" in character — that is, "produce many types of tumors in the host." For example, even though antibodies against BK virus are widespread in the population, human tumors and human malignant cell lines were negative when analyzed for BK virus-specific DNA sequences. Because the probes used could detect one copy of BK virus DNA if only 10 per cent of the cells were tumor cells, the results are very strong evidence that the tumors analyzed did not have a BK virus etiology. The tumors tested represent about 50 per cent of all types of cancers in the United States; thus there is no evidence that this member of papovavirus group B is involved in production of these tumors in the host.

TABLE 2. DNA-Containing Viruses With Envelopes or Complex Coats

Capsid symmetry	Virion: naked or enveloped	Site of capsid assembly	Site of nucleocapsid envelopment	Reaction to ether	No. of capsomeres	Diam. of virion (nm)	Mol. wt. of NA in virion ($\times 10^6$)	Genus name	Subfamily name	Family name
Cubic	Envel.	Nucl.	Nuclear membrane	Sens.	162	100	54–92	(Many genera not yet named)	ALPHAHERPESVIRINAE BETAHERPESVIRINAE GAMMAHERPESVIRINAE	HERPESVIRIDAE
		Cyto-plasm	Cytopl. membrane	Sens.	1500	130	130	IRIDOVIRUS		IRIDOVIRIDAE
Complex	Complex coats	Cyto-plasm		Resist.		230 × 300	160	ORTHOPOXVIRUS AVIPOXVIRUS CAPRIPOXVIRUS LEPORIPOXVIRUS PARAPOXVIRUS ENTOMOPOXVIRUS		POXVIRIDAE

ADENOVIRIDAE
(Norrby et al., 1976)

The adenovirus virion is a nonenveloped isometric particle with 252 capsomeres, each 7 to 9 nm in diameter. Vertex capsomeres are antigenically distinct from the others and carry one or two filamentous projections. The adenovirus genome is a single linear molecule of double-stranded DNA. At least 33 serotypes infect man, and there are distinct serotypes for a number of other species. Adenoviruses have a predilection for mucous membranes and may persist for years in lymphoid tissue. Some cause acute respiratory diseases, febrile catarrhs, pharyngitis, and conjunctivitis. Human adenoviruses rarely cause disease in laboratory animals, but certain serotypes do produce tumors in newborn hamsters. Common antigens are shared by all mammalian adenoviruses (*Mastadenovirus* genus); these antigens are different from the corresponding antigens of members of the genus *Aviadenovirus*.

HERPESVIRIDAE
(O'Callaghan and Randall, 1976)

The herpesvirus family is a heterogeneous group of viruses identified by their structure. As shown in Table 2, the particle consists of a DNA-containing core enclosed by an icosahedral capsid with 162 hollow cylindric capsomeres. A lipid membrane containing virus-specific proteins surrounds the inner structures. The double-stranded DNA of various herpesviruses differs considerably in size (92 to 102 $\times$ 10^6 molecular weight), cytosine and guanine content (44 to 74 per cent), and structural complexity. The DNA of herpesviruses is sufficiently large to carry the genetic code for 80 to 100 proteins, of which about 50 have been observed.

Herpesviruses are noteworthy for their ability to establish latent and/or persistent infections, which may last for the lifetime of the host, even in the presence of circulating antibodies. Special interest has been generated by the association of EB herpesvirus with human Burkitt lymphomas and nasopharyngeal carcinomas and by the possible role of the genital herpesvirus, herpes simplex type 2, in cancer of the uterine cervix. Several simian herpesviruses have been shown to be oncogenic in experimentally infected animals. Herpesvirus infections of heterologous species are in many cases very serious; examples are the fatal infection of man by one of the simian herpesviruses, so-called B virus, and the infection of cattle by swine pseudorabies virus. Human diseases include oral and genital herpes; chickenpox and shingles due to varicella/zoster virus; cytomegalic inclusion disease; and infectious mononucleosis. Subfamilies have now been established to include most of the members of this large virus family (Matthews, 1979).

IRIDOVIRIDAE

The best-known members of this family are the members of the insect iridescent virus group (for example, *Tipula* iridescent virus), now placed in the genus *Iridovirus*. However, other members of this family include important pathogens of vertebrates: African swine fever virus and many viruses of frogs and fish. No human iridovirus is known.

POXVIRIDAE
(Fenner, 1978)

These large viruses are brick-shaped or ovoid with a complex virion structure. The virion contains more than 30 structural proteins and several viral enzymes including a DNA-dependent RNA polymerase. This is the major DNA-containing virus family whose members replicate entirely within the cytoplasm. The genus *Orthopoxvirus* includes smallpox virus and the other poxviruses of man. Some of the animal poxviruses (for example, monkeypox) can infect humans, and with the eradication of smallpox from the world it has been speculated that human infections by these agents might be detected more frequently.

RNA VIRUSES

PICORNAVIRIDAE
(Melnick et al., 1974b; Cooper et al., 1978)

These are shown at the top of Table 3. Members of this family — the smallest of the viruses with RNA genomes — are classed into four genera and into several hundred species. At least 70 members of the *Enterovirus* genus infect man; these include polioviruses, coxsackieviruses, echoviruses, and in recent years, new enterovirus serotypes that are assigned sequential numbers, such as enterovirus 68. Well over 100 viruses that

TABLE 3. RNA-Containing Viruses With Cubic Capsid Symm

Capsid symmetry	Virion: naked or enveloped	Site of Capsid assembly	Site of nucleocapsid envelopment	Reaction to ether	No. of capsomeres	Diam. of virion (nm)	Mol. wt. of NA in virion (× 10⁶)	Genus name	
Cubic	Naked	Cyto-plasm		Resistant (or relatively resistant)	32	24–30	2.5	ENTEROVIRUS	PICORNAV.
							2.3–2.8	RHINOVIRUS	
							2.5	CARDIOVIRUS	
							2.3–2.8	APHTHOVIRUS	
					32	60–80	12	ORBIVIRUS	REOVIRIDAE
					92	75	15	REOVIRUS	
					32	70	12	ROTAVIRUS	
	Enve-loped	Cyto-plasm	Surface membrane	Sensitive	32 or 42	60	4	ALPHAVIRUS	TOGAVIRIDAE
								RUBIVIRUS	
			Intracyto-plasmic membrane	Sensitive	?	40	3–4	PESTIVIRUS	
								FLAVIVIRUS	

infect human beings belong to the genus *Rhinovirus*. Large numbers of agents from both of these genera are indigenous to other hosts. The two other genera are *Cardiovirus*, a rodent agent that may also infect man, and *Aphthovirus*, which includes the economically important foot-and-mouth disease viruses of cattle. The caliciviruses, once thought to belong to the Picornaviridae, have recently been shown to differ in both virus structure and method of replication.

The picornavirus genome is one piece of linear, single-stranded RNA of low molecular weight (about 2.5×10^6). The RNA is infectious and serves as its own messenger for protein translation. The enteroviruses and cardioviruses are acid-stable and have a buoyant density in CsCl of about 1.34 g/cm^3; in contrast, the rhinoviruses and aphthoviruses are acid-labile and have a higher buoyant density, about 1.4 g/cm^3.

The diseases caused by picornaviruses range from severe paralysis (paralytic poliomyelitis) to aseptic meningitis, pleurodynia, myocarditis, skin rashes, and common colds; inapparent infection is very common. Different viruses may produce the same syndrome; on the other hand, the same picornavirus may cause more than a single syndrome. After decades of investigation it now seems that hepatitis A virus is an enterovirus.

REOVIRIDAE
(Joklik et al., 1980)

Members of this virus family share a property unique among the RNA-containing viruses of vertebrates — the possession of a double-stranded, rather than single-stranded, RNA genome. The genome consists of several segments. The capsid has a double shell, and the structure of the outer capsid layer is indistinct, but icosahedral symmetry has been demonstrated in the inner capsid layers of all three recognized groups of reoviruses that infect vertebrates, the genera *Reovirus, Orbivirus,* and *Rotavirus*. The capsomeres of the orbiviruses are unusually large (10 to 15 nm wide) and appear ring-shaped. The human reoviruses are found in the bowel but their association with disease is not clear. Some orbiviruses have been considered to be arboviruses. The diseases caused by orbiviruses include Colorado tick fever of man, blue-tongue of sheep, African horsesickness, and epizootic hemorrhagic disease of deer. The members of the genus *Rotavirus* that infect human beings are increasingly recognized as major pathogens, responsible for a large share of nonbacterial infantile diarrhea. The gastroenteritis syndrome they cause is clinically very severe and is one of the common-

ood illnesses throughout the world; in ng countries it is a leading cause of death. f the initial study of the rotaviruses was plished by electron microscopy and immune scopy, and rotaviruses that infect man have been isolated in cell cultures.

TOGAVIRIDAE
(Porterfield et al., 1978)

Members of this family include most arboviruses* of antigenic groups A and B, now classed in the genus *Alphavirus* (group A) and the genus *Flavivirus* (group B), and in newly designated genera, which include nonarbo togaviruses, rubella *(Rubivirus)*, and the mucosal disease virus group *(Pestivirus)*. The virions are spherical, 40 to 70 nm in diameter, and have a lipoprotein envelope with lipid and virus-specified glycopeptide tightly applied to an icosahedral nucleocapsid. The genome is a single molecule of single-stranded RNA. The alphaviruses and flaviviruses include many of the major human arboviral pathogens: the viruses of Venezuelan, eastern, and western equine encephalitis are alphaviruses, and the viruses of yellow fever, dengue, Japanese encephalitis, St. Louis encephalitis, Omsk hemorrhagic fever, and Russian spring-summer encephalitis are flaviviruses. Rubella virus thus far is the only member placed in the genus *Rubivirus*. Members of the *Pestivirus* genus include the viruses of hog cholera, bovine virus diarrhea, and other animal viruses.

*One important and well-known virus group name that does not appear in the diagrams is a category based on ecologic properties, the *arbovirus group* (Berge, 1975). The more than 350 arthropod-borne viruses survive through a complex cycle involving vertebrate hosts and arthropods that serve as vectors, transmitting the viruses by their bites. This grouping, based on transmission, remains a useful one despite the wide diversity of its members in regard to properties of the virion. The vast majority of arboviruses now have been sufficiently well-characterized to permit their taxonomic placement. Their classic serologic interrelationships previously delineated by arbovirologists are paralleled by morphologic similarities, and have provided vital clues that can speed the taxonomic location of large numbers of viruses. Once some of the members of a classic serologic group have been characterized in terms of biophysical and biochemical properties, attention of taxonomists can be focused on their antigenic relatives. Arboviruses now are included in a number of families, chiefly Togaviridae, Bunyaviridae, Rhabdoviridae, Arenaviridae, and Reoviridae.

ORTHOMYXOVIRIDAE
(Dowdle et al., 1975) (See Table 4)

All orthomyxoviruses recognized to date are influenza viruses. The virions may be spherical, elongated, or filamentous. For most members of the family, there are "spikes" projecting from the surface of the envelope; these are glycosylated protein peplomers 10 to 14 nm long and 4 nm in diameter, consisting of two types, the hemagglutinin and the neuraminidase. (In the nomenclature of variants of influenza type A, which arise frequently either as minor variants or as major new strains, the designation of these two antigens as H_1N_1, H_3N_2, and so on, supplies an important part of the information needed about new strains — their relatedness, if any, to strains that circulated previously — as a guide to the probable degree of immunity in the population.) During replication the helical nucleocapsid is first detected in the nucleus, whereas the hemagglutinin and neuraminidase are formed in the cytoplasm. The virus matures by budding at the cell surface

TABLE 4. RNA-Containing Viruses With Helical Symmetry

Capsid symmetry	Virion: naked or enveloped	Site of capsid assembly	Site of nucleocapsid envelopment	Reaction to ether	Diam. of helix (nm)	Diam. of virion (nm)	Mol. wt. of NA in virion ($\times 10^6$)	Genus name	Family name
Helical	Enveloped	Cytoplasm	Surface membrane	Sens.	9–15	80–120	4	INFLUENZAVIRUS	ORTHOMYXOVIRIDAE
					12–15	150–300	6–8	PNEUMOVIRUS	PARAMYXOVIRIDAE
					18	150–300	6–8	PARAMYXOVIRUS	
								MORBILLIVIRUS	
					50	60 × 180	3–4	VESICULOVIRUS	RHABDOVIRIDAE
								LYSSAVIRUS	

TABLE 5. RNA-Containing Viruses With Architecture Unsymmetric or Unknown

Capsid symmetry	Virion: naked or enveloped	Site of capsid assembly	Site of nucleocapsid envelopment	Reaction to ether	Diam. of virion (nm)	Mol. wt. of NA in virion ($\times 10^6$)	Genus name	Subfamily name	Family name
Unsymmetric or unknown	Enveloped	Cytoplasm	Surface membr.	Sens.	~100	10–12		ONCOVIRINAE	RETROVIRIDAE
								SPUMAVIRINAE	
								LENTIVIRINAE	
					50–300	3–5	ARENAVIRUS		ARENAVIRIDAE
			Intracytoplasmic membr.	Sens.	80–130	9?	CORONAVIRUS		CORONAVIRIDAE
					100	6–7	BUNYAVIRUS		BUNYAVIRIDAE
							(Other genera)		

membrane. The genus *Influenzavirus* has been established and includes viruses of type A and type B; type C is considered a "probable genus." Antigenic variation is common, particularly among members of type A. Recombination occurs with high frequency within types but not between types or genera. Type A influenza viruses include agents of human, equine, and swine influenza and of fowl plague. Only human strains are known for types B and C.

PARAMYXOVIRIDAE
(Kingsbury et al., 1978)

Usually spherical, these virions may also be pleomorphic; filamentous forms may be several micrometers long. On the lipid bilayer envelope are surface projections. Virions are formed in the cytoplasm by budding from the plasma membrane. Paramyxovirus infectivity is sensitive to ether, acid, and heat. The genera include the following: *Paramyxovirus* (parainfluenza viruses, mumps virus, Newcastle disease virus, Yucaipa and other avian paramyxoviruses); *Morbillivirus* (the viruses of measles, canine distemper, rinderpest, and peste de petite ruminant); and *Pneumovirus* (respiratory syncytial viruses of man and of cattle, and pneumonia virus of mice). Paramyxoviridae are genetically stable and genetic recombination does not occur.

RHABDOVIRIDAE
(Fenner, 1976)

Members of this family have enveloped virions that are rod-shaped, resembling a bullet (with one end rounded and the other flattened) or bacilliform. Enclosed within the lipoprotein envelope and membrane protein is the long tubular nucleocapsid with helical symmetry. Members of some genera multiply in arthropods as well as in vertebrates or higher plants; others multiply only in insects. Infectivity is sensitive to ether, acid, and heat. The genera that infect vertebrates are *Lyssavirus* (including rabies virus, Duvenhage virus, and Mokola virus, all of which infect man; Lagos bat virus; and several agents isolated as yet only from insects), and *Vesiculovirus* (including vesicular stomatitis virus and a number of antigenically interrelated viruses from various animal species). Marburg virus, a simian virus highly pathogenic for man, is rhabdovirus-like in most properties but has very elongated forms.

RETROVIRIDAE
(Dalton et al., 1974; Vogt, 1976)
(See Table 5)

Much remains to be settled concerning this family, but a good deal of progress has been made. The members include all of the RNA tumor viruses ("oncornaviruses"; "leukoviruses"), which are now assigned to a subfamily, Oncovirinae; other subfamilies are Lentivirinae (the slow viruses of the maedi/visna group) and Spumavirinae (the foamy virus group — agents that form syncytia in cell cultures). Members of the Retroviridae characteristically have a reverse transcriptase (RNA-dependent DNA polymerase) within the virion. For the most thoroughly studied members, the lipoprotein envelope encloses an inner shell with icosahedral symmetry and a central core or nucleocapsid with helical symmetry. Infectivity is ether, acid, and heat sensitive. Replication of the viral RNA involves a DNA provirus that is integrated into host cell DNA. In the case of the subfamily Oncovirinae, all normal cells of several animal species contain integrated copies of genes of the endogenous species of oncovirus. The oncovirus genes may be unexpressed but can be activated by physical and chemical agents, by superinfection with other oncoviruses, and even by herpesviruses. According to certain morphologic, antigenic, and enzymatic differences, oncoviruses also have been divided into A, B, and C (and possibly D) types of viruses. With some exceptions, oncoviruses fall into host-species–specific groups of agents inducing either leukemias or sarcomas — that is, leukemia-sarcoma complexes of avian, murine, feline, or hamster oncoviruses; other groups are murine mammary tumor virus and primate oncoviruses. One of the primate oncoviruses is the monkey mammary tumor virus (MoMTV).

ARENAVIRIDAE
(Pfau et al., 1974)

Members of this family have spherical or pleomorphic virions with a dense lipid bilayer membrane that bears surface projections; within the virion core are electron-dense RNA-containing granules about 20 to 30 nm in diameter that resemble ribosomes. Most member viruses have a single restricted rodent host in which persistent infection occurs; spread to other mammals and to man can take place, but is unusual. Members include lymphocytic choriomeningitis virus (LCM virus), which infects mice but may spread to man; Lassa virus; and members of the Tacaribe complex (Junin and Machupo viruses of South American hemorrhagic fevers, Pichinde virus, and several other viral agents that have been isolated as yet only from arthropods). Some arenaviruses (for example, Junin and Machupo, and Lassa) are very serious pathogens when they do spread to man.

CORONAVIRIDAE
(Tyrrell et al., 1975)

The family is named for unique petal-shaped or club-shaped peplomers that project from the envelope; in negatively stained electron micrographs these projections form a fringe resembling the solar corona. The interior structure of the virion is believed to be a loosely-wound helically symmetric nucleocapsid. The genome consists of one large molecule of single-stranded RNA. Infectivity is sensitive to ether, acid, and heat. Nucleocapsids develop in the cytoplasm and mature by budding through intracytoplasmic membranes. Several serotypes of human coronaviruses have been isolated from patients with acute upper respiratory tract illnesses, primarily through the use of human embryonic tracheal and nasal organ cultures. There are distinct coronaviruses that infect a number of animal species.

BUNYAVIRIDAE
(Porterfield et al., 1975/76)

This family is the largest and most recently recognized taxonomic grouping assigned to an antigenically interrelated set of arboviruses. There are at least 200 members, more than 100 belonging to the Bunyamwera supergroup of arboviruses. The virions are spherical; they develop in the cytoplasm and mature by budding through intracytoplasmic membranes. Members of the family produce a number of important diseases of man and of domestic animals — for example, California encephalitis, Crimean hemorrhagic fever, sandfly fever, Rift Valley fever, and Nairobi sheep disease. Most Bunyaviridae members are mosquito-transmitted, but some are tickborne.

EMERGING PROBLEMS IN VIRUS CLASSIFICATION

Some of the present and developing problems that viral taxonomists will have to meet are those posed by the recently discovered forms of life called viroids, and also by viral hybrids (between unrelated viruses), pseudovirions, and recombinant DNA.

Viroids constitute a recently discovered class of infectious agents smaller than viruses. They are known to cause several diseases of plants (for example, potato spindle tuber disease), and may ultimately be found to cause disease in man and higher animals. (For example, the agent of scrapie disease of sheep — one of the puzzling "slow viruses" not yet placed taxonomically — may, according to recent findings, prove to be a viroid.) Viroids exhibit the characteristics of nucleic acids in crude extracts; that is, they are insensitive to heat and to organic solvents but are sensitive to nucleases. They do not appear to possess a protein coat. Viroids known at present consist solely of a short strand of RNA with a molecular weight of 75,000 to 100,000.

VIRUS HYBRIDS

The fact that virus hybrids can exist in nature should be more widely recognized. If the simian papovavirus SV40 had not already been known as a virus before the discovery of SV40-adenovirus hybrid particles, these particles would have presented viral taxonomists with a very confusing puzzle. The hybrid particles, in which portions of SV40 genome material are covalently linked to adenovirus genetic material and encased within an adenovirus coat, would have seemed to be a new and very strange virus that reacted antigenically like an adenovirus (of the serotype from which its coat was derived) but whose progeny had many properties altogether different from other members of the adenovirus group.

PSEUDOVIRION

This is another viral form that is difficult to classify. During viral replication the capsid sometimes encloses *host* nucleic acid rather than viral nucleic acid. Such particles look like ordinary virus particles when observed by electron microscopy, but they do not replicate. Pseudovirions contain the "wrong" nucleic acid. These particles present the taxonomist with problems based on natural events, but future laboratory manipulations will probably add to these problems of classification.

RECOMBINANT DNA

Recently developed techniques allow DNA to be cleaved into specific pieces by using enzymes from bacteria called restriction endonucleases. These distinct fragments have importance in two areas: (1) the physical mapping of genes in large, complicated DNA genomes, and (2) genetic engineering. In addition to the overriding concern for safety precautions to ensure that new genetic combinations thus produced do not result in new organisms with dangerous properties, virologists must also give attention to how the new recombinant organisms should be classified. Classification of these new forms of life needs to be developed in ways that will reflect their origin and relatedness to each other and to other living things.

References

Bachmann, P. A., Hoggan, M. D., Melnick, J. L., Pereira, H. G., and Vago, C.: Parvoviridae. Intervirology 5:83, 1975.

Berge, T. O.: International Catalogue of Arboviruses Including Certain Other Viruses of Vertebrates. DHEW Publication No. (CDC) 75-8301, U. S. Dept. of Health, Education and Welfare, 1975.

Cooper, P. D., et al.: Picornaviridae: Second Report. Intervirology 10:165, 1978.

Dalton, A. J., Melnick, J. L., Bauer, H., Beaudreau, G., Bentvelzen, P., Bolognesi, D., Gallo, R., Graffi, A., Haguenau, F., Heston, W., Huebner, R., Todaro, G., and Heine, U. I.: The case for a family of reverse transcriptase viruses: Retraviridae. Intervirology 4:201, 1974.

Davis, B. D., Dulbecco, R., Eisen, H. N., Ginsberg, H. S., and Wood, W. B., Jr.: Microbiology. New York, Hoeber Medical Division, 1967.

Dowdle, W. R., Davenport, F. M., Fukumi, H., Schild, G. C., Tumova, B., Webster, R. G., and Zakstelskaja, L. Ya.: Orthomyxoviridae. Intervirology 5:245, 1975.

Fenner, F.: Classification and Nomenclature of Viruses: Second Report of the International Committee on Taxonomy of Viruses. Intervirology 7:1, 1976.

Fenner, F.: Portraits of viruses: The poxviruses. Intervirology 11:137, 1979.

Fenner, F., and White, D. O.: Medical Virology. 2nd ed. New York, Academic Press, 1976.

Joklik, W. K., et al.: Reoviridae. Intervirology, 1980.

Kilham, L., and Margolis, G.: Problems of human concern arising from animal models of intrauterine and neonatal infections due to viruses. Progr Med Virology 20:113, 1975.

Kingsbury, D. W., Bratt, M. A., Choppin, P. W., Hanson, R. P., Hosaka, Y., ter Meulen, T., Norrby, E., Plowright, W., Rott, R., and Wunner, W. H.: Paramyxoviridae. Intervirology 10:137, 1978.

Matthews, R. E. F.: Classification and Nomenclature of Viruses. Third Report of the International Committee on Taxonomy of Viruses. Intervirology 12:129, 1979.

Mayor, H. D., and Melnick, J. L.: Small deoxyribonucleic acid-containing viruses (picodnavirus group). Nature 210:331, 1966.

Melnick, J. L.: Papovavirus group. Science 135:1128, 1962.

Melnick, J. L.: Taxonomy of viruses, 1978. Progr Med Virology 25:160, 1979.

Melnick, J. L., Allison, A. C., Butel, J. S., Eckhart, W., Eddy, B. E., Kit, S., Levine, A. J., Miles, J. A. R., Pagano, J. S., Sachs, L., and Vonka, V.: Papovaviridae. Intervirology 3:106, 1974a.

Melnick, J. L., et al.: Picornaviridae. Intervirology 4:303, 1974b.

Norrby, E., Bartha, A., Boulanger, P., Dreizin, R. S., Ginsberg, H. S., Kalter, S. S., Kawamura, H., Rowe, W. P., Russell, W. C., Schlesinger, R. W., and Wigand, R.: Adenoviridae. Intervirology 7:117, 1976.

O'Callaghan, D. J., and Randall, C. C.: Molecular anatomy of herpesviruses: Recent studies. Progr Med Virology 22:152, 1976.

Pfau, C. J., Bergold, G. H., Casals, J., Johnson, K. M., Murphy, F. A., Pedersen, I. R., Rawls, W. E., Rowe, W. P., Webb, P. A., and Weissenbacher, M. C.: Arenaviruses. Intervirology 4:207, 1974.

Porterfield, J. S., Casals, J., Chumakov, M. P., Gaidamovich, S. Ya., Hannoun, C., Holmes, I. H., Horzinek, M. C., Mussgay, M., Oker-Blom, N., and Russell, P. K.: Bunyaviruses and Bunyaviridae. Intervirology 6:13, 1975/76.

Porterfield, J. S., Casals, J., Chumakov, M. P., Gaidamovich, S. Ya., Hannoun, C., Holmes, I. H., Horzinek, M. C., Mussgay, M., Oker-Blom, N., Russell, P. K., and Trent, D. W.: Togaviridae. Intervirology 9:129, 1978.

Tyrrell, D. A. J., Almeida, J. D., Cunningham, C. H., Dowdle, W. R., Hofstad, M. S., McIntosh, K., Tajima, M., Zakstelskaya, L. Ya., Easterday, B. C., Kapikian, A., and Bingham, R. W.: Coronaviridae. Intervirology 5:76, 1975.

Vogt, P. K.: The oncovirinae — a definition of the group. In First Report of the WHO Collaborating Centre for Collection and Evaluation of Data on Comparative Virology (Proceedings of the Official Opening of the Centre, in Munich). Published by the WHO Centre, Munich, 1976.

Wildy, P.: Classification and Nomenclature of Viruses: First Report of the International Committee on Nomenclature of Viruses. Monographs in Virology, Vol. 5, Basel, S. Karger, 1971.

9 VIRAL REPLICATION

Samuel C. Silverstein, M.D.

All viruses share a common problem: namely, to maintain and propagate their genetic information. To this end viruses parasitize the biosynthetic machinery of their host cells, availing themselves of one or more of the many subcellular compartments, synthetic capacities, processing pathways and substrates available in animal cells. Even the picornaviruses, which are among the smallest and apparently least complex of the animal viruses, require dozens of host-cell cytoplasmic functions and structures to replicate their genomes. The larger DNA-containing viruses, such as herpesviruses, require for their replication the biosynthetic capacities of both the nucleus and cytoplasm of their cellular hosts. Some tumor-forming viruses, such as the papova and oncornaviruses, integrate their genomes into their host cells' DNA. Locked together in this way, papovavirus or oncornavirus genes and host-cell genes replicate coordinately.

The number and variety of replicative strategies utilized by the different classes of RNA and DNA viruses suggest that they have left no potential replicative or biosynthetic pathway unexplored. Even the central dogma of molecular biology, "DNA makes RNA," must be qualified in the case of the oncornaviruses in which DNA is synthesized on an RNA template. Viewed in this context, each virus group occupies a specific ecologic niche among intracellular parasites and identifies one of the many pathways that have evolved for the faithful and efficient transfer of genes from cell to cell.

INITIATION OF A VIRAL INFECTION

The initial problem facing any potential pathogen, be it bacterial, protozoan, or viral, is to enter the host. This is an especially demanding task for viruses since, unlike schistosomules or salmonellae, they cannot crawl through the skin or break open the junctions that join the cells of the intestinal epithelium. Viruses enter their hosts by infecting a susceptible cell.

Viral pathogens that enter without the assistance of an insect or animal vector do so by infecting cells of the respiratory, intestinal, and genital tracts. Although epithelial cells represent the major targets in these locations, viruses also appear to have access to the lymphoid tissues that surround the oropharynx. Once infected, the epithelial cells may release progeny virions into the bloodstream as occurs with smallpox and polioviruses, or the recirculating cells of the lymphoid tissues may themselves become infected and carry the virus to tissues elsewhere in the body. For viruses that utilize an insect (arboviruses), animal (rabies), or human (hepatitis B) vector, entry into the host is not synonymous with infection; these agents must still find and penetrate a susceptible cell to initiate the infectious process.

CELL SURFACE RECEPTORS AND TISSUE TROPISM

The precise mechanism(s) that regulate the entry of viral nucleic acids or nucleoproteins into the cytoplasm or nucleus are poorly understood. It is clear that different viruses adsorb to unique "receptor" molecules on the host cells' surface and that these plasma membrane receptors determine the susceptibility of cells bearing them to infection with a given virus. For instance, both mouse and human cells produce poliovirus when they are infected with RNA extracted from poliovirus Type 1. That is, both human and mouse cells are biosynthetically capable of translating and replicating poliovirus RNA and encapsidating it into virions. However, human cells, unlike mouse cells, have plasma membrane receptors for poliovirus. As a consequence, only human cells bind and uncoat the intact virus, thereby allowing the viral RNA to gain access to the cell's cytoplasm and to initiate an infection.

Why have viral receptors been retained on the surface of cells? If the presence of these receptors served only to enhance the susceptibility of cells to a viral pathogen, then cells retaining these receptors would have a selective disadvantage in terms of survival. If, on the other hand, these surface receptors perform important roles in the maintenance of normal cellular function, then selective pressures for their elimination would be balanced by pressures for their retention. Although no data are available concerning the evolutionary history of cellular receptors for viruses, we do know that many cells have membrane receptors that mediate the binding and internalization of hormones, vitamins, immunoglobulins, and other physiologically important macromolecules. Viruses may utilize one or more of these receptor systems to penetrate their host cells.

Specific viral capsid proteins determine cell and tissue tropisms of closely related viruses. For instance, reovirus Type 1 infects ependymal cells in suckling mice and causes hydrocephalus; reovirus Type 3 infects neurons and causes a fatal encephalitis in these animals. The viral protein that controls these type-specific effects is the outer capsid protein responsible for the hemagglutinating properties of the virus. Each reovirion contains less than 30 copies of Type 1 or Type 3 hemagglutinin protein on its surface. Recombinant viruses containing all Type 3 genes, except for the single Type 1 gene encoding the hemagglutinin protein, behave like reovirus Type 1. That is, this recombinant virus infect ependymal cells, not neurons, and causes hydrocephalus, not encephalitis.

PENETRATION OF THE VIRAL GENOME

Studies of avian RNA tumor viruses and of reoviruses suggest that the interaction of cell membrane receptors with viral capsid proteins regulates the penetration of a viral nucleoprotein into the cytoplasm, and not merely the attachment of the virus to the cell's surface. We currently recognize two mechanisms by which penetration occurs: fusion of the virus with the cell surface and endocytosis of the intact virion.

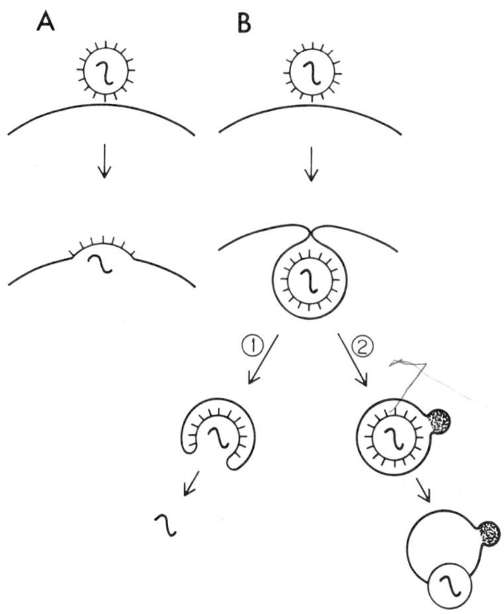

FIGURE 1. *Pathways of viral uncoating. A, The viral envelope fuses with the cell surface, resulting in release of viral nucleoprotein into the cytoplasmic matrix. B, The virus is endocytosed. Encircled 1, The interaction of the viral envelope with the membrane of the endocytic vacuole causes fusion or dissolution of the vacuolar membrane and partial or complete uncoating of the virus. Encircled 2, The virus-containing vacuole fuses with lysosomes. The outer viral capsids are degraded by lysosomal enzymes, and the viral nucleoprotein is released into the cytoplasm.*

Fusion

Herpes, pox, and paramyxoviruses are agents whose outer envelope is composed of a lipid bilayer in which viral glycoproteins are embedded. These viruses appear to enter by fusion of this envelope with the host cell's plasma membrane (Fig. 1*A*). Fusion of a viral envelope with the cell membrane has been shown most convincingly for the paramyxoviruses. A glycoprotein named F (fusion factor) has been identified on the surface of paramyxoviruses, and activity of the F protein has been proved to be required for viral infectivity. As a consequence of its fusion with the cell's surface, viral envelope proteins are incorporated into the host cell's plasma membrane, and the viral nucleoprotein is released directly into the cytoplasm. It is of interest that most of the viruses that enter by fusion also have the capacity to fuse infected cells with their uninfected neighbors. Presumably the same viral glycoproteins that promote viral penetration are inserted into the plasma membranes of infected cells and are responsible for the fusion of these cells with neighboring cells.

Endocytosis

Fusion factors have not been identified in several classes of enveloped viruses (myxo, toga, rhabdo, and oncorna), or in viruses (picorna, reo, papova, adeno) containing only proteins in their outer shells. These agents appear to be engulfed by the host cell in a phagocytic vacuole. The interaction of the viral capsids with the membrane of the phagocytic vacuole is presumed to result in the modification of the surface structure of the virus and the lysis of the phagocytic vacuole (Fig. 1*B*, pathway I). In this way the picorna, toga, myxo, rhabdo, and oncornaviruses are thought to release their infectious nucleic acids or nucleoproteins into the cytoplasm. Papova and adenoviruses require an additional uncoating step within the cytoplasm. These viruses are partially uncoated during their escape from the phagocytic vacuole. Subsequently, these partially uncoated virions are transported to the nuclear membrane at which site their infectious DNA is released and transported via the nuclear pores into the nucleoplasm.

Reoviruses illustrate a unique variation of this uncoating pathway. These viruses enter by phagocytosis but are transported within endocytic vacuoles to the host cell's lysosomes. Proteolytic enzymes within the lysosomal compartment degrade the outer viral capsids; the resulting nucleoprotein core is resistant to further proteolysis and appears to be released from the lysosomes into the cytoplasmic matrix (Fig. 1*B*, pathway 2).

THE ROLE OF ANTIBODIES AND COMPLEMENT IN PREVENTING INFECTION

Antibodies directed against viral surface determinants inhibit viral infection by several mechanisms. First, antibodies block the interaction of viral capsids with cell-surface receptors and thereby markedly decrease the efficiency of viral adsorption to cells. Second, antibodies stimulate the fixation of serum complement to the virus. In the case of some enveloped viruses, complement causes lysis of the viral particle. Third, antibodies enhance the binding of viruses to the antibody and complement receptors of monocytes and macrophages and promote phagocytosis of neutralized viruses by these cells. Mononuclear phagocytes exhibit a reduced capacity to replicate a variety of viruses and thereby may slow the spread of an infection. Fourth, antibodies promote the segregation of antibody-coated virus into cellular lysosomes. Most viruses are destroyed by the degradative enzymes contained in this cellular compartment.

Although serum and/or secretory antibodies generally protect against infections with viruses, several exceptions have been noted. Infants with low antibody titers against dengue virus may develop an exceptionally severe disease on reexposure to the virus, presumably because at low levels the antibodies promote adsorption to and infection of mononuclear phagocytes. Individuals experiencing recurrent herpes infections have high titers of neutralizing antibody against herpes virus. Reactivation of the latent virus leads to virus spread by cell-cell fusion, a process that is controlled poorly by humoral immunity.

REPLICATION

Having successfully gained access to the cell's biosynthetic machinery, viruses have the opportunity to subvert this machinery for their own purposes. Lytic animal viruses quickly shut off host-cell synthetic processes while temperate and transforming viruses rely on the integrity and longevity of the host cell for their own survival. Regardless of the replicative strategy to be employed, however, the initial problem facing all viruses is that of synthesizing messenger RNA (mRNA); it is only by forming new proteins that the virus can take command of the cell. In the sections that follow I summarize the steps leading to mRNA formation, genome replication, and synthesis of progeny virus for each of the major virus groups. Throughout, single-stranded polynucleotides of the same polarity as messenger RNA are termed plus (+) strands and single-stranded polynucleotides of complementary base sequence are termed minus (−) strands.

SYNTHESIS AND STRUCTURE OF MESSENGER RNA

Polynucleotide chains are synthesized by the sequential addition of nucleotides. The direction of polymerization is 3′→5′ with respect to the template and 5′→3′ with respect to the newly formed polynucleotide. For example, an RNA template that has the sequence 3′ HO-UpGpApCp, etc.* at its 3′ end directs the formation of a complementary RNA strand with ATP at its 5′ terminus

5′ pppA-OH 3′
(3′ HO-UpGpApCp---etc.).

The next nucleotide, cytidylic acid, will join to the 3′ hydroxyl of ATP to give

5′ pppApC-OH 3′
3′ HO-UpGpApCp---etc.

Polymerization continues by the stepwise addition of nucleotides until the polynucleotide chain is completed. In the case of the RNA chain illustrated above, the next two nucleotides specified by the template RNA are uridylic and guanylic acids, giving the structure

5′ pppApCpUpGp-OH 3′
3′ HO-UpGpApCp---etc.

As the chain is synthesized, it is released from the template.

Most cellular and viral messenger RNAs have distinctive chemical modifications at their 5′ and 3′ ends. These modifications enhance the efficiency of messenger RNA translation by the ribosomes. The 5′ end is modified by a guanylic acid residue, the so-called "capping" nucleotide, joined by an unusual 5′ to 5′ linkage to the 5′ end of the RNA chain; both the capping guanylic acid residue and the nucleotide to which it is joined are methylated. Assuming the polynucleotide illustrated above is to be capped and methylated, the reactions proceed in the sequence pppApCpUpG-OH → GpppApCpUpG-OH → methyl-GpppAmethylpCpUpG-OH. The enzymes that carry out the capping and methylating reactions may be viral, as occurs in the formation of reovirus mRNAs, or

*p represents the phosphates that join the ribonucleosides in phosphodiester linkages. pppA-OH is adenosine triphosphate. The OH is the hydroxyl group at the 3′ position of the ribose of the nucleotide. Gppp is guanosine triphosphate.

cellular, as occurs in the formation of influenza virus mRNAs.

The 3′ end of most messengers is modified by a sequence of 50 to 200 adenylic acid residues. By and large these adenylic acid residues are not determined by a sequence of complementary nucleotides in the template. Rather, they are added to the 3′ end of the nascent mRNA chain by a viral or a cellular polyadenylic acid polymerase according to the reaction sequence - - -pU-OH 3′+(ATP)$\rightarrow_n$ - - -pU(pA)$_n$-OH 3′.

POSITIVE-STRAND RNA VIRUSES

Picornaviruses

The entero and rhinoviruses, the smallest of the RNA-containing animal viruses, contain a single molecule of single-stranded RNA of molecular weight 2.6×10^6 daltons. The genome RNA has the same size, base sequence, and polarity as mRNA. Upon entry into the cytoplasm, the genome RNA binds to ribosomes and initiates the synthesis of viral proteins (Fig. 2, pathway I). One or more of these proteins forms an RNA polymerase, and this enzyme catalyzes the synthesis of a negative RNA strand, using as its template the parental RNA plus (+) strand (Fig. 2, pathway II). The RNA minus strands act as templates for the synthesis of single-stranded RNA plus strands. Some of these progeny plus strands function as mRNA and direct the synthesis of viral proteins. Other plus strands are enclosed in viral capsids and thus become the genomes of the newly formed progeny virus.

The initial translation product of the viral

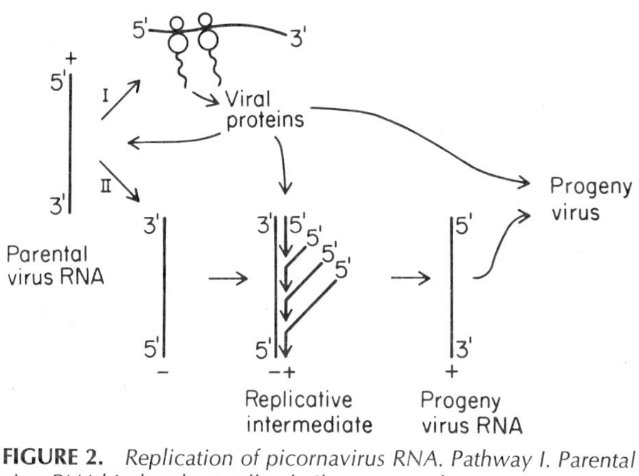

FIGURE 2. *Replication of picornavirus RNA. Pathway I. Parental virus RNA binds to host cell polyribosomes and serves as messenger RNA for the synthesis of viral proteins. Pathway II. Viral proteins form an RNA-dependent RNA polymerase that makes a negative-strand RNA copy of the parental positive-strand RNA. This negative-strand serves as a template for the synthesis of progeny positive strands.*

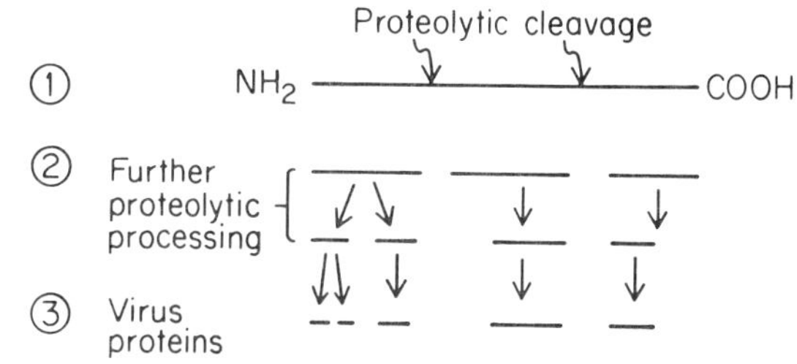

FIGURE 3. *Processing of the large precursor protein, encircled 1, that is the primary translation product of poliovirus messenger RNA. Proteolytic enzymes in the cytoplasm of the host cell perform the initial cleavage steps, encircled 2. Viral encoded enzymes may be involved in later cleavages. The proteins derived from the left (amino) half of the large protein form the viral coat proteins. The proteins derived from the right (carboxyl) half of the precursor protein form the viral RNA polymerases.*

mRNA is a large precursor protein (Fig. 3). Proteases present in the cytoplasm of the host cell catalyze the cleavage of this "polyprotein" into the smaller proteins that form the viral polymerase and capsid proteins. The capsid proteins self-assemble with one another, and with viral RNA, to form the completed virus. Progeny virus particles accumulate at an exponential rate within the cytoplasm until, for unknown reasons, the biosynthetic capacity of the cell is exhausted. The cell then lyses, releasing virus into the surrounding tissues.

Togaviruses or Arthropod-borne Viruses

These agents have a slightly larger RNA genome (4×10^6 daltons) than the picornaviruses, but replicate their genomes in essentially the same manner. However, in addition to genome-sized RNA, togavirus-infected cells contain an additional species of mRNA that is about one-third the length of viral genome RNA. This mRNA directs the synthesis of the viral core protein, and the three viral envelope glycoproteins E_1, E_2, and E_3. Viral genome RNA also serves as mRNA and is presumed to encode nonstructural proteins in addition to the envelope glycoproteins described above. Both genome-sized and shorter RNA plus strands are formed on genome length complementary RNA minus strands. Transcription of the less-than-genome length mRNA on the minus-strand RNA template is initiated independently of genome-length RNA.

Togaviruses are composed of an RNA-containing core wrapped within a lipoprotein envelope. The envelope is composed of host-cell lipids and viral glycoproteins. The core is assembled in the cytoplasmic matrix from RNA and soluble proteins while the envelope glycoproteins are synthesized in the rough endoplasmic reticulum and transported to the cell surface in membrane vesicles in a manner similar to that shown for vesicular stomatitis glycoprotein (Fig. 6). Formation of the virus is completed by the encapsidation of cores within their envelopes (see Fig. 6), and their release from the cell surface. Mature togaviruses do not accumulate in the cell but are released into the surrounding tissues as they are completed.

Coronaviruses

In comparison with the picorna and togaviruses, our information about the molecular biology of coronavirus replication is less complete. Human coronaviruses contain single-stranded RNA of ~ 6×10^6 daltons molecular weight. This RNA, like picornavirus RNA and most messenger RNAs, contains a polyadenylic acid sequence at its 3′ terminus. RNA extracted from coronaviruses is infectious. These findings, and the lack of a nucleic acid polymerase within the virus, suggest that coronaviruses are positive-strand viruses and that they replicate their RNA by a mechanism similar to the one described for picornaviruses.

Coronaviruses are named for the prominent spikes that radiate from their lipid envelopes. They replicate in the cytoplasm. Their helical viral ribonucleoprotein is encapsidated by budding into the cisternae of the endoplasmic reticulum. The virus is presumably released by lysis of infected cells; transport of coronaviruses from the rough endoplasmic reticulum to the cell surface in membrane vesicles may also occur.

Oncornaviruses

These RNA-containing viruses have been identified as tumor-causing agents in a variety of vertebrates. Although they are not presently associated with disease in man, it seems unlikely that the species *Homo sapiens* has been excluded as a host for these ubiquitous viruses.

Oncornaviruses are diploid; they contain duplicate copies of single-stranded RNA of the same polarity as viral messenger RNA. In addition, they contain a specific transfer RNA (tryptophanyl tRNA in the case of avian oncornaviruses) hydrogen bonded to genome RNA, and an RNA-dependent DNA polymerase that synthesizes a double-stranded DNA copy of the viral RNA. The mechanism of synthesis of this double-stranded DNA copy is not completely understood, but the following steps have been identified (Fig. 4). Upon release of the virus core into the cytoplasm, synthesis of a DNA minus strand is initiated using viral RNA as template and the transfer RNA that is bound to viral RNA as primer (Fig. 4, *encircled 1*). Minus-strand synthesis is followed by plus-strand formation (Fig. 4, *encircled 2*), and the resulting linear DNA duplex is circularized

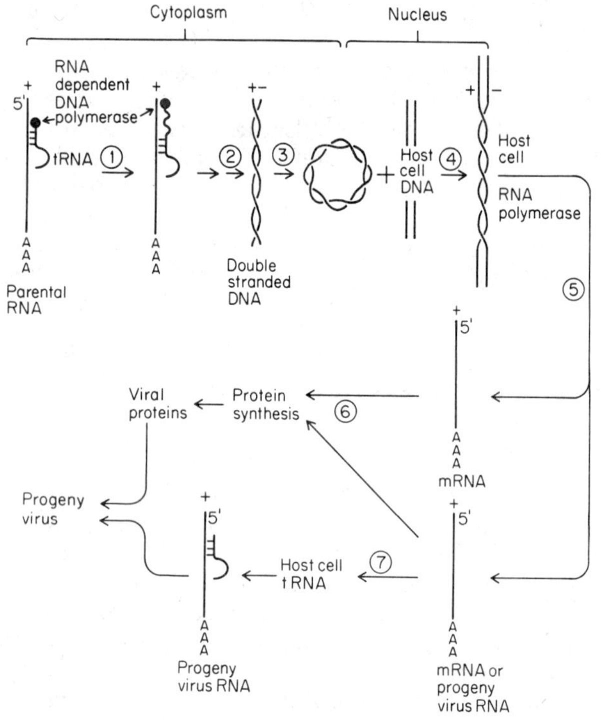

FIGURE 4. *Replication of oncornavirus RNA. Reactions shown on the left side of the diagram occur in the cytoplasm; those on the right side occur in the nucleus. Synthesis of a DNA copy of viral RNA begins (encircled 1) as soon as the viral ribonucleoprotein is released into the cytoplasm. Following its covalent integration into host cell DNA (encircled 4), the double-stranded DNA copy of the viral genome becomes a template for host cell RNA polymerases (encircled 5). These enzymes synthesize viral messenger RNAs (encircled 6) and progeny virus RNAs (encircled 7), using the integrated viral DNA as template. Proteins encoded by these messengers form the envelope and internal proteins of progeny virions. See text and Figure 6 for further details of oncornavirus morphogenesis.*

and integrated into host-cell DNA (Fig. 4, *encircled 3 and 4*). Host-cell RNA polymerases now synthesize messenger RNAs of less than genome length, and single-stranded RNA plus strands of genome length (Fig. 4, *encircled 5*). RNAs of both sizes are capped and polyadenylated at their 5′ and 3′ ends, respectively. They are exported from the nucleus into the cytoplasm where they direct the synthesis of viral proteins (Fig. 4, *encircled 6*). Genome-length RNA, together with host cell transfer RNA, is incorporated into progeny virions (Fig. 4, *encircled 7*). Progeny virions are formed by the envelopment of their ribonucleoprotein core within a lipoprotein envelope at the surface of the infected cell. Synthesis of oncornavirus envelope glycoproteins and their insertion into the nascent virus particle as it buds from the cell's surface occurs by a mechanism that is similar to the one described in Figure 6 for the formation of the envelope of vesicular stomatitis virus. The host cell is not destroyed by an oncornavirus infection; it continues to divide and to release virus. Thus oncornavirus-infected cells express viral envelope glycoproteins on their plasma membranes and viral proteins in their cytoplasm indefinitely.

A 60,000 molecular-weight protein that is encoded by the viral sarcomagenic ("sarc") gene has been identified as the molecule responsible for oncogenic transformation of cells infected with avian oncornaviruses. This protein has kinase activity; that is, using ATP as a cofactor, it catalyzes the phosphorylation of cellular and/or viral proteins. Whether this enzymatic activity is responsible for the malignant behavior of cells transformed by avian sarcoma viruses is unknown.

NEGATIVE-STRAND RNA VIRUSES

These viruses contain RNA genomes that are complementary in nucleotide sequence to messenger RNA. Since messenger RNAs are by definition plus strands, complementary RNAs are termed minus or negative strands. Three virus groups of major significance are classified under this heading. In order of increasing genome complexity, they are rhabdoviruses (rabies, vesicular stomatitis virus); paramyxoviruses (measles, mumps, parainfluenza, and respiratory syncytial viruses); and orthomyxoviruses (influenza viruses). These viruses share several design features. They have envelopes whose lipids are derived from the surfaces of their host cells; they have helical nucleocapsids containing RNA of negative polarity; and they contain RNA-dependent RNA transcriptases, enzymes that make positive-strand mRNA using the viral genome as template.

The mechanisms of messenger RNA transcription by negative-strand viruses have been examined in great detail. Nevertheless, key issues regarding mRNA synthesis are unresolved. The intermediate steps in the replication of genome RNA are even less clearly defined. Thus the steps in mRNA synthesis and viral genome replication shown in Figures 5 and 7 should be taken as representative of the overall process of progeny-negative RNA strand formation but not as a precise description of mechanisms or of the intermediates involved.

Rhabdoviruses

Rhabdoviruses are bullet-shaped; they contain five proteins and a single strand of RNA of molecular weight 4.5×10^6 daltons. The RNA is not by itself infectious. Within the virus, the RNA is associated throughout its length with multiple copies of the nucleocapsid protein. Also associated with the ribonucleoprotein are two proteins that form an RNA transcriptase. A matrix protein lies on the under or internal surface of the lipid bilayer that forms the viral envelope. A single glycoprotein spans the lipid bilayer; this glycoprotein forms the spikes that project from the virus surface. Information concerning rhabdovirus replication is chiefly derived from studies of vesicular stomatitis viruses. These viruses replicate in the cytoplasm.

Entry of the viral nucleoprotein into the cytoplasm activates the nucleoprotein-associated RNA transcriptase and initiates the synthesis of viral messenger RNA (Fig. 5, pathway I). Messenger RNA molecules are synthesized sequentially from the 3′ end of the viral genome. They are presumably cleaved from a large RNA transcript that is formed as the transcriptase proceeds down the genome from the 3′ to the 5′ end. These RNA messages are modified by the addition of methylated guanylic acid "caps" at their 5′ ends and of polyadenylic acid at their 3′ ends. Each messenger RNA directs the synthesis of one of the five viral proteins. Synthesis of viral proteins is required for the formation of full-length RNA plus strands and of progeny virus minus-strand RNA. Progeny minus strands serve as templates for amplification of messenger RNA synthesis and as the genomes of newly formed virions.

The viral proteins that form the internal components of the virus (in the case of vesicular stomatitis virus, these are the nucleocapsid, RNA transcriptase, and matrix proteins) are synthesized on polyribosomes that are free in the cy-

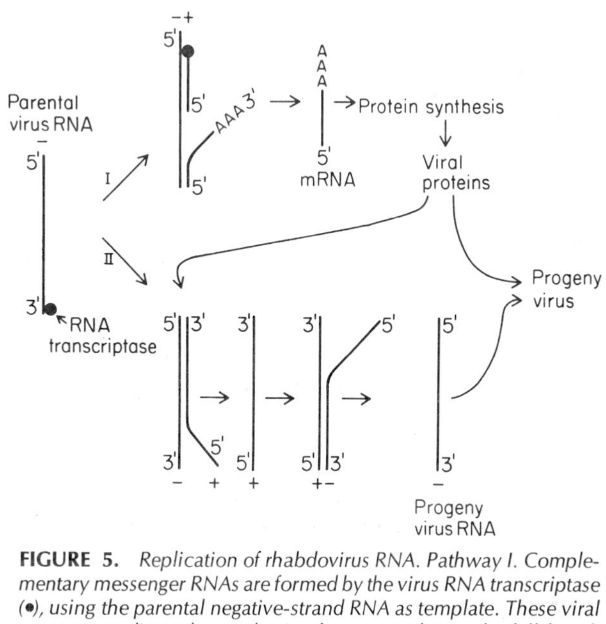

FIGURE 5. *Replication of rhabdovirus RNA. Pathway I. Complementary messenger RNAs are formed by the virus RNA transcriptase (•), using the parental negative-strand RNA as template. These viral messengers direct the synthesis of enzymes that make full length positive-strand RNA copies of the parental negative-strand RNAs (pathway II). The positive strands are the templates on which progeny virus negative-strand RNA is formed.*

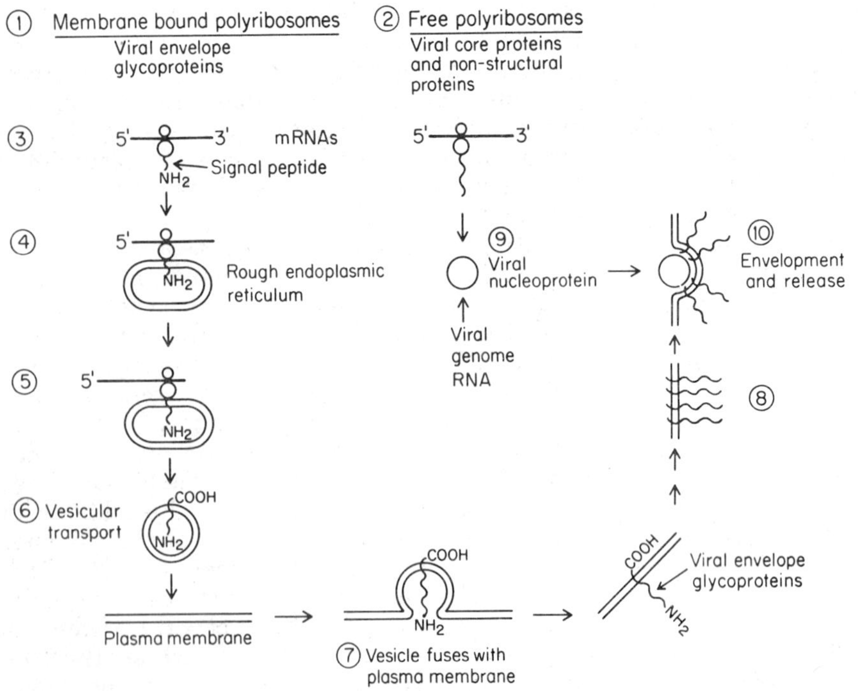

FIGURE 6. *Formation of viral envelope glycoproteins and the assembly of enveloped viruses. See text for details.*

toplasmic matrix (Fig. 6, *encircled 2*). The viral envelope glycoprotein (G) is formed on the ribosomes of the rough endoplasmic reticulum (Fig. 6, *encircled 1*). A "signal" peptide at the amino terminus of the nascent G protein promotes ribosome binding to the membranes of the rough endoplasmic reticulum and is responsible for the incorporation of the G protein into these membranes during its synthesis (Fig. 6, *encircled 3, 4, 5*). In this way, information contained in the viral messenger RNA specifies both the primary structure of the G protein and the cellular compartments through which it is to be processed. Glycosyl transferases within the rough endoplasmic reticulum add oligosaccharide chains to the nascent viral glycoprotein (Fig. 6, *encircled 5*). The viral glycoprotein is then transported via the smooth endoplasmic reticulum to the Golgi apparatus in which the final sugar residues are added. The composition and sequence of sugars in viral glycoproteins are determined by the host cell. The glycoproteins are thought to be transported from the smooth endoplasmic reticulum to the Golgi apparatus and from there to the cell surface as part of membrane-bound vesicles (Fig. 6, *encircled 6*), and to be inserted into the cell surface by fusion of these vesicles with the plasma membrane (Fig. 6, *encircled 6, 7, 8*).

The viral glycoprotein has the same orientation in the membranes of the rough endoplasmic reticulum as it has in the plasma membrane and in the completed virus. The glycoprotein spans the lipid bilayer. Its amino terminus and carbohydrate lie on the outside of the viral envelope while its carboxyl terminus projects into the interior of the virus. Viral glycoproteins in the plasma membrane can be thought of as the nucleation sites for the assembly of matrix proteins on the cytoplasmic side of the membrane. The interaction of the ribonucleoprotein with membrane-bound matrix protein leads to the formation of the completed virus and its release from the cell (Fig. 6, *encircled 9, 10*).

The mechanisms governing the synthesis, glycosylation, vesicular transport, and insertion of G protein into the cell surface are generally applicable to the formation of viral envelope glycoproteins. Thus, it seems likely that these pathways and principles are employed in the morphogenesis of most enveloped viruses.

Paramyxoviruses

Paramyxoviruses are pleomorphic, filamentous, and spherical particles containing single-stranded RNA of molecular weight 5.5 to 6×10^6 daltons. They have two externally disposed glycoproteins, each of which forms spikes on the surface of the virion. One of these glycoproteins (termed HN) is both a hemagglutinin and a neuraminidase. It is responsible for attaching the virus to sialic acid residues on cell surfaces during penetration and for the virus's hemagglutinating activity. The neuraminidase removes sialic acid from cellular glycolipids and glycoproteins, thereby enhancing virus release from the surfaces of infected cells. The second viral glycoprotein (called F) promotes fusion of the viral envelope with cell surfaces, and fusion of infected cells with neighboring uninfected cells. Virus particles with an inactive fusion protein are not infectious.

Aside from these differences, paramyxoviruses resemble rhabdoviruses in overall replication strategy. Paramyxovirus ribonucleoprotein contains one major nucleocapsid protein and two other proteins that form the RNA transcriptase. A matrix protein lines the inner surface of the viral envelope. Release of the paramyxovirus ribonucleoprotein into the cytoplasm activates its RNA transcriptase and initiates the synthesis of viral messenger RNAs. Subsequently, viral proteins and progeny genome RNA are formed as outlined in Figure 5. Synthesis and insertion of viral glycoproteins HN and F into the cell surface occur in a manner similar to that described in Figure 6 for vesicular stomatitis G proteins. Matrix protein and ribonucleoprotein assemble at the cytoplasmic face of plasma membrane segments that contain HN and F proteins. Envelopment of the nucleoprotein and release of mature virions from the cell surface complete the replicative cycle.

Orthomyxoviruses

Influenza virions occur in spheric and filamentous forms. Both forms have lipid envelopes containing two glycoproteins, a hemagglutinin and a neuraminidase. The hemagglutinin is responsible for virus attachment to sialic acid residues on the surface of the host cell, while the neuraminidase promotes release of progeny virus from infected cells. Unlike the paramyxoviruses, myxoviruses lack a fusion protein and do not cause cell fusion.

The influenza viruses contain eight segments of single-stranded RNA of negative polarity. Each RNA segment has a unique nucleotide sequence; each encodes a single messenger RNA and thereby a single protein. The genome contains 5.6 to 6 $\times 10^6$ daltons of RNA.

Each RNA segment forms a helical coil around the nucleocapsid proteins. Four other viral proteins are bound to the viral nucleoprotein; the entire complex has RNA transcriptase activity. The RNA transcriptase catalyzes the synthesis of single-stranded messenger RNA copies of all eight segments of the viral genome.

After penetration, the influenza ribonucleoproteins are transported into the host cell's nucleus. Formation of viral messenger RNA requires host-cell RNA synthesis. The capped and methylated 5′ ends of host-cell messenger RNAs are used as "primers" for the viral RNA transcriptase. These primers are joined to RNA sequences encoded by the parental RNA templates. The host's polyadenylic acid polymerase adds adenylic acid residues to the 3′ ends of viral RNA transcripts (Fig. 7, pathway I). These messenger RNAs are exported into the cytoplasm where they direct the synthesis of viral proteins (Fig. 7, pathway I) and serve as templates for genome (minus strand) RNA formation (Fig. 7, pathway II). Viral envelope glycoproteins (neuraminidase and hemagglutinin) are synthesized on membrane-bound ribosomes; viral nucleocapsid* and matrix proteins* are formed on free polysomes. The envelope glycoproteins are inserted into the plasma membrane by vesicular transport; the final steps of myxovirus assembly are similar in principle to those described for paramyxo- and rhabdoviruses (Fig. 6); that is, matrix protein is bound to the

*See Chapter 7 for a description of the location of these proteins in the virion.

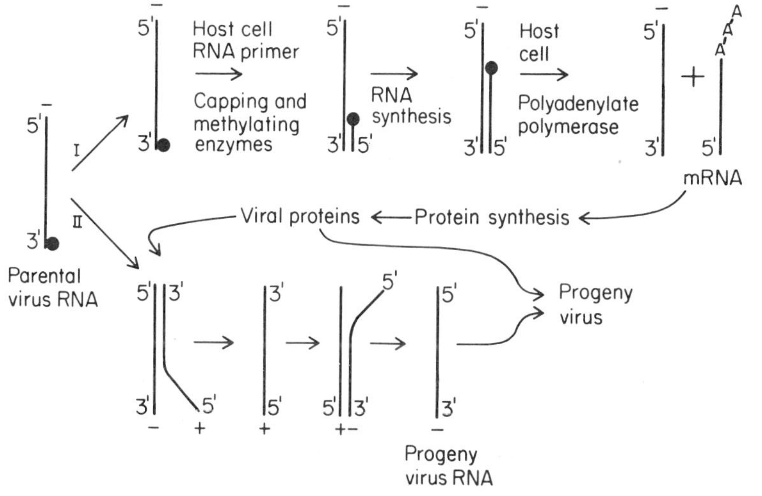

FIGURE 7. *Replication of influenza virus RNA. Pathway I. Viral messenger RNAs are synthesized by the virion-contained RNA transcriptase (•), using the parental genome RNAs as templates. Proteins encoded by these messengers form an RNA polymerase. This enzyme synthesizes positive-strand RNA copies of each of the segments of the parental genome (pathway II). These positive strands serve as templates for progeny negative-strand RNAs.*

cytoplasmic face of the plasma membrane. It is assumed, but not proved, that matrix protein binds preferentially to plasma membrane segments containing viral envelope glycoproteins. The eight pieces of viral ribonucleoprotein may bind to the matrix proteins; they are enveloped as the segment of membrane to which they are attached evaginates around them.

Sialic acid residues are removed from the host cell's surface by viral neuraminidase. Removal of those sialic acid groups prevents readsorption of progeny virions to the host cell and promotes the release of virus into the surrounding tissues. The negative-strand viruses are not stored within the cells they infect. They are released throughout the infectious cycle. Viral replication is terminated by death of the host cell.

The process by which viruses containing segmented genomes package one copy of each RNA segment is not known. It is possible, albeit unlikely, that packaging is a random event.

VIRUSES CONTAINING DOUBLE-STRANDED RNA

Reo and rotaviruses contain 10 and 12 separate segments, respectively, of double-stranded RNA. A single messenger RNA encoding a single viral protein is transcribed from each segment of the double-stranded RNA genome. These viruses contain no lipid. They are composed of two concentric protein shells; the RNA is contained within the inner shell or core.

Reoviruses enter their host cells within endocytic vacuoles and are subsequently concentrated within lysosomes (Fig. 1*B*, pathway 2). Lysosomal proteases degrade the outer layer of viral proteins and thereby activate the RNA transcriptase contained within the core. This enzyme synthesizes single-stranded RNA plus strands using the complementary minus strands of the viral double-stranded RNAs as templates. The mechanism of RNA transcription on genome double-stranded RNA is analogous to the mechanism of RNA transcription on double-stranded DNA. That is, the parental double strands are conserved, and the newly synthesized single-stranded RNA transcripts are released (Fig. 8, *encircled 2*). In the case of reoviruses, the parental double-stranded RNAs are conserved within the virus's core. The parental virions synthesize multiple single-stranded RNA copies of each segment of genome double-stranded RNA. These newly formed RNA transcripts are exported from the particle (Fig. 8, *encircled 2*) and serve as messengers for the synthesis of viral proteins (Fig. 8, *encircled 3*) or as templates for the formation of progeny double-stranded RNA (Fig. 8, *encircled 4*). Synthesis of double-stranded RNA occurs within the cores of nascent progeny virions (Fig. 8, *encircled 5*). The mechanism(s) responsible for segregating one copy of each of the ten or twelve single-stranded RNA templates into each nascent core particle are unknown. Cores containing newly formed double-stranded RNA (Fig. 8, *encircled 6*) synthesize additional RNA single strands and thereby amplify the rate of messenger and genome template RNA formation. RNA synthesis is terminated by the assembly of the outer layer of capsid proteins on the viral core (Fig. 8, *encircled 7*). Progeny virus accumulates in the cytoplasm and is released by lysis of the host cell.

VIRUSES CONTAINING DNA GENOMES

Papovaviruses

Papova is an acronym that stands for *pa*pilloma, *po*lyoma, and simian *va*cuolating viruses. Of

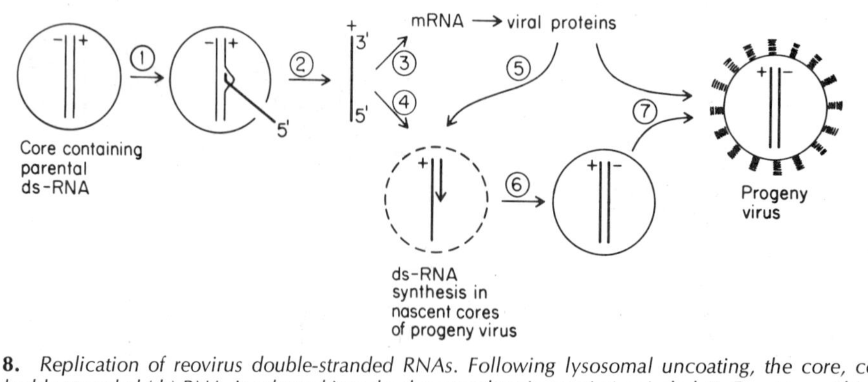

FIGURE 8. *Replication of reovirus double-stranded RNAs. Following lysosomal uncoating, the core, containing parental double-stranded (ds) RNA, is released into the the cytoplasmic matrix* (encircled 1). *Enzymes within the core synthesize and export single-stranded RNA plus strands* (encircled 2). *Progeny plus strands serve as messengers* (encircled 3) *for the synthesis of viral proteins* (encircled 5), *or as templates* (encircled 4) *for double-stranded RNA formation. Completed cores* (encircled 6) *containing all ten segments of double-stranded RNA combine with outer capsid proteins* (encircled 7) *to form progeny virions. See text for further details.*

these, wart virus (a papilloma virus) and JC virus (a vacuolating virus that is the putative causative agent of progressive multifocal leukoencephalopathy) are human pathogens. Interest in these viruses derives principally from the finding that simian virus 40 (a vacuolating virus) and polyoma virus cause oncogenic transformation of hamster and mouse cells, respectively. Cells transformed by SV40 or polyoma virus have one or more copies of the viral genome integrated into cellular DNA.

SV40 virions are small, icosahedral particles about 500 Å in diameter containing only protein and DNA. The genome is a circular molecule of double-stranded DNA of molecular weight 3×10^6 daltons. The DNA is infectious.

Bacterial restriction endonucleases, enzymes that cleave double-stranded DNA at sites containing specific nucleotide sequences, have been used to map the SV40 genome. One of these endonucleases, EcoRI, produces a single cut in the SV40 genome, converting it from a circular to a linear molecule. This EcoRI cleavage site provides a point of reference on SV40 DNA, and all functions of the DNA are mapped with respect to it. For instance, the origin of viral DNA replication begins roughly two-thirds of the way around the circle from the EcoRI cleavage site, and DNA replication occurs bidirectionally from this point of origin (Fig. 9).

After viral penetration into a permissive host cell, SV40 DNA is found in the nucleus. Host-cell RNA polymerase initiates transcription of "early" messenger RNAs using viral DNA as template. "Early" viral functions are messenger RNAs and their protein products that are synthesized prior to the replication of the viral genome. "Late" functions are messenger RNAs and proteins whose

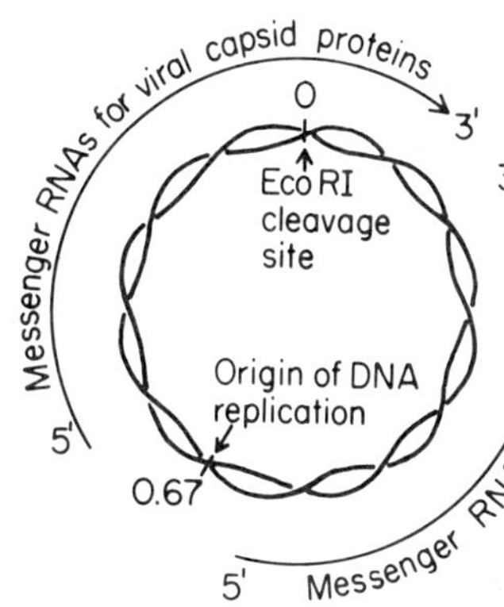

FIGURE 9. *Map of SV40 DNA. Early messenger RNAs are encoded by one strand of the DNA duplex. These messenger RNAs direct the synthesis of viral T-antigens. Late messengers are encoded by the complementary DNA strand on the opposite half of the circular molecule. These messengers direct the synthesis of viral capsid proteins. Messenger transcription begins near the origin of DNA replication (0.67) and proceeds in opposite directions on the two DNA strands.*

synthesis is initiated after replication of the viral genome. In general, early functions are enzymes required for genome replication, while late functions are viral capsid proteins. This is so for SV40-infected cells.

SV40 produces a single class of early messenger RNAs. These RNAs are exported to the cytoplasm where they direct the synthesis of tumor (T) antigens. T-antigens are a group of proteins, the largest of which has a molecular weight of 90,000. Animals with SV40-induced tumors have antibodies directed against T-antigens. These antibodies show that T-antigens are located principally in the nucleus of SV40-infected and transformed cells.

T-antigens stimulate the production of host-cell enzymes involved in DNA replication, and they induce host-cell DNA synthesis. T-antigens also regulate the replication of SV40 DNA. In cells that are transformed by SV40 the viral DNA is integrated into host-cell DNA early in the replicative cycle, and later viral genes encoding capsid proteins are not expressed. Monkey cells support a lytic infection with SV40. In these cells, the late genes that encode the three viral capsid proteins are transcribed and translated, and SV40 virions are formed within the nucleus. Virus is released when the cell lyses.

Adenoviruses

Adenoviruses are icosahedral particles containing linear double-stranded DNA of about 25×10^6 daltons molecular weight. Like SV40, they replicate in the nucleus, use host-cell RNA polymerase to initiate their RNA transcription, produce a unique T-antigen, and cause transformation in some cell lines (rat) and lytic infections in others (human). Cells transformed by adenovirus contain a specific segment of adenovirus DNA integrated into cellular DNA. This "transforming" DNA comprises less than 10 per cent of the adenovirus genome.

Adenoviruses enter their host cells by endocytosis. Fibers projecting from the 12 vertices of these icosahedral particles bind to the cell surface and are required for successful penetration. These fibers are removed from the particle as it enters the cytoplasmic matrix. The particle is then transported to the nucleus. The final stage of adenovirus uncoating occurs at the pores of the nuclear envelope and results in the release of viral DNA into the nucleus. Within the nucleus, early adenovirus messenger RNAs are transcribed by host-cell RNA polymerase. These messengers direct the synthesis of viral proteins required for DNA replication and late RNA transcription. Late messenger RNAs direct the synthesis of adenovirus capsid proteins.

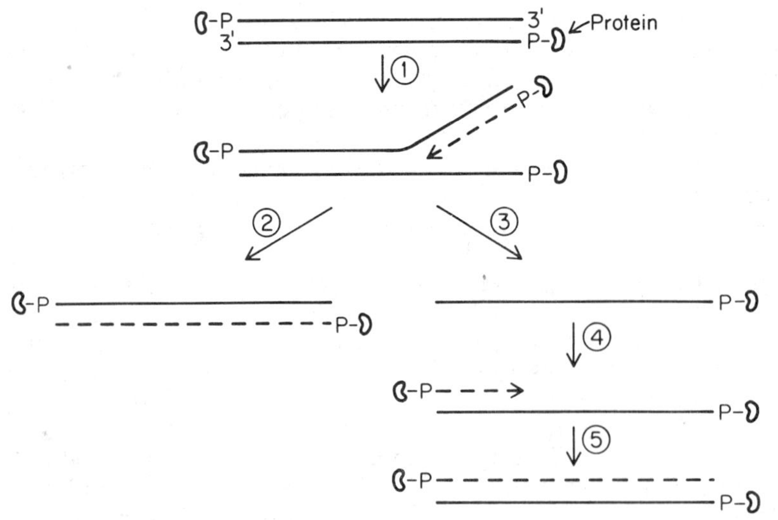

FIGURE 10. *Replication of adenovirus DNA. Solid lines (———) signify parental DNA strands. Dotted lines (---) indicate newly synthesized DNA. DNA synthesis is initiated at the 3′ end of one strand of the double-stranded DNA molecule (encircled 1). An adenovirus protein-nucleotide complex may be required to initiate DNA replication. Completion of DNA elongation produces one double-stranded progeny DNA molecule (encircled 2) and one single-stranded parental DNA molecule (encircled 3). A complementary DNA strand is formed on the parental single strand (encircled 4), giving a second molecule of double-stranded DNA (encircled 5).*

Both strands of adenovirus DNA have a protein covalently linked to their 5′ termini. Nucleotides linked to this protein may act as primers for progeny DNA synthesis. Adenovirus DNA is replicated by displacement of a parental DNA strand by a progeny DNA strand (Fig. 10). This yields one double-stranded progeny DNA molecule and one single-stranded DNA molecule. Synthesis of a complementary DNA strand using this single-stranded DNA as template yields a second molecule of double-stranded DNA (Fig. 11). Although the resultant DNA molecules contain one parental and one progeny DNA strand, this mechanism of DNA replication differs markedly from that observed for papovavirus or cellular DNA. In the latter cases, the DNA replication forks grow bidirectionally from a single initiation site.

Incubation of adenovirus DNA with a bacterial endonuclease, such as EcoRI, results in the formation of a small number of DNA fragments of limited size. The order of these fragments along

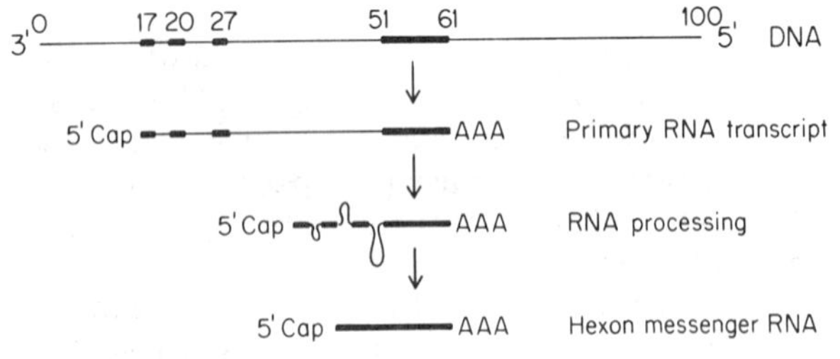

FIGURE 11. *Formation of adenovirus messenger RNA. Adenovirus DNA is shown at the top. Numbers along the DNA are map coordinates representing the percentage of the distance from the left to the right ends of the DNA of any specific nucleotide sequence. For instance, the bulk of the hexon messenger RNA is encoded by the segment of DNA that lies between 51 and 61 per cent of the distance from the 3′ to the 5′ ends of the DNA. Heavy lines (———) = DNA or RNA sequences that appear in the completed messenger RNA. Light lines (—) = intervening nucleotide sequences. See text for further explanation.*

the genome has been determined and the fragments have been used to map the position on the viral chromosome of specific messenger RNAs. In this way, the nucleotide sequences encoding a major viral capsid protein, the hexon, were shown to lie between 51 and 61 per cent of the distance from the left to the right ends of the viral DNA (Fig. 11).

By separating the two strands of viral DNA from one another, it has been possible to identify the strand encoding each messenger RNA. Experiments of this kind show that both DNA strands contain nucleotide sequences encoding the various early and late messenger RNAs. Thus, unlike SV40 (see Fig. 9, legend), the switch from early to late adenovirus messenger RNA synthesis is not a consequence of a switch in the DNA strand that is transcribed.

Studies of the regulation of adenovirus messenger RNA transcription led to the unexpected observation that the nucleotide sequences that specify the messenger RNA for the hexon protein are located in several noncontiguous segments of the viral chromosome (Fig. 11). Further studies have shown that hexon messenger RNA is derived from a large RNA precursor. This precursor, or primary RNA transcript, is an exact copy of all the DNA sequences lying in the interval between the sites of initiation and termination (17 to 61 per cent) of hexon messenger RNA. Thus, hexon messenger RNA is formed by splicing together specific segments of the primary RNA transcript (Fig. 11). Nucleotide sequences in the primary RNA transcript that are not required in the messenger RNA are cleaved from the primary transcript and degraded within the nucleus.

The presence of intervening DNA sequences and the formation of messenger RNAs by splicing of noncontiguous RNA segments are now recognized as general features of gene structure and function of all eukaryotic cells and of the DNA viruses that parasitize them. The discovery of RNA splicing has added a new dimension to our understanding of the genetics of eukaryotic cells.

Herpes Viruses

Herpes viruses are relatively large (2000 Å in diameter), spherical particles composed of a lipoprotein envelope wrapped around a protein shell. The shell contains 162 regularly spaced capsomeres arranged in the shape of an icosahedron.* Within this shell lies the linear double-stranded DNA genome (~ 10^8 daltons molecular weight) of the virus. The DNA appears to be wound around a central protein hublike thread on a spool.

Although they have similar anatomic features, the viruses of the herpes group are distinguishable from one another by the antigens they bear, by the structure of their DNA genomes, and by the diseases they cause. In terms of human pathology, herpes simplex virus (oral and genital herpes), Epstein-Barr (EB) virus (mononucleosis), cytomegalovirus (respiratory disease), and varicella-zoster virus ("chickenpox," "shingles") are the most frequently encountered members of this group. Of these viruses, herpes simplex is the prototypic strain. It is the easiest to grow to high titer in tissue culture. Its replicative cycle is summarized below and in Figure 13.

The structure of herpes simplex DNA is complex. Four DNA isomers have been identified in all virus stocks. Each arm has an inverted repeated nucleotide sequence at its ends. The two arms differ in their nucleotide sequences and in the proteins they encode. It is evident from their structures that the DNA sequences of the long and short arms can be joined together in four different orientations with respect to one another (Fig. 12). Presumably, randomization of the orientation of the long and short segments is a consequence of the mechanisms of herpes DNA replication, a process that is not well understood at present.

Penetration of herpes simplex virus into a suitable host cell occurs by the fusion of the lipoprotein envelope of the virus with the plasma membrane or with the membrane of an endocytic vacuole. In either case, the DNA-containing icosahedral shell of the virus is released into the cytoplasm; the viral DNA is subsequently concentrated in the nucleus (Fig. 13, *upper left*). Cellular RNA polymerases initiate transcription of RNA messengers on the herpes DNA. Like cellular messenger RNAs, these viral messengers are capped and polyadenylated and exported from the nucleus into the cytoplasm in which they direct the synthesis of herpes virus proteins. With the exception of viral glycoproteins, the newly synthesized herpes proteins return to the nucleus in which they regulate the synthesis of additional classes of viral messenger RNAs, shut off cellular RNA and DNA synthesis, initiate herpes DNA replication, and provide structural proteins for the assembly of virions.

Herpes-encoded glycoproteins are inserted into intracellular membranes (nuclear envelope, rough endoplasmic reticulum) and plasma membrane, presumably by a mechanism similar to that described for the insertion of vesicular stomatitis virus glycoprotein into the internal and surface membranes of vesicular stomatitis virus-infected cells (Fig. 6). Virus-encoded glycoproteins in the nuclear envelope form the outer envelope of progeny virions during the final stages of particle maturation (see below and Figure 13). Similar glycoproteins are inserted into the cell surface and may regulate spread of the virus by promoting fusion of infected cells with their uninfected neighbors.

Herpes DNA replicates in the nucleus. Newly replicated viral DNA occurs in concatemers, i.e., DNA molecules of greater than unit (10^8 MW) length. These multimeric genomes are then cleaved to molecules of unit length, condensed, and packaged within the icosahedral shell of the virus. Virus morphogenesis is completed when the DNA-containing icosahedron buds through the inner membrane of the nuclear envelope (Fig. 13). By this means the virus acquires its lipoprotein envelope and enters the cisternae of the endoplasmic reticulum. From this site it is transported to the extracellular environment, presumably by a process that is analogous to the transport of secretory proteins from the exocrine pancreas. That is, the virus is contained in smooth surfaced vesicles and carried within them to the plasma membrane. Fusion of these vesicles with the plasma membrane results in release of virus from the cell. In this way herpes virus-infected cells release virus throughout the replicative cycle. Ultimately, however, the cell is lysed as a result of the viral infection.

Several aspects of herpes virus-host cell in-

*See Chapter 7 for a definition of icosahedral symmetry.

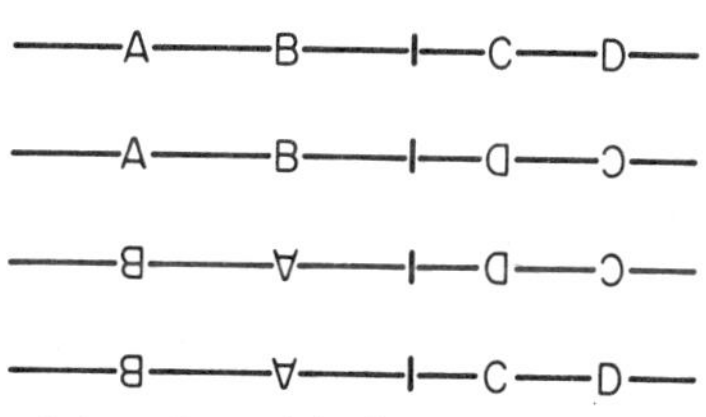

FIGURE 12. *Orientations of the long (A–B) and short (C–D) arms of herpes simplex DNA. "Stick" figures represent double-stranded DNA molecules.*

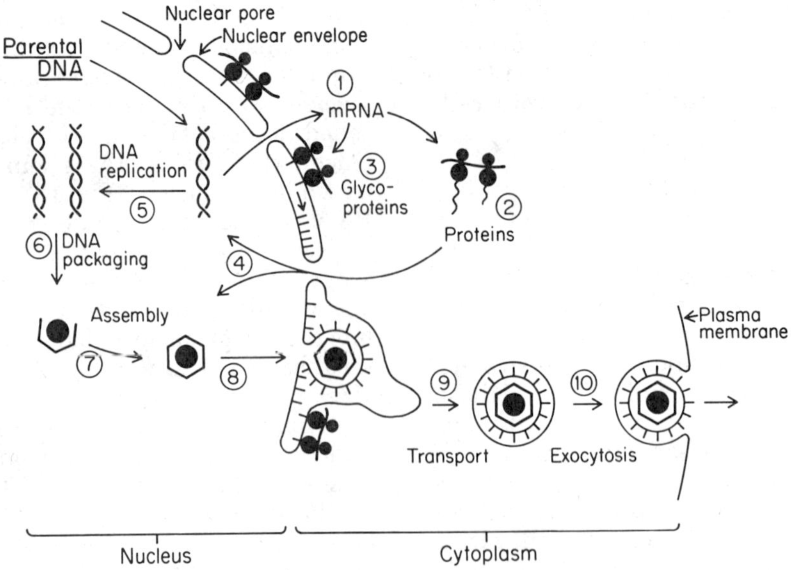

FIGURE 13. *Herpes simplex virus morphogenesis. Diagram begins with parental DNA (upper left) in nucleus of host cell.* Encircled 1, *Viral messenger RNAs are synthesized by host cell RNA polymerase on the viral DNA template. Messengers are exported from the nucleus and initiate the synthesis of viral proteins (free polyribosomes,* encircled 2) *and glycoproteins (membrane-bound polyribosomes,* encircled 3). *Viral proteins enter the nucleus* (encircled 4), *where they promote synthesis of additional classes of viral messenger RNAs. Viral DNA polymerase enters the nucleus and initiates viral DNA replication* (encircled 5). *DNA is cut into unit lengths, packaged into nucleoids* (encircled 6) *and encapsidated* (encircled 7) *within an icosahedral shell. The icosahedral shell buds through the nuclear membrane, acquiring a lipoprotein envelope containing viral glycoproteins* (encircled 8). *The mature virus is transported in vesicles to the plasma membrane* (encircled 9). *Fusion of the vesicle with the plasma membrane results in the release of the virus into the extracellular space* (encircled 10).

teractions have direct medical relevance and deserve further emphasis here.

Cell Fusion

Cells infected with herpes simplex viruses fuse with their neighbors, forming multinucleated giant cells, presumably as a consequence of the insertion of viral glycoproteins into the cell surface. Fusion of infected with uninfected cells may allow spread of an infection even when virus-neutralizing antibodies are present in the extracellular fluids.

Biosynthetic Requirements

Herpes viruses encode a thymidine kinase. The virus specified enzyme has a less stringent substrate specificity then the host enzyme and has been identified, together with the virus-induced DNA polymerase, as a target for nucleoside analogs exhibiting potent antiviral effects. For instance, 5′-iodo-5′-amino 2′,5′dideoxyuridine (AIdU), and 9-(2-hydroxyethoxymethyl) guanine (acyclo-guanine) are efficiently converted to their respective triphosphates by viral, but not by cellular, thymidine kinase. Both compounds efficiently inhibit DNA synthesis in herpes-infected cells but have little effect on DNA synthesis in uninfected cells.

Biosynthesis of viral glycoproteins is essential for formation of the viral envelope. Glucose and mannose analogs such as 2-deoxyglucose have proved to be effective inhibitors of viral glycoprotein biosynthesis in tissue culture. The potential therapeutic efficacy of these compounds is now being evaluated.

Latent Infections

Herpes simplex, varicella-zoster, and EB virus cause latent infections in man. The recurrent nature of herpes infections, even in the presence of high titers of circulating antibodies, the recovery of infectious virus from ganglion cells of man and animals (simplex, zoster), and the identification of EB viral antigens and DNA in lymphoid cell lines from "normal" individuals indicate that the genomes of these viruses have the capacity to lie dormant for decades within the cells of the host. Efforts to prove that the entire genome of any one of these agents is integrated into host-cell DNA have as yet been unsuccessful. Multiple copies of EB virus DNA are found in human lymphoblasts and Burkitt's lymphoma cells. Although the DNA extracted from EB virus is linear, the EB DNA molecules found in transformed cells are closed circular duplex DNAs. These circular molecules resemble bacterial episomes; that is, they replicate in step with host-cell DNA without becoming part of it.

Oncogenic Agents

A number of herpes-type viruses have been shown to be oncogenic agents in frogs, chickens, and monkeys, and herpes simplex and cytomegaloviruses, whose DNA has been damaged by ultraviolet light, have the capacity to transform cells in culture. Most important is the demonstration that EB virus causes lymphomas in monkeys and that EB virus can be recovered from the lymphoma cells of these primates. At present EB virus is the best candidate for a human tumor virus.

EB virus selectively infects and transforms human bone marrow-derived (B) lymphocytes. Its narrow host-cell range may be a consequence of the close association of EB virus receptors with the receptors for the cleaved third component of complement (C3d) on B-lymphocytes. The entire EB genome appears necessary for transformation. All EB transformed cells express an EB-specific nuclear antigen (EBNA). This antigen appears analogous to the T (tumor) antigens induced in the nuclei of animal cells transformed by papovaviruses or by adenoviruses. The vast majority of lymphocytes transformed by EB virus do not produce progeny virions. Viral capsid antigen (VCA) is found in the small proportion of cells that do synthesize infectious virus. These cells are destroyed as a result of the infection.

Poxviruses

Poxviruses, the largest of all animal viruses, are brick shaped particles roughly 3500 by 2500 Å in size. They contain a single linear molecule of double-stranded DNA of about 1.3×10^8 molecular weight. Like cellular chromosomal DNA, the two strands of poxvirus DNA are covalently cross-linked at or near the ends of the DNA molecule.

In the virus, the DNA lies within a protein core. Also packed within this core are a DNA-dependent RNA transcriptase, a polyadenylic acid polymerase, a nucleotide phosphohydrolase, and RNA-capping and methylating enzymes. Flanking the core on two sides are protein-containing structures called lateral bodies; their functions are unknown. A lipoprotein envelope containing viral glycoproteins surrounds the entire particle.

Unlike other DNA-containing viruses, poxviruses replicate in the cytoplasm. However, host-cell nuclear functions are required late in the replicative cycle for the formation of infectious virus from immature virions (Fig. 14, *encircled 11*).

Entry of the viral core into the cytoplasm is accomplished by fusion of the viral envelope with the cell surface or by endocytosis of the virus and release of the core from the endocytic vacuole. Both processes have been shown to occur. Once liberated from its lipoprotein envelope and exposed to substrates in the cytoplasm, the core becomes biosynthetically active. Enzymes within the core synthesize and export single-stranded RNA messengers (Fig. 14, *encircled 2*) that direct the synthesis of proteins such as the thymidine kinase, DNA polymerase, and uncoating enzyme; these enzymes are used in subsequent biosynthet-

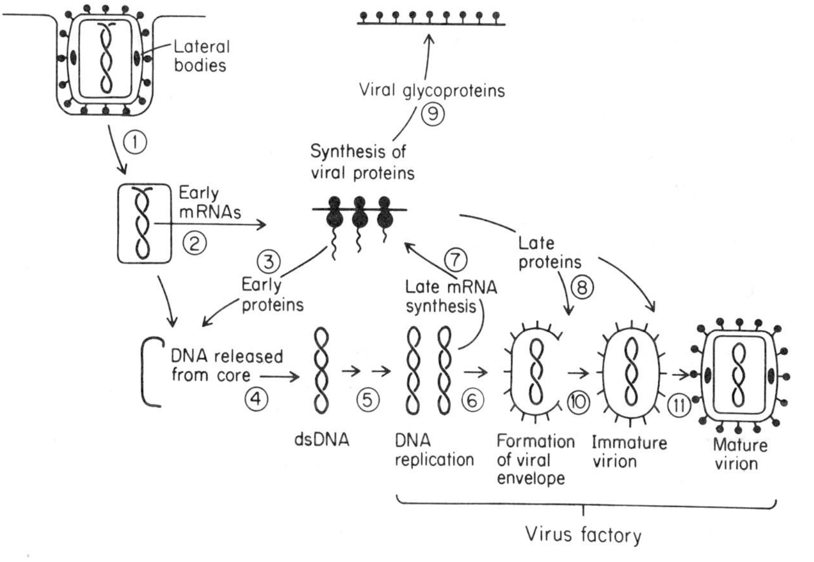

FIGURE 14. *Poxvirus replication. Infecting virus is shown at upper left. Progeny virus at lower right. See text for an explanation of the numbered intermediate steps.*

ic steps. Since these proteins are the products of the first RNA transcripts formed by the virus, they are termed "early" proteins.

Uncoating enzyme opens the viral core (Fig. 14, *encircled 4*), thereby releasing poxvirus DNA into the cytoplasmic matrix. Poxvirus DNA replicates semiconservatively; progeny double-stranded DNAs are composed of one parental and one progeny DNA strand. However, the precise mechanisms of DNA duplication and of breakage and synthesis of the bonds that cross-link the two DNA strands are not known.

Viral morphogenesis takes place in factories within the cytoplasm (Fig. 14, *encircled 5, 6, 10, 11*). These factories can be though of as "extranuclear nuclei." That is, viral DNA (Fig. 13, *encircled 5*) and viral messenger RNAs (Fig. 14, *encircled 7*), are synthesized within the factories. The DNA is retained in this location while the messenger RNAs are exported into the surrounding cytoplasm where they direct the synthesis of viral proteins (Fig. 14, *encircled 7 and 8*. Viral proteins return to the factories where they are assembled into virions.

In the factories the newly formed viral envelope appears to engulf a small portion of the matrix of the factory, thereby forming an immature particle (Fig. 14, *encircled 6 and 10*). The DNA-containing core and lateral bodies condense and assemble within these immature particles. Subsequently, other proteins are added to the envelope. Maturation of envelope and internal components completes the replicative cycle (Fig. 14, *encircled 11*). A small number of completed virions are released continuously from the infected cell throughout the replicative cycle. However, the major portion of virions is not released until the cell lyses.

Viral envelope glycoproteins are inserted into the host cell's plasma membrane. These glycoproteins alter the properties of the cell surface so that virus-infected cells fuse with nearby uninfected cells. In this way, poxviruses spread from cell to cell without entering the extracellular compartment. Mutant poxviruses that encode glycoproteins that inhibit cell-cell fusion have been identified.

HEPATITIS VIRUS TYPE B

Hepatitis B viruses form a unique taxonomic group; they cannot be clasified with any of the recognized viruses. Two types of particles are found in the serum of individuals carrying the virus: (1) The Dane particle, a 420-Å diameter DNA-containing particle, is presumed to be the infectious agent of hepatitis B; (2) Australia antigen, a 200-Å diameter lipoprotein particle that is similar in composition to the outer envelope of the Dane particle. No nucleic acid has been found in Australia antigen, and it is presumed to be noninfectious.

The outer lipoprotein envelope of the Dane particle contains Australia or hepatitis B surface antigen. Within this envelope is core-containing circular DNA of molecular weight 1.6×10^6. Each DNA molecule is composed of a single-stranded covalently closed circular DNA and a complementary linear DNA strand that is about two-thirds the length of the circular DNA. Thus, the intact molecule is a circle that is double-stranded for two-thirds of its circumference and linear for the remaining one-third.

The Dane particle contains less genetic information than any known DNA or RNA virus. There is presently no cell culture system for in vitro growth of the virus. Thus the number of proteins encoded by the viral genome is not known. Since individuals infected with hepatitis B produce antibodies to Australia antigen, core antigen, and viral DNA polymerase, it is assumed that these proteins are viral gene products. Immunofluorescent studies indicate that the virus replicates in the hepatocyte nucleus.

Viroids

Viroids are naked pieces of infectious circular single-stranded RNA of defined nucleotide sequence and weighing approximately 120,000 daltons. This is barely enough genetic information to code for a single protein of moderate molecular weight. However, no viroid-encoded protein has yet been found.

Viroids have been identified as the agents of plant diseases whose names range from the mundane (potato tuber spindle disease) to the exotic (Cadang-Cadang, a disease of coconut trees). They have not been identified as the agents of human disease, but they are candidate agents for Kuru, a progressive degenerative neurologic disease of New Guinea cannibals, transmissible dementias such as Creutzfeldt-Jakob disease, and scrapie, a progressive neurologic disease of sheep and goats. All three diseases have been transmitted to primates, and scrapie has been passaged in mice as well. The agents of these diseases appear to be smaller than any known virus and exhibit physical properties that suggest that they are viroids.

DEFECTIVE INTERFERING PARTICLES

Defective interfering (DI) particles, as their name suggests, are defective viruses that interfere with the replication of their parent viruses.

DI particles contain insufficient genomic information to replicate themselves. In some cases, they contain less nucleic acid than standard virus. DI reo or influenza viruses lack the largest segment of viral RNA, while DI polio, vesicular stomatitis, and papovaviruses contain nucleic acids that are less than unit length. In other cases, DI particles contain gene duplications or segments of host DNA that have been substituted for viral DNA.

DI particles share four essential properties: (1) They are unable to propagate themselves. (2) They multiply only in the presence of a helper virus. (3) They interfere with the replication of homologous standard virus but not with the replication of viruses of another type (vesicular stomatitis DI particles interfere with VSV but not with poliovirus replication). (4) They have the capacity to compete with the standard virus and thereby to enrich their numbers at the expense of standard virus.

DI particles have been identified in stocks of virtually every virus type. They appear in increasing frequency when viruses are repeatedly passaged at high titer.

Persistent infections have been established in cell cultures with mixtures of DI particles and standard virus. Mice inoculated with a mixture of DI and standard vesicular stomatitis viruses develop a slowly progressive paralytic disease, while mice infected with standard vesicular stomatitis virus alone are rapidly killed. Whether DI particles promote chronic or persistent virus infections in man is not known.

DI particles may also provide protection against viral infection. Animals immunized with DI particles are protected against challenge with the homologous standard virus. Some strains of polio vaccine have been shown to contain DI particles.

CONCLUSION

It is evident that the animal viruses have explored every possible combination of nucleic acids for the transfer of their genetic information. Negative strands encode complementary positive strands, positive strands encode complementary negative strands, and information shuttles back and forth between single and double strands and between RNA and DNA. The viruses have been no less adventurous in exploring cellular pathways for penetration of macromolecules into the cytoplasm, insertion of proteins into membranes, and assembly of complex structures than they have for information transfer. They highlight the major biosynthetic and morphogenetic pathways of eucaryotic cells. Where a pathway exists, a virus has evolved to exploit it.

References

General

Luria, S. E., Darnell, J. E., Jr., Baltimore, D., and Campbell, A.: General Virology. New York, John Wiley and Sons, 1978.

Fraenkel-Conrat, H., and Wagner, R. R.: Comprehensive Virology. Vols. 1–10. New York, Plenum Press.

Virus Penetration

Dales, S.: Early events in cell-animal virus interactions. Bacteriol Rev 37:103, 1973.

Picornaviruses

Dasgupta, A., Baron, M. H., and Baltimore, D.: Poliovirus replicase. A soluble enzyme able to imitate copying of poliovirus RNA. Proc Natl Acad Sci USA 76:2679, 1979.

Nomoto, A., Kitemmura, N., Golini, F., and Wimmer, E.: The 5′ terminal structures of poliovirion RNA and poliovirus mRNA differ only in the genome-linked protein VPg. Proc Natl Acad Sci USA 74:5345, 1977.

Oncornaviruses

Bishop, J. M.: Retroviruses. Ann Rev Biochem 47:35, 1978.

Collett, M. S., and Erickson, R. L.: Protein kinase activity associated with the avian sarcoma virus Src gene product. Proc Natl Acad Sci USA 75:2021, 1978.

Rhabdoviruses

Banerjee, A. K., Abraham, G., and Colonno, R. J.: Vesicular stomatitis virus: Mode of transcription. J Gen Virol 34:1, 1977.

Formation of Membranes of Enveloped Viruses

Katz, F. N., Rothman, J. E., Knipe, D. M., and Lodish, H. F.: Membrane assembly: Synthesis and intracellular processing of the vesicular stomatitis viral glycoprotein. J Supramol Struct 7:353, 1977.

Blobel, G., and Lingappa, V. R.: Transfer of proteins across intracellular membranes. In Silverstein, S. C. (ed.): Transport of Macromolecules in Cellular Systems. Berlin, Abakon Verlagsgesellschaft, 1978, pp. 289–298.

Paramyxoviruses

Scheid, A., and Choppin, P. W.: Two disulfide-linked polypeptide chains consitute the active F protein of paramyxoviruses. Virology 80:54, 1977.

Myxoviruses

Palese, P., Plotch, S. J., Bouloy, M., and Krug, R. M.: The genes of influenza virus. Transfer of 5′-terminal cap of globin mRNA to influenza viral complementary RNA during transcription *in vitro*. Proc Natl Acad Sci USA 76:1618, 1979.

Reoviruses

Silverstein, S. C., Christman, J. C., and Acs, G.: The reovirus replicative cycle. Ann Rev Biochem 45:375, 1976.

Weiner, H. L., Drayna, D., Averil, D., Jr., and Fields, B. N.: Molecular basis of reovirus virulence: Role of the S1 gene. Proc Natl Acad Sci USA 74:5744, 1978.

Papova and Adenoviruses

Nathans, D.: Dissecting the genome of a small tumor virus (SV-40). (The Harvey Lectures, vol. 70.) New York, Academic Press, 1976, pp. 111–157.

Griffin, J. D., Spangler, G., and Livingston, D. M.: Protein kinase activity associated with SV40 T-antigen. Proc Natl Acad Sci USA 76:2610, 1979.

Sussenbach, J. S.: The mechanism of replication of adenovirus DNA. Virology 84:509, 1978.

Restriction Endonucleases

Nathans, D., and Smith, H. D.: Restriction endonucleases in the analysis and restructuring of DNA molecules. Ann Rev Biochem 44:273, 1975.

Gene Splicing

Abelson, J.: RNA processing and the intervening sequence problem. Ann Rev Biochem 48:1035, 1979.
Crick, F.: Split genes and RNA splicing. Science 204:264, 1979.

Herpes Viruses

Jacob, R. J., Morse, L. S., and Roizman, B.: Anatomy of herpes simplex virus DNA. J Virol 29:448, 1979.
Klein, R. J.: Pathogenetic mechanisms of recurrent herpes simplex virus infection. Arch Virol 51:1, 1976.
Henle, W., Henle, G., and Lennette, E. T.: The Epstein-Barr virus. Sci Am 241:48, 1979.

Poxviruses

Stern, W., and Dales, S.: Biogenesis of vaccinia. Virology 75:232, 1976.
Esteban, M., Flores, L., and Holowaczak, J. A.: Model for vaccinia virus DNA replication. Virology 83:467, 1977.

Hepatitis B Virus

Blumberg, B. S.: Australia antigen and the biology of hepatitis B. Science 197:17, 1977.
Charnay, R., Pourcel, C., Louise, A., Fritsch, A., and Tiollais, P.: Cloning in *E. coli* and physical structure of hepatitis B virion DNA. Proc Natl Acad Sci USA 76:2222, 1979.
Robinson, W. S.: The genome of hepatitis B virus. Ann Rev Microbiol 31:351, 1977.
Summer, J., Smolec, J. M., and Snyder, R.: A virus similar to human hepatitis B virus associated with hepatitis and hepatoma in woodchucks. Proc Natl Acad Sci USA 75:4533, 1978.

Viroids

Gajdusek, D. C.: Slow infections with unconventional viruses. (The Harvey Lectures.) New York Academic Press, 1978, pp. 283–353.
Randles, J. W., and Hatta, T.: Circularity of the ribonucleic acids associated with Cadang-Cadang disease. Virology 96:47, 1979.

Defective Interfering Particle

Huang, A. S., and Baltimore, D.: Defective Interfering Animal Viruses. New York, Planum Press, 1977, pp. 73–116.

10 THE GENETICS OF VIRUSES

C. R. Pringle, B.Sc., Ph.D

INTRODUCTION: THE SIGNIFICANCE OF VIRUS GENETICS

The science of virus genetics is concerned with the origin and mechanism of variation of viruses. The variability of viruses and the extent to which it is heritable have great importance for human medicine. The surface antigens of some viruses are inherently stable, and the diseases associated with these infectious agents can be controlled by vaccination (e.g., rubella, measles, rabies) or in exceptional circumstances even eradicated (e.g., smallpox). Other viruses exist as a range of antigenic types (e.g., the enteroviruses and rhinoviruses) or appear to change progressively in response to the development of immunity in their host (e.g., the antigenic drift of influenza virus). Control of these agents by vaccination is more difficult and ultimately may depend on accurate prediction of the potential evolution of the antigens of these viruses. Indeed, the design and development of appropriate influenza virus vaccines in anticipation of epidemics and even pandemics could become one of the principal applications of virus genetics in human medicine.

Viruses that cause diseases characterized by short incubation period, recurrent infection, and aviremia may be difficult to combat by vaccination and may be controllable only by chemotherapy. Suitable antiviral drugs have not been developed yet, and progress towards successful chemotherapy of virus infection depends on clear discrimination of virus-specific processes from those processes that are part of the biosynthetic apparatus of the host cell and essential for the viability of both the virus and its host. The isolation of virus mutants helps to identify these virus-specific processes, defining appropriate targets for pharmacologic attack and allowing a rational approach to antiviral chemotherapy.

Assessment of the oncogenic potential of a virus requires knowledge of the nature of its genome. To be oncogenic, the genome of the virus must exist as DNA at some stage to permit integration of viral information into the genome of the host cell and to produce a heritable change (transformation) in the growth properties of the cell. However, integration by itself does not invariably cause transformation and may not be obligatory in all cases. Determination of what is critical for virus-induced transformation and oncogenesis requires additional analysis of the organization and function of the viral genome.

THE NATURE AND CODING CAPACITY OF VIRAL GENOMES

The viruses of man and animals exhibit great diversity of genome structure. From a genetic standpoint, viruses fall into three distinct categories: the *DNA-containing viruses*, the *RNA-containing viruses*, and the viruses endowed with reverse transcriptase (the *retroviruses*), whose genome alternates between RNA in the virion and proviral DNA in the host cell.

The DNA viruses, with the exception of the poxviruses, multiply in the nucleus of the host cell, and messenger RNA (mRNA) is transcribed by the host cell transcriptase. Hence, the naked genome of these viruses is infectious. The RNA viruses multiply predominantly in the cytoplasm, and since the host cell is unable to replicate RNA molecules, an RNA polymerase (or the information for its synthesis) must be introduced with the virus. It is an accepted convention to call a single-stranded RNA molecule the positive strand if it functions as the messenger for protein synthesis, and to call its complement the negative strand (or antimessage). The genome of RNA viruses may be represented by a positive or negative strand or both. The genome of positive-strand RNA viruses is infectious because it functions as mRNA for synthesis of the entire complement of viral proteins including the RNA replicase. The genome of negative-strand RNA viruses, on the other hand, is not infectious because it cannot replicate or be translated directly into protein. Consequently, the nucleocapsid of these viruses includes an RNA transcriptase to initiate synthesis of mRNA in the infected cell. The genome of retroviruses functions as mRNA and is therefore the positive strand; however, this RNA is not infectious, and the reverse transcriptase enzyme is a part of the structure of the virion. The proviral DNA form of these viruses is infectious, however.

The diverse nature of the viral genome can be rationalized by regarding the genome as a particular stage in the cycle of replication of nucleic acids sequestered in an extracellular particle. This concept is illustrated in Figure 1. The only

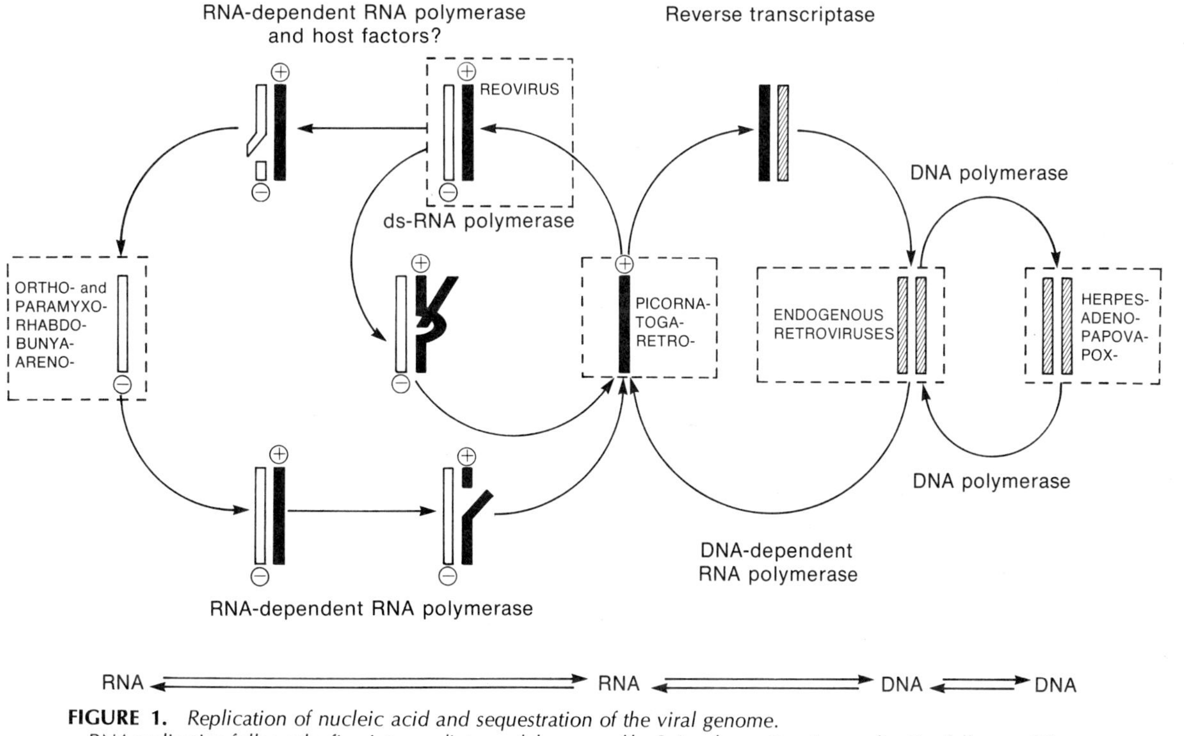

FIGURE 1. *Replication of nucleic acid and sequestration of the viral genome.*

RNA replication follows the five-intermediate model proposed by Spiegelman. Reovirus replication follows a different path because there is no displacement of strands, and the parental duplex RNA is conserved. Positive strands are shown as black; negative strands are shown as white. DNA strands are indicated by cross-hatching. The stages of the nucleic acid replication cycles sequestered in extracellular virus particles are indicated by dashed lines.

The segmentation of the genome of some RNA viruses is not considered here, because subdivision of the genome is probably a form of transcriptional control. The single-strand DNA viruses have been omitted for simplicity. The only form among known viruses not represented is the RNA:DNA intermediate in the reverse transcription cycle.

TABLE 1. The Physical and Genetic Characteristics of the Genomes of Animal Viruses

			NUCLEIC ACID					GENETIC PROPERTIES			
Group[a]	Example	Site of Replication	Type	Strand in Virion	Form	Infectivity	MW × 10^6	Relative Coding Capacity[b]	Complementation Groups	Recombination Frequency and Grouping	Special Features
Parvo	Murine minute	Nucleus	SS-DNA	−	Linear	?	1.7	3–4	N.D.	N.D.	Replicates in unison with host; mitosis-dependent
Parvo	Adeno-associated	Nucleus	SS-DNA	−and+	Linear	Yes	1.8	3–4	N.D.	N.D.	Helper and mitosis-dependent. Inverted terminal repeat
Papova	SV40	Nucleus	DS-DNA	+/−	Super-coiled helix	Yes	3.0	3	3 (or 5)	Low	Induces host DNA synthesis
Adeno	Adeno 5	Nucleus	DS-DNA	+/−	Linear	Yes	23.0	23	12	Low-high linear map	Inverted terminal repetition. Covalently bonded protein at 5′ terminus
Herpes	HSV-2	Nucleus	DS-DNA	+/−	Linear	Yes	100.0	100	18	Low-high linear map	Terminal and internal inverted repeats
Pox	Vaccinia	Cytoplasm	DS-DNA	+/−	Linear	No	160.0	160	N.D.	Low-high linear map	Cross-linked ends. DNA-dependent RNA polymerase and other virion enzymes
Picorna	Polio-1	Cytoplasm	SS-RNA	+	Linear	Yes	2.6	5	Nil	Low linear map	One polycistronic message, post-translational cleavage; covalently bonded protein at 5′ terminus
Alpha	Sindbis	Cytoplasm	SS-RNA	+	Linear	Yes	4.0	8	6	None	Two polycistronic messages; post-translational cleavage
Flavi	Dengue	Cytoplasm	SS-RNA	+	Linear	N.D.	4.0	8	N.D.	None	Polycistronic (with internal initiation of translation?)
Bunya	Snowshoe hare	Cytoplasm	SS-RNA	−	3 unique subunits	No	6.4	13	2	High 2 groups	Virion RNA-dependent RNA polymerase
Areno	Pichinde	Cytoplasm	SS-RNA	−	3 unique subunits	No	3.2	6	N.D.	High 2 groups	Virion RNA-dependent RNA polymerase
Orthomyxo	Influenza	Cytoplasm (and nucleus?)	SS-RNA	−	8 unique subunits	No	5.0	10	N.D.	High 8 groups	Virion RNA-dependent RNA polymerase
Paramyxo	Sendai	Cytoplasm	SS-RNA	−	Linear	No	7.0	14	7	None	Virion RNA-dependent RNA polymerase
Rhabdo	Vesicular stomatitis	Cytoplasm	SS-RNA	−	Linear	No	4.0	8	6	None	Virion RNA dependent RNA polymerase
Diplorna	Reo	Cytoplasm	DS-RNA	+/−	10 unique subunits	No	15.0	15	N.D.	High 10 groups	Virion DS-RNA-dependent RNA polymerase
Retro	Rous sarcoma	Nucleus and cytoplasm	SS-RNA	+	2 identical subunits	RNA no DNA Yes	6.0	6	N.D.	High	Virion RNA-dependent DNA polymerase

[a]The papilloma, corona, irido, orbi, and metamyxovirus have been omitted because of lack of information.
[b]The number of possible gene products, assuming that an average sized protein contains 500 amino acids.

stage, excluding the replicative intermediates, not represented in known viruses, is the RNA:DNA intermediate in the reverse transcription pathway.

The *information capacity* of the genome is determined by the universal coding assignment of three nucleotides per amino acid. Therefore, 1500 nucleotides (or nucleotide pairs in the case of double-stranded nucleic acid) are required for an average-sized protein of 500 amino acids. The information content of different viruses is indicated in Table 1, together with other characteristics that define the properties of the genome. The genomes of DNA viruses differ by as much as fiftyfold in coding capacity, whereas the RNA viruses vary only within a threefold range (or fourfold if the retroviruses are included). A large proportion of the genome of the large DNA viruses, e.g., herpes simplex virus, is concerned with regulation of DNA synthesis, since nearly half of all *ts* mutants (see Mutation and Mutants) have DNA-negative phenotypes. This is probably reflected in the complex biology of herpes viruses and their tendency to cause latent infection.

These estimates of the information content of the viral genome are minimum values and do not take account of proteolytic cleavage and other modifications of primary gene products. Furthermore, recent analysis of the genome of the small single-stranded DNA bacteriophage ΦX174 has shown that polypeptides of entirely different amino acid sequence can be transcribed from the same polynucleotide by displacement of the triplet reading frame; thus gene E of ΦX174 lies entirely within gene D, and gene B within gene A. Similarly, functionally distinct gene products can be obtained by translation of mRNA from a fixed point with occasional read-through of a specific stop signal resulting in two polypeptides with overlapping sequences (e.g., the A and A′ proteins of bacteriophage Qβ). Also, among the mammalian viruses, polyomavirus conserves the coding potential of its genome by utilizing host proteins as structural compounds of the virion (Table 1).

MUTATION AND MUTANTS

Mutation can be defined as a discontinuous event that results in a change in information content. Virus mutants arise spontaneously or can be induced by chemical or physical agents (mutagens). *Mutagens* that have proven effective with human and animal viruses include base analogues (5-fluorouracil and 5-azacytidine for RNA viruses and 5-bromode oxyuridine for DNA viruses), alkylating agents (ethyl methane sulfonate, among others), intercalating agents (proflavine and NTG), deaminating agents (nitrous acid), hydroxylamine and ionizing irradiation (ultraviolet light). The base analogues are incorporated into the genome, and mutations are produced by miscoding during replication. The other mutagens induce mutation by direct chemical change of the nucleic acid. Frameshift and nonsense (polypeptide chain-terminating) mutants of human or animal viruses have not been unequivocally identified, and most induced mutants appear to be missense mutants in which substitution of an amino acid modifies the functional activity of a gene product.

Mutants may have specific phenotypes such as resistance to heat inactivation or an inhibitory drug; for instance, mutants of poliovirus resistant to guanidine hydrochloride have played an important role in mapping the poliovirus genome. In general, however, mutants with specific phenotypes are less useful than conditional lethal mutants, in which a single common phenotype (e.g., inability to multiply at high temperature) is sufficient to provide mutants with lesions in any gene that is indispensable for normal replication. This is because a change in amino acid sequence of any polypeptide can produce a conformational change that affects the stability of a protein at high temperature. *Conditional lethal mutants* of animal viruses are exclusively temperature-sensitive (*ts*) mutants. Host-restricted mutants, analogous to the amber-ochre mutants of prokaryotes, are unknown probably because mammalian cells with the characteristics of the suppressor strains of prokaryotes have not yet been identified.

Extensive changes in polynucleotide sequence of the genome can occur by rearrangement, duplication, and deletion. Spontaneous *deletion mutants* of viruses are common and occur in all virus groups. Particles with defective genomes frequently interfere specifically with the replication of the undeleted genome and are known as DI (defective interfering) particles. It has been suggested that the generation of DI particles is responsible for the self-limitation of many virus diseases by moderating the course of infection until an immune response is developed. Figure 2 illustrates the extent of the deletions observed in DI particles of the rhabdovirus vesicular stomatitis virus (VSV). Deletion mutants can only multiply with the assistance of a helper virus. This complex relationship is exemplified best in the leukemia-sarcoma viruses, in which the mammalian sarcoma viruses lack genetic information for envelope protein (and internal proteins according to the strain), and these viruses are entirely dependent on the "helper" leukemia virus for normal maturation.

The deletion of sequences in DI particles of

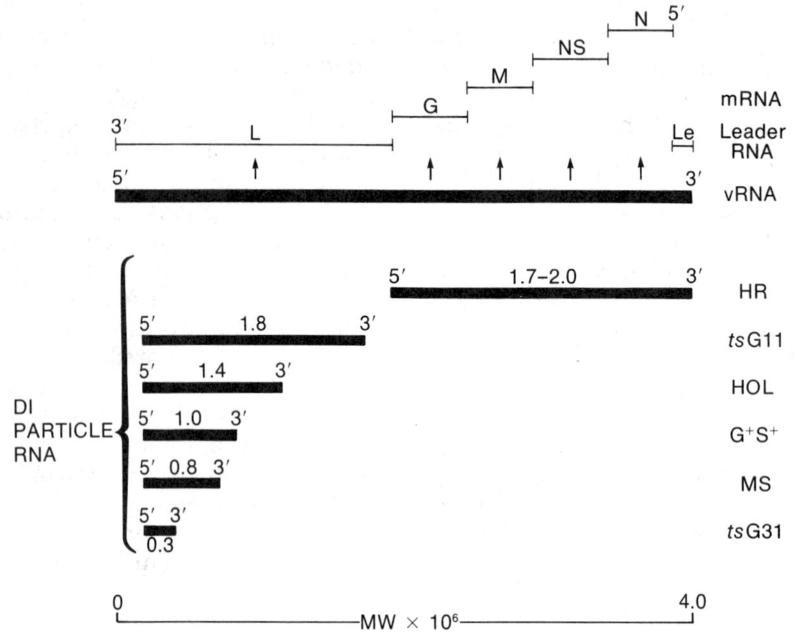

FIGURE 2. *Physical maps of DI particle RNA of vesicular stomatitis virus (VSV).*

The DI particles of VSV are deletion mutants. The physical location of the RNA from six different DI particles (HR, tsG11, HOL, Gts⁺, MS and tsT31) are illustrated. All these DI particles have interfering activity with homotypic virus. The HR DI particle is unique in its ability to interfere heterotypically. It is also the only DI particle that contains complete viral genes.

papovaviruses, on the other hand, may be accompanied by rearrangements, reiteration of sequences, and substitutions of extraneous host DNA. Indeed, *duplications* and rearrangements are characteristic features of the normal genome of the herpes viruses and presumably have arisen by mutational events during the evolution of these viruses. The location and orientation of the redundant (duplicated) sequences in the herpes simplex virus genome are illustrated in Figure 3. The four forms of the genome are present simultaneously in any population of molecules and are thought to be continually generated by recombinational events within the redundant sequences.

GENETIC INTERACTIONS BETWEEN VIRUSES

Recombination in RNA Viruses and Subunit Reassortment

Since informational RNA is unique to viruses, so also is recombination between RNA molecules. The genome of several RNA viruses (influenza, reo, bunya, and arenoviruses) is segmented, and high frequency recombination is a characteristic feature of all these viruses. Recombination occurs by *reassortment* of the subunits of the genome. The mechanism that ensures that the correct complement of subunits is assembled in the mature virion is still unknown. Each of the eight subunits of the influenza virus genome codes for a single protein, and hybrid viruses can be selected from the progeny of mixed infections. For example, if the PR8 and Hong Kong strains are crossed, a recombinant with the Hong Kong hemagglutinin and the PR8 neuraminidase can be selected by exposure to antiPR8 hemagglutinin and anti–Hong Kong neuraminidase sera. The genomic subunits of different strains of influenza virus (and similarly their gene products) have different electrophoretic mobilities in polyacrylamide gel, and by analysis of the subunits present in specific recombinants, it has been possible to map completely the influenza virus genome. The coding assignment of the subunits of the genome of the PR8 and Hong Kong strains, and a recombinant with the hemagglutinin of Hong Kong (HK) and the neuraminidase of PR8 are illustrated in Figure 4. From the relative mobilities of the subunits, it is apparent that only the hemagglutinin subunit has been exchanged in this particular recombinant. In any cross of two influenza viruses, 254 different recombinants (i.e., $[^1/_2]^8 - 2$) are possible by subunit rearrangement. This particular recombinant was obtained because one of the parents had been irradiated by ultraviolet light, so that most viable progeny viruses would derive all their genomic segments from the unirradiated parent with the exception of the hemagglutinin

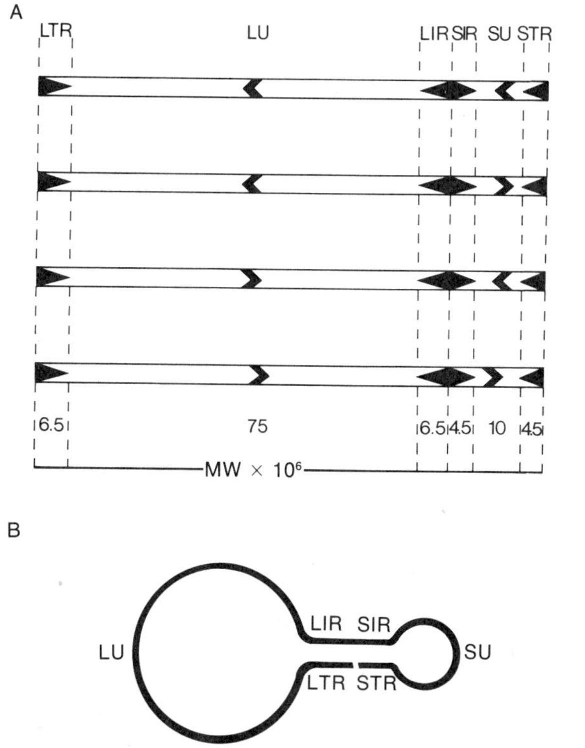

FIGURE 3. *The herpes simplex virus genome. A, The four arrangements of the genome of herpes simplex virus. LTR = long terminal repeat; LU = long unique region; LIR = long internal repeat; SIR = short internal repeat; SU = short unique region; STR = short terminal repeat. Inversion of the long and short regions is indicated by arrows and could occur as a result of internal recombination within the redundant sequences. B, The homology of the internal and terminal repeat regions. This double ring structure is observed in electronmicrographs of single-strand DNA following self-annealing.*

This arrangement of the genome is characteristic for herpes simplex virusesType 1 and Type 2 but not for other members of the herpes group. The least complex herpesvirus genome is that of channel catfish virus, which lacks a short unique region and has no internal redundancy.

gene, which was selected against by exposure to antiPR8 hemagglutinin.

The retroviruses also exhibit high frequency recombination. However, although the genome exists as two subunits, the mechanism is not independent reassortment because the two components of the genome are identical, and therefore the particles are genetically *diploid*. The subunits of the retrovirus genome are linked by base pairing at their 5′ end as an inverted dimer, and transcription is initiated from primers (tryptophan tRNA in Rous sarcoma virus) located close to the 5′ ends. Synthesis of a complete progeny strand necessitates a jump from the 5′ end of one strand to the 3′ end of the same or the other strand of the dimer. Hence, each round of transcription provides an opportunity for intermolecular recombination, which could account for the high frequency of recombination.

Recombination has not been detected in any negative-strand RNA virus with an unsegmented genome, but it does occur at low frequency with at least some positive-strand RNA viruses (see Table 1). *Genetic maps* of poliovirus and foot-and-mouth disease virus (FMDV) have been constructed from crosses of *ts* mutants. In both cases, the mutants were oriented relative to the locus for resistance to the inhibitor guanidine and then arranged in a linear order on the basis of the recombination frequencies observed in crosses of pairs of mutants. The genetic map of poliovirus is illustrated in Figure 5. Analysis of the properties of the constituent *ts* mutants defines the different coding regions. The organization of the genome of FMDV is probably similar to that of poliovirus, but the gene order in other RNA viruses differs from the picornavirus pattern. For example, in Sindbis virus and VSV the coat protein genes are located centrally or towards but not at the 5′ end. These differences probably reflect different modes of transcription and translation. In fact, the genome of picornaviruses is polycistronic and not partitioned into genes, since the entire genome is translated directly into a single giant polypeptide, which is subsequently cleaved into smaller polypeptides, which are the functional gene products. *Post-translation cleavage* is probably a feature of all positive-strand viruses, because in eukaryotic cells, initiation of translation occurs at only a single site in all mRNA. (A recent study of Kunjin virus suggests, however, that the flavi-

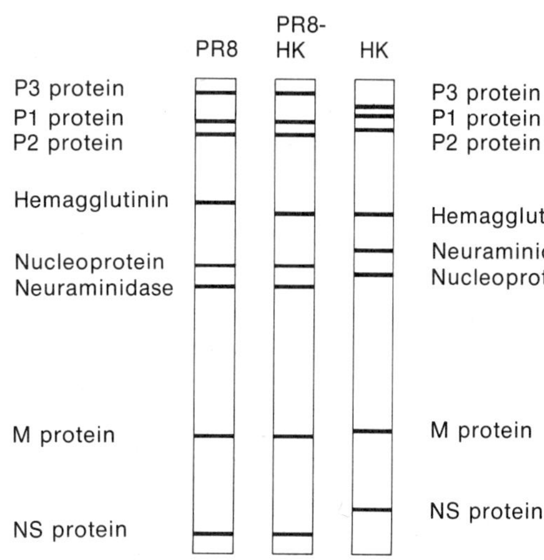

FIGURE 4. *The influenza virus genome.*

The eight subunits of the influenza virus genome can be separated by polyacrylamide gel electrophoresis. The relative mobilities of the RNA subunits of the PR8 (A/PR/8/34) and HK(A/HK/8/68) strains of influenza virus, and the assignments of gene function are illustrated in the left and right tracks, respectively. The center track is a recombinant clone, and it can be seen by inspection that it contains the hemagglutinin gene of the HK strain together with seven subunits from the PR8 strain.

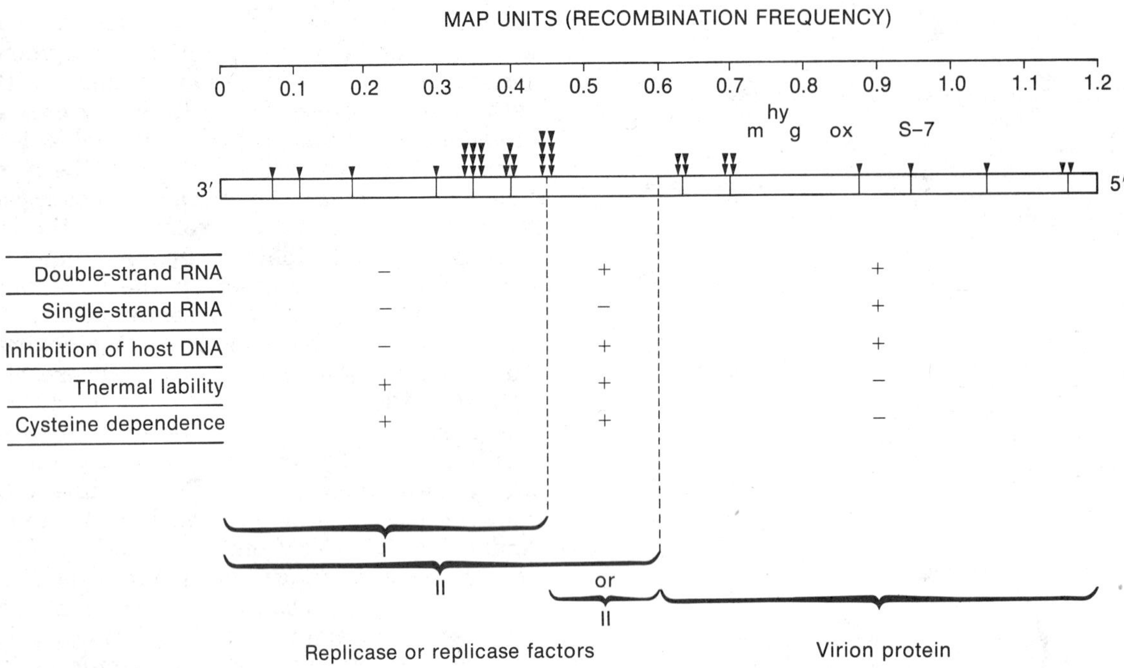

FIGURE 5. *The picornavirus genome.*

Recombination map of 39 ts mutants of poliovirus Type 1 and the assignment of gene function. The symbol indicates the map position of a ts mutant. Identification of gene function is summarized in the lower part of the figure. The loci of four inhibitors of multiplication all fall within the coat protein region. The inhibitors are guanidine (g), dextran sulfate (m), and derivatives of hydantoin (hy), oxadiazole (ox), and pyrimidine carboxylate (S–7).

The genetic map of foot-and-mouth disease virus, an unrelated picornavirus, closely resembles that of poliovirus. Thirty-five mutants have been arranged in a linear sequence extending over 0.7 map units. Again, the locus of the guanidine inhibitor falls in the presumptive coat protein region.

viruses may prove exceptions to this generalization.)

The gene order of avian sarcoma virus has been established by a combination of genetic and physical methods to be the following: 5′-*gag-pol-env-src*-3′, in which *gag* is the gene for the precursor of the internal proteins, *pol* is the gene for viral polymerase, *env* is the gene for the surface glycoproteins, and *src* is the sarcomagenic sequence. In avian leukemia viruses, the *src* gene is deleted.

Recombination in DNA Viruses and Correlation of Physical and Genetic Maps

Recombination between DNA viruses does not require any exceptional mechanisms, since the enzymes capable of recombining DNA molecules preexist in the host cell. Genetic maps of the genomes of polyomavirus, SV40 virus, adenovirus, and herpes simplex virus have been derived from crosses of *ts* mutants and correlated with physical maps of the genome. In adenovirus, for example, the *physical mapping* of cross-over events was achieved by isolating recombinants between *ts* mutants of two serologically distinct strains of adenovirus, whose DNA differed in their pattern of cleavage by restriction endonucleases. (Restriction endonucleases are bacterial enzymes that recognize particular nucleotide sequences in double-stranded DNA and cleave specific phosphodiester bonds within these sites to produce unique DNA fragments that can be separated by electrophoresis in polyacrylamide gel). Since the genome of the recombinant contains sequences from each parent, the position of the cross-over event can be determined by comparison of the types of DNA fragment produced. The location of mutants to particular regions of DNA can also be determined by *marker rescue* experiments in which *ts* mutant-infected cells are coinfected with individual fragments of the DNA of wild-type virus obtained by restriction endonuclease treatment. The genetic and physical maps of the genome of adenovirus and the two papovaviruses, SV40 and polyoma, have been aligned by these methods, and mapping of the more complex herpes simplex virus is progressing rapidly. Genetic mapping is an important step towards the goal of defining the specific features of viral biosynthesis. Analysis of the properties of the different mutants locates the regions of the molecule that code for particular proteins. Figure

6 illustrates the alignment of the genetic and physical maps of adenovirus. Perhaps one of the most significant features of this map is that the viral genes that are responsible for transformation of the growth properties of cells are confined to one region of the molecule (the 5′ end).

Hybrid Viruses

Genetic recombination between two completely unrelated DNA viruses can occur in certain special circumstances. Adenovirus Type 7 has been adapted to grow in nonpermissive monkey-kidney cells by serial passage. This modified adenovirus was found to induce SV40-type tumors in hamsters, however, and it was evident that the adaptation was due to the "helper" effect of contaminating SV40 virus. Attempts to eliminate the SV40 component by antiserum treatment led instead to isolation of a true hybrid of adenovirus and SV40, in which 10 per cent of the adenovirus genome had been deleted and replaced by 75 per cent of the SV40 genome. The hybrid was defective and only able to replicate in the presence of "helper" adenovirus, presumably because essential genes had been deleted. The inserted fragment of the SV40 genome included the early region (with a small repetition) and part of the late region, and consequently the hybrid virus expressed SV40 T-antigen in infected cells and induced SV40-type tumors in hamsters. *Defective adenovirus Type 2-SV40 hybrids* have been isolated that have up to one half of the adenovirus genome deleted and more than one genome equivalent of SV40 inserted.

Nondefective adenovirus Type 2-SV40 hybrids have also been isolated with 4.5 to 7.1 per cent of the adenovirus genome deleted and substituted by 7 to 59 per cent of the SV40 genome. Again, it is the early region of the SV40 genome that is inserted. Evidently the region of the adenovirus genome between map positions 0.79 and 0.86 (see Fig. 6), where all the deletions occur, is not essential for replication. The generation of these nondefective hybrids was a unique event, since it has not been possible to produce this hybrid again. Nevertheless, it illustrates that under rare circumstances DNA viruses at least have the potential to acquire genes from other DNA viruses, and how a nononcogenic virus might become oncogenic.

NONGENETIC INTERACTIONS

Complementation and Genetic Analysis in the Absence of Recombination

When cells are simultaneously infected by two *ts* mutants and incubated at high temperature, an enhanced yield may result if the mutations are in different viral genes. This is because the normal gene product of one virus compensates for the defective gene product of the other virus and vice versa. This phenomenon is termed *complementation*. Complementation defines the functional units of the genome and involves interaction of gene products only. Progeny virus produced by complementation retains the *ts* phenotype of the parents. In some cases, complementation may be unidirectional or even result in a depressed yield if a defective protein is produced in excess and incorporated into the majority of virions. The operational distinction of complementation and recombination is illustrated in Figure 7. Mutants can be grouped according to their ability to complement other mutants, and the number of complementation groups gives a minimum estimate of the number of genes in the genome of any virus (Table 1). In addition, complementation facili-

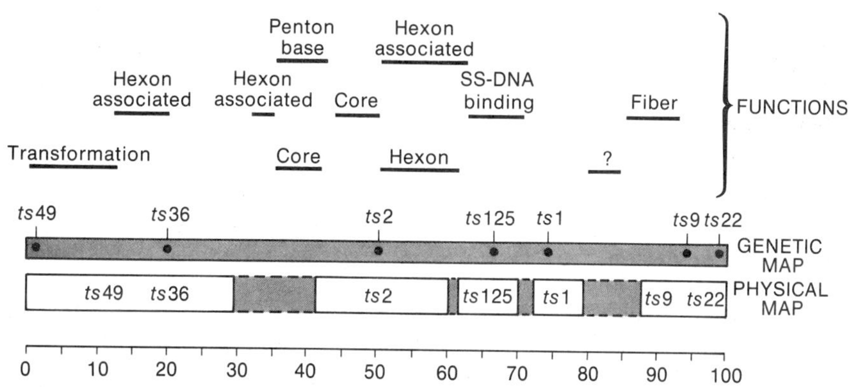

FIGURE 6. *The adenovirus genome.*

The genetic map of seven ts mutants of adenovirus Type 5 obtained by crosses of ts mutants is compared with the location of the same seven mutants obtained by physical mapping where fragments of adenovirus DNA generated by restriction endonuclease treatment are used to rescue ts mutants. There is no discrepancy between the two maps. The presumptive gene functions of the various regions are also illustrated.

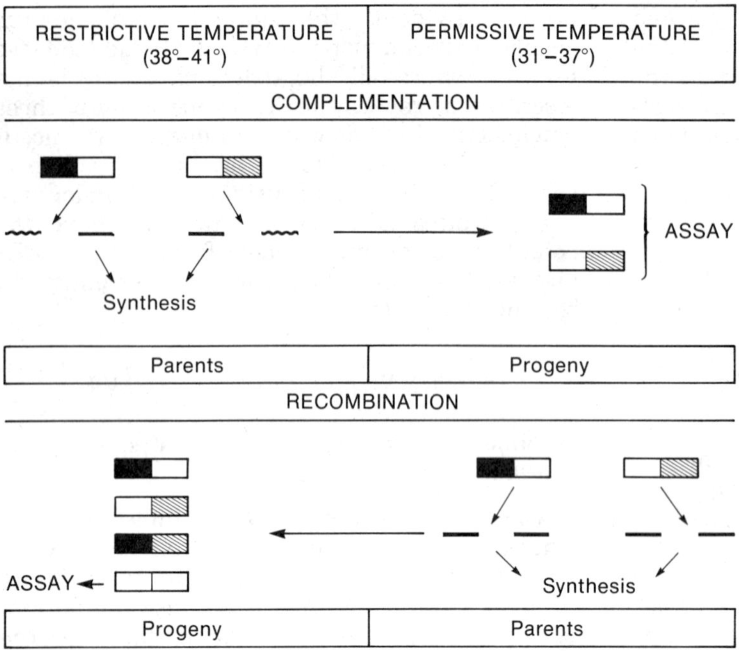

FIGURE 7. *The operational distinction between complementation and recombination using temperature-sensitive mutants.*

In complementation the mixed infection is conducted at restrictive temperature, and the yield is assayed at permissive temperature. The progeny are temperature-sensitive like the parents. In recombination the mixed infection is conducted at permissive temperature, and the yield is assayed at restrictive temperature. Thus, only one of the recombinants (the nonmutant) is assayed, since the parents and the double mutant recombinant are restricted. ▭ *= normal gene;* ■ *= independent;* ▨ *= ts mutants;* —— *= normal gene product;* ╱╲╱╲ *= defective gene product.*

tates analysis of the organization of the genome of viruses (such as the unsegmented negative-strand RNA viruses) that do not undergo recombination. The example of the rhabdovirus VSV illustrates the value of this approach. Five proteins are present in the virion of VSV, and no other virus-specified proteins have been detected in infected cells. The existence of six groups of complementing mutants points to the existence of another VSV protein, as yet undetected. Further study of the properties of *ts* mutants may lead to identification of this putative VSV protein and its function. So far four of the complementation groups of VSV Indiana have been correlated with viral proteins. Identification of gene functions in this way has confirmed that the RNA-dependent RNA polymerases of negative-strand viruses and the RNA-dependent DNA polymerase (reverse transcriptase) of retroviruses are virus-specified enzymes.

Phenotypic Mixing and Pseudotypes

Two different viruses multiplying in the same cell may produce progeny with the phenotypic characteristics of both parents. This phenomenon is particularly evident among enveloped viruses that mature by budding from membranes, but it is also prevalent among nonenveloped viruses. For instance, influenza virus can acquire antigens from serologically related strains and also from quite unrelated parainfluenza viruses and rhabdoviruses. The majority of phenotypically mixed particles contain antigens from both parents and can be neutralized by antisera to both viruses. A minority may be entirely invested by the envelope of the heterologous virus. This can have important epidemiologic consequences, since it would allow a virus to survive and persist in an apparently immune population.

The defective avian and mammalian sarcoma viruses exist only as *pseudotypes*, i.e., as phenotypically mixed particles whose envelope proteins are donated by the nondefective helper leukemia virus. Pseudotypes are also produced when cytolytic viruses (VSV is often used experimentally) multiply in cells chronically infected with oncornaviruses (the oncogenic retroviruses). Virions with the envelope antigens of avian leukosis virus and the nucleocapsid of VSV, designated VSV (ALV), and the reciprocal pseudotype ALV (VSV), have been identified among the progeny following VSV infection of chick embryo cells carrying avian leukosis virus. Pseudotype formation is also known to occur between DNA viruses (herpes simplex) and RNA viruses (VSV).

Production of pseudotypes of VSV can be used as a sensitive probe for detection of cryptic RNA tumor virus agents, as an aid in the study of membranes and host restriction, or as a rapid method of assay of such agents as mouse mammary tumor virus in which the alternative in vivo assay is lengthy and difficult. However, pseudotypes of the ALV (VSV) type constitute a potential biohazard. The leukemogenic and sarcoma genic viruses of animals are characteristically host-restricted, and envelopment in a membrane provided by an unrestricted virus like VSV could enable the oncogenic virus to reach cells that are normally inaccessible, with unpredictable consequences.

Phenotypic mixing is not confined to envelope proteins. The α mutants of Rous sarcoma virus, for example, have part of the polymerase gene deleted (as well as the entire envelope gene), and they can only be propagated in the presence of a helper virus that contributes the polymerase and envelope functions.

Host-Induced Modification

Host-induced modification, i.e., nonheritable phenotypic variation, can occur in certain circumstances. The lipid and the carbohydrate portions of the glycolipids and glycoproteins of the viral envelope are contributed by the host cell and therefore vary according to the cell of origin. For instance, the glycoproteins of arthropod-borne viruses released from insect cells lack sialic acid, while the same virus propagated in mammalian cells may possess sialic acid. However, in general this variability does not appear to have important biologic consequences.

DETERMINATION OF VIRULENCE, DISEASE POTENTIAL, AND ONCOGENICITY

The factors that determine the pathogenic potential of viruses can be defined by study of the effects of mutations in specific genes. For instance, study of temperature-sensitive mutants of reovirus revealed that mutation in a late function of the viral genome resulted in a change of disease pattern. Wild-type reovirus is lethal for newborn rats, whereas *ts* mutants exhibit greatly reduced virulence. In the case of one mutant (*ts* B1), although the majority of animals survived infection with high doses of virus, later in life progressive neurologic disease developed, which culminated in death. The assembly of the outer capsid is defective in *ts* B1-infected cells, and it was proposed that intracellular accumulation of viral nucleocapsid is responsible for the altered disease pattern.

Mutations affecting *neurovirulence* have been described frequently. One mutant of measles virus is less neurovirulent than wild-type measles in suckling hamsters. Hydrocephalus, a response to infection normally masked by the dominant neurologic signs, is a consequence of infection with this mutant. On the other hand, *ts* mutants of VSV often exhibit enhanced neurovirulence for newborn hamsters. However, there is no specific association of neurovirulence with a particular gene. The alteration in disease pattern is related more to the increased survival time of infected animals.

The virulence of avian influenza (fowl plague) virus appears to be influenced by many factors; so far no one gene has been implicated as a major determinant of virulence, and probably the same is true for human influenza virus. However, the virulence of the avian parainfluenza virus (Newcastle disease virus) for chickens is dependent on extracellular cleavage of an envelope glycoprotein (the F or fusion protein). Mutation affecting the sensitivity of this protein to proteolytic cleavage influences host range in vitro and virulence in vivo. Furthermore, there are known mutants of a murine parainfluenza virus that differ in the sensitivity of the F protein to cleavage by specific proteases and, as a consequence, their ability to infect different cells.

The role of viruses in *oncogenesis* can also be approached by study of the properties of mutants. Transformation defective (*td*) mutants of Rous sarcoma virus occur spontaneously and are predominantly deletion mutants in which part or all of the sarcoma determinant (the *src* gene) is missing. These mutants are still oncogenic, however, because they retain the ability to induce leukosis. The genome of *td* deletion mutants and the naturally occurring avian leukosis viruses are similar in size. The *src* gene sequences of different avian sarcoma viruses are homologous, indicating that they have been conserved in evolution. *Src* sequences can be detected also in the normal cells of several avian species, including Japanese quail (which has no inducible endogenous retrovirus). The prevalence of these *src*-related sequences suggests that the *src* gene of avian sarcoma viruses, which is a function not required for virus multiplication, was acquired from the cell at some time in the past.

The origin of the *src* gene in mammalian sarcoma viruses is more complex. The Harvey and Kirsten strains of murine sarcoma virus appear to have acquired rat genome sequences and transforming ability during passage in rats. The Moloney strain of murine sarcoma virus, on the other hand, acquired transforming ability during passage in mice, and there is no sequence homology between the *src* gene of the Moloney virus and the *src* genes of the Harvey or Kirsten viruses.

Almost all *ts* mutants of avian sarcoma virus that affect transformation are late mutants. Temperature-shift experiments with *ts* mutants have shown that transformation is dependent on the continuous activity of a viral gene. However, the nature of the *src* gene product is still unknown.

Cells transformed by SV40 virus carry viral

DNA sequences integrated into their chromosomes. All or only part of the SV40 genome may be present. Study of cells transformed by *ts* mutants of SV40 virus showed that expression of an early gene function of SV40 only was necessary for initiation and maintenance of the transformed state. However, there seems to be no single chromosomal site of integration of SV40 (in rat cells at least), and the number of insertions of the SV40 genome or its early genes is variable in different cell lines.

Similarly, the isolation of *ts* mutants of adenovirus that failed to transform rodent cells at nonpermissive temperature showed that a viral gene product was required at least for initiation of transformation by adenovirus. These mutants map at one end of the adenovirus genome, and only those fragments of the adenovirus genome obtained by restriction endonuclease digestion that contain the left-hand end are able to transform cells; no more than 8 per cent of the genome is required to cause transformation.

EVOLUTION IN ACTION

Human influenza A virus undergoes dramatic changes in its surface antigens (*antigenic shift*) at intervals, causing severe pandemics. Between pandemics, the virus is constantly changing in a more progressive manner (*antigenic drift*). The available evidence supports the hypothesis that antigenic drift occurs by accumulation of small mutational changes, whereas antigenic shift is the result of hybridization between a human influenza and a virus maintained in a reservoir species. The pandemic of 1918 was caused by a strain that may have acquired an antigenically novel hemagglutinin by recombination with a swine influenza virus. Reconstruction experiments in animals have confirmed that hybridization between different influenza virus strains can occur in vivo. For example, reciprocal recombinants of swine influenza and fowl plague virus (FPV) have been obtained from pigs, despite the inability of FPV to multiply efficiently in this host animal.

The history of the introduction of *myxomatosis* into Australia illustrates that evolutionary changes may occur in both the virus and its host. Myxomavirus was extremely lethal for European rabbits when it was liberated in Australia in 1950 as a pest control measure: mortality rates were in excess of 99 per cent. However, attenuated variants soon made their appearance and within 3 to 4 years they became dominant, because reduced virulence favored survival of animals from one season to the next and allowed a susceptible population to be reestablished each year. Following appearance of the initial attenuated variants, there was a corresponding increase in the genetic resistance of the host animal, so that the mortality induced by a standard virulent strain fell from over 90 per cent to 25 per cent in 5 years. Thus, a milder disease and stable endemicity were evolved by changes in the disease potential of the virus and the genetic resistance of the host animals. It is possible that similar factors have had a similar influence in the evolution of endemic virus disease in man.

References

The Nature and Coding Capacity of Viral Genomes

Baltimore, D.: Expression of animal virus genomes. Bacteriol Rev 35:235, 1971.

Watson, J. D.: Molecular Biology of the Gene, 2nd ed. New York and Amsterdam, W. A. Benjamin Inc., 1970.

Weisbeek, P. J., Borrias, W. E., Langeveld, S. A., Baas, P. D., and van Arkel, G. A.: Bacteriophage ΦX174: gene A overlaps gene B. Proc Natl Acad Sci USA 74:2504, 1977.

Mutation and Mutants

Fenner, F.: Conditional lethal mutants of animal viruses. Curr Topics Microbiol Immunol 48:1, 1969.

Fraenkel-Conrat, H., and Wagner, R. R.(eds.): Comprehensive Virology, Vol. 9: Regulation and Genetics, The Genetics of Animal Viruses. New York and London, Plenum Press, 1977.

Huang, A. S., and Baltimore, D.: Defective viral particles and viral disease processes. Nature (London) 226:325, 1973.

Pringle, C. R.: Conditional lethal mutants of vesicular stomatitis virus. Curr Topics Microbiol Immunol 68:85, 1975.

Genetic Interactions Between Viruses

Fenner, F., McAuslan, B. R., Mims, C. A., Sambrook, J., and White, D. O.: The Biology of Animal Viruses, 2nd ed. New York and London, Academic Press, 1974.

Kelly, T. J., Jr., and Nathans, D.: The genome of simian virus 40. Adv Virus Res 21:86, 1977.

Palese, P.: The genes of influenza virus. Cell 10:1, 1977.

Tooze, J. (ed.): The Molecular Biology of Tumour Viruses. Cold Spring Harbor Laboratory, 1973.

Westway, E. G.: Strategy of the flavivirus genome; evidence for multiple internal initiation of translation of proteins specified by Kunjin virus in mammalian cells. Virology 80:320, 1977.

Williams, J., Grodzicker, T., Sharp, P., and Sambrook, J.: Adenovirus recombination: physical mapping of cross-over events. Cell 4:113, 1975.

Nongenetic Interactions

Fields, B. N.: Genetic manipulation of reovirus — a model for modification of disease? N Engl J Med 287:1026, 1972.

Stehelin, D., Guntaka, R. C., Varmus, H. E., and Bishop, J. M.: Purification of DNA complementary to nucleotide sequences required for neoplastic transformation of fibroblasts by avian sarcoma viruses. J Mol Biol 101:349, 1976.

Determination of Virulence, Disease Potential, and Oncogenicity

Fenner, F., and Ratcliffe, F. N.: Myxomatosis. London and New York, Cambridge University Press, 1965.

Kilbourne, E. D. (ed.): The Influenza Viruses and Influenza. New York, Academic Press, 1975.

PERSISTENT VIRAL INFECTIONS 11

Julius S. Youngner, Sc.D.

The attention of medical virologists has been focused primarily on acute febrile diseases. Acute infections, such as influenza, smallpox, and measles, have incubation periods of days to weeks, well-defined symptoms, and closely associated pathologic manifestations. In individuals who recover from acute viral infections, both cellular and humoral mechanisms operate during the incubation period and the virus usually disappears from the body within several weeks after the first symptoms appear. The period in which virus can be isolated from clinical specimens is very limited; in most cases virus is demonstrable only briefly before and after the onset of clinical disease. In virus infections, such as poliomyelitis and mumps, the pattern of acute disease and recovery is often more difficult to study, since the infection may be subclinical and produce no clearly recognizable or characteristic symptomatology.

Increasing attention is being devoted to persistent viral infections in which the infectious agent is present for long periods of time. Infections of this type are diverse and difficult to classify with precision. A great variety of viruses, patterns of pathogenesis, and clinical entities contribute to the complexity of this group of infections (Table 1).

CLASSIFICATION OF PERSISTENT VIRAL INFECTIONS IN HUMANS AND ANIMALS

Persistent viral infections can be classified into three general categories: (1) latent infections, (2)

TABLE 1. Persistent Viral Infections of Humans

VIRUS Genome	CLASSIFICATION Group	AGENT	DISEASE	VIRUS PRESENCE*	MECHANISM OF PATHOGENESIS†
DNA	Herpesvirus	Herpes simplex virus (Types 1 and 2)	Herpes	Latent	Cytolysis
		Varicella-Zoster virus	Zoster	Latent	Cytolysis
		Cytomegalovirus	Cytomegalo-inclusion disease	Latent	Cytolysis
			Mononucleosis	Latent	Transformation
		EB virus	Mononucleosis	Latent	Transformation
			Lymphoma	Latent	Transformation
DNA	Papovavirus	JC virus	Progressive multifocal leukoencephalopathy (PML)	Minimal	Cytolysis
		BK virus	?	Latent	?
		Human papilloma virus	Warts	Overt	Transformation
DNA	Adenovirus	Adenovirus (many types)	?	Latent	?
DNA	Unclassified	Hepatitis B virus	Hepatitis	Overt	Immunopathology(?)
?		Non A/B hepatitis virus	Hepatitis	Overt	Immunopathology(?)
RNA	Paramyxovirus	Measles	Subacute sclerosing panencephalitis (SSPE)	Minimal	Cytolysis
RNA	Togavirus	Rubella	Fetal anomalies	Overt	Inhibition of cell division(?)
			Panencephalitis		Cytolysis(?)
?	?	Creutzfeldt-Jakob agent	Spongiform encephalopathy	Overt	Cytolysis
		Kuru agent	Spongiform encephalopathy	Overt	Cytolysis

*Latent = infective virus cannot be demonstrated, except during active or recrudescent symptoms of disease.
Minimal = infective virus difficult to demonstrate; may be absent or present in exceedingly small numbers.
Overt = infective virus usually demonstrable with or without presence of symptoms.

†Cytolysis = lesions result from cell destruction by virus.
Transformation = lesions due to cell proliferative response to infection in the absence of cytolysis.
Immunopathology = lesions due to specific virus antigen-antibody complex reactions in affected tissues.

chronic infections, and (3) slow infections (Fenner and White, 1976).

Latent Infections

In this type of persistence, the virus disappears after acute primary disease and reappears during recurrences of disease. There may be one or more such recurrences after long intervals. The best understood diseases in this class are caused by members of the herpesvirus group, herpes simplex, and varicella-zoster viruses.

The most common manifestations of herpes simplex is a recurrent vesicular eruption ("cold sores"), which appears most commonly at the mucocutaneous junctions around the lips or nostrils. These eruptions progress to pustules and crust formation and then disappear. The virus remains latent in the sensory cells of the trigeminal nerve ganglion between attacks (Stevens, 1975). The mechanism that is responsible for reactivation of virus is not fully understood. Nonspecific stimuli, such as respiratory infection, fever, or sunlight cause the virus to spread from the trigeminal ganglion via the sensory nerves to the areas of the skin supplied by the nerves. The relapses of herpes simplex occur despite the presence of circulating antibody in patients with recurring disease. The virus probably passes directly from cell to cell without release; this protects the virus from exposure to antibody in the body fluids.

A similar pattern of latency is seen in herpes zoster, a disease due to reactivation of virus latent in dorsal root or cranial nerve ganglia, many years after childhood varicella. In herpes zoster, which mainly affects adults, painful vesicles erupt in an area of skin supplied by a sensory nerve fiber from a single dorsal root. Most commonly affected are the thoracic nerves; less frequently the ophthalmic nerve of the trigeminal ganglion may be involved.

Disseminated infection with cytomegalovirus, a ubiquitous member of the herpesvirus group, is seen when immunosuppressive therapy, neoplasms, or other debilitating diseases lower resistance of the patient. Reactivation of latent cytomegalovirus is frequently seen in renal transplant recipients; these reactivations are often not associated with any overt signs or symptoms of disease. The site of latency of cytomegalovirus is not clearly established. In some cases, cytomegalovirus infection may be transmitted during the transfusion of large volumes of fresh blood, and reflects the widespread occurrence of healthy carriers of cytomegalovirus.

Chronic Infections

Persistent infections in this class have a common characteristic: virus is always demonstrable and often is shed. Disease symptoms may be absent or associated with immunopathologic manifestations. A good example of chronic human infection is hepatitis B, a disease associated with varying degrees of hepatic necrosis, and usually contracted by transmission of virus in blood from a chronic carrier (Committee on Viral Hepatitis, 1975). A large amount of a virus-associated antigen (HB-Ag) is commonly present in the serum. This antigen usually disappears rapidly after an acute clinical or subclinical infection. However, in about 5 per cent of infected individuals, HB-Ag may persist in serum in high concentration for years. Although the concentration of infective virus is lower than the HB-Ag, sera from such carriers are highly infectious for recipients of blood transfusion, staff of renal dialysis units, and drug addicts who use unsterilized syringes.

Sera from patients acutely or chronically ill with hepatitis B infection contain antigen-antibody complexes that may play a role in the pathogenesis of acute and chronic hepatic disease. In addition, there is evidence that the glomerulonephritis and polyarteritis nodosa seen in some HB-Ag carriers may be caused by antigen-antibody complexes.

Another example of a chronic infection in humans is the rubella syndrome seen in infants infected in utero with rubella virus during the first 16 weeks of pregnancy. Such infants have various disorders and defects due to generalized infection with rubella virus, and it is usually possible to isolate virus from almost any of their organs. Maternal antibody may inhibit the spread of virus but does not eliminate virus-producing cells. These cells divide at a slower rate than normal uninfected cells, probably giving rise to some of the developmental abnormalities common in infants with rubella syndrome.

Slow Infections

Diseases in this group have many of the characteristics of the chronic infections described in the previous section. However, they differ from chronic infections in several important characteristics. Slow infections have very long incubation periods ranging from months to years; the disease has a long, chronic, and progressive course, and the outcome is usually fatal (Kimberlin, 1976).

The viruses that cause slow infections are heterogeneous. Nononcogenic retroviruses cause visna, maedi, and progressive pneumonia, which are slowly progressive infections of sheep. Another group of agents is poorly defined and difficult to isolate and assay — namely, the agents that produce subacute spongiform encephalopathies in sheep (scrapie), in mink (mink encephalopathy), and in humans (kuru and Creutzfeldt-Jakob disease). The agents of these diseases resemble viruses in that they pass filters with small pore diameters and can be transmitted to experimental animals. The resistance of these agents to irradiation, chemicals, and heat, and the failure of the agents to elicit a detectable antibody response in infected hosts indicate that they are not typical viruses. A third group of agents is made up of conventional viruses that produce slow, degenerative diseases of the central nervous system in humans: subacute sclerosing panencephalitis (measles and rubella viruses) and progressive multifocal leukoencephalopathy (JC virus, a papovavirus). Table 2 summarizes the characteristics of slow virus infections of humans.

Spongiform Encephalopathies. Kuru was the first chronic degenerative disease of the central nervous system of humans proved to have a viral etiology. This disease occurred only among a group of about 50,000 highland New Guineans. The classic studies of Gajdusek (1977) revealed that the kuru agent was spread by ritualistic cannibalism and had an incubation period of from four to 20 years. The incidence of the disease has declined since the early 1960s, when the practice of cannibalism was discouraged. All attempts to cultivate the agent of kuru in cell cultures have been unsuccessful. However, transmission experiments have shown that intracerebral inoculation of brain tissue from kuru victims into chimpanzees and monkeys causes the animals to develop symptoms of kuru after an incubation period of about two years.

The importance of the studies of kuru has been illuminated by the demonstration that a rare, progressive, fatal neurologic disease of humans, Creutzfeldt-Jacob disease, is due to an agent with many of the characteristics of the kuru virus. Creutzfeldt-Jakob disease is characterized by presenile dementia with symptoms due to lesions in the cerebral cortex (status spongiosus) and lesions in the spinal cord. The disease is reproduced in chimpanzees and other monkeys by intracerebral inoculation of brain tissue from victims of the disease; the incubation period in experimental animals is about one year. The natural route of infection is not known, but the disease has been accidentally transmitted via a cornea graft (Duffy et al., 1974). Transmission has also been reported via electrodes used for electroencephalography from a patient with the disease to other patients in whom the electrodes were subsequently implanted (Bernoulli et al., 1977).

Subacute Sclerosing Panencephalitis (SSPE). This severe chronic neurologic disease is seen in children and young adults several years after primary, and usually uncomplicated, measles (ter Meulen et al., 1972). Affected children have high titers of measles antibody in their serum and spinal fluid. The earliest signs of illness are personality and behavioral changes with intellectual impairment. The disease progresses to convulsions, myoclonal spasms, and increasing neurologic deterioration leading to coma and death. At postmortem there is electron microscopic evidence of measles virus infection, but measles virus cannot be recovered directly. Only by cocultivation of brain tissue with permissive indicator cells can infective measles virus be recovered. Despite the presence of high levels of specific antibody in serum and cerebrospinal fluid, the progress of the disease is not arrested. The factors involved in determining the occurrence of this slow complication of measles in rare individuals are not known. Chronic infection with rubella virus infrequently produces a progressive neurologic disease resembling SSPE (Weil et al., 1975).

Progressive Multifocal Leukoencephalopathy (PML). This is a rare disease that occurs only in patients whose immunologic responsiveness has been severely compromised by such preexisting conditions as leukemia, reticulosis, or immunosuppressive therapy. There are varied neurologic signs, such as dementia, incoordination, and impaired vision; the disease is usually fatal in three or four months. On autopsy, multiple foci of demyelination are found in the cerebral hemi-

TABLE 2. Slow Virus Infections of Humans

DISEASE	SITE	PATHOLOGY	VIRUSES
Kuru	Brain	Spongiform encephalopathy, especially in cerebellum	Filterable agent transmissible to chimpanzees and monkeys
Creutzfeldt-Jakob disease	Brain and spinal cord	Spongiform encephalopathy	Filterable agent transmissible to chimpanzees and monkeys
Subacute sclerosing panencephalitis (SSPE)	Brain	Neuronal degeneration	Measles and rubella viruses
Progressive multifocal leukoencephalopathy (PML)	Brain	Multiple foci of demyelination in cerebral hemispheres and cerebellum	Human papovavirus (JC virus)

spheres and cerebellum; the brain stem and basal ganglia may also be affected. Most cases are due to a papovavirus (JC virus) but some are caused by a virus almost identical to SV40, another papovavirus. These viruses can be grown in human fetal glial cell cultures. Another serotype of papovavirus (BK virus) has been recovered from the urine and genitourinary tract of patients undergoing immunosuppression to prevent rejection of transplanted kidneys. BK virus has not yet been associated with any specific clinical disease (Padgett and Walker, 1976).

Infections with both JC and BK viruses appear to be widespread, since serologic surveys show that many people have antibody to these agents. However, the type of disease caused by primary infection with these viruses during childhood or early adolescence has not been identified. The site of viral latency in the body before reactivation is also not known.

In spite of the increased understanding of the etiology of the slow virus diseases described above, the causes of the most important demyelinating diseases of the central nervous system — that is, multiple sclerosis, the common presenile dementias, and Parkinson's disease — are all still obscure. In the case of multiple sclerosis, common viruses such as measles and parainfluenza type 1 have been suggested as the causative agent in the recent past. However, these associations have so far not stood up to thorough and critical investigation.

PERSISTENT VIRAL INFECTION AT THE CELLULAR LEVEL

The complexities involved in studying persistent infections in humans and the difficulties in establishing model systems in animals have led to intensive efforts to use cell culture models of persistent viral infection. Infection of permissive cells with lytic viruses in culture usually leads to productive infection, the release of large numbers of progeny viruses, and cell death. However, many animal viruses that are considered to be highly cytocidal can establish infections of cell cultures that result in long-term multiplication of the virus while at the same time the cells continue to grow and divide. These stable virus-host cell relationships can be classified into several general categories.

Cell Culture Models of Persistent Viral Infections

The classification that follows is oversimplified because, in many systems, cells may be protected by several mechanisms that may function simultaneously or sequentially. Despite these complexities, the following classification is useful in understanding the model systems that have been established in cell cultures (Walker, 1968). The salient characteristics of these model systems are summarized in Table 3.

Class I. Carrier State in Genetically Resistant Cells. In this type of infection a majority of the cells is genetically resistant. However, permissive (susceptible) cells continually appear and these cells, when infected by free virus in the medium, permit a normal cycle of virus replication and the release of progeny. Characteristically, viral specific antibody or other antiviral factors, such as interferon, are not required to maintain the stable carrier state. Only a small proportion of the cells is infected; when persistently infected cells are cloned, infected cells are destroyed and the clones obtained are free of virus. The clearest examples of this type of carrier culture are infections of a human cell line (HeLa) by poliovirus and Coxsackie virus, members of the picornavirus group.

Class II. Carrier State in Permissive Cells Protected by Antibody in the Medium. In this type of carrier culture only a small fraction of the genetically susceptible cells is infected; most cells are protected from infection by the presence of antiviral factors, usually antibody, in the me-

TABLE 3. Classification of Persistently Infected Carrier Cultures

Class of Carrier Culture	Antiviral Antibody Required	All or Most Cells Infected	Infected Cells Can Divide and Form Colonies	Infection Cured by Antiviral Antibody	Interfering Factors Produced in Medium
I. Genetically resistant cells	No	No	No	Yes	Yes or No
II. Permissive cells protected by antibody in medium	Yes	No	No	Yes	Yes or No
III. Permissive cells protected by interference or interferon	No	No	No	Yes	Yes
IV. Regulated infections	No	Yes	Yes	No	Yes or No
V. Virus DNA integrated into the host cell genome	No	Yes	Yes	No	No

dium. Removal of the antibody from the medium results in the spread of the infection and destruction of the cell culture. Infected cells do not divide and grow into colonies except in the presence of antibody in the medium. In the absence of antibody, the cell clones obtained are virus-free. The best documented example of this type of carrier state is the persistent infection of various human cell lines by herpes simplex virus.

Class III. Carrier Cultures of Permissive Cells Protected by Interference or Interferon. The general characteristics of this class of persistently infected cells are the following: only a small fraction of the cell population is infected at any time; infected cells do not divide and cannot be cloned; cell clones, when obtained, are free of virus; endogenously produced interfering factors are always present in the culture medium; therefore, antibody or other antiviral factors do not have to be provided for establishment or maintenance of the carrier state. The relative importance of different endogenously produced interfering factors, such as defective-interfering virus particles and interferon, will be considered below.

An outstanding feature of cultures of this type is that the cells are resistant to superinfection by the carried virus or by heterologous viruses, even though only a small fraction of the cells is infected and producing virus at any given time. However, when the infections are cured by prolonged exposure to antibody, the cured cells are just as susceptible to the homologous and heterologous viruses as the original cell line, indicating that genetically resistant cells do not arise during the carrier phase. There are many examples of Class III carrier cultures involving paramyxoviruses and togaviruses.

Class IV. Carrier Cultures in Which the Infections Are "Regulated." The first three classes of persistently infected cells that were described above have important common characteristics. Only a small fraction of the cell population is infected at any given time, infected cells do not divide and grow into colonies but are killed by virus, and clones of cells that are obtained form the carrier culture are free of virus. In contrast, the "regulated" infections of Class IV have the following properties: (1) a high proportion, sometimes all, of the cells in the culture are infected; (2) infected cells can divide and grow into clones that consist of infected cells; (3) antibody or other antiviral factors are not required to establish or maintain the persistently infected state and the culture is not cured of virus by the addition of antibody to the medium; (4) although carrier cultures of this type are fully resistant to superinfection by the homologous virus, they exhibit no resistance to unrelated viruses.

Little is known about the mechanisms by which the "regulated" infections prevent the virus from going through the usual replicative events and cell destruction characteristic of cytocidal infections. The designation "regulated" infection seems appropriate because there seems to be some sort of intracellular regulation or control not yet understood. A large variety of enveloped, budding, RNA viruses can establish this type of persistence — for example, the paramyxoviruses (mumps, measles, parainfluenza viruses 1 and 3, Sendai), rubella, and rabies viruses.

Class V. Carrier State Due to Virus DNA Integration into the Host Cell Genome. Another model for persistent infections involves the integration of the viral genome into the DNA genome of the host cell, an event that would permit the long-term maintenance of viral genetic information in infected cells. There is strong evidence in cell culture systems that persistence by some members of the papovavirus, adenovirus, and herpesvirus groups is maintained by this mechanism. However, in most instances rigorous proof is lacking that genomes of these DNA viruses are integrated into the host cell chromosome in humans or animals. Even in cell culture models, the mechanisms by which the viral genomes are regulated and the phenomena associated with activation are mostly unknown.

There is convincing evidence that visna virus, an RNA C-type retrovirus, which causes a persistent inflammatory demyelinating disease of the central nervous system of sheep, replicates via a DNA intermediate that is integrated into the host cell genome (Haase, 1977). The reverse transcriptase present in the visna virion is responsible for the transcription of a DNA copy of the RNA genome, and it is this DNA intermediate that is inserted into the host DNA.

The possibility has been raised that some enveloped cytolytic RNA viruses, such as paramyxoviruses and togaviruses, which do not contain reverse transcriptase, have the choice of alternate pathways of nucleic acid synthesis and can employ an integrated DNA intermediate for maintaining long-term persistence in infected cells. This mechanism, which is yet to be proved, will be considered in more detail below.

Virus-Specific Factors That May Be Involved in the Establishment and Maintenance of Persistent Infections

Mechanisms of Persistence of DNA Viruses. Herpes simplex, varicella-zoster, and cytomegaloviruses, all members of the herpes-virus group, probably establish persistence in which the virus remains latent; the DNA genomes are present as cytoplasmic or nuclear plasmid forms in either a completely regulated or a nonreplicating state. Also likely is the possibility that in some instances a less tightly controlled, very slowly replicating state may be involved. In either case there is no a priori need to postulate integration as a means of perpetuating the infection.

In regard to the papovaviruses, the best-studied example is SV40, a virus of simian origin that is related to human papovavirus (JC virus) associated with progressive multifocal leukoencephalopathy. SV40 virus maintains a latent state in cell cultures by integration of the viral DNA into the host genome; such cells may contain up to 10 copies of SV40 DNA per cell (Doerfler, 1975). In the case of human papilloma virus, a papovavirus that causes warts, the mechanism by which the viral DNA persists is not yet known.

There is strong circumstantial evidence that persistence of adenovirus in certain cell systems is maintained by the integration of the viral genome into the host cell DNA (Doerfler, 1975). There is, however, no rigorous proof of this mechanism, and the manner in which adenoviruses persist in adenoids and tonsils of humans is unknown. Also unknown is the mechanisms of persistence of the SV40-related viruses, JC and BK, which are isolated from patients with progressive multifocal leukoencephalopathy or from immunosuppressed patients shedding these viruses in their urine.

The only definitive information concerning integration of viral DNA of herpesviruses comes from studies of the Epstein-Barr (EB) virus that is associated with infectious mononucleosis, Burkitt's lymphoma, and nasopharyngeal carcinoma (Klein, 1973). There is evidence that both in lymphoblastoid cell lines and in the tumors, the EB virus DNA is maintained in two physical states; a linear form integrated into cellular DNA, and a circular, extrachromosomal plasmid.

In summary, although there is some understanding of the state of the viral DNA in cell culture systems latently infected with some papovaviruses (except papilloma virus), adenoviruses, and EB virus, there is little definitive information concerning other herpesviruses. With the exception of EB virus, little is known about the mechanisms of DNA virus latency in man or animals. The mechanism by which the viral genomes are controlled and regulated during latent infection, and more particularly, the phenomena associated with reactivation are, at this time, poorly understood.

Mechanisms of Persistence of RNA Viruses. A number of different mechanisms, which are not necessarily mutually exclusive, have been proposed to explain the persistence of ordinarily cytolytic RNA viruses in cell cultures. Although there is some evidence that these mechanisms may operate in animal models, there is no direct support at this time for their involvement in human disease (Rima and Martin, 1976).

ROLE OF DEFECTIVE-INTERFERING (DI) PARTICLES. Evidence is accumulating that defective particles capable of interfering with the replication of homologous standard virus occur spontaneously in almost every virus system that has been studied (Huang, 1973). Serial passage of viruses at high multiplicities of infection (high virus per cell ratios in the inoculum) results in the production of DI particles that have the following general characteristics: (1) they contain normal structural capsid proteins; (2) a portion of the viral genome is missing (in reality they are deletion mutants); (3) they can replicate only in cells coinfected with the homologous virus helper; and (4) they can interfere specifically with the replication of homologous standard virus.

It has been postulated that the production of DI particles may regulate the synthesis of virus in persistent infections by specifically inhibiting production of the homologous standard virus, thereby damping down infection (Huang and Baltimore, 1970). There is some evidence that this concept is plausible even for a highly cytolytic virus. When vesicular stomatitis virus (VSV), a rhabdovirus, is passaged serially at high multiplicity in susceptible Chinese hamster ovary cells, and the relative concentrations of standard VSV and DI particles are determined at each passage, there is a cyclical overlapping pattern of production of complete virus and production of DI particles. This pattern, based on serial high multiplicity passages, serves as a theoretical model for persistent infection.

Evidence concerning the possible role of DI particles in persistent infection is provided by a model system involving a line of hamster cells (BHK-21) persistently infected with VSV (Holland and Villarreal, 1974). A mixture of infective standard virus and large numbers of DI particles is required to establish the initial persistent infection. During the course of the infection, more DI particles are generated.

Synthesis of DI particles has been implicated in maintenance of measles virus persistence in human cell lines, and in cultured cells persistently infected with such diverse viruses as reovirus, rabies, several togaviruses, and lymphocytic choriomeningitis virus. The precise relationship of DI particles to autointerference and to persistence is not well understood. Although it is evident that DI particles play an important role in many carrier cell cultures, much more work is needed to prove a role for DI particles in persistent infections in animals and humans.

ROLE OF TEMPERATURE-SENSITIVE (TS) MUTANTS. A body of information has accumulated that shows that viruses recovered from persistent infections often differ in biologic properties from the standard virus used to initiate the infection. Quite commonly, when compared to wild-type viruses, the agents recovered from persistently infected cells produce smaller plaques in cell cultures and are less able to infect experimental animals. In many model systems, there is a natural selection of ts mutants* that replace the standard virus population during the evolution of the persistently infected state (Preble and Youngner, 1975). Such ts viruses occur in a variety of host cells persistently infected with agents such as mumps, measles, rubella, Sendai, Newcastle disease, Sindbis, Western equine encephalomyelitis, and vesicular stomatitis viruses. Selection of ts mutants in persistent infections is not limited to any particular class of virus or type of cultured cell. In fact, one report (Valentine et al., 1969) describes a commonly occurring spontaneous ts mutant of bacteriophage Qβ that causes a persistent infection of its bacterial host, *Escherichia coli*.

The reasons why ts mutants are selected under conditions of persistent infection are not fully understood, but the following is known. In the cases of such viruses as VSV, Newcastle disease virus, and Sindbis, ts mutants interfere strongly with the replication of standard virus at both permissive and nonpermissive temperatures, interference occurring before or at the level of RNA transcription. Second, ts mutants of VSV are rescued by standard virus at nonpermissive temperatures; in effect, ts mutants act as conditionally defective interfering (DI) viruses at the nonpermissive temperature. As a result, the replication of ts virus tends to be dominant at the nonpermissive or partially restrictive temperature, thereby explaining why ts mutants are selected. Also, since under these conditions ts mutants tend to be less cytocidal and the replication of standard virus is suppressed, an explanation is provided of why ts mutants tend to be elected and why this selection leads to the maintenance of persistent infections.

Despite these observations, the presence of multiple genetic alterations in viruses isolated from persistently infected cells mandates caution in attributing to any one defect a key role in the establishment and maintenance of a persistent infection. Any genetic alteration that converts a lytic virus-cell interaction to a noncytocidal one has the potential to cause a persistent infection.

There is much evidence that ts mutation plays a role in the establishment and maintenance of many persistent infections in cell cultures. However, in animals and humans the relationship of these observations to persistence is not yet established.

ROLE FOR DEFECTS IN HOST CONTRIBUTIONS TO VIRUS REPLICATION. It is likely that increasing evidence will appear dealing with the importance of defects or alterations in virus assembly and release in persistent infections, particularly those involving enveloped viruses. The yield of these viruses from persistently infected cells is often less than in cytocidal infections, and frequently no infective virus is produced. Several mechanisms could operate in such a system.

Specific protein cleavages, especially of viral envelope glycoproteins, are important steps in virus maturation (Scheid and Choppin, 1976). In some paramyxoviruses, cleavage is accomplished by a host enzyme and activates virus activities such as cell fusion, hemolysis, and viral penetration of host cells. The host range and tissue tropisms in large measure depend on the presence in host tissue of the appropriate activating protease. Since virus mutants have been isolated in the laboratory that require different proteases for activation, the possibility exists that infection of a tissue not normally susceptible to a given virus may result from the appearance of a mutant that can be activated by a protease present in that tissue. In addition, infection by a cytocidal virus of a cell that cannot cleave the progeny viral glycoprotein could result in persistent infection without the production infective virus. If proper cleavage reactions fail in paramyxoviruses, mature but noninfective virus is produced, whereas failure of cleavage in togaviruses prevents virus assembly. It is possible that specific proteolysis of viral proteins by host enzymes is important in many persistent infections.

POSSIBLE ROLE OF INTEGRATED VIRAL GENOMES. DNA copies of RNA virus genomes are integrated into chromosomes of cells persistently infected by tumorigenic retroviruses and by visna virus, a retrovirus that causes persistent central nervous system disease in sheep. There have been several reports that DNA copies of usually cytocidal single-stranded RNA viruses have been integrated into the genome of persistently infected host cells (Simpson and Iinuma, 1975; Zhdanov, 1975). The viruses involved include respiratory syncytial virus, SV5, measles, Sindbis, and tick-borne encephalitis virus. A mechanism proposing an integrated DNA intermediate of usually cytolytic RNA viruses raises questions concerning (1) the source of the RNA-dependent DNA polymerase

*Temperature-sensitive (ts) conditional lethal mutants have been described for many different viruses (Fenner, 1969). These ts viruses occur in low frequency spontaneously and in higher frequency after exposing standard virus to mutagens. Ts viruses replicate well at low temperatures (31 to 33° C) and poorly at higher temperatures (38 to 40° C); in contrast, standard wild-type viruses replicate almost equally as well at both permissive and nonpermissive (restrictive) temperatures.

that initially transcribes the RNA of the infecting virus into a DNA copy, (2) the nature of the integration of this DNA copy into the host cell DNA, and (3) the regulation of the expression of the integrated DNA copy of the viral RNA. Some workers have not found DNA intermediates in cells persistently infected with several cytolytic RNA viruses (Haase et al., 1977). The implications and potential of this mechanism are so important that the original observations concerning integrated DNA copies of RNA virus genomes should be confirmed.

There is abundant evidence, however, that a DNA integration mechanism is not necessary for persistence of RNA viruses. In many instances the viral genome persists as RNA, replicates independently of the host cell genome, and passes from one daughter cell to another for generations, with viral products demonstrable at any stage.

Other Mechanisms. The intact animal probably has more mechanisms for establishing and maintaining persistent infections than cultured cells possess. Both nonspecific resistance factors (fever; interferon production) and specific immunity complicate the situation. Much more information is needed to understand the mechanisms by which virus-infected cells are eliminated in the body. The role of antibody modulation of viral antigens from the cell surface in persistent infection in the animal must also be clarified (Joseph and Oldstone, 1975).

References

Bernoulii, C., Siegfried, J., Baumgartner, G., Regli, F., Rabinowtiz, T., Gajdusek, D. C., and Gibbs, C. J.: Danger of accidental person-to-person transmission of Creutzfeldt-Jakob disease by surgery. Lancet 1:478, 1977.
Committee on Viral Hepatitis: Symposium on Viral Hepatitis. Am J Med Sci 270:2, 1975.
Doerfler, W.: Integration of viral DNA into the host genome. Curr Top Microbiol Immunol 71:1, 1975.
Duffy, P., Wolf, J., Collins, G., De Voe, A. G., Streeten, B., and Cowen, D.: Possible person-to-person transmission of Creutzfeldt-Jakob disease. N Engl J Med 290:692,1974.
Fenner, F.: Conditional lethal mutants of animal viruses. Curr Top Microbiol Immunol 48:1, 1969.
Fenner, F., and White, D. O.: Medical Virology, 2nd ed. New York, Academic Press, Inc., 1976.
Gajdusek, D. C.: Unconventional viruses and the origin and disappearance of kuru. Science 197:943, 1977.
Haase, A. T., Stowring, L., Ventura, P., Traynor, B., Johnson, K., Swoveland, P., Smith, M., Britten-Darnall, M., Faras, A., and Morayan, O.: Role of DNA intermediates in persistent infections caused by RNA viruses. In Schlessinger, D. (ed.): Microbiology—1977, Washington, American Society for Microbiology, 1977, p. 478.
Holland, J. J., and Villarreal, L. P.: Persistent noncytocidal vesicular stomatitis virus infections mediated by defective T particles that suppress virion transcriptase. Proc Natl Acad Sci USA 71:2956, 1974.
Huang, A. S.: Defective interfering viruses. Ann Rev Microbiol 27:101, 1973.
Huang, A. S., and Baltimore, D.: Defective viral particles and viral disease processes. Nature 226:325, 1970.
Joseph, B. S., and Oldstone, M. B. A.: Immunologic injury in measles virus infection. II. Suppression of immune injury through antigenic modulation. J Exp Med 142:864, 1975.
Kimberlin, R. H. (ed.): Slow virus diseases of animals and man. Frontiers in Biology. Amsterdam, North-Holland Publishing Co., 1976.
Klein, G.: The Epstein-Barr virus. In Kaplan, A. S. (ed.): The herpesviruses. New York, Academic Press, 1973, p. 521.
Padgett, B. L., and Walker, D. L.: New human papovaviruses. Prog Med Virol 22:1, 1976.
Preble, O. T., and Youngner, J. S.: Temperature-sensitive viruses and the etiology of chronic and inapparent infections. J. Infect Dis 131:467, 1975.
Rima, B. K., and Martin, S. J.: Persistent infection of tissue culture cells by RNA viruses. Med Microbiol Immunol 162:89, 1976.
Scheid, A., and Choppin, P: Protease activation mutants of Sendai virus. Activation of biologic properties by specific proteases. Virology 69:265, 1976.
Simpson, R. W., and Iinuma, M.: Recovery of infectious proviral DNA from mammalian cells infected with respiration syncytial virus. Proc Natl Acad Sci USA 72:3230, 1975.
Stevens, J. G.: Latent herpes simplex virus and the nervous system. Curr Top Microbiol Immunol. 70:31, 1975.
ter Meulen, V., Katz, M., and Müller, D.: Subacute sclerosing panencephalitis: a review. Curr Top Immunol 57:1, 1972.
Valentine, R. C., Ward, R., and Strand, M.: The replication cycle of RNA bacteriophages. Adv Virus Res 15:1, 1969.
Walker, D. L.: Persistent viral infection in cell cultures. In Sanders, M., and Lennette, E. H. (eds.) Medical and Applied Virology. St. Louis, Missouri, Warren H. Green, Inc., 1968, p. 99
Weil, M. L., Itabashi, H. H., Cremer, N. E., Oshers, L. S., Lennette, E. H., and Carnay, L.: Chronic progressive panencephalitis due to rubella virus simulating subacute sclerosing panencephalitis. N Engl J Med 19:994, 1975.
Zhdanov, V. M.: Integration of viral genomes. Nature 256:471, 1975.

12 INTERFERON AND INTERFERENCE

Monto Ho, M.D.

The story of interferon illustrates how an obscure phenomenon originally of interest only to laboratory scientists can become important for the whole field of biomedical and clinical sciences (Finter, 1973; Ho and Armstrong, 1975; Metz, 1975). In the 1930s and 1940s, virologists were already familiar with "viral interference." They found that, for example, if virus A was injected into an experimental animal at the same time or shortly before virus B, the replication of virus B would be inhibited. Virus A, the "interfering virus," frequently interfered even if it was inactivated or made noninfectious by heat or ultraviolet irradiation. No satisfactory explanation for this phenomenon was provided until 1957 when Isaacs and Lindenmann did a classic experiment, described in Figure 1.

The experiment formed the basis of a simple but important discovery. This was that viruses can induce in cells an antiviral substance, interferon, which in turn inhibited the replication of the challenge virus. In many ways, interferon is an ideal antiviral substance. It is relatively nontoxic, and it is broadly effective against a large number of RNA and DNA viruses. Further, since interferon is produced by cells under "natural" conditions of viral infections in animals, it was thought that a novel antiviral defense system was discovered.

INDUCTION AND INDUCERS

Interferon is a cellular protein. In this respect it is like hormones, enzymes, antibodies, and other cellular proteins. Such proteins are usually made by specialized cells, but as far as we know all cells can make interferon if given the appropriate stimulus ("induced"). The interferon-making capacity of cells is ordinarily not expressed; it is "repressed." It is widely believed that when cells are "induced" to make interferon, this capacity is "derepressed." However, there is a curious aspect of "derepression" of interferon production. Usually in induced or derepressed protein synthesis, the presence of the inducer or derepressor guarantees continued production. In the case of interferon production, it is a "one-shot affair" no matter how much inducer is applied or for how long the inducer is present, as shown in Figure 2 (Finter, 1973).

This suggests that another regulatory mechan-

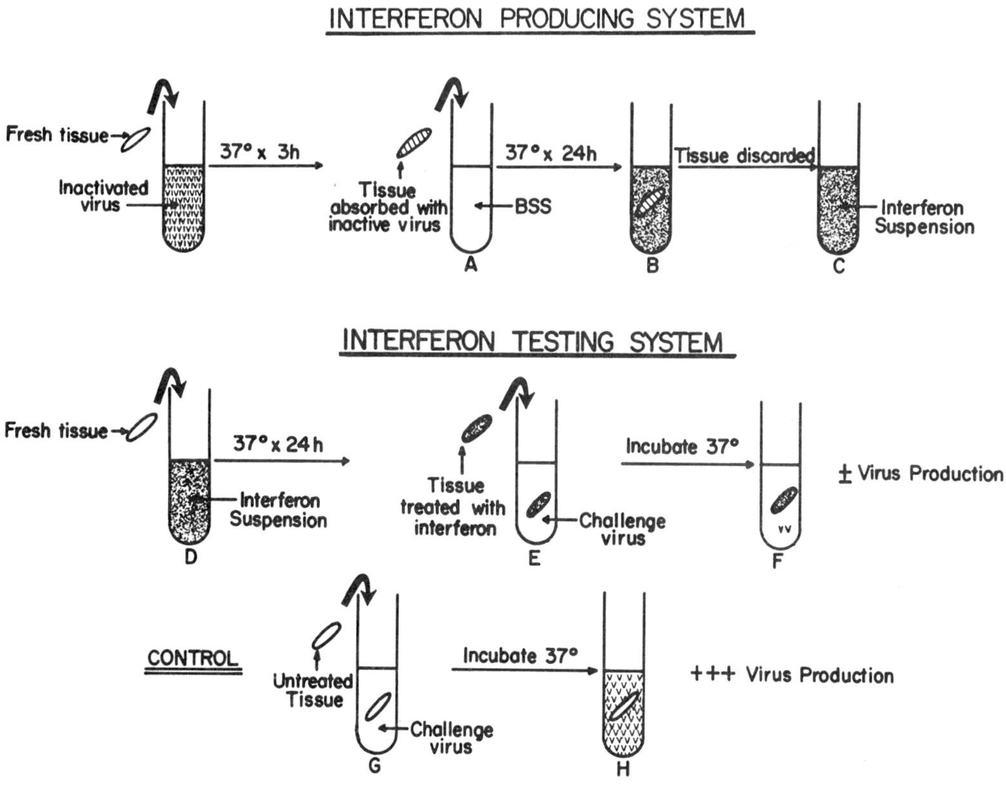

FIGURE 1. *Scheme of a system for producing and testing interferon (From M. Ho: Interferons N Engl J Med 266:1260, 1962).*

Fragments of chick chorioallantoic tissue are bathed in a suspension of inactivated Type A influenza virus (interfering virus). The tissue absorbed with the inactivated virus was incubated for 24 hours (A, B). A substance called "interferon" was released (C). Its presence was demonstrated in the "Interferon Testing System." Fresh tissue fragments are treated in the suspension containing interferon (D). The treated tissue (E) along with untreated controls (G) is inoculated with a "challenge virus." Virus production in treated and control tissue is measured. The degree of reduction in (F) as compared with (H) represents the potency of interferon. This testing system is still the conceptual basis for all methods of interferon assay.

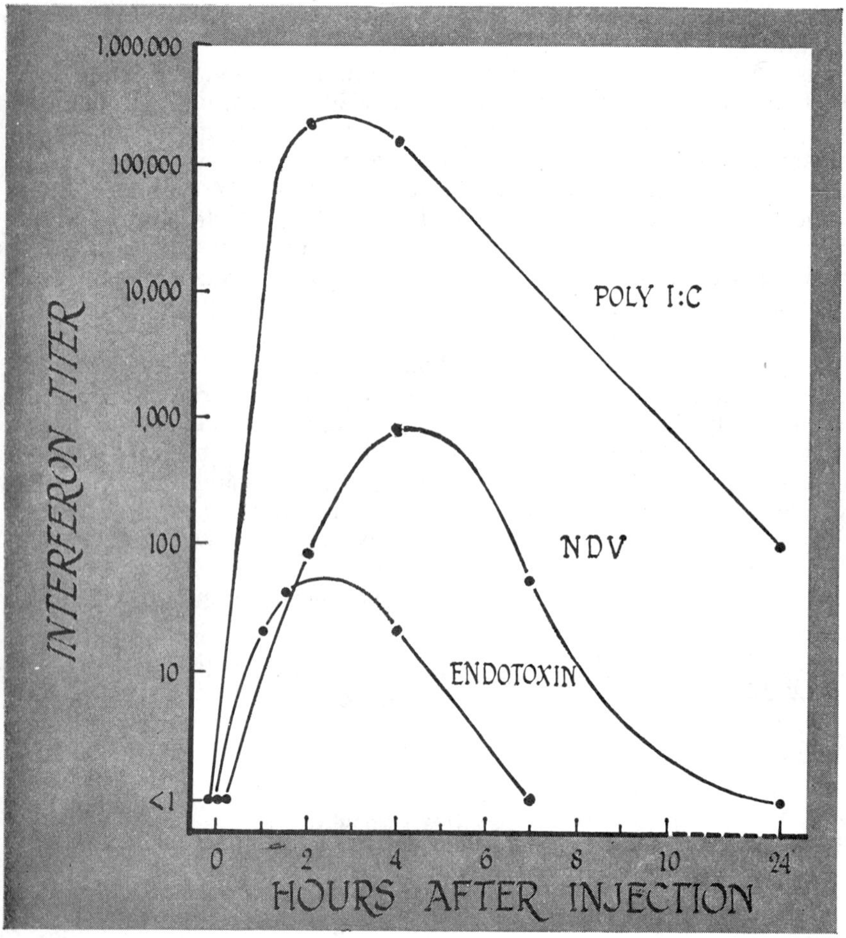

FIGURE 2. *The appearance of interferon in the serum of a rabbit after injection of* E. coli *lipopolysaccharide (endotoxin), Newcastle disease virus (NDV), or a complexed synthetic double-stranded ribonucleic acid, polyriboinosinic-polyribocytidylic acid (poly I:C). Note the marked variation in the effectiveness of these three inducers, and that maximum interferon levels are only transiently maintained. (From Ho, M.: Factors influencing the interferon response. Arch Int Med 126:136, 1970.)*

ism in cells that inhibits interferon synthesis is triggered off by the inducer. Cells in culture as well as animals cannot respond to an interferon inducer for varying intervals after being induced. This "hyporeactive" or "refractory" period varies from hours to days. In animals it may be as long as one week and is a serious deterrent to repeated administrations of inducers to achieve steady interferon concentrations in tissues.

There are many substances that induce interferon. They are classified in Table 1 (Ho and Armstrong, 1975). Two types of interferon production may be distinguished. Type 1 follows stimulation with specific, known inducers. These inducers are divided into two classes, A and B.

Class A inducers are good inducers that induce the production of 1000 units/ml or more of interferon in cell culture or in the bloodstream of animals. They contain double-stranded RNA, or may form such RNA in the course of cell infection. Newcastle disease virus and poly I:C (see Fig. 2) are examples.

Class B inducers are moderate to poor inducers that may act only when injected in animals. They are a motley group of macromolecular substances often without effect in fibroblast or epithelial cell cultures. Some are microorganisms that infect intracellularly. It is assumed that uptake by macrophages is necessary in induction by many of

TABLE 1. Types of Interferon Production and Inducers

I. Type I Interferon Production: Stimulated by Inducers
   A. Class A Inducers (good inducers, double-stranded RNA):
      1. RNA animal viruses
      2. DNA animal viruses
      3. Plant, insect fungus and bacterial viruses
      4. Natural and synthetic double-stranded RNA's
   B. Class B Inducers (moderate to poor inducers):
      1. Intracellular organisms (bacteria, rickettsia, protozoa)
      2. Bacterial products (lipopolysaccharides, polysaccharides)
      3. Polymers (polycarboxylic, polysulfates, polyphosphates)
      4. Low molecular weight substances (tilorone, cycloheximide, etc.)

II. Type II Interferon Production: Specific and Nonspecific Immune Induction
   A. Immune Specific Induction
   B. Nonspecific Stimulation of T or B Lymphocytes by Mitogens

these substances. Some members of this class may eventually relate to class A inducers or to Type II production.

Type II is the production of interferon by specific and nonspecific immunologic activation of lymphocytes and macrophages. Although this important phenomenon produces only moderate amounts of interferon, it may be an effector limb of cell-mediated immunity.

During immune specific induction, interferon is produced by sensitized lymphocytes upon exposure to specific antigen. The sensitizing antigen may be a viral antigen, or nonviral antigens such as PPD, diphtheria, or tetanus toxoid. Immune induction may occur in vitro or in the animal. A most effective method is to stimulate circulating interferon in mice infected with BCG by an intravenous injection of BCG or PPD.

Interferon is also produced when lymphocytes are stimulated by nonspecific mitogens such as phytohemagglutinin, pokeweed antigen, and concanavalin A. The responding lymphocyte is the one specifically stimulated by the mitogen.

PROPERTIES OF INTERFERON

Interferon is a heterogeneous class of proteins with unique biologic properties. Biochemically it is still inadequately characterized.

Its definitive biologic effect is its ability to inhibit intracellular replication of viruses. In addition it has a host of other less apparent but rather bewildering biologic properties. Interferon can inhibit the multiplication of normal and cancerous cells, act as an immunologic adjuvant or depressant, enhance or inhibit the production of interferon itself, enhance the toxicity of polyribonucleotides for cells, and enhance the phagocytic effect of phagocytes. The significance of these heterogeneous effects is still unclear, but at this time they appear insufficient to render interferon toxic. In any case, interferon can no longer be assumed to be a pristine antiviral substance. Eventually, its other effects may explain its efficacy or toxicity, depending on whichever action one may be observing. The antitumor effect of interferon is potentially of great therapeutic interest.

Another cardinal property of interferon is its so-called "species," or actually genus, specificity. That is, its antiviral effect is observed only in cells of the same species or genus as the cells from which the interferon was made. For example, chick cell interferon acts only on chick cells or other fowl cells but not human cells. To obtain interferon effective for human cells or for human beings, it is necessary to produce interferon in human or primate cells. This implies interferons from different species vary biochemically, but such chemical differences have not been worked out.

The purification of interferon has been an elusive objective. Almost every physicochemical method used to study interferon has revealed a new dimension of heterogeneity. Interferon, even from one species, is heterogeneous with respect to molecular weight, isoelectric point, carbohydrate content, physical stability, and polyacrylamide gel electrophoresis. That is, more than one type of molecule is always obtained.

Immunologic analyses have recently revealed another dimension of the heterogeneity of interferon. Antibody affinity column chromatography has also been a powerful tool in interferon purification. For example, it has been shown that antibody made against human fibroblast interferon neutralizes the homologous interferon but it has no effect on leukocyte interferon. Conversely, antibody made against leukocyte interferon neutralizes both fibroblast and leukocyte interferon. This suggests that human leukocyte interferon possesses both "L" and "F" antigens, whereas fibroblast interferon possesses only "F" antigen. Apparently the type of cell producing interferon is one of the determinants of the characteristics of the interferon produced. Other work suggests that the mode of induction also plays a role. Interferon produced by the Type I method (Table 1) is antigenically different from interferon production by Type II, or "immune," induction. At this time we still do not understand why there are so many different interferons, or how it is explained genetically or developmentally. Apparently, in the case of human interferon, at least two chromosomes, 2 and 5, are involved in interferon production. But we have no idea whether the different interferons are variations of one basic gene product or of many.

MECHANISM OF ACTION OF INTERFERON

Very little is understood about the actions of interferon, except its antiviral effect. It was realized soon after its discovery that interferon does not directly inactivate viruses, nor does it affect their adsorption on cells. The antiviral property of interferon was found to be eliminated if cell RNA or protein synthesis was inhibited. The interpretation was that interferon, in order to be active, must induce in the cell another protein that is the active antiviral substance. This explains why interferon must first incubate with the cell to create an "antiviral state." Once this state is created, the presence of interferon is no longer required, in striking contrast to other antimicrobial substances, whose activity usually requires their constant presence. It is also possible that the full expression of the antiviral state induced by interferon requires a virus infection.

Another piece of indirect evidence suggesting that interferon itself is not antiviral, and that an intermediary substance is needed comes from somatic hybrid studies. In cultures of fused hybrid human and mouse cells, in which specific human chromosomes are lost, the sensitivity to interferon (or the ability of interferon to induce the antiviral state) is lost in cells without chromosome G 21. Presumably this chromosome has a gene that facilitates the expression of interferon, perhaps by coding for the antiviral protein.

Little is known about the "actual" antiviral substance or how it acts. Indeed, its existence is still largely inferential, although some studies suggest its isolation in cell-free systems. Interferon has been thought to act at the transcriptional or translational level, and there are experiments to support each thesis. These two views are not necessarily exclusive. It is possible that with further advances in the state of the art both in molecular virology and in the study of interferons, these two views may be reconciled or subsumed by a more basic mechanism of action.

Inhibition of transcription was demonstrated by two types of experiments. First, interferon was found to inhibit the synthesis of early RNA in SV 40–infected cells. Such early RNA includes transcripts of the SV 40 genome, which code for T antigen and other early viral proteins, also found to be inhibited by interferon. Second, a number of RNA viruses (such as vesicular stomatitis virus) contain, as part of the virus, an RNA polymerase. After infection, transcription of the viral RNA by the polymerase occurs. This transcript, a product complementary to the viral RNA, has also been reported to be inhibited in interferon-treated cells.

A number of experiments suggest that translation of viral messenger RNA is inhibited in interferon-treated cells. In cell-free systems containing ribosomes, transfer RNA's, elongation factors, messenger RNA, and labeled amino acids extracted from interferon-treated cells, translation of viral messenger is not supported as well as it is in extracts from untreated cells (Fig. 3). In some studies, certain nonviral messengers, such as rabbit globin messenger RNA, may not be translated either. The question arises as to how specific is the antiviral effect of interferon. Perhaps interferon, in view of its other effects, such as inhibition of normal cell multiplication, is a metabolic regulator of broader significance.

ROLE OF INTERFERON IN HOST DEFENSE

Interferon is a novel antiviral substance produced in the course of natural virus infections, but its relative importance among specific and nonspecific antiviral defenses is uncertain. The missing link in our understanding is that there is no "experiment in nature" in which a viral infection may be observed in an animal whose capacity to make interferon or react to it is genetically deficient. Analogous congenital deficiencies in humoral and cellular immunity are well known and highly instructive. The best evidence that interferon probably plays some role in host defense is that viral infections in mice run a more lethal course if interferon formed is neutralized by specific antiserum against interferon (Gresser et al., 1976).

The discovery that lymphocytes activated by mitogens and sensitized lymphocytes exposed to specific antigens produce interferon adds another, although uncertain, dimension to the role of interferon in host defense. Interferon may be regarded as one of the many lymphokines, such as migration inhibition factor, which help explain how cellular immunity works. A number of viruses, particularly the Herpetoviridae, produce more frequent and more severe infections when

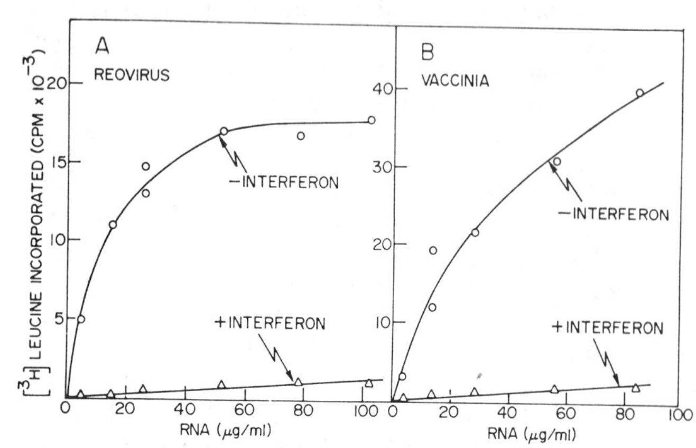

FIGURE 3. *The translation of reovirus and vaccinia virus messenger RNA by cell-free fraction (S10) of Krebs II ascites tumor cells (a form of suspended mouse tumor cell) pretreated or untreated with interferon. Interferon pretreatment inhibits translation of these two viral proteins, but not of normal cell mRNA. (From Samuel, C. E., and Joklik, W. K., Virology 58: 483, 1974.)*

cellular immunity is suppressed. One method by which cellular immunity is assumed to work is by release of interferon from sensitized T lymphocytes that come in contact with viral antigen. This mechanism should be particularly effective in the lesions, such as the vesicles of herpes simplex or herpes zoster, where there is an abundance of viral antigens and inflammatory lymphocytes. Interferon is produced in such lesions, but what would happen if interferon were not produced is less clear. Herpes zoster is reported to disseminate more frequently when the interferon content of vesicles is low in patients who have lymphomas and are on chemotherapy. But immunosuppression by disease or chemotherapy has not been shown to reduce interferon formation.

THERAPEUTIC ROLE AS AN ANTIVIRAL SUBSTANCE

There are two ways to use the interferon system therapeutically. One may administer either an inducer of interferon or interferon itself. In animals the administration of "inducer" may be effective against viral infections. "Hyporeactivity" to repeated induction is an intrinsic problem of this method. It limits the efficacy of repeated administration of inducer and places a limit on the amount of interferon that can be induced, but it is not insurmountable. A more serious problem has been the toxicity of practically every inducer that has been carefully studied. Some of the best and chemically purest inducers, such as synthetic double-stranded polyribonucleotides, are prohibitively toxic. The search for nontoxic inducers, either by molecular manipulation of polyribonucleotides or by search for new types of inducers, continues.

Although interferon has properties of an ideal antiviral substance, it is still not available in the physician's armamentarium, more than 20 years after its discovery, for two main reasons. First, interferon does not eradicate viral infections. It is unusually potent, since as little as 0.001 μg possesses detectable antiviral activity, but its antiviral activity at any dose is incomplete so that it cannot cure or prevent a viral infection even with relatively large amounts. Any clinical effect requires large amounts of interferon. Second, interferon is species specific and difficult to prepare in large amounts, especially human interferon, but this technical problem can be solved.

Studies in experimental animals suggest that interferon should be more effective as a prophylactic than a curative agent. In humans it has been shown that when relatively large amounts are sprayed in the nose, subsequent infection by a rhinovirus may be ameliorated. The severity of established herpes zoster in Hodgkin's disease can also be lessened by interferon. Eventually, interferon may be useful in prevention of a widespread epidemic viral disease, or in treatment of severe infections.

Two observations suggest other unexpected possibilities for the use of interferon. One is the report that interferon reduced hepatitis B virus activity in patients with chronic aggressive hepatitis (Greenberg et al., 1976). Thus, interferon may even have a place in the therapy of chronic viral infection. Second, interferon has been reported to retard the progression of osteogenic sarcoma in humans. It is unclear whether this effect is due to the antiviral or some other effect of interferon. It is also possible that the antitumor effect is due to some impurity from the white cells from which the interferon was made. These questions can be answered only by further studies.

OTHER TYPES OF VIRAL INTERFERENCE

There are three types of viral interference that are not mediated by interferon.

Receptor Interference

One type of interference is based on competition for specific viral receptors. Infection by a myxovirus, such as Newcastle disease virus, may be interfered with if cell receptors are occupied or blocked by a prior infection. Another example of this type of interference may be found in the avian leukoviruses. It is used to study the relatedness of these agents and their titration. For example, inapparent infection of chick cells by one leukovirus will abrogate cellular receptors for this and related agents, so that a superinfection by a related transforming or cytocidal agent will be interfered with.

Intracellular Homologous Interference

Another type of interference involves competition for limiting factor *within cells* among homologous or homotypic viruses. Examples are known among many viruses, such as influenza virus, vesicular stomatitis virus, and poliovirus. The best studied example is the ability of defective particles to interfere. These defective interfering particles (DIP) explain "autointerference" and may have a modifying role on viral infections. They are discussed in Chapter 11.

Interference by Heterologous Agents

This is also intracellular and may be based on repression of transcription or translation. One example is so-called "intrinsic interference," which depends on a protein (not interferon) specified by the interfering virus. A number of viruses, such as rubella virus or measles virus, may interfere in chick embryo fibroblast cultures with the replication of Newcastle disease virus. This type of interference may be used to titer noncytopathic or nonreplicating virus.

References

Finter, N. B.: Interferons and Interferon Inducers. New York, American Elsevier Pub. Co., Inc., 1973, p. 598.

Greenberg, H. B., Pollard, R. B., Lutwick, L. I., Gregory, P. B., Robinson, W. S., and Merigan, T. C.: Effect of human leukocyte interferon on hepatitis B virus infection in patients with chronic active hepatitis. N Engl J Med 295:517, 1976.

Gresser, I., Tovey, M. G., et al.: Role of interferon in the pathogenesis of virus diseases in mice as demonstrated by the use of anti-interferon serum. I. Rapid evolution of encephalomyocarditis virus infection. J Exper Med 144:1305, 1976.

Ho, M., and Armstrong, J. A.: Interferon. Ann Rev Micro 29:131, 1975.

Isaacs, A., and Lindenmann, J.: Virus interference. I Interferon Proc Roy Soc B, 147:258, 1957.

Metz, D. H.: Interferon and interferon inducers. Adv Drug Res 10:101, 1975.

Samuel, C. E., and Joklik, W. K.: A protein synthesizing system from interferon-treated cells that discriminates between cellular and viral messenger RNAs. Virology 58:476, 1974.

Mycology

13 MORPHOLOGY AND STRUCTURE OF FUNGI

Vladimir Farkaš, Ph.D.

The fungi (Latin: *fungus,* mushroom) can be defined as eukaryotic, thallus-forming organisms *(Thallobiota)* that lack both chlorophyll and chemolithotrophic machinery. Consequently, they can produce structural elements neither from carbon dioxide via photosynthesis nor from inorganic matter. For this reason they depend on external sources of organic carbon. The fungi can live as saprophytes on dead organic matter or as parasites on other organisms, mainly plants, but also animals and man. (For the taxonomic position of fungi, see Chapter 14).

The vegetative body of the fungus, the *thallus,* is never differentiated into roots, stem, and leaves, and the fungi have no specialized vessels for transport of water and nutrients as in vascular plants. The fungi can form only a plectenchymatic type of "tissue" consisting of aggregates of cells largely retaining their individuality. In certain cases, as in sexual processes, the cells can communicate by means of hormone-like compounds.

In the vegetative phase the fungal population grows essentially as an undifferentiated group of similar cells. The differentiation occurs in the second developmental phase, known as the reproductive or fruiting phase. The hyphae that penetrate into the growth medium or spread over its surface absorbing the nutrients are called *vegetative mycelium.* The vegetative mycelium projects above the surface of the medium as *aerial mycelium* bearing reproductive structures. Some parasitic fungi produce special branches of the vegetative mycelium called *haustoria* (s. haustorium), which penetrate into the cytoplasm of host cells and absorb the necessary nutrients.

REPRODUCTION

The fungi usually reproduce by means of motile or nonmotile *spores;* however, in many cases even small fragments of viable mycelium are sufficient to initiate growth of a new individual. The spores can be one-celled, two-celled, or many-celled. They are derived from the parent thallus and contain all the genetic information necessary for the development of a new thallus. The shapes of spores and the mode of sporulation are important characteristics in classification and identification of fungi (Fig. 1).

The spores can arise either asexually by differentiation of the thallus or sexually as a result of fusion of parent haploid nuclei. This is followed by formation of a sexual zygote and by restoration of the haploid state by meiosis. The sexual process is initiated by fusion of pairs of motile or nonmotile sexual cells *(gametes)* of opposite sexes or by somatic copulation of undifferentiated vegetative hyphae (Fig. 2).

MORPHOLOGY

Morphologically, the fungi represent a very heterogeneous group of microorganisms. Some fungi grow as single cells *(yeasts),* others as multinuclear filaments *(molds).* Certain fungi, including many human pathogens, exhibit dimorphism — that is, they can grow as yeast or as mold (Fig. 3). In deep infections the yeast form predominates, whereas in superficial infections of keratinized skin layers the mycelial form of fungi is more common. The biochemical nature of dimorphism is poorly understood. The yeast–mold transitions of the fungal phenotype can be influenced by temperature, nutrition, redox potential, partial pressure of carbon dioxide, and other environmental factors. The two morphologic types of the same organism are distinctly different in the chemical composition of their cell walls.

Yeasts

Yeasts are unicellular fungi that reproduce by budding or fission; the latter can be considered as a broad-base type of budding. Yeast cells are spherical or ellipsoidal, usually about 3 to 10 μm in diameter and 5 to 30 μm in length. Different species of yeast differ in their mode of budding (apical, lateral, bipolar, multipolar) and in other details. Yeast cells grow by a somewhat uniform extension of the cellular surface of the growing bud. During budding the nucleus of the mother cell divides, and one daughter nucleus passes into the bud. The two cells are then separated by a crosswall, called a *septum,* and after a certain period the bud breaks away. A birth scar is visible on the daughter cell, and a prominent bud scar remains on the mother cell's wall surface. The chemical composition of the bud scar with the septum usually differs from that of the rest of the cell wall. The bud scars remain as permanent

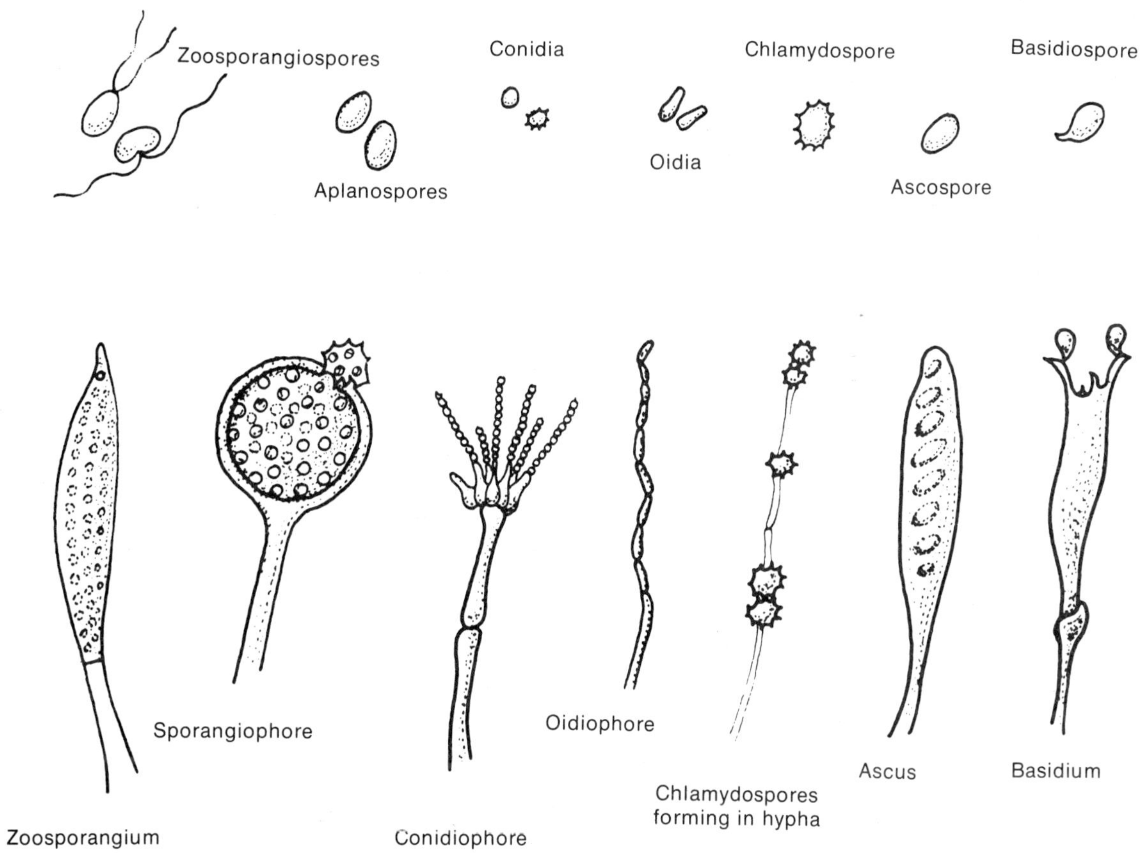

FIGURE 1. *Different types of reproductive structures and spores in fungi. (After Encyclopedia of Science and Technology. Vol. 5. New York, McGraw-Hill Book Company, 1971, p. 117.)*

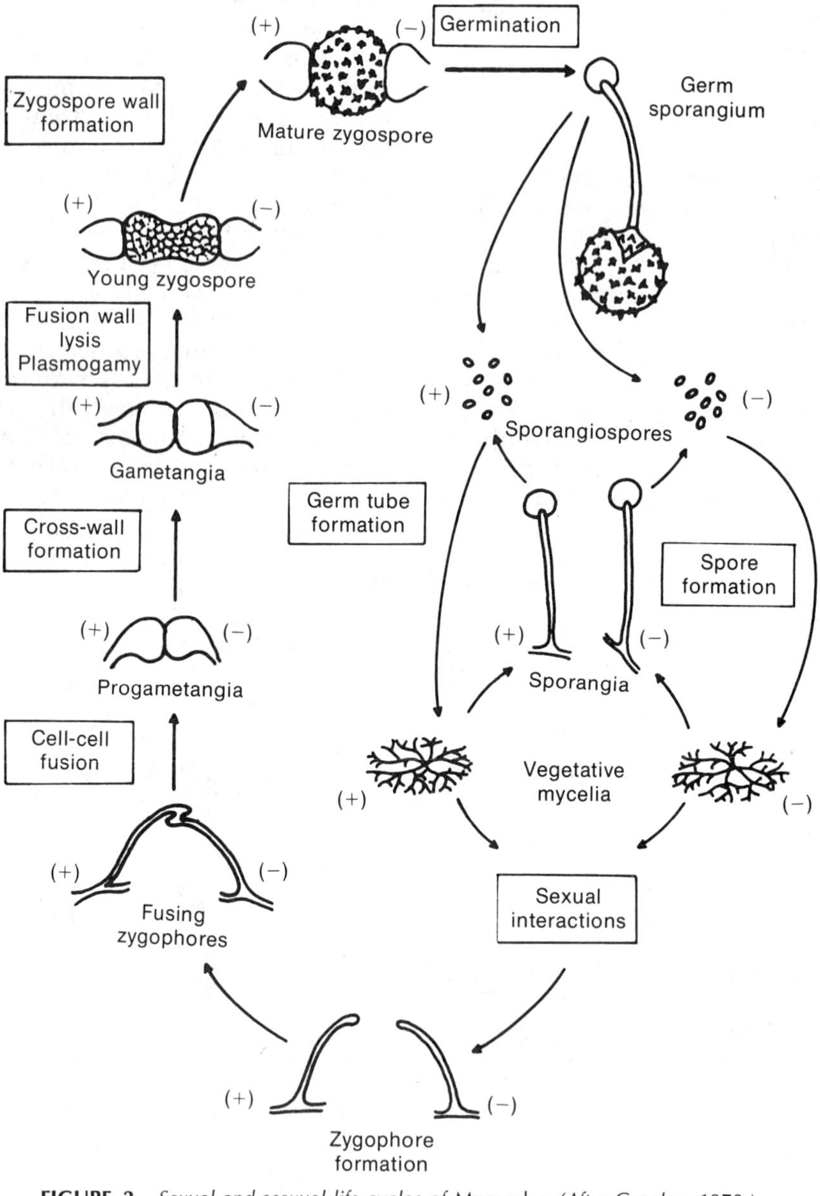

FIGURE 2. *Sexual and asexual life cycles of* Mucorales. *(After Gooday, 1973.)*

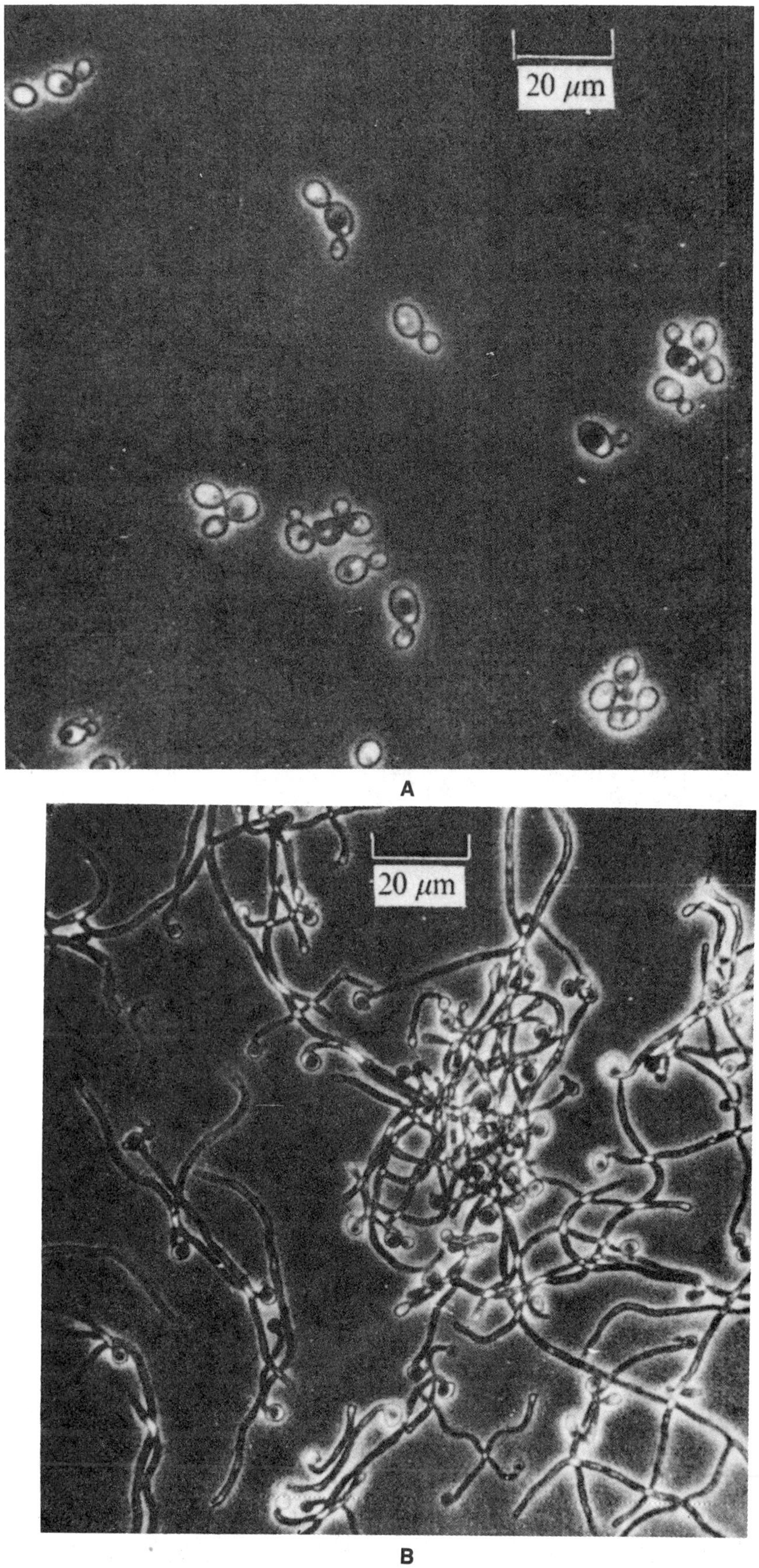

FIGURE 3. *Yeast* (A) *and mycelial* (B) *forms of* Candida albicans. *(From Mariott, M. S.: J Gen Microbiol 86:115, 1975.)*

structures on the surface of the walls, and their number indicates the age of the cells (Fig. 4).

Molds

The principal living unit of a mold is a *hypha*. Hyphae are about 3 to 12 μm diameter and can reach several centimeters in length. The hyphae grow by elongation at the tip *(apex)*. In older portions of the hypha lateral branches arise that then develop like the main hypha. The densely packed, interwoven hyphae constitute web-like *mycelia*, which are visible as macroscopic fungal *colonies* (Fig. 5).

In lower fungi the hyphae are usually not divided by septa into individual cells, but rather the whole mycelium is coenocytic — that is, it contains many nuclei in nonseparated cytoplasm. Regular septation observed in higher fungi does not change the coenocytic character of the hyphae because the septa have fine pores that ensure the continuity of the cytoplasm.

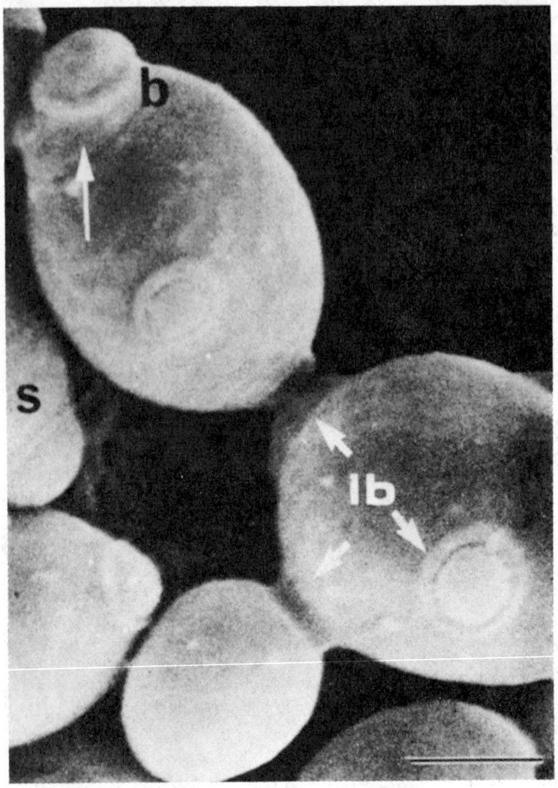

A

FIGURE 4. A, *Scanning electronmicrograph of the cells of* Candida slooffii *showing lateral budding (1b), bud scar (bs) and a birth scar (s). Bar indicates 1 μm. (From Watson, K., and Arthur, H.: J Bacteriol 130:312, 1976.)*

B, *Electronmicrograph of an ultrathin section through the cell of* Cryptococcus neoformans. *C, capsule; CW, cell wall; M, mitochondria; N, nucleus; V, vacuole; B, bud. Bar indicates 1 μm. (From Peterson, E. M., Hawley, R. J., and Calderone, R. A.: Can J Microbiol 22:1518, 1976.)*

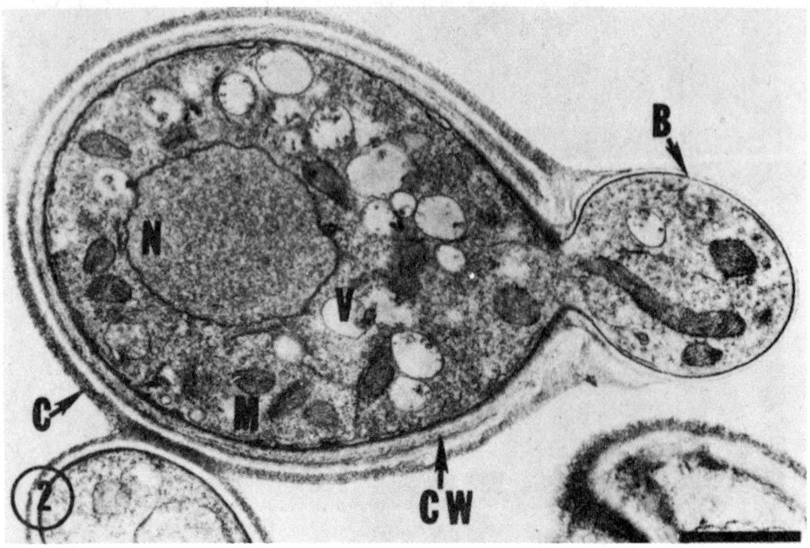

B

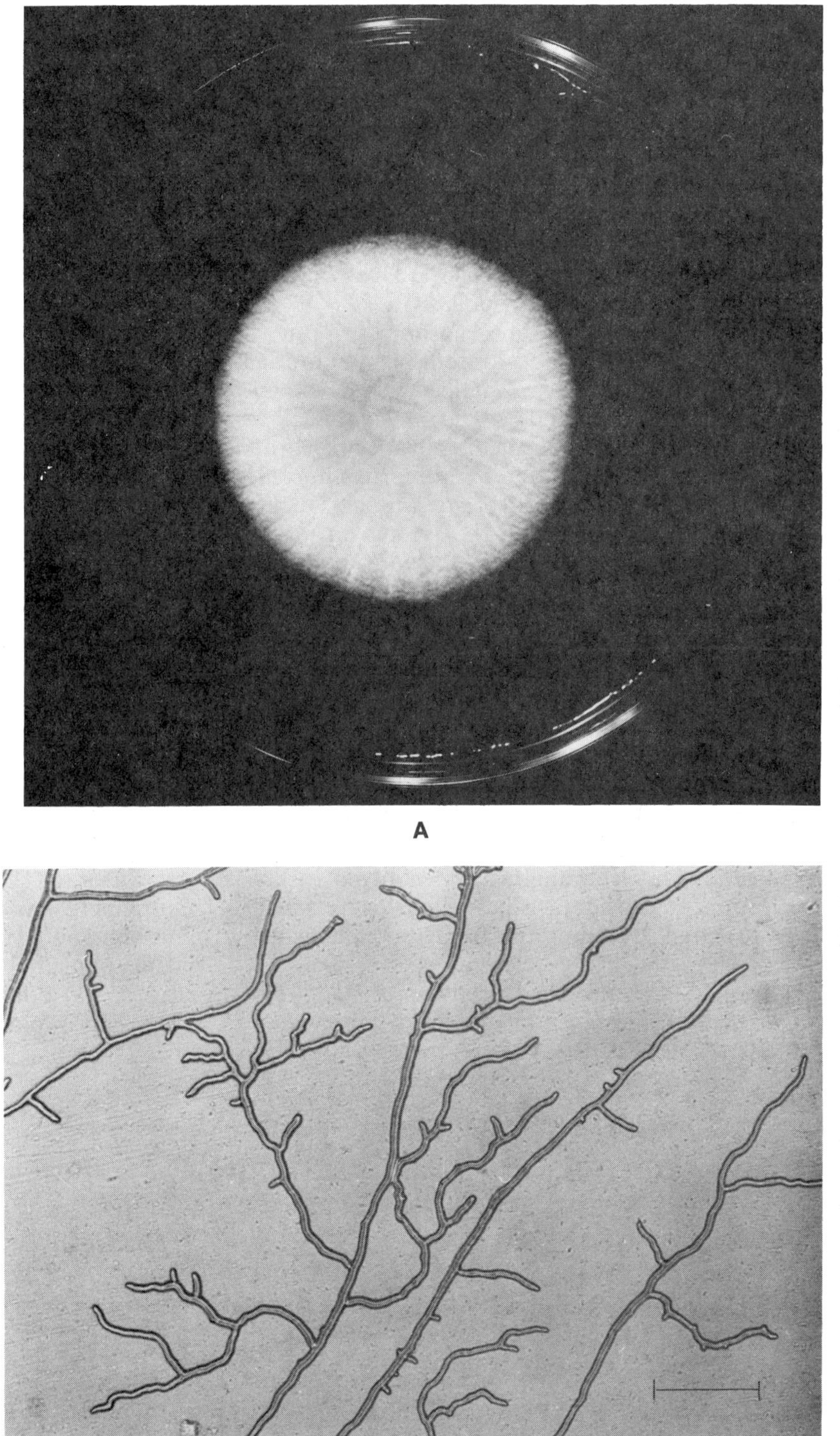

FIGURE 5. A, *Colony of the mold* Aspergillus niger. B, *Microphotograph from the margin of the same colon showing individual hyphae. Bar represents 50 μm.*

ULTRASTRUCTURE

In the anatomic complexity of their cells the fungi resemble plants and animals. Unlike bacteria, the fungi possess a "true" nucleus that is typical of eukaryotic cells; it contains different chromosomes separated from the cytoplasm by a nuclear envelope and a mitotic apparatus within the nucleus that ensures the separation of duplicated chromosomes into two identical parts during mitosis. The quantity of DNA per cell varies from four to ten times the amount present in bacterial cells. This figure can be compared with the 1000 to 10,000-fold increase of DNA content observed in plants and animals.

Electron microscopic observations on fungal cells reveal a complicated system of membranes and membrane-bound organelles within the cytoplasm (see Figs. 4B and 6). The *nuclear envelope* consists of two parallel unit membranes perforated at their mutual contacts by numerous *nuclear pores*. Nuclear pores assure communication between the nucleoplasm and the cytoplasm.

Closely associated with the nuclear membrane is the complicated system of membranes called the *endoplasmic reticulum*. The membranes of the endoplasmic reticulum serve as the attachment sites for *ribosomes* that are involved in protein synthesis. The endoplasmic reticulum also channels different components within the cytoplasm and divides the cytoplasm into compartments. The closely packed membranes of the endoplasmic reticulum that are free of ribosomes form special cisternae termed *dictyosomes* or the *Golgi apparatus*. These organelles are involved in glycosylation of peptide acceptors and excretion of prefabricated glycoproteins from the cell (exocytosis).

Vacuoles are spherical vesiculoid structures enclosed by a unit membrane. They can contain water with dissolved solutes, gases, polymetaphosphate granules (volutine), waste products of cellular metabolism, and various hydrolytic enzymes used for breakdown of polymeric substances. Small vacuolar bodies derived from the membranes of the endoplasmic reticulum or from the cisternae of the Golgi apparatus can convey prefabricated glycoproteins to the cell exterior.

Metabolic energy is produced in *mitochondria* in the form of adenosine 5′-triphosphate (ATP) through respiration-mediated oxidative phosphorylation of adenosine 5′-diphosphate (ADP). The mitochondria, about 1 μm in diameter and several micrometers in length, are ubiquitous organelles of all eukaryotic cells.

Besides the principal organelles described above, the fungal cytoplasm can contain granules of reserve materials (lipids and glycogen) and various other inclusions.

The cytoplasm is surrounded by a single unit membrane, about 8 to 9 nm thick, called the *cytoplasmic membrane* or *plasmalemma*. The cytoplasmic membrane fulfills many functions in mediating communication between the cytoplasm and the cell exterior. Among these functions are the diffusion of solutes and nutrients and the passive and active carrier transport of amino acids and sugars. It is also an attachment site of enzymes involved in the biosynthesis of the cell wall. The cytoplasmic membrane, like other cellular membranes, is composed of phospholipids, proteins (or glycoproteins) and sterols. The presence of sterols distinguishes the membranes of eukaryotic cells from bacterial membranes. By association of their lipophilic portions, the phospholipids form a lipid bilayer or sandwich, whereby the hydrophilic portions of the participating phospholipid molecules face out on both sides of the bilayer. The phospholipid bilayer is broken up in places by globular protein (or glyco-

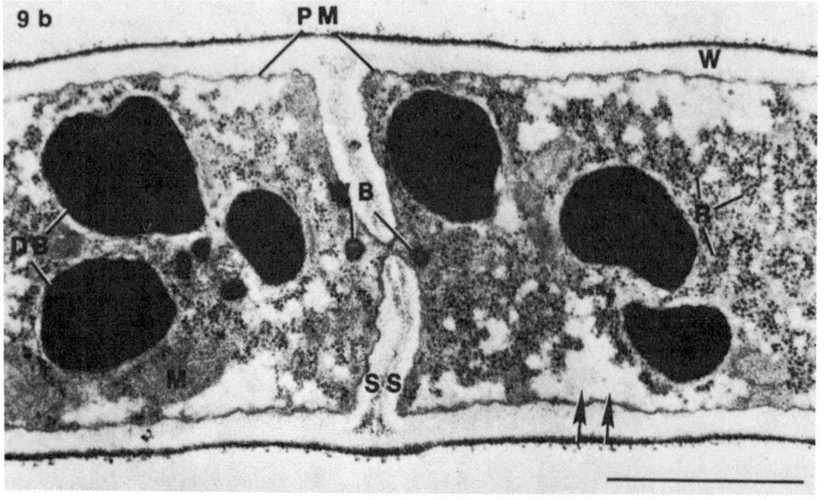

FIGURE 6. *Electronmicrograph of sectioned hyphal cell of* Phialophora dermatitis. *W, cell wall; R, ribosomes; DB, dense bodies; WB, Woronin bodies; M, mitochondria; SS, single septum. Double arrows indicate polysaccharide storage areas. Bar indicates 1 μm. (From Oujezdsky, K. B., Grove, S. N., and Szaniszlo, P. J.: J Bacteriol 113:468, 1973.)*

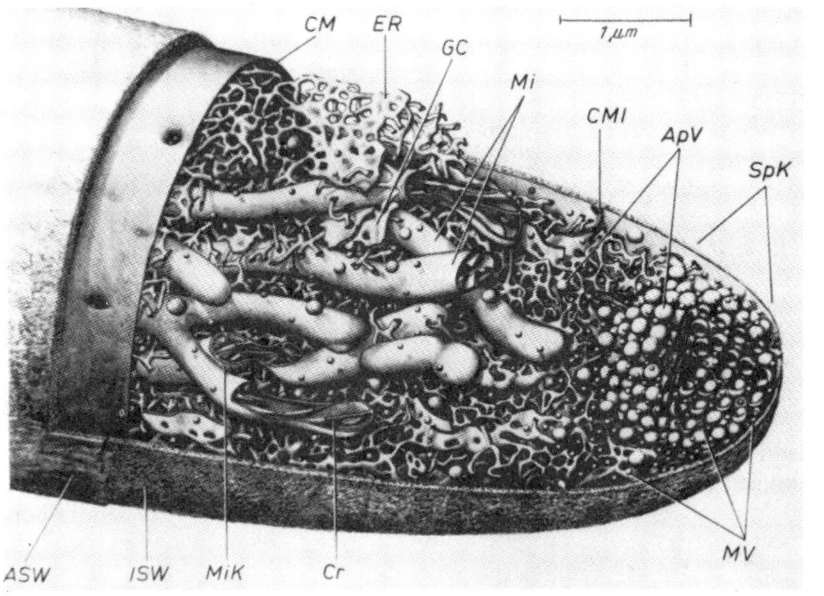

FIGURE 7. *A three-dimensional model of the apical region of a hypha of* Polystictus versicolor. *ApV, apical vesicles; ASW, outer wall layer; CM, plasmalemma; CMI, invaginations of the plasmalemma (lomasome); Cr, mitochondrial cristae; ER, endoplasmic reticulum; GC, Golgi cisternae; ISW, inner fibrillar wall layer; Mi, mitochondria; MiK, mitochondrial membrane; MW, microvesicle; SpK, Spitzenkörper. (From Girbardt, M.: Protoplasma 67:413, 1969.)*

protein) molecules, which are completely or partially immersed in the lipids. The "fluid mosaic" model of the cytoplasmic membrane (Singer and Nicolson, 1972) presumes in extreme cases a free lateral movement of the protein globules in the lipid bilayer.

The space between the cytoplasmic membrane and the cell wall, called the *periplasmic space,* contains soluble macromolecular precursors of the cell wall and various extracellular hydrolases that cleave the oligomeric substrates to monomers before their transport into the cell. A three-dimensional model of a fungal cell is shown in Figure 7.

CELL WALLS

In the great majority of fungi the outer surface of the cytoplasmic membrane is covered by a rigid cell wall. The primary role of the cell wall is to determine and to maintain cellular morphology. Dissolution of the cell walls by special lytic enzymes liberates osmotically fragile protoplasts that are shaped spherically by hydrostatic forces. Under suitable conditions (osmotic stabilizers, nutrition) the protoplasts can regenerate the cell wall on their surface. Lowering the osmolarity of the medium causes lysis of the protoplasts and kills them.

Besides their morphogenetic role, the cell walls mediate the interactions of the fungus cells with their environment, other cells, and the infected host.

Chemical Composition

The fungal walls are composed of polysaccharides, protein-polysaccharide complexes, variable amounts of lipids, and lesser quantities of other components.

Polysaccharides, which represent about 80 to 90 per cent of the dry matter of the cell walls, are composed of amino sugars, hexoses, hexuronic acids, methylpentoses, and pentoses (Bartnicki-Garcia, 1970). Owing to their distinctive physicochemical properties, the different polymers fulfill specific functions in the cell walls. The crystalline, insoluble polysaccharides, such as cellulose, chitin, and beta-glucans, form the wall skeleton that is responsible for the mechanical strength and morphology of the walls. The amorphous homo- and heteropolysaccharides, often in association with proteins, act as cementing substances, constitute the carbohydrate moieties of extracellular enzymes, and form the immunodeterminant groups of wall antigens.

Wall Architecture

The organization of individual components within the wall can be different in various genera of fungi and can vary with the age of the cells. The older parts of the walls are usually thicker, more rigid, and more resistant to hydrolytic enzymes.

Refined electron microscopic techniques coupled with cytochemical staining and selective action of purified polysaccharide-hydrolases have revealed that several layers of building material exist within the fungal cell wall. The general picture is that the outer surface of the wall is smooth or slightly granular in texture and is composed of amorphous glycoprotein material, whereas the skeletal, microcrystalline component is prominent in the layer adjacent to the cytoplasmic membrane (Fig. 8). The spaces between the fibrils of the microcrystalline layer, the *wall matrix,* are filled with an amorphous component that also penetrates the periplasmic space. Some yeasts (e.g., *Cryptococcus* sp.) produce viscous polysaccharide capsules on the outer surface of the cell wall. The chemical nature of the capsular slime closely resembles that of the protein-polysaccharides from the cell walls. A great portion of the capsular material is released into the medium during growth.

STRUCTURE AND BIOSYNTHESIS OF CELL-WALL COMPONENTS

The polysaccharides of the fungal cell walls are polymerized from their activated monomers, nucleoside 5′-diphosphate sugars (NDP-sugars; Fig. 9) under the catalytic action of corresponding glycosyl transferases. The general equation for this process is:

$$\underset{\text{donor}}{\text{NDP–sugar}} + \underset{\text{acceptor}}{(\text{sugar})_n} \xrightarrow{\text{glycosyl transferase}} \underset{\text{product}}{(\text{sugar})_{n+1}} + \text{NDP}$$

The result of the transglycosylic reaction is the lengthening of the saccharidic chain of the acceptor molecule by one glycosyl unit. A single carbohydrate unit as well as the product from the preceding reaction can serve as the acceptor. The multiple repetition of this process leads to formation of long chains of glycosyl units linked by glycosidic bonds.

In the biosynthesis either of heteropolysaccharides containing different carbohydrate units, or of polymers containing different glycosidic bonds, any of several glycosyl transferases may catalyze the transfer of a specific sugar and the formation of a specific glycosidic bond. The following are examples of biosynthesis of some principal cell-wall constituents in fungi.

FIGURE 8. *Electronmicrograph of an isolated empty hyphal wall of* Pythium accanthicum *showing granular outer surface (OS) and microfibrillar inner surface (IS) of the cell wall. (From Siestma, J. H., Child, J. J., Nesbitt, L. R., and Haskins, R. H.: J Gen Microbiol 86:29, 1975.)*

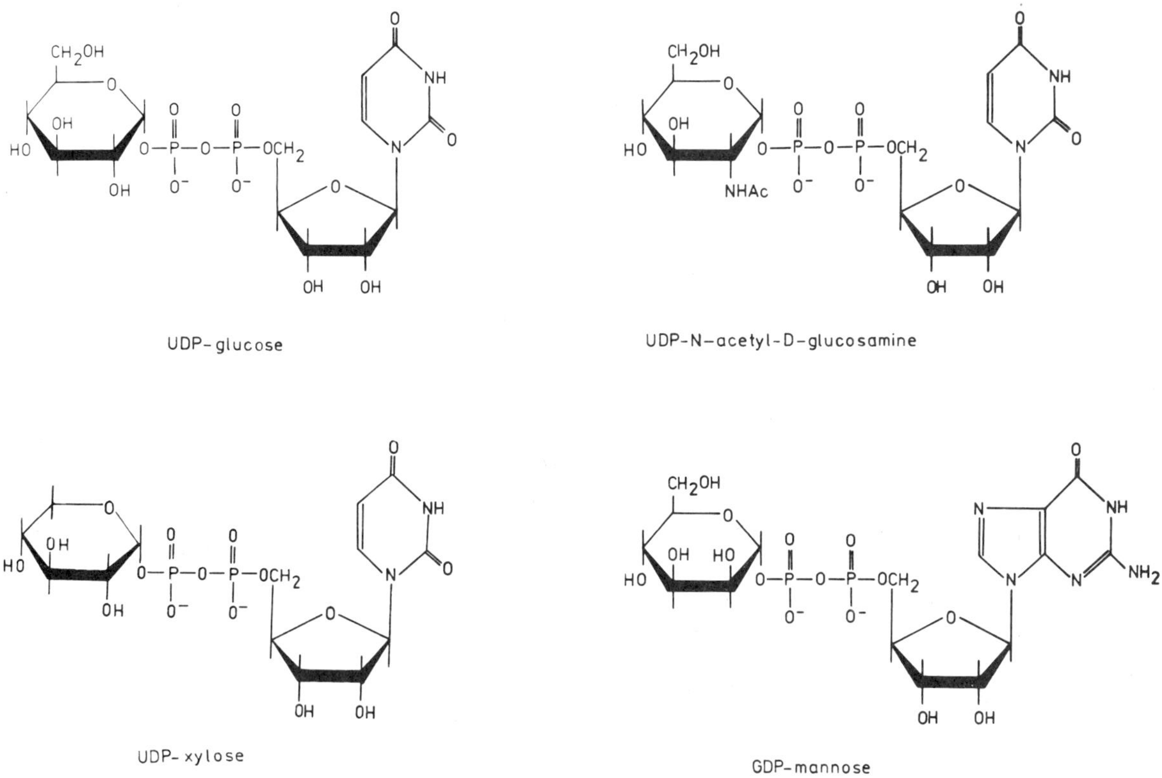

FIGURE 9. *Structural formulae of some nucleoside 5′-diphosphate sugars serving as precursors in biosynthesis of fungal wall polysaccharides.*

Chitin

Chitin is a linear polysaccharide composed exclusively of *N*-acetyl-D-glucosamine units linked by beta-1,4 glycosidic bonds. Owing to low solubility the linear chitin molecules tend to form microcrystalline aggregates in water. Chitin is polymerized from UDP-*N*-acetyl-D-glucosamine under the catalytic action of the enzyme chitin synthetase. The enzyme is located in the cytoplasmic membrane, where it can exist in two interconvertible forms, as active enzyme or as temporarily inactive zymogen. The interconversion of the two forms of chitin synthetase is supposed to play a decisive role in regulation of biosynthesis of the cell wall and, consequently, of fungal morphogenesis.

Glucans

Various polymers of glucose represent important cell-wall constituents in many fungi. The insoluble cellulose and beta-glucans (i.e., those containing only beta-glycosidic bonds) form, together with chitin, the skeletal portion of the walls, whereas the alpha-glucans are usually amorphous and are located in the wall matrix. Most glucans are polymerized by transglycosylic reactions from UDP-glucose. The exact cellular location of the reactions is, so far, not known. It is supposed that the skeletal wall glucans are synthesized by enzymes closely associated with the plasmalemma.

Protein-Polysaccharide Complexes

Protein-polysaccharides, or glycoproteins, are the center of much interest because they are both structural components of the cell walls and determinants of biologic specificity on cell surfaces. Structurally, the fungal wall glycoproteins exhibit many similarities with surface protein-polysaccharide complexes from higher organisms. For this reason the glycoproteins of fungi are used as models to study various aspects of glycoprotein structure, function, and biosynthesis.

So far, the most systematic studies on fungal glycoproteins have been performed with yeast mannan (Ballou, 1976). The polysaccharide moiety of yeast mannan consists of an alpha-1,6-linked polymannose backbone to which short side chains of mannosyl units linked by alpha-1,2 and alpha-1,3 glycosidic bonds are attached. The whole polysaccharide is linked via a diacetylchi-

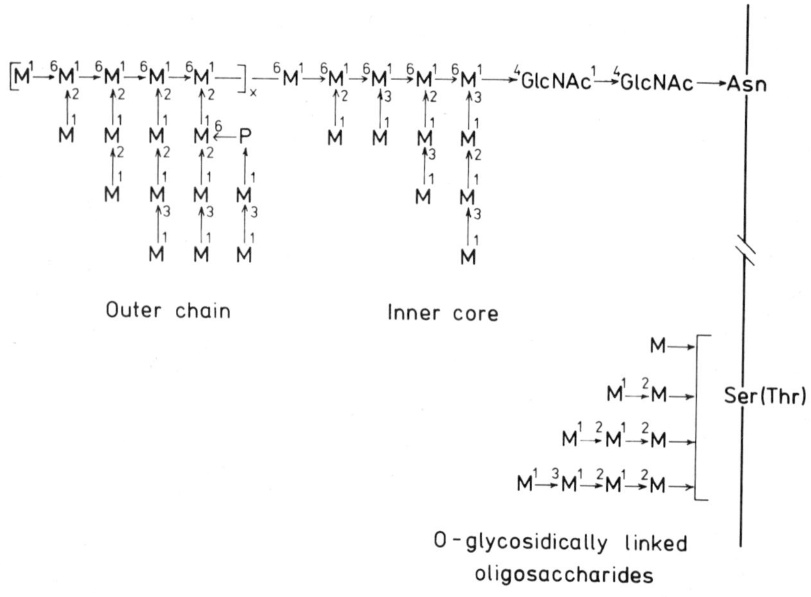

FIGURE 10. *Structure of mannan from the yeast* Saccharomyces cerevisiae. *M, mannopyranose residue; GlcNAc, N-acetyl-D-glucosamine; P, phosphate; Asn, asparagine; Ser, serine; Thr, threonine. (From Farkaš, V.: Microbiol Rev 43:117, 1979.)*

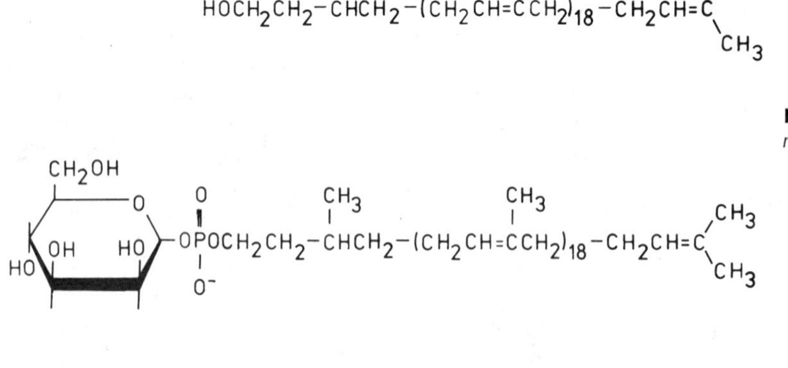

FIGURE 11. *Structural formula of dolichol monophosphate-mannose.*

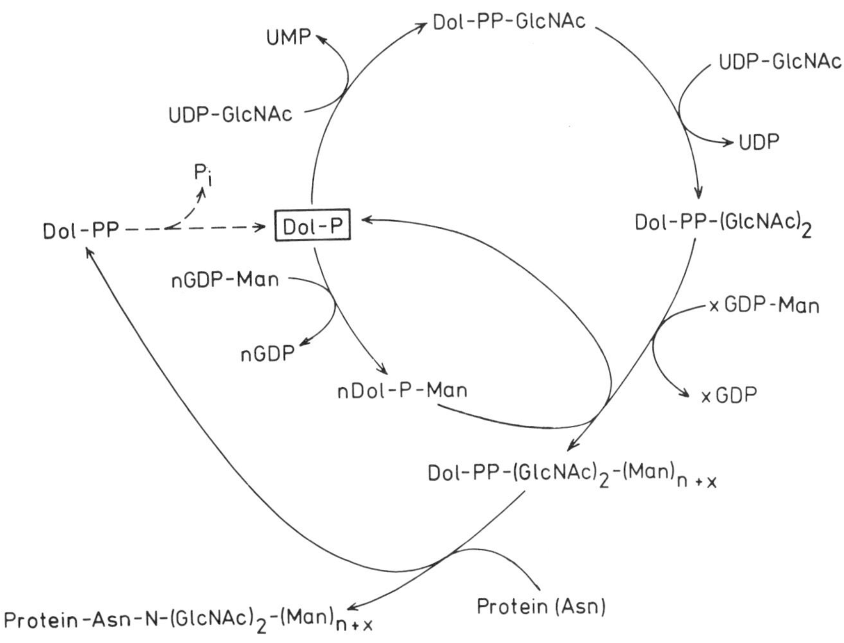

FIGURE 12. *Schematic representation of possible steps in biosynthesis of the "inner core" of yeast mannan. Dol-P, dolichol phosphate; Dol-PP, dolichol pyrophosphate; P_i, inorganic phosphate. x + n = 0 to 17. (From Farkaš, V.: Microbiol Rev 43:117, 1979.)*

tobiose bridge to an asparaginyl residue in the protein part of the molecule. Besides that, short manno-oligosaccharides containing alpha-1,2 and alpha-1,3 glycosidic links are attached directly to the hydroxyaminoacids serine and threonine by O-glycosidic bonds (Fig. 10).

Biosynthesis of the relatively complex structure of yeast mannan requires participation of a whole set of mannosyltransferases, each of them catalyzing the formation of a specific linkage. The precursor of mannosyl units in the biosynthesis of yeast mannan is guanosine 5′-diphospho mannose (GDP-Man). The transfer of mannosyl units from their donor to the acceptor can proceed either in a single step or through lipophilic intermediates such as dolichol-monophosphate mannose (Fig. 11). It is assumed that the involvement of dolichols is necessary in those reactions that take place in a lipophilic environment, as for example on the surface or inside the membranes of the endoplasmic reticulum. The binding of hydrophilic sugar residues to a lipophilic carrier facilitates

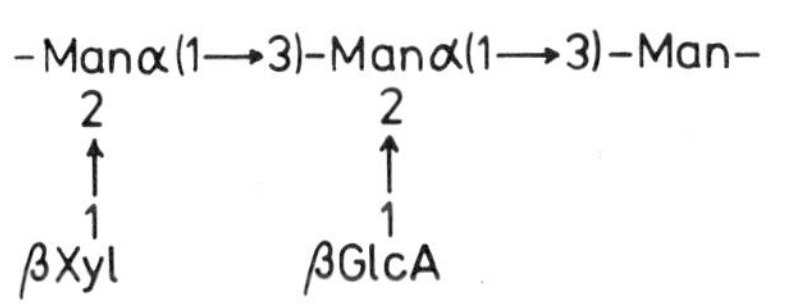

FIGURE 13. *Structure of acidic polysaccharide from capsules of* Cryptococcus neoformans. *Man, mannopyranose residue; Xyl, xylopyranose residue; GlcA, glucuronic acid residue.*

their solubility in the membrane and, consequently, their translocation within the cell. The biosynthesis of the whole "inner core" of yeast mannan probably proceeds while the product is attached to the dolichol carrier (Fig. 12).

The mannan of *Candida albicans* has a structure very similar to that depicted in Figure 10. The difference is that the mannan from *Candida albicans* contains a much higher proportion of unsubstituted mannosyl units in the polymannose backbone and, in addition, contains longer side branches that reach the size of a heptasaccharide. The side chains containing terminal alpha-1,3 linked mannosyl units are immunodeterminant groups in the mannan antigens, and the intensity of the precipitin reaction with antibodies induced by whole cells increases with the length of the side chains in the mannan.

Surface antigens from other fungi have usually more complex structures than the yeast mannans. The polysaccharide portions of these antigens can be chemically characterized as neutral or acidic heteropolysaccharides, composed of hexoses, pentoses, methylpentoses, and, in acidic heteropolysaccharides, uronic acids. For example, the capsular glucurono-xylo-mannan from *Cryptococcus neoformans* has a branched structure in which a linear alpha-1,3 linked polymannose chain is substituted at positions C-2 by beta-glycosidically linked glucuronic acid or xylose, with the xylose predominating (Fig. 13).

Glycosylation of nascent proteins takes place in

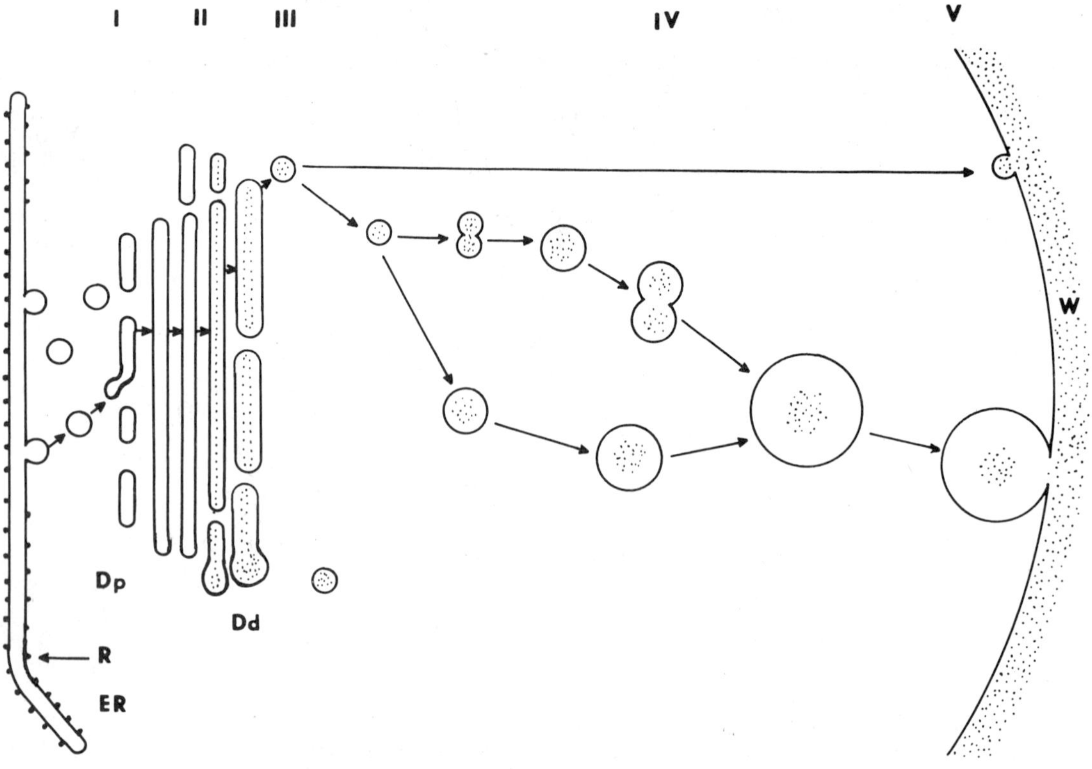

FIGURE 14. *Diagrammatic representation of biosynthesis and excretion of cell wall glycoproteins. Stage I: Formed proteins are transferred from ER to dictyosome by blebbing of ER and refusion of vesicles to form a cisterna at the proximal pole of the dictyosome (Dp). Stage II: Cisternal contents and the membranes are transformed as the cisterna is displaced to the distal pole (Dd) by the continued formation of new cisternae. Stage III: Cisternae vesiculate to form secretory vesicles as they approach and reach the distal pole. Stage IV: Secretory vesicles migrate to the cell wall. Some may increase in size or fuse with other vesicles to form large secretory vesicles, while others are carried directly to the cell surface. Stage V: Vesicles accumulate at the growth region of the wall and fuse with the cytoplasmic membrane, liberating their contents into the wall region (W). (After Grove et al., 1970.)*

dictyosomes and in the smooth endoplasmic reticulum. In the course of this process the formed glycoproteins are packed into vesicles derived from the original membrane systems and carried toward the cytoplasmic membrane. The carrier vesicles fuse with the membrane and discharge their contents into the cell exterior (Fig. 14). A portion of the supplied glycoproteins remains entrapped in the periplasm or anchored to the cell wall, and the other part diffuses into the surrounding medium. Some of the extracellular glycoproteins exhibit enzymatic activity; they act as hydrolases of various kinds. The hydrolytic action of these enzymes facilitates the growth and penetration of the fungus into the solid substrate or the invaded tissue.

References

Ballou, C. E.: Structure and biosynthesis of the mannan component of the yeast cell envelope. Adv Microbiol Physiol 14:93, 1976.

Bartnicki-Garcia, S.: Cell wall composition and other biochemical markers in fungal phylogeny. In Hasborne, J. B. (ed.): Phytochemical Phylogeny. New York and London, Academic Press, 1970, p. 81.

Farkaš, V.: Biosynthesis of cell walls of fungi. Microbiol Rev 43:117, 1979.

Girbardt, M.: Die Ultrastruktur der Apikalregion von Pilzhyphen. Protoplasma 67:413, 1969.

Gooday, G. W.: Differentiation in *Mucorales*. Symp Soc Gen Microbiol 23:269, 1973.

Grove, S. N., Bracker, C. E., and Morré, D. J.: An ultrastructural basis for hyphal tip growth in *Pythium ultimum*. Am J Botany 57:245, 1970.

Mariott, M. S.: Isolation and chemical characterization of plasma membranes from the yeast and mycelial forms of *Candida albicans*. J Gen Microbiol 86:115, 1975.

Oujezdsky, K. B., Grove, S. N., and Szaniszlo, P. J.: Morphological and structural changes during the yeast-to-mold conversion of *Phialophora dermatitis*. J Bacteriol 113:468, 1973.
Peterson, E. M., Hawley, R. J., and Calderone, R. A.: An ultrastructural analysis of protoplast-spheroplast induction in *Cryptococcus neoformans*. Can J Microbiol 22:1518, 1976.
Siestma, J. H., Child, J. J., Nesbitt, L. R., and Haskins, R. H.: Chemistry and ultrastructure of the hyphal walls of *Pythium accanthicum*. J Gen Microbiol 86:29, 1975.
Singer, S. J., and Nicolson, G. L.: The fluid mosaic model of the structure of cell membranes. Science 175:720, 1972.
Watson, K., and Arthur, H.: Cell surface topography of *Candida* and *Leucosporidium* yeasts as revealed by scanning electron microscopy. J Bacteriol 130:312, 1976.

CLASSIFICATION OF FUNGI 14

Willard A. Taber, Ph.D.

Fungi live in soil, on plants, in water, and occasionally on man. Most plant diseases are caused by fungi; most diseases of animals are caused by bacteria and viruses. For this reason fungi are usually associated with botany, whereas bacteria are associated with zoology. For an overview of fungi and their impact on man, see Taber and Taber, 1967.

The term fungus refers to nonphotosynthetic plant-like forms that resemble plants in possessing cell walls and the glyoxylate bypass of the Krebs cycle. Fungi resemble both plants and animals in being eukaryons — that is, they possess nuclei, mitochondria, endoplasmic reticula, more than one chromosome, protoplasmic streaming, and the capacity to synthesize steroids. They differ from bacteria, which are prokaryons, and which lack the above structures and activities. Whether microscopic or macroscopic in size (Fig. 1), fungi are united by the possession of hyphae (Fig. 2) or hypha-like structures. Fungi vary greatly in structural detail, and consequently a single all-inclusive definition is difficult to formulate. Fungi are nonphotosynthetic forms that consist of multinucleated cytoplasm within a much-branched system of tubes (the hyphae of filamentous fungi), of naked multinucleated cytoplasm (the plasmodium of Myxomycota), or of single vegetative cells that multiply either by formation of buds (the yeasts) or of motile spores (certain primitive aquatic phycomycetes).

There are a minimum of 50,000 species of fungi (Ainsworth, 1968b) distributed in approximately 4000 genera. Enumeration of fungal taxa is made difficult because some species have two generic names. This results from the fact that fungal taxa are based on sexual structures even though many fungi exist that produce only asexual spore structures. They are given a form-type name. If such a fungus is found to produce a sexual stage, it is given a second name to relate it to similar fungi of the classification. As an example, recent crossings of various strains of the deadly asexual fungus *Cryptococcus neoformans* showed it to be capable of producing a sexual stage characteristic of the Basidiomycetes and according to the provisions of the International Code of Botanical Nomenclature was given a second name, *Filobasidiella neoformans*. The species name is always retained although the ending may have to be changed to conform to the new generic name. Either name can be used, but the name based on the sexual state takes preference taxonomically. For definition of mycological terms, see the *Dictionary of Fungi* (Ainsworth, 1971).

SOME CHARACTERISTICS OF REPRESENTATIVE FUNGI

Myxomycota

The Myxomycota as represented by the common myxomycetes, or slime molds, possess a vegetative or assimilative phase that consists of a naked mass of multinucleated cytoplasm, the plasmodium. This plasmodium increases in size as it crawls over the substrate consuming small forms of life and soluble nutrients (Fig. 3). When food or water supply becomes limited, the plasmodium is transformed into sporangia-bearing spores. Meiosis precedes either formation or germination of these spores. A spore breaks open and approximately four amoeboid or flagellated cells escape and feed on surrounding nutrients. Pairs of such cells functioning as gametes fuse (Fig. 3) and develop into the 2N plasmodium. Myxomycete plasmodia or sporangia can be found on logs, fallen leaves, or grass after rains. None is known to be pathogenic to animals although the related Plasmodiophoromycetes are pathogens of certain

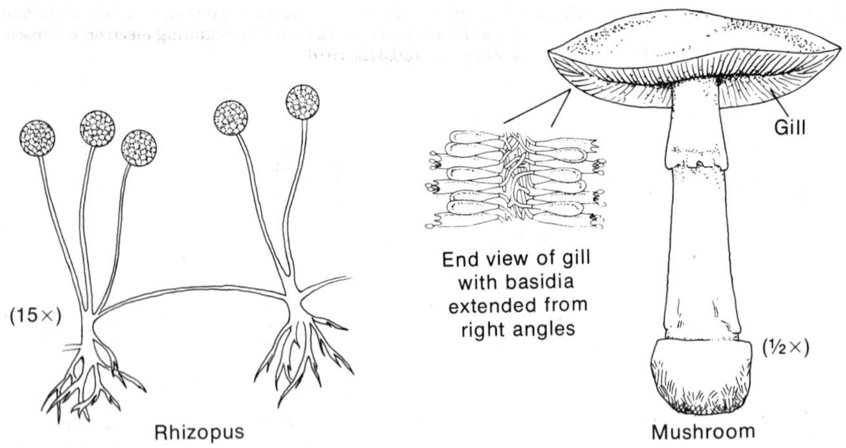

FIGURE 1. *Asexual sporangiospores are borne in sporangia that are supported on sporangiophores. Rhizopus also possesses rhizoids at the base of the sporangiophore. (Left) Basidiocarp of Amanita sp. Basidiospores are borne on basidia. (Right)*

FIGURE 2. *Vegetative or assimilative forms of growth of Fungi. (Left) Hyphal form with branches. (Center) Yeast form of growth in which proliferation occurs by budding. (Right) Unicellular growth characteristic of certain simple Phycomycetes. Approximate size of a bacterium is shown for comparison. Bacteria are always very narrow.*

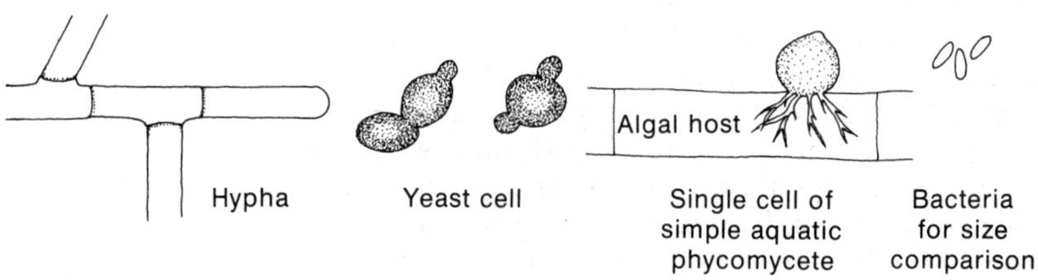

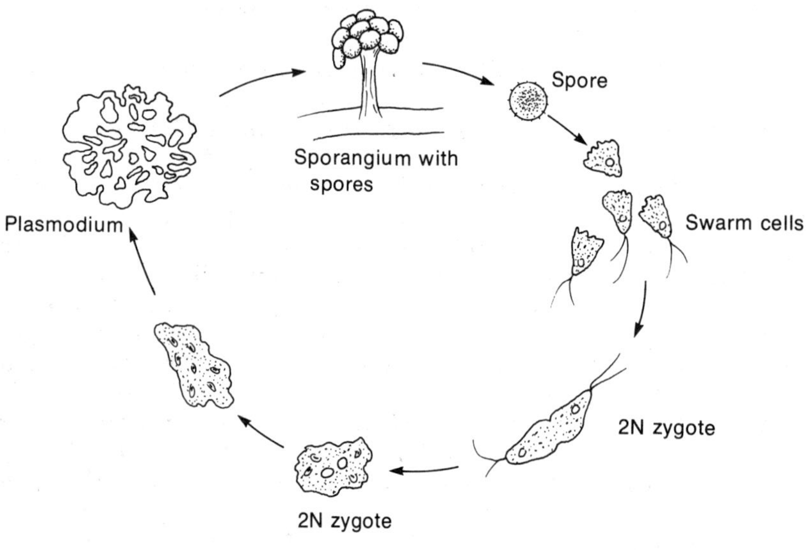

FIGURE 3. *Life cycle of myxomycete.*

plants. For descriptions of species, see Martin and Alexopoulos, 1969.

Eumycota

Mastigomycotina

Mastigomycotina are nonplasmodial fungi that produce motile cells. Motility is by flagella of the eukaryotic type. Many species, such as some of the Chytridiomycetes, consist of a single cell with rhizoids (Fig. 2), which becomes transformed into a sporangium containing motile spores. Chytridiomycetes possess one posterior flagellum, Hypochytridiomycetes possess one anterior flagellum, and the Oomycetes possess two flagella and in addition are usually mycelial (Table 1). Oomycetes also possess a female oogonium and a male antheridium (Table 1). The fertilized oosphere in the oogonium becomes an oospore. The oospore germinates by germ tube, which develops into mycelium or hyphae. Some species of Saprolegnia can parasitize aquatic animals, producing a woolly halo of mycelium around the host. The Oomycetes consist of four orders, Saprolegniales, Leptomitales, Lagenidiales, and Peronosporales. The Peronosporales embrace many plant pathogens, such as the downy mildews, which cause severe crop losses (Fig. 4). The asexual spores of some can produce germ tubes as well as flagella, and this group represents a transition to nonmotile terrestrial forms of fungi.

Zygomycotina

Zygomycetes are nonmotile phycomycetes that possess coenocytic hyphae, thick-walled sexual spores called zygospores (Table 1), and sporangia containing asexual spores (Hesseltine, 1973). Most species produce woolly aerial mycelium when grown on spoiled foods or agar in a Petri plate. The aerial hyphae may fill the air space of a Petri plate culture; other fungi do not. Some species are homothallic. That is, hyphae developing from one spore can produce the sexual spore. Others are heterothallic, and hyphae of different mating types must fuse before sexual spores can be produced. Hormones are involved in hyphal attraction. Several species are pathogenic to man (Table 2) and their invasiveness is associated with diabetes. These fungi grow rapidly on sugars and are frequently referred to as sugar fungi. Members of the Entomophthorales attack insects, grow in association with amphibians, or grow in ground litter. Dead flies fixed to window panes and surrounded by a halo have been killed by *Entomophthora muscae*. The halo consists of sporangia shot off from the fungus growing on the fly.

Ascomycotina

There are approximately 15,000 species of ascomycetes distributed in 1950 genera. Ascomycetes are those fungi that produce the sexual spore, the ascospore, in an ascus (Table 1, Fig. 5). The vegatative or assimilative phase can consist either of coenocytic hyphae compartmentalized by regularly placed simple perforated septa (crosswalls) or yeast cells. An ascus usually contains eight ascospores but certain species can produce four and others more than eight ascospores.

Ascomycetes are divided into three major categories: Euascomycetes, which possess a unitunicate ascus and usually an ascocarp (Table 1) that forms around the developing asci; Loculoascomycetes, which possess a bitunicate ascus (Fig. 5) that is formed within a cavity or locule hollowed out of a stroma that may superficially resemble an ascocarp; and Laboulbeniomycetes, which are minute nonmycelial ascomycetes living exclusively on the exoskeletons of insects and mites (Benjamin, 1973). Certain species of Euascomycetes can also possess a stroma, but if so, the ascocarp develops in or on top of it. The stroma is composed of pseudoparenchymatous tissue (see Taber and Taber 1973 for a review of structures). Unexplainably, the concept Euascomycete is not employed taxonomically.

Euascomycetes

Hemiascomycetes. Yeast and yeast-like fungi that produce unitunicate asci but not in an ascocarp are placed in this class. At least one, *Pichia guilliermondii*, is pathogenic to man.

Plectomycetes. Plectomycetes are normally filamentous ascomycetes that produce a simple closed ascocarp, the cleistothecium, containing globose dissolving asci at various heights in the ascocarp (Table 1). Many of the human pathogenic fungi are members of this class and most are in the Family Gymnoascaceae. *Histoplasm capsulatum, Blastomyces dermatitidis, Paracoccidioides brasiliensis*, and the unrelated *Sporothrix schenkii* grow as true yeasts in human tissue and as hyphae elsewhere in nature. Note that the first three names represent the asexual forms of these fungi, the forms generally observed. Names of the sexual and asexual forms are listed in Table 2. Many of the dermatophytes are members of this group. Note that dermatophytes are housed in three asexual genera: Trichophyton, Microsporum, and Epidermophyton; and note that those producing a sexual state are also members of either Arthroderma or Nannizzia (Table 2). Microascus produces a perithecium (Table 2) but here is placed in the Plectomycetes because the

TABLE 1. Overview of the Fungi

OLD	NEW	CHARACTERISTICS
Myxomycetes (Class)	MYXOMYCOTA (Division) Myxomycetes (Class) "Slime molds" Plasmodiophoromycetes Plant Pathogens Acrasiomycetes "Labyrinthulales"	Vegetative phase is a plasmodium that consists of fused or single amoeba. Creeps over surface of substrate. Some produce dry spore-bearing structures that resemble fungal structures.
	EUMYCOTA (Division) All other fungi.	Do not creep over substrate. Spread by growing. Vegetative phase a hypha, yeast cell, or single cell.
Phycomycete-Oomycete Aquatic phycomycetes, many are plant pathogens	Mastigomycotina (Subdivision) Chytridiomycetes (Class) Hyphochytridiomycetes Oomycetes	Asexual Spores are motile owing to flagella. Hyphae, when present, multinucleate (coenocytic) without regular cross walls (septa). 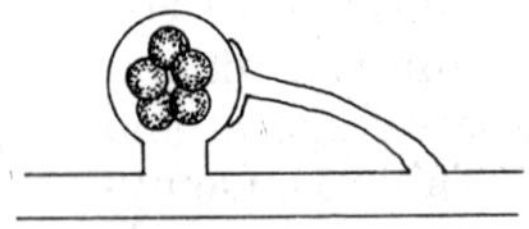 Sexual phase 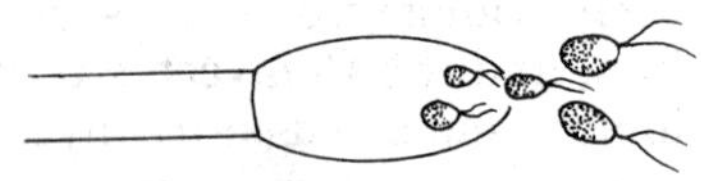 Asexual phase
Phycomycete-Zygomycete	Zygomycotina (Subdivision) Zygomycetes (Class) Sexual spore is a zygospore and asexual spore is usually a sporangiospore (borne within a sporangium). Human pathogens are found among: Rhizopus, Absidia, Mucor, and Basidiobolus. Trichomycetes (on or in arthropods)	Coenocytic hyphae without regular perforated septa but no motile spores. 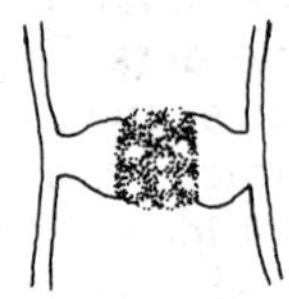 Sexual spore 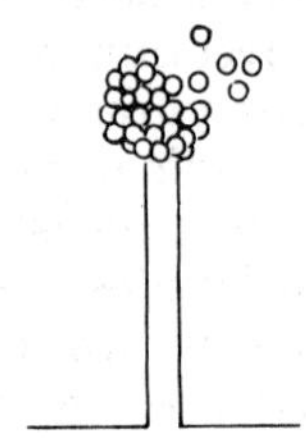 Asexual spore
Ascomycete Ascocarp is a structure housing or supporting asci. It forms around the developing asci of the Euascomycetes. Loculoascomycetes possess a hollowed-out stroma in which asci develop.	Ascomycotina (Subdivision) 1. Euascomycetes. Unitunicate ascus (no classification) Hemiascomycetes. (Class). Yeast or yeast-like. No ascocarp; no ascogenous hyphae. Yeasts usually have no hyphae. Plectomycete. Round asci scattered throughout cleistothecium type ascocarp; asci dissolving. (Human pathogens: dermatophytes, Histoplasma, Aspergillus are members of this group.) Pyrenomycetes. Ascocarp usually a perithecium bearing oblong asci from one layer.	Coenocytic hyphae with simple perforated septa. Usually have an ascocarp. Always have ascus and ascospore.
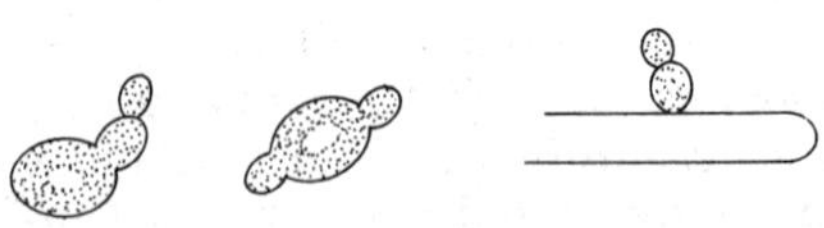	 Ascus	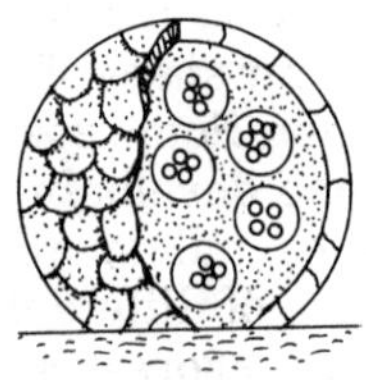 Cleistothecium 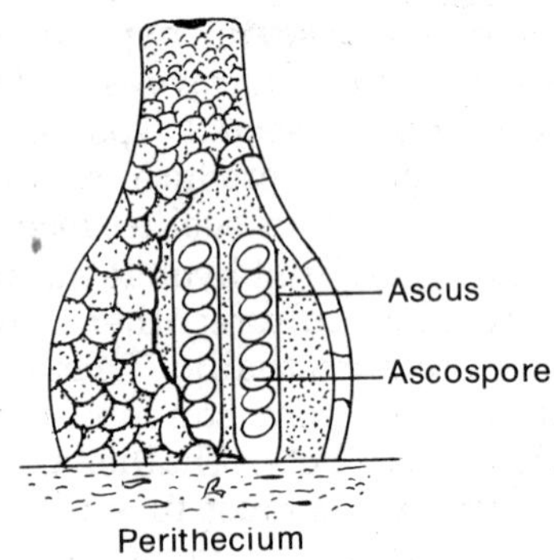 Perithecium
	Discomycetes. Oblong ascus with or without operculum in an (open) apothecium	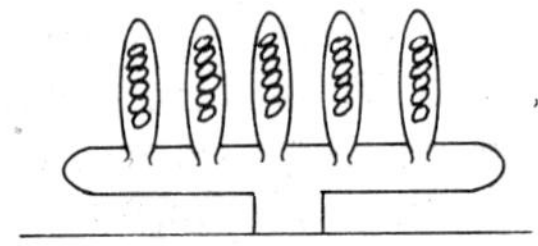

TABLE 1. Overview of the Fungi *(Continued)*

OLD	NEW	CHARACTERISTICS
	2. Loculoascomycetes (Class) Bitunicate (two-layered) ascus in a hollowed-out stroma. May resemble perithecium or black tar spots. Piedraia hortae on man. 3. Laboulbeniomycetes. Minute, on arthropod exoskeletons.	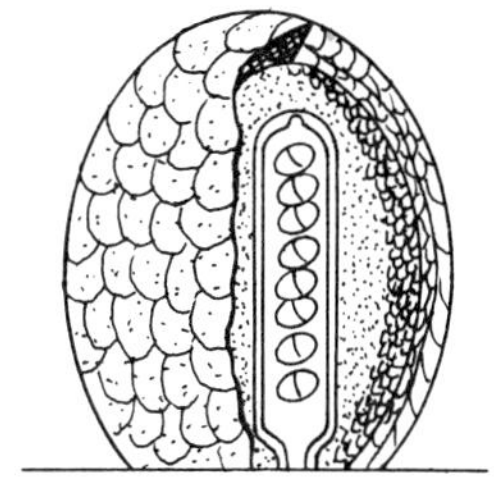
Fungi Imperfecti (Most fungi growing on laboratory Petri plate cultures are Deuteromycotina or Zygomycetes).	Deuteromycotina (Subdivision) (asexual fungi) Blastomycetes (Class) yeasts Hyphomycetes. Spores borne in open on hyphae Coelomycetes. Spores borne in (enclosed) pycnidium or on acervulus cushion breaking plant tissue epidermis	 Yeast or hyphal fungi without a sexual phase. If a sexual phase is found, both names can be used. Many are now known to be ascomycetes or to have ascomycetous affinities. Hyphae are coenocytic with perforated septa. Candida, Cryptococcus, Torulopsis, Pityrosporum.
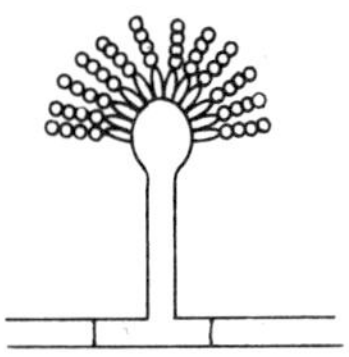 Aspergillus 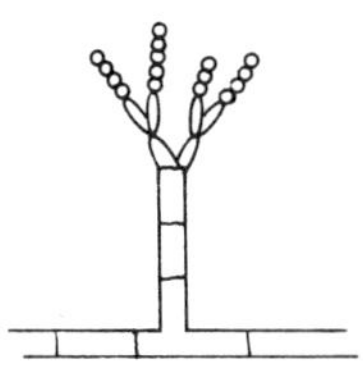 Penicillium	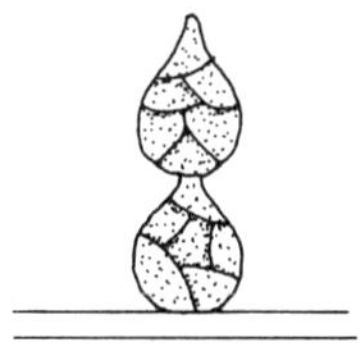 Alternaria 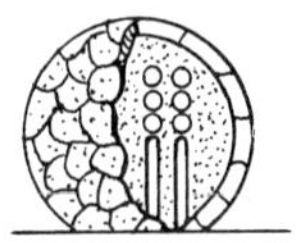 Phoma	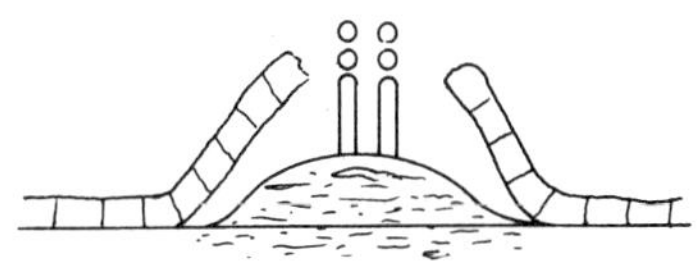 Acervulus in leaf
Basidiomycetes	Basidiomycotina (Subdivision) Coenocytic hyphae with complex perforated septa and often with two nuclei per compartment. Basidium and basidiospore, and often clamp connection characterize the Subdivision. The human pathogen *Cryptococcus neoformans* recently found to be asexual phase of the basidiomycete *Filobasidiella neoformans*. Teliomycetes (Class) Rusts and smuts. No basidiocarp. Hymenomycetes. Spores shot off. Mushrooms. Gasteromycetes. Spores not shot off; in an enclosed basidiocarp.	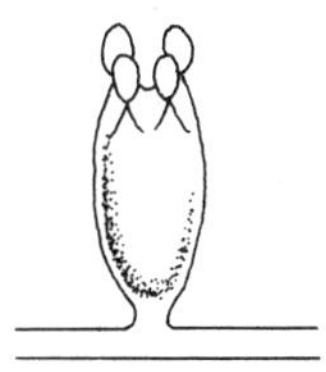 Basidium with spores
	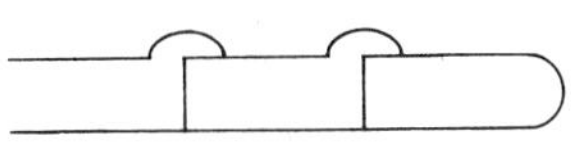 Hypha with clamp connection  Mushroom	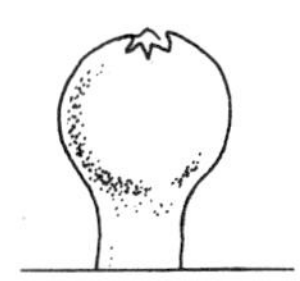 Puffball

TABLE 2. Fungal Taxa That Cause Diseases of Man

- Eumycota
  - Mastigomycotina
    - Chytridiomycete (?)
      - *Rhinosporidium seeberi*
      - *Emmonsia parva* and *Emmonsia crescens ?*
  - Zygomycotina
  - *Coccidioides immitis?* coccidioidomycosis. Taxonomic location uncertain
  - Zygomycetes
    - Mucorales
      - *Absidia corymbifera*
      - *Absidia ramosa*
      - *Hyphomyces destruens ?*
      - *Mortierella* sp. (questionable)
      - *Mucor circinelloides*
      - *Mucor pusillus*
      - *Mucor ramosissimus*
      - *Rhizopus arrhizus*
      - *Rhizopus oryzae*
    - Entomophthorales (usually grow on insects or in ground litter)
    - *Basidiobolus haptosporus* (tropical)
    - *Entomophthora coronata*
  - Ascomycotina
  - Euascomycete (unitunicate ascus; ascocarp wall built around developing asci)
    - Hemiascomycete (yeast or yeast-like)
      - Endomycetales
        - *Endomyces geotrichum,* ascomycetous state of *Geotrichum candidum*
        - *Pichia guilliermondii* (yeast). Sexual state of *Candida guilliermondii*
    - Plectomycete (globose asci borne at different levels in ascocarp; asci dissolving)
      - Eurotiales (cleistothecium)
        - Gymnoascaceae
          - *Ajellomyces dermatitidis*—ascomycetous state of *Blastomyces dermatitidis.* Blastomycosis
          - *Arthroderma benhamiae*—ascomycetous state of the dermatophyte *Trichophyton mentagrophytes*
          - *Arthroderma gertleri*—ascomycetous state of dermatophyte *Trichophyton vanbreuseghemii*
          - *Arthroderma simii*—ascomycetous state of dermatophyte (?) *Trichophyton simii*
          - *Allescheria boydii* (location ?)—ascomycetous state of *Monosporium apiospermum,* mycetoma
          - *Emmonsiella capsulata*—ascomycetous state of *Histoplasma capsulatum,* histoplasmosis
          - *Nannizzia cajetani*—ascomycetous state of dermatophyte (?) *Microsporum cookei*
          - *Nannizzia fulva*—ascomycetous state of dermatophyte *Microsporum fulvum*
          - *Nannizzia grubyia*—ascomycetous state of dermatophyte *Microsporum vanbreuseghemii*
          - *Nannizzia gypsea*—ascomycetous state of dermatophyte *Microsporum gypseum*
          - Nannizzia incurvata—ascomycetous state of dermatophyte *Microsporum gypseum* (same asexual state as above).
          - *Nannizzia obtusa*—ascomycetous state of dermatophyte *Microsporum nanum*
          - Nannizzia persicolor—ascomycetous state of dermatophyte *Microsporum persicolor*
        - Eurotiaceae (Trichocomataceae?)
          - *Sartorya fumigata* —ascomycetous state of *Aspergillus fumigatus,* aspergillosis
      - Microascales (perithecium; asci at various levels, usually spheric, asci dissolving)
        - Microascaceae
          - *Microascus cinereus* —ascomycetous state of *Scopulariopsis cinerea.* Dermal. This genus placed by von Arx in Sphaeriales (see Muller & von Arx 1973).
          - *Microascus manganii* — ascomycetous state of *Scopulariopsis albo-flavescens.* Dermal.
          - *Microascus trigonosporus*—ascomycetous state of a Scopulariopsis
  - Loculoascomycete (asci borne in locules of hollowed-out ascostroma; bitunicate ascus)
    - *Leptosphaeria senegalenis.* Mycetoma
    - *Neotestudina rosatii.* Mycetoma
    - *Piedraia hortae.* Black piedra (tropical) nodules as ascostroma on hair
- Deuteromycotina (asexual state, often, of ascomycetes; if sexual and asexual states are known, fungus will have two names with sexual name taking preference)
  - Blastomycetes (asexual yeasts)
    - Cryptococcaceae
      - *Cryptococcus neoformans.* Cryptococcus. Asexual state of Basidiomycete *Filobasidiella neoformans.*
      - *Pityrosporum* sp. *(Malassezia furfur).* In the stratum corneum; pityriasis versicolor
      - *Candida albicans.* Candidiasis
      - *Candida guilliermondii*
      - *Candida parapsilosis*
      - *Candida stellatoidea*
      - *Candida tropicalis*
      - *Trichosporon beigelii.* White piedra
      - *Trichosporon cutaneum*

TABLE 2. Fungal Taxa That Cause Diseases of Man (*Continued*)

Geotrichum candidum (white, moist colony but not a yeast—no buds). See below.
Torulopsis glabrata. Secondary invader
Hyphomycetes (filamentous, spores borne in open)
Moniliales
Moniliaceae (hyaline or light colored except for Aspergillus niger)
Trichophyton (see also Arthroderma in Ascomycetes) dermatophyte
T. schoenleinii; T. rubrum; T. mentagrophytes; T. concentricum; T. verrucosum; T. violaceum; T. tonsurans; T. equinum; T. simii (monkeys); *T. ajelloi* (?), *T. megninii*
Microsporum (see also Nannizzia in Ascomycetes) dermatophyte
M. canis; M. gypseum; M. audouini; M. ferrugineum; M. fulvum; M. nanum; M. obtusa; M. vanbreuseghemii; M. persicolor; M. gallinae; M. distortum (?); *M. cookei* (?)
Epidermophyton floccosum. Dermatophyte
Geotrichum candidum (wet, white growth with yeast-like arthrospores formed by fragmentation of hyphae
Cepahalosporium (now *Acremonium*) *mycetoma*
Sporothrix schenkii (white at first, becoming black with age) sporotrichosis (a yeast when growing in the body; hyphal otherwise)
Coccidioides immitis (Phycomycete?) coccidioidomycosis. (Spherule in the body, hyphal otherwise)
Histoplasma capsulatum (see Ascomycetes) (Yeast in the body; otherwise hyphal)
Blastomyces dermatitidis (see Ascomycetes) (Yeast in the body; otherwise hyphal)
Paracoccidioides brasiliensis (Chrysosporium?) (Yeast in the body; otherwise hyphal)
Penicillium commune. Pulmonary
Aspergillus fumigatus (see Ascomycetes) pulmonary, aspergillosis
Aspergillus niger. Otomycosis, fungus balls
Scopulariopsis brevicaulis (see Ascomycetes) in nails
Fusarium solani, F. oxysporum, F. nivale. Keratitis
Madurella mycetomi, M. grisea. Mycetoma
Monosporium apiospermum (see Ascomycetes) mycetoma
Dematiaceae (hyphae black or black-green)
Cercospora apii (also pathogen of celery and other crops) subcutaneous
Torula bantiana. Banti's mycosis
Cladosporium carrionii
Cladosporium trichoides. Cladosporiosis
Cladosporium werneckii. On hands
Curvularia lunata and *C. geniculata*. Mycetoma
Phialophora (at least one species pathogenic to both man and certain trees)
P. compacta chromomycosis
P. dermatitidis chromomycosis
P. gougerotii phaeosporotrichosis
P. jeanselmei mycetoma
P. parasitica. subcutaneous
P. pedrosoi. chromomycosis
P. repens. mycetoma
P. richardsiae. phaeosporotrichosis
P. spinifera. granulomatous lesion
P. verrucosa. chromomycosis
Coelomycetes (asexual spores in an enclosed pycnidium or on an acervulus)
Sphaeropsidales
Macrophoma sp. keratitis
Pyrenochaeta romeroi. Mycetoma
Loboa loboi (never cultured) Lobo's disease
Prototheca zopfii. Probably a fungus but may be an algal mutant

spheric asci are produced scattered throughout the ascocarp. Muller and von Arx (1973) place Microascus in the Sphaeriales to be considered below.

Pyrenomycetes. This class houses those ascomycetes that produce oblong unitunicate asci in a perithecium that is an ascocarp containing an ostiole, or opening, at the top. Asci originate from one layer of cells rather than being distributed throughout the ascocarp as with Plectomycetes. The spores are usually one celled. Some cleistothecial "analogs" or perithecial forms do exist, however. The ascospores of most pyrenomycetes are shot out of the ascus, but in a very few genera, such as Chaetomium and Melanospora, the asci dissolve and spores ooze out. Virtually all live on plant tissue or in the soil and probably none is a human pathogen if Microascus is considered to be a Plectomycete.

Discomycetes. Cylindric unitunicate asci are

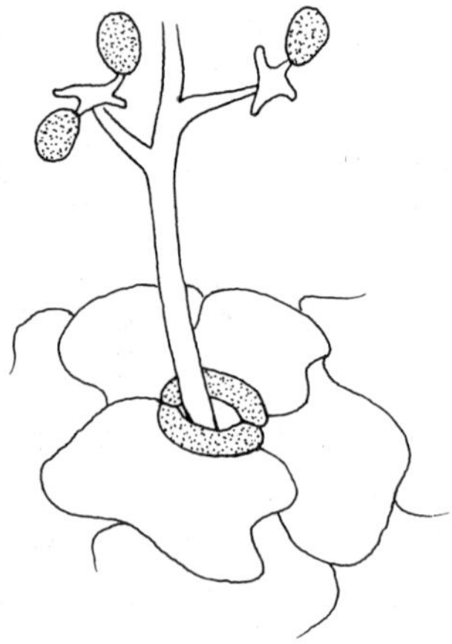

FIGURE 4. *Sporophore of a downy mildew protruding out of a stoma of an infected leaf.*

produced on, or supported on, an open saucer-like ascocarp, the apothecium (Table 1). None is known to be a human pathogen.

Loculoascomycetes

Ascomycetes possessing bitunicate asci in the locule or cavity of a "hollowed-out" ascostroma are placed in this class (Luttrell, 1973). The asci are usually thick walled with an apical dimple and a distinct shoulder (Fig. 5). Ascopores are usually multiseptate and dark colored. *Testudina rosatti* causes mycetoma (Table 2) and *Piedraia hortae*, which grows as ascostroma on hair, causes black piedra.

Laboulbeniomycetes

These are minute, nonmycelial ascomycetes that grow on the exoskeletons of insects. Benjamin (1973) reviews this unique group of fungi that may or may not be pathogens of insects and certain mites.

Basidiomycotina

Basidiomycetes are those fungi that produce sexual spores, basidiospores, on a basidium (Table 1, Fig. 1). As with the ascus, nuclear fusion and meiosis takes place in this cell. Mating types occurring at one or two loci exist. These segregate as a result of meiosis, and the four basidiospores that are the direct product of meiosis contain the two or four different mating types. Germinated basidiospores are haploid and contain as a rule one nucleus per compartment. There are no sexual organs in the basidiomycetes (except rusts) and the nuclei differing in mating type are brought together as a result of fusion of hyphae derived from basidiospores of different mating types. These secondary hyphae usually contain two nuclei per compartment. The perforated septa dividing the coenocytic mycelium into compartments often possess ornamentations that distinguish them from the simple septa of the ascomycetes.

Teliomycetes

This class unites basidiomycetes not having basidiocarps. The plant pathogens, rusts and smuts, and the sexual stage of *Cryptococcus neoformans* and *Filobasidiella neoformans* are members of this class.

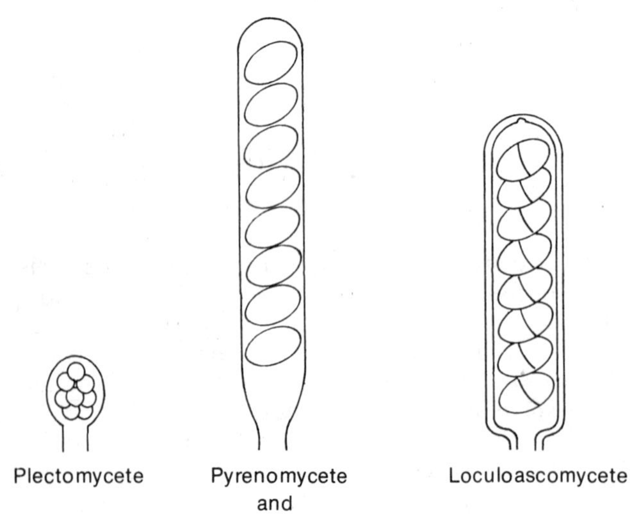

FIGURE 5. *Asci and ascospores.*

Hymenomycetes

Mushrooms (Fig. 1; Table 1) and other fleshy basidiomycetes bearing basidia on gills, pores, smooth surfaces, or spines are allocated to this class. In all cases, spores are shot off of the basidium.

Gasteromycetes

Basidiomycetes producing spores that are not shot off are placed in this class. Basidia are contained within an enclosed basidiocarp, such as that of the puffball (Table 1).

Deuteromycotina

The Class Deuteromycotina (Fungi Imperfecti) unites those fungi producing asexual spores or no spores at all. As indicated earlier, many are now known to produce sexual spores under conditions such as certain environmental conditions or presence of two mating types, and such fungi are assigned two names. Most fungi observed growing on Petri plate cultures will be Deuteromycotina or Zygomycotina. Probably the vast majority of Fungi Imperfecti are ascomycetes or were derived from ascomycetes. The pathogenic yeast *Cryptococcus neoformans* and the yeast *Candida scottii* along with a few others are asexual yeasts of basidiomycetes, however.

Blastomycetes

Asexual yeasts are placed in this class. There are approximately 17 genera of asexual yeasts, 17 genera of ascomycete yeasts, and five genera of basidiomycete yeasts (Kreger-van Rij, 1973). This publication provides keys to genera of all yeasts. Identification keys are also given in the following publications (Lodder, 1970; Barnett and Pankhurst 1974; and DeHoog, 1977).

Hyphomycetes

Most Fungi Imperfecti observed in the laboratory are members of this class, which embraces those asexual fungi producing spores on hyphae not contained within sporocarps (Table 1). Approximately 595 genera are recognized in the latest treatment (Kendrick and Carmichael, 1973). They do not assign genera to families and they base their classification of genera on four properties: 1) Saccardoan spore group, which indicates shape and number of compartments of the spore, 2) arrangement of conidia, 3) color of conidia, and 4) type of conidiogenous cell.

The Hyphomycetes can be divided into five form families: Moniliaceae — mycelia and spores colorless or bright colored (except for the black *Aspergillus niger,* which otherwise conforms to the characteristics of the moniliaceous Aspergillus); Dematiaceae — mycelia and spores black, dark green or dark brown; Stilbaceae — conidiophores united into vertical fascicles called synnemata or coremia; Tuberculariaceae — sporophores borne on an acervulus-like cushion of hyphae infecting a plant; Mycelia Sterilia — spores not produced.

Aspergillus flavus and *Aspergillus fumigatus* cause pulmonary diseases and are unique in being the only non-Zygomycete fungi that grow deep in tissue in mycelial rather than yeast form. The asexual forms of the dermatophytes are also placed in the Moniliaceae or Hyphomycetes. Dark-spored Phialophora and Cladosporium are dematiaceous fungi.

Coelomycetes

Those fungi producing asexual spores in an enclosed sporocarp, the pycnidium, or in the cushion-like structure, the acervulus, are placed in this class (Table 1). There are approximately 500 genera (Sutton, 1973). *Macrophoma* sp. and *Pyrenochaeta romeroi* cause diseases in man.

HABITAT OF HUMAN PATHOGENIC FUNGI

Fungal diseases can be divided into systemic mycoses, subcutaneous mycoses, and superficial mycoses. Of those causing systemic mycoses and subcutaneous mycoses (histoplasmosis, coccidioidomycosis, sporotrichosis, blastomycosis, aspergillosis, and candidiasis), only *Candida albicans*, the causal agent of candidiasis, normally lives in man. The rest live in soil or on plant products. Some of the dermatophytes live in soil and it is now believed (Emmons et al., 1977) that some of the dermatophytes have become so well adapted to man that they now normally live on him. This conclusion is based on the observation that some of the human dermatophytes have not been isolated from soil. Ainsworth (1968a) discusses the fungi parasitizing vertebrates.

References

Ainsworth, G.C.: Fungal parasites of vertebrates. In Ainsworth, G.C., and Sussman, A.S. (eds): The Fungi. Volume III. New York, Academic Press, 1968a, p. 211.

Ainsworth, G.C.: The number of fungi. In Ainsworth, G.C., and Sussman, A.S. (eds): The Fungi. Volume III. New York, Academic Press, 1968b, p. 505.

Ainsworth, G.C.: The Dictionary of Fungi. 7th ed. Kew, Surrey, England, Commonwealth Mycological Institute, 1971.

Barnett, J.A., and Pankhurst, R.J.: A new key to the yeasts. A key for identifying yeasts based on physiological tests only. Amsterdam, North Holland Publishers, 1974.

Benjamin, R.K.: Laboulbeniomycetes. In Ainsworth, G.C., Sparrow, F.K., and Sussman, A.S. (eds): The Fungi. Volume IVA. New York, Academic Press, 1973, p. 223.

DeHoog, G.S., and Hermanides-Nijhof, E.J.: The black yeasts and allied genera. Baarn, Holland, Centraalbureau voor Schimmelcultures, 1977.
Emmons, C.W., Binford, C.H., Utz, J.P., and Kwon-Chung, K.J.: Medical Mycology. 3rd ed. Philadelphia, Lea & Febiger, 1977.
Hesseltine, C.W., and Ellis, J.J.: Mucorales. In Ainsworth, G.C., Sparrow, F.K., and Sussman, A.S. (eds): The Fungi. Volume IVB. New York, Academic Press, 1973, p. 187.
Kendrick, W.B., and Carmichael, J.W.: Hyphomycetes. In Ainsworth, G.C. Sparrow, F.K., and Sussman, A.S. (eds): The Fungi. Volume IVA. New York, Academic Press, 1973, p. 323.
Kreger-van Rij, N.J.W.: Endomycetales, basidiomycetous yeasts and related fungi. In Ainsworth, G.C., Sparrow, F.K., and Sussman, A.S. (eds): The Fungi. Volume IVA. New York, Academic Press, 1973, p. 11.
Lodder, J. (ed): The yeasts, a taxonomic study. Amsterdam, North Holland press, 1970.
Luttrell, E.S.: Loculoascomycetes. In Ainsworth, G.C., Sparrow, F.K., and Sussman, A.S. (eds): The Fungi. Volume IVA. New York, Academic Press, 1973, p. 135.
Martin, G.W., and Alexopoulos, C.J.: The Myxomycetes. Iowa City, The University of Iowa Press, 1969.
Muller, E., and von Arx, J.A.: Pyrenomycetes. In Ainsworth, G.C., Sparrow, F.K., and Sussman, A.S. (eds): The Fungi. Volume IVA. New York, Academic Press, 1973.
Sutton, B.C.: Coelomycetes. In Ainsworth, G.C., Sparrow, F.K., and Sussman, A.S. (eds): The Fungi. Volume IVA. New York, Academic Press, 1973, p. 513.
Taber, W.A., and Taber, R.A.: The Impact of Fungi on Man. Boulder, Col., Educational Programs Improvement Corporation, 1967.
Taber, W.A., and Taber, R.A.: Ascomycetes. In Lechevalier, H.A., and Laskin, A.I. (eds): Handbook of Microbiology. 2nd ed. West Palm Beach, Fl., CRC Press, 1978, p. 97.

15 *FUNGAL TOXINS*

Makoto Enomoto, M.D.

Mycotoxicosis is a disease caused by fungal toxins (mycotoxins), and should be distinguished from *mycosis* caused by opportunistic infections of fungi. Fungi include mushrooms, molds, mildews, blights, rusts, and yeasts. Generally, fungi are valued for the organic fermentation they produce and as a source of many antibiotics. In addition, edible mushroms are used as human food. Toxin-producing fungi were occasionally encountered during exploration for antibiotics, but were given little attention. Except for griseofulvin, cancer production was usually found in animals only after parenteral administration of antibiotics, including mitomycin C, streptozotocin, penicillin G, azaserine, daunomycin, elaiomycin, and actinomycin D, L, and S. The hepatocarcinogenicity of griseofulvin was proved in mice by the oral administration of a daily dosage of 5000 to 10,000 ppm. In the bone marrow and small intestine there was no tumor, but cytotoxicity was observed in rats given a high dose of griseofulvin.

MUSHROOM POISONS

Mycotoxicoses known to be related to human disease are shown in Table 1. Among them are ergotism and mushroom poisoning *(mycetismus)*, which are known in many countries. The poisonous principles of mushrooms of the genus Amanita are cell-destroying cyclic peptides, such as hepatotoxic "phalloidin" and "amanitin," each representing two groups of lethal sulfur-containing cyclopeptides, the phallotoxins and amatoxins. The phallotoxins act on the endoplasmic reticulum of liver cells, whereas the amatoxins act on the nucleus. After the isolation from *Amanita muscaria* of muscarine, which excites the parasympathetic nervous system, centrally active substances of isoxazole type, like "muscimol," "ibotenic acid," and "tricholomic acid," have been discovered in it and in other mushrooms (Wieland, 1968).

ERGOT

Widespread epidemics of ergotism, popularly known as St. Anthony's fire, were caused by a fungus, *Claviceps purpura,* infecting rye and other cereal grains during the Middle Ages. When ingested, the ergotized grain produced either cardiovascular effects, such as vasoconstriction and gangrenous symptoms, or neurologic changes, including convulsions and confusion. It is an interesting fact that the ravages of ergotism during the Middle Ages in Europe were largely alleviated by the introduction of new dietary staples such as potatoes and maize.

The first chemically pure alkaloid exhibiting the typical biologic properties of ergot was named ergotamine by Stoll and Hofmann (1970). Ergot alkaloid is a highly variable mixture. The ergot alkaloids have been used as therapeutic agents because of their ability to cause vasoconstriction and contraction of the uterus, and to affect the central nervous system. One of the alkaloids, lysergic acid diethylamide (LSD), is a well-known prototype of hallucinogens.

TABLE 1. Mycotoxicosis in Man

DISEASE (Date of outbreaks)	LOCATION (Reported)	FUNGAL TOXINS	SPECIES OF FUNGUS	SYMPTOMS
Mushroom poisoning (sporadic)	Distributed	Phallotoxin, Amatoxin	Amanita (*A. phalloides, verna, bisporigera, tenuifolia, A. muscaria*)	Vomiting, diarrhea, jaundice
		Muscarin		Hallucination, excites parasympathic nervous system
Ergotism, "St. Anthony's fire" (Middle Ages)	Central Europe, U.S.A., India	Ergot alkaloids (ergotamine, ergostine, lysergic acid derivative)	*Claviceps purpurea, pasali*	Lassitude, muscular pain, gangrenous symptoms, convulsion, confusion
Alimentary toxic aleukia or specific angina (1913, 1942–1955)	USSR, Japan	Trichothecenes	Fusaria (*F, poae, sporotrichoides, tricinctum, gramine-arum,* etc.) Cephalosporium, Trichothecium, Stachybotrys, Myrothecium, Trichoderma	Vomiting, diarrhea, fever, hemorrhagic rash, necrotic angina, leukopenia, sepsis, etc.
Onyalai (1904–1975)	Africa	Not determined	*Phoma sorghina*	Hemorrhagic bullae in the mouth, thrombocytopenia
Aflatoxicosis	Africa, India, Thailand	Aflatoxins	*Aspergillus flavus, parasiticus*	Edema, anorexia, abdominal pain and distention (ascites), jaundice, Reye's syndrome (Thai children)
Luteoskyrin poisoning	Japan*	Luteoskyrin	*Penicillium islandicum,* Mycelia Sterilia	Anorexia, edema, jaundice

*Accidental case

CARCINOGENIC MYCOTOXINS

Mycotoxicosis received attention in recent years upon discovery of a number of carcinogenic mycotoxins contaminating human and animal foods. The infrequent reports of human intoxication attributable to fungal toxins in foods may be explained by the absence of dramatic acute effects. Acute mycotoxicosis is common in livestock, fish, birds, and experimental animals that eat moldy feed polluted by mycotoxins. A series of outbreaks of unexplained hemorrhage and/or hepatic injury in domestic animals and "spontaneous" hepatic tumors in hatchery-raised fish and laboratory animals resulted from mycotoxin contamination of feed. Of major importance in this regard is a group of hepatocarcinogenic aflatoxins produced by *Aspergillus flavus* (Detroy et al., 1971; Butler, 1974). Luteoskyrin and cyclochlorotine isolated from culture of *Penicillium islandicum* were also shown to be responsible for acute hepatic damage, cirrhosis, and tumors of the liver in mice and rats. The type of injury depended on the amount of mold metabolites produced (Uraguchi, 1971; Enomoto and Ueno, 1974). Table 2 shows the animals susceptible to both the carcinogenic and the toxic actions of several important mycotoxins.

The *aflatoxins* are a group of secondary metabolites produced by *Aspergillus flavus* and *A. parasiticus*. They are a mixture of chemically related compounds, derivatives of difuranocoumarin. Aflatoxins B_2 and G_2 are dihydro derivatives of the parent aflatoxins B_1 and G_1. Aflatoxins P_1, M_1, and Q_1 are hydroxylated derivatives of aflatoxin B_1. Aflatoxin M_1 has been recovered from milk, feces, and urine of various mammals, including humans who ingested aflatoxin B_1. In monkeys and chickens, conjugates of aflatoxin P_1 have been detected in the urine in addition to M_1. Aflatoxin B_1 is the most potent carcinogen, followed by G_1, B_2, and M_1 in order of decreasing carcinogenic potency.

Aflatoxin B_1 suppresses synthesis of DNA, RNA, and protein and inhibits the polymerase in rat liver cells. These changes are reflected in the segregation of nucleolar granules and fibrils, capping of the nucleolus, and disruption of the cytoplasmic rough endoplasmic reticulum in liver cells. Aflatoxin B_1 also causes increased activity of DNase and other lysosomal enzymes in the liver or pancreas of rats. The active metabolite is thought to be an epoxide such as B_1-2,3-epoxide.

Sterigmatocystin was first isolated from a culture of *Aspergillus versicolor* and was also found to be a metabolite of *A. nidulans* and Bipolaris species. Sterigmatocystin is a derivative of difurano-xanthone and has a difuranmethoxybenzene ring in its structure, just as aflatoxins have. Consequently, these two mycotoxins can be derived from a common intermediary in their biosynthesis (Van der Watt, 1974). Sterigmatocystin is carcinogenic to the rat and newborn mouse. A high incidence of liver cell carcinoma was observed in rats surviving a dose of 0.75, or 1.5 to 2.25 mg/rat/day (Van der Watt, 1974; Fujii et al., 1976). The relatively high carcinogenic potency of sterigmatocystin and the wider distribution of this mycotoxin in Japan and South Africa suggest that it may present a more formidable danger than aflatoxins. The metabolic fate of sterigmatocystin in mammals is unknown. The biologic activity of derivatives of sterigmatocystin has not been extensively studied, but *O*-acetylsterigmatocystin is more toxic and carcinogenic than sterigmatocystin (Terao, 1977). Various derivatives of sterigmatocystin are mutagenic for bacteria, including sterigmatocystin itself, demethylsterigmatocystin, stergmatin, O-acetylsterigmatocystin, 6, 8-dimethylversicolin A and B, 5-methoxysterigmatocystin, and versicolin A.

Luteoskyrin and *rugulosin* are anthraquinone pigments. Both induce acute hepatic centrilobular toxic lesions in mice and rats and liver cell adenoma or carcinoma in mice. Rugulosin has less carcinogenic activity than luteoskyrin, but rugulosin is produced by a number of common contaminants of foodstuffs, such as *Penicillium rugulosum, P. brunneum,* and *P. tardum* (Enomoto and Ueno, 1974).

MYCOTOXINS AND HUMAN LIVER CANCER

Data from Thailand, Kenya, Mozambique, and Swaziland provide strong circumstantial evidence of a causal relationship between aflatoxin ingestion and liver cancer incidence in man (Wogan, 1976). The evidence for human aflatoxicosis has also been increasing. Outbreaks of hepatitis in both man and dogs in West India were traced to consumption of spoiled maize heavily contaminated with *Aspergillus flavus*. Aflatoxins were detected in these contaminated samples and in serum samples of a few patients (Krishnamachari et al., 1975). The symptoms of high fever and rapidly progressive jaundice and ascites and the morphologic changes in the liver suggested toxic hepatitis. There was centrilobular scarring with varying degrees of occlusion of the hepatic veins, severe cholangiolar proliferation with cholestasis, and syncytial giant cells of liver-cell origin (Tandon et al., 1977). In Thai children afla-

TABLE 2. Mycotoxins Known to Produce Cancer in Animals by Oral Administration

CARCINOGENIC MYCOTOXIN	ANIMAL SPECIES SUSCEPTIBLE TO BOTH THE CARCINOGENIC AND TOXIC EFFECTS (target organs)	REGULAR* DOSAGE (ppm in diet)	ANIMAL SPECIES SENSITIVE ONLY TO THE TOXIC EFFECTS
Aflatoxin B_1	Rat (liver, kidney, colon); trout, duck, guppy, ferret, newborn mouse, salmon, monkey, marmoset (liver); sheep (liver, nose)	0.5–1.5 (rat)	Turkey, mink, cattle, guinea-pig, swine, dog, hamster, rabbit, pheasant, quail, chicken, cat, frog
Aflatoxin G_1	Rat (liver, kidney); duck (liver)	1–3 (rat)	
Aflatoxin B_2	Rat, duck (liver)		
Aflatoxin M_1	Rat, trout (liver)		Duckling
Sterigmatocystin†	Rat, velvet monkey, newborn mouse (liver)	10–50 (rat)	Guinea pig
Luteoskyrin	Mouse, rat (liver)	30–100 (mouse)	Rabbit, chicken, monkey
Rugulosin	Mouse (liver)	200 (mouse)	Rat
Griseofulvin	Mouse (liver)	5000–10,000 (mouse)	Rat

*Oral carcinogenic dosage per day in the representative animal shown in parentheses.
†O-Acetylsterigmatocystin is more carcinogenic than sterigmatocystin in rats. Hepatocellular carcinomas were produced in 53 per cent of rats fed 10 ppm O-acetylsterigmatocystin for 52 weeks.
(From Enomoto, M.: Carcinogenicity of mycotoxin. In Uraguchi, K., and Yamayaki, M. (eds): Toxicology, Biochemistry and Pathology of Mycotoxins. Tokyo, Kodansha International, Ltd., 1977.)

toxin presents the picture of Reye's syndrome, i.e., fatty liver and encephalopathy (Shank et al., 1971). The chronic and carcinogenic effect of aflatoxins in man is potentially more important than acute toxicosis but is difficult to prove. Aflatoxin has been detected in human urine, milk, biopsy specimens of liver, and autopsied liver or other tissues by improved chemical methods of aflatoxin analysis (Pong et al., 1974) or immunoassay with antibody against aflatoxin B_1. Yet its role in neoplastic transformation of the liver is not proved by its presence in liver tissue any more than that of viruses that are frequently found in human livers with carcinomas.

In evaluating the danger from fungal toxins after long-term exposure, one must take into account the fact that experimental evidence linking cirrhosis and liver cancer has not been convincing, especially with respect to aflatoxins or sterigmatocystin. Liver cirrhosis and cancer were consistently associated with high incidence rates only in rats and mice fed moldy rice infested with *Penicillium islandicum*. The metabolites of this fungus contain both hepatocarcinogenic luteoskyrin and cirrhogenic cyclochlorotine (Miyake et al., 1960; Enomoto and Ueno, 1974). These facts suggest that in man liver cirrhosis itself is attributable to the combined effects of other factors, such as hepatitis virus, cirrhogenic mycotoxin, or nutritional agents. Further studies are needed to give definitive evidence of the relationship between liver cell cancer and carcinogenic mycotoxins other than aflatoxins. If prevention of mycotoxin contamination in human foodstuffs results in a decrease in human liver cancer it would be convincing evidence for the positive role of carcinogenic mycotoxins in human carcinogenesis (Fig. 1).

ALIMENTARY TOXIC ALEUKIA (ATA)

Alimentary toxic aleukia has been endemic in Siberia and the Amur region of the U.S.S.R. since 1913. The causal agents are the Fusarium species, which grow on overwintered moldy cereals and other crops and produce burning sensations of the mouth, anorexia, vomiting, and diarrhea

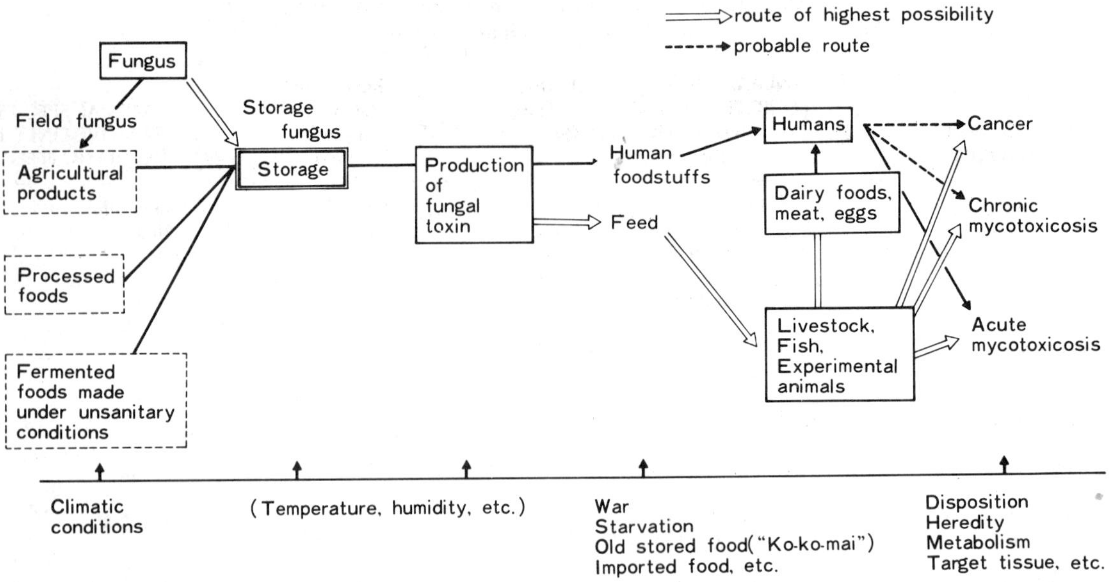

FIGURE 1. *Mycotoxin-induced diseases of man and animals: related conditions and causal relationships. (After Enomoto, M.: In Yamamura, Y., and Sugimura, T., (eds.): Cancer. Vol. 15. Tokyo, Iwanami Shoten, 1976. [Japanese].)*

(Stage I), followed by depression of all bone marrow and thrombopoiesis (Stage II). Fever, hemorrhagic rash, bleeding from gums and nose, necrotic angina, extreme leukopenia, and agranulocytosis with general sepsis occur in the final stage (Stage III) (Joffe, 1971). Death occurs six to eight weeks after ingestion of large amounts of toxin. Skin testing is reliable for toxic assay of overwintered grains. The only prophylactic measure against ATA is the elimination of polluted grain from food.

After the isolation of diacetoxyscirpenol from *Fusarium scirpi,* a number of trichothecene compounds, including T2-toxin, trichothecolon, nivalenol, fusarenon-X, verrucarin A, roridin A, and neosolaniol, have been isolated from imperfect fungi, including Cephalosporium, Fusarium, Trichoderma, Myrothecium, and Stachybotrys species (Bamburg and Strong, 1971; Smalley and Strong, 1974). At present these trichothecenes are considered to be responsible for human ATA because of their characteristic biologic activities in animals. Mice exhibit decreased activity shortly after administration of these trichothecene samples, followed by diarrhea, loss of reaction against skin stimulation, lowered body temperature, and lethargy. The animals usually die from 12 to 72 hours after administration of toxins. Histologic examination revealed marked cytotoxic changes in tissues with actively dividing cells, such as the mucosa of the gastrointestinal tract, lymph follicles, thymus, and bone marrow (Saito et al., 1969, 1974).

ONYALAI

Onyalai is an acute purpuric disease endemic in Africa since 1904. It is attributed to the mycotoxin of *Phoma sorghina* isolated from millet and grain sorghum (Rabie et al., 1975).

CARDIAC BERIBERI

Acute cardiac beriberi was an enigmatic toxic disease prevalent in Japan until 1950. Many patients still suffer from this disease in Southeast Asia. Although the disease is considered a vitamin B_1 deficiency, Japanese investigators attribute it to a fungal toxin of rice and other staple foods. Citroviridin isolated from *Penicillium citroviride* Biourge causes an acute poisoning in the cat and monkey, characterized primarily by an ascending paralysis like that in human beriberi (Uraguchi, 1971; Ueno, 1974). Cardiac damage occurs in mice given xanthoascin (Takahashi et al., 1976) from *Aspergillus candidus,* a frequent contaminant of human foodstuffs in Southeast Asia. Cardiac muscle undergoes vacuolar degeneration and necrosis followed by cardiac dilatation (Ohtsubo et al., 1976).

References

Bamburg, J. R., and Strong, F. M.: 12,13-Epoxytrichothecenes. In Kadis, S., Ciegler, A., and Ajl, S. J. (eds.): Microbial Toxins. Vol. 7. New York, Academic Press, 1971, p. 207.

Butler, W. H.: Aflatoxin. In Purchase, I. F. H. (ed.): Mycotoxins. Amsterdam, Elsevier Scientific Publishing Company, 1974, p. 1.

Detroy, R. W., Lillehoj, E. B., and Ciegler, A.: Aflatoxin and related compounds. In Ciegler, A., Kadis, S., and Ajl, S. J. (eds.): Microbial Toxins. Vol. 6. New York, Academic Press, 1971, p. 4.

Enomoto, M.: Carcinogenicity of mycotoxin. In Uraguchi, K., and Yamazaki, M. (eds.): Toxicology, Biochemistry and Pathology of Mycotoxins. Tokyo, Kodansha International, Ltd., 1977, p. 239.

Enomoto, M., Mabuchi, M., Miyata, K., Naoe, S., Takada, N., and Yamazaki, M.: Development of liver cell carcinoma in rats fed toxic rice culture *(Aspergillus versicolor))* containing 1 ppm of sterigmatocystin. St Marianna Med J 5:101, 1977.

Enomoto, M., and Ueno, I.: *Penicillium islandicum* (toxic yellowed rice) — luteoskyrin-islanditoxin-cyclochlorotine. In Purchase, I. F. H. (ed.): Mycotoxins. Amsterdam, Elsevier Scientific Publishing Company, 1974, p. 303.

Fujii, K., Kurata, H., Odashima, N., and Hatsuda, Y.: Tumor induction by a single subcutaneous injection of sterigmatocystin in newborn mice. Cancer Res 36:1615, 1976.

Joffe, A. Z.: Alimentary toxic aleukia. In Kadis, S., Ciegler, A., and Ajl, S. J. (eds.): Microbial Toxins. Vol. 7. New York, Academic Press, 1971, p. 139.

Krishnamachari, K. A. V., Baht, R. T., Nagarajan, V., and Tilak, T. B. G.: Hepatitis due to aflatoxicosis. Lancet 1:1061, 1975.

Miyake, M., Saito, M., Enomoto, M., Shikata, T., Ishiko, T., Uraguchi. K., Sakai, F., Tatsuno, T., Tsukioka, M., and Sakai, F.: Toxic liver injuries and liver cirrhosis induced in mice and rats through long term feeding with *Penicillium islandicum* Sopp-growing rice. Acta Path Jap 10:75, 1960.

Ohtsubo, K., Horiuchi, T., Hatanaka, Y., and Saito, M.: Hepato- and cardiotoxicity of xanthoascin, a new metabolite of *A. candidus* Link, to mice. Japan J Exp Med 46:277, 1976.

Pong, R. T. L., Husaini, and Karyadi, D.: Aflatoxin and primary hepatic cancer in Indonesia. Presented at the V World Congress of Gastroenterology, 1974.

Rabie, C. G., Van Rensburg, S. J., Van der Watt, J. J., and Lübben, A.: Onyalai — the possible involvement of a mycotoxin produced by *Phoma sorghina* in the aetiology. S A Med J 49:1647, 1975.

Richer, C., Paccalin, J., Larcebeau, S., Faugeres, J., Morard, J-L, and Lamant, M.: Présence d'aflatoxine B_1 dans le foie humain. Biologie Générale 223, 1975.

Saito, M., Enomoto, M., and Tatsuno, T.: Radiomimetic biologic properties of the new scirpenol metabolites of *Fusarium nivale.* Gan 60:599, 1969.

Saito, M., and Ohtsubo, K.: Trichothecene toxins of *Fusarium* species. In Purchase, I. F. H. (ed.): Mycotoxins. Amsterdam, Elsevier Scientific Publishing Company, 1974, p. 263.

Shank, R. C., Bourgeois, C. H., Keshamra, N., and Cahndavimol, P.: Aflatoxins in autopsy specimens from Thai children with an acute disease of unknown aetiology. Food Cosmet Toxicol 9:501, 1971.

Smalley, E. B., and Strong, F. M.: Toxic trichothecenes. In Purchase, I. F. H. (ed): Mycotoxins. Amsterdam, Elsevier Scientific Publishing Company, 1974, p. 199.

Stoll, A., and Hofmann, A.: The chemistry of ergot alkaloids. In Pelletier, S. W. (ed.): Chemistry of the Alkaloids. New York, Van Nostrand Reinhold, 1970, p. 267.

Takahashi, C., Yoshihira, K., Natori, S., and Umeda, M.: Xanthoascin, a new metabolite isolated from *Aspergillus candidus.* Chem Pharm Bull 24:613, 1976.

Tandon, B. N., Krishnamurthy, L., Koshy, A., Tandon, H. D., Ramalingaswami, V., Mathur, M. M., and Mathur, P. D: Study of an epidemic of jaundice, presumably due to toxic hepatitis, in Northwest India. Gastroenterology 72:488, 1977.

Terao, K., and Ueno, Y.: Morphological and functional damage to cells and tissues. In Uraguchi, K., and Yamazaki, M. (eds.): Toxicology, Biochemistry and Pathology of Mycotoxins. Tokyo, Kodansha International, Ltd., 1977, p. 189.

Ueno, Y.: Citrioviridin from *Penicillium citreo-viride* Biourage. In Purchase, I. F. H. (ed.): Mycotoxins. Amsterdam, Elsevier Scientific Publishing Company, 1974, p. 283.

Uraguchi, K.: Pharmacology of mycotoxins. In Raskova, H. (ed.): International Encyclopedia of Pharmacology and Therapeutics. Section 71. Oxford, Permagon, 1971, p. 143.

Van der Watt, J. J.: Sterigmatocystin. In Purchase, I. F. H. (ed.): Mycotoxins. Amsterdam, Elsevier Scientific Publishing Company, 1974, p. 369.

Wieland, T.: Poisonous principles of mushrooms of the genus Amanita. Science 159:946, 1968.

Wogan, G. N.: The induction of liver cell cancer by chemicals. In Cameron, H. M., Linsell, D. A., and Warwick, G. P. (eds.): Liver Cell Cancer. Amsterdam, Elsevier Scientific Publishing Company, 1976, p. 121.

Parasitology

16 CLASSIFICATION AND ANATOMY OF PARASITES

D. L. Lee, B.Sc., Ph.D.

The parasites that will be mentioned in this chapter belong to the Protozoa, Platyhelminthes, Nematoda, and Acanthocephala.

PHYLUM PROTOZOA

The Protozoa are regarded as a phylum within the animal kingdom but consist of a single cell, unlike the rest of the animal kingdom, in which the cells are grouped to form tissues and organs. The taxonomic position of the Protozoa is controversial, but the most commonly accepted scheme is that proposed by the Society of Protozoologists, as modified by Baker in 1973. Those groups that contain parasites of man are given below.

SUBPHYLUM I. SARCOMASTIGOPHORA

Locomotor organs are flagella, pseudopodia, or both.

Superclass 1. Mastigophora (The Flagellates)

One or more flagella present on trophozoite; asexual reproduction by binary fission; sexual reproduction unknown in many groups. Free-living and parasitic classes.

Class Zoomastigophorea

Order Kinetoplastida. One to four flagella; kinetoplast.

SUBORDER TRYPANOSOMATINA. One flagellum, free or attached to body by undulating membrane (Fig. 1); all species parasitic (*Trypanosoma, Leishmania*).

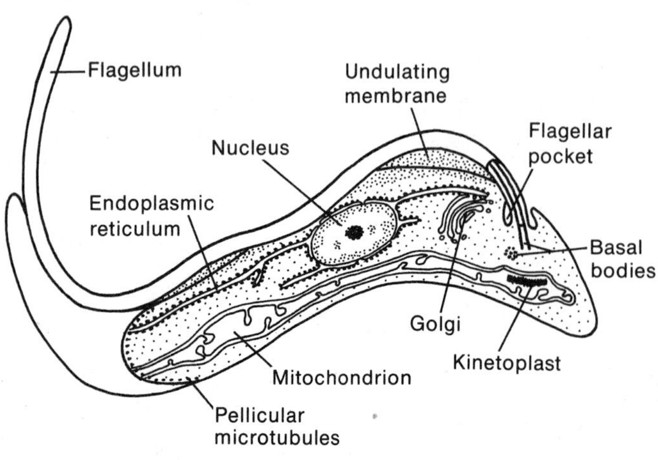

FIGURE 1. *Diagram to show the internal structure of a* Trypanosoma *species, as revealed by electron microscopy. Note the elongate mitochondrion with tubular cristae, the undulating membrane and the kinetoplast.*

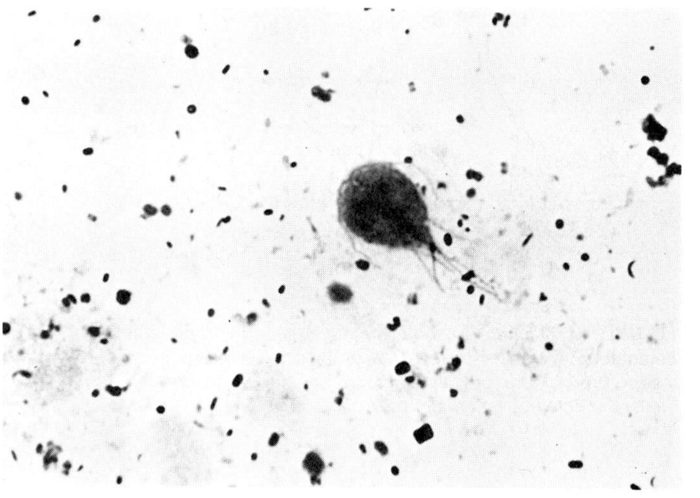

FIGURE 2. *Photomicrograph of Giardia. Note the two nuclei and four pairs of flagella.*

Order Retortamonadida. One to four flagella, one runs posteriorly and is associated with a cytostome; harmless parasites of man (*Retortamonas, Chilomastix*).

Order Diplomonadida. Four pairs of flagella; body bilaterally symmetric; two similar nuclei (*Giardia*) (Fig. 2).

Order Trichomonadida. Four to six flagella, one trailing and associated with undulating membrane if present; axostyle and parabasal body; no cystic stage (*Trichomonas, Pentatrichomonas*) (Fig. 3).

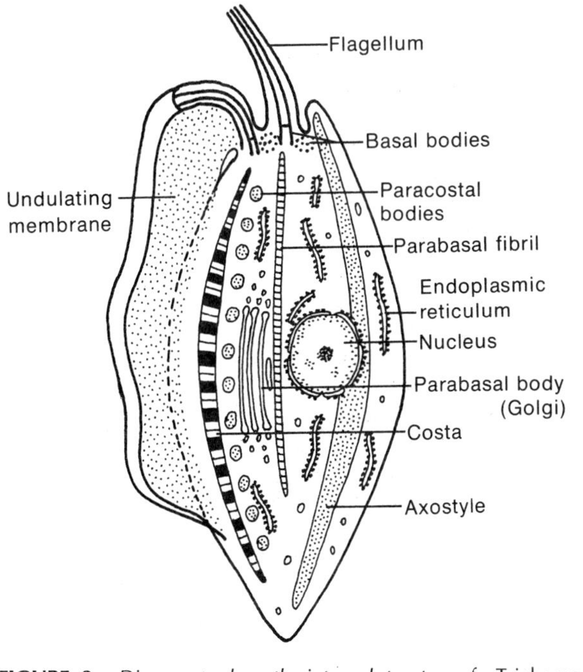

FIGURE 3. *Diagram to show the internal structure of a* Trichomonas *species as revealed by electron microscopy. Note the various rodlike structures (costa, parabasal fibril, axostyle), the anterior flagella (4 in* Trichomonas vaginalis*), the recurrent flagellum, which is associated with the undulating membrane, and the lack of mitochondria.*

Superclass 2. Opalinata

Parasites of amphibia, fish, and snakes; numerous cilia in rows over body; two to many nuclei of one type.

Superclass 3. Sarcodina

The amebae; locomotion by means of pseudopodia.

Class Rhizopodea

Subclass Lobosia

Pseudopodia broad and blunt (lobopods), rarely long and thin (filiform).

Order Amoebida. Naked; single nucleus; multiply by binary fission; many produce resistant cysts (*Entamoeba, Hartmanella, Naegleria*) (Fig. 4).

SUBPHYLUM II. SPOROZOA

Produce simple spores containing one to many sporozoites (Fig. 5); cilia and flagella absent except on microgametes of some; single nucleus.

Class 1. Telosporea

Reproduction sexual and asexual; spores and sporozoites usually present; move by gliding or flexing body.

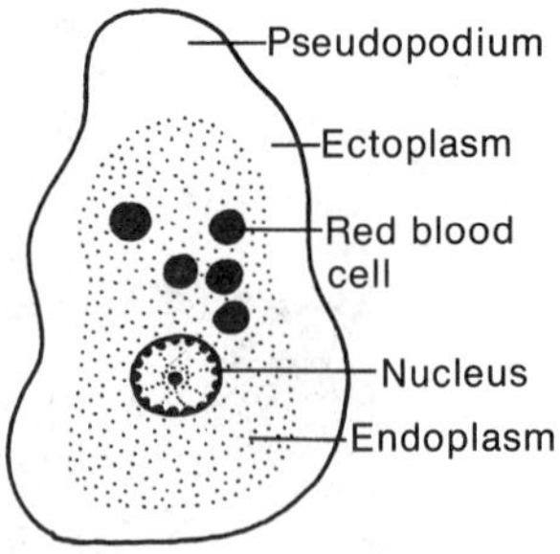

FIGURE 4. *Diagram to show the structure of a trophozoite of* Entamoeba histolytica *as revealed by light microscopy. Note the ingested red blood cells, the characteristic appearance of the nucleus, the ectoplasm, and the endoplasm.*

Subclass Coccidia

Mature trophozoites are intracellular parasites.

Order Eucoccidia. Schizogony (a type of asexual reproduction) present; asexual and sexual phases in life cycle; parasitic in epithelial and blood cells of vertebrates and invertebrates.

SUBORDER EIMERIINA. Micro- and macrogametes develop independently (no syzygy); zygote nonmotile; oocyst present; sporozoites (Fig. 5) enclosed in sporocyst. (*Eimeria, Isospora, Toxoplasma.*)

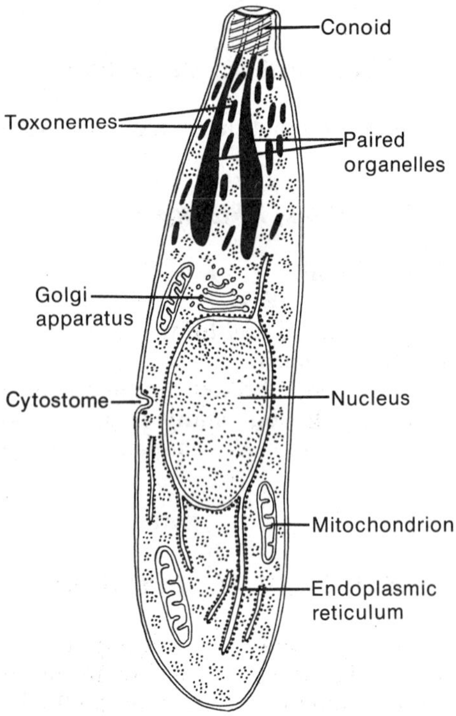

FIGURE 5. *Diagram to show the internal structure of a sporozoite, as revealed by electron microscopy. Note the paired organelles and associated taxonemes or dense bodies (the function of these is unknown but they may be concerned with penetration of the host cell, the conoid with its reinforcing rings, the mitochondria, and the cytostome.*

SUBORDER HAEMOSPORINA. Micro- and macrogamete develop independently (no syzygy); zygote motile; oocyst present; sporozoites naked; schizogony in vertebrate host and sporogony in invertebrate host (*Plasmodium, Haemoproteus, Hepatocystis*) (Fig. 6).

Class 2. Piroplasmea

No spores; small, nonpigmented parasites of erythrocytes of vertebrates, occasionally in other cells; transmitted by an invertebrate host, usually a tick.

Order Piroplasmida. (*Babesia, Theileria.*)

SUBPHYLUM III. CNIDOSPORA

Spores possess one or more polar filaments; parasites of fish and invertebrates, including the protozoa.

SUBPHYLUM IV. CILIOPHORA

Cilia are present in at least one stage of life cycle; two different types of nucleus (Fig. 7); sexual reproduction by conjugation.

Class 1. Ciliatea

Subclass Holotrichia

Cilia simple, uniform over body; buccal cilia usually absent.

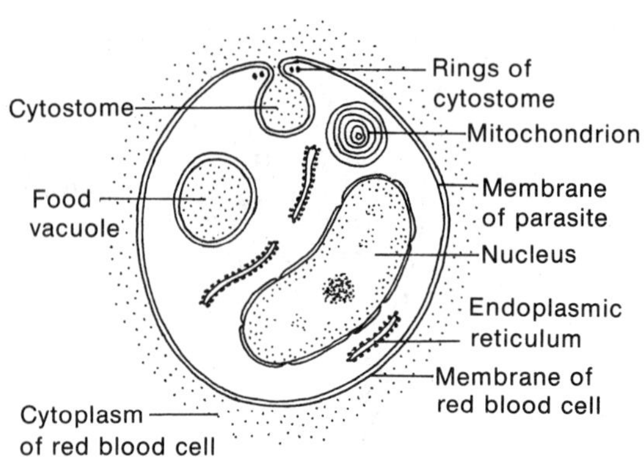

FIGURE 6. *Digram to show the internal structure of a trophozoite of* Plasmodium *as revealed by electron microscopy. Note the membrane of the red blood cell around the parasite, the cytostome, the food vacuole containing cytoplasm from the red blood cell, and the so-called "mitochondrion," which lacks cristae but sometimes contains whorls of membranes.*

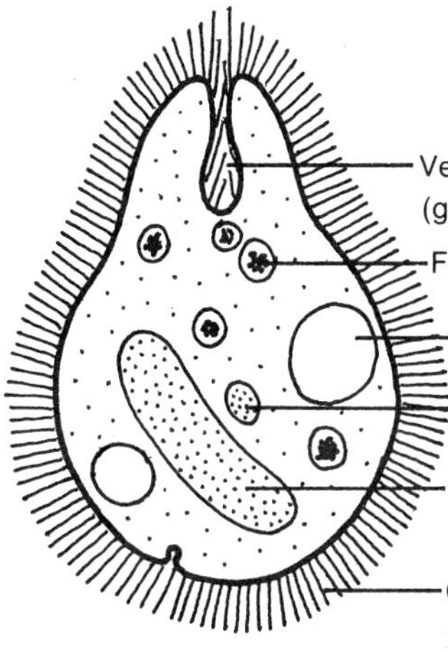

FIGURE 7. *Diagram to show the structure of* Balantidium coli *as revealed by light microscopy. Note the uniform cilia over the surface, the micro- and the macronuclei, the two contractile vacuoles, the vestibule or gullet at the base of which lies a cytostome, and the food vacuoles.*

Order Trichostomatida. Cytostome at base of a vestibule; cilia normally uniform but asymmetric in some (*Balantidium*) (Fig. 7).

Order Hymenostomatida. Cilia of buccal cavity fused to form membranelles; cilia uniform over body (*Tetrahymena*). (Most species free living but much used in study of protozoan physiology.)

For general references to the classification of protozoa see Honigberg et al., 1964; Corliss, 1961; Kudo, 1966; Levine, 1961; Garnham, 1966; and Baker, 1973.

ANATOMY OF PROTOZOA

The protozoan cell resembles cells of higher animals in that it may contain some, or all, of the structures (organelles) found in the metazoan cell. The protozoon consists of a mass of cytoplasm enclosed by a limiting membrane (plasmalemma or pellicle) and contains one or more nuclei, mitochondria, endoplasmic reticulum, ribosomes, Golgi apparatus (dictyosome, parabasal body), centrioles, flagella, cilia, microtubules, and fibrils. Some protozoa, and specific stages in the life cycle of others, lack certain of these organelles. In some groups — for example, the Sarcodina — the cytoplasm is divided into an outer, clearer ectoplasm and an inner, denser endoplasm that contains most of the organelles (Fig. 4). Some protozoa contain contractile vacuoles (Fig. 7) and some contain skeletal rods (Fig. 3), which do not occur in metazoan cells.

The nucleus is similar to that of the metazoan cell. The ciliates have two different types of nucleus; one is the sexual nucleus (micronucleus), and the other is the asexual nucleus (macronucleus) (Fig. 7). In *Entamoeba* the nucleoprotein is arranged around the periphery of the nuclear membrane in a characteristic pattern (Fig. 4).

The mitochondria are similar to those found in the cells of higher organisms, but the cristae (infoldings of the inner membrane) are usually tubular rather than plate-like (Figs. 1, 8). Certain species are apparently anaerobic and lack mitochondria (*Trichomonas* (Fig. 3), *Entamoeba*). Some species of malaria parasites lack true mi-

FIGURE 8. *Electron micrograph of a section through a macrogamete of* Eimeria *to show the structure of the mitochondria. Note the tubular nature of the cristae.*

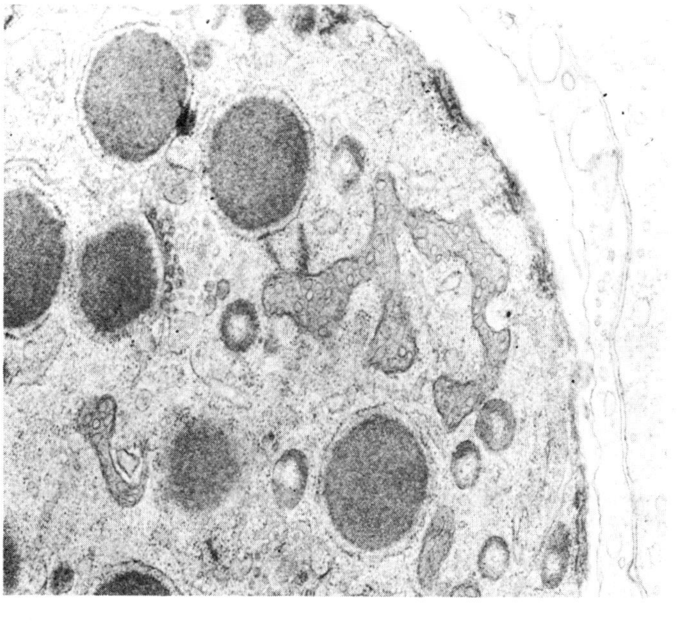

tochondria in their asexual stages but contain membranous organelles that may perform a similar function (Fig. 6). Members of the Trypanosomatina possess a spherical or rod-shaped structure called the kinetoplast; this contains DNA and lies within a single, large mitochondrion (Fig. 1).

Rough and smooth endoplasmic reticulum, ribosomes, and Golgi apparatus (dictyosome) are similar to those found in metazoan cells. The parabasal body found in the Mastigophora is apparently a type of Golgi apparatus (Figs. 1, 3).

Many protozoa possess contractile flagella or cilia on their body surface (Figs. 3, 7). These perform a locomotory, and sometimes an attachment, function. Cilia are usually shorter and more numerous than flagella, but they have a similar internal structure consisting of nine pairs of microtubules arranged around the periphery of the flagellum and two single fibrils in the center (Fig. 9); this is called the axoneme. The whole flagellum is enclosed by an extension of the limiting membrane of the protozoon. Cilia and flagella arise from a basal body or kinetosome within the cell (Figs. 1, 3). This basal body is a hollow cylinder, about 1.5 μm in diameter and is composed of nine outer double or triple microtubules that connect with the outer microtubules of the flagellum; the two central microtubules of the flagellum are not present in the basal body. A basal plate marks the junction of the basal body with the axoneme of the flagellum. In certain species (*Trypanosoma, Trichomonas*) an undulating membrane is associated with the flagellum and is an extension of the plasmalemma of the body surface (Figs. 1, 3).

In many protozoa the outer, limiting membrane is a typical membrane (plasmalemma), but in certain groups it is a more complicated structure and is called the pellicle. The pellicle consists of two or more unit membranes in some species (malaria parasites) and may have a sculptured appearance, as in many ciliates.

Cysts are produced by many parasitic protozoa, and the cyst wall forms a protective covering for the parasites when they spend part of their life cycle outside of their host (*Entamoeba, Eimeria, Isospora*). Granules arise inside the protozoan and are secreted to the exterior, where their contents coalesce and harden to form the cyst wall. Cysts are produced by some species of parasitic protozoa that use a vector as a means of transmission to another host (oocysts of malaria parasites) and by other species as a means of protecting the parasite from the defenses of the host (tissue cysts of *Toxoplasma*). These cysts usually have a softer and more flexible wall.

Skeletal structures, fibers, and microtubules are found in several groups of parasitic protozoa. Fibrils are present in the cytoplasm of many protozoa and may have a skeletal or contractile function. Microtubules are often found beneath the limiting membrane (Fig. 1) as well as in the flagellum (Figs. 1, 3, 9) and may have a skeletal or locomotory function. Some may also play a role in the transport of materials within the cell. Larger skeletal structures, such as the costa and the axostyle of *Trichomonas* (Fig. 3), are also found in certain groups of protozoa.

Contractile vacuoles are unusual in parasitic protozoa but they do occur in parasitic ciliates (*Balantidium*) (Fig. 7) and in the amebae *Naeg-*

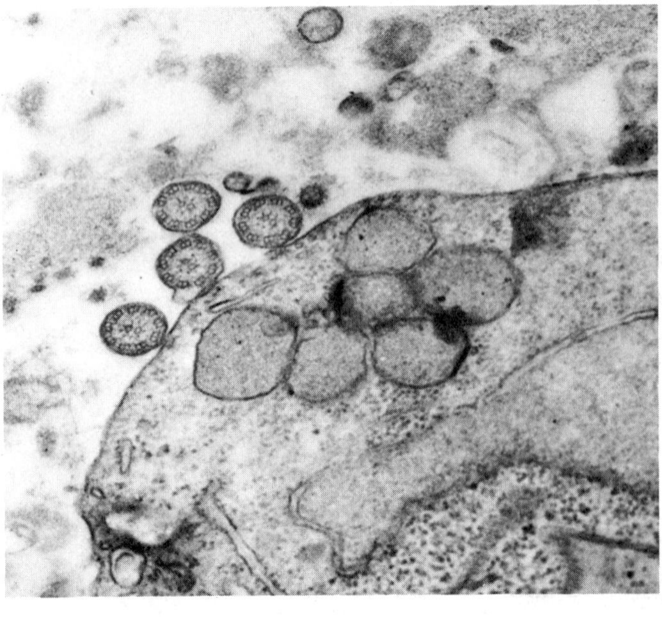

FIGURE 9. *Electron micrograph of a section through the flagella of* Trichomonas *to show the internal structure of the flagellum.*

leria and *Hartmanella*, which occasionally infect man. The contractile vacuole is a vesicle or vacuole that pulsates in a rhythmic manner and opens to the outside through a small pore in the limiting membrane.

Many parasitic protozoa possess a mouth, or cytostome, through which food particles are ingested. This cytostome may be a small depression on the surface of the body (Sporozoa) (Figs. 5, 6) or may lie at the base of a groove or cytopharynx, where it is usually associated with one or more flagella (Fig. 7).

Further reading on the structure of protozoa: (Baker, 1973; Dogiel, 1965; Mackinnon and Hawes, 1961; Adam *et al.*, 1971; Sleigh, 1973; Vickerman and Cox, 1967).

PHYLUM PLATYHELMINTHES

The platyhelminthes are worms that are usually flattened dorsoventrally, bilaterally symmetric, and without true segmentation. The digestive tract is incomplete or absent. The excretory system is the flame cell type and there is no body cavity, circulatory system, or respiratory system. The nervous system consists of a pair of anterior ganglia from which extend longitudinal nerves connected by transverse commissures. Individuals are hermaphroditic, with a few exceptions. The body is covered by a cytoplasmic epidermis or tegument (formerly thought to be a secreted cuticle in the Trematoda and Cestoidea).

The classification given here is incomplete because particular attention was given to parasites that attack man. For details of the complete classification see the references at the end of each section.

Class Trematoda (Flukes)

Subclass Digenea

Adult worms are typically leaf-shaped, but cylindric forms exist. A cup-shaped sucker surrounds the mouth (oral sucker) and a second sucker (acetabulum) is situated on the ventral surface. Almost all species are endoparasitic as adults and live in the alimentary tract, bile duct, blood vessels, lungs, bladder, or other organs of the vertebrate host.

The classification of the Digenea is controversial. The system described here is based on the system proposed by La Rue (1957) and used by Erasmus (1972) and by Noble and Noble (1976), but reference should be made to Yamaguti (1958) and to Dawes (1956).

Superorder Anepitheliocystidia

Cercaria with thin excretory bladder; tails simple or forked; stylet absent.

Order Strigeatoiida. Miracidia with one or two pairs of flame cells; cercaria with forked tail.

SUBORDER STRIGEATA

Superfamily Schistosomatoidea (=Bilharziidae). Sexes separate; adults live in blood vessels of mammals and birds. (*Schistosoma.*)

Order Echinostomida

SUBORDER ECHINOSTOMATA

Superfamily Echinostomatoidea. (Fasciola hepatica, Fasciolopsis buski, Echinostoma ilocanum.)

SUBORDER PARAMPHISTOMATA

Superfamily Paramphistomatoidea. (Gastrodiscoides = Gastrodiscus hominis.)

Superorder Epitheliocystida

Cercaria with thick walled excretory bladder; tail of cercaria single, reduced in size, or absent; miracidium with a single pair of flame cells.

Order Plagiorchiidae

SUBORDER PLAGIORCHIATA

Superfamily Plagiorchioidea. (Dicrocoelium dendriticum; Paragonimus westermani.)

Order Opisthorchiida

SUBORDER OPISTHORCHIATA

Superfamily Opisthorchioidea. (Opisthorchis = Clonorchis sinensis; Heterophyes heterophyes; Metagonimus yokogawai.)

ANATOMY OF THE DIGENEA

The life cycle of the digeneans consists of some or all of the following stages: egg, miracidium, mother sporocyst, daughter sporocyst, redia, cercaria, metacercaria, and adult.

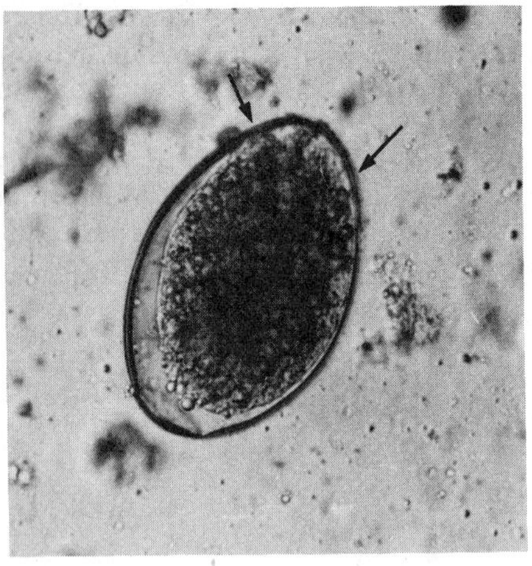

FIGURE 10. *Photomicrograph of the egg of* Fasciola hepatica. *Note the operculum at one end of the egg shell (arrows).*

The egg shell should correctly be referred to as an egg capsule. It is usually oval, light to dark brown, and operculated at one end (Fig. 10). The egg capsule of the human blood flukes (*Schistosoma*) is nonoperculate and bears a terminal spine (*S. haematobium*), a lateral spine (*S. mansoni*) (Fig. 11), or a lateral knob (*S. japonicum*).

The miracidium is a ciliated, bullet-shaped, free-swimming larva in most species (Figs. 12, 13). Ciliated epidermal plates on the outer surface are separated from each other by extensions of the subepidermal layer that lie beneath the plates (Fig. 13). The anterior end of the miracidium bears an apical papilla that lacks cilia but carries the openings of the apical and penetration glands. Most miracidia possess a pair of eyespots, each of which is a pigmented cup containing a lens. The cerebral mass, which lies behind the eyespots, is connected by nerves to the eyespots, the sense organs, the apical papilla, and the posterior end of the larva. Sense organs occur on the apical papilla and also laterally on the body of the miracidium. The posterior part of the miracidium contains the germinal cells that may form clusters of cells called germ balls. The miracidium possesses one or more pairs of flame cells (Figs. 13, 21).

In most species the miracidium develops into a sporocyst (Fig. 14) after penetration of a suitable snail host. It loses its ciliated epidermal plates, the subepidermal layer develops into the new outer layer or tegument, and the larva becomes an elongated sac containing germinal cells. These germinal cells subsequently give rise to daughter sporocysts (as in the schistosomes) or to rediae that escape through a birthpore.

The redia is not present in all species. It is an elongate developmental stage that possesses a mouth, muscular pharynx, and a simple sac-like gut (Fig. 15). It usually bears a ridge-like collar around its anterior end and a pair of lobe-like lappets near the posterior end. The outer covering is a thin tegument (Fig. 16) that covers circular and longitudinal muscles and a cellular layer. The body cavity contains germinal cells, germinal masses, and developing daughter rediae or cercariae (Fig. 15). The excretory system consists of flame cells that open into two lateral excretory canals.

The cercaria arises from germinal masses within the daughter sporocysts (as in the schistosomes) or within the rediae. They vary in appear-

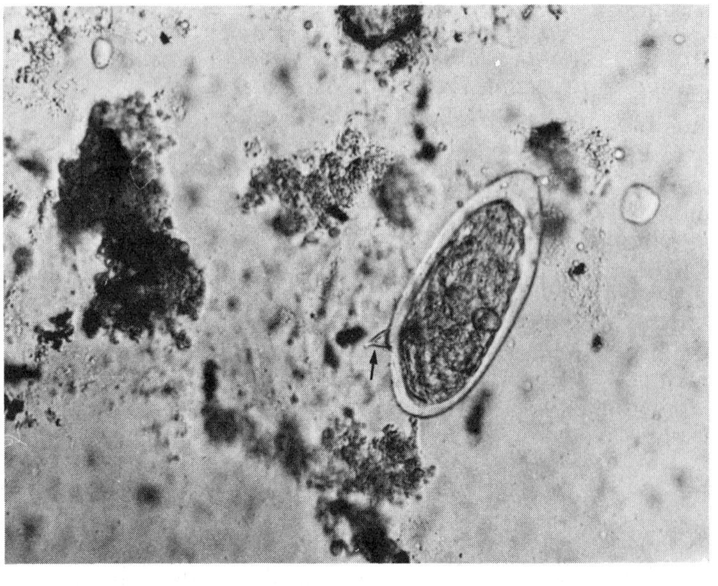

FIGURE 11. *Photomicrograph of the egg of* Schistosoma mansoni. *Note the lack of an operculum and the characteristic spine on the egg shell (arrow).*

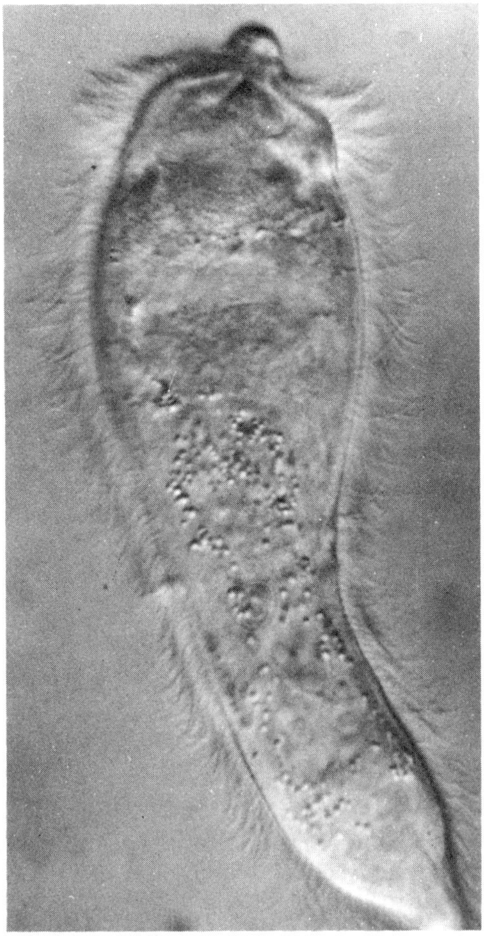

FIGURE 12. *Photomicrograph of the miracidium of* Schistosoma mansoni *(Nomarski interference microscopy). Note the ciliated epidermis and apical papilla.*

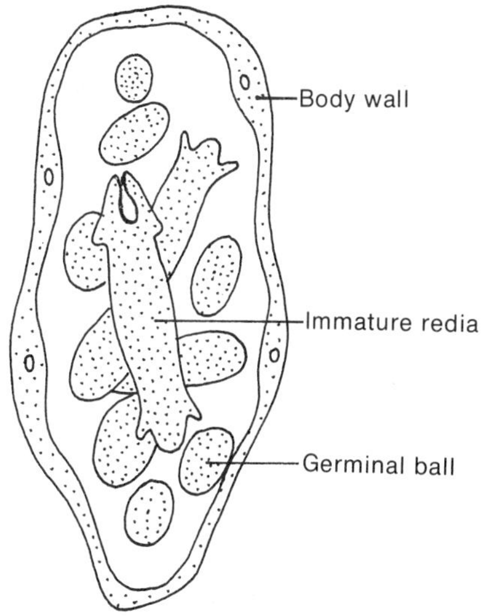

FIGURE 14. *Diagram of a sporocyst. Note the sac-like appearance and the germinal balls, which develop into rediae (or into daughter sporocysts or cercariae in those species that lack a redia stage).*

ance according to the species (Figs. 17 *A, B*) but have a cylindric or globular body and a muscular tail that is used in locomotion. The tail may be simple or forked (Figs. 17 *A, B*); both the body and the tail are covered by a cytoplasmic tegument. An oral sucker surrounds the mouth, and most species also have a ventral sucker. The rudimentary digestive system consists of a mouth, muscular pharynx, esophagus, and a pair of blind-ended ceca. A cerebral ganglion lies just behind the oral sucker and supplies nerves to various parts of the

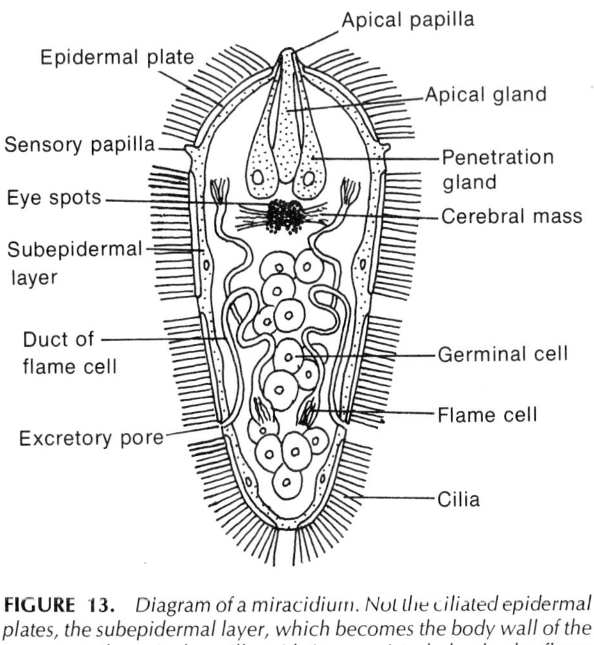

FIGURE 13. *Diagram of a miracidium. Not the ciliated epidermal plates, the subepidermal layer, which becomes the body wall of the sporocyst, the apical papilla with its associated glands, the flame cells with their collecting ducts, and the germinal balls.*

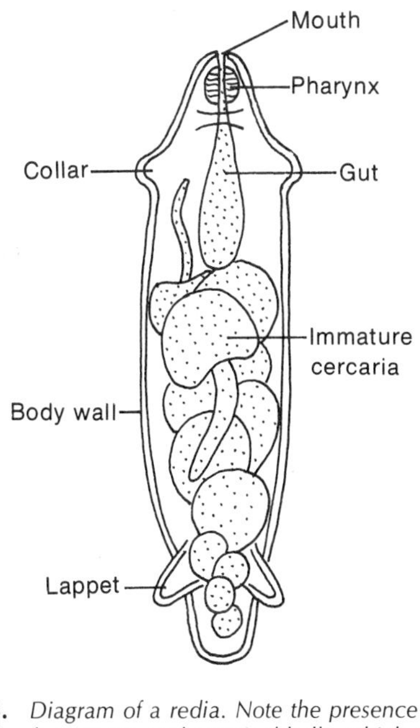

FIGURE 15. *Diagram of a redia. Note the presence of a simple gut, the collar, lappets (2), and germinal balls, which develop into cercariae.*

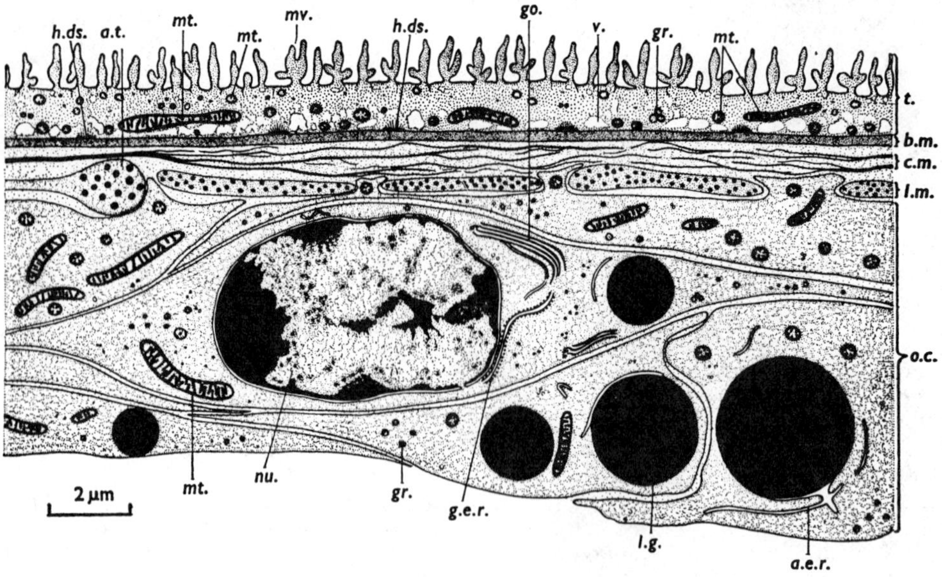

FIGURE 16. *Diagram of the ultrastructure of the body wall of a redia. Note the extension of the surface into branched microvilli.* a.e.r., *agranular endoplasmic reticulum;* a.t., *axon terminal;* b.m., *basement membrane;* c.m., *circular muscle;* g.e.r., *granular endoplasmic reticulum;* go., *Golgi complex;* gr., *granule;* h.ds., *half-desmosome;* l.g., *lipid globule;* l.m., *longitudinal muscle;* mt., *mitochondrion;* mv., *microvillus;* nu., *nucleus;* o.c., *overlapping cell;* t., *tegument;* v., *vacuole. (Reproduced with permission from G. Rees:* Parasitology *56, 1966.)*

cercaria. Some species possess pigmented eyespots. The cercaria also possesses a well-developed excretory system of flame cells and collecting ducts.

Gland cells secrete the metacercarial cyst in those species that encyst, or secrete materials such as an adhesive substance or enzymes if the cercaria penetrates a second intermediate host or the final host. Penetration glands (Fig. 17 *A*) are a characteristic feature of the cercariae of schistosomes and they secrete several substances. Cystogenous gland cells, which are usually more widely scattered around the body of the cercaria, give rise to the wall of the metacercarial cyst (Fig. 17 *B*) in those species that form a cyst (*Fasciola hepatica*). The tail of the cercaria, which is well-supplied with muscles and mitochondria, is shed when the cercaria penetrates another host or encysts.

The cercaria of many species (*Fasciola hepati-*

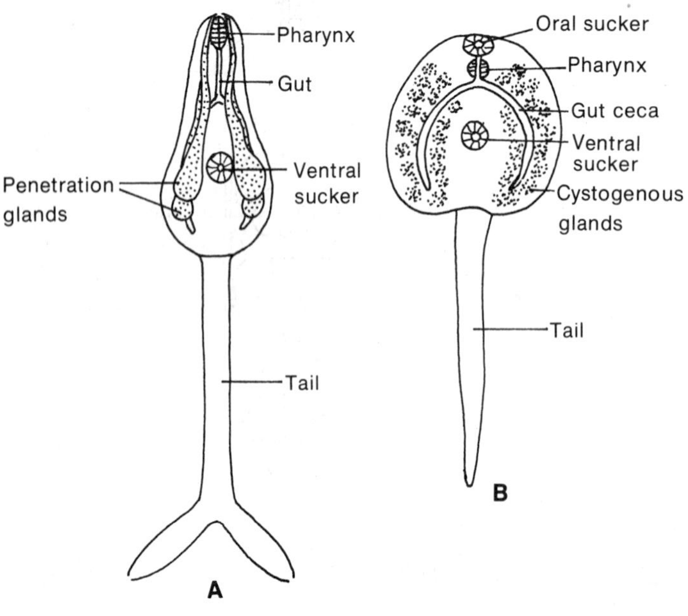

FIGURE 17. A, *Diagram of a schistosome-type cercaria. Note the bifurcate tail, the penetration glands, and the sucker.* B, *Diagram of a* Fasciola-*type cercaria. Note the unbranched tail, the lack of penetration glands, the cystogenous glands, which produce the wall of the metacercaria, and the suckers.*

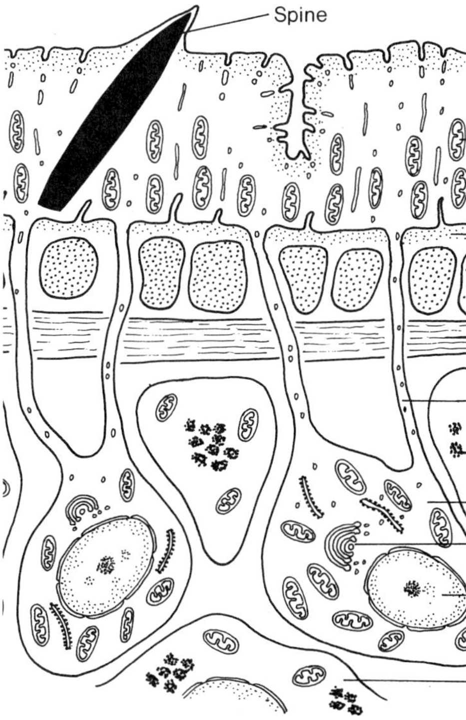

FIGURE 18. *Diagram to show the ultrastructure of the body wall of* Fasciola hepatica. *Note the syncytial cytoplasmic nature of the tegument, the tegumental cells, which lie below the tegument but connect with it by processes, the muscle layers, and the spine. The nuclei of the tegument lie in the tegumental cells and not in the tegument itself. (Based on Threadgold; Quart J Microscopical Sci. 104, 1963.)*

ca) encyst upon vegetation, whereas others penetrate a second intermediate host (*Opisthorchis sinensis*) and encyst inside the tissue; these stages are called the metacercaria. The cercaria of schistosomes penetrates the skin of the final host and becomes an active schistosomule; the schistosomule lacks the tail of the cercaria and has discharged its penetration glands.

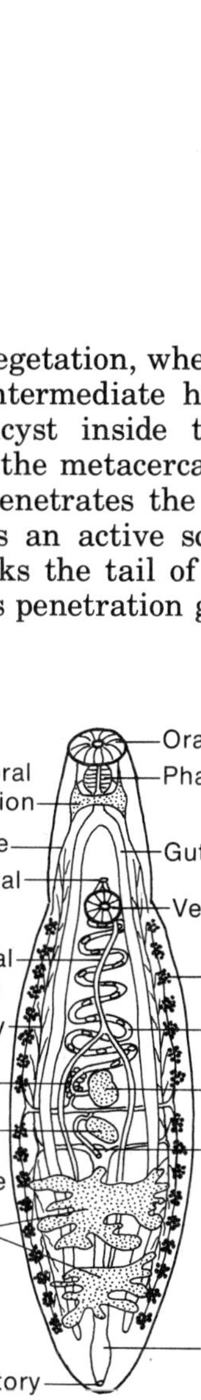

FIGURE 19. *Diagram to show the morphology of a digenetic trematode* (Opisthorchis-Clonorchis sinensis).

The adult digenean is covered with a syncytial cytoplasmic epidermis called the tegument, the nucleated portions of which lie below the main outer layer (Fig. 18). The tegument contains mitochondria, dense rod-shaped bodies of unknown function, endoplasmic reticulum, vacuoles, and, in some species, spines that extend from the surface of the tegument (Fig. 18). Nuclei and lateral cell walls are not present in the outer covering. Nucleated cells make contact with the main tegument by means of cytoplasmic tubes (Fig. 18). These cells contain endoplasmic reticulum, mitochondria, Golgi apparatus, and ribosomes. The tegument of adult schistosomes is covered by a series of membranes that may be important in evasion of the immune response of the host.

The terminal mouth of the adult leads to a muscular pharynx, esophagus, and a pair of blind-ending intestinal ceca in most species (Fig. 19). The wall of the ceca consists of a single layer of epithelial cells lying upon a basement lamella; a thin layer of muscle surrounds the cecal wall. The surface of the cells that line the cecal lumen is covered with long microvilli (Fig. 20). These intestinal cells are capable of secretion and absorption.

The excretory system consists of flame cells (Fig. 21) and collecting ducts, which open either directly to the outside from two large lateral excretory tubules or into a bladder that opens to the exterior (Fig. 19). A system of lymph channels occurs in some species but little is known about them.

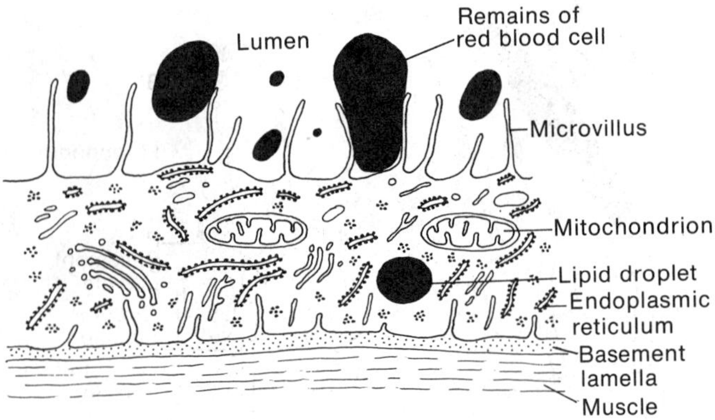

FIGURE 20. *Diagram to show the ultrastructure of part of the gut ceca of* Schistosoma mansoni. *Note the long, widely spaced microvilli and the partly broken-down red blood cells. (Drawn from an electron micrograph in Morris, A. P.* Experientia, *24, 1968).*

The nervous system consists of a pair of cerebral ganglia from which longitudinal nerves arise (Fig. 19). Transverse commissures connect these longitudinal nerves. Nerves extend from the ganglia and from the longitudinal nerves to supply the sense organs, suckers, muscles, and other organs of the body.

A layer of circular and a layer of longitudinal muscles lie beneath the tegument (Fig. 18). The muscle fibers contain thick and thin myofilaments, and an extension of the sarcoplasm contains the nucleus, mitochondria, and endoplasmic reticulum of the muscle cell. The muscle fibers appear nonstriated in the adult worm, but the muscles in the tail of the cercaria are striated.

The parenchyma consists of large cells that lie in contact with the various tissues and organs of the body; there is no body cavity in the adult worm.

Most digeneans are hermaphroditic but the sexes are separate in the schistosomes. Figure 19 shows a diagram of the reproductive system of a typical digenean. There is usually a single ovary and one or more testes, which may be either compact or branched organs. Yolk for the eggs is produced by paired vitelline glands that may be branched or compact. A characteristic chamber called the ootype lies between the oviduct and the uterus. A series of small cells that form the Mehlis' gland are associated with the ootype. This combined structure is important in the formation of the egg capsule. The uterus, which is usually long and full of eggs, opens to the exterior at the genital atrium. Spermatozoa pass from the testes along the vas deferens to a seminal vesicle and an eversible cirrus.

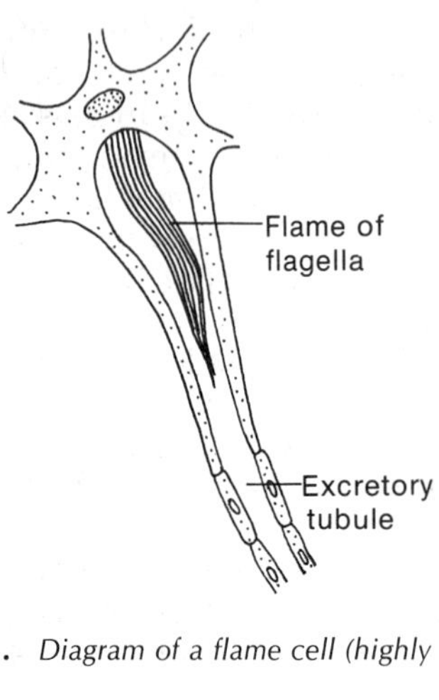

FIGURE 21. *Diagram of a flame cell (highly magnified).*

For further reading on the structure and anatomy of trematodes see Hyman, 1951a; Erasmus, 1972; Smyth, 1966; Dawes, 1956; and Yamaguti, 1958.

Class Cestoidea

The class Cestoidea consists of hermaphroditic worms, which, as adults, are parasitic in the alimentary tract of vertebrates. They are commonly called cestodes or tapeworms.

Subclass Eucestoda (= Cestoda)

The long, ribbon-like body consists of 4 to 4000 proglottids and a scolex with adhesive organs at the anterior end of the worm.

Order Pseudophyllidea. Scolex with two bothria (Fig. 22*B*); length varies from a few millimeters to 25 meters. Mainly parasites of fish except for *Dibothriocephalus latus (= Diphyllobothrium latum)*, which is parasitic in man.

Order Cyclophyllidea (= Taenoidea). Scolex with four muscular suckers (acetabula) and often with one or more rows of hooks on the tip (rostellum) of the scolex (Fig. 22*C*); proglottids well defined. Includes most tapeworms of higher animals and man (*Taenia, Echinococcus, Hymenolepis*).

For further reading on the classification of tapeworms see Wardle and McLeod, 1952; Wardle et al., 1974; Smyth, 1969; Voge, 1969; Hyman, 1951; Joyeux and Baer, 1961; and Yamaguti, 1959.

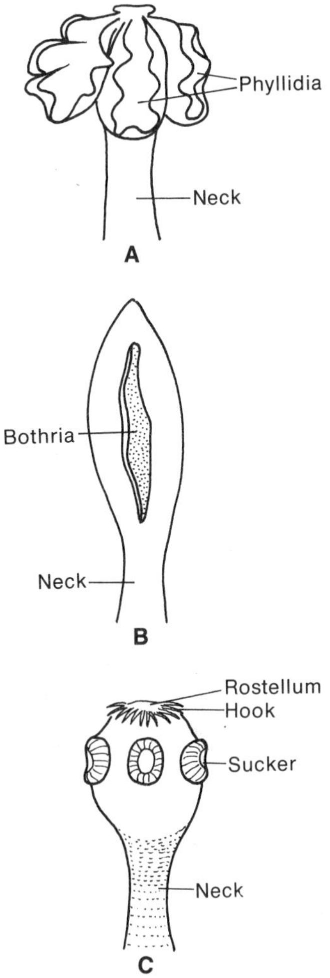

FIGURE 22. *Diagrams of scoleces of some cestodes.* A, *Tetraphyllidea-type, note the trumpet-like attachment organs (phyllidia);* B, *Pseudophyllidea-type* (Dibothriocephalus latus), *note the slit-like bothria, one of two;* C, *Cyclophyllidea-type* (Taenia solium), *note the rostellum bearing hooks (these are absent in some species) and the muscular suckers (acetabula).*

ANATOMY OF CESTOIDEA

All tapeworms that parasitize man belong to the Eucestoda. They are mostly slender, elongate, white or cream-colored worms with a body consisting of few to many segments (proglottids). Each proglottid contains a complete set of male and female reproductive organs at some stage in development. There are no circulatory, respiratory, or skeletal organs and there is no alimentary system. Larval stages develop in one or more intermediate hosts.

The head is an attachment organ (scolex) that is armed with suckers of various sorts (bothria are slit-like grooves, Fig. 22*B*; phyllidia are trumpet-like or leaf-like structures, Fig. 22*A*; acetabula are muscular suction cups, Fig. 22*C*, 24) and may or may not also possess hooks. The anterior end of the scolex may be formed into a rostellum (Figs. 22*C*, 24). The neck lies behind the scolex and is a zone of proliferation from which immature proglottids are formed. These proglottids pass through a phase when they increase in size, develop gonads (mature proglottids), and fill the uterus with eggs (gravid proglottids) (Fig. 24). These proglottids form the thin, tape-like body of the worm and are constantly renewed from the neck region. Gravid proglottids drop off the end of the tape and are excreted in the feces of the host.

The adult tapeworm is covered by a cytoplasmic epidermis called a tegument (Figs. 23, 26) and not by a cuticle. The outer surface of the tegument is covered with numerous microvillus-like extensions (microtriches) that vary in length from species to species; each microthrix is tipped with an electron-dense, spine-like structure. The microtriches are covered with a membrane that is continuous with the limiting membrane of the tegument. The cytoplasm of the tegument contains mitochondria, small electron-dense bodies of unknown function, and vesicles, but no lateral cell walls or nuclei. Nucleated tegumental cells lie in the parenchyma beneath the tegument and con-

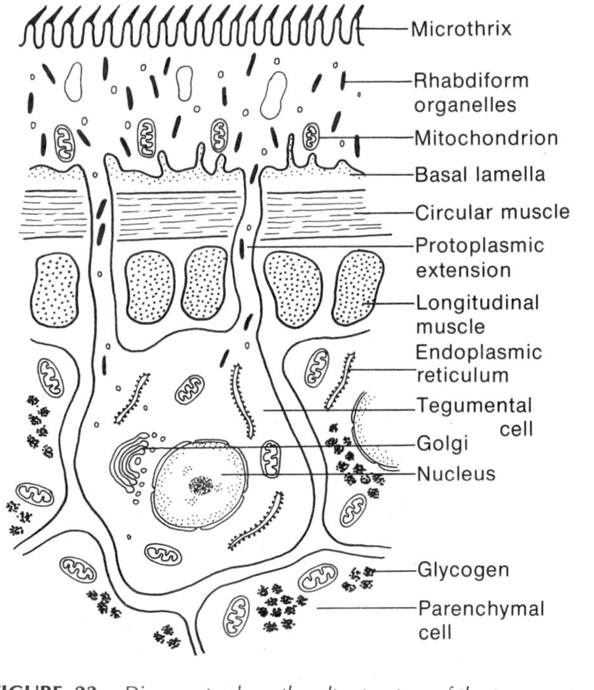

FIGURE 23. *Diagram to show the ultrastructure of the tegument of a typical cestode. Note the syncytial cytoplasmic outer layer which bears microvilli called microtriches (plural) or microtrix (singular), with spine-like tips; the nucleated tegumental cells, which are sunk in the parenchyma; and the muscles of the body wall.*

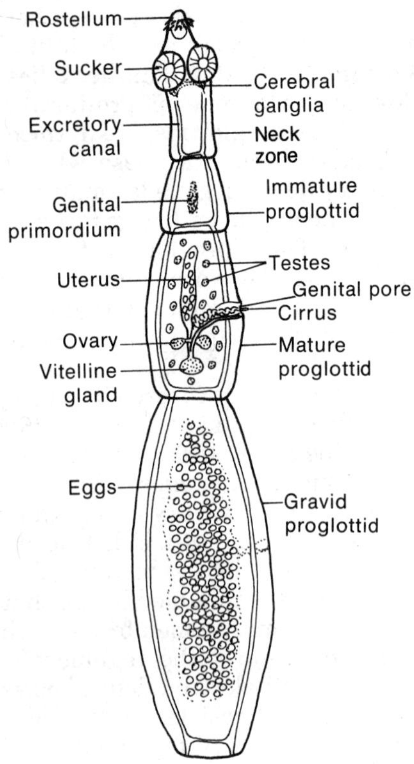

FIGURE 24. *Diagram of adult* Echinococcus granulosus *to show various anatomic features of a cyclophyllidean cestode.*

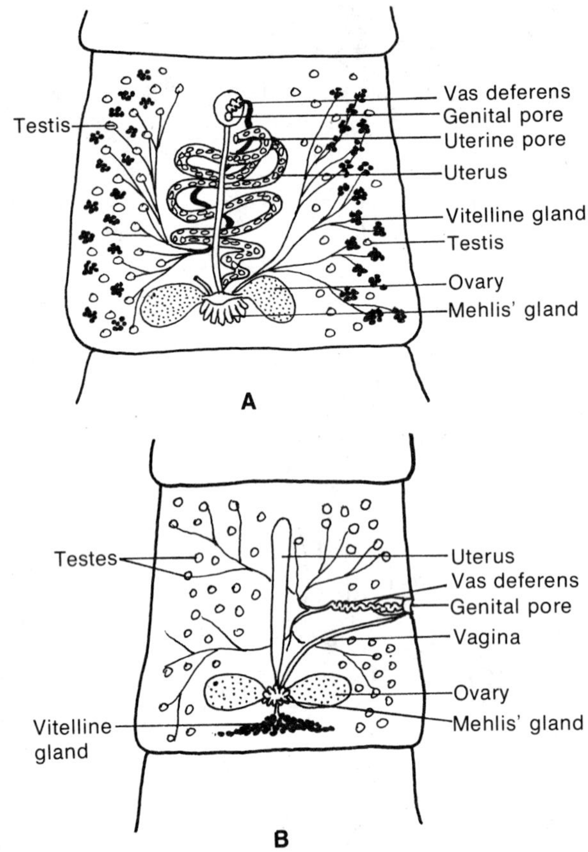

FIGURE 25. *A, Diagram of a mature proglottid of* Dibothriocephalus latus *to show the reproductive system. Note the scattered nature of the vitelline glands. Excretory canals and nerves omitted from the diagram. B, Diagram of a mature proglottid of* Taenia saginata *to show the reproductive system. Note the compact vitelline gland. Excretory canals and nerves omitted from the diagram.*

nect with it by means of cytoplasmic extensions (Fig. 23).

A layer of circular and a layer of longitudinal muscle lie beneath the tegument (Fig. 23), and transverse, diagonal, and dorsoventral muscles traverse the proglottid. The scolex is often very muscular. The nervous system consists of paired cerebral ganglia in the scolex and longitudinal nerves that run along the lateral margins of the worm. Sensory receptors are present on the scolex and on the proglottids. The excretory system consists of flame cells (Fig. 21), collecting vessels, and longitudinal excretory canals (dorsal and ventral) located along the lateral margins of the worm (Figs. 24, 26). Transverse canals link the longitudinal canals in each proglottid (Fig. 24).

All of the organs and tissues of the body are embedded in a cellular parenchyma (Fig. 26); there is no body cavity.

The worms are usually hermaphrodites (Figs. 25, 26), but in many species the male organs ripen before the female. The reproductive organs can be divided into two groups: the first group (Fig. 25*A*) has the vitellaria scattered throughout the proglottid or in laterally situated masses (Pseudophyllidea); the second group has compact vitelline glands usually situated toward the center of the proglottid (Cyclophyllidea) (Fig. 25*B*).

The egg of cestodes varies in different groups, but essentially it has four layers, or envelopes, enclosing the embryo. These layers are the capsule, outer envelope, inner envelope (which may be subdivided to form the embryophore), and oncospheral membrane (Fig. 27). The pseudophyllidean egg (Fig. 27) has an operculum, and a thick capsule composed of tanned protein. It usually

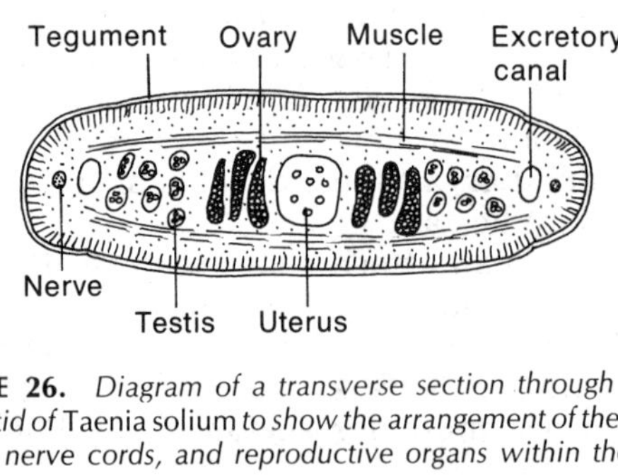

FIGURE 26. *Diagram of a transverse section through a mature proglottid of* Taenia solium *to show the arrangement of the excretory canals, nerve cords, and reproductive organs within the section.*

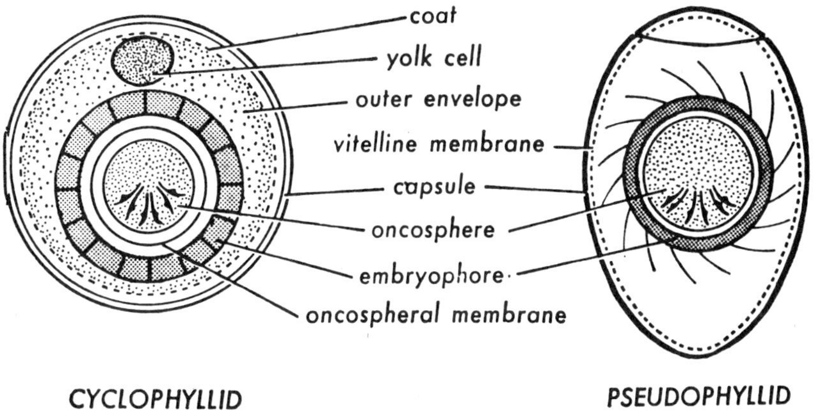

FIGURE 27. *Diagrams of a cyclophyllidean egg* (Taenia) *and a pseudophyllidean egg* (Dibothriocephalus). *Note the thick, toughened egg capsule with its operculum, and the ciliated embryophore (coracidium larva) in the pseudophyllidean egg, and the lack of a thick egg capsule in the egg of the cyclophyllidean but the presence of a thick, striated embryophore. (Reproduced, with permission, from Smyth: The Physiology of Cestodes, Oliver & Boyd 1969.)*

hatches in water to release a ciliated, free-swimming larva (coracidium). Embryonation of the egg occurs in water. The eggs of other tapeworms are usually embryonated when laid and do not have a free-living aquatic stage in the life cycle. They fall into three types: type 1 has a thin capsule and a thin embryophore *(Hymenolepis);* type 2 has an outer tanned protein capsule and a thick, striated embryophore (Fig. 27) *(Taenia, Echinococcus);* and type 3 has a shell formed by the egg and not by the vitellaria (*Stilesia*).

In the Pseudophyllidea, of which *Dibothriocephalus latus* is the most important species to man, the egg hatches in water to release a ciliated larva (coracidium). This coracidium contains an oncosphere or hexacanth embryo (Fig. 27) that carries three pairs of hooks. The cilia are shed when the coracidium is eaten by a copepod and the oncosphere migrates to the body cavity, where it develops into a proceroid larva. This larval stage is a small worm-like creature that possesses a tail-like appendage containing the oncospheral hooks. When the infected copepod is eaten by a suitable fish, the larva migrates to the muscles of the fish and develops into a plerocercoid larva. The plerocercoid of *Dibothriocephalus latus* is an undifferentiated worm with a poorly developed

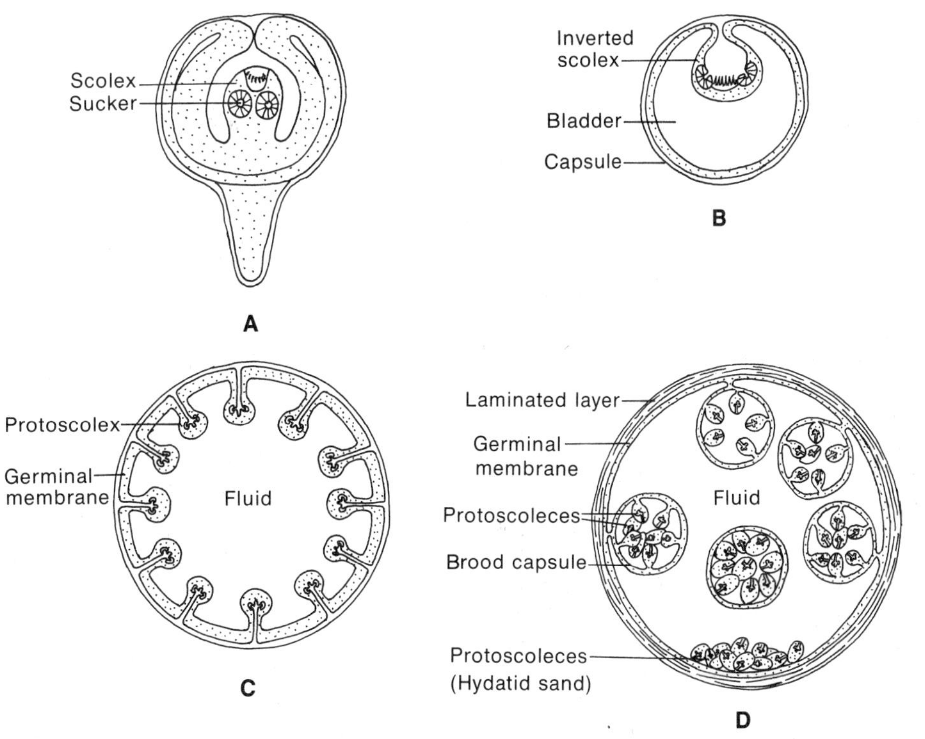

FIGURE 28. *Diagrams of A, cysticercoid larva; B, cysticercus larva; C, coenurus larva; and D, hydatid cyst.*

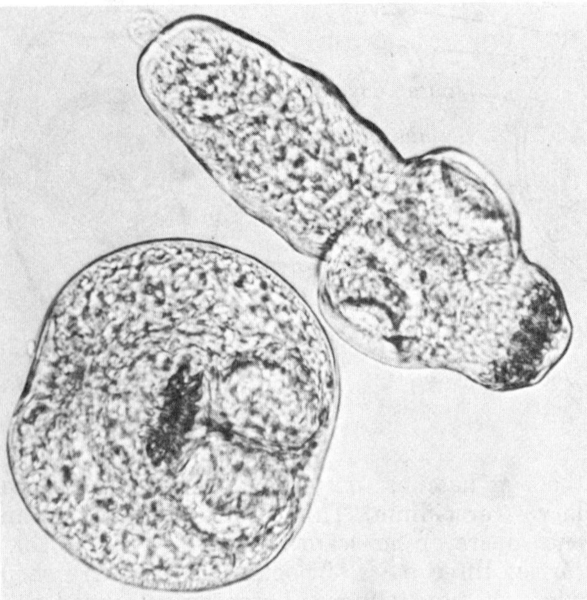

FIGURE 29. *Photomicrograph of protoscoleces of* Echinococcus granulosus, *one is evaginated and the other invaginated.*

scolex at the anterior end. When ingested by the final host, this larva differentiates into the adult worm.

In the order Cyclophyllidea, which includes most of the tapeworms parasitic in man, there are three types of larvae. The oncosphere (Fig. 27) hatches from the egg and is a six-hooked embryo that penetrates the tissues of the intermediate host. The oncosphere then develops into a cysticercoid larva or into a cysticercus, depending upon the species. The cysticercoid larva develops as a small cyst or bladder in which the scolex of the future adult is retracted (Fig. 28*A*) as in *Hymenolepis nana.* The cysticercus (often called the bladderworm) consists of a rounded, fluid-filled cyst or bladder within which the scolex of the future adult is invaginated (Fig. 28*B*) as in *Taenia solium.* A coenurus larva is a larger cysticercus in which several inverted scoleces (protoscoleces) develop from the inner (germinal) layer of the bladder wall (Fig. 28*C*) as in *Multiceps.* A hydatid cyst is a form of coenurus and is a large bladder that buds off numerous daughter cysts (brood capsules) from the germinal layer of the bladder wall. Each of these brood capsules contains several inverted scoleces (protoscoleces) (Figs. 28, 29).

For further reading on the structure and anatomy of cestodes see Wardle and McLeod, 1952; Wardle et al., 1974; Smyth, 1969; Voge, 1969; Hyman, 1951a; Joyeux and Baer, 1961; Yamaguti, 1959; Noble and Noble 1976.

PHYLUM ASCHELMINTHES

Class Nematoda (Roundworms)

Nematodes are found in most habitats and there are many species. They are important parasites of man, animals, and plants, but many species are free living. This classification will concentrate on those groups that are important in human disease. Some authorities regard the nematodes as a separate phylum (Maggenti, 1976), whereas others regard them as a class in the phylum Aschelminthes (Hyman, 1951b; De Coninck, 1965; Anderson et al., 1974).

Nematodes are slender, cylindric worms, usually tapered at both ends (Fig. 30), that vary in length from less than a millimeter to 9 meters. The body is covered with a collagenous cuticle, is unsegmented, has only longitudinal muscles in the body wall, has a body cavity (pseudocoelom) (Fig. 31), and has a straight alimentary tract that is usually complete. The excretory system, when present, does not contain flame cells. The sexes are usually separate. The male has a cloaca that usually contains one or two copulatory spicules.

Subclass Adenophorea (= Aphasmidia)

Phasmids (caudal sensory organs) (Fig. 30) absent; excretory system without lateral canals; esophagus (should more correctly be called a pharynx) cylindric, esophageal glands may be free in body cavity and form a stichosome or a trophosome; eggs of some species have polar plugs (Fig. 32*A*).

Order Enoplida. Esophagus cylindric, usually with anterior muscular and posterior glandular parts.

Superfamily Trichuroidea. Stichosome or trophosome present; intestine and rectum present; no caudal sucker in male; females with one set of gonads. *(Trichuris, Trichinella, Capillaria).* Eggs of *Trichuris* and *Capillaria* with polar plugs (Fig. 32*A*).

Subclass Sercenentea (= Phasmidia)

Phasmids present (Fig. 30); excretory system usually with paired lateral canals (Fig. 34); sti-

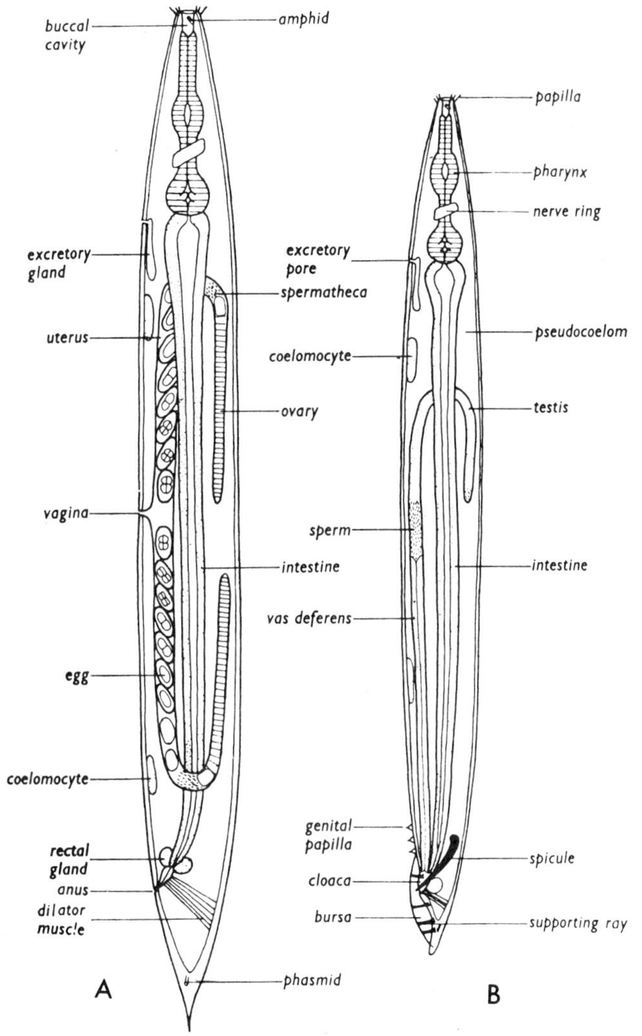

FIGURE 30. *Diagram of a generalized female (A) and male (B) nematode to show their anatomy. (Reproduced, with permission, from Lee & Atkinson: Physiology of Nematodes, 2nd ed. Macmillan 1976.)*

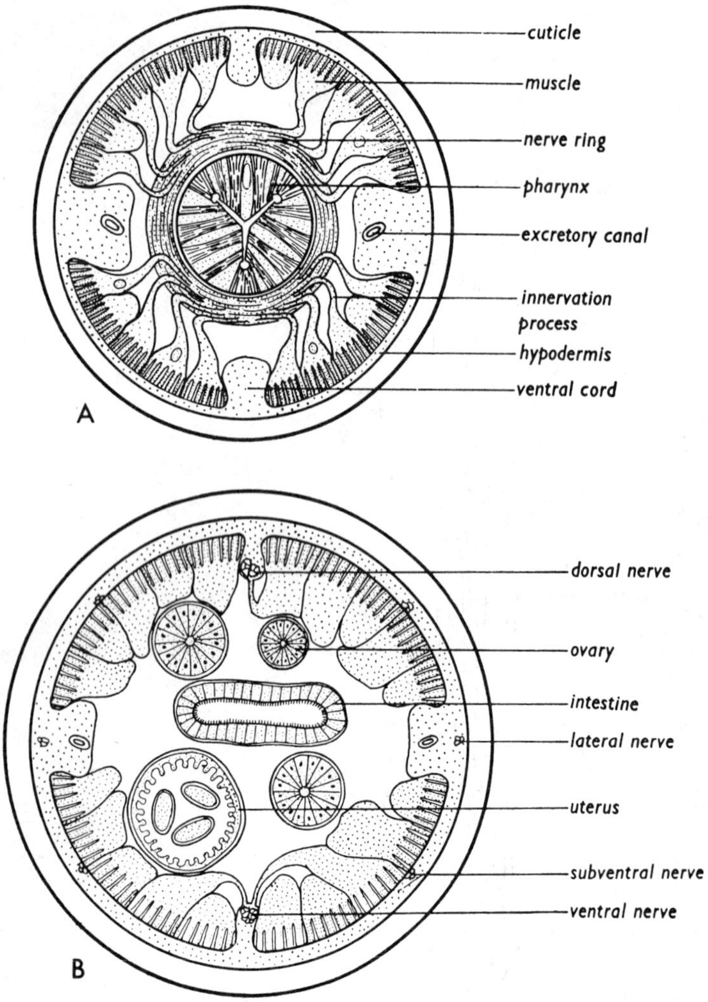

FIGURE 31. *Diagram of transverse sections through the esophageal* (A) *(should more correctly be called pharynx) and middle* (B) *regions of a female nematode to show the arrangement of the tissues and organs. (Reproduced with permission from Lee & Atkinson: Physiology of Nematodes, 2nd ed. Macmillan 1976.)*

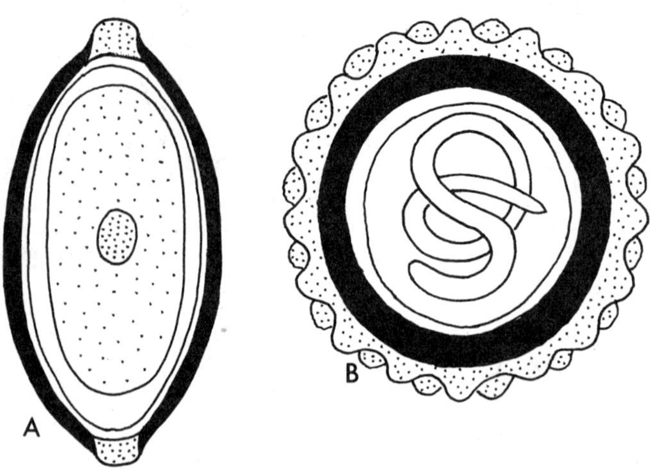

FIGURE 32. *Diagram of an egg of* Trichuris (A) *(note the polar plugs at each end), and* Ascaris (B).

chosome or trophosome absent; eggs without polar plug.

Order Rhabditida. Parasitic generation (females only) alternating with free-living generations (males and females); parasitic in lungs of amphibia or small intestine of vertebrates. (*Strongyloides stercoralis*).

Order Tylenchida. Stylet present in buccal cavity. Feed on plants or fungi; same species parasitic in body cavity of insects.

Order Strongylida. Male has copulatory bursa with supporting rays (Fig. 33); usually two spicules; ovejector of female with well-developed sphincter muscles. Mouth and buccal cavity variable in size and shape. Excretory system H-shaped and with two large glands (Fig. 34*A*, *B*). Parasitic in vertebrates.

Superfamily Diaphanacephalodea. Intestinal parasites of snakes and lizards.

Superfamily Ancylostomatoidea. Buccal capsule large, strongly cuticularized; lips absent; mouth with teeth and cutting plates (Fig. 35) or unarmed. Parasitic in intestine of mammals. (*Necator americanus, Ancylostoma duodenale.*)

Superfamily Strongyloidea. Buccal cavity variable, sometimes surrounded by ring of projections (corona radiata); no teeth or cutting plates. Parasites of intestine, respiratory tract, or urinary tract of mammals and birds. (*Strongylus, Oesophagostomum.*)

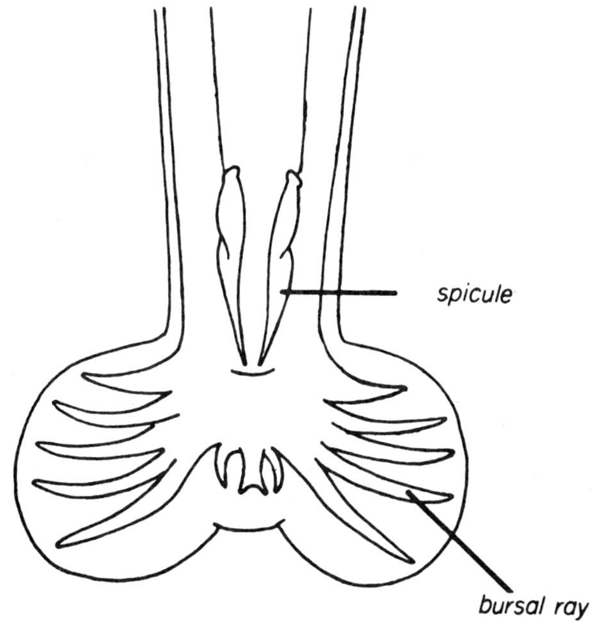

FIGURE 33. *Diagram of tail of a male trichostrongyle to show the copulatory bursa and spicules. (Reproduced with permission from Lee & Atkinson: Physiology of Nematodes, 2nd ed. Macmillan 1976.)*

Superfamily Trichostrongyloidea. Buccal cavity small (Fig. 36); cuticle around head often inflated; longitudinal cuticular ridges often present.

FIGURE 34. *Representative types of excretory system found in nematodes. A, Rhabditoid-type, an H-shaped system with two subventral glands, the lateral canals are embedded in the lateral cords (see Fig. 31); found in the Strongylida. B is a variant of A found in* Oesophagastomum. *C, Tylenchoid type. D, Oxyuroid type, an H-shaped system lacking subventral gland cells. E, Ascaroid type. F, Cephaloboid type. G, Anisakid type, a reduced form of E. H, Single ventral cell type present in Chromadorina, Manhysterina, and Enoploidea. (Reproduced with permission from Lee & Atkinson: Physiology of Nematodes, 2nd ed. Macmillan 1976. Based on originals by Chitwood and Chitwood, 1950.)*

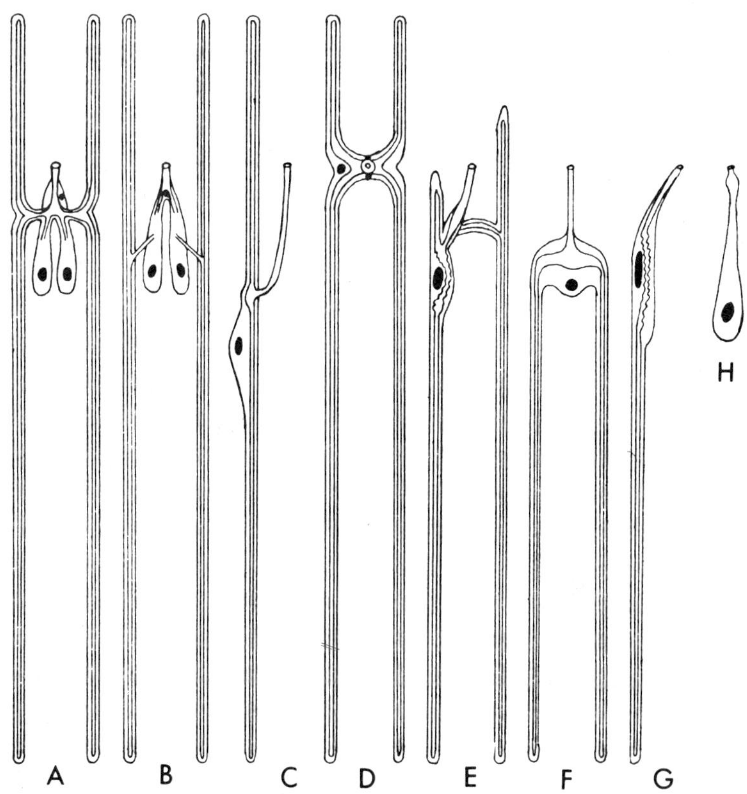

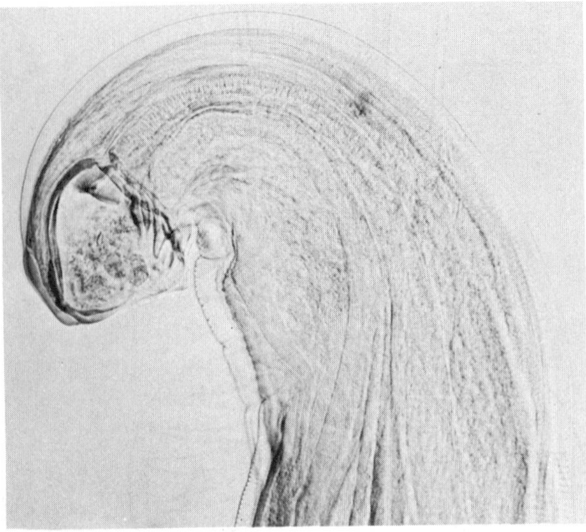

FIGURE 35. Photomicrograph of the head end of Ancylostoma duodenale. Note the cutting plates at the opening of the mouth and the teeth at the base of the buccal cavity.

Parasites of alimentary tract or respiratory tract of mammals. (*Trichostrongylus colubriformis.*)

Superfamily Metastrongyloidea. Buccal cavity small; copulatory bursa often reduced; vulva near anus; longitudinal cuticular ridges absent. Mainly parasites of respiratory tract of mammals. Intermediate host usually a gastropod, oligochaete, crustacean, or occasionally a vertebrate. (*Metastrongylus* spp., *Angiostrongylus cantonensis.*)

Order Oxyurida. Male without copulatory bursa, one or two spicules; excretory system H-shaped (Fig. 34*D*); esophageal bulb present; body usually short and stout. Parasitic in lower intestine, colon, or rectum of vertebrates or hindgut of insects. (*Enterobius vermicularis.*)

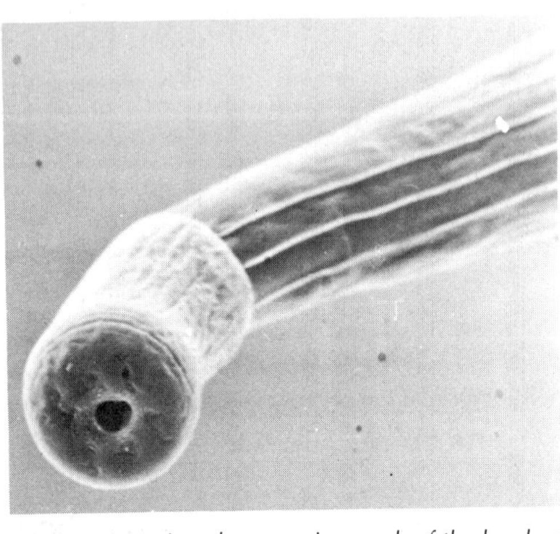

FIGURE 36. Scanning electron micrograph of the head end of a trichostrongyle nematode. Note the small mouth.

Order Ascaridida. Usually three or six lips; esophagus variable in form but not divided into short muscular and long glandular parts; excretory system a modified "H" system (Fig. 34*E*). Usually parasitic in intestine of vertebrates.

Superfamily Cosmocercoidea

Superfamily Seuratoidea

Superfamily Heterakoidea

Superfamily Ascaridoidea. Mouth surrounded by three lips (Fig. 37); buccal cavity and esophagus simple. Tail of male usually curved or coiled. (*Ascaris lumbricoides, Toxocara canis.*)

Superfamily Heterocheiloidea (= Anisakoidea). Esophageal ventriculus (glandular modification of esophagus, which extends posteriorly beside the intestine) present; intestinal cecum often present. Parasitic in stomach of mammals (*Anisakis*).

Order Spirurida. Usually two lateral lips surround mouth; esophagus with anterior muscular and posterior glandular part; most males have two unequal spicules.

SUBORDER CAMALLANINA. Larvae without cephalic hooks; esophageal glands usually uninucleate.

Superfamily Camallinoidea. Buccal cavity well developed. Parasites of gut.

Superfamily Dracunculoidea. Buccal cavity weakly developed; six conspicuous labial papillae present. Usually parasitic in tissues of host. (*Dracunculus medinensis.*)

SUBORDER SPIRURINA. Larvae with cephalic hooks; esophageal glands multinucleate.

Superfamily Gnathostomatoidea. Cuticular outgrowths (pseudolabia) overlie and replace lips; cuticle on inner face folded to form projections that interdigitate with projections on adjacent pseudolabium. Head sometimes bulbous. Parasites of gut and tissues. (*Gnathostoma spinigerum.*)

Superfamily Physalopteroidea

Superfamily Rictularioidea

Superfamily Thelazioidea. Buccal cavity variable, sometimes long and thin; mouth usually without definite lips. Parasites of eye cavity of mammals and birds, lungs of mammals, or intestine of fishes. (*Thelazia.*)

Superfamily Spiruroidea. Buccal cavity never long and cylindric, but well cuticularized; two lateral lips surround mouth. Vulva near middle of body; males usually have two unequal spicules. Parasites of alimentary tract, respiratory tract, eye cavity, nasal cavity, or sinuses of vertebrates. (*Gongylonema pulchrum.*)

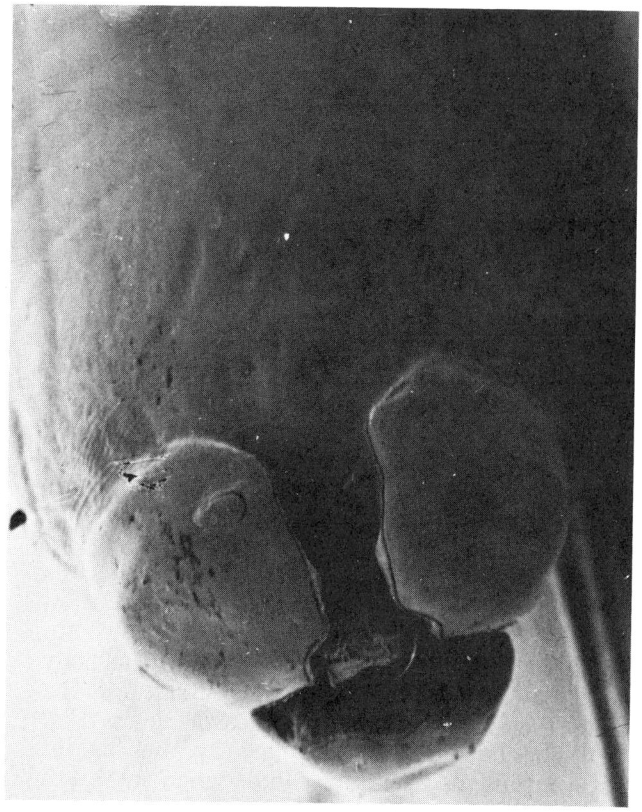

FIGURE 37. *Scanning electron micrograph of the head end of Ascaris. Note the three lips.*

Superfamily Habronematoidea

Superfamily Acuarioidea

Superfamily Filarioidea. Lips absent; ovoviviparous; first larva a microfilaria. Parasites of tissues and tissue spaces of vertebrates; transmitted by arthropod vectors. (*Wuchereria bancrofti, Onchocerca volvulus, Loa loa.*)

Superfamily Aproctoidea

Superfamily Diplotriaenoidea

For further information on the classification of nematodes see Hyman, 1951b; de Coninck, 1965; Chitwood, 1969; Maggenti, 1976; Anderson, Chabaud, and Willmott, 1974; Yamaguti, 1961.

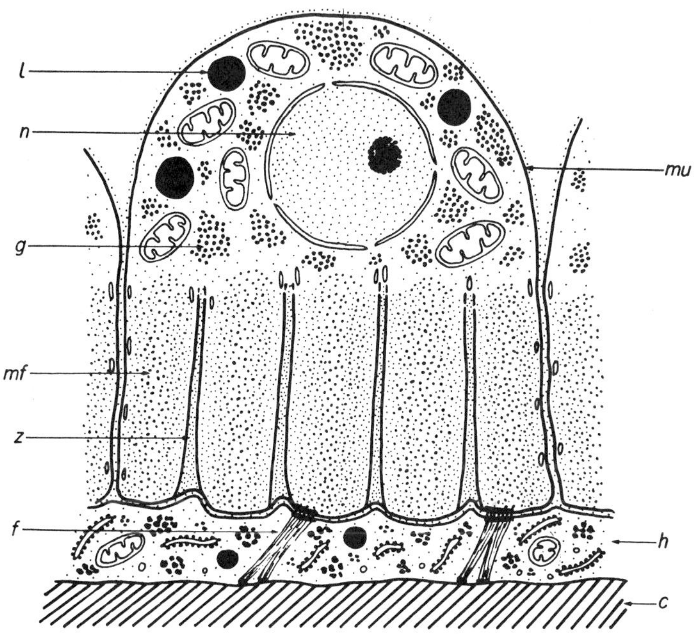

FIGURE 38. *Diagram of a section through a longitudinal muscle of the body wall of a nematode to show the internal structure of the muscle and the association with the hypodermis and the cuticle. c, cuticle; f, fibres connecting muscle to cuticle; g, glycogen; h, hypodermis; l, lipid; mf, myofilaments; mu, muscle; n, nucleus; z, z-band region of muscle. (Reproduced with permission from Lee & Atkinson: Physiology of Nematodes. 2nd ed. Macmillan 1976.)*

ANATOMY OF NEMATODA

The typical nematode is a spindle-shaped, unsegmented, and bilaterally symmetric worm (Fig. 30) that is round in cross-section (Fig. 31). The internal organs of the male and of the female are shown in Figures 30 and 31.

The body wall has an outer collagenous cuticle, a cellular or syncytial hypodermis, and a layer of longitudinal muscle (Fig. 31). The cuticle lines the buccal cavity, esophagus (more correctly called the pharynx), excretory pore, rectum, cloaca, and vulva. Teeth, cutting plates, stylets, and spicules are formed from toughened and hardened cuticle. The cuticle is basically three layered, but further subdivisions often occur. The outer surface of the cuticle is superficially annulated and may also be formed into fin-like structures (alae) or ridges that run along the length of the body. The hypodermis lies between the cuticle and the longitudinal muscles of the body wall. It is normally a thin layer of cytoplasm but it projects into the body cavity along the middorsal, midventral, and the lateral lines to form four ridges or cords (Fig. 31). The lateral cords are usually the largest and contain the excretory canals when these are present (Fig. 31).

The muscles of the body wall are spindle shaped, longitudinal cells that have a contractile and a noncontractile portion (Fig. 38). There are no circular muscles in the body wall. The muscles are innervated by arms that extend from the muscle to the nerves (Fig. 31).

The nervous system consists of a ganglionated circumesophageal ring with a large ventral, a smaller dorsal, and two or more lateral nerves running from it along the length of the hypodermal cords (Figs. 30, 31). Nerves extend forward from the nerve ring to innervate the sense organs around the mouth (Fig. 39). Sense organs occur elsewhere on the body, particularly around the tail of the male.

The body cavity is not a true coelom and is called a pseudocoelom. It is filled with fluid under pressure and forms the hydrostatic skeleton of the nematode.

The alimentary system is usually complete. There is a mouth surrounded by lips (Fig. 39) (these may be reduced in number from the normal six, or may be absent), a buccal cavity of varying shape and size, a muscular and glandular esophagus (pharynx), which usually has a triradiate lumen, a straight intestine, a rectum (cloaca in the male), and anus (Figs. 30, 31). The intestine consists of a single layer of epithelial cells that carry microvilli on the lumen side (Fig. 31*B*).

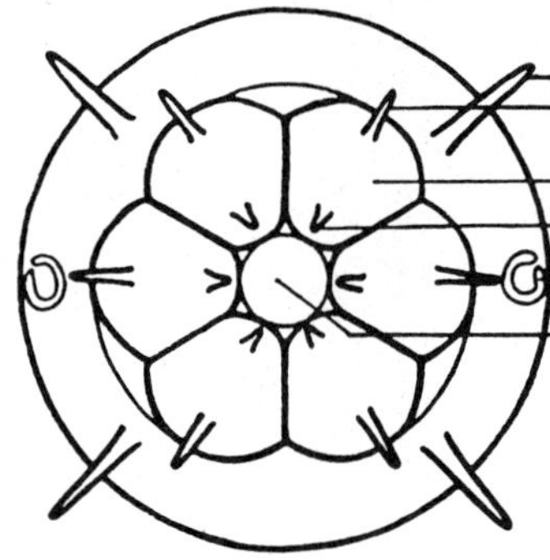

FIGURE 39. *En face view of a nematode head showing the position of the mouth, lips, amphids, labial sensillae, and cephalic sensilla (usually called papilla). (Redrawn from Jones:* Plant Nematology. *H.M.S.O. 1959.)*

The excretory system varies in structure from group to group and is absent in some species. There are two basic types — a glandular system and a tubular system (Fig. 34). The glandular system is found in many free-living nematodes and consists of a ventral gland cell situated in the body cavity near the base of the esophagus. It usually has a terminal ampulla that opens to the exterior on the ventral surface. The tubular system varies in structure but is basically an H-shaped system with a lateral canal in each lateral cord. It is united by a transverse canal in the anterior end of the nematode (Fig. 34*D, E*). This opens to the exterior through a common excretory duct and pore on the ventral surface. The Strongylida also contain a pair of glands in the pseudocoelom and these open into the transverse excretory canal just behind the excretory pore (Fig. 34*A, B*); although these are called excretory glands they appear to have a secretory rather than an excretory function. The excretory system contains no flagella or cilia.

The sexes are usually separate, and males are frequently smaller than females. The males have one or two testes, each opening into a seminal vesicle and then into a common vas deferens. The vas deferens open into a cloaca (Fig. 30). Many males possess one or two copulatory spicules that lie in pouches connected to the cloaca (Fig. 33). In some groups of nematode (Strongylida) the area around the cloaca is expanded to form a copulatory bursa (Fig. 33). Females have one or two ovaries that open into an oviduct(s) and a uterus or uteri (Fig. 30). The uterus often ends in an ovejector that is usually very muscular and opens to the exterior via the vagina and vulva. The eggs of nematodes are essentially ovoid and have three main layers (Fig. 32), the middle layer containing a chitin-protein complex. The sperm of nematodes are ameboid.

Nematodes have no circulatory or respiratory organs and lack true flagella.

The larval stages are really juveniles, since they are usually miniature adults with similar structure but without sex organs. All nematodes moult four times, and any structural changes that occur during the life cycle, such as modifications to the buccal cavity or to the cuticle, occur during the moult.

For further reading on the structure and anatomy of nematodes see Bird, 1971; Chitwood and Chitwood, 1950; Hyman, 1951b; Croll, 1976; and Lee and Atkinson, 1976.

PHYLUM ACANTHOCEPHALA

These worm-like parasites live as adults in the intestine of vertebrates and as larvae in arthropods. They are distinguished by their spined anterior proboscis, which may be retracted into the main body of the worm. A digestive system is completely absent and there are no circulatory or respiratory structures. The sexes are separate; the males possess a posterior bell-like bursa.

Order Archiacanthocephala. Proboscis spines are concentric; contain protonephridia, eight cement glands in male. Parasitic in terrestrial hosts (*Moniliformis, Macracanthorhynchus*).

Order Palaeacanthocephala. Proboscis spine in alternate radial rows; no protonephridia; usually six cement glands in males. Parasitic mainly in aquatic hosts.

Order Eoacanthocephala. Proboscis spines arranged radially; no protonephridia; cement gland syncytial. Parasitic in aquatic hosts.

ANATOMY OF ACANTHOCEPHALA

Adult acanthocephala are unsegmented worms in which the sexes are separate (Figs. 40*A, B*). Most species are 1 to 2 cm long, but a few, notably *Macracanthorhynchus hirudinaceus* from pigs, are longer and can reach 45 cm. The body is divided into an anterior praesoma that contains the retractable spiny proboscis together with its associated structures, and a larger posterior portion called the metasoma that includes the other organs and tissues of the body. The body wall consists of about five layers. There is a thin outer epicuticle, a tough cuticle penetrated by numerous pores that lead into canals of the striped layer. This striped layer merges into the fibrous felt layer. The innermost layer of the body wall is the thickest layer and is called the radial layer. It contains nuclei, mitochondria, ribosomes, and other cellular constituents, but has no lateral cell walls. The radial layer is separated from the circular and longitudinal muscles of the body wall by a basement lamella. There is a body cavity but it is not a true coelom. There is no alimentary tract in this group of worms. Food is taken up across the body wall. There are no circulatory or respiratory structures; nephridia are present in some groups but not in others. A nerve ganglion associated with the proboscis sheath sends nerves to various regions of the body. The male reproduction system consists of a copulatory bursa, a number of cement glands (usually six to eight), a pair of testes, and a sperm duct that leads to the genital pore. The female system has no persistent ovary; ova develop on a ligament in the body cavity and are set free in the body cavity, where they are fertilized. The eggs are sorted in a uterine bell and pass through a uterus to the genital pore.

For further reading on the Acanthocephala see Hyman, 1951b, and Crompton, 1970.

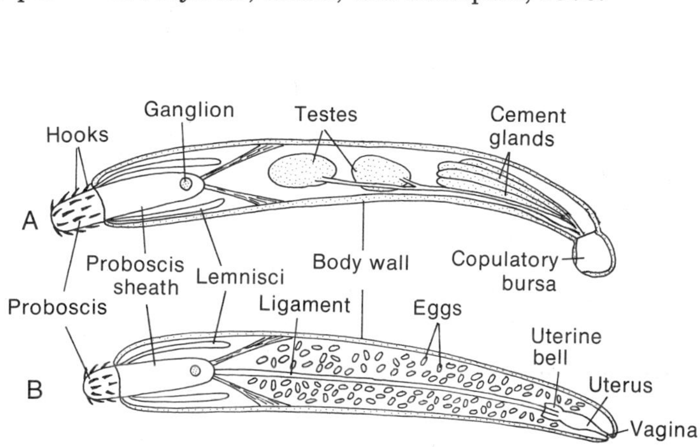

FIGURE 40. *Diagram of a male (A) and female (B) acanthocephalan to show their internal anatomy. Note the spiny proboscis and the lack of an alimentary system.*

References

Adam, K. M. G., Paul, J., and Zaman, V.: Medical and Veterinary Protozoology. Edinburgh, Churchill Livingstone, 1971.
Anderson, R. C., Chabaud, A. G., and Willmott, S: CIH keys to the nematode parasites of vertebrates. No. 1. Commonwealth Agricultural Bureaux, Farnham Royal, Slough, U. K., 1974.
Baker, J. R.: Parasitic Protozoa. London, Hutchinson, 1973.
Bird, A. F.: The Structure of Nematodes. New York, Academic Press, 1971.
Chitwood, B. G., and Chitwood, M. B.: An Introduction to Nematology. Baltimore, Maryland, Monumental Printing Co., 1950.
Corliss, J. O.: The Ciliated Protozoa. Oxford, Pergamon Press, 1961.
Croll, N. A.: The Organization of Nematodes. New York, Academic Press, 1976.
Crompton, D. W. T.: An Ecological Approach to Acanthocephalan Physiology. Cambridge, Cambridge University Press, 1970.
Dawes, B.: The Trematoda. Cambridge, Cambridge University Press, 1956.
de Coninck, L.: Classe des nématodes. In Grassé, P-P (ed.): Traité de Zoologie. IV Némathelminthes (Nématodes). Paris, Masson et Cie, 1965.
Dogiel, V. A.: General Protozoology. Oxford, Clarendon Press, 1965.
Erasmus, D. A.: The Biology of Trematodes. London, Edward Arnold, 1972.
Garnham, P. C. C.: Malaria Parasites and Other Haemosporidea. Oxford, Blackwell Scientific Publications, 1966.
Honigberg, B. M., Balamuth, W., Bovee, E. C., Corliss, J. O., Gojdics, M., Hall, R. P., Kudo, R. R., Levine, N. D., Loeblich, A. R., Weiser, J., and Wenrich, D. H.: A revised classification of the Phylum Protozoa. Protozool, 11:7, 1964.
Hyman, L. H.: The Invertebrates. Vol. II. New York, McGraw-Hill, 1951a.
Hyman, L. H.: The Invertebrates. Vol. III. New York, McGraw-Hill, 1951b.
Joyeux, C., and Baer, J. G.: Classe des Cestodaires. In Grassé, P-P. (ed.): Traité de Zoologie, IV. Paris, Masson et Cie, 1961.
Kudo, R. R.: Protozoology. Springfield, Illinois, Charles C Thomas, 1966.
La Rue, G. A.: The classification of digenetic trematodes: a review and a new system. Exp. Parasitol 6:306, 1957.
Lee, D. L., and Atkinson, H. T.: Physiology of Nematodes. London, Macmillan Press, 1976.
Levine, N. D.: Protozoan Parasites of Domestic Animals and of Man. Minneapolis, Burgess Publishing Co., 1961.
Llwellyn, J.: The evolution of parasitic platyhelminths. Symposia of the British Society for Parasitology 3:47, 1965.
Mackinnon, D. L., and Hawes, R. S. J.: An Introduction to the Study of Protozoa. Oxford, Clarendon Press, 1961.
Maggenti, A. R.: Taxonomic position of Nematoda among the Pseudocoelomate Bilateria. In Croll, N. A. (ed.): The Organization of Nematodes. New York, Academic Press, 1976.
Noble, E. R., and Noble, G. A.: Parasitology. The Biology of Animal Parasites. Philadelphia, Lea and Febiger, 1976.
Sleigh, M.: The Biology of Protozoa. London, Edward Arnold, 1973.
Smyth, J. D.: The Physiology of Trematodes. Edinburgh, Oliver & Boyd, 1966.
Smyth, J. D.: The Physiology of Cestodes. Edinburgh, Oliver & Boyd, 1969.
Vickerman, K., and Cox, F. E. G.: The Protozoa. London, John Murray, 1967.
Voge, M.: Systematics of cestodes — present and future. In Schmidt, G. D. (ed.): Problems in Systematics of Parasites. Baltimore, University Park Press, 1969.
Wardle, R. A., and McLeod, J. A.: The Zoology of Tapeworms. Minneapolis, University of Minnesota Press, 1952.
Wardle, R. A., McLeod, J. A., and Radinovsky, S.: Advances in the Zoology of Tapeworms 1950–1970. Minneapolis, University of Minnesota Press, 1974.
Yamaguti, S.: Systema Helminthum. I. The Digenetic Trematodes of Vertebrates. New York, Interscience Publishers Inc., 1958.
Yamaguti, S.: Systema Helminthum II. Cestodes of Vertebrates. New York, Interscience Publishers Inc., 1959.
Yamaguti, S.: Systema Helminthum III. The Nematodes of Vertebrates. New York, Interscience Publishers Inc., 1961.

17 BIOCHEMISTRY OF PARASITES: HELMINTHS

Howard J. Saz, Ph.D.

In order to understand the mode of action of anthelmintic agents, it must be realized that the metazoan parasites differ from other disease-causing organisms in several important pharmacologic as well as biochemical respects. Pathogenesis of bacterial, viral, protozoan, and mycotic diseases depends on the replication or multiplication of the invading pathogens. Therefore, agents that inhibit pathogen replication will arrest or cure the disease. In helminth infections, the picture is quite different. Generally, the problem is to remove the adult worm, which is not reliant upon multiplication for its survival in a host. Hence, antibiotics that work effectively against bacteria by inhibiting macromolecular synthesis have little or no value in the chemotherapy of helminth infections.

What are the vulnerable sites for drug action in helminths? Anthelmintics are inhibitory primarily at one of two sites. Many, but not all, anthelmintics inhibit either muscle contraction or energy-generating processes within the parasites. Table 1 lists some anthelmintic compounds and their reported sites of inhibition. Although the inhibitions noted may account for the chemotherapeutic effect of each of these compounds, it is not established that these inhibitions represent their primary modes of chemotherapeutic action. At any rate, the importance of studying the biochemistry of the metazoan parasites now becomes more obvious. A better understanding of parasite biochemistry would help to explain the modes of action of anthelmintics. It would also aid in pointing out possible sites of inhibition that are unique to the parasite metabolism and therefore suggest the design of drugs that are more toxic for the parasite than they are for the host. In the remainder of this chapter we will discuss the biochemistry of the metazoan parasites with major emphasis on their energy metabolisms, since this area is quite susceptible to drug action.

Before 1950, most biochemists were concerned

TABLE 1. Reactions Inhibited by Anthelmintic Compounds

ANTHELMINTIC AGENT	SITE OF INHIBITION
Trivalent antimonials	Phosphofructokinase
p-Rosaniline	Glycogen accumulation Acetylcholine esterase
Cyanine dyes (Dithiazanine)	Respiration Glucose uptake
Chlorsalicylamide (Yomesan) Dichlorophen Desaspidin 2,4-Dihexanoyl-6-methyl phloroglucinol	Electron transport phosphorylation
Tetramisole	Neurotransmission Fumarate reductase
Thiabendazole Cambendazole	Fumarate reductase
Mebendazole	Glucose and amino acid uptake
Dibenzylamines	Glucose transport

primarily with elucidating the concept of unity in biochemistry. They were struck by the biochemical similarities of all living tissues. However, small groups of scientists, particularly the pharmacologists and the immunologists, began to stress the fact that tissues even from related species were biochemically distinct in some respects. These considerations gave rise to studies of comparative biochemistry. It is now readily accepted that differences exist both at the subtle level of protein or enzyme structure and kinetics as well as at the more obvious metabolic level.

Krebs and Najjar (1948) were the first to report that enzymes carrying out the same reaction in different animal species could be antigenically different. Antisera to purified yeast glyceraldehyde-3-phosphate dehydrogenase inhibited the yeast enzyme, but not the corresponding enzyme isolated from rabbit muscle. Similarly, certain enzymes of *Schistosoma mansoni* were found to be immunologically distinct from the corresponding enzymes of rabbit muscle. It may be concluded, therefore, that the protein structure of the host and parasite enzymes is different.

Such structural differences between corresponding host and parasite enzymes may account for the specificity of some anthelmintics. For example, the trivalent organic antimonials, such as stibophen and potassium antimony tartrate, have been used to treat both schistosomiasis and filariasis. As with numerous other anthelmintics, the antimonials destroy the parasites by inhibiting energy metabolism; in this case by inhibiting glycolysis. Specifically, in schistosomes, filariids, *Ascaris,* and the rat tapeworm *Hymenolepis diminuta,* the antimonials are known to be potent inhibitors of the glycolytic enzyme phosphofructokinase. Most important, and presumably because of differing protein structures, the mammalian phosphofructokinase is not affected by therapeutic concentrations of antimonials. Only slight inhibition of the mammalian enzyme occurs even at levels of drug 80- to 100-fold higher than those that inhibit the parasite enzyme almost completely (Mansour and Bueding, 1953; Saz and Dunbar, 1975).

In addition to the differences in protein structure, striking differences also exist between the metabolic pathways of the parasites and their hosts. This first became obvious at the level of overall metabolism. Although all helminths examined can consume oxygen, none can catalyze the complete oxidation of substrates to carbon dioxide and water. All helminths examined accumulate end products of metabolism other than, or in addition to, carbon dioxide and water. The same is true of the parasitic protozoans. Thus, in contrast to their mammalian hosts, terminal aerobic respiration in parasites is either absent or rate limiting. Similarly, it appears that many helminths surprisingly have lost their ability to synthesize long chain fatty acids de novo and must rely on their host to supply these important nutritional components (Meyer et al., 1966). Such an abbreviated lipid metabolism occurs in the cestodes *Spirometra mansonoides* and *Hymenolepis diminuta;* the nematode *Ascaris lumbricoides;* and the trematode *Schistosoma mansoni.*

Although the oxidative capacity of all helminths is limited, some are obligate aerobes and are destroyed rapidly by anaerobiosis. In contrast, others require no oxygen for energy metabolisms and survive equally well in the presence or absence of oxygen. Relatively few parasites have been cultured through their life cycles in vitro, making it difficult to assess their oxygen requirements for growth. Investigators have determined oxygen requirements mainly from survival time in air of a given stage of the parasite. Among worms that have been completely or partially cultured, a gradation of oxygen requirements exists ranging from highly aerobic to anaerobic. It has been suggested that oxygen may even inhibit the normal development of some parasites. In all helminths and in all stages of their development, however, oxygen can be consumed even though it may be deleterious to survival. In many instances, when the energy metabolism appears to be anaerobic, neither the physiologic significance nor the mechanism of this oxygen consumption is understood. It must be borne in mind that we are dealing with complex developmental stages in the life cycles of the helminths. In a given parasite, some of these stages may be aerobic while others are not. Shifts or changes in the metabolism of the parasites occur upon development from one stage to another. Almost nothing is known of the mechanisms by which these "switches" in metabolism occur. Examples of this "switching" phenomenon will be discussed below.

In many respects, the helminths resemble the facultative and obligate anaerobic bacteria. A wide range of fermentation products is formed, indicating the utilization of terminal electron acceptors other than oxygen. More simply stated, rather than oxygen being reduced to H_2O, other organic compounds take the place of oxygen and accept electrons, resulting in an accumulation of fermentation products. Some of the helminths are homolactate fermenters, accumulating lactate as the sole fermentation product. This also occurs in certain bacteria, rapidly contracting muscle, mammalian red blood cells, and numerous other organisms. Most helminths accumulate an array of fermentation products that might include volatile fatty acids, succinate, lactate, and neutral volatile compounds such as ethanol and acetylmethylcarbinol (acetoin). In general, all of these products arise from either pyruvate or succinate, or from both of these acids. Therefore, the carbohydrate metabolisms of most helminths studied might be divided into two categories. The first group would comprise those parasites, such as the schistosomes and filarial worms, that rely entirely on glycolysis and a few adjunct reactions for their energy metabolisms. Products of these fermentations would include lactate or ethanol and possibly acetate and acetylmethylcarbinol (acetoin). The second group of parasites, as exemplified by *Ascaris* and *Hymenolepis diminuta,* rely on a similar series of reactions, but can also fix carbon dioxide into phosphoenolpyruvate leading to the formation of a C_4 dicarboxylic acid, which in turn could give rise to succinate, propionate, and other volatile fatty acids. A representative sampling of a few parasites and their type of metabolism are listed in Table 2 together with the products formed by each worm. In the remainder of this chapter, examples of the "primarily glycolytic helminths" and the "CO_2-fixing helminths" will be discussed. A limited number of examples of each type of parasite will be examined.

TABLE 2. Glycolytic and CO_2 Fixing Helminths and Products Formed

PARASITE	FERMENTATION PRODUCTS
Glycolytic	
Schistosoma mansoni	Lactate
Schistosoma haematobium	Lactate
Schistosoma japonicum	Lactate
Brugia pahangi	Lactate
Dipetalonema viteae	Lactate
Litomosoides carinii	Lactate, Acetate, CO_2
Dirofilaria uniformis	Lactate
Dracunculus insignis	Lactate
Angiostrongylus cantonensis	Lactate
CO_2 Fixing	
Ascaris lumbricoides	Succinate, volatile fatty acids
Heterakis gallinae	Succinate, propionate
Trichuris vulpis	Succinate, volatile fatty acids
Trichinella spiralis (larvae)	Acetate, propionate, volatile fatty acids
Hymenolepis diminuta	Succinate, acetate, lactate
Moniezia expansa	Succinate, lactate
Echinococcus granulosus (cysts)	Succinate, lactate, acetate, ethanol
Taenia taeniaformis (adults and larvae)	Succinate, lactate, pyruvate, acetate, ethanol, glycerol
Spirometra mansonoides (spargana and adults)	Propionate, acetate, succinate, lactate
Echinostoma liei	n-Valerate, propionate, acetate, n-hexanoate
Fasciola hepatica	Acetate, propionate, lactate
Paragonimus westermani	Acetate, formate, propionate, n-valerate, 2-methylbutyrate, n-caproate
Moniliformis dubius	Succinate, lactate, acetate, ethanol

PRIMARILY GLYCOLYTIC HELMINTHS

Schistosomes

The schistosomes constitute one of the most prevalent of the parasitic helminths that infect man. It is not surprising, therefore, that *Schistosoma mansoni* was one of the first helminths whose pathway of carbohydrate metabolism was elucidated (Bueding, 1950). This blood fluke was found to utilize glucose very rapidly. In one hour an amount of glucose equivalent to approximately one fifth of its dry weight was dissimilated by a schistosome. Essentially all of the glucose carbon that disappeared from the medium was recoverable as lactate carbon, indicating that these organisms were homolactate fermenters, almost quantitatively converting the glucose utilized to lactate by the reactions of glycolysis (Fig. 1).

It is of particular interest that these blood flukes, living free in the aerobic environment of the bloodstream, should still employ the anaerobic reactions of glycolysis as their sole source of carbohydrate energy metabolism. Presumably, glucose concentrations in the blood are always sufficiently high to allow the parasite the extravagance of such a wasteful metabolism. Further evidence in support of the anaerobic nature of this energy metabolism comes from several sources. First, the rates of glucose utilization, as well as the rates of lactate formation, by the schistosomes are the same under either aerobic or anaerobic conditions of incubation. Second, a group of compounds referred to as the cyanine dyes (of which the anthelmintics dithiazanine and pyrivinium chloride are members) inhibits oxygen uptake almost completely in the schistosomes, but has no effect upon their rate of glycolysis or their surviv-

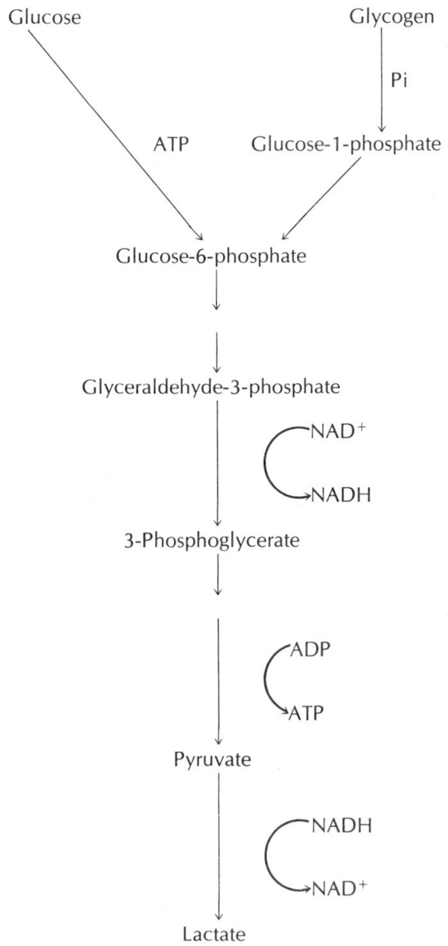

FIGURE 1. *The homolactate fermentation pathway in Schistosomes.*

al. Third, during in vitro incubation of adult schistosomes for 12 days, their gross morphology, motor activity, frequency of sex pairings, rates of glucose utilization, and lactate formation were the same in the presence or absence of oxygen. However, egg production was essentially stopped when oxygen was omitted. It is not known if oxidative metabolism is required for energy to produce eggs, or whether another non-energy-yielding reaction is essential for the developmental processes within the egg. Preliminary reports indicate that one non–energy-yielding reaction, tanning of the egg shell, is needed for maturation. Tanning requires the oxidation of some phenolic compounds with oxygen. There is little question, however, that for survival the adult parasite requires little or no oxygen.

The importance of carbohydrate metabolism to the schistosomes is indicated by more evidence than their rapid utilization of exogenous glucose and endogenous glycogen. As mentioned, the trivalent antimonials impair carbohydrate utilization by inhibiting phosphofructokinase, which, in turn, destroys the parasite. Glycolysis, therefore, is vital for schistosomes. Another example concerns the antischistosomal drug niridazole. Its first detectable effect is to lower glycogen levels of the male *S. mansoni.* Glycogen is depleted through accelerated utilization. Glycogen phosphorylase catalyzes this degradation of glycogen in tissues, including those of the schistosomes. This enzyme has two rapidly interconvertible forms, the "inactive" b form and the "active" a form (Fig. 2). Phosphorylase a is formed by the phosphorylation of phosphorylase b. Normally, when glycogen utilization is no longer necessary, glycogen phosphorylase a is converted back to the "inactive" phosphorylase b by the removal of phosphate as catalyzed by phosphorylase phosphatase. Niridazole inhibits phosphorylase phosphatase, thereby preventing the removal of the "active" glycogen phosphorylase a. As a consequence, in the presence of niridazole the phosphorylase continues to break down glycogen, depletes the parasite of stored glycogen, and kills it. Unfortunately, niridazole also lowers muscle glycogen in patients and causes muscular weakness. These findings further indicate the importance of carbohydrate metabolism to schistosomes.

Relatively little is known of the metabolism of other developmental stages of the schistosomes. The cercariae appear to differ from the schistosomules and adults. Cercariae oxidize all three carbons of pyruvate to CO_2, indicating (but not proving) the operation of a tricarboxylic acid cycle. In support of this possibility, some of the intermediates of the tricarboxylic acid cycle also are utilized by cercariae. Schistosomules, on the other hand, show a much reduced pyruvate catabolism and appear to more closely resemble the adults. The factors that regulate this shift from an apparent aerobic metabolism of the cercariae to the more nearly anaerobic metabolism of the adults (and presumably the schistosomules) are obscure.

Filarial Worms

Most filarial parasites that infect man are not amenable to biochemical studies, since they have not been reared in sufficient quantities in laboratory animals. This difficulty has been overcome partially by utilizing other filarial systems as models. Of these models, the three that have received most attention from the biochemical point of view are these: *Litomosoides carinii,* the adult form of which invades the pleural cavity of cotton rats or jirds; *Dipetalonema viteae,* which matures in the subcutaneous tissues of jirds or hamsters; and *Brugia pahangi,* which normally invades lymphatics but for experimental purposes is

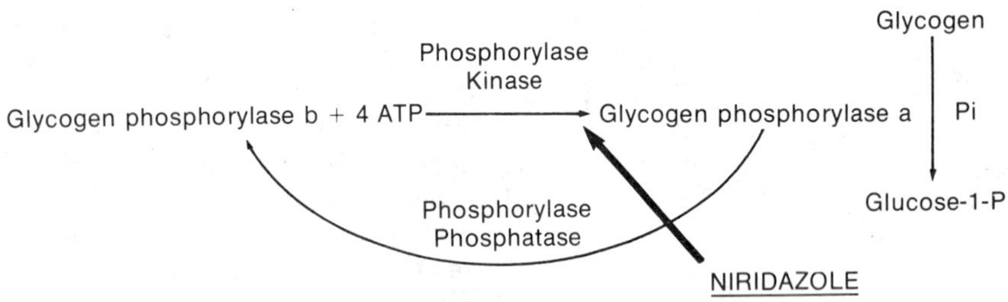

FIGURE 2. *The mechanism of the stimulation of glycogen breakdown by niridazole.*

much more readily recovered from the peritoneal cavity of jirds when they are infected intraperitoneally. Of these three species, only *B. pahangi* can infect man.

L. carinii was the first of this group of parasites to be reared in small animals, studied biochemically, and employed routinely for the screening of antifilarial compounds. Diethylcarbamazine (Hetrazan) was discovered by using this screen. Subsequently, another series of compounds, the cyanine dyes (for example, dithiazanine) was found to eliminate *L. carinii* infections in cotton rats. Cyanines strongly inhibit oxygen uptake of the parasite, which requires this gas for survival. Unfortunately, dithiazanine was without effect on those filariids that infect man. This disparity between the effectiveness of the cyanine dyes on these closely related species was disappointing and raised the question of the ways *L. carinii* differed from the other filariids.

A probable answer to this question has been arrived at by comparing the energy metabolisms of *L. carinii, B. pahangi,* and *D. viteae*. The motility of *L. carinii* adults ceases almost completely within one hour of anaerobic incubation, whereas aerobically, the parasites maintain good motility for several days. Therefore, *L. carinii* is an obligate aerobe. In contrast to this, both the motility and survival of *B. pahangi* and *D. viteae* is the same aerobically or anaerobically. Thus, oxygen does not appear to be required for motility or maintenence in either of these two parasites.

In accord with the apparent lack of an oxygen requirement for the motility of *B. pahangi* and *D. viteae,* subsequent studies have demonstrated that, in vitro, both of these parasites are homolactate fermenters, as was found with the schistosomes. All of the glucose carbons utilized could be accounted for as lactate. No other products were detected. If the filarial worms that infect man are similar in their energy metabolism to *B. pahangi* and *D. viteae* then they too would have no need for oxygen for survival, and depression of oxygen consumption by the cyanine dyes would be expected to have little or no effect upon them. On the other hand, the trivalent antimonials that inhibit the anaerobic reactions of glycolysis at the level of phosphofructokinase were once used for treating filarial infections but later abandoned because of their high toxicity.

In contrast with *B. pahangi* and *D. viteae,* adult *L. carinii* are obligate aerobes. This pleural cavity-invading parasite is a heterolactate fermenter. That is, it accumulates other fermentation products in addition to lactate. Approximately 50 per cent of the glucose carbon it dissimilates aerobically is recovered as lactate. Acetate, CO_2 acetylmethylcarbinol (acetoin), and one or more unidentified products also accumulate. Presumably energy is derived from this further metabolism, since under anaerobic incubation the fermentation shifts toward a homolactate fermentation but the parasite dies. Studies employing various species of ^{14}C-glucose as substrates indicate that essentially all of the respiratory CO_2 formed arises from the 3 and 4 carbons of glucose with very little arising from the other carbon atoms of glucose (Wang and Saz, 1974). According to the glycolytic sequence of reactions, the 3 and 4 carbons of glucose give rise to the carboxyl carbons of pyruvate, which, in turn, would be lost as CO_2 in the oxidative decarboxylation to acetate. Since almost no CO_2 arises from the 1,2 or 5,6 carbons of glucose, presumably acetate is not further oxidized and the tricarboxylic acid cycle is not a quantitatively significant energy-yielding pathway in *L. carinii.* The findings suggest that the aerobic requirement of *L. carinii* may reside completely in one system, the oxidative decarboxylation of pyruvate to acetate and CO_2. A summary of the metabolic energy pathways of the three filariids is outlined in Figure 3.

As might be expected, the metabolisms of the adult and microfilarial stages differ from each other. The microfilariae lose their motility in the absence of oxygen, but even after seven days of anaerobic incubation the reintroduction of air results in full restoration of motility. Therefore, although oxygen is required for motility, it does not appear to be required for survival of the microfilariids.

Under aerobic conditions, *B. pahangi* microfilariae switch to a heterolactate fermentation that resembles the process described for the adult *L. carinii,* forming lactate, acetate, and CO_2. Most of the pyruvate decarboxylated accumulates as the one step oxidation product, acetate. However, a small quantity of the acetate may be oxidized completely to CO_2, presumably via a tricarboxylic acid cycle mechanism. When the microfilariae are incubated anaerobically they become nonmotile and concurrently shift their metabolism toward total lactate accumulation (Rew and Saz, 1977).

A further indication of the coupling of motility with the aerobic metabolism of *B. pahangi* microfilariae is obtained when they are incubated with the anthelmintic levamisole. Levamisole inhibits neuromuscular transmission (possibly by ganglionic blockade) so that the parasites are paralyzed. Aerobic incubation in vitro in the presence of low concentrations of levamisole results in a rapid loss of motility. After this loss of motility, the energy metabolism shifts toward lactate production, mirroring an anaerobic metabolism. The effects of levamisole on the metabolism of the microfilariid appear to be secondary to the paralysis because loss of motility from tetramisole blockade reduces the energy requirement. The lowered energy requirement may cause a secondary shift toward an anaerobic lactate-forming metabolism. These findings indicate a tight coupling between motility and aerobiosis in these microfilariae.

CO_2-FIXING HELMINTHS

Ascaris lumbricoides

Although all helminths examined dissimilate carbohydrates to triose via the glycolytic pathway, many of them have evolved another sequence of reactions leading to the accumulation of succinate or products derived from succinate. This succinate-forming pathway requires the fixation of carbon dioxide and has the potential of yielding additional energy (ATP) for the worm. *A. lumbricoides*, the parasitic intestinal roundworm, has served as the model system in elucidating this pathway. Increasing numbers of species of organisms that accumulate succinate under reduced oxygen tension and appear to obtain energy for survival by employing this "*Ascaris* pathway" are being reported. In addition to many of the parasitic helminths and protozoans, numerous free-living metazoans, such as the intertidal molluscs, which spend part of their life cycle under anoxic conditions burrowing in the sand, diving mammals such as seals and porpoises, and ischemic or anoxic rat hearts are all reported to use this pathway of succinate formation. Therefore, the pathway reported initially for *Ascaris* represents an important sequence in many organisms, including some mammals, in which it is a means of obtaining additional energy during oxygen deprivation to the tissues.

It has long been known that *Ascaris* survives equally well under aerobic or anaerobic conditions. The presence of CO_2 in the environment enhances survival. Adult *Ascaris*, therefore, can obtain energy for survival via anaerobic pathways. Accordingly, the organism is not sensitive to cyanide and levels of cytochrome c oxidase are so low as to be of doubtful physiologic significance. Accumulation of hydrogen peroxide in the presence of air indicates the existence of a flavin terminal oxidase rather than the usual cytochrome system, although it has been postulated that cytochrome o, which can react directly with oxygen to give hydrogen peroxide, may be involved in terminal respiration. Due to a deficiency of catalase, hydrogen peroxide accumulates and is toxic to the tissues of the worm.

Succinate and a mixture of volatile fatty acids

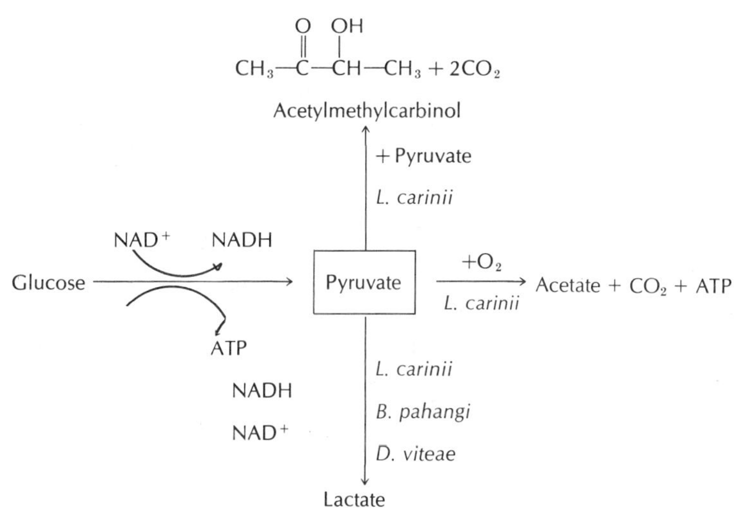

FIGURE 3. *Pathways of glucose dissimilation in* L. carinii, B. pahangi, *and* D. viteae.

comprise the major fermentation products of *Ascaris* metabolism. Volatile acids formed include acetate, propionate, traces of butyrate, pentanoate (valerate), tiglate (2-methylcrotonate), 2-methylbutyrate, and 2-methylvalerate. Lactate is not a fermentation product of intact *Ascaris* adults. The current concept of the carbohydrate fermentation pathway in *Ascaris* muscle is illustrated in Figure 4 (Saz, 1971). According to this pathway, the glycolytic enzymes of *Ascaris* muscle, which are present in the cytosol portion of the cell, function similarly to those of the host tissues up to the point of phosphoenolpyruvate (PEP) accumulation. At this point the metabolism of the parasite diverges from that of the host. In *Ascaris,* cytoplasmic pyruvate is not formed from PEP, since pyruvate kinase activity is barely detectable and of doubtful physiologic significance. Instead, the cytoplasmic PEP carboxykinase catalyzes the fixation of carbon dioxide into PEP to form oxalacetate according to the following reaction:

$$\text{PEP} + CO_2 + \text{IDP(GDP)} \rightleftharpoons \text{Oxalacetate} + \text{ITP(GTP)}$$

In this reaction a substrate level phosphorylation of inosine diphosphate or of guanosine diphosphate takes place to form the corresponding energy-rich triphosphate. This conserves the energy of the phosphate bond of PEP as is accomplished by the pyruvate kinase reaction of the host tissues. Therefore, up to this point, identical amounts of energy are recovered from both pathways.

In order for glycolysis to continue, the NADH formed at the glyceraldehyde dehydrogenase level of glycolysis must be reoxidized back to NAD^+ so that another molecule of glyceraldehyde may be oxidized. In mammalian tissues this would be accomplished by the lactate dehydrogenase reaction, wherein pyruvate is reduced to lactate with NADH as the electron donor, thus reforming NAD^+. In *Ascaris,* however, cytoplasmic pyruvate is not formed and lactate is not an end product. Instead of employing lactate dehydrogenase, the parasite uses its cytoplasmic malate dehydrogenase for this function. Oxalacetate formed from the PEP carboxykinase reaction as illustrated above is reduced with NADH to form malate and regenerate NAD^+ as follows:

$$\underset{\text{(Oxalacetate)}}{HOOC{-}CH_2{-}CO{-}COOH} + NADH + H^+ \rightleftharpoons \underset{\text{(Malate)}}{HOOC{-}CH_2{-}CHOH{-}COOH} + NAD^+$$

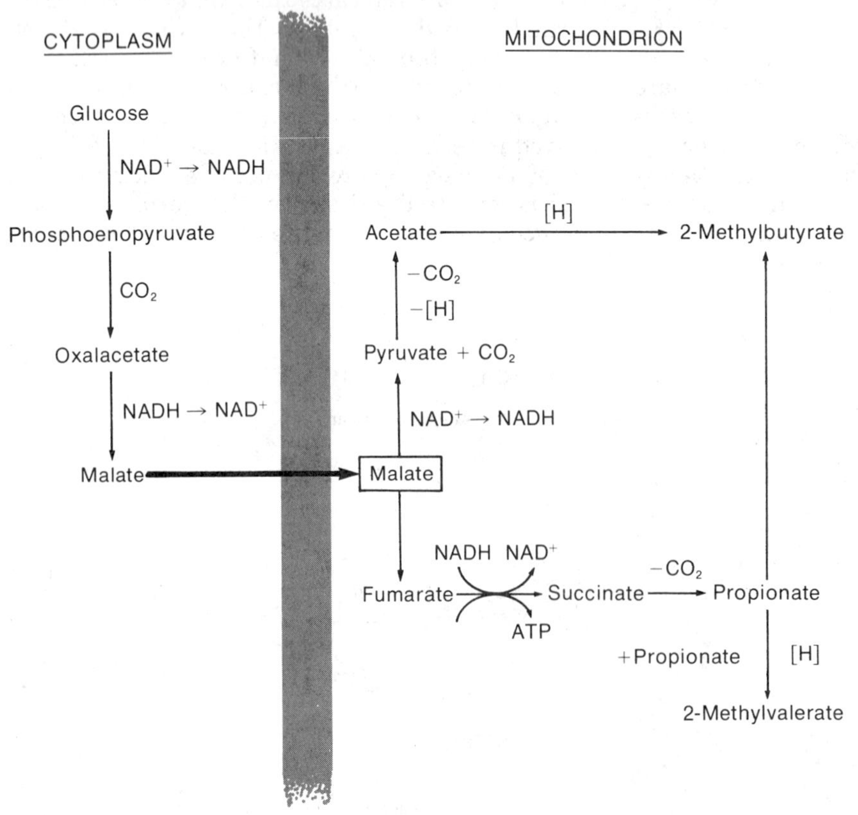

FIGURE 4. *Pathway for the dissimilation of glucose and the formation of succinate and volatile acids in* Ascaris lumbricoides *muscle.*

Malate then permeates through the mitochondrial membrane and serves as the substrate for the mitochondrion of the nematode. In contrast with mammalian mitochondria, evidence indicates that a functional tricarboxylic acid cycle is absent from the *Ascaris* organelles. The roundworm mitochondria function and generate energy anaerobically. Under these conditions, the malate that enters the mitochondrion cannot be oxidized directly, but rather must undergo a dismutation reaction. That is, for each mole of malate oxidized, a corresponding amount must be reduced simultaneously.

As illustrated in Figure 4, the oxidative leg of this dismutation is provided by the NAD^+-linked oxidation of malate to pyruvate and CO_2 catalyzed by the mitochondrial "malic" enzyme (malate dehydrogenase decarboxylating). This reaction proceeds as follows:

$$\text{1-Malate} + NAD^+ \rightleftharpoons \text{Pyruvate} + CO_2 + NADH + H^+$$

The "malic" enzyme reaction then serves to generate intramitochondrial reducing power in the form of NADH. This NADH will be subsequently employed in the ATP-generating reductive leg of the dismutation reaction.

An equivalent quantity of malate must now be reduced through fumarate to succinate as catalyzed by reactions (1) and (2):

(1) Fumarase:

$$\underset{\text{Malate}}{HOOC\text{—}CH_2\text{—}CHOH\text{—}COOH} \rightleftharpoons \underset{\text{Fumarate}}{HOOC\text{—}CH\text{=}CH\text{—}COOH} + H_2O$$

(2) Fumarate Reductase:

$$\underset{\text{Fumarate}}{HOOC\text{—}CH\text{=}CH\text{—}COOH} + NADH + H^+ \xrightarrow{ADP \rightarrow ATP} \underset{\text{Succinate}}{HOOC\text{—}CH_2\text{—}CH_2\text{—}COOH} + NAD^+$$

NADH formed from the "malic" enzyme reaction donates electrons to fumarate, thereby accumulating succinate. The fumarate reductase reaction differs dramatically from the corresponding mammalian enzyme, succinate dehydrogenase. It does not catalyze a simple reversal of the succinate dehydrogenase system found in host tissues. The mammalian succinate dehydrogenase passes electrons directly to flavins, bypassing NAD^+. In contrast, the fumarate reductase of *Ascaris* couples with NADH, which, in turn, passes the electrons to a presumed flavin. This transfer of electrons from NADH to flavin would be analogous to a site I phosphorylation of mammalian electron transport and, as would be expected, electron-transport–associated ATP is generated by the *Ascaris* fumarate reductase reaction. Therefore, this reaction provides energy for the cell in excess of that which would be obtained by the homolactate fermentation, providing an advantage to the succinate-forming organisms. For each glucose molecule fermented, two malates would enter the mitochondrion, but one half of the malate would be oxidized and one half would be reduced to succinate. Thus, 1 mole of ATP would be formed in the fumarate-reductase reaction for each mole of glucose dissimilated over this pathway.

The fumarate reductase and its associated reactions of electron transport are of great importance for chemotherapeutic and other reasons. A number of anthelmintic agents such as thiabendazole, cambendazole, and tetramisole inhibit the reduction of fumarate, although it is not clear whether this inhibition would constitute the primary site of action. A number of other anthelmintics uncouple the phosphorylation system associated with this reaction. Chlorsalicylamide (Yomesan), desaspidin, dichlorophen, and phloroglucinol derivatives constitute some examples. The fumarate reductase pathway of metabolism is by no means limited to *Ascaris,* but appears to be a means of anaerobic energy generation in numerous other parasitic and free living forms (Table 2).

Many helminths, including *Ascaris* adults, accumulate additional products that arise from succinate; that is, succinate serves as the precursor for these products. The most common is the three-carbon volatile acid propionate, which is formed by an overall decarboxylation of succinate (Fig. 4). The reaction occurs in two steps as it does in mammalian tissues and requires the coenzyme A derivatives as illustrated by reactions (A) and (B) below:

$$\text{(A) Succinyl-CoA} \underset{\text{Vitamin } B_{12}}{\overset{\text{Mutase}}{\rightleftharpoons}} \text{Methylmalonyl-CoA}$$

$$\text{(B) Methylmalonyl-CoA} \overset{\text{Propionyl-CoA Carboxylase}}{\rightleftharpoons} \text{Propionyl-CoA} + CO_2$$

Mammalian tissues generally employ these reactions in a direction opposite to that shown. Propionate is utilized by means of CO_2 fixation to form succinate. As a consequence, propionate is a glycogenic fatty acid in mammals. That is, it can give rise to succinate that, in turn, can go on to form glycogen. In *Ascaris,* however, the reaction operates primarily in the reverse direction, toward propionate accumulation. What controls the direction of this sequence in *Ascaris* is not understood completely. However, the fact that propionyl-CoA is further metabolized by the nematode, as will be discussed below, may pull the reaction in the indicated direction. Since the methylmalonyl-CoA mutase requires vitamin B_{12}, those parasites that form propionate contain high levels of this vitamin (Tkachuck et al., 1977). In contrast, those helminths that form primarily succinate, lactate, or products unrelated to propionate do not contain appreciable amounts of vitamin B_{12}.

Pyruvate, formed as a product of the "malic" enzyme reaction (Fig. 4), is a precursor of acetate that arises within the mitochondrion. In mammalian tissues the oxidative decarboxylation of pyruvate to acetate and CO_2 results in the generation of ATP, since the reaction is linked to the electron transport system with oxygen acting as the terminal electron acceptor. Pyruvate utilization by *Ascaris* is particularly interesting in that acetate is formed anaerobically. What substitutes for oxygen as the electron acceptor is still not known, but it is possible that the oxidation of pyruvate may be coupled to the formation of the volatile fatty acids, 2-methylbutyrate and 2-methylvalerate. Whether or not ATP is generated in the nematode by this anaerobic oxidative decarboxylation is unknown.

The mechanism by which *Ascaris* muscle forms its major fermentation products, 2-methylbutyrate and 2-methylvalerate, is not completely understood. However, acetate and propionate, presumably as the coenzyme A derivatives, are precursors to these volatile fatty acids. Incubating various radioactive substrates with *Ascaris* muscle preparations, then isolating the fermentation products formed and chemically degrading these products to determine the distribution of isotope in the carbon atoms of each have led to a postulation of the pathways for the formation of the two branched chain volatile acids (Fig. 5). Observations are consistent with the hypothesis that 2-methylbutyrate is formed after a condensation of the carboxyl carbon of acetyl-CoA with the number-two carbon of propionyl-CoA. The condensation product would be the coenzyme A derivative of methylacetoacetate, which possesses the appropriate carbon skeleton and could be reduced to 2-methylbutyrate. A similar series of reactions appears to account for the formation of the six carbon branched chain volatile acid, 2-methylvalerate. In this case, two molecules of propionyl-CoA condense to form the corresponding C_6 keto acid, which would then be reduced to 2-methylvalerate. Although it was thought that these reactions were unique to *Ascaris,* it was

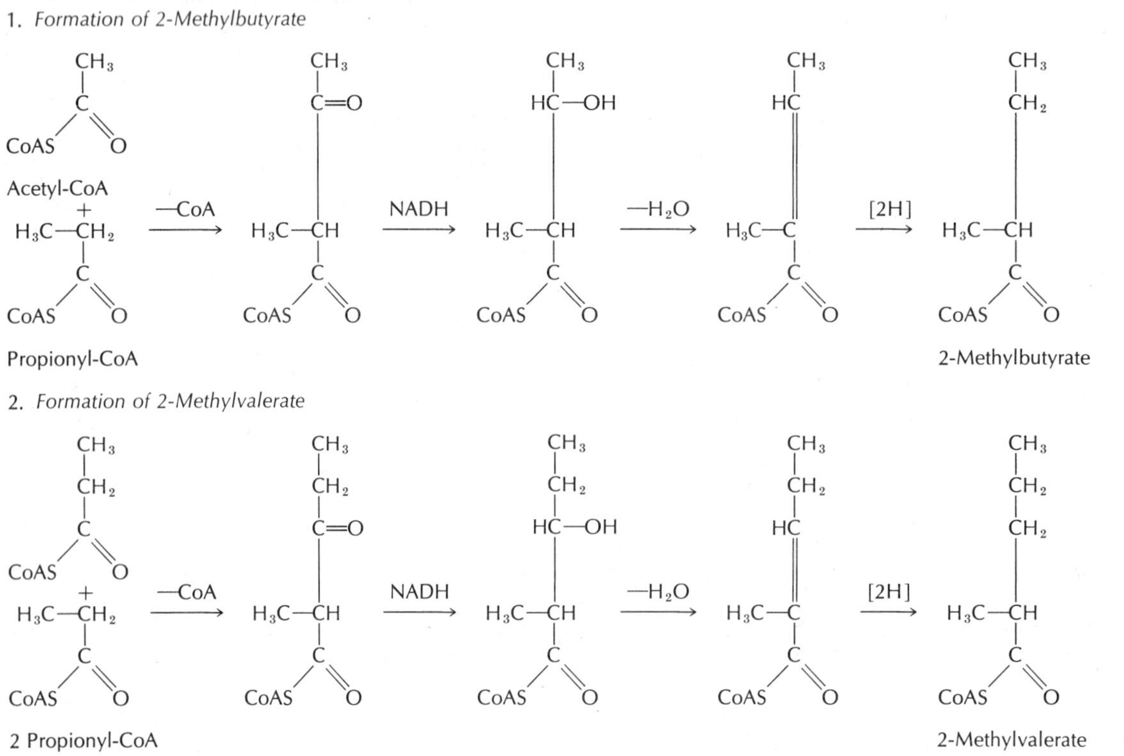

FIGURE 5. *Proposed pathways for the formation of 2-methylbutyrate and 2-methylvalerate by* Ascaris lumbricoides *muscle.*

reported recently that the clinically important lung fluke *Paragonimus westermani* also accumulates 2-methylbutyrate as a fermentation product.

Although many of the reactions in the energy pathways of *Ascaris* and its host are similar or almost identical, others are distinct and therefore vulnerable to chemotherapeutic attack. The fumarate reductase system is vital to the adult ascarids, but of minor, if any, physiologic importance to the mammalian host. Thus, interference with the fumarate reductase reaction should be detrimental to the parasite, but not to the host. As stated above, many anthelmintics inhibit some component of the fumarate reductase electron transport system. Accordingly, succinate formation and the fumarate reductase pathway are common among other parasites and invertebrates (Table 2). On the other hand, the formation of the branched-chain volatile acids seems to be of great importance in the overall economy of *Ascaris* and *Paragonimus,* but may be unique to these two helminths. Therefore, each parasite may have unique as well as common features in its biochemical arsenal of reactions. Each parasite is in some ways similar to and in some ways different from other parasites and from the host tissues, indicating that each worm should be examined independently.

In contrast to the adult *Ascaris,* fertilized eggs laid by the female in the intestines cannot develop past the single cell stage until they are passed out and gain access to oxygen. The eggs require oxygen to initiate and maintain development. At any stage, growth and differentiation may be arrested by removing oxygen. Yet, the potential for continued development remains even after extended periods of anaerobiosis, and normal maturation may be resumed merely by the introduction of air. Unlike adults, respiration of the eggs is highly sensitive to the cytochrome oxidase inhibitors, cyanide, azide, and carbon monoxide in the dark. Cytochrome oxidase activity is not detectable in unembryonated eggs. After exposure to air, however, the cytochrome oxidase activity increases continuously throughout development and high levels of activity are attained. Metabolic pathways in the egg are switched from those of the adult to one more closely resembling the aerobic metabolism of the host. The eggs are cleidoic; that is, only gases and water can permeate the vitelline membrane. As a consequence, all of the substrates for development are contained and utilized within the egg. Interestingly, *Ascaris* eggs were the first cells unequivocally shown to catalyze the net conversion of lipid into carbohydrate (Passey and Fairbairn, 1957). Most tissues cannot accomplish this transformation, since acetate formed from long chain fatty acids is oxidized by the mechanism of the tricarboxylic acid cycle, which results in the complete oxidation of the equivalent of both carbons of acetate to CO_2, leaving no carbons for a net synthesis of glycogen. Hence, no *net* synthesis of carbohydrate can take place from lipid. However, the glyoxylate cycle pathway in certain bacteria, plant cells, and in *Ascaris* eggs allows for this conversion of lipid into carbohydrate (Fig. 6), thereby allowing the egg to utilize its stored lipids to replenish depleted carbohydrate stores for subsequent use.

The first molt in the *Ascaris* life cycle occurs within the eggs, resulting in formation of the second stage larvae that are released from the eggs on hatching in the upper intestine. The second molt occurs in the lungs of the host and the third stage larvae make their way up the bronchial tree to the esophagus to be swallowed back down to the intestine. Before or shortly after entering the intestine the third molt to the fourth stage occurs. Cytochrome c oxidase activity is present in each stage from the developing egg through the third (lung stage) larvae. When the larvae molt to the fourth stage, the cytochrome c oxidase activity is lost, indicating a shift from aerobic to an anaerobic metabolism. Factors that control the "switchovers" from anaerobic to aerobic metabolism in the developing egg and from aerobic to anaerobic in going from the third to the fourth larval stages are unknown.

Hymenolepis diminuta

The large intestinal tapeworm of the rat, *Hymenolepis diminuta,* also has served as a biochemical model for studying parasitic helminths. There is little doubt that this cestode is an anaerobe, since it has been cultivated from larva through adult in a CO_2-containing anaerobic environment. Oxygen may be detrimental to normal development. Another cestode, *Hymenolepis nana,* which infects humans, has also been cultivated. This cultivation also requires a CO_2-containing anaerobic gas phase. It is noteworthy that both the adult and larval stages were cultivated anaerobically. This is different from the larval stages of *Ascaris,* most of which require oxygen.

Hymenolepis diminuta adults accumulate succinate as the major fermentation product. Smaller quantities of acetate and lactate also are produced. Succinate is formed by a pathway involving CO_2 fixation, a "malic" enzyme, and fumarate reductase, as was described in *Ascaris.* The tapeworm also is similar to *Ascaris* in other respects. The tricarboxylic-acid cycle does not appear to be of physiologic significance as an energy

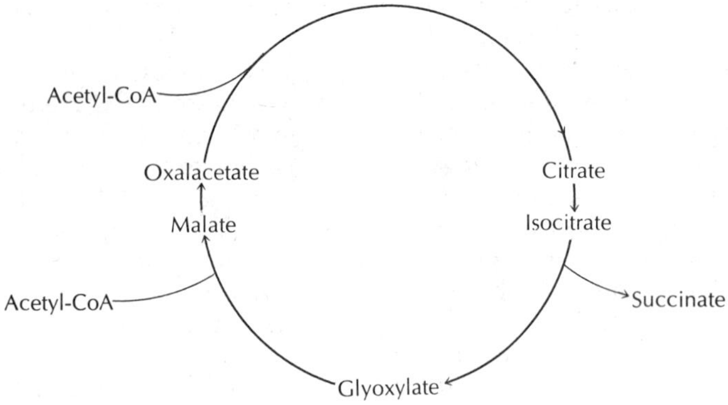

(a) ATP + CoA + Acetate ⟶ Acetyl CoA + AMP + P-P

(b) Isocitrate ⟶ Glyoxylate + Succinate

(c) Glyoxylate + Acetyl CoA ⟶ Malate + CoA

(d) = (a) + (b) + (c)

ATP + Isocitrate + Acetate ⟶ Malate + Succinate + AMP + P-P

(e) From TCA cycle reactions:

ATP + Malate + Acetate ⟶ Isocitrate + 2(H) + AMP + P-P

(f) = (d) + (e)

2 ATP + 2 Acetate ⟶ Succinate + 2(H) + 2 AMP + 2 P-P

FIGURE 6. *The glyoxylate cycle for the net synthesis of carbohydrates from lipids in Ascaris eggs.*

yielding pathway in *H. diminuta*. Mitochondrial ATP generation from the fumarate reductase reaction is inhibited by chlorosalicylamide (Yomesan), desaspidin, dichlorophen, and some of the phloroglucinol derivatives, and the cestode has lost its ability to synthesize long-chain fatty acids de novo.

Unlike *Ascaris* metabolism, succinate is a terminal fermentation product in *H. diminuta* in that it is not further utilized. The adult cestode does not form propionate from succinate, and in accordance with this does not contain appreciable levels of vitamin B_{12}. Another interesting difference between *Ascaris* and *H. diminuta* was uncovered upon comparing their respective "malic" enzymes. In the case of the former, the oxidative decarboxylation of malate is coupled with NAD^+ (see text above for reaction). In the cestode, the reaction is coupled specifically with $NADP^+$. This is important because the reducing power generated by these mitochondrial reactions must be further employed in the specific NADH-requiring fumarate reductase reaction, which presumably will not accept directly from NADPH. To overcome this discrepancy, *H. diminuta* adults contain an active $NADP^+$:NAD^+ transhydrogenase as a constituent of their inner mitochondrial membranes. This activity is not detectable in *Ascaris*. The $NADP^+$:NAD^+ transhydrogenase provides the *H. diminuta* fumarate reductase with NADH by reaction with the NADPH formed in the "malic" enzyme reaction as follows:

1. $\text{1-Malate} + NADP^+ \rightleftharpoons \text{Pyruvate} + CO_2 + NADPH + H^+$

2. $NADPH + NAD^+ \underset{}{\overset{\text{Transhydrogenase}}{\rightleftharpoons}} NADH + NADP^+$

3. $\text{Fumarate} + NADH + H^+ \xrightarrow{\text{ADP} \curvearrowright \text{ATP}} \text{Succinate} + NAD^+$

Fasciola hepatica

Although it seems clear that some of the helminths possess primarily, if not solely, an anaerobic energy metabolism as in the cases of the homolactate fermenters and the cestodes that have been cultured anaerobically, it is difficult to assess the physiologic significance of oxygen in the energy metabolisms of many other worms. One of the indications that oxygen plays a role in the metabolism of an organism is when the presence of oxygen alters the rate or type of metabolic products formed (for example, Pasteur effect). It has been known for many years that the liver fluke *Fasciola hepatica* exhibits such an effect of oxygen, but it has only recently been indicated that oxygen may serve as a terminal electron acceptor in the energy metabolism of the parasite.

During the adult phase of its life cycle, *F. hepatica* resides in a habitat of low oxygen tension. Therefore, the major part of the trematode's energy metabolism may be expected to be anaerobic. The major end products of either aerobic or anaerobic incubation are acetate and propionate, with considerably smaller quantities of lactate. Incubation of the fluke in air, however, decreases the amount of propionate formed but does not alter acetate formation. Therefore, oxygen may play a role in the energy metabolism of this parasite even though it appears to be predominantly anaerobic in this respect. Recent evidence indicates that either fumarate or oxygen may serve as the terminal respiratory acceptor of electrons. In both cases, phosphorylation of ADP to ATP is associated with the transport of electrons. If oxygen competes with fumarate as terminal acceptor, then aerobically less fumarate and hence less propionate would accumulate. Therefore, although the liver fluke is similar to *Ascaris* in most of its energy-forming pathway, aerobic oxidations may be of considerably greater physiologic significance to this trematode than they are to the nematode.

SUMMARY

Chemotherapy of most helminth infections of man is still inadequate. Agents employed in the therapy of other infectious diseases are almost universally without effect on helminth infestations. Therefore, the therapy of the metazoan diseases has lagged far behind other therapies and must rely upon its own unique arsenal of drugs. Of those anthelmintics whose mode of action has been examined, most appear to act by inhibiting either neurotransmission or interrupting the energy-yielding reactions of the parasites. These inhibitions are possible because many of the components of the parasite tissues differ from those of the host. It is of both academic and practical importance to understand the biochemistry of the parasitic helminths.

It should be clear that no two helminths will have completely identical biochemical reactions. Differences exist not only between parasite and host, but also between any two parasites. These differences may be at the subtle level of enzyme or protein structure and kinetics or at a more obvious level of pathways, reactions, and products. The key, in examining the energy metabolism of the parasites, is that each organism is different and is a separate biochemical entity. Some helminths, such as *Hymenolepis diminuta* and *Hymenolepis nana,* appear to be anaerobic in all stages of development. Others are capable of a complete or major anaerobic energy metabolism in one or more developmental stages, but require aerobiosis during other parts of the life cycle. Still other parasites are obligate aerobes in one or more of their developmental forms.

Most parasitic worms possess one of two types of energy metabolism; either they are primarily glycolytic, or they are glycolytic but also depend, for part of their energy supply, on the fixation of CO_2 and the formation of succinate or products arising from succinate. Many worms employ both pathways and accumulate lactate as well as succinate. Generally, where this is found, the succinate pathway is dominant. Schistosomes and the filarial worms *Brugia pahangi* and *Dipetalonema viteae* are the best examples of glycolyzing helminths, since in vitro they are all homolactate fermenters and do not appear to require oxygen for energy generation. *Litomosoides carinii,* another filarial worm, is an example of a primarily glycolyzing parasite that forms mostly lactate but also requires oxygen for survival, presumably for the single-step energy-yielding oxidative decarboxylation of pyruvate to acetate and CO_2. Many other parasites accumulate acetate; some require air for this reaction, others accomplish this oxidative decarboxylation anaerobically. Little is known about the mechanism or energetics of the anaerobic oxidative decarboxylation of pyruvate to acetate in the helminths.

Ascaris and to some extent *Hymenolepis diminuta* have served as models for parasites that require CO_2 fixation and succinate formation for their energy needs. Many anthelmintics act by inhibiting either the fumarate reductase reaction or the ATP-generating electron transport system associated with it. The role of oxygen in the parasites that form succinate is still not clear. In some — for example, *F. hepatica* — oxygen appears capable of partially competing with fumarate as the terminal electron acceptor. Other parasites, such as *Ascaris* and *Hymenolepis,* will take up oxygen, but the parasites survive equally well without air.

Progress has been slow in the field of parasite biochemistry for many reasons, but some of the biochemical relationships that exist between the host and parasite systems are emerging. Pathways and reactions of greater physiologic significance to the parasite than to the host are being recognized, thereby designating areas that may be amenable to chemotherapeutic attack. However, the giant void between recognizing vulnerable pathways and designing specific inhibitors of these pathways must still be bridged.

References

Bueding, E.: Carbohydrate metabolism of *Schistosoma mansoni*. J Gen Physiol 33:475, 1950.

Krebs, E. G., and Najjar, V. A.: The inhibition of d-glyceraldehyde 3-phosphate dehydrogenase by specific antiserum. J Exp Med 88:569, 1948.

Mansour, T. E. and Bueding, E.: The actions of antimonials on glycolytic enzymes of *Schistosoma mansoni*. Br J Pharmacol 9:459, 1954.

Meyer, F., Kimura, S., and Mueller, J. F.: Lipid metabolism in the larval and adult forms of the tapeworm *Spirometra mansonoides*. J Biol Chem 241:4224, 1966.

Passey, R. F., and Fairbairn, D.: The conversion of fat to carbohydrate during embryonation of *Ascaris* eggs. Can J Biochem Physiol 35:511, 1957.

Rew, R. S., and Saz, H. J.: The carbohydrate metabolism of *Brugia pahangi* microfilariae. J Parasitol 63:123, 1977.

Saz, H. J.: Facultative anaerobiosis in the invertebrates: Pathways and control systems. Am Zoologist 11:125, 1971.

Saz, H. J. and Dunbar, G. A.: The effects of stibophen on phosphofructokinases and aldolases of adult filariids. J Parasitol 61:794, 1975.

Tkachuck, R. D., Saz, H. J., Weinstein, P. P., Finnegan, K., and Muller, J. F.: The presence and possible function of methylmalonyl CoA mutase and propionyl CoA carboxylase in *Spirometra mansonoides*. J Parasitol 63:769, 1977.

Wang, E. J., and Saz, H. J.: Comparative biochemical studies of *Litomosoides carinii, Dipetalonema viteae,* and *Brugia pahangi* microfilariae. J Parasitol 60:361, 1974.

18 ARTHROPODS OF MEDICAL IMPORTANCE

A. L. Dohany, Ph.D.
John J. S. Burton, Ph.D.

The animals within the phylum Arthropoda are characterized by bilateral symmetry, a chitinous exoskeleton, and paired jointed appendages (Borror et al., 1976). This phylum, the largest of all animal phyla, contains species that transmit some of the most important diseases known to man, including malaria, yellow fever, plague, dengue, typhus, and the encephalitides. The majority of the arthropods of medical importance are contained within two large classes: Insecta (insects) and Arachnida (mites, ticks, spiders, and scorpions) (Furman and Catts, 1970).

The purpose of this chapter is to provide an overview of the arthropods of medical importance. Specific relationships between the arthropods and their animal hosts will be covered in chapters pertaining to the diseases they cause or transmit.

CLASS: INSECTA

Order: Diptera

Flies and Mosquitoes

Adult Diptera or true flies have only a single pair of wings. The hind wings are modified into a pair of small stalked knobs called halteres (balancing organs). The mouth parts of adult biting Diptera are broadly classed as piercing-sucking, and those of filth flies as sponging-lapping.

The Diptera are the most important arthropods medically, and Culicidae (mosquitoes) is the most important family within the group. The activities of Diptera in relation to man can be placed in three categories. (1) As *disease vectors*, they are capable of rapid dissemination of infection either by biologic transmission (e.g., malaria, filariasis) or by mechanical transmission (e.g., typhoid, anthrax) (Chamberlain and Sudia, 1961). (2) As *pests*, the persistence of flies can interrupt and interfere with human activity over broad geographic areas. (3) As *allergens*, biting Diptera sometimes produce rather severe reactions in man, depending on the immune system of the individual (Feingold, 1973). Scratching the site of the reaction can lead to secondary infection.

Family: Culicidae

Mosquitoes

Mosquitoes can be recognized by their relatively long and slender wings, with scales along the wing veins and posterior margin, and by their long, forward-projecting sucking proboscis. Only the females suck blood. The males can be distinguished by the much greater plumosity of the antennae.

There are a number of medically important genera. Living specimens of *Anopheles* (subfamily Anophelinae) are readily distinguishable by their resting and biting posture (Fig. 1). Their body plane is generally at a considerable angle to the substrate, giving the impression that they are standing on their heads. The palpi of both sexes are long like the proboscis. The eggs float singly on the water surface, and the larvae lie parallel to and just below the surface. Certain species in this genus are exclusively responsible for the biologic transmission of the four species of human malaria (*Plasmodium*) as well as simian malaria. They are also known to transmit filariasis (both *Wuchereria* and *Brugia*), as well as some viruses of man, notably that producing O'nyong-nyong fever in Africa.

The genera *Aedes* and *Culex* (subfamily Culicinae) hold their thoracic-abdominal plane essentially parallel to the substrate during resting and biting (Figs. 2 and 3). Palpi of the male are long like the proboscis; those of the female are short. The eggs of *Aedes* are laid singly, most commonly on moist substrate above the water line; those of *Culex* are laid in groups (rafts) on the water. The larvae of this subfamily hang head down into the water and contact the surface film only with an elongated respiratory siphon. *Aedes* mosquitoes are best known for transmitting viral diseases, and *A. aegypti* is clearly the most important disease vector in this genus. Yellow fever and dengue are the major viral diseases transmitted by *Aedes*, but they are also known to transmit chikungunya and encephalitis viruses. Several species also transmit filariasis. *Culex* mosquitoes may be most important as vectors of filariasis (*Wuchereria*), but they also have significant roles in the transmission of viral encephalitis, including Japanese encephalitis, western and eastern equine encephalitides, and St. Louis encephalitis.

Some other genera are important vectors of *specific* diseases; e.g., *Mansonia* mosquitoes are the principal vectors of *Brugia*. Larvae of this genus obtain oxygen peculiarly by attaching to the submerged parts of aquatic plants rather than at the water surface, a characteristic that complicates control measures.

Toxorhynchites mosquitoes (subfamily Toxorhynchitinae) are beneficial. The adults do not suck blood; instead they live on plant juices. The larvae are predaceous and may partially control the larvae of other mosquito genera locally.

Family: Simuliidae

Black Flies

Black flies are small (1 to 5 mm) and have stout bodies that are usually dark in color (Fig. 4).

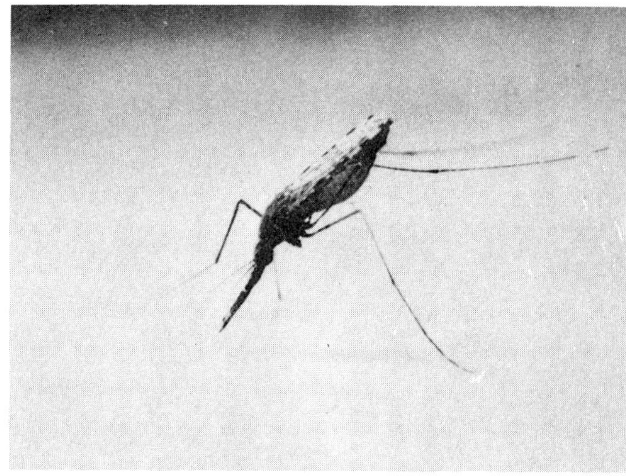

FIGURE 1

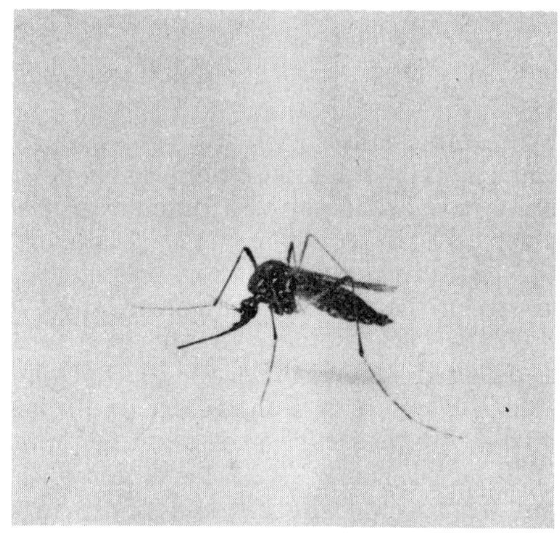

FIGURE 2

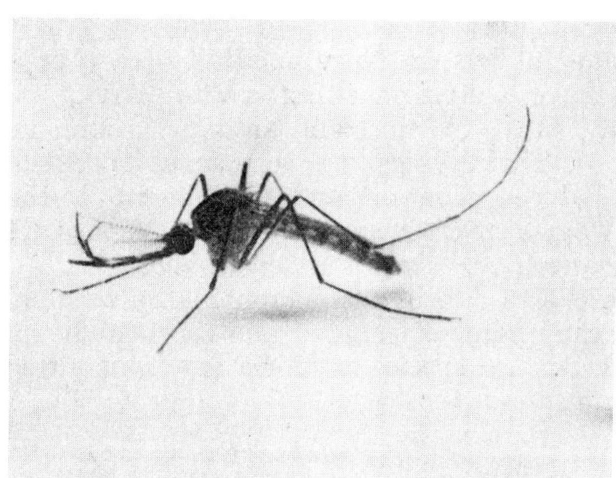

FIGURE 3

FIGURES 1–3. Anopheles, Aedes *and* Culex *mosquitoes in typical resting positions.*

FIGURE 4. *A slide-mounted specimen of the family Simuliidae, black flies.*

Their antennae contain 9 to 12 segments (usually 11) and are without arista (long sensory hairs). The wings are broad and are folded over the body when at rest. The anterior wing veins are strong and the legs are relatively short and stout.

The genus *Simulium* transmits the filarial worm *Onchocerca volvulus* in Africa and in Central and South America (Smith, 1973). *Onchocerca* typically causes subcutaneous swellings and may cause blindness when the organism migrates into the eyes. Black fly attacks may also be serious. Irritation, allergic reactions, toxemia, and secondary infections may complicate the bites of simulia (Smith, 1973).

Family: Ceratopogonidae

Biting Midges

Biting midges are very small (0.6 to over 4 mm), have piercing-sucking mouth parts, and usually have 14 apparent antennal segments (Fig. 5). Their wings are sometimes speckled, have reduced venation, and are folded over the body when at rest. The most important genera are *Culicoides*, *Lasiohelea*, and *Leptoconops*.

These flies can be a major nuisance to man. Bites may become irritated and secondarily infected. Several species of filaria are transmitted by *Culicoides*.

Family: Psychodidae

Sand Flies and Moth Flies

The Psychodidae are small flies (1.5 to 4 mm) with hairy bodies (Fig. 6). They usually have rounded wings that are pointed at the tip. The wings have numerous longitudinal veins but few if any basal cross veins.

Moth flies are important only for their nuisance effects. They are often found around sewage treatment plants or in kitchen or bathroom drains. They hold their wings rooflike over the body when at rest.

Phlebotomine sand flies, of which *Phlebotomus* is the important genus, are best known as vectors of visceral and dermal leishmaniasis throughout Central and South America, Africa, southern Europe, and eastward through India and China. Bartonellosis *(Bartonella)* is transmitted by sand flies in South America. Sand fly fever, a nonfatal viral disease, occurs from the Mediterranean area eastward to Sri Lanka and China.

Family: Tabanidae

Horse Flies and Deer Flies

The Tabanidae have medium to stout bodies and large convex heads. The overall length of Tabanidae varies from 5 to 30 mm. Their piercing-sucking mouth parts have broad, bladelike stylets. Their wing venation is very characteristic (Figs. 7 and 8).

Horse flies and deer flies are vicious biters of

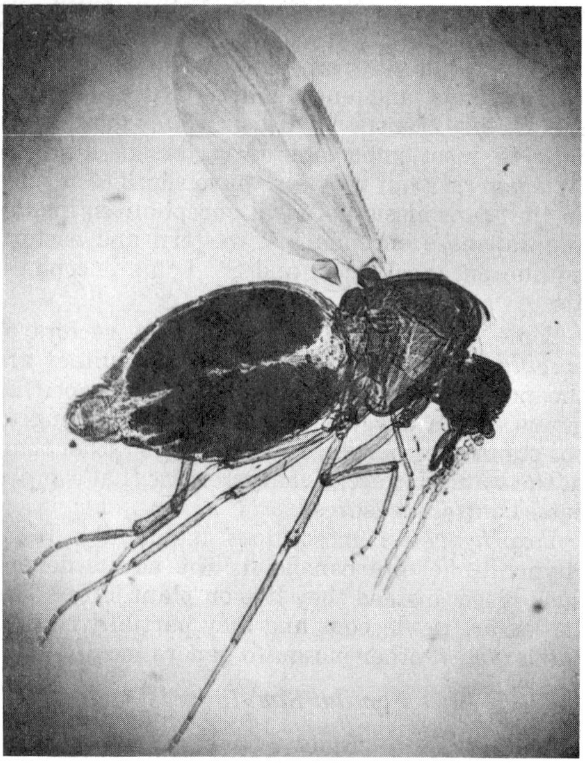

FIGURE 5. *A slide-mounted specimen of a biting midge,* Culicoides.

FIGURE 6. *A slide-mounted specimen of a phlebotomine sand fly.*

FIGURE 7

FIGURE 8

FIGURES 7–8. *Pinned specimens of a horse fly,* Tabanus, *and of a deer fly,* Chrysops, *showing their stout bodies and large convex heads.*

man and animals, producing a large wound that can become secondarily infected. They can be efficient mechanical vectors of disease when their mouth parts have been recently contaminated with pathogenic organisms. Anthrax (*Bacillus anthracis*) and tularemia (*Pasteurella tularensis*) are diseases known to be transmitted in that manner (Krinsky, 1976). In Africa, several species of deer flies (*Chrysops*) are responsible for the biologic transmission of loiasis caused by the filaria *Loa loa*.

Family: Muscidae

House Flies, Tsetse Flies, and Relatives

The Muscidae are usually dull-colored and medium to small in size. Their mouth parts vary considerably.

Most species of the genus *Musca,* including *M. domestica*, the common house fly (Fig. 9), have sponging-lapping mouth parts. The larvae breed in filth and the adults frequent feces and garbage. Their sponging-lapping mouth parts and regurgitation of food make them efficient transmitters of various diseases by food contamination. The diseases that house flies are known to transmit include: many bacterial diseases such as dysenteries, typhoid, and paratyphoid; protozoan diseases including amebic dysentery; helminthic dis-

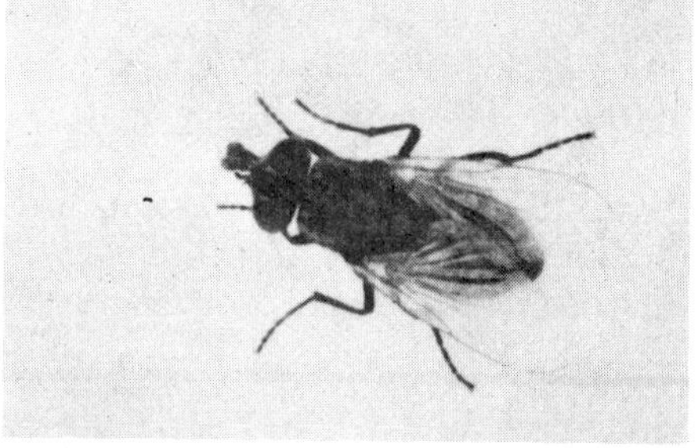

FIGURE 9. *A specimen of the common house fly,* Musca domestica.

FIGURE 10. *A pinned specimen of* Stomoxys calcitrans, *the stable fly.*

eases; and viral infection such as poliomyelitis.

Piercing-sucking mouth parts suitable for blood sucking are found throughout the family. *Stomoxys calcitrans* (Fig. 10), the stable fly, is a blood sucker that may transmit pathogens to man. Members of the genus *Glossina* (Fig. 11), the tsetse flies, are well known for their ability to transmit African sleeping sickness (*Trypanosoma rhodesiense* and *T. gambiense*).

FIGURE 11. *A pinned specimen of* Glossina, *a tsetse fly, showing a distinct hatchet-shaped wing cell.*

Various members of this family and the following two familes are known to cause myiasis, with the fly larvae living, at least temporarily, in man. Larvae may enter the body by several routes. For example, they may be ingested in contaminated food, or the adult fly may deposit eggs at body orifices or in wounds.

Families: Calliphoridae and Sarcophagidae

Blow Flies and Flesh Flies

Many of the blow flies (Calliphoridae) are metallically green or blue (Fig. 12) and have an antennal arista that is plumose nearly to the tip. The flesh flies (Sarcophagidae) (Fig. 13) have a black and gray variegated pattern (not metallic). If the antennal arista is plumose, the plumosity does not extend to the tip of the antenna.

In both families, the larvae generally feed on carrion and excrement. Some are known to cause myiasis. The larvae of some species are able to pierce the skin. Adult flies are able to transmit dysentery by contaminating food.

Order: Anoplura

Sucking Lice

The sucking lice are small (1.5 to 3mm) wingless insects (Figs. 14 and 15) with piercing-sucking mouth parts. The tarsi of the legs are usually reduced. The single tarsal claw is opposed by a toothed projection on the tibia. These two structures provide an excellent means of grasping hairs of their hosts.

FIGURE 12. *A pinned specimen of a blow fly, Calliphoridae.*

FIGURE 13. *A pinned specimen of a flesh fly, Sarcophagidae, showing a non-metallic black and gray color-pattern.*

Two varieties of the species, *Pediculus humanus*, the head and the body louse, and *Pthirus pubis*, the crab louse, are important to man.

Louse infestation, pediculosis, may cause extreme irritation and discomfort. Red papules, swelling, and sensitization may result from the feeding of sucking lice. Often the egg or nit may be easily found (Fig. 16).

Epidemic relapsing fever (*Borrelia recurrentis*) is transmitted by crushing an infected body louse on the skin. Epidemic typhus or louse-borne typhus (*Rickettsia prowazekii*) and trench fever (*R. quintana*) may be transmitted to humans either by the feces of the louse or by crushing of the body of the louse (Hunter et al., 1976). Transmission from the salivary glands of the louse is not involved in any of these diseases.

Order: Siphonaptera

Fleas

Fleas, which are well known for their jumping ability, are wingless insects (Fig. 17) with laterally compressed bodies. The usual size varies from 1.5 to 4 mm. Their mouth parts are adapted for piercing-sucking. Only adult fleas feed on mammals. The larvae and pupae are normally found in soil or bedding.

Fleas are the vector of the bacterium *Yersinia pestis*, which causes plague. Plague, an infection of rodents, is transmitted to humans when conditions become favorable. Fleas that have fed on an infected animal often develop a blocked proventriculus. Then, when they try to feed on another host, the plague bacilli are forced into the uninfected host. Plague bacilli have also been shown to be transmitted by the flea's feces. The major flea involved in urban plague is *Xenopsylla cheopis*; a large number of other species have been incriminated, however, especially in sylvatic or campestral plague (Hunter et al., 1976).

Xenopsylla cheopis is the main vector of murine typhus (*Rickettsia typhi*), although *Nosopsyllus*

FIGURE 14. *A slide-mounted specimen of the human body louse,* Pediculus humanus.

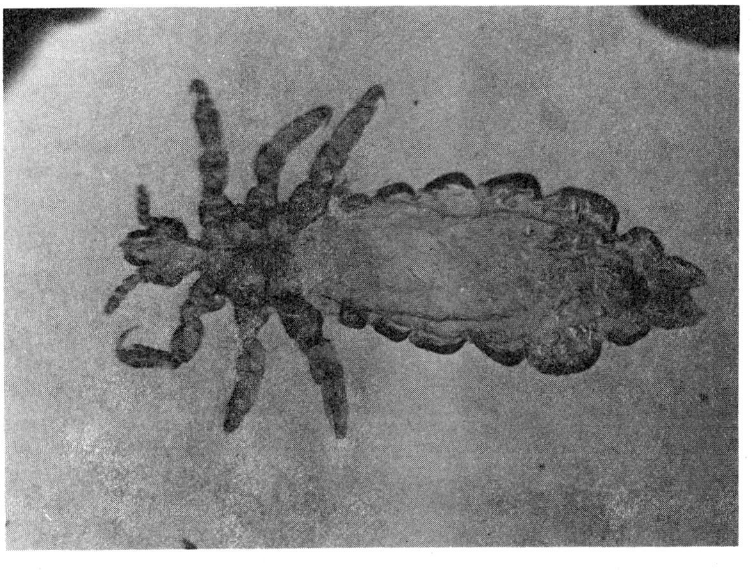

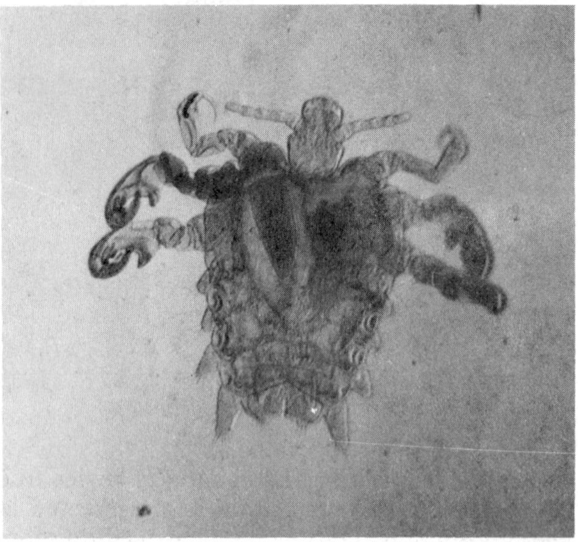

FIGURE 15. *A slide-mounted specimen of the pubic louse,* Pthirus pubis, *showing the distinctive tarsal claws used for grasping hairs.*

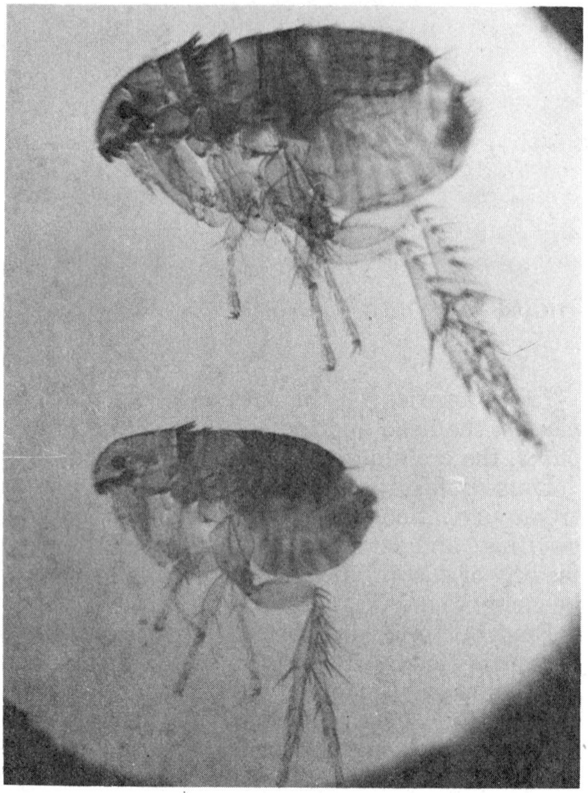

FIGURE 17. *Male (top) and female (bottom) adult Siphonoptera, showing their lack of wings and their strong hind legs.*

fasciatus and *Leptopsylla segnis* are also known to be vectors (James and Harwood, 1969). The mode of infection is scratching of the vector's feces into the skin.

Order: Orthoptera

Cockroaches

Cockroaches are normally flattened dorsoventrally and have a smooth but tough integument (Fig. 18). Adults are characterized by leathery outer wings and membranous inner wings. Their color varies from green to orange in tropical species to dark brown to black in the more common household species. The size of adults varies from 12 to 40 mm. Most species remain in dark locations during the day and can be seen feeding only during the night, usually in kitchens and food storage areas when lights are suddenly turned one.

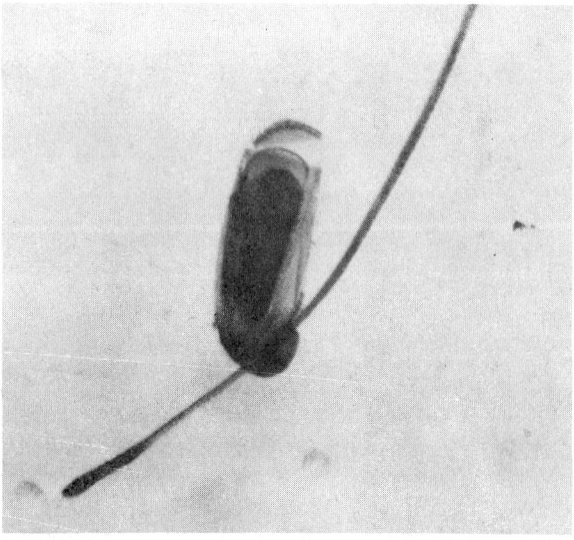

FIGURE 16. *Egg of a sucking louse cemented to a hair.*

FIGURE 18. *Specimen of the cockroach,* Periplaneta americana.

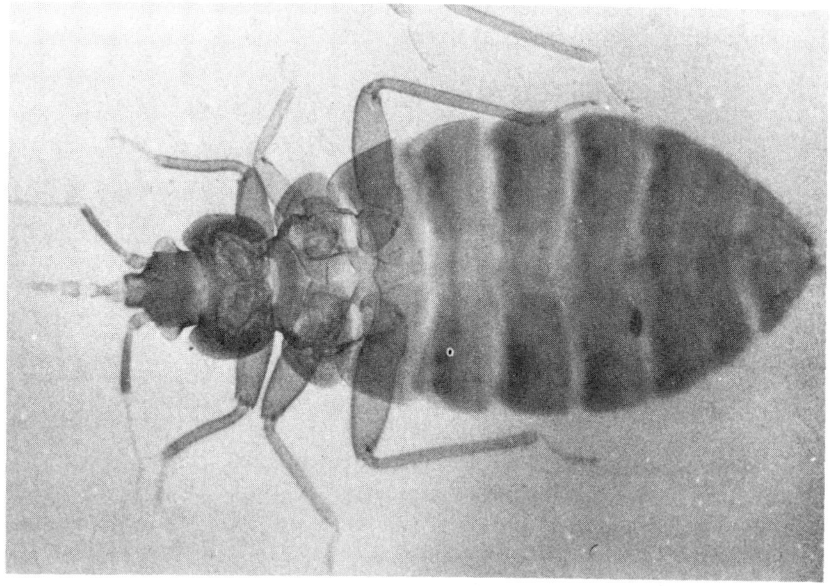

FIGURE 19. *A slide-mounted specimen of* Cimex, *showing rudimentary wings.*

Circumstantial and experimental evidence indicates that cockroaches could be involved in the mechanical transmission of several organisms. They are omnivorous feeders and have the habit of disgorging partially digested food and feces as they feed. Thus, the medical importance of cockroaches lies in their ability to contaminate food and foodstuffs.

Numerous viruses, pathogenic bacteria, pathogenic fungi, and protozoa have been isolated from cockroaches. Additionally, the list of different organisms that cockroaches may harbor experimentally is quite long (James and Harwood, 1969). Cockroaches are also known to be intermediate hosts for several rat nematodes.

Order: Hemiptera

True Bugs

The fore wing of the adult Hemiptera is usually divided into a leathery basal portion and a membranous apical portion. The hind wing is totally membranous. The bed bug, which has only rudimentary wings, is an exception to this general rule. The mouth parts of the order are adapted for piercing and sucking.

Bed bugs, family Cimicidae (4 to 5 mm) (Fig. 19), are not known to be vectors of human pathogens, but their bite does cause extreme irritation and swelling in individuals who are allergic to the saliva that is introduced at the time of feeding. Bed bugs are nocturnal. They spend the daylight hours hidden in cracks, crevices, or bedding. *Cimex lectularius* is normally found in the temperate regions, whereas *C. hemipterus* is usually restricted to the tropics. A third species, *Leptocimex boueti*, is limited to Africa (James and Harwood, 1969).

The bite of the assassin bugs (Fig. 20), family Reduviidae, can cause extreme local irritation and swelling. Systemic reactions may also occur as a result of the injection of foreign protein.

Chagas' disease (*Trypanosoma cruzi*), normally found in South and Central America, is transmitted by several genera of reduviids. This disease is transmitted when the reduviid defecates while feeding (Faust et al., 1968). The feces are then transported by the fingers to the conjunctiva of the eye or the mucosa of the mouth or nose. Although these bugs are large (up to 25 mm) and formidable in appearance, they are seldom dis-

FIGURE 20. *A specimen of a reduviid,* Triatoma.

turbed during feeding because their bite is painless and they usually feed at night.

Order: Coleoptera

Beetles

The adults of the order Coleoptera have two pairs of wings. The outer pair is thickened to form wing covers called elytra, while the inner pair is membranous (Fig. 21). The mouth parts are formed for chewing.

Some families of beetles are known for their ability to produce blisters by discharging their body fluids onto the skin. The "Spanish fly" (*Lytta vesicatoria*) (10 to 15 mm) is perhaps the best known vesicating beetle. Blisters are formed when this beetle is crushed on the skin.

Order: Lepidoptera

Butterflies and Moths (Caterpillars)

The adults of the Lepidoptera (butterflies and moths) may be separated from other insects by their two pairs of membranous wings that are covered with overlapping scales. Their mouth parts are adapted for siphoning. Larvae and caterpillars are usually cylindrical in shape and have three pairs of thoracic legs and two to five pairs of abdominal legs or prolegs (Fig. 22).

Contact with poison hairs on the external surface of caterpillars causes a stinging dermatitis. The urticating fluid is released when the tips of the hairs are broken upon contact.

Order: Hymenoptera

Bees, Wasps, Hornets, and Ants

The Hymenoptera usually have two pairs of membranous wings with hind wings that are

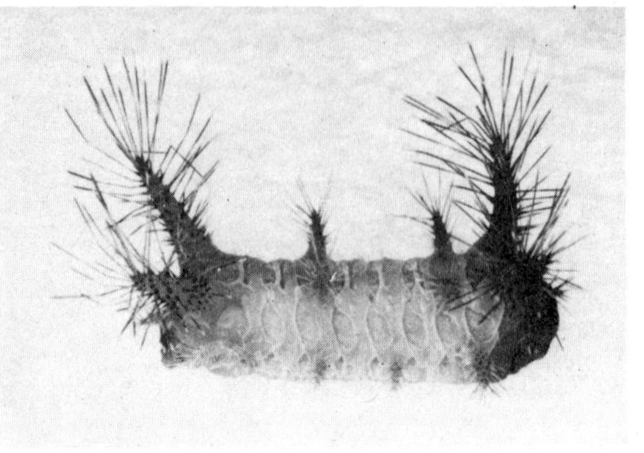

FIGURE 22. *A specimen of a lepidopterous larva, showing its urticating hairs.*

smaller than the fore wings (Fig. 23), but some members of the order, notably the ants, are wingless. The abdomens of the females are usually provided with a stinging apparatus.

The effects of a Hymenoptera sting may range from moderate, temporary pain and slight local swelling to shock and death.

CLASS: ARACHNIDA

Subclass: Acari

Mites and Ticks

The subclass Acari is composed of the arthropods that are commonly referred to as ticks and mites. These arthropods are distinctive in that they have two body regions, an idiosoma (fused abdomen and cephalothorax) and a gnathosoma (mouth parts) (Krantz, 1978). Adult Acari usually have four pairs of legs, but larvae have only three.

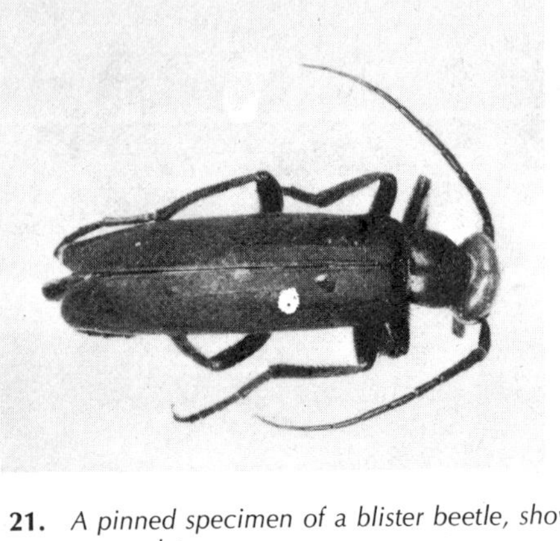

FIGURE 21. *A pinned specimen of a blister beetle, showing its thick wing covers or elytra.*

FIGURE 23. *A pinned specimen of an adult hymenopteran, showing a stinger on the terminal abdominal segment.*

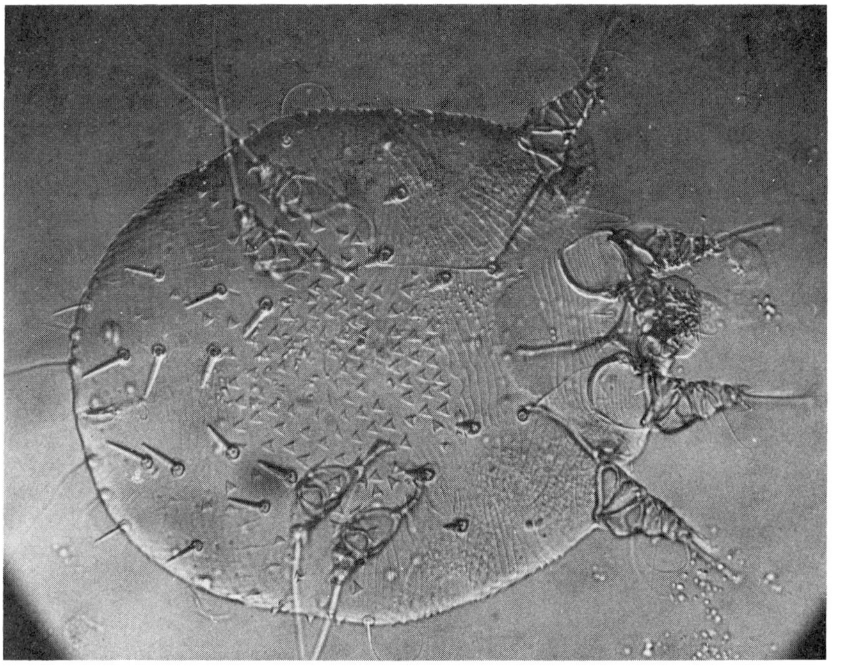

FIGURE 24. *A slide-mounted specimen of* Sarcoptes scabiei.

Only two species of mites are parasites of man: the skin mite, *Sarcoptes scabiei* (0.3 to 0.5 mm) (Fig. 24), and the follicle mite, *Demodex folliculorum* (0.4 mm) (Fig. 25). *Demodex folliculorum* may not cause any disease in man; it may only be a commensal. However, *Sarcoptes scabiei* can cause extreme itching and keratotic crusts covering large numbers of mites. *Demodex* and *Sarcoptes* cause the condition commonly known as mange in animals.

A large number of mites and ticks can parasitize man if their normal host is not available. Additionally, another large group of mites may affect many by biting and causing dermatitis (Krantz, 1978). This group includes many mites that feed on plants and organic matter and that come into contact with man only accidentally.

Scrub typhus or chigger-borne rickettsiosis (*Rickettsia tsutsugamushi*) is transmitted by the larvae of the trombiculid mite or chigger (Fig.

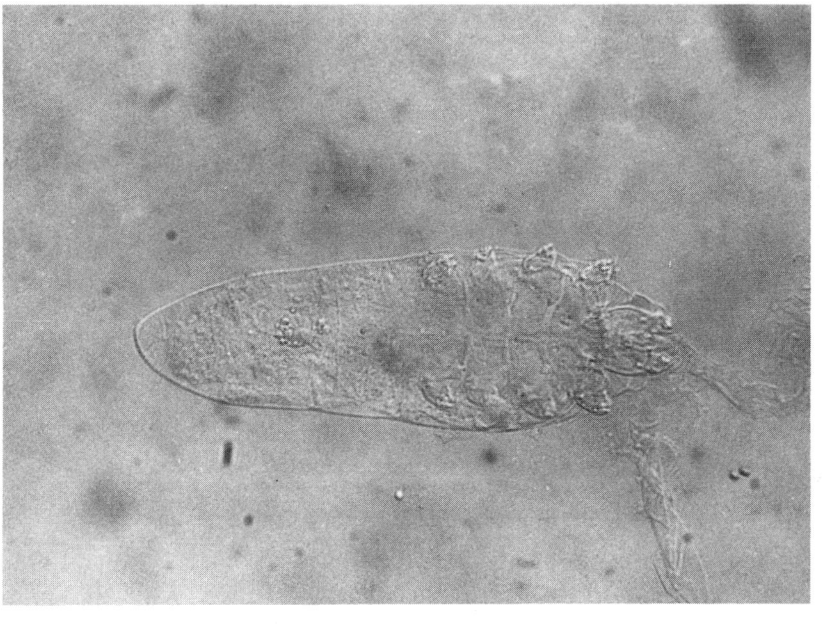

FIGURE 25. *A slide-mounted specimen of* Demodex folliculorum.

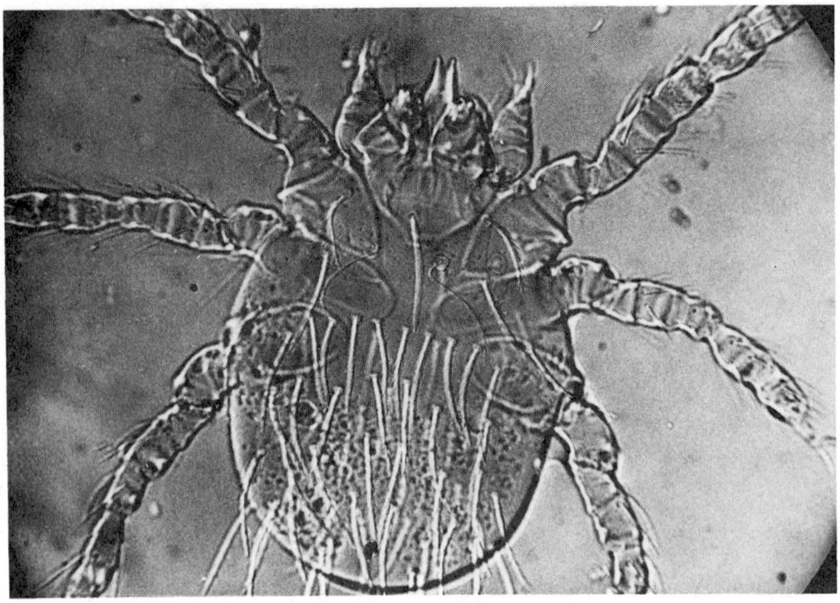

FIGURE 26. *A slide-mounted specimen of a* Leptotrombidium *chigger.*

26). The disease occurs throughout Asia in a triangle roughly bound by India, northern Australia, and northern Japan. All known vectors are in the genus and subgenus *Leptotrombidium* (Traub and Wisseman, 1974). The small (0.2 to 0.4 mm), six-legged larvae normally feed on small rodents but will attack man if he is in the chiggers' habitat. Feeding by these mites is often inapparent until the onset of the disease.

Rickettsialpox (*Rickettsia acari*) reported from the United States and Russia is transmitted by nymphs and adults of the house mouse mite, *Liponyssoides sanguineus*. Although this disease is apparently uncommon, it may be more widely spread than has been reported, since a similar disease has been described from Africa and the causative organism has been isolated from wild rodents in Korea.

A number of mites commonly referred to as house dust mites have been associated with the

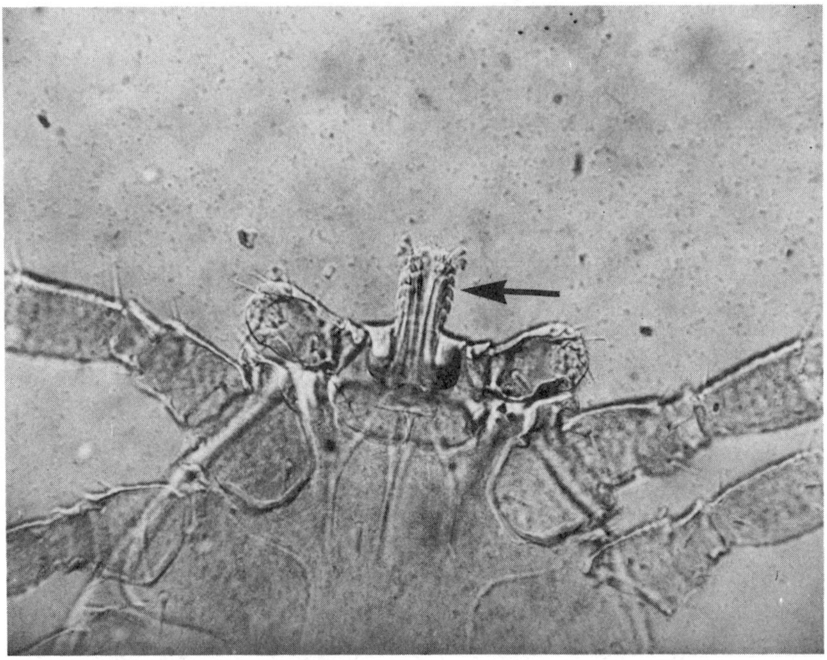

FIGURE 27. *A slide-mounted specimen showing the hypostome (arrow) of a tick.*

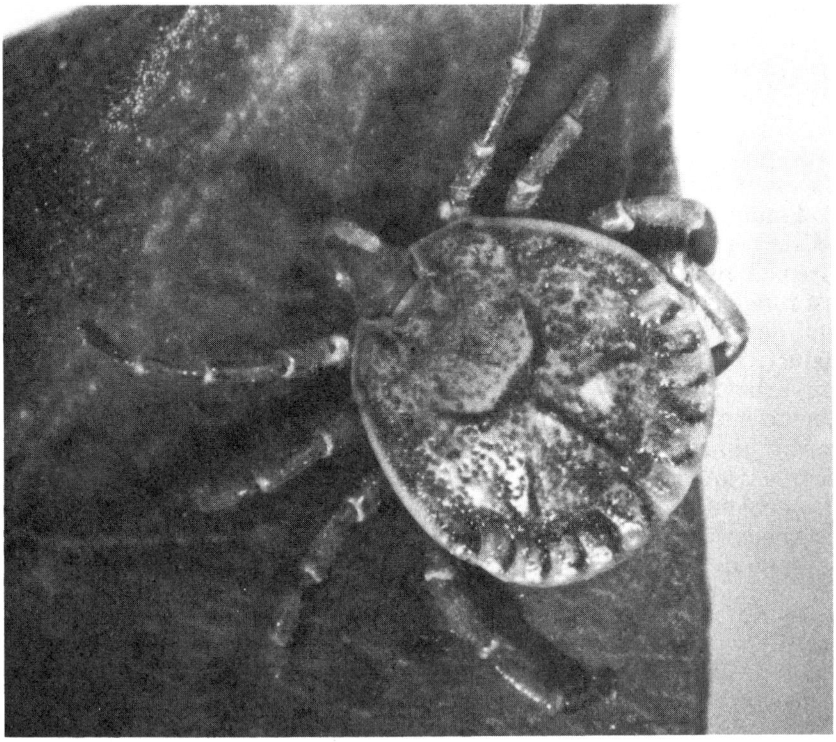

FIGURE 28. *A specimen of an adult tick, Ixodidae.*

production of allergens that cause bronchial asthma and rhinitis (Wharton, 1976). Although a large number of different genera may occur in dust samples, species of the genus *Dermatophagoides* (0.2 to 0.4 mm) have proved to be the most important in relation to allergies. Dust mites are normally found in beds and furniture and under beds, feeding on human dander.

Ticks (superfamily Ixodoidea) can be differentiated from the other acari by the presence of a sensory structure known as Haller's organ on the first pair of tarsi. Also, many species have a toothed hypostome (Fig. 27). Generally, adult ticks are relatively large (0.3 to 5 mm) in comparison to other acari and have a leathery appearance (Fig. 28). Ixodoidea is composed of three

FIGURE 29. *Scorpion, showing the bulbous sac and prominent stinger.*

families: Ixodidae, or hard ticks; Argasidae, or soft ticks; and Nuttalliellidae, an intermediate represented by a single species. Of the vast number of tick-borne diseases, by far the largest number are transmitted to man by the Ixodidae (Balashov, 1972).

James and Harwood (1969) list eight factors that make ticks such good vectors of mammalian diseases. Ticks are persistent bloodsuckers, slow feeders, highly sclerotized, and free of natural enemies. They also have a wide host range, a long life span, ability to transmit the disease agents transovarially, and a great reproductive potential. Ticks are known to transmit arboviruses, rickettsiae, bacteria, and piroplasms to man. Additionally, ticks may cause paralysis in man.

By far the largest number of diseases transmitted to man by ticks are caused by arboviruses. The majority fall into the category known as the Russian spring-summer complex (flavivirus or antigenic group B).

Subclass: Scorpiones

Scorpions

Scorpions are easily recognizable by their long, five-segmented, tail-like post-abdomen that terminates in a bulbous sac with a prominent stinger. Anteriorly, the pedipalps are enlarged and resemble an additional pair of legs, and the last two segments are modified to form pincers (Fig. 29).

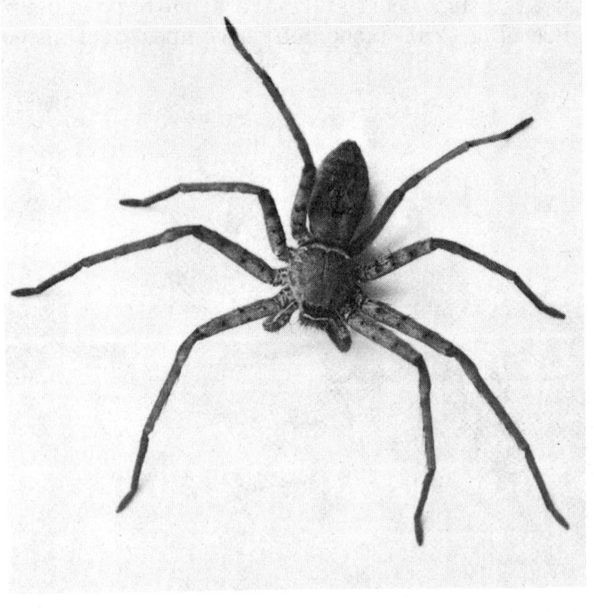

FIGURE 30. *Spider, characterized by the prominent body division of the cephalothorax and abdomen.*

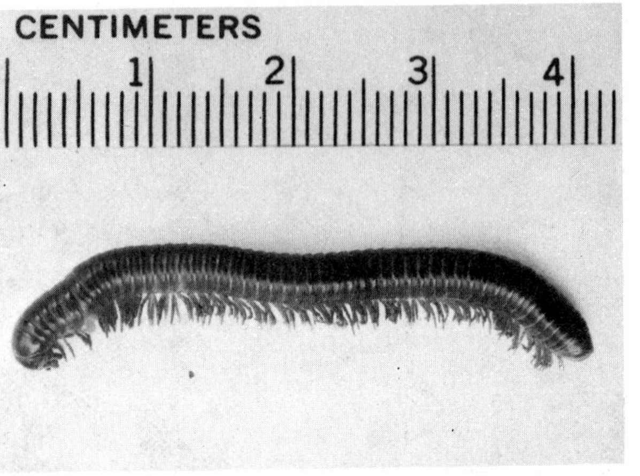

FIGURE 31. *Millipede, bearing 2 pairs of legs per body segment.*

Scorpions are nocturnal by nature, remaining hidden in the daytime under stones, litter, and lumber. The sting usually results from accidental contact with a scorpion, such as accidentally stepping on it, placing one's foot in its hiding place (e.g., a boot), or disturbing its habitat. The sting of the scorpion can be very dangerous, depending upon the species. Areas such as Mexico, southwestern United States, parts of South America, and North Africa west of Egypt are known to have particularly dangerous species. The severity of the sting is not necessarily related to the size of the scorpion; the smaller species are often more dangerous. Children are more seriously affected. Pain is usually associated with the sting, but swelling and discoloration at the site may or may not occur, depending upon the species of scorpion. The sting of dangerous species can be fatal, particularly to children.

Subclass: Araneae

Spiders

The spiders have two body parts, cephalothorax and abdomen, and four pairs of legs (Fig. 30). The cephalothorax is separated from the abdomen by a narrow constriction, the pedicel. The eyes are simple and antennae are absent. All spiders have silk glands that produce silk through spinnerets located near the rear of the abdomen. The legs are borne on the cephalothorax.

All spiders feed by injecting venom into their prey, but relatively few species are of medical importance. Two types of effects are commonly noted from spider bites: the formation of a necrotizing ulcer and systemic symptoms. The bite of *Loxosceles* spp., brown recluse or fiddle-back spiders, causes necrotic ulcers on the skin, a condition termed loxoscelism. In mild cases the necrotic ulcer may be slow to heal, leaving a disfiguring scar. In severe cases erosion of all mucous membranes, hemorrhage, and death may occur (Southcott, 1976). Systemic reactions from spider bites are particularly well known from the hairy spiders known as tarantula, Theraphosidae, and the "black widows," *Latrodectus* spp. Swelling and pain may occur at the site of the bite, followed by burning and aching in the general area of the bite. Circulating venoms may induce abdominal cramps. Untreated bites may cause convulsions, shock, and death.

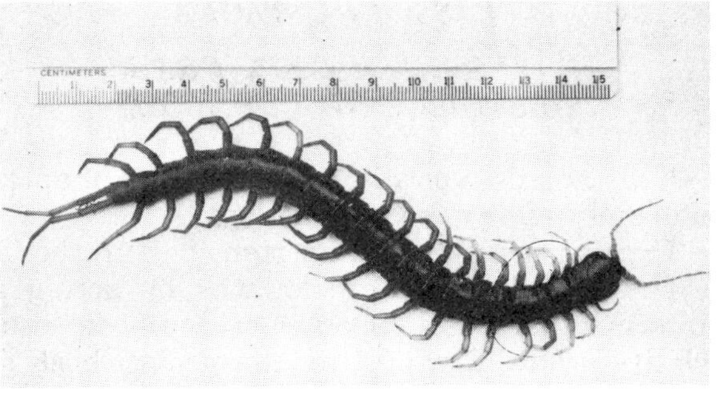

FIGURE 32. *Centipede, bearing a single pair of legs per segment.*

Erythema and edema may occur and persist for several weeks.

CLASS: DIPLOPODA

MILLIPEDES

Members of the class Diplopoda are multi-segmented and multi-legged arthropods. Each body segment bears two pairs of legs (Fig. 31). Most species of millipedes feed on decaying organic matter and many species are equipped with offensive stink glands.

The stink glands of the millipede may have the capability of forcefully ejecting secretions. This liquid may cause severe pain and staining of the skin at the site of contact. The lining epithelium of the eye and mouth are particularly susceptible.

CLASS: CHILOPODA

CENTIPEDES

The Chilopoda have a single pair of antennae and a multi-segmented trunk. Each segment bears a single pair of legs (Fig. 32). Paired poison claws arise from the first trunk segment.

The venom of the poison claws of centipedes is used to kill prey, usually insects, but occasionally small reptiles, mammals, and birds. Some species of centipedes are able to pierce human skin and cause severe pain and localized swelling. The severity of a centipede bite is often compared with that of a bee or wasp sting.

FIGURE 33. *A slide-mounted specimen of a copepod,* Cyclops.

CLASS: CRUSTACEA

CRAYFISH, SHRIMP, CRABS, LOBSTERS, AND COPEPODS

Crustaceans typically have two pairs of antennae and five or more pairs of legs.

Crustaceans play an important part in the life cycle of a number of helminths by acting as intermediate hosts. Species of copepods, especially *Cyclops* (Fig. 33), are intermediate hosts for the broad tapeworm *Diphyllobothrium latum* and for the guinea worm of Africa, *Dracunculus medinensis*.

References

Balashov, Y.S.: Bloodsucking ticks (Ixodoidea) — vector of diseases of man and animals (1968). Misc Pub Entomol Soc Am 8:161, 1972.

Borror, D.J., DeLong, D.M., and Triplehorn, C.A.: An Introduction to the Study of Insects. 4th ed. New York, Holt, Rinehart & Winston, 1976.

Chamberlain, R.W., and Sudia, W.D.: Mechanism of transmission of viruses by mosquitoes. Ann Rev Entomol 6:371, 1961.

Faust, E.C., Beaver, P.C., and Jung, R.C.: Animal Agents ;and Vectors of Human Disease. 3rd ed. Philadelphia, Lea & Febiger, 1968.

Feingold, B.F.: Introduction to Clinical Allergy. Springfield, Ill., Charles C Thomas, 1973.

Furman, D.P., and Catts, E.P.: Manual of Medical Entomology. 3rd ed. Palo Alto, Calif., National Press Books, 1970.

Hunter, G.W., III, Swartzwelder, J.C., and Clyde, D.F.: Tropical Medicine. 5th ed. Philadelphia, W.B. Saunders Co., 1976.

James, M.T., and Harwood, R.F.: Herms's Medical Entomology. 6th ed. New York, Macmillan, 1969.

Krantz, G.W.: A Manual of Acarology. 2nd ed. Corvallis, Ore., State University Book Stores, Inc., 1978.

Krinsky, W.L.: Animal disease agents transmitted by horse flies and deer flies (Diptera: Tabanidae). J Med Entomol 13:225, 1976.

Smith, K.G.V. (Ed.): Insects and Other Arthropods of Medical Importance. London, Trustees of the British Museum (Natural History), 1973.

Southcott, R.V.: Arachnidism and allied syndromes in the Australian Region. Records Adelaide Children's Hosp 1:97, 1976.

Traub, R., and Wisseman, C.L., Jr.: The ecology of chigger-borne rickettsiosis (scrub typhus). J Med Entomol 11:237, 1974.

Wharton, G.W.: House dust mites. J Med Entomol 12:577, 1976.

Principles of Antimicrobial Chemotherapy of Infections

19 DESCRIPTION OF ANTIMICROBIAL DRUGS

Abraham I. Braude, M.D., Ph.D.

CLASSIFICATION

Antimicrobial drugs may no longer be classified according to the organisms they inhibit because their spectra can be broadened by minor adjustments in structure or dosage. Classification of antibiotics and synthetic antimicrobial drugs is based instead on chemical structure or biochemical effects. Since the biochemical effects are considered in the next chapter, "Mechanisms of Action of Antimicrobial Drugs," a chemical classification will be presented here. All antimicrobial drugs are ring compounds; otherwise, the various groups have little structural similarity.

THE PEPTIDES

The penicillins, cephalosporins, bacitracin, and polymyxins are the most important peptide antibiotics. Their molecular weight seldom exceeds 3000. For this reason they are too small to be antigenic and stimulate neutralizing antibodies that would cause them to lose activity with continued administration. Small nonantigenic peptide antibiotics are a lucky evolutionary byproduct in the molds and bacteria that produce these antimicrobial agents. They are small because they are synthesized by a pathway that does not depend on the elaborate system for coding amino acids and their sequences required for protein synthesis. It is likely that microbial peptides, including peptide antibiotics, evolved before the process of protein synthesis via nucleic-acid coding on ribosomes. Since the functions performed by peptides were later taken over by microbial proteins, the peptides can be regarded as "fossils."[1] The fact that the majority of these antibiotics have a cyclic structure is also consistent with the idea that they once performed a functional role in microbial metabolism because cyclic structures are found in enzymes and other functional proteins.

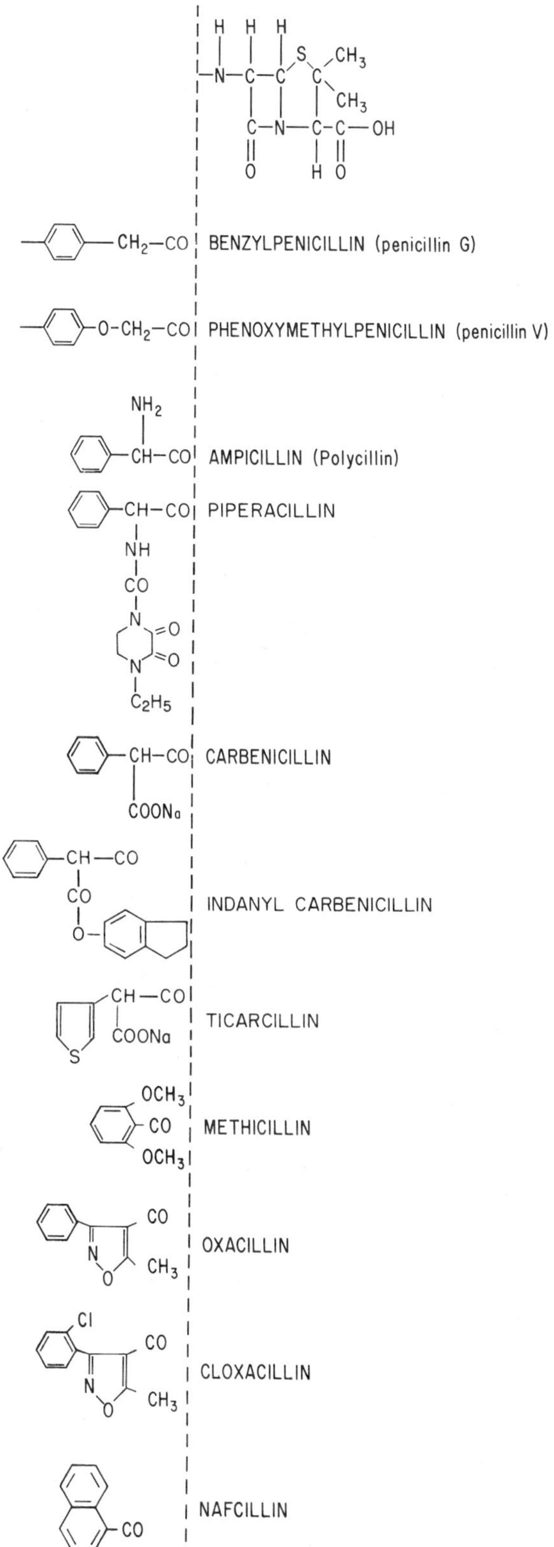

FIGURE 1. *The side chains responsible for the biologic differences in each of the penicillins are shown to the left of the dotted line, and 6-aminopenicillanic acid is on the right. The last four side chains protect the β-lactam ring from the action of penicillinase.*

The Penicillins

The common nucleus of the penicillins is 6-aminopenicillanic acid, a cyclic dipeptide of L-cysteine and D-valine (Fig. 1). These are arranged in a basic structure consisting of a thiazolidine ring joined to a β-lactam ring. Individual penicillins differ with respect to the side chains attached to the common nucleus. The most important of the naturally occurring penicillins is penicillin G, containing a benzyl side chain (Fig. 1). Penicillin V is obtained when phenoxyacetic acid is added as a precursor to the fermentation medium so that a phenoxymethyl side chain becomes attached to the penicillin nucleus. This compound is well suited for oral use because of its resistance to gastric acid. Other penicillins are prepared by the synthetic addition of various groups as side chains to 6-aminopenicillanic acid after it has been isolated from *Penicillium* fermentation media. Depending on their chemical structure, these side chains can broaden the antimicrobial spectrum, protect the penicillin nucleus from acid hydrolysis, or protect it against penicillinase. In ampicillin, for example, the presence of an amino group in the phenyl radical of benzylpenicillin produces a compound both resistant to acid and more active against gram-negative bacteria than penicillin G. If a carboxyl group is introduced instead, as in carbenicillin, the spectrum is altered so that the compound becomes the only penicillin derivative with activity against *Pseudomonas aeruginosa.* None of these modifications, however, confers resistance to staphylococcal penicillinase. This was first accomplished in the synthesis of methicillin by the introduction of methoxy groups at positions 2 and 6 on the benzyl ring so that affinity of the substrate site for penicillinase is reduced 10,000 times.[13] This modification did not achieve acid resistance, but the synthesis of isoxazolyl penicillin introduced a series of products with combined resistance to penicillinase and gastric acid. Oxacillin and dicloxacillin are the most widely used of these doubly resistant penicillins.

These and other penicillin derivatives have been modified further to improve their blood levels, broaden their antimicrobial spectrum and reduce side effects. The conversion of ampicillin to amoxicillin, by the introduction of a hydroxyl radical at the para position in the benzyl side chain, improves absorption. Absorption is also improved by conversion of amoxicillin to pivampicillin, the pivaloyloxymethyl ester of ampicillin. After absorption, pivampicillin is hydrolyzed first to pivalic acid and the hydroxymethyl ester of ampicillin; these are then hydrolyzed within 15 minutes to ampicillin and formaldehyde. Either

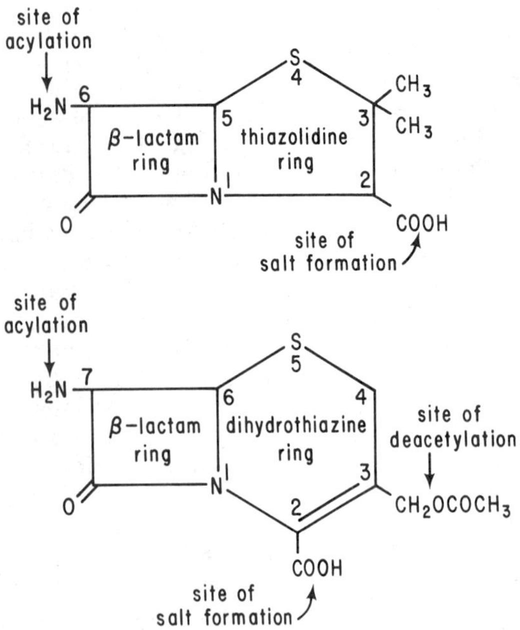

FIGURE 2. *Comparison of penicillin nucleus, 6-aminopenicillanic acid (upper figure), with cephalosporin nucleus, 7-aminocephalosporanic acid (lower diagram). The main difference in the two compounds is the thiazolidine ring in penicillin and the dihydrothiazine ring in the cephalosporins. Most penicillin derivatives are produced by addition of side chains at the acylation site. Cephalosporin derivatives vary in the chemical groups at the acylation site, and at the deacetylation site as well.*

modification provides blood levels after oral administration that are approximately twice that obtained with ampicillin.[12]

The acid lability of carbenicillin in the stomach has been overcome by converting it to the 5-indanyl ester, as indanyl-carbenicillin. In this compound the α-carboxylic acid moiety of carbenicillin is bound through ester linkage to 5-indanol and hydrolyzed after ingestion to active carbenicillin. The carbenicillin is then well absorbed. If the benzyl group in carbenicillin is replaced with the 5-carbon thienyl ring, the resulting compound, ticarcillin, is two to four times more active against *Pseudomonas aeruginosa,* but remains acid labile and must be given parenterally. Even greater activity against *P. aeruginosa* is obtained with piperacillin. In this antibiotic, ampicillin is bound through its amino group to a derivative of piperazine.

The structural modifications of semisynthetic penicillins are illustrated in Figure 1.

The Cephalosporins

The common nucleus of the cephalosporins resembles 6-aminopenicillanic acid but differs by having a dihydrothiazine ring, instead of a thiazolidine ring, attached to the β-lactam ring (Fig. 2). The active nucleus, known as 7-aminocephalosporanic acid, can also be manipulated chemically to yield more active and useful derivatives.[6] The basic advantages of the cephalosporin nucleus are its innate resistance to staphylococcal penicillinase and its safety in patients with allergy to penicillin. These properties are retained in cephalothin (Fig. 3), in which the side chain at the 7-position increases its potency and range of activity. Another derivative, cephalexin (Fig. 3), has a different side chain at the 7-position that confers resistance to gastric acid and allows good absorption from the alimentary tract. Cefazolin (Fig. 3), with substitutions at both the 7- and 3-positions, can be injected intramuscularly without pain and possesses greater activity than cephalothin against certain bacteria.

Two newer cephalosporins, cefamandole and cefoxitin, are more resistant to the β-lactamase of *Bacteroides fragilis* and more active against that organism than cephalothin and cefalexin. Cefoxitin is derived from cephamycin C, which has a methoxy group at position 7 in 7-aminocephalosporanic acid. The methoxy group may be responsible for the resistance of this drug to the β-lactamase of *B. fragilis* and other gram-negative bacilli. Cefoxitin also appears to be able to reach higher levels in the cerebrospinal fluid than other cephalosporins.[11]

Bacitracin

This antibiotic is produced by a strain of *Bacillus subtilis,* which was isolated from the dirty compound fracture of a girl named Tracy and called "bacitracin" out of joint deference to the bacillus and patient.[8] The antibiotic consists of a mixture of polypeptides, the most important of which is bacitracin A. Like the penicillins, it contains a thiazolidine ring but does not have their β-lactam ring. In bacitracin A the thiazolidine ring is a condensation product of isoleucine and a cysteine residue and is attached through L-leucine to a peptide composed of D- and L-amino acids (Fig. 4).

The Polymyxins

These cyclic polypeptides are also produced by a spore-forming aerobic bacillus, *Bacillus polymyxa.* Their detergent activity on bacteria can be attributed to two unique components: an amino acid, α,γ-diaminobutyric acid (DAB), and a C9 fatty acid, 6-methyloctanoic acid (Fig. 5). The cationic α-amino groups of DAB and the hydrophobic side chain of the fatty acid give the polymyxins the surface-active properties of a cationic detergent. Only two of the polymyxins, B and E, are used in clinical medicine. Polymyxin E is generally known as colistin.

THE AMINOGLYCOSIDES

The aminoglycosides are derived from different species of *Streptomyces* and are composed of

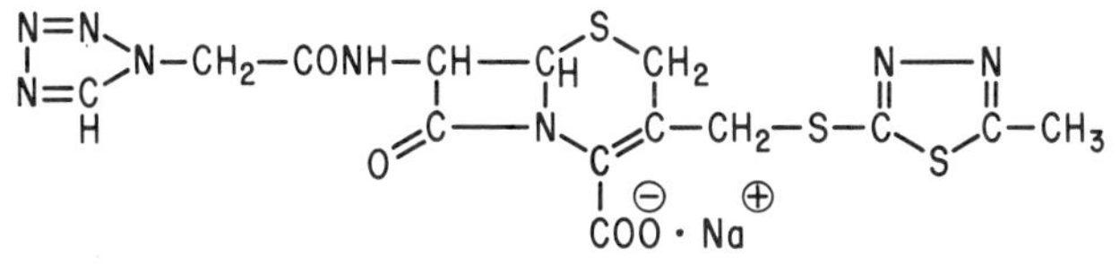

Cefazolin

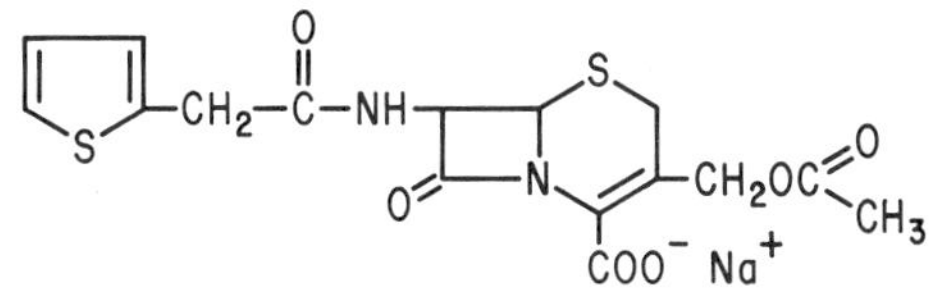

Cephalothin

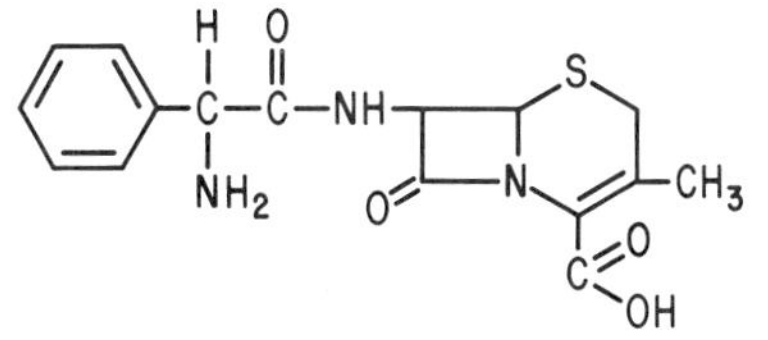

Cephalexin

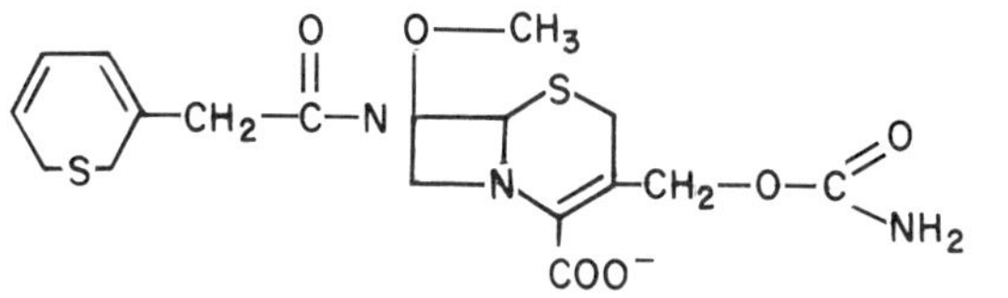

Cefoxitin

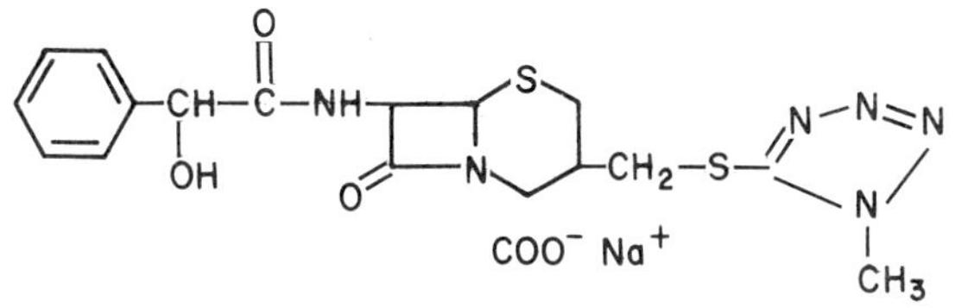

Cefamandole

FIGURE 3. *Structural formulas of five important cephalosporin derivatives illustrate differences in side chains at acylation and deacetylation sites.*

amino sugars. In contrast to penicillin, they are organic bases rather than acids. Streptomycin, neomycin, kanamycin, and paromomycin — listed in order of discovery — are all members of this group. They resemble each other because of the inositol residue and the prominent basic groups, ranging from three in streptomycin to six in neomycin. They differ in the sugars attached to inositol and in the guanidine groups of streptomycin (Fig. 6). This similarity in structure accounts for common toxic effects on patients and bacteria.

Gentamicin is structurally similar to these aminoglycosides but is produced by a species of *Micromonospora* rather than *Streptomyces.* It has an inositol residue with two amino sugars and is a complex of three antibiotics (designated C_1, C_2, and C_{1A}) which differ in structure only by one or two methyl groups.

THE TETRACYCLINES

The tetracyclines are so named because their common hydronaphthacene nucleus contains four

BACITRACIN A

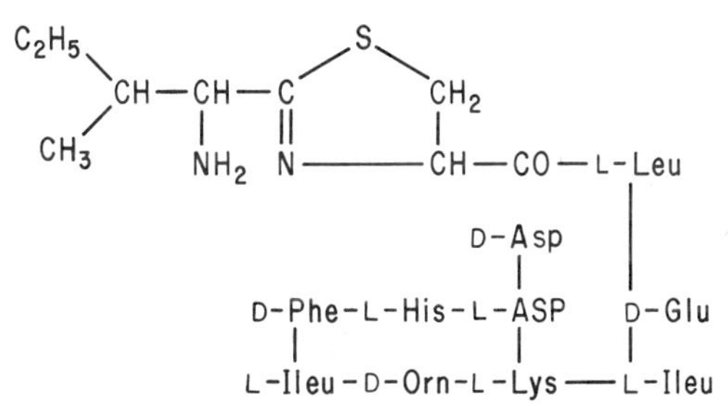

FIGURE 4. *Like penicillin, bacitracin A contains a thiazolidine ring. It is a condensation product of isoleucine and cysteine, in contrast to that of penicillin which contains* D*-valine and* L*-cysteine. Bacitracin lacks the β-lactam ring of penicillin.*

POLYMYXIN B

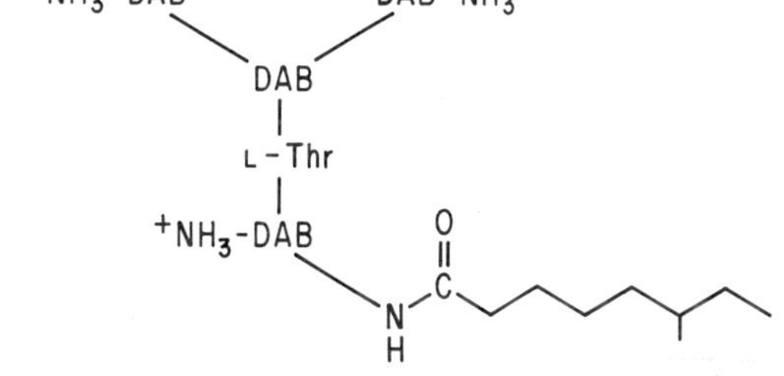

FIGURE 5. *Diaminobutyric acid (DAB) is present in all polymyxins, including colistin (polymyxin E). The terminal residue of DAB is acetylated by 6-methyloctanoic acid, the C9 fatty acid at the bottom of the structural formula. The cationic α-amino group of DAB and the hydrophobic fatty acid are responsible for the surface-active properties of the cationic detergent.*

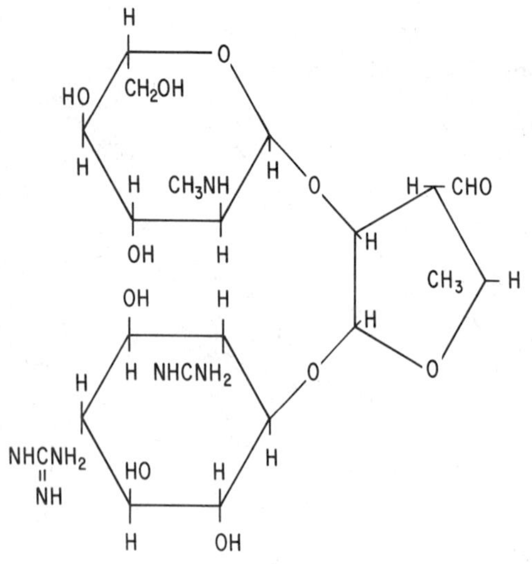

STREPTOMYCIN

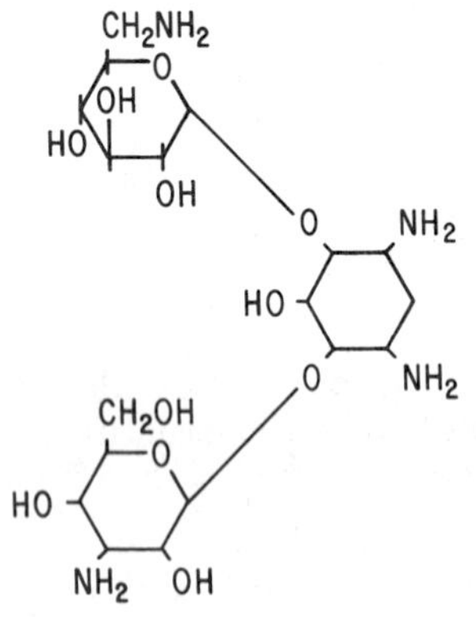

KANAMYCIN

FIGURE 6. *The basic structure of the aminoglycosides, illustrated by kanamycin and streptomycin, is that of a polycationic compound composed of amino sugars and connected by glycosidic linkages. In streptomycin, the inositol residue common to both compounds is in the form of streptidine (the bottom sugar with two guanido groups), and in kanamycin it is deoxystreptamine (the central sugar with two amino groups). Amikacin and tobramycin are related structurally to kanamycin. In amikacin the lower amino group in the diagram of the deoxystreptamine nucleus (position 1) is replaced with a side chain consisting of:*

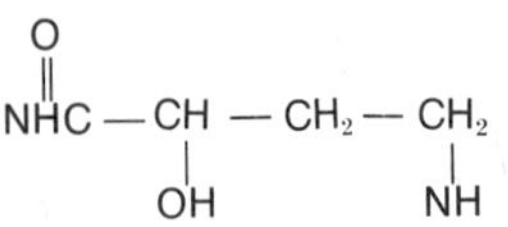

in order to increase its resistance to bacterial inactivating enzymes. In tobramycin the main differences from kanamycin are the absence of the first two hydroxyl groups in the upper ring and the presence of an amino group at position 1.

fused rings (Fig. 7). Chlortetracycline (Aureomycin) was the first tetracycline compound described, and oxytetracycline (Terramycin) was discovered 2 years later, in 1950. Both compounds were produced by strains of *Streptomyces.* Although tetracycline was made by the catalytic reduction of chlortetracycline, all other compounds in this group are named in relation to the basic structure of tetracycline. In chlortetracycline a hydrogen is replaced by chlorine, and in oxytetracycline it is replaced by a hydroxyl ion. Demeclocycline (Declomycin) is chlortetracycline without the 6-methyl group, and doxycycline is oxytetracycline without the 6-hydroxyl group. These minor structural differences have less effect on their antimicrobial spectra than on stability in solution and pharmacologic properties. Chlortetracycline not only is less stable than the other tetracyclines but is the least stable of any important antibiotic.

CHLORAMPHENICOL

Chloramphenicol is the only naturally occurring antibiotic with nitrobenzene in its structure (Fig. 8). This chemical grouping probably accounts for its toxicity to both bacteria and patients. Its tendency to cause aplastic anemia is explained by its benzene ring, a component of most organic substances involved in that disorder. Its ability to compete with messenger RNA for ribosomal binding is explained by its spatial similarity to uridine-5-phosphate.

THE MACROLIDES

Erythromycin, isolated from *Streptomyces erythreus,* is the only important member of the

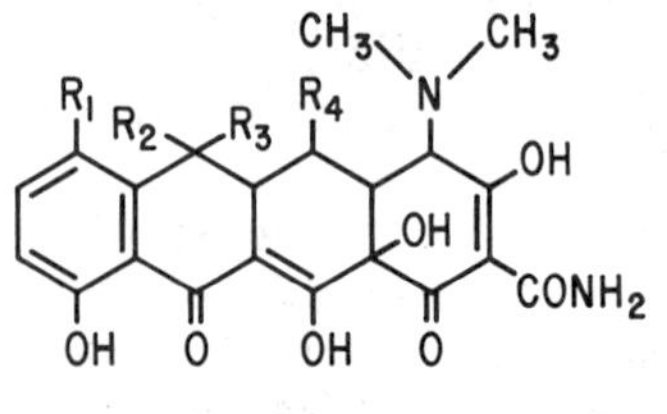

	R_1	R_2	R_3	R_4
Tetracycline	H	CH_3	OH	H
Oxytetracycline	H	CH_3	OH	OH
Doxycycline	H	CH_3	H	OH
Methacycline	H	CH_2		OH
Chlortetracycline	Cl	CH_3	OH	H
Demethylchlor-tetracycline	Cl	H	OH	H
Minocycline	$N(CH_3)_2$	H	H	H

FIGURE 7. *The compounds of this group are named in relation to the basic structure of tetracycline and on the basis of substituted chemical groups at one or more of the four R positions.*

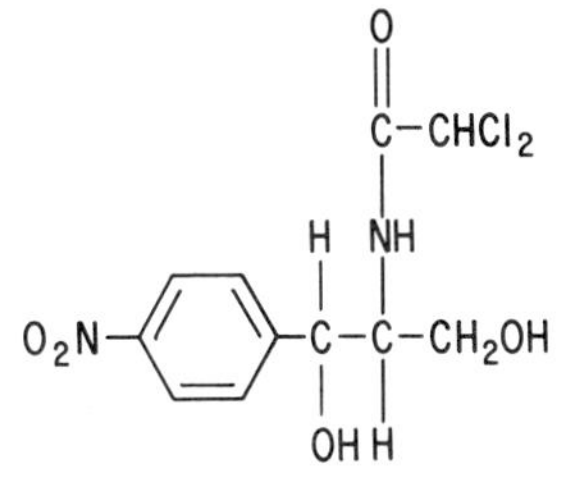

CHLORAMPHENICOL

FIGURE 8. *Chloramphenicol is the only naturally occurring antibiotic with nitrobenzene in its structure, a property that may account for its tendency to cause aplastic anemia.*

macrolides. The basic structure is a large lactone ring to which unusual sugars are attached (Fig. 9). The term "macrolide" refers to the large ring, formed from a chain of 14 to 20 carbon atoms by lactone condensation of a carboxyl and hydroxyl group. The other 37 macrolides, such as oleandomycin, spiramycin, kitasamycin, and carbomycin, differ from erythromycin both in the structure of the lactone ring and in the attached sugars. Since the other macrolides have the same spectrum but are less potent than erythromycin, they are not widely used.

LINCOMYCIN

Despite striking similarity in the biologic effects of lincomycin and erythromycin, their chemical structures are totally dissimilar. In contrast to the macrolides, lincomycin consists of an amino acid joined to a sulfur-containing amino sugar. The amino acid is trans-L-4-*n*-propyl hygric acid,

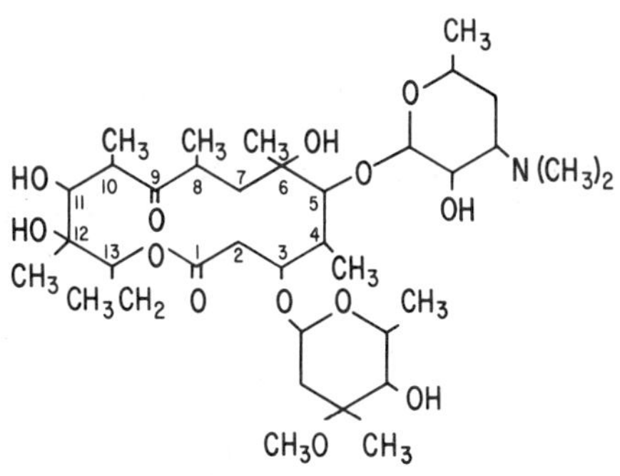

ERYTHROMYCIN

FIGURE 9. *Erythromycin is the only important member of a group of 37 different compounds known as macrolides. The term "macrolide" refers to the large lactone ring formed from a chain of 14 to 20 carbon atoms by lactone condensation.*

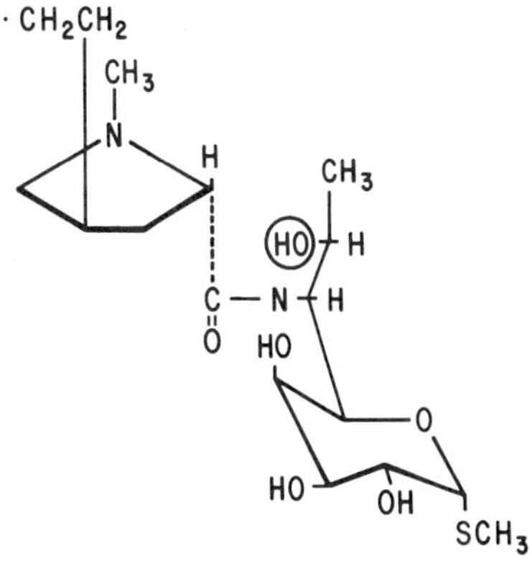

LINCOMYCIN

FIGURE 10. *The sulfur-containing amino acid, methyl-α-thiol lincosamide, is named for the parent compound which was obtained from a mold growing in Lincoln, Nebraska. Note total dissimilarity of lincomycin from macrolides (Fig. 9). The circle indicates the 7-hydroxyl group where a chloro group is substituted in clindamycin.*

and the amino sugar is methyl-α-thiol lincosaminide (Fig. 10).

Clindamycin is a synthetic modification of lincomycin. As indicated by its chemical name, 7-chloro-7-deoxylincomycin, clindamycin is produced by a 7-chloro substitution of the 7(R)-hydroxyl group of the parent compound, lincomycin. These modifications in structure appear to increase absorption, blood levels, and antibacterial activity.

THE RIFAMYCINS

The rifamycin antibiotics are fermentation products of *Streptomyces mediterranei.* Their basic structure is an aromatic ring compound spanned by an aliphatic bridge. The most active of the original compounds, rifamycin B, was not well absorbed after ingestion. After the chemical structure of various rifamycins was determined in 1963, it was possible to synthesize a great many semisynthetic derivatives.[16] At the time of this writing, rifampin is the most important of these because it is orally effective in tuberculosis and leprosy, and in infections by various gram-negative and gram-positive bacteria. The special features of rifampin, shown in Figure 11, are the double ring compound, the long aliphatic bridge, and the side chain: CH = N − N ◯ N − CH_3. The ring is a naphthohydroquinone and is the chemical grouping responsible for the red color of the antibiotic. The importance of the rifamycins lies not only in their currently important derivative, rifampin, but also in the great potential for new derivatives with activity against many different microorganisms, including viruses and fungi.

RIFAMPIN

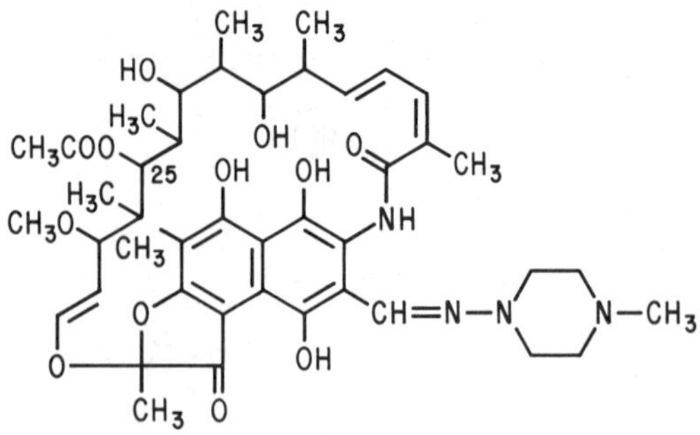

FIGURE 11. *The basic structure of rifampin is the double ring compound spanned by a long aliphatic bridge, and the side chain.*

PARTIAL STRUCTURE FOR AMPHOTERICIN B

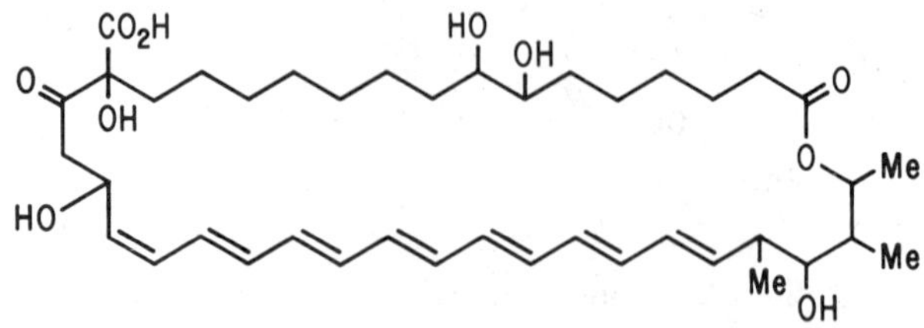

FIGURE 13. *Amphotericin B is called a polyene because it contains a series of carbon atoms with multiple conjugated double bonds.*

VANCOMYCIN

Vancomycin is a bactericidal antibiotic produced by *Streptomyces orientalis.* The structure of vancomycin has recently been determined by x-ray analysis.[15] The principal units in the structure are a disaccharide linked to three aromatic rings: N-methylleucine, aspartic acid, and a biphenyl system. These units are connected by secondary amide bonds to form a tricyclic molecule containing N-terminal N-methylleucine and two free carboxyl groups.

GRISEOFULVIN

Although derived from *Penicillium* molds, griseofulvin is active against fungi rather than bacteria. Discovered in 1939 in London, griseofulvin was used against plant fungi before its value against human dermatophytes was demonstrated.[14] It has a spirocyclic structure formed from acetate units (Fig. 12).

THE POLYENES

Amphotericin B, nystatin, and pimaricin are polyenes. These antifungal antibiotics are classified as polyenes because the molecules contain a series of carbon atoms with four or five conjugated double bonds. Nystatin and amphotericin also possess the aminodeoxyhexose, mycosamine (Fig. 13), which is not present in many of the polyenes that are unsuitable for medical use.[3] Their extensive unsaturation makes them unstable compounds, especially in acid or alkaline solutions. They are also unstable in light and air. Although the exact formula has not been worked out for either one, amphotericin B is known to be a conjugated heptaene lactone linked with mycosamine and has the tentative formula $C_{46}H_{73}O_{20}N$ (Fig. 14).

SYNTHETIC ANTIMICROBIAL DRUGS

The Sulfonamides

The term "sulfonamide" refers to derivatives of sulfanilamide, or π-aminobenzenesulfonamide (Fig. 15), the first antimicrobial shown to be effective systemically for the treatment of human bacterial infection.[10] Thousands of sulfonamides have been synthesized, but only a dozen are of value for patients.

Most derivatives are made by substitutions on the sulfonamide group (SO_2NH_2), since these increase the antimicrobial activity. The para-NH_2 group, on the other hand, must remain free, or become free, after hydrolysis. Succinylsulfathiazole (Sulfasuxidine) is a good example of how the

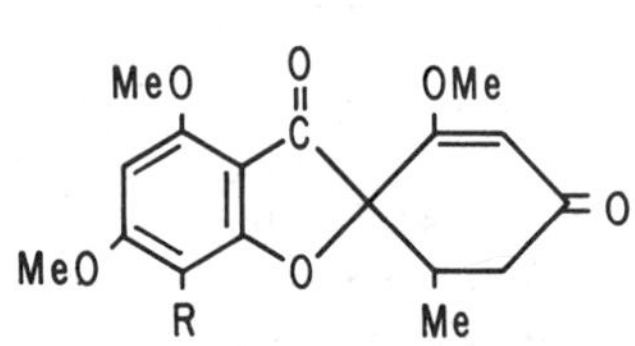

GRISEOFULVIN

FIGURE 12. *Griseofulvin is a spirocyclic compound formed from acetate units.*

MYCOSAMINE

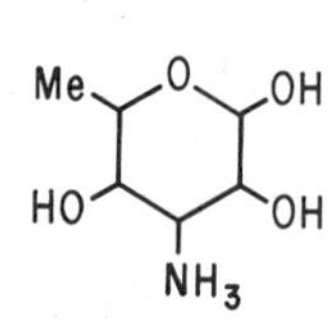

FIGURE 14. *This amino sugar, mycosamine, is linked to a polyene in both amphotericin B and nystatin.*

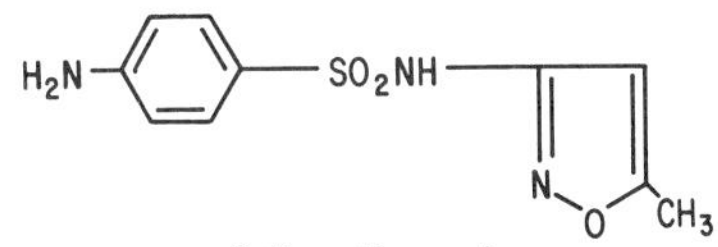

Sulfamethoxazole

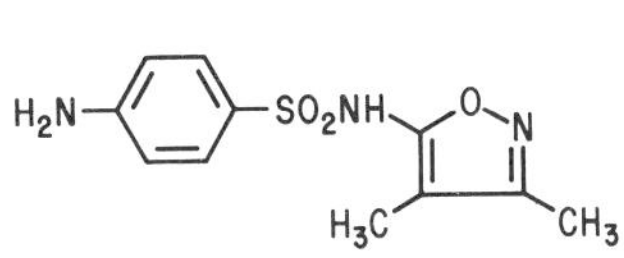

Sulfisoxazole, U.S.P.

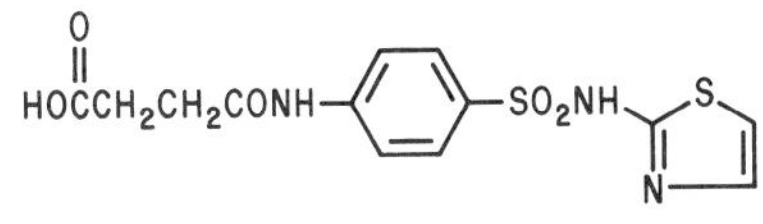

Succinylsulfathiazole, U.S.P.

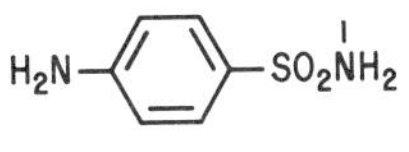

Sulfanilamide

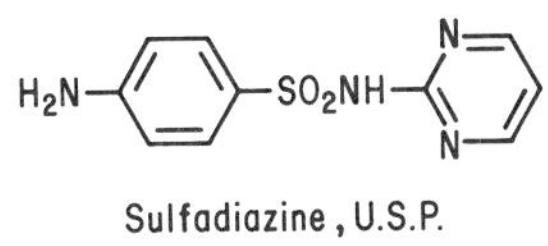

Sulfadiazine, U.S.P.

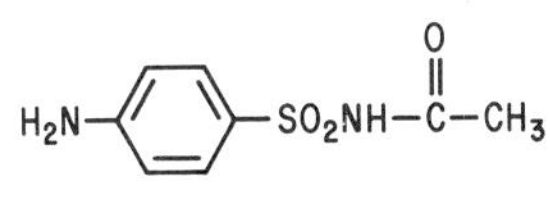

Sulfacetamide, N.F.

FIGURE 15. *These and other sulfonamide derivatives are made by substitutions on the SO_2NH_2 group of the parent compound sulfanilamide. The para-NH_2 group must remain free.*

para-NH_2 group becomes free after slow hydrolysis from its inactive form to the active sulfathiazole. The substitutions on the SO_2NH_2 group of other sulfonamides are shown in Fig. 15.

The Sulfones

As is evident from the structural formula of dapsone in Figure 16, sulfones are related to sulfonamides but lack the sulfonamide group and its broad antibacterial spectrum. Instead, the sulfones are limited in antibacterial activity to the leprosy bacillus.

Isoniazid and Ethionamide

Isoniazid was the first modern chemotherapeutic agent synthesized (1912), but its value in tuberculosis was not appreciated until 1952.[4] It is the hydrazide of isonicotinic acid (Fig. 17). Ethionamide is also a derivative of isonicotinic acid.

The Diaminopyrimidines

The diaminopyrimidine compounds were first synthesized as analogs of the nitrogenous bases found in DNA. Pyrimethamine, for example, was prepared as a thymine analog. This drug and trimethoprim are the only two members of this group that are of practical medical importance. Their formulas are given in Figure 18. Both are useful in protozoal infections, and trimethoprim in bacterial infections.[7]

The Fluorinated Pyrimidines

The substitution of fluorine for hydrogen in the 5-position on pyrimidines was first made in order to produce the pyrimidine analog 5-fluorouracil, a potent antimetabolite for mammalian cells and useful in cancer chemotherapy. Another fluorinated pyrimidine, 5-fluorocytosine (5FC), does not seem to be metabolized in mammalian cells and has no activity against cancer cells. The activity of 5FC against microorganisms seems to be related to its conversion to 5-fluorouracil by deamination[9] (Fig. 19).

The Nitrofurans

The nitrofurans are derivatives of the 5-membered ring sugars known as furans and possess a nitro group in the 5-position. The chief nitrofuran available for clinical application is shown in Figure 20 and is the only sugar derivative with clinically important antibacterial properties.

The Aminoquinolines

The most important member of the aminoquinolines is chloroquine, a 4-aminoquinoline with

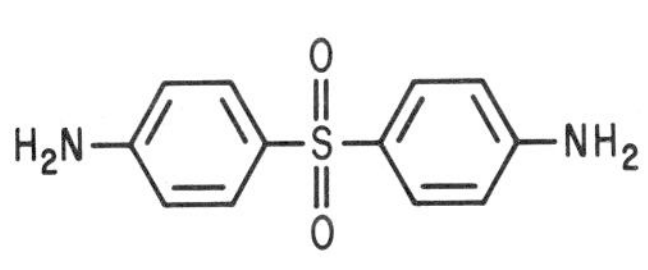

Dapsone, U.S.P.

FIGURE 16. *Note structural relationship of dapsone to sulfonamides.*

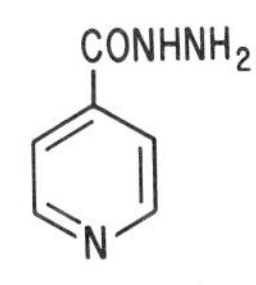

ISONIAZID

FIGURE 17. *The drug isoniazid is a derivative of isonicotinic acid.*

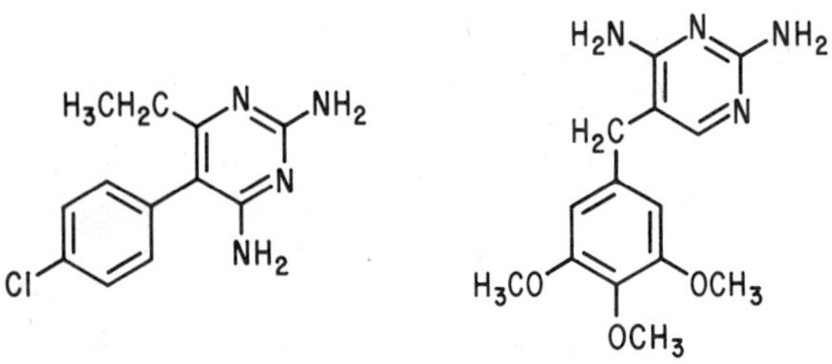

Pyrimethamine Trimethoprim

FIGURE 18. *These diaminopyrimidines were first synthesized as analogs of the nitrogenous bases found in DNA.*

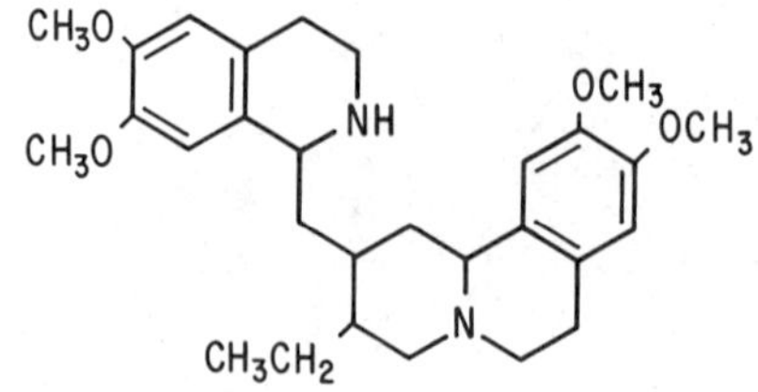

EMETINE

FIGURE 22. *Emetine. This is the structure of the active principle of ipecac.*

DEAMINATION OF
5-FLUOROCYTOSINE TO 5-FLUOROURACIL

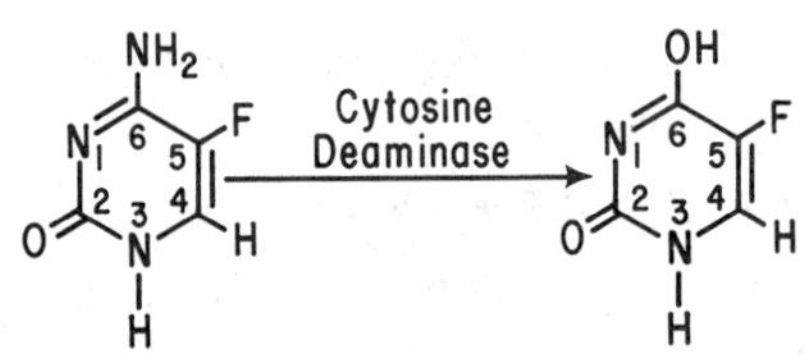

5-Fluorocytosine 5-Fluorouracil

FIGURE 19. *The activity of 5-fluorocytosine against microorganisms seems to be related to its conversion to 5-fluorouracil.*

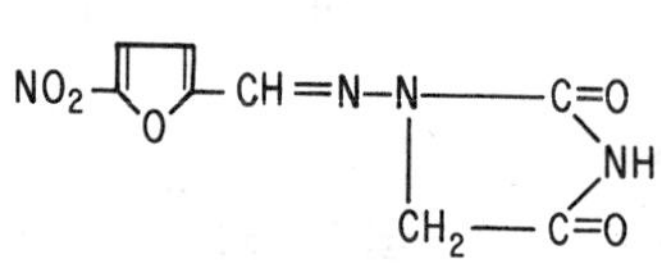

FURADANTIN

FIGURE 20. *The nitrofurans are the only synthetic sugar derivatives with clinically useful antibacterial properties.*

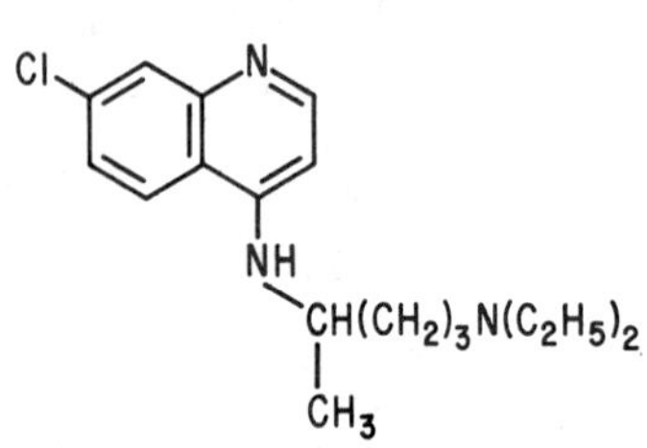

Chloroquine

FIGURE 21. *The Cl atom in position 7 is necessary for maximum antimalarial activity of chloroquine.*

the structure shown in Figure 21. The main parts of the molecule are the quinoline nucleus composed of a double ring and the alkyl side chain. The chlorine atom in position 7 of the quinoline nucleus appears necessary for maximum antimalarial activity.

Although first used as an antimalarial, chloroquine was later found to be valuable in the treatment of amebic liver abscess and giardiasis.

Emetine

Emetine is the oldest amebicidal drug and the active principle chiefly responsible for the clinical efficacy of ipecac in amebic infections. Emetine hydrochloride is obtained from ipecac as a hydrated hydrochloride[5] (Fig. 22).

Nitroimidazoles

Metronidazole (Fig. 23) is the newest amebicidal drug in clinical use. It is a nitroimidazole derivative active against two groups of anaerobic organisms, the anaerobic bacteria and the pathogenic protozoa, including amebae, *Trichomonas, Giardia,* and *Balantidium.* Trinidazole, another promising nitroimidazole drug, has been effective in amebic dysentery and trichomonas infections.

Imidazoles

Three antifungal drugs, clotrimazole, miconazole, and ketoconazole, are imidazoles that are related to the nitroimidazoles used for anaerobic bacteria and protozoa (Fig. 23). The benzimidazoles (see Fig. 27) are potent anthelmintics. Clotrimazole is limited to topical use for ringworm and moniliasis of the skin, but miconazole can be given intravenously for generalized fungus infections. Ketoconazole, the newest imidazole, is well absorbed orally and is being examined for its clinical effectiveness.

The Anthelmintics[2]

The simplest structure of the anthelmintics, a group of nitro ring compounds, is that of pipera-

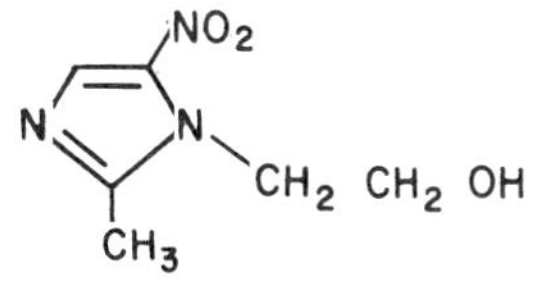

METRONIDAZOLE

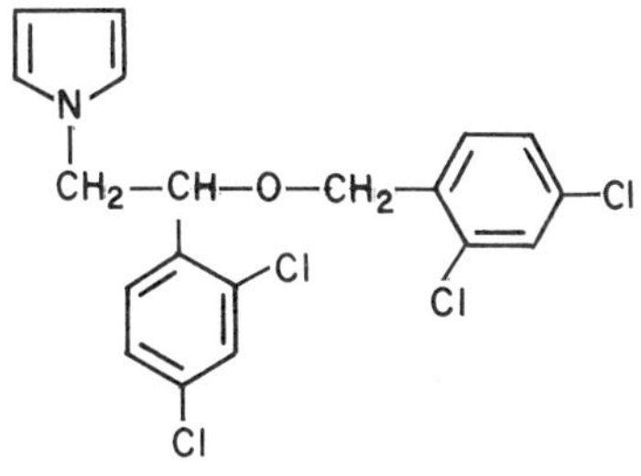

MICONAZOLE

FIGURE 23. *Comparison of the nitroimidazole, metronidazole, with the imidazole, miconazole. Metronidazole attacks anaerobic organisms only (bacteria and protozoa) after the nitro group is reduced, whereas miconazole acts against fungi, which are strictly aerobic. Compare also the benzimidazoles (Fig. 27), which are also active against anaerobes—the intestinal worms.*

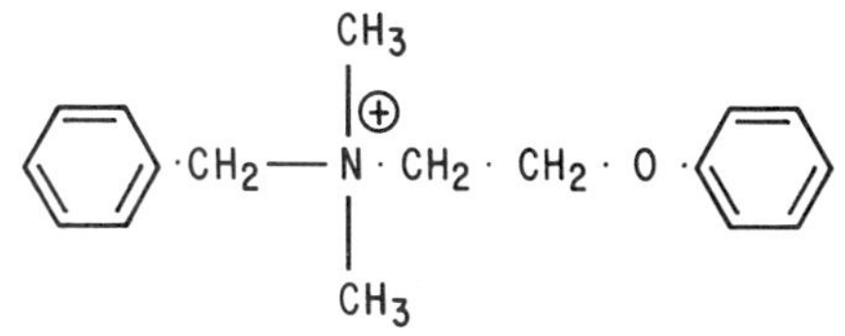

BEPHENIUM

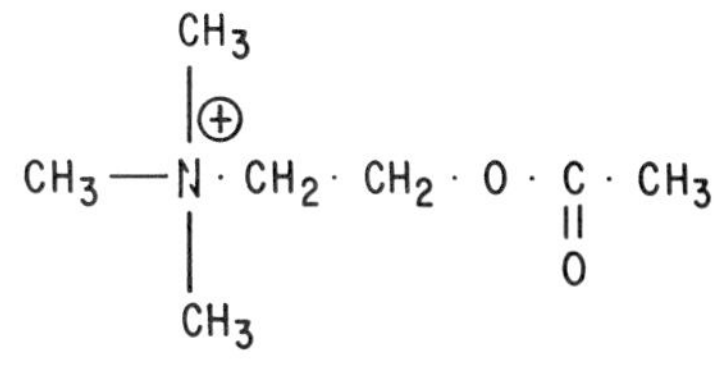

ACETYLCHOLINE

FIGURE 25. *The anthelmintic bephenium resembles acetylcholine in structure and action. Bephenium paralyzes nematodes by causing membrane depolarization of muscle in the worms.*

zine, a drug used unsuccessfully at first for the treatment of gout but later found to be highly effective against *Ascaris* and *Enterobius* worms. Modification of piperazine (Fig. 24) by substituting a diethylcarbamyl group at the 1-position and a methyl group at the 4-position created diethylcarbamazine, a drug that can kill adult filaria and microfilaria. Bephenium is a quaternary amine and resembles acetylcholine in both structure and function (Fig. 25). The anticestodal drug, niclosamide (chlorsalicylamide), resembles closely the structure of desaspidin, the active principle of oleoresin of aspidium, the oldest remedy for tapeworms (Fig. 26). Another important group of anthelmintics are the substituted benzimidazole compounds. One of these, thiabendazole, is effective against various round worms, while mebendazole is the only anthelmintic effective against both tapeworms and round worms. The formula of thiabendazole is shown in Figure 27. The only compound with a close structural resemblance to antibacterial drugs is niridazole, which is derived from a nitrothiazole nucleus (Fig. 28). Its structure is related to that of nitrofurantoin and metronidazole, two agents with antibacterial activity.

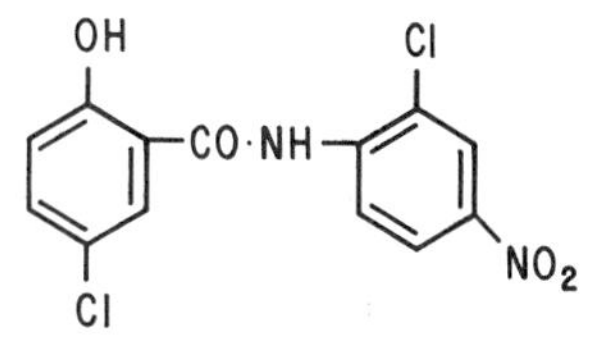

NICLOSAMIDE

FIGURE 26. *Niclosamide resembles in structure desaspidin, the active principle of oleoresin of aspidium, an old remedy for tapeworms.*

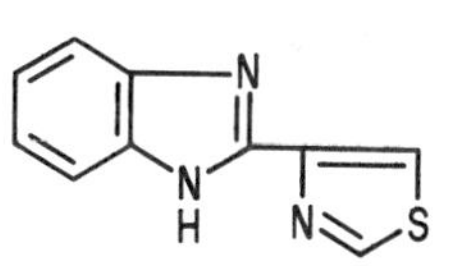

Thiabendazole

FIGURE 27. *Thiabendazole is the best anthelmintic among several hundred substituted benzimidazole compounds.*

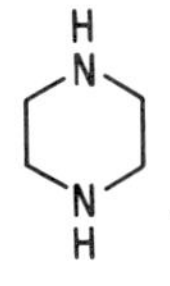

PIPERAZINE

FIGURE 24. *Piperazine has the simplest structure of the anthelmintics.*

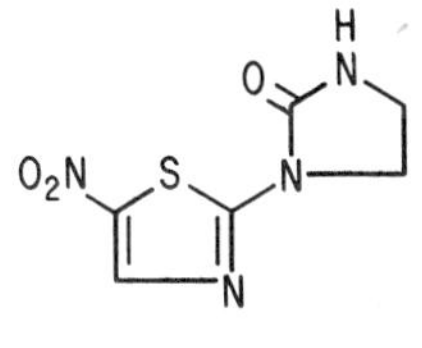

NIRIDAZOLE

FIGURE 28. *Niridazole. Note structural similarity of this anthelmintic to the antibacterial drug nitrofurantoin.*

MICROBIAL SUSCEPTIBILITY TO DRUGS

The antimicrobial activity of a drug is generally expressed as its minimum inhibitory concentration in nutrient broth or agar. The test conditions would seem far removed from those in infected tissues or body fluids, but there is a remarkable correlation between successful treatment of the patient and inhibitory concentrations in the laboratory.

Microbial inhibition is measured either by serial dilution or by diffusion of antibiotics.[41] In the dilution tests the antibiotic is diluted serially in broth or agar, and results are expressed as the lowest concentration that inhibits growth of a standard bacterial inoculum at 37° C. Three diffusion tests are used: the disk test, the filter strip, and the gradient plate. In the disk test, antibiotics are incorporated into filter paper disks, and the minimum inhibitory concentration is calculated indirectly from the diameter of the inhibitory zone.[5, 17] (Fig. 29). Long filter paper strips can be used instead of disks. The strips have the advantage of measuring inhibitory concentrations against multiple strains of bacteria that are streaked perpendicularly to the strip.

The gradient plate is prepared by pouring agar with a given concentration of antibiotic in square Petri dishes that are tilted so that the agar solidifies on a slant. The plate is then placed on a flat surface and an equal volume of agar without antibiotic is poured over the slanted agar to provide a horizontal layer, as shown in Figure 30. The antibiotic diffuses into the upper layer to give a perfect linear gradient ranging from zero to the concentration present in the original layer. Multiple strains can be streaked on the agar surface along the gradient. The minimum inhibitory concentration is directly related to the distance bacterial growth extends from the zero end of the gradient.

The gradient plate method is the best method for laboratories that need to perform sensitivity on many strains because multiple organisms can be examined on one plate and there is little room for error.[7] The disk test is the most widely used (despite a number of disadvantages) because it lends itself to commercial distribution and has been publicized. The most serious problem encountered with the disk method is the occasional failure to detect strains of staphylococci that are resistant to penicillin. These resistant staphylococci are easily recognized by tests for penicillinase. The simplest is a rapid capillary tube assay

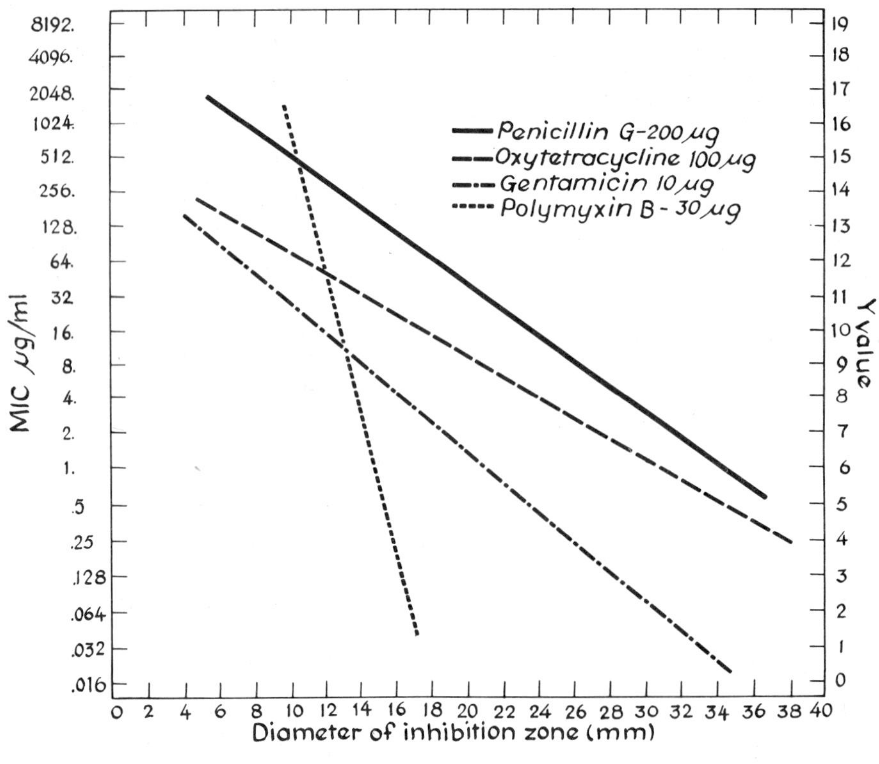

FIGURE 29. *Determination of antibiotic sensitivity by disk method is based on correlation between zone diameter and minimum inhibitory concentration. Regression lines for minimum inhibitory concentrations and inhibitory zone diameter are shown for penicillin G (200 μg), oxytetracycline (100 μg), gentamicin (10 μg), and polymyxin B sulfate (30 μg). (From Stamey, T. A.: Urinary Infections. © 1972, The Williams and Wilkins Co., Baltimore, p. 45.)*

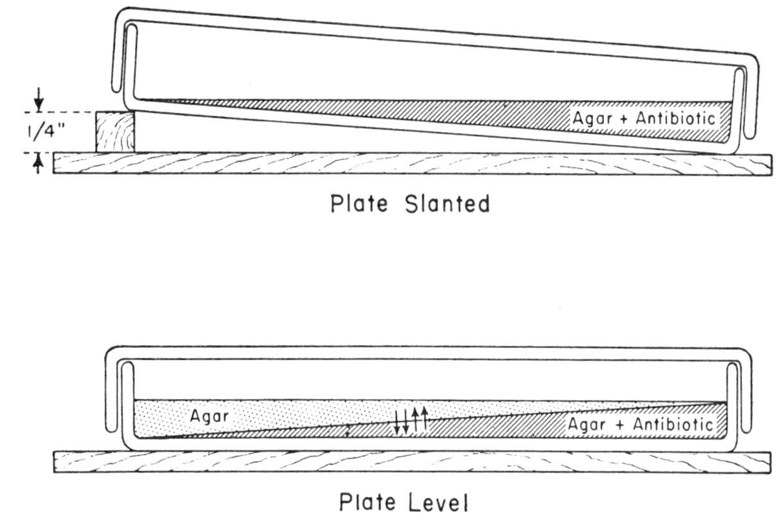

FIGURE 30. *Diagram of procedures used for preparing gradient plates. The extreme right end of the completed plate (bottom) is where the gradient begins; there is initially no dilution of antibiotic at this point. Halfway across the plate (arrows) there is a twofold dilution of antibiotic. At the extreme left there is virtually no antibiotic and its concentration is zero. Thus a uniform linear gradient is formed by the vertical diffusion of antibiotic. (From Braude, A. I., Banister, J., and Wright, N.: Use of the gradient plate for routine clinical determinations of bacterial sensitivities to antibiotics. Antibiot Annu p. 1134, 1954–1955.)*

that should be carried out routinely in all laboratories.[44] It is based on the change in pH that occurs when staphylococcal penicillinase hydrolyzes penicillin to penicilloic acid.

The values given below for antibiotic sensitivity are derived from either dilution methods or the gradient plate and are a composite of results from our laboratory and several other representative laboratories.

THE COCCI

With the exception of certain staphylococci, gonococci,[11, 42] and enterococci, cocci remain exquisitely sensitive to penicillin G (benzylpenicillin). Pneumococci, hemolytic streptococci, meningococci, anaerobic streptococci, and *Streptococcus viridans* are usually inhibited by less than 0.1 microgram (μg) per ml and frequently by as little as 0.01 μg/ml[18, 34, 46] (Table 1). Gonococci and staphylococci are no longer uniformly sensitive to penicillin. Staphylococci that produce penicillinase cannot be treated with benzylpenicillin at any dosage. The number of staphylococci that produce penicillinase varies among hospitals. At the University Hospital in San Diego 90 per cent do so. Before 1960, 95 per cent of gonococci were sensitive to 0.1 μg/ml benzylpenicillin, but now less than half are inhibited by this concentration, and 1.0 μg/ml is necessary to inhibit 95 per cent of gonococcal strains in many communities.[32]

In contrast to those gonococci and staphylococci which have lost initial sensitivity, enterococci have always been relatively resistant to penicillin G. Enterococci generally grow readily in 1.0 μg/ml and usually require 3 to 6 μg/ml for inhibition.[18]

Modification of penicillin G to acid-resistant or penicillinase-resistant compounds causes loss in antibacterial power. In penicillin V and ampicillin the loss of activity against most cocci is usually slight; in fact, ampicillin is more active than penicillin G against the enterococci. *Neisseria* show the biggest differences in sensitivity between penicillin V and G, with the meningococcus and penicillin-sensitive gonococci requiring concentrations of penicillin V four times greater than those of penicillin G for inhibition.

The penicillinase-resistant penicillins show greater loss of activity against penicillin-sensitive cocci. Methicillin is 35 to 40 times less active than penicillin G against streptococci and penicillinase-negative staphylococci. Cloxacillin, oxacillin, and dicloxacillin are approximately 8 to 12 times less active than penicillin G against pneumococci, various streptococci, and penicillinase-negative staphylococci.[22, 58] The same, except for a little more activity against the pneumococcus, is true for nafcillin. All four inhibit penicillinase-producing staphylococci in a range of 0.15 to 1 μg/ml.

The related group of penicillinase-resistant antibiotics, the cephalosporins, are also less active than penicillin against all cocci (Table 2). Cephaloridine approaches benzylpenicillin in potency in vitro against the pneumococcus and group A streptococcus but otherwise the cephalosporins are considerably less active against cocci than is benzylpenicillin.[25] The cephalosporins are so inactive against the enterococcus and meningococcus that they are virtually useless in infections with those organisms.[6, 47]

Tetracyclines are losing some of their potency against cocci.[18, 34] In all species of cocci, strains have appeared that are too resistant for treatment with this group of antibiotics. Resistance to tetracyclines has been most marked among the

TABLE 1. Usual Minimum Inhibitory Concentrations (μg/ml) of Penicillin Derivatives Against Cocci

	BENZYL-PENICILLIN	PHENOXYMETHYL-PENICILLIN	AMPICILLIN*	METHICILLIN	OXACILLIN	DICLOXACILLIN	NAFCILLIN
Pneumococcus	0.01	0.03	0.02	0.1	0.5	0.15	0.04
Group A streptococcus	0.005	0.015	0.02	0.2	0.02	0.05	0.02
Staphylococcus aureus (penicillinase-negative)	0.03	0.03	0.05	1.0	0.30	0.15	0.40
S. aureus (penicillinase producer)	R	R	R	1.0	0.40	0.10	0.50
Streptococcus faecalis	3.6	3.2	1.6	>25.0	>25.0	—	25.0
Streptococcus viridans	0.01	0.01	0.1	0.1	—	—	0.06
Gonococcus	0.01–3.0	0.03–>3.0	0.3	12.0	12.0	0.06–4.0	—
Meningococcus	0.03	0.25	0.05	6.0	—	6.0	6.0
Peptostreptococcus	0.2	0.5	0.2	2.0	0.6	2.0	<25.0

*The activity of amoxicillin is very similar to that of ampicillin, except for *Streptococcus faecalis*. Amoxicillin inhibits most strains of *S. faecalis* at 0.6 μg/ml.

R = Resistant of all concentrations.

— = Inadequate data.

TABLE 2. Usual Minimum Inhibitory Concentrations (μg/ml) of Cephalosporin Antibiotics Against Cocci[29, 47]

	CEPHALOTHIN	CEPHALEXIN	CEFACLOR	CEFAZOLIN	CEFOXITIN
Pneumococcus	0.4	3.1	1.0	0.1	2.0
Group A streptococcus	0.1	1.0	0.8	0.2	0.8
Staphylococcus aureus	0.6	6.0	4.0	0.6	2.5
Staphylococcus epidermidis	0.6	12.0		1.0	3.0
Streptococcus faecalis	50.0	50.0	60.0	50.0	>50.0
Streptococcus viridans	2.0	6.25		0.3	12.0
Gonococcus	4.0	6.0	0.02–2.0	4.0	0.5–1.0
Meningococcus	1.6	100.0		–	0.5–1.5

– = Inadequate data.

enterococci, but many staphylococci have also developed resistance beyond the limits of clinical efficacy. In certain hospitals as many as 50 per cent of staphylococci show such resistance while staphylococci from patients in outpatient clinics remain sensitive. So much resistance has developed among strains of pneumococci (5 to 23 per cent) and group A streptococci (20 to 40 per cent)[34] that tetracyclines cannot be relied on for the treatment of pneumonia or sore throat unless sensitivity tests establish their susceptibility. Among anaerobic gram-positive cocci, tetracycline resistance is a lesser problem. In contrast to the aerobic gram-positive cocci, over 90 per cent of *Neisseria* remain sensitive to the tetracyclines, which can cure most cases of gonorrhea. One tetracycline, minocycline, has also been somewhat effective in treating pharyngeal carriers of meningococci.[21] The in vitro susceptibility of sensitive strains of cocci to the tetracyclines is shown in Table 3.

Certain tetracycline analogues seem to be more active in vitro against cocci that are resistant to other tetracyclines.[57] Minocycline is the best example of this phenomenon since it inhibits in low concentrations staphylococci, group A streptococci, and enterococci that are resistant to tetracycline.

Lincomycin and clindamycin inhibit most of the gram-positive cocci, but not the *Neisseria.* Clindamycin is 4 to 16 times more active than lincomycin against staphylococci, and the minimum inhibitory concentration (MIC) against hemolytic streptococci (groups A, B, and C), pneumococci, and *Streptococcus viridans* is less than 0.05 μg/ml.[12] Among gram-positive cocci, only enterococci are resistant. Anaerobic cocci are almost all sensitive to 1.0 μg/ml or less of clindamycin.

The spectrum of erythromycin resembles that of lincomycin and clindamycin except for the *Neisseria.*[12, 20] Most strains of gonococci and meningococci are inhibited by 1.0 μg/ml of erythromycin.[46] The pyogenic streptococci and pneumococci are inhibited by 0.04 μg/ml, and sensitive staphylococci by 0.4 μg/ml. Staphylococci sometimes become resistant to erythromycin, especially in the hospital, and may show cross resistance to lincomycin. Strains of group A streptococci[48] and pneumococci are rarely resistant, but in some large hospitals over half the strains of enterococci are resistant to erythromycin.[60] Other enterococci range in sensitivity from 0.1 to 1.5 μg/ml erythromycin. Most anaerobic streptococci also fall in this range of sensitivity.[33]

The cocci do not show the exquisite sensitivity to chloramphenicol often found with other antibiotics, but most cocci are inhibited by 1 to 4 μg/ml and acquired resistance is unusual. The average pneumococcus or group A streptococcus, for example, is inhibited by 3.0 μg/ml of chloramphenicol. Meningococci tend to be more sensitive, with the average strain inhibited by 1.0 μg/ml. Staphylococci, anaerobic cocci, enterococci, gonococci, and *Streptococcus viridans* are generally sensitive to 4 μg/ml or less.[18, 46]

The aminoglycosides are less active against cocci than other antibiotics that interfere with protein synthesis. Streptococci and pneumococci are naturally resistant, and the *Neisseria* generally show only marginal sensitivity. The staphylococci are an exception, because kanamycin and gentamicin are both highly active against them, and some strains are sensitive to streptomycin.[52]

TABLE 3. Usual Minimum Inhibitory Concentrations (μg/ml) of Tetracyclines Against Sensitive Strains of Cocci

	TETRA-CYCLINE	DOXY-CYCLINE	MINO-CYCLINE
Staphylococcus aureus	0.3	0.8	0.8
Group A streptococcus	0.3	0.2	0.2
Pneumococcus	0.3	0.1	0.1
Peptostreptococcus	1.6	0.8	0.8
Gonococcus	0.8	0.8	0.8
Meningococcus	0.8	0.8	0.8

TABLE 4. Usual Minimum Inhibitory Concentrations (μg/ml) of Benzylpenicillin Against Sensitive Bacilli

Gram-positive rods	
Listeria monocytogenes	0.2
Actinomyces israelii	0.06
Clostridium perfringens	0.16
Bacillus anthracis	0.02
Corynebacterium diphtheriae	0.08
Erysipelothrix rhusiopathiae	0.03
Bordetella pertussis	0.5
Gram-negative rods	
Hemophilus influenzae	0.8
Hemophilus ducreyi	1.0
Pasteurella multocida	0.4
Streptobacillus moniliformis	0.01
Bacteroides oralis	1.6
Bacteroides melaninogenicus	1.0
Fusobacterium nucleatum	0.8

Staphylococci range in sensitivity to gentamicin from 0.1 to 1.0 μg/ml, and to kanamycin from 0.4 to 4.0 μg/ml. Spectinomycin is the only aminoglycoside with enough activity against the gonococcus to warrant its use for the routine treatment of gonorrhea. Gonococci range in sensitivity from 6.2 to 25 μg/ml.[24]

One of the most active antibiotics against cocci is rifampin. The MIC for group A streptococci, staphylococci, and pneumococci is 0.02 μg/ml or less, while meningococci are inhibited by 0.2 μg/ml, gonococci by 0.5 μg/ml, and anaerobic streptococci by 1.6 μg/ml.[28, 49] Only enterococci have too much innate resistance for clinical effectiveness of the drug. Acquired resistance to rifampin among staphylococci and other cocci occurs readily and limits its usefulness.

Only one other antibiotic, vancomycin, deserves brief mention for its use against cocci. Its main value is for enterococcal endocarditis in patients with severe penicillin allergy.[19] The MIC for vancomycin against enterococci is 0.3 to 3.0 μg/ml.

Most other antimicrobials have little place in the treatment of coccal infections. The polymyxins are inactive against cocci, and antimicrobials other than antibiotics are of such limited value that they need not be discussed here. An exception is trimethoprim, whose antimicrobial spectrum is described later.

THE BACILLI

Penicillin G is highly effective against all gram-positive bacilli and many gram-negative bacilli. Most enteric bacilli are relatively resistant but the penicillin derivatives, ampicillin, carbenicillin, and pipericillin usually inhibit these gram-negative rods in concentrations that can be reached in infected body fluids. The gram-positive rods, *Listeria monocytogenes, Actinomyces israelii,* the clostridia, *Erysipelothrix rhusiopathiae, Bacillus anthracis,* and *Corynebacterium diphtheriae,* are all very sensitive to penicillin, and most are inhibited by less than 0.1 μg/ml. Certain important gram-negative rods such as *Hemophilus influenzae, Pasteurella multocida, Streptobacillus moniliformis,* and most *Bacteroides* other than *Bacteroides fragilis* are also sensitive to penicillin in a range of 0.5 to 2.0 μg/ml (Table 4). Despite impressions to the contrary, ampicillin is not superior to penicillin G against *H. influenzae.* Most careful studies in the United States have shown that the MIC of both antibiotics against *H. influenzae* ranges from 0.2 to 1.6 μg/ml, with a median near 0.8 μg/ml.[35, 46] A few strains have developed resistance, but since this is mediated by a β-lactamase, both drugs are ineffective against them.

TABLE 5. Comparison of Usual Minimum Inhibitory Concentrations (μg/ml) of Penicillin G, Ampicillin, and Carbenicillin for Enteric Gram-Negative Rods

	BENZYLPENICILLIN	AMPICILLIN*	CARBENICILLIN	PIPERICILLIN
E. coli	100.0	1.6–500	12.0	3.1–500
P. mirabilis	50.0	1.0–32.0	1.6	0.8
K. pneumoniae	>100.0	25.0–400.0	>200.0	>200.0
Enterobacter species	>500.0	20.0–250.0	6.0	50.0
S. marcescens	>500.0	40.0–100.0	12.0–400.0	3.1–500
P. aeruginosa	>500.0	>200.0	50.0–100.0	12.0
Proteus vulgaris	>100.0	3.0–100.0	2.0–25.0	–
Other indole + *Proteus*	>100.0	3.0–50.0	1.0–12.0	1.6–500
Salmonella species	12.0	6.0	12.0	1.6–500
Pseudomonas pseudomallei	25.0	10.0	>100.0	–

*Amoxicillin activity against gram-negative rods is very similar to that of ampicillin except that amoxicillin is approximately four times as active against *Salmonella* strains.[38]
– = Inadequate data.

TABLE 6. Usual Minimum Inhibitory Concentrations (μg/ml) of Cephalosporins for Pathogenic Bacilli

	CEFAZOLIN	CEPHALOTHIN	CEFACLOR	CEPHALEXIN	CEFOXITIN
Gram-positive rods					
Listeria monocytogenes	–	2.0	–	64.0	–
Actinomyces israelii	–	2.0–>100	–	–	0.5–32.0
Clostridium perfringens	–	0.6	12.0	1.2	1.2
Bacillus anthracis	–	0.6	–	2.0	–
Gram-negative rods					
Hemophilus influenzae	25.0	6.0–10.0	1.0	6.0–20.0	7.0
Pasteurella multocida	–	0.6	–	2.0	–
Bacteroides fragilis	>200.0	25.0	>100	25.0	10.0
Bacteroides oralis	–	0.1	–	–	–
Bacteroides melaninogenicus	–	0.2–6.0	–	–	1.0
Fusobacterium nucleatum	2.0	1.0	–	–	2.5
E. coli	0.8	6.0	3	12.0	2.0
P. mirabilis	3.0	6.0	6	20.0	2.0
K. pneumoniae	3.0	6.0	12	20.0	2.5
E. aerogenes	6.0–400.0	50.0–>400.0	>100	>100.0	>75.0
S. marcescens	>400.0	>100.0	–	>100.0	>40.0
P. aeruginosa	>400.0	>400.0	–	>100.0	>100.0
Indole + *Proteus*	200.0	100.0–400.0	>100	>100.0	5.0
Salmonella spp.	1.0	2.0	0.8	4.0	3.0
Shigella spp.	1.0	8.0	0.8	12.0	–
Pseudomonas pseudomallei	–	>1000.0	–	–	–

– = Inadequate data. The antibacterial activity of cephradine is nearly identical to that of cephalexin.

Enteric bacilli are considerably less sensitive to penicillin G than the rods listed in Table 4.[18, 47, 56] Table 5 shows the greater susceptibility of these organisms to ampicillin and carbenicillin. Ampicillin inhibits sensitive strains of *Proteus mirabilis* and *E. coli* at a concentration of 3.0 μg/ml or less, and salmonellae at 6μg/ml. Carbenicillin is active in a range achieved clinically against all gram-negative bacilli listed except for certain strains of *Serratia*.[10, 53]

The cephalosporins differ from the penicillins in the poor activity of the former against *H. influenzae* (Table 6). Another notable difference is the greater activity of cephalothin, cefoxitin, cefamandole, and cefazolin against *Klebsiella pneumoniae* than that shown by ampicillin and penicillin G. *Enterobacter aerogenes, Serratia,* indole + *Proteus,* and *Pseudomonas* are resistant to the older cephalosporins,[15] but indole + *Proteus* is inhibited by 5 μg/ml of cefamandole or cefoxitin, and many strains of *Enterobacter* are inhibited by 2.5 μg/ml cefamandole.

Erythromycin, lincomycin, and clindamycin are effective in vitro against *Bacteroides*[33] and the gram-positive rods, but not against enteric gram-negative bacilli. Erythromycin also shows activity against *H. influenzae* and *B. pertussis,* but lincomycin does not. Table 7 compares the activi-

TABLE 7. Usual Minimum Inhibitory Concentrations (μg/ml) of Erythromycin, Lincomycin, and Clindamycin Against Susceptible Bacilli

	ERYTHROMYCIN	LINCOMYCIN	CLINDAMYCIN
A. israelii	0.12	0.25	–
L. monocytogenes	2.0	–	–
C. diphtheriae	1.6	–	–
C. perfringens	1.5	2.0	0.8
B. anthracis	0.6	6.0	–
H. influenzae	3.0	20.0	6.0
P. multocida	3.1	–	–
B. fragilis	2.0	2.0	0.2
B. oralis	0.1	0.1	0.1
B. melaninogenicus	0.4	0.1	0.01
F. nucleatum	1.6	3.1	0.4
C. fetus (jejuni)	3.1	12.5	0.8

– = Inadequate data.

TABLE 8. Usual Minimum Inhibitory Concentrations (μg/ml) of Chloramphenicol and Tetracycline Against Gram-Positive and Gram-Negative Rods

	CHLORAMPHENICOL	TETRACYCLINE
A. israelii	3.0	3.0
L. monocytogenes	5.0	1.0
C. diphtheriae	0.5	0.3–1.0
C. perfringens	3.0	3.0
B. anthracis	1.5	4.0
H. influenzae	2.0	2.0
H. ducreyi	<1.0	<1.0
B. pertussis	2.0	2.0
P. multocida	1.5	3.0
B. fragilis	6.0	1.0–25.0
B. oralis	1.5	0.5
B. melaninogenicus	1.5	1.0
F. nucleatum	3.1	6.2
E. coli	6.0–12.0	6.0–50.0
E. aerogenes	20.0	50.0–>100.0
K. pneumoniae	10.0	50.0–>100.0
S. marcescens	25.0	100.0
P. mirabilis	6.0–12.0	50.0–100.0
Indole + *Proteus*	50.0	25.0
P. aeruginosa	>100.0	>100.0
Y. enterocolitica[64]	3.0	6.0
B. abortus[63]	3.0	1.0
P. pseudomallei[16]	6.4	1.6
Salmonella spp.	2.0	1.0
Shigella spp.	2.0	8.0
Vibrio cholerae[30]	1.25	1.05
Vibrio parahemolyticus	3.1	3.1

ties of these three antibiotics against bacilli. The activity of clindamycin against *Campylobacter fetus* subsp. *jejuni* is noteworthy.

All the rods, both gram-positive and gram-negative, were initially sensitive to the tetracyclines and chloramphenicol, but some resistance has developed among the enteric bacilli. Table 8 shows that *Pseudomonas aeruginosa, Yersinia enterocolitica,* indole + *Proteus,* and most *Serratia* are beyond the reach clinically of chloramphenicol and tetracycline.

The aminoglycosides differ from other antibiotics in the weakness of their inhibitory action against anaerobes[33] but show potent in vitro activity against nearly all other bacilli. The usual MIC of streptomycin, kanamycin, and gentamicin against aerobic gram-negative bacilli is given in Table 9. Two MIC values are given for a number of enteric bacteria because of differences in susceptibility between "street" and "hospital" strains and because of variations from one community or hospital to another. The in vitro sensitivities of *Salmonella* and *Brucella* are misleading because streptomycin and kanamycin are not effective in treating brucellosis or *Salmonella* infections.

The polymyxins — polymyxin B and colistin methane sulfonate — are also impotent against the anaerobes, as well as the gram-negative bacilli, *Proteus, Serratia, Brucella,* and *P. pseudomallei.* Most strains of *E. coli, K. pneumoniae,* and *P. aeruginosa* are inhibited by 2.5 μg/ml or less of either drug. *Salmonella, Shigella,* and *H. influenzae* are even more sensitive (0.2 to 0.4 μg/ml), but the polymyxins are seldom used to treat infections by these three groups of organisms.

Rifampin is effective to some degree against all the enteric rods, *H. influenzae,*[2, 26, 49] and *H. ducreyi* (Table 10).

Among the sulfonamides, sulfadiazine and sulfisoxazole will inhibit most enteric bacilli in con-

TABLE 9. Sensitivity in vitro of Gram-Negative Bacilli to Aminoglycosides

	USUAL MINIMUM INHIBITORY CONCENTRATIONS (μg/ml)			
	Streptomycin	Kanamycin	Gentamicin*	Amikacin
H. influenzae	8.0	4.0	2.0	–
B. pertussis	4.0	2.0	1.0	–
P. multocida	25.0	–	–	–
E. coli	4.0–25.0	3.0	2.0	2.0–4.0
E. aerogenes	4.0–25.0	2.0–10.0	0.5–2.0	1.4–2.0
K. pneumoniae	6.0–100.0	2.0–10.0	1.0–4.0	1.8
S. marcescens	>100.0	10.0	4.0	3.2
P. mirabilis	6.0–>100.0	2.0–>100.0	0.3–5.0	1.2–2.0
Indole + *Proteus*	6.0–>100.0	2.0–>100.0	2.0	–
P. aeruginosa	50.0	>100.0	1.0–4.0	3.0–6.0
P. pseudomallei	>200.0	25.6	50.0	–
Y. enterocolitica[64]	6.0–>100.0	6.0	–	–
B. abortus[63]	2.0	1.5	0.3	–
Salmonella spp.	4.0–16.0	3.0	0.8	–
Shigella spp.	3.0–10.0	5.0	2.0	–
Francisella tularensis[3]	0.4	–	–	–
Vibrio cholerae	20.0	–	–	–

*Tobramycin activity against gram-negative bacilli is similar to that of gentamicin except that tobramycin is twice as active against *P. aeruginosa* and less active against some *Serratia*.

– = Inadequate data.

TABLE 10. Sensitivity of Gram-Negative Bacilli to Rifampin

ORGANISM	USUAL MINIMUM INHIBITORY CONCENTRATIONS (μg/ml)
E. coli	2.5–10.0
P. mirabilis	1.0
Indole + *Proteus*	3.0
P. aeruginosa	10.0
Salmonella spp.	7.5
Shigella spp.	2.5
K. pneumoniae	10.0
P. pseudomallei	50.0
H. influenzae	0.2–0.8
H. ducreyi	0.02

centrations of 8 to 64 μg/ml in the absence of acquired resistance. Of the gram-negative rods, *P. aeruginosa* is most likely to be innately resistant to these concentrations. Acquired resistance to sulfonamides is so common among all bacteria, however, that sensitivity tests are usually needed to predict results.

Many gram-negative bacilli are also susceptible to nalidixic acid in concentrations of 20 to 50 μg/ml. *Klebsiella, E. coli, P. mirabilis,* and indole + *Proteus* organisms are often sensitive, while *P. aeruginosa* is uniformly resistant. Resistance tends to develop so rapidly, however, that it can appear overnight during treatment with nalidixic acid.[43, 54]

Metronidazole, a drug with interesting potential activity in anaerobic infections, can inhibit *B. fragilis* and other species of the family *Bacteroidaceae.* Concentrations of 3.1 μg/ml inhibit most *B. fragilis.*[59]

OTHER PATHOGENIC MICROORGANISMS

Mycobacteria and Nocardia

Mycobacteria show a different pattern of drug susceptibility than that of most bacteria (Table 11). Among antibiotics in general use in various bacterial infections, only the aminoglycosides and rifampin are active enough to warrant extensive use in tuberculosis and other mycobacterial infections.[28, 40] In fact, the most important antituberculosis drug, isoniazid, has no significant inhibitory effect on any other group of bacteria. The same is true of ethambutol and pyrazinamide. Despite its spectacular activity against human and bovine tubercle bacilli, isoniazid is inactive against most mycobacteria. Some other mycobacteria, especially *Mycobacterium kansasii,* are very sensitive to rifampin (but not rifamycin) and are moderately sensitive to streptomycin.[31] This organism is also sensitive in vitro to erythromycin. The major difference in susceptibility between the human and bovine tubercle bacillus is that *M. tuberculosis* var. *bovis* is resistant to pyrazinamide while *M. tuberculosis* var. *hominis* is sensitive. The slight sensitivity of the human tubercle bacillus to tetracycline has had some practical implications in combined drug therapy for preventing resistance to another antituberculous drug, but not for primary treatment.

Nocardia asteroides, the only important acid-fast bacillus other than the mycobacteria, is resistant to most antimycobacterial drugs. In general, strains of *Nocardia* show considerable variation in susceptibility to a given drug but are usually sensitive in vitro to the sulfonamides, tetracyclines, and cycloserine. The most consistently active antibiotic in vitro is minocycline, which inhibits 90 per cent of nocardial strains at a concentration of 3.1 μg/ml.[4] Erythromycin inhibits 40 per cent of strains at 0.8 μg/ml, but the others are resistant to >100 μg/ml. These in vitro results, based on standard testing methods, are probably not applicable to clinical therapy because the sulfonamides have been the most consistently successful drug for treating human nocardiosis even though *Nocardia* strains are highly resistant (>1600 μg/ml) to sulfonamides in vitro unless tiny inocula (<100 organisms) are used. Likewise, a number of patients have not responded to antibiotics that inhibited the infecting strain of *Nocardia* in low concentrations in vitro.

Mycoplasma

Mycoplasma pneumoniae, the cause of primary atypical pneumonia, is the only mycoplasma which has been proved to cause human infection, although Ureaplasma has been implicated in nongonococcal urethritis. Both organisms are resistant to drugs that affect the mucopeptide of bacteria because this structure is not present in mycoplasma. Their in vitro sensitivities are given in Table 12.[8, 39] Erythromycin and tetracycline are active against them, while lincomycin and clindamycin are ineffective.

Yeasts and Fungi

Amphotericin B, 5-fluorocytosine, nystatin, miconazole, and griseofulvin are the major drugs given for treating fungus infections. Amphotericin B, 5-fluorocytosine, and miconazole are active against the fungi causing deep mycoses, and griseofulvin against dermatophytes. Nystatin is used only topically. The sulfonamides, clotrimazole, and hydroxystilbamidine have some antifungal properties, but toxicity (clotrimazole,

TABLE 11. Drug Sensitivity of Mycobacteria

	USUAL MINIMUM INHIBITORY CONCENTRATIONS (μg/ml)				
	M. tuberculosis var. *hominis*	*M. kansasii*	*M. marinum*	*M. intracellulare*	*M. fortuitum*
Isoniazid	0.2	>25.0	>25.0	>25.0	>25.0
Rifampin	0.1	0.1–0.5	0.6	10.0	>20.0
Ethambutol	1.0	5.0	—	—	>20.0
Streptomycin	0.3–6.0	12.0–25.0	2.0–10.0	12.0–25.0	>100.0
Kanamycin	8.0	>5.0	5.0	—	—
Erythromycin	>32.0	1.0–2.0	—	32.0	>32.0
Tetracycline	10.0	—	—	—	—
PAS	1.0	>10.0	>2.0	—	>10.0
Pyrazinamide	18.0–20.0	>100.0	50.0	>100.0	—
Cycloserine	5.0–20.0	>50.0	50.0	—	>100.0
Gentamicin	3.0–6.0	3.0	—	3.0	25.0

— = Inadequate data.

TABLE 12. In Vitro Sensitivity of Mycoplasmas to Antibiotics

	USUAL MINIMUM INHIBITORY CONCENTRATIONS (μg/ml)	
	M. pneumoniae	*Ureaplasma*
Tetracycline	1.6	0.4
Erythromycin	0.025	1.6
Lincomycin	20.0	200.0
Chloramphenicol	12.0	1.6
Streptomycin	1.0	1.6
Polymyxin	—	500.0
Kanamycin	—	3.1
Gentamicin	—	6.2
Clindamycin	—	6.2–50.0

— = Inadequate data.

hydroxystilbamidine) or narrow spectrum (sulfonamides) greatly limit their use. Their activity against sensitive strains in vitro[1, 13, 14, 23, 27, 50, 51, 55, 61] is summarized in Table 13.

Griseofulvin inhibits virtually all dermatophytes at a concentration of less than 0.5 μg/ml. *Microsporum canis, M. gypseum, M. audouinii, Epidermophyton floccosum, Trichophyton mentagrophytes, T. rubrum, T. tonsurans, T. versicolor,* and all other clinically significant dermatophytes are susceptible in tube dilution tests to this concentration, or less, of griseofulvin.[45] Miconazole is less active against dermatophytes than griseofulvin, inhibiting most of these skin fungi at 32 μg/ml or less; it is especially active against *E. floccosum* (<0.125 μg/ml). Miconazole is useful when applied topically in ringworm.

SYNERGISM

A few important drugs are more active in the presence of another. The second drug can potentiate by increasing permeability. Thus, the aminoglycosides are more active with penicillin, while rifampicin, 5-fluorocytosine, and tetracycline are potentiated against fungi by amphotericin B. By injuring the mucopeptide of the cell wall, penicillin facilitates entry so that more of the aminoglycoside can reach the ribosomes.[37, 65] Amphotericin B alters permeability by binding with sterols so that more 5-fluorocytosine can enter the cells and inhibit nucleic acid synthesis through its action on pyrimidine biosynthesis.[36] This idea is the basis for the combined treatment of cryptococcosis with amphotericin B plus 5-fluorocytosine, and of enterococcal endocarditis with streptomycin and penicillin. More than 50 μg/ml streptomycin and more than 100 μg/ml

TABLE 13. Usual Minimum Inhibitory Concentrations (μg/ml)

	AMPHOTERICIN	5-FLUOROCYTOSINE	CLOTRIMAZOLE	NYSTATIN	MICONAZOLE
Cryptococcus neoformans	0.4–3.0	0.5–4.0	1.6	1.5	0.5
Coccidioides immitis	0.5	>100.0	0.4	1.5	3.0
Aspergillus fumigatus	>40.0	50.0–>100.0	1.6	3.0	8.0–16.0
Blastomyces dermatitidis	0.4–0.8	25.0	0.2	0.8	1.0
Candida albicans	2.0–4.0	0.5–500.0	4.0–16.0	3.0	30.0
Torulopsis glabrata	0.5–2.0	0.5–5.0	2.0–8.0	—	16.0–64.0
Histoplasma capsulatum	0.1–0.8	>100.0	3.0	1.5	0.5
Sporothrix schenckii	0.1–0.6	>100.0	—	12.0	2.0
Mucoraceae	0.03–>2.5	>100.0	0.8	12.0	1000.0

— = Inadequate data.

TABLE 14. Synergism between Trimethoprim and Sulfamethoxazole[9]

	USUAL MINIMUM INHIBITORY CONCENTRATIONS (μg/ml)*			
	Sulfamethoxazole Alone	Trimethoprim Alone	Sulfamethoxazole in Combination	Trimethoprim in Combination
Streptococcus pyogenes	100.0	1.0	1.0	0.05
Pneumococcus	30.0	2.0	2.0	0.1
S. aureus	3.0	1.0	0.3	0.015
H. influenzae	10.0	1.0	0.3	0.015
K. pneumoniae	100.0	3.0	4.0	0.2
E. coli	3.0	0.3	1.0	0.05
Salmonella typhimurium	10.0	1.0	0.3	0.05
Shigella sonnei	>10.0	1.0	0.3	0.05
N. gonorrhoeae	5.0	18.3	4.0	2.3
B. fragilis	6.0	25.0	4.0	0.2
Fusobacterium spp.	40.0	100.0	15.0	1.0
Peptostreptococcus spp.	>100.0	>100.0	30.0	2.0

*The levels of sulfamethoxazole usually reached in serum are 16 to 32 μg/ml and those of trimethoprim are 0.8 to 1.6 μg/ml.

penicillin are required to kill most enterococci, but many of these are killed by 6.25 μg of streptomycin in the presence of 6.25 μg penicillin or less.[62] In other words, one-fourth or less of the minimum bactericidal concentration of each drug alone killed enterococci in combination. Similar synergism against enterococci occurs with penicillin plus kanamycin and with penicillin plus gentamicin.

A second drug can also potentiate by reinforcing the metabolic disturbance of the first. This approach is used for synergism between trimethoprim and sulfonamides, two drugs that block two sequential steps in folinic acid synthesis.[9] Maximum potentiation occurs when the two drugs are present in proportions corresponding to their respective MIC when acting singly. Trimethoprim is usually about 20 times more active than sulfamethoxazole, the sulfonamide with which it is usually combined. When the two drugs are mixed according to this ratio (20 sulf:1 trimethoprim), they potentiate each other against many organisms, including streptococci, pneumococci, *S. aureus, H. influenzae, B. pertussis, K. pneumoniae, E. coli, Salmonella, Shigella, Proteus,* and gonococci. Examples are given in Table 14 and Figure 31, which give the minimum inhibitory concentrations of each drug when they are combined in a ratio of 1 part trimethoprim and 20 parts sulfamethoxazole.[9] The results show a po-

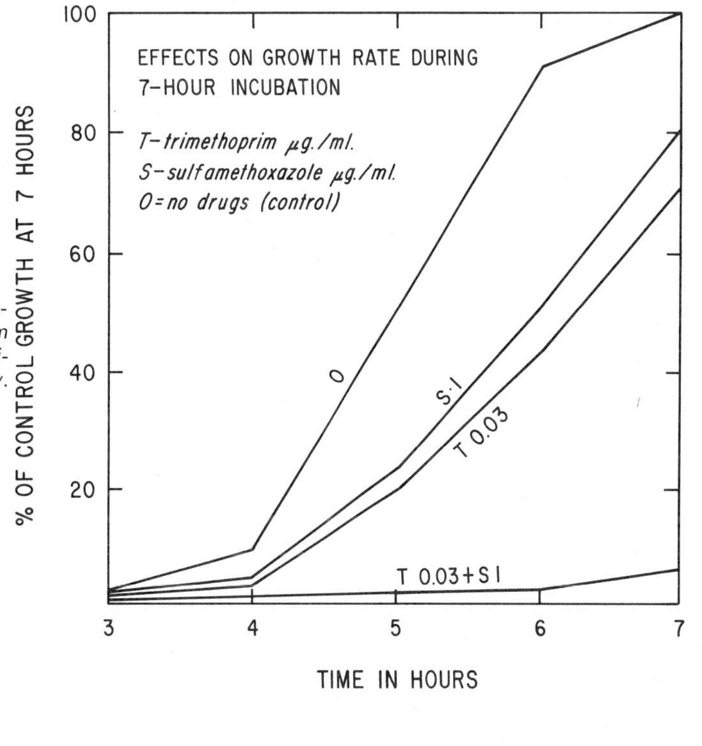

FIGURE 31. *Synergism between trimethoprim and sulfamethoxazole on growth inhibition of* E. coli. *(Modified from Bushby, S. R. M.: Combined antibacterial action* in vitro *of trimethoprim and sulphonamides. Postgrad Med J 45:17, Nov. 1969.)*

TABLE 15. Average Serum Concentrations After Single Oral Doses in Fasting Subjects

	DOSE (mg)	AVERAGE PEAK SERUM LEVEL (μg/ml)	AVERAGE SERUM HALF-LIFE (hours)	AVERAGE PERCENTAGE EXCRETED IN URINE IN 6 HOURS
Benzylpenicillin (penicillin G)	500	0.5	4.0	10.0
Phenoxymethyl penicillin (penicillin V)	500	2.9	2.0	35.0
Ampicillin	500	3.0	5.0	30.0
Amoxicillin	500	7.5	1.0	60.0
Oxacillin	500	2.6	1.5	20.0
Dicloxacillin	500	11.5	3.0	40.0
Cloxacillin	500	8.0	2.0	20.0
Indanyl carbenicillin	500	12.0	3.5	22.0
Cephalexin	500	17.0	2.0	95.0
Erythromycin base	500	1.0	3.0	0.5
Erythromycin base	250	0.35	5.5	0.5
Erythromycin stearate	500	1.1	5.0	0.5
Erythromycin stearate	250	0.4	6.0	0.5
Erythromycin estolate	500	2.5	3.5	0.9
Lincomycin	500	3.5	5.0	5.0
Clindamycin	450	4.5	3.5	5.0
Tetracycline	500	3.0	8.5	17.0
Tetracycline	250	2.2	8.0	25.0
Oxytetracycline	500	2.2	5.6	–
Chlortetracycline	500	1.4	8.0	–
Minocycline	300	4.0	15.0	2.0
Doxycycline	100	1.6	18.0	10.0
Doxycycline	200	2.8	20.0	10.0
Chloramphenicol	500	11.0	3.5	<5.0 (as active drug)
Rifampin	600	7.0	3.3	<5.0 (as active drug)
Sulfadiazine	–	–	–	7.0 (as active drug)
Sulfamethoxazole	800	53.8	12.0	10.0
Trimethoprim	160	1.58	12.0	16.0
Chloroquine	500	0.2	72.0	–
Isoniazid	600	8.5	2.6	30.0 (as active drug in 24 hr)
5-Fluorocytosine	2000	25.0	6.0	70.0 (24 hr)
Griseofulvin	500	1.25	–	<0.5
Neomycin	2000	0.95	8.0	0.85 (48 hr)
Ethambutol	1750	3.5	–	50.00 (24 hr)

– = Inadequate data.
See references 11, 21, 26, 35–38, 57, 60.

TABLE 16. Average Serum Concentrations After Single Intramuscular Dose

	DOSE (gm)	AVERAGE PEAK SERUM LEVEL (μg/ml)	AVERAGE SERUM HALF-LIFE (hours)	AVERAGE PERCENTAGE EXCRETED IN URINE IN 6 HOURS
Benzylpenicillin	0.6	12.0	0.5	70.0
Procaine penicillin G	0.2	1.0	18.0	70.0
Benzathine penicillin G	0.75	0.01	—	—
Ampicillin trihydrate	0.5	4.3	4.0	—
Sodium methicillin	1.0	16.0	1.5	66.0
Sodium oxacillin	0.5	10.9	—	—
Sodium nafcillin	0.5	—	—	10.0
Carbenicillin disodium	1.0	19.0	3.5	65.0
Sodium cephalothin	0.5	10.0	2.0	—
Sodium cefazolin	0.5	42.2	1.8	60.0
Cephaloridine	0.5	18.0	1.1	85.0 (24 hr)
Erythromycin ethylsuccinate	0.1	0.6	2.5	—
Clindamycin	0.6	5.1	8.0	15.0 (24 hr)
Lincomycin hydrochloride	0.6	9.5	8.0	30.0 (24 hr)
Streptomycin	0.5	18.0	5.0	60.0 (12 hr)
Kanamycin sulfate*	0.5	21.0	4.0	60.0
Gentamicin*	0.06	4.7	4.0	60.0 (24 hr)
Chloramphenicol sodium succinate	1.0	5.0	6.0	1.0 (as active drug)
Polymyxin B sulfate	0.05	3.0	7.0	—
Sodium colistimethate	0.15	3.0	3.0	—
Spectinomycin	2.0	100.0	5.0	—
Pentamidine isethionate	0.280	0.5	>24.0	15.0 (24 hr)

— = Inadequate data.
*The peak level, half-life, and urinary excretion of tobramycin are approximately the same as those of gentamicin, whereas these properties of amikacin are similar to those of kanamycin.
See references, 8, 10, 16, 21, 28, 32, 37, 44, 46, 48–50.

TABLE 17. Average Serum Concentrations After Single Intravenous Dose

	DOSE (gm)	AVERAGE PEAK SERUM LEVEL (μg ml)	AVERAGE SERUM HALF-LIFE (hours)	PERCENTAGE OF ACTIVE DRUG EXCRETED IN URINE IN 6 HOURS
Benzylpenicillin	0.5	5.5	0.15	90.0
Ampicillin	0.5	9.0	1.5	70.0
Sodium methicillin	0.5	16.2	0.7	80.0
Sodium oxacillin	0.5	60.0	0.4	45.0
Sodium nafcillin	0.5	40.0	0.5	35.0
Carbenicillin disodium	1.0	125.0	1.0	70.0
Sodium cephalothin	1.0	70.0	0.5	52.0 (24 hr)
Sodium cefazolin	0.5	118.0	1.8	95.0 (24 hr)
Sodium cephapirin	1.0	36.7	0.6	—
Erythromycin lactobionate	0.2	3.5	1.0	8.0
Clindamycin	0.6	8.5	3.0	20.0
Streptomycin	0.6	32.8	3.0	60.0 (12 hr)
Gentamicin sulfate	0.08	11.0	2.0	—
Amikacin	1.0	20.0	2.5	60.0
Sodium colistimethate	0.15	20.0	2.5	75.0
Doxycycline hyclate	0.2	4.0	18.0	71.4 (48 hr)
Tetracycline hydrochloride	0.5	8.5	8.5	45.0 (24 hr)
Oxytetracycline hydrochloride	0.5	7.0	9.2	60.0 (24 hr)
Chlortetracycline hydrochloride	0.5	9.0	5.6	15.0 (24 hr)
Amphotericin B	0.7	5.0	24.0	—
Chloramphenicol sodium succinate	1.0	11.0	3.5	5.0
Rifampin	0.6	7.0	3.3	5.0 (as active drug)
Vancomycin	0.5	33.0	6.0	80.0 (24 hr)
Sulfadiazine	—	—	—	7.0 (as active drug)
Sulfamethoxazole	0.8	53.8	10.0	10.0
Trimethoprim	0.16	1.58	10.0	16.0
Chloroquine	0.5	0.2	72.0	—
Isoniazid	0.6	8.5	6.0	30.0 (as active drug in 24 hr)
5-Fluorocytosine	2.0	25.0	6.0	70.0 (24 hr)
Griseofulvin	0.5	1.25	—	0.5
Miconazole	1.0	1.5	1.0	2.0

— = Inadequate data.
See references 21, 28, 30, 31, 44, 48, 50, 62.

tentiation of at least 20-fold for each drug against most organisms when used in combination.

SERUM CONCENTRATIONS OF ANTIMICROBIAL DRUGS

The aim of chemotherapy is to get enough drug into the infected tissues to inhibit or kill the pathogenic organism. In other words, the drug level must exceed the minimum inhibitory concentration. The levels of drug that can be anticipated in serum after different routes of administration are listed in Tables 15, 16, and 17. By comparing these levels to the minimum inhibitory concentrations given in the preceding tables, an estimate can then be made of appropriate drugs and dosage for a given infection. These estimates will be modified, in turn, by the pharmacologic considerations discussed in the next chapter. Renal failure, diffusion barriers, and the type of inflammatory reaction, if any, must be taken into account in selecting a drug and determining its dose. The interval between doses is a function of half-life, and the effect of renal failure can be predicted by the percentage of drug excreted in the urine.

References

Classification

1. Bodanszky, M., and Perlman, D.: Peptide antibiotics. Science 163:352, 1969.
2. Bueding, E.: Some biochemical effects of anthelmintic drugs. Biochem Pharmacol 18:1541, 1969.
3. Dutcher, J. D., Young, M. B., Sherman, J. H., et al.: Chemical studies on amphotericin B. I. Preparation of the hydrogenation product and isolation of mycosamine, an acetolysis product. Antibiot Annu 866, 1956–1957.
4. Fox, H.: The chemical attack on tuberculosis. Trans. NY Acad Sci 15:234, 1953.
5. Grollman, P.: Structural basis for inhibition of protein synthesis by emetine and cycloheximide based on an analogy between ipecac alkaloids and glutarimide antibiotics. Proc Nat Acad Sci USA 56:1867, 1966.
6. Hewitt, L.: The cephalosporins — 1973. J Infect Dis 128: Suppl:S312, 1973.
7. Hitchings, H.: Species differences among dihydrofolate reductases as a basis for chemotherapy. Postgrad Med J 45(Suppl):7, 1969.
8. Johnson, A., Anker, H., and Meleney, L.: Bacitracin; new antibi-

otic produced by member of *B. subtilis* group. Science 102:376, 1945.

9. Lacroute, F.: Regulation of pyrimidine biosynthesis in Saccharomyces cerevisiae. J. Bacteriol 95:824, 1968.
10. Long, H., and Bliss, A.: Para-amino-benzene-sulfonamide and its derivatives; experimental and clinical observations on their use in treatment of beta-hemolytic streptococcic infection: preliminary report. JAMA 108:32, 1937.
11. Nair, S., and Cherubin, C.: Use of cefoxitin, new cephalosporin-like antibiotic, in the treatment of aerobic and anaerobic infections. Antimicrob Agents Chemother 14:866, 1978.
12. Neu, H. D.: Antimicrobial activity and human pharmacology of amoxicillin. J Infect Dis 129 (Suppl):123, 1974.
13. Novick, P.: Staphylococcal penicillinase and the new penicillins. Biochem J 83:229, 1962.
14. Oxford, E., Raistrick, H., and Simonart, P.: XXIX. Studies in the biochemistry of microorganisms: LX: Griseofulvin, $C_{17}H_{17}O_6Cl$, a metabolic product of *Penicillium griseofulvum* Dierckx. Biochem J 33:240, 1939.
15. Sheldrick, G., Jones, P., Kennard, O., Williams, D., and Smith, G.: Structure of vancomycin and its complex with acetyl-D-alanyl-D-alanine. Nature 271:223, 1978.
16. Wehrli, W., and Staehelin, M.: Actions of the rifamycins. Bacteriol Rev 35:290, 1971.

Microbial Susceptibility to Drugs

1. Artis, D.,and Baum, G. L.: In vitro susceptibility of 24 strains of *Histoplasma capsulatum* to amphotericin B. Antibiot Chemother 11:373, 1961.
2. Atlas, E. and Turck, M.: Laboratory and clinical evaluation of rifampicin. Am J Med Sci 256:47, 1968.
3. Avery, F. W., and Barnett, T. B.: Pulmonary tularemia. A report of five cases and consideration of pathogenesis and terminology. Am Rev Resp Dis 95:584, 1967.
4. Bach, M.C., Sabath, L. D., and Finland, M.: Susceptibility of *Nocardia asteroides* to 45 antimicrobial agents in vitro. Antimicrob Agents Chemother 3:1, 1973.
5. Bauer, A. W., Kirby, W. M., Sherris, J. C., et al.: Antibiotic susceptibility testing by a standardized single disk method. Am J Clin Pathol 45:493, 1966.
6. Benner, E. J.: The cephalosporin antibiotics. Pediatr Clin North Am 15:31, 1968.
7. Braude, A. I., Banister, J., and Wright, N.: Use of the gradient plate for routine clinical determinations of bacterial sensitivites to antibiotics. Antibiot Annu 1133, 1954–1955.
8. Braun, P., Klein, J. O., and Kass, E. H.: Susceptibility of genital mycoplasmas to antimicrobial agents. Appl Microbiol 19:62, 1970.
9. Bushby, R. S.: Combined antibacterial action in vitro of trimethoprim and sulphonamides. The in vitro nature of synergy. Postgrad Med J 45 (Suppl):10, 1969.
10. Butler, K., English, A. R., Ray, V. A., et al.: Carbenicillin: chemistry and mode of action. J Infect Dis 122:(Suppl):51, 1970.
11. Cave, V. G., Hurdle, E. S., and Catelli, A. R.: Sensitivity of *Neisseria gonorrhoeae* to penicillin and other drugs. NY State J Med 70:844, 1970.
12. Chadwick, P.: Bacteriological assessment of clindamycin, a new lincomycin derivative. J Med Microbiol 4:529, 1971.
12a. Dixon, D., Shadomy, S., Shadomy, H., Espinel-Ingroff, A., and Kerkering, T.: Comparison of the *in vitro* antifungal activities of miconazole and a new imidazole, R 41,400. J Infect Dis 138:245, 1978.
13. Drouhet, E.: Basic mechanisms of antifungal chemotherapy. Mod Treat 7:539, 1970.
14. Drutz, D. J., Spickard, A., Rogers, D. E., et al.: Treatment of disseminated mycotic infections. A new approach to amphotericin B therapy. Am J Med 45:405, 1968.
15. Edmondson, E. B., and Sanford, J. P.: The Klebsiella-Enterobacter (Aerobacter)-Serratia group. A clinical and bacteriological evaluation. Medicine (Balt) 46:323, 1967.
16. Eickhoff, T. C., Bennett, J. V., Hayes, P. S., et al.: *Pseudomonas pseudomallei:* susceptibility to chemotherapeutic agents. J Infect Dis 121:95, 1970.
17. Ericsson, H., Hogman, C., and Wickman, K.: Paper disc method for determination of bacterial sensitivity to chemotherapeutic and antibiotic agents. Scand J Clin Lab Invest 6 (Suppl):21, 1954.
18. Finland, M.: Changing patterns of susceptibility of common bacterial pathogens to antimicrobial agents. Ann Intern Med 76:1009, 1972.
19. Friedberg, C. K., Rosen, K. M., and Bienstock, P. A.: Vancomycin therapy for enterococcal and *Streptococcus viridans* endocarditis. Successful treatment of six patients. Arch Intern Med (Chicago) 122:134, 1968.
20. Griffith, R. S., and Black, H. R.: Erythromycin. Med Clin North Am 54:1199, 1970.
21. Guttler, R. B., Counts, G. W., Avent, C. K., et al.: Effect of rifampin and minocycline on meningococcal carrier rates. J Infect Dis 24:199, 1971.
22. Hammerstrom, C. F., Cox, F., McHenry, M. C., et al.: Clinical laboratory, and pharmacological studies of dicloxacillin. Antimicrob Agents Chemother 69, 1966.
23. Hildick-Smith, G.: Antifungal antibiotics. Pediatr Clin North Am 15:107, 1968.
24. Judson, F. N., Allaman, J., and Dans, P. E.: Treatment of gonorrhea — Comparison of penicillin G procaine, doxycycline, spectinomycin, and ampicillin. JAMA 230:705, 1974.
25. Kayser, F. H.: In vitro activity of cephalosporin antibiotics against gram-positive bacteria. Postgrad Med J 47 (Suppl):14, 1971.
26. Kunin, C. M., Brandt, D., and Wood, H.: Bacteriologic studies of rifampin, a new semisynthetic antibiotic. J Infect Dis 119:132, 1969.
27. Larsh, H. W., Hinton, A., and Silberg, S. L.: The use of the tissue culture method in evaluating antifungal agents against systemic fungi. Antibiot Annu 988, 1957–1958.
28. Lester, W.: Rifampin: a semisynthetic derivative of rifamycin — a prototype for the future. Am Rev Microbiol 26:85, 1972.
29. Levison, M.E., Johnson, W. D., Thornhill, T. S., et al.: Clinical and in vitro evaluation of cephalexin. A new orally administered cephalosporin antibiotic. JAMA 209:1331, 1969.
30. Lindenbaum, J., Greenough, W. B., and Islam, M. R.: Antibiotic therapy of cholera. Bull WHO 36:871, 1967.
31. Lorian, V., and Finland, M.: In vitro effect of rifampin on mycobacteria. Appl Microbiol 17:202, 1969.
32. Martin, J. E., Jr., Lester, A., Kellogg, D. S., Jr., et al.: In vitro susceptibility of *Neisseria gonorrhoeae* to nine antimicrobial agents. Appl Microbiol 18:21, 1969.
33. Martin, W. J., Gardner, M., and Washington, J. A. II: In vitro antimicrobial susceptibility of anaerobic bacteria isolated from clinical specimens. Antimicrob Agents Chemother 1:148, 1972.
34. Matsen, J. M., Blazevic, D. J., and Chapman, S. S.: In vitro susceptibility patterns of beta-hemolytic streptococci. Antimicrob Agents Chemother 485, 1969.
35. McLinn, S. E., Nelson, J. D., and Haltalin, K. C.: Antimicrobial susceptibility of *Hemophilus influenzae.* Pediatrics 45:827, 1970.
36. Medoff, G., Comfort, M., and Kobayashi, G. S.: Synergistic action of amphotericin B and 5-fluorocytosine against yeast-like organisms. Proc Soc Exp Biol Med 138:571, 1971.
37. Moellering, R. C., Jr., Wennersten, C., and Weinberg, A. N.: Studies on antibiotic synergism against enterococci. I. Bacteriologic studies. J Lab Clin Med 77:821, 1971.
38. Neu, H.: Antimicrobial activity and human pharmacology of amoxicillin. J Infect Dis 129 (Suppl):S123, 1974.
39. Niitu, Y., Hasegawa, S., Suetake, T., et al.: Resistance of *Mycoplasma pneumoniae* to erythromycin and other antibiotics. J Pediatr 76:438, 1970.
40. Raleigh, J. W.: Rifampin in treatment of advanced pulmonary tuberculosis. Report of a VA cooperative pilot study. Am Rev Respir Dis 105:397, 1972.
41. Ribble, J. C.: Laboratory assistance in the treatment of bacterial infections. Pediatr Clin North Am 18:115, 1971.
42. Ronald, A. R., Eby, J., and Sherris, J. C.: Susceptibility of *Neisseria gonorrhoeae* to penicillin and tetracycline. Antimicrob Agents Chemother 431, 1968.
43. Ronald, A. R., Turck, M., and Petersdorf, R. G.: A critical evaluation of nalidixic acid in urinary-tract infections. N Engl J Med 275:1081, 1966.
44. Rosen, I. G., Jacobsen, J., and Rudderman, R.: Rapid capillary tube method for detecting penicillin resistance in *Staphylococcus aureus.* Appl Microbiol 23:649, 1972.
45. Roth, F. J., Sallman, B., and Blank, H.: In vitro studies of the antifungal antibiotic griseofulvin. J Invest Dermatol 33:403, 1959.
46. Sabath, L. D., Stumpf, L. L., Wallace, S. J., et al.: Susceptibility

of *Diplococcus pneumoniae, Hemophilus influenzae,* and *Neisseria meningitidis* to 23 antibiotics. Antimicrob Agents Chemother 53, 1970.
47. Sabath, L. D., Wilcox, C., Garner, C., et al.: In vitro activity of cefazolin against recent clinical bacterial isolates. J Infect Dis 128 (Suppl):S320, 1973.
48. Sanders, E., Foster, M. T., and Scott, D.: Group A beta-hemolytic streptococci resistant to erythromycin and lincomycin. N Engl J Med 278:538, 1968.
49. Sensi, P., Maggi, N., Füresz, S., et al.: Chemical modifications and biological properties of rifamycins. Antimicrob Agents Chemother 699, 1966.
50. Shadomy, S.: Further in vitro studies with 5-fluorocytosine. Infect Immun 2:484, 1970.
51. Shadomy, S., Kirchoff, C. B., and Ingroff, A. E.: In vitro activity of 5-fluorocytosine against *Candida* and *Torulopsis* species. Antimicrob Agents Chemother 3:9, 1973.
52. Simon, H. J.: Streptomycin, kanamycin, neomycin, and paromomycin. Pediatr Clin North Am 15:73, 1968.
53. Smith, C. B., Wilfert, J. N., Dans, P. E., et al.: In-vitro activity of carbenicillin and results of treatment of infections due to *Pseudomonas* with carbenicillin singly and in combination with gentamicin. J Infect Dis 122 (suppl):S14, 1970.
54. Stamey, T. A., Nemoy, N. J., and Higgins, M.: The clinical use of nalidixic acid. A review and some observations. Invest Urol 6:582, 1969.
55. Sterr, P. L., Marks, M. I., Klite, P. D., et al.: 5-Fluorocytosine: an oral antifungal compound. A report on clinical and laboratory experience. Ann Intern Med 76:15, 1972.
56. Steigbigel, N. H., McCall, C. E., Reed, C. W., et al.: Antibacterial action of "broad spectrum" penicillins, cephalosporins and other antibiotics against gram-negative bacilli isolated from bacteremic patients. Ann NY Acad Sci 145:224, 1967.
57. Steigbigel, N. H., Reed, C. W., and Finland, M.: Susceptibility of cillin, a new isoxazolyl penicillin, compared with oxacillin, in vitro. Am J Med Sci 255:179, 1968.
58. Sutherland, R., Croydon, E. A., and Rolinson, G. N.: Flucloxacilin, a new isoxazolyl penicillin, compared with oxacillin, cloxacillin, and dicloxacillin. Br Med J 4:455, 1970.
59. Tally, F. P., Sutter, V. L., and Finegold, S. M.: Metronidazole versus anaerobes. In vitro data and initial clinical observations. Calif Med 117:22, 1972.
60. Toala, P., McDonald, A., Wilcox, C., et al.: Comparison of antibiotic susceptibility of group D streptococcus strains isolated at Boston City Hospital in 1953–54 and 1968–69. Antimicrob Agents Chemother 479, 1969.
61. Watson, D. C., and Neame, P. B.: In vitro activity of amphotericin B on strains of *Mucoraceae* pathogenic to man. J Lab Clin Med 56:251, 1960.
62. Wilkowske, C. J., Facklam, R. R., Washington, J. A., II., and Geraci, J.: Antibiotic synergism: enhanced susceptibility of group D streptococci to certain antibiotic combinations. Antimicrob Agents Chemother 195, 1970.
63. Yow, E. M., and Spink, W. W.: Symposium on antibiotics; experimental studies on the action of streptomycin, aureomycin, and chloromycetin on brucella. J Clin Invest 28:871, 1949.
64. Zen-Yoji, H., and Maruyama, T.: The first successful isolations and identification of *Yersinia enterocolitica* from human cases in Japan. Jap J Microbiol 16:493, 1972.
65. Zimmermann, R. A., Moellering, R. C., Jr., and Weinberg, A. N.: Mechanism of resistance to antibiotic synergism in enterococci. J Bacteriol 105:873, 1971.

20 MECHANISMS OF ACTION OF ANTIMICROBIAL DRUGS

Abraham I. Braude, M.D., Ph.D.

The key to antibiotic action is a selective toxicity for the infecting organism but not for the patient. Antibiotics can hit at least four targets in bacteria and other organisms that are either missing or less vulnerable in human cells: the cell wall, the cytoplasmic membrane, the ribosomes, and the molecules involved in transcription of genetic information.

CELL WALL

The cell wall is where the penicillins and cephalosporins do their damage. It is the most important target for antibiotics because it is absent in human cells. The cell wall of bacteria is a thick rigid envelope that surrounds the cell membrane, maintaining the shape of the bacteria and keeping them from osmotic damage in water and body fluids. The internal pressures of pathogenic bacteria are somewhat higher than serum and other extraceullar fluids so that the organisms would imbibe water, swell, and burst if the cell wall became defective. The rigid portion of the cell wall, known as the sacculus, is reminiscent of grape hulls in cocci and balloons in bacilli. The rigid material in the sacculus is a material composed of sugar and peptides and called mucopeptide or murein. The mucopeptide is at least four times thicker in gram-positive cells than in gram-negative cells. In *E. coli* the layer of mucopeptide is made up of polysaccharide chains linked together by peptides. The polysaccharide chains are composed of repeating units of two sugars, muramic acid and *N*-acetylglucosamine, as indicated in Figure 1. The cross-linking peptide strands are attached to the polysaccharide chains by a peptide bond between the carboxyl group in each muramic acid unit and the amino group of L-alanine. Another peptide bond connects L-alanine to D-glutamic acid, followed in turn by α, ϵ-diaminopimelic acid, and finally D-alanine to form a tetrapeptide. The free carboxyl groups on the D-alanine residues are also attached to the NH_2 group of the α, ϵ-diaminopimelic acid in adjacent tetrapeptides to produce a cross-linked

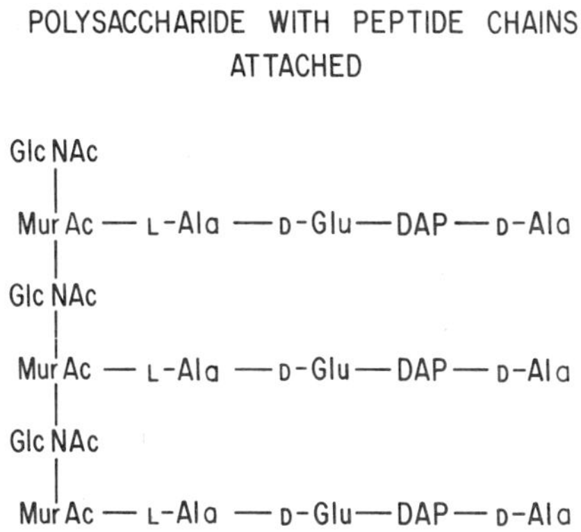

FIGURE 1. *The rigid structure in bacterial cell walls is a sugar peptide called mucopeptide or murein. The sugar is a polysaccharide of muramic acid (MurAc) and glucosamine (GlcNAc). The peptides are composed of four amino acids: L-alanine (L-Ala), D-glutamic acid (D-Glu), α, ε-diaminopimelic acid (DAP), and D-alanine.*

structure (Fig. 2). In staphylococci a second chain composed of five glycine molecules is used to connect neighboring peptides so that the structure is more like that in Figure 3. The peptide group in staphylococci is also different from that in *E. coli,* as shown in Figure 4. The staphylococcal peptide lacks diaminopimelic acid and, instead, has L-lysine as the third amino acid. This is followed by D-alanyl-D-alanine at the end of the chain. The murein network in the *Staphylococcus* is completed when D-alanine is split from D-alanyl-D-alanine so that a peptide bond can form between the carboxyl group of the residual D-alanine and the terminal amino group of the pentaglycine chain (Fig. 4).

Penicillin is thought to prevent the final peptide bond between D-alanine and glycine.[30] It has been suggested that penicillin combines with the enzyme responsible for this final cross linkage (the cross-linking enzyme). Since the stearic configuration of penicillin is like that of D-alanyl-D-alanine, penicillin might react with the cross-linking enzyme and inactivate it so that it could not complete the transpeptidation reaction (Fig. 5). By preventing this final step in murein synthesis, penicillin seems to have at least two deleterious effects on bacteria: (1) it inhibits multiplication; and (2) it creates weak points in the cell wall through which the growing cytoplasm can bulge. There is also evidence that penicillin binds to proteins that suppress murein hydrolases. It has been suggested that binding of penicillin by the proteins releases hydrolases from suppression, allows them to destroy the mucopeptide, and thus kills the bacterium by autolysis.[29]

An important feature of penicillin action is its ability to kill growing bacterial cells, but not stationary organisms. During growth of certain bacteria hydrolases seem to produce gaps in the mucopeptide that are filled with new structural units. These units are tied into the mucopeptide by the transpeptidation (cross-linking) reaction and can therefore be blocked by penicillin so that open gaps remain in the mucopeptide. The cell membrane extends through these gaps and ruptures under osmotic stress, and the cell dies (Fig. 6). Cells that are not undergoing multiplication, or cells that have no murein hydrolases ("penicillin-tolerant"), can survive in the presence of penicillin because their mucopeptide is unbroken and there is no reparative cross-linking activity for the penicillin to block. These cells are known as "persisters" and may be responsible for recurrence of infection after penicillin treatment has been stopped.[12] Penicillin and cephalosporin derivatives attack bacteria by similar mechanisms.

While penicillins block the terminal cross-linking reaction of mucopeptide formation, it is also theoretically possible for antibiotics to prevent the synthesis or transfer of mucopeptide precursors. Such action against precursors by certain antibiotics does, in fact, take place but these are of minor clinical importance. Cycloserine is a structural analog of D-alanine and competitively inhibits the enzyme responsible for synthesis of D-alanyl-D-alanine, an essential component of the pentapeptide.[22] Bacitracin[28] and vancomycin[1] both block the stages of cell wall construction which involve the transfer of the sugar pentapeptide from the site of synthesis in

FIGURE 2. *Cross linking of peptide chains in gram-negative bacilli by bonds between the free carboxyl groups of D-alanine and the NH_2 groups of α,ε-diaminopimelic acid. (Modified from Rogers, H. J.: The mode of action of antibiotics. In Bittar, E. E. (ed.): The Biological Basis of Medicine. Vol. 2. New York, Academic Press, 1968, p. 427.)*

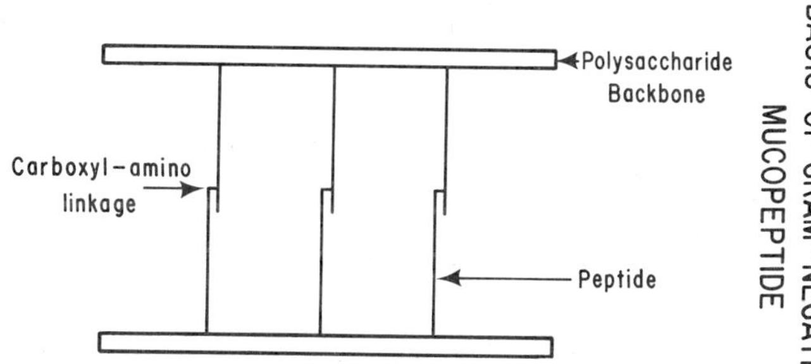

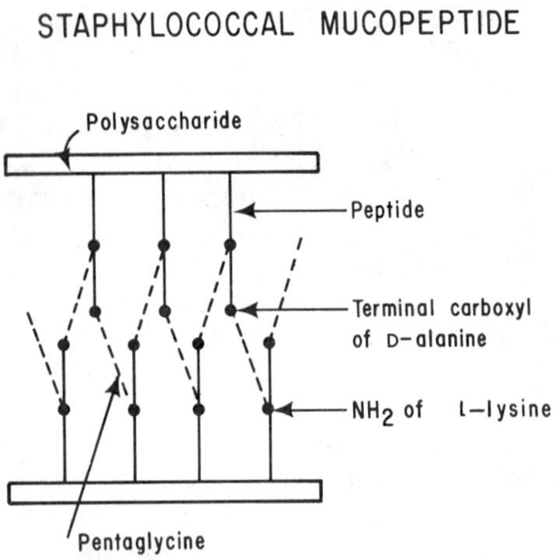

FIGURE 3. *In staphylococci a second chain composed of 5-glycine (pentaglycine) molecules connects the neighboring peptides. (Modified from Rogers, H. J.: The mode of action of antibiotics. In Bittar, E. E. (ed.): The Biological Basis of Medicine. Vol. 2. New York, Academic Press, 1968, p. 428.)*

the cytoplasm to its attachment to a lipid in the cell membrane.

CYTOPLASMIC MEMBRANE

Beneath the rigid cell wall is a membrane that totally encloses the cytoplasm (Fig. 7*A*). This cytoplasmic membrane resembles that of human cells in possessing lipid and protein structural elements. Bacterial lipids are mainly phospholipids. Fungi contain sterols in their membranes that are not present in bacteria.

The lipoproteins in the cytoplasmic membrane of all cells account for selective permeability to water, ions, and nutrients. The polymyxins are cationic detergents that react with the phosphate groups of cell envelope phospholipids and disorganize the lipoproteins in the bacterial cytoplasmic membrane by inserting the lipophilic portion of their molecule into the membrane lipid. This causes leakage of amino acids, purines, pyrimidines, and other small molecules from inside the cell so that nucleic acids and proteins break down and the cell dies.[11, 23]

Amphotericin and other polyene antibiotics also alter the permeability of sensitive cells, but only in organisms whose membranes contain sterols. For this reason they do not affect bacteria and are limited in their action against pathogenic organisms to yeasts, fungi (Fig. 7*B*), and certain amebae. After reacting with polyenes, sterols probably become reoriented within the membranes so that permeability is altered.[18] If sterol synthesis is blocked in fungi, the polyenes lose their target and are no longer effective. This occurs in fungi exposed to miconazole, which inhibits ergosterol synthesis and thus interferes with the antifungal action of amphotericin B. For this reason amphotericin is of no value in patients under treatment with miconazole.[32] By itself, however, miconazole and other imidazoles impair membrane structure and function of fungal membranes by depriving them of ergosterol. Electron microscopy in vitro and in vivo discloses that the first effect of imidazole on fungi is a disturbance in the cell membrane structure.[33]

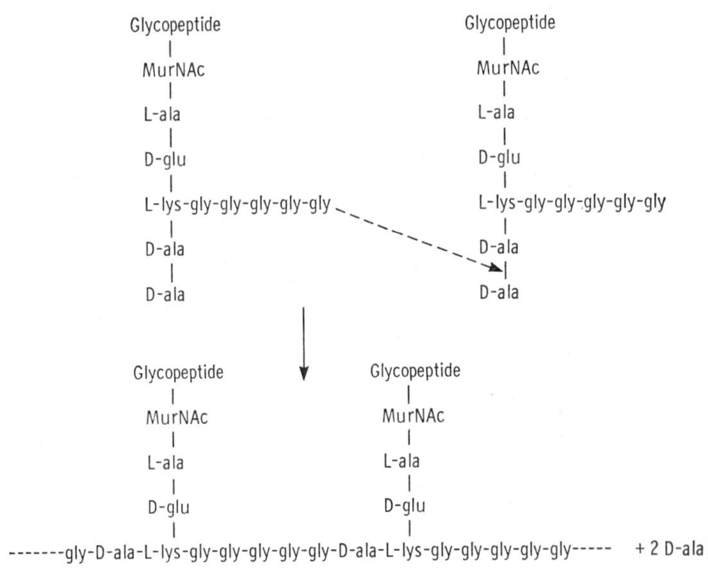

FIGURE 4. *Mucopeptide from staphylococci. The peptide chain differs from that in* E. coli *(Fig. 1) in that staphylococcal peptide lacks diaminopimelic acid and instead has L-lysine as the third amino acid. The murein network is completed when D-alanine is split from D-alanyl-D-alanine so that a peptide bond can form between the carboxyl group of the residual D-alanine and the terminal group of the pentaglycine chain. Penicillin is thought to act by inhibiting the enzyme (transpeptidase) responsible for this cross linkage. (From Strominger, J. L.: Enzymatic reactions in bacterial cell wall synthesis sensitive to penicillins and other antibacterial substances. In Guze, L. B. (ed.): Microbial Protoplasts, Spheroplasts and L-Forms. © 1968, The Williams & Wilkins Co., Baltimore, p. 57.)*

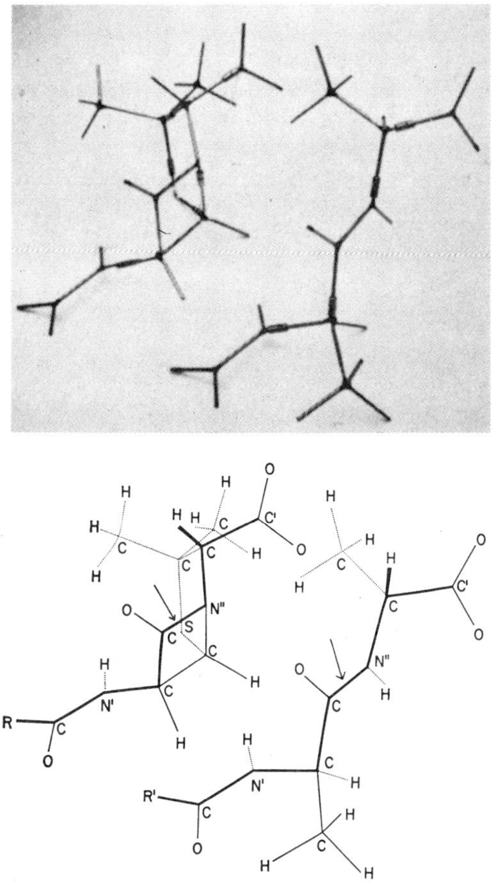

FIGURE 5. *Stereomodels of penicillin (left) and the end of the peptide* D-alanyl-D-alanine *have suggested that penicillin is a structural analog of the* D-alanyl-D-alanine *end of the peptide and thereby can react with the transpeptidase to prevent the transpeptidation reaction required for closure of the glycine bridges between peptide chains. (From Strominger, J. L.: Enzymatic reactions in bacterial cell wall synthesis sensitive to penicillins and other antibacterial substances. In Guze, L. B. (ed.): Microbial Protoplasts, Spheroplasts and L-Forms. ©1968, The Williams & Wilkins Co., Baltimore, p. 58.)*

RIBOSOMES

Ribosomes act as an assembly line where amino acids are strung together in peptide chains and proteins. The process is directed by messenger RNA (mRNA), which carries the code for protein synthesis from nuclear DNA. The message in the code is *transcribed* from DNA to RNA (Fig. 8*A*) and translated into the appropriate amino acid sequence by the four types of ribonucleotides in mRNA. These four ribonucleotides are prearranged in 4^3 (or 64) different triplet combinations that are able to specify different amino acids. Since there are 64 triplets and only 20 amino acids used in protein synthesis, many amino acids are selected by more than one triplet (or codon). Three of the codons (UAA, UAG, and UGA) code for chain termination only and do not select amino acids. Each amino acid specified by the triplet is carried to the ribosome for incorporation into the growing peptide chain by a second type of RNA, transfer RNA (tRNA). The triplet AUG (adenylyl-uridylyl-guanylyl) or GUG initiates peptide chain formation by directing tRNA carrying methionine (as N-formylmethionine) to attach to the ribosome.

The bacterial ribosomes are spherical particles with a molecular weight of nearly 3 million. They sediment in the Svedberg ultracentrifuge at a rate expressed as 70S (Svedberg units). Before protein synthesis is started, the ribosome dissociates into a 30S subunit and a 50S subunit (Fig. 8*B*). Protein synthesis starts when mRNA attaches to the 30S subunit and tRNA carrying methionine is bound (Fig. 8*C*). This is followed by recombination with the 50S subunit to form the functioning 70S ribosome (Fig. 8*D*). The complete 70S ribosome has another binding site for tRNA, so that a second tRNA molecule carrying another

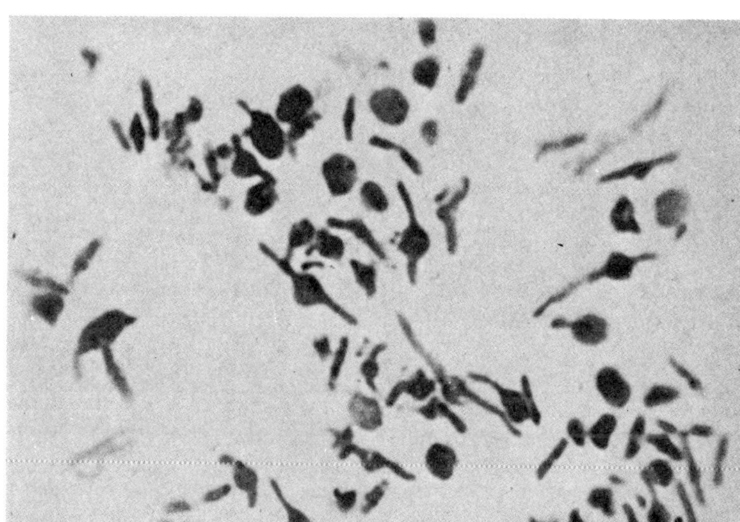

FIGURE 6. *Effect of penicillin on cell wall of* Klebsiella pneumoniae. *Defective portions of the mucopeptide allow the high internal osmotic pressure to cause swellings in both the central and terminal portions of the rods. (From Braude, A. I., Siemienski, J., and Jacobs, I.: Protoplast formation in human urine. Trans Assoc Am Physicians 74:238, 1961.)*

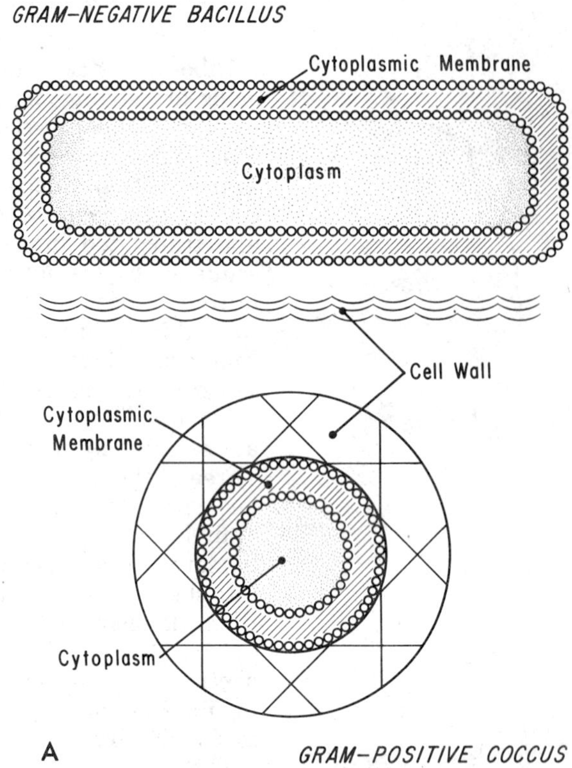

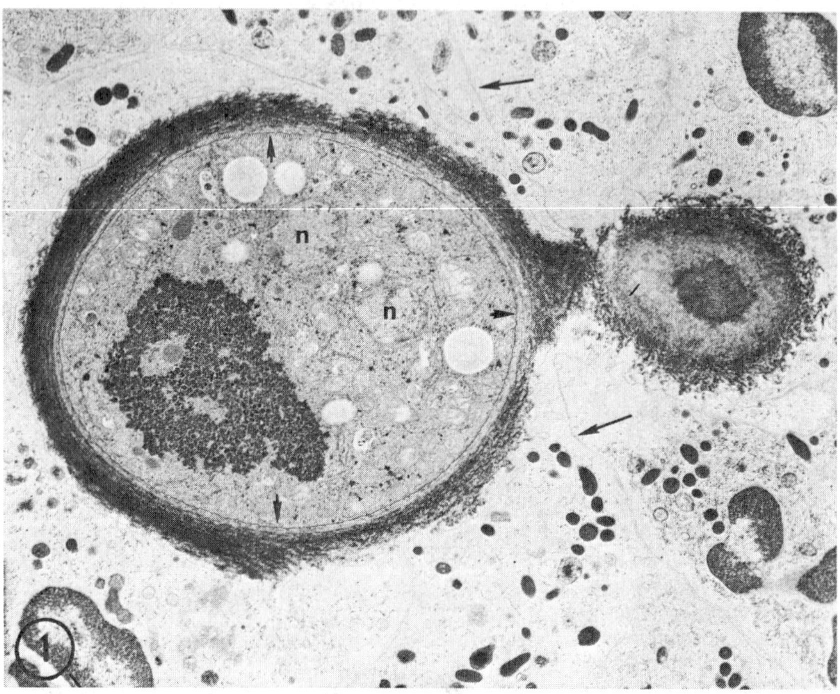

FIGURE 7. A, *Bacterial cell wall and underlying cytoplasmic membrane. The circles indicate the protein, and the adjacent diagonal lines represent the lipid in the lipoprotein cytoplasmic membrane. The cytoplasmic membrane is drawn out of proportion to its true relationship to the rest of the cell for purposes of illustration. In gram-negative bacilli the cytoplasmic membrane is the site of action of the polymyxins, but in gram-positive cocci it is not damaged by those drugs in usual doses. Differences in accessibility to the membrane, related to differences in cell wall, may account for these differences in susceptibility between gram-positive and gram-negative organisms.*

B, *Effect of amphotericin B on cytoplasmic membrane of fungi. 1, 2, 3, and 4 show* in vivo *effect of amphotericin during treatment of human North American blastomycosis. (Courtesy Dr. Henry C. Powell.)*

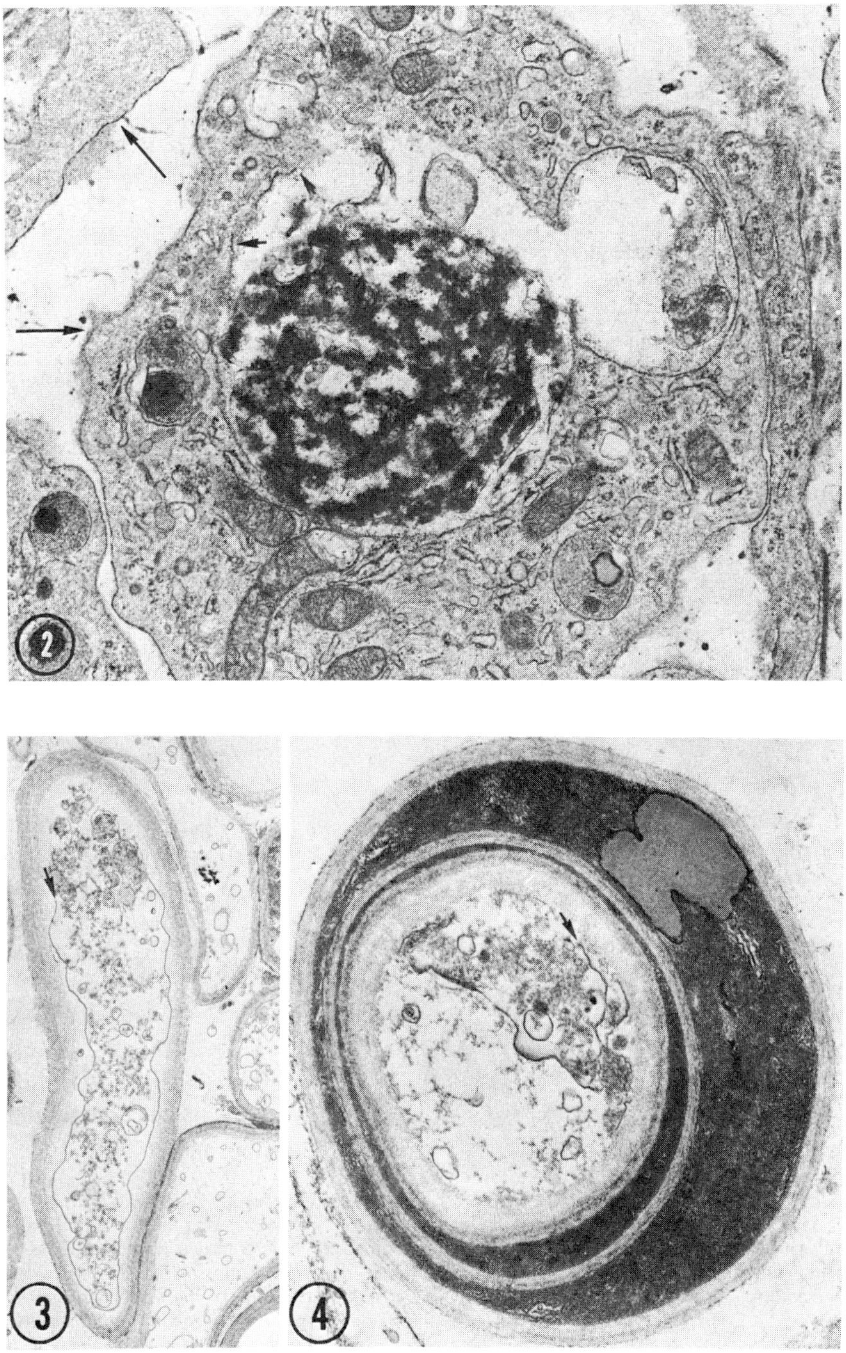

FIGURE 7 continued.

1. Skin biopsy, North American blastomycosis. Budding organism: the parent cell shows two nuclei (n), numerous mitochondria, and aggregates of electron-dense material (probably glycogen) enclosed by cell membrane (arrowheads) and the rough outer coat. The surfaces of surrounding macrophages are closely apposed (arrows). ×20,000. 2. Skin biopsy, North American blastomycosis, after amphotericin treatment. Rupture of the cell membrane (arrowhead) and cytolysis. *×20,000. 3. North American blastomycosis. Fungal cells* in vitro *after incubation with amphotericin and 5-fluorocytosine. Note lysis of the cell membranes (arrow) and cytoplasmic degeneration. ×7000. 4. Degenerating fungal cell with electron-dense deposits in the cell wall. ×20,000.*

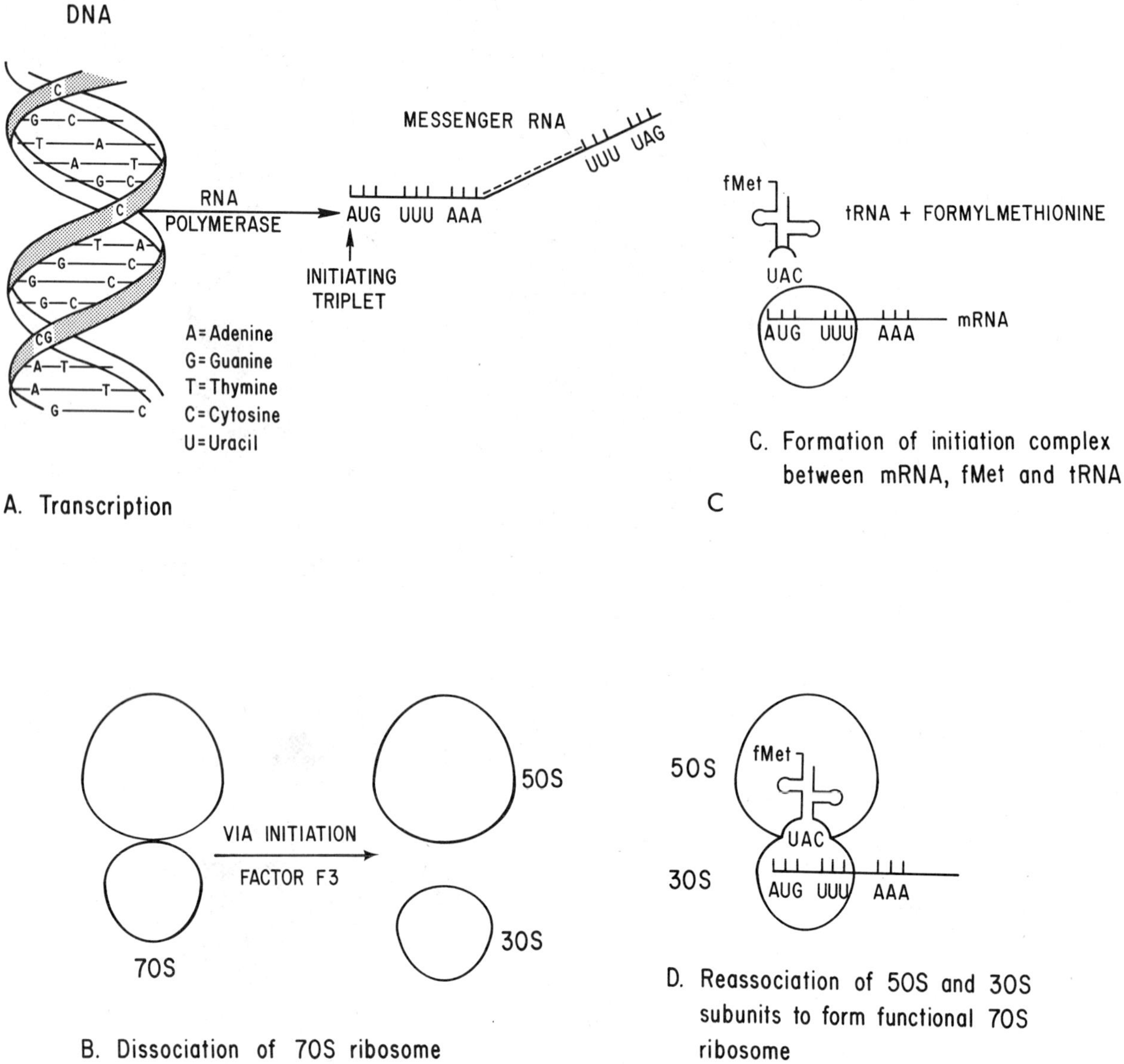

FIGURE 8. A, *Messenger RNA receives the code for amino acid sequence (and thus protein synthesis) from DNA in a process known as transcription. The message is carried to the ribosomes in the form of nucleotide triplets. Since there are four nucleotides in mRNA, they can be arranged in* 4^3*, or 64, different triplet combinations, and each can specify one of the 20 amino acids. As each triplet reaches the ribosome it directs the attachment of a specific amino acid. Rifampin, chloroquine, 5-iodo-2′-deoxyuridine, 5-fluorocytosine, sulfonamides, pyrimethamine, and trimethoprim all interfere with transcription by one or more mechanisms as described in the text.* B, *Dissociation of the 70S ribosome is the first step in protein synthesis on ribosomes.* C, *Protein synthesis starts when mRNA attaches to the 30S subunit and then tRNA + formylmethionine are bound. The 50S subunit then reassociates and the initiation complex is completed. Streptomycin binds the 30S subunit and inactivates the initiation complex so that it cannot form peptide bonds. The tetracyclines also bind to the 30S subunit and prevent binding of tRNA.* D, *Reassociation of 50S and 30S subunits to form functional 70S ribosome.*

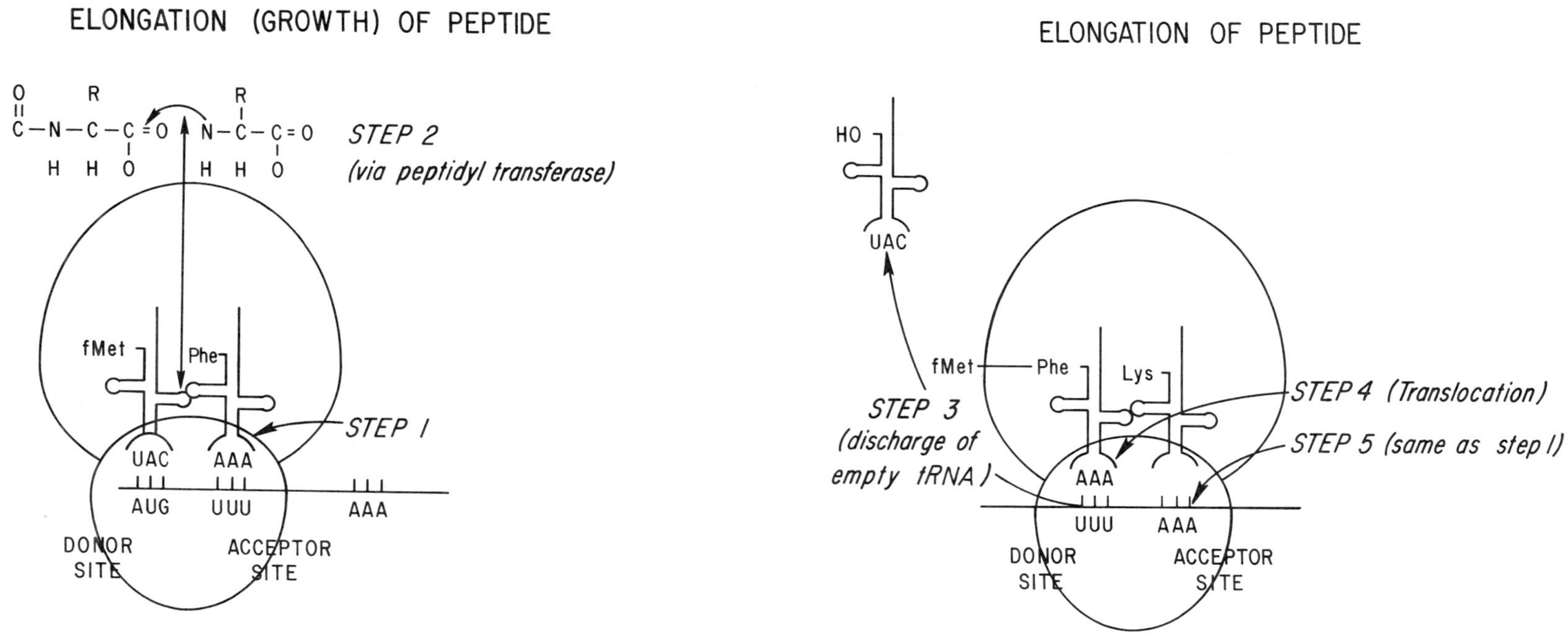

STEP I: Binding of tRNA + 2nd amino acid (e.g. phenylalanine) to acceptor site of 70S ribosome, as directed by 2nd triplet on mRNA (e.g. UUU).

STEP 2: Formation of peptide bond by reaction of amino group of newly bound amino acid (e.g. phenylalanine) with carboxyl group of methionine. The enzyme peptidyl transferase in the 50S subunit catalyzes this reaction.

STEP 3: Discharge of empty tRNA (formerly carrying formylmethionine) from donor site on 70S ribosome.

STEP 4: Translocation of mRNA and tRNA with new peptide bond (between methionine and e.g. phenylalanine) to donor site on 70S ribosome.

E

FIGURE 8 Continued. E, *Step 1: The second amino acid is brought in by tRNA after the initiation complex is completed. Step 2: This binds the two amino acids by a peptide bond and is blocked by chloramphenicol and lincomycin. Steps 3 and 4: The first tRNA is ejected from the ribosome and the second tRNA with its dipeptide is translocated to the site vacated by the first tRNA. The ribosome then moves along the chain of mRNA to the next triplet codon for instruction on the identity of the third amino acid to be introduced into the growing peptide chain. Erythromycin blocks the translocation step.*

amino acid becomes attached (Fig. 8*E*, step 1). The first amino acid (methionine) is linked to the second amino acid by the enzyme peptidyltransferase, the two being joined by a peptide bond between the carboxyl group of formylmethionine and the amino group of the second amino acid (Fig. 8*E*, step 2). This beginning chain with its tRNA is translocated to the first site (donor site) after it has been vacated by the release of tRNA that had formerly carried formylmethionine (Fig. 8*E*, steps 3 and 4). The ribosome then moves along the chain of mRNA to the next triplet codon for further instruction. Here directions are given for the attachment of the unoccupied acceptor site of a third tRNA with its amino acid. The dipeptide on the donor site is transferred to the third amino acid, and the process of chain elongation is continued until a termination code triplet in mRNA announces that the protein chain is complete.

Since the mRNA strand is "read" by several ribosomes simultaneously, multiple proteins are synthesized simultaneously. A connecting mRNA fiber between adjacent ribosomes, which forms an assembly of as many as 100 ribosomes on a single mRNA strand, can be seen in the electron microscope. This arrangement of multiple ribosomes, each producing its own protein, is called a polyribosome.

Antibiotics that bind to ribosomes cure infections by interfering at certain points with peptide chain formation in bacteria. Thus, they may interfere with initiation of the peptide chain, the attachment of tRNA after initiation, peptide-bond formation, translocation, and the movement of ribosomes along mRNA. Among antimicrobials important in clinical medicine, the mechanism of such interference has been worked out best for five groups: the aminoglycosides, the tetracyclines, chloramphenicol, erythromycin, and emetine.

The Aminoglycosides

The mode of action of streptomycin has been examined far more than that of other aminoglycosides. Streptomycin binds to the 30S subunit of the ribosome by irreversibly combining with a specific ribosomal protein, designated P10. At this site it has three effects on protein synthesis: (1) It permits formation of the initiation complex but blocks its normal activity. When streptomycin attaches to ribosomes, they fall off the "assembly line"; i.e., they leave mRNA prematurely. These ribosomes dissociate into 30S and 50S subunits which subsequently reassociate at the normal initiation sites on mRNA, but they remain irreversibly inactivated initiation complexes that cannot form peptide bonds. (2) It interferes with the attachment of tRNA. (3) It distorts the triplet codons of mRNA so that the message is misread, the wrong amino acids are inserted into the peptide chain, and faulty proteins are produced. Of these three effects, the first is the most important cause of bacterial killing by streptomycin. Cells are killed by the accumulation of aberrant inactive initiation complexes.[19, 24] The third occurs only at borderline inhibitory concentrations of streptomycin.[6]

Other aminoglycosides such as kanamycin, neomycin, and gentamicin probably act similarly, but more research is necessary to establish their mode of action. There is some evidence that amikacin, neomycin, and gentamicin have binding sites on both 30S and 50S ribosomes.

The Tetracyclines

These drugs also bind to the 30S subunit of bacterial ribosomes and block the binding of tRNA to the mRNA 30S ribosomal subunit.[5] In other words, tetracyclines prevent the introduction of new amino acids into the peptide chain so that protein synthesis cannot proceed.

Chloramphenicol

In contrast to the aminoglycosides and tetracyclines, chloramphenicol attaches exclusively to the larger (50S) moiety of the ribosome. The drug prevents peptide-bond formation by inhibiting the enzyme peptidyltransferase. This enzyme is located in the 50S subunit, so that it is blocked when chloramphenicol binds to that portion of the ribosome.[25]

Erythromycin

Like chloramphenicol, erythromycin binds to 50S ribosomal subunits and can compete with chloramphenicol for binding sites on 50S ribosomes. Like chloramphenicol, it interferes with peptidyltransferase activity.[20]

Lincomycin

This antibiotic resembles erythromycin in its antibacterial spectrum and also acts like chloramphenicol in inhibiting protein synthesis. Thus, lincomycin binds to 50S ribosomes (but not ribosomes of *E. coli,* organisms whose growth is not inhibited by lincomycin) and also appears to block the peptidyltransferase reaction necessary for peptide-bond formation.

Emetine

This ancient amebicidal drug has only recently been found to be an inhibitor of protein synthesis. It inhibits the transfer of amino acids from tRNA to the polypeptide on the ribosome and prevents elongation, rather than initiation, of peptide chains. Emetine also inhibits protein synthesis in bacteria and mammalian cells. This lack of selectivity may explain the toxicity of emetine for patients during treatment of amebic dysentery or liver abscess.[13]

The Thiosemicarbazones (Methisazone)

If polyribosomes are disrupted, protein synthesis stops. At least one group of antimicrobials, the thiosemicarbazones, seem to interfere with protein synthesis of smallpox virus by breaking up the mRNA into smaller fragments and disrupting the ribosomes.[2]

TRANSCRIPTION MECHANISMS

The information that determines the sequence of amino acids in a given protein is coded in the DNA and *transcribed* into messenger RNA (see Fig. 8*A*). Messenger RNA then becomes attached to ribosomes, where the code is translated into protein synthesis. Antibiotics interfering with translation act on the ribosomes, as described in the preceding section. Drugs acting on transcription may interfere either with separation of DNA strands or with the synthesis of RNA. DNA consists of two polynucleotide chains twisted about each other in the form of a double helix. During transcription these strands of the double helix separate, and one of them serves as a specific surface or *template* upon which a complementary strand of RNA is synthesized through the action of RNA polymerase. The complementary RNA strand and its template DNA differ in only two respects: the presence of deoxyribose in DNA in place of ribose, and of thymine, in place of uracil, as one of the four major bases.

A drug can interfere with the transcription process by preventing strand separation of DNA, by breaking a strand, by introducing an improper component into the replicating RNA strand, or by blocking access of RNA polymerase to the template strand. Only one clinically valuable antibiotic, rifampicin, interferes with transcription. Other antimicrobial drugs, however, active against protozoa, fungi, and viruses act by interfering with transcription.

Rifampicin

This antibiotic is the most potent inhibitor of DNA-dependent RNA polymerase in bacteria. Human DNA-dependent RNA polymerase, on the other hand, is resistant to rifampicin, so that the drug is selectively toxic for bacterial but not human cells. By binding to RNA polymerase, rifampicin inhibits the formation of all forms of RNA in bacteria.[34]

Chloroquine

This important antiprotozoal drug inhibits nucleic acid synthesis by interfering with the ability of DNA to act as a template.[15] Chloroquine is inserted (intercalated) between the stacked base pairs of the double helix. This drug inhibits nucleic acid synthesis in mammalian cells as well, but protozoa concentrate chloroquine so that their intracellular level is much higher than that in the body fluids of the patient.

Nitroimidazoles

The selective toxicity of metronidazole and other 5-nitroimidazoles against anaerobic bacteria and protozoa probably involves reduction of the nitro group to a nitrosohydroxyl amino group. This is carried out by a reduced electron transport protein similar to ferredoxin, which is important in the terminal energy metabolism of susceptible organisms. The intracellular concentration of the unreduced nitroimidazole is kept low by its metabolism so that the drug is taken up by simple diffusion to reach intracellular concentrations 50 to 100 times that in the environment.[14] The reduced drug is lethal to anaerobic organisms because it causes strand breaks in DNA.[9] Mammalian cells are unharmed because they lack the enzymes required for reduction of the nitro group.

5-Iodo-2′deoxyuridine (IUDR)

This nucleoside analog is widely used in the treatment of herpes simplex infections of the cornea. The drug is incorporated into *viral* DNA instead of thymidine.[17] Normally, deoxuridilic acid is converted to thymidilic acid by the enzyme thymidilic acid synthetase. IUDR inhibits this enzyme so that insufficient thymidine phosphate is available for DNA synthesis. Strands of DNA containing IUDR in place of thymidine are more easily broken. In addition, the presence of IUDR in DNA could lead to abnormal base pairing and the consequent production of nonfunctional proteins that are not assembled into virus particles.[2]

5-Fluorocytosine

Like IUDR, this halogenated (fluorinated) pyrimidine probably acts by eventually inhibiting the action of the enzyme thymidilic acid synthetase. Cytosine is first deaminated to uracil by cytosine deaminase. 5-Fluorouracil is then converted to 5-fluorodeoxyuridylate. This fluorinated compound causes a lethal thymine deficiency by blocking the conversion of the normal deoxyribonucleotides to deoxythymidylate by thymidylate synthetase. 5-Fluorouracil is also incorporated into mRNA so that errors are produced in translation of information from DNA into protein.[7]

The Sulfonamides

This group of drugs is placed here among transcription inhibitors because sulfonamides block the synthesis of thymidine and all purines. Thymidine is necessary for DNA synthesis, and the purines for nucleic acid synthesis. This action of sulfonamides is accomplished by preventing the synthesis of folic acid (pteroylglutamic acid) by microbial cells. Sulfonamides are structural analogs of para-aminobenzoic acid (PABA), an essential ingredient of folic acid (Fig. 9). Competitive inhibition of PABA utilization by sulfonamides interferes with folic acid synthesis. Since folic acid functions as a coenzyme for transporting 1-carbon units from one molecule to another, the sulfonamides block these reactions which are necessary for the synthesis of thymidine, purines, methionine, and serine. Folic acid metabolism in patients is not affected by sulfonamides because human cells cannot synthesize folic acid. Instead, patients must obtain folic acid in their diets. Since bacteria cannot transport exogenous folic acid into their cells, dietary folate does not interfere with the action of sulfonamide drugs. In pus, however, the breakdown of cells may cause a considerable accumulation of thymidine, purines, methionine, and serine which reverse the inhibitory effect of sulfonamides on bacteria by replenishing the end products of folic acid metabolism. In this way, the sulfonamides may lose therapeutic effectiveness.[10]

In addition to the sulfonamides, para-aminosalicylic acid (PAS) and the sulfones (both active against certain mycobacteria) are PABA analogs and block folic acid synthesis by competitive inhibition.

The Diaminopyrimidines

Both pyrimethamine and trimethoprim, the two important members of this group, are folic acid antagonists ("antifols"). Their site of action is different from that of the sulfonamides, however. The diaminopyrimidines are structurally similar to the pteridine portion of dihydrofolate; therefore, instead of blocking PABA utilization, they prevent the conversion of folic acid to tetrahydrofolic acid by depression of the enzyme dihydrofolic reductase, as indicated in the following reactions:

$$\text{PABA} \rightarrow \text{folic acid} \xrightarrow[\text{reductase}]{\text{folic}} \text{dihydrofolic acid}$$

$$\xrightarrow[\text{reductase}]{\text{dihydrofolic}} \text{tetrahydrofolic acid } (H_4FA)$$

$$\text{Precursors} \xrightarrow{H_4FA} \text{components of nucleic acids}$$

The dihydrofolic acid reductase of protozoa and certain other pathogenic organisms is far more sensitive to trimethoprim than that of man, so that folic acid deficiency in patients given trimethoprim is not a serious problem.[16]

In order to take advantage of the two vulnerable metabolic sites in protozoa and other pathogens, the diaminopyrimidines are usually given in conjunction with sulfonamides. This combination has a much greater antifol action than a simple summation of the two, and, since the two sequential depression steps are present only in the parasite, the combination has markedly increased activity against infection without an increased toxicity for patients.

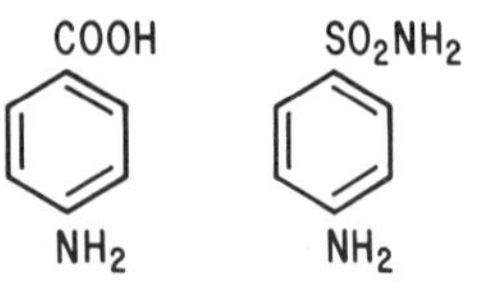

Structural relationship between p-aminobenzoic acid (left) and sulfanilamide (right)

FIGURE 9. *The close structural relationship of sulfonamides to para-aminobenzoic acid (PABA) is used to explain their antibacterial effect. The competitive inhibition of PABA utilization by sulfonamides interferes with folic acid synthesis in bacteria but not in man.*

MECHANISM OF ACTION OF ANTHELMINTIC DRUGS

Chemotherapy for worms is based on physiologic damage rather than protein inhibition. This is because pathogenic worms are fully grown when treatment is needed. Drugs that interfere with growth through inhibition of protein synthesis can stop egg production, but egg production is not needed for worm survival.

Anthelmintics kill worms by blocking energy metabolism or by paralysis.[3] Pyrvinium and mebendazole, for example, interfere with energy metabolism by blocking uptake of glucose, while thiabendazole inhibits fumaric reductase, a key enzyme in fermentation of glucose.[31] Like other intestinal microflora, worms are anaerobic and do not have the enzymes that mammalian cells use for terminal oxidation of glucose.[26] In the generation of energy-rich phosphate during glucose metabolism in worms, electron transfer is characterized by the reduction of fumarate to succinate through the action of fumarate reductase, which serves as an electron carrier from flavoproteins to fumarate. In other words, fumarate, instead of O_2, becomes the ultimate electron acceptor in worms as in the following schema of electron transfer in worms and man:

Worms: Flavoprotein → fumaric reductase → fumarate
Man: Flavoprotein → cytochrome oxidase → O_2

This selective inhibition of fumaric reductase explains the toxicity of thiabendazole for worms, but not man.

Niclosamide, another important anthelmintic, interferes with energy metabolism by blocking the phosphorylation of adenosine diphosphate (ADP) and thus the formation of adenosine triphosphate (ATP) during electron transport. In other words, the high-energy phosphate required for energy by the worm is not generated in the presence of niclosamide. Niclosamide would, no doubt, show the same effect on human ATP formation, but fortunately the drug is not absorbed from the intestine. Niridazole, a powerful new remedy for schistosomiasis, affects energy metabolism by depleting glycogen reserves. It blocks the inhibitor of glycogen phosphorylase so that glycogenolysis becomes excessive.[4]

Two important paralyzing anthelmintics are piperazine and bephenium hydroxynaphthoate.[27] Piperazine produces flaccid paralysis of *Ascaris* so that the worm can be expelled from the patient by intestinal peristalsis.[8, 21] It stabilizes the membrane potential of *Ascaris* muscle by hyperpolarization, but has no effect on human muscle. Hence its toxicity is fully selective for the parasite. Bephenium, which resembles acetylcholine in

TABLE 1. Mechanisms of Action of Antimicrobial Drugs

NATURE OF INJURY	ANTIMICROBIAL DRUG	MODE OF ACTION
Defective cell wall mucopeptide	Penicillins and cephalosporins	Prevent final peptide bond between D-alanine and glycine
	Cycloserine	As structural analogue of D-alanine, it inhibits enzymes responsible for synthesis of D-alanyl-D-alanine, an essential component of mucopeptide
	Bacitracin and vancomycin	Block transfer to cell membrane of sugar pentapeptide from site of synthesis in cytoplasm
Damaged cytoplasmic membrane	Polymyxins	Disorganize lipoproteins by inserting lipophobic moiety into membrane lipid
	Polyenes	React with steroids in fungal membranes so that permeability is altered
Impaired function of ribosomes	Aminoglycosides	Bind to 30S ribosomal unit, causing ribosomes to leave mRNA prematurely; also interfere with attachment of tRNA and distort triplet codons so that message is misread
	Tetracyclines	Bind to 30S unit and block binding of tRNA so that new amino acids cannot be introduced into peptide chain
	Chloramphenicol	Attaches to 50S subunit of ribosomes and prevents peptide-bond formation by inhibiting enzyme peptidyltransferase
	Erythromycin and lincomycin	Same as chloramphenicol
	Emetine	Prevents elongation of peptide chain by inhibiting transfer of amino acids from tRNA to polypeptide on ribosome
	Thiosemicarbazones	Disrupt polyribosomes
Impaired nucleic acid function	Rifampicin	Blocks bacterial RNA formation by inhibiting DNA-dependent RNA polymerase
	Chloroquine	Inserted between stacked base pairs in double helix and thus interferes with ability of DNA to act as a template for nucleic acid synthesis
	5-Iodo-2′deoxyuridine	Incorporated into viral DNA instead of thymidine so that nonfunctional proteins are synthesized
	5-Fluorocytosine	Converted to 5-fluorouracil which blocks thymidylate synthetase so that lethal thymine deficiency results
	Sulfonamides and diaminopyrimidines	By preventing synthesis of folic acid, they block formation of thymidine and purines needed for nucleic acid synthesis
	Metronidazole	After partial reduction of the nitro group, the activated drug causes strand breaks in DNA
Impaired energy metabolism	Pyrvinium and mebendazole	Block glucose uptake
	Thiabendazole	Inhibits fumaric reductase so that glucose fermentation is impaired
	Niclosamide	Blocks phosphorylation of ATP
	Niridazole	Depletes glycogen reserves
Paralysis	Piperazine	Stabilizes membrane potential of *Ascaris* muscle by hyperpolarization
	Bephenium	Depolarizes membranes (i.e., reverse of piperazine)

structure and function, has the reverse effect on nematodes. Instead of hyperpolarization and flaccid paralysis, it causes depolarization and contraction. Since the cuticle of many worms is impervious to bephenium, the drug is effective against only a few nematodes of clinical importance, including *Necator americanus, Ancylostoma duodenale,* and *Ascaris lumbricoides.* The mucosa of the human gastrointestinal tract is also impervious to bephenium, as it is to acetylcholine, presumably because these two compounds possess a quarternary nitrogen, i.e., a central nitrogen attached to four methyl groups. The poor absorption explains the low toxicity of this anthelmintic.

References

1. Anderson, J. S., Meadows, P. M., Haskin, M. A., et al.: Biosynthesis of the peptidoglycan of bacterial cell walls. I. Utilization of uridine diphosphate acetylmuramyl pentapeptide and uridine diphosphate acetylgucosamine for peptidoglycan synthesis by particulate enzymes from *Staphylococcus aureus* and *Micrococcus lysodeikticus.* Arch Biochem 116:487, 1966.
2. Appleyard, G.: Chemotherapy of viral infections. Br Med Bull 23:114, 1967.
3. Bueding, E.: Some biochemical effects of anthelmintic drugs. Biochem Pharmacol 18:1541, 1969.
4. Bueding, E., and Fisher, J.: Biochemical effects of niridazole on *Schistosoma mansoni.* Molec Pharmacol 6:532, 1970.
5. Craven, G. R., Gavin, R., and Fanning, T.: The transfer RNA binding site of the 30 S ribosome and the site of tetracycline inhibition. Sympos Quant Biol 34:129, 1969.
6. Davis, B. D.: Streptomycin resistance and the study of ribosomal structure and function. N Engl J Med 83:1405, 1970.
7. De Kloet, S. R.: Effects of 5-fluorouracil and 6-azauracil on the synthesis of ribonucleic acid and protein in *Saccharomyces carlsbergensis.* Biochem J 106:167, 1968.
8. Del Castillo, J., De Mello, W. C., and Morales, T.: Mechanism of the paralysing action of piperazine on ascaris muscle. Br J Pharmacol 22:463, 1964.
9. Edwards, D. I.: The action of metronidazole on DNA. J Antimicrob Therap 30:43, 1977.
10. Feingold, D. D.: Antimicrobial chemotherapeutic agents: the nature of their action and selective toxicity. N Engl J Med 269:957, 1963.
11. Few, A. V.: Interaction of polymyxin E with bacterial and other lipids. Biochim Biophys Acta 16:137, 1955.
12. Greenwood, D.: Mucopeptide hydrolases and bacterial "persisters." Lancet 2:465, 1972.
13. Grollman, A. P.: Structural basis for inhibition of protein synthesis by emetine and cycloheximide based on an analogy between ipecac alkaloids and glutarimide antibiotics. Proc Nat Acad Sci USA 56:1867, 1966.
14. Gutteridge, W., and Coombs, G.: Biochemistry of Parasitic Protozoa. Baltimore, University Park Press, 1977.
15. Hahn, F. E., O'Brien, R. L., Ciak, J., et al.: Studies on modes of action of chloroquine, quinacrine, and quinine and on chloroquine resistance. Milit Med 131(Suppl):1071, 1966.
16. Hitchings, G. H.: Species differences among dihydrofolate reductases as a basis for chemotherapy. Postgrad Med J 45(Suppl):7, 1969.
17. Kaplan, A. S., and Ben-Porat, T.: Differential incorporation of iododeoxyuridine into the DNA of pseudorabies virus-infected and noninfected cells. Virology 31:734, 1967.
18. Kinsky, S. C.: Nystatin binding by protoplasts and a particulate fraction of *Neurospora crassa,* and a basis for the selective toxicity of polyene antifungal antibiotics. Proc Nat Acad Sci USA 48:1049, 1962.
19. Luzzato, L., Apirion, D., and Schlessinger, D.: Mechanism of action of streptomycin in *E. coli*: interruption of the ribosome cycle at the initiation of protein synthesis. Proc Nat Acad Sci USA 60:873, 1968.
20. Mao, J. C., and Robishaw, E. E.: Erythromycin, a peptidyltransferase effector. Biochemistry 11:4864, 1972.
21. Mansour, T. E.: Chemotherapy of parasitic worms: New Biochemical Strategies 205:462, 1979.
22. Neuhaus, F. C., and Lynch, J. L.: The enzymatic synthesis of D-alanyl-D-alanine. 3. On the inhibition of D-alanyl-D-alanine synthetase by the antibiotic D-cycloserine. Biochemistry (Wash) 3:471, 1964.
23. Newton, B. A.: The properties and mode of action of the polymyxins. Bacteriol Rev 20:14, 1956.
24. Ozaki, M., Mizushima, S., and Nomura, M.: Identification and functional characterization of the protein controlled by the streptomycin-resistant locus in *E. coli.* Nature (London) 222:333, 1969.
25. Pongs, O., Bald, R., and Erdmann, V. A.: Identification of chloramphenicol-binding protein in *Escherichia coli* ribosomes by affinity labeling. Proc Nat Acad Sci USA 70:2229, 1973.
26. Saz, H. J.: Comparative energy metabolisms of some parasitic helminths. J Parasitol 56:634, 1970.
27. Saz, H. J., and Bueding, E.: Relationships between anthelmintic effects and biochemical and physiological mechanisms. Pharmacol Rev 18:871, 1966.
28. Siewert, G., and Strominger, J. L.: Bacitracin: an inhibitor of the dephosphorylation of lipid pyrophosphate, an intermediate in biosynthesis of the peptidoglycan of bacterial cell walls. Proc Nat Acad Sci USA 57:767, 1967.
29. Thomasz, A.: The mechanism of the irreversible antimicrobial effects of penicillins. How the beta-lactam antibiotics kill and lyse bacteria. Ann Rev Microbiol 33:113, 1979.
30. Tipper, D. J., and Strominger, J. L.: Mechanism of action of penicillins: a proposal based on their structural similarity to acyl-D-alanyl-D-alanine. Proc Nat Acad Sci USA 54:1133, 1965.
31. Van den Bossche, H.: Biochemical effects of the anthelmintic drug mebendazole. In Van den Bossche, H. (ed.): Comparative Biochemistry of Parasites. New York, Academic Press, 1972, pp. 139–157.
32. Van den Bossche, H., Willemsens, G., Couls, W., Lauwers, W., and Le June, L.: Inhibition of ergosterol biosynthesis in *Candida albicans* by miconazole. Curr Chemother 3:228, 1978.
33. Voigt, W.: On the mode of action of antimycotics, especially of clotrimazole (canesten) on the ultrastructural level of human pathogenic fungi. Scand J Infect Dis Suppl 16:51, 1978.
34. Wehrli, W., and Staehelin, M.: Actions of the rifamycins. Bacteriol Rev 35:290, 1971.

RESISTANCE TO ANTIMICROBIAL DRUGS 21

Microbial resistance is acquired after a change in DNA. This change may occur by alteration in the structure of chromosomal DNA or by acquisition of extrachromosomal DNA. The alteration of chromosomal DNA is called mutation, and the acquisition of extrachromosomal DNA is the result of genetic exchange. These changes lead to the formation of enzymes or other proteins that inactivate drugs or hinder access to their site of action.

GENETIC EXCHANGE: R FACTORS

Susumu Mitsuhashi

GENERAL PROPERTIES OF R FACTORS

Genetic exchange is the more important cause of clinical drug resistance because it produces epidemic resistance to multiple drugs. The extrachromosomal DNA that is responsible for such resistance can reproduce itself within the bacterial cell and then spread to other bacteria by transduction or mating (conjugation) (Mitsuhashi et al., 1965; Novick, 1969). Transduction is a type of gene transfer in which DNA from one bacterial cell is introduced into another bacterial cell by bacteriophage infection. The transmission of resistance from one bacterial cell to another is known as "infectious" drug resistance, because sensitive bacteria become "infected" with resistance determinants. By this process, resistance to one or multiple antibiotics may spread from one bacterium to another during mating, even in the absence of antibiotics. In other words, bacteria may become resistant to multiple drugs without ever having been exposed to them.

The genetic elements that control infectious drug resistance are a form of DNA known as plasmids. Plasmids are not part of the large circular bacterial chromosome but exist as small cyclic DNA molecules that are capable of reproducing themselves independently. The extrachromosomal resistance factors in gram-negative bacteria are called R factors (Mitsuhashi, 1977B). R factors are composed of two kinds of genes: those that carry determinants for antibiotic resistance and those that promote transfer from one cell to another. The first are called resistance determinants (RD) and the second are called resistance transfer factors (RTF). The transfer of the R factor during mating depends on external hair-like appendages — the sex pili. They facilitate plasmid transfer from male bacteria (donors) to female bacteria (recipients), which have no pili (Fig. 1). The number of RD attached to the RTF determines the number of drugs to which the bacteria become resistant. One R factor may have many genes, each of which is responsible for resistance to a different antibiotic (Fig. 2).

FIGURE 1. *Sex pili in* E. coli *06. The sex pili are the long, hairlike appendages that extend from one bacillus to another. The sex pilus is about 90 Å across, and of variable length (mean 1 to 2 μm). The first step in conjugation is the formation of a mating pair by interaction between the tip of one pilus and a receptor site on the surface of the recipient cell. According to one concept, the pilus then serves as the bridge along which R-factor DNA passes. In an alternative postulate for DNA transfer, the pilus may retract by sequential depolymerization of its subunits into the donor cell membrane. This draws the surfaces of the 2 bacterial cells together and allows formation of a conjugation bridge not involving the pilus (× 8000 scanning electron micrograph made by Dr. R. Weller, Univ. Calif., San Diego).*

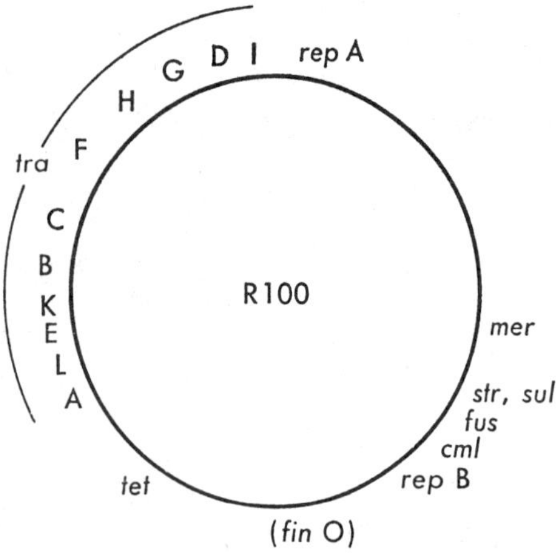

FIGURE 2. *Circular map of an R factor (R100 factor, a conjugative plasmid in* E. coli *K12). Distance between each marker is drawn arbitrarily. The following genetic determinants are represented:* tet = *tetracycline resistance;* str = *streptomycin resistance;* sul = *sulfonamide resistance;* mer = *mercury resistance;* cml = *chloramphenicol resistance;* fus = *fusidic acid resistance;* tra *(A→L inclusive are each required for the different steps in this complex process. Thus 9 of the 12 cistrons are required for synthesis of the sex pilus and others for penetration of plasmid DNA during conjugation;* rep *determines the ability to replicate in the host cell;* fin = *fertility inhibition by repressing piliation.*

When a bacterium becomes infected with an R factor, the cell develops sex pili and becomes a donor cell. Donor competence, or the capacity to transfer resistance by conjugation, is greatest in bacteria that have recently acquired an R factor. The number of cells that are competent donors declines after a few generations as the capacity to produce sex pili is repressed.

DISCOVERY OF R PLASMID

The appearance and rapid increase in number of multiply resistant *Shigella* strains in Japan attracted the attention of microbiologists — from the standpoint of epidemiology especially, and later from the standpoint of genetics (Mitsuhashi, 1977B).

Evidence for the existence of drug resistance genes in the cytoplasm of resistant bacteria was first obtained in Japan (Mitsuhashi, 1977B) by demonstrating the properties of resistant bacteria that would be indicative of conjugative plasmids: 1) transmission of drug resistance by mixed cultivation of resistant and sensitive strains; 2) interruption of the transmission with a fritted glass disk to separate the two parental cultures; and 3) spontaneous and induced loss of drug resistance from drug-resistant cells. The artificial elimination of R plasmids is carried out with drugs (ethylenediaminetetraacetate [EDTA], dodecyl sodium sulfate) that kill bacteria carrying such plasmids, or with DNA intercalating agents (acriflavine) that inhibit DNA synthesis. Drug resistance was transmitted by mixed cultivation from drug-resistant *Escherichia coli* K12 F^- or Hfr to drug-sensitive *Shigella,* regardless of the polarity of the F agent. Between substrains of *E. coli* K12, drug resistance was found to be transmitted similarly by mixed cultivation without regard to the polarity of the F agent. These facts indicated that transferable drug resistance is transmitted independently of the chromosomal transmission of the donor strain and that this agent is different from the F factor. Direct proof of the cytoplasmic existence of drug resistance genes was obtained by physical isolation of plasmid DNA. The R plasmids were isolated as a satellite DNA different from the chromosomal DNA by centrifugation, and were demonstrated by electron microscopy as covalently closed circular or open circular DNA (Falkow, 1975). A mixed incubation of a small number of bacterial cells carrying transmissible drug resistance results in the rapid acquisition of multiple drug resistance by a majority of the recipient cells. R factors are transferable among all species of the family Enterobacteriaceae, as well as among *Neissaria gonorrhoeae, Hemophilus influenzae,* the *Vibrio* group, *Pasteurella pestis, Bordetella bronchiseptica,* and the *Pseudomonas* group (Iyobe et al., 1974). They can also be transferred from *Bacteroides* to *E. coli* (Guiney and Davis, 1978). These properties of the R factor (i.e., autonomous replication and a wide range of transmission among various species of bacteria) are of great importance in public health and animal husbandry.

From these studies, and others, it can be concluded that R factors are self-replicating genetic elements that 1) confer drug resistance on the host; 2) possess plasmid characteristics, including autonomous replication, self-regulation, conjugal transferability, and stable residence within the host; and 3) interfere with other genetic elements, such as the F, R, and Col factors, and bacteriophages.

RESISTANCE MARKERS

Transfer between gram-negative enteric bacteria of multiple drug resistance — i.e., resistance to tetracycline (TC), chloramphenicol (CM), streptomycin (SM), and sulfanilamide (SA) — was first described in 1959. An R factor carrying kanamycin (KM) resistance was isolated in 1963 and later was found also to confer resistance to both neomycin (NM) and paromomycin (PM). In 1965, ampicillin (APC) resistance conferred by an R factor was reported, APC inactivation resulting from the production of a β-lactamase. R factors have been found that are capable of conferring resistance to aminoglycosidic antibiotics, such as the gentamicin C (GM) complex. In addition to conferring resistance to antibacterial agents, R factors are capable of causing resistance to heavy metal ions, such as Ni^{2+}, Hg^{2+}, and Co^{2+}.

R FACTOR PROPERTIES OTHER THAN DRUG RESISTANCE

R factors can control many genetic properties other than those responsible for drug resistance, e.g., conjugal transferability, autonomous replication, stable maintenance in the host, maleness of the host (R mating), and interference with other plasmids and bacteriophages. It has been demonstrated that R factors, like the F factor, carry structural genes for pili formation, which govern conjugation, R factor transfer, and R mating. However, R factors, unlike the F factor, have a regulator gene that can repress piliation due to R and F factors; thus, when R factors and F factor coexist within the same cell, F mating is inhibited. At the present time, the locus that governs R mating (the ability of R factors to promote the transfer of the host chromosome) has not been separated genetically from the locus *tra,* which is responsible for the transfer of R factors.

One R factor interferes with superinfection with another R factor by the phenomena of entry exclusion and incompatibility. Superinfection inhibition occurs when the frequency of R factor transfer is decreased by the presence of an R factor within the cell, i.e., in an R^+ recipient. Mutual exclusion occurs in cells that contain two incompatible R factors. The factors become labile, and either one or both are easily lost during subsequent growth of the bacteria. R factors can also suppress plaque formation by bacteriophages. This trait is designated as *spp.* The suppression is not caused by an inhibition of phage adsorption, but is the result of degradation of the injected phage DNA.

The genetic properties of R factors are summarized in Table 1, and the genetic structure of an R factor is shown in Figure 2.

BIOCHEMICAL MECHANISMS OF R-MEDIATED RESISTANCE

Drug resistance that is conferred by R factors is achieved in some instances by the production of inactivating enzymes, as has been demonstrated in *E. coli* strains that are resistant to CM, KM, and SM. Chloramphenicol is inactivated by an *o*-acetyltransferase that in the presence of acetyl CoA and CM forms 3-0-acetyl CM and 1-3-0-0-diacetyl CM (Fig. 3A). Streptomycin can be inac-

TABLE 1. Genetic Properties of R Factors

GENETIC PROPERTY		MECHANISM
Drug resistance	Tetracycline	Impermeability
	Chloramphenicol	Acetylation Impermeability
	Aminoglycoside antibiotics	Phosphorylation Adenylylation Acetylation Impermeability
	Sulfonamides	Impermeability, production of sulfonamide-resistant dihydropteroate synthetase
	Ampicillin	β-Lactamase
Resistance to metal ions	Hg^{2+} Co^{2+} Ni^{2+}	Mercury vaporization
Conjugal transfer(*tra*)	R transfer Chromosome transfer	Pili formation
	Repression	Repressor of pili formation
Replication		
Stable inheritance		
Superinfection inhibition	Entry exclusion Incompatibility	
Phage interference		Restriction enzyme

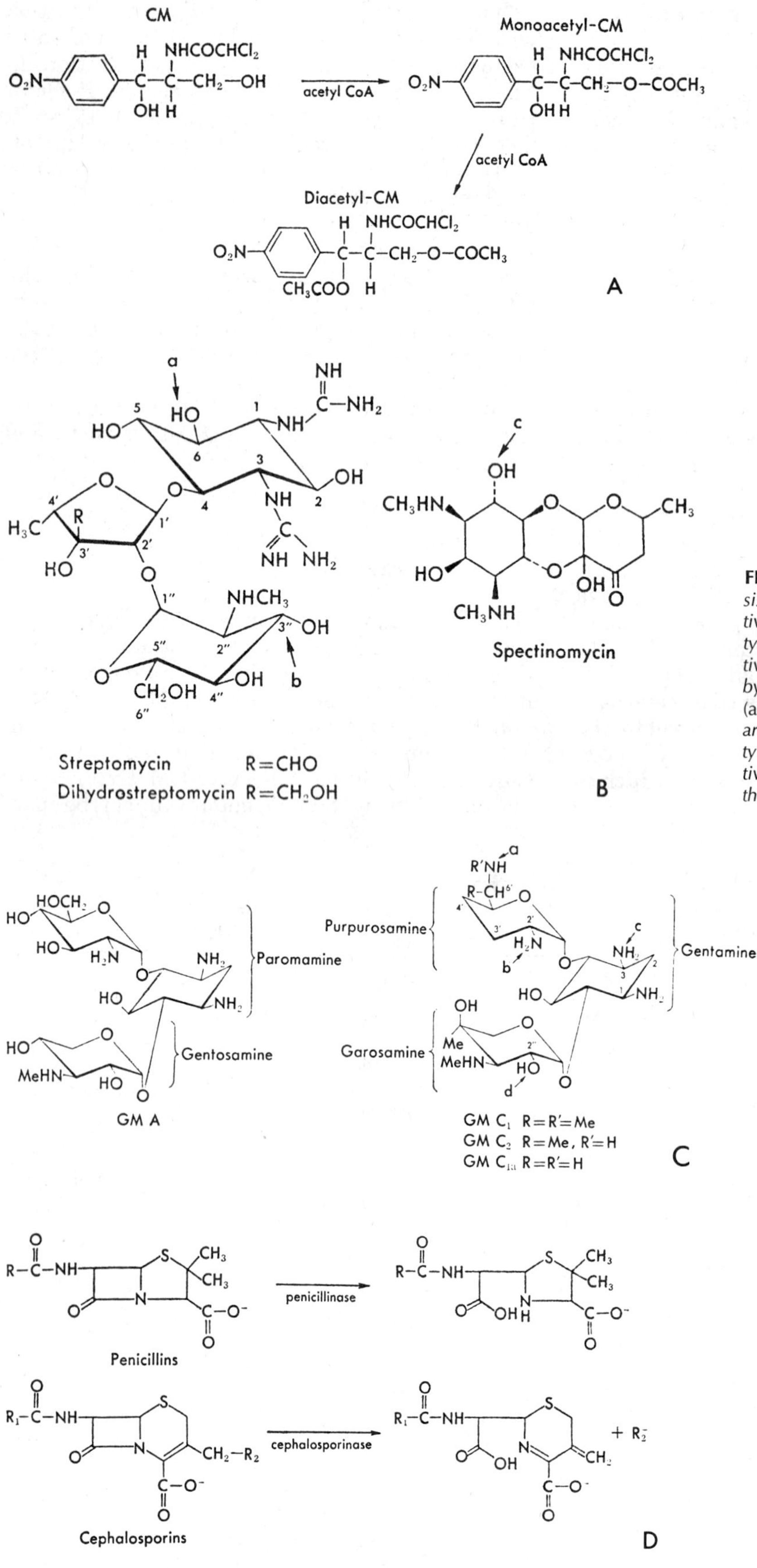

FIGURE 3. *Mechanisms responsible for resistance in bacteria carrying R factors. A, Inactivation of chloramphenicol (CM) by CM acetyltransferase. B, Arrows indicate site of inactivation of streptomycins and spectinomycin by adenylylation (a,b,c), and phosphorylation (a,b). C, Chemical structures of gentamicins; arrows indicate the site of inactivation, acetylation (a,b,c), and adenylylation (d). D, Inactivation of penicillins and cephalosporins by the beta-lactamases.*

tivated by adenylylating or phosphorylating enzymes in the presence of adenosinetriphosphate (ATP), resulting in the formation of adenylylstreptomycin (Fig. 3B) or phosphorylstreptomycin. Kanamycin can also be inactivated either by acetylation in the presence of acetyl CoA, which produces 6′-*N*-acetylkanamycin, or by phosphorylation in the presence of ATP, which produces 3′-phosphorylkanamycin.

The newly introduced aminoglycoside gentamicin C lacks a 3-hydroxyl group in its amino sugars and consequently is not inactivated by the KM or SM phosphorylating enzyme previously isolated from resistant R factor–carrying strains. Enzymes isolated from bacterial cells harboring new types of R factors can inactivate gentamicin C components either by acetylation or by adenylylation (Fig. 3C).

Hydrolytic enzymes can also mediate R factor resistance. Penicillin, ampicillin, and cephalosporin resistance, for example, are produced by enzymes that hydrolyze the C-N bond of the β-lactam ring (Fig. 3D). There are two general types of β-lactamases that differ in antigenic and substrate specificity but have similar molecular weights in the range of 20,000 to 25,000 (Table 2). Type I is more active against cephaloridine than is type II, and type II is more active against ampicillin, cloxacillin, methicillin, and oxacillin than is type I. These penicillinases elaborated by R factors thus differ from staphylococcal penicillinase, which is relatively inactive against the cephalosporins, oxacillin, cloxacillin, and methicillin. In fact, the clinical success of the cephalosporins and these three semisynthetic penicillins is primarily related to their continued effectiveness against serious infections by β-lactamase-producing staphylococci. In addition to the gram-negative enteric bacteria, the R plasmids mediating type I penicillinase have been demonstrated in *Pseudomonas aeruginosa*, gonococci, and *H. influenzae*. These R factor penicillinases are found between the cell wall and the cell membrane. They are all constitutive.

Drug resistance due to a decrease in permeation of the drug into the bacterial cell has also

TABLE 2. Summary of Comparison between Types I and II Penicillinases Mediated by R Plasmids

PROPERTIES	PENICILLINASE		
	Type Ia Penicillinase (R_{GN14}, Renamed R_{ms212})	Type Ib Penicillinase (R_{GN823})	Type II Penicillinase (R_{GN238}, Renamed R_{ms213})
Penicillinase activity* per mg of dry weight of bacteria	0.6	16.7	0.025
Specific penicillinase activity per mg of enzyme protein	1330	1670	20
pH optimum	6.5–7.0	6.5–7.0	7.6
Temperature optimum (°C)	45	40–45	30
Inhibition of activity by chloride ion	No	No	Yes
Inhibition of activity by anti-serum (Type Ia penicillinase)	Yes	Yes	No
Molecular weight†	20,600	24,000	25,400
Isoelectric point‡	5.1 (5.4)	6.9 (5.6)	8.3 (7.4)
$s_{20,w}$	—	2.45	2.66
Secondary structure§	α25%	α25%	α + β
Substrate specificity◗ and *Km*			
Substrate			
Benzylpenicillin	100 (27 μM)	100 (24 μM)	100 (5 μM)
Phenethicillin	33 (32 μM)	27 (14 μM)	155 (6 μM)
Ampicillin	115 (30 μM)	112 (32 μM)	450 (16 μM)
Cloxacillin	2	1	292 (13 μM)
6-Aminopenicillanic acid	87 (222 μM)	89 (200 μM)	363 (26 μM)
Cephaloridine	130 (400 μM)	111 (500 μM)	36 (111 μM)

*Penicillinase activity is expressed in units; one unit of the enzyme activity is defined as the activity which hydrolyzes 1 μmole of benzylpenicillin per min at 30 °C.

†The molecular weight was estimated by the gel filtration method.

‡Isoelectric points were determined by the use of agar-gel electrophoresis. The values in parenthesis were taken from the data of Matthew, which were determined by the use of analytical isoelectric focusing.

§α-Helix content was determined by ORD spectrum.

◗Substrate specificity is expressed as the percentage of hydrolysis of benzylpenicillin.

been reported. Tetracycline and its derivatives have been shown to inhibit, to the same degree, protein synthesis in cell-free systems prepared from TC-sensitive and R factor–carrying TC-resistant strains. Strains resistant to TC by virtue of an R factor also have been shown to be incapable of inactivating tetracycline. This suggests that the mechanism of TC resistance involves a decrease in the permeability of the drug. R factors causing CM resistance by decreasing CM penetration into the cell have also been found. Similarly, the mechanism of sulfonamide (SA) resistance in strains of *E. coli* containing an R factor has been explained by a decrease in the permeability of the cell membrane to sulfanilamide. Plasmid-mediated SA resistance is also achieved by the formation of SA-resistant dihydropteroate synthetase. Resistance to Hg^{2+} in R-carrying strains has been attributed to mercury vaporizing activity (Mitsuhashi et al., 1977).

CLASSIFICATION OF R PLASMIDS

The R factors of the *Enterobacteriaceae* are diverse and distributed all over the world — in hospitals, livestock farms, and fish ponds. Classification of R plasmids is useful in epidemiologic studies (Mitsuhashi, 1977A). Resistance patterns of R plasmids and their biochemical mechanisms of resistance can be used for their classification. Another classification of these plasmids is based on their compatibility properties. The closely related plasmids are incompatible; i.e., they cannot coexist in their bacterial host. The unknown R factor to be tested is transferred from its wild host to an auxotrophic strain of *E. coli* K12. The R^+ strain is used as the donor in matings with a set of K12 cultures, each of which carries a plasmid of known compatibility group (Datta, 1977).

Most R plasmids from *P. aeruginosa* strains are nontransferable to *E. coli* strains and are conjugally transferred only between *P. aeruginosa* strains. Therefore, the R factors of *P. aeruginosa* strains can be classified by an intraspecies conjugation system (Sagai et al., 1976).

NONCONJUGATIVE RESISTANCE (r) PLASMIDS

Nonconjugative resistance (r) plasmids were discovered after R plasmids in multiply resistant *Staphylococcus aureus* strains by the irreversible loss of resistance (Mitsuhashi, et al., 1963) and by genetic analysis of staphylococcal resistance (Novick, 1963). The r plasmids are self-replicating genetic elements that confer drug resistance to the host and possess plasmid characteristics, including autonomous replication and stable residence within the host. A big difference from the conjugative R plasmids is a lack of conjugal transferability with r plasmids because staphylococci do not mate. The r plasmids are usually smaller than R plasmids, because a *tra* (transferability) region is absent from the r molecule (Fig. 4). The transmission of r plasmids in staphylococci is mediated by transduction with bacteriophages, by which staphylococci are lysogenized (Mitsuhashi et al., 1965). It has been found that the drug resistance of *Streptococcus pyogenes* is mostly due to the presence of a nonconjugative plasmid (Mitsuhashi, 1979).* Further studies have disclosed that nonconjugative resistance (r) plasmids are widely distributed even in gram-negative bacteria, and drug resistance of naturally occurring resistant strains is mostly caused by the presence of conjugative (R) or nonconjugative (r) plasmids, or both. Nonconjugative plasmids confer single resistance on *S. aureus* and other host bacteria, and those carrying multiple resistance are very exceptional in contrast to R plasmids, which confer multiple resistance.

In staphylococci with r factors, penicillin resistance is caused by an inducible β-lactamase. The β-lactamases of enteric bacteria are not inducible. The enzyme responsible for chloramphenicol resistance in r-bearing staphylococci is also inducible. Chloramphenicol resistance in these staphylococci is produced by an acetyltransferase like that in enteric bacteria, but the chloramphenicol-acetylating enzyme of enteric bacteria is constitutive (noninducible).

MOLECULAR NATURE OF R FACTOR DNA

The relatively small size of plasmids permits their isolation as intact DNA molecules in both the resting and replicating states, and allows a definitive analysis of their molecular structure. Fragmentation during this manipulation is much less of a problem than in the case of bacterial chromosomes whose large size makes them vulnerable to shearing. The DNA of R plasmids has a molecular weight in the range of 60×10^6 and a buoyant density of 1.710 to 1.712 g/ml (Rownd and Womble, 1977). This is similar to the density of chromosomal DNA in *E. coli* and *Shigella*, so that the plasmid DNA cannot be distinguished by a satellite band in CsCl gradients. The molecular weight of nontransmissible r plasmids is only 5 to 10×10^6 because they have only one or two

*In group D streptococci (enterococci) conjugative plasmids confer resistance to erythromycin, kanamycin, tetracycline, and streptomycin.

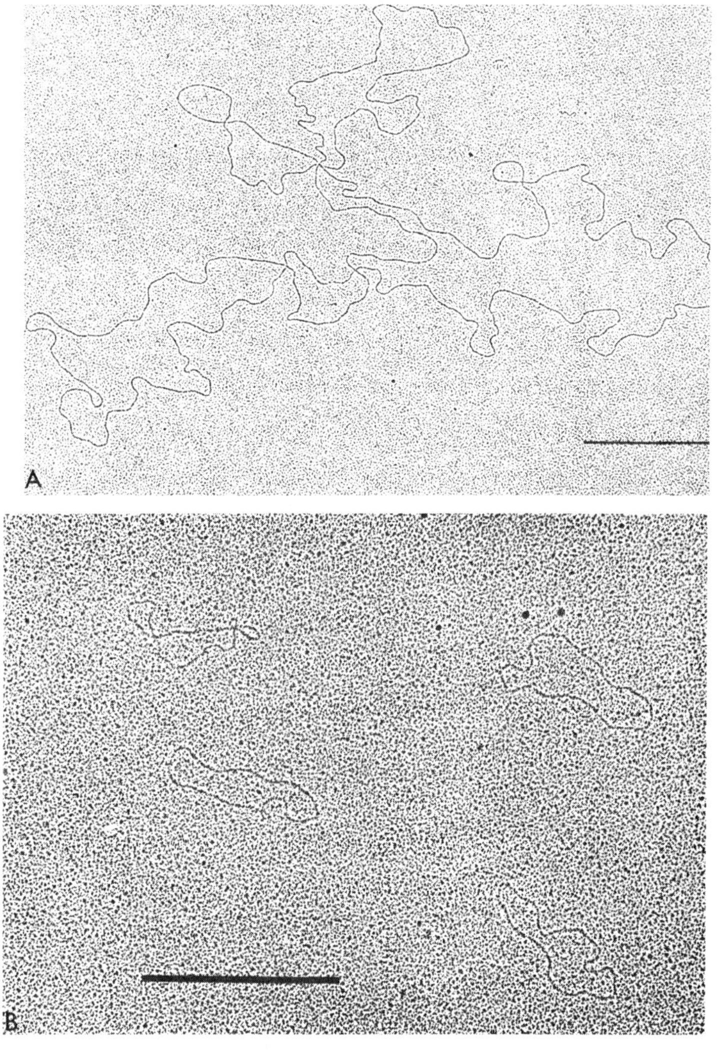

FIGURE 4. *Electron micrograph of the open circular molecules of plasmid DNA from a conjugative (A) and nonconjugative (B) plasmid to illustrate that conjugative plasmids are larger because they have more resistance genes and transfer genes. Bar represents 1 μm. (A) a conjugative RMS201 plasmid DNA (about 31 μ) encoding resistance to TC, CM, SM, SA and APC; (B) a nonconjugative rMS31 plasmid DNA (about 2 μ) encoding resistance to SA.*

resistance determinants and no transfer factor (Fig. 5).

Like the bacterial chromosome, R factors are circular molecules of DNA. The circular R factors are twisted (super coiled) double strands closed by covalent bonding. Through various physicochemical manipulations (e.g., pH 12.5) the covalently bonded R factor becomes more compact and can be separated in density gradients from the separated chromosomal strands by its faster sedimentation. Such maneuvers overcome the problem of separating the R factor from the chromosome that results from their uniform buoyant density under normal conditions.

CHROMOSOMAL DRUG RESISTANCE

Abraham I. Braude

In contrast to infectious drug resistance, resistance after mutation involves chromosomal genes. Spontaneous mutation to drug resistance is infrequent, generally occurring in only 1 of 10 million to 1 trillion bacterial cells. The resistant mutants are difficult to detect unless the drug is present as a selective agent to suppress the overwhelming number of nonmutated cells and to allow the multiplication of resistant ones. The multiplication of resistant organisms during antibiotic therapy usually occurs on mucous membranes, in urine, or in lung cavities, where no immune mechanisms can check the growth of the organisms. Streptomycin has been the most important antibiotic responsible for selecting resistant mutants during treatment. Resistance develops in *E. coli* mutants because of a change in the gene that specifies the P10 protein in 30 S ribosomes (Ozaki et al., 1969).

The P10 protein of sensitive strains allows the attachment of streptomycin to the 30 S ribosome, but in resistant strains the attachment sites for streptomycin are masked so that streptomycin is not bound to the ribosome. The loss of binding to the ribosome produces more or less total resistance to streptomycin and occurs in a single mutational step. Hence, there can be an abrupt appearance of resistant organisms during treatment with streptomycin after the original sensitive organisms have been rapidly eliminated. In urinary tract infections, for example, massive resistance to streptomycin can develop after five days of treatment. High-level resistance also occurs after exposure to erythromycin, apparently due to the modification of 23 S ribosomal RNA by mutation so that binding of the drug is reduced (Lai et al., 1973). High-level resistance is not confined to drugs acting on ribosomes, since it develops to antituberculous drugs such as isoniazid and rifampin (Guttler et al., 1971) that do not interfere directly with ribosomal function.

Mutation appears to be responsible for one type of gonococcal resistance to penicillin, but the level of such resistance is much lower than that seen in gonococci possessing R factors. The reason for this is that chromosomal resistance in gonococci to penicillin increases slowly by succeeding mutational steps. Similar low-level resistance to tetracycline has occurred in gonococci, in occasional strains of pneumococci, and in group A streptococci. A higher level of resistance of pneumococci to penicillin has recently been reported in South Africa (see later discussion).

Drug resistance can also occur in fungi and protozoa. Among fungi, resistance to 5-fluorocytosine in *Cryptococcus neoformans, Candida albicans,* and other species of *Candida* has limited the usefulness of this important new drug. High-level resistance of cryptococci and *Candida* to 5-fluorocytosine can develop by single-step mutation, so that a strain that was sensitive to 3.0 μg/ml is no longer inhibited by 1000 μg/ml after short periods of treatment (Weese, 1972).

The most important example of drug resistance in protozoa is that of *Plasmodium falciparum* (the cause of malignant subtertian malaria) to chloroquine (Peters, 1970). The major difference between chloroquine-sensitive and chloroquine-resistant malarial parasites is the loss of high-affinity binding of chloroquine by red cells infected with chloroquine-resistant parasites. Chloroquine resistance is thus attributed to a decrease in the number, affinity, or accessibility of chloroquine receptors at the site of action within the malaria parasite. The nature and subcellular location of substances that bind chloroquine are still under study.

The mechanism of acquired resistance of cryptococci or *Candida* to 5-fluorocytosine can be inferred from studies of the nonpathogenic yeast, *Saccharomyces cerevisiae* (Jund and Lacroute 1970). By such an analogy it has been suggested that resistance could occur through one of three processes: 1) failure of the deamination of cytosine to 5-fluorouracil — the product responsible for blocking nucleic acid synthesis; 2) loss of the permeases required for permeation of the drug into the yeast cell; or 3) the absence of uracil-binding sites. The first of these mechanisms, loss of cytosine deaminase activity, was responsible for acquired resistance to 5-fluorocytosine in a strain of *Candida parapsilosis,* causing fatal endocarditis (Hoeprich et al., 1974).

PRACTICAL IMPLICATIONS OF DRUG RESISTANCE

Gram-negative bacteria carrying R factors are the most serious problem among antibiotic-resistant organisms because their resistance to multiple drugs can be spread in epidemic proportion throughout hospitals and whole communities. In Japan the situation became so bad that in 1966 to 1967, 79 per cent of *Shigella* strains were resistant to one or more drugs, and at least half of these strains carried R factors (Mitsuhashi, 1969). Similar outbreaks have occurred in other parts of the world and with other enteric pathogens (Jonsson, 1972; Lowbury et al., 1972; Anderson and Smith, 1972). In Mexico, for example, multiresistant *Salmonella typhi* recently produced the worst epidemic of typhoid fever in modern history (Gangarosa et al., 1972).

Shigella and *Salmonella* resistance is primarily an epidemiologic problem outside the hospital. In hospitals, as many as 60 to 70 per cent of all enteric bacteria (other than *Salmonella-Shigella*) carry R factors for multiple antibiotics. The high prevalence of infectious drug resistance can be traced directly to the increased use of antibiotics both in and out of the hospital. The importance of selective pressure by antibiotics in promoting infectious drug resistance in hospitals can be shown by the parallel rise in resistance with drug use and by the disappearance of gram-negative bacilli transferring drug resistance when antibiotic therapy is sharply restricted (Lowbury et al., 1972). Outside the hospital, when antibiotics are used in animal feed and sold without restriction, selective antibiotic pressure encourages the spread of R factors. Feed containing a penicillin or tetracycline for growth stimulation of livestock has led to the widespread distribution of enteric bacteria containing R factors. In Great Britain, the extensive use of antibiotics for preventing *Salmonella* infections in cattle was followed by a human epidemic of antibiotic-resistant *Salmonella typhimurium* infection (Anderson, 1968). In Mexico, the recent epidemic of chloramphenicol-resistant typhoid fever can be related to the unrestricted sale of chloramphenicol in pharmacies to patients without prescriptions.

Although antibiotics have generated the problem, they did not create the R factor. Davis and Anandan (1970) have shown that R factors existed in communities that had never been exposed to commercial antibiotics (Fig. 5). Their evolutionary development can be explained as a survival mechanism in the presence of antibiotics produced by other organisms in the bowel or soil. After the introduction of commercial antibiotics, these bacteria with R factors possessed a selective advantage and became predominant. They can transfer multiple resistance in the bowel to pathogenic enteric bacteria. Multiple resistance in one organism may involve as many as seven antibiotics. Determinants for resistance to kanamycin, for example, tend to occur on R factors in conjunction with determinants for resistance to sulfonamides, streptomycin, tetracycline, and chloramphenicol. The use of kanamycin, therefore, could promote the simultaneous spread of resistance to itself and to four other antibiotics (Cohen, 1969).

Penicillinase plasmids in staphylococci have created a less serious clinical problem than the R factors in gram-negative bacteria, because multiple resistance is not usually transmitted by penicillinase plasmids. Antibiotics that are resistant to penicillinase can therefore successfully treat infections caused by penicillinase-producing staphylococci. An exception to this is the resistance to methicillin in 10 to 17 per cent of staphylococcal infections in England, France, and Switzerland occurring in hospitalized patients with chronic debilitating diseases (Benner and Kayser, 1968). This resistance may cross over to other penicillinase-resistant penicillins and is not caused by inactivation of the drug. Because methicillin-resistant staphylococci are confined to extremely debilitated patients, they have probably lost virulence, even though they cause serious and even fatal infections. They are becoming a problem in the United States, where such strains have been isolated recently in growing numbers.

Chromosomal drug resistance is a major clinical problem in the treatment of tuberculosis, cryptococcosis, candidiasis, and gram-negative urinary infections. In cavitary pulmonary or genitourinary infections, any of the antituberculous drugs can select resistant mutants, which then produce relapses during treatment. In contrast to

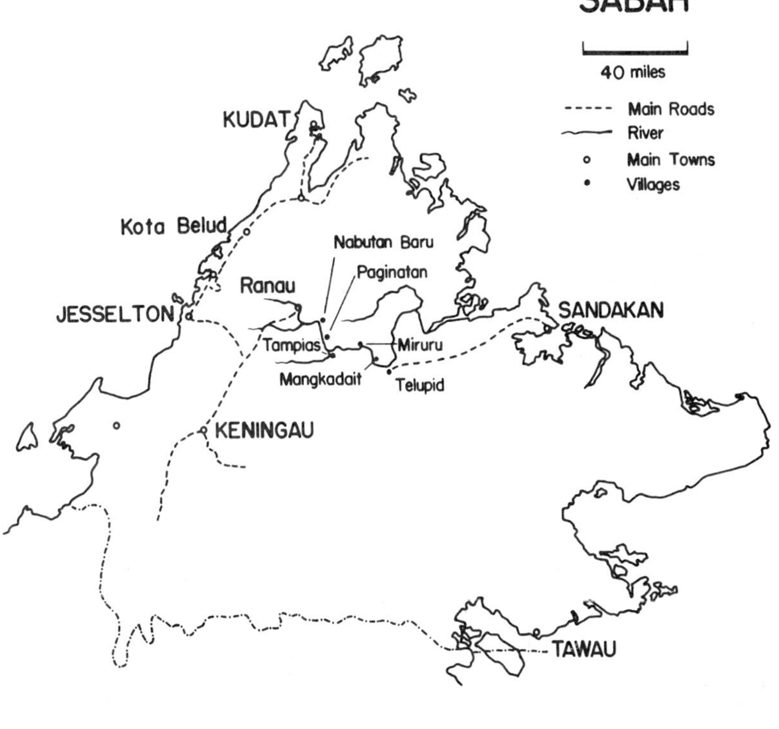

FIGURE 5. *A remote locale in Borneo was used for exploring the existence of R factor before the introduction of commercial antibiotics. Davis found that R factors existed in Miruru, a community that had never been exposed to commercial antibiotics. From his discovery, it is reasonable to infer that the evolutionary development of R factor can be explained as a survival mechanism in the presence of antibiotics produced by other organisms in the bowel or soil. (From Davis, C. E., Anandan, J.: The evolution of R factor. A study of a preantibiotic community in Borneo. N Engl J Med 282:118, 1970.)*

plasmids, which produce multiple drug resistance, chromosomal mutations affect susceptibility to only one drug. Clinical resistance can, therefore, be prevented by combination treatment with two drugs. Since the mutation rate of the tubercle bacillus to streptomycin resistance is approximately 10^{-10}, and 10^{-6} to isoniazid, the chance of developing resistance to both is only one in 10^{16} cell divisions. This probability is so small that the combined use of streptomycin and isoniazid in treating cavitary pulmonary tuberculosis is almost never followed by the development of resistance to either drug. This principle is also used in cryptococcosis by giving amphotericin with 5-fluorocytosine to prevent resistance to 5-fluorocytosine; in protozoal infections by giving sulfonamides with trimethoprim to prevent resistance to trimethoprim; and in *E. coli* urinary infections by giving small doses of tetracycline with streptomycin to prevent resistance to streptomycin.

Except for the foregoing examples, chromosomal drug resistance has not been a serious clinical problem. Clinical resistance to penicillin, the most important antimicrobial, probably never occurs during the course of treating a specific infection with penicillin. Despite the high frequency of infection by penicillinase-producing staphylococci in hospitals, these organisms do not arise from penicillin-sensitive staphylococci in patients undergoing treatment with penicillin. Their exact origin is unknown, but it is more likely that penicillinase-producing staphylococci are selected out in the nasopharynx of staphylococcal carriers among hospital personnel who are exposed to low concentrations of penicillin in aerosols that develop from open vials or syringes. Penicillin resistance rarely occurs among pyogenic streptococci, but it is becoming a serious problem among pneumococci in certain parts of the world. The resistance was first observed in a type 23 pneumococcus isolated in 1967 from Australian patients with hypogammaglobulinemia and in type 4 pneumococci recovered from 15 asymptomatic carriers in remote areas of New Guinea where penicillin was in frequent use (Hansman et al., 1971). Although these strains apparently lost the extreme sensitivity to penicillin that characterizes pneumococci, they were still sensitive to 0.5 μg/ml — a level easily reached in treatment. More recently, high-level resistance to penicillin led to fatal infection from pneumococcal bacteremia in Johannesburg, South Africa. Carriers of types 6A and 19A penicillin-resistant pneumococci were found in 29 per cent of pediatric patients and in 2 per cent of hospital staff members in the Johannesburg Hospital. These pneumococci and others isolated in Durban, Minneapolis, and London have ranged in resistance from 2.5 to 4 μg/ml. However, the new South African strains were multiply resistant to other β-lactam antibiotics as well as erythromycin, clindamycin, tetracycline, and chloramphenicol. The Durban strains of resistant pneumococci caused 14 deaths among 29 patients with systemic infections. These episodes emphasize the importance of testing the sensitivity of pneumococci in certain parts of the world. The mechanism of this resistance is unknown, but it is probably chromosomal because the resistant pneumococci do not produce penicillinase.

References

Anderson, E. S.: The ecology of transferable drug resistance in the enterobacteria. Annu Rev Microbiol 22:131, 1968.

Anderson, E. S., and Smith, H. R.: Chloramphenicol resistance in the typhoid bacillus. Br Med J 3:329, 1972.

Benner, E. J., and Kayser, F. H.: Growing clinical significance of methicillin-resistant *Staphylococcus aureus*. Lancet 2:741, 1968.

Cohen, S.: A decade of R factors. J Infect Dis 119:104, 1969.

Datta, N.: R factors in Enterobacteriaceae. In Mitsuhashi, S. (ed.): Transferable Drug Resistance Plasmid—R Factor. Tokyo, University of Tokyo Press; Baltimore, University Park Press, 1977, p. 109.

Datta, N., and Richmond, M. H.: The purification and properties of a penicillinase whose synthesis is mediated by an R-factor in *Escherichia coli*. Biochem J 98:204, 1966.

Davis, C. E., and Anandan, J.: The evolution of R factor. A study of a "preantibiotic" community in Borneo. N. Engl J Med 282:117, 1970.

Falkow, S.: Infectious Multiple Drug Resistance, London, Pion Limited Publishing Company, 1975.

Gangarosa, E. J., Bennett, J. V., Wyatt, C., et al.: An epidemic-associated episome? J Infect Dis 126:215, 1972.

Guiney, D., and Davis, C.: Identification of a conjugative R plasmid in *Bacteroides ochraceus* capable of transfer to *Escherichia coli*. Nature 274:181, 1978.

Guttler, R. B., Counts, G. W., Avent, C. K., et al.: Effect of rifampin and minocycline on meningococcal carrier rates. J Infect Dis 24:199, 1971.

Hansman, D., Glasgow, H., Sturt, J., et al.: Increased resistance to penicillin of pneumococci isolated from man. N Engl J Med 284:175, 1971.

Hoeprich, P. D., Ingraham, J. L., Kleker, E., et al.: Development of resistance to 5-fluorocytosine in *Candida parapsilosis* during therapy. J Infect Dis 130:112, 1974.

Iyobe, S., Hasuda, K., Fuse, I., and Mitsuhashi, S.: Demonstration of R factors from *Pseudomonas aeruginosa*. Antimicrob Ag Chemother 5:547, 1974.

Jonsson, M.: Antibiotic resistance and R factors in gram-negative bacteria. A study from Sweden. Scand J Infect Dis 5(Suppl):1, 1972.

Jund, R., and Lacroute, F.: Genetic and physiological aspects of resistance to 5-fluoropyrimidines in *Saccharomyces cerevisiae*. J Bacteriol 102:607, 1970.

Lai, C. J., Weisblum, B., Fahnestock, S. R., et al.: Alteration of 23 S ribosomal RNA and erythromycin-induced resistance to lincomycin and spiramycin in *Staphylococcus aureus*. J Mol Biol 74:67, 1973.

Lowbury, E. J., Babb, J. R., and Roe, E.: Clearance from a hospital of gram-negative bacilli that transfer carbenicillin-resistance to *Pseudomonas aeruginosa*. Lancet 2:941, 1972.

Mitsuhashi, S.: The R factors. J Infect Dis 119:89, 1969.

Mitsuhashi, S.: Epidemiology of bacterial drug resistance. In Mitsuhashi, S. (ed.): Transferable Drug Resistance Plasmid — R Factor. Tokyo, University of Tokyo Press; Baltimore, University Park Press, 1977A, P. 3.

Mitsuhashi, S.: Discovery of R factors. In Mitsuhashi, S. (ed.): Transferable Drug Resistance Plasmid — R Factor. Tokyo, University of Tokyo Press; Baltimore, University Park Press, 1977B, p. 18.

Mitsuhashi, S.: Drug Resistance Plasmids. Mol Cell Biochem 26:135, 1979.

Mitsuhashi, S., Morimura, M., Kono, M., and Oshima, H.: Elimination of drug resistance of *Staphylococcus aureus* by treatment with acriflavine. J Bacteriol 86:162, 1963.

Mitsuhashi, S., Oshima, H., Kawarada, U., and Hashimoto, H.: Drug resistance of staphylococci. I. Transduction of tetracycline resistance with phage lysates obtained from multiply resistant staphylococci. J Bacteriol 89:988, 1965.

Mitsuhashi, S., Yamagishi, S., Sawai, S., and Kawabe, H.: Biochemical mechanisms of plasmid-mediated resistance. In Mitsuhashi, S. (ed.): Transferable Drug Resistance Plasmid — R Factor. Tokyo, University of Tokyo Press; Baltimore, University Park Press, 1977, p. 195.

Novick, R. P.: Analysis by transduction of mutation affecting penicillinase formation in *Staphylococcus aureus*. J Gen Microbiol 33:121, 1963.

Novick, R. P.: Extrachromosomal inheritance in bacteria. Bacteriol Rev 33:210, 1969.

Ozaki, M., Mizushima, S., and Nomura, M.: Identification and functional characterization of the protein controlled by the streptomycin-resistant locus in *E. coli*. Nature 222:333, 1969.

Peters, W.: Chemotherapy and Drug Resistance in Malaria. New York and London, Academic Press, 1970.

Rownd, R. H., and Womble, D. D.: Molecular nature and replication of R factors. In Mitsuhashi, S. (ed.): Transferable Drug Resistance Plasmid — R Factor. Tokyo, University of Tokyo Press; Baltimore, University Park Press, 1977, p. 161.

Sagai, H., Hasuda, K., Iyobe, S., et al.: Classification of R plasmids by incompatibility in *Pseudomonas aeruginosa*. Antimicrob Ag Chemother 10:573, 1976.

Watanabe, T.: The origin of R factors. Ann NY Acad Sci 182:126, 1971.

Weese, W. C.: 5-Fluorocytosine therapy. Ann Intern Med 77:1003, 1972.

22 THE PHARMACOLOGY AND TOXICOLOGY OF ANTIMICROBIAL AGENTS

Harold C. Neu, M.D.

There has been great progress in understanding the pharmacology of antibiotics. Unfortunately, it is not always possible to apply data about the biologic half-life, plasma and renal clearance, and bioavailability to clinical practice in a way that improves on the results of empirical antimicrobial therapy. Unlike the situation with anesthetics and anticonvulsants, in which serum concentrations of an agent correlate well with therapeutic effect, there has been little substantiation that serum or tissue concentrations of antibiotics correlate well with successful therapy. The reason for this is that response to therapy with antimicrobials is the result of many disparate factors in addition to the antimicrobial effect of the drugs, including the infectious and pathogenic abilities of the particular microorganism and the resistance of the patient.

Knowledge of the pharmacology of antimicrobials can, however, be used to minimize drug toxicity. Close attention to certain pharmacokinetic properties of antibiotics results in fewer adverse side effects, since many of the toxic reactions of antimicrobials are due to the accumulation of drugs in patients who have renal or hepatic dysfunction, which can be avoided by adjustments in dosage programs.

GENERAL CONSIDERATIONS

Antibiotics can be administered by oral, intramuscular, intravenous, or topical routes. After absorption, they dissolve in the plasma water. They are bound to plasma proteins and occasionally are absorbed into erythrocytes. In the plasma, they are distributed to various extravascular tissues and fluids in which they may be free or bound. As the antibiotic is distributed into extravascular compartments, there is an initial rapid fall in plasma concentration. This initial fall in the plasma level occurs at the end of an I.V. infusion. After intramuscular injection and after oral ingestion, the initial distribution phase is obscured by the combination of slow absorption and simultaneous excretion. The continued decrease in serum levels of antibiotics is related to renal and biliary excretion as well as to biotransformation in the case of some drugs. The amount of drug that reaches extravascular tissue depends not only upon the concentration gradient from serum to tissue fluid, but also on the degree of protein binding in serum and in tissues, as well as on the diffusibility of the agent. Diffusibility of a drug is a function of its molecular size, dissociation constant, and lipid solubility.

Although blood and tissue levels both decline after each dose, the decline in the two compartments is not usually parallel. Some tissues may even avidly bind the drug, although the amount of drug bound in relation to the total dose is usually small, and the rate of decline of drug concentration is not usually affected. Locally bound drug may be extremely important from a toxicologic viewpoint, however.

The pharmacokinetics of most antibiotics are first or second order. Thus, a plot of antibiotic concentration in serum on a logarithmic scale versus time on a linear scale yields a straight line after the initial phase of absorption and distribution. The slope of the linear phase is a measure of biologic life of the antibiotic. The most commonly used measure is the time required for a 50 per cent decrease from the peak value; this is the half-life of the antibiotic. For most antibiotics, the half-life is independent of the dose, initial concentration, or route of administration.

ABSORPTION

Penicillins

There has not been any demonstration of an active transport of penicillins across the lipid protein barrier of the intestinal mucosal cell. Penicillin G is not stable in the presence of gastric acid at pH 1-2, and 50 per cent of the drug is destroyed in 20 minutes. In contrast, at pH 4-5, penicillin G is stable for four hours. Other forms of penicillin are more acid-stable, such as penicillin V, and they are well absorbed by mouth (McCarthy and Finland, 1960). Oral absorption of the semi-synthetic penicillinase-resistant penicillins (cloxacillin, dicloxacillin, and flucloxacillin) is excellent. Oral absorption of nafcillin is erratic, and the oral absorption of oxacillin is so much less than the other penicillins that it probably should not be used orally. Methicillin is not orally absorbed owing to its acid liability. Wide variations in oral absorption exist among the aminopenicillins. Amoxicillin is well absorbed, approximately twice as well as ampicillin. The acetone derivative of ampicillin, hetacillin, is no better absorbed orally than is the parent compound. Pivampicillin, talampicillin, and bacampicillin are esters of ampicillin that are covered in the intestinal mucosa and in serum to ampicillin. These compounds produce blood levels twice those achievable with ampicillin. Cyclacillin, azidocillin, and epicillin are adequately absorbed after oral ingestion. Pivmecillinam, a pivolyl ester of mecillinam, is absorbed and converted immediately to the parent compound. None of the extended spectrum or anti-*Pseudomonas* penicillins (carbenicillin, ticarcillin, azlocillin, mezlocillin, and piperacillin) are absorbed orally. An indanyl ester of carbenicillin, indanylcarbenicillin, is well absorbed (Butler et al., 1972), but the serum and tissue levels are too low to treat systemic *Pseudomonas* infections. It does achieve adequate urinary levels. Ingestion of food at the same time as the pencillin decreases the absorption of ampicillin but not of amoxicillin, pivampicillin, or pivmecillinam; and it may increase the absorption of indanylcarbenicillin.

Cephalosporins

The orally absorbed cephalosporins are cephalexin, cephradine, cephaloglycin, cefaclor, cefatrizine, and cefadroxil. Cephalexin is extremely well absorbed even in the presence of food (Kirby and Regamy, 1973). Cephaloglycin is poorly absorbed, which results in inadequate serum levels, although urinary concentrations are adequate for the treatment of infections with susceptible bacteria. Cephradine appears to behave almost identically to cephalexin. None of the other cephalosporins are absorbed after oral ingestion.

Tetracyclines

With the exception of chlortetracycline, the tetracyclines are well absorbed when taken by mouth (Fabre et al., 1973). Food, divalent cations (such as those found in antacids and milk), and iron interfere with absorption of the tetracyclines. If a patient ingests a tetracycline on an empty stomach, about 75 per cent of tetracycline hydrochloride, 55 per cent of oxytetracycline, 30 per cent of chlortetracycline, 65 per cent of demeclocycline, and 95 per cent of minocycline and doxycycline are absorbed. The absorption of both of the latter compounds is less affected by food, but iron and calcium decrease their absorption.

Macrolides and Lincinoids

Because erythromycin base is destroyed by acid it must be coated with an acid-resistant coating if it is taken by mouth. Since absorption of the coated antibiotic occurs in the duodenum and ileum, peak levels occur later than with erythromycin estolate or stearate, which are absorbed by the stomach. Food does not appreciably alter the absorption of erythromycin estolate, but absorption of erythromycin stearate is decreased if it is given with food. Spiramycin, oleandomycin, and triacetyloleandomycin (all macrolides) are all absorbed orally, as are rosamicin and kitasamycin. Lincomycin is absorbed after oral ingestion but not as well as clindamycin, which produces blood levels five-fold greater. Furthermore, although the presence of food markedly impairs lincomycin absorption, food does not decrease clindamycin absorption.

Other Agents

Chloramphenicol is well absorbed orally, yielding levels equal to those after intravenous injection (Dupont et al., 1970). However, in the form of palmitate ester, unless particle size is well controlled, absorption may be erratic.

Fusidic acid is usually administered by the oral route. There is considerable variation in the amount of absorption of fusidic acid from one individual to another.

None of the aminocyclitol-aminoglycoside antibiotics yield levels adequate for therapeutic purposes after oral ingestion, although ototoxic serum levels of neomycin can be reached after chronic oral ingestion or after rectal dosing. Some aminoglycosides are, in fact, now used primarily to suppress microflora, e.g., neomycin and paromomycin. The same is true of vancomycin, which is not an aminoglycoside.

All the sulfonamides (except sulfaguanidine, succinylsulfathiazole, and phthalylsulfathiazole) are well absorbed after oral administration. Trimethoprim and pyrimethamine are also well absorbed orally. All of the urinary "antiseptics" (such as methenamine, furadantoin, nalidixic acid, oxolinic acid, and cinoxacin) are absorbed when taken orally but yield inadequate serum levels to treat infections outside the urinary tract. Urinary levels are, however, adequate to treat urinary tract infections.

Among the antituberculosis drugs (isoniazid, para-amino salicylic acid, ethambutol, pyrazinamide, ethionamide, thiacetazone, and rifampin), all are absorbed orally.

Antifungal agents that can be given orally are 5-fluorocytosine, clotrimazole, and griseofulvin. However, the orally absorbed clotrimazole is rapidly converted to an inactive form. There are two antiviral agents that are taken orally: amantadine and methisazone.

Parenterally Administered Agents

In general, intramuscular (I.M.) administration of antibiotics is safe and effective. However, a number of the antibiotics cannot be given frequently by intramuscular administration because of pain on injection, and others probably should not be used this way since oral administration yields levels greater than those achieved by I.M. injection. Intramuscular routes should not be relied on in individuals in shock nor in obese individuals or in diabetics because there may be altered absorption due to poor perfusion from the injection sites.

Among the penicillins, crystalline penicillin G, given I.M., is so rapidly cleared from the body that it yields inadequate serum levels unless given frequently. Indeed, when used in prophylaxis for dental procedures to prevent endocarditis in individuals with valvular heart disease, it should be given no more than 30 minutes before dental work is begun. Penicillin combined with procaine produces low levels for 12 hours and is satisfactory to treat hemolytic streptococcal or pneumococcal infections. Doubling the dose of procaine penicillin does not double the serum level unless the dose is given at two sites. Benzathine penicillin G is a repository salt of penicillin that provides tissue and serum levels for 15 to 30 days (depending upon the size of the dose used) that will treat syphilis or streptococcal pharyngitis and will prevent recurrences of rheumatic fever. In general, in adults the levels of penicillinase-resistant penicillins produced by I.M. injection are below those needed for serious systemic staphylococcal illness. Ampicillin, carbenicillin, ticarcillin, and the newer broad spectrum penicillins yield serum levels after I.M. injection that are adequate to treat gram-positive infections or urinary infections but not to treat systemic infections with gram-negative bacteria such as *E. coli, Pseudomonas, Klebsiella,* and *Enterobacter*. Indeed, relatively high serum levels are needed to treat most gram-negative infections with agents such as carbenicillin, so they should be administered every four hours by intravenous infusion.

Cephalothin, cephapirin, cephacetrile, and cefoxitin are too painful to be given by the I.M. route. Cefamandole can be given by I.M. injection, but cephaloridine and cefuroxime yield slightly higher serum levels. Cefazolin yields the highest serum levels of all the cephalosporins after either intravenous or intramuscular administration (Kirby and Regamey, 1973). Cefotaxime, ceftizoxime, and moxalactam all can be given by the I.M. route but it is preferable to administer cefoperazone intravenously.

Most of the tetracyclines can be given by either intramuscular or intravenous routes, although the acid pH of tetracycline solutions causes phlebitis. Chloramphenicol as an ester with succinate can be used intravenously, but, given intramuscularly, the ester is not hydrolyzed, and blood levels are inadequate. Lincomycin and clindamycin can be given by either the I.M. or the I.V. route, but rapid infusions should be avoided. Erythromycin is usually too painful to use by the I.M. route, but, as lactobionate or gluceptate, it can be given intravenously. Vancomycin can be given only intravenously since I.M. injection produces sterile abscesses.

All of the aminoglycosides (amikacin, dibekacin, gentamicin, kanamycin, netilmicin, sisomicin, and tobramycin) yield effective serum levels after I.M. injection or I.V. infusion. Streptomycin and spectinomycin are given only by the I.M. route. Rifamide is administered by I.M. injection. It is also possible to give isoniazid by I.M. injection. Amphotericin is given only by I.V. infusion. Pentamidine isoethenate is given I.M. since there is an increased risk of hypotension when it is given intravenously. Adenosine arabinoside is given by intravenous infusion.

Nearly all antibiotics that can be administered I.M. can be given intravenously (some exceptions are mentioned above). In general, higher peak blood levels are produced by equal doses of intravenously administered antibiotics. If a drug can be safely administered by either route, intravenous administration is preferred when: (1) absorption from I.M. sites cannot be relied upon; (2) the volume of the required dose precludes I.M. injection; or (3) there is a risk of hemorrhage from the trauma of the infection.

Topical Absorption of Antibiotics

Antibiotics can exert their antibacterial effect when used as topical agents. The degree of absorption of antibiotics from the skin varies widely, but in the presence of denuded or burned skin, even poorly absorbed agents may be absorbed sufficiently to accumulate and produce toxic reactions. Topical administration of penicillins and cephalosporins is probably unwise, since not only does it promote the selection of resistant bacterial species, but it is also a means of sensitization of the patient by coupling the beta-lactam to body proteins. Aminoglycosides can be absorbed through burned skin to a degree sufficient to produce renal toxicity and ototoxicity if combined with parenteral use of the agents. In contrast, topical use of silver sulfadiazine, mafenide, and silver nitrate has been effective and safe as a therapeutic maneuver in reducing bacterial colonization of the wound eschar and thereby preventing wound sepsis. Compounds such as bacitracin, vancomycin, and aminoglycosides, which can be incorporated into fibrin material but are not bound to the protein, are able to inhibit entrapped bacteria. In general, however, the topical use of antibiotics should be avoided.

ELIMINATION OF ANTIBIOTICS

Antibiotics may be removed from the body by renal mechanisms (glomerular filtration, tubular secretion, or both), by secretion into the bile, or by enzymatic inactivation in the liver.

Renal Excretion

Most antimicrobials are excreted by the kidney as active compounds (Table 1). The renal clearance of compounds within the same class of drug may vary widely. All penicillins, which are weak anions, are cleared from the body by proximal tubular secretion, but the rate of clearance of the penicillins varies somewhat. Thus the half-lives of the penicillins vary from a half-life for penicillin G of 40 minutes to that of carbenicillin, 66 minutes. Renal clearance of oxacillin is half that of dicloxacillin, accounting to some extent for differences in serum levels. Amoxicillin and ampicillin are cleared equally by the kidney, and the clearances of carbenicillin, ticarcillin, azlocillin, mezlocillin, and piperacillin are similar. Some of the difference in the half-lives of penicillin is due not only to differences in renal clearance but to differences in metabolism. For example, only 5 per cent of carbenicillin is metabolized to penicilloic acid versus 17 per cent of penicillin G.

The renal clearance of cephalosporins is accomplished by tubular secretion for the most part, although some are also metabolized. Cefazolin is cleared most slowly, followed by cephaloridine and cefuroxime. Cephaloridine differs from the other cephalosporins because it is both excreted by glomerular filtration and actively transported into the proximal tubule cell, where it can accumulate to toxic levels. The newer cephalosporins (cefamandole, cefuroxime, cefoxitin, cefotaxime, and ceftizoxime) all are rapidly secreted by renal tubular cells. Moxalactam and cefoperazone are excreted more slowly with $T^{1/2}$ of about 2h compared with 1h or less for the previous agents.

The excretion of beta-lactam compounds, both penicillins and cephalosporins, can be blocked by probenecid, which seems to bind to the transport protein in the renal tubule and thereby competitively interferes with the tubular transport of the antibiotics.

All the aminoglycosides are excreted by glomerular filtration. There may be a minor role for tubular secretion. In the presence of decreased renal function, the aminoglycosides accumulate. Even in normal individuals, not all of a dose is excreted, and variable amounts, depending on the particular aminoglycoside, bind to renal cortical tissue. Vancomycin is also excreted by the kidney via glomerular filtration. Polymyxins, polymyxin B and colistimethate, are excreted by the kidney, but the precise mechanism is unclear. Polymyxins also bind to renal tissue.

That amount of tetracycline not eliminated in the feces is excreted by the kidney with the exception of chlortetracycline, which is metabolized, and doxycycline, which is excreted into the intestine.

Renal excretion accounts for only a small fraction of the elimination of erythromycin, other macrolides, chloramphenicol, lincomycin, and clindamycin.

Sulfonamides are partly filtered and partly reabsorbed and secreted, depending on the nature of the individual drug. The solubility of sulfonamides in the urine depends on the drug concentration, which is a function of the plasma concentration and state of hydration, the urinary pH, and the temperature and inherent solubility of the drugs and their acetylated derivatives. Trimethoprim appears to be handled by both glomerular filtration and tubular secretion, and renal clearance increases with acidity.

If the hepatic clearance of rifampin is exceeded, the excess drug, as well as the desacetyl derivative, is cleared by the kidney. Isoniazid, both the free form and the acetylated derivative, is cleared by the kidney.

Hepatic Elimination

Although the liver can secrete weak anions such as the beta-lactam compounds, hepatic clearance of penicillins (except for nafcillin and dicloxacillin) is not significant. In the presence of combined hepatic and renal insufficiency, however, carbenicillin and ticarcillin do accumulate.

Many antibiotics are converted in the liver to compounds that are either less active or inactive. Cephalothin, cephapirin, and cephacetrile are all converted by hepatic esterases to desacetyl derivatives, which are much less active than the parent compounds. Isoniazid is also modified in the liver by acetylation, as are sulfonamides (Hughes, 1953). Chloramphenicol is converted by glucuronyltransferase to the inactive glucuronide, which is then excreted by renal tubular secretion. Although rifampin is converted in the liver to a desacetyl derivative, this metabolite retains significant activity against bacteria and mycobacteria. The desacetyl rifampin is excreted in bile but is reabsorbed in the intestine to reenter the enterohepatic circulation and thereby maintain prolonged serum levels (Cohn, 1969). Rifampin induces the microsomal liver enzymes responsible for its excretion, thereby resulting in a shorter half-life and lower peak blood levels after chronic therapy. Cefotaxime is converted to a desacetyl derivative that is as active as the parent compound except against *Morganella* and *Pseudomonas*. Cefoporazone is excreted to a major degree by the liver.

Although erythromycin, lincomycin, clindamycin, fusidic acid, and doxycycline are excreted in bile, this route does not account for all of the drug excreted, and it seems probable that there are other mechanisms of inactivation.

Some drugs may be bound to tissues, as appears to be the case with amphotericin, aminoglycosides, and certain quinolines that disappear from serum but can be detected in urine for months after the last dose has been administered.

The metabolism of antibiotics can be altered by other drugs that diminish the microsomal degradation system of the liver or by intrinsic liver disease. Thus, toxic concentrations of free drug may develop in patients with liver failure. There is no way to predict the blood level of hepatically excreted or metabolized antibiotics in patients with liver disease.

DISTRIBUTION

The tissue distribution of antibiotics can be considered as consisting of three major areas: (1) highly perfused lean tissues, such as the heart, lung, and hepatoportal system; (2) less well perfused tissues, such as muscle and skin, representing lean tissue mass; and (3) adipose tissues of negligible perfusion, such as ligaments, cartilage, and some areas of bone. Concentrations of antibiotics in the heart, lung, and liver are equivalent to the levels in serum. However, the level of antibiotic found in normal tissues may not be representative of the amount in an area of infection since the inflammation alters blood flow. Certain "barriers" in the body markedly alter the concentration of antibiotic in a compartment such as the brain and CSF (blood-brain and blood-cerebrospinal fluid barriers) and the aqueous and vitreous barriers in the eye between the ocular fluid and plasma water). Some agents, because of ionization, cannot cross tissue barriers such as the prostatic acini.

Many of the commonly used antibiotics are bound to serum proteins, principally to albumin. The amount of an antibiotic bound to protein usually represents only a small fraction of the antibiotic in the body. If a drug is 50 per cent bound to albumin, only 10 per cent of the total drug is bound. The precise effect that protein binding of antibiotics has on biologic activities is poorly understood. Only the free antibiotic has antibacterial action. Most antibiotic receptor sites lie below the outer membrane of the bacterial cell wall and, if an antibiotic is bound to a protein the size of albumin, it cannot reach its receptor, e.g., penicillin receptors of the inner wall or aminoglycoside receptors on ribosomes in the cytoplasm of the bacterium. On the other hand, protein binding of some antibiotics acts as a temporary store of the agents and thereby prevents large fluctuations in the concentrations of free drug in the body fluids. Indeed, some penicillins have never been used clinically since, as a result of low protein binding, there is rapid tubular secretion, and they are cleared from the body too rapidly to be effective.

Antibiotics differ greatly in protein binding from one class to another and markedly within a class, as illustrated in Table 1. For example, aminoglycosides are not protein-bound, while penicillins range in protein binding from 17 per cent for ampicillin and amoxicillin to 97 per cent for dicloxacillin. The values given in the literature for the protein binding of antibiotics show a marked variation from one study to another, depending upon the technique used to measure protein binding. Protein concentration, pH, and presence of other substances alter the binding and probably do so in vivo since the serum binding of antibiotics in uremic patients is markedly less for certain agents, and uremic patients have higher levels of free drug. If other competing

TABLE 1. Pharmacokinetic Properties of Antibiotics

ANTIBIOTIC	ORAL ABSORP-TION	MAJOR EXCRETION ROUTE	PROTEIN BINDING (%)	SERUM HALF-LIFE Normal (h)	SERUM HALF-LIFE $C_{cr} < 10$* (h)	HEPATIC METAB-OLISM EXCRE-TION	DOSE ADJUSTMENT IN RENAL FAILURE	SERUM LEVEL AFFECTED BY Hemo-dialysis	SERUM LEVEL AFFECTED BY Peritoneal Dialysis
Amikacin	No	R–G	0	2	35–50	No	Major	Yes	Yes
Amoxicillin	Yes	R–T	17	1	6–18	No	Yes	Yes	Yes
Amphotericin	No	Nonrenal	90	24	24	No	Minor	No	No
Ampicillin	Yes	R–T	17	1	6–18	No	Yes	Yes	Yes
Azlocillin	No	R–T	50	1	10–15	Yes	Major	Yes	Yes
Carbenicillin	No	R–T	50	1	10–15	Yes	Major	Yes	Yes
Cefaclor	Yes	R–T	15	1	20	No	Yes	Yes	Yes
Cefamandole	No	R–T	70	0.5	9	No	Yes	Yes	Yes
Cefazolin	No	R–T	85	1.9	30–50	No	Major	Yes	Yes
Cefoperazone	No	R, L	90	2	3–6	Yes	Minor	Yes	
Cefotaxime	No	R–T	50	1	2–4	Yes	Minor	Yes	Yes
Cefoxitin	No	R–T	70	0.5	9	No	Yes	Yes	Yes
Ceftizoxime	No	R–T	50	1.5		No	Minor	Yes	
Cefuroxime	No	R–T	50	1		No	Yes	Yes	Yes
Cephacetrile	No	R–T	70	0.5	3–8	Yes	Yes	Yes	Yes
Cephalexin	Yes	R–T	15	1	8–20	No	Yes	Yes	Yes
Cephaloridine	No	R–G, T	50	1.5	25	No	Avoid	Yes	Yes
Cephalothin	No	R–T	70	0.5	3–8	Yes	Yes	Yes	Yes
Cephapirin	No	R–T	70	0.5	3–8	Yes	Yes	Yes	Yes
Cephradine	Yes	R–T	15	1	8–20	No	Yes	Yes	Yes
Chloramphenicol	Yes	L	25	1–2	3–5	Yes	Minor	Yes	Yes
Chlortetracycline	Yes	L	55	6	7–11	Yes	Avoid	No	No
Clindamycin	Yes	L	94	2–2.5	6–10	Yes	Minor	No	No
Cloxacillin	Yes	R–T	94	0.5	1–2	Yes	Minor	Yes	No
Colistimethate	No	R	< 10	5	50–70	No	Avoid	No	No
Demethylchlortetracycline	Yes	R	50	10	50	Yes	Avoid	Yes	Yes
Dicloxacillin	Yes	R–T	97	0.5	1–2	Yes	Minor	No	No
Doxycycline	Yes	L	90	15–20	15–20	Yes	No	No	No
Erythromycin	Yes	L	18	1.5	4–6	Yes	No	No	No
Ethambutol	Yes	R–G	< 10	4	7–10	No	Major	Yes	Yes
Flucloxacillin	Yes	R–T	96	0.5	1–2	Yes	No	No	No
5-Fluorocytosine	Yes	R–G	< 10	3–6	70	No	Yes	Yes	Yes
Gentamicin	No	R–G	0	2	35–50	No	Major	Yes	Yes
Isoniazid	Yes	L, R	< 10	1–4	1–4	Yes	Minor	Yes	Yes

Kanamycin	No	R–G	0	2	35–50	No	Major	Yes	Yes
Lincomycin	Yes	L	70	4	10–15	Yes	No	No	No
Methenamine	Yes	R		4		No	Avoid	—	—
Methicillin	No	R–T	35	0.5	4	No	Minor	No	No
Metronidazole	Yes	R, L	20	6–14		No	Major	Yes	
Mezlocillin	No	R–T	50	1	10–15		Major	Yes	Yes
Minocycline	Yes	R, L	90	15	35	Yes	Major	No	No
Moxalactam	No	R–T	60	2	19	No	Major	Yes	
Nafcillin	Yes	R–T	90	0.5	1–2	Yes	Minor	No	No
Nalidixic acid	Yes	R	93	1–2	6	No	Avoid		
Neomycin	No	R–G	0	2	50	No	Avoid	Yes	Yes
Netilmicin	No	R–G	0	2	35–50	No	Major	Yes	Yes
Nitrofurantoin	Yes	R		0.3	1	No	Avoid		
Oxacillin	Yes	R–T	94	0.5	1–2	No	Minor	Yes	
Oxytetracycline	Yes	R	35	10	50	Yes	Avoid	No	No
Para-aminosalicylic acid	Yes	R	70	1	25	Yes	Avoid	No	No
Paromomycin	No	R–G	0	2	50	No	Avoid	Yes	
Penicillin G	Yes	R–T	60	0.5	2–6	No	Minor	Yes	Yes
Penicillin V	Yes	R–T	80	1	1–2	No	Minor	No	No
Pentamidine	No	R		—	—	No	—		
Piperacillin	No	R–T	50	1	10–15	Yes	Major	Yes	Yes
Polymyxin B	No	R	< 10	5	40–70	No	Avoid	No	No
Pyrimethamine	Yes	Nonrenal		2–5	20	No	Major		
Quinine	Yes	Nonrenal	70	4	—	Yes	Major	Yes	No
Rifampin	Yes	L	70	1.5–5	1.5–5	Yes	Minor	No	No
Sisomicin	No	R–G	0	2	35–50	No	Major	Yes	Yes
Spectinomycin	No	R–G	0	2		No	Avoid	—	—
Streptomycin	No	R–G	0	2	25	No	Major	Yes	Yes
Sulfadiazine	Yes	R	50	2–5	12–25	Yes	Major	Yes	Yes
Sulfamethoxazole	Yes	R	50	5–7	12	Yes	Major	Yes	Yes
Sulfisoxazole	Yes	R	50	5–7	12	Yes	Major	Yes	Yes
Tetracycline-HCl	Yes	R	55	6–8	50	Yes	Avoid	No	No
Thiamphenicol	Yes	L	25	1–2	3–5	Yes	Minor	Yes	No
Ticarcillin	No	R–T	50	1	12–17	Yes	Major	Yes	Yes
Trimethoprim	Yes	R	60	10	25	No	Major	Yes	Yes
Tobramycin	No	R–G	0	2	35–50	No	Major	Yes	Yes
Vancomycin	No	R	< 10	6	240	No	Major	No	No

*Creatinine clearance less than 10.
R = renal, T = tubular, G= glomerular
L = liver

compounds are present in the serum, the amount of each substance bound to protein depends on the relative affinity of each agent for the particular protein receptor site.

Florey and associates studied the distribution of penicillin into tissues in 1946 and showed that there was penicillin in wound exudates eight hours after an intramuscular injection of penicillin. This observation has been confirmed and extended by recent studies utilizing a variety of model systems that have shown that antibiotics that are not highly protein-bound reach peak levels in tissue one to two hours after administration, but highly protein-bound antibiotics reach peak levels later and remain in the tissue for longer periods. Using a fibrin clot model, Barza and colleagues (1974) found that there was an inverse correlation between penetration of beta-lactam antibiotics into the clot and the degree of protein binding. With a chamber implanted in an animal, it has been shown that highly protein-bound agents penetrate the chambers best and remain for the longest period.

Although protein-bound drug is also carried to tissue sites, the amount of free drug in serum governs the level of free drug in tissues. This is important since it might be thought that leakage of serum into an area of inflammation would provide a higher concentration of drug than one could achieve on the basis of diffusion of unbound drug. However, it is unlikely that inhibitory levels of an antibiotic will be achieved at sites of infection if the concentration of free drug in serum never reaches inhibitory levels.

Highly protein-bound agents tend to be excluded from areas of the body in which a complex carrier system across a lipid-protein barrier is needed to transport antibiotics. Protein binding also delays the removal of drugs handled by glomerular filtration, but, in the case of tubular secretion, even a highly protein-bound antibiotic can be eliminated rapidly without eliminating a volume of plasma fluid that contains the drug.

TISSUE BINDING

Many body tissues can inactivate antibiotics by binding them and releasing such small amounts so slowly that the agent is rendered ineffective. This is particularly true of the polymyxin class and the antifungal polyene antibiotics. Polymyxin B and E bind to acid phospholipids present in mammalian tissues. Polyene antibiotics are taken up in the reticuloendothelial system. Recently it has been shown that intracellular ligands in liver tissue bind penicillins, cephalosporins, tetracyclines, and chloramphenicol. Nucleic acid debris, acid proteins, and high concentrations of divalent cations, such as calcium and magnesium, decrease the activity of aminoglycosides. Inert materials, such as fecal material, can inactivate aminoglycosides; and calcium, iron, or magnesium salts in the intestine chelate tetracyclines, rendering them inert.

DISTRIBUTION OF ANTIBIOTICS TO SPECIFIC BODY AREAS

Pulmonary

Concentrations of most antibiotics within the lung are satisfactory, provided there is some blood flow. There is a wide variation in the concentrations of penicillins and tetracyclines in sputum. Peak sputum levels occur about two hours after the drugs are administered and are dose-dependent, although disproportionately higher sputum concentrations are achieved at higher doses. Levels of carbenicillin and ticarcillin are only 5 per cent of simultaneous serum levels. Appreciable levels of the newer cephalosporin antibiotics, cefoxitin and cefuroxime, are reached in bronchial secretions within one hour after an intravenous dose. Cephalothin concentrations in sputum are approximately 25 per cent of the serum level. Tetracycline concentrations in sputum are generally low and do not seem to vary with the degree of purulence. Although sputum concentrations of tetracyclines are low, levels in bronchial epithelium and lung are close to serum levels. Aminoglycoside levels in bronchial secretions are appreciable, but, because of charge properties, they may be less effective in sputum than other agents. Chloramphenicol, because of its lipid solubility and small size, achieves high concentrations in sputum and bronchial secretions. Trimethoprim also has chemical properties that cause it to achieve effective levels in sputum and bronchial secretions. All of the antituberculosis agents (isoniazid, ethambutol, rifampin, and others) achieve appreciable levels in pulmonary tissue.

It has not been established that either sputum or bronchial concentrations of antibiotics influence therapy of pneumonia.

Pleural, Pericardial, and Ascitic Fluid

Most of the penicillins, cephalosporins, sulfonamides, macrolides, clindamycin, chloramphenicol, fusidic acid, and antituberculosis drugs diffuse into serous cavities. Aminoglycosis, such as tobramycin and gentamicin, diffuse slowly into the peritoneal cavity, but, once equlibrium with extracellular fluid is achieved, the serum levels are lower than predicted because of the increased volume of distribution. Peritonitis increases the rate at which aminoglycosides enter ascites. In most cases, it is unnecessary to inject antibiotics into the peritoneum, but penicillins, cephalosporins, and aminoglycosides can be added safely to peritoneal dialysis fluids.

Bone

Penicillins, tetracyclines, cephalosporins, and the lincomycin-clindamycin antibiotics penetrate bone and bone marrow. Levels of these antibiotics in infected bone are greater than in normal bone. Administration of semisynthetic penicillinase-resistant penicillins or cephalosporins at the time of surgery yields bone hematoma levels adequate to inhibit most staphylococci. Further injection over the next 24 to 48 hours causes continued incorporation of antibiotic into the wound site hematoma, which, during this period, continues to develop. Tetracycline binds to bone in areas in which bone is being laid down. Tetracyclines temporarily depress normal skeletal growth, but permanent effects on the skeleton have not been seen. It is not established whether the high calcium environment of bone depresses aminoglycoside activity, but these drugs have been used successfully to treat osteomyelitis due to *Pseudomonas* and *Serratia*. Antituberculosis drugs (isoniazid, rifampin, and ethambutol) all achieve therapeutic concentrations in bone.

Synovial Fluid

Most antibiotics used in the treatment of joint infections enter inflamed joints adequately so that intra-articular instillation of antibiotic is unnecessary (Parker and Schmid, 1971). The penicillins, cephalosporins, chloramphenicol, tetracycline, lincomycin, and clindamycin reach peak levels in joint fluid about one to two hours after administration and are present in the synovial fluid at levels equal to or in excess of those in serum for the next four to six hours. Aminoglycosides (kanamycin, gentamicin, tobramycin, and amikacin) all reach therapeutic levels in synovial fluid in the presence of inflammation. Measurement of antibiotic activity in joint fluid should be done and the drugs instilled only if patients do not respond to treatment. Polymyxins are not transported into synovial fluid in adequate concentrations and must be injected into the joint.

Ear, Sinuses, and Tears

Most of the penicillins, including penicillin G, penicillin V, ampicillin, and amoxicillin, reach levels in middle ear fluid in acute otitis adequate to eradicate the major organisms involved in this infection. Amoxicillin and bacampicillin yield higher middle ear fluid levels than ampicillin. Cephalosporins reach the middle ear but, with the exception of cefaclor, their concentrations are too low to inhibit most strains of *Haemophilus influenzae*. Erythromycin, sulfonamides, and trimethoprim achieve adequate middle ear fluid levels. In chronic otitis in which there is a great deal of scarring, middle ear fluid levels of penicillins are often inadequate, and perhaps in these situations large doses of the drugs are needed. In such situations benzathine penicillin does not yield adequate levels.

Concentrations of antibiotics in sinuses have been shown to be adequate for ampicillin, amoxillin (and the other aminopenicillin esters), tetracyclines, erythromycin, sulfonamides, and trimethoprim. High concentrations of sulfonamides, minocycline, and rifampin are present in the lacrimal secretions, which bathe the posterior pharynx where meningococci are harbored, explaining the value of these agents for the prophylaxis of meningococcal disease.

Eye

Very few antibiotics penetrate the eye well. Measurements in patients undergoing cataract extraction or rabbit models to determine tissue levels in the eye have been disparate. In general, the levels of penicillins and cephalosporins in the aqueous are less than 10 per cent of the peak serum level and so inhibit only highly sensitive bacteria such as pneumococci or streptococci (Records, 1968). The penicillinase-resistant penicillins (methicillin, oxacillin, and others) do not penetrate the aqueous of noninflamed eyes, but, in the presence of infection, achieve levels adequate to inhibit staphylococci, provided large doses are given intravenously. Cefamandole can reach levels in the aqueous high enough to inhibit streptococci and staphylococci. Aminoglycosides do not achieve adequate aqueous levels when given by I.M. or I.V. injection. Chloramphenicol given orally or I.V. yields measurable levels in the aqueous, but not if injected subconjunctivally as the succinate salt (McPherson et al., 1968). Amphotericin concentrations are not measurable when the drug is administered parenterally.

Subconjunctival instillation of aminoglycosides, gentamicin, or tobramycin has produced levels adequate to treat experimental *Pseudomonas* infection in the rabbit eye. Amphotericin has also been administered by this technique. No significant penetration of any antibiotic into the vitreous has been demonstrated, and intracameral injection of aminoglycosides has been used on rare occasions.

Skin

Highly protein-bound penicillins and cephalosporins have not achieved high concentrations in skin windows. Tetracyclines and clindamycin concentrate in skin tissue, accounting for their effectiveness in the treatment of acne. Minocycline, because of its high lipid solubility, has the greatest concentration in skin among the tetracyclines. Chloramphenicol also achieves excellent levels in skin tissues but obviously should not be used for skin infections unless other agents are not available. In second and third degree burns, antibiotics do not penetrate the subeschar level, and even sensitive bacteria can proliferate in this area although serum levels are very high.

Kidney

Renal parenchymal concentrations of antibiotics differ in relation not only to renal blood flow but also to the state of hydration and the presence of other drugs within the kidney that compete for transport mechanisms. In general, it has been stated that cure of urinary tract infection is related to urinary levels rather than to serum levels (Whelton et al., 1972). This is well substantiated by compounds such as nalidixic acid or the nitrofurantoins that are effective in the treatment of urinary tract infections in spite of inadequate serum levels. In the treatment of pyelonephritis, high concentrations within medullary, cortical, and papillary interstitial fluid may be important. Penicillin G and ampicillin levels are two to eight times the serum level. Similar findings have been made for cephalosporins. Aminoglycosides concentrate in the renal cortical tissue, which undoubtedly contributes to the development of proximal tubular cell damage. Bound gentamicin may also function as an antibiotic, however. The renal cortical concentrations of tobramycin and netilmicin are considerably less than those of gentamicin.

In the presence of markedly decreased renal function, oral ampicillin, cephalexin, and trimethoprim-sulfamethoxazole still give urinary concentrations adequate to treat infections by most urinary bacteria. In contrast, nitrofurantoins (Sachs et al., 1968), methenamine, and nalidixic acid do not yield urine levels in renal failure necessary to treat urinary infection unless toxic serum levels are achieved. The parenterally administered penicillins (carbenicillin and ticarcillin) and parenteral cephalosporins (such as cephalothin, cefazolin, cefamandole, and cefoxitin) yield adequate intrarenal and urinary concentrations even in the presence of markedly decreased renal function.

Prostate

The high degree of ionization and protein binding of most antibiotics excludes most antibiotics from the prostate (Stamey, 1972). In acute inflammatory prostatic infections, this is not a consideration since the acute inflammatory process allows the antibiotics to enter the infected gland. Although some sulfonamides, erythromycin, and rifampin enter the normal gland, these agents are usually not effective against the microorganisms that cause chronic prostatitis. Trimethoprim achieves prostatic fluid levels eightfold greater than serum levels, and doxycycline (Garnes, 1973) and minocycline also achieve high prostate levels.

Placenta

Penicillin G, ampicillin, amoxicillin, tetracyclines, clindamycin, lincomycin, erythromycin, cephalothin, and cephaloridine reach effective levels in fetal blood within a short time after administration to the mother. Penicillins cross the placental barrier poorly if they are very highly bound to proteins. Aminoglycosides, trimethoprim, and metronidazole cross the placenta, as do the antimalarial quinolines. Penicillin G, ampicillin, erythromycin, and clindamycin reach therapeutic concentrations in amniotic fluid readily, but only negligible amounts of streptomycin, tetracycline, and chloramphenicol are found there (Duignan et al., 1973).

Central Nervous System

Only lipid-soluble compounds reach the brain readily because of the blood-brain barrier created by tight intracellular junctions and specialized cells surrounding the brain capillaries (Rall, 1971). Chloramphenicol achieves high brain tissue levels. In the presence of inflammation, such as a brain abscess, penicillin G, ampicillin, methicillin, oxacillin, and nafcillin achieve levels of antibiotic high enough to be measured. Lincomycin and clindamycin do not produce adequate brain tissue levels nor do the aminoglycosides. Vancomycin does achieve levels in the brain in the presence of inflammation. Trimethoprim and many of the sulfonamides achieve adequate levels in brain tissue. Unfortunately, with most of the antibiotics, the kinetics of entry into the brain and the rate of decline of brain levels are unknown.

The level of antibiotics in the cerebrospinal fluid (CSF) is a function of lipid permeability, protein binding, and the secretion and removal of weak anions by the transport systems of the choroid plexus and other sites in which CSF is formed. CSF levels of most agents except chloramphenicol, trimethoprim, some sulfonamides, isoniazid, and 5-fluorocytosine are low in the absence of inflammation. Penicillins not only have difficulty passing through the blood-brain barrier into the CSF but are removed from the CSF by active transport. Penicillin G and ampicillin can achieve adequate CSF levels in the presence of inflammation (Barrett et al., 1966). Oxacillin, nafcillin, and methicillin administered at high doses (12 g/day) yield CSF levels adequate for the treatment of staphylococcal meningitis. Vancomycin also achieves adequate CSF levels, but most cephalosporins do not. Relapses of meningitis due to pneumococcus and meningococcus have occurred in patients being treated with cephalothin. Recent studies suggest that cefamandole and cefuroxime may yield CSF levels adequate to treat selected organisms, but the data are not firm enough to advocate their use. The tetracyclines enter CSF but to a lesser extent than does chloramphenicol. The concentration of a tetracycline is about 10 per cent of the serum level. Minocycline penetrates normal CSF more readily than the other tetracyclines. Erythromycin does not enter the CSF in the absence of inflammation. Although chloramphenicol produces CSF levels adequate to treat pneumococci, meningococci, and *Haemophilus*, the levels needed to eradicate some gram-negative bacteria may be higher than those that are usually achieved. Aminoglycosides do not penetrate the CSF adequately, and administration via the intrathecal route or intraventricular injection via an Omaya reservoir is necessary to treat *Pseudomonas* or multi-resistant Enterobacteriaceae. Polymyxins do not enter the CSF and must be given intrathecally if they are to be used in the treatment of central nervous system infections. Cefotaxime, moxalactam, ceftizoxime and cefoperazone enter the CSF in the presence of inflammation in concentrations adequate to inhibit *S. pneumoniae, H. influenzae, N. meningitidis,* and even many *E. coli* and *K. pneumoniae.*

Amphotericin achieves concentrations in CSF that are sufficient to treat cryptococcal meningitis but not meningitis due to *Coccidioides immitis* or amebas. Rifampin and ethambutol enter the CSF with inflammation of levels adequate to treat most mycobacteria. Quinine and pyrimethamine enter CSF and brain tissue in concentrations that inhibit Plasmodia. Adenosine arabinoside enters both brain tissue and CSF in concentrations adequate to inhibit growth of herpes simplex virus.

Intracellular

Many antibiotics do not enter phagocytic cells, such as polymorphonuclear cells and macrophages. Chloramphenicol, tetracycline, trimethoprim, rifampin, and isoniazid all enter these cells, and this property may be important in the eradication of intracellular bacteria. The uptake of amphotericin by the reticuloendothelial system may contribute to its destruction of fungi within these cells.

MODIFICATION OF ANTIBIOTIC PROGRAMS IN THE PRESENCE OF RENAL FAILURE

For some antibiotics, little or no dosage adjustment is required in the presence of renal failure, either because they are eliminated by extrarenal mechanisms or because the margin of safety is so great that accumulation does not result in toxicity. With the aminoglycosides, vancomycin, anti-*Pseudomonas* penicillins, sulfonamides, and trimethoprim, major adjustments in dosage are necessary. There are a number of methods available to adjust dosage in the presence of renal failure. These are based on the assumption that the drugs are eliminated in an exponential fashion and that the rate of elimination is proportional to the glomerular filtration rate. In order to reach a therapeutic level in critically ill patients with a decreased renal function, it may be necessary to administer a loading dose of antibiotics. Dosage adjustments may be made by giving an initial dose and then subsequent doses at prolonged intervals, which are calculated from the serum creatinine or creatinine clearance. The alternative method is to give the loading dose and then a fraction of the usual dose at the usual interval for the drug. This method avoids the high peak and low trough concentrations that result from the former, but, in some animal studies, aminoglycosides administered this way are more nephrotoxic.

Aminoglycosides require dose adjustment most commonly even when there is only moderate renal insufficiency because their toxic to therapeutic ratio is so narrow. Since, in hospitalized patients, medications are conveniently administered at certain regular intervals, Table 2 shows a method for estimating the dose to be given at fixed intervals to provide effective but safe peak blood levels for all degrees of renal insufficiency. The creatinine clearance can be estimated from the serum creatinine as follows:

$$C_{cr} = \frac{(140 - \text{Age})\ \text{Wt}^*}{72\ \text{Cr}}$$

This method provides only an estimate of the serum levels that will be obtained, and these levels should be measured. If this method is followed, the blood levels may fall below the minimal inhibitory concentration toward the end of the interval between doses, but there is no evidence that sustained levels above the MIC are necessary.

All techniques for dosage adjustment are guides to therapy and should be confirmed by

*Weight in kilograms.

TABLE 2. Aminoglycoside Dosing Chart

1. Select loading dose in mg/kg [ideal weight] to provide peak serum levels in range listed below for desired aminoglycoside.

AMINOGLYCOSIDE	USUAL LOADING DOSES	EXPECTED PEAK SERUM LEVELS
Tobramycin Gentamicin	1.5 to 2.0 mg/kg	4 to 10 μg/ml
Amikacin Kanamycin	5.0 to 7.5 mg/kg	15 to 30 μg/ml

2. Select maintenance dose (as percentage of chosen loading dose) to continue peak serum levels indicated above according to desired dosing interval and the patient's corrected creatinine clearance.

PERCENTAGE OF LOADING DOSE REQUIRED FOR DOSAGE INTERVAL SELECTED

C(c)cr (ml/min)	Half Life† (hr)	8 hr	12 hr	24 hr
90	3.1	84%	–	–
80	3.4	80	91%	–
70	3.9	76	88	–
60	4.5	71	84	–
50	5.3	65	79	–
40	6.5	57	72	92%
30	8.4	48	63	86
25	9.9	43	57	81
20	11.9	37	50	75
17	13.6	33	46	70
15	15.1	31	42	67
12	17.9	27	37	61
10	20.4	24	34	56
7	25.9	19	28	47
5	31.5	16	23	41
2	46.8	11	16	30
0	69.3	8	11	21

†Alternatively, one half of the chosen loading dose may be given at an interval approximately equal to the estimated half life.
Modified from Sarubii, F.A., and Hull, J.H.: Amikacin serum concentrations: prediction of levels and dosage guidelines. Ann Intern Med 89:612, 1978.

measuring serum levels. Checking the serum level at the time of anticipated peak concentrations (just after an I.V. dose or 30 to 60 minutes after an I.M. dose) and just before the next dose establishes both whether the levels are adequate and if the agents are accumulating unduly.

Effects of Peritoneal Dialysis and Hemodialysis

Because patients who are treated by dialysis have minimal or no renal function, no antibiotic is eliminated in the urine. However, many antibiotics are removed by dialysis, and during that procedure they are eliminated at a constant rate. The exact rate of elimination depends both on the characteristics of the drug (i.e., protein binding, charge, and molecular weight) and on the conditions of dialysis (such as the machine that is used, the flow-pressure relationships, and the duration of dialysis). Table 1 provides data on which antibiotics can be removed by dialysis. If a drug is removed, it is necessary to administer an additional dose after dialysis. It is possible to add the antibiotic to the peritoneal dialysate at the same concentration that is desired in the blood. If that is done, it is not necessary to administer additional doses of the antibiotic. If the toxic to therapeutic ratio is narrow, it should be measured just before and just after dialysis to determine whether safe but effective concentrations have been attained. If that is not possible, then drugs with a wider margin of safety should be used.

TOXICOLOGY OF ANTIMICROBIAL AGENTS

Antimicrobial agents cause direct toxicity, can interact with other drugs to influence the toxicity of the other agent, or can by alteration of microbial flora result in infection with organisms that are normally saprophytic (Table 3).

Hypersensitivity Reactions

Almost any antimicrobial agent has the potential of being an antigenic stimulus and provoking an immunologic reaction due either to immuno-

TABLE 3. Major Toxicities of Selected Antimicrobial Agents

AGENT	MECHANISM	SIGNS
Hematologic		
Chloramphenicol	Inhibits protein synthesis	Reversible anemia, leukopenia
Chloramphenicol	Damages stem cell	Aplastic anemia
Sulfonamides	G-6-PD deficiency	Hemolytic anemia
Carbenicillin and ticarcillin	Platelet aggregation inhibited	Bleeding
Nervous System		
Aminoglycosides	Binds hair cells of organ of Corti	Deafness
Aminoglycosides	Binds vestibular cells	Vertigo
Aminoglycosides	Competitive neuromuscular blockade	Respiratory paralysis
Polymyxins	Noncompetitive neuromuscular blockade	Respiratory paralysis
Penicillins and cephalosporins	Cortical stimulation	Myoclonic seizures
Gastrointestinal		
Rifampin, isoniazid and tetracyclines	Liver cell damage	Hepatitis
Neomycin	Villous damage	Malabsorption
Clindamycin, lincomycin	Clostridium difficile	Diarrhea
All agents	Altered bowel flora	Diarrhea

globulins or to sensitized lymphocytes, but in fact the most common agent to do this has been penicillin G or one of the other penicillins. Penicillins are widely dispersed in nature and contaminate milk products and other foods so that individuals who have never received penicillin for medical reasons have been otherwise exposed to them. Penicillin G and its derivatives act as haptens that can combine either with proteins contaminating the penicillin solution or with human proteins. The most important antigenic component of penicillin is the penicilloyl determinant. This hapten develops after the beta-lactam ring opens and allows formation of an amide linkage with cellular and serum proteins. Penicillenic acid is produced by normal breakdown in solution of penicillin G, ampicillin, and carbenicillin, and this material also makes amide linkages with serum proteins. Both compounds are the major determinants of penicillin allergy. Minor determinants of allergy are benzylpenicillin itself and benzylpenicilloate (Levine et al., 1966). Both major and minor determinants may be involved in anaphylactic reactions and urticaria. These reactions are due to IgE antibodies. Immediate anaphylactic reactions are uncommon, occurring in 0.04 to 0.004 per cent of patients treated with penicillin.

Positive skin tests with benzylpenicilloyl polylysine and with penicillin G, benzylpenicilloic acid, and benzylpenicilloate predict who will have immediate reactions to penicillins (Voss et al., 1966). An immediate wheal and flare reaction to the major and minor determinants after skin pricks or intradermal injection indicates a high probability of immediate (2 to 30 minutes) anaphylactic or accelerated (1 to 72 hours) urticarial reactions (with wheezing, pharyngeal edema, or local inflammation) to penicillins. Negative skin tests virtually rule out an anaphylactic reaction, although they do not exclude late cutaneous reactions. These later skin reactions probably are mediated by IgM antibodies. Indeed, continued administration of penicillin during the development of the skin eruption may elicit production of blocking IgG antibodies that prevent allergic reactions. Serum sickness from immune complexes containing IgG antibodies can develop, but it appears to be rare now that purer preparations of penicillin, made with improved production methods, are being used.

Ampicillin produces rashes twice as frequently as do other penicillins (7 per cent versus 3 per cent). The mechanism of this reaction is not understood. The rash occurs in almost all individuals with Ebstein-Barr virus infection (infectious mononucleosis) and cytomegalovirus infections. Rechallenge of these individuals with penicillin does not elicit a reaction. Allergic cross-reactions with cephalosporins in patients allergic to penicillins are uncommon.

Anaphylaxis and serum sickness have been associated with most of the other antimicrobial agents but are uncommon for erythromycin, clindamycin, and chloramphenicol. Serum sickness due to antigen-antibody complexes has complicated treatment with sulfadiazine and sulfathiazole.

Drug fever mediated via sensitized lymphocytes has been associated with all antimicrobials. Indeed, allergy to cephalosporins is manifest more often as fever than as a rash. Isoniazid can produce a syndrome indistinguishable from systemic lupus erythematosus that is presumably on an allergic basis. Both penicillins and sulfonamides have produced a nonspecific vasculitis similar to Henoch-Schönlein purpura.

Skin Reactions

Cutaneous reactions ranging from urticaria, fixed drug eruptions, and photodermatitis to exfoliative dermatitis, toxic epidermal necrolysis, and erythema nodosum have been reported with antimicrobial agents of every class. Sulfonamides and penicillins are the most common offenders. Tetracyclines and antituberculosis medications (such as isoniazid, ethambutol, rifampin, and particularly para-aminosalicylic acid) have produced dermatologic reactions. Antimalarials have produced alopecia. Nalidixic acid, griseofulvin, chlortetracycline, and demethylchloretetracycline (Saslaw, 1961) cause photosensitivity. This reaction is probably phototoxic rather than photoallergic and apparently results from conversion of the drug by light to a noxious agent. It can range from intense sunburn and loosening of the nails with tetracyclines to bullous skin eruptions with nalidixic acid. For unknown reasons, chlorination makes the tetracyclines phototoxic.

Hematologic Reactions

Pancytopenia and aplastic anemia have been produced by chloramphenicol. The incidence of aplastic anemia is estimated at one in 60,000. There are two types of marrow reaction to chloramphenicol. In one type, which occurs in nearly everyone after prolonged serum levels over 25 μg/ml, there is gradual development of anemia and mild thrombocytopenia. Serum iron increases, and the reticulocyte count falls. Bone marrow cells show maturation arrest with vacuolization of the cytoplasm of erythroid cells. These changes reverse if chloramphenicol is discontinued. The second type of marrow depression is not reversible or dose-related. It has occurred in individuals who have had multiple exposures to the drug and is characterized by fatal aplastic anemia. Its occurrence in identical twins raises the question of genetic predisposition (Nagao and Mauer, 1969).

Hemolytic anemia occurs in individuals with glucose-6-phosphate dehydrogenase deficiency who take eight aminoquinolines, such as quinacrine, sulfonamides, nitrofurans, sulfones, nalidixic acid, and chloramphenicol. Immune hemolysis has developed during treatment with PAS, penicillins, and cephalosporins. With cephalosporins, most positive Coombs' tests are false-positives, and hemolysis is very rare. Hemolysis due to membrane damage occurs with amphotericin, but most of the anemia seen with amphotericin results instead from marrow suppression (Brandiss et al., 1964).

Leukopenia and agranulocytosis have followed use of chloramphenicol, penicillins, cephalosporins, sulfonamides, dapsone, trimethoprim, and 5-fluorocytosine. In most instances, the reaction is reversible. In some cases leukopenia due to penicillins and cephalosporins is dose-related, and reduction in dose causes the white count to become normal. Sulfisoxazole has produced a pseudo Pelger-Huët anomaly of white cells.

Thrombocytopenia can result from an immunologic reaction with penicillins, sulfonamides, cephalosporins, and rifampin. Direct damage to platelets has occurred with other antibiotics. Trimethoprim produces thrombocytopenia, leukopenia, and megaloblastic anemia by interference with the dihydrofolate reductase enzyme (Kahn et al., 1968). A serious platelet injury is caused by carbenicillin and other penicillins through unknown mechanisms. When carbenicillin is given in therapeutic doses, all recipients develop defective platelet function within 12 to 24 hours. The drug or a metabolite appears to impair the function of platelets by decreasing their sensitivity to aggregation by adenosine diphosphate (ADP). Since ADP is the physiologic agent responsible for aggregation of platelets into plugs that repair blood vessels, it is not surprising that carbenicillin frequently causes prolonged bleeding and that it is responsible for bleeding problems after surgery (Brown et al., 1974).

Cardiovascular Toxicity

Most cardiovascular toxicity occurs with antiparasitic drugs. Emetine and pentamidine can cause hypotension and electrocardiographic changes. Quinine depresses conduction velocity. Rapid injection of large doses of a potassium salt of penicillin can produce cardiac arrest. Miconazole has produced hypotension and arrhythmias.

Gastrointestinal Toxicity

Many antibiotics produce gastrointestinal complaints that vary from hairy tongue to enterocolitis. Severe diarrhea progressing to pseudomembranous enterocolitis has occurred with ampicillin, indanylcarbenicillin, tetracyclines, and chloramphenicol, but more frequently with lincomycin and clindamycin. The mechanism for this toxicity appears to be alteration of intestinal flora. Proliferation of *Clostridium difficile* and accumulation of toxin within the colon has been established as one cause of this syndrome (Bartlett et al., 1978). Nonabsorbable aminoglycosides, such as neomycin, produce malabsorption of fat, protein, and carbohydrates. Oral neomycin causes striking villous shortening, round cell infiltration of the upper small bowel, and crypt-cell damage. This mucosal injury, plus an interaction of neomycin with bile salts, are responsible for malabsorption.

HEPATIC TOXICITY

Cholestatic jaundice has followed use of tetracyclines, erythromycin estolate, oxacillin, nitrofurans, and sulfonamides. Tetracyclines, taken in doses above 2 g/day during the third trimester of pregnancy or postpartum, have produced hepatic damage with extensive fatty changes in the liver and a fatality rate of 80 per cent (Schultz et al., 1963).

Antituberculosis agents are the antimicrobials most often associated with hepatocellular toxicity. Isoniazid produces hepatitis, but the incidence of this side effect varies with the age of the patient (Brummer et al., 1971). The risk of individuals below 20 years of age developing hepatitis due to isoniazid is 0.03 per cent, while those above 35 years, the risk is over 1 per cent. Approximately 10 per cent of children receiving isoniazid develop abnormal liver chemistry values, but clinical signs of hepatitis are extremely rare in children. Toxicity due to isoniazid is not related to previous hepatic disease but is more common in rapid acetylators of isoniazid, suggesting that the toxic component is the acetyl derivative. Para-aminosalicylic acid, ethionamide, and pyrazinamide have produced fulminant hepatic necrosis. Rifampin is infrequently associated with liver toxicity but seems to increase the likelihood of isoniazid toxicity.

Sulfonamides may produce both hepatocellular and cholestatic liver damage. Penicillins (penicillin G, ampicillin, oxacillin, and carbenicillin) rarely produce a hepatitis and usually in association with a general allergic reaction. Erythromycin, many of the penicillins, and cephalosporins produce elevations in serum glutamic-oxaloacetic transaminase by mechanisms that are unexplained.

Respiratory Toxicity

Nitrofurantoin produces two forms of pulmonary reaction (Hailey et al., 1969). In the acute form, chills, fever, cough, and dyspnea begin 2 to 10 days after the antibiotic is started. X-rays reveal diffuse alveolar infiltrates, and there usually is an eosinophilia. This reaction resolves when the drug is stopped. The second reaction to nitrofurans is a chronic one that occurs after the patient has taken the drug for a prolonged period, usually months. Dyspnea, cough, and cyanosis develop gradually. The x-ray shows diffuse basilar interstitial infiltrates, and pulmonary function studies show restrictive lung disease. Partial recovery occurs if the drug is stopped.

Metabolic Toxicity

Hypoglycemia can follow pentamidine administration. Use of large doses of penicillins, particularly the anti-*Pseudomonas* agents like carbenicillin, present such a large load of nonreabsorbable anion to the distal tubule that hypokalemia results. Carbenicillin also can produce sodium overload since it contains 4.7 mEq/g and is given in doses of 24 to 40 g/day. Demethylchlortetracycline produces nephrogenic diabetes insipidus (Singer and Rotenberg, 1973). Amphotericin B and stale (outdated) tetracyclines produce renal tubular acidosis and hypokalemia.

Renal Toxicity

Many factors intrinsic to the kidney make it particularly vulnerable to nephrotoxic reactions to antimicrobial agents (Fig. 1). The most common form of renal damage due to penicillins is an acute interstitial nephritis that has been associated with methicillin therapy but can occur with any penicillin. Fever, eosinophilia, and rash commonly antedate or accompany the renal lesion. Urinalysis may reveal white cells, casts, and proteinuria, but microscopic hematuria is most common. Urinary output is normal at first, although the serum creatinine and urea nitrogen increase. With discontinuation of the drug, the renal function of most patients returns to normal. Renal biopsy shows patchy tubular damage with interstitial edema, accumulation of lymphocytes, monocytes, eosinophils, and plasma cells. Glomeruli are normal. Most of the evidence suggests that this is a hypersensitivity reaction. Antitubular basement membrane antigens have been demonstrated in the serum in methicillin nephritis; and IgG, C_3, and a methicillin antigen were present in a linear pattern along the tubular basement membrane (Border et al., 1974).

Nephrotoxicity due to cephaloridine is dose-related (Silverblatt et al., 1973). There is proximal tubular damage due to accumulation of high

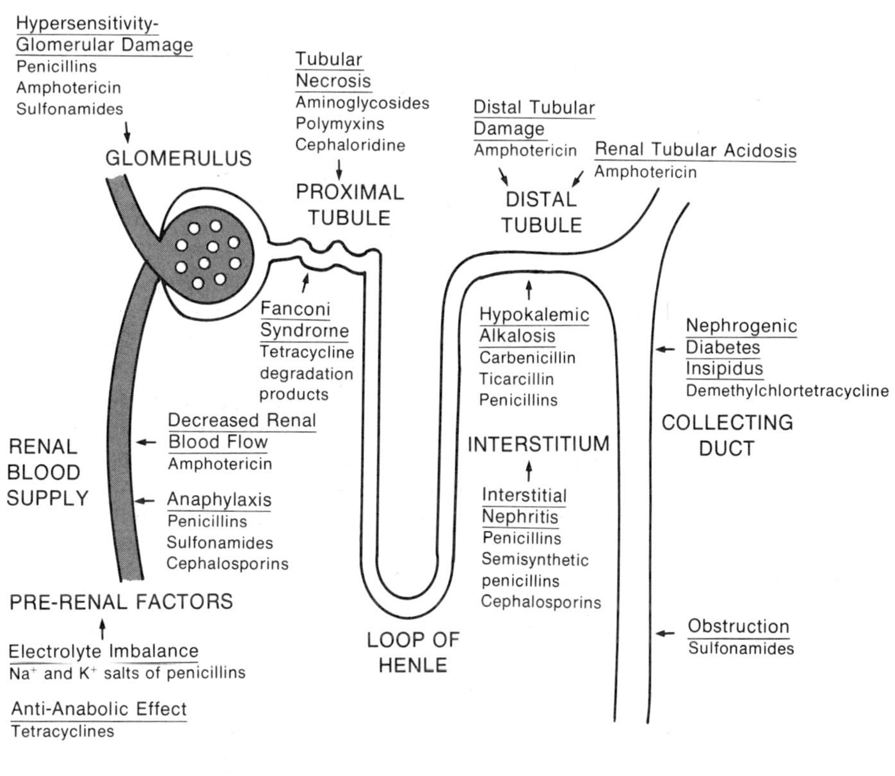

FIGURE 1. *Nephrotoxic sites of antimicrobial agents.*

intracellular concentrations of the drug in cortical tubular cells. The other cephalosporins have not been convincingly shown to be nephrotoxic.

Polymyxins bind to proximal tubular cells and damage cellular membranes. The aminoglycosides vary in their nephrotoxic potential ranging from neomycin, the most toxic agent, to streptomycin, the least toxic. Gentamicin and amikacin appear to be more nephrotoxic in animals than tobramycin, but clinical differences among the agents are difficult to establish. Gentamicin damages proximal tubules with loss of renal tubular enzymes, concentrating ability, and the development of intracytoplasmic bodies that seem to be injured cellular organelles. Aminoglycoside nephrotoxicity is dose-related and seems to be increased by simultaneous administration of cephalosporins. The overall incidence of nephrotoxicity due to aminoglycosides is about 5 to 10 per cent.

Crystallization-producing obstructive nephropathy, allergic and hypersensitivity reactions, precipitation of hemoglobin casts secondary to drug-induced hemolysis, and an intrinsic toxic effect have all been reported as secondary to sulfonamide therapy. Whether trimethoprim produces nephrotoxicity is unclear, but it does seem to alter handling of creatinine by the kidney.

Amphotericin can cause renal vasoconstriction and damage to proximal and distal tubular cells with intratubular calcium deposits. Renal abnormalities occur in about 25 per cent of patients who receive the drug. Most of the toxicity due to amphotericin is reversible, provided less than 5 g have been given. The primary disturbance is distal tubular acidosis (Burgess and Birchall, 1972). The drug seems to produce a permeability defect that allows the tubular cells to leak potassium and become permeable to hydrogen ions. Hydrogen ions then pass from tubular urine into the cells. The overall picture of amphotericin toxicity to the kidney is one of azotemia, acidosis, hypokalemia, and polyuria. Daily supplements of 100 mg of potassium are often needed after even small doses of amphotericin.

Neurologic Toxicity

Peripheral neuropathy has been produced by isoniazid most often in patients who acetylate the drug slowly. It is dose-related and can be reversed by pyridoxine (La Du, 1972). Ethionamide also produces peripheral neuropathy. High concentrations in the blood of nitrofurans, polymyxins, and emetine cause peripheral neuropathy in patients with decreased renal function. Polymyxins and aminoglycosides produce a reversible neuromuscular blockade resembling myasthenia gravis (Warner and Sanders, 1971). This toxicity is dose-related and has followed use of the agents as a peritoneal lavage. The drug produces a competitive blockade at the myoneural end-plate. Aminoglycosides probably compete with acetylcholine for receptor sites so that the end-plate is not depolarized. Neostigmine reverses this paralysis (Ream, 1963). Polymyxins also produce a noncompetitive prolonged depolarization due to calcium depletion, which can be reversed by calcium (Zauder et al., 1966).

Central nervous system excitation and seizures can be induced by isoniazid, amantadine, and penicillin. Penicillin-induced seizures are a myoclonic type and are more frequent in elderly patients and following cardiopulmonary by-pass (Seamans et al., 1968). Direct instillation of methicillin, oxacillin, or cephalosporins into the ventricles when CSF shunts are installed usually does not produce seizures. Cycloserine also causes seizures.

Ototoxicity

Aminoglycosides vary in their potential to produce deafness. Neomycin is the most ototoxic, followed by kanamycin, gentamicin, and amikacin. Tobramycin and netilmicin are less ototoxic. Deafness is the result of damage to the hair cells of the organ of Corti (Friedman et al., 1966). The mechanism for this reaction is unknown. Sensory cells of the vestibular system are also damaged by aminoglycosides, and, although it was originally thought that certain aminoglycosides produced only vestibular damage and other agents produced only auditory damage, it is clear that individual drugs can injure both vestibular and auditory function. Streptomycin and gentamicin also seem to have a greater propensity to produce vestibular toxicity than auditory toxicity, but kanamycin, neomycin, tobramycin, and any of the agents can produce either or both forms of toxicity. Both forms of ototoxicity seem to occur most frequently in older people, those who have received repeated doses of aminoglycosides, and individuals with decreased renal function in whom the drugs accumulate. The exact incidence of ototoxicity is not known, since hearing loss first affects perception of sounds above 2000 cycles/sec, which is outside the conversational range and therefore overlooked by patients and physicians, and the body adjusts for minor damage to the vestibular system, making its recognition difficult.

The other antibiotic that causes vestibular dysfunction is minocycline (Williams et al., 1974). This occurs more often in females. This toxicity can often be avoided by giving the drug in divided

TABLE 4. Drug Interactions of Antimicrobial Agents With Other Compounds

INTERACTING DRUGS	ADVERSE EFFECT
Aminoglycoside antibiotics with:	
Cephaloridine	Nephrotoxicity
Cephalothin and other cephalosporins	Nephrotoxicity
Curariform drugs	Neuromuscular blockade
Digoxin	Possible decreased digoxin effect
Ethacrynic acid	Ototoxicity
Furosemide	Ototoxicity
Methoxyflurare	Nephrotoxicity
Polymyxins	Nephrotoxicity
Aminosalicyclic acid (PAS) with:	
Probenecid	Aminosalicylic acid toxicity
Amphotericin B with:	
Curariform drugs	Curare effect
Digitalis	Digitalis toxicity
Cephaloridine with:	
Aminoglycoside antibiotics	Nephrotoxicity
Ethacrynic acid	Nephrotoxicity
Furosemide	Nephrotoxicity
Cephalosporins with:	
Aminoglycoside antibiotics	Nephrotoxicity
Chloramphenicol with:	
Dicoumarol	Excessive anticoagulation
Phenytoin	Phenytoin toxicity
Hypoglycemics, oral	Hypoglycemia
Griseofulvin with:	
Anticoagulants, oral	Decreased anticoagulant effect
Isoniazid with:	
Aluminum antacids	Decreased isoniazid effect
Disulfiram	Psychotic episodes, ataxia
Phenytoin	Phenytoin toxicity
Lincomycin with:	
Kaolin-pectin	Decreased lincomycin effect
Metronidazole with:	
Alcohol	Antabuse-like reaction
Disulfiram	Psychosis
Nalidixic acid with:	
Dicoumarol	Prolonged prothrombin time
Polymyxins with:	
Aminoglycoside antibiotics	Nephrotoxicity
Curariform drugs	Neuromuscular blockade
Rifampin with:	
Anticoagulants, oral	Decreased anticoagulant effect
Contraceptives, oral	Decreased contraceptive effect
Corticosteroids	Decreased corticosteroid effect
Hypoglycemics, oral	Possible decreased effect
Methadone	Methadone withdrawal symptoms
Sulfonamides with:	
Anticoagulants, oral	Anticoagulant effect
Hypoglycemics	Sulfonylurea hypoglycemia
Tetracyclines with:	
Antacids, oral	Decreased effect of tetracyclines
Barbiturates	Decreased doxycycline effect
Carbamazepine	Decreased tetracycline effect
Iron, oral	Decreased effect of tetracyclines
Methoxyflurane	Nephrotoxicity

doses rather than at a single time. Vancomycin can also produce deafness, and erythromycins at very high concentrations can cause temporary decrease of hearing.

ALTERATION OF MICROBIAL FLORA

Alteration of intestinal flora by use of antibiotics allows overgrowth of *Candida*, which often results in thrush (Solomon, 1961) or vaginitis and may cause disseminated candidiasis in patients with depressed resistance. Prolonged use of oral penicillins as prophylaxis against recurrent rheumatic fever has been complicated by colonization of the oropharynx with streptococci that are relatively resistant to penicillin. Administration of antibiotics in *Salmonella* gastroenteritis results in prolongation of the carrier state and occasionally increases the risk of bacteremia (Aserkoff and Bennett, 1969). Pseudomembranous colitis due to overgrowth of *C. difficile* has been discussed.

DRUG INTERACTIONS

Penicillins can inactivate aminoglycosides when the two are mixed in the same solution. Drug interactions within the patient are numerous and may result in increased toxicity of an agent or a decreased effect of a drug (Table 4). These antibiotic reactions with other agents in vivo involve a change in the metabolism of a compound. Microsomal enzymes that destroy more drug may be induced or there may be competition for enzymes involved in inactivation of both drugs. Binding to receptor sites on proteins or tissues, so that the second drug is blocked, may make more drug available. Competition for excretory pathways in the kidney or liver may result in accumulation of a drug.

References

Aserkoff, B., and Bennett, J. V.: Effect of antibiotic therapy in acute salmonellosis on the fecal excretion of salmonellae. N Engl J Med 281:636, 1969.

Barrett, F. F., Eardley, W. A., Yow, M. D., et al.: Ampicillin in the treatment of acute suppurative meningitis. J Pediatr 69:343, 1966.

Bartlett, S., Chang, T., Gururth, M., Gorback, S. L., and Guderdonk, S.: Antibiotic-associated pseudomembranous colitis due to toxin-producing clostridia. N Eng J Med 298:531, 1978.

Barza, M., Brusch, J., Bergeron, M., et al.: Penetration of antibiotics into fibrin loci in vivo. III. Intermittent vs. continuous infusion and the effect of probenecid. J Infect Dis 129:73, 1974.

Border, W. A., Lehman, D. H., Egan, J. D., et al.: Antitubular basement-membrane antibodies in methicillin-associated interstitial nephritis. N Engl J Med 291:381, 1974.

Brandiss, M. U., Wolff, S. M., Moores, R., et al.: Anemia induced by amphotericin B. JAMA 189:663, 1964.

Brown, C. H., Natelson, E. A., Bradshaw, W., et al.: The hemostatic defect produced by carbenicillin. N Engl J Med 291:265, 1974.

Brummer, D. L., et al.: Summary of the report of the Ad Hoc Advisory Committee on isoniazid and liver disease. Morbidity and Mortality Weekly Report, 20:231, 1971.

Burgess, J. L., and Birchall, R.: Nephrotoxicity of amphotericin B, with emphasis on changes in tubular function. Am J Med 53:77, 1972.

Butler, K., English, A., Knirsch, A., and Korst, J.: Metabolism and laboratory studies with indanylcarbenicillin. In Holloway, W. J.

(ed.): Infectious Disease Reviews. Wilmington, Futura Publishing Company, 1972, pp. 157–166.

Cohn, H. D.: Clinical studies with a new rifamycin derivative. J Clin Pharmacol 9:118, 1969.

Duignan, N. M., Andrews, J., and Williams, J. D.: Pharmacological studies with lincomycin in late pregnancy. Br Med J 3:75, 1973.

DuPont, H. L., Hornick, R. B., Weiss, C. F., et al.: Evaluation of chloramphenicol acid succinate therapy of induced typhoid fever and Rocky Mountain spotted fever. N Engl J Med 282:53, 1970.

Fabre, J., Milek, E., Kalfopoulos, P., et al.: The kinetics of tetracyclines in man. I. Digestive absorption and serum concentrations. In Doxycycline, A Compendium of Clinical Evaluations. New York, Pfizer Laboratories Division, Pfizer Inc., 1973, pp. 13–18.

Florey, T. E., Turton, E. C., and Duthe, E. S.: Penicillin in wound exudates. Lancet 2:405, 1946.

Friedmann, I., Dadswell, J. V., and Bird, E. S.: Electronmicroscope studies of the neuroepithelium of the inner ear in guinea-pigs treated with neomycin. J Pathol Bacteriol 92:415, 1966.

Garnes, H. A.: Doxycycline levels in serum and prostatic tissue in man. Urology 1:205, 1973.

Hailey, F. J., Glascock, H. W., Jr., and Hewitt, W. F.: Pleuropneumonic reactions to nitrofurantoin. N Engl J Med 281:1087, 1969.

Hughes, H. B.: On the metabolic fate of isoniazid. J Pharmacol Exp Ther 109:444, 1953.

Kahn, S. B., Fein, S. A., and Brodsky, I.: Effects of trimethoprim on folate metabolism in man. Clin Pharmacol Ther 9:550, 1968.

Kirby, W. M., and Regamey, C.: Pharmacokinetics of cefazolin compared with four other cephalosporins. J Infect Dis 128: Suppl. S341–346, 1973.

La Du, B. N.: Pharmacogenetics: Defective enzymes in relation to reactions to drugs. Annu Rev Med 23:453, 1972.

Levine, B. B., Redmond, A. P., Fellner, M. J., et al.: Penicillin allergy and the heterogeneous immune responses of man to benzylpenicillin. J Clin Invest 45:1895, 1966.

McCarthy, C. G., and Finland, M.: Absorption and excretion of four penicillins: penicillin G, penicillin V, phenethicillin, and phenylmercaptomethyl penicillin. N Engl J Med 263:315, 1960.

McPherson, S. D. Jr., Presley, G. D., and Crawford, J. R.: Aqueous humor assays of subconjunctival antibiotics. Am J Ophthalmol 66: 430, 1968.

Nagao, T., and Mauer, A. M.: Concordance of drug-induced aplastic anemia in identical twins. N Engl J Med 281:7, 1969.

Parker, R. H., and Schmid, F. R.: Antibacterial activity of synovial fluid during therapy of septic arthritis. Arthr Rheum 14:96, 1971.

Rall, D. P.: Drug entry into brain and cerebro-spinal fluid. In Brodie, B. B., and Gillette, J. R. (eds.): Handbook of Experimental Pharmacology: Concepts in Biochemical Pharmacology, Part. 1. New York, Springer-Verlag, 1971, pp. 240–248.

Ream, C. R.: Respiratory and cardiac arrest after intravenous administration of kanamycin with reversal of toxic effects by neostigmine. Ann Intern Med 59:384, 1963.

Records, R. E.: Intraocular penetration of cephalothin. Am J Ophthalmol 66:436, 1968.

Sachs, J., Geer T., Noell, P., et al.: Effect of renal function on urinary recovery of orally administered nitrofurantoin. N Engl J Med 278:1032, 1968.

Saslaw, S.: Demethylchlortetracycline phototoxicity. N Engl J Med 264:1301, 1961.

Schultz, J. C., Adamson, J. S., Jr., Workman, W. W., et al.: Fatal liver disease after intravenous administration of tetracycline in high dosage. N Engl J Med 269:999, 1963.

Seamans, K. B., Gloor, P., Dobell, R. A. R., et al.: Penicillin-induced seizures during cardiopulmonary bypass. A clinical and electroencephalographic study. N Engl J Med 278:861, 1968.

Silverblatt, F., Harrison, W. O., and Turck, M.: Nephrotoxicity of cephalosporin antibiotics in experimental animals. J Infect Dis 128: Suppl. S367–377, 1973.

Singer, I., and Rotenberg, D.: Demeclocycline-induced nephrogenic diabetes insipidus. In vivo and in vitro studies. Ann Intern Med 79:679, 1973.

Solomon, P.: Oral moniliasis complicating combined broad-spectrum antibiotic and antifungal therapy. N Engl J Med 265:847, 1961.

Stamey, T.: Urinary infections. Baltimore, The Williams & Wilkins Company, 1972, pp. 161–200.

Voss, H. E., Redmond, A. P., and Levine, B. B.: Clinical detection of the potential allergic reactor to penicillin by immunologic tests. JAMA: 196:679, 1966.

Warner, W. A., and Sanders, E.: Neuromuscular blockade associated with gentamicin therapy. JAMA 215:1153, 1971.

Whelton, A., Sapir, D. G., Carter, G. G., et al.: Intrarenal distribution of ampicillin in the normal and diseased human kidney. J Infect Dis 125:466, 1972.

Williams, D. N., Laughlin, L. W., and Lee, Y.: Minocycline: Possible vestibular side-effects. Lancet 2:744, 1974.

Zauder, H. L., Barton, N., Bennett, E. J., et al.: Colistimethate as a cause of postoperative apnoea. Canad Anaesthiol Soc J 13:607, 1966.

B. SPECIFIC MICROBIAL AGENTS OF DISEASE

1 *AEROBIC BACTERIA OR FACULTATIVELY ANAEROBIC BACTERIA*

Gram-Positive Cocci

STAPHYLOCOCCI 23

*Stephen I. Morse, M.D.**

The staphylococci make up the medically most important genus in the family *Micrococcaceae*. There are three species: *Staphylococcus aureus* is responsible for most cases of staphylococcal disease in man; *S. epidermidis* usually causes minor skin lesions; and the newly recognized species *S. saprophyticus* may produce bladder infection.

MORPHOLOGY

Staphylococci are nonmotile, nonspore forming, gram-positive cocci; in old cultures, or after ingestion by phagocytes, they may appear gram negative. Individual cells have a diameter of 0.7 to 1.2 μm and are characteristically grouped in irregular aggregates that resemble clusters of grapes, hence the name *staphylo* from the Greek *staphylē* — bunch of grapes (Fig. 1). Cell division takes place in successive perpendicular planes, but there is incomplete separation of the daughter cells and, instead of residual attachment along the division plane, the attachment point is usually eccentric to the division plane, resulting in irregular aggregates (Tzagaloff and Novick, 1977). Moreover, daughter cells may shift from their original site of attachment. Although clusters are usually seen in pathologic material and in growth on solid media, staphylococci often form short chains in liquid culture.

Some *S. aureus* strains possess distinct capsules, as demonstrated by the India ink method, electron microscopic techniques, or the Quellung reaction with specific capsular antiserum (Morse, 1978). Encapsulation may be more frequent in vivo than after cultivation in vitro.

The peptidoglycan of staphylococcal cell walls is characterized by unique pentaglycine bridges that link the tetrapeptides attached to the muramic acid residues (see Chapter 20). The pentaglycine bridges are specifically susceptible to the action of the enzyme lysostaphin, thereby providing a useful means of identifying the genus.

Osmotically unstable L forms, deficient in cell wall material, can be induced by agents that affect cell wall synthesis or structure — for example, penicillin or lysostaphin — and can be propagated in vitro in hypertonic medium. Small colony variants of *S. aureus* (G forms) occur when microbial growth is inhibited by a variety of agents, including antimicrobials, but also may appear spontaneously. Both L and G forms are resistant to many drugs, especially those affecting the cell wall, and it has been postulated that recrudescent *S. aureus* disease is sometimes due to reversion of persisting dormant L or G forms to invasive parental organisms.

On solid media, most strains of *S. aureus* produce a characteristic golden-yellow (*aureus*) carotenoid pigment; however, colonial coloration may vary from white to orange. *S. epidermidis* colonies usually are white and those of *S. saprophyticus* are white to greyish-white. Colonies of staphylococci are sharply defined, round, con-

*Deceased.

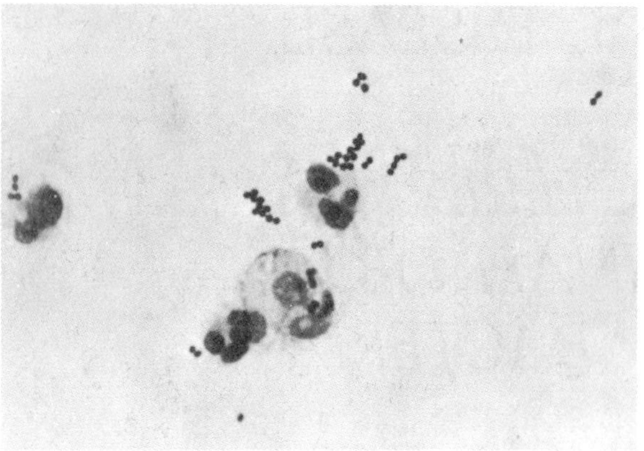

FIGURE 1. *Gram stain of exudate containing intracellular and extracellular staphylococci × 650. (From White, A., and Brooks, G. F.: Furunculosis, pyoderma, and impetigo. In Hoeprich, P. D. (ed.): Infectious Diseases. 2nd ed. Hagerstown. Harper and Row, 1977, p. 786.)*

vex, and measure 4 mm in diameter. On blood agar, *S. aureus* is usually surrounded by a zone of clear (β) hemolysis. Strains of *S. aureus* may produce one or more of four distinct hemolysins with different hemolytic specificities; therefore, the occurrence and extent of hemolysis depends upon both the strain and the source of blood. Some *S. epidermidis* strains are also hemolytic, whereas *S. saprophyticus* is nonhemolytic.

ANTIGENIC COMPOSITION

Species Antigens

The species-specific antigens of staphylococci are cell wall teichoic acids. In *S. aureus*, the antigen is a ribitol teichoic acid composed of a linear backbone of ribitol linked by phosphodiester bridges. *N*-acetylglucosamine is attached to the C4 position of the ribitol residues, and ester-linked *D*-alanine is attached to approximately 50 per cent of the C2 residues. The antigenic determinant is the glucosamine moiety, which may be in either α or β glycosidic linkage to ribitol. In most strains both types of glycosidic linkage are found, but in some only one is present. Therefore, antisera that will react with both anomers must be used for immunochemical determination of the species. Similarly, teichoic acids with both specificities must be used when testing for the presence of serum antibodies.

The *S. epidermidis* species-antigen is a glycerol teichoic acid in which glucose residues, in either α or β glycosidic linkage, are attached to a glycerol phosphate backbone and constitute the antigenic determinant. Cell walls of *S. saprophyticus* contain ribitol teichoic acid of two types. One has *N*-acetylglucosamine residues, whereas the other has glucose residues. Both may be present in the same strain.

Cellular Antigens of S. Aureus

Capsules. The few strains of *S. aureus* that are encapsulated in vitro tend to be more virulent in animals, and anticapsular antibodies protect against experimental disease. The capsule of one group of strains is a polymer of glucosaminuronic acid; that of another contains mannosaminuronic acid; and components characteristic of the peptidoglycan, e.g., glycine, alanine, and glucosamine, are found in the capsules of other strains.

Protein. Protein A (agglutinogen A) is a surface component of most strains of *S. aureus*. Depending upon the method of isolation, the molecular weight of protein A ranges between 13,000 and 42,000. Although the bulk of protein A is covalently linked to the peptidoglycan, approximately one third is released extracellularly. The most striking property of protein A is its nonspecific interaction with the Fc portion of the IgG of a wide range of mammalian species. The result of the interaction may be either precipitation or the formation of soluble complexes. In the case of human IgG, protein A reacts with subclasses IgG_1, IgG_2, and IgG_4, but not IgG_3. Protein A also binds to some samples of human IgM and IgA_2. The interaction with the Fc portion of IgG produces a variety of biologic effects, including: activation of complement by both the classical and the alternative pathways; local wheal and flare reaction; the Arthus phenomenon; local and systemic anaphylaxis; inhibition of phagocytosis of opsonized particles through competition with the Fc receptors of phagocytes for the Fc portion of opsonic antibody (Peterson et al., 1977); and in vitro induction of proliferation of both human T lymphocytes and B lymphocytes (Sakane and Green, 1978). Protein A is also a true antigen and reacts with the Fab portion of specific antibody.

Clumping Factor (Bound Coagulase). Unencapsulated strains of *S. aureus* clump when suspended in fibrinogen-containing solutions. There are apparently specific receptors on the bacterial surface for fibrinogen and the clumping is due to cross-linkage of the cells by fibrinogen. The receptors have not been characterized.

Extracellular Antigens and Products of S. aureus

Coagulase. *S. aureus* has the unique ability to clot a variety of mammalian plasmas. Clotting is caused by an extracellular product, coagulase (or free coagulase), which exists in several antigenically different forms, all having the same mechanisms of action. Virtually all naturally occurring strains of *S. aureus* produce coagulase, and since tests for coagulase are simple to perform, the *aureus* species is usually identified by coagulase production rather than by seriologic, physiologic, or biochemical tests. Coagulase-positive staphylococci are by definition *S. aureus*.

Coagulase does not clot fibrinogen directly but first reacts with a plasma constituent, coagulase-reacting factor (CRF), which is most probably prothrombin, to form a thrombin-like substance. The fibrinopeptides formed during clotting are identical to those produced by the action of thrombin on fibrinogen and, as in the case of thrombin-mediated clotting, diisopropylfluorophosphate (DFP) blocks the reaction.

Hemolysins. Four distinct hemolysins are produced by *S. aureus* and many strains manufacture more than one type (Wiseman, 1975). All cause β (clear) hemolysis. They have different lytic spectra with respect to the susceptibility of various erythrocyte species and are also cytotoxic for cells other than erythrocytes. The hemolysins are antigenic proteins; activity is neutralized by specific antiserum.

α-Hemolysin (α-toxin) is the most commonly encountered hemolysin in clinical isolates of *S. aureus*. Rabbit erythrocytes are highly susceptible, whereas human erythrocytes are unaffected. In experimental animals, α-hemolysin causes dermonecrosis at the site of local injection and is lethal when injected intravenously. The mode and site of action of α-hemolysin are unknown.

β-Hemolysin is found in many strains of animal origin but is produced by less than 20 per cent of human strains. It is a "hot-cold" hemolysin; lysis is maximal only after blood agar plates or toxin-erythrocyte mixtures are held at low temperatures after incubation at 37°C. β-hemolysin is a sphingomyelinase, which, in the presence of magnesium ions, catalyzes the breakdown of sphingomyelin to *N*-acylsphingosine and phosphorylcholine. Erythrocyte susceptibility is directly correlated with sphingomyelin content. Sheep, human, and guinea pig erythrocytes contain decreasing amounts of sphingomyelin and are decreasingly susceptible to β-hemolysin. β-Hemolysin is toxic for experimental animals only in high doses.

γ-Hemolysin consists of two components that act in concert to induce hemolysis. γ-Hemolysin is inhibited by sulfonated polymers, including agar, and hence its activity is not readily seen on blood agar plates. Rabbit, human, and sheep erythrocytes are susceptible to lysis.

δ-Hemolysin consists of aggregates of low molecular weight subunits and has a broad range of lytic and cytotoxic activity probably due to a nonspecific detergent-like action. Since serum phospholipids inhibit δ-hemolysin, it is not clear whether it functions in vivo.

Leucocidin. In addition to the leukocytoxic effects of some of the hemolysins, *S. aureus* also produces a distinct nonhemolytic leukocytotoxic substance, Panton-Valentine (P-V) leucocidin. P-V leucocidin contains two components, F (electrophoretically fast moving) and S (slow moving). Both are required for activity, and antibody to either component neutralizes toxicity. Human and rabbit polymorphonuclear leukocytes and macrophages are the only susceptible cells.

Enterotoxins. Five chemically and immunologically related enterotoxins (designated a-e) are produced by *S. aureus*. These proteins have a molecular weight of 3.5×10^4, are relatively heat stable, and only moderately susceptible to trypsin digestion (Bergdoll et at., 1974). They are responsible for many cases of food poisoning, but the mechanism of action has not been completely elucidated. It appears that both the bowel itself and neural innervation are affected.

Exfoliatin. Exfoliatin is an extracellular product that cleaves the stratum granulosum layer of the epidermis and causes several clinical syndromes. It is produced by many *S. aureus* strains in phage group II, and it is probable that more than one antigenic type exists.

A number of other extracellular substances with biologic activity are also produced by *S. aureus*. These include staphylokinase (fibrinolysin). hyaluronidase, and phospholipase.

The wide variety of antigenic components and extracellular products found in *S. aureus* is not seen in either *S. epidermidis* or *S. saprophyticus*. Some strains of *S. epidermidis* cause β-hemolysis owing to a hemolysin termed ϵ-hemolysin, which is similar to the *S. aureus* δ-hemolysin. *S. saprophyticus* is not hemolytic.

METABOLISM

S. aureus and *S. epidermidis* are facultative anaerobes that ferment sugars with the formation of large amounts of lactic acid, as occurs in lactic acid bacteria (Baird-Parker, 1974). They also resemble the spherical lactic acid bacteria in the low G + C content (30 to 35 moles per cent) of their DNA, but staphylococci possess heme-containing enzymes that allow normal respiratory metabolism, which is absent in lactic acid bacteria. The most important of these respiratory enzymes in staphylococci are cytochromes a, b, and o, which contain firmly bound prosthetic groups capable of donating or accepting electrons as they undergo oxidation or reduction. They are bound to the cell membrane of the staphylococcus and, with a group of quinones (menaquinones), form the membrane-bound electron transport system. The quinones are nonprotein carriers of relatively low molecular weight. Catalase, another heme enzyme possessed by staphylococci, is important for splitting hydrogen peroxide and preventing accumulation of this highly toxic compound. In the diagnostic laboratory, catalase activity helps distinguish staphylococci from streptococci, which are catalase deficient. Of particular usefulness in species differentiation is the capacity of *S. aureus* to ferment mannitol and to grow in concentrations of sodium chloride that are inhibitory for other microorganisms — for example 7.5 per cent NaCl. *S. aureus* also produces a heat-stable nuclease, not found in the other species. In contrast to *S. aureus* and *S. epidermidis*, *S. saprophyticus* grows very poorly under anaerobic conditions.

Although capable of anaerobic growth, *S. aureus* and *S. epidermidis* grow better aerobically. Moreover, they exhibit different metabolic activities under aerobic and anaerobic conditions. For example, oxygen is necessary for pigment production, and nonpigmented colonies of *S. aureus* grown anaerobically will develop their characteristic yellow color upon exposure to air (aerochromogens; compare with photochromogens among mycobacteria, Chapter 40). Glucose, lactose, maltose, and mannitol are energy sources for staphylococci anaerobically. These sugars are also metabolized in air, and so are other hexoses, pentoses, disaccharides, and sugar alcohols. The products of glucose metabolism change from lactic acid during fermentation to acetate and CO_2 in air. *S. epidermidis* produces nitrite from nitrate when grown in air but may not do so under anaerobic conditions because nitrate reductase does not function in the absence of air. For reduction of nitrate, oxygen is required not as a terminal electron acceptor but for the biosynthesis of a cytochrome, which mediates in electron transfer to nitrate.

GENETICS

Many of the properties and products of *S. aureus* are genetically controlled by plasmids. Plasmid regulation of antibiotic resistance, especially through the production of penicillinase (β-lactamase), is of profound clinical importance (Lacey, 1975).

In contrast to R-factors in gram-negative bacteria, penicillinase plasmids do not tend to carry other antibiotic resistance markers, with the exception of erythromycin. Hence, multiple drug resistance mediated by plasmids is unusual in *S. aureus*. There appear to be two molecular classes of staphylococcal resistance plasmids. The larger of these has a molecular weight of 20×10^6, may be isolated as a covalently closed ring molecule, and its replication appears to depend on cell division so that only one (or at most a few) copy occurs in each cell. The other type is much smaller and has a molecular weight in the range of 3×10^6. The smaller type may carry resistance determinants for tetracycline or chloramphenicol and multiply independently of cell division so that multiple copies are produced in each cell. As with gram-negative bacteria, chloramphenicol resistance is mediated by staphylococcal plasmids through the enzyme chloramphenicol acetyltransferase. In contrast, however, to the constituent enzymes of gram-negative R-factors, the penicillinase and acetyltransferase of resistant staphylococci are inducible (see Chapter 4). Transfer of genetic material between strains of *S. aureus* occurs by transduction. Most strains of *S. aureus* are lysogenic, and lysogenic bacteriophages belonging to serologic group B are capable of generalized transduction, in which bacterial DNA (rather than phage DNA) is incorporated with the phage particle. Spontaneous prophage induction and subsequent transduction occurs in mixed cultures in vitro and more importantly transduction occurs in vivo in animals and humans carrying an appropriate lysogenic donor and a suitable recipient strain of *S. aureus* (Novick and Morse, 1968). In addition to transduction, bacteriophages also exert genetic control by the process of phage conversion as in the production of staphylokinase and β-hemolysin. Under highly specialized circumstances, transformation reactions can occur, but conjugation as a means of genetic transfer in staphylococci is not seen.

PATHOGENIC PROPERTIES

Two types of disease are produced by *S. aureus*, invasive and toxinogenic. The hallmark of invasive disease is abscess formation. Most often the abscesses are superficial (furuncle), but in some cases the furuncles develop into burrowing lesions consisting of a number of interconnecting abscesses (carbuncle). Serious deep-seated disease usually is not seen in healthy people, but may occur in those debilitated by disease, malnutrition, extensive surgical procedures, and immunosuppression. In these instances, abscesses may occur in almost any organ; bacteremia is frequent with metastatic lesions appearing most commonly in the lung, bones, and kidney. Approximately two-thirds of patients with bacteremia may develop endocarditis due to *S. aureus* and the course is often fulminating. Inhalational staphylococcal pneumonia generally is a complication of viral pneumonia, particularly influenza.

No single component or product of *S. aureus* has been shown to be the primary determinant of either the initiation or the progression of invasive lesions. It appears, therefore that a number of factors may be involved. These include the intraphagocytic survival of a few cells; inhibition of phagocytosis by capsules and protein A (Peterson et al., 1977); and the toxic effects on cells and tissues of the hemolysins, leucocidin, and enzymatic products. Some strains of *S. aureus* appear to be more virulent for man than others, but the reasons for this difference are unknown.

There are two clinical syndromes produced by *S. aureus* that are not the consequence of invasion of host tissues. *S. aureus* is a major cause of food poisoning due to ingestion of preformed enterotoxin. Typically, staphylococcal food poisoning follows consumption of food such as custards, meats, pastries, or salad dressings that have been contaminated by enterotoxin-producing organisms from a food handler. If the food is kept at a temperature that permits bacterial multiplication, enterotoxin production occurs. If the foods are then heated, the organisms may be killed but the heat-stable enterotoxin persists. Alternatively, contamination occurs after cooking, in which case live organisms may also be ingested. Nausea, vomiting, and sometimes diarrhea appear one to six hours after ingestion of enterotoxin. Prompt recovery without therapy is the rule except in the case of infants and the elderly, where replacement of fluid and electrolytes may be required.

In patients with syndromes caused by exfoliation, viable phage group II organisms are present, with or without overt lesions. There are three basic syndromes that together are termed the staphylococcal scalded skin syndrome (Melish and Glasgow, 1970). The term derives from the most serious manifestation that occurs in infants and young children. Wide areas of skin are denuded and generalized bullous formation occurs (toxic epidermal necrolysis or Ritter's disease) (Fig. 2). In older children or adults the disorder is characterized by the formation of local bullae (Lyell's disease) or by a scarlatina-like rash that mimics streptococcal scarlet fever except that the tongue and palate are spared. The different clinical manifestations may be related to the individual's state of immunity to exfoliatin. Fatalities are unusual, since the lesions are at the stratum granulosum and severe exudation and transudation do not occur.

S. epidermidis is often responsible for minor skin abscesses, but in the compromised host and after major surgery, particularly involving the insertion of an intravascular prosthetic device, bacteremic disease may occur. *S. saprophyticus* may cause infection of the urinary bladder in man.

IMMUNITY

The roles of humoral antibodies and cell-mediated immunity in invasive *S. aureus* disease are controversial. In experimental animals, except for the uncommon encapsulated organisms, there is little evidence for protection against invasive disease by humoral anticellu-

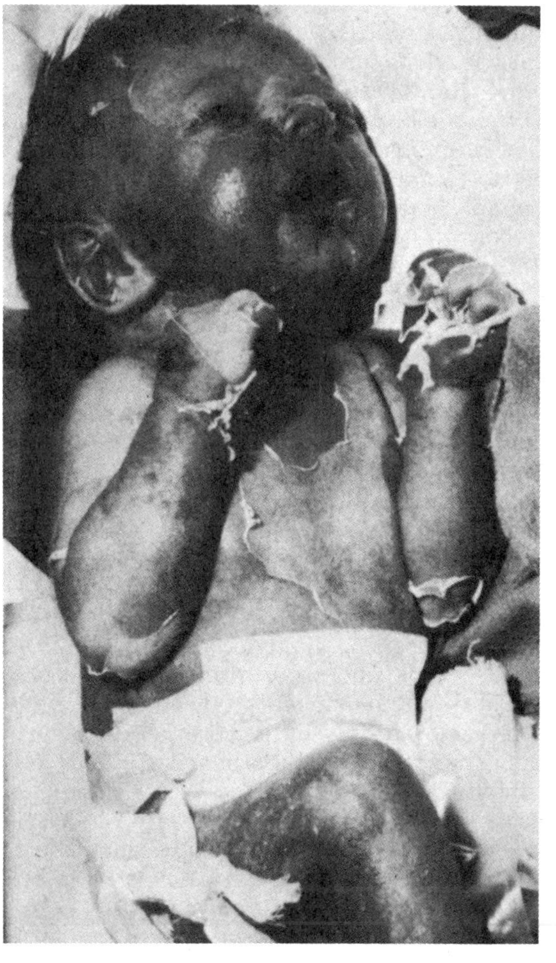

FIGURE 2. *Staphylococcal scalded skin syndrome in an infant. (From Melish, M. E., and Glasgow, L. A.: N Engl J Med 282:1114, 1970. Reprinted by permission from the N Engl J Med.)*

lar antibodies. Although toxic effects such as those produced by α toxin are diminished in experimental animals by antitoxin, toxic effects in man are usually not primary, and antitoxins do not appear to protect. Moreover, antibodies to a variety of components and products of *S. aureus* are found in normal humans, but there is no correlation between the presence of these antibodies and the occurrence and progression of disease.

In experimental animals, cell-mediated immunity may restrict the spread of organisms from the inoculation site but, as in the case of tuberculosis, local reactions are often more severe. Thus, the characteristic morphology and confinement of the staphylococcal abscess may be related to the effects of cell-mediated immunity.

Man has a high innate resistance to *S. aureus* disease, but resistance can be readily decreased by local factors. For example, intradermal injection of as many as 10^8 to 10^9 organisms usually causes no abscess, but if a suture containing as few as 10^2 organisms is passed through the skin and tied, marked abscess formation occurs, often requiring surgical and antimicrobial therapy. Other local factors, particularly those that impeded circulation, enhance susceptibility to staphylococcal disease.

There is considerable evidence that colonization of normal human skin or mucosa with one strain of *S. aureus* inhibits colonization with another strain, a phenomenon known as bacterial interference. This has been used therapeutically in the case of epidemics in newborn nurseries. The neonate is purposefully colonized in the nares with an avirulent strain of *S. aureus*, and suprainfection with an epidemic virulent strain is thereby prevented. The nature of the inhibition is unknown but may be based upon competition for nutrients, or, more likely, the production of bactericidins.

LABORATORY DIAGNOSIS

The diagnosis of staphylococcal disease is suggested by the finding of gram-positive cocci in clumps in pathologic material, but final diagnosis is achieved only by culture and appropriate tests. The characteristic β-hemolysis and yellow pigmentation strongly point to *S. aureus*, and selective media are useful. But, a positive coagulase test is required for identification. The test is performed by adding an aliquot of a broth culture or an inoculum from agar to 0.5 ml of rabbit plasma (either undiluted or diluted 1:10). Most strains produce clotting within three hours of incubation at 37°C, but cultures should be held overnight before a negative result is reported.

Identification of individual strains of *S. aureus* is often epidemiologically useful. Serologic tests are complex and typing sera are not readily available for routine use. Determining the spectrum of antibiotic resistance (resistogram) is simple but imprecise. Phage typing is the most useful technique and is based upon the susceptibility of *S. aureus* to lytic phages. The phages used for typing are divided into groups that have similar, but not identical, host range. An agar plate, subdivided into sections, is heavily swabbed with a suspension of the test strain and a standard drop of typing phage is placed onto each section. After overnight incubation, the areas of lysis are noted and the strain is designated by the identification number of the typing phages to which it is susceptible and by the phage group of the lytic phages. On the basis of phage susceptibility *S. aureus* strains are divided into four groups. Phage group II is distinctive in that production of bacteriocins and exfoliatin are limited to this group.

Typing phages for *S. epidermidis* strains are at present not available for general use and strains are usually distinguished by their patterns of resistance to antimicrobials, or their metabolic and biochemical properties (biotypes).

Strains of coalgulase-negative staphylococci, isolated from urine, can be identified as *S. saprophyticus* by their poor growth anaerobically and their resistance to novobiocin (M.I.C. > 2.0 μg/ml, Baird-Parker, 1974).

DRUG SUSCEPTIBILITY

When penicillin was first introduced, the vast majority of strains of *S. aureus* were susceptible. Depending upon the geographic area, as many as 50 per cent or more of strains occurring in the population-at-large are now penicillin-resistant, and 80 per cent of strains responsible for hospital-acquired disease may be resistant. Resistance is due to the plasmid-mediated enzyme penicillinase, which is a β lactamase that splits that β lactam ring of the penicillin nucleus (Lacey, 1975). In severe staphylococcal disease, unless the strain is sensitive to penicillin, penicillinase-resistant penicillins such as oxacillin and methicillin are the drugs of choice. In patients allergic to penicillin, cephalosporin derivatives may be used. Combined therapy with an aminoglycoside such as gentamycin is often employed, but there is no clear evidence that such combinations are more effective than the use of the bactericidal penicillins or cephalosporins alone. Vancomycin is used in patients who are allergic to the penicillins and cephalosporins.

Infections due to *S. epidermidis* are often caused by strains resistant to a variety of antimicrobial agents, and therapy of deep-seated disease may be very difficult. Strains of *S. saprophyticus* are generally sensitive to penicillin.

Superficial staphylococcal abscesses generally do not require antimicrobial therapy. Local application of moist heat, immobilization, and incision and drainage usually suffice. In both serious and minor staphylococcal disease, infected foci such as foreign bodies or necrotic bone or tissue must be removed.

EPIDEMIOLOGY

S. aureus is ubiquitous. Approximately 30 to 40 per cent of adults are asymptomatic carriers. Colonization begins in the neonatal period and approximately 90 per cent of infants discharged from hospital nurseries are carriers. The carrier rate in infants delivered at home is less but still substantial. The prevalence drops in the ensuing several years and then rises again to the adult level. The nares are the usual carrier site, but organisms can be found on skin and mucous membranes, and occasionally in the gut. *S. aureus* is extremely hardy and survives in the air and on inanimate objects and surfaces for long periods of time, but it is likely that person-to-person transmission is more important than environmental contamination.

Because of the widespread occurrence of *S. aureus*, it is inappropriate to consider methods to prevent acquisition of strains in a normal population. However, in "at risk" patients, such as newborns and patients in hospitals with serious underlying disorders, it is imperative that individuals with open lesions be isolated from the patient. In the face of hospital epidemics, which appear to be much less frequent than a decade ago, personnel who are carriers of the epidemic strain also should be removed from contact with such patients.

References

Baird-Parker, A.C.: *Micrococcaceae*. In Buchanan, R.E. and Gibbons, N.W. (eds.): Bergey's Manual of Determinative Bacteriology. 8th ed. Baltimore, Williams and Wilkins, 1974, p. 483.

Bergdoll, M.S., Huang, I.Y., and Schantz, E.J.: Chemistry of the staphylococcal enterotoxins. J Agr Food Chem 22:9, 1974.

Bernheimer, A.W.: Interactions between membranes and cytolytic bacterial toxins. Biochim et Biophysica Acta 344:27, 1974.

Cohen, J.O.: The Staphylococci. New York, John Wiley and Sons Incorporated, 1972.

Forsgren, A.: Immunological aspects of protein A. In Schlessinger, D., (ed.): Microbiology-1977. Washington, D.C., American Society for Microbiology, 1977, p.353.

Jeljaszewiez, J. (ed.): Staphylococci and Staphylococcal Diseases. New York, Gustav Fischer Verlag, 1976.

Lacey, R.W.: Antibiotic resistance plasmids of *Staphylococcus aureus* and their clinical importance. Bacteriol Rev 39:1, 1975.

Melish, M.E., and Glasgow, L.A.: The staphylococcal scalded-skin syndrome: development of an experimental model. N Engl J Med 282:1114, 1970.

Morse, S.I.: Staphylococci. In Davis, B.D., Dulbecco, R., Eisen, H.N., and Ginsberg, H.S. (eds.): Microbiology. 3rd ed. Hagerstown, Harper and Row, 1978.

Novick, R.P., and Morse, S.I.: In vivo transmission of drug resistance factors between strains of *Staphlococcus aureus*. J Exp Med 125:45, 1968.

Peterson, P.K., Verholf, J., Sabbath, L.D., and Quie, P.G.: Effect of protein A on staphylococcal opsonization. Infect Immun 15:760, 1977.

Sakane, T., and Green, I.: Protein A from *Staphylococcus aureus* — a mitogen for human T lymphocytes and B lymphocytes but not L lymphocytes. J Immun 120:302, 1978.

Tzagaloff, H., and Novick, R.P.: Geometry of cell division in *Straphylococcus aureus*. J Bact 129:343, 1977.

Wiseman, G.M.: The hemolysins of *Staphylococcus aureus*. Bacteriol Rev 39:317, 1975.

STREPTOCOCCUS 24

Isaac Ginsburg, M. Sc., Ph.D.

Since the discovery of the streptococcus by Billroth and Ehrlich in 1877, it has received special attention because it is involved in numerous diseases of humans and lower animals. Few other microorganisms of medical importance can elaborate as many exotoxins and enzymes and also produce serious infections in virtually any tissue. Certain streptococci are notorious for inducing nonsuppurative sequelae that affect the heart, kidney, joints, and brain, and take the form of rheumatic fever, acute glomerulonephritis (AGN), and chorea.

Although certain streptococci are human pathogens, others are integral parts of the normal flora of the mouth and intestinal tracts. The normal streptococcal flora may, however, on occasion invade the bloodstream and cause septicemia, usually associated with endocarditis. Certain streptococci that inhabit the mouth are involved in the pathogenesis of dental caries.

The genus *Streptococcus* belongs to the family Streptococcaceae. The most important species of streptococci with pathogenic properties for man are *Streptococcus pyogenes* (group A streptococci),

Streptococcus pneumoniae, Streptococcus agalactiae (group B), *Streptococcus faecalis* (enterococci — group D) and *Streptococcus viridans* (a heterogenous group of different species). The guanine + cystosine content of streptococcal DNA ranges from 33 to 44 moles per cent. Analysis of DNA-RNA homology among streptococci (including pneumococci) discloses a close relationship except for *S. faecalis* (Weissman et al., 1966).

GROUP A STREPTOCOCCI

MORPHOLOGY

Streptococci are gram-positive, usually nonmotile cocci that are arranged in chains. Growth occurs by elongation on the axis parallel to the chain and division is at right angles in the equatorial plane. Whereas certain streptococci — that is, pneumococci and the oral streptococci — grow in pairs or relatively short chains in liquid medium, the hemolytic streptococci grow in longer chains. These do not elongate indefinitely, since in some cases the streptococci produce a "dechaining factor."

The most typical colonies of *S. pyogenes* are disk-like, matt (dull), and 0.5 to 2.0 mm in diameter. The matt colonies are derived from younger mucoid colonies whose hyaluronic acid capsules become dehydrated. If plate cultures are sealed to prevent dehydration, the colonies of group A streptococci remain mucoid. With loss of virulence the matt form reverts to small, smooth colonies with a dry, glistening surface or a typical rough colony.

When grown on blood agar, various streptococcal groups can modify hemoglobin. Thus, most *S. faecalis* species do not modify hemoglobin and no discrete zones are formed around colonies on blood agar. Most oral streptococci and pneumococci modify hemoglobin to green pigments (biliverdin and other heme compounds) and are designated α-hemolytic, whereas most Lancefield groups A, B, C, and G streptococci (see below) produce a discrete clear zone of true hemolysis around the colonies; this is known as β-hemolysis. These characteristics greatly aid in a rapid preliminary screening of different streptococcal groups, but final classification depends on detailed biochemical, serologic, and genetic analysis of streptococci.

L-forms

Protoplasts may be induced by penicillin or by the group C streptococcus phage-associated lysin and may be propagated on hypertonic media to produce L-forms (see Chapter 54). Removal of penicillin usually allows reversion to the parent strain. However, there are many stable L-form strains that are no longer capable of reversion. Certain L-forms can also be propagated in isotonic media. The role played by L-forms in disease or in the persistence of streptococci in tissues is not known (Ginsburg, 1972).

METABOLISM

Streptococci are facultatively anaerobic, acidogenic, and acidophilic. Growth in media that lack body fluids is usually poor but it may be greatly enhanced by the addition of serum proteins and reducing agents in the form of sulfhydryl compounds. Most streptococci virulent for humans show an absolute demand for at least 12 amino acids, 3 purines and pyrimidines, 4 vitamins, certain divalent metals, and an energy source like glucose. Growth is sometimes enhanced by certain peptides (strepogenin).

Luxurious growth in fluid media always forms aggregates (floccules) that fall to the bottom, leaving a relatively clear supernatant fluid. Growth of bacteria in the submerged floccules is inhibited by lactic acid, which is formed from glucose by glycolysis. Excellent growth is usually obtained on brain heart infusion media or trypticase-soy broth. Growth is also facilitated by constant neutralization of the lactic acid and rich growth is achieved in a chemostat (steady-state culture) in a complete synthetic medium.

All streptococci ferment glucose, maltose, and most of them ferment lactose and sucrose. They all give a negative catalase reaction, since they all lack cytochrome systems. The major end product of fermentation is lactic acid. A most important characteristic of certain viridan streptococci (normal flora of the oral cavity) is the synthesis of polysaccharides from sucrose (dextrans) and fructose (levans). The mucoid appearance of streptococcal colonies grown on 5 per cent sucrose is used to classify these streptococci (Table 1). Certain streptococci liquify gelatin and about half of the *S. pyogenes* strains hydrolyze casein. Production of ammonia from arginine and hydrolysis of hippuric acid and esculin characterize certain streptococci that grow in the colon. Strong reducing activity (for example, early reduction of litmus

TABLE 1. Distinguishing Feature Among Members of Viridans Group of Streptococci

SPECIES	LANCEFIELD GROUP ANTIGENS	LEVAN (L) OR DEXTRAN (D) FROM SUCROSE	FERMENTATION OF				
			Inulin	Raffinose	Salicin	Lactose	Trehalose
S. sanguis	H	+ (D)	±	0	+	+	+
S. salivarius	K(?)	+ (L)	+	+	+	±	±
S. mitis	O(?),N(?)	∓	0	∓	±	±	∓
S. milleri	A,C,F,G	0	?	0	+	+	+
S. anginosus	F,G,L	0	?	±	+	+	+
S. mutans	none	+ (D)	?	±	+	+	+

± = usually +
∓ = rarely +
0 = never
? = questionable

milk), black colonies on tetrazolium agar, and growth in media with high NaCl content and at high pH help identify *S. faecalis* species.

Cell Wall Structure

The rigid cell of streptococci has several layers of a linear polymer that is composed of *N*-acetylglucosamine-*N*-acetylmuramic acid (the peptidoglycan, PPG) held together by peptide bridges (Wannamaker and Matsen, 1972) and forming a three-dimensional lattice. The PPG accounts for about 40 to 80 per cent of the wall. In most β-hemolytic strains the PPG is attached covalently to a surface polysaccharide, the C carbohydrate.

Many strains of hemolytic streptococci produce a prominent capsule in young cultures and is composed of hyaluronic acid in Group A streptococci. It is not immunogenic and is susceptible to streptococcal hyaluronidase.

Streptococci also possess characteristic surface polymers known as teichoic (TA) and teichuronic acids. These acids contain polyglycerol phosphate and alanine and are covalently linked to the PPG. Certain strains belonging to Lancefield's groups A, B, C, D, and G possess an immunoglobulin (Ig) receptor responsible for the interaction with the Fc fragment and possibly with the CH2 domain. This Ig receptor is heat labile and trypsin sensitive, and is probably responsible for the agglutination of streptococci by sera devoid of specific antistreptococcal antibodies. The IgG receptor is different from many known surface components of streptococci and its role in the biology of the streptococcus is unknown.

The cellular membrane, which is triple layered (75 to 90 Å), is composed of lipoprotein that differs in structure and content from that of L-forms. The cell membrane is attached by hydrophobic interaction with the glycolipid moiety of lipoteichoic acid (LTA). LTA consists of chains of polyglycerol phosphate units that may contain sugar or amino sugar substituents covalently linked to a glycolipid. The LTA extends through the width of the wall and appears on its surface. The LTA is haptenic when extracted from the streptococci by phenol, but is highly immunogenic when bound to the streptococcal cells. Although only a fraction of the LTA is spontaneously released into the medium during growth, the bulk of LTA may be released by phenol, lysozyme, extracts of human leukocytes, and a variety of cationic polyelectrolytes. Although LTA's function is not fully known, it is possible that, like M-protein, it is involved in the attachment of the streptococci to cell surfaces. LTA of all streptococci cross-react immunologically among themselves and with LTA of other gram-positive bacteria.

The cytoplasm of streptococci contains proteins (including enzymes) and nucleic acids. At least 11 distinct components have been detected in the cytoplasm. A high degree of antigenic heterogeneity is a particular characteristic of the nucleoprotein fraction.

ANTIGENIC COMPOSITION

Group-Specific Carbohydrate

The work of Rebecca Lancefield laid the ground for the serologic classification of streptococci into at least 18 groups (A–R) on the basis of the group-specific carbohydrate, the "C carbohydrate." Groups A, B, and C "C carbohydrates" are composed of branched polymers. The carbohydrate is covalently linked to the PPG. They can be extracted with either hot formamide, hot TCA autoclaving, or enzymatic digestion with lysozyme and *Streptomyces albus* enzymes. The carbohydrate probably protects the peptidoglycan against lysozyme. It accounts for approximately 30 to 50 per cent of the dry weight of the cell wall and about 10 per cent of the intact cell, and may be found on both sides of the PPG. The polysaccharide of group A streptococci, which is composed of a terminal *N*-acetyl glucosamine (NAGA) and rhamnose, may be cleaved by *N*-acetylglucosaminidase of human macrophages. Upon cleavage of the terminal NAGA a structure known as the A-variant antigen is left; rhamnose is the antigenic determinant.

The group-specific antigen of Group D streptococci is a teichoic acid, not a carbohydrate, and it is a component of the cytoplasmic membrane, not of the cell wall.

Type-Specific M-Proteins

These proteins are almost exclusively found in group A streptococci. M-proteins are surface antigens (Fox, 1974). They are responsible for the virulence of group A streptococci by virtue of their unique antiphagocytic characteristic and are lost when virulence is lost. The large M-protein complex contains hair-like structures (fimbria or pili) on the cell wall that are probably attached to it by covalent bonds. They are attachment structures for colonization of the susceptible patient. As protein antigens, they confer type specificity upon Group A streptococci; about 80 types are known. M-protein has been released from the streptococci by extraction with HCl, sonic oscillation, pepsin digestion; phage lysis, extraction with guanidine, and by nonionic detergents. The M-protein molecule has been purified. It is a heat- and acid-stable protein, soluble in alcohol, has a molecular weight of 40,000, and an isoelectric point of 5.3. It possesses multiple subunits, all having identical serologic activity. M-protein is susceptible to proteases, including streptococcal proteinase, which can prevent typing by digesting this antigen.

Until recently, all preparations of M-protein were contaminated with non–type-specific polypeptides. These strongly associated proteins give rise to non–type-specific antibodies and also elicit local and systemic hypersensitivity reactions. Beachey and his colleagues have separated M-protein from these contaminants and their preparation is both immunogenic and nontoxic.

T Antigen

In addition to the type-specific M-protein, streptococci possess another set of protein antigens known as the T-system. The T-protein can be isolated by peptic, tryptic, or pancreatic digestion of heat-killed streptococci. It is protease-resistant, insoluble in alcohol, antigenic in the cell, and antigenic when cell free. It is not related to virulence, and antibodies against it are not protective. Some T-antigens are restricted to a single M-type whereas others may be shared by several M-types. The T-system is employed for typing those group A strains that either fail to synthesize M-protein or to which no anti-M-type serum is available (see *Laboratory Diagnosis*).

R Antigen

Certain types of group A streptococci possess R antigens that can also be found among groups B, C, G, and L streptococci. The antigen is a nonprotective protein that cross-reacts with grouping and typing antisera that contain antibodies to the R protein. This antigen is not commonly employed for classification and its biologic significance is still not known.

PATHOGENETIC PROPERTIES

The most prominent pathogen for humans is *Streptococcus pyogenes* belonging to Lancefield's groups A and C (See also *Infections by Other Streptococci*). Pyogenic streptococci (Groups A and C) produce large amounts of extracellular products during growth in vitro in defined media or in human tissues. At least 25 different antigens are detected in concentrated culture supernatants by immunoelectrophoresis with pooled human γ-globin or rabbit sera hyperimmune to streptococcal extracellular products (Ginsburg, 1972). Only a few of these products have been identified, and a small number purified. Of the numerous factors elaborated by the Group A streptococcus, few have been implicated in the pathogenesis of tissue damage in animal models.

The major virulence factor of group A streptococci is the type-specific M-protein that deters phagocytosis. Whether the hyaluronic acid capsule also has significant antiphagocytic activity in natural infections is less certain.

Purified M-protein precipitates fibrinogen and interacts with fibrin lysates and fibrinomonomer complexes. It clumps platelets and leukocytes, it lyses PMN, and inhibits the migration of leukocytes in capillary tubes. The cytotoxic effect of M-protein requires both heat-stable and heat-labile serum factors, probably related to complement. It is highly immunogenic, and complexes of M-protein and antibodies can localize in the glomeruli and may be associated with glomerulonephritis (see below). M-positive strains can adhere to mucosal epithelial cells in vitro, whereas M-negative streptococci do not. The adherence of virulent streptococci to mucosal surfaces can be inhibited with type-specific antibody (including secretory IgA).

Extracellular Toxins

Streptococcal infection is probably facilitated by extracellular products, some of which possess cytopathic properties (Ginsburg, 1972).

Streptolysins. The majority of group A, C, and G streptococci elaborate two distinct hemolysins called Streptolysins O and S. The hemolytic zones around streptococcal colonies on blood agar, under aerobic conditions, are primarily due to streptolysin S (SLS). Both hemolysins are cytopathic for mammalian cells and block phagocytosis, presumably by impairing chemotaxis and ingestion by leukocytes and disrupting their lysosomes.

Streptolysin O (SLO). This oxygen-labile protein antigen has a molecular weight of approximately 70,000. It is synthesized only by growing streptococci. It possesses a labile SH-group and is hemolytic or cytolytic only under reducing conditions. Its activity is strongly inhibited by cholesterol and other sterols. SLO is also cardiotoxic. It may cause interstitial myocarditis in experimental animals and systolic arrest of perfused mammalian hearts. Its cardiotoxicity is probably caused by inducing the release from atria of acetylcholine, which poisons the ventricles.

Streptolysin S (SLS). This is an oxygen-stable nonantigenic peptide with a molecular weight of approximately 2800. It is synthesized by both growing and resting cells and is found on the surface of washed streptococcal cells as a cell-bound hemolysin. This loosely bound peptide can, however, be released into the surrounding medium by a variety of carrier molecules (for example, serum albumin, α-lipoproteins, yeast RNA, and nonionic detergents like tweens and tritons). Upon release by the carriers, the hemolysin becomes recognized as SLS. Extracellular SLS is thus a complex between a nonspecific carrier molecule and the specific hemolytic peptide. The peptide can be transferred among the various carriers and finally to the surface of mammalian cells. After the interaction with membrane phospholipids the hemolytic peptide is inactivated. SLS activity is strongly inhibited by lecithin and β-lipoproteins but not by cholesterol. RBC treated with SLO show distinct "lesions" similar to those caused by complement; no such lesions have been shown to be induced by SLS. Upon intravenous injection, SLS causes necrosis of liver and kidney tubuli and massive intravascular hemolysis. It can induce a chronic arthritis when injected intraarticularly. Group A streptococci also possess an intracellular hemolysin with properties similar to SLS. This hemolysin can be released only after prolonged bacterial sonication (Ginsburg, 1972).

Erythrogenic Toxin (ET). Most group A streptococci produce one (or more) of three immunologically distinct erythrogenic toxins (ET) known today as Streptococcal Pyrogenic Toxin (SPT), which were first described by the Dicks. In addition to causing fever and rash, SPT enhances susceptibility to lethal endotoxin shock, injures macrophages, alters antibody response to red blood cells, and is mitogenic for lymphocytes. SPT has been purified and crystallized from culture supernatant fluid of certain streptococcal strains. It has 1×10^7 rabbit skin doses per 20/mg N and a molecular weight of approximately 29,000. It contains 80 per cent protein and hyaluronic acid; the latter probably acts as a carrier and is not associated with toxicity. The symptoms of scarlet fever are believed due to both the primary toxicity of SPT and the secondary effects of hypersensitivity; the typical scarlatinal rash is due to the latter. The role, if any, played by SPT in the pathogenesis of streptococcal infections and their sequelae is not established. SPT is probably produced under the direction of a temperate phage (Wannamaker and Matsen, 1972).

Proteinase. Most group A streptococci elaborate an SH-dependent proteinase precursor (M.W. = 44,000), which can be converted to an active proteinase by reducing agents and by proteases. Proteinase paradoxically destroys M-protein. This protease acts on several naturally-occurring proteins as well as on some synthetic substrates. The intravenous injection of reduced proteinase causes massive myocardial and skeletal muscle necrosis in rabbits. It is immunogenic and the specific antibodies that develop neutralize the proteolytic activity of this enzyme. Its role in the pathogenesis of streptococcal infections is not known.

Streptokinase (SK). Most group A and C strains produce SK, an activator of the fibrinolytic system of human blood. SK is a protein with a molecular weight of approximately 47,000. It converts a proactivator present in plasma to an activator, which converts plasminogen to the proteolytic enzyme, plasmin. Plasmin splits fibrinogen, fibrin, and other proteins. A small amount of plasmin activates more plasminogen. SK may be a virulence factor of streptococci by lysing blood clots and fibrin precipitates. Together with hyaluronidase and the nucleases it may function as a "spreading factor" (see below). Preparations of SK have been widely employed for debriding surface infections, for enhancement of wound healing, for treatment of fibrinous exudates, and for the lysis of intravascular thrombi.

The Spreading Factors. Several extracellular products of streptococci are regarded as "spreading factors" because they promote spread of streptococci in tissues by liquifying and reducing the viscosity of inflammatory exudates. Group A strains elaborate hyaluronidase, four serologic varieties of deoxyribonucleases (A, B, C, and D), and a ribonuclease. All these enzymes are immunogenic.

Nicotinamide adenine dinucleotide glycohydrolase (NADG). Some strains belonging to groups A, C, and G elaborate NADG. At one time this enzyme was thought to be associated with leukotoxicity of streptococci, presumably because of its ability to affect leukocyte metabolism. Now it is believed that leukotoxicity is due to cell-bound SLS. The role of NADG in pathogenicity of streptococci is not known. Production of NADG by streptococci is sometimes used in serologic differentiation of certain strains (see below).

Cardiohepatic Toxin (CHT). Injection of viable group A streptococci into tonsils is followed within 24 hours by interstitial myocarditis of rabbits with granulomatous lesions in the myocardium of some animals that resemble Aschoff bodies.

An active principle isolated from supernatant culture fluids of group A streptococci, called "cardiohepatic toxin," causes myocardial and diaphragmatic lesions in rabbits similar to those induced by tonsilar infection. It also produces giant cell granulomas in the liver. Animals injected with CHT develop hyperlipemia and high blood levels of glutamicoxaloacetic transaminase, sorbitol-dehydrogenase, and creatine phosphokinase. The chemical nature of CHT is not known.

Cell-Associated Toxins

Peptidoglycan-Polysaccharide Complexes (PPGPS). Although streptococci are killed after phagocytosis by PMN and macrophages, intracellular digestion (degradation) of their cell walls is extremely slow. It is claimed that mammalian cells lack specific enzymes capable of degrading streptococcal wall components. The extreme resistance to degradation is due either to the C-polysaccharide or to the presence on the cell surface of teichoic acid, which hinders the interaction of lysozyme with the PPG.

Cell-wall components of group A streptococci persist for many months within macrophages in granulomatous lesions in the heart, synovium, and liver of laboratory animals. Since the persistence of cell wall components within macrophages leads to secretion of lysosomal enzymes, it was postulated that enzyme secretion initiated these chronic lesions (Davies et al., 1974; Ginsburg and Sela, 1976).

Undegraded streptococcal cell-wall components have also been shown to localize selectively in tissue sites remote from the original focus of infection after transport within phagocytic cells. Several investigators suggest that the pathogenesis of rheumatic fever may be associated with the localization and persistence of undegraded PPGPS in the cardiac muscle (Ginsburg and Sela, 1976). Wall components of streptococci may also activate complement.

BIOLOGIC PROPERTIES OF PEPTIDOGLYCANS (PPG). Isolated and partially purified PPG can cause the following: fever, the localized Shwartzman reaction, nonspecific resistance to infections, enhancement of lesions induced by pyogenic bacteria, enhanced immunologic response to other antigens (adjuvant effect), inhibition of macrophage migration, tumor inhibition, impaired phagocytosis by PMN, activation of complement, and lysis of platelets. Undegraded complexes of PPG and other polysaccharide moieties have induced granulomatous inflammation in the skin, heart, joints, and liver of laboratory animals. It appears that the size of the PPG molecule and its association with the naturally occurring polysaccharide moiety of the cell wall is of utmost importance for the expression of toxicity, immunogenicity, and persistence in tissues. The probability that different individuals may degrade streptococcal cell walls to different entities may explain the enigma that poststreptococcal sequelae affect only certain patients (see below).

Lipoteichoic Acid (LTA). This membrane-associated haptene can sensitize mammalian cells to lysis in the presence of antistreptococcal IgM antibodies and complement. LTA can cause chronic arthritis after intraarticular injection into animals preimmunized with group A streptococci. It can cause resorption of bone (via cyclic AMP and prostaglandins) in culture and nephrocalcinosis in laboratory animals. LTA is the adherence factor that accounts for the affinity of streptococci for mucous membranes. Both LTA and M-protein, which had been thought to be the adherence factor, are components of the hairy fibrils on the surface of the streptococcus. The nature of the epithelial cell receptor is unknown except that it is trypsin sensitive.

SEQUELAE OF ACUTE STREPTOCOCCAL INFECTIONS

A unique characteristic of group A streptococci is their ability to induce nonsuppurative sequelae in susceptible individuals. These are rheumatic fever (RF), chorea, and acute glomerulonephritis (AGN) and they may follow overt or inapparent infections. The onset of these complications may start two to three weeks after infection. The common denominator among these diverse late complications of Group A streptococcal infections is the failure to isolate living streptococci from the affected tissue. The failure to duplicate these sequelae in lower animals has hampered research into the pathogenesis of these syndromes. Our present concepts are based on circumstantial evidence and no mechanism can be considered proved or accepted by most investigators in this field.

There are several absolute requirements for the development of rheumatic carditis and arthritis.

First, group A streptococci must have been present; second, the infection must be localized in the upper respiratory tract; third, the bacteria must persist for complication to occur; and fourth, a streptococcal antibody response indicative of a recent infection must be known.

The importance of the site of infection for the development of RF has been stressed; many observations suggest that RF does not develop after streptococcal infection of the skin. The factors that localize streptococci to the throat are not fully understood, but M-protein and LTA are both implicated in anchoring streptococci to the surface of epithelial cells.

Current Theories of Pathogenesis

The numerous theories proposed to explain how group A streptococci initiate sequelae fall under several headings: (Wannamaker and Matsen, 1972; Ginsburg, 1972; Stollerman, 1975; Glynn, 1975) (a) The "rheumatic" toxin; (b) autoimmunity and immunopathology; (c) genetic and anatomic abberations; (d) a combination of these theories.

Is There a "Rheumatic Toxin"? Several extracellular substances elaborated by group A streptococci can induce myocardial, endocardial, hepatic, articular, and renal lesions in laboratory animals (see above). As a rule, none of the lesions fully resemble the human lesions of RF, and no clinical manifestations similar to those seen in humans have been duplicated in the animals. Undegraded peptidoglycan-polysaccharide complexes (PPGPS) of the cell wall have produced rheumatic-like granulomatous lesions in mice and rabbits. Another approach suggests that in early streptococcal pharyngitis certain susceptible patients develop acute myocardial and articular injuries explainable by diffusible "toxins." None of the "toxic" theories have been based on firm ground and all data have been obtained from experimental models (Wannamaker and Matsen, 1972; Ginsburg, 1972; Glynn, 1975; Ginsburg and Sela, 1976).

Immunopathology and Autoimmunity. Several investigations have proposed that complexes formed between streptococcal products and autologous antibodies (for example, SLO–anti-SLO complexes) may localize in tissues and damage them by activating complement (serum sickness type), or through cell-mediated immunity (CMI). Sensitized T lymphocytes may then interact with streptococcal antigens to cause the release of lymphokines. Cell-sensitizing agents (lipoteichoic acids) have also been implicated in tissue damage. These agents can sensitize mammalian cells to passive lysis in the presence of antistreptococcal antibodies and complement. This theory suffers from lack of evidence for the localization of streptococcal antigens in the heart or in joint lesions.

The theory of autoimmunity gained favor because substances in mammalian cardiac tissue can act as autoantigens and because certain streptococcal toxins can release such antigens. However, the theories implicating autoimmunity in RF have been stated only in general terms, with no focus on precise mechanisms, and have been overshadowed by theories implicating immunologic cross-reaction between streptococcal antigens and components of mammalian tissues.

The phenomenon of crossed immunity between mammalian and bacterial constituents (the concept of molecular mimicry) is well established (Kaplan, 1976; Zabriskie, 1976). Several antigens in group A streptococcal cell walls and protoplast membrane have been isolated. These cross-react with glycoproteins of the heart valves, sarcolemma of cardiac and skeletal muscle, bundle of His, glycoprotein of glomerular basement membrane, human fibroblasts, endothelial cells, astrocytes, cytoplasm of subthalamic and caudate nuclei, and histocompatibility antigens (Kaplan 1976; Zabriskie, 1976). Despite the presence in human sera of cross-reactive antibodies and their elution from myocardium of patients with RF there is no evidence that such antibodies have any relationship to the lesions of rheumatic fever. It is interesting that no bound cross-reactive globulin is found in Aschoff bodies. There is no evidence that the cross-reactive antibodies are cytotoxic to mammalian cells.

Genetic and Anatomic Abberations. Certain investigators claim that the "rheumatic patient" may possess direct anatomic routes (probably lymphatic channels) that connect the tonsils with the heart. It is proposed that through such channels streptococci, streptococcal L-forms, and their diffusible extracellular products (toxins) and undegraded cell-wall components may freely migrate and localize in the heart to cause tissue damage (Wannamaker, 1973). Others favor the theory that the rheumatic patient is predetermined genetically to develop abnormal cellular and humoral responses to streptococcal antigens (Zabriskie, 1976).

Arthritis

Migratory arthritis is part of the rheumatic syndrome. The similarity of the tissue alterations in the joints of rheumatic patients to those induced experimentally by anaphylactic and delayed hypersensitivity reactions suggest that the human joint disease may be caused by such mechanisms. As a rule, neither viable streptococci nor their extracellular or cellular constituents are found within the inflamed joints. A decrease in both early and late components of complement within the synovial fluid of RF patients suggest local activation by immune complexes. Although the streptococcus antigen that could be present in these complexes has not been identified, this phenomenon may point to the mechanism of arthritis in rheumatic fever. Arthritic lesions can be induced in laboratory animals by the intra-articular injection of Streptolysin S, by a pool of extracellular products, by soluble cell wall components, by L-forms, and by lipoteichoic acid (LTA) (Ginsburg et al., 1977).

Acute Glomerulonephritis (AGN)

One of the most common complications of upper respiratory and skin infections with group A streptococci is AGN. It is an old observation that acute RF and AGN rarely, if ever, occur in the same person at the same time (Stollerman, 1975). AGN is probably caused by special nephritogenic strains. The pharyngeal M-types 12, 14, and 49 and a family of "impetigo" streptococcal strains belonging to the following serologic complexes: T-3/13/B3264/12 (related to the M serotypes (22, 33, 39, 41, 43, 52, 56); T8/25/Dmp19(M-types 2, 8, 25, 55, 57, 58); and T5/11/12/27/44 (M-types 11, 59, 61) are all associated with AGN. The latter pyodermia strains were previously unrecognized owing to lack of specific antisera for typing them. The factors elaborated by these skin strains, which endow them with nephritogenicity, is not understood. Furthermore, in spite of the detailed serologic, epidemiologic, and renal morphologic studies, the pathogenesis of nephritis remains unknown. The theories regarding pathogenesis of AGN can be summarized under four main headings and are based primarily on experimental models (Ginsburg, 1972).

The Nephrotoxin. Streptococcal agents like Streptolysin S, M-protein, autolysates of nephritogenic streptococci, extracts of streptococcal cell walls, solubilized protoplast membranes, and undefined diffusible agents released by certain nephritogenic streptococci (nephrotoxin) can cause renal disease in laboratory animals but not the full spectrum of clinical and pathologic changes seen in humans.

Immunologic Cross-Reactivity. Cross-reactive antigens between soluble components of glomerular basement membranes are chemically and immunologically similar to antigens of protoplast membranes of nephritogenic streptococci. Antibodies to protoplast membranes can cause glomerular lesions.

Autoantibody. Extracts of kidneys that had been mixed with streptococci were claimed to give rise to nephrotoxic antibodies. Although this theory was favored by several investigators, the results were controversial and could not be duplicated in several other laboratories.

Immune Complex Disease. Immunofluorescent and electron microscopic analysis of kidneys obtained from patients with AGN showed localized immune complexes and complement in the glomeruli. The nature of the antigenic substance in these complexes and its possible relationship to streptococci is still controversial. It has also been suggested that immune complexes can be obtained when streptococcal enzymes modify the chemical composition of IgG; the altered IgG then interacts with autoantibodies that recognize both native and altered globulin. Such complexes can cause glomerular lesions.

Despite the numerous models established, none are satisfactory analogs to the human disease (Wannamaker and Matsen, 1972).

Immunity. Infections with group A streptococci that possess M-antigen are usually followed by a type-specific humoral immunity. If the anti-M antibodies disappear, they can be recalled by small booster doses of homologous M antigen. The anti-M antibodies are mainly IgG. They appear slowly (over 30 to 60 days) and are opsonins (Stollerman, 1975). None of the antibodies that are regularly produced against C-polysaccharide, T antigen, teichoic acid, or the numerous extracellular enzymes and other factors can prevent infection. Type-specific antibody can be detected by opsonization, mouse protection, or chain elongation in vitro. These tests are difficult to perform in a routine bacteriologic laboratory.

Streptococcal Vaccine

Since the immunity to Group A streptococci is type-specific, and since about 80 types are known to be present in different populations, the rationale for preparing an adequate vaccine depends on the distribution of the major M-types in a particular population. Immunization, if available, should aim to prevent both suppurative and nonsuppurative complications. The preparation of a vaccine should also take into consideration the risk of inducing delayed hypersensitivity reactions, the elimination from the M-protein of antigens cross-reactive with heart, and the elimination from M-protein of covalently bound lipoteichoic acid (LTA). Preliminary field trials have used M-protein in alum given either subcutaneously (Sc) or intrapharyngeally. The patients were then challenged by smearing the homologous M-types on their tonsils. Seventy-five per cent of the subjects immunized revealed a primary opsonic response to the vaccine with no reactions. Although the parenteral immunization did not prevent colonization of the throat, it protected against clinical illness. The main questions to be answered are these: how many serologic types warrant inclusion in the vaccine; how often do prevalent M-types change under natural circumstances; and how much variation in prevalence of specific M-types is there in different parts of the world.

Recent experiments have been conducted on a vaccine against dental caries by employing strains of *S. mutans*. Monkeys injected with this streptococcus developed antibodies that could be detected in gingival fluid. Such antibodies (mostly IgG) may be responsible for the substantial decrease in the incidence of cervical caries but had no effect on the incidence of carious lesions on smooth surfaces (Lehner et al., 1976).

Pneumococcal Vaccine

See section of *S. pneumoniae*.

LABORATORY DIAGNOSIS

Tests for Streptococcal Antibodies

Antibodies against streptolysin O, streptokinase, hyaluronidase DNase-B, NADase, and proteinase are routinely measured to diagnose recent streptococcal infections. Proof of a streptococcal infection is based, in part, upon an increase in one or more of these antibodies. Although antistreptolysin O (ASO) titration in patients' sera is the most common test, it is often advantageous to measure other antistreptococcal antibodies. The streptozyme test is a two-minute slide hemagglutination procedure that quantitatively measures multiple antibody (A-STZ) to streptococcal extracellular products. The reagents are sheep RBC sensitized simultaneously with streptolysin-O, deoxyribonuclease-B, hyaluronidase, streptokinase, and nicotinamide adenine dinucleotide glycohydrolase. A-STZ correlates well with ASO in the sera of rheumatic fever patients. Moreover, it has been found that the A-STZ test detects a higher incidence of elevated antistreptococcal antibodies than does one ASO test. The specificity of the test (no antibodies to cellular elements of streptococci are detected with this reagent), its high reproducibility, the ease of performance, and the minimal quantities of patient serum required indicate that the test is helpful in estimating antibody response to streptococcal infection whenever the performance of a battery of tests for individual antibodies is impractical and expensive.

Culture of Group A Streptococci

For the diagnosis of a group A streptococcal infection, the Streptococcus must be isolated and identified. Specimens must be transported to the laboratory without delay, because cotton swabs may contain substances that inhibit group A streptococci. Viability is well maintained in swabs transported in a silica-gel. Use of enrichment media (for example, Pike's broth) increases the isolation rate of Group A streptococci compared with direct plating on blood agar. The growth on blood agar is screened for complete (β) hemolysis. Sheep blood is preferred because it inhibits the growth of *Hemophilus hemolyticus*, a β-hemolytic saprophyte commonly found in the throat. Overgrowth of other microorganisms can be prevented by various selective agents in the media. Group A streptococci can be tentatively identified, rapidly and easily, with bacitracin disks. About 90 per cent of Group A streptococci are inhibited by disks containing 0.04 unit of bacitracin, whereas most Group B, C, and G streptococci are resistant. Definitive identification is made by demonstrating the C carbohydrate of Group A by immunofluorescence with group-specific antiserum, or by precipitin tests with C carbohydrate extracted from the bacteria.

DRUG SUSCEPTIBILITY

Unlike staphylococci, no group A streptococcus strain has been resistant to penicillin. This makes penicillin the drug of choice in the treatment of streptococcal infections and in prevention of rheumatic fever. The MIC for penicillin G is usually 0.001 μg/ml. About 40 per cent of strains are resistant to tetracycline and sulfadiazine. Resistance to erythromycin is uncommon and is mediated by a plasmid. Streptococci are resistant to polymyxins, kanamycin, and streptomycin.

EPIDEMIOLOGY

Upper Respiratory Infections

Understanding of the pathogenesis of streptococcal infections, its nonsuppurative sequelae, and its epidemiology depends on the distinction between a "carrier" state and acute streptococcal upper respiratory infection. Group A streptococci cultured from acute sore throats contain large amounts of M-protein. Pharyngitis and tonsillitis occur most frequently in children from 5 to 15 years of age. It is unlikely for a child to reach the age of ten without having encountered the Group A streptococcus.

Transmission occurs via droplets from respiratory secretion of patients or from healthy carriers. Both food and milk may be sources for occasional outbreaks. Organisms recovered from clothing, bed, or house dust are usually noninfective. Control of hemolytic streptococcal disease is difficult because many infections are either exceedingly mild or inapparent, and persons with subclinical infections can disseminate streptococci. Overcrowding allows the explosive spread of single serotypes, especially among civilians.

Peak incidence occurs between December and May in temperate zones.

Pyoderma

Occurrence is usually confined to the summer and early fall in hot and tropical climates. It mostly affects preschool children and infants. Transmission may be aided by insects. Poverty, filth, and overcrowding are major predisposing factors. The same streptococcus type is often found in throats and in pyoderma lesions.

PREVENTION

Infections with Group A streptococci may be prevented by therapeutic intervention during epidemics or by prophylactic drugs given to those at high risk, such as in boarding schools, orphanages, and military camps in which infections are endemic. Immunization with M-protein has been tested on a relatively small scale and was found to be effective, but the use of vaccine in general or in selected population needs to be explored (see above). Impetigo may be prevented by improved skin hygiene. Epidemics are best halted by antibiotic treatment of all cases.

OTHER β-HEMOLYTIC STREPTOCOCCI PATHOGENIC FOR HUMANS

ANTIGENIC STRUCTURE

Group B Streptococci (Streptococcus agalactiae)

Group B streptococci contain as a major constituent of the cell wall, a group-specific carbohydrate (C-substance) composed of rhamnose, *N*-acetylglucosamine, and galactose. L-rhamnose is the significant component of the antigenic determinant. The capsule antigen (S-substance) can serve to classify further the group into four serologic separate and distinct types, Ia, Ib, II, and III. A fifth type, Ic, contains the Ia carbohydrate antigen plus a protein antigen. Both human and bovine strains are found in all these types but with different frequencies.

Groups C and G streptococci (S. equisimilis, S. zooepidemicus, S. equi, S. dysgalactiae)

These differ from group A by causing neither glomerulonephritis nor rheumatic fever and by possessing a group-specific carbohydrate polymer of L-rhamnose and NAGA (in group C) and galactose, galactosamine, and rhamnose (group G). Neither group contains type-specific protein antigens like the M-protein of group A.

Group D streptococci (S. faecalis-enterococci, S. faecium, S. durans, S. bovis, S. equinus)

The Group D antigen is unique among the streptococci in that it is a glycerol teichoic acid containing D-alanine and glucose, is probably located in the region of the wall, and may not be linked through a primary bond with the wall components. The type-specific antigen is a cell-wall carbohydrate that is composed of D-glucose and NAGA and that also contains rhamnose. In order to preserve the type-specific antigens of Group D streptococci, they must be released from the cells not by the routine hot HCl method but by lysozyme or by lytic or autolytic enzymes.

METABOLISM

Carbohydrate metabolism is chiefly homofermentative in all these groups. Glucose is fermented to lactic acid. An oxidative pathway also can be utilized by *S. agalactiae,* which yields lactic acid, acetic acid, acetylmethylcarbenol, and carbon dioxide. A noncytochrome iron-containing chromophore appears to be utilized, and oxygen is the terminal electron receptor in this reaction.

PATHOGENIC PROPERTIES

S. agalactiae (Group B) has long been recognized as an important bovine pathogen. Recently it has been recognized as a frequent cause of meningitis and pneumonia in neonates (Patterson and Hafeez, 1976). Meningitis is caused almost exclusively by type III, and pneumonia by type Ia. Sporadic infections in adults show no type predilection. The explanation for the organotropism in infants of these two serotypes is obscure. Nevertheless, the type-specific antigens are the most important virulence factors known. *S. agalactiae* produces a hemolysin distinct from streptolysins O and S and a hyaluronidase, but their pathogenic significance is unknown. Group C streptococci have a hyaluronic acid capsule that deters phagocytosis. Many Group C and G strains produce streptolysin O, but its pathogenic significance is as uncertain for these groups as it is for Group A streptococci.

IMMUNITY

It has been shown in mice that antibody to the type-specific antigens of *S. agalactiae* is protective. Furthermore, transplacental passage of antibody to capsular type III protects newborns from Group B streptococcal meningitis. Both type-specific antibody and a heat labile component (presumably complement) are opsonic for type Ia strains. Most maternal sera and cord sera lack opsonins for that strain. Leukocytes from neonates have diminished bactericidal activity for Group B streptococci but this defect is probably not specific for these bacteria.

LABORATORY IDENTIFICATION

Non–Group A hemolytic streptococci are usually distinguished from Group A by their resistance to an 0.04 unit bacitracin disk. However, because up to 5 per cent of Group A streptococci are also resistant and a like percentage of other Groups are sensitive, this is not a definitive test. β-Hemolytic streptococci are identified with specific antibody against the C carbohydrate. It is often difficult to identify the Group D strains by this reaction and they are identified by their hydrolysis of esculin in the presence of 40 per cent bile (Table 2). Group D streptococci are divided into enterococci (including *S. faecalis, S. faecium, S. zymogenes*) and nonenterococcal species *(S. bovis, S. equinis)*. This distinction, which is based on their ability to grow in hypertonic salt media, has therapeutic implications. Group B strains hydrolyze hippurate and produce a soluble hemolysin that acts synergistically with the β-hemolysis of *Staphylococcus aureus* (CAMP phenomenon).

DRUG SUSCEPTIBILITY

β-Hemolytic streptococci, other than Group D, are uniformly susceptible to less than 1.0 unit/ml of penicillin G. Ninety per cent are also susceptible to erythromycin and clindamycin, but up to 50 per cent of recently isolated strains of Group B are resistant to tetracycline. Nonenterococcal Group D streptococci are also susceptible to penicillin, but enterococci are much less so. Enterococci are somewhat more susceptible to ampicillin than to penicillin G and show variable susceptibilities to chloramphenicol, tetracycline, and erythromycin. They are susceptible to vancomycin. Synergism between penicillin and aminoglycoside antibiotics is important in the treatment of enterococcal endocarditis. The combination is bactericidal in concentrations of each drug that are only one tenth that required to kill the enterococcus when either acts alone.

EPIDEMIOLOGY

Group B, C, and G streptococci are often isolated from the nasopharynx. Group D predominates in the gastrointestinal tract, but as many as 30 per cent of normal people have Group B streptococci in their feces. Group B streptococci can also be isolated from the genital tract of 30 per cent of normal women. Neonatal colonization with Group B streptococci is common, but the risk of disease is probably less than 5 per cent. Serotype III Group B streptococci are often acquired by infants in the nursery as well as in utero or at delivery. These bacteria may spread from infant to infant or from nurses to infants, because many nurses may be colonized.

TABLE 2. Reactions That are Useful for the Differentiation of Several Groups of Streptococci

GROUP	HEMOLYSIS	BACITRACIN SENSITIVITY	HIPPURATE	ESCULIN	6.5% NaCl	CAMP
A	β	+	−	−	−	−
B	β	−	+	−	+	+
D enterococci	α, β	−	−	+	+	−
D nonenterococci	−	−	−	+	−	−

THE VIRIDANS GROUP

This is a heterogenous group of very poorly defined α-hemolytic organisms that are commonly found in the mouth and pharynx. About 50 per cent of bacterial endocarditis is caused by these bacteria. Since they do not have a defined group carbohydrate antigen, no correlation exists between serologic and physiologic characteristics of these strains.

MORPHOLOGY

Viridans streptococci usually form a narrow zone of α-hemolysis around their pinhead colonies. No exotoxins or hydrolytic enzymes similar to those produced by pyogenic streptococci are found in culture supernatants. Several of the streptococci elaborate a constitutive extracellular glycosyltransferase and fructosyltransferase responsible for the synthesis from sucrose of large capsules of dextran and levan respectively. Certain of the streptococci also accumulate intracellular polysaccharides of the amylopectin type.

Their cell wall structure is essentially that of the pyogenic streptococci. They possess peptidoglycan, wall teichoic acid, and membrane-associated lipoteichoic acid.

METABOLISM

Viridans streptococci resemble pneumococci both culturally and in Gram stains, but unlike pneumococci they do not possess well-developed autolytic systems and are not lysed by bile salts. The viridans streptococci are fastidious and grow luxuriantly only in heart infusion broth or in media rich in protein. They are microaerophilic and acidophilic, and the major product of fermentation is lactic acid. This heterogenous group can, however, be classified by its ability to ferment inulin, raffinose, salicin, lactose, and trehalose, and to release ammonia from arginine. The most common species recognized are *S. salivarius, S. milleri, S. M. G. (anginosus), S. mitis, S. sanguis,* and *S. mutans* (Table 1).

PATHOGENIC PROPERTIES

Dental Caries

Four recognized species, *S. sanguis, S. mitis, S. salivarius,* and *S. mutans,* are associated with dental caries in humans, rats, and in monkeys. *S. mutans* appears to be the most virulent and is the only species recognized that consistently initiates decay affecting smooth enamel surfaces. This pathogenic potential is probably related to its ability to adhere and accumulate on the surface of the teeth and to form large bacterial plaque deposits. It can synthesize high molecular weight dextrans and other glucans from sucrose but not from other sugars. The extracellular polysaccharides enable them to adhere to surfaces. It has the unique property of aggregating in the presence of very small amounts of high molecular weight dextran, indicating the presence on its surface of specific receptors. Small fragments of dextran prevent plaque formation and tooth decay in laboratory animals presumably by functioning as glucosyl acceptors for dextransucrase, thereby inhibiting synthesis of high molecular weight polymers.

Dental caries may be looked upon as an infectious disease and studies with monoinfected gnotobiotic animals show that *S. mutans* can spread among members of hamster or rat colonies, colonize the mouth, and induce carious lesions in animals fed high sucrose diets. Thus, the unique combination of streptococci, sucrose, and tooth surface are prerequisite for the initiation of caries.

Bacterial Endocarditis

Most strains of viridans streptococci can adhere to human heart valves. The chemical nature of the receptors has not been identified.

STREPTOCOCCUS PNEUMONIAE (PNEUMOCOCCUS)

The pneumococcus is a *Streptococcus* that forms α-hemolytic zones on blood agar and readily undergoes spontaneous autolysis. It causes pneumococcal pneumonia, septicemia, and meningitis.

MORPHOLOGY

The cell is characteristically ovoid or lanceolate and is usually arranged in pairs, but chains are also present in infected secretions and in cultures.

Pneumococci grow well on meat infusions supplemented with serum or blood under reducing conditions. Pneumococcal colonies are 1 mm in diameter, raised, smooth, and circular. Type 3 strains grow as larger mucoid colonies. When autolysis occurs, the center of the pneumococcal colony becomes depressed. Growth in the presence of antiserum may yield chains of unencapsulated variants that produce small, granular, rough colonies.

ANTIGENIC STRUCTURE

About 82 pneumococcal types are recognized. Typing is based on variation of the chemical composition of the capsular polysaccharide. Eighty per cent of strains isolated from cases of lobar pneumonia in adults fall into types 1 to 8.

The different types of capsular polysaccharides differ serologically and chemically. Thus, the polysaccharide antigen of Type 1 pneumococci contains galacturonic acid, galactose, fucose, and glucosamine; that of type 2 contains rhamnose, glucose, and glucuronic acid; type 3 is a polymer of cellobiuronic acid that consists of alternate residues of glucose and glucuronic acid. The isolated capsular polysaccharides are immunogenic in man and mice, but not in horses or in rabbits unless they are injected with a protein carrier. Human beings injected with 30 to 60 μg of pneumococcal polysaccharide produce antibodies that protect them against pneumonia.

Some capsular surface polysaccharides cross-react immunologically with other α-hemolytic streptococci, with capsular polysaccharides of *Klebsiella* or *Salmonella,* and with blood group substances. The capsule swells when reacting with specific antibodies (the Quellung reaction). The cell wall of pneumococci is composed of a peptidoglycan similar in structure to the group A streptococci. In addition, a carbohydrate C-substance is present that reacts with C-reactive protein, a protein that appears in the blood during acute inflammatory conditions. The pneumococcal cell wall also contains choline, which forms part of the wall teichoic acid. The key role of choline in susceptibility to lysis by bile salts and penicillin is evident when pneumococci are grown in the presence of ethanolamine instead of choline. With ethanolamine they form chains and are no longer solubilized by bile salts nor lysed by penicillin. The choline-containing lipoteichoic acid is a powerful inhibitor of the homologous autolytic enzyme *N*-acetylmuramyl-L-alinine amidase and is lost from the cell in the presence of penicillin so that autolysis is no longer inhibited. The C-substance may be identical with the wall teichoic acid. Pneumococci also possess protein antigens similar to the streptococcal M-proteins, but unlike the streptococcus this pneumococcal protein antigen is not associated with virulence.

LABORATORY DIAGNOSIS

Streptococcus pneumoniae is bile soluble and inhibited by optochin (ethylhydrocuprine). The nutritional requirements are complex, but they can be grown readily on infusion agars supplemented with 5 per cent blood. The white mouse is exquisitely sensitive to many pneumococcal types (type 14 is a notable exception) and pneumococci are isolated from sputum by injecting it intraperitoneally into a mouse. The mouse will eliminate other bacteria but die from pneumococcal septicemia and its heart blood will contain only pneumococci. This is the most sensitive way of isolating the pneumococcus from sputums, but it is rarely necessary if blood agar is inoculated and the growth examined with a dissecting microscope. Acid is formed from lactose, sucrose, trehalose, and raffinose and the fermentation of inulin is considered a diagnostic characteristic.

Pure cultures are identified as *S. pneumoniae* by capsular swelling. A diagnostic reagent,

"Omni-serum," which contains antibodies to all 82 capsular polysaccharides, has been employed for rapid identification (obtained from Statens Serum Institut, Copenhagen, Denmark). This serum can also be used to identify by counterimmunoelectrophoresis (CIE) capsular polysaccharide in the blood and urine of patients with pneumonia and the cerebrospinal fluid (CSF) of patients with meningitis. The CIE technique gives rapid diagnosis and identifies pneumococcal infection in patients whose CSF has been sterilized by antibiotics.

PATHOGENICITY

The capsular polysaccharide makes the pneumococcus virulent. Like the M-protein of group A streptococci, this surface polysaccharide deters phagocytosis by both polymorphonuclear leukocytes and macrophages. On the other hand, decapsulated pneumococci or encapsulated organisms opsonized with specific antibodies are readily taken up by phagocytes. Complement may also be opsonic for some types, even in the absence of specific antibody. Solubilized polysaccharide may overcome host resistance by binding with circulating antibodies. The polysaccharide, being resistant to degradation by host enzymes, may also contribute to immune paralysis by overwhelming the mononuclear phagocytic system. The polysaccharide may thus remain within phagocytic cells for the life of the animal without inciting an inflammatory response. "Surface phagocytosis" is a phenomenon in which encapsulated pneumococci can be phagocytosed when trapped by phagocytes in vitro on rough surfaces. Its role, if any, in the phagocytosis of encapsulated pneumococci in the alveoli has not been established.

No pneumococcal toxin is known to be definitely related to pathogenicity. Upon washing and autolysis, pneumococci release a cell-bound lysin that is antigenically related to streptolysin O. A neuraminidase, which cleaves terminal *N*-acetyl-neuraminic acid from glycoprotein substrates on cell membranes, may be involved in the neurotoxicity of meningeal infections. In the presence of immunoglobulin, platelets are aggregated by heat-killed bacteria; this may play a role in the pathogenesis of disseminated intravascular coagulation seen in severe infections. It is believed that lethality depends entirely on the ability of the organism to grow extracellularly and that toxins are not significant lethal factors.

DRUG SUSCEPTIBILITY

Although most pneumococci are susceptible to less than 0.03 μg/ml penicillin G, acquired resistance to this antibiotic has been described in South Africa, New Guinea, and elsewhere. Tetracycline, erythromycin, and vancomycin resistance have also been reported. Multiple drug resistance, not associated with β-lactamase production, has also been described (see Chapter 21).

EPIDEMIOLOGY

Infection with pneumococci occurs through droplets released from infected patients. Whether or not soiled clothing and blankets can serve as infective material has not been established. There is no animal reservoir. Intrafamily spread and asymptomatic infection occurs frequently, especially in association with viral upper respiratory infections.

References

Davies, P., Page, R. C., and Allison, A. C.: Changes in cellular enzyme levels and extracellular release of lysosomal acid hydrolases in macrophages exposed to group A streptococcal cell wall substance. J Exp Med 139:1962, 1974.

Fox, E. N.: M Proteins of group A streptococci. Bact Rev 38:57, 1974.

Ginsburg, I.: Mechanism of cell and tissue injury induced by group A Streptococci: Relation to poststreptococcal sequelae. J Infect Dis 126:294, 1972.

Ginsburg, I., and Sela, M. N.: The role of leukocytes and their hydrolases in the persistence, degradation and transport of bacterial constituents in tissues: Relation to chronic inflammatory processes in staphylococcal, streptococcal and mycobacterial infections and in chronic periodontal disease. Critical Reviews in Microbiology, 4:249, 1976.

Ginsburg, I., Zor, U., and Floman, Y.: Experimental Models of Streptococcal arthritis: Pathogenetic role of streptococcal products and prostaglandins and their modification by anti-inflammatory agents. In Glynn, L. E., and Schlumberger, H. D. (eds.): Experimental Models in Chronic Inflammatory Diseases. Berlin, Springer Verlag, 1977, 256.

Glynn, L. E.: Rheumatic Fever. In Gell, P. G. H., Combs, R. R. A., and Lachmann, P. J. (eds.): Clinical Aspects of Immunology. 3rd ed. Oxford, London, Blackwell Scientific Publications, 1975, p. 1079.

Kaplan, M. H.: Antoimmunity in Rheumatic Fever: Relationship to Streptococcal antigens cross-reactive with valve fibroblasts, myofibers and smooth muscle. In Dumonde, D. C. (ed): Infection and Immunity in the Rheumatic Disease. Oxford, Blackwell, Scientific Publication, 1976, p. 113.

Lehner, T., Challacombe, S. J., and Caldwell, J.: Immunologic basis for vaccination against dental caries in rhesus monkeys. J Dent Res 55:c166, 1976.

Patterson, M. J., and Hafeez, A. El.B. Group B streptococci in human disease. Bact Rev 774:40, 1976.

Stollerman, G. H.: Rheumatic fever and streptococcal infection. New York, Grune and Stratton, 1975.

Wannamaker, L. W., and Matsen, J. M.: Streptococci and Streptococcal Diseases. Recognition, Understanding and Management. New York, Academic Press, 1972.

Wannamaker, L. W.: The chain that links the heart and the throat. Circulation 48:9, 1973.

Weissman, S. M., Reich, P. R., Somerson, N., and Cole, R.: Genetic differentiation by nucleic acid homology. IV. Relationships among Lancefield groups and serotypes of Streptococci. J Bacteriol 92:1372, 1966.

Wilson, G. S., and Miles, A. A. (eds.): The Streptococci. In Topley and Wilson's Principles of Bacteriology, Virology and Immunity. 6th ed. London, Edward Arnold, 1975, p. 712.

Zabriskie, J. B.: Rheumatic Fever: A streptococcal-induced autoimmune disease? In Dumonde, D. C. (ed.): Infection and Immunology in the Rheumatic Diseases. Oxford, Blackwell Scientific Publications, 1976, p. 97.

Gram-Positive Rods

25 *DIPHTHERIA BACILLI AND OTHER CORYNEBACTERIA*

Lane Barksdale, Ph.D.

Diphtheria is a contagious disease of man in which *Corynebacterium diphtheriae* colonizes the mucous membranes of the fauces and pharynx (sometimes extending to the larynx and the trachea). This is *faucial (laryngeal,* etc.) *diphtheria.* When colonization occurs on the subcutaneous tissue (of the skin) *cutaneous diphtheria* may result.

THE DISCOVERY OF DIPHTHERIA, CORYNEBACTERIUM DIPHTHERIAE, *DIPHTHERIAL TOXIN, AND ANTITOXIN*

Just one century ago, in most countries of the world, diphtheria was a dreaded disease with a case fatality rate of about 8 per cent (see Table 1).

TABLE 1. Decline in the Case Fatality Rate for Diphtheria in England According to Returns from the Hospitals of the Metropolitan Asylums Board*

YEAR	CASE MORTALITY PER CENT	YEAR	CASE MORTALITY PER CENT
1889	40.7	1909	9.4
1890	33.5	1910	7.8
1891	30.6	1911	8.4
1892	29.3	1912	6.2
1893	30.4	1913	6.2
1894	29.3	1914	7.9
1895	22.8	1915	8.4
1896	21.2	1916	6.8
1897	17.7	1917	6.7
1898	15.4	1918	7.7
1899	13.9	1919	9.3
1900	12.3	1920	8.6
1901	11.1	1921	8.8
1902	11.0	1922	8.7
1903	9.7	1923	6.8
1904	10.0	1924	7.0
1905	8.3	1925	5.0
1906	8.8	1926	4.9
1907	9.6	1927	4.0
1908	9.7		

*Antitoxin came into general use in the treatment of diphtheria in 1895. During the same period the diagnosis of diphtheria began to pick up mild cases that previously may have fallen into some other clinical category. There was probably a slow but general improvement in living standards including personal hygiene. Mandatory immunization with toxoid was not begun until more than 20 years after the gathering of the data in this table. (Data from Wilson, G. S., and Miles, A. A.: Topley and Wilson's Principles of Bacteriology and Immunity. 4th ed. Baltimore, The Williams and Wilkins Company, 1955.)

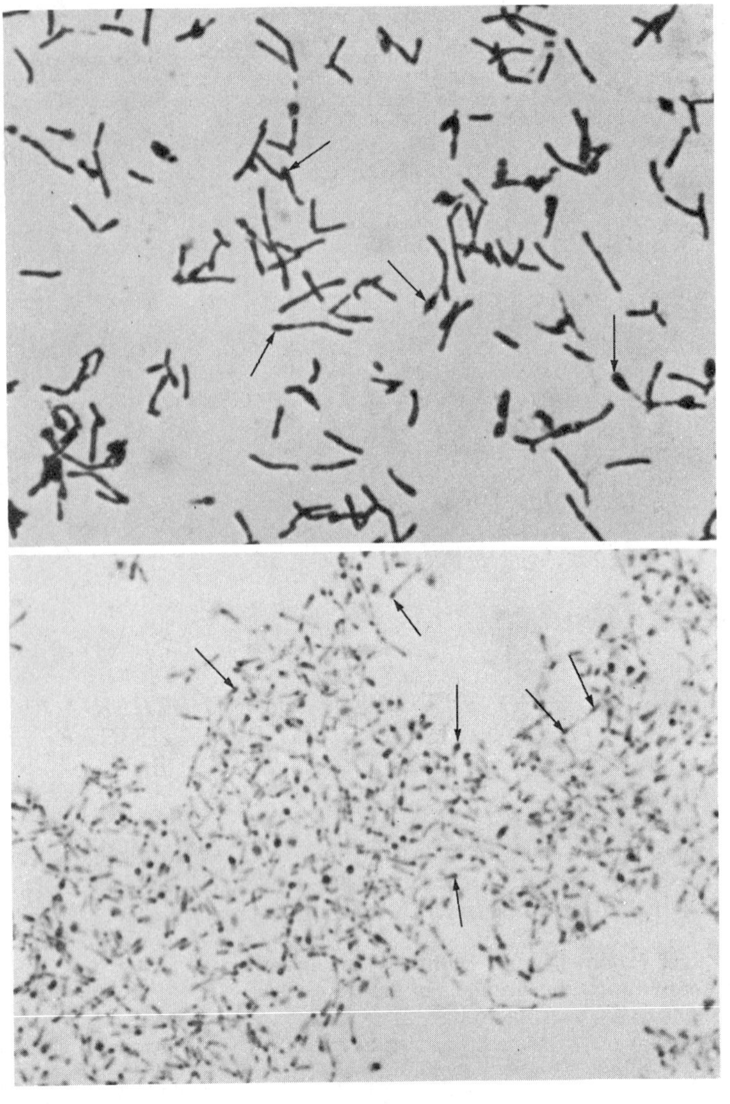

FIGURE 1. Corynebacterium diphtheriae, *strain* $C7_S(-)^{tox-}$ *grown on Loeffler's Medium with added phosphate.* Upper, *Cells stained by the method of Gram. Arrows indicate swollen areas, "club shapes." See text.* Lower, *Cells stained with alkaline methylene blue. Arrows indicate metachromatic granules. Compare these shapes with those found in tissue culture (Fig. 2) and those found under conditions ideal for maximal growth (Fig. 3, upper). ×2000. Courtesy K.-S. Kim.*

The history of our coming to understand diphtheria and to devise a means of combating *diphtheritic death* stands as a model investigation of the etiology of an infectious disease and as a lasting tribute to the clinicians and microbiologists concerned (Andrewes et al., 1923). It was the astute French physician Pierre Fidèle Bretonneau who (from 1821 to 1826) conceived the clinical entity diphtheria to be characterized by the formation of a tightly adhering membranous growth (the pseudomembrane) on the mucous membranes of the throat, sometimes extending into the trachea. At that time, before bacterial agents of disease were known, Bretonneau's clear delineation of the specific clinical condition of diphtheria was the beginning of separation of one throat infection from another, for example, diphtheria from streptococcal pharyngitis. The question "can diphtheria be transmitted from one human subject to another?" was asked by Bretonneau and answered by Trendelenburg (1879), who showed that injection of pseudomembranous material from human cases of diphtheria into pigeons and rabbits gave rise to pseudomembranes. Thus the pseudomembrane of diphtheria was transmissible. With these important facts to go on, Friedrich Loeffler began a painstaking investigation into the cause(s) of diphtheria. Being a student in the laboratory of Robert Koch (see *Mycobacterium tuberculosis*), he was disposed to think that diphtheria might have a bacterial etiology. He also accepted the idea of Bretonneau that diphtheria was a singular disease, in the sense that physicians of Bretonneau's day considered smallpox to be singular.

Loeffler's approach to the microscopic examination of material from diphtheritic lesions was essentially that employed today. The staining of smears from some pseudomembranes with alkaline methylene blue revealed to him club-shaped bacilli containing reddish, spheroidal inclusions (metachromatic granules), as shown in Figure 1. Loeffler also knew that diphtheria bacilli growing in vitro often assumed a unique shape (Fig. 2). In smears from 22 cases of clinical diphtheria *he was able to demonstrate bacilli in only 13*. From 6 of these he isolated diphtheria bacilli in culture (on inspissated serum slants). Thus, it was not possible for Loeffler to isolate the causal agent in every case of diphtheria. To the present day, such has been a common experience of bacteriologists. He also isolated diphtheria bacilli from one normal child. Thus, he realized that *the isolation of diphtheria bacilli from the human throat does not necessarily mean that that throat is a diphtheritic one*. He observed that diphtheria and streptococcal pharyngitis *could occur together*. To this day *streptococcal infections sometimes occur simultaneously with diphtheria*. Thus, a diagnosis of diphtheria may require sophisticated clinical judgment backed up with careful bacteriology.

Loeffler's experimental infections in animals led to the discovery that diphtheria bacilli tended to remain at the site where they had been injected, although autopsy of those animals revealed damage to organs far from that site. The *connection* between diphtheria bacilli at a superficial location (in experimentally infected animals) and damage to distant organs with subsequent death was shown by Roux and Yersin (1888) to be the filterable poison diphtherial toxin. It remained for Behring and Kitasato (1890) to discover that antibodies prepared against diphtherial toxin could neutralize its toxicity, thus providing a means of

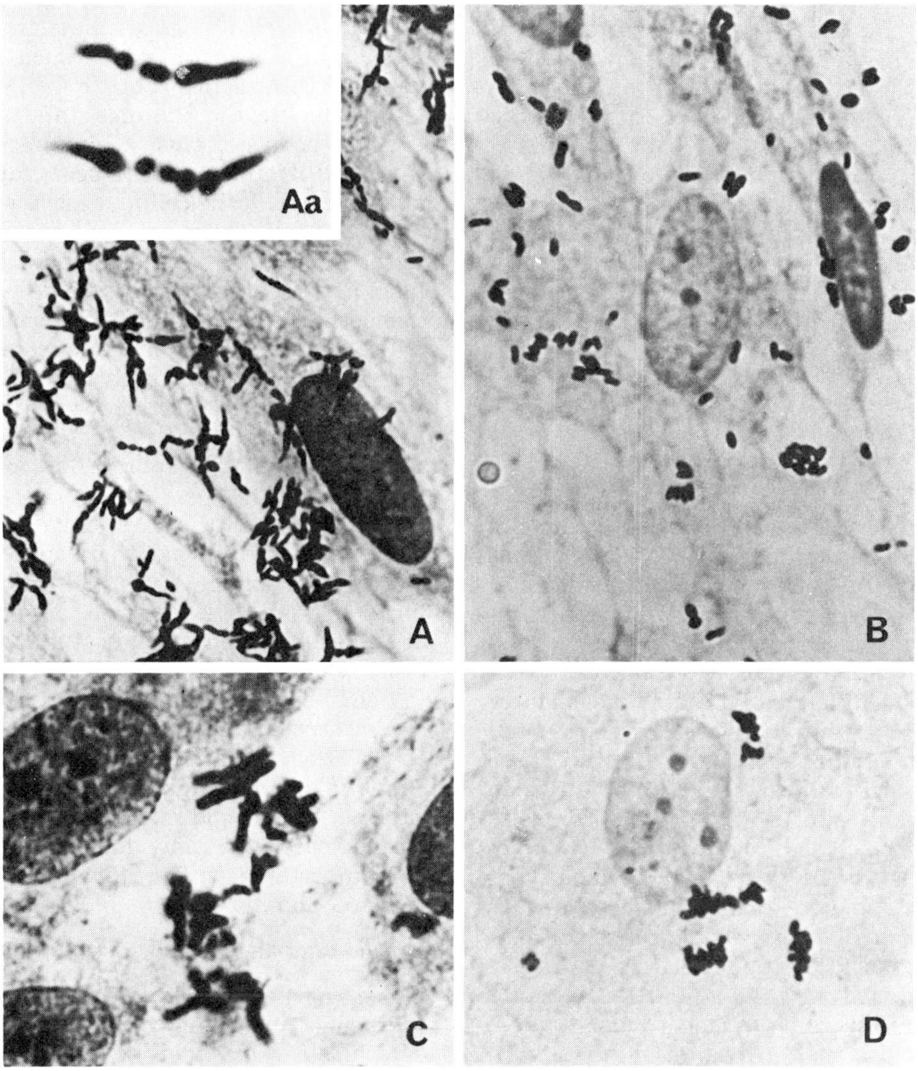

FIGURE 2. Corynebacterium diphtheriae *strain* $C7_S(-)^{tox-}$ *growing in tissue culture,* A *and* Aa. B *and* C, diphtheriae gravis *growing as in* A. C, Bacterium sp. *strain 22M growing as in* A. D *and* C, xerosis *growing as in* A. ×*3,000. Courtesy K.-S. Kim.*

rescuing patients from diphtheritic death by *passively* immunizing them against diphtherial toxin.

CHEMICAL BASIS OF TAXONOMY OF THE GENUS CORYNEBACTERIUM *AND* C. DIPHTHERIAE

Corynebacterium is the Latin name applied to a variety of tapered, gram-positive, nonmotile, nonspore-forming, rod-shaped bacteria. Their closest relatives are in the genus *Nocardia* and the genus *Mycobacterium,* collectively known as the CMN group (Barksdale, 1970). This group of organisms share a basic cell wall peptidoglycan of cross-linked subunits containing *meso*diaminopimelic acid, L-alanine, and D-alanine to which are covalently bonded polysaccharides consisting of arabinose and galactose (arabinogalactan) and some mannose as arabinomannan. In addition, the cell walls of organisms of each of the three genera contain ester-linked, β-hydroxylated, α-branched, long-chain fatty acids of characteristic carbon lengths ranging from about C_{28} to C_{90} (corynomycolic, C_{28} to C_{40}, nocardomycolic, C_{40} to C_{56} and mycolic (mycobacterial), C_{60} to C_{90}, acids). These mycolic acids are important constituents of the outer cell walls and such cell-surface-associated conjugates as trehalose dimycolates (cord factors) (Fig. 3). Members of the CMN group can be used as adjuvants in experimenal immunization.

MORPHOLOGY

Diphtheria bacilli may be separated primarily into different kinds on the basis of the size, shape, and texture of the colonies they form (colonial types). They can be further characterized according to the various biochemical properties discussed under The Laboratory Identification of *C. diphtheriae*. The difference in colony morphology reflects certain essential differences in the cell surfaces of individual diphtheria bacilli. These surface peculiarities affect the manner in which individual bacilli pile up to form a colony. To identify diphtheria bacilli, knowledge of the colonial type is useful. The three most common types of colony are (1) smooth, (2) dwarf-smooth, and (3) semirough. (Rough strains of *C. diphtheriae* lack characteristic surface antigens present in semirough organisms.)

When the strains that are smooth also produce diphtherial toxin, they are called *mitis* (Fig. 4) (McLeod, 1943; Robinson, 1934). Similarly, toxin-producing strains of the dwarf-smooth variety have been termed *intermedius* and certain starch-fermenting, toxin-producing rough strains have been designated *gravis* (Fig. 4). Each of these colonial categories can be further subdivided on the basis of phage types (specific patterns of sensitivity to selected corynebacteriophages) and antigenic types identifiable with specific antisera (serotypes). Some public health laboratories have also typed strains according to their sensitivity to bacteriocins (see Chapter 4). Individuals cells of semirough strains of *C. diphtheriae* tend to be short and stubby. Cells of smooth strains are much longer, whereas the cells of dwarf-smooth strains are intermediate in length.

Common to cells of diphtheria bacilli and most other corynebacteria are certain properties which can be observed with the light microscope: (1) When growing in tissue their cell walls have in them thin spots that are leaky to the Gram stain; they are gram-variable. (2) Old cells store phosphate as polymetaphosphate, localized as phosphate glass and known as metachromatic granules. When bacilli containing such granules are stained with a metachromatic dye, e.g., alkaline methylene blue, the granules stand out as reddish refractile bodies against a blue background (see Fig. 1). (3) Bacilli with thin spots tend to balloon out at one end of the cell and assume a shape like a club (coryne-, coryneform, corynebacterium). Misshapenness of this sort is called pleomorphism. In smears from tissue, the pleomorphism of smooth strains is more exaggerated than that of dwarf-smooth and rough strains. Thus, the bizarre morphology of diphtheria bacilli from a pseudomembrane associated with infection by a *mitis* strain could be more readily recognized than that from a membrane associated with a *gravis* infection.

Since in each of the colonial categories (*gravis, mitis,* and *intermedius*) there are a number of serologic and phage types, the species *C. diphtheriae* has the potentiality for accommodating to the defenses of the human host much as do Group A *streptococci* and *pneumococci*. Thus, persons who have experienced infection with one type of *C. diphtheriae* may later be infected with another type. When this happens in young adults or older individuals having adequate levels of circulating antitoxin, the second infection is apt to be of little consequence to the individual though it may be of some importance in the epidemiology of the disease (see Chapter 90).

C. *diphtheriae* and Salts of Tellurium

Most corynebacteria share with certain other organisms, including some staphylococci and some yeasts, the capacity to grow in the presence of 100 μg/ml potassium tellurite, to reduce the tellurite to tellurium and to concentrate the tel-

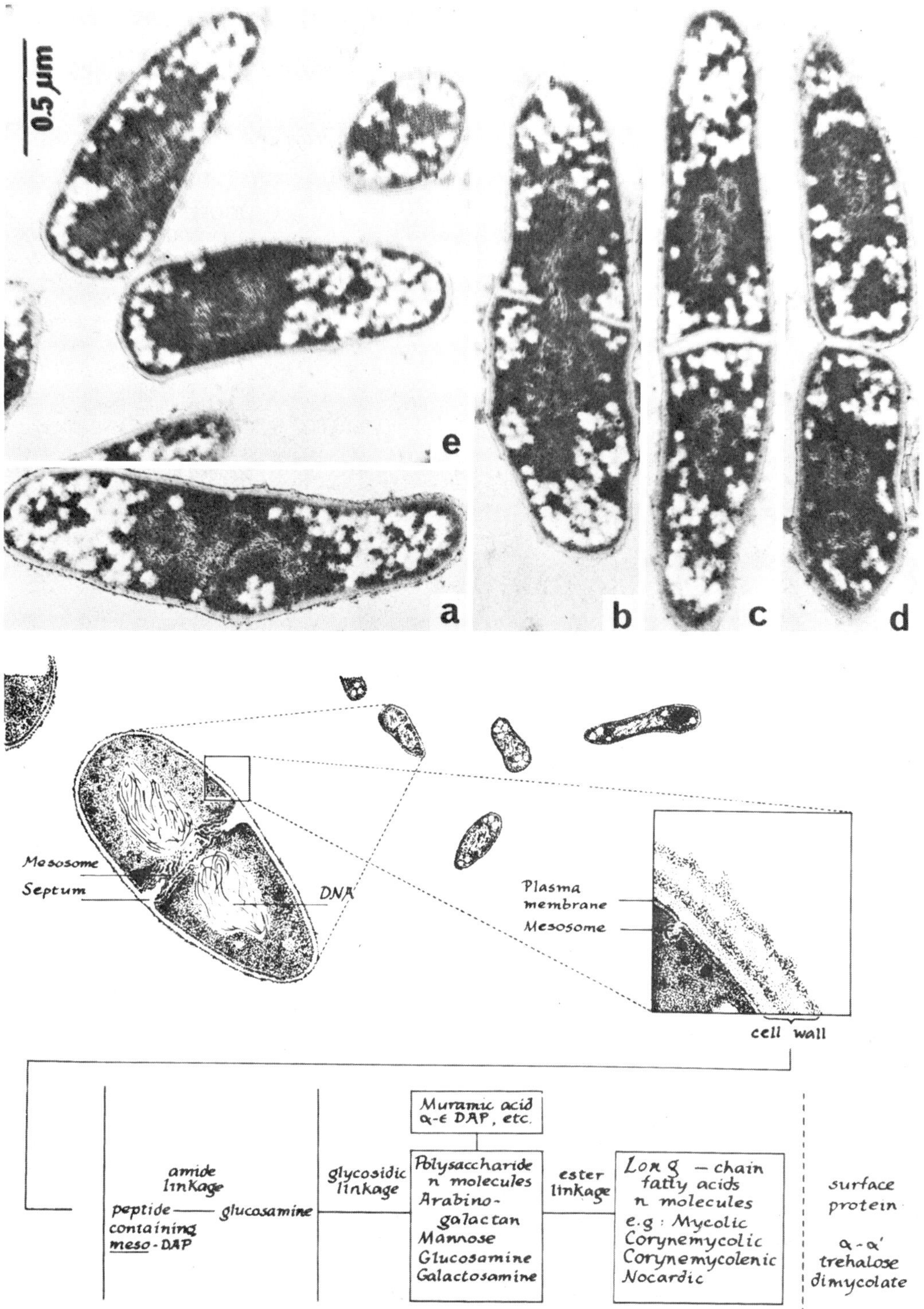

FIGURE 3. Upper, *An electron micrographic record of cell division as it occurs in logarithmically growing* C. diphtheriae, $C7_S(-)^{tox-}$. *Division time = 60 minutes. A, Initiation of septum formation by ingrowth of membrane. B, Well-developed septal initials showing "layers" of the components of the cell envelope. C, Two cells still connected showing that the septum consists of two full complements of membrane and envelope components. D, Beginning of separation of cell doublets. E, The "snapping" involved in the pulling apart of two corynebacterial cells, showing the characteristic taper from septal to distal end. Electronopaque areas, peculiar to actively growing cells, seem not to be glycogen but may represent lipid associated with loci of intense biosynthetic activity. ×51,000. Bar = 0.5 µm. From data of Sheila Heitner.*

Lower, *Diagrammatic sketch of an actively growing bacterial cell representing a composite of the CMN group. A "resting" cell with metachromatic granule is shown. A portion of the envelope of an actively growing cell has been expanded to show the relation of the complex envelope to the cytoplasmic membrane. A portion of the wall is shown to consist of murein, arabinogalactan-mannan linked to species of long-chain, α-branched, β-hydroxylated fatty acids, the mycolic acids, and to dimycolates of trehalose and to a surface protein antigen. The mureins and arabinogalactans are distinctive from genus to genus as are the mycolic acids—for example, mycolic* (Mycobacterium), *corynemycolic and corynemycolenic* (Corynebacterium), *and nocardomycolic* (Norcardia). *In general terms, the murein-arabinogalactan is a heat-stable O antigen; the heat-labile surface protein antigen is the K antigen. Drawing by James E. Ziegler. Rearrangement by Kwang-Shin Kim. Reproduced with permission of Bacteriological Reviews.*

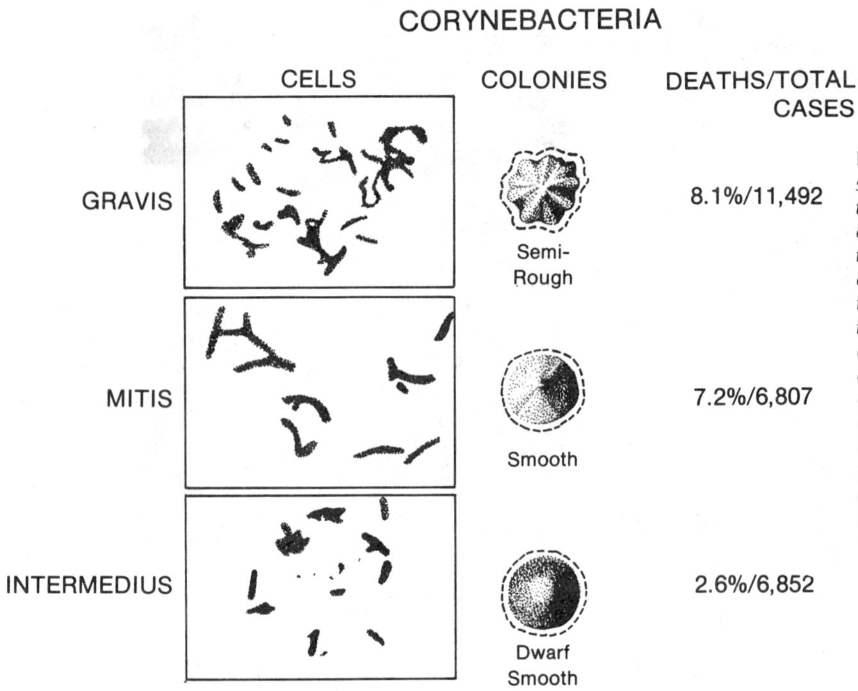

FIGURE 4. *Separation of toxinogenic strains of* Corynebacterium diphtheriae *on the basis of the morphology of their colonies. In a series of 25,000 cases of diphtheria studied in England, outbreaks of the disease could be related to specific colony types* (gravis, mitis, *and* intermedius), *and there was a rough correlation between colony type and clinical severity of the disease. It should be noted that once these colonial types are cultivated in complex media, a number of nutritional and colonial mutants grow out. Thus, in order to keep a stock culture of a model of any one of the colony types, repeated reisolation of that type is necessary.*

lurium (as the metal or its sulfide) in the cells, so that the bacterial colonies that develop on tellurite agar appear grayish black to jet black. The growth of most streptococci, some staphylococci, and other members of the throat flora is inhibited on tellurite agar. For over 60 years a variety of tellurite media have been used for the selective cultivation of *C. diphtheriae* (see The Laboratory Identification of *C. diphtheriae*).

PATHOGENIC PROPERTIES

Diphtherial Toxin

Diphtherial toxin is a simple protein with a molecular weight of about 62,000 daltons. It is lethal for man in amounts of 130 ng/kg of body weight. Toxin is liberated into the extracellular milieu by growing cells of *C. diphtheriae* and is therefore called an exotoxin. It consists of a single polypeptide chain cross-connected by two disulfide bridges, as shown in Figure 5. Mild treatment of toxin with trypsin (termed "nicking"), followed by reduction of its disulfide bridges with dithiothreitol, yields two fragments: an N-terminal fragment A (m.w. 21,000) and a C-terminal fragment B (m.w. 39,000); see Pappenheimer (1977).

Although the only biologic activity directly associated with the intact toxin molecule is the ability to inactivate a variety of animal cells, fragment A behaves as a diphosphopyridine nucleotidase (NADase) as well as an adenosine diphosphoribosyl transferase. This latter function of fragment A stops protein synthesis in the test tube by covalently linking the adenosine diphosphoribose moiety of nicotinamide adenine dinucleotide to eukaryotic elongation factor 2 (EF2) (Collier, 1975). This *enzymatic activity of fragment A* (liberated from diphtherial toxin) operationally places toxin in the category of a proenzyme. There is good evidence to suggest that toxin fixes to the sensitive animal cells by its carboxy terminal end (Fig. 5) and that proteolytic and reductive steps occurring near the cell surface (or infolds related to the process of endocytosis) liberate fragment A, which, once inside the cell, inactivates EF2. Enough such inactivation would stop protein synthesis, leading to cell death.

The Genetic Control of Toxin Production. The genetic information required for the synthesis of diphtherial toxin is carried in the genomes of certain temperate corynebacteriophages as the gene *tox*. This information gains expression in strains of *C. diphtheriae* undergoing lysis by the phage as well as in lysogenic strains that carry tox^+ in the prophage state (Fig. 6). Integration of *tox*-containing prophages into the corynebacterial nucleoid assures perpetuation of the toxinogenic character from one generation of toxinogenic corynebacteria to the next.

Toxoid

The early efforts at immunizing against diphtherial toxin with graded doses of toxin and, later, with mixtures of toxin and antitoxin (horse) sometimes led to unfortunate accidents. A satisfactory nontoxic antigen was obtained by Ramon

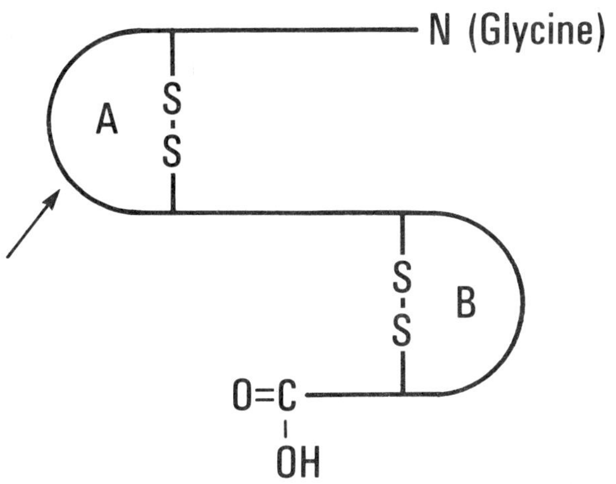

FIGURE 5. *A molecule of diphtherial toxin schematically represented as a single polypeptide chain. At one terminus is the amino group of glycine; the amino acid functioning as the carboxy* ($-\overset{O}{\overset{\|}{C}}-OH$) *terminal is not known. The molecule is interconnected by disulfide bridges between cystine residues. At the bend shown by the arrow is a sequence of three arginines. Trypsin acts to nick toxin at the arrow point. When such nicking action is followed by exposure of the molecule to dithiothreitol, the disulfide bridges are broken and two fragments of toxin result. The smaller piece (from the arrow point back to the terminal glycine residue) has been designated fragment A. The larger piece (from the arrow point to the carboxy terminus) has been designated fragment B. Fragment A is the enzymically active fragment responsible for the following group transfer reactions:*

1. $NAD^+ + EF2 \leftrightarrows$ *ADP-ribosyl EF2 + nicotinamide +* H^+
2. $NAD^+ + HOH \leftrightarrows$ *ADP-ribose + nicotinamide +* H^+
3. NAD^+ *+ toxin* $\leftrightarrows$ *ADP-ribosyl-toxin + nicotinamide +* H^+

EF-2 = elongation factor 2. See text. Fragment B is thought to be important for the fixing of diphtherial toxin to receptors on toxin-sensitive cells.

and Glenny, who found that prolonged incubation of diphtherial toxin with formalin under alkaline conditions converted it into a nontoxic antigen which, when injected into laboratory animals, stimulated the production of antitoxin. Formalin interacts with the tyrosine residues of toxin, effecting their cross-linkage with the ε-amino groups of constituent molecules of lysine. Toxoid is unable to fix to animal cells and it cannot be split into A and B fragments. It lacks, therefore, all of the properties of toxin except antigenicity and the capacity to interact with antitoxic antibody. This remarkably stable antigen has served for more than 50 years as satisfactory prophylactic agent against death from diphtheritic intoxication (see also Schick Test).

In the early days of production of toxoid for immunization, it was necessary to find an exceptional strain of *C. diphtheriae* that produced large amounts of toxin. Park and Williams discovered an *avirulent* diphtheria bacillus, the Park-Williams number 8 strain (PW8), which is used throughout the world in the "manufacture" of diphtherial toxoid. In suitable media the PW8 strain can produce 300 μg toxin protein/10^9 bacilli/ml, or around 300 mg per liter.

Invasiveness of Virulence of *C. diphtheriae*

In Figure 6 are listed the properties of a smooth strain of *C. diphtheriae*, $C7_S(-)^{tox^-}$, carrying no tox^+ prophage and the properties of $C7_S(\beta)^{tox^+}$, a lysogenic strain carrying prophage β^{tox^+}. Antigenically, these two strains are identical except

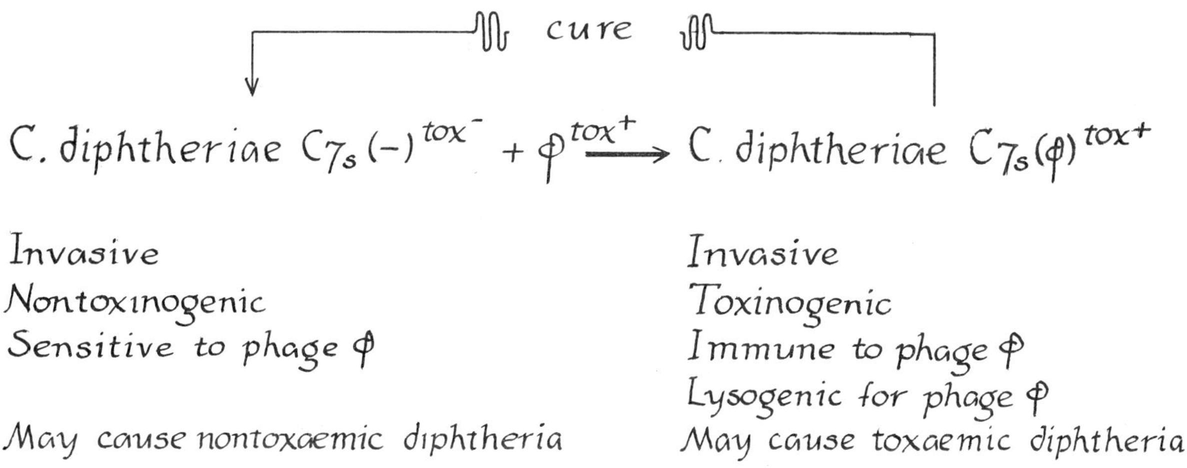

FIGURE 6. *Changes brought to* Corynebacterium diphtheriae, *strain* $C7_s(-)^{tox-}$, *following integration (into its genome) of a probacteriophage that carries the tox gene. Presumably, the indicator strain* $C7_s$ *is nonlysogenic, hence the designation (−), and is nontoxinogenic, tox.⁻ When nontoxinogenic, nonlysogenic C7 is lysed by a phage carrying the* tox^+ *marker, such as* ϕ^{tox-} *or* β^{tox-}, *toxin is produced during the course of phage multiplication and lysis of the cell. When lysogenized by such phages, the genome of* $C7_s(\phi)^{tox+}$ *includes phage genes that endow it with immunity to homologous phage (lysogenic immunity = synthesis of specific repressor) and the ability to synthesize diphtherial toxin. The subscript s refers to the smooth (surface) antigen of the strain. See text for those corynebacterial products that play a role in invasiveness. From Barksdale, L.:* Corynebacterium diphtheriae *and its relatives. Bacteriological Reviews 34:378, 1970. Reproduced with permission of the publisher.*

for the synthesis of diphtherial toxin by the toxinogenic strain. Each of the strains can colonize the pharyngeal mucous membranes of susceptible people and produce a pseudomembrane. Invasiveness in *C. diphtheriae* is associated with production of a dimycolate of trehalose, a so-called cord factor (see Chapter 40), which can inactivate the mitochondria of mammalian cells. *C. diphtheriae* can survive in pharyngeal mucus by producing a neuraminidase (sialidase) capable of cleaving residues of *N*-acetylneuraminic acid, NANA (sialic acid), from parent molecules such as mucins, the glycoproteins of cell surfaces, and gangliosides. In addition, these corynebacteria liberate an *N*-acetylneuraminic acid lyase which can split NANA into its constituents, *N*-acetyl mannosamine and pyruvate. Pyruvate markedly stimulates growth of corynebacteria and other members of the CMN group.

THE LABORATORY IDENTIFICATION OF C. DIPHTHERIAE

At present, *C. diphtheriae* is not sought in the average diagnostic laboratory *unless* a death from diphtheria has occurred. In many countries there are few if any laboratory technicians who have had first-hand experience with *C. diphtheriae*. Yet diphtheria occurs in at least one major city of the world each year. The problem of diagnostic inadequacy of laboratories is compounded by deficiencies in contemporary textbooks regarding the characterization of diphtheria bacilli. The deficiencies include (1) inadequate definitions of "pleomorphism" in general and with regard to *C. diphtheriae* in particular; (2) absence of a clear statement of the relationship existing between nontoxinogenic and toxinogenic *C. diphtheriae* (see Fig. 6); and (3) erroneous fermentation patterns (repeated from earlier textbooks) by which various corynebacteria are identified.

Procedures

Smears are prepared from diphtheritic membrane, and media are inoculated for enrichment of numbers of diphtheria bacilli and for primary isolation. *Note*: Corynebacteria thrive in a CO_2-rich environment, e.g., 10 per cent in air. A candle jar provides satisfactory levels of CO_2.

The diphtheritic membrane (pseudomembrane, see Figs. 1 and 2 in Chapter 90) is usually so tenacious that rubbing it with a cotton swab will remove only some of its outermost material. More of the fibrinous exudate may be obtained by using a bacteriologic loop (steel) that has been opened into a hook. Pulling away bits of pseudomembrane with such a device often results in slight bleeding. This behavior is in contrast to that of the easily removable exudate found in uncomplicated streptococcal pharyngitis. Some of the pseudomembrane so obtained is used for inoculating (1) a slant of Loeffler's medium supplemented with 2.5 gm K_2HPO_4/liter, (2) a blood agar plate, and (3) a Mueller-Miller tellurite or Tinsdale tellurite agar plate. From the remainder, smears should be prepared for future reference and for (1) the Gram stain and (2) staining with alkaline methylene blue. The former should provide an assessment of the different kinds of bacteria present in the exudate; the latter indicates whether or not any of the tapered, pleomorphic rods present contain polyphosphate (metachromatic) granules. Corynebacteria growing in a low phosphate environment produce very few metachromatic granules. This probably explains the limited numbers of granules found in *C. diphtheriae* in smears from tissues. The tapered ends of *C. diphtheriae* (see Figs. 1 and 2) will appear more exaggerated with the methylene blue stain than with the Gram stain. The exaggeration of the taper is most striking with *mitis* strains and less so with *intermedius* and *gravis* strains. Tapering of these rods is stressed here because it represents a particular kind of misshapenness or pleomorphism. Pleomorphism per se is not uncommon among gram-positive bacteria such as α-hemolytic streptococci, certain group G streptococci, propionibacteria, lactobacilli, and corynebacteria. In addition, fusobacteria, ordinarily gram-negative, in mixed cultures sometimes do not decolorize properly and may be mistaken for pleomorphic gram-positive rods. Any one of these bacteria might grow well in and about a streptococcal sore throat or a diphtheritic membrane. Since the normal pharyngeal flora varies as to numbers and kinds of bacteria from one individual to another, the secondary organisms in a pseudomembrane vary greatly as to both kind and absolute numbers. Their numbers more often than not obscure the presence of *C. diphtheriae*. However, *C. diphtheriae*, particularly *mitis* strains, when present may be presumptively singled out on the basis of the taper of the cells and the presence of metachromatic granules in a few cells.

A presumptively positive set of smears should correlate with the growth of similar bacteria on the inoculated Loeffler's slant, in use all of these years because it favors the outgrowth of *C. diphtheriae*, and the morphologic characteristics of the organisms growing on it are closer to those seen in vivo (e.g., in smears from diphtheritic exudates). Compare cells in Figure 3 with those in Figures 1 and 2. From rich inocula *C. diphtheriae* may show considerable increase in numbers as early as 7 hours after incubation.

Final Identification

Colonies of suspected corynebacteria are picked, suspended in sterile broth, and restreaked onto sterile plates. Once isolated colonies have grown up, one (or more) is selected for the preparation of stock cultures for further propagation and identification. Identification is made possible by determining, for the strain under study, the reactions given in Table 2. The way in which the reactions obtained identify the organism under study is shown in Table 3.

Methods for Detecting Diphtherial Toxin and Antitoxin

Agar Gel Diffusion (The Ouchterlony-Elek Plate). Elek and Ouchterlony independently found that precipitates of diphtherial toxin and antitoxin, formed in agar gels, could be used as a means for the in vitro detection of toxinogenic strains of *C. diphtheriae*. A horse serum agar plate (Petri dish) is poured and, before the agar has hardened, a strip of filter paper (1.6 × 8 cm),

TABLE 2. Ten Properties Useful for Separating Members of the Genus *Corynebacterium**

1. Fermentation of lactose
2. Production of catalase†
3. Formation of iodinophilic polymer from glucose-1-phosphate
4. Production of disulfide reductase
5. Fermentation of dextrose
6. Fermentation of maltose
7. Fermentation of starch
8. Production of urease
9. Production of pyrazinamidase
10. Hydrolysis of gelatin

*The author is much indebted to M. C. Pollice and Ioan T. Sulea for data supporting this Table.
†In the absence of exogenous hemin.

TABLE 3. Formulation of Species of *Corynebacterium* According to the 10 Properties Listed in Table 2*

Species	Above the line	Below the line
C. diphtheriae mitis:	2 3 4 5 6	1 7 8 9 10
C. diphtheriae gravis:	2 3 4 5 6 7	1 8 9 10
C. diphtheriae intermedius:	2 3 4 5 6	1 7 8 9 10
C. diphtheriae variety *ulcerans:*	2 3 4 5 6 7 8 10	1 9
C. kutscheri:	2 3 4 5 6 8 9	1 7 10
C. minutissimum:	2 3 4 5 6 9	1 7 8 10
C. pseudotuberculosis (ovis):	2 3 4 5 6 8	1 7 9 10
C. pseudodiphtheriticum: (hofmanni)	2 3 4 8 9	1 5 6 7 10
C. renale:	2 3 4 5 ⑨	1 6 7 8 10
C. xerosis:	2 3 4 5 9	1 ⑥ 7 8 10

*Numbers above the line indicate positive traits. Numbers below the line indicate negative traits. Circled numbers below the line indicate that a very occasional strain may be positive; circled numbers above the line indicate that a very occasional strain may be negative. Sucrose has not been included. Most *C. xerosis* are sucrose positive; some *C. diphtheriae mitis* are sucrose positive, as are some strains of *C. minutissimum* and *C. kutscheri*. Note that the pattern $\frac{2\ 3\ 4}{1}$ is common to each of the 10 species. The three types of *C. diphtheriae* differ only with regard to starch fermentation. However, when this pattern is supplemented with colonial morphology on tellurite agar and cellular morphology (under the microscope), as shown in Figure 4, precise identification can readily be accomplished. Those laboratories using sheep's blood agar plates with a concentration of erythrocytes low enough to reveal feeble hemolysis will be able to separate hemolytic *mitis* strains from nonhemolytic *gravis* and *intermedius* strains. *Note:* To obtain the most prompt and reliable test reactions, actively growing corynebacteria should be washed with saline, resuspended as a thick slurry and inoculated into test media so that each tube receives 100 million or more bacteria.

The author is much indebted to M. C. Pollice and Ioan T. Sulea for data supporting this table.

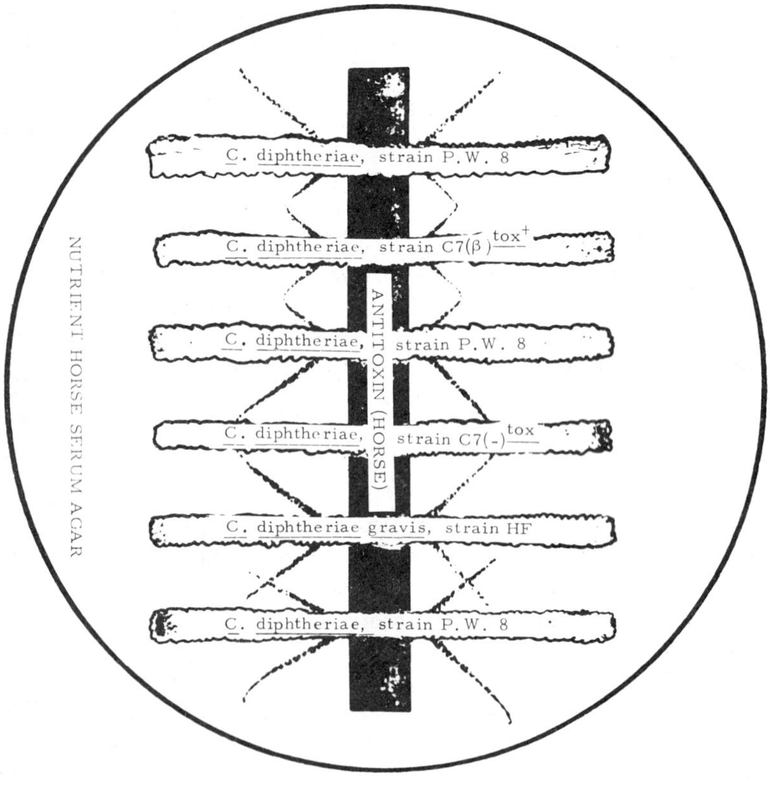

FIGURE 7. *Diagram of an Elek/Ouchterlony plate (see text for details) for the in vitro detection of diphtherial toxin. Cultures of strains of* C. diphtheriae *have been streaked at right angles to the central strip of filter paper soaked in antitoxin. See text for information concerning each of the different strains of diphtheria bacilli and the antigen-antibody precipitates formed in their vicinities.*

previously dipped into horse antitoxin (500 au/ml) and drained of excess liquid, is placed across the center of the agar surface. The Petri dish, its lid ajar, is kept in a 37° C incubator until dry. Onto the dry surface of the plate in well separated parallel lines (Fig. 7) are streaked known toxinogenic and nontoxinogenic strains of *C. diphtheriae* and the unknown organisms being tested. Cotton swabs, soaked in heavy bacterial cultures and then freed of excess liquid by expression against the sides of the culture tubes, are useful for introducing each of the bacterial cultures onto the plate. A satisfactory positive control is the toxinogenic avirulent PW8 strain, and a suitable negative control is the nontoxinogenic mutant of *C. diphtheriae gravis,* HF. Consider as unknown organisms to be tested strains $C7_s(-)^{tox-}$ and $C7_s(\beta)^{tox+}$. Each is streaked so that it is growing next to the known toxinogenic strain, PW8. If, during growth on the antitoxin-containing agar plate, toxin is produced, then a line of toxin-antitoxin precipitate will become visible in the agar, as can be seen to the left of the PW8 strain. When gradients of identical precipitates of antigen-antibody are formed from different loci in agar gels, they will fuse, producing an arc of identity. Thus, between C7 (β) and PW8 the diffusing toxins interacting with antitoxin in the agar have produced arcs between toxin (C7 (β))· antitoxin from one side to toxin (PW8)· antitoxin from the other. Diphtheria bacilli produce various proteins other than toxin. In most preparations of antitoxin there are antibodies to some of these proteins. In the case of $C7(-)^{tox-}$ (Fig. 7) an anomalous protein has formed a precipitate with its antibody. Between strain PW8 and strain HF anomalous antigen-antibody precipitates have formed showing no arc of identity.

Intradermal Skin Tests

Necrosis by diphtherial toxin can readily be seen following the introduction of minute amounts of toxin into the skin of shaved rabbits or guinea pigs. Neutralization of toxicity by antitoxin specifically identifies the toxin. Bacterial cultures may similarly be tested intradermally for their capacity to produce toxin. For example, at numbered multiple sites on the back of a rabbit, *known* and *test cultures* are injected at time zero, antitoxin is given intravenously at time *t* (usually 3½ hours later), and at time *t* + 30 minutes *the same cultures* are injected at new (prime) sites. The reactions are read daily up to 96 hours. When diphtherial toxin is present, necrosis will occur at the sites inoculated at time *t* but not at the sites inoculated 30 minutes after the administration of antitoxin. Necrosis at both sites indicates the production of a toxin not neutralized by diphtherial antitoxin, i.e., a toxin immunologically distinct from diphtherial toxin.

IMMUNITY: THE SCHICK TEST

In 1913 Bela Schick described a skin test for the detection of circulating antitoxin in human subjects. Once in use, it became clear that a modified Schick test would also detect delayed hypersensitivity to other products of *C. diphtheriae.* In the modified Schick test used at New York University Medical School for over 30 years, minute amounts of toxin are injected at one site (on the forearm) and toxoid at a second site. There are five pairs of reactions possible. (1) If at 48 to 96 hours there is no reaction at either site, the subject has sufficient circulating antitoxin to neutralize the dose of toxin and shows no allergy to corynebacterial products. This rather positive capacity has been designated as *the Schick-negative reaction.* (2) If there is *necrosis* at the toxin site and no reaction at the toxoid site, the subject has insufficient circulating antitoxin to neutralize the toxin. This relatively negative condition is designated as *the Schick-positive reaction.* (3) An *immediate wheal* and *erythematous reaction* at both sites at about 35 minutes post injection indicates circulating IgE specific for the injected products. (4) An *erythematous reaction* at *48* to *96 hours* at both sites indicates that there has been enough circulating antitoxin to neutralize the toxin administered and that in addition the subject shows delayed allergy to corynebacterial products. This has been called *the pseudoreaction.* (5) *Necrosis* at the test site and *delayed allergy* at both sites indicate insufficient circulating antitoxin to neutralize the toxin and an allergy to corynebacterial products. This has been called *the combined reaction.*

Over the years the examination of human populations with the Schick test has given various insights into immunobiology (Barksdale, 1980). It has revealed that carriers of *C. diphtheriae* do not necessarily have detectable levels of circulating antitoxin. It has revealed that circulating antitoxin per se does not prevent the onset of a diphtheritic infection. And, most recently, it has led to the discovery that about 1.5 to 2 per cent of the human population cannot respond to immunizing doses of toxoid with the production of detectable levels of circulating antitoxin. As long as these people are around and as long as diphtheria bacilli continue to be carried in populations, there will probably be occasional deaths from diphtheria.

SENSITIVITY OF DIPHTHERIA BACILLI TO COMMONLY AVAILABLE ANTIBIOTICS

Although *C. diphtheriae* is sensitive to a variety of antibiotics active on gram-positive organisms, including chloramphenicol, erythromycin, kanamycin, methicillin, penicillin G, rifampin, streptomycin, and tetracycline, the antibiotics that have been most thoroughly examined in relation to cases of diphtheria and carriers of diphtheria bacilli are penicillin G and erythromycin. In a recent study of the antibiotic sensitivity of 337 strains of *C. diphtheriae* isolated from cases of diphtheria, 90 per cent were sensitive to the aforementioned antibiotics, and 100 per cent of the strains were sensitive to gentamicin. The minimal inhibitory concentrations of antibiotics for *C. diphtheriae* are as follows (μg/ml): penicillin G, 0.08; erythromycin, 1.6; chloramphenicol, 0.5; and tetracycline, 1.0.

OTHER CORYNEBACTERIA

Listed in Table 3 are most of the species of *Corynebacterium* encountered in material from human and animal sources. Those that are of significance in diphtheritic infections are *C. diphtheriae* and sometimes *C. ulcerans.* Rare infections with *C. pseudotuberculosis (ovis)* have been reported in human subjects whose occupation put them into intimate contact with animals. *C. pseudotuberculosis (ovis)* causes infections in sheep and horses. *C. xerosis,* originally isolated from the conjunctiva, is one of those "background" corynebacteria from which *C. diphtheriae* must be differentiated. Aside from its distinguishing properties given in Table 3, *C. xerosis* tenaciously retains crystal violet, in contrast to the leaky, gram-variable reaction given by *C. diphtheriae* to the Gram stain. *C. pseudodiphtheriticum (C. hofmanni)* (Table 3) is another species of *Corynebacterium* that was originally isolated from normal throats and sometimes is encountered in routine bacteriologic study of the throat and upper respiratory tract. Among corynebacteria associated with genitourinary infections is a group recently designated *C. genitalium. C. kutscheri* is found in overt and latent infections of mice. *C. renale* is the cause of pyelonephritis in cattle.

Organisms that have been wrongly listed as belonging in the genus *Corynebacterium* include the group G streptococcus originally designated *C. pyogenes,* and the propionibacterium called *C. acnes.* The group G streptococcus (*C. pyogenes*), which under natural conditions produces suppurative lesions in cattle, sheep, pigs, and goats, has been associated with pharyngitis and skin ulcers occurring in humans. In these latter cases, the bacillary shape assumed by this streptococcus has led to its confusion with *Corynebacterium.* However, the absence of an obvious taper to its cells and the absence of metachromatic granules in them readily suggests that these *lactose-fermenting* streptococci are not corynebacteria.

References

Andrewes, F. W., Bulloch, W., Douglas, S. R., Dreyer, G., Gardner, A. D., Fildes, P., Ledingham, J. C. G., and Wolf, C. G. L.: Diphtheria, Its Bacteriology, Pathology and Immunology. London, Medical Research Council, His Majesty's Stationery Office, 1923.

Barksdale, L.: *Corynebacterium diphtheriae* and its relatives. Bacteriol Rev 34:378, 1970.

Barksdale, L.: The immunobiology of diphtheria. In Nahmias, A. J., and O'Reilly, R. J. (eds.): Immunology of Human Infection. Part I in Comprehensive Immunology, Vol. 8. R. A. Good and S. B. Day (eds.) New York, Plenum Medical Book Company, 1980.

Collier, R. J.: Diphtheria toxin: mode of action and structure. Bacteriol Rev 39:54, 1975.

McLeod, J. W.: The types mitis, intermedius and gravis of *Corynebacterium diphtheriae,* a review of observations during the past ten years. Bacteriol Rev 7:1, 1943.

Pappenheimer, A. M., Jr.: Diphtheria toxin. Ann Rev Biochem 46:69, 1977.

Robinson, D. T.: Further investigations on the *gravis, mitis* and "intermediate" types of *C. diphtheriae*: type stability. J Path Bact 39:551, 1934.

26 *LISTERIA MONOCYTOGENES*

H. P. R. Seeliger, M.D.

MORPHOLOGY

Listeria monocytogenes (Murray, et al., 1926) represents a bacterial species consisting of gram-positive to gram-variable asporogenous, acapsular, aerobic to microaerophilic rods. At temperatures of around 20° C the cells show peritrichous flagellation. At 37° C usually only one polar flagellum is apparent. Motility is best at around 20° C. The growth range is between +4° C and 42° C with most rapid growth of delicate colonies at around 36° C on ordinary media at slightly alkaline pH preferably in the presence of 5 per cent sheep or rabbit blood. Dissociation of colonies occurs with formation of S− and R− and several intermediary forms. Culturally, *L. monocytogenes* may be confused with streptococci, notably motile enterococci.

ANTIGENIC COMPOSITION

By means of antigenic analysis, various somatic and flagellar antigens allow the differentiation of several serogroups with one to several serovarieties in each of these groups. This forms the basis for serologic identification (Table 1). It should be noted that the former serovariants 4f and 4g with numerous subvarieties do not belong to *L. monocytogenes* proper, but rather to a related apathogenic species of *Listeria* that has been named *L. innocua* (Seeliger and Schoofs, 1978).

METABOLISM

Glucose and other carbohydrates are attacked by production of acid, but not of gas. Contrary to the biovarieties of *Murraya grayi,* mannitol is not acidified (see Table 2). NO_3 is not reduced to NO_2. All strains that have been isolated from pathologic specimens of human and animal origin produce a β-hemolysin. β-hemolysis is most pronounced with strains representing serovar 5. Nonhemolyzing strains usually belonging to *L. innocua* and related subvarieties (cf., Table 2) are otherwise biochemically identical with *L. monocytogenes,* but, because of complete lack of virulence, absent β-hemolysis, and failure to produce an experimental monocytosis in rabbits, do belong to a different species. *L. monocytogenes,* including this closely related group of organisms, is differentiated from similar gram-positive rods classified with the genus *Murraya* by a set of biochemical tests as outlined in Table 2, although their G + C ratio of 38 to 40 moles per cent is identical.

PATHOGENIC PROPERTIES

Strains of human and animal origin do not show any differences in pathogenic properties. Smooth virulent cultures can stimulate in rodents an experimental monocytosis through the action of extractable lipids. Instillation of *L. monocytogenes* into the conjunctiva of rabbits provokes a characteristic keratoconjunctivitis. In warm-blooded animals, parenteral inoculation causes a septic disease with varying clinical and organic manifestations. Liver, spleen, central nervous system, and the reproductive organs are most commonly attacked. Infections via the oral route usually are successful only in young or pregnant animals. Typical focal granulomas (listerioma) are prominent; suppuration and abscess formation occur less frequently.

TABLE 1. Serovars of Listeria monocytogenes, Murraya grayi, and Related Species

DESIGNATION Paterson	Seeliger-Donker-Voet	O-ANTIGENS	
1	1/2 a	I II (III)	
	1/2 b	I II (III)	A B C
2	1/2 c	I II (III)	B D
3	3 a	II (III) IV	A B
	3 b	II (III) IV	A B C
	3 c	II (III) IV	B D
4	4 a	(III) (V) VII IX	A B C
	4 ab	(III) V VI VII IX	A B C
	4 b	(III) V VI	A B C
	4 c	(III) V VII	A B C
	4 d	(III) (V) VI VIII	A B C
	4 e	(III) V VI (VIII)(IX)	A B C
	5	(III) (V) VI VIII X	A B C
	7?	(III) XII XIII	A B C
Listeria innocua*	6a (4f)	(III) V VI VII IX XV	A B C
	6b (4g)	(III) V VI VII IX X XI	A B C
M grayi (ssp. grayi)		(III) XII XIV	E
(spp. murrayi)		(III) XII XIV	E

*nonhemolytic strains, additional antigen combinations are known but not listed.
†nonhemolytic serovar, strongly related to serovars 4 f and 4 g.

TABLE 2. Nitrate Reduction and Acid Production from Carbohydrates by Species of Genus Listeria* and Murraya

	L. MONO-CYTOGENES†	MURRAYA subspec. *grayi*	MURRAYA subpec. *murrayi*
$NO_3 \rightarrow NO_2$	–	–	+
L-arabinose	–	–	–
D-galactose	d	+	(+)
glycogen	–	–	–
lactose	(d)	+	+
mannitol	–	+	+
melezitose	d	–	–
melibiose	–	–	–
rhamnose	d	–	d
sucrose	(d)	–	–
xylose	d	–	–

Key: + = acid produced 24-48 hrs (90% or more strains)
– = no acid produced 21 days (90% or more strains)
d = some strains positive, some negative
(+) = acid produced slowly (3-7 days)
(d) = some strains produce acid slowly (3-7 days), other strains negative.

*All species produce acid but no gas in 24-48 hours from amygdalin, esculin, cellobiose, dextrin, fructose, glucose, maltose, mannose, salicin, starch and trehalose. No acid in 21 days from adonitol, dulcitol, erythritol, inositol, inulin or raffinose.
†Including nonhemolyzing, serologically identifiable strains that represent a different nonpathogenic species named *L. innocua*.

L. monocytogenes is the cause of sporadic enzootic and epizootic disease of numerous small and large animals. It is globally spread. The infection may follow the course of a septicemia or may localize in certain organs, especially the central nervous system.

Septicemia occurs mainly in rodents, poultry, and young animals. The central nervous system is involved predominantly in large adult animals. Atypical and silent forms of the disease are not rare. The maturity grade, the resistance of the animals, and the portal of entry probably play a decisive role in the course of the disease. In general, young animals are known to have an increased susceptibility to listeriosis. In older animals, *L. monocytogenes* often is a secondary invader after primary damage from another disease. Only sheep seem to be an exception, since the disease in them often runs a fulminating course with a high mortality.

L. monocytogenes sometimes invades the reproductive organs of wild and domestic animals. Infection of the pregnant animal may result in damage and death of the fetus or abortion. The extent of listeriosis as a cause of infectious abortion in cattle, sheep, and other animals needs more study. Permanent damage to the reproduc-

... has been noted only infrequently. ...tural conditions infections occur sporad... ...nd in epizootics; however, they seem to be ...e frequent than is generally appreciated.

Mortality is high in clinically manifest cases. Epizootics in breeding stocks and live stocks may cause severe economic losses.

In human listeriosis there are many symptoms that may easily lead to confusion with diseases of different etiology. The symptomatology comprises "grippe-like" infections, sore throat with glandular swelling and monocytosis, severe septicemia with predominant involvement of the liver, disease of the central nervous system, and localized purulent inflammations of the mucous membranes. The infected tissues develop granulomatous reactions.

All age groups are affected, but predominantly the newborn and infants, and to a lesser degree the aged. Inapparent infections during pregnancy are transmitted transplacentally to the fetus and may result in abortion or stillbirth owing to infection of the fetus. In other cases, sick infants die shortly after birth. The newborn may also be infected during delivery either by infected vaginal secretions or by swallowing infected amniotic fluid. The immaturity of the neonatal tissues, especially of the intestinal tract, facilitates the penetration of the causative organisms into the body of the newborn. A purulent acute meningitis frequently kills the baby. Apparently, listeriosis is a more frequent cause of fetal damage, abortions, and neonatal death than is generally recognized. In adults, meningitis and meningoencephalitis compose the main clinical syndrome. The disease has also been observed as a secondary infection during corticosteroid therapy or repeated hemodialysis.

In newborns and the aged listeriosis has a fatality rate of 70 per cent and higher. Other forms of the disease are somewhat more benign; asymptomatic infections and a carrier state (fecal, genital, and oropharyngeal) have been observed.

IMMUNITY

Little is known about immunity to listeriosis. Passive protection has not been achieved experimentally by treatment of animals with homologous antiserum. Vaccination with killed suspensions of *L. monocytogenes* has not afforded protection. However, experiments with living vaccines have given more promising results.

Present research indicates that protection may depend on cell-mediated immunity. Although in rare cases circumstantial evidence points to repeated attacks of *Listeria* infection in subsequent pregnancies endangering the fetus, no second clinical *Listeria* infection has been observed in individuals or animals who had been cured of proven listeriosis.

LABORATORY DIAGNOSIS

The diagnosis of listeriosis is made primarily by culture. With certain restrictions, serologic procedures are applicable for case finding and retrospective diagnosis. Blood, sternal bonemarrow, spinal fluid, amniotic fluid, meconium, placenta, throat swabs, and all kinds of discharge should be examined by smear and culture. *L. monocytogenes* may also be found postmortem in the liver, spleen, lymph nodes, cerebrum, cerebellum, and medulla.

Primary culture of *Listeria* is occasionally difficult. Most of the difficulties may be overcome by cultivating the specimens in tryptose or sodium thioglycollate broth, and by use of Gray's enrichment technique — that is, by several weeks' storage at +4° C with repeated subcultures on blood agar or tryptose agar. Media containing nalidixic acid and trypaflavine are helpful in the isolation of *Listeria* from contaminated samples. The recognition of suspicious colonies is facilitated by use of Henry's illumination technique. Owing to morphologic, cultural, biochemical, and serologic similarities with other bacterial species, accurate identification of the organism is required in every suspected case. There are many sources of error and often *Listeria* is overlooked or wrongly classified. Similar organisms may be mistaken for *L. monocytogenes*. This applies particularly to nonhemolytic *L. innocua* and to certain strains of enterococci and corynebacteria. Hemolysis, characteristic tumbling (end over end) motility, and ability to produce keratoconjunctivitis in rabbits are among the most useful diagnostic features.

During and after Listeria infections serum antibodies are found in animals and human patients by agglutination reactions with somatic and flagellar antigens. However, Listeria agglutinins are frequently found in sera of healthy animals and man without a history of listeriosis. Whether those agglutinins are "specific" is still an unsolved problem. Newborn and infants up to the sixth month may lack serum antibodies against *L. monocytogenes,* even in proven infections. The distribution of agglutinin titers varies with age and reaches its maximum between the third and fifteenth year of life. Whether this is due to subclinical infections is a matter of conjecture.

Cross-reactions with enterococcal and especially with staphylococcal antigen create a serious handicap. Therefore, the interpretation of agglutination reactions requires great caution and a critical approach. Only titers of 1:200 and above, or rises in titer by at least two dilutions are considered of some diagnostic significance. Recently the serodiagnosis of *Listeria* infections has been improved by a complement-fixation test. Positive findings with this test must, however, in many instances be checked for specificity by absorption of the serum samples with staphylococcal antigens. About two thirds of patients above the age of six months show rises in antibody titers during the disease and in convalescence. Serologic tests are inferior to the bacteriologic methods for establishing diagnosis. Skin tests have been tried repeatedly. A "polypeptide" skin test antigen has been developed that has given specific results in some suspected human cases.

DRUG SUSCEPTIBILITY

Chemotherapy with sulfonamides, and recently with certain antibiotics, may be successful. Ampicillin, tetracyclin, chloromycetin, and erythromycin have reduced the case fatality rate of the severe forms of human listeriosis. But treatment is usually successful only if it is instituted early in the disease. Occasional relapses require a second course of antibiotic therapy.

Although the sensitivity pattern of *L. monocytogenes* as found with several thousand isolates from human and animal sources is rather uniform, in vitro sensitivities should be determined for all strains isolated from clinical material. *L. monocytogenes* is usually inhibited by the following concentrations of antibiotics (μg/ml): benzylpenicillin 0.2, cephalothin 2.0, erythromycin 2.0, chloramphenicol 5.0, tetracycline 1.0.

EPIDEMIOLOGY

The epizootology remains largely unknown. Direct contact, aerogenous transmission, mating infection, and rhinogenous as well as otogenous infections have been considered as essential modes of infection. Susceptibility is definitely influenced by the state of general resistance and by many environmental and climatic factors. Inade-

FIGURE 1.

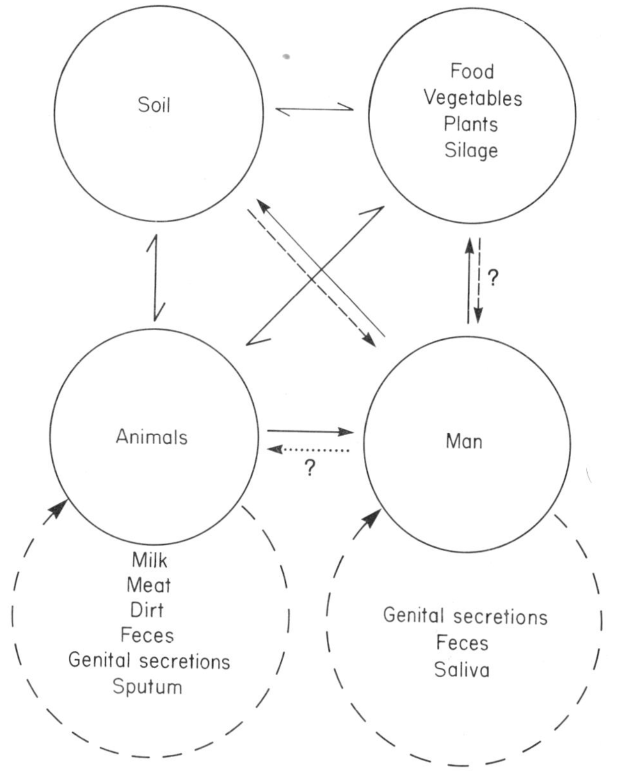

Seasonal prevalence

Animal	Man
Autumn–Winter	Spring–Summer

quate food seems to be essential in lowering resistance. In sheep and cows the consumption of poorly fermented silage has been shown to be the main source of infection during the stabling period. Most cases among domestic animals occur in late autumn and winter.

A risk exists for anyone dealing with animals or animal products, meat, game, and nonpasteurized milk and milk products in particular. Like other zoonoses listeriosis is transmissible from animals to humans. The chains of infection usually end "blind"; that is, in most cases the organisms are not known to be transmitted from a human host to others, with the exception of neonatal listeriosis, which is transmitted from mother to fetus. Transmission has taken place by direct contact with diseased animals or their excretions, possibly by consumption of contaminated food and by inhalation of infected dust. In most cases, however, the origin of the infection and the mode of transmission are unknown.

In view of the increased isolation of *L. monocytogenes* from environmental sources and its psychrophile nature, soil, plants, and certain vegetables would seem to be an important reservoir. The present view on the epizootic and epidemiologic pattern of listeriosis is outlined in Figure 1. In humans, listeriosis occurs more frequently in spring and summer than in the other seasons.

Prophylactic measures comprise a general improvement in sanitation with particular reference to an enforced control of raw material products destined for human use and consumption. Warnings and instruction are necessary for exposed individuals, particularly the pregnant, the obstetrician and his staff, including midwives and nurses, the veterinary profession, and all animal keepers and animal breeders. If listeriosis is suspected during pregnancy, prophylactic short-term antibiotic treatment is advised.

References

Gray, M. L., and Killinger, M. H.: Listeria monocytogenes and listeric infections. Bacteriol Rev 30:309, 1966.

Murray, E. G. D., Webb, R. A., and Swann, M. B. R.: A disease of rabbits characterized by a large mononuclear leukocytosis caused by a hitherto undescribed bacillus: Bacterium monocytogenes (n.sp.). J Path Bact 29:407, 1926.

Paterson, J. St.: The antigenic structure of organisms of the genus Listerella. J Path Bact 48:25, 1940.

Seeliger, H. P. R., and Finger, H.: Analytical serology of Listeria. In Kwapinski, J. B. G. (Ed.): Analytical serology of microorganisms. J. S. Wiley and Sons, Inc., vol. 2, 1969.

Seeliger, H. P. R., and Höhne, K.: Determination of serovars of Listeria. In Bergan, T. (ed.): Methods in microbiology. vol. 10 (in press).

27 *ERYSIPELOTHRIX RHUSIOPATHIAE*

Charles E. Davis, M.D.

Erysipelothrix rhusiopathiae is a facultatively anaerobic, gram-positive bacillus that shares many characteristics with *Listeria monocytogenes* (Chapter 26), corynebacteria, lactobacilli, and certain streptococci. The technique of numerical taxonomy, and the guanine plus cytosine content of its DNA indicate a closer relationship to *Listeria,* lactobacilli, and some streptococci than to coryneform bacteria but no definite relationship to any family of bacteria. In the eighth edition of Bergey's Manual of Determinative Bacteriology (1974), *Erysipelothrix* and *Listeria* have been removed from the chapter on coryneform bacteria and placed in the *Lactobacillus* chapter as genera of uncertain affiliation.

Erysipelothrix was first recovered from mice in 1880 by Koch. In 1882, Löffler and Pasteur and Thuillier independently isolated *Erysipelothrix* from pigs with erysipelas. Only two years later, Rosenbach isolated the same organism from an erysipeloid lesion of a patient and established it as the cause of erysipeloid by inoculating himself with the isolate. Although *Erysipelothrix* is still isolated from zoonotic skin infections of patients and from rare cases of septicemia and endocarditis, its major importance is the economic loss associated with outbreaks of swine erysipelas.

MORPHOLOGY

Erysipelothrix rhusiopathiae is a gram-positive, nonmotile, nonsporulating bacillus that may form either smooth or rough colonies on primary isolation. Smooth colonies are entire, convex, transparent, and about 1.0 mm in diameter. There may be a blue sheen by reflected light. Broth cultures are uniformly turbid. Rough colonies are larger and granular with a matt appearance on agar and flocculent with hair-like projections in broth. Cells from smooth colonies are small, slender, straight or slightly curved bacilli that measure from 0.2 to 0.4 μm in width by 0.5 to 2.5 μm in length. Cells from rough colonies vary from short forms to long chains of bacilli. Some are filamentous and beaded like the actinomycetes. Others may have large, fundus-like swellings.

Fully developed colonies on blood agar are surrounded by a zone of greening that may clear on further incubation. Colonies on tellurite agar are pinpoint and gray at 24 hours, but they become larger and jet black by 48 to 72 hours.

ANTIGENIC COMPOSITION

Erysipelothrix has been divided into at least 20 serotypes according to cross-agglutination studies against heat-resistant, acid-soluble antigens (Dedie, 1949; Kucsera, 1973; Wood et al., 1978). These antigens have not been further characterized but are thought to be peptidoglycans of the cell wall.

Although Pasteur and Thuillier successfully immunized swine against *Erysipelothrix* in 1883 with live organisms attenuated by passage through rabbits, little was known about the protective antigen until recently. Traub (1947) and Gledhill (1952) showed that at least part of the protective activity of the vaccine was a result of antibody directed at a soluble antigen found in the supernatants of cultures. White and Verway (1970 and 1970a) extended these observations and found that the soluble protective antigens contained a glycolipoprotein that was probably derived from the cell wall, since 75 per cent of the protective activity was destroyed by pretreatment of the antigen with muramidase. Trypsin and heat also destroyed part of the protective activity of this antigen, which was soluble in butanol but unaffected by lipase or ribonuclease.

Animals immunized with either the vaccine of Pasteur and Thuillier or cell-free culture supernatants develop agglutinins to whole bacteria.

METABOLISM

Erysipelothrix rhusiopathiae is microaerophilic. It will grow when exposed to atmospheric concentrations of oxygen but grows best in reduced oxygen tension with 5 to 10 per cent carbon dioxide. *E. rhusiopathiae* multiplies at temperatures of 16 to 41° C. Optimum growth occurs at 33 to 37° C. It obtains the energy for growth by glycolysis and does not produce catalase, even when grown aerobically. All strains produce acid but no gas from glucose, galactose, fructose, and lactose.

Riboflavin and oleic acid are required for growth. Acid production from carbohydrates is poor in 1 per cent peptone water. Five per cent rabbit serum or yeast autolysate should be added to the sugar in 1 per cent peptone water.

One of the unique metabolic properties of *Erysipelothrix* is the production of hydrogen sulfide. Among gram-positive organisms, only a few species of streptococci and *Bacillus* show this characteristic. *Erysipelothrix* acidifies the butt of Kligler's iron agar by producing organic acids from glucose by the process of anaerobic respiration with thiosulfate as the electron acceptor. In this acid environment, thiosulfate is broken down to sulfite and H_2S gas by the enzyme thiosulfate reductase, which is produced in most bacteria that generate H_2S.

PATHOGENIC PROPERTIES

Erysipelothrix is a natural pathogen primarily of mice, fish, swine, and man. Sheep and turkeys are also susceptible to natural infections. In swine, *Erysipelothrix* produces three clinical types of disease: acute fatal septicemia, red rhomboid-shaped skin lesions ("diamond skin disease"), and chronic disease with endocarditis of the mitral valves. Arthritis is a frequent complication of all forms of swine erysipelas and may also occur independently of the other manifestations. Laboratory mice die of overwhelming septicemia after intraperitoneal inoculation.

Erysipelothrix causes skin infections in man that are called erysipeloid because they resemble streptococcal erysipelas. These lesions usually occur on the hand after contact with infected animals or animal products. Arthritis of nearby joints is common (see Chapter 245).

The virulence of different strains of *Erysipelothrix* for mice varies, but most strains kill mice. Variations in virulence are independent of serotype, biochemical properties, and protein composition as determined by patterns on electrophoresis and electrofocusing (White and Mirikitani, 1976). The virulence factors have not been determined, but virulence is related to rapid growth and the production of neuraminidase (Krasemann and Muller, 1975). Inoculation of a partially purified glycoprotein from culture filtrates into the skin of rabbits causes necrosis. Intravenous inoculation of this product also causes high fevers in rabbits (Leimbeck et al., 1975).

The pathogenesis of *Erysipelothrix* arthritis has been extensively studied because it is a prominent feature of swine erysipelas and erysipeloid in man (Chapter 245). It has been used as a laboratory model of relapsing, erosive arthritis because its histopathology and course are similar

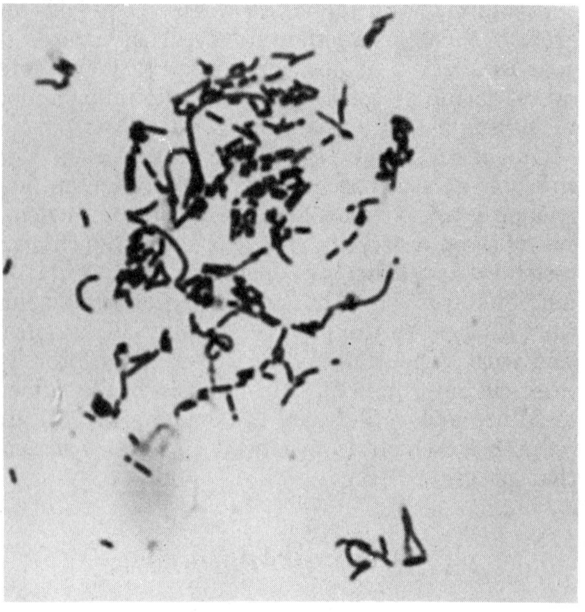

FIGURE 1. *Gram stain appearance of rough colony of* E. rhusiopathiae. *Note predominance of filamentous beaded bacilli.*

to rheumatoid arthritis (Hadler, 1976). Intravenous inoculations of whole live organisms causes polyarthritis in swine, dogs, and rabbits. Rabbits develop heterologous rheumatoid factor (Astorga, 1969) and polyarthritis after injection of dead organisms and cell-free culture extracts (White et al., 1971). This cell-free extract contains murein and many proteins. Some of these products bind rapidly to the synovium, persist for as long as six months, and cause cytopathic changes in synovial cell cultures after a single brief exposure (White et al., 1976). The exact antigen or antigens responsible for this effect have not been isolated, but it is clear that the pathogenic factors do not require the continuing presence of the organism.

IMMUNITY

Active immunization with the live attenuated vaccine of Pasteur and Thuillier protects against swine erysipelas. This protection is probably mediated by antibody because passive protection with antiserum provides effective prophylaxis for at least two weeks. The protective antigen in this preparation is a glycolipoprotein that is destroyed by lysozyme (White and Verway, 1970 and 1970a). The immune response after vaccination with either whole cells or cell-free culture extracts may be monitored by the titer of agglutinins to *Erysipelothrix*. Immunization is effective after a single inoculation. Swine do not develop clinical arthritis after immunization. Polyarthritis from culture extracts develops only after multiple intravenous inoculations. This arthritis does not seem to be primarily an antibody-mediated autoimmune phenomenon, since persistence of bacterial antigens in the synovium has been demonstrated (White et al., 1976).

Natural human infection with *E. rhusiopathiae* also stimulates the production of agglutinins, but patients undergo relapses and reinfections. Immunization of swine probably does not prevent colonization of the tonsils and gastrointestinal tract. It is possible that swine are more effectively immunized by vaccination than man is by natural infection because swine are reimmunized by colonization with *Erysipelothrix* (see Epidemiology, below).

Crude culture extracts are cytopathic for synovial cells and probably mitogenic for lymphocytes (White and Mirikitani, 1976), but the role of cellular immunity against this bacterium has not been determined.

LABORATORY DIAGNOSIS

Gram stains of aspirates or biopsies from erysipeloid or blood cultures from patients with septicemia and endocarditis may reveal coccobacillary, coryneform, or filamentous gram-positive bacilli. Forms that are almost indistinguishable from streptococci are not uncommon either from infected material or from primary isolation plates (Seeliger, 1961). Growth is enhanced by carbon dioxide, reduced oxygen tension, and 5 per cent serum. Both smooth and rough colonies (see Morphology, above) usually produce a zone of greening on blood agar. This zone may clear on further incubation. Gram stains of smooth colonies usually reveal coccobacillary or coryneform bacilli, while smears of the larger rough colonies often show a predominance of filamentous, beaded bacilli that may resemble actinomycetes or lactobacilli (Fig. 1).

Two unique cultural features of *Erysipelothrix* should prevent confusion with any other gram-positive rod. Except for a few species of *Bacillus* and a few streptococci, *E. rhusiopathiae* is the only gram-positive bacterium that will produce H_2S in Kligler's iron agar incubated in air (Fig. 2). It also produces a unique appearance in gelatin stab cultures if the tube is incubated at temperatures low enough to keep the gelatin in a solid state. Bead-like colonies with lateral filamentous growth develop and resemble a test-tube brush (Fig. 3).

Erysipelothrix is nonmotile, and is negative for catalase, oxidase, and indole production; it does not reduce nitrate to nitrite. It makes acid but not gas

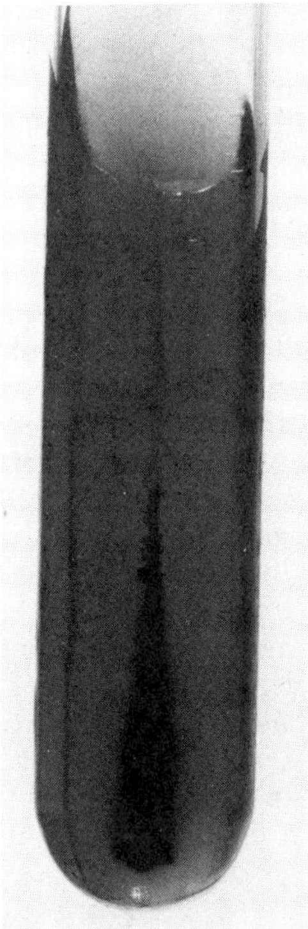

FIGURE 2. *Twenty-four-hour culture of* E. rhusiopathiae *in Kliger's iron agar. Note H_2S production along stab line.*

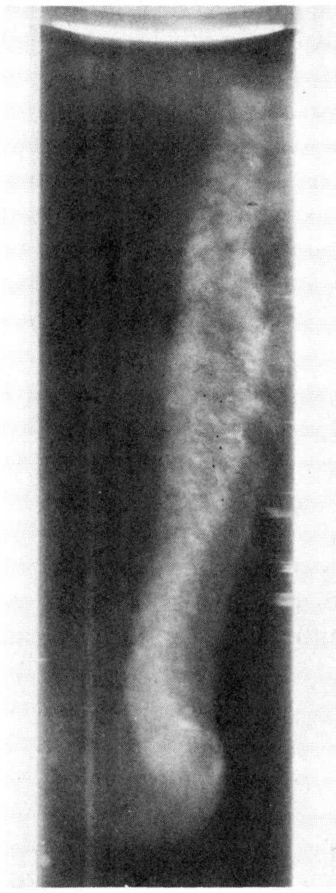

FIGURE 3. *Forty-eight-hour gelatin stab culture of* E. rhusiopathiae *with lateral filamentous growth that resembles a test tube brush.*

from glucose, lactose, galactose, and fructose. Some strains make acid from mannose (10/14) and xylose (4/14) (White and Mirikitani, 1976). It makes an acid slant and butt in Kligler's iron agar and triple sugar iron agar with enough H_2S formation to blacken the butt. Key biochemical reactions are shown in Table 1.

Erysipelothrix can be easily differentiated from *Listeria* (Chapter 26) because *Listeria* is motile and beta-hemolytic and does not produce H_2S (Table 1). *Erysipelothrix* is resistant to neomycin and kanamycin, but *Listeria* is not. Intraperitoneal inoculation of mice with a 24- to 48-hour culture of *Erysipelothrix* kills mice within two to three days but will not cause keratoconjunctivitis when dropped into the eye of a rabbit. *Listeria* is

TABLE 1. Characteristics of *Erysipelothrix rhusiopathiae*

TEST OR COMPOUND	REACTION	COMMENT
Alpha-hemolysis	+	
Beta-hemolysis	−	*Listeria, C. pyogenes,* and *C. haemolyticum* are +
Motility	−	*Listeria* is +
H_2S in KIA or TSI	+	*Listeria,* corynebacteria and lactobacilli are −
Nitrate reduction	−	*Listeria* is +
Catalase	−	*Listeria* and most corynebacteria are +
Esculin hydrolysis	−	*Listeria* is +
Glucose	A[a]	*Erysipelothrix* produces acid without gas from carbohydrates
Lactose	A	
Fructose	A	
Galactose	A	
Mannose	A(10/14)	(White and Mirikitani, 1976)
Xylose	−(10/14)	(White and Mirikitani, 1976)
Resistance to neomycin and kanamycin	+	*Listeria* is sensitive
Pathogenic for mice	+	*Listeria* is −
Keratoconjunctivitis in rabbits	−	*Listeria* is +
Per cent G + C	36.5%	*Listeria* is 38%

[a]A few drops of rabbit serum or yeast autolysate should be added to carbohydrate tests if the base is 1 per cent peptone water.

not pathogenic for mice but causes keratoconjunctivitis in rabbits. *Erysipelothrix* can be differentiated from all lactobacilli and corynebacteria by the production of H_2S in KIA or TSI slants. Corynebacteria also produce catalase except for *C. pyogenes* and *C. haemolyticum*, which are beta-hemolytic.

DRUG SUSCEPTIBILITY

E. rhusiopathiae is exquisitely sensitive to penicillin (0.0025 to 0.02 μg/ml), ampicillin, the semisynthetic penicillins, the cephalosporins, erythromycin, and clindamycin. It is also susceptible to readily achievable levels of tetracycline. Some strains are only moderately sensitive to chloramphenicol. It is resistant to the aminoglycosides, vancomycin, sulfonamides, and polymyxin.

EPIDEMIOLOGY

Erysipelothrix lives on dead and dying organic matter in the environment. It can be isolated in large numbers from the slime covering the bodies of fish and has been isolated from the soil of pig pens as long as five years after an outbreak of swine erysipelas. The organism tolerates high salt concentrations and develops as well in sea water as it does in fresh water.

Erysipelothrix also colonizes many animals including fish, crustaceans, wild and domestic birds, and rodents, especially mice and rats. Its primary economic importance is in agriculture, in which significant losses of poultry and swine are common. *Erysipelothrix* colonizes a high percentage of apparently normal swine. In one recent study (Stephenson and Berman, 1978), the tonsils of 62 out of 63 normal pigs were colonized by *Erysipelothrix* at slaughter. The organism is widely distributed throughout Europe and the United States. Tissue (primarily spleens) of hogs from 2633 herds in 46 states of the United States were contaminated with *E. rhusiopathiae* in another recent study (Harrington and Ellis, 1975). In this study, the swine came from herds in which hog cholera was suspected, and 21.1 per cent were infected with *Erysipelothrix*.

Hogs and other wild and domestic animals can probably be infected either through the skin or through the alimentary canal. The tonsils may be one of the major foci of infection in hogs. Hogs and poultry probably acquire Erysipelothrix primarily from the soil of their pens, but uncooked garbage and rodents may also be a major source of swine infection. The peak incidence of swine erysipelas occurs in the summer. The peak incidence of human disease occurs in the fall and winter shortly after outbreaks of swine erysipelas.

E. rhusiopathiae infection is primarily an occupational disease in humans. Farmers, abbatoir workers, poultry workers, fishermen, and housewives acquire the infection through skin abrasions while handling infected meat. Occasionally, cases of septicemia and endocarditis have no apparent preceding skin infection. Some investigators believe that these infections may be acquired through the gastrointestinal tract after ingestion of improperly prepared infected meat or fish. Human-to-human transmission has never been proved.

Disease in domestic animals and humans can be prevented by immunization of animals with the vaccine of Pasteur and Thuillier and by improved practices of animal husbandry. Housewives and those at occupational risk should handle meat and fish cautiously and wear gloves when possible.

References

Astorga, G.P.: Immunologic studies of an experimental chronic arthritis resembling rheumatoid arthritis. Arthritis Rheum. 12:589–596, 1969.

Dedie, K.: Die Säurelöslichen Antigene von *Erysipelothrix rhusiopathiae* Mh. Vet. Med. 4:7, 1949.

Gledhill, A. W.: The immunizing antigen of *Erysipelothrix*-the role of the L-antigen. J. Gen. Microbiol. 7:179–191, 1952.

Hadler, Nortin H.: A pathogenetic model for erosive synovitis. Lessons from animal arthritidies. Arthritis and Rheumatism 19:256–266, 1976.

Harrington, R., and Ellis, E. M.: Salmonella and Erysipelothrix infection in swine: A laboratory survey. Am. J. Vet. Res. 36:1379–80, 1975.

Jones, D.: The taxonomic position of *Listeria*. *In* M. Woodbine (ed.): *Problems of Listeriosis*. Leicester University Press, Leicester, 1975, pp. 4–17.

Krasemann, C., and H.E. Muller: The virulence of *Erysipelothrix rhusiopathiae* strains and their virulence production. Zentrabl. Bakteriol. [ORIG A] 231 (1-3), 206–213, 1975.

Kucsera, G.: Proposal for standardization of the designations used for serotypes of *Erysipelothrix rhusiopathiae* (Migula) Buchanan. Int. J. Syst. Bacteriol. 23:184–188, 1973.

Leimbeck, R., K.H. Bohm, H. Ehard, and L.-Cl Schulz: Studies of the toxic components of *Erysipelothrix rhusiopathiae*: Detailed characterization of an extracted endotoxin. Zentrabl. Bakteriol. [ORIG A] 232 (2–3), 266–286, 1975.

Seeliger, H.P.R.: *Listeriosis*. Karger Co., Basle, Switzerland, 1961.

Stephenson, E. H., and D.T. Berman: Isolation of *Erysipelothrix rhusiopathiae* from tonsils of apparently normal swine by 2 methods. Am. J. Vet. Res. 39:187–188, 1978.

Traub, E.: Immunisierung gegen Schweinerotlauf mit konzentrientem Adsorbatimpstoffen. Monatsh. Veterinaermed. 10:165–173, 1947.

White, R.R., and W.F. Verway: Isolation and characterization of a protective antigen-containing particle from culture supernatant fluids of *Erysipelothrix rhusiopathiae*. Infect. Immun. 1:380–386, 1970.

White, R.R., and W.F. Verway: Solubilization and characterization of a protective antigen of *Erysipelothrix rhusiopathiae*. Infect. Immun. 1:387–393, 1970(a).

White, T.G., J.L. Puls, and F.K. Mirikitani: Rabbit arthritis induced by cell-free extracts of *Erysipelothrix*. Infect. Immun. 3:715–722, 1971.

White, T.G., F.K. Mirikitani, and Patricia Hargrove: The effect of bacterial extracts on synovial cells in tissue culture. *In Vitro* 12:702–707, 1976.

White, T.G., and F.R. Mirikitani: Some biological and physical-chemical properties of *Erysipelothrix rhusiopathiae.* Cornell Vet. 66:152–163, 1976.

Wood, R.L., and R.A. Packer: Isolation of *Erysipelothrix rhusiopathiae* from soil and manure of swine-raising premises. Am. J. Vet. Res. 33:1610–1620, 1972.

Wood, R.L., D.R. Haubrich, and R. Harrington: Isolation of previously unreported serotypes of *Erysipelothrix rhusiopathiae* from swine. Am. J. Vet. Res. 39:1958–1961, 1978.

28 BACILLUS ANTHRACIS AND OTHER AEROBIC SPORE-FORMING BACILLI

Robert P. Williams, Ph.D.

MORPHOLOGY

The genus *Bacillus* is a member of the family Bacillaceae. The bacteria are rod-shaped, and the majority are motile by means of lateral flagella. An important exception is *Bacillus anthracis,* which lacks flagella and is not motile. Formation of a single endospore in the vegetative bacterium is a dominant feature of the genus. Spores may be oval or spherical, and may be located centrally, subterminally, or terminally. The presence of spores causes the vegetative cell to swell in some species. Capsules are present in a few species, particularly *Bacillus anthracis,* whose virulent strains synthesize a polypeptide capsular material composed of repeating units of D-glutamic acid.

The bacteria can be divided into two groups depending on the size of the vegetative cell (Table 1). In small cell species, the bacterium has a width of 0.6 to 0.8 μm and a length of 1.5 to 3 μm; in large cell species, the width is 1.0 to 1.5 μm and the length 2 to 5 μm. Although the Gram reaction is usually positive, some species show a variable reaction, particularly if the stain is made from samples taken in the later stages of growth. Sometimes globules of a reserve metabolic material, poly-beta-hydroxybutyrate, are seen in vegetative bacteria of *B. anthracis, B. cereus,* and related species.

The structure of the endospore and its location within the vegetative cell are characteristics useful for taxonomy (Table 1). In group I, which includes *B. anthracis, B. cereus,* and *B. subtilis,* the spore is oval, has a central location, and does not distend the vegetative cell. Vegetative cells of *B. subtilis* do not contain globules of poly-beta-hydroxybutyrate. Spores of group II, including *B. macerans* and *B. polymyxa,* cause the vegetative cell to swell. These spores usually are located centrally but also may be found subterminally or terminally. When thin sections of these spores are examined with the electron microscope, they show a thick spore coat with several prominent longitudinal, parallel ridges on the surface. The ridges are not seen on spores of other species. *B. sphaericus* is included in Group III, in which the spores are round, are located terminally, and cause the vegetative cell to swell. The endospores of *B. cereus, B. anthracis,* and *B. thuringiensis* are enclosed within a loose outer coat, the exosporium.

The different species have an enormous range of colonial morphology that can vary according to the composition of the medium. Isolated colonies may be quite rough and may spread over the surface, particularly if the agar is moist. Pigments sometimes color colonies yellow, pink, red, or even black.

B. anthracis grows in vitro in long chains that look like a jointed bamboo rod. In vivo, the chains are shorter, and single bacilli or pairs of rods may be seen (Fig. 1). When grown in culture, no capsule can be seen around anthrax bacteria unless they are grown under an increased tension of carbon dioxide. Capsules are readily apparent around the bacteria when smears from infected animals are examined, particularly if a polychrome stain is used (Fig. 1). When grown on conventional media in air, encapsulated anthrax bacteria, the virulent form, grow as rough colonies. But when the same bacteria are incubated on media containing bicarbonate in an atmosphere of increased carbon dioxide, the colonies are mucoid and smooth (Fig. 2), and capsules are readily seen around the rods (Fig. 3). No other species of *Bacillus* grows in this fashion under these conditions. *B. anthracis* forms spores only under aerobic conditions; thus, spores are not seen in the circulating blood of infected animals.

TABLE 1. Characteristics Useful to Differentiate Between Some Species of *Bacillus*[a]

CHARACTERISTIC	SPECIES OF *BACILLUS*[b]														
	anthracis	*alvei*	*brevis*	*cereus*	*circulans*	*coagulans*	*firmus*	*laterosporus*	*licheniformis*	*macerans*	*megaterium*	*polymyxa*	*pumilus*	*sphaericus*	*subtilis*
Morphologic															
Gram stain	+	v	v	+	v	+	+	v	+	v	+	v	+	v	+
Size[c]	l	s	s	l	s	s	s	s	s	s	l	s	s	s	s
Motility	−	+	+	v	v	+	v	+	+	+	v	+	+	+	+
Spore shape[d]	o	o	o	o	o	o	o	o	o	o	o	o	o	r	o
Spore location[e]	c	cts	cts	c	cts	cts	c	c	c	t	c	cts	c	t	c
Swelling of vegetative cell	−	+	+	−	+	v	−	+	−	+	−	+	−	+	−
Capsule	+	−	−	−	−	−	−	−	+	−	v	+	−	−	−
Biochemical															
Catalase	+	+	+	+	+	+	+	+	+	+	+	+	+	+	+
Fermentations (acid)[f]															
Arabinose	−	−	−	−	+	v	v	−	+	+	v	+	+	−	+
Glucose	+	+	+	+	+	+	d	+	+[g]	+[g]	+	+[g]	+	−	+
Mannitol	−	−	v	−	+	v	+	+	+	+	v	+	+	−	+
Xylose	−	−	−	−	+	v	v	−	+	+	v	+	+	−	+
Citrate utilization	v	−	v	+	v	v	−	−	+	v	+	−	+	v	+
Nitrate reduction	+	−	v	+	v	v	v	+	+	+	v	+	−	−	+
Gelatin liquefaction	+	+	+	+	+	−	+	+	+	+	+	+	+	v	+
Hemolysis on sheep blood agar[h]	−	n	n	+	n	n	n	n	n	n	n	n	n	n	±
Lecithinase activity	+	−	−	+	−	−	−	+	−	−	−	−	−	v	−
Starch hydrolysis	+	+	−	+	+	+	+	−	+	+	+	+	−	−	+
Voges-Proskauer reaction	+	+	−	+	−	v	−	−	+	−	−	+	+	−	+
Anaerobic growth[i]	+	+	−	+	v	+	−	+	+	+	−	+	−	−	−
Growth at pH <6	+	−	v	+	v	+	−	−	+	+	+	+	+	v	+
Growth in 7% NaCl	+	−	−	v	v	−	+	−	+	−	+	−	+	v	+
Growth at 50° C	−	−	+	−	+	+	−	v	+	v	−	−	+	−	+

[a]Data obtained principally from Bergey's Manual of Determinative Bacteriology, 8th ed., and Cowan and Steel, Manual for the Identification of Medical Bacteria, 2nd ed.
[b]Variable reactions indicated by v; d indicates delayed reaction; n indicates reaction not reported in manuals.
[c]l represents large (width 1.0 to 1.5 μm and length 2 to 5 μm); s represents small (width 0.6 to 0.8 μm and length 1.5 to 3 μm).
[d]o represents oval; r, round.
[e]c represents central; t, terminal; and s, subterminal.
[f]In media containing $(NH_4)_2HPO_4$ and the appropriate sugar.
[g]Gas is also produced.
[h]Presence of hemolysis eliminates possibility of *B. anthracis,* but lack of hemolysis may be characteristic of other species also.
[i]In broth containing glucose.

ANTIGENIC COMPOSITION

The genus *Bacillus* forms an antigenically heterogeneous group. However, data about the definitive antigenic composition of the various species are scant. Vegetative cells and spores contain different antigens. Since the bacteria usually are motile, flagellar or H antigens are also present. Unfortunately, except for *B. cereus* and *B. thuringiensis,* in which serotyping schemes have been developed based upon agglutinins formed against the H antigens, the antigenic composition of the genus has been of little help in distinguishing species or strains from one another.

Spores and vegetative cells are antigenically distinct, although some investigators claim that vegetative antigens can be found in spores. Spore antigens can distinguish among the small cell species of the genus, but differentiation among large cell species is less successful. Injection of autoclaved spores into animals produces agglutinins and precipitins. If live spores are injected, antibodies against both the spore and the vegetative cell develop. The precipitinogens seem to be mainly species-specific, while the agglutinogens show subspecies distribution.

Vegetative cell, spore, and flagellar antigens show different levels of specificity. Considerable cross-reaction occurs among species with respect to vegetative cell antigen. Spore antigens seem to provide the highest species specificity, whereas flagellar antigens show the greatest strain specificity within species.

A serotyping scheme has been developed based

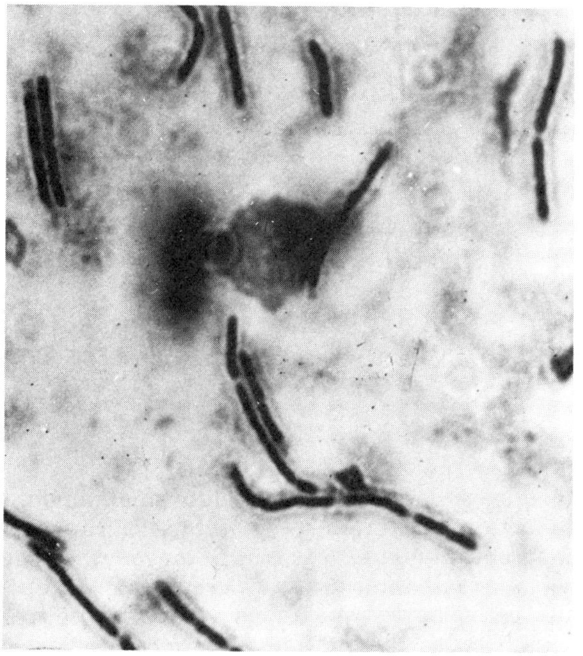

FIGURE 1. *Blood smear from a moribund guinea pig infected with* B. anthracis. *Smear was stained with polychrome methylene blue.*

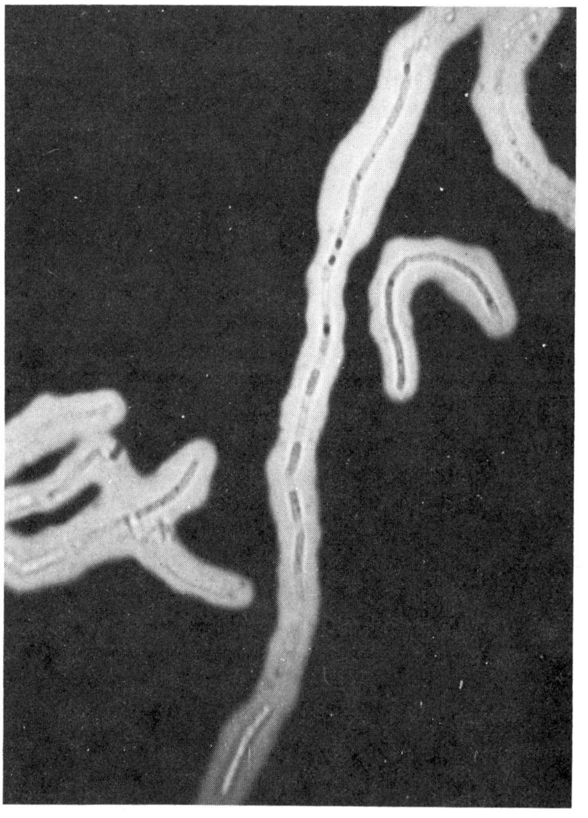

FIGURE 3. *Smear of bacteria stained with India ink from colonies of* B. anthracis *grown in a candle jar on bicarbonate medium. Note the abundant capsule around the bacteria.*

on agglutination with the H antigen of *B. cereus* and has been used in epidemiologic studies of outbreaks of food poisoning caused by the bacteria. Strains of *B. cereus* were classified into 23 serotypes, but about half of the strains isolated could not be typed. Strains isolated from infections other than food poisoning often were not typable.

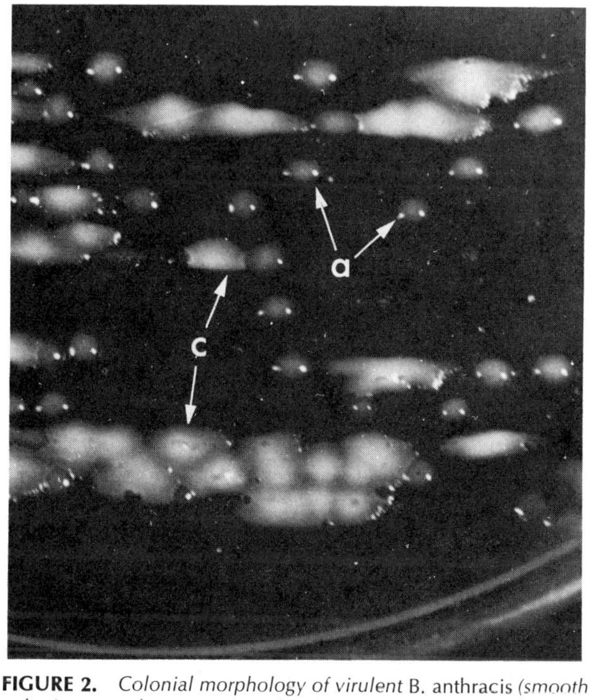

FIGURE 2. *Colonial morphology of virulent* B. anthracis *(smooth colonies, a) and* B. cereus *(rough colonies, c) when grown on nutrient agar containing 0.7 per cent of sodium bicarbonate and incubated at 35° C in a candle jar.*

The capsular polypeptide of *B. anthracis* is antigenic, although antibodies against it are not protective against infection. Only one capsular type of *B. anthracis* is known. An antigenic polysaccharide is a component of the cell wall of *B. anthracis*. Antibodies to the polysaccharide cross-react with sera prepared from vegetative cells of *B. cereus*. Whereas *B. anthracis* seems to possess only one type of cell-wall antigen, several types are present in *B. cereus* and *B. megaterium*. However, little success has been achieved in the serologic identification of strains of *B. cereus* on the basis of cell-wall antigens. The complex protein toxin produced by *B. anthracis* also is antigenic (Table 2).

METABOLISM

The great diversity among species of *Bacillus* is exemplified by their metabolism. The bacteria are classified as aerobes, but most species are facultative anaerobes, although growth occurs best aero-

TABLE 2. Anthrax Toxin: Composition and Biological Effects

FACTOR[a]	CHEMICAL NATURE	TOXIC EFFECT[b] Edema	TOXIC EFFECT[b] Lethal	ANTIGENIC ACTIVITY	IMMUNOGENIC ACTIVITY
Edema factor (EF); I	Protein, carbohydrate	–	–	+	–
Protective antigen (PA); II	Protein	–	–	+	+
Lethal factor (LF); III	Protein	–	–	+	–
EF + PA		+	+	+	+
EF + LF		–	–	+	+
PA + LF		–	+	+	+
EF + PA + LF[c]		+	+	+	+

[a]The descriptive terms and acronyms are the American terminology. The Roman numerals are the British terms.
[b]Edema tested in guinea pig skin; lethal effect in mice.
[c]Complete toxin that would simulate the substances present in infected animals.

bically. All are chemoheterotrophs and can dissimilate a variety of organic substrates such as amino acids, organic acids, and sugars to obtain carbon and energy by aerobic respiration, anaerobic respiration, or fermentation. Adenosine triphosphate (ATP) usually is generated from an aerobic electron transport chain.

Growth requirements range from simple to complex. Most species grow readily on nutrient agar or peptone media. *B. subtilis* grows on a minimal medium containing glucose, citrate, ammonium phosphate, and the usual mineral salts. *B. cereus* requires the addition of certain amino acids, whereas *B. anthracis* needs in addition thiamine, and its growth is stimulated by adenine, guanine, and uracil. Some of the insect pathogens have a more fastidious nutrition and require factors for growth that have not been identified.

Most species grow optimally at pH 7, but some species can grow under alkaline conditions at pH 9, while a few tolerate acidic conditions as low as pH 2. The optimal temperature for growth of the majority of species is 30 to 45° C. *B. stearothermophilus* is thermophilic and grows vigorously at 65° C. Other extreme thermophiles occur in the genus, but these species have not been well characterized. They grow in hot, moist environments such as hay stacks, where they decompose plant material. A few species are psychrophiles and grow at temperatures below 25° C.

Fermentation of glucose can yield a mixture of end products depending upon the species. Initial dissimilation of the sugar is through the Embden-Meyerhof pathway. *B. cereus* and other large cell species of group I convert 2-glyceraldehyde-3-phosphate (triose phosphate) into pyruvate that then is metabolized through acetoin to 2,3-butanediol and carbon dioxide. The reduced nicotinamide adenine dinucleotide (NADH) remaining from the latter reaction is used to reduce another molecule of triose phosphate to glycerol. In addition to 2,3-butanediol, glycerol, and carbon dioxide, these species also produce small amounts of lactate and ethanol. *B. anthracis* can grow anaerobically on sugars, but *B. megaterium* cannot. A small cell variety of group I, *B. subtilis,* also cannot grow anaerobically on glucose, perhaps because the enzyme for reduction of triose phosphate to glycerol is missing, but in air this bacterium does produce large quantities of 2,3-butanediol. *B. licheniformis* is unique in the genus because it can carry out denitrification. Under anaerobic conditions, this bacterium can grow on nonfermentable organic substrates if furnished with nitrate. Another unique member of group I, *B. fastidiosus,* a species of no medical importance, utilizes uric acid as the only source of energy, carbon, and nitrogen, producing ammonia, carbon dioxide, and cell material.

Good growth of species in group II occurs only on utilizable carbohydrates. Although *B. polymyxa* produces 2,3-butanediol from glucose, the fermentation is different from that carried out by bacilli in group I because no glycerol is formed. *B. macerans* ferments glucose with the formation of ethanol, acetone, acetate, formate, carbon dioxide, and H_2. Both species can dissimilate starch and pectins, and they participate in the retting of flax. When grown anaerobically, both species can fix atmospheric nitrogen.

Species of group III such as *B. sphaericus* do not use carbohydrates effectively as sources of energy because they lack fermentative ability. They utilize amino and organic acids as oxidizable substrates. Many strains produce urease, which hydrolyzes urea to form ammonia, and during growth of these bacteria the pH of the medium rapidly becomes alkaline. Growth of these strains usually does not occur unless the medium is initially alkaline and contains urea or ammonium salts. Actively growing cultures of these bacteria can decompose three g of urea per liter of medium per hour.

The different species produce an abundance of

extracellular products including antimicrobial substances, enzymes, pigments, and in some cases toxins. With few exceptions, primarily among the insect pathogens and thermophiles, catalase is formed. Amylase, collagenase, hemolysin, lecithinase, phospholipase, protease, and urease are some of the other enzymes that can be found in cultures. The hemolysin of *B. cereus* can be separated into two components: one, termed cereolysin, has a molecular weight of 52,000; the other has a molecular weight of 31,000. These hemolysins appear similar to those produced by *B. thuringiensis*. *B. anthracis* does not synthesize hemolysin, and this deficiency is used to separate this pathogen from other members of the genus. The enzyme activities, fermentation reactions, and other characteristics of metabolism form the basis for the differentiation of the several species of *Bacillus* (Table 1, Fig. 4).

Several species produce pigments, two of which have been identified. Pulcherrimin, the ferric salt of the pyrazine compound pulcherriminic acid, is synthesized by strains of *B. cereus, B. licheniformis,* and *B. subtilis*. An aromatic compound, protocatechuic acid, is formed by *B. anthracis,* probably as a derivative of shikimic acid. When complexed with iron in various proportions, protocatechuic acid causes liquid cultures to vary in color from brown to pink to red. Virulent strains of *B. anthracis* reportedly produce more protocatechuic acid than avirulent strains. Unidentified pigments formed by some species are black, tan, or yellow.

B. cereus and *B. anthracis* synthesize toxins that cause disease in humans (Tables 2 and 3). Cultural conditions, particularly the tension of carbon dioxide, are critical for synthesis of anthrax toxin. If cultures of *B. cereus* are rapidly transferred in broth, they produce an uncharacterized toxin that is extremely lethal for guinea pigs and mice. Toxins from *B. lentimorbus, B. popilliae,* and *B. thuringiensis* are lethal for insects.

Production of antimicrobial substances prior to the onset of sporulation is characteristic of the genus. These compounds are linear or cyclic polypeptides with a molecular weight of about 1400. Many of the amino acids in the peptides are of the D configuration. Biosynthesis of the peptides does not involve transfer ribonucleic acid (tRNA) and

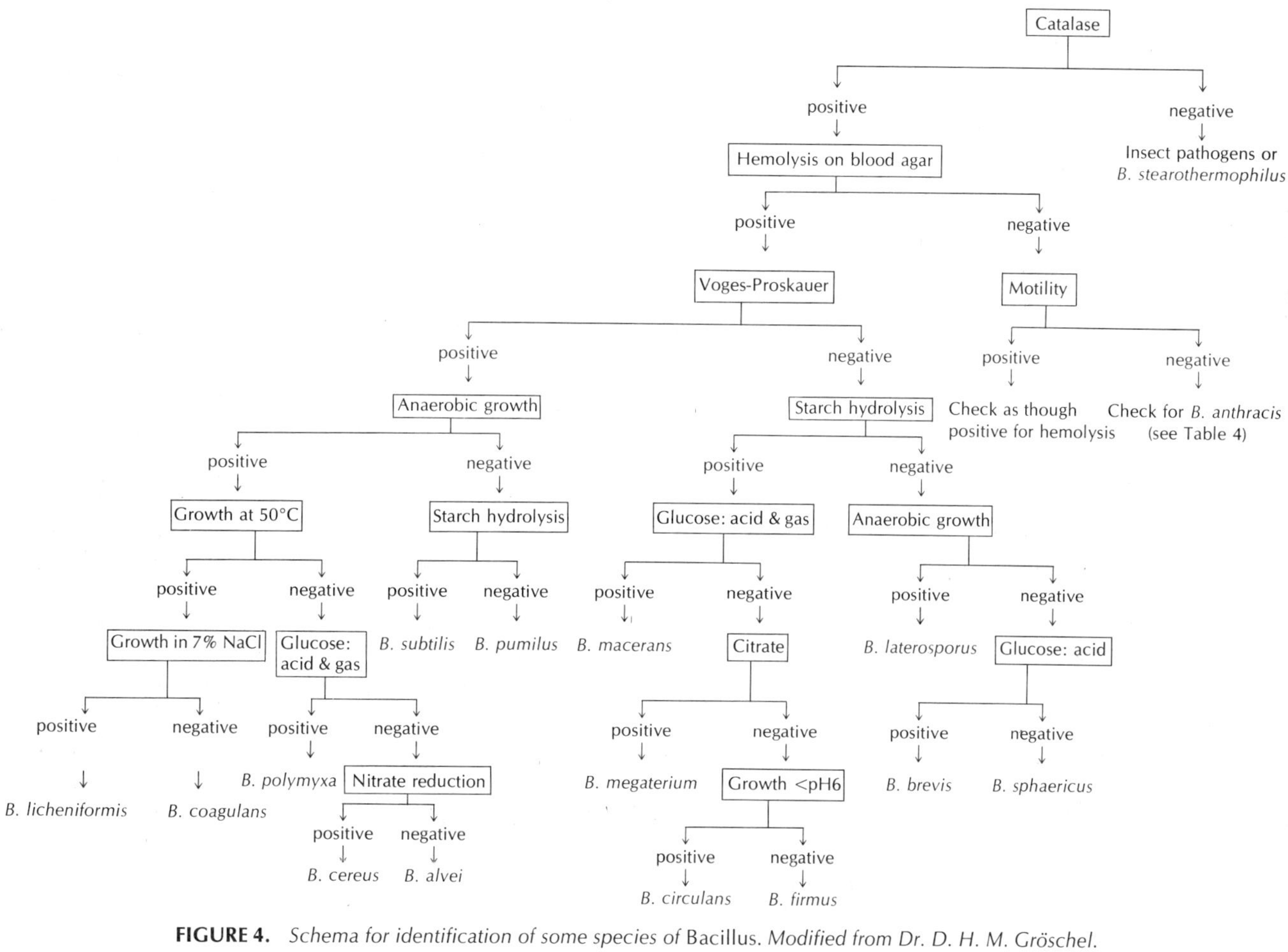

FIGURE 4. *Schema for identification of some species of* Bacillus. *Modified from Dr. D. H. M. Gröschel.*

TABLE 3. Comparison of Two Enterotoxins Produced by *Bacillus cereus*[a]

CHARACTERISTIC	TYPE OF ENTEROTOXIN	
	Diarrheal	**Emetic**
Clinical syndrome		
Incubation period	8 to 16 hr	1 to 5 hr
Diarrhea	Very common	Fairly common
Vomiting	Occasional	Very common
Duration of illness	12 to 24 hr	6 to 24 hr
Foods implicated	Meat products, soups, vegetables, puddings, sauces	Fried or boiled rice
Serotypes of *B. cereus* involved	1, 2, 6, 8, 9, 10, 12, 19	1 (majority) 3, 4, 5, 8, 12, 19
Enterotoxin[b]		
Molecular weight	ca. 50,000	<5,000
Stability to heat	−	+
Fluid accumulation in ligated rabbit ileal segment	+	−
Increased vascular permeability in guinea pig or rabbit skin	+	−
Lethal for mice after intravenous injection	+	−
Stimulation of adenylate cyclase-cAMP system in intestinal epithelial cells	+	−
Response when fed to rhesus monkeys	Diarrhea	Vomiting

[a]From Terranova and Blake, 1978; Gilbert, 1979.
[b]Except for molecular weight, data obtained from cell-free culture filtrates.

ribosomes; their assembly instead is directed by a protein enzyme. Examples of these antimicrobial compounds are bacitracin, gramicidin, polymyxin, and tyrocidine.

The peptidoglycan of the vegetative cell wall usually is similar to the type found in gram-negative bacteria. The diamino acid, *meso*-diaminopimelic acid, is present and is cross-linked to D-alanine to form the interpeptide bridge. In some species, L-lysine replaces diaminopimelic acid, and additional amino acids are present in the interpeptide bridge. The cortex of spores contains a unique peptidoglycan that contains three repeating subunits: a muramic acid unit without attached amino acids; a muramic acid unit attached to an alanine residue; and a muramic acid unit attached to L-alanine, D-glutamic acid, *meso*-diaminopimelic acid, and D-alanine.

The guanine plus cytosine content of the DNA of species ranges from 32 to 53 moles per cent. The DNA base composition of *B. cereus* and *B. anthracis* is similar but not the same, indicating that these two species are related but not identical. Relationships between some species have been studied by use of genetic recombination mediated by transformation or by transduction, or by use of DNA-to-DNA hybridization. Because the DNA of *B. subtilis* is readily transferred, and the bacteria grow well on a minimal medium, this species has been used extensively for studies of bacterial physiology and genetics. Mutants, both natural and induced, of various species can differ in nutritional requirements, sensitivity to drugs and bacteriophage, heat resistance, and, in the case of *B. anthracis,* virulence and production of capsule and toxin.

Formation of endospores is one of the most salient characteristics of *Bacillus*. The process is complex. Essentially, the phenomenon involves cell differentiation in which morphologic, biochemical, genetic, and physical changes occur in the bacteria. Endospores are a normal stage in the life history of *Bacillus,* but their biologic function is unknown. Their high resistance to inimical environments has survival value, and the mode of life of these bacteria depends upon the accumulation of spores in nature. Asporogenous mutants have little capacity to survive.

The initiation of endospore formation begins as the exponential growth of vegetative cells ends and the cultures approach the stationary phase as a result of limitation of nutrients, particularly sources of nitrogen and carbon. Synthesis of polypeptide antimicrobial substances commences, and cytologic changes begin in the vegetative cell. The two nuclear bodies in a vegetative cell condense to form a single filament. Near the terminal end of the bacterium, a transverse septum forms from the cytoplasmic membrane, probably in conjunction with development of a mesosome. This septum divides the nuclear material and cytoplasm to form a smaller cell, which will become the spore, from a larger cell that is destined eventually to lyse. The cytoplasmic membrane and the cytoplasm of the larger cell engulf the smaller cell and its membrane so that the forespore is surrounded by two cytoplasmic membranes, one from the larger cell and the other from growth of the septum that originally developed to separate the smaller cell. The process is irreversible at this stage, and the vegetative cell is committed to sporulation.

Synthesis and deposition of new structures distinctive to the spore then ensue. A cortex composed of a unique peptidoglycan develops from the two membranes. Exterior to the cortex, a spore coat is synthesized. The coat is a protein like keratin that is resistant to the usual procedures for solubilization of protein. It represents about 80 per cent of the spore protein. In some species, such as *B. cereus,* the exosporium, a loose, thin layer composed of lipoprotein and carbohydrate, surrounds the spore coat. As the spore coat forms, the spores become more refractile. Heat resistance then develops and is associated with an uptake of calcium ions by the spores and synthesis of dipicolinic acid, 2,6-dicarboxylpyridine, which is synthesized by spores from aspartic acid as an offshoot of the lysine pathway. The calcium salt of dipicolinic acid accounts for 10 to 15 per cent of the dry weight of spores and probably is a major factor in heat resistance. After development through the above stages, the mature spore is liberated from the parent cell by autolysis.

Morphologically, the mature endospore consists of a core, the spore protoplast, which contains the nucleus and the enzymes needed for subsequent germination. A complex envelope surrounds and protects the core. The innermost layer is the spore membrane, the original cytoplasmic membrane of the forespore, which contains normal peptidoglycan. External to this membrane is the cortex, the thickest layer of the envelope. The relatively impermeable spore coat is the most superficial layer of most spores, although it may be enclosed by an exosporium in some species.

Mature endospores are less easily stained than vegetative cells and are highly refractile. They have no detectable metabolism, are highly dehydrated, and can survive for decades in a state of dormancy. Spores are highly resistant to deleterious treatment by chemical agents, enzymes, heat, ionizing radiation, and ultraviolet light. Boiling for 10 minutes, dry heat at 140° C for three hours, or autoclaving for 20 to 30 minutes at 15 pounds of steam pressure kills most spores. Alcohols, phenols, heavy metal ions, and detergents are poor agents for killing spores.

When an endospore is placed in an environment with favorable nutritional conditions, it germinates to produce a single vegetative cell. Germination takes place in three stages. First, the spores must be activated by chemical or mechanical agents that damage the spore coat. Abrasion with glass beads, a low pH, or contact with compounds that contain free sulfhydryl groups activates spores. A common laboratory procedure for activation is called "heat shock," in which a suspension of spores is heated to 65° C for a few minutes to activate them. After the dormancy of spores is broken by activation, initiation of germination commences if environmental conditions are favorable. Water and an effector compound that activates spore enzymes are required. The effector agent varies with different species. L-alanine or adenosine or other amino acids or compounds can serve. Enzymes in the spore are activated to degrade spore constituents, in particular the peptidoglycan of the cortex. Water is taken up by the spores, and calcium dipicolinate is released. Respiration commences, and the spores lose their refractility and heat resistance. If the nutrients required for growth of vegetative cells are present, outgrowth of the spore occurs. Cell wall is synthesized from the spore membrane, and other biosyntheses rapidly develop, culminating in cell division.

During sporulation, the genes for spore formation are activated, while the genes for vegetative cell growth are repressed. The pattern of gene transcription is shifted, and the specificity of RNA polymerase is altered so that different regions of DNA are selected for transcription into messenger RNA. The new gene products direct the formation and composition of the spore. Linkage maps constructed for *B. subtilis* show that the genes for sporulation are clustered in about five widely dispersed regions of the chromosome. A variety of asporogenous mutants exist that are blocked at different stages in the development of endospores. Study of these mutants has contributed enormously to knowledge about sporulation. At least 20 genes are known to be involved in sporulation, and the total number may be several times that. Sporulation provides a relatively simple model in which cell differentiation and morphogenesis can be studied.

PATHOGENIC PROPERTIES

Susceptible animals injected with *B. anthracis* develop a massive septicemia. Even in the natural infection, the number of bacilli in the blood is so great that they can be seen in stained smears. These observations led to the early conjecture that death from anthrax was caused by blockade of capillaries that prevented oxygenation of tissues. In 1955, British investigators demonstrated that sterile plasma obtained from guinea pigs dying of anthrax was lethal when injected into guinea pigs or mice. The presence of masses of bacteria in the blood was not required, because termination of bacteremia by administration of streptomycin did not prevent death of guinea pigs once the critical number of *B. anthracis* was reached. This number was about 1/300 of the number of bacteria present at death in untreated animals.

Anthrax toxin is a complex substance that contains three factors (Table 2): edema factor (EF), protective antigen (PA), and lethal factor (LF), or factors I, II, and III respectively. Various combinations cause edema or are lethal. The most lethal is a combination of the three. Problems arise in precise interpretations of the effects of mixtures because of the difficulty of recombining three relatively unstable proteins in the appropriate proportions. Purified protective antigen is immunogenic and has been used as a vaccine for humans. The toxin complex is produced by virulent, encapsulated and avirulent, nonencapsulated *B. anthracis*. For production of protective antigen, cultures of avirulent, nonencapsulated *B. anthracis* are used.

The purified toxin complex interferes with phagocytosis and the bacteriocidal activity of serum. The toxin also increases vascular permeability. The decrease in circulating blood volume plus capillary thrombosis produced by the toxin may account for death from anthrax. The virulence of *B. anthracis* apparently depends upon both the ability to produce the toxin complex and the possession of a capsule. The presence of bicarbonate or serum is required in the medium for production of both the capsule and the toxin. Although the capsule is antigenic, it is not immunogenic. However, the capsule does interfere with phagocytosis and probably plays a role in the pathogenicity of anthrax.

The complex nature of anthrax toxin has left unresolved the precise nature of its pathogenic effects. Paradoxically, some animals such as mice are very susceptible when challenged with spores but relatively resistant when toxin is given intravenously. In contrast, dogs and rats are resistant when injected parenterally with spores but succumb to small doses of toxin. In addition, various animal species show different pathophysiologic responses to injection of the substance. Other factors with pharmacologic activity are produced by *B. anthracis,* such as a material with epinephrine-like activity and a cardioactive substance. The bacteria also produce protocatechuic acid that combines with iron to form pigment in liquid cultures of *B. anthracis*. The role, if any, of these substances in the pathophysiology of anthrax is unknown. Other species of *Bacillus* produce a variety of extracellular metabolites with antibiotic, hemolytic, lethal, lipolytic, or proteolytic activity, but their association with the occasional pathogenicity of the bacteria is unknown.

Two enterotoxins are produced by *B. cereus* during exponential growth (Table 3), one that causes diarrhea and the other vomiting. The diarrheal toxin is produced readily on media such as brain-heart infusion broth containing added glucose. Detection of emetic toxin was difficult at first because ordinary media did not support its production, but growth on a medium prepared from rice permitted its isolation. The diarrheal toxin causes fluid accumulation in rabbit ileal loops, promotes vascular permeability in rabbits, and is lethal for mice when injected intravenously. Purulent, pyrogenic, lethal, and necrotic activities are associated with strains of *B. cereus* isolated from patients with severe infections not associated with the gastrointestinal tract. These various activities and the diarrheal toxin may be manifestations of a single substance. Hemolytic and phospholipolytic activities are not associated with pathogenicity of *B. cereus* or with the enterotoxins. Strains of *B. cereus* produce amounts of the diarrheal toxin that vary from undetectable to considerable. Loss of the capacity to produce toxin after transfer in the laboratory led to the speculation that elaboration of the toxin might be under the control of plasmids or bacteriophages. This speculation has not been confirmed.

Some species of *Bacillus* are pathogenic for insects but not for vertebrates. *B. lentimorbus* and *B. popilliae,* which grow readily and produce large numbers of spores in the hemolymph of Japanese beetles, cause milky disease that is lethal for the insects. During sporulation, *B. thuringiensis* forms a crystalline protein that lies adjacent to the spore. After ingestion by larvae of *Lepidoptera,* this parasporal protein dissolves in the gut and effects paralysis of the caterpillar. A specific strain of *B. thuringiensis* also is lethal for larvae of *Diptera,* and its use as a bacterial pesticide may provide means to control mosquitoes and flies.

IMMUNITY

Louis Pasteur in 1881 dramatically established that herbivorous animals developed immunity to anthrax. He immunized a group of sheep with a vaccine made from an attenuated strain of *B. anthracis*. When later challenged with virulent *B. anthracis,* all 25 vaccinated animals survived, while an equal number of unvaccinated controls did not. Because the attenuated Pasteur vaccine sometimes caused severe disease in vaccinated animals, its use has been supplanted by a suspension of viable spores prepared from a nonencapsulated, avirulent strain of *B. anthracis*. This vaccine is not approved for human use, and occasional cases of disease still occur in vaccinated animals. Both of these vaccines probably produce immunity by release of toxin by growing bacteria.

Second attacks of anthrax in man are rare, and humans probably are relatively resistant to the disease. Manufacturing plants that process wool have large numbers of spores in the air and dust that is inhaled by workers, yet few cases of inhalation anthrax occur in spite of the fact that some of the particles in the dust are of a size that could reach the alveoli. Nor is cutaneous anthrax common in such plants. Antibodies to protective antigen were found in workers who were not immunized and who had no history of anthrax. This serologic evidence suggests that inapparent or mild forms of anthrax occur in humans to provide an antibody response and immunity.

Vaccines of killed bacilli evoke no significant immunity probably because they cannot grow and produce anthrax toxin. Antibody against the polypeptide capsule is not protective except in mice. The complex, purified anthrax toxin is both antigenic and immunogenic, but isolated edema and lethal factors are only antigenic (Table 2). Protective antigen is immunogenic. A vaccine prepared by precipitation and concentration of protective antigen by addition of aluminum potassium sulfate to sterile culture filtrates was calculated to be 92 per cent effective in an exposed, susceptible population of humans. Antibody can be measured by complement fixation, agar gel precipitation, or indirect microhemagglutination tests. The latter procedure is the most sensitive.

Acquired immunity to anthrax apparently involves primarily antibody to the toxin, although the antiphagocytic nature of the capsule suggests that it also may be important. Some animals have a high natural immunity to anthrax. Rats and dogs are only slightly susceptible to infection, but the nature of their immunity is unknown. Guinea pigs, mice, and rabbits are very susceptible, and reportedly one virulent *B. anthracis* organism is lethal after injection into a mouse. The natural disease occurs primarily in herbivorous animals such as cattle, sheep, and horses, who possess little immunity.

Whether other species of *Bacillus* evoke an immune response is unknown. Except for food poisoning, primarily by *B. cereus,* other species cause no well-defined infection. Apparently there is little or no immunity to the enterotoxin produced by *B. cereus* (Table 3).

LABORATORY DIAGNOSIS

An isolated, hemolytic colony of an aerobic, spore-forming, gram-positive, motile bacillus often is dismissed by the microbiology laboratory as an unimportant contaminant. But the increasing frequency with which severe infections caused by *Bacillus* is being reported mandates that the laboratory consult with the clinician before discarding such cultures. If the specimen originated from a seriously ill patient or from an unusual source or was isolated in large numbers in repeated specimens, the physician should be notified so that the results can be interpreted in terms of the clinical findings.

Identification of the species of *Bacillus* usually can be made by evaluation of several morphologic and biochemical characteristics (Table 1, Fig. 4), although speciation may not be made quickly because of the number of species and the fact that variant strains are common. The bacteria most often encountered in humans grow readily on common nutrient media. Specimens should be obtained from patients prior to administration of antimicrobial therapy and inoculated onto 5 per cent horse or sheep blood agar. The cultures are incubated at 35 to 37° C in air for 24 to 48 hours. Hemolytic, catalase-positive colonies of gram-positive bacilli that contain spores can then be identified by a series of commonly used biochemical tests that include the Voges-Proskauer reaction, starch hydrolysis, glucose fermentation, and anaerobic growth (Table 1, Fig. 4). Experienced technicians may make a preliminary identification from the morphologic appearance of the colonies.

When *B. cereus* is suspected as a cause of food poisoning, the food can be cultured to isolate the bacteria, or isolation of *B. cereus* from the stool of patients can be attempted. If more than 10^5 bacteria per gram of food are identified as *B. cereus,* the diagnosis is confirmed. Serologic typing of isolated strains can be done for epidemiologic purposes, but the antisera developed by the Food Hygiene Laboratory in London are available only in a few centers. *B. licheniformis* and *B. subtilis* also may be isolated from foods and incriminated as causes of food poisoning.

Although *B. anthracis* is seldom encountered in an average hospital or public health laboratory, it can be confused with the more common *B. cereus* or encapsulated *B. megaterium.* The three species can be differentiated easily (Table 4). The "string of pearls" phenomenon is observed when *B. anthracis* is inoculated onto agar containing penicillin (0.05 to 0.5 units per ml). After incubation for about three to six hours, the areas where growth might occur are examined microscopically for the presence of large spherical bacilli in chains, the "string of pearls." Animals injected subcutaneously with *B. anthracis* develop an overwhelming infection within 36 to 48 hours, and smears from the blood or spleen show numerous encapsulated bacteria. Suspensions should be diluted and made in saline to avoid nonspecific deaths.

TABLE 4. Characteristics Useful for Distinguishing *B. anthracis*, *B. cereus*, and *B. megaterium*

CHARACTERISTIC	*B. anthracis*	*B. cereus*	*B. megaterium*
Hemolysis on sheep's blood agar	−	+	−
Peptonization of litmus milk	−	+	±
Fermentation of salicin	−	±	−
Specific fluorescent antibody stain	+	−	−
Appearance of colonies grown on bicarbonate (0.7%) nutrient agar[a]			
In air	(Rough)[b]	Rough	No growth
In 5% CO_2 or candle jar	Smooth	Rough	Rough
Growth in penicillin (10 μ/ml) agar	−	+	−
String of pearls reaction	+	−	−
Pathogenicity for mice	+	−	−
Positive capsular stain for bacteria from spleen[c]	+	−	−
Susceptibility to specific gamma bacteriophage	+	−	−

All cultures incubated at 35 to 37° C.
[a]Bicarbonate must be sterilized separately by filtration and added to agar.
[b]*B. anthracis* may not grow under these conditions.
[c]Stained by Loeffler's methylene blue stain (polychrome methylene blue stain).

Samples from cutaneous anthrax lesions should be obtained carefully. Fluid can be taken from the vesicles with a Pasteur pipette, or less preferably, with a syringe and needle or swab, and used for smear and culture. Tentative identification can be made if a smear shows short chains of large gram-positive rods, perhaps with a few spores. Pulmonary anthrax cannot be diagnosed by culture because the sputum rarely contains bacteria; a blood culture usually is positive. Blood cultures also should be obtained to diagnose intestinal anthrax. Specimens for diagnosis in animals are best taken from the jugular vein.

The fluorescent antibody stain for *B. anthracis* is presumptive because *B. megaterium* and *B. cereus* also fluoresce with the antiserum. Cultures for fluorescent staining must be grown in carbon dioxide to permit capsule production. This test and the test for susceptibility to gamma bacteriophage are obtainable only in certain laboratories such as the Center for Disease Control in the United States. Serologic diagnosis for anthrax is not valuable because the disease is of short duration, although procedures such as an agar gel precipitation test, complement fixation, and indirect microhemagglutination have been used to detect antibody. Selective media for isolation have been unsatisfactory because they inhibit growth of *B. anthracis*.

B. anthracis can be isolated from contaminated wool and hair. The material should be soaked in detergent or dilute potassium hydroxide. After thorough soaking, the sample is teased, heated to 65° C for five minutes, and centrifuged. The sediment is cultured, and suspicious colonies are identified.

Any competent microbiology laboratory can handle *B. anthracis* safely. Aerosols should be avoided, and use of a biological safety hood is advisable, particularly if contaminated materials are examined. Animals should be inoculated only when adequate facilities are available for isolation, but reasonable identification of *B. anthracis* can be made without use of animals. Benches should be washed with 5 per cent carbolic acid (phenol) or 5 per cent hypochlorite, and instruments should be autoclaved to reduce the possibility of contamination.

DRUG SUSCEPTIBILITY

With one exception, broad spectrum antimicrobial agents are usually effective against species of *Bacillus*. The exception is penicillin. *B. cereus* in particular produces a beta-lactamase that renders penicillin ineffective for treatment. Chloramphenicol, erythromycin, penicillin, tetracyclines, and sulfonamides have been used to treat anthrax. About 24 to 48 hours after treatment with penicillin, *B. anthracis* disappears from cutaneous lesions. The bacteria are reported to persist longer after treatment with chloramphenicol or tetracycline. However, some patients with cutaneous anthrax recover without treatment, another indication that humans are relatively resistant to the disease. *B. anthracis* is resistant to neomycin and polymyxin. Other species of *Bacillus* are sensitive to chloramphenicol, gentamicin, kanamycin, and tetracyclines. Treatment of infections caused by these organisms usually involves initial treatment with a broad spectrum antibiotic, followed by laboratory tests on the isolated organism to determine sensitivities. *B. subtilis* is susceptible, *B. pumilus* less so, to ampicillin, cephalothin, methicillin, and penicillin G. *B. cereus* is resistant to these drugs as well as to sulfonamides and trimethoprim.

EPIDEMIOLOGY

Gram-positive, aerobic, spore-forming bacilli are ubiquitous in the environment. They are found in decaying organic matter, dust, soil, vegetables, and water, and some species are part of the normal flora. From these sources, the bacteria can readily cause infections in debilitated, immunosuppressed, or traumatized patients. However, except for *B. anthracis,* members of the genus usually are not considered pathogenic. But the opportunity for infection with other species is increasing with the prolonged survival of patients with severe illnesses. When the bacteria are isolated from blood, body fluids, or infections of closed spaces, they should not be dismissed as contaminants, particularly if patients are receiving immunosuppressive or myelosuppressive drugs, are undergoing hemodialysis, or are drug addicts. Although the true prevalence is unknown, infection caused by the bacteria occurs more frequently than is appreciated. They have been documented as causes of bronchopneumonia, meningitis, ocular infections, septicemia, osteomyelitis, endocarditis, abscesses, and several other types of infections.

The diarrheal type of gastroenteritis caused by *B. cereus* is associated with a wide range of foods, whereas boiled and fried rice is primarily implicated in the emetic disease (Table 3). Several serotypes are found in outbreaks of diarrhea, but Serotype 1 predominates in cases of vomiting, perhaps because the spores of this type are more resistant to heat, and they survive in cooked rice. Outbreaks of food poisoning caused by *B. cereus* have been reported from Asia, Australia, Europe, and North America. The diseases may be more common than reported because they are easily confused with the more frequent outbreaks of food poisoning caused by staphylococci and *C. perfringens*. Thorough cooking and refrigeration of foods help control gastroenteritis effected by *B. cereus*.

B. anthracis is a saprophytic bacterium that can become a facultative parasite. The bacteria probably survive in soil in a dynamic state in which they undergo cycles of germination and sporulation depending upon conditions in the microenvironment. Contamination of the soil originally occurs when infected animals soil the ground with blood and excreta that contain vegetative bacteria. Sporulation ensues in the aerobic environment outside of the body. In an alkaline, calcareous soil that is poorly drained, the spores germinate in decayed vegetation. When the weather becomes dry, spores form, and, if grazing animals inadvertently ingest them, an outbreak of anthrax can occur. Such focal areas may remain contaminated indefinitely, with *B. anthracis* undergoing cycles of germinatio
tion. The dependency of the cycle
environment may account for the seasona
rence of anthrax in the late summer and earl
subsequent to previous rain, as well as for th
sporadic outbreaks of the disease.

Anthrax primarily afflicts herbivorous animals. It is one of the major livestock diseases in the world and affects thousands of cattle, goats, horses, and sheep annually. Carnivorous animals can be infected secondarily if they feed on infected carcasses, but the disease is not spread from animal to animal. About a third of the cases in animals occur in Europe, with the major portion of infections occurring in Asia and Africa. The disease is enzootic in Asia Minor, Central and South America, Africa, and Southeast Asia, from whence epizootics periodically are reported in wild and domestic animals. In the United States, sporadic cases occur in the Gulf coast and midwest states and in California. The disease may be more prevalent than reported because many sudden deaths in animals, the typical outcome of an infection, may not be recognized as anthrax.

As many as 20,000 to 100,000 cases of anthrax in humans are estimated to occur worldwide per year, but such figures are unreliable because many countries do not report the disease. Anthrax was a formidable scourge in England in the mid-19th century, when fatal outbreaks of "woolsorter's disease" (inhalation anthrax) among textile workers led to legislation to control the disease. Today in industrial countries anthrax still is predominantly an occupational disease in the textile and tanning industries. Wool, hair, hides, and bone meal imported from countries where the disease is prevalent in animals contain large numbers of spores. Workers are exposed to the spores in the air and dust of plants and while handling the contaminated materials. However, although the bacteria are widespread, the number of cases of disease is few. Cutaneous infections are most common when spores invade the skin of the head, neck, or arms through cuts or abrasions, but if properly treated, the mortality is nil. Inhalation anthrax is rare but almost uniformly fatal. Although industrial anthrax is the most commonly reported form of the disease, agricultural anthrax may be widespread among farmers in countries where the disease is epizootic. Intestinal anthrax that has a high mortality occurs when meat from diseased animals is eaten. Veterinarians occasionally develop anthrax from contact with infected animals or carcasses. Cases of disease now occur in amateur craftsmen who use raw wool contaminated with anthrax spores. There are no age or sex differences in cases. Industrial anthrax has no seasonal incidence, but agricultural cases are

summer and early fall. reported in the United

quires elimination or re- h infected animals. Car- died from anthrax must be with lime deep in the soil. Autop... be done if anthrax is suspected to avoid ... tamination of the soil. Potential focal areas for growth of *B. anthracis* can be eliminated by proper drainage or tillage. If anthrax does occur, infected animals must be treated, and the exposed herd quarantined and vaccinated with the viable spore vaccine. This vaccine protects for about one year. Immunization and improved working conditions in industry seem to be responsible for the progressive decline of human cases of anthrax, so that the disease now is rare in the United States and Europe. Vaccine prepared from protective antigen can be used to immunize exposed individuals who handle materials contaminated with anthrax spores. Antibody titer declines rapidly; booster doses of the vaccine are suggested at six months or one year to maintain immunity.

References

Gilbert, R. J.: *Bacillus cereus* gastroenteritis. In Riemann, H., and Bryan, F. L. (eds.): Foodborne Infections and Intoxications. 2nd ed. New York, Academic Press, 1979.

Gilbert, R. J., Turnbull, P. C. D., Parry, J. M., and Kramer, J. M.: Food poisoning and other clinical infections associated with *Bacillus* species with particular reference to *B. cereus*. In Berkeley, R., and Goodfellow, M. (eds.): The Aerobic Endospore-Forming Bacteria: Classification and Identification. Soc Gen Microbiol Special Publications Series. New York, Academic Press, 1980.

Nungester, W. J.: Proceedings of the conference on progress in the understanding of anthrax. Fed Proc 26:1485, 1967.

Terranova, W., and Blake, P. A.: *Bacillus cereus* food poisoning. N Engl J Med 298:143, 1978.

Tuazon, C. U., Murray, H. W., Levy, C., Solny, M. W., Curtin, J. A., and Sheagreen, J. N.: Serious infection from *Bacillus* species. JAMA 241:1137, 1979.

Turnbull, P. C. B., Jorgensen, K., Kramer, J. M., Gilbert, R. J., and Parry, J. M.: Severe clinical conditions associated with *Bacillus cereus* and the apparent involvement of exotoxins. J Clin Pathol 32:289, 1979.

Van Ness, G. B.: Ecology of anthrax: Anthrax undergoes a propagation phase in soil before it infects livestock. Science 172:1303, 1971.

Wright, G. G.: Anthrax toxin. In Schlessinger, D. (ed.): Microbiology — 1975. Washington, D. C., American Society for Microbiology, 1975, p. 272.

Gram-negative Cocci and Coccobacilli

29 *NEISSERIA*

Stephen A. Morse, Ph.D.

NEISSERIA

The genus *Neisseria* is one of four genera included in the family *Neisseriaceae*. The other genera in the family are *Branhamella, Moraxella,* and *Acinetobacter* (see Table 1). Members of the genus *Neisseria* inhabit the mucosal surfaces of warm-blooded animals. The genus includes two species that are pathogenic for man: *Neisseria gonorrhoeae* (the gonococcus) and *Neisseria meningitidis* (the meningococcus). It also includes several nonpathogenic species (*N. flavescens, N. mucosa, N. perflava, N. sicca,* and *N. lactamica*) that may be part of the normal flora and therefore can be confused with gonococci and meningococci.

Descriptions of a condition resembling gonococcal urethritis can be found in the earliest recorded histories of man. The disease, gonorrhea, was named by Galen in about A.D. 130 (*gonos*, seed; *rhoia*, flow). Galen had the mistaken impression that the disease was due to a morbid loss of semen. Gonorrhea was known to be of venereal origin in the 13th century, but was thought to be an early symptom of syphilis. It was not until the middle of the 19th century that gonorrhea and syphilis were clearly differentiated. *N. gonorrhoeae*, the causative agent of gonorrhea, was first described by Neisser in 1879 in smears of purulent exudates from patients with acute gonorrhea and from newborn infants with conjunctivitis. The gonococcus was first cultivated by Leistikow and Loeffler in 1882. Bumm showed that the gonococcus fulfilled Koch's postulates in 1885.

N. meningitidis is the causative agent of men-

TABLE 1. Taxonomy and Nomenclature of the Neisseriaceae

Genus I.	Neisseria
Species:	*N. gonorrhoeae*
	N. meningitidis
	N. lactamica
	N. sicca
	N. flavescens
	N. subflava (includes *N. flava, N. perflava*)
	N. mucosa
Genus II.	Branhamella
Species:	*B. catarrhalis* (formerly *N. catarrhalis*)
	B. ovis (formerly *N. ovis*)
	B. caviae (formerly *N. caviae*)
Genus III.	Moraxella
Species:	*M. lacunata*
	M. nonliquifaciens
	M. bovis
	M. osloensis
	M. phenylpyruvica
	M. kingii
	*M. urethralis** (formerly *Mima polymorpha* var *oxidans*)
Genus IV.	Acinetobacter
Species:	*A. calcoaceticus* (formerly *Bacillus anithratum, Herrelea vaginicola,* or *M. lwoffi*)

*Tentatively placed in this genus.

ingococcal meningitis (formerly called epidemic cerebrospinal meningitis). The disease has the potential for occurring in epidemic form. It was first recognized as a contagious disease early in the 19th century. Outbreaks were eventually described on all continents, and the prevalence of the disease, particularly among military personnel, became apparent. The causative organism was first described by Marchiafava and Celli (1884) in meningeal exudate. In 1887, Weichselbaum isolated the organism in pure culture and described it in detail as the characteristic bacterium found in six cases of acute cerebrospinal meningitis.

Morphology

Neisseria are gram-negative cocci, 0.6 to 1.0 μm in diameter. The organisms are usually seen in pairs with their adjacent sides flattened. Tetrads or clusters are occasionally seen. Negatively stained cells display a rugose outer cell surface that is characteristic of many gram-negative bacteria. Fresh isolates of most *N. meningitidis* serogroups are encapsulated. Gonococci also possess a capsule; however, the capsule is most evident on recent isolates (Hendley et al., 1977; Richardson and Sadoff, 1977). *Neisseria* do not possess flagella; however, twitching motility may sometimes be observed with piliated organisms.

Thin sections of gonococci or meningococci grown in vitro exhibit cell structures in electron micrographs similar to those of other gram-negative bacteria (Fig. 1). An undulating outer membrane, approximately 7.5 to 8.5 nm in thickness, appears as a bilayered structure. The periplasmic space, between the cytoplasmic membrane and the outer membrane, contains a thin, electron-dense layer, approximately 6.0 nm in diameter, which corresponds to the peptidoglycan layer of the bacterial cell envelope. The chemical composition of the purified peptidoglycan from *N. gonorrhoeae* consists of muramic acid, glutamic acid, alanine, *meso*-diaminopimelic acid, and glucosamine in approximate molar ratios of 1:1:2:1:1 respectively (Morse, 1978; Roberts, 1977). The peptidoglycan layer and the outer membrane adhere to each other at regular intervals around the periphery of the cell.

Cell wall blebs produced by budding of the outer membrane have been seen both in lag phase broth- or agar-grown cultures of *N. meningitidis* and *N. gonorrhoeae*. The blebs consist of outer membrane components, including lipopolysaccharide. The production of these blebs occurs on rapidly growing cells since no blebs are seen on stationary phase cells.

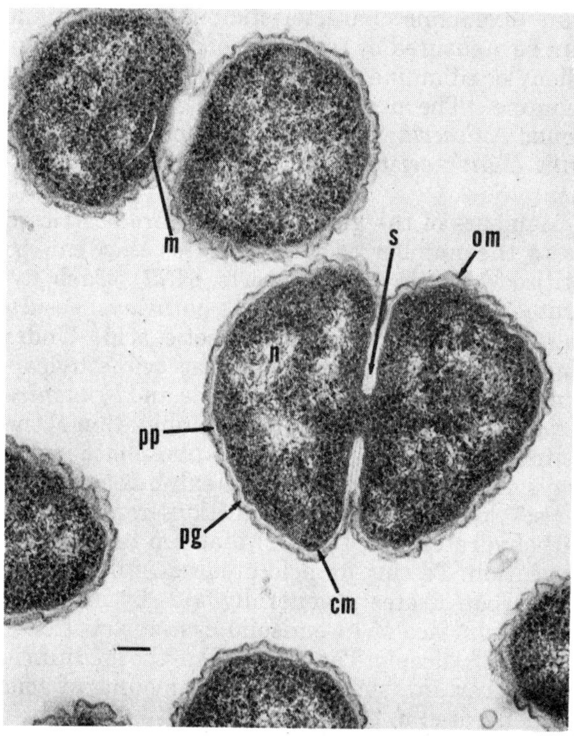

FIGURE 1. *Electron micrograph of a thin section of* N. gonorrhoeae *stained with uranyl acetate and lead citrate. Symbols: om, outer membrane; pg, peptidoglycan; pp, periplasmic space; cm, cytoplasmic membrane; s, septum; n, nucleoid zone; m, mesosome. Bar + 0.1 μm. (Used with permission of the American Society for Microbiology.)*

TABLE 2. Differential Characteristics of the Genus Neisseria*

	N. gonorrhoeae	*N. meningitidis*	*N. lactamica*	*N. sicca*	*N. subflava*†	*N. flavescens*	*N. mucosa*
Acid from:							
Glucose	+	+	+	+	+	−	+
Maltose	−	+	+	+	+	−	+
Lactose	−	−	+	−	−	−	+
Fructose	−	−	−	+	v	−	+
Sucrose	−	−	−	+	v	−	+
Starch	−	−	−	v	v	−	+
Glycogen synthesis from 5% sucrose	∅	∅	?	+‡	d	+	+
Growth at 22° C	−	−	−	d	d	+	+
Presence of capsule	v	d	?	v	+	−	+
Production of H_2S	−	−	−	+	+	+	+
Pigment	−	−	−	d	+	+	−
Reduction of nitrate	−	−	−	−	−	−	+
Reduction of nitrite	=	d	?	+	+	+	+

*+, most strains positive (≧ 90%); −, most strains negative (≧ 90%); d, some strains positive, some negative; v, character inconstant within single strain; ∅, no growth on medium with 5% sucrose.
†New species consisting of *N. subflava*, *N. flava*, and *N. perflava*.
‡*N. sicca* forms an iodine-positive product when grown on trypticase soy agar without 5% sucrose. This reaction, which does not occur with *N. subflava*, may be used as a differentiating characteristic.

Metabolism

Neisseria are aerobic and contain high levels of cytochrome *c* oxidase activity (Morse, 1978; Roberts, 1977). Cytochrome *c* oxidase is an important taxonomic characteristic of this genus and can be measured by reduction of *N*,*N*-dimethyl-*p*-phenylenediamine (or tetramethyl-*p*-phenylenediamine). The principal species included in the genus *Neisseria* are listed in Table 2, together with characteristics used in species differentiation.

Members of the genus *Neisseria* are restricted as to the number of carbohydrates that can be utilized (Morse, 1978; Roberts, 1977). Carbohydrates are utilized by oxidative pathways, resulting in the accumulation of acetic acid. Under some conditions, lactic acid and acetylmethylcarbinol are produced. *N. gonorrhoeae* and *N. meningitidis* dissimilate glucose by a combination of the Entner-Doudoroff and pentose phosphate pathways. Each species of *Neisseria* exhibits a characteristic pattern of acid production from specific carbohydrates (see Table 2) that can be used for speciation. Testing for acid production from various carbohydrates is generally done by inoculating the surface of a semisolid cystine trypticase agar (CTA) deeply. The initial pH of the medium is very important, since only small amounts of acid are generated in positive reactions.

Both the gonococcus and meningococcus have a functional tricarboxylic acid cycle (Morse, 1978; Roberts, 1977). However, the utilization of this pathway is markedly influenced by growth conditions.

N. gonorrhoeae and *N. meningitidis* are fastidious organisms (the gonococcus more so than the meningococcus) with complex nutritional growth requirements. Both organisms are sensitive to growth inhibition by free fatty acids present in various peptones or in agar. The toxic effect of the free fatty acids can be eliminated by adding a binding agent, such as soluble starch, serum, or charcoal, to the medium. Gonococci and meningococci are unique among gram-negative bacteria because they are inhibited by low concentrations of gonadal steroids (Morse, 1978; Roberts, 1977).

Many strains of *N. gonorrhoeae* require increased atmospheric carbon dioxide concentrations (ca 4 to 8 per cent) for initial isolation. This requirement is often lost after repeated subculture. An exogenous source of carbon dioxide is required during the lag phase. This requirement may be met by either the addition of bicarbonate ions or gaseous carbon dioxide. During exponential growth, the exogenous source is replaced by metabolically generated carbon dioxide. Meningococci are less stringent in their requirement for carbon dioxide.

Gonococci and meningococci typically undergo autolysis in older cultures or when suspended in appropriate buffers. In *N. gonorrhoeae*, autolysis results from a combination of peptidoglycan hydrolysis and destabilization of the cell membranes (Morse, 1978; Roberts, 1977). The presence of divalent cations (Mg^{2+} or Ca^{2+}) will prevent cellular lysis but will not inhibit peptidoglycan hydrolysis.

The pathogenic *Neisseria* have an optimum growth temperature that ranges from 36 to 39°C.

The maximum growth temperature is ca 41°C and the minimum growth temperature is ca 24°C. The inability of *N. gonorrhoeae* and *N. meningitidis* to grow at 22°C distinguishes these species from most of the nonpathogenic species of *Neisseria*. The pH range for growth varies from strain to strain. Many strains will grow in media buffered over a pH range of 6.0 to 8.0.

NEISSERIA GONORRHOEAE

Colonial Types

As early as 1904 differences were noted in the colonial appearance of *N. gonorrhoeae*. More recently, an association was observed between virulence and colonial morphology. Four morphologically distinct colonial types were initially described and designated 1, 2, 3, and 4 (T1, T2, T3, T4). Later, other colonial types, such as T5, were reported. The properties of these colonial variants are summarized in Table 3.

Freshly isolated strains consist of type 1 and type 2 colonies. The type 1 and type 2 colonial morphology can be maintained by selective subculture at daily intervals. Organisms from type 1 and type 2 colonies are piliated and retain virulence when they are inoculated experimentally into the urethra of the human or chimpanzee male, or intravenously into chicken embryos. During nonselective transfer, the predominant colonial types (T1 or T2) may be rapidly replaced by the less virulent colonial types T3, T4, and T5. Organisms from these colonial forms lack pili.

Antigenic Composition

Strains of *N. gonorrhoeae* differ not only when freshly isolated, but also change antigenically on continued subculture. Recent evidence suggests that some antigens may be expressed only in vivo or in response to specific environmental factors. The gonococcus also shares antigens with the meningococcus and other gram-negative bacteria. Most current knowledge concerns antigens associated with the cell envelope.

Capsules. Encapsulation of gonococci is suggested by the finding that meningococci have a polysaccharide capsule that provides a basis for grouping. It has been demonstrated by negative staining and by electronmicroscopy of cells exposed to hyperimmune serum that freshly isolated strains of *N. gonorrhoeae* possessed a capsule (Hendley, 1977; Richardson and Sadoff, 1977). Environmental and nutritional factors are important in capsule synthesis. The capsule is loosely associated with the cell surface and can be easily removed by mild shearing. Although the capsule is most evident on recent isolates or on organisms recently grown in vivo, it is also present on the surface of laboratory strains. Cells from all colony types produce capsules. The capsules are antiphagocytic; however, anticapsular antibodies promote phagocytosis. The chemical nature of the capsule and homo- or heterogeneity of capsular types are still unknown (Richardson and Sadoff, 1977).

Pili. Metal-shadowed electronmicrographs of *N. gonorrhoeae* reveal hairlike filaments, known as pili, extending from the surface. Pili have a

TABLE 3. Morphology and Other Properties of Colonial Types of N. gonorrhoeae

	COLONIAL TYPE				
	1	2	3	4	5
Average diameter (mm)	0.5–0.7	0.5–0.7	1.0–1.4	1.0–1.4	1.4
Elevation (convexity)	++	++	+	+	+
Reflection of light	+	++	−	−	+
Surface granularity	−	+	+	−	++
Edge	Entire	Entire	Entire	Entire	Slightly crenated
Opacity	Translucent	Translucent	Translucent	Translucent	Opaque
Color*	Gray-Gold	Dark Gold	Light Brown	Colorless	Brown
Structure	Amorphous	Amorphous	Granular	Amorphous	Granular
Consistency	Viscid	Friable	Viscid	Viscid	?
Autoagglutinability in saline	+	+	−	−	++
Hemadsorption with rabbit erythrocytes	+	+	−	−	−
Hemagglutination	+	+	−	−	+/+
Clumping during growth in liquid medium	Slight or moderate	Marked	Moderate	None	?
Presence of visible pili	+	+	−	−	−
Presence of rhamnose in LPS	−	−	+	+	+
Competent for transformation	+	+	−	−	?
Cryptic 2.6×10^6 dalton plasmid	+	+	+	+	?
Self-transmissible 24.5×10^6 dalton plasmid	+	+	+	+	−
Approximate LD_{50} for chick embryos inoculated intravenously (colony-forming units)	$<10^4$	$<10^2$	$<10^7$	$<10^7$	$<10^7$

*Light variants of colonial types 1 and 2 have been reported.

diameter of ca 7 nm and are found on the surface of gonococci from colony types 1 and 2; gonococci from colony types 3 and 4 are devoid of visible pili. The presence of pili on *N. gonorrhoeae* is not unique. Nonpathogenic *Neisseria* spp and *N. meningitidis* also possess pili. Short pili (175 to 210 nm in length) are seen only on nonpathogenic species, whereas long pili (up to 4300 nm) are seen on organisms of both nonpathogenic and pathogenic species. There is no consistent relationship between colonial morphology and pili among the nonpathogenic *Neisseria* spp and *N. meningitidis*.

Gonococcal pili have been purified and are primarily protein in nature. Neutral sugar and phosphate residues have also been reported in them. Pilar subunit molecular weights vary among strains and range between 18,000 and 20,000. Pili also exhibit antigenic heterogeneity. The amino-terminal amino acid sequence is identical for the first 24 residues in pili purified from three strains of *N. gonorrhoeae* and one strain of *N. meningitidis*. This finding was unexpected in that the three sources of gonococcal pili exhibited less than 1 per cent shared antigenicity.

A potential antiphagocytic role for pili has been proposed to explain the decreased association of piliated gonococci with human polymorphonuclear leukocytes as compared with nonpiliated gonococci. This role has been challenged, and the association of the gonococcus with polymorphonuclear leukocytes has been ascribed to another protein surface factor (leukocyte association (LA) factor). On the other hand there is considerable evidence that gonococcal pili play an important role in the attachment of the gonococcus to host cells. Pili appear to recognize specific binding sites on host cell surfaces. It is noteworthy that the greatest attachment of purified pili is to cells that are histologically the most similar to the actual sites of gonococcal infection. It has been postulated that pili overcome the long-range electrostatic repulsion between the negatively charged gonococcal and host cell surfaces.

Outer Membrane Proteins. The outer membrane of *N. gonorrhoeae* contains relatively few proteins, with one protein accounting for as much as 66 per cent of the total outer membrane protein. This protein varies antigenically among strains and can be used to divide the gonococcus into serotypes. At least 16 serotypes have been isolated. Antibodies against the principal outer membrane protein are bactericidal in a complement-dependent assay.

Lipopolysaccharides. The lipopolysaccharide (LPS) from *N. gonorrhoeae* has been isolated and characterized by several groups of investigators. There is some controversy as to whether the gonococcus possesses a smooth-type or rough-type LPS. A smooth-type LPS has been reported in organisms from type 1 and type 2 colonies. However, growth conditions may alter the type of LPS synthesized. The LPS from types 3 and 4 organisms contains rhamnose. The gonococcus also produces an enzyme that degrades the polysaccharide portion of its LPS. The core region of the LPS is antigenic and antibodies are bactericidal.

Genetics

The chromosome of the gonococcus has approximately 1.5×10^6 nucleotide pairs and a molecular weight of 9.8×10^8. Most strains of *N. gonorrhoeae* contain a 2.4×10^6 dalton plasmid. There has been no function ascribed to the presence of this plasmid and it remains phenotypically cryptic. A smaller percentage of gonococcal strains contains a 24.5×10^6 dalton plasmid. It has been recently shown that this plasmid is a sex factor capable of promoting the transfer of chromosomal genes and R-plasmids via conjugation.

Genetic transformation readily occurs with *N. gonorrhoeae*. Unlike most bacteria, the gonococcus is competent for transformation throughout the growth cycle. All strains of *N. gonorrhoeae* examined have been found competent for genetic transformation. Transformation frequencies are significantly higher ($>10^3$-fold higher) with piliated cells from type 1 or type 2 colonies than with nonpiliated cells from type 3 or type 4 colonies.

Conjugation has recently been described in *N. gonorrhoeae*. Only strains harboring the 24.5×10^6 dalton plasmid serve as donors. There is no relationship between colony type (and presumably the presence of pili) and the ability to serve as donor.

No bacteriophage has been described for *N. gonorrhoeae*, and transduction has not been reported in the gonococcus.

Epidemiology

The only natural host for *N. gonorrhoeae* is the human. An estimated 3 million cases of gonorrhea were treated in the United States during 1977. The annual incidence of reported gonorrhea in the United States tripled between 1963 and 1975. This increase in incidence has apparently leveled off in the last few years. Gonorrhea is a worldwide problem. Significant increases in incidence have been reported from Argentina, Austria, Bolivia, Canada, Chad, Colombia, Costa Rica, Denmark, El Salvador, Ecuador, Finland, Gabon, Germany, Hong Kong, Israel, Iran, Iraq, Jamaica, Mali, Norway, Panama, the Philippines, Poland, Rwanda, Swaziland, Trinidad and Tabago, the United Kingdom, and Venezuela.

The incidence has leveled off in the United Kingdom and is apparently falling in the Scandinavian countries.

The gonococcus is susceptible to drying and will die within one to two hours. Because it cannot survive for long away from its host, and because it can produce primary infection only on certain specific epithelial surfaces, the gonococcus is almost always venereally transmitted.

The highest attack rate for gonorrhea in both men and women occurs in the 20 to 24 year old age group. The second highest attack rate is in the 15 to 19 year old age group. Because of numerous biases in case detection and reporting, the true male-female ratio for new cases of gonorrhea is unknown but is probably close to 1:1 in large populations.

There are several factors that contribute to the incidence of uncomplicated gonorrhea. Some host-related factors are as follows: the number of sexual partners; patterns of contraceptive use; population mobility; homosexuality; recidivism; and individuals with blood group B. An increase in the resistance of the gonococcus to therapeutic antibiotics may also be an important factor.

Behavioral factors predisposing to gonococcal infection also increase the risk of acquiring other sexually transmitted diseases. Such diseases are prevalent in groups with a high incidence of gonorrhea, and populations or individuals with any sexually transmitted disease should be screened for gonorrhea.

Pathogenic Properties

Not everyone exposed to gonorrhea acquires the disease. It is uncertain whether this is due to variations in virulence or in the size of the inoculum, to nonspecific resistance, or to specific immunity. Several potential virulence factors (pili, capsule, lipopolysaccharide) have been discussed in a preceding section. The gonococcus elaborates a cytotoxic factor that damages ciliated epithelial cells in fallopian tube cultures. In addition, gonococci and meningococci produce an extracellular protease that cleaves a pro-thr bond on the heavy chain of the IgA_1 subclass of IgA. Cleavage of this bond results in a loss of antimicrobial activity. Because IgA is the antibody subclass that predominates in secretions, this enzyme may be important in pathogenesis of infection.

There is considerable controversy as to the intracellular survival and multiplication of the gonococcus. Gonococci are present both on and within PMN leukocytes soon after phagocytosis. Although most of the ingested gonococci are killed, gonococci may survive and multiply within phagocytes. Acquired IgA and IgG serum antibody enhances the association of gonococci with PMN leukocytes in vitro. However, piliated cells tend to remain extracellular, whereas nonpiliated cells are readily ingested and killed.

Drug Susceptibility

Sulfonamides were used extensively to treat gonorrhea until the end of World War II, when sulfonamide resistance of gonococci became extensive. Fortunately, penicillin was introduced at about the same time as the sulfa drugs were losing effectiveness. Initially, gonococci were extremely sensitive to penicillin with minimum inhibitory concentrations (MIC) ranging from 0.003 to ca 0.03 units/ml. Treatment with a single injection of 150,000 units of penicillin produced cure rates of 90 per cent or more.

By the mid-1950s, gonococci with reduced sensitivity to penicillin were isolated from various parts of the world. This low-level resistance was not due to a β-lactamase but necessitated higher doses of penicillin. At the same time, gonococci developed low-level resistance to several other antibiotics, including tetracycline, chloramphenicol, and erythromycin. The use of streptomycin for therapy has resulted in resistance to levels of streptomycin far in excess of achievable serum levels of this drug.

β-Lactamase-producing strains of *N. gonorrhoeae* were first reported in 1976. The β-lactamase was active against penicillin G, ampicillin, and cephalosporins. The β-lactamase-producing strains possess either a 3.2×10^6 or 4.4×10^6 dalton plasmid. Loss of this plasmid results in loss of β-lactamase production. Some strains which possess the 4.4×10^6 dalton plasmid also possess a 24.5×10^6 dalton sex factor capable of mediating the conjugal transfer of the resistance factor. It is generally believed that the 4.4×10^6 and 3.2×10^6 dalton plasmid are genetically related and originated in an enteric bacterium. Penicillin-resistant strains of *N. gonorrhoeae* may be successfully treated with other drugs, including spectinomycin. However, spectinomycin resistance has also been reported.

Immunity

The natural route of infection by *N. gonorrhoeae* brings the organism into direct contact with mucosal surfaces. However, it is still not clear to what extent gonococcal infection stimulates local immunity and whether any subsequent secretory immune response will protect against subsequent infection (Griffiss, 1976).

Repeated infections with *N. gonorrhoeae* are common. The development of serologic and biochemical typing schemes will allow differentiation between reinfection and infection with a new strain. It is also not clear whether disseminated infection elicits protective antibody. There is no vaccine against gonorrhea.

Gonococci in urethral exudates are resistant to the bactericidal activity of the patients' serum, but lose this resistance upon subculture on laboratory medium. Serum resistance can be maintained by subculture on a medium resembling the chemical environment of the host. Virulent colonial types are generally more resistant to killing by serum than the avirulent colonial types. After exposure to killing by serum, virulent colonial types can acquire complete resistance to serum. This acquired resistance is rapidly lost upon subculture on serum-free medium. The loss and gain of serum resistance is due to phenotypic variation.

TABLE 4. Chemical Composition of the Capsular Polysaccharides of N. meningitidis

SEROGROUP	COMPOSITION
A	Partially 0-acetylated comopolymer of 2-acetamido-2-deoxy-D-mannose-6-phosphate
B	N-acetyl neuraminic acid (2→8-α-linked)
C	0-acetylated N-acetylneuraminic acid (2→9-α-linked)
X	2-acetamido-2-deoxy-D-glucose-4-phosphate
Y	Partially 0-acetylated alternating sequence of D-glucose and N-acetylneuraminic acid
Z	Not determined
29E	Equimolar amounts of 3-deoxy-D-*manno*-octulosonic acid and 2-acetamido-2-deoxy-galactosamine
W-135	Alternating sequence of D-galactose and N-acetylneuraminic acid

Laboratory Diagnosis

Men. Gram-stained smears of urethral exudate will often reveal gram-negative intracellular diplococci. This finding is sufficient for a diagnosis of gonorrhea. Culture specimens should be obtained from the anterior urethra and inoculated onto a selective medium (modified Thayer-Martin). Plates should be incubated in a candle jar or CO_2 incubator. Oxidase-positive colonies composed of gram-negative diplococci are presumed to be *N. gonorrhoeae*. Fluorescent antibody staining or the production of acid from glucose, but not from maltose, sucrose, or fructose, may be used for confirmation. In homosexuals, an additional specimen should be obtained from the anal canal and pharynx and processed as above.

Women. Gram-stained smears of exudate are not recommended for diagnosis of gonorrhea in women. Culture specimens should be obtained from the cervix and anal canal and inoculated on modified Thayer-Martin plates. The criteria for presumptive and confirmatory identification are as above. Serologic tests for gonorrhea have not proved satisfactory for either women or men.

NEISSERIA MENINGITIDIS

Antigenic Composition

Capsule. The capsular polysaccharides of *N. meningitidis* provide the basis for grouping these organisms (Bhattacharjee and Jennings, 1974; Bhattacharjee et al., 1976; Gotschlich et al., 1969). The serogroups and the chemical composition of their capsular polysaccharide, where known, are listed in Table 4. Meningococcal capsular polysaccharides are immunogenic when the intact bacterium is presented either to man or to rabbits. Highly specific anticapsular antibodies produced by rabbits are used in an agglutination test to group meningococci.

Outer Membrane Proteins. Serogroups B and C meningococci have been subdivided into serotypes based upon the presence of outer membrane proteins. The majority of both group B and group C disease is caused by one serotype, type 2. The type 2 antigens from groups B and C meningococci are chemically and serologically identical. Antibodies against the serotype antigens are bactericidal in the presence of complement.

Lipopolysaccharide. Purified meningococcal LPS is highly toxic, and is as lethal for mice as is the LPS from *Escherichia coli* or *Salmonella typhimurium*. However, meningococcal LPS is five- to tenfold more effective than enteric LPS in eliciting a dermal Schwartzman reaction in rabbits (Davis and Arnold, 1974). Humans develop antibodies against meningococcal LPS during systemic disease. However, a role for these antibodies in immunity to meningococcal infection has yet to be demonstrated.

Pathogenic Properties

Untreated meningococcal meningitis has a mortality rate of approximately 85 per cent. The virulence of *N. meningitidis* depends in part upon the antiphagocytic properties of its capsule. Antibodies directed against the capsular polysaccharide are bactericidal. In the animal, meningococci behave as extracellular parasites. Although they are often visible within PMN leukocytes, there is no evidence that they can multiply intracellularly. The toxic properties of the meningococcus reside in its LPS (Davis and Arnold, 1974).

Epidemiology

The meningococcus usually inhabits the human nasopharyngeal area without causing disease (Griffiss, 1976). This carrier state may last for days or months and is important because it provides a reservoir for meningococcal infections and enhances the immunity of the host. From 3 to 30 per cent of normal individuals are carriers at any given time, yet few develop meningococcal disease. Even during epidemics of meningococcal meningitis in military recruits, when the carrier rate may reach 95 per cent, the incidence of systemic disease is less than 1 per cent (Goldschneider et al., 1969, 1969a). Carriers of meningococci are usually over 21 years of age, but the attack rates of disease are highest in children: with group B, under 5 years of age; with group C, 4 to 14 years of age. The low incidence of disseminated disease after colonization suggests that host rather than bacterial factors play an important determining role.

Meningococcal meningitis occurs sporadically and in epidemics, with highest incidence during the late winter and early spring. The disease appears to be declining in frequency in the United States and in Great Britain, whereas there has been an increased frequency in Finland and in Brazil. Most epidemics are caused by group A strains, but small outbreaks have occurred with both group B and group C strains. Sporadic cases are generally caused by group B, group C, and group Y strains, with groups B and C alternating in predominance. Whenever group A strains become prevalent, the incidence of meningitis increases markedly.

Drug Susceptibility

The usual minimal inhibitory concentrations of antibiotics against the meningococcus are as follows (μg/ml): benzylpenicillin 0.03, ampicillin 0.05, methicillin 6.0, dicloxacillin 6.0, cephalothin 1.6, cephalexin 10.0, tetracycline 0.8, minocycline 0.8, rifampin 0.2, chloramphenicol 1.0, and erythromycin 1.0. Penicillin is the most potent and is the drug of choice in treatment, although useless in prophylaxis (see Chapter 180). Chloramphenicol has less in vitro activity than penicillin but penetrates the cerebrospinal fluid so well that it is an acceptable alternative in meningococcal meningitis for patients who cannot take penicillin. Rifampin, minocycline, and sulfonamides are sometimes used for eliminating meningococci from pharyngeal carriers. During World War II sulfonamides were remarkably effective in eliminating the carrier state but many strains of meningococci are now resistant to that drug.

Immunity

Group A and group C capsular polysaccharide vaccines are available (Gotschlich et al., 1969, 1969a). These vaccines are highly protective against strains of these serogroups. Protection is provided through complement-mediated bactericidal antibodies that persist for at least five years after a single parenteral injection of vaccine. The capsular polysaccharide from serogroup B organisms is not immunogenic for humans.

Laboratory Diagnosis

Specimens of blood, spinal fluid, and nasopharyngeal secretions should be examined for the presence of *N. meningitidis* in cases of suspected meningococcal disease. Specimens should be collected before treatment with antimicrobial agents because they reduce cultural isolation. However, smears may still show gram-negative diplococci, and counter-immunoelectrophoresis will detect meningococcal capsular polysaccharide. The CSF should be cultured on chocolate or blood agar in either a candle jar or a CO_2 incubator. The presence of oxidase-positive colonies consisting of gram-negative diplococci provides a presumptive identification of *N. meningitidis*. Production of acid from glucose and maltose, but not from sucrose, lactose, or fructose, may be used for confirmation. The serologic group may be determined by a slide agglutination test, using first polyvalent then monovalent antisera.

Nasopharyngeal specimens must be obtained from the posterior nasopharyngeal wall behind the soft palate. Specimens should be inoculated onto selective medium (modified Thayer-Martin) and processed as above.

Gram-stained smears of CSF may be diagnostic. However, organisms in CSF smears are often more difficult to find than in pneumococcal meningitis. Quellung tests may be of value.

References

Bhattacharjee, A.K., and Jennings, H.J.: Characterization of 3-deoxy-*D-manno*-octulosonic acid as a component of the capsular polysaccharide antigen from *Neisseria meningitidis* serogroup 29-e. Biochem Biophys Res Commun 61:489, 1974.

Bhattacharjee, A.K., Jennings, H.J., Kenny, C.P., Martin, A., and Smith, I.C.P.: Structural determination of the polysaccharide antigens of *Neisseria meningitidis* serogroups Y, W-135, and BO. Canad J Biochem 54:1, 1976.

Davis, C. E., and Arnold, K.: Role of meningococcal endotoxin in meningococcal purpura. J Exp Med 140:159, 1974.

Goldschneider, I., Gotschlich, E.C., and Artenstein, M.S.: Human immunity to the meningococcus. I. The role of humoral antibodies. J Exp Med 129:1307, 1969.

Goldschneider, I., Gotschlich, E.C., and Artenstein, M.S.: Human immunity to the meningococcus. II. Development of natural immunity. J Exp Med 129:1327, 1969a.

Gotschlich, E.C., Goldschneider, I., and Artenstein, M.S.: Human im munity to the meningococcus. IV. Immunogenicity of group A and group C meningococcal polysaccharides in human volunteers. J Exp Med 129:1367, 1969.

Gotschlich, E.C., Goldschneider, I., and Artenstein, M.S.: Human immunity to the meningococcus. V. The effect of immunization with meningococcal group C polysaccharide on the carrier state. J Exp Med 129:1385, 1969a.

Gotschlich, E.C., Liu, T.Y., and Artenstein, M.S.: Human immunity to the meningococcus. III. Preparation and immunochemical properties of the group A, group B, and group C meningococcal polysaccharides. J Exp Med 129:1349, 1969.

Griffiss, J.M., and Artenstein, M.S.: The ecology of the genus *Neisseria*. Mt Sinai J Med 43:746, 1976.

Hendley, J.O., Powell, K.R., Rodewald, R., Holzgrefe, H.H., and Lyles, R.: Demonstration of a capsule on *Neisseria gonorrhoeae*. N Engl J Med 296:608, 1977.

Morse, S.A.: The biology of the gonococcus. Crit Rev Microbiol 7:000, 1978.

Richardson, W.P., and Sadoff, J.C.: Production of a capsule of *Neisseria gonorrhoeae*. Infec Immun 15:663, 1977.

Roberts, R. B. (ed.): The Gonococcus. New York, John Wiley and Sons, 1977.

30 MORAXELLA AND ACINETOBACTER

Herman Baer, M.D.

Organisms of *Moraxella* and *Acinetobacter* are gram-negative, strictly aerobic, coccobacillary rods that do not produce flagella or spores. Their classification has changed repeatedly and there are numerous synonyms for some of the species. The most recent reclassification in the eighth edition of *Bergey's Manual,* 1974, lists five species of *Moraxella: M. lacunata* (the diplobacillus of Morax and Axenfeld), *M. bovis, M. nonliquefaciens, M. phenylpyruvica,* and *M. osloensis.* The validity of these species, which are very similar phenotypically, has been confirmed genetically by transformation studies. *M. kingii* and *M. urethralis* (formerly *Mima polymorpha,* variety oxidans) have been tentatively grouped with the five valid species of *Moraxella. Acinetobacter* is considered in *Bergey's Manual* to consist of one species, *A. calcoaceticus.* This species includes both saccharolytic (formerly *Herellea vaginicola)* and nonsaccharolytic strains (formerly *Mima polymorpha,* variety nonoxidans).

Both genera are currently included in the family Neisseriaceae (see Chapter 29, Table 1). *Moraxella* is genetically and phenotypically related to the other members of this family, especially *Branhamella,* but *Acinetobacter* is a hardier, nutritionally versatile organism that shows no significant genetic homology with the other Neisseriaceae. Although the G + C content of both *Moraxella* and *Acinetobacter* is in the range of 40 to 47 per cent, DNA hybridization studies show very limited homology (Henriksen, 1976).

MORAXELLA

MORPHOLOGY

The chief reason for considering *Moraxella* and *Acinetobacter* as closely related bacteria is their microscopic morphology. In stained smears of wild organisms obtained directly from patients or the environment, *Moraxella* and *Acinetobacter* are short, plump bacteria that tend to occur in pairs and to be so coccoid that they may be confused with *Neisseria.* Careful examination almost always reveals some bacillary forms that increase in number with repeated passages in culture. *Moraxella* are usually fastidious and produce small colonies. Colonies from freshly isolated strains of *M. nonliquefaciens, M. bovis,* and *M. kingii* spread and corrode agar surfaces after incubation for several days in a humid environment. This type of growth is associated with polar fimbriae of about 5 nm diameter. Fimbriated cells show twitching motility that appears to cause the spreading phenomenon. Nonfimbriated bacteria neither corrode agar surfaces nor exhibit twitching motility (Henriksen, 1976). Because the twitching motility and corrosion are difficult to demonstrate by routine procedures, *Moraxella* are still considered nonmotile for purposes of differentiation from flagellated bacteria in the diagnostic laboratory. The polar fimbriae appear to determine competence in transformation.

Capsules may or may not be present. The cell walls of *Moraxella* have not been extensively studied.

ANTIGENIC COMPOSITION

Because *Moraxella* are not important pathogens, few studies of antigenic composition have been undertaken. Instead, the purpose of serologic studies has been to confirm taxonomic relationships. Studies of a collection of *Moraxella* species and a few Acinetobacter strains showed serologic relationships between *M. lacunata, M. bovis,* and *M. nonliquefaciens*. There was no cross-reactivity with the other *Moraxella* species or *Acinetobacter.*

METABOLISM

All *Moraxella* are relatively fastidious and grow best on rich media such as blood agar. In the case of *M. lacunata,* this fastidiousness is due in part to inhibition by fatty acids and other media components that can be neutralized by the addition of binding agents such as starch, charcoal, or serum. *Moraxella* are mesophilic and obligately aerobic. They grow poorly at room temperature and not at all at 5° C. They do not attack sugars. All *Moraxella* are oxidase positive when tested by *N,N*-dimethyl-*p*-phenylenediamine, a characteristic demonstrated by Baumann et al. (1968) to be associated with the presence of cytochrome C in the electron transport system. *Moraxella* species are chemoorganotrophs that utilize only a limited number of organic acids, alcohols, and amino acids as carbon and energy sources. Little is known about specific growth requirements. *M. osloensis,* but not the other species, will grow in mineral medium with acetate and ammonium salts. Only *M. bovis* produces a β-hemolysin. It is also the only species that does not produce detectable catalase activity.

Nitrogen can be obtained from reduction of nitrate by *M. lacunata, M. nonliquefaciens,* most strains of *M. phenylpyruvica,* and occasional strains of *M. osloensis. M. bovis* and *M. lacunata* are proteolytic and digest blood serum in Loeffler's medium. The other species are nonproteolytic. Only *M. phenylpyruvica* degrades urea. *M. osloensis* can utilize ammonium as the sole nitrogen source.

PATHOGENIC PROPERTIES

Moraxella are obligate parasites of the mucous membranes of animals. They rarely cause systemic disease even when defense mechanisms are impaired. In the last century, *M. lacunata* was described by Morax and Axenfeld as a cause of keratoconjunctivitis in man, but this infection has essentially disappeared. *M. bovis* causes epizootic keratoconjunctivitis in cattle. *M. caprae* and *M. equi,* which are indistinguishable bacteriologically from *M. lacunata,* are isolated from keratoconjunctivitis of goats and horses, respectively. *M. phenylpyruvica, M. osloensis,* and *M. nonliquefaciens* are part of the indigenous flora of mammals and are virtually nonpathogenic. *M. urethralis,* which includes some of the strains formerly called *Mima polymorpha* var. oxidans, is a rare cause of local and systemic infection in compromised patients.

Because *Moraxella* rarely cause disease, there have been few studies of possible pathogenic properties. Possession of fimbriae is a prerequisite for the ability of *M. bovis* to colonize mucous membranes and cause bovine disease. Fimbriae are probably also responsible for the adhesiveness of *M. nonliquefaciens* and *M. kingii,* but the nonfimbriated *M. osloensis* and *M. lacunata* also adhere readily to mucosal surfaces. Because *Moraxella* possess a typical gram-negative wall, they presumably contain LPS but its toxicity has not been studied.

IMMUNITY

Immune mechanisms do not prevent colonization of mucous membranes by *Moraxella.* If the limited pathogenicity of these bacteria were related to immunity against invasion stimulated by colonization, *Moraxella* should be a more common cause of infection in the immunocompromised patient. It seems likely, therefore, that *Moraxella* rarely cause infection because they lack pathogenic properties, not because they are limited to the mucous membranes by acquired immunity.

LABORATORY DIAGNOSIS

Moraxella should be suspected when the Gram stain reveals deep-staining, gram-negative diplococci accompanied by definite bacillary forms. Colonies are small and delicate, and some strains may grow only on agar enriched with blood or serum *(M. lacunata* and *M. bovis).* All species are oxidase positive, nonmotile by standard tests, indole negative, nonsaccharolytic, and penicillin sensitive. They are differentiated from *Acinetobacter* with 100 per cent accuracy by their oxidase positivity and penicillin sensitivity. They are more difficult to differentiate from several unnamed bacteria that resemble *Moraxella* phen-

TABLE 1. Diagnostic Characteristics of Moraxella and Acinetobacter

TEST OR SUBSTRATE	MORAXELLA *lacunata*	*bovis*	*nonlique-faciens*	*osloensis*	*phenyl-pyruvica*	ACINETOBACTER *var. anitratus (Herellea)*	*var. lwoffi (Mima)*
Enrichment[1]	[required]	±	0	0	0	0	0
β-Hemolysis	0	[+]	0	0	0	partial[2]	partial[2]
Oxidase	+	+	+	+	+	[0]	[0]
Catalase	+ (w)	0	+ (w)	+ (w)	+ (w)	+	+
CO_2 enhancement	+	0	0	0	0	0	0
MacConkey agar	NG	NG	NG	+ (w) or NG	+ (w) or NG	[+]	[+]
H_2S (Pb ac paper)	0 or + (w)	+ (w)	+ (w) or −	+ (w) or −	− or + (w)	0	0
Citrate	0	0	0	0	0	+ (90%)	+ (50%) or −
Nitrate reduction	+	0	+	− or + (28%)	+ (64%) or −	[0]	0
Phenylalanine deaminase	0	0	0	0	[+]	0	0
Urease (Christensen's)	0	0	0	0	[+ (95%)]	− or + (40–50%)	− or + (15%)
Acid from glucose	0	0	0	0	0	[+[3]]	0
xylose	0	0	0	0	0	[+]	0
lactose	0	0	0	0	0	[+]	0
OF media	−	−	−	−	−	[oxid.]	−
Penicillin sensitivity	S	S	S	S	S	[R]	[R]

[1]with blood or 3% rabbit serum

[2]Most strains of Acinetobacter cause indeterminate, partial hemolysis. About 12.5% of each variety produce complete hemolysis. These strains differ from other Acinetobacter in that they grow on SS agar, liquefy gelatin, and peptonize peptone milk.

[3]Reactions clearer in OF media than in liquid peptone media; (w) = weak; NG = no growth; 0 = negative; K = alkaline; N = no change; S = sensitive; R = resistant; TSI = triple sugar iron agar. [ ] = reaction of major importance in identification.

otypically (M-3, M-4, M-4f, M-5, and M-6). Careful attention to yellow pigment production (M-4f, M-5, and M-6), and alkalinization of citrate media (M-4) will usually separate these unusual isolates. *Moraxella* are differentiated from one another and from *Acinetobacter* by the diagnostic characteristics shown in Table 1. *M. lacunata* is recognized by its requirement for enrichment with blood or serum and enhancement of growth by CO_2; *M. bovis* by its β-hemolysis and catalase negativity; and *M. phenylpyruvica* by production of urease and phenylalanine deaminase. *M. nonliquefaciens* can usually be separated from *M. osloensis* by failure to grow on MacConkey agar and by production of nitrate reductase. *M. osloensis* is a hardier organism than the other *Moraxella* and produces colonies that approach those of *Acinetobacter* in size and character. *M. phenylpyruvica* is a very fastidious organism that grows poorly even in complex media.

Both the species tentatively grouped with the *Moraxella* are easy to separate from the valid members of this genus. Except for oxidase positivity, the biochemical reactions of *M. urethralis* are identical to those of *A. lwoffi* (Mima). *M. kingii* is catalase negative and β-hemolytic, grows without enrichments, and makes acid from glucose and maltose on ascites agar medium.

DRUG SUSCEPTIBILITY

Moraxella species are uniformly sensitive to penicillin and most other antibiotics.

EPIDEMIOLOGY

Moraxella species are part of the indigenous flora of the mucosal surfaces of man and other warm-blooded animals. Colonization probably occurs after transmission by droplets or direct contact. Fimbriae are a prerequisite for colonization of mucous membranes by *M. bovis* and are probably important colonization factors for *M. nonliquefaciens* and *M. kingii.* The other *Moraxella* colonize successfully without fimbriae, however. *M nonliquefaciens* and *M. osloensis* colonize the upper respiratory tract, and *M. urethralis* the urogenital tract. These and the other species also inhabit other mammalian mucosal surfaces including the conjunctivae.

ACINETOBACTER

Acinetobacter calcoaceticus is the only species recognized in *Bergey's Manual,* but it is customary to divide the species into strains that oxidize carbohydrates and strains that fail to attack carbohydrates. The oxidative strains have been known under a variety of names, including *Herellea vaginicola, Moraxella glucidolytica, Bacterium anitratum,* and *Acinetobacter anitratus.* Synonyms for the nonoxidizing strains include *Moraxella lwoffi, Mima polymorpha* var. nonoxidans, *B5W*-organism, and *Acinetobacter lwoffi.*

MORPHOLOGY

The microscopic morphology of *Acinetobacter* strains is indistinguishable from that of *Moraxella.* The nonsaccharolytic members of *A. calcoaceticus* were presumably named *Mima* because they mimicked *Neisseria.* Larger cells, typical bacilli, and filamentous forms can be found in most cultures. *Acinetobacter* strains generally grow more vigorously than *Moraxella* and produce butyrous colonies resembling those of the Enterobacteriaceae.

Electron microscopy of thin sections of Acinetobacter has revealed a multilayer cell envelope typical of gram-negative bacilli. Chemical studies of the LPS and peptidoglycan confirm their similarity to the equivalent cell wall structures of Enterobacteriaceae. Capsules have been seen in electron micrographs and have been characterized chemically and serologically. Some strains of *Acinetobacter* display twitching motility that is dependent on the production of polar fimbriae. The relationship of genetic competence to fimbriae has not been studied.

ANTIGENIC COMPOSITION

Fluorescein-labeled antisera to capsular antigens have identified 28 serotypes of saccharolytic acinetobacters and suggested that there are different serotypes of nonsaccharolytic acinetobacters (Marcus et al., 1969). Juni (1978) has studied the capsules of two strains of *Acinetobacter* and shown that one was composed of L-rhamnose and D-glucose and the other of glucose and galactose. The capsule that was composed of rhamnose and glucose cross-reacted in precipitin studies with antisera to groups B and G streptococci and type XIII pneumococci.

The peptidoglycan of saccharolytic and nonsaccharolytic strains is similar and contains muramic acid, glucosamine, alanine, D-glutamic acid, and *m*,-diaminopimelic acid, a composition characteristic of peptidoglycans of chemotype I (Horisberger, 1977). Neither the LPS nor the polar fimbriae of *Acinetobacter* have been characterized antigenically.

METABOLISM

Acinetobacters are strictly aerobic chemoorganotrophs that are versatile in the utilization of organic compounds for carbon and energy sources. None has specific growth requirements and most can grow in a mineral medium containing a single organic carbon and energy source. Nitrate is not reduced to nitrite or other reduced compounds but ammonium and nitrate salts can be used as nitrogen sources. Urease-producing strains can substitute urea for these compounds.

Acinetobacters capable of using glucose as a source of carbon and energy degrade this compound exclusively via the Entner-Doudoroff pathway. Acinetobacters lack enzymes for the direct phosphorylation of hexoses, and the first step for entry into the Entner-Doudoroff pathway is the oxidation of glucose to gluconic acid (Juni, 1978).

Acinetobacters that produce acid from glucose, lactose, mannose, arabinose, and xylose produce a glucose dehydrogenase that oxidizes D-glucose to D-gluconolactone. Oxygen is the ultimate electron acceptor for the particulate form of this enzyme which is tightly complexed to cytochrome b (Hauge, 1960). Nonsaccharolytic acinetobacters lack glucose dehydrogenase (Juni, 1978).

All the enzymes of the tricarboxylic and glyoxylate cycles have been demonstrated in either cell-free extracts or ethanol-grown cells of *Acinetobacter*. Acinetobacters contain functional electron transport pathways consisting of cytochrome a_1, a_2, and b as well as flavin. All can be reduced with NADH or succinate and re-oxidized by oxygen. As expected for an oxidase negative organism, *Acinetobacter* does not contain cytochrome C.

Pathways used by this versatile organism for biosynthesis and for the degradation of aromatic and alicyclic compounds, hydrocarbons, and 2,3-butanediol are similar to those of the pseudomonads and have recently been extensively reviewed by Juni (1978).

Most strains grow well at mesophilic temperature ranges of 35 to 37° C but have an optimal temperature range of 30 to 32° C. Some strains grow well at 5° C. The optimal pH for growth is about 7.

PATHOGENIC PROPERTIES

Acinetobacter is common in the environment and has been isolated from a nearly unlimited variety of clinical specimens. In most cases *Acinetobacter* can be dismissed as a contaminating or colonizing strain, but it is occasionally isolated from patients in whom its pathogenic role is indisputable. When *Acinetobacter* is isolated in mixed culture, the clinical situation and the culture results must be carefully evaluated before the diagnosis of *Acinetobacter* infection is accepted.

Acinetobacter behaves essentially like any other gram-negative, aerobic, opportunistic pathogen. Because of its low pathogenicity, the organism rarely, if ever, causes infection in otherwise healthy individuals. The number of nosocomial infections by *Acinetobacter* has increased in recent years. Several outbreaks of nosocomial *Acinetobacter* septicemia have been reported, and indwelling intravenous catheters were suspected as the portal of entry of the organisms. Nosocomial *Acinetobacter* pneumonia most often affects patients who are severely debilitated and have undergone major surgery or trauma. The pneumonia may be necrotizing. Abscesses, pleural effusions, and septicemia are occasional complications. *Acinetobacter* urinary tract infection is usually associated with the presence of an indwelling bladder catheter. When postoperative infections of the surgical area have been observed, they usually involve complicated procedures in severely ill patients or orthopedic procedures with implantation of a prosthetic device (Glew et al., 1977).

Some strains of *Acinetobacter* produce fimbriae. The role of these structures in the usual habitat of this free-living bacterium is unknown. It is not clear whether fimbriae promote nosocomial colonization of severely ill, hospitalized patients who are colonized more frequently than healthy people.

Since *Acinetobacter* causes disease uncommonly, there have been few studies of possible pathogenic properties. The heteropolysaccharide capsules may prevent phagocytosis by the severely ill patient who may be deficient in opsonins or phagocytes.

Acinetobacter contains a typical trilaminar, gram-negative cell wall, and LPS has been isolated and partly characterized chemically. Biologic studies of the endotoxic potency of this LPS have not been reported.

IMMUNITY

Healthy people resist colonization and infection by *Acinetobacter*. Since severely ill, immunocompromised patients are frequently colonized and occasionally infected, it is likely that normal resistance mechanisms play a role in preventing *Acinetobacter* infection. The relative importance of natural versus acquired immunity is unknown. The heteropolysaccharide capsule is antigenic in animals and it is likely that both the *Acinetobacter* capsule and LPS stimulate antibody production in human beings. It is possible that natural resistance to infection is partly related to the known cross-reactions between *Acinetobacter* capsules and those of Streptococcus B and G and *S. pneumoniae* type XIII.

LABORATORY DIAGNOSIS

Although the microscopic morphology of *Acinetobacter* is indistinguishable from that of *Moraxella,* the colonial morphology is usually quite different. Acinetobacters produce larger, butyrous colonies on most standard laboratory media; *Moraxella* produce smaller, more fragile colonies that may require media enriched with blood or serum. Unlike *Moraxella,* acinetobacters are strongly catalase positive, oxidase negative, and resistant to penicillin. They are easily differentiated from *Pseudomonas* by the oxidase reaction and from Enterobacteriaceae by their failure to reduce nitrate and their lack of reactivity in triple sugar iron agar or Kligler's iron agar. Acinetobacters also give uniformly negative results in the following tests: indole, H_2S, decarboxylation of lysine, ornithine, and arginine, and deamination of phenylalanine. The remainder of the laboratory characteristics of the saccharolytic (var. anitratus) and the nonsaccharolytic (var. lwoffi) strains are given in Table 1.

ANTIBIOTIC SUSCEPTIBILITY

Acinetobacter is resistant to penicillin, ampicillin, the cephalosporins, and chloramphenicol. The susceptibility to tetracycline varies. Most strains are sensitive to the aminoglycosides, colistin, and carbenicillin. However, R-factor-mediated resistance to carbenicillin, kanamycin, and gentamicin is becoming increasingly prevalent.

EPIDEMIOLOGY

Acinetobacter is not part of the indigenous flora of man or animals, but it is commonly found in soil and water. The organism is occasionally recovered in specimens obtained from nonhospitalized patients, but it is much more prevalent in hospitalized patients. *Acinetobacter,* like many other aerobic gram-negative organisms, is a frequent contaminant of moist environments in the hospital, e.g., respirators, humidifiers, and sinks (Smith and Massanari, 1977). Patients acquire the organism either by direct contact with an environmental reservoir of the organism, or indirectly from members of the hospital staff who carry the organism on their hands. Nosocomial colonization by *Acinetobacter* may involve the respiratory tract, intestinal tract, urogenital tract, and the moist intertriginous areas of the skin. The severity of illness seems to be the most important risk factor for nosocomial colonization.

Fluorescein-labeled antisera to capsular antigens can be used as an epidemiologic marker for tracing the source of nosocomial infections.

References

Baumann, P., Doudoroff, M., and Stanier, R. Y.: Study of the Moraxella group. I. Genus Moraxella and the Neisseria catarrhalis group. J Bacteriol 95:58, 1968.

Buchanan, R. E., and Gibbons, N. E. (eds.): Bergey's Manual of Determinative Bacteriology. 8th ed. Baltimore, The Williams and Wilkins Company, 1974, pp. 426–438.

Glew, R. H., Moellering, R. C., and Kim, L. J.: Infections with Acinetobacter calcoaceticus (Herellea vaginicola): Clinical and laboratory studies. Medicine 56:79, 1977.

Hauge, J. G.: Purification and properties of glucose dehydrogenase and cytochrome b from Bacterium anitratum. Biochim Biophys Acta 45:250, 1960.

Henriksen, S. D.: Moraxella, Neisseria, Branhamella, and Acinetobacter. Annu Rev Microbiol 30:63, 1976.

Horisberger, M.: Structure of the peptidoglycans of Moraxella glucidolytica and Moraxella lwoffi grown on hydrocarbons. Arch Microbiol 112:297, 1977.

Juni, E.: Genetics and physiology of Acinetobacter. Annu Rev Microbiol 32:349, 1978.

Marcus, B. B., Samuels, S. B., Pittman, B., and Cherry, W. B.: A serologic study of Herellea vaginicola and its identification by immunofluorescent staining. Am J Clin Pathol 52:309, 1969.

Smith, P. W., and Massanari, M.: Room humidifiers as the source of Acinetobacter infections. JAMA 237:795, 1977.

Gram-Negative Rods

31 ENTEROBACTERIACEAE

Frits Ørskov, M.D., and Ida Ørskov, M.D.

Enterobacteriaceae are a family of five tribes (Table 1) of closely related, nonspore-forming gram-negative bacilli that ferment glucose and, except for some strains of *Erwinia,* reduce nitrates to nitrites. Many genera are important normal inhabitants of the intestinal flora of mammals. Others, like the typhoid bacillus, the Shiga bacillus, and the plague bacillus, are historically and medically among the most important bacterial pathogens encountered by man. The plague bacillus, *Yersinia pestis,* and the other *Yersinia* species are discussed more extensively in Chapter 37 for historical reasons and because they cause clinical syndromes similar to those caused by *Francisella* and *Pasteurella.*

MORPHOLOGY

Enterobacteriaceae are relatively short, 2 to 3 μ by 0.4 to 0.6 μ, rod-shaped, gram-negative bacteria. Most genera are so similar microscopically and colonially that they cannot be differentiated from one another reliably without the use of differential media or a battery of biochemical tests. Most Enterobacteriaceae are motile, and motility is always dependent upon the production of peritrichous flagella. Many strains produce common pili or fimbriae. Strains that can serve as donors of plasmid DNA during conjugation also produce sex pili. Several members of the family produce capsules or microcapsules that are composed of repeating units of acidic polysaccharides. The capsules of *Klebsiella* and the microcapsules that are more typical of *Escherichia coli* and some other Enterobacteriaceae can be recognized by electron-microcapsules of Enterobacteriaceae are referred glutination or capsular swelling. The capsules and microcapsules of Enterbacteriaceae are referred to as K antigens except for the microcapsule of *Salmonella typhosa,* which is called the Vi (virulence) antigen.

TABLE 1. Family Enterobacteriaceae

TRIBES	GENERA
I Escherichieae	Escherichia Edwardsiella Citrobacter Salmonella Shigella
II Klebsielleae	Klebsiella Enterobacter Hafnia Serratia
III Proteeae	Proteus Providence (*Proteus inconstans*)
IV Yersinieae[a]	Yersinia
V Erwinieae	Erwinia

[a]This tribe is discussed in depth in Chapter 37.

Enterobacteriaceae contain the typical, lipid-rich multilayered cell wall of gram-negative bacteria. The outer layer contains the lipopolysaccharide (LPS) that is responsible for many of the toxic and biologic properties of these organisms. LPS also contains complex oligosaccharide side chains that determine O antigen specificity. The peptidoglycan forms a distinct layer of the cell wall between the LPS and the cytoplasmic membrane but is neither as thick nor as complex as that of gram-positive bacteria. Enterobacteriaceae contain 5 to 10 per cent lipoprotein in the cell wall. This substance is intermixed with the LPS and covalently linked to the peptidoglycan. It does not form a distinct layer. Unlike gram-positive bacteria, Enterobacteriaceae do not produce teichoic acid.

Enterobacteriaceae grow on most common media and on media with concentrations of bile salts and aniline dyes that inhibit most gram-positive bacteria. Colonies are circular, convex, entire, and more or less glistening and mucoid, depending partly on the production of polysaccharide surface structures. Unencapsulated forms, which have also lost the hexose epimerases necessary for synthesis of the O antigenic polysaccharide side chains of the LPS, produce flat, granular, irregular colonies that are referred to as "rough." The highly motile *Proteus* spp. may move away from the original point of inoculum on solid media and produce the unique, poorly understood swarming phenomenon. Only *E. coli* frequently produce hemolysins on blood agar. Some strains of *Serratia* produce a nondiffusible red pigment called prodigiosin. *Enterobacter agglomerans* (classified by some taxonomists as the Herbicola group of *Erwinia*) produces yellow colonies. Other *Erwinia* strains that are infrequently isolated from man produce blue or pink pigments. The rest of the Enterobacteriaceae are unpigmented.

ANTIGENIC COMPOSITION

General

Most genera of Enterobacteriaceae can be subdivided into serotypes by antibody raised against their antigenic surface components. The O, K, and H antigens are the fundamental serotyping antigens because of the great variety of their chemical composition and their antigenic stability. The properties of these antigens are relatively stable because they are chromosomally determined. Serotyping is used to establish the precise identification of strains isolated during outbreaks and epidemics. As certain serotypes have special pathogenic properties, antigenic analysis may also be important diagnostically.

O Antigens

The O antigens that are the polysaccharide side chains of the LPS are the backbone of analytic antigenic systems. LPS is a highly complex molecule (see Chapter 6), which consists of three regions (Luderitz et al., 1971): (1) Lipid A, which is buried in the outer membrane of the cell wall, is responsible for many of the biologic properties of LPS. (2) The LPS core, which is linked to lipid A and expresses R (rough) specificity, is partly hidden in S (smooth) forms by the attachment to the core of O antigens. (3) The O-specific polysaccharides of the bacterial S forms are the chemical basis for the serologic classification of Enterobacteriaceae into hundreds of complex O antigen groups, e.g., *Salmonella* into 60 and *E. coli* into more than 160 such groups (Ørskov et al., 1977).

O antigenic analysis is carried out by bacterial agglutination.

K Antigens

K antigens are polysaccharides that usually contain acidic groups. When well developed, they form microscopically distinct capsules. K antigens are determined by the capsular swelling technique, in which the capsule is made visible by the addition of specific antiserum or by the development of precipitin lines when tested against specific antisera by countercurrent immunoelectrophoresis (CIE). Bacterial agglutination can also be used, but the results may be difficult to interpret because of the many other surface antigens (Ørskov and Ørskov, 1978). Capsules interfere with phagocytosis and serum bactericidal activity.

H Antigens

Flagella are protein organelles that determine the H antigen specificity of motile members of the family. Many *Salmonella* strains show phase variation, i.e., the H antigens of a single strain may occur in either one or both of two serologically different phases called Phase 1 and Phase 2. Serologic determination is performed by agglutination techniques.

Fimbrial or Pili Antigens

Most Enterobacteriaceae may develop fimbriae (or pili). These will usually be the so-called type 1 fimbriae. General serologic analysis of fimbrial antigens is not yet available.

The K88 and K99 antigens of *E. coli* are fimbrial, proteinaceous antigens that are important colonization factors of strains associated with diarrhea in young pigs and calves. They received the K label at a time when their true nature was not recognized and may, therefore, soon receive a new designation that will also be an appropriate term for other fimbrial antigens.

Serotypes

Serotyping schemes, such as the Kauffmann-White scheme (Kauffmann, 1966), are based on the systematic examination of the antigenic specificities and cross-reactions between the surface antigens. Complete analysis may require numerous cross-absorption and cross-reaction tests.

The table below shows the basic agglutination-absorption experiment used in Enterobacteriaceae serology. Two strains, A and B, react mutually in antisera A and B. By reciprocal absorption followed by agglutination, it is shown that both strains contain a common and a specific antigenic factor. The two absorbed sera are called factor sera. Strain A might be given the antigenic designation of 1,2 and Strain B the designation of 1,3, where 1 represents the common factor and 2 and 3 the specific factors.

	ANTISERUM A		ANTISERUM B	
	Unabsorbed	Absorbed by B	Unabsorbed	Absorbed by A
Antigen A	+	+	+	0
Antigen B	+	0	+	+

Some typical examples of serotypes are:

Salmonella 1,4,5,12:b:1,2 (*S. paratyphi* B), where 1,4,5,12 represents the O antigen and b:1,2 the two phases of the H antigen.

Klebsiella 01:K1, or just K1, because the 0 antigens of *Klebsiella* are not usually determined.

Escherichia coli O1:K1:H5 or 1:1:5.

Shigella II:1,3,4 (*Shigella flexneri* 2a), where both II and 1,3,4 stand for O antigen determinants, II for the so called type-specific O factor and 1,3,4 for the group factors.

Simplified serologic labels can be found in the literature, e.g., when *E. coli* O111 and O75 are decribed as serotypes, even though they are O groups. The serotype of strains belonging to these O groups could be O111:H2 or O75:K100:H5.

Special

Escherichia Coli

A recent list of all the antigenic test strains that have been published contains 164 O antigens, 103 K antigens, and 75 H antigens. Most of the K antigens are polysaccharides, but a few proteinaceous, fimbrial antigens, which confer adhesiveness on enterotoxigenic *E. coli*, are presently listed as K antigens. They may be established as a special group of antigens (F antigens). Since many O, K, and H antigens can be combined in different ways, the number of possible serotypes is very high. Fortunately, some are more common than others, both among the special serotypes associated with disease and among the flora of the normal colon. Consequently, most routine laboratories use only a limited number of antisera that correspond to certain "pathogenic" serotypes. More detailed serologic analysis should be carried out in specialized national or international centers.

Salmonella

Salmonella strains are serotyped by the Kauffmann-White scheme, which has divided this genus into more than 1500 serotypes (Kauffmann, 1966). The O and H antigens in this highly complex scheme were defined only after extensive cross-absorption and cross-agglutination tests. More than 65 O-antigenic groups have been described. Several strains contain more than one O antigen factor. This means that more than one determinant has been found on the O-specific side chain of the LPS molecule, e.g., O factors 1,2,12 in *Salmonella paratyphi* A. Lysogenization by converting bacteriophages can bring about modification of certain O antigens.

Only one K antigen, the Vi antigen, is important for *Salmonella* serotyping, and it is usually associated only with *S. typhi*.

H antigens, the flagellar antigens, can be found in two alternative phases in many *Salmonella* cultures. One is called "specific phase," or Phase 1, and the other "non-specific phase," or Phase 2. During subculture of the two types, the population of each gives rise to cells of the alternative phase. The Kauffmann-White scheme first divides *Salmonella* into O groups. Strains of the same O group are then subdivided according to their H antigen.

Fimbriae can interfere with serotyping by blocking antigenic sites of the other surface antigens.

Shigella

Shigella is closely related to *E. coli* phenotypically and genetically. When these groups were given taxonomic status, *Shigella* strains had been isolated primarily from patients with severe dysentery, while *E. coli* was considered to be a harmless, normal inhabitant of the mammalian intestinal tract. The fact that otherwise typical *Escherichia* strains may cause dysentery has underscored the similarities between the two groups. By agglutination with O antiserum and minor biochemical differences *Shigella* is subdivided into four subgroups: *S. dysenteriae, S. flexneri, S. boydii,* and *S. sonnei.* The first three of these can be subdivided into several serotypes based on further analysis of O antigens. Capsular polysaccharide antigens have not been demonstrated in *Shigella.* Fimbrial antigens have been described but are not used for serotyping.

Edwardsiella and *Citrobacter*, the other two genera of Tribe I, are usually not serotyped.

Klebsiella

This genus is characterized by a rich variety of K antigens but a small number of different O antigens. Thus, antigenic analysis of *Klebsiella* strains is based on testing the organism against about 80 antisera to different K types.

Enterobacter, Hafnia, Serratia, and *Proteus* are usually not serotyped.

Yersinia

Yersinia enterocolitica has attracted much interest in recent years because of its possible relationship to certain enteric diseases and to rheumatic conditions. Serotyping is based on O and H antigens.

METABOLISM

Enterobacteriaceae organisms are facultative anaerobes. They possess a respiratory electron transport system that enables them to grow aerobically at the expense of a rich variety of oxidizable organic compounds. Organic acids, amino acids, and carbohydrates are utilized by all members of the family, and some can use aromatic compounds. They are oxidase negative because their electron transport system does not include cytochrome C. All except one species of *Shigella* produce catalase. Under anaerobic conditions, growth becomes strictly dependent upon carbohydrate fermentation. Many monosaccharides, disaccharides, and polyalcohols can be fermented by members of this group. Polysaccharides are utilized less commonly, but some plant pathogens (*Erwinia*) do attack pectin. All, except some species of *Erwinia*, reduce nitrates to nitrites.

Enterobacteriaceae ferment sugars by the Embden-Meyerhof pathway (Chapter 3). The pyruvic acid produced by this fermentation can be catabolized through several different pathways. All Enterobacteriaceae cleave pyruvate to formic acid. This unique cleavage is not encountered in any other bacterial fermentation. Formic acid does not always accumulate because many Enterobacteriaceae synthesize formic hydrogenylase, which converts formic acid to hydrogen and carbon dioxide. The remainder of the pyruvic acid is broken down either by mixed-acid fermentation or by butylene glycol (butanediol) fermentation. Enterobacteriaceae that metabolize glucose by the mixed-acid fermentation produce succinic acid by one pathway, lactic acid by another, and acetic acid and ethanol by the third. Enterobacteriaceae that use the butylene glycol fermentation convert a major portion of the pyruvate to a different end product, 2,3-butanediol, which is formed by an additional independent pathway. Although bacteria that use the butylene glycol pathway also produce succinic, lactic, and acetic acids, the neutral end products (butanediol and ethanol) predominate, and the total amount of acid formed per mole of glucose is much less than in the mixed-acid fermentation. Two simple tests on cultures in glucose-peptone broth divide the Enterobacteriaceae into two groups according to which fermentation pathway they use. The methyl red test detects those that use the mixed-acid fermentation because they produce enough acid from glucose to convert this indicator to its red, acidic form. The Voges-Proskauer test detects the butylene-glycol fermentation because acetoin, an intermediate in this pathway, is oxidized to a diacetyl form at an alkaline pH and complexes with creatine in the media to form a pink compound. The mixed-acid pattern is typical of the Escherichieae and Yersinieae tribes, and the butylene glycol pattern is typical of the Klebsielleae tribe.

All Enterobacteriaceae produce formic acid, but only those that synthesize formic hydrogenylase produce gas. The production of butanediol also results in a net production of CO_2, but the CO_2 is very soluble in water and does not cause observable gas in the medium. Gas formation is a property of differential value in identification of Enterobacteriaceae, since it differentiates the gas formers of the genus *Escherichia* from *Salmonella typhosa* and *Shigella* spp. and the gas forming *Enterobacter* from *Serratia*.

The ability to ferment lactose, a disaccharide, depends on the possession of β-galactosidase and is a characteristic of considerable diagnostic importance in this family. Utilization of lactose also depends upon a specific galactoside permease, which facilitates the entry of lactose into the cell. Lactose fermentation is characteristic of *E. coli* and *Enterobacter* and is absent from *Shigella, Salmonella,* and *Proteus*. Some *Shigella* species produce β-galactosidase but cannot ferment lactose because they lack a permease.

PATHOGENIC PROPERTIES

General

The natural habitat of many Enterobacteriaceae is the mammalian intestine. Some, like *E. coli,* are found in the normal healthy intestine; others, like *Salmonella, Shigella,* and enteropathogenic *E. coli,* can be considered to be pathogenic.

Although many Enterobacteriaceae are part of the normal intestinal flora, they are also "opportunistic" pathogens that may cause fatal infections when the normal host defenses are insufficient, e.g., in newborns, in patients in terminal stages of disease or in patients receiving immunosuppressive therapy. Enterobacteriaceae are the most common causes of infection of the obstructed biliary and urinary tracts and, along with the intestinal anaerobes, the major cause of infection of the soiled peritoneum. Enterobacteriaceae, especially *E. coli* and *Klebsiella,* are now the most common cause of septicemia in hospitalized patients.

Endotoxin

All Enterobacteriaceae synthesize endotoxin. Lipid A seems to be responsible for most of the potent biologic activities of endotoxin. Endotoxin is extensively discussed in Chapter 6, but it is important to point out in this chapter that many of the symptoms of invasive disease caused by Enterobacteriaceae can be ascribed to endotoxin.

Among these effects are: (1) pyrogenicity: microgram doses of LPS given intravenously cause a diphasic rise of the temperature of humans and many experimental animals; (2) consumption of complement: the survival rate of patients with enterobacterial septicemia is inversely related to the complement level; (3) consumption of coagulation factors: hemorrhage and fibrin deposition occur in patients and are equivalent to the Shwartzman phenomenon in experimental animals; and (4) shock: injection of larger doses of LPS causes hypotension, irreversible shock, and death. Similar symptoms can be caused by the presence of large numbers of Enterobacteriaceae in the blood.

Enterotoxins

Enterotoxin-producing *E. coli* strains have been the subject of intensive investigation in recent years (Sack, 1975). These protein exotoxins, which are genetically determined by transferable plasmids, play a prominent role in the pathogenesis of diarrhea in animals and man. Enterotoxin-positive strains colonize but do not invade the small intestine. Toxin-producing strains of other members of the family Enterobacteriaceae have been reported recently. There at least two types of enterotoxins; one is called ST (thermostable toxin) and the other LT (thermolabile toxin). Diarrhea occurs as a result of water and electrolyte loss caused by enterotoxin-induced adenyl cyclase activity of the epithelial cells. LT (the thermolabile toxin) is immunogenic but ST is not. Enterotoxigenic *E. coli* also affects infant pigs and calves, causing great economic loss to farmers.

Colonization Factors

Many enterotoxigenic strains produce special, plasmid-determined fimbriae (or pili) that play an important pathogenic role as adhesive structures. Experimental evidence suggests that these structures make it possible for enterotoxin-producing *E. coli* to deliver their toxin close to the intestinal epithelial cells.

The best known colonization factor is the K88 antigen of the *E. coli* strains that cause diarrhea in piglets. Enterotoxin-producing strains are not virulent for piglets if they have lost the K88 antigen or if the pigs have been immunized with K88. Studies of the K99 antigen suggest that it plays a similar role in diarrhea of calves. Similar colonization factors have been identified in human enterotoxigenic strains. K88 and K99 are both mediated by transferable plasmids.

K Antigens

K antigens are acidic polysaccharides that probably promote pathogenicity primarily by interfering with phagocytosis.

E. coli K1 strains cause a very high percentage of neonatal meningitis without an equivalent increase in either gastrointestinal colonization or septicemia. The sialic acid-containing K1 capsule cross-reacts immunologically with the capsule of serogroup B meningococci (Robbins et al., 1975). The reasons for this unique predilection for the meninges are not known. K1, K2, K3 K5, K12, and K13 are the most common K types associated with infections of the urinary tract. Certain O types — O1, O2, O4, O6, O7, O8, O9, O11, O18, O22, O25, and O75 — are also more commonly associated with urinary tract and other extraintestinal infections. These O and K groups are commonly found in the intestine, and it is uncertain whether they are especially virulent or are just more prevalent.

Special

E. Coli Diarrhea. Three groups of special-*Escherichia* strains are associated with three types of intestinal disease (Table 2). The enterotoxigenic *E. coli* species (ETEC), which were discussed above, cause diarrhea in piglets and calves, traveler's diarrhea in man, and a cholera-like disease in the people living in the warm climates of developing countries.

Another group of special *Escherichia* serotypes (EIEC), which are related to certain serotypes of *Shigella,* is invasive into the epithelial cells of the human colon and can cause dysentery-like disease. These serotypes have also been associated with food poisoning.

A third group of special serotypes is associated with diarrhea in institutionalized infants in developed countries. These special serotypes, which are not commonly found in the normal intestine, have been called the enteropathogenic *E. coli* (EPEC). Because they do not invade the epithelial cells of the small intestine and do not produce readily detectable enterotoxins, there has been doubt about their pathogenic role. However, the close association of these highly specialized serotypes with severe outbreaks of diarrhea cannot be disputed. Only future research will reveal their actual role in the pathogenesis of infantile diarrhea.

Salmonella

More than 1500 different serotypes have been

TABLE 2. *Escherichia Coli* and Acute Diarrhea

	ENTEROPATHOGENIC EPEC	ENTEROTOXIGENIC ETEC	ENTEROINVASIVE EIEC
Pathogenic mechanism	Unknown	Enterotoxin (LT, ST)	Epithelial cell invasion
Age groups affected	Infants	Infants and adults	Adults and infants
Epidemiology	Sporadic cases and outbreaks	Sporadic cases and outbreaks	Sporadic cases and outbreaks
O groups associated	26, 55, 86, 111, 114, 119, 125, 126, 127, 128, 142	6, 8, 15, 25, 27, 78, 148, 159	28ac, 112ac, 124, 136, 143, 144, 152, 164

From Rowe, B.: *Escherichia coli* in acute diarrhoea. Lab-Lore 7:449, 1977.

described, all of which should be considered as potentially pathogenic to man. Several cause generalized infection in specific animals, such as *S. typhimurium* in mice and *S. typhi* in man. Most serotypes can cause enteritis in many different animals when large numbers of organisms are ingested.

Enteric Fever. In humans, enteric fever is caused by *S. typhi* or *S. paratyphi* A, B, and C, depending on the geographic location. Other *Salmonella* serotypes are occasionally isolated from cases of enteric fever. A relatively small oral inoculum of the strains of *Salmonella* that cause enteric fever can initiate disease. The organisms invade the mucosa and the mucosal macrophages of the small intestine, multiply in local lymph nodes, spread through the lymphatics to the bloodstream, and disseminate to many organs. *S. typhi* and *S. paratyphi* C (*S. hirschfeldii*) frequently produce the Vi antigen, which is composed of a simple polymer of *N*-acetyl galactosaminuronic acid. This substance is elaborated as a thin, outermost layer and is equivalent to K antigen.

Salmonella septicemia may be caused by some other serotypes, notably *S. choleraesuis*. This infection is characterized by high remittent fever and bacteremia usually without involvement of the intestinal tract. Focal suppuration, meningitis, osteomyelitis, pneumonia, and endocarditis may occur.

Enteritis (Food Poisoning). Most other *Salmonella* serotypes can cause enteritis in man when ingested in large numbers. Typically, these strains cause disease after ingestion of food in which the *Salmonella* strain has had the opportunity to multiply. The disease is often labeled food poisoning, but invasion and multiplication are probably a necessary part of the pathogenesis. The incubation period ranges from 8 hours to 3 days. The disease is usually limited to the local intestinal lymph nodes. Some serotypes, e.g., *S. typhimurium, S. enteritidis, S. heidelberg,* and *S. adelaide,* are more frequently associated with this disease, but almost all strains have caused enteritis.

Shigella

Shigella causes bacillary dysentery in man and higher apes. This enteritis is localized to the mucosa of the terminal ileum and colon. The bacteria invade the colonic epithelial cells, where they multiply and cause mucosal ulcerations. One of the serotypes, *Shigella dysenteriae* Type 1 (the Shiga bacillus), also produces a special exotoxin that was earlier described as a neurotoxin.

Yersinia

Yersinia pestis, the cause of plague, is discussed in Chapter 37. *Y. pseudotuberculosis* and *Y. enterocolitica* may cause dysentery, mesenteric adenitis and an interesting syndrome consisting of dysentery and inflammatory noninfectious arthritis. The pathogenesis of this disease is not known, but it appears to be more frequent in certain geographic areas, and perhaps in certain HLA types (Chapter 129).

Other Enterobacteriaceae

The other members of the family are primarily causes of opportunistic infections. Several reports have suggested an association of the *Klebsiella* group with enterotoxic diarrhea, but there is no current evidence to suggest that there is a regular association of the enterotoxin plasmid with any serotype of *Klebsiella*.

Proteus species are found regularly in the mammalian intestine. These urease-producing bacteria, which break down urea to ammonium, are so frequently associated with stones of the urinary tract that the isolation of *Proteus* from the urine alerts clinicians to this possibility.

IMMUNITY

The intestinal tract of the newborn mammal is colonized shortly after birth by several different groups of bacteria including the Enterobacteriaceae. Of the many bioserotypes or clones of *E. coli,* only a limited number are highly prevalent. Immune reactions take place, and low titers of anti-Escherichia antibodies develop against the more common serogroups. The *E. coli* flora undergoes changes with time, age, and diet and is also dependent on the bacteria that are continuously taken in by mouth. Thus, some of the so-called natural antibodies are determined by immune reactions to the gram-negative intestinal bacteria. Antigenic cross-reactions that occur between different species of Enterobacteriaceae may be of importance for protection against intestinal diseases, and similar cross-reactions also exist between Enterobacteriaceae and pathogenic organisms from quite different bacterial groups. Thus, immune reactions to normal intestinal *E. coli* may protect against the invasion of many different bacteria; for example, *Haemophilus influenzae* Type b, which has a capsule that is serologically similar to the capsular antigen K100 of *E. coli* (Robbins et al., 1975).

Escherichia Coli

People living in an area in which enterotoxigenic *E. coli* strains are widely distributed are more resistant to disease from this type of organism than visitors coming from areas with few enterotoxigenic strains. Experimental evidence suggests that this immunity may be mediated by antibody production either to the heat-labile toxin or to colonization factors such as pili or K antigens. Experimental vaccines against enterotoxigenic *E. coli* are being tried in animals and may eventually be available for human beings.

Salmonella

Enteric fever provides some protection against the infecting serotype. Vaccination with TAB vaccine against *S. typhi* and *S. paratyphi* A and B has been given for more than 50 years to travelers going to parts of the world in which these diseases are endemic. The vaccine provides some protection against exposure to smaller doses of *S. typhi*.

Shigella

People living in geographic areas in which *Shigella* species are endemic acquire some immunity to the disease. There are no effective vaccines, probably because most *Shigella* vaccines have not been able to increase the local intestinal defenses.

LABORATORY DIAGNOSIS

Enterobacteriaceae are easy to cultivate either directly from feces, urine, or blood or from swabs of feces held in Stuart's medium. In addition to blood agar or nutrient agar, specimens that may contain Enterobacteriaceae are plated onto media that are both selective and differential. These media incorporate bile salts (MacConkey's agar, deoxycholate agar) or dyes (eosin-methylene blue or bromthymol blue) to inhibit gram-positive bacteria, lactose as the sole fermentable substrate, and a pH indicator (e.g., methylene blue), so that colonies that make acid from lactose are colored. Strains that do not ferment lactose are able to grow on the peptone provided in the media but are colorless because they do not produce acid.

Salmonella and *Shigella* are very resistant to bile salts and certain dyes and can be selected by plating directly on media (e.g., Salmonella-Shigella agar, Hektoen agar) that partly inhibit *E. coli* and the other Enterobacteriaceae. Further selection is accomplished by culturing feces into selenite or tetrathionate broth for 18 to 24 hours before subculturing onto the selective solid media.

Final identification of Enterobacteriaceae is established by a battery of biochemical tests, agglutination with appropriate O, K (Vi), and H antisera, and occasionally by phage typing. A limited number of biochemical tests (Table 3) separates the five tribes. In practice, identification to the species level is usually accomplished easily in one step by selecting certain differential tests from the major biochemical characteristics of Enterobacteriaceae listed in Table 4. A typical battery of these tests, which will differentiate between the major species of the *Escherichia* and *Klebsiella* groups, is given in Table 5. Tables 6 and 7 highlight the limited number of tests necessary to separate the species of *Klebsiella* and *Enterobacter*, respectively. Most *Klebsiella* isolates are typical strains of *K. pneumoniae; K. ozaenae* and *K. rhinoscleromatis* are unusual isolates. *Enterobacter aerogenes* is the most commonly isolated *Enterobacter* species, but *E. cloacae* and *E. agglomerans* are isolated fairly frequently. *E. agglomerans (Erwinia herbicola)* is still classified with the *Erwinia* in the 8th edition of Bergey's Manual (1974), but many medical diagnostic laboratories have followed the suggestion of Ewing and Fife (1972) and classified it with *Enterobacter*.

There are three species or biotypes of *Citrobacter (Escherichia* tribe) that can be separated from one another by a limited number of biochemical tests (Table 8). *Citrobacter freundii* is by far

TABLE 3. Distinguishing Characteristics of the Five Primary Groups (Tribes)

	TRIBE I ESCHERICHIEAE	TRIBE II KLEBSIELLEAE	TRIBE III PROTEEAE	TRIBE IV YERSINIEAE	TRIBE V ERWINIEAE
Fermentation pattern	Mixed acid	2,3-Butanediol		Mixed acid	Mixed acid and 2,3-butanediol
M.R.	+	D	+	+	
V.P.	−	D	D	−	D
Phenylalanine deamination	−	−	+	−	D
Nitrate reduction	+	+	+	+	D
Urease	−	D	D	D	−
KCN, growth in	D	+	+	−	D
Optimal temp. for growth	37 C	37 C	37 C	30–37 C	27–30 C
G + C, %	50–53	52–59	39–42	45–47	50–58

From Buchanan, C. E., and Gibbons, N. E. (eds.): Bergey's Manual of Determinative Bacteriology, 8th ed. Baltimore, The Williams & Wilkins Company, 1974.
G + C = Guanine + cytosine
D = different reactions from different genera or species

TABLE 4. Main Biochemical Characters of Primary Groups I to IV

	GROUP I					GROUP II				GROUP III PROTEUS	GROUP IV YERSINIA
	Escherichia	*Edwardsiella*	*Citrobacter*	*Salmonella*	*Shigella*	*Klebsiella*	*Enterobacter*	*Hafnia*	*Serratia*		
Catalase	+	+	+	+	D[a]	+	+	+	+	+	+
Oxidase	−	−	−	−	−	−	−	−	−	−	−
β-Galactosidase	+	−	+	D	d	+	+	+	+	−	+
Gas from glucose at 37 C	+	+	+	+	−	d	+	+	d	D	−
KCN (growth on)	−	−	+	D	−	+	+	+	+	+	−
Mucate (acid)	+	−	+	D	−	d	d	−	−		
Nitrate reduced	+	+	+	+	+	+	+	+	+	+	+
G + C, moles %	50–51			50–53		52–56	52–59	52–57	53–59	39–42 (one species = 50)	45–47
Carbohydrates (acid from)											
Adonitol	−	−	−	−	−	d	+	−	d	D	D
Arabinose	+	−	+	+		+	+	+	−	−	+
Dulcitol	d	−	d	D	d	d	−	−	−	−	−
Esculin	d	−	d	−			D	−	−	d	D
Inositol	−	−	−	d	−	+	D	−	d	D	−
Lactose	+ or ×	−	+ or ×	D	D	D	+	−	−	−	−
Maltose	+	+	+	+		+	+	+	+	D	+
Mannitol	+	−	+	+	D	+	+	+	+	D	+
Salicin	d	−	d	−	−	+	+	−	+	d	D
Sorbitol	+	−	+	+		+	+	−		−	D
Sucrose	d	−	d	−	D	+	+	− or ×	+	D	D
Trehalose	+	−	+	+		+	+	+	+	d	+
Xylose	d	−	+	+	D	+	+	+	d	D	D
Related C sources											
Citrate	−	−	+	+	−	d	+	+	+		−
Gluconate		−		−	−	+	+	+	+		
Malonate	−	−	d	D	−	D	+		−		−
d-Tartrate	d	−	+	D		d	−	−			D
M.R.	+	+	+	+	+	D	−	−	−	+	+
V.P.	−	−	−	−	−	D	+	+	D	d	−
Protein reactions											
Arginine	d	−	d	+	−	−	D	−	−	−	−
Gelatin hydrolysis	−	−	−	D	−	(d)	(+)	−	+	D	−
H_2S from TSI	−	+	D	+	−	−	−	−	−	D	D
Indole	+	+	D	−	D	d	−	−	−	D	D
Lysine decarboxylated	+	+	−	+	−	d	D	+	+	d	−
Ornithine	d	+	d	+	d	−	+	+	+	D	D
Urea hydrolyzed	−	−	(+)	−	−	d	(d)	−	−	D	D
Glutamic acid	−	−	−	−	−	−				+	
Phenylalanine	−	−	−	−	−	−	−	−	−	+	−

[a]D = different reactions given by different species of a genus; d = different reactions given by different strains of a species or serotype; × = late and irregularly positive (mutative).

From Buchanan, C. E., and Gibbons, N. E. (eds.): Bergey's Manual of Determinative Bacteriology, 8th ed. Baltimore, The Williams & Wilkins Company, 1974.

TABLE 5. Differentiation of Major Genera of Escherichieae and Klebsielleae

	E. COLI	EDWARD-SIELLA	SHIGELLA	SAL-MONELLA	CITRO-BACTER	KLEBSIELLA PNEU-MONIAE	ENTERO-BACTER AEROGENES	HAFNIA	SERRATIA MAR-CESCENS
KIA	A/AG	K/AG	K/A	K/AG	K or A/AG	A/AG	A/AG	K/AG	K/A or AG
H_2S (KIA)	−	+	−	+	+ or −	−	−	−	−
Indole	+	+	− or +	−	−	− or +	−	−	−
Methyl red	+	+	+	+	+	−	−	+ or −	− or +
Voges-Proskauer	−	−	−	−	−	+	+	+ or −	+
Citrate (Simmons)	−	−	−	d	+	+	+	(+) or −	+
Motility	+	+	−	+	+	−	+	+	+
Urease	−	−	−	−	−	+	−	−	−
KCN	−	−	−	−	+ or −	+	+	+	+
Lysine decarboxylase	d	+	−	+	−	+	+	+	+
Ornithine decarboxylase	d	+	d	+	d	−	+	+	+
Gas from glucose	+	+	−	+	+	+	+	+	− or +
Lactose	+	−	−	−	d	+	+	− or (+)	− or (+)
Salicin	d	−	−	−	d	+	+	d	+
Sorbitol	+	−	d	+	+	+	+	−	+
Raffinose	d	−	d	−	d	+	+	−	−
Rhamnose	d	−	d	+	+	+	+	+	−
Gelatin liquefaction (22°C)	−	−	−	−	−	−	− or (+)	−	+

+ = 90% or more positive in 1 to 2 days; − = 90% or more negative; d = different biochemical types; (+) delayed positive; − or + = majority of cultures negative; + or − = majority positive; KIA = Kligler's iron agar; K = alkaline or no change; A = acid; G = gas.
Modified from Edwards, P. R., and Ewing, W. H.: Identification of Enterobacteriaceae, 3rd ed. Minneapolis, Burgess Publishing Co., 1972.

the most common isolate. Most laboratories do not recognize *Citrobacter intermedius* (not listed in Table 8 — see 8th ed. Bergey's Manual of Determinative Bacteriology) and regard strains that are either H_2S negative, indole positive, or ornithine-positive as variants of *C. freundii*. *C. diversus* is the biotype of Citrobacter that combines these 3 divergent characteristics with the inability to grow in KCN and inability to decarboxylate lysine.

The biochemical reactions used to differentiate between *Salmonella* and *Shigella* species are listed in Tables 9 and 10. *Shigella* can be differentiated to the species level by a combination of the indicated biochemical tests and simple agglutination tests with four O antisera. Speciation of *Salmonella* is an entirely different matter. The Salmonella species *S. typhi*, *S. paratyphi* A, *S. arizona* (formerly Arizona), and *S. choleraesuis* can be identified biochemically but have distinct serotypes. More than 1500 other bioserotypes are now grouped together as *S. enteritidis* and differentiated mainly by serologic methods (see below).

Biochemical differentiation of the *Proteus* tribe is usually easily accomplished by the tests listed in Table 11. The Providencia species have been renamed *Proteus inconstans*, which appropriately points out their close relationship to the other *Proteus* spp.

Table 12 lists the major differential charac-

TABLE 6. Differentiation of Klebsiella Species

	K. PNEU-MONIAE	K. OZAENAE	K. RHINO-SCLEROMATIS
Urease	+	− or +	−
Methyl red	− or +	+	+
Voges-Proskauer	+	−	−
Citrate (Simmons)	+	− or +	−
Malonate	+	−	+ or −
Lysine decarboxylase	+	− or +	−
Gas from glucose	+	+ or −	−
Lactose	+	(+) or −	(+) or −

+ = 90% or more positive in 1 to 2 days; − = 90% or more negative; − or + = majority of cultures negative; + or − = majority positive; (+) = delayed positive.
Modified from 8th ed., Bergey's Manual of Determinative Bacteriology, 1974; and Lannette et al., Manual of Clinical Microbiology, 2nd ed., Am Soc Microbiol, Wash., D.C., 1974. Some bacteriologists recognize the oxytoca variant of *K. pneumoniae* as a separate species. These strains differ from other Klebsiella because they produce indole, liquefy gelatin, and produce a brown pigment when grown on defined medium containing gluconate and ferric citrate (Korth et al., 1969).

TABLE 7. Differentiation of Enterobacter Species

	E. AEROGENES	E. CLOACAE	E. AGGLO-MERANS
Urease	−	+ or −	− or +
Lysine decarboxylase	+	−	−
Ornithine decarboxylase	+	+	−
Adonitol	+	− or +	−
Inositol (acid)	+	− or +	− or +
Inositol (gas)	+	−	
Esculin	+	− or +	d
Yellow pigment	−	−	+ or −

+ = 90% or more positive in 1 to 2 days; − = 90% or more negative; − or + = majority of cultures negative; + or − = majority positive; d = different biochemical types.
Modified from Edwards, P. R., and Ewing, W. H.: Identification of Enterobacteriaceae. 3rd ed., Minneapolis, Burgess Publishing Co.; and Ewing, W. H., and Fife, M. A.: *Enterobacter agglomerans* (Baijerinck) Comb. Nov. (The Herbicola-hathyri Bacteria.) Int J Syst Bacteriol *22*(1):4–11, 1972.

TABLE 8. Differentiation of Citrobacter Species

	C. FREUNDII	C. DIVERSUS
Citrate (Simmon's)	+	+
Indole	−	+
H_2S (KIA)	+	−
Ornithine decarboxylase	− or +	+
Lysine decarboxylase	−	−
KCN	+	−

+ = 90% or more positive in 1 to 2 days; − = 90% or more negative; d = different biochemical types; − or + = majority negative; KIA = Kligler's iron agar.

Modified from 8th ed. Bergey's Manual of Determinative Bacteriology, 1974; and Ewing, W. H.: Differentiation of Enterobacteriaceae by Biochemical Reactions, Revised. DHEW Publication No. (CDC) 74-8270, May, 1974.

teristics of the *Yersinia* species, which are discussed more extensively in Chapter 37. Although *Yersinia* has recently been reclassified in the family Enterobacteriaceae, it is discussed separately for historic reasons and because it causes clinical syndromes similar to those of *Francisella* and *Pasteurella*.

Erwinia, another recent addition to the family of Enterobacteriaceae, also cause classification problems. These organisms are plant pathogens or commensals and are isolated infrequently from human sources except for the Herbicola group *(Enterobacter agglomerans)*. *Erwinia* are usually divided into three groups: the Amylovora group (sometimes referred to as "true *Erwinia*"), the Herbicola group (*E. agglomerans*), and the Carotovora group (pectobacteria). The Amylovora group is frequently nitrate-negative, attacks glucose primarily oxidatively, fails to grow at 37°C, and is isolated very infrequently from humans. Since this group fits the definition of Enterobacteriaceae poorly and is rarely isolated from humans, we will not deal with it further in this chapter. *E. agglomerans* (Herbicola group) is characterized in Table 7. The Carotovora group (pectobacteria) is occasionally isolated from man and is biochemically similar to the *Klebsiella* tribe but also grows better at room temperature than at 37° C; reactions are given at both temperatures.

As indicated in the tables, most Enterobacteriaceae can be differentiated to the species level by biochemical tests. Serologic methods are used most frequently for the identification of *Shigella, Salmonella,* and certain serotypes of *E. coli*. Although *Shigella* can frequently be speciated by the tests shown in Table 10, the identification is always confirmed by agglutination with O antisera. Shigellae have no special capsular antigens and no H antigens since they are nonmotile. Salmonellae are speciated by the Kaufmann and White scheme described previously under Antigenic Composition. Most laboratories use only O and Vi antisera for preliminary identification of salmonellae and send their isolates to reference laboratories for final identification. If *S. typhi* is suspected in the clinical laboratory, it is important to attempt agglutination with Vi antisera, because the Vi antigen may block access to the O antigenic sites by O antisera. Blocking of O agglutination by either Vi or H antigens may be removed by boiling. *S. typhi* should be suspected when a *Salmonella*-like organism (Table 10) fails to utilize citrate as the sole carbon source, produces no gas from glucose, produces little or no H_2S, and fails to decarboxylate ornithine. A *Salmonella* with these biochemical reactions that agglutinates with Vi antisera before boiling and with O antisera after boiling should be identified as *S. typhi*. A *Salmonella* that agglutinates with C_1 antisera and has typical *Salmonella* reactions, except for failure to produce H_2S, is identified as *S. choleraesuis*. All other salmonellae are customarily identified as *S. enteritidis*, Group — according to the group of O antisera with which they agglutinate, and are sent to reference laboratories for final identification.

Enteropathogenic E. Coli

The *E. coli* bioserotypes associated with diarrhea constitute a special problem because they cannot be differentiated from other *E. coli* on existing media.

Enterotoxigenic *E. coli* isolated from diseases of pigs are usually hemolytic, so that only hemolytic colonies are tested with agglutinating antisera against the serotypes that cause piglet diarrhea. No similar characteristic trait is known for strains that are enteropathogenic to human beings. Recent studies suggest that some serotypes harbor enterotoxin plasmids more frequently than others. This finding may simplify the screening for enterotoxigenic strains, but it will not represent a definitive test.

The laboratory identification of enterotoxigenic strains is very time-consuming and is ordinarily a research procedure. Single colonies of each specimen are tested for LT (heat-labile) enterotoxin by determining their influence on the morphology of adrenal (Y1) or Chinese hamster ovary cells (CHO) in tissue culture. ST (heat-stable) enterotoxin is detected in infant mice by peroral or intragastric injections of suspected colonies followed by measurement of fluid accumulation (weight gain) of the total intestinal tract. Both toxins can be detected by fluid accumulation in the "rabbit ileal loop," which is even more forbidding as a routine test. The recent isolation in several laboratories of pure LT has brought simple serologic tests for LT within reach. A similar approach for ST will not be possible because it is nonimmunogenic.

TABLE 9. Differentiation of Major Species of Salmonella

	S. TYPHI	S. CHOLERAESUIS[a]	S. ENTERITIDIS	S. PARATYPHI A	S. ARIZONAE
H_2S (KIA or TSI)	$+^w$	d	+	− or +	+
Citrate (Simmons)	−	(+)	+	− or (+)	+
Lysine decarboxylase	+	+	+	+	+
Ornithine decarboxylase	−	+	+	+	+
Gas from glucose	−	+	+	+	+
Dulcitol	− or(+)	− or +	+	+	−
Trehalose	+	−	+	+	
Arabinose	−	−	+	+	+
Rhamnose	−	+	+		+
Malonate	−	−	−	−	+
O group	D	C_1	All groups	A	Arizona[b]

[a] H_2S positive strains are referred to as *S. choleraesuis* var. *Kunzendorf.*

Salmonella are now named either *S. typhi, S. choleraesuis,* or *S. enteritidis,* serogroup—. *S. paratyphi* A and *S. arizonae* are included in this table because they can be identified biochemically and because *S. arizonae* was considered a separate species until recently.

Precise identification is made in reference centers by extensive cross agglutination tests and occasionally by phage typing.

+ = 90% or more positive in 1 to 2 days; − = 90% or more negative; d = different biochemical types; (+) = delayed positive; − or + = majority of cultures negative; + or − = majority positive; A = acid, KIA = Kligler's iron agar; TSI = triple sugar iron agar.

[b] *S. arizona* is agglutinated by commercial Arizona antisera and by some specific salmonella antisera.

Modified from 8th ed. Bergey's Manual of Determinative Bacteriology, 1974; and Edwards, P. R., and Ewing, W. H.: Identification of Enterobacteriaceae, 3rd ed., Minneapolis, Burgess Publishing Co., 1972.

Invasive *E. coli* can be detected by the capacity of a few drops of an overnight culture to cause keratoconjunctivitis in guinea pigs or rabbits. The so-called enteropathogenic serotypes (EPEC) from infantile diarrhea can be detected by screening colonies with appropriate antisera.

DRUG SENSITIVITY

Except for benzyl penicillin, Enterobacteriaceae are intrinsically sensitive to achievable serum and tissue concentrations of most antimicrobials when they are first introduced. Resistant strains are selected quickly, however, because the Enterobacteriaceae in the normal intestinal tract are frequently exposed to antibiotics directed at other, more invasive bacteria. Furthermore, some genera of Enterobacteriaceae are intrinsically resistant to certain antimicrobials. Some examples of this property include the resistance of *Proteus* spp. and *Serratia* to polymyxins, the resistance of *Proteus* spp. to tetracycline, and the resistance of *Enterobacter* to cephalothin. *Klebsiella* is usually resistant to aminobenzylpenicillin (ampicillin) but sensitive to the cephalosporins. *Proteus mirabilis* is unique among the Enterobacteriaceae species for its sensitivity to benzylpenicillin. Rapid selection of resistant mutants may occur when Enterobacteriaceae are exposed to nalidixic acid, rifampin, or the aminoglycosides as single antimicrobial therapy. All Enterobacteriaceae are intrinsically resistant to achievable levels of erythromycin and the other macrolides.

Most of the genetic determinants of resistance, however, are carried on transferable plasmids that have been responsible for the appearance of

TABLE 10. Differentiation of Shigella Species

	S. DYSENTERIAE	S. FLEXNERI	S. BOYDII	S. SONNEI
Indole	d	d	d	−
Lactose (acid)	−	−	−	(+)
Mannitol	−	d	+	+
Sucrose	−	−	−	(+)
Dulcitol	d	d	d	−
Xylose	d	d	d	−
Ornithine decarboxylase	−	−	−	+
Rhamnose	d	d	d	+
Raffinose	−	d	−	d
O groups	A	B	C	D

+ = 90% or more positive in 1 to 2 days; − = 90% or more negative; d = different biochemical types; (+) = delayed positive.

Modified from 8th ed. Bergey's Manual of Determinative Bacteriology, 1974; and Edwards, P. R., and Ewing, W. H.: Identification of Enterobacteriaceae, 3rd ed., Minneapolis, Burgess Publishing Co., 1972.

TABLE 11. Differentiation of Proteus Species[a]

	P. VULGARIS	P. MIRABILIS	P. MORGANII	P. RETTGERI	P. INCONSTANS[a] A	B
Phenylalanine deaminase	+	+	+	+	+	+
Urease	+	+ or (+)	+	+	–	–
Indole	+	–	+	+	+	+
H_2S (KIA or TSI)	+	+	–	–	–	–
Lysine decarboxylase	–	–	–	–	–	–
Ornithine decarboxylase	–	+	+	–	–	–
Moles % G + C	39	39	50	39	41	41
Gas from glucose	+	+	+ or –	– or +	+ or –	–
Esculin	d	–	–	+	–	–
Inositol	–	–	–	+	–	+
Maltose	+	–	–	–	–	–
Mannitol	–	–	–	+ or –	–	– or +
Mannose	–	–	+	+	+	+
Rhamnose	–	–	–	+ or –	–	–
Adonitol	–	–	–	+ or –	+	–
Spontaneous swarming	+	+	–	–	–	–
Penicillin sensitivity	–	+	–	–	–	–

[a]*P. inconstans* subgroups A and B (8th ed., Bergey's Manual of Determinative Bacteriology) were formerly known as *Providencia alcalifaciens* and *Providencia stuartii*.
+ = 90% or more positive in 1 to 2 days; – = 90% or more negative; d = different biochemical types; (+) = delayed positive; – or + = majority of cultures negative; + or – = majority positive.
Modified from Edwards, P. R., and Ewing, W. H.: Identification of Enterobacteriaceae. 3rd ed. Minneapolis, Burgess Publishing Company, 1972, and 8th ed., Bergey's Manual of Determinative Bacteriology, 1974.

an increasing number of multiresistant Enterobacteriaceae. The use of antibiotics for treatment of disease and as food additives in animal breeding has exerted a powerful selection pressure for antibiotic-resistant strains. When the use of antibiotics is restricted, the number of resistant strains in that area decreases (Falkow, 1975).

Thus, the antimicrobial sensitivities of individual species of this important family of bacteria are unpredictable. They vary from broad sensitivity to complete resistance to the tetracyclines, aminoglycosides, the sulfonamides, cephalosporins, polymyxins, chloramphenicol, ampicillin, and carbenicillin. For this reason, antimicrobial sensitivity tests must be conducted quickly and accurately to ensure effective therapy against enterobacterial infections.

EPIDEMIOLOGY

Escherichia Coli

The source of infection with the three types of *E. coli* diarrhea described above is human. Few cases exist that point to direct transfer of pathogenic strains from domestic animals to human beings. The enteropathogenic types (EPEC), are transferred nosocomially in nurseries and wards directly or indirectly from one child to another. The ultimate source of human enterotoxigenic strains (ETEC) is probably the infected patient, but transmission is indirect, possibly through contaminated food. The finding of the same highly defined serofermentative types among enterotoxigenic strains isolated all over the world indicates that the same clones of bacteria have spread around the world. Enterotoxigenic piglet diarrhea strains spread from country to country within a short period of time and demonstrated how easily special pathogenic strains can be globally disseminated.

The epidemiologic characteristics of *E. coli* strains from dysentery-like disease (EIEC) are the same as those of *Shigella* (see below).

TABLE 12. Differentiation of Yersinia Species

	Y. PESTIS	Y. PSEUDO-TUBERCULOSIS	Y. ENTERO-COLITICA
Motility			
22°C	–	+	+
37°C	–	–	–
Urease	–	+	+
Esculin	+	+	– or +
Rhamnose	– or +	+	–
Salicin	+ or –	+	–
Sucrose	–	–	+
Ornithine decarboxylase	–	–	+
Indole	–	–	+ or –

+ = 90% or more positive in 1 to 2 days; – = 90% or more negative; – or + = majority negative; + or – = majority positive.
Modified from 8th ed., Bergey's Manual of Determinative Bacteriology; and Lennette et al., Manual of Clinical Microbiology, 2nd ed., Am Soc Microbiol, Wash., D.C., 1974.

Salmonella

Typhoid and paratyphoid bacteria are primarily human parasites, and the corresponding diseases originate from human sources, i.e., either sick patients or carriers. Some geographic differences occur, e.g., *S. paratyphi* C is found primarily in Eastern Europe and Asia, while *S. paratyphi* A is found in the Americas and India.

Typhoid fever in developing countries is often waterborne in contrast to paratyphoid fevers. Shellfish, especially oysters, are an important source of typhoid fever but are rarely the source of paratyphoid fever. Milk and other diary products are often vehicles for typhoid and paratyphoid infections. A small percentage of clinical cases of typhoid fever become chronic carriers and excrete the organism indefinitely. In developed countries, it is often possible to trace nonimported cases of typhoid fever to such carriers.

Salmonella enteritis, associated with many different serotypes, such as *S. typhimurium, S. newport, S. adelaide,* and others, is often described under the heading of food poisoning. It is true that large infectious doses are generally necessary to cause disease, but the pathogenesis includes inflammation of the mucosal tissues and often involves the local lymph nodes. Human *Salmonella* enteritis is associated with consumption of food contaminated either directly or indirectly by an infected animal, e.g., pig, fowl (eggs), cattle, and others. *S. typhimurium* is the most widespread of the serotypes causing enteritis, but the predominant *Salmonella* serotype varies in different geographic areas. Patterns of prevalence change with time as new serotypes are introduced and spread within certain geographic locations. The present close contact between countries and the extensive exchange of travelers and food products have also increased the exchange of pathogenic *Salmonella* serotypes. The same applies, of course, to other pathogenic (and nonpathogenic) Enterobacteriaceae.

Shigella

Human cases are always the source of infection, so the prevalence of bacterial dysentery is highly dependent on sanitary conditions. Bacillary dysentery is endemic in many developing countries. In countries with better conditions of public health, this disease is often confined to institutions or to outbreaks caused by contamination of dairy products. In the highest developed countries, shigellosis is rare and often imported. Chronic carriers are uncommon. *Shigella dysenteriae* Type 1 (the Shiga bacillus) causes the most severe disease.

PREVENTIVE MEASURES

Sanitary measures are all important to control enteric diseases caused by *Salmonella, Shigella,* pathogenic *E. coli,* and other Enterobacteriaceae, e.g., control of water, food, milk, and sewage disposal. Carriers (of *Salmonella*) must not be allowed to work as food handlers.

References

Buchanan, C. E., and Gibbons, N. E. (ed.): Bergey's Manual of Determinative Bacteriology. 8th ed. Baltimore, The Williams & Wilkins Company, 1974, pp. 290–340.

Edwards, P. R., and Ewing, W. H.: Identification of *Enterobacteriaceae*. 3rd ed. Minneapolis, Burgess Publishing Company, 1972.

Ewing, W. H.: Differentiation of Enterobacteriaceae by Biochemical Reactions, Revised. DHEW Publication No. (CDC) 74-8270, May, 1974.

Ewing, W. H., and Fife, M. A.: Enterobacter agglomerans (Beijernick) Comb. Nov. (the Herbicola-Lathyri bacteria). Int J. System Bacteriol 22(1):4, 1972.

Falkow, S.: Infectious Multiple Drug Resistance. London, Pion Limited, 1975.

Kauffmann, F.: The Bacteriology of *Enterobacteriaceae*. Copenhagen, E. Munksgaard, 1966.

Korth, H., Ørskov, I., and Pulverer, G.: Farbstoffbildende *Klebsiella*-Stamme. Zentrabl. Bacteriol Parasitenk Infectionskr Abt Orig I 211 (1):105–107, 1969.

Lennette, E. H., Spaulding, E. H., and Truant, J. P.: Manual of Clinical Microbiology, 2nd ed. American Society of Microbiology. Washington, D.C. 1974.

Lüderitz, O., Westphal, O., Staub, A. M., and Nikaido, H.: Isolation and chemical and immunological characterization of bacterial lipopolysaccharides. In Weinbaum, G., Kadis, S., and Ajl, S. J. (eds.): Microbial Toxins. Vol. 4. New York and London, Academic Press, 1971, pp. 145–233.

Ørskov, F., and Ørskov, I.: Serotyping of *Enterobacteriaceae* with special emphasis on K determination. In Norris, J. R., and Bergan, T. (eds.): Methods in Microbiology. Vol. 11. London and New York, Academic Press, 1978.

Ørskov, I., Ørskov, F., Jann, B., and Jann, K.: Serology, genetics and chemistry of O and K antigens of *Escherichia coli*. Bacterial Rev 41:667, 1977.

Robbins, J. B., Schneerson, R., Liu, T., Schiffer, M. S., Schiffman, G., Myerowitz, R. L., McCracken, G. H., Jr., Ørskov, I., and Ørskov, F.: Cross reacting bacterial antigens and immunity to disease caused by encapsulated bacteria. In Neter, E., and Milgrom, F. (eds.): The Immune System and Infectious Disease. Buffalo, Fourth International Convocation on Immunology, 1975, pp. 218–241.

Rowe, B.: *Escherichia coli* in acute diarrhoea. Lab-Lore 7:499, 1977.

Sack, R. B.: Human diarrhea disease caused by enterotoxigenic *Escherichia coli*. Ann Rev Microbiol 29:333, 1975.

VIBRIOS 32

Barun Deb Chatterjee, M.B.B.S., Ph.D., Dip. Bact.

Curved bacteria with polar flagellation are called *vibrios*. These chem-organotrophic bacteria are grouped as follows: I. Family Vibrionaceae (respiratory and fermentative metabolism) — (a) *Vibrio cholerae,* (b) Noncholera vibrios (NCV) or Nonagglutinating (NAG) vibrios, (c) *Aeromonas hydrophila,* (d) *Plesiomonas shigelloides;* and II. Family Spirillaceae (respiratory metabolism) — *Campylobacter fetus*. The earliest designation, *Vibrio gindha,* Pfeiffer 1896 (Chalmers and Waterfield, 1916),. is preferred to NCV or NAG vibrios, since these organisms produce cholera-like disease and agglutinate in homologous antisera (although not in anticholera serum). The taxonomic position of *Vibrio parahaemolyticus* is unsettled: being peritrichous, it is not a vibrio. Although related to different genera, as shown in Table 1, *V. parahaemolyticus* and *C. fetus* are provisionally described here until their taxonomy is clarified.

In the tropics, cholera vibrios are the principal but not the sole agents of acute dehydrating rice water diarrhea. Cholera syndrome may be related to a range of pathogens of widely different taxons — for example, NCV, *V. parahaemolyticus, A. hydrophila, Pl. shigelloides,* and enterotoxigenic *E. coli* (Chatterjee and Neogy, 1972).

TABLE 1. Alliance of *V. parahaemolyticus* to different Fermentative, Gram-Negative Bacteria

Characters	*V.parahaemolyticus*	*Pasteurella*	*Yersinia*	*Lucibacterium*	*Aeromonas*	*Vibrio* ‡
Cells : coccoid	+	+	+	−	−*	−
Capsule	+	+	−	−	+	−
Flagella : polar and peritrichous	+	−	−	+	−	−
Swarming	+	−	+	+	−	−
Growth in 0% NaCl conc.	−	+	+	−	+	+
Oxidase	+	+	−	+	+	+
β-galactosidase	−	−	+	?	+	+
Lysine decarboxylase	+	d	−	+	−	+
Zoonosis	+	+	+	−	+	−
Tissue invasion	+	+	+	?	+	−
Luminescence	−	−	−	+	−	−
GC ratio : 46	+	+†	+	+	−	−

\+ = more than 90% strains positive
− = more than 90% strains negative
*occasionally positive
† top range
‡ type species
d = 11-89% strains positive

VIBRIO CHOLERAE

(*vibrio,* which vibrates; *cholerae,* an intestinal disease)

MORPHOLOGY

Cells are slightly curved, gram negative, 0.5 by 1.5 to 3.0 μm, and occur singly (Fig. 1). Bacilli rarely appear straight, spherical, and, if joined, S- or C-shaped and as spirilla. The cholera vibrios are fimbriated and show fast linear motility by a long wavy polar flagellum (Fig. 2). They are essentially noncapsulated, but occasionally a cell may look encapsulated. In smears from mucus flakes of rice water stool, bacilli lie parallel to each other like "fish in a stream," presumably

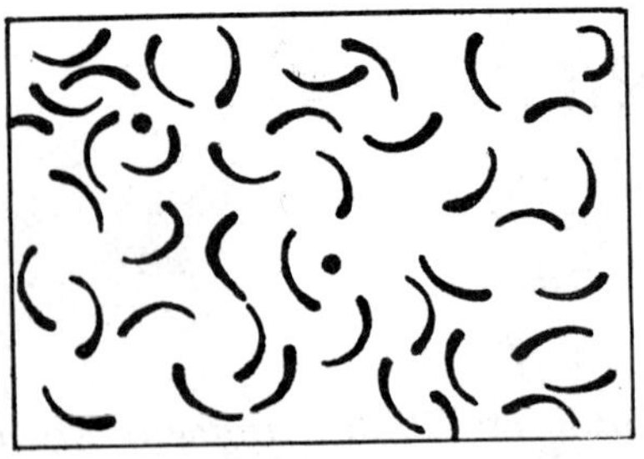

FIGURE 1. *Cell morphology of cholera vibrios.*

because of the degenerating spirilla forms. Cells are vividly stained by dilute carbol fuchsin.

On nutrient agar, the organism produces semitransparent, low convex, 2 to 3 mm round colonies that suspend uniformly in saline. In one week, the growth is overlaid by minute secondary colonies and develops needle-shaped crystals underneath. Colonies may undergo variations to (1) dwarf, (2) entirely opaque, (3) centrally opaque with smooth periphery, and (4) reversible rugose colonies.

After primary isolation in bile salt agar (BSA) and lactose teepol agar (LTA), the conspicuously transparent, flat colonies of cholera vibrios are readily differentiated from the opaque raised ones of coliforms. Under stereo-plate microscopy, the cholera vibrios on BSA exhibit greenish to red-bronze iridescence with fine granular transparency.

Colonies appear opaque-yellow on thiosulfate citrate bile salts-sucrose (TCBS) agar, translucent-yellow on sucrose teepol tellurite (STT) agar (Chatterjee et al., 1977), translucent bluish gray on thiosulfate citrate bile salts-sucrose lauryl sulfate agar (Vibrio agar), grayish translucent on lauryl sulfate tellurite agar (Cholera medium), and gelatin taurocholate tellurite (GTT) agar. While BSA and LTA inhibit gram-positive bacilli, others suppress coliforms as well, thus permitting vibrios to grow in pure form. The slide agglutination with antisera can be done from isolates of all media except TCBS and vibrio agar, for which, because of the stickiness, subculture on nutrient agar is necessary.

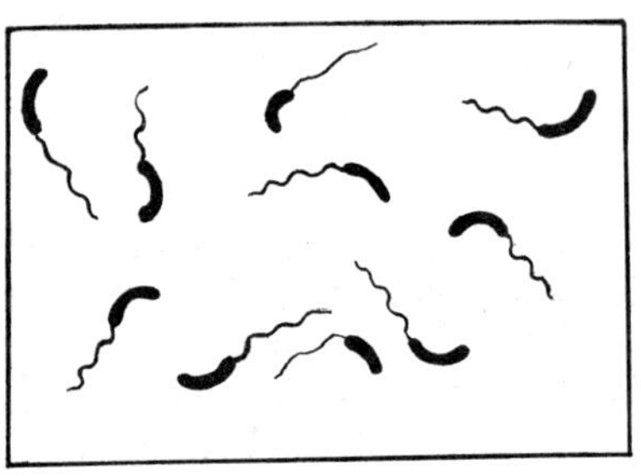

FIGURE 2. *Long, wavy polar flagella of cholera vibrios.*

ANTIGENIC COMPOSITION

Cholera vibrios possess heat labile flagellar (H) and heat stable somatic (O) antigens. The former is a protein, the latter a lipopolysaccharide. Three fractions of the O antigen: A, B, and C determine Ogawa (AB), Inaba (AC), Hikojima (ABC), and one unusual (A) serotypes. In vivo conversion from Ogawa to Inaba and the reverse is a rare possibility. The isolates from chronic carriers even if appearing smooth may lack O antigen because of S-R variation. Detection of such a variant is facilitated by the "rough" antiserum, made against O antigen-deficient strain. Minor proportions of cholera H antigen are present in certain strains of NCV and *S. enteritidis*. A common O antigenic factor being shared by cholera vibrios, brucella species, *Y. enterocolitica* serotype IX, and *C. terrigena* may confuse serologic diagnosis.

METABOLISM

Cholera vibrios are facultative anaerobes and can multiply at 16 to 42°C and pH 6.4 to 9.6. Optimum growth occurs aerobically, at 37°C, pH 8.0 to 8.2, in ordinary media and even on minimal synthetic medium containing sources of carbon, nitrogen, and salts. Higher pH (8.5 to 9.5) is beneficial but a lower one (pH 4.0) is lethal, so that strains grown in media containing glucose do not survive. In Table 2, the various distinctive characters of *V. cholerae* are presented. The positive oxidase, catalase, and nitrate reduction tests should be noted.

V. cholerae differs from *V. cholerae* biotype El Tor (El Tor vibrio, named for its discovery in El Tor lazeret of Sinai Peninsula). The *V. cholerae* (classic vibrio) is sensitive to 50 IU polymyxin B disk (or 15 μg/ml in nutrient agar) and 1 RTD group IV cholera phage, whereas the El Tor is resistant to these. Only El Tor produces acetoin and agglutinates with chicken, sheep, and human erythrocytes. Further, El Tor causes distinct opacity and a pellicle in nutrient broth, whereas the *V. cholerae* creates minimum turbidity and no pellicle. Until 1962, El Tor strains were hemolytic, but subsequent isolates have not been.

With four lytic phages and standard laboratory strains, classic vibrios exhibit five lytic patterns and El Tor is distributed into six types by five phages. However, phage typing provides limited information in tracing the source of infection, because in actual practice, wild strains of the classic and El Tor strains comprise only two phage types, and the El Tor phages are unstable. Strains of El Tor vibrio isolated from clinical cases from the year 1961 onward, almost invariably show ly-

TABLE 2. Distinguishing Features of Vibrio and Allied Pathogens (All single polar flagellated, rapidly motile, positive to catalase, oxidase and nitrate reduction tests)

Characters	*V.cholerae*	*NCV*	*A.hydrophila*	*Pl.shigell-oides*	*V.para-haemolyticus*	*C.fetus*
Cells:						
curved	+	+	−	−	−	+
coccoid	−	−*	−*	−	+	−
Flagella						
lophotrichous	−	−	+	+	+	−
peritrichous	−	−	−	−	+	−
Capsule	−	+	+	+	+	−
Motility: corkscrew-like	−	−	−	−	−	+
Facultative anaerobic	+	+	+	+	+	−
Microaerophilic	−	−	−	−	−	+
Growth:						
TCBS agar	+	+	−	−	+	NT
SS agar	−	−	+	+	−	−
0% NaCl in gelatin agar	+	+	+	+	−	NT
7% NaCl in gelatin agar	−	−	−	−	+	NT
KCN broth	d	d	+	−	−	NT
O-F test: fermentative	+	+	+	+	+	−
Gas from: glucose	−	−	d	−	−	−
Acid from						
glucose	+	+	+	+	+	−
mannitol	+	+	+	−	+	−
inositol	−	−	−	+	−	−
Current Heiberg groups	I	(I II V)	(I III IV V)	VI	VII	−
Acid from						
mannose	+	+ −	+ ++ − +	−	+	−
sucrose	+	+ +	− ++ + −	−	−	−
arabinose	−	− −	− −+ + −	−	+	−
β-galactosidase	+	+	+	+	−	NT
Arginine dihydrolase	−	−	+	+	−	−
Lysine decarboxylase	+	+	−	+	+	−
Ornithine decarboxylase	+	+	−	+	+	−
Methyl red	d	d	+	+	−	−
Voges-Proskauer	−	d	d	−	−	−
Indole	+	+	+	+	d(weak)	−
Citrate utilization	d	+	d	−	−	−
Hydrolysis						
gelatin	+	+	+	−	+	−
casein	+	+	+	−	+	−
Tween 80	+	+	?	−	+	−
starch	+	+	+	−	+	−
esculin	−	−	d	−	+	−
Hemolysis						
sheep cell suspension	−	d	d	−	−	−
0/129 sensitivity	+	+	−	+	+	−
Moles per cent GC	47	46–48	57–63	51	46	32–35
Mice: s.c. injection bacteremia	−	+	+	+	+	NT
Saprophytic existence	−	+	+	−	+	−

+ = more than 90% strains positive
d = 11–89% strains positive
− = more than 90% strains negative
NT = not tested
*occasionally positive
S.C. = subcutaneous

sogenization by a specific temperate phage called the "kappa" phage. Such strains are identified by their capacity to elicit turbid plaques on *V. cholerae* H 218. By detecting "kappa" phage lytic mutants in their feces, bacteriologically negative cases of El Tor cholera and their contacts can be identified. The *V. cholerae* is resistant to "kappa" phage; El Tor acquires susceptibility after eliminating the temperate phage.

PATHOGENIC PROPERTIES

Evidence suggests that low gastric acid predisposes to attacks of cholera, a finding in keeping with the susceptibility in vitro of *V. cholerae* to low pH. The infective dose in natural infections is probably less than that causing disease in volunteers (10^6), since the organism is short-lived in soil, impure water, saprophytic contaminants, and chlorinated water, and perishes when desiccated, gently heated, or exposed to sunlight.

After bypassing the acid gastric barrier, the organisms reach the favorable alkaline medium in the small bowel, where they anchor to the gut mucosa and colonize, and liberate enterotoxin to cause acute rice water diarrhea. The toxin, irreversibly fixed to the intestinal mucosa, stimulates adenyl cyclase, which in its turn builds up cAMP to trigger outpouring of isotonic fluid. Cholera toxin is a protein of molecular weight 84,000 and comprises synergistically acting toxin and toxoid moieties. Experimentally, the toxin produces diarrhea in nine day old rabbits and mongrel dogs. It causes dilatation and fluid accumulation in the ligated loops of adult rabbits, induces rounding of cells and steroidogenesis in clonal cell lines of Y-1 and OS-3 adrenal tumor cells in tissue culture, and elongation of Chinese hamster ovarian (CHO) cell line.

The toxin moiety consists of six light subunits (L) and one heavy subunit (H). While the former helps adherence to the receptor (ganglioside) of the cell membrane, the latter performs the toxic activity.

However, other metabolites of cholera vibrios, like vascular permeability factor, cytolysin, hemolysin, and mucinase do not appear to be related to the pathogenesis of cholera.

IMMUNITY

An attack of cholera confers short-lived immunity to reinfection. Within a few weeks after onset of the disease, there is rise of both serum antibodies (agglutinins [IgM], vibriocidal antibody [IgM], antitoxin [IgG]) and coproantibody (IgA), which persists for three months, but shortly afterward falls to low titer. It is well known that the IgM and IgG lose their normal functions within the gut lumen because of enzymatic inactivation, and because complement-mediated bactericidal activity as well as complement-dependent phagocytosis are not operative in the anticomplementary intestinal contents. However, it is believed that the antibacterial antibody prevents attachment of the organism to the gut mucosa, whereas the antitoxic immunity inhibits fixation of the toxin.

Likewise, the cholera vaccines, killed parenteral (with or without adjuvant) and live or killed oral varieties, without exception impart short-term (three to nine months) protection. The efficacy of vaccines is further limited by their failure to prevent epidemics, death, and inapparent cases and carriers.

LABORATORY DIAGNOSIS

Collection

Fecal samples are collected aseptically in sterile containers with soft rubber catheters (No. 22 or 24), previously lubricated in sterile liquid paraffin. Rectal swabs are not recommended because they often become dry if sampling is inadequate or if the interval between collection and inoculation is too long. If cotton swabs are used they should be soaked in Sorensen's buffer, pH 7.4, before autoclaving. A sample that cannot be inoculated soon after collection should be kept in Cary-Blair's transport medium, which is ideal for vibrios, shigella, salmonella, and *E. coli*.

Culture

Since a range of pathogens are expected in a rice water stool, a semiquantitative approach sheds light on the role of organisms, especially in mixed infections (Chatterjee and Neogy, 1972). Usually one of these agents is found in high numbers (10^6 to 10^9/ml), whereas others constitute a minor proportion (10^1 to 10^4/ml). A calibrated loop (0.01 ml; 4 mm internal diameter) is filled with neat, 1/10 and 1/100 dilutions of feces in sterile saline containing 0.1 per cent peptone, and streaked on selective media as for a urine culture.

Six hours of enrichment in alkaline peptone water (with 3 per cent NaCl), pH 9.2 enhances isolation of cholera vibrios, NCV, and *V. parahaemolyticus*. For the salmonella, aeromonas, and plesiomonas species, 18 hour enrichment in selenite F broth is recommended. Subculture on selective media after preliminary enrichment allows identification of pathogens.

Media

Colony counts of cholera vibrios, NCV, and *V. parahaemolyticus* are done from TCBS or STT agar. The SS agar provides similar information on salmonella, aeromonas, plesiomonas, and pseudomonas; the desoxycholate citrate agar (DCA) is primarily used for salmonella and shigella species, and the MacConkey agar is used for *E. coli.*

Identification

Oxidase-positive isolates on TCBS or STT agar that ferment sucrose are tested in cholera polyvalent and (if necessary) "rough" antisera. Cholera vibrios showing agglutination are divided into Inaba, Ogawa, or (rarely) Hikojima serotypes by specific sera and classic or El Tor biotypes after noting sensitivity to phage IV and polymyxin B (and/or hemagglutination in sheep cells). Contrarily, the sucrose fermenters not reacting with cholera sera are identified as NCV on the basis of decarboxylase-dihydrolase tests. Of the sucrose nonfermenting colonies, the Heiberg group VII strains, showing growth at 7 per cent but not in 0 per cent NaCl concentrations and a negative β-galactosidase reaction are identified as *V. parahaemolyticus* biotype 1; whereas those of Heiberg group V, lacking the halophilic property with a positive β-galactosidase test are regarded as NCV (Table 2). *V. parahaemolyticus* is subjected to serotyping and the Kanagawa test (see below). The oxidase-positive colonies of MacConkey agar and DCA are treated in a similar manner.

The nonlactose-fermenting, oxidase-positive colonies from SS agar are tested for their oxidation-fermentation (O-F) reaction: *C. terrigena* is negative, whereas *P. aeruginosa* is oxidative. The fermentative breakdown of glucose suggests aeromonas and plesiomonas species. They are identified from the results of mannitol and inositol fermentation and decarboxylase tests (Table 2, Fig. 3).

DRUG SUSCEPTIBILITY

The mean inhibitory concentrations (MIC) in μg/ml for cholera vibrios are as follows: tetracycline 2.5, chloramphenicol 1.5, streptomycin 25, paromomycin 25, viomycin 200, and neomycin 20 to 30. The cholera vibrios are also sensitive to disks of ampicillin (2 μg), sulfisoxazole (4 μg), and polymyxin B (100 IU). Only 2.5 per cent of the global strains are not susceptible to one of these agents. The resistance is chromosomal in origin and seldom, if ever, mediated by plasmids.

EPIDEMIOLOGY

From remote times, *V. cholerae,* the pathogen of Asiatic cholera had been endemic in the vicinity of the Ganges and Brahmaputra rivers in Eastern India and Bangladesh. Over the 19th and early part of the present century, a series of six pandemics originated from the endemic focus in which areas of Asia, Europe, Africa, and America were devastated. During 1923 to 1960, it produced seasonal epidemics within India and Bangladesh with occasional extensions to the neighboring countries.

In 1961, the biotype El Tor was implicated in the seventh pandemic, which, since 1937 had been causing bouts of mild cholera in the Sulawesi island of Indonesia. It spread to all states of Southeast Asia in 1961 to 1962, states of mainland Asia within 1963 to 1969, and, almost replaced classic vibrio. Lately, it also invaded Africa, Europe, and South USSR, and, to a minor extent, Canada and Australia. There have been cases of imported cholera in nonendemic areas owing to rapid air travel.

Under natural conditions, cholera vibrios produce attacks of diarrhea, inapparent cases, and carrier states in no other animals than man. Incubation period ranges from one to five days, and, the infectivity 5 to 14 days. The bacterium is transmitted from man to man by the fecal oral route through food and drink. It is particularly prevalent in overcrowded and unsanitary conditions with sharing of common water sources and latrines.

In endemic areas, such as India, Philippines, and Bangladesh, the outbreak occurs before, during, and after the monsoon. In the interepidemic periods, the endemicity continues through a sort of "silent epidemic," moving from the carriers to inapparent cases or vice versa.

In comparison to the classic biotype, El Tor proliferates faster, is more resistant to environmental conditions, and produces many more cases, carriers, and newer foci of infections. An El Tor epidemic runs a protracted course rather than the explosive pattern of the Asiatic cholera. The Inaba and Ogawa serotypes, respectively, predominate among the isolates of *V. cholerae* and *V. cholerae* biotype El Tor. Quite recently, the latter has revealed more strains of Inaba type.

Cost benefit analyses point to the superiority of hygiene and improved sanitation over cholera vaccines. Vaccination is economically justified only when the case incidence exceeds 8 per 1000 population, which is hardly seen in practice. Although the immunity is serotype specific, one biotype induces nearly equal protection against the other. However, vaccination is indicated in persons moving from a cholera-free area to the endemic zone.

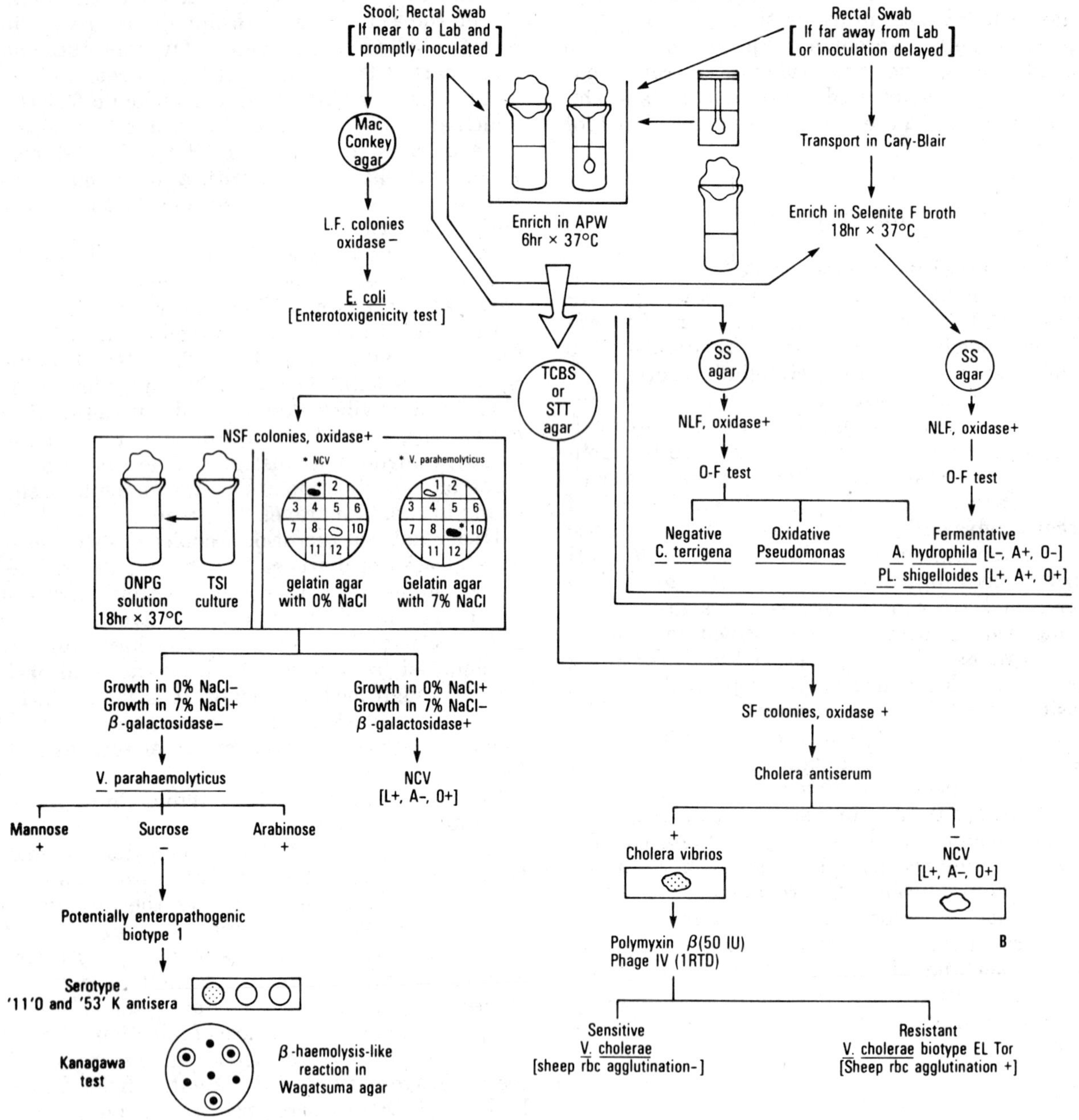

[FOOT NOTES]

SF → Sucrose fermenting; LF-Lactose fermenting
O-F Test → Oxidation-fermentation
NSF → Non-sucrose fermenting test
APW → Alkaline peptone water with 3% NaCl, pH 9.2
NLF-non lactose fermenting
A → Arginine, L → Lysine, O → Ornithine; Growth ●, No growth ○

FIGURE 3. *Laboratory diagnosis of cholera syndrome.*

NONCHOLERA VIBRIOS (NCV)

MORPHOLOGY

The features are identical with *V. cholerae,* except that all NCV are thinly capsulated and a few are coccobacillary. Fimbriated forms of NCV are not known. Colonies on BSA and nutrient agar may mimic those of cholera vibrios.

In TCBS and STT agar, the colonies are slightly mucoid and yellow or bluish green. Rough colonies have not been recognized.

ANTIGENIC COMPOSITION

On the basis of heat stable somatic antigens, the strains of human diarrhea fall under 40 O-groups. The flagellar (H) and capsular (K) antigens are heat labile and heterogenous. Smith (1974) disclosed major differences between the H antigens of cholera vibrios and NCV, although previously they were thought to be identical.

METABOLISM

Characteristics do not differ from *V. cholerae* (Table 2), but NCV in order of prevalence belong to Heiberg groups I, II and V.

PATHOGENIC PROPERTIES

The NCV cause mild gastroenteritis or illness simulating cholera and dysentery. Some strains elaborate enterotoxin that produce fluid accumulation and dilatation in the ligated loops of adult rabbits and diarrhea in 9 day old rabbits.

DRUG SUSCEPTIBILITY

The MIC in μg/ml: tetracycline 1.5, chloramphenicol 1.5, streptomycin 75, paromomycin 50, viomycin 500, and neomycin 30. It is also sensitive to disks of polymyxin B (100 IU) and ampicillin (25 μg). Sometimes drug resistance is found.

EPIDEMIOLOGY

NCV can be isolated usually from natural waters, sewage, flies, and the feces of cold-blooded and hot-blooded animals, but not from normal human stool. These organisms produce sporadic human diarrhea in different regions of the world. The incidence commonly rises before and after the cholera outbreak in eastern India. NCV do not produce epidemics.

AEROMONAS HYDROPHILA

(*aer,* gas; *monad,* unit; *aeromonas,* gas-producing one; *hydrophila,* water loving)

MORPHOLOGY

Cells are gram-negative, thinly capsulated, straight rods, or coccobacilli, 1.0 to 4.4 μm, arranged singly, in pairs and short chains, and rapidly motile by one or more terminal flagella.

On nutrient agar, the colonies are 2 to 3 mm, translucent, smooth, round, and convex after 18 hrs at 30° to 37° C. Broth culture shows diffuse turbidity with or without pellicle.

METABOLISM

A. hydrophila is a facultative anaerobe and is nonfastidious. Its generation time is short. Growth occurs at 21° to 40° C, pH 5.5 to 9.0, but some biochemical characters (Table 2) are better elicited at 22°C than 37°C — for example, gas from carbohydrates, production of acetoin, and utilization of citrate. Whereas the aerogenic biotype produces acetoin and H_2S from cysteine, the anaerogenic one is negative to these tests.

PATHOGENIC PROPERTIES

A. hydrophila causes mild gastroenteritis, choleraic diarrhea, and dysentery in man. Its enterotoxin is heat labile, neutralized by the cholera antitoxin and gives positive reaction in rabbit loops and adrenal tumor cells (Wadström et al., 1976). The enteropathogenic serotypes remain uncharacterized.

The organism may also be associated with bacteremia, osteomyelitis, cutaneous ulcers, and urinary tract infection. It is pathogenic to amphibians and reptiles.

DRUG SUSCEPTIBILITY

A. hydrophila is sensitive to disks of streptomycin (10 μg), tetracycline (30 μg), chloramphenicol (30 μg), erythromycin (15 μg), nalidixic acid (100 μg), furadantin (100 μg), kanamycin (5μg), and polymyxin B (100 IU), but resistant to ampicillin (25 μg). Resistance is rarely noted to streptomycin and tetracycline. The organism may possess drug-resistant plasmids.

PLESIOMONAS SHIGELLOIDES

(Plesios, nearer to *(aeromonas); shigelloides,* like *shigella)*

MORPHOLOGY

Cells are rod shaped with rounded ends, 0.8 to 1.0 by 3.0 μm, thinly capsulated, arranged singly, in pairs, short chains, and are rapidly motile with single polar or lophotrichous (tufted) flagella. Colonies appear small, 1 to 1.5 mm, almost transparent, convex, and glossy.

ANTIGENIC COMPOSITION

Strains are classified into 16 O-groups, of which one reveals major sharing with *Sh.sonnei.* The H antigen is divided into four types.

METABOLISM

Pl. shigelloides is a facultative anaerobe and grows on ordinary media at 18 to 37°C. Unlike the vibrios and aeromonas, it lacks diastase, lipase, DNase, proteinase, hemolysin (Table 2), and has no biotypes.

PATHOGENIC PROPERTIES

The organism is suspected as an enteropathogenic agent in Japan, India, and Africa, but enteric effects have yet to be reproduced in laboratory models. It may be isolated from blood, cerebrospinal fluid, and feces of animals. Its significance for man is not known.

DRUG SUSCEPTIBILITY

It is usually sensitive to disks of streptomycin (10 μg), tetracycline (10 μg), chloramphenicol (25 μg), kanamycin (30 μg), gentamycin (10 μg), paromomycin (30 μg), furazolidone (100 μg), but resistant to ampicillin (25 μg).

VIBRIO PARAHAEMOLYTICUS

(para, like; *haema,* blood; *parahaemolyticus,* like *haemolytica)*

MORPHOLOGY

Cells are ellipsoidal, gram-negative, capsulated rods, 0.5 to 1.4 μm by 0.4 to 0.6 μm, slightly pleomorphic, arranged singly, in pairs, short chains and small clusters; filaments are rare but consistent (Fig. 4). Adverse conditions or aging produces marked pleomorphism. Strains grown on nutrient agar containing 3 per cent NaCl after two or three days show involution forms — for example, bipolar staining, spheroplasts, slender rods with tapering ends, and clubs. Broth culture at 37° C reveals polar flagellated cells with rapid linear motility. Contrarily, 18-hr growth at 22 to

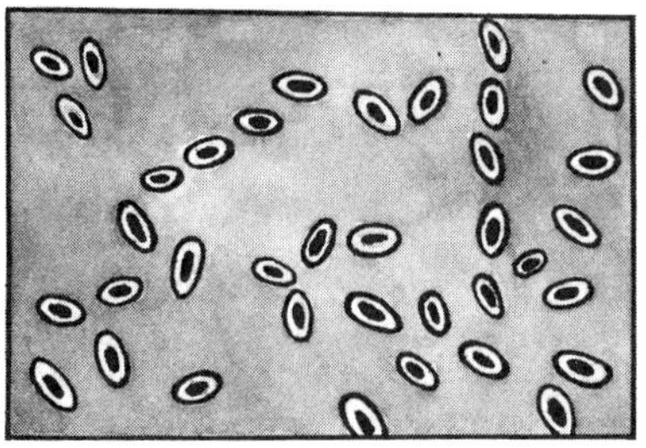

FIGURE 4. *Well formed capsule of* V. parahaemolyticus.

28°C on semisolid agar and 5-hr nutrient agar culture at 37°C disclose feebly motile peritrichous and polar flagellated cells (Fig. 5). The polar flagella are thick and wavy but the lateral ones are thin, fragile, and curly (Chatterjee, 1974).

The organism produces dome-shaped, off-white, sticky colonies in meat extract agar that measure 2 to 3 mm in diameter. With time, the culture is slowly covered by minute daughter colonies. Colonies of the enteropathogenic variety are bluish green on TCBS and STT agar but others are yellow. Sometimes a smooth colony exhibits wrinkled opaque transformation at the center with unaltered periphery; this is reversed on subculture at 22°C. An old culture in agar slant may show brown coloration of the medium and a bunch of needle shaped crystals underneath. Growth on semisolid agar at 22°C may show swarming.

ANTIGENIC COMPOSITION

V. parahaemolyticus possess flagellar (H) capsular (K) and somatic (O) antigens, each having 2, 53, and 11 types respectively. Flagellar antigenicity and agglutinability are better developed at 22 to 28°C than 37°C. Since these H antigens are shared by all strains, they do not help in classification. Serotyping is based on O and K antigens. The enteropathogenic serotypes are: 01:K38, 01:K56, 02:K3, 03:K4, 03:K33, 04:K8, 04:K9, 04:K11, 04:K12, 04:K13, 04:K55, 05:K12, 05:K15, 05:K17, 05:K30, 05:Cal/Ka, 05:K47, 08:K20, 08:K21, 08:K22, 09:K23, 010:K19, and 010:K24; they are rarely distributed in the fish and water.

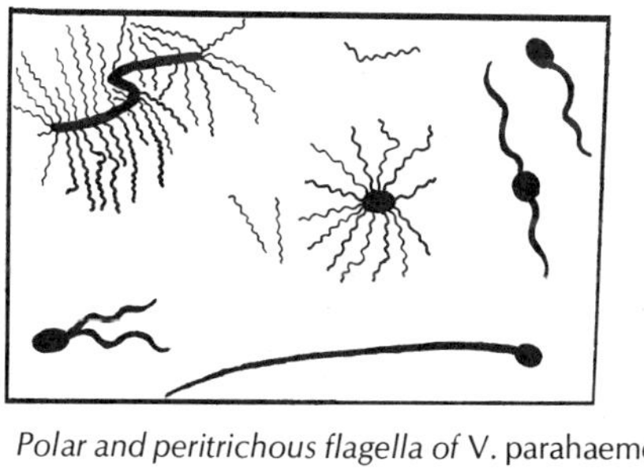

FIGURE 5. *Polar and peritrichous flagella of* V. parahaemolyticus.

METABOLISM

V. parahaemolyticus differs sharply from the other members in that it gives a negative test for β-galactosidase and fails to grow in NaCl-deficient media. In order to cultivate the organism in gelatin agar, at least 0.1 per cent NaCl is needed. Table 2 shows characteristics of enteropathogenic strains grouped under the biotype 1. It differs from the biotype 2 (*V. parahaemolyticus* biotype *alginolyticus),* which ferments sucrose, produces acetoin, and grows in 10 per cent NaCl. There are intermediate forms between the biotypes 1 and 2. *V. parahaemolyticus* is proteolytic, saccharolytic but essentially nonhemolytic for suspended red cells of different species. Yet, most (> 90 per cent of strains from human cases of diarrhea and their contacts show the Kanagawa phenomenon — that is, a reaction resembling β-hemolysis in Wagatsuma agar during overnight incubation at 37°C. On the other hand, the isolates of fish, food, and water are Kanagawa-negative after overnight incubation and positive only after 48 hours. Growth on Wagatsuma agar, with its high pH, excess NaCl, and fermentable carbohydrate-like mannitol is a test for hemodigestion rather than true hemolysis.

The media for methyl red (MR), V–P, and decarboxylase-dihydrolase tests being salt free are supplemented with 0.5 per cent NaCl. Results of the MR test, KCN tolerance, and citrate utilization are negative in media having 0.5 per cent NaCl (Table 2), but can be reversed if the same media are modified with 3 per cent NaCl (Chatterjee, 1974).

PATHOGENIC PROPERTIES

V. parahaemolyticus causes attacks of food poisoning, sporadic rice water diarrhea, and dysentery. It may also cause fulminating septicemia that is manifested by hemolytic anemia, hypotension, thrombophlebitis, erythema multiforme, and vasculitis leading to gangrene. Live cultures fed orally to infant rabbits cause bacteremia. Factors determining pathogenicity are not known.

An "aberrant biotype" of *V. parahaemolyticus* ("group F vibrio") ferments sucrose but reacts negatively to lysine and ornithine decarboxylase reactions and V-P test. Like *A. hydrophila,* it reveals arginine dihydrolase activity. This organism can be isolated from cold blooded animals and water. (Chatterjee, 1974; Furniss *et al,* 1977). It caused sporadic and epidemic forms of gastroenteritis in Bangladesh.

LABORATORY DIAGNOSIS

The steps of investigating *V. parahaemolyticus* food poisoning are shown below in addition to those already indicated in connection with the cholera syndrome.

Food Bacteriology

Demonstration of numerous organisms in the food strongly supports the diagnosis. Accordingly, 50 g of food are homogenized and diluted 1/10 and 1/100 in sterile saline with 0.1 per cent peptone. The neat and diluted samples are streaked on TCBS or STT agar with a calibrated loop for quantitation. A similar quantity of homogenized food is also put up for enrichment in alkaline peptone water (pH 9.2) containing 3 per cent NaCl.

Isolation and identification of *V. parahaemolyticus* from feces are already discussed under *V. cholerae*. The isolates of food and feces are compared by serotyping and the Kanagawa test; they may reveal multiple serotypes, of which one remains the dominant type.

DRUG SUSCEPTIBILITY

MIC in μg/ml are as follows: streptomycin 15 to 25, tetracycline 2.5 to 6, and chloramphenicol 0.5 to 2.5. It is also sensitive to disks of kanamycin (5 μg), polymyxin B (50 IU), colistin (10 μg), cephalothin (30 μg), furazolidone (100 mg), but resistant to ampicillin (25 μg). The organism produces β-lactamase.

EPIDEMIOLOGY

V. parahaemolyticus is a free-living bacterium. It is distributed all over the world in offshore sea water and marine animals. In eastern India where the inland water is slightly brackish it is found in pond water and its fish; it is also recovered from flies and sewage.

The organism produces sporadic diarrhea and food poisoning in Japan, Southeast Asia, Australia, and the United States. The food poisoning may break out suddenly in a community after sharing common meals. It is occasionally lethal in Japan. It may also be confined within a family.

Fish are most often responsible for infection. In Japan more than 70 per cent of summer food poisoning cases are produced by *V. parahaemolyticus* after eating uncooked or undercooked fish. Since adequate cooking reduces the risk of infection, the incidence of food poisoning is much lower in other countries. However, the food may be recontaminated by the pathogen after cooking by exposure to kitchen utensils or food handlers. Its short generation time (ten minutes) allows it to attain infective numbers (10^5 to 10^7 viable cells in human volunteers) rapidly at ambient temperatures.

CAMPYLOBACTERS

(*campylo,* curved; *bacter,* rod)

MORPHOLOGY

Cells are curved, S- or gull-shaped, gram-negative, slender rods, 1.5 to 5 μm by 0.2 to 0.5 μm and occur in chains with three to five spirals. The organism bears a polar flagellum at one or both ends and displays rapid linear or corkscrew-like motility (Smibert, 1978). It is not encapsulated.

On primary isolation, the colonies are small (0.5 to 1 mm) and vary in appearance from smooth to rough forms and in color from white to cream. Upon further incubation, the smooth colonies become mucoid. Subculture may produce "cut glass" colonies with reflecting facets. Broth culture shows uniform turbidity with a powdery deposit.

ANTIGENIC COMPOSITION

On the basis of heat-stable somatic antigens, *C. fetus* and *C. fetus* biotype *intestinalis* have two serotypes each, whereas *C. fetus* biotype *jejuni* shows one serotype. *C. fetus* may cross-react among its biotypes and with *Brucella abortus*.

METABOLISM

The organism needs a microaerophilic condition for growth in solid medium at 37° C, which is provided by an atmosphere of 5 per cent O_2, 10 per cent CO_2, and 85 per cent N_2. Aerobic incubation in 10 per cent CO_2 yields growth on semisolid media (0.16 per cent agar), such as thiol agar, yeast extract agar, *Brucella* agar, and also in a liquid medium like *Brucella* broth. It is not very active biochemically (Table 2).

Lately, the organism is grown in 10 per cent CO_2 and 90 per cent H_2 or air. The tests for characterizing the species and the biotypes of campylobacters are shown in Table 3.

PATHOGENIC PROPERTIES

C. fetus biotype *jejuni* causes acute diarrhea in children (5.8 per cent) and in adults (2.3 per cent). In children below 8 months, it can account for 30 per cent of the cases of acute diarrhea in developing countries. It is sometimes related to bacteremia. Debilitated patients are prone to infections with *C. fetus* biotype *intestinalis,* which produces septicemia (without splenomegaly), meningitis, meningoencephalitis, abortion, enteritis, thrombophlebitis, septic arthritis, and jaundice with hepatomegaly. Other species and biotypes are not pathogenic to man (Table 3). *C. fetus* biotype *jejuni* and *C. sputorum* may be rarely isolated respectively from the feces and mouth of normal persons.

IMMUNITY

In older children (above 8 months) and adults, *C. fetus* biotype *jejuni* appears to impart immunity localized to the gut. Information from animal diseases may help us to understand the mechanism of protection of man. It is well known that *C. fetus* infection remains confined to the reproductive organs of cattle. On the other hand, *C. fetus* biotype *intestinalis* in sheep produces bacteremia before localization in the placenta. The cattle develop a local immune response (usually IgA antibody against 0 antigen), whereas the infection of sheep produces a generalized immune response. Live or heat-killed vaccine prevents bovine infection, but the ovine disease is averted by the purified, heat-labile surface protein or the live bivalent-type vaccine.

LABORATORY DIAGNOSIS

A spiral organism in blood culture should arouse suspicion of *Campylobacter,* since *Spirillum minus,* the pathogen of relapsing fever fails to grow

TABLE 3. Differential Features of Campylobacters

SPECIES CHARACTERIZED BY TESTS	BIOTYPES KNOWN TO GROW ON Glycine 1%	NaCl 3.5%	Bile 1%	25° C	RESERVOIR HOSTS	PATHOGENICITY
Catalase + H_2S nitrate reduction *C. fetus*	−	−	+	+	Bulls: genital tract.	Cattle: abortion, infertility, venereally transmitted.
C. fetus biotype *intestinalis*	+	−	+	+	?	Man: infections (see text) orally transmitted; cattle, sheep: abortion.
C. fetus biotype *jejuni*	+	−	+	−	Man, cattle, sheep, goats, pigs, birds: gut lumen.	Man: infections (see text) orally transmitted; sheep: abortion; turkeys: bluecomb disease, enteritis; pigs: dysentery.
Catalase H_2S + nitrate reduction + : *C. sputorum*	+	−	+	+	Man: oral cavity.	?
C. sputorum biotype *bubulus*	+	+	−	+	Cattle, sheep: genital tract.	?
C. sputorum biotype *mucosalis*	−	−	−	?	Pigs: oral cavity.	Pigs: intestinal adenomatosis, necrotic enteritis, regional ileitis, proliferative enteropathy.
Catalase + H_2S + nitrate reduction *C. fecalis*	+	V	V	−	Sheep: genital tract, gut.	?

v, variable; ? unknown.

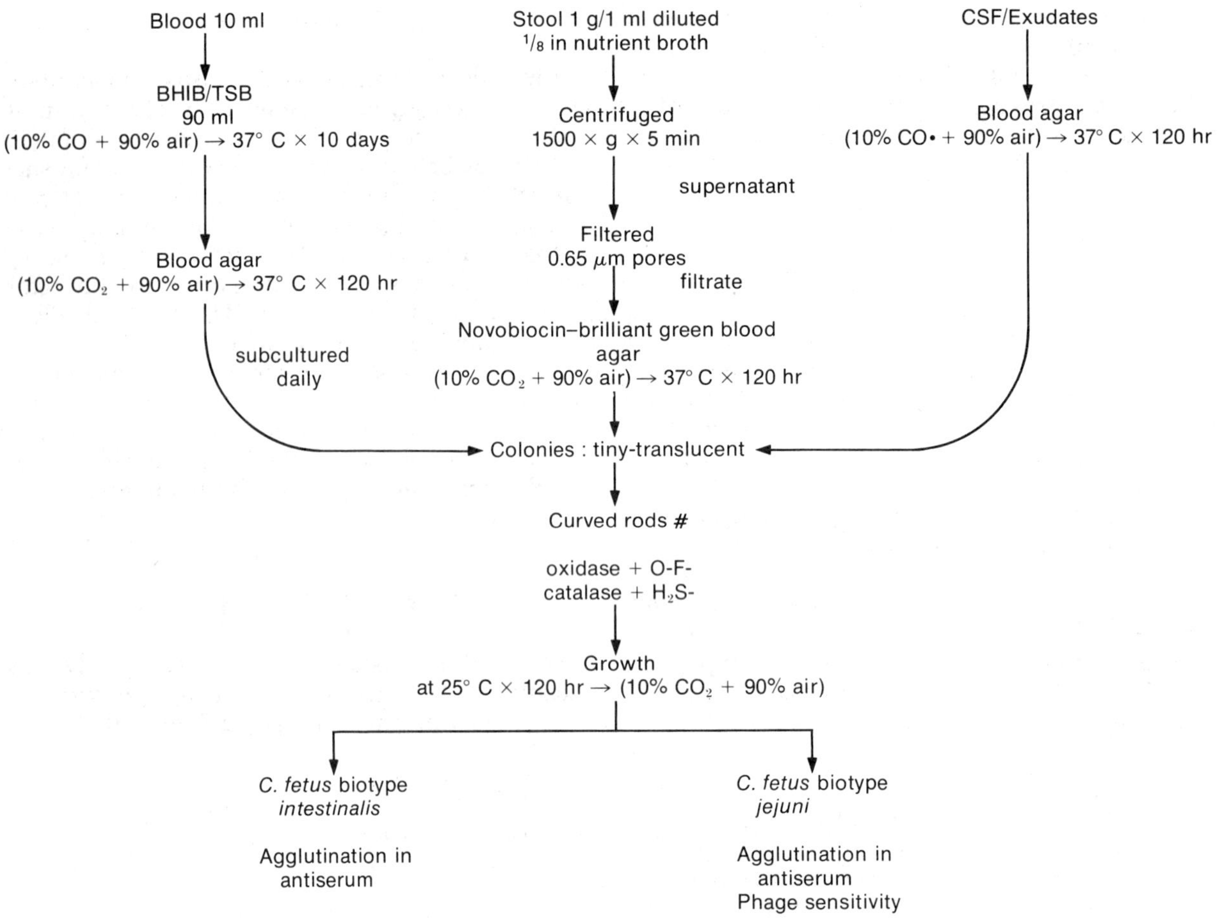

BHIB, brain heart infusion broth; TSB, trypticase soy broth; O-F, oxidation-fermentation. Five to 7 days' old culture show coccoid cells.

FIGURE 6. *Laboratory diagnosis of human campylobacteriosis.*

on artificial medium. *Campylobacter* can be easily isolated from blood or body fluids under microaerophilic conditions. In case of stools, the normal fecal flora are first partially removed by millipore filtration (0.65 μm pores) and then suppressed on selective-inhibitory media for *Campylobacter* (Fig. 6). Rising titers of complement-fixation and indirect hemagglutination tests suggest infection.

DRUG SUSCEPTIBILITY

The MIC in μg/ml for *C. fetus* biotypes *intestinalis* and *jejuni* are: clindamycin, 0.4 to 1.6; tetracycline, 0.8; streptomycin, 2 to 4; neomycin, 8; kanamycin, 0.01 to 3.12; gentamicin, 0.19 to 1.56; ampicillin, 1.56 to 25; erythromycin, 2 to 8; chloramphenicol, 4; cephalothin, 50; and carbenicillin, 200. They are resistant to penicillin (5 units), polymyxin B (1024 units), and bacitracin (120 units). Tetracycline and trimethoprim resistance is sometimes encountered.

EPIDEMIOLOGY

Human campylobacteriosis until lately was reported from the United States, Europe, and Africa. Awareness of the disease has resulted in increasing recognition from many other places in the world. It is suspected that the organism is transmitted from animals to man, but the exact mode of acquisition is obscure.

References

Chalmers, A. J., and Waterfield, N. E.: Paracholera caused by *Vibrio gindha*. Pfeiffer 1896. J Trop Med Hyg 19:165, 1916.

Chatterjee, B. D.: Epidemiologic and taxonomic status of *Vibrio parahaemolyticus*. In Fujino, T., Sakaguchi, G., Sakazaki, R., and Takeda, Y. (eds.): International Symposium on *Vibrio parahaemolyticus*. Tokyo, Saikon Publ. Co., 1974, p. 177.

Chatterjee, B. D., and Neogy, K. N.: On the etiology of choleraic diarrhoea. Indian J Med Res 60:531, 1972.

Chatterjee, B. D., De, P. K., and Sen, T.: Sucrose teepol tellurite agar: A new selective indicator medium for isolation of *Vibrio* species. J Infect Dis 135:654, 136:716.

Furniss, A. L., Lee, J. V., and Donovan, T. J.: Group F, a new vibrio? Lancet ii:565, 1977.

Smith, H. L., Jr.: Antibody responses in rabbits to injections of whole cell, flagella and flagellin preparations of cholera and non-cholera vibrios. J Appl Microbiol 27:375, 1974.

Wadström, T., Ljungh, A., and Wretlind, B.: Enterotoxin, haemolysin and cytotoxic protein in *Aeromonas hydrophila* from human infections. Acta Path Microbiol Scand Sec B 84:112, 1976.

GENUS 33 *PSEUDOMONAS* *

M. J. Pickett

DEFINITION

Pseudomonads are nonsporulating, gram-negative rods. Most species are polarly flagellated, and a few also have lateral flagella. The cells of some species may contain granules of poly-β-hydroxybutyrate (PHB). These PHB granules may be evident in Gram-stained preparations and, by special staining procedures, can be used as an aid in identification. Most pseudomonads are oxidase-positive and noncapsulated. Excluding *P. vesicularis,* which may be negative, all are catalase-positive. With few exceptions, they grow readily on nutrient and blood agar media at both room (22 ± 2° C) and incubator (35 to 37° C) temperatures. They are obligatively aerobic and nonfermentative. Most species oxidatively acidify carbohydrates. All are anaerogenic. One species, *P. putrefaciens,* produces hydrogen sulfide in Kligler iron agar (KIA) and triple sugar iron agar (TSIA) media. None forms indole or gives a positive Voges-Proskauer test for acetoin. With few exceptions, they are resistant to penicillin. The moles percentage of guanine plus cytosine (percentage GC) in their chromosomal DNA ranges from 48 to 70.

Habitat, Resistance

Many of the species, including those associated with human disease, are commonly present in soil and in the hospital environment and may be a part of the normal flora of the body. Plants and both wild and domestic animals may harbor pseudomonads, but rarely play a significant role as reservoirs and vectors for transmission to humans. One species, *P. fluorescens,* can grow at 4° C and has been associated with endotoxemia following administration of contaminated intravenous fluids. Others, such as *P. cepacia,* can grow in commercial distilled water. Several have been recovered from cosmetics and other toiletries. Most are relatively resistant to several common germicides, particularly quaternary ammonium compounds such as Zephiran. They display both species and strain variation in respect to susceptibility to antimicrobials.

More than 25 different species of pseudomonads are associated with man and his immediate environment, but only two of these, *P. aeruginosa* and *P. maltophilia,* account for more than 75 per cent of those recovered from clinical specimens (Tables 1 and 2).

PATHOGENICITY

Only two species of pseudomonads are uniquely associated with specific diseases of man; *P. mallei* is the etiologic agent of glanders and *P. pseudomallei* of melioidosis. However, many of the species, notably *P. aeruginosa,* are opportunistic pathogens. They may be etiologically significant, usually in previously compromised patients, in wounds and postburn sepsis, postsurgical infections, septicemia, and respiratory and urinary tract infections, among others.

TABLE 1. Distribution of Aerobic Gram-Negative Bacilli in Clinical Specimens[a]

GROUP	% (APPROX.)
Enterobacteriaceae	73
Haemophilus	12
Nonfermentative bacilli	14
Other[b]	1

[a]Unpublished data from Pickett and Martin, 1032 strains, June, 1976, UCLA Hospital and Clinics, Los Angeles, California, and Blachman, 768 strains, March-April, 1976, Olive View Medical Center, Sylmar, California.

[b]E.g., *Aeromonas, Cardiobacterium, Eikenella.*

*This chapter is not documented with an exhaustive list of references. Readers wishing such documentation should consult the references given in the manual of Lennette et al. (1974), the textbook of Wilson and Miles (1975), and the monograph of Gilardi (1978a).

TABLE 2. Distribution of Nonfermentative Bacilli in Clinical Specimens[a]

	% (APPROX.)		% (APPROX.)
Pseudomonas aeruginosa	73	*P. alcaligenes*	<1
(pyocyanin-positive)	(64)	*P. denitrificans*	<1
(pyocyanin-negative)	(9)	*P. diminuta*	<1
Acinetobacter	9	*P. pseudoalcaligenes*	<1
P. maltophilia	6	*P. putrefaciens*	<1
Flavobacterium	3	*P. vesicularis*	<1
P. cepacia	1	*Pseudomonas*, IIk group	<1
P. fluorescens	1	*Pseudomonas*, Ve group	<1
P. pickettii	1	*Achromobacter*	<1
P. putida	1	*Alcaligenes*	<1
P. stutzeri	1	*Bordetella*	<1
Moraxella	1	*Kingella*	<1
P. acidovorans	<1		

[a]Unpublished data from Pickett and Pedersen, 486 strains, 1968, and Greenwood and associates, 525 strains, 1978, UCLA Hospital and Clinics, Los Angeles, California.

TABLE 3. Differential Features for Fluorescent Pseudomonads[a,b]

	P. AERUGINOSA	P. FLUORESCENS	P. PUTIDA
Moles % GC	67	59–64	60–63
Flagella	1	>1	>1
Growth at 42°C*[c]	+[d]	0	0
Pyocyanin	(+)	0	0
Acidification of			
Lactose	0	(+)	(−)
Sorbitol	0	(+)	0
Sucrose	0	(+)	−
Alkalinization of			
Acetamide*	+	0	−
Tartrate	0	0	(+)
Nitrate → gas	(+)	(−)	0
Gelatin*	(+)	+	0
Hippurate	0	0	(+)

[a]*P. aeruginosa* occasionally fails to show fluorescence at 20 to 25°C; *P. fluorescens* and *P. putida* frequently fail to show fluorescence at 35 to 37°C.
[b]All three species are oxidase-positive and glucose-positive, grow on MacConkey agar medium, and are susceptible to polymyxin; none contains PHB granules.
[c]Asterisk indicates features that are most useful for identification.
[d]+, 90 per cent (or more) of strains are positive; (+), usually positive; (−), usually negative; −, 10 per cent (or less) of strains are positive; 0, never positive.

DETECTION AND IDENTIFICATION

Initial detection is similar to that for Enterobacteriaceae, namely, in broth medium for blood specimens or on a blood agar plate for other specimens. Since they are obligatively aerobic, most strains fail to grow when these media are incubated anaerobically. Most species, including the two most commonly encountered, grow rapidly and present 0.5 to 2 mm colonies on 24-hour blood agar plates. Confirmation that a clinical isolate is an obligatively aerobic, nonfermentative bacillus (NFB) is commonly effected by demonstrating its failure to grow in a tube of oxidation-fermentation (OF) medium overlayed with petrolatum. Alternatively, its identity as an NFB may be confirmed by demonstrating growth on the slant but neither growth nor acidification in the butt of KIA or TSIA medium.

Many of the features used for identification of enteric bacteria are equally applicable to the pseudomonads, but since the latter not only are obligate aerobes but also usually alkalinize media containing peptone, special methods are usually used for biochemical tests (Tatum et al., 1974; Gilardi, 1978b). Features for identification of the more commonly encountered pseudomonads include (1) colonial and microscopic morphology including pigmentation, if any; (2) production of fluorescein; (3) production of gas from nitrate; (4) growth at 42° C; (5) acidification of carbohydrates, particularly glucose, lactose, mannitol, and sucrose; (6) decarboxylation of lysine (the LDC test); and (7) alkalinization of acetamide and urea. Tables 3 through 5 present the principal differential features of these bacilli. "Antibiograms" are also useful guides for identification (Table 6).

PSEUDOMONAS AERUGINOSA

P. aeruginosa (*Pseudomonas pyocyanea*) is more frequently encountered and more important as an agent of human disease than all the other nonfermentative bacilli. Understandably, many of its features have been studied in considerable detail.

MORPHOLOGY

Microscopically, *P. aeruginosa* is a slender, nonsporulating, usually noncapsulated rod bearing a single polar flagellum. The cells do not contain PHB inclusions. This bacterium presents three types of colonies. The most common, on a 24-hour blood agar plate, is low convex to flat and 1 to 5 mm in diameter with a rough or frosted glass surface and an undulate or erose periphery. It may be β-hemolytic on a 24-hour plate and usually shows diffuse β-hemolysis on a 48-hour plate. Most strains form a water- and chloroform-soluble phenazine pigment, pyocyanin (from the Greek, blue pus), which usually imparts a green or blue-green color to the medium surrounding

TABLE 4. Differential Features for Nonfluorescent, Glucose-Positive Pseudomonads

	P. CEPACIA	P. MALTOPHILIA	P. PAUCIMOBILIS (BIOGROUP IIk-1)	P. PICKETTII (BIOTYPE Va-1)	P. PICKETTII (BIOTYPE Va-2)	P. PSEUDOMALLEI	P. PUTREFACIENS	P. STUTZERI	P. VESICULARIS	PSEUDOMONAS (BIOGROUP Ve-1)	PSEUDOMONAS (BIOGROUP Ve-2)
Moles % GC	68	67	65	63	64	70	48–59	65	66	57	67
Flagella	>1	>1	1	1	1	>1	1–2	1–2	1	>1	1
PHB granules	+	−	+	+	+	+	ND[d]	−	+	ND	ND
Pigmented growth[a]	(−)	(−)	+	0	0	(−)	(−)	(−)	(−)	+	+
Oxidase*[b]	+	−	+	+	+	+	+	+	+	−	−
MacConkey, growth	(+)	+	(+)	+	+	+	+	+	(+)	+	+
Growth at 42°C	(+)	(+)	−	(−)	(+)	+	(+)	+	−	(+)	(−)
Polymyxin susceptible[c]	−	+	(+)	−	−	−	+	+	+	+	+
Acidification of											
Arabinose*	+	−	+	+	+	+	(+)	+	+	+	+
Cellobiose	+	+	+	+	0	+	(−)	0	+	0	(−)
Ethanol	(+)	−	+	0	0	(+)	−	+	+	+	+
Lactose*	+	+	+	+	0	+	(−)	0	0	−	−
Maltose*	+	+	+	+	0	+	(−)	+	+	+	+
Mannitol*	(+)	0	0	0	0	+	0	(+)	0	+	+
Sorbitol	+	0	0	0	0	+	−	0	0	−	(+)
Sucrose*	(+)	+	+	0	0	(+)	(−)	0	0	−	(−)
Amylase	0	0	(+)	0	0	(+)	0	+	−	(−)	(+)
Citrate (Simmons)	+	(−)	(−)	+	+	+	(−)	+	−	+	(+)
Esculin	(+)	+	+	0	0	(+)	(−)	0	+	+	−
H_2S (in KIA or TSIA)*	0	0	0	0	0	0	+	0	0	0	0
Lysine (LDC)*	+	+	0	0	0	0	0	0	0	0	0
Nitrate → gas*	0	0	0	+	+	+	0	+	0	0	0
Nitrate → nitrite	(−)	(−)	−	+	+	(+)	(+)	(+)	−	(+)	(−)
Urea (Christensen)*	(−)	−	−	+	+	(−)	−	(−)	−	(+)	(+)

[a]Growth tan, yellow, or orange on a 24- to 48-hr blood agar plate. Other notations as for Table 3.
[b]Asterisk indicates features that are most useful for identification.
[c]300 unit disk.
[d]ND, no data.

TABLE 5. Differential Features for Glucose-Negative Pseudomonads[a]

	P. ACIDOVORANS	P. ALCALIGENES	P. DIMINUTA	P. PSEUDO-ALCALIGENES	P. TESTOSTERONI
Moles % GC	67	66	67	63	62
Flagella	>1	1	1	1	>1
PHB granules	+	−	+	(+)	+
Growth at 42°C	0	+	(+)	+	0
Polymyxin susceptible	(+)	+	(−)	+	(+)
Acidification of					
Arabinose	0	0	(−)	(−)	0
Ethanol*[b]	+	−	(+)	+	(−)
Fructose*	+	0	0	+	0
Mannitol	(+)	0	0	0	0
Alkalinization of					
Acetamide*	+	0	0	0	0
Allantoin	+	0	0	0	(+)
Formate*	+	(+)	0	+	+
Nicotinamide	+	0	0	0	0
Tartrate	+	(−)	0	0	−
Citrate (Simmons)*	+	(+)	−	(+)	(+)
Nitrate → nitrite	+	+	(−)	+	(+)

[a]All are oxidase-positive and grow on MacConkey agar medium; none is pigmented, fluorescent, or forms gas from nitrate. Other notations as for Table 3.
[b]Asterisk indicates features most useful for identification.

TABLE 6. In vitro Antimicrobial Susceptibility of Pseudomonads[a]

SPECIES/GROUP	NUMBER OF STRAINS	CARBENI-CILLIN	GENTAMICIN	NOVO-BIOCIN	POLYMYXIN	STREPTO-MYCIN	TRIMETHO-PRIM/SULFA-METHOX-AZOLE
P. acidovorans[b]	81	37[e]	7	39	70	4	99
P. aeruginosa[c]	72	63	93	0	97	36	76
P. alcaligenes[d]	71	75	73	10	94	37	62
P. cepacia	132	10	13	48	0	2	100
P. diminuta	33	97	100	100	49	15	94
P. fluorescens	137	4	96	0	96	53	47
P. maltophilia	390	38	43	1	95	20	95
P. paucimobilis	61	90	80	77	70	13	95
P. pickettii							
biotype Va-1	15	20	53	0	0	20	100
biotype Va-2	22	36	0	0	0	0	100
P. pseudomallei	7	0	0	86	0	0	0
P. putida	180	13	96	1	99	58	9
P. putrefaciens	44	80	100	0	100	89	100
P. stutzeri	116	89	100	0	100	90	70
P. vesicularis	19	95	100	100	90	90	90
Biogroup Ve-1	16	100	100	0	100	100	60
Biogroup Ve-2	39	100	100	0	100	54	39

[a]Data adapted from Gilardi (1979, personal communication).
[b]Includes closely related *P. testosteroni*.
[c]Apyocyanogenic strains.
[d]Includes closely related *P. pseudoalcaligenes*.
[e]Per cent of strains susceptible.

the colony. Other brown to black, phenazine pigments are also occasionally formed. Fluorescein (pyoverdin), a water-soluble, pale yellow-green pigment, is formed by nearly all strains of *P. aeruginosa* and is an important feature for identification of this species. Fluorescein is not apparent on blood agar medium; a special medium enriched with magnesium and phosphate should be used, and detection is by fluorescence under ultraviolet light.

A second type of colony, commonly displayed by strains recovered from the sputum of patients suffering from cystic fibrosis, is relatively large and markedly mucoid, particularly on MacConkey agar medium. Such strains, when repeatedly subcultured, may yield a third small and smooth type colony. This third phenotype, in turn, may yield the familiar large rough colony upon repeated subculture. Qualitatively identical biochemical features obtain for the three colonial varieties of *P. aeruginosa*, although the reactions (e.g., acidification of sugars and alkalinization of acetamide) are frequently weak and delayed with mucoid strains.

The characteristic rough or mucoid colony, β-hemolysis, blue-green pigmentation, and fruity odor represent tentative identification of this species in the clinical laboratory. Approximately 12 per cent of strains are nonpigmented, but these, too, can easily be identified by their fluorescence and alkalinization of acetamide (Table 3).

NUTRITION AND GROWTH

P. aeruginosa, like most pseudomonads, is not nutritionally fastidious. It grows in an inorganic basal medium supplemented with a single organic compound such as acetate or glucose (Stanier et al., 1966). As noted above, it is nonfermentative. Electron acceptors other than oxygen (e.g., nitrate) permit growth in an otherwise anaerobic environment. It grows readily at 20 to 42° C; in contrast, another fluorescent species, *P. fluorescens*, grows at 4° C but not at 41 to 42° C (Table 3). All strains of *P. aeruginosa* grow readily on MacConkey agar medium. Most strains also grow on a selective medium containing hexadecyltrimethylammonium bromide ("cetrimide agar").

PHYSIOLOGY

All of the three fluorescent pseudomonads (*P. aeruginosa, P. fluorescens,* and *P. putida*) are oxidase-positive. All are relatively versatile in their oxidative attack of amides, amino acids, carbohydrates, and organic salts (Stanier et al., 1966; Gilardi, 1978b). They rarely hydrolyze starch and esculin; never decarboxylate lysine and ornithine; and never produce acetoin, indole, and H_2S (in KIA or TSIA medium). Most strains of the fluorescent group, but not other pseudomonads, oxidize gluconate to 2-ketogluconate. All alkalinize Simmons citrate medium. Most strains of *P. aeruginosa,* but not of the other two species, produce gas from nitrate and rapidly digest gelatin, casein (skim milk medium), and serum (Loeffler's slant). *P. aeruginosa* acidifies glucose, arabinose, and xylose but not lactose, maltose, and sucrose.

PATHOGENICITY

P. aeruginosa is etiologically significant in many diseases and is particularly associated with postburn sepsis and other nosocomial infections, cystic fibrosis, and septicemia in patients with immunodeficiencies, including those common to neoplastic disease. Neonates are particularly prone to *P. aeruginosa* infection, although this may be referable more to the high humidity and contaminated equipment of the nursery than to high susceptibility. Infection of neonates commonly occurs as otitis media, meningitis, septicemia, and infection of the umbilicus.

P. aeruginosa may be a primary infectious agent but may also emerge following eradication of other bacteria by antibiotics. The normal eye is usually resistant, but following physical trauma may become infected with this bacterium, frequently with resultant blindness. Infection of the eye may also arise from contact lenses that have been stored in contaminated solutions. Normal skin, too, is relatively resistant, but a soggy dermatitis may occur on chronically moist surfaces of the body. Mucoid strains are particularly associated with patients suffering from cystic fibrosis and bronchiectasis. Such strains are only rarely recovered from other patients; indeed, there is suggestive evidence that the in vivo environment of these patients favors the mucoid phenotype of *P. aeruginosa.*

ANTIGENS, TOXINS, AND PATHOGENESIS

A broth culture of *P. aeruginosa* contains not only bacteria but also extracellular factors proliferated by this bacterium. Peripheral to the bacterial cell envelope, at least with some strains, are a capsule-like slime layer, pili, and a flagellum. In vivo, flagella elicit antibodies, and these may be useful for serotyping this species, but there is no evidence that flagella are significant in host-parasite interactions. Though pili (fimbriae) are frequently present on strains recently recovered from clinical specimens, there is no evidence that these are significant for attachment to mammalian cells or for phage-mediated genetic transfer, thus differing from observations on *Escherichia coli.* The extracellular slime surrounding the bacterial cell contains polysaccharide that is both toxic and antigenic (see Sadoff in Artenstein and Sanford, 1974). In experimental animals, both active and passive immunization against slime protect against lethal challenge with live bacteria.

The cell envelope, consisting of cytoplasmic membrane, the intermediate peptidoglycan layer, and the outer cell wall, is typical of that associated with other gram-negative bacteria. Endotoxicity, O-antigen, and lipopolysaccharide (LPS) are resident in the outer cell wall. As with Enterobacteriaceae, advantage is taken of the O-antigens in serotyping. At present there are several nomenclatures for these O-antigens, and a universal system has been proposed (Serotyping Committee, 1979). It is not clear whether O-specificity resides in the polysaccharide or in the lipid moiety of LPS. Opsonizing antibody directed against LPS appears to play an important role in immunologic defense against *P. aeruginosa,* but high titer antibody in patients with cystic fibrosis usually does not lead to its eradication from the respiratory tract (see Neter in Artenstein and Sanford, 1974). The LPS is similar to that of the Enterobacteriaceae, but there are differences in the hydroxy acids of its lipid A moiety (see Eagon in Artenstein and Sanford, 1974).

Among the several extracellular substances associated with *P. aeruginosa* are proteases, phospholipase, hemolysin, and exotoxin A (see Liu in Artenstein and Sanford, 1974). According to Liu, the "collagenase" of *P. aeruginosa* appears to be a nonspecific activity of proteases. In vitro, purified proteases cause clumping and vacuolization of rabbit alveolar macrophages (Leake et al., 1978). Injected intratracheally into rabbits, they elicit hemorrhage histopathologically resembling that occurring in *P. aeruginosa* pneumonia of man. Exotoxin A is a protein of more than 10,000 daltons. Injected intravenously, the LD_{50} for mice is approximately 7 μg (see Pavlovskis et al. in Artenstein and Sanford, 1974). Injected intradermally, it causes edema and necrosis. Injected intraperitoneally, 5 to 10 LD_{50} causes leukopenia and death; autopsy discloses necrosis of the liver, edema and hemorrhage of the lungs, and necrosis and hemorrhage in the kidneys. Its mode of action appears to be similar to that of diphtheria toxin (Chung and Collier, 1977; Leppla et al., 1978).

To summarize, all of the following are associated with *P. aeruginosa,* and all may be significant in the pathogenesis of disease referable to this bacterium: lipopolysaccharide, a slime layer consisting of polysaccharide and protein, proteases, phospholipase, and exotoxin A.

THERAPY AND CONTROL

Most strains of *P. aeruginosa* are susceptible to one or more of the aminoglycosides, amikacin, gentamicin, kanamycin, netilmicin, sisomycin, and tobramycin (Artenstein and Sanford, 1974; Neu, 1978). Resistance to one aminoglycoside is not necessarily accompanied by resistance to others. Resistance, when present, may be attributed either to failure of the agent to enter the bacterium or to enzymatic inactivation. Both chromosomal and plasmid DNA are involved in the mediation of resistance.

The uniform resistance to penicillin appears referable to both exclusion by the cell membrane and chromosomally mediated β-lactamase. Penicillin derivatives such as carbenicillin and ticarcillin presently are effective therapeutic agents since they do cross the membrane and are relatively resistant to the β-lactamases. Mucoid strains are exceptional in that they frequently are resistant to both carbenicillin and tobramycin but are susceptible to tetracycline (Govan and Fyfe, 1978). Most strains of *P. aeruginosa* are susceptible to polymyxin B and E (colistemethate) but are resistant to other antimicrobials.

Attempts to control *P. aeruginosa* infections call to mind similar problems with another ubiquitous bacterium, *Staphylococcus aureus,* and have been met with limited success. Routine monitoring and rigorous control of the entire hospital environment suppress but seldom eradicate nosocomial infections. Particular attention to epidemic strains should be a significant factor in controlling *P. aeruginosa* but has not been extensively practiced. Phage-, pyocin-, and serotyping do permit delineation of strains within this species (Zierdt, 1978). Nevertheless, none of these methods has disclosed types with unique antibiograms, epidemic propensities, or anatomic specificities. Additionally, and although there has been one recent favorable report (Shivananda et al., 1978), there is little evidence that either prophylactic antimicrobials or vaccines significantly control nosocomial infections.

DIAGNOSTIC PROCEDURES

As in the search for other pathogenic bacteria, body fluids such as blood and aspirates are transferred to a fluid medium. *P. aeruginosa* requires an aerobic environment for optimal growth, and hence commercial blood-broth bottles containing an inert gas are "vented" to admit air if this bacterium is anticipated. Secondary culture of such specimens and primary culture of swabs, exudates, and urine are made on blood agar plates, and frequently also on a differential medium such as MacConkey agar. A selective medium such as a cetrimide agar plate is also used in environmental and epidemiologic surveys when only this species is being sought.

In the routine clinical laboratory most isolates of *P. aeruginosa* are, as noted above, pyocyanin-positive; this feature, along with characteristic colonial morphology (large, flat, frosty) and odor permit tentative identification. When the pigment is not obvious and the colony is mucoid or smooth, thus mimicking enteric bacilli, a colony should be transferred to KIA or TSIA medium. Failure to grow in, or acidify, the butt of this medium and good growth on the slant is confirmatory of a nonfermentative bacterium. The following tests, along with appropriate microscopic morphology as seen in Gram stain, confirm its identification as *P. aeruginosa:* fluorescence +, acetamide +, and growth at 41 to 42° C (Table 3). Identification to the subspecies level may be desirable for epidemiologic purposes; this can be effected by sero-, phage-, or pyocin-typing (Zierdt, 1978).

SUMMARY

P. aeruginosa is widely distributed in nature and the hospital environment. This obligatively aerobic, gram-negative, oxidase-positive, nonsporulating motile rod is more common than all other nonfermentative bacilli in clinical specimens, and is etiologically significant as both a primary and a secondary opportunistic pathogen. Although *P. aeruginosa* is resistant to many antimicrobials, it is usually susceptible to one or more of the aminoglycosides. Easily grown and identified, its salient features are blue-green pigmentation, colonial morphology, odor, fluorescence, growth at 42° C, and alkalinization of acetamide.

PSEUDOMONAS MALTOPHILIA

Approximately 7 per cent of the pseudomonads recovered from hospital specimens are *P. maltophilia (Alcaligenes bookeri)*. This species is strongly proteolytic and only weakly saccharolytic and hence is frequently misidentified as *Alcaligenes faecalis*. As the "maltose-loving" name suggests, it readily acidifies this sugar.

MORPHOLOGY

Microscopically, *P. maltophilia* is a slender, gram-negative rod bearing a tuft of polar flagella. Colonies on a 24-hour blood agar plate are small (0.2 to 0.5 mm), convex, smooth, colorless, and entire. On a 48-hour plate, the colonies are 1 to 2 mm in diameter, slightly brown, and surrounded by a zone of diffuse β-hemolysis.

NUTRITION AND GROWTH

Like *P. aeruginosa, P. maltophilia* grows luxuriantly on nonselective media routinely used in the clinical laboratory. Though most strains require methionine for growth, this amino acid is already present in such media. Growth occurs on MacConkey agar but is frequently less luxuriant than on nonselective media. *P. maltophilia* does not grow on salmonella-shigella and cetrimide agar and, like the fluorescent pseudomonads, rarely grows on media containing 6.5 per cent sodium chloride. Good growth and motility are obtained in OF basal medium, and such cultures remain viable for months at 4° C.

PHYSIOLOGY

P. maltophilia is usually oxidase-negative, but nearly 5 per cent of strains are weakly to moderately oxidase-positive. Though it is a saccharolytic pseudomonad, acidification of carbohydrates is masked in media containing 0.5 per cent or more of organic nitrogen. However, in suitable basal medium containing 0.2 per cent or less of peptone, not only maltose but usually also glucose, lactose, and sucrose are acidified (Otto and Pickett, 1976; Gilardi, 1978b). It hydrolyzes esculin and deoxyribonucleic acid but not starch and urea. It is one of only two pseudomonads that are LDC-positive. These two, *P. cepacia* and *P. maltophilia,* are easily distinguished by other features, particularly their differential acidification of mannitol and arabinose (Table 4).

HABITAT

P. maltophilia has been recovered from an impressive variety of sites both within and outside the hospital environment (Hugh and Gilardi, 1974), and therefore it appears to be readily available to colonize compromised hosts. Among these sites are deionized water, eggs, feces, fish, milk, normal nasopharynges, sewage, soil, and therapeutic baths.

PATHOGENICITY AND THERAPY

This species has been etiologically associated, in both mixed and pure cultures, with lymphadenopathy, infected wounds, pulmonary disease, septicemia, superficial ulcers, and urinary tract infections. Most strains are susceptible in vitro to chloramphenicol, colistin, nalidixic acid, polymyxin B, and trimethoprim/sulfamethoxazole but are resistant to ampicillin, cephalothin, erythromycin, nitrofurantoin, and novobiocin. The trimethoprim-sulfamethoxazole combination has been favorably reported as a therapeutic agent (Moody and Young, 1975; von Graevenitz, 1978).

SUMMARY

P. maltophilia is widely distributed in nature. This obligatively aerobic, gram-negative, usually oxidase-negative, nonsporulating, noncapsulated rod grows well on blood, infusion, and Kligler iron agar media. The small, smooth, weakly pigmented colonies are strongly proteolytic but weakly saccharolytic and LDC-positive. Maltose is acidified, but arabinose and mannitol are not. Among the pseudomonads, *P. maltophilia* is second only to *P. aeruginosa* as an opportunistic pathogen. Although it is resistant to many antimicrobials, it is usually susceptible to chloramphenicol and trimethoprim/sulfamethoxazole.

OTHER PSEUDOMONADS

Pseudomonas (*Bacillus, Actinobacillus, Loefflerella, Malleomyces, Pfeifferella*) *mallei* is almost never encountered today in the Western world. It is the etiologic agent of glanders in ungulates, particularly equines, and only rarely in other animals, but this disease is largely controlled today in most areas of the world. Human infections, frequently fatal, were acquired from the nasal discharge of diseased animals, commonly horses and donkeys (Wilson and Miles, 1975). In man, glanders is usually an acute, though sometimes chronic respiratory disease. In the acute form there is usually fever, prostration, a mucopurulent discharge, and death within 2 weeks of onset. The chronic form may persist for months or years, usually with lymphadenopathy.

P. mallei differs from other pseudomonads in its parasitic habitat, slow growth in both defined and complex media, and absence of flagella, but many of its features, including oxidase positivity, aerobic metabolism, and moles per cent GC (69) in its chromosomal DNA are those of the genus *Pseudomonas* (Redfearn et al., 1966; Mandel, 1966). It resembles *P. pseudomallei* in its production of gas from nitrates, moles per cent GC, antigenic composition, oxidative assimilation of organic substrates, and sensitivity to temperate *P. pseudomallei* phages. Features that serve to distinguish these two species are shown in Table 7.

Pseudomonas (*Loefflerella, Malleomyces, Pfeifferella*) *pseudomallei,* formerly called Whitmore's bacillus, is the etiologic agent of melioidosis. Its normal habitat appears to be tropical soils throughout the world, but subclinical colonization of man may occur months prior to overt disease. Hence cases of melioidosis, often fatal, have been reported from temperate and subtropical as well as tropical areas of the world. There is suggestive evidence that both overt and covert infections exist in rodents and domestic animals in the tropics, but there is little evidence that these serve as a source of human infections. In man, pulmonary infection is the more common form of melioidosis, and the x-ray picture may mimic that of tuberculosis or mycotic disease. It may occur as a rapidly progressive and fatal septicemia, despite antimicrobial therapy. Chloramphenicol, tetracycline, and trimethoprim/sulfamethoxazole have been reported as effective therapeutic agents (Howe et al., 1971; John, 1976; Girard et al., 1976). It is usually not susceptible to penicillin, its derivatives, and the aminoglycosides.

TABLE 7. Differential Features for Pseudomonas mallei and P. pseudomallei[a]

	P. MALLEI	P. PSEUDO-MALLEI
Flagella	0	>1
Colonies, rugose	−	V[b]
Growth, MacConkey agar	−	+
Amylase	V	+
Assimilation of		
L-Arabinose	V	0
Erythritol	0	+
D-Ribose	0	+
D-Xylose	+	0

[a]Adapted from Redfearn et al., 1966, and Wilson and Miles, 1975.
[b]V, variable.

P. pseudomallei is a slender, relatively short, noncapsulated, gram-negative rod bearing two or more polar flagella. When grown on a suitable medium, the rods contain PHB granules, visible by Sudan black staining or phase microscopy. On a 24-hour blood agar plate, colonies are 0.2 to 0.6 mm in diameter, translucent, convex, entire, nonpigmented, and oxidase-positive. At 48 to 72 hours the colonies are usually 3 to 4 mm in diameter, opaque, and yellow-brown with a uniquely corrugated appearance. Note that this unique, rugose, "textbook" description refers to aged, not 24-hour, colonies. Strains may vary from rough to mucoid and in color from cream to orange; this heterogeneity in colonial morphology is in contrast to the homogeneity of other features (Redfearn et al., 1966). Useful features for identification of this species, in addition to unique colonial morphology when present, are gas from nitrate; growth at 42° C, acidification of mannitol, sorbitol, and sucrose; and hydrolysis of gelatin, casein (milk), and serum proteins (Loeffler's slant). Denitrification may not occur in peptone-rich media and with stock cultures, and such cultures may also fail to acidify sucrose (Table 4).

Pseudomonas cepacia (*P. multivorans,* biogroup EO-1) was first associated with skin rot of the onion, *Allium cepa.* Later EO-1, associated with humans, and *P. multivorans,* recovered from soil, were shown to be synonymous with *P. cepacia.* Though strains of both plant and clinical origin are now given a common species designation, they do have distinguishing features, particularly their sensitivity to bacteriocins (Gonzalez and Vidaver, 1979). This species is particularly important as an opportunistic pathogen in that it can cause fulminating septicemia. It has also been implicated in endocarditis, necrotizing vasculitis, pneumonitis, and urinary tract and wound infections. Differing from *P. aeruginosa,* it is usually highly resistant to the aminoglycosides. Trimethoprim/sulfamethoxazole, trimethoprim/sulfisoxazole, and chloramphenicol have been effective therapeutic agents.

This species appears to be widely distributed in nature. It has been recovered from cannulas, catheters, distilled water, the eye, feces, nose, plants, soil, solutions of disinfectants (chlorhexidine, Zephiran), throat, and the urinary tract. The two synonyms for *P. cepacia* reflect its features. It grows luxuriantly (eugonically) on routine laboratory media and oxidatively acidifies carbohydrates; "multivorans" indicates its utilization of many substrates, particularly sugars. It is a slender, gram-negative, noncapsulated rod bearing two or more polar flagella and, like *P. pseudomallei,* forms PHB inclusions. Many strains are yellow to orange on iron-rich media such as KIA, but pigmentation is rare on blood agar plates. On such plates, 24-hour colonies are 0.2 to 0.8 mm in diameter, translucent, convex, smooth, entire, and oxidase-positive (frequently weak). At 48 hours, the colonies are 1 to 2 mm in diameter, opaque, and occasionally pigmented. Though sharing many features with *P. pseudomallei,* it differs in acidifying adonitol, never producing gas from nitrate, never attacking arginine (ADH-negative), and always decarboxylating lysine (Table 4).

Pseudomonas vesicularis (*Corynebacterium vesiculare*), initially recovered from a medicinal snail, has recently been implicated in genitourinary tract infections, particularly cervicitis (Otto et al., 1978). Its normal habitat is not known. Like *P. diminuta,* it differs from other pseudomonads in having specific nutritional requirements (biotin, cyanocobalamin, and pantothenate) and bearing a polar flagellum of exceptionally short wavelength (Ballard et al., 1968; Gilardi, 1978b). It has been recovered from cervical and urethral swabs inoculated onto Thayer-Martin medium and grows poorly on initial isolation. Salient features for identification of this species are orange pigmentation; acidification of glucose and maltose but not fructose, lactose, mannitol, or sucrose; and hydrolysis of esculin (Table 4).

OTHER SACCHAROLYTIC PSEUDOMONADS

P. fluorescens and *P. putida,* like *P. aeruginosa,* are fluorescent pseudomonads (Table 3). As noted above, *P. fluorescens* has been implicated in endotoxemia following intravenous administration of contaminated solutions. *P. putida* is present in the hospital environment but, excepting an occasional genitourinary tract infection is rarely implicated in human disease. Like *P. putida, P. pickettii* appears to be present in the hospital environment but only rarely is etiologically significant. Among its salient features (Table 4) are hydrolysis of urea and production of gas from nitrate (at 20 to 30° C but usually not at 35 to 37° C). *P. putrefaciens,* a frequent cause of spoilage in marine foods, is occasionally recovered from clinical specimens and may have etiologic significance (Vandepitte and Debois, 1978). Presently it is a heterogeneous species but nevertheless is easily identified, since it is the only H_2S-positive pseudomonad and also decarboxylates ornithine (ODC-positive), a feature shared with only one other species, *P. cepacia. P. stutzeri,* an occasional opportunistic pathogen in compromised patients, particularly drug addicts, is well known as a resident in the soil and on plants. Though commonly described as presenting a unique colonial morphology and producing gas from nitrate, neither of these features may be evident with clinical isolates. Other features permit identification of this species (Table 4); among these are hydrolysis of starch and acidification of glucose, maltose, and mannitol but not lactose and sucrose.

Pseudomonads presenting lemon-yellow colonies are occasionally recovered from clinical specimens. These are usually biogroups IIk and Ve. Some investigators have equated IIk with genus *Xanthomonas,* a phylogenetic group of plant pathogens, but this synonymy remains controversial (Gilardi, 1978b). Two biotypes of the IIk group are recognized. IIk-1 has the features of a pseudomonad (65 moles per cent GC, oxidative metabolism, one polar flagellum) and has recently been named *P. paucimobilis* (Holmes et al., 1977). IIk-2 differs in having 40 moles per cent GC; this appears to exclude it from genus *Pseudomonas* (48 to 70 moles per cent GC). IIk-2 also differs from IIk-1 in not being strongly pigmented and in several biochemical features. The Ve group ("*Chromobacterium*" *typhiflavum*) similarly contains two biotypes that can easily be distinguished (Table 4). Both appear to be pseudomonads (Gilardi et al., 1975), but neither has been assigned a species epithet.

GLUCOSE-NEGATIVE PSEUDOMONADS

These organisms (Table 5) are infrequently encountered in clinical specimens, are rarely etiologically significant, and are commonly misidentified as *Alcaligenes faecalis;* indeed, the latter species is almost never present in such specimens.

Although they almost never acidify glucose, other features permit their identification. *P. acidovorans* acidifies fructose and usually mannitol; some strains weakly acidify glucose. *P. pseudoalcaligenes* acidifies fructose and, rarely and weakly, arabinose, glucose, and xylose. Additionally, growth at 42° C and alkalinization of acetamide are useful features for identification of these species. These two species and *P. alcaligenes* are the more commonly encountered glucose-negative pseudomonads; *P. denitrificans* (gas from nitrate), *P. diminuta,* and *P. testosteroni* are far less commonly encountered.

References

Artenstein, M. S., and Sanford, J. P. (eds.): Symposium on *Pseudomonas aeruginosa.* J Inf Dis 130(Supplement):51, 1974.

Ballard, R. W., Doudoroff, M., Stanier, R. Y., and Mandel, M.: Taxonomy of the aerobic pseudomonads: *Pseudomonas diminuta* and *P. vesiculare.* J Gen Microbiol 53:349, 1968.

Chung, D. W., and Collier, R. J.: Enzymatically active peptide from the adenosine diphosphate-ribosylating toxin of *Pseudomonas aeruginosa.* Infect Immun 16:832, 1977.

Gilardi, G. L. (ed.): Glucose Nonfermenting Gram-Negative Bacteria in Clinical Microbiology. West Palm Beach, CRC Press, Inc., 1978a.

Gilardi, G. L: Identification of *Pseudomonas* and related bacteria. In Gilardi, G. L. (ed.): Glucose Nonfermenting Gram-Negative Bacteria in Clinical Microbiology. West Palm Beach, CRC Press, 1978b.

Gilardi, G. L., Hirschl, S., and Mandel, M.: Characteristics of yellow-pigmented nonfermentative bacilli (groups VE-1 and VE-2) encountered in clinical bacteriology. J Clin Microbiol 1:384, 1975.

Girard, D. E., Nardone, D. A., and Jones, S. R.: Pleural melioidosis. Am Rev Resp Dis 114:1175, 1976.

Gonzalez, C. F., and Vidaver, A. K.: Bacteriocin, plasmid and pectolytic diversity in *Pseudomonas cepacia* of clinical and plant origin. J Gen Microbiol 110:161, 1979.

Govan, J. R. W., and Fyfe, J. A. M.: Mucoid *Pseudomonas aeruginosa* and cystic fibrosis: Resistance of the mucoid form to carbenicillin, flucloxacillin and tobramycin and the isolation of mucoid variants *in vitro.* J Antimicrob Chemother 4:233, 1978.

Holmes, B., Owen, R. J., Malnick, A. E. H., and Willcox, W. R.: *Pseudomonas paucimobilis,* a new species isolated from human clinical specimens, the hospital environment, and other sources. Int J System Bacteriol 27:133, 1977.

Howe, C., Sampath, A., and Spotnitz, M.: The pseudomallei group: A review. J Inf Dis 124:598, 1971.

Hugh, R., and Gilardi, G. L.: *Pseudomonas.* In Lennette, E. H., Spaulding, E. H., and Truant, J. P. (eds.): Manual of Clinical Microbiology. 2nd ed. Washington, D.C., American Society for Microbiology, 1974.

John, J. F., Jr.: Trimethoprim-sulfamethoxazole therapy of pulmonary melioidosis. Am Rev Resp Dis 114:1021, 1976.

Leake, E. S., Wright, M. J., and Kreger, A. S.: *In vitro* effect of purified proteases of *Pseudomonas aeruginosa* on rabbit lung macrophages. Exp Mol Pathol 29:241, 1978.

Lennette, E. H., Spaulding, E. H., and Truant, J. P. (eds.): Manual of Clinical Microbiology. 2nd ed. Washington, D.C., American Society for Microbiology, 1974.

Leppla, S. H., Martin, O. C., and Muehl, L. A.: The exotoxin of *P. aeruginosa:* A proenzyme having an unusual mode of activation. Biochem Biophys Res Commun 81:532, 1978.

Mandel, M.: Deoxyribonucleic acid base composition in the genus *Pseudomonas.* J Gen Microbiol 43:273, 1966.

Moody, M. R., and Young, V. M.: In vitro susceptibility of *Pseudomonas cepacia* and *Pseudomonas maltophilia* to trimethoprim and trimethoprim-sulfamethoxazole. Antimicrob Agents Chemother 7:836, 1975.

Neu, H. C.: Clinical role of *Pseudomonas aeruginosa.* In Gilardi, G. L. (ed.): Glucose Nonfermenting Gram-Negative Bacteria in Clinical Microbiology. West Palm Beach, CRC Press, Inc., 1978.

Otto, L. A., Deboo, B. S., Capers, E. L., and Pickett, M. J.: *Pseudomonas vesicularis* from cervical specimens. J Clin Microbiol 7:341, 1978.

Otto, L. A., and Pickett, M. J.: Rapid method for identification of gram-negative, nonfermentative bacilli. J Clin Microbiol 3:566, 1976.

Redfearn, M. S., Palleroni, N. J., and Stanier, R. Y.: A comparative study of *Pseudomonas pseudomallei* and *Bacillus mallei.* J Gen Microbiol 43:293, 1966.

Serotyping Committee, Japan *Pseudomonas aeruginosa* Society: Proposal of an International Standard for the infra-specific serologic classification of *Pseudomonas aeruginosa.* Jpn J Exp Med 49:89, 1979.

Shivanananda, P. G., Nayak, M. N., Mohan, M., Rao, K. N. A., and Singh, Y. I.: Therapeutic trial of pseudomonas vaccine — a preliminary report. Indian J Med Res 68:580, 1978.

Stanier, R. Y., Palleroni, N. J., and Doudoroff, M.: The aerobic pseudomonads: A taxonomic study. J Gen Microbiol 43:159, 1966.

Tatum, H. W., Ewing, W. H., and Weaver, R. E.: Miscellaneous gram-negative bacteria. In Lennette, E. H., Spaulding, E. H., and Truant, J. P. (eds.): Manual of Clinical Microbiology. 2nd ed. Washington, D.C., American Society for Microbiology, 1974.

Vandepitte, J., and Debois, J.: *Pseudomonas putrefaciens* as a cause of bacteremia in humans. J Clin Microbiol 7:70, 1978.

Von Graevenitz, A.: Clinical role of infrequently encountered nonfermenters. In Gilardi, G. L. (ed.): Glucose Nonfermenting Gram-Negative Bacteria in Clinical Microbiology. West Palm Beach, CRC Press, Inc., 1978.

Wilson, G. S., and Miles, A.: Topley and Wilson's Principles of Bacteriology, Virology and Immunity. 6th ed. Baltimore, The Williams & Wilkins Company, 1975.

Zierdt, C. H.: Systems for typing *Pseudomonas aeruginosa.* In Gilardi, G. L. (ed.): Glucose Nonfermenting Gram-Negative Bacteria in Clinical Microbiology. West Palm Beach, CRC Press, Inc., 1978.

34 BRUCELLA

Wesley W. Spink, A.B., M.D.

MORPHOLOGY

Members of the genus *Brucella* occur as small, nonmotile, noncapsular, gram-negative bacilli or coccobacilli that do not form spores and are not acid-fast. They vary from 0.4 to 1.5 μm in length and 0.4 to 0.8 μm in width. Young colonies are pinpoint in size, moist, translucent, and glistening. Six species of *Brucella* are recognized, some with several biotypes. The most significant species in human disease are *Brucella melitensis, Brucella abortus,* and *Brucella suis,* which cause contagious abortion in goats (melitensis), cattle (abortus), and swine (suis), and are shed in their milk. *Brucella neotomae* has been identified in the desert wood rat *(Neotoma lepida)* in the western United States. *Brucella ovis* is an important pathogen in sheep, causing chiefly ram epididymitis. *Brucella canis* is found in dogs, especially in the beagle family, and causes epidemic abortions in kennels. Neither *B. neotomae* nor *B. ovis* has been known to cause human illness, and *B. canis* has resulted in only a few recognized human cases. *Brucella neotomae* grows as smooth colonies on solid media and is probably a derivative of *B. abortus*. *Brucella ovis* and *B. canis* reproduce as rough colonies, and both are probably variants of *B. suis*. Further discussion centers on the three classic species: *melitensis, abortus,* and *suis,* and some of their biotypes. These and other members of the genus are closely related as determined by DNA hybridization. The G & C content of their DNA ranges from 56 to 58 moles per cent (Brinley-Morgan and McCullough, 1974).

Brucella cells contain nucleoprotein, protein, lipopolysaccharide, and polysaccharide antigens. A soluble polysaccharide fraction elicits an immediate localized reaction when injected intradermally into an infected subject, whereas the nucleoprotein material produces a delayed response.

METABOLISM

Although *Brucella* may sometimes be cultivated on chemically defined media containing amino acids, glucose, and B vitamins (nicotinic acid, thiamine, and biotin), the fastidious nutritional requirements of *Brucella* are ordinarily met by tryptose or trypticase broth and agar. Aerobic growth at 37° C requires several days, and for some strains (*abortus*) added carbon dioxide is essential. Neither acid nor gas is produced from carbohydrates in peptone media, catalase activity is variable, and urease is present in *B. suis* and *B. melitensis*. Hydrogen sulfide is produced by all three species, but in variable amounts. All species except *B. neotomae* and *B. ovis* are usually oxidase-positive (Brinley-Morgan and McCullough, 1974). The differential metabolic requirements of the three species and their subtypes are discussed later under Laboratory Diagnosis.

Brucella organisms are killed in 3 minutes at 143° to 145° F, that is, at temperatures used for pasteurization. They are also killed by gastric juice so that persons who ingest the organism are infected by *Brucella* less easily than those who are exposed through skin abrasions (Garrod, 1939; Morales-Otero, 1929).

PATHOGENIC PROPERTIES

The three classic species of *Brucella* are highly invasive, gaining entrance into the body through the oral or ocular mucosae or through abrasions of the skin. Controversy exists as to whether the organisms invade the respiratory tract. A state of bacterial hypersensitivity much like that in tuberculosis is induced. Strains having attenuated virulence such as those used in immunization programs in cattle (*B. abortus* strain 19) can cause human illness when accidentally injected through the skin during the procedure of vaccination. When this occurs in veterinarians having a previous history of brucellosis, an immediate localized reaction takes place at the site of entry of the organisms, which is accompanied by a systemic response with chills and fever. This is due to the acquired *Brucella* hypersensitivity caused by the previous infection (Spink, 1956).

ANTIGENIC COMPOSITION

The basic antigenic components of *Brucella* as defined by Wilson and Miles (1932) with quantitative agglutinin-absorption tests distinguish *B. melitensis* from *B. abortus* and *B. suis*. These species all possess A and M antigen in different proportions; in *B. melitensis* the ratio of A and M is 1:20 and in *B. abortus* it is 20:1. This permits the production of specific monovalent sera for agglutination tests with *melitensis* or *abortus*. From a practical point of view a standardized *Brucella* antigen for diagnostic agglutination tests can be prepared from cultures of any one of the three species.

Brucella organisms produce no exotoxins but, as in all gram-negative bacteria, the lipopolysaccharide (LPS) in their cell wall is an endotoxin. *Brucella* endotoxins have the same chemical and biologic properties as the endotoxins of enteric bacilli (see Chapter 6). The O-specific side chains of *Brucella* LPS are lost when smooth (S) strains undergo dissociation to rough (R) strains. This S→R transformation occurs more readily than in enteric bacteria and is accompanied by loss in virulence and serologic specificity. The loss in virulence implies that the complete LPS is a virulence factor. This implication is supported by the fact that patients with brucellosis uniformly develop hypersensitivity to *Brucella* LPS, and that certain important clinical manifestations can be reproduced in these patients with LPS (Spink, 1956).

A prominent feature of *Brucella* infection is intracellular parasitism of the macrophages and histiocytes of the reticuloendothelial system. Focal aggregates of the parasitized cells produce granulomas of the liver, spleen, and bone marrow. In experimental animals the pathologic response varies with the species (Braude, 1951). In infections of guinea pigs caused by *B. abortus,* there is marked splenomegaly, little if any suppuration, and no discernibly ill health. Guinea pigs infected with *B. suis,* on the other hand, develop extensive suppurating granulomas of the liver, bones, and testicles. *B. melitensis* infections are intermediate in their tissue damage, but more disabling than the other two.

IMMUNITY

Recovery from brucellosis is accompanied by an effective, but not total, acquired immunity to subsequent attacks of the disease. The humoral immune response can be evaluated by measuring bacteriolysins, precipitins, and agglutinins in the serum. Cellular immunity is associated with acquired hypersensitivity, as it is in tuberculosis, and can be detected with the use of *Brucella* antigens for skin tests.

Evidence for immunity against brucellosis has been demonstrated by the remarkable protection afforded cattle, sheep, and goats through the use of vaccines containing living organisms with attenuated virulence.

LABORATORY DIAGNOSIS

Bacteriologic

A definitive diagnosis of brucellosis is made through the isolation of *Brucella* from body fluids or tissues (Alton et al., 1975). For this purpose the culture media of tryptose or trypticase broth and agar is recommended. Liquid media are employed for primary isolation and subcultures on agar slants and plates are used for identification. For the majority of specimens such as blood, cerebrospinal fluid, urine, and minced tissue suspensions, the double medium of Castaneda is recommended. The medium is prepared under sterile conditions and stored in 4 oz rubber-stoppered glass bottles until ready for use. Twenty ml of broth are added to each bottle and then 10 ml of melted agar are layered on one of the narrow edges. Ten per cent of the bottled air should be displaced with CO_2. After the bottle is inoculated with suspected material, it is incubated at 37° C for a minimum of 2 to 3 weeks. After 48 to 72 hours minute colonies of *Brucella* may be detected on the agar surface. Subcultures should be made on agar plates by withdrawing material through the rubber stopper with a needle and syringe every 4th, 7th, 15th, and 21st day. The agar plates should be placed in a jar containing 10 per cent CO_2. Usually organisms can be detected for identification within a week after the primary culture is made.

The isolation of *Brucella* from heavily contaminated substances is possible by the intraperitoneal injection of the material into guinea pigs.

Colonies appearing on the agar surface should be transferred to agar plates and to broth suspensions. A suspected culture of *Brucella* reveals small gram-negative bacilli or coccobacilli. Discrete colonies appearing within 48 to 72 hours should be examined by indirect light. Freshly isolated smooth colonies are gray, translucent, sharply edged, and about 0.5 mm in size. *Brucella* organisms tend to dissociate, especially in broth cultures, from smooth to rough forms, revealing morphologic changes and a decrease in virulence. Rough colonies are flat, dull, opaque, and granular. Mucoid colonies are slimy in consistency.

The observations made by direct examination of the colonial morphology can be supplemented quickly by exposing the organisms to acriflavine and to crystal violet. After a drop of fresh acriflavine solution in a concentration of 1:1000 is added to a loopful of suspended smooth colonies on a slide, examination under low-power magnification shows that the colonies remain in suspension, but rough colonies are immediately agglutinated, and mucoid variants form threads. When a fresh culture of *Brucella* on a plate is flooded with a fresh 1:40 dilution of crystal violet, the smooth colonies are not stained, whereas dissociated rough forms appear red and purple with radial cracks on the surface. Only smooth forms should be used for identification with *Brucella* antiserum because rough variants not only lose their antigenic specificity but also become autoagglutinable owing to loss of their O side chains.

A presumptive identification of *Brucella* can be readily made by mixing a drop of high titer human or rabbit anti-*Brucella* serum with a suspended drop of smooth organisms on a slide when agglutination of *Brucella* occurs. Further substantiation is obtained with the tube-agglutination test and by using appropriate controls of a known *Brucella* culture, anti-*Brucella* serum, and a serum without *Brucella* antibody.

Having identified a culture as belonging to the genus *Brucella,* it is necessary to determine the species of *Brucella*. For practical purposes the aim is to distinguish between the three principal species: *melitensis, abortus,* and *suis*. Methods that are used are indicated in Table 1.

TABLE 1. Tests for Distinguishing Among the Species of Brucella

1. Requirement of added CO_2 for growth
2. Inhibitory action of basic fuchsin and thionin on growth
3. Production of H_2S
4. Urease activity
5. Inhibitory action of bacteriophage on growth
6. Agglutination with monospecific serum
7. Oxidative-metabolic tests

TABLE 2. Inhibiting Action of Fuchsin and Thionin on Brucella

B. melitensis	Growth on all six plates
B. abortus	Growth only in CO_2 and with fuchsin
B. suis	Growth with and without CO_2 but only with thionin

CO_2 Requirement. The vast majority of the initial isolates of *B. abortus* strains require the addition of 10 per cent CO_2 for optimal growth. After many subsequent subcultures, particularly in broth, this requirement is reduced or lost. Strain 19 *B. abortus,* used widely in the United States for vaccinating cattle, does not require added CO_2.

Inhibitory Action of Fuchsin and Thionin. Freshly prepared agar plates with either of these dyes from a stocked solution should be used. The test is performed by dividing a plate into four segments and streaking suspensions of each of the known three species on three segments, and the unknown on the fourth. One set of two plates, one with fuchsin and one with thionin, is placed in a jar with 10 per cent CO_2 added and incubated at 37° C. A second set of plates is incubated without added CO_2. Growth is checked at 48 and 72 hours. The following results in Table 2 distinguish among the three species.

H_2S Production. The three species can be differentiated by the amount of S liberated from agar slants by reproducing organisms. This is determined by suspending in tubes dried filter strips that have been soaked in a 10 per cent solution of lead acetate and then noting the intensity of the black precipitated sulfide on the paper. Each of three tubes contain one of the known species of *Brucella* and the fourth has the unknown culture. The tubes are incubated at 37° C for 4 days with 10 per cent CO_2 added to each tube. Every 24 hours the strips are examined and replaced by fresh ones. The degree of blackening is measured from 0 to 4+. A typical series of observations is as follows:

DAY	ABORTUS	MELITENSIS	SUIS	UNKNOWN
1	+	0	++	+
2	++	0	+++	+
3	±	0	++++	±
4	0	0	+++	0

The "unknown" is *B. abortus.* H_2S production is most intense with *B. suis* and is not observed with *B. melitensis.*

Urease Activity. Most *Brucella* strains split urea to form ammonia. This activity can be measured by adding urea to the media along with phenol red as an indicator. A loopful of known *Brucella* organisms and an unknown culture is added to the surface of an agar slant and observed at room temperature. *Brucella suis* produces a pink color of the medium almost immediately, reaching a maximum intensity within 15 to 30 minutes, whereas the color change with *B. abortus* takes several hours. The color change with *B. melitensis* occurs more slowly than with *B. abortus* and the overall results are variable.

Thus, the H_2S and urease tests are most specific in identifying strains of *B. suis.*

Inhibitory Action of Brucella Bacteriophage. Tb (Tbsili) phage, originally isolated in the U.S.S.R., is specific for most *abortus* strains. This test is rarely necessary in a routine diagnostic laboratory. If it is necessary, assistance should be sought from one of the FAO/WHO Research Centers for Brucellosis.

Agglutination with Monospecific Serum. For practical purposes the foregoing procedures for the differentiation of *Brucella* species can be adapted to a diagnostic laboratory. Thus, *B. suis* can be readily distinguished from *abortus* and *melitensis* cultures. The vast majority of fresh isolates of *B. abortus* can be differentiated from *melitensis* and *suis* cultures because of the requirement of added CO_2 for optimal growth of *abortus.* However, on rare occasions some difficulty may be encountered in differentiating *abortus* cultures from those of *melitensis,* especially if the CO_2 requirement for the former is not definite. For this reason the use of monospecific absorbed agglutinating serum can be helpful. An example from our laboratory using monospecific anti-*abortus* and anti-*melitensis* sera is shown in Table 3.

The "unknown" in Table 3 is not *B. melitensis,* but it still could be *B. suis* on the basis of this single procedure. The H_2S and urease tests will make the final distinction.

Although many *Brucella* isolates have been studied intensively on a world-wide basis, the taxonomy of the *Brucella* species is focused on three basic species. Nevertheless, on the basis of biologic and metabolic studies, some strains cannot be strictly assigned to *abortus, suis,* or *melitensis.* These strains have not been classified as new species but rather as biotypes of one of the three. Three biotypes of *B. melitensis* are recognized, two of which have antigenic properties of *abortus* strains as determined with monospecific agglutination sera. Some isolates of *abortus* strains do not require additional CO_2 for growth; in others agglutinins for *melitensis* predominate. Nine types of *abortus* are known. There are four biotypes of *suis.* Although the classic strain forms H_2S readily, three do not, including type 4, which is found in reindeer, and type 2, found in the Danish hare and swine.

Table 3. Differentiation of Abortus and Melitensis Cultures Using Monospecific Serum

	MONOSPECIFIC ANTI-ABORTUS SERUM				MONOSPECIFIC ANTI-MELITENSIS SERUM			
	1:20	1:40	1:80	1:160	1:20	1:40	1:80	1:160
B. abortus	+++[a]	+++	+++	+++	0	0	0	0
B. melitensis	0	0	0	0	++++	+++	+++	+++
Unknown	+++	+++	+++	+++	0	0	0	0

[a]Degree of agglutination ranges from 0 to ++++.

Additional information useful for differentiating the strains has been derived from oxidative-metabolic studies. These studies involve a measurement of the utilization of oxygen with individual sugars or amino acids used as substrates (Meyer, 1974). This procedure is performed in brucellosis research centers such as the U.S. Department of Agriculture, Veterinary Services Laboratories, Box 70, Ames, Iowa 50010, U.S.A.; FAO/WHO Collaborative Centre for Brucellosis, Central Veterinary Laboratory, Weybridge, England; and the Pan American Zoonoses Center, Casella 23, Ramos Mejía, Buenos Aires, Argentina.

Serologic

If *Brucella* organisms are not isolated, the diagnosis can sometimes be made by testing the serum of patients for antibody. The most reliable serologic test is the tube-dilution, saline-agglutination test using a standardized *B. abortus* antigen. The test has minor limitations as far as specificity is concerned. Cross-agglutination with *Vibrio cholera* and *Pasteurella tularensis* occurs in the sera of patients who have either had these diseases or have been vaccinated for them. The antigen-serum mixtures should be incubated for 48 hours at 37° C and then examined. Titers of 1:100 and above are significant, though on occasion healthy individuals who have had the disease months and years previously may have titers of 1:100 or below, rarely higher. "Blocking antibodies" may interfere with a clear-cut reading so that the lower dilutions reveal little or no agglutination, and the higher dilutions show incomplete agglutination. When such blocking is present, centrifuging the mixtures at 3000 rpm may reveal further clumping. The presence of blocking antibody can also be detected by incubating the suspect serum with a known serum of high antibody content. The known serum is then diluted with antigen and incubated, and if blocking is occurring the titer will be lower than in the untreated serum.

Although agglutinins are rarely absent in patients with active disease, especially in those with blood cultures that reveal *Brucella* organisms, the interpretation of agglutinin titers of 1:100 or less is often difficult when the cultures remain sterile. The *Brucella* agglutinin in active disease is IgG antibody, whereas in those who have recovered from the disease it is IgM. The type of antibody present in the serum may be differentiated by adding 2-mercaptoethanol to the antigen-antibody mixtures. Under these circumstances IgM dissociates and no agglutination occurs.

Other serologic tests carried out in some laboratories include the complement-fixation, the Coombs'-antiglobulin, and the fluorescent antibody tests.

Drug Susceptibility

It is not necessary to carry out routinely in vitro drug sensitivity tests on isolated strains of *Brucella* for the purposes of human therapy. Extensive clinical experience in several areas of the world has shown that tetracycline is the drug of choice for the treatment of brucellosis. In severe illness tetracycline can be supplemented with streptomycin. An alternate drug for tetracycline is chloramphenicol.

Tests in vitro also show that the tetracycline antibiotics are the most effective. Tetracycline, for example, inhibits 95 per cent of strains in a concentration of 0.02 μg/ml, and is more bactericidal for *Brucella* than is generally appreciated (Hall and Manion, 1970). Chloramphenicol is approximately 100 times less active against *Brucella* than tetracycline. Streptomycin is as active in vitro as chloramphenicol but the aminoglycoside is not effective in clinical therapy, presumably because it does not effectively enter the cells in which *Brucella* organisms are living.

EPIDEMIOLOGY

Brucellosis is one of the zoonoses — that is, a disease transmitted directly or indirectly from animals to man. The disease is very rarely transmitted from one person to another. The most common animal reservoirs are sheep, goats, cattle, hogs, and caribou, although other domestic animals can harbor the organisms. The prevalence of human disease will depend on the concentration of the animal populations just mentioned, the incidence of animal disease, and the efficiency of disease eradication programs in animals.

Where disease is prevalent in sheep and goats (*B. melitensis*) there will be a high incidence of human infection transmitted through the milk and cheese of these animals. Likewise, brucellosis in cattle (*B. abortus*) results in human disease by ingestion of unpasteurized dairy products. Hogs and caribou infected with *Brucella (B. suis)* transmit disease to veterinarians, farmers, and packing-house workers who handle infected tissues and vaginal discharges of aborting animals. The frequent association of *B. abortus* with fistulous withers in horses provides another source of potential infection. *Brucellae* will survive in dry soil for 40 to 60 days after it is soiled by animal discharges or tissues, and in milk for 10 days at 10° C. Unpasteurized cheese preserves the vitality of these organisms for up to 2 months, but they do not survive in ripened, aged cheeses.

Eradication programs in domestic animals include testing and slaughtering infected animals. Vaccination has been successful in sheep and goats, and in cattle. Prophylactic vaccines in humans are not feasible. The successful eradication program of bovine brucellosis in the United States has markedly reduced the incidence of human brucellosis from 3500 cases reported in 1950 to several hundred in 1970. In other parts of the world, however, animal and human brucellosis is still widespread. The disease is still prevalent in Mexico, Central and South America, much of Africa, France, Italy, Spain, the Middle East, India, Australia, Polynesia, and New Zealand.

References

Alton, G. G., Jones, L. M., and Pietz, D. E.: Laboratory Techniques in Brucellosis. 2nd ed. Geneva, World Health Organization, 1975.

Braude, A. I.: Studies on the pathology and pathogenesis of experimental brucellosis. I. A comparison of the pathogenicity of *Brucella abortus, Brucella melitensis,* and *Brucella suis* for guinea pigs. J Inf Dis *89*:76, 1951.

Brinley-Morgan, W. J., and McCullough, N. B.: *Brucella. In* Bergey's Manual of Determinative Bacteriology. 8th ed. Baltimore, The Williams & Wilkins Company, 1974, p. 278.

Garrod, L. P.: A study of the bactericidal power of hydrochloric acid and of gastric juice. St. Barth Hosp Rep 72:145, 1939.

Hall, W., and Manion, R.: In vitro susceptibility of *Brucella* to various antibiotics. Appl Microbiol 20:600, 1970.

Meyer, M. E.: Advances in research on brucellosis, 1957–1972. In Bradley, C. A., and Cornelius, C. E. (eds.): Advances in Veterinary Science and Comparative Medicine. Vol. 18. New York, Academic Press, 1974, p. 231.

Morales-Otero, P.: Experimental infection of *Brucella abortus* in man: Preliminary report. Puerto Rico J Public Health and Trop Med 5:144, 1929.

Spink, W. W.: The Nature of Brucellosis. Minneapolis, University of Minnesota Press, 1956.

Wilson, G., and Miles, A. A.: The serologic differentiation of smooth strains of the *Brucella* group. Br J Exp Pathol *13*:1, 1932.

BORDETELLA 35

Jean M. Dolby, Ph.D.

Bordetella pertussis is virulent for man and is the cause of whooping cough. *B. parapertussis* also causes whooping cough in children, and *B. bronchiseptica* is an animal pathogen of the upper respiratory tract rarely transmissible to man.

MORPHOLOGY

All three species are minute, gram-negative, nonacidfast coccobacilli 1 × 0.3 to 0.5 μm; *B. pertussis* is shown in Figure 1. *B. bronchiseptica* is the only motile species, and *B. pertussis* is the only one that is capsulated. Pili have been demonstrated on *B. pertussis* (Morse and Morse, 1976).

ANTIGENIC COMPOSITION

The three members of the species are more closely related antigenically to each other than to *Haemophilus* and *Brucella,* with which one or the other has been classified. Some of the antigens are common to the genus *Bordetella,* some are shared by only two of the three species, and some are species- or even strain-specific. Most antigens are protein or lipopolysaccharide components of the bacterial cell wall.

When freshly isolated, *B. pertussis* organisms are nutritionally exacting; they can be slowly adapted to grow on less complex media. The derivative strains, growing on nutrient agar or in nutrient broth, are called Phase IV; they have lost most of the Phase I antigens (Leslie and Gardner, 1931; Standfast, 1951). Phase IV organisms can still be identified, however, because the lipopolysaccharide or their cell walls retains the ability to elicit "bactericidal antigen" as do Phase I cells (Dolby and Ackers, 1975).

More rapid loss of Phase I properties occurs in cells grown in a medium containing (a) magnesium sulfate, to give C-model cells (Lacey, 1960) or

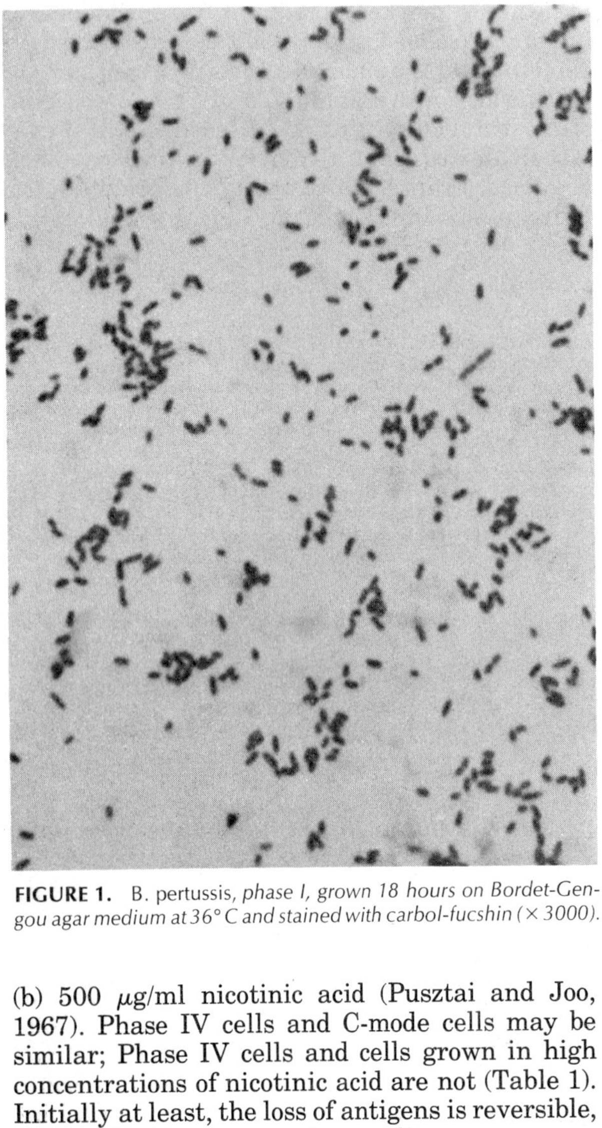

FIGURE 1. B. pertussis, *phase I, grown 18 hours on Bordet-Gengou agar medium at 36° C and stained with carbol-fucshin (× 3000).*

(b) 500 μg/ml nicotinic acid (Pusztai and Joo, 1967). Phase IV cells and C-mode cells may be similar; Phase IV cells and cells grown in high concentrations of nicotinic acid are not (Table 1). Initially at least, the loss of antigens is reversible, and cultivation on complex media causes reversion to typical Phase I cells.

By simple immunodiffusion techniques, about 20 precipitin lines denoting antigen-antibody systems of Phase I antigens can be demonstrated. By chemical methods, 35 proteins can be demonstrated in the cell walls of Phase I *B. pertussis;* Phase IV cells have only two less protein bands (Fig. 2), which cannot account for all the antigens lost (Wardlaw et al., 1975). Electronmicroscope photographs show more differences between Phases I

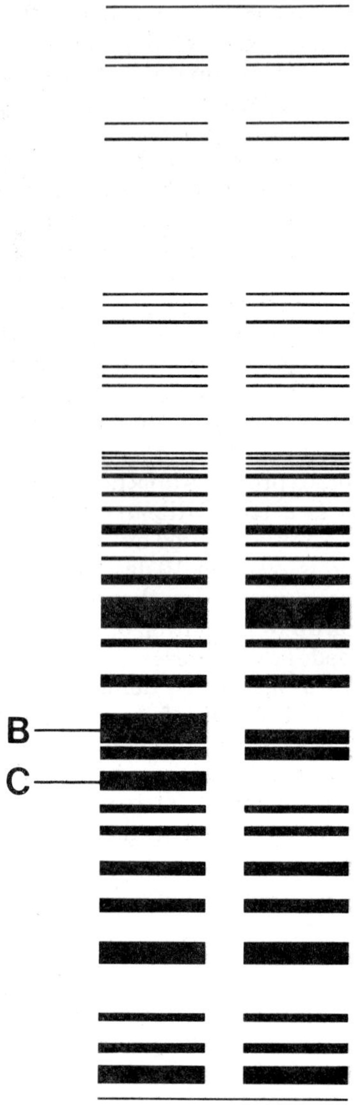

FIGURE 2. *Line drawing of polypeptide band patterns of SDS-polyacrylamide gel electrophoresis preparations of the cell envelopes of phase I and phase IV* B. pertussis. *(From Parton, R., and Wardlaw, A. C.: J Med Microbiol 8:47, 1975.)*

TABLE 1. Antigenic Relationship of *B. pertussis* and Derivatives

ANTIGEN	PHASE I	PHASE IV	NICOTINIC-ACID-GROWN	C-MODE
K agglutinogens 1–6	+	−	−	−
HSF, LPF[a]	+	−?	−	
"Bactericidal antigen"	+ or ±	as ph I	as ph I	
"Protective antigen"	+	−	−	−
"Adhesion antigen"	+	−	+	
Antigen-stimulating opsonin	+	+	−	
Hemagglutinin (pili)	+	−	−	

[a]HSF = histamine-sensitizing factor; LPF = lymphocytosis-promoting factor.

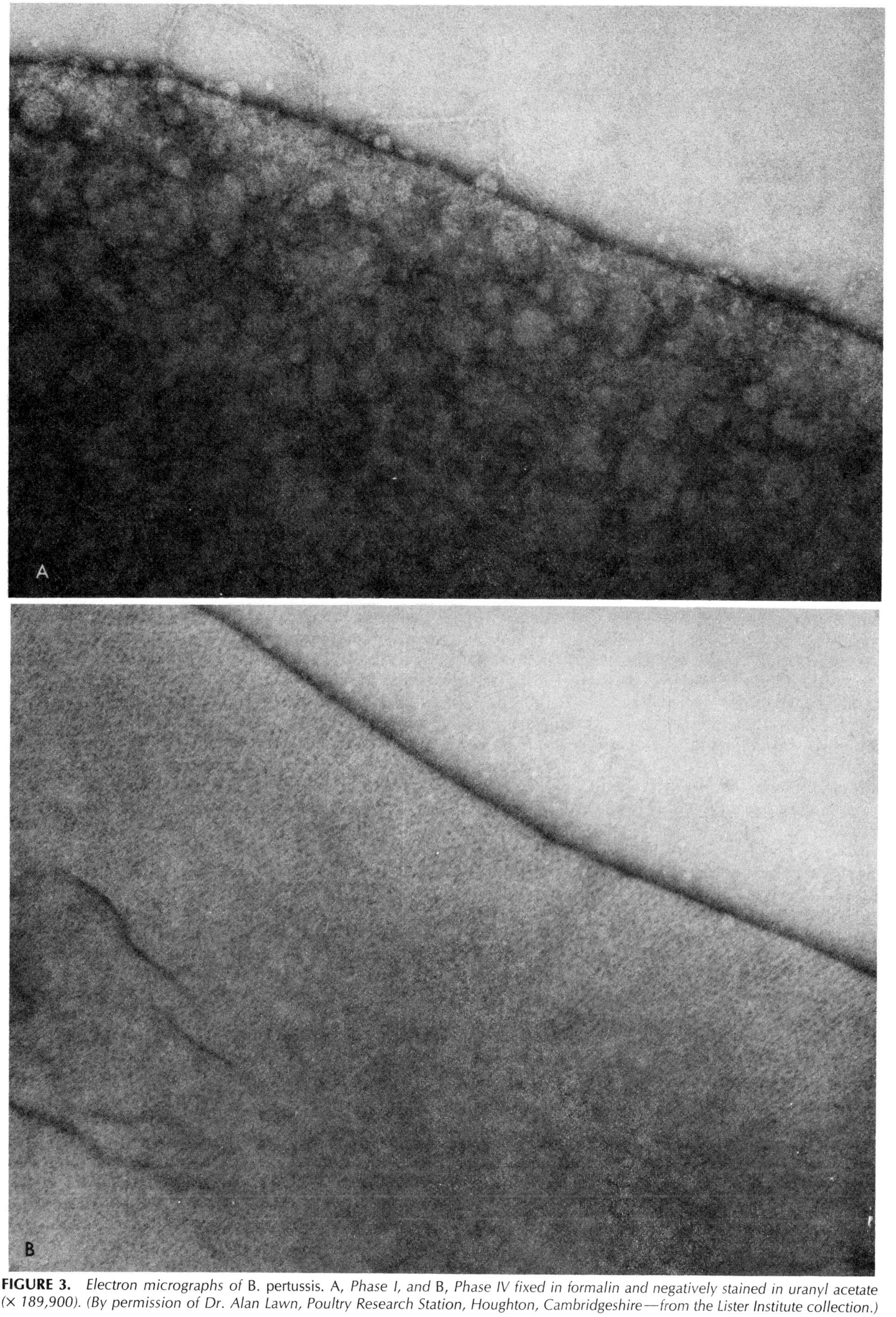

FIGURE 3. *Electron micrographs of* B. pertussis. *A, Phase I, and B, Phase IV fixed in formalin and negatively stained in uranyl acetate (× 189,900). (By permission of Dr. Alan Lawn, Poultry Research Station, Houghton, Cambridgeshire—from the Lister Institute collection.)*

TABLE 2. Antigenic Relationships in the Genus *Bordetella*

	ANTIGEN	B. PERTUSSIS	B. PARAPERTUSSIS	B. BRONCHISEPTICA
	K agglutinogens 7–14	7, 13	7–11, *14*	7–11, *12*, 13
	Toxin	+	+	+
	Antigen-stimulating opsonin	+	+	+
	Pertussis "bactericidal antigen"	+ or ±	−	−
Precipitinogen	Pertussis LPS[a]	+	−	variable
	Parapertussis LPS	−	+	variable
	Bronchiseptica LPS	variable	variable	+
Protection by Vaccine	Pertussis	+	−	variable
against	Parapertussis	−	+	variable
	Bronchiseptica	variable		+

[a]LPS = lipopolysaccharide.

and IV than are apparent from protein analysis (Fig. 3).

B. parapertussis and *B. bronchiseptica* grow on simple media upon isolation, and cultivation on complex media does not alter their antigenic make-up.

Table 1 lists some of the defined antigens of *B. pertussis,* and Table 2 shows those shared by other *Bordetella* organisms. A few are discussed below.

(1) The agglutinogens are a heterogeneous group of antigens that stimulate antibodies including the relatively heat-labile K antigens, protein components of the wall, and the less specific heat-stable lipopolysaccharide components. The K antigens are demonstrable in cells heated at 56° C and in vaccine but are destroyed at 100° C; they are numbered 1 through 14, of which 1 through 7 and 13 are found in *B. pertussis,* 1 through 6 being specific (Andersen, 1953; Eldering et al., 1957). Agglutinogen 1 is found in all *B. pertussis* organisms together with one or more of all of the others, usually in the combinations 3 and 6, 2 and 4. Serotypes are distinguished on the basis of agglutinogens (K antigens), which are stable characteristics with the exception of variable strains of serotypes 1,2,3 (Bronne-Shanbury and Dolby, 1976).

(2) The histamine-sensitizing factor (HSF) and lymphocyte-promoting-factor (LPF) appear to be one antigenic moiety, distinct from the hemagglutinating pili and responsible for several different biologic effects. The responsible moiety appears to be a protein in the cell wall and is composed of four subunits. When injected into mice, it produces leukocytosis and lymphocytosis (LPF) and sensitizes the mice so that they are more susceptible to killing by histamine (HSF). This moiety also prevents the hyperglycemia that follows epinephrine injection and causes fasting hypoglycemia. The four phenomena may be mediated through effects on adenylate cyclase. This protein may also promote the formation of IgE (reaginic antibody) by eliminating the suppressor activity of T-lymphocytes (Munoz and Bergman, 1968; Morse, 1976).

(3) Hemagglutinin has been isolated as pili (Morse and Morse, 1976) and separated from the LPF with which it was thought to be associated (Sato et al., 1974). Others, however, report two hemagglutinins, one associated with LPF and one with fimbriae (Arai and Sato, 1976; Irons and MacLennan, 1978).

(4) Protective antigen (PA) is part of the cell wall and is so called because it induces immunity to infection. It was reported to be a protein 22S antigen (Sato and Nagase, 1967), but this has not been confirmed. This is the most important antigen from the standpoint of conferring immunity to whooping cough.

By the standard assay for PA in mice (see below) it has been shown not to be agglutinogen 1 (Ross and Munoz, 1971), agglutinogens 2 through 6 (Dolby and Bronne-Shanbury, 1975), HSF (Dolby, 1958), LPF (Morse, 1976), hemagglutinin (Pillemer, 1950; Masry, 1952), or the cell wall lipopolysaccharide (Ackers and Dolby, 1972). Evidence that agglutinogens may be involved in protection has been taken from experiments using other assays (Preston and Evans, 1963; Stanbridge and Preston, 1974). Recent work has suggested that a protective property is associated with the fimbrial hemagglutinin.

The antigens of *B. parapertussis* and *B. bron-*

chiseptica that are shared with *B. pertussis* and the species-specific ones are shown in Table 2. *B. bronchiseptica* is antigenically heterogeneous.

METABOLISM

Phase I *B. pertussis* requires a complex medium; the original solid one of Bordet and Gengou (1906) and still used in a modified form contains starch, peptone glycerol, and 33 per cent blood. Increasingly defined liquid ones have been suggested (Cohen and Wheeler, 1946; Lacey, 1954; Stainer and Scholte, 1970) on which *B. pertussis* may be grown for a vaccine. *B. parapertussis* and *B. bronchiseptica* present no problems and can be grown on nutrient agar, broth, and citrate agar. None of them ferment carbohydrates or reduce nitrates to nitrites. *B. pertussis* does not possess a urease, but the other two do.

All three species derive energy from oxidative deamination so that suitable buffering is necessary to retain the optimum pH at about 7.6. Amino acids essential to the metabolism of *B. pertussis* are cysteine and glutamic acid or proline, without which the others cannot be utilized. Growth factors nicotinic acid, glutathione, and ferrous salt are necessary (Rowatt, 1955).

The blood in the agar medium absorbs inhibiting substances and allows growth from small inocula of *B. pertussis.* Albumin or charcoal can be used instead.

The subject has been reviewed by Rowatt (1957) and Parker (1976).

PATHOGENIC PROPERTIES

B. pertussis is virulent for humans (Bordet and Gengou, 1906) and so is *B. parapertussis* (Bradford and Slavin, 1937; Eldering and Kendrick, 1938). The organisms are not invasive but attach themselves to the ciliated epithelium of the upper respiratory tract (Rich, 1932) causing catarrh, lasting about 2 weeks, and then paroxysmal coughing and "whooping," which may last 4 to 6 weeks. *B. bronchiseptica,* mainly a disease of animals, attaches to ciliated epithelium in the same way (Bemis et al., 1977). In exposed humans, it causes chronic catarrh and spasmodic cough (McGowan, 1911; Brown, 1926). In experimental situations, ciliated epithelium elsewhere can become the target of *B. pertussis,* as shown in Figure 4. The surface component of the bacterial cell upon which this attachment depends is obviously of great importance in pathogenicity. It may be the adherence antigen (Holt, 1972), hemagglutinin, or some other antigen. Toxin, common to all three species, that paralyzes the cilia

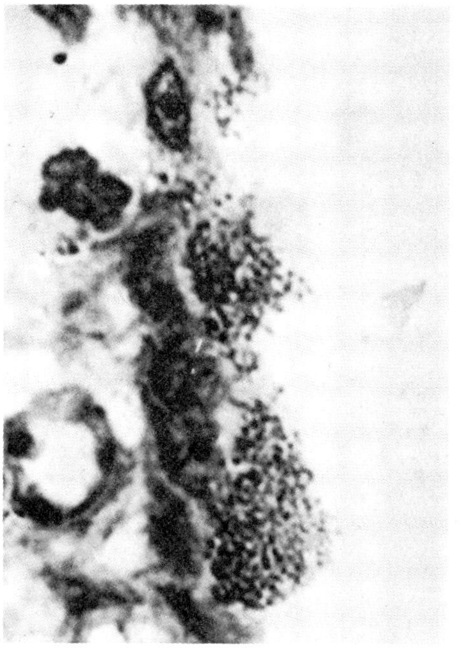

FIGURE 4. B. pertussis *on the ependymal cells of mouse brain 24 hours after infection, Giemsa stain (× 1500). (From Iida, T., et al.: Jap J Exp Med 32:490, 1962.)*

may be only a minor factor in aiding lodgment (Standfast, 1958a).

Toxin may be responsible for the cough and lung tissue damage either directly or indirectly and perhaps also brain damage (Munoz, 1971; Lehrer et al., 1977).

IMMUNITY

Successful immunization has been achieved with parenteral vaccination of children with a whole cell *B. pertussis* vaccine (Medical Research Council, 1951; 1956), and the incidence of whooping cough has been reduced to a very low level in communities with a high acceptance of vaccine (Kendrick, 1975). A cell-free but not purified absorbed bacterial extract (Pillemer, 1950) has also been effective (Medical Research Council, 1959). The effect of vaccination on pertussis in the United Kingdom is shown in Figure 5. *B. parapertussis* vaccine is similarly used with success in areas in which parapertussis is prevalent. An experimental study on vaccination in dogs with *B. bronchiseptica* has been reported (Bemis et al., 1977).

The ability of vaccine to protect mice against an intracerebral but not an intranasal challenge was correlated with the success of the vaccine in children (Standfast, 1958b), as shown in Figure 6. This correlation is the basis of the international assay of vaccine in mice (World Health Organization, 1964).

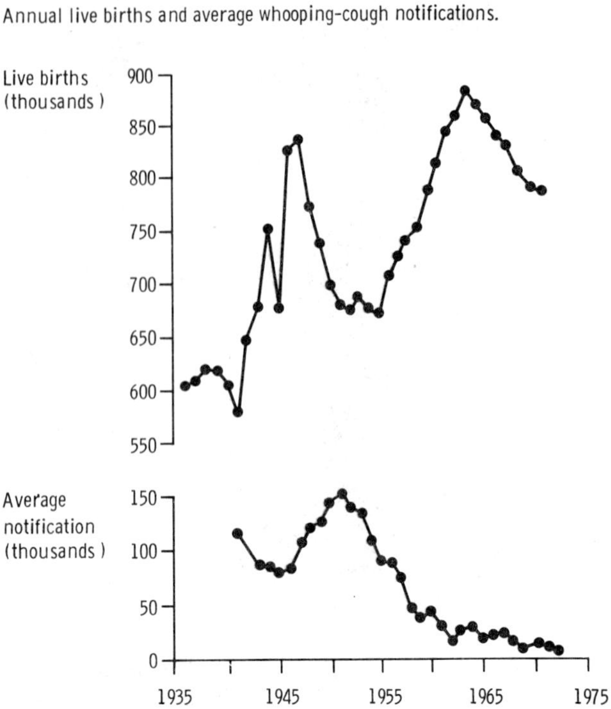

FIGURE 5. *Annual live births and average whooping cough notifications in the United Kingdom from 1940 to 1975. Widespread vaccination was introduced in 1957 at the time that the number of births increased — whooping cough notifications continued to decrease. (From Miller, C. L., et al.: Lancet 2:510, 1974.)*

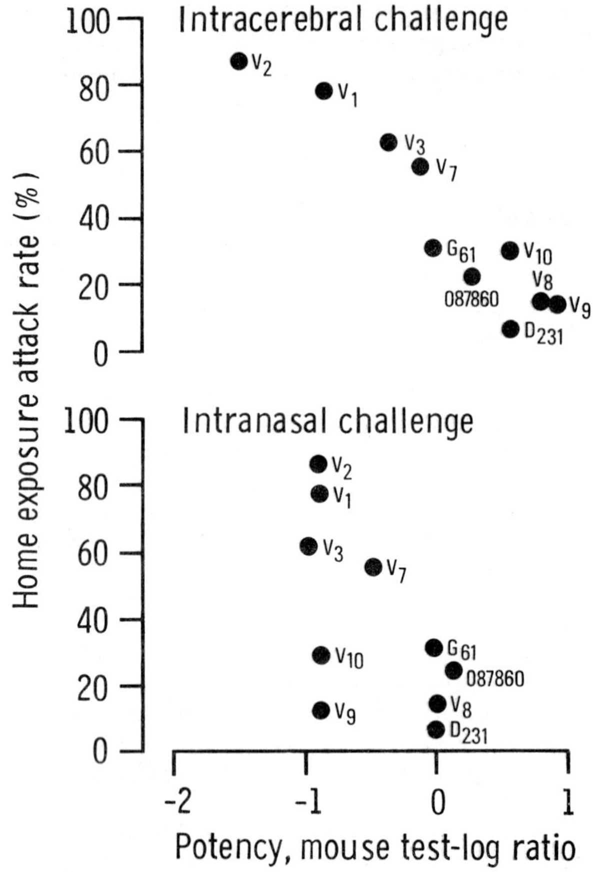

FIGURE 6. *The relationship between mouse protection tests and home exposure attack rate for ten pertussis vaccines:* (top), *close relationship by intracerebral mouse challenge* (bottom), *no relationship by intranasal mouse challenge. (From Standfast, A. F. B.: Immunology 1:135, 1958.)*

The vaccine is a suspension of inactivated cells of *B. pertussis* in Phase I, which is free from unacceptable amounts of endotoxin, heat-labile toxin, and LPF. A course consisting of a minimum of 12 units in three doses is recommended. The addition of alum, which increases antibody response and potency, is not recommended in areas in which poliomyelitis is prevalent.

Protection of the mouse against an intranasal challenge follows vaccination by all *B. pertussis* organisms, including Phase IV (Logan et al., 1959) and is due to antibody against lipopolysaccharide. Protection against an intracerebral challenge, which is correlated with child protection, can be achieved only with Phase I vaccine (Fig. 6). Both 7S and 11S immunoglobulins protect mice against intracerebral infection. Antibody in the 19S fraction is not effective unless injected intracerebrally along with the bacteria. Passive transfer of lymph node cells from immune animals is also protective, but the responsible cell type is unknown (Adams and Hopewell, 1970; Dolby et al., 1975). These antibodies may function as opsonins (Holt, 1972), but not all opsonic antibodies give protection. Mice can be immunized by repeated oral vaccine. This route may offer a new approach to human immunization (Hof et al., 1976). Host-parasite relationships in pathogenesis and immunity to the disease have been reviewed by Pittman (1970).

LABORATORY DIAGNOSIS

Organisms can be cultured in the 1st or 2nd week of the disease from the posterior nasopharynx by using swabs made from 16-cm lengths of thin wire wrapped with cotton wool. Unless plated immediately, the swabs should be held in a tube containing 0.2 ml of 1 per cent casamino acids and plated within 4 to 6 hours onto solid media (Bordet-Gengou or blood-charcoal medium) containing 0.25 to 0.3 unit of benzyl penicillin per ml and, if possible, 2 μg/ml of M&B 938 (4:4 diamidino phenylamine dihydrochloride) (Lacey, 1954). The identity of the isolated *Bordetella* can be checked by agglutination and growth characteristics: *B. parapertussis* grows on nutrient agar with brown pigment formation; *B. bronchiseptica* grows on nutrient agar without forming pigment.

Fluorescent-labeled antipertussis serum to a strain of 1,2,3,4 serotype has given good diagnostic results on smears from nasopharyngeal swabs (Kendrick et al., 1961). A greater success rate has been obtained by using a combination of culture and labeled antibody than by either method alone (Chalvardjian, 1966).

DRUG SUSCEPTIBILITY

B. pertussis is sensitive in vitro to several antibiotics; erythromycin is best, but chloramphenicol, kanamycin, and oxytetracycline are moderately effective (Bass et al., 1969a). None of these has any effect on the course of the disease, but they reduce the infectivity of the patients (Bass et al., 1969b).

EPIDEMIOLOGY

Pertussis and parapertussis are spread by droplet infection, particularly during the catarrhal stage. *B. parapertussis* may account for 0.5 to 5 per cent of all whooping cough (Linneman, 1977). *B. bronchiseptica* is very rare in man.

B. pertussis infection occurs sporadically the world over (Raska, 1970), mainly in children up to 8 to 10 years old, and an attack probably confers immunity of 15 to 20 years, or the same duration as vaccination. The disease incidence undulates (Fig. 7). Newborns are particularly susceptible, but it is still not known whether transfer of immunity from the mother to her baby is impossible, or whether transmissible immunity at child-bearing age is low (Kendrick et al., 1945). Adult vaccination is not recommended, even though the adult population may be susceptible in countries with successful vaccination campaigns (Linneman et al., 1975). Partial immunity decreases the severity of the disease (Miller and Fletcher, 1976), which may pass undiagnosed and may be a source of subsequent infection (Kurt et al., 1972). There is probably no completely asymptomatic carrier state.

A high vaccination rate (greater than 80 per cent) can protect a community. At the beginning of a vaccination campaign, babies should be inoculated at 1, 2, and 3 months of age, but, as these grow up to be immune siblings, the vaccination of the newborn can be postponed to begin at 3 months. Later vaccination produces less general reaction to the vaccine and enables pertussis to be combined with diphtheria and tetanus toxoids (Dudgeon, 1976).

There has been concern in Northern Europe about the rare (1:100,000) brain damage to vaccinated children (Strom, 1967; Kulenkamphff et al., 1974). Such reactions are idiosyncratic and can occur after vaccine or infection (Byers and Moll, 1948). Less rare reactions from which the child recovers are shock and screaming. Decline in the acceptance of vaccine because of the ensuing reactions is an obstacle to any vaccination program. The balance of benefits and hazards in the United Kingdom has been reviewed (Department of Health and Social Security, 1977).

Over the last 30 years, there has been a change in the infecting serotype of *B. pertussis* in some countries, from predominantly 1,2,4 to 1,3,6 (Blaskett et al., 1971; Bronne-Shanbury et al., 1976). It has been suggested that this may be influenced by vaccination (Preston, 1976), particularly

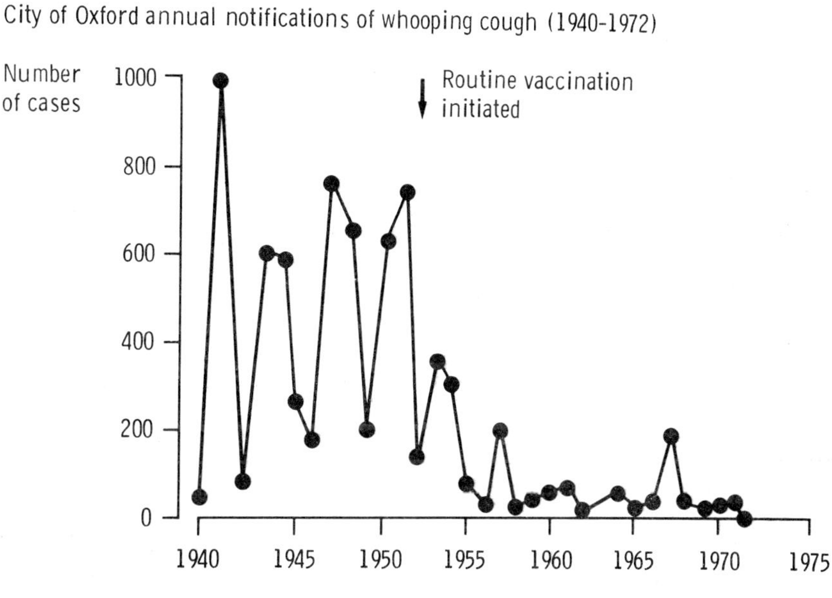

FIGURE 7. *The seasonal fluctuation of* B. pertussis *in Oxfordshire, England. (From Warin, J. F.: Proc Roy Soc Med 67:374, 1974.)*

by vaccine from which one serotype is missing. Vaccine containing the main serotypes is now made, and the consequences are awaited with interest.

References

Ackers, J. P., and Dolby, J. M.: The antigen of *Bordetella pertussis* that induces bactericidal antibody and its relationship to protection in mice. J Gen Microbiol 70:371, 1972.

Adams, G. J., and Hopewell, J. W.: Enhancement of intracerebral infection of mice with *Bordetella pertussis*. J Med Microbiol 3:15, 1970.

Andersen, E. K.: Serological studies on *H. pertussis, H. parapertussis* and *H. bronchisepticus*. Acta Pathol Microbiol Scand 33:202, 1953.

Arai, H., and Sato, Y.: Separation and characterization of two distinct haemagglutinins contained in purified leukocytosis-promoting factor from *Bordetella pertussis*. Biochem Biophys Acta 444:765, 1976.

Bass, J. W., Crast, F. W., Kotheimer, J. B., and Mitchell, I. A.: Susceptibility of *Bordetella pertussis* to 9 antimicrobial agents. Am J Dis Child *117*:276, 1969a.

Bass, J. W., Klenk, E. L., Kotheimer, J. B., Linneman, C. C., Smith, M. H. D.: Antimicrobial treatment of pertussis. J Pediatr *75*:768, 1969b.

Bemis, D. A., Greisen, H. A., and Appel, M. J. G.: Pathogenesis of canine Bordetellosis. J Infect Dis 135:753, 1977.

Blaskett, A. C., Gulasekharam, L. M. S., and Fulton, L. C.: The occurrence of *Bordetella pertussis* serotypes in Australia, 1950–70. Med J Aust 1:781, 1971.

Bordet, J., and Gengou, O.: Le microbe de la coquelucheuse. Ann Inst Pasteur 23:415, 1906.

Bradford, W. C., and Slavin, B.: An organism resembling *Hemophilus pertussis*. Am J Public Health *27*:1277, 1937.

Bronne-Shanbury, C. J., and Dolby, J. M.: The stability of serotypes of *Bordetella pertussis* with particular reference to serotypes 1,2,3,4. J Hyg (Camb) 76:277, 1976.

Bronne-Shanbury, C. J., Miller, D., and Standfast, A. F. B.: The serotypes of *Bordetella pertussis* isolated in Great Britain between 1941 and 1968 and a comparison with serotypes observed in other countries over this period. J Hyg (Camb) 76:265, 1976.

Brown, J. H.: *Bacillus bronchisepticus* infections in a child with symptoms of pertussis. Bull Johns Hopkins Hosp 38:147, 1926.

Byers, R. K., and Moll, F. C.: Encephalopathies following prophylactic pertussis vaccine. Pediatrics 1:437, 1948.

Chalvardjian, N.: The laboratory diagnosis of whooping-cough by fluorescent antibody and by culture methods. Can Med Assoc J 95:263, 1966.

Cohen, S. M., and Wheeler, M. W.: Pertussis vaccine prepared with Phase-I cultures grown in fluid medium. Am J Public Health 36:371, 1946.

Department of Health and Social Security: Whooping Cough Vaccination. London, Her Majesty's Stationery Office, 1977.

Dolby, J. M.: The separation of the histamine-sensitising factor from the protective antigens of *Bordetella pertussis*. Immunology 1:328, 1958.

Dolby, J. M., and Ackers, J. P.: Taxonomic distribution of the antigen eliciting bactericidal antibody for *Bordetella pertussis*. J Gen Microbiol 87:239, 1975.

Dolby, J. M., and Bronne-Shanbury, C. J.: The use of spheroplast-derived strains to differentiate between *Bordetella pertussis* heat-labile agglutinogens and protective antigen for mice. J Biol Stand 3:89, 1975.

Dolby, J. M., Dolby, D. E., and Bronne-Shanbury, C. J.: The effects of humoral, cellular and non-specific immunity on intracerebral *Bordetella pertussis* infections in mice. J Hyg (Camb) 74:85, 1975.

Dudgeon, J. A.: Immunising procedures in childhood. In Hull, D. (ed): Recent Advances in Paediatrics. London, Churchill-Livingstone, 1976, p. 169.

Eldering, G., Hornbeck, C., and Baker, J.: Serological study of *Bordetella pertussis* and related species. J Bacteriol 74:133, 1957.

Eldering, G., and Kendrick, P. L.: Bacillus para-pertussis: A species resembling both *Bacillus pertussis* and *Bacillus bronchisepticus* but identical with neither. J. Bacteriol 35:561, 1938.

Hof, H., Finger, H., Korner, L., and Milke, L.: Effectiveness of orally administered *Bordetella pertussis* vaccine in mice. Dev Biol Stand 33:47, 1976.

Holt, L. B.: The pathology and immunology of *B. pertussis* infections. J Med Microbiol 5:407, 1972.

Iida, T., Kusano, N., Yamamoto, A., and Shiga, H.: Studies on experimental infection with *Bordetella pertussis*. Bacteriological and pathological studies on the mode of infection in the mouse brain. Jpn J Exp Med 32:471, 1962.

Kendrick, P. L.: Can whooping cough be eradicated? J Infect Dis 132:707, 1975.

Kendrick, P. L., Eldering, G., and Eveland, W. C.: Fluorescent antibody techniques. Am J Dis Child 101:149, 1961.

Kendrick, P. L., Thompson, M., and Eldering, G.: Immunity response of mothers and babies to injections of pertussis vaccine during pregnancy. Am J Dis Child 70:25, 1945.

Kulenkampff, M., Schwartzmann, J. S., and Wilson, J.: Neurological complications of pertussis inoculation. Arch Dis Child 49:46, 1974.

Kurt, T. L., Yeager, A. S., Guenette, S., and Dunlop, S.: Spread of pertussis by hospital staff. J Am Med Assoc 221:264, 1972.

Lacey, B. W.: A new selective medium for *Haemophilus pertussis* containing a diamidine, sodium fluoride and penicillin J. Hyg (Camb) 52:273, 1954.

Lacey, B. W.: Antigenic modulation of *Bordetella pertussis*. J Hyg (Camb) 58:57, 1960.

Lehrer, S. B., Nakamura, R. M., and Tan, E. M.: Histopathological changes in mice treated with extracts of histamine sensitising factor of *Bordetella pertussis*. Int Arch Allergy Appl Immunol 54:129, 1977.

Leslie, P. H., and Gardner, A. D.: The phases of *Haemophilus pertussis*. J Hyg (Camb) 31:423, 1931.

Linneman, C. C.: *Bordetella parapertussis*. Recent experience and review of literature. Am J Dis Child 131:560, 1977.

Linneman, C. C., Ramundo, N., Perlstein, P. H., Minton, S. D., Englender, G. S., McCormick, J. B., and Hayes, P. S.: Use of pertussis vaccine in an epidemic involving hospital staff. Lancet 2:540, 1975.

Logan, J. E., Griffiths, B. W., and Mason, M. A.: Specificity of the clearance phenomenon in the lungs of pertussis-immunised mice following intratracheal administration of radio-iodinated pertussis vaccine. Can J Microbiol 5:405, 1959.

McGowan, J. P.: Some observations on a laboratory epidemic principally among dogs and cats in which animals affected presented the symptoms of the disease called 'distemper.' J Pathol Bact 15:372, 1911.

Masry, F. L. G.: Production, extraction and purification of the haemagglutinin of *Haemophilus pertussis*. J Gen Microbiol 7:201, 1952.

Medical Research Council: Vaccination against whooping cough. Investigation. Br Med J 1:1464, 1951; Relation between protection of children and laboratory tests. Ibid 2:454, 1956; Final report. Ibid 1:994, 1959.

Miller, C. L., and Fletcher, W. B.: Severity of notified whooping cough. Br Med J 1:117, 1976.

Miller, C. L., Pollock, T. M., and Clewer, A. D. E.: Whooping cough vaccine, an assessment. Lancet 2:510, 1974.

Morse, S. I.: Biologically active components and properties of *Bordetella pertussis*. Adv Appl Microbiol 20:9, 1976.

Morse, S. I., and Morse, J. H.: Isolation and properties of the leukocytosis- and lymphocytosis-promoting factor of *Bordetella pertussis*. J Exp Med *143*:1483, 1976.

Munoz, J. J.: Protein toxins from *Bordetella pertussis*. In Kadis, S., Montie, T. C., and Ajl, S. J. (eds.): Microbial Toxins. Vol. IIA. New York, Academic Press, 1971, p. 271.

Munoz, J. J., and Bergman, R. K.: Histamine-sensitising factors from microbial agents, with special reference to *Bordetella pertussis*. Bacteriol Rev 32:103, 1968.

Parker, C.: Role of the genetics and physiology of *Bordetella pertussis* in the production of vaccine and the study of host-parasite relationships in pertussis. Adv Appl Microbiol 20:27, 1976.

Parton, R., and Wardlaw, A. C.: Cell envelope proteins of *Bordetella pertussis*. J Med Microbiol 8:47, 1975.

Pillemer, L.: Adsorption of protective antigen of *Hemophilus pertussis* on human red cell stromata. Proc Soc Exp Biol Med 75:704, 1950.

Pittman, M.: *Bordetella pertussis* — bacterial, and host factors in the pathogenesis and prevention of whooping cough. In Mudd, S. (ed.): Infectious agents and host reactions. W. B. Saunders Co., p. 239, 1970.

Preston, N. W.: Prevalent serotypes of *Bordetella pertussis* in nonvaccinated communities. J Hyg (Camb) 77:85, 1976.

Preston, N W., and Evans, P.: Type specific immunity against intracerebral pertussis infection in mice. Nature 197:508, 1963.

Pusztai, S., and Joo, I.: Influence of nicotinic acid on the antigenic structure of *Bordetella pertussis.* Ann Immunol Hung 10:63, 1967.
Raska, K.: Whooping cough: Epidemiological situation in the world. Symposia Series in Immunobiological Standardisation 13:4, 1970.
Rich, A. R.: On the aetiology and pathogenesis of whooping cough. Bull Johns Hopkins Hosp 51:346, 1932.
Ross, R. F., and Munoz, J. J.: Antigens of *Bordetella pertussis.* V. Separation of agglutinogen 1 and mouse-protective activity. Infect Immun 3:243, 1971.
Rowatt, E.: Amino-acid metabolism in the genus *Bordetella.* J Gen Microbiol 13:552, 1955.
Rowatt, E.: Growth requirements of *Bordetella pertussis:* A review. J Gen Microbiol 17:279, 1957.
Sato, Y., Arai, H., and Suzuki, K.: Leucocytosis-promoting factor of *Bordetella pertussis.* III. Its identity with protective antigen. Infect Immun 9:801, 1974.
Sato, Y., and Nagase, K.: Isolation of protective antigen from *Bordetella pertussis.* Biochem Biophys Res Commun 27:195, 1967.
Stainer, D. W., and Scholte, M. J.: A simple, chemically defined medium for the production of phase I *Bordetella pertussis.* J Gen Microbiol 63:211, 1970.
Stanbridge, T. N., and Preston, N. W.: Experimental pertussis infection in the marmoset: Type specificity of active immunity. J. Hyg (Camb) 72:213, 1974.
Standfast, A. F. B.: The phase I of *Haemophilus pertussis.* J Gen Microbiol 5:531, 1951.
Standfast, A. F. B.: Some factors influencing the virulence in mice of *Bordetella pertussis* by the intracerebral route. Immunology 1:123, 1958a.
Standfast, A. F .B.: Comparison between field trials and mouse protection tests against intranasal and intracerebral challenges with *Bordetella pertussis.* Immunology 1:135, 1958b.
Strom, J.: Further experience of reactions, especially of a cerebral nature in conjunction with triple vaccination. Br Med J 4:320, 1967.
Wardlaw, A. C., Parton, R., and Hooker, M. J.: Loss of protective antigen, histamine sensitising factor and envelope polypeptides in cultural variants of *Bordetella pertussis.* J Med Microbiol 9:89, 1975.
Warin, J. F.: Immunisation: Balancing the risks. Proc R Soc Med 67:374, 1974.
World Health Organisation: Requirements for pertussis vaccine. Technical Reports Series No. 274, 25–40, 1964.

HAEMOPHILUS 36

Mogens Kilian, D.D.S., Ph.D.

The organisms of the genus *Haemophilus* are small, gram-negative, nonmotile, and nonspore-forming, facultatively anaerobic rods, which require accessory factors for in vitro growth. The name of the genus refers to the fact that these growth factors are contained in blood cells and may be supplied to ordinary growth media by the addition of blood or its derivatives.

HISTORY

The German bacteriologist Robert Koch was the first to observe bacteria of the genus *Haemophilus* during his examinations of conjunctivitis exudates in Egypt (Koch, 1883). However, the isolation and description of another Haemophilus organism in 1892 by Koch's colleague Richard Pfeiffer caused more sensation (Pfeiffer, 1892). Pfeiffer's bacillus was isolated from purulent sputum and postmortem lung cultures of patients succumbing to influenza during the pandemic of 1889 through 1892. The discovery was published as "the exciting cause of influenza," and the organism was accordingly designated *Haemophilus influenzae* by a committee of the American Society of Bacteriologists. Following the discovery in 1933 of the influenza virus, the actual etiologic agent of epidemic influenza, interest in *H. influenzae* vanished for several years.

The precise role played by *H. influenzae* in the pandemic of 1890 and again in that of 1918 through 1919 is still not clear. The view was held by a number of investigators of the 1918 pandemic that the unprecedented severity of the disease reflected the concurrent interplay of a virus and Pfeiffer's bacillus. This thesis was strengthened by the subsequent demonstration by Shope (1931) that the synergistic effect of a Haemophilus organism (*H. suis*) and swine influenza virus is essential for both natural and experimental swine influenza. Because *H. influenzae* has clearly not played a similar part in more recent pandemics, the importance of viral and bacterial synergism in the pathogenesis of the human disease needs further study (Michaels et al., 1977).

MORPHOLOGY

Bacteria of the genus *Haemophilus* tend to be pleomorphic (Fig. 1). In samples of pathologic material such as cerebrospinal fluid, the bacteria are predominantly coccobacillary, sometimes simulating diplococci and erroneously suspected of being meningococci or pneumococci. In older cultures, or when grown under unfavorable conditions, the rods become elongated and frequently appear filamentous. The pleomorphic appearance is most prominent among the X-factor independent species.

Strains of *H. influenzae* may have capsules that are readily demonstrable only in young cultures. After 24 hours of incubation, the capsules often

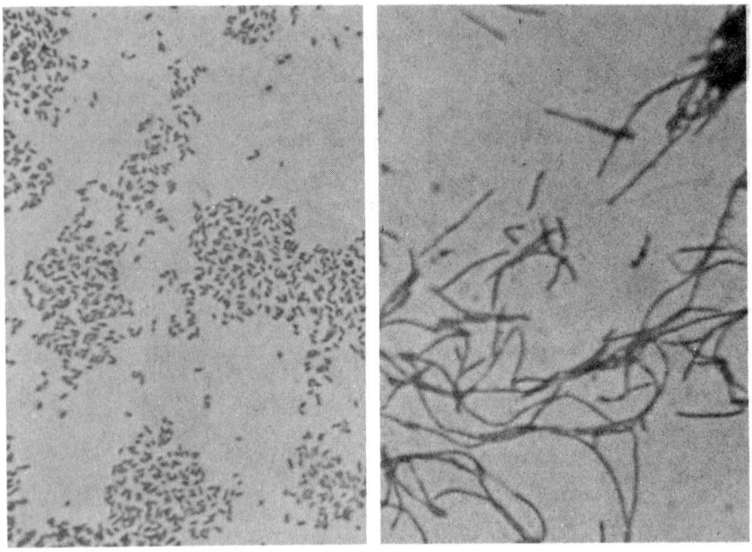

FIGURE 1. *Variations in microscopic appearance of Hemophilus organisms (Gram-stained smears × 800).*

disappear. Capsular material is shed into liquid culture media during growth.

Colonies formed on solid media are 1 to 3 mm long after incubation for 24 hours. Young cultures (8 to 18 hours) of encapsulated strains on translucent solid media are characteristically iridescent (produce a rainbow effect) when examined in obliquely transmitted light. Strains of some species are β-hemolytic (Table 1).

ANTIGENIC COMPOSITION

Virtually all strains of *H. influenzae* isolated from invasive infections possess capsules. The capsular material evokes antibodies that protect against infection. Six distinct antigenic types, designated a through f, have been described by Pittman (1931). The capsular material of all six types are polysaccharides, and all except those of Types d and e contain phosphorus. Strains causing meningitis and other acute illnesses virtually all belong to serotype b. The capsule of this serotype is a polymer of ribose- and ribitolphosphate. It is structurally and serologically closely related to the cell-wall teichoic acid of gram-positive bacteria and the capsular polysaccharides of certain serotypes of pneumococci and certain enterobacteria (Table 2) (Alexander, 1958; Argaman et al., 1974; Bradshaw et al., 1971).

Very little is known about the somatic antigens of *H. influenzae.* A protein common to all types has been described, as well as an endotoxin that resembles those from other gram-negative bacteria.

METABOLISM

Haemophilus organisms grow very poorly on ordinary laboratory media. The hemophilic nature of these bacteria is due to their inability to carry out the biosynthesis of either one or two substances present in blood cells (as well as in most other cells). The two growth factors were first recognized as a heat-stable substance referred to as the *X-factor* and a heat-labile substance referred to as the *V-factor*. The requirement for these factors has been used for the definition of the genus as well as for its internal subdivision into species. *H. influenzae* thus requires both X- and V-factors for growth, whereas most other species require only one (Table 1).

Both X- and V-factors are of vital importance in the metabolism and growth of living cells. The X-factor is used in the biosynthesis of respiratory cytochromes and has been identified as hematin or certain precursors. The V-factor, which is involved in oxidation-reduction processes, can be replaced by one of the two co-dehydrogenases, NAD or NADP (Lwoff and Lwoff, 1937a, b).

Optimal growth occurs in rich media to which the two growth factors have been added. The growth on ordinary blood agar is very poor, since only the X-factor is directly available to the bacteria in satisfactory quantities. Both growth factors may be liberated into the medium either by peptic digestion of the blood (Fildes medium) or by briefly heating the medium (chocolate agar and Levinthal's medium).

Various microorganisms (e.g., staphylococci) excrete the V-factor and may support the growth of haemophili on media deficient in this growth

TABLE 1. Principal Differential Characteristics and Primary Habitat of the Common Species of the Genus *Haemophilus*

	X-FACTOR REQUIRED[a]	V-FACTOR REQUIRED	HEMOLYSIS	INDOLE	GLUCOSE, ACID	SUCROSE, ACID	LACTOSE, ACID	XYLOSE, ACID	NITRATE REDUCTION	PRIMARY HABITAT
H. influenzae	+	+	–	d[b]	+	–	–	+	+	pharynx
H. haemolyticus	+	+	+	d	+	–	–	+	+	pharynx
H. ducreyi	+	–	–	–	–	–	–	–	+	(genital organs)
H. parainfluenzae	–	+	d	–	+	+	–	–	+	oral cavity and pharynx
H. aphrophilus	–	–	–	–	+	+	+	–	+	oral cavity
H. paraphrophilus	–	+	–	–	+	+	+	–	+	oral cavity

[a]X-factor requirement as determined by the porphyrin test (Kilian, 1974). Ref.: Kilian, 1976.
[b]d = differences observed

factor. This satellite phenomenon, which is illustrated in Figure 2, is widely used for the demonstration of this growth requirement.

All *Haemophilus* species reduce nitrate, and all, except for *H. ducreyi*, are heterofermentative. In addition to the requirement for growth factors, fermentation reactions and other biochemical characteristics are useful for the identification of the *Haemophilus* species (Table 1). Virtually all strains of *H. influenzae* isolated from meningitis produce indole and ornithine decarboxylase, characters that are rarely encountered concurrently among nasopharyngeal commensals of the same species (Kilian, 1976).

PATHOGENIC PROPERTIES

H. Influenzae

The infections in which *H. influenzae* is a primary etiologic agent are almost invariably due to encapsulated strains of serotype b, and they are diseases usually associated with bacteremia. The most important of these are meningitis and epiglottitis.

H. influenzae (serotype b) is among the three leading causes of bacterial meningitis, the two others being meningococci and the pneumococci. *Haemophilus* meningitis is, however, more strictly a disease of early childhood than are the two other forms. About 80 per cent of all cases occur between the age of 2 months and the end of the 3rd year, and most of the rest occur during the next few years (Fig. 3). Adults are rarely affected.

Epiglottitis is another serious disease caused by *H. influenzae* serotype b, although less common. As with meningitis, the patients are children but usually somewhat older.

Invasion of the bloodstream by *H. influenzae* (usually serotype b) frequently results in suppurative arthritis, osteomyelitis, and pericarditis. Primary *H. influenzae* pneumonia occurs in children as well as in adults but is an uncommon condition.

H. influenzae is a common and important cause of surface infection of mucous membranes of the respiratory tract and attached sinuses. Such in-

TABLE 2. Bacterial Species Possessing Antigens Cross-Reactive with the Capsular Material of *H. influenzae* Type b

Bacillus alvei
Bacillus pumilus
Escherichia coli K 100
Lactobacillus plantarum
Staphylococcus aureus
Staphylococcus epidermidis
Streptococcus faecalis
Streptococcus pneumoniae types 6, 15a, 29, 35a
Streptococcus pyogenes (group A)

Refs.: Alexander, 1965; Argaman *et al.*, 1974; Bradshaw *et al.*, 1971.

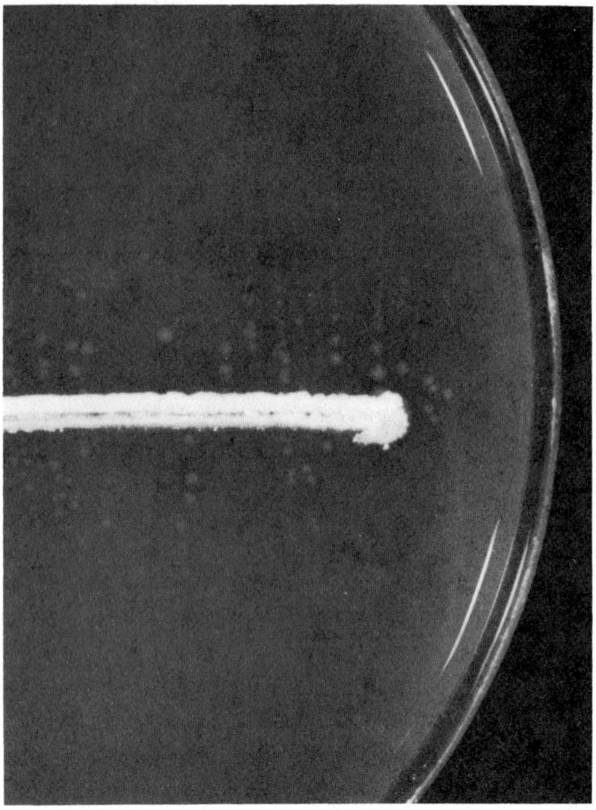

FIGURE 2. *Satellite phenomenon: Colonies of* H. influenzae *growing on an agar plate cross-inoculated with a staphylococcus. The agar medium is autoclaved blood agar that lacks the V-factor.*

fections are secondary to the breakdown of local resistance, as in bronchitis, bronchiectasis, cystic fibrosis, otitis media, and chronic sinusitis. The responsible organisms are usually noncapsulated nasopharyngeal commensals, which enter the normally sterile locations after viruses, allergic reactions, or dysfunctions have initiated the inflammatory processes. The conditions, which appear as acute exacerbations of chronic inflammation, usually respond to chemotherapy. Infection often recurs owing to recolonization with the same or other commensals.

Likewise, lowered resistance of the conjunctival mucosae due to trachomatous or viral infection or obstruction of the lacrimal duct may lead to colonization with noncapsulated *H. influenzae* organisms from the respiratory tract. A hemagglutinating variety of *H. influenzae,* the Koch-Weeks bacillus ("*H. aegyptius*") is associated with a more acute, purulent, and contagious form of conjunctivitis. This disease occurs as seasonal endemics, especially in hot climates.

Capsulation has a bearing upon pathogenicity in the invasive diseases caused by *H. influenzae.* As with pneumococci and meningococci, the polysaccharide capsule protects the organism against phagocytosis. However, the properties that determine the ability of *H. influenzae* serotype b strains to colonize and invade the mucosa of the respiratory tract are largely unknown. There is no evidence that haemophili produce an exotoxin.

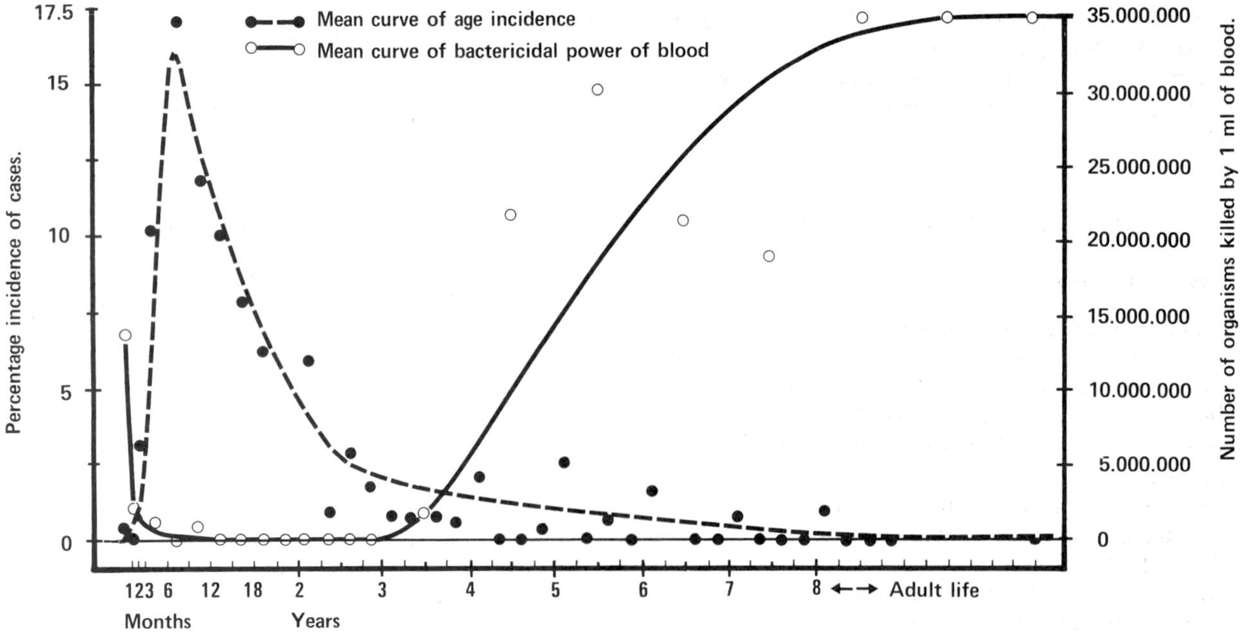

FIGURE 3. *Relation of the age incidence of* H. influenzae *meningitis to bactericidal activity in the blood. (From Fothergill, L. D., and Wright, J.: J Immunol 24:281, 1933. © (1933) The Williams & Wilkins Co., Baltimore.)*

The endotoxin of some strains may cause lethal toxic injury in animals when it is injected in large doses, but there is no conclusive evidence that it plays a significant role in the pathogenicity for humans.

It has recently been discovered that strains of *H. influenzae,* meningococci, and pneumococci produce an enzyme that specifically cleaves immunoglobulin A (subclass 1). This property, which is lacking in nonpathogenic species of the same genera, may prove to be an important virulence factor.

H. ducreyi

This organism was first described by Ducrey (1889) in purulent discharge from the venereal disease "soft chancre" or "chancroid." The organism is recognized as the etiologic agent of the disease. However, it is often isolated together with poorly staining, gram-positive rods, which have been misidentified as *H. ducreyi.* In purulent material collected from ulcerative lesions or bubonic lymph nodes, the organism appears as chains of small gram-negative coccobacilli often located intracellularly.

Other Haemophilus Species

The species found indigenously in the oral cavity, *H. parainfluenzae, H. aphrophilus,* and *H. paraphrophilus,* like other oral bacteria, occasionally cause subacute endocarditis and brain abscesses in predisposed patients. *H. haemolyticus* occurs in the healthy upper respiratory tract and is nonpathogenic. Its β-hemolytic colonies may be mistaken for those of hemolytic streptococci on blood agar.

IMMUNITY

Both *H. influenzae* meningitis and epiglottitis are caused by strains possessing a capsule of serotype b, and anticapsular antibodies, which promote phagocytosis, are protective against disease. Apart from showing distinctive clinical and epidemiologic features, patients who develop meningitis and epiglottiditis also show different serum antibody responses (Whisnant et al., 1976). Children with *H. influenzae* meningitis develop surprisingly low levels of antibodies against the capsular material (Norden et al., 1976). The convalescent-phase levels of antibody after meningitis are even lower than levels found in healthy, age-matched siblings. In patients less than 2 years of age, an antibody response is even undetectable (Anderson et al., 1972; Sell and Karzon, 1973). However, this apparent failure of some children to respond to the invasion of this organism is not due to a general deficiency in immunoglobulin production and may, therefore, suggest some limited and genetically determined specific deficiency in immune response to this particular antigen.

The characteristic age distribution of *H. influenzae* meningitis has been shown by Fothergill and Wright (1933) to be inversely related to that of a serum factor bactericidal to the causative organism (Fig. 3). This finding was used to support the hypothesis that immunity in the newborn is passively acquired from the mother transplacentally, is lost within the first months of life, and then is gradually restored by natural immunization during childhood. However, the serum bactericidal activity detected by these authors does not seem to be caused only by anticapsular antibodies but also by antibodies directed against somatic antigens (Mpairwe, 1971). Although the evaluation of the importance of these latter antibodies is still indirect, resistance to *H. influenzae* Type b meningitis is probably determined by a combined action of anticapsular antibodies, which promote phagocytosis, and complement-dependent bacteriolytic antibodies directed against one or more somatic antigens.

In contrast to meningitis patients, children with epiglottiditis react with high levels of antibody to the capsular antigen of *H. influenzae* Type b. The magnitude of the antibody response correlates with increasing age of the patients. Although the pathogenesis of the disease is largely unknown, the dramatic changes taking place in the epiglottis are likely to be a result of a local allergic reaction.

Acute sinusitis, otitis media, or bronchial infection due to noncapsulated *H. influenzae* strains results in significantly raised serum levels of complement-fixing antibodies against somatic antigens. Detection of this antibody response may be used as a diagnostic tool (Turk and May, 1967).

The contribution of secretory antibodies of the IgA type to resistance against *H. influenzae* infection is still unknown.

Because of the relative rarity of *H. influenzae* Type b in the normal population, the wide prevalence of anticapsular antibodies in normal adults has been difficult to explain. However, there is growing evidence that "natural" antibodies, and hence protective immunity to this organism, is due in part to antigenic stimulation by cross-immunogenic substances in the normal bacterial flora of the body (Table 2).

LABORATORY DIAGNOSIS

Haemophili do not survive well in clinical specimens. Whenever possible a suitable transport medium should be used for transportation of swabs for throat or nose cultures or scrapings from the conjunctivae or chancroid lesions.

If gram-negative rods resembling *H. influenzae* are found in normally sterile material, such as cerebrospinal fluid or pleural fluid, an immediate diagnosis can be made by exposing the organisms in the fluid to a specific anticapsular antiserum. The diagnosis is established if capsular swelling (positive quellung-reaction) is observed. Even in the absence of visible bacteria, a precipitin test (counter-immunoelectrophoresis) for free capsular material may give the diagnosis. Blood culture is important in establishing the diagnosis in the invasive diseases caused by *H. influenzae*, especially in epiglottiditis, since recovery of *H. influenzae* by swabbing the epiglottis is unreliable.

Likewise, smears of scrapings from conjunctivae or chancroid lesions, although not diagnostic, are clinically useful, since recovery of the infectious agent of these diseases is often unsuccessful.

Cultivation of clinical samples for *H. influenzae* is performed on chocolate (heated blood) agar, (Fildes enrichment agar), or ordinary blood agar cross-inoculated with a staphylococcus (Fig. 2). For primary isolation from sources such as the respiratory tract, advantage may be taken of the selective action of bacitracin (300 mg/l). *H. ducreyi* may be cultivated on an agar medium containing 30 per cent freshly drawn rabbit blood or chocolate agar supplemented with 10 per cent Iso-Vitalex plus vancomycin 5 μg/ml (Hammond et al., 1978). Incubation of agar plates is performed in a mixture of air plus 5 to 10 per cent extra carbon dioxide (candle jar).

Identification of isolated strains is done by determining the growth factor requirements and by performing selected biochemical tests (Table 1). The V-factor requirement can be demonstrated by observing the satellite phenomenon, preferably on autoclaved blood agar. The X-factor requirement is difficult to determine on agar media, since most media contain small quantities of hematin. It is most clearly established by a biochemical detection of the in vitro ability of strains to carry out some of the steps involved in the biosynthesis of hematin (porphyrin test) (Kilian, 1974).

DRUG SUSCEPTIBILITY

Most Haemophilus strains are susceptible to penicillin and its derivatives, chloramphenicol, sulfonamides, trimethoprim, and the tetracycline group. These antibiotics have been extensively used to treat infections caused by these organisms. For the treatment of *Haemophilus* meningitis, chloramphenicol and ampicillin are particularly effective. However, since 1974 an increasing number of *Haemophilus* isolates have been found to be resistant to ampicillin due to plasmid-mediated β-lactamase production (Katz, 1975; Sykes et al., 1975). In areas of the world with extensive use of antibacterial drugs, the prevalence of ampicillin resistance among *Haemophilus* strains is approaching 10 per cent. More recently, plasmid-mediated resistance against chloramphenicol and tetracycline have likewise been recorded. The choice of initial therapy in acute *Haemophilus* infections should, at present, depend on the prevalence of resistance to ampicillin and chloramphenicol in the particular area. Subsequent revision of the chemotherapy is done according to the results of the bacteriologic examination.

EPIDEMIOLOGY

Haemophili form part of the normal flora of the upper respiratory tract and oral cavity soon after birth. The predominant species in the pharynx is *H. influenzae*, which is virtually absent from the oral cavity (Table 1). The majority of *H. influenzae* strains found in the indigenous flora are noncapsulated and differ biochemically from strains producing invasive diseases. The carriage rate of capsulated strains of *H. influenzae* seldom exceeds 5 per cent, although it may be considerably higher in closed communities of children (Kilian, 1976).

Over a recent 25-year period, *H. influenzae* Type b infection has shown both relative and absolute increases in incidence (Sell and Karzan, 1973). Statistics from the United States indicate that, before the age of 5 years, one in 400 to 500 children will contract *Haemophilus* meningitis. The attack rate in blacks appears to be higher.

Haemophilus meningitis occurs almost always sporadically. The infectious organism spreads from healthy carriers, but case-to-case spread of *Haemophilus* meningitis is very rare. Asymptomatic nasopharyngeal carriage of *H. influenzae* Type b is common in homes in which *Haemophilus* meningitis occurs. It seems that the organism must pass through one or two partially immune contacts before it can cross the meningeal barrier.

The unpredictability of *Haemophilus* meningitis deprives us of any practicable means of preventing it. Prophylactic chemotherapy is indicated only in special situations in which younger children are at high risk. The effect of an experimental vaccine prepared from purified polyribophosphate from the capsule of *H. influenzae* Type b has been studied. Immunization is well tolerated, but a significant antibody response is only achieved in adults and children older than 2 years (Anderson et al., 1972).

References

Alexander, H. E.: The Hemophilus group. In Dubos, R. J. (ed.): Bacterial and Mycotic Infections in Man. 3rd ed. Philadelphia, J.B. Lippincott Company, 1958, p. 470.

Anderson, P., Peter, G., Johnston, R., Jr., Wetterlow, L. H., and Smith, D. H.: Immunization of humans with polyribophosphate, the capsular antigen of *Hemophilus influenzae*, type b. J Clin Invest 51:39, 1972.

Argaman, M., Liu, T-Y., and Robbins, J. B.: Polyribitolphosphate: An antigen of four gram-positive bacteria cross-reactive with the capsular polysaccharide of *Haemophilus influenzae* type b. J Immunol 112:649, 1974.

Bradshaw, M. W., Schneerson, R., Parke, F. C., and Robbins, F. B.: Bacterial antigens cross-reactive with the capsular polysaccharide of *Haemophilus influenzae* type b. Lancet 1:1095, 1971.

Fothergill, L. D., and Wright, J.: Influenzal meningitis: The relation of age-incidence to the bactericidal power of blood against the causal organism. J. Immunol 24:273, 1933.

Hammond, G. W., Lean, C. J., Wilt, S. C., and Ronald, A. R.: Comparison of specimen collection and laboratory techniques for the isolation of *Haemophilus ducreyi*. J Clin Microbiol 7:39, 1978.

Katz, S. L.: Ampicillin-resistant *Haemophilus influenzae* type b: A status report. Pediatrics 55:6, 1975.

Kilian, M.: A rapid method for the differentiation of *Haemophilus* strains. The porphyrin test. Acta Pathol Microbiol Scand [8] 82:835, 1974.

Kilian, M.: A taxonomic study of the genus *Haemophilus*, with the proposal of new species. J Gen Microbiol 93:9, 1976.

Koch, R.: Bericht über die Thätigkeit der deutschen Cholerakommision in Aegypten und Ostindien. Wien Med Wochenschr 33:1548, 1883.

Lwoff, A., and Lwoff, M.: Role physiologique de l'hémine pour *Haemophilus influenzae*. Pfeiffer. Ann Inst Pasteur 59:129, 1937a.

Lwoff, A., and Lwoff, M.: Studies on codehydrogenases. I. Nature of growth factor "V". Proc Royal Soc B 122:352, 1937b.

Michaels, R. H., Myerowitz, R. L., and Klaw, R.: Potentiation of experimental meningitis due to *Haemophilus influenzae* by influenza A virus. J Infect Dis 135:641, 1977.

Mpairwe, Y.: Immunity to *Haemophilus influenzae* type b: The nature of the bactericidal antibody in human blood. J Med Microbiol 4:43, 1971.

Norden, C. W., Michaels, R. H., and Melish, M.: Serologic responses of children with meningitis due to *Haemophilus influenzae* type b. J Infect Dis 134:495, 1976.

Pfeiffer, R.: The influenza bacillus. I. Preliminary communication on the exciting causes of influenza. Br Med J:i:128, 1892.

Pittman, M.: Variation and type specificity in the bacterial species *Haemophilus influenzae*. J Exp Med 53:471, 1931.

Sell, S. H. W., and Karzon, D. T. (eds.): *Hemophilus influenzae*. Nashville, Vanderbilt University Press, 1973.

Shope, R. E.: Swine influenza: Experimental transmission and pathology. J Exp Med 54:349, 1931.

Sykes, R. B., Matthew, M., and O'Callaghan, C. H.: R-factor mediated β-lactamase production by *Haemophilus influenzae*. J Med Microbiol 8:437, 1975.

Turk, D. C., and May, J. R.: *Hemophilus influenzae*. Its clinical importance. London, The English Universities Press Ltd., 1967.

Whisnant, J. K., Rogentine, G. N., Gralnick, M. A., Schlesselman, J. J., and Robbins, J. B.: Host factors and antibody response in *Haemophilus influenzae* type b meningitis and epiglottitis. J Infect Dis 133:448, 1976.

37 YERSINIA, PASTEURELLA, AND FRANCISELLA

T. H. Chen, M.D., and
Sanford S. Elberg, Ph.D.

The new genus *Yersinia*, of the family Enterobacteriaceae, comprises the species *Y. pestis*, *Y. pseudotuberculosis*, and *Y. enterocolitica* (Bergey's Manual of Determinative Bacteriology, 1974). *Pasteurella* is now restricted to the type species *P. multocida* and a number of other closely related animal pathogens. *Francisella tularensis* is a species of another new genus, *Francisella*. These microorganisms are the causes of plague (*Y. pestis*), yersiniosis (*Y. pseudotuberculosis* and *Y. enterocolitica*), pasteurellosis (*P. multocida* et al.), and tularemia (*F. tularensis*). Although they are primarily rodent, other mammalian, and fowl parasites, they are also responsible for severe human infections.

YERSINIA PESTIS AND PLAGUE

MORPHOLOGY

Yersinia pestis, the causative agent of plague, was independently discovered by Kitasato and Yersin during the great epidemic in Hong Kong in 1894 (Bibel and Chen, 1976). The bacilli (small, oval, pleomorphic rods with rounded ends) are nonmotile and form no spores. They vary considerably in size (average 1.5 × 0.7 μm), stain readily with the basic aniline dyes, and are gram-

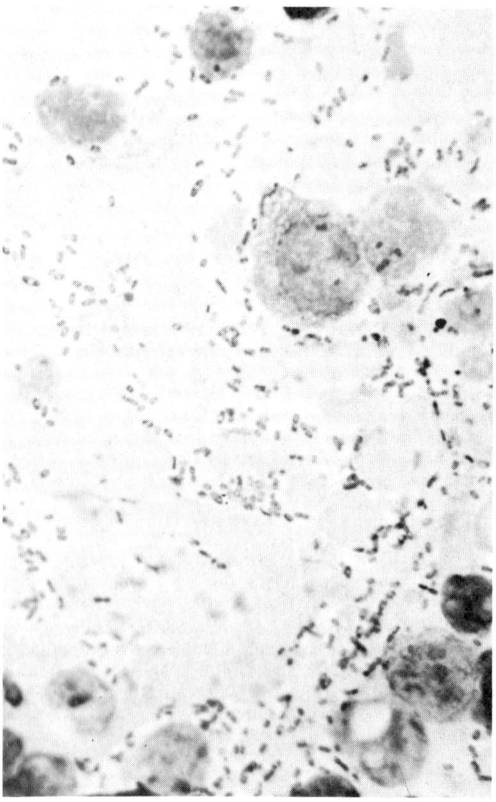

FIGURE 1. Y. pestis *from impression smear of spleen of experimental guinea pig. Wayson's stain.* × *1050.*

negative. With the use of such stains as Wayson's, Giemsa, and Gram's, the central part of the bacillus is often left colorless, yielding the so-called bipolar staining. Thus, the overall morphology of plague bacilli in a smear resembles a scattering of various sized "safety pins" (Fig. 1). In cultures grown at 37° C, envelopes can be demonstrated by negative staining with India ink (Fig. 2). When the organisms are cultivated under suboptimal conditions (with media containing 2 to 4 per cent sodium chloride, i.e., "salt agar"), these involuted forms become very large and take on a striking diversity of shapes, becoming large globular, oval, or pyriform yeastlike bodies (Fig. 3).

On blood agar plates *Y. pestis* grows well but slowly. The pinpoint, transparent colonies become gray and grossly visible after 48 hours and, following 72 hours of incubation, enlarge up to 3 to 4 mm in diameter. If growth is prolonged, particularly at lower temperatures, the colonies have a "fried egg" appearance. Those grown at 37° C for 72 hours, because of the gelatinous surface antigen (envelope) of this organism, are more abundant and have a stringy consistency when tested with a wire loop. In broth *Y. pestis* causes little turbidity, appears slightly granular, and collects at the bottom and sides of the container, illustrating its stalactiform growth property. A feathery vortex can be seen when the culture is gently shaken. In smears from young broth cultures, the bacilli generally emerge in chains, sometimes of considerable length, exhibiting a streptobacillary configuration (Fig. 4).

METABOLISM

Although biochemical tests are not used extensively to identify this microorganism, they may help to distinguish it from other species of *Yersinia* (Buchanan and Gibbons, 1974) (Table 1). The two most important distinctive characteristics of *Y. pestis* are (1) slower alkali production from nitrogenous constituents than other species of *Yersinia* and (2) no hydrolysis of urea. Thus this strain, cultivated on deoxycholate citrate agar plates for 48 hours at 37° C, grows rather scantily in reddish, pinpoint colonies while the medium retains its original pink hue. *Yersinia pseudotuberculosis* bacilli, in contrast, grow plentifully within this period of time in large opaque colonies that (like the medium in general) become yellow. *Yersinia pestis* growth occurs in urea medium without hydrolyzing it. The urease of *Y. pseudotuberculosis* strains reddens the urea medium within 24 hours (Thal and Chen, 1955).

FIGURE 2. Y. pestis *grown on blood agar at 37° C for 72 hours. India ink preparation. Note the envelope antigen around the organisms (arrows).* × *1050.*

FIGURE 3. Y. pestis *grown on 3 per cent salt agar. Wayson's stain. × 1050. Note the involution forms.*

ANTIGENIC COMPOSITION

Yersinia pestis contains at least 19 different antigenic components recognizable by gel diffusion and/or biochemical methods (Albizo and Surgalla, 1970; Chen, 1965; Chen, 1972), 15 of which are shared by the closely related *Y. pseudotuberculosis*. Some of these 19 antigenic constituents of *Y. pestis* have been isolated in pure form, and five components (Fraction 1, the V and W antigens, exotoxin, and endotoxin) correlate well or at least somewhat with immunogenic activity and virulence (see Pathogenic Properties later in this chapter), whereas the remainder bear little or no relationship to plague immunity and infection.

Fraction 1

Fraction 1 or "envelope antigen," which was first isolated in a highly purified state by Baker et al. (Baker et al., 1974), is a principal antigen involved with both virulence and immunity. Envelope antigen consists of two serologically identical components, 1A and 1B, which appear to differ chemically only in respect to the presence of a carbohydrate moiety in the former and its absence in the latter (Bennett and Tornabene, 1974). Bennett and Tornabene reported that the soluble Fraction 1 antigen of *Y. pestis* exists in the form of aggregates more than 300,000 in molecular weight, and that each aggregate can be further separated into a single antigenic subunit of 15,000 to 17,000 MW by dissociation treatment with 0.1 per cent mercaptoethanol in 0.25 per cent sodium dodecyl sulfate (SDS) at 95° C for 5 minutes. These subunits reaggregate into a variety of large molecular weight structures upon removal of the SDS (Bennett and Tornabene, 1974). Chen and Elberg have confirmed these physicochemical observations by scanning electronmicroscopy (Chen and Elberg, 1977).

V and W

In 1956, Burrows and Bacon demonstrated that the virulent *Y. pestis* strain contains two additional antigens, designated V and W, which are not produced by most avirulent strains of *Y. pestis* (Burrows and Bacon, 1956). It was their belief that these components were responsible for the virulence of a strain. V antigen is a protein with a molecular weight of 90,000, and the W antigen is a lipoprotein of 145,000 MW. The titers of both antigens fall upon prolonged storage at 5° C or after lyophilization, but not upon storage at −20° C. Thus, V and W are unstable antigens.

PATHOGENIC PROPERTIES

In experimental plague, the term "virulence" is used to define the comparative lethality of different strains as measured in LD_{50}. Virulent strains have lower LD_{50} levels than avirulent strains. Low virulence resulting from repeated cultivation in artificial media will be restored to full virulence by passage through susceptible animals (guinea pigs), in which selection occurs and the nonvirulent organisms are killed by phagocytes.

FIGURE 4. Y. pestis *grown in broth for 24 hours at 37° C. Wayson's stain. × 1050.*

TABLE 1. Comparison of the Physical and Metabolic Characteristics of *Yersinia* and *P. multocida*[a]

	YERSINIA			
	pestis	enterocolitica	pseudo-tuberculosis	PASTEURELLA MULTOCIDA
Growth in MacConkey's	+	+	+	−
Growth at 5 °C	−	⊕	−	−
Motile	⊖	+ (22°)	+ (22°)	−
Nitrate	+	+	+	+
Citrate	−	−	−	−
Oxidase	−	−	−	v
Indole	−	d	−	+
Urease	−	⊕	⊕	−
β-Galactosidase	+	+	+	−
Ornithine decarboxylase	−	⊕	−	+
Phenylalanine deaminase	·	−	−	·
Fermentations				
lactose	−	−	−	−
glucose	·	·	·	+
galactose	(+)	+	+	+
sucrose	−	⊕	−	+
rhamnose	v	−	+	−
salicin	v	−	+	−
inositol	·	·	·	−
sorbose	−	−	+	−
adenitol	−	−	+	−
esculin	+	⊖	+	−

[a]v, variable reaction; >50% of stains are +; circled reactions are most helpful in differentiating between species.

The virulence determinants of a strain are complex and are associated with its specific antigens (see Antigenic Composition) and nutritional factors. In *Y. pestis*, the five generally established determinants of virulence consist of the capacity to (1) develop an elaborate envelope of Fraction 1 antigen, (2) produce V and W antigens, (3) synthesize a surface structure that permits absorption of hemin from the medium and pigments, (4) synthesize purines, and (5) generate toxins. The major functions of the five virulence determinants are as follows:

Fraction 1, V, and W

Fraction 1, a surface envelope antigen, protects organisms grown in vivo against phagocytosis (Chen and Meyer, 1954) so that they can multiply and induce fatal infection. However, certain nonenveloped strains that still produce V and W antigens retain considerable virulence. Indeed, it has been contended that these components are also responsible for virulence (Burrows and Bacon, 1956).

In an effort to clarify the relationship of Fraction 1 and V and W antigens to the virulence of *Y. pestis*, Donovan et al. (Donovan et al., 1961) studied the infectivity and lethality of two antigenically different strains, enveloped MP6 (wild strain, F+, V and W+), and nonenveloped M23 (mutant strain, F−, V and W+). They found that a single bacterium of the enveloped strain, intradermally injected, could cause fatal infection in the guinea pig. Nonenveloped strains possessing the ability to produce V and W antigens lose their lethality but maintain the capacity to establish infection (skin lesions, buboes, and fever); death does not ensue if the organism is injected in nontoxic dosages. Thus the strain possessing the envelope antigen would appear to be more virulent.

Further studies of the Fraction 1 and V and W antigens by Cavanaugh and Randall (1959) have elucidated the process of plague infection. Following the injection of bacilli in the phagocytosis-sensitive state from the blocked flea into a host (plague-blocked fleas contain nonenveloped [F−, V and W+] virulent *Y. pestis* because the organisms are grown at a temperature lower than 37° C), the majority of organisms were ingested by polymorphonuclear leukocytes (PMN) and destroyed, while the minority, engulfed by monocytes, survived and multiplied. Later the pathogens were released in a well-enveloped form (F+, V and W+) that was resistant to phagocytosis by both PMN and monocytes. These phagocytosis-resistant, enveloped organisms account for the establishment of fatal infection.

Janssen et al. (1963) found no strict correlation between virulence and the envelope antigen's ability to resist phagocytosis. The well-enveloped avirulent strains (F+, V and W−) were more resistant to ingestion by nonsessile phagocytes than nonenveloped strains that produced V and W antigens. Hence these authors concluded that the primary factors in determining the virulence of *Y. pestis* were, rather, the V and W antigens, which permit survival and multiplication of the organisms within the phagocytic cells (monocytes). Apparently the factors determining virulence are Fraction 1, associated with resistance to phagocytosis, and the V and W antigens, associated with the capacity for intracellular multiplication. In contrast, an avirulent (F+, V and W−) strain of *Y. pestis* cannot multiply extensively and is destroyed by phagocytic activity.

Pigmentation

Virulent strains grown on a defined medium ("pigmentation medium" containing hemin) (Jackson and Burrows, 1956) form dark brown (P+) colonies; others produce whitish or straw-colored (P−) colonies in the same medium. To demonstrate high virulence, both for guinea pigs and mice, it is insufficient for a strain to produce only Fraction 1 and V and W antigens (F+, V and W+); it must also be capable of absorbing hemin (P+) from the medium. Nonpigmented colonies consist of avirulent organisms. Loss of pigmentation will result in the loss of virulence. For example, the avirulent, living vaccine strain EV76 is F+, V and W+ but P−.

PURINE

Many nutritional factors can differentiate virulent from avirulent *Y. pestis*. As noted above, fully virulent strains possessing the properties of F+, V and W+, and P+ must also synthesize purine (Pu+) to grow in the host. If a mutant strain derived from a fully virulent one loses the ability to synthesize purine (Pu−), a considerable loss of virulence and failure to produce disease result. However, if purine is injected into mice simultaneously with the bacilli, disease develops (Burrows, 1955).

TOXINS

All of the fully virulent strains have produced toxins (Ts+), which are doubtless responsible for death from plague, suggesting that toxigenicity is a significant virulence determinant.

Two classes of toxins have been isolated and characterized from the plague bacillus. One, a soluble, heat-labile, formalin-sensitive exotoxin, is composed of two active proteins of toxin A (associated with the cytoplasmic membrane) and toxin B (within the cytoplasm) with molecular weights of 240,000 and 120,000, respectively (Montie et al., 1966). These two proteins are highly toxic for rats and mice: the LD_{50} is less than 1 μg in the mouse, as determined via the intravenous route (Ajl et al., 1958). The other, an insoluble, heat-stable lipopolysaccharide endotoxin contained in the cell wall (Albizo and Surgalla, 1970), has an intraperitoneal LD_{50} of about 500 μg in guinea pigs and mice and an intravenous LD_{50} of 32 μg in rabbits. Both the exotoxin and the endotoxin contribute to or account for death in plague, acting respectively on the peripheral vascular system to produce hemoconcentration and shock and on the liver to produce gross pathologic changes and endotoxic shock.

Yet many avirulent strains produce about the same amount or more toxin than the virulent strains (Englesberg et al., 1954); none are fatal, however, if the number of injected bacilli is below the toxic death level. Toxicity alone, then, is inadequate to render an organism virulent.

In summary, the pathogenic properties of *Y. pestis* are complex. To cause fatal infection, a strain should possess all virulent features, i.e., F+, V and W+, P+, Ts+ and Pu+. Absense of one or more factors will cause loss of virulence. Still, there are some strains with all of the known determinants of virulence that, surprisingly, remain not fully virulent. Perhaps the final determinant of virulence is the *quantity* in which these substances are present.

IMMUNITY

Those who recover from plague appear to be protected from a second attack or at least incur a milder form on reinfection. Antibodies to three of nineteen recognized antigens correlate well with immunity, and the remainder have no bearing on plague immunity. The three most important immunogens are Fraction 1, exotoxin, and endotoxin.

A variety of antibodies present in immunized animals cannot eradicate *Y. pestis* in the absence of phagocytic cells. In the blood of immune hosts, plague bacilli are more readily phagocytized than in the blood of the nonimmune host. This is because Fraction 1 antibodies react to the surface envelope of the organisms and act as opsonins. Antibody to Fraction 1 is the principal protective factor against plague in animals as well as in man, as concluded from the positive correlation found between Fraction 1 antibody (CF and PHA titers), the opsonocytophagic index, and resistance to virulent infection following immunization by Fraction 1. Phagocytosis by neutrophils and macrophages of virulent *Y. pestis* directly parallels the development of Fraction 1 antibodies. Immunity to *Y. pestis* infection ultimately depends upon the ability of the macrophage to inactivate and to resist cytotoxic effects of the intracellular organism. Factors other than antibody that are produced by a subpopulation of lymphocytes from spleens of specifically immunized animals are essential for this macrophage function (Wong and Elberg, 1977).

Albizo and Surgalla (1970) isolated and purified the endotoxin and showed that endotoxin can contribute significantly to death in plague. Rabbits can be immunized and develop monospecific antibody that reacts with *Y. pestis* endotoxin. Although we do not understand the neutralizing effect of this antitoxin, it has a curative value, despite the fact that the antitoxic antibodies of exotoxin and endotoxin are not considered essential for serotherapy. Only in near-terminal cases, or when excessive antibiotic therapy may have caused the lysis of numerous bacilli, is antitoxin needed to neutralize the toxin.

Prevention of plague by vaccination has been practiced since the advent of the Haffkine vaccine in 1897. Thereafter, many investigators developed various vaccines, which were prepared as (1) killed broth cultures of *Y. pestis*, (2) chemically inactivated, agar-grown cultures of *Y. pestis*, (3) certain antigenic fractions of *Y. pestis*, or (4) live, attenuated plague strains. At present, the recommended regimen of immunization consists of administering two intramuscular injections, one month apart, of 1×10^9 formalin-killed plague bacilli preserved with phenol in physiologic saline (0.5 ml), followed by a booster dose of 4×10^8 bacilli (0.2 ml) 4 to 12 weeks after the second dose. Measurable levels of antibody can be maintained by booster inoculations given at intervals of not less than six months apart. When administered parenterally, attenuated live *Y. pestis* in repeated revaccinations has been effective during interepidemic periods (Payne et al., 1956). Severe local and systemic reactions have occurred, however, and there is no proof of the efficacy of this vaccine in preventing pneumonic plague. Oral administration of live attenuated EV76 (Paris) F vaccine is efficacious against both bubonic and pneumonic plague in the very susceptible vervet (*Cercopithecus aethiops*) (Chen et al., 1976, 1977). Thus the prospect that an oral, live, avirulent vaccine will be effective for human immunization is promising.

LABORATORY DIAGNOSIS

The clinical diagnosis of plague must be confirmed by conventional bacteriologic procedures. Smears from bubo aspirates, sputum, and blood should be stained with Gram's stain and also with Wayson's reagent. The direct smear should show gram-negative bipolar staining bacilli. A pure culture may be obtained after 48 to 72 hours from such material, with the use of the blood agar plate and infusion broth tube. *Yersinia pestis* grows slowly on blood agar, requiring 48 hours for macroscopic visibility, especially at 37° C. Touched with a wire loop, the 72-hour-growth colonies exhibit some stickiness. *Yersinia pestis* is nonmotile and never displays turbidity in standing broth culture tubes. A feathery vortex can be seen when the tube is gently shaken. Confirmation can also be achieved by specific bacteriophage lysis at 20° C. At a temperature of 37° C, but not 20° C, this phage can also lyse *Y. pseudotuberculosis*. Passive hemagglutination is the most specific and sensitive serologic procedure (Chen and Meyer, 1954, 1966). Positive results can usually be obtained after about five to seven days of illness.

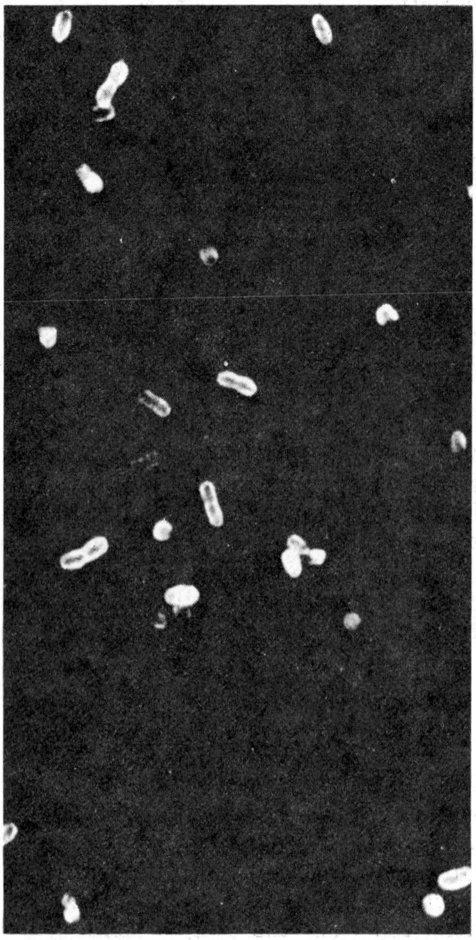

FIGURE 5. *Smear of* Y. pestis *strain 195/P stained with fluorescent Fraction I specific antibody. × 1000. (From Quan, S. F., et al., Am J Trop Med Hyg 14:424, 1965.)*

Inoculation of the plague specimens into mice and guinea pigs produces bacteremia and death within two to seven days. The appearance of the bacilli in these animals, combined with the pathologic changes induced by the infection, helps identify *Y. pestis*. Direct fluorescent antibody techniques allow screening of the cultures and detection of the organisms in infected material (Moody and Winter, 1959) (Fig. 5).

DRUG SUSCEPTIBILITY

Yersinia pestis is susceptible to sulfonamides and antibiotics, including streptomycin, tetracycline, kanamycin (Cantey, 1974), and chloramphenicol. The in vitro and in vivo effects of trimethoprim-sulfamethoxazole on *Y. pestis* are at least equal to those of streptomycin and chloramphenicol (Nguyen Van Ai et al., 1972a, b).

EPIDEMIOLOGY

Plague, primarily a naturally occurring disease of rodents and mammals, infects over 200 species of wild rodents, which are frequently incriminated hosts for *Y. pestis* (Pollitzer and Meyer, 1961). Humans may become victim to the disease through an aberrant interruption of the rodent-flea-rodent sequence or through the handling of infected animals. Usually bubonic plague in humans is transmitted by a bite from fleas (*Xenopsylla cheopis*, rat flea; *Pulex irritans*, human flea) that have previously sucked blood from a plague-infected animal or human. Plague bacilli (as many as 10^7/ml) are present in rat blood during the acute stage, and the flea becomes infected by feeding on the diseased rat (Douglas and Wheeler, 1943). As ingested bacilli multiply in the midgut of the flea, massive infection eventually blocks the proventriculus to the extent that little or no blood can pass. During feeding, some bacilli mix with the blood of the new host and are regurgitated into the wound of the bite, transmitting about 25,000 to 100,000 bacilli. Thus, under natural conditions, transmission from rodent to rodent or human is always mediated by infected fleas. Generally the bubonic form proceeds to the secondary pneumonic form and may lead to person-to-person "droplet" transmission of primary pneumonic plague without the insect vector.

The most prevalent vector species for transmitting plague is *Xenopsylla cheopis*, the rat flea, because of its longevity as a reservoir of plague bacilli and the more frequent impediment of its proventriculus than in other types of fleas. Infected fleas can sometimes survive for 396 days (Pollitzer and Meyer, 1961), suggesting that possibly fleas carry infection from one epizootic season to another, even in the absence of definitive hosts. The human flea, *Pulex irritans*, is a relatively ineffective vector but can transmit plague from person to person under certain circumstances. *Diamanus montanus*, the common squirrel flea, is also an important vector and is a far more serious threat in introducing plague into rat populations in rural and perhaps urban areas (Meyer and Holdenride, 1949). The danger of human plague in an area can be measured by the "flea index" (the average number of fleas per rat). An *X. cheopis* index of at least 3 appears to prevail during epidemics.

Plague now occurs in Central and South Africa, South America, and Asia. In the United States, rural plague has been discovered in ground squirrels, prairie dogs, jack rabbits, voles, pack rats, and other animals in 15 states. A minimum of 50 rodent species is involved in enzootic and epizootic plague (Meyer, 1965). Under natural conditions, camels, sheep, coyotes, deer, dogs, and cats can contract plague infection, either by feeding on infected carcasses or, as in the case of dogs and cats, by acting as mechanical conveyances transporting the infected fleas to their owners. Seventy-eight cases were recognized in the United States from 1966 to 1976, all acquired from areas of wilderness in the enzootic western states (Center for Disease Control, 1975, 1977).

The incidence of plague continues to decline only because the epidemics are under control, rather than because the disease has been eradicated. It is now, in fact, firmly established in endemic foci. Plague is still a potential danger, due to the vast areas of wilderness in which infection persists and the increasing use of wilderness areas for recreational purposes.

YERSINIA PSEUDOTUBERCULOSIS AND YERSINIOSIS

MORPHOLOGY

Yersinia pseudotuberculosis, a pleomorphic, coccoid-appearing, oval or rod shaped microorganism, is gram-negative and measures 0.8 to 6.0 × 0.8 μm. Some strains disclose bipolar staining with Wayson's stain. Neither spores nor definite capsules are detectable, but at 22° C a viscous layer may be visible in India ink preparations. When cultured between 20 and 30° C, the organism is motile, and examination by scanning electronmicroscopy reveals peritrichous flagella (Fig. 6) (Chen and Elberg, 1977).

METABOLISM

A facultative anaerobe, *Yersinia pseudotuberculosis* ferments rhamnose, esculin, salinigrin, adonitol, arabinose, arbutin, dextrin, galactose, and melibiose, producing acid (but no gas). It has both a urease and a β-galactosidase. It can be distinguished from *Y. enterocolitica* and *Y. pestis* by its fermentation of adonitol (Table 1).

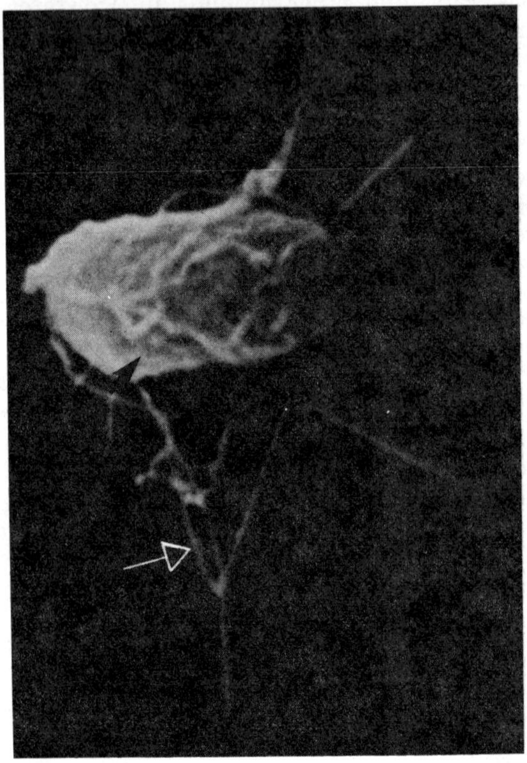

FIGURE 6. Y. pseudotuberculosis *Type I bacillus cultured at 22° C for 48 hours. Note the prominent flagella (arrows). Scanning Electron Micrograph. × 43,000. (From Chen, T. H., and Elberg, Sanford S.: Infect Immun 15:972, 1977.)*

ANTIGENIC COMPOSITION

The multiplicity of antigens in *Y. pseudotuberculosis* is exemplified by the variety of somatic (O) antigens present in different strains. Fifteen somatic and five flagellar (H) antigens have been defined by Thal and Knapp (1969), and a diagnostic antigenic scheme of ten serotypes has been proposed. In diagnostic work it is sufficient to identify the O group.

Many of the antigenic components of *Y. pseudotuberculosis* are shared with those of *Y. pestis*. Indeed, extensive investigation of the antigenic structure of both *Yersinia* species has revealed 17 antigenic components, 15 of which are common to both (Burrows and Bacon, 1956, 1960; Lawton et al., 1960; Larrabee et al., 1965). A protein-lipopolysaccharide complex, protective factor (PF), has been isolated from avirulent *Y. pseudotuberculosis* type IV and *Y. pestis* strains (Lawton and Surgalla, 1963). Burrows and Bacon (1960) demonstrated that only newly isolated *Y. pseudotuberculosis* strains also produce V and W antigens identical to those of *Y. pestis*. (The importance of these two antigens is discussed in the preceding section on the antigenic composition of *Y. pestis*.)

Exotoxins have been isolated mainly in group III strains of *Y. pseudotuberculosis*, although they may also be found in some other groups. Despite the fact that each of these exotoxins is unique, they are nevertheless immunologically identical (Thal, 1954, 1966).

Certain O antigens of *Y. pseudotuberculosis* are related to *Salmonella* antigens, and group II and IV strains can be agglutinated by sera from patients infected with *Salmonella*.

PATHOGENIC PROPERTIES

In humans, yersiniosis, which is characterized by diarrhea and lymphadenopathy with necrosis and septicemia, may be severe and even fatal. Generally, however, the human disease is localized to the mesenteric lymph nodes and commonly has a benign course. The capacity of *Y. pseudotuberculosis* for intracellular survival and reproduction represents a significant factor in the virulence of the organism (Besednova et al., 1975).

IMMUNITY

Immunization against yersiniosis caused by *Y. pseudotuberculosis* has been of wide interest because this pathogen is intimately related antigenically to *Y. pestis* (see Antigenic Composition). Guinea pigs recovered from yersiniosis become immune to reinfection, and active immunity in these animals can be induced by inoculation with suspensions of avirulent, live Type IV strain 32 *Y. pseudotuberculosis* (Thal, 1962). Such immunization will also protect them from infection with *Y. pestis*. Yet guinea pigs so immunized or recovered from *Y. pestis* remain susceptible to yersiniosis. The relationship of the PF antigen to plague and yersiniosis immunity is still unclear. Injection of PF antigen protects guinea pigs as early as the first day by increasing nonspecific resistance to infection. Anti-*Y. pestis* serum usually agglutinates a variety of strains of *Y. pseudotuberculosis*, and anti-*Y. pseudotuberculosis* R serum also reacts with nonenveloped *Y. pestis* strains.

LABORATORY DIAGNOSIS

The diagnosis of yersiniosis is confirmed by isolating *Y. pseudotuberculosis* from the affected blood, liver, spleen, lymph nodes, or feces, or by demonstrating the presence of specific antibodies in the serum. *Y. pseudotuberculosis* is easily cultured in ordinary media. The smooth phase produces a uniform turbidity in broth and a sediment after 24 hours. The rough phase strains grow without generating turbidity. On solid media, transparent colonies reach diameters of 0.5 to 1.0 mm in 24 hours. The pathogen grows abundantly in large opaque colonies on deoxycholate citrate agar, whereas *Y. pestis* grows scantily in reddish pinpoint colonies (Thal and Chen, 1955).

Biochemical and growth characteristics (discussed under Metabolism) also should be assayed (Table 1). The commercial specific typing sera containing agglutinins for types of *Y. pseudotuberculosis* will not agglutinate *Y. pestis*. Likewise, *Y. pseudotuberculosis* bacteriophage will lyse all strains of this organism but will not lyse *Y. pestis*. Further, *Y. pestis* phage will lyse *Y. pseudotuberculosis* at 37° C but not at 22° C. For these reasons, two sets of cultures should be prepared for incubation at two different temperatures (Gunnison et al., 1951). The two simple media that are used to differentiate plague and *Y. pseudotuberculosis* bacilli, namely urea and deoxycholate citrate agar, are extremely useful (Thal and Chen, 1955). *Yersinia pseudotuberculosis* grown below 30° C, but not at 37° C, is motile.

DRUG SUSCEPTIBILITY

Yersinia pseudotuberculosis is sensitive to chloramphenicol, kanamycin, tetracycline, and streptomycin, and some strains are sensitive to penicillin or ampicillin.

EPIDEMIOLOGY

Rodents and birds constitute the primary reservoirs of yersiniosis. Mice and rats, although apparently resistant to natural infection, shed the organisms for weeks in their urine and feces. Thus the resistant rodents may be the carriers and would seem to play an important epidemiologic role in spreading the disease (Mair, 1973).

Yersinia pseudotuberculosis inhabits a diversity of mammals, which are probably the main sources of most human infections. Evidence indicates that the strains isolated from man and animals possess the same cultural, biochemical, and pathologic characteristics. The majority of human cases can be attributed to contact with or ingestion of contaminated food and drinking water, and the consistent involvement of the abdominal organs in this disease also suggests that the mode of transmission may be via the oral route.

YERSINIA ENTEROCOLITICA AND YERSINIOSIS

MORPHOLOGY, METABOLISM, AND ANTIGENIC COMPOSITION

Yersinia enterocolitica was originally isolated by Hässing et al. in 1949 from abscesses at autopsy in two cases of human septicemia. The microorganism is a facultative anaerobe, grows well in air, is gram-negative and oval or coccoid-shaped, and has dimensions of 0.5 to 1.0 × 1.0 to 3.0 μm. It is easily grown on both blood agar and selective media and demonstrates motility by peritrichous flagella when cultivated at less than 30° C. The optimal temperature for the growth of *Y. enterocolitica* is 30 to 37° C, and it has a somewhat slower growth rate at this temperature than other enteric pathogens. The outstanding multiplication feature of the organism is its ability to increase at lower temperatures (±5° C) (other kindred organisms do not). Refrigeration in 2 per cent peptone water of the specimen (from feces, lymph nodes, etc.) therefore results in selective enrichment. Upon primary isolation this bacterium produces smooth colonies, 0.5 to 2.0 mm in diameter, that often exhibit R forms in subculture. When cultivated in broth, either turbidity or a pellicle with clear supernatant and sediment is visible. Arbutin, cellobiose, galactose, sorbitol, and sorbose are fermented, forming acid but no gas. Urease, β-galactosidase, and ornithine decarboxylase are all produced by this organism. *Yersinia enterocolitica* can be distinguished from *Y. pestis* and *Y. pseudotuberculosis* by its fermentation of sucrose, cellobiose, and sorbose as well as by its production of ornithine decarboxylase, which the other organisms do not produce; *Y. enterocolitica* in some instances also forms indole.

On the basis of experimental evidence of heat-stable O-antigen types, a total of 34 different O antigens has lately been established in *Y. enterocolitica* (Wauters et al., 1972). In 1973 Knapp and Thal proposed a revision of the serologic scheme of *Y. enterocolitica* on the basis of six O-antigen groups. Serotype O9 (now known as O-group V) cross-reacts with *Brucella abortus*. The cross-reacting antigens were isolated by hot phenol water extraction of the respective bacteria and identified as lipopolysaccharides.

PATHOGENIC PROPERTIES AND IMMUNITY

O-antigen Types 3 and 9 of *Y. enterocolitica* are pathogenic for humans and are the most common types isolated from humans and swine in Europe. In the United States, Serotype 8 has been isolated from patients, and Canadian strains contain Type 3. Clinical infections with this organism take the form of either acute mesenteric lymphadenitis or enterocolitis. Usually enterocolitis provokes nonbloody diarrhea and abdominal pain due to enlargement of the lymph nodes, while mesenteric lymphadenitis or acute, terminal ileitis elicits appendicitis-like symptoms. Maximal antibody titers are demonstrable in patients by the agglutination test within one or two weeks, declining within one or two months.

LABORATORY DIAGNOSIS AND DRUG SUSCEPTIBILITY

Yersinia enterocolitica has been confirmed as the causative infection by isolation of the organisms from the affected blood, urine, feces, lymph nodes, wounds, or abscesses, or by demonstrating the presence of specific antibodies in the serum. The *Yersinia* species, *enterocolitica* and *pseudotuberculosis*, can be differentiated by their biochemical reactivity, bacteriophage susceptibility, and pathogenicity for laboratory animals. Whereas *Y. enterocolitica* is nonpathogenic for mice, guinea pigs, and rabbits by subcutaneous, intraperitoneal, or intravenous injection (Knapp and Thal, 1963), *Y. pseudotuberculosis* is pathogenic for all these animals. Both species are motile at 22° C, but not at 37° C.

Generally *Y. enterocolitica* and *Brucella abortus* infections result in cross-reacting antibodies, so that yersiniosis and brucellosis are indistinguishable by the agglutination and complement fixation tests, a situation that has caused great confusion in brucellosis eradication work in cattle. However, the two diseases can now be differentiated by electroimmunoassay (Hurwell, 1975).

DRUG SUSCEPTIBILITY

Most strains of *Y. enterocolitica* are sensitive to polymyxin, gentamicin, kanamycin, chloramphenicol, and streptomycin.

EPIDEMIOLOGY

Apparently *Y. enterocolitica* is worldwide in distribution. Extensive animal reservoirs of this organism exist, and it has been isolated from both sick and healthy animals throughout Europe and the United States. Such animals include horses, swine, sheep, goats, cows, dogs, deer, monkeys, and chinchilla. A common source of human infection is food, especially fluids (milk, ice cream, water), contaminated by feces and urine from infected animals. Yersiniosis in both animals and humans is more prevalent in the colder months, possibly because it grows at lower temperature.

PASTEURELLA MULTOCIDA AND PASTEURELLOSIS

MORPHOLOGY

Pasteurella multocida appears as an ellipsoid gram-negative rod or a coccobacillus, ranging from 1.0 to 1.8 × 0.3 to 0.5 μm in size with bipolar-staining capsules (when present), which are best demonstrated by negative staining, as with India ink or Giesma stain. *Pasteurella multocida* is nonsporogenous and nonmotile at both 22 and 37° C. Its optimal growth temperature is 37° C. Small, nonhemolytic, gray colonies form on blood agar, but most nonhemolytic strains create a brownish discoloration in the blood medium. Broth cultures of smooth or mucoid strains are evenly turbid, while the rough strain produces floccular or granular deposits.

METABOLISM

Pasteurella multocida is an aerobe (or facultative anaerobe) that will not grow in MacConkey's agar. It produces acid from galactose, mannose, and sucrose, and most strains ferment mannitol, sorbitol, and xylose (lactose is not fermented). Indole and H_2S are also usually produced, as well as ornithine decarboxylase. The organism is urease-negative and catalase-positive.

ANTIGENIC COMPOSITION

Pasteurella multocida has both capsule polysaccharide and somatic lipopolysaccharide antigens, which can be used to serotype the pathogen. The antigenic structure correlates with strain virulence and host susceptibility (Collins, 1977). Bain (1955) and Dhanda (1960) both isolated a mouse-protective protein antigen from some of their strains of *P. multocida.* On the basis of the capsular antigen, Carter (Carter and Annau, 1953; Carter, 1972a), using the indirect hemagglutination (IHA) test, separated *P. multocida* into four serotypes (A to D). Later, a new type (E) was added (Carter, 1961, 1963) and type C was discarded, narrowing the serotypes to A, B, D, and E.

P. multocida also contains a number of smooth, somatic, lipopolysaccharide antigens. The rapid agglutination test (Carter, 1972b) has been used for subtyping strains classified in the same capsular serotype category. The arabic numeral is utilized for the specific somatic antigen, followed by a capital letter designating the capsular antigen. So far, 11 specific O antigens have been identified, e.g., 1:A, 3:A, 11:B (Namioka and Bruner, 1963).

Endotoxins and their component lipopolysaccharides have been isolated from whole cells and cell walls. They are toxic for mice and pyrogenic in rabbits (Bain and Knox, 1961). Type E lipopolysaccharide is antigenic for rabbits, and mice can be passively immunized with this antiserum (Perreau and Petit, 1963).

PATHOGENIC PROPERTIES AND IMMUNITY

P. multocida represents an animal parasite of considerable economic and veterinary importance, especially to cattlemen and fowl ranchers (Collins, 1977). This microorganism is so highly pathogenic for mice and rabbits that only one to ten cells of a smooth or mucoid variant are usually fatal in the mouse. Infrequently, it may be found in the throats of healthy humans (Smith, 1959), and under conditions of stress or debilitation of the host, the parasite, previously benign, may penetrate beneath the mucous membrane and result in frank disease. Most infections take one of three clinical patterns: local infection by cat scratch or animal bite, respiratory tract infection, or systemic infections such as meningitis and bacteremia.

Agglutinins in the sera of patients with generalized infection can be detected by the hemagglutination test. (Normally, no antibodies are present in the patient with a localized wound infection.) It is not known if these antibodies are bactericidal or antitoxic, although evidence indicates that the immunity is humorally mediated (Collins, 1973). The immunization of domestic animals (cattle) and birds (poultry) with a killed vaccine conveying protection against pasteurellosis is often practiced. Viable, attenuated, oral vaccines have also proved effective in turkeys and chickens. No vaccine exists for man.

LABORATORY DIAGNOSIS

Specimens from infected wounds, sinuses, sputum, pleural or spinal fluid, blood and postmortem tissues should be placed on 5 per cent blood agar. Pure cultures can be further identified as *P. multocida* by their inability to grow in MacConkey's agar, biochemical reactions, lack of motility at both 22 and 37° C, sensitivity to penicillin, and high pathogenicity for mice. Bacteriophage can also be used for identification of *P. multocida* (Gadberry and Miller, 1977).

DRUG SUSCEPTIBILITY

P. multocida is inhibited by drugs at the following average concentrations (mg/ml): chloramphenicol (1.5), tetracycline (3.0), penicillin (0.4), and erythromycin (3.1). Chang and Carter (1976), studying the multiple drug resistance of *P. multocida* and *P. hemolytica,* observed that penicillin and tetracycline may not always be effective. Chloramphenicol proved most efficacious; only 4 of 403 isolates were resistant to this agent.

EPIDEMIOLOGY

P. multocida can survive for only a short time outside the host; thus, the reservoir of infection is probably a healthy carrier. Mammals, birds, and people have been found to be reservoirs for *P. multocida.* In general, human infections are acquired from animal bites, and 15 per cent of cat bites and dog bites treated in United States and Russian hospitals have yielded positive cultures of *P. multocida* (Francis et al., 1975). Most human respiratory and septicemic infections occur in farmers. Most human isolates are Serotype A or D (Carter, 1962). On occasion, the organism is discovered in the throats of healthy individuals exposed to infected animals (Smith, 1959), so the possibility of interhuman spreading by nasopharyngeal excretions, urine, and feces must be considered (Hubbert and Rosen, 1970). *P. multocida* may remain in the host for a long time, later asserting its pathogenic properties when resistance to the infection diminishes. In animals, pasteurellosis has been characterized as a disease of debility. In fact, stress in livestock during transportation is responsible for the illness known as "shipping fever."

FRANCISELLA TULARENSIS AND TULAREMIA

In 1911, McCoy, exploring plague in California ground squirrels, discovered a disease characterized by lesions similar to those of plague. The following year, he and Chapin isolated the causative organism of this plaguelike disease by means of coagulated egg yolk medium and named it *Bacterium tularense,* after Tulare County, California. After previous assignment to several genera, this organism is now known by its own genus name, *Francisella,* in honor of Francis, a pioneer investigator of the disease in humans (Francis, 1921a, b), who infelicitously termed it "tularemia" to indicate its ability to cause septicemia.

MORPHOLOGY

Francisella tularensis is small and aerobic, measuring 0.2 × 0.2 to 0.7 μm. A gram-negative, nonmotile, noncapsular, nonspore-forming bacillus, it is highly pleomorphic, often appearing as a rod or coccus and always singly. The bacterium forms minute, transparent droplike colonies on blood-glucose-cystine agar. Colonies may reach a diameter of 4 mm after two to five days of incubation and are readily emulsified. They require more cystine than is present in the ordinary nutrient medium.

METABOLISM

The optimal growth temperature of *F. tularensis* is 37° C. Most strains ferment glucose, maltose, mannose, fructose, and dextrose, creating acid without gas. The organism is grown in medium containing cystine or cysteine and produces H_2S (Table 1). Enriched media with incorporated antibiotics (penicillin, polymyxin B, and cycloheximide) may be of value when clinical specimens harboring normal flora are cultured for *F. tularensis.*

ANTIGENIC COMPOSITION

Several strains of *F. tularensis* have been studied by serologic methods, but no immunologic types have emerged. Agglutinins appear during the 2nd and 3rd weeks of illness and reach a peak two or three months after infection. Cross-reactions with *Brucella melitensis* and *B. abortus* do occur, but rising titers for *F. tularensis* in the convalescent phase will confirm this infection.

Jellison (1970) and Russian scientists have observed and discussed the fact that two basic types of *F. tularensis* occur in nature; they are distinguishable by differences in virulence and biochemical reactions but cannot be antigenically differentiated. The highly virulent strains, mainly isolated in North America, can cause fatality in mice, guinea pigs, and rabbits (Bell et al., 1955). They are usually isolated from tick-borne tularemia in rabbits. They utilize glycerol and produce citrulline ureidase. In doses of 1 to 100 organisms, these highly virulent strains result in necrotic nodules of the lung, spleen, and liver. The low-virulent strains are lethal for rabbits only when the dose is at least 5×10^9 organisms; these strains do not utilize glycerol or generate citrulline ureidase. This type is primarily associated with rodents; it may also be transmitted by ticks but is frequently associated with water and occurs in both the western and eastern hemispheres.

Treatment of aqueous suspensions of *F. tularensis* with ethyl ether yields an ether-soluble, immunogenic antigen that protects mice from death following challenge with more virulent strains of *F. tularensis* (Larson, 1945; Bell et al., 1952). This ether-soluble antigen has been further purified by salt precipitation, differential centrifugation, and dialysis, resulting in a uniform immunogenic activity. The immunizing antigens appear to be composed of pure cell wall material (Ormsbee and Larson, 1955). Ether-water extracts of strain SCHU S4 from stationary-phase, liquid-grown, saline suspensions have been demonstrated to stimulate better antibody formation in rabbits than do antigens obtained by other chemical procedures. Thus they provide a good crude antigenic mixture for further purification and the production of related monospecific antisera (Nutter, 1971). It has been reported that *F. tularensis* yields both a toxic filtrate and an endotoxin (Meyer, 1965).

PATHOGENIC PROPERTIES

F. tularensis is a facultative, intracellular parasite that may persist for many years in the organs (Carr and Kadull, 1957). Direct human contact with infected vertebrates and the discharges or bites of arthropods (deerflies and ticks) generally evoke tularemia. The persistent immune response and the occasional tendency of the disease toward relapse and chronicity are probably attributable to the prolonged intracellular survival of the microorganisms.

IMMUNITY

An attack of tularemia confers solid lifelong immunity. Agglutinins appear during the 2nd or 3rd week of illness and persist for years. The tendency of the disease to progress (and sometimes relapse) despite high antibody titers is probably due to the ability of the organism to survive and multiply intracellularly. Although the inactivated vaccines do not offer complete resistance, they modify the course of the infection favorably. A live, attenuated strain of *F. tularensis* has been used to immunize both animals and people and has afforded significant protection against the pathogen (Eigelsbach and Downs, 1961). In the United States, live vaccine in volunteers protected 10 of 14, whereas the cell wall antigen and the killed vaccine protected few or none against intracutaneous or respiratory-route challenge infection (McCrumb, 1961; Saslaw et al., 1961a, b). Burke (1977), in an effort to determine the incidence of tularemia in laboratory workers, compared the number of individuals vaccinated with the phenol-killed Foshay vaccine and those given the live, avirulent vaccine and demonstrated that the live vaccine was more protective; the incidence of typhoidal tularemia fell from 5.70 to 0.27 cases per 1000, whereas the incidence of ulceroglandular tularemia remained constant, although the clinical signs and symptoms were much milder than in those vaccinated with the killed vaccine. At present, vaccination with the live, attenuated vaccine is indicated only for persons at high risk, such as laboratory workers.

LABORATORY DIAGNOSIS

A definitive diagnosis of tularemia is usually achieved by a combination of bacteriologic, serologic, and animal inoculation methods. *F. tularensis* cannot be grown in ordinary media but thrives on coagulated egg yolk, enriched blood glucose-cystine agar, and thioglycollate broth. It can be isolated from sputum, pharyngeal exudate, scrapings, and biopsy material from local lesions or gastric washings when it is cultured on specific media, such as blood-glucose-cystine agar, or when guinea pigs are inoculated. Colonies may appear in two to four days and should be identified by the rapid slide agglutination test with specific antiserum. For fast screening of cultures and detection of the organism in infected material, it is possible to use fluorescent antibody techniques (Yager et al., 1960). Cultures should be incubated for two weeks before being discarded as negative, and culture plates should be enclosed in polyethylene bags to reduce desiccation during the prolonged incubation period. It is also of value to determine the dependence of the organism upon special media (the possible requirement for extra cystine or sulfhydryl compound) for growth.

Guinea pigs, mice, and hamsters are susceptible to *F. tularensis* and die with characteristic lesions two to seven days after subcutaneous or intraperitoneal inoculation. Animal inoculation is particularly valuable when the specimen is contaminated with other organisms that tend to overgrow *F. tularensis* in culture. At autopsy there is hemorrhagic edema at the site of inoculation and whitish necrotic lesions of the liver, spleen, lung, and bone marrow, each capable of yielding positive culture.

Berdal and Søderlund (1977) observed that specimens grown on selective chocolate agar (containing 7.7 μg/ml colistin, 12.5 μg/ml nystatin, 0.5 μg/ml lincomycin, and 5 μg/ml trimethoprim lactate) at 37° C in an atmosphere of 5 per cent CO_2 for two to three days disclosed smooth colonies of *F. tularensis,* 0.1 to 1.0 mm in diameter, cultures that were either pure or only slightly contaminated. Thus, the selective chocolate agar medium might be a practical supplement to nonselective media.

Serologic tests are also of value in diagnosis. Agglutinins appear during the 2nd or 3rd week of illness and reach peak titer at two to three months. Only rising titers are indicative of recent infection, since previous infections with the same or other organisms can cause detectable antibody titers which may persist for years. Cross-reactions with *Brucella* immune sera may occur, but they can be distinguished by comparative agglutination tests performed with antigens of both species, since the homologous titers are much higher than the heterologous ones.

Delayed hypersensitivity reactions to the tularemia skin test may be an early diagnostic aid because the positive reaction may be obtained during the first week of illness (Buchanan et al., 1971).

DRUG SUSCEPTIBILITY

F. tularensis is susceptible to streptomycin, kanamycin, tetracycline, and chloramphenicol.

EPIDEMIOLOGY

Tularemia is a sporadic disease, largely limited to lagomorphs (rabbits, hares) and rodents in certain areas of enzootic infection. Humans are an accidental end host. Human infections have been traced to a broad variety of sources, including ectoparasites, vertebrates, aerosols and contaminated water. *F. tularensis,* originally isolated from California ground squirrels in 1912, has since been found in innumerable species of wild animals (muskrats, beavers, woodchucks, water rats, voles, skunks, deer, dogs, cats, and rabbits) throughout North America and primarily in Europe, Russia, Turkey, and Japan. The insects that are most significant in the transmission of tularemia and that keep the infection active in vertebrates are lice, mites, and ticks (Meyer, 1965).* The propagation of *F. tularensis* in nature depends upon the wood tick and dog tick, feeding on cottontail and jack rabbits and many other rodents. Animals naturally infected with *F. tularensis* can therefore serve as reservoirs. The tick efficiently maintains the chain of infection by transmitting the organisms via the ovary to the eggs, through the larva and nymph, and finally to the adult, which is then capable of infecting rodents.

In the United States, there are two distinct types of tularemia, the highly virulent and the low-virulent forms. The former is tick-borne from the rabbit; the low-virulent, rodent form is contracted from water contaminated by water rats. Low-virulent strains growing in water and mud can infect muskrats and then trappers. Tularemia in humans, however, is largely transmitted by direct contact with the blood or other tissue fluids of infected rodents. Vector-borne tularemia in the United States, mainly conveyed by ticks and deerflies (Boyce, 1975), has a mortality rate of 7.5 per cent in untreated cases.

References

Yersinia pestis

Ajl, J., Jr., Rust, J., Jr., Hunter, D., Woebke, J., and Bent, D. F.: Preparation of serologically homogeneous plague murine toxin and its reactions with physical, chemical and enzymatic agents. J Immun 80:435, 1958.

Albizo, J. M., and Surgalla, M. J.: Isolation and biological characterization of *Pasteurella pestis* endotoxin. Infect Immun 2:229, 1970.

Baker, E. E., Somer, H., Foster, L. E., Meyer, E., and Meyer, K. F.: Antigenic structure of *Pasteurella pestis* and the isolation of a crystalline antigen. Proc Soc Exp Biol Med 64:139, 1947.

Bennett, L. G., and Tornabene, T. G.: Characterization of the antigenic subunits of the envelope protein of *Yersinia pestis.* J Bacteriol 117:48, 1974.

**Haemaphysalis leporispalustris,* rabbit tick; *Amblyomma americanum,* Lone Star tick; *Dermacentor andersoni,* common wood tick; *D. variabilis,* dog tick; *Chrysops discalis,* bloodsucking deerfly.

Bibel, D. J., and Chen, T. H.: Diagnosis of plague: An analysis of the Yersin-Kitasato controversy. Bacteriol Rev 40:633, 1976.

Buchanan, R. E., and Gibbons, N. E. (eds.): Bergey's Manual of Determinative Bacteriology. 8th ed. Baltimore, The Williams & Wilkins Company, 1974.

Burrows, T. W.: The basis of virulence for mice of *Pasteurella pestis*. *In* Mechanisms of Microbial Pathogenicity. Fifth Symposium of the Society for General Microbiology. Cambridge, Cambridge University Press, 1955, p. 152.

Burrows, T. W., and Bacon, G. A.: The basis of virulence in *Pasteurella pestis*. An antigen determining virulence. Br J Exp Pathol 37:481, 1956.

Burrows, T. W., and Bacon, G. A.: V and W antigen in strains of *Pasteurella pseudotuberculosis*. Br J Exp Pathol 41:38, 1960.

Cantey, J. R.: Plague in Vietnam. Clinical observation and treatment with kanamycin. Arch Intern Med 133:280, 1974.

Cavanaugh, D. C., and Randall, R.: The role of multiplication of *Pasteurella pestis* in mononuclear phagocytes in the pathogenesis of fleaborne plague. J Immunol 83:348, 1959.

Center for Disease Control: Annual Supplement Summary 1975. Morbid Mortal Weekly Rep 24(54):1, 1975.

Center for Disease Control: Morbid Mortal Weekly Rep 26:159, 1977.

Chen, T. H.: The antigenic structure of *Pasteurella pestis* and its relationship to virulence and immunity. Acta Trop (Basel) 22:97, 1965.

Chen, T. H.: The immunoserology of plague. In Kwapinski, J. B. (eds.): Research in Immunochemistry and Immunobiology. Vol. 1. Baltimore, University Park Press, 1972, p. 223.

Chen, T. H., and Elberg, S. S.: Scanning electron microscopic study of virulent *Yersinia pestis* and *Yersinia pseudotuberculosis* Type I. Infect Immun 15:972, 1977.

Chen, T. H., Elberg, S. S., and Eisler, D. M.: Immunity in plague: Protection induced in *Cercopithecus aethiops* by oral administration of live, attenuated *Yersinia pestis*. J Infect Dis 133:302, 1976.

Chen, T. H., Elberg, S. S., and Eisler, D. M.: Immunity in plague: Protection of the vervet *(Cercopithecus aethiops)* against pneumonic plague by the oral administration of live attenuated *Yersinia pestis*. J Infect Dis 135:289, 1977.

Chen, T. H., and Meyer, K. F.: Studies on immunization against plague. VII. A hemagglutinin test with the protein fraction of *Pasteurella pestis:* A serologic comparison of virulent and avirulent strains with observations on the structure of the bacterial cells and its relationship to infection and immunity. J Immunol 72:282, 1954.

Chen, T. H., and Meyer, K. F.: An evaluation of *Pasteurella pestis* Fraction-I-specific antibody for the confirmation of plague infections. Bull WHO 34:911, 1966.

Donovan, J. E., Han, D., Fukui, G. M., and Surgalla, M. J.: Role of the capsule of *Pasteurella pestis* in bubonic plague in the guinea pig. J Infect Dis 169:154, 1961.

Douglas, J. R., and Wheeler, C. M.: Sylvatic plague studies. II. The fate of *Pasteurella pestis* in the flea. J Infect Dis 72:18, 1943.

Englesberg, E., Chen, T. H., Levy, J. B., Foster, L. E., and Meyer, K. F.: Virulence in *Pasteurella pestis*. Science 119:413, 1954.

Jackson, S., and Burrows, T. W.: The pigmentation of *Pasteurella pestis* on a defined medium containing hemin. Br J Exp Pathol 37:570, 1956.

Janssen, W. A., Lawton, W. D., Fukui, G. M., and Surgalla, M. J.: The pathogenesis of plague: A study of the correlation between virulence and relative phagocytosis resistance of some strains of *Pasteurella pestis*. J Infect Dis 113:139, 1963.

Meyer, K. F.: *Pasteurella* and *Francisella*. In Dubos, R. J., and Hirsch, J. G. (eds.): Bacterial and Mycotic Infections. 4th ed. Philadelphia, J. B. Lippincott Company, 1965, p. 659.

Meyer, K. F., and Holdenride, R.: Rodents and fleas in a plague epizootic in a rural area of California. Puerto Rico J Public Health Trop Med 24:201, 1949.

Montie, T. C., Montie, D. B., and Ajl, S. J.: A comparison of the characteristics of two murine-toxic proteins from *Pasteurella pestis*. Biochim Biophys Acta 130:406, 1966.

Moody, M. D., and Winter, C. C.: Rapid identification of *Pasteurella pestis* with fluorescent antibody. III. Staining *Pasteurella pestis* in tissue impression smears. J Infect Dis 104:288, 1959.

Nguyen Van Ai, Nguyen Duc Hanh, Pham Van Dien, and Nguyen Van Le: Action *in vitro* and *in vivo* of trimethoprim-sulfamethoxazole on *Yersinia pestis*. Bull Soc Pathol Exot 65:759, 1972a.

Nguyen Van Ai, Nguyen Duc Hanh, Pham Van Dien, and Nguyen-Van Le: Successful treatment of bubonic and septicaemic plague by trimethoprim-sulfamethoxazole. Preliminary note. Bull Soc Pathol Exot 65:770, 1972b.

Payne, F. E., Smadel, J. E., and Courdurier, J.: Immunologic studies on persons residing in a plague epidemic area. J Immunol 77:24, 1956.

Pollitzer, R., and Meyer, K. F.: The ecology of plague. In May, J. M. (ed.): Studies in Disease Ecology. New York, Hafner Publishing Company, 1961, p. 433.

Thal, E., and Chen, T. H.: Two simple tests for the differentiation of plague and pseudotuberculosis bacilli. J Bacteriol 69:103, 1955.

Wong, J. F., and Elberg, S. S.: Cellular immune response to *Yersinia pestis* modulated by product(s) from thymus-derived lymphocytes. J Infect Dis 135:67, 1977.

Yersinia pseudotuberculosis

Besednova, N. N., Timchenko, N. F., Gorshknova, R. P., and Somov, G. P.: An interaction of the causative agent of pseudotuberculosis with the peritoneal macrophages of the immune and nonimmune organism. Zh Mikrobiol Epidemiol i Immunobiol (10):39, 1975.

Burrows, T. W., and Bacon, G. A.: The basis of virulence in *Pasteurella pestis*. An antigen determining virulence. Br J Exp Pathol 37:481, 1956.

Burrows, T. W., and Bacon, G. A.: V and W antigens in strains of *Pasteurella pseudotuberculosis*. Br J Exp Pathol 41:38, 1960.

Chen, T. H., and Elberg, S. S.: Scanning electron microscopic study of virulent *Yersinia pestis* and *Yersinia pseudotuberculosis* Type I. Infect Immun 15:972, 1977.

Gunnison, J. B., Larson, A., and Lazarus, A. S.: Rapid differentiation between *Pasteurella pestis* and *Pasteurella pseudotuberculosis* by action of bacteriophage. J Infect Dis 88:254, 1951.

Larrabee, A. R., Marshall, J. D., and Crozier, D.: Isolation of antigens of *Pasteurella pestis*. I. Lipopolysaccharide-protein complex and R and S antigens. J Bacteriol 90:116, 1965.

Lawton, W. D., Fukui, G. W., and Surgalla, M. J.: Studies on the antigens of *Pasteurella pestis* and *Pasteurella pseudotuberculosis*. J Immunol 84:476, 1960.

Lawton, W. D., and Surgalla, M. J.: Immunization against plague by a specific fraction of *Pasteurella pseudotuberculosis*. J Infect Dis 113:39, 1963.

Mair, N. S.: Yersiniosis in wildlife and its public health implications. J Wildlife Dis 9:64, 1973.

Thal, E.: Untersuchungen über *Pasteurella pseudotuberculosis* und besonderer Berücksichtigung ihres immunologischen Verhaltens. Nord Vet Med 6:829, 1954.

Thal, E.: Oral immunization of guinea pigs with avirulent "*Pasteurella pseudotuberculosis*." Nature 194:490, 1962.

Thal, E.: Immunobiologische Studien an *Yersinia pseudotuberculosis* (Syn. *Pasteurella pseudotuberculosis*). Schweiz Arch Tierheilkd 108:372, 1966.

Thal, E., and Chen, T. H.: Two simple tests for the differentiation of plague and pseudotuberculosis bacilli. J Bacteriol 69:103, 1955.

Thal, E., and Knapp, W.: The revised antigenic scheme of *Yersinia pseudotuberculosis*. Symposia Series in Immunobiological Standardization, Bern 15:219, 1969.

Yersinia enterocolitica

Hässig, A., Karrer, J., and Pusterla, F.: Über Pseudotuberkulose beim Menschen. Schweiz Med Wochenschr 79:971, 1949.

Hurwell, B.: Differentiation of cross-reacting antibodies against *Brucella abortus* and *Yersinia enterocolitica* by electroimmunoassay. Acta Vet Scan 16:318, 1975.

Knapp, W., and Thal, E.: Untersuchungen über die kulturell-biochemischen, serologischen, tierexperimentellen und immunologischen Eigenschaften einervorläufig "*Pasteurella* X" benannten Bakterienart. Zbl Bakt (Abt I, orig.) 190:472, 1963.

Knapp, W., and Thal, E.: Die biochemische Charakterisierung von *Yersinia enterocolitica* (Syn. "*Pasteurella* X") als Grundlage eines vereinfachten O-Antigenschemas. Zbl Bakt (Abt I, orig.) 223:88, 1973.

Wauter, G., Le Minor, L., Chalon, A. M., and Lassen, J.: Supplément au schéma antigénique de "*Yersinia enterocolitica*." Ann Inst Pasteur Lille 122:951, 1972.

Pasteurella multocida

Bain, R. V. S.: Studies on hemorrhagic septicemia of cattle. IV. A preliminary examination of antigens of *Pasteurella multocida* Type I. Br Vet J 111:492, 1955.

Bain, R. V. S., and Knox, K. W.: The antigens of *Pasteurella multocida* Type I. II. Lipopolysaccharides. Immunology 4:122, 1961.
Carter, G. R.: A new serological type of *Pasteurella multocida* from central Africa. Vet Rec 73:1052, 1961.
Carter, G. R.: Animal serotypes of *Pasteurella multocida* from human infections. Can J Public Health 53:158, 1962.
Carter, G. R.: Immunological differentiation of Type B and E strains of *Pasteurella multocida.* Can Vet J 4:61, 1963.
Carter, G. R.: Improved hemagglutination test for identifying Type A strains of *Pasteurella multocida.* Appl Microbiol 24:162, 1972a.
Carter, G. R.: Agglutinability of *Pasteurella multocida* after treatment with hyaluronidase. Vet Rec 91:150, 1972b.
Carter, G. R., and Annau, E.: Isolation of capsular polysaccharides from colonial variants of *Pasteurella multocida.* Am J Vet Res 14:474, 1953.
Chang, W. H., and Carter, G. R.: Multiple drug resistance in *Pasteurella multocida* and *Pasteurella hemolytica* from cattle and swine. J Am Vet Med Assoc 169:710, 1976.
Collins, F. M.: Growth of *Pasteurella multocida* in vaccinated and normal mice. Infect Immun 8:868, 1973.
Collins, F. M.: Mechanisms of acquired resistance to *Pasteurella multocida* infections: A review. Cornell Vet 67:103, 1977.
Dhanda, M. R.: Purification and properties of the soluble antigen of *Pasteurella septica,* Type I. Indian J Pathol Bacteriol 2:59, 1960.
Francis, D. P., Holmes, M. A., and Brandon, G.: *Pasteurella multocida.* Infections after domestic animal bites and scratches. JAMA 233:42, 1975.
Gadberry, J. L., and Miller, N. G.: Use of bacteriophages as an adjunct in the identification of *Pasteurella multocida.* Am J Vet Res 38:129, 1977.
Hubbert, W. T., and Rosen, M. N.: II. *Pasteurella multocida* infection in man unrelated to animal bite. Am J Public Health 60:1109, 1970.
Namioka, S., and Bruner, D. W.: Serological studies on *Pasteurella multocida.* IV. Type distribution of organisms on the basis of their capsular and O groups. Cornell Vet 53:41, 1963.
Perreau, P., and Petit, J. P.: Lipopolysaccharide antigens of *Pasteurella* Type E. Rev Élèvage Med Vet Pays Trop 16:5, 1963.
Smith, J. E.: Studies on Pasteurella septica. III. Strains from human beings. J Comp Pathol 69:231, 1959.

Francisella tularensis

Bell, J. F., Larson, C. L., Wight, W. C., and Ritter, S. S.: Studies on the immunization of white mice against infections of *Bacterium tularense.* J Immunol 69:515, 1952.
Bell, J. F., Owen, C. R., and Larson, C. L.: Virulence of *Bacterium tularense.* I. A study of the virulence of *Bacterium tularense* in mice, guinea pigs, and rabbits, J Infect Dis 97:162, 1955.
Berdal, B. P., and Søderlund, E.: Cultivation and isolation of *Francisella tularensis* on selective chocolate agar as used routinely for the isolation of gonococci. Acta Pathol Microbiol Scand (B) 85:108, 1977.
Boyce, J. M.: Recent trends in the epidemiology of tularemia in the United States. J Infect Dis 131:197, 1975.
Buchanan, T. M., Brooks, G. F., and Branchman, P. S.: The tularemia skin test. 325 skin tests in 210 persons: Serologic correlation and review of the literature. Ann Intern Med 74:336, 1971.
Burke, D. S.: Immunization against tularemia: Analysis of the effectiveness of live *Francisella tularensis* vaccine in prevention of laboratory acquired tularemia. J Infect Dis 135:55, 1977.
Carr, E. A., Jr., and Kadull, P. J.: Persistence of *Bacterium tularense* in man in the absence of serious clinical illness. Arch Pathol 64:382, 1957.
Eigelsbach, H. T., and Downs, C. M.: Prophylactic effectiveness of live and killed tularemia vaccines. I. Production of vaccine and evaluation in the white mouse and guinea pigs. J Immunol 87:415, 1961.
Francis, E.: Tularemia Francis 1921. I. The occurrence of tularemia in nature as a disease of man. Public Health Rep 36:1731, 1921a.
Francis, E.: Tularemia Francis 1921. A new disease of man. JAMA 78:1015, 1921b.
Jellison, W. L.: Tularemia in Montana. *In* Montana Wildlife, Series 2 of Montana Animals, Montana Fish and Game Commission. Butte, McKee Printing Company, 1970.
Larson, C. L.: Immunization of white rats against infections with *Pasteurella tularensis.* Public Health Rep 60:725, 1945.
McCoy, G. W.: A plague-like disease of rodents. Public Health Bull 43:53, 1911.
McCoy, G. W., and Chapin, G. W.: Further observations on a plague-like disease of rodents with a preliminary note on the causative agent, *Bacterium tularense.* J Infect Dis 10:61, 1912.
McCrumb, F. R.: Aerosol infection of man with *Pasteurella tularensis.* Bacteriol Rev 25:262, 1961.
Meyer, K. F.; *Pasteurella* and *Francisella.* In Dubos, R. J., and Hirsch, J. G. (eds.): Bacterial and Mycotic Infections of Man. Philadelphia, J. B. Lippincott Company, 1965, p. 659.
Nutter, J. E.: Antigens of *Pasteurella tularensis:* Preparative procedures. Appl Microbiol 22:44, 1971.
Ormsbee, R. A., and Larson, C. L.: Studies on *Bacterium tularense.* II. Chemical and physical characteristics of protective antigen preparations. J Immunol 74:359, 1955.
Saslaw, S., Eigelsbach, H. T., Wilson, H. E., Prior, J. A., and Carhart, S.: Tularemia vaccine study. I. Intracutaneous challenge. Arch Intern Med 107:689, 1961a.
Saslaw, S., Eigelsbach, H. T., Prior, J. A., Wilson, H. E., and Carhart, S.: Tularemia vaccine study. II. Respiratory challenge. Arch Intern Med 107:702, 1961b.
Yager, R. H., Spertzel, O. R., Yaeger, R. F., and Tigertt, W. D.: Domestic fowl: Source of high titer *P. tularensis* serum for the fluorescent antibody technique. Proc Soc Exp Biol Med 105:651, 1960.

38 *STREPTOBACILLUS MONILIFORMIS*

Joshua Fierer, M.D.

Streptobacillus (Actinobacillus) moniliformis is a pleomorphic bacterium. It is nonmotile, unencapsulated, and nonacid-fast. Gram-stained smears of exudate show small rods or short filaments. When grown in serum-enriched broth, they are branching filaments that form interwoven masses. There usually is marked pleomorphism. Many filaments contain cells with rounded or fusiform swellings. After prolonged culture, tiny curved and ring forms appear.

Much of the pleomorphism of this bacterium is due to the spontaneous appearance of variants with defective cell walls, called L-phase variants (after the Lister Institute where this observation was first made). These are stable variant bacteria that lack a rigid cell wall. As a consequence, in broth they grow as small round cells, and on agar the colonies are flat and grow into the agar like *Mycoplasma* colonies. For many years this similarity in colonial morphology and in osmotic fragility between L-forms and mycoplasmatales obscured the large and fundamental differences between these disparate microorganisms (see Chapters 53 and 54. Stable L-forms lack the mucopeptide that confers on the parent bacteria their rigidity and rod-like shape. The L-forms divide and propagate by binary fission as cell-deficient forms. Unstable L-forms revert back to parental bacteria. Protoplasts, in contrast to L-forms, are cell wall-deficient forms that are created artificially by exposing gram-positive bacteria to enzymes (e.g., lysozyme) or antibiotics such as penicillin that directly interfere with cell wall synthesis. Sometimes a stable protoplast is produced by these manipulations that can multiply without synthesizing a cell wall in the absence of the inducing agent.

ANTIGENIC COMPOSITION

The L-phase variant and the bacillus share a common antigen, but the L-phase lacks antigens present in the latter. There appears to be only one serotype of the bacillus, but the chemical nature of the antigen has not been established.

METABOLISM

Growth can occur aerobically but is equally good or better under anaerobic conditions (Wittler, 1974). A 10 per cent CO_2 atmosphere probably enhances growth. Optimum temperature for growth is 37° C. The metabolic characteristics of the organism have not been studied in detail. It is both catalase and oxidase negative and does not reduce nitrate to nitrite, nor produce idole. Urea is not hydrolyzed nor is phenylalanine deaminated. Acid but not gas is produced from dextrin, fructose, galactose, glucose, maltose, mannose, starch, and glycogen. No acid is produced from glycerol, inositol, inulin, sorbitol, rhamnose, or mannitol. Tetrazolium salt and tellurite are reduced aerobically and anaerobically though the tellurite reduction test may be only weakly positive.

IMMUNITY

The pathogenic properties of this organism have not been studied in detail. The stable L-forms do not produce disease in mice whereas bacilli do. If injected into mice, unstable L-forms revert to the bacterial form and cause disease. Vaccination with the L-form does not protect mice against challenge with the bacterial form, but a killed bacterial vaccine is protective.

LABORATORY DIAGNOSIS

Diagnosis of infection is made by isolating the bacteria from a lymph node, pus, joint fluid, or blood. Bacteria grow in ordinary blood culture media (Rogosa, 1974). This organism does not grow on nutrient agar, but on sheep blood agar it produces very small, gray, translucent, nonhemolytic colonies with entire edges after 48 to 72 hours of incubation. In serum or ascitic fluid broth, it grows as white granules or flocculent balls that sediment rapidly or stick to the side of the glass tube. Nutrient broth supplemented with 20 per cent horse serum is useful for primary isolation. Injection of pus into mice can also be used to isolate the bacterium, but care must be taken not to use naturally infected animals for such isolation.

The tube-agglutination test with whole bacterial antigen can be used for serologic diagnosis. Demonstration of a rising titer is significant. A negative test does not exclude the diagnosis.

DRUG SUSCEPTIBILITY

The organism is susceptible to penicillin (0.05 unit/ml), streptomycin, and tetracycline. The L-phase variant is resistant to penicillin, and persistent symptoms and signs may be associated with the presence of L-phase organisms in the blood during penicillin therapy. Tetracycline or streptomycin can be used in that circumstance.

EPIDEMIOLOGY

The distribution of this organism is worldwide, although human infections are rare in Europe and North America. The reservoir is infected rats or, rarely, other rodents. It may cause an epizootic disease in mice characterized by arthritis and lymphadenitis. Most human infections are acquired as the result of a rat bite (Cole et al., 1979). Consequently, it is an occupational hazard of research laboratory workers. After a bite, the incubation period is 3 to 10 days. Infection can occur as a consequence of ingesting contaminated milk (Haverhill fever or erythema arthriticum epidemicum). Some patients who live in rat-infested environments have had no history of a bite, and infection may have been food-borne.

References

Cole, J. S., Stoll, R. W., and Bulger, R. J.: Rat bite fever, report of three cases. Ann Intern Med 71:979, 1969.

Rogosa, M.: Streptobacillus moniliformis and Spirillum minus. In Manual of Clinical Microbiology, 2nd ed. American Society for Microbiology, Washington, 1974.

Wittler, R. G.: In Bergey's Manual of Determinative Bacteriology, 8th ed. Baltimore, The Williams & Williams Company, 1974.

39 ACTINOBACILLUS AND CARDIOBACTERIUM

Irving J. Slotnick, Ph.D.

ACTINOBACILLUS

Only one species in the genus *Actinobacillus*, *Actinobacillus actinomycetemcomitans*, has been consistently associated with human infection. The other species are pathogenic in cattle, horses, and pigs, but only rarely in man.

MORPHOLOGY

Members of the genus *Actinobacillus* are gram-negative rods, coccoid to coccobacillary in shape, 0.3 to 0.5 μm wide and 0.5 to 1.5 μm long, nonspore-forming and nonacid-fast. They often present pleomorphic coccal and long bacillary and filamentous forms. They are not distinguishable morphologically from organisms in the genera *Brucella*, *Haemophilus*, and *Pasteurella* (Figs. 1 and 2).

In infections of animals, grayish white granules may appear in pus that are similar to but smaller (0.4 mm diameter) than the "sulfur" granules of actinomycosis. The granules contain clublike bodies emanating radially from the center, which is made up of the gram-negative bacilli and cell deposits. Care must be taken to differentiate these from granules caused by *Actinomyces* spp, *Nocardia* spp, *staphylococci*, *Monosporium* spp, fragments of caseous material, and clumps of pus cells and fibrin. A Ziehl-Neelsen stain colors the granule interior blue whereas the clubs are weakly acid-fast.

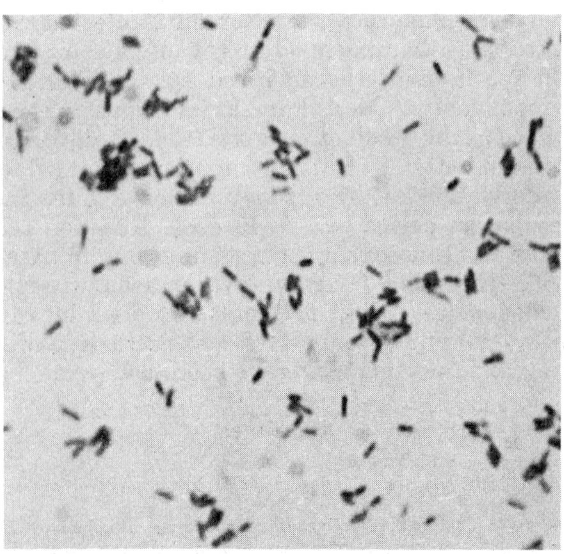

FIGURE 1. Actinobacillus lignieresi. *A T C C #10811. Gram stain. Blood agar. 48-hours growth. 1800 ×.*

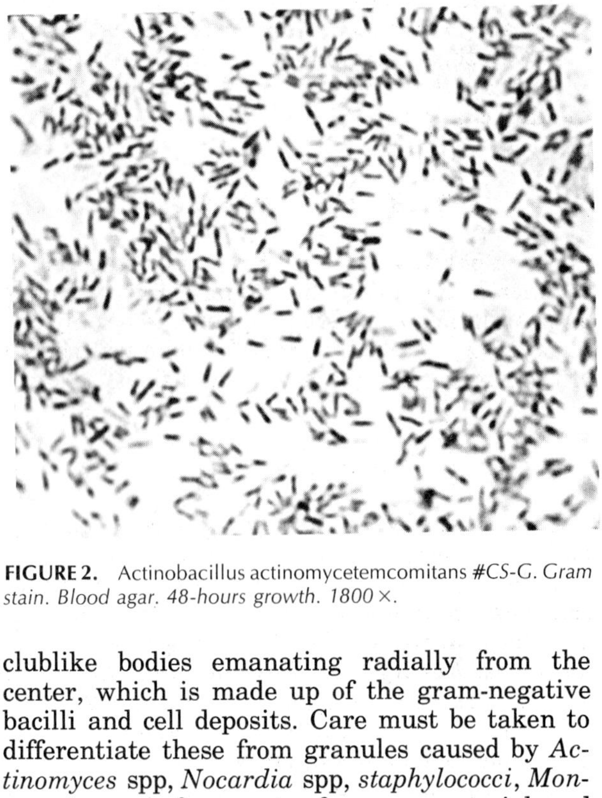

FIGURE 2. Actinobacillus actinomycetemcomitans *#CS-G. Gram stain. Blood agar. 48-hours growth. 1800 ×.*

Growth is improved in the presence of increased carbon dioxide and by adding blood (5 per cent) or serum (10 per cent) to any of the common agar media. No growth occurs on *Salmonella-Shigella* (SS) or eosin-methylene blue (EMB) agar. *A. lignieresii* and *A. equuli* grow on MacConkey agar, whereas *A. actinomycetemcomitans* grows only slightly at best. Both rough and smooth types of colonies occur in primary cultures. Rough colonies are tiny at 24 hours, ranging up to 1.0 mm in diameter, they are raised, grayish, opaque, convex, tenacious, and firmly adherent to the medium. The surface appears dry and dull. In 72 hours the colony increases to 3.4 mm in diameter.

Smooth colonies are the same size as rough forms. They are small, discrete, convex, and semi-opaque, with entire edges and glistening surfaces. The colonies appear gray but may have a blue to yellowish hue when viewed by transmitted light. After four to five days of incubation, the colony has a domed appearance when examined microscopically. The formation of smooth colony variants is favored on continued subculture (Alexander, 1974).

The central area of the colonies may, after three to five days of incubation, appear to contain a four- to six-pointed star formation. This portion of the colony grows into the agar, and when the colony is scraped away, the star formation remains. This is more readily observed on clear media than on blood agar.

The consistency of bacterial growth on agar is strain-dependent. On blood agar growth may be butyrous and can be scraped up easily, or it may be adherent and difficult to remove with a loop.

Serum enriched thioglycollate, soy broths, cooked meat, and other infusion media show growth at the end of 24 hours in air. Granular growth may follow uniform turbidity. The granules tend to adhere to the walls of the tubes, leaving the broth fairly clear. A surface film and a heavy tenacious sediment also occur in old broth cultures.

ANTIGENIC COMPOSITION

Heat-stable polysaccharide agglutinating antigens associated with extracellular slime and several agglutinogens are formed by the actinobacilli. Antigenic relationships exist between *A. lignieresii* and *A. equuli*. Cross-reactions between *Pseudomonas mallei, Pseudomonas pseudomallei,* and *A. equuli* are well known. Phillips (1967) defined six types (1 through 6) and two subtypes (1a and 4c) of *A. lignieresii* on the basis of heat-stable somatic antigens. Some host specificity in the distribution of serotypes seems to be involved, because most cattle strains belong to Type 1 and most sheep strains belong to Types 2, 3, and 4. Heat-labile antigens common to different antigenic types also occur.

Pulverer and Ko (1972) examined 100 strains of *A. actinomycetemcomitans* by a variety of serologic procedures and delineated 6 heat-stable agglutinating antigens and 24 agglutinating patterns. Ouchterlony agar-gel diffusion tests proved to be inferior to the agglutination procedures for typing purposes.

METABOLISM

The actinobacilli are representative chemoorganotrophs, using only organic compounds as their source of energy. Growth is aerobic and facultatively anaerobic. Carbohydrates are fermented with production of acid and usually no gas. Fermentation reactions are weak, variable, and often delayed. Glucose, maltose, mannitol, sucrose, xylose, galactose, trehalose, and dextrin can be used as substrates. Lactose fermentation is variable. Dulcitol, rhamnose, and inositol are not attacked. The OF medium of Hugh and Leifson is not recommended for fermentation reactions because it does not support the growth of most strains of actinobacilli. Carbohydrate broths used to test reactions should be supplemented with 5 to 10 per cent sterile horse serum and not incubated more than 14 days to minimize nonspecific reactions.

Pulverer and Ko (1970) classified eight fermentative biotypes of *A. actinomycetemcomitans* on the basis of galactose, mannitol, and xylose reaction patterns. Biotypes I through IV make up the majority of strains isolated from cases in the United States reported to the Center for Disease Control.

Optimal temperature for growth is 37° C, with a range of 20 to 42° C. Optimal pH is 7.6, and no growth occurs at pH below 6.5. The guanine plus cytosine (G+C) content of the DNA ranges from 40.6 to 42.0 moles per cent. Capsules are not produced, but some strains produce extracellular slime. Flagella are not synthesized, and the bacilli are nonmotile.

PATHOGENIC PROPERTIES

Localized purulent granulomatous lesions, abscess formation in soft tissues, septicemia, endocarditis, and occasionally meningitis characterize the disease spectrum of the actinobacilli. No exotoxins have been demonstrated. Subcutaneous inoculation of a pure culture of *A. lignieresii* or *A. equuli* into cattle leads to the formation of an abscess similar to that occurring in the natural disease, but not all strains are virulent. Experimental studies have been limited due to absence of another animal model and lack of pathogenicity for rabbits, mice, or guinea pigs.

IMMUNITY

Definitive data concerning the immune state following an overt case of actinobacillosis are meager. Repeated attacks have been singularly rare, although chronic cases can occur. Serum antibodies are formed in animals and humans, but it is not known if they are protective. Horse antiserum has been employed in the treatment of animals without conspicuous success.

Nonspecific agglutination reactions with sera from normal hosts and seronegative reactions in proven cases make serodiagnostic testing unsatisfactory (Phillips, 1967). Agglutination tests with known organisms or known antisera can be used to identify antibody or species, but commercial sources for either are not available. However, it is possible to follow antibody development in individuals from whom an isolate has been obtained (Pathak and Ristic, 1962). The serum of animals suffering from actinobacillosis frequently agglutinates the causative organism agent to a titer of 1:160 or higher. We observed a transitory rise to a titer of 1:320 in a human case of *A. actinomycetemcomitans* endocarditis and found no rise in a second case.

LABORATORY DIAGNOSIS

Identification of culture isolates is based on biochemical reactions in Table 1.

DRUG SUSCEPTIBILITY

Streptomycin, tetracycline, and chloramphenicol inhibit all strains tested in reasonably low concentrations (Page and King, 1966). Less than 50 per cent of *A. actinomycetemcomitans* organisms are inhibited by 12.5 μg/ml of penicillin, only 35 per cent are inhibited by 3.12 μg/ml of ampicillin, and all are resistant to methicillin.

Potassium iodide is very effective and is the drug of choice in the treatment of spontaneous actinobacillosis. The response is dramatic and permanent. It has been used with some success in the more chronic cases of disease in humans. The iodides have little bactericidal effect against *A. lignieresii*. The sulfonamides, penicillin, streptomycin, and broad-spectrum antibiotics are used to treat *A. lignieresii* and *A. equuli* infection.

EPIDEMIOLOGY

Even though actinobacilli are recognized obligate parasites of animals and humans, their eco-

TABLE 1. Identification Characteristics of Suspected Actinobacilli and Cardiobacterium Strains

	ACTINOBACILLUS ACTINOMYCETEMCOMITANS	ACTINOBACILLUS LIGNIERESII	ACTINOBACILLUS EQUULI	CARDIOBACTERIUM HOMINIS
Oxidase activity	−/weak	+	+	+
Growth on MacConkey's agar	−/slight	+	+	−
Acetyl-methylcarbinol (VP test)	−	v	−	−
Catalase activity	+	−/weak	−/weak	−
Indole production	−	−	−	+/weak
Citrate utilization	−	−	−	−
Nitrate reductase	+	+	+	−
H_2S production (TSI)[a]	−	−	−	−
H_2S: lead acetate paper	+/weak	+/weak	+/weak	+/weak
β-galactosidase (ONPG)[a]	−	+	+	−
Urease activity	−	+	+	−
LDC/ODC[a]	−	−	−	−
TSI, acid butt	+	+	+	+ or NG[a]
Gelatin hydrolysis	−	−	+	−
Hippurate hydrolysis	−	−	+	−
Carbohydrates attacked				
arabinose	−	d	−	−
lactose	−	+	+	−
maltose	+	+	+	+
mannitol	d[a]	+	+	d
raffinose	−	−	+	·
salicin	−	−	−	−
sorbitol	−	d	−	+
sucrose	−	+	+	+
trehalose	−	−	+	−
xylose	d	+	+	−
galactose	d			−
glucose	+	+	+	+

Data adapted from Weaver et al. (1972) and Cowan and Steele (1974). Both references cite excellent sets of key reactions to differentiate other fastidious gram-negative fermentative bacteria.

[a]TSI, triple sugar iron agar; LCD, lysine decarboxylase; ODC, ornithine decarboxylase; ONPG, *O*-nitrophenyl−β−D-galactopyranoside (test for β-galactosidase); NG, no growth; d, some strains positive but may be delayed.

logic habits are not fully known. They are recovered with ease from the normal mouth, tonsillar area, and intestinal tract of all hosts. The genital tract may also be colonized. The oral cavity and upper respiratory tract are primary portals of entry and serve as a focal endogenous habitat.

Human infection caused by *A. actinomycetemcomitans* develops as endocarditis in rheumatic or congential heart disease or as opportunistic invaders in malignant lymphoma and acute leukemia. Infections have followed dental manipulations, but soft tissue *A. actinomycetemcomitans* abscesses have also been seen without evidence of trauma to the area of infection (Burgher et al., 1973). A human urinary tract infection has been reported (Townsend and Gillenwater, 1969).

A. lignieresii and *A. equuli* are chiefly the cause of actinobacillosis in animals and can cause epidemics. The disease can spread rapidly in a herd, attacking up to 50 per cent of the animals in a matter of weeks. When human infections have been caused by *A. lignieresii* or *A. equuli*, they occurred in persons whose occupations required heavy contact with animals, such as farmers, animal husbandmen, veterinarians, and abbatoir personnel (Custus et al., 1944).

CARDIOBACTERIUM

MORPHOLOGY

The only species in this genus is *Cardiobacterium hominis*. It is a gram-negative rod, 0.5 μm wide and 1.0 to 2.2 μm long, arranged singly, in pairs, short chains, or clusters. The rods may be uniform or pleomorphic. The pleomorphic forms have one or both ends enlarged, tear-drop forms, or rosette clusters (Figs. 3, 4, and 5). There is evidence that pleomorphism is related to the yeast extract content of the medium. Organisms appear homogeneously stained and of *uniform dimensions* on blood agar containing yeast extract. The organism is nonspore-forming and nonacid-fast. Sudanophilic bodies and metachromatic inclusions are demonstrable.

Thin sections of *C. hominis* show a cell wall of gram-negative type with a cytoplasm containing numerous intrusive membranes disposed around the periphery of the cells, especially at the poles (Reyn et al., 1971). The cell wall consists of a unit membrane sandwiched between dense outer and inner layers. Polar caps are seen on the ends of cells. The substructure of the surface layer is unusual for a gram-negative bacterium.

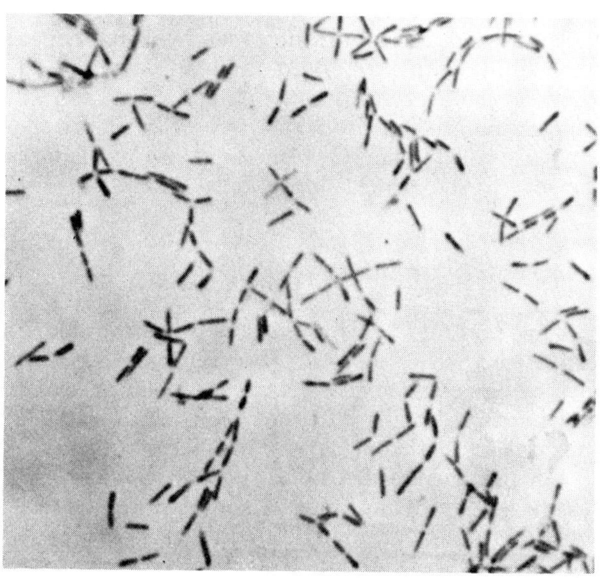

FIGURE 3. Cardiobacterium hominis. Strain *#6518. Blood agar culture illustrative of a culture with singular lack of pleomorphic cell-types. Rod shape cells predominate. 1800 ×.*

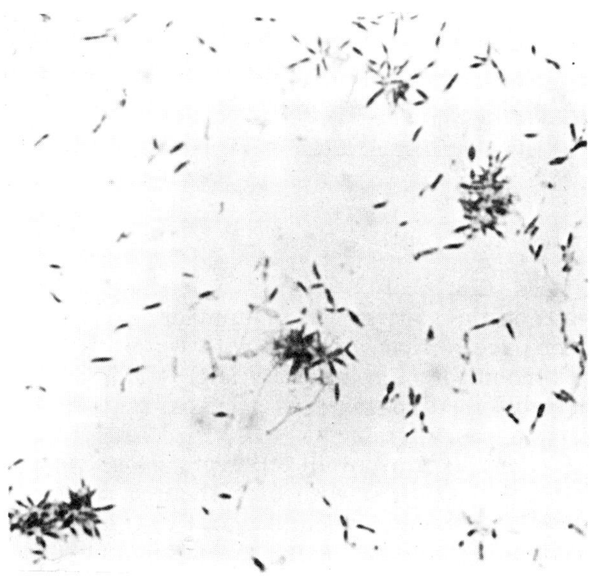

FIGURE 4. Cardiobacterium hominis. Strain *#6573. Rosette-clusters in a Gram stain preparation of a blood agar culture. 1800 ×.*

C. hominis colonies on blood agar are punctiform after 24 hours, and attain a maximum size of one to two mm in 48 to 72 hours. Colonies are circular, convex, smooth, entire, glistening, opaque, and butyrous. In older cultures, the center of the colony appears more opaque and grayish white in contrast to the edges. A slight greening of blood agar, which develops around dense areas of growth in two to three days, becomes brownish with further incubation. Clear hemolytic zones are not observed.

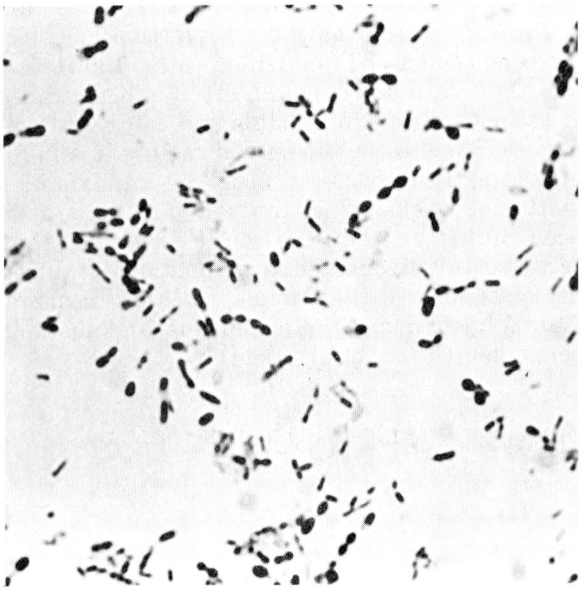

FIGURE 5. Cardiobacterium hominis. *Strain #6573. Gram stain of blood agar culture illustrating tear-drop morphology and chaining tendency. 1800 ×.*

Good growth of *C. hominis* is easily obtained when cultures are grown in humidors containing filter paper strips saturated with water. The requirement for high humidity is growth-limiting. In addition, atmospheres of 3 to 5 per cent carbon dioxide are preferable.

C. hominis grows well on trypticase soy agar, tryptose blood agar with or without 5 per cent blood enrichment, chocolate agar, PPLO medium, cystine-heart agar, Casman's medium, and heart infusion agar. Moderate growth occurs on nutrient agar and Lowenstein-Jensen's medium. The organism does not grow on MacConkey's agar, Simmon's citrate, SS agar, tellurite medium, potato medium, Sabouraud's dextrose agar, phenylethylalcohol medium, EMB medium, or endo agar medium.

ANTIGENIC COMPOSITION

Antigenically, *C. hominis* is relatively homogeneous. The organisms cross-agglutinate to high titer in unabsorbed immune antisera. Antiserum absorption with heterologous strains gives residual titers indicative of varied as well as common antigenic groupings. Fluorescent-isothiocyanate-labeled antibody detects the same antigenic relationships (Slotnick and Dougherty, 1964).

METABOLISM

C. hominis, like the actinobacilli, is an aerobic, facultative anaerobic, chemoorganotroph with a fermentative, albeit weak, type of metabolism.

C. hominis ferments carbohydrates without gas production. Fermentable substrates include glucose, sucrose, levulose, mannose, and sorbitol. Maltose and mannitol breakdown varies with individual strains. Fermentation is late and irregular in raffinose, dextrin, glycogen, and starch. Xylose, lactose, glycerol, salicin, inulin, arabinose, adonitol, dulcitol, galactose, rhamnose, trehalose, inositol, cellobiose, erythritol, melibiose, and melezitose are not attacked. Glucose is converted primarily to lactate and smaller amounts of pyruvate, formate, and propionate.

Optimal temperature for growth is 37° C. No growth occurs at 42° C or above, and only sporadic light growth occurs at 25° C. Optimal pH range 7.0 to 7.2. The G+C content of the DNA is 61.7 moles per cent (Hill and LaPage, 1969). Neither capsules nor flagella are produced.

PATHOGENIC PROPERTIES

Infections with *C. hominis* have been primarily associated with human endocarditis (Tucker et al., 1969; Midgley et al., 1970; Snyder and Ellner, 1969; Weiner and Werthamer, 1975; Savage et al., 1977; Perdue et al., 1974). Their low intrinsic virulence is in keeping with the subacute course of the endocarditis. Infection occurs most frequently in patients with pre-existing cardiovascular defects or disease. However, an ever increasing number of diseased sites are being encountered, including the cervix, vagina, cheek, mandible, empyema fluid, and spinal fluid, as well as isolates from sputum, nose, and throat. Exotoxins are not produced. Other virulence factors are not known. *C. hominis* is not pathogenic for any of the several laboratory animals.

IMMUNITY

The immune mechanisms operative in hosts infected with *C. hominis* are unknown. Second attacks or chronic cases have not been observed. Humoral antibodies are not regularly formed in normal adult throat carriers of *C. hominis*.

Serum of endocarditis patients may give high titers in agglutination and complement fixation tests. Midgley et al. (1970) report a case with admission titers of 320 and 640 respectively in agglutination and complement fixation tests. Serum obtained nine months later gave an agglutination titer of 160, showing little change, whereas the complement fixation titer had fallen considerably to 20. Neither of these sera produced agglutination with a variety of other organisms isolated in the authors' laboratory, and conversely, numerous sera from other patients failed to agglutinate *C. hominis* in slide agglutination tests.

LABORATORY DIAGNOSIS

Conventional blood culture media containing 10 per cent carbon dioxide are suitable for recovering *C. hominis.* The biochemical identification of this organism is based on reactions given in Table 1 and is confirmed serologically with specific agglutinating or immunofluorescent antisera.

DRUG SUSCEPTIBILITY

C. hominis is susceptible to a wide range of antibiotics. Penicillin and streptomycin, either singly or in combination, should be considered the drugs of choice. Drug resistance in treatment has not been a problem.

EPIDEMIOLOGY

C. hominis is part of the indigenous commensal human respiratory flora. A large proportion of normal individuals of all ages and both sexes harbor these organisms. Their presence in the stool is easily detectable with direct fluorescent antibody staining. An extremely low incidence in cervical and vaginal cultures coupled with complete absence in urine supports the conclusion that they are only transient contaminants of the genitourinary tract.

In contrast to actinobacillosis, *C. hominis* infection does not occur in epidemics. It is primarily an infection of compromised persons.

No animal host or reservoir is known, and extensive sampling of several soil and hospital environments has failed to detect these organisms.

References

Alexander, A. D.: Actinobacillus. In Lennette, E. H., Spaulding, E. H., and Truant, J. P. (eds.): Manual of Clinical Microbiology. 2nd ed. Washington, D.C., American Society for Microbiology, 1974, p. 320.

Burgher, L. W., Loomis, G. W., and Ware, F.: Systemic infection due to *Actinobacillus actinomycetemcomitans.* Am J Clin Pathol 60:412, 1973.

Custus, D. L., Halley, H., and Bacon, C. M.: *Actinobacillus lignieresii* endocarditis. Arch Pathol 38:332, 1944.

Hill, L. R., and LaPage, S. P.: The DNA base compositions and taxonomy of *Bacteroides corrodens* and *Cardiobacterium hominis.* Brno Conference on the Taxonomy of Bacteria, 24–27 September, 1969.

Midgley, J., LaPage, S. P., Jenkins, B. A. G., Barrow, G. I., Roberts, M. E., and Buck, A. G.: *Cardiobacterium hominis* endocarditis. J Med Microbiol 3:91, 1970.

Page, M. I., and King, E. O.: Infection due to *Actinobacillus actinomycetemcomitans* and *Haemophilus aphrophilus.* N Engl J Med 275:181, 1966.

Pathak, R. C., and Ristic, M.: Detection of an antibody to *Actinobacillus lignieresii* in infected human beings and the antigenic characterization of isolates of human and bovine origin. Am J Vet Res 23:310, 1962.

Perdue, C. D., Dorney, E. R., and Ferrier, F.: Embolomycotic aneurysm associated with bacterial endocarditis due to *Cardiobacterium hominis.* Am Surg 34:901, 1974.

Phillips, J. E.: The incidence of agglutinating antibodies to *Actinobacillus lignieresii* in the sera of normal and infected cattle. J Pathol Bacteriol 90:557, 1965.

Phillips, J. E.: Antigenic structure and serological typing of *Actinobacillus lignieresii.* J Pathol Bact 93:463, 1967.

Pulverer, G., and Ko, H. L.: *Actinobacillus actinomycetemcomitans:* Fermentative capabilities of 140 strains. Appl Microbiol 20:693, 1970.

Pulverer, G., and Ko, H. L.: Serological studies on *Actinobacillus actinomycetemcomitans.* Appl Microbiol 23:207, 1972.

Reyn, A., Birch-Anderson, A., and Murray, R. G. E.: The fine structure of *Cardiobacterium hominis.* Acta Pathol Microbiol Scand [B]72:51, 1971.

Savage, D. D., Kagan, R. L., Young, A. N., and Horvath, A. E.: *Cardiobacterium hominis* endocarditis: Description of two patients and characterization of the organism. J Clin Microbiol 5:75, 1977.

Slotnick, I. J., and Dougherty, M.: Further characterization of an unclassified group of bacteria causing endocarditis in man: *Cardiobacterium hominis* gen. et sp.n. Antonie van Leeuwenhoek 30:261, 1964.

Snyder, A. I., and Ellner, P. D.: *Cardiobacterium hominis* endocarditis. NY State J Med 69:704, 1969.

Townsend, T. R., and Gillenwater, J. Y.: Urinary tract infection due to *Actinobacillus actinomycetemcomitans.* JAMA 210:558, 1969.

Tucker, D. N., Slotnick, I. J., King, E. O., Tynes, B., Nicholson, J., and Crevasse, L.: Endocarditis caused by *Pasteurella*-like organism: Report of four cases. N Engl J Med 267:913, 1962.

Weiner, M., and Werthamer, S.: *Cardiobacterium hominis* endocarditis: Characterization of the unusual organisms and review of the literature. Am J Clin Pathol 63:131, 1975.

Acid-Fast Rods

40 MYCOBACTERIA

Donald W. Smith, Ph.D.

The genus *Mycobacterium* includes the causative agents of tuberculosis and leprosy, two diseases that, according to the World Health Organization, are among the main public health priorities of many of the developing countries of the world. For example, the global tuberculosis problem is estimated to be the following: 1500 million persons infected with tubercle bacilli; 20 million sputum-positive persons capable of disseminating the disease; 3 to 5 million new cases each year; and 600,000 deaths per year.

The principal species of mycobacteria that cause disease in man and other animals are listed in Table 1, along with the most common saprophytic (free-living) mycobacteria.

TABLE 1. Pathogenic and Saprophytic Mycobacteria

PATHOGENIC MYCOBACTERIA	
Microorganism	**Disease Produced**
M. tuberculosis	Tuberculosis in man and subhuman primates
M. bovis	Tuberculosis in cattle, man and subhuman primates
*M. avium intracellulare**	Tuberculosis in birds and swine; tuberculosis-like disease in man
*M. kansasii**	Tuberculosis-like disease in man
*M. fortuitum complex** (including *M. chelonei*)	Wound infection in man
*M. marinum**	Tuberculosis in fish and cutaneous disease in man
*M. ulcerans**	Ulcerative lesions in man
M. leprae	Leprosy in man
SAPROPHYTIC MYCOBACTERIA	
Microorganism	**Source in Nature**
M. gordonae	Soil, water Not responsible for human disease but may be isolated from sputum, gastric washings, etc.
M. terrae complex (includes *M. terrae, M. triviale, M. novum,* and *M. nonchromogenicum*)	Soil, water
M. flavescens	Soil, water
M. gastri	Soil, water

*Atypical mycobacteria.

MYCOBACTERIUM TUBERCULOSIS AND ATYPICAL MYCOBACTERIA

Morphology

Mycobacteria are slender, rod-shaped organisms that cannot be distinguished from each other on the basis of morphology. Tubercle bacilli in tissue and in sputum smears frequently have an irregularly staining, beaded appearance. The unstained regions of the cell are presumably areas containing inclusion bodies such as glycogen and polymetaphosphate. Electron microscopic examination of mycobacteria shows a thick cell wall, the presence of mesosomes, and inclusions of lipid (Fig. 1). Mycobacteria do not form spores. They have an unusually high lipid content, greater than 25 per cent in contrast to 0.5 per cent lipid in gram-positive and 3 per cent lipid in gram-negative bacteria. The high lipid content is presumed to be responsible for their characteristic resistance to drying, alcohol, acids, alkali, and certain germicides. The high lipid content also renders the Gram stain invalid for mycobacteria. On the other hand, mycobacteria are acid-fast; this property is shared with certain species of only one other genus of bacteria, *Nocardia*. Once stained with basic fuchsin dyes, acid-fast organisms resist decolorization even with alcohol containing 3 per cent mineral acid. Acid-fastness is an important characteristic useful in the laboratory diagnosis of all mycobacterial diseases. Although the lipids extracted from mycobacteria are weakly acid-fast and the lipid-free cell residue is not acid-fast, tubercle bacilli also become non-acid-fast after such damage as crushing the cells between glass slides or exposure to ultrasound. According to Barksdale and Kim (1977), the acid-fastness of intact mycobacterial cells depends on trapping of intracellular fuchsin. A barrier results from mycolate-fuchsin complexes that form in the cell wall.

Growth

Most pathogenic mycobacteria grow unusually slowly. The doubling time for *M. tuberculosis* is

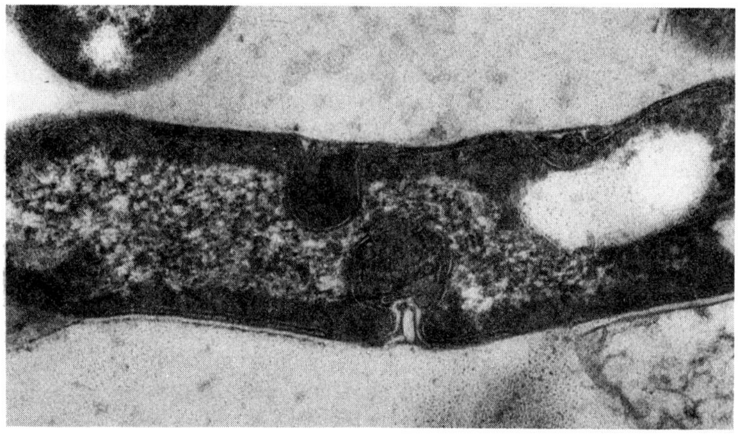

FIGURE 1. *Section of actively growing cell of* M. tuberculosis *strain H37Rv. Note the presence of mesosomes at the initiation of cross-wall formation and the presence of vacuole-like fatty inclusion bodies. ×78,000. (From Barksdale, L., and Kim, K. S.: Mycobacterium. Bacteriol. Rev. 41:217, 1977.)*

12 to 18 hours, in contrast to 15 minutes for the Enterobacteriaceae. Although tubercle bacilli can grow on simple synthetic media, their isolation from clinical specimens requires complex media. The organism is a strict aerobe, an attribute that appears to play an important role in the pathogenesis of tuberculosis.

Virulent (but not avirulent) strains of *M. tuberculosis* grown on the surface of liquid or solid media characteristically form strands or cords, and this is reflected in a difference in the appearance of the colonies on solid media (Fig. 2).

Drug Susceptibility

Tubercle bacilli are inhibited in vitro and in vivo by isonicotinic acid hydrazide (INH), streptomycin (ST), para-aminosalicylic acid (PAS), thiacetazone (TH), ethambutol (EMB), and rifampin (RIF), as well as by several other more toxic drugs (e.g., ethionamide, kanamycin, cycloserine) that are used primarily in patients whose organisms are resistant to the first-line drugs (INH, ST, PAS, TH, EMB, RIF). RIF inhibits DNA-dependent RNA polymerase in prokaryotic cells; ST inhibits function of the 30 S ribosomal subunit. The mode of action of INH is unclear but may involve suppression of mycolic acid synthesis. Although antituberculosis drugs administered singly to experimentally infected animals bring about bacteriostasis, the combination of RIF and INH has been demonstrated to have bactericidal activity in vivo. Tubercle bacilli isolated from the sputum of inadequately treated patients may be resistant to one or more of the drugs used. This resistance is not associated with plasmids (no R-factor) and is due to the selection and the eventual predominance of naturally occurring resistant mutants present in small numbers among the high populations of bacilli that develop in cavitary tuberculosis.

Lipids

Mycobacteria contain a number of unusual high molecular weight complex lipids, for example, mycolic acid, mycosides, waxes D, trehalose-6,6′-dimycolate, and sulfolipid. A number of the complex lipids of mycobacteria contain mycolic acid, which has the general formula

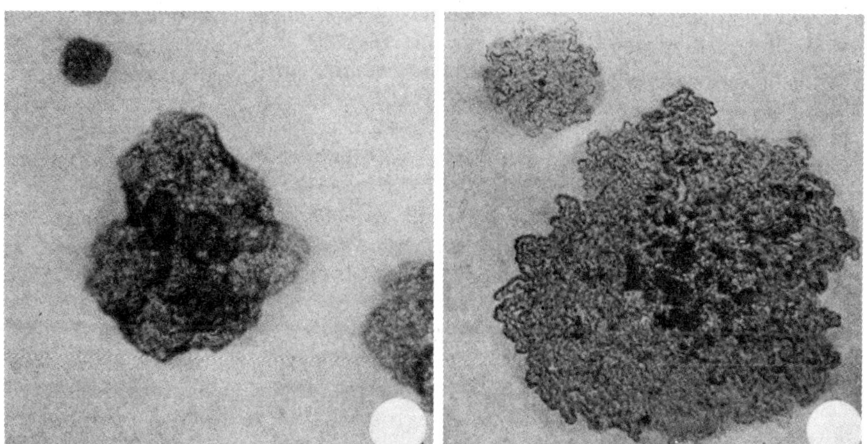

FIGURE 2. *Colonies of virulent (H37Rv) and attenuated (H37Ra)* M. tuberculosis *grown on the surface of agar medium. A, H37Ra: 12-day-old culture illustrating the nonoriented heaped-up structure of the colony. B, H37Rv: the colonies are flat with a serpentine structure. Cording is visible at the edge of the colony as thin strands and loops. (From Middlebrook, G., Dubos, R. J., and Pierce, C.: Virulence and morphological characteristics of mammalian tubercle bacilli. J. Exp. Med. 86:175, 1947.)*

$C_{88}H_{176}O_4 \pm 5CH_2$. The general structure of mycolic acid from *M. tuberculosis* is:

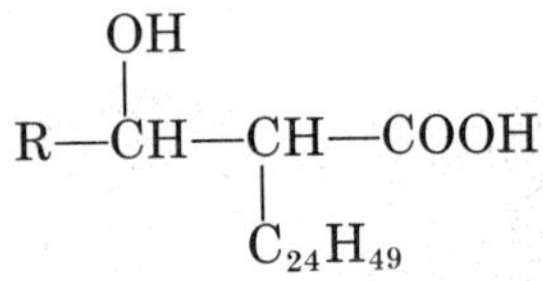

The R group in the formula contains about 60 carbon atoms and an undetermined oxygen function and occurs in the molecule in three chains.

The mycosides, a series of mycolic acid–containing glycolipids or glycolipid peptides, are uniquely distributed among the different species of mycobacteria, a chemically distinct mycoside associated with each species (Randall and Smith, 1964). Some of the mycosides have been shown to occur on the outer surface of the cell and to act as mycobacteriophage receptors. Although determination of the specific mycoside present in an isolate could contribute to its identification, the steps involved are too complex for routine application.

Waxes D are a family of closely related substances composed of mycolic acid, peptides, and polysaccharides. When extracted from *M. tuberculosis*, these substances have unique adjuvant properties in that they not only enhance antibody production against a protein antigen incorporated in a wax D oil emulsion, but also induce a cell-mediated immune (CMI) response against the protein (White et al., 1958). Because of this attribute, waxes D may contribute to the pathogenesis of tuberculosis through enhancement of the CMI response (specifically, delayed-type hypersensitivity) against mycobacterial proteins. Recent work from several laboratories indicates that the adjuvant-active component of waxes D (and the mycobacterial cell wall) is an N-acetylmuramyl dipeptide (see Barksdale and Kim, 1977).

Cord factor, so named because it was thought to be responsible for the cord-forming tendency of virulent tubercle bacilli grown on the surface of liquid or solid media, can be extracted from the cells with hexane. The purified material has been reported by various investigators to have the following properties: lethality for mice; inhibition of migration of polymorphonuclear leukocytes; induction of protection against virulent infection; and induction of granuloma formation. Despite this impressive array of potential contributors to virulence, the role played by cord factor in the pathogenesis of tuberculosis is unknown.

Binding of the dye neutral red, another property reported to distinguish between virulent and avirulent mycobacteria, has been shown to be due to the presence of a sulfonated glycolipid characterized as a tetraester of trehalose; again, the role of this substance in the disease process has not been elucidated.

Antigenic Structure

As is true of all other microbes, mycobacteria contain many antigens and antigenic determinants. Antibody-mediated immunity (AMI) and cell-mediated immunity (CMI) develop against many of these determinants during the course of infection; however, no association between a particular antigenic structure and virulence has been established. Various species of mycobacteria share antigenic determinants, and this is responsible for the cross-reactions observed in tuberculin skin test responses.

Pathogenic Properties

Tuberculosis epitomizes the pathogenesis of chronic infectious diseases. Virulent mycobacteria produce no potent toxins or tissue-destructive enzymes. The principal basis for their virulence lies in the fact that, though readily phagocytosed, the organisms resist the destructive properties of normal macrophages and are capable of multiplying intracellularly. This process damages tissue once CMI is activated.

The principal features in the initiation of tuberculous infection and its evolution to tuberculous disease, as revealed by studies of the disease in man and in an experimental model of tuberculosis in guinea pigs, can be summarized as follows. The organism (usually one to three), airborne as droplet nuclei coming from an individual with cavitary tuberculosis (or in the case of the guinea pig model, from an aerosol-generating device) is inhaled by a susceptible individual. Droplet nuclei less than 5 μm in diameter escape the defense mechanisms of the upper respiratory tract, and usually one such particle is carried into a terminal alveolar space in the well-ventilated mid to lower lung. The small number of organisms present in the particle is readily ingested by an alveolar macrophage.

After a lag period of about three days, the organism begins to multiply slowly, with one generation occurring each 12 to 18 hours. The bacilli, accumulating intracellularly, kill the initial alveolar phagocyte, and the organisms released are readily ingested by macrophages (transformed from blood monocytes) carried to the site and replicating in response to the developing inflammatory process. This slowly evolving local lesion continues to enlarge, and about two weeks after initiation of the infection, bacilli are transported via the lymphatics to the lymph node draining that region of the lung. The fixed macrophages in the lymph node ingest the organisms, but again they continue to multiply intracellularly.

A few days later, organisms leave the lymph node and are carried via the efferent lymphatics and thoracic duct to the bloodstream. This bacillemic phase of the infection leads to the dissemination of organisms throughout the body, where they are again trapped by macrophages already present, or they are ingested by macrophages carried to the sites of deposition. Again, evidence suggests that the bacilli likely multiply inside cells at each of these sites.

About three to four weeks after infection, the CMI response has been initiated, and immune (sensitized) lymphocytes coming from lymph nodes and spleen are carried by the bloodstream to the sites of the developing microscopic lesions. The interaction between lymphocytes sensitive to mycobacterial antigens and the corresponding specific antigen presumably leads to the release of mediators (lymphokines) and in turn to the activation of macrophages and to the initiation of caseation necrosis. The joint action of the activated macrophages and the unfavorable oxygen tension at most of the sites of hematogenous dissemination are presumably responsible for the death of the mycobacteria. In contrast, the organisms that by chance lodge in the apex or the subapical area of the lungs, the kidney, or the growing ends of long bones (in children) are favored by the high pO_2 in these sites (the high pO_2 in the apical and subapical areas of the lungs results from a high ventilation-perfusion ratio). The tubercle bacillus, a strict aerobe, may survive in small numbers in these sites of high pO_2 in a slowly metabolizing, relatively dormant state.

These events can all occur without the development of symptoms; that is to say, this is tuberculous infection, but not tuberculous disease. The host-parasite interaction terminates at this stage in about 90 per cent of infected persons. The only evidence that infection has taken place is conversion of the tuberculin reaction from negative to positive (tuberculin conversion) and perhaps the later development of a calcified focus at the site of the primary respiratory implantation and perhaps at the site of the draining lymph node in the hilus of the lung. These calcified sites are called the primary complex or Ghon complex. There may also be radiologic abnormalities and even calcification in the apical and subapical areas of the lung. These are sometimes referred to as Simon's foci.

Paradoxically, if tuberculous infection is to progress to tuberculous disease, the subsequent critical events usually take place at the apical or subapical site seeded during the bacillemia (Fig. 3), rather than at the site of the primary respiratory implantation (primary complex). This progression occurs during the first two years of infection in about 5 per cent (recent experience in India suggests a lower percentage) of those infected and much later in life in another 5 per cent. One can postulate the following events.

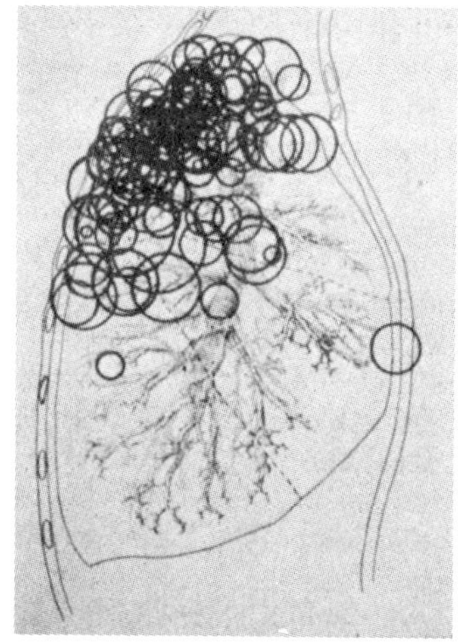

FIGURE 3. *Location of 268 cavities in the lungs of 204 tuberculosis patients as revealed by stereoscopic radiographs. Note the tendency for cavities to occur in the apical and subapical areas of the lungs. (From Sweaney, H. C., Cook, C. E., and Kegerreis, R.: A study of the position of primary cavities in pulmonary tuberculosis. Am. Rev. Tuberc. 24:558, 1931.)*

There is an antecedent temporary suppression of the CMI response, perhaps due to another infection or therapy for another disease. Activated macrophages are no longer produced, and the organisms that had been dormant in the apical focus begin to multiply and within several weeks are present in relatively high numbers. Later, when the CMI response is restored, sensitive lymphocytes reappear at the site of the developing lesion, lymphokines are released, and macrophages are activated. However, because of the high level of bacillary antigen, the entire process is exaggerated, and although the population of bacilli is temporarily reduced, caseation necrosis is extensive. Some change in the lesion, which is not yet fully understood, causes the caseous mass to undergo liquefaction. This creates a favorable environment for the bacillus, and massive growth ensues, more lymphocytes infiltrate, more lymphokines are released, and the lesion enlarges. As the developing lesion erodes through a bronchus, the liquid caseum is released into the bronchial tree, coughed up, and expectorated or swallowed. This leaves a cavity in the lung, and bacilli continue to multiply in the lining of the cavity. This process is usually accompanied by low-grade fever, cough, and malaise — that is, the symptoms of tuberculous disease.

Immunity to Tuberculosis

Substantial evidence from work with animal models of tuberculosis, as well as work with other diseases caused by facultative intracellular pathogens, indicates that immunity to these microbial agents is most likely due to the CMI response. Macrophages, activated as a result of the lymphokines released from the interaction of immune lymphocytes and specific mycobacterial antigen, are apparently altered so that they become competent to control the intracellular organism. The process of macrophage activation results in increases in the following: metabolic activity; membrane activity; phagocytic activity; and granule (lysosomal) enzymes (Dannenberg, 1968). It has not been established, however, which if any of these changes is responsible for the ability of the activated macrophage to limit the intracellular multiplication of tubercle bacilli. The role of CMI in the tissue-destructive aspects of the disease and in acquired resistance to tuberculosis is a clear example of the immune response as a "double-edged sword."

Human field trials have revealed that injection of bacille Calmette-Guérin (BCG) vaccine, a viable attenuated strain of *M. bovis*, leads to as much as 70 to 80 per cent reduction in the incidence of tuberculosis in vaccinated subjects (Hart, Sutherland, and Thomas, 1967). It is theorized that the protective effect of BCG stems from an interference with the bacillemic phase of the infection in vaccinated persons. This hypothesis is directly supported by studies in a guinea pig model of tuberculosis constructed so as to mimic the conditions under which tuberculosis is initiated in humans (Fok et al., 1976). Many reports indicate that killed mycobacterial vaccines or bacillary fractions induced protective immunity in experimental animals (Smith, Wiegeshaus, and Grover, 1968); however, none of these substances has been shown yet to have application for the prevention of human disease.

Epidemiology of Tuberculosis (Prevalence, Mode of Spread, Prevention)

The *prevalence* of tuberculous infection is estimated by tuberculin test surveys of sample populations. The *incidence* of tuberculous disease is obtained from data giving the number of new cases of tuberculosis, as revealed by laboratory studies and clinical findings, or from the frequency of individuals shown to have tubercle bacilli in their sputum. The incidence of tuberculosis varies greatly from one country to another. In several of the developing countries, tuberculous infection rates are as high as 85 to 90 per cent of the general population. In these same areas the prevalence of tuberculous disease may be as frequent as one to two per hundred. In contrast, in some of the developed countries, rates of tuberculous infection have fallen to as low as 10 to 15 per cent, and of tuberculous disease as low as 10 to 15 per 100,000.

As was indicated earlier, transmission of tuberculosis occurs when an individual with tubercle bacilli present in his or her sputum (sputum-positive tuberculosis) coughs, sings, or otherwise creates an aerosol and a few organisms are inhaled by a nearby susceptible individual. This situation, referred to as a primary aerosol, is one in which droplet nuclei containing bacilli are transmitted via the air directly from the diseased person to a susceptible person. It is this primary aerosol that is most likely to be associated with transmission of the infection. In contrast, droplets of sputum that fall to the floor, bedding, or furniture are less likely to result in transmission even if they become airborne once again. This situation is referred to as a secondary aerosol, and successful transmission is less likely for several reasons. The most important is that the bacilli in the secondary aerosol are almost always associated (by electrostatic forces) with particles of dust or lint, and the particle size is thereby increased. Because the defense mechanisms of the upper respiratory tract are very effective in trapping and eliminating particles greater than 5 μm in diameter, the organisms in the secondary aerosol are removed before they reach susceptible tissues deep in the lung. Moreover, bacilli in the secondary aerosol usually have had longer exposure to the inactivating influence of ultraviolet light. It has been shown experimentally that tubercle bacilli in an aerosol are rapidly killed upon exposure to ultraviolet light. The importance of the contrast between primary and secondary aerosol is that it focuses attention on the most important source of transmission, namely, the infectious person. It is possible to reduce transmission of tuberculosis to direct contacts of patients with positive sputum by lowering the concentration of infectious particles in the air. This can be done by means of an adequate number of air changes and by ultraviolet irradiation of the upper air in the patient's room. The concept of primary aerosol indicates that less attention need be given to the handling of bedding and objects used by the patient.

The most effective means of preventing tuberculosis is to block the transmission of the organism. This is best accomplished by early detection and treatment of patients with tubercle bacilli in their sputum. Treatment for only several weeks with isoniazid alone (despite the need for two or more drugs to prevent drug resistance) will greatly reduce the number of secondary cases among close contacts of a patient under treatment, even though organisms may still be recovered from his sputum.

A second means of reducing the tuberculosis problem in a given area is vaccination with BCG. As was indicated earlier, field trials have shown that use of an effective BCG vaccine is accompanied by a 70 to 80 per cent reduction in the incidence of tuberculous disease.

Laboratory Diagnosis

Direct Microscopy

Sputum collected from the diseased person should be examined after the preparation of an acid-fast stain. A loopful of material from the sediment in the sputum container (the maximal chance of finding bacilli is realized from a study of the solid particles) is smeared on a glass slide, air-dried, and heat-fixed. Some modification of the Ziehl-Neelsen acid-fast stain is usually employed. The smear is flooded with carbol-fuchsin and then heat is applied to cause the stain to steam. After five minutes of steaming, the carbol-fuchsin is washed off with water and the smear is decolorized for up to two minutes with a reagent consisting of 3 ml of hydrochloric acid in 97 ml of 95 per cent ethanol or with 25 per cent sulfuric acid. The preparation is then washed with water and counterstained for one minute with methylene blue. After a final water wash, the slide is air-dried and examined under the oil-immersion lens of a microscope. Tubercle bacilli appear as deep-red–staining, thin-beaded, often curved rods in a blue background. The number of tubercle bacilli in a positive sputum smear could vary from a few in 20 to 30 oil-immersion fields to many in each oil-immersion field. *Fluorochrome staining* may be superior to Ziehl-Neelsen staining for the detection of tubercle bacilli in sputum. The staining reagent is tetramethyldiaminodiphenyl ketoimine (auramine O), and although most laboratories examine auramine-stained smears under ultraviolet light with a special microscope, techniques have been developed that permit use of a standard microscope and an illuminator with a special filter (Runyon et al., 1974). The fluorochrome staining procedure has the advantage of permitting the survey of a sputum smear at a lower magnification, and this speeds up the examination and increases the sensitivity of the detection of acid-fast organisms.

Digestion and Concentration

If the laboratory has resources for culture of the specimen, acid-fast stains would be made after the specimen has been subjected to digestion and concentration. The primary objectives of the digestion-concentration step are to break up tissue debris, to kill nonmycobacterial organisms present as contaminants, and to concentrate tubercle bacilli in a smaller volume of specimen. Digestion can be accomplished by vigorous shaking of a mixture of equal parts of specimen and 3 per cent sodium hydroxide for 30 minutes. The process can be accelerated by incorporating 0.5 per cent acetyl cysteine in 2 to 4 per cent sodium hydroxide, mixing, and allowing the suspension to stand for 15 minutes. The shorter contact time increases the number of surviving mycobacteria and thus increases the sensitivity of the technique for detection of a positive specimen. After digestion, the specimen is centrifuged, the supernatant fluid is decanted, and the sediment is used for smears and stains and for culture.

Culture

Recovery of mycobacteria from clinical specimens requires an enriched medium. Ordinarily this is some modification of Löwenstein-Jensen's (L-J) medium containing egg, potato extract, glycerine, and an inhibitory dye (often malachite green) added to retard the growth of contaminants that have survived the decontamination step. Slants of L-J medium are inoculated with the concentrated sediment. The cultures are incubated at 37° C for as long as eight weeks and are examined weekly for the appearance of colonies. If viable *M. tuberculosis* organisms were present in the specimen, small buff-colored, rough colonies could appear in about three weeks but may take as long as six to eight weeks to appear. Acid-fast stains made from these colonies will reveal a morphology like that described above. If growth appears within one week, the organism could be *M. fortuitum* or one of the saprophytic mycobacteria. Although a slow-growing, nonpigmented, acid-fast organism may be *M. tuberculosis*, it could also be one of the atypical mycobacteria.

Atypical Mycobacteria

Depending on the region of the world, mycobacteria other than *M. tuberculosis* are recovered from less than 0.1 per cent to as many as 50 per cent of individuals with tuberculosis-like diseases.* These organisms, collectively called "atypical mycobacteria," are indicated by an asterisk

*The high levels are based on recent reports from the low-prevalence areas of developed countries. In contrast, in a recent report from Madras, only one case of disease due to atypical mycobacteria was found among many thousands of cases of tuberculosis.

in the list of mycobacteria given in Table 1. Recent reports suggest that atypical mycobacteria more frequently cause disease in patients whose lung or immune response is already damaged by another disease. Atypical mycobacteria and the disease they produce differ in several important respects from classic tuberculosis. Of primary significance is that examination of close contacts of patients with disease due to an atypical mycobacterium reveals *no evidence of human to human transmission* (atypical mycobacteria have been recovered from the soil and from water). Atypical mycobacteria are more likely to be *resistant to the first-line antituberculous drugs* than is *M. tuberculosis*. Patients infected with atypical mycobacteria usually give a *negative tuberculin test* to the 5-TU test dose of tuberculin prepared from *M. tuberculosis*. Finally, atypical mycobacteria can be distinguished from *M. tuberculosis* by a series of laboratory tests, primarily biochemical tests.

Identification of the Species of Mycobacteria (Speciation)

When laboratory resources are available, the identification of the species of mycobacteria causing disease in a given patient is important because of the implications for the treatment regimen employed and for the need to search for other diseased persons among the contacts of the patient. Speciation is accomplished by examining the results of a series of differential tests. Although complete speciation requires a comprehensive series of tests, the 12 characteristic tests shown in Table 2 permit a satisfactory identification of most species. (Note, for example, that *M. avium* and *M. intracellulare* differ primarily in the proportion of strains showing photo-inducible pigmentation and the proportion that grow on MacConkey agar. This is the reason why these species were listed in Table 1 as the *M. avium intracellulare* complex.)

Drug Susceptibility Tests

As many as 5 to 10 per cent or even more of cultures isolated from human specimens may contain bacilli resistant to one or more of the first-line antituberculosis drugs. Therefore, pathogenic mycobacteria recovered from clinical material should be examined for susceptibility to the first-line antituberculosis drugs. If acid-fast organisms are still grown from the sputum six months after the initiation of therapy, susceptibility tests should be repeated to look for drug-resistance. The drugs to be tested are added to slants of L-J medium, and the drug-containing media and suitable control media are then inoculated with standardized suspensions of organisms prepared from the primary L-J culture. The appearance of colonies on drug-containing media is indicative of the presence of drug-resistant mycobacteria. In this event, it may be necessary to determine the susceptibility of the organism to the second-line (more toxic, less effective, and often more expensive) antituberculosis drugs.

MYCOBACTERIUM LEPRAE

Mycobacterium leprae, the causative agent of leprosy, was discovered by Hansen in 1874. In honor of his discovery and in part to circumvent the stigma associated with leprosy, the affliction is sometimes referred to as Hansen's disease and the organism as Hansen's bacillus.

Morphology

M. leprae is morphologically indistinguishable from the other mycobacteria; they are rod-shaped organisms approximately 0.3 to 0.5 μm wide and 2 to 5 μm in length. *M. leprae* is acid-fast and can be stained by the Ziehl-Neelsen method.

Metabolism and Drug Susceptibility

Of paramount importance is that *M. leprae* is an obligate intracellular parasite. The microbe has not been cultured in the absence of living tissues and is therefore referred to as a noncultivable mycobacterium. *M. leprae* appears to require something in the parasitized host cell in order to replicate. Even when multiplying intracellularly, the generation time of the organism is exceptionally long. Estimates based on experimental infection of mouse foot pads (Shepard, 1971) indicate a generation time of 13 days. *M. leprae* is susceptible to diaminodiphenylsulfone (dapsone) and to rifampin. Inoculation tests in the mouse food pad indicate that *M. leprae* present in the circulation of lepromatous patients is killed faster after rifampin therapy than dapsone therapy.

Antigenic Composition

M. leprae contains no unusual antigens. Patients infected with *M. leprae* may undergo both an antibody-mediated immune (AMI) response and a cell-mediated immune (CMI) response. Because the organism is noncultivable, the skin test reagent, called lepromin, is prepared from organisms concentrated from diseased tissue.

Pathogenic Properties

Like the tubercle bacillus, *M. leprae* produces no potent toxins or extracellular factors contrib-

TABLE 2. Characterization of Clinically Significant Mycobacteria According to Twelve Key Properties*

TWELVE PROPERTIES[a]

(1) Rate of growth (S = slow; F = fast)
(2) Secretion of niacin
(3) Reduction of nitrate ($NaNO_3$)
(4) Semiquantitative test for hyperproduction of catalase (column of gas bubbles, >45 mm)
(5) Stability of catalase to 68°C, 20 min
(6) Carotenogenesis constitutive (scotochromogenic)
(7) Carotenogenesis photoinducible (photochromogenic)
(8) Hydrolysis of Tween 80 after 10 days
(9) Reduction of tellurite ($KTeO_2$), 3 days
(10) Growth on media containing 5% (wt/vol) NaCl
(11) Hydrolysis of tripotassium phenolphthalein disulfate by arylsulfatase, 3 days
(12) Growth on MacConkey agar

FORMULAE FOR 17 MYCOBACTERIAL TAXA AND 1 SPECIES COMPLEX[a]

Taxon	Above the line	Below the line
M. avium:	S ⑤ [7] [9]	3 4 6 8 10 12
M. bovis:	S	2 3 4
M. chelonei:	F ④⑤[8][9]⑩[b] 11 ⑫	2
M. flavescens:	S 3 ④ 5 ⑥ 8 [9] ⑩	
M. fortuitum[c]*:*	F ③④ 5 [8][9] 10 ⑪ ⑫	2
M. gastri:	S [12]	
M. gordonae:	S ④⑤ 6 ⑧	3 9 12
M. intracellulare:	S ⑤ ⑨ [12]	3 4 6 7 8 10 11
M. kansasii:	S ③④ 5 ⑦ 8	2 6 12
M. marinum:	S [2] [4][5]⑦⑧	3 10 11
M. phlei:	F ③ 4 5 6 8 [9] 10	
M. scrofulaceum:	S [3]④ 5 6	9
M. smegmatis:	F 3 ④⑤ 8 ⑨ 10	11
M. terrae complex[d]*:*	S ③ 4 ⑤ ⑧	6 7 10 12
M. triviale:	S 3 ④ 5 8 10 [11]	
M. tuberculosis:	S ②③ [8]	5 6 9 10
M. vaccae:	F ③ 4 5 [6] 7 8 ⑨ [10] [11]	
M. xenopi:	S ⑤ 6 [11]	9

*From Barksdale, L., and Kim, K. S.: Mycobacterium. Bacteriol Rev *41*:217, 1977.

[a]Key for reading formulae. When number N is unboxed, *N* (number above the line) = 100% strains tested was positive (e.g., *11* in *M. chelonei*); Ⓝ = 70 to 99% strains tested, positive (e.g., ⑤ in *M. avium*); [N] = 15 to 60% strains tested, positive (e.g., [7] in *M. avium*); N (number below the line) = 0.4 to 14% strains tested, positive (e.g., 3 in *M. avium*). Absence if N = 100% strains tested, negative (e.g., 11 absent in *M. avium*).

[b]*M. chelonei* subsp. chelonei fails to grow in 5% NaCl.

[c]*M. fortuitum* includes strains designated as *M. peregrinum*.

[d]*M. terrae* complex includes *M. terrae, M. nonchromogenicum* and *M. novum*.

uting to the disease process. The pathogenesis of leprosy appears to derive from the ability of the microbe to survive and replicate in macrophages and other host cells, especially nerve cells, and the consequent immune response to the invader. In one polar form of leprosy, lepromatous leprosy, the CMI response appears to be suppressed, and consequently the organisms accumulate in the tissues in large numbers. On the other hand, the AMI response in the lepromatous patient is not suppressed (and perhaps not even controlled), and excessive quantities of antibody are produced. The sera of lepromatous patients often contain circulating immune complexes composed of antibacillary antibody and anti-immunoglobulin. These immune complexes, which precipitate in the cold (+4° C) and therefore are called cryoglobulins, contribute to the pathogenesis of the disease. In patients with the other polar form of leprosy, tuberculoid leprosy, the CMI response appears to be intact and the bacillary population in the tissues is very low or even nondemonstrable.

Experimental infections have been produced in the foot pads of mice and in the nine-banded armadillo. Mice that have been thymectomized and x-irradiated develop a more rapidly progressing infection upon inoculation with *M. leprae*. A recent report suggests that a naturally occurring infection in armadillos in the United States is caused by an organism indistinguishable from *M. leprae* (Walsh et al., 1975).

Immunity to Leprosy

The patient with tuberculoid leprosy appears to be making an effective immune response against *M. leprae*, and bacillary populations are reduced to a very low level. On the other hand, the patient with lepromatous leprosy appears to have either a specific or a general defect in CMI responsiveness, and bacillary populations in the tissues reach very high levels. The increased frequency of leprosy in certain families in endemic areas and the studies of Jamison and Vollum (1968) suggest that the CMI defect in lepromatous leprosy has a genetically determined component. They compared tuberculin-negative children with a family history of leprosy and tuberculin-negative children with no family history of leprosy for their response to a vole bacillus vaccine (a live attenuated antituberculosis vaccine about equal in potency to BCG). Whereas 90 per cent of the children with no family history of leprosy responded to the vaccine by converting to a positive tuberculin test, tuberculin conversion was observed in less than 20 per cent of children with a family history of leprosy.

Considering all lines of evidence, it appears that the protective host response in leprosy is the CMI response. No vaccine has been developed from *M. leprae*. Field trials with BCG vaccine have yielded discouraging results in reducing the extent of leprosy (only 20 per cent over a nine year follow-up — Bechelli et al., 1974).

Laboratory Diagnosis of Leprosy

The bacteriological diagnosis of leprosy is based on finding acid-fast organisms in scrapings from nasal mucosa and skin lesions or in biopsy specimens. Leprosy is differentiated from other mycobacterial diseases by demonstrating the lack of cultivability and by showing nerve involvement in histologic examination of tissue biopsies.

Epidemiology of Leprosy

Leprosy appears to be transmitted by direct contact. The disease is not highly contagious, and prolonged contact with infected persons may be required before transmission occurs. It is estimated that approximately 35 million persons have leprosy, most of them living in states in central Africa and certain regions in Asia. Prevention of leprosy in endemic areas is based on diagnosis, isolation, and treatment of infected persons.

References

Barksdale, L., and Kim, K.S.: Mycobacterium. Bacteriol Rev 41:217, 1977.

Bechelli, L.M., Lwin, K., Garbajosa, P.G., Gyi, M.M., Uemura, K., Sundaresan, T., Tamondong, C., Matejka, M., Sansarricq, H., and Walter, J.: BCG vaccination of children against leprosy: Nine-year findings of the controlled WHO trial in Burma. Bull WHO 51:93, 1974.

Dannenberg, A. M., Jr.: Cellular hypersensitivity and cellular immunity in the pathogenesis of tuberculosis: Specificity, systemic and local nature, and associated macrophage enzymes. Bacteriol Rev 32:85, 1968.

Fok, J.S., Ho, R.S., Arora, P.K., Harding, G.E., and Smith, D.W.: Host-parasite relationships in experimental airborne tuberculosis. V. Lack of hematogenous dissemination of *Mycobacterium tuberculosis* to the lungs in animals vaccinated with bacille Calmette-Guerin. J. Infect Dis 133:137, 1976.

Hart, P.D., Sutherland, I., and Thomas, J.: The immunity conferred by effective BCG and vole bacillus vaccines, in relation to individual variations in induced tuberculin sensitivity and to technical variations in the vaccines. J Brit Tuberc Assoc 48:201, 1967.

Jamison, D.G., and Vollum, R.L.: Tuberculin conversion in leprous families in northern Nigeria. Lancet 11 (no.7581):1271, 1968.

Middlebrook, G., Dubos, R.J., and Pierce, C.: Virulence and morphological characteristics of mammalian tubercle bacilli. J Exp Med 86:175, 1947.

Randall, H.M., and Smith, D.W.: Characterization of mycobacteria by infrared spectroscopic examination of their lipid fractions. Zentralbl Bakteriol 194:686, 1964.

Runyon, E.H., Karlson, A.G., Kubica, G.P., and Wayne, L.G.: Mycobacterium. In Lennette, E.H., Spaulding, E.H., and Truant, J.P. (eds.): Manual of Clinical Microbiology, 2nd ed. Washington, D.C., American Society for Microbiology, 1974, pp. 148–174 (pertinent section, p. 155).

Shepard, C.C.: The first decade in experimental leprosy. Bull WHO 44:821, 1971.

Smith, D.W., Grover, A.A., and Wiegeshaus, E.: Nonliving immunogenic substances of mycobacteria. Adv Tuberc Res 16:191, 1968.

Sweany, H.C., Cook, C.E., and Kegerreis, R.: A study of the position of primary cavities in pulmonary tuberculosis. Am Rev Tuberc 24:558, 1931.

Walsh, G.P., Storrs, E.E., Burchfield, H.P., Cottrell, E.H., Vidrine, M.F., and Binford, C.H.: Leprosy-like disease occurring naturally in armadillos. J Reticuloendothel Soc 18:347, 1975.

White, R.G., Bernstock, L., Johns, R.G.S., and Lederer, E.: The influence of components of *M. tuberculosis* and other mycobacteria upon antibody production to ovalbumin. Immunology 1:54, 1958.

NOCARDIA AND STREPTOMYCES 41

Saroj K. Mishra
and Ruth E. Gordon

Nocardia and *Streptomyces* are important aerobic actinomycetes known mainly for the causation of serious diseases of man and animals, for the production of potent antibiotics as well as toxic metabolites, and for their decomposition of organic matter in the soil (Waksman, 1959). The infections caused by members of the two genera are broadly classified into two categories: 1) Nocardiosis, a generalized or systemic disease due to members of the genus *Nocardia,* principally *N. asteroides,* and occasionally to *N. caviae* (Causey, 1974) and *N. brasiliensis* (Berd, 1973, see Chapter 112). The disease is usually characterized by primary pulmonary involvement that may be pneumonic or subclinical, chronic or transitory (Emmons et al., 1977). A hematogenous dissemination may occur from the lungs or other parts of the body, especially the brain, for which the pathogen seems to have a predilection. 2) Actinomycetoma, a localized swollen lesion, usually on the foot, hand, or back, and involving the skin, subcutaneous tissues, fascia, and bone (Mariat et al., 1977, see Chapter 238). The pus draining through the sinuses contains granules (compact colonies of the causative microorganism surrounded by reactive cells), the size, shape, and color of which may suggest specific etiology. The chief etiologic agents are strains of *N. madurae,** *N. pelletieri, N. brasiliensis, N. caviae, N. asteroides,* and *Streptomyces somaliensis.* In addition, strains belonging to other species (Table 1), namely *N. dassonvillei, N. autotrophica, S. griseus, S. albus,* etc., are isolated from clinical specimens. But their actual role in the causation of disease is uncertain owing to the lack of supporting clinical, experimental, and histologic evidence. With increasing use of immunosuppressants in modern medicine, their role as "possible opportunistic invaders" should be judiciously considered.

MORPHOLOGY

Cultures of nocardiae and streptomycetes are composed of branching filaments (hyphae) approximately 1 μ in width. In stained smears the growth appears as gram-positive filaments (Fig. 1) or as fragmented forms (rods and/or coccoid bodies). Motile forms have not been reported.

In infections caused by some species of *Nocardia* (mainly *N. asteroides, N. caviae,* and *N. brasiliensis*), the microorganisms observed in clinical specimens are generally acidfast. Acidfastness (See *Appendix* in this chapter) of cultures of the same species grown in vitro is, however, a variable characteristic. In some cultures of these few species of *Nocardia,* usually those recently isolated, a high percentage (85 to 100 per cent) of the filaments and fragments retain the carbol fuchsin. Smears of other cultures reveal a decreasing range of acidfastness with only portions of the filaments resisting decolorization (Fig. 1), and many nonacidfast cultures have been observed.

Routinely, the best observation of morphologic characteristics of the nocardiae and streptomycetes is by microscopic examination (100, 200, and 400 ×) of individual, mechanically undisturbed colonies. After three to five days of incubation at 28° C, individual colonies of cultures streaked on plates of tap water agar (see *Appendix*) display a network of branching filaments (vegetative or substrate hyphae) that spread over the surface of the agar. (Caution: The microorganisms must be killed by formalin fumigation before opening the plate.) The vegetative hyphae normally give rise to aerial hyphae that project into the air and, therefore, appear wider and darker than the substrate hyphae (Fig. 2). Junctures of a substrate hypha and an aerial hypha can usually be clearly seen at the periphery of the colony. At the center of the colony, longer and more abundant aerial hyphae hide the substrate hyphae. After longer incubation (7 to 14 days), the aerial hyphae generally are more extensive and branching. Some are straight; some flexuous; and some form loose spirals or a variety of other forms. Under these conditions of growth, the aerial hyphae of cultures of nocardiae and streptomycetes may or may not segment into bead-like conidia (spores).

In the routine diagnostic laboratory, the microscopic appearance of the individual colonies of nocardiae and streptomycetes offers the best

*The authors prefer the retention of *N. madurae, N. pelletieri,* and *N. dassonvillei* in the genus *Nocardia* to their assignment to *Actinomadura.*

TABLE 1. Some Properties of Typical Strains of Species of Nocardia and Streptomyces of Medical Importance

PROPERTY	*N. asteroides*	*N. carnea*	*N. caviae*	*N. brasiliensis*	*N. transvalensis*	*N. autotrophica*	*N. orientalis*	*N. aerocolonigenes*	*N. dassonvillei*	*N. madurae*	*N. pelletieri*	*S. somaliensis*	*S. lavendulae*	*S. albus*	*S. rimosus*	*S. griseus*
Acidfastness	v*	v	v	v	v	–	–	–	–	–	–	–	–	–	–	–
Decomposition of																
Adenine	–	–	–	–	v	+	–	–	+	–	–	–	–	–	+	+
Casein	–	–	–	+	–	–	+	+	+	+	+	+	+	+	+	+
Hypoxanthine	–	–	+	+	+	+	+	+	+	+	v	–	+	+	+	+
Tyrosine	–	–	–	+	+	+	+	+	+	+	+	+	+	+	+	+
Urea	+	–	+	+	+	+	+	+	v	–	–	–	v	+	v	+
Xanthine	–	–	+	–	v	v	v	–	+	–	–	–	v	v	+	+
Resistance to																
Lysozyme	+	+	+	+	+	–	–	+	–	–	–	–	+	v	+	–
Rifampin	+	+	+	+	+	–	v	–	–	v	v	–	–	+	+	–
Hydrolysis of																
Hippurate	v	–	v	–	–	v	v	v	+	–	–	–	–	v	v	v
Starch	v	v	v	v	+	+	+	+	+	+	–	v	+	–	+	+
Nitrite from nitrate	+	+	+	+	+	v	v	v	+	+	+	–	v	v	v	v
Survival at 50°C, 8 hr	+	+	+	–	+	v	v	v	+	+	+	v	–	+	+	+
Acid from																
Adonitol	–	–	–	–	+	+	+	v	–	+	–	–	–	+	+	v
1(+)Arabinose	–	–	v	–	–	v	+	+	v	+	–	–	v	–	v	v
Cellobiose	–	–	–	–	–	v	+	+	+	+	–	–	v	+	+	+
1-Erythritol	–	–	–	–	+	+	+	–	–	–	–	–	–	+	+	–
Glucose	+	+	+	+	+	+	+	+	+	+	+	v	+	+	+	+
Glycerol	+	+	+	+	+	+	+	+	+	+	–	–	+	+	+	+
i-Inositol	–	v	+	+	v	v	+	+	–	v	–	–	v	v	+	v
d(+)Lactose	–	–	–	–	–	–	+	+	–	v	–	–	v	+	+	+
d(+)Maltose	–	–	–	–	–	+	v	+	+	v	–	v	+	+	+	+
d(–)Mannitol	–	+	v	+	v	+	+	+	+	+	–	–	–	+	+	+
d(+)Melezitose	–	–	–	–	–	+	v	–	–	–	–	–	–	v	–	–
α-Methyl-D-glucoside	–	–	–	–	–	v	+	–	v	–	–	–	–	+	–	+
d(+)Raffinose	–	–	–	–	–	–	v	v	–	–	–	–	–	–	+	v
1(+)Rhamnose	v	–	–	–	–	–	v	v	v	+	–	–	–	–	–	v
d-Sorbitol	–	+	–	–	v	+	v	–	–	–	–	–	–	–	v	–
d(+)Trehalose	v	+	v	+	v	+	v	+	v	+	+	–	v	+	+	+
d(+)Xylose	–	–	–	–	–	+	+	+	v	+	–	–	v	+	v	+

*v, variable property; +, positive property; –, negative property of typical strains.

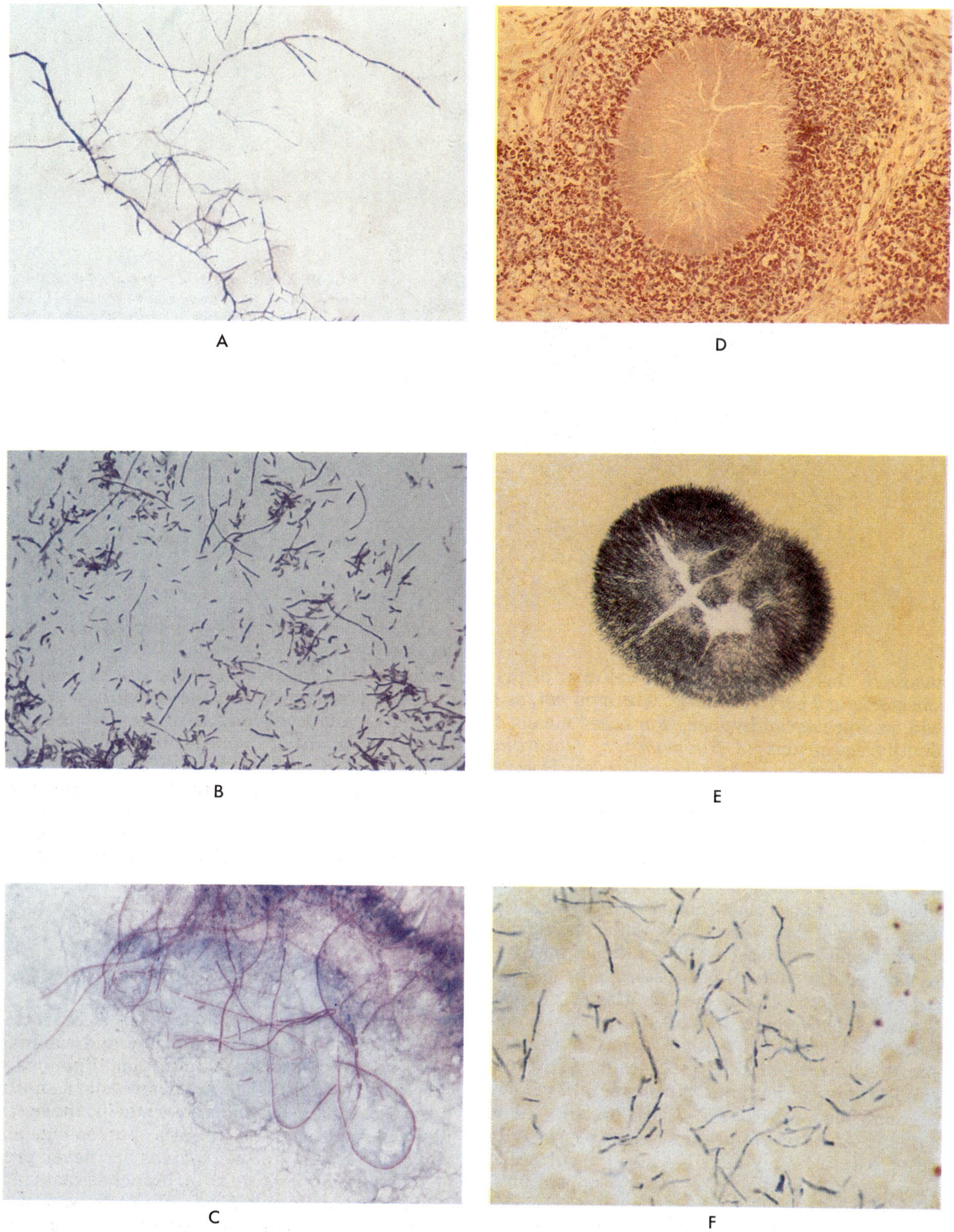

FIGURE 1. *Microscopic morphology of some Nocardiae in culture (A–C) and in tissue (D–F).*

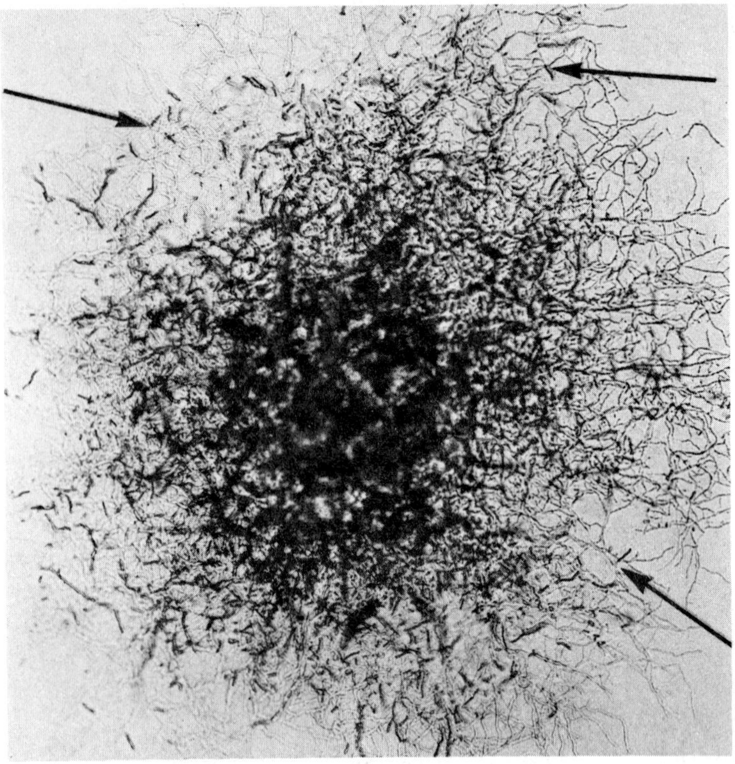

FIGURE 2. *Colony grown on tap water agar for 4 days at 28° C, showing vegetative and aerial hyphae. Arrows point to juncture of a vegetative and an aerial hypha. ×700.*

means of their separation from cultures of mycobacteria and corynebacteria, which do not, as a rule, produce aerial hyphae. (For other means of identifying strains of *Streptomyces, Nocardia, Mycobacterium,* and *Corynebacterium* see Species Identification.)

On the whole, the gross appearance of cultures of strains of *Nocardia* and *Streptomyces* belonging to the same species varies considerably. Figures 3 and 4 are presented to illustrate not only the diversity in appearance among cultures of the same species but also the similarity among strains of different species, for example, the likeness of certain cultures of *N. dassonvillei* to cultures of *S. griseus* (Fig. 3). The growth of some strains of *N. asteroides* (Fig. 4) is thick and abundant, whereas growth of other strains is thin and sparse. Cultures of different strains of *N. asteroides* may be red, pink, purple, peach, yellow, or cream, and some form a dark brown soluble pigment. The growth of some cultures is heavily coated with aerial hyphae that give the surface of the growth a powdery or chalky appearance. Other cultures are less thickly coated with aerial hyphae and varying amounts of the substrate growth can be seen. The aerial hyphae of other strains cannot be seen without the aid of a microscope.

Despite this quantitative variation, the formation of aerial hyphae, with few exceptions, is a taxonomically reliable characteristic of the nocardiae and streptomycetes; the aerial hyphae may be sparse and even rudimentary, but they are there. Exceptions are some strains of *N. madurae, N. pelletieri, N. aerocolonigenes,* and *S. somaliensis*. Strains that do not produce aerial hyphae are, however, identifiable by a distinctive pattern of physiologic, chemotaxonomic, and other morphologic properties (see section on *Laboratory Diagnosis*).

PATHOGENIC PROPERTIES

Nocardiosis due to *N. asteroides* is characterized by the formation of multiple and confluent abscesses and intense suppuration (Emmons et al., 1977). In contrast to actinomycosis, a similar disease caused by *Actinomyces israelii,* the nocardial lesions show less fibrosis, burrowing, and sinus formation; sulfur granules are never present. The lesions contain thin, branched filaments, as well as bacillary bodies, which can be demonstrated only when the sections are properly stained. Loose conglomeratory growth may sometimes be observed in the tissues, but granules and the giant cell granulomatous tissue reaction, as seen in actinomycetoma or in organs of mice experimentally infected with nocardiae (Fig. 1), are

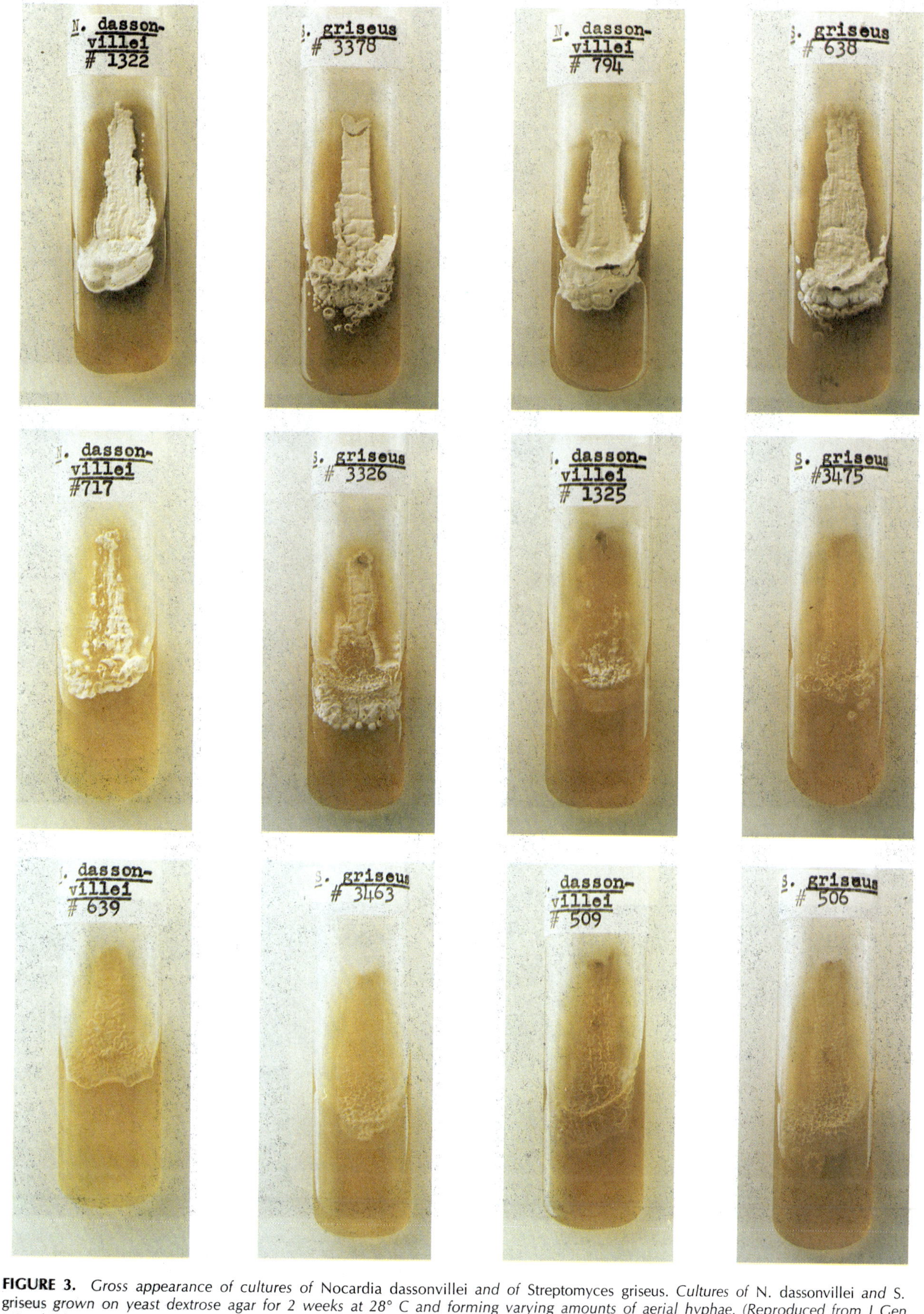

FIGURE 3. *Gross appearance of cultures of* Nocardia dassonvillei *and of* Streptomyces griseus. *Cultures of* N. dassonvillei *and* S. griseus *grown on yeast dextrose agar for 2 weeks at 28° C and forming varying amounts of aerial hyphae. (Reproduced from J Gen Microbiol 50:235–240, 1968, with the kind permission of the editors).*

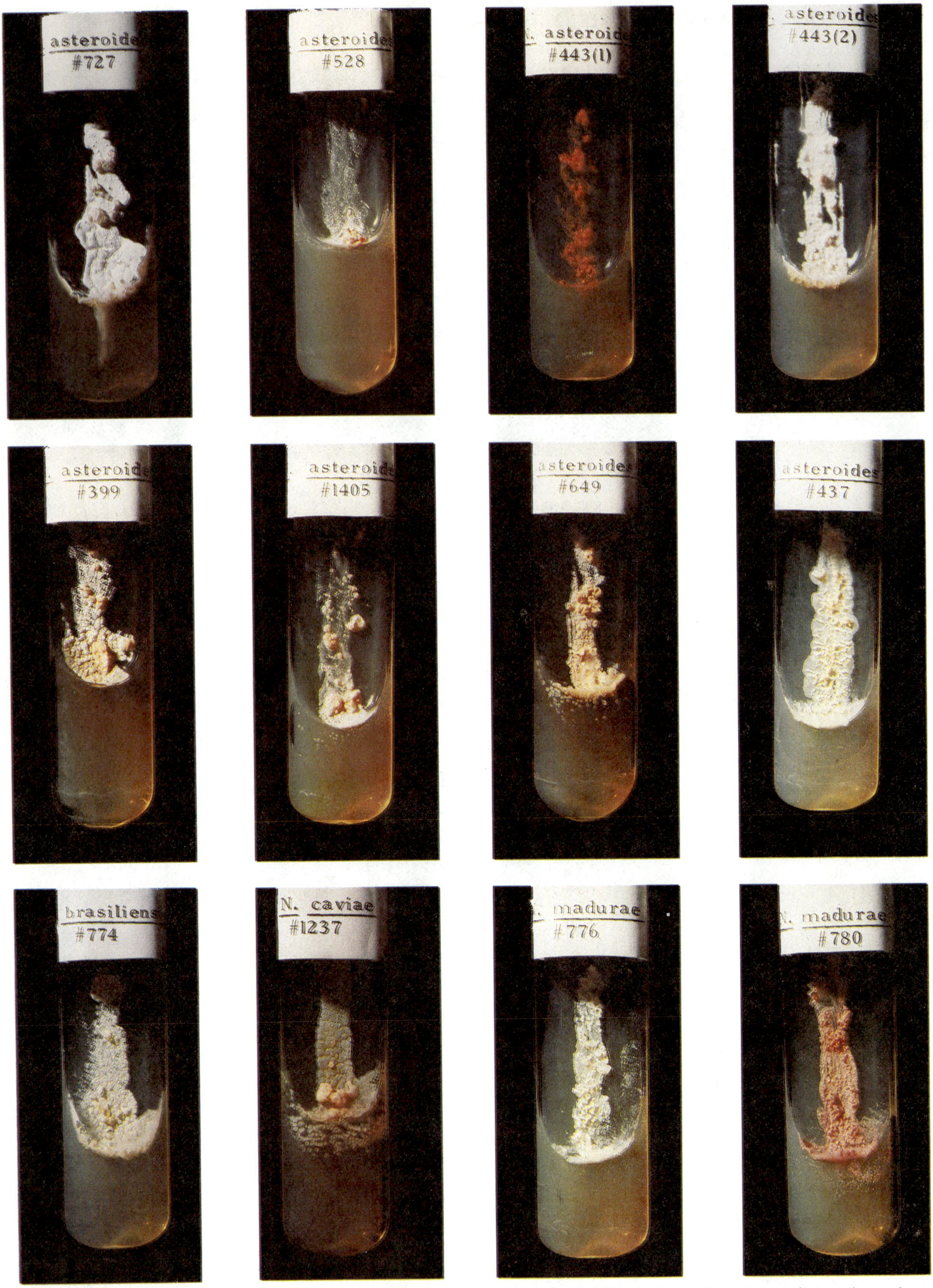

FIGURE 4. *Gross appearance of some cultures of* Nocardia asteroides, Nocardia brasiliensis, Nocardia caviae, *and* Nocardia madurae. *Cultures grown on yeast dextrose agar for 1 week at 28° C.* N. asteroides: *no. 727, heavy growth thickly covered with aerial hyphae; no. 528, thin growth covered with aerial hyphae; no. 443(1), red-pigmented growth without macroscopically visible aerial hyphae; no. 443(2), variant of red strain with abundant aerial hyphae; nos. 399, 1405, 649, and 437, cultures forming decreasing amounts of aerial hyphae; nos. 399 and 649, cultures producing dark brown soluble pigment.* N. brasiliensis: *no. 774.* N. caviae: *no. 1237.* N. madurae: *nos. 776 and 780, cultures with different pigments.*

seldom present in generalized or systemic nocardiosis in man.

For demonstrating nocardiae in the tissue, best results are obtained when the sections are stained by the method of Brown and Brenn or McCallum-Goodpasture-Gram. They can also be demonstrated by Gomori's methenamine-silver stain by extending the staining time. Periodic acid-Schiff or Gridley stains are not satisfactory, and the microorganisms cannot be detected by hematoxylin and eosin stain. Although most strains of *N. asteroides, N. caviae,* and *N. brasiliensis* possess a certain degree of acidfastness, this property is hard to demonstrate in sections of formalin-fixed tissue.

Generalized nocardiosis can be successfully produced in laboratory animals, although suitability of animal species varies and the degree of virulence may differ from microbial strain to strain. In our experience, albino mice are the most suitable for testing pathogenicity of nocardiae. Usually a heavy inoculum prepared in 5 per cent, sterile, gastrin mucin and injected intraperitoneally can cause multiple abscess formation involving most of the peritoneal organs. Dissemination to the lungs and brain may occur and the majority of animals die within one to two weeks. Alternatively, the infection may heal spontaneously with indefinite survival of most of the animals. The age of the culture (three to five days), amount of inoculum, and use of adjuvant are important factors. *N. brasiliensis* is reportedly more virulent than *N. asteroides* when injected intraperitoneally (Kurup et al., 1970A) or in the footpads of the mice (Gonzalez Ochoa, 1973). When injected intravenously, however, *N. asteroides* and *N. caviae* cause much higher mortality and morbidity than *N. brasiliensis* (Mishra et al., 1973B). Mishra and coworkers also demonstrated that cortisone administration can significantly lower the host resistance to the infection by the three species. The histopathologic picture in experimental nocardiosis is similar to that observed in generalized nocardiosis in man and animals. Attempts to infect mice or other laboratory animals with *N. madurae, N. pelletieri, N. dassonvillei,* and *S. somaliensis* have been, so far, unsuccessful.

LABORATORY DIAGNOSIS

Due to the lack of a characteristic clinical syndrome and specific immunologic tests, the laboratory diagnosis of nocardiosis demands greater attention and effort than most of the other common bacterial and fungal infections. The problem is also aggravated by the in vitro sensitivity of cultures of *Nocardia* and *Streptomyces* to most of the commonly used antibiotics that are incorporated into culture media to suppress the fast-growing saprophytic or commensal bacteria and fungi. On the other hand, filaments of *N. asteroides* may occasionally survive the concentration procedure commonly used for the isolation of *Mycobacterium tuberculosis* and grow on Lowenstein-Jensen medium. Because the filaments may be very fragile and may show strong acidfastness, nocardiae may easily be mistaken for *M. tuberculosis.*

Collection of Clinical Specimens

Samples of sputum, bronchial aspirates, gastric lavage (in the case of young children who tend to swallow their sputum), surgically removed tissues (biopsies and resected lungs), cerebrospinal fluid, and midstream urine are collected under aseptic conditions. A fresh morning sample of sputum, collected in a wide mouthed, glass stoppered bottle, after the patient has thoroughly cleaned his teeth and mouth, is more productive than a 24-hour specimen. The specimen is promptly brought to the laboratory and processed immediately. Sputum is homogenized by vigorous shaking with sterile glass beads. The biopsied tissues may be homogenized with the help of glass tissue grinders. Samples of cerebrospinal fluid and urine must be concentrated by centrifugation.

Direct Examination

Smears prepared from the homogenized or concentrated specimen are stained and examined microscopically. Several fields should be searched for the presence of filaments and coccobacillary bodies. The filaments are gram positive but often take the stain irregularly, look beaded, and may sometimes be 50 μ in length. Branching is usually at right angles and tends to be at long intervals. Acidfastness, if demonstrable, is particularly helpful in differentiating *N. asteroides, N. caviae,* and *N. brasiliensis* from *N. madurae, N. pelletieri,* and *S. somaliensis.*

Direct Culture

After the initial processing, the specimens are streaked upon neutral Sabouraud's dextrose agar and yeast dextrose agar plates (see *Appendix*) and incubated at 37° C (anaerobic or microaerophilic conditions are not required for growth). These media are useful for isolating nocardiae and streptomycetes from relatively less contaminated clinical specimens such as biopsied tissues, washed grains from actinomycetoma, and cerebrospinal fluid. If the specimens are suspected of harboring a wide spectrum of contaminating microbial flora, for example, gastric lavage and sputa of patients with chronic pulmonary diseases, the chances of isolating some nocardiae are enhanced by the use of the paraffin bait technique (see *Appendix*) as described by Mishra and Randhawa (1969) and Mishra et al. (1973A). It depends on the ability of nocardiae to use as energy sources certain organic compounds, such as paraffin, which are not attacked by other aerobic bacteria.

N. asteroides and other pathogenic nocardiae and streptomycetes, although widespread in nature, are surprisingly seldom encountered as laboratory contaminants or as constituents of normal microbial flora of the human body. Isolation of *N. asteroides* from a clinical specimen, therefore, has a definite diagnostic value. An unequivocal diagnosis of systemic nocardiosis depends upon repeated demonstration of gram-positive and partially acidfast branching filaments in the direct smear and isolation of the pathogen in culture.

Laboratory diagnosis of actinomycetoma is relatively easier. Biopsied tissue or the pus collected from the draining sinuses should be examined carefully for the presence of granules. If present, the granules should be separated from the pus and washed carefully in sterile physiologic saline. In the case of *N. brasiliensis, N. caviae,* and *N. asteroides,* the granules are soft, white to cream colored, and range from 0.5 to 1 mm in size. The grains produced by *N. pelletieri* in tissue are hard, red, pink, or occasionally yellowish in color and measure about 1 mm. The granules produced by *N. madurae* are soft, big in size, mostly white or cream, rarely pink. *S. somaliensis* produces yellowish to brown, hard granules, 0.5 to 2 mm in diameter. After gross examination, the granules are crushed, examined microscopically, and inoculated on Sabouraud's dextrose agar and yeast dextrose agar. (The paraffin bait technique is not suitable for the isolation of *N. madurae, N. pelletieri,* and *S. somaliensis* because they do not utilize paraffin as the sole source of carbon.) Although morphology of the granules and staining characteristics of the filaments present therein are fairly distinctive of some of the etiologic species, the organism must be grown in culture for its specific identification.

Single colonies, morphologically compatible with the cultures of *Nocardia* and *Streptomyces,* should be picked from the plates or tubes directly

A KEY TO THE TENTATIVE IDENTIFICATION OF NOCARDIAE AND STREPTOMYCES OF MEDICAL IMPORTANCE

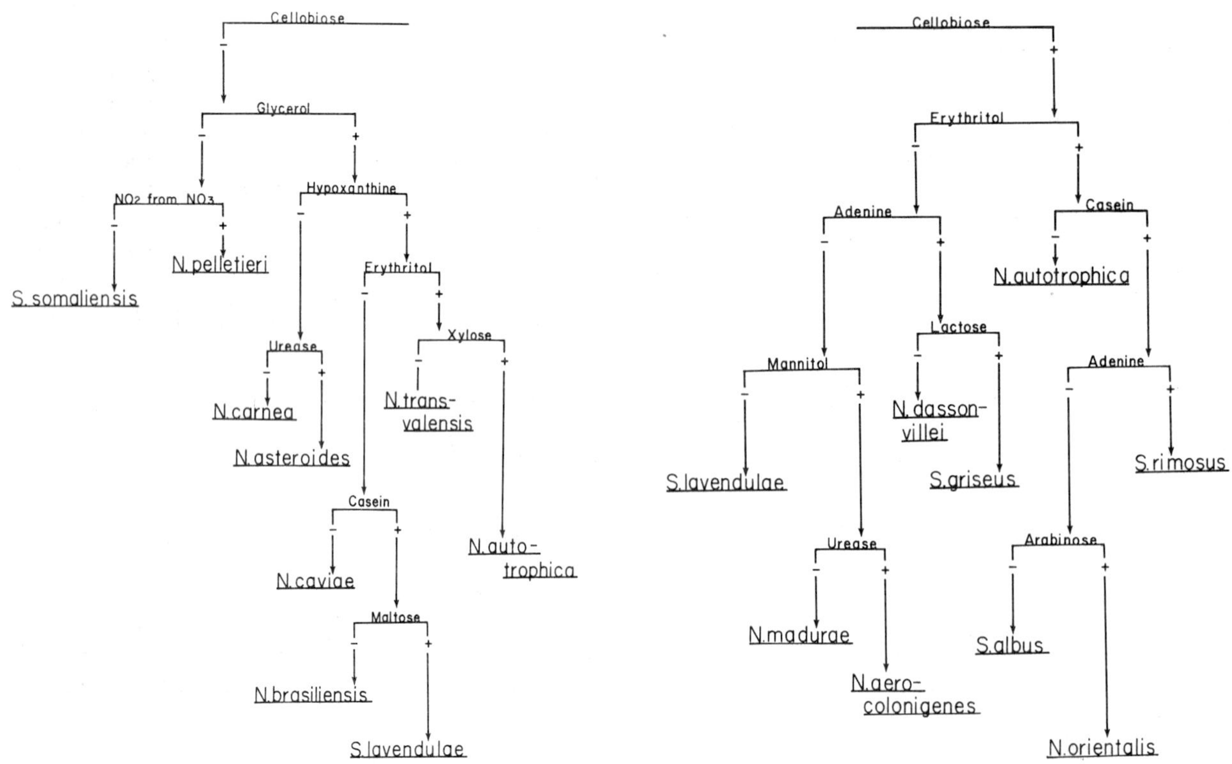

FIGURE 5.

inoculated with the clinical specimens or isolated from the growth on paraffin-coated rods. After purifying the culture by serial dilution or by repeated streaking on yeast dextrose agar plates, the growth, microscopic morphology, and physiologic characteristics are studied.

Species Identification

In the absence of a reliable morphologic feature, or features, for the separation of the genera *Nocardia* and *Streptomyces,* the form of diaminopimelic acid (DAP) in the cell wall or in whole-cell hydrolysates is used to divide the genera (Lechevalier, 1977). Strains of *Nocardia* species have the *meso*-form of DAP and strains of *Streptomyces* species, the L-form.

Other properties — pigment formation, fragmentation of substrate hyphae, development of conidia, and production of an earthy odor — are possessed by some strains of nocardiae and by some strains of streptomycetes. Nor have we found any physiologic test that differentiates strains of the two genera. The key (Fig. 5) presented here for the tentative identification of strains of nocardiae and streptomycetes includes the species represented by the strains received here as isolations from patients. It is designed to permit the user, in whose laboratory chromatographic determination of DAP is not feasible, to arrive at species identifications of unknown isolations with the morphologic characteristics of nocardiae and streptomycetes as though the species belonged to one genus (Gordon, 1976). If the 13 tests employed in the key (see *Appendix*) are applied to an unknown isolate and the strain is found to be, for example, cellobiose negative, glycerol positive, hypoxanthine negative, and urease positive, the strain may be tentatively identified as one of *N. asteroides*. If the remaining properties of the strain are as follows: acid formation negative from arabinose, erythritol, lactose, maltose, mannitol, and xylose; inability to decompose adenine and casein; and positive reduction of nitrate to nitrite, the tentative identification of the strain as one of *N. asteroides* is partially confirmed.

Further confirmation may be obtained by studying additional physiologic characteristics listed in Table 1 and described in detail by Gordon et al. (1974, 1977, 1978).

If chemotaxonomic examinations are feasible in the laboratory, the presence of *meso*-DAP, arabinose, and galactose in the cell walls or in hydrolysates of whole cells will provide further evidence of the strain's identity as one of *N. asteroides*. The lipids (mycolic acids) of strains of mycobacteria and of some nocardiae offer another valuable chemotaxonomic property (Lechevalier, 1976). Species lacking mycolic acids are usually nonacidfast. Among the species described in Table I, cultures of *N. asteroides, N. carnea, N. caviae, N. brasiliensis,* and *N. transvalensis* contain LCN-A (lipid characteristic of *Nocardia* species-asteroides type) that distinguishes them from the other species in Table I and from strains of mycobacteria and corynebacteria. Mycolic acids apparently have no significant role in the pathogenicity of nocardiae and streptomycetes. Certain nonpathogenic species, e.g., *N. amarae,* are known to have LCN-A, whereas some of the well known pathogenic species, such as *N. madurae, N. pelletieri,* and *S. somaliensis,* do not have LCN-A or any other mycolic acids.

The importance of comparing unknown isolates with as many named strains of the various species as possible cannot be overstressed to anyone attempting to identify strains of nocardiae and streptomycetes for the first time. By applying the tests to known strains, the investigator soon learns whether each test as applied in his laboratory yields results comparable with those given here or elsewhere in the literature (Ajello et al., 1963; Goodfellow, 1971). If only a tentative identification of an isolate can be made or if the strain cannot be identified by the information provided, it should be sent to a reference laboratory (our Institute, the Waksman Institute of Microbiology, Piscataway, New Jersey; the Center for Disease Control, Atlanta, Georgia; or any other laboratory actively engaged in the identification of aerobic actinomycetes).

The name *N. farcinica,* the type species of the genus *Nocardia,* is not used here because its status is uncertain. Diversity of interpretations of Nocard's description (1888) of his strain isolated from a case of bovine farcy in Guadeloupe, the different species identity of strains purported to be Nocard's original isolation, and the assignment of causal agents of African cases of bovine farcy to the genus *Mycobacterium* (Asselineau et al., 1969) confuse present-day significance of the name.

ANTIGENIC COMPOSITION, SEROLOGY AND IMMUNITY

The immunology of *Nocardia* and *Streptomyces* is still poorly understood. Cross-reaction with mycobacteria and members of related genera is frequently observed. Strains of *Nocardia, Mycobacterium,* and *Corynebacterium,* which have identical cell wall sugars, seem to have some common antigenic components. Crude and purified polysaccharides, protein derivatives, and polypeptide preparations are immunologically active but both homologous and heterologous reactions may occur. Pier and Fichtner (1971), using a gel diffusion precipitin test and soluble extracellular antigens, distinguished four serotypes in nocardiae. Type I, II, and III antigens were confined to *N. asteroides;* the type IV antigen occurred in strains of *N. caviae* and *N. brasiliensis* as well as in several strains of *N. asteroides.* By comparative reciprocal intradermal sensitin tests in guinea pigs, some success has been reported in delineating *Nocardia* at the species level but not at the generic level (Magnusson, 1976). Despite the promising value of immunologic methods in demonstrating homologous antigenicity in experimental models, the diagnostic worth of sensitin tests in demonstrating immediate or delayed hypersensitivity is limited. Skin test with purified protein derivatives (nocardin-PPD) derived from *N. brasiliensis,* is reported to elicit a positive reaction in patients with actinomycetoma caused by *N. brasiliensis* and to be negative in healthy individuals and patients with tuberculosis and various other diseases (Bojalil and Zamora, 1963). Similar convincing results, however, are lacking in cases of actinomycetoma caused by other species, and in cases of systemic or generalized nocardiosis.

The relationship of antibodies to immunity against infection by nocardiae and streptomycetes is not clear. Since members of the two genera are widely distributed in nature, it is logical to assume that more people are frequently exposed to these microorganisms than acquire clinical disease. The immunity may develop after a very mild or subclinical infection that diminishes spontaneously if host-defense mechanisms are not impaired.

DRUG SUSCEPTIBILITY

Strains of *Nocardia* and *Streptomyces* are sensitive to a number of antibiotics in vitro, but there is considerable strain variation and lack of correlation between the in vitro and in vivo effects. Since the first successful use of sulphadiazine in the treatment of subcutaneous nocardiosis (Lyons et al., 1943) and the demonstration of its in vitro inhibitory effect on aerobic actinomycetes (Cutting and Gebhardt, 1941), sulphonamides have remained the drug of choice in the treatment of nocardial infections. *N. asteroides* is demonstrably more sensitive to chlortetracycline, oxytetracycline, and chloramphenicol than to sulphadiazine in vitro, but sulphadiazine is more effective in vivo (Strauss et al., 1951). Sensitivity to penicillin is poor, whereas ampicillin, streptomycin, erythromycin, neomycin, gentamycin, and many other antibiotics and chemotherapeutics are quite active against nocardiae and streptomycetes (Bach et al., 1973; Pridham and Tresner, 1974). Some of these, alone or in combination with sulphonamides, have been used in the treatment of nocardiosis and actinomycetoma. A combination of trimethoprim and sulphamethoxazole has in vitro inhibitory effect on strains of *N. asteroides* and *N. brasiliensis* at a relatively low concentration and appears to be more useful in the treatment of actinomycetoma. Minocycline and fusidic acid have been recently reported to inhibit growth in vitro of all strains of *N. asteroides* tested at an impressively low concentration of 6.2 μg/ml and 3.1 μg/ml, respectively (Schaal and Heimerzheim, 1974; Wallace et al., 1977). There are not enough data, however, on the in vivo effect of these two drugs, as well as amikacin and tobramycin, which are also very active in vitro. Antifungal antibiotics, such as nystatin, cycloheximide, griseofulvin, and amphotericin B, are by and large not effective against *Nocardia* and *Streptomyces* infections, although occasional clinical success and in vitro inhibitory action have been reported.

EPIDEMIOLOGY

Nocardia and *Streptomyces* are soil-inhabiting, ubiquitous microorganisms, commonly present in the human environment. After the first report on the isolation of *N. asteroides* from soil by Gordon and Hagan (1936), several investigators have demonstrated the occurrence of *N. asteroides, N. caviae, N. brasiliensis,* and other medically important aerobic actinomycetes in the soil in many different parts of the world (Cross et al., 1976). Since their filaments are delicate and highly fragile, they are easily airborne. Systemic or primary pulmonary infection is usually acquired by inhalation of the airborne mycelial fragments or spores. Occasionally, primary lesions may develop in the gastrointestinal tract through ingestion of food contaminated with dust or water harboring nocardiae. In the case of actinomycetoma or subcutaneous manifestations, traumatic implantation of the pathogen directly into the skin is the commonest mode of infection. These microorganisms have no special affinity for any particular race, sex, age, or geographic region. Cases of systemic nocardiosis are reported from all over the world (Kurup et al., 1970B) although more frequently from America and Europe than from Asia and Africa. The differences are mainly due to the greater availability of highly specialized laboratory diagnostic facilities and an increased awareness among the investigators in the more developed countries. Occupational relationship is not established, but farmers and laborers heavily exposed to the soil and dust are more likely to contract the infection, especially in the presence of debilitating metabolic diseases and poor nutritional and working conditions.

The etiologic agents of actinomycetoma also have a worldwide distribution, although the disease is more frequently observed in Sudan, Mexico, and certain other arid regions lying between the equator and the Tropic of Cancer. Incidentally, *N. pelletieri* and *S. somaliensis* are more commonly incriminated in the etiology of actinomycetoma in Sudan and other African countries, whereas *N. brasiliensis* and *N. madurae* are more frequently reported in Mexico. Strains of almost all the species have been isolated from cases in other countries but no etiologic or regional significance could be attached to any particular species.

APPENDIX

Acidfastness

Slides with air-dried smears of the cultures are immersed in carbolfuchsin (saturated ethanol solution of basic fuchsin, 10 ml; 5 per cent aqueous solution of phenol, 90 ml). Heat is applied and the fuchsin boiled for five minutes. The slides are washed in water, quickly dipped once in acid alcohol (concentrated HCl, 3 ml; 95 per cent ethanol, 97 ml), immediately washed again in water, and counterstained with methylene blue (saturated solution of methylene blue in 95 per cent ethanol, 30 ml; 0.01 per cent aqueous solution of KOH, 100 ml). Use of 0.5 per cent H_2SO_4 instead of acid alcohol was abandoned in this laboratory because filaments of some strains of *S. griseus* were not decolorized.

Tap Water Agar

This medium contains only 1.5 g of agar in 100 ml of tap water.

Neutral Sabouraud's Dextrose Agar

The ingredients of this agar are: glucose, 20 g; neopeptone, 10 g; agar, 20 g; distilled water, 1000 ml; pH 7.0.

Yeast Dextrose Agar

This agar comprises 10 g of yeast extract, 10 g of glucose, 15 g of agar, 1000 ml tap water, and the pH is adjusted to 7.0.

Paraffin Bait Technique

Mix 2 ml of the homogenized specimen with 5 ml of carbon-free broth ($NaNO_3$, 2 g; K_2HPO_4, 0.8 g; $MgSO_4 \cdot 7H_2O$, 0.5 g; $FeCl_3$, 10 mg; $MnCl_2 \cdot 4H_2O$, 8 mg; $ZnSO_4$, 2 mg; distilled water, 1000 ml; pH 7.0). Introduce into the tube a paraffin-coated glass rod (sterilized by storing overnight in 95 per cent alcohol), after the alcohol has drained off. Incubate the tubes at 37° C and observe regularly up to four weeks before discarding as negative. Cream to orange growth on the paraffin rod, just above the surface of the medium, is removed and streaked on Sabouraud's dextrose agar (pH 7.0). Single colonies morphologically compatible with those of nocardiae are picked and processed further for specific identification. Better results can be obtained if, after 1 week, the paraffin rod is carefully removed from the inoculated tube and transferred to another tube containing only 6 ml of carbon-free broth.

Acid from Carbohydrates

The basal inorganic nitrogen medium contains $(NH_4)_2HPO_4$, 1 g; KCl, 0.2 g; $MgSO_4 \cdot 7H_2O$, 0.2 g; agar, 15 g; distilled water, 1000 ml. The pH value of the medium is adjusted to 7.0 before the addition of 15 ml of a 0.04 per cent solution (w/v) of bromcresol purple. After this agar is tubed and sterilized by autoclaving, 0.5 ml of a 10 per cent solution (w/v) of each carbohydrate (autoclaved separately) is added aseptically to the tubes. Cultures on slants of these carbohydrate agars are observed for acid color of the indicator after 7 and 28 days of incubation at 28° C. Tubes showing a negative reaction at 28 days are retained 4 more weeks and observed again.

Decomposition of Adenine and Hypoxanthine

Adenine (0.5 g), suspended in 10 ml of distilled water, is autoclaved, mixed carefully with 100 ml of sterile nutrient agar (peptone, 5 g; beef extract, 3 g; agar, 15 g; distilled water, 1000 ml; adjusted to pH 7.0), cooled to 45° C, mixed again, and poured into sterile plates (60 mm diameter). Care must be taken to obtain an even distribution of the crystals of adenine throughout the solidified agar. Each culture is streaked once across a plate, incubated at 28° C, and observed at 14 and 21 days for the disappearance of the crystals underneath and around the growth. Plates of hypoxanthine are prepared, inoculated, incubated, and observed in the same way.

Decomposition of Casein

A suspension of 5 g of skim milk powder in 50 ml of distilled water and a suspension of 1 g of agar in 50 ml of distilled water are autoclaved separately and cooled to 45° C. The two suspensions are then mixed and poured into plates. Each culture is streaked once across a plate and incubated at 28° C. At 7 and 14 days the plates are examined for clearing of the casein underneath and around the growth.

A heavy inoculum is necessary in testing for the decomposition of casein, adenine, and hypoxanthine. Some cultures grow well but do not dissolve the casein or the crystals of adenine and hypoxanthine except around the larger clumps of inoculum.

Decomposition of Urea

A 10 ml amount of 15 per cent solution of urea (w/v), sterilized by filtration, is added to 75 ml of sterile urease broth (KH_2PO_4, 10 g; Na_2HPO_4, 9.5 g; yeast extract, 1 g; 0.04 per cent solution of phenol red (w/v), 20 ml; distilled water, 1000 ml; adjusted to pH 7.0). The mixture is pipetted aseptically in 2.5 ml amounts into sterile, capped tubes and inoculated with actively growing cultures. An alkaline reaction after 28 days of incubation at 28° C demonstrates the presence of urease.

Reduction of Nitrate to Nitrite

Cultures in nitrate broth (peptone, 5 g; beef extract, 3 g; KNO_3, 1 g; distilled water, 1000 ml; pH 7.0) are tested for nitrite at 5, 10, and 14 days by mixing 1 ml of culture with three drops of each of the following solutions: 1) sulfanilic acid, 8 g; 5 N acetic acid (glacial acetic acid and water 1:2.5), 1000 ml; and 2) dimethyl-α-naphthylamine, 6 ml; 5 N acetic acid, 1000 ml. (Caution: Dimethyl-α-napthylamine is carcinogenic and should be used with care.) Development of red or yellow (high concentration of nitrite) is proof of the presence of nitrite. In the absence of a positive reaction for 14 days, 4 to 5 mg of zinc dust are added to the tube previously tested for nitrite. The presence of nitrate (absence of reduction) is demonstrated by the development of a red color.

We thank the Foundation for Microbiology for providing color Figures 1 and 4 and Dr. A. H. McIntosh and Mrs. R. Shamy for Figures 2 and 5. Permission of the *Journal of General Microbiology* to reproduce Figure 3 is gratefully acknowledged.

References

Ajello, L., Georg, L. K., Kaplan, W., and Kaufman, L.: Laboratory Manual for Medical Mycology. Public Health Service Publication No. 994. Washington, U. S. Government Printing Office, 1963, p. G65.

Asselineau, J., Lanéele, M. A., and Chamoiseau, G.: De l'étiologie du farcin de zébus tchadiens: nocardiose ou mycobactériose? II. Composition lipidique. Rev. Élev Méd Vét Pays Trop 22:205, 1969.

Bach, M. C., Sabath, L. D., and Finland, M.: Susceptibility of *Nocardia asteroides* to 45 antimicrobial agents *in vitro*. Antimicrob Agents Chemother 3:1, 1973.

Berd, D.: *Nocardia brasiliensis* infection in the United States: A report of nine cases and a review of the literature. Am J Clin Path 60:254, 1973.

Bojalil, L. F., and Zamora, A.: Precipitin and skin tests in the diagnosis of mycetoma due to *Nocardia brasiliensis*. Proc Soc Exp Biol Med 113:40, 1963.

Causey, W. A.: *Nocardia caviae:* a report of 13 new isolations with clinical correlation. Appl Microbiol 28:193, 1974.

Cross, T., Rowbotham, T. J., Mishustin, E. N., Tepper, E. Z., Antoine-Portaels, F., Schaal, K. P., and Bickenbach, H.: The ecology of nocardioform actinomycetes. In Goodfellow, M., Brownell, G. H., and Serrano, J. A. (eds.): The Biology of the Nocardiae. London, Academic Press, 1976, p. 337.

Cutting, W. C., and Gebhardt, L. P.: Inhibitory effect of sulphonamides on cultures of *Actinomyces hominis*. Science 94:568, 1941.

Emmons, C. W., Binford, C. H., Utz, J. P., and Kwon-Chung, K. J.: Medical Mycology, 3rd ed. Philadelphia, Lea and Febiger, 1977.

Gonzalez Ochoa, A.: Virulence of nocardiae. Canad J Microbiol 19:901, 1973.

Goodfellow, M.: Numerical taxonomy of some nocardioform bacteria. J. Gen Microbiol 69:33, 1971.

Gordon, R. E.: A taxonomist's obligation. In Goodfellow, M., Brownell, G. H., and Serrano, J. A. (eds.): The Biology of Nocardiae. London, Academic Press, 1976, p. 66.

Gordon, R. E., and Barnett, D. A.: Resistance to rifampin and lysozyme of strains of some species of *Mycobacterium* and *Nocardia* as a taxonomic tool. Int J Syst Bacteriol 27:176, 1977.

Gordon, R. E., Barnett, D. A., Handerhan, J. E., and Pang, C. H-N.: *Nocardia coeliaca, Nocardia autotrophica,* and the nocardin strain. Int J Syst Bacteriol 24:54, 1974.

Gordon, R. E., and Hagan, W. A.: A study of some acid-fast actinomycetes from soil with special reference to pathogenicity for animals. J Infect Dis 59:200, 1936.

Gordon, R. E., Mishra, S. K., and Barnett, D. A.: Some bits and pieces of the genus *Nocardia: N. carnea, N. vaccinii, N. transvalensis, N. orientalis,* and *N. aerocolonigenes*. J. Gen Microbiol 109:69, 1978.

Kurup, P. V., Randhawa, H. S., Sandhu, R. S., and Abraham, S.: Pathogenicity of *Nocardia caviae, N. asteroides,* and *N. brasiliensis*. Mycopathologia 40:133, 1970A.

Kurup, P. V., Randhawa, H. S., and Gupta, N. P.: Nocardiosis: a review. Mycopathologia 40:193, 1970B.

Lechevalier, H. A.: The actinomycetales: Soil or oxidative actinomycetes. In Laskin, A. I., and Lechevalier, H. A. (eds.): Handbook of Microbiology, 2nd ed. Cleveland, CRC Press, 1977, p. 363.

Lechevalier, M. P.: The taxonomy of the genus *Nocardia:* Some light at the end of the tunnel? In Goodfellow, M., Brownell, G. H., and Serrano, J. A. (eds.): The Biology of the Nocardiae. London, Academic Press, 1976, p. 1.

Lyons, C., Owen, C. R., and Ayers, W. B.: Sulphonamide therapy in actinomycotic infections. Surgery 14:99, 1943.

Magnusson, M.: Sensitin tests as an aid in the taxonomy of *Nocardia* and its pathogenicity. In Goodfellow, M., Brownell, G. H., and Serrano, J. A. (eds.): The Biology of the Nocardiae. London, Academic Press, 1976, p. 236.

Mariat, F., Destombes, P., and Segretain, G.: The mycetomas: clinical features, pathology, etiology and epidemiology. Contr Microbiol Immunol 4:1, 1977.

Mishra, S. K., and Randhawa, H. S.: Application of paraffin bait technique to the isolation of *Nocardia asteroides* from clinical specimens. Appl Microbiol 18:686, 1969.

Mishra, S. K., Randhawa, H. S., and Sandhu, R. S.: Observations on paraffin baiting as a laboratory diagnostic procedure in nocardiosis. Mycopathologia 51:147, 1973A.

Mishra, S. K., Sandhu, R. S., Randhawa, H. S., Damodaran, V. N., and Abraham, S.: Effect of cortisone administration on experimental nocardiosis. Infect Immun 7:123, 1973B.

Nocard, E.: Note sur la maladie des boeufs de la Guadeloupe connue sous le nom de farcin. Ann Inst Pasteur 2:293, 1888.

Pier, A. C., and Fichtner, R. E.: Serologic typing of *Nocardia asteroides* by immunodiffusion. Am Rev Resp Dis 103:698, 1971.

Pridham, T. G., and Tresner, H. D.: Family VII. *Streptomycetaceae* Waksman and Henrici. In Buchanan, R. E., and Gibbons, N. E. (eds.): Bergey's Manual of Determinative Bacteriology. 8th ed. Baltimore, Williams and Wilkins Company, 1974, p. 747.

Schaal, K. P., und Heimerzheim, H.: Mikrobiologische Diagnose und Therapie der Lungennocardiose. Mykosen 17:313, 1974.

Strauss, R. E., Kligman, A. M., and Pillsbury, D. M.: Chemotherapy of actinomycosis and nocardiosis. Am Rev Tuberc 63:441, 1951.

Waksman, S. A.: The Actinomycetes. Vol. 1. Baltimore, Williams and Wilkins Company, 1959.

Wallace, R. J., Jr., Septimus, E. J., Musher, D. M., and Martin, R. R.: Disk diffusion susceptibility testing of *Nocardia* species. J Infect Dis 135:568, 1977.

SPIROCHETES

LEPTOSPIRA 42

A. D. Alexander, Ph.D.

GENERAL CHARACTERISTICS AND MORPHOLOGY

The genus *Leptospira* consists of a large number of serologically heterogeneous strains that can be separated into two large complexes, the biflexa, or "saprophytic," leptospires and the parasitic leptospires. The biflexa leptospires are omnipresent in natural waters and wet soils, where apparently they are free-living. They are not known to produce infections in man or other mammalian hosts except for a few rare reports. The parasitic leptospires include the pathogenic strains known to infect man and other mammals. At present only one species, *Leptospira interrogans,* is recognized, although separate speciation has been proposed —*L. biflexa* for the saprophytic leptospires and *L. interrogans* for the parasitic types (International Committee on Systematic Bacteriology, 1974).

Biflexa strains are distinguishable from pathogenic strains in the following respects: inability to infect laboratory animals; relative resistance to growth-inhibitory effects of bivalent copper ions, 8-azaguanine and aniline dyes; less fastidious nutritional growth requirements; ability to grow at low temperatures; and serologic and genetic characteristics (Turner, 1976). Members of the two complexes share no nucleotide sequences as determined by DNA-DNA annealing tests. However, leptospires within each complex can be further separated into three groups that have partial DNA homology. A leptospire designated *illini* has been isolated which resembles the biflexa leptospires phenotypically but is genetically unrelated to any of the known genetic types (Brendle et al., 1974).

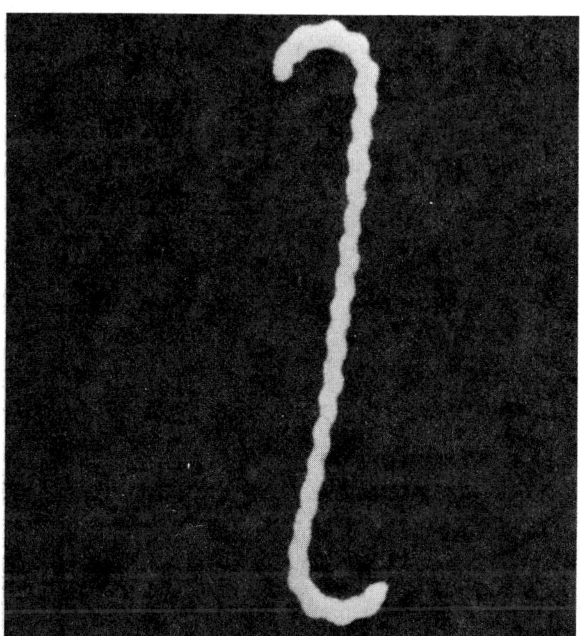

FIGURE 1. *Leptospira. Dark-ground illumination. (Courtesy C. D. Cox, University of Massachusetts, Amherst.)*

Morphologically the saprophytic and parasitic leptospires are indistinguishable. Leptospires are helicoidal, flexuous organisms with semicircular hooked ends (Fig. 1). Occasional strains are straight-ended or have only one hook. The organisms usually measure 6 to 20 μm in length (range 3 to 40 μm) and approximately 0.1 μm in diameter. The coils are tightly wound with an overall diameter of 0.2 to 0.3 μm and a wavelength of approximately 0.5 μm. When examined by electron microscopy the structure of leptospires consists of a helicoidal protoplasmic cylinder that is wound about two independent axial filaments (Fig. 2). The filaments are inserted subterminally at each end with their free ends positioned toward the middle, where they usually do not overlap. The axial filaments resemble bacterial flagella structurally, in chemical composition, and in method of attachment. The helicoidal body is delineated by a cytoplasmic membrane–cell wall complex similar to that of gram-negative bacteria. The flagella and protoplasmic cylinder are enclosed by a common outer envelope or sheath (Johnson, 1977).

In fluid media, leptopsires appear to rotate alternately along their longitudinal axis, moving forward or backward without polar differentiation. The movement is characteristic because of the spinning hooked ends. In semisolid media, serpentine, boring, and flexing movements are seen. Leptospires can be seen by dark-field or phase contrast but not by bright-field microscopy. They are not readily visualized when stained with aniline dyes but can be demonstrated by the

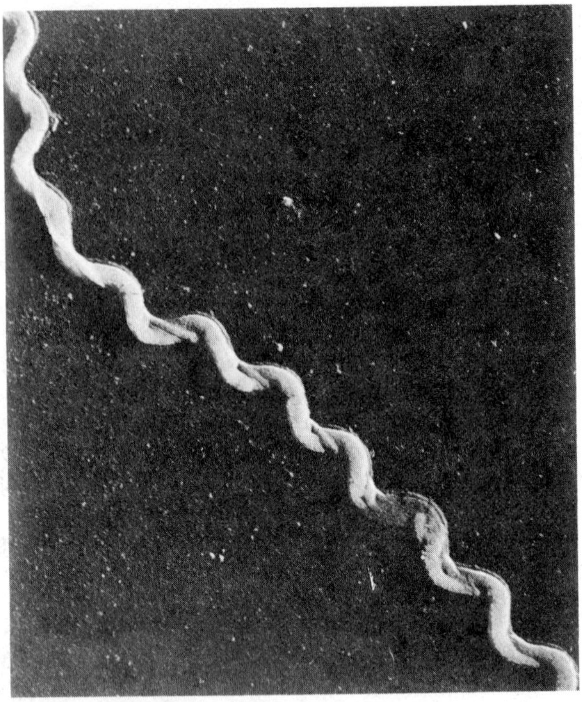

FIGURE 2. *Electron micrograph of a portion of a leptospire. (Courtesy Armed Forces Institute of Pathology, Washington, D.C.)*

use of silver deposition techniques. Leptospires are capable of penetrating bacterial retaining filters. These attributes coupled with the fact that they do not grow on conventional media have frequently led to the mistaken identification of leptospirosis as a viral disease.

ANTIGENIC COMPOSITION

The pathogenic leptospires have distinct agglutinogenic properties that provide the basis for their differentiation. Otherwise they are indistinguishable by morphologic, cultural, and physiologic properties. The basic taxonomic group is the serovar (syn., serotype), which is included within a species. Currently, approximately 160 different serovars have been recognized on the basis of cross-agglutination and agglutinin-adsorption tests with serovar-specific rabbit antiserum (Turner, 1976). The serovars have been assembled into 18 serogroups on the basis of shared major agglutinogens. The serogroup has no taxonomic status but is used as an expedient for selection of antisera or antigens to identify isolates and to test sera. The agglutinogenic relationships do not necessarily correlate with genetic groupings. Each genetic group contains serologically diverse serovars. However, strains with major antigenic affinities appear to be genetically homologous (Brendle et al., 1974).

The agglutinogen, not surprisingly, has been associated with the enveloping sheath, which is composed of lipids, carbohydrates, and proteins (Johnson, 1977). Flagellar antigens have been isolated and shown to fall into eight antigenic groups by immunodiffusion techniques. They appeared to correlate best with genetic groupings (Chang et al., 1974). A variety of other antigenic substances have been extracted from leptospires by various chemical or physical treatments and have been shown to have specific or generic activity by the use of various serologic techniques. Their exact location and role in immunity and virulence needs additional study. An extract from biflexa serovars that is soluble in 50 per cent but insoluble in 95 per cent ethanol has proved to be particularly useful as a genus-specific, erythrocyte-sensitizing antigen in indirect hemagglutination and hemolytic tests (Cox et al., 1957; Sulzer et al., 1975).

METABOLISM

The pathogenic leptospires are aerobic organisms that can be grown in media containing serum, or serum components and fatty acids, at pH 6.8 to 7.8 (optimally at pH 7.2 to 7.4). Long-chain fatty acids (not carbohydrates or amino acids) are used as a source of carbon and energy. They can utilize ammonium salts or urea but not amino acids for their nitrogen requirements. Exogenous purines but not pyrimidines can be incorporated. They require thiamin and vitamin B_{12} in media. Cultures are usually incubated at 30° C. Incubation time for optimal growth ranges from a few days to 4 weeks or longer but is usually 6 to 14 days. At time of optimal growth the concentration of organism ranges from 1 to 4 $\times$ 10^8/ml. Cytochrome enzymes, catalase, and enzymes of the citric acid cycle, glycolytic, and pentose pathways, and an acyl-coenzyme A dehydrogenase for β-oxididation of fatty acids have been demonstrated (Smibert, 1973).

PATHOGENIC PROPERTIES

The clinical and histopathologic findings in leptospirosis are consistent with the action of a toxin. However, attempts to isolate and identify a toxic component or product of leptospires has been elusive. Endotoxic properties of leptospires have been reported but not confirmed, nor were findings of toxic substances in lysed cultures or culture filtrates consistent (Johnson, 1976). Toxic substances have been demonstrated in extracts of infected tissues of experimental animals (Arean et al., 1964; Hanson, 1976). The nature of these toxins has not been defined. A hot-cold heat-labile, oxygen-stable hemolysin with specific activity for ruminant red blood cells is produced by some pathogenic leptospires and may be a significant factor for the hemolytic manifestations in bovine and other ruminant species (Alexander, 1976b). Association of virulence with the outer envelope is suggested by findings that it is the site of the cidal action of antibody and complement and that virulent leptospires are more resistant to the antibody complement system than avirulent cells (Bey et al., 1974; Johnson and Harris, 1967). Pathogenicity has also been associated with production of catalase (Baseman and Cox, 1969). Pathogenic leptospires produce hyaluronidase (Yanagihara et al., 1973), which may contribute to the high invasiveness of these organisms.

IMMUNITY

Immunity in leptospirosis is best correlated with specific agglutinogenic characteristics of serovars (Alexander, 1976b). Individuals who recover from infection with a specific serovar develop immunity against the same serovar but are susceptible to frank infection with antigenically different serovars. Where cross-protection was observed, it was related to common major agglutinogenic components. In accordance with this observation, the outer sheath has been found to be highly immunogenic and to have the same serologic specificity as agglutinogens (Auran et al., 1978). Intertype immunity against disease but not against infection has been demonstrated for certain serologically heterologous strains in experimental infections in hamsters and guinea pigs, and appeared to occur only between genetically homologous serovars (Alexander, 1976b).

The protection afforded by specific immunoglobulins reflects their agglutinating and opsonizing properties, which enhance destruction of the organisms by phagocytosis. Cell-mediated immunity does not appear to be important in combating acute disease but its role in the development or prevention of the renal carrier state and other persistent forms of infection cannot be ruled out (Adler and Faine, 1977). Immunologic factors may be important in some of the pathogenetic features of leptospirosis. Antigen-antibody complex reactions have been suggested to explain neurologic, renal, and ocular complications that occur after the first week of disease. Autoimmune reactions have been postulated to explain lesions in the kidney and eye. Various immediate and delayed-type hypersensitivity reactions have been elicited in experimental animals (Alexander, 1976b).

Vaccines are usually prepared from fixed cells. In man, vaccines have been extensively used in Japan, Italy, USSR, and other Eastern European countries. Vaccine prophylaxis for livestock and pets is carried out in the United States and in many other countries throughout the world. Vaccines are serovar specific and usually incorporate two or more serovars, depending on the types of prevailing leptospiral infections.

LABORATORY DIAGNOSIS

Various microscopic, cultural, and serologic procedures are available for the laboratory diagnosis of leptospirosis. The selection and use of a suitable test depend on knowledge of the course of infection. After an incubation period, usually of 10 to 12 days, the disease appears abruptly, ushering in a septicemic stage persisting approximately 1 week. During this period the organisms may be found in the cerebrospinal fluid as well as in blood. The septicemic period usually ends with the appearance of detectable antibodies. Maximal antibody levels occur the third or fourth week of disease. Thereafter antibody levels gradually decline but may be detectable for years. Leptospires may be found in the urine after the first week of disease. Shedding is more pronounced during the first few weeks of convalescence; thereafter it may persist intermittently for two to three months in many cases and longer in some.

Culture

Leptospires are best demonstrated during the acute stage by culture of blood on such media as Stuart's liquid or Fletcher's semisolid, which contain rabbit serum, or EMJH medium, which contains albumin and fatty acid in lieu of serum (Vera and Dumoff, 1974). Media are usually dispensed in test tubes in 5 ml amounts and inoculated with one or two drops of blood. At least four tubes of media, preferably from two different lots, should be used for each sample. Repeated blood cultures during the acute phase of disease are recommended. If culture medium is not immediately available at time of collection, blood may be defibrinated or mixed with anticoagulants, and subsequently cultured. Triturated clotted blood can also be used as inoculum. Spinal fluid, if available, may be cultured during the acute stage.

After the first week of disease, cultural recovery of leptospires from urine is possible in many cases. A midstream urine sample is carefully collected from cleansed genitalia and diluted 1:10 and 1:100 with media. Undiluted and diluted samples are cultured by using one to two drops of inoculum. In fatal cases leptospires may be isolated from suspensions of triturated liver and kidney. Minimal inocula of blood, urine, and tissue samples are recommended to dilute out growth-inhibitory substances. To inhibit the growth of contaminating organisms that may be present in urine or other samples, media may be supplemented with 5-fluorouracil, or with various antibiotics such as neomycin, furazolidone, cyclohexamide, or sulfathiazole, used singly or in combination (Alexander, 1974; Sulzer and Jones, 1976).

Cultures are incubated at 30° C or in the dark at room temperature, examined at five- to seven-day intervals, and discarded if negative after six weeks of incubation.

In semisolid media a linear disk of growth is usually seen 1 to 3 cm below the surface. Fluid media becomes slightly turbid. On plating media containing 1 per cent agar, many strains form subsurface colonies. Plating media are useful for purifying cultures but are less satisfactory than tubed media for primary isolation. To examine cultures, a drop is placed on a slide, covered with a coverslip, and examined by dark-ground microscopy at high dry magnification (450 times). Leptospires are recognized by their characteristic morphology and motility. Cultures can be stored in screw-capped or other airtight tubes. Transfers are made at two- or three-month intervals from semisolid media, and at three to six weeks if kept in fluid media. Cultures are typed by the use of microscopic agglutination techniques. Initially the isolate employed as antigen is tested against a battery of different serovar antisera to determine serogroup affinities. They are then tested with additional antisera to elucidate further relationships. For definitive identification, agglutinin-adsorption tests are carried out. Culture typing is usually carried out in state or national laboratories, or in special WHO Leptospirosis Reference Laboratories (Sulzer and Jones, 1976).

Microscopic Examination

The concentration of leptospires in the blood of human patients is low and may not be demonstrated by dark-ground microscopic examination. In a small proportion of cases the chance of demonstrating leptospires may be increased by differential centrifugation of whole unclotted blood, first to remove cellular elements, and then to concentrate organisms in plasma (Wolff, 1954). Although the method may be valuable in establishing a rapid diagnosis, it is not recommended as a routine diagnostic procedure because artifacts such as cellular extrusions and fibrils are frequently mistaken for spirochetes. Direct microscopic methods are of value in examination of specimens in which a profuse number of leptospires may be present, such as blood and liver suspensions of hamsters or guinea pigs infected with clinical material or urine or kidney suspensions of natural animal hosts. Fluorescent-antibody staining techniques and silver deposition techniques have been used to demonstrate organisms in postmortem tissues and are especially useful if cultural and serologic tests are not possible (Alexander, 1974; Sulzer and Jones, 1976).

Animal Inoculation

The use of laboratory animals offers no greater chance than direct culture for recovery of leptospires from blood or from other materials that can be obtained aseptically. Weanling hamsters and young guinea pigs are most commonly preferred because of their greater susceptibility to frank infection with a broad range of serovars (Wolff, 1954). The course of disease in laboratory animals varies with different serovars and also with different strains of the same serovar, and range from inapparent to lethal. Material is inoculated intraperitoneally. Heart blood for culture is taken on the fourth and sixth day after inoculation or whenever disease signs appear, and if necessary, at periodic intervals up to the twentieth day. On the twentieth day kidneys should also be cultured.

Serologic Diagnosis

The microscopic agglutination test has been most widely used. It is highly sensitive for diagnosis of recent as well as past infections and can be applied to animal as well as human sera. However, it has high specificity. Consequently, to ensure detection of antibodies that may be provoked by any of the large number of different serovars, it is necessary to use multiple serovar antigens that would cross-react with most of the known serovars that may be present. The number of test serovars may range from a few to 15 or more, depending on the number of serovars and their relative occurrence in specific geographic regions (Alexander, 1976a). Young cultures of leptospires in fluid media are used as antigen. These may be used live or be treated with Formalin. Test sera are serially diluted two- to four-fold with physiologic salt solution to provide serum dilutions from 1:50 to 1:3200. Two-tenths-milliliter amounts of serum in each dilution are distributed in a series of agglutination tubes, to which is added an equal volume of antigen. The reaction mixtures are shaken, incubated at room temperature for two to three hours, shaken again, and examined for agglutination by dark-ground microscopic examination. The test has been adapted for use with microtitration techniques (Sulzer and Jones, 1976). Titers of 1:100 are considered to be significant, and may range as high as 1:25,600 or greater.

The laboriousness of the microscopic agglutination test limits its usefulness by the small laboratory. To circumvent this drawback, macroscopic-agglutination slide tests incorporating single or pooled antigens were developed and are commercially available (Alexander, 1976a). These tests are simple to perform and have good sensitivity for detecting recently or previously formed antibodies. "Genus-specific" tests have been proposed, which entail the use of specific biflexa strains known to cross-react frequently with leptospiral antibodies elicited by a large variety of pathogenic serovars (Alexander, 1976a; Turner, 1968). A hemolytic test with a "genus-specific" biflexa erythrocyte-sensitizing antigen has been used advantageously in areas of multiple leptospirosis (Cox et al., 1957; Tan, 1974). The test has been simplified by use of glutaraldehyde-fixed sensitized erythrocytes (Sulzer et al., 1975). Complement-fixation tests with biflexa antigens are in use in some European countries (Turner, 1968). Various fluorescent-antibody techniques have been reported for diagnosing leptospirosis but have not been widely adopted.

DRUG SUSCEPTIBILITY

Penicillin, streptomycin, tetracycline, and macrolide antibiotics are all active against leptospires in experimental animals and in vitro. Leptospires are resistant to chloramphenicol, sulfa drugs, neomycin, actidione, and isoniazid (Stahlheim, 1973).

EPIDEMIOLOGY

Leptospiroses are zoonoses that are globally distributed in a large variety of rodents and other feral and domestic mammals. Approximately 160 different mammalian species have been found to harbor leptospires (Leptospire Serotype Distributions Lists, 1966, 1975). Pathogenic leptospires have occasionally been isolated from birds, amphibians, reptiles, and ticks. The epidemiologic importance of nonmammalian hosts for leptospires has not been demonstrated. The distribution of the various serovars in mammalian hosts varies according to geographic region. The host range of selected serovars that are frequent causes of human disease in many parts of the world is shown in Table 1. Many serovars appear to be preferentially adapted to select mammalian hosts. For example, serovar *icterohaemorrhagiae* is primarily associated with the Norway rat, *canicola* with dogs, and *pomona* with swine and cattle. The occurrence of a specific serovar in a select host is not exclusive. The same mammalian species may be the primary reservoir for several serovars. In addition, this species may be infected and serve as a carrier of serovars that occur predominantly in other hosts. Infections in natural hosts are usually inapparent, but may be morbid. Naturally occurring disease in animals has been principally observed in dogs, swine, cattle, and other livestock. Cats appear to be relatively resistant to disease. In many countries leptospirosis is an important veterinary medical problem.

In natural hosts, leptospires nest in the lumen of convoluted tubules within the kidneys, whence they are shed in the urine. The degree and duration of leptospiruria vary with the host and infecting serovar. The Norway rat infected with *ic-*

TABLE 1. Host Range* for Some Widely Distributed, Commonly Occuring Serovars Causing Human Leptospirosis†

SEROVAR	HOSTS: Domestic	HOSTS: Wild
Icterohaemorrhagiae	dog, pig, horse, cattle	rats,‡ mice, mongoose, raccoon, muskrat, foxes, opossum, skunk, woodchuck, nutria, cavy, hedgehog, civet, apes
canicola	dog,‡ pig, cattle, cat, horse	skunk, raccoon, armadillo, mongoose, hedgehog, jackal, bandicoot, nutria, rats, vole
pomona	pig,‡ cattle,‡ goat, dog, cat, horse, sheep, water buffalo	skunk, opossum, foxes, raccoon, deer, hedgehog, mice, civet, cavy, woodchuck, vole, souslik, wolf, rabbit, sea lion
grippotyphosa	cattle, goat, sheep, swine, dog, cat	vole,‡ mice, rats, shrew, hedgehog, raccoon, skunk, fox, opossum, cavy, muskrat, weasel, mole, bobcat, leopard cat, bandicoot, rabbit, gerbil, squirrel, polecat
bataviae	dog,‡ cat, cattle	rats,‡ mice,‡ bandicoot, leopard cat, armadillo, vole, hedgehog, shrew
autumnalis	cattle, dog	mice,‡ rat, bandicoot, raccoon, opossum
australis	cattle, dog, horse	rats,‡ hedgehogs, mice, skunk, nutria, raccoon, vole, opossum, weasel, fox, bandicoot

*Based on cultural isolation.
†Source: Leptospiral Serotype Distribution Lists, 1966, 1975.
‡Known major maintenance host.

terohaemorrhagiae sheds profuse numbers of organisms, apparently for the remainder of its natural life. Strains of *canicola* are less efficient in colonizing rat kidneys. On the other hand, in the dog *canicola* is better able to establish leptospiruria than is *icterohaemorrhagiae* (Babudieri, 1958). Shedding in infected dogs, swine, cattle, and other domestic animals may be heavy for several months after infection but is usually sparse or absent after six months. The prevalence of infection in natural hosts can be remarkably high. Carrier rates of 50 per cent or greater in Norway rats and other feral mammals, and of 10 to 50 per cent in domestic animals, are common findings in many parts of the world.

Leptospires are transmitted through contact with urine of animal carriers, either directly or by contact with damp soil or natural waters soiled by carriers. Pathogenic leptospires can survive for three months or more in neutral or slightly alkaline waters. They do not persist in brackish or acid waters. Organisms enter hosts through abrasions of the skin or through mucosal surfaces of the mouth, nasopharynx, eye, or esophagus. Individuals of all ages and both sexes are susceptible. Infection in man is accidental and is related to occupational or avocational exposure, or to contact with infected pets. Man is usually a dead-end host. Sewer workers, butchers, fish and poultry processors, miners, ditch diggers, and other individuals who work or live in rat-infested areas are at risk. Cases in dairymen, swineherds, and in abattoir workers are usually epidemiologically related to infection in livestock. Sporadic cases or epidemics have occurred in various parts of the world in agricultural workers engaged in raising flax, rice, vegetables, or sugar cane, in rubber plantation workers, and in soldiers exposed to natural waters contaminated by carriers. Common-source epidemics have occurred repeatedly in children and young adults who bathe or swim in ponds or streams located in pasturelands.

References

Adler, B., and Faine, S.: Host immunological mechanisms in the resistance of mice to leptospiral infections. Infect Immun 17:67, 1977.

Alexander, A. D.: *Leptospira*. In Lennette, E. H., Spaulding, E. H., and Truant, J. P. (eds.): Manual of Clinical Microbiology. 2nd ed. Washington, D. C., American Society for Microbiology, 1974, p. 347.

Alexander, A. D.: Serological diagnosis of leptospirosis. In Rose, N. R., and Friedman, H. (eds.): Manual of Clinical Immunology. Washington, D. C., American Society for Microbiology, 1976a, p. 352.

Alexander, A. D.: Immunity in leptospirosis. In Johnson, R. C. (ed.): The Biology of Parasitic Spirochetes. New York, Academic Press, 1976b, p. 339.

Arean, V. M., Sarasin, G., and Green, J. H.: The pathogenesis of leptospirosis: Toxin production by *Leptospira icterohaemorrhagiae*. Am J Vet Res 25:836, 1964.

Auran, N. E., Johnson, R. C., and Alexander, A. D.: Chemical composition and serological activity of leptospiral outer envelope. In Proceedings of the National Symposium on Leptospirosis, *Leptospira*, and other *Spirochaeta*, 25-27 September, 1975. Cantacuzino Institute, Bucharest. Ilexum Editure Medicala, 1978, p. 277.

Babudieri, B.: Animal reservoirs of leptospires. Ann NY Acad Sci 70:393, 1958.

Baseman, J. B., and Cox, C. D.: Terminal electron transport in Leptospira. J Bacteriol 97:1001, 1969.

Bey, R. F., Auran, N. E., and Johnson, R. C.: Immunogenicity of whole cell and outer envelope leptospiral vaccines in hamsters. Infect Immun 10:1051, 1974.

Brendle, J. J., Rogul, M., and Alexander, A. D.: Deoxyribonucleic acid hybridization among selected leptospiral serotypes. Int J Systematic Bacteriol 24:205, 1974.

Chang, A., Faine, S., and Williams, W. T.: Cross-reactivity of the axial filament antigen as a criterion for classification of *Leptospira*. Aust J Exp Biol Med Sci 52:549, 1974.

Cox, C. D., Alexander, A. D., and Murphy, L. C.: Evaluation of the hemolytic test in the serodiagnosis of human leptospirosis. J Infect Dis 101:203, 1957.

Hanson, H. E.: Pathogenesis of leptospirosis. In Johnson, R. C. (ed.): The Biology of Parasitic Spriochetes. New York, Academic Press, 1976, p. 295.

van der Hoeden, J.: Epizootiology of leptospirosis. Adv. Vet Sci 4:277, 1958.

International Committee on Systematic Bacteriology. Subcommittee on the Taxonomy of *Leptospira*. Minutes of the meetings, 30 August–4 September 1973, Jerusalem, Israel. Int J. Systematic Bacteriol 24:381, 1974.

Johnson, R. C.: Comparative spirochete physiology and cellular composition. In Johnson, R. C. (ed.): The Biology of Parasitic Spirochetes. New York, Academic Press, 1976, p. 39.

Johnson, R. C.: The spirochetes. Annu Rev Microbiol 31:89, 1977.

Johnson, R. C., and Harris, V. G.: Antileptospiral activity of serum. II. Leptospiral virulence factor. J. Bacteriol 93:513, 1967.

Leptospiral Serotype Distribution Lists According to Host and Geographic Area. Atlanta, Georgia, U.S. Dept. Health, Education and Welfare, Center For Disease Control, 1966 (Supplement 1975).

Smibert, R. M.: *Spirochaetales*, a review. CRC Critical Reviews in Microbiol 2:491, 1973.

Stahlheim, O. H. V.: Chemical aspects of leptospirosis, CRC Critical Reviews in Microbiol 2:423, 1973.

Sulzer, C. R., Glosser, J. W., Roger, F., Jones, W. L., and Tiex, M.: Evaluation of an indirect hemagglutination test for the diagnosis of human leptospirosis. J Clin Microbiol 2:218, 1975.

Sulzer, C. R., and Jones, W. L.: Leptospirosis. Methods in Laboratory Diagnosis. Revised ed. U.S. Dept. Health, Education, and Welfare, Public Health Service, Center for Disease Control, HEW Publ. No. (CDC) 76–8275, 1976.

Tan, D. S. K., and Welch, Q. B.: Evaluation of *Leptospira biflexa* antigens for screening human sera by the microscopic agglutination (MA) test in comparison with the sensitized-erythrocyte-lysis (SEL) test. Southeast Asian J Trop Med Publ Hlth 5:12, 1974.

Turner, L. H.: Leptospirosis. II. Serology. Trans R Soc Trop Med Hyg 62:880, 1968.

Turner, L. H.: Classification of spirochaetes in general and of the Genus *Leptospira* in particular. In Johnson, R. C. (ed.): The Biology of Parasitic Spirochetes. New York, Academic Press, 1976, p. 95.

Vera, H. D., and Dumoff, M.: Culture media. In Lennette, E. H., Spaulding, E. H., and Truant, J. P. (eds.): Manual of Clinical Microbiology. 2nd ed. Washington, D. C., American Society for Microbiology, 1974, p. 881.

Wolff, J. W.: The Laboratory Diagnosis of Leptospirosis. Springfield, Ill., Charles C Thomas, 1954.

Yanagihara, Y., Taniyama, T., and Mifuchi, I.: Separation and purification of hyaluronidase of Leptospira (in Japanese). Medicine and Biology 86:83, 1973.

2. ANAEROBIC BACTERIA

GRAM-POSITIVE COCCI: PEPTOCOCCUS, PEPTOSTREPTOCOCCUS, STREPTOCOCCUS (ANAEROBIC), and SARCINA 43

Howard Robert Attebery, D.D.S.

The anaerobic gram-positive cocci are a controversial and difficult group of organisms in clinical microbiology. Their pathogenicity is doubted by some (Facklam and Smith, 1976) and respected by others (Finegold, 1977; Lambe et al., 1973; Pien et al., 1972). Their classification has been difficult.

Gram-positive cocci that upon isolation grow both in a mixture of 10 per cent carbon dioxide gas in air and in an anaerobic environment are microaerophilic cocci and are not classified with anaerobic cocci. However, some strains of anaerobic cocci can become tolerant of air upon subculturing and confuse classification efforts. Many other strains of anaerobic cocci, on the other hand, are fastidious in their anaerobic requirements, and good transport and anaerobic methods are needed for their recovery.

MORPHOLOGY

It is not possible to classify anaerobic cocci to genera by using only cellular morphology. Though peptococci are usually clumped, they do occur singly or in pairs and are frequently found as short chains. Peptostreptococci and anaerobic streptococci cannot be differentiated by a Gram stain. Peptostreptococci may cluster as do peptococci with only short chains being evident. Chaining is best demonstrated in liquid cultures.

Many strains of anaerobic gram-positive cocci destain easily and have a gram-negative appearance. However, *Veillonella,* an anaerobic gram-negative coccus, is much smaller than the anaerobic gram-positive cocci, with the exception of *Peptostreptococcus micros.* Also, the anaerobic gram-positive cocci may be gram-variable during their life cycle. Anaerobic gram-positive cocci may elongate, so that they appear as short rods. This is common with *Peptostreptococcus productus* but may occur with other species. Cells may appear to be of unequal size in a culture. This size differential may be so great that it gives the appearance of budding.

The cellular morphology of *Sarcina ventriculi* is distinctive from that of the other anaerobic gram-positive cocci. The cells are usually about 2.0 mm in diameter and nearly spherical with their adjacent sides usually flattened. These cells occur in packets of eight or more.

Peptococcus (Pc.), *Peptostreptococcus* (Ps.) and the anaerobic members of the genus *Streptococcus* (S.) have no distinguishing colonial morphology. On blood agar plates, colonies of these anaerobic cocci are small (0.5 mm after 48 hours of incubation), convex, opaque, have an entire edge, and are either shiny or dull. Most colonies are gray to white in color. However, some strains of *Ps. micros* and *Ps. anaerobius* produce black colonies.

S. constellatus and *S. morbillorum* usually show alpha hemolysis. *S. intermedius* is usually not hemolytic; however, some strains are alpha-hemolytic. *Ps. productus* may be nonhemolytic or may give alpha or beta reactions. The other peptococci and peptostreptococci are not hemolytic except for some strains of *Ps. micros* and *Pc. prevotii* that may produce a beta hemolysin.

Surface colonies of *S. ventriculi* are 1 to 3 mm after 24 hours of incubation, colorless, circular, and opaque with an irregular edge. Most strains are not hemolytic; a few are alpha hemolytic. Subsurface colonies are frequently star-shaped or irregular in appearance.

ANTIGENIC COMPOSITION

Few studies have been done on the antigens of the anaerobic gram-positive cocci. Agglutination and precipitation tests for the immunologic classification of anaerobic cocci were not successful (Stone, 1940). Another study showed no indication of a common *Peptococcus* or a *Peptostreptococcus* antigen but did show that antigens of *Pc. magnus* may be distinct from the other anaerobic gram-positive cocci. Identification of this species may therefore be practical by fluorescence technique (Porschen and Spaulding, 1974). A recent study of three strains of *Pc. magnus* and one strain of *Ps. anaerobius* using counterimmunoelectrophoresis also indicated that *Pc. magnus* isolates were immunologically distinct (Markowitz and Lerner, 1977).

TABLE 1. Clinically Important Anaerobic Gram-Positive Cocci and Their Major Metabolic End-Products from Amino Acid and Glucose Metabolism[a]

NAME	MAJOR END PRODUCTS		MINOR END PRODUCTS
Pc. asaccharolyticus	A	H	B(SLF)
Pc. magnus	A	H	(LSF)
Pc. prevotii	A	H	B(L)
Pc. saccharolyticus	FA2		
Ps. anaerobius	A	H	(2BL, et al.)
Ps. micros	A		(LSF)
Ps. parvulus	LA		(S)
Ps. productus	A	(H)	S(LP)
St. constellatus	L		
St. intermedius	L		
St. morbillorum	L		
S. ventriculi	A2	H,CO_2	

[a]*Pc., Peptococcus; Ps., Peptostreptococcus; St., Streptococcus; S., Sarcina;* A, acetic acid; B, butyric acid; F, formic acid; L, lactic acid; P, propionic acid; S, succinic acid; H, hydrogen gas; CO_2, carbon dioxide gas; ( ), variable; 2, ethyl alcohol.

METABOLISM

The clinically important species of the anaerobic gram-positive cocci are listed in Table 1, together with their major end-products of amino acid and glucose metabolism.

Peptococci and peptostreptococci are heterofermentative, not producing lactic acid as the sole major end-product from glucose metabolism. They utilize amino acids as their main sources of nitrogen and energy. The anaerobic streptococci are homofermentative, producing only lactic acid as the major end product of carbohydrate fermentation.

Table 2 lists some of the metabolic activities of the anaerobic gram-positive cocci. Peptococci are not very active biochemically. *Pc. asaccharolyticus* is the only clinically important anaerobic gram-positive coccus that produces indole.

Sarcina ventriculi has a fermentative metabolism; the principal products from glucose metabolism are carbon dioxide, hydrogen, ethyl alcohol, and acetic acid. This organism has available two pathways for conversion of pyruvate to ethyl alcohol. One is the same as that used by yeast and the other is similar to that of *Escherichia coli* (Canale-Parola, 1970).

PATHOGENICITY

Evidence for the pathogenicity of anaerobic gram-positive cocci comes from clinical studies that show that only pure cultures of anaerobic gram-positive cocci are obtained from some infections. Pure cultures have been reported from

TABLE 2. Metabolic Reactions of Clinical Gram-Positive Anaerobic Cocci[a]

	FERMENTATION								
	Glucose	Lactose	Maltose	Sucrose	INDOLE	NITRATE REDUCTION	GELATIN HYDROLYSIS	ESCULIN HYDROLYSIS	CATALASE
Pc. asaccharolyticus	–	–	–	–	+	–	–	–	–
Pc. magnus	–^{w}	–	–	–	–	–	V	–	–
Pc. prevotii	–^{w}	–	–	–	–	–$^{+}$	–	–	–$^{+}$
Pc. saccharolyticus	A^{w}	–	–	–	–	+$^{-}$	–	–	+
Ps. anaerobius	W^{a-}	–	–^{w}	–	–	–$^{+}$	–^{w}	–	–$^{+}$
Ps. micros	–^{w}	–	–	–	–	–	–	–	+$^{-}$
Ps. parvulus	A	A	A	–	–	–	–	+	–
Ps. productus	A	A	A	A	–	–	–	+	–
St. constellatus	A	–	A	A	–	–	–	+	–
St. intermedius	A	A	A	A	–	–	–	+	–
St. morbillorum	A	–	A^{w}	A	–	–	–	+	–
S. ventriculi	A	A	A^{w}	A	–	+$^{-}$	–	+	–

[a]–negative reaction; +, positive reaction; A, acid; W, weak acid; V, variable.

From Holdeman, L. V., Cato, E. P., and Moore, W. E.: Anaerobe Laboratory Manual. Blacksburg, Va., Virginia Polytechnic Institute and State University, 1977.

breast abscess, prostatitis, liver abscess, pleuropulmonary infections, epidermoid cysts, finger infections, osteomyelitis, and other infections. *Pc. magnus* was found in pure culture in four postoperative wound infections, a thigh abscess, and one case of septicemia (Lambe et al., 1973). Reports also show a high incidence of *Peptostreptococcus* isolated in pure cultures from brain abscess, meningitis, and extradural and subdural empyema. *Peptococcus* is not isolated as frequently in pure culture from clinically significant infections as are peptostreptococci.

The anaerobic gram-positive cocci are more commonly found in infections together with other anaerobes or with facultative bacteria. The pathogenic role of the anaerobic gram-positive cocci in such conditions is unproven. About one third of intraabdominal infections have a mixed flora in which anaerobic cocci are found together with either *Bacteroides fragilis* or clostridia or both. Anaerobic streptococci and clostridia are the most frequent anaerobes isolated from biliary tract infections. The pathogenic potential of the anaerobic cocci in these infections is unknown.

Certain bacteria act together in a synergistic manner to produce infections. For example, a synergistic gangrene is produced by a microaerophilic *Streptococcus* together with a hemolytic *Staphylococcus aureus*. If animals are monoinoculated with individual cultures no disease is produced, but if the organisms are combined and then inoculated, typical necrotic ulcers are formed (Meleney, 1949). The synergy involves hyaluronidase production by the *Staphylococcus* plus a growth factor that makes the microaerophilic *Streptococcus* invasive.

Little has been done to determine the mechanisms of pathogenicity in the anaerobic cocci. The hemolytic activity of the group has been mentioned. Most anaerobic cocci produce ammonia that causes inflammation.

Sarcina ventriculi is not considered a pathogen. It is a free-living form that enters the stomach frequently. Prolonged gastric retention allows *Sarcina* to grow rapidly and produces so much gas from fermentable substrates that abdominal discomfort results.

S. ventriculi is occasionally found in infections, especially of the gallbladder and bile duct, but never in pure culture. *S. ventriculi* is a minor member of the normal flora in humans.

IMMUNITY

The role of immunologic mechanisms against specific anaerobic gram-positive cocci is unknown. Intact epithelium affords a high degree of resistance to the endogenous peptococci, peptostreptococci, and anaerobic streptococci. Also, the essential exogenous *Sarcina ventriculi* is usually denied entrance to the body surface. Leukocytes can ingest and kill *Pc. magnus* and *Ps. anaerobius* in the absence of oxygen.

LABORATORY DIAGNOSIS

Differentiation between the genera *Peptococcus* and *Peptostreptococcus* is difficult because the species are variable in morphologic, cultural, and metabolic properties and are relatively inert biochemically. Criteria established by the Anaerobe Laboratory of the Virginia Polytechnic Institute (Holdeman et al., 1977) are used by most laboratories for the identification of the anaerobic gram-positive cocci. Their method requires determination of the end-products of metabolism of glucose and amino acids in the base medium by gas-liquid chromatography.

Peptococci can be differentiated from peptostreptococci on the basis of the guanine plus cytosine content of their DNA. Peptostreptococci have 33.5 moles per cent, peptococci 35.7 to 36.7 moles per cent. However, anaerobic streptococci range from 33 to 44 moles per cent, making it necessary to use fermentation end-product analysis for the differentiation of anaerobic streptococci.

Anaerobic streptococci and peptostreptococci give negative catalase tests. Peptococci may give weak and variable catalase reactions. Therefore, a negative catalase test does not separate this group of organisms. A positive test does suggest the genus *Peptococcus*.

Members of the *Peptococcus* group usually do not form long chains. Peptostreptococci and anaerobic streptococci both have a strong chaining tendency. Absence of chaining, then, suggests peptococci. If long chains are present, end-product analysis of glucose fermentation is needed to differentiate peptostreptococci from anaerobic streptococci and to confirm members of the genus *Peptococcus*.

Two species common in clinical specimens may be identified by simple tests. *Pc. asaccharolyticus* is the only gram-positive anaerobic coccus found in clinical infections that produces indole. *Ps. anaerobius* is the only anaerobic gram-positive coccus that is susceptible to sodium polyanethol-sulfonate (SPS), which can be used easily to differentiate it from the other anaerobic gram-positive cocci (Wideman et al., 1976).

Sarcina ventriculi can easily be distinguished by microscopic morphology. Large round cells with adjacent sides flattened and in packets of eight or more are characteristic of this species, the only anaerobic *Sarcina* found in man.

A key for the identification of the clinically

TABLE 3. Identification Key to The Clinically Important Anaerobic Gram-Positive Cocci

Lactic acid the sole major end-product	
Esculin hydrolysis +	
Lactose +	Streptococcus intermedius
Lactose −	Streptococcus constellatus
Esculin hydrolysis −	Streptococcus morbillorum
Lactis acid not the sole major end-product	
Indole +	Peptococcus asaccharolyticus
Indole −	
Sensitive to SPS[a]	Peptococcus anaerobius
Resistance to SPS	
Lactose	
Maltose +	Peptococcus productus
Maltose −	Peptococcus parvulus
Lactose −	
Glucose +	Peptostreptococcus micros
Glucose −	
Butyric acid produced	Peptococcus prevotii
Butyric acid not produced	Peptococcus magnus

[a]Sodium polyanethol sulfonate.

important anaerobic gram-positive cocci is given in Table 3. This is one of several schemes that have been proposed for the identification of these bacteria.

Not all anaerobic gram-positive cocci isolates can be identified. On occasion, what are usually facultative gram-positive cocci grow only anaerobically upon isolation and initial subculture (Yatabe et al., 1977). Facultative cocci may also mutate to form obligate anaerobes that retain all their characteristics except anaerobiosis.

DRUG SUSCEPTIBILITY

Tables 4 and 5 show the minimum inhibitory concentrations of ten antibiotics for the genus *Peptococcus* and *Peptostreptococcus*. Published information is lacking about the bactericidal activity of these drugs against the anaerobic gram-positive cocci.

The anaerobic gram-positive cocci are susceptible to the drugs listed except the aminoglycoside kanamycin and the structurally related antibiotic gentamicin. Many strains of anaerobic gram-positive cocci are resistant to tetracycline and erythromycin. Penicillin G is the drug of choice for most anaerobic gram-positive cocci infections. Clindamycin and cephalothin are alternate choices (Braude, 1976).

EPIDEMIOLOGY

Peptococci, peptostreptococci, and anaerobic streptococci are part of the indigenous microbiota of the human oral cavity, gastrointestinal tract, genitourinary system, and skin. These opportunist bacteria are commonly found in infections that are closely associated with these organ systems.

Sarcina ventriculi is commonly found in soil and mud, and on grain, but only infrequently and in low numbers in the normal human fecal flora.

A 14-year study on the incidence of anaerobes in clinical specimens showed that 58.5 per cent yielded anaerobes (Holland et al., 1977). Anaerobic gram-positive bacteria were present in 66 per cent of these specimens and were second only to *Bacteroides,* which occurred in 70 per cent of the specimens. This is a major clinical concern, especially since 15 per cent of the specimens contained anaerobic cocci as the only anaerobe present.

Pc. asaccharolyticus is found in about 10 per cent of human infections. *Ps. anaerobius* occurs in about 9 per cent of the cases, and the third most numerous isolate, *S. intermedius,* in about 5 per cent of infections. *Pc. prevotti* and *Pc. magnus* are both found in about 4 per cent of infections.

In a study of 100 anaerobic pleuropulmonary infections, peptostreptococci were recovered 19

TABLE 4. In Vitro Susceptibility of 145 Clinical Isolates of Peptococcus to 10 Antibiotics[a]

	CUMULATIVE PERCENTAGE SUSCEPTIBLE AT INCREASING CONCENTRATION (μg/ml)									
	0.1	**0.2**	**0.4**	**0.8**	**1.6**	**3.1**	**6.2**	**12.5**	**25**	**>25**
Penicillin G	48	91	95	96	97	99		100		
Erythromycin	9	10	16	24	58	79	80	86	87	100
Cephalothin	22	38	56	78	88	89	97	98	99	100
Tetracycline	12	15	29	52	56	59	62	74	89	100
Lincomycin	29	45	69	92	95	96			97	100
Clindamycin	62	76	87	94	95	96	97			100
Kanamycin	1				2		4	13	44	100
Chloramphenicol	5	6	8	25	67	97	98		99	100
Gentamicin	2	3		5	7	11	30	62	95	100
Rifampin	47	50	62	76	96	98	99			100

[a]From Pien, F. D., Thompson, R. L., and Martin, W. J.: Mayo Clin Proc 47:251, 1972.

TABLE 5. In Vitro Susceptibility of 72 Clinical Isolates of Peptostreptococcus to 10 Antibiotics

	CUMULATIVE PERCENTAGE SUSCEPTIBLE AT INCREASING CONCENTRATION (μg/ml)									
	0.1	**0.2**	**0.4**	**0.8**	**1.6**	**3.1**	**6.2**	**12.5**	**25**	**>25**
Penicillin G	58	91	97	98	100					
Erythromycin	37	39	46	56	73	88	92	99		100
Cephalothin	40	57	70	78	96		99		100	
Tetracycline	18	26	44	55	65	73	77	83	96	100
Lincomycin	39	58	72	85	96	100				
Clindamycin	81	85	90	98	100					
Kanamycin				1	5	11	12	23	44	100
Chloramphenicol		3	11	37	63	96	100			
Gentamicin	1	2	9	13	78	20	33	58	78	100
Rifampin	68			72			94		96	100

[a]From Pien, F. D., Thompson, R. L., and Martin, W. J.: Mayo Clin Proc 47:251, 1972.

times, 3 times in pure culture. Peptococci were recovered 15 times, always in a mixed culture.

Anaerobic cocci are found in about 1 per cent of blood cultures. They make up approximately 7 per cent of the anaerobic isolates from blood cultures. The foci of infection in anaerobic gram-positive cocci bacteremias are usually infections of the soft tissue, gastrointestinal tract, liver, female genital tract, and oropharynx.

Anaerobes have been found in about 3 per cent of bacterial endocarditis. Anaerobic and microaerophilic streptococci make up over 90 per cent of these anaerobes.

Anaerobic cocci have been reported in about one third of all abdominal infections. The anaerobic gram-positive cocci are associated with *Bacteroides fragilis* and clostridia. Anaerobic cocci and clostridia are the dominant anaerobes in biliary tract infections, but overall they represent a minor cause of these infections.

The anaerobic gram-positive cocci are frequently found in liver abscesses and brain abscesses, and 40 per cent of abscesses of the female pelvis contain anaerobic streptococci. In septic abortions, *B. fragilis, B. melaninogenicus,* and anaerobic streptococci are isolated in greatest number. Bacteremias from septic abortions most frequently yield anaerobic streptococci and *Bacteroides*. These two anaerobic groups are also the most frequent isolates from postpartum endometritis.

One study of brain abscess reported recovery of anaerobes in 89 per cent of the cases, with peptostreptococci being the predominant organism (Heinemann and Braude, 1963). The most common isolates from chronic sinusitis are peptostreptococci.

The bacteria seen in synergistic necrotizing cellulitis are gram-negative bacilli, peptostreptococci, and *Bacteroides*. Peptostreptococci are the anaerobes most frequently found in necrotizing fasciitis.

References

Braude, A. I.: Antimicrobial Drug Therapy. Philadelphia, W. B. Saunders Company, 1976.

Canale-Parola, E.: Biology of the sugar-fermenting sarcinae. Bacterial Rev 34:82, 1970.

Facklam, R. R., and Smith, P. B.: The gram positive cocci. Human Pathol. 7:187, 1976.

Finegold, S. M.: Anaerobic Bacteria in Human Disease. New York, Academic Press, 1977.

Heineman, H. S., and Braude, A. I.: Anaerobic infection of the brain: Observations on eighteen consecutive cases of brain abscess. Am J Med 35:682, 1963.

Holdeman, L. V., Cato, E. P., and Moore, W. E.: Anaerobe Laboratory Manual. Blacksburg, Va., Virginia Polytechnic Institute and State University, 1977.

Holland, J. W., Hill, E. O., and Altemeier, W. A.: Numbers and types of anaerobic bacteria isolated from clinical specimens since 1960. J Clin Microbiol 5:20, 1977.

Lambe, D. W., Vroon, D. H., and Rietz, C. W.: Infections due to anaerobic cocci. In Ballows, A., et al. (eds.): Anaerobic Bacteria: Role in Disease. Springfield, Ill., Charles C Thomas, 1973.

Markowitz, A., and Lerner, A. M.: Differentiation of several isolates of *Peptococcus magnus* by counterimmunoelectrophoresis. Infect Immun 16:152, 1977.

Meleney, F. L.: Clinical Aspects and Treatment of Surgical Infections. Philadelphia, W. B. Saunders Company, 1949.

Pien, F. D., Thompson, R. L., and Martin, W. J.: Clinical and bacteriologic studies of anaerobic gram-positive cocci. Mayo Clin Proc 47:251, 1972.

Porschen, R. K., and Spaulding, E. H.: Fluorescent antibody study of the gram-positive anaerobic cocci. Appl Microbiol 28:851, 1974.

Stone, M. L.: Studies on the anaerobic streptococcus. I. Certain biochemical and immunological properties of anaerobic streptococci. J Bacteriol 39:559, 1940.

Wideman, P. A., Vargo, V. L., et al.: Evaluation of the sodium polyanethol sulfonate disk test for the identification of *Peptostreptococcus anaerobius*. J Clin Microbiol 4:330, 1976.

Yatabe, J. H., Baldwin, K. L., and Martin, W. J.: Isolation of an obligately anaerobic *Streptococcus pneumoniae* from blood culture. J Clin Microbiol 6:181, 1977.

44 ACTINOMYCES AND MICROAEROPHILIC ACTINOMYCETES

Leo Pine, Ph.D.

Perhaps the earliest description of the actinomycotic organism was made by von Graefe (1854), who described fungal masses he observed in granules taken from infections of the lacrimal canal. Cohn (1875) also observed a fine branching fungus in lacrimal concretions, which he named *Streptothrix foersteri.* In 1877, Bollinger published the first description of actinomycosis in cattle. This organism, resembling *Streptothrix,* was described in the tissues by Hartz (1879), who named it *Actinomyces bovis.* Israel described actinomycosis in man in 1878, and Ponfick (1880) compared human and bovine actinomycosis. Although Bujwid (1889) first isolated *Actinomyces* from human actinomycosis, it was the classic description of the morphology and physiology of this bacterium by Wolff and Israel (1891) that permitted subsequent workers to identify and establish this organism as the cause of actinomycosis in man.

Silberschmidt first cultured the anaerobic actinomycete from lacrimal concretions; he compared the organisms isolated from maxillary actinomycosis of a cow, from human lacrimal concretions, and from maxillary and thoracic infections in man (Silberschmidt, 1901). Because all strains appeared identical, Silberschmidt concluded that *Streptothrix* was the correct name on the basis of taxonomic priority. Thus *Streptothrix* and names such as *Leptothrix, Cladothrix,* and *Leptotrichia* were all commonly used in early ophthalmologic and dental literature to refer to long filamentous or branching organisms. These names were used so commonly that some investigators no longer associated the *Streptothrix* or *Leptothrix* of lacrimal canaliculitis or of dental plaque with the *Actinomyces* of thoracic infections. To add to the confusion, Bostroem (1891) had earlier isolated aerobic forms of filamentous bacteria not only from human and bovine cases of actinomycosis but also from plant materials and soil. Although years later these aerobic forms were given the name *Streptomyces,* the generic name *Actinomyces* is still used today in certain countries to describe the aerobic nonpathogenic streptomycetes.

The source of the infections became apparent when Naeslund (1925) identified the bacterium as part of the normal flora of the mouth; Emmons (1935) later found *Actinomyces* in granules in tonsillar crypts. Although Erikson (1940) believed that human and bovine strains were sufficiently different to warrant separation as *Actinomyces israelii* and *Actinomyces bovis,* respectively, these species were not accepted by other workers until after World War II (Thompson, 1950; Pine et al., 1960). *A. naeslundii,* originally isolated from the mouth by Thompson and Lovestedt (1951) and Howell et al. (1959), was later found to be pathogenic for animals and man (Colemen et al., 1969).

Of great importance for taxonomy and for laboratory diagnosis was the observation that the pathogenic *Actinomyces* had no catalase, whereas coryneforms were catalase positive (Suter, 1956). Analysis of fermentation products and cell wall composition helped delineate those actinomycetes that showed true branching or mycelial microcolonies (Cummins, 1962).

The 8th edition of *Bergey's Manual of Determinative Bacteriology* (Buchanan and Gibbons, 1974) describes five genera in the family. Five species of *Actinomyces* are accepted and an additional three species are described, although their present taxonomic status is uncertain (Table 1). All species of *Actinomyces* were originally isolated from patients with actinomycosis, actinomycosis-like disease, periodontal plaque, or dental caries. The exception is *Actinomyces humiferus,* which was isolated from the soil and is nonpathogenic. Similarly, *Bacterionema matruchotii,* a commensal in the mouth of animals, has not been associated with disease. The most definitive discussion of these organisms appears in *Actinomyces, Filamentous Bacteria: Biology and Pathogenicity* (Slack and Gerencser, 1975).

MORPHOLOGY

The species of bacteria in the families Actinomycetaceae and Propionibacteriaceae have gram-positive, straight or curved, branching, pleomorphic rod-like cells that are nonacid-fast, nonmotile, and nonspore-forming (Figs. 1 and 2). The organisms are generally 0.2 to 0.3 μm in

TABLE 1. Taxonomic Relationships of Fermentative Actinomycetes and Similar-Form Organisms Associated with Human or Animal Disease

FAMILY[a]	GENUS (SPECIES)	SYNONYM	DISEASE	HABITAT	MORPHOLOGY IN DISEASE TISSUE	CELLULAR MORPHOLOGY[c]	PATHOGENICITY IN LABORATORY ANIMALS
Actinomycetaceae (microaerophilic, fermentative)	*Actinomyces*[b] *bovis, A. israelii, A. naeslundii, A. suis, A. viscosus, A. odontolyticus*	*Streptothrix, Cladothrix, Leptothrix, Actinobacterium*	Actinomycosis, lacrimal canaliculitis	Oral cavity, tonsillar crypts, dental plaque, calculus	Long branching mycelium or filamentous cells, sulfur granules, mycelial masses, streptococcal-like chains	Filaments, long rods, branching rods, clubs, small rods, diphtheroids	Pathogenic for mice and hamsters; forming limited or progressive abscesses in most animals when injected intraperitoneally
	Arachnia propionica	*Actinomyces propionicus*	Actinomycosis, lacrimal canaliculitis	Oral cavity, dental plaque, tonsillar crypts	Long threadlike filaments, sulfur granules; branching mycelial masses	Long branching rods, nonbranching filaments, coccoid bodies (spheroplasts)	Same as *Actinomyces*
	Bifidobacterium adolescentis[d]	*Actinomyces eriksonii*	Lung abscess, pleural fluid	Oral cavity, tonsils, intestinal tract	Small bifid rods; club-shapes; curved rods	Small singly branched rods, clubs, unbranched rods	Nonpathogenic; forms limited abscesses
	Bacterionema matruchotii	*Leptotrichia dentium, L. buccalis*	None	Dental plaque, calculus	None	Branching filaments, long filaments with terminal bacillary head	Nonpathogenic
	Rothia dentocariosa	*Nocardia salivae*	None	Dental plaque, calculus	Beaded filaments, short branching rods	Short or long branching rods, cocci, filaments	Nonpathogenic; forms limited abscesses
Propionibacteriaceae (microaerophilic, fermentative)	*Propionibacterium acnes*	*Corynebacterium acnes*	Skin abscess	Gingival plaque, skin abscesses, intestinal contents, lacrimal canal	"Diphtheroid" rods, curved rods, branched rods, irregular shapes	Diphtheroid rod-like cells	Forms limited abscesses when inoculated into mice subcutaneously or intraperitoneally
Nocardiaceae (aerobic, oxidative)	*Nocardia asteroides, N. brasiliensis, Actinomadura madurae, Actinomadura pelletiera*	*Streptothrix, Proactinomyces, Actinomyces, Streptomyces, Nocardia*	Maduromycosis	Soil	Long branching mycelial elements, granules	Nonsporulating fragmenting mycelium gives rise to rod-shaped and coccoidal elements	Pathogenic for laboratory animals depending on route of inoculation and species
Streptomycetaceae (aerobic, oxidative)	*Streptomyces somaliensis, S. paraguayensis*	*Actinomyces*	Maduromycosis	Soil	Long branching mycelial elements, granules	Nonfragmenting surface mycelium; may form chains of spores	May cause limited progressive abscesses with intraperitoneal inoculation

[a]Based primarily on the data in Bergey's Manual of Determinative Bacteriology (Buchanan and Gibbons, 1974) and Slack and Gerencser (1975). *Actinomyces humiferus*, not listed above, is isolated from the soil and is not pathogenic.
[b]*A. odontolyticus* is nonpathogenic.
[c]Grown for 24 to 72 hours in semisolid thioglycolate or Trypticase Soy Broth.
[d]Mitsuoka et al. (1974).

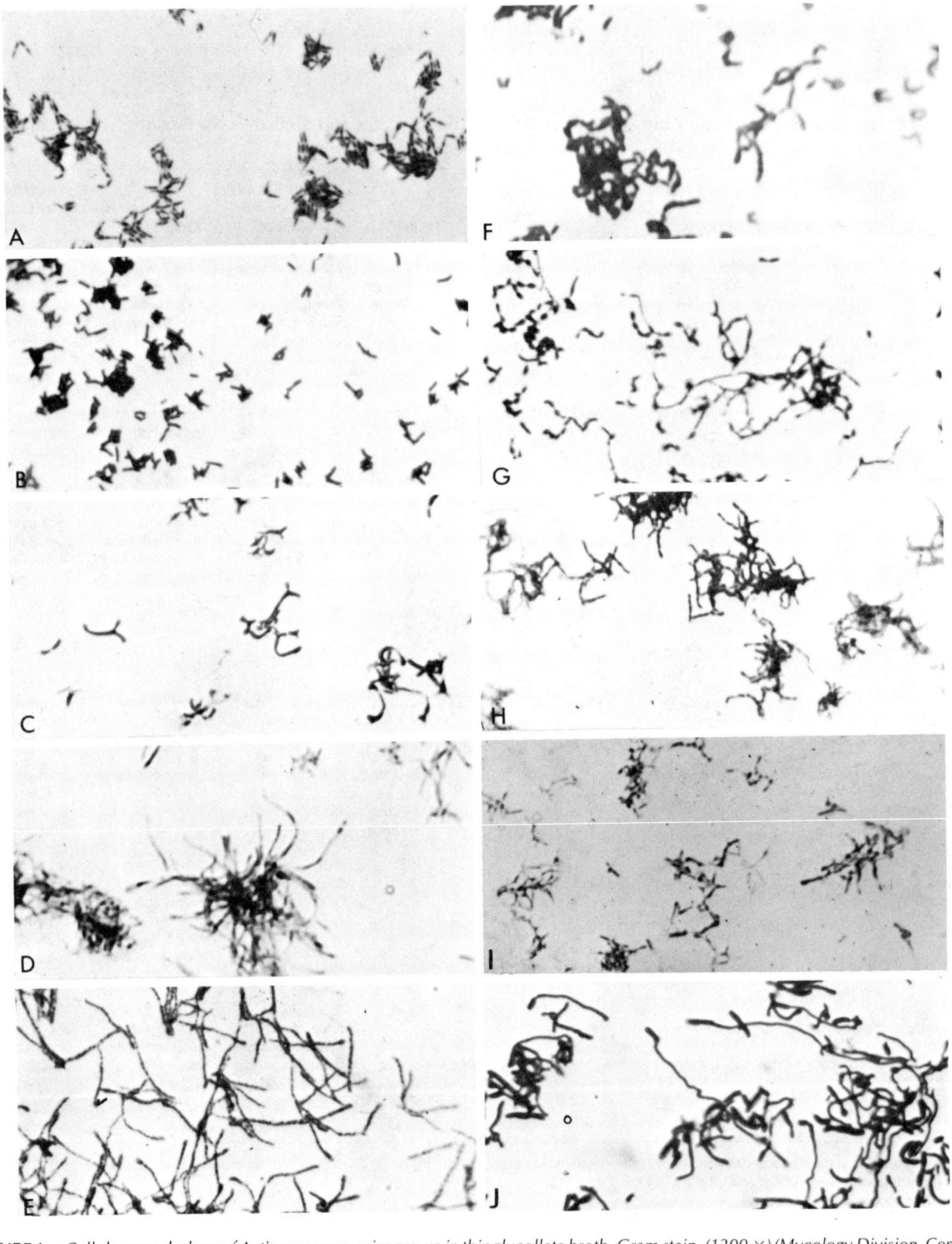

FIGURE 1. *Cellular morphology of* Actinomyces *species grown in thioglycollate broth, Gram stain, (1200 ×) (Mycology Division, Center for Disease Control, Atlanta, Georgia). (A)* Actinomyces bovis, *(B)* A. odontolyticus, *(C)* Bifidobacterium adolescentis (A. eriksonii), *(D)* A. naeslundii, *(E)* A. israelii, *(F)* A. israelli, *(G)* Arachnia propionica, *(H)* Propionibacterium acnes, *(I)* Rothia dentocariosa, *(J)* Bacterioma matruchotii. *Most species exhibit the range of cellular morphologies seen in Figure 2. However, the long filamentous form terminating in a bacillary head is characteristic only of* Bacterionema matruchotti *(J), whereas the bifid branching of* B. adolescentis *(C) is characteristic of the genus* Bifidobacterium. *Within the remaining species, short cells, often in pairs as V forms (A, F, and G) to long filamentous branching cells are often seen (E and F). The morphology of* P. acnes *cannot be distinguished from that of the* Actinomyces *species.*

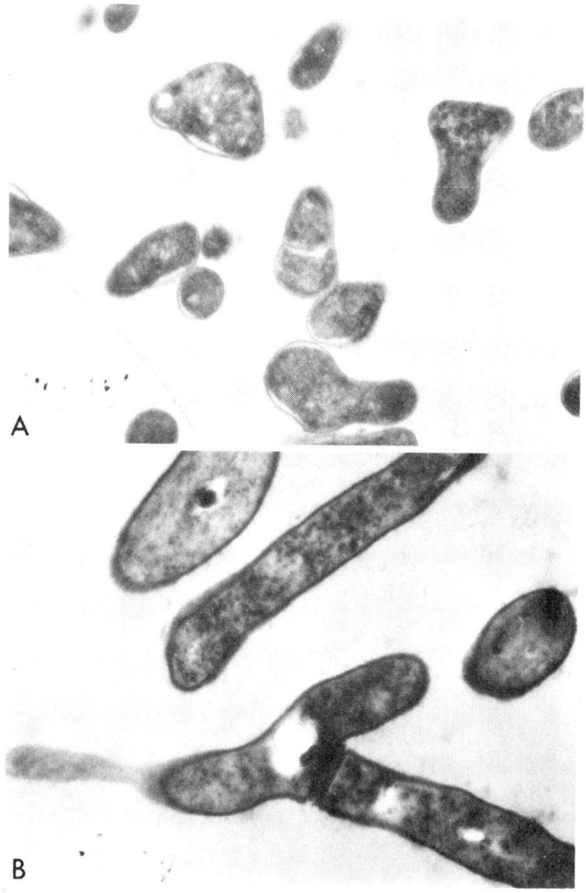

FIGURE 2. *Comparative cellular morphology of "diphtheroid" and filamentous forms (23,000 ×) (J. L.Overman).*
A. Actinomyces bovis
B. Arachnia propionica

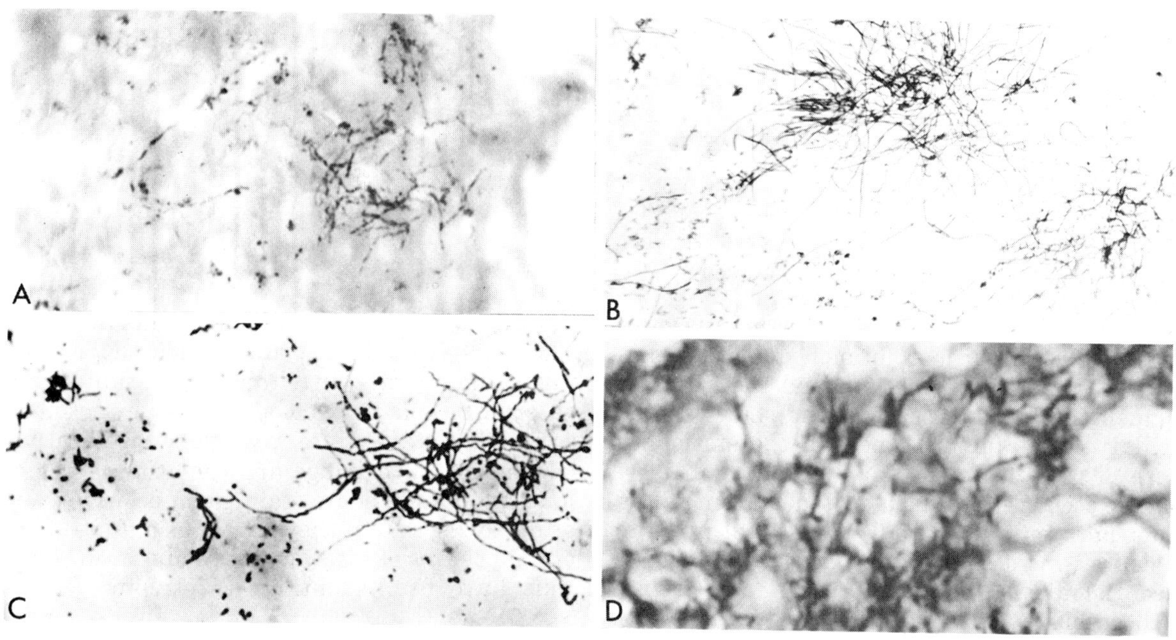

FIGURE 3. *Morphologic aspects of microaerophilic actinomycetes in tissues.*
A. Actinomyces bovis *in pus from inoculated hamster. Gram stain (900 ×). Note streptococcal-like chains.*
B. Arachnia propionica *in pus from inoculated hamster. Gram stain (900 ×). (Pine and Hardin, 1959.)*
C. Actinomyces bovis *in pus from inoculated hamster. Gram stain (900 ×). (Pine et al., 1960.) Note coccal forms and filamentous forms without branching.*
D. Actinomyces israelii *in lung, human infection. Brown and Brenn stain (1000 ×).*

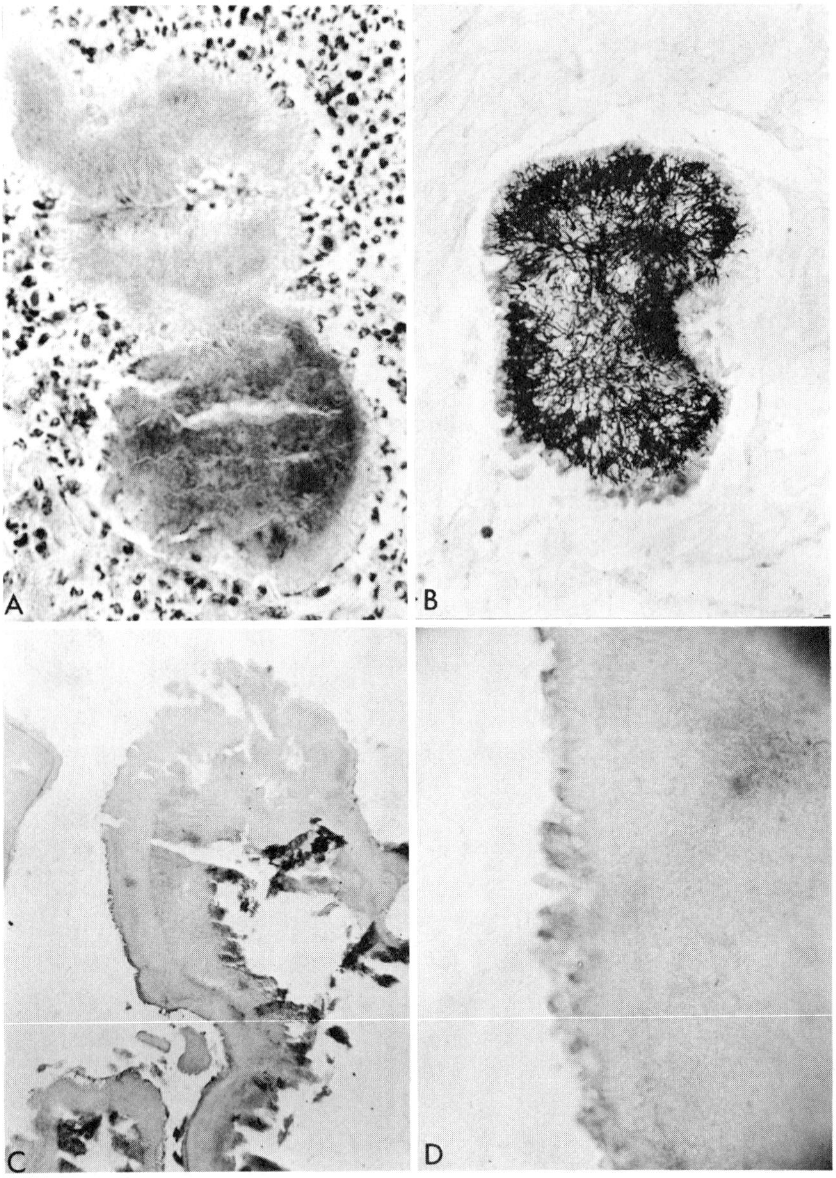

FIGURE 4. *Morphologic aspects of the sulfur granule.*

A. Sulfur granule from human infection, H & E stain. (Mycology Division, Center for Disease Control, Atlanta, Ga.)

B. Sulfur granule from human infection. Brown and Brenn (Gram) stain (315 ×). (Mycology Division, Center for Disease Control, Atlanta, Ga.)

C. Sulfur granule from bovine lumpy jaw. Colloidal iron stain (100 ×)). (Pine and Overman, 1963.)

D. As C above (800 ×) showing peripheral arrangement of clubs.

diameter and may vary from 0.4 to many μm in length. Cells appear more branched and mycelial in the early stages of growth and then become fragmented, forming coccoidal and short tapered bacillary forms (diphtheroids) in the stationary phase of growth. Important differences in morphology occur (1) in infected tissues (2) in young liquid cultures, and (3) on agar plate within 24 to 72 hours (microcolonies) and after 3 to 14 days (macrocolonies).

All species that cause deep infections (except *Bifidobacterium adolescentis*) are seen in pus or infected tissues as long, nonbranching filaments or as complexes of loosely or tightly interwoven branching mycelial elements (Fig. 3). Masses of *B. adolescentis* with a morphology similar to that seen in culture have been found in human tissues. The Gram stain is the primary method for detecting actinomycetes in clinical material and for observing their morphology; however, actinomycetes may stain gram positively or negatively depending on their age and condition. Often the long hyphal elements stain gram positively only at intervals, so that they appear as masses of cocci or long chains of coccoidal elements. Such stained cells have been identified as streptococci (Fig. 3).

The various *Actinomyces* or *Arachnia* species form special structures, called drusen, sulfur granules, or mycelial granules within the tissues. Only trace amounts of sulfur are present in sulfur granules, so the term probably derived from their

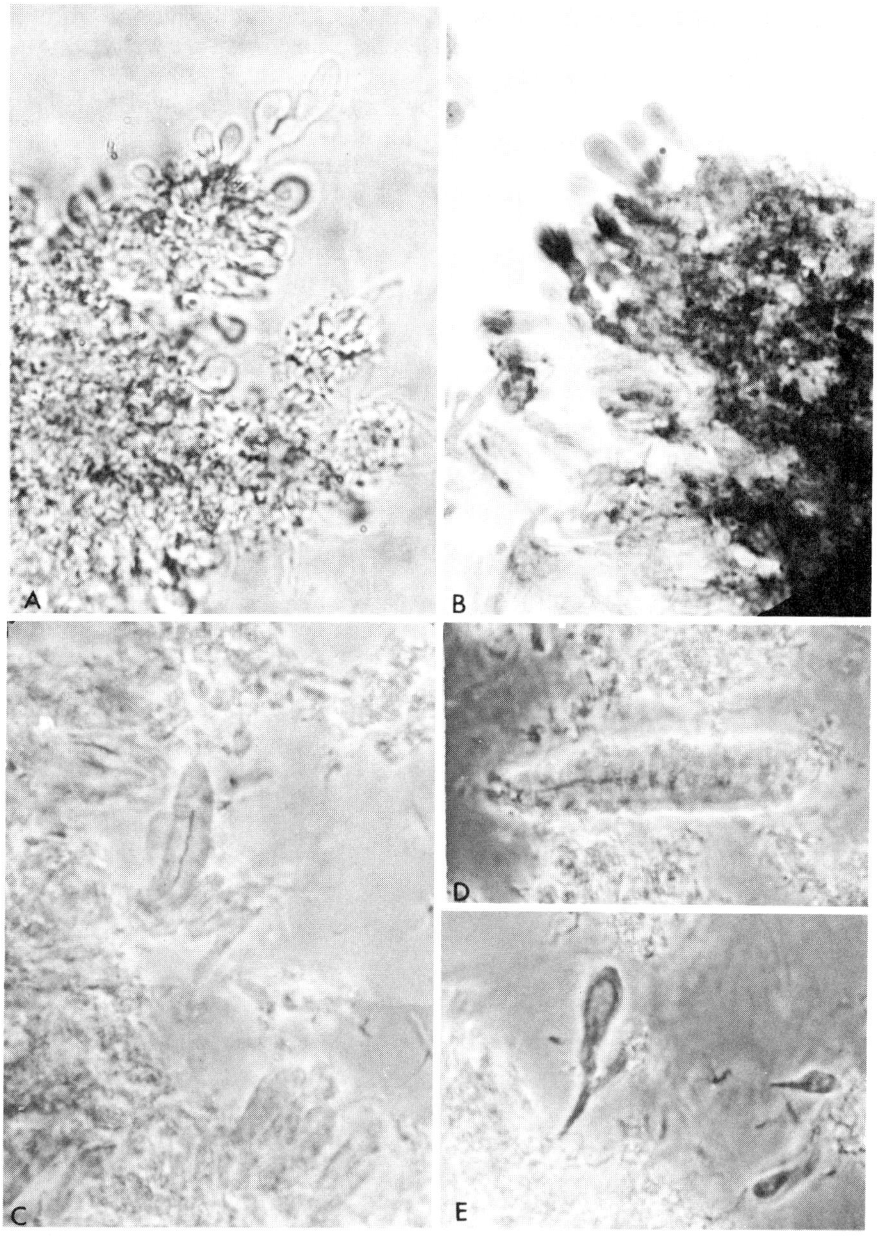

FIGURE 5. *Sulfur granule —structure of clubs.*

A. Crushed granule mounted in 10 per cent KOH showing multibranched hyaline clubs (970 ×).

B. Periodic acid-Schiff stain of sectioned granule (970 ×). Note hyaline nature of clubs and absence of clearly defined mycelium.

C. Crushed granule, digested in 10 per cent KOH; phase contrast, (970 ×). Note presence of filaments within clubs. (Pine and Overman, 1966.)

D. Large intact club showing internal filament after successive treatments with trichloracetic acid and proteolytic enzymes (970 ×). (Pine and Overman, 1963.)

E. Crushed granule, lactophenol mount; phase contrast (970 ×). Note successive layers of capsular material.

yellow color in pus. *A. odontolyticus* does not invade the tissues and does not make granules. *B. adolescentis* does not form granules, although it does form compact masses of bacterial cells in human infections (Georg et al., 1965). Characteristically, sulfur granules are seen in the pus or tissue as discrete yellow grains of relatively hard consistency (Fig. 4). They are 40 to 400 μm in diameter and can be seen by eye, trapped in the gauze of surgical dressings.

When placed in a water or 5 per cent potassium hydroxide mount and crushed between the coverslip and the glass slide, the mycelial and cellular elements can be seen directly with phase optics or with the use of Gram, methylene blue, periodic acid-Schiff, or silver methenamine stains. The sulfur granules generally have peripheral clubs that with proper chemical treatment can be shown to contain an internal mycelial filament (Fig. 5). The sulfur granules of *A. bovis* have been analyzed by a combination of histochemical stains of thin sections, chemical analyses, and electron microscope studies (Pine and Overman, 1963, 1966). Whereas the overall chemical composition of the granules was similar to that of the organism grown on culture medium, they contained some 40 to 60 per cent $Ca_3(PO_4)_2$. Some of the $Ca_3(PO_4)_2$ was poorly crystallized apatite (Frazier and Fowler, 1967). The mycelial elements of the granule are cemented together by mucopolysaccharide with the clubs forming a rosette around the periphery (Fig. 4). The clubs may measure from 3 to 20 times the diameter of the hyphae embedded within them. The $Ca_3(PO_4)_2$ may completely saturate the embedding material, simply fill certain individual cells, or form a layer at their internal periphery. Coupled with Wright's (1905) demonstration of the production of typical sulfur granules in cultures containing serum, these findings suggest that $Ca_3(PO_4)_2$ deposition within the cells is caused by the biochemical activities of the bacterium itself. The normal hyphal elements within the center of a granule are readily see with Gram or silver methenamine stains (Fig. 5). However, heavily calcified granules or clubs do not stain intensely, and clear observation of the cellular elements that are either within the club or embedded in the granule requires extraction of the granule with trichloracetic acid or treatment by other chemical means (Fig. 5).

Although the presence of sulfur granules strongly suggests *Actinomyces* infection, *Staphylococcus, Nocardia, Streptomyces,* and *Actinobacillus lignieresii* may also form "granules." The bacterium of the granule must therefore be identified by histopathologic and bacteriologic procedures.

Morphology of strains even within the same species may vary greatly (Fig. 6). In one medium a strain may be rod-shaped, and in another it may be a longer branched cell or even a coccus. Simi-

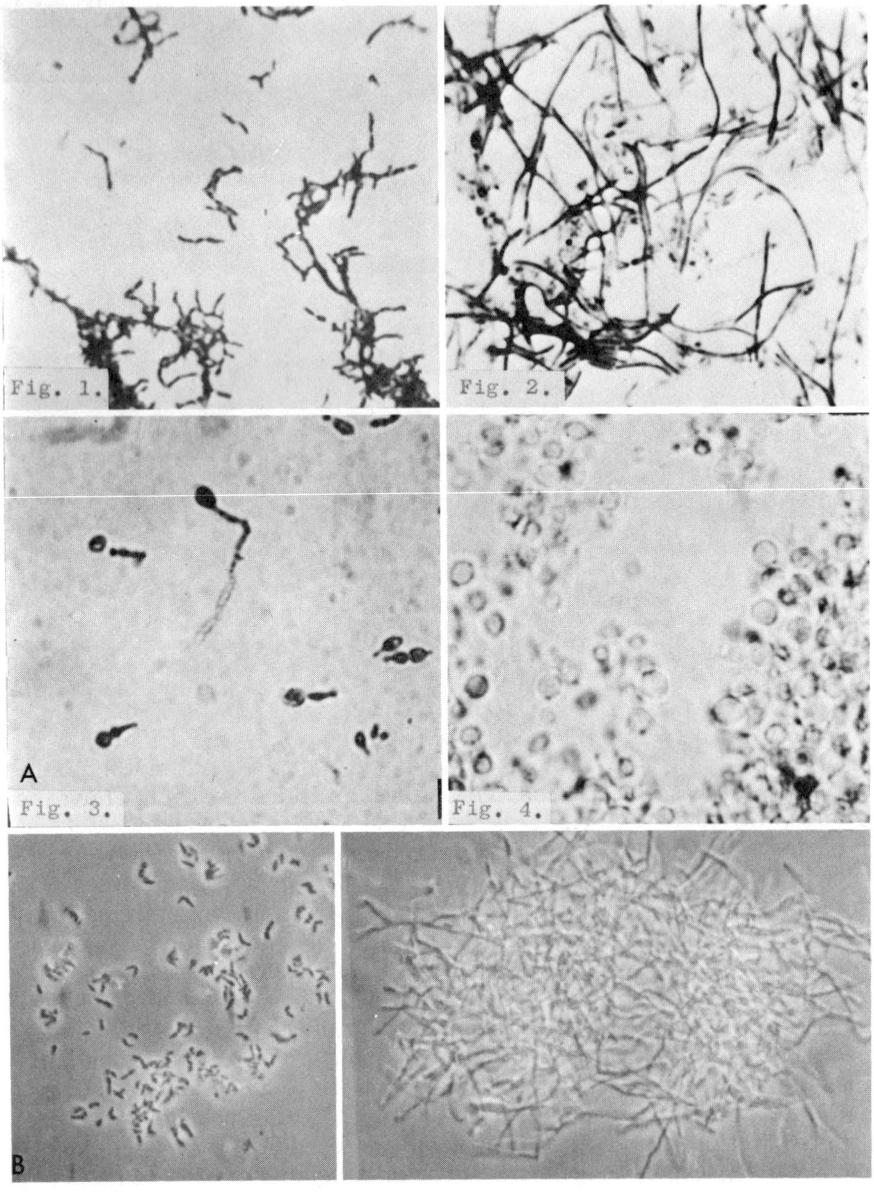

FIGURE 6. *Variation in cellular morphology during growth.*

A. Variation of Arachnia propionica *during growth in broth media. Changes from diphtheroidal morphology to filament and coccoidal forms, and formation of spheroplasts in glucose casitone broth (Figs. 1–3); formation of spheroplasts in rhamnose broth (Fig. 4), phase contrast (970 ×).*

B. Actinomyces bovis *diphtheroid and mycelial forms; phase contrast, (970 ×)). The mycelial form upon transfer may give rise to the diphtheroid morphology observed on the left.*

larly, morphology may change radically as the bacterium goes through its various phases of growth and the medium becomes depleted (Fig. 6).

The various strains of a given species may have cellular characteristics in common with all members of that species (Fig. 1; Table 2). Cellular morphology is best characterized in fluid (0.075 per cent agar) thioglycolate medium with 0.1 per cent sugar. Long incubation periods lead to fragmentation of branched hyphae, conversion to atypical forms, and loss of gram-positivity.

The two basic cell morphologies are: (1) diphtheroids, which are small branched or unbranched forms of comma-shaped cells or V- or Y-forms with two or three cells connected, and (2) long rods, sometimes branching (Table 2). *A. bovis*, and *A. odontolyticus* are most often diphtheroids or short rods. *B. adolescentis* has the characteristic Y-branched form (Fig. 1). *A. israelii, A. naeslundii, A. viscosus,* and *A. suis* are V-forms, whereas *Arachnia propionica* and *Rothia dentocariosa* are mixed rods and long filaments with numerous coccoid forms. *Bacterionema matruchotii* has a long filament that gradually increases in diameter and terminates in a bacillary-shaped body (Fig. 1). *Propionibacterium acnes* and *Actinomyces* species are often found together in clinical materials and cannot be distinguished morphologically.

TABLE 2. Morphologic Aspects of the Microaerophilic Actinomycetes

SPECIES	Morphology in Pus or Infected Tissue	CELLULAR AND COLONY MORPHOLOGY IN CULTURE: Sulfur Granules	Liquid Medium[a]	Solid Medium[b]: Microcolony	Solid Medium[b]: Macrocolony
Bifidobacterium adolescentis	Short filaments	None	Short rods, branched V and Y shape	Smooth glistening entire edge, no mycelium	Large, shiny convex colonies; circular or irregular; surface smooth or granular
Actinomyces odontolyticus	None	None	Diphtheroid, very short, pleomorphic	Smooth circular granular colonies, no mycelium	Large, circular, glistening, white, opaque colonies; red colonies on blood
A. bovis	Long filaments, branching mycelium	Typical sulfur granules, clubs calcified	Short (rarely long) branching rods	Round streptococcal-like, smooth glistening colony, no mycelium or small compact mycelium	Colonies of a wide range of morphology (see Figs. 8 and 9); white to yellow; generally adherent to agar; roughly circular with smooth, granular, raspberry or molar tooth morphology
A. suis	Long filaments, branching mycelium	Typical sulfur granules, clubs	Short branching rods	Like *A. bovis*	As above
A. israelii	Long filaments, branching mycelium	Typical sulfur granules, clubs calcified	Long branching rods	Large mycelial colonies, spider colonies	As above
A. naeslundii	Long filaments, branching mycelium	Granules, no clubs reported	Short to long branching rods	Mycelial colonies with dense centers	As above
A. viscosus	Filamentous coccoidal	Granules, no clubs reported	Short branching rods	Like *A. bovis*	As above
Arachnia propionica	Long filaments, branching mycelium, short rods, coccoidal cells	Typical sulfur granules, clubs	Short branching rods, coccoidal cells, long unbranched filaments	Mycelial colonies, spider colonies	As above
Rothia dentocariosa	None	None	Short to long branched filaments, coccoidal forms	Diffuse mycelial colonies, branching filaments	As above
Bacterionemia matruchotii	None	None	Long filaments with bacillary or club-shaped terminal ends, short rods	Small diffuse clumps of branching filamentous cells	As above
Propionibacterium acnes	None	None	Short pleomorphic rods, resembles *Actinomyces*	Smooth on entire edges, glistening colony, streptococcal or micrococcal-like	Small, circular, white to pink colonies

[a]Thioglycolate medium, 10 ml per tube.
[b]Brain Heart Infusion agar, incubated anaerobically or under microaerophilic conditions.

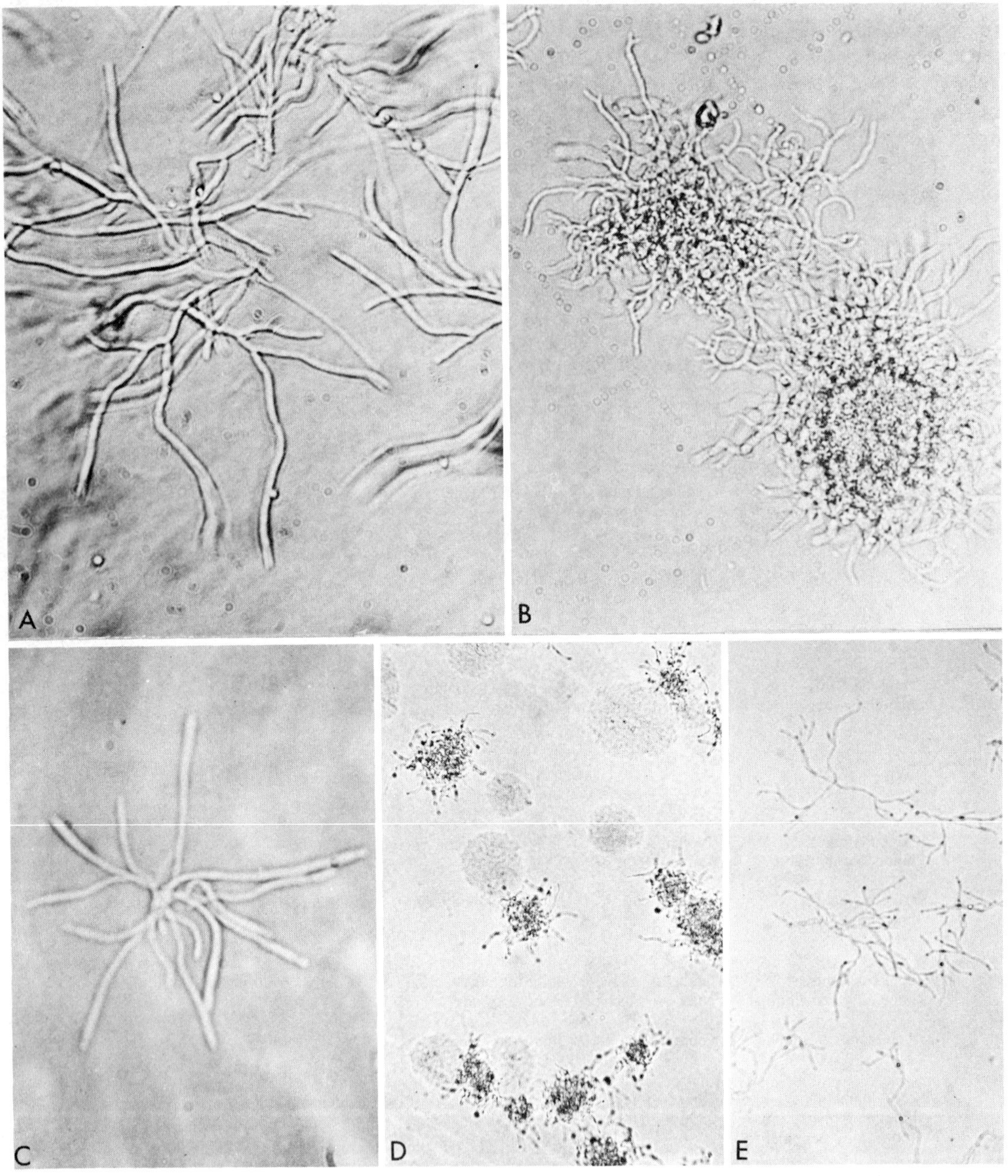

FIGURE 7. *Microcolonies of the microaerophilic actinomycetes.*

A. Actinomyces israelii *(Howell et al., 1959). The diffuse branching colony is characteristic of the species.*

B. A. naeslundii *(Howell et al., 1959). The tightly interwoven compact mycelium of the central portion of the colony is characteristic of this species.*

C. A. viscosis *(Howell, 1963).*

D. A. bovis *(Pine et al., 1960). Spider-like and smooth circular colonies isolated from the same case of bovine lumpy jaw. Transmitted light (290 ×).*

E. Arachnia propionica *(Buchanan and Pine, 1962). Web-like mycelial colonies (430 ×).*

Colonial morphology is best seen on agar plates at 24 to 48 hours by transmitted light. Young colonies have two basic forms (Fig. 7): (1) small mycelial colonies, and (2) smooth streptococcal-like colonies without true filaments. In general, microcolonies of *A. bovis* are small, circular, translucent or opaque, and have an entire edge with a smooth or granular surface. They cannot be distinguished from similar colonies of *Streptococcus pyogenes* (Fig. 7D). Although colonies of *A. odontolyticus* and *B. adolescentis* have characteristics very similar to those of *A. bovis* described above, some isolates of *A. bovis* are petite mycelial microcolonies (Fig. 7). No mycelial colonies have been reported for *A. odontolyticus* or *B. adolescentis*. Mycelial-like colonies may be formed by *Propionibacterium* species, although *P. acnes* forms only a smooth circular colony similar to that of *A. bovis*.

After incubation for three to seven days the microcolonies of the Actinomycetaceae develop into 1- to 3-mm diameter macrocolonies, which are often described as raspberry, molar tooth, or rough colonies. The forms vary considerably (Figs. 8 and 9) with no one morphologic pattern specific for a given species. Furthermore, the so-called smooth and matt strains of *Streptococcus pyogenes* may be confused with the macrocolonies formed by *Actinomyces* and *Arachnia* species (Fig. 8). In summary, various features of cellular or colonial morphology in tissues or in media that strongly suggest *Actinomyces* or *Arachnia* may instead indicate the genera *Streptococcus, Propionibacterium, Nocardia,* and *Streptomyces* (Table 2).

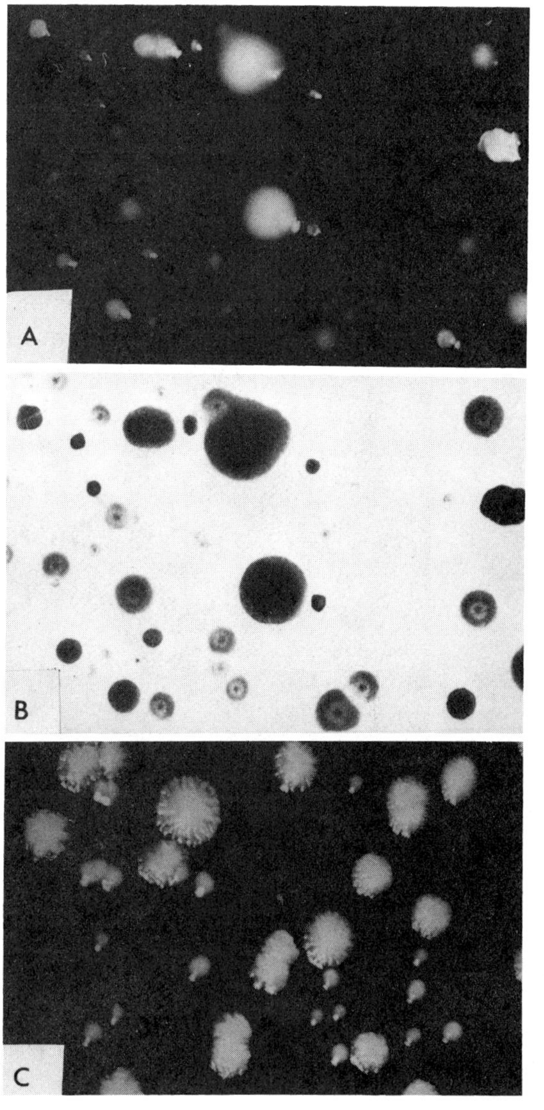

FIGURE 8. *Macrocolony morphology of* Actinomyces bovis *(Pine et al., 1960).*

A. Rough and smooth colonies; reflected light (13 ×).

B. Same field as A, viewed by transmitted light. Note "doughnut" aspect of smooth colonies of A. bovis.

C. A. bovis, *strain P3, smooth and rough colonies; reflected light, (13 ×). Both colony forms are indistinguishable from the smooth and matt colonies of some strains of* Streptococcus pyogenes.

ANTIGENIC STRUCTURE

Actinomyces antigens prepared from formalized whole cells, cell walls, soluble semipurified cell walls, or culture supernate precipitates have been used for taxonomy and serotyping (Bowden and Hardie, 1973). Also, formalized whole cells have been used to prepare hyperimmune fluorescein-tagged rabbit serum which very effectively demonstrates actinomyces in infected tissue, human saliva, and dental plaque. Because of cross-reactions with sera from cases of streptococcal infections, nocardiosis, and tuberculosis (Holm and Kwapinski, 1959; Georg et al., 1968), no satisfactory routine complement fixation, immunodiffusion, or hemagglutination test has been developed.

Species differentiation has relied principally on immunodiffusion, whole cell or cell wall agglutination, and fluorescent antibody staining of whole cells. Agar gel immunodiffusion has used soluble antigens either precipitated from the culture supernate by acetone or extracted from whole cells with formamide and trichloracetic acid. The protein associated with carbohydrate seems to be the serologically reactive component. Identification of species by agar gel immunodiffusion has not been entirely successful. Brock and Georg (1969) clearly separated *A. naeslundii* and *A. israelii* by agar gel diffusion, and Cummins (1968) distinguished the two serotypes of *A. israelii* using trichloracetic acid-acetone precipitated antigens from strains of *A. israelii*. These sero-

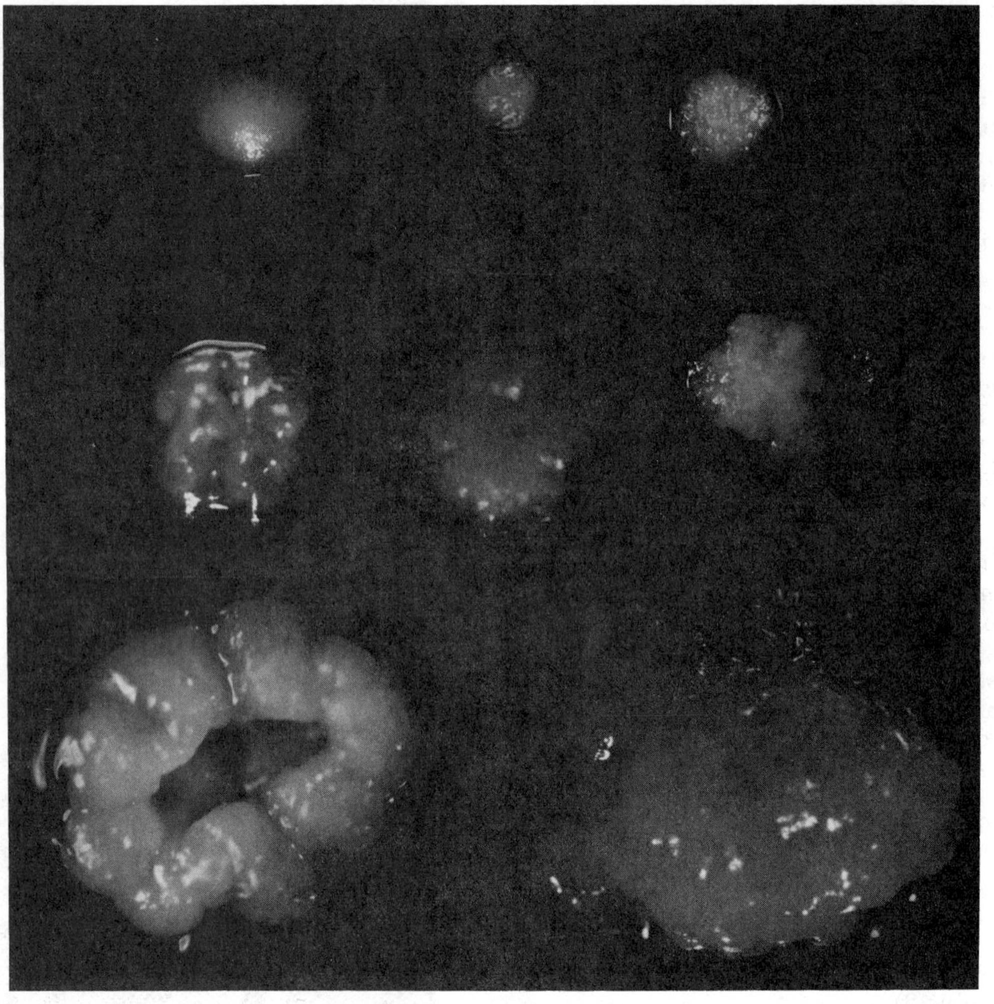

FIGURE 9. *Macrocolonies of* Actinomyces israelii; *reflected light (23 ×) (Howell et al., 1959). The diverse colony forms shown may be formed by strains of all the species of* Actinomyces *and* Arachnia.

types were confirmed with fluorescent antibody (Brock and Georg, 1969).

Chemical differentiation of purified cell walls (Cummins and Harris, 1958) combined with the use of whole cells for cell wall agglutination studies (Cummins, 1962) provided a definitive basis for species identification. What was once an extremely complex heterogeneous mixture of overlapping morphologic and physiologic groups was now organized primarily on the basis of a definitive chemical cell wall composition (Table 3). After delineation of these groups, their serologic relationships became clear.

The fluorescent antibody technique, which became the most fruitful and least complex serologic procedure for identifying species and serotypes, was first used by Slack and his colleagues (1966) to identify *A. israelii* from two patients with actinomycosis. The technique, tested with pure cultures and specific sera, has led to serologic separations and identification of species that correlate with species designations determined by other means. Nevertheless, cross-reactions occur, and sera specific for the species *A. israelii, A. naeslundii,* and *A. viscosus* interact in a complex manner (Slack and Gerencser, 1975). All *Actinomyces* species and *Ar. propionica* have had two serotypes described; the exceptions are *A. naeslundii,* which has one, and *R. dentocariosa,* which has four or more.

METABOLISM

All members of the family Actinomycetaceae are fermentative, and their substrates are largely converted to volatile and nonvolatile acids in anaerobic, microaerophilic, or strongly aerobic conditions. Alcohols are not formed. The fermentative actinomycetes do not contain a sequence of

TABLE 3. Cell-Wall Composition of the Pathogenic *Actinomyces* and Related Species[a]

	AMINO ACIDS[b]													
	Aspartic Acid	Lysine	Ornithine	Glycine	LL Diamino-pimelic Acid	DL Diamino-pimelic Acid	GALACTOSE	GLUCOSE	MANNOSE	RHAMNOSE	DEOXYHEXOSE	DEOXYTALOSE	FUCOSE	ARABINOSE
Actinomyces														
A. bovis ATCC13683	+	+	−	−	−	−	−	+	+	+	+	+	+	−
A. bovis P2R	−	+	−	−	−	−	−	+	+	+	+	+	+	−
A. humiferus	+	+	+	−	−	−	−	−	−	+	−	−	−	−
A. odontolyticus	−	+	+	−	−	−	+	+	+	−	−	−	−	−
A. suis	−	+	+	−	−	−	−	−	−	+	−	−	−	−
A. israelii/1	−	+	+	−	−	−	+	−	−	−	−	−	−	−
A. israelii/2	−	+	+	−	−	−	+	−	−	+	−	−	−	−
A. naeslundii	−	+	+	−	−	−	−	+	+	+	+	+	−	−
A. viscosus/1	−	+	+	−	−	−	+	+	−	+	+	+	−	−
A. viscosus/2	−	+	+	−	−	−	+	+	−	+	+	+	−	−
Bifidobacterium adolescentis	+	+	+	−	−	−	+	+	−	−	−	−	−	−
Rothia dentocariosa	−	+	−	−	−	−	+	−	−	−	−	−	−	−
Bacterionemia matruchotii	−	−	−	+	−	+	+	+	−	−	−	−	−	+
Arachnia														
A. propionica/1	−	−	−	+	+	−	−	+	−	−	−	−	−	−
A. propionica/2	−	−	−	+	+	−	+	+	−	−	−	−	−	−
Propionibacterium														
P. acnes/1	−	−	−	+	+	−	+	+	±	−	−	−	−	−
P. acnes/2	−	−	−	+	+	−	−	+	±	−	−	−	−	−

[a]Based on data and references in Cummins (1968), Johnson and Cummins (1972), Schleifer and Kandler (1972), Pine (1973), and Cummins (personal communication).
[b]All strains have alanine and glutamic acid.

cytochromes to oxidize the substrate for utilizable energy. A carbohydrate energy source such as sugar, alcohol, pentose, hexose, di- or trisaccharide, or starch is required for growth, but amino acids or proteins are not used. In contrast with this family, which accumulates acids in the medium, strictly aerobic genera such as *Mycobacterium, Nocardia, Actinomadura,* and *Streptomyces* oxidize their substrates mainly to carbon dioxide and water; in addition, these genera may also use amino acids and proteins for growth once the carbohydrate is completely metabolized.

Within the family Actinomycetaceae, the fermentations of glucose are best described as homolactic, heterolactic, or propionic acid (Pine, 1970). All species of *Actinomyces* are homolactic acid organisms because, given the correct conditions of growth and medium, they will ferment glucose primarily to lactic acid. Therefore, they have fructose diphosphate aldolase and use primarily the Embden-Meyerhof glycolytic scheme for catabolizing sugar. However, under anaerobic conditions when carbon dioxide is present, all species form acetic, lactic, and succinic acids, and traces of formic acid (Table 4). Some species use large amounts of carbon dioxide to synthesize succinic acid and form aspartic acid internally. Since these carbon dioxide-fixing strains lack a permease for the uptake of aspartic acid (Buchanan and Pine, 1965), carbon dioxide is a factor required for growth. But once growth is initiated, some members of the genus can grow aerobically, converting glucose primarily to acetic acid and carbon dioxide.

The genus *Bifidobacterium,* however, does not have fructose diphosphate aldolase or glucose-6-phosphate dehydrogenase, and glucose is degraded by the fructose-6-phosphate shunt to acetic and lactic acids (DeVries and Stouthamer, 1967). Since two moles of lactic acid cannot be formed from one mole of glucose, this fermentation is heterolactic. Acetate and lactate are formed in molar ratios approximating 3:2; only traces of other products such as formic or succinic acid are formed.

Although they contain fructose diphosphate aldolase, the species of the third group *(Ar. propionica* and *B. matruchotii)* ferment glucose to form propionic and acetic acids as the major products. Small amounts of lactic and succinic acids with considerable amounts of carbon dioxide are also formed in the single anaerobic fermentation within the family in which carbon dioxide is a major product. Although the fermentations resemble those of the propionobacteria, they can differ significantly in the relative quantities of the products formed.

There is an excellent correlation of the cell wall

TABLE 4. Fermentative Characteristics Differentiating Genera of *Actinomycetaceae* and *Propionibacterium*

GENUS	OXYGEN RELATIONSHIP	CELL-WALL MUCOPEPTIDE[a]	FERMENTATION	MAJOR ACIDS FORMED FROM GLUCOSE	
				Anaerobic	Aerobic
Actinomyces	Anaerobe, facultative anaerobe	Lysine, ornithine, glutamic acid (1:1:2)	Homolactic	(Formic), acetic, lactic, succinic	CO_2, acetic
Rothia	Aerobe	Lysine	Homolactic	Lactic	Acetic, lactic
Bifidobacterium	Anaerobe	Alanine, lysine (ornithine): glutamic acid (1:1)	Heterolactic	Acetic, lactic	No growth
Arachnia	Facultative anaerobe	Glycine, alanine, glutamic acid (2:1:1) plus LL diaminopimelic acid	Propionic	CO_2, acetic, propionic, lactic, succinic	CO_2, acetic
Propionibacterium	Anaerobe	Glycine, alanine, glutamic acid (1:2:1) plus LL diaminopimelic acid	Propionic	CO_2 acetic, propionic (lactic, succinic)	CO_2, acetic
Bacterionemia	Aerobic, facultative anaerobe	Glycine, DL diaminopimelic acid	Propionic	CO_2, propionic, lactic	CO_2, acetic, propionic

[a]Data based on Schleifer and Kandler (1972).

mucopeptide components to the fermentation. All *Actinomyces* species except *A. bovis* have lysine, ornithine, and glutamic acid in their cell walls in a ratio of 1:1:2, and all ferment glucose homolactically. However, in *B. adolescentis (A. eriksonii)* ornithine or lysine occurs in a ratio of 1:1 with glutamic acid, replacing one or the other within the peptide subunit (Schleifer and Kandler, 1972). The propionic acid-forming species, however, have diaminopimelic acid instead of lysine or ornithine as the dibasic component of the cell wall. These relationships of fermentation to cell wall composition are summarized in Table 4.

The general identifying physiologic characteristics of the various species of Actinomycetaceae are listed in Table 5. Basically, all organisms are anaerobic or facultatively anaerobic; the latter can grow in oxygen when growth is initiated with larger inocula. Carbon dioxide stimulates growth and increases the fermentation products of glucose from only lactic acid to volatile and nonvolatile acids that are more characteristic of the species. Under strongly aerobic conditions, certain species ferment glucose to acetate and carbon dioxide; the cell yields under these conditions are doubled, indicating that the use of oxygen is energy yielding and linked to growth. *Ar. propionica*, *B. matruchotii*, *A. viscosus*, and *R. dentocariosa* are aerobic or facultatively aerobic and may not require carbon dioxide to initiate growth. In general, the presence of carbon dioxide improves the primary isolation from small numbers of cells whether the incubation is anaerobic or aerobic.

Although numerous primary colonies of species defined as aerobic may grow on isolation plates, incubated anaerobically, they cannot be transferred under anaerobic conditions. Conversely, particularly when the inoculum is heavy, anaerobic species may sometimes grow under aerobic conditions. On occasion microcolonies of even the aerobic *Nocardia asteroides* may grow on primary isolation plates incubated anaerobically. Thus these results suggest that for primary isolation of an actinomycete, both aerobic and anaerobic isolations should be made in the presence of carbon dioxide.

The definition of a species as aerobic is strongly supported by the presence of catalase or of cytochromes, as shown by the benzidine test. *R. dentocariosa* is aerobic and catalase-positive. *P. acnes* and *A. viscosus* are anaerobic and facultatively anaerobic respectively, but both have catalase. This distinguishes them from all other facultative or anaerobic species within the family Actinomycetaceae.

The growth requirements of this family of organisms are complex and resemble those of lactobacilli and streptococci. A few strains of *A. israelii* require cysteine, glutamic acid, lysine, leucine, isoleucine, and tryptophan (Keir and Porteus, 1962).

Of the physiologic characteristics in Table 5, acid is formed by certain strains, listed as acid-negative, only with prolonged incubation of 14 days or more. The results given in Table 5 should be apparent within seven or eight days. Thus, hydrolysis of starch by *A. bovis* and *B. adolescen-*

TABLE 5. Physiologic Characteristics of Species of *Actinomycetaceae* and *Propionibacterium acnes* [a]

ORGANISM	CATALASE	NITRATE REDUCTION	NITRATE REDUCED	CASEIN AND GELATIN HYDROLYSIS	STARCH HYDROLYSIS	GLUCOSE	STARCH	MANNITOL	XYLOSE	ARABINOSE	GLYCEROL	OXYGEN REQUIREMENTS	CO_2 REQUIRED	%G+C COMPOSITION OF DNA
Actinomyces														
A. humiferus	−	−	−	+	+	A	A	A	A	A	A(−)	Ae	−	73 −
A. bovis	−	−	nr	−	+	A	A	−	−(A)	−	−	An	+	57–63
A. suis	−	nr	nr	−	+	A	A	A(−)	nr	A(−)	−(A)	An	+	nr
A. odontolyticus	−	+(−)	−	−	−	A	−	−	A	−(A)	(A)	FacAn	+	62
A. israelii	−	+(−)	−	−	−	A	−	A	A	A−	−	An	+	57–60
A. naeslundii	−	+	−	−	−	A	−	−	−	−	A,−	FacAn	+	63–64
A. viscosus	+	+	+−	−	+	A	A−	−	−	A−	A,−	FacAn	+	63–70
Rothia dentocariosa	+	+	+	−	−	A	−	−	−	−	A,−	FacAn	+	63–70
Bifidobacterium adolescentis	−	−	−	−	+	A	A	A	A	A	−	An	−	62
Arachnia propionica	−	+	nr	−	−	A	−	A	−	−	−(A)	FacAn	−	63–65
Bacterionemia matruchotii	+−	+	+	−	+(−)	A	nr	−	−	−	−	Ae	−	50–56
Propionibacterium acnes	+	+	nr	+	−	A	−	A(−)	−	−	A(−)	An	−	57–60

[a]+, positive reaction; −, negative reaction; +−, positive and negative reactions among strains; A, acid formed but no gas; nr, not reported; Ae, aerobic; An, anaerobic FacAn, facultative anaerobe; ( ), occurs in a few strains.

Data compiled from Pine (1973), and Slack and Gerencser (1975).

tis is rapidly apparent. Although a few strains of *Ar. propionica* may ferment glycerol or lactic acid after prolonged incubation, they do so less readily than *P. acnes* or other propionibacteria. Gelatin liquefaction or casein hydrolysis differentiates *P. acnes* and *A. humiferus* from all other species. Catalase production may require a short period of aerobic induction. A major source of confusion in identification of pathogenic species in actinomycosis is *P. acnes,* because it is often present as a contaminant or commensal. However, *P. acnes* can be differentiated from the other species by the catalase test and gelatin liquefaction.

PATHOGENIC PROPERTIES

A. humiferus, which was isolated from the soil, is the only *Actinomyces* species that is not resident in the animal body. *A. bovis* has been isolated only from infected cows and *A. suis* has been described only in pigs, primarily as a cause of mammary actinomycosis. In the cases of bovine and swine actinomycosis, it would appear that both species inhabit the oral cavity of their host as do the other species in this genus. *A. israelii* is reported to be the major cause of actinomycosis in man; although actinomycosis has been reported in dogs, deer, and other animals, species identification has been minimal. *A. israelii* causes lacrimal canaliculitis, which is not a true actinomycosis because the deeper tissues are not penetrated. In this infection, large mycelial masses of the organism block the lacrimal canal and may erode the surface membrane without invading the underlying tissues. *Ar. propionica* also shares these characteristics with *A. israelii.* Typical actinomycosis, i.e., progressive chronic inflammatory disease, suppuration, and sinus and sulfur granule formation, is rarely caused by the other species of the family (Table 6).

Transmission of these diseases from man to man or animal to animal is unknown except in cases of human or animal bite; this mode of transmission is suspected in cases of swine actinomycosis. Bovine actinomycosis is attributed to injury of the tissues surrounding the mandible during the chewing of roughage by the cow. In lacrimal canaliculitis the assumption is that the organisms spread from the mouth or nasopharynx to the lacrimal duct by a stifled sneeze or by transfer of saliva to the eye from which it drains into the lacrimal canal. There is no evidence that *Actinomyces* normally inhabits the lacrimal canal. All other cases of actinomycosis are attributed to the direct introduction of the organism from its normal habitat in the mouth into the deeper tissues or bloodstream as a result of some physical injury. Thus, tooth extraction or dental surgery may be followed by actinomycosis of the jaw. Appendectomy, cholecystectomy, or other operations were the major factors preceding actinomycosis in most patients reported by Brown (1973). Disseminated infections are best explained by spread of the organism from the mouth or bowel into the bloodstream (Bowden and Hardie, 1973).

Of the species examined for pathogenicity

TABLE 6. Pathogenicity of Microaerophilic Actinomycetes in Man and Experimental Animals

	NATURAL INFECTION	EXPERIMENTAL INFECTION		
ORGANISM	Number of Cases and Characteristics in Man	Animal	Lesions	Abscesses Formed with Mycelium, Sulfur Granules[h]
Actinomyces				
A. bovis	None	Hamster, mouse	Progressive	6/10
A. israelii	Numerous cases[a] (true actinomycosis)	Hamster, mouse	Progressive	52/66 341/397
A. naeslundii	Three cases[b] (atypical actinomycosis)	Mouse	Progressive	111/129
A. odontolyticus	One case[a] (atypical actinomycosis)	Mouse	Regressive	6/68
A. viscosus	One case[d] (atypical)	Mouse	Progressive	23/24
Arachnia propionica	Numerous cases[e] (true actinomycosis)	Mouse	Progressive	28/28
Bifidobacterium adolescentis	Several cases[f] (abscess formation)	Mouse	Regressive	34/49
Rothia dentocariosa	One case[g]	Mouse	Regressive	0/32
Bacterionema matruchoti	None	Hamster	Regressive	0/6

[a]Brown, J. R. (1973); Bowden and Hardie (1973).
[b]Coleman et al. (1969).
[c]Morris and Kilbourn (1974).
[d]Adeniyi-Jones et al. (1973).
[e]Slack and Gerencser (1975).
[f]Georg et al. (1965).
[g]Sharfen, J.: Zentralbl Bakterial [Orig A] 233:80, 1975.
[h]Data taken in part from Slack and Gerencser (1975); Animals positive/animals inoculated.

Data partially compiled from Table 11–3 in Slack and Gerencser (1975).

(Table 6), *A. israelii, A. naeslundii, A. viscosus,* and *Ar. propionica* were equally effective in causing actinomycotic lesions in mice. These four species usually produced progressive lesions with mycelial filaments, mycelial clumps, and occasionally sulfur granules, unlike the abscesses formed by other species. From the table and the results of population studies in dental calculus it is evident that *A. israelii* and *Ar. propionica* are the most pathogenic species for man regardless of their relative population in the oral cavity or on calculus. In mice and other animals all four species appear to be equally infectious and pathogenic.

The mechanisms by which species of *Actinomyces* establish a progressive infection in humans are unknown. In humans the disease is degenerative and may be accompanied by destruction of tissue and dissolution of bone. In cattle, lumpy jaw may destroy bone, although it most often stimulates bone growth in the upper or lower jaw. No toxin has been implicated in the pathogenesis of actinomycosis, nor is there any evidence that sulfur granules protect the bacterium from the defensive mechanisms of the host. There are sufficient reports of successful vaccine therapy to support the contention that high levels of antibody may limit the progress of the disease (Buchs, 1963). It is often stated but not proved that other organisms contribute to the pathogenesis of *A. israelii* infections.

The oral actinomycetes are strongly associated with the formation of cavities and periodontal disease. *A. naeslundii* and *A. viscosus* are implicated in gingivitis in man and cause extensive periodontal disease in hamsters and germ-free rats (Jordan et al., 1972).

From studies with *Streptococcus mutans,* it is believed that dental plaque forms as a result of bacterial adherence to the tooth surface, where the bacteria colonize the salivary glycoproteins coating the enamel. The bacterium has both a soluble and a cell-associated glycosyltransferase that converts sucrose to insoluble glucans and oligosaccharides of low molecular weight. The insoluble polysaccharide precipitates on the tooth surface and, because of the affinity of the cell-bound enzyme with its substrate, the streptococcus adheres to the polysaccharide. Other bacteria also adhere to the insoluble polysaccharides, which may be used for further bacterial growth and colonization. The final result is the formation of plaque.

Within the plaque the fermenting bacteria form their characteristic volatile and nonvolatile acids that cause localized areas of low pH and subsequent dissolution of enamel; this latter process initiates the formation of dental caries. When the plaque becomes calcified by mechanisms not fully understood, it is converted to calculus, and both are colonized by all the *Actinomyces* species, which become the predominant populations. At this point, intimate contact or abrasion of the periodontal tissue by the calculus exposes the tissue to both the organism and its products, without being invaded by them.

Baker et al. (1976) summarized the immunologic events leading to the pathogenesis of gingivitis or periodontosis of 11 common oral bacteria. They strongly implicated *A. israelii, A., naeslundii, A. viscosus, Ar. propionica, P. acnes,* and a gram-negative rod as the primary causes of lymphocyte transformation. The immunologic responses to the organisms or their products result in the production of sensitized lymphocytes that release cell products. These products enhance antibody production, attract phagocytic cells, and initiate processes that resorb bone and damaged tissue. Ivanyi and Lehner (1970) showed a direct correlation between the severity of periodontal disease and the degree of blastogenesis of the host's lymphocytes when the host is exposed to oral bacteria. Baker et al. (1976) showed that the lymphocytic transformations caused by the above species were the result of antigenic sensitivity and not of the bacterial walls acting as mitogens. Thus, in general, leukocyte transformation induced by *Actinomyces* requires the sensitized leukocyte of the diseased host or immune serum. Leukocytes of patients with periodontal disease respond to sonicated dental plaque, whole *Actinomyces* cells and cell walls, and soluble cytoplasmic components. These components contain protein and neutral sugars and act as specific antigens for the immune cellular response (Reed et al., 1976). These data strongly implicate the four species above as major causes of periodontal disease by mechanisms linked with bacterial hypersensitivity. This mechanism may also be involved in pathogenesis of actinomycosis when the living actinomycete is introduced into the tissues.

IMMUNITY

In the early 1920s, opsonins and agglutinins were observed in actinomycosis; more recently agar gel precipitins, hemagglutinins, and complement-fixing antibodies have been described. In a few cases, the serum precipitin response corresponded with the course of the disease in patients with actinomycosis. Thus recovery after treatment was accompanied by a decline in precipitins.

There is also a suggestion that vaccine therapy can control the disease when used alone or in combination with surgical procedures or antibiotic therapy. Colebrook (1921) found that vaccine therapy was particularly successful in treating cervico-facial actinomycosis. There are no reports of immunity to reinfection.

DRUG SUSCEPTIBILITY

The susceptibility of pathogenic *Actinomyces* and *Arachnia* species to drugs is shown in Table 7. The correlation of such results with treatment of actinomycosis is summarized by Lerner (1974). The effectivenss of penicillin is well-established in human actinomycosis (Brown, 1973; Weese and Smith, 1975). The antifungal agents amphotericin B, nystatin, and griseofulvin are not effective.

Clinical failures with penicillin and other drugs active in vitro against *A. israelii* are attributed to secondary infections with Gram-negative organisms. Weese and Smith (1975) reported that 65 per cent of 57 cases over a 36-year period had *Actinomyces* in association with other bacteria, the majority of which were Gram-positive. There have been no extensive reports of antibiotic resistance of *Actinomyces* developing during therapy. Garrod (1952) found eightfold and twentyfold increases respectively in the resistance of two strains during treatment.

LABORATORY DIAGNOSIS

Media

For primary isolation of *Actinomyces* and related organisms, Brain Heart infusion agar plates both without blood and with 5 per cent whole defibrinated animal blood should be used. Rabbit, sheep, or horse blood is recommended; human blood is not. However, *Actinomyces* broth (BBL Division of BioQuest, Becton, Dickinson and Company) with 1.5 per cent agar can be used for isolation plates or with 0.7 per cent agar for dilution tubes and for maintenance. The recom-

TABLE 7. In Vitro Sensitivity of Pathogenic Species in the Family Actinomycetaceae

	A. ISRAELII	*A. NAESLUNDII*	*A. ODONTOLYTICUS*	*A. VISCOSUS*	*A. BOVIS*	*A. ERIKSONII* [a]	*ARACHNIA PROPIONICA*
No strains tested	32	5	5	15	4	6	7
			μg/ml for total inhibition of colony growth				
Penicillin G	0.5	1.0	0.15	0.5	1.0	0.5	0.5
Ampicillin	2.0	2.0	2.0	1.0	2.0	2.0	1.0
Cephaloridine	0.5	0.5	0.5	1.0	2.0	0.5	1.0
Cephalothin	2.0	2.0	0.25	0.5	1.0	2.0	1.0
Cephalexin	8.0	16.0	4.0	8.0	8.0	16.0	4.0
Minocycline	1.0	1.0	0.5	1.0	0.5	0.5	0.5
Doxycycline	2.0	4.0	0.5	2.0	1.0	4.0	4.0
Tetracycline	2.0	8.0	8.0	2.0	2.0	4.0	2.0
Clindamycin	0.5	8.0	0.5	2.0	0.5	0.25	4.0
Lincomycin	1.0	4.0	0.5	2.0	2.0	0.5	2.0
Oxacillin	8.0	4.0	4.0	4.0	16.0	8.0	8.0
Dicloxacillin	3.2	16.0	8.0	16.0	32.0	16.0	16.0
Erythromycin	0.25	0.12	0.06	0.12	0.06	0.06	0.12
Chloramphenicol	8.0	8.0	4.0	8.0	32.0	4.0	8.0
Vancomycin	20.0	10.0	10.0	20.0	10.0	20.0	10.0
Rifampin	>0.5	>0.5	0.03	>0.5	0.6	>0.5	>0.5
Fusidic acid	20.0	>20.0	5.0	>20.0	>20.0	>20.0	>20.0
Novobiocin	8.0	>25.0	16.0	>25.0	16.0	>25.0	16.0

[a]*A. eriksonii* = *Bifidobacterium adolescentis.*

Data from Lerner (1974).

mended formula for the *Actinomyces* broth is: potassium dihydrogen phosphate, 15.0 g/L; ammonium sulfate, 1.0 g/L; magnesium sulfate, 0.2 g/L: calcium chloride (anhydrous), 0.01 g/L; heart infusion broth, 25.0 g/L; dextrose, 5.0 g/L; cysteine hydrochloride, 1.0 g/L; casitone, 4.0 g/L; yeast extract, 5.0 g/L; insoluble potato starch (preferred) or soluble starch, 1.0 g/L; final pH adjusted to 6.5 to 6.8 with potassium hydroxide.

Fluid thioglycolate broth may also be used for dilution tubes. However different commercial products vary in sugar concentrations and in the presence or absence of cystine or phosphate as buffer, and also in their basic complex protein extracts. Media with cystine and phosphate are preferred. For best results the commercial fluid thiglycolate medium should be fortified with 0.2 per cent sterile rabbit serum just before use.

Although the Brain Heart infusion media are excellent for isolation, they are not satisfactory for maintenance. For maintenance, deep cultures are made in *Actinomyces* broth with 0.7 per cent agar. Tubes of the medium are sealed with the standard pyrogallol-sodium carbonate seal described below. For the maintenance of the aerobic species *R. dentocariosa,* 1.5 per cent agar slants of Brain Heart Infusion agar or Trypticase Soy Agar (BBL, BioQuest) are used.

The medium used for fermentation tests is Thiglycolate Fermentation Medium without added dextrose (BBL, BioQuest) or indicator to which 2.0 g of yeast extract (glucose free) and 2.0 ml of 1.0 per cent Brom cresol purple have been added (Slack and Gerencser, 1975). This medium is used for all species except *R. dentocariosa,* for which meat-extract peptone base is used. The formula for this medium is: meat extract, 3.0 g/L; bactopeptone, 10.0 g/L; sodium chloride, 5.0 g/L; Andrade's indicator, 10 ml/L, pH adjusted to 7.4.

All media may be prepared and stored; before use, however, they should be fresly melted or heated to eliminate absorbed oxygen. For isolation or stock cultures, the *Actinomyces* maintenance medium with 0.7 per cent agar can be tubed in 10 ml quantities and stored in anaerobic containers or with pyrogallol-sodium carbonate seals at 5°C for one year.

Steps in Laboratory Diagnosis

1. Examination of Clinical Materials. All materials should be examined under wet mounts in dilute KOH, and by Gram and acid-fast stains of thinly prepared smears. Acid-fast cells should be regarded as potential *Nocardia* or *Mycobacteria.*

Pus is examined for mycelium or sulfur granules by adding a few drops of it to 5 ml of sterile distilled water and shaking the mixture well to make a fine suspension and to lyse tissue cells. The sediment is examined in KOH mounts and by Gram stains. Granules are washed several times by passage through 4 or 5 ml of sterile distilled water and then crushed in 0.5 ml of liquid. This suspension is used to streak plates or make dilution tubes. Sulfur granules or mycelial colonies are characteristic of *Actinomyces* and *Arachnia*.

2. Preparation and Examination of Primary Isolation Cultures Two plates each of Brain Heart Infusion agar with and without blood should be made; a series of dilution tubes in the 0.7 per cent agar-*Actinomyces* medium or in thioglycolate broth are also made. One plate each, with and without blood, should be incubated anaerobically with carbon dioxide, and one set should be incubated aerobically with carbon dioxide. A candle jar or Gaspak (BBL, BioQuest) is adequate for the anaerobic carbon dioxide systems. However, anaerobic chambers that can be evacuated and flushed with the nitrogen-5 per cent carbon dioxide are preferred. A desiccator to which a few pea-sized granules of dry ice are added can be used for aerobic incubation with carbon dioxide. Dilution tubes made with *Actinomyces* medium are extremely effective for primary isolation, particularly if the tubes are sealed with pyrogallol-sodium carbonate. Once the tubes are inoculated, the top half of the cotton plug is cut off, and the plug is pushed into the tube, care being taken to keep the plug several inches above the agar. A second snug fitting stopper of absorbent cotton is inserted; five drops of a saturated pyrogallol solution and five drops of a 10 per cent sodium carbonate solution are added in that sequence, and the tube is rapidly sealed with a tight rubber stopper. Dilution tubes made in fluid thioglycolate medium may be used without the anaerobic seals. All cultures are incubated at 37° C.

All plates should be examined after 24 to 48 hours by transmitted or reflected light to observe the spider-like microcolonies of *A. israelii,* or the round entire microcolonies characteristic of *A. bovis, A. odontolyticus,* and *B. adolescentis.* Beta hemolysis or strong alpha hemolysis is not produced by the actinomycetes. After four to five days the macrocolonies are examined by Gram stain and for catalase.

To test for catalase one drop of a cell suspension is added to one drop of 1.5 per cent hydrogen peroxide on a microscope slide and immediately covered with a coverslip. If catalase is present, bubbles will form beneath the coverslip within 30 seconds and stream continuously to the edges. Blood agar media must not be transferred with the colony, since the red cells themselves contain catalase. Also, metallic ions such as iron can catalyze the breakdown of peroxide. In a positive test gas formation continues for at least 20 to 30 seconds after bubbling begins. A positive catalase test suggests *A. viscosus, R. dentocariosa,* or contaminating bacteria.

Primary cultures should be restreaked before a stock culture is made. Cultures that appear pink and are catalase-positive may contain *P. acnes.* After moderate growth in *Actinomyces* agar deeps, the cultures may be stored at 5° C for 6 to 12 months.

3. Differential Physiologic Characteristics. The identification of species rests on the characteristics summarized in Tables 3, 4, and 5. The physiologic and fermentation tests of Table 5 are the minimum required for generic if not species identification. Identification of species may also be made by the fluorescent antibody test (Slack and Gerencser, 1975). Volatile and nonvolatile acids are identified by chromatography; cell wall sugars and amino acids are identified by paper chromatography after enzymatic digestion or sodium hydroxide extraction of cell walls followed by acid hydrolysis (Cummins, 1962; Boone and Pine, 1968).

EPIDEMIOLOGY

There have been no recorded epidemics of *Actinomyces* infections. In humans the disease occurs throughout the world and shows no propensity for any given racial group. One clinic reported an average incidence of from one to three cases per year over a 36-year period, with one three-year period averaging five cases per year (Weese and Smith, 1975). Another reported about five cases per year over a ten-year period (Buchs, 1963). Distribution does not seem to be age-related, although approximately 80 per cent of the patients were over 20 years old. In a study of 181 cases, Brown (1973) observed that two thirds of the patients were between 30 and 60 years old. A greater number of cases are reported in males, with only 20 to 40 per cent reported in females.

The incidence of human infections ranges from 35 to 55 per cent in the tissues of the head and neck, 14 to 33 per cent in the thorax, and 23 to 28 per cent in the pelvic and abdominal region, of which 10 per cent may occur in the pelvic organs alone (Brown, 1973). In cattle, where 0.03 to 0.5 per cent of the herds may have actinomycosis, the disease manifests itself primarily as lumpy jaw or as infections of the softer tissues of the head.

Because the organisms are commensals prevalent in the mouth of the adult, it is assumed that the young are colonized by direct oral contact or through some other physical vector. Once the organism is established in the host, an infection is most often associated with some physical trauma by which the organism gains entrance to the deeper tissues. Supporting the data on the effect of physical trauma are the reports of Henderson (1973), Schiffer et al. (1975), and Gupta et al. (1976), relating the presence of *Actinomyces* and pelvic actinomycosis to the use of intrauterine devices by women. These reports represent only a few of those published recently in America, England, and Canada on this subject. In recent observations, the description of abscesses with pus containing sulfur granules, the typical aspect of the sulfur granule stained with hematoxylin and eosin, and the typical gram-positive mycelium in sections stained by the Brown and Brenn stain all support the conclusion that *Actinomyces* is associated with infections that occur during the use of intrauterine devices. Even more significant, Spence et al. (1977) have observed organisms consistent with actinomycetes in 350 pancervicovaginal (Fast) smears of patients who used intrauterine devices. In 35 of these, *A. israelii* was identified by specific fluorescent antibody. Similar stains for *A. naeslundii* were negative. These reports of IUD-associated infections have been supported by isolations of *A israelii* from three cases by Luff and Gupta (1977) and by one fatality associated with *A. israelii* (Serotype 1) infection (Hager and Majmudar, 1979). The overall results strongly implicate *A. israelii* as a causal organism in infections resulting from the use of intrauterine devices and show that women who use IUDs or pessaries have an increased risk of infection by *A. israelii* (Christ and Haja, 1978).

REFERENCES

Baker, J. J., Chan, S. P., Socransky, S. S., Oppenheim, J. J., and Mergenhagen, S. E.: Importance of *Actinomyces* and certain gram-negative anaerobic organisms in the transformation of lymphocytes from patients with periodontal disease. Infect Immun 13:1363, 1976.

Bollinger, O.: Ueber eine ne Pilzkrankheit beim Rinde. Zentralbl Med Weiss. 15:481, 1877.

Boone, C. J., and Pine, L.: Rapid method for characterization of actinomycetes by cell wall composition. Appl Microbiol 16:279, 1968.

Bostroem, E.: Untersuchungen über die Aktinomykose des Menschen. Beitr Pathol 9:1, 1891.

Bowden, G. H., and Hardie, J. M.: Commensal and pathogenic *Actinomyces* species in man. In Sykes, G., and Skinner, F. A. (eds.): Actinomycetales: Characteristics and Practical Importance. New York, Academic Press, 1973, p. 277.

Brock, D. W., and Georg, L. K.: Determination and analysis of *Actinomyces israelii* serotypes by fluorescent-antibody procedures. J Bacteriol 97:581, 1969.

Brown, J. R.: Human actinomycosis. A study of 181 subjects. Hum Pathol 4:319, 1973.

Buchanan, B. B., and Pine, L.: Relationship of carbon dioxide to aspartic acid and glutamic acid in *Actinomyces naeslundii.* J Bacteriol 89:729, 1965.

Buchanan, R. E., and Gibbons, N. E.: Bergey's Manual of Determinative Bacteriology. 8th ed. Baltimore, The Williams & Wilkins Company, 1974.

Buchs, H.: Zur klinik und therapei der cervicofaciaten aktinomykose. Dtsch. Zahnaerztl Z 18:1069, 1963.

Bujwid, O.: Ueber die Reinkulter des *Actinomyces.* Zentralbl Bakteriol [Orig. B] 6:630, 1889.

Cohn, F.: Untersuchungen über Bacterien. Beitr. Biol Pflanzen 1:141, 1875.

Colebrook, L.: A report on 25 cases of actinomycosis with special reference to vaccine therapy. Lancet 1:893, 1921.

Coleman, R. M., George, L. K., and Rozzell, A. R.: *Actinomyces naeslundii* as an agent of human actinomycosis. Appl Microbiol 18:420, 1969.

Christ, M. L., and Haja, J.: Cytologic changes associated with vaginal pessary use with special reference to the presence of *Actinomyces.* Acta Cytol 22:146, 1978.

Cummins, C. S.: Chemical composition and antigenic structure of cell walls of *Corynebacterium, Mycobacterium, Nocardia, Actinomyces* and *Arthrobacter.* J Gen Microbiol 28:35, 1962.

Cummins, C. S.: *Actinomyces israelii* type 2. In Prouser, H. (ed.): The Actinomycetales. Jena, Gustav Fischer Verlag, 1968.

Cummins, C. S., and Harris, H.: Studies on the cell wall composition and taxonomy of Actinomycetales and related groups. J. Gen. Microbiol 18:173, 1958.

DeVries, W., and Stouthamer, A. H.: Pathway of glucose fermentation in relation to the taxonomy of bifidobacteria. J Bacteriol 93:574, 1967.

Emmons, C. W.: *Actinomyces* and actinomycosis. Puerto Rico J Public Health Trop Med 11:63, 1935.

Erikson, D.: Pathogenic anaerobic organisms of the *Actinomyces* group. Med Res Council Spec Rep Ser 240:1, 1940.

Frazier, P. D., and Fowler, B. O.: X-ray diffraction and infrared study of the "sulphur granules" of *Actinomyces bovis.* J Gen Microbiol 46:445, 1967.

Garrod, L. P.: The sensitivity of *Actinomyces israelii* to antibiotics. Br Med J 1:1263, 1952.

Georg, L, K., Coleman, R. M., and Brown, J. M.: Evaluation of an agar gel precipitin test for the serodiagnosis of actinomycosis. J Immunol 100:1288, 1968.

Georg, L. K., Robertstad, G. W., Brinkman, S. A., and Hicklin, M. D.: A new pathogenic anaerobic *Actinomyces* species. J Infect Dis 115:88, 1965.

von Graefe, A.: Koncretionen in unteren Thränenröhrchen durch Pilzbildung. Arch Ophthalmol 1:284, 1854.

Gupta, P. K., Hollander, D. H., and Frost, J. K.: Actinomycetes in cervicovaginal smears: An association with IUD usage. Acta Cytol 20:295, 1976.

Hager, W. D., and Majmudar, B.: Pelvic actinomycosis in women using intrauterine devices. Am J Obstet Gynecol 156:60, 1979.

Hartz, C. O.: *Actinomyces bovis:* ein neuer Schimmel in dem Geweben des Rindes. Dtsche Z Tier-Med 5:125 (Suppl.) 1879.

Henderson, S. R.: Pelvic actinomycosis associated with an intrauterine device. Obstet Gynecol 41:726, 1973.

Holm, P., and Kwapinski, J. B.: Studies on the detection of *Actinomyces* antibodies in human sera by use of pure antigenic fractions of *Actinomyces israelii.* Acta Pathol Microbiol Scand [B] 45:107, 1959.

Howell, A., Jr., Murphy, W. C., Paul, F., and Stephan, R. M.: Oral strains of *Actinomyces.* J Bacteriol 78:82, 1959.

Israel, J.: Neue Beobachtungen auf dem Gebiete der Mykosen des Menschen. Arch Pathol Anat 74:15, 1878.

Ivanyi, L., and Lehner, T.: Stimulation of lymphocyte transformation by bacterial antigens in patients with periodontal disease. Arch Oral Biol 15:1089, 1970.

Jordan, H. V., Keyes, P. H., and Bellack, S.: Periodontal lesions in hamsters and gnotobiotic rats infected with *Actinomyces* of human origin. J Periodont Res 7, 21, 1972.

Keir, H. A., and Porteus, J. W.: The amino acid requirements of a single strain of *Actinomyces israelii* growing in a chemically defined medium. J Gen Microbiol 28:581, 1962.

Lerner, P. I.: Susceptibility of pathogenic actinomycetes to antimicrobial compounds. Antimicrob Agents Chemother 5:302, 1974.

Luff, R. D., and Gupta, P. K.: Actinomycetes-like organisms in wearers of intrauterine contraceptive devices. Am J Obstet Gynecol 129:477, 1977.

Morris, J. F., and Kilbourn, P.: Systemic actinomycosis caused by *Actinomyces odontolyticus.* Ann Int Med 81:700, 1974.

Naeslund, C.: Studies of *Actinomyces* from the oral cavity. Acta Pathol Microbiol Scand 2:110, 1925.

Pine, L.: Classification and phylogenetic relationship of microaerophilic actinomycetes. Int J System Bacteriol 20:445, 1970.

Pine, L.: Parasitic or fermentative actinomycetes. In Laskin, A. I., and Lechevalier, H. A. (eds.): Handbook of Microbiology. Cleveland, CRC Press, 1973, p. 212.

Pine, L., Howell, A., Jr., Watson, S. J.: Studies of the morphological, physiological and biochemical characters of *Actinomyces bovis.* J Gen Microbiol 23:403, 1960.

Pine, L., and Overman, J. R.: Determination of the structure and composition of the "sulphur granules" of *Actinomyces bovis.* J Gen Microbiol 32:209, 1963.

Pine, L., and Overman, J. R.: Differentiation of capsules and hyphae in clubs of bovine sulphur granules. Sabouraudia 5:141, 1966.

Ponfick, E.: Ueber Actinomykose. Berl Klin Wchnehr 17:660, 1880.

Reed, M. J., Potters, M. R., Mashimo, P. A., Genco, R. J., and Levine, M. J.: Blastogenic response of human lymphocytes to oral bacterial antigens: Characterization of bacterial sonicates. Infect Immun 14:1202, 1976.

Schiffer, M. A., Elguezabal, A., Sultana, M., and Allen, A. C.: Actinomycosis infections associated with intrauterine contraceptive devices. Obstet Gynecol 45:67, 1975.

Schleifer, K. H., and Kandler, O.: Peptidoglycan types of bacterial cell walls and their taxonomic implications. Bacteriol Rev 36:407, 1972.

Silberschmidt, W.: Über Actinomykose. Z Hyg 37:345, 1901.

Slack, J. M., and Gerencser, M. A.: *Actinomyces,* Filamentous Bacteria: Biology and Pathogenicity. Minneapolis, Burgess Publishing Company, 1975.

Slack, J. M., Moore, D. W., and Gerencser, M. A.: Use of the fluorescent antibody technique in the diagnosis of actinomycosis. W Va Med J 62:228, 1966.

Spence, M. R., Gupta, P. K., Frost, J. K., and King, T. M.: Cytological detection and clinical significance of *Actinomyces israelii* in women employing intrauterine contraceptive devices. Am J Obstet Gynecol 131(3):295, 1978.

Suter, V. L.: Evaluation of criteria used in the identification of *Actinomyces bovis* with particular reference to the catalase reaction. Mycopathologica 6:220, 1956.

Thompson, L.: Isolation and comparison of *Actinomyces* from human and bovine infections. Proc Staff Meet Mayo Clin 25:81, 1950.

Thompson, L., and Lovestedt, S. A.: An *Actinomyces*-like organism obtained from the human mouth. Proc Staff Meet Mayo Clin 26:169, 1951.

Weese, W. C., and Smith, I. M.: A study of 57 cases of actinomycosis over a 36-year period. Arch Intern Med 135:1562, 1975.

Wolff, M., and Israel, J.: Ueber Reincultur des *Actinomyces* and seine Uebertragbarkeit auf Theire. Virchow's Arch [Path Anat] 126:11, 1891.

Wright, J. J.: The biology of the microorganism of actinomycosis. J Med Res 13:349, 1905.

THE CLOSTRIDIA 45

Betty C. Hobbs, Ph.D., D.Sc., F.R.C.Path., Dip. Bact.

GENERAL CHARACTERISTICS

The *Clostridium* species are gram-positive, spore-forming bacilli, motile except for *C. perfringens* and obligate anaerobes. They vary in their requirements for reduced oxygen, and some species tolerate concentrations of oxygen not far below those found in the atmosphere.

The clostridia are biochemically active and may be saccharolytic, fermenting carbohydrate, such as *C. perfringens, C. septicum, C. oedematiens,* and *C. fallax,* and proteolytic, decomposing protein, such as *C. sporogenes, C. histolyticum,* and *C. botulinum.*

Most species are saprophytic, others exist as commensals in the human and animal intestine, and many are pathogenic. The pathogenic clostridia cause serious diseases such as tetanus *(C. tetani),* gas gangrene *(C. perfringens, C. oedematiens,* and *C. septicum),* and botulism *(C. botulinum). C. perfringens* also causes gastroenteritis, usually mild but sometimes fatal in elderly and debilitated persons, and also enteritis necroticans, which is a more serious disease and often fatal. In veterinary medicine clostridia are important intestinal pathogens. They cause enterotoxemia in sheep, calves, lambs, goats, and piglets. Dysentery in lambs (*C. perfringens* Type B) and piglets (*C. perfringens* Type C) may necessitate extensive vaccination programs. Toxemias in adult animals may follow a change of diet. High carbohydrate levels may encourage the growth of clostridia and toxin formation, as in the production of epsilon toxin by *C. perfringens* Type D in pulpy kidney disease. Similarly, it is suggested by Lawrence and Walker (1976) that in Papua New Guinea the sudden intake of pork during feasting encourages the growth of *C. perfringens* Type C and the production of beta toxin, which cannot be destroyed because of the low level of digestive proteases in the intestine; the staple diet is sweet potato that possesses a heat-stable trypsin inhibitor. The use of a *C. perfingens* Type C toxoid has given encouraging results in children (Leader, 1977).

The artificial feeding of 12 hospital-born neonates was followed by necrotizing enteritis of varying severity within ten days to six weeks. Evidence of the presence of *C. butyricum* was found in the blood of nine to ten babies examined; this organism was thought to be the invader responsible (Howard et al., 1977). Rifkin et al. (1977) described *C. sordellii* infection in adults with antibiotic-induced colitis. Thus it seems that clostridia can grow and produce toxin under con-

TABLE 1. Industrial Uses of the Clostridium Group

C. pasteurianum	Nitrogen fixation in the soil
C. beijerinckii	Anaerobic digestion
C. cellobiofavum	Cellulose degradation
C. butyricum	Butyric acid production in butter and cheese
C. tyrobutyricum	Butyric acid production in cheese
C. butylicum	Acetic and butyric acid production from glucose; also ethyl, butyl and propyl alcohols and acetone
C. iodophilum	Acetic and butyric acid production from glucose; also ethyl and butyl alcohols and acetone
C. toanum	Butyric acid; also ethyl, butyl and isopropyl alcohols and acetone
C. acetobutylicum	Acetic and butyric acids; also ethyl and butyl alcohols and acetone. Anaerobic digestion
C. amylosaccharobutylpropylicum	Butyl and propyl alcohols from starch and sugar

ditions favorable to them and in the absence of the usual flora or mechanisms that control them (Leader, 1977). It is suggested also that clostridia may be involved in the breakdown of bile salts to steroids during digestion and thus indirectly produce conditions that predispose toward cancer.

Aside from the medical and veterinary significance of the clostridia, many species are used commercially in the production of acids and alcohols in fermentative or digestive processes; some have been and could be used in the chemical industry when fossil fuels expire. Many species break down nitrogen and cellulose (denitrifying and degrading cellulose in the soil) and many fix nitrogen in the soil. They are prominent in decomposition processes such as silage maturation. Table 1 lists the names of some clostridia and their uses.

In food technology, the presence of mesophilic clostridia in canned nonacid foods indicates that heating has been insufficient to destroy spores of *C. botulinum,* which are the greatest hazard in sterilized or pasteurized canned food. In chilled or dried foods, an unusually large number of viable clostridia might indicate a hazard from *C. perfringens* or *C. botulinum,* either in the food as preserved or in its future use. Spores of the more heat-resistant clostridia are also used to test the sterility of medical equipment.

C. perfringens is one of the more important members of the group, and its characteristics are described in detail.

CLOSTRIDIUM PERFRINGENS

MORPHOLOGY

The organism is a strongly gram-positive, nonmotile rod with blunt or square ends; it is not a strict anaerobe. On some media the bacilli may be longer and more slender. Human Type C strains may be larger with filaments and swollen forms. Sometimes there are short, almost coccoid, forms. Spores are rarely seen in laboratory media or in cooked foods. Sporulation may be encouraged in special media described by Ellner (1956), Duncan and Strong (1968), and Clifford and Annellis (1971); sporulating mutants have been induced. The spores are large, oval, and central and distend the organism. Capsules are mostly seen in the animal body.

The colonies are low, convex, semiopaque, and shiny; they may be round and entire, or somewhat irregular with a vine leaf appearance. Hemolysis is variable according to the strain and the animal source of the blood. It may be clear beta or partially clear alpha, sometimes with a double zone.

ANTIGENIC COMPOSITION

C. perfringens is divisible into five serologic types, A to E, according to the kinds and proportions of exotoxins (soluble antigens) produced. Antitoxic sera are used in the routine typing of strains. The α-toxin or lecithinase is produced by all types; it is antigenically related to the lecithinase of *C. bifermentans* and *C. sordellii.* Although agglutination is of little value in subdividing the species because of its heterogeneity, Type A strains nevertheless are further divided into a large number of serotypes by simple agglutination. This is of value for the epidemiologic study of food poisoning and gas gangrene.

METABOLISM

Glucose, lactose, maltose, and sucrose are fermented, and gelatinase is produced. H_2S is formed, but not indole. *C. perfringens* may be differentiated from most other species of *Clostridium* by lactose fermentation, nitrate reduction, tests for motility and sporulation, and lecithinase activity; the Nagler reaction is a simple in vitro test for this substance.

PATHOGENIC PROPERTIES

Types A, C, and D are pathogenic for man, Type A is responsible for gas gangrene and food poisoning, and Type C for enteritis necroticans; Type A has also been incriminated in necrotizing colitis. Types B, C, D, E, and possibly A affect animals. More than 80 per cent of environmental isolates have been shown to be Type A. Infection of experimental animals and patients causes extensive blood-stained edema fluid, often with gas in the tissues, and invasion of the bloodstream. A number of exotoxins are involved.

TOXICOLOGY

There are at least 12 different soluble antigens described for *C. perfringens.* Alpha, beta, epsilon, and iota are the major lethal toxins in mice.

TABLE 2. The Soluble Antigens of Clostridium perfringens[a]

	α	β	γ	δ	ϵ	θ	ι	κ	λ	μ	ν
A Gas gangrene, puerperal infection, septicemia, food poisoning:											
Classical	++++	−	−	−	−	++−	−	++	−	−	+
Atypical	+−	−	−	−	−	−	−	+−	−	−	−
B Lamb dysentery, foal enterotoxemia, goats and sheep: hemorrhagic enteritis	+	+++	+	++	++	+	−	−	+++	+	+
C Sheep toxemia, calves and lambs: hemorrhagic enteritis,	+	+++	+	++	−	+	−	+	−	−	+
Man: enteritis necroticans	+	+	+	−	−	−	−	−	−	−	+
D Sheep, lambs, goats, cattle: enterotoxemia	+++	−	−	++	+++	++	−	++	++	++−	++−
E Sheep, cattle (? pathogenic)	+++	−	−	−	−	++	+++	++	++	(+)	+−

[a]+++, produced by all strains; +, ++, produced in increasing quantities; ++−, +−, present in some strains only; −, not produced. Adapted from Willis (1977).

α-toxin is common to all types of *C. perfringens*, but it is produced in largest amount by strains of Type A. It is lethal for mice 6 to 12 hours after intravenous inoculation. It also causes necrosis after intradermal injection in guinea pigs. The lecithinase activity of alpha toxin splits lecithin to phosphorylcholine and a diglyceride and can be demonstrated by production of opalescent clostridia on egg yolk agar (Nagler reaction) and by lysis of sheep or mouse erythrocytes.

β-toxin is produced by Types B and C. It is necrotizing and lethal, but not hemolytic; it is heat-labile. It is identified by the intradermal inoculation of depilated albino guinea pigs, in which it produces a purplish necrotic area.

δ-toxin is lethal and hemolytic. It lyses sheep erythrocytes but not those of the horse or rabbit. It is produced by Types B and C.

θ-toxin is produced in greatest amount by Type C strains but can be found also in culture filtrates of Types A, B, D, and E. It is an oxygen-labile hemolysin, active against sheep and horse red cells but not very active against those of the mouse.

ϵ-toxin is produced by Types B and D. ι-toxin is produced by Type E strains. Both toxins are lethal and necrotizing. They are detected in culture filtrates by intradermal tests in guinea pigs. Trypsinization must be used to ensure activation of the prototoxins. They are both absorbed from the intestinal tract.

κ-toxin is produced by Types A, C, E, and some strains of Type D. It is a collagenase and gelatinase and is also lethal and necrotizing.

λ-toxin is produced by Types B and E and some strains of Type D. It is a proteolytic enzyme that attacks gelatin and azocoll (commercial hide powder coupled to a dye) and also casein and hemoglobin (unlike κ-toxin).

γ- and η-toxins are lethal toxins without other demonstrable activity.

μ- and ν-antigens are a hyaluronidase and deoxyribonuclease; they may be estimated by the ACRA (acid-congo red-alcohol) test using horse synovial fluid (hyaluronic acid) (μ) and sodium deoxyribonuclease (ν) as the substrates (Oakley and Warrack, 1951).

Other soluble substances found in filtrates have included neuraminidase and fibrinolysis. The toxin patterns in relation to pathogenicity in man and animals are shown in Table 2 (Oakley and Warrack, 1953).

Enterotoxin

This component of the sporulating cells of Types A and C is responsible for food poisoning and is produced and released in the large intestine during sporulation (Skjelkvåle and Duncan, 1975). Sporulation in cooked food as in laboratory media is poor, so that the toxin is unlikely to be detected either in the laboratory or by clinical manifestations. By means of ligated loop experiments, the association between a large intake of *C. perfringens*, sporulation, and enterotoxin production in the intestine was confirmed (Duncan and Strong, 1971).

Hauschild (1973, 1975) described the purified toxin as a protein of approximately 36,000 MW and isoelectric point 4.3, containing 19 amino acids with a predominance of aspartic and glutamic acids, serine, and leucine. The toxicity for mice was 2000 MLD/mg N. The enterotoxin is heat-labile with a decimal reduction time of 4 minutes at 60° C. The toxin has lethal and emetic properties and causes cutaneous erythema in the guinea pig and rabbit.

Antienterotoxin antibodies have been found in the blood of a high proportion of persons who yet remain sensitive to the enterotoxin (Torres-Anjel and Riemann, 1975).

Biologic assays are carried out with suckling mice and the ileum loop technique; the fluorescent antibody technique may be used also. Torres-Anjel et al. (1975) described a fluorescent antibody technique for the detection of enterotoxin-producing cells of *C. perfringens* Type A. Enterotoxin appeared at the end of the cell after four hours of growth and gradually spread throughout the cell (Niilo, 1977). Enterotoxin was demonstrated in the stools of patients by reversed-passive hemagglutination (Dowell et al., 1975).

LABORATORY DIAGNOSIS

Pathologic material may be inoculated directly onto plates of blood agar and egg yolk agar and incubated anaerobically. Neomycin sulfate may be incorporated into the medium to suppress aerobic organisms. Cooked meat broth cultures may be subcultured onto the same two agar media after eight to ten hours growth. As the organism does not form spores in the tissue or in cooked foodstuffs, heat treatment is not required. The organism produces spores readily in feces, and heat treatment of fecal suspensions is an advantage for isolation of heat-resistant strains. Direct inoculation from a fecal sample onto blood agar (with neomycin sulfate) and subculture from cooked meat broth without heat treatment should also be carried out (Sutton et al., 1971).

C. perfringens is a minor component of the fecal flora in most, if not all, persons. The median count of *C. perfringens* in fecal samples from 50 healthy adults was 7.5×10^3 per g. The ratio of the spore count to the count of unheated material was generally of the order of 1/10 to 1/100 (Sutton, 1969). In outbreaks of *C. perfringens* food poisoning, counts of the organism in feces from patients are higher, usually 10^6 to 10^7, and the heat-resistant spore/vegetative cell count was found to be 1/2000 to greater than 1/60,000 in one outbreak reported by Sutton (1966). Thus quantitation is useful in the assessment of the significance of *C. perfringens* in the stool. The Miles and Misra (1938) technique may be used on blood agar spread with neomycin sulfate; an approximately 1/10 suspension of the fecal sample may be used as the first dilution (Sutton et al., 1971). Heavily contaminated foods responsible for outbreaks may be cultured onto blood agar with and without neomycin sulfate and incubated anaerobically. For the enumeration of small numbers of *C. perfringens* for specification purposes, black colonies in pour plates of sulfite cycloserine agar may be preferred. The isolation of small numbers of *C. perfringens* through cooked meat or liver enrichment media may be of value in epidemiologic studies and when typing is carried out, but care must be taken in the interpretation of the results.

The simplest means of identification is by the reaction described by Nagler in 1939 and developed as a diagnostic test by McClung and Toabe (1947). Cultures growing on media incorporating lecithin, egg yolk agar with or without mannitol, and an indicator show an opalescent precipitate around colonies. The opalescence is due to the interaction between the lecithin in the medium and lecithinase (α-toxin) produced by the organism. It is inhibited by antiserum to the α-toxin. The antitoxin is usually spread over half the plate (half antitoxin egg yolk agar), so that single streaks of cultures across the plate show the precipitate on one half only. Several strains can be tested on one plate. Lecithinase-negative strains occur from time to time. Stringer et al. (1978) described five outbreaks due to lecithinase-negative strains, and found that in four, the serotype of the strains isolated was identical.

Other tests for identification include motility, proteolysis, nitrate reduction, and biochemical reactions; the action on litmus milk (stormy clot) is not always typical. Laboratory reactions used in differentiating *C. perfringens* from other pathogenic clostridia are summarized in Table 3.

EPIDEMIOLOGY

The organism is widely distributed in nature in soil, sewage, water, and the intestinal tract of man and animals. Wounds may be infected from spores or cells in the environment or from contaminants on the skin surrounding the lesion. Serologic studies have indicated that *C. perfringens* in postoperational gangrene frequently originates from the intestine of the patient (Parker, 1967; Ayliffe and Lowbury, 1969).

C. perfringens is an important agent of food

TABLE 3. Differential Laboratory Reactions of Clostridia Pathogenic for Man

	NAGLER REACTION (LECITHINASE)	SPORES	MOTILITY	β-HEMOLYSIS	LACTOSE FERMENTATION	GLUCOSE FERMENTATION	H_2S	INDOLE	UREASE	NITRATE REDUCTION
C. perfringens	+	C	0	+	+	+	+	0	0	+
C. tetani	0	T	+	+	0	0	0	+	0	0
C. botulinum										
Types A, B, E, F	0	ST	+	+	0	+	+	0	0	0
Types C, D	0	ST	+	+	0	+	+	0	0	0
Type G	0	ST	+	+	0	0	+	0	0	0
C. novyi										
Type A	+	ST	+	+	0	+	+	0	0	+
Type B	+	ST	+	+	0	+	+	0	0	+
C. histolyticum	0	ST	+	+	0	0	+	0	0	0
C. septicum	0	ST	+	+	+	+	+	0	0	+
C. sporogenes	+	ST	+	0	0	+	+	0	0	+
C. sordelli	+	C	+	+	0	+	+	+	+	+

[a]T, terminal; C, central; ST, subterminal.

poisoning in the United States and the United Kingdom and perhaps (unrecognized) in many other countries also. It was reported as a cause of intestinal disturbances in hospital patients as early as 1895 (Klein, 1895). Knox and MacDonald (1943), McClung (1945), Zeissler and Rassfeld-Sternberg (1949), Hobbs et al. (1953), and many subsequent workers described the role of *C. perfringens* Type A in food poisoning. *C. perfringens* Type C was described by German investigators in a more serious and often fatal food-borne disease known as enteritis necroticans. Murrell et al. (1966) described the same disease in Papua, New Guinea. The β-toxin was thought to be responsible for the more severe intestinal lesions. The enterotoxin, which is distinct from the other toxins of *C. perfringens*, was later proved to be rsponsible for the milder type of food poisoning due to the Type A strains.

In humans symptoms of food poisoning from Type A strains develop 8 to 24 hours after eating food heavily contaminated with the organism. Symptoms include abdominal pain, nausea, and acute diarrhea lasting 12 to 24 hours; fever is rare. Occasionally there are fatalities among elderly debilitated patients. Deaths of some malnourished children with diarrhea may be due to terminal septicemia with *C. perfringens*. There were approximately 6500 cases of *C. perfringens* food poisoning in England and Wales during the years 1973 to 1975 (Vernon, 1977), representing between 15 and 32 per cent of the annual numbers reported from all bacterial causes. The outbreaks occurred mostly in hospitals, schools, and institutions and tended to be larger (50 persons average) than outbreaks due to other organisms. Outbreaks resulting from faults in the cooling and storage of foods cooked in bulk occurred repeatedly in hospitals and schools. Means for rapid cooling and adequate cold storage of meat and poultry, and other foods cooked in bulk ahead of requirements, should be incorporated into the design of canteen kitchens and other establishments preparing food on a large scale. The importance of this measure needs emphasis in teaching.

Statistics for enteritis necroticans due to the Type C strains are not available. The disease has been reported from Germany and Papua, New Guinea ("Pigbel"), but it is rarely noted in other countries. The factors leading to illness are similar to those for the Type A strains: the survival of spores resistant to heat and the multiplication of vegetative cells in the slow cooking and cooling of masses of meat. In New Guinea the slow spit roasting of pig carcasses eaten at feasts and the "cairn" cooking of meat by the heat of the sun have encouraged the multiplication of the organism to large numbers. Laurence and Walker (1976) suggest that the low protein diet of the people of the Highland of Papua, New Guinea, and the presence of heat-stable trypsin-inhibitor from sweet potatoes results in low levels of digestive proteases in the intestinal lumen. Thus there is no proteolytic activity, and the β-toxin from *C. perfringens* Type C that is normally susceptible to proteolysis is not destroyed and causes necrosis.

The heat resistance of the spores of *C. perfringens* Type A varies from strain to strain and also within strains. Some spores can survive for an hour or more in boiled meats, in which they may be protected by the meat or fat; others live for a few minutes only. Cooking stimulates (heat shock) most spores of *C. perfringens* to germinate; the activation of germination is detected at about 75 to 80° C. The optimum temperatures for growth in cooked meat are 43 to 47° C (Collee et al., 1961).

The rate of division or generation time at optimum temperatures is 10 to 12 minutes (Mead, 1969) in food favorable for growth, principally a meat medium. Cooking drives off dissolved oxygen and leaves an anaerobic environment suitable for the growth of *C. perfringens*. The resistance to heat and activation by heat of the spores during cooking, and the fast multiplication of the vegetative cells during cooling result in the food becoming a medium for an almost pure culture of *C. perfringens*. When many (millions per g) vegetative cells in cooked foods are swallowed, they survive passage through the gastric acid and become established in the intestine. In the United Kingdom and United States, outbreaks of *C. perfringens* food poisoning occur throughout the year with no seasonal prevalence. The fault lies in the prolonged storage at 30 to 50° C between cooking and eating. Without proper cooling and adequate cold storage facilities, the preparation of food in large catering establishments often leads to this type of food poisoning. In addition to outbreaks associated with hospitals, schools, and factories, there are many that follow banquets and meals prepared for touring groups. Kitchens designed to prepare a few meals are often expected to cater for many travelers, despite inadequate space and facilities.

Vegetative cells and spores of *C. perfringens* are common on raw meat and on and within poultry and fish. Their origin is assumed to be feces from man and animals. Because the spores are resistant to dehydration, they can survive for long periods in soil and dust. The ubiquitous nature of the organism makes epidemiologic investigation difficult without finer methods of typing the strains isolated. The serologic differentiation of Type A strains was originally described by Henderson (1940) and Hobbs et al. (1953). It has now been expanded widely into an international typing scheme (Hughes et al., 1976; Stringer et al., 1975; Stringer et al., 1978). Approximately 56 per cent of strains may be typed by simple slide agglutination method using a battery of 50 or more sera made up into polyvalent groups, each containing about five sera of different types. The antigenic specificity is thought to originate from the capsular material.

Serotyping is useful also for tracing the origin of infection in gas gangrene cases (Parker, 1967; Ayliffe and Lowbury, 1969).

CLOSTRIDIUM TETANI

MORPHOLOGY

The characteristic bacilli are slender, gram-positive, motile, and strictly anaerobic. The mature spores are terminal and spherical so that the bacillus has a "drumstick" appearance; immature spores are oval and terminal. Sporefree forms occur, and older cultures may be gram-negative. The organism grows as a fine rhizoidal film with a finely filamentous edge of growth; the extremely fine growth may be missed. Nonmotile variants form discrete colonies with no swarming. Antitoxin will prevent swarming. There is alpha-to-beta hemolysis on horse blood agar.

ANTIGENIC COMPOSITION

There are at least ten types of flagellar antigens. Antigenically identical neurotoxin is formed by all types. Nontoxigenic variants may be isolated from wounds of patients.

METABOLISM

No sugars are fermented except glucose rarely. *C. tetani* is nonproteolytic but produces gelatinase. H_2S is not produced, but most strains produce indole. A rennin-like enzyme gives zones of opacity around colonies on milk agar. Some strains produce deoxyribonuclease. Metabolic products include acetic, propionic, and butyric acids and ethanol and butanol.

PATHOGENIC PROPERTIES

The pathogenicity of *C. tetani* for man arises from its neurotoxic exotoxin. The neurotoxin first causes local peripheral nerve and muscle spasms. The toxin (tetanospasmin) then spreads to the anterior horn cells via the bloodstream and causes generalized convulsions. A hemolysin (tetanolysin) is also produced but is less important; it is not related to the neurotoxin and may be produced by nontoxigenic strains. Tetanospasmin and *C. botulinum* neurotoxin are the most potent poisons known. In culture, tetanospasmin is produced after the phase of active growth; it is heat-labile and is destroyed at 65° C in five minutes.

IMMUNITY

The disease may be prevented by active immunization with tetanus toxoid or passive immunization with antitetanus hyperimmune human periumglobulus. Recovered patients must be immunized because the disease is produced by amounts of toxin that are too small for immunization. Toxoid given to pregnant women prevents tetanus of the newborn.

LABORATORY DIAGNOSIS

Pathologic material should be placed directly onto culture media and incubated in cooked meat broth for two to four days under anaerobic conditions. Fresh or heated blood agar is used for subculture, and the inoculum is placed near the edge of the plate so that the spreading or swarming growth may leave the contaminants behind. For purity, subcultures may be made from the edge of the growth. Neomycin sulfate and cautious heat treatment may be used to inhibit contaminants in material containing nonmotile variants of *C. tetani*.

DRUG SUSCEPTIBILITY

The organism is sensitive to penicillin and its derivatives and to tetracycline.

EPIDEMIOLOGY

C. tetani is widely distributed in soil, especially cultivated soil, and in the intestinal tract of man and animals. Infection of wounds and umbilical stumps of the newborn occurs after contamination with soil and human and animal feces. Military wounds, trauma among farm workers, and injuries from auto accidents are especially subject to soil contamination and thus to tetanus intoxication.

CLOSTRIDIUM BOTULINUM

MORPHOLOGY

The bacillii are large stout rods, motile and gram-positive, especially in young cultures. Sporulation may be sparse; the spores are oval and central or subterminal.

The colonies are irregular or circular and translucent with a granular surface; they may be lobular with an indefinte spreading edge. Growth may spread over the plate. There is hemolysis on horse blood agar.

ANTIGENIC COMPOSITION

There are seven known toxicologic types, A to G, which produce antigenically distinct neurotoxins. There are minor antigenic relationships between the neurotoxins of Types C and D and between Types E and F. Walker and Batty (1964) found that the toxicologic types fell into three distinct serologic groups. Types A, B, and F are serologically related and are distinct from Types C, D, and F. Types C and D are related and are distinguishable from Type E. Walker and Batty also noted limited cross-reactions between strains of *C. botulinum* and *C. sporogenes*.

METABOLISM

Types A, B, E, and F ferment glucose, maltose, and sucrose; Types C and D ferment glucose and maltose; Type G is nonsaccharolytic. All types produce gelatinase and H_2S; indole is not produced.

PATHOGENIC PROPERTIES

C. botulinum Types A, B, E, and F cause botulism in man. Types C and D and more rarely A and B are pathogenic for mammals and birds. Type G has not been reported from outbreaks in man or animals. The exotoxin is responsible for the disease; the organisms do not multiply easily in the body. *C. botulinum* in wounds can give rise to botulism in a few cases (Mersam and Dowell, 1973). Toxin elaborated by *C. botulinum* in the bowel of infants can cause sudden death (Midura and Arnon, 1976).

The potent neurotoxins, although antigenically distinct, are pharmacologically similar; they are absorbed from the alimentary tract through the gastric and upper intestinal mucosa. They can be demonstrated in the blood, and they reach the peripheral nervous system. The toxin acts at the tips of the motor nerve endings at the neuromuscular junction and interferes with the release of acetylcholine. To be effective, antitoxin should be administered before the toxin becomes fixed to the tissues. The incubation period is usually less than 24 hours. Vomiting, thirst, and pharyngeal and ocular paresis occur. Voluntary muscles weaken, and death may occur within 24 hours from paralysis of the respiratory muscles. In poultry and wildfowl the condition known as limberneck gives rise to paralysis of the neck and leg muscles.

A fatal dose of toxin for man has been estimated to be between 0.1 and 1.0 μg (Schantz and Sugyama, 1974). The toxicity of the neurotoxin is about 10^3 that of *C. perfringens* α-toxin, 10^5 that of crotactin (rattlesnake venom), and 10^6 that of strychnine (van Heyningen, 1968, 1970). The toxin is described as a simple protein consisting of 19 amino acids, molecular weight about 150,000, and isoelectric point 5.6. The toxin is rapidly destroyed by boiling. Toxigenic strains may lose their toxigenicity, while naturally occurring nontoxigenic strains can revert to a toxigenic state after infection with certain phages.

IMMUNITY

Natural immunity is unlikely to occur because the lethal dose of toxin is less than that required to elicit an antibody response. Laboratory workers may be immunized with polyvalent antitoxin (A, B, and E).

LABORATORY DIAGNOSIS

The toxin in supernatants of food can be detected in mice injected with the extracts and identified by protection with antitoxin. The clinical diagnosis can be confirmed by demonstrating the toxin in blood and stool specimens. Reverse passive hemagglutination, immunofluorescence and agar plates containing specific antitoxins are also used for identification. The organism may be isolated from cooked meat cultured on to blood agar plates incubated anaerobically. Shaking the cooked meat culture with a little alcohol or benzol before subculture inhibits contaminants. Neomycin sulfate in culture media helps to repress aerobic organisms. However, low concentrations of neomycin sulfate may be inhibitory to *C. botulinum* Type E. Likewise, heat treatment should be used with caution. The selection of colonies is aided by the use of egg yolk agar; with the exception of type G, *C. botulinum* produces opalescence and a pearly layer. Culture on lactose egg yolk milk agar distinguishes proteolytic and nonproteolytic variants. Types C, D, and E produce nonlactose, nonproteolytic fermenting lipolytic colonies. Also there are nonproteolytic variants of Types B and F.

EPIDEMIOLOGY

C. botulinum is unevenly but widely distributed throughout the world. The natural habitat is the soil. Types A and B are the most common around the world. Type E has been found in Japan, Alaska, Canada, Russia, and also near the Baltic Sea. Type G was first isolated by Gimenez and Cicarrelli (1966, 1970) from Argentinian soil. Type E occurs in fish and in mud and shore waters; toxin production may increase to levels that affect fish. Intensive growth of *C. botulinum* may occur in overstocked fishfarms in which the fish are overfed. The infection is more difficult to control when the floor of the pond is earth rather than concrete.

Drought appears to encourage proliferation of the organism in mud and wildlife, such as fish, ducks, and seagulls. In foods, the germination of spores and the growth of vegetative cells are necessary for toxin production. The spores survive cooking, smoking, and salting; nitrite is required in the curing of salt meats to prevent the outgrowth of spores. Home-preserved foods are most frequently responsible for botulism. Examples are home-canned or salted low-acid vegetables, smoked or pickled fish, and pork products cured with salt without the addition of nitrite. Cheese can become toxigenic when acid production is deficient. Processes for commercially canned foods are usually calculated to destroy the spores of *C. botulinum* that are particularly resistant to heat. Outbreaks of botulism attributable to commercially canned foods have involved both underprocessing and postprocessing contamination. Micro-leaks in cans are significant even when chlorinated water is used for cooling, because vegetative cells may be destroyed by chlorine, but spores are not. Thus, adequate methods of food preservation are important in the prevention of botulism in both commercial and home processing of foods. Nonacid foods may be particularly dangerous.

CLOSTRIDIUM NOVYI (C. OEDEMATIENS)

Type D is sometimes called *C. hemolyticum.*

MORPHOLOGY

The organism is a fairly large gram-positive bacillus that is motile and strictly anaerobic. It sporulates freely, and the spores are large and subterminal and expand the rod. In young culture it resembles *C. perfringens.* Colonies are irregular or circular and semitranslucent with a finely lobulated or crenelated edge. The surface is finely granular. Growth may spread.

ANTIGENIC COMPOSITION

There are four types: A, B, C, and D. All strains of *C. novyi* Types A, B, and C share two somatic antigens in varying proportions. Fluorescent-labeled antiserum prepared against Type B organisms stains strains of all types.

METABOLISM

Glucose and maltose are fermented by Types A and B, but only glucose is fermented by Types C and D. All types produce gelatinase but do not attack more complex proteins. Type D produces large amounts of indole, and H_2S is produced by all strains and especially by Type D. Products of metabolism include propionic and butyric acids and small amounts of acetic and valeric acids.

PATHOGENIC PROPERTIES

Types A and B cause gas gangrene in man and animals; Types B and D are responsible for other diseases in animals. Type A is the most common clinical pathogen.

TOXICOLOGY

Eight different soluble toxic antigens have been demonstrated in culture filtrates. Types A and B produce necrotizing lethal α-toxin, which increases capillary permeability and causes the characteristic gelatinous edema of muscle seen in *C. novyi* gas gangrene. Production of α-toxin is apparently bacteriaphage-dependent. Types A and B also produce lecithinase C enzymes that are hemolytic, necrotizing, and lethal (Type B).

LABORATORY DIAGNOSIS

Isolation is often difficult and slow, particularly for Type D. Media should be freshly poured or prereduced. The cysteine dithiothreitol medium of Moore (1968) is recommended for Types B, C, and D. Before culture, the material may be heated at 80° to 100° C for 10 to 15 minutes. Preliminary enrichment in cooked meat broth is of value. Reactions on half-antitoxin lactose egg yolk milk agar is species specific and type specific for Type A strains; there is a diffuse lecithinase C opalescence and a restricted pearly layer due to lipolysis. Strains of Types B and D produce opalescence only, while Type C strains are egg yolk-negative. Animal inoculation and protection tests may be used. Neomycin sulfate may be used in culture media to suppress aerobic organisms.

EPIDEMIOLOGY

The organism is widely distributed in soil and (Types A and B) in the livers of healthy animals. Approximately 42 per cent of the cases of gas gangrene in World War II were due to *C. novyi.*

CLOSTRIDIUM HISTOLYTICUM

MORPHOLOGY

The bacillus is gram-positive, motile, and not an exacting anaerobe. Spores are formed readily except under aerobic conditions. They are large and oval, subterminal, and distend the organism.

The colonies are small, circular, opaque, shiny with an entire edge, and grayish white. There is a narrow zone of hemolysis on horse blood agar.

ANTIGENIC COMPOSITION AND TOXICOLOGY

There is no division into types. Oakley and Warrack (1950) demonstrated three separated soluble antigens in culture filtrates as follows:

α-toxin, lethal and necrotizing.

β-antigen, collagenase attacking azocoll and gelatin.

γ-antigen, proteinase, attacks azocoll, gelatin, and casein also.

δ-antigen, an elastase (Oakley and Banerjee, 1963).

METABOLISM

The organism ferments no sugars. It is strongly proteolytic and attacks gelatin and more complex proteins. H_2S is produced but not indole. Acetic acid is produced as the only volatile product.

PATHOGENIC PROPERTIES

The organism is pathogenic for man and animals. In man it is associated, together with other anaerobes, with gas gangrene infections.

LABORATORY DIAGNOSIS

The organism grows slowly under aerobic conditions when it is atypical and pleomorphic without spores; growth is improved under anaerobic conditions. Colonies are easily recognised on lactose egg yolk milk agar. There is partial clearing, lactose is not attacked, and there is no typical egg yolk reaction. Neomycin sulfate may be used to eliminate aerobic organisms, and differential heating may be used for the same purpose. In cooked meat broth, there is vigorous proteolysis of the meat particles.

EPIDEMIOLOGY

The organism is widely but sparsely distributed in soil and probably also in the intestinal tract of man and animals.

CLOSTRIDIUM SEPTICUM AND CLOSTRIDIUM CHAUVOEI

Because the characteristics of these two organisms are similar, they are described together. Sometimes they are considered as a single species, *C. septicum* Types A and B (*C. chauvoei*).

MORPHOLOGY

They are gram-positive bacilli, particularly in young culture. Spores are oval and subterminal, and they distend the organism. They are strict anaerobes. Colonies of *C. septicum* are small with a coarse rhizoidal edge; they may spread. There is beta hemolysis on horse blood agar. *C. chauvoei* colonies are small, umbonate, irregular or circular, shiny, and semitranslucent.

ANTIGENIC COMPOSITION

C. septicum is divisible into two groups on the basis of the O antigen; neither group cross-reacts with *C. chauvoei*, which possesses a common O antigen (Moussa, 1959). Batty and Walker (1963) found that both organisms could be differentiated by the use of fluorescent antibodies. The swarming growth of *C. septicum* may be prevented by treating plates with a polyvalent *C. septicum* O antiserum before inoculation.

METABOLISM

Both organisms ferment glucose, maltose, and lactose; in addition, *C. septicum* ferments salicin and *C. chauvoei* ferments sucrose. Both produce H_2S but not indole. Both produce gelatinase and deoxyribonuclease. Acetic, butyric, and formic acids are formed.

PATHOGENIC PROPERTIES

C. septicum is pathogenic for man and animals and is associated with gas gangrene in man. *C. chauvoei* is pathogenic for animals only, especially ruminants.

TOXICOLOGY

C. septicum produces three exotoxins. The α-toxin is lethal, necrotizing, and hemolytic; the β-toxin is deoxyribonuclease; and the γ-toxin is hyaluronidase. A neuraminidase and hemagglutinin are also produced.

C. chauvoei also produces α- β- and γ-toxins and an oxygen-labile hemolysin; all appear to be related to the corresponding antigens produced by *C. septicum*.

IMMUNITY

C. septicum antitoxin provides homologous protection and also protection against *C. chauvoei*. *C. chauvoei* antitoxin does not protect against *C. septicum* toxemia, because *C. septicum* produces additional toxins, in particular, the α-toxin, which is unrelated to the α-toxin of *C. chauvoei*.

LABORATORY DIAGNOSIS

Both organisms are easy to grow; *C. chauvoei* is encouraged by the presence of liver in the medium. Preliminary enrichment is helpful and neomycin sulfate may be used to discourage growth of aerobic contaminants; initial heating is also useful. The organisms are egg yolk–negative. In clinical medicine, it is important to distinguish *C. septicum* from *C. chauvoei*. Animal protection tests may be used. In cooked meat medium, both organisms produce small amounts of gas, and the meat particles may become pink.

EPIDEMIOLOGY

Both organisms are found chiefly in soil and also in the intestinal tract of herbivorous animals.

CLOSTRIDIUM SPOROGENES

MORPHOLOGY

The bacillus is strongly gram-positive. Sporulation is profuse, and the spores are oval and subterminal, and they distend the cell. Free spores are common. The colonies are umbonate, opaque with a grayish white center. They are flat, irregular or circular with a rhizoidal edge, and they may spread. There is no hemolysis.

ANTIGENIC COMPOSITION

Soluble antigens of clinical significance are not produced.

METABOLISM

Glucose and mannitol are fermented but not lactose or sucrose. Gelatin and more complex proteins are broken down. H_2S is formed but not indole. A wide range of acids and alcohols are produced, particularly acetic, propionic, isobutyric, isovaleric, and isocaproic acids.

PATHOGENIC PROPERTIES

In mixed clostridial infections *C. sporogenes* increases the virulence of pathogenic species such as *C. perfringens* and *C. septicum*. There are local putrefactive changes in inoculated muscles, especially after damage has occurred; occasionally guinea pigs may die after inoculation.

LABORATORY DIAGNOSIS

Heat treatment and neomycin sulfate may be used to suppress contaminating aerobic organisms. On lactose milk yolk agar, there is restricted opalescence, and a pearly layer develops owing to lipase activity. Growth is surrounded by wide zones of clearing due to proteolysis of the milk. In cooked meat, broth meat particles are attacked and partly digested. A scum of fatty acids may develop on the surface. Cultures have an odor of skatol.

EPIDEMIOLOGY

C. sporogenes is widely distributed in nature, both in soil and in the intestinal tract of man and animals. It is not easily distinguishable from nontoxic sporulating strains of *C. botulinum*.

CLOSTRIDIUM BIFERMENTANS AND C. SORDELLII

MORPHOLOGY

The organisms are strongly gram-positive bacilli, motile, and not exacting in their requirements. They spore readily, and the spores are large and cylindrical, usually central and slightly distending the cell. Chains of sporulating organims are formed, and free spores are common.

The colonies of both organisms are low, convex with an entire but irregular edge, and grayish white; they may swarm. There is a narrow zone of hemolysis on horse blood agar.

ANTIGENIC COMPOSITION

Walker (1963) differentiated the two species on the basis of spore agglutinogens but not spore precipitinogens. There was no cross-agglutination or precipitation between the spores of the *C. bifermentans-sordellii* group and those of *C. sporogenes, C. histolyticum, C. sphenoides,* and *C. perfringens* Types A to D.

METABOLISM

Glucose and maltose are attacked but not lactose or sucrose. In addition, *C. bifermentans* ferments mannose, sorbitol, and salicin; and *C. sordellii* is urease positive. Gelatin and more complex proteins are broken down by both organisms, and H_2S and indole are produced.

Large amounts of acetic, isobutyric, and isovaleric and small amounts of propionic and isocaproic acids are formed.

PATHOGENIC PROPERTIES

Intramuscular inoculation of a pathogenic strain of *C. sordellii* kills guinea pigs in one to two days. Proteolysis of tissues may progress in the same way as that of *C. histolyticum* infections; there is hemorrhage and gas production. Rifkin et al. (1977) describe infection by *C. sordellii* in colitis induced by antibiotics.

TOXICOLOGY

Pathogenic strains of *C. sordellii* produce a lethal toxin for which antitoxic sera are available. All strains of *C. sordellii* and *C. bifermentans* produce a lecithinase C that is antigenically related to, but not identical with, the lecithinase C (α-toxin) of *C. perfringens.* There are indications that at least one other soluble antigen is produced by these organisms.

LABORATORY DIAGNOSIS

Heat treatment and neomycin sulfate in or on solid media help to inhibit aerobic organisms. On lactose egg yolk milk agar there is extensive opalescence that is inhibited by *C. perfringens* Type A antitoxin, but the inhibition is seldom complete. There is no pearly layer; lactose is not fermented; zones of partial clearing indicate that milk is attacked. In cooked meat broth, the meat particles are partially digested, and a viscous mucoid deposit is produced that is peculiar to these organisms.

DRUG SUSCEPTIBILITY

C. sordellii is apparently resistant to clindamycin and sensitive to vancomycin. Thus, clindamycin treatment of patients may induce overgrowth of the bowel with *C. sordellii* and consequent clostridial colitis. This in turn can be treated successfully with orally administered vancomycin.

EPIDEMIOLOGY

Both organisms are widely distributed in the soil and in the large bowel of man and animals.

Acknowledgements

All those whose work and research have enabled this information to be assembled are acknowledged with gratitude. The author and editors are particularly grateful to Dr. A. T. Willis for his clear exposition of the facts in his book *Anaerobic Bacteriology: Clinical and Laboratory Practice,* third edition, and for his willingness to allow some of his material to be abstracted. We are grateful also to Butterworth and Co. (Publishers) Ltd. for their permission.

References

Ayliffe, G. A. J., and Lowbury, E. J. L.: Sources of gas gangrene in hospital. Br Med J 2:333, 1969.

Batty, I., and Walker, P. D.: Differentiation of *Clostridium septicum* and *Clostridium chauvoei* by the use of fluorescent labelled antibodies. J Pathol Bacteriol 85:517, 1963.

Clifford, W. J., and Annellis, A.: *Clostridium perfringens*. I. Sporulation in a biphasic glucose-ion-exchange resin medium. Appl Microbiol 22:856, 1971.

Collee, J. G., Knowlden, J. A., and Hobbs, B. C.: Studies on the growth, sporulation, and carriage of *Clostridium welchii* with special reference to food poisoning strains. J Appl Bacteriol 24:326, 1961.

Dowell, V. R., Jr., Torres-Anjel, M. J., Reimann, H. P., Merson, M., Whaley, D., and Darland, G.: A new criterion for implicating *Clostridium perfringens* as the cause of food poisoning. Rev Lat-Am Microbiol 17:137, 1975.

Duncan, C. E., and Strong, D. H.: Improved medium for the sporulation of *Clostridium perfringens*. Appl Microbiol 16:82, 1968.

Duncan, C. E., and Strong, D. H.: *Clostridium perfringens* type A food poisoning. Response of the rabbit ileum as an indication of enteropathogenicity of strains of *Clostridium perfringens* in monkeys. Infec Immun 3:167, 1971.

Ellner, P. D.: A medium promoting rapid sporulation in *Clostridium perfringens*. J Bacteriol 71:495, 1956.

Gimenez, D. F., and Ciccarelli, A. S.: A new type of *Clostridium botulinum*. In Ingram, M., and Roberts, T. A. (eds.): Botulism. Proceedings of the 5th International Symposium on Food Microbiology, July 1966. London. Chapman and Hall, 1966, p. 455.

Gimenez, D. F., and Ciccarelli, A. S.: Another type of *Clostridium botulinum*. Zentralbl Bakteriol Parasitol Infektionskrankh Hyg Abt I [Orig] 215:221, 1970.

Hauschild, A. H. W.: Criteria and procedures for implicating *Clostridium perfringens* in food-borne outbreaks. Can J Public Health 66:388, 1975.

Hauschild, A. H. W.: Food poisoning by *Clostridium perfringens*. Can Inst Food Sci Technol J 6:106, 1973.

Henderson, D. W.: The somatic antigens of the *Cl. welchii* group of organisms. J Hyg Camb 40:501, 1940.

Hobbs, B. C., Smith, M. E., Oakley, C. L., Warrack, G. H., and Cruickshank, J. C.: *Clostridium welchii* food poisoning. J Hyg Camb 51:75, 1953.

Howard, F. M., Bradley, J. M., Flynn, D. M., Noone, P., and Szawatkowski, M.: Outbreak of necrotizing enterocolitis caused by *Clostridium butyricum*. Lancet 11:1099, 1977.

Hughes, J. A., Turnbull, P. C. B., and Stringer, M. E.: A serotyping system for *Clostridium welchii (Clostridium perfringens)* type A, and studies on the type-specific antigens. J Med Microbiol 9:475, 1976.

Klein, E.: On a pathogenic anaerobic intestinal bacillus, *Bacillus enteritidis sporogenes*. Zantralbl Bakteriol 1 Abt 18:737, 1895.

Knox, R., and MacDonald, E. J.: Outbreaks of food poisoning in certain Leicester institutions. Med Off 69:21, 1943.

Lawrence, G., and Walker, P. D.: Pathogenesis of enteritis necroticans in Papua New Guinea. Lancet 1:125, 1976.

Leader.: Clostridia as intestinal pathogens. Lancet ii:1113, 1977.

McClung, L. S.: Human food poisoning due to growth of *Clostridium perfringens (Cl. welchii)* in freshly cooked chickens: Preliminary note. J. Bacteriol 50:229, 1945.

McClung, L. S., and Toabe, R.: Egg-yolk plate reaction for presumptive diagnosis of *Clostridium sporogenes* and certain species of gangrene and botulinum groups. J. Bacteriol 53:139, 1947.

Mead, C. C.: Growth and sporulation of *Clostridium welchii* in breast and leg muscle of poultry. J. Appl Bacteriol 32:86, 1969.

Mersom, M. H., and Dowell, V. R.: Epidemiologic clinical and laboratory aspects of wound botulism. N. Engl J Med 289:1005, 1973.

Midura, T. F., and Arnon, S. S.: Infant botulism. Identification of *Clostridium botolinum* and its toxins in feces. Lancet 11:934, 1976.

Miles, A. A., and Misra, S. S.: The estimation of the bactericidal power of blood. J Hyg Camb 38:732, 1938.

Moore, W. B.: Solidified media suitable for the cultivation of *Clostridium novyi* type B. J Gen Microbiol 53:415, 1968.

Moussa, R. S.: Antigenic formulae for *Clostridium septicum* and *Clostridium chauvoei*. J Pathol Bacteriol 77:341, 1959.

Murrell, T. G. C., Egerton, J. R., Rampling, A., Samuels, J., and Walker, P. D.: The ecology and epidemiology of the pig-bel syndrome in man in New Guinea. J Hyg Camb 64:375, 1966.

Nagler, F. P. O.: Observations on a reaction between the lethal toxin of *Cl. welchii* (Type A) and human serum. Br J Exp Pathol 20:473, 1939.

Niilo, L.: Enterotoxin formation by *Clostridium perfringens* type A studied by the use of fluorescent antibody. Can J Microbiol 23:908, 1977.

Oakley, C. L., and Banerjee, N. G.: Bacterial elastases. J Pathol Bacteriol 85:489, 1963.

Oakley, C. L., and Warrack, G. H.: The ACRA test as a means of estimating hyaluronidase deoxyribonuclease and their antibodies. J Path Bact 63:45, 1951.

Oakley, C. L., and Warrack, G. H.: Routine typing of *Clostridium welchii*. J Hyg Camb 51:102, 1953.

Parker, M. T.; Clostridial sepsis. Br Med J 2:698, 1967.

Rifkin, G. D., Fekety, F. R., Silva, J., and Sack, R. B.: Antibiotic-induced colitis: Implications of a toxin neutralized by *Clostridium sordellii* antitoxin. Lancet ii:1103, 1977.

Schantz, E. J., and Sugyama, H.: Toxic proteins produced by *Clostridium botulinum*. Agric Food Chem 22:26, 1974.

Skjelkvåle, R., and duncan, C. L.: Enterotoxin formation by different toxigenic types of *Clostridium perfringens*. Infect Immun 11:563, 1975.

Stringer, M. F., Shah, N., and Gilbert, R. J.: Serological typing of *Clostridium perfringens* and its epidemiological significance in the investigation of food poisoning outbreaks. Proceedings of IAMS meeting in Szczecin, Poland, 1978.

Sutton, R. G. A.: The pathogenesis and Epidemiology of *Clostridium welchii* Food Poisoning. Thesis (Ph.D), University of London, 1969.

Sutton, R. G. A., Ghosh, A. C., and Hobbs, B. C.: Isolation and enumeration of *Clostridium perfringens*. In Shapton, D. A., and Board, R. G. (eds.): Isolation of Anaerobes. Soc Appl Bact Tech Ser No. 5. London, Academic Press, 1971, p. 39.

Torres-Anjel, M. J., Reimann, H. P., and Tsai, Che C.: A Fluorescent-Antibody Technique for the Detection of Enterotoxin-Producing Cells of *Clostridium perfringens* Type A. Scientific Publication No. 305. New York, Pan American Health Organization (World Health Organization), 1975.

Torres-Anjel, M. J., and Reimann, H. P.: Enterotoxic *Clostridium perfringens* in selected humans. II. A cohort study. Rev Lat Am Microbiol 17:199, 1975.

van Heyningen, W. E.: The pathogenic actions of exotoxins. Zentralbl Bacteriol Parasitol Infect Hyg Abt I [Orig] 212:191, 1970.

van Heyningen, W. E.: Tetanus. Sci Am 218:69, 1968.

Vernon, E.: Food poisoning and salmonella infections in England and Wales 1973–1975. An analysis of reports to the Public Health Laboratory Service. Public Health 91:225, 1977.

Walker, P. D.: The spore antigens of *Clostridium sporogenes, Cl. bifermentans* and *Cl. sordellii*. J Pathol Bacteriol 85:41, 1963.

Walker, P. D., and Batty, I.: Fluorescent studies in the genus *Clostridium*. II. A rapid method for differentiating *Clostridium botulinum* types A, B, and F, types C and D, and type E. J Appl Bacteriol 27:140, 1964.

Willis, A. T.: Anaerobic Bacteriology: Clinical and Laboratory Practice. Third edition. Butterworth & Co. (Publishers) Ltd. 1977.

Willis, A. T., and Williams, K.: Prevention of swarming of *Clostridium septicum*. J Med Microbiol 5:493, 1972.

Zeissler, J., and Rassfeld-Sternberg, L.: Enteritis necroticans due to *Clostridium welchii* type F. Br Med J 1:267, 1949.

46 BACTEROIDES

Tor Hofstad, M.D.

The genus *Bacteroides* comprises a heterogeneous group of nonspore-forming, obligately anaerobic rods that often are very pleomorphic. Their habitat is the mucous membranes of man and animals. They can be differentiated from *Fusobacterium* by their inability to produce substantial amounts of butyric acid from glucose.

The 8th edition of *Bergey's Manual of Determinative Bacteriology* (1974) lists 22 species of *Bacteroides.* A few of these, namely *B. fragilis, B. melaninogenicus,* and *B. corrodens,* are pathogenic in man. Strains of *Bacteroides* species of uncertain pathogenicity have been isolated from clinical specimens together with other anaerobic or facultative bacteria. At present, *B. fragilis* is subdivided into the subspecies ss. *fragilis,* ss. *distasonis,* ss. *vulgatus,* ss. *thetaiotaomicron,* and ss. *ovatus.* The different subspecies are, however, phenotypically and genotypically distinct. *B. melaninogenicus* is a heterogene species containing saccharolytic and nonsaccharolytic strains. *Bacteroides* strains classified as *B. oralis* are nonpigmented, saccharolytic *B. melaninogenicus strains.*

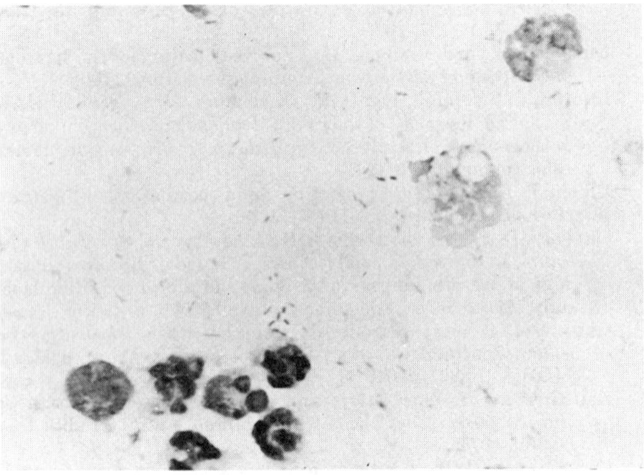

FIGURE 1. B. fragilis *in a smear from a liver abscess.*

MORPHOLOGY

Cells of *B. fragilis* are nonmotile, pale-staining, gram-negative straight or slightly curved rods with rounded ends, 0.5 to 1.0 μm in diameter and 0.5 to 5.0 μm long, occurring singly and in pairs (Fig. 1). Irregular staining, which is often bipolar, is frequently seen. Cells grown in fluid media containing a fermentable carbohydrate tend to be more pleomorphic than cells grown in solid media. Short filaments and, in particular, vacuolated cells may be seen in such cultures. Ovoid forms are common in some strains. *B. melaninogenicus* is a nonmotile, even-staining, gram-negative coccobacillus, 0.3 to 0.4 μm wide and 0.6 to 2.0 μm long (Fig. 2). Fluid cultures of some strains show larger, pale, and highly vacuolated cells, with some densely stained areas. Cells of *B. corrodens* are nonmotile, pale-staining, gram-negative rods, 0.5 to 0.7 μm in width and 1.0 to 3.0 μm long, occurring singly or in pairs. Filaments are occasionally seen. Most other *Bacteroides* species are characterized by short to medium-sized pleomorphic rods.

The ultrastructure of *Bacteroides,* as seen in the electron microscope, is that of a gram-negative bacterium. The cell wall has an electron-dense, solid layer containing the peptidoglycan and an outer, trilaminar membrane made up of protein, lipid, and polysaccharide.

ANTIGENIC COMPOSITION

The best studied *Bacteroides* antigens are the cell wall lipopolysaccharides of *B. fragilis* and *B. melaninogenicus* (Hofstad, 1974). These are high molecular weight complexes composed of polysaccharide, lipid, and small amounts of protein. The lipopolysaccharides exhibit serologic specificity

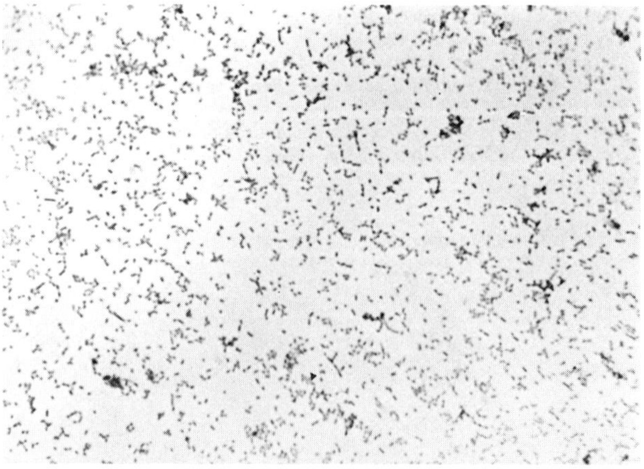

FIGURE 2. B. melaninogenicus *from blood agar.*

analogous to the O antigens of Enterobacteriaceae. Thus each *B. fragilis* strain shares several O-antigenic specificities with other *B. fragilis* strains and, occasionally, with other species of *Bacteroides*. All antigenic determinants are carried by the polysaccharide part of the lipopolysaccharide macromolecule. Beyond this, the chemical basis of the antigenic specificity is unknown. The lipopolysaccharides of *B. fragilis* and *B. melaninogenicus* seem to contain the same sugars, i.e., glucose, galactose, rhamnose, fucose, glucosamine, and galactosamine. Heptose and 2-keto-3-deoxyoctonate, which are essential constituents of most other bacterial lipopolysaccharides, are not present in these two or in any other species of *Bacteroides*. The O antigens are best examined by indirect hemagglutination and hemagglutination inhibition.

A group-specific capsular antigen from *B. fragilis* ss. *fragilis* has a molecular weight of 800,000 and is composed of hexose, hexosamine, methylpentose, and sialic acid (Kasper, 1976).

The outer membrane of *B. fragilis* contains a protein antigen that reacts with precipitation and is apparently group-specific. Several other precipitating antigens, the nature of which is unknown, have been observed in extracts of *B. fragilis* and *B. melaninogenicus* strains.

GROWTH AND METABOLISM

The clinically important *Bacteroides* species grow well anaerobically on blood agar and in enriched media containing peptone and yeast extract. Surface colonies of *B. fragilis* on sheep or horse blood agar are after two days of incubation, 1 to 3 mm in diameter, circular, entire, convex, semiopaque, grayish to whitish, and nonhemolytic. Colonies of some strains may be soft and shiny. *B. melaninogenicus* organisms grow on blood agar with entire, circular, convex to pulvinate colonies, which after incubation for two days measure 0.5 to 2 mm in diameter. Old colonies are often surrounded by flattened margins. Colonies of some strains are shiny or slightly mucoid. Colonies are whitish or slightly gray in color when young, but after two or three days of incubation they take on a slightly tan to brownish pigment, and soon become jet black. The pigment is formed sooner on agar that contains laked blood rather than intact erythrocytes. The tan and light brown colonies fluoresce pink or orange, and the darker colonies show red under ultraviolet light (365 nm). Entirely black colonies do not fluoresce. The colonies of *B. melaninogenicus* are β-hemolytic or nonhemolytic. Surface colonies of *B. corrodens* range from barely visible to 0.5 mm in diameter after two days of incubation. Upon continued incubation they grow to about 1 mm in diameter and are gray-white, convex or umbonate circular colonies with slightly undulating or entire margins that seem to disappear into the agar. Their ability to corrode, or pit, the agar may be lost after continued cultivation in fluid media. Colonies of most other *Bacteroides* species resemble those of *B. fragilis* but are usually smaller.

In fluid media growth of bacteroides causes turbidity and, frequently, a sediment that may be smooth, ropy or stringy. A small amount of gas is produced by most *B. fragilis* strains.

The bacteroides grow best at 37° C and at a pH of about 7.0. With some exceptions they are moderate anaerobes that grow maximally in oxygen tensions of up to 3 per cent. *B. fragilis* seems to be the most oxygen-tolerant species. *B. ochraceus,* a member of the normal oral microflora, grows in 5 to 10 per cent carbon dioxide in air.

The energy metabolism of *Bacteroides* is virtually unknown. There is evidence that *B. fragilis* has a primitive type of electron transport system with cytochrome B that is linked to the reduction of fumarate. The asaccharolytic *B. corrodens* seems to gain energy by the transfer of electrons from hydrogen or formate to fumarate.

B. fragilis has simple nutritional requirements. The organism has an obligate need of vitamin B_{12}, and the growth of most strains is stimulated by hemin and bile. Ammonium is the preferred nitrogen source rather than amino acids. A fermentable carbohydrate is essential for good growth. *B. melaninogenicus* and most other *Bacteroides* species are more fastidious anaerobes. *B. melaninogenicus* has an obligate requirement for hemin and for peptides, and many strains need vitamin K. The organism grows well in media based on tryptone and tryptose. Serum or L-asparagine often stimulates growth. Bile is growth-inhibitory. Fumaric acid has been found to stimulate growth of *B. corrodens*. This organism grows best in a liver digest medium with added yeast extract and blood. Strains of several *Bacteroides* species require carbon dioxide.

Most bacteroides metabolize carbohydrates. The fermentation products include combinations of acetic, lactic, succinic, propionic, isobutyric, isovaleric, and formic acids. Butyric acid is formed by a few species but not as a major product. *B. fragilis* is highly saccharolytic, giving a final pH in glucose broth of about 5.0 to 5.4. *B. corrodens* does not ferment carbohydrates. Asaccharolytic strains of *B. melaninogenicus* are highly proteolytic. The organism can hydrolyze casein, gelatin, plasma proteins, and collagen. Indole is produced by some *Bacteroides* species or subspecies. Catalase, lecithinase, and lipase are usually not produced. *B. fragilis* and most other bacteroides hydrolyze aesculin. Dehydrogenases specific for different amino acids and other organic acids are present in *Bacteroides* and have been used for differentiation between members of the species. β-Lactamase (Pinkus et al., 1968) and neuraminidase are produced by *B. fragilis*.

PATHOGENIC PROPERTIES

The pathogenic *Bacteroides* species are part of the normal microflora of mucous membranes, particularly those of the gastrointestinal and upper respiratory tracts. They are harmless inhabitants unless the mucous membrane barrier is broken. The bacteroides, like other microorganisms on the mucosa, may then invade the adjacent tissue and cause localized infection or spread hematogenously to the liver, lung, or brain. Following aspiration they may be carried from the mouth into the lung. These endogenous infections are not communicable from one person to another.

A requisite for *Bacteroides* infection is lowering of the redox potential by tissue necrosis, impaired blood supply, and growth of facultative microorganisms in an infected area. The importance of underlying disease was illustrated in a study of 112 patients with bacteroidaceae bacteremia, caused mainly by *B. fragilis* (Chow and Guze, 1974). Forty-three per cent of the patients had either malignancy, arteriosclerosis, alcoholic liver disease, diabetes mellitus, or end-stage renal disease. Underlying disease was the overriding determinant of mortality. The most significant factors in nonfatal infections were old age and a history of trauma or surgery.

The most common serious anaerobic infections encountered are caused by members of the family Bacteroidaceae, especially *B. fragilis*. This organism may be implicated in suppurative lesions of almost every part of the human body, and is the anaerobic microorganism isolated most often from microbial infections. It is also the anaerobic gram-negative rod found most frequently in pure culture. Usually, however, *B. fragilis* and other bacteroides are accompanied by facultative or other anaerobic bacteria such as peptostreptococci. Other bacteria are almost universally present in infections with *B. melaninogenicus*, probably because they satisfy its need for vitamin K.

B. fragilis is now isolated from two thirds or more of intra-abdominal infections. Other *Bacteroides* species are isolated less often. *B. fragilis* and *B. melaninogenicus* are each found in 25 per cent of patients with anaerobic pleuropulmonary disease (empyema, lung abscess, pneumonitis), and in 25 per cent of those with infections of the female genital tract (Ledger et al., 1971). *Bacteroides* species are often isolated from abscesses of the liver and the brain. *B. fragilis* is the anaerobic organism isolated most often from soft tissue infections. In septicemic patients *B. fragilis* is isolated in 5 to 10 per cent of all positive blood cultures. The common portals of entry are the gastrointestinal and female genital tract. *Bacteroides* species may also be implicated in decubital ulcers (Rissing et al., 1974) and gangrene, in dental infections, and in chronic infections of the paranasal sinuses and the middle ear. Occasionally *Bacteroides* species are isolated from osteomyelitis, purulent arthritis, infective endocarditis, and urinary tract infections. *B. corrodens* has been found in oral and upper respiratory infections and in blood cultures.

Approximately 80 per cent of the *B. fragilis* strains isolated from anaerobic infections belong to subspecies ss. *fragilis*. The other four *B. fragilis* subspecies are less important as human pathogens. *B. fragilis* ss. *ovatus* does not seem to occur in systemic infections.

Pure cultures of *B. fragilis* or other *Bacteroides* species are noninfective or only slightly infective in experimental animals. Little is known, therefore, about the pathogenesis of *Bacteroides* infections. Synergistic mechanisms are thought to play a role in mixed infections. Two such mechanisms are known: production of a low oxidation-reduction potential by facultative organisms, and supply of an essential growth factor. A capsule in *B. fragilis,* particularly ss. *fragilis* and in some *B. melaninogenicus* strains may protect them from phagocytosis. The cell wall lipopolysaccharides of *B. fragilis* and *B. melaninogenicus* have low toxicity compared to endotoxins of the facultative enteric bacilli (Sveen et al., 1977). Certain enzymes, like neuraminidase, deoxyribonuclease, and heparinase may also contribute to pathogenicity. Heparinase may contribute to the septic thrombophlebitis observed in serious *B. fragilis* infections (Mitre and Rotheram, 1974). *B. melaninogenicus* is one of the few microorganisms that produce collagenase.

IMMUNITY

Low titers of IgM antibodies against O antigens from *B. fragilis* and *B. melaninogenicus* are present in normal human serum, but their protective function is not known. Normal human serum kills *B. fragilis* isolated from stools, and polymorphonuclear leukocytes can phagocytose it with the help of heat-labile opsonins (Bjornson and Bjornson, 1978). A rise in IgM and IgG antibodies in patients has been observed after *B. fragilis* septicemia and local suppurative lesions. The significance of these humoral antibodies in recovery from infection is unknown.

LABORATORY DIAGNOSIS

Pus tissue biopsies from normally sterile sites, transtracheal aspirates, pleural and ascitic fluids, aspirates from the nasal sinuses, carefully collected materials from the root canals and periapical tissues of teeth, cervical material collected by direct visualization, and specimens of blood should be cultured routinely for *Bacteroides* and other anaerobes. It is imperative to avoid contamination with normal flora. Whenever possible specimens should be taken by aspiration in large volumes. Purulent exudate is in itself a good transport medium.

Processing of Specimens

A Gram-stained smear should always be examined for characteristic gram-negative rods that do not appear in aerobic culture. The specimen is streaked on a nonselective solid medium, preferably Brucella blood agar (tryptose agar with glucose added), and on laked blood agar containing 75 μg/ml of kanamycin and 7.5 μg/ml of vancomycin, which is selective for *Bacteroides*. The plates are incubated for two days in GasPak jars or in ordinary anaerobic jars in an atmosphere of nitrogen and hydrogen with 5 to 10 per cent carbon dioxide. An enriched liquid medium, for instance chopped-meat glucose, is also inoculated. If prereduced anaerobically sterilized (PRAS) media are not available, semifluid media can be used.

Identification

A presumptive identification of *B. fragilis* and *B. melaninogenicus*, the *Bacteroides* species encountered most frequently, can be made on the basis of cellular morphology, colonial growth, antibiotic susceptibility patterns, and ability to grow in a medium containing 20 per cent ox bile. A scheme for the identification of *B. fragilis* subspecies and *B. melaninogenicus* is shown in Table 1. The scheme, which permits the distinction between the two *Bacteroides* species and the most common *Fusobacterium* species, can be used in any bacteriologic laboratory. Fermentation of carbohydrates and hydrolysis of esculin should be performed in media containing peptone and yeast extract. The indole reaction can be carried out as a spot test or in a medium containing tryptone or tryptose, which are peptones rich in tryptophan. Rapid fermentation of the *B. fragilis* subspecies can be tested by using large inocula in small volumes of carbohydrate-containing media. Also, micromethod multitest systems for identification of *Bacteroides* species and other anaerobic bacteria pathogenic in man are sold commercially. *B. corrodens* produces urease, reduces nitrate to nitrite, and pits the agar. The pitting may take days to develop or may not occur.

By means of indirect fluorescence, *B. fragilis* ss. *fragilis* may be identified with antiserum from rabbits immunized with capsular material from

TABLE 1. Distinguishing Characters of the Most Common *Bacteroides* and *Fusobacterium* Species

SPECIES	MORPHOLOGY: Cellular	MORPHOLOGY: Colony	GROWTH INHIBITION: Penicillin[a] (2 units/ml)	GROWTH INHIBITION: 20% Bile	BIOCHEMICAL TESTS: Indole Production	Esculin Hydrolysis	Fermentation of: *Glucose*	*Rhamnose*	*Mannitol*	*Trehalose*
B. fragilis	Short rods with rounded ends, may be ovoid or vacuolated, may show irregular staining	Convex, white to gray, glistening. May be slightly hemolytic								
ss. fragilis			−	−	−	+	+	−	−	−
ss. vulgatus			−	−	−	+	+	+	−	−
ss. thetaiotaomicron			−	−	+	+	+	V	−	+
ss. distasonis			−	−	−	+	+	V	−	+
ss. ovatus			−	−	+	+	+	V	+	V
B. melaninogenicus	Coccobacilli or short rods with rounded ends	Tan to black, or gray-white (*B. oralis*), hemolytic	+	+	V	−	V	−	−	−
B. corrodens	Slender rods with rounded ends	Pin-point, corroding, nonhemolytic	+	+	−	−	−	−	−	−
F. nucleatum	Slender, spindle-shaped bacilli with tapered ends (fusiform)	Convex, slightly irregular, grayish, glistening	+	+	+	−	−	−	−	−
F. necrophorum	Pleomorphic, often beaded, rods, with rounded to tapered ends	Convex, often raised opaque center and translucent entire edge. Hemolytic	+	+	+	−	−	−	−	−

[a]Growth inhibition = zones <10 mm; +, positive; −, negative; V, variable.

Modified from Bartlett et al.: Anaerobic Bacteria and Disease. Kalamazoo, Mich., The Upjohn Company, 1975.

that organism. Most sera from rabbits immunized with whole bacterial cells appear to be species specific but do not discriminate among the subspecies. Double diffusion in agar and whole cell agglutination can differentiate only incompletely among the subspecies of *B. fragilis*.

DRUG SUSCEPTIBILITY

All *B. fragilis* subspecies have a similar susceptibility pattern to antibacterial drugs. *B. fragilis* is the anaerobic microorganism most resistant to antimicrobial agents. Of particular importance is the fact that *B. fragilis* is resistant to the aminoglycoside antibiotics, and is resistant to penicillins and first generation cephalosporins. Clindamycin usually inhibits *B. fragilis* in a concentration of 0.2 μg/ml, and for the time being is the drug of choice for treatment of anaerobic infections with *B. fragilis*. The antibiotic is active also against most other anaerobes and has produced good results in the treatment of anaerobic infections. It is about ten times more active against *B. fragilis* than the parent compound lincomycin, but is not consistently bactericidal and does not penetrate well into the central nervous system. Serious colitis is seen sometimes in patients treated with clindamycin. Metronidazole, which inhibits *B. fragilis* at 3 μg/ml or less, is bactericidal in the same concentraction against anaerobic bacteria including *B. fragilis* (Finegold and Nastro, 1972) but inactive against aerobic bacteria. The drug has a low toxicity and crosses the blood-brain barrier, but prolonged administration in large doses has caused tumors in an animal model. It is also mutagenic in bacteria. *B. fragilis* is regularly inhibited by chloramphenicol at a level of 5 to 8 μg/ml and by erythromycin in concentrations of 2.0 μg/ml. The β-lactamase resistant cephalosporin, cefoxitin, is active against *B. fragilis* (Dailand and Birnbaum, 1977). A relatively small percentage of *B. fragilis* isolates, varying from one geographic area to another, is susceptible to tetracycline in a concentration of as low as 1.0 μg/ml.

The susceptibility pattern of other *Bacteroides* species is insufficiently known. Most isolates have been highly susceptible to the commonly used antibiotics, including penicillin. A few isolates of *B. melaninogenicus* and *Bacteroides* species other than *B. fragilis* have proved to be relatively or highly resistant to penicillin. Development of resistance to antimicrobial agents may occur among the *Bacteroides* and may be due to R-factors.

EPIDEMIOLOGY AND ECOLOGY

In human subjects on a Western diet, *B. fragilis* is the predominant organism of the lower intestinal tract, amounting to 10^{11} or more organisms per gram dry weight of feces. The great majority is made up of ss. *vulgatus* and ss. *thetaiotaomicron*. Subspecies *fragilis* is present in a comparatively low number. *Bacteroides* is normally found in a small number in the terminal ileum. In patients with a continent ileostomy or with intestinal blind loops, there is a bacterial overgrowth, a substantial part of which is caused by *B. fragilis*. The overgrowth is regularly associated with vitamin B_{12}, malabsorption, and steatorrhea (stagnant loop syndrome) (Farrar et al., 1975). *B. fragilis* is present also in the normal urethral, vaginal, and cervical flora, and in the upper respiratory tract.

B. melaninogenicus constitutes 4 to 8 per cent of the cultivable microflora of the gingival crevices and is the predominating *Bacteroides* species of the mouth. It becomes established in the mouth upon eruption of the permanent teeth. The organism is a member of the normal flora of the upper respiratory tract, vagina, external male and female genitalia, and skin between the toes.

B. corrodens inhabits the mouth and the upper respiratory tract. The principal habitat of other *Bacteroides* species occasionally isolated from anaerobic infections seems to be the lower intestinal tract.

Bacteroides is frequently involved in hospital infections. Infection may be facilitated by radiation and surgery, and treatment with cytotoxic drugs. A substantial portion of *Bacteroides* infections is associated with surgery. The use of antibiotics that select resistant anaerobes such as *B. fragilis* is thought to predispose to infection.

A significant morbidity and mortality attend infections with *Bacteroides*. Prophylaxis would require avoidance of any condition that reduces the normal redox potential of the tissues and the introduction of the normal mucosal microflora into wounds, closed cavities, or other sites that are normally sterile. Prophylactic antimicrobial therapy may be of value in surgery involving the colon and the female genitalia.

References

Bjornson, A., and Bjornson, H.: Participation of immunoglobulin and the alternative complement pathway in opsonization of *Bacteroides fragilis* and *Bacteroides thetaiotaomicron*. J Infect Dis 138:351, 1978.

Chow, A., and Guze, L.: Bacteroidaceae bacteremia: Clinical experiences with 112 patients. Medicine 53:93, 1974.

Dailand, G., and Birnbaum, J.: Cefoxitin resistance to beta-lactamase: A major factor for susceptibility of *Bacteroides fragilis* to the antibiotic. Antimicrob Agents Chemother 11:725, 1977.

Farrar, W., O'Dell, N., Achord, J., and Greer, H.: Intestinal microflora and absorption in patients with stagnation-inducing lesions of the small intestine. Am J Dig Dis 17:1065, 1975.

Finegold, S., and Nastro, L.: Bactericidal activity of five antimicrobial agents. J Infect Dis 126:104, 1972.

Hofstad, T.: Endotoxins of anaerobic gram-negative microorganisms. In Balows, A. (ed.): Anaerobic Bacteria. Springfield, Ill., Charles C Thomas, 1974, pp. 295–305.

Kasper, D. L.: The polysaccharide capsule of *Bacteroides fragilis* subspecies *fragilis:* Immunochemical and morphologic definition. J Infect Dis 133:79, 1976.

Ledger, W., Sweet, R., and Headington, J.: Bacteroides species as a cause of severe infections in obstetric and gynecologic patients. Surg Gynecol Obstet 133:837, 1971.

Mitre, R., and Rotheram, E.: Anaerobic septicemia from thrombophlebitis of the internal jugular vein. JAMA 230:1168, 1974.

Pinkus, G., Veto, G., and Braude, A. I.: Bacteroides penicillinase. J Bacteriol 96:1437, 1968.

Rissing, J., Crowder, J., Dunfee, T., and White, A.: Bacteroides bacteremia from decubitus ulcers. S Med J 67:1179, 1974.

Sveen, K., Hofstad, T., and Milner, K. C.: Lethality for mice and chick embryos, pyrogenicity in rabbits and ability to gelate lysate from amoebocytes of *Limulus polyphemus* by lipopolysaccharides from *Bacteroides, Fusobacterium* and *Veillonella.* Acta Pathol Microbiol Scand (B) 85:388, 1977.

FUSOBACTERIA 47

Howard Robert Attebery, D.D.S.

Thirteen recognized species in the genus *Fusobacterium* have been found to be associated with human clinical infections (Moore and Holdeman, 1974). Four of these species, *F. mortiferum, F. necrophorum, F. nucleatum,* and *F. varium*, are of primary concern because they are the major isolates recovered from clinical infections. The other species are rarely or infrequently found and will not be discussed individually. For more information, refer to Moore and Holdeman, 1974; Holdeman, Cato, and Moore, 1977; and Finegold, 1977.

MORPHOLOGY

Fusiform is the term used to describe thin, gram-negative anaerobic rods with pointed ends. Fusiforms are not all members of the genus *Fusobacterium*, and all fusobacteria are not fusiform shaped. *Fusobacterium nucleatum* organisms are fusiform, slender, and spindle shaped with tapered ends. *F. varium* organisms are coccobacillary to bacillary with rounded ends. The variability of size and shape may make one suspect a mixed culture. *F. necrophorum* and *F. mortiferum* are heavy bodied bacteria with rounded ends. Pleomorphic cells may be short or elongated. Terminal or central swellings are common. *F. mortiferum* is slightly larger than *F. necrophorum* and more pleomorphic.

The colonies and microscopic morphology of *F. nucleatum* offer a unique combination that assures recognition of this species. The colonies are circular, convex, and translucent with a "flecked" appearance. Mature colonies on blood agar produce a greenish discoloration of the surrounding agar when exposed to air for 15 minutes.

Colonies of *F. varium* are minute to 1 mm in diameter, circular with entire edges, flat to low convex, and translucent with gray-white centers and colorless edges. Colonies of *F. necrophorum* are 1 to 2 mm in diameter, circular with irregular to erose edges, convex to umbonate, frequently with an even surface. A mosaic interior can be seen with transmitted light.

Colonies of *F. mortiferum* are 1 to 2 mm in diameter, circular with entire to irregular edges, convex to umbonate, translucent, and smooth.

ANTIGENIC COMPOSITION

There are few reports on the immunochemistry and the antigenic structure of *Fusobacterium.* Purified lipopolysaccharides with endotoxic activity exhibit a high degree of serologic specificity so that several serologic types can be distinguished (De Araujo et al., 1963). A carbohydrate, probably the O antigen, determines the serologic specificity (Kristoffersen et al., 1971). Protein containing antigens that may be associated with the outer membrane are responsible for broad cross-reactivity among strains of fusobacteria. The sera of a majority of normal people contain antibody to this antigen (Kristoffersen and Hofstad, 1970).

METABOLISM

Fusobacteria obtain their energy for growth by metabolizing carbohydrates or peptones. Large amounts of butyric acids with small amounts of acetic and lactic acids are produced. Small amounts of propionic, succinic, and formic acids may be produced as well as short-chained alcohols. Pyruvate is converted predominantly to butyrate and acetate. Some strains convert threonine to propionate. All strains are strong ammonia producers. Some strains produce indole. Catalase is usually not produced. Nitrates are rarely reduced. The production of hydrogen sulfide gas, volatile amines, aldehydes, and large amounts of butyric acid give cultures a malodorous character and infections a putrid odor.

Fusobacteria are obligate anaerobes. Although they do not require an extremely low oxidation-reduction (redox) potential, some strains are very sensitive to peroxides and cannot survive exposure to air for even a few minutes. Carbon dioxide gas or bicarbonate stimulates growth. Some strains require purines, pantothenic acid, or tryptophan.

Some fusobacteria synthesize and store intracellular polysaccharides.

PATHOGENIC PROPERTIES

Because fusobacteria are part of the normal mammalian flora, they are often isolated from mixed infections with facultative bacteria or other anaerobes such as *Bacteroides melaninogenicus, B. fragilis,* or anaerobic cocci. Although it is difficult to assess the pathogenic role of normal flora in mixed infections, it is clear that fusobacteria are capable of causing disease. Fusobacteria have been isolated many times in pure culture from the blood as well as from infections of the skin and mucous membranes. *F. nucleatum* was isolated from 18 per cent of anaerobic pleuropulmonary infections in one hospital series and was recovered four times in pure culture (Bartlett and Finegold, 1972). Fusobacteria are also one component of the fusospirochetal synergistic infection that has been demonstrated in experimental animals (Lewis and Barenberg, 1929; Smith, 1930), and *F. nucleatum* is one of the components of acute necrotizing ulcerative gingivitis (see Chapter 91, Vincent's infection). Another indicator of the pathogenicity of fusobacteria is the occurrence of metastatic lesions at sites distant from the primary lesion.

The only pathogenic factor of fusobacteria that has been studied is its lipopolysaccharide. Endotoxins have been isolated from several species of fusobacteria, and a recent study of the endotoxin of *F. necrophorum* has shown it to be very potent in all endotoxin assays (Garcia et al., 1975). The endotoxins of fusobacteria have been implicated in human periodontal disease and in fatal cases of septic shock in man and animals.

Hemolytic strains of *F. necrophorum* are isolated more frequently from serious infections than nonhemolytic strains and are more pathogenic in experimental infections of mice. It is not known whether hemolysin is directly responsible for the increased virulence.

The mechanism of symbiosis between fusobacteria and spirochetes in mixed infections is unknown, but these synergistic infections usually induce necrotic lesions. Purulent tissue reactions are more common when fusobacteria are isolated in pure culture or in association with other bacteria from infections such as periapical dental abscess, brain abscess, or mastoiditis.

IMMUNITY

Little is known about specific immune responses to colonization or infection with fusobacteria. The sera of most people contain antibody to protein-containing antigens of fusobacteria (Kristoffersen and Hofstad, 1970), and the oligosaccharide side chains of fusobacterial endotoxins induce antibody production in experimental animals (De Araujo et al., 1963). The capacity of these antibodies to protect against invasion of the tissues by fusobacteria is unknown. Fusobacteria are probably sensitive to killing by serum complement and to ingestion and killing by phagocytes.

The most important defense against invasion of the tissues by all anaerobes is the redox potential of well-oxygenated tissue (about 150 microvolts). Obligate anaerobes like the fusobacteria cannot propagate at this redox potential. Consequently, death of tissue after trauma, surgery, arterial insufficiency, and infection by aerotolerant bacteria are important predisposing factors to fusobacterial infection.

The growth of facultative bacteria in the tissues may promote fusobacterial infection by lowering the redox potential, whereas other members of the normal flora may limit the growth of fusobacteria. The production of hydrogen peroxide and the superoxide ion during aerobic growth of competing microflora inhibits fusobacteria, since they have little or no catalase or superoxide dismutase.

LABORATORY DIAGNOSIS

Fusobacteria are gram-negative obligate anaerobes that produce large amounts of butyric acid from fermentable carbohydrates or metabolized peptones (with little or no isobutyric or isovaleric acids). The differential reactions of the four major species are listed in Table 1.

F. necrophorum is the only *Fusobacterium* species that produces lipase. Not all *F. necrophorum* strains produce lipase, however. A lipase-

TABLE 1. Differential Culture Reactions of The Major Fusobacteria[a]

CULTURE REACTION	F. NECROPHORUM	F. NUCLEATUM	F. MORTIFERUM	F. VARIUM
Microscopic morphology	L-form like	spindle-shaped	L-form like	small, variable
Beta-hemolysis	+	−	−	−
Indole	+	+	−	$+^{-}$
Esculin hydrolysis	−	−	+	−
Nitrate	−	−	−	−
Growth in 20% bile	$-^{2+}$	$-^{2+}$	4+	4+
Gas in glucose agar deeps	4+	$-^{2+}$	4+	4+
Threonine → propionate	+	+	+	+
Lactate → propionate	+	−	−	−
Lipase	+	−	−	−
Fermentation				
glucose	$-^{w}$	$-^{w}$	$+^{w}$	w^{+}
lactose	−	−	w^{+}	−
fructose	$-^{w}$	w^{-}	W^{+}	w
maltose	−	−	w^{-}	−

[a]Numbers refer to intensity of reaction. Superscripts refer to reactions of 10 to 40% of strains. +, 90% or more of strains positive; −, 90% or more of strains negative; w, weak reaction.

negative fusobacterium that is indole-positive and produces propionic acid from lactate is *F. necrophorum*. Most strains of *F. necrophorum* are hemolytic on horse and rabbit blood agar. Sheep red blood cells are less susceptible to this hemolysin.

F. nucleatum is the only *Fusobacterium* species with fusiform cells from colonies that show internal flecking. This species is indole-positive and does not produce propionic acid from lactate. Mannose is not fermented. Propionic acid is not produced from threonine.

F. varium is indole-variable. No propionic acid is produced from lactate. Mannose is fermented to a weak acid. Esculin is not hydrolyzed. Fructose is fermented. Propionic acid is produced from threonine.

F. mortiferum is indole-negative. Esculin is hydrolyzed. Acid is produced from lactose.

DRUG SUSCEPTIBILITY

Penicillin G is the drug of choice in the treatment of fusobacterial infections (Table 2). However, an occasional strain of *F. varium* and *F. mortiferum* may be highly resistant to this drug. Occasional strains of *F. varium* are also highly resistant to clindamycin.

TABLE 2. Susceptibility of Fusobacteria to Antimicrobial Agents[a]

	F. VARIUM	OTHER SPECIES
Penicillin G	1^{R}	1
Lincomycin	3	2
Clindamycin	3^{R}	2
Metronidazole	3	2
Chloramphenicol	2	2
Tetracycline	3	2
Erythromycin	4	4
Vancomycin	4	4

[a]1, excellent activity, drug of choice; 2, good activity; 3, moderate activity; 4, poor activity; R, occasional strains may be resistant.

Erythromycin is relatively inactive against fusobacteria. At a level of 4 μg/ml only 30 to 50 per cent of the strains of fusobacteria tested were shown to be susceptible to erythromycin (Sutter and Finegold, 1976).

EPIDEMIOLOGY

Fusobacteria live in the normal mucous membranes of the mouth, bowel, and urogenital tract, and often invade underlying tissue that has been damaged by biting, other accidental trauma, surgery, tumors, or another infection. They make up about 3 per cent of the anaerobes recovered from clinical infections (Finegold et al., 1977).

Fusobacteria in the mouth and throat frequently infect the teeth and throat. They may also infect any adjacent area such as the head, eyes, ears, mastoids, and sinuses. They may be aspirated into the terminal bronchioles, causing lung abscess, or spread by local or systemic circulation to cause brain abscess, meningitis, or pleuropulmonary disease. A recent literature survey showed that fusobacteria were the most common organisms isolated from brain abscesses (50 out of 142 cases), anaerobic meningitis (24 out of 104 cases), and from pleuropulmonary disease (Finegold, 1977).

Fusobacteria from the gastrointestinal tract, especially *F. necrophorum* and *F. varium*, may cause intra-abdominal and perineal infections and septicemia. They are the third most common isolate from these sites after *Bacteroides* and Peptostreptococci.

Fusobacteria are among the most common isolates from the 70 per cent of normal postpartum uterine cavities that are culturally positive for anaerobes. They frequently colonize the genital tract of women and account for up to 20 per cent of serious pelvic infections (Gorbach and Barttlett, 1974).

Recent evidence suggests that the gastrointestinal tract is colonized with fusobacteria that are ingested in the diet. Japanese ingesting a Japanese diet have fusobacteria in their stools in concentrations of 10^8 to 10^{10} (Ueno et al., 1974). *F. necrophorum* was present in ten out of ten studied, *F. varium* in two out of ten, and other fusobacteria in about five out of ten. Of 11 Americans (two black and nine Caucasian), only six had fusobacteria isolated from stool cultures. *F. varium* was present in three out of six and *F. nucleatum* in three out of six. None of the Americans had *F. necrophorum* in their stools. Two of the patients with *F. varium* ate rice regularly. Furthermore, the number of fusobacteria steadily declined in those Japanese who switched to American diets. Thus, diet determines whether fusobacteria appear in the stool and which species are present.

References

Bartlett, J. G., and Finegold, S. M.: Anaerobic pleuropulmonary infections. Medicine 51:413, 1972.

De Araujo, W. C., Varan, E., and Mergenhagen, S. E.: Immunochemical analysis of human oral strains of *fusobacterium* and *leptotrichia*. J Bacteriol 86:837, 1963.

Finegold, S. M.: Anaerobic Bacteria in Human Disease. New York, Academic Press, 1977.

Finegold, S. M., Shepherd, W. E., and Spaulding, E. H.: Practical anaerobic bacteriology. Cumitech 5. American Society of Microbiology, April, 1977, p 14.

Garcia, M. M., Charlton, K. M., and McKay, K. A.: Characterization of endotoxin from *Fusobacterium necrophorum*. Infect Immun 11:371, 1975.

Gorbach, S. L., and Bartlett, J. G.: Anaerobic infections. N Engl J Med 290:1177, 1974.

Holdeman, L. V., Cato, E. P., and Moore, W. E.: Anaerobe Laboratory Manual. Blacksburg, Va., Virginia Polytechnic Institute and State University, 1977.

Kristoffersen, T., and Hofstad, T.: Antibodies in humans to an isolated antigen from oral fusobacteria. J Periodontol Res 5:110, 1970.

Kristoffersen, T., Maelund, J. A., and Hofstad, T.: Serologic properties of lipopolysaccharide endotoxins from oral fusobacteria. Scand J Dent Res 79:105, 1971.

Lewis, J. M., and Barenberg, L. H.: Pulmonary gangrene due to spirochetes and fusiform bacilli. Am J Dis Child 37:351, 1928.

Moore, W. E. C., and Holdeman, L. V.: Fusobacterium. In Buchanan, R. E., and Gibbons, N. E. (ed.): Bergey's Manual of Determinative Bacteriology, 8th ed. Baltimore, The Williams & Wilkins Co., 1974, p. 1268.

Smith, D. T.: Fuso-spirochetal disease of the lungs produced with cultures from Vincent's angina. J Infect Dis 46:303, 1930.

Sutter, V. L., and Finegold, S. M.: Susceptibility of anaerobic bacteria to 23 antimicrobial agents. Antimicrob Agents Chemother 10:736, 1976.

Ueno, K., Sugihara, P. T., Bricknell, K. S., Attebery H. R., Sutter, V. L., and Finegold, S. M.: Comparison of characteristics of gram-negative anaerobic bacilli isolated from feces of individuals in Japan and the United States. In Balows, A. (ed.): Anaerobic Bacteria, Role in Disease. Springfield, Ill., Charles C Thomas, Publisher, 1974.

48 VEILLONELLA

Howard Robert Attebery, D.D.S.

Veillonella are gram-negative, obligately anaerobic cocci. They are universally present as part of the indigenous oral microbiota of man and are commonly found in the respiratory tract, intestine, and vagina of humans. By reason of their usual occurrence and high numbers as parasites on the mucous membranes of man, they can be found in infected sites when the indigenous flora are transported from the protective mucous membrane barrier into deeper tissues that can also support their growth.

Veillonella are not regarded as overt pathogens, although they do exhibit some pathogenic properties. They are not found in pure culture in any infections but are always mixed with facultative bacteria or other anaerobes or both. Veillonella are spoken of as "opportunists" because they share an infected site with more aggressive and more pathogenic microorganisms, and their role in the infectious process is considered minor.

Two species of *Veillonella* are commonly seen in the literature, *Veillonella parvula* and *Veillonella alcalescens*. These species have been found to be genetically homologous, and now only one species is recognized for the genus, *Veillonella parvula* (Holdeman et al., 1977).

MORPHOLOGY

Veillonella are small cocci that measure 0.3 to 0.5 μm in diameter. They appear usually in masses or as diplococci and on occasion singly or as short chains.

On blood agar plates, colonies appear as small, convex, translucent, and glistening with an entire edge. Colony size is small unless pyruvate or lactate and carbon dioxide gas are added to the growth system.

ANTIGENIC COMPOSITION

Rogosa found seven major serologic groups for the strains he tested (Rogosa, 1965). He could clearly distinguish 21 oral strains that were isolated from hamsters. These group I strains all possessed a distinctive antigen and were singularly homologous with each other. He placed 28 strains in group II, all oral isolates from hamsters, rats, and rabbits. Group III consisted of two oral strains isolated from rats. Group IV, *Veillonella alcalescens*, comprised 15 members, which included 3 from the rat, 1 from a rabbit, 4 from human bacteremias that followed dental operations, and 5 from the human mouth. Group V consisted of only two strains, one each from man and the rat. Group VI, *Veillonella parvula,* consisted of 28 strains. Most were oral strains from humans, but four were isolated from the intestinal tract of man. Group VII consisted of only three strains, two from the human mouth and one from the human respiratory tract.

A major antigenic component is associated with an endotoxin that induces pyrogenic and Shwartzman reactions and is structurally associated with an outer three-layered cell membrane (Bladen and Mergenhagen, 1964).

METABOLISM

Veillonella has complex nutritional requirements. Carbohydrates are not fermented. D-ribose is, however, utilized biosynthetically and incorporated into nucleic acid (Kafkewitz and Delwiche, 1967).

For energy requirements, *Veillonella* utilizes pyruvate best but also lactate, oxaloacetate, malate, and fumarate. Veillonella also obtains energy by reducing nitrates to nitrites.

Lactate is metabolized to acetic and propionic acids and to carbon dioxide and hydrogen.

Hydrogen sulfide is produced from a number of sulfur-containing compounds such as thioglycolate, thiocyanate, thiosulfate, cysteine, cystine, and glutathione. Carbon dioxide is required for growth.

PATHOGENICITY

Veillonella can produce disease in experimental animals. Its lipopolysaccharide is an endotoxin that produces fever and Shwartzman reactions in rabbits.

The role of *Veillonella* in human clinical infections is not clear. It appears that *Veillonella*, the parasite of mucous membranes, is an opportunist in human infection and does not play a major role in any infection. This view has evolved because of the relative infrequency of involvement of *Veillonella* in clinical infections and the fact that *Veillonella* is not present in pure culture in infections.

LABORATORY DIAGNOSIS

Three genera of anaerobic, gram-negative cocci that belong to the family Veillonellocae may be found in infected material. These genera are easily differentiated. *Megasphaera* is fermentative and produces abundant acid from glucose, fructose, and maltose. *Veillonella* and *Acidaminococcus* are nonfermentative. In addition, *Megasphaera* produces butyric acid from carbohydrate growth media, while the others do not. *Megasphaera* has a large cell, between 1.7 and 2.5 μm, which also distinguishes this genus from *Veillonella* and *Acidaminococcus*. *Veillonella* reduces nitrates, whereas *Acidaminococcus* and *Megasphaera* do not.

The criteria separating *Veillonella* from the other anaerobes are as follows: small, gram-negative cocci, 0.3 to 0.5 μm, that reduce nitrate but do not ferment. Growth is stimulated by pyruvate and lactate, and carbon dioxide is required for growth.

One species is now recognized, *Veillonella parvula.* Commonly seen in the literature are two species that were separated by their catalase activity, *V. alcalescens,* which gives a positive catalase reaction, and *V. parvula,* which gives a negative reaction. Now both are classified as *V. parvula.*

DRUG SUSCEPTIBILITY

Veillonella organisms are exquisitely sensitive to penicillin and clindamycin. Table 1 lists varying susceptibilities to ten antibiotics. Poor in vitro activity against *Veillonella* is seen with erythromycin, kanamycin, and gentamicin. In addition, *Veillonella* organisms are resistant to streptomycin, neomycin, and vancomycin.

EPIDEMIOLOGY

Veillonella is part of the indigenous microbiota of the mouth, respiratory system, intestinal tract, and vagina of humans. The organisms are universally present in high numbers in the human mouth and might be expected in infections in which the oral flora is the primary infecting

TABLE 1. In Vitro Susceptibility of 13 Clinical Isolates of The Genus Veillonella to 10 Antibiotics*

ANTIBIOTIC	CUMULATIVE PER CENT SUSCEPTIBLE AT INCREASING CONCENTRATIONS (μg/ml)									
	0.1	0.2	0.4	0.8	1.6	3.1	6.2	12.5	25	>25
Penicillin G	23	77	100							
Erythromycin		8	16		23	31	38	69	77	100
Cephalothin	23	69		84	100					
Tetracycline	23		31	69	76	85			100	
Lincomycin	46	77			92		100			
Clindamycin	100									
Kanamycin	8						23	38	61	100
Chloramphenicol	15		23	46	85	100				
Gentamicin					8		31	46	85	100
Rifampin	23			31	100					

*Adapted from Martin, W. J.: In Linnette, E. H., Spaulding, . H., and Trauant, J. P. (eds.): Manual of Clinical Microbiology. Washington, D.C., American Society for Microbiology, 1974.

agent. *Veillonella* can on occasion be found in infections associated with the microbiota of the intestinal tract and vagina.

Veillonella can be found in transient bacteremias following dental extraction, in sinusitis, tonsillar and peritonsillar abscesses, otitis media, mastoiditis, and brain abscesses. The organisms are occasionally found in intra-abdominal infections and genitourinary infections. They are commonly found in the mixed infections in pneumonitis, necrotizing pneumonia, lung abscess, aspiration pneumonia, empyema, and pleuropulmonary infections. They can be seen in female genital tract infections, intrauterine infections, tubo-ovarian abscesses, and postoperative genitourinary infections.

Veillonella has been reported in bile. Some strains can dehydroxylate bile acids, and *Veillonella* organisms have the potential to be involved in the chain of microorganisms producing chemical carcinogens from bile acids. Veillonella was found in 5 of 25 subjects with colonic polyps but not in 25 matched control subjects without polyps (Finegold et al., 1975).

References

Bladen, H. A., and Mergenhagen, S. E.: Ultrastructure of *Veillonella* and morphological correlation of an outer membrane with particles associated with endotoxic activity. J Bacteriol. 88:1482, 1964.

Finegold, S. M., Flora, D. J., Attebery, H. R., and Sutter, V. L.: Fecal bacteriology of colonic polyp patients and control patients. Cancer Res 35:3407, 1975.

Holdeman, L. V., Cato, E. P., and Moore, W. E.: Anaerobe Laboratory Manual. Blacksburg, Va., Virginia Polytechnic Institute and State University, 1977.

Kafkewitz, D., and Delwiche, E. A.: Utilization of D-ribose by *Veillonella*. J Bacteriol 98:903, 1967.

Martin, W. J.: Anaerobic cocci. In Lennette, E. H., Spaulding, E. H., and Truant, J. P. (eds.): Manual of Clinical Microbiology. Washington, D.C., American Society for Microbiology, 1974, p. 970.

Rogosa, M.: The genus *Veillonella*. IV. Serological groupings, and genus and species emendations. J Bacteriol 90:704, 1965.

49 SPIROCHETES: TREPONEMA AND BORRELIA

Daniel M. Musher, M. D.

DEFINITION

The order Spirochaetales includes the following five genera: *Treponema, Borrelia, Spirochaeta, Leptospira,* and *Cristispira*. This chapter will deal with the first two. The genus *Treponema,* which is pathogenic for man, includes *T. pallidum, T. pertenue,* and *T. carateum,* which cause syphilis, yaws, and pinta, respectively. These organisms have not been cultivated successfully in vitro, although recent studies have demonstrated that the viability and virulence of *T. pallidum* can be maintained for days to weeks in artificial media, and the organism may even have shown some degree of replication. Our inability to cultivate these bacteria in vitro is, to a great extent, responsible for our incomplete understanding of their properties. No differences between *T. pallidum, T. pertenue,* and *T. carateum* have been detected by microscopic, electronmicroscopic, or immunologic techniques, although the infections they produce in animals or humans are readily distinguishable (Turner and Hollander, 1957). *T. cuniculi* and *T. hyodysenteriae* cause dermal and systemic disease in rabbits and diarrheal disease in swine, respectively; *T. cuniculi* also produces experimental infection in primates. Neither of these organisms has been shown to infect people. *T. macrodentium* and *T. vincentii* may contribute to infections of the mouth and gums. Except for *T. cuniculi,* these other treponemes can all be cultivated in artificial media. This chapter will be confined to discussing the three treponemes that are pathogenic for man.

The genus *Borrelia* includes about 18 species or strains that are human pathogens. Some of these may be distinguished serologically, although formal classification is usually based on their arthropod vectors. Classification of these organisms is imprecise also, in part because in vitro cultivation of most strains has not been achieved.

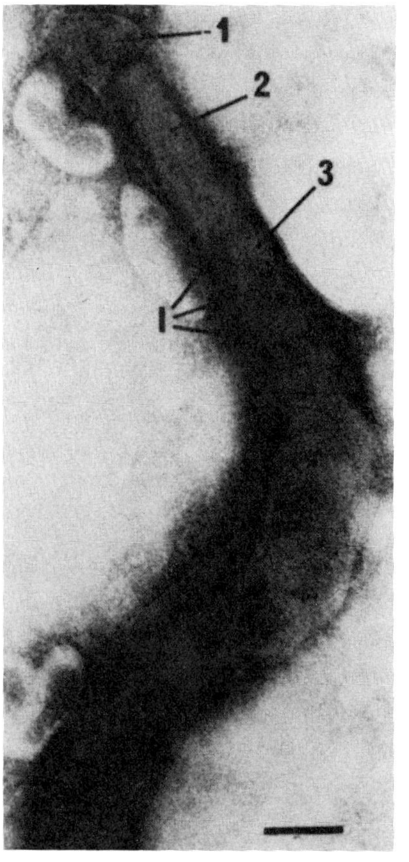

FIGURE 1. *Electron micrograph of* Treponema pallidum *(Nichols strain). Zone 1 is the outermost tip, which comes to a point. Zone 2 is the distal portion of the cytoplasm and tapers at its end beneath the outermost tip. Zone 3 is the region where flagella are inserted. Pathogenic treponemes have three flagella inserted in a row subterminally at each end. (From Hovind-Hougen, K.: Acta Pathol Microbiol (Scand) 255:1, 1976.)*

MORPHOLOGY

The pathogenic treponemes are so narrow ($\leq 0.15\ \mu$) that they are below the resolution of the light microscope; they are, therefore, considered gram-indeterminate. Electron microscopy shows them to be wavelike organisms with tapered ends (Fig. 1). They have tight, regular spirals with a wavelength of 1.1 μm and an amplitude of 0.2 to 0.3 μm; their length ranges from 6 to 15 μm, and is usually 10 to 13 μm. Examination of *T. pallidum* obtained from syphilitic rabbit testis reveals an amorphous outer layer. Recent electronmicroscopic studies of tissues have shown that *T. pallidum* in situ has an outermost layer that stains with ruthenium red and that may be the same as the amorphous external layer. The relation of these findings to a long-hypothesized slime layer that prevents phagocytosis and inhibits early antigen processing is not known. The virulent treponemes have three-sheathed flagella (also called axial fibrils) that emerge from electron-dense areas near each end of the organism; these flagella are contained between the three-layered outer membrane and an electron-dense layer (peptidoglycan) that is thought to be responsible for the wavelike configuration of treponemes. The cytoplasmic membrane is within and is adherent to the electron-dense layer. Although it might seem reasonable to assume that, by analogy to bacterial flagella, the axial fibrils are responsible for motility, their location within the outer membrane and the failure of antiflagellar antibody to immobilize certain spirochetes makes this assumption less certain. Intracytoplasmic tubules are present, but their function is unknown (Hovind-Hougen, 1976).

Borreliae are composed of pointed, helical organisms that usually are 10 to 20 μm in length. They have coarse irregular coils with variable amplitude. Because their width at the center is 0.4 to 0.5 μm, they are easily detectable in stained preparations as gram-negative, wavelike organisms. Wright's and Giemsa's stains also delineate them nicely, especially if prolonged staining is used. As is the case with *Treponema,* an outermost amorphous surface layer surrounds an outer membrane that has three distinct layers (Fig. 2). Between the outer and the cytoplasmic membrane may be found 15 to 20 flagella that lack sheaths; they are inserted at each end of the organism by means of basal knobs.

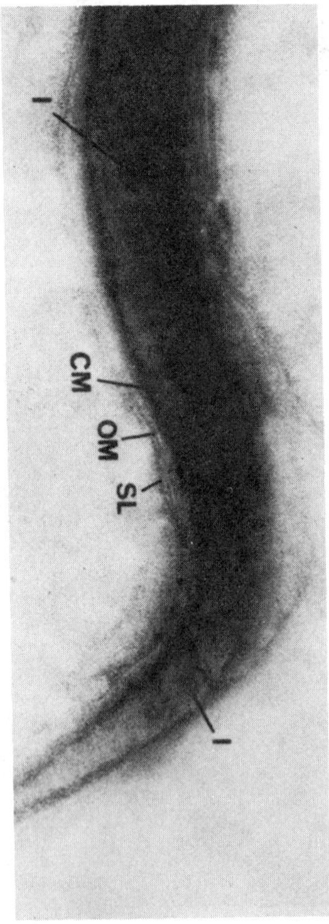

FIGURE 2. *Electron micrograph of* Borrelia recurrentis. *Note amorphous surface layer (SL), three-layered outer membrane (OM), and cytoplasmic membranes (CM). Flagella are inserted at I. The end is sharply pointed. (From Hovind-Hougen, K.: Acta Pathol Microbiol (Scand) 255:1, 1976.)*

ANTIGENIC COMPOSITION

Little is known about the antigenic properties of the virulent treponemes because, as noted above, they must be obtained from infected tissue and because the problem of differentiating *Treponema* from mammalian antigens precludes precise analysis. It is not even known with certainty whether cardiolipin, the antigen that stimulates production of Wassermann (VDRL [Veneral Disease Research Laboratory]) antibody (see below), is a component of the treponeme or whether it is released from mammalian tissues during active treponemal infection. Favoring the former possibility is the observation that VDRL antibody follows clinically inapparent infection of primates with *T. cuniculi*. Favoring the latter is the association in experimental animals and humans between active lesions and the detection of this antibody, as well as its appearance in autoimmune disease states such as systemic lupus erythematosus. A recent study has shown that cardiolipin represents 13 per cent of the total lipid composition of *T. pallidum* (Matthews et al., 1979). However, since treponemes are unable to synthesize some fatty acids, their lipid composition may simply reflect that of the tissue in which they have grown. The outermost layer of *T. pallidum* has been shown to consist of mucopolysaccharides (Fitzgerald and Johnson, 1979) to which mammalian proteins readily adhere (Alderete and Baseman, 1979), further complicating studies of antigenicity of these organisms. Patients who have been infected with any of the pathogenic treponemes have antibodies against the others. Finally, because of cross-reactivity with nonpathogenic treponemes, it is necessary to absorb human serum with *T. phagedenis* (previously called *T. reiteri*) before performing tests to detect antibody to *T. pallidum*.

Antigenic properties of borreliae have not been studied thoroughly. Antiserum to one strain of *Borrelia* agglutinates other borreliae as well as *T. pallidum*. Typhus-immune serum may also contain complement-fixing antibodies that react with *Borrelia*. However, infection with one strain of *Borrelia* does not necessarily protect against others. Evidence related to "relapse strains" (see below) suggests that the antigens against which the host makes protective antibodies are unstable. Antibody to cardiolipin appears in a small percentage of cases of borreliosis. Both *Treponema* and *Borrelia* have ornithine and lack muramic and diaminopimelic acid in their cell walls, in contrast to *Leptospira*.

METABOLISM

T. pallidum has traditionally been regarded as an obligate anaerobe because it best retains motility in vitro under anaerobic conditions. Recent studies, however, have shown that *T. pallidum* takes up oxygen and degrades glucose aerobically to carbon dioxide and acetate, as well as anaerobically to pyruvate and lactate, although of 22 carbon sources studied, only glucose and pyruvate were metabolized aerobically (Baseman et al., 1976). *T. pallidum* is better able to incorporate amino acids into proteins in the presence of 10 per cent oxygen than anaerobically. These studies are difficult to interpret because the organisms do not actively replicate in the in vitro systems utilized (Lysko and Cox, 1977).

The metabolic characteristics of most borreliae are unknown. *Borrelia duttoni* metabolizes glucose anaerobically via the Embden-Meyerhof pathway, with pyruvate acting as the terminal electron acceptor and reducing to lactate; aerobic respiration does not appear to play a role, and the borreliae are generally considered to be anaerobic or microaerophilic. Borreliae have been shown to metabolize lysolecithin (but not lecithin) to fatty acids and glycerol, and to incorporate cholesterol from culture medium in vitro.

PATHOGENIC PROPERTIES

The means by which *Treponema* or *Borrelia* produces infection are unknown.

IMMUNITY

By the time primary syphilis is diagnosed most patients have antibodies to *T. pallidum* that can be detected by immunofluorescent and hemagglutination techniques. Also present are antibodies that, together with complement, inhibit motility and eliminate infectivity of the organism (*T. pallidum* immobilizing [TPI] antibodies). Despite these antibodies, syphilis progresses and secondary lesions develop in virtually all patients unless specific therapy is given. Passive immunization with serum from rabbits that have recovered from experimental infection and are immune to rechallenge with *T. pallidum* delays and attenuates but does not prevent the development of syphilitic lesions (Bishop and Miller, 1976). An opsonizing effect of immune globulin has not been found, and phagocytosis of *T. pallidum* by polymorphonuclear leukocytes or macrophages has not been shown to contribute to resistance. Antibody to cardiolipin does not appear to contribute to immunity. Delayed hypersensitivity to treponemal antigens appears late in secondary syphilis and might, therefore, appear to be related to the onset of latency. Infection with *T. pallidum* stimulates cellular immune mechanisms, as demonstrated by enhanced ability of macrophages from syphilitic rabbits to suppress the growth of *Listeria monocytogenes;* this reaction is mediated by thymus-dependent lymphocytes (Schell et al., 1975). However, stimulation of acquired cellular resistance with unrelated organisms such as *Mycobacterium bovis* (BCG) or *Propionibacterium acnes* does not protect animals against challenge with *T. pallidum*. In addition, adoptive transfer of spleen cells from syphilis-immune rabbits fails to protect recipients against syphilitic infection. Several findings that suggest that thymus-dependent immune responses may be impaired early in syphilis might support the hypothesis that the failure of the host to control infection results from immune suppression: (1) in vitro blastogenic transformation of lymphocytes following stimulation with a variety of antigens is depressed, (2) paracortical (thymus-dependent) areas of lymph nodes are depleted, (3) delayed hypersensitivity to several antigens may be depressed, and (4) ability to be sensitized to sheep red blood cells with consequent production of IgG is markedly inhibited (Baughn and Musher, 1978). It is entirely possible that there is no complete immunity to *T. pallidum*. Active lesions may be brought under control and animals may be resistant to rechallenge with *T. pallidum,* but the host is unable to rid itself of the infecting organism, which continues to persist in lymph nodes. Human subjects have also been found to be resistant to rechallenge with *T. pallidum* during latency, although a number of observations indicate that they have not succeeded in eradicating the infecting organism (Magnuson, et al., 1956).

Humoral immunity is thought to be responsible for arresting borreliosis. Three to five days after the onset of symptoms, circulating antibodies appear that agglutinate, immobilize, and/or lyse borreliae; their appearance corresponds with clinical improvement and disappearance of these organisms from the bloodstream. Although these antibodies can be shown by some techniques to cross-react with other borreliae, immunity itself is species- and strain-specific. After the initial humoral response, and as a result of mechanisms that are not understood, the infecting organism undergoes a series of antigenic shifts, each one leading to the emergence of a new "relapse" serotype that is not immediately subject to humoral control. These serotypes cause a clinical relapse, which is then arrested within four days by a newly emerging set of specific antibodies. The shorter duration and the reduced intensity of the relapses suggest antigenic cross-reactivity with the original infecting strain. It is interesting that relapse strains appear to revert to parental forms on passage through insect vectors but not through mammalian hosts. Immunity to rechallenge with a particular strain of *Borrelia* may persist up to several years in experimental animals. In clinical reports that have described reinfection within a few months, the infecting strains have not been specifically identified. The physiologic role of phagocytosis in infection due to borreliae is thought to be minimal. Cell-mediated immunity has not been shown to play a role (Felsenfeld, 1965; Southern and Sanford, 1969).

LABORATORY DIAGNOSIS

Three kinds of laboratory tests are important in diagnosing syphilis, yaws, or pinta: (1) Treponemes in exudates from suspicious lesions can be detected by darkfield microscopy. This method is sensitive if a good specimen has been obtained and is highly specific if care has been used to prevent contamination from oral or fecal sources. (2) Antibody to cardiolipin can be detected by a variety of flocculation tests. The most widely used is the VDRL reaction. The rapid plasma reagin (RPR) test or a modification thereof is particularly useful for screening large numbers of specimens. It is still considered best to verify and quantify positive RPR reactions by measuring VDRL antibodies. The VDRL test is reactive in about 80 per cent of patients at the time primary syphilis is diagnosed and in virtually all patients with secondary syphilis. Although VDRL reactivity is closely associated with active infection, this reaction may be negative in patients with cardiovascular or neurosyphilis. (3) Antibody to virulent treponemes can be detected by using an immunofluorescent technique (FTA-ABS, an acronym for *f*luorescent *t*reponemal *a*ntibody following *abs*orption with nonvirulent treponemes to remove nonspecific reactivity) or hemagglutination (TPHA, *T. pallidum h*emagglutination *a*ssay). Of these, the FTA-ABS technique is slightly more sensitive, being positive in more than 90 per cent of patients with primary syphilis. Both are positive in virtually all patients with secondary, latent, and late infections. The same antibodies are also detectable in yaws and pinta. These tests are highly specific for treponemal infection, although serum from patients with autoimmune diseases is rarely reactive. Despite the advantages of sensitivity and specificity, detection of antitreponemal antibody may not be useful in diagnosing active infection in an individual case because, once present, the antibody persists indefinitely; the FTA-ABS is most useful in excluding a treponemal cause for a chronic lesion (Rudolph, 1977).

Borreliosis is usually diagnosed by finding loosely coiled spirochetes during microscopic examination of Wright- or Giemsa-stained peripheral blood, or during darkfield microscopic examination of the whole fresh blood. This is highly specific but is relatively insensitive, being positive in one quarter to one half of patients who are found to have borreliosis. Injecting mice with blood from infected patients is helpful diagnostically because it produces an obvious spirochetemia, often within two to three days. Results in a recent outbreak suggested that agglutination of the *Proteus* Ox-K antigen is useful diagnostically because it was positive in at least two thirds of cases. Antibodies that agglutinate or kill *Borrelia* have been identified but have not been shown to be of practical clinical value. The VDRL test is negative in borreliosis.

ANTIMICROBIAL SUSCEPTIBILITY

The virulent treponemes are highly susceptible to penicillins, cephalosporins, macrolides (erythromycin and clindamycin), tetracyclines, and chloramphenicol. They are resistant to sulfonamides and aminoglycosides. Borreliae are susceptible to penicillin, aminoglycosides, tetracycline, and chloramphenicol.

EPIDEMIOLOGY

T. pallidum, T. pertenue, and *T. carateum* naturally infect only humans, although experimental infection has been produced in a variety of animals. *T. pallidum* infections have a world-wide distribution. Infection is spread primarily by venereal contact but may be acquired in utero. Yaws (*T. pertenue*) is limited to tropical areas and has largely been eradicated in recent years thanks to efforts of the World Health Organization. Infection appears to spread by direct contact, although flies may act as vectors. Pinta (*T. carateum*) is prevalent in areas of Central and South America and the Carribbean. It also appears to spread by contact from person to person. There is no known animal reservoir. Infection occurs sporadically in all age groups, although yaws generally begins in childhood.

In the United States borreliosis has almost always been acquired in areas above sea level where pine and fir trees predominate. The north rim of the Grand Canyon has become recognized as a highly endemic area with tick-infected log cabins playing the predominant role (Boyer et al., 1977). The reservoir of *B. recurrentis,* the classic cause of louse-borne relapsing fever, is unknown. Only human beings are clinically infected, and there is no transovarian transmission by the insect vector (*Pediculus humanus* var *humanus*). All the other borreliae infect rodents and small mammals; infection is transmitted to humans by *Ornithodorus* ticks. Elimination of rodent nests adjacent to human habitations and use of insecticide and insect repellents appear to control the spread of infection.

References

Alderete, J. F., and Baseman, J. B.: Surface-associated host proteins on virulent *Treponema pallidum.* Infect Immun 26:1048, 1979.

Baseman, J. B., Nichols, J. C., and Hayes, N. S.: Virulent *Treponema pallidum:* Aerobe or anaerobe. Infect Immun 13:704, 1976.

Baughn, R. E., and Musher, D. M.: Altered immune responsiveness associated with experimental syphilis in the rabbit: elevated IgM and depressed IgG response to sheep erythrocytes. J Immunol 120:1691, 1978.

Bishop, N. H., and Miller, J. N.: Humoral immunity in experimental syphilis. I. The demonstration of resistance conferred by passive immunization. J Immunol 117:191, 1976.

Boyer, K. M., Munford, R. S., Maupin, G. O., Pattison, C. P., Fox, M. D., Barnes, A. M., Jones, W. L., and Maynard, J. E.: Tick-borne relapsing fever: an interstate outbreak originating at Grand Canyon National Park. Am J Epidemiol 105:469, 1977.

Canale-Parola, E.: Physiology and evolution of spirochetes. Bacteriol Rev 41:181, 1977.

Felsenfeld, O.: Borreliae, human relapsing fever, and parasite-vector-host relationships. Bacteriol Rev 29:46, 1965.
Fitzgerald, T. J., and Johnson, R. C.: Surface mucopolysaccharides of *Treponema pallidum*. Infect Immun 24:244, 1979.
Hovind Hougen, K.: Determination by means of electron microscopy of morphological criteria of value for classification of some spirochetes, in particular treponemes. Acta Pathol Microbiol (Scand) [B] Suppl 255:1, 1976.
Lysko, P. G., and Cox, C. D.: Terminal electron transport in *Treponema pallidum*. Infect Immun 16:885, 1977.
Magnuson, H. J., Thomas, E. W., Olansky, S., Kaplan, B. I., Demello, L., and Cutler, J. C.: Inoculation syphilis in human volunteers. Medicine 35:33, 1956.
Matthews, H. M., Yang, T.-K., and Jenkin, H. M.: Unique lipid composition of *Treponema pallidum* (Nichols virulent strain). Infect Immun 24:713, 1979.
Rudolph, A. H.: Laboratory diagnosis of syphilis. In Demis, D. J., Dobson, R. L., and McGuire, J. (eds.): Clinical Dermatology, vol. 3. Hagerstown, Md., Harper and Row, 1977, pp. 1–17.
Schell, R. F., Musher, D. M., Jacobson, K., and Schwethelm, P.: Induction of acquired cellular resistance following transfer of thymus-dependent lymphocytes from syphilitic rabbits. J Immunol 114:550, 1975.
Southern, P. M., Jr., and Sanford, J. P.: Relapsing fever: A clinical and microbiological review. Medicine 48:129, 1969.
Turner, T. B., and Hollander, D. H.: Biology of the Treponematoses. Geneva, World Health Organization, Monograph series No. 35, 1957.

3 UNIQUE INTRACELLULAR GRAM-NEGATIVE BACTERIA

RICKETTSIA 50

Garrison Rapmund, M.D.

INTRODUCTION

Rickettsiae are small bacteria that are natural parasites of certain arthropods. Some of these arthropods can transmit rickettsiae to animals, including man. Most of our knowledge of rickettsiae come from studies of the few species that are pathogenic for man.

Rickettsiae are named after the American physician, H. T. Ricketts, who first recognized rickettsiae as the cause of typhus and spotted fever and who died of typhus in 1910 while investigating the disease in Mexico.

Rickettsiae are distinct from other bacteria because of their small size, their transmission by arthropods, and their obligate intracellular parasitism. Four groups of rickettsiae are pathogenic for man: the typhus group, of which the type species is *Rickettsia prowazekii;* the spotted fever group, of which the type species is *Rickettsia rickettsii;* the scrub typhus group, composed of one species complex, *Rickettsia tsutsugamushi*; and the Q fever group, *Coxiella burnetii*. The first three groups are similar in their growth and metabolism, fine structure, and pathogenic properties in man. They differ significantly, however, in size, intracellular location, extracellular behavior, and antigenic composition. These differences distinguish the species of rickettsiae and separate the organism that causes Q fever into a different genus, *Coxiella*. Several organisms that cause disease in man have been called rickettsiae in the past but have now been reclassified. *Rickettsia quintana,* the cause of louse-borne trench fever in man, is now placed in a separate genus, *Rochalimaea,* because of its ability to grow on artificial media. *Rickettsia sennetsu,* the cause of an infectious mononucleosis-like syndrome in southern Japan and perhaps elsewhere, is now known to be related serologically to organisms of the genus *Ehrlichia,* which are pathogens of dogs and domestic ruminants. For reasons to be discussed later, *R. sennetsu* may require reclassification with the *Ehrlichia.*

MORPHOLOGY

Rickettsiae are typically short rods 0.3 to 0.7 μm by 1.0 to 2.0 μm, a size just visible by light microscopy. Pleomorphism is common. Coccobacilli, diplobacilli, and individual rods are seen in infected tissue. A developmental cycle has never been identified in rickettsiae. Some investigators have postulated that morphologic variation represents a developmental cycle, but this view is not generally accepted. It is true that rickettsiae in the logarithmic growth phase are typically rod-shaped, while those in the stationary phase are often coccoid, filamentous, or swollen, but there is no evidence that this pleomorphism reflects a developmental cycle. *C. burnetii* is smaller than pathogenic rickettsiae and is filterable through certain pore filters that trap other bacteria.

Rickettsiae are gram-negative but stain poorly by the Gram technique. Rickettsiae stain red by the Macchiavello (Fig. 1) and Giminez techniques (Giminez, 1964) and purple by the Giemsa stain. Only specific immunofluorescent antisera can differentiate rickettsiae definitely from other microorganisms by means of a staining reaction.

The ultrastructure of rickettsiae is typical of gram-negative bacteria (Anacker et al., 1967). A five-layered cell wall-plasma membrane complex (Fig. 2) surrounds the cytoplasm, which contains a large array of ribosomes and fibrouslike strands

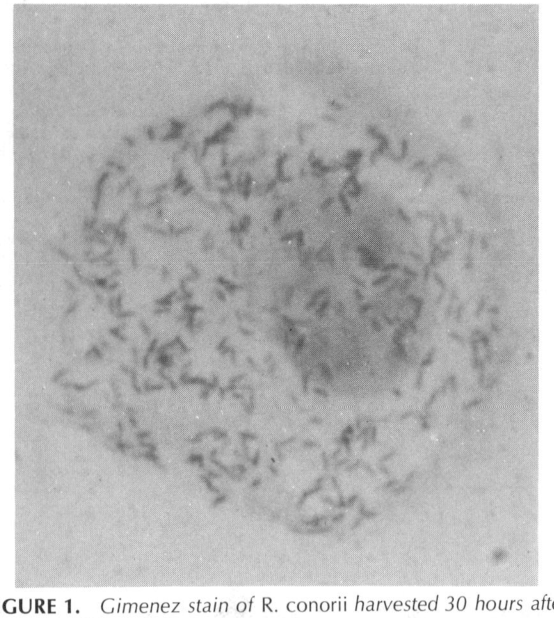

FIGURE 1. *Gimenez stain of* R. conorii *harvested 30 hours after infection of L929 cells in tissue culture.*

indicative of deoxyribonucleic acid. Intracytoplasmic vacuoles and invaginations of the plasma membrane are common (Figs. 2 and 3). Discrete, sharply defined nuclear structures have not been observed. *C. burnetii* contains a central fibrous body with radiating fibrils that are susceptible to treatment with ribonuclease. This suggests an organized DNA center that may be physically connected with RNA-containing ribosomes in the cytoplasm.

Several workers have demonstrated a zone on the outer surface of rickettsiae (Fig. 4). Silverman and Wisseman (1978) called this a slime layer rather than a capsule because of its tendency to slough away. This extracellular layer corresponds to an electron-lucent zone surrounding *R. prowazekii* and *R. rickettsii* in infected host cells and appears to be a complex of three distinct zones: an inner layer at the cell surface of small projections with a periodicity of 13 nm radiating outward, an intermediate clear zone, and an outer fibrous slime layer that varies in width from 25 to 130 nm. The morphologic similarities to gram-negative bacteria are striking, but there seem to be significant functional differences. The presence of a capsule in other bacteria is associated with virulence, but the avirulent (Madrid E) strain of *R. prowazekii* contains the same extracellular structures as the fully virulent Breinl strain. Furthermore, the ultrastructure of rickettsiae in arthropod tissue is identical to that in mammalian tissue, and the ultrastructure of viable extracellular *R. prowazekii* in dried louse feces is identical to its intracellular morphology in the louse gut epithelium (Silverman et al., 1974). Thus, the ultrastructure of rickettsiae has not yet provided clues to the important functional

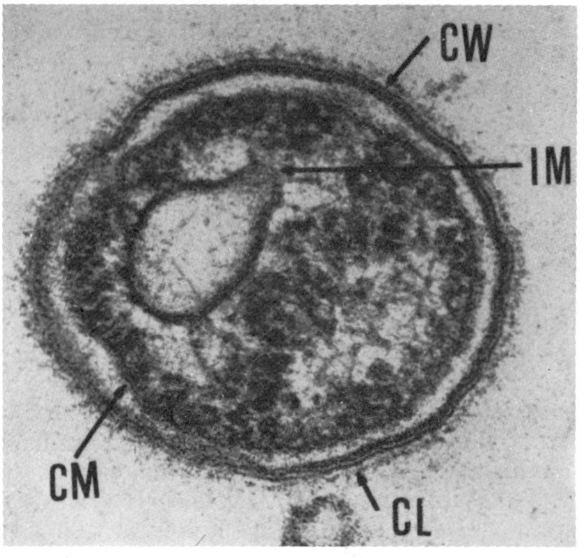

FIGURE 2. *Thin section of* R. prowazekii. *Note capsule-like layer (CL), five-layered cell wall (CW), and intracytoplasmic membrane (IM). × 185,000 (From Anacker, R. L., et al.: J Bacteriol 94:260, 1967.)*

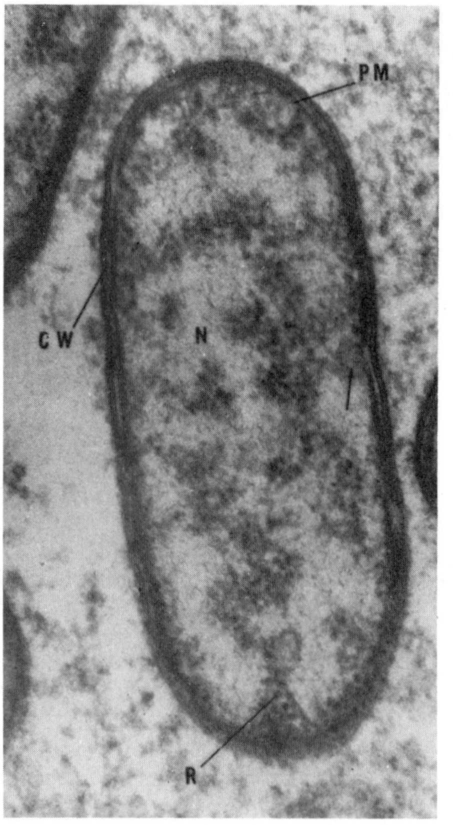

FIGURE 3. *Thin section through several* R. quintana *cells from the intestine of an infected louse. Cell wall (CW) and plasma membrane (PM) both appear trilaminar. Arrow indicates an invagination of the plasma membrane. R indicates granules that are presumed to represent ribosomes. The "nuclear" component (N) is composed of fine filaments between the granules. × 125,000. (From Ito, S., and Vinson, J. W.: J Bacteriol 89:481, 1965.)*

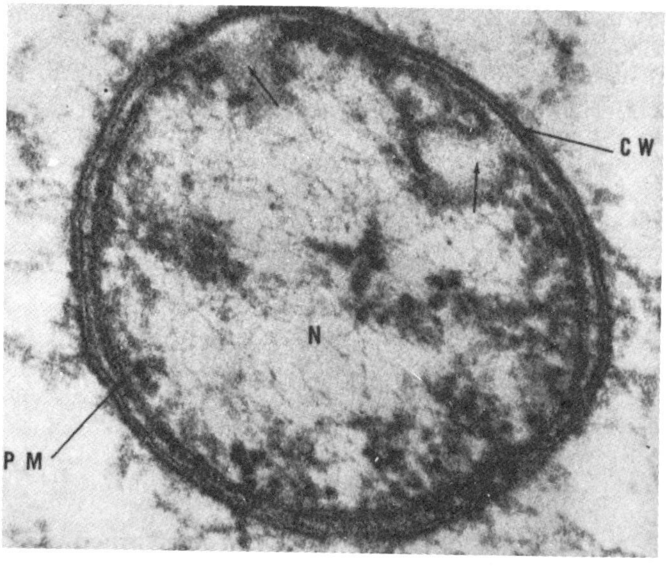

FIGURE 4. *Transverse section of* R. quintana. *Arrows designate multiple invaginations of the plasma membrane. Symbols are the same as those in Figure 3. × 190,000. (From Ito, S., and Vinson, J. W.: J Bacteriol 89:481, 1965.)*

properties of virulence and intracellular or extracellular survival.

The chemical composition of the rickettsial cell wall is similar to other gram-negative bacteria. It contains muramic acid, diaminopimelic acid, other amino acids, sugars, and amino sugars. Teichoic acid, a characteristic component of gram-positive bacteria, has not been found. Some years ago, Wood and Wisseman (1967) reported that *R. typhi* had endotoxic activity that suggested the possibility that it contained a lipopolysaccharide layer. More recently, Schramek and co-workers (1976) isolated lipopolysaccharide from the typhus group, the spotted fever group, and purified cells and soluble antigen preparations of the scrub typhus group. These workers have not yet reported a full chemical analysis of this rickettsial lipopolysaccharide, but they have analyzed similar preparations of lipopolysaccharide from purified phase 1 cells of *C. burnetii* (Schramek and Brezina, 1976). The lipopolysaccharide contains 17 fatty acids, including hydroxymyristic acid, which is a common constituent of most gram-negative bacterial lipopolysaccharides. The sugar moiety of *C. burnetii* lipopolysaccharide contains mannose, glucose, galactose, xylose, heptose, and 2-keto-3-deoxyoctonate (KDO). The presence of the amino sugars, glucosamine, galactosamine, and possibly fucosamine is further evidence of the similarity of this lipopolysaccharide to that of other gram-negative bacteria.

Rickettsiae contain both RNA and DNA. The RNA species include transfer RNA and ribosomal RNA. The size of the DNA genome is about one third that of *E. coli* and the same or slightly smaller than that of pathogenic *Neisseria*. Myers and Wisseman (1980) reported the genome size to be about 1×10^9 daltons. Tyeryar et al. (1973) analyzed the base composition of rickettsial DNA and found a small but distinct difference in the molar percentage of guanine plus cytosine (G + C) between the typhus group (30 per cent) and the spotted fever group (32.5 per cent). Strains within each group show remarkable homogeneity of G + C content, even when isolated from widely separated geographic areas. *Rickettsia canada,* an organism that cross-reacts immunologically with members of both the typhus and the spotted fever groups, has a G + C content of 30 moles per cent, which suggests that it evolved from the typhus group. *C. burnetii* has a G + C content of 43 to 45 per cent (Schramek, 1968), distinctly different from other rickettsiae. *R. quintana's* G + C content of 38.6 moles per cent is also quite distinct. DNA-base composition can be identical in quite different organisms, however. The DNA-base ratio of typhus rickettsiae is identical to that of *Wolbachia persica,* a rickettsia-like symbiont of ticks, which has significant metabolic differences from typhus rickettsiae (Weiss, 1978).

ANTIGENIC COMPOSITION

The typhus group of rickettsiae includes *R. prowazekii,* the cause of louse-borne epidemic typhus; *R. typhi,* also referred to as *R. mooseri,* the cause of flea-borne murine typhus; and *R. canada,* isolated from ticks in Canada. Members of this group possess common soluble antigens that are released into the aqueous phase during ether extraction of infected yolk sacs. These antigens cross-react in the complement fixation (CF) test. Members of the group can be distinguished from each other in the CF test by using washed whole rickettsial cells as the CF antigen.

The spotted fever group of rickettsiae includes representatives from all continents and Australia. Members of this group also possess common soluble antigens that are released into the aqueous phase by ether treatment of infected yolk sacs. The type species for the group, *R. rickettsii,* causes spotted fever in man in North America. Other members of this group that are pathogenic for man include Siberian tick typhus *(Dermacentroxenus sibericus),* North Queensland tick typhus *(R. australis),* fievre boutonneuse in the Mediterranean region *(R. conorii),* and the cause of rickettsialpox *(R. akari).* Washed rickettsial cells of each species are species-specific in the CF test. All spotted fever rickettsiae except *R. akari* cause acute toxic death of mice when inoculated live intravenously in high titer. In the so-called toxin neutralization test, rickettsiae and immune guinea pig sera are incubated together before inoculation. Neutralization of the toxic death is species-specific and distinguishes between members of the goup. A number of strains of spotted fever rickettsiae that are generally avirulent for guinea pigs have been isolated from different tick species in the West, Southwest, and Southeast regions of the United States. Some, like *R. parkeri* and *R. rhipicephalus,* have been given species designation (Burgdorfer et al., 1975). Other workers (Tarasevich et al., 1976; Urvolgyi and Brezina, 1978) have isolated from ticks spotted fever rickettsiae that are also distinguishable from the type species of their respective regions. A number of the tick and human strains of North American origin have been examined by Philip et al. (1978) by immunofluorescence with mouse antisera. They reported that all human isolates were identical to *R. rickettsii* and that the isolates from ticks can be sorted into one of eight other serotypes. None cross-reacted with *R. akari.* Rickettsiae of one serotype were not limited to one geographic locality. The serotypes may be grouped epidemiologically according to association with either one or more than one species of arthropod. Some serotypes were recovered from only one tick species and are apparently host-specific. They include *R. parkeri* from the Gulf Coast tick *Amblyomma maculatum;* an unclassified rickettsia from the rabbit tick, *Dermacentor parumapertus;* an unclassified rickettsia isolated many times from the Pacific Coast tick, *Ixodes pacificus;* and an unclassified rickettsia from another Pacific Coast tick, *Dermacentor occidentalis.* Rickettsiae of the five other serotypes were recovered from more than one species of ticks that were widely distributed in the United States. The tick species involved were *Haemaphysalis leporispalustris, Dermacentor andersoni, Dermacentor variabilis,* and *Rhipicephalus sanguineus.* The serotype to which all human isolates belonged was also recovered from the first three of the foregoing tick species. The stability of these antigenic characteristics is not known yet, but, if they prove to be genetically stable, they offer an interesting opportunity to search for antigenic markers of virulence for man among the spotted fever rickettsiae.

The scrub typhus group of rickettsiae is antigenically much more heterogeneous than the typhus and spotted fever groups. Yet, only one species, *R. tsutsugamushi,* is recognized. Soluble antigens extracted by ether treatment of infected yolk sacs are not group-reactive in the CF test. Therefore, members of the group are identified by a mouse vaccination-challenge procedure. A laboratory mouse that survives infection with one strain is immune to all other strains. Such broad heterologous immunity is unfortunately not seen in man, and an extraordinary amount of unsuccessful effort has been expended in attempts to overcome this vulnerability with a broadly protective vaccine. Three strains (Karp, Gilliam, and Kato), all isolated from human infections, have evolved into reference strains to which all other strains are compared. The full number of immunotypes of *R. tsutsugamushi* is still not known. Using the immunofluorescence procedure, Elisberg and co-workers (1968) identified six additional serotypes (as well as Karp, Gilliam, and Kato) from an array of human, larval mite, and wild small mammal isolates from Thailand. An immunofluorescence study of 74 isolates from Pakistan (Shirai and Wisseman, 1975) found almost all to be related to Karp. The remainder cross-reacted with Gilliam and/or Kato. Antigenic heterogeneity was also found by Russian workers in the Soviet Far East and in Tadjikistan and by Japanese workers throughout Japan. A single immunologic strain does not appear to be restricted to one species of mite vector. Implications of this strain diversity for the ecology of scrub typhus have been considered in a comprehensive review by Traub and Wisseman (1974).

METABOLISM

Although rickettsiae produce some energy from independent metabolic activity, they are obligate intracellular parasites because their essential substrates must come from the host cell. Rickettsial respiration is stimulated most by glutamate, but less so by glutamine, which the rickettsia must deaminate to glutamate. Pyruvate and the dicarboxylic acid intermediates of the citric acid cycle support respiration to a much lesser extent. Glucose, glucose-6-phosphate, lactate, sucrose, and naturally occurring amino acids other than glutamate do not stimulate respiration. End products of glutamate metabolism include ammonia and carbon dioxide, but most of the amino group of glutamate is transaminated to form aspartate by glutamate-oxaloacetate transminase. Rickettsiae have a tricarboxylic acid cycle but not a complete glycolytic sequence and pentose shunt (Coolbaugh et al., 1976). In comparison with *Chlamydia, R. typhi* can generate appreciable amounts of energy. It has also been shown that typhus rickettsiae synthesize levels of monophosphate and diphosphate kinases comparable to those of *Salmonella typhimurium.* The fact that *C. burnetii* does catabolize glucose is further evidence of the great difference between this organism and other rickettsiae. All metabolic studies are hindered by the difficulties of harvesting viable undamaged whole cells or extracts free of host-cell material. The recent development of renografin density gradient centrifugation (Weiss et al., 1975) provides, for the first time, highly purified viable rickettsiae. Like other bacteria, rickettsiae have an electron-transport system, including nicotinamide adenine dinucleotide (NADPH) and NADPH-dependent enzymes. Oxidative energy can be converted into high energy phosphate bonds by formation of ATP. Rickettsiae can synthesize small amounts of protein as measured by incorporation of radiolabeled amino acids and the inhibition of this incorporation by chloramphenicol. The phosphorylation and stimulation of endogenous levels of adenine nucleotides by glutamate metabolism indicates that rickettsiae have the potential for sustained biosynthetic capabilities. However, the paucity of other nucleotides places severe restrictions on nucleic acid synthesis. Rickettsiae also synthesize lipids, but on a minute scale, similar to their synthesis of protein. These synthetic activities require the donation from outside the rickettsiae of complex energy-yielding substrates, including ATP.

One of the chief objectives of the studies of the metabolic pathways of rickettsiae is the definition and synthesis of an artificial medium that supports extracellular rickettsial growth. To date, this has not been achieved. Given the chemical composition and structural similarity between rickettsiae and other gram-negative bacteria, there is obviously some fundamental difference between the two that precludes extracellular growth of rickettsiae. Much attention has focused on the integrity and function of the rickettsial cell wall as a possible explanation for this difference. Moulder (1966) propounded the concept of "leakiness" of the cell wall. He suggested that rickettsial cell walls were nonspecifically porous and allowed passage of host-cell nutrients into intracellular rickettsiae, but allowed loss of essential nutrients from extracellular rickettsiae. Winkler (1976) disproved this thesis in a series of elegant experiments showing that rickettsial cell membranes are highly specialized and active rather than passive with respect to nutrient transport. He defined a carrier-mediated transport system that is very specific for ATP and ADP and the analog β, γ-methylene-ATP. The system is an obligatory exchange system in that a molecule can enter a cell only if another molecule can exit. The pool of ATP and ADP remains the same, but ATP can be acquired in exchange for ADP. Smith and Winkler (1977) defined a second system that actively transports lysine against a gradient. The system is highly specific only for L-lysine, D-lysine, and aminoethyl cysteine.

Rapid death of extracellular rickettsiae can be retarded by suspension in a medium composed of sucrose and glutamate in potassium phosphate buffered to pH 7.0 (SPG). This medium, devised years ago by Bovarnick et al. (1950), has been widely used to sustain rickettsiae during purification, metabolic studies, and titrations. Its performance was improved by the addition of serum albumin. Rickettsiae can be preserved for long periods frozen at minus 60° C, in whole yolk sac membranes or membranes triturated in SPG. In an effort to improve plaquing of rickettsiae, Wike et al. (1972) showed that extracellular infectivity and stability could be maintained more satisfactorily for short periods if rickettsiae were suspended in brain-heart infusion broth.

Rickettsiae multiply by transverse binary fission. Wisseman and Waddell (1975) have studied the kinetics of growth in irradiated chick embryo cells. As described by Weiss (1978), growth of *R. prowazekii* proceeded immediately if the inoculum was derived from rickettsiae harvested in the logarithmic phase of growth. There was a lag period of about 7.5 hours if rickettsiae were harvested in the stationary phase. The generation time of rickettsiae was slightly less than 9 hours at 34° C, and this rate continued for about 36 to 48 hours. After this period, heavily infected cells released their rickettsiae, and the cycle of growth became irregular.

Well-adapted strains of rickettsiae grow luxuriously in cell cytoplasm, often pushing the nucleus to one side, but they do not aggregate in the cytoplasm to form inclusion bodies like *Ehrlichia.* Spotted fever rickettsiae are distinguished by their ability to grow in the cell nucleus as well as the cytoplasm. *R. canada,* a member of the typhus group, also grows in the nucleus occasionally. Rickettsiae form plaques in monolayer cultures, and it is possible to clone strains. *R. rickettsii* forms plaques larger and faster than other rickettsiae. This is consistent with the independent observation that *R. rickettsii* is released from the cell and infects other cells at a much earlier stage of the growth cycle than either *R. prowazekii* or *R. tsutsugamushi.*

Rickettsiae require well-nourished host cells for growth, but the host cells do not need to multiply. It is possible, therefore, to stop host-cell metabolism with irradiation or colchicine and use these stationary cells as temporary microenvironments for the culture and study of rickettsiae. In such cells, rickettsiae may grow until the host cell bursts, but release of rickettsiae from actively metabolizing cells appears to occur also by extrusion. Schaechter et al. (1957) observed rickettsiae trapped in microfibrillar structures protruding from the edge of the cell. When the microfibrillae retracted, they either carried the rickettsiae back into the cytoplasm or released them to the outside. Stork and Wisseman (1976) have observed that *R. prowazekii* can infect and multiply in enucleated L and chicken embryo cells, but the rickettsiae do not achieve true exponential growth. This finding is consistent with the observations of Weiss (1973) that *R. typhi* continued to grow in cells that could not synthesize protein because of treatment with cycloheximide.

PATHOGENIC PROPERTIES

Rickettsiae enter most mammalian cells by first attaching to the cell membrane and then being phagocytized. Attachment is not a casual process. Close contact is required; centrifugation of mixtures of entodermal cell cultures and rickettsiae increases penetration. Active participation of rickettsiae is also required. Ramm and Winkler (1973) have studied rickettsial red blood cell interactions as a possible model of cell penetrance. They showed that the lysis of sheep red blood cells required live rickettsiae. The erythrocyte was found to have a receptor for rickettsiae, a neutral lipid complex of cholesterol and palmitic acid. Energy poisons, such as potassium cyanide, inhibited both adsorption and lysis, and addition of ATP reactivated this rickettsial activity (Winkler, 1974). Scanning electronmicrographs revealed that rickettsiae adsorb end-on almost exclusively. Winkler and Ramm (1975) have also observed that rickettsiae harvested after lysing red blood cells can adsorb to other red blood cells. The mechanism of red blood cell lysis is unknown, but Winkler has observed in negative-stained preparations that lysed erythrocytes appear to be full of discrete holes. Hyperimmune rabbit serum inhibits adsorption but only by aggregation of rickettsiae. Hyperimmune serum enhances phagocytosis of rickettsiae by professional phagocytes, but it does not improve phagocytosis by such nonprofessional cells as endothelial cells and fibroblasts.

Once inside the cell, virulent rickettsiae can escape from the phagosome before it fuses with the lysosome and thus survive to multiply freely in the cell cytoplasm. By contrast, the avirulent Madrid E strain of *R. prowazekii* remains in the phagosome and is destroyed when the phagosome forms and the rickettsiae are exposed to lysosomal enzymes. Phagocytosed rickettsiae coated with immune serum are destroyed in professional phagocytes but survive in nonprofessional phagocytes. In contrast, *C. burnetii* remains in the phagosome and multiplies, protected by unknown means from the lysosomal hydrolases.

For experimental work, rickettsiae are usually cultivated in fertile hen eggs, tissue cultures, and small laboratory animals. The minimum infectious dose for eggs has been estimated to be one infectious rickettsia for the Madrid E strain of *R. prowazekii.* Ley et al. (1952) also estimated that only one *R. tsutsugamushi* organism could cause human disease. The optimum growth temperature for *R. rickettsii* is about 33.5° C and for other rickettsiae about 35° C. Large numbers of rickettsiae can be harvested from infected yolk sacs of chicken embryos (Cox, 1938), and this is usually the source of rickettsiae for a wide variety of investigations and for preparation of diagnostic antigens. Many host cell-parasite relationships have been worked out in explant and continuous-cell-line tissue cultures. Small laboratory animals are used chiefly for isolation of rickettsiae from patients. Guinea pigs are the animal of choice for typhus and the spotted fever group, and the laboratory mouse for scrub typhus (see *Laboratory Diagnosis*).

The basic lesion caused by rickettsiae is a vasculitis, localized to the endothelium and smooth muscle of the vessel wall. The lesion contains rickettsiae. There is endothelial swelling, thrombosis, and vascular and perivascular necrosis. Vascular permeability is increased. The infected host displays varying degrees of hemorrhage, tissue edema, and peripheral circulatory failure. Depending on the distribution of the vasculitis, the patient develops a petechial rash, frank hemorrhage into the skin, gangrene of appendages, interstitial inflammation of major viscera, lymph node hyperplasia, and meningoencephalitis. The circulatory failure of severe rickettsial disease has been studied by Harrell and Aikawa (1949). Plasma volume is decreased, and the extravascular fluid spaces are increased. The pathophysiology has been studied in experimental hosts including the guinea pig, rabbit, and monkey. Different species of rickettsiae have typical pathophysiologic manifestations, as shown by variations in the diseases they cause in man. Each pathogenic species causes some infections in man and experimental animals, however, that are entirely inapparent and others that are so severe as to be lethal. In some geographic regions, inapparent infection is the rule. As yet, there are no defined markers that distinguish virulent from avirulent rickettsiae in terms of host response to infection, nor is there clear evidence for any mechanism of rickettsiae-induced pathophysiology other than direct cell invasion and injury by live rickettsiae. De Brito et al. (1968) found gamma globulin and complement in late vascular lesions of guinea pigs infected with *R. rickettsii*, which suggests that immune complex vasculitis occurs in spotted fever. More recently, Walker and Henderson (1978) studied *R. rickettsii* infection in the immunosuppressed guinea pig. They found as severe a vasculitis in the immunosuppressed as in control animals and argued that rickettsiae damage cells without the participation of immune mechanisms.

Schramek, Brezina, and Tarasevich (1976) have reported isolation of lipopolysaccharide antigens from *R. prowazekii*, *R. typhi*, *R. canada*, *R. conorii*, and *R. tsutsugamushi*. From hypothermic reactions in intraperitoneally inoculated white rats and chemical analysis, they concluded that these antigens had endotoxic properties. Full characterization of a rickettsial endotoxin has not been reported. Nevertheless, the endotoxin hypothesis is attractive as an explanation of some aspects of rickettsial pathophysiology.

Yamada and co-workers (1978) reported activation of the kallikrein-kinin system in five patients with American spotted fever. Four had petechial rashes characteristic of vasculitis and then developed disseminated intravascular coagulation. They concluded that even if kinins do not initiate the early vascular changes, they may promote progression of the lesions.

Despite these studies of the basic pathology and pathogenic properties of rickettsiae, the virulence factors are still largely unknown. The attenuated Madrid E strain of *R. prowazekii* (Perez, 1963), in contrast to fully virulent strains, disappears quickly from the tissues of guinea pigs, so that serial passage is usually unsuccessful. It does not differ substantially from virulent strains, however, in a variety of other properties, including toxicity for mice, hemolytic activity, infectivity for body lice, antigenic characteristics, and quantity of slime layer. There is no difference in its ability to infect and grow in chicken embryo cells and a variety of other cell lines. It does differ in one important respect: in human macrophages, the Madrid E strain cannot escape from the phagosome and is killed by lysosomal hydrolases. This may be the mechanism of its attenuation.

The stability of the E strain has been questioned by Russian (Balayeva and Nikolskaja, 1972) and Slovak (Kaźar et al., 1973) workers who noted reversion to full virulence for guinea pigs after prolonged serial passage in the mouse lung and the guinea pig peritoneum. It is unclear whether this phenomenon represents selection of a virulent strain from a mixture that is predominantly avirulent or is a true back-mutation. Studies with cloned E strains have not been reported.

There have been many studies of the virulence of spotted fever rickettsiae. Long ago it was observed that inoculation of rickettsiae into guinea pigs from unfed adult ticks *(Dermacentor andersoni)* would produce immunity but no disease. If the ticks were given a blood meal before harvest and inoculation of rickettsiae, the guinea pigs became sick. Later, Spencer and Parker reexamined this phenomenon by inoculating a virulent strain of *R. rickettsii* into adult ticks and storing them for several months at refrigerator temperature. Inoculation at this point produced, as before, immunity but no disease in guinea pigs. If the ticks were warmed first or fed a blood meal, virulence for guinea pigs was restored. Close examination of this phenomenon showed that it was not caused by an increase in the number of rickettsiae (Gilford and Price, 1955). Addition of para-aminobenzoic acid (PABA), parahydroxybenzoic acid, or PABA plus coenzyme A to rickettsiae held for 60 hours at 25° C restored virulence to the stored rickettsiae. Virulence could also be restored by incubating the rickettsiae with recently fed and ground-up ticks.

Generally, the virulence of rickettsiae for laboratory animals does not parallel virulence for man. However, there does appear to be a consistent geographic difference in the virulence of scrub typhus rickettsiae for man. A strain from the Pescadores Islands, where disease is characteristically mild, was used as a live vaccine in Japan. Disease contracted during the winter months in Japan is said to be much milder than disease contracted in the warmer months. It is interesting that different mite vectors are involved in the Pescadores and during different seasons in Japan. Certainly, scrub typhus is commonly very mild in Southeast Asia, where the predominant vector mite is still another species. Because a good marker for virulence is lacking, it is not possible to examine the stability of rickettsial virulence during passage through different species of mites.

IMMUNOLOGY

Natural host resistance to infection must be considered in assessments of rickettsial virulence. Groves and Osterman (1978) examined genetic resistance to lethal infections with *R. tsutsugamushi* in over 30 inbred strains of mice. They found some strains fully resistant, some fully susceptible, and a few selectively resistant to lethal infection with the Gilliam strain of scrub typhus rickettsiae. Resistance was genetically dominant and controlled by a single gene or a closely linked cluster of autosomal genes. Resistance was not due to an inability of host cells to support rickettsial growth. Susceptibility, on the other hand, was not due to inability to mount an immune response. Russian workers (Kokorin et al., 1976) analyzed the same phenomenon and noted a marked difference in the reaction of the macrophages of resistant and suceptible mice. Fatal infection was accompanied by death of macrophages and peritoneal necrosis. The resistant mice showed no clinical signs of infection but mounted an intensive macrophage reaction at the inoculation site. Most of the inoculated rickettsiae died at the inoculation site, and the macrophages remained viable. Some rickettsiae survive and persist in resistant mice for a long time without producing overt clinical disease. Recently, Anderson and Osterman (1980) have also observed a genetic basis for natural resistance to lethal infection with *R. akari* (rickettsialpox).

Rickettsiae stimulate both humoral and cellular immunity. The role of each in response to infection of man and animals is not completely understood. Humoral antibody appears in response to rickettsial infections between the seventh and fourteenth days of disease. Passive transfer of immune serum provides protection against clinical disease in animals challenged with the same strain. Persistence of rickettsiae in immune people and animals without recrudescence of disease may be the result of protection by humoral antibody, but humoral antibody is clearly not the sole factor in immunity to rickettsiae. Cellular immune processes also occur, as shown by the development of dermal hypersensitivity, lymphocyte transformation, and elaboration of migration-inhibiting factor.

Recent studies of *R. tsutsugamushi* and *C. burnetii* revealed some details of the immune process. Shirai and co-workers (1976) have shown that passive transfer of spleen cells from infected mice protects the recipient against challenge with homologous and heterologous strains of *R. tsutsugamushi*. The protective spleen cells were thymus-dependent lymphocytes. Subsequently, these investigators (Catanzaro et al., 1977) demonstrated that peritoneal exudate lymphocytes (PEL) were the mediators of this cellular protection. It was not possible to determine from their experiments whether PELs were the primary mediators, as they are in other bacterial infections. Subsequently, Nacy and Osterman (1979) determined that protective splenic lymphocytes elaborated lymphokines that activate the killer macrophages.

LABORATORY DIAGNOSIS

Isolation of rickettsiae should ordinarily be attempted only in a research laboratory. Isolation is accomplished preferably by immediate (within one hour of collection) intraperitoneal inoculation of the specimen into a suitable experimental animal. The isolation host of choice for the typhus and spotted fever groups of rickettsiae is the guinea pig and for scrub typhus the laboratory mouse. The most appropriate isolation specimen for human diagnosis is blood clot prepared as a 10 per cent suspension in skim milk or brain-heart infusion broth. Other diluents adversely affect rickettsiae. Several passages of tissue from inoculated animals at approximately 14-day intervals may be necessary before symptomatic infection develops. In guinea pigs, the only indication may be an increase in rectal temperature, which must be recorded daily from the time of inoculation. On the first passage of typhus group rickettsiae, fever does not appear until the third week after inoculation. Either brain or spleen should be harvested from the febrile animal and inoculated intraperitoneally into other guinea pigs. If no fever or other evidence of infection occurs, animals in the second or third passage are bled on day 28 for antibody studies. Antibody may be the only evidence that rickettsiae have been isolated. The only gross evidence of *R. prowazekii* infection in the guinea pig is a fibrous exudate on the spleen surface. Examination of smears taken from scrapings of the spleen surface may show rickettsiae in serosal cells. If the inoculum of rickettsiae is large, the scrotum may enlarge, and the testes may adhere to the scrotal sac. This lesion is called the Neill-Mooser reaction after the investigators who discovered it. The lesion is much more often found in *R. typhi* infections and occurs at the time of fever. In other respects, *R. typhi* infections are like those of *R. prowazekii*, except that the incubation period of established infections is shorter (three to seven days), and the Neill-Mooser reaction is more severe. The isolated rickettsiae are identified by using the rickettsiae as antigen in the CF or other serologic test against various antisera and by cross-vaccination and challenge against known rickettsiae.

The response of guinea pigs to virulent spotted fever group rickettsiae is more severe. After several passages to establish the isolate, the incubation period is shortened to two to five days, the fever course is often prolonged, and the animals may die on the sixth to eighth day. A severe scrotal reaction usually begins on the third or fourth day of fever and first consists of edema and erythema. The scrotum becomes necrotic with eventual tissue sloughing. Gangrene of the ears and footpads may also occur. As in infections with the typhus group, necropsy findings consist chiefly of a fibrinous exudate on the spleen surface and the scrotal reaction.

Scrub typhus rickettsiae are isolated by intraperitoneal inoculation of the white laboratory mouse. Several passages of infected liver, spleen, or brain may be required to establish the agent. Gross signs of infection include ruffled fur and a swollen abdomen. Necropsy reveals an enlarged spleen and mucoid peritoneal fluid. Scrapings of the peritoneal lining yield mesothelial cells containing rickettsiae in the cytoplasm.

Rickettsiae can be isolated by inoculation either into the yolk sac of the hen eggs or into tissue cultures. Adaptation to these substrates is often slow, and small inocula may be lost. Laboratory acquired infection is a constant hazard, even in the most sophisticated laboratory (Oster et al., 1977), so isolation of rickettsiae should be attempted only by experienced personnel in well-equipped facilities. Identification of rickettsiae or rickettsial antigen in the first week of disease would greatly assist clinicians to diagnose atypical cases. DeShazo et al. (1976) diagnosed Rocky Mountain spotted fever by the fourth day of disease in rhesus monkeys upon identification of *R. rickettsii* by immunofluorescence in primary monocyte culture. Shirai et al. (1978) achieved similar results in *R. tsutsugamushi* infections of monkeys and dogs. Coolbaugh and colleagues (1978) have used this procedure for the early diagnosis of a human case of scrub typhus. For pathologic specimens, Walker and Cain (1978) have described a method allowing retrospective analysis of fixed, paraffin-embedded tissue by *R. rickettsii*-specific immunofluorescence. The diagnosis of *R. rickettsii* infection may also be suggested by examination of ticks taken from patients. Burgdorfer (1970) identified *R. rickettsii* by immunofluorescent examination of the hemolymph of live ticks. In desiccated ticks, rickettsial antigen may be detected by immunofluorescence for several weeks (Kurz and Burgdorfer, 1978).

Human rickettsial antibodies are detectable by many serologic procedures. Complement fixation has been the most widely used specific test because antigens for the typhus and spotted fever groups have been available commercially or from central reference laboratories. Group-specific soluble antigen is prepared by ether extraction of infected yolk sacs of embryonated hen eggs. The antigen is inactivated. Typhus group antigen is prepared from *R. typhi,* while spotted fever is usually prepared from *R. akari.* The CF test can be made species-specific by using washed rickettsial cell suspensions as antigen. For diagnosis of spotted fever infections in various parts of the world, it is best to use the prevalent rickettsial species as antigen since *R. akari,* while broadly group-reactive, does not detect CF antibody to all the spotted fevers. Scrub typhus-soluble CF antigens are principally strain-specific, and the CF test is unsuitable for group diagnosis.

The indirect microimmunofluorescence procedure (Bozeman and Elisberg, 1963) is increasingly employed for serologic diagnosis. Concentrated suspensions of rickettsiae from infected yolk sacs or cell culture are used as antigen. In the micro-technique, many antigen spots can be applied to one slide. After air drying and fixation in acetone, the slides may be stored at −70° C for later use. Philip and co-workers (1977) have compared this procedure with others for the diagnosis of both typhus and spotted fever group infections. In their hands, it was superior to the CF test for identifying disease.

Rickettsial agglutination has been used for many years for the diagnosis of a variety of rickettsial infections by French workers and has been widely used by others for the diagnosis of Q fever. A microagglutination procedure (Fiset et al., 1969) conserves antigen (washed rickettsial suspensions).

Chang et al. (1954) performed hemagglutination with an erythrocyte-sensitizing substance (ESS) from typhus and spotted fever group rickettsiae. This antigen, which depends for its activity on a carbohydrate moiety (Osterman and Eiseman, 1978), has not been isolated from *R. tsutsugamushi.* The hemagglutination and microagglutination procedures compare favorably with the microimmunofluorescence procedure for the diagnosis of North American spotted fever (Philip et al., 1977).

Because CF antibodies are not found until the third week of disease, much effort has been expended to develop procedures for early diagnosis. The indirect immunofluorescence procedure, the indirect hemagglutination test, and the ELISA (enzyme-linked immunosorbent assay) all detect antibodies as early as the sixth to seventh day of disease. Woodward et al. (1976) demonstrated *R. rickettsii* in skin biopsy specimens from two patients on days 4 and 8 of illness by direct immunofluorescence. While confirming this experience in some patients, other workers have failed to identify rickettsiae in skin biopsies from patients already treated with tetracycline or chloramphenicol. Years ago, Fleck et al. (1960) found murine typhus antigen in the urine by indirect hemagglutination, but this early diagnostic technique has not been evaluated by others.

Specific rickettsial antibody appears by the end of the first week of disease in human and animal infections and persists in some cases for years. Rickettsemia and circulating specific antibody are detectable in the same blood specimens during disease and convalescence. Humans respond early to primary typhus infection with IgM antibodies, but this class of antibody is not recalled in recrudescent typhus (Murray et al., 1965). Philip and co-workers (1976) found IgM antibodies by microimmunofluorescence early in American spotted fever patients but noted its absence in 15 per cent of them. IgM antibody usually disappears within three to four months. IgG antibody usually appears by the third week of disease and persists for one or more years.

The antibody response can be suppressed by antibiotics early in the disease. Patients with American spotted fever that were treated early, especially during the first two days of illness, were less likely to be seropositive and had lower titers of antibody by the CF test during convalescence (Philip et al., 1977). Smadel (1954) also observed delayed or diminished antibody production by patients who were treated for scrub typhus by the third day of disease but there was no effect when therapy was started on the sixth day or later. It is assumed that early treatment suppresses antibody production by reducing the antigenic mass.

Antigenic cross-reactions occur between the typhus and spotted fever groups. The degree of cross-reactivity reflects not only the extent of shared antigens but also the species of host involved. Ormsbee et al. (1978) found that most patients who were acutely ill with epidemic typhus produced IgG and IgM antibodies to purified antigens of *R. prowazekii*, *R. typhi*, and *R. canada*. They also produced IgG, but seldom IgM, antibodies to spotted fever rickettsiae. Similarly, patients acutely ill with spotted fever produced IgG and IgM antibodies to *R. rickettsii*, *R. conorii*, and *R. akari*, as well as both classes of antibody to the typhus group. By contrast, the mouse produces species-specific rickettsial antibody. Ormsbee et al. (1978) have speculated that the more specific rodent antibody response reflects the greater antiquity of this host-parasite relationship. Scrub typhus and Q fever rickettsiae do not share antigens with other rickettsiae. The degree of heterologous antibody response in spotted fever and typhus infections is usually less than the homologous response, which helps clarify the serologic diagnosis of these infections. Prior immunization with epidemic typhus or spotted fever vaccine can cause confusion in the serologic findings because the anamnestic response of vaccine-induced antibody may exceed that stimulated by infection.

Rickettsiae share antigens with other bacteria. The antigens shared with *Proteus* organisms form the basis of the Weil-Felix agglutination test. For many years, this test was the only serologic test widely available for the diagnosis of rickettsial disease. The shared antigens have never been fully characterized, but years ago Castaneda (1934) described a carbohydrate antigen shared by *Proteus vulgaris* OX-19 and *R. prowazekii*. In human typhus, agglutinins occur chiefly against OX-19 and, to a much lesser extent, to the OX-2 strain. These agglutinins are rarely found in recrudescent typhus (Brill-Zinsser disease), although Ormsbee and co-workers (1977) found a brisk agglutinin response in two of five cases in Ethiopia. The cross-reactive antibody response stimulated by the spotted fever groups varies widely. Some patients produce chiefly OX-19 antibody, some chiefly OX-2, and some both. Therefore, the Weil-Felix agglutinins do not distinguish spotted fever from typhus group infections. Scrub typhus may stimulate production of agglutinins to *Proteus mirabilis* strain OXK. A titer of 1:200 or greater suggests infection. Because these antigens are commercially available, and the procedures are simple, physicians have relied on the Weil-Felix agglutination tests for rickettsial diagnosis. With the development of other procedures that use strain-specific antigens that are also noninfectious, the Weil-Felix test should be discarded. It is unreliable because agglutinins are often absent in proven rickettsial disease, especially scrub typhus. Furthermore, the presence of agglutinins may only indicate infection with *Proteus* or other bacteria (Hechemy et al., 1970). Infections with *R. akari* and *C. burnetii* do not produce Weil-Felix agglutinins.

DRUG SUSCEPTIBILITY

The growth of rickettsiae is inhibited by chloramphenicol and the tetracyclines that are given for treatment of rickettsial diseases. Penicillin and streptomycin are of no clinical value but are slightly inhibitory to rickettsiae. Sulfonamides have no effect on rickettsiae. On the other hand, para-aminobenzoic acid, an antagonist of sulfonamides, inhibits most rickettsiae, but *R. tsutsugamushi* is affected only slightly and *C. burnetii* not at all. Ormsbee et al. (1955) studied the relative inhibitory effect of various antibiotics in infected chicken embryos and found that the inhibitory dose of oxytetracycline for typhus and spotted fever rickettsiae was about one-eighth that of chlortetracycline, which was one third as active as chloramphenicol. Erythromycin was highly inhibitory for *R. prowazekii,* less so for spotted fever rickettsiae, and almost not at all for *C. burnetii.* Chloramphenicol is also relatively inactive against *C. burnetii. R. tsutsugamushi* is as sensitive to chloramphenicol and the tetracyclines in laboratory culture as the spotted fever rickettsiae. Doxycycline, a long-acting tetracycline, is highly effective clinically against the typhus group and *R. tsutsugamushi.*

Growth-inhibiting antibiotics do not eliminate viable rickettsiae from infected people. *R. tsutsugamushi* can persist in lymph nodes for years after recovery from infection. Persistence of *R. prowazekii* in patients who had survived an attack of epidemic typhus in the preantibiotic era was dramatically illustrated by recrudescent disease long after they had emigrated from endemic areas of Europe to America. Recrudescent typhus is called Brill-Zinsser disease. In laboratory animals that survive a rickettsial infection, recrudescent disease can be provoked by administration of cortisone. It is assumed that cortisone in animals mimics stress in man, which seems to modify the steady-state host-parasite relationship and permit rickettsiae to proliferate. Clinical recrudescence of other rickettsial diseases long after recovery has not been observed, but persistence of viable rickettsiae after treatment indicates that these drugs are rickettsiostatic. Continous exposure of *R. tsutsugamushi* in cultured L-cells to high concentrations of chloramphenicol eventually sterilized the culture, but only after three weeks.

Antibiotic-resistant strains of rickettsiae have not been isolated from nature. Many serial passages of the Madrid E strain of *R. prowazekii* in the presence of increasing amounts of chloramphenicol did yield a substrain resistant to chloramphenicol (Weiss and Dressler, 1962). Resistance was not lost during 10 drugless egg passages. Weiss and Dressler (1960) isolated an erythromycin-resistant strain by the same procedure but could not induce resistance to tetracycline.

HOST RANGE

The human pathogens among the rickettsiae are distributed widely in many arthropods and mammals. Among the typhus group, *R. prowazekii* parasitizes both the head and body varieties of the human louse, *Pediculus humanis.* The human body louse is the major vector of classic epidemic typhus. *R. prowazekii* infects only the gut epithelium of the louse, which acquires the rickettsiae by feeding on the blood of a rickettsemic human. After a few days, proliferating rickettsiae burst the epithelial cells of the gut and appear in the louse feces. The louse usually dies from the infection within 7 to 10 days. Its salivary glands are not infected. Man acquires the infection from the louse by scratching the skin that has become contaminated with rickettsiae-containing louse feces (autoinoculation). Until recently, it was believed that the only cycle of *R. prowazekii* in nature was man-louse-man. Man was thought to be the reservoir because the louse does not pass the rickettsiae from generation to generation transovarially like other rickettsiae of ticks and mites. Recently, Bozeman and co-workers (1975), in a remarkable series of studies, demonstrated the first nonhuman mammalian host of *R. prowazekii,* the American flying squirrel *(Glaucomys volans).* They found enzootic foci in flying squirrels from Florida, North Carolina, Virginia, and Maryland and also recovered rickettsiae from the squirrel lice and fleas. The squirrel flea is known to feed on man. The strains of rickettsiae isolated from flying squirrels have been identified as *R. prowazekii* (Dasch et al., 1978). McDade and coworkers (1980) have just reported that examination of sera from 7 of 1575 persons for rickettsial antibodies from 1976 through 1978 suggested recent infection with *R. prowazekii.* Two of the seven patients had had contact with flying squirrels. Reiss-Gutfreund (1966) reported infection with *R. prowazekii* in domestic animals and their ticks in Ethiopia that seemed to confirm earlier reports of epidemic typhus antibodies in domestic animals as well as man in the Upper Volta, Ruwanda-Burundi, and the Central African Republic. But other workers are skeptical (Burgdorfer et al., 1972).

R. typhi is found naturally in rats and mice. It is transmitted from rat to rat by the rat louse and the rat flea, *Xenopsylla cheopis,* which is usually the vector to man. Many other wild animals are also susceptible to *R. prowazekii.* The host range of *R. canada* is not well defined. It has been isolated in nature only from *Haemophysalis leporispalustris* ticks in Ontario, Canada (McKiel et al., 1967).

The host ranges of the spotted fever group and scrub typhus group are discussed in the chapters on each disease.

Q FEVER RICKETTSIAE

Coxiella burnetii has been placed in a separate genus because it differs in many ways from other rickettsiae (Table 1). The differences have been summarized well by Ormsbee (1969). *C. burnetii* stains gram-positive when alcoholic iodine is used as mordant instead of aqueous iodine. Under these conditions, rickettsiae stain gram-negative. Rickettsiae release a soluble antigen when aqueous suspensions of cells are shaken in ether. *C. burnetii,* like other gram-negative bacteria, does not. Immunizing antigens of *C. burnetii* are very resistant to heat, whereas those of rickettsiae are heat-sensitive. The guanine plus cytosine molar per cent content of *C. burnetii* DNA is about 43 per cent, whereas rickettsiae have a G + C content in the range of 30 to 32.5 per cent. *C. burnetii* is remarkably resistant to physical and chemical agents. It survives in wool for seven to nine months at 20° C, at least two years in skim milk and tap water at 15 to 20° C, and for about six months in dried blood at room temperature. Disinfection of *C. burnetii* requires exposure to 5 per cent H_2O_2 or 2 per cent formaldehyde. Because of the extraordinary survival of *C. burnetii* in the extracellular environment, it probably infects man more often by the aerosol route than by tick transmission. Rickettsiae are generally transmitted to man by arthropods. The energy-producing metabolism of *C. burnetii* is quite different from that of rickettsiae. *C. burnetii* metabolizes pyruvate, synthesizes enzymes of the Embden-Meyerhof pathway, and contains many enzymes generally associated with the catabolism of glucose. Rickettsiae synthesize none of these enzymes.

TABLE 1. Differences Between *Coxiella burnetii* and the Rickettsiae

	C. burnetii	RICKETTSIAE
Gram stain (alcohol mordant)	Gram-positive	Gram-negative
Soluble antigen released with ether	No	Yes
Immunizing antigens	Heat-resistant	Heat-sensitive
Guanine plus cytosine molar per cent	43	30 to 32.5
Usual route of infection	Aerosol	Arthropods
Embden-Meyerhof pathway	Yes	No
Persistence in phagosomes	Yes	No
Susceptibility to chloramphenicol and erythromycin	±	++++
Acute toxicity for mice	No	Yes
Phase variation	Yes	No

The extraordinary extracellular resistance of *C. burnetii* may simply reflect its intracellular behavior, which is quite distinct from that of rickettsiae. *C. burnetii* enters the cell in phagosomes, remains there even as the phagosomes fuse with lysosomes, and can protect itself from the harsh action of lysosomal hydrolases. Rickettsiae quickly escape from the phagosome and exist free in the cell cytoplasm. Growth of *C. burnetii* in experimental yolk sac infections is only slightly affected by chloramphenicol and erythromycin, which are highly rickettsiostatic. *C. burnetii* is sensitive to the tetracyclines, however. Unlike certain rickettsiae, live *C. burnetii* do not cause acute toxic death of mice when inoculated intravenously in high titer. Finally, *C. burnetii* exhibits a unique phase variation, which is host-controlled. *C. burnetii* exists in nature only in Phase I. When cultivated in chick embryos, the organism shifts to Phase II but reverts to Phase I upon passage through laboratory animals. The significance of phase variation has been a topic of intense study since it was originally recognized by Stoker et al. (1955). Shift in phase is accompanied by changes in major antigenic components, pathogenicity, immunogenicity, buoyant density, agglutinability, staining properties, resistance to phagocytosis, and pyrogenicity. Phase variation appears to represent a phenotypic rather than a genotypic change. There is no discernible morphologic difference between phases of *C. burnetii.*

ROCHALIMAEA QUINTANA

This organism causes trench fever in man. It was originally recognized as a natural inhabitant of the gut lumen of the human body louse in the course of studies undertaken to elucidate the etiology of louse-borne epidemic typhus during World War I. Although it was called a rickettsia for many years, this organism has been reclassified into its own genus because it is usually located extracellularly and can be cultivated on cell-free media. The organism is named for the Brazilian scientist H. da Rocha-Lima, who was one of the early investigators of rickettsial diseases, and for the form of the disease, in which febrile episodes recur after every fifth (quintana) day without fever.

The organisms resemble rickettsiae in size and

cell structure. The cell wall is similar to that of rickettsiae in chemical composition (Osterman et al., 1974) and in structure by electron microscopy (Ito and Vinson, 1965). It multiplies by binary fission and, like rickettsiae, metabolizes glutamine and succinate but not glucose. It is gram-negative and stains like rickettsiae in the Giminez and Macchiavello stains. It is susceptible to inactivation by standard disinfectants, but it survives for months in dried louse feces, like *R. prowazekii.*

R. quintana is distinct from all other rickettsiae in its G + C content of 38.5 moles per cent. This characteristic contributed to the recent identification of Baker's vole agent as *R. quintana* (see Chapter 191). This was the first isolation of *R. quintana* in nature from a nonhuman mammalian source (Weiss et al., 1978).

The second major difference of *R. quintana* from rickettsiae is its capacity to grow on cell-free media. It was first cultivated by Vinson (1966) on a blood agar base enriched with 6 per cent horse serum inactivated at 56° C for 30 minutes and 4 per cent washed, hemolyzed horse erythrocytes incubated at 37° C in a moist atmosphere of 5 per cent CO_2 in air. After 12 to 14 days incubation, colonies are round, lenticular, translucent, mucoid, and 65 to 200 μ in diameter. Colonies develop only on the surface of the agar. Organisms do not multiply without added carbon dioxide nor under anaerobic conditions. Mason (1970) developed a liquid culture medium for study of the growth cycle of *R. quintana* and noted that fetal calf serum could be substituted for the erythrocyte lysate. The growth cycle was characterized by a lag phase of approximately 24 hours, an exponential growth phase of 72 hours, and a doubling time of approximately 4.5 hours. Myers and co-workers (1972) demonstrated a growth requirement for hematin and concluded that the hematin was required as a precursor in the synthesis of various heme-proteins. *R. quintana* would not grow on hematin substitutes that satisfied other hematin-requiring bacteria. Apparently, it absolutely requires hematin for the synthesis of heme-proteins. *R. quintana* does not appear to produce H_2O_2 and is catalase-negative (Myers et al., 1972).

The only experimental host for *R. quintana* is the rhesus monkey (Mooser and Weyer, 1953), but specific antibody can be produced in rabbits and guinea pigs immunized with soluble or whole cell antigens in Freund's adjuvant (Herrmann et al., 1977). A number of serologic procedures detect *R. quintana* antibody in man, including complement fixation, (Vinson and Campbell, 1968); passive hemagglutination (Cooper et al., 1976); enzyme immunoassay; and radioimmunoprecipitation tests (Herrmann et al., 1977). Hollingdale and co-workers (1978) examined sera by the enzyme immunoassay method from patients with various rickettsial infections. Nine of 15 sera from patients with typhus group infections, 2 of 7 sera from *R. rickettsii* infections, and 3 of 6 sera from scrub typhus infections produced antibody to *R. quintana* antigen. Cross-reactions also occurred in guinea pigs infected with scrub typhus. These findings suggest that *R. quintana* shares antigens with rickettsiae. Seroepidemiologic studies, which have not yet been undertaken on a broad scale, must take these cross-reactions into account.

RICKETTSIA SENNETSU

In 1954, two Japanese groups (Fukuda, 1954; Misao and Kobayashi, 1954) reported the isolation of a rickettsia-like agent from patients with an infectious mononucleosis-like syndrome on the island of Kyushu. Subsequent studies established *R. sennetsu* to be immunologically distinct from *R. tsutsugamushi;* from *Neorickettsia helminthoeca,* the cause of salmon poisoning in dogs; and from Elokomin fluke fever agent, the cause of a similar lymphadenopathy in bears and raccoons in the Pacific Northwest of the United States (Kitao et al., 1973). Korb and co-workers (1966) in Czechoslovakia examined sera from 17 patients with clinical infectious mononucleosis. They, and other European workers, concluded that there was no good evidence to associate *R. sennetsu* with infectious mononucleosis in central Europe. Electronmicrographs show that *R. sennetsu* resides within an intracytoplasmic vesicle (Fig. 5) in contrast to other rickettsiae, which lie free in the cytoplasm (Anderson et al., 1965). Huxsoll recognized that this property was similar to that of organisms of the genus *Ehrlichia,* which is taxonomically close to *Rickettsia* but pathogenic for animals. Recent studies by Huxsoll, Ristic, and colleagues (personal communication) suggest an immunologic relationship between *R. sennetsu* and *Ehrlichia canis,* a tick-borne disease agent of dogs found in many parts of the world. It is not yet known whether *R. sennetsu* is tick-transmitted to man, like *Ehrlichia,* or is transmitted by eating raw fish, like salmon poisoning in dogs. The discovery of this interesting disease agent and its possible relationship to *Ehrlichia*

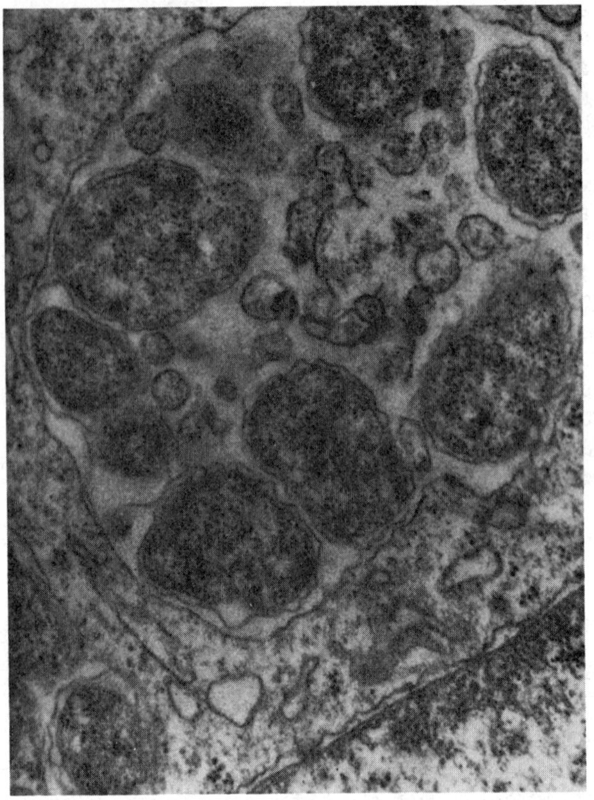

FIGURE 5. R. sennetsui *in a distended vacuole. Several organisms are enclosed in a membrane-lined vacuole. Note that each organism is enclosed by a cell wall and a plasma membrane. ×30,000. (From Anderson, D. R., et al.: J Bacteriol 90:1387, 1965.)*

underscores the importance of considering agents of animal disease when searching for the specific etiology of obscure human febrile diseases. These studies will not always be as rewarding as those on *R. sennetsu* mononucleosis. An example of one of these febrile diseases that still remains obscure is Kawasaki disease (or mucocutaneous lymph node syndrome of infants and children), first reported by Kawasaki on the basis of 50 cases in Tokyo (see Chapter 201). Hamashima et al. (1973) reported rickettsia-like bodies by electron microscopy of skin and lymph node biopsies, but others cannot confirm these results. The unresponsiveness of the disease to antibiotics suggests an alternate etiology.

REFERENCES

Anacker, R. L., Pickens, E. G., and Lackman, D. B.: Details of the ultrastructure of *Rickettsia prowazekii* grown in the chick yolk sac. J Bacteriol 94:260, 1967.

Anderson, D. R., Hopps, H. E., Barile, M. F., and Bernheim, B. C.: Comparison of the ultrastructure of several rickettsiae, ornithosis virus, and mycoplasma in tissue culture. J Bacteriol 90:1387, 1965.

Anderson, G. W., and Osterman, J. V.: Host defenses in experimental rickettsialpox: Genetics of natural resistance to infection. Infect Immun 28:132, 1980.

Balayeva, N. M., and Nikolskaya, V. N.: Enhanced virulence of the vaccine strain E of *Rickettsia prowazekii* on passaging in white mice and guinea pigs. Acta Virol 16:80, 1972.

Bovarnick, M. R., Miller, J. C., and Snyder, J. C.: The influence of certain salts, amino acids, sugars, and proteins on the stability of rickettsiae. J Bacteriol 59:509, 1950.

Bozeman, F. M., and Elisberg, B. L.: Serological diagnosis of scrub typhus by indirect immunofluorescence. Proc Soc Exp Biol Med 112:568, 1963.

Bozeman, F. M., Masiello, S. A., Williams, M. S., and Elisberg, B. L.: Epidemic typhus rickettsiae isolated from flying squirrels. Nature 255:545, 1975.

Burgdorfer, W.: Hemolymph test. A technique for detection of rickettsiae in ticks. Am J Trop Med Hyg 19:1010, 1970.

Burgdorfer, W., Ormsbee, R. A., and Hoogstraal, H.: Ticks as vectors of *Rickettsia prowazekii* — a controversial issue. Am J Trop Med Hyg 21:989, 1972.

Burgdorfer, W., Sexton, D. J., Gerloff, R. K., Anacker, R. L., Philip, R. N., and Thomas, L. A.: *Rhipicephalus sanguineus*: Vector of a new spotted fever group rickettsia in the United States. Infect Immun 12:205, 1975.

Castaneda, M. R.: The antigenic relationship between proteus X-19 and typhus rickettsia. II. A study of the common antigenic factor. J Exp Med 60:119, 1934.

Catanzaro, P. J., Shirai, A., Agniel, L. D., and Osterman, J. V.: Host defenses in experimental scrub typhus: Role of spleen and peritoneal exudate lymphocytes in cellular immunity. Infect Immun 18:118, 1977.

Chang, R. S. M., Murray, E. S., and Synder, J. C.: Erythrocyte-sensitizing substances from rickettsiae of the Rocky Mountain spotted fever group. J Immunol 73:8, 1954.

Coolbaugh, J. C., Progar, J. J., and Weiss, E.: Enzymatic activities of cell-free extracts of *Rickettsia typhi*. Infect Immun 14:298, 1976.

Coolbaugh, J. C., Ho, C-M., and Fang, R. C. Y.: Diagnosis of scrub typhus by culture of blood monocytes. In Tan, D. S. K. (ed.): Proceedings of the 18th South East Asia Ministers of Education Organization Tropical Medicine Seminar. Bangkok, SEAMEO Regional Tropical Medicine Publishing Health Project, 1978.

Cooper, M. D., Hollingdale, M. R., Vinson, J. W., and Costa, J.: A passive hemagglutination test for diagnosis of trench fever due to *Rochalimaea quintana*. J Inf Dis 134:605, 1976.

Cox, H. R.: Use of yolk sac of developing chick embryo as medium for growing rickettsiae of Rocky Mountain spotted fever and typhus groups. Public Health Rep 53:2241, 1938.

Dasch, G. A., Samms, J. R., and Weiss, E.: Biochemical characteristics of typhus group rickettsiae with special attention to the *Rickettsia prowazekii* strains isolated from flying squirrels. Infect Immun 19:676, 1978.

deBrito, T., Tiriba, A., Godoy, C. V. F., Penna, D. O., and Jordao, F. M.: Glomerular response in human and experimental rickettsial disease (Rocky Mountain spotted fever group). A light and electron microscopy study. Pathol Microbiol (Basel) 31:365, 1968.

DeShazo, R. D., Boyce, J. R., Osterman, J. V., and Stephenson, E. M.: Early diagnosis of Rocky Mountain spotted fever. Use of primary monocyte culture technique. JAMA 235:1353, 1976.

Elisberg, B. L., Campbell, J. M., and Bozeman, F. M.: Antigenic diversity of *Rickettsia tsutsugamushi*: Epidemiologic and ecologic significance. J Hyg Epidemiol Microbiol Immunol 12:18, 1968.

Fiset, P., Ormsbee, R. A., Silberman, R., Peacock, M., and Spielman, S. H.: A microagglutination technique for detection and measurement of rickettsial antibodies. Acta Virol 13:60, 1969.

Fleck, L., Porat, S., Evenchik, Z., Klingberg, M. A.: The renal excretion of specific microbial substances during the course of infection with murine typhus rickettsiae. Am J Hyg 72:351, 1960.

Fukuda, T., Kitao, T., and Keida, Y.: Study on the causative agent of "Hyuga netsu" disease (infectious mononucleosis). Med Biol 32:200, 1954.

Gilford, J. H., and Price, W. H.: Virulent-avirulent conversions of *Rickettsia rickettsii* in vitro. Proc Nat Acad Sci 41:870, 1955.

Giminez, D. F.: Straining rickettsiae in yolk-sac cultures. Staining Technol 39:135, 1964.

Groves, M. G., and Osterman, J. V.: Host defenses in experimental scrub typhus: Genetics of natural resistance to infection. Infect Immun 19:583, 1978.

Hamashima, Y., Kishi, K., and Tasaka, K.: Rickettsia-like bodies in infantile acute febrile mucocutaneous lymph-node syndrome. Lancet 2:42, 1973.

Harrell, G., and Aikawa, J.: Pathogenesis of circulatory failure in Rocky Mountain spotted fever. Medicine 28:333, 1949.

Hechemy, K. E., Stevens, R. W., Sasowski, S., Michaelson, E. E.,

Casper, E. A., and Philip, R. N.: Discrepancies in Weil-Felix and microimmunofluorescence test results for Rocky Mountain spotted fever. J Clin Microbiol 9:292, 1979.

Herrmann, J. E., Hollingdale, M. R., Collins, M. F., and Vinson, J. W.: Enzyme immunoassay and radioimmunoprecipitation tests for the detection of antibodies to *Rochalimaea (Rickettsia) quintana*. Proc Soc Exp Biol Med 154:285, 1977.

Hollingdale, M. R., Herrmann, J. E., and Vinson, J. W.: Enzyme immunoassay of antibody to *Rochalimaea quintana*: Diagnosis of trench fever and serologic cross-reactions among other rickettsiae. J Inf Dis 137:578, 1978.

Ito, S., and Vinson, J. W.: Fine structure of *Rickettsia quintana* cultivated in vitro and in the louse. J Bacteriol 89:481, 1965.

Kazar, J., Brezina, R., and Urvolgyi, J.: Studies on the E strain of *Rickettsia prowazekii*. Bull WHO 49:257, 1973.

Kitao, T., Farrell, R. K., and Fukuda, T.: Differentiation of salmon poisoning disease and Elokomin fluke fever: Fluorescent antibody studies with *Rickettsia sennetsu*. Am J Vet Res 34:927, 1973.

Kokorin, I. N., Chyong, D. K., Kekcheeva, N. G., and Miskarova, E. D.: Cytological investigation on *Rickettsia tsutsugamushi* infection of mice with different allotypic susceptibility to the agent. Acta Virol 20:147, 1976.

Korb, J., Kouba, K., and Kulková, H.: *Rickettsia sennetsu* and the etiology of infectious mononucleosis. Čas lék Čes 105:975, 1966.

Kurz, J., and Burgdorfer, W.: Detection of the Rocky Mountain spotted fever agent, *Rickettsia rickettsii*, in dead ticks, *Dermacentor andersoni*. Infect Immun 20:584, 1978.

Ley, H. L., Smadel, J. E., Diercks, F. H., and Paterson, P. Y.: Immunization against scrub typhus. V. The infective dose of *Rickettsia tsutsugamushi* for men and mice. Am J Hyg 56:313, 1952.

Mason, R. A.: Propagation and growth cycle of *Rickettsia quintana* in a new liquid medium. J Bacteriol 103:184, 1970.

McDade, J. E., Shephard, C. C., Redus, M. A., Newhouse, V. F., and Smith, J. D.: Evidence of *Rickettsia prowazekii* infections in the United States. Am J Trop Med Hyg 29:277, 1980.

McKiel, J. A., Bell, E. J., and Lackman, D. B.: *Rickettsia canada:* A new member of the typhus group of rickettsiae isolated from *Haemophysalis leporispalustris* ticks in Canada. Can J Microbiol 13:503, 1967.

Misao, T., and Kobayashi, Y.: Studies on infectious mononucleosis: On the isolation of causative agents from blood, bone marrow fluid, and lymph gland with mice. Tokyo Med J 71:683, 1954.

Mooser, H., and Weyer, F.: Experimental infection of *Macacus rhesus* with *Rickettsia quintana*. Proc Soc Exp Biol Med 83:699, 1953.

Moulder, J. W.: The relation of the psittacosis group (chlamydiae) to bacteria and viruses. Ann Rev Microbiol 20:107, 1966.

Murray, E. S., Gaon, J. A., O'Connor, J. M., and Mulahasanovic, M.: Serologic studies of primary epidemic typhus and recrudescent typhus (Brill-Zinsser disease). 1. Differences in complement-fixing antibodies: High antigen requirement and heat lability. J Immunol 94:723, 1965.

Myers, W. F., and Wisseman, C. L., Jr.: Genetic relatedness among the typhus group of rickettsiae. Int J Syst Bacteriol 30:143, 1980.

Myers, W. F., Osterman, J. V., and Wisseman, C. L., Jr.: Nutritional studies of *Rickettsia quintana*: Nature of the hematin requirement. J Bacteriol 109:89, 1972.

Nacy, C. A., and Osterman, J. V.: Host defenses in experimental scrub typhus: Role of normal and activated macrophages. Infect Immun 26:744, 1979.

Ormsbee, R. A., Parker, H., and Pickens, E. G.: The comparative effectiveness of aureomycin, terramycin, chloramphenicol, erythromycin, and thiocymetin in suppressing experimental rickettsial infections in chick embryos. J Inf Dis 96:162, 1955.

Ormsbee, R. A.: Rickettsiae (as organisms). Ann Rev Microbiol 23:275, 1969.

Ormsbee, R., Peacock, M., Philip, R., Casper, E., Plorde, J., Gabre Kidan, T., and Wright, L.: Serologic diagnosis of epidemic typhus fever. Am J Epidemiol 105:261, 1977.

Ormsbee, R., Peacock, M., Philip, R., Casper, E., Plorde, J., Gabre Kidan, T., and Wright, L.: Antigenic relationships between the typhus and spotted fever groups of rickettsiae. Am J Epidemiol 108:53, 1978.

Oster, C. N., Burke, D. S., Kenyon, R. H., Ascher, M. S., Harber, P., and Pedersen, C. E., Jr.: Laboratory-acquired Rocky Mountain spotted fever. The hazard of aerosol transmission. N Engl J Med 297:859, 1977.

Osterman, J. V., Myers, W. F., and Wisseman, C. L., Jr.: Chemical composition of the cell envelope of *Rickettsia quintana*. Acta Virol 18:151, 1974.

Osterman, J. V., and Eisemann, C. S.: Rickettsial indirect hemagglutination test: Isolation of erythrocyte-sensitizing substance. J Clin Microbiol 8:189, 1978.

Perez Gallardo, F.: Study of the E stain isolated in Madrid in 1941. Its antigenic properties, its avirulence. Bull Soc Pathol Exot 56:805, 1963.

Philip, R. N., Casper, E. A., Ormsbee, R. A., Peacock, M. G., and Burgdorfer, W.: Microimmunofluorescence test for the serological study of Rocky Mountain spotted fever and typhus. J Clin Microbiol 3:51, 1976.

Philip, R. N., Casper, E. A., MacCormack, J. N., Sexton, D. J., Thomas, L. A., Anacker, R. L., Burgdorfer, W., and Vick, S.: A comparison of serologic methods for diagnosis of Rocky Mountain spotted fever. Am J Epidemiol 105:56, 1977.

Philip, R. N., Casper, E. A., Burgdorfer, W., Gerloff, R. K., Hughes, L. E., and Bell, E. J.: Serologic typing of rickettsiae of the spotted fever group by microimmunofluorescence. J Immunol 121:1961, 1978.

Ramm, L. E., and Winkler, H. H.: Rickettsial hemolysis: Adsorption of rickettsiae to erythrocytes; effect of metabolic inhibitors upon hemolysis and adsorption. Infect Immun 7:93, 550, 1973.

Reiss-Gutfreund, R. J.: The isolation of *Rickettsia prowazekii* and *mooseri* from unusual sources. Am J Trop Med Hyg 15:943, 1966.

Schaechter, M., Bozeman, F. M., and Smadel, J. E.: Study on the growth of rickettsiae. II. Morphologic observations on living rickettsiae in tissue culture cells. Virology 3:160, 1957.

Schramek, S.: Isolation and characterization of deoxyribonucleic acid from *Coxiella burnetii*. Acta Virol 12:1822, 1968.

Schramek, S., and Brezina, R.: Characterization of an endotoxic lipopolysaccharide from *Coxiella burnetii*. Acta Virol 20:152, 1976.

Schramek, S., Brezina, R., and Tarasevich, I. V.: Isolation of a lipopolysaccharide antigen from *Rickettsia* species. Acta Virol 20:270, 1976.

Shirai, A. and Wisseman, C. L.: Serologic classification of scrub typhus isolates from Pakistan. Am J Trop Med Hyg 24:145, 1975.

Shirai, A., Catanzaro, P. J., Phillips, S. M., and Osterman, J. V.: Host defenses in experimental scrub typhus: Role of cellular immunity in heterologous protection. Infect Immun 14:39, 1976.

Shirai, A., Sankaran, V., Gan, E., and Huxsoll, D. L.: Early detection of *Rickettsia tsutsugamushi* in peripheral monocyte cultures derived from experimentally infected monkeys and dogs. Southeast Asian J Trop Med Public Health 9:11, 1978.

Silverman, D. J., Boese, J. L., Wisseman, C. L., Jr.: Ultrastructural studies of *Rickettsia prowazekii* from louse midgut cells to feces: Search for "dormant" forms. Infect Immun 10:257, 1974.

Silverman, D. J., and Wisseman, C. L., Jr.: Comparative ultrastructural study on the cell envelopes of *Rickettsia prowazekii*, *Rickettsia rickettsii*, and *Rickettsia tsutsugamushi*. Infect Immun 21:1020, 1978.

Smadel, J. E.: Influence of antibiotics on immunologic response in scrub typhus. Am J Med 17:246, 1954.

Smith, D. K., and Winkler, H. H.: Characterization of a lysine-specific active transport system in *Rickettsia prowazekii*. J Bacteriol 129:1349, 1977.

Stoker, M. G. P., and Fiset, P.: Phase variation of the Nine Mile and other strains of *Rickettsia burneti*. Can J Microbiol 2:310, 1956.

Stork, E., and Wisseman, C. L., Jr.: Growth of *Rickettsia prowazekii* in enucleated cells. Infect Immun 13:1743, 1976.

Tarasevich, I. V., Plotnikova, L. F., Fetisova, N. F., Makarova, V. A., Jablonskaja, V. A., Rehacek, J., Zupacico, M., Kovacova, E., Urvolgyi, J., Brezina, R., Zakarjan, A. V., and Kocinjan, M. E.: Rickettsioses studies. I. Natural foci of rickettsioses in the Armenian Soviet Socialist Republic. Bull WHO 53:25, 1976.

Traub, R., and Wisseman, C. L., Jr.: The ecology of chigger-borne rickettsioses (scrub typhus). J Med Entomol 11:237, 1974.

Tyeryar, F. J., Jr., Weiss, E., Millar, D. B., Bozeman, F. M., and Ormsbee, R. A.: DNA base composition of rickettsiae. Science 180:415, 1973.

Urvolgyi, J., and Brezina, R.: *Rickettsia slovaca*: A new member of the spotted fever group rickettsiae. In Kazar, J., Ormsbee, R. A., and Tarasevich, I. V. (eds.): Rickettsiae and Rickettsial Diseases. Bratislava, VEDA, 1978, pp. 299–306.

Vinson, J. W.: *In vitro* cultivation of the rickettsial agent of trench fever. Bull WHO 35:155, 1966.

Vinson, J. W., and Campbell, E. S.: Complement fixing antigens from *Rickettsia quintana*. Acta Virol 12:54, 1968.

Walker, D. H., and Cain, B. G.: A method for specific diagnosis of Rocky Mountain spotted fever on fixed paraffin-embedded tissue by immunofluorescence. J Inf Dis 137:206, 1978.

Walker, D. H., and Henderson, F. W.: Effect of immunosuppression on

Rickettsia rickettsii infection in guinea pigs. Infect Immun 20:221, 1978.

Weiss, E., and Dressler, H. R.: Selection of an erythromycin-resistant strain of *Rickettsia prowazekii*. Am J Hyg 71:292, 1960.

Weiss, E., and Dressler, H. R.: Increased resistance to chloramphenicol in *Rickettsia prowazekii* with a note on failure to demonstrate genetic interaction among strains. J Bacteriol 83:409, 1962.

Weiss, E.: Growth and physiology of rickettsiae. Bacteriol Rev 37:259, 1973.

Weiss, E., Coolbaugh, J. C., and Williams, J. C.: Separation of viable *Rickettsia typhi* from yolk sac and L cell host components by renografin density gradient centrifugation. Appl Microbiol 30:456, 1975.

Weiss, E.: In Kazar, J., Ormsbee, R. A., and Tarasevich, I. V. (eds.): Rickettsiae and Rickettsial Diseases. Bratislava, VEDA, Slovak Academy of Sciences, 1978.

Weiss, E., Dasch, G. A., Woodman, D. R., and Williams, J. C.: Vole agent identified as a strain of the trench fever rickettsia, *Rochalimaea quintana*. Infect Immun 19:1013, 1978.

Wike, D. A., Ormsbee, R. A., Tallent, G., and Peacock, M. G.: Effects of various suspending media on plaque formation by rickettsiae in tissue culture. Infect Immun 6:550, 1972.

Winkler, H. H.: Inhibitory and restorative effects of adenine nucleotides on rickettsial adsorption and hemolysis. Infect Immun 9:119, 1974.

Winkler, H. H., and Ramm, L. E.: Adsorption of typhus rickettsiae to ghosts of sheep erythrocytes. Infect Immun 11:1244, 1975.

Winkler, H. H.: Rickettsial permeability. An ADP-ATP transport system. J Biol Chem 251:389, 1976.

Wisseman, C. L., Jr., and Waddell, A. D.: In vitro studies on rickettsia-host cell interactions: Intracellular growth cycle of virulent and attenuated *Rickettsia prowazekii* in chicken embryo cells in slide chamber cultures. Infect Immun 11:1391, 1975.

Wood, W. H., Jr., and Wisseman, C. L., Jr.: Studies of *Rickettsia mooseri* cell walls. II. Immunologic properties. J Immunol 98:1224, 1967.

Woodward, T. E., Pedersen, C. E., Oster, C. N., Bagley, L. R., Romberger, J., and Synder, M. J.: Prompt confirmation of Rocky Mountain spotted fever: Identification of rickettsiae in skin tissues. J Inf Dis 134:297, 1976.

Yamada, T., Harber, P., Pettit, G. W., Wing, D. A., and Oster, C. N.: Activation of the kallikrein-kinin system in Rocky Mountain spotted fever. Ann Int Med 88:764, 1978.

51 *BARTONELLA BACILLIFORMIS*

Manuel Cuadra, M.D.

MORPHOLOGY

The morphology of *Bartonella bacilliformis* varies greatly with environmental conditions. In the red cells of blood smears from patients with Oroya fever, it appears as a pleomorphic organism (Fig. 1), varying from slender bacilli to coccoid forms. Although the bacillary forms usually have a smooth surface, there are also rosary-like forms that are considered to be produced by multisegmentation of the smooth forms; later the beaded forms disintegrate into coccoid forms. During the 1st week in typical cases of Oroya fever, when the rate of red cell parasitism increases geometrically, the bacillary (vegetative) forms predominate greatly over the coccoid forms (Fig. 1, top). In the 2nd week, both forms are found in approximately equal numbers. When the fever subsides in the 3rd week (usually by lysis), the coccoid forms greatly predominate over the bacillary forms. During convalescence (an afebrile stage with profound anemia), only coccoid forms are seen (Fig. 1, bottom). Electron microscopy shows that *B. bacilliformis* lies within the red cells (Fig. 2) and that its internal structure, showing a defined cell wall, resembles that of bacteria (Cuadra and Takano, 1969). Compact masses of organisms filling the cytoplasm of capillary endothelial cells have been found in the stage of Oroya fever (Strong, 1945; Aldana, 1929; Pinkerton and Weinman, 1937–38; Alzamora Castro, 1940; Urteaga, 1948).

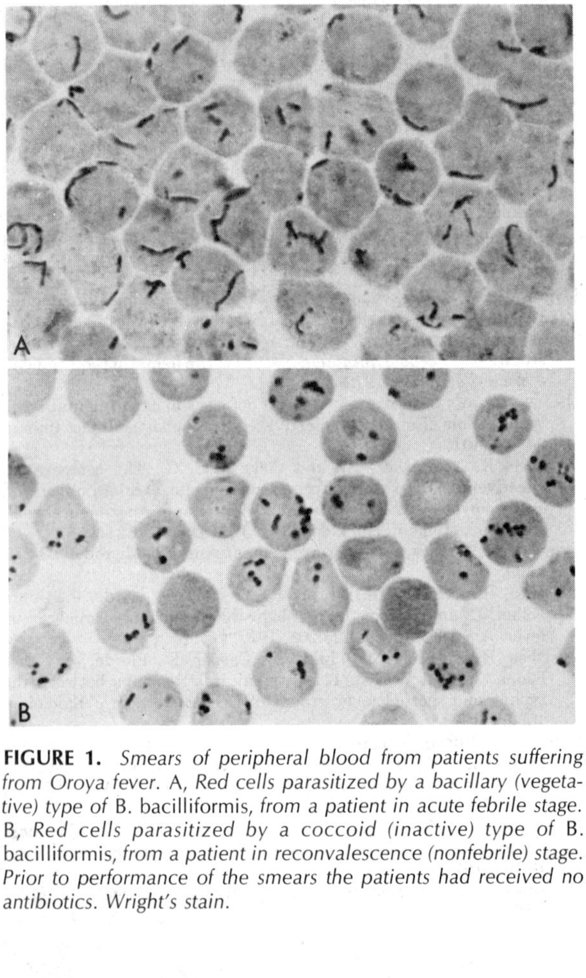

FIGURE 1. *Smears of peripheral blood from patients suffering from Oroya fever. A, Red cells parasitized by a bacillary (vegetative) type of* B. bacilliformis, *from a patient in acute febrile stage. B, Red cells parasitized by a coccoid (inactive) type of* B. bacilliformis, *from a patient in reconvalescence (nonfebrile) stage. Prior to performance of the smears the patients had received no antibiotics. Wright's stain.*

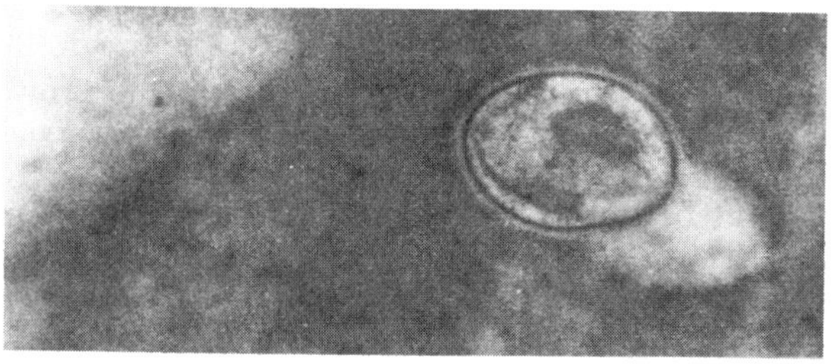

FIGURE 2. *Thin section of a red cell parasitized by* B. bacilliformis. *The organism lies within a cavity in the red cell and shows a dense cell wall. Lead hydroxide stain. ×50,000.*

B. bacilliformis has been demonstrated within verruga nodules and experimental monkey nodules by light microscopy (Mackehenie and Weiss, 1926; Noguchi, 1926a, b; Noguchi, 1927a; Marquez da Cunha and Muniz, 1928; Weiss, 1932; Weinman and Pinkerton, 1937a; Alzamora Castro, 1945; Urteaga and Calderón, 1965) and electron microscopy (Recavarren and Lumbreras, 1972; Takano, 1970). Both extracellular (Recavarren and Lumbreras, 1972) and intracellular (Takano, 1970) organisms occur; they resemble those forms parasitizing red blood cells.

Inclusion-like bodies, consisting of aggregated organisms that are presumably in stages of intracellular digestion, have also been found in the cytoplasm of the verruga cells (angioblasts) (Mayer et al., 1913; Marquez da Cunha and Muniz, 1928; Weiss, 1932; Pinkerton and Weinman, 1937–38; Noguchi, 1927b; Kikuth, 1931). Both extracellular and intracellular bacterial forms and the inclusion-like bodies develop in cultured cells (Pinkerton and Weinman, 1937). The assumption that the inclusions are composed of a viral form of *B. bacilliformis* (Aldana, 1947) is speculative.

The morphology of *B. bacilliformis* in blood agar cultures is shown in Figure 3. During active growth (3rd to 6th day at 28° C), the organisms appear in Giemsa-stained touch preparations as small rods with parallel sides, straight or slightly curved axis, slightly rounded ends, moderate variation in length (monomorphism), and arranged as single elements or in palisades. Later, pleomorphism becomes marked and increases with time of incubation. Long and short bacilli, coccobacilli, coccoid forms, short chains, diplococcoid forms, diplobacilli, and bacilli with bipolar staining closely resembling *Pasteurella* are seen. In cultures more than seven days old, coccoid forms predominate over bacillary forms, and by the 9th day, ring-shaped organisms are predominant. The coccoid forms are derived either by segmentation of the rods or by dissolution of the midportion of bipolar rods, so that the poles become freestanding. Under electron microscopy,

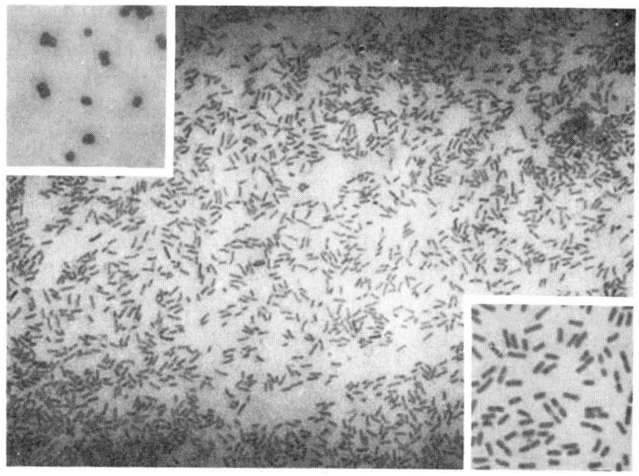

FIGURE 3. *Touch preparation from a blood agar culture of* B. bacilliformis *(4th day of incubation at 28° C) showing monomorphic rods. For comparison see insets: touch preparations of* Staphylococcus epidermidis *(top) and* Escherichia coli *(bottom). Giemsa stain. The three organisms are shown at the same magnification (×840).*

many organisms show an envelope secondary to retraction of the cytoplasm (Fig. 4), and a defined cell wall is shown by thin sections (Cuadra and Takano, 1969) (Fig. 2). Most bacilli have single or multiple (up to 30 μm long and up to 10 in number) flagella (Fig. 4) at one pole (cephalotrica) (Peters and Wigand, 1951–1952; Perez Alva and Giustini, 1957). If transverse septation occurs centrally, two identical daughter cells result; if septation occurs eccentrically, the daughter cells are pleomorphic.

Growth of *B. bacilliformis* in liquid media produces no morphologic differences from the forms grown on solid media. Flagellated and nonflagel-

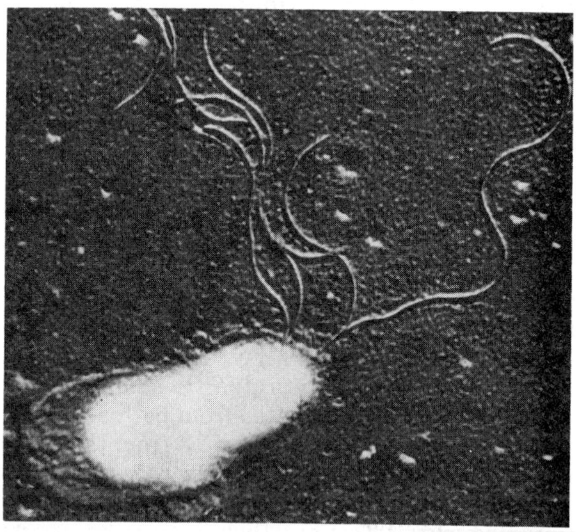

FIGURE 4. B. bacilliformis *as seen by electronmicroscope. The organism shows a pronounced retraction of the cytoplasm and flagella emerging from a pole. Coated with palladium. ×19,000. (From Peters, D., and Wigand, R.: Z. Tropenmed Parasitol 3:313, 1951.)*

lated forms appear in solid, semisolid, and liquid media.

MOTILITY

Free forms in plasma (wet preparations) of patients suffering from Oroya fever have not been reported, but free forms lying in the spaces between the red cells of stained blood smears are occasionally seen. Most, however, lie in ghosts of erythrocytes (Cuadra, 1957). Old reports (Barton, 1909) that the organisms move inside the red cells have not been confirmed. Both motile and nonmotile forms appear in cultures. Nonmotile forms predominate markedly over motile ones, and paradoxically, flagellated forms predominate over nonflagellated ones. Hence, the flagellated organisms are not necessarily motile. This deduction is supported by the fact that *Bartonella* organisms growing in the semisolid medium of Noguchi may not be motile (Peters and Wigand, 1951–1952; Noguchi and Battistini, 1926) but are flagellated. Moreover, the yield of flagella increases with the time of incubation (Peters and Wigand, 1951–1952) as motility decreases.

Bartonellae can attach to red blood cells in cultures. The "parasitized" erythrocytes appear to rotate (dance) in a characteristic fashion (Cuadra, 1978). Dance is prevented, specifically, by antisera.

B. bacilliformis is gram-negative and readily takes the Romanowsky stains (Giemsa, Wright, Leishman) in smears from patients' blood or cultures and in histologic sections of organs and verruga nodules after fixation with Ragoud's fixative (Noguchi and Battistini, 1926; Pinkerton and Weinman, 1937–1938). Flagella stain by the method of Zettnow-Fontana (Noguchi and Battistini, 1926; Pinkerton and Weinman, 1937–1938).

ANTIGENIC COMPOSITION

There have been no reports of flagellar or somatic antigens of *Bartonella*. Different clinical patterns in two different areas of Peru (Callejon de Huaylas and Rimac Valley) suggest that there are two antigenic variants of *Bartonella*. Morphologically, the strains isolated from both areas are indistinguishable. Strains isolated from the blood of Oroya fever patients cross-react in complement-fixation reaction with strains from verruga nodules (Noguchi, 1927c), and there is cross-immunity between Oroya fever and verruga (Noguchi, 1927d). No cross-reaction occurs with *Hemobartonella* or related organisms (Wigand, 1956, 1958).

CULTURAL CHARACTERISTICS

B. bacilliformis, a nonsporing organism, grows slowly in different bacteriologic media. On blood-broth of Hercelles (1927), the first medium devised for cultivation of *B. bacilliformis,* the organism does not produce turbidity; the change in color of the medium from red to dark reddish violet is the only indicator of growth. Both single elements and clusters of them can be seen under the microscope from the 3rd to the 4th day of incubation. The organisms have a marked tendency to attach to red blood cells (Cuadra, 1978). Attachment is evidenced by) the "dance" of "parasitized erythrocytes" and by clumped red cells containing clusters of organisms arranged in a framework. We use both signs as indicators of growth. On blood agar, the colonies are too small for study by the naked eye. Under the stereoscopic microscope, the colonies appear from the 3rd to the 4th day of incubation to be variable in size, circular, transparent, and smooth. The colonies can also be studied microscopically on stained touch preparations (Fig. 3). A square piece of blood agar culture is removed from a Petri dish culture and placed, surface-to-surface, on a slide; when the piece is suddenly removed, a layer of organisms, arranged in colonies or singly, remains on the slide. On leptospirasemisolid medium of Noguchi (Noguchi and Battistini, 1926), which also contains blood factors essential for growth, *Bartonella* forms clusters; the organism, predominantly coccobacilli in form, are less motile than those on blood agar slants (Peters and Wigand, 1951-1952; Noguchi and Battistini, 1926). On liquid, semisolid, or solid media of Geiman (1941), which contain ascorbic acid and glutathione as key substances with animal serum, *Bartonella* grows well; in the liquid medium, it causes a fine dustlike deposit but neither turbidity nor clusters. The individual organisms are dispersed and are mainly nonmotile.

All media that support growth of *Bartonella* contain either whole blood or serum. Attempts to determine if the *Haemophilus influenzae* growth promoting V and X blood factors (Thjötta and Avery, 1921) are involved have given confusing results (Jimenes, 1940; Colichón and Bedón, 1973). In comparison with *H. influenzae* or any other current bacteria, *Bartonella* slowly reaches a moderate amount of growth.

Metabolism

B. bacilliformis is an obligate aerobe (Noguchi and Battistini, 1926), but in blood broth the organism grows partly at the bottom on the mass of sedimented red cells and partly as a thin pellicle on the surface of the medium (Aldana,

1929, 1946). A difference in oxygen or carbon dioxide requirements therefore exists between the two populations. Growth occurs between about 20° and 37° C, but the optimal temperature is 28° C, and no growth occurs at 40° C. The pH range for growth is between 6.8 and 8.4, with an optimum level of 7.8 (Noguchi and Battistini, 1926). The organism ferments no carbohydrates (Noguchi and Battistini, 1926). No hemolysis occurs on blood agar, but in blood broth the erythrocytes attached to clusters of organisms become hemolyzed, and red cell ghosts can be seen by phase-contrast microscopy (Cuadra, 1978).

Resistance and Viability

B. bacilliformis, a nonsporing organism, dies within seven minutes at 56° C. Formalin, phenol, mercurochrome, neosalvarsan, and neutroflavine inhibit growth in cultures. The inhibitory concentration varies. Thus, formalin is inhibitory at 1:100,000 and phenol at 1:100 concentrations (Aldana, 1929; Noguchi, 1928a).

Viability varies with the specimen, the cultural conditions, and the strain. Thus, in citrated blood from patients with Oroya fever, the organism remains viable for four months (Aldana, 1929), and in blood clots, for nine months (Herrer, 1948) at room temperature. In the excised verruga nodule from monkeys, *B. bacilliformis* survives for about 56 days at 4° C and 28 days at room temperature; on blood agar slants or leptospira semisolid medium, it remains viable for about 50 days at 37° C and four months at 4° C or room temperature (Noguchi, 1926c). On blood broth or blood agar slants, it survives for nine months at room temperature (Aldana, 1929). Protection of the cultures from drying by adequate stoppers, without impairing the oxygen supply (since *Bartonella* is an obligate aerobe), is essential to maintain viability. Under storage at −20° to −70° C, after freezing the cultures in the period of optimal growth (abundance of motile forms), *B. bacilliformis* remains viable for years and resists the process of lyophilization (Perez Alva et al., 1957).

PATHOGENIC PROPERTIES

B. bacilliformis is pathogenic only for man and causes Oroya fever and verruga peruana, which represent two successive stages of the same disease (Carrión's disease).

Inoculations of *Bartonella* into human beings have given variable results. Carrión, a medical student, was inoculated intracutaneously with juice from a verruga nodule. After 21 days of incubation, he developed Oroya fever with severe anemia and died on the 16th day of disease (Rebagliati, 1940). Strong et al. (1945) inoculated a man intracutaneously with verruga material, and a local verruga nodule developed but no Oroya fever. Garcia Rosell, a physician, inoculated himself accidentally with the blood of an Oroya fever patient while giving a blood transfusion (Rebagliati, 1940). A febrile illness without anemia lasted for two weeks, and then a florid verruga eruption appeared. Another physician, Kuchinsky-Godard (Mackehenie, 1937), who inoculated himself voluntarily with a culture of bartonellae, developed verruga nodules locally and at distant sites but no Oroya fever. Urteaga (1950) could induce Oroya fever in splenectomized individuals, but not in normal ones, by inoculation of parasitized blood from Oroya fever patients.

Inoculations carried out in monkeys (*Macacus rhesus* is the animal of choice) have also given variable results (Jadassohn and Seiffert, 1910; Mayer et al., 1913; Arce et al., 1913; Noguchi and Battistini, 1926; Noguchi, 1926a, b; Marquez da Cunha and Muniz, 1928; Aldana, 1929; Weinman and Pinkerton, 1937a; Wigand and Weyer, 1952). In an effort to reproduce human bartonellosis with its two stages, infected blood from Oroya fever patients, juice from verruga nodules, infected wild sandflies, and cultures have been inoculated by all possible routes (intravenously, cutaneously by scarification, and intra- or subdermally). Local verruga nodules occur with relative ease after cutaneous inoculation of these materials. A mild Oroya fever characterized by fever, anemia, and parasitism of the red blood cells occurs in a limited number of cases after intravenous inoculation with cultures. With parasitized blood from Oroya fever patients as the inoculum, negative results are obtained as a rule. The high degree of parasitism that usually occurs in Oroya fever patients, involving nearly 100 per cent of red cells (Fig. 1) with associated severe anemia (one million erythrocytes/mm^3), has never been obtained in monkeys (Weinman and Pinkerton, 1937a). The discrete Oroya fever induced in monkeys is very exceptionally followed by a generalized (endogenous) eruption (Noguchi, 1926a). However, it seems that anemia is produced more easily in splenectomized monkeys (Weinman and Pinkerton, 1937a).

According to old reports from Peruvian doctors, verruga peruana but not Oroya fever can be induced by inoculation of various animal species (Tamayo, 1899; Arce et al., 1913; Aldana, 1929). Noguchi confirmed the results in dogs and donkeys (Noguchi et al., 1929).

IMMUNITY

Oroya fever is an acute disease. In most nonfatal cases, recovery occurs spontaneously within two to three weeks. It is followed first by an asymptomatic "intercalary period" of variable duration (one or more weeks) and then by the verruga peruana stage, which lasts about one to three months. Chronic Oroya fever with persistent anemia and *B. bacilliformis* in peripheral red blood cells has never been reported. After spontaneous remission of the verrucous eruption, a solid lifelong immunity is established. A second attack of Oroya fever has never been reported in patients who returned to endemic areas. However, two, three, or more episodes of verruga peruana, usually accompanied by rheumatoid manifestations (verrucous rheumatism) and mild fever may occur in some individuals (chronic verruga peruana). The fact that some subjects with or without a past history of Oroya fever or verruga peruana give positive blood cultures (Howe, 1943; Weinman and Pinkerton, 1937b; Herrer, 1959) supports the existence of a state of infection-immunity (i.e., immunity depending on existing but not prior infection).

LABORATORY DIAGNOSIS

Blood smears stained with Giemsa, Wright, or Leishman stain reveal the organisms in the red cells of patients suffering from Oroya fever throughout the course of the fever and convalescent period (Fig. 1). The percentage of parasitized erythrocytes and the relative proportions of bacillary and coccoid forms should be estimated because bacillary forms, whatever the number of associated coccoid forms, are always related to the presence of fever. For this reason, the bacillary form is considered to be the active form. In contrast, in the convalescent period, when fever is absent, only coccoid forms, may be encountered, although they may be parasitizing 100 per cent of red cells (Fig. 1). Thus, the coccoid form, originating from the bacillary form, is considered to be an inactive (stationary) form. Parasitized erythrocytes disappear gradually during convalescence, which, in cases with a high rate of parasitism, may last up to three weeks. By this time, the residual red cells harboring coccoid forms are spherocyte-like microcytes. The finding of pure coccoid forms in febrile patients indicates that the fever is not caused by *Bartonella* but by an associated infection, usually caused by *Salmonella* (Cuadra, 1956).

In an atypical form of the disease characterized by little or no anemia, the organisms are found with difficulty. In these cases, the thick-blood technique is of great aid. Blood cultured in broth always produces growth of *B. bacilliformis* after a variable incubation period at 28° C. The incubation time is usually three to five days and depends on the degree of parasitism of red blood cells. Blood cultures are frequently positive for *Salmonella* after a shorter incubation time. *S. typhimurium* is encountered most frequently (Cuadra, 1956).

No bartonellae are found in blood films of patients suffering from verruga peruana, although the blood cultures, especially the cultures of excised verruga nodules, may be positive. The nodules have a characteristic histologic appearance (Rocha-Lima, 1913; Mackehenie, 1938).

The agglutination reaction was thoroughly studied by Howe (1942, 1943) in a group of 203 individuals. The titer of agglutinins appeared to be low but was of diagnostic significance. Practically all patients with Oroya fever had measurable agglutinins; in contrast, agglutinins were found in only 28 per cent (15 out of 54) of patients in the verruga stage. In both Oroya fever and verruga peruana, the titer showed a tendency to decline and to disappear within a relatively short time after remission of the symptoms. In many individuals, blood cultures were positive (with or without symptoms), and agglutination tests were negative; the reverse was uncommon. In a group of 21 apparently healthy residents of endemic areas, both blood cultures and agglutination tests were negative.

The complement fixation reaction (Strong et al., 1915; Noguchi, 1928b; Aldana, 1929; Reese et al., 1950) has not been evaluated by extensive clinical trials.

DRUG SUSCEPTIBILITY

Penicillin, streptomycin, chloramphenicol, tetracycline, and erythromycin are highly effective against Oroya fever (Larrea, 1958). They produce defervescence within 24 hours, in parallel with morphologic changes of bartonellae and their disappearance from the peripheral blood cells. In vitro, *Bartonella* is highly susceptible to antibiotics and relatively resistant to sulphonamides (Wigand, 1952). Penicillin induces extraordinarily large coccoid forms in the cultures (Aldana and Tisnado Munoz, 1945; Wigand, 1952). Such forms correspond to L forms of *Bartonella* (Sharp, 1968), Resistance develops to streptomycin during treatment, whereupon the fever returns, the bacillary forms of bartonellae coincidentally reappear in the peripheral erythrocytes (Aldana et al., 1948; Cuadra, 1957).

The effectiveness of antibiotics on verruga peruana is doubtful. A comparative in vitro assay on drug susceptibility of strains isolated from Oroya fever patients with those isolated from verruga peruana patients is needed.

EPIDEMIOLOGY

Bartonellosis occurs in certain Andean valleys of Colombia, Ecuador, and Peru between 2 degrees north and 13 degrees south latitude. In Peru, the most afflicted country, the endemic foci lie at altitudes of between 800 and 3500 meters above sea level. The most important foci lie in valleys watered by rivers that flow into the Pacific Ocean at about 30 km. distance from the sea Foci of minor importance are scattered in valleys with rivers that flow into the Amazon River (Rebagliati, 1940).

The inhabitants of the endemic areas constitute the reservoir. A high proportion of blood cultures of healthy individuals in those areas, with or without a past history of bartonellosis, is positive (Weinman and Pinkerton, 1937b; Howe, 1943; Herrer, 1959). Among animals, field mice have occasionally been found to harbor *Bartonella* in their blood (Hertig, 1948). The fact that some persons have contracted Oroya fever after spending the night in the forest supports the concept of the existence of an animal reservoir. Infection is transmitted by certain species of *Phlebotomus: Ph. verrucarum, Ph. pescei,* and *Ph. bicornutus* (Noguchi et al., 1929; Hertig, 1942; Herrer et al., 1959-1960). The bite of wild sandflies in the endemic areas can cause infection in monkeys (Battistini, 1931). However, the crucial experiment of transmission to man by *Phlebotomus* fed on Oroya fever patients failed. The cycle of bartonellae within *Phlebotomus* is unknown. In an epidemic area, Guaytara (Colombia), no *Phlebotomus* species were found, and only lice were present as a possible vector (Patino, Camargo, 1939). *Bartonella* can multiply within the celomic cavity of lice (Wigand and Weyer, 1952). Experimentally, the infection could be transmitted from infected to normal rhesus monkeys by the bite of the tick *Dermacentor andersoni* (Noguchi, 1926d).

References

Aldana, L. D.: Bacteriologia de la enfermedad de Carrión. Cron Med Lima 46:237, 1929.

Aldana, L.: Estado actual del tratamiento de la enfermedad de Carrión por la penicilina. Rev Med Per Lima 19:617, 1946.

Aldana, L.: Estados biológicos de la Bartonella en la enfermedad de Carrión. Rev Sanid Polic Lima 7:391, 1947.

Aldana, L., and Tisnado Munoz, S.: Penicilina y enfermedad de Carrión. Rev Med Per Lima 18:343, 1945.

Aldana, L., Zubiate, P., and Contreras, F.: Un caso de verruga peruana resistente a la estreptomicina. Arch Per Pat Clin Lima 2:553, 1948.

Alzamora Castro, V.: Contribución al estudio de la bartonellosis humana o enfermedad de Carrión. Grac Med Lima 2:78, 1945.

Alzamora Castro, V.: Enfermedad de Carrión: Ensayo de etiopatogenia. Anales Fac Cien Med 23:1, 1940.

Arce, J., Mackehenie, D., and Ribeyro, R.: Estudio experimental de la enfermedad de Carrión. I. Inoculabilidad de la verruga peruana a los animales. Cron Med Lima 30:394, 1913.

Barton, A.: Descripción de elementos endoglobulares en los enfermos de fiebre de verruga. Cron Med Lima 26:7, 1909.

Battistini, T.: La verrue péruviene. (Sa transmission par le phlébotome.) Rev Sud Amér Med Chirug 2:719, 1931.

Colichón, H., and Bedón, C.: Factores que la *Bartonella bacilliformis* requiere para su crecimiento. Rev Lat Am Microbiol 15:11, 1973.

Cuadra, M.: The ability of *Bartonella bacilliformis* growing in culture to attach to red blood cells. Abstracts of the Twelfth International Congress of Microbiology. Sept 3-8, 1978, München, p. 178.

Guadra, M.: Mecanismo de destrucción de los eritrocitos. La hemolisis intravascular. Anal Fac Med lima 40:872, 1957.

Cuadra, M.: Salmonellosis complication in human bartonellosis. Texas Rep Biol Med 14:97, 1956.

Cuadra, M.: Tratamiento con cloranfenicol de casos de bartonellosis aguda (enfermedad de Carrión) en periodo de inicio. Anal Fac Med Lima 40:147, 1957.

Cuadra, M., and Takano, J.: The relationship of *Bartonella bacilliformis* to the red blood cell as revealed by electron microscopy. Blood 33:708, 1969.

Geiman, Q. M.: New media for the growth of *Bartonella bacilliformis.* Proc Soc Exp Biol Med 47:329, 1941.

Hercelles, O.: El germen de la verruga peruana. Anal Fac Med Lima 10:231, 1927.

Herrer, A.: Estudios sobre la enfermedad de Carrión en el valle interandino del Mantaro. II. Incidencia de la infección bartonellósica en la población humana. Rev Med Exper Lima 23:47, 1959.

Herrer, A.: Supervivencia de la *Bartonella bacilliformis* en la sangre coagulada de los enfermos de verruga. Rev Med Exp Lima 7:70, 1948.

Herrer, A., Blancas, F., Cornejo Ubillús, J., Lung, J., Espejo, L., and Flores, M.: Estudios sobre la enfermedad de Carrión en el valle interandino del Mantaro. I. Observaciones entomólógicas. Rev Med Exper Lima 13:27, 1959.

Hertig, M.: Phlebotomus and Carrión's disease. Am J Trop Med (suppl) 22:1942.

Hertig, M.: Sandflies of the genus *Phlebotomus.* A review of their habits, disease relationships, and control. Proceedings of the 4th International Congress of Tropical Medicine and Malaria, 1948, Washington, DC, USA, p. 1609.

Howe, C.: Carrión's disease. Immunologic studies. Arch Int Med 72:147, 1943.

Howe, C.: Demonstration of agglutinins for *Bartonella bacilliformis.* J Exp Med 75:65, 1942.

Jadassohn, W. E., and Seiffert, G.: Ein Fall von Verruga peruviana: Gelungene Übertragung auf Affen. Z Hyg Infec Krankh, 6:247, 1910.

Jimenes, J. F.: Carrión's disease. I. Some growth factors necessary for cultivation of *Bartonella bacilliformis.* Proc Soc Exp Biol Med 45:402, 1940.

Kikuth, W.: Experimentelle Untersuchungen über Oroyafieber und Verruga peruana. Z Immunforschg 73:1, 1931.

Larrea, P.: Los antibioticos en la bartonellemia humana. Arch Per Pat Clin 12:1, 1958.

Mackehenie, D.: Un caso de verruga humana por autoinoculación experimental. Ref Med Lima 23:741, 1937.

Mackehenie, D.: Estudio del noduloma verrucoso. Ref Med Lima 24:50, 1938.

Mackenhenie, D., and Weiss, P.: Contribución al estudio de la verruga peruana. Gac Med Per Lima 4:51, 1926.

Marquez de Cunha, A., and Muniz, J.: Pesquisas sobre la verruga peruana. Med Inst Oswaldo Cruz 21:161, 1928.

Mayer, M., Rocha-Lima, H., and Werner, H.: Untersuchungen über verruga peruviana. Münch Med Wochenschr 60:739, 1913.

Noguchi, H.: Comparative studies of different strains of *Bartonella bacilliformis* with special reference to the relation between the clinical types of Carrion's disease and the virulence of the infecting organism. J Exp Med 47:219, 1928b.

Noguchi, H.: Etiology of Oroya fever. II. Viability of *Bartonella bacilliformis* in cultures and in the preserved blood and an excised nodule of Macacus rhesus. J Exp Med 44:533, 1926c.

Noguchi, H.: Etiology of Oroya fever. III. The behavior of *Bartonella bacilliformis* in Macacus rhesus. J Exp Med 44:697, 1926a.

Noguchi, H.: Etiology of Oroya fever. IV. The effect of inoculation of anthropoid apes with *Bartonella bacilliformis*. J Exp Med 44:715, 1926b.

Noguchi, H.: Etiology of Oroya fever. VI. Pathological changes observed in animals experimentally infected with *Bartonella bacilliformis*. The distribution of the parasites in the tissues. J Exp Med 45:437, 1927a.

Noguchi, H.: Etiology of Oroya fever. VII. The response of the skin of Macacus rhesus and anthropoid apes to inoculation with *Bartonella bacilliformis*. J. Exp Med 45:455, 1927b.

Noguchi, H.: Etiology of Oroya fever. VIII. Experiments on cross-immunity between Oroya fever and verruga peruana. J Exp Med 45:781, 1927d.

Noguchi, H.: Etiology of Oroya fever. XIII. Chemotherapy in experimental *Bartonella bacilliformis* infection. J Exp Med 48:619, 1928a.

Noguchi, H.: The etiology of verruga peruana. J Exp Med 45:175, 1927c.

Noguchi, H.: The experimental transmission of *Bartonella bacilliformis* by ticks (*Dermacentor andersoni*). J Exp Med 44:729, 1926d.

Noguchi, H., Shannon, R. C., Tilden, E. B., and Tyler, J. R.: Etiology of Oroya fever. XIV. The insect vectors of Carrión's disease. J Exp Med 49:993, 1929.

Noguchi, H., Muller, H. R., Tilden, E. B., and Tyler, J. R.: Etiology of Oroya fever. XVI. Verruga in the dog and the donkey. J Exp Med 50:455, 1929.

Noguchi, H., and Battistini, T.: Etiology of Oroya fever. I. Cultivation of *Bartonella bacilliformis*. J Exp Med 43:851, 1926.

Patino Camargo, L.: Batonellosis del guaytara o fiebre verrucosa del guaytara. Ref Med Lima 24:649, 689, 1939.

Perez Alva, S., and Giustini, J.: La maladie de Carrión. Etude morphologique de *Bartonella bacilliformis* au microscope électronique. Bull Soc Path Exot 50:188, 1957.

Perez Alva, S., Roger, F., and Roger, A.: La maladie de Carrión. Resistance de *Bartonella bacilliformis* an processus de lyophilisation. Bull Soc Path Exot 50:243, 1957.

Peters, D., and Wigand, R.: Neue Untersuchungen über *Bartonella bacilliformis*. I. Morphologie der Kulturform. Z Tropenmed Parasitol 3:313, 1951.

Pinkerton, H., and Weinman, D.: Carrion's disease. I. Behavior of the etiological agent within cells growing or surviving in vitro. Proc Soc Exp Biol Med 37:587, 1937.

Pinkerton, H., and Weinman, D.: Carrión's disease. II. Comparative morphology of the etiological agent in Oroya fever and verruga peruana. Proc Soc, Exp Biol Med 37:591, 1937.

Rebagliati, R.: Verruga Peruana (Enfermedad de Carrión). Imprenta Lima, Torres Aguirre, 1940.

Recavarren, S., and Lumbreras, H.: Pathogenesis of the verruga of Carrión's disease. Am J Pathol 66:461, 1972.

Reese, J. D., Morrison, M. E., and Fowler, E. M.: Complement fixation and Weil-Felix reactions in rabbits inoculated with *Bartonella bacilliformis*. J Immunol 65:355, 1950.

Rocha-Lima, H.: Zur Histologie der Verruga peruviana. Verh Dtsch Path Ges 16:409, 1913.

Sharp, J. T.: Isolation of "L" forms of *Bartonella bacilliformis*. Proc Soc Exp Biol Med 128:1072, 1968.

Strong, R. P.: Verruga peruana and Oroya fever. In Stitt's Diagnosis, Prevention and Treatment of Tropical Disease, 7th ed. vol 2. Philadelphia, The Blakiston Company, 1945, pp. 997–1014b.

Strong, R. P., Tyzzer, E. E., and Sellards, A. W.: Differential diagnosis of verruga peruana. Fifth report. J Trop Med 18:122, 1915.

Takano, J.: Enfermedad de Carrión (bartonellosis humana). Estudio morfológico de la fase hemática y del periodo eruptivo con el microscopio electrónico. Doctoral thesis. Faculty of Medicine of the National University of San Marcos, 1970.

Tamayo, M. O.: Inoculabilidad de la verruga. Cron Med Lima 16:81, 1899.

Thjötta, T., and Avery, O. T.: Studies on bacterial nutrition. II. Growth accessory substances in the cultivation of hemophilic bacilli. J Exp Med 34:97, 1921.

Urteaga, O.: Histopatogenia de la anemia en la verruga peruana. Arch Per Pat Clin 2:355, 1948.

Urteaga, O.: Lecture at the Cathedra of Infectious, Parasitic and Tropical Diseases of the Faculty of Medicine, Lima, 1950.

Urteaga, O., and Calderón, J.: Ciclo biológico de reproducción de la *Bartonella bacilliformis* en los tejidos de pacientes de verruga peruana o enfermedad de Carrión. Arch Per Pat Clin Lima 19:1, 1965.

Weinman, D., and Pinkerton, H.: Carrion's disease. III. Experimental production in animals. Proc Soc Exp Biol Med 37:594, 1937a.

Weinman, D., and Pinkerton, H.: Carrión's disease. IV. Natural source of *Bartonella* in the endemic zone. Proc Soc Exp Biol Med 37:596, 1937b.

Weiss, P.: Contribución al estudio de la verruga peruana o enfermedad de Carrión. Rev Med Per Lima 4:5, 58, 1932.

Wigand, R.: Morphologische, biologische und serologische Eigenschaften der Bartonellen. Stuttgart. Georg Thieme Verlag, 1958.

Wigand, R.: Neue Untersuchungen über *Bartonella bacilliformis*. 2. Verhalten gegenüber Sulfonamide und Antibiotica in vitro. Z Tropenmed Parasitol 3:453, 1952.

Wigand, R.: Serologische Reaktionen an Haemobartonella muris und Eperythrozoon coccoides. Z Tropenmed 7:322, 1956.

Wigand, R., und Weyer, F.: Neue Untersuchungen über *Bartonella bacilliformis*. 3. Übertragungsversuche auf rhesus Affen und auf Kleiderläuse. Z Tropenmed Parasitol 4:243, 1952.

52 THE CHLAMYDIAE

G. Philip Manire, Ph.D.

No other pathogenic microorganisms are better adapted for persistence and survival than the chlamydiae and no others are more widespread. Epidemiologic surveys suggest that some members of almost every species of birds and mammals may be infected with these organisms. Their ability to produce inapparent infections in these hosts is unexcelled, and individuals may be infected for extremely long periods of time without apparent harm. Estimates have been made that 10 to 20 per cent of the human population of the world is infected with chlamydiae. These organisms are ubiquitous, rarely kill their hosts, are generally highly infectious and easily transferred to new hosts, and have a remarkable ability to escape normal host immune mechanisms.

CLASSIFICATION

The genus *Chlamydia* is composed of two species, *C. trachomatis* and *C. psittaci* (Page, 1974). Although trachoma has been recognized as a specific disease for many centuries, and the first microscopic evidence of a specific causative agent was obtained early in the twentieth century, laboratory studies on these organisms began only in 1930 following the first successful isolation of

TABLE 1. Differentiation of *C. trachomatis* and *C. psittaci*

C. TRACHOMATIS	C. PSITTACI
Sensitive to sulfonamide	Resistant to sulfonamide
Rigid micro colonies	Diffuse micro colonies
Glycogen in inclusion	No glycogen in inclusion
Guanine + cytosine content of DNA ~ 44%	Guanine + cytosine content of DNA ~ 41%

C. psittaci from infected humans and birds. Isolation of *C. trachomatis* was not achieved until 1957.

Since it was shown very quickly that chlamydiae are strict intracellular parasites, they were considered for many years to be the largest known viruses. As evidence regarding their growth and biochemical characteristics accumulated, it became obvious that these organisms are bacteria with unique characteristics. Like other bacteria they (1) contain both DNA and RNA, (2) possess a continuous cell envelope similar in most respects to that of gram-negative bacteria, (3) possess ribosomes similar to those of other bacteria and synthesize their own proteins and nucleic acids, (4) possess limited but definite numbers of metabolic systems, and (5) are susceptible to a wide range of antibiotics.

The characteristics separating the two species are shown in Table 1. This classification was further justified by the finding that there is a significant DNA homology within each species but very little homology between the two, suggesting a long-standing evolutionary separation.

MORPHOLOGY

All members of the genus *Chlamydia* share a unique developmental cycle in which two distinct forms occur. These two forms are especially adapted for extracellular survival and transfer from cell to cell, and for intracellular growth. The extracellular form, generally referred to as an elementary body (EB), is a spherical bacterium 0.25 to 0.3 μm in diameter. It is surrounded by a rigid trilaminar cell envelope similar in composition to those of other gram-negative bacteria, except that the cell walls do not exhibit characteristic endotoxic properties. The inner side of the envelope is composed of a continuous layer of small subunit structures in a hexagonal array, and these subunits apparently serve to maintain envelope rigidity. The cytoplasm of the elementary body is surrounded by a typical plasma membrane. The DNA and ribosomes are condensed in the center of the organisms, and in electron micrographs of shadow-cast preparations these assume a "derby hat" or "fried egg" appearance. Thin-section electron microscopy shows a similar central condensation.

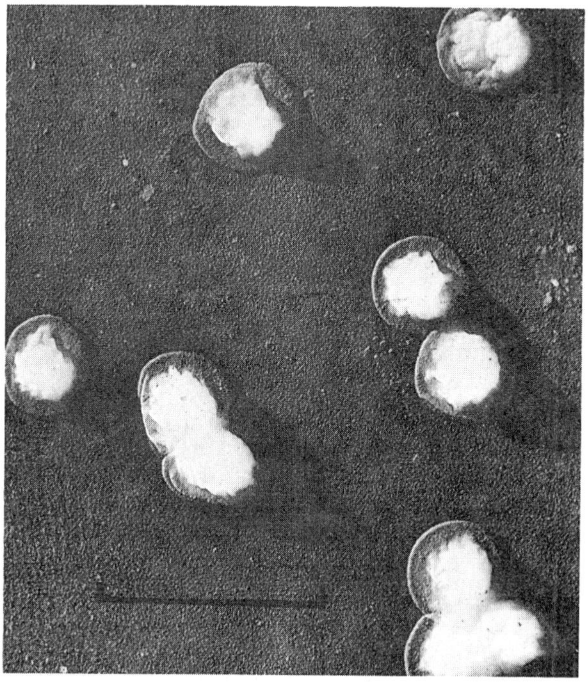

FIGURE 1. *Shadowcast preparation of purified elementary bodies showing typical central condensation of cytoplasmic contents.*

The intracellular form of chlamydiae, referred to here as the reticulate body (RB), is much larger than the extracellular form and differs from it in many respects. It is surrounded by a trilaminar envelope that is very fragile and flexible so that pleomorphism results. The subunit layer found on the inside of the elementary body envelope is missing, and no sulfur-containing amino acids have been detected in the outer cell wall. Ribosomes and other cytoplasmic constituents are distributed homogeneously throughout the cytoplasm.

The genome of the organism has a molecular weight of 660 × 10^6 as estimated by electron-microscopic measurements. As in other bacteria, the chlamydiae contain numerous proteins, which comprise 60 per cent of dry weight of the organisms, and lipids, about 30 per cent of dry weight.

A summary of the differing characteristics of elementary bodies and reticulate bodies is shown in Table 2.

In a typical growth cycle, the elementary body attaches to a susceptible cell by a process that involves a heat-sensitive component on the surface of the organism and a trypsin-sensitive site on the host cell (Byrne, 1976). The organism enters the cell by phagocytosis, and the resulting

FIGURE 2. *Cell envelope fragments prepared from disrupted suspensions of C. psittaci. Fragment A illustrates the hexagonal array of subunit structure found on the inner side of the cell envelope. Fragment B illustrates the outer granular side of the envelope. The structure C is an intact envelope.*

phagosome membrane separates the developing colony of chlamydiae from the host cell cytoplasm throughout the active growth cycle. The developing phagosome membrane contains no protein synthesized by chlamydiae, but the membranes continue to increase in size during colony development even when the host cell synthesis of protein is largely inhibited by cycloheximide. This suggests that chlamydiae play a role in control of phagosome membrane synthesis (Stokes, 1974).

By an unknown process the elementary bodies within the phagosome prevent lysosome activation and fusion of lysosomes and phagosomes, whereas when opsonized or heat-treated elementary bodies enter susceptible cells, there is rapid fusion of lysosomes and phagosomes, and the noninfectious chlamydiae are rapidly digested.

Within 6 to 8 hours after phagosome formation, the elementary bodies undergo a conversion to form reticulate bodies, and by 12 hours binary fission begins. In 20 to 24 hours some of the reticulate bodies have begun to show central condensation and conversion to elementary bodies, but most continue to undergo binary fission until about 40 hours, when large numbers of elementary bodies begin to appear. Following this the host cells begin to die and chlamydiae are released by lysis. It has been suggested that host cell lysosomes release into the cytoplasm hydrolytic enzymes that digest host cell constituents with consequent membrane lysis and release of chlamydiae (Todd and Storz, 1975).

ANTIGENIC COMPOSITION

As with all bacteria, the chlamydiae have a complex antigenic structure. However, there are

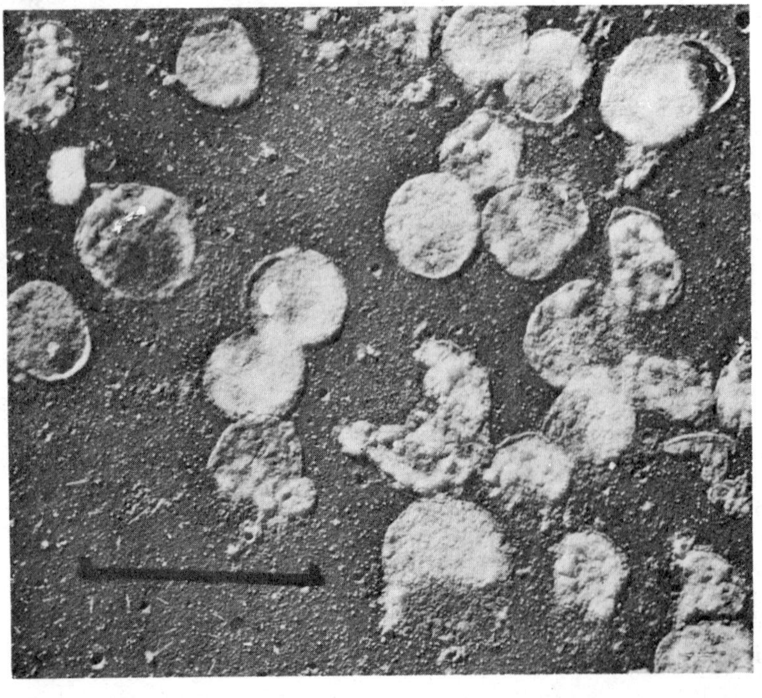

FIGURE 3. *Shadowcast preparation of purified reticulate bodies.*

TABLE 2. Characteristics of Elementary Bodies (EB) and Reticulate Bodies (RB)

CHARACTERISTICS	EB	RB
Morphology	Small, dense centered	Large, homogeneous
RNA:DNA	1:1	3:1
Sonication	Resistant	Sensitive
Effect of trypsin	Resistant	Sensitive, lysis
Infectivity	+	−
Toxicity	+	−
Hemagglutinin	Present	Absent
Permeability	Slight	Marked
Envelope subunit	Present	Absent
Location	Extracellular	Intracellular

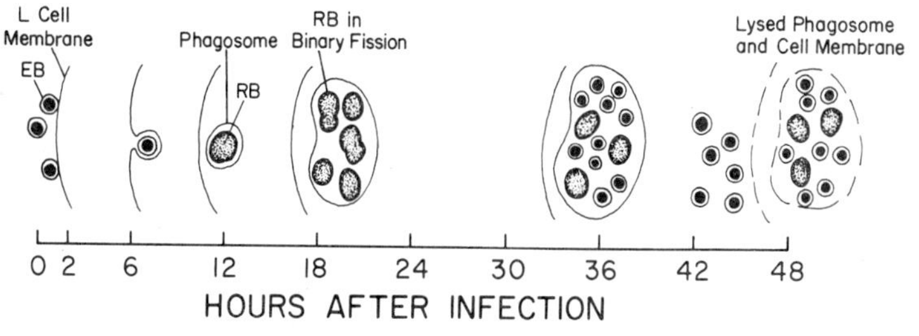

FIGURE 4. *Schematic representation of the reproductive cycle of* C. psittaci *in L cells. (EB = elementary body, RB = reticulate body).*

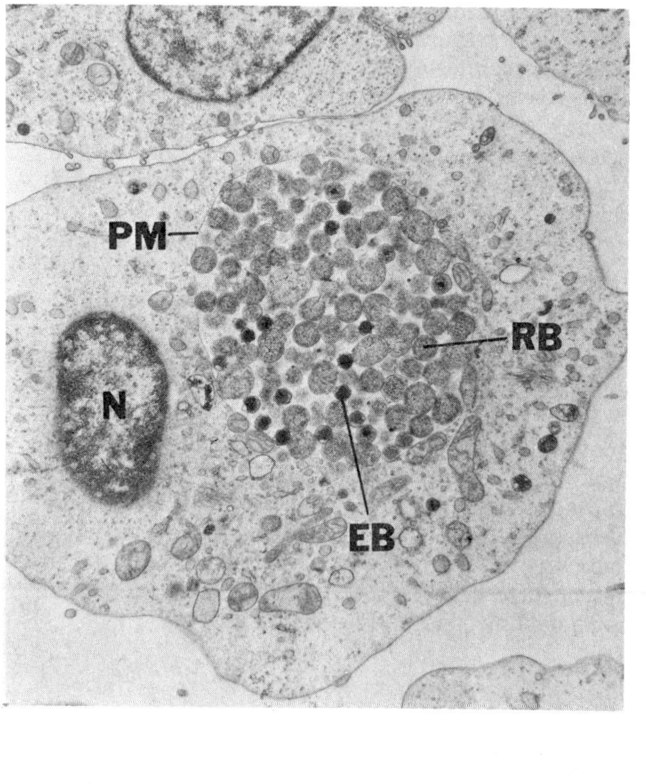

FIGURE 5. *Thin section of L cell 30 hours after infection with* C. psittaci. *Typical reticulate bodies (RB), newly formed elementary bodies (EB) and the continuous phagosome membrane (PM) are labeled.*

two general types of antigens, first recognized by Bedson in 1936. The first of these is the group antigen that is found in all chlamydiae and is usually demonstrated by complement fixation test. This antigen is a lipid-carbohydrate complex, stable when heated at 100° C for 30 minutes, and its antigenic determinant appears to be an acidic polysaccharide that is destroyed by periodic acid or acid hydrolysis.

There are a variety of specific antigens on or near the surface of the envelope of chlamydiae. In *C. psittaci* a large number of these have been demonstrated by toxin neutralization and plaque reduction tests. Among *C. trachomatis* strains, there are three serologically distinct types of lymphogranuloma venereum (LGV) agents, and 12 serotypes of the trachoma–inclusion conjunctivitis–urethritis group. The serotypes of *C. trachomatis* have been determined mostly by use of the microimmunofluorescence test (Wang and Grayston, 1970). Little is known about these antigens and none have been isolated for study, but they appear to be protein in nature.

METABOLISM

The adaptation of chlamydiae for intracellular parasitism is well illustrated by their apparent total dependency on host cells for energy. They do synthesize their own macromolecules including DNA, RNA, and proteins, and they inhibit to varying degrees macromolecular synthesis in host cells, although the mechanism of this inhibition is not known. The chlamydiae contain numerous enzymes and can complete some metabolic processes, but they cannot complete the pentose cycle and do not utilize pyruvate by way of the tricarboxylic acid cycle. There is no evidence that they can generate high energy phosphate bonds, and they are completely dependent on the host cell for ATP. This dependence gives added significance to the microscopic evidence that mitochondria accumulate around the phagosome membrane of developing inclusions and are among the last cell organelles to distintegrate in the dying infected cell.

Essentially all of the in vitro studies done on metabolism of chlamydiae have been conducted with partially purified organisms prepared by methods that tend to destroy the reticulate bodies. Since the reticulate bodies are the metabolically active form of chlamydiae, further studies need to be done using purified and stabilized suspensions of reticulate forms.

PATHOGENIC PROPERTIES

The chlamydiae do not possess demonstrable exotoxin and their cell envelopes, while similar in most characteristics of those of gram-negative bacteria, do not exhibit typical endotoxic properties. They do have several properties of significance in pathogenesis, one of which is toxicity of live elementary body preparations. Toxicity is usually measured by the intravenous injection of elementary body preparations into young mice. The component responsible for toxicity cannot be separated from the elementary body, and only live infectious elementary bodies exhibit this phenomenon. Toxicity can be neutralized by specific antisera, although the organisms may remain infectious. However, both live elementary bodies and their cell envelopes will absorb antitoxin, indicating that the antigen responsible for toxin neutralization lies on the surface of the envelope.

In contrast, the reticulate body is nontoxic, and antitoxin is not absorbed by suspensions of reticulate bodies, indicating the absence of this specific antigen during the growth of chlamydiae (Christoffersen and Manire, 1969).

The relatively harmless or nontoxic nature of the developing form may well contribute to another property of chlamydiae — the ease with which these organisms establish chronic inapparent carrier states in man and animals. This lack of specific antigenicity may allow reticulate bodies to reside intracellularly for long periods of time protected from host antibody.

Hypersensitivity may also play a role in *C. trachomatis* infection in that prior vaccination or infections often produce increasingly severe clinical symptoms.

In general, *C. trachomatis* tends to produce localized infections in mucous membranes with some extension in the case of genital infections. It may also produce pneumonitis in newborn infants. *C. psittaci*, in contrast, almost always produces generalized infections involving many organs in different animal hosts.

Little information is available to explain these differences or to answer other questions about the ability to chlamydiae to damage cells, establish latent or inapparent infections, or overcome immune mechanisms.

IMMUNITY

Diseases produced by chlamydiae in animals and man tend to be chronic, and carrier states frequently develop following recovery from clinical disease. Neutralization of these organisms by specific antisera can be demonstrated in vitro, but the intracellular site of their growth and the absence of specific antigens on the surface of the dividing reticulate bodies apparently protect the organisms from specific antibody.

IgM, IgG, and IgA against chlamydiae are all detectable in infected animals and man, but the role of these antibodies and of cell-mediated immune mechanisms in recovery from infection is not known. Severe infection often occurs in healthy avian, and probably mammalian, carriers when they are subjected to stress, nutritional deprivation, and other traumatic events, indicating a delicate balance between host defense and chlamydial pathogenicity.

Since 1960 many trachoma vaccines produced by a variety of techniques have been tested in primates and humans. Short-term protection has been achieved in numerous instances, but subsequent susceptibility and development of hypersensitivity have been consistent long-term consequences.

LABORATORY DIAGNOSIS

In *C. psittaci* infections, the complement fixation test using the heat-resistant group antigen has been the most commonly used serologic procedure. The organisms can be isolated without difficulty but with considerable hazard by injection of specimens into mice by various routes and into the chick embryo yolk sac, and many laboratory infections have occurred.

In *C. trachomatis* infections, recovery of the organisms by inoculation of specimens into irradiated or otherwise treated McCoy cells or by injection into chick embryo yolk sacs are well-established techniques. The examination of iodine- and/or Giemsa-stained smears has long been used to detect organisms in patients with trachoma. The more recently devised microimmunofluorescence test is rapidly replacing other serologic tests both for identification of specific antibody and for recognition of the organisms in cells removed from infected areas. A comparison of diagnostic methods currently available is shown in Table 3.

DRUG SUSCEPTIBILITY

The first specific therapy for chlamydial infections followed the discovery that these organisms are sensitive to sulfonamides and penicillin. In general, *C. trachomatis* strains are sensitive to sulfonamides and these drugs have long been used in treatment of trachoma and LGV. *C. psittaci* strains are generally resistant to sulfonamides.

Penicillin has been used successfully in psittacosis infections, but its effect is to prevent conversion of reticulate forms to elementary bodies; it does not inhibit cell infection, conversion of elementary bodies to reticulate bodies, or growth of reticulate bodies.

Both of these drugs have been largely replaced by chloramphenicol, tetracycline, and rifamycin. The action of these drugs on chlamydiae is the same as with other bacteria.

A very useful observation in the studies on metabolism of chlamydiae is that cycloheximide will almost completely shut down host cell protein synthesis without effect on chlamydiae.

TABLE 3. Comparative Sensitivities of Diagnostic Laboratory Tests for *C. trachomatis* Infections

	ANATOMIC SITE		
TEST	Conjunctiva (130)[a]	Male Urethra (200)[a]	Female Cervix (300)[a]
Isolation			
Tissue culture	95%[b]	100	95
Yolk sac	38	26	40
Stain of cell scrapings			
Giemsa	45	15	41
Fluorescent antibody	85	60	66
Serum antibody			
Complement fixation	50	15	40
Microimmunofluorescence	100	90	99

[a]Number tested
[b]% positive
From Schachter, J.: N Engl J Med 298:545, 1978.

EPIDEMIOLOGY

In 1967 Meyer presented a comprehensive review on the distribution of chlamydiae in humans and in wild and domestic mammals and birds. At that time more than 130 species of birds had been found to be infected, and it is probable that almost all bird species are carriers of *C. psittaci*. There is a wide range of mammals that are naturally infected. These include mice, cats, dogs, cattle, sheep, seals, and many other species. Various arthropods associated with infected animals also carry chlamydiae, but there is no evidence that arthropods play a role in disease transmission. The extent of infection in laboratory animals is illustrated by Meyer's finding that none of the colonies of mice and hamsters tested between 1940 and 1944 were free from chlamydiae.

Although clinical trachoma appears to have been diminishing since World War II, there are hundreds of millions of infected humans, primarily in Africa and Asia. Recent evidence that *C. trachomatis* is an important cause of nongonococcal urethritis and of cervicitis, and the finding of chlamydiae in 5 to 10 per cent of cervical specimens from asymptomatic women undergoing routine examinations, further confirm the extraordinarily widespread occurrence of these organisms and their role as a cause of infectious disease.

C. psittaci organisms spread from birds to man only sporadically, and outbreaks of human psittacosis occur mostly in persons processing carcasses of domestic turkeys, ducks, and other birds. There are several recorded instances of direct person-to-person transmission primarily involving hospital personnel.

Humans have been frequently infected following contact with apparently healthy birds. The organisms may be introduced from dust in bird cages and in other material contaminated with fecal materials from domestic or wild birds. Infection in humans most commonly results from inhalation of contaminated droplets or dust.

C. trachomatis is spread principally by three methods. The most common is direct contact by uninfected persons with eye secretions from infected carriers. In populations where hygiene is little practiced and where the carrier rate is very high, essentially every child is infected by contact with older carriers. The second method of transmission is sexual intercourse, resulting in acute, subacute, or inapparent infections of the genital tract of females and the genitourinary tract of males. The third method of transmission occurs during birth when the infant becomes infected during passage through the birth canal, with resulting conjunctivitis and/or pneumonitis.

References

Bryne, G. I.: Requirements for ingestion of *C. psittaci* by mouse fibroblasts. Infect Immun 14:645, 1976.

Christoffersen, G., and Manire, G. P.: The toxicity of meningopneumonitis organism (*C. psittaci*) at different stages of development. J Immunol 103:1085, 1969.

Meyer, K. F.: The host spectrum of psittacosis-lymphogranuloma venereum agents. Am J Ophthalmol 63:1225, 1967.

Page, L. A.: Chlamydiales. In Buchanan, R. E., and Gibbons, N. E. (eds.): *Bergey's Manual of Determinative Bacteriology,* 8th ed. Baltimore, The Williams & Wilkins Company, 1974.

Schachter, J.: Chlamydial infections. N Engl J Med 298:428, 490, 540, 1978.

Stokes, G. V.: Cycloheximide-resistant glycosylation in L-cells infected with *C. psittaci*. Infect Immun 9:497, 1974.

Storz, J., and Spears, P.: Chlamydiales: Properties, cycle of development and effect on eukaryote host cells. Curr Topics Microbiol Immunol 76:167, 1977.

Todd, W., and Storz, J.: Ultrastructural cytochemical evidence for the activation of lysosomes in the cytocidal effect of *C. psittaci*. Infect Immun 12:638, 1975.

Wang, S. P., and Grayston, J. T.: Immunological relationship between genital TRIC, lymphogranuloma venereum and related organisms in a new microtiter indirect immunofluorescence test. Am J Ophthalmol 70:367, 1970.

4 MYCOPLASMAS AND L-FORMS

53 MYCOPLASMAS

Zell A. McGee, M.D.
David Taylor-Robinson, M.D.

Mycoplasmas, originally called pleuropneumonia-like organisms (PPLO), are the smallest of free-living organisms. In addition to their characteristic small size (smaller than some viruses), they all lack a cell wall and therefore are resistant to penicillin and other cell-wall active antimicrobials. The individual organisms are bounded by a pliable unit membrane that encloses the cyto-

TABLE 1. Characteristics of Mycoplasmas Compared to Those of Bacteria and Viruses

CHARACTERISTIC	MYCOPLASMAS	BACTERIA	VIRUSES
Size (diameter)	0.3 μm[a]	1–2 μm	<0.5 μm
Lack a cell wall	Yes	No	Yes
Propagate on cell-free media	Yes	Yes	No
Usually require sterol and native protein for growth	Yes	No	No
Intrinsic energy metabolism	Yes	Yes	No
Usually narrow range of host specificity	Yes	No	Yes
Growth inhibited by specific antibody	Yes	No	Yes
Resistant to cell wall-active antibotics (e.g., penicillins)	Yes	No	Yes
Resistant to antibiotics that inhibit metabolism (e.g., tetracycline)	No	No	Yes

[a]Smallest organism capable of propagation.

plasm, DNA, RNA, and other metabolic components necessary for propagation on cell-free media. Mycoplasmas have a number of other characteristics that distinguish them from bacteria and viruses (Table 1). They are not related to L-phase variants of bacteria (see Chapter 54) (Hayflick, 1969; McGee et al., 1967).

Despite the superficial similarities among mycoplasmas, they compose a heterogeneous assemblage of microorganisms that differ from one another in DNA composition, nutritional requirements, metabolic reactions, antigenic composition, and host species specificity. Taxonomically, the mycoplasmas are divided into two families: the Mycoplasmataceae, which require cholesterol for growth, and the Acholeplasmataceae, which do not. All the mycoplasmas commonly isolated from humans belong to the Mycoplasmataceae. This family comprises the genus *Mycoplasma*, which contains organisms that do not hydrolyze urea, and the genus *Ureaplasma*, the organisms of which do hydrolyze urea. The latter were originally termed T strains or T mycoplasmas because of the tiny colonies they form.

Although the small size of their genome restricts the metabolic capabilities of mycoplasmas, certain species are important because they cause pneumonia, arthritis, keratoconjunctivitis, and mastitis among livestock and poultry in Africa, Australia, and other parts of the world. Eleven species constitute the normal flora or are pathogens of humans (Table 2). Many mycoplasma species are a laboratory nuisance as occult contaminants of tissue cultures. In addition, some mycoplasmas infect plants and insects (spiroplasmas), and others have been isolated from environments with a pH of less than 2 or with a temperature of 60° C (thermoplasmas).

TABLE 2. Mycoplasmas of Humans

	SPECIES	FREQUENCY OF ISOLATION	USUAL SITE OF ISOLATION	NATURE OF DISEASE
Nonpathogenic				
	M. orale	frequent	oropharynx	none
	M. salivarium	frequent	oropharynx	none
	M. buccale	rare	oropharynx	none
	M. faucium	rare	oropharynx	none
	M. lipophilum	rare	oropharynx	none
	A. laidlawii	very rare	oropharynx	none
	M. fermentans	infrequent	genitourinary tract	none[a]
	M. primatum	rare	genitourinary tract	none
Occasionally pathogenic				
	M. hominis	frequent[b]	genitourinary tract	Septicemia, abscess, endometritis-salpingitis, reproductive disorders (sterility, abortion, prematurity)?
	U. urealyticum	frequent[b]	genitourinary tract	Nongonococcal urethritis, septicemia? endometritis-salpingitis? reproductive disorders
Frequently pathogenic				
	M. pneumoniae	frequent[c]	oropharynx, respiratory tract	Tracheobronchitis, pneumonia, hemorrhagic bullous myringitis

[a]Reports of isolation from joint fluids in rheumatoid arthritis remain unconfirmed.
[b]Colonization frequent, disease infrequent.
[c]During disease but otherwise infrequent.

MORPHOLOGY

The morphology of individual organisms varies from one mycoplasma species to another, from time to time during the growth cycle, and from one environmental condition to another. Most mycoplasmas are spherical (0.3 to 0.8 μm in diameter) and divide by binary fission. Some mycoplasmas develop branching filaments 0.3 to 0.4 μm in diameter and up to 150 μm long. These divide to produce new spherical bodies.

The morphology of three species of mycoplasmas, *Mycoplasma pneumoniae, M. pulmonis* and *M. gallisepticum*, pathogens of the human, murine, and avian respiratory tracts respectively, is characterized by specialized structures at one or both ends of the organisms. *M. pneumoniae* has a tapered, filamentous tip that contains a dense, central rodlike core (Fig. 1). A specialized structure also has been observed in *M. pulmonis. M. gallisepticum* has a pear-shaped bleb at one or both ends of the organism. These structures may help attach the organism to the respiratory tract mucosa; they usually are the point of intimate contact of the mycoplasmas with the host cell membrane (Fig. 1). They also may be involved in motion, since these three mycoplasmas exhibit gliding motility in which the organisms move tip first.

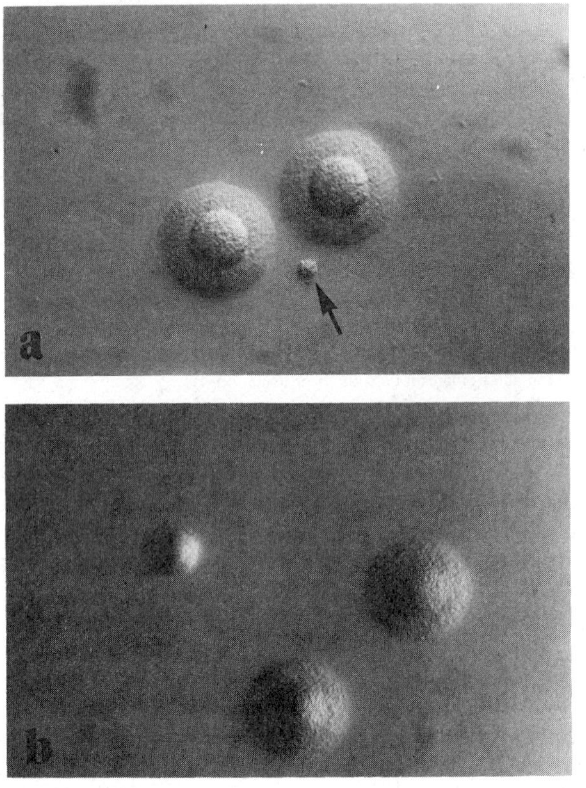

FIGURE 2. *Colonial morphology of mycoplasmas. (a) Two colonies (90 μm diameter) of* Mycoplasma hominis *and one colony (15 μm diameter) of* Ureaplasma urealyticum *or T-strain (tiny colony) mycoplasmas (arrow) from urethral exudate. (b) Three colonies of* Mycoplasma pneumoniae *from the sputum of a patient with pneumonia. Two colonies (80 μm diameter) have a poorly defined periphery and the third, more typical of* M. pneumoniae, *lacks a periphery.*

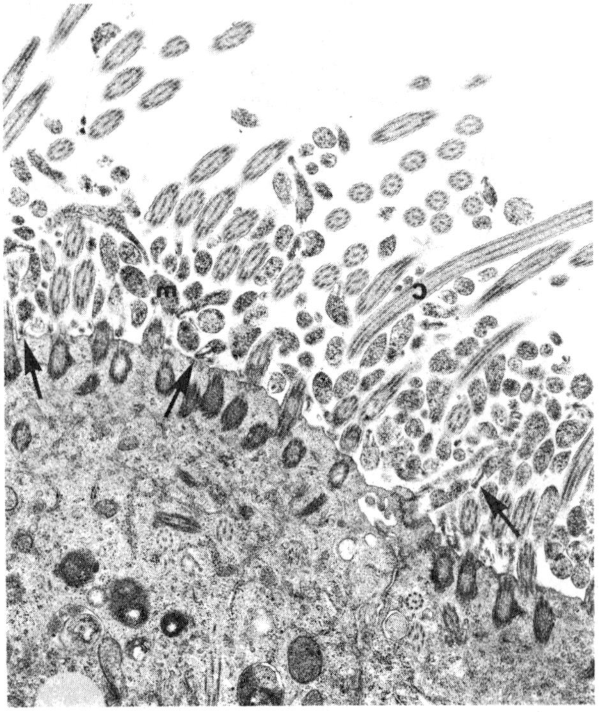

FIGURE 1. *Electron micrograph of ciliated epithelial cells in tracheal mucosa infected with* Mycoplasma pneumoniae. *Note cilia (c) and individual organisms of* M. pneumoniae *(m) with specialized terminal structure oriented toward the membrane of the host cell (arrows). (×13,000).*

Colonies of most mycoplasma species vary in size from 50 to 600 μm and are therefore most easily viewed through clear media with a dissecting microscope. The characteristic "fried egg" morphologic appearance of mycoplasma colonies on agar medium (Fig. 2*a*, large colonies) is caused by an opaque central zone of growth into the agar and a translucent peripheral zone on the surface. Such typical colonies do not always develop, since development depends on the species of mycoplasma, the constituents and degree of hydration of the medium, and atmospheric conditions. Ureaplasmas were originally called T-strains or T-mycoplasmas (T for tiny) because of the very small colonies they produce (15 to 30 μm in diameter—Fig. 2*a*, small colony); however, on medium buffered to pH 6, typical colonies up to 300 μm may develop. On primary isolation, *M. pneumoniae* usually produces mulberry-like colonies with no translucent peripheral zone, but larger colonies with a periphery sometimes develop (Fig. 2*b*).

ANTIGENIC COMPOSITION

The major antigenic determinants of mycoplasmas are membrane proteins and glycolipids. The glucose- and galactose-containing glycolipids of *M. pneumoniae* are haptens, which are antigenic only when bound to membrane protein. They induce antibodies that react in tests of complement fixation, metabolism inhibition, and growth inhibition. Glycolipids of similar structure and activity are found in some other mycoplasmas, in most plants, and in human brains. The cross-reactivity of human brain tissue antigens with *M. pneumoniae* antibodies could account for the neurologic manifestations of *M. pneumoniae* infection. In contrast, a glycoprotein fraction of *M. pneumoniae*, rather than the glycolipid moiety, is involved in the development of a cell-mediated immune response to this mycoplasma species.

Several mycoplasma species, such as *M. mycoides* (bovine), *Acholeplasma laidlawii*, and *M. meleagridis* (avian), possess extracellular carbohydrates or capsule-like material. Although the function of these substances is not clear, the galactan of *M. mycoides,* which is similar to a substance found in bovine lung, may play an immunologic role in the production of pneumonia.

Antigenic relationships among mycoplasmas infecting humans have been defined by relatively specific serologic procedures (such as metabolism inhibition, growth inhibition, and immunofluorescence) that measure antibodies directed against mycoplasma membrane antigens. By means of these tests, mycoplasma "species" have been established. Species isolated from humans are distinct from each other and from those isolated from other animals. In contrast, tests employing mycoplasma cell extracts are less specific and show group relationships; thus, by immunodiffusion, common antigens are demonstrable among all the human mycoplasma species that have been tested except *M. pneumoniae.* Ureaplasmas of human origin comprise a single species, *U. urealyticum*, which has been subdivided into eight serotypes by such tests as metabolism and growth inhibition.

NUTRITION AND METABOLISM

Mycoplasmas require lipids and lipid precursors for synthesis of the plasma membrane (Hayflick, 1969; Razin, 1978). Animal sera provide a complex of lipoprotein and cholesterol, the latter being incorporated into the lipid bilayer membrane of sterol-requiring mycoplasmas and functioning as a regulator of the bilayer fluidity. Acholeplasmas do not need exogenous cholesterol because they synthesize saturated long-chain fatty acids and carotenoids that substitute for cholesterol in the membrane.

TABLE 3. Metabolic Properties of Common Human Mycoplasmas

	ARGININE HYDROLYZED	GLUCOSE FERMENTED TO ACID	UREA HYDROLYZED
M. orale	+	−	−
M. salivarium	+	−	−
M. hominis	+	−	−
M. pneumoniae	−	+	−
U. urealyticum	−	−	+

Mycoplasmas generally multiply at a slower rate than bacteria. The mean generation time for many mycoplasmas, including ureaplasmas, is one to three hours, and for some, six to nine hours. Most mycoplasmas are facultative anaerobes, but they will grow aerobically, some preferring this atmosphere. They utilize either glucose or arginine as a major source of energy. The carbohydrate-fermenting species catabolize glucose by glycolytic pathways, mainly to lactic acid. In most but not all mycoplasmas the respiratory pathways are flavin-terminated so that the heme compounds, cytochromes, and catalase are absent. Mycoplasmas that metabolize arginine use a three-enzyme system that converts it via ornithine to ammonia, and in so doing, supplies the organisms with ATP. A few species utilize both glucose and arginine, but ureaplasmas metabolize neither substrate. They convert urea to ammonia by means of a urease, although the need for urea as an energy source has not been established. Some metabolic properties of common human mycoplasmas are summarized in Table 3.

PATHOPHYSIOLOGY

Although *M. hominis* and *U. urealyticum* have been associated with perinatal sepsis, endometritis-salpingitis, and other genitourinary infections, their pathogenic role in many cases is unclear, and little is known of their pathogenic mechanisms (McCormack et al., 1973; Bowie et al., 1976). The pathogenicity of *M. pneumoniae*, which has been studied in humans, hamsters, and tracheal organ cultures, seems to depend on attachment of the specific terminal structure to neuraminic acid receptor sites on the surface of mucosal cells in the trachea and bronchi (Fig. 1). Proteins on the membrane of *M. pneumoniae* play a role in the attachment (Hu et al., 1977). The organisms do not invade the mucosal cells but appear to damage them by a toxic factor. *M. neurolyticum,* a pathogen of mice, produces a soluble protein exotoxin, but the only toxic factor definitively identified in *M. pneumoniae* so far is hydrogen peroxide. Attachment is important in pathogenesis; disease is produced by strains of *M. pneumoniae* that produce peroxide and attach, but no disease is caused by strains that produce peroxide but do not attach. *M. pneumoniae* can also activate the classic and alternate pathways of complement and inhibit phagocytosis by alveolar macrophages.

Because fatal cases of *M. pneumoniae* pneumonia are rare, the histopathologic picture of this disease is derived mainly from experimental infection of hamsters and natural mycoplasmal disease in other animals. The pneumonic infiltrate is predominantly a peribronchiolar and perivascular cuffing by lymphocytes (Fig. 3*a*). This tissue response is followed by a change of the exudate in the bronchioles from a predominance of lymphocytes to a predominance of polymorphonuclear leukocytes and macrophages. These latter cells may, in association with antibody, help clear mycoplasmas by immune phagocytosis. The rather slow development of these events in a primary infection contrasts with the markedly accelerated and often more intense response that occurs with reinfection. Some of the lymphocytes in the peribronchial infiltrate contain immunoglobulin, but most of these cells appear to be thymus-dependent. Ablation of the thymus in hamsters by various procedures prevents the development of peribronchial cuffing when these animals are challenged with *M. pneumoniae* (Fig. 3*b*). To at least some extent, therefore, the pneu-

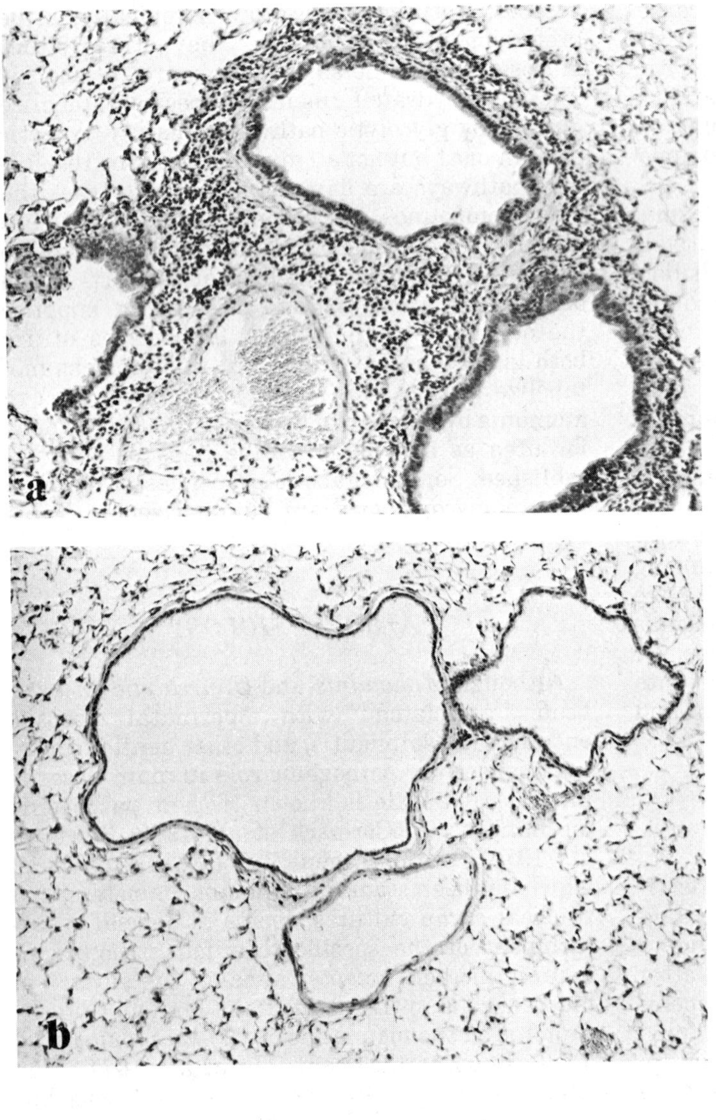

FIGURE 3. *Pathology of pneumonia caused by* Mycoplasma pneumoniae. *Pneumonia 2 weeks after intranasal inoculation with* M. pneumoniae *of (a) immunologically normal hamsters (note peribronchiolar and perivascular infiltrate of mononuclear cells, predominantly lymphocytes) and (b) hamsters previously depleted of T-lymphocytes (note virtual absence of infiltrate). (Hematoxylin and eosin, ×176).*

monia caused by *M. pneumoniae* is an immunopathologic process (Taylor and Taylor-Robinson, 1975). Since many young children possess *M. pneumoniae* antibody, it is possible that the pneumonia that occurs in older age groups is an immunologic overresponse to reinfection, the lung being infiltrated by previously sensitized lymphocytes.

The pathophysiology of the Stevens-Johnson syndrome (erythema multiforme), meningoencephalitis, and other conditions that sometimes complicate *M. pneumoniae* infections is not understood. No firm evidence supports a pathogenic role for the antiheart and antilung antibodies that are observed in some patients with *M. pneumoniae* pneumonia.

The localization of specific receptor sites, and thus the organisms of *M. pneumoniae*, in the trachea and bronchi with subsequent peribronchial infiltrates, contrasts with the localization of pneumococi in fluid-filled alveoli that are adjacent to the pleura. This may explain why pneumonia caused by *M. pneumoniae* is characterized by substernal burning and scanty production of sputum on coughing, whereas pneumococcal pneumonia is characterized by lateral pleuritic chest pain and generous production of sputum on coughing.

IMMUNITY

Antibodies in serum may be detected during mycoplasma infections by a variety of techniques, the most sensitive of which are radioimmunoprecipitation, complement-dependent mycoplasmacidal and metabolism inhibition tests. Generally, the first antibodies produced are of the IgM class; later IgG predominates. Humoral antibodies, however, do not fully protect against infection or disease, since infection and development of pneumonia may occur despite high levels of, for example, mycoplasmacidal antibodies to *M. pneumoniae* in the serum. Furthermore, natural or experimental respiratory tract infection of animals may produce only low levels of serum antibodies, yet such local infection provides greater resistance to reinfection than does parenteral inoculation of organisms, which produces high levels of serum antibody. This suggests that local factors are crucial in resistance. Indeed, resistance of adult volunteers to *M. pneumoniae* has been related to the presence of specific IgA antibody in respiratory secretions. This antibody may function as the first line of defense by preventing the attachment of organisms to the respiratory epithelial cells. On the other hand, development of a cell-mediated immune response, which is detectable by lymphocyte transformation and macrophage migration inhibition, does not seem to be important in resistance.

In field trials, inactivated *M. pneumoniae* vaccines have prevented disease in only about half of those vaccinated. The inability of these and other killed mycoplasma vaccines to protect completely may be due to their failure to stimulate local antibody production. With this in mind, live but attenuated temperature-sensitive mutants of *M. pneumoniae* have been developed that multiply at the temperature of the upper respiratory tract but not at that of the lower tract. Some of the mutants have produced pulmonary infection both in the hamster and in human volunteers without causing disease and have induced resistance to challenge with virulent wild-type *M. pneumoniae*. Whether such mutants can be made sufficiently stable and attenuated for large-scale use as a human vaccine remains to be seen. However, permanent protection seems unlikely because naturally acquired *M. pneumoniae* infection does not provide lifelong immunity. Reinfection occurs with appreciable frequency, and second attacks of *M. pneumoniae* pneumonia also occur.

LABORATORY DIAGNOSIS

Mycoplasmas grown in vitro are gram-negative. However, individual mycoplasma organisms are too small and do not take up the safranin counterstain well enough to be recognized on microscopic examination of gram-stained clinical specimens. The laboratory diagnosis depends, therefore, on cultural identification and on serologic tests. Details of these techniques have recently been outlined by Gump and Forsyth (1977).

The medium for primary isolation of *M. pneumoniae* consists of PPLO agar or broth, 20 per cent horse serum and 10 per cent vol/vol fresh yeast extract (25 per cent wt/vol). To enhance detection of *M. pneumoniae*, the medium is supplemented with thallium acetate and penicillin to inhibit bacteria and fungi, methylene blue to inhibit other human mycoplasmas, and glucose with phenol red as a pH indicator to detect acid produced by *M. pneumoniae*. A vial of diphasic medium (broth over agar) is inoculated with sputum or a pharyngeal swab and incubated at 37° C. A color change from purple to yellow-green, which usually occurs within 4 to 21 days, signals the fermentation of glucose with production of acid and the consequent change in color of the phenol red to yellow. This presumptive positive identification is confirmed by subculturing to agar medium and demonstrating inhibition of colony development with commercial disks impregnated with specific antiserum. Hemadsorption and rapid hemolysis of guinea pig red cells by colonies of *M. pneumoniae* on agar are also used to help in presumptive identification. The hemadsorption is attributed to neuraminic acid receptors on red cells.

A similar strategy, utilizing the metabolism of arginine or urea, has been employed to design media that select for and presumptively identify *M. hominis* and *U. urealyticum* in specimens from the genitourinary tract.

The most readily available specific serologic test for *M. pneumoniae* is the complement fixation test, for which reagents can be obtained commercially. It is positive in about 80 per cent of cases as indicated by a fourfold or greater rise in antibody titers between acute and convalescent specimens. The cold agglutinin test depends on nonspecific agglutination of O Rh-negative red blood cells at 4° C by antibodies to I antigen that arise in *M. pneumoniae* infection. The test is positive in only about 50 per cent of patients, usually those with more severe disease. Serologic tests that detect antibody to *M. hominis* and *U. urealyticum* are still research procedures.

DRUG SUSCEPTIBILITY

Since the mycoplasmas lack a cell wall, they are resistant or indifferent to the penicillins, cephalosporins, and other antimicrobials that act on the cell wall. In contrast, mycoplasmas are generally sensitive to antimicrobials that inhibit protein synthesis. At clinically achievable concentrations, tetracycline is inhibitory for *M. pneumoniae, M. hominis,* and *U. urealyticum*. Whereas erythromycin has marked inhibitory and mycoplasmacidal activity against *M. pneumoniae*, it is only moderately active against *U. urealyticum* and is inactive against *M. hominis*. Both genital mycoplasmas, *U. urealyticum* and *M. hominis*, are sensitive to spectinomycin. Despite the sensitivity of *M. pneumoniae* to tetracycline and erythromycin and the good clinical responses to therapy that occur with these drugs, about 50 per cent of treated patients continue to harbor *M. pneumoniae* in the posterior pharynx for one to three months.

EPIDEMIOLOGY

Infections with *M. pneumoniae* apparently occur world-wide and account for approximately 20 per cent of all cases of pneumonia in some cities. The disease is endemic with periodic epidemics. Spread of the organism, which occurs by aerosol transmission, is slow. Therefore, most infections occur in small groups of people who have frequent close contact such as families, military units, and college fraternities. The incubation period is two to three weeks. The disease is usually introduced into the family by a small child and moves from person to person over a period of months with an attack rate of greater than 50 per cent. Pneumonia is most frequent in persons aged 5 to 20. Thereafter, pneumonia occurs less frequently but is generally more severe as the patient's age increases.

Ureaplasmas are acquired by up to 30 per cent of infants at birth. Colonization of the genital tract is most common but is usually transient. The organisms are found again after puberty, and antibody is first detectable at this time. Thereafter, the frequency of occurrence is related to sexual activity; recovery rates of about 60 per cent and 80 per cent have been observed for men and women, respectively, attending venereal disease clinics (McCormack et al., 1973). The occurrence of ureaplasmas in apparently healthy persons has posed a problem in attributing a pathogenic role to them. However, the results of several studies, including the intraurethral inoculation of human volunteers with ureaplasmas, suggest that they cause some cases of nongonococcal urethritis (Bowie et al., 1976).

M. hominis is acquired at birth by about 6 per cent of infants. Most of these individuals lose these strains but become colonized later in life when sexual activity commences. *M. hominis* is unlikely to play a role in nongonococcal urethritis, but it has been associated with reproductive failure, low birth weight, recurrent abortions, and a variety of pelvic infections. It has been isolated from the blood in the absence of other microorganisms in 8 per cent of women with febrile abortions.

The frequent causation of arthritis and neurologic diseases by different mycoplasmas in various animal species has suggested that they might cause such diseases in humans. Although the search for mycoplasmas in rheumatoid arthritis and other diseases of unknown etiology continues, these studies have not yet yielded definitive evidence.

References

Bowie, W. R., Floyd, J. F., Miller, Y., Alexander, E. R., Holmes, J., and Holmes, K. K.: Differential response of chlamydial and ureaplasma-associated urethritis to sulphafurazole (sulfisoxazole) and aminocyclitols. Lancet 2:1276, 1976.

Gump, D. W., and Forsyth, B. R.: Mycoplasma. In von Graevenitz, A. (ed.): Clinical Microbiology, vol. I. (CRC Handbook Series in Clinical Laboratory Science, Sec. E) Cleveland, CRC Press, 1977, p. 357.

Hayflick, L. (ed.): The Mycoplasmatales and the L-Phase of Bacteria. New York, Appleton-Century-Crofts, 1969.
Hu, P. C., Collier, A. M., and Baseman, J. B.: Surface parasitism by *Mycoplasma pneumoniae* of respiratory epithelium. J Exp Med 145:1328, 1977.
McCormack, W. M., Braun, P., Lee, Y. H., Klein, J. O., and Kass, E. H.: The genital mycoplasmas. N Engl J Med 288:78, 1973.
McGee, Z. A., Rogul, M., and Wittler, R. G.: Molecular genetic studies of relationships among mycoplasma, L-forms and bacteria. Ann N Y Acad Sci 143:21, 1967.
Razin, S.: The mycoplasmas. Microbiol Rev 42:414, 1978.
Taylor, G., and Taylor-Robinson, D.: The part played by cell-mediated immunity in mycoplasma respiratory infections. Develop Biol Stand 28:195, 1975.

54 CELL WALL-DEFECTIVE BACTERIA

Zell A. McGee, M.D.

Susceptible bacteria that are exposed to penicillin or other substances that damage the bacterial cell wall are usually killed. Under certain conditions, however, bacteria can survive with damaged cell walls and multiply as organisms with markedly altered morphology, physiology, and cultural characteristics (Dienes and Weinberger, 1951; Gutman et al., 1967; Hayflick, 1969). These wall-defective bacteria are undetectable by routine cultural techniques, are not affected by penicillin and other cell wall-active antimicrobials, and may revert to the intact bacterium if the cell wall-active antibiotic is removed. These characteristics of elusiveness, indifference to the most frequently administered antimicrobials, and morphologic versatility have stimulated interest in the possibility that wall-defective bacteria may play a role in various chronic, recurrent, or culture-negative infections of humans (Clasener, 1972; Guze, 1967).

MORPHOLOGY AND PHYSIOLOGY

The pliable bacterial cytoplasmic membrane is encased in a rigid peptidoglycan layer, which gives the bacterial cell its coccal or rod shape. In gram-negative bacteria, the intracellular osmotic pressure is 300 to 400 milliosmoles per kilogram water (mOsm/kg), and in gram-positive bacteria, it is 900 mOsm/kg or greater. When bacteria are suspended in serum or most bacteriologic media (270 to 300 mOsm/kg), the higher internal osmotic pressure of the bacteria favors the flow of water into their cytoplasm, but there cannot be a net gain of water because the peptidoglycan layer restricts any increase in volume. If the peptidoglycan layer is removed, water moves freely into the cytoplasm with increasing distention of the cytoplasmic membrane and dilution of the cytoplasm until the concentration of molecules inside the membrane equals the concentration outside. Before this equalization occurs, however, the cytoplasmic membrane usually ruptures, and the organism is killed. This seems to be the way that bacteria die when their cell walls are damaged by penicillin.

If the concentration of molecules in the environment is increased so that it balances the concentration within the bacterium, removal of the peptidoglycan may result in a change of shape (e.g., from a rod to a spherical shape), but the organism will not rupture (see Fig. 1*b*). Bacteria with damaged or absent cell walls can survive in environments with osmolalities lower than their intracellular osmotic pressure, but their cytoplasmic membranes must be stabilized intrinsically by divalent cations or polyamines, or by an adaptive change to a higher ratio of saturated to unsaturated fatty acids in the membrane (Leon and Panos, 1976).

When bacteria are exposed to penicillin in a hypertonic medium (strictly, a medium with an osmolality greater than that of human serum [290 mOsm/kg] but conventionally, about 900 mOsm/kg), they may simply round up to become *spheroplasts* (which have partial loss of the cell wall) or *protoplasts* (which have complete loss of the cell wall) (Fig. 1*a, b*). Since bacteria in the spheroplast or protoplast phase cannot multiply and are lysed at physiologic osmolalities, they are unlikely to play a role in human infections. Bacteria may also undergo a more fundamental change, however, either directly or by further transition from spheroplasts or protoplasts to L-phase variants or L-forms (Fig. 1*c,d*) (L stands for the Lister Institute in London where they were first described in 1935). Some bacteria, such as *Neisseria gonorrhoeae* and *Bacteroides fragilis,* can undergo this transition to the L-phase spontaneously; others require the presence of penicillin, lysozyme, or other cell wall-active substances.

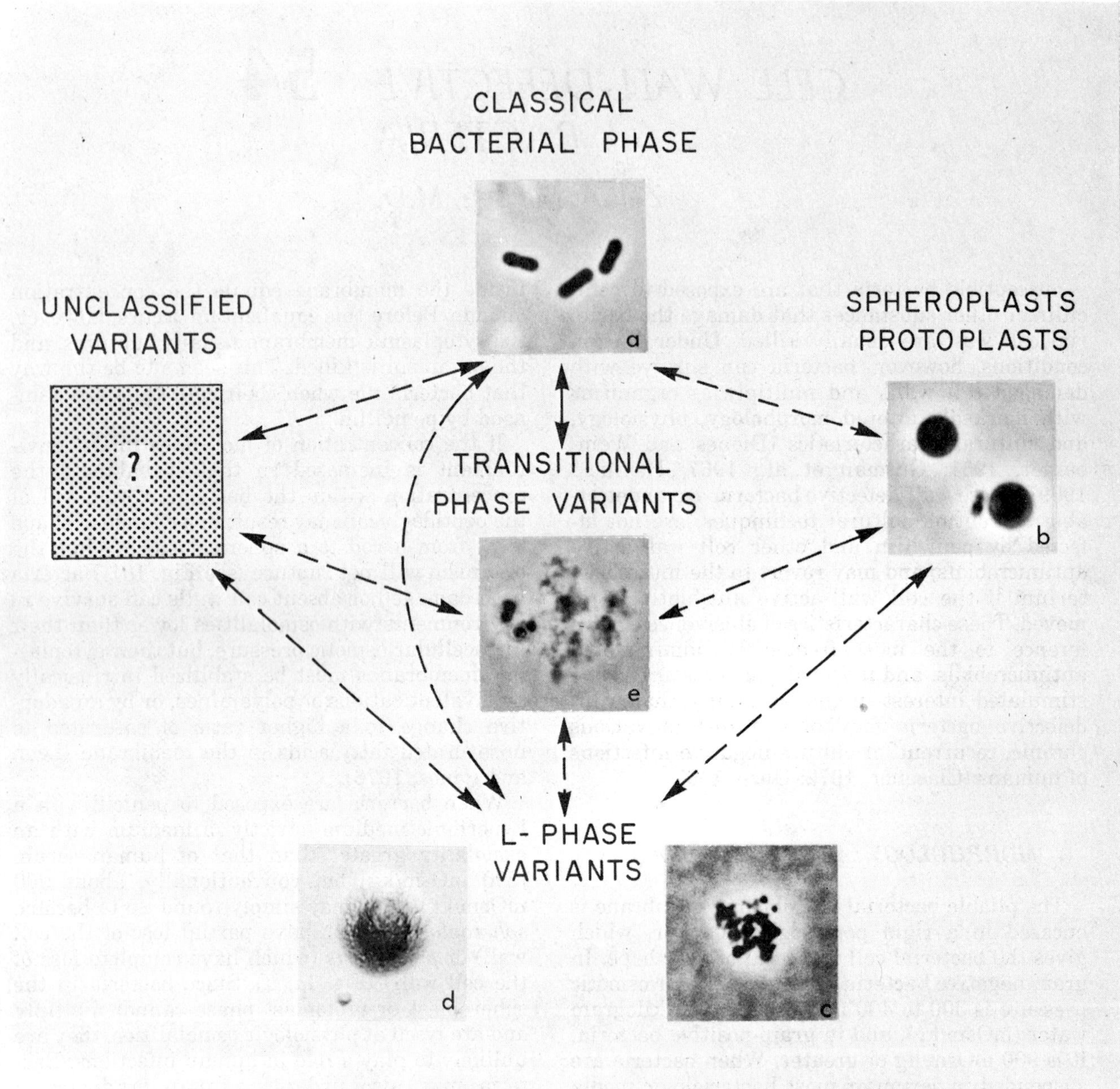

FIGURE 1. *Schematic representation of the transition of a bacterium to different variant phases. Photos* a, b, *and* c *demonstrate the contrasting morphology of the classic bacterial phase, spheroplast phase, and a clump of L-phase variants of* Proteus mirabilis *in identical media (900 mOsm/kg) and at the same magnification (phase contrast, × 3400). Photo* d *shows a "fried egg" colony of L-phase variants of* Proteus mirabilis *(× 150). Photo* e *shows a clump of transitional phase variants of* Staphylococcus albus *from the blood culture of a patient with an infected prosthetic heart valve (phase contrast, × 1600). (From McGee, Z.A., Wittler, R.G., Gooder, H., and Charache, P.: J Infect Dis 123:433, 1971.)*

Bacteria in the L-phase can replicate serially as nonrigid cells that are spherical or pleomorphic and vary in size from "large bodies," larger than some human cells, to elementary bodies, which may be as small as some viruses. On solid media, L-phase variants produce tiny colonies composed of a central core, which grows into the agar, and a superficial growth around the core, which gives them a "fried egg" appearance (Fig. 1*d*). L-phase variants that cannot revert to the classic bacterial phase are called "stable" L-phase variants. Bacteria that have reverted from the L-phase may be indistinguishable from the original bacteria, except that when exposed to the same or different inducing agents, they enter the L-phase more readily. L-phase variants resemble mycoplasmas in their individual and colonial morphology, but studies have failed to show any immunologic, biochemical or genetic relationship between these groups of microorganisms (Hayflick, 1969; McGee et al., 1971).

L-phase variants and bacteria undergoing transition to or from L-phase variants (*transitional phase variants*, Fig. 1*e*) are the types of wall-defective bacteria most likely to play a role in human disease.

ANTIGENIC COMPOSITION

The antigenic composition of a bacterium in the L-phase is similar to that of its bacterial phase. Even L-phase variants that completely lack cell-wall antigens on their external surface may continue to elaborate them into the environment. Although L-phase variants sometimes show antigenic determinants that are different from those of the parental bacteria, these may not represent new antigens but merely greater accessibility of certain determinants to antibody-producing cells.

One type of antigen-antibody reaction that occurs with the L-phase but not with the classic bacterial phase is inhibition of growth by specific antibody (Braude and Siemienski, 1968). These growth-inhibiting antibodies, which may also be elicited by the bacterial phase, do not require complement and seem to be directed at antigens of the cytoplasmic membrane.

METABOLISM

Transition to the L-phase is sometimes accompanied by loss of the ability to carry out certain metabolic reactions such as fermentation of a particular sugar or production of a hemolysin. The L-phase variants of a few bacterial species, such as *Salmonella typhi,* require anaerobic conditions, whereas the parental bacterium can grow aerobically. Nevertheless, L-phase variants generally carry out metabolic reactions and generate products characteristic of the bacteria from which they were derived, although they may do so at a slower rate.

HOST DEFENSES

A number of factors make the human body a potentially adverse environment for L-phase variants: (1) the osmolality of human serum and tissue fluid is approximately 290 mOsm/kg, well below the osmolality at which wall-defective variants of some bacteria undergo osmotic rupture; (2) normal human serum contains an antibody-complement system that rapidly disrupts the cytoplasmic membranes and kills L-phase variants of a variety of gram-positive and gram-negative pathogens; (3) growth-inhibiting antibodies may be elicited by L-phase variants or their parental bacteria; and (4) L-phase variants can be phagocytized by polymorphonuclear and mononuclear phagocytes.

PATHOGENIC PROPERTIES

None of the defenses mentioned above precludes the survival and multiplication of L-phase variants in humans. L-phase variants of both gram-negative and gram-positive bacteria can survive at osmolalities comparable to those of serum or purulent exudates. The hypertonic milieu of the kidney provides osmotic protection and inactivates the antibody-complement system in serum that might otherwise be lethal to L-phase variants. These variants can be formed and can persist within certain host cells under experimental conditions in vitro and in vivo.

L-phase variants of toxigenic bacteria, such as *Clostridium tetani* or *Vibrio* species, also produce toxins and might thereby be pathogenic. The L-phase variants of some gram-negative bacteria elaborate endotoxin and can evoke the localized Shwartzman reaction. Infection of the urinary tract of experimental animals with L-phase variants of *Proteus mirabilis,* which produce urease, has been associated with formation of bladder stones (Braude and Siemienski, 1968). L-phase variants of group A beta-hemolytic streptococci have been shown to adapt to osmotic conditions like those in human tissues and to invade and kill human heart cells in tissue culture (Leon and Panos, 1976). However, L-phase variants have persisted for prolonged periods in the tissues of experimental animals with little evidence of damage. The greatest pathogenic potential of L-phase variants may reside in their ability to maintain a pathogenic organism in a host under conditions such as antibiotic therapy that might kill the bacterial phase. The L-phase variant may then revert to the pathogenic bacterial phase when conditions become favorable for its survival.

ROLE IN INFECTIONS

Despite the varied methods of formation and survival of wall-defective variants in vivo under experimental conditions, it has been most difficult to prove that wall-defective variants play a role in human diseases. Koch's postulates have not been helpful in approaching this problem because they do not allow for a significant change in the nature of an organism during the course of an infection.

There are many reports of pyelonephritis, osteomyelitis, endocarditis, and other acute and chronic infections in which wall-defective variants were isolated on hypertonic but not on routine media. However, only a few of these studies provide evidence that L-phase variants were actually present in the patient and not merely induced on the media in the laboratory.

Microscopic examination of the urine from patients with pyelonephritis has revealed spheroplasts and other atypical bacteria during periods of antimicrobial therapy but only classic rod-shaped bacteria when therapy was not being administered. In one series of patients, the presence of viable cell-wall defective bacteria in the urine of 11 (16 per cent) of 57 patients with chronic pyelonephritis was documented by filtrating the specimens to remove bacteria and isolating L-phase variants from the filtrates on hypertonic but not on isotonic media. In three patients, who had only wall-defective variants in their urine and were not receiving antibiotics, the variants were apparently either induced by autolytic bacterial enzymes or by host factors.

One patient, a 10-year-old girl (described by Gutman et al., 1967), had recurrent bouts of urinary tract infection characterized by urgency, chills, fever, and more than 10^5 *P. mirabilis* per ml. of urine. During ampicillin therapy, the patient's symptoms decreased and routine cultures became sterile, but L-phase variants of the *Proteus* were recovered from filtered urine cultured on hypertonic media. After therapy was discontinued, bacterial phase *P. mirabilis* returned, and cultures for L-phase variants became negative. The bacterial phase was sensitive to ampicillin but resistant to erythromycin. The L-phase variant was resistant to ampicillin but sensitive to erythromycin. After ampicillin therapy was begun, routine cultures became negative, but cultures for L-phase variants became positive. When erythromycin was added to the regimen, cultures for L-phase variants became negative within three days. After completion of 12 days of combined therapy, all cultures remained negative and the patient was free of urinary tract infection for a follow-up period of 18 months.

Wittler and colleagues (1960) described the five-year course of a young girl with similar changes of symptoms and culture results during periods of penicillin therapy for *Corynebacterium* endocarditis. When penicillin was administered, she became asymptomatic and cultures on routine media became negative, but wall-defective variants were recovered on hypertonic media. The bacterial phase returned when penicillin was discontinued. This study was more convincing because the variants were visualized microscopically in peripheral blood cells when variants were recovered from hypertonic culture media and because penicillinase was used in the cultures to decrease the likelihood of producing variants in vitro.

These and a few other careful clinical studies as well as data from experimental animals strongly suggest that L-phase variants may be produced and persist in vivo. It is not clear whether such organisms are of clinical significance in more than rare cases.

LABORATORY DIAGNOSIS

Cultures for wall-defective organisms can be performed by the laboratory on two different levels: (1) the laboratory may routinely include media that are appropriate for recovering wall-defective variants but will not determine whether organisms isolated were present in the patient as typical bacteria or as L-phase variants; or (2) the laboratory may perform on highly selected patients the specialized procedures necessary to isolate and identify wall-defective variants that do not revert on their first passage in vitro, determine their antimicrobial susceptibility, and establish whether the organism existed in a wall-defective form in the patient. The details of these procedures have recently been reviewed (McGee, 1977). Either or both approaches may be adopted, but the second requires experienced personnel, complex media, and a large commitment of time.

DRUG SUSCEPTIBILITY

Bacteria undergoing transition to the L-phase always lose susceptibility to penicillins, cephalosporins, and other agents that act on the cell wall. Conversely, gram-negative bacilli, such as *Pseudomonas aeruginosa* become more susceptible to erythromycin, tetracycline, and other agents that act within the bacterial cytoplasm but are normally excluded by the gram-negative cell wall. These changes in susceptibility provide a rationale for the therapy of infections involving bacterial variants.

Therapy of infections with cell wall-defective variants has generally consisted of a combination of two drugs: (1) a cell wall-active agent such as penicillin that either kills residual bacteria or makes them wall-defective, and (2) an antimicrobial such as erythromycin or gentamicin that acts within the bacterial cell. Although this approach has been effective in controlled studies of infected experimental animals, reports of its success in human infections are anecdotal. Currently, the use of such therapy seems justified only in highly selected and closely monitored patients whose clinical specimens contain wall-defective variants (McGee, 1977).

References

Braude, A. I., and Sieminski, J.: Production of bladder stones by L-forms. Trans Assn Amer Phys 81:323, 1968.

Clasener, H.: Pathogenicity of the L-phase of bacteria. Ann Rev Microbiol 26:55, 1972.

Dienes, L., and Weinberger, H. J.: The L-forms of bacteria. Bacteriol Rev 15:245, 1951.

Gutman, L. T., Schaller, J., and Wedgwood, R. J.: Bacterial L-forms in relapsing urinary-tract infection. Lancet 1:464, 1967.

Guze, L. B. (ed.): Microbial protoplasts, spheroplasts and L-forms. Baltimore, Williams and Wilkins, 1967.

Hayflick, L. (ed.): The mycoplasmatales and the L-phase of bacteria. New York, Appleton-Century-Crofts, 1969.

Leon, O., and Panos, C.: Adaptation of an osmotically fragile L-form of *Streptococcus pyogenes* to physiological osmotic conditions and its ability to destroy human heart cells in tissue culture. Infect Immun 13:252, 1976.

McGee, Z. A.: Cell-wall deficient bacteria. In von Graevenitz, Alexander (ed.): CRC Handbook Series in Clinical Laboratory Science, Section E: Clinical Microbiology, Vol. I. Cleveland, CRC Press, Inc., 1977, p. 367.

McGee, Z. A, Wittler, R. G., Gooder, H., and Charache, P.: Wall-defective microbial variants: terminology and experimental design. J Infect Dis 123:433, 1971.

Wittler, R. G., Malizia, W. F., Kramer, P. E., Tuckett, J. D., Pritchard, H. N., and Baker, H. J.: Isolation of a corynebacterium and its transitional forms from a case of subacute bacterial endocarditis treated with antibiotics. J Gen Microbiol 23:315, 1960.

5 VIRUSES

DNA Viruses

ADENOVIRUS 55

Lennart Philipson, M.D., Dr. Med. Sci.

Adenoviruses constitute a family of viruses originally isolated from the respiratory tract of man and other animals. They were discovered in 1953. They are nonenveloped icosahedral viruses with DNA genomes of intermediate size (20 to 30 × 10^6 daltons). The adenovirus family has been subdivided in two genera, mastadenovirus and aviadenovirus, referring to viruses isolated from mammalian and avian hosts respectively (Norrby et al., 1976). There is no cross-reacting antigen between these two genera. The mastadenovirus comprises at least 80 different serotypes, of which 33 have been isolated from human sources. The aviadenovirus contains 14 distinct serotypes.

Precise biochemical tools to characterize adenovirus genomes have become available during recent years, including separated strands of the DNA, restriction enzyme fragments, and liquid hybridization methods to detect the viral transcription products. The structural and nonstructural proteins encoded by the adenovirus genome have also been extensively characterized with regard to both structure and function. The adenoviruses have therefore become a model virus to investigate the complex control mechanism involved in macromolecular synthesis in mammalian cells. Studies of the molecular biology of these viruses may help to elucidate mechanisms for transcription, processing of the transcripts to messenger RNA, and control mechanisms for protein synthesis in mammali-

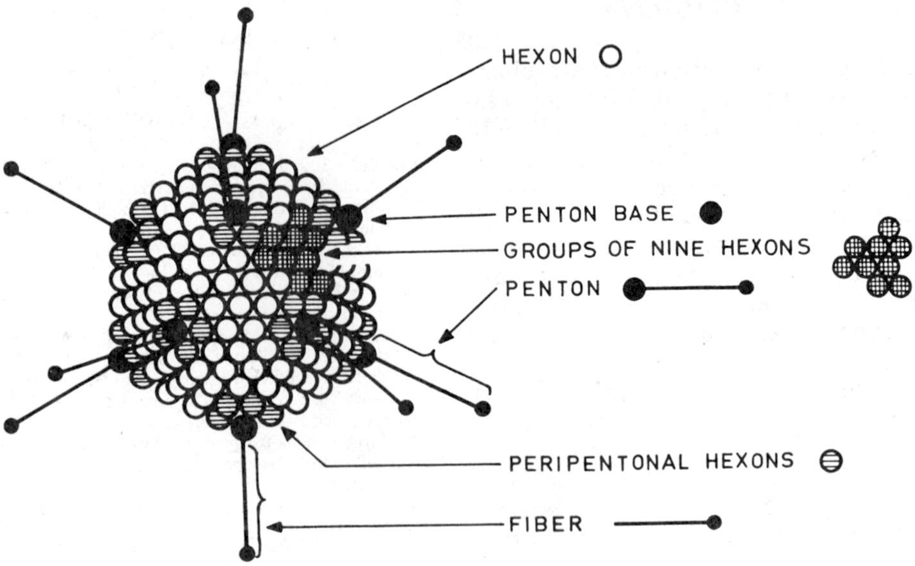

FIGURE 1. *Structure of the adenovirus capsid. Schematic drawing showing the icosahedral outline of the adenovirus capsid and the location of various components. (Reprinted from Philipson et al., 1975.)*

an cells. The molecular biology of the adenoviruses has recently been reviewed (Philipson et al., 1975; Wold et al., 1978), but the role of these viruses in human infections has not been evaluated during recent years. Several reviews may provide a more detailed insight into the clinical aspects of adenovirus infections (Sohier et al., 1965; Potter, 1967; and Rose, 1969). This chapter will focus on some properties of adenoviruses and their role in infectious diseases in man.

THE STRUCTURE OF THE VIRION

Morphology

The adenoviruses are nonenveloped viruses with a diameter of 65 to 80 nm. The capsid has 252 capsomers arranged into an icosahedron with 20 triangular facets and 12 vertices, as schematically shown in Figure 1. Two hundred forty of the 252 capsomers have six neighbors, and they are called *hexons;* whereas the 12 capsomers at the vertices have five neighbors, and are therefore called *pentons.* Each penton unit consists of a penton base anchored in the capsid and a rod-like projection with a knob attached at the distal end. The rod-like portion is referred to as the *fiber.* Two types of hexons may be defined: (1) those 60 located in juxtaposition to the pentons, *peripentonal hexons* that have the penton base as one of their six neighbors, and (2) those 180 that form the triangular facets and the edges of the icosahedron. The latter may be released from the virions in aggregates of nine hexons, *ninemers,* which form a defined structure with threefold rotational symmetry. Inside the capsid there is a core with a diameter of 40 to 45 nm, which contains the DNA and additional proteins. It can be revealed by electron microscopy of thin sections of particles stained with uranyl acetate. A circular protein-DNA complex was recently observed after degradation of virions with guanidine, suggesting that the two termini of the DNA in the virion are linked to each other by proteins.

Chemical Composition

The human adenovirus type 2 (ad2), which has been studied in detail, has an estimated particle weight of 175 × 10^6 daltons. It contains 13 per cent DNA corresponding to 23 × 10^6 daltons, and the additional components are all proteins. Unlike most other viruses the adenoviruses have capsid units that are soluble in nondenaturing solvent, which has facilitated the purification and characterization of the virus components. Methods have been developed for sequential disintegration of the virion. The pentons alone or together with peripentonal hexons can be selectively removed by dialysis at low pH. The release of pentons is accompanied by the release of additional antigens probably located in the peripentonal region. After treatment of the virus particle with sodium dodecyl sulfate (SDS), urea, or pyridine, the capsid is disrupted and the hexons from the triangular facets are released as ninemers. The polypeptide composition of adenoviruses has been studied by SDS-polyacrylamide electrophoresis. The virion of ad2, ad7, and ad12 contains a minimum of nine polypeptides that range in size from 7500 to 120,000 daltons. Eight of these are antigenically distinct and reside in different structures after sequential degradation of the virion. Five of the polypeptides are integral parts of the capsomers or the core. Thus, polypeptide II is the only polypeptide that is detected in purified hexons from infected cells. Polypeptide III resides in the penton base, IV, in the fiber, and polypeptide V and VII are associated with the core. The remaining polypeptides have been tentatively localized in the virion structure as indicated in Figure 2. The structural units observed in the virion as hexon, penton, and fiber are multimers of the integral polypeptide. Hexon contains three polypeptides, penton probably contains five, and fiber contains three polypeptides, and each unit must therefore be assembled independently.

The DNA of human adenoviruses has been extensively characterized. There is a difference in base composition between the different subgroups of the human adenoviruses (Table 1). In addition, there is considerable homology of the DNA sequence within each subgroup but only 10 to 20 per cent homology between DNA of members of different subgroups, regardless of whether filter hybridization or electron microscopic analysis of heteroduplexes was used to measure homology. The DNA for several types is infectious but the specific infectivity is at least 10^6 times lower than for virions measured as number of infectious units/μg of DNA. The protein associated with the termini of the viral DNA appears to enhance the infectivity of viral DNA. Specific fragments of DNA representing around 7.5 per cent of the total genome can transform cells. The DNA has been cleaved by several restriction enzymes and fragment maps have been established (Fig. 3). The fragments have been used to map the position of tran-

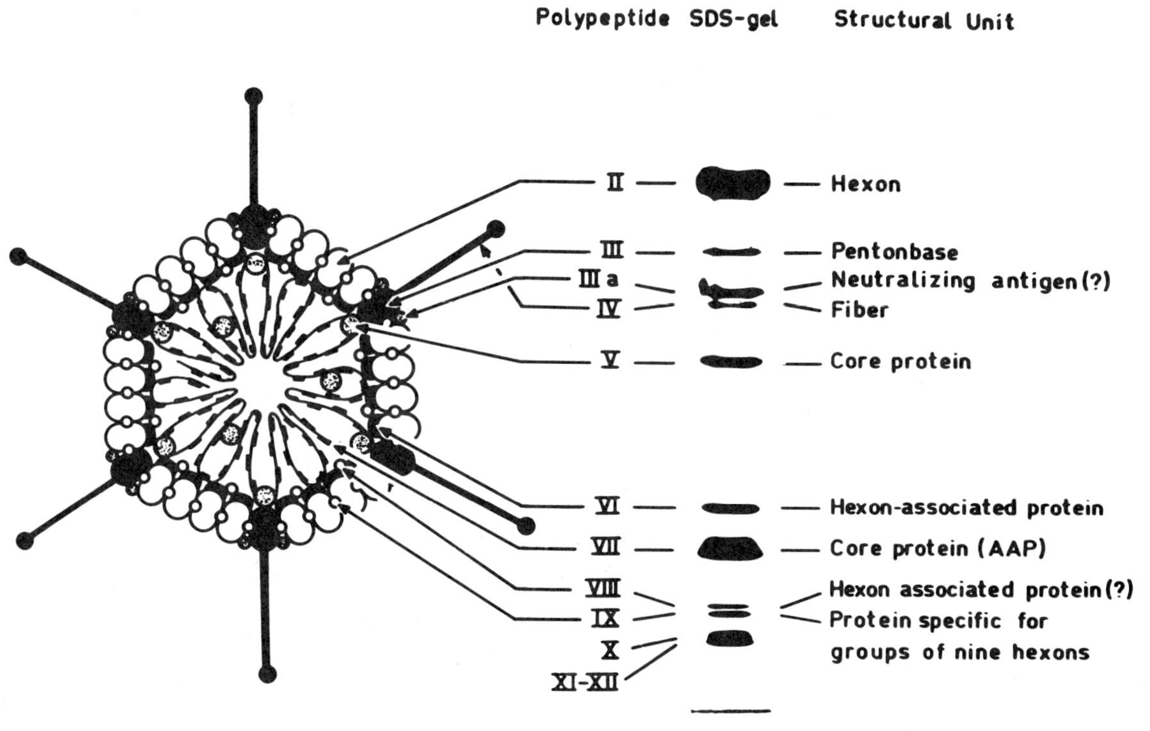

FIGURE 2. *A tentative model of the location of different proteins in the ad2 virion. The core protein V may be located inside at the vertices, since it is partially released with the peripentonal region. It has been estimated that core protein VII may neutralize about 50 per cent of all phosphate residues in the DNA. The molar ratio between polypeptide VI and the hexon polypeptide is about 2, and the native protein VI exists as a dimer. Protein VI is not iodinated in intact virions, which suggests that protein VI is located at the inner surface of the hexons. Peripentonal hexons also possess this protein. Protein IX appears to be the cementing substance between hexons from the facets, since it is associated with groups of nine hexons. Protein VIII is also associated with the hexons and may reside at the inner surface of the triangular facets, since it is not iodinated in intact virions. Polypeptide IIIa is probably located in the peripentonal region. The localization of protein X is unknown. The polypeptide composition of the virion proteins as identified in a stained exponential (10 to 16 per cent) SDS-polyacrylamide gel are also shown. (From Philipson et al., 1975.)*

TABLE 1. Classification of Human Adenoviruses*

SUBGROUP	SEROTYPES	PER CENT GC IN DNA	PER CENT DNA HOMOLOGY BETWEEN TYPES	HEMAGGLUTI-NATION SUBGROUP	LENGTH OF FIBER (nm)	ONCOGENICITY†
A	12,18,31	48–49	80–85	IV	28–31	High
B	3,7,11,14,16,21	50–52	70–95	I	9–11	Weak
C	1,2,4,5,6	57–59	85–95	III	23–31	None but transform
D	8,9,10,13,15,17,19, 20,22,23–28	57–61	Not known	II	12–13	None but transform

*Reprinted from Philipson et al., 1975, where the original references have been cited.
†Highly oncogenic types cause tumors within 2 months; weakly oncogenic types cause tumors in some animals in 4 to 18 months.

scripts and messenger RNA by nucleic acid hybridization. Isolated messenger RNAs have also been analyzed by in vitro translation in order to map the position of most of the gene products on the viral genome.

CLASSIFICATION OF HUMAN ADENOVIRUSES

Human adenoviruses have been separated into four groups based on their ability to agglu-

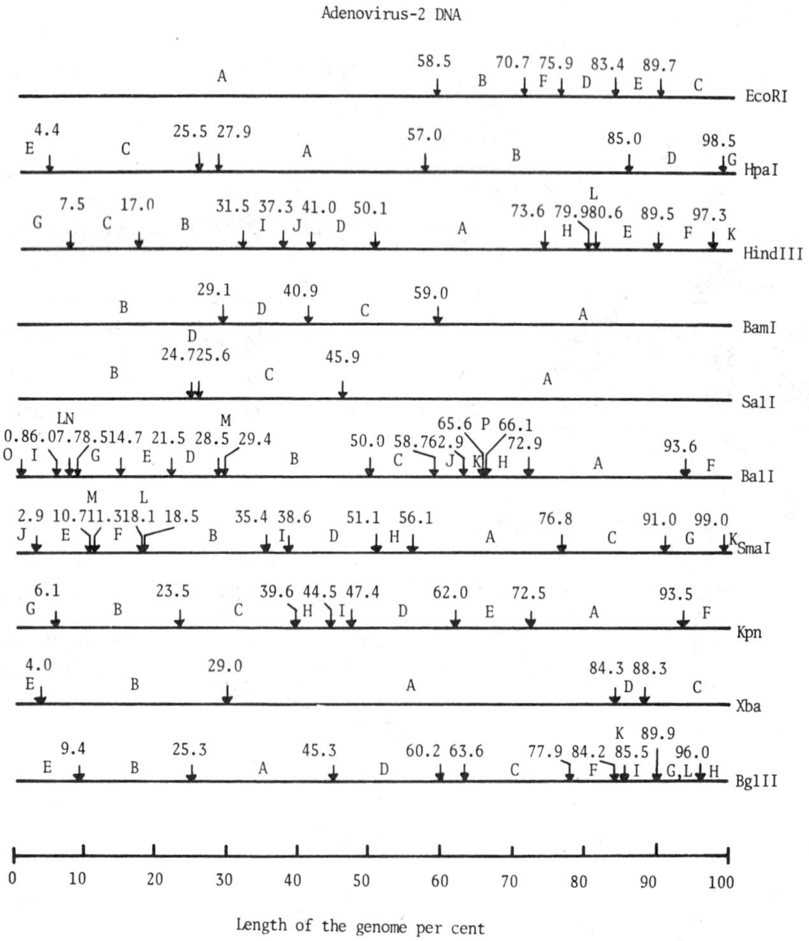

FIGURE 3. *Restriction enzyme maps of adenovirus type 2 DNA. The restriction enzymes have been described by Roberts (1976) and the map positions of the different enzymes have been determined by several investigators at the Cold Spring Harbor Laboratory (Tooze, 1979). The presented maps were compiled by Dr. Marc Zabeau and distributed at the Cold Spring Harbor Tumor Virus Meeting, 1977.*

tinate monkey or rat erythrocytes. Group I viruses cause complete agglutination of monkey erythrocytes. group II complete agglutination of rat erythrocytes, group III incomplete agglutination of rat erythrocytes, and group IV viruses, which were originally reported to lack hemagglutinin, agglutinate rat erythrocytes with an incomplete pattern similar to group III viruses. Members of each subgroup have fibers of a characteristic length (Table 1). A classification based on oncogenicity into subgroups A to D has also been proposed, which also correlates well with the DNA homology. Not all types within each subgroup have been studied with regard to DNA homology or biologic parameters and the proposed classification in Table 1 remains tentative.

ANTIGENIC COMPOSITION

Most of the antigens of adenoviruses reside in the outer capsid (Table 2). The hexons contain at least two different antigenic determinants, one is group specific (α) and the other type specific (ϵ). The penton base contains a dominant group-specific antigenic determinant (β). The fiber contains a type-specific determinant (γ) residing in the distal knob, which probably serves to anchor the virus to the cell at infection. An intrasubgroup determinant (δ) is also present in the proximal part of the fiber of subgroups A, C, and D.

Hemagglutination by adenoviruses was first demonstrated in 1958. All human types display hemagglutinating capacity. The type-specific part of the fiber interacts with the red cell surface. Since the intact virion carries several vertex projections, they can establish a bridge between the erythrocytes and give a complete hemagglutination pattern. Multimers of pentons and fibers can in the same way give rise to hemagglutination. Monomers of penton and fiber can establish only a monovalent link with the cell and agglutination can be detected only when antibodies are used to bridge the fibers anchored with their type-specific moiety on the erythrocyte. Heterotypic antibodies must be used to link the group-specific determinants available on the surface of the cells. The short fibers of subgroup B viruses therefore cannot agglutinate, since they only contain type-specific determinants. This method to reveal adenovirus hemagglutination by heterotypic antisera has been called *hemagglutination enhancement.* Antibodies directed toward the fiber can be assayed by *hemagglutination inhibition.*

Adenoviruses carry a mosaic of antigenic specificities on their surface. Since, by definition, all types are distinct by neutralization, type-specific antigens should induce neutralizing antibodies. Both the fiber (γ) and the hexon (ϵ) contain type-specific antigenic determinants and both have been claimed to induce neutralizing antibodies. The mechanism for adenovirus neutralization is still a matter of controversy. Antisera against crude hexon preparations induce neutralizing antibodies, and purified hexons from subgroup B and D viruses give rise to neutralizing antibodies, but hexons against subgroup C virions contain no or few exposed such determinants. Antisera against fibers from subgroup C viruses do not contain neutralizing activity when measured with the plaque assay. On the other hand, when neutralization is scored by the fluorescent focus assay, fiber anti-

TABLE 2. Antigens Associated with the Major Structural Proteins

PROTEIN	CORRESPONDING POLYPEPTIDE (See Fig. 2)	DESIGNATION	ANTIGEN SPECIFICITY	REMARKS
Hexon	II	α	Group	Oriented toward the inside of the virion
		—	Inter- and intrasubgroup	
		ϵ	Type	Available at the surface of the virion from serotypes belonging to subgroups B and D
Penton base	III	β	Group	Carries toxin activity
		—	Inter- and intrasubgroup	
Fiber	IV	γ	Type	Reacts with HI-antibody
		—	Intersubgroup	Shared between members of subgroups C and D
		δ	Intrasubgroup	At the proximal part of the fiber present only in subgroups A, C and D

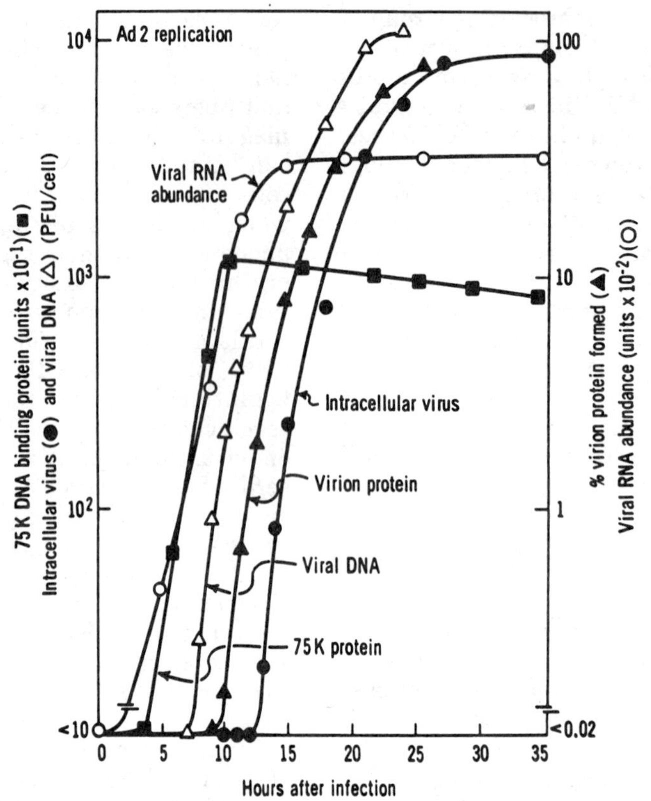

FIGURE 4. *The replication cycle of adenovirus type 2. Time course of synthesis of viral RNA, early viral 75K protein, viral DNA, virion protein, and intracellular virus. (From Wold, W.S.M., Green, M., and Büttner, W.: Adenoviruses. In Nayak, D.P. (ed.): The Molecular Biology of Animal Viruses. New York, Marcel Dekker, Inc., 1978. With permission.)*

sera contain high titers of neutralizing activity. In general antisera prepared against disrupted virions are more efficient in neutralization than antisera against purified capsid components. The neutralizing antigen of adenoviruses may therefore have escaped detection. It may be necessary to purify additional proteins in the peripentonal region to assess their capacity to induce neutralizing antibodies.

SUSCEPTIBLE HOST CELLS

All adenoviruses of human origin produce cytopathic changes (CPE) when propagated in primary or continuous cell lines of human origin, such as diploid cells of human embryonic lung or kidney, HeLa, KB, or Hep-2 cells. Characteristic CPE manifests itself as rounding, enlargement, increased opacity, and aggregation of the cells into irregular clusters. The rapidity with which CPE occurs is a function of the dose of virus in the inoculum but the lower numbered types ad1 to ad7 produce CPE faster than do the other types. Monkey cells do support the early and most of the late expression of human adenovirus genomes but virions fail to assemble. The block may be overcome by concurrent infection with SV 40 virus. Hamster cells are permissive for ad2 and ad5. Ad12, on the other hand, can express early products but fails to replicate its DNA and express late genes. Ad12, nevertheless, can induce CPE. Rat cells are semipermissive for ad2 in the sense that single cells may support replication but the majority of the cells fail to do so.

The lytic cycle of adenoviruses has been studied mostly in suspension cultures of KB or HeLa cells. The time course of the infection with ad2 is shown in Figure 4. When the cells are infected at high multiplicity (500 to 1000 PFU/cell), a synchronous response is observed with two functionally different phases of the infectious cycle. During the early phase that precedes viral DNA replication, about 40 per cent of the viral genome is expressed, exemplified with the 75K DNA-binding protein, which is one of the early proteins. In the late phase, beginning with the onset of viral DNA synthesis about six hours after infection, approximately, 90 per cent of the genome is expressed, and the amount of viral RNA increases about ten-fold. At about 14 hours postinfection the synthesis of host cell protein is, to a large extent, replaced by fabrication of viral products, mostly viral structural proteins. New virus particles begin to appear at 15 hours, and the infectious cycle for ad2 and ad5, both members of subgroup C, is completed within 20 to 25 hours. Similar time courses have been obtained with other serotypes although the growth cycle of group A viruses is longer than for ad2. The growth cycle in primary cells is prolonged at least for 12 to 15 hours even for group C viruses.

One system for subdivision of human adenovirus into subgroups A to D (Table 1) was originally based on the frequency with which newborn hamsters develop tumors after inoculation with virus. Several human adenoviruses as well as adenoviruses from other species can, in addition, transform cells in vitro, regardless of their oncogenic capacity in vivo. Cells that are nonpermissive for virus replication are more susceptible to transformation. The events leading to transformation may therefore occur also in permissive cells but it escapes detection because infected cells do not survive. The frequency of transformation is in all cases low for adenoviruses with only one transforming focus per 10^6 cells or more. Only 4 to 8 per cent of the genome of ad5 is required for transformation, as revealed by transformation with DNA fragments. Accordingly, several cells transformed with ad2 and ad5 contain a limited amount of the viral DNA integrated into the host cell genome. With ad12, on the other hand, most of the viral genome appears to be integrated into the host genome of the transformed cell. Whether this difference could explain the varying oncogenic capacity of these viruses is unsettled.

PATHOGENIC PROPERTIES

The human adenoviruses are almost exclusively pathogenic for man. A single type of adenovirus may cause different clinical syndromes and, conversely, more than one type may be responsible for the same clinical pattern. Four different syndromes of respiratory infection have been associated with adenoviruses.

1. *Acute febrile pharyngitis* is perhaps the most common clinical manifestation, especially in infants and children. It is difficult to distinguish the disease in individual cases from infections caused by influenza and parainfluenza viruses, respiratory syncytial virus, certain enteroviruses, and members of the rhinovirus group. It appears that subgroup C viruses are mainly responsible for infection in early life. The majority of children show serologic evidence of prior exposure to one or more of these antigenic types by the time they reach school age.
2. *Pharyngeal-conjunctival fever* is an infection usually observed in children and it may occur in epidemic form. It manifests itself as an acute febrile pharyngitis with concurrent conjunctivitis. Ad3 has been most frequently implicated as the etiologic agent, although ad7 and ad14 have also been involved. All belong to subgroup B. Subgroup C viruses may cause this syndrome in exceptional cases.
3. *Acute respiratory disease (ARD)* is prevalent in military recruits, although the same illness may be observed among other institutionalized young adults of similar age. Ad4 and ad7 are most frequently associated with this syndrome in recruit populations, but other types within subgroup B may periodically be involved as the responsible agent.
4. *Adenovirus pneumonia* is usually seen as a complications of ARD in patients infected with ad4 or ad7. The clinical course resembles that of pneumonia caused by *Mycoplasma pneumoniae* but the disease does not respond to treatment with antibiotics. Pneumonia may also be observed occasionally in infants and several adenovirus types have been isolated from fatal cases in infants.

Adenovirus may also cause acute follicular conjunctivitis. It is mainly seen in adults and principally is caused by ad3 and ad7, although other types have also been incriminated. Ad8 is the major cause of epidemic keratoconjunctivitis of the classic type with corneal infiltrates, but conjunctivitis may be caused by several adenovirus types. Adenoviruses may also be associated with gastroenteritis without respiratory disease and intussusception in infants and young children. Acute mesenteric lymphadenitis is thought to be caused by adenoviruses, although a causal relationship has not been established. Systemic infection with adenoviruses with involvement of several organs has mostly been observed in infants or young children.

IMMUNITY

Susceptibility or resistance to most clinical infections by adenovirus appears to be correlated directly to the absence or the presence of circulating neutralizing antibodies. The adenoviruses are exceptionally good antigens for development of vaccines. One dose of vaccine apparently gives almost maximal antibody response. Several studies have demonstrated a reduction of 85 to 90 per cent in the attack rate of adenovirus infection in immunized groups. The low incidence of adenovirus infection in the civilian population, however, raises questions concerning the advisability of the indiscriminate use of adenovirus vaccines. It has been estimated from the incidence of adenovirus infections among civilians that a vaccine that completely prevented infection would lower by only 6 per cent the number of common respiratory illnesses experienced by an average child during the first 10 years of life. Since adenoviruses can transform cells and can be oncogenic to hamsters, adenovirus vaccines were abandoned. More recently, adenovirus vaccines that only contain purified protein components from infected cells have been reintroduced. Purified hexon or fiber preparations induce high levels of neutralizing antibodies in volunteers and a vaccine produced from these components was effective in a field trial. A vaccine of this type could be used for high risk military recruits. Successful vaccination has also been achieved by oral administration of live virus in capsules that decompose in the intestine.

LABORATORY DIAGNOSIS

For virus isolation, throat and fecal samples are inoculated into primary cell cultures of human origin. Rapid diagnosis by immunofluorescence has more recently been introduced (Gardner and McQuillin, 1974). Aspirated cells from the nasopharynx are fixed and stained with group-specific immunofluorescent antibodies.

The complement fixation test is the most useful single procedure for serologic diagnosis of infection. One antigen detects antibodies to all human adenovirus strains and these antibodies increase significantly in titer (four-fold or greater) in the sera of adult patients within two weeks after onset of infection. The antigen may be prepared by propagating any virus type in cultures of human embryonic kidney, HeLa or KB cells, and purifying the hexon component. After primary isolation adenovirus is classified into groups by determining if the isolate will agglutinate monkey or rat red blood cells. Hemagglutination inhibition and virus neutralization tests with type-specific rabbit antisera can establish the virus type. To establish the type-specific serologic response in the patient, both hemagglutination inhibition and neutralization tests should be performed on acute phase and convalescent sera. In general, heterologous reactions occur more frequently in hemagglutination inhibition than in neutralization. The heterologous reactions observed with the neutralization test appear to be based on the presence of overlapping intertypic antigenic determinants more than on an anamnestic antibody response.

EPIDEMIOLOGY

Adenoviruses exist in nearly all parts of the world. At least 33 antigenic types have been recovered from human sources, but not all are proven agents of disease. Serologic surveys indicate that most children become infected with one or more of the adenovirus types before the age of six years, but adenoviruses are probably responsible only for 2 or 3 per cent of acute febrile respiratory illness. The incidence may rise to 5 to 10 per cent of cases that require hospitalization. There is a general problem with estimating the incidence or significance of infections with adenoviruses, since infection leading to antibody response is not necessarily associated with clinical overt disease. Therefore, the recovery of the adenovirus from the respiratory tract or intestinal tract and a serologic response does not in itself permit the assumption that the virus is related to the observed disease. In general, adenoviruses do not cause more than 5 to 8 per cent of all cases of acute respiratory disease. Only sporadic cases have been observed in children and adults, but localized outbreaks may be encountered in boarding schools, summer camps, or colleges. Interestingly enough, the highest attack rates have been observed among military recruits where ad3, ad4, and ad7 have been responsible for major outbreaks. Ad14 and ad21 may also cause outbreaks among this population. The exceptional susceptibility of military recruits to these viruses is unexplained.

References

Gardner, P. S., and McQuillin, J.: Rapid Virus Diagnosis. Application of Immunofluorescence. London, Butterworth and Co., Ltd., 1974, p. 181.

Norrby, E., Bartha, A., Boulanger, P., Dreizin, R. S., Ginsberg, H. S., Kalter, S. S., Kawamura, H., Rowe, W. P., Russell, W. C., Schlesinger, R. W., and Wigand, R.: Adenoviridae. Intervirology 7: 117, 1976.

Philipson, L., Pettersson, U., and Lindberg, U.: Molecular Biology of Adenoviruses. Virology Monographs, Vol. 14. Vienna and New York, Springer Verlag, 1975.

Potter, C. W.: Modern Trends in Medical Virology. Vol. 1., 1967, p. 162.

Roberts, R. J.: Restriction endonucleases. CRC Crit Rev Biochem 4:123, 1976.

Rose, H. M.: Adenoviruses. In Lennette, E. H., and Schmidt, N. J. (eds.): Diagnostic Procedures. 4th ed. American Public Health Association Inc., 1969, p. 205.

Sohier, R., Chardonnet, Y., and Prunieras, M.: Adenoviruses. Status of current knowledge. In Melnick, J. L. (ed.): Progress in Medical Virology, Vol. 7., New York, Karger Basel, 1965, p. 1.

Wold, W. S. M., Green, M., and Büttner, W.: Adenoviruses. In Nayak, D. P. (ed.): The Molecular Biology of Animal Viruses. Marcel Dekker, Inc., 1978, p. 673.

THE HERPESVIRUSES 56

Joseph S. Pagano, M.D.
Stanley M. Lemon, M.D.

Few viruses have greater medical importance than the herpesviruses. These similar yet very different viruses are ubiquitous, infecting virtually everyone; they are all potentially pathogenic; and together they are responsible for a wide variety of diseases. Some of these viruses, especially the Epstein-Barr virus, have a special importance for developing nations. Once infection has taken place, these viruses or their genomes persist for life, usually silently. However, these latent infections are subject to reactivation, and they may cause serious illness, especially in immunosuppressed patients who receive modern therapies ranging from organ transplants to cancer chemotherapy.

Although over 70 viruses of the herpes group are known to infect many different animal species, there are only five distinct herpesviruses that commonly infect humans: herpes simplex virus (HSV) Types 1 and 2, varicella-zoster virus (VZV), cytomegalovirus (CMV), and Epstein-Barr virus (EBV). As a group these viruses are marked by both unity and diversity. All the herpesviruses resemble each other structurally and have biologic properties in common, particularly the hallmarks of latency and reactivation. However, pathogenetically and clinically they exhibit a spectrum of effects. On the molecular level the arrangement of genetic information is structured in herpesvirus genomes in a unique way.

All of these viruses may reactivate at any time during life upon natural or iatrogenic provocation. Curiously, the clinical manifestations of reactivated infection may be quite different from disease caused by primary infection. Finally, a most important property of these viruses is their oncogenic potential: Epstein-Barr virus is well known for its ability to "transform" or alter the growth potential of infected cells, and it is closely associated with two different human malignancies.

VIRUS STRUCTURE

HSV, VZV, CMV, and EB virus particles are indistinguishable by electron microscopy. They are among the largest of viruses, with a diameter of approximately 150 to 200 nm. An inner core consisting of DNA intertwined with protein is surrounded by a protein capsid of symmetrical structure composed of 162 capsomeres. This nucleocapsid is covered by a loose amorphous envelope that is derived from the nuclear membrane of the host cell (Fig. 1). The nucleocapsid is assembled in the cell nucleus and is enveloped as it exits from the nucleus. Loss of this envelope greatly reduces virus infectivity. There are at least 33 polypeptides involved in the structure of the HSV virion including some that are included in the envelope. Other polypeptides are in the core of the virion intimately associated with the viral DNA and may affect regulation of viral gene expression. These viruses do not appear to contain polypeptides with enzymatic activities as part of their structure. They do code for several virus-specific enzymes, including DNA polymerases and thymidine kinases that are distinct from those found in mammalian cells.

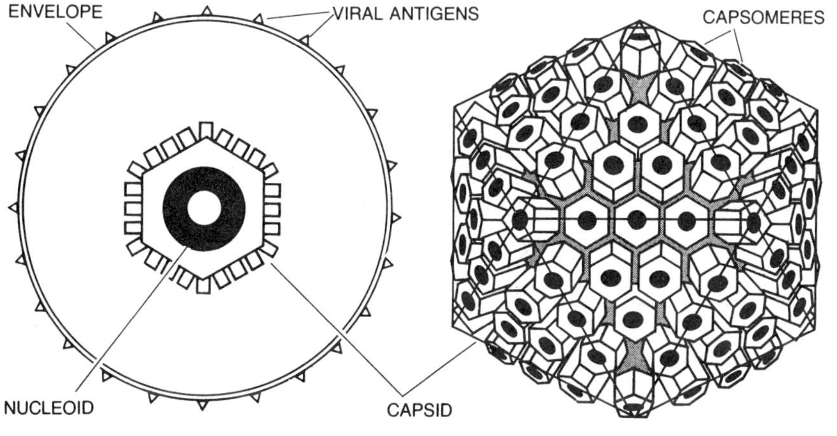

FIGURE 1. *Structure of herpes group viruses. The virion consists of a central core, or nucleoid, which contains the viral DNA; a capsid, which is icosahedral in shape and made of tubular protein subunits called capsomeres; and an envelope derived from cellular membranes. The envelope contains viral proteins or antigens. (From Henle, W., Henle, G., and Lennette, E. T.: The Epstein-Barr virus. Sci Am 24:48, 1979.)*

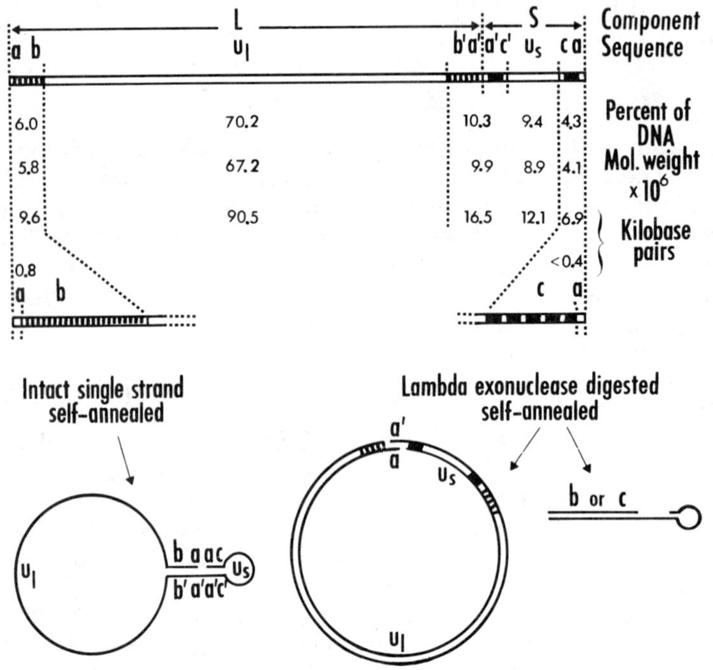

FIGURE 2. *Sequence arrangement in herpes simplex type 1 DNA. The HSV genome is divided into a long (L) and a short (S) segment. At the ends of the genome are nucleotide sequences that are also repeated within the genome at the junction of the L and S segments. The terminal sequences ab and ca of single strands anneal to their inverted repeats (b' a' and a' c') to produce the barbell structure on the lower left. The digestion of the DNA with a progressive exonuclease allows it to circularize. Most if not all members of the herpes group exhibit similar although not identical structural arrangements. (From Roizman, B.: The structure and isomerization of herpes simplex genomes. Cell 16:481, 1979.)*

Herpesvirus genomes consist of linear double-stranded DNA ranging from approximately 90 × 10^6 daltons for certain EBV strains to 150 × 10^6 daltons for CMV. Herpesvirus genomes possess terminal and internal stretches of repeated sequences. That is, in addition to the unique sequences, some of the nucleotide sequences found at the ends of the genome are also situated internally in the genome. The general structure of the herpesvirus genome is shown in Figure 2. A variety of such arrangements, ranging from simple to complex permuted structures, exist in the herpes group of viruses. Such sequence arrangements, the significance of which is unknown, appear to exist only in herpesvirus genomes and not in other genomes, viral or nonviral. These novel arrangements may relate to the generation of defective genomes.

Defective or incomplete virus particles are common among the herpesviruses. These virions possess less than the full complement of viral DNA. For example, in the case of cytomegalovirus many more virions contain only 100 × 10^6 daltons of DNA than the full genome complement of 150 × 10^6 daltons. Herpes simplex virions may possess a genome that is of normal unit length but is still defective in that many of the unique sequences are missing and are replaced by reiterations of a single segment of the genome. These defective forms may be important in some of the biologic effects of the herpesviruses such as persistent infection.

In addition to the linear form, the EBV genome can assume an entirely novel closed circular form within the cell. This supercoiled genome, called the EBV episome, is not encapsulated by virion structural protein (Fig. 3). It is found in the nuclei of latently infected cells within a nucleosomal structure similar to that of cellular chromatin. During latency the EBV episome is replicated as if it were a cellular constituent by host-cell DNA polymerase. This is the only episomal or plasmid DNA form known to exist in eucaryotic cells save for certain yeasts. Other herpesvirus genomes may be able to assume a similar episomal form, but thus far such closed circular genomes have been detected only in cells infected with an oncogenic herpesvirus of monkeys, *Herpesvirus saimiri*. In the laboratory, however, linear HSV genomes can be circularized by exonuclease digestion of the ends of the DNA strands (Fig. 2). This digestion exposes homologous nucleotide sequences at either end of the genome, thereby permitting circularization. Some EBV DNA may also be directly integrated by covalent bonding into cellular DNA in latently infected cells.

With the exception of HSV Type 1 and HSV Type 2, the genomes of the human herpesviruses are completely or almost completely different as shown by cross-hybridization experiments. HSV Type 1 and HSV Type 2 share approximately 50 per cent of their genome content. A lesser degree of homology between, for example, the Epstein-Barr virus and cytomegalovirus — in the range of 5 per cent or less — has not yet been excluded. With the exception of certain simian EBV-like agents, the many herpes-group viruses of other animal species bear little if any genetic homologous relation to human herpes-group viruses. A few of the animal herpesviruses are capable of infecting man — herpes simiae and perhaps simian CMV — but do so rarely.

Sequence homology of DNA virus genomes can be analyzed by digestion with restriction endonucleases. These nucleases digest only at certain sites on different genomes, and thus generate specific DNA fragments which when examined by electrophoresis show characteristic patterns of sizes. Such patterns are a direct and precise reflection of virus strain differences that may be difficult to detect otherwise. Herpes simplex virus genomes and also CMV and EBV genomes isolated from various sources exhibit great diversity, especially in some regions of the genomes. Such analyses have demonstrated epidemiologic usefulness in tracing transmission of specific strains of virus.

ANTIGENIC COMPOSITION

The relatively large herpesvirus genome contains sufficient genetic information to encode probably more than 50 different proteins, placing the herpesviruses among the more complex viruses of man. Many of these proteins, some of which are nonstructural (that is, not present in the intact virion), serve as efficient antigens. However, despite the identical morphologic appearance of the herpesviruses, there is little antigenic similarity among most members of this group.

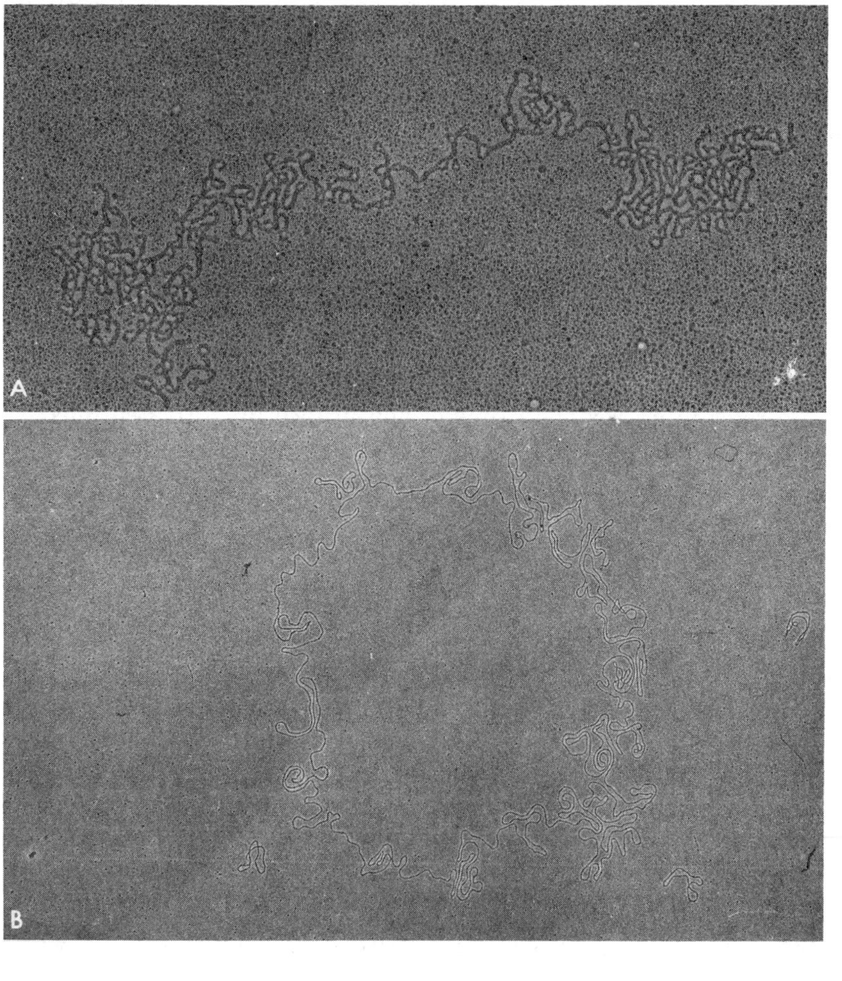

FIGURE 3. *The Epstein-Barr Virus Episome. The episomal or plasmid form of the EBV genome is a closed-circular, supercoiled double-stranded DNA molecule of molecular weight 100 × 10⁶ daltons. The EBV episome is found intracellularly, organized as part of the nucleosomal structure of chromosomes. It is replicated by host-cell enzymes. A, the closed-circular form; panel B, the open-circular form of a similar molecule. (Courtesy of J. E. Shaw, B. Colby, C. Moore, and J. Griffith; from J. S. Pagano, The Epstein-Barr Virus Plasmid, ICN-UCLA Symposium on Extrachromosomal DNA, Academic Press, 1979.)*

Among the human herpesviruses only HSV Type 1 and HSV Type 2 share a significant degree of antigenic relatedness. Although these two viruses are clearly separable by several different serologic techniques, antigenically related polypeptides common to both HSV Type 1 and HSV Type 2 have been identified by immunoprecipitation in gel-diffusion studies. This is not surprising in view of the high degree of nucleic acid homology (approximately 50 per cent) between these viruses. Heterologous neutralizing antisera are cross-reactive, which indicates that related antigenic sites exist on the envelope surface. Cross-reacting internal and nonstructural antigens have been identified as well. Antigenic diversity in HSV is not restricted to differences between Type 1 and Type 2 viruses; there is also considerable variation among strains of the same type. Such variation sometimes leads to the identification of an HSV strain as "intermediate" between Type 1 and Type 2.

Unlike strains of HSV, VZV isolates are relatively homogeneous in their antigenic make-up. There is no demonstrable cross-reactivity between VZV and HSV antigens despite the heterologous antibody rises that are occasionally observed following infection with one of the viruses.

Human cytomegalovirus strains are antigenically unrelated to other human herpesviruses, although they are related to CMV-like agents isolated from nonhuman primates. As with HSV, strains of CMV demonstrate considerable antigenic diversity when analyzed either by neutralization kinetics or complement fixation. However, no definitive grouping of strains on the basis of antigenic differences has yet emerged. Recent studies have identified "early" nonstructural CMV antigens that are distinct from "late" CMV antigens including virion components. Early and late refer to approximate time of appearance in the viral replicative cycle.

Several distinct antigens have been associated with EBV. None of these cross-reacts in any way with the other human herpesviruses, reflecting the general lack of nucleic acid homology among these agents. By fluorescent antibody technique, viral capsid antigen (VCA) activity has been recognized as part of the virion structure. However, additional antigenic activities have been identified that are nonstructural. EBV "early" antigen (EA) is found in cells that are abortively infected with virus and unable to produce mature infectious virions. "Membrane" antigen (MA) appears on the surface of some malignant cells grown in vitro from neoplasms associated with EBV. Such cells also contain EBV nuclear antigen (EBNA), the function of which is still unknown. Because of the superficial similarity between EBNA and the T-antigens of the smaller DNA-containing tumor viruses, SV40 and adenovirus, there is a suspicion that EBNA may be involved in the transforming function of EBV. The appearance of antibodies to each of these distinct EBV antigens follows a characteristic temporal sequence after primary infection with the virus, and they are therefore of diagnostic significance.

SUSCEPTIBLE HOST CELLS AND PATHOGENESIS

Many different types of cells become infected with herpesviruses. The type of cell susceptible to infection is to some extent dependent upon age, since infants seem to have a broader cellular susceptibility than adults. Primary infection, latency, and reactivation may involve different cell types even with the same virus. To add to the complexity, despite a common tendency to infect epithelial cells, each of the herpesviruses exhibits its own unique cellular interactions. Disease production may be linked directly to virus-caused cell death, or it may occur secondarily in relation to the immune response.

Both HSV Type 1 and Type 2 infect epithelial cells primarily. Type 1 virus is classically associated with oropharyngeal lesions, whereas Type 2 infects the genital mucosa and adjacent skin sites. Such territorial or organ-specific localization, while not absolute, is typical of herpes-group viruses. In severely affected infants and adults, HSV is found in the respiratory and gastrointestinal epithelium, the skin, and the parenchymal organs such as the liver. Such patients often have underlying immunocompromising conditions. HSV secondarily infects nervous tissue; Type 1 causes a latent infection in the trigeminal ganglion, and Type 2 infects the sacral ganglia. Reactivated infection occurs in mucosa and skin that the virus reaches from the ganglion by axonal spread. In addition to these silent ganglionic infections both viruses cause neurologic disease. In the United States, HSV Type 1 is the leading cause of sporadic and often fatal encephalitis. Whether the principal route of infection of the brain is directly from the trigeminal ganglion or through a cell-associated viremia is uncertain. However, it does seem likely that encephalitis can be produced not only by primary infection but also by reactivated infection. This view is supported by the occurrence of the condition at all ages. HSV Type 2 more commonly causes a benign aseptic meningitis. Herpes neonatorum is usually caused by Type 2 infections contracted during birth from active genital lesions in the mother. The majority of primary and secondary infections with HSV are probably asymptomatic.

Varicella-zoster virus primarily infects the respiratory epithelium, in which replication of the virus is for the most part asymptomatic. In adults who are infected for the first time with VZV, however, there is a high probability that a serious viral pneumonia will develop. The virus spreads to the skin by way of a cell-associated viremia. Virus may also spread to organs such as the liver in severe reactivated infection as well as in primary infection. Large amounts of infectious virus are replicated in the vesicles of chickenpox, but the virus seems to be transmitted primarily in respiratory aerosols. In the course of the infection peripheral sensory nerves in most parts of the body become infected asymptomatically. Here the virus persists for life, later to be reactivated in the form of herpes zoster, or "shingles." By all available criteria the viruses associated with varicella and with zoster are indistinguishable.

Cytomegalovirus replicates in the epithelial cells of the respiratory tract, salivary glands, and kidney; in the kidney the tubular cells shed virus for prolonged periods. In addition, CMV is frequently present in cervical secretions, especially late in pregnancy. Virus may also be found in the semen in CMV mononucleosis after its disappearance from other sites; the exact site of replication in the male genital tract is uncertain. As with other herpesviruses, spread via the bloodstream is accomplished by a cell-associated viremia, the virus being associated with both lymphocytes and polymorphonuclear leukocytes. Cell susceptibility to human CMV is strikingly affected by age. Infections in utero result in devastating destruction of the central nervous system, whereas encephalitis from postnatal CMV infection is exceedingly rare. CMV infection acquired later in life is usually asymptomatic, but occasionally it results in a mononucleosis-like syndrome with liver involvement. Cytomegalovirus infection exemplifies the different clinical forms infection may take upon reactivation. In patients who have received bone-marrow transplants CMV appears to cause an interstitial pneumonitis, whereas pneumonic involvement in primary CMV infection is virtually unknown. However, in recipients of renal allografts fever, leukopenia, and mild hepatitis may occur both in primary and in reactivated infection. Such infections appear to be more severe if they are primary.

Epstein-Barr virus probably replicates initially in the epithelial cells of the oropharynx and parotid gland. Subsequently, bone marrow-derived (B) lymphocytes are infected. Infected B lymphocytes do not appear to be a source of freely replicating virus in vivo, but they do harbor the episomal form of EBV DNA. Although symptoms of EBV infection arise from many organs, including the throat, lymphatic tissue, liver, and both the peripheral and central nervous systems, there is no direct evidence that the virus actually infects cells in parenchymal organs or the nervous system. Involvement of these organs may result from an intense immunologic response to the presence of EBV-bearing B lymphocytes and not from cellular invasion and destruction by virus itself.

In East and West Central Africa, EBV infection is closely associated with the development of Burkitt's lymphoma, a B-lymphocytic malignancy. Cells taken from African but not usually from the rarer North American Burkitt's lymphomas uniformly contain latent EBV DNA in its episomal form. In these areas of Africa, EBV infection takes place early in life. During the first year of life virtually 90 per cent of the children become infected, apparently asymptomatically, with EBV. This early exposure may be an important factor in the development of Burkitt's lymphoma. Persistently high EBV antibody titers early in life are predictive of the later development of Burkitt's lymphoma. Other factors, such as chronic infection with malaria, may be involved.

Several lines of evidence associate EBV with a second human malignancy, undifferentiated nasopharyngeal carcinoma. From a pathogenetic standpoint this association is especially interesting because this malignancy is epithelial in origin and points to the primary target cell of EBV replication. Although certain ethnic groups, especially Cantonese Chinese, are at high risk of developing this tumor, the association of the tumor with the virus is world-wide. EBV remains the prime candidate human tumor virus.

LABORATORY DIAGNOSIS

HSV infection produces multinucleate giant cells and Cowdry type A intranuclear inclusion bodies. In cervical cytologic smears, such changes correlate well with isolation of virus. Similar cytologic changes occur with VZV infection (see Fig. 2 in Chap. 222), and the examination of Giemsa-stained smears of cells scraped from the floor of fresh vesicles is helpful in the diagnosis of vesicular disease due to these two herpesviruses. Intracytoplasmic as well as intranuclear inclusions occur with CMV infection. Cytologic examination of exfoliated cells in urine may lead to the diagnosis of cytomegaloviruria, but it is much less sensitive than virus isolation. Infection with other viruses, adenovirus and measles, for example, produce inclusion bodies and may cause confusion in the diagnosis.

HSV strains are easily isolated in a variety of cell cultures and can be recovered from the oropharynx, cervix, vesicular and ocular lesions, brain, and other tissues. Cytopathic effects caused by these virus strains develop rapidly and are usually obvious within 24 to 48 hours after inoculation of cell monolayers. VZV is less readily recovered but can be isolated from fresh vesicles (less than four days old) when fluid is inoculated into primary or diploid human cells. Cytopathic effects develop more slowly, and, unlike HSV, the virus remains tightly cell-associated. CMV can be recovered from urine, the cervix, semen, the oropharynx, and peripheral blood leukocytes. Replication of this virus in vitro is restricted to human diploid fibroblasts, despite the obvious presence of CMV within epithelial cells in vivo. Characteristic cytopathic effects may take several weeks to develop, and blind passage of infected cells may even be necessary. Final identification of these agents rests on fluorescent antibody tests or other immunologic procedures.

An in vitro cell system fully permissive for EBV replication has yet to be found. EBV biologic activity can, however, be detected by in vitro transformation assays with human cord-blood lymphocytes. Transforming EBV, found in oropharyngeal washings and parotid gland secretions, alters the growth potential of these lymphocytes so that they proliferate indefinitely in culture. Such transformed or "immortalized" lymphocytes contain EBNA. Transformation assays for EBV are complex and laborious, and are restricted to research laboratories. Clinical diagnosis depends on serologic techniques.

Antibodies to the herpesviruses can be detected by several methods. Serum neutralization tests have been developed for each virus and are both sensitive and specific. Following infection, neutralizing antibodies generally persist for the life of the individual. Neutralization tests are, however, time-consuming and expensive and are not suited for general clinical use. For detecting HSV antibodies the complement fixation technique is an excellent alternative because it is widely available, and complement-fixing antibodies to HSV are generally long-lasting. Significant antibody titer changes take place following severe recurrent infections as well as during primary infections. Because of the strong degree of antigenic relatedness between HSV Type 1 and HSV Type 2, however, neither of these techniques is able to distinguish type-specific antibodies unless special modifications are made.

Complement-fixing antibodies to VZV are short-lived, usually declining to undetectable levels within a year of infection. Antibodies to VZV antigen expressed on the membrane of infected cells may be detected by a fluorescent antibody technique. This assay is preferable for indicating immunity.

As with VZV, complement-fixing antibodies to CMV may disappear following primary infection only to reappear after reactivation or reinfection. Antibody detected by fluorescent antibody technique is more persistent and correlates better with neutralization test results. Detection of IgM antibody to CMV is important in the diagnosis of congenital infection.

Serodiagnosis of EBV infection depends on the detection of both virus-specific and heterophile antibodies. Paul-Bunnell-Davidsohn heterophile antibodies are predominantly IgM and are directed to surface antigens of erythrocytes from sheep and certain other animal species. These antigens do not cross-react with known EBV virion antigens; the reason for their occurrence in up to 90 per cent of cases of infectious mononucleosis due to EBV is unknown. In any case, such antibodies are highly suggestive of primary EBV infection. Indirect fluorescent antibody techniques may detect antibody to each of the several EBV-associated antigens. Antibodies to VCA and EA develop shortly after primary infection. VCA antibodies persist for life, whereas EA antibodies usually fall to low or undetectable levels within several months of infection. EBNA antibodies do not appear until one to three months after infection, but then they persist for life. This kind of antibody interplay is helpful in diagnosing and roughly dating EBV infection.

The promise of effective chemotherapy for some herpesvirus infections makes critical the need for rapid diagnostic procedures. The success of adenine arabinoside (Vidarabine) treatment of HSV encephalitis is highly dependent on how promptly therapy is begun. The most promising methods for rapid diagnosis of herpesvirus infections are based on the detection of virus antigen. A variety of techniques can be used to demonstrate virus antigens directly in fresh specimens. The techniques make use of virus-specific antibody labeled with fluorescein (direct and indirect IF tests), with radioisotopes (radioimmunoassays), or with enzymes (ELISA tests). Detection of HSV membrane antigens in the cerebrospinal fluid is an example of a new technique under evaluation. There appears to be a fairly good correlation with herpes encephalitis, although VZV infections sometimes give false-positive reactions for HSV antigens. A reliable test of this type would be welcome inasmuch as HSV is not shed into the cerebrospinal fluid. Fluorescent antibody techniques may demonstrate HSV antigens in brain-biopsy material and may shorten the time required to document cytomegaloviruria. Rapid diagnosis of congenital CMV infection also is possible with electronmicroscopy of urine specimens. However, the enzyme-linked (ELISA) tests appear to be potentially the most widely applicable.

A major problem in diagnosis is the tendency of herpesvirus to cause latent infection and the frequency with which reactivation of virus occurs. Many stimuli, including stress, fever, and immunosuppression, may lead to reactivation. For example, in patients with meningoencephalitis, the recovery of HSV from superficial mucosal sites does not correlate with its presence or absence in the brain. In severe or atypical herpesvirus infections the actual demonstration of virus in the affected organ is essential. Even then, the isolation of a herpesvirus must fit with the histopathologic findings and overall clinical pattern before an etiologic association can be made.

IMMUNITY

Immunity to the herpesviruses is the result of a complex interplay between humoral and cell-mediated immune systems. Infection with the herpesviruses does not necessarily produce permanent immunity either to exogenous reinfection or to reactivation of endogenous virus. By and large the presence of circulating antibodies to EBV and VZV prevents full-fledged reinfection with these viruses, whereas the antibodies conferred by previous infection with CMV and HSV may modify disease but do not prevent reinfection, which may recur repeatedly. Reinfection from exogenous sources with HSV Type 1, HSV Type 2, and cytomegalovirus is common. Reinfection may in part be permitted by virus-strain differences inasmuch as some strains of herpes simplex virus and cytomegalovirus differ enough antigenically that there is suboptimal cross-protection. Probably more important, however, is the ability of these viruses to establish infection in local areas that are sheltered from immunologic response mechanisms because of their superficial location. There is some epidemiologic evidence to suggest that exogenous reinfection with VZV may occasionally occur as well, although it is clear that most cases of zoster are due to reactivation of latent virus. There is no evidence of reinfection with EBV.

Cell-mediated immune mechanisms are crucial for preventing or limiting reactivated infection and probably also limit primary infection once it is underway. Depressed or absent cellular immune responses to HSV and VZV correlate with the risk of reactivation of these viruses in recipients of organ transplants. Also, CMV reactivation occurs in transplant recipients with defective cellular immunity despite the presence of high complement-fixing or neutralizing antibody titers. Asymptomatic reactivation of EBV can also be detected in such patients. Thus, with each of the herpesviruses, intact humoral immunity is not by itself a deterrent to reactivation of endogenous virus. Although depressed cellular immunity certainly predisposes to reactivation, the common syndrome of reactivated oral or genital herpes is probably linked more to environmental effects that act as local triggers than to immunologic factors. Exposure to ultraviolet light, stress, fever, and local irritation are provocative factors. Pregnancy also seems to increase susceptibility to HSV and CMV infections.

Cellular immune mechanisms probably limit the extent of primary herpesvirus infections. For example, the proliferation of infected B lymphocytes that occurs during acute infectious mononucleosis appears to be held in check by a cytotoxic thymus-derived (T) lymphocytic response; this response is most intense shortly after primary infection. Cytotoxic cell-mediated response may be broadly reactive and may play a significant role in the production of disease. Thus, the atypical lymphocytosis that characterizes infectious mononucleosis is composed mainly of T lymphocytes responding to new antigens generated on the surface of B lymphocytes infected with EBV, although a nonspecific so-called "killer cell" (lymphocytic) response is also evoked.

EPIDEMIOLOGY

Virtually everyone is infected with HSV Type 1, CMV, VZV, and EBV before middle life. There is a significant difference in the epidemiology of these viruses in developing and industrialized countries, however, in that infections occur considerably earlier in life in poorly developed areas. This difference in age-related incidence of infection is striking in the case of EBV in Kenya and Uganda, where virtually every inhabitant has acquired antibodies to EBV by the end of the first year of life. In contrast, only about 40 per cent of students entering college in the United States have antibodies to EBV, although lower socioeconomic groups living in crowded conditions are more likely to be infected with EBV at an earlier age. Infectious mononucleosis is a manifestation of EBV infection when acquired in the second or third decade of life, and it is therefore a disease that is virtually limited to the more highly developed nations.

Another epidemiologic difference is the striking incidence of Burkitt's lymphoma in endemic fashion in the so-called lymphoma belt of Africa. The pattern of occurrence of this tumor, including such features as its age-related incidence, geographic distribution according to elevation, and case-clustering in families and in districts, suggests a transmissible etiology. African Burkitt's lymphoma appears to be declining in incidence for unknown reasons. In the United States Burkitt's lymphoma is much rarer and is not invariably associated with EBV infection.

The very consistent association of EBV with nasopharyngeal carcinoma holds true both in populations with a genetic predisposition to this malignancy and in other groups without such a predisposition. There appears to be an association of nasopharyngeal carcinoma and an HLA-2 subtype in genetically susceptible groups.

Among the herpesviruses, only cytomegalovirus can infect in utero. It differs also in causing rather frequent perinatal infections. Thereafter the incidence of infection is related to age and socioeconomic status much like EBV infection. HSV Type 1 and VZV infections largely occur in childhood. HSV Type 2 and genital CMV infections occur mostly in late adolescence and early adulthood.

HSV Type 1 is spread primarily by exchange of oral secretions from person to person but also by direct contact with skin or mucous membranes infected with the virus. The disposition of HSV Type 1 to infect the mouth and lips and of HSV Type 2 to infect the genital region is by no means absolute. Both HSV Types 1 and 2 may be sexually transmitted. HSV is found not only in lesions of the external genitalia but also in cervical secretions where it may or may not be associated with cervicitis. The virus may be transmitted to infants during vaginal delivery and may cause disseminated neonatal infections. Obstetricians recommend cesarean section when there is active herpes genitalis.

Varicella is one of the most contagious of all human diseases. VZV infects children (or the rare adult who has escaped infection earlier in life) primarily by the respiratory route, although vesicular fluid is also infectious. Children incubating varicella are infectious for several days before the rash develops. Patients with herpes zoster are also a source of infection for susceptible children and adults.

Cytomegalovirus resembles HSV in that it is probably spread primarily from person to person by direct contact and exchange of secretions such as saliva. Venereal transmission is thought to occur because the virus is found in both semen and vaginal fluids, and because women with other venereal infections are much more likely to have CMV antibodies. Direct contact with virus results during the passage of an infant through an infected birth canal. Transplacental transfer of virus early in parturition is unfortunately all too common and frequently results in severe damage to the fetus. The presence of antibodies to CMV does not prevent reactivation of infection, nor do antibodies prevent reinfection with exogenous strains of CMV. Consequently, a mother who has given birth to one infant with congenital CMV infection may, rarely, deliver a second child who has been infected in utero. Virus may be present in milk and be transmitted to the infant via breast feeding. Last, leukocyte-associated CMV can be transmitted by transfusion of fresh blood and result in mild hepatitis with or without a mononucleosis-like syndrome.

EBV infection is also thought to spread primarily by close contact with exchange of oral secretions containing infectious virus. The virus is shed from the throat intermittently for long periods — certainly for several months and possibly for as long as a year. The period of contagiousness is unknown; estimates of the incubation period are imprecise and range from two to four weeks.

PREVENTION AND TREATMENT

There are as yet no accepted herpesvirus vaccines. The use of varicella hyperimmune globulin to ameliorate symptoms of VZV infection is well established, particularly in leukemic children who incur high mortality and morbidity. Experimental vaccines against cytomegalovirus and varicella-zoster virus, which are presumed to be attenuated by virtue of passage in vitro in tissue cultures, are now under study. These are live virus vaccines, and the virus is expected to persist in a latent form just as in natural infection. Inactivated vaccines, including so-called herpes simplex virus "subunit" vaccines, are also under investigation; these preparations consist of disrupted virus with the viral DNA removed. Preparations of this type require large amounts of virus, which are hard to obtain with the other herpes-group viruses, but they will not produce infection, either active or latent. Although herpesvirus vaccines present complicated and novel problems, their continued development is important, especially for the prevention of infections such as congenital CMV infection that elude early diagnosis and treatment.

Several specific HSV infections can be modified by available antiviral compounds. Iododeoxyuridine (IUdR) is of proven efficacy in herpes simplex keratitis. This is the only topical use for which IUdR and other such antiviral compounds is indicated. There is no consistent effect when such compounds by themselves are applied to lip or skin lesions, presumably because of failure of the drug to penetrate to the site of viral replication. Many nonspecific therapies have been proposed for the treatment of oral or genital herpes, but all have been unsuccessful when subjected to rigorous testings.

Although both IUdR and cytosine arabinoside

are either ineffective or even harmful in the treatment of HSV encephalitis, adenine arabinoside (Vidarabine) given intravenously early in the course of the illness appears to reduce mortality. The drug is now licensed for this use within the United States. Adenine arabinoside may also cause improvement in VZV infections in immunocompromised adults. Exogenous human interferon is of potential benefit in such cases of VZV infections because it tends to decrease the risk of dissemination. One striking feature of studies of these agents in VZV infection, however, is the surprisingly benign course of untreated disease in immunocompromised individuals. Interferon may also suppress CMV infection when administered prophylactically after renal transplantation.

New experimental drugs include phosphonoacetic acid and phosphonoformate, which are extremely effective in vitro against all herpesviruses and act by inhibiting the virus-induced DNA polymerases.

A new drug, acyclovir (9-(2-hydroxyethoxymethyl guanine), represents a novel class of antiviral agents. The mechanism of action of acyclovir in HSV infections requires that it first be phosphorylated in vivo to its mono-, di-, and triphosphorylated forms before it will act. The target of action of the drug appears to be confined to the infected cell where the drug is first phosphorylated and then interferes with virus-induced DNA polymerase activity and also may become inserted into DNA, acting as a terminator of DNA chain elongation. In laboratory tests the drug is highly specific for infected cells; it appears to derive this specificity from the requirement that the drug be phosphorylated by a virus-induced enzyme—in the case of herpes simplex virus infection, the virus-coded thymidine kinase. The second level of specificity may come in its selective inhibition of virus-induced DNA polymerase and lack of effect on normal host cellular DNA polymerases.

The drug is undergoing active clinical trials in herpes simplex encephalitis, disseminated herpes simplex infections, and disseminated varicella-zoster infections. In vitro the drug does not appear to be as effective against human cytomegalovirus; this relative insensitivity may be because CMV does not seem to induce the appearance of a new viral thymidine kinase in infected cells. In vitro the drug also has an effect against the freely replicating form of EBV but not against the latent episomal form of EBV. EBV DNA polymerase is especially sensitive to the phosphorylated drug in vitro. Preliminary studies in animals indicate that acyclovir is effective when applied topically or administered systemically against genital and oral herpetic lesions in animal models; the drug is soon to undergo tests in humans for these maladies.

References

Epstein, M. A., and Achong, B. G.: Recent progress in Epstein-Barr virus research. Ann Rev Microbiol 31:421, 1977.

Henle, W., Henle, G. E., and Horwitz, C. A.: Epstein-Barr virus specific diagnostic tests in infectious mononucleosis. Human Pathol 5:551, 1974.

Huang, E. S., and Pagano, J. S.: Comparative diagnosis of cytomegaloviruses: New approach. In Kurstak, E. (ed.): Comparative Diagnosis of Viral Diseases. New York, Academic Press, 1977.

Kaplan, A. S. (ed.): The Herpesviruses. New York, Academic Press, 1973.

Nahamias, A. J., and Roizman, B.: Infection with herpes-simplex viruses 1 and 2. N Engl J Med 289:667, 719, 781, 1973.

Pagano, J. S.: Diseases and mechanisms of persistent DNA virus infection: Latency and cellular transformation. J Infect Dis 132:209, 1975.

Pagano, J. S., and Shaw, J. E.: Molecular probes and genome homology. In Epstein, M. A., and Achong, B. G. (eds.): The Epstein-Barr Virus. New York, Springer-Verlag, 1979.

Weller, T. H.: The cytomegaloviruses: Ubiquitous agents with protean clinical manifestations. N Engl J Med 285:203, 267, 1971.

Weller, T. H.: Varicella-herpes zoster virus. In Evans, A. S. (ed.): Viral Infections of Humans. New York, Plenum Medical Book Company, 1976, p. 457.

POXVIRUSES 57

Abraham I. Braude, M.D., Ph.D.

The poxviridae are a family of viruses composed of six genera. Five of the genera produce pox lesions or benign tumors of invertebrates and the sixth infects insects (Table 1). They have a similar morphology, a common nucleoprotein antigen, and the ability to recombine genetically. With the exception of ectromelia and myxoma viruses, the poxviruses are one of the most resistant groups of viruses. They are stable in storage under adverse conditions and remain infective in dried exudates and tissues for remarkably long periods.

VIRAL STRUCTURE

The poxviruses have the greatest size and the most complicated structure of any animal virus. The double-stranded DNA virus has a molecular weight of approximately 150×10^6, the largest of any animal virus. It is enclosed in a dense, dumbbell-shaped core, which is covered by a closely adherent lipoprotein inner membrane. Two oval masses of obscure function known as lateral bodies fill the concavity of the dumbbell. A second outer membrane covers the entire struc-

TABLE 1. The Genera of Pox Viridae

GENUS	PRIMARY HOSTS	DISEASE
Orthopoxvirus	Man Cows Mice Rabbits Monkeys	Variola, vaccinia, alastrim Cowpox[a] Ectromelia Rabbitpox Monkeypox[a]
Capripoxvirus	Ungulates	Sheep-pox, goatpox
Leporipoxvirus	Rabbits and squirrels	Myxoma, fibroma
Parapoxvirus	Sheep Cows Cows	Orf[a] Papular stomatitis Milkers nodules[a]
Avipoxvirus	Birds	Fowlpox, canarypox
Entomopoxvirus	Arthropods	

[a]Also produce human infections

ture. The resulting particle is oblong, and the surface is covered with ridges (see Chapter 7, Fig. 8). In contrast to other viral envelopes, the outer membrane of the virion is not derived from the cytoplasmic membrane of the infected cell but is synthesized by the virus. This difference in origin of the poxvirus envelopes accounts for a difference in sensitivity of the viruses to ether. Many poxviruses (ortho and poxviruses) are resistant to ether, whereas other enveloped viruses are inactivated by ether. Vaccinia virions measure 300 × 240 nm, whereas the parapoxvirus particles (Orf, papular stomatitis, and milkers nodes) are 260 × 160 nm.

In addition to DNA, poxviruses contain neutral lipid, phospholipid, protein, and carbohydrate. There are at least 75 proteins, but the function of most has not been identified. The core contains at least 17 structural proteins or polypeptides, and the inner membrane has two structural glycoproteins. Five other proteins are located at the surface of the viral particle. Seven enzymes have been identified in the core that are involved in RNA synthesis. One of these is a DNA-dependent RNA polymerase, the first to be demonstrated in any virion. All the phospholipid in the virion is located in the outer membrane. The predominant phospholipid is lecithin. Poxvirus DNA ranges from 80 to 100 nm in length. Its guanine plus cytosine content of 35 to 40 per cent is the lowest of any vertebrate virus.

ANTIGENIC COMPOSITION

In addition to the nucleoprotein antigen common to all members of the family, poxviruses contain a number of antigens derived from the structural components of the virion (structural antigens) and others known as soluble antigens (Table 2). The complex nucleoprotein antigen (NP), which is obtained by extraction with dilute alkali, constitutes over half the substance of the viral particle. NP contains 6 per cent DNA as well as certain core proteins. The structural antigens are demonstrated by chemical disruption of the virion or by neutralization of infectivity. At least 20 antigens are found in immunodiffusion tests after virion disruption, of which four have been localized to the core and one outside the core beneath the outer membrane. Two surface antigens have been identified by neutralization of infectivity. The antigens that produce neutralizing antibody are the basis for division of the poxvirus family into the genera listed in Table 1.

The soluble antigen known as LS is identified by complement fixation. It is found in virus-free extracts of tissue infected by vaccinia virus and appears to be a complex protein with a molecular weight of 240,000. The protein has both a heat-labile and a heat-stable antigen designated L and S, respectively. The L antigen is inactivated at 60 ° C and the S antigen is resistant to 100° C. The LS antigen seems to be derived from the viral surface because it stimulates neutralizing antibodies; it does not, however, immunize animals against vaccinia infection of the skin, and absorption with LS does not remove neutralizing antibody from protective sera obtained after vaccinia infection. Thus, it does not appear to be related to the protective structural surface antigens that delineate the orthopoxviruses. Another soluble antigen agglutinates chicken red cells. This hemagglutinin is a lipoprotein and is inactivated by

TABLE 2. Major Poxvirus Antigens

CATEGORY	ANTIGEN	COMPOSITION	SIZE	PROTECTIVE
Structural	NP Surface antigens	6% DNA + core proteins Probably protein	50% of virion mass ?	0 +
Soluble	LS Hemagglutinin	Protein Lipoprotein	MW 240,000 65 nm	± 0

lecithinase. In contrast to hemagglutinin inhibition antibodies against influenza virus, antibodies that block hemagglutination of a poxvirus cannot neutralize infection by that virus. For this reason, the poxvirus hemagglutinin does not appear to be a structural unit of the viral surface.

SUSCEPTIBLE HOST CELLS

Besides the primary hosts listed in Table 1, each poxvirus can infect other animals. The most extensive studies on the range of susceptibility of different animal tissues have been carried out with vaccinia virus. Calves, sheep, and rabbits can be infected with vaccinia virus and are used for the production of a smallpox vaccine composed of virus harvested from the skin lesions. Monkeys, rats, mice, guinea pigs, and hamsters are susceptible to vaccinia infection but less so than rabbits. Chick embryos are used to prepare smallpox vaccine by propagation of virus on the chorioallantoic membrane. Chick embryos can also be infected by inoculation of the amniotic cavity or yolk sac. Repeated propagation of vaccinia strains on the chorioallantoic membrane or in chick embryo tissue culture tends to lessen the virulence of vaccinia strains for man (Downie, 1965).

In contrast to their susceptibility to vaccinia virus, most animals other than primates are resistant to variola virus. It has not been possible to propagate variola virus serially in rabbits, guinea pigs, calves, sheep, and goats. Although smallpox virus can produce in rabbits a keratitis with the formation of Guarnieri bodies (intracytoplasmic spherical eosinophilic inclusions) and a local skin lesion, the virus cannot be propagated further in the rabbit skin. Variola virus is also less virulent than vaccinia virus upon intranasal inoculation of adult mice, but it kills suckling mice after intraperitoneal inoculation. On the cholioallantoic membrane, the pocks produced by variola virus are smaller and less heat-tolerant than those of vaccinia virus; variola pocks do not occur in embryos incubated at 39° C or above, whereas vaccinia pocks develop even at 41° C with some strains. Monkeypox virus shows the restricted host range of variola virus (Ho and Wenner, 1973), whereas the host range of cowpox virus resembles that of vaccinia virus.

Serial passage of poxviruses establishes an affinity for certain tissues, a phenomenon known as tropism. Passage of vaccinia virus, for example, in rabbit or calf skin creates dermotropic strains that have scant neurotropic properties. On the other hand, certain poxviruses are naturally neurotropic or become so after cerebral passage. Rabbitpox, cowpox, and vaccinia virus have all become neurotropic. The dermotropism of fowlpox virus is evident from the extensive benign hyperplasia it produces in epithelial cells of the skin. Fibroma viruses have connective tissue tropism and produce benign connective tissue tumors in rabbits and squirrels.

Poxviruses multiply in cultured cells from many different tissues derived from rabbits, mouse, monkey, man, fowl, and cow. Variola virus, for example, can be propagated in HeLa, Hep 2, bovine kidney, bovine skin, human embryo skin, human embryo muscle, and L-cell cultures.

PATHOGENIC PROPERTIES

In poxvirus infections, the cell dies after its ability to synthesize proteins is blocked during viral replication. From studies with vaccinia virus, it has been learned that the virion enters the cell by phagocytosis. The outer viral membrane and lateral basal bodies are degraded within the phagocytic vacuole so that the core is released into the cytoplasm. It is clear that this early step of removing the outer coat is carried out by constitutive enzymes of the cell because all of the phospholipid in the virion and half of the virion protein are released in a soluble form (Joklik, 1966). The inner viral core membrane, on the other hand, is degraded by viral enzymes. The DNA-dependent RNA polymerase within the core synthesizes mRNAs which are extruded into the cytoplasm by a mechanism that requires ATP (McAuslan, 1979). The mRNAs are translated on the ribosomes of the infected cell into proteins that direct the release of virus DNA from the core. One of these proteins is an uncoating enzyme that degrades the viral core membrane and allows the naked DNA to leave the core and enter the cytoplasm. The DNA-dependent RNA polymerase is heat-sensitive, so that poxviruses can be inactivated by heat. The heat-inactivated virus can be reactivated, however, by a viable poxvirus of any genus because all poxviruses appear to contain the RNA polymerase in their cores.

After the viral DNA leaves the core, viral growth begins with synthesis of viral DNA, virion enzymes, and structural proteins of the virion. This late phase of synthesis begins about 90 minutes after infection and lasts for about four hours. In contrast to other DNA viruses, the synthesis of poxvirus DNA takes place entirely in the cytoplasm in foci that are composed of dense granules and random fine threads and are independent of mitochondria or other cytoplasmic organelles. During this phase of viral growth, the virus blocks the ability of the host cell to transport RNA from its nucleus or to manufacture

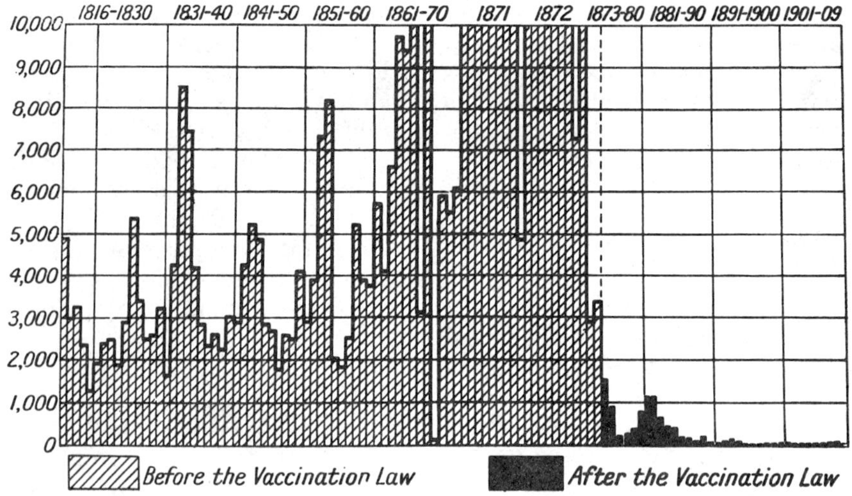

FIGURE 1. *The chart shows smallpox mortality during a period of a century in Germany. Vaccination was made compulsory in April, 1875, and in that year the total deaths from smallpox dropped to about 1500 compared with totals exceeding 10,000 in each of the years 1871 and 1872. Thereafter, the death totals fell nearly to zero. (After Jockmann: Pocken und Vaccinationlehre, Berlin, 1913.)*

protein. The synthesis of early virus proteins is also prevented.

Viral assembly begins after replication of viral DNA is completed and takes place on the cytoplasmic foci where viral DNA is synthesized. The virus synthesizes its own membrane, which encloses the viral DNA to form an "immature particle." These particles mature into the complex virion with its dumbbell-shaped core, lateral bodies, and outer membrane.

An interesting fluctuation in cell metabolism is seen in infections with rabbit fibroma virus. When cells are infected with this virus, cell DNA synthesis may be stopped for long periods, but the cells survive and DNA synthesis resumes after the virion matures (Tompkins et al., 1969).

From studies with other poxviruses, it has been proposed that in human infections variola virus first replicates at the portal of entry in the cells of the upper respiratory tract (Lancaster et al., 1966; Roberts, 1962). After spreading through the lymphatics, it multiplies in the regional lymphatic tissue and then enters the bloodstream. Virus is cleared from the blood by the reticuloendothelial system, where futher multiplication occurs. A second viremia occurs about 12 days after infection begins, marking the end of the incubation period. This viremia carries the infection to the skin and mucous membranes, where the virus first infects capillary endothelium and then epithelial cells. It has been suggested that the virus localizes in the skin because its lower temperature during fever is more conducive to viral replication than the temperature of visceral organs. Multiplication of virus in epithelial cells causes cytolysis, intradermal fluid accumulation, release of lysosomal enzymes, hydrolysis of matrix proteins, and the development of vesicles. An inflammatory reaction to the infection and to cell necrosis delivers polymorphonuclear leukocytes into the lesions, converting the vesicles to sterile pustules.

IMMUNITY

Observations on immunity were made in antiquity, when it was known that recovery from smallpox prevented another attack. Much later, a popular notion developed in Gloucestershire, England, that dairy maids who caught cowpox were afterwards immune to smallpox. The validity of this folklore was established by Jenner, who proved that vaccination of people with living cowpox virus produced solid immunity to smallpox (Fig. 1). Jenner had no idea of the mechanism of protection, let alone the nature of the material he was using, but he managed to discover the phenomenon of delayed hypersensitivity and its connection to smallpox immunity. This discovery was recorded in his description of the accelerated inflammatory reaction that occurred when smallpox material was injected into the skin of Mary Barge, who had once had cowpox. He wrote: 'It is remarkable that variolous matter, when the system is disposed to reject it, should excite inflammation more speedily than when it produces the smallpox. It seems as if a change which endures had been produced in the action, or disposition to action, in the vessels of the skin" (Jenner, 1798).

This observation that immunity to smallpox is related to a reaction that we now recognize as delayed hypersensitivity has been substantiated and elaborated by current research.

Contemporary observations on the response to smallpox vaccination have also given important insight into the mechanism of immunity in poxvirus infections. The relative importance of humoral and cellular immunity has been examined after vaccinating patients who have immune-deficiency diseases. Most patients with severe deficiencies in gamma globulin respond normally to smallpox vaccination and recover from it completely. On the other hand, vaccination of patients who have depressed cellular immunity but normal humoral antibody (autosomal recessive lymphopenia, or Nezelof's syndrome) usually causes disseminated vaccinia (Fulginiti et al., 1968). This finding in patients is supported by experiments that showed that administration of antithymocyte serum to mice markedly increased mortality from mousepox, even though humoral antibody levels were the same as those in controls (Blanden, 1970). Furthermore, immunity against mousepox was restored by transfer of immune T cells (Blanden, 1971).

These observations by themselves would imply that cellular immunity is more important than humoral immunity in vaccinia and mousepox. Other observations indicate that antibody may also contribute to protection against poxviruses. The most compelling of these is simply that passive immunity to pox infection can be transferred with virus-neutralizing antibody that develops after natural infection or after immunization with live virus. Neutralizing antibody can prevent skin lesions in animals and fatal encephalitis in mice.

LABORATORY DIAGNOSIS

Now that smallpox has virtually disappeared and vaccination is no longer a standard practice, the laboratory diagnosis of smallpox or vaccinia should be confirmed in one of the few centers at which rapid reliable tests are available. Specimens from a suspected case of smallpox, vaccinia, and monkeypox should be examined by the electron microscope, inoculated into chick embryos and tissue cultures, and tested for pox antigens against specific antiserum by immunodiffusion in agar (Nakano and Bingham, 1974). Virus can be isolated earliest from heparinized blood during the pre-eruptive stage, and later from vesicular fluid and crusts. The vesicular fluid should be collected in capillary tubes and the crusts in screw-capped vials. Preparations of vesicle fluid or ground crusts are scanned in the electron microscope at a magnification of 10,000 or 20,000, and the virion is examined in detail at 50,000 to 150,000. The brick-shaped smallpox virus is readily distinguishable from the enveloped icosahedrons of varicella virus. Variola virus must also be distinguished from nonviral particles and from herpes viruses, which can produce similar vesicles and crusts.

Examination of smears by light microscopy is less reliable. Smears are made by scraping material from the base of a vesicle or from papules with a scalpel blade and spreading it on a glass slide. After washing carefully with distilled water and ether, the smears are fixed with 95 per cent methyl alcohol and stained with a mixture of equal parts of 1 per cent gentian violet and 2 per cent sodium bicarbonate. The staining mixture is filtered onto the smeared slide and steamed for five minutes. The stained preparation is then examined for the characteristic elementary bodies of smallpox.

Isolation of the virus is carried out in fertile chicken eggs that have been incubated at 38 to 39° C for 11 to 13 days (World Health Organization, 1969). The chorioallantoic membrane (CAM) can be inoculated with plasma or buffy coat from blood samples or with vesicular fluid or scab material diluted in a buffered solution containing penicillin and streptomycin. The eggs are incubated at 35° C for 48 to 72 hours and the CAM examined for pocks. The pocks from vaccinia, variola, and monkeypox are differentiated from each other, and from those of herpesviruses by their appearance (Fig. 2).

Tissue cultures are also used for poxvirus isolation because the CAM method fails occasionally. Variola, vaccinia, and monkeypox viruses can all produce cytopathogenic effects in monkey kidney cells within 24 hours after inoculation.

Human poxvirus antigens in crusts or vesicle fluid can be detected in nearly 80 per cent of cases by agar gel precipitation. Antivaccinia antiserum is placed in the central well and the test specimens in outer wells. Upon incubation at 35° C, precipitin bands appear in a few hours between the wells containing antiserum and those containing pox antigen (Fig. 3). Antigens of smallpox, vaccinia, and human monkeypox viruses all react with vaccinia antiserum.

Serologic tests are of no practical value in smallpox because the antibody response is too slow for diagnostic help during the illness.

EPIDEMIOLOGY

Man is the only host or reservoir of smallpox. He transmits the infection only when he has an outspoken illness in the active stage of the dis-

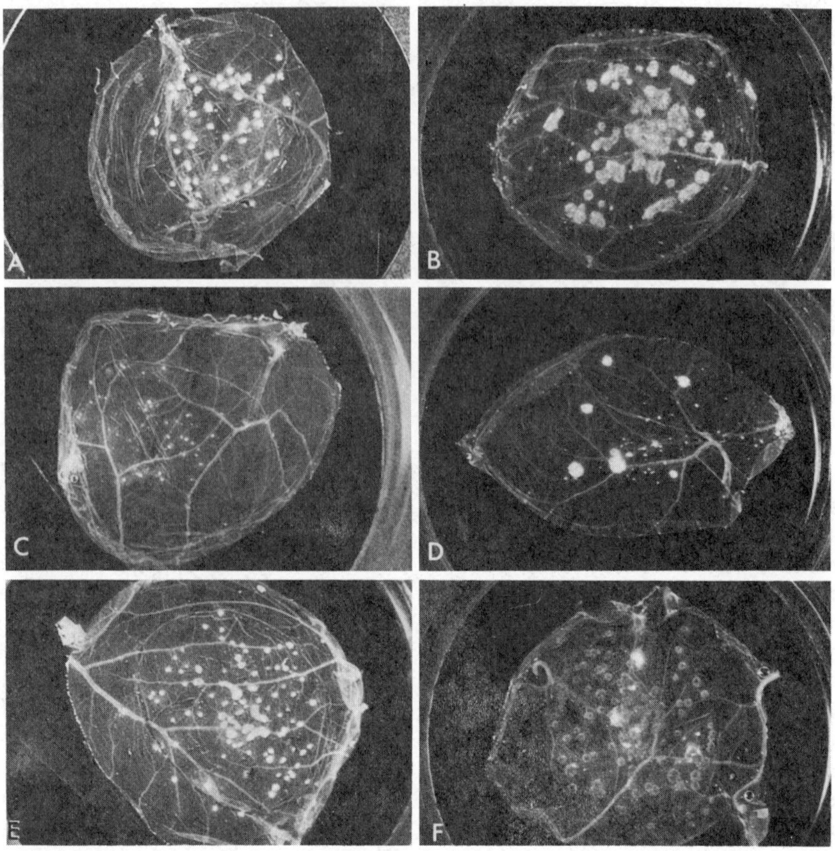

FIGURE 2. A, *Variola major lesions on chick chorioallantois (CAM) after three days at 35° C. Variola minor lesions are not distinguishable from those of variola major.* B, *Vaccinia lesions on CAM after three days at 35° C. Lesions are significantly larger than those of variola and are ulcerated.* C, *Variola major lesions on CAM after three days at 38.2° C. Inoculum was the same as that on CAM shown in Figure 1. Note that lesions are reduced in size and number.*

D, *Vaccinia lesions on CAM after three days at 35° C. Same virus as shown in Figure 3 but inoculated at a lower dose. Note the emergence of small secondary lesions.* E, *Herpes simplex lesions on CAM after three days at 35° C. Most strains of this virus give much smaller lesions, but strains isolated from genital lesions may have this appearance. The lesions in CAM resemble those of variola but are more irregular in size and some of them are ulcerated. After further incubation these lesions continue to increase in size, whereas those of variola do not (Barton, personal communication).* F, *Cowpox lesions on CAM after three days at 35° C. Most lesions are hemorrhagic, but a small portion of white variants are seen.*

All chorioallantoic membranes shown were taken from eggs inoculated when the embryos were 12 days old. (From Dumbell, K. R.: Laboratory aids to the control of smallpox in countries where the disease is not endemic. Progr Med Virol 10:395, 1968.)

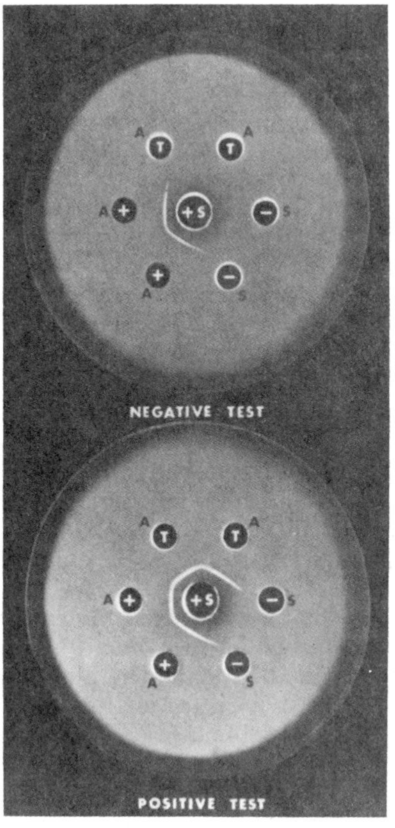

FIGURE 3. *+S (center wells): Positive rabbit antivaccinial serum; TA (top wells): test antigen made from material taken from the skin lesions of a case of smallpox; +A (left): positive antigen prepared from the crusts of a known case of smallpox; −S (right): Normal rabbit serum from an animal known to be devoid of antibodies to variola.*

Above, A negative test. There is no precipitation between the test antigen (TA) and the positive serum (+S). The only lines of precipitation are between the positive serum and the positive control antigen (+A).

Below, A positive test. The lines of precipitation between the positive serum (+S) and the test antigen (TA) join the lines between the positive control antigen (+A) and the positive serum (+S). (From Swain, R., and Dodds, T.: Clinical Virology. Baltimore, The Williams & Wilkins Company, 1967, p. 106.)

ease. Even in patients with partial immunity after vaccination, the attenuated infection still produces enough symptoms to identify the patient as a sick person so that direct contact with him is avoided. These features, plus the absence of asymptomatic carriers, have made smallpox easier to control than most other epidemic infections.

At first the eradication of smallpox was attempted through universal vaccination and revaccination. This approach was not entirely successful, however, because in many countries it was possible to vaccinate only three-fourths of the population. This might be enough with other epidemic infections such as poliomyelitis, but not with smallpox because its relatively low infectivity does not generate herd immunity. Small pockets of smallpox can persist in villages and spread by close contact in families to susceptible individuals who have escaped natural immunization or vaccination. These characteristics make smallpox susceptible to control by surrounding each new case with a ring of immunized subjects (World Health Organization, 1968).

This change in approach from universal to selective (or strategic) vaccination came about in 1967 and involved two stages: First, the source of each outbreak was determined, and the chain of transmission was traced. Then the remaining susceptibles along the route of transmission were identified and vaccinated. This surveillance-containment program has been so successful that smallpox has been eradicated from all parts of the world.

References

Blanden, R. V.: Mechanisms of recovery from a generalized viral infection: Mousepox. I. The effects of antithymocyte serum. J Exp Med 132:1035, 1970.

Blanden, R. V.: Mechanisms of recovery from a generalized viral infection: Mousepox. II. Passive transfer of recovery mechanisms with immune lymphoid cells. J Exp Med 133:1074, 1971.

Downie, A. W.: Poxvirus group. In Horsfall, G., and Tamm, I. (eds.): Viral and Rickettsial Diseases of Man. Philadelphia, J. B. Lippincott Company, 1965, p. 932.

Dumbell, K. R.: Laboratory aids to the control of smallpox in countries where the disease is not endemic. Progr Med Virol 10:388, 1968.

Fulginiti, V., Kempe, C., Hathaway, W., Pearlman, D., Seeber, O., Eller, J., Joyner, J., and Robinson, A.: Progressive vaccinia in immunologically deficient individuals. In Bergema, D. (ed.): Immunologic Deficiency Diseases in Man. Washington, D.C., The National Foundation, 1968, p. 128.

Ho, T., and Wenner, H. A.: Monkeypox virus. Bacterial Rev 37:1, 1973.

Jenner, E.: An inquiry into the causes and effects of the variolae vaccinae, a disease discovered in some of the western countries of England, particularly Gloucestershire, and known by the name of the cowpox. London, Sampson Low, 1798.

Jochmann, G.: Pocken und Vaccinationslehre, Vienna, Alfred Hölder, 1913.

Joklik, W. K.: The poxviruses. Bacteriol Rev 30:33, 1966.

Lancaster, M., Boulter, E., Westwood, J., and Randles, J.: Experimental respiratory infection with poxviruses. II. Pathological studies. Br J Exp Pathol 47:466, 1966.

McAuslin, B. R.: The biochemistry of poxvirus replication. In Levy, H. B. (ed.): Biochemistry of Viruses. New York, Dekker Publishing Company, p. 360, 1969.

Nakano, J., and Bingham, P.: Smallpox, vaccinia, and human infections with monkeypox viruses. In Lennette, E., Spaulding, E., and Truant, J. (eds.): Manual of Clinical Microbiology. Washington, D.C., American Society of Microbiology, 1974, p. 782.

Roberts, J. A.: Histopathogenesis of mousepox. I. Respiratory infection. Br J Exp Pathol 43:451, 1962.

Swain, R., and Dodds, T.: Clinical Virology. Baltimore, The Williams & Wilkins Company, 1967, p. 106.

Tompkins, G., Gelehter, T., Granner, D., Martin, D., Samuels, H., and Thompson, E.: Control of specific gene expression in higher organisms. Science 166:1474, 1969.

World Health Organization: Smallpox Eradication. Report of a WHO Scientific Report Group. Technical Report Series No. 393, Geneva, World Health Organization, 1968.

World Health Organization: Guide to the Laboratory Diagnosis of Smallpox for Smallpox Eradication Programmes. Geneva, World Health Organization, 1969.

58 *PAPOVAVIRUSES*

Michael N. Oxman, M.D.

Papovaviruses are small DNA-containing viruses that characteristically produce latent and chronic infections in their natural hosts. When adequately tested, all papovaviruses appear to be capable of inducing neoplasms in at least some animal species. The name papovavirus is derived from the first two letters of the names of the viruses first placed in the group: *pa*pilloma virus, mouse *po*lyoma virus, and simian *va*cuolating virus (Melnick, 1962). Papovaviruses have been divided into two groups or genera, papillomaviruses and polyomaviruses (Table 1), primarily on the basis of the size of the viral genome (Melnick et al., 1974; Melnick, 1977).

The papillomavirus genus includes the human papilloma (wart) viruses, the Shope rabbit papilloma virus, and a number of other species-specific viruses that usually cause benign skin papillomas or warts in their natural hosts (Rowson and Mahy, 1967). Human warts were one of the first neoplasms shown to be caused by a transmissible agent with the properties of a virus (Rowson and Mahy, 1967; Ciuffo, 1907). Nevertheless, little is known about the multiplication of papillomaviruses or about their interactions with animal cells because there is no satisfactory tissue culture system for their propagation (Rowson and Mahy, 1967; Butel, 1972; zur Hausen, 1977.)

Simian vacuolating virus (simian virus 40, SV40) and mouse polyoma virus, members of the polyomavirus genus that can cause tumors and cell transformation, have been intensively studied as models of viral oncogenesis. As a consequence, we have an enormous amount of information about their structure, replication, genetics, and interactions with animal cells (Enders, 1965; Black, 1968; Benjamin, 1972; Sambrook, 1972; Levine, 1974; Eckhart, 1977; Salzman and Khoury, 1974; Kelly and Nathans, 1977; Fareed and Davoli, 1977; Fried and Griffin, 1977; Howley, 1980). In fact, the elucidation of the complete nucleotide sequence of SV40 DNA (Fiers et al., 1978; Reddy et al., 1978) is a major triumph of modern virology.

Interest in the potential role of papovaviruses in human disease was aroused in the early 1960s by the discovery that SV40 produced a latent infection of rhesus monkey kidneys and was a common but unrecognized contaminant of viral vaccines administered to millions of people in the 1950s (Sweet and Hilleman, 1960; Melnick and Stinebaugh, 1962; Shah and Nathanson, 1976). In 1965, particles resembling papovavirus virions were found within glial nuclei in the brains of patients with a rare demyelinating disease, progressive multifocal leukoencephalopathy (PML) (Zu Rhein and Chou, 1965; Silverman and Rubinstein, 1965; Howatson et al., 1965). In 1971, Padgett and his colleagues (Padgett et al., 1971) isolated a papovavirus, JCV, from the brain of a patient with PML, and Gardner and his associates (Gardner et al., 1971) isolated a papovavirus, BKV, from the urine of a renal allograft recipient. These two viruses, which are distinct from each other and from any previously recognized papovavirus (Padgett and Walker, 1976; Takemoto, 1978), commonly cause unrecognized infections in children. In 1972, Weiner et al. isolated papovaviruses SV40-PML$_1$ and SV40-PML$_2$ from the brains of two patients with PML. Surprisingly, these isolates were nearly indistinguishable from SV40 (Fareed and Davoli, 1977; Takemoto, 1978; Weiner et al., 1972). The almost exclusive relationship of JCV infection to PML and the prevalence of asymptomatic BKV infections in immunosuppressed patients are now established, and considerable progress has been made in characterizing these two human papovaviruses (Howley, 1980).

VIRAL STRUCTURE

Papovaviruses are small, nonenveloped, icosahedral DNA viruses that replicate in the nuclei of susceptible vertebrate cells. Their virions

TABLE 1. Papovaviruses

PAPILLOMAVIRUS GENUS
Human papilloma (wart) viruses
Shope rabbit papilloma virus
Rabbit oral papilloma virus
Bovine papilloma virus
Canine papilloma virus
Hamster papilloma virus
Papilloma viruses of horses, sheep, deer, and other species
POLYOMAVIRUS GENUS
Mouse polyoma virus
Simian vacuolating virus (SV40)[a]
Mouse K virus
Rabbit kidney vacuolating virus
Human BK virus
Human JC virus
Stumptailed macaque virus (STMV)
Papovavirus of chacma baboons (SA12)

[a]Includes two SV40-like viruses isolated from patients with progressive multifocal leukoencephalopathy (PML) that appear to be SV40 variants (Weiner et al., 1972).

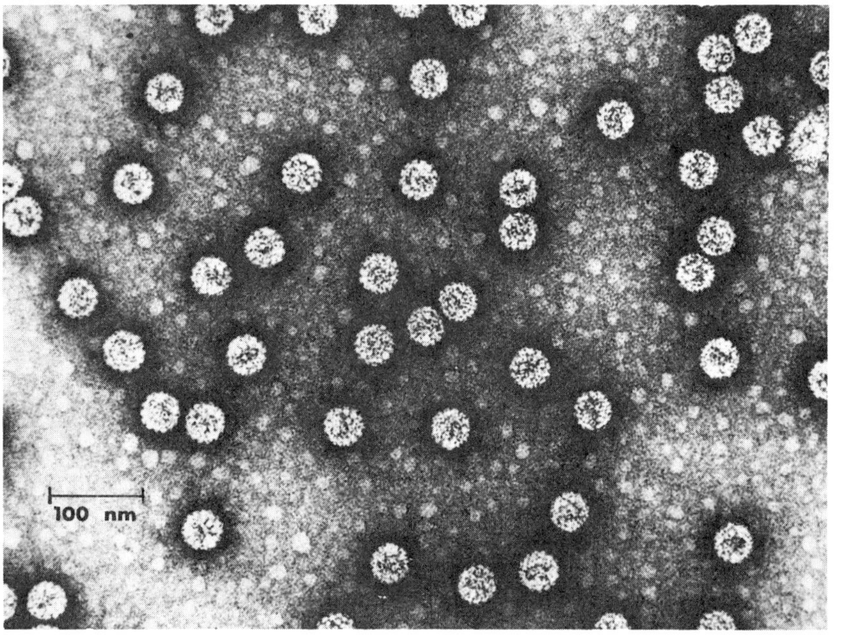

FIGURE 1. *Electron micrograph of purified SV40 virions (full particles: P = 1.34 g/cm³ in CsCl) negatively stained with sodium phosphotungstate (Bar = 100 nm).*

consist of a molecule of circular double-stranded DNA enclosed in a spherical protein coat or capsid composed of 72 morphologic subunits or capsomeres (Fig. 1). Many empty capsids are observed when unfractioned papovavirus preparations are examined electron-microscopically following negative staining with sodium phosphotungstate. The dense core of DNA is missing from these capsids, and they are thus penetrated by the stain. Filamentous or tubular forms may also be seen in the nuclei of infected cells and in preparations of extracellular virus (Finch and Klug, 1965; Mattern et al, 1967). The diameter of papillomaviruses is approximately 55 nm; they contain about 5 × 10^6 daltons of DNA. The polyomaviruses have diameters of 40 to 45 nm and contain approximately 3.3 × 10^6 daltons of DNA. The capsids of human papillomaviruses appear to be composed of one major protein that represents 60 to 80 per cent of the total viral protein and has an estimated molecular weight of between 54,000 and 63,000 (Gissman et al., 1977). The papillomavirus DNA within the capsid consists of a double-stranded, supercoiled, circular molecule closely associated with cellular histones.

In the case of the polyomaviruses, the major capsid protein VP-1 accounts for more than 75 per cent of the total virion protein and has an estimated molecular weight of 43,000 to 48,000. Two minor virus-coded proteins, VP-2 and VP-3, have estimated molecular weights of 32,000 to 39,000 and 23,000 to 31,000, respectively (Salzman and Khoury, 1974; Kelly and Nathans, 1977; Fareed and Davoli, 1977; Fried and Griffin, 1977). Within the polyomavirus virion the viral DNA is a double-stranded, supercoiled, circular molecule closely associated with four cellular histones, H2A1, H2A2, H2B, and H3, in a structure resembling the chromatin of eukaryotic cells (Kelly and Nathans, 1977; Fareed and Davoli, 1977; Fried and Griffin, 1977; Howley, 1980).

PHYSICAL AND CHEMICAL PROPERTIES

DNA accounts for 10 to 12 per cent of the weight of papovavirus virions, the remainder being protein. There is no essential lipid or carbohydrate. Complete virions have a buoyant density of 1.34 g/cm³ in CsCl; incomplete virions (empty capsids) have a buoyant density of 1.29 g/cm³. The infectivity of the polyomaviruses is resistant to lipid solvents (e.g., ether and chloroform) because they have no envelope; they are relatively resistant to irradiation, and are stable over a wide range of pH. These properties are exploited in extracting and purifying papovaviruses from tissues. Infectivity also resists heating at 50° C for one hour in distilled water but not at 50° C for one hour in 1M $MgCl_2$. Treatment with 0.1 per cent beta-propriolactone inactivates the infectivity of BK virus without destroying its hemagglutinating properties (Pitko et al., 1974), so that noninfectious BK virus hemagglutinin can be used in serologic tests.

When viral DNA is extracted from purified papovavirus virions or from infected cells, two species are generally found. Most of the viral DNA is in the form of a covalently closed circular

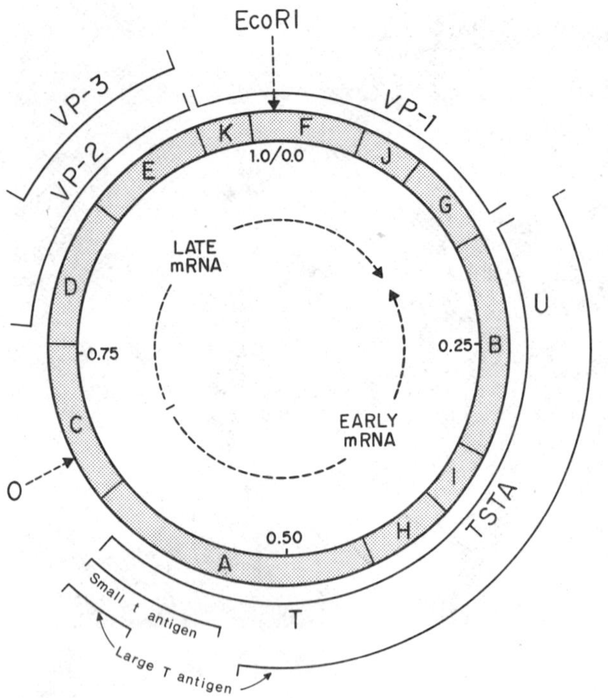

FIGURE 2. Hind *II/III restriction endonuclease map of SV40 DNA. Map coordinates are oriented clockwise from the* Eco *RI restriction endonuclease cleavage site at 1.0/0.0. Letters refer to the* Hind *II/III fragments. Inner dashed lines indicate regions transcribed into early and late viral mRNA and the direction of transcription. Outer arcs indicate the approximate location of the genetic information for virus-coded proteins. The "O" indicates the approximate location of the origin of DNA replication in* Hind *II/III fragment C (prepared from information summarized in Fareed and Davoli, 1977; Fried and Griffin, 1977; Kelly and Nathans, 1977; Fiers et al., 1978; Reddy, et al., 1978; and Howley, 1980).*

duplex molecule (DNA I) that, in SV40, consists of 5227 purine and pyrimidine bases and contains 26 negative superhelical turns. A small portion of the viral DNA is in the form of an open circular duplex molecule (DNA II), which is generated from DNA I by the rupture of one phosphodiester bond in either strand. Both forms of DNA are infectious. The degree of base sequence homology between various papovavirus isolates has been studied by nucleic acid hybridization. Bacterial restriction endonucleases, which recognize specific nucleotide sequences in duplex DNA and cleave both strands at these sites, have been used to construct physical maps of papovavirus genomes (Salzman and Khoury, 1974; Kelly and Nathans, 1977; Fareed and Davoli, 1977; Fried and Griffin, 1977; Padgett and Walker, 1976; Takemoto, 1978; Howley, 1980). With these techniques it is possible to compare the DNAs of different papovaviruses. An example of such a physical map of the SV40 genome is shown in Figure 2.

VIRUS REPLICATION

In contrast to papillomaviruses, the replication of polyoma virus and SV40 has been extensively studied (Enders, 1965; Black, 1968; Benjamin, 1972; Sambrook, 1972; Levine, 1974; Eckhart, 1977; Salzman and Khoury, 1974; Kelly and Nathans, 1977; Fareed and Davoli, 1977; Fried and Griffin, 1977). When SV40 or polyoma virus infects the cells of its natural host (permissive cells), the usual result is a lytic (productive) infection characterized by extensive virus replication and cell lysis. These viruses can also infect cells of other species (termed nonpermissive cells) which do not support efficient virus replication. This results in an incomplete or abortive infection, in which only a part of the viral genome is expressed. Some abortively infected cells may be transformed to a malignant phenotype, a process that requires the persistence and expression of at least one viral gene.

Productive infection begins with the adsorption of virus to receptors on the cell surface. In the case of polyoma virus, and probably other hemagglutinating viruses such as BKV and JCV, adsorption can be prevented by treating cells with neuraminidase, which destroys sialic acid-containing cell receptors. After adsorption, the virus penetrates the cell membrane and is transported to the cell nucleus, where it is uncoated (Salzman and Khoury, 1974; Diacumakos and Gershey, 1977).

The replication cycle that follows can be divided into two phases: early and late. During the early phase, which precedes viral DNA replication (Fig. 3*A*), early viral mRNA is transcribed from approximately one half of one strand of the viral DNA, i.e., from the E strand of *Hind* fragments A, H, I, and B, counterclockwise from 0.65 to 0.17 on the map of the SV40 genome (Fig. 2). Transcription of early mRNA, which can be detected as early as six hours after infection, is followed by synthesis of virus-specific early proteins. Two of these, SV40 T antigen and SV40 U antigen, accumulate in the cell nucleus, where they can be detected by fluorescent antibody (FA) staining beginning about eight hours after infection (Fig. 4). A third early antigen, tumor specific transplantation antigen (TSTA), is transported to the cell membrane. The late phase begins with the onset of viral DNA replication 12 to 15 hours after infection (Fig. 3*B*). Late viral mRNA is transcribed from the remaining half of the genome (Fig. 2), its template being the opposite strand of the viral DNA from that involved in early mRNA synthesis (i.e., the L strand). Late structural viral proteins, VP-1, VP-2, and VP-3, are then synthesized and transported to the nu-

PRODUCTIVE PAPOVAVIRUS INFECTION

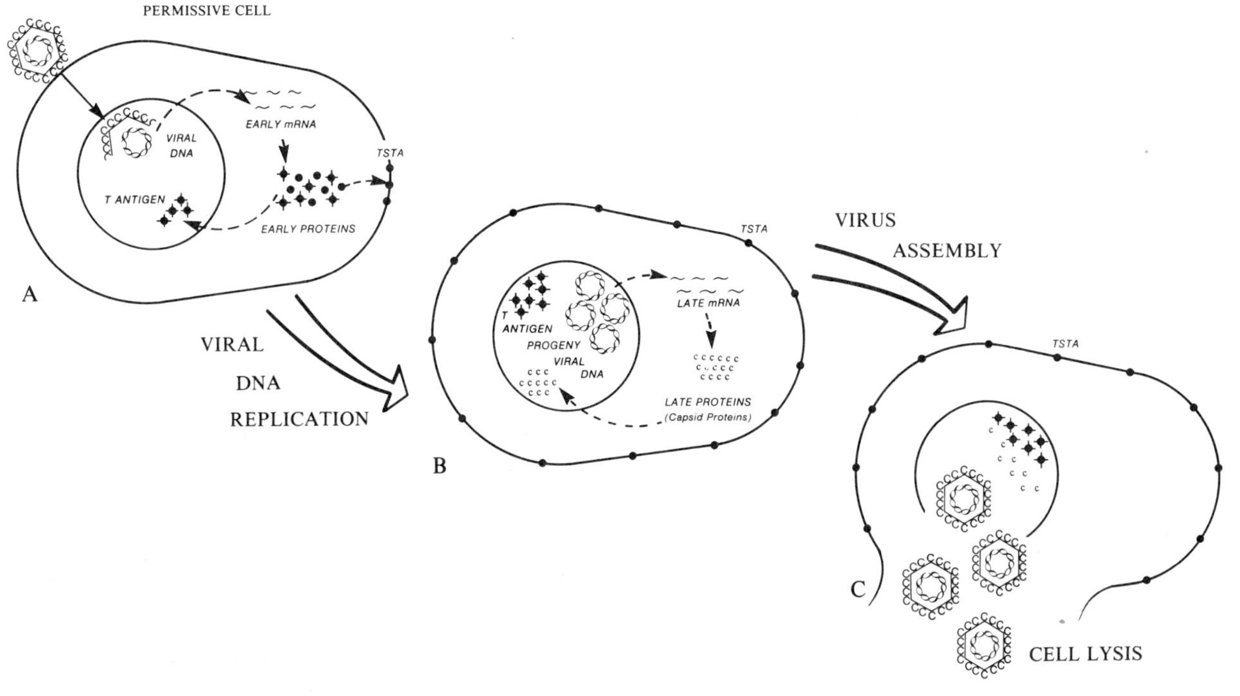

FIGURE 3. *Productive papovavirus infection as exemplified by SV40 infection of primary African green monkey kidney cells.*

cleus, where they can be detected by FA staining by about 15 hours after infection. Progeny virions assemble in the nucleus beginning about 18 hours after infection and accumulate for a prolonged period, usually for at least 40 hours, before cell lysis occurs.

It has recently been shown that there are actually two SV40 T antigens: large T antigen with a molecular weight of approximately 100,000 and small t antigen with a molecular weight of 17,000. The large T and small t proteins are immunologically related, and they have the same NH_2-terminal amino acids as well as several common methionine-tryptic peptides (Pauca et al., 1978; Simmons and Martin, 1978). The synthesis of each of these early proteins is directed by a distinct cytoplasmic 19S mRNA, but both mRNAs are derived from the same primary early RNA transcript by a process of splicing. The mRNAs for large T and small t both span almost the entire early region from approximately 0.65 counterclockwise to 0.15 on the map of the SV40 genome (Fig. 2), but they differ in the size and location of deleted intervening sequences (Berk and Sharp, 1978; Crawford et al., 1978). The mRNA coding for large T antigen lacks a 346-nucleotide intervening sequence from 0.59 to 0.535 map units (Fig. 5). Since

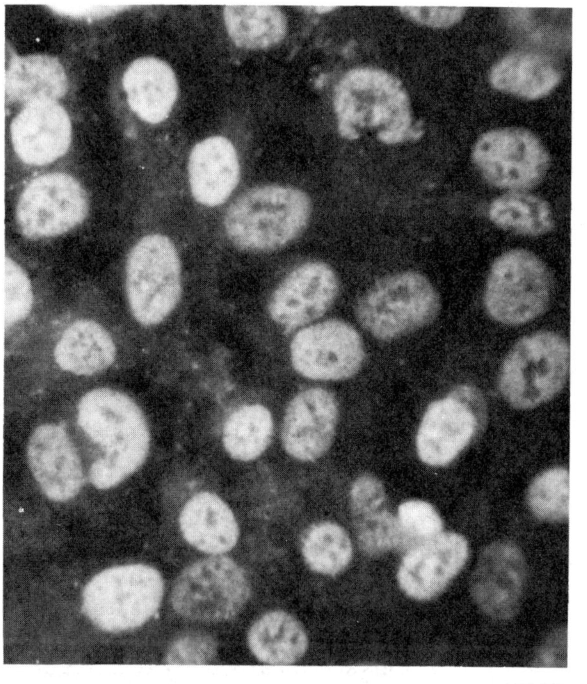

FIGURE 4. *SV40 T antigen demonstrated by fluorescent antibody staining of SV40 transformed cells. Note the finely granular staining confined to the nucleus and exhibiting characteristic nucleolar sparing (1000 ×).*

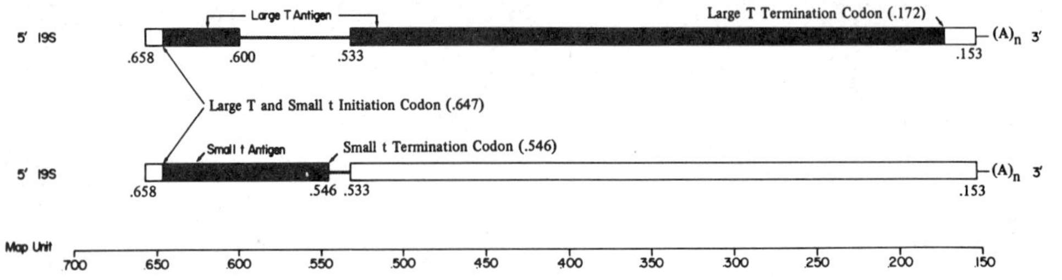

FIGURE 5. *The structure of SV40 early mRNAs. The map of the early region of the SV40 genome is shown counterclockwise below. The regions coding for the large T antigen and the small t antigen are indicated by the shaded areas within the bars that define the two early 19S mRNAs. The thin lines indicate regions of the SV40 genome that are spliced out of the mature poly-adenylated cytoplasmic mRNAs (from Howley, 1980).*

termination codons located at about 0.54 map units are deleted, the entire mRNA can be translated into the 100,000 molecular weight large T protein. The mRNA for small T antigen lacks a 65-nucleotide intervening sequence from 0.54 to 0.53 map units (Fig. 5). However, a termination codon remains just before the site of the deleted sequences of 0.54, and thus only the portion of the mRNA from 0.65 to 0.54 can be translated. This yields the 17,000 molecular weight small t protein, which is identical to large T antigen in the region encoded by sequences from 0.65 to 0.59 in the SV40 genome (Howley, 1980). The stable cytoplasmic late mRNAs, which direct the synthesis of VP-1, VP-2 and VP-3, are similarly formed by the splicing out of intervening sequences present in primary transcripts of the late region of the viral genome (Howley, 1980).

In contrast to many other viruses, papovavirus infection of permissive cells does not result in the early inhibition of host cell synthetic functions. Rather, there is an induction of host cell DNA, RNA, and protein synthesis during the early phase of infection. There are also changes in the surface properties of infected cells. These include increased uptake of sugars, enhanced agglutinability by plant lectins, and the appearance of new surface antigens, including TSTA. Most of these changes appear to be dependent upon the expression of the early portion of the papovavirus genome, and particularly upon the synthesis of large T antigen, a virus-coded early protein with a molecular weight of approximately 100,000. Large T antigen is required for productive infection. It is responsible for the initiation of viral DNA replication and late viral mRNA synthesis, as well as for the induction of host cell DNA replication (Howley, 1980). The large T protein contains an ATPase activity and binds specifically to SV40 DNA at the origin of replication. It also binds to host cell chromosomes. It may act by locally unwinding a portion of the duplex DNA, allowing polymerases to bind and initiate replication. Large T antigen is also required for the initiation and maintenance of transformation in nonpermissive host cells.

Productive SV40 infection results in characteristic cytopathic effects (Sweet and Hilleman, 1960; Hsiung and Gaylord, 1961). In unstained monolayer cultures these cytopathic effects (CPE) consist of cell rounding and enlargement with the development of extensive cytoplasmic vacuolization (Fig. 6). This extensive cytoplasmic vacuolization led to the original designation of SV40 as "simian vacuolating virus." Papovavirus CPE is clearly visible when infected monolayer cultures are fixed and stained with hematoxylin and eosin. At about 24 hours after infection there is nuclear swelling, some clumping of chromatin, and the appearance of irregular patches of eosinophilic material within the nucleus. By about 48 hours after infection the nuclei show margination of chromatin and large basophilic Feulgen-positive intranuclear inclusions that, when examined by electron microscopy, are found to be composed of papovavirus virions. At about this time the extensive cytoplasmic vacuolization develops (Fig. 6). The fully developed intranuclear inclusions produced by papovaviruses of the polyomavirus genus may be basophilic, amphophilic, or eosinophilic, depending upon the virus strain, cell type, and method of fixation. Although productive infection results in cell lysis, progeny papovavirus virions are not always efficiently released. This is especially true of the hemagglutinating human papovaviruses, BKV and JCV, which remain associated with cells and cell debris. To achieve good yields of these viruses it is necessary to disrupt the infected cells mechanically and use receptor-destroying enzyme (RDE) to free the virions from cell debris.

Perhaps the most significant biologic property of the papovaviruses is their capacity to transform normal cells into tumor cells. Thus, while members of the polyomavirus genus are not associated with spontaneous tumors in their natural hosts, they regularly produce tumors experimentally. Moreover, they can transform normal cells in tissue culture into cells that closely resemble those derived from virus-induced tumors. Since transformation implies the survival and subsequent multiplication of the infected cell, it can occur only in the setting of abortive rather than lytic infection. Such abortive infections can result either from the infection of a nonpermissive cell or from the infection of a permissive cell with a defective viral genome (Black, 1968; Sambrook, 1972; Eckhart, 1977; Salzman and Khoury, 1974; Kelly and Nathans, 1977; Fareed and Davoli, 1977; Oxman, 1967). The common denominator appears to be the persistence and continued expression of early viral genetic information in the absence of virus replication.

When papovaviruses of the polyomavirus genus infect nonpermissive cells, e.g., when SV40 infects mouse 3T3 cells (Black, 1968; Sambrook, 1972; Eckhart, 1977; Salzman and Khoury, 1974; Kelly and Nathans, 1977; Fareed and Davoli, 1977; Oxman, 1967), the early phase of infection is similar to that described for infection of permissive cells. Virus adsorption, penetration, and uncoating occur, and the early region of the viral genome is transcribed into early mRNA. The early viral proteins (T antigens, U antigen, and TSTA) are synthesized, the surface properties of the infected cell are altered, and cellular DNA, RNA, and protein synthesis are stimulated (Fig. 7*A*). However, the late phase of the replication cycle does not occur. There is little or no viral DNA replication, no synthesis of viral capsid proteins (V antigens), and no production of papovavirus virions. Such abortively infected cells are stimulated to divide and behave for a few generations like transformed cells, i.e., they become insensitive to the controls that regulate the multiplication of normal cells. This transient abnormal behavior, called "abortive transformation," is coincident with and dependent upon the expression of the early region of the viral genome, as evidenced by the synthesis of T antigens.

Gradually, and over several cell generations, most abortively transformed cells return to their original state because, in the absence of viral DNA replication, cell division and enzymatic destruction result in the loss of the early viral

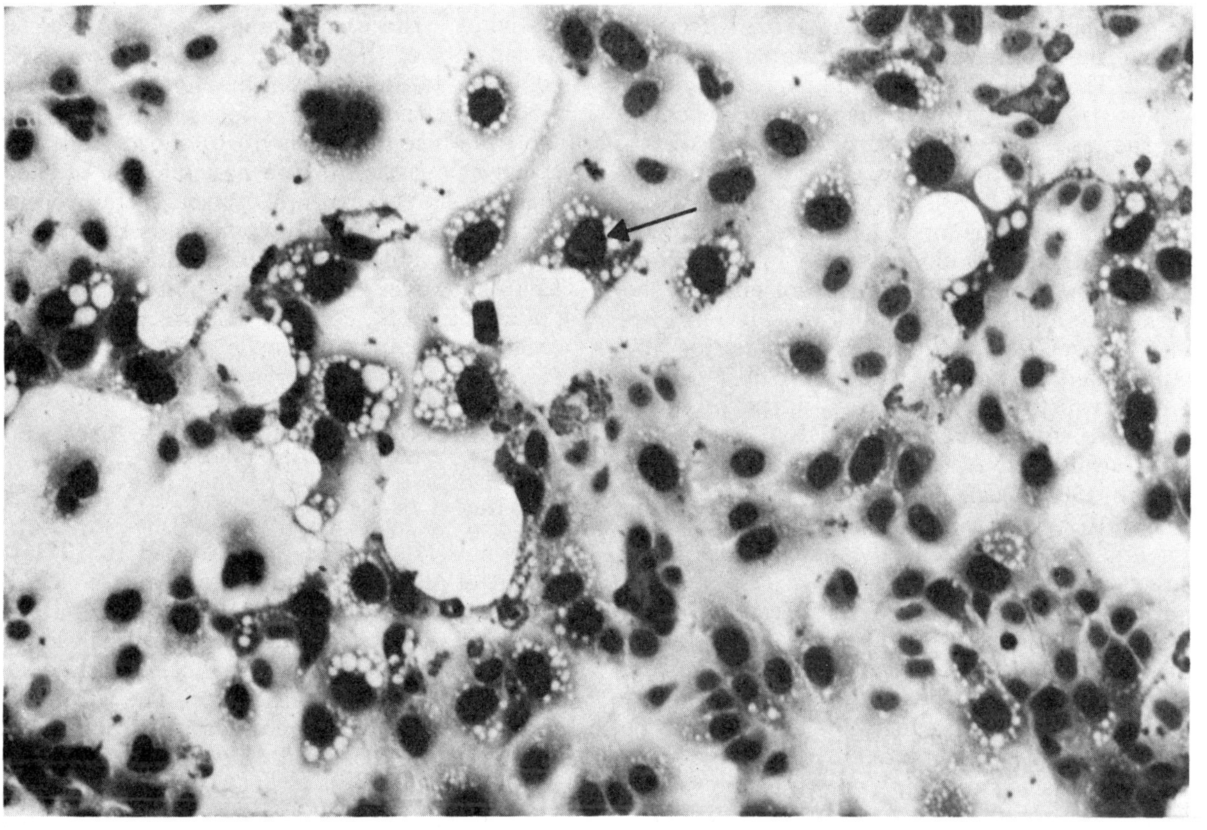

FIGURE 6. *SV40 infection of permissive (AGMK) cells. Monolayer cultures were fixed and stained with hematoxylin and eosin 3 days after infection with a multiplicity of approximately 5 plaque-forming units per cell. Note the extensive cytoplasmic vacuolization. The arrow indicates a cell with a typical basophilic intranuclear inclusion body (350 ×).*

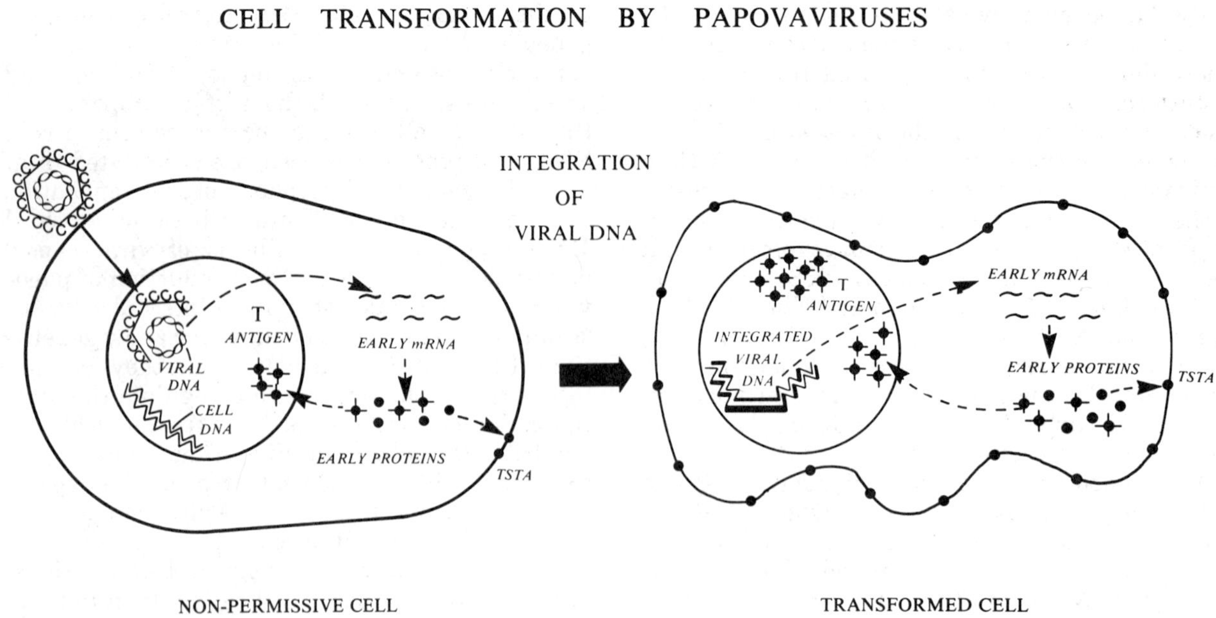

FIGURE 7. *Cell transformation by papovaviruses as exemplified by SV40 transformation of mouse 3T3 cells.*

gene(s) and gene product(s) responsible for the transformed phenotype (Sambrook, 1972; Oxman, 1967). However, abortive transformation is not the only possible outcome of abortive infection. Some of the infected cells are permanently transformed. These cells and their progeny continue to express the early papovavirus gene(s) and never lose the transformed phenotype. The critical difference between the permanently and the abortively transformed cells is the integration of the papovavirus genome (or at least the early region of the viral genome) into cell DNA in the permanently transformed cells. This event ensures the persistence and expression of the early viral genetic information that is responsible for the transformed phenotype (Fig. 7*B*). Thus, permanently transformed cells contain papovavirus DNA covalently linked to cell DNA, as well as early viral mRNA and early viral antigens (Enders, 1965; Black 1968; Benjamin, 1972; Sambrook, 1972; Levine, 1974; Eckhart, 1977; Salzman and Khoury, 1974; Kelly and Nathans, 1977; Fareed and Davoli, 1977; Oxman, 1967). Moreover, although they contain no infectious virus, transformed nonpermissive cells often contain one or more copies of the whole viral genome, and infectious virus can be rescued from them by cocultivation with permissive cells. It should be noted that papovavirus transformation (i.e., permanent transformation) is a very inefficient process; 10^4 to 10^5 infectious virions are required per transforming event (Sambrook, 1972).

Papovavirus transformation is of considerable importance to students of human disease. Because they contain early viral proteins, cells transformed by papovaviruses are extremely useful sources of antigens for serologic tests to detect antibodies to these viral gene products. Moreover, the capacity of the human papovaviruses, BKV and JCV, to induce malignancies in animals, together with the high frequency with which both agents produce infections in childhood, raises the important question of the possible role of these viruses in human cancers, especially those occurring in immunologically compromised individuals (Padgett and Walker, 1976; Takemoto, 1978). To date, however, there is no convincing evidence implicating SV40, BKV, or JCV in the etiology of human malignancies (Howley, 1980).

VIRUS STRAINS AND ANTIGENIC PROPERTIES

JC Virus

JC virus (JCV) was first isolated in 1971 by Padgett et al. from brain tissue of a case of PML that occurred in a 38-year-old man with Hodgkin's disease. Papovavirus virions were observed by electron microscopy in the nuclei of oligodendrocytes in the diseased tissue. The virus was isolated in primary human fetal glial (PHFG) cell cultures that had been inoculated with extracts of brain tissue.

JCV grows and produces CPE only in PHFG cell cultures that contain a high proportion of spongioblasts. Even in these cells the CPE is subtle, and the majority of the progeny virions remain cell-associated. In view of the restricted host range of JCV it is not surprising that earlier attempts to isolate virus from brain tissue of patients with PML failed in spite of the inoculation of multiple cell types and animals (Gardner et al., 1971; Weiner and Narayan, 1974). Recently, replication of JCV has been reported in primary human amnion and vascular endothelial cells in vitro (Howley, 1980). The amnion cells are readily available, and this should facilitate biologic and biochemical studies of JCV.

Since the original isolation of JCV, this virus has been identified in the brain tissue of more than 25 patients with PML (Padgett and Walker, 1976; Weiner and Narayan, 1974; Gardner, 1977; Takemoto, 1978; Padgett et al., 1977). JCV has also been isolated from patients without neurologic disease, including renal transplant recipients and a healthy 38-year-old pregnant woman (Gardner, 1977; Coleman et al., 1977; Hogan et al., 1980).

JCV agglutinates human, guinea pig, and chicken erythrocytes at 4° C, but not those of sheep, African green monkeys, rhesus monkeys, or hamsters (Padgett and Walker, 1976; Padgett et al., 1977). The hemagglutinin is associated with the virion, which has a buoyant density in CsCl of 1.34 g/cm^3 (Osborne et al., 1977). The hemagglutination inhibition (HI) test detects antibody to JCV.

BK Virus

BK virus (BKV) was first isolated in 1971 by Gardner et al. from the urine of a 39-year-old man receiving immunosuppressive therapy following a renal allograft. His urine contained abnormal epithelial cells with basophilic intranuclear inclusions. Electron-microscopic examination revealed papovavirus virions in the nuclei of these inclusion-bearing cells, as well as extracellular virions in the urine. Papovavirus virions were also detected in the nuclei of epithelial cells bordering the lumen of the donor ureter after its removal because of obstruction. BKV was isolated in the VERO monkey cell line, but primary human cells are considerably more susceptible to BKV than monkey cells, and thus human cells have been used to obtain many of the more recent isolates of BKV (Padgett and Walker, 1976; Gardner, 1977; Takemoto et al., 1974; Dougherty and Di Stefano, 1974; Takemoto and Mollarkey, 1973; Lecatsas et al., 1974). BKV grows very well in human embryonic kidney (HEK), human fetal fibroblast, and human fetal brain cultures, and produces a CPE like that produced by SV40 in AGMK cells (Fig. 6).

There is no serologic cross-reactivity between BKV and the human wart viruses or mouse polyoma virus, and only a very minor antigenic relationship between BKV and SV40. Since SV40 does not hemagglutinate and BKV agglutinates human O, chicken, and guinea pig cells to high titer, there is little likelihood of confusing BKV with SV40. Specific antibody to BKV inhibits hemagglutination, and thus the HI test provides a rapid and sensitive method for the detection and titration of antibody to BKV.

Since 1971 a number of investigators have isolated and identified BKV in the urine of renal allograft recipients, children and adults receiving chemotherapy for malignancies, and children with Wiskott-Aldrich syndrome, a congenital disorder involving defects in both cellular and humoral immunity (Padgett and Walker, 1976; Gardner, 1977; Takemoto, 1978; Hogan et al., 1980).

Although BKV is less oncogenic in vivo than JCV (see below), hamster and rat cells have been successfully transformed in vitro by BKV, as well as by BKV DNA (Padgett and Walker, 1976; Takemoto, 1978; Portolani et al., 1975; van der Noordaa, 1976; Takemoto and Martin, 1976; Howley and Martin, 1977). The transformed cells resemble cells transformed by SV40, and contain BKV T antigens and BKV DNA. In some cases, BKV can be rescued from them by fusion with permissive cells (Howley, 1980).

Antigenic Relationships

Virion Antigens. Hyperimmune rabbit antisera to SV40, BKV, and JCV all show minor reciprocal cross-reactions at low dilutions. No cross-reactions have been observed between either BKV or JCV and other members of the polyomavirus genus (including other primate papovaviruses) or with human papilloma viruses. Neutralization and FA staining appear to be the most sensitive techniques for detecting these cross-reacting virion antigens. There is a genus-specific antigenic determinant on the major capsid protein, VP-1, which is shared by all members of the polyomavirus genus. It is not located on the surface of the virion, and antibodies to it are only produced after immunization with disrupted virions or purified VP-1.

Nonvirion Antigens. Like SV40, BKV and JCV induce T antigens that are detectable by FA staining in the nuclei of acutely infected and transformed cells; hamsters bearing tumors induced by these viruses develop anti-T antibodies (Padgett and Walker, 1976; Takemoto, 1978; Dougherty, 1976). The T antigens induced by all three viruses have been shown to be immunologically similar (Padgett and Walker, 1976; Takemoto, 1978; Dougherty, 1976). This is not unexpected, because they have some, but not all, of their methionine-tryptic peptides in common.

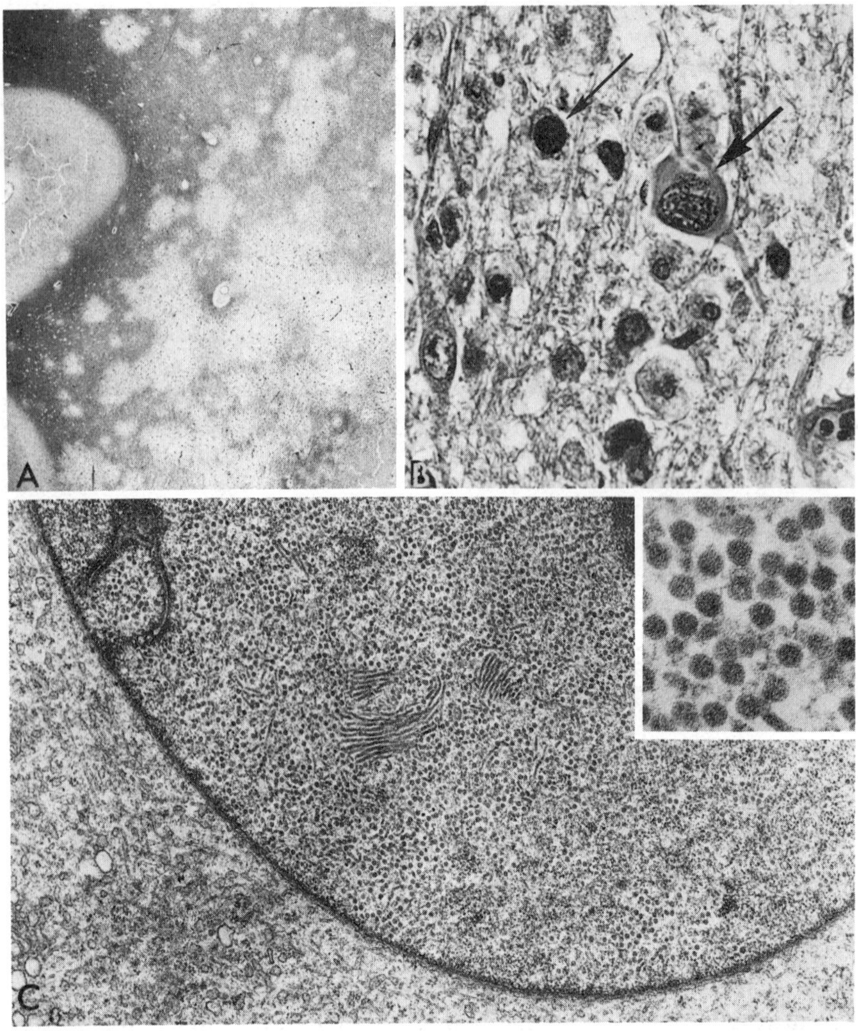

FIGURE 8. *The brain in a typical case of progressive multifocal leukoencephalopathy. A, Patchy demyelination unrelated to vessels merging into confluent lesions. Stained with Luxol fast blue (8 ×). B, The enlarged oligodendroglial nucleus showing margination of chromatin and effacement of nuclear structure by an intranuclear inclusion body (small arrow) and the hypertrophic astrocyte with a bizarre, giant nucleus (large arrow) are characteristic light microscopic findings. Stained with hematoxylin and eosin (350 ×). C, Oligodendrocyte with enlarged nucleus filled with spherical and filamentous virus particles (20,000 ×) that are enlarged in inset (80,000 ×). (Courtesy of Dr. Peter W. Lampert)*

EPIDEMIOLOGY

JC Virus

The consistent finding of JCV in the diseased brains of patients with progressive multifocal leukoencephalopathy (PML) and the localization of JCV virions in the abnormal oligodendrocytes that are the hallmark of the disease provide strong evidence that JCV is the cause of PML. PML is a rare disease that typically presents as a rapidly evolving neurologic illness in a patient with a long-standing systemic disease such as Hodgkin's disease, chronic lymphocytic leukemia, or lymphosarcoma that is associated with an impairment of immunity (Astrom et al., 1958; Richardson, 1961). The main clinical features are those of a diffuse but asymmetric disease of the cerebral hemispheres: mental aberrations, blindness and disturbances of ocular motility, abnormalities of speech, ataxia, hemiparesis, and other focal neurologic deficits. The disease is relentlessly progressive in the absence of fever or abnormalities in the cerebrospinal fluid, and death usually occurs within three to six months of onset. Occasionally a patient may live for two years or longer.

The lesions in the central nervous system constitute the truly distinctive feature of PML (Zu Rhein and Chou, 1965; Silverman and Rubinstein, 1965; Howatson et al., 1965). These consist of multiple foci of demyelination of varying sizes and stages of evolution, which become confluent and may eventually involve the white matter of entire lobes (Fig. 8*A*). The lesions show no relationship to blood vessels and there is no inflammatory response. They do, however, show two characteristic cytopathologic abnormalities. The pathognomonic feature of PML is the presence of altered oligodendrocytes in and around the early foci of demyelination and at the periphery of the more advanced lesions. These oligodendrocytes are enlarged with markedly enlarged nuclei that are intensely basophilic and show effacement of their nuclear structure (Fig. 8*B*). These changes are followed by necrosis, and thus oligodendrocytes are generally absent from the central portions of more advanced lesions.

The other distinctive cytopathologic feature of PML is the presence, in the more advanced lesions, of giant astrocytes with bizarre hyperchromatic nuclei that resemble the malignant astrocytes of pleomorphic glioblastomas. These cells also resemble astrocytes that have been transformed in vitro by SV40. These changes, and especially the presence of inclusions in the nuclei of oligodendrocytes, led Cavanagh et al (Cavanagh et al., 1959) and Richardson (Richardson, 1961) to suggest that PML might represent an opportunistic viral infection involving oligodendrocytes in patients with impaired immunity. The demonstration in 1965 (Zu Rhein and Chou, 1965; Silverman and Rubinstein, 1965; Howatson et al., 1965) that nuclei of the affected oligodendrocytes in lesions of PML were packed with viral particles resembling papovavirus virions (Fig. 8*C*) lent strong support to this concept and set the stage for the subsequent isolation of JCV by Padgett et al. in 1971 (Padgett et al., 1971). It now seems likely that PML is due to an opportunistic infection by JCV of persons with impaired immunity, and that demyelination occurs because JCV produces a lytic infection of the oligodendrocytes.

Although PML is a rare disease, serologic studies indicate that JCV is widely disseminated in the human population (Padgett and Walker, 1976; Gardner, 1977; Padgett and Walker, 1973; Brown, Tsai, and Gajdusek, 1975). Infection is most frequently acquired in childhood, with HI antibody to JCV already present in 65 per cent of children aged 10 to 14 years (Padgett and Walker, 1973). Approximately 75 per cent of adults in the United States and England have antibody to JCV. There is no information about the route of transmission or the characteristics of the primary infection. However, excretion in the urine and infection via the oral or respiratory route seems likely in view of the isolation of JCV from urine and the epidemiology of other papovaviruses, such as polyoma virus in mice or SV40 in monkeys. Most members of the polyomavirus genus appear to be inhabitants of the urinary tract.

The highly restricted host range of JCV makes a nonhuman reservoir of infection very unlikely. Furthermore, Padgett et al. (Padgett et al., 1977) examined sera from a wide variety of animals for antibody to JCV and found that no species other than man had antibody to this virus. Thus, JCV appears to be a strictly human virus.

BK Virus

BKV can be isolated from the urine of 15 to 40 per cent of renal allograft recipients, as well as from other immunosuppressed individuals, but there is no evidence of any significant disease resulting from BKV infection. There is no serologic evidence to suggest an association between BKV infection and malignancy. Furthermore, extensive studies have not confirmed early reports that BKV DNA could be detected by nucleic acid hybridization in certain human tumors (Howley, 1980).

Serologic surveys indicate that BKV infection is common throughout the world. Approximately 70 per cent of children have already acquired HI antibody to the virus by age 4 to 6 years. Thus, BKV is readily transmitted to children at an earlier age than JCV infection. As in the case of JCV, there is no information on the route of transmission or the characteristics of the primary infection. Furthermore, isolation of BKV has not yet been reported from normal children or adults. Like JCV, BKV is likely to be a strictly human virus.

SV40

The isolation of papovaviruses indistinguishable from SV40 from brain tissue of two patients with PML (Weiner et al., 1972), the exposure of millions of individuals to SV40 in viral vaccines administered during the 1950s, the capacity of this virus to induce tumors in animals and transform human cells in vitro, and the presence of low titers of antibody to SV40 in a small proportion of human sera (Shah and Nathanson, 1976; Shah, 1972) have raised the question of the potential role of SV40 in human disease, especially cancer. However, no additional cases of PML have been associated with SV40-like viruses, and follow-up studies have not revealed any significant excess of malignancies in recipients of SV40-contaminated vaccines (Shah and Nathanson, 1976). Furthermore, it now appears that most or all of the antibody to SV40 that is not accounted for by exposure to SV40 in vaccines or contact with monkeys is due to cross-reacting antibody induced by the human papovavirus, BKV. Thus, there is no evidence that SV40 is a significant cause of disease in humans or that SV40 infects humans under natural conditions.

Human Papillomaviruses

The human papillomaviruses cause a variety of hyperplastic lesions of the skin and mucous membranes. These papillomas or warts are generally benign and tend to spontaneously regress, but they may occasionally undergo malignant degeneration. Transmission of human papillomaviruses is by contact, either direct or indirect; and, since papillomaviruses appear to be highly species specific, humans are likely to be the only reservoir of infection. The peak incidence of warts occurs during the second decade of life, and most adults have been afflicted at least once. Cellular immunity is important in controlling papillomavirus infections, and patients treated with immunosuppressive drugs often have persistent and widely disseminated warts (Spencer and Anderson, 1970; Ingelfinger et al., 1977).

Although human papillomaviruses have not been propagated in cell culture, the ready availability of virus-containing papilloma tissue has permitted the extraction of sufficient quantities of virus to allow an analysis of human papillomavirus (HPV) DNA and capsid proteins. Restriction endonuclease analysis and nucleic acid hybridization of HPV DNA, together with immunologic and electrophoretic analysis of HPV proteins, has led to the identification of at least six distinct human papillomaviruses (Pass and Maizel, 1973; Favre et al., 1975; Gissmann et al., 1977; Orth et al., 1977; zur Hausen, 1977; Orth et al., 1978; Coggin and zur Hausen, 1979). Human papilloma virus type 1 (HPV-1) and HPV-4 are associated mainly, but not exclusively, with deep plantar warts. HPV-2 is associated with common hand warts and mosaic plantar warts. HPV-3 is associated with flat warts in normal individuals and in patients with epidermodysplasia verruciformis, a rare familial disease characterized by generalized warty lesions that persist and exhibit a tendency to become malignant. HPV-5 has been isolated from benign lesions in patients with epidermodysplasia verruciformis. HPV-6 is associated with condyloma acuminata (genital warts) and perhaps also with juvenile laryngeal papillomas. DNA hybridization under stringent conditions has revealed little if any nucleotide sequence homology between the DNAs of these different human papillomaviruses. This suggests that, in evolutionary terms, the different HPV types may have diverged as much from each other as they have from papillomaviruses of other species.

LABORATORY DIAGNOSIS

JC Virus

In view of our current lack of knowledge of the natural history of JCV infections, it would be desirable to attempt the isolation of virus from a variety of tissues and body fluids of patients with PML and of immunosuppressed patients without neurologic disease. Padgett and his associates (Padgett and Walker, 1976; Padgett et al., 1977) have examined a wide variety of cells of human, nonhuman primate, and nonprimate origin, including primary human fetal cells derived from virtually every organ, but have been able to cultivate JCV only in cultures of primary human fetal glial (PHFG) cells. The isolation of JCV requires an enormous investment of time and energy, as well as the use of a cell culture that is not widely available. Thus the recent report that JCV can be propagated in primary human amnion cells is especially significant (Howley, 1980).

Fortunately, tissues and body fluids suspected of containing JCV can be examined directly by one or more of the following techniques. Tissue sections or cytologic preparations from urine, cerebrospinal fluid, or other body fluids may be stained with hematoxylin and eosin and examined by light microscopy for the presence of cells with typical intranuclear inclusion bodies (Rand et al., 1977; Coleman, 1975; Coleman et al., 1977). Tissue sections, cytologic preparations, and urine or other body fluids may also be examined by electron microscopy in search of papovavirus virions. When virions are observed in such specimens, they may be extracted and serotyped with monospecific rabbit sera by immune electron microscopy (Narayan et al., 1973). FA staining with specific antisera may be used to detect and identify virion antigens in tissue sections and cytologic preparations (Narayan et al., 1973). Finally, radiolabeled JCV DNA probes may be used to detect JCV DNA sequences in tissue specimens or cultured cells.

The HI test has been extensively used to detect and quantitate antibodies to JCV in human serum (Padgett and Walker, 1976; Padgett and Walker, 1973). Consideration should be given to the use of FA tests for the detection of antibody to JCV T antigens. However, the greatest need at present is to employ the HI test more widely in order to better define the natural history of JCV infections.

BK Virus

As in the case of JCV, it would be desirable to attempt the isolation of BKV from a variety of tissues and body fluids of immunosuppressed patients. It would also be useful to screen children under 6 years of age for excretion of BKV in urine, feces, saliva, and respiratory secretions to determine the mode of transmission of BKV infection. Unfortunately, although the host range of BKV is not as restricted as that of JCV, this is still a formidable undertaking.

When VERO cells have been employed in the isolation of BKV from urine, the interval from inoculation to the development of CPE has generally ranged from one to five months, and the CPE has not been very distinctive (Gardner et al, 1971; Coleman et al, 1973; Takemoto and Mullarkey, 1973). Primary human cells, including primary HEK, fetal fibroblasts, and PHFG, are considerably more susceptible to BKV than VERO cells and exhibit a much more easily recognized CPE. However, CPE may not appear for five or more weeks following inoculation. Thus, it would be reasonable to employ a series of four blind passages, each of 21 days duration and initiated with sonicated RDE-treated cells from the preceding passage. Moreover, cultures in the final passage should be examined by electron microscopy, FA staining for BKV V antigens, and testing of supernatant fluids and sonicated RDE-treated cell extracts for hemagglutination before being discarded as negative. BKV may be assayed by CPE end-point dilution, plaque formation in HEK monolayers, or the formation of foci of infected cells detectable by FA staining with antibody to V antigen (Padgett and Walker, 1976; Gardner, 1977; Takemoto, 1978). Although the isolation of BKV is not as difficult as that of JCV, it is still a costly and time-consuming endeavor.

Fortunately, tissues and body fluids suspected of containing BKV can be examined directly. Tissue sections, or cytologic preparations from urine, cerebrospinal fluid, or other body fluids may be stained with hematoxylin and eosin and examined by light microscopy for the presence of cells with typical basophilic intranuclear inclusion bodies. Tissue sections, cytologic preparations, and urine or other body fluids may also be examined by electron microscopy in search of papovavirus virions. When virions are observed in such specimens, they may be extracted and serotyped with monospecific rabbit sera by immune electron microscopy. FA staining with specific antisera may detect and identify virion antigens in tissue sections and cytologic preparations. Similarly, sera from tumor-bearing hamsters may be employed to detect BKV T antigen, but, as already noted, this test is not virus-specific. Finally, BKV DNA sequences in suspect tissues may be detected by hybridization with radiolabeled BKV DNA probes (Howley, 1980). The HI test has been used extensively to detect and quantitate antibodies to BKV in human serum.

PATHOGENICITY IN EXPERIMENTAL ANIMALS

JC Virus

JCV appears to be a strictly human virus with a very restricted host range, and antibodies to it have not been detected in any species of animal except man. Furthermore, with the exception of tumors that arise after the inoculation of newborn hamsters and owl monkeys (see below), no acute or chronic disease attributable to JCV has been found in any of a variety of species inoculated by multiple routes and observed for months to years (Padgett and Walker, 1976). The animals tested include rabbits, guinea pigs, hamsters, mink, and rhesus monkeys. Newborn rhesus monkeys inoculated by several routes apparently became infected because they developed antibody to T antigens, as well as to V antigen. However, no disease or tumors have been recognized during a three-year period of observation (Padgett and Walker, 1976).

JCV is highly oncogenic in newborn Syrian hamsters, especially following intracerebral inoculation, when it produces tumors, which are usually multiple malignant gliomas. Subcutaneous and intraperitoneal inoculation of JCV produces multiple sarcomas in visceral organs (Padgett and Walker, 1976). Recently, malignant astrocytomas have been reported in two owl monkeys inoculated intracerebrally, intravenously, and subcutaneously 16 and 25 months earlier. JCV T antigen was detected in the cells of one of the tumors by FA staining (London et al., 1978). This observation is particularly interesting in the light of a recent report (Castaigne et al., 1974) of multiple gliomas in the brain of an 18-year-old immunodeficient patient with PML.

BK Virus

Large doses of BKV administered by various routes have produced no recognizable disease or tumors in newborn mice, rats, guinea pigs, or rabbits (Takemoto, 1978). Five newborn rhesus monkeys inoculated intracerebrally and intraperitoneally with large doses of purified BKV remained healthy during a 30-month period of observation, although infection was present, as evidenced by the development of antibodies to both BKV T and V antigens (Takemoto, 1978). In contrast to JCV, BKV is only weakly oncogenic for hamsters.

Relevance of Human Papovaviruses to Human Disease

Except for the etiologic role of JCV in PML, there is no evidence linking infections with JCV or BKV to human disease. However, the oncogenic potential of these agents and the ubiquity of infections in childhood would appear to justify a careful examination of their potential roles in human cancer. Although several reports concerning the finding of papovavirus virions, antigens, or nucleic acids in human tumors have appeared (Takemoto, 1978), they have not been confirmed by subsequent studies, and there is no convincing evidence in favor of any etiologic role for these agents in human cancer (Howley, 1980). Since none of the members of the polyomavirus genus appears to cause malignancies in their natural hosts under normal conditions, it seems likely that if the human papovaviruses cause any human tumors, the most probable candidates are tumors occurring in immunosuppressed or immunodeficient individuals.

References

Astrom, K. E., Mancall, E. L., and Richardson, E. P., Jr.: Progressive multifocal leuko-encephalopathy: A hitherto unrecognized complication of chronic lymphatic leukaemia and Hodgkin's disease. Brain 81:93–111, 1958.

Benjamin, T.: Physiological and genetic studies of polyoma virus. Curr Top Microbiol Immunol 59:107, 1972.

Berk, A. J., and Sharp, P. A.: Spliced early mRNAs of simian virus 40. Proc. Natl. Acad. Sci. U.S.A., 75:1274, 1978.

Black, P. H.: The oncogenic DNA viruses. A review of *in vitro* transformation studies. Annu Rev Microbiol 22:391, 1968.

Brown, P., Tsai, T., and Gajdusek, D. C.: Seroepidemiology of human papovaviruses: Discovery of virgin populations and some unusual patterns of antibody prevalence among remote peoples of the world. Am J Epidemiol 102:331, 1975.

Butel, J. S.: Studies with human papilloma virus modeled after known papovavirus systems. J Natl Cancer Inst 48:285, 1972.

Castaigne, P., Rondot, P., Escourolle, R., Dumas, J. L. R., Cathala, F., and Hauw, J. J.: Leucoencephalopathie multifocale progressive et "gliomes" multiples. Rev Neurol 130:379, 1974.

Cavanagh, J. B., Greenbaum, D., Marshall, A. H. E., and Rubinstein, L. J.: Cerebral demyelination associated with disorders of the reticuloendothelial system. Lancet 2:525, 1959.

Ciuffo, G.: Innesto positivo con filtrato di verruca volgare. Giorn Ital Mal Venereol 48:12, 1907.

Coggin, J. R. Jr., and zur Hausen, H.: Workshop on papillomaviruses and cancer. Cancer Res 39:545, 1979.

Coleman, D. V.: The cytodiagnosis of human polyomavirus infection. Acta Cytol 19:93, 1975.

Coleman, D. V., Daniel, R. A., Gardner, S. D., Field, A. M., and Gibson, P. E.: Polyoma virus in urine during pregnancy. Lancet 1:709, 1977.

Coleman, D. V., Gardner, S. D., and Field, A. M.: Human polyomavirus infection in renal allograft recipients. Br Med J 3:371, 1973.

Coleman, D. V., Russell, W. J. I., Hodgson, J., Pe, T., and Mowbray, J. F.: Human papovavirus in Papanicolaou smears of urinary sediment detected by transmission electron microscopy. J Clin Pathol 30:1015, 1977.

Crawford, L. V., Cole, C. N., Smith, A. E., Paucha, E., Tegtmeyer, P., Rundell, K., and Berg, P.: Organization and expression of early genes of simian virus 40. Proc. Natl. Acad. Sci. U.S.A., 75:117, 1978.

Diacumakos, E. G., and Gershey, E. L.: Uncoating and gene expression of simian virus 40 in CV-1 cell nuclei inoculated by microinjection. J Virol 24:903, 1977.

Dougherty, R. M.: A comparison of human papovavirus T antigens. J Gen Virol 33:61, 1976.

Dougherty, R. M., and Di Stefano, H. S.: Isolation and characterization of a papovavirus from human urine. Proc Soc Exp Biol Med 146:481, 1974.

Eckhart, W.: Genetics of polyoma virus and simian virus 40. In Fraenkel-Conrat, H., and Wagner, R. R. (ed.): Comprehensive Virology, vol 9. New York, Plenum Press, 1977, pp. 1–26.

Enders, J. F.: Cell transformation by viruses as illustrated by the response of human and hamster renal cells to simian virus 40. Harvey Lectures 59:113, 1965.

Fareed, G. C., and Davoli, D.: Molecular biology of papovaviruses. Ann Rev Biochem 46:471, 1977.

Favre, M., Breitburd, F., Croissant, O., and Orth, G.: Structural polypeptides of rabbit, bovine, and human papillomaviruses. J Virol 15:1239, 1975.

Fiers, W., Contreras, R., Haegeman, G., Rogiers, R., Van de Voorde, A., Van Heuverswyn, H., Van Herreweghe, J., Volckaert, G., and Ysebaert, M.: Complete nucleotide sequence of SV40 DNA. Nature, 273:113, 1978.

Finch, J. T., and Klug, A.: The structure of viruses of the papilloma-polyoma type. III. Structure of rabbit papilloma virus. J Mol Biol 13:1, 1965.

Fried, M., and Griffin, B. E.: Organization of the genomes of polyoma virus and SV40. Adv Cancer Res 24:67, 1977.

Gardner, S. D.: New human papovaviruses: Their nature and significance. In Waterson, A. P. (ed.): Recent Advances in Clinical Virology, vol 1. Edinburgh, Churchill-Livingstone, 1977, pp. 93–115.

Gardner, S. D., Field, A. M., Coleman, D. V., and Hulme, B.: New human papovavirus (B.K.) isolated from urine after renal transplantation. Lancet 1:1253, 1971.

Gissmann, L., Pfister, H., and zur Hausen, H.: Human papilloma viruses (HPV): Characterization of four different isolates. Virology 76:569, 1977.

Hogan, T. F., Borden, E. C., McBain, J. A., Padgett, B. L., and Walker, D. L.: Human polyomavirus infections with JC virus and BK virus in renal transplant patients. Annals Int Med 92:373, 1980.

Howatson, A. F., Nagal, M., and Zu Rhein, G. M.: Polyoma-like virions in human demyelinating brain disease. Can Med Assoc J 93:379, 1965.

Howley, P. M.: Molecular biology of SV40 and the human polyomaviruses BK and JC. In Klein, G. (ed.): Viral Oncology. New York, Raven Press, 1980, pp. 489–549.

Howley, P. M., and Martin, M. A.: Uniform representation of the human papovavirus BK genome in transformed hamster cells. J Virol 23:205, 1977.

Hsiung, G. D., and Gaylord, W. H., Jr.: The vacuolating virus of monkeys. I. Isolation, growth characteristics, and inclusion body formation. J Exp Med 114:975, 1961.

Ingelfinger, J. R., Grupe, W. E., and Topor, M.: Warts in a pediatric renal transplant population. Dermatologica 155:7, 1977.

Kelly, T. J., Jr., and Nathans, D.: The genome of simian virus 40. Adv Virus Res 21:85, 1977.

Lecatsas, G., Prozesky, O. W., and Scheepers, F.: The cytopathology and development of a human polyoma virus (BK). Arch Ges Virusforsch 44:319, 1974.

Levine, A. J.: The replication of papovavirus DNA. Prog Med Virol 17:1, 1974.

London, W. T., Houff, S. A., Madden, D. L., Fucillo, D. A., Gravell, M., Wallen, W. C., Sever, J. L., Padgett, B. L., Walker, D. L., Zu Rhein, G. M., and Ohashi, T.: Brain tumors in owl monkeys following inoculation with a human polyomavirus (JC virus). Science, 201:1246, 1978.

Mattern, C. F. T., Takemoto, K. K., and DeLeva, A. M.: Electron microscopic observations on multiple polyoma virus-related particles. Virology 32:378, 1967.

Melnick, J. L.: Papova virus group. Science 135:1128, 1962.

Melnick, J. L.: Taxonomy of viruses. Prog Med Virol 23:196, 1977.

Melnick, J. L., Allison, A. C., Butel, J. S., Eckhart, W., Eddy, B. E., Kit, S., Levine, A. J., Miles, J. A. R., Pagano, J. S., Sachs, L., and Vonka, V.: Papovaviridae. Intervirology 3:106, 1974.

Melnick, J. L., and Stinebaugh, S.: Excretion of vacuolating SV-40 virus (papovavirus group) after ingestion as a contaminant of oral poliovaccine. Proc Soc Exp Biol Med 109:965, 1962.

Narayan, O., Penney, J. B., Jr., Johnson, R. T., Herndon, R. M., and Weiner, L. P.: Etiology of progressive multifocal leukoencephalopathy. N Engl J Med 289:1278, 1973.

Orth, G., Favre, M., and Croissant, O.: Characterization of a new type of human papillomavirus that causes skin warts. J Virol 24:108, 1977.

Orth, G., Jablonska, S., Favre, M., Croissant, O., Jarzabek-Chorzelska, M., and Rzesa, G.: Characterization of two types of human papillomaviruses in lesions of epidermodysplasia verruciformis. Proc Natl Acad Sci USA 75:1537, 1978.

Osborn, J. E., Robertson, S. M., Padgett, B. L., Zu Rhein, G. M., Walker, D. L., and Weisblum, B.: Comparison of JC and BK human papovaviruses with simian virus 40: Restriction endonuclease digestion and gel electrophoresis of result fragments. J Virol 13:614, 1977.

Oxman, M. N.: Some behavioral studies of simian virus 40 (SV40). Arch Ges Virusforsch 22:171, 1967.

Padgett, B. L., Rogers, C. M., and Walker, D. L.: JC virus, a human polyomavirus associated with progressive multifocal leukoencephalopathy: Additional biological characteristics and antigenic relationships. Infect Immun 15:656, 1977.

Padgett, B. L., and Walker, D. L.: New human papovaviruses. Prog Med Virol 22:1, 1976.

Padgett, B. L., and Walker, D. L.: Prevalence of antibodies in human

sera against JC virus, an isolate from a case of progressive multifocal leukoencephalopathy. J Infect Dis 127:467, 1973.
Padgett, B. L., Walker, D. L., Zu Rhein, G. M., Eckroade, R. J., and Dessel, B. H.: Cultivation of papova-like virus from human brain with progressive multifocal leucoencephalopathy. Lancet 2:1257, 1971.
Padgett, B. L., Walker, D. L., Zu Rhein, G. M., Hodach, A. E., and Chou, S. M.: JC papovavirus in progressive multifocal leukoencephalopathy. J Infect Dis 133:686, 1976.
Pass, F., and Maizel, J. V., Jr.: Wart-associated antigens. II. Human immunity to viral structural proteins. J Invest Dermatol 60:307, 1973.
Pauca, E., Mellor, A., Harvey, R., Smith, A. E., Hewick, R. M., and Waterfield, M. D.: Large and small tumor antigens from simian virus 40 have identical amino termini mapping at 0.65 map units. Proc Natl Acad Sci USA 75:2165, 1978.
Pitko, V. M., Pyokari, P., Nase, L., and Mantyjarvi, R.: Effect of β-propiolactone on infectivity and haemagglutinin of the BK virus. Acta Pathol Microbiol Scand [B] 83:141, 1975.
Portolani, M., Barbanti-Brodano, B., and La Placa, M.: Malignant transformation of hamster kidney cells by BK virus. J Virol 15:420, 1975.
Rand, K. H., Johnson, K. P., Rubinstein, L. J., Wolinsky, J. S., Penney, J. B., Walter, D. L., Padgett, B. L., and Merigan, T. C.: Adenine arabinoside in the treatment of progressive multifocal leukoencephalopathy: Use of virus-containing cells in the urine to assess response to therapy. Ann Neurol 1:458, 1977.
Reddy, V. B., Thimmappaya, B., Dhar, R., Subramanian, K. N., Zain, B. S., Pan, J., Ghosh, P. K., Celma, M. L., and Weissman, S. M.: The genome of simian virus 40. Science 200:494, 1978.
Richardson, E. P., Jr.: Progressive multifocal leukoencephalopathy. N Engl J Med 265:815, 1961.
Rawson, K. E. K., and Mahy, B. W. J.: Human papova (wart) virus. Bacteriol Rev 31:110, 1967.
Salzman, N. P., and Khoury, S.: Reproduction of papovaviruses. In Fraenkel-Conrat, H., and Wagner, R. R. (eds.): Comprehensive Virology, vol 3. New York, Plenum Press, 1974, pp. 63–141.
Sambrook, J.: Transformation by polyoma virus and simian virus 40. Adv Cancer Res 16:141, 1972.
Shah, K. V.: Evidence for an SV40-related papovavirus infection of man. Am J Epidemiol 95:199, 1972.
Shah, K., and Nathanson, N.: Human exposure to SV40: Review and comment. Am J Epidemiol 103:1, 1976.
Silverman, L., and Rubinstein, L. J.: Electron microscopic observations on a case of progressive multifocal leukoencephalopathy. Acta Neuropathologica 5:215, 1965.
Simmons, D. T., and Martin, M. A.: Common methionine-tryptic peptides near the amino-terminal end of primate papovavirus tumor antigens. Proc Natl Acad Sci USA 75:1131, 1978.
Spencer, E. S., and Anderson, H. K.: Clinically evident, nonterminal infections with herpes viruses and the wart virus in immunosuppressed renal allograft recipients. Br Med J 3:251, 1970.
Sweet, B. H., and Hilleman, M. R.: The vacuolating virus, SV40. Proc Soc Exp Biol 105:420, 1960.
Takemoto, K. K.: Human papovaviruses. Int Rev Exp Pathol 18:281, 1978.
Takemoto, K. K., and Martin, M. A.: Transformation of hamster kidney cells by BK papovavirus DNA. J Virol 17:247, 1976.
Takemoto, K. K., and Mullarkey, M. G.: Human papovavirus. BK strain: Biological studies including antigenic relationship to simian virus 40. J Virol 12:625, 1973.
Takemoto, K. K., Rabson, A. S., Mullarkey, M. F., Blaese, R. M., Garon, C. F., and Nelson, D.: Isolation of papovavirus from brain tumor and urine of a patient with Wiskott-Aldrich syndrome. J Natl Cancer Inst 53:1205, 1974.
van der Noordaa, J.: Infectivity, oncogenicity and transforming ability of BK virus and BK virus DNA. J Gen Virol 30:371, 1976.
Weiner, L. P., Herndon, R. M., Narayan, O., Johnson, R. T., Shah, K., Rubinstein, L. J., Preziosi, T. J., and Conley, F. K.: Isolation of virus related to SV40 from patients with progressive multifocal leukoencephalopathy. N Engl J Med 286:35, 1972.
Weiner, L. P., and Narayan, O.: Virologic studies of progressive multifocal leukoencephalopathy. Prog Med Virol 18:229, 1974.
zur Hausen, H.: Human papillomaviruses and their possible role in squamous cell carcinomas. Curr Top Microbiol Immunol 78:1, 1977.
Zu Rhein, G. M., and Chou, S. M.: Particles resembling papova viruses in human cerebral demyelinating disease. Science 148:1477, 1965.

RNA Viruses

59 *ORTHOMYXOVIRUSES AND PARAMYXOVIRUSES*

Douglas D. Richman, M.D.

MYXOVIRUSES

DEFINITION

The term myxovirus (*Gr. myxo,* mucus) has been applied to two families of enveloped RNA viruses, the *Orthomyxoviridae* and *Paramyxoviridae* (Table 1). All myxoviruses contain a single-stranded genome enclosed in a helical array of nucleoprotein. The viral RNA polymerase associated with this ribonucleoprotein structure transcribes viral messenger RNA from the virion RNA. Myxovirus virions are pleomorphic ellipsoids or filaments possessing a lipid envelope with surface glycoprotein projections that bind to host cell receptors.

Myxoviruses are transmitted via the respiratory tract, where influenza viruses, parainfluenza

TABLE 1. Human Myxoviruses

	VIRION DIAMETER (nm)	GENOME SIZE (daltons × 10^6)	NUMBER OF GENE SEGMENTS	DIAMETER OF RIBONUCLEO-PROTEIN*	NUCLEAR REQUIREMENT FOR REPLICATION	ENZYME ACTIVITIES RNA Polymerase
Orthomyxoviridae						
Influenza A & B	80–170	5	8	9	+	+
Influenza C	80–170	5	7 (?)	9	?	+
Paramyxoviridae						
Parainfluenza 1–4b	150–250	6	1	18	−	+
Mumps	150–250	6	1	18	−	+
Measles (Rubeola)	150–250	6	1	18	−	+
Respiratory syncytial virus	90–180	6	1	14	−	+

*Three size classes of ribonucleoprotein apparently exist, although their precise measurements vary with techniques of preparation.

viruses, and respiratory syncytial virus produce their predominant disease. Measles, mumps, and the veterinary myxovirus infections also have a respiratory portal of entry, although disease may be manifested primarily in extrarespiratory sites. In spite of these common myxovirus characteristics, the genome of the orthomyxoviruses is segmented and the genome of the paramyxoviruses is not. These two virus families will be considered separately.

ORTHOMYXOVIRUSES: THE INFLUENZA VIRUSES

STRUCTURE AND REPLICATION

Influenza virions are most frequently 80 to 120 nm in diameter (Figs. 1 and 2). Although generally spheroid, the virions may exhibit great variation in length; long filamentous forms are frequently observed. The virions have a bouyant density of 1.23 g/cm^3 and a composition of approximately 1 per cent RNA, 73 per cent protein, 20 per cent lipid, and 6 per cent carbohydrate. There are seven structural proteins of the influenza virion. Three large proteins (P1, P2, P3), associated with the virion RNA, have molecular weights between 85,000 and 95,000. Their functions probably include RNA transcription and replication. A 60,000-dalton nucleoprotein (NP) polypeptide is associated with the virion RNA in the ribonucleoprotein structures. The 25,000-dalton membrane or matrix (M) protein underlies the lipid envelope of the virion. The virion surface is studded with numerous glycoprotein projections that consist of two types, the hemagglutinin (H) and the neuraminidase (N). The H is a trimer of a polypeptide with a molecular weight of 75,000 to 80,000; the N is a tetramer of a 60,000-dalton polypeptide. The most remarkable structural feature of influenza virus is its segmented genome. The genomes of influenza types A and B consist of eight separate segments of single-stranded RNA, each of which is the gene for one viral polypeptide. When a cell is coinfected by two different influenza viruses, gene segments are assembled into progeny virions with an apparently random distribution with regard to parental origin (Fig. 3). This property, termed *genetic reassortment,* permits major single step changes in surface antigens and host susceptibility. This property carries major implications, both for the epidemiology of the disease and for vaccine development.

The total molecular weight of the influenza virion RNA is 5×10^6. The segments of RNA are each associated with NP to form eight separate ribonucleoprotein structures with a diameter of 9 nm and a bouyant density of 1.34 g/cm^3. Influenza viruses are "negative stranded" RNA viruses; the virion RNA is complementary to mRNA. A virus

TABLE 1. Human Myxoviruses *(Continued)*

ENZYME ACTIVITIES				CULTIVATION OF PRIMARY ISOLATES		
Hemagglutinin	Neuraminidase	Hemolysin	Membrane Fusion	Embryonated Egg	Primary Human or Monkey Kidney	Continuous Cell Lines
+	+	−	−	+	+	MDCK
+	−	−	−	+	−	MDCK
+	+	+	+	−	+	Some types only: MDCK, HeLa, HEp-2, WI-38
+	+	+	+	+	+	Vero, HeLa, LLC-MK2
+	−	+	+	+	+	HeLa, HEp-2, LLC-MK2
−	−	−	+	−	+	Hep-2, HeLa WI-38

coded RNA polymerase is associated with the influenza RNA within the virion and transcribes mRNA from the (−) strand virion RNA. Less than 10 per cent of most influenza virions are infectious, perhaps because they lack a complete set of eight gene segments. Influenza virus C may contain fewer segments, thus differing from types A and B in this characteristic, in the lack of N activity, and in several other respects. It appears to be medically insignificant and will not be considered further.

A single proteolytic cleavage of the H glycoprotein by a cell or serum protease is necessary to activate the H and thus to render the influenza virion infectious. The replicative cycle of influenza is initiated by attachment of the virion H to an *N*-

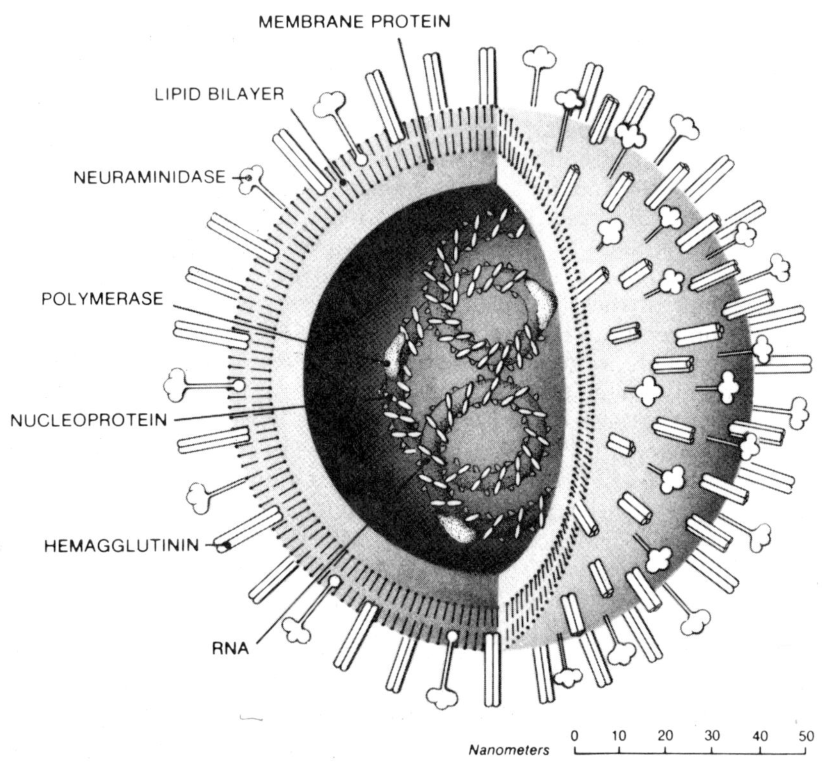

FIGURE 1. *Cutaway diagram of influenza virion structure.*

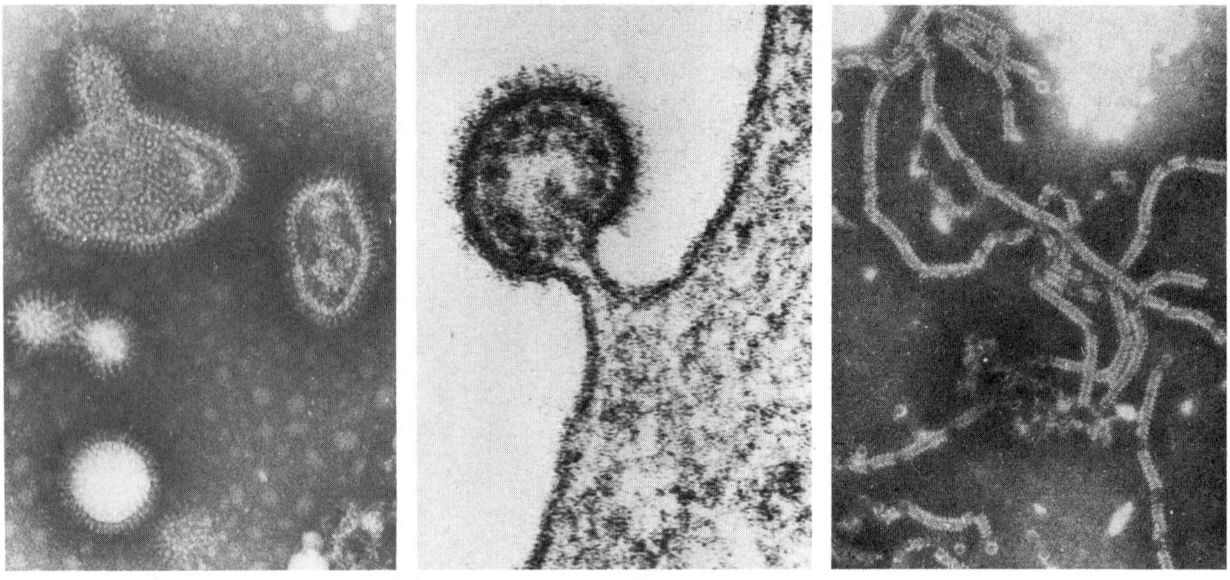

FIGURE 2. Left, *Purified influenza A/Hong Kong/68 (H3N2). Note pleomorphic shapes, variable size, and glycoprotein projections covering the virion surface. Electron-photomicrograph kindly provided by A. R. Kalica, Ph.D. Laboratory of Infectious Diseases, NIAID National Institutes of Health, Bethesda, Maryland. (182,000 ×).*

Center, *Respiratory syncytial virus budding from the surface of an infected HeLa cell in culture. Electron-photomicrograph from Kalica, Wright, Hetrick, and Chanock electron microscopic studies of respiratory syncytial temperature-sensitive mutants. Arch fur die ges Virusf 41:248, 1973, with permission. (132,000 ×).*

Right, *Purified nucleocapsids from mumps virus. Note helical array of ribonucleoproteins to form the nucleocapsid structure with an 18 nm diameter. This nucleocapsid contains an apparent hollow "core," which is apparent on both transverse and end-on views. Electron-photomicrograph from Huppertz, Hall and ter Meulen. Polypeptide composition of mumps virus. Med Microbiol Immunol 163:25, 1977, with permission. (142,025 ×).*

acetylneuraminic acid–containing receptor on a host cell membrane. Two mechanisms of penetration have been proposed: fusion of viral and cell membranes, and endocytotic ingestion of the virus by the cell (viropexis). The precise location, sequence, and control of influenza RNA and protein synthesis are still being investigated; nevertheless, the unusual requirement for the participation of the cell nucleus in influenza replication is well recognized. In contrast to most other RNA viruses, such as picornaviruses and paramyxoviruses, whose cytoplasmic replication is independent of the cell nucleus, influenza virus replication requires the presence and function of the cell nucleus, which contains both viral RNA and protein during influenza replication. Enucleation of the host cell or treatment of it with sublethal doses of UV irradiation, actinomycin D, or α-amanitin will prevent influenza virus replication. None of these inhibits paramyxovirus replication. Studies using cells resistant to the action of α-amanitin suggest that the host cell RNA polymerase II enzyme is essential for influenza replication.

RNA synthesis in the early hours of infection represents messenger RNA that is complementary to the input virion RNA. Subsequent RNA synthesis represents the production of (−) strand RNA that will be utilized in the assembly of new virions. The viral RNA and proteins are transported to the cell periphery by an undetermined mechanism. Also indefinite are the processes of assembly of these viral components and of budding of virions from the host cell. The first evidence of a budding site is the accumulation of the H and N glycoprotein projections in a limited region of the cell outer membrane. A layer of M protein then appears subjacent to the host membrane lipid bilayer in this region, which is to be incorporated into the budding virion. Ribonucleoprotein segments can then be seen by electron microscopy attached to these localized regions of altered cell membrane, which then progressively pouch out and finally bud off from the cell surface. Release of new virions from the host cell surface requires hydrolysis by the virion N of the *N*-acetylneuraminic acid–containing receptors that normally are present in the host cell membrane.

ANTIGENIC COMPOSITION

Two internal structural proteins, the NP and the M protein are used in typing influenza viruses. Each of these two proteins are antigenically identical or highly similar within each of the three influenza types. Moreover, these proteins exhibit no antigenic cross-reactivity among the types. Consequently, whole virus or purified preparations of NP (previously called the soluble [S] antigen) or M protein are used in complement fixation or immunodiffusion tests to type influenza virus isolates as A, B, or C. In addition, these tests are used in the type-specific, but not subtype- or antigenic strain–specific, serodiagnosis of infection. Antibody to the NP or M proteins, which are not exposed on the virion, provides no substantial protection from infection.

The two surface glycoproteins, H and N, are the important antigens for host immunity, for antigenic variation of influenza viruses, and for serodiagnosis of infection. The H binds to *N*-acetylneuraminic acid residues, which are present in all eukaryotic cell membranes. Thus, the virus will probably bind to any cell, including both susceptible host cells and erythrocytes. Attachment to erythrocytes is responsible for hemagglutination, and inhibition of this reaction with antibody, hemagglutination inhibition (HI), constitutes the most utilized assay for measuring antibody against influenza viruses. This technique is also used to characterize antigenic variants of influenza virus. HI antibody neutralizes the infectivity of influenza viruses in vitro and in vivo.

The N enzyme facilitates the release of free virus during the budding process and may also help virions reach the respiratory epithelium, which is protected by mucus and its rich content of glycoproteins containing *N*-acetylneuraminic acid that bind to the virus. N activity is measured by a colorimetric assay of *N*-acetylneuraminic acid released by the enzymatic hydrolysis of a substrate like fetuin. Antibody to N is measured by inhibition of this activity when serum and virus are mixed. This neuraminidase-inhibiting (NI) antibody appears to restrict the spread of virus from infected cells. In monolayer cell cultures infected with influenza virus, HI antibody reduces the number of plaques formed (neutralization of inoculum virus), but the plaques that do appear are of normal size; NI antibody does not reduce plaque number, but does reduce plaque size (restriction of spread of virus from infected to adjacent susceptible cells). In both animal models and in human studies, NI antibody tends to modify the severity of disease expression rather than prevent the acquisition of infection.

Influenza virus has the unique capability of changing the antigenic identity of its H and N. Antigenic changes occur in the H and N polypeptides with a frequency not observed in other viruses or in other influenza virus proteins. The resulting antigenic variants have a selective advantage over the parental antigenic strain in the

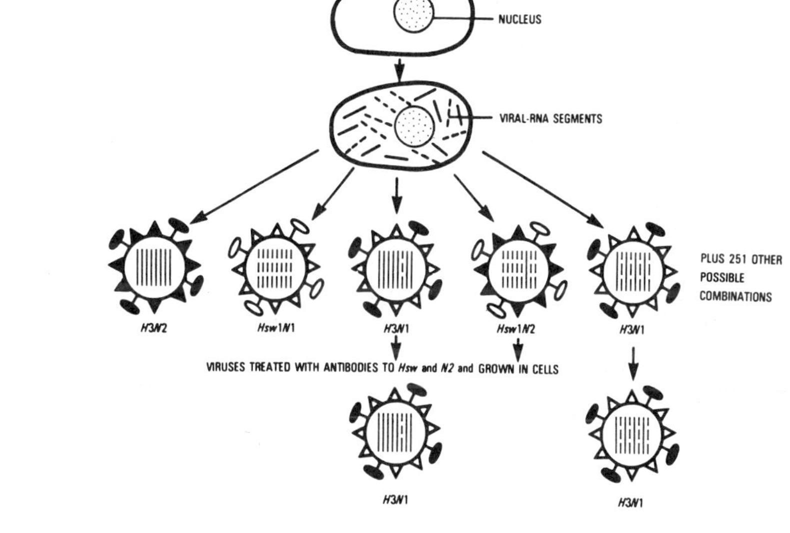

FIGURE 3. *Recombination (reassortment) of genetic material from two different type A influenza viruses is represented schematically in this diagram. Here the two influenza viruses, one from man (H3N2) and the other from swine (HswineN1), are shown being inoculated into the nose of a pig. The inoculation results in the simultaneous infection of a single lung cell with the eight separate RNA segments from each virus. Once inside the cell, the viruses multiply, and the 16 different RNA segments can be recombined in many ways during the "packaging" of the new virus particles. In the presence of antibodies to the neuraminidase of the human strain and to the hemagglutination of the swine strain, all the resulting viruses will be neutralized except for those possessing the H3N1 combination of surface antigens. The H3N1 influenza virus may contain many different combinations of RNA that include those coding for the two surface proteins, but their composition with respect to the remaining six segments can vary and some of the recombinants may cause infection in pigs. Such recombination has been demonstrated experimentally to occur in cell culture, in swine, and in fowl. (Adapted from reference 3).*

TABLE 2. Subtypes of Human Type A Influenza Viruses*

SUBTYPE	PERIOD OF PREVALENCE	REPRESENTATIVE ANTIGENIC VARIANTS
H0N1	1933–1947	A/PR/8/34
H1N1	1946–1957 1977–	A/FM/1/47 A/USSR/90/77
H2N2	1957–1968	A/Japan/305/57
H3N2	1968–	A/Hong Kong/1/68 A/England/42/72 A/Texas/1/77
HswineN1	None, sporadic infections documented	A/New Jersey/6/76

*Table summarizes experience since first isolation in 1933. Prior experience indicated by serologic recapitulation not included.

presence of antibody to the original strain. All three types of influenza virus undergo minor annual changes in their H and N glycoprotein antigens termed *antigenic drift.* These minor changes do not eliminate the basic antigenic relationship to the original strain. When the antigenic change in either H or N is of sufficient magnitude to render them not immunologically cross-reactive, they are given a distinct subtype designation. Such H and N subtypes have been identified only with influenza type A. The appearance of a new subtype is termed *antigenic shift.* Antigenic drift and shift have profound epidemiologic implications as will be considered below.

NOMENCLATURE

In recognition of the vast array of antigenic variants, combinations of H and N subtypes, and hosts of origin, a World Health Organization committee has devised a systematic classification for influenza viruses. The schema includes *(1) the type; (2) the host species, if not human; (3) geographic origin; (4) strain number for the year; and (5) year of isolation (H and N subtype, if type A).* Consequently, the human isolate of the swine-like influenza virus was designated A/New Jersey/6/76 (H swine N1). One of the earliest swine isolates of influenza virus is designated A/swine/Wisconsin/15/30 (H swine N1). This nomenclature supercedes the old terminology of A_0, A_1, and A_2, which ignored the independent variation of the H and N proteins. The influenza A subtypes that have circulated since the first human isolation in 1933 by Smith, Laidlaw, and Andrewes are indicated in Table 2. In addition, over two dozen different subtype combinations have been recognized in swine, horses, and fowl.

PATHOGENESIS AND CLINICAL MANIFESTATIONS

Deposition of droplet nuclei containing influenza virus on almost any portion of the respiratory epithelium can initiate infection. Virus shedding precedes symptoms, which are first noticed one to three days after exposure. Typical symptoms are sneezing, chills, fever, headache, myalgia, nasal obstruction and discharge, sore throat, and nonproductive cough. These symptoms usually subside in two to four days with little residual after a week. Coincident with the onset of symptoms the ciliated epithelium desquamates. Symptoms and local pathology usually lag out of phase with virus shedding, which is usually rapidly diminishing within a day of the onset of symptoms. The involved epithelium regenerates, passing through a series of stages from a transitional basal layer, to stratified squamous to hyperplastic cuboidal until the normal ciliated columnar epithelium is reestablished in one to two weeks.

The pneumonia that can occur in the course of influenza infection can be viral, secondary bacterial, or a mixture of the two. Pure viral pneumonia causes focal alveolar exudation, denuded alveolar walls, capillary thrombosis, and necrosis. The increased susceptibility of lungs infected with influenza virus to secondary bacterial invasion has been well documented and is attributed to loss of ciliary clearance, a rich bacterial growth medium in the alveolar exudate, and injury to leukocytes by influenza. *Staphylococcus aureus, Streptococcus pneumoniae,* and *Haemophilus influenzae* are the most frequent secondary pathogens. The bacterial infection may often be difficult to distinguish from the viral on clinical grounds. The course may be biphasic with an interval of partial recovery between the viral syndrome and the bacterial pneumonia, or it may present as a bacterial pneumonia with only epidemiologic, serologic, or cultural evidence to indicate viral infection.

Severe myalgia is common in influenza, and serum creatinine phosphokinase is often elevated. The pathogenesis of this myalgia, which can be severe enough to cause myoglobinuria and renal failure, has not been elucidated. Myocarditis and encephalitis have also been associated with epidemic influenza. In addition, Reye described a syndrome in 1963 of hepatic failure and encephalopathy in children shortly after one of several viral infections—including influenza A on occasion, but most frequently type B. In spite of sporadic claims of virus isolation from blood, there has been as yet no systematic documentation of extrarespiratory viral infection in humans with influenza.

It is important to note that only one half of infected individuals, as determined by antibody responses or viral isolation, develop clinical disease. It is not clear to what extent site of inoculation, inoculum size, and host susceptibility factors contribute to disease severity. Nevertheless, the frequency of lethal complications is significantly increased at the extremes of age, in pregnancy, and by underlying cardiopulmonary disease such as mitral stenosis or emphysema.

IMMUNITY

Antibody to the H and N prevents reinfection or modifies the severity of the disease. HI antibody plays the major role, although NI antibody alone in high titer can also prevent disease. Protection is correlated with both serum antibodies and IgA secretory antibodies in the respiratory tract. At the same level of serum HI antibody, live respiratory infection provides the greatest degree of protection. Active parenteral immunization with killed virus provides less protection, and passive parenteral immunization provides the least protection of the three methods. Multiple factors are thus important in immunity. The relative contributions of serum antibody, secretory antibody, cell-mediated immunity, and other undefined mechanisms are unknown.

Immunity to a newly emerged antigenic variant is a function both of the degree of immunity to the previously circulating variant and the antigenic relatedness of the two strains. The three types of influenza share no antigenic relatedness and thus induce no cross-protection. Cross-protection becomes more complicated within a type, however. Thus, with a minor antigenic drift a person with immunity sufficient for full protection against the older strain might develop a subclinical infection or mild disease with the newer strain; a person with partial protection against the older strain might develop full-fledged influenza with the newer strain. Antibody levels correlate with protection on a statistical basis in a population, but not with the severity of individual responses to virus infection, which exhibit tremendous variation, in keeping with the multifactorial basis of resistance, including perhaps, nonimmune factors in disease response such as genetics and nutrition. The multiple infections a population experiences with influenza viruses that possess varying antigenic coats often make vaccine studies or immunodiagnosis insecure and difficult to interpret. Many interesting phenomena have been recognized, including cross-reactive and non–cross-reactive antigenic sites on the H and N proteins, and *original antigenic sin,* a term referring to the stronger memory for and the increased responsiveness to the influenza antigen to which an individual was first exposed.

Induction of immunity against mankind's last great infectious plague is a prime goal of modern medicine. Intramuscular injection of the standard vaccine, composed of inactivated egg-grown virus protects 70 per cent of adults against infection with an antigenically identical strain. Children respond with less antibody and more adverse effects to killed vaccine than do adults. With current purification of inactivated vaccine, adverse effects appear to correlate directly with the dose of virion components.

One contribution to influenza vaccine production has been Kilbourne's proposal to recombine high yielding egg-adapted laboratory strains of influenza with each newly circulating epidemic virus. Recombinants possessing the new surface antigens and the high yield growth characteristics have cut cost and increased speed of production of killed vaccines. Nevertheless, the relative inefficacy of the inactivated influenza vaccine, when compared with polio or measles virus vaccines, for example, demands improvement. Among new approaches to immunoprophylaxis is the use of purified glycoprotein subunits. Another proposal is to immunize against the currently circulating N antigen in order to permit a subclinical infection with the circulating strain that would give the complete protection of live virus infection via the respiratory route.

Three approaches to live attenuated vaccine have been pursued. One approach proposed recombination between an older strain attenuated by multiple passages in eggs and an antigenically new epidemic strain. Unfortunately, the recombinants with the new antigens appear to have unpredictable levels of attenuation. A second method that may be combined with the first method selects for mutants that are resistant to nonspecific inhibitors to the H3 hemagglutinin that are present in guinea pig or horse serum, and that, for some reason, correspond to reduced pathogenicity for man.

Chanock has proposed the use of a master strain of donor virus that contains temperature-sensitive mutations that specify a desired level of attenuation and that are not located on the gene segments corresponding to the H and N. An attenuating gene could also be derived from an animal influenza A virus that is not virulent for man. By reassortment, a specified level of attenuation could then be transferred by a rapid, single-step process to a virus containing the prevalent antigens. The precise genetic composition of the recombinant with regard to contribution from the two parental strains can now be ascertained with high resolution electrophoretic or hybridization techniques. The efficacy, safety, and feasibility of producing a live attenuated vaccine virus with the glycoproteins of the most recent strain are being actively investigated.

LABORATORY DIAGNOSIS

Because the clinical characteristics of viral respiratory syndromes can be produced by dozens of different agents, the diagnosis of influenza requires isolation of the virus from the patient, identification of specific viral antigens in a patient's tissue, or documentation of a specific serologic response by the patient. The standard technique for influenza virus isolation is growth in embryonated hen's eggs after allantoic or amniotic inoculation. Types A and B virus can also be isolated in primary kidney cell cultures, especially those from human embryos and rhesus monkeys. With the diminishing availability of these sources of primary kidney tissue, a reliable tissue culture substrate that can be propagated continuously has been seriously needed. Laboratory strains of virus have been adapted to continuous cell lines; however, as is common with many viruses, the growth requirements of primary isolates are often fastidious. The Madin Darby canine kidney cell line along with trypsin in the culture medium to cleave and activate the H glycoprotein now appears to be a readily available continuous cell line that is as sensitive as eggs or primary kidney cell cultures for the isolation of influenza viruses.

The best specimens for isolation are nasal washings obtained within the first few days of the onset of symptoms. The presence of virus in allantoic or tissue culture fluid is identified by hemagglutinating activity. The viral antigens can then be determined with known antisera in an HI test. Alternatively, known infected tissue culture monolayers can be identified by the hemadsorption of erythrocytes and the antigenic specificity of the agent can be determined by inhibiting the hemadsorption reaction with specific antisera. More recently, specific fluorescent antibody straining has been used to identify infected tissue culture.

The fluorescent antibody method is also showing great promise in the rapid and specific identification of viral antigens in exfoliated nasopharyngeal cells or in expectorated sputum. Caution is warranted, however, because nonspecific staining can lead to false positive diagnoses unless great care has been observed in the preparation of the specific antibody conjugates and in the antigen and serum controls.

Although a fourfold rise in a virus-specific antibody has been used as a diagnostic standard of infection, this approach contains many pitfalls, especially for influenza virus infections. Any of the various influenza antibody measurements available may not reveal a rise in titer in some individuals with documented infection. Consequently, the failure to demonstrate an immune response by using appropriate acute and convalescent sera and a single reliable assay like the HI test does not exclude the diagnosis of influenza.

The complement-fixation test measures antibody to the RNP and M proteins. This antibody usually falls to undetectable levels over a period of months so that an elevated convalescent serum complement-fixation titer has been considered indicative of a recent infection. Unfortunately, children, especially upon primary exposure, usually develop little or no complement-fixing antibody. With repeated influenza exposure this antibody response increases both in magnitude and duration so that elderly adults will often sustain elevated complement-fixation antibody levels for years.

The HI antibody response is generally considered the most reliable serologic indicator of infection. Caution is warranted, however, in using the HI antibody response to determine the antigenic identity of the infecting influenza virus because of the possibility of minor antigenic cross-reactions between sub-types and the existence of immune responses to unrelated subtype antigens based on the phenomenon of original antigenic sin. Consequently, knowledge of the antigenic composition of the prevalent viruses and use of potentially cross-reacting antigens as controls are helpful when performing and interpreting HI tests. Additional and potentially more rapid or less complicated methods, such as radial hemolysis, radial immunodiffusion, radioimmunoassay, indirect immunofluorescence, and immunoenzymatic assays, are continually under investigation.

EPIDEMIOLOGY

Although structurally similar, the three influenza types vary immensely in epidemiologic impact. Influenza C causes sporadic infections with mild upper respiratory tract disease; it does not cause epidemic influenza. Influenza B causes epidemics and can produce serious pneumonitis in adults. More importantly, it is associated with Reye syndrome in children and is responsible for pneumonitis and croup in infants. The notoriety of influenza A stems from its capacity to produce pandemics.

Because numerous agents can cause similar syndromes, influenza must be detected either by laboratory surveillance for virus isolates and seroconversions, by absenteeism associated with respiratory symptoms, or by excess pneumonia and mortality. By comparing the number of deaths or cases of pneumonia in a selected population with the baseline levels for previous years, the magnitude of the excess can be calculated. This measurement correlates well with the circulation of influenza in the population. Infants, the elderly, and those with underlying cardiopulmonary impairment are most susceptible to the complications of influenza.

The incidence of influenza peaks in the winter months. In tropical climates influenza remains an important disease, but it tends to occur over longer periods with dampened epidemic peaks. The mechanism of "overwintering" — that is, the maintenance of the agent in the population between epidemics has not been completely determined; nevertheless, surveillance of large populations during the summer indicates that the virus remains endemic in the population with sporadic infrequent isolates and subconversions often associated with subclinical infection.

Influenza infection usually induces protective immunity to rechallenge with the homologous virus. The periodic recurrence of influenza in populations with previous experience with the virus results from the capacity of influenza virus to change surface antigens. Both in vitro and in vivo, the presence of partially protective levels of antibody gives a selective advantage to viruses with surface antigens that react less avidly with the antibody. Since a majority of the population develops protective immunity to the circulating influenza strain, an *antigenic drift* occurs in either or in both of the surface antigens, resulting in a relatively more susceptible population. Peptide analysis ("fingerprinting") of the hemagglutinin polypeptide reveals that one or a few amino acid substitutions are responsible for an antigenic drift. Thus, antibody pressure provides a selective advantage for the growth of certain missense mutants.

Although all three types of influenza virus exhibit antigenic drift, influenza A is the most virulent and influenza C the least virulent at given levels of immunity in the population. The progressively increasing sensitivity of the replication of types A, B, and then C to temperatures above body temperature has been proposed as an explanation for the observed gradient in virulence. Regardless of the intrinsic virulence of the different types, one additional property provides influenza A with the capacity to produce pandemic disease. Every 10 to 30 years, one or both of the surface antigens of influenza A changes subtype, rendering a previously immune population completely susceptible. This *antigenic shift* is qualitative rather than the quantitative one characteristic of antigenic drift. Rather than a limited amino acid change typical of antigenic drift, a new subtype polypeptide shows little resemblance to its predecessor on peptide mapping. An accumulating body of circumstantial evidence suggests that antigenic shifts occur as a consequence either of rapid adaptation to man by an animal influenza A virus or of genetic reassortment during coinfection with influenza A viruses derived from man and animal.

Influenza A occurs in animals and man, whereas types B and C appear to be restricted to man. The influenza A strains circulating in animals may provide a source of new antigenic subtypes for human influenza A viruses, thus accounting for antigenic shift. Experimental studies with pigs and turkeys have demonstrated that genetic reassortment occurs in vivo after coinfection with two viruses (Fig. 3). Moreover, the selection of particular combinations of antigenic subtypes can be manipulated by adjusting the preinfection antibody levels present in the animal. These recombinants are then transmissible. One species can transmit its influenza A virus to susceptible members of another species. The phenomenon of influenza virus infection with virus from another host species has been well described on swine farms with the exchange going in either direction between man and pig. Such exchange has also been observed on several occasions between migratory and domestic fowl.

Various combinations of H and N subtypes have been described, suggesting reassortment and interchange. For example, the N1 has been associated with the H0, H1, Hswine, Havian 1, Havian 5, and Havian 6 subtypes. In addition, the human H2 and H3 have been shown by Webster and Laver not only to be antigenically related to several avian and equine subtypes, but to be remarkably similar in their peptide fingerprints to the antigenically related animal hemagglutinins.

Although the first subtype of influenza virus (H0N1) was isolated in 1933, the period of circulation and the antigenic specificity of the three previous subtypes have been deduced with a technique termed *immunologic recapitulation,* by which HI titers against numerous H subtypes are determined in numerous age cohorts. These studies demonstrate that HI antibody to Hswine not based on cross-reaction is present in almost everyone born before 1918 and disappears in people born after 1927. Thus the 1918–1919 pandemic has been attributed to the Hswine subtype of influenza. This process of immunologic recapitulation indicates that the 1890 epidemic was probably caused by the H3 or related Hequine2 subtypes, and the 1877 pandemic by the H2 subtype. The accumulation of such circumstantial evidence supports the currently accepted working hypothesis that there are a finite number of subtype antigens that recirculate, probably by means of infrequent exchange, between human and animal species.

PARAMYXOVIRUSES

CLASSIFICATION

The paramyxoviruses are enveloped, single-stranded, RNA viruses that are transmitted via the respiratory route. They are among the most important respiratory viruses of infants (respiratory syncytial virus and the parainfluenza viruses) and two of the most important contagious diseases of childhood (measles and mumps). Properties of these viruses are delineated in Table 1. The family *Paramyxoviridae* has been divided into three genera. The genus *Paramyxovirus* includes parainfluenza virus types 1, 2, 3, 4a, 4b, and mumps virus. These viruses possess two surface glycoproteins, one with both hemagglutinating and neuramidinase enzyme activities and one with cell fusion and hemolysin activities. These viruses are all antigenically related. Sendai virus (mouse parainfluenza virus type 1), simian virus 5, and Newcastle disease virus of fowl belong to this genus. The genus *Morbillivirus* includes measles, canine distemper, and bovine rinderpest viruses, which are antigenically related, but do not cross-react immunologically with paramyxoviruses of the other two genera. These viruses lack neuramidinase activity and they hemagglutinate only red blood cells of old world monkeys. The genus *Pneumovirus* includes respiratory syncytial virus (RSV) and pneumonia virus of mice. These agents have a smaller nucleocapsid and lack the hemagglutinating, neuramidinase, and hemolysin activities characteristic of the surface glycoproteins of the members of the other two genera.

STRUCTURE

The most complete biophysical, biochemical, and genetic information regarding the paramyxoviruses has been derived from the following animal agents: Sendai virus, simian virus 5, and Newcastle disease virus. Information regarding the human agents will be specified only when it is known and distinctive. *Paramyxovirus* virions are usually spheroids 150 to 250 nm in diameter, although preparations often show variant forms with dimensions ranging from 100 to 700 nm. These virions have a buoyant density of 1.19 to 1.22 g/cm^3 and a sedimentation coefficient of 1000 to 1100S. Their genetic material is a linear, single-stranded RNA molecule that has a sedimentation coefficient of 50 to 57S and weighs approximately 6×10^6 daltons. This RNA represents 0.91 per cent of the total virion weight. The remainder of the virion is composed of 73 per cent protein, 20 per cent lipids, and 6 per cent carbohydrate.

This RNA genome has a coding capacity for one or two more proteins than the slightly smaller influenza virus genome, which codes for seven structural and one nonstructural protein. The protein number and function of the paramyxoviruses have not yet been completely defined. The RNA is surrounded by a helical array of nucleocapsid protein resulting in a ribonucleoprotein structure 18 nm in diameter with a central core 5 nm in diameter (Fig. 2). This structure has a buoyant density of 1.27 to 1.31 c/cm^3 and a length of 1 μm. Associated with the nucleocapsid is the RNA transcriptase, which synthesizes viral messenger RNA with the virion RNA serving as template and host cell providing synthetic machinery and energy. This, the paramyxoviruses, like the influenza viruses, are (−) strand RNA viruses. The ribonucleoprotein is enclosed in an envelope composed of a layer of membrane or matrix (M) protein that underlies the lipid bilayer derived from the host cell during the budding process.

Attached to the envelope are projections consisting of two different glycoproteins, the activities of which distinguish the three genera of the paramyxovirus family. The parainfluenza and mumps viruses possess a glycoprotein (HN) with both hemagglutinin (H) and neuramidinase (N) activities and a glycoprotein (F) with both membrane fusion and hemolysin activities. The HN protein alone presumably serves the two functions of host cell attachment and release served by the two glycoproteins (H and N) of the influenza viruses. As in the case of the influenza viruses, *N*-acetylneuraminic acid present in host cell and red blood cell membranes serves as the receptor and the substrate for the HN activities. This glycoprotein has a molecular weight of 67,000. The F glycoprotein is derived from a precursor polypeptide (F_0) with a molecular weight of 65,000. The proteolytic cleavage of an 8000 to 10,000 dalton polypeptide is required to convert the F_0 to F and activate the fusion and hemolytic functions. This renders the virion infectious. The cleavage step is accomplished by a host cell protease and that represents one, if not the major, determinant of cell permissiveness in vitro. This host cell characteristic may also represent a critical factor in host range and in tissue tropism in vivo.

Measles virus possesses a similar F glycoprotein; however, the second glycoprotein does not contain the HN activities. Measles virus does, however, agglutinate old world monkey red blood cells. This characteristic has some diagnostic utility, but presumably has no biologic relevance. This glycoprotein is responsible for host cell attachment although the receptor site is not *N*-acetylneuraminic acid. Antibody to this glycoprotein neutralizes measles virus. The glycoproteins of respiratory syncytial virus presumably contribute to host cell attachment and release, but the activities of these proteins have not been delineated.

The replication of the paramyxoviruses is unaffected by inhibitors of cellular DNA synthesis such as actinomycin D or by irradiation or enucleation of the host cell. This requirement only for intact host cell cytoplasmic synthetic machinery is another characteristic that distinguishes the paramyxoviruses from the orthomyxoviruses; however, in other respects, the synthesis and assembly of the virion components closely resemble that of the influenza viruses (Choppin and Compans, 1975). The only recognized difference is the requirement of the orthomyxoviruses to assemble multiple, different ribonucleoprotein structures rather than a large single ribonucleoprotein structure into a virion.

ANTIGENIC COMPOSITION

Parainfluenza Viruses and Mumps Virus

There are four serotypes of the human parainfluenza viruses, which are distinct by neutralizing, hemagglutination inhibiting (HI), and complement-fixing antibody (CF) assays. Types 1, 2, and 3 are each antigenically homogeneous and stable as is mumps virus. Type 4 is composed of two subtypes, 4a and 4b, which are distinguishable by neutralization and HI, but not by CF. Antibody to the HN protein has both neutralizing and HI activity. The presence and function of antibody to the F glycoprotein has not been evaluated. CF utilizes the soluble (S) antigen associated with the nucleocapsid. Although these antigens are distinct for each type, hyperimmunization will stimulate cross-reactive antibody to the other types, to mumps virus, and to Newcastle disease virus. This phenomenon occurs with neutralizing, HI, and CF antibody responses. Humans are exposed to most of these agents during childhood and are repeatedly infected with the parainfluenza viruses; this accounts for the heterotypic antibody responses to infection often observed in humans, especially in older people.

Measles Virus

Measles virus is antigenically homogeneous and stable. It is antigenically related to distemper and rinderpest viruses. CF antibody is detected by utilizing the nucleocapsid protein (soluble [S] antigen). Neutralizing and HI antibody is directed against the monkey red cell hemagglutinating protein. Antibody to the fusion-hemolysin has been rarely studied except in relation to vaccine responses.

Respiratory Syncytial Virus

Several minor antigenic variants have been detected by neutralization tests with early postinfection sera; however, convalescent human sera appear to be unable to make this distinction. This antigenic variation has no known biologic significance. Analogous to the other paramyxoviruses, CF and neutralizing antibodies can be measured; however, HI is obviously not possible.

SUSCEPTIBLE HOST CELLS

These agents all initiate their infections on the respiratory epithelium. The susceptibility of epithelial cells to these agents probably accounts for the value of primary human and primate kidney cell cultures for their laboratory isolation (Table 1). The respiratory pathogens (the parainfluenza viruses and RSV) are restricted to the respiratory tract. Measles and mumps disseminate from the respiratory tract via the bloodstream to various target organs as discussed below.

PATHOGENIC PROPERTIES

All paramyxoviruses are transmitted by inhalation of droplet aerosols or by inoculation of respiratory secretions (by hand, for example) onto the respiratory mucosa. The nature of the resulting infection then depends both on host factors, such as age and immune status, and on the specific agent.

Parainfluenza Viruses

The replication of these agents is restricted to the respiratory tract. No extrapulmonary replication has been well documented, although viremia has been reported. These agents cause only upper respiratory infections (the common cold syndrome) except in a proportion of infants who develop laryngotracheobronchitis (croup), bronchiolitis or pneumonitis during their primary infection.

Mumps

Mumps infection causes parotitis after an incubation period of 15 to 18 days after exposure to an infected individual. The incubation period has been reported to range from the extremes of 7 to 23 days. Early in the incubation period, after replication in the respiratory epithelium, viremia disseminates virus throughout the body. Virus is usually detectable in blood, saliva, and parotid duct secretions for two to three days before onset of symptoms. The presence of virus in saliva before symptoms, the variable incubation period, and the numerous asymptomatic but still infectious cases, account for the difficulty in controlling transmission. Fever, malaise, myalgia, and coryza usually initiate the clinical disease. Parotitis usually follows within a day. Virus can be isolated from Stensen's duct or from the saliva for another five to seven days. Parotitis may be bilateral, unilateral, or absent. Other organs are involved, either singly or in a variety of combinations, even in the absence of parotitis. Most frequently recognized is infection of other salivary glands, testes, ovaries, pancreas, breasts, and meninges. Rare reports of thyroiditis and arthritis need confirmation. Myocarditis has been documented but fetal mumps infection as a cause of endocardial fibroelastosis remains controversial. Mumps virus is shed in the urine for ten days or longer after the onset of symptoms, even in the absence of orchitis. In addition, hematuria, proteinuria, and a reversible reduction in creatinine clearance also suggest that nephritis may be common, although neither viral replication in the kidneys nor immune complex disease has been documented.

Mumps meningoencephalitis causes about 10 per cent of all cases of aseptic meningitis, and ranges in presentation from aseptic meningitis to encephalitis. The frequency of subclinical meningitis is unknown; however, an abnormal electroencephalogram (EEG) is uncommon with mumps parotitis in contrast to its frequency in uncomplicated measles. The cerebrospinal fluid in mumps meningitis tends to have lower sugars and higher polymorphonuclear leukocyte counts than in most other viral meningitides. Parotitis is absent in 25 to 50 per cent of cases of mumps meningoencephalitis, but when present, usually occurs three to ten days before CNS symptoms. Although parotitis occurs with equal frequency in both sexes, CNS mumps occurs three times more frequently in males. Sequelae are unusual after CNS mumps; the mortality from encephalitis is 1 per cent. Deafness due to auditory neuritis has been described, and may even occur without encephalitis.

Measles

After replication in the respiratory mucosa, measles virus is transported to lymphoid tissues, where it multiplies further, and then spreads via the bloodstream in leukocytes. During this asymptomatic incubation period, cytolysis, tissue necrosis, and inflammation are not manifest; the only histologic sign of local replication is multinucleate giant cells with intranuclear inclusions (Warthin-Finkeldey cells). Just before or during the prodrome, these are seen in the lymph nodes, tonsils, appendix, and exfoliated nasal mucosa. The usual incubation period of 10 to 12 days terminates with the onset of coryza, conjunctivitis, dry cough, sore throat, lymphadenopathy, headache, fever, and Koplik's spots. Koplik's spots are tiny erythematous patches with bluish-white apices on the buccal mucosa along the molar bite line. These spots contain the giant cells with eosinophilic intranuclear inclusions that are composed of measles virus-specific proteins as determined by immunofluorescence and viral nucleocapsids as determined by electron microscopy. The exanthem usually appears on the second day, beginning on the head, then spreading to the trunk, and finally to the extremities. The incubation period is usually two to three days shorter when the usual respiratory portal of entry is bypassed by parenteral inoculation with either the wild-type virus experimentally or the live attenuated vaccine. The incubation period is prolonged up to 3 weeks, especially in older people or in recipients of measles antiglobulin.

During the prodrome, virus is present in tears, nasal secretions, throat, urine, and blood. The viremia disappears with the onset of the rash, which is followed within two to three days by the appearance of serum antibody. A causal relationship between the appearance of immunoglobulins and the onset of symptoms has been proposed. Although the cutaneous and upper respiratory involvement is most easily recognized, viremia carries the virus elsewhere throughout the body. Respiratory complications include otitis media, croup, and pneumonitis with or without secondary bacterial pneumonia. An acute abdomen, easily mistaken for appendicitis, may precede the exanthem. Myocarditis and thrombocytopenic purpura are other rare complications.

Although EEG aberrations are detectable in most patients with measles, clinically apparent encephalomyelitis occurs in only one of every 1000 cases. The encephalomyelitis results in a 15 per cent mortality and 25 per cent residual morbidity. Most cases of encephalitis occur on about day six and usually within two weeks of the onset of the rash. Despite its temporal relationship to the rash, the pathology in encephalitis differs from that in other organs, in that inclusion bodies and giant cells are not common. Instead, a perivascular mononuclear infiltrate, demyelination, petechial hemorrhages, and microglial proliferation are seen. The overall mortality from measles can approach 10 per cent in malnourished populations, primarily because of secondary bacterial pneumonia. Infants and the elderly are most susceptible and mortality in them may exceed 25 per cent. In affluent societies the mortality rate approximates 0.1 per cent.

Subacute sclerosing panencephalitis (SSPE) is a rare (1 per 100,000 cases) but intriguing expression of measles infection. It causes an insidious intellectual, behavioral, and motor deterioration, which progresses to convulsions, coma, and death usually within a year of onset. Victims of SSPE manifest a characteristic EEG pattern, markedly elevated serum and CSF measles antibodies, and typical histologic findings distinct from the acute encephalitis of measles. Measles antigen is readily demonstrable in SSPE brains, although virus isolation usually requires extensive effort, including cocultivation techniques. SSPE occurs with a median latent period of 7 years after typical measles in a typical child, but occurring at a younger age and in a more rural area than average. Immune defects and virus variants have been proposed to explain the pathogenesis of SSPE, but documentation for either proposal is inconclusive.

Respiratory Syncytial Virus

The incubation period of RSV infection is four to five days. RSV replication appears to be restricted to the respiratory mucosa. It produces bronchiolitis, pneumonitis, and, less frequently, croup and apnea. These severe infections usually occur in children less than 6 months old who are experiencing their first RSV infection. The occurrence of severe RSV disease in the first 6 months, when maternal antibodies are still present, or in older children who have received inactivated vaccines suggests an immunopathologic basis. Occurrence of bronchiolitis during primary infection does not correlate with neutralizing or CF antibody levels of either the mother or the infant.

IMMUNITY

Parainfluenza Viruses

Severe life-threatening infection occurs in infants experiencing their first infection. IgA-neutralizing antibody appears in the nasal secretions after infection. The presence of this secretory antibody correlates well with protection from reinfection but it disappears in one to six months, leaving the child susceptible to reinfection despite substantial levels of serum neutralizing antibody. Past experience with the virus, however, does provide some protection; repeat infections are limited to the upper respiratory tract.

Life-threatening bronchiolitis and pneumonitis due to parainfluenza type 3, as with RSV, occur in infants, many of whom have maternally derived serum antibody. In contrast to RSV, administration of an inactivated vaccine (in one study) did not predispose to more severe disease, although it also failed to provide protection. In contrast to both parainfluenza type 3 and RSV, maternal antibody seems to give some protection from types 1 and 2 infection to infants in their first 4 months. Croup is seen after infection by these two viruses in some infants but also occurs in primary infection of older children.

Mumps Virus

CF antibody to the nucleocapsid protein (S antigen) is usually apparent at 1 week and peaks at 2 weeks postinfection; it is thus useful for laboratory diagnosis of recent infection. Antibody to the hemagglutinin glycoprotein (V antigen), as measured by hemagglutination inhibition or neutralization of viral infectivity, appears at two to three weeks and peaks at three to four weeks. This antibody is the best indicator of immunity because it is neutralizing antibody and because it is measurable for much longer than CF antibody. Reinfection is extremely rare; however, because asymptomatic infections are common and because other viruses may also cause parotitis, neither a history of parotitis nor its absence is useful in determining previous experience with mumps virus. In addition to mumps virus, parainfluenza viruses, enteroviruses, and influenza A virus have all been implicated as etiologic agents of parotitis.

The live attenuated mumps vaccine is grown in chick embryo cells. The vaccine causes an inapparent infection, seroconversion, and protection from mumps infection in over 95 per cent of seronegative recipients. The vaccine, which is inoculated subcutaneously, stimulates all commonly measured antibody responses but with lower mean titers than does natural infection. This vaccine does not produce detectable virus in saliva, viremia, viruria, or spread to contacts; however, vaccine virus has been isolated from an aborted placenta 10 days postinoculation of the mother. The vaccine was licensed in the United States in 1967, too recent for the long-term duration of immunity to be determined. However, studies to date are very encouraging. Killed mumps vaccines have been shown to protect only for short durations. The efficacy of passive immunization with convalescent serum or pooled γ-globulins has not been demonstrated.

Measles

Infection confers lifelong protection. The presence of CF, HI, or neutralizing antibody indicates immunity. Passive immunization with pooled IgG is usually effective if given within 1 week of exposure. It is indicated for pregnant women, neonates, and immunosuppressed patients who are exposed and susceptible. Although it prevents reinfection, serum antibody does not appear to be essential for recovery from primary infection or even constitute the sole defense against reinfection because immunoglobulin-deficient patients resolve a measles infection normally and they resist reinfection. In contrast, patients with defective cellular immunity suffer serious complications of measles infection. Enders demonstrated that the syndrome of giant cell pneumonia without the usual morbilliform rash was a consequence of measles infection in such patients.

Modified measles is an abbreviated and milder form of ordinary measles, often with a prolonged incubation period. Modified measles is seen in children passively immunized with prophylactic IgG or with transplacental IgG. *Atypical measles* results from infection with either wild type or vaccine measles virus at least 2 years after immunization with formalin-inactivated measles vaccine. It is characterized by fever, pneumonitis, pleural effusion, edema of extremities, and an atypical rash. The rash has a predilection for the extremities and tends to be discrete, often with an urticarial, petechial, or vesicular component in contrast to the typical blanching, erythematous, maculopapular, morbilliform rash of ordinary measles. Norrby has proposed that atypical measles infection is a consequence of the inability of the killed vaccine to stimulate antibody to the F (hemolysin-fusion) protein as does natural or live attenuated vaccine infection.

The live attenuated vaccine, grown in chick embryo fibroblasts at reduced temperature, has supplanted the killed vaccine. The live vaccine is >95 per cent effective. Failures have been attributed to vaccine inactivation and to administration to infants with residual transplacental immunity, usually in infants less than 1 year old. Ten per cent of vaccine recipients have a fever or a mild rash 10 days postinoculation. There is little, if any, vaccine virus excretion and no transmission has been documented. Mean antibody titers tend to be lower after vaccination than with natural infection; however, many vaccinees subsequently exposed to measles appear to have asymptomatic antibody boosts. The measles vaccine is administered in one dose subcutaneously. It can be given in combination with the mumps and rubella live attenuated vaccines with no apparent reduction in the efficacy of any of the three components.

Respiratory Syncytial Virus

Newborns enjoy relative freedom from infection for two to four weeks, perhaps because exposure is limited or immunity is passed from the mother. For the next six months, with a peak in month two, susceptibility to life-threatening infection increases with no clear relation to levels of passively acquired antibody. Active immunity from primary infection does not prevent reinfection during epidemics, but reinfections are limited to the upper respiratory tract. The role of the secretory immune system has not been evaluated in infants. Adults are all seropositive, and elevated neutralizing antibody in their nasal secretions is correlated with resistance to infection.

Formalin-inactivated virus not only fails to

protect, but places children as old as two years at risk for life-threatening bronchiolitis, a syndrome usually not seen after the sixth month in unimmunized children. Temperature-sensitive mutants of respiratory syncytial virus are being evaluated in the hope of developing an attenuated, live virus vaccine.

LABORATORY DIAGNOSIS

Parainfluenza Viruses

When respiratory tissue or secretions are inoculated into cell cultures, the presence of parainfluenza virus may be recognized by cell death or syncytium formation; however, hemadsorption or specific immunofluorescence is more sensitive because isolates in early passage may produce little or no visible cell injury. The parainfluenza virus isolate can be typed by hemadsorption inhibition with type-specific antiserum, by HI or CF characterization of progeny virus from the tissue culture supernatant, or by immunofluorescence of infected tissue culture cells. Rapid diagnosis by identification of specific viral antigen by immunofluorescence in exfoliated nasopharyngeal cells is possible; however, highly specific immune reagents and appropriate controls are required. Persistent infection with simian virus 5 (related to type 2) in simian cell cultures, and of that virus or Sendai virus (related to type 1) in rodent cultures is a common source of error. Primary human embryonic or simian kidney cell cultures are the most sensitive cell lines for isolation of human parainfluenza viruses; HeLa, HEp2, and WI-38 cell lines are also effective for types 2 and 3. Recently a continuous monkey kidney cell line, LLC-MK2, along with trypsin in the culture medium to cleave and activate the F glycoprotein, has been shown to be a sensitive line for the isolation and propagation of parainfluenza viruses. CF antibody responses may not occur in mild disease; consequently, the neutralization or HI tests are more sensitive. In addition, cross-reactive rises in antibody to different types may occur in reinfection.

Mumps

Although the typical syndrome of parotitis has also been attributed to parainfluenza viruses, coxsackieviruses, and influenza A virus, laboratory diagnosis is indicated when complications occur. Virus can be isolated in the first few days of illness from saliva, parotid duct secretions, urine, or cerebrospinal fluid if meningitis is present. Viruria may persist for over two weeks. Although the virus was originally isolated by inoculation into the amniotic cavity of eight day old embryonated chicken eggs, primary or continuous human and primate cell lines are more sensitive. Cell rounding and syncytia are seen in infected cell monolayers, but hemadsorption and immunofluorescence are more sensitive indicators of infection. Prompt inoculation of a specimen is important because mumps virus infectivity is unstable both at room temperature and at 4° C. As discussed in the section on immunity, the CF antibody response is more prompt but less durable than the HI and neutralization response.

Measles

Measles usually produces a distinct syndrome even when the typical exanthem cannot be easily recognized — for example, in dark-skinned people. Confirmation is indicated in the atypical or complicated case and can be accomplished by isolation of virus from respiratory secretions, blood, or urine with embryonated eggs or cell culture. The virus is excreted in the respiratory secretions, tears, and urine during the prodrome and for about two days after the appearance of the rash. The virus is also present in blood leukocytes, lymph nodes, spleen, kidney, skin, and lungs. Measles isolation in cell culture, first described by Enders and Peebles in 1954, provided the technology for the most sensitive isolation technique, and thus enabled us to obtain the data that underlie our current concepts of measles immunology, pathogenesis, and epidemiology. Primary human embryo or rhesus monkey kidney cell cultures are the most sensitive; cytopathic effects become apparent in one to two weeks. Isolation may require holding cell cultures for a month with "blind" passaging to permit adaptation of the isolate to cell culture. The classic cytopathology — multinucleate giant cells and intranuclear inclusions — may be apparent only with passaging. Earlier detection of infected cell cultures is possible with specific immunofluorescence or hemadsorption with monkey erythrocytes.

Respiratory Syncytial Virus

RSV can be isolated from the respiratory secretions by using susceptible host cell cultures, such as primary monkey or bovine kidney, HeLa, HEp-2, or WI-38. Syncytia and eosinophilic cytoplasmic inclusions are produced, but specific confirmation is accomplished by CF of progeny virus or by immunofluorescence of infected cells. Immunofluorescence of exfoliated respiratory epithelial cells is an effective method for the rapid diagnosis of RSV infection. Because RSV is an extremely labile virus, isolation is maximized by collecting good quantities of respiratory secretions — for example, nasopharyngeal washings rather than swabs, and by promptly inoculating cell cultures without prior storage of the specimen. Infants suffering primary infection may shed virus for four to six weeks, even after apparent recovery.

CF is the most readily available and least cumbersome serologic test. However, the neutralization test is more sensitive, with plaque reduction being slightly more sensitive than end-point neutralization of cytopathic effect.

EPIDEMIOLOGY

Parainfluenza

As with most respiratory agents, the risk of infection increases in those with contact with young children. Thus, high risk situations include large families, child care centers, schools, and pediatric wards of hospitals. Type 3 is the most prevalent of the parainfluenza types. Fifty per cent of children have serologic evidence of infection during their first year and 95 per cent by age six. Type 3 infections occur endemically throughout the year, whereas type 1 and, to a lesser extent, type 2 occur in epidemics during the fall or winter, often in a biannual pattern. Only respiratory syncytial virus exceeds type 3 in frequency as a cause of bronchiolitis and pneumonitis in infants. Type 1 and, to a lesser extent, type 2 are largely responsible for croup (acute laryngotracheobronchitis) in infants.

Mumps

Mumps is not as contagious as measles or varicella; consequently, there is a fair proportion of seronegative young adults, perhaps 10 per cent. This percentage is increased in isolated areas; for example, more than 80 per cent of certain populations of Alaskan Eskimos may be seronegative. Infection is endemic throughout the year with a peak in the late winter and early spring in nontropical climates. Epidemics occur in congregations of children and young adults, especially schools and military situations.

The incidence of encephalitis is 2.5 per 1000 cases of reported mumps; the incidence of fatal mumps is approximately 0.25 cases per 1000 cases of reported mumps. Mumps infections are probably underreported, however, because approximately one third of all infections are subclinical. The age incidence of encephalitis correlates well with the age incidence of parotitis; however, fatalities are disproportionately increased in infants and older adults.

Measles

Measles represents an ideal subject for epidemiologic study because it is highly contagious, there is a low proportion of subclinical infection, infection confers immunity to reinfection, and there is no evidence to suggest that the virus is heterogeneous with regard to either antigens or virulence. There is no apparent animal reservoir; thus, maintenance requires transmission to susceptible, nonimmune people. Consequently, the virus is best maintained in populous, developing areas, where annual late winter and early spring epidemics are seen, with cases detectable throughout the year. Epidemics occur with lower attack rates and at longer intervals in less populous and more economically developed areas. The virus disappears periodically from sparsely populated and isolated areas, which are subject to outbreaks when the virus is reintroduced from the outside after a critical number of nonimmune individuals have accumulated.

Measles is usually contracted by preschool children in populous, developing areas. In developed countries the disease is most common in elementary school children. In a remote Eskimo village, measles occurs at the age at which the virus is introduced from the outside. Thus, before measles vaccine, children rarely reached adulthood without measles infection except in remote and sparsely populated areas. Since the introduction of measles vaccine — in the United States, for example — the incidence of measles has diminished to 1 per cent of its previous incidence and the disease is occurring in somewhat older individuals. Infection is occurring in older individuals because unvaccinated children tend to be older and because there are fewer susceptibles to maintain annual epidemics.

Respiratory Syncytial Virus

RSV consistently produces annual winter epidemics between January and March in temperate climates in the northern hemisphere. Epidemics during the rainy summer season have been described in Trinidad. These epidemics attack approximately one third of seronegative children, so that by age five, 95 per cent of children are seropositive. The mechanism of "overwintering," the maintenance of the virus between epidemics, is an enigma. Low rates of endemic infection between epidemics have not been documented as has been done with influenza. In primary infection, viral excretion often persists for one or two months after the resolution of symptoms; however, prolonged excretion after reinfection of either children or adults has not been documented. Reinfection occurs in as many as 20 per cent of children and adolescents annually, diminishing to an annual rate of 3 to 5 per cent in adults. The highest risk of infection occurs in nurseries, institutions, and pediatric wards of hospitals where there are high concentrations of infants.

References

Stuart-Harris, C. H., and Schild, G. C.: Influenza: The Viruses and the Disease. Littleton, Massachusetts, Publishing Sciences Group, 1976.

Kilbourne, E. D. (ed.): The Influenza Viruses and Influenza. New York, Academic Press, 1975.

Kaplan, M. M., and Webster, R. G.: The epidemiology of influenza. Sci Am 237:88, 1977.

Choppin, P. W., and Compans, R. W.: Reproduction of paramyxoviruses. In Fraenkel-Conrat, H., and Wagner, R. R. (eds.): Comprehensive Virology. New York, Plenum Publishing Corp., p. 95, 1975.

Evans, A. S. (ed.): Viral Infections of Humans, Epidemiology and Control. New York, Plenum Medical Book Company, 1976.

PICORNAVIRUSES 60

Nathaniel A. Young, M.D.

PHYSICAL STRUCTURE AND BIOCHEMICAL PROPERTIES

Picornaviridae is a large family of animal viruses named for their diminutive size, 20 to 30 nm, and nucleic acid type, RNA (Cooper et al., 1978). Picornavirus particles (virions) are icosahedral and consist of two structural moieties: (1) an outer protein shell, or capsid, comprising 70 per cent of the mass, and (2) an inner strand of RNA, which is 30 per cent of the mass. Lacking a lipid envelope, the "naked" virions are not affected by ether, alcohol, or other lipid solvents, although they are readily inactivated by phenol, formaldehyde, or ionizing radiation, which damages the capsid or the enclosed nucleic acid.

The capsid consists of 60 identical subunits, or capsomeres, which are synthesized from equimolar amounts of four different virion polypeptide chains, VP1–4. Three of these polypeptides are large (molecular mass 20,000 to 40,000 daltons, depending on the polypeptide and the particular picornavirus), and the fourth is small (5000 to 10,000 daltons). The capsid not only protects the enclosed viral genetic apparatus but also specifies which cells the virus can infect, since attachment of virus to specific receptors on the cell surface is a consequence of the chemical structure and surface configuration of the capsomeres. This "lock and key" fit of viruses with receptors is probably responsible for the host and tissue tropisms of individual picornaviruses.

The RNA of picornaviruses is linear, single-stranded, and nonsegmented, and has a molecular mass of 2.6 million daltons. After virions adsorb to susceptible cells, the RNA is uncoated and released into the cytoplasm, where it serves as a template for the synthesis of additional RNA. Newly synthesized RNA in turn may be either translated into viral proteins or encapsidated to form progeny virions. A distinctive feature of picornavirus RNA is its function as a monocistronic messenger; protein synthesis on the polyribosome is initiated at a single point on the RNA, from which the entire genome is translated into one large "polyprotein" of molecular mass greater than 200,000 daltons. Nascent chains of the polyprotein are then enzymatically cleaved to form individual polypeptides (Jacobson and Baltimore, 1968).

Also encoded by the viral genome are other noncapsid polypeptides. These include (1) enzyme(s), which catalyze the synthesis of viral RNA, and (2) VPg, a small polypeptide of uncertain function, which is apparently unique to picornaviruses (Cooper et al., 1978). VPg is covalently linked to the 5′ end of genomic RNA, but it is not attached to viral messenger RNA. It perhaps functions as a primer for the synthesis of new viral RNA strands, or, alternatively, it may regulate the processing of viral RNA molecules. According to the latter hypothesis, VPg must be removed from RNA molecules before they can be translated into protein and must be added to those RNA molecules destined to become encapsidated as progeny virions (Nomoto et al., 1977).

CLASSIFICATION OF PICORNAVIRUSES

Picornaviruses are subdivided into four genera* — *Enterovirus, Cardiovirus, Rhinovirus,* and *Aphthovirus* — primarily on the basis of sen-

*The four genera have been proposed by the Study Group on Picornaviridae (Cooper et al., 1978) but have not yet been formally recognized by the International Committee on Taxonomy of Viruses.

sitivity to acid, buoyant density of the virion in CsCl, and clinical manifestations (Cooper et al., 1978) (Table 1). Only enteroviruses and rhinoviruses are major human pathogens.

Enteroviruses multiply throughout the alimentary tract and cause a variety of infections, which range in severity from asymptomatic to fatal. Among their distinctive properties are a buoyant density in CsCl of 1.33 to 1.35 g/cm^3 and maintenance of infectivity at a pH as low as 3. Because of their acid stability, enteroviruses that have undergone limited replication in the oropharynx survive transit through the stomach and implant in the lower intestinal tract, where they undergo more extensive multiplication. Enteroviruses are also relatively resistant to thermal inactivation at ambient temperatures, especially in the presence of 1 M $MgCl_2$. This chemical is therefore added to live poliovirus vaccines to enhance their stability, especially under field conditions or in tropical regions lacking refrigeration.

Rhinoviruses inhabit the upper respiratory tract and are the principal recognized etiologic agents of the common cold. Unlike enteroviruses, rhinoviruses are acid labile. They begin to lose infectivity at pH 6 and are completely unstable at pH 3. They are further distinguished from enteroviruses by their optimal temperature of replication (33° C) and by their higher buoyant density in CsCl. Preferential replication at lower than body temperature probably reflects their adaptation to the nasal passages.

Cardioviruses, examples of which include encephalomyocarditis virus and Mengo virus, have been recognized only rarely as causes of human diseases (Gajdusek, 1955; Tesh, 1978). Although they have been recovered chiefly from rodents, their natural host(s) is uncertain (Tesh, 1978; Tesh and Wallace, 1978). Their buoyant density and optimal temperature of replication are similar to those of enteroviruses. However, although they are usually stable over the broad range of pH 4 to 10, they differ from enteroviruses in being unstable at pH 6 in the presence of 0.1 M chloride ions.

Aphthoviruses, named for the vesicular lesions that they produce in cloven-footed animals, are an important cause of economic loss in the cattle industry. Their buoyant density is intermediate between those of rhinoviruses and enteroviruses/-cardioviruses. They resemble rhinoviruses in their instability in acid, but at pH 5 to 6 in solutions of high ionic strength they are relatively stable.

Species of the four picornavirus genera are distinguished immunologically, usually by the ability of specific antiserum to neutralize only the

TABLE 1. A Classification of the Family Picornaviridae

GENUS	SUBGENUS[a]	NUMBER OF IMMUNOTYPES (SPECIES)	DISTINCTIVE PROPERTIES: Infectivity at pH 3	Infectivity at pH 6	Density in CsCl g/cm^3	Optimal Growth Temperature
Enterovirus	Poliovirus	3	+	+	1.33–1.35	37° C
	Coxsackievirus A	23[b]				
	Coxsackievirus B	6				
	Echovirus	30[c]				
	Enterovirus	4				
Cardiovirus	–	1	+	–[d]	1.34	37° C
Rhinovirus	–	>110	–	±[e]	1.38–1.41	33° C
Aphthovirus	–	7	–	±[f]	1.43–1.45	37° C

[a]The designation subgenus has not been proposed by the International Committee on Taxonomy of Viruses but is used here for convenience simply to indicate a taxon smaller than a genus but larger than a species.

[b]Coxsackieviruses A1–A24; coxsackievirus A23 has been reclassified as echovirus 9.

[c]Echoviruses 1–34; echoviruses 1 and 8 are identical; echovirus 10 has been reclassified as reovirus 1, and echovirus 28 as rhinovirus 1A; echovirus 34 is a variant of coxsackievirus A24.

[d]Cardioviruses are stable at pH 4–10, but in the presence of chloride ions they are unstable in the region of pH 6.

[e]Rhinoviruses are borderline stable at pH 6 and unstable at lower pH.

[f]Aphthoviruses are unstable below pH 7 at low ionic strength but stable at pH 5–6 in high ionic strength.

homotypic virus. Sixty-six human enterovirus species (immunotypes) and more than 100 human rhinoviruses have been identified. Not included in Table 1 are many immunotypes of simian, porcine, bovine, and insect enteroviruses, as well as bovine, equine, and feline rhinoviruses.

ANTIGENIC STRUCTURE OF PICORNAVIRUSES

Enteroviruses and rhinoviruses exhibit two distinct antigenic reactivities, C and D, associated with different physical forms of the virus particle (Roizman et al., 1958; Hughes et al., 1974). Empty capsids are C antigenic and give rise to group-specific reactions detectable by complement fixation (CF) or immunoprecipitin reactions. In intact virions, the C antigen is ordinarily masked; instead, the D antigen, which elicits type-specific neutralizing antibodies, is expressed. After infection, serum contains antibodies to both C and D antigens, but when the serum is absorbed with empty capsids it retains only D-specificity. Mild denaturing conditions, such as heating at 50° C or ultraviolet irradiation, convert the reactivity of virions from D to C. When enteroviruses are completely disrupted or when rhinoviruses are subjected to acidity (pH 3), the integrity of both C and D antigens is lost.

D and C antigenic reactivities are explained by changes in the conformation and molecular architecture of the virion. VP4 is essential for D reactivity (Breindl, 1971). When virions are heated, they lose VP4 and RNA and become noninfectious. Empty capsids, consisting of VP1 and VP3 (and some VP2), retain C antigenicity, but this reactivity is also lost when capsids are further disrupted into their constituent capsomeres (Katagiri et al., 1971).

SUBCLASSIFICATION AND HOST RANGE OF PICORNAVIRUSES

Enteroviruses

Historically, human enteroviruses have been subclassified into polioviruses, coxsackieviruses, and echoviruses based on antigenic relationships, difference in host range, and types of disease that result. *Polioviruses,* the first to be recognized, produce characteristic lesions when inoculated into the central nervous system of primates (see Chapter 173) and replicate only in primates or primate cell cultures. In contrast, most *coxsackieviruses* are not readily isolated in cell cultures and are only occasionally neuropathogenic for monkeys. However, they cause paralysis and death when inoculated into suckling mice. This property was responsible for their detection and differentiation from polioviruses in 1958, when they were first recovered from the feces of children with a poliomyelitis-like syndrome in the village of Coxsackie, New York (Dalldorf and Sickles, 1948). Shortly after their discovery, it was recognized that some coxsackieviruses, designated group A, produced widespread lesions and flaccid paralysis in skeletal muscle of mice, whereas others, designated group B, produced only focal myositis but more widespread visceral lesions involving the heart, fat, pancreas, and central nervous system, causing spastic paralysis. Moreover, the group B coxsackieviruses, unlike most of those of group A, could be propagated in primate cell cultures (Melnick, 1954). Further attempts to recover viruses from the feces of healthy children led to the discovery in 1949–1950 of still other agents that produced cytopathic effects in primate cell cultures but failed to cause disease in suckling mice or in the central nervous system of primates (Melnick and Agren, 1952; Melnick, 1954; Ramos-Alverez and Sabin, 1954). These agents, initially considered "orphan" viruses because they were unrelated to any disease, were called *echoviruses* (*e*nteric *c*ytopathic *h*uman *o*rphan).

During the quarter century that has elapsed since techniques for detecting enteroviruses were developed, three immunotypes of polioviruses have been recognized, as well as 23 group A coxsackieviruses, 6 group B coxsackieviruses, and 30 echoviruses. Many enteroviruses cannot be subclassified unambiguously. For example, some have the host range of a coxsackievirus but are antigenically related to echoviruses. Therefore, new agents are now designated simply as *enteroviruses* with serial numbers (Rosen et al., 1970). Since adoption of this simplified taxonomic scheme four new immunotypes, enteroviruses 68 to 71, have been recognized. However, because of established precedent, the first 67 immunotypes are still subclassified as polioviruses, coxsackieviruses, or echoviruses.

Rhinoviruses

There are 89 numbered species of human rhinoviruses. Additional immunotypes awaiting formal recognition have brought the total to at least 110. Human rhinoviruses are infectious only for primates or primate cell cultures, and many can be propagated in vitro only in human embryonic fibroblasts. A few fastidious strains grow only in organ cultures of embryonic human tracheal epithelium. Rhinoviruses have been classified as H or M strains according to their growth in cells of human or monkey origin, but this subdivision has little clinical or epidemiologic significance.

RELATIONSHIPS AMONG PICORNAVIRUSES

Although each immunotype is distinct, human enteroviruses and rhinoviruses are related genetically and immunologically. Genetic relationships among enteroviruses are shown by reciprocal hybridization of their nucleic acids (Young, 1973) or by oligonucleotide "fingerprinting" (Frisby et al., 1976). Different immunotypes within a major group, such as the three species of poliovirus, share 30 to 50 per cent of their nucleotide sequences, whereas homologies among viruses of different groups, such as group A coxsackieviruses and echoviruses, are generally less than 20 per cent. At least 5 per cent of the RNA sequences are common to all human enteroviruses. On the other hand, less than 4 per cent homology has been found among the RNA of three rhinovirus species, suggesting that they are genetically very diverse (Yin et al., 1973).

D antigens of each enterovirus immunotype stimulate type-specific neutralizing antibodies, but immunologic cross-reactions are elicited by the C antigens of related viruses. Heterotypic reactions are most readily demonstrated in the serum of individuals who have previously experienced sequential infections with other enteroviruses. For example, a child whose first coxsackievirus infection is with immunotype B5 usually elaborates neutralizing and CF antibodies that react only or predominantly with homotypic antigens; on the other hand, if he has previously been infected by another group B coxsackievirus, titers of CF antibodies against all six group B coxsackieviruses may rise, even though neutralizing antibody titer rises will be greatest between certain pairs of viruses, such as polioviruses 1 and 2, coxsackieviruses A3 and A8, and echoviruses 6 and 30. Furthermore, a distant immunologic relationship between human enteroviruses and rhinoviruses can be demonstrated in immunoprecipitin reactions or by immune electron microscopy with undiluted human serum (Hughes et al., 1977). The weak cross-reaction of enteroviruses and rhinoviruses probably reflects the broadened antibody response resulting from multiple picornavirus infections, because no cross-reaction can be demonstrated with serum from animals hyperimmunized with a single picornavirus antigen.

PATHOGENESIS

Enteroviruses

The pathogenesis of poliovirus infections has been extensively studied and is discussed in Chapter 173. Data for other enteroviruses, meager because there are no readily available experimental models, indicate a basic similarity of the major pathogenetic events, excluding those involving the central nervous system. Briefly recapitulated, these events are as follows.

Infection is primarily a result of ingesting fecally contaminated material. In individuals lacking type-specific antibodies, enteroviruses implant and replicate in susceptible tissues of the gut, which probably include mucosal epithelial cells and lymphoid tissue of the lamina propria. Pathologic lesions are not apparent in these sites. Although most enteroviruses multiply in both the oropharynx and the distal small bowel or colon, evidence favors the view that replication in the lower gut is far more efficient (Ogra and Karzon, 1971): (1) Large doses of oral polio vaccines result in virus shedding in both oropharyngeal secretions and feces, whereas with small doses only fecal shedding occurs; (2) virus titers are substantially higher in feces, often 10^6 infectious units per gram; (3) virus can be recovered from the oropharynx only for several days, rarely as long as three weeks, whereas fecal excretion of virus is more sustained, often five to six weeks or longer; (4) although some enteroviruses — for example, group B coxsackieviruses — are often recovered from both ends of the alimentary canal, echoviruses are commonly recovered only in the feces (Kogon et al., 1969).

After multiplying in the submucosal lymphatic tissues, enteroviruses reach the cervical and mesenteric lymph nodes. From these sites small quantities of virus escape into the bloodstream ("minor viremia") and are disseminated to the reticuloendothelial tissues, such as the liver, spleen, and bone marrow. All the replicative events up to this point produce no symptoms, and in most individuals the infection is contained by host defense mechanisms without further progression. However, in a few infected persons, additional replication in reticuloendothelial tissues produces heavy, sustained shedding of virus into the blood ("major viremia"). This event is temporally associated with the "minor illness" of poliomyelitis and probably the nonspecific febrile illnesses caused by other enteroviruses. Sustained viremia is also responsible for dissemination of virus to target organs, such as the meninges, skin, and heart. In these tissues the virus produces inflammatory lesions, often accompanied by necrosis.

Rhinoviruses

Rhinoviruses implant and replicate in epithelial cells of the nasal mucosa, producing hyperemia and edema of the mucous membrane and a seromucinous exudate. A scant infiltrate of neutrophils, eosinophils, and mononuclear inflammatory cells is present beneath the epithelium, and mucin secretion by goblet cells is increased. Columnar epithelial cells exhibit impaired mucociliary clearance, become necrotic, and slough. Rhinovirus replication can be detected within 24 hours after implantation, often several days before the onset of symptoms, and reaches a peak in two to three days. Excretion generally ceases by seven days, but in a few individuals it persists up to two weeks (Douglas et al., 1966; Kettler et al., 1969). Rhinoviruses are shed principally from the nose, and the concentration of virus in the nasal mucus is 10- to 100-fold higher than it is in saliva or oropharyngeal secretions (Hendley et al., 1973). The precise anatomic range of infected tissues has not been defined; it has not been established whether infection extends beyond the nose or whether virus in the oropharynx comes principally from contamination from the nose. In normal individuals, as well as in patients with chronic bronchitis, rhinovirus infections have been associated with abnormalities of pulmonary function (Cate et al., 1973), but this may be a consequence of reflex bronchoconstriction rather than of direct viral invasion of the lower respiratory tract.

Although rhinoviruses usually implant in the nose, infection can be initiated experimentally by dropping virus on the conjunctiva. Rhinoviruses introduced into the oropharynx cause infection only with relative difficulty (Hendley et al., 1973), and they are unable to replicate in the lower intestinal tract, even when the acidic environment of the stomach is bypassed and virus is artificially introduced into the small bowel (Cate et al., 1967). Among the possible reasons for failure to replicate in the intestine are lack of appropriate viral receptors on mucosal epithelial cells and the higher temperature at this site compared with the nasal passages. Neither viremia nor dissemination of rhinoviruses beyond the respiratory tract has been documented.

IMMUNITY

Type-specific neutralizing antibodies in the serum and in mucosal secretions are the factors most responsible for acquired immunity against picornaviruses. Cell-mediated immunity has not been thoroughly studied.

In the nasopharynx and the intestinal tract, secretory IgA antibodies directed against specific immunotypes of polioviruses and rhinoviruses are elicited by the initial infection with these agents (Ogra and Karzon, 1971; Cate et al., 1966; Gwaltney, 1976). Depending on the titer of secretory antibodies, either reinfection by the same immunotype is prevented or the outcome of infection is modified. High titers of IgA antibodies prevent implantation and replication of virus in the gut and pharynx. In contrast, reinfection is possible when antibody titers are low or when the virus challenge is massive. Reinfections are unaccompanied by illness and give rise only to brief excretion of virus (Kogon et al., 1969). IgA antibodies to poliovirus may also be acquired passively in milk by breast-fed infants whose mothers are immune; these antibodies interfere with a "take" by live, oral poliovirus vaccines (OPV) and may also protect against natural infection during early infancy (Warren et al., 1964).

IgM and IgG neutralizing antibodies to enteroviruses begin to appear in the serum as early as one to three days after infection (Ogra and Karzon, 1971). IgA antibodies are usually not detected in serum until two to six weeks; they are generally of low titer and do not develop in all individuals. IgM antibodies usually persist for two to three months, but serum-neutralizing antibodies of the IgG and IgA classes are present for many years, possibly for life. Although serum-neutralizing antibodies do not prevent alimentary infection by enteroviruses, they effectively block viremia and hematogenous dissemination to target organs such as the meninges, skin, and heart. Parenterally administered pooled gamma globulin produces trace levels of serum antibodies to poliovirus that protect against paralysis if given before exposure (Stevens, 1959). Protection against paralytic poliomyelitis is also afforded by either OPV or parenterally administered inactivated polio vaccine (IPV) because both types of vaccine stimulate serum-neutralizing antibodies in a high proportion of recipients. However, OPV has the additional advantage of interrupting the transmission of wild polioviruses in the community, because the attenuated viruses replicate in the alimentary tract and stimulate secretory immunity, thereby preventing subsequent reinfection as well as central nervous system illness (Melnick, 1978).

Rhinoviruses, like enteroviruses, stimulate neutralizing antibodies in both the serum and the nasopharyngeal secretions of up to 80 per cent of infected individuals. Antibody responses at these sites are closely linked, even after nasal or parenteral immunization with inactivated virus. It has not been established unequivocally, therefore, whether secretory or circulating antibodies are more important in type-specific immunity to the common cold (Gwaltney, 1976). Because antibodies in either serum or nasal secretions are not detected until approximately two weeks after infection, when virus shedding has already ceased, it is likely that recovery from acute infection is mediated by other mechanisms, possibly interferon.

Complement-fixing antibodies stimulated by enteroviruses or rhinoviruses are detectable for one to five years after infection. They do not appear in all individuals, and it is unclear whether they are protective.

LABORATORY DIAGNOSIS

Appropriate specimens for isolation of enteroviruses include, depending on the clinical syndrome, throat washings or swabs, feces, cerebrospinal fluid, vesicle fluid from skin lesions, pericardial fluid, conjunctival swabs (acute hemorrhagic conjunctivitis), and autopsy tissues such as brain, spinal cord, and myocardium. Isolation of an enterovirus from one or more of these sites is usually convincing evidence of an etiologic relationship to illness, but isolation from feces or throat secretions, especially in infants during the summer months when enteroviruses are prevalent, may be only circumstantial evidence of such a relationship because etiologically unrelated intercurrent infection may be present. Isolation of a rhinovirus from nasopharyngeal secretions likewise must be interpreted with regard to the possibility that other, unidentified respiratory pathogens may be present.

Specimens for enterovirus isolation optimally should be transported on wet ice (4° C) to the virus laboratory, where they should be inoculated into human or primate cell cultures, such as rhesus monkey kidney or human embryonic kidney. For the diagnosis of illness due to group A coxsackieviruses, infant mice must also be inoculated. Rhinoviruses can usually be isolated in human diploid fibroblast cell cultures, such as WI 38, although the expense of this procedure is rarely warranted in patients with the common cold. Specimens that cannot be inoculated until many hours after their collection should be frozen, preferably at −70° C or lower. However, the characteristic stability of enteroviruses sometimes permits their recovery even from unrefrigerated specimens that have been mailed to the laboratory. Characteristic cytopathic effects in cell cultures are commonly observed after two to five days and permit a presumptive diagnosis of picornavirus infection. The specific immunotype is identified by the use of polyvalent immune sera in an "intersecting pool" scheme (Lim and Benyesh-Melnick, 1960).

For serodiagnosis, the acute phase serum should be collected as soon after onset of illness as possible, and the convalescent serum should be collected approximately two to three weeks later.

In the individual patient, the diagnosis of enterovirus or rhinovirus infection is made most conveniently by virus isolation. Rising titers of neutralizing antibodies in paired acute and convalescent sera are confirmatory, but serologic diagnosis generally cannot substitute for virus isolation. The reason is that group-reactive antigens encompassing all immunotypes are lacking, and it is not practical to perform serologic tests with type-specific antigens individually for more than 60 enteroviruses or 100 rhinoviruses. The physician should therefore not expect the virus diagnostic laboratory to perform antibody "screens" unless a virus has been isolated from the patient, the particular epidemic immunotype is known or suspected, or the type of clinical illness suggests only a limited number of probable immunotypes. For example, serologic diagnosis of sporadic cases of aseptic meningitis or the common cold is not feasible, but it may be reasonable to measure antibodies to the three poliovirus immunotypes in paralyzed individuals or to coxsackieviruses, B1–B5 in patients with myocarditis.

EPIDEMIOLOGY

Most enteroviruses and rhinoviruses are distributed worldwide. Seasonal variation in prevalence and patterns of transmission of these two virus groups are dissimilar (Melnick, 1976; Gwaltney, 1976).

Enteroviruses

In temperate climates, enteroviral infections are most common in the summer and early autumn, whereas in tropical climates infection occurs year round. Virus is spread chiefly by the fecal-oral route directly from person to person or by fomites, although spread by respiratory secretions may play a lesser role. In economically developed countries, usually only one or two immunotypes are prevalent in the community each season; in contrast, children living in urban areas with low standards of sanitation are often multiply infected. Approximately 50 to 80 per cent of enteroviral infections are asymptomatic, and many of the remainder are characterized by undifferentiated febrile illnesses lasting only a few days, often accompanied by upper respiratory tract symptoms. The so-called "characteristic" enterovirus syndromes — such as aseptic meningitis, myopericarditis, and exanthems — are in fact unusual manifestations of infection. Enteroviruses are most efficiently disseminated by infected children less than 2 years old. Introduction of virus into the household by one family member results in high rates of infection among others lacking type-specific neutralizing antibodies; secondary attack rates of approximately 92 per cent for polioviruses, 75 per cent for coxsackieviruses, and 50 per cent for echoviruses have been observed in family surveillance studies in the city of New York (Kogon et al., 1969).

Although the epidemiology of most enteroviruses is basically similar, patterns of infection with some immunotypes are distinctive. Improved standards of hygiene and widespread use of poliovirus vaccines have resulted in striking changes in the epidemiology of poliomyelitis, discussed in Chapter 173. Enterovirus 70 and coxsackievirus A24, the principal etiologic agents of acute hemorrhagic conjunctivitis, are probably transmitted from hand to eye by conjunctival secretions; replication of these viruses in the alimentary tract, if it exists at all, appears to be limited. Coxsackievirus A21 also is shed primarily from the upper respiratory tract, where it produces rhinovirus-like illness.

Control of nonpolio enterovirus infections and illness is best effected by hygienic measures and improvements in sanitation. Isolation of patients with enteroviral illness is generally not helpful in controlling epidemics because simultaneously there is a large reservoir of unidentified, asymptomatically infected individuals who excrete virus. Because nonpolio enteroviral illness is rarely life-threatening and because there are so many immunotypes, control by vaccines is not practical.

Rhinoviruses

The peak incidence of rhinovirus infections in temperate climates occurs in the colder months, especially the early autumn and spring. In the tropics, peak activity occurs in the rainy season. Crowding indoors and the return of children to school are believed to be among the factors that account for this periodicity. Contrary to popular belief, there is no evidence that susceptibility to the common cold is enhanced by chilling of the body during the winter months. Rhinovirus infections experimentally induced at cold ambient temperatures do not occur with greater frequency of severity (Douglas et al., 1968).

Rhinovirus infections spread primarily in the home and in the classroom. As with enteroviruses, secondary attack rates of 50 to 80 per cent are common among susceptible family members, especially young children. Transmission of the common cold occurs with two to five day intervals between cases and is most efficient when there is close contact between individuals. Although it has long been assumed that rhinoviruses are disseminated by respiratory droplet nuclei, recent data suggest that spread may occur more commonly by transfer of nasal mucus to the hands of infected individuals, who transmit virus via fomites or hand contact to others, who in turn autoinoculate the nose or conjunctiva (Gwaltney et al., 1978).

Multiple rhinovirus immunotypes often circulate concurrently. Antibody prevalence rates increase with age, beginning in early childhood and peaking in young adults, who commonly have serum antibodies to approximately half of the recognized rhinovirus immunotypes. Most infections are symptomatic, the ratio of apparent to inapparent infections being approximately 3:1.

Practical measures for the control of rhinovirus infections have not been established. If additional studies confirm that transmission occurs primarily from hand to nose rather than by inhalation of respiratory droplets, then handwashing or avoidance of hand contact with individuals with colds may reduce contagion. Prophylaxis with large doses of ascorbic acid is controversial, but this agent appears to be marginal at best in reducing the incidence of severity of colds. Specific chemotherapeutic agents are not available. Control by vaccines is not practical because of the large number of immunotypes.

References

Breindl, M.: VP4, the D-reactive part of poliovirus. Virology 46:962, 1971.

Cate, T. R., Douglas, R. G., Jr., Johnson, K. M., Couch, R. B., and Knight, V.: Studies on the inability of rhinovirus to survive and replicate in the intestinal tract of volunteers. Proc Soc Exp Biol Med 124:1290, 1967.

Cate, T. R., Roberts, J. S., Russ, M. A., and Pierce, J. A.: Effects of common colds on pulmonary function. Am Rev Respir Dis 108:858, 1973.

Cate, T. R., Rossen, R. D., Douglas, R. G., Jr., Butler, W. T., and Couch, R. B.: The role of nasal secretion and serum antibody in the rhinovirus common cold. Am J Epidemiol 84:352, 1966.

Cooper, P. D., Agol, V. I., Bachrach, H. L., Brown, F., Ghendon, Y., Gibbs, A. J., Gillespie, J. H., Lonberg-Holm, K., Mandel, B., Melnick, J. L., Mohanty, S. B., Povey, R. C., Rueckert, R. R., Schaffer, F. L., and Tyrrell, D. A. J.: Picornaviridae: Second report. Intervirology 10:165, 1978.

Dalldorf, G., and Sickles, G. M.: An unidentified, filtrable agent isolated from the feces of children with paralysis. Science 108:61, 1948.

Douglas, R. G., Jr., Cate, T. R., Gerone, P. J., and Couch, R. B.: Quantitative rhinovirus shedding patterns in volunteers. Am Rev Respir Dis 94:159, 1966.

Douglas, R. G., Jr., Lindgren, K. M., and Couch, R. B.: Exposure to cold environment and rhinovirus common cold: Failure to demonstrate effect. N Engl J Med 279:743, 1968.

Frisby, D. P., Newton, C., Carey, N. H., Fellner, P., Newman, J. F. E., Harris, T. J. R., and Brown, F.: Oligonucleotide mapping of picornavirus RNAs by two-dimensional electrophoresis. Virology 71:379, 1976.

Gajdusek, D. C.: Encephalomyocarditis virus infection in childhood. Pediatrics 16:902, 1955.

Gwaltney, J. M., Jr.: Rhinoviruses. In Evans, A. S. (ed.): Viral Infections of Humans. New York, Plenum Medical Book Company, 1976, pp. 383–408.

Gwaltney, J. M., Jr., Moskalski, P. B., and Hendley, J. O.: Hand-to-hand transmission of rhinovirus colds. Ann Intern Med 88:463, 1978.

Hendley, J. O., Wenzel, R. P., and Gwaltney, J. M., Jr.: Transmission of rhinovirus colds by self-inoculation. N Engl J Med 288:1361, 1973.

Hughes, J. H., Chema, S., Lin, N., Conant, R. M., and Hamparian, V. V.: Acid lability of rhinoviruses: Loss of C and D antigenicity after treatment at pH 3.0. J Immunol 112:919, 1974.

Hughes, J. H., Gnau, J. M., Hilty, M. D., Chema, S., Ottolenghi, A. C., and Hamparian, V. V.: Picornaviruses: Rapid differentiation and identification by immune electronmicroscopy and immunodiffusion. J Med Microbiol 10:203, 1977.

Jacobson, M. F., and Baltimore, D.: Polypeptide cleavages in the formation of poliovirus proteins. Proc Natl Acad Sci USA 61:77, 1968.

Katagiri, S., Aikawa, S., and Hinuma, Y.: Stepwise degradation of poliovirus capsid by alkaline treatment. J Gen Virol 13:101, 1971.

Kettler, A., Hall, C. E., Fox, J. P., Elveback, L., and Cooney, M. K.: The virus watch program: A coninuing surveillance of viral infections in metropolitan New York families. VIII. Rhinovirus infections: Observations of virus excretion, intrafamilial spread and clinical response. Am J Epidemiol 90:244, 1969.

Kogon, A., Spigland, I., Frothingham, T. E., Elveback, L., Williams, C., Hall, C. E., and Fox, J. P.: The virus watch program: A continuing surveillance of viral infections in metropolitan New York families. VII. Observation on viral excretion, seroimmunity, intrafamilial spread and illness association in coxsackievirus and echovirus infections. Am J Epidemiol 89:51, 1969.

Lim, K. A., and Benyesh-Melnick, ML Typing of viruses by combinations of antiserum pools. Applications to typing of enteroviruses (Coxsackie and ECHO). J Immunol 84:309, 1960.

Melnick, J. L.: Application of tissue culture methods to epidemiological studies of poliomyelitis. Am J Public Health 44:571, 1954.

Melnick, J. L.: Enteroviruses. In Evans, A. S. (ed.): Viral Infections of Humans. New York, Plenum Medical Book Company, 1976, pp. 163–207.

Melnick, J. L.: Advantages and disadvantages of killed and live poliomyelitis vaccines. Bull W H O 56:21, 1978.

Melnick, J. L., and Agren, K.: Poliomyelitis and Coxsackie viruses isolated from normal infants in Egypt. Proc Soc Exp Biol Med 81:621, 1952.

Nomoto, A., Kitamura, N., Golini, F., and Wimmer, E.: The 5′-terminal structures of poliovirion RNA and poliovirus mRNA differ only in the genome-linked protein VPg. Proc Natl Acad Sci USA 74:5345, 1977.

Ogra, P. L., and Karzon, D. T.: Formation and function of poliovirus antibody in different tissues. Progr Med Virol 13:157, 1971.

Ramos-Alvarez, M., and Sabin, A. B.: Characteristics of poliomyelitis and other enteric viruses recovered in tissue culture from healthy American children. Proc Soc Exp Biol Med 87:655, 1954.

Roizman, B., Mayer, M. M., and Rapp, H. J.: Immunochemical studies of poliovirus. III. Further studies on the immunological and physical properties of poliovirus particles produced in tissue culture. J Immunol 81:419, 1958.

Rosen, L., Melnick, J. L., Schmidt, N. J., et al.: Subclassification of enteroviruses and ECHO virus type 34. Arch Ges Virusforsch 30:89, 1970.

Stevens, K. M.: Estimate of molecular equivalent of antibody required for prophylaxis and therapy of poliomyelitis. J Hyg (Camb) 57: 198, 1959.

Tesh, R. B.: The prevalence of encephalomyocarditis virus neutralizing antibodies among various human populations. Am J Trop Med Hyg 27:144, 1978.

Tesh, R. B., and Wallace, G. D.: Observations on the natural history of encephalomyocarditis virus. Am J Trop Med Hyg 27:133, 1978.

Warren, R. J., Lepow, M. L., Bartsch, G. E., and Robbins, F. C.: The relationship of maternal antibody, breast feeding, and age to the susceptibility of newborn infants to infection with attenuated polioviruses. Pediatrics 34:4, 1964.

Yin, F. F., Lonberg-Holm, K., and Chan, S. P.: Lack of a close relationship between three strains of human rhinoviruses as determined by their RNA sequences. J Virol 12:108, 1973.

Young, N. A.: Polioviruses, coxsackieviruses, and echoviruses. Comparison of the genomes by RNA hybridization. J Virol 11:832, 1973.

61 TOGAVIRUSES

Marian C. Horzinek, D.V.M., Ph.D.

DEFINITION

The family Togaviridae includes the genera *Alphavirus, Flavivirus, Rubivirus,* and *Pestivirus*. Members of the family are characterized by spherical virions, 40 to 70 nm in diameter, that consist of a lipoprotein envelope with cellular lipids and virus-specified glycopeptides surrounding a spherical nucleocapsid of icosahedral symmetry. The virus contains a single molecule of single-stranded RNA (molecular weight about 4 $\times 10^6$) with positive polarity that is infectious when extracted and assayed under appropriate conditions. Togaviruses multiply in the cytoplasm and mature by budding (Fenner, 1976).

All species of the genus *Alphavirus* and most flaviviruses multiply in arthropods as well as in vertebrates and constitute the classic arthropod-borne viruses (formerly termed arbo-A and arbo-B viruses, respectively). Arboviruses are viruses that are maintained in nature principally, or to an important extent, through biologic transmission between susceptible vertebrate hosts by hematophagous arthropods (WHO Study Group, 1967). Essential to this ecologic definition is the term "biologic" (as opposed to mechanical) transmission, by which is meant that an "extrinsic incubation period" (5 to 12 days) elapses between the moments when the arthropod vector becomes infected after a blood meal and when it can transmit the virus to a new vertebrate host. Biologic transmission is not confined to the above mentioned togavirus genera: Members of the Reoviridae and Rhabdoviridae families and all Bunyaviridae are also arthropod-borne. On the other hand, members of the Rubivirus (Rubellavirus) and Pestivirus genera (hog cholera, bovine diarrhea viruses) as well as mouse lactic dehydrogenase virus and equine arteritis virus are nonarbotogaviruses (Horzinek, 1973). The following chapters will cover only togaviruses of medical importance—namely, a number of alphaviruses and flaviviruses (Tables 1 and 2) and rubella virus (RUV).

STRUCTURE

Togaviruses are the smallest enveloped animal viruses; alphaviruses and RUV are about 60 nm in diameter, and flaviviruses are smaller, about

TABLE 1. Alphaviruses

		ISOLATED FROM												ISOLATED IN						HUMAN DISEASE	
		Arthropods				Vertebrates															
		Mosq.																			
VIRUS	ABBREVIATION	Culicine	Anopheline	Culicoides	Other	Man	Other Primates	Rodents	Birds	Bats	Marsupials	Other	Sentinels	Africa	Asia	Australasia	Europe	North America	South America	Natural Infection	Lab Infection
Aura	AURA	+																	+		
Bebaru	BEB	+													+						
Chikungunya	CHIK	+				+	+		+	+			+	+	+					+	+
Eastern equine enc.	EEE	+	+	+	+	+	+	+	+		+	+			+		+	+	+	+	+
Everglades	EVE	+	+					+			+							+		+	
Getah	GET	+	+									+			+	+					
Mayaro	MAY	+			+	+						+							+	+	+
Middleburg	MID	+												+							
Mucambo	MUC	+				+		+	+		+		+						+	+	+
Ndumu	NDU	+												+							
O'nyong-nyong	ONN		+			+							+	+						+	
Pixuna	PIX	+	+					+											+		
Ross River	RR	+				+			+		+					+				+	
Sagiyama	SAG	+												+							
Semliki Forest	SFV	+	+						+			+	+								+
Sindbis	SIN	+	+		+	+			+				+	+	+	+				+	
Una	UNA	+	+									+	+					+	+		
Venezuelan equine enc.	VEE	+	+			+		+		+	+	+	+					+	+	+	+
Western equine enc.	WEE	+	+		+	+		+	+			+	+				?	+	+	+	+
Whataroa	WHA	+														+					

Berge, T. O. (ed.): International Catalogue of Arboviruses. Washington, D.C., U.S. Department of Health, Education, and Welfare, 1975.

45 nm. The size differences as determined by negative staining and thin section electron microscopy are reflected by the sedimentation coefficients, which range between 240 and 300S for the former group and 170 to 220S for the latter group. Gradient analysis has shown that the buoyant densities of infectious virions depend on the substance used for building the gradient; values vary between < 1.18 g/ml for sucrose and 1.24 for cesium chloride and other salts.

In the electron microscope togavirions appear as spherical particles consisting of an envelope and an isometric core. The envelope shows unit membrane characteristics and carries glycoprotein surface projections that are organized in a regular (icosahedral) surface lattice (Fig. 1); they are the substrate for a pH-dependent hemagglutinating activity. Exposure of virions to organic solvents or to detergents inactivates their infectivity by disintegrating the envelope, leading to hemagglutinating membrane fragments and to liberation of the nucleocapsid (30 to 40 nm in diameter for alphaviruses and RUV, and 20 to 30 nm for flaviviruses). The togavirion has a shape and substructure suggestive of icosahedral symmetry. In contrast to naked animal viruses, however, isolated togavirus capsids are nonrigid and susceptible to RNase, which degrades the viral genome to small fragments. Since this treatment has no effect on intact virions, it is the function of the envelope to protect the viral RNA from enzymatic attack.

Consistent data of overall chemical analysis have been obtained so far only for alphaviruses, since these grow to high titers in tissue culture and can be obtained as suspensions of sufficient purity and concentration. The reported proportions of RNA: protein: lipid: carbohydrate (in glycoprotein) are about 6:61:27:6. The RNA of togaviruses is a single-stranded, colinear molecule that has an average molecular weight of 4.2 $\times 10^6$, which corresponds to about 12,000 nucleotides, and sedimentation coefficient of 42 to 49S in sucrose gradients. Studies with alphaviruses have indicated that there is a significant degree of secondary structure in extracted RNA, which is necessary for infectivity. Polyadenylic acid sequences located near the 3′ terminus have been demonstrated; these, together with the absence of a virion-associated polymerase, indicated that togaviral RNA is plus-stranded and might serve as the initial messenger molecule during infection. These assumptions have been confirmed by the demonstration of parental virion RNA in

TABLE 2. Mosquito-borne Flaviviruses

		ISOLATED FROM													ISOLATED IN						HUMAN DISEASE	
		Arthropods					Vertebrates															
		Mosq:		Ticks																		
VIRUS	ABBREVIATION	*Culicine*	*Anopheline*	*Ixodid*	*Argasid*	*Other*	*Man*	*Other Primates*	*Rodents*	*Birds*	*Bats*	*Marsupials*	*Other*	*Sentinels*	*Africa*	*Asia*	*Australasia*	*Europe*	*North America*	*South America*	*Natural Infection*	*Lab Infection*
Alfuy	ALF	+								+							+					
Bagaza	BAG	+													+							
Banzi	BAN	+					+							+	+						+	
Boubyui	BOU	+	+					+							+							
Bussuquara	BSQ	+					+		+					+					+	+	+	
Dengue-1	DEN-1	+					+								+	+	+				+	+
Dengue-2	DEN-2	+					+	+							+	+	+			+	+	+
Dengue-3	DEN-3	+					+									+	+			+	+	+
Dengue-4	DEN-4	+					+									+	+				+	
Edge Hill	EH	+	+														+					
Ilheus	ILH	+					+			+				+					+	+	+	+
Japanese enceph.	JE	+	+				+			+	+		+			+					+	+
Jugra	JUG	+									+					+						
Kokobera	KOK	+															+					
Kunjin	KUN	+					+			+							+					+
Murray Valley enceph.	MVE	+					+										+				+	
Ntaya	NTA	+													+							
Sepik	SEP	+															+				+	
St. Louis enceph.	SLE	+	+			+	+			+	+		+						+	+	+	+
Spondweni	SPO	+					+								+						+	+
Stratford	STR	+															+					
Tembusu	TMU	+	+													+	+					
Uganda S	UGS	+								+				+	+							
Usutu	USU	+							+	+					+							
Wesselbron	WSL	+	+				+		+				+		+	+					+	+
West Nile	WN	+	+	+	+		+		+	+	+		+	+	+	+		+			+	+
Yellow fever	YF	+					+	+				+			+					+	+	+
Zika	ZIKA	+					+						+	+	+						+	+

Berge, T. O. (ed.): International Catalogue of Arboviruses. Washington, D.C., U.S. Department of Health, Education, and Welfare, 1975.

polysomes of infected cells and by its translation into virus-specific polypeptides in cell-free systems. Like other mRNAs from eukaryotic systems, viral RNAs contain methylated "capped" structures at their 5′ termini, which in alphaviruses are of the form mG (5′) pppAp.

A minimum of three structural polypeptides has been identified in togaviruses, one of them associated with the nucleocapsid; in alphaviruses and RUV this protein has a molecular weight of 30,000 to 35,000, and in flaviviruses the molecular weight ranges from 13,000 to 14,000. Amino acid analyses have shown that the alphavirus capsid protein has an *N*-terminal lysine and is relatively hydrophilic, a property consistent with its extensive and rapid interaction with viral RNA.

Two glycosylated envelope proteins have been identified in RUV (50,000 and 63,000 mw) and in most alphaviruses (E1 and E2, 50,000 to 53,000 mw) by SDS polyacrylamide gel electrophoresis. Using chromatographic techniques, three envelope proteins (52,000, 49,000, and 10,000 mw) were isolated from SFV (for abbreviations see Tables 1 and 2). In flaviviruses only one envelope protein (50,000 to 60,000 mw) is glycosylated, whereas the other (8000 to 9000 mw) seems to be buried in the lipid bilayer and may have a bridging function between the peripheral glycoprotein and the capsid. In alphaviruses both the E1 and the E2 proteins occupy a superficial position because they can be stripped away by treatment with proteolytic enzymes and can be demonstrated by enzymatic iodination of undisrupted virions. They probably penetrate the lipid bilayer and interact directly with the capsid.

Togaviral lipid is confined to the envelope; in alphaviruses it has been found that 25 to 31 per

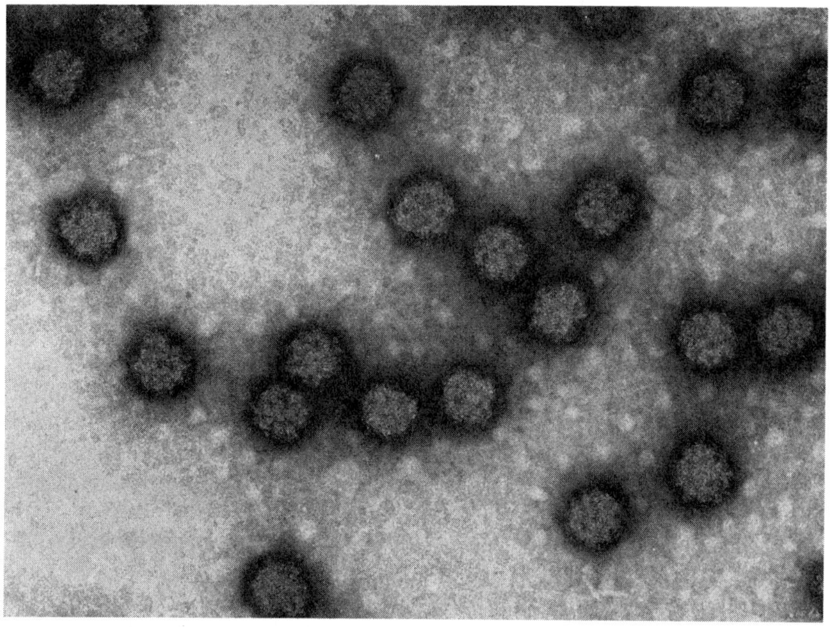

FIGURE 1. *Sindbis virus, a member of the genus* Alphavirus *of the* Togaviridae. *Note the regular pattern on the virion surface. Negatively stained preparation (uranyl acetate, after fixation with glutaraldehyde). (Courtesy of Dr. D. J. Ellens, CDI Lelystad, The Netherlands.)*

cent (by weight) is neutral lipid, predominantly cholesterol. Most of the remaining lipids are represented by phosphatidylethanolamine, phosphatidylserine, sphingomyeline, and phosphatidylcholine, which is the principal phosphatide. Of the viral fatty acids, oleic, palmitic, and stearic acids predominate. In general, the composition of togaviral lipids reflects that of the host cell membranes, as shown by a comparative analysis of SFV grown in BHK (baby hamster kidney) and *Aedes albopictus* cells, respectively (Pfefferkorn and Shapiro, 1974).

ANTIGENIC COMPOSITION

Assignment of a togavirus to the alphagenus or flavigenus is made on the basis of immunologic cross-reactions. Virion surface antigens responsible for adsorption to susceptible cells and erythrocytes can be identified and compared by neutralization (NT) and hemagglutination inhibition (HI) tests. By using solvent-detergent-extracted antigens in complement fixation, gel diffusion, or radioimmunoassays, the total of antigen-antibody reactions can be measured, including those with the capsid protein. Capsid protein has been described as carrying group-reactive determinants, similar to the common nucleoprotein (gs) antigen of the orthomyxoviruses. These capsid determinants may mean that all alphaviruses are related (Dalrymple et al., 1976).

By using sera from animals that have been immunized by one inoculation of virus, it is possible to divide alphaviruses into subgroups and to make specific serologic identifications. Within the alphavirus genus the viruses BEB, CHIK, GET, MAY, ONN, RR, SFV, and UNA form one cluster of antigenically related viruses, AUR, SIN, WEE, and WHA a second one, and MUC, PIX, and VEE a third one. NDU, EEE, and MD are not related to each other nor to any other alphavirus (Casals and Reeves, 1959; Karabatsos, 1975). The following flavivirus subgroups have been recognized by using plaque neutralization: (1) the tick-borne flaviviruses (Table 2) with the exceptions of KAD and POW; (2) APOI, CR, DB, ENT, KAD, MOD, and RB; (3) ALF, JE, KOK, KUN, MVE, SLE, STR, USU, and WN; (4) SPO and ZIKA; (5) IT, NTA, and TMU; (6) BAN, EH, and UGS; and (7) the four DEN type viruses. No relationships to any other flavivirus could be demonstrated with BUS, ILH, MML, POW, WSL, and YF (de Madrid and Porterfield, 1974). RUV is unrelated to any other animal virus and has been classified as a togavirus exclusively for structural reasons. It is the only representative of the rubivirus genus.

It has been possible to determine the antigenic function of the respective alphavirion surface polypeptides. The E1 glycoprotein of SIN carries the complete hemagglutinating activity, whereas E2 appears to be important to infectivity because only antiserum directed against this molecular species effectively neutralized the virus. Glycoprotein E1 appears to be cross-reactive with antiserum to the closely related WEE and may therefore carry the determinant(s) responsible for defining the serologic complexes; E2 appears to be virus-specific (Dalrymple et al., 1976). By studying nucleotide sequence homologies between the nucleic acids of different alphaviruses it was demonstrated that the closely related viruses of CHIK and ONN have only 13 per cent of the base sequences in common, whereas ≤ 1 per cent of RNA–RNA homologies occur between the other viruses of this genus. These rather low homologies can be reconciled with the established serologic relationship by assuming that the antigenic sites are composed of only a small number of amino acids and/or that the degeneracy of the genetic code allows a much larger homology to be present in the protein sequences than in corresponding RNA base sequences (Wengler et al., 1977).

Because of the simplicity of the test system the hemagglutinating antigen has been studied most extensively (Clarke and Casals, 1958). For diagnostic purposes alkaline aqueous extracts, fluorocarbon, ether-acetone, sucrose-acetone, or Tween 80-ether extracts of infected tissues (mostly mouse brain material for alpha- and flaviviruses) are used. Treatment with protamine sulfate results in precipitation of host material and improvement in the HA patterns. In contrast to most other viruses, the demonstration of HA by togaviruses is dependent on the ionic environment; in general, alphaviruses show optimal activity at pH 5.8 to 6.2 and flaviviruses at 6.2 to 6.4. RUV requires Ca^{++} ions for HA. Dependence on the ionic environment indicates that a specific conformation of the envelope glycoproteins is essential; for example, it has been shown that the isoelectric point of SIN virus E1 protein is the same as the pH that is optimum for HA.

REPLICATION

Adsorption of alphaviruses to susceptible cells occurs within a few minutes. Because it is dependent on the salt concentration (divalent cations inhibit attachment) and pH but independent of temperature, electrostatic forces are probably involved. The enhancing effect of DEAE-dextran, a polycation, further supports this assumption. Little is known about the requirement for cellular receptors. Since alphaviruses are capable of growing in a variety of cells from phylogenetically unrelated species including mammalian, avian, reptilian, amphibian, piscine, and arthropod, these viruses either have a broad range of receptors or they do not require specific receptors. Penetration of adsorbed virus is dependent on temperature (optimum at 37° C) but independent of the ionic environment and follows first order kinetics. The virion is engulfed by a pinocytotic vacuole and subsequently uncoated.

Replication of viral RNA requires a new protein, a virus-specific RNA-dependent RNA polymerase. For its synthesis, the input parental RNA serves as a messenger. The polymerase is bound to smooth cytoplasmic membranes, where it catalyses the synthesis of so-called replicative intermediates. These consist of partially double-stranded RNA molecules containing one polynucleotide species serving as a template, and a second species of complementary strandedness made up of nascent chains of varying length that are partially base-paired to the template. The major species of alphaviral RNA synthesized in infected cells is not virion RNA (4.2×10^6) but a subgenomic RNA species of 26S (1.6×10^6) that has the same polarity as virion RNA but contains only one-third of its nucleotide sequences. It serves as the mRNA for the structural proteins of the virion. Studies in SFV-infected cells and with cell-free synthesizing systems have shown that the $5' \rightarrow 3'$ gene order in 26S RNA is C, E3, E2, and E1 and that these proteins are synthesized as a polyprotein of about 130,000 daltons from a single initiation site. By a sequence of nascent and post-translational proteolytic cleavages, this molecule is processed to give the structural virion polypeptides.

The remaining coding capacity corresponding to protein of about 300,000 daltons is used for nonstructural viral polypeptides. Since the 42S RNA is infectious, some, if not all, of these are likely to be components of the viral polymerase. The nucleotide sequence of the 26S RNA is located inward from the 3′ end of the 42S RNA, which implies that the genes coding for the nonstructural proteins must be situated in the two thirds of the genome near the 5′ terminal. Their synthesis is initiated at a single site near the 5′ end and is terminated internally before the structural protein genes. Also, the nonstructural polypeptides are synthesized as a giant precursor molecule that is subsequently processed by proteolytic cleavage.

The alphavirus capsid is assembled in the cytoplasm of the infected cell around the RNA and then attaches to the inner surface of the host cell plasma membrane after that membrane has been modified by the insertion of the virus-specified glycoproteins. The membrane at this stage contains the definite E1 protein and a precursor to the second glycoprotein. The envelopment of the nucleocapsid takes place as it is progressively wrapped into the modified membrane while moving from the cytoplasm to a position physically outside the cell. During the terminal stages of this "budding" process the precursor polypeptide is cleaved to form the E2 glycoprotein.

Intracytoplasmic vacuoles are the first evidence of subcellular alteration associated with virus replication. Alphaviruses and flaviviruses differ with respect to their morphogenesis; the former envelop their capsids by budding preferentially from the marginal membrane, whereas the latter emerge from internal vacuolar membranes. The general picture for RUV resembles that of alphaviruses. Only in alphavirus-infected cells may assembled capsids be detected around cytoplasmic vacuoles.

Different togaviruses vary in the intensity of their cytopathic action, which may depend on the cell species on the one hand and on environmental conditions (medium, pH, temperature, virus dose) on the other. In vertebrate cells growth of alphaviruses is rapid. Only a few hours after infection, newly formed virus is released. Virus production continues at a constant linear rate for up to 10 or 12 hours, approaching 1000 infectious units per cell. Flaviviruses show a latent period of about 12 hours, and a further 10 to 20 hours are required to achieve maximal titers of extracellular virus (Mussgay et al., 1975; Pfefferkorn and Shapiro, 1974).

PATHOGENIC PROPERTIES

Alphaviruses and Flaviviruses

The animal classes from which the arthropod-borne togavirus types have been isolated are listed in Tables 1 to 4. Additional information on susceptibility to infection of experimental hosts can be obtained from the *International Catalogue of Arboviruses* (Berge, 1975). In man, infections range in severity from completely subclinical but immunizing ones to rapidly fatal illness. Similar syndromes may be produced by immunologically different viruses, and completely different disease pictures can be caused even by strains of a virus that are immunologically identical.

A tendency to a biphasic temperature curve has been noted, in which the peaks may follow each other without an afebrile interval; this "saddle back" curve is considered typical for YF and DEN infections. However, in infections caused by the tick-borne encephalitis viruses (Table 3) there may be a period of 6 to 14 afebrile days between the two fever peaks. With the first rise in temperature—which may be mild and is often overlooked (in infections with WEE, SLE, and JBE, for example)—systemic infection occurs and virus can be isolated from the blood. The second febrile period is associated with more serious clinical manifestations. During this phase it is often impossible to isolate virus from the blood, and antibodies are usually already demonstrable. During the first febrile attack virus multiplies in the internal organs and leukopenia is usually observed. Leukocytosis is often associated with later, more severe manifestations such as en-

TABLE 3. Tick-borne Flaviviruses

		ISOLATED FROM											ISOLATED IN						HUMAN DISEASE	
		Arthropods			Vertebrates															
		Ticks																		
VIRUS	ABBREVIATION	*Ixodid*	*Argasid*	*Other*	*Man*	*Other Primates*	*Rodents*	*Birds*	*Bats*	*Marsupials*	*Other*	*Sentinels*	*Africa*	*Asia*	*Australasia*	*Europe*	*North America*	*South America*	*Natural Infection*	*Lab Infection*
[a]Absettarov	ABS	+			+						+					+			+	+
[a]Hanzalova	HAN	+			+											+			+	
[a]Hypr	HYPR	+			+		+	+	+		+					+			+	+
Kadam	KAD	+											+							
Karshi	KSI		+											+						
[a]Kumlinge	KUM	+			+		+	+			+					+			+	+
[a]Kyasanur Forest dis.	KFD	+	+	+	+	+	+	+	+		+			+					+	+
Langat	LGT	+												+						
[a]Louping ill	LI	+			+		+	+			+					+			+	+
[a]Negishi	NEG	+			+									+					+	+
[a]Omsk hem. fev.	OMSK	+			+		+				+			+		?			+	+
[a]Powassan	POW	+			+		+				+						+		+	
Royal Farm	RF		+											+						
[a]Russian spring summer enceph.	RSSE	+			+		+	+						+		+			+	+
Tyuleniy	TYV	+												+			+			

[a]Encephalitis in man has been observed after infection with the indicated viruses.

From Berge, T. O. (ed.): International Catalogue of Arboviruses. Washington, D.C., U.S. Department of Health, Education, and Welfare, 1975.

TABLE 4. Other Flaviviruses, No Arthropod Vector Demonstrated

		ISOLATED FROM								ISOLATED IN						HUMAN DISEASE	
		Vertebrates															
VIRUS	ABBREVIATION	Man	Other Primates	Rodents	Birds	Bats	Marsupials	Other	Sentinels	Africa	Asia	Australasia	Europe	North America	South America	Natural Infection	Lab Infection
Apoi	APOI			+							+						+
Batu Cave	BC					+					+						
Carey Island	CI					+					+						
Cowbone Ridge	CR			+										+			
Dakar bat	DB	+				+				+							
Entebbe bat	ENT					+				+							
Israel turkey meningo.	IT				+						+						
Jutipa	JUT			+										+			
Koutango	KOU			+						+							
Modoc	MOD			+										+			
Montana myotis leuko.	MML					+								+			
Phnom-Penh bat	PPB					+					+						
Rio Bravo	RB					+								+		+	+
Saboya	SAB			+						+							
Sokuluk	SOK					+					+						

From Berge, T. O. (ed.): International Catalogue of Arboviruses. Washington, D.C., U.S. Department of Health, Education and Welfare, 1975.

cephalitis. No consistent clinical picture is associated with arbo-togavirus infections. An influenza-like syndrome is observed in mild infections with accompanying headache, nausea, fever, gastrointestinal disturbances, myalgia, and joint pains. Encephalitis in humans and domestic animals is known to be caused by the alphaviruses EEE, VEE, and WEE; the flaviviruses that at times produce central nervous conditions are SLE, JBE, MVE, WN, members of the tick-borne flaviviruses, and possibly ILH. Hepatitis is a typical manifestation of YF in man and has also been observed after infection with KFD. Maculopapular rashes with lymphadenopathy are common in DEN infections and have also been reported for WN, CHIK, and ONN infections. Hemorrhages of varying severity are observed in some DEN epidemics, and in OMSK and KFD infections. The black vomit ("vomito negro") of YF is the result of gastric hemorrhages.

The DEN hemorrhagic fever and a newly recognized DEN shock syndrome (which occurs with significant mortality especially in children in Asia) are known to be immunologic diseases. The four DEN serotypes do not confer a high level of cross-protection; nevertheless, there is a rapid serologic response when an individual who has recovered from an attack of DEN caused by one serotype is infected with another. The secondary antibody response coincides with viremia due to the superinfecting DEN type; circulating immune complexes are then formed, and complement depletion occurs. The immune complexes are deposited in the vessel walls, and hemorrhagic fever and shock result (Theiler and Downs, 1973).

Rubella Virus

Primates appear to be the natural hosts of RUV; in monkeys (rhesus and African monkeys) infection occasionally causes symptoms but is mostly inapparent. Ferrets, hamsters, guinea pigs, rabbits, and newborn mice also have been found to support multiplication of RUV in vivo and display no clinical signs. Infection in pregnant rats can cause teratogenic damage to the offspring that is similar to that seen in human infants with congenital rubella.

Postnatal infection in adult man frequently starts with mild fever, headache, malaise, and respiratory symptoms some days before the appearance of a maculopapular rash associated with lymphadenopathy. The latter symptom is a prominent feature of the disease, whereas the rash may be absent, especially in younger individuals. Although postnatal infection usually follows a benign course, arthritis, thrombocytopenic purpura, and encephalitis may occur as complications.

The pathogenic significance of RUV lies in its teratogenic properties. About 10 to 15 per cent of living infants born to mothers with apparent or inapparent rubella during the first trimester of pregnancy show evidence of infection that is recognizable at birth or during the first year of life. The consequences of in utero infection include spontaneous abortion, stillbirth, and live birth with moderate to severe abnormalities, as well as completely normal infants. The congenital rubella syndrome consists of neurologic and developmental defects such as hearing loss, eye lesions (pearly cataract, glaucoma, chorioretinitis, microphthalmia, corneal clouding), cardiovascular defects, hepatosplenomegaly, thrombocytopenic purpura, anemia, interstitial pneumonia, metaphyseal bone lesions, and intrauterine growth retardation. Any combination of anomalies may occur in an individual infant.

IMMUNITY

In infected or vaccinated individuals togaviral antigens cause a pronounced immune response that can be measured in vitro; like most viruses that produce systemic disease with a viremia, togavirus infection is controlled primarily by circulating antibody. Lifelong immunity after a togavirus attack has been observed. For instance, yellow fever antibodies have been found persisting for 75 years in the absence of reexposure. A specific aspect of alphavirus and flavivirus infections is the antibody response of individuals after multiple exposures to antigenically related viruses. In general, with repeated immunizations there is a broadening of the specificity of the antibodies produced and a heightened reaction to the original virus rather than to the virus responsible for reinfection. Thus, sequential infection of humans with two flaviviruses induced neutralizing antibodies that cross-reacted with a third flavivirus to which the subjects had never been exposed. Unfortunately, the existence of cross-reactive antibodies does not necessarily provide resistance to infection. Otherwise, vaccination with a series of carefully selected antigenically related viruses would produce a broad immunity that might be effective against all viruses of this genus. However, this cannot be achieved because of the "original antigenic sin" phenomenon: The immunologic response of a person to a togavirus is dominated throughout his life by his first experience with a virus of that type.

Vaccines against the epidemic and endemic arthropod-borne togavirus infections have been developed for human (medical personnel, laboratory workers, military forces) and veterinary use. Of the inactivated vaccines, formalin-treated TBE virus preparations have been in wide use in the U.S.S.R. Experimental vaccines against other flaviviruses (JBE, SLE, and KFD) have been tested. The accumulated evidence suggests that potent inactivated vaccines can be prepared. The substrate for virus production must be carefully selected, first for its ability to produce virus in high titer and second for the absence of undesirable components, such as the encephalitogenic factor in brain tissue. A vaccination regime of two initial doses followed by a yearly booster injection has been recommended (Smith, 1969). The immunizing potency of an experimental SFV virion subunit vaccine obtained after different solvent/detergent treatments has been tested in animals. Its efficiency depended on the size of the envelope fragments produced; large fragments (e.g., after Tween 80-ether treatment) offered the best protection (Mussgay et al., 1973).

Of the attenuated live virus vaccines, the YF strains 17D (prepared in chick embryo) and FN (mouse embryo) strains are in common use. The seed virus has to be strictly defined, since loss of immunogenicity has been observed at higher passage levels. The 17D vaccine has been one of the most innocuous and effective vaccines; the first fatality (in a 3-year-old child) was reported after the administration of some 34 million doses and 25 years of use. A number of other live attenuated viruses have given satisfactory antibody responses in human volunteers and experimental animals. Heterotypic immunization using LGT virus, a naturally occurring tick-borne virus from Malaya, against tick-borne viruses highly pathogenic to man (see Table 3) has been successful. When heterotypic vaccinations are used the above-mentioned hazard of immunologic aggravation of diseases must be recognized.

After natural RUV infection, persistent levels of nasopharyngeal-specific IgA, which play an important role in preventing reinfection of adults, have been detected at the portal of entry; cell-mediated immunity has also been demonstrated in tonsillar lymphoid tissue. If fetal infection is to be prevented, adequate and persistent levels of serum antibody, especially IgG, must be present in the mother's plasma. Several attenuated vaccine strains have been developed by serial passages of RUV in vervet monkey kidney cells (HPV 77) followed by passage in primary rabbit kidney (Cendehill) and primary guinea pig kidney cells (the Japanese strain To-336) or in human embryonic kidney cells and W1 38 fibroblasts (RA27/3). Administration of these strains resulted in seroconversion of 95 per cent of susceptible recipients, although antibody levels were often lower than those following naturally acquired disease. Reinfection of reconvalescents or vaccinated individuals does occur and may in fact be responsible for maintaining adequate levels of immunity in a community. In contrast to measles and poliomyelitis, however, the aim of RUV vaccination is not so much to protect the vaccinee herself against a relatively mild disease but to prevent infection of the fetus. Fetal infection is likely to occur only if maternal viremia develops, and viremia has not been observed in immune individuals after either natural or experimental challenge (Banatvala, 1977).

LABORATORY DIAGNOSIS

Alphaviruses and Flaviviruses

Suckling mice are considered the universal host system for virus isolation; they are inoculated intracerebrally with sera from patients, postmortem specimens, or extracts from pools of trapped arthropods. The mice become ill, and frequently have hind limb paralysis, usually two to three days or longer after injection. Affected mice are then sacrificed. Isolation of virus is confirmed by passage of brain suspension material to other suckling mice. For primary isolation, cell cultures of chicken fibroblasts and hamster kidney cells are commonly used; as for continuous lines, Vero (green monkey) and BHK-21 (hamster) cells are routinely employed. Not all cell-virus systems show cytopathic effects after viral multiplication. To isolate virus it is important to inoculate unknown specimens at two or more dilutions, since high concentrations of defective interfering particles may restrict or completely mask the symptoms of infection.

Identification of an isolated agent is usually done by neutralization tests employing mice or cell cultures. The most sensitive animal method is the intraperitoneal inoculation of infant mice with serum-virus mixtures. The reproducibility of this assay is improved by the addition of complement, which in the presence of antibody causes lysis of the viral membrane and inactivation of its genome by RNase. In a neutralization test for virus identification the amount of serum is kept constant and incubated with serial tenfold virus dilutions. By comparing the neutralization indices (the difference in log virus titers in the presence and absence of antibodies) of the virus in question with homologous virus-serum mixtures, type identification can be made. By using this procedure it is also possible to determine the extent of antigenic cross-relationship between viruses within one genus.

In general, cross-reactions between related viruses are more pronounced by hemagglutination inhibition (HI) and complement fixation (CF) tests than by neutralization. From studies on the HI antibody response of experimental animals the following conclusions can be drawn:

1. Each virus, after primary infection, produces a distinctive HI pattern. The reaction may be entirely specific, such that the serum reacts only with the homologous antigen. When cross-reactions occur, they are always significantly lower in titer than the homologous reactions. Cross-reactions are highest with antigens prepared from viruses that are closely related to the immunizing antigen.
2. When animals are inoculated repeatedly with the same virus, the sera are more broadly cross-reactive and titers are higher, but the basic pattern remains the same.
3. Animals immunized to two different viruses of the same genus produce sera that are very widely cross-reactive; their cross-reactivity is always more marked than that observed with sera obtained from animals after two inoculations of the same virus. Anti-flavivirus sera tend to show more pronounced intragenus cross-reactivity than anti-alphavirus sera.

Sera contain nonspecific inhibitors for togavirus HA in high titer; they are lipoproteins and are removed by acetone extraction or adsorption to acid-washed kaolin. In addition, many sera show hemagglutinating activity for goose red blood cells (RBC), which is eliminated by adsorption to erythrocytes before the serum is titrated. Goose RBC are used in many laboratories to test HA, but chicken and pigeon cells (RUV) are also suitable. Optimal temperature is 25° to 37° C for alphaviruses; some flavivirus and RUV require 4° C. Spontaneous elution does not occur.

From the foregoing it is clear that serologic tests for determining the type identity of alphavirus or flavivirus antibodies can be inconclusive, particularly when the sera are collected from patients in endemic areas. In these cases, the mere demonstration of a rise in antibody to a given virus by whatever test is not sufficient evidence that the infection is attributable to the virus used in the test. Only primary infection in individuals not previously exposed to any alphavirus or flavivirus can be reliably diagnosed by means of serology (Theiler and Downs, 1973).

Rubella Virus

Throat swabs or washings, acute phase blood, cerebrospinal fluid, urine, amniotic fluid, and placental or fetal tissues can be used for isolation in cultures of primary African green monkey kidney, human amnion, RK13 (rabbit), BHK-21 cells, and others. In established cell lines rubella virus may produce a CPE that is markedly influenced by environmental variables (e.g., serum concentration), whereas in primary cells CPE is often absent and virus multiplication must be demonstrated by interference with a challenge virus. For instance, primary African green monkey kidney cells show no evidence of CPE after infection with rubella virus but lose their ability to support growth of added echovirus type 11, coxsackievirus A9, or Newcastle disease virus. From fetal specimens RUV can be best isolated by explanting tissue fragments or trypsinizing fetal tissue; infected cells and media thus obtained can be used for virus identification. Since no antigenic cross-reactions occur between RUV and other togaviruses, unequivocal identification can be made by neutralization of the interfering or cytopathic effect by using a specific antiserum. A more rapid method consists of the application of indirect immunofluorescent antibody staining. In this method, acetone-fixed infected cells on coverslips are incubated with known immune rabbit sera and subsequently with fluorescein-conjugated goat anti-rabbit globulin preparations. The specificity of the cytoplasmic fluorescence must be verified by including preimmune rabbit serum and uninfected cells as controls.

For the determination of RU antibodies, HI, CF, and indirect immunofluorescence are used mostly; the reagents are commercially available. The suggested serologic tests and their interpretations are summarized in Table 5 (Plotkin, 1969). In general, the HI test gives the highest titer levels; CF antibodies tend to fade quickly and are often absent when other tests are still positive. For indirect immunofluorescence a fluorescein isothiocyanate-conjugated anti-human globulin preparation must be used as a second antibody. If this preparation is IgM-specific it can be employed for the diagnosis of congenital infection (see case E in Table 5). After disease, precipitating antibodies directed against two viral antigens (ϑ and ι) may be detected in gel double-diffusion tests. Anti-ϑ antibodies usually parallel those detected by HI, whereas anti-ι antibodies appear to correlate with resistance to reinfection.

EPIDEMIOLOGY

For an epidemiologic cycle to occur in nature the following conditions must be met: A critical minimal amount of virus is necessary to establish infection of the arthropod vector. This threshold value is characteristic for every arthropod-virus combination and implies that the virus must circulate in the vertebrate host in sufficient quantities. The susceptibility of a given vector to a given virus can be either inferred from material collected in nature under endemic or epidemic conditions or determined by experimental infection of the arthropod (feeding on or inoculation with virus-containing material). For the virus to be transmitted by the vector, it must pass from the lumen of the arthropod's intestinal tract to susceptible cells, multiply, and reach the salivary glands in quantities adequate to inoculate an infectious dose by bite. This process determines the transmitting efficiency of the vector. Epidemiologically, climate conditions are critical. Thus, the extrinsic incubation period in the vector is shortened by increased temperature. Rainfall

TABLE 5. Suggested Testing for Rubella Antibodies in Different Situations

	SOURCE OF SERUM	CHOICE OF SEROLOGIC TEST 1st	2nd	3rd	INTERPRETATION
A	Patient with rash less than 1 week previously. Convalescent serum 14 to 21 days posteruption	HI	FA	CF	Fourfold rise proves recent infection
B	Patient with rash more than 1 week previously. Convalescent serum 21 to 28 days posteruption	CF	FA	HI[a]	Fourfold rise proves recent infection
C	Subject in whom immunity is to be determined or pregnant women exposed to rubella	HI	Nt	FA	Presence of antibody at any titer shows immunity
D	Possible rubella-syndrome infant older than 6 months	HI	Nt	FA	Presence of antibody suggests congenital infection if child has not been exposed to rubella
E	Possible rubella-syndrome infant younger than 6 months	HI	Nt	FA	Presence of antibody suggests congenital infection if titer significantly higher than mother's or if contained in IgM fraction

[a]Determination of IgM rubella antibody may assist in the diagnosis.

From Plotkin, S. A.: In Lenette, E. H., and Schmidt, N. J. (eds.): Diagnostic Procedures for Viral and Rickettsial Infections. New York, American Public Health Association, 1969.

may favor breeding conditions for mosquitoes. Climate may control the geographic distribution of arboviruses by providing the ecologic conditions for vectors and hosts. Since each species of blood-sucking arthropod has an affinity for a vertebrate species or group of species, detailed knowledge of vector-host preferences or dependencies is essential to the epidemiology and control of arthropod-borne virus diseases. Infection chains and reservoirs (animal species in which virus is maintained and can be disseminated for a prolonged period) have been established for many arthropod-borne togaviruses (Theiler and Downs, 1973).

Rubella virus transmission occurs through droplet infection. The portal of entry is the lymphoid tissue of the pharynx and possibly the conjunctiva. Sources of infection are virus-excreting individuals with or without clinical symptoms. Children with chronic infection are an important source of infection. Rubella has a worldwide distribution. Infections in the northern hemisphere increase during March to May, and more extensive epidemics occur at regular intervals of 5 to 10 years (Norrby, 1969).

References

Banatvala, J. E.: Rubella vaccines. In Recent Advances in Clinical Virology, Edinburgh, London, New York, Churchill-Livingstone, vol.1, 1977, p. 171.

Berge, T. O. (ed.): International Catalogue of Arboviruses. Washington, D. C., U. S. Department of Health, Education and Welfare, 1975.

Casals, J., and Reeves, W. C.: Arthropod-borne animal viruses. In Rivers, T. M., and Horsfall, F. L. (eds.): Viral and Rickettsial Infections of Man. London, Pitman Medical Publishing Company, 1959, p. 269.

Clarke, D. H., and Casals, J.: Techniques for hemagglutination and hemagglutination-inhibition with arthropod-borne viruses. Am J Trop Med Hyg 7:561, 1958.

Dalrymple, J. M., Schlesinger, S., and Russell, P. K.: Antigenic characterization of two Sindbis envelope glycoproteins separated by isoelectric focusing. Virology 69:93, 1976.

Fenner, F.: Classification and nomenclature of viruses. Intervirology 7:44, 1976.

Horzinek, M. C.; The structure of togaviruses. Progr Med Virol 16:109, 1973.

Karabatsos, N.: Antigenic relationship of group A arboviruses by plaque reduction neutralization testing. Am J Trop Med Hyg 24:527, 1975.

de Madrid, A. T., and Porterfield, J. S.: The flaviviruses (group B arboviruses): A cross-neutralization study. J Gen Virol 23:91, 1974.

Mussgay, M., Weiland, E., Strohmaier, K., Ueberschär, S., and Enzmann, P. J.: Properties of components obtained by treatment of Semliki Forest virus with Tween-80 and tri (N-butyl) phosphate. J Gen Virol 19:89, 1973.

Mussgay, M., Enzmann, P. J., Horzinek, M. C., and Weiland, E.: Growth cycle of arboviruses in vertebrate and arthropod cells. Progr Med Virol 19:257, 1975.

Norrby, E.: Rubella Virus. New York, Springer Verlag, 1969, p. 115.

Pfefferkorn, E. R., and Shapiro, D.: Reproduction of togaviruses. In Fraenkel-Conrat, H., and Wagner, R. R. (eds.): Comprehensive Virology, vol. 2. New York, Plenum Publishing Corporation, 1974, p. 171.

Plotkin, S. A.: Rubella virus. In Lenette, E. H., and Schmidt, N. J. (eds.): Diagnostic Procedures for Viral and Rickettsial Infections. New York, American Public Health Association, 1969, p. 364.

Smith, C. E., Gordon: Arbovirus vaccines. Br Med Bull 25:142, 1969.

Theiler, M., and Downs, W. G.: The Arthropod-Borne Viruses of Vertebrates. New Haven, Yale University Press, 1973.

Wengler, G., Wengler, G., and Filipe, A. R.: A study of nucleotide sequence homology between the nucleic acids of different alphaviruses. Virology 78:124, 1977.

World Health Organizaiton Study Group: Arboviruses and human disease. WHO Technical Report Series No 369, 1967.

62 ARENAVIRUSES

Fritz Lehmann-Grube, M.D.

ARENAVIRUSES

The group of arenaviruses comprises at present ten members, of which four are pathogenic for man, causing lymphocytic choriomeningitis, Lassa fever, Argentine hemorrhagic fever, and Bolivian hemorrhagic fever. The latter three illnesses are severe and of great public health importance. Arenaviruses are of special interest for investigating virus-host relationships such as pathologic immune phenomena in virus diseases and persistent virus infections.

CLASSIFICATION

The lymphocytic choriomeningitis (LCM) virus was the first arenavirus to be discovered (Table 1). During serial transfers in monkeys of materials from a fatal case of the 1933 St. Louis epidemic of encephalitis, a virus was encountered that differed from the other isolates of that outbreak. The true origin of this first strain of LCM virus was never determined.

In 1958 the Junin virus was obtained from people suffering from Argentine hemorrhagic fever. The Tacaribe virus was isolated next. Tacaribe virus and Junin virus were antigenically related, as was Machupo virus, recovered from cases of Bolivian hemorrhagic fever in 1965. These three agents formed the Tacaribe complex, soon to be joined by Amapari virus, Tamiami virus, Parana virus, Pichinde virus, and Latino virus.

Similarities between Machupo virus and LCM virus were noted early, but it was their striking

TABLE 1. Isolation History and Natural Occurrence of Viruses Presently Recognized as Arenaviruses

VIRUS	FIRST ISOLATION Reference	Source	NATURAL OCCURRENCE Principal Host	Geographic Distribution
LCM	Armstrong and Lillie (1934)	Monkey (?)	*Mus musculus*	America, Europe
Lassa	Buckley and Casals (1970)	Man	*Mastomys natalensis*	Africa (Sierra Leone, Nigeria; other parts)
Amapari	Pinheiro et al. (1966)	*Oryzomys* sp. *Neacomys guianae*	*Oryzomys capito* *Neacomys guianae*	South America (Brazil)
Junin	Parodi et al. (1958)	Man	*Calomys laucha** *Akodon azarae*	South America (Argentina)
Latino	Webb et al. (1973)	*Calomys callosus*	*Calomys callosus*	South America (Bolivia, Brazil)
Machupo	Johnson et al. (1965)	Man	*Calomys callosus*	South America (Boliva)
Parana	Webb et al. (1970)	*Oryzomys buccinatus*	*Oryzomys buccinatus*	South America (Paraguay)
Pichinde	Trapido and Sanmartin (1971)	*Oryzomys albigularis* *Thomasomys lugens*†	*Oryzomys albigularis*	South America (Colombia)
Tacaribe	Downs et al. (1963)	*Artibeus lituratus*‡ *Artibeus jamaicensis* Mosquitoes	?‡	Central America (Trinidad)
Tamiami	Calisher et al. (1970)	*Sigmodon hispidus*	*Sigmodon hispidus*	North America (Florida)

*Predominantly *C. laucha musculinus.*
†Subspecies *T. lugens fuscatus.*
‡Isolated between March 1956 and December 1958 from 11 bats and one pool of mosquitoes, but never since.

morphologic resemblance that led to the proposal that the Tacaribe complex viruses and LCM virus should form a taxonomic group. In 1970 the agent causing Lassa fever was isolated from patients with the disease in Nigeria. Its relatedness with the LCM virus was soon recognized, and in the same year a new virus group was defined containing LCM, Lassa, and the Tacaribe complex viruses. The proposed name was arenoviruses (from the Latin word *arenosus*, meaning sandy) — later changed to arenaviruses — which was to reflect the characteristic granules seen by electron microscopy in ultrathin sections of infected cells (Rowe et al., 1970). By approval of the International Committee on Taxonomy of Viruses this virus group has been given the status of a genus named *Arenavirus* belonging to the family *Arenaviridae*; type species is the LCM virus (Fenner, 1976).

MORPHOLOGY AND BIOCHEMICAL PROPERTIES

Of the many features that the arenaviruses have in common, the morphology of virus particles is salient (Murphy and Whitfield, 1975). Overall shape as well as structural details are identical. Arenavirus particles are round or pleomorphic with diameters of 50 to 300 nm (average 120 nm). In vivo and in vitro they are released from the infected cell by budding (Fig. 1). The viral envelopes are formed from plasma membranes: the density of both layers as well as the width of the intermediary zone increase, and club-shaped surface projections, approximately 10 nm long, are inserted often forming well-delineated margins. In their interior, arenavirus particles characteristically contain variable numbers of electron-dense granules that resemble cellular ribosomes, although morphologic, cytochemical, and functional differences have been noted (Dalton et al., 1968; Mannweiler and Lehmann-Grube, 1973; Welsh et al., 1975). They are usually randomly distributed, but sometimes they are arranged beneath the envelope in a circular pattern. If they are indeed host cell ribosomes, they are not essential for infectivity (Leung and Rawls, 1977). Perhaps they are accidentally incorporated during the budding process as are glycogen granules that have occasionally been observed inside arenavirus particles (Kajima and Majde, 1970).

Nucleocapsids have never been observed in arenavirus particles by thin section electron microscopy of infected cells. Such structures have been visualized, however, in purified preparations after removal of the viral envelopes with detergent (Palmer et al., 1977; Vezza et al., 1977; Gard et al., 1977). Their form varied among different arenaviruses, which may represent dissimilar architectures of individual members but, more likely, resulted from technical differences.

Many determinations of the size of the infec-

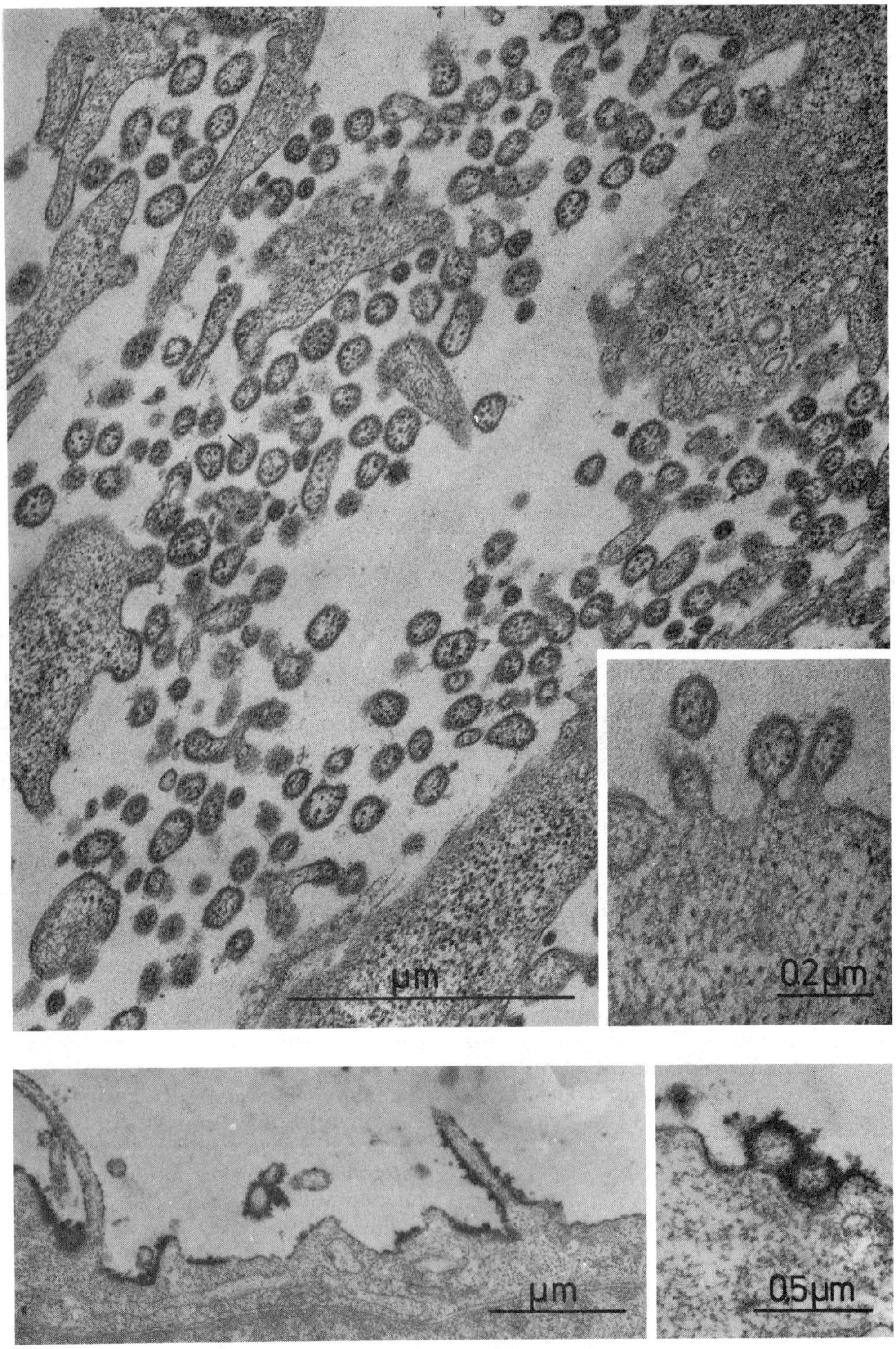

FIGURE 1. *Thin section electron micrographs of LCM virus-infected cultivated mouse (L) cells.*

Above. Multiple arenavirus particles are seen either released from the plasma membrane or still budding.

Below. Demonstration by means of the immunoperoxidase method of LCM virus-specific antigen(s) in association with budding virus particles and in distinct areas of the plasma membrane that are not otherwise visibly altered. Photographs kindly supplied by Dr. K. Mannweiler, Hamburg.

tious unit by various physical methods have given greatly divergent results, not only between different arenaviruses but also for one virus with different methods. Immuno-electron microscopy has shown (Fig. 1) that the characteristic particles budding from infected cells contain viral antigen(s). However, whether they are all virions and potentially infectious is unknown.

The nucleic acid of arenaviruses is single stranded RNA (Pfau, 1974) consisting of four components that sediment with 31 to 33S, 28S, 22 to 25S, and 18S, respectively. Small molecules with 4 to 6S have also been identified, and a 15S species has been resolved in Pichinde virus, but not yet in other arenaviruses. Present evidence indicates that the 31 to 33S ("L") and the 22 to 25S ("S") RNA classes are coded for by the virus and contain the viral genetic information. In contrast, the 28S and 18S RNA components originate from host cell ribosomes. They probably correspond to the ribosome-like granules seen inside arenavirus particles by electron microscopy (see above). The small RNA species resemble host cell transfer RNA with respect to both size and methylation ratio (Pedersen, 1973; Farber and Rawls, 1975; Vezza et al., 1978).

The virus-specific RNA (L and S) from the virions of Pichinde virus does not have properties of messenger RNA. Polysomal RNA, however, from virus-infected cells does have properties of messenger RNA and, furthermore, is complementary to the virion RNA (Leung et al., 1977). These findings, together with the demonstration of an association of RNA-dependent RNA polymerase with the virion (Carter et al., 1974), strongly suggest that the Pichinde virus RNA — and by analogy the RNA of the other arenaviruses — is negative stranded. This conclusion would explain why attempts to prepare infectious RNA from virions of LCM virus were unsuccessful (Welsh et al., 1975).

The protein composition of the arenaviruses is less complex than their morphology suggests. Analyses have disclosed three major proteins for LCM and Pichinde virions and only two for Tacaribe and Tamiami virions (Table 2). The place of Junin virus within the two groups is undecided. In view of the divergent results obtained for one virus (Pichinde) by different investigators, the significance of the differences of molecular weights between individual arenaviruses is uncertain. However, the difference of the number of structural polypeptides is real, and is one noteworthy exception to the rule that these viruses possess similar properties. In addition to the (two or three) major proteins, several minor ones have been found associated with the virions (Table 2), but whether all these are of viral origin is not certain.

The arenaviruses disintegrate rapidly in the presence of lipid solvents and detergents, indicating that lipids are needed for their structural integrity. By analogy with other budding viruses, these lipids probably are derived from the host cell membrane forming part of the viral envelope. The glycoproteins are situated on or close to the surface of the virion, probably in the spikes. The one nonglycosylated major protein has always been found in association with the RNA and is probably part of the ribonucleoprotein complex.

The buoyant densities of the infectious units of different arenaviruses vary between 1.17 and 1.18 when determined in sucrose and between 1.18 and 1.20 when determined in CsCl (Pfau, 1974). Gschwender et al. (1975) arrived at values for LCM virus of 1.22 in CsCl and 1.14 in a gradient formed of iodinated organic compounds.

During replication of LCM virus, in addition to

TABLE 2. Structural Proteins of Arenaviruses

VIRUS	MAJOR PROTEINS: Glycosylated 1	Glycosylated 2	Nonglycosylated	REFERENCE
LCM	"GP-1" 44*	"GP-2" 35	"NP" 63	Buchmeier and Oldstone (1978)
Pichinde	"V_{II}" 72	"V_{III}" 34	"V_I" 72†	Ramos et al. (1972)
	"G1" 64	"G2" 38	"N" 66†,‡	Vezza et al. (1977)
Tacaribe	None	"G" 42	"N" 68‡	Gard et al. (1977)
Tamiami	None	"G" 44	"N" 66‡	Gard et al. (1977)
Junin		"5" 38§	"3" 64§	Martínez Segovia and De Mítri (1977)

*Designation and approximate molecular weight ($\times 10^3$).

†A minor nonglycosylated protein with molecular weight 12×10^3 was not reproducibly found.

‡In addition, one minor nonglycosylated protein ("P") with molecular weight 77×10^3 (Pichinde virus and Tamiami virus) and 79×10^3 (Tacaribe virus).

§In addition, three minor glycoproteins ("1," "2," "4") with molecular weights 91×10^3, 72×10^3, 52×10^3, and one minor nonglycosylated protein ("6") with molecular weight 25×10^3.

fully infectious (standard) virus, particles are produced that cannot multiply by themselves but reduce the infectious yield and, if the standard virus is cytolytic, abolish cytopathic effects. These interfering particles are slightly less dense and relatively resistant to ultraviolet light, neutral red, and heat (Welsh et al., 1975; Popescu et al., 1976). Although these properties are characteristic for defective interfering (DI) particles, there is one essential difference. "By definition DI particles are defective virions, which can grow only in the presence of a helper standard virus" (Huang, 1973). In contrast, LCM virus-interfering particles multiply predominantly in cells infected by standard virus alone rather than in cells dually infected with standard virus and interfering particles (Lehmann-Grube et al., 1975). Interfering particles have also been detected in association with several other arenaviruses (Pfau, 1977).

ANTIGENIC COMPOSITION

Serologic relationships between arenaviruses have been demonstrated by complement-fixation tests (Casals et al., 1975) and immunofluorescence procedures (Wulff et al., 1978). In the case of Tacaribe complex viruses, the viral component responsible for cross-reactivity is the viral nucleoprotein (Buchmeier and Oldstone, personal communication). As a rule, serologic relationships cannot be demonstrated by use of the neutralization test with one notable exception: guinea pigs infected with Tacaribe virus develop protective immunity against Junin virus and also Junin virus-neutralizing antibody (Weissenbacher et al., 1975).

The immunogen that induces complement-fixing antibody has been characterized for LCM and Pichinde viruses (Gschwender et al., 1976; Buchmeier et al., 1977). It is produced in abundance in infected cells and is also a structural component of the virion; it is not, however, represented on the surface of either the virion or the infected cell and is distinct from the antigen combining with neutralizing antibody. Although serologically indistinguishable, the complement-fixing antigen from infected cells differs chemically from the antigen that is associated with the virion. Possibly the cell-associated polypeptides with complement-fixing activity are cleavage products of the nucleocapsid protein, which is the carrier of the complement-fixing activity residing inside the virion.

By necessity, the immunogen leading to the formation of neutralizing antibody is on the surface of the virion, but its chemical nature is unknown; presumably it is a glycoprotein.

In double immunodiffusion tests in agar using extracts of LCM or Pichinde virus-infected cells and the respective antisera one line of precipitation is consistently formed. Its antigenic component is heat-stable and pronase-resistant, and corresponds to the complement-fixing antigen just mentioned. A second line is weaker and less regularly observed; its antigenic component is heat-labile and susceptible to the activity of the enzyme (Bro-Jørgensen, 1971; Buchmeier et al., 1977).

In the cytoplasm of LCM virus-infected cells two types of antigens can be distinguished by immunofluorescence (Rutter and Gschwender, 1973). One consists of coarse granules with bright fluorescence; serologically it is identical with the complement-fixing antigen mentioned above. The other antigen has a fine dust-like distribution; its relationship with viral structures is not known.

Several methods detect new antigens on the surfaces of LCM virus-infected cells. Letting the viable cell interact with fluorescein-labeled antibody and inspecting it with an immunofluorescence microscope is one, and incubating the infected cell with antiserum plus complement and watching for lysis is another (Rutter and Gschwender, 1973). New antigens on the surface of LCM virus-infected cells have also been localized by immuno–electron microscopy (Mannweiler and Lehmann-Grube, 1973). Not surprisingly, budding viruses are thus labeled. In addition, virus-specific antigen is found in areas of the plasma membrane where morphologic alterations are not apparent (Fig. 1). Whether they are sites in which budding of virions is imminent or whether some other viral antigen is labeled that is not part of the surface of the viral particles is not known.

The LCM virus does not damage vital tissues of the mouse, at least not enough to impair function (Lehmann-Grube, 1971). Nonetheless, infection of an adult mouse with LCM virus results in severe, sometimes lethal, disease. The explanation is that the immune response may kill the host; the LCM disease of the adult mouse is a pathologic immune response to virus-induced antigens (Hotchin, 1962). The antigens (allergens) that induce the allergic response must be on the surface of infected cells, but whether they are the budding virions or other virus-induced alterations of the cell membrane or a combination of both has not been decided. The immune response of the mouse leading to disease and even death is entirely cell-mediated; T lymphocytes play an essential role (Cole and Nathanson, 1974; Johnson et al., 1978). It is postulated that they damage virus-infected cells of mouse tissues just as the cytotoxic T cells damage virus-infected target cells in vitro (Zinkernagel and Doherty, 1977).

Similar observations have been made with other arenavirus-host combinations — for instance, LCM virus-adult rat, Junin virus-suckling mouse, Tamiami virus-suckling mouse, Lassa fever virus-adult mouse — but in none of these has the immunopathologic nature of the disease been as convincingly demonstrated as in the LCM virus-infected adult mouse.

VIRUS-HOST RELATIONSHIPS

The host range of LCM virus is so wide that no mammalian species is known to resist infection. Virus strains differ in pathogenic properties. As a rule, disease signs are moderate or not detected; but there are exceptions. One has already been mentioned: the adult mouse responds with a severe illness to infection with LCM virus. Clinical signs also develop in humans and monkeys. One widely employed virus strain ("WE") is absolutely lethal for guinea pigs; in other words, one guinea pig infectious dose of WE strain virus is identical with one lethal dose. In vitro, too, cells from most mammalian species can be infected with LCM virus and respond with replication of infectious virus. Whether virus multiplication is accompanied by cytopathic effects depends on the strain of virus and its previous passage history as much as on origin and properties of the host cells. For instance, passage of LCM virus through mouse brains selects for cytolytic variants, whereas prolonged replication in the mouse spleen results predominantly in noncytopathogenic virus (Popescu and Lehmann-Grube, 1976). Cytopathic effects are blocked by interfering particles (see above).

The principal hosts of the other arenaviruses are listed in Table 1. Infection with Lassa virus of squirrel monkey (*Saimiri scirreus*) and guinea pig (Walker et al., 1975), with Machupo virus of rhesus and cynomolgus monkeys (*Macaca mulatta* and *M. fascicularis*), suckling mouse and suckling hamster, strain C-13 guinea pig, and marmoset (*Saguinus geoffroyi*) (Eddy et al., 1975; Webb et al., 1975), and with Junin virus of guinea pig and suckling mouse (Weissenbacher et al., 1975) leads to disease signs that often resemble human Lassa fever, Bolivian hemorrhagic fever, and Argentine hemorrhagic fever, respectively. Most arenaviruses multiply with cytopathic effects in Vero cells; other cell lines widely employed are L and BHK21.

Arenaviruses have attracted much attention by their propensity to establish prolonged infections in rodents but there is no reason to assume that prolonged infection is restricted to these animals. The best studied example is the persistent infection of *M. musculus* with LCM virus. When a mouse is infected with LCM virus in utero or shortly after birth, it becomes a lifelong carrier, meaning that the virus is not eliminated and that it multiplies in all organs without causing overt disease (Fig. 2). Although carrier mice respond normally to most antigens, LCM virus-specific antibodies are formed at only low concentrations or cannot be detected (Volkert et al., 1975), and LCM virus-specific cell-mediated immunity is absent (Cihak and Lehmann-Grube, 1978). Since immune elimination of LCM virus is readily accomplished by mice infected as adult animals, failure of virus elimination by a mouse infected congenitally or neonatally is attributed to LCM virus-specific immunologic tolerance (Volkert et al., 1975).

In carrier mice, replication of the virus is regulated so as to replenish losses due to natural decay, but not allow uninhibited increase. The mechanism of virus control is probably very complex (Lehmann-Grube, 1977) and presumably in-

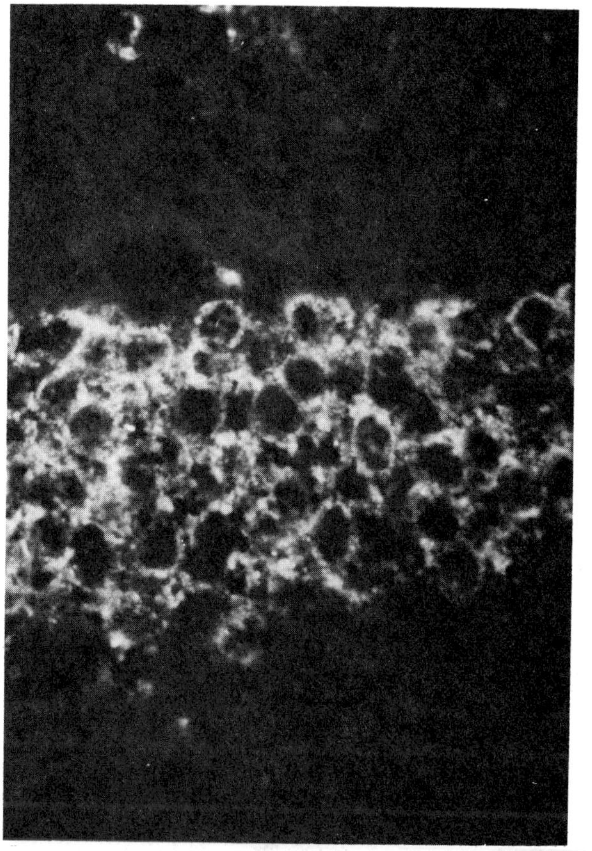

FIGURE 2. *Brain of an LCM virus carrier mouse (stratum granulosum areae dentatae of the hippocampal region). Infected cells, primarily neurons, are visualized by immunofluorescence procedure. Photograph kindly supplied by Dr. J. Löhler, Hamburg.*

volves interfering particles that are generated in LCM virus-carrier mice (Popescu and Lehmann-Grube, 1977).

The type of virus-host relationship exemplified by LCM virus *M. musculus* has been termed persistent tolerant infection; its salient features are lifelong duration, lack of a specific immune response, and absence of disease signs. Vertical transmission is also characteristic of the LCM virus-mouse relationship, but not a *conditio sine qua non* for designating an infection persistent tolerant.

The statement that mice persistently infected with LCM virus remain healthy requires qualification. In certain mouse strain-virus strain combinations aging carrier mice develop a "late onset disease" (Hotchin and Collins, 1964) characterized by runting and shortened life span. Many organs are affected, but a mesangioproliferative glomerulonephritis is prominent (Fig. 3). The evidence that the pathogenic mechanism is formation and deposition of immune complexes (Fig. 4) is impressive (Oldstone, 1975). However, the immunologic specificity of the deposits has not been unequivocally determined.

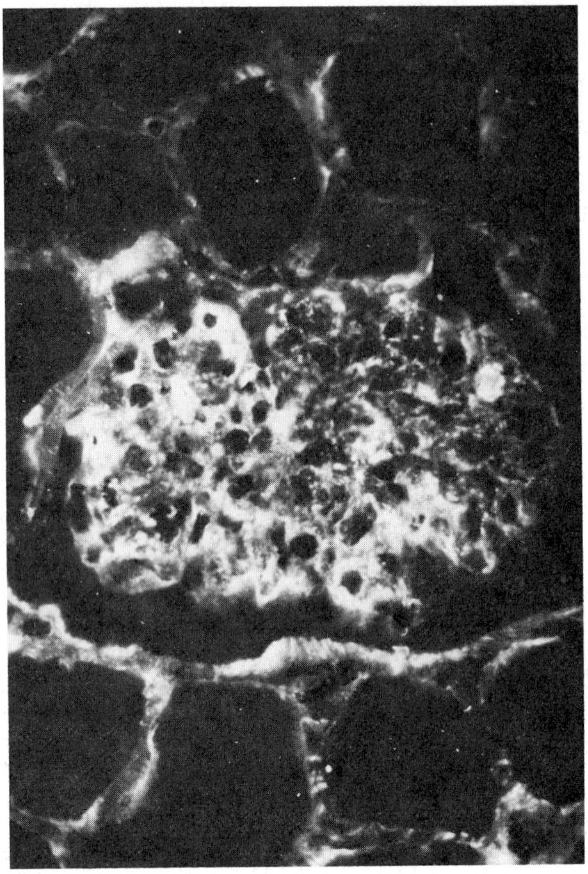

FIGURE 4. *Deposition of immune complexes in the kidney of an LCM virus carrier mouse. Host immunoglobulin in the subendothelial space of capillaries and basement membrane of Bowman's capsule of a glomerulus and in the basement membranes of tubuli is visualized by immunofluorescence procedure. Photograph kindly supplied by Dr. J. Löhler, Hamburg.*

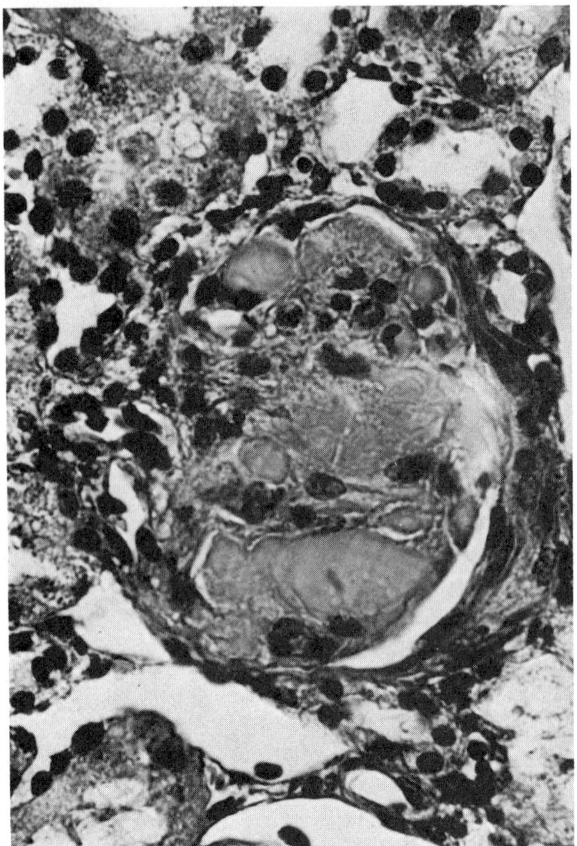

FIGURE 3. *Final stage of glomerulonephritis of an LCM virus carrier mouse as part of late onset disease. Note the marked hyalinization of the glomerulum. Photograph kindly supplied by Dr. J. Löhler, Hamburg.*

With the exception of Tacaribe virus all arenaviruses have been isolated from rodents with prolonged infections. However, only for LCM virus *Mus musculus* has the pattern of infection been fully established. Reduced virus-specific immune responsiveness has been documented in several instances, especially well for *C. callosus* infected with Machupo virus (Justines and Johnson, 1969), but it is uncertain whether other arenavirus-host combinations (listed in Table 1) are persistent tolerant infections or merely unusually prolonged, in which case they should be termed chronic.

A better understanding of prolonged infection has been obtained by analyzing persistent infections of cell cultures with LCM virus in L cells and LCM virus, Parana virus, and Pichinde virus in BHK21 cells (Pfau, 1977). It appears that regulation of these infections, which go on for years, depends on an intricate interplay involving cells, virulent standard virus, attenuated standard virus, and interfering particles (Lehmann-Grube et al., 1975).

EPIDEMIOLOGY

The LCM virus has been found throughout Europe and America; whether it is distributed worldwide, as often stated, is doubtful (Lehmann-Grube, 1971). All other members of this group occur in small areas in South and Central America and in Africa. With the exception of the Tacaribe virus, arenaviruses are maintained in nature by rodents, and rodents are the principal source for infections of other animals and man. The mode of spread is not known with certainty, but the high virus concentrations in nasal secretions, saliva, and urine suggest that the virus is disseminated either by direct contact or via fomites and food. Airborne transmission is a further possibility. Arthropod vectors have never been convincingly incriminated.

The principal host for LCM virus is the persistently infected grey house mouse (*Mus musculus*) (Armstrong et al., 1940). More recently, the Syrian hamster (*Mesocricetus auratus*) has also been identified as an important source for human infections (Ackermann et al., 1972; Gregg, 1975). Probably this animal is not a natural host of LCM virus but is infected from house mice that invade the hamsteries (Skinner et al., 1976). Transmission among members of mammalian species other than *M. musculus* and *M. auratus* are rare. In only one instance has man-to-man infection been documented; the circumstances were unusual (Smadel et al., 1942). Lassa virus is indigenous to certain areas of Africa where it is maintained by *Mastomys natalensis* (Monath, 1975). Unlike infections with other arenaviruses, Lassa virus is easily transmitted from person to person and several hospital-centered outbreaks with usually one index case and secondary infections of medical personnel and visitors have been described. Outbreaks have also occurred in communities. Sources for human infections with Machupo and Junin viruses, etiologic agents of Bolivian and Argentine hemorrhagic fevers, are rodents of the genera *Calomys* and *Akodon* (Johnson et al., 1973; Sabattini et al., 1977). As the names indicate, these diseases occur in Bolivia and Argentina, where they cause considerable health problems.

References

Ackermann, R., Stille, W., Blumenthal, W., Helm, E.B., Keller, K., und Baldus, O.: Syrische Goldhamster als Überträger von Lymphozytärer Choriomeningitis. Dtsch med Wsch 97:1725, 1972.

Armstrong, C., and Lillie, R.D.: Experimental lymphocytic choriomeningitis of monkeys and mice produced by a virus encountered in studies of the 1933 St. Louis encephalitis epidemic. Public Health Rep 49:1019, 1934.

Armstrong, C., Wallace, J.J., and Ross, L.: Lymphocytic choriomeningitis. Gray mice, *Mus musculus*, a reservoir for the infection. Public Health Rep 55:1222, 1940.

Bro-Jørgensen, K.: Characterization of virus-specific antigen in cell culture infected with lymphocytic choriomeningitis virus. Acta path microbiol scand, Sect B 79:466, 1971.

Buchmeier, M.J., Gee, S.R., and Rawls, W.E.: Antigens of Pichinde virus. I. Relationship of soluble antigens derived from infected BHK-21 cells to the structural components of the virion. J Virol 22:175, 1977.

Buchmeier, M.J., and Oldstone, M.B.A.: Virus-induced immune complex disease: identification of specific viral antigens and antibodies deposited in complexes during chronic lymphocytic choriomeningitis virus infection. J. Immun 120:1297, 1978.

Buckley, S.M., and Casals, J.: Lassa fever, a new virus disease of man from West Africa. III. Isolation and characterization of the virus. Am J Trop Med 19:680, 1970.

Calisher, C.H., Tzianabos, T., Lord, R.D., and Coleman, P.H.: Tamiami virus, a new member of the Tacaribe group. Am J Trop Med 19:520, 1970.

Carter, M.F., Biswal, N., and Rawls, W.E.: Polymerase activity of Pichinde virus. J Virol 13:577, 1974.

Casals, J., Buckley, S.M., and Cedeno, R.: Antigenic properties of the arenaviruses. Bull WHO 52:421, 1975.

Cihak, J., and Lehmann-Grube, F.: Immunological tolerance to lymphocytic choriomeningitis virus in neonatally infected virus carrier mice: evidence supporting a clonal inactivation mechanism. Immunology 34:265, 1978.

Cole, G.A., and Nathanson, N.: Lymphocytic choriomeningitis. Pathogenesis. Progr Med Virol 18:94, 1974.

Dalton, A.J., Rowe, W.P., Smith, G.H., Wilsnack, R.E., and Pugh, W.E.: Morphological and cytochemical studies on lymphocytic choriomeningitis virus. J Virol 2:1465, 1968.

Downs, W.G., Anderson, C.R., Spence, L., Aitken, T.H.G., and Greenhall, A.H.: Tacaribe virus, a new agent isolated from *Artibeus* bats and mosquitoes in Trinidad, West Indies. Am J Trop Med 12:640, 1963.

Eddy, G.A., Scott, S.K., Wagner, F.S., and Brand, O.M.: Pathogenesis of Machupo virus infection in primates. Bull WHO 52:517, 1975.

Farber, F.E., and Rawls, W.E.: Isolation of ribosome-like structures from Pichinde virus. J Gen Virol 26:21, 1975.

Fenner, F.: Classification and nomenclature of viruses. Second report of the International Committee on Taxonomy of Viruses. Intervirology 7:1, 1976.

Gard, G.P., Vezza, A.C., Bishop, D.H.L., and Compans, R.W.: Structural proteins of Tacaribe and Tamiami virions. Virology 83:84, 1977.

Gregg, M.B.: Recent outbreaks of lymphocytic choriomeningitis in the United States of America. Bull WHO 52:549, 1975.

Gschwender, H.H., Rutter, G., and Lehmann-Grube, F.: Lymphocytic choriomeningitis virus. II. Characterization of extractable complement-fixing activity. Med Microbiol Immunol 162:119, 1976.

Gschwender, H.H., Rutter, G., and Popescu, M.: Use of iodinated organic compounds for the density gradient centrifugation of viruses. Arch Virol 49:359, 1975.

Hotchin, J.: The biology of lymphocytic choriomeningitis infection: virus-induced immune disease. Cold Spr Harb Symp Quant Biol 27:479, 1962.

Hotchin, J., and Collins, D.N.: Glomerulonephritis and late onset disease of mice following neonatal virus infection. Nature 203:1357, 1964.

Huang, A.S.: Defective interfering viruses. Ann Rev Microbiol 27:101, 1973.

Johnson, E.D., Monjan, A.A., and Morse, H.C.: Lack of B-cell participation in acute lymphocytic choriomeningitis disease of the central nervous system. Cell Immunol 36:143, 1978.

Johnson, K.M., Webb, P.A., and Justines, G.: Biology of Tacaribe-complex viruses. In Lehmann-Grube, F. (ed.): Lymphocytic Choriomeningitis Virus and Other Arenaviruses. Berlin, Heidelberg, New York, Springer-Verlag, 1973, p. 241.

Johnson, K.M., Wiebenga, N.H., Mackenzie, R.B., Kuns, M.L., Tauraso, N.M., Shelokov, A., Webb, P.A., Justines, G., and Beye, H.K.: Virus isolations from human cases of hemorrhagic fever in Bolivia. Proc Soc Exper Biol Med 118:113, 1965.

Justines, G., and Johnson, K.M.: Immune tolerance in *Calomys callosus* infected with Machupo virus. Nature 222:1090, 1969.

Kajima, M., and Majde, J.: LCM virus as a carrier of non-viral cellular components. Electron microscopic evidence. Naturwissenschaften 57:93, 1970.

Lehmann-Grube, F.: Lymphocytic Choriomeningitis Virus. Virology Monographs Vol. 10. Wien and New York, Springer-Verlag, 1971.

Lehmann-Grube, F.: Lymphocytic choriomeningitis virus carrier mice. Factors determining virus persistence. Medicina (B.A.) 37 Suppl 3:78, 1977.

Lehmann-Grube, F., Popescu, M., Schaefer, H., and Gschwender, H.H.: LCM virus infection of cells *in vitro*. Bull WHO 52:443, 1975.

Leung, W.C., Ghosh, H.P., and Rawls, W.E.: Strandedness of Pichinde virus RNA. J Virol 22:235, 1977.

Leung, W.C., and Rawls, W.E.: Virion-associated ribosomes are not required for the replication of Pichinde virus. Virology 81:174, 1977.

Mannweiler, K., and Lehmann-Grube, F.: Electron microscopy of LCM virus-infected L cells. In Lehmann-Grube, F. (ed.): Lymphocytic Choriomeningitis Virus and Other Arenaviruses. Berlin, Heidelberg, New York, Springer-Verlag, 1973, p. 37.

Martínez Segovia, Z.M. de, and De Mitri, M.I.: Junin virus structural proteins. J Virol 21:579, 1977.

Monath, T.P.: Lassa fever: review of epidemiology and epizootiology. Bull WHO 52:577, 1975.

Murphy, F.A., and Whitfield, S.G.: Morphology and morphogenesis of arenaviruses. Bull WHO 52:409, 1975.

Oldstone, M.B.A.: Virus neutralization and virus-induced immune complex disease. Progr Med Virol 19:84, 1975.

Palmer, E.L., Obijeski, J.F., Webb, P.A., and Johnson, K.M.: The circular, segmented nucleocapsid of an arenavirus-Tacaribe virus. J Gen Virol 36:541, 1977.

Parodi, A.S., Greenway, D.J., Rugiero, H.R., Frigerio, M., de la Barrera, J.M., Mettler, N., Garzón, F., Boxaca, M., Guerrero, L., y Nota, N.: Sobre la etiología del brote epidémico de Junín. Día méd (B.A.) 30:2300, 1958.

Pedersen, I.R.: LCM virus: its purification and its chemical and physical properties. In Lehmann-Grube, F. (ed.): Lymphocytic Choriomeningitis Virus and Other Arenaviruses. Berlin, Heidelberg, New York, Springer-Verlag, 1973, p. 13.

Pfau, C.J.: Biochemical and biophysical properties of the arenaviruses. Progr Med Virol 18:64, 1974.

Pfau, C.J.: The role of defective interfering (DI) virus in arenavirus infections. Medicina (B.A.) 37 Suppl 3:32, 1977.

Pinheiro, F.P., Shope, R.E., Paes de Andrade, A.H., Bensabath, G., Cacios, G.V., and Casals, J.: Amapari, a new virus of the Tacaribe group from rodents and mites of Amapa Territory, Brazil. Proc Soc Exper Biol Med 122:531, 1966.

Popescu, M., and Lehmann-Grube, F.: Diversity of lymphocytic choriomeningitis virus: variation due to replication of the virus in the mouse. J Gen Virol 30:113, 1976.

Popescu, M., and Lehmann-Grube, F.: Defective interfering particles in mice infected with lymphocytic choriomeningitis virus. Virology 77:78, 1977.

Popescu, M., Schaefer, H., and Lehmann-Grube, F.: Homologous interference of lymphocytic choriomeningitis virus: detection and measurement of interference focus-forming units. J Virol 20:1, 1976.

Ramos, B.A., Courtney, R.J., and Rawls, W.E.: Structural proteins of Pichinde virus. J Virol 10:661, 1972.

Rowe, W.P., Murphy, F.A., Bergold, G.H., Casals, J., Hotchin, J., Johnson, K.M., Lehmann-Grube, F., Mims, C.A., Traub, E., and Webb, P.A.: Arenoviruses: proposed name for a newly defined virus group. J Virol 5:651, 1970.

Rutter, G., and Gschwender, H.H.: Antigenic alteration of cells in vitro infected with LCM virus. In Lehmann-Grube, F. (ed.): Lymphocytic Choriomeningitis Virus and Other Arenaviruses. Berlin, Heidelberg, New York, Springer-Verlag, 1973, p. 51.

Sabattini, M.S., González de Ríos, L.E., Díaz, G., y Vega, V.R.: Infección natural y experimental de roedores con virus Junín. Medicina (B.A.) 37 Suppl 3:149, 1977.

Skinner, H.H., Knight, E.H., and Buckley, L.S.: The hamster as a secondary reservoir host of lymphocytic choriomeningitis virus. J Hyg 76:299, 1976.

Smadel, J.E., Green, R.H., Paltauf, R.M., and Gonzales, T.A.: Lymphocytic choriomeningitis: two human fatalities following an unusual febrile illness. Proc Soc Exper Biol Med 49:683, 1942.

Trapido, H., and Sanmartín, C.: Pichindé virus. A new virus of the Tacaribe group from Colombia. Am J Trop Med 20:631, 1971.

Vezza, A.C., Clewley, J.P., Gard, G.P., Abraham, N.Z., Compans, R.W., and Bishop, D.H.L.: Virion RNA species of the arenaviruses Pichinde, Tacaribe, and Tamiami. J Virol 26:485, 1978.

Vezza, A.C., Gard, G.P., Compans, R.W., and Bishop, D.H.L.: Structural components of the arenavirus Pichinde. J Virol 23:776, 1977.

Volkert, M., Bro-Jørgensen, K., and Marker, O.: Persistent LCM virus infection in the mouse. Immunity and tolerance. Bull WHO 52:471, 1975.

Walker, D.H., Wulff, H., Lange, J.V., and Murphy, F.A.: Comparative pathology of Lassa virus infection in monkeys, guinea-pigs, and *Mastomys natalensis*. Bull WHO 52:523, 1975.

Webb, P.A., Johnson, K.M., Hibbs, J.B., and Kuns, M.L.: Parana, a new Tacaribe complex virus from Paraguay. Arch ges Virusforsch 32:379, 1970.

Webb, P.A., Johnson, K.M., Peters, C.J., and Justines, G.: Behavior of Machupo and Latino viruses in Calomys callosus from two geographic areas of Bolivia. In Lehmann-Grube, F. (ed.): Lymphocytic Choriomeningitis Virus and Other Arenaviruses. Berlin, Heidelberg, New York, Springer-Verlag, 1973, p. 313.

Webb, P.A., Justines, G., and Johnson, K.M.: Infection of wild and laboratory animals with Machupo and Latino viruses. Bull WHO 52:493, 1975.

Weissenbacher, M.C., de Guerrero, L.B., and Boxaca, M.C.: Experimental biology and pathogenesis of Junin virus infection in animals and man. Bull WHO 52:507, 1975.

Welsh, R.M., Burner, P.A., Holland, J.J., Oldstone, M.B.A., Thompson, H.A., and Villarreal, L.P.: A comparison of biochemical and biological properties of standard and defective lymphocytic choriomeningitis virus. Bull WHO 52:403, 1975.

Wulff, H., Lange, J.V., and Webb, P.A.: Interrelationships among arenaviruses measured by indirect immunofluorescence. Intervirology 9:344, 1978.

Zinkernagel, R.M., and Doherty, P.C.: Major transplantation antigens, viruses, and specificity of surveillance T cells. Cont Top Immunobiol 7:179, 1977.

63 REOVIRIDAE PATHOGENIC FOR MAN

Neville F. Stanley, D. Sc.

Reoviridae viruses are of interest to molecular biologists because all of them have a double-stranded RNA genome. Because of their great ubiquity in infecting plants, arthropods, and vertebrates, there still exist taxonomic problems that have not yet been resolved owing to paucity of data with some of these viruses. Our concern in this chapter will be with human infections with viruses belonging to three genera within this family. It is convenient on morphologic, biochemical, and biophysical properties to include two recognized (Reovirus and Orbivirus) and two 'tentative' (Rotavirus and Plant-insect) genera within this family (see Table 1) whose nomenclature is still under consideration by an international committee. The table sets out some of the better known viruses that have been sufficiently well described to include them in the family, but only those appropriately marked infect humans and will form the basis for discussion in this brief review.

Most of the orbiviruses and plant-insect viruses replicate in arthropods, whereas there is no evidence to show that reoviruses or rotaviruses replicate in arthropods, although the former have been isolated from mosquitoes (Stanley, 1977).

REOVIRUSES

The properties, comparative biology, pathogenesis, and diagnostic procedures have been described in a number of reviews (Hassan and Cochran, 1966; Stanley, 1967, 1974, 1977; Rosen, 1968) that are useful for further reference. The viruses were originally discovered by isolation in infant mice when they were called *hepatoencephalomyelitis* viruses (Stanley et al., 1953, 1954). Later these agents were named *reoviruses*, which derives from *r*espiratory *e*nteric *o*rphan (Sabin, 1959). The double-stranded RNA core is surrounded by an inner and an outer protein shell. The outer shell has a diameter of between 60 and 75 nm and is icosahedral (5:3:2 symmetry).

Antigenic Structure

All mammalian strains of reovirus fall into three serotypes, known as 1, 2, and 3. Hemagglutination-inhibition and neutralization tests have been primarily used for this differentiation, although a common antigen is detected by complement-fixation or by immuno-diffusion. It has not been possible to distinguish antigenic differences between human and animal isolates. All three serotypes produce hemagglutinins to erythrocytes of humans and other animals.

Susceptible Cell Lines

All three types produce a distinctive cytopathic effect in many cell lines. Those cells that have been used for isolation or replication studies comprise *Macaca* kidney, *Cercopithecus* kidney, primary human kidney, human amnion, HELA and mouse L, as well as cells from marsupials, dogs, cats, guinea pigs, rabbits, dolphins, calves, and chick embryos (Stanley, 1977). The reovirus cytoplasmic inclusions are recognized in the perinuclear areas by staining or electron microscopy. Persistent infection with some cells has been reported and plaques useful for assays and replication studies have been produced with murine and simian cell lines. The main effects on host cell function are inhibition of DNA and protein synthesis and interferon induction.

TABLE 1. Reoviridae

PROPERTIES OF WELL-DEFINED MEMBERS

Capsid of 60–80 nm diameter
Icosahedral symmetry (with inner and outer capsid)
Genomes of 10–12 molecules of ds rna
Virion Contains Transcriptase
Cytoplasmic replication
Resistant to lipid solvents

REOVIRUS	ORBIVIRUS*	ROTAVIRUS	PLANT-INSECT VIRUSES*
3 Mammalian†	Many members, including:	Infantile diarrhea†	Several members in at least two genera including:
5 Avian	Bluetongue	Nebraska calf scours	Wound tumor
Simian	African horse sickness	Simian virus SA 11	Rice dwarf
Canine	Colorado tick fever†	EDIM§	Maize rough
	Corriparta	Pig enteritis	Sugarcane gall
	Eubenangee	"O" virus	
	EHD‡	Foal diarrhea	
	Ibaraki		

*Natural arthropod transmission.
†Infects humans.
‡Epizootic hemorrhagic disease of deer.
§Epizootic diarrhea of infant mice.

Clinical Disease

Human. The association of any of three serotypes with disease in man is uncertain because of great virus ubiquity and the absence of clinical manifestations after infection. Outbreaks of mild clinical illness, supported by virus isolation, occur mainly with children with respiratory and gastrointestinal tract involvement. Only four fatal cases have been recorded, and all of these appear to have had a disseminated infection with lesions in brain, heart, lung, or liver (Stanley, 1977). There have been many other associations that need further confirmation — one of particular interest being keratoconjunctivitis (Jarudi et al., 1973).

Animal. Ubiquity of the reoviruses also poses similar difficulties with natural animal infections, although there is little doubt that reovirus 3 produces spontaneous disease in some mouse colonies, and reovirus 2 causes upper respiratory tract infection in laboratory-housed chimpanzees. A considerable amount of useful information has been derived from experimental murine models exploited by Australian workers (Stanley, 1974, 1977). These have developed our knowledge of virus pathogenesis in three areas: reovirus 1 — hydrocephalus; reovirus 3 — active chronic hepatitis and lymphoma; and reovirus 3 — chronic biliary obstruction. In addition to mice, experimental infections have been established in rats, guinea pigs, hamsters, rabbits, ferrets, dogs, swine, cattle, and nonhuman primates. Little is known about humoral or cell-mediated immunity in reovirus infections. Recent studies suggest the association of temperature-sensitive mutants with persistent infection and possibly specific clinicopathologic syndromes (Raine and Fields, 1974).

Laboratory Diagnosis

This is based on the isolation and identification of the virus by cell cultures or infant mouse inoculation, the detection of the virus or its antigens, the presence of specific cellular changes and/or the development of a specific, significant, serologic response. Standard and detailed techniques have been published (Lennette et al., 1974). When cell cultures are used, primary *Macaca* kidney or human kidney are adequate. Intraperitoneal inoculation of newborn mice is a useful adjunct to cell culture for virus isolation, but the mouse colony should be demonstrated to be free from reovirus. Immunofluorescence and immunoperoxidase techniques have been developed for the detection of reovirus antigens. Serologic diagnosis is not helpful because of the early acquisition of antibody in most populations.

Four serologic tests are currently used for antibody estimation: hemagglutination-inhibition, complement-fixation, neutralization, and immunodiffusion.

Epidemiology

Antibodies to the three serotypes are widely distributed and antibody conversion rates show that infection is acquired early in life (Stanley et al., 1954, 1964; Stanley, 1961, 1967, 1974, 1977; Stanley and Leak, 1963; Rosen, 1968). The virus is usually isolated from the feces of children. Although the three mammalian types of reovirus have been isolated from large numbers of many vertebrates, an even greater ubiquity is indicated by the widespread occurrence of naturally occurring antibody. It would appear that the following, in addition to man, are frequently infected: nonhuman primates, cattle, horses, sheep, dogs, cats, rats, mice, rabbits, hares, guinea pigs, marsupials, reptiles, poultry, wild birds, and bats (Stanley, 1977).

The epidemiologic patterns indicate fecal-oral transmission, but infection via the respiratory tract has yet to be substantiated. It is possible that animal strains infect man and that human strains may infect animals. Their physical properties enable reoviruses to survive in the environment and they may, therefore, contaminate some ecosystem. It has been suggested (Stanley, 1977) that they be used as a virus marker for bird, animal, or human fecal pollution of water. All types have been isolated from stagnant water, river water, or raw sewage, and sometimes in higher concentrations than entero-viruses. The high rate of isolation from raw sewage (34 per cent) does not correlate with the low isolation rate (0.2 per cent) from human feces. Reoviruses may be recovered more consistently from sewage than most other viruses (England, 1972). Other evidence for gross contamination of the environment is the isolation of reoviruses from mosquitoes.

Although reoviruses then are not to be considered as serious producers of human disease, their involvement in congenital virus infection should be considered (Stanley, 1977).

ORBIVIRUSES

Borden et al. (1971) suggested the name *orbivirus* (from *orbis* [L], ring or circle) for a distinctive group of viruses equal in hierarchy to the reoviruses. They may be divided into antigenic subgroups, but do not share a common antigen (see Table 1). Although reoviruses are stable at pH 3.0, orbiviruses are inactivated at this pH. By thin section and negative-stain electron microscopy, Murphy et al. (1971) were able to distinguish orbiviruses from togaviruses and reoviruses. The surface architecture of bluetongue virus suggests that this virus with 32 morphologic units has an icosahedral structure, distinct from reoviruses. The virion is smaller than that of reoviruses and possesses large capsomeres with the appearance of rings. The outer layer covering the inner icosahedral capsid is frequently indistinct (see Fig. 1).

Although most members are transmitted by and multiply in arthropods they frequently cause viremia in the vertebrate hosts. The best known and perhaps the most important of the animal pathogens are bluetongue (20 serotypes) and African horse sickness (9 serotypes). The only known human pathogen in this subdivision is Colorado tick fever, which is spread to man through the bite of an infected tick. Only one antigenic type is known and an infection usually produces effective long-lasting immunity. With the exception of one fatal case, the disease is benign with headache, ocular pain, muscle and joint pains, lumbar pains, nausea, and vomiting. The pathogenesis is not clearly understood. Leukopenia is common. Patients have a persistent circulating erythrocyte-associated viremia and the disease has been transmitted by blood transfusion. Electron microscopy studies suggest that the virus infects early-stage hematopoietic cells with movement of virus to the circulating blood cells (Emmons et al., 1972).

Diagnosis may be made accurately and rapidly by fluorescent antibody staining of peripheral blood smears. The virus can also be isolated from whole blood by inoculation of newborn mice or hamsters, or by cell cultures. Antibodies are usually detected by a complement-fixation test.

The major distribution of the human disease confined to North America has followed the distribution of the wood tick, *Dermacentor andersoni*, which is primarily found in Colorado, Oregon, Utah, Idaho, Montana, and Wyoming. The tick is clearly the true reservoir and the virus may be transmitted transovarially by the adult female. Infected rodents act as reservoirs for the immature tick. Recent studies have indicated endemic areas in California, where 205 cases have occurred since 1954. It has been estimated that more than 2000 cases occur annually in Colorado.

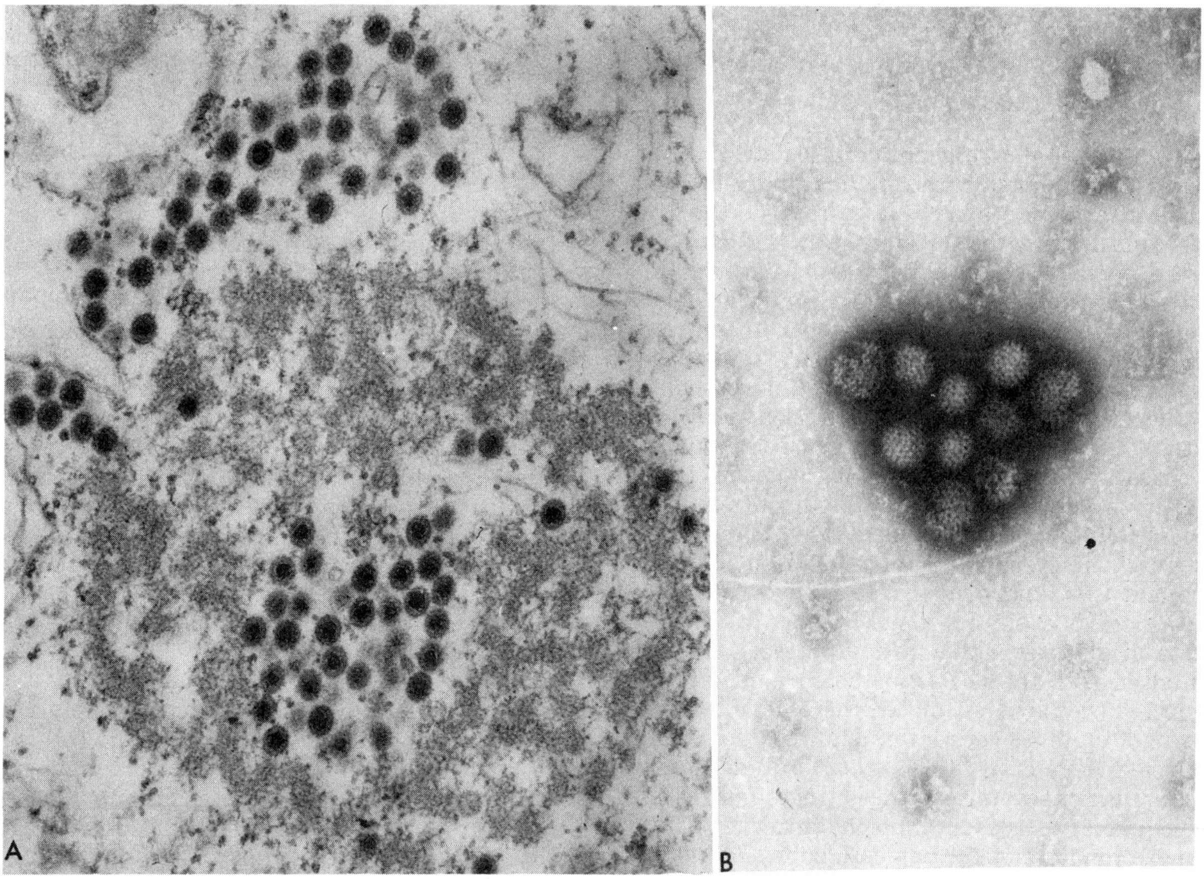

FIGURE 1. *Electron micrographs of bluetongue (an orbivirus) grown in baby hamster kidney cells. (Supplied through the courtesy of Dr. Frederick A. Murphy, Center for Disease Control, Atlanta, Georgia, USA.)*
(A) × 109,000 (B) × 174,000

The virus has also been found in an area where *D. andersoni* is absent. This suggests a maintenance cycle different from the chipmunk-ground squirrel *D. andersoni* cycle. The virus has recently been isolated from Ixodid and Argarid ticks (for further reference see Emmons et al., 1972; Oshiro, et al., 1978).

ROTAVIRUSES

These viruses are the most common cause of infantile enteritis — a disease now known to be widespread throughout the world and usually with a seasonal winter peak. Until agreement has been reached on nomenclature the term *rotavirus* will be used. It is synonymous with *duovirus* used by the Australian workers and some others (Holmes et al., 1975; Davidson et al., 1975). The disease and the virus are currently under extensive study and many of the observations relate to direct electron microscopy of human fecal material. New methods are now being employed and a chapter of this nature will need consistent modification as more precise information becomes available, particularly with the use of antisera against other nonhuman rotaviruses such as calf diarrhea.

Clinical and Epidemiologic Characteristics

Vomiting and significant dehydration are consistently associated with rotavirus diarrhea. The highest incidence is in children between the ages of 7 and 24 months. Most hospitalized children present with high fever between the second and fifth days after the diarrhea and vomiting. There may be involvement of the upper respiratory tract. The mean duration of uncomplicated cases is between one and two weeks, but coincident pathogenic enterobacterial infection may increase the severity and extend the time of the disease. The incubation period appears to be about two to three days. There is little doubt about the high winter incidence but no completely satisfactory explanation has yet been offered to explain this epidemiology. This epidemiology is not worldwide, since it is not commonly observed in tropical areas. Although adults associated with children with rotavirus diarrhea are usually not infected (serologic estimate), there have been reports of up to 41 per cent of adult contacts being infected (Kim et al., 1977). Nearly all adult contact infections were subclinical. In an interesting epidemiologic study by Murphy et al. (1977), in an examination of 628 newborn babies in Sydney hospitals in Australia, it was found that 49 per cent had rotavirus in their stools. Although none of the one day neonates were positive, many commenced excreting virus within three or four days after birth. Most of these were symptom free. Since there was no seasonal variation observed, it appears that rotaviruses may persist in some Sydney hospital nurseries and are spread between neonates.

Diagnosis

It is important to differentiate rotavirus infantile diarrhea from enteritis produced by other microorganisms. Although direct electron microscopy, after differential centrifugation, has been the main and initial technique, other effective methods and variations have been introduced. These may be virus or virus antigen detection (by antibody rise) or both. The following methods are currently being explored:

1. Free viral immunofluorescence assay in human stools (Yolken et al., 1977b)
2. Enzyme-linked immunosorbent assay (ELISA) (Yolken et al., 1977a)
3. Fluorescent virus precipitation test (Peterson et al., 1976)
4. Cell culture plus immunofluorescence (Woode et al., 1974; Wyatt et al., 1974, 1976; Banatvala et al., 1975; Albrey and Murphy, 1976; Bryden et al., 1977)
5. Complement-fixation (Kapikian et al., 1975; Tufvesson and Johnsson, 1976a)
6. Electrophoresis:
   a. Immunoelectroosmophoresis (Tufvesson and Johnsson, 1976b)
   b. Polyacrylamide gel (Kalica et al., 1976)
   c. Counter-immunoelectrophoresis (Spence et al., 1977)

Transmission

Human rotaviruses have been successfully transmitted to lambs, monkeys, dogs, and pigs. This is not surprising in view of the antigenic relationship between rotaviruses of animals and infants.

Antigenicity

Some of the animal rotaviruses, notably calf rotavirus and the simian SA 11, grow in cell cultures. These viruses cross-react with human rotavirus by complement-fixation, immunofluorescence, and immuno-electron microscopy. In addition the rotaviruses derived from pigs and mice (EDIM) also cross-react with the human strains. So far, all known members share a common antigen, but antigenic differences have yet to be explored. As the simian rotavirus (SA 11) replicates easily in primary cell culture it is being examined for use in the serodiagnosis of human infections (Schoub et al., 1977).

Morphology

It is impossible to distinguish between rotaviruses from simian, porcine, equine, murine, or human origin by electronmicroscopy. Particles with and without outer capsids have been found in each animal species and their diameter varies between 50 and 68 nm depending on the presence of the outer capsid. When the outer layer is lost the inner capsomeres projecting from the surface give the particles a rough appearance around the edge. Sometimes flattened tubules with hexagonally packed subunits are observed.

For further reading see Flewett et al., 1973; Rodriguez et al., 1977; Middleton, 1977).

References

Albrey, M. B., and Murphy, A. M.: Rotavirus growth in bovine monolayers. Lancet 1:753, 1976.

Banatvala, J. E., Totterdell, B., Chrystie, I. L., and Woode, G. N.: In-vitro detection of human rotaviruses. Lancet 2:821, 1975.

Borden, E. C., Shope, R. E., and Murphy, F. A.: Physicochemical and morphological relationships of some arthropod-borne viruses to bluetongue virus — a new taxonomic group. Physicochemical and serological studies. J. Gen Virol 13:261, 1971.

Bryden, A. S., Davies, H. A., Thouless, M. E., and Flewett, T. H.: Diagnosis of rotavirus infection by cell culture. J Med Microbiol 10:121, 1977.

Davidson, G. P., Bishop, R. F., Townley, R. R. W., Holmes, I. H., and Ruck, B. J.: Importance of a new virus in acute sporadic enteritis in children. Lancet 1:242, 1975.

Emmons, R. W., Oshiro, L. S., Johnson, H. N., and Lennette, E. H.: Intra-erythrocytic location of Colorado tick fever virus. J Gen Virol 17:185, 1972.

England, E.: Concentration of reovirus and adenovirus from sewage and effluents by protamine sulfate (salmine) treatment. Appl Microbiol 24:510, 1972.

Flewett, T. H., Bryden, A. S., and Davies, H. A.: Virus particles in gastroenteritis. Lancet 2:1947, 1973.

Hassan, S. A., and Cochran, K. W.: Teratogenicity of reo and poliovirus in mice. Bact Proc 42:115, 1966.

Holmes, I. H., Ruck, B. J., Bishop, R. F., and Davidson, G. P.: Infantile enteritis viruses: morphogenesis and morphology. J. Virol 16:937, 1975.

Jarudi, N. I., Huggett, D. O., and Golden, B.: Reovirus keratoconjunctivitis. Canad J Ophthal 8:371, 1973.

Kalica, A. R., Garon, C. F., Wyatt, R. G., Mebus, C. A., van Kirk, D. H., Chanock, R. M., and Kapikian, A. Z.: Differentiation of human and calf reoviruslike agents associated with diarrhea using polyacrylamide gel electrophoresis of RNA. Virology 74:86, 1976.

Kapikian, A. Z., Cline, W. L., Mebus, C. A., Wyatt, R. G., Kalica, A. R., James, H. D., Jr., van Kirk, D., and Chanock, R. M.: New complement-fixation test for the human reovirus-like agent of infantile gastroenteritis. Nebraska calf diarrhea virus used as antigen. Lancet 1:1056, 1975.

Kim, H. W., Brandt, C. D., Kapikian, A. Z., Wyatt, R. G., Arrobio, J. O., Rodriguez, W. J., Chanock, R. M., and Parrott, R. H.: Human reovirus-like agent infection. Occurrence in adult contacts of pediatric patients with gastroenteritis. JAMA 238:404, 1977.

Lennette, E. H., Spaulding, E. H., and Truant, J. P. (eds.): Manual of Clinical Microbiology. 2nd ed. Washington, American Society for Microbiology, 1974.

Middleton, P. J.: Rotavirus: clinical observations and diagnosis of gastroenteritis. In Kurstak, E., and Kurstak, C. (eds.): Comparative Diagnosis of Viral Diseases, vol. 1, Human and Related Viruses, part A. New York, San Francisco, London, Academic Press, 1977, p. 423.

Murphy, A. M., Albrey, M. B., and Crewe, E. B.: Rotavirus infections of neonates. Lancet 2:1149, 1977.

Murphy, F. A., Borden, E. C., Shope, R. E., and Harrison, A.: Physicochemical and morphological relationships of some arthropod-borne viruses to bluetongue virus — a new taxonomic group. Electron microscopic studies. J Gen Virol 13:273, 1971.

Oshiro, L. S., Dondero, D. V., Emmons, R. W., and Lennette, E. H.: J Gen Virol, 39:73, 1978.

Peterson, M. W., Spendlove, R. S., and Smart, R. A.: Detection of neonatal calf diarrhea virus, infant reovirus-like diarrhea virus, and a coronavirus using the fluorescent virus precipitin test. J Clin Microbiol 3:376, 1976.

Raine, C. S., and Fields, B. N.: Neurotropic-virus host relationship alterations due to variation in viral genome as studied by electron microscopy. Am J Path 75:119, 1974.

Rodriguez, W. J., Kim, H. W., Arrobio, J. O., Brandt, C. D., Chanock, R. M., Kapikian, A. Z., Wyatt, R. G., and Parrott, R. H.: Clinical features of acute gastroenteritis associated with human reovirus-like agent in infants and young children. J Pediat 91:188, 1977.

Rosen, L.: Reoviruses. In Monographs in Virology, 1. Wien, New York, Springer-Verlag, 1968, p. 73.

Sabin, A. B.: Reoviruses. Science 130:1387, 1959.

Schoub, B. D., Lecatsas, G., and Prozesky, O. W.: Antigenic relationship between human and simian rotaviruses. J Med Microbiol 10:1, 1977.

Spence, L., Fauvel, M., Petro, R., and Bloch, S.: Comparison of counterimmunoelectrophoresis and electron microscopy for laboratory diagnosis of human reovirus-like agent-associated infantile gastroenteritis. J Clin Microbiol 5:248, 1977.

Stanley, N. F.: Reovirus — a ubiquitous orphan. Med J Aust 2:815, 1961.

Stanley, N. F.: Reoviruses. Br Med Bull 23:150, 1967.

Stanley, N. F.: The reovirus murine models. In Hotchin, J. (vol. ed.), Melnick, J. L. (ser. ed.): Progress in Medical Virology: Slow Virus Diseases, vol. 18. Basel, Karger, 1974, p. 257.

Stanley, N. F.: Diagnosis of reovirus infections: comparative aspects. In Kurstak, E., and Kurstak, C. (eds.): Comparative Diagnosis of Viral Diseases, vol. 1, Human and Related Viruses, part A. New York, San Francisco, London, Academic Press, 1977, p. 385.

Stanley, N. F., Dorman, D. C., and Ponsford, J.: Studies on the pathogenesis of a hitherto undescribed virus (hepato-encephalomyelitis) producing unusual symptoms in suckling mice. Aust J Exp Biol Med Sci 31:147, 1953.

Stanley, N. F., Dorman, D. C., and Ponsford, J.: Studies on the hepato-encephalomyelitis virus (HEV). Aust J Exp Biol Med Sci 32:543, 1954.

Stanley, N. F., and Leak, P. J.: The serologic epidemiology of reovirus infection with special reference to the Rottnest Island quokka (*Setonix brachyurus*). Am J Hyg 78:82, 1963.

Stanley, N. F., Leak, P. J., Walters, M.N-I., and Joske, R. A.: Murine infection with reovirus. II. The chronic disease following reovirus type 3 infection. Br J Exp Path 45:142, 1964.

Tufvesson, B., and Johnsson, T.: Occurrence of reo-like calf viruses in young children with acute gastroenteritis. Diagnoses established by electron microscopy and complement fixation, using the reo-like virus as antigen. Acta Path Microbiol Scand(B) 84:22, 1976a.

Tufvesson, B., and Johnsson, T.: Immunoelectroosmophoresis for detection of reo-like virus: methodology and comparison with electron microscopy. Acta Path Microbiol Scand(B) 84:225, 1976b.

Woode, G. N., Bridger, J. C., Hall, G., and Dennis, M. J.: The isolation of a reovirus-like agent associated with diarrhoea in colostrum-deprived calves in Great Britain. Res Vet Sci 16:102, 1974.

Wyatt, G. B., Hocking, B., Bishop, R., and Wyatt, J. L.: Duovirus infection as a cause of infantile gastro-enteritis in Port Moresby. Papua New Guinea Med J 19:134, 1976.

Wyatt, R. G., Kapikian, A. Z., Thornhill, T. S., Sereno, M. M., Kim, H. W., and Chanock, R. M.: In vitro cultivation in human fetal intestinal organ culture of a reovirus-like agent associated with nonbacterial gastroenteritis in infants and children. J Infect Dis 130:523, 1974.

Yolken, R. H., Kim, H. W., Clem, T., Wyatt, R. G., Kalica, A. R., Chanock, R. M., and Kapikian, A. Z.: Enzyme-linked immunosorbent assay (ELISA) for detection of human reovirus-like agent of infantile gastroenteritis. Lancet 2:263, 1977a.

Yolken, R. H., Wyatt, R. G., Kalica, A. R., Kim, H. W., Brandt, C. D., Parrott, R. H., Kapikian, A. Z., and Chanock, R. M.: Use of a free viral immunofluorescence assay to detect human reovirus-like agent in human stools. Infect Immun 16:467, 1977b.

64 *RHABDOVIRUS*

Bosko Postic, M.D.
and Tadeus J. Wiktor, D.V.M.

The rhabdoviruses are a group of bullet-shaped viruses that infect vertebrates, invertebrates, and plants. An electron micrograph of a rabies virion, a member of the rhabdovirus group, is shown in Figure 1. Most rhabdoviruses measure 180 × 75 nm. Short particles measuring 70 to 100 nm and longer ones measuring up to 400 nm are also seen. Plant rhabdoviruses appear to have a similar conformation but are usually longer and bacilliform in shape.

Out of some twenty animal rhabdoviruses only the vesicular stomatitis virus (VSV) and the rabies virus subgroup are pathogenic for humans and domestic animals. In addition, several rhabdoviruses can cause economically important diseases of fish.

The central helical core of a rhabdovirus consists of a single-stranded RNA genome and various structural proteins forming the nucleocapsid, which is surrounded by a lipid envelope and an outer surface of glycoprotein spikes. The chemical composition of purified rabies virus consists of 22 per cent lipids, 3 per cent carbohydrate, 1 per cent RNA, and 74 per cent protein. Four major polypeptide components of different molecular sizes form the protein moiety of rabies virus; a nucleoprotein, a glycoprotein, and two membrane proteins. Lipid solvents can inactivate the infectivity of rhabdoviruses.

ANTIGENIC COMPOSITION

Using immunologic techniques (virus neutralization, complement fixation, antibody binding, or fluorescent antibody staining) rhabdoviruses can be classified into several antigenic groups.

In the VSV group containing some seven known strains, two major serotypes (VSV Indiana and VSV New Jersey) can be distinguished; other strains of VSV are related to major serotypes.

The rabies group contains rabies virus and five other viruses that share antigens with rabies virus. These are the Lagos bat, Mokola, Duvenhage, Obodhiang, and Kotonkan viruses, but only two, the Mokola and Duvenhage viruses, may be associated with human disease.

Two antigens of rabies virus have been obtained in pure form: (1) The glycoprotein induces virus-neutralizing antibody and is therefore important in immunity. Antibody to glycoprotein plus complement can produce lysis of cultured cells infected by rabies virus. (2) Antibody stimulated by the viral nucleocapsid does not neutralize virus or cause lysis, but does fix complement (CF). The CF antibody can identify the intracytoplasmic inclusions known as Negri bodies, which consist mostly of rabies virus ribonucleoprotein.

Using cross-neutralization, a limited relationship has been detected between rabies, Mokola, and Lagos bat viruses. In CF tests and fluorescent antibody (FA) assays, there is considerable antigenic similarity between these viruses (Shope et al., 1970). Immunization of mice with Mokola or Lagos virus was not protective against challenge with rabies virus. On the other hand, mice immunized with attenuated rabies virus (high egg passage [HEP], Flury strain) resisted challenge with Lagos bat virus and, to a lesser degree, with Mokola virus (Tignor and Shope, 1972).

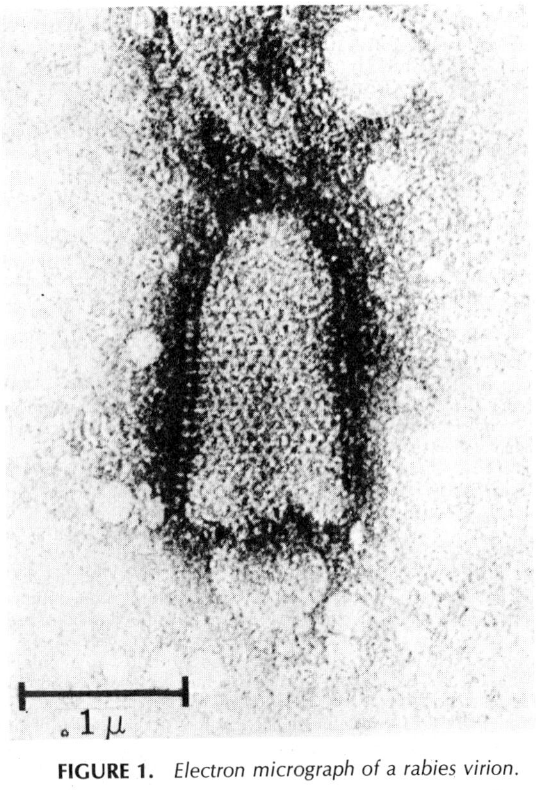

FIGURE 1. *Electron micrograph of a rabies virion.*

SUSCEPTIBLE HOST CELLS

Vesicular stomatitis virus possesses great virulence for many cell cultures, causing a rapid cytopathic effect that is usually complete by 48 hours after infection. Plaques are also formed in cell cultures of several species. This virus is extremely sensitive to the action of interferon and is used for titration of interferon. Under certain experimental conditions, truncated virus particles are produced. These can interfere with the production of standard VSV particles in vivo and in vitro (see the section on Interference in Chapter 12).

In the last century, Louis Pasteur found that the agent of rabies replicates in the nerve tissue. He and his co-workers injected saliva from rabid dogs submeningeally into rabbits (Pasteur et al., 1884). The disease was then serially transmitted with suspensions of brain tissue from inoculated rabbits.

Seventy years later, Kissling and co-workers (Kissling, 1958) propagated rabies virus in cultured cells of non-nervous origin. A wide variety of cells can be infected, but generally no cytotoxicity is observed. The cytoplasm of cultured cells was shown to contain rabies virus antigen by staining with fluorescent antibody. This observation is the basis for diagnostic tests in tissue culture, which replace animal inoculations (see Chapter 167). A continuous line of hamster cells BH-21, is particularly supportive of rabies virus growth and can be used for plaque formation. In the authors' laboratories, rabies virus-infected cell cultures have produced interferon, and conversely, interferon added before inoculation of virus suppressed its replication. In other words, rabies virus not only induces interferon but is also suppressed by it. Interferon also protects laboratory animals from rabies virus (Postic and Fenje, 1971). (See Chapter 12.)

PATHOGENIC PROPERTIES

Both the Indiana and the New Jersey serotypes of VSV affect many species of animals but have a special predilection for livestock. In these animals vesicular stomatitis is characterized by vesicles on the oral mucosa, especially on the tongue. Vesicles and ulcers may also occur on the udders of cows and mares, and linear necrotic lesions have been observed on the corium and heels of horses and cattle. Over 95 per cent of animals survive the natural infection. The illness in cattle resembles foot-and-mouth disease, but vesicular stomatitis is milder. A related coccal virus was isolated from small rodents but not livestock. This agent is serologically close but not identical with the Indiana strain of VSV. Both types of VSV can produce illness in humans. Most outbreaks occur through professional exposure to animals and are due to the New Jersey serotype. The illness in humans is acute, self-limited, and similar to influenza. Fields and Hawkins (1967) described eight cases of vesicular stomatitis in persons exposed to animals naturally infected with the Indiana serotype. Three of these patients had oral lesions: vesicles on the gum, buccal, and pharyngeal mucosa, or a herpes-like lip lesion. One adult patient experienced a 9-kg weight loss over a period of two weeks. All patients were febrile, and half of them complained of myalgia. Headache, nausea, vomiting, and pharyngitis were also prominent. Recovery occurred in all patients after a course of treatment lasting from two to seven days. In other outbreaks, including those in laboratory workers, most patients became febrile after an incubation period of two to six days, and developed pharnygitis, conjunctivitis, and oral and buccal vesicles with submaxillary and submandibular lymphadenitis. Occasionally, the febrile course was diphasic, and transient leukopenia occurred.

Rabies can affect all warm-blooded animals either naturally or experimentally. It is discussed in Chapter 167. Rabies-related viruses were all isolated in Africa. They appear to play a minor role in human disease. Their pathogenicity for animals is suggested by the sources of isolation. Lagos bat virus was isolated from the pooled brains of an African frugivorous bat. Experimentally, mice, dogs, and monkeys could be infected by the intracerebral route. Negri bodies were not seen in these animals. Mokola virus was isolated from shrews and from two Nigerian patients experiencing a central nervous system infection. One had a fatal poliomyelitis-like disease, and the second patient had a nonfatal febrile illness with convulsions. The pathogenicity of Mokola virus for animals is similar to that of Lagos bat virus. Negri bodies were found in infected monkey brains but not in mice or dogs. Duvenhage virus was isolated in South Africa from the brain of a patient bitten by a bat.

Obodhiang and Kotonkan viruses were isolated from mosquitoes by inoculating homogenized pools into baby mice. According to serologic results, infection by Kotonkan virus may be prevalent in livestock and rodents.

IMMUNITY

The reader is referred to Chapter 167 for the immunologic aspects of rabies.

Antibodies to VSV appear in infected animals in the 1st week after infection. Antibody determinations to the Indiana or New Jersey serotypes are carried out by a plaque neutralization technique using the homologous virus as indicator. Neutralizing antibodies mediate immunity to the homologous virus in experimental infections.

The immunologic relationship of the rabies-related viruses to rabies virus was discussed previously under Antigenic Composition.

LABORATORY DIAGNOSIS

As in all virus diseases, the laboratory methods used for rhabdovirus can be subdivided into: (1) tissue diagnosis (see Chapter 167), (2) virus isolation, and (3) serology.

VSV can be cultured from a variety of samples (serum, vesicle fluid, throat swab) by the intracerebral inoculation of baby mice, but only if the time of sampling coincides with viral replication in vivo. Since this event appears to be brief in humans, most cases are diagnosed serologically by the plaque neutralization test.

Rabies-related viruses are isolated by the mouse inoculation method from the sources discussed in Pathogenic Properties. In the brains of infected mice, the effect of these viruses may be hard to distinguish from that produced by rabies virus. Identification of isolates is accomplished by virus neutralization, CF, FA, and vaccination challenge tests. The vaccination challenge appears to be most discriminating; intraperitoneal immunization of mice with Mokola and Lagos bat viruses does not protect against intracerebral challenge with rabies virus. Likewise, immunization with Lagos bat virus does not protect against Mokola virus challenge. Immunization with Mokola virus, however, protects against both homologous and Lagos bat virus challenge. Mice immunized with rabies HEP Flury vaccine resisted challenge with Lagos bat virus and, to a lesser extent, with Mokola virus.

EPIDEMIOLOGY

VSV has been isolated from a variety of naturally infected animals, mostly livestock. The virus has also been isolated from arthropods, such as *Phlebotomus* sandflies and *Aedes* sp mosquitoes. The role of these arthropods in transmitting the disease is not clear. A prolonged vesicular stomatitis viremia has not been observed naturally or experimentally in a vertebrate. Thus, an infective blood meal for arthropods is probably not readily available in nature.

In humans VSV infection is most commonly acquired in the virus laboratory. Humans who processed animals from an epizootic have also been infected. The route by which VSV infects man is not known in most instances. Inhalation of aerosol, inoculation into the conjunctiva, and entry through abrasions of the skin are equally plausible portals of entry. Human-to-human transmission of vesicular stomatitis is unlikely.

Rabies is primarily a disease of domestic and wild animals. In areas were animal control programs are not extensively developed, dogs or cats account for most rabid animals and cause most human exposures to rabies. In such areas 90 per cent or more of human cases result from exposure to rabid dogs or cats. Effective domestic animal rabies control programs in these areas reduced sharply the numbers of rabid dogs and cats as illustrated by the decrease in the United States in the period 1950 to 1960 (Fig. 2). In the United States since 1960 the vast majority of cases of animal rabies occurred in wild animals, and most of the human rabies cases were secondary to bites by rabid wild animals. Where no wild animal reservoirs exist, as in the United Kingdom, eradication of dog rabies can eliminate rabies.

The epidemiology of human rabies closely follows the epizootiology of animal rabies. Human rabies has been reported from all continents except Australia and the Antarctic, although most cases occur in countries where control of domestic animal rabies has not been well developed. About 700 rabies deaths are reported each year to the World Health Organization, a number that is probably only a fraction of the actual number of cases. Human rabies is most common in persons under age 15, with about 40 per cent of cases occurring in children age 5 to 14. All age groups are susceptible. Most rabies victims are male, since they are more likely to be in contact with infected animals.

Rabies epidemics in humans follow two patterns. In one, a single rabid dog or wolf may attack and bite several people. In the other type of outbreak, human cases result from an increase of rabies in the domestic or wild animal population. Thus Central and Eastern Europe is currently experiencing a threat from rabies "fallout" from a major epizootic of rabies in foxes. The geographic distribution of rabies in wildlife is variable. In the USA, rabies reservoirs are found in wildlife: skunks, foxes, raccoons, and insectivorous bats. In South and Central America, dog and cattle rabies is prevalent; vampire bats there are an important reservoir as well as vectors. In Southeast Asia, dog rabies is enzootic except in Japan and Taiwan, which are rabies-free. European fox rabies has already been noted. North Africa abounds in dog rabies, whereas in sub-Saharan regions other *Canidae* (jackal) and *Felidae* genera also participate in forming the reservoir of enzootic rabies.

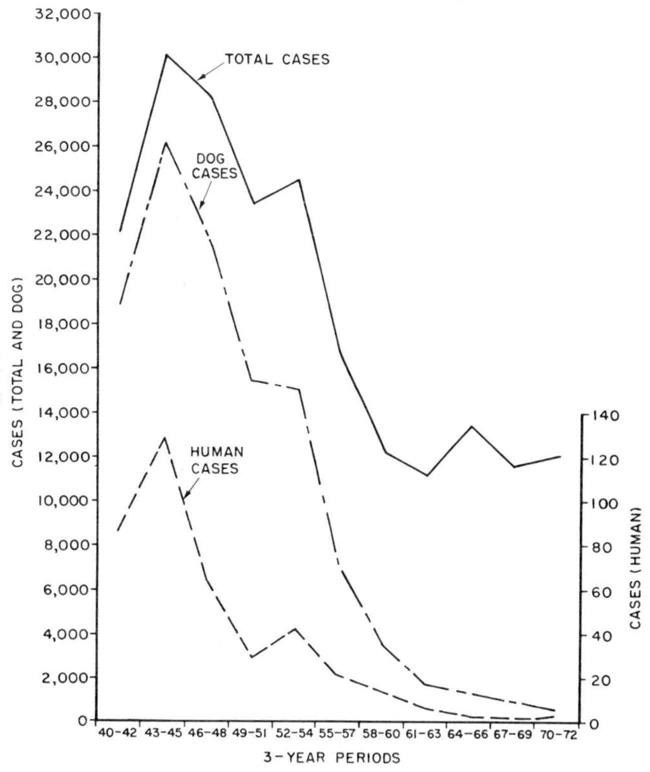

FIGURE 2. *Rabies cases by 3-year periods, United States, 1940-1972. (Source: Wiktor and Hattwick, New York, Academic Press, Rhabdoviruses, In Kurstak, E. and C., Comparative diagnosis of viral diseases, Vol. I, Part A, p. 775, 1977.)*

The epidemiology of rabies-related viruses is largely unexplored. As noted previously, these viruses only occasionally involve man. Their main medical significance lies in differentiation of them from rabies virus upon isolation from animals, or rarely, a human.

References

Fields, B. N., and Hawkins, K.: Human infection with the virus of vesicular stomatitis during an epizootic. N Engl J Med 277:989, 1967.

Kissling, R. E.: Growth of rabies virus in non-nervous tissue culture. Proc Soc Exp Biol Med 98:223, 1958.

Pasteur, L., Chamberland, M. M., and Roux, E.: Nouvelle communication sur la rage. C. R. Acad Sci (D) (Paris) 98:457, 1884.

Postic, B., and Fenje, P.: Effect of administered interferon on rabies in rabbits. Appl Microb 22:428, 1971.

Shope, R. E., Murphy, F. A., Harrison, A. K., Causey, O. R., Kemp, G. E., Simpson, D. I. H., and Moore, D. L.: Two African viruses serologically and morphologically related to rabies virus. J Virol 6:690, 1970.

Tignor, G. H., and Shope, R. E.: Vaccination and challenge of mice with viruses of the rabies serogroup. J Infect Dis 125:322, 1972.

Wiktor, T. J.,, and Hattwick, M. A. W.: Rhabdoviruses: Rabies and rabies related viruses. In Kurstak, E., and Kurstak, C. (eds.): Comparative Diagnosis of Viral Diseases. New York, Academic Press, Inc., 1977.

CORONAVIRUS 65

D. A. J. Tyrrell

The family of viruses known as the coronaviridae affects primarily various species of animals, but there are also coronaviruses that affect humans and cause mild but frequent diseases. Laboratory diagnosis is usually impossible, but it is important for the physician to know about these organisms so that he may understand better the nature of the disease he encounters clinically.

VIRUS STRUCTURE

The virus particles vary in size and shape (diameter 60 to 200 nm) but are generally spherical. The internal nucleic acid is RNA, which functions as messenger for the translation of protein; this means that the virus is "positive stranded," and thus in the strategy of its multiplication it is quite unlike influenza and parainfluenza viruses, which it superficially resembles. There is a lipid-containing envelope that carries rather widely spaced characteristic projections. These are described as resembling little "petals" or "clubs" and form a ring from which the name of the virus group was derived (Latin, *corona*, crown or halo, Fig. 1).

Virus particles develop in the cytoplasm of infected cells by budding through the membrane into the endoplasmic reticulum. They are released when the cell degenerates and the cytoplasm shatters into many small fragments.

Because of its structure the virus is easily destroyed by detergents and other common antiseptics. It is also easily inactivated by ordinary room temperatures and by drying.

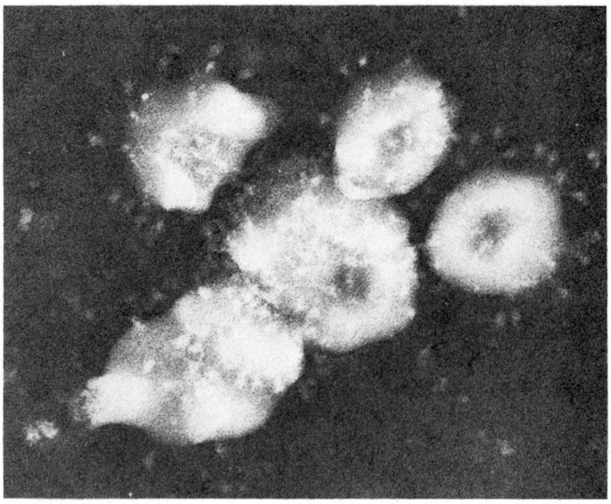

FIGURE 1. *A group of five coronavirus particles visualized by negative contrast electron microscope. The particles are pleomorphic and show the typical surface projections. (Courtesy of Mrs. Heather Davies.)*

ANTIGENIC COMPOSITION

Antigenic composition has not been adequately studied, but there are probably three distinct antigens, one of which is the internal nucleoprotein. The surface glycoproteins are apparently the antigens that determine immunity — antibody attaches to them and neutralizes the virus. For technical reasons, mentioned below, we know nothing of the antigens of most human coronaviruses. Some resemble the strain 229E and others OC43 (or OC38) for which antigens are available. Others are apparently distantly related to one or the other of these.

SUSCEPTIBLE HOST CELLS

The viruses seem to grow in a very limited range of human cells; in fact, many of them were originally isolated only by inoculating them into organ cultures of human fetal trachea or nasal epithelium. Of these viruses, two (OC43 and OC38), which proved to be identical, were found to produce disease in suckling mice and were then adapted to monkey cells in culture, but this has not been done with other viruses. A number of other coronaviruses have been isolated by inoculating clinical specimens into cultures of human embryo fibroblasts; these viruses have all proved to be antigenically related to the first one cultivated in this way, strain 229E. The virus may cause little obvious damage in organ cultures of respiratory epithelium, but it destroys susceptible strains of fibroblasts. It is not as obviously cytopathic as rhinoviruses, which may explain why the type of colds it causes clears up rather quickly.

PATHOGENIC PROPERTIES

As might be expected, the human coronavirus is infectious when inoculated into the upper respiratory tract. After an incubation period of two to five days it produces a disease characterized by profuse nasal discharge, with nasal obstruction and sneezing. Constitutional upset is mild. There is little or no evidence of an effect on the lower respiratory tract, and the illness clears up rapidly and completely. It may be characterized as a typical streaming cold. In some cases, particularly children, there may be lower respiratory symptoms including wheezing. The virus seems to have no tendency to cause pneumonia or to predispose to bacterial infections.

IMMUNITY

There is little detailed information on immunity to coronavirus infection. Volunteers may resist infection, in some cases because they possess circulating antibody. Secretory antibody is probably important, but this has not been proved. There is no exact information on the duration of immunity. After infection there may be no detectable antibody response, and it seems likely that reinfection may be possible after a period of months or a few years.

LABORATORY DIAGNOSIS

The virus is found in respiratory secretions and can be isolated from nasal or throat swabs or from nasal washings. At present the test is strictly a research procedure and requires facilities for organ culture and electron microscopy. Infections can be more easily diagnosed by serologic procedures. There is a complement fixation test for the 229E serotype and a hemagglutination-inhibition test for the OC43 serotype. Paired sera must be collected, the first at onset and the second two to three weeks later; a fourfold rise in titer indicates infection with the virus used in the test or one related to it. The antigens are difficult to obtain, and the antibody titers are often low. Rapid diagnosis by immunofluorescence has been attempted but is not available at the moment.

EPIDEMIOLOGY

Reliable figures are difficult to obtain, but coronaviruses are probably the second most frequent cause of the common cold, after rhinoviruses, and most likely cause about one case in five. They infrequently precipitate attacks of wheezing and asthma in children. They are usually prevalent during the coldest months of the year, January to April, in the northern temperate zones. There appear to be epidemic waves that can be recognized by laboratory tests but not by any distinctive clinical feature of the disease. These waves may affect cities and sometimes a whole country in a short period of time.

Although some of the coronaviruses of animals are related antigenically to those that infect man — mouse hepatitis virus is an example — there is no evidence that the viruses actually cross from one species to another.

HUMAN ENTERIC CORONAVIRUSES

Coronavirus-like particles have been seen by electron microscopy in the feces of patients with diarrheal disease and of others without such symptoms. The virus seems to be extremely difficult to grow but has been propagated in a limited way in organ cultures of human intestine. It has not been proved to cause disease, and nothing is known of its antigens. Future research may show these particles to be important, since there are several coronaviruses that cause significant diarrhea in animals, especially the young.

References

Bradburne, A. F., and Tyrrell, D. A. J.: Coronaviruses of man. Prog Med Virol 13:373, 1971.

McIntosh, K.: Coronaviruses: A comparative review. Current Topics in Microbiology and Immunology, vol. 63. New York, Springer Verlag, 1974.

Monto, A. S.: Coronaviruses. Yale J Biol Med 47:234, 1974.

66 MARBURG VIRUS AND EBOLA VIRUS

Rudolf Siegert, M.D.

Marburg Virus

VIRAL STRUCTURE

Marburg virus is named after a German city in which it was isolated in patients suffering from an unknown hemorrhagic fever (Siegert et al., 1967). Like the closely related Ebola virus, Marburg virus exhibits unique morphologic features, and thus has not yet been classified. Electron microscopy of infected organs and tissue cultures shows cylindrical particles with bizarre shapes, the most frequently found being staffs and long filaments with one rounded end and the other often coiled (Peters et al., 1971; Almeida et al., 1971). In addition, U-shaped, 6-shaped, and annular structures are found (Fig. 1 *A-C*).

The length of Marburg virus is also unusual. According to electron microscope measurements the average length is 665 nm but can be as much as 12,000 nm. The particles consist of a core with helical symmetry and an envelope with surface projections. The membranous material surrounds the inner cylindrical structure, which is considered to be the nucleocapsid.

The physicochemical composition of Marburg virus is largely unknown; from a few indirect tests, it was concluded that the genetic material is RNA. Experiments with lipid solvents and enzymes indicate that the infectious virion contains lipoprotein.

The morphogenesis of the virus takes place in

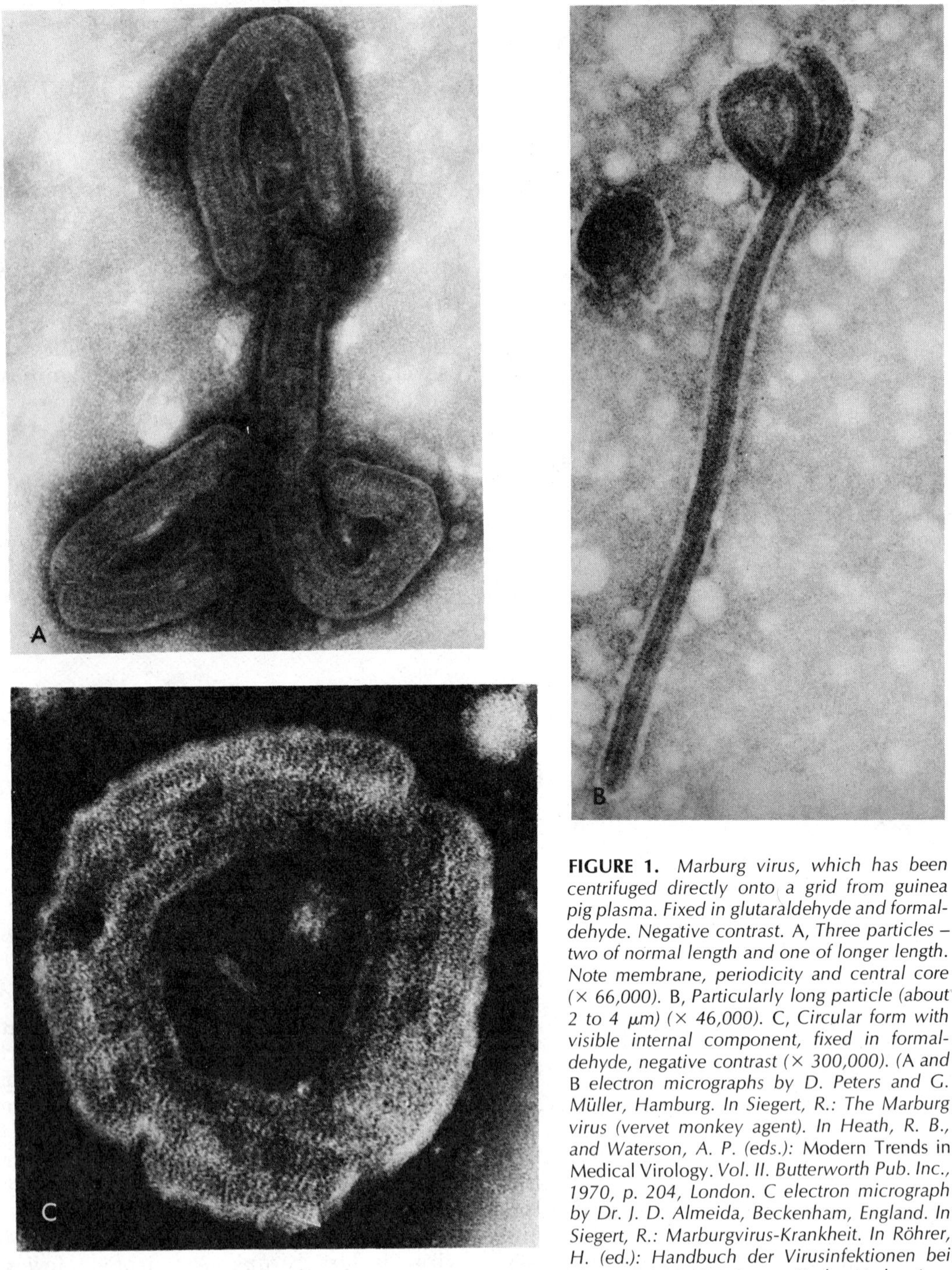

FIGURE 1. *Marburg virus, which has been centrifuged directly onto a grid from guinea pig plasma. Fixed in glutaraldehyde and formaldehyde. Negative contrast.* A, *Three particles – two of normal length and one of longer length. Note membrane, periodicity and central core (× 66,000).* B, *Particularly long particle (about 2 to 4 μm) (× 46,000).* C, *Circular form with visible internal component, fixed in formaldehyde, negative contrast (× 300,000). (*A and B *electron micrographs by D. Peters and G. Müller, Hamburg. In Siegert, R.: The Marburg virus (vervet monkey agent). In Heath, R. B., and Waterson, A. P. (eds.):* Modern Trends in Medical Virology. *Vol. II. Butterworth Pub. Inc., 1970, p. 204, London. C electron micrograph by Dr. J. D. Almeida, Beckenham, England. In Siegert, R.: Marburgvirus-Krankheit. In Röhrer, H. (ed.): Handbuch der Virusinfektionen bei Tieren, Bd. VI, VEB Gustav Fischer Verlag, Jena 1978, p. 579.)*

cytoplasmic inclusion bodies in the infected cells. They consist of an aggregation of tubular structures (which may be nucleocapsids). These structures are provided with a coat from the cytoplasmic membrane during a budding process.

ANTIGENIC COMPOSITION

Virus-specific antigen has been demonstrated in infected cells by immunofluorescence and immunoelectron microscopy. Infectiousness can be neutralized by specific antisera. A specific antigen for complement fixation has been prepared from infected Vero-cell cultures. It is antigenically distinct from other viruses.

SUSCEPTIBLE CELLS

Many cell-culture systems have been examined for their suitability for growing Marburg virus (Siegert, 1972). Primary monkey kidney cells, human amnion, and guinea pig fibroblasts have been found to be most suitable. Although in some cases considerable virus concentrations have been obtained, complete cytopathic effects have developed only in vervet monkey kidney cells after serial passages. In the absence of cell destruction, the propagation of the virus can be recognized by the development of cytoplasmic inclusion bodies as revealed by immunofluorescence (Fig. 2 *A*, *B*).

Of the established cell lines examined so far, Vero cells derived from vervet-monkey kidneys and kidney cells from baby hamsters have been best for virus propagation. The cell sheet could not usually be destroyed.

PATHOGENIC PROPERTIES

Marburg virus is pantropic. Its pathogenicity depends upon virulence and upon the nature of the host (Siegert, 1978).

Experimentally infected monkeys always died, regardless of the virus dose or the route of infection. Observable symptoms are scanty but resemble those found in humans suffering from the same infection (fever, anorexia, petechial exanthema, hemorrhagic diathesis).

Guinea pigs infected intraperitoneally developed fever for several days, as well as certain noncharacteristic symptoms (anorexia, loss of weight, dyspnea, conjunctivitis). The animals almost always survived the first passage, but after several passages they usually died.

We were able to adapt Marburg virus to hamsters and mice, and we found that newborn animals were more susceptible than adults. In contrast, chick embryos were not susceptible.

IMMUNITY

Infection with Marburg virus induced antibodies in all patients (Slenczka et al., 1970). With

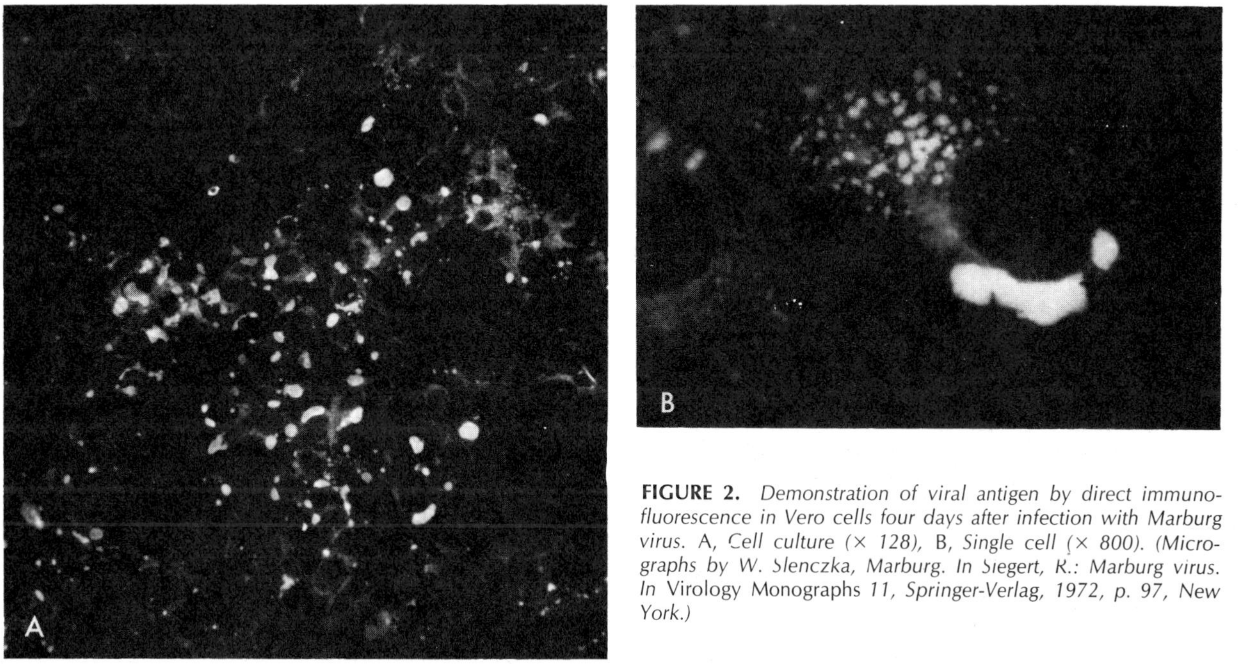

FIGURE 2. *Demonstration of viral antigen by direct immunofluorescence in Vero cells four days after infection with Marburg virus.* A, *Cell culture (× 128),* B, *Single cell (× 800). (Micrographs by W. Slenczka, Marburg. In Siegert, R.: Marburg virus. In* Virology Monographs *11, Springer-Verlag, 1972, p. 97, New York.)*

indirect immunofluorescence, antibodies could be demonstrated during the second week of the disease. They reached a peak one to two weeks later, then declined slowly, and were clearly demonstrable eight years later. Complement-fixing antibodies developed more slowly and declined faster, so that only very low titers were present two years later. The persistence of the antibodies indicates that immunity is long-lasting. Maternal antibodies transmitted transplacentally disappeared within three months in newborn infants. Nothing is known about cellular immunity during the disease.

LABORATORY DIAGNOSIS

Samples for laboratory examination should be transported in unbreakable plastic containers with screw-on lids. The World Health Organization recommends sending the material to the high-security laboratories (Class III facilities) of the Center for Disease Control in Atlanta, Georgia, or to the Microbiological Research Establishment in Porton, England.

Marburg virus can be demonstrated most reliably in whole blood, serum, and organ specimens obtained during the febrile period (Siegert, 1969). Virus concentrations in throat washes or urine were much lower. In one case the virus was demonstrated in semen, and in another it was found in the anterior chamber of the eye, but all the stool samples were negative.

The high concentrations of virus in the blood and organs in the acute phase of the disease permit rapid demonstration under the electron microscope (Siegert and Slenczka, 1971). Viral particles in serum or plasma can be sedimented directly upon the grid. Their size and shape are sufficiently characteristic to allow a diagnosis. Virus-specific antigen can be demonstrated in autopsy specimens by immunofluorescence.

Virus isolation should always be attempted in case the concentration of virus is too low to be seen by the electron microscope. Vero cells are most suitable for this purpose. Virus multiplication can be recognized a few days after inoculation by the appearance of a virus-specific antigen. Likewise, under an electron microscope, the virus can be demonstrated intracellularly and in the culture fluid. Lassa and yellow fever viruses, which are important for purposes of differential diagnosis, can also be propagated in Vero cells. Antigenic differentiation from Ebola virus is important.

Guinea pigs that become febrile after infection are also suitable for virus isolation. During the febrile phase, the virus can be demonstrated in the serum by electron microscopy and the antigen can be revealed in cell inclusions of organ specimens by immunofluorescence.

For serodiagnosis, blood from the acute and convalescent phases of the disease must be examined in order to detect a decisive rise in the antibody titer. The complement-fixation reaction provides useful results with purified cell-culture antigens. However, the method of choice is immunofluorescence, owing to its greater sensitivity.

EPIDEMIOLOGY

Only two outbreaks of Marburg virus disease have been reported, with a total of 34 patients. The two episodes differed considerably in basic epidemiology. In 1967, 31 persons in Marburg and Frankfurt (Federal Republic of Germany) and in Belgrad (Yugoslavia) became ill almost simultaneously; seven of them died. The epidemiologic situation was puzzling at first, but it was soon noted that the first victims all worked in institutes in which live poliomyelitis vaccines are produced and tested (Hennessen et al., 1966). It was then learned that all the primary cases had had contact with blood and organs of African green or vervet monkeys (*Cercopithecus aethiops*), which had been imported from Uganda by air freight via London. Those persons who ran the greatest risk were individuals involved in the actual surgical removal of organs from the animals. This is not surprising, since the virus concentrations in the blood and tissues of infected monkeys are particularly high and the measures taken to protect the workers were incomplete. The incidence was lower in persons working with cultures of monkey kidney cells. The remaining cases were persons in contact with patients, either in the family or in hospitals. Later serologic examinations of the environment produced no evidence of inapparent infections.

The most severe courses of the disease were observed in patients with primary infections, and all the fatalities occurred in this group. Patients with secondary infections had a milder course. In these cases, infection took place mainly via lesions of the skin and mucous membranes. In the hospital personnel who contracted the disease while attending infected patients, the infectious medium was blood from the febrile phase. The only infection that occurred in a family was in a woman whose husband was a carrier; she was infected by his semen. As to other possible transmission mechanisms, there is no reason for assuming the existence of oral, aerogenic, or conjunctival portals of entry. The chain of infection ended after at most two human passages.

A second outbreak of the disease, with only three cases, took place in 1975 in Johannesburg

(South Africa) (Gear et al., 1975). The first victim was an Australian who had traveled through Rhodesia shortly before becoming ill. The woman accompanying him then contracted the disease, and the third case was the nurse who had attended the Australian before his death. During the latter's journey, he had had no direct contact with animals. The two women became infected through contact with him while attending him during his illness.

In spite of extensive field studies, the natural reservoir of the virus has not yet been ascertained. Its presence in Uganda and Rhodesia indicates that it is more widespread than was originally assumed. It is probable that monkeys, like humans, were only coincidental hosts. Although Marburg virus has been propagated in intrathoracically inoculated *Aedes aegypti* mosquitoes, no attempt has yet been made to transmit the virus from infected mosquitoes.

Ebola Virus

VIRAL STRUCTURE

Ebola virus was isolated in 1976, following an epidemic of hemorrhagic fever in central Africa; it is named after a river in Zaire. In size, shape, and ultrastructure the virus is strikingly analogous to Marburg virus (Bowen et al., 1977; Johnson et al., 1977; Pattyn et al., 1977). At first the two were considered to be identical, but careful comparisons have shown that the normal and double lengths of Ebola virus are 1.25 times those of Marburg virus. Particles with a length of up to 14,000 nm have been found (Slenczka and Peters, 1977). The diameter is listed as being 70 to 100 nm, with an inner helix of 40 nm.

Because of their special morphologic position, these two viruses may be considered the first representatives of an undesignated new group (Simpson and Zuckerman, 1977).

The physicochemical composition of Ebola virus has not yet been examined because it is too dangerous to work with. It is assumed that the genome, like that of Marburg virus, consists of RNA and that the envelope contains lipoprotein.

Ebola virus is constructed of preformed nucleocapsids that develop in the cytoplasm of the host cells and are coated by the plasma membrane during the budding process.

ANTIGENIC COMPOSITION

The Ebola virus strains from the Sudan and Zaire are identical not only morphologically but also antigenically. They share no antigen with Marburg virus, in spite of their similar structure. Antigens were compared by direct and indirect immunofluorescence.

SUSCEPTIBLE CELLS

Of the numerous primary and permanent cells of varying origin that have been tested, Vero cells are the most suitable for propagation of Ebola virus. A cytopathic effect develops according to the culture medium used and increases with the number of passages made. The virus can be demonstrated in the cell-culture fluid before the cytopathic effect is apparent. Numerous intracytoplasmic antigen inclusions appear in the infected cells.

PATHOGENIC PROPERTIES

The range of the natural hosts of Ebola virus is not known. As with Marburg virus, monkeys are highly susceptible in experiments. Rhesus (*Macaca mulatta*) and vervet (*Cercopithecus aethiops*) monkeys become ill a few days after intraperitoneal infection, with symptoms of fever, maculopapular rash, diarrhea, and hemorrhages; they die after a short time.

In addition, Ebola virus is pathogenic for guinea pigs, which develop fever for several days but no other striking symptoms. Mortality is low after the first passages. Newborn and suckling mice die after intracerebral or intraperitoneal infection.

IMMUNITY

In patients and infected guinea pigs, antibodies were demonstrated by means of indirect immunofluorescence, the complement fixation reaction, and the neutralization test. It is still too early to evaluate the persistence of the antibodies. There is no cross-immunity with Marburg virus.

LABORATORY DIAGNOSIS

Diagnostic examinations are based on the same principles as those used for diagnosis of Marburg virus (Emond et al., 1977). In every case, a rapid diagnosis should be attempted, with direct electron microscope demonstration of the virus in serum or plasma during the febrile phase and in ultra-thin sections of autopsy liver tissue. In addition, virus isolation should be attempted in Vero cells and guinea pigs. The isolated virus must then be identified by means of immunofluorescence or immunoelectron microscopy, and differentiated from Marburg virus. Ebola virus was found regularly in serum samples taken during the febrile phase and also in the semen of one patient, but could not be demonstrated in throat washings, urine, or feces.

The method of choice for serodiagnosis is immunofluorescence, since it is more sensitive than the complement fixation reaction in demonstrating antibodies.

EPIDEMIOLOGY

From August to November of 1976 there were epidemics of a hemorrhagic fever with high mortality in the southern Sudan and simultaneously in northern Zaire, around 1000 km away (Brès, 1977). The disease was similar to that caused by Marburg virus. In the Sudanese towns of Nzara and Maridi there were 70 and 229 cases, respectively; of these almost one half died. In Maridi, 76 of 230 hospital employees became ill, and 41 of them died. After transfer of the patients to other hospitals, contact infections occurred in these personnel as well. In Zaire, at least 43 villages within a radius of 50 km around the missionary station at Yambucu were affected; 237 patients were registered, and 211 of them died. Here too there were contact infections among the hospital personnel. In Porton, England, a laboratory technician infected himself accidentally with an injection needle while working with guinea pigs.

The morbidity rates during the two epidemics varied considerably. In the Sudan, the rate was listed as 3.5 to 15.3/1000, and as less than 1 to 8/1000 in Zaire. The number of secondary cases is said to have been approximately 15 per cent in Zaire, whereas in the Sudan there were reports of 13 per cent secondary, 14 per cent tertiary, and 9 per cent quaternary cases. Longer infection chains were rare.

It is still not known how the virus is eliminated from the organism or transmitted. Person-to-person infections require extremely close contact. Exposure was greatest among hospital personnel. In them, contamination with blood and other body fluids played an important part, and transmission was favored by skin and mucous lesions as well as by insufficiently sterilized instruments. Airborne droplet infections appear to play only a minor role, if any.

No arthropods or mammals have been demonstrated to be carriers or hosts.

References

Marburg Virus

Almeida, J. D., Waterson, A. P., and Simpson D. I. H.: Morphology and morphogenesis of the Marburg agent. In Martini, G. A., and Siegert, R. (eds.): Marburg Virus Disease. New York, Springer Verlag, 1971, p. 84.

Gear, J. S. S., Cassel, G. A., Gear, A. J., Trappler, B., Clausen, L., Meyers, A. M., Kew, M. C., Bothwell, T. H., Sher, R., Miller, G. B., Schneider, J., Koornhof, H. J., Gomperts, E. D., Isaacson, M., and Gear, J. H. S.: Outbreak of Marburg virus disease in Johannesburg. Br Med J. 1:489, 1975.

Hennessen, W., Bonin, O., and Mauler, R.: Zur Epidemiologie der Erkrankung von Menschen durch Affen. Dtsch Med Wochenschr 93:582, 1968.

Peters, D., Müller, G., and Slenczka, W.: Morphology, development, and classification of the Marburg virus. In Martini, G. A., and Siegert, R. (eds.): Marburg Virus Disease. New York, Springer Verlag, 1971, p. 68.

Siegert, R.: Probleme der Virusdiagnostik, dargestellt am Beispiel des Marburg-Virus. Verh Dtsch Ges Inn Med 75:565, 1969.

Siegert, R.: Marburg virus. In Virology Monographs 11. New York, Springer Verlag, 1972, p. 97.

Siegert, R.: Marburgvirus-Krankheit. In Röhrer, H. (ed.): Handbuch der Virusinfektionen bei Tieren, Bd. 6. Jena, Gustav Fischer Verlag, 1978, p. 579.

Siegert, R., and Slenczka, W.: Laboratory diagnosis and pathogenesis. In Martini, G. A., and Siegert, R. (eds.): Marburg Virus Disease. New York, Springer Verlag, 1971, p. 157.

Siegert, R., Shu, H. L., Slenczka, W., Peters, D., and Müller, G.: Zur Ätiologie einer unbekannten, von Affen ausgegangenen menschlichen Infektionskrankheit. Dtsch Med Wochenschr 92: 2341, 1967.

Slenczka, W., Siegert, R., and Wolff, G.: Nachweis komplementbindender Antikörper des Marburg-Virus bei 22 Patienten mit einem Zellkultur-Antigen. Arch Virusforsch 31:71, 1970.

Ebola Virus

Bowen, E. T. W., Lloyd, G., Harris, W. J., Platt, G. S., Baskerville, A., and Vella, E. E.: Viral haemorrhagic fever in southern Sudan and northern Zaire. Lancet 1:571, 1977.

Brès, P.: Report of the Informal Consultation on the Marburg Virus-like Disease Outbreaks in the Sudan and Zaire in 1976. Held at the London School of Hygiene and Tropical Medicine, 4 and 5 January 1977. New York, World Health Organization, 1977, p. 569.

Emond, R. T. D., Evans, B., Bowen, E. T. W., and Lloyd, G.: A case of Ebola virus infection. Br Med J 2:541, 1977.

Johnson, K. M., Lange, J. V., Webb, P. A., and Murphy, F. A.: Isolation and partial characterisation of a new virus causing acute haemorrhagic fever in Zaire. Lancet 1:569, 1977.

Pattyn, S., van der Groen, G., Jacob, W., Piot, P., and Courteille, G.: Isolation of Marburg-like virus from a case of haemorrhagic fever in Zaire. Lancet 1:573, 1977.

Simpson, I. H., and Zuckerman, A. J.: Marburg and Ebola: Viruses in search of a relation. Nature 266:217, 1977.

Slenczka, W., and Peters, D.: Ebola-Virus, ein neuer Vertreter der Marburg-Virus-Gruppe. Tropenmed Parasitol 28:260, 1977.

Unclassified Viruses

HEPATITIS VIRUSES 67

Shalom Z. Hirschman, M.D.

Many viruses, such as cytomegalovirus, measles virus, Epstein-Barr virus, and yellow fever virus, attack the liver during systemic infection. However, in this chapter we will examine the viruses whose only known organ of tropism is the liver. These viruses are the heaptitis A virus, the agent causing infectious hepatitis (short incubation hepatitis), and the hepatitis B virus, the cause of serum hepatitis (transfusion hepatitis, long incubation hepatitis).

Two classic epidemiologic patterns of transmission of viral hepatitis were recognized early and formed the basis for classification of viral hepatitis into two major clinical types. The more common variety of viral hepatitis, which often involved many people and exhibited a shorter incubation period, was termed infectious hepatitis. Hepatitis following blood transfusions or needle wounds, which had a longer incubation period, was called serum hepatitis. Only recently have the virologic differences between these two types of viral hepatitides been established in the laboratory.

The morphologic, biophysical, and biochemical properties of the newly identified hepatitis A and B viruses will be discussed here. The salient characteristics of the two viruses are summarized in Table 1.

TABLE 1. Characteristics of Hepatitis A and B Viruses

CHARACTERISTIC	VIRUS A	VIRUS B	Non-A, Non-B
Epidemiology	Endemic and epidemic, water-borne and food-borne epidemics	Endemic	Endemic
Transmission	Fecal-oral	Direct inoculation, ? venereal	Direct inoculation ? other
Incubation period	2–7 weeks	4 weeks to 6 months	2–8 weeks
Disease	Acute	Acute and chronic	Acute and chronic
Size of virus	27 nm	42 nm (27 nm core)	?
Coat protein	–	+	?
Nucleic acid	RNA	Circular DNA (mainly double-stranded; molecular weight $\sim 2.1 \times 10^6$)	?
DNA polymerase	–	+	?
Cell culture system	+	–	–
Animal infection	Marmosets, chimpanzees	Chimpanzees	Chimpanzees
Chronic carrier	–	+	+
Vaccine	–	?Purified coat protein	–
Passive immunity with immune globulin	+	+	?+

HEPATITIS A VIRUS

EPIDEMIOLOGY

Early epidemiologic studies conducted almost three decades ago and studies with stool filtrates fed to human volunteers suggested a fecal-oral route of transmission for the virus causing infectious hepatitis. Indeed, many water- and food-borne outbreaks of disease have been reported. The infection is more common in countries with low standards of living, and in these countries it infects the population at a younger age.

ISOLATION AND DETECTION OF VIRUS

The modern study of hepatitis A virus (HAV) began a decade ago with attempts to transmit this virus to marmosets (*Sanguinus mystax*) (Dienhardt et al., 1967). These studies were only recently brought to practical fruition when strain CR326 of HAV was isolated in Costa Rica from the blood of a child with typical infectious hepatitis by serial passage in marmosets. Neutralizing antibody to CR326 developed in patients with hepatitis A infection. Use of viral antigen (now called hepatitis A antigen [HAAg]) derived from marmoset liver allowed the development of specific complement fixation, immune adherence, and radioimmunoassay tests for detection of antibody, thus providing sensitive serologic tests for viral infection (Krugman et al., 1975). At the same time, the virus was identified by immune electron microscopy in stool extracts of volunteers infected with virus by the oral route (Feinstone et al., 1973). Viral particles found in stool extracts were clumped by antibody present in convalescent serum.

HAV has been purified from human and chimpanzee stool extracts, from infected marmoset and chimpanzee livers, and from bile of infected chimpanzees (Dienstag et al., 1975). The purification techniques have included precipitation by polyethylene glycol (PEG), gel filtration on Sepharose 2B columns, and isopyknic ultracentrifugation on cesium chloride density gradients in sundry combinations (Bradley et al., 1976).

MORPHOLOGY AND BIOCHEMISTRY

The immunoreactive particles have a diameter of 27 nm (Fig. 1) (Provost et al., 1975). Electron-microscopic observations with positive staining of the particles showed that many particles have a dense core that is presumably composed of nucleo-

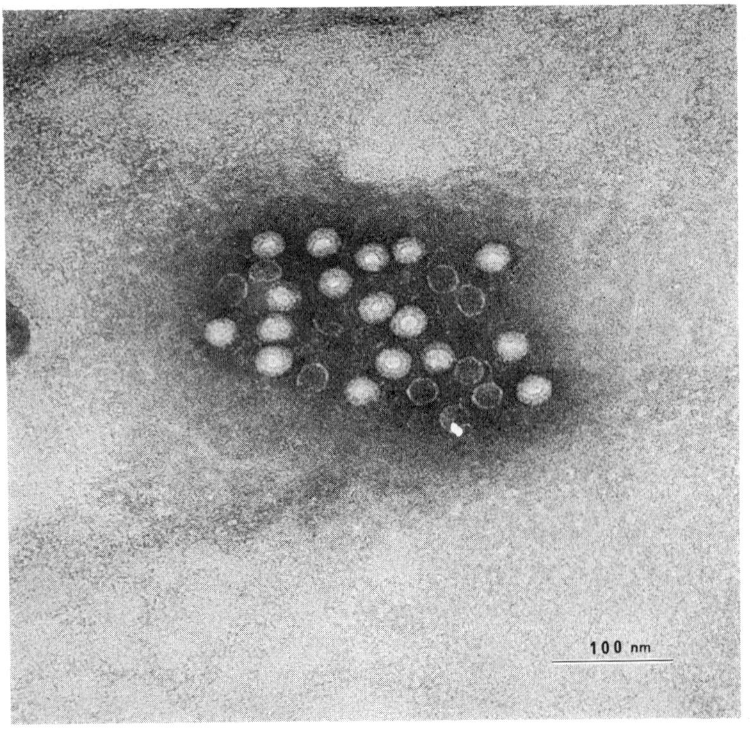

FIGURE 1. *Electron micrograph of HAV particles extracted from HAV-infected marmoset liver showing electron dense cores (× 328,600; Prepared by Mr. E.H. Cook, Jr., and provided by courtesy of Dr. Daniel W. Bradley).*

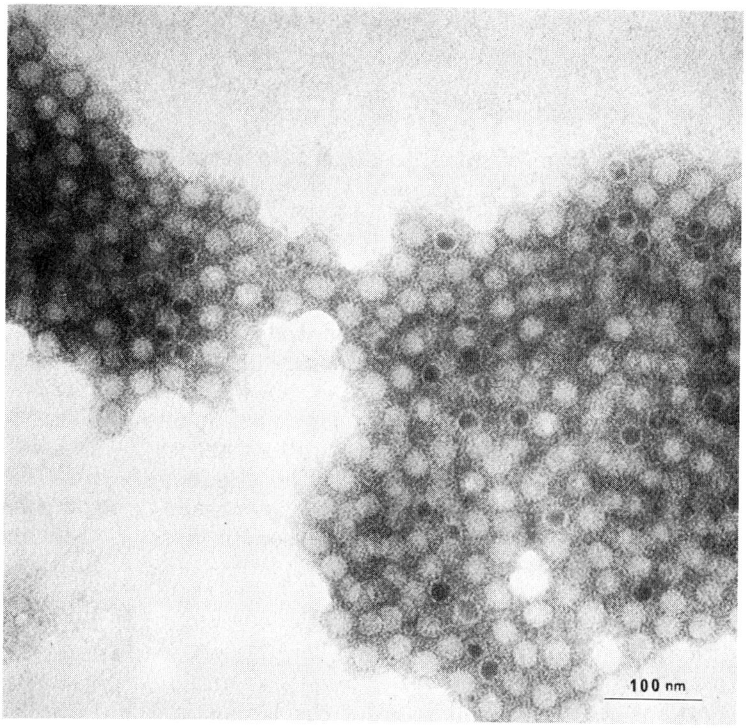

FIGURE 2. *Highly purified HAV from pre-acute phase chimpanzee stool aggregated by anti-HAV (× 256,300; Prepared by Mr. E.H. Cook, Jr., and provided by courtesy of Dr. Daniel W. Bradley).*

protein (Fig. 1). The viral particles are clumped by specific antibody (Fig. 2). Infectious virus bands at a density of 1.32 to 1.34 g/ml on cesium chloride density gradients. The infectivity of purified virus is destroyed by heating at 100° C for five minutes but is only partially reduced by heating at 60° C for one hour. Infectivity can be reduced by exposure to ultraviolet light. Purified virus is totally inactivated by treatment with 1:4000 formalin for three days at 37° C. The virus is acid stable. Electronmicroscopic study of marmoset liver showed that the virus was localized in small vesicles in cytoplasm but was absent in the nucleus. Preliminary studies suggested that hepatitis A virus may contain RNA. Acridine orange staining of purified concentrates of CR326 spread on slides gave an orange-red color characteristic of RNA staining, and partial inactivation of viral infectivity, beyond that produced by heat alone, was obtained by reacting the virus with pancreatic RNAse for one hour at 60° C.

ANIMAL TRANSMISSION AND IMMUNITY

Recently, hepatitis A virus has been successfully transmitted to chimpanzees (Pantroglodytes) by fecal extracts from infected patients administered either orally or intravenously. The chimpanzees developed lassitude and anorexia, accompanied by biochemical abnormalities such as elevation of serum glutamic oxaloacetic transaminase (SGOT) within two to five weeks. Viral 27-nm particles appeared in stools during the second week after inoculation and disappeared by the fifth week. Antibody appeared by the third week and remained elevated. Although IgG and IgA immunoglobulins remained unchanged, there was a rise in the level of IgM with the onset of clinical illness. Highly purified HAV derived from extracts of acute phase feces of infected chimpanzees has transmitted typical clinical hepatitis when inoculated into previously uninfected chimpanzees. Thus, it has been established that the 27-nm particles containing HAAg derived from extracts of acute phase feces from both humans and chimpanzees are infectious and provoke a specific antibody response in inoculated chimpanzees. A tissue culture system for propagation of HAV has recently been developed.

Infection with HAV provokes a prolonged antibody response, and immunity appears to be lifelong. Passive protection with immunoglobulin appears to both prevent and ameliorate disease. There is still no vaccine for hepatitis A infection.

HEPATITIS B VIRUS

EPIDEMIOLOGY

A second form of viral hepatitis is associated with blood transfusions or direct needle inoculation and has a rather prolonged incubation period. In Western Europe and North America cases of serum hepatitis usually appear singly. However, in Asia large segments of the population have been found to be infected. Indeed, infection seems to be acquired even during the neonatal period. Recent evidence has shown that mosquitoes can carry the virus. However, it has not been established whether mosquitoes are important in transmitting the infection. Bedbugs also can harbor the virus and may possibly transmit disease from person to person in close family quarters.

BIOCHEMISTRY

The first biochemical trace of the hepatitis B virus was discovered when a new antigen was found in the blood of an Australian aborigine. This antigen formed a precipitin line with serum from a patient with hemophilia who had received multiple blood transfusions (Blumberg et al., 1967). It was quickly appreciated that this new antigen, termed Australia antigen (also called SH-antigen and hepatitis-associated antigen), was associated with serum hepatitis. The antigen was found to exist in particulate form in the blood of patients with serum hepatitis (Fig. 3) (Bayer et al., 1967). The predominant form was a disk-shaped particle 18 to 20 nm in diameter. Tube forms with diameters of 20 nm and lengths of 50 to 500 nm, and showing cross-striations with a periodicity of 3 nm also were observed in sera from most patients. A larger particle 40 to 42 nm in diameter, with a double shell and an inner core resembling a viral nucleoid, was also found in sera of patients with antigenemia (Dane et al., 1970). It was suggested that the larger particles represented the complete virion and that the other morphologic forms were excess coat protein. This suggestion was given experimental foundation when the internal component of the larger particle, also called the Dane particle, was released by treating the particles with 0.5 per cent Tween-80 (polysorbate) in phosphate-buffered saline (PBS) (Almeida et al., 1971). The internal component had a diameter of 27 nm and seemed to be morphologically identical to particles found in the nuclei of hepatocytes of patients with serum hepatitis. Thus, the coat protein of the Dane particle was called hepatitis B surface antigen (HBsAg), and the core particle was called hepatitis B core antigen (HBcAg). The long tube forms of HBsAg could be converted to the disk forms by exposure to mildly acidic buffers. Antibody directed against the internal component of the Dane particle, which was different from antibody to HBsAg, was found during acute attacks of hepatitis.

DNA polymerase is present in preparations of hepatitis B antigen containing Dane particles. The enzyme is associated with the core of the Dane particle. The core of the Dane particle contains covalently closed, open circular, mainly double-stranded DNA (Robinson et al., 1974), but has single-stranded regions comprising 30 to 60 per cent of the nucleotides, of molecular weight of 2.1 × 10^6 (Fig. 4). The base composition of this DNA is approximately 49 per cent guano-

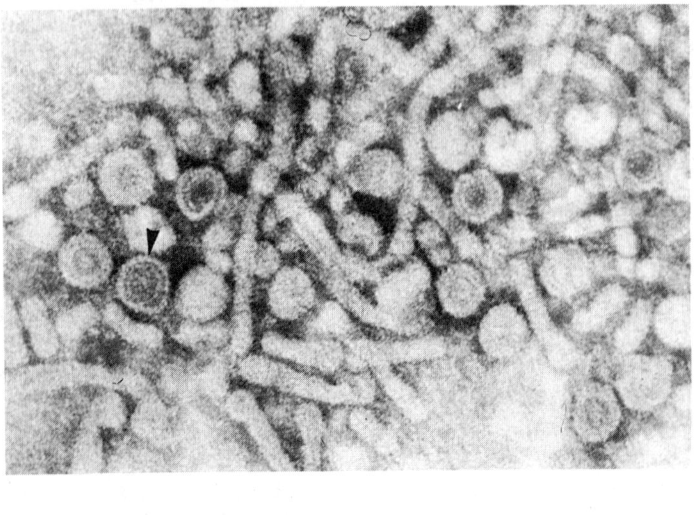

FIGURE 3. *Electron micrograph of particulate Australia antigen (hepatitis B surface antigen) showing 18 to 20 nm discs, long tubular forms and 42 nm Dane particles (arrow) (× 250,000; electron micrograph taken by Dr. Michael Gerber).*

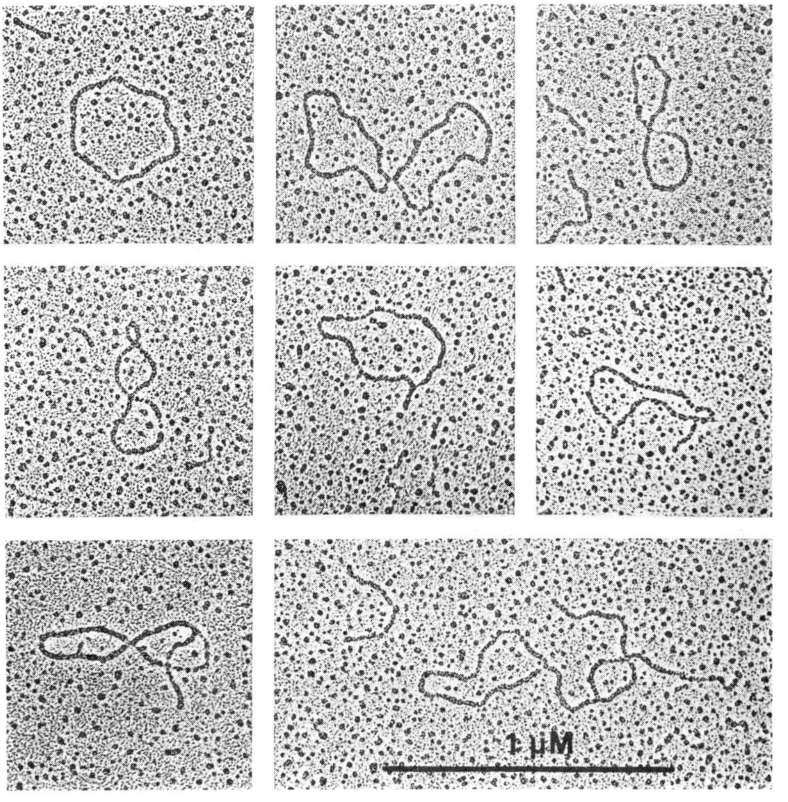

FIGURE 4. *Electron micrograph of DNA extracted from Dane particles purified from human plasma showing open circles. (Provided by courtesy of Dr. Lacy R. Overby.)*

sine plus cytosine. In the test tube, the DNA polymerase appears to function as a repair enzyme for the region of single-stranded DNA.

Two populations of Dane particles can be identified in the electron microscope by using positive stains. One population, usually the minor one, shows densely staining cores, whereas the other population, usually the major one, shows lightly staining cores. The densely staining population of Dane particles contains very high DNA polymerase activity, while much less enzyme activity is found in the lightly staining particles. The densely staining particles sediment faster in sucrose. Core particles isolated from the heavier Dane particles band at densities of 1.36 to 1.34 gm/ml in cesium chloride, while core particles from the lighter Dane particles band at 1.33 to 1.30 gm/ml. It has been suggested that the densely staining core particles are complete, containing a full complement of both DNA and DNA polymerase, while the lightly staining core particles may not contain DNA. Core particles of density 1.30 to 1.32 gm/ml have been isolated from purified nuclei of hepatocytes infected by HBV (Fig. 5); although they contained linear double-stranded DNA, these particles did not show DNA polymerase activity. Core particlcs with DNA polymerase activity have been isolated from total homogenates of both human and chimpanzee liver infected with HBV.

The core particle-associated DNA polymerase is active in very high concentrations of both salts of monovalent cations and magnesium chloride in which other known bacterial, mammalian, and viral DNA polymerases are inhibited (Hirschman and Garfinkel, 1977). This singular property of the core particle DNA polymerase forms the basis for a differential assay of the hepatitis B enzyme in reaction mixtures with $\geq$0.4 M KCl and $\geq$0.04 M $MgCl_2$. Hepatitis B DNA polymerase is inhibited by $MnCl_2$ and most other divalent cations.

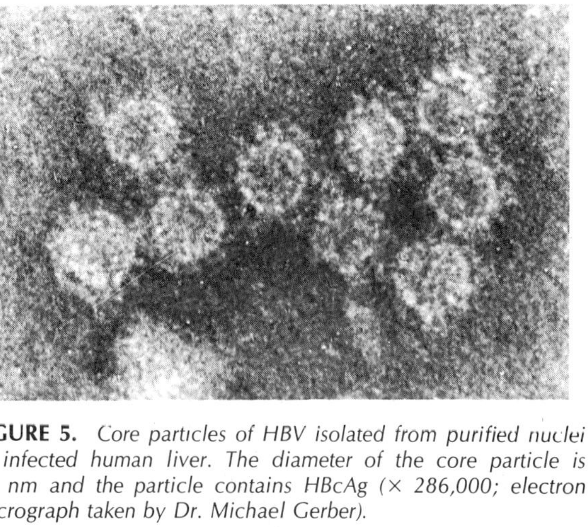

FIGURE 5. *Core particles of HBV isolated from purified nuclei of infected human liver. The diameter of the core particle is 27 nm and the particle contains HBcAg (× 286,000; electron micrograph taken by Dr. Michael Gerber).*

Nonionic detergents such as Nonidet P-40 enhance the DNA polymerase reaction of Dane particles, perhaps by removing the coat of HBsAg, thus allowing freer access of precursor deoxyribonucleotides to the viral core. The activity of core particle-associated DNA polymerase increases with rising concentrations of both salts of monovalent cations and $MgCl_2$. Hepatitis B DNA polymerase is not activated by linear or circular single- or double-stranded exogenous DNAs, suggesting specificity of the enzyme for its own DNA or inability of the exogenous templates to reach the active site of the enzyme.

Particles 27 nm in diameter containing HBcAg have been purified from infected human and chimpanzee liver. Double-stranded DNA has been isolated from these particles. Hepatitis B DNA appears to be integrated into the nuclear DNA of infected liver cells, and free circulating DNA hybridizing to Dane particle DNA has also been found in plasma of infected patients.

A soluble antigen, called e antigen (HBeAg), is also found in sera of patients with hepatitis B virus infection. The antigen appears during acute infection and then usually disappears, but can be carried chronically in patients with chronic hepatitis B antigenemia and chronic hepatitis. The e antigen appears to be a protein of approximately 17×10^3 daltons, but the exact nature of this antigen is unknown. At present, three major antigenic subtypes, Hbe_1Ag, HBe_2Ag, and HBe_3Ag are recognized. The presence of e antigen appears to be associated with the presence of Dane particles. Along with hepatitis B core antigen, persistence of HbeAg usually portends chronic hepatitis. HBeAg also appears to be a marker for infectivity of HBsAg positive blood. The major component antigens of HBV are summarized in Table 2.

ASSAY OF HBAg

The first method used to detect hepatitis B surface antigen in the laboratory was the rather insensitive method of immunodiffusion; this test was followed by the more sensitive counterimmunoelectrophoresis. Newer, more sensitive procedures for detecting HBsAg include the techniques of complement fixation, hemagglutination, and radioimmunoassay. The most commonly used test is the radioimmunoassay, which is now commercially available. Hepatitis B core antigen can be detected both by complement fixation and radioimmunoassay, but the latter is the preferred and more widely used technique. Unfortunately, HBeAg is still determined by immunodiffusion although radioimmunoassays have recently been developed. The wide use of radioimmunoassay, with its greater accuracy in detecting HBsAg, has certainly contributed to the great decrease of transmission of hepatitis B from the blood of commercial donors.

TABLE 2. Major Antigens Comprising Hepatitis B Virus

ANTIGEN	CHARACTERISTIC
Surface; HBsAg	Particulate; viral coat protein; has major and minor antigenic subdeterminants; found in cytoplasm
Core; HBcAg	Particulate; viral nucleoid; found in nucleus
e; HBeAg	Soluble; present in viral core several subtypes

SUBTYPES OF HBsAg

HBsAg contains a common immunologic determinant, a, and several major subdeterminants that are specified by the viral genome (LeBouvier, 1971). The subdeterminants can be detected by the presence of spurs in immunodiffusion tests with various antisera. Eight distinct categories and two of mixed subtype have been recognized (Table 3). These consist of various combinations of the subdeterminants d/y and w/r. The subdeterminants appear to comprise two groups composed of d/y on the one hand and w1, w2, w3, w4, and r on the other. The two mixed subtypes are very rare and may have been due to phenotypic or genotypic mixing of immunologic markers during simultaneous infection with viruses associated with more than one subtype of HBsAg. Several minor antigenic subdeterminants have been described, including q, x, f, p, j, n, and g; g has been found with w2. Antigenic types ayw2 and ayw3 appear to be more common in Africa and the Middle East. The subdeterminant r appears to predominate in the Far East and is very common in Japan. The antigenic type adw2 is common in the United States. In experiments in which chimpanzees were infected with different strains of HBV carrying distinct antigenic determinants it was found that the viral strains breed true with respect to the antigenic determinants. Purified hepatitis B surface antigen appears to share cross-antigenicity with some normal human proteins. Both lipid and carbohydrates are associated with purified HBsAg. The carbohydrate content appears to vary from 3.6 to 6.5 per cent, and antigenic activity is lost after mild periodate oxidation. The lipids associated with HBsAg include phosphatidylcholine, sphingomyelin, lysophosphatidylcholine, and glycosphingolipid.

TABLE 3. Antigenic Subdeterminants of Hepatitis B Surface Antigen

MAJOR	MINOR
aywl	q
ayw2	x
ayw3	f
ayw4	t
ayr	j
adw2	n
adw4	g
adr	
adwy	
adyr	

HBsAg of adw and ayw subtypes appears to differ in both biophysical and biochemical characteristics. The buoyant density of HBsAg in cesium chloride is 1.21 g/cc. The molecular weight of adw preparations of HBsAg is 3.7×10^6. HBsAg of ayw subtype has an average molecular weight of 4.6×10^6. The acidic isoelectric point (pI) of HBsAg is due to a high proportion of acidic amino acid residues relative to basic amino acids. HBsAg contains high levels of cysteine, proline, leucine, and phenylalanine. Reduction and alkylation destroy the antigenic activity of HBsAg. Nine separate polypeptides have been found in preparations of the ayw subtype of HBsAg and only seven in the preparations of the adw subtype. The polypeptides appear to vary in antigenic activity. Much less is known about the chemical composition of hepatitis B core antigen, but preliminary studies indicate that this antigen has a much simpler composition than that of the surface antigen.

It is thus evident that HBV is a unique virion that contains not only partly single-stranded circular DNA but also a DNA polymerase. The virus contains many antigens that may not all be coded for by the small DNA in the virus (unless one postulates proteins composed of similar subunits and perhaps also overlapping DNA codons). Much further work will be required to delineate the replicative cycle, especially the role of the curious DNA polymerase, of this fascinating virus in the hepatocyte.

ANTIBODIES TO HBAg

After infection with hepatitis B virus, antibodies to HBcAg (anti-HBc) usually appear when liver disease becomes evident. Antibodies to HBsAg (anti-HBs) usually appear during convalescence and coincide with disappearance of circulating HBsAg. Both animal and human studies indicate that anti-HBc tends to decrease gradually and may become undetectable after one to two years. Anti-HBs lasts much longer and indeed may be present throughout life. Antibody to HBsAg seems to bestow immunity to further infection with HBV. In the chronic carrier state, with persistence of HBsAg, a high level of anti-HBc, but not anti-HBs, is usually found in serum. Anti-HBe appears with onset of overt liver disease in acute hepatitis and is considered a good prognostic sign.

VACCINES

The thesis that HBsAg represents excess coat protein of the complete hepatitis B virion led to the conclusion that vaccines of purified HBsAg would be protective but not infectious. Preliminary studies with such vaccines composed of purified HBsAg appear to bear out this hypothesis (Melnick et al., 1976). In attempts to avoid possible cross-reactions of HBsAg vaccines with normal human proteins, much effort is being expended on the study of protein subunit vaccines. Small antigenic peptides have been purified from HBsAg.

OTHER HEPATITIS VIRUSES

Evidence has been accumulating from epidemiologic studies that there may be at least two hitherto unrecognized viral agents causing acute hepatitis. The first such evidence came from epidemiologic and serologic studies of transfusion-associated hepatitis. In these studies it was found that half of the cases that occurred following transfusion were not due to either hepatitis A or B virus and were not caused by other recognizable viral agents such as Epstein-Barr virus or cytomegalovirus. Similar evidence for the presence of other hepatitis viruses has been deduced from a study of patients with multiple attacks of acute viral hepatitis. Some of the patients had four attacks of acute hepatitis, two of which could be accounted for by hepatitis A and hepatitis B infection. Thus it was presumed that the other two attacks may have been caused by other hitherto unrecognized hepatitis viruses. Now that infections of HAV and HBV can be determined serologically, much evidence is accumulating that non-A, non-B hepatitis is a prevalent infection (Feinstone et al., 1975).

References

Almeida, J. D., Rubinstein, D., and Stott, E. J.: New antigen-antibody system in Australia-antigen-positive hepatitis. Lancet 2:1225, 1971.

Barker, L. F., Chisari, F. V., McGrath, P. P., Dalgard, D. W., Kirschstein, R. L., Almeida, J. D., Edgington, T. S., Sharp, D. G., and Peterson, M. R.: Transmission of type B viral hepatitis to chimpanzees. J Infect Dis 127:648, 1973.

Bayer, M. E., Blumberg, B. S., and Werner, B.: Particles associated with Australia antigen in the sera of patients with leukemia, Down's syndrome and hepatitis. Nature 318:1057, 1968.

Blumberg, B. S., Gerstley, B. J. S., Hungerford, D. A., London, W. T., and Sutnick, A. I.: A serum antigen (Australia antigen) in Down's syndrome, leukemia and hepatitis. Ann Int Med 66:924, 1967.

Bradley, D. W., Hollinger, F. B., Hornbeck, C. L., and Maynard, J. E.: Isolation and characterization of hepatitis A virus. Am J Clin Pathol 65:876, 1976.

Dane, D. S., Cameron, C. H., and Briggs, M.: Virus-like particles in serum of patients with Australia antigen associated hepatitis. Lancet 1:695, 1970.

Dienhardt, F., Holmes, A. W., Capps, R. B., and Popper, H.: Studies on the transmission of human viral hepatitis to marmoset monkeys. J Exp Med 125:673, 1967.

Dienstag, J. L., Feinstone, S. M., Purcell, R. H., Hoofnagle, J. H., Barker, L. E., London, W. T., Popper, H., Peterson, J. M., and Kapikian, A. Z.: Experimental infection of chimpanzees with hepatitis A virus. J Infect Dis 132:532, 1975.

Feinstone, S. M., Kapikian, A. Z., and Purcell, R. H.: Hepatitis A: Detection by immune electron microscopy of a viruslike antigen associated with acute illness. Science 182:1026, 1973.

Feinstone, S. M., Kapikian, A. Z., Purcell, R. H., Alter, H. J., and Holland, P. V.: Transfusion-associated hepatitis not due to viral hepatitis type A or B. N Engl J Med 292:767, 1975.

Hirschman, S. Z.: Integrator enzyme hypothesis for replication of hepatitis B virus. Lancet 2:436, 1975.

Hirschman, S. Z., and Garfinkel, E.: Ionic requirements of the DNA polymerase associated with serum hepatitis B antigen. J Infect Dis 135:897, 1977.

Krugman, S., Friedman, H., and Lattimer, C.: Viral hepatitis, type A. Identification by specific complement fixation and immune adherence tests. N Engl J Med 292:1141, 1975.

LeBouvier, G. L.: The heterogeneity of Australia antigen. J Infect Dis 123:671, 1971.

Melnick, J. N., Dreesman, G. R., and Hollinger, F. B.: Approaching the control of viral hepatitis type B. J Infect Dis 133:210, 1976.

Provost, P. J., Wolanski, B. S., Miller, W. J., Ittensohn, O. L., McAleer, W. J., and Hilleman, M. R.: Physical, chemical and morphologic dimensions of human hepatitis. A virus strain. Proc Soc Exp Biol Med 148:532, 1975.

Robinson, W. S., Clayton, D. A., and Greenman, R. L.: DNA of a human hepatitis B virus candidate. J Virol 14:384, 1974.

6 FUNGI

68 CRYPTOCOCCUS

Charles E. Davis, M.D.

Cryptococcus neoformans is a round, yeast-like fungus that produces urease and forms a large heteropolysaccharide capsule in infected tissues. It is abundant in the feces of healthy pigeons but occurs throughout the world, even in areas where pigeons are rare or absent. While it is ordinarily a saprophyte, *C. neoformans* is an important cause of fatal meningitis in certain susceptible individuals. It was classified for many years as an asporogenous yeast in the Cryptococcaceae family of the Fungi Imperfecti, but Kwon-Chung (1975, 1976) has recently demonstrated two perfect (sexual) states among different serotypes of *C. neoformans*. The perfect forms, which appear to be related to the rust and smut pathogens of higher plants, were named *Filobasidiella neoformans* and *F. bacillospora*.

Several other species of encapsulated, round, yeast-like fungi that produce urease also belong to the cryptococcal genus. They look and behave biochemically like *C. neoformans,* except that they grow poorly or not at all at 37° C. This growth limitation probably explains why they cannot cause natural or experimental disease in mammals. Little is known about their activities in the environment, but they are important to the medical mycologist, who must differentiate them from *C. neoformans*.

MORPHOLOGY

Colors of cryptococci are white or creamy and typical of yeasts. They become yellowish or light tan with age. The colonies are entire without pseudomycelia, and convex and mucoid because of the capsular polysaccharide. Colonies that produce abundant capsule on artificial media may flow to the bottom of agar slants. All strains of *C. neoformans* form capsules in mammalian tissue, but some produce little or no capsule on artificial media. Others lose part or all of their capsule on serial passage in culture. Colonies of unencapsulated cryptococci are dry and glabrous.

Individual cryptococci are round and 3.5 to 7 μ in diameter. They are surrounded by a capsule of 1 to 30 μ. Reproduction is accomplished by the formation of one or two daughter cells connected to the parental cell by a narrow isthmus. Buds may break off when quite small and cause considerable variation in the size of these round yeasts. Part of the parental wall is broken by the emerging bud.

The cell wall of *C. neoformans* has at least two layers (Fig. 1): an outer, electron-transparent layer and an inner electron-dense, lamellar layer (Cassone et al., 1974). The capsule is just outside the electron-transparent layer, which contains 20-nm particles (Takeo et al., 1973). These particles, which have not been observed in unencapsulated yeast, are excreted into the capsule when cryptococci are grown in vivo (Fig. 2) and are thought to be involved in assembly and transport of capsular polysaccharide (Takeo et al., 1973; Edwards et al., 1967).

When *C. neoformans* reproduces, part of the bud wall (the electron-transparent layer) is synthesized in the parental cell wall, while the electron-dense, lamellar wall is a direct continuation of the parental wall. This process breaks up the parental wall, which is repaired by formation of a septum (Fig. 3). Shadomy (1971) has shown clamp connections between the parental and daughter cells in two strains of *C. neoformans*.

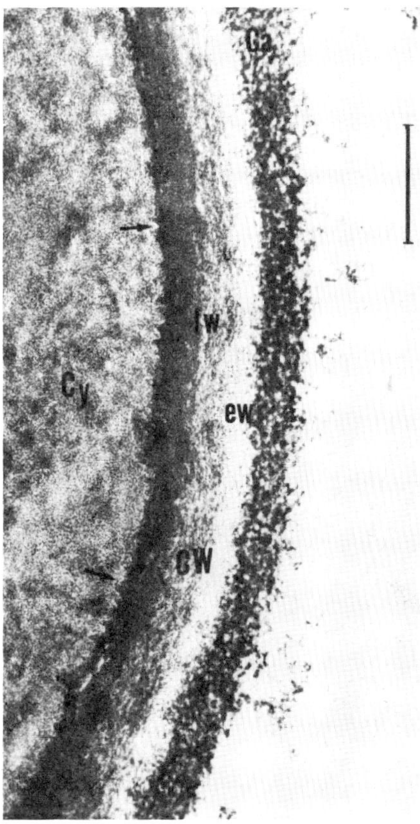

FIGURE 1. *Cell wall of* C. neoformans. *Ca, capsule; CW, cell wall; Cy, cytoplasm; ew, electron-transparent layer; lw, electron-dense lamellar layer. The arrows point to the cytoplasmic membrane. The bar corresponds to 0.2 μ. (From Cassone, A., Simonetti, N., and Strippoli, V.: Arch Microbiol 95:205, 1974.)*

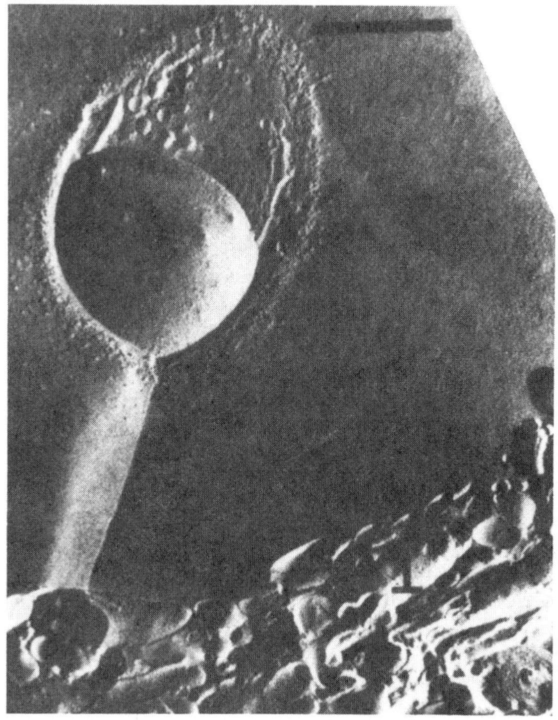

FIGURE 2. Cryptococcus *in brain tissue. Wall particles appear to be excreted into inner portion of capsule. Bar corresponds to 1 μm. (From Takeo, K., Ueasaka, I., Uehina, K., and Nishiura, M.: J Bacteriol 113:1449, 1973.)*

The electron-dense wall structure, the rupture of the parental wall by the newly emerging bud, and the presence of clamp connections all support the classification of the cryptococci among

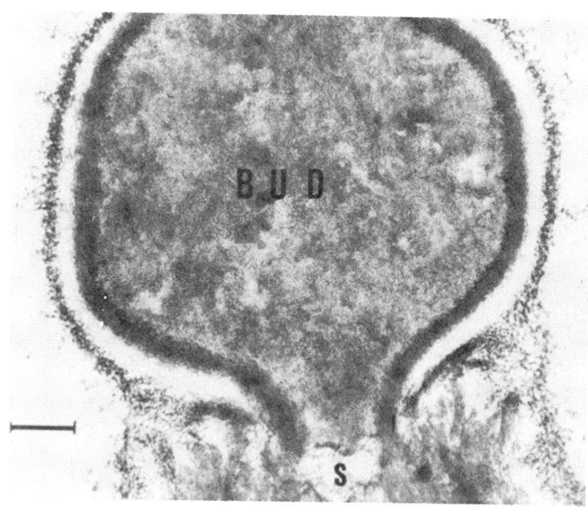

FIGURE 3. *Late cryptococcal bud with septum. Note the broken remnants of the electron-dense layer of the parental wall covered by the electron transparent layer and the capsule. S, septum; bar, 0.2 μ. (From Cassone, A., Simonetti, N., and Strippoli, V.: Arch Microbiol 95:205, 1974.)*

the basidiomycetes. In the basidiomycetes, two nuclei of mated cells travel to the newly formed cell by different routes, fuse, and form hyphae that bear a club-shaped structure called a basidium. These characteristics were the basis for Kwon-Chung's discovery (1975, 1976) that the sexual stage of *C. neoformans* was related to the rust and smut pathogens of plants and strengthened the suspected analogy between the budding of basidiomycetous yeasts and the germination of some fungal spores (Marchant and Smith, 1968). Figure 4 shows the hyphal form of *C. neoformans* (*Filobasidiella*) with clamp connections and a basidium. Basidiospores (Fig. 5) that are produced from the basidium complete the sexual cycle and become budding yeasts (Erke, 1976).

During the yeast (asexual) cycle, cryptococci do not form germ tubes, chlamydospores, or hyphae, either in vitro or in vivo. Although they are always encapsulated during infection, cryptococci in the environment are often shrunken and unencapsulated. Hyperosmolarity has been shown to be at least one of the factors that suppress capsule formation (Dykstra et al., 1977). There appear to be surface receptors for capsular polysaccharide in the outer wall of *C. neoformans,* since purified capsule adheres to unencapsulated *C. neoformans* but not to other yeasts or to heat-killed *C. neoformans* (Kozel, 1977).

The nonpathogenic species of cryptococci cannot grow at or above 37° C because an apparent imbalance of growth develops between the cytoplasmic contents and the cell wall. Growth of the cell wall is inhibited and bizarre-shaped spheroplasts and protoplasts emerge, probably from the site of bud formation (Dabbagh et al., 1974). These cells lyse almost immediately unless the experiment is carried out in hypertonic media.

ANTIGENIC COMPOSITION

The capsule of *C. neoformans,* which is a polymer of mannose, xylose, and glucuronic acid, can be divided into four serotypes (A, B, C, or D) by agglutination or fluorescent antibody testing with hyperimmune sera. The capsular polysaccharide usually elicits only a minimal humoral antibody response in man or experimental animals during infection, possibly because most of the antibody is adsorbed by the abundant capsular polysaccharide in the tissues. During the treatment of cryptococcosis, a good prognosis is associated with falling levels of cryptococcal polysaccharide in the serum; these falling levels are sometimes accompanied by rising levels of cryptococcal antibody (Bardana et al., 1968).

Another possible explanation for the poor or absent antibody response to *C. neoformans* is that the capsule inhibits the immune response. Cryptococcal capsular polysaccharide causes immunologic paralysis in mice (Murphy and Cozad, 1972), unless the dose is carefully monitored (Kozel and Cazin, 1974). Mice form good levels of antibody if given small doses of polysaccharide along with adjuvants or protein carriers, and rabbits produce high titers of antibody if they are hyperimmunized. Antisera to *C. neoformans* cap-

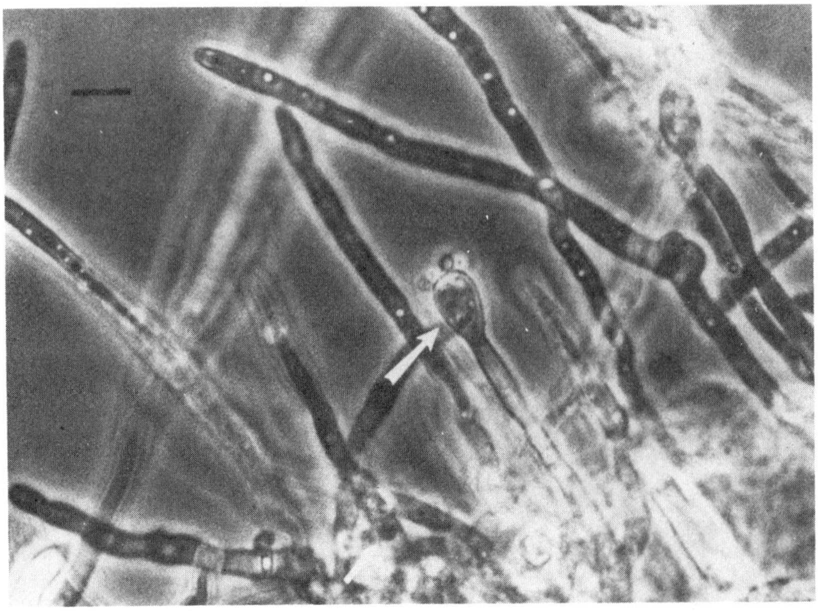

FIGURE 4. *Sexual stage of C. (Filobasidiella) neoformans. The arrow points to a basidium with four basidiospores still attached. A clamp connection can be seen four bar lengths to the right bar length 12 μm. (From Erke, K. H.: J Bacteriol 128:445, 1976.)*

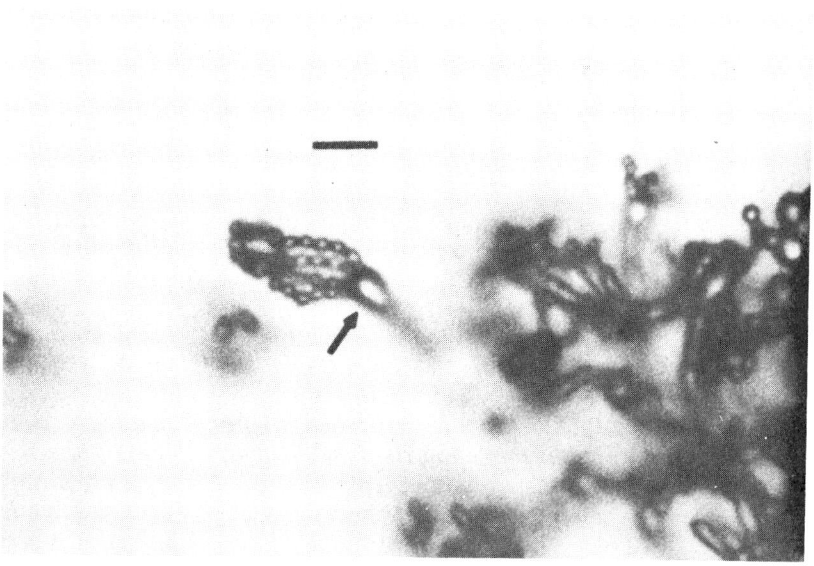

FIGURE 5. *Basidium of F. neoformans with four chains of basidiospores. Bar 16 μm. (From Erke, K. H.: J Bacteriol 128:445, 1976.)*

sule cross-reacts with the capsule of Types 2 and 14 pneumococci. Like bacteria, cryptococci undergo a Quellung reaction (capsular swelling) when suspended in homologous antiserum.

The physicochemical differences responsible for the antigenic variations in the four capsular serotypes are unknown, but this grouping appears to have important taxonomic implications. Kwon-Chung (1976) noted that *Filobasidiella neoformans* was produced only by mating Serotype A or D strains, and *F. bacillospora* only by mating Serotypes B and C. The serotypes differ in geographic distribution, rate of isolation from pigeon droppings, and biochemical activities (Bennett et al., 1977, 1978). These differences suggest that *C. neoformans* may consist of two species, *C. neoformans* of Serotypes A and D and *C. neoformans* of Serotypes B and C. There is no evidence that these two "species" differ in pathogenicity.

Because of the prominence of the cryptococcal capsule, the cell-wall glycopeptides of cryptococci have received little attention. Most of the work has been done with *C. laurentii*, a nonpathogenic *Cryptococcus* species. Raizada et al. (1975) have isolated a glycopeptide that consists of three oligosaccharide side chains bound by O-glycosidic linkages to threonyl and seryl residues in a common polypeptide core. The oligosaccharides are α and β linked polymers of mannose, galactose, and xylose. Like *Candida* species, cryptococci also contain glucans in their cell wall (Meyer and Phaff, 1977). *C. neoformans* can be converted to protoplasts or spheroplasts by the action of certain glucanases. These polymers of mannose, glucose, and other sugars in the cell wall of cryptococci may be responsible for protective antibody that can be raised in mice by immunization with either live or killed unencapsulated *C. neoformans*.

PHYSIOLOGY

Cryptococci are nonfermentative aerobes that attack carbohydrates and other substrates oxidatively. Components of the electron-transport chain (Yamada and Kondo, 1973) and alpha-keto components of the Krebs cycle (Norkrans and Tumblad-Johannson, 1977) have been demonstrated in cryptococci. The complete electron-transport chain has not been worked out, but detailed studies of the benzoquinone coenzymes (coenzyme Q) have been used as taxonomic keys.

The type of coenzyme (categorized according to the number of isoprene units attached to the benzoquinone nucleus) in cryptococci is homogeneous and differs from *Candida*. All species of cryptococci except one saprophytic variant contain coenzyme Q_{10} (Yamada and Kondo, 1973). This coenzyme generally occurs in urease-positive yeasts with a high guanine plus cytosine content.

C. neoformans can acquire nitrogen from peptones, urea, and creatinine but cannot reduce nitrate. Among the saprophytic species, only *C. albidus* and *C. terreus* can obtain nitrogen from nitrate, but all can utilize peptones and urea as nitrogen sources.

C. neoformans synthesizes phenol oxidases that produce melanin-like pigments from phenols. The color of the pigment depends on the substrate. Staub (1962) found that colonies of *C. neofor-*

mans, but not nonpathogenic cryptococci or *Candida,* were brown if extracts from seeds of *Guizotia abyssinica,* a common weed, were included in the growth medium (nigerseed agar). Caffeic acid is the phenolic compound in *Guizotia* that is converted to a brown, melanin-like pigment. The phenoloxidase activity of *C. neoformans* is located in the cell wall and produces pigment in the colonies that does not diffuse into the medium (Shaw and Kapica, 1972). This observation has led to the production of several selective and differential media that have been useful for the isolation and identification of *C. neoformans.* Other species of cryptococci can produce pigment from certain aminophenol and diaminobenzene compounds, and can be presumptively identified by the substrates from which they produce pigment (Chaskes and Tyndall, 1978).

Immunity

The natural resistance of people to *Cryptococcus neoformans* is so strong that cryptococcosis is sometimes referred to in Europe as "malade signal" because it is frequently a sign of underlying disease. About 50 per cent of patients with cryptococcal meningitis are found to have another disease, most frequently a lymphoid malignancy such as Hodgkin's disease. This association and the persistence of abnormal cell-mediated immunity to the cryptococcus in otherwise normal individuals after successful treatment suggests that cell-mediated immunity is the critical line of defense against cryptococcosis. Skin reactivity, lymphocyte transformation, and production of migratory-inhibition factor often show impaired response to cryptococcal antigens and sometimes to other antigens as well (Graybill and Alford, 1974; Schimpff and Bennett, 1975).

Abnormal phagocytes and humoral reactions have not been definitely linked to susceptibility to cryptococcosis. There is no question, however, that human neutrophils and macrophages can ingest and kill *C. neoformans* in vitro (Diamond et al., 1972). In pulmonary infections of mice and other experimental animals, neutrophils clear most of the cryptococci from early infiltrates, and monocytes remove the remainder in the later stages. Although human macrophages do not kill *C. neoformans* very efficiently (Diamond and Bennett, 1973), macrophages that have been activated by sensitized lymphocytes appear to make experimental animals resistant to cryptococcosis. However, two observations suggest that the quantity of the inflammatory response may be more important than the capacity of individual cells to ingest and kill cryptococci. First, many encapsulated cryptococci are probably too large for phagocytes to ingest. Second, there is evidence that white cells can surround cryptococci and kill them by nonphagocytic mechanisms. Monocytes and macrophages may kill cryptococci by release of hydrolases (Kalina et al., 1974); neutrophils, monocytes, and lymphocytes may be cryptococcicidal in the presence of specific antibody (Diamond and Allison, 1976).

Natural humoral substances that inhibit the growth of cryptococci are present in normal serum, saliva, and spinal fluid (Howard and Bolande, 1966; Reiss et al., 1975). Estrogens and other steroids may damage *C. neoformans* (Mohr et al., 1972) and promote phagocytosis (Mohr et al., 1974). There is little evidence that acquired antibody prevents infection. Immunization of mice with an appropriate concentration of capsule plus adjuvant or live unencapsulated cryptococci provides some protection against experimental infection (Gadebusch and Johnson, 1966; Abrahams, 1966), but the mechanism of this resistance is unknown.

It is surprising that the relative importance of the various aspects of immunity against *C. neoformans* have not been defined because it is the only fungus with a definite pathogenic factor. The capsule is essential to the pathogenicity of *C. neoformans.* Strains of *C. neoformans* that cannot form a capsule do not cause disease. When the capsule of cryptococci is lost through manipulation of growth media, they apparently regain the capsule after inoculation because they kill mice and have capsules in the tissues (Dykstra et al., 1977). The capsular polysaccharide has so many effects on the immune response that it is hard to determine the most important. It suppresses antibody-forming cells (Murphy and Cozad, 1972; inhibits phagocytosis (Bulmer and Sans, 1972) and leukocyte migration (Drouhet and Segretan, 1951); decomplements serum by activation of the alternative pathway (Diamond et al., 1974; Macher et al., 1978); and adsorbs or neutralizes opsonins and other protective antibodies. These various effects of the capsule and the numerous immune responses that are activated by experimental cryptococcosis suggest that a normal person has several lines of resistance against the cryptococcus. Nevertheless, the great predilection for cryptococcal infections in patients with depressed cell-mediated immunity implies that this immunologic defense is of critical value in resisting this infection.

The other property of *C. neoformans* that enables it to infect humans and other animals is its ability to grow at 37° C. The other encapsulated cryptococci grow poorly if at all at 37° C and do not cause disease. Rabbits with a body temperature of 39.6° C are much more resistant than mice to infection, and the only recorded infection of a pigeon (body temperature 42° C) was in superficial tissues that were cooler than the core temperature (Ensley et al., 1979). Cryptococci do not grow above 39° C in vitro but survive temperatures up to 44° C. A fascinating and important corollary to these observations is that cryptococcal meningitis does not cause fever.

LABORATORY DIAGNOSIS

Identification of *C. neoformans* in wet mounts and smears of infected material and cultivation from these specimens are the cornerstones of the diagnosis of cryptococcal infection. This large, round, encapsulated yeast is most easily identified in gram-stained smears of spinal fluid, respiratory secretions, or other infected material. The yeast cell (3.5 to 7.0 μ) retains the crystal violet and is easily differentiated from surrounding leukocytes because of the large (up to 30 μ), lightly safranin-stained capsular halo (Fig. 6). Virtually the entire slide can be examined under $\times$ 100 power, and suspicious structures can be visualized in more detail under the oil immersion lens. The India ink preparation, prepared by mixing a few drops of the spinal fluid, spinal fluid sediment, or other specimens with a drop of India ink, is also a popular means of detecting cryptococci (Chapter 155), but there is a danger of confusing lymphocytes, tissue cells, fat droplets, aggregated particles of India ink, or even yeast contaminants of the India ink with cryptococci. If CSF (cerebrospinal fluid) is to be examined by India ink, the presumptive diagnosis of cryptococcosis should not be made without observing the typical small single or double bud and confirming the presence of the organism by Gram stain. Specimens of body fluids may be sedimented by centrifugation at 5,000 g or more before microscopic examination and cultivation, but care must be taken to avoid contamination during the process. Alternatively, a small portion of the fluid can be sedimented for microscopic examination and the remainder cultivated on as many different slants of media as required. Either a portion of the whole fluid or the supernatant from the sedimented fluid should be retained for antigen detection. The larger the volume of spinal fluid cultivated, the greater the chance of isolating the organism, and clinicians should be encouraged to submit as much spinal fluid as possible. We request 15 to 20 ml if the patient does not have increased CSF pressure.

If pulmonary or brain tissue is submitted for culture, impression smears should be made for Gram and special stains and the tissue carefully macerated with a scalpel. Before culture, smears should be made from the macerated tissue, and a portion should be mixed with 1 or 2 ml of 15 per cent potassium hydroxide and incubated at 37° C for 30 minutes before examination as a wet mount with and without India ink. The potassium hydroxide digests cells and artifacts that could be confused with cryptococci.

The histologic examination of tissue should include special stains as well as the routine hematoxylin and eosin stain. Cryptococci can be readily seen on hematoxylin and eosin stain, but the capsule does not stain and can be confused with a space containing a round yeast. Periodic-acid Schiff and Gomori methenamine silver stains are excellent screening stains and often permit visualization of the capsule as an unstained halo surrounding the round budding yeast, but Mayer's mucicarmine technique is the definitive stain because it stains the cryptococcal capsule. In addition to their large capsule, cryptococci differ from *Blastomyces* because the cryptococcal wall is thinner, and the buds are attached to the parent cell by a narrow isthmus. They can be

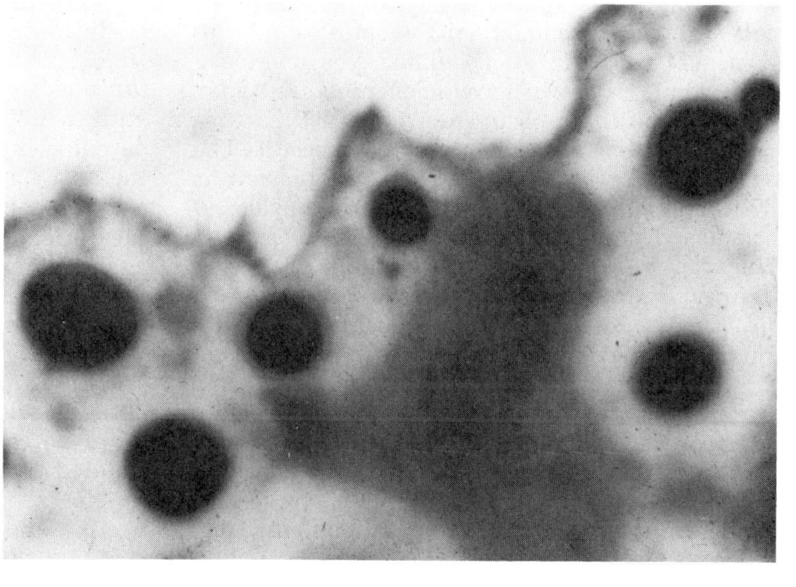

FIGURE 6. *Gram stain of* C. neoformans *in cerebrospinal fluid at* $\times$ *1000. Note the large capsules in this stain.*

distinguished from *Candida* because cryptococci are round instead of oval and form no pseudohyphae or hyphae. The spherules of *Coccidioides immitis* contain endospores and do not bud.

In patients with meningitis, the respiratory secretions, CSF, urine, blood, and any tissue obtained by biopsy should be examined microscopically and cultured, at least on Sabouraud's agar. It is helpful, especially if the specimen is contaminated, to cultivate a portion of the specimen on nigerseed agar and on media that contain antibacterial antibiotics. Cryptococci are sensitive to cycloheximide (Actidione), and media containing this antibiotic should be avoided. As many slants as possible should be inoculated and incubated at room temperature or 25 to 30° C. Some tubes may also be placed at 37° C for quick differentiation from saphrophytic cryptococci. Growth often appears within 48 to 72 hours but may be delayed for seven days or more if the inoculum is small. For this reason, cultures should be kept for six weeks before being discarded as negative, and tubed media should be used to avoid drying.

When colonies appear, they are usually convex, mucoid, cream-colored to light tan, or very rarely a light pink. Individual cells are more round than oval, have at least a small capsule in India ink, and do not produce hyphae. In practice, every round yeast, encapsulated yeast, or urease-producing yeast should be subjected to a complete work-up to exclude *C. neoformans*. The cultural identification of *C. neoformans* can be definitely established by three properties: growth of subcultures at 37°C, mouse pathogenicity, and urease production. If mice are inoculated intracerebrally, intravenously, or intraperitoneally with 10^5 to 10^7 *C. neoformans*, a rapid, progressively fatal infection follows within one to three weeks. One or two mice should be sacrificed at weekly intervals up to one month and the brain cultured and examined microscopically for the presence of round encapsulated yeast. Mouse pathogenicity, brown growth on nigerseed agar (Fig. 7), and the capacity to grow at 37° C separate *C. neoformans* from other cryptococci. The saprophytic cryptococci that occasionally grow at 37° C usually produce small, poorly developed colonies. We have also found the *C. neoformans*, but not the saprophytes, grows at 39° C.

In addition to these major characteristics, *C. neoformans* does not produce pseudohyphae, chlamydospores, or germ tubes; does not reduce nitrate; does not ferment (produce gas from) any sugars; and always assimilates inositol. Although these properties can separate *C. neoformans* from any other fungus, the assimilation tests shown in Table 1 are usually performed. Mouse pathogenicity, failure to produce deep salmon-pink pigment, and assimilation of inositol separate *C. neoformans* from *Rhodotorula* species, which have similar sugar assimilation patterns.

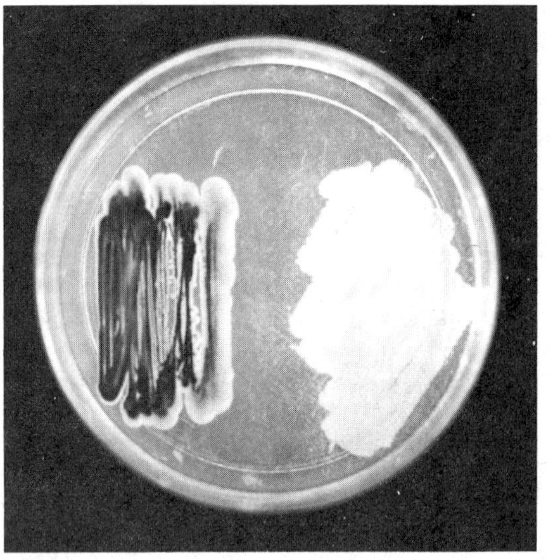

FIGURE 7. *The darkly pigmented growth of* C. neoformans *on nigerseed agar clearly differentiates it from the white or cream-colored growth of* Candida albicans *on the other half of the agar plate.*

Although the perfect state of *C. neoformans* has been described (see earlier sections on Morphology and Epidemiology), and several strains have produced hyphae in the laboratory, this is an unusual phenomenon and does not ordinarily occur without special techniques and media (Kwon-Chung, 1975, 1976). If hyphae are observed, they resemble the forms shown in Figures 4 and 5. *Filobasidiella (Cryptococcus) bacillospora* (Serotypes B and C) assimilates malic, fumaric, and succinic acids, while *F. (Cryptococcus) neoformans* (Serotypes A and D) does not (Bennett et al., 1978).

The only useful serologic technique for identifying patients with cryptococcal infection is the agglutination of latex particles that are coated with antibody raised against the cryptococcal capsule. This test measures cryptococcal capsular antigen in the serum or spinal fluid of infected patients; the capsular polysaccharide reacts with the antibody on the latex particle and agglutinates them. False-positive tests are unusual unless the patient produces rheumatoid factor that also agglutinates the antibody-coated latex. A control to rule out rheumatoid factor should be performed in every test and is provided in the commercial test kits. The latex-agglutination test is very sensitive and is positive in the spinal fluid of most patients with cryptococcal meningitis but in a lower percentage of the serum of patients with isolated pulmonary disease. Although false-positive tests are rare, *C. neoformans* is usually easy to grow, and the diagnosis of cryptococcal infection

TABLE 1. Characteristics of *Cryptococcus neoformans**

SPECIES	MOUSE VIRULENCE	GROWTH AT 37° C	PIGMENT ON NIGERSEED AGAR	UREASE ACTIVITY	KNO_3	ASSIMILATIONS					
						Inositol	Sucrose	Lactose	Melibiose	Cellobiose	Raffinose
C. neoformans	+	+	+	+	−	+	+	−	−	+	+
C. laurentii	−	+ or −	−	+	−	+	+	+	+	+	+
C. terreus	−	+ or −	−	+	+	+	−	rare	−	+	−
C. gastricus	−	−	−	+	−	+	−	−	−	+	−
C. uniguttulatus	−	−	−	+	−	+	+	−	−	− or +	+
C. luteolus	−	−	−	+	−	+	+	−	+	+	−
C. albidus var. *albidus*	−	rare	−	+	+	+	+	+ or −	− or +	+	+
C. albidus var. *diffluens*	−	− or +	−	+	+	+	+	−	+ or −	+	+

*Modified from Widra and Long, 1970, and Bennett et al., 1978.

should always be confirmed by culture. Tests for detection of cryptococcal antibody are not useful in the diagnosis of cryptococcal infection because some normal people have antibody and most patients with cryptococcal infection do not. During the course of successful treatment, the titer of cryptococcal polysaccharide should fall, but this may or may not be accompanied by the development of cryptococcal antibody. While this combination of serologic events is probably a good prognostic sign, some patients recover without developing antibody. Furthermore, dead cryptococci may be seen by microscopic examination of the spinal fluid for many weeks or months after successful treatment. If these organisms cannot be cultivated, their presence should not be taken as a bad prognostic sign.

Cryptococcal skin test preparations are useful for epidemiologic surveys but are of no value for diagnosis.

EPIDEMIOLOGY

Unlike many other fungi that can cause life-threatening infections in apparently normal, nonhospitalized patients, *C. neoformans* is widely distributed throughout the world. It can be isolated from fruit skins and juice, milk, and soil but is most often associated with pigeon dung. It is not part of the normal flora of man or other animals. Although the most common manifestation of cryptococcal infection is meningitis, the primary lesion almost certainly occurs in the lung from inhalation of infective particles. Only 200 to 300 cases of cryptococcal meningitis are reported each year in the United States, but it is estimated that more than 15,000 subclinical respiratory infections occur annually in New York City alone (Ajello, 1969; Littman and Walter, 1968).

Although the capsule of *C. neoformans* is an essential virulence factor, its large size would appear to prevent cryptococci from reaching the terminal bronchioles and causing parenchymal disease. This paradox has probably been solved by two observations. First, in the hyperosmolar, alkaline environment of pigeon feces, *C. neoformans* is unencapsulated and shrunken in size, so that a small percentage of the yeast population has a diameter of 1 μ or less and could easily reach the alveoli (Powell et al., 1972). Second, the sexual stage of *C. neoformans*, *Filobasidiella*, produces basidiospores that are also about 1 μ in diameter. A role for the sexual stage in dissemination of cryptococci has not been proven but is an attractive possibility.

Cryptococci can reach numbers as high as 5×10^7 in pigeon feces and can be recovered in large numbers from the accumulated filth of pigeon roosts, attics, barn lofts, and cornices. It is often the predominant organism in this alkaline, hyperosmolar environment, with its rich content of creatinine and other nitrogen-containing compounds, but may disappear when the bird droppings are mixed with soil. It is infrequently isolated from organically enriched soil (Emmons, 1960). Direct exposure to sunlight also inhibits its growth. In shaded, moist, or desiccated pigeon feces, it may persist for as long as two years or more. It survives better if relative humidity is increased (Ishaq, et al., 1968).

Since *Cryptococcus neoformans* is a stable, soil-dwelling organism in its sexual stage, it is possible that *C. neoformans* was infrequently isolated from soil because it converted to *Filobasidiella* and was not recognized. In fact, only *C. neoformans* Serotypes A and D are isolated from pigeon droppings. The ecologic site of Types B and C is unknown. Almost all clinical infections in the United States are caused by Serotype A strains except in southern California, where 51 per cent of infections are caused by Serotypes B and C (Bennett et al., 1977). The ecologic characteristics of the organism undoubtedly explain this interesting difference in the geographic distribution of infections.

C. neoformans infects many animals, but transmission from animal to man or man to man has never been documented. Although Green and Bulmer (1979) have shown that a small number of mice could be infected by gastrointestinal inoculation, ingestion of cryptococci has never been implicated as the portal of entry in human infection, even after exposure to large numbers of cryptococci in unpasteurized milk from cows with cryptococcal mastitis.

The reasons why cryptococci do not compete well with other organisms in the soil are unknown but may involve biological control mechanisms. Bunting et al. (1979) have shown that *Acanthamoeba polyphaga,* a free-living ameba, can phagocytose and kill up to 99 per cent of *C. neoformans* in mixed cultures. One ameba can ingest and kill as many as 84 cryptococci per day. Interestingly, some of the surviving yeast cells developed into colonies containing hyphae. These forms may be a biological "escape hatch" and suggest the possibility that *C. neoformans* survives the hostile environment of the soil by conversion into the sexual stage.

References

Abrahams, J.: Further studies on acquired resistance to murine cryptococcosis. Enhancing effect of *Bordetella pertussis.* J Immunol 98:914, 1966.

Ajello, L.: A comparative study of the pulmonary mycosis of Canada and the United States. Publ Health Rep 84:869, 1969.

Bardana, E. J., Kauffman, L., and Benner, E. J.: Amphotericin B and cryptococcal infections. Arch Intern Med 122:517, 1968.

Bennett, J. E., Kwon-Chung, K. J., and Howard, D. H.: Epidemiological differences among serotypes of *Cryptococcus neoformans.* Am. J. Epidemiol 10:582, 1977.

Bennett, J. E., Kwon-Chung, J. K., and Theodore, T. S.: Biochemical. differences between serotypes of *Cryptococcus neoformans.* Sabouraudia 16:167, 1978.

Bulmer, G. S., and Sans, M. D.: *Cryptococcus neoformans.* III. Inhibition of phagocytosis. J Bacteriol 95:5, 1968.

Bunting, L. A., Neilson, J. B., and Bulmer, G. S.: *Cryptococcus neoformans*: Gastronomic delight of a soil amoeba. Sabouraudia 17:225, 1979.

Cassone, A., Simonetti, N., and Strippoli, V.: Wall structure and bud formation in *Cryptococcus neoformans.* Arch Microbiol 95:205, 1974.

Chaskes, S., and Tyndall, R. L.: Pigment production by *Cryptococcus neoformans* and other cryptococcus species from aminophenols and diaminobenzenes. J Clin Microbiol 7:146, 1978.

Dabbagh, R., Conant, N. F., Nielsen, H. S., and Burns, R. O.: Effect of temperature on saprophytic cryptococci: Temperature-induced lysis and protoplast formation. J Gen Microbiol 85:177, 1974.

Diamond, R. D., Root, R. K., and Bennett, J. E.: Factors influencing killing of *Cryptococcus neoformans* by human leukocytes *in vitro.* J Infect Dis 125:367, 1972.

Diamond, R. D., and Bennett, J. E.: Growth of *Cryptococcus neoformans* within human macrophages *in vitro.* Infect Immun 7:231, 1973.

Diamond, R. D., May, J. E., Kane, M. A., Frank, M. M., and Bennett, J. E.: The role of the classic and alternate complement pathways in host defenses against *Cryptococcus neoformans* infection. J Immunol 112:2260, 1974.

Diamond, R. D., and Allison, A. C.: Nature of the effector cells responsible for antibody-dependent cell-mediated killing of *Cryptococcus neoformans.* Infect Immun 14:716, 1976.

Drouhet, E., and Segretain, G.: Inhibition de la migration leukocytaire in vitro par un polyoside capsulaire de *Torulopsis* (*Cryptococcus*) *neoformans.* Ann Inst Pasteur 81:674, 1951.

Dykstra, M. A., Friedman, L., and Murphy, J. W.: Capsule size of *Cryptococcus neoformans*: Control and relationship to virulence. Infect Immun 16:129, 1977.

Edwards, M. R., Gordon, M. A., Lapa, E. W., and Ghiorse, W. C.: Micromorphology of *Cryptococcus neoformans.* J Bacteriol 94:766, 1967.

Emmons, C. W.: Prevalence of *Cryptococcus neoformans* in pigeon habitats. Publ Health Rep 75:362, 1960.

Ensley, P. K., Davis, C. E., Anderson, M. P., and Fletcher, K. C.: Cryptococcosis in a male Beccari's Crowned Pigeon. J Am Vet Med Assoc 175:992, 1979.

Erke, K. H.: Light microscopy of basidia, basidiospores, and nuclei in spores and hyphae of *Filabasidiella neoformans* (*Cryptococcus neoformans*). J Bacteriol 128:445, 1976.

Gadebusch, H. H., and Johnson, A. G.: Natural host resistance to infection with *Cryptococcus neoformans.* J Infect Dis 116:551, 1966.

Graybill, J. R., and Alford, R. H.: Cell-mediated immunity in cryptococcosis. Cell Immunol 14:12, 1974.

Green, J. R., and Bulmer, G. S.: Gastrointestinal incubation of *Cryptococcus neoformans* in mice. Sabouraudia 17:233, 1979.

Howard, J. I., and Bolande, R. P.: Humoral defense mechanisms in cryptococcosis. Substances in normal human serum, saliva, and cerebral spinal fluid affecting the growth of *Cryptococcus neoformans.* J Infect Dis 116:75, 1966.

Ishaq, C. M., Bulmer, G. S., and Felton, F. G.: An evaluation of various environmental factors affecting the propagation of *Cryptococcus neoformans.* Mycopathol Mycol Appl 35:81, 1968.

Kalina, M., Kletter, Y., and Aronson, M.: The interaction of phagocytes and the large-sized parasite *Cryptococcus neoformans*: Cytochemical and ultrastructural study. Cell Tiss Res 152:165, 1974.

Kozel, T. R., and Cazin, J., Jr.: Induction of humoral antibody response by soluble polysaccharide of *Cryptococcus neoformans.* Mycopathol Mycol Appl 54:21, 1974.

Kozel, T. R.: Non-encapsulated variant of *Cryptococcus neoformans* II. Surface receptors for cryptococcal polysaccharide and their role in inhibition of phagocytosis by polysaccharide. Infect Immun 16:99, 1977.

Kwon-Chung, J. K.: A new genus, *Filobasidiella,* the perfect state of *Cryptococcus neoformans.* Mycologia 67:1197, 1975.

Kwon-Chung, J. K.: A new species of *Filobasidiella,* the sexual state of *Cryptococcus neoformans* B and C serotypes. Mycologia 68:942, 1976.

Littman, M. L., and Walter, J. E.: Cryptococcosis. Current issues. Am. J. Med. 45:922, 1968.

Macher, A., Bennett, J., Gadek, J., and Frank, M. M.: Complement depletion in cryptococcal sepsis. J Immunol 120:1686, 1978.

Marchant, R., and Smith, D. G.: Bud formation in *Sacchoromyces cerevisiae* and a comparison with the mechanism of cell division in other yeasts. J Gen Microbiol 53:163, 1968.

Meyer, M. T., and Phaff, H. J.: Survey for alpha-(1–3) glucanase activity among yeasts. J Bacteriol 131:702, 1977.

Mohr, J. A., Long, H., McKown, B. A., et al.: *In vitro* susceptibility of *Cryptococcus neoformans* to steroids. Sabouraudia 10:171, 1972.

Mohr, J. A., Muchmore, H. G., and Tacker, R.: Stimulation of phagocytosis of *Cryptococcus neoformans* in human cryptococcal meningitis. J. Reticuloendothel Soc 15:149, 1974.

Murphy, J. W., and Cozad, G. C.: Immunological unresponsiveness induced by cryptococcal capsular polysaccharide assayed by the hemolytic plaque technique. Infect Immun 5:896, 1972.

Norkrans, B., and Tumblad-Johannson, I.: Cellular contents of the Krebs cycle keto acids in yeasts grown on different nitrogen sources including hydroxylamine. Arch Microbiol 115:127, 1977.

Powell, K. E., Bernhoff, A., Dahl, A., Weeks, R. J., and Tosh, F. E.: Airborne *Cryptococcus neoformans:* Particles from pigeon excreta compatible with alveolar deposition. J Infect Dis 125:412, 1972.

Raizada, M. K., Schutzbach, J. S., and Ankel, H.: *Cryptococcus laurentii* cell envelope glyco-protein. J Biol Chem 250:3310, 1975.

Reiss, F., Szilagyi, G., and Mayer, E.: Immunological studies of the anti-cryptococcal factor of normal human serum. Mycopathologia 55:175, 1975.

Schimpff, S. C., and Bennett, J. E.: Abnormalities in cell-mediated immunity in patients with *Cryptococcus neoformans* infection. J Allergy Clin Immunol 43:430, 1975.

Shadomy, H. J.: Clamp connections in two strains of *Cryptococcus neoformans.* Recent trends of yeast research. Atlanta Research Articles, School of Arts and Sciences, Georgia State University, 1971.

Shaw, C. E., and Kapica, L.: Production of diagnostic pigment by phenoloxidase activity of *Cryptococcus neoformans.* Appl Microbiol 24:824, 830, 1972.

Staub, F.: *Cryptococcus neoformans* and *Guizotia abyssinica* (syn. *G.* oleifera D.C.(Z Hyg 148:446, 1962.

Takeo, K., Ueasaka, I., Uehina, K., and Nishiura, M.: Fine structure of *Cryptococcus neoformans* grown *in vivo* as observed by freeze etching. J Bacteriol 113:1449, 1973.

Widra, A., and Long, I.: Taxonomic studies on cryptococci and related yeasts. Mycopathol Mycol Appl 40:89, 1970.

Yamada, Y., and Kondo, K.: Coenzyme Q system in the classification of the yeast genera *Rhodotorula* and *Cryptococcus,* and the yeast-like genera *Sporobolomyces* and *Rhodosporidium.* J Gen Appl Microbiol 19:59, 1973.

CANDIDA AND TORULOPSIS 69

Abraham I. Braude, M.D., Ph.D.

Members of the genera *Candida* and *Torulopsis* are indigenous human yeasts that colonize the skin and mucous membranes of normal people but produce infection only when natural resistance is compromised. The two genera resemble each other morphologically and are related antigenically, but only members of the genus *Candida* commonly cause serious human illness. The genus comprises seven species of medical interest. The most important of these is *Candida albicans,* but *Candida tropicalis* has become a serious cause of disseminated infection in patients undergoing treatment for hematologic malignancies or receiving bone marrow transplantation (Table 1) (Wingard et al., 1979). The other five species encountered in human disease are designated *C. krusei, C. guilliermondi, C. parapsilosis, C. pseudotropicalis,* and *C. stellatoidea.*

CANDIDA

MORPHOLOGY

Members of the genus *Candida* characteristically develop both as yeast cells and as pseudohyphae. *C. albicans,* the principal pathogenic member of the genus, also produces true hyphae. Pseudohyphae are chains of budding cells that fail to detach, so that they develop into a branching network that resembles true hyphae. Colonies composed of pseudohyphae have the soft, white character of yeasts in contrast to the cottony or woolly growth of true mycelia. The yeast cells of *C. albicans* are oval and gram-positive; they vary in size from $2 \times 3\ \mu$ to $8.5 \times 14\ \mu$. Less frequently, elongated cells may be seen after incubation at room temperature. On cornmeal agar, a nutritionally deficient medium composed only of cornmeal and water, *C. albicans* reproduces in four different forms: (1) true mycelia, (2) pseudomycelia, (3) blastospores, and (4) chlamydospores. The blastospores grow in round clusters at intervals along the pseudomycelia. Terminal chlamydospores (macroconidia) growing at the end of the hyphae are the most distinctive feature of *C. albicans* and are rarely produced by any other species of *Candida.* The chlamydospores are large (8 to 12 μ) and round and have a thick wall (Fig. 1). They are particularly adapted for maintaining vitality during starvation and other adverse conditions. Their large size is due to the storage of

TABLE 1. Frequency of Isolation of Different Species of *Candida* and *Torulopsis* from Blood Cultures

SPECIES	NUMBER OF PATIENTS WITH POSITIVE BLOOD CULTURES
C. albicans	68
C. tropicalis	18
C. parapsilosis	12
Torulopsis glabrata	2

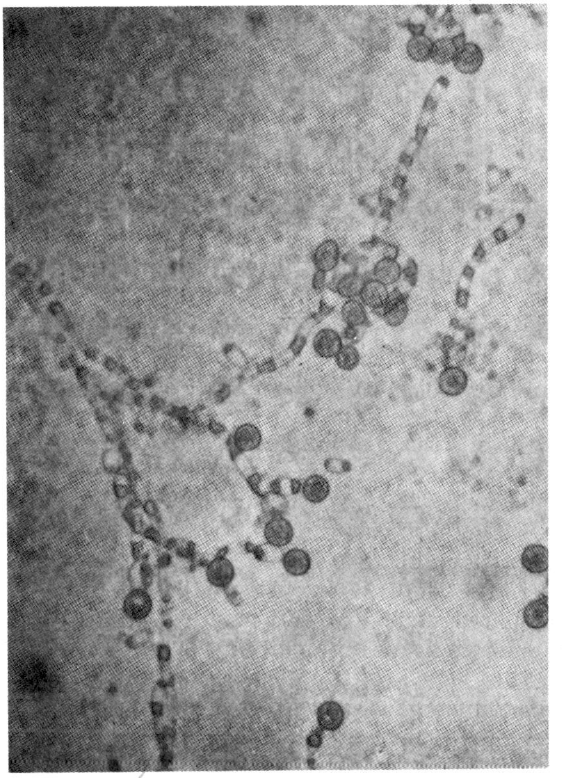

FIGURE 1. Chlamydospores of *C. albicans* after growth on cornmeal agar. The chlamydospores are the sharply defined large round bodies at the tips of the hyphae.

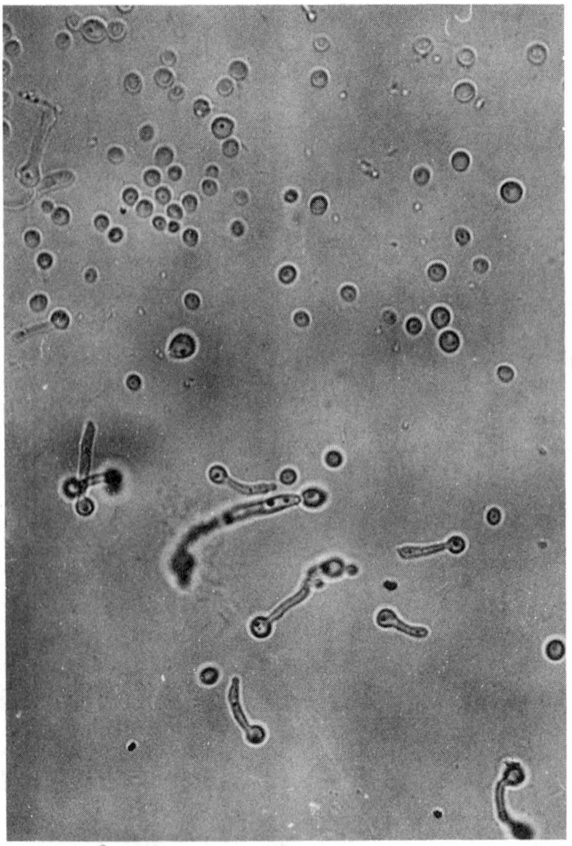

FIGURE 2. *Germ-tube formation in* C. albicans. *Germ tubes are short filaments that protrude from yeast cells within two to four hours after incubation in serum at 37° C. (Reynolds, R., and Braude, A. I.: Clin Res Proc 4:40, 1956.)*

reserve nutritional substance, and their thick wall protects them from an unfavorable environment. The thick wall has two layers, of which the outer is polysaccharide (beta 1:3 glucan) and the inner, protein. There is also a high lipid content (Jansons and Nickerson, 1970). The poor nutritional quality of cornmeal agar causes the yeast cells to differentiate into the well-provisioned chlamydospores. Formation of chlamydospores is a test for distinguishing between *C. albicans* and those species of *Candida* that do not develop into these forms.

Perhaps the most interesting form is the germ tube (Reynolds and Braude, 1956). When yeast cells of *C. albicans* are suspended in serum and incubated at 37° C, they produce within two to four hours a short filament measuring $1.5 \times 15\ \mu$ and having the appearance of a bean sprout (Fig. 2). Because germ tubes develop so quickly, they are also used as a rapid test for *C. albicans*. Some of the morphologic features of *C. albicans* are sometimes seen with *C. stellatoidea,* which can also produce blastospores, pseudohyphae, germ tubes, and rarely, chlamydospores. Since the two are also similar in antigenic structure, *C. stellatoidea* is regarded by some students of yeast taxonomy as an avirulent variant of *C. albicans* (Pollack and Benham, 1957). *C. tropicalis* also produces pseudohyphae, true mycelia and chlamydospores rarely, but not germ tubes. In broth cultures, *C. tropicalis* tends to form a thin film that can trap gas bubbles. The film of *C. tropicalis* is in contrast to that of *C. krusei,* which is much thicker and crawls up the side of the broth culture tube. The flat, dry, dull colonies of *C. krusei* on agar and its cylindrical cell morphology in broth are additional distinctive characteristics. These cells vary considerably in size, and elongated forms may look like crossed matchsticks. The cells of *C. pseudotropicalis* are also elongated in broth cultures but fall apart and lie parallel to each other.

The morphologic differences and species of *Candida* are summarized in Table 2.

ANTIGENIC COMPOSITION

The various species of *Candida* have been classified into six antigenic groups on the basis of slide agglutination tests with monospecific absorbed rabbit sera (Table 3). Each of 22 species of *Candida* were grouped according to their content of seven heat-stable and three heat-labile antigens. In this system of classification, both *C. stellatoidea* and *C. tropicalis* are closely related antigenically to *C. albicans.* Although all *Candida* species share a common antigen, the antigenic structures of *C. pseudotropicalis, C. krusei, C. parapsilosis,* and *C. guilliermondi* are distinctly different from *C. albicans.* These antigenic relationships have been largely substantiated by means of immunoelectrophoresis of soluble yeast extracts (Biguet et al., 1965). The important antigenic determinants in *Candida* appear to be surface polysaccharides, such as mannans and glucans. Mannan forms the outer layer and glucan the inner layer of the cell wall of *C. albicans*. The two sugars appear to occur naturally as complexes of polysaccharide-protein linked together by *N*-acetylglucosamine. The antigenic specificity of the mannans depends on the lengths of the polysaccharide side branches and the type of glycosidic linkages present in them. The mannans are polymers of mannose that, in the main chain or backbone of *C. albicans,* are connected by alpha 1 to 6 linkages. In the side chains, the linkages are alpha 1 to 2 or rarely, alpha 1 to 3, and there are six mannose units or less. Two serotypes of *C. albicans* have been described whose antigenic distinction may depend on the number and position of the linkages.

TABLE 2. Gross and Microscopic Appearance of the Different Species of *Candida*

SPECIES	COLONIES ON AGAR[a]	FILM[b]	YEAST CELL MORPHOLOGY	GERM TUBES	CHLAMYDOSPORES
C. albicans	Creamy	0	Ovoid	+	+
C. stellatoidea	Creamy	0	Ovoid	±	Rare
C. tropicalis	Creamy	Thin, traps bubbles	Ovoid	0	0
C. krusei	Flat, dull, dry	Thick, climbs tube	Cylindrical, elongated cells arranged as crossed matchsticks	0	0
C. pseudotropicalis	Soft, smooth, white	0	Elongated cells lying parallel like logs in a stream	0	0
C. guilliermondi	Thin, flat, glossy	0	Ovoid or cylindrical	0	0
C. parapsilosis	Creamy	0	Ovoid, sometimes giant cells	0	0

[a]Three days' growth on Sabouraud's dextrose agar at 25° C.
[b]Three days' growth on Sabouraud's dextrose broth at 25° C.

PHYSIOLOGY

The metabolism of *Candida* cells is the same as that of other aerobic eukaryotic cells. They are capable of both aerobic glycolysis via the hexose monophosphate pathway and of anaerobic glycolysis through the Embden-Meyerhof pathway. They also have Krebs cycle enzymes and the mitochondrial enzymes for oxidative phosphorylation. Oxidative phosphorylation in *Candida* involves mainly cytochromes a, a_3, b, c, and c_1.

Protein synthesis also resembles that in eukaryotes. *C. albicans* has 80S ribosomes, which dissociate into 60S and 38S subunits (Yamaguchi and Iwata, 1970).

Little is known of the physiologic changes associated with or responsible for the development of hyphae or chlamydospores. Temperature is one important factor; relatively high temperatures (37° C) promote hyphae and blastospores, whereas lower temperatures (<25° C) seem to favor chlamydospores. Mycelial formation is accompanied by a suppression of the pentose phosphate pathway and the diversion of hexoses for cell wall biosynthesis.

TABLE 3. Antigenic Grouping of *Candida* Species of Medical Importance

GROUP	SPECIES	THERMOSTABLE ANTIGENS	THERMOLABILE ANTIGENS
I	*C. albicans*	1, 2, 3, 4, 5, 6, 7	
	C. tropicalis	1, 2, 3, 4, 5, 6	
II	*C. stellatoidea*	1, 8, 10	a
	C. pseudotropicalis	1, 2, 3, 4, 5, 10	
III	*C. krusei*	1, 2, 5, 11	b
IV	*C. parapsilosis*	1, 2, 3, 5, 13, 14, 15	c
V	*C. guilliermondi*	1, 2, 3, 4, 9	
VI	None of medical importance		

IMMUNITY

Cell-mediated immunity is thought to be more important than humoral immunity in resistance to *Candida* infection. At least in chronic infections of the skin due to *C. albicans*, there are often disturbances of T-cell function. Since there is a familial tendency for mucocutaneous candidiasis to occur in children who have abnormal T lymphocytes it would seem that the disturbance in these cells may sometimes be genetic in origin (see Chapter 211). There is some evidence that candidiasis occurs because the T cell cannot recognize the antigen or produce migration inhibition factor (MIF). There is also suggestive evidence that macrophage chemotaxis is depressed. Despite universal delayed sensitivity to *Candida* antigens in healthy people, those with chronic candidiasis are often anergic to its antigens, and transformation of their lymphocytes by *Candida* antigen is sometimes depressed. B lymphocytes and circulating antibodies, on the other hand, are normal. The importance of abnormal T cells in the cause of impaired immunity to *Candida* is also suggested by the tendency of mucocutaneous candidiasis to occur in patients who have frank thymic disorders such as thymomas and congenital thymic aplasia (DiGeorge syndrome).

Abnormalities in humoral immunity, such as agammaglobulinemia, hypogammaglobulinemia, and multiple myeloma have not displayed a special susceptibility to *Candida* infection. Furthermore, the ability of humoral antibody to prevent experimental candidiasis after active or passive immunization has been inconsistent and unimpressive (Hurd and Drake, 1953).

The natural immunity of healthy persons to *Candida* infection is probably generated early in life when the alimentary tract becomes colonized with *C. albicans*. The surface glycoproteins (mannoproteins and glucoproteins) are thought to stimulate both humoral and cellular immunity.

Thus, normal people develop antibodies and delayed hypersensitivity to *Candida* culture filtrates, which contain glycoprotein and polysaccharide antigens.

LABORATORY DIAGNOSIS

Morphologic and Cultural Methods.

C. albicans infection of the skin, urinary tract, mouth, vagina, esophagus, and other tissues is easily recognized by microscopic examination of wet mounts (Fig. 3) or Gram stains (Fig. 4) of the lesions. The material is taken for examination by scraping the skin or mucous patches (thrush) and smearing the infected scrapings on a microscope slide for Gram staining. Yeast and mycelial forms of *Candida* are strongly gram-positive. Both superficial and deeper hematogenous *Candida* infections of the skin can be recognized by microscopic examination of smears made from skin scrapings. *C. albicans* is also cultured without difficulty from lesions on any conventional solid or liquid medium. It may appear overnight at 37° C on blood agar or nutrient agar and seldom needs more than 48 hours to grow out. Although less important for diagnosis than smears, culture is necessary for identifying the species of *Candida* and for testing sensitivity to 5-fluorocytosine, a drug to which *C. albicans* can become resistant.

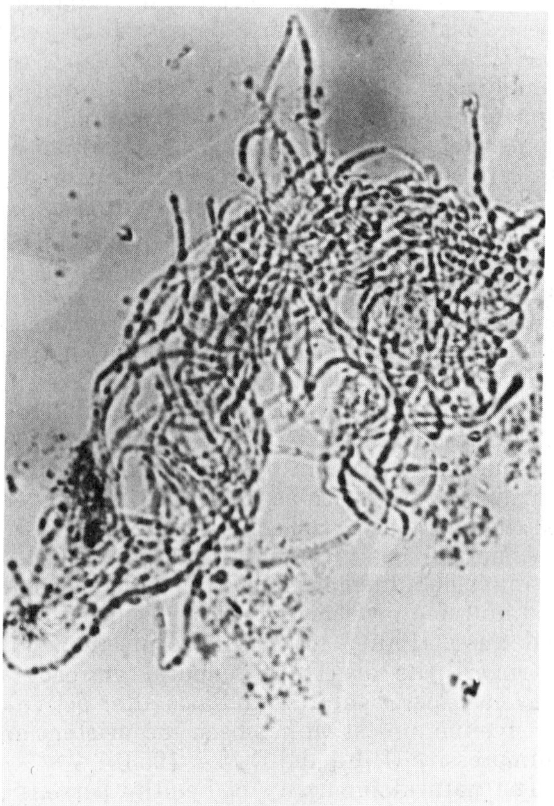

FIGURE 3. *Wet mount of* C. albicans *in skin lesion from girl with chronic mucocutaneous candidiasis. The skin scraping had been treated with potassium hydroxide to digest away the epithelial cells and debris before microscopic examination. Note predominance of hyphae.*

Culture of *Candida* from blood is more difficult than direct culture from infected tissues. The organism may not appear in culture for several days after onset of candidemia, if at all. Blood must be cultured aerobically to get an optimum recovery of *Candida* organisms. The speed of cultural isolation is also accelerated if blood is cultured in pour plates (1.0 ml heparinized blood added to 9 ml warm molten nutrient agar). If candidemia is secondary to contaminated intravenous lines, the diagnosis can also be made by smear and culture of the catheter tip after it is removed from the vein.

The first step in species identification is to test for germ tubes by suspending the yeast in serum and incubating it at 37° C for two to four hours. A presumptive diagnosis of *C. albicans* can be made if germ tubes are produced, and a final diagnosis is made by sucrose assimilation. *C. stellatoidea,* which can also produce germ tubes, can be excluded if sucrose is fermented and assimilated. The other species of *Candida* are also identified by fermentation and assimilation tests.

Assimilation tests are used to determine the ability of a yeast to use a given substance as the sole source of carbon or nitrogen. The test can be done either with solid or liquid media. In the test on solid medium, organisms are dispersed throughout warm agar, which is then poured into a plate before it solidifies. The seeded agar is prepared so that it lacks the specific sugar or nitrogen source under test, and small amounts of the test substance are placed at various points on the surface. The ability to assimilate a given substance is determined by the presence of growth at room temperature in the region where the test substance is applied. The test can also be done by streaking the surface and applying filter paper disks containing test substrate, or by inoculating the surface of agar slants containing each of the individual sugars to be examined. Bromcresyl blue is used as a pH indicator, and each tube is watched for growth and for a change in color from blue to yellow as acid is produced by fermentation. The same media can be used in liquid form for broth assimilation tests. Each broth tube contains one sugar as the test substrate and is inoculated with a suspension of organisms grown for 48 hours in the basal medium to deplete the cells of stored substrate. The tubes are incubated at room temperature, and

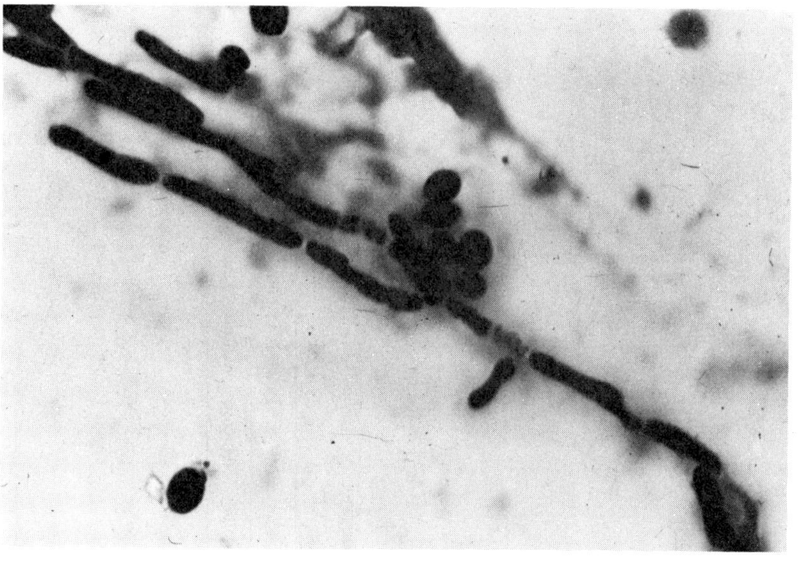

FIGURE 4. *Gram stain of smear made from embolus to femoral artery in a patient with Candida endocarditis. C. albicans is strongly gram-positive. Both hyphae and blastospores appear to be growing in the embolus.*

growth is determined by the development of turbidity. In the nitrate assimilation test, a carbon base is inoculated with 0.078 per cent potassium nitrate in order to determine if this substrate can be used as the sole source of nitrogen.

Fermentation tests determine the ability of yeasts to ferment individual sugars in media that support their growth even in the absence of the test sugar. These media contain peptone, yeast extract (the water-soluble portion of autolyzed bakers' or brewers' yeast that is rich in B vitamins), and a pH indicator. Observations are made on production of both acid and gas. As recorded in Table 4, the yeasts cannot ferment to gas all sugars that they can assimilate. Examples of sugars that are assimilated but not fermented to gas are galactose by *C. stellatoidea* and sucrose by *C. parapsilosis.* It should also be noted that all species of *Candida* of medical importance assimilate and ferment glucose, and all except *C. krusei* assimilate galactose and xylose. Lactose, on the other hand, is assimilated and fermented only by *C. pseudotropicalis,* and inositol by none of them.

In contrast to certain cryptococci, no species of *Candida* in Table 4 can use nitrate as a sole source of nitrogen. Even more important, no species of *Candida,* except certain strains of *C. krusei,* has urease activity. The urease test is thus important for distinguishing *C. albicans* and other species of *Candida* from *Cryptococcus neoformans,* which does have urease activity.

Serologic Methods

Attempts to diagnose candidiasis by tests for antibody or circulating antigen are not yet reliable. Attempts have been made to find antibody in sera from patients by testing them for their ability to agglutinate *C. albicans* yeast cells, or to precipitate antigens from disrupted *C. albicans* in tubes or in gel diffusion plates. These tests have seldom been useful clinically because patients who develop disseminated candidiasis are often suppressed immunologically and cannot generate a specific immune response. Moreover, disseminated candidiasis is a disorder that requires recognition long before antibody can make its appearance. For this reason, efforts are being made to detect circulating *Candida* antigens by immunologic or chemical methods in order to make a diagnosis of systemic candidiasis in patients with negative blood cultures (Miller et al., 1974).

EPIDEMIOLOGY

In contrast to the other fungi that cause human infection, members of the genus *Candida* are not inhabitants of the soil but rather indigenous members of the normal flora of the human alimentary tract and skin. *C. albicans* is also widely distributed among animals and is responsible for oral, esophageal, genital, skin, and disseminated hematogenous infections in domesticated mammals or fowl. For example, *C. albicans* has caused epidemic venereal disease among geese and many cases of esophagitis in pigs (Beemer et al., 1973; Kadel et al., 1969). Despite its wide prevalence in these animal reservoirs, the source of human infections by *C. albicans* is endogenous, arising from carriage sites on the patient himself. About 10 per cent of healthy people carry *C. albicans* in their mouth, and 15 per cent carry it in the

TABLE 4. Assimilation and Fermentation Reactions of Different Species of *Candida* and *T. glabrata*

	ASSIMILATION RESULTS												FERMENTATION (GAS PRODUCTION)				
	Glucose	Maltose	Sucrose	Lactose	Galactose	Melibiose	Cellobiose	Inositol	Xylose	Raffinose	Trehalose	Dulcitol	Glucose	Maltose	Sucrose	Lactose	Galactose
C. albicans	+	+	+	0	+	0	0	0	+	0	+	0	+	+	A*	0	+
C. stellatoidea	+	+	0	0	+	0	0	0	+	0	+	0	+	+	0	0	0
C. tropicalis	+	+	+	0	+	0	+	0	+	0	+	0	+	+	+	0	+
C. parapsilosis	+	+	+	0	+	0	0	0	+	0	+	0	+	0	0	0	+
C. krusei	+	0	0	0	0	0	0	0	0	0	0	0	+	0	0	0	0
C. pseudotropicalis	+	0	+	+	+	0	+	0	+	+	0	0	+	0	+	+	+
C. guilliermondi	+	+	+	0	+	+	+	0	+	+	+	+	+	0	+	0	+
T. glabrata	+	0	0	0	+	0	0	0	0	0	+	0	+	0	0	0	0

*A, acid, no gas

rectum. Carriage in the mouth and rectum can reach 40 per cent or more in patients and tends to increase with age and antibiotic treatment (Berdon and Seita, 1971; Chitale and Bhende, 1965). Vaginal carriage is highest normally in the childbearing age, when it may reach 10 per cent. *C. albicans* is found less frequently on the skin than on mucous membranes; the skin isolation rate seldom exceeds 5 to 8 per cent even in hospitalized patients. *C. tropicalis* is found occasionally in the mouth and feces, and *C. parapsilosis* in feces but in much fewer numbers than *C. albicans.*

The normal residence of *C. albicans* in the vagina, bowel, mouth, and skin is significant in at least two respects. First, cultural isolation of this yeast from materials passing through the mouth or vagina does not constitute proof of infection. This must be kept in mind especially in the case of sputum, because pulmonary infection due to *C. albicans* is extremely rare, if it ever occurs, and positive candidal cultures of sputum almost always represent contamination of pulmonary secretions by the oral resident yeast flora. The second significant point about the normal carriage sites of *C. albicans* is that they represent foci for transmission. For example, *C. albicans* can spread from the vagina or rectum to cause intertrigenous candidiasis of the groin in diabetics; from the vagina to the infant during passage through the birth canal to cause oral thrush in the newborn; and from the skin via intravenous catheters or from the gut via mucosal ulcers to the circulating blood. Transmission of infection from environmental sources of *C. albicans* occurs in exceptional circumstances but is far less important than transmission from normal carriage sites in the patient (Malmatinis et al., 1968; Cremer and DeGroot, 1967).

TORULOPSIS

The yeast *Torulopsis glabrata* shares with *C. albicans* all of the first seven heat-stable antigens of Tsuchiya except antigen 4. Like *C. albicans,* it is also a member of the resident flora and forms ovoid cells. In contrast to members of the genus *Candida, T. glabrata* does not form mycelia or pseudomycelia. In this respect, it resembles four other yeasts (*Torulopsis pintolopesii, Saccharomyces tellustris, Candida bovina,* and *Candida sloofii*), from which it can be differentiated by the fact that *T. glabrata* can ferment trehalose, whereas the other four yeasts cannot. This one fermentation is important because all five species have otherwise identical fermentation and assimilation patterns.

It is also important to differentiate this yeast from *Cryptococcus neoformans,* which also forms no mycelia. The quickest way to differentiate them is to test for urease activity, which is present in *C. neoformans* but not in *T. glabrata.* Two other differential points are that *T. glabrata* grows at room temperature (25° C) but *C. neoformans* does not, and *C. neoformans* has a capsule but *T. glabrata* has none. Failure to assimilate inositol distinguishes *T. glabrata* from other cryptococci. The most common clinical source of *T. glabrata* is the urine, but occasionally it has been cultured from the blood (Table 1) and spinal fluid. When inoculated intraperitoneally in mice, it produces small granulomas and has an intracellular appearance in them that resembles that of *Histoplasma capsulatum.*

T. glabrata is sensitive to nystatin, amphotericin B, and 5-fluorocytosine in concentrations that can be achieved in human infections.

References

Beemer, A., Kutten, E., and Katz, Z.: Epidemic venereal disease due to *Candida albicans* in geese in Israel. Avian Dis 17:639, 1973.

Berdon, J., and Seita, C.: The incidence of *Candida albicans* in hospital patients. J Oral Med 26:123, 1971.

Biquet, J., Tran Van Ky, P., Andrieu, S., and Legaey, R.: Etude immunoelectrophorétique de la nature et de l'ordre d'apparition des anticorps précipitants du sérum de lapins en fonction de leur mode d'immunisation contre *Candida albicans.* Sabouraudia 4:148, 1965.

Chitale, A., and Bhende, Y.: The incidence of *Candida albicans* in throat and feces of healthy persons and patients on antibiotic therapy. J Postgrad Med 11:30, 1965.

Cremer, G., and DeGroot,: An epidemic of thrush in a premature nursery. Dermatologia 135:107, 1967.

Hurd, R., and Drake, C.: *Candida albicans* infections in actively and passively immunized animals. Mycopathologia 6:290, 1953.

Jansons, V., and Nickerson, W.: Chemical composition of chlamydospores of *Candida albicans.* J Bacteriol 104:922, 1970.

Kadel, W., Kelly, D., and Coles, E.: Survey of yeastlike fungi and tissue changes in esophagogastric region of stomachs of swine. Am J Vet Res 30:401, 1969.

Malamatinis, J., Mattniller, E., and Westphal, J.: Cutaneous moniliasis affecting varsity athletes. J Am Coll Health Assoc 16:294, 1968.

Miller, G., Witwer, M., Braude, A., and Davis, C.: Rapid identification of *Candida albicans* septicemia by gas-liquid chromatography. J Clin Invest 54:1235, 1974.

Pollack, J., and Benham, R.: The chlamydospores of *Candida albicans*: Comparison of three media for their induction. J Lab Clin Med 50:313, 1957.

Reynolds, R., and Braude, A. I.: The filament-inducing property of blood for *Candida albicans*: Its nature and significance. Clin Res Proc 4:40, 1956.

Rippon, J. W.: Medical Mycology, 1st ed. Philadelphia, W. B. Saunders Co., 1974.

Tsuchiya, T., Fukazawa, Y., and Kawakita, S.: Serologic classification of the genus *Candida.* In Tokyo Research Committee of Candidiasis: Studies of Candidiasis in Japan. Tokyo, Education Ministry of Japan, 1961, p. 34.

Wingard, J., Nierz, W., and Sarah, R.: *Candida tropicalis*: A major pathogen in immunocompromised patients. Ann Intern Med 91:539, 1979.

Yamaguchi, H., and Iwata, K.: *In vitro* and *in vivo* protein synthesis in *Candida albicans.* 2. Dissociation properties in ribosomes. Sabouraudia 8:189, 1970.

70 SPOROTHRIX SCHENCKII

F. Mariat, Dr. ès Sciences;
and R.G. Garrison, Ph.D.

Sporothrix schenckii is the etiologic agent of sporotrichosis, a chronic mycotic disease of man and animals. The fungus is a dimorphic, pathogenic hyphomycete that occurs naturally on vegetation or as a soil saprophyte. The disease develops after the traumatic introduction of the fungus into the dermis. Classic clinical sporotrichosis usually is seen as an ascending lymphangitis of the extremities after development of the primary lesion at the site of inoculation. Sporotrichosis may occur in other diverse clinical forms and at times it may be severe.

The morphologic and cultural characteristics of this fungal pathogen were first described by Schenck in 1898. In the same publication, E.F. Smith suggested its inclusion in the genus *Sporotrichum*. In 1900, Hektoen and Perkins described a new human case of sporotrichosis and proposed the name *S. schenckii*. A black variety of the fungus was isolated in Europe in 1903 by de Beurmann and Ramond and was described under the name *Sporotrichum beurmanni*. In less than ten years, these authors observed more than 200 cases of sporotrichosis in France and published two classic studies: "Les sporotrichum pathogènes" in 1911 and "Les sporotrichoses" in 1912. Other interesting studies of the fungus were those describing epidemics of sporotrichosis in South Africa in 1927 and especially that of 1941 to 1943. Extensive literature reviews on *S. schenckii* are quoted by Nicot and Mariat (1973), Rippon (1974), and Mariat (1977).

TAXONOMY

For a long time the binomial *Sporothrix schenckii,* as proposed by Hektoen and Perkins, was ignored and the fungus was classified in the genus *Sporotrichum.* At first, two species were recognized on the basis of pigment production: *Sprotorichum beurmanni* (brown to black colonies) and *S. schenckii* (non-pigmented colonies). Other species were created on the basis of differences in cultural characteristics. Later, these species were consolidated into a single taxon, *S. schenckii.* The observations of Carmichael and Nicot and Mariat have established the validity of the binomial *S. schenckii*. The fungus is provisionally classed as Deuteromycotina (Fungi Imperfecti), Hyphomycetes, sympodulosporae, amerosporae.

The perfect state of *S. schenckii* has not yet been precisely established. On the basis of ecologic, morphologic, physiologic, immunologic, and biochemical properties, it has been speculated that *S. schenckii* is closely related to the ascomycetous fungi of the genus *Ceratocystis*. Mariat (1977) has reviewed the taxonomic problems concerned with the *Sporothrix-Ceratocystis* complex and pointed out the possible phylogenetic relationship of *S. schenckii* to *C. stenoceras* as well as other species of *Ceratocystis*.

Synonymy

Sporothrix schenckii Hektoen and Perkins 1900 = Sporotrichum sp. = *Sporotrichum beurmanni* = *Sporotrichum asteroides* = *Sporotrichum equi* = *Sporotrichum jeanselmei* = *Sporotrichum councilmani* = *Sporotrichum grigsby* = *Sporotrichum oculare* = *Sporotrichum biparasiticus* = *Sporotrichum tropicale* = *Rhinocladium beurmanni*.

Possible perfect stage: *Ceratocystis stenoceras* (Robak) Moreau or another *Ceratocystis* species with a *Sporothrix* mode of conidiogenesis.

MORPHOLOGY

The Saprophytic Mycelial Phase

The colonial morphology of *S. schenckii* varies with the strain when grown at 20 to 30°C on Sabouraud dextrose agar. It may be smooth, glabrous or velvety, folded or wrinkled, and sometimes moist with or without erect coremia. The color of the colony varies from yellowish white to black. The consistency is always membranous and resistant. Sectorial variation is frequently observed.

The microscopic characters are relatively constant (Fig. 1). The thin, septate, mycelial filaments (1.5 μm in diameter) branch and bear mononucleate, hyaline conidia, which measure 1.5 to 2.5 by 2.5 to 5.5 μm. The ovoid to elongate conidia are borne directly on the vegetative filament by short sterigmata (radulaspores), or in clusters or

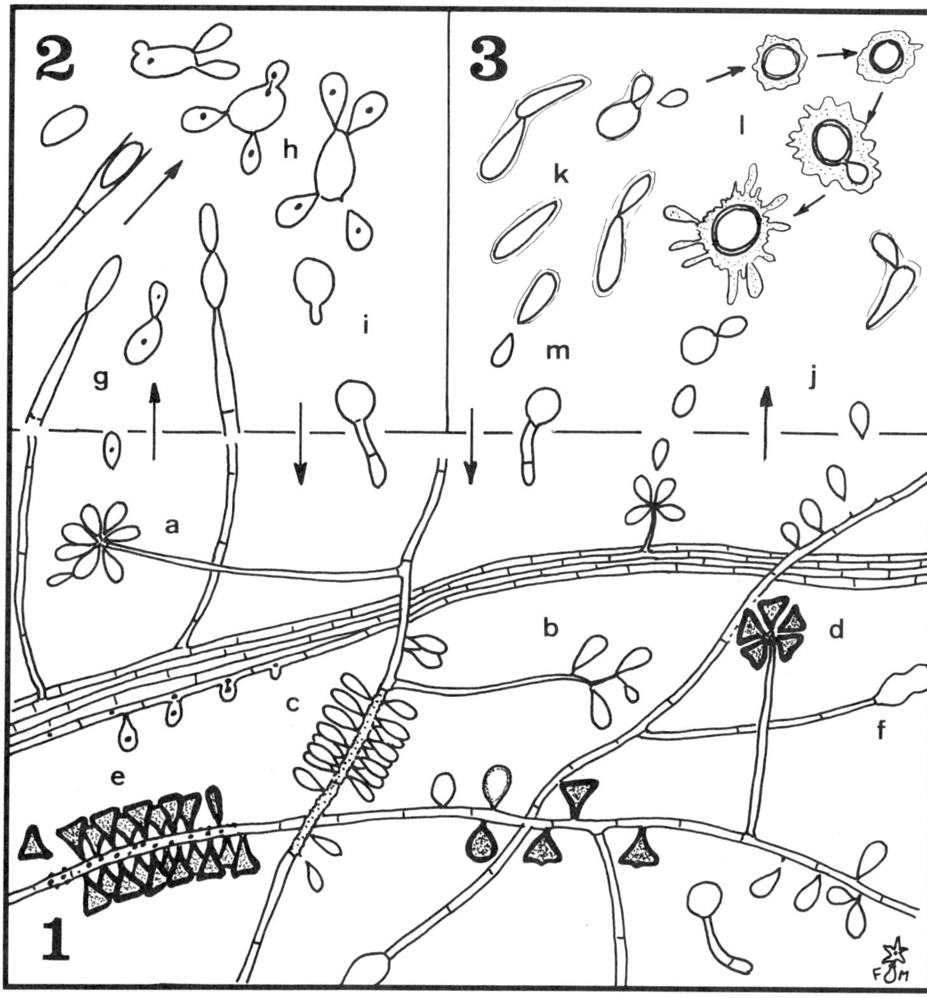

FIGURE 1. *Diagrammatic representation of the life cycle of* Sporothrix schenckii. *1, Saprophytic mycelial phase. 2, Yeastlike phase in vitro. 3, Parasitic yeastlike phase in vivo. a, Hyaline conidia in clusters. b, Conidia in sympodium. c, Conidia in radula. d, Triangular pigmented macrospores in clusters. e, Macrospores in radula. f, Chlamydospores. g, Mycelial (M) to yeast (Y) transformation in vitro. h, Yeastlike budding cells in vitro. i, Y→M transformation. j, M→Y transformation in vivo. k, Yeastlike budding cells in vivo. l, Asteroid bodies. m, Y→M transformation.*

bouquets at the top of fertile erect branches. In the latter case, the spores appear successive (sympodulospores). In addition, many strains of *S. schenckii* produce a thick-walled, pigmented spore that is spherical to conical and characteristically triangular in contour. These spores measure from 2.5 to 4.0 μm in diameter. They are borne on the filament in the same fashion as the hyaline conidia but the sterigmata are larger. These conidia may be interpreted as "macrospores" and are considered specific for *S. schenckii*.

Yeast Phase in vitro

S. schenckii is a dimorphic fungus. The saprophytic state is filamentous when grown at 20 to 30°C, whereas the parasitic form is yeastlike when grown under special conditions at 37° C. The unicellular yeastlike form is obtained when grown at 37°C under CO_2 on a blood agar base (brain heart infusion agar, for instance) or in liquid shaker culture. Yeastlike colonies grown on a blood agar base appear similar to bacterial colonies and are of a creamy consistency. The surface is moist, smooth, and off-white in color.

Yeastlike cells obtained from in vitro cultures differ in morphology from those seen in vivo. The cells are ovoid to globose to elongate and measure 2.5 to 5.0 by 3.5 to 6.5 μm. Mycelial to yeast transformation occurs by direct budding from the mycelium or by the formation of oidial yeastlike cells within the interior of the same filament (Garrison et al., 1975).

Ultrastructure of S. schenckii

Ultrastructural studies of yeastlike and mycelial phase cells of *S. schenckii* indicate that the cytoplasm of both forms contains typical fungal organelles (Lane *et al.*, 1969). The septal area of the mycelial phase cell is typical of that of an ascomycetous fungus. Cells of both phases are usually mononucleate and contain numerous mitochondria, an endoplasmic reticulum with ribosomes, and scattered vacuoles of varying size. Membrane-bound, osmiophilic structures thought to be involved in lipid storage may be observed in young cells of both phases (Garrison et al., 1977). A thickened, electron-dense layer of microfibrillar material is seen at the outer surface of the yeastlike cell wall. This material may be composed in part of an acid mucosubstance and represent portions of an extracellular capsular or slime layer.

PHYSIOLOGY

Physiologic characteristics are of little importance for the recognition of the species. All cultures isolated from cases of sporotrichosis require the pyrimidine moiety of thiamine as a growth factor. A study of the carbon nutrition of 15 strains showed that xylose, glucose, fructose, galactose, maltose, starch, and glycerol are readily utilized, but that acetate, citrate, and tartrate are not. Utilization of sucrose, lactose, and arabinose is strain variable.

The precise physiologic conditions necessary for the development of the yeastlike form are not known, although the presence of CO_2 appears important. It is interesting that the RNA/DNA ratio of mycelial phase cells is significantly higher than that of yeastlike cells.

LABORATORY DIAGNOSIS

Direct Examination

Smears of pus or sputum may be examined by the Gram stain or Giemsa stain. In most cases of sporotrichosis, the parasitic yeastlike form of *S. schenckii* is observed very rarely or only with difficulty on staining of clinical materials. However, in some cases they may be abundant. The yeastlike cells are small (2 to 3 by 3 to 6 μm) and spherical to ovoid to cigar-shaped, and may possess one or two buds. The cells are gram variable and are often surrounded by a clear halo. Staining by the fluorescent antibody method of Kaplan and González Ochoa or pretreatment of smears with diastase as proposed by Fetter may aid in demonstration of the yeastlike forms.

Histopathology

The cutaneous lesion is an irregular or papillomatous acanthosis with intradermal microabscesses containing lymphocytic and plasmocytic infiltrates. There are frequent giant cells and a peripheral fibrosis. The granuloma may appear with three concentric zones: the central suppurative, the intermediate tuberculoid, and the peripheral syphiloid. However, these zones are often indistinguishable. The pathologic aspects of the lesion are not specific.

The parasitic yeastlike forms are extremely difficult to demonstrate on direct microscopic examination of stained histologic sections of clinical materials. When present, the yeastlike cells are readily stained by the PAS or Gomori-Grocott methods. The cellular morphology is similar to that seen in cells from pure cultures.

In numerous cases, and especially in chronic sporotrichosis, a cellular form known as an asteroid body may be observed. It is readily seen on staining with hematoxylin-eosin. These structures are spherical to ovoid, large (3 to 10 μm in diameter), thick-walled cells that may show budding. The cell is surrounded by an eosinophilic substance that may have an arrangement of projections in a rather irregular pattern. This eosinophilic material is thought to be the result of precipitation of an antigen-antibody complex (the Hoeppli-Splendore phenomenon). The diameter of the complete asteroid body may reach about 30 μm.

Cultures

S. schenckii may be isolated readily from clinical materials on a variety of routine culture media. The mycelial phase of the fungus is obtained in three to five days when grown on Sabouraud dextrose agar at room temperature. The fungus is resistant to both cycloheximide and chloramphenicol. Thus, the addition of 0.5 g/liter cycloheximide and 0.1 g/liter chloramphenicol to the isolation medium before sterilization aids the isolation of the fungus in pure culture. The yeastlike form may be obtained by inoculation on brain-heart infusion blood agar and then incubating at 37°C under CO_2. Mycelial to yeastlike cell transformation is important for the specific identification of *S. schenckii*.

Laboratory Animal Inoculations

Intraperitoneal inoculations may be made in male mice or male rats, although male hamsters are the animals of choice. Characteristic disease is produced in two to three weeks, with involvement of the spleen, liver, and testes as target organs. Sections of these organs stained by the PAS, Gomori-Grocott or Gram method reveal abundant intracellular yeastlike forms. Asteroid bodies are usually frequent in histologic sections taken from the testes. It is interesting to note that yeastlike forms are much easier to demonstrate in experimental animal infections than in naturally acquired disease. They are gram positive and resemble large gram-positive rods.

ANTIGENIC COMPOSITION

Protein and lipid fractions from extracts of *S. schenckii* have been studied, but cell-wall glycoproteins are the most active antigenic components.

Comparative studies done by Toriello and Mariat have shown that polysaccharides synthesized by *S. schenckii* contain mannose and rhamnose as the main sugars, although appreciable amounts of galactose and traces of glucose and hexosamine occur. It is noteworthy that rhamnose is not frequently found in fungi. A protein moiety containing 1.1 to 1.9 per cent nitrogen is bound strongly to the polysaccharides.

The structure of the polysaccharides as peptido-rhamno-mannans has been defined by Lloyd et al., Travassos et al., and Gorin et al., quoted by Mariat (1977).

IMMUNITY

S. schenckii is ubiquitous in the environment, and natural resistance to infection is high. In endemic areas, hypersensitivity is noted in individuals with no clinical signs of disease. In such individuals, this state of hypersensitivity may affect the course of the infection.

Hypersensitivity may be determined by intradermal skin tests with cellular sporotrichin (a suspension of heat and/or chemically killed spores or yeastlike cells containing about 5×10^7 cells/ml). A positive skin test to cellular sporotrichin occurs in persons having immunosensitizing contact with *S. schenckii*. Clinical signs may or may not occur. Sporotrichin prepared from crude polysaccharide or purified peptido-rhamnomannan extracts from the culture fluid of yeastlike phase *S. schenckii* is epidemiologically less interesting, but has significant diagnostic value. It is strongly reactive and highly specific.

Humoral immunity in sporotrichosis is not easily detected, the possible reason being the insufficient quality of the antigens employed. Even in cases demonstrated by culture of *S. schenckii*, the titers of the reactions are often very low and are not diagnostic.

Whole yeastlike cell and latex particle agglutination tests seem the most sensitive serologic reactions for study of the disease. Fresh or lyophilized yeastlike cells, crude cultural extracts, or purified polysaccharide extracts of the yeastlike cultural medium may be employed as antigens (Rippon, 1974). Immunodiffusion tests using the same antigens and complement-fixation tests using disrupted yeastlike cell suspensions are diagnostically equivocal.

ECOLOGY AND EPIDEMIOLOGY

Ecology of the Fungus

S. schenckii is a normal inhabitant of the soil. The fungus occurs in soils rich in organic matter or vegetable debris. It is frequently associated with wood fragments or the bark of trees such as the eucalyptus or pine. It occurs often in association with *Ceratocystis stenoceras*, which *S. schenckii* closely resembles. Thus, both plants and soil may be considered as the natural reservoir of *S. schenckii*. At the same time, the fungus has been isolated from apparently healthy animals, air, water, and a variety of other substrates; these should be considered as possible vectors of the disease. Mammals may develop sporotrichotic infections comparable to that of the human and occasionally are active vectors of the disease.

Mycelial phase elements of the fungus may infect man by inhalation, thereby initiating a primary pulmonary complex that is no longer considered rare. However, the most frequent mode of contamination is through trauma of the skin. Various forms of trauma have been implicated. The most common is a prick from a thorn or plant fragment. Other types of trauma are animal bites, pecks from chickens or other birds, and stings of insects. The handling of fish from a volcanic lake in Guatemala resulted in 43.3 per cent of 53 infections reported from this endemic area of sporotrichosis.

The presence of *S. schenckii* and *C. stenoceras* in the same ecologic site is to be emphasized. *C. stenoceras* is widespread in nature, and its morphologic, biochemical, and antigenic characteristics are very similar, if not identical, to those of *S. schenckii*. It may play an important role in the epidemiology of sporotrichosis.

Epidemiology of Sporotrichosis

The disease occurs worldwide, but it is found mainly in warm temperate and tropical zones.

The majority of patients are under 30 years of age, and children under 10 years of age are frequently affected. Infection is equally distributed between the sexes. Sporotrichosis is found mainly in persons who are in active contact with plants or soil. It is an occupational disease of nursery workers, florists, potters, workers packing earthenware, and masons working with raw bricks.

S. schenckii is only slightly pathogenic, and healthy individuals generally resist infection. The absence of direct contagion from man to man, even under conditions most favorable for the fungus, is indicative of low virulence. To produce lethal sporotrichosis experimentally, male hamsters must receive at least 5×10^6 cells of a known pathogenic strain by intraperitoneal injection. This number is far higher than that introduced spontaneously by accidental trauma. If traumatic inoculation of a few infecting particles is to induce spontaneous sporotrichosis, then favorable host conditions must be present. Dietary deficiencies, primary pathologic defects, and resensitization are believed to be predisposing factors.

References

Garrison, R.G., Boyd, K.S., and Mariat, F.: Ultrastructural studies of the mycelial to yeast transformation of *Sporothrix schenckii*. J Bact 124:959, 1975.

Garrison, R.G., Mariat, F., Boyd, K.S., and Fromentin, H.: Ultrastructural observations of an unusual osmiophilic body in the hyphae of *Sporothrix schenckii* and *Ceratocystis stenoceras*. Ann Microbiol (Inst Pasteur) 128B:275, 1977.

Lane, J.W., Garrison, R.G., and Field, M.F.: Ultrastructural studies of the yeast-like and mycelial phases of *Sporotrichum schenckii*. J Bact 100:1010, 1969.

Mariat, F.: Taxonomic problems related to the fungal complex *Sporothrix schenckii/Ceratocystis spp.* In Iwata, K. (ed.): Recent Advances in Medical and Veterinary Mycology. Baltimore, University Park Press, 1977, p. 265.

Nicot, J., and Mariat, F.: Caractères morphologiques et position systématique de *Sporothrix schenckii*, agent de la sporotrichose humaine. Mycopath Mycol Applic 49:53, 1973.

Rippon, J.W.: Medical Mycology. The Pathogenic Fungi and the Pathogenic Actinomycetes. Philadelphia, W.B. Saunders Company, 1974.

71 *HISTOPLASMA CAPSULATUM*

Howard W. Larsh, Ph.D.
and Nancy K. Hall, Ph.D.

MORPHOLOGY

Histoplasma capsulatum is a dimorphic fungus. In its natural habitat in soils and at room temperature on agar, it grows as a mold. The colony on agar is usually fluffy and either white or buff-brown. Two types of spores are formed by most strains: small spherical aleuriospores (microconidia) that are 2 to 4 μ in diameter and larger macroaleuriospores (macroconidia) that measure 8 to 14 μ in diameter. Both types of spores are found at the end of narrow aleuriophores that arise at right angles from vegetative mycelia. With time, the surface of the macroaleuriospores may become covered with evenly spaced spines ("tubercles"), which gives them a characteristic appearance (Fig. 1). These structures are not absolutely diagnostic, however, since a saprophytic mold *Sepedonium* also produces macroaleuriospores.

When *H. capsulatum* grows in tissue, it transforms into a small oval yeast. The yeast phase can also be propagated in vitro on blood agar media. Colonies are white and moist. Individual yeast are 2 to 4 μ long and reproduce by budding at their narrow end (Fig. 2). The pore that connects mother and daughter cells is narrow and fragile and therefore may break prematurely, producing marked variation in the size of individual cells. The cell wall is thin. When stained, the cytoplasm shrinks away from the cell wall, creating a halo that has been mistaken for a capsule. Each cell contains only one nucleus, which may be hard to see in stained preparations.

The perfect stage of *H. capsulatum* has been described. It is heterothallic and is therefore classified in the family Gymnoascaceae of the Ascomycetes.

The etiologic agent of histoplasmosis, *H. capsulatum*, was identified in 1906 by Darling as a protozoan (*Leishmania*). Later, the infectious agent was proved to be a diphasic fungus with a pathogenic yeast stage and a saprophytic mycelial stage.

ANTIGENIC COMPOSITION

Four major antigenic preparations are now used for investigative, clinical, and epidemiologic

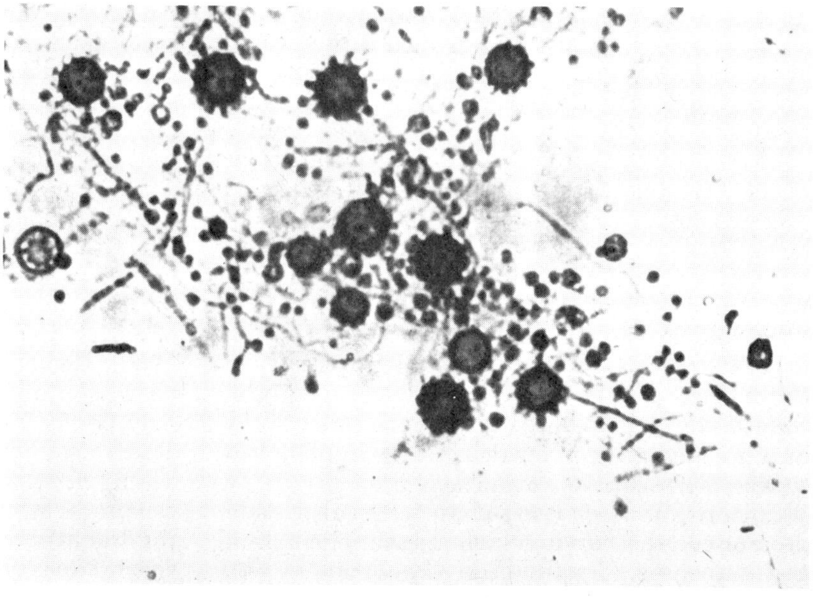

FIGURE 1. *Macroaleuriospores (macroconidia) of* Histoplasma capsulatum *covered with spines or "tubercles."*

studies of histoplasmosis: extracts of mycelial and yeast-phase cells of *H. capsulatum* and culture filtrates of the two morphologic types. The classic mycelial culture filtrate antigen (histoplasmin) is prepared from a synthetic broth culture incubated at room temperature. Filtration of the broth after a period of growth and standardization of the activity by intradermal injection is required. Comparison of potency with that of a standard histoplasmin is also necessary (Goodman et al., 1968).

The active components of mycelial culture filtrates have been identified by using column chromatography and polyacrylamide disk gel electrophoresis (Sprouse et al., 1969). Two antigenic fractions were found in a five-month asparagine medium preparation, the more potent being a high molecular weight protein that elicits a skin-test reaction in guinea pigs with experimental histoplasmosis.

By using an ammonium sulfate precipitation and Pevikon block electrophoresis, crude histoplasmin has also been separated into two main fractions, a protein that elicits a positive skin test and a less reactive fraction rich in polysaccharide (O'Connell et al., 1967). No increase in skin sensitivity was observed with either fraction.

Many techniques have been used to isolate and characterize a reactive component from the mycelial phase cells themselves. Alkali-extracted mycelium yielded a galactomannan protein complex that is separable into polysaccharide and glycoprotein complexes by ion-exchange chromatography (Reiss et al., 1974). The galactomannan did not elicit a skin-test response, but it elicited a cell-mediated reaction (production of migration-inhibition factor) from sensitized lymphocytes.

Polysaccharide antigens of mycelial *H. capsulatum* have been studied after using enzymatic digests to eliminate protein antigens (Kobayashi, 1971). The mannose-glucose polysaccharide was active in serologic tests but was not tested for its ability to elicit skin responses.

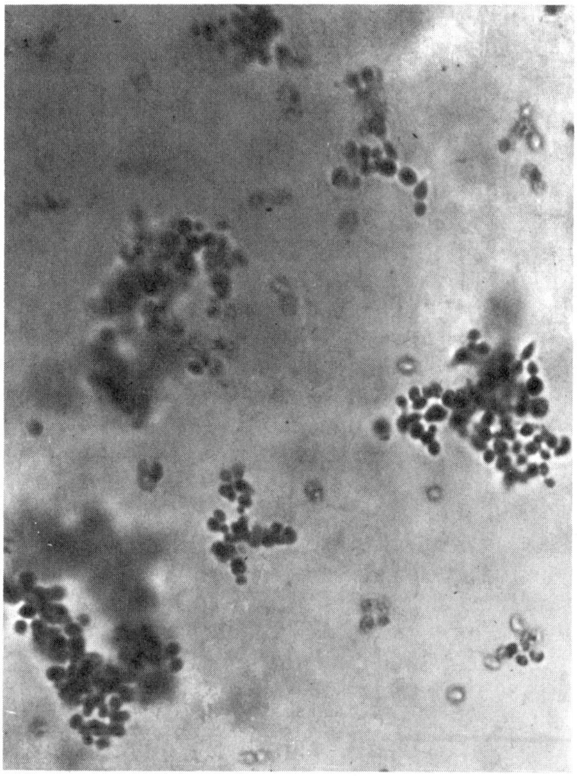

FIGURE 2. *Yeast phase of* H. capsulatum *grown in vitro. Individual yeast are 2 to 4 μ long and reproduce by budding.*

Since the yeast phase is the tissue form, it seems likely that the in vivo response would probably be to yeast-cell constituents. As with the mycelial preparations, most of the antigenicity resides in the protein-rich moiety of yeast cells.

Yeast cells are fractionated in sonicators or tissue disintegrators to separate cell-wall and protoplasmic fractions. Salvin and Ribi (1955) demonstrated that the cell wall contained both protective antigens and antigens used for complement fixation tests. Domer and Ichinose (1977) used soluble cytoplasmic substances and ethylenediamine extractions of cell walls in assays of cell-mediated immunity. The protein moiety of the isolated cell-wall glycoprotein is under study as a possible antigen for detecting pure delayed-type hypersensitivity to *H. capsulatum*.

Culture fluids of yeast-phase cells also yield protein-polysaccharide complexes that elicit skin-test responses. Treatment of the complex with chloroform eliminated the skin reactivity of the antigen (Dyson and Evans, 1955). An antigen from yeast-phase filtrates was isolated and shown by electrophoresis to be composed of two components, a skin-test reactive protein and a less reactive polysaccharide. Increased specificity has not been achieved.

A purified, specific antigen capable of standardization is being sought in mycelium, mycelial culture filtrates, yeast, and yeast-phase culture filtrates.

IMMUNITY

Immunity to *H. capsulatum* correlates with the development of delayed hypersensitivity. Antibody to the fungus is produced regularly in response to infection but does not result in immunity. The antibody response is of limited value in the diagnosis of histoplasmosis, because available antigens cross-react with antibodies to other fungi. To maximize the usefulness of serologic testing, two or more samples of sera should be taken for examination during the acute, convalescent, and recovery stages of the infection. Also, tests with a battery of antigens should be used (Campbell, 1960). Interpretation of serologic results should take into account not only cross-reacting antibodies but also antibodies induced with intradermal histoplasmin. Precipitation, agglutination, and complement fixation are the three basic serologic procedures.

The antibody response to *H. capsulatum* is complex. Immunodiffusion of histoplasmin against *Histoplasma* antisera results in six distinct precipitin bands (Heiner, 1958). These bands correlate with the clinical condition of the patient: h-band is associated with infection; m-band appears after skin tests with mycelial antigen or infection; c-band indicates exposure to skin testing with histoplasmin, coccidioidin, or blastomycin; n-band is nonspecific, and x- and y-bands are of unknown significance.

The complement fixation antigens in common use are whole yeast-phase cells or histoplasmin. Due to the unpredictability of the individual patient response, both antigens should be used; for example, sera may react with only one antigen or with both antigens in variable titer (Campbell, 1960). Complement fixation titers normally appear soon after onset of clinical symptoms and may persist for months or years after infection. Titers usually parallel the course of the disease. Titers are low or negative at first (and may remain so in asymptomatic infections), rise during the acute illness, and disappear slowly after several months. In the progressive form of the disease, titers remain high for a long time.

A particulate agglutination test is also used to demonstrate antibodies to *H. capsulatum*. Histoplasmin is adsorbed to collodion particles, to sheep erythrocytes, or to latex particles to serve as antigen. Whole yeast-phase cells can also be used successfully (Cozad and Larsh, 1960).

LABORATORY DIAGNOSIS

A definitive diagnosis of histoplasmosis can be made by culturing the fungus from tissue secretions or excretions, or by demonstrating the yeast cells histologically. *H. capsulatum* stains very poorly with hematoxylin-eosin and is usually evident only as a protoplasmic mass surrounded by a halo. The yeast cell walls stain red with periodic acid-Schiff stains, although the cytoplasm stains poorly. With silver stains (e.g., Gomori) the cell wall is black. Organisms appear larger with cell-wall stains because the artifactual shrinkage of the cell cytoplasm that occurs during fixation is not evident.

In primary histoplasmosis, sputum is often not produced for culture, but *H. capsulatum* may be cultured from urine. In cavitary pneumonias, sputum or gastric lavage specimens should be submitted for culture. In disseminated disease, scrapings from surface lesions, tissue biopsies (especially liver, bone marrow, and lymph nodes), blood, urine, and sputum may yield positive cultures.

Specimens for microscopic examination of secretions are fixed and air-dried on a microscope slide and then stained using either the Giemsa or Wright method. The yeast appear in histiocytes and macrophages as small oval cells, 2 to 5 μ in diameter. If rupture of the phagocytic cells occurs, it may be difficult to differentiate the extracellular *H. capsulatum* from other yeasts. Nevertheless, their small size, the budding from the narrow end, and the narrow attachment to the mother cells are distinguishing characteristics.

In pulmonary histoplasmosis, an early morning specimen of sputum should be collected in a sterile glass jar and transported to the laboratory within an hour. A delay in processing sputum increases the degree of contamination by faster growing bacteria and fungi.

The use of test tubes or Petri dishes for inoculation of sputum is debatable. The organism is an obligate aerobe, and Petri dishes allow more inoculum to be spread on the surface of the medium, but unsealed Petri dishes may present a hazard to laboratory personnel. The specimens suspected of harboring pathogenic fungi should be cultured in a bacteriologic safety hood.

Sputum as well as all other secretions and excretions should be treated with antibiotics to suppress contaminating bacteria. To ensure that the specimens are well homogenized, they may be diluted with sterile cysteine-saline and shaken in a mechanical shaker for 30 minutes to an hour. One-half milliliter of sputum, urine, gastric, or bronchial lavage is added to the following media, and the plates incubated for at least four weeks: (1) Emmon's modified Sabouraud's agar pH 7 without antibiotics; (2) Emmon's modified Sabouraud's agar (1 per cent Neopeptone, 2 per cent glucose, 2 per cent agar) pH 7 with penicillin, streptomycin, and Actidione or the same medium with chloramphenicol and Actidione; (3) blood agar base with 5 per cent defibrinated blood with gentamicin and Actidione, or blood agar base with 5 per cent defibrinated blood with gentamicin, carbenicillin, and Actidione. At least one medium should contain no Actidione, since *Cryptococcus neoformans, Petriellidium boydii, Aspergillus* sp., and other fungi that may cause pulmonary disease are inhibited by the antifungal agent.

Sterile tissue grinders are used to prepare tissues for plating. The homogenates are diluted and one-half milliliter is spread on the media listed above. Spinal fluids and other aseptic specimens are plated directly on the media without addition of antibiotics.

A policy of not incubating specimens at 37° C has been followed successfully in our laboratory for years. Most isolates of *H. capsulatum* do not grow well at this temperature , since it is near the maximal growth range of this fungus. If the yeast stage is to be obtained from infected tissue, an enriched medium containing blood should be used and incubated at 30° C. Mycelium can be converted without difficulty to the pathogenic yeast stage by inoculating tissue cultures (e.g., Hela cells).

Periodic examination of the cultures after 7 to 10 days usually reveals growth. From "typical" cultures, lactophenol preparations can be made of the mycelia that reveal microaleuriospores and macroaleuriospores. The former are spherical conidia 2 to 4 μm in diameter that are borne singly at the tip of short conidiophores. The walls of these spores may be smooth or have small spines. The macroaleuriospores are spherical but on occasion may be pear-shaped. These spores may be 8 to 15 μm in diameter and are firmly attached to narrow conidiophores. They may have smooth walls, but characteristically there are evenly spaced spines or finger-like appendages.

In the patient with chronic histoplasmosis, an isolate of *H. capsulatum* frequently is "atypical." Neither microaleuriospores nor macroaleuriospores are produced, and the cultures may be neither white nor brown. These isolates must be inoculated into experimental animals or tissue culture to obtain the yeast form of the fungus. Mice are useful for this purpose. Mice often survive the infection and should be autopsied two to four weeks after inoculation to demonstrate the yeast cells in liver and spleen. Sputum, or other material contaminated with bacteria, should be treated with antibiotics before injection.

In ecologic studies, mice are usually used to isolate *H. capsulatum* from soils, bird feces, and other highly contaminated specimens. The samples must be treated with antibiotics before inoculation intraperitoneally. In addition, antibiotics may be added to the drinking water and a second injection of antibiotic after 24 hours administered.

EPIDEMIOLOGY

The growth of *H. capsulatum* in soils enriched with bird or bat droppings is the source of human and animal exposure. Disturbances of the environment in which the fungus is growing produce aerosols containing aleuriospores that may be inhaled and cause primary infections. The specific role of feces in large bird roosts in the epidemiology of the disease is clear. The birds themselves do not contact histoplasmosis as a natural infection, since the blood temperature of birds is too high to support growth of the fungus. The fungus is not communicable from person to person, animal to person, or animal to animal. Each species acquires a natural infection by inhalation of spores. The fungus is also associated with bat guano, and some caves are heavily contaminated. Bats can develop histoplasmosis. The urban brown bat may be another source of histoplasmosis in urban areas.

H. capsulatum is an aerobic organism that grows in the upper 1 to 3 centimeters of soil or other enriched materials. Although there is controversy about whether or not this fungus has been isolated from dry pigeon droppings, it is conceded that pigeon droppings will enrich the ability of the soil to support the growth of *H. capsulatum* (Ajello, 1964). It is important to understand that although birds are not naturally infected, their droppings provide an enriched medium for the growth of *H. capsulatum,* whereas certain bats have the disease and their droppings can likewise support the growth of the fungus (Emmons, 1958).

The distribution of *H. capsulatum* in the soil is not uniform, even in highly endemic areas. Frequently *H. capsulatum* can be isolated from soil at the site of a point-source epidemic, but the soil only a few inches away will be negative. It is very difficult to disinfect contaminated soil.

The fungus has a worldwide distribution. Highly endemic areas have been identified by serologic and skin-test surveys with histoplasmin. In highly endemic areas such as the Ohio and Mississippi River valleys in the United States, up to 90 per cent of residents react to histoplasmin. However, the fungus does exist outside these areas where it can be cultured from soil and from native animals (Ajello, 1958; Edwards, 1971). In these areas, such as the eastern seaboard of the United States, skin-test surveys do not reveal the presence of the fungus because only isolated areas of soil are contaminated, and the rate of infection is consequently low.

References

Ajello, L.: Geographic distribution of *Histoplasma capsulatum.* Mykosen 1:147, 1958.

Ajello, L.: Relationship of *Histoplasma capsulatum* to avian habitats. Pub Health Rep 79:266, 1964.

Campbell, C. C.: The accuracy of serologic methods in diagnosis. Ann NY Acad Sci 89:163, 1960.

Cozad, G. C., and Larsh, H. W.: A capillary tube agglutination test for histoplasmosis. J Immunol 85:387, 1960.

Darling, S. T. A.: A protozoon general infection producing pseudotubercles in the lungs and focal necroses in the liver. JAMA 46:1283, 1906.

Domer, J. E., and Ichinose, H.: Cellular immune responses in guinea pigs immunized with cell walls of *Histoplasma capsulatum* prepared by several different procedures. Infect Immun 16:293, 1977.

Dyson, J. E., and Evans, E. E.: Delayed hypersensitivity in experimental fungus infection. J Lab Clin Med 45:449, 1955.

Edwards, P. Q.: Histoplasmin sensitivity patterns around the world. In Histoplasmos. Proceedings of the Second National Conference. Springfield, Ill., Charles C Thomas, 1971.

Emmons, C. W.: Association of bats with histoplasmosis. Pub Hlth Rept 73:590, 1958.

Goodman, N. L., Sprouse, R. F., and Larsh, H. W.: Histoplasmin potency as affected by culture age. Sabouraudia 6:273, 1968.

Heiner, D. C.: Diagnosis of histoplasmosis using precipitin reactions in agar gel. Pediatrics 22:616, 1958.

Kobayashi, G. S.: Isolation and characterization of polysaccharide of *Histoplasma capsulatum.* In Histoplasmosis Proceedings of the Second National Conference. Springfield, Ill., Charles C Thomas, 1971.

O'Connell, E. J., Hermans, P. E., and Markowitz, H.: Skin reactive antigens of *Histoplasma capsulatum.* Proc Soc Exp Biol Med 124:1015, 1967.

Reiss, E., Mitchell, W. O., Stone, S. H., and Hasenclever, H. F.: Cellular immune activity of a galactomannan-protein complex from mycelia of *Histoplasma capsulatum.* Infect Immun 10:802, 1974.

Salvin, S. B., and Ribi, E.: Antigens from yeast phase of *Histoplasma capsulatum.* Proc Soc Exp Biol Med 108:498, 1955.

Sprouse, R. F., Goodman, N. L., and Larsh, H. W.: Fractionation, isolation and chemical characterization of skin test active components of histoplasmin. Sabouraudia 7:1, 1969.

Thor, D. E., and Dray, S.: A correlation of human delayed hypersensitivity: Specific inhibition of capillary tube migration of sensitized human lymph node cells by tuberculin and histoplasmin. J Immunol 101:51, 1968.

72 *COCCIDIOIDES IMMITIS*

Henry A. Walch, Ph.D.

MORPHOLOGY

Dimorphic Growth Cycle

Coccidioides immitis is a fungus that is a part of the microbial population of the soil in certain geographic regions. In the soil it grows as a mycelium, and from branches and strands of hyphae forms square to rectangular arthrospores about 2 to 4 μ in width and 3 to 10 μ in length. These chains of arthrospores are arranged with intervening cytoplasmically empty cells and are easily detached and dispersed into the air (Figs. 1 and 2).

Infection of man and other animals usually occurs through inhalation of arthrospores into the lungs. Fungus growth in the body occurs as a rounding up of the arthrospores (Fig. 3) to form spherules that may enlarge to 20 to 100 μ in diameter (Fig. 4). The one to several nuclei originally present in the arthrospore undergo division and the multinucleate cytoplasm becomes differentiated into numerous, predominantly uninucleate 2 to 3 μ diameter endospores (Fig. 5), each of which can develop into a mature spherule and continue the parasitic cycle. A clinical specimen containing spherules and endospores that is cultured to a routine isolation medium will pro-

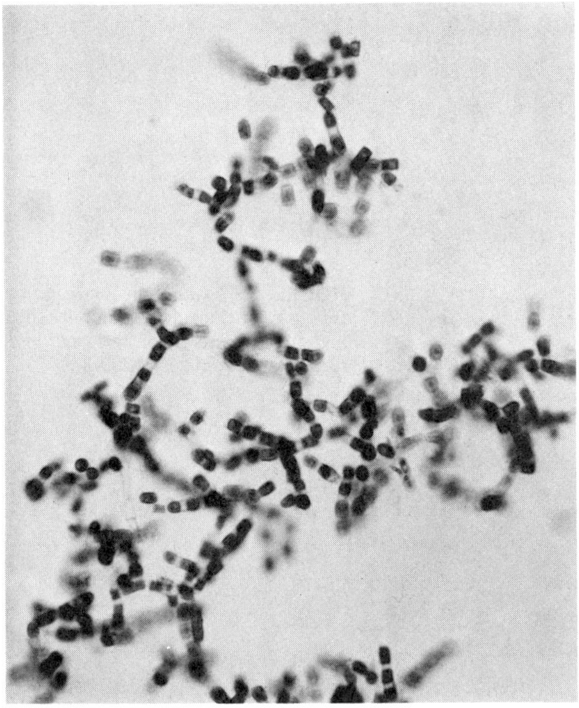

FIGURE 1. *Arthrospores of* Coccidioides immitis *(approximately 600 ×).*

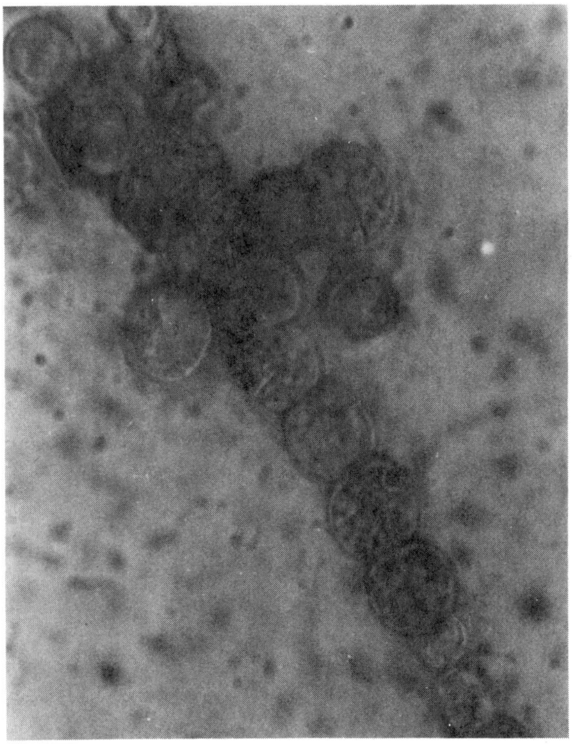

FIGURE 3. *Chain of arthrospores of* Coccidioides immitis *undergoing spherulation* in vitro *(approximately 1200 ×).*

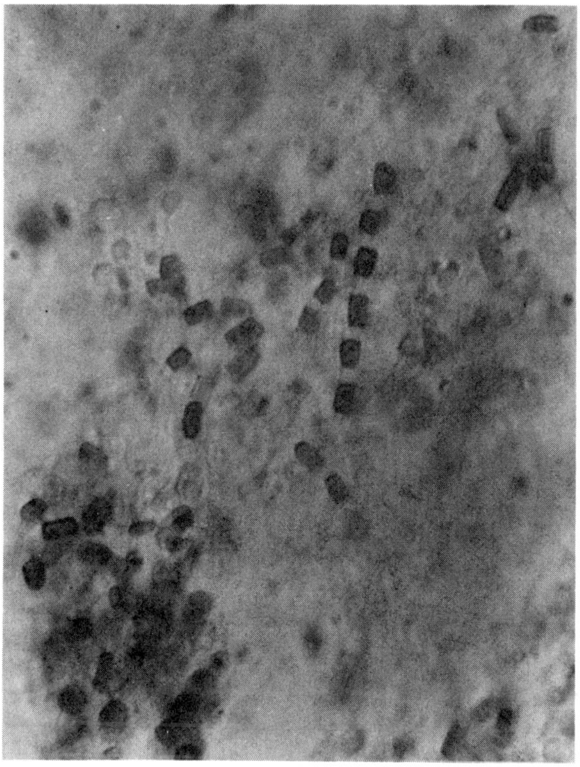

FIGURE 2. *Arthrospores of* Coccidioides immitis *(approximately 1200 ×).*

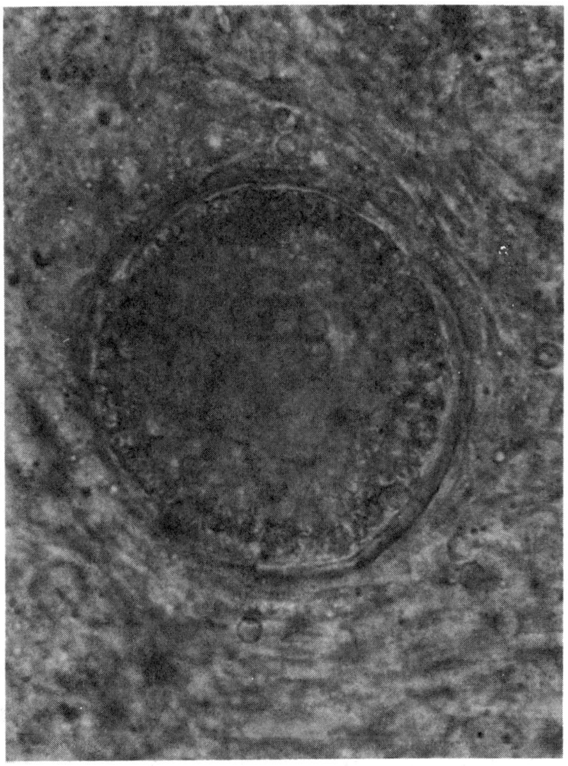

FIGURE 4. *Large, mature spherule of* Coccidioides immitis, *unstained in tissue cleared with sodium hydroxide solution (approximately 1200 X).*

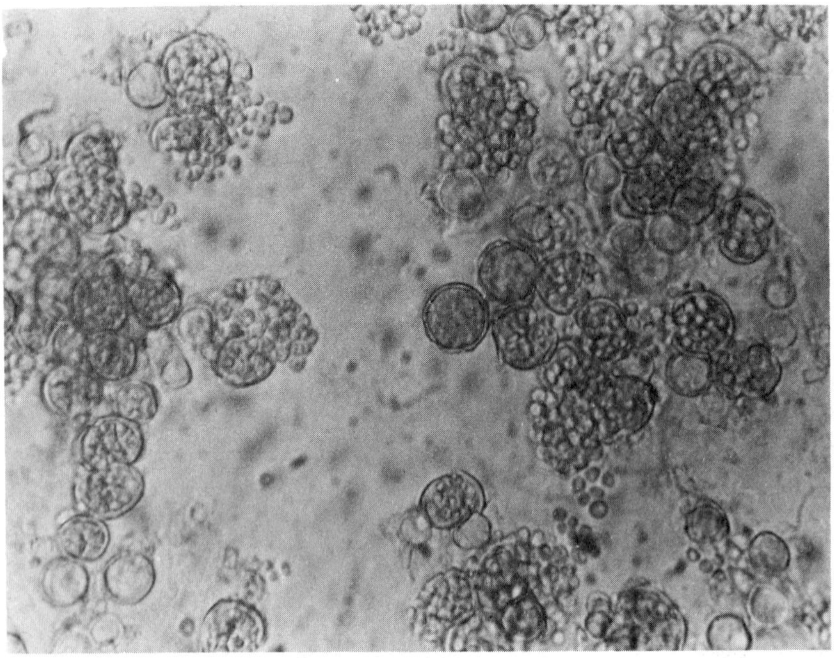

FIGURE 5. *Mature* in vitro *formed spherules of* Coccidioides immitis *showing large numbers of released endospores (approximately 1200 X).*

duce the mycelium-arthrospore colony stage (Fig. 6) through the germination and growth of endospores and immature spherules.

It is historically pertinent to note that the spherule-endospore tissue form led the original students of this disease (1892–1896) to conclude that it was of protozoan (coccidioidal) etiology and to speciate it as *Coccidioides immitis* (not mild), since the few known human cases were all fatal.

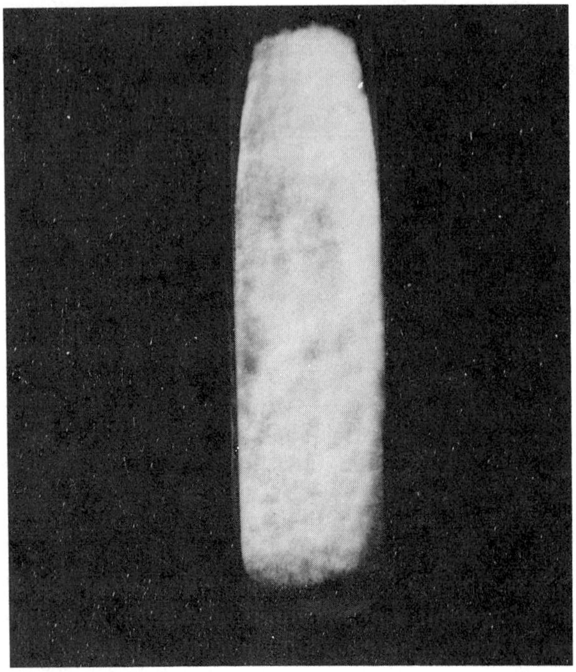

FIGURE 6. *Mature culture of* Coccidioides immitis *showing typical aerial mycelium.*

Classification

The classification of the fungus is still to be resolved, since no sexual stage of growth has been demonstrated. However, cytologic studies of the formation of arthrospores and the ultrastructural characteristics of hyphae (Kwon-Chung, 1969) and studies of arthrospore to spherule morphogenesis (Sun et al., 1976), are mutually supportive of a tentative classification as an ascomycete.

Cell Wall Composition

The chemical composition of the cell wall is apparently similar to that of other fungi. Analysis of the cell wall has shown chitin to be present in the wall of the three morphologic forms. Other carbohydrate polymers are also present such as B 1–3 glucans, mannans, or possible glucomannans as major cell wall components. Protein is present also, and differences in amino acid composition have been found in the different morphologic forms (Wheat et al., 1977). The cell wall of the mature spherule has five times more chitin than that of the endospore. The spherule wall also differs from the endospore in containing considerable glucose, which may be present as a B 1–3 glucan. Such results, as noted by Collins (1977), indicate that major differences occur in the polysaccharide composition of the wall during endospore to spherule maturation.

ANTIGENIC COMPOSITION

Immunologic Test Antigens

Antigens that are used in testing for humoral antibodies or cellular reactivity are called coccidioidins. Two kinds of coccidioidins are available — mycelial and spherule derived.

Mycelial coccidioidin is made from a filtrate of an aging (two months) and autolyzing stationary broth culture, or a pool of broth cultures of different strains to allow for possible antigenic variability. This type of coccidioidin was originally used as a successful serologic and skin test antigen beginning in the early 1940 period (Smith et al., 1950). Because of the extensive past and continued record of use, it serves as a standard of comparison for test antigens. Another type of mycelial coccidioidin is a toluene-induced lysate of a young (two to three day) shaken culture (Pappagianis et al., 1961a). This has been used in immunodiffusion testing and in experimental applications. These coccidioidins can be dialyzed without loss of activity. Heat treatment by autoclaving does not impair the activity of antigens involved in precipitin testing or skin test reactivity. Complement-fixing antigen activity is destroyed, however, by heating at 60° C for 30 minutes.

Spherule coccidioidin (spherulin) is a lysate of a washed, water suspension of in vitro cultured, mature, endosporulating spherules (Levine et al., 1969). In common with the mycelial coccidioidins, the active portion is nondialyzable. In contrast to mycelial coccidioidin, its skin testing potency is definitely impaired by heating at 60° C for 20 minutes.

Analysis of these two complex test antigens by two-dimensional immunoelectrophoresis methods demonstrated at least 26 antigens for mycelial coccidioidin and 12 for spherulin. Ten antigens were in common, two were unique to the spherule, and 16 to the mycelial derived antigens (Huppert et al., 1978).

Spherulin has only recently become available for use as a routine test antigen. It is more reactive and sensitive for cutaneous hypersensitivity testing than is mycelial coccidioidin, and probably more specific (Levine et al., 1973; Stevens, 1974; Levine et al., 1975). It has also been more effective as an antigen when used in the lymphocyte transformation test (Deresinski et al., 1974). Use as a complement-fixing antigen, in comparison with the mycelial coccidioidin, has shown that both were equally sensitive. The results of comparing the reactivity of the two antigens with sera from noncoccidioidal mycoses, however, showed that spherulin was considerably less specific (Huppert et al., 1977).

Antigen Standardization

Coccidioidins, as mixtures of antigens, are subject to variability of reactivity between different lots produced at different times even though conditions of culture are uniformly maintained. Because of this antigenic heterogeneity and variability, standardization by weight or chemical analysis is not useful. Also, no comparative laboratory animal testing system for biologic reactivity and potency testing is available. Therefore, standardization for serologic or cutaneous hypersensitivity testing must be done by different procedures. Details of these procedures are considered by Huppert (1970), Kobayashi and Pappagianis (1970) and Huppert et al. (1974). In summary, skin test standardization requires human subjects of known cutaneous hypersensitivity and a reference antigen. Serologic standardization requires both a reference serum and antigen and strict adherence to the prescribed test procedure.

Chemical Composition

Various attempts have been made to chemically separate antigenically active components from mycelial coccidioidin. Ethanol precipitation was used to separate out a skin test reactive and precipitating polysaccharide composed primarily of mannose, which was associated with three to four per cent nitrogen in amino acids. Immunodiffusion and quantitative tube precipitin testing, however, showed that it contained different antigens (Pappagianis et al., 1961b).

Gel molecular filtration and ultrafiltration techniques were used in the course of later studies of the chemical composition. These results showed mannose, 3-O-methylmannose, and peptide components to be consistently present in skin test active fractions (Anderson et al., 1971).

IMMUNITY

Infection and Immunity

The extensive human case studies of Smith and his associates (1957) provided evidence that immunity resulted from primary coccidioidomycosis. In addition, asymptomatic infection with conversion to skin test reactivity also seems to result in immunity.

Infection with *C. immitis* begins with a transitory exposure to the infecting arthrospores, which, after a period of days, begin a transformation to the spherule form. The resulting spherule-to-endospores-to-spherules parasitic cycle involves a continuing multiple antigenic experience for the host. Incidental exposure to the cytoplasm of these structures, in turn, provides additional antigenic stimulation. Some of these antigens will engender the formation of humoral antibodies and others will be active in stimulating a protective cell-mediated immune response.

Cell-Mediated Immunity

Immunity to *C. immitis* is of the cell-mediated type. Infection is accompanied by cutaneous delayed hypersensitivity, lymphocyte transformation, and macrophage migration inhibition through the use of mycelial and spherule coccidioidins. The role of T cells is implied by the resistance of normal mice to infection after receiving an intravenous inoculation of spleen cells from spherule-vaccinated mice (Beaman et al., 1977), and transfer of delayed hypersensitivity in mice (Rifkind et al., 1976).

In disseminated coccidioidomycosis, the skin test is frequently negative and coccidioidin does not stimulate lymphocyte transformation. Whether this condition represents a nonspecific, preexisting and predisposing immune deficiency, or specific anergy induced by disseminated fungus growth is unsettled.

Immunization

Formalin-killed vaccines consisting of either mycelium-arthrospores or mature spherules have been compared in mice. The mycelium-arthrospore vaccine provided some protection, whereas the spherule-vaccinated group showed notably better survival and reduced pathologic involvement (Levine et al., 1960).

After mechanical fractionation of ruptured spherules, the spherule wall was an effective immunogen in mice but the soluble cell lysate and the prematurely released endospores were not (Kong et al., 1963). Optimum protection followed immunization with the intact, mature spherule and naturally released endospores were immunogenic, but less so (Levine et al., 1965). Thus, a cycle of immunogen development and decline is associated with the morphogenetic alterations of the parasitic phase of growth.

The monotype spherule-endospore vaccine protects against heterologous strains. When groups of vaccinated mice were subsequently challenged with typical and atypical strains of established virulence, equal protection was demonstrated among all groups, indicating a similarity of immunogenic antigens among different isolates of the fungus (Huppert et al., 1967).

LABORATORY DIAGNOSIS

Fungus Growth and Confirmation

C. immitis will grow on a wide range of culture media. Commercially available glucose-peptone based selective media containing cycloheximide and antibacterial antibiotics will, generally, provide better growth and sporulation if supplemented with about 0.5 per cent yeast extract. Incubation temperatures of 25 to 30° C are recommended. Although the fungus will grow as the mycelial form on these media at 37° C, the sporulation of some strains is inhibited.

In general, methods that are used for preparation of clinical specimens for bacterial culture will not be inhibitory for *C. immitis* isolation. The processing of sputum, however, must not employ methods designed for isolation of the tubercle bacillus. If it is desirable to digest sputum for fungus culture, equal volumes of either of the mucolytic agents, N-acetyl-l-cysteine or dithiothreitol at a 0.5 per cent concentration made up in 2.94 per cent sodium citrate solution, as per the method of Reep and Kaplan (1972), is suitable.

Confirmation of the presumptively identified culture can be done by spherule demonstration in animals or by in vitro methods. Mice are commonly inoculated intraperitoneally, guinea pigs intratesticularly, in order to demonstrate spherules in tissues. Virulence varies markedly in mice for different strains of the fungus. Also, the mere culture of a fungus from the tissues of an injected animal is not confirmatory since many nonpathogenic fungi will persist without any pathologic effect.

Two convenient in vitro confirmatory identification methods are available. With the method of Sun et al. (1976) arthrospores convert to spherules in three to five days. The method of Standard and Kaufman (1977) uses immunodiffusion for detecting antigens in a fluid culture supernatant and is particularly useful for atypical nonsporulating cultures.

Immunologic Diagnosis

The two serologic tests that have been most extensively documented and serve as the standards for comparison with other methods are the tube precipitin and the complement-fixation tests. Strict adherence to prescribed procedures and the use of a standardized antigen are required for diagnostic use. Details for the performance of the precipitin test are provided by Kaufman (1976). The complement-fixation procedure using the full volume test is described in the Pan American Health Organization Manual, part II (1974); for the micro complement-fixation test technique consult the United States Public Health Service Publication Number 1228 (1965). Critical evaluations of these complement-fixation methods are contained in the publications of Huppert et al. (1970) and of Kaufman et al. (1970).

Precipitating antibodies occur transiently in the serum and may appear as early as the first week of illness during primary disease. Serum complement-fixing antibodies generally are not evident as early as precipitin antibodies, but they remain detectable for a longer period of time.

Serologic diagnosis and prognosis depends on the testing of subsequent serum samples to determine changes of the complement-fixing antibody titer. The titer usually rises in proportion to severity of disease and declines with improvement of the patient. The generally accepted interpretation of these titers in relation to the disease process is based on the studies of Smith and his associates (1950, 1956). Titers above 1:16 indicate possible dissemination. Lower titers, 1:8 or less, may indicate early, stable residual or meningeal coccidioidomycosis or may be false positives. Negative results do not exclude disease, however, since about 40 per cent of chronic residual pulmonary disease cases are nonreactive. About 25 per cent of patients with coccidioidal meningitis give negative results when the spinal fluid is used for complement fixation but a positive complement-fixation test of cerebrospinal fluid is a valid test for the diagnosis of coccidioidal meningitis. If, however, the spinal fluid is allowed to bind complement overnight in the cold (4° C), only 5 per cent of the meningitis cases tend to be negative (Pappagianis and Crane, 1977).

Alternate immunologic procedures have been studied for routine screening or supplementary test systems. Huppert and his associates (1968) have recommended a combination of the latex particle agglutination test (in place of the tube precipitin test) and the immunodiffusion test that correlates with the complement-fixation test) (IDCF) as a rapid and reliable primary test system. The IDCF test in conjunction with the microcomplement-fixation test is recommended by Kaufman and Clark (1974), since a positive immunodiffusion test and a low complement-fixation titer indicates active or recent disease.

EPIDEMIOLOGY

Geographic Distribution

Coccidioidomycosis is endemic only in limited areas of North, Central, and South America. In Southwestern United States, annual infection rates have been estimated at 35,000 to 100,000. These have been predominantly from Arizona and California, with fewer from Texas. New Mexico is of low endemicity and in Nevada and Utah skin test surveys give the only indication of the disease.

Mayorga and Espinoza (1970) reviewed the distribution of coccidioidomycosis in Mexico and Central America. Although most of the reported cases in Mexico are from hospitals in the states of Sonora, Neuvo Leon, and Coahuila, a more extensive occurrence of the disease is indicated by the skin test surveys of Gonzalez-Ochoa (1967). In Central America disease has been reported only from Guatemala, with eight, and Honduras, with two cases. The Guatemala endemic area, however, has a skin test reactor rate of 42.5 per cent.

At the time of Campins' report (1970) Argentina and Venezuela had 27 and 35 cases respectively, and Paraguay and Colombia 2 each. Recent soil isolations have been reported from Argentina (Borghi et al., 1977).

Transmission to Humans

The incidence of infection is directly related to occupations that break up soil. Incidental infection in an endemic area depends on proximity to areas of soil disturbance, prevailing and seasonal wind conditions, and weather patterns affecting growth of the fungus in the soil.

About 60 per cent of humans infected will be asymptomatic and the remainder will have symptoms ranging from those of a mild respiratory condition to prolonged severe illness. A small number of the primary infections will disseminate to bones, joints, internal organs, or meninges. Disseminated disease occurs in approximately 0.5 to 1.0 per cent of white males, 5 to 10 per cent of black males, and in an even higher percentage of Filipinos.

Ecology

In the United States and elsewhere the fungus has been most frequently isolated from the soil in semiarid areas from sea level to a few hundred feet elevation, which have mild winters and hot summers, with a yearly rainfall total of 5 to 20 inches occurring in one or two seasonal periods. There are exceptions to these usual geographic areas of disease endemicity. In Mexico, two small tropical areas are apparently endemic. One of these, comprising Colima and Michoacan border areas, has recorded two disseminated cases in children. The other is a limited area in the state of Guerrero where skin test reactivity indicates low endemicity (González Ochoa, 1967). In the United States, from California, soil isolations have been made from a mediterranean woodland region, from areas as high as 3200 feet and from sites where occasional snow and freezing temperatures occur during winter months (Swatek, 1975).

There is no association with the macroflora (Lacy and Swatek, 1974). The fungus has been isolated from the environs and even the burrows of rodents, but there is no evidence that such sites are regular ecologic niches. The chemical and physical conditions in the soil that favor the fungus are not known beyond the general association with alkaline, sandy soil types with a high content of salts.

References

Anderson, K.L., Wheat, R.W. and Conant, N.F.: Fractionation and composition studies of skin test active components of sensitins from *Coccidioides immitis*. Appl Microbiol 22:294, 1971.

Beaman, L., Pappagianis, D., and Benjamini, E.: Significance of T cells in resistance to experimental murine coccidioidomycosis. Infect Immun 17:580, 1977.

Borghi, A.L., Rossi de Benetti, M.S., and Corallini de Bracalenti, B.J.: *Coccidioides immitis:* Su Aislamiento de Muestras de Suelos de las Provincias de San Luis Y Mendoza. Sabouraudia 15:51, 1977.

Campins, H.: Coccidioidomycosis in South America. A review of its epidemiology and geographic distribution. Mycopath et Mycologia Appl 40:1, 1970.

Collins, M.S., Pappagianis, D., and Yee, J.: Enzymatic solubilization of precipitin and complement fixing antigen from endospores, spherules and spherule fraction of *Coccidioides immitis*. In Ajello, L. (ed.): Coccidiomycosis. Miami, Symposia Specialists, 1977, p. 429.

Deresinksi, S.C., Levine, H.B., and Stevens, D.A.: Soluble antigens of mycelia and spherules in the in vitro detection of immunity to *Coccidioides immitis*. Infect Immun 10:700, 1974.

González Ochoa, A. Coccidioidomycosis in Mexico. In Ajello, L. (ed.): Proceedings of Second Coccidioidomycosis Symposium. Phoenix, The University of Arizona Press, 1967, p. 293.

Huppert, M.: Standardization of immunological reagents. Proceedings — International Symposium on Mycoses. Scientific Publication PAHO No. 205, 1970, p. 243.

Huppert, M., Chitjian, P.A., and Gross, A.J.: Comparison of methods for coccidioidomycosis complement fixation. Appl Microbiol 20:328, 1970.

Huppert, M., Krushow, I., Vukovich, K.R., Sun, S.H., Rice, E.H., and Kutner, L.J.: Comparison of coccidioidin and spherulin in complement fixation tests for coccidioidomycosis. J Clin Microbiol 6:33, 1977.

Huppert, M., Levine, H.B., Sun, S.H., and Peterson, E.T.: Resistance of vaccinated mice to typical and atypical strains of *Coccidioides immitis*. J Bact 94:924, 1967.

Huppert, M., Peterson, E.T., Sun, S.H., Chitjian, P.A., and Derrevere, W.J.: Evaluation of a latex particle agglutination test for coccidioidomycosis. The Am J Clin Path 49:96, 1968.

Huppert, M., Spratt, N.S., Vukovich, K.R., Sun, S.H., and Rice, E.H.: Antigenic analysis of coccidioidin and spherulin determined by two-dimensional immunoelectrophoresis. Infect Immun 20:541, 1978.

Huppert, M., Sun, S.H., and Vukovich, K.R.: Standardization of mycological reagents. Proceedings: International Conference on Standardization of Diagnostic Materials. U.S. Dept. HEW, CDC, Atlanta, 1974, p. 187.

Kaufman, L.: Serodiagnosis of fungal diseases. In Manual of Clinical Microbiology. American Society of Microbiology, 1976, p. 363.

Kaufman, L., and Clark, M.J.: Value of the concomitant use of complement fixation and immunodiffusion tests in the diagnosis of coccidioidomycosis. Appl Microbiol 28:641, 1974.

Kaufman, L., Hall, E.C., Clark, M.J., and McLaughlin, D.: Comparison of macrocomplement and microcomplement fixation techniques used in fungus serology. Appl Microbiol 20:579, 1970.

Kobayashi, G.S., and Pappagianis, D.: Preparation and standardization of antigens of *Histoplasma capsulatum* and *Coccidioides immitis*. Mycopath et Mycolog Appl 41:139, 1970.

Kong, Y.M., Levine, H.B., and Smith, C.E.: Immunogenic properties of undisrupted and disrupted spherules of *Coccidioides immitis* in mice. Sabouraudia 2:131, 1963.

Kwon-Chung, K.J.: *Coccidioides immitis*: Cytological study on the formation of the arthrospores. Canad J Genet Cytol 11:43, 1969.

Lacy, G.H., and Swatek, F.E.: Soil ecology of *Coccidioides immitis* at Amerindian middens in California. Appl Microbiol 27:379, 1974.

Levine, H.B., Cobb, J.M., and Scalarone, G.M.: Spherule coccidioidin in delayed dermal sensitivity reaction of experimental animals. Sabouraudia 7:20, 1969.

Levine, H.B., Cobb, J.M., and Smith, C.E.: Immunity to coccidioidomycosis induced in mice by purified spherule, arthrospore and mycelial vaccines. Trans N Y Acad Sci 22:436, 1960.

Levine, H.B., González-Ochoa, A., and Ten Eyck, D.: Dermal sensitivity to *Coccidioides immitis*. Am Rev Resp Dis 107:379, 1973.

Levine, H.B., Kong, Y.M., and Smith, C.E.: Immunization of mice to *Coccidioides immitis*: dose regimen and spherulation stage of killed spherule vaccines. The J Immun 94:132, 1965.

Levine, H.B., Restrepo, M.A., Ten Eyck, D.R., and Stevens, D.A.: Spherulin and coccidioidin: cross reactions in dermal sensitivity to histoplasmin and paracoccidioidin. Am J Epidem 101:512, 1975.

Mayorga, R.P., and Espinoza, H.: Coccidioidomycosis in Mexico and Central America. Mycopath et Mycologia Appl 40:13, 1970.

Pan American Health Organization. Manual of standardized serodiagnostic procedures for systemic mycoses. Part II. Complement fixation tests. Washington, D.C., Pan American Health Organization, 1974.

Pappagianis, D., and Crane, R.: Survival in coccidioidal meningitis since introduction of amphotericin B. In Ajello, L. (ed.) Coccidioidomycosis. Miami, Symposia Specialists, 1977, p. 223.

Pappagianis, D., Smith, C.E., Kobayashi, G.S., and Saito, M.T.: Studies of antigens from young mycelia of *Coccidioides immitis*. J Infect Dis 108:35, 1961a.

Pappagianis, D., Putman, E.W., and Kobayashi, G.S.: Polysaccharide of *Coccidioides immitis*. J Bact 82:714, 1961b.

Reep, B.R., and Kaplan, W.: The use of N-acetyl-L-cysteine and dithiothreitol to process sputa for mycological and fluorescent antibody examinations. Health Lab Sci 9:118, 1972.

Rifkind, D., Frey, J.A., Davis, J.R., and Petersen, E.A.: Delayed hypersensitivity to fungal antigens in mice. I. Use of the intradermal skin and foot pad swelling tests as assays of active and passive sensitization. J Infect Dis 133:50, 1976.

Smith, C.E., Pappagianis, D., and Saito, M.: The public health significance of coccidioidomycosis. U.S. Public Health Service Publication No. 575, 1957, p. 3.

Smith, C.E., Saito, M.T., Beard, R.R., Kepp, R.M., Clark, R.W., and Eddie, B.V.: Serological tests in the diagnosis and prognosis of coccidioidomycosis. The Amer J Hyg 52:1, 1950.

Smith, C.E., Saito, M.T., and Simons, S.A.: Pattern of 39,500 serologic tests in coccidioidomycosis. JAMA 160:546, 1956.

Standard, P.G., and Kaufman, L.: Immunological procedure for the rapid and specific identification of *Coccidioides immitis* cultures. J Clin Microbial 5:149, 1977.

Stevens, D.A., Levine, H.B., and Ten Eyck, D.R.: Dermal sensitivity to different doses of spherulin and coccidioidin. Chest 65:530, 1974.

Sun, S.H., and Huppert, M.: A cytological study of morphogenesis in *Coccidioides immitis*. Sabouraudia 14:185, 1976.

Sun, S.H., Huppert, M., and Vukovich, K.R.: Rapid in vitro conversion and identification of *Coccidioides immitis*. J Clin Microbiol 3:186, 1976.

Swatek, F.E.: The epidemiology of coccidioidomycosis. In Al-Doory (ed.): The Epidemiology of Human Mycotic Diseases. Springfield, Charles C Thomas Publishers, 1975, p. 74.

United States Public Health Service. Standardized diagnostic complement fixation method and adaption to micro test. U.S. Public Health Service Publication No. 1228, 1965.

Wheat, R.W., Tritschler, C., Conant, N.F., and Lowe, E.P.: Comparison of *Coccidioides immitis* arthrospore, mycelium and spherule cell walls, and influence of growth medium on mycelial cell wall composition. Infect Immun 17:91, 1977.

73 BLASTOMYCES AND PARACOCCIDIOIDES

Smith Shadomy, Ph.D.,
and Dennis M. Dixon, Ph.D.

MORPHOLOGY

Blastomyces (Ajellomyces) dermatitidis and *Paracoccidioides brasiliensis* are dimorphic fungi that grow in tissue as budding yeasts and in culture at room temperature as molds. The two fungi are morphologically similar and share antigens. Despite its name, *P. brasiliensis* is now considered to be in the genus *Blastomyces*.

Yeast cells of *B. dermatitidis* range from 8 to 15μ in size. In wet mounts or when stained with lactophenol cotton blue, the cell wall appears as a highly refractile structure (Fig. 1). The thick nature of the cell wall has led various authors to describe the fungus as doubly rounded or doubly contoured as if delimited by two cell walls. Yeast phase cells of *B. dermatitidis* produce single buds that are characterized by a wide pore, approximately one-half the width of the parent cell, between the parent cell and bud. The wall of the bud is initially thinner than that of the parent cell. These features distinguish *B. dermatitidis* from the other human fungal pathogens.

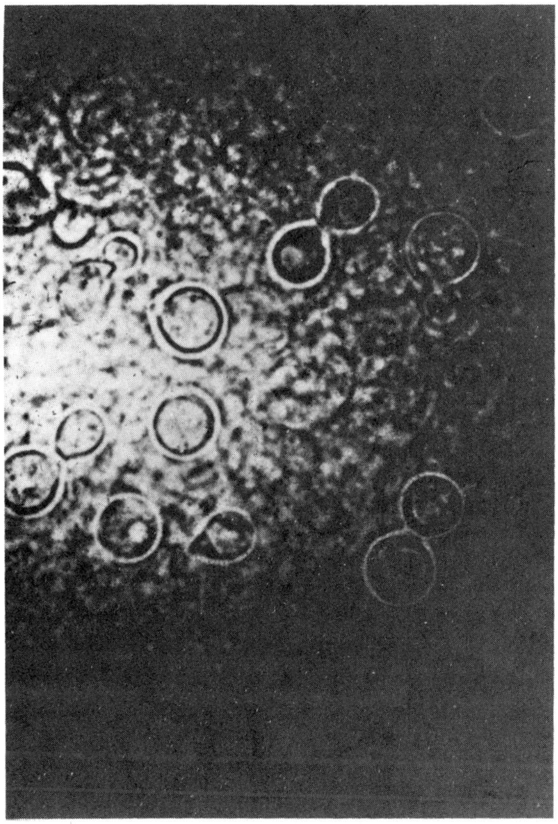

FIGURE 1. *Yeast cells of* Blastomyces dermatitidis *in pus. Note thick cell wall and single buds. Compare with multiple buds of* Paracoccidioides brasiliensis *in Figure 2. (× 1200)*

Yeast cells of *P. brasiliensis* are from 2 to 30μ in diameter and have thinner cell walls than *B. dermatitidis*. Additionally, *P. brasiliensis* is differentiated by the production of multiple buds from a single cell. This gives rise to the familiar "Mickey Mouse hat" appearance when three such buds are involved, or the "pilot's wheel" when multiple buds are produced around the circumference of the parent cell (Fig. 2). Also, pores between the parent cell and buds are narrower than those with *B. dermatitidis*.

In culture, *B. dermatitidis* grows as a whitish buff-colored mold with a yellow-brown underside. Colonies may be floccose or glabrous, furrowed or smooth. Microscopically, one sees fine, septate hyphae that have characteristic pyriform conidia measuring 2 to 10μ. The conidia are borne on short lateral or terminal branches (conidiophores) and typically have a truncated base that can be more clearly seen after their separation from the parent conidiophore. Such conidia are commonly termed aleurioconidia or aleuriospores and may be confused with the smooth-walled conidia of *Histoplasma capsulatum* and certain saprophytic *Chrysosporium* species. *B. dermatitidis* and *H. capsulatum* can be differentiated on the basis of yeast phase morphology. *Chrysosporium* species do not grow as a yeast in vitro.

P. brasiliensis is essentially indistinguishable from *B. dermatitidis* in the mycelial phase and often does not produce conidia. When present, the conidia are aleurioconidia measuring 2 to 4μ in diameter. Arthroconidia, produced by disarticulation of the vegetative hyphae, and intercolony chlamydospores, or thick-walled swollen cells, have been reported. For both fungi, media such as potato dextrose agar or yeast extract agar usually enhance the production of conidia while limiting the growth of vegetative hyphae.

On Sabouraud's agar at room temperature, both *B. dermatitidis* and *P. brasiliensis* appear as

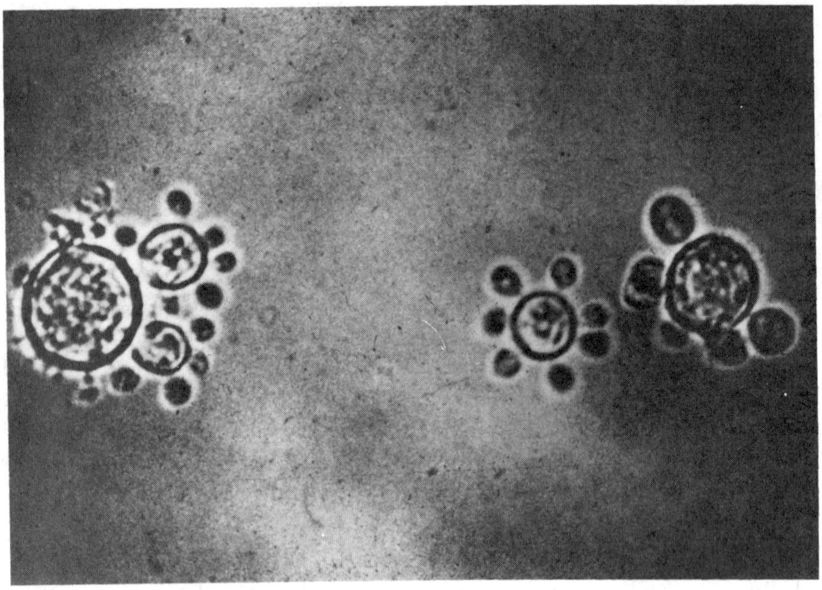

FIGURE 2. *Yeast cells of* Paracoccidioides brasiliensis. *Note thinner cell wall than those of* B. dermatitidis *in Figure 1, and multiple buds in a single cell (pilot's wheel). (×1500)*

slow-growing molds and are indistinguishable from each other. Colonies are generally apparent after one week of incubation, but maximum growth may not be attained for three or more weeks. Colonies typically are white-buff to brown with a yellow-brown reverse and may be floccose to glabrous, furrowed or smooth. *B. dermatitidis* and *P. brasiliensis* occasionally produce tufts of aerial hyphae (coremia) that are most distinctive in young cultures; otherwise, the colonies are similar in appearance to *H. capsulatum*. Microscopically, *B. dermatitidis* consists of fine, septate hyphae and characteristic pyriform conidia measuring from 2 to 10μ.

On brain-heart infusion blood agar at 37° C, both *B. dermatitidis* and *P. brasiliensis* grow as cream-colored, wrinkled, heaped colonies. Growth rates are similar to those of mycelial phase cultures, but the colonial growth is more restricted than the mycelial growth that may cover the agar surface.

B. dermatitidis has a sexual stage, *Ajellomyces dermatitidis*. The sexes are separate (heterothallism), and opposite mating types (+ and −) are required to obtain fertile cultures of *A. dermatitidis*. When the two mating types are cultured at 25° C on appropriate media, specialized rounded structures (cleistothecia) are produced. Within these are sac-like structures, or asci, each of which contains eight ascospores. The fungus is therefore classified as an *Ascomycete*. The ascospores are somewhat smaller than the asexual conidia and are infectious. A sexual stage has not been described for *P. brasiliensis*.

ANTIGENIC STRUCTURE

Blastomycin and paracoccidioidin are concentrates that have been prepared from filtrates of mycelial phase cultures. They are used as the antigens for serodiagnosis of infection with these fungi as well as antigens for in vivo skin tests. These materials are heterogenous mixtures of antigens that lack both sensitivity and specificity, and, to compound the problem, they are poorly standardized. Work is under way to isolate antigens that will be both specific and sensitive for serodiagnosis (Cox and Larsh, 1974a, b). Cell-wall proteins are believed to be the antigens responsible for skin-test reactions. An alkaline-stable, water-soluble extract of yeast cell walls appears to contain a skin-test antigen that is both sensitive and specific (Lancaster and Sprouse, 1976). Restrepo and others have isolated specific antigens from yeast cells that can be used for serodiagnosis of paracoccidioidomycosis (Restrepo, 1966; Kaufman, 1972).

IMMUNITY

Cell-mediated immunity is a more significant factor than humoral immunity in most of the mycoses, and blastomycosis and paracoccidioidomycosis are no exceptions. Although circulating antibodies can generally be detected at some point in the course of the diseases, they do not appear to be protective. In contrast, cell-mediated immune responses show a positive correlation with the host's ability to overcome the infections. Poor clinical progress is generally correlated with impairment in some aspect of cell-mediated immunity (Mok and Greer, 1977; Musatti et al., 1976). Whether there is a causal relationship between lymphocyte responses and immunity remains to be determined; however, it has been suggested that paracoccidioidomycosis leads to impairments in cell-mediated immunity. Various parameters of cell-mediated immunity need be examined closely in order to assess their roles in determining the clinical manifestations and outcome in these and other systemic fungal infections.

LABORATORY DIAGNOSIS

B. dermatitidis and *P. brasiliensis* bear striking mycologic similarities. Both are dimorphic, existing as molds at room temperature and as yeasts in infected tissues and in vitro at 37° C on enriched media. While *B. dermatitidis* and *P. brasiliensis* are virtually indistinguishable in the mycelial phase, both have certain characteristic morphologic differences in the yeast phase.

While unequivocal diagnosis of any mycosis requires isolation of the causative fungus in vitro, a presumptive diagnosis can be made by direct microscopic observation of the fungus in clinical materials. Samples of cerebrospinal fluid and urine are first sedimented by centrifugation and the sediment taken for direct examination and culture. Tissue must be homogenized using a small amount of physiologic saline and a glass tissue grinder or mortar and pestle. The tissue homogenate is then subject to both culture and direct microscopic examination. Viscous materials are digested by adding a drop of 10 per cent KOH to the material on a microscope slide and covering it with a coverslip while still wet.

The morphologies of *B. dermatitidis* and *P. brasiliensis* in infected tissues are the same as those of the yeast phases when grown in vitro. Histopathologic staining of infected tissue can therefore be a useful adjunct to laboratory diagnosis. Sections of involved tissue should be stained with methenamine silver, periodic-acid Schiff (PAS), and hematoxylin and eosin (H & E). Yeasts that were viable at the time of fixation are discernible with H & E stain, but they are most easily located with methenamine silver, which stains them black. With the Gomeri stain, they appear slightly enlarged due to the impregnation effect of the silver. More details of cell morphology can be resolved with PAS and H & E.

The presence of thick-walled cells producing single buds with a broad base of attachment represents presumptive histopathologic evidence of infection with *B. dermatitidis*. There may be superficial resemblance to *Histoplasma duboisii* (the etiologic agent of African histoplasmosis), but this latter fungus buds with narrow pores. Also, *H. duboisii* is uninucleate, whereas *B. dermatitidis* is typically multinucleate. With PAS and H & E stain, the cytoplasm of *B. dermatitidis* shrinks away from the cell wall, leaving a clear space. This appearance should not be confused with the capsule of *Cryptococcus neoformans*, which stains red with both PAS and Mayer's mucicarmine stains and which produces buds through narrow pores. (*B. dermatitidis* stains poorly at best with mucicarmine.) Single cells of *B. dermatitidis* have been mistaken for spherules of *Coccidioides immitis*. In some cases, it may be difficult to distinguish small intracellular yeast cells of *B. dermatitidis* from *H. capsulatum*. The former may have multiple nuclei that are visible with H & E stain, while the latter has only a single nucleus per yeast cell. In some cases, it may not be possible to exclude a dual infection by histologic criteria alone.

The morphology of *P. brasiliensis* in tissues is like that of the yeasts grown in vitro. Small forms of *P. brasiliensis* with single buds can be differentiated from *B. dermatitidis* on the basis of pore size. Single yeast cells of *P. brasiliensis* can be differentiated from *C. neoformans* and *C. immitis* by using the same criteria that are used for *B. dermatitidis*.

Materials for culture are prepared as described above for direct observation and are streaked in duplicate on both Sabouraud's agar and brain-heart infusion blood agar. Agar media may be used that contain both chloramphenicol, to inhibit bacterial contaminants, and cycloheximide (Actidione), to inhibit saprophytic fungal contaminants. However, such media should not be the sole media employed because certain opportunistic fungal pathogens as well as *Cryptococcus neoformans* are also inhibited by cycloheximide. One set of plates should be incubated at room temperature, or 30° C if possible, and one set at 37° C. Plates should be kept for four weeks before discarding, but growth should be observable after one week of incubation.

B. dermatitidis and *P. brasiliensis* can be grown in vitro on a variety of media. The nutritional requirements of both fungi are quite similar, and no special growth factors or vitamins are required. Both fungi are slow growing, with mean generation times of 12 hours or more, and both fungi grow as molds at room temperature as well as at 30° C. Temperature is the primary stimulus

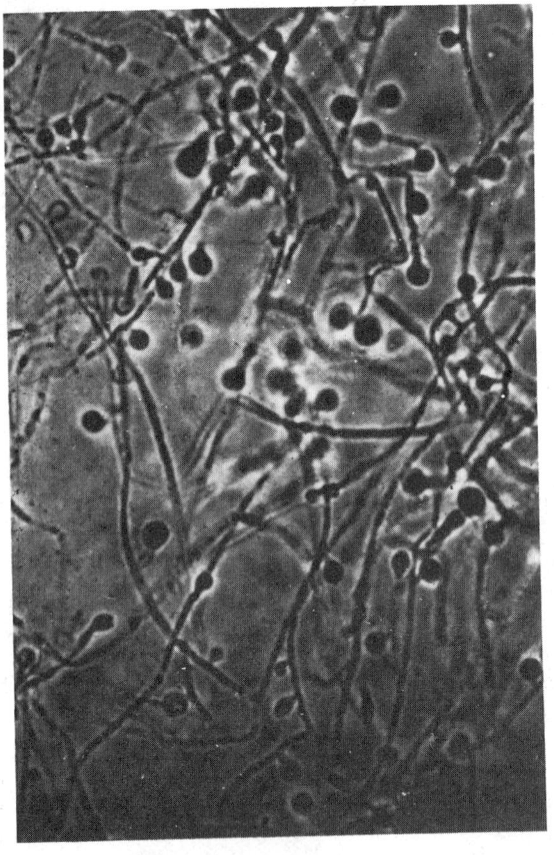

FIGURE 3. *Filamentous growth of* Blastomyces dermatitidis *showing spherical or pyriform spores. Note short pedicles that attach spores to hyphae.*

for dimorphism, and culturing on an enriched medium such as brain-heart infusion blood agar with incubation at 35° or 37° C will result in conversion of the fungi to their parasitic yeast phases. Colonies of yeast cells should appear within five days; if they do not, subsequent transfers may be necessary.

The preparations are stained with lactophenol cotton blue and examined microscopically for the charactistic aleurioconidia. For confirmation, yeast phase conversion should be demonstrated. Production of the characteristic budding yeasts differentiate *B. dermatitidis* from *P. brasiliensis* and from the saprophytic *Chrysoporium* species. Cottonseed agar may be more efficient for conversion of *B. dermatitidis* mold to its yeast phase than is brain-heart infusion blood agar. Not all isolates convert to the yeast phase. Inoculation of mice intraperitoneally with mycelia of *B. dermatitidis* produces (after four weeks) local infection that will contain yeasts.

EPIDEMIOLOGY

Recent studies of the epidemiology of blastomycosis have shown that the disease is not limited to North America as believed earlier. Rather, autochthonous infections have been reported in Europe and throughout Africa. The majority of cases, however, do come directly or indirectly from North America, where the disease is endemic to several geographic regions, including the Mississippi, Ohio, and Missouri River basins, as well as the Great Lakes region on the western shore of Lake Michigan.

Paracoccidioidomycosis is limited to Central and South America. In the case of infections reported from other countries, travel to known endemic areas usually can be documented. Highly endemic regions include both humid tropical forests and humid or very humid subtropical forest zones. Most cases have been reported from the state of São Paulo, Brazil.

Both blastomycosis and paracoccidiodomycosis are reported more commonly from rural than from urban areas. With both infections, similar occupational patterns are seen among the patients who are often laborers, farmers, and tree cutters. A unique epidemic of blastomycosis occurred in Bigfork, Minnesota, where members of four families were infected during the clearing of forest undergrowth and construction of a cabin (Tosh et al., 1974).

Neither blastomycosis nor paracoccidioidomycosis is naturally transmissible from man to man or from animal to man. The infections are acquired from some as yet poorly defined sources in nature, probably by inhalation of infectious conidia. The natural habitats of these fungi are assumed to be soil; however, only a few reports of isolation of these organisms from soil have been recorded. Isolation studies employing bird and bat manure have ruled out these materials as natural sources.

Blastomycosis and paracoccidioidomycosis occur in all age groups, but the preponderance of cases occurs in the 20- to 50-year-old age group. Males are more commonly affected than females in ratios of approximately 9 to 1. This is probably a reflection of hormonal as well as sociologic and occupational differences. There is no clear-cut predilection with respect to race as is seen in coccidioidomycosis.

REFERENCES

Cox, R. A., and Larsh, H. W.: Isolation of skin test-active preparations from yeast-phase cells of *Blastomyces dermatitidis*. Infect Immun 10:42, 1974a.

Cox, R. A., and Larsh, H. W.: Yeast and mycelial-phase antigens of *Blastomyces dermatitidis*: Comparison using disc gel electrophoresis. Infect Immun 10:48, 1974b.

Kaufman, L.: Evaluation of serological tests for paracoccidioidomycosis: Preliminary report. In Pan American Health Organization: Paracoccidioidomycosis. Proceedings of the First Pan American Symposium. Geneva, World Health Organization, 1972, pp. 221–223.

Lancaster, M. V., and Sprouse, R. F.: Isolation of a purified skin test antigen from *Blastomyces dermatitidis* yeast-phase cell wall. Infect Immun 14:623, 1976.
Mok, P. W. Y., and Greer, D. L.: Cell-mediated immune responses in patients with paracoccidioidomycosis. Clin Exp Immunol 28:89, 1977.
Musatti, C. C., Rezkallah, M. T., Mendes, E., and Mendes, N. F.: *In vivo* and *in vitro* evaluation of cell-mediated immunity in patients with paracoccidioidomycosis. Cell Immunol 24:365, 1976.
Restrepo, A.: La prueba de immunodiffusion en el diagnostico de la paracoccidioidomicosis. Sabouraudia 4:223, 1966.
Tosh, F. E., Hammerman, K. J., Weeks, R. J., and Sarosi, G. A.: A common source epidemic of North American blastomycosis. Am Rev Resp Dis 109:525, 1974.

THE ASPERGILLI 74

Abraham I. Braude, M.D., Ph.D.

The aspergilli are a group of versatile fungi belonging to the class Hyphomycetes and the family Moniliaceae. As noted in Chapter 14, *Aspergillus fumigatus* and *Aspergillus flavus* are unique in being the only nonzygomycete fungi that grow deep in tissue in mycelial form rather than in yeast form.

The name for this group of fungi is taken from the word aspergillum, which means a brush used for sprinkling holy water. This brush has a remarkable resemblance to the microscopic appearance of the fungus *Aspergillus niger* (Fig. 1).

MORPHOLOGY

In order to appreciate the structure of the brushlike portion of the aspergilli, it is necessary to be familiar with *four* components: the conidiophore, the vesicle, the sterigmata, and the spores. These are identified in Figure 2, where it can be seen that the handle of the brush consists of the unbranched conidiophore, which has a swollen end, the vesicle. A number of little stalks, shaped like a bottle, radiate from the vesicle and are arranged in one or two rows (Fig. 2*A* or 2*B*). Chains of spores arise from the tips of these stalks, or sterigmata, in the outer row.

The spores (or conidia) develop from the tips of the fertile sterigmata. The first stage in spore formation is a constriction of an elongated portion of the sterigma; the second stage is the develop-

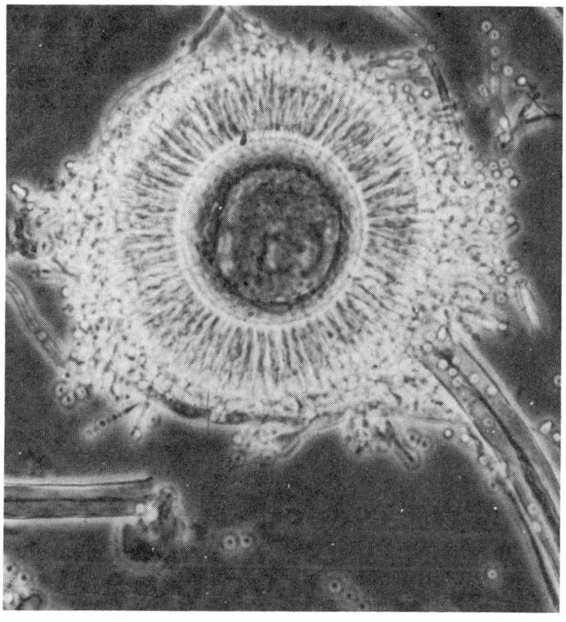

FIGURE 1. Aspergillus niger *bears a remarkable resemblance to the aspergillum, a brush used for sprinkling holy water.*

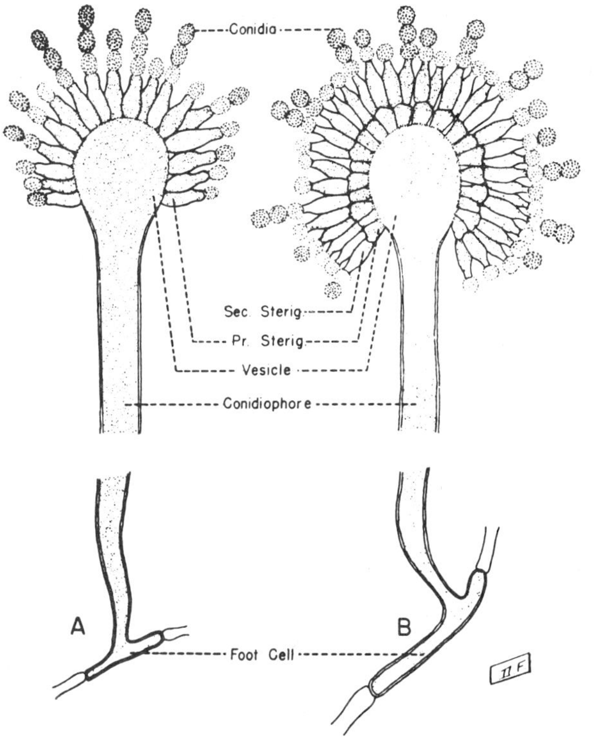

FIGURE 2. *Diagram of two different morphologic types of aspergilli showing one row of sterigmata (uniseriate) in A and two rows (biseriate) in B. (From Roper, K. B., and Fennel, D. I.: The Genus Aspergillus. Baltimore, The Williams & Wilkins Company, 1965.)*

ment of a septum that separates the spore from the sterigma. Chains of attached spores evolve as the cutting process continues at the tip of the sterigma. The compact mass of spores around the vesicle is called a conidial head and corresponds to the portion of a brush in which the bristles are set.

At its opposite end, the conidiophore arises from a foot cell, an enlarged thick-walled cell of the segmented mycelium. The conidiophore is borne as a stalk perpendicular to the long axis of the foot cell (Fig. 2).

The morphologic classification of all species of the genus *Aspergillus* is enormously complex and is beyond the scope of this chapter. Those members of medical importance, however, are relatively few and can generally be recognized by their color, the shape of their vesicle, and the arrangement of their sterigmata and spores. The groups of *Aspergillus* species that cause human disease are listed in Table 1 and can be arranged according to the following classification.

I. Green colonies
   A. Sterigmata arranged in single rows (uniseriate)
      1. *Conidial heads are split* (clavate) into two or more divergent columns of compact chains of spores; they have swollen ends. Vesicles are also split and appear as an elongated, gradually swelling portion of the end of the conidiophore. They are fertile over their entire surface and are blue-green (*A. clavatus*).
      2. *Conidial heads are compact, not split*; vesicles are fertile in their upper half so that sterigmata arise from the upper half only. Sterigmata are arranged parallel to each other and to the long axis of the conidiophore, pointing upward; they are green to dark green in color (*A. fumigatus* Fig. 3).
      3. *Conidial heads are not split*; interspersed among the green conidial heads are abundant bright yellow cleistothecia (round bodies containing multiple asci, which are sacs of sexual spores). Cleistothecia are about equal in size to conidial heads (*A. glaucus*).
   B. Sterigmata are arranged in two rows (biseriate): purple cleistothecia and orange-red ascospores (*A. nidulans*).

II. Black colonies. Sterigmata are in either one row or two rows, depending on the species, and arise from the entire surface of the globular vesicle (*A. niger*, Fig. 1).

III. Yellow colonies. A mixture of uniseriate or biseriate sterigmata occur in the same culture or even on the same vesicle; vesicles are globular and are sometimes split. They are fertile over most of their surface (*A. flavus*).

IV. Orange-brown or cinnamon colonies. There are two rows of sterigmata; the vesicles are hemispherical *(A. terreus)*.

TABLE 1. Diseases Caused by Different Species of Aspergilli

TYPE OF ASPERGILLOSIS	SPECIES OF ASPERGILLUS
Invasive pulmonary	*fumigatus*
Allergic bronchopulmonary	*fumigatus*
	flavus
	niger
	terreus
	clavatus
Pulmonary aspergilloma	*fumigatus*
	niger
Otomycosis	*niger*
Sinusitis	*flavus*
Disseminated	*fumigatus*
	flavus
Mycetoma	*glaucus*
	nidulans

ANTIGENIC COMPOSITION

Most immunologic reactions in *Aspergillus* infections have been studied with crude filtrates of *Aspergillus* cultures grown for several weeks at room temperature in Sabouraud broth. These contain countless unidentified antigens that can give up to 20 precipitin lines when examined by immunodiffusion against hyperimmune rabbit serum. Both protein and carbohydrate antigens have been identified in these *Aspergillus* culture filtrates, and some of the proteins have been characterized as enzymes (oxoreductases, hydrolases, proteases). One of the polysaccharides reacts with C-reactive protein. C-reactive protein appears during the acute phase of various infections and other inflammatory processes and reacts with the somatic polysaccharide (C substance) of the pneumococcus. Its reaction with an antigen from *Aspergillus* suggests that this fungal polysaccharide may have properties that are similar to the pneumococcal C polysaccharide. Since the reaction with C-reactive protein clouds the specificity of immunodiffusion reactions with *Aspergillus* filtrates, it is necessary to eliminate the C-reative substance by incorporating sodium citrate in the gel (Longbottom and Pepys, 1964).

The antigens in different species of *Aspergillus* cross-react in serologic tests with each other and

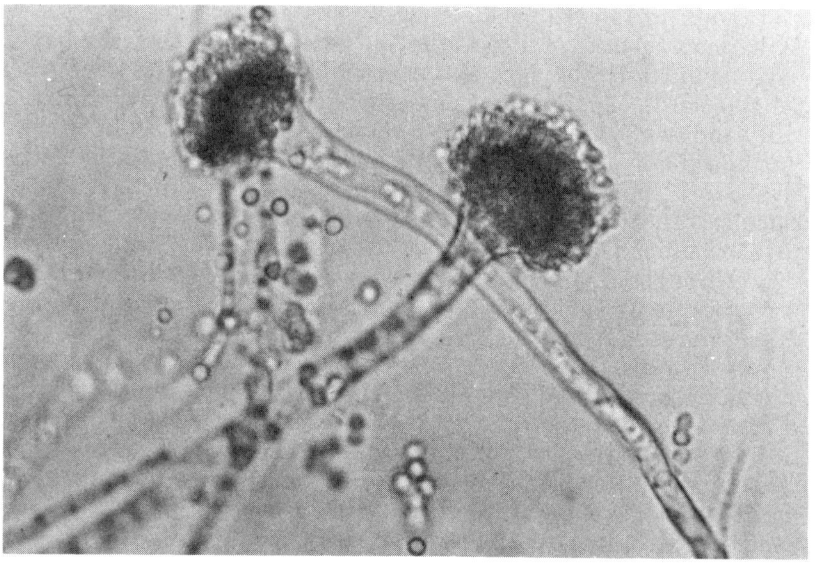

FIGURE 3. *Wet mount of* Aspergillus fumigatus *cultured from ethmoid sinus (× 400). Note following diagnostic features typical of* A. fumigatus: *(1) compact (unsplit) conidial head; (2) single row of sterigmata; (3) sterigmata arise from upper half only; (4) sterigmata are arranged parallel to each other and to the long axis of the conidiophore.*

with antigens of other fungi. These cross-reactions occur mainly between polysaccharide antigens and may occur not only with dermatophytes and *Cladosporium* but also with nematodes, trematodes, cestodes, and vegetable dusts. All of these cause immediate reactions to skin tests in patients with *Aspergillus* hypersensitivity and precipitin reactions with sera from some patients with pulmonary aspergilloma (Pepys, 1969). Galactomannans are prominent polysaccharide antigens in aspergilli as in other fungi, and they may be partly responsible for the cross reactions observed between different species of *Aspergillus* and between aspergilli and other fungi.

IMMUNITY

At least two types of antibody are important in the pathogenesis of aspergillosis. IgE antibody is responsible for bronchospasm and asthma, whereas IgG seems to cause Arthus-type reactions in the walls of bronchi that progress to bronchiectasis. If the respiratory tract becomes colonized with aspergilli, it releases the multiple antigens described in the previous section, and these sensitize certain patients with inherited hypersensitivity to aspergilli. These atopic patients respond by producing IgE antibody to *Aspergillus* antigens and immediate hypersensitivity reactions in the respiratory tree when they are re-exposed to inhaled *Aspergillus* spores. They also develop high serum levels of IgG antibody, which precipitates *Aspergillus* antigens around the vessels in the alveoli and bronchial walls. The ensuing Arthus inflammatory reaction (Chapter 83) is thought to cause transient pneumonitis and a progressive injury to the bronchial wall that culminates in the characteristic bronchiectasis of allergic bronchopulmonary aspergillosis (Golbert and Patterson, 1970).

There is no evidence that antibody can protect against aspergillosis; on the contrary, passive transfer of antibody can reproduce allergic bronchopulmonary aspergillosis (Golbert and Patterson, 1970). There are also no concrete data to indicate that cellular immunity is involved in resistance to *Aspergillus*. From observations of patients and experimental animals, it would appear that neutropenia, steroid administration, and loss of mechanical antimicrobial clearance mechanisms are factors in lowering resistance to aspergillosis. In patients who become neutropenic from leukemia or drugs, aspergilli invade pulmonary vessels and spread unopposed to the brain, kidney, liver, thyroid, and other organs (Young et al., 1970).

Some patients who develop aspergillosis do so while receiving adrenal steroids, but their importance in causing loss of immunity is obscured because many other debilitating factors are associated with steroid use in these cases. In mice, however, the ability of steroids to lower resistance to *Aspergillus* is dramatic; one injection of cortisone allows fatal aspergillosis to develop in animals that are otherwise completely resistant to aspergilli (Sidransky and Friedman, 1959). The importance of mechanical clearing processes, which prevent colonization of the respiratory or alimentary tract by pathogenic organisms, is illustrated by the types of aspergillosis that develop in the lung and esophagus when these mechanisms are lost. Structural damage to the lung and bronchopulmonary tree by any form of bronchiectasis or lung cavity interferes with the ability of cilia and cough to remove inhaled aspergillus spores and allows them to proliferate in the defenseless cavities in which they grow as aspergillomas (or fungus balls). Similarly, when esophageal motility and swallowing is disturbed by severe debilitating illness, aspergilli are propelled less effectively into the acid gastric juice. Instead, they colonize the esophagus and can produce severe invasive esophagitis. Thus, loss of natural immunity through impaired protective mechanical processes may be among the most important disturbances responsible for defective resistance to aspergillosis.

Acquired resistance after infection with aspergillus has been observed experimentally only in limited studies in mice. Mice are normally resistant to aspergillus infection but become susceptible to pulmonary, renal, and fatal disseminated aspergillosis after treatment with cortisone. If they are injected first with viable aspergillosis but not cortisone, they are no longer susceptible to progressive aspergillosis when rechallenged with aspergilli plus cortisone. The mechanism or specificity of this immunity is unknown; neither antibodies nor cellular immunity has been examined carefully enough to indicate its relative contribution to this important phenomenon. Although patients with disseminated aspergillosis have no antibodies to *A. fumigatus* antigens, a loss of humoral immunity cannot be blamed for susceptibility to this infection because the patients are also subject to suppression of cellular immunity by the chemotherapy given for their hematologic malignancies (Young and Bennett, 1971).

LABORATORY DIAGNOSIS

Mycologic Examination

In spite of the fact that aspergilli are frequently cultured as contaminants in specimens from patients, they may be difficult to recover from patients with certain forms of invasive aspergillosis. Thus, it is unusual to recover aspergilli from the sputum of patients with pulmonary aspergilloma or invasive *Aspergillus* pneumonias, and the fungus is rarely found in blood cultures of patients with disseminated aspergillosis. The organism can be isolated most readily in invasive disease of the lung, brain, sinuses, orbit, or skin when the infected tissue is obtained by biopsy (Fig. 4). In such tissues, the mycelial elements are usually found easily in histologic sections stained with hematoxylin and eosin or the methenamine-silver method; they appear in 24 to 48 hours in culture after aerobic incubation at 35° C on any standard medium. The only nonbiopsied specimens from which aspergilli are easily cultured are the sputa of patients with allergic bronchopulmonary aspergillosis, the pleural fluids in *Aspergillus* empyema, the exudates of burns infected with aspergilli, and the discharges from ears of patients with *A. niger* infection of the external auditory canal. Characteristic hyphae can also be seen in these specimens but may require special treatment before microscopic examination. Such treatment usually consists of mounting the material in 20 per cent potassium hydroxide under a coverslip and incubating at 37 or 56° C until there is dissolution of the organic material in the bronchial plug, ear wax, or burn crust. In some materials, the conidial head may be recognizable and will allow identification of *A. fumigatus*. For the most part, however, exact identification depends on cultural isolation.

Serologic Tests

Serologic tests for antibodies to aspergilli are valuable in certain forms of aspergillosis that cannot easily be identified by culture. Antibody determinations are also useful in determining whether or not the isolation of the fungus from specimens represents contamination or infection with aspergilli. Although antibodies against *Aspergillus* can be found by immunodiffusion, complement fixation, immunoelectrophoresis, and indirect immunofluorescence, the most widely used test is immunodiffusion (ID) because of its simplicity, sensitivity, and relative reliability. The ID test is most useful for the diagnosis of fungus ball and allergic bronchopulmonary aspergillosis. Patients with fungus balls due to one of the aspergilli have positive reactions. The diagnosis of fungus ball is suspected by characteristic lesions in chest radiographs in patients with underlying cavitary lung disease due to healed tuberculosis, sarcoidosis, bronchiectasis, or bronchogenic carcinoma, and an ID test should be done regardless of a negative sputum culture for aspergilli. Standardized antigen separations should be made from *A. fumigatus, A. niger*, and *A. flavus*. Each of these three species is grown separately in Sabouraud broth cultures at 31° C for five weeks. Reproducible test antigens can be obtained by precipitating them out of culture with acetone, concentrating them eightfold and adjusting the carbohydrate content to 1000 μg/ml by the anthrone test (Coleman and Kaufman, 1972). If simultaneous ID tests are done in separate plates with each of the three antigens, precipitins will be found in over 90 per cent of patients with fungus balls and 70 per cent of those with allergic bronchopulmonary aspergillosis. The test is controlled with positive reference sera, and the diagnostic precipitin lines are recog-

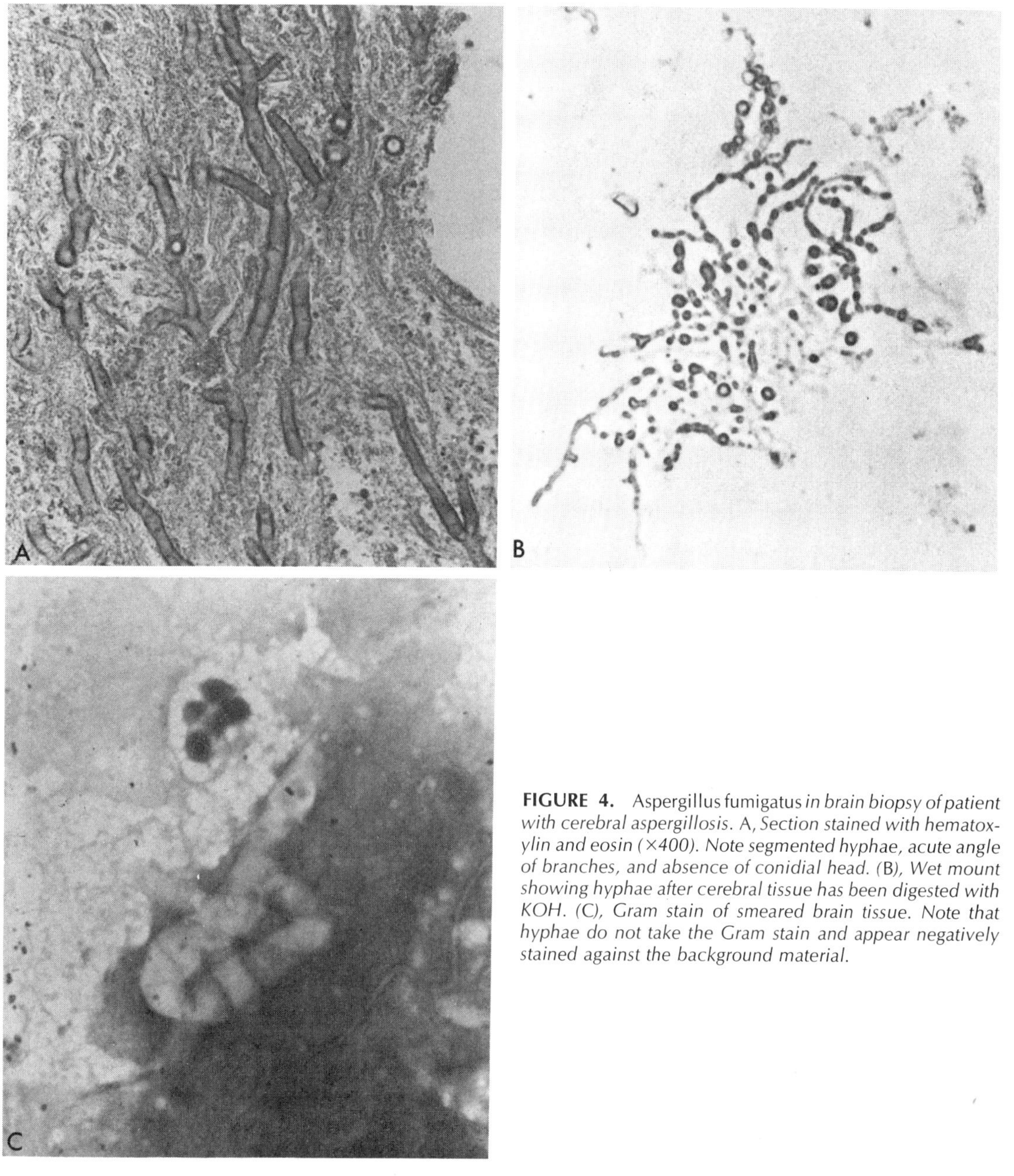

FIGURE 4. Aspergillus fumigatus *in brain biopsy of patient with cerebral aspergillosis. A, Section stained with hematoxylin and eosin (×400). Note segmented hyphae, acute angle of branches, and absence of conidial head. (B), Wet mount showing hyphae after cerebral tissue has been digested with KOH. (C), Gram stain of smeared brain tissue. Note that hyphae do not take the Gram stain and appear negatively stained against the background material.*

nized by their fusion with the reference lines (Fig. 5). Nonspecific lines produced by C-reactive substance are identified by their failure to fuse with the reference lines and their fuzzy appearance. They are greatly minimized by using antigens prepared from five week old cultures and eliminated by soaking the agar with 5 per cent sodium citrate for 45 minutes before reading the test. At least three specific precipitin bands are produced by antibody to *A. fumigatus* and one or more by antibody to *A. niger* or *A. flavus*. The precipitating antibodies belong to the IgG class of immunoglobulin (Warnock and Eldred, 1975). When multiple lines of identity occur with those of reference sera, a diagnosis of *Aspergillus* infection should be very seriously considered. Cross-reactions with antibody to fungi other than aspergilli are not likely to give multiple lines of identity, and clinical experience indicates that precipitins against *Aspergillus* antigens do not occur in normal sera or in those from other systemic fungus infections, bacterial infections, or cancer.

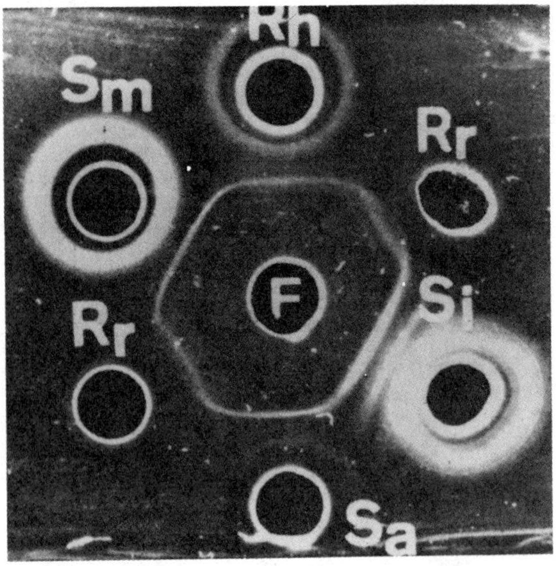

FIGURE 5. *Immunodiffusion reactions obtained with* A. fumigatus *against sera of patients with aspergilloma* (S_m) *and invasive* (S_i) *and allergic bronchopulmonary disease* (S_a). *Reference positive human serum is* R_r *(From Coleman, R., and Kaufman, L.: Appl. Microbiol 23:301, 1972.)*

EPIDEMIOLOGY

Aspergilli are found everywhere "from the winds of the Sahara to the snows of the Antarctic." They grow on soil, dead vegetation, any kind of organic debris, and laboratory reagents. They are extremely hardy and can withstand extremes of temperature, pH, and salt concentration. They are frequently encountered in farm houses, stables, barns, and grains, and infections have been acquired through inhalation of spores by farmers, millers, and gardeners. The most pathogenic species, *A. fumigatus*, is thermophilic —i.e., it can grow at the high temperatures that develop in compost piles from bacterial decomposition of wet leaves and wood chips. Enormous numbers of spores are produced, and, if inhaled by susceptible gardeners, they can produce pulmonary aspergillosis. The same phenomenon accounts for the heavy growth of *A. fumigatus* in stored hay or grains. Grain is an important source of infection in herds and livestock. Epidemics of aspergillosis involving the air sacs or lungs of birds have been traced to feeding on moldy grain. Horses, cattle, and sheep also get pulmonary aspergillosis by inhaling spores from their feed. Many years ago, infected squabs were thought to be a source of human aspergillosis. These pigeons were fattened by persons known as "gaveurs de pigeons" who used their tongues to force chewed grain into the bird's esophagi (Renon, 1897). It is also possible that the infection was acquired from the grain rather than from the squab.

Among patients newly admitted to sanitariums in the south central United States, the incidence of aspergillosis was reported in 1970 to be nearly as great as that of chronic pulmonary histoplasmosis (Parker et al., 1970). Since then, it is likely that more cases of aspergillosis have developed in patients after admission to the hospital. These are debilitated patients who have been given immunosuppressive drugs for treatment of hematologic malignancy or for prevention of renal transplant rejection. Air ducts contaminated by bird droppings have been incriminated as a source of exposure for immunosuppressed patients (Kyriakides et al., 1976), and some observations suggest that the acquisition of pulmonary aspergillosis in the hospital might be eliminated if all incoming hospital air were filtered, properly vented, and not recirculated (Rose, 1972).

Aflatoxins, the mycotoxins produced by *A. flavus*, are a potential epidemiologic problem because they are found in the ground peanuts that are heavily consumed by certain African tribes afflicted with a high incidence of hepatoma. A direct etiologic relationship between cancer and aflatoxin has not been established, however, although aflatoxin can cause hepatomas in experimental animals. The epidemiologic problems related to mycotoxin are fully covered in Chapter 15.

REFERENCES

Coleman, R., and Kaufman, L.: Use of the immunodiffusion test in the serodiagnosis of aspergillosis. Appl Microbiol 23:301, 1972.

Golbert, T., and Patterson, R.: Pulmonary allergic aspergillosis. Ann Intern Med 72:395, 1970.

Kaufman, L.: Value of immunodiffusion tests in the diagnosis of systemic mycotic diseases. Ann Clin Lab Sci 3:141, 1973.

Kyriakides, G., Zinneman, H., Hall, W., Arora, V., Lifton, J., Dewolf, W., and Miller, J.: Immunologic monitoring and aspergillosis in renal transplant patients. Am J Surg 131:246, 1976.

Longbottom, J. L., and Pepys, J.: Pulmonary aspergillosis: Diagnostic and immunologic significance of antigens and C-substance in *Aspergillus fumigatus*. J Pathol Bacteriol 88:141, 1964.

Parker, J., Sarosi, G, Doto, I., and Tosh, F.: Pulmonary aspergillosis in sanitoriums in the South Central United States. Am Rev Resp Dis 101:551, 1970.

Pepys, J.: Hypersensitivity diseases of the lung due to fungi and organic dusts. In Kallos, P., Goodman, H., Hasek, M., Inderbitzin, T., and Waksman, B. (eds.): Monographs in Allergy, Vol. 4. Basel, S. Karger, 1969.

Renon, L.: Étude sur l'Aspergillose Chez les Animaux et Chez l'homme. Paris, Masson et cie, 1897.

Rose, H. D.: Mechanical control of hospital ventilation and aspergillus infections. Am Rev Resp Dis 105:306, 1972.

Sidransky, H., and Friedman, L.: The effect of cortisone and antibiotic agents on experimental pulmonary aspergillosis. Am J Pathol 35:169, 1959.

Warnock, D., and Eldred, G.: Immunoglobulin classes of antibodies to *Aspergillus fumigatus* in patients with pulmonary aspergillosis. Sabouraudia 13:204, 1975.

Young, R., Bennett, J., Vogel, C., Carbone, P., and DeVita, V.: Aspergillosis. The spectrum of the disease in 98 patients. Medicine 49:147, 1970.

Young, R., and Bennett, J.: Invasive aspergillosis. Absence of detectable antibody response. Am Rev Resp Dis 104:710, 1971.

THE ZYGOMYCETES 75

Abraham I. Braude, M.D., Ph.D.

The Zygomycetes are the only medically important class belonging to the group of primitive fungi known as Phycomycetes. The various members of the class Zygomycetes have in common the production of sexual spores known as zygospores. The Zygomycetes produce two entirely different human infections: (1) Mucormycosis is a malignant infection of cerebral, pulmonary, and abdominal blood vessels due to Mucorales, the bread molds. The Mucorales are unique among the pathogenic fungi because they have no septa in their hyphae. (2) Tropical subcutaneous phycomycosis and nasal phycomycosis (or rhinoentomophthoromycosis) are benign chronic infections due to the insect fungi known as Entomophthorales. These curious fungi can shoot off their conidia forcibly.

MORPHOLOGY

The three pathogenic genera of the order Mucorales are known as *Mucor, Rhizopus*, and *Absidia*. All three are characterized by a white or gray woolly growth on food or agar. Under the microscope, the woolly growth consists of mycelium that is not divided into individual cells by septa, so that the entire mass of mycelium forms one large cell containing multiple nuclei. This nonseptate mycelium is called coenocytic, which means *common cell* in Greek.* In all genera of Mucorales, the mycelia give rise to specialized branches known as conidiophores. At the tip of the conidiophores are sacs filled with asexual spores. These sacs, called sporangia, measure 20 to 115 μ and are seen with the naked eye as dark dots scattered throughout the woolly colony (Fig. 1). In *Rhizopus* and *Absidia*, other specialized branches arise from the mycelium that allow them to spread rapidly over the surface of the agar, climb the sides of the culture plate, and fill it with aerial mycelium. One of these structures is the stolon (or "runner"). It is a filament of mycelium that extends out into the air from the point at which the sporangiophore branches from the mycelium. The newly extended runner then produces a "hold-fast," or rhizoid, which anchors the filament to the surface of the agar, the side of the plate, or its cover. In *Rhizopus*, the rhizoid is found at a node opposite the origin of the conidiophore. In *Absidia*, on the other hand, the sporangiophore arises from the arch of the stolon at a point between the nodes where the rhizoids branch off. In contrast to *Rhizopus* and *Absidia*, *Mucor* has neither stolons nor rhizoids and consequently fills the culture dish less rapidly and does not adhere to the lid. These relationships are shown in Figures 2, 3, and 4.

In addition to the asexual spores in the sporangia, the Mucorales also produce sexual spores, but they are rarely seen on routine agar cultures. The sexual spores are designated zygospores and result from the fusion of hyphae from two neighboring thalli, a process known as heterothallous conjugation. Despite the implication that two different sexes participate in this conjugation, the conjugating hyphae cannot be distinguished from each other. When two hyphae conjugate, their tips come together and become separated from the rest of the mycelium by a cross-wall, the only septum that characteristically occurs in these

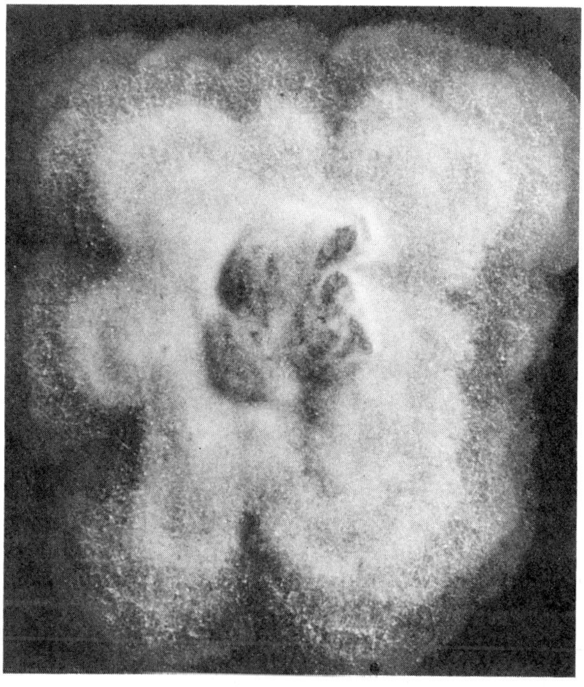

FIGURE 1. Mucor pusillus *isolated from blood of patient. Note dots scattered throughout the woolly colony. These dots represent sporangia.*

*Since certain green algae are also coenocytic, these nonseptate fungi are also called Phycomycetes, or algae-like fungi.

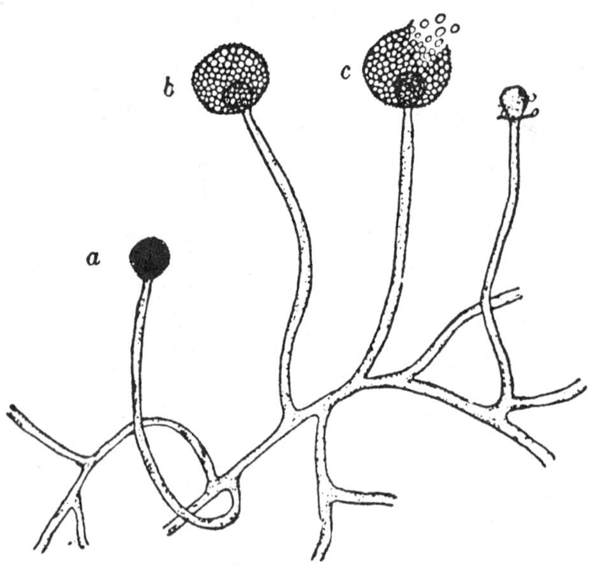

FIGURE 2. *Diagram of Mucor sp showing young sporulating head (a), mature sporangium (b and c), and columella after scattering of spores (d). Note that in contrast to* Rhizopus *and* Absidia *(Fig. 4) the sporangiophores arise from all parts of the thallus, and no stolons or rhizoids are produced. (From Waksman, S. A., and Starkey, R. L.: The Soil and the Microbe. New York, John Wiley and Sons, 1931.)*

otherwise nonseptate fungi. As the tips fuse, the walls separating them dissolve, so that the multinucleated masses of protoplasm in the two cells can run together, and the nuclei fuse. The wall of the fused cell becomes very thick.

In contrast to the prolific woolly colonies of the Mucorales, those of the Entomophthorales are flat and are covered by short, white mycelial fuzz. In further contrast, the mycelia of the Entomophthorales are septate. The two main pathogenic genera of Entomophthorales are known as *Entomophthora* and *Basidiobolus*.

Basidiobolus haptosporus, the organism that causes tropical subcutaneous phycomycosis, grows rapidly at room temperature and produces three kinds of spores: conidia, zygospores, and chlamydospores. The conidia are actually uninucleate sporangia that are carried at the end of a short sporangiophore. These sporangia (or conidia) are discharged forcibly onto the glass surface of the culture tube or lid of the petri dish. The fragment of sporangiophore (or "basidium") immediately adjacent to the ejected sporangium is carried along with it. The sporangium is a smooth, globular structure measuring 16 to 45 μ in length. The zygospores are the most distinctive spores. They are spherical, have a thick, smooth wall, and measure 20 to 40 μ in diameter. Two "conjugation breaks" protrude from the zygospore (Fig. 5) as remnants of the hyphal ends (copulation tubes) that fuse during conjugation. The third spore, the chlamydospore, differentiates from other cells within the septate mycelium by developing a markedly thickened cell wall. Production of chlamydospores continues to increase as the culture ages and the relative number of zygospores and sporangia declines.

Entomophthora coronata, which causes African nasal phycomycosis (rhinoentomophthoromycosis), also produces many chlamydospores. In contrast to *Basidiobolus*, however, *E. coronata* rarely has any zygospores and has only sparse septa in the mycelium. The organisms contain numerous conidia, which are shot off from their short individual conidiophores, but the expelled conidia do

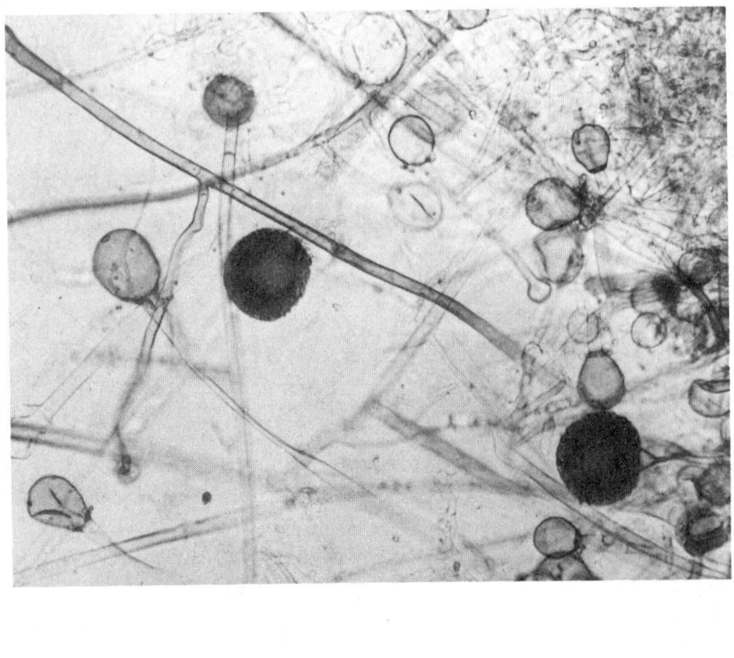

FIGURE 3. *Culture mount of* Mucor pusillus *from blood of fatal case of pulmonary mucormycosis. Note nonseptate mycelium, sporangiophore, and sporangium filled with asexual spores. The tip of the sporangiophore appears within the sporangium as a round swelling known as the columella. (Identification confirmed by J. J. Ellis, Research Mycologist, United States Department of Agriculture Research Service, Peoria, Illinois, Accession #NRRLA 12, 299.)*

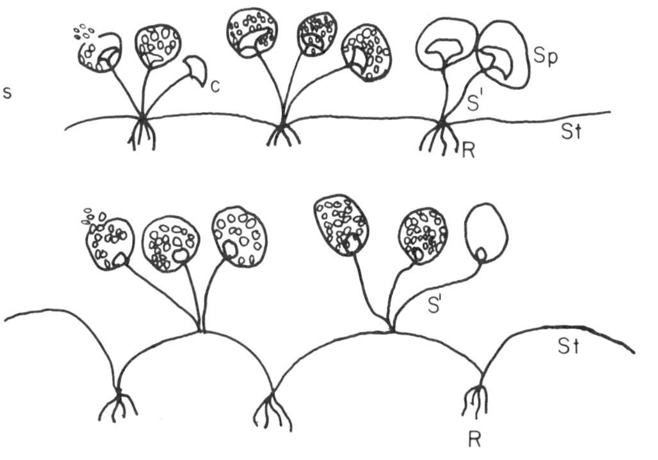

FIGURE 4. *Diagram of structural relationships of stolons (ST), rhizoids (R), sporangiophore (S), columella (C), and sporangium (SP) in* Rhizopus *and* Absidia. *Note that the sporangiophores of* Rhizopus *arise from the point at which the rhizoids are formed, whereas in* Absidia *the sporangiophores come off from the arched stolons. Also note the hemispherical shape of the columella in* Rhizopus *and the pear-shaped columella in* Absidia.

not carry a fragment of conidiophore (or "basidium") as do the conidia of *Basidiobolus*. The discharged conidia of *E. coronata* have the unique property of producing small secondary conidia that may form a crown around the parent (hence the name "coronata"). These secondary conidia are carried on short conidiophores that emerge from the primary conidium. Instead of secondary conidia, degenerating conidia may produce at their periphery many hair-like appendages that resemble flagella.

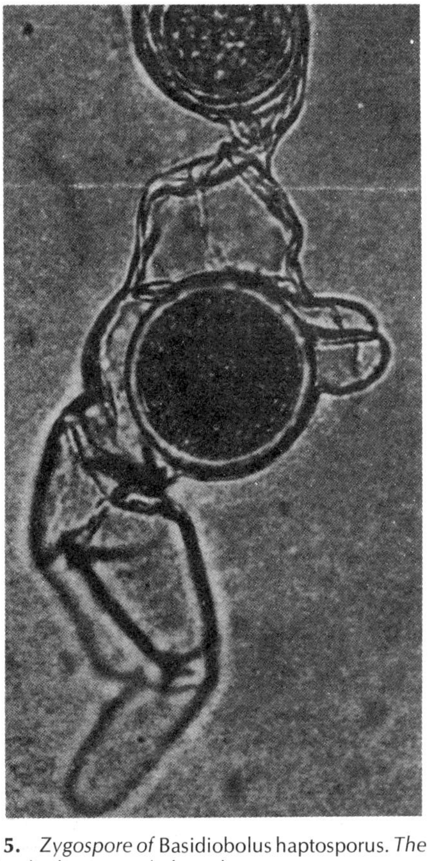

FIGURE 5. *Zygospore of* Basidiobolus haptosporus. *The two conjugation beaks that protrude from the zygospore are remnants of the hyphal ends that fuse during conjugation. (From Burkitt, D. P., Wilson, A., and Jelliffe, D.: Br Med J 1:1669, 1964.)*

ANTIGENIC COMPOSITION

The Zygomycetes have not been subject to the extensive antigenic analysis carried out with other pathogenic fungi, and no serologic test has been standardized for the diagnosis of mucormycosis, subcutaneous phycomycosis, or rhinoentomophthoromycosis. In preliminary studies using immunodiffusion, common antigens shared by *Absidia, Mucor*, and *Rhizopus* have been found in culture filtrates, and antigens specific for each of these genera were found in mycelial homogenates (Jones and Kaufman, 1978). The homogenate antigens showed some diagnostic promise when used in tests with sera from patients with zygomycosis, but a great deal of study is needed with this technique before it can be introduced for clinical purposes.

Limited analyses of the Entomophthorales have also found evidence of both common and specific antigens in species belonging to this order (Greer and Friedman, 1966).

IMMUNITY

Mucormycosis does not characteristically occur in patients who have normal immunity to infection. Instead, it is a disease of severely ill patients who lose their resistance because of hematologic malignancies, immunosuppressive drugs, or metabolic disorders. Uncontrolled diabetes mellitis is often the underlying disease, and an infection closely resembling the disease in man can be produced in rabbits with acute alloxan diabetes, but not in metabolically normal rabbits. The hyperglycemia in both human and experimental diabetes probably stimulates the growth of the fungus and at the same time impairs phagocytosis. In experimental diabetes, the polymorphonuclear (PMN) leukocytes respond vigorously to invasion by *Rhizopus oryzae*, but the PMN leukocytes uniformly show nuclear pyknosis and karyorrhexis (Bauer et al., 1956). It is also known that high glucose levels impair phagocytosis of other pathogenic organisms (Chernew and Braude, 1962). The importance of normal leukocytes in preventing mucormycosis is evident when normal rabbits are inoculated with spores of *R. oryzae*; the leukocytes in the acute inflammatory exudate restrain the fungal proliferation and destroy the spores. Moreover, in neutropenic animals or patients, there is less resistance to mucormycosis. Most reported cases of phycomycosis in nondiabetics have occurred in leukemic patients with severe neutropenia (Parkhurst and Vlahides, 1967; Meyer et al., 1972; McBride et al., 1960). Other cases have occurred in burn wounds in which phagocytosis is impaired.

If leukocytes and other immune factors cannot restrain the fungus, it invades vessels and infarcts tissues. The usual portal of entry appears to be the nasal turbinates or paranasal sinuses; from there, the organism extends along the invaded vessels to the retroorbital tissues and cerebrum. Thrombosis of arteries and veins causes multiple infarcts throughout the brain, but only a minimal inflammatory response is found because the inflammatory cells are depleted or their function is impaired. If spores of the Mucorales are inhaled or ingested, they may invade the walls and vessels of bronchi, stomach, or intestine of neutropenic patients.

In contrast to the Mucorales, the Entomophthorales produce infection in apparently normal people. Like mucormycosis, rhinoentomophthoromycosis also invades the nasal mucosa but stops short of the vessels, and there is no tendency to infarct the tissues. The affinity of spores from both Mucorales and Entomophthorales for the human nasal mucosa is a striking phenomenon, but its mechanism is mysterious. Normally, the cilia of the nose carry foreign particles to the posterior pharynx, where they are swallowed. Since other fungal spores are not arrested in the nose, it is possible that those of the Zygomycetes can impair ciliary action and inactivate this important means of natural resistance to nasal infection. A second possibility is that the nasal mucosa has receptors that hold the spores to the cell surfaces. Finally, it has been suggested that the *N*-acetyl neuraminic acid in nasal mucin has a specific affinity for spores of the Zygomycetes. Although none of these speculations on natural resistance has been verified, there is more solid information on the mechanism of acquired immunity. It appears that immunity to *E. coronata* involves eosinophils through a mechanism similar to that operating in schistosomiasis, filariasis, and other parasitic infections. The inflammatory exudate around the fungi may be composed almost entirely of eosinophils, and eosinophil granules are aggregated in close apposition to the fungal cell walls (Williams et al., 1969). Charcot-Leyden crystals, which are derived from eosinophils, are also present. In addition to the eosinophil granules and crystals, lysosomes, mitochondria, vacuoles, and pinocytic vesicles are released by cell necrosis. For this reason it has been proposed that the fungi lyse the eosinophils and other inflammatory cells so that their cytoplasmic and nuclear contents are released. These form an eosinophilic precipitate or "sleeve" around the fungus, called the Splendore-Hoeppli phenomenon, like that seen around schistosome ova. The adjacent granulomatous infiltrate is composed of macrophages, lymphocytes, and plasma cells. If current concepts on cellular immunity to worms can be translated into immunity against fungi, these histologic observations in enterophthoromycosis imply that IgE interacts with macrophages and eosinophils so that both cells can kill *E. coronata*. The tissue reaction in subcutaneous phycomycosis due to *Basiliobolus haptosporus* is identical, producing an eosinophilic sheath around the hyphae and showing no invasion of vessels. The cellular immune reaction of eosinophils and macrophages appears to be effective because both infections are benign and may even regress spontaneously (Burkitt et al., 1964). Since the infection by *B. haptosporus* occurs almost exclusively in childhood, it would seem that immunity acquired then protects adults.

LABORATORY DIAGNOSIS

In cerebral mucormycosis, the fungus can be found in sections and cultures of the infarcted nasal turbinate or paranasal sinus. In pulmonary mucormycosis, it can be found in lung biopsies or in metastatic skin lesions. For mysterious reasons, the etiologic agent of cerebral mucormycosis has rarely if ever been cultured from the brain or spinal fluid, even when fresh tissues known to harbor the characteristic mycelium were subjected to expert mycologic techniques. The fungus most frequently isolated in mucormycosis is *Rhizopus oryzae*. Other species have been identified as *Rhizopus arrhizus*, *Rhizopus rhizopodiformis*, *Absidia corymbifera*, and *Mucor pusillus*. Two other Mucorales, *Cunninghamella elegans* and *C. mortierella*, have also caused human infection on rare occasions. These are easily separated from the other Mucorales because they have no columellas. In addition, *C. elegans* has no sporangium and its conidia arise from a vesicle like that found in *Aspergillus*.

It is impossible to distinguish among the species of Mucorales in the tissues of patients with mucormycosis, but they can be differentiated from the fungi that produce hyphae in tissues. In contrast to the hyphae of *Aspergillus* and *Candida*, those of the Mucorales are nonseptate, branch irregularly, and stain deeply with hematoxylin and eosin or silver-methenamine but poorly with the periodic acid-Schiff (PAS) stain. *Aspergillus* is septate, stains better with PAS than hematoxylin, and branches with great regularity. The various genera of Mucorales are identified by the characteristics in Figures 2, 3, and 4. Special guides must be consulted for identification of species (Ellis and Hesseltine, 1966; Zycha, 1935).

The Entomophthorales, *B. haptosporus* and *E. coronata*, are also identified by biopsy and culture. As noted in the previous section, the two produce identical histologic changes that strongly suggest the diagnosis without culture. In contrast to those of the Mucorales, the hyphae of the Entomophthorales take up hematoxylin poorly and are usually septate. The poorly stained hyphae may appear as circular spaces surrounded by a cuff of amorphous eosinophilic material representing the Splendore-Hoeppli phenomenon. Large numbers of eosinophils are also present in acute exudates. The hyphae may be surrounded by epithelioid cells and ingested by giant cells. The poorly staining walls of the hyphae and the absence of cytoplasm in subcutaneous phycomycosis give the impression of tunnels through subcutaneous inflammatory tissue. Because of the prominent eosinophils, these tunnels were thought at one time to be the track of an unidentified worm larva (Burkitt et al., 1964).

If tiny fragments of tissue are incubated on Sabouraud's medium at 37° C, the fungi grow rapidly, and the characteristic radially folded, wax-colored colonies appear in three days. The diagnostic zygospores are seen after one or two weeks in cultures of *B. haptosporus* but do not develop in *E. coronata*.

EPIDEMIOLOGY

The Mucorales are ubiquitous thermotolerant molds that thrive in any organic material. *Rhizopus* is an important cause of spoilage in fruits and sweet potatoes. The disease in strawberries is known as leak because of drippings from the softened fruit. The mucors are also found frequently on stale, moist breads ("bread molds") and can hydrolyze starch to sugar. Yet despite their wide prevalence in food and other organic materials, the Mucorales are not commonly encountered in hospital environments where susceptible diabetics, leukemics, and burn patients are likely to be exposed. Recent outbreaks of skin and subcutaneous mucormycosis were reported in hospitals using contaminated commercial elastic dressings (Elastoplast tapes), but cultures of the environment over a ten year period in the same institution have not grown Mucorales (Gartenberg et al., 1978). Their absence in the hospital despite their frequent presence in food is puzzling and might indicate the Mucorales in food can infect debilitated patients without spreading to the hospital environment. This puzzle, and the mysterious failure to culture the fungus from the brains of patients with histologically proven cases of mucormycosis, imply that pathogenic Mucorales may sometimes lose their ability to grow on environmental or laboratory substrates but not in certain tissues.

The epidemiology of infection by the Entomophthorales is also puzzling because infections by these fungi are remarkably infrequent for such ubiquitous fungi. It has been suggested that environmental temperatures limit the period during which *B. haptosporus* can be infectious. The fungus grows poorly at 15° C, so that in temperate climates it can be disseminated only during the few warm months from its reservoir in the gastrointestinal tracts of insectivorous reptiles (Clark, 1968). This factor also explains why nearly all cases of subcutaneous phycomycosis occur in tropical areas such as Uganda, Kenya, Sudan, the Cameroons, Indonesia, India, and parts of Southeast Asia. It has not been reported from the western hemisphere. Infections with *B. haptosporus* occur mainly in children, who probably acquire them from minor trauma or insect bites. It is a fungus of low pathogenicity and is seldom able to cause infection in adults, who seem to acquire immunity from clinical or subclinical exposure earlier in life.

The epidemiology of rhinoentomophthoromycosis due to *E. coronata* differs in two main respects from that of subcutaneous phycomycosis caused by *B. haptosporus*. First, *E. coronata* infections are probably transmitted by spore inhalation rather than insect bites, since the portal of entry is the nose. Second, it is a disease of adult agricultural workers, especially men exposed to the fungus growing in the soil or vegetation of tropical rain forests in Africa. It has also been reported from similar regions such as Columbia, Brazil, India, and Puerto Rico.

References

Bauer, H., Flanagan, J., and Sheldon, W.: The effects of metabolic alterations on experimental *Rhizopus oryzae* (mucormycosis) infection. Yale J Biol Med 29:23, 1956.

Burkitt, D. P., Wilson, A., and Jelliffe, D.: Subcutaneous phycomycosis: A review of 31 cases seen in Uganda. Br Med J 1:1669, 1964.

Chernew, I., and Braude, A.: Depression of phagocytosis by solutes in concentrations found in the kidney and urine. J Clin Invest 41:1945, 1962.

Clark, B.: The epidemiology of phycomycosis. In Wolstenholme, G., and Porter, R. (eds.): Symposium on Systemic Mycoses. Boston, Little, Brown & Company, 1968, p. 179.

Ellis, J., and Hesseltine, C.: Species of *Absidia* with ovoid sporangiospores. Sabouraudia 5:59, 1966.

Gartenberg, G., Bottone, E., Keusch, G., and Weitzman, I.: Hospital-acquired mucormycosis (*Rhizopus rhizopodiformis*) of skin and subcutaneous tissues: Epidemiology, mycology and treatment. N Engl J Med 299:1115, 1978.

Greer, D, and Friedman, L.: Antigenic relationships between the fungus causing subcutaneous phycomycosis and saprophytic isolates of *Basidiobolus meristosporus* and *B. ranarum*. Sabouraudia 5:7, 1966.

Jones, K., and Kaufman, L.: Development and evaluation of an immunodiffusion test for diagnosis of systemic zygomycosis (mucormycosis): Preliminary report. J Clin Microbiol 7:97, 1978.

McBride, R., Corson, J., and Dammin, G.: Mucormycosis: Two cases of disseminated disease with cultural identification of *Rhizopus*: Review of literature. Am J Med 28:832, 1960.

Meyer, R., Rosen, P., and Armstrong, D.: Phycomycosis complicating leukemia and lymphoma. Ann Intern Med 77:871, 1972.

Parkhurst, G., and Vlahides, G.: Fatal opportunistic diesase. JAMA 202:279, 1967.

Waksman, S. A., and Starkey, R. L.: The Soil and the Microbe. New York, John Wiley and Sons, 1931.

Williams, A., Lichtenberg, F., Smith, J., and Martinson, F.: Ultrastructure of phycomycosis due to entomophthora, basiodiobolus, and associated "Splendore-Hoeppli" phenomenon. Arch Pathol 87:495, 1969.

Zycha, H.: Mucorineae, Kryptogamenflora. Leipzig, Mark Brandenberg, Berlin, Gebrüder Borntraeger, 1935.

76 DEMATIACEAE: AGENTS OF CHROMOMYCOSIS

Yousef Al-Doory, Ph.D.

The dematiaceous fungal agents of chromomycosis comprise a group of genetically related fungi. Five species have been isolated repeatedly from confirmed cases of this disease. The clinical appearance of this chronic disease of the skin and subcutaneous tissues does not indicate which species is the causative agent.

The five fungi can be classified into two types: as a group, *Fonsecaea pedrosoi, F. compactum, Phialophora verrucosa,* and *Cladosporium carrionii* are characterized by a moldy appearance and an abundance of dry aerial hyphae on culture media; whereas *F. dermatiditis,* the other type, has soft, dark, moist colonies that undergo a yeastlike phase during the course of development.

A few other dematiaceous fungi have been reported occasionally as causative agents of chromomycosis. As an example, *Cladosporium bantianum* has been reported to be the agent for cladosporiosis, which is considered by some workers as cerebral chromomycosis (Al-Doory, 1972).

MORPHOLOGY

All of the fungi of chromomycosis are slow growing and have dark-colored colonies that show black color through the reverse of the agar. However, individual strains do reveal great variations in color, rates of growth, and gross colony characteristics. *F. pedrosoi* colonies are flat, covered with feltlike, short, aerial hyphae that are dark green to brown or black. *F. compactum* colonies are heaped, brittle, and covered with a coarse aerial hyphae. Colonies of *C. carrionii* may appear smooth or irregular with a defined margin bordered by a darker submerged hyphae. *P. verrucosa* shows dark brown to black colonies with olive to gray aerial hyphae. *F. dermatitidis* colonies mature from moist, glistening, olive to black colonies to those showing feathery strands of submerged, tightly compressed hyphae radiating outward from the margin. These hyphae eventually give rise to olive-gray aerial mycelia (Fig. 1).

Microscopically, these five species of fungi show dark conidia and mycelia with three types of conidiophores (Fig. 2). The phialophora-type is found mainly in *P. verrucosa.* A distinct conidiophore is formed terminally or along the mycelium. It is generally flask-shaped with a rounded oval or elongated base, a constricted neck, and an opening with or without a flaring collarette and lip. The conidia are formed semiendogenously and extrude through the neck.

The second type of conidiophore is the cladosporium-type that is found mainly in *C.*

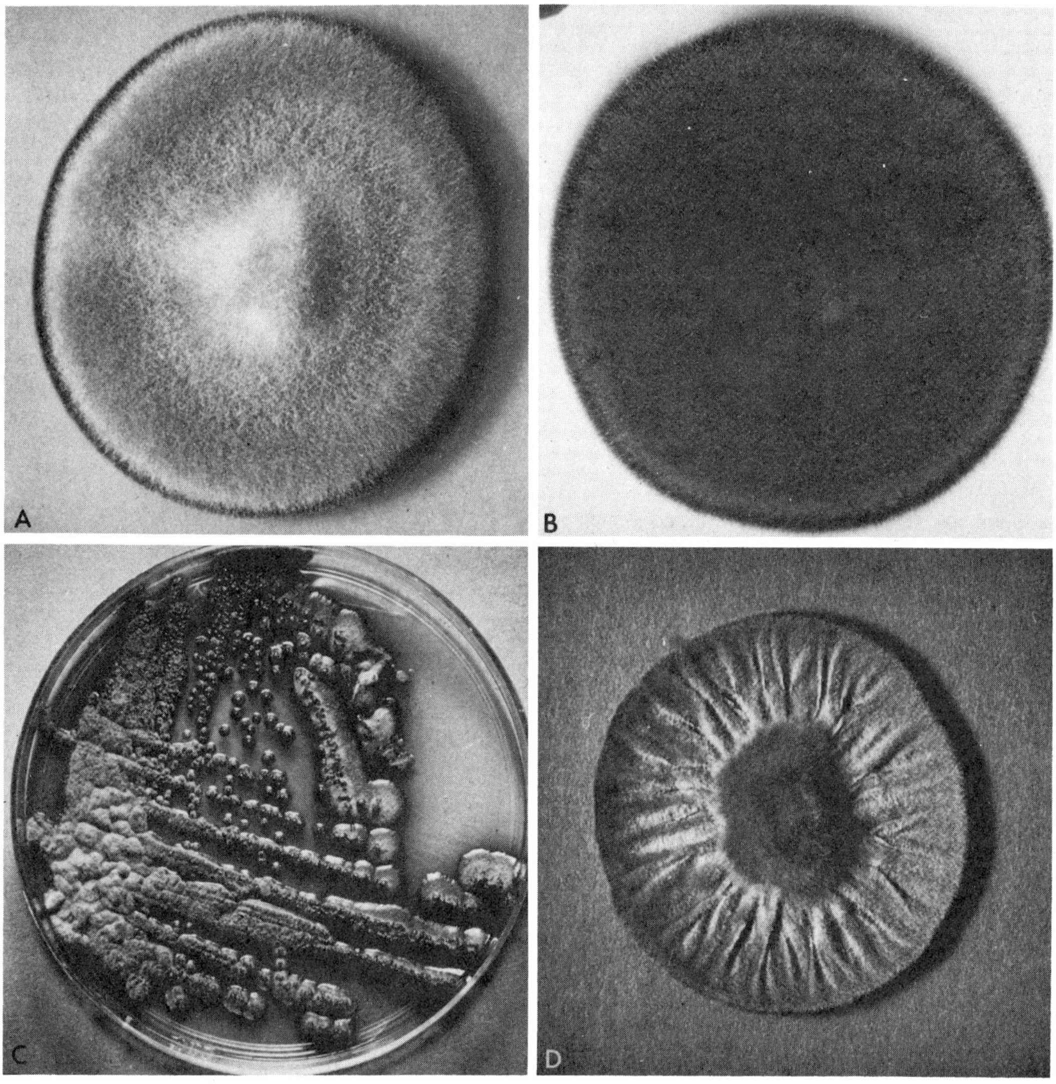

FIGURE 1. *Colonies of three causative agents of chromomycosis:* A, *Fluffy, black-gray colony of* Fonsecaea pedrosoi. B, *Fuzzy, black colony of* F. compactum. C,D, *Glistening, yeast colony* (C) *transferring to fuzzy, mycelial form* (D) *of* F. dermatitidis. *(Courtesy of Rippon, J. W.: Medical Mycology. W. B. Saunders Co., 1975.)*

carrionii. The conidia in this type are formed at the tip of a simple conidiophore that is slightly enlarged at the distal end. A few conidia are formed first at the tip and then bud, forming secondary conidia at their distal poles, and the process continues. This produces long branching chains of conidia.

The third type of conidiophore is the fonsecaea-type (acrotheca-type) found in the three species of *Fonsecaea* in addition to the previous two types. The conidiophore is similar to the vegetative hyphae, in which conidia are formed irregularly at the tip and along the sides of the conidiophore.

ANTIGENIC COMPOSITION

Few studies of the antigenic composition and relationships of the various etiologic agents of chromomycosis have been reported. Chemical analysis of the cell walls of *F. pedrosoi*, *P. verrucosa*, and *C. carrionii* reveals a composition of 17 to 31 per cent glucose, 29 to 42 per cent protein components, 8 to 14 per cent mannose, and 6 to 8 per cent glucosamine (Szaniszlo et al., 1972).

Complement fixation, agglutination, agglutination-absorption, and precipitation tests have been used and described (Martin et al., 1936; Seeliger, 1959, 1968). Furthermore, patients with

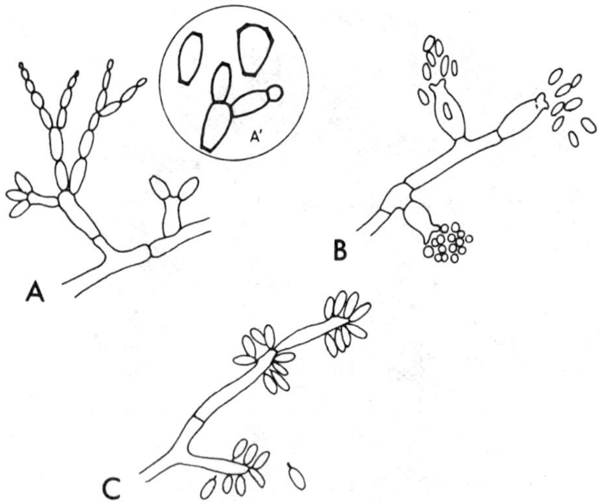

FIGURE 2. *Sporulation types in the causative agents of chromomycosis. A, Cladosporium-type. Note the thickened sacs or disjunctors (A'). B, Phialophora type. C, Acrotheca type. (Courtesy of Rippon, J. W.: Medical Mycology. W. B. Saunders Co., 1975.)*

chromomycosis were found to have circulating antibodies (Buckley and Murray, 1966; Smith et al., 1968). By cross-precipitation tests, it was shown that *F. pedrosoi, F. compactum,* and *P. verrucosa* are closely related and are antigenically distinct from *C. carrionii* (Buckley and Murray, 1966). By the use of the fluorescent antibody technique, it was shown that *C. carrionii* can be differentiated from *C. bantianum,* which is the causative agent of cerebral chromomycosis (Al-Doory and Gordon, 1963; Gordon and Al-Doory, 1965).

So far, no consistent antigenic patterns have been found to describe firmly and delineate antigenically the various species of chromomycotic fungi. Similarly, there is no serologic test with diagnostic significance for patients in whom the disease is suspected.

METABOLISM

There are no major studies on the metabolism of chromomycotic fungi in the literature. It is known that these fungi produce neither exotoxins nor endotoxins. At the same time, these fungi lack any of the proteolytic enzymes, which liquify Loeffler's coagulated serum medium or gelatin, or which peptonize milk (Fuentes and Bosch, 1960).

PATHOGENIC PROPERTIES

The chromomycotic fungi are found in soil, certain plants, and debris. Therefore, infection from these fungi is exogenous, and actual implantation of the fungus occurs through a break in the skin and/or subcutaneous tissues. It is believed that these fungi are not highly pathogenic because they require repeated exposure to overcome the normal resistance of the human body. Accordingly, chromomycosis is a noncontagious disease, and transfer from man to man has never been reported.

The chromomycotic fungi can infect tissues and organs beyond the portal subcutaneous tissues to cause dissemination or cerebral involvement (Al-Doory, 1972). From the medical histories in the available cases of chromomycosis, there is evidence that, in addition to foci of infection in localized extremity lesions, an aerosol-type of infection may cause a primary pulmonary focus, with subsequent lymphangitic or hematogenous seeding of other anatomic sites (Jotisankasa et al., 1970).

The pathogenicity of the chromomycotic fungi has been tested in various animals (Al-Doory, 1972; Wilson et al., 1933) and has been found capable of producing lesions. The clinical appearance of infected animals was not similar to that of chromomycotic patients, and histopathologically, abscesses in animals contained mycelial elements rather than the regular round sclerotic bodies (MacKinnon, 1934).

IMMUNITY

Natural resistance to infection with agents of chromomycosis is normally high. It is higher both in preteen individuals than in adults and in females than in males.

No race seems immune. However, a review of 124 patient histories revealed the following: 60 per cent were Caucasian, 30 per cent were black, and 10 per cent were of mixed racial background; among these: six cases were Mongolian; two were Malayan; two were Hindu, and one was a Mexican and one a Jamaican (Carrion and Silva, 1947).

Hypersensitivity has been demonstrated in chromomycotic patients and is diminished following clinical recovery. The relationship of this finding to resistance against reinfection is not known, however (Rippon, 1974).

LABORATORY DIAGNOSIS

The appearance of the chromomycotic fungi in pus as round, thick-walled, chestnut-brown, sclerotic cells is a significant diagnostic feature in histopathology (Fig. 3). Microscopically, the

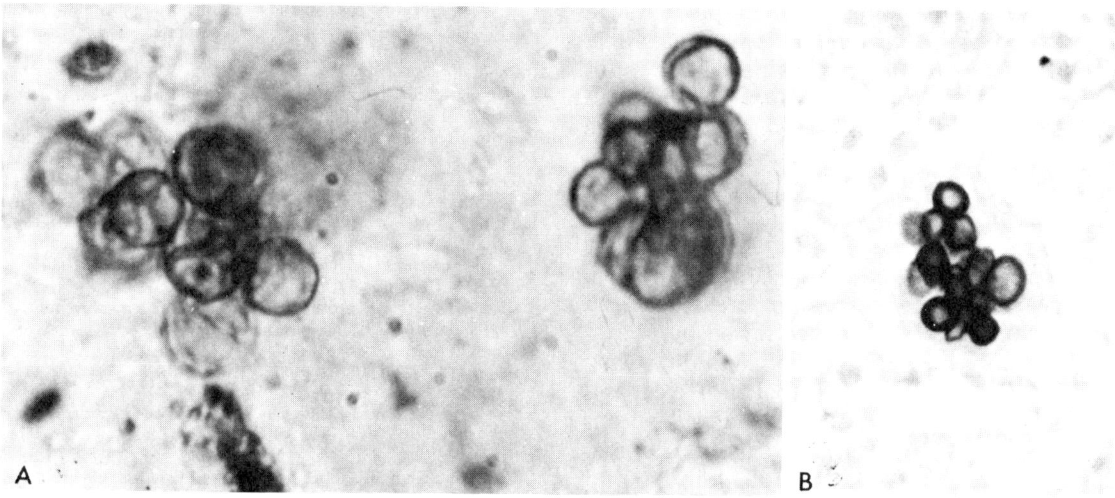

FIGURE 3. *Clusters of sclerotic fungal cells from skin of patients with chromomycosis. Note internal septa in A and thick walls in B. (Courtesy of Al-Doory: Chromomycosis. The Mountain Press.)*

presence of the three types of conidiophore formation is an important feature of the culture of the three species of *Fonsecaea* (conidiophores show various ratios depending on species, strain, and growth environment). In addition, the appearance of the moist, glistening yeast-like colonies is an indication of *F. dermatitidis,* while the appearance of subspherical conidia borne tightly in a compact chain is an indication of *F. compactum.* The presence of only phialophora-type conidiophores in the culture is the characteristic of *P. verrucosa,* while the cladosporium-type conidiophore with its extralong chains of conidia is indicative of *C. carrionii.*

Aging cultures of most strains of these fungi develop masses of pigmented chlamydospores, which have characteristic coarse, dark walls surrounding yellowish-brown, granular protoplasm.

Macroscopically, these fungi feature slow growth of colonies compared with the relatively fast growth of the saprophytic *Cladosporium sp.,* and they are unable to hydrolyze starch, coagulate milk, or liquify gelatin, Loeffler's medium, or any substrate containing protein. The saprophytic species are proteolytic (Fuentes and Bosch, 1960).

DRUG SENSITIVITY

Chromomycotic fungi are insensitive to various bacterial antibiotics. However, they show various degrees of sensitivity to other chemotherapeutic agents in vitro. The fungal growth of these fungi was found to be inhibited with 0.75 μg/ml of thiabendazole (2-[4′-thiazolyl]benzimidazole) (Blank and Rebell, 1965). The growth of *F. pedrosoi* was inhibited by sodium sulfamerazine (Keeney et al., 1944), stilbamidine, propamidine, and pentamidine (Bocobo et al., 1953), and deamidines (Al-Doory, 1972).

Amphotericin B is the drug of choice in most mycotic diseases. Chromomycotic fungi are killed by huge doses of this drug in vitro — i.e., a culture of *F. dermatitidis* containing 5 μg/ml of amphotericin B was nonviable at the end of 24 hours; with 2.5 μg/ml of the drug, it was nonviable after 48 hours. Even 0.5 μg/ml caused a reduction in the colony count (Keeney et al., 1944). These fungi are too resistant for use of amphotericin in patients, however. The MIC is greater than 40 μg/ml in vitro, while the maximum dose tolerated by humans is 1 to 1.5 mg per kg per day, which produces a blood level of the drug no greater than 1.8 μg/ml (Littman et al., 1958).

In vivo these fungi show sensitivity to potassium iodide, sodium iodide, copper sulfate, and 5-fluorocytosine. However, currently no drug can be considered the drug of choice for these fungi, either in vitro or in vivo.

EPIDEMIOLOGY

Although chromomycosis has been reported from every continent and is therefore, of worldwide distribution, the fungi of this disease are more abundant in the tropical and subtropical regions.

It appears that certain species are localized more in certain geographic areas than others. *C. carrionii* has been isolated only from patients in South Africa, Venezuela, and Australia. *F. pedro-*

soi has been isolated mostly from patients in tropical or subtropical regions (Brazil, Costa Rica, Cuba), while *P. verrucosa* is considered the main causative agent of the disease in colder climates, mainly the United States. *F. dermatitidis* has been reported as the causative agent of cases in Japan (Al-Doory, 1972).

For a number of reasons, including completeness and accuracy, the reported cases do not fully represent the distribution of the fungi; they are, however, the only guide available. Brazil leads in the number of cases reported, followed by Madagascar, Costa Rica, Dominican Republic, Australia, Cuba, Colombia, Mexico, South Africa, and Paraguay. It is known that the individual's occupation, continued exposure to environmental hazards or trauma, personal hygienic habits, and nutritional and environmental conditions are among the major factors in the infectivity of these fungi. Therefore, these factors play a large role in the distribution of reported cases and may influence the geographic distribution of the fungal agents.

References

Al-Doory, Y.: Chromomycosis. Missoula, Mont., Mountain Press, 1972.

Al-Doory, Y.: The Epidemiology of Human Mycotic Diseases. Springfield, Ill., Charles C Thomas, 1975.

Al-Doory, Y., and Gordon, M.: Application of fluorescent-antibody procedures to the study of pathogenic dematiaceous fungi. I. Differentiation of *Cladosporium carrionii* and *Cladosporium bantianum*. J Bacteriol 86:332, 1963.

Blank, H., and Rebell, G.: Thiabendazole activity against the fungi of dermatophytosis, mycetomas and chromomycosis. J Invest Dermatol 44:219, 1965.

Bocobo, F. C., et al.: *In vitro* fungistatic activity of stilbamidine, propamidine, pentamidine and diethylstilbestrol. J Invest Dermatol 21:149, 1953.

Buckley, H. R., and Murray, I. G.: Precipitating antibodies in chromomycosis. Sabouraudia 5:78, 1966.

Carrion, A. L., and Silva, M.: Chromoblastomycosis and its etiologic fungi. Ann Cryptogamici Phytopathol 6:20, 1947.

Fuentes, C. A., and Bosch, Z. E.: Biochemical differentiation of the etiological agents of chromoblastomycosis from nonpathogenic *Cladosporium* species. J Invest Dermatol 34:419, 1960.

Gordon, M. A., and Al-Doory, Y.: Application of fluorescent-antibody procedures to the study of pathogenic dematiaceous fungi. II. Serological relationships of the genus *Fonsecaea*. J Bacteriol 89:551, 1965.

Jotisankasa, V., et al.: *Phialophore dermatitidis:* Its morphology and biology. Sabouraudia 8:98, 1970.

Keeney, E. I., et al.: Studies on common pathogenic fungi and on *Actinomyces bovis*. Bull Johns Hopkins Hosp 75:393, 1944.

Littman, M. L., et al.: Coccidioidomycosis and its treatment with amphotericin B. Am J Med 24:568, 1958.

MacKinnon, J. E.: Estudio del primer casa Uruguayo de chromoblastomycosis y revista critica sobra el enfermedad. Arch Urug Med Cir Esep 5:201, 1934.

Martin, D. S., et al.: A case of verrucous dermatitis caused by *Hormodendrum pedrosoi* (chromoblastomycosis) in North Carolina. Am J Trop Med 16:593, 1936.

Rippon, J. W.: Medical Mycology: The Pathogenic Fungi and the Pathogenic Actinomycetes. Philadelphia, W. B. Saunders Company, 1974.

Seeliger, H. P. R.: In The Fungi, vol. 3. New York, Academic Press, 1968.

Seeliger, H. P. R., et al.: Identification of fungi by serologic tests. Further serologic studies with dematiaceous fungi. Proceedings of the Sixth International Congress of Tropical Medicine. Malaria 4:636, 1959.

Smith, D. T., et al. (eds.): Zinsser's Microbiology, 14th ed. New York, Appleton-Century-Crofts, 1968.

Szaniszlo, P. J., et al.: Chemical composition of the hyphal walls of three chromomycosis agents. Sabouraudia 10:94, 1972.

Wilson, S. J., et al.: Chromoblastomycosis in Texas. Arch Dermatol 27:107, 1933.

77 MISCELLANEOUS FUNGI: THE AGENTS OF MYCETOMA AND RHINOSPORIDIUM

Abraham I. Braude, M.D., Ph.D.

AGENTS OF MYCETOMA

Mycetoma is a clinical syndrome (Chapter 238) produced either by soil bacteria, the *Actinomycetes,* or by certain soil fungi that enter the tissues after trauma to the skin. The most common site of infection is the foot (Taralakshmi et al., 1977; Reddy et al., 1972), and the most common fungi causing mycetoma are *Petriellidium boydii, Madurella mycetomi* and *Acremonium* sp. Among the less common causes are *Madurella grisea, Leptosphaeria senegalensis, Pyenochaeta rosneroi, Phialophora jeanselmei,* and *Neotestudina rosatti*. Descriptions are provided here of *P. boydii, M. mycetomi,* and *Acremonium* sp. The properties of the others are summarized in Chapter 237.

Petriellidium boydii

This fungus, also known as *Allescheria boydii,* is the most common cause of mycetoma in the United States and Europe. It produces on any

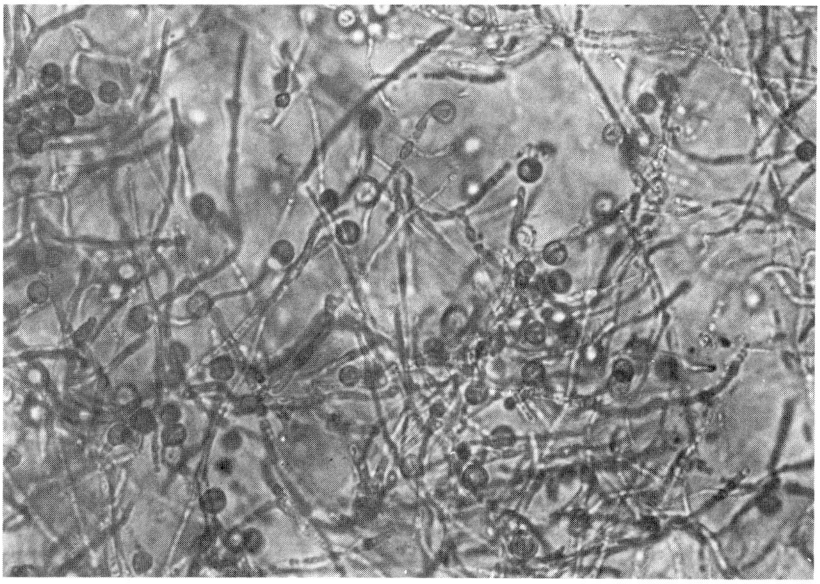

FIGURE 1. *Hyphae and asexual spores of* Petriellidium boydii. *The asexual spores are borne singly at the tips and sides of single conidiophores.*

medium a rapidly spreading, fluffy white colony that turns gray and resembles the fur of mice. The most distinctive feature is the manner in which the conidia are attached to the mycelium. These asexual spores are usually borne singly at the tips and sides of single conidiophores but may sometimes occur in small groups (Fig. 1). The spores are egg- or lemon-shaped, measure 4 to 8 by 5 to 15 μ, and have a brown color.

Sexual reproduction is homothallic, i.e., the sexual spores, or ascospores, result from the fusion of cells from the same mycelial mat or thallus. Groups of eight ascospores are contained in sacs known as asci, which measure 8 to 20 μ in diameter. After a couple of weeks, the asci become enclosed in a protective hollow globe known as an ascocarp or perithecium. The ascocarps are dark brown and have a thin wall composed of one layer of modified hyphal cells. Ascocarps are not seen in many cultures but when present can be found in the agar or in the mycelium near the edge of the colony. When they reach maturity, the thin wall of the ascocarp ruptures and releases the ascospores (Fig. 2).

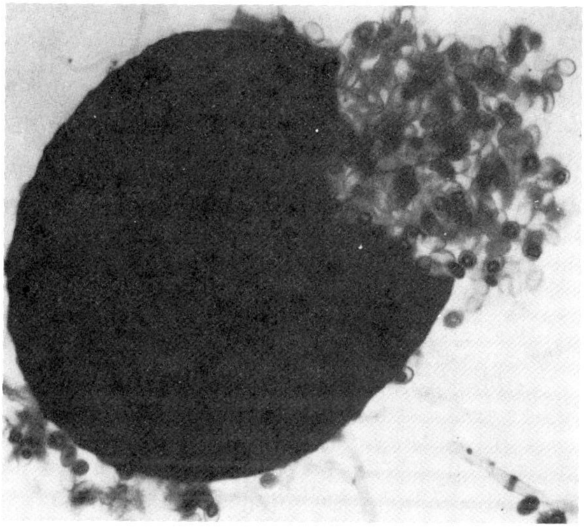

FIGURE 2. *Ascospores of* P. boydii *escaping from perithecium (ascoscarp). (From Parker, J. C., and Klintworth, G. F.: In Baker R. D. (ed.): Human Infections with Fungi, Actinomycetes, and Algae. New York, Springer-Verlag, 1971.)*

In infected fistulas, colonies of the fungus take the form of white or yellow grains. In wet preparations or stained smears of crushed granules, a mycelial mass is seen. The mycelium measures 2 to 4 μ in diameter and swells peripherally to 10 μ. The periodic acid-Schiff stain or the Gomori methenamine silver stain may sometimes bring out details of the organism that are not visible in wet mounts.

In addition to mycetomas, the fungus can sometimes be cultured from *Petriellidium* infections of the lung, cornea, ear, and meninges. The infection can be introduced or disseminated by surgery and other invasive therapeutic procedures. Thus, chronic meningitis due to *P. boydii* has followed spinal anesthesia (Benham and George, 1948), and disseminated infection to the spine and long bones has occurred after lung biopsy performed at the time of heart surgery. The fungus is usually but not always resistant to amphotericin B and may be sensitive (<1.0 μg/ml) to miconazole and ketoconazole.

M. mycetomi

This fungus is probably the most common cause of eumycotic mycetoma and is found in Asia, Africa, and South America. In India, for example, it is isolated at least ten times more often from mycetoma than *P. boydii* (Taralakshmi et al., 1977). *M. mycetomi* varies in colonial appearance from strain to strain, from case to case, and from one geographic area to another. In general, *M. mycetomi* grows well at 37° C, producing in several days a white mycelium that soon turns brown. The growth is compact and has a wrinkled leathery surface, which becomes covered with a powdery down. Diffusible brown pigments stain the reverse side of the agar.

The hyphae are 3 to 4 μ in width and segmented. Spores are rare but, when present, take two forms: pyriform conidia and phialospores. The pyriform conidia are carried on short conidiophores, and the phialospores are extruded from short bottle-shaped stalks known as phialides.

In tissues, the hyphae form round masses that issue from draining fistulas as hard, black granules measuring 0.5 to 5 mm. The hyphal mass in the dense type is held together by a brownish cement that is homogenously distributed throughout the granule. In the vesicular type of granule, the cement is located around its periphery.

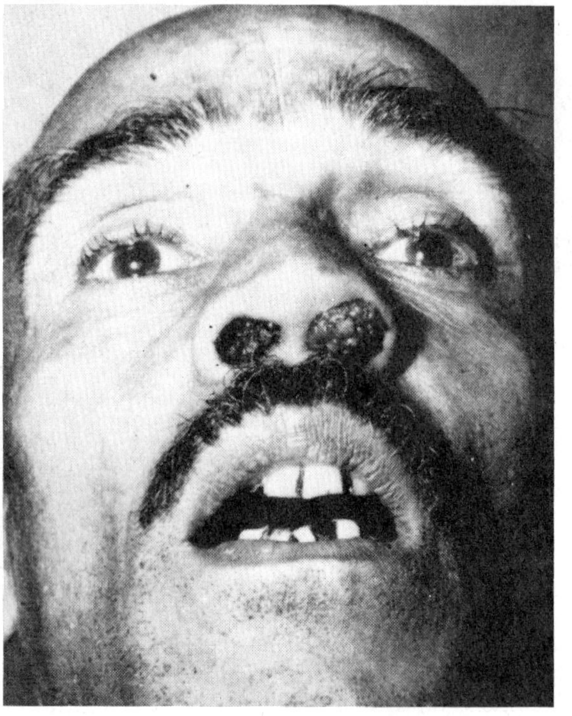

FIGURE 4. *Rhinosporidiosis with tumor-like masses in nares. (From Karunaratne, W. A. E.: Rhinosporidiosis in Man. London, The Athlone Press, 1964.)*

Acremonium Species (Cephalosporium Species)

The characteristic feature of this genus is the balls of conidia (Fig. 3). The conidia are borne one at a time at the tips of short erect conidiophores that branch from the aerial hyphae. Successive conidia push the others aside so that they form clusters or small balls held together by a sticky exudate. The conidia are nonpigmented and elongated.

The fungus grows rapidly at room temperature and produces cottony white colonies with mycelia measuring 2 to 3 μ in diameter. In the tissues, it produces white oval granules less than 1 mm in size. The granules contain in their center a dense matt of mycelium with eosinophilic clubs at the periphery. There is no cement. They resemble the granules of *A. boydii*.

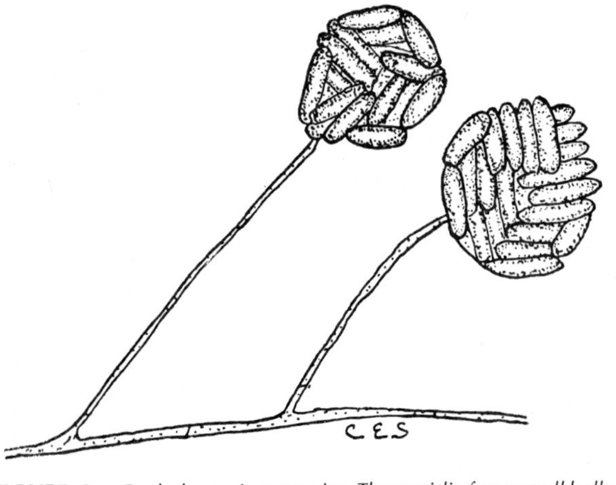

FIGURE 3. Cephalosporium *species. The conidia form small balls at the tips of the conidiophores. (Drawing by C. E. Skinner from Skinner, C. E., Emmons, C. W., and Tsachiya, H. M.: Molds, Yeasts, and Actinomycetes. New York, John Wiley and Sons, 1947.)*

RHINOSPORIDIUM

The fungus *Rhinosporidium seeberi* produces small tumor-like masses usually in the nose (Fig. 4) and nasopharynx, but sometimes in the eye. The endosporulating fungus is seen in the tissues but cannot be cultured on laboratory media. It is placed in the class Phycomycetes (Ashworth, 1923) and in the family Coccidioidaceae because characteristic giant sporangia develop in the tissues (Figs. 4 and 5). These thick-walled, endospore-filled sporangia resemble the spherules of *Coccidioides immitis*.

The nasal disease is probably acquired by bathing or diving in infected water. In India, where rhinosporidiosis reaches epidemic proportions, its rarity in women is attributed to social taboos that prohibit women from bathing in open places. The characteristic lesion is a vascularized papillomatous proliferation of the nasal or pharyngeal mucous membrane containing sporangia in vari-

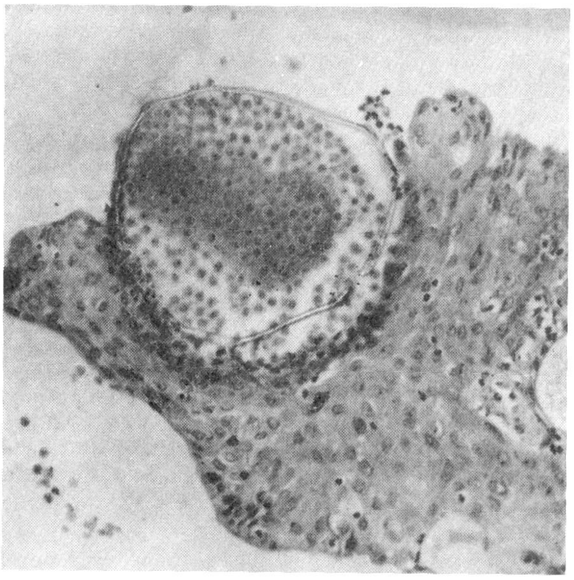

FIGURE 5. *A sporangium is embedded in the epithelium, which is attempting to encircle it; sporangium is filled with endospores. (From Karunaratne, W. A. E.: Rhinosporidiosis in Man. London, The Athlone Press, 1964.)*

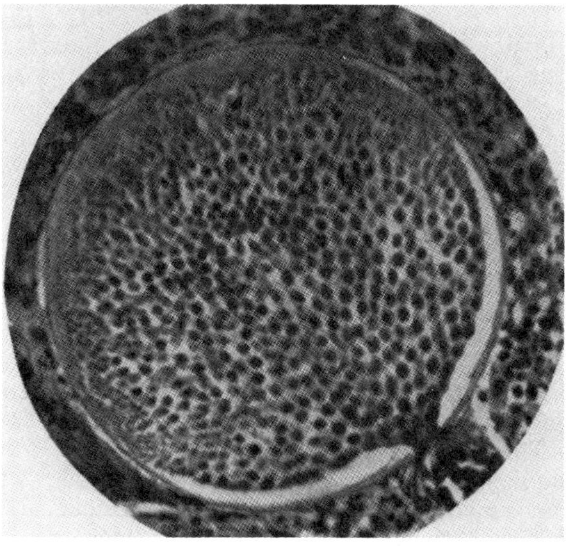

FIGURE 6. *Giant sporangium bursting and releasing endospores. (From Karunaratne, W. A. E.: Rhinosporidiosis in Man. London, The Athlone Press, 1964.)*

ous stages of maturity. Red blood cells, inflammatory pus cells, and extruded endospores fill the interstitial tissues. As sporangia enlarge, they compress the columnar epithelium of the pharynx and allow endospores to escape and reinoculate the adjacent tissue. Ocular infection occurs in dry, dusty regions and may be transmitted in dust storms. In Texas, half the cases of rhinosporidiosis are ocular.

The vascularized papillomas produce single or multiple pedunculated, fleshy, red masses in the nares or pharynx and cause rhinitis, epistaxis, and nasal obstruction. In exceptional cases, hoarseness may develop from laryngeal infection. The conjunctiva of the lids is predominantly affected in the ocular form, which tends to be unilateral and remarkably free of pain. Although usually confined to the eye or nose, the infection may infrequently be seen in the vagina, rectum, male urethra, and bronchi. Generalized rhinosporidiosis and visceral involvement have also been reported.

The lesions can be completely removed by surgery or electrocautery. Electrocautery is preferred, because surgery leaves open incisions in which spores can be implanted and produce recurrences.

References

Ashworth, J. H.: On *Rhinosporidium seeberi* (Wernicke, 1903) with special reference to its sporulation and affinities. Trans R Soc Edin 53:301, 1923.

Benham, R., and George, L.: *Allescheria boydii,* causative agent of a meningitis. J Invest Dermatol 10:99, 1948.

Karunaratne, W. A. E.: Rhinosporidiosis in Man. London, The Athlone Press, 1964.

Parker, J. C., and Klintworth, G. K.: Miscellaneous uncommon diseases attributed to fungi and actinomycetes. In Baker, R. D. (ed.): Human Infections with Fungi, Actinomycetes, and Algae. New York, Springer-Verlag, 1971, p. 958.

Reddy, C., Sundareschwar, B., Pattabni Rama Rao, A., and Reddy, S.: Mycetoma: Histopathologic diagnosis of causal agents in 50 cases. Indian J Med Sci 26:733, 1972.

Skinner, C. E., Emmons, C. W., and Tsuchiya, H. M.: Molds, Yeasts, and Actinomycetes. New York, John Wiley and Sons, 1947, p. 109.

Taralakshmi, V. V., Pankajalakshmi, V. V., Paramaswan, C. N., Shetty, B. M. V., and Subramanian, S.: Mycetomas in Madras. Sabouraudia 15:17, 1977.

7 PARASITES

78 THE PROTOZOA

Joseph H. Miller, M.S., Ph.D.

THE INTESTINAL AND ATRIAL PROTOZOA OF MAN

Parasitic Amebas of Man

Members of this group include free-living soil/water amebas that are opportunistic parasites of man and the amebic parasites of man's large intestine and mouth.

Three genera of soil amebas — *Naegleria*, *Hartmannella*, and *Acanthamoeba* — are thought to cause meningoencephalitis in man. In most cases, however, the organisms have been identified in fixed tissue specimens from necropsy and exact identification could have been confused by morphologic similarity. When culture was possible, *Naegleria fowleri* was isolated.

The intestinal amebas, on the other hand, are a well-defined group and consist of five species: (1) *Entamoeba histolytica*, (2) *Entamoeba coli*, (3) *Dientamoeba fragilis*, (4) *Endolimax nana*, and (5) *Iodamoeba bütschlii*. Another species, *Entamoeba gingivalis*, is found in the mouth. Only *Entamoeba histolytica* is an important pathogenic species, although *Dientamoeba fragilis* may cause chronic but mild intestinal symptoms. The comparative morphology of these six parasitic species in humans is presented in Figure 1.

Soil/Water Amebas

Morphology

Naegleria fowleri, the species most often associated with disease in man, is distributed worldwide in water and moist soil. In an ameboid stage, pseudopodia form eruptively. The endoplasm and ectoplasm are well defined, and there is one contractile vacuole. The nucleus has a large central karyosome. A biflagellate form develops in water. A cystic stage may also occur.

Epidemiology

Most cases have occurred in young adults after vigorous swimming, leading to the theory that these water-borne amebas may traverse the nasal mucosa, penetrate the cribriform plate, and multiply in the gray matter of the brain.

Laboratory Diagnosis

The cerebral spinal fluid should be examined for the presence of motile amebas. Stained preparations will show the characteristic large central karyosome. The organism may also be cultured. *Naegleria* grow out from a central inoculum of *Escherichia coli* on plates containing 1.5 per cent agar without added nutrient or sodium chloride.

Drug Susceptibility

Amphotericin B is effective against experimental *Naegleria* infections of mice and may be useful in humans. Sulfadiazine is very effective against experimental acanthamebiasis.

Immunity

Experimental animals seem to be immune to reinfection, but these studies are preliminary and incomplete. Most natural infections have occurred in young, healthy individuals.

Entamoeba histolytica

Morphology

E. histolytica may exist as a trophozoite, precyst, or cyst in feces (Fig. 2). The trophozoite is usually 15 to 25 μ in length. The clear ectoplasm is sharply demarcated from the granular endoplasm, which may contain ingested red cells. Its single nucleus appears ringlike when stained with iodine in fresh feces. Iron hematoxylin or other permanent staining techniques demonstrate a fine nuclear membrane encrusted with a layer of minute chromatin granules. A small punctiform karyosome is centrally located. The trophozoite is usually extremely active and exhibits progressive motility until it is cooled below the temperature of the human body. The precystic stage is a rounded, nonmotile trophozoite, devoid of inclusions. It is smaller than a trophozoite but usually larger than a cyst. The cysts are round-to-oval hyaline bodies, 5 to 20 μ in size, with smooth, refractile walls. Immature cysts are uninucleate with a large ringlike nucleus. As nuclear division occurs, individual nuclei decrease in size until the quadrinucleate mature cyst forms. In immature cysts glycogen is usually diffuse, and sausage-shaped chromatoid bodies may be seen. In unstained preparations chromatoid bodies are highly refractile, whereas in iron hematoxylin stains, they are uniformly dark. The chromatoid bodies, which contain ribonucleic acid, disappear as nuclear division occurs in the cyst.

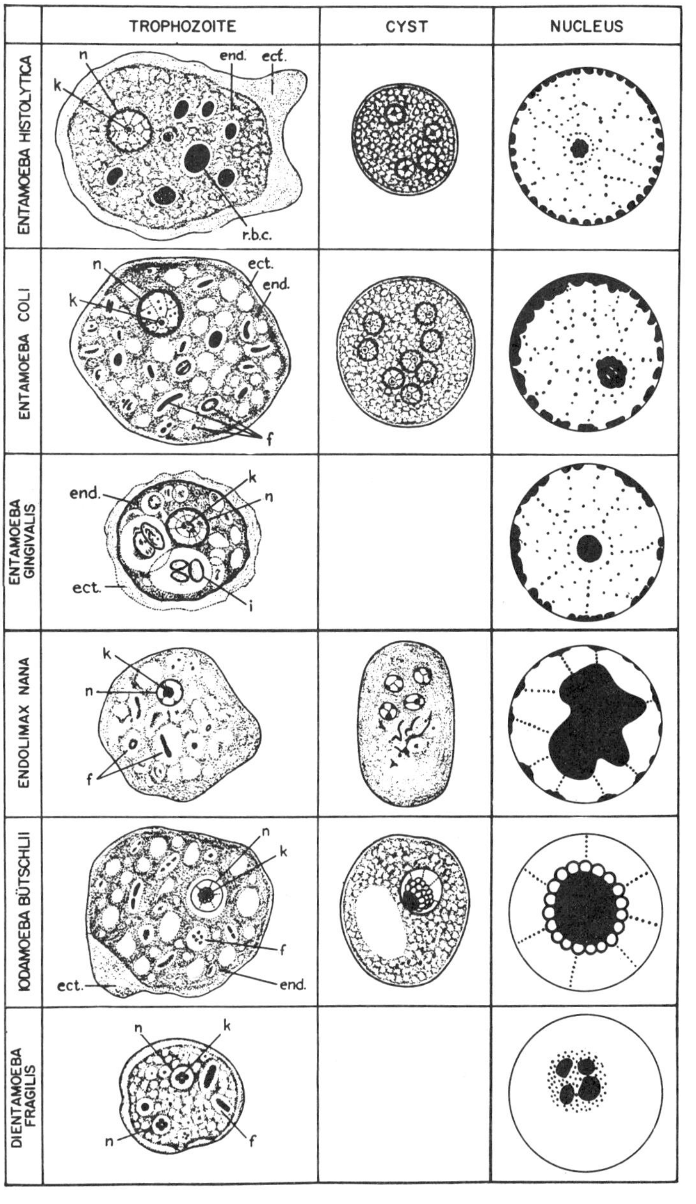

FIGURE 1. *Comparative morphology of the amebae of man and schematic representation of their nuclei as seen in iron-hematoxylin stains. ect., ectoplasm; end., endoplasm; f, food vacuoles; i, inclusions; k, karyosome; n, nucleus; r.b.c., red blood cells. (From Brown, H. W.: Basic Clinical Parasitology. Appleton-Century-Crofts, New York, 1975.)*

Life Cycle/Epidemiology

Cysts in feces are the infective stage. Persons are infected through food or drink contaminated by infective feces, flies, or the unwashed hands of food handlers. Cross-connections between sewage and water lines have been responsible for waterborne epidemics. Excystation occurs when the mature cyst reaches the lower small intestine. A four-nucleate ameba is released, which ultimately divides into eight small trophozoites. These amebas move downward and establish themselves at areas of stasis in the large intestine. They may feed in the lumen and invade the mucosa. Multiplication occurs by binary fission. Cysts are produced and passed in the feces. Trophozoites are seen in diarrheic and dysenteric stools, but they are not resistant to environmental conditions and do not spread the infection. Man is the principal host and source of infection, although natural infections of monkeys, dogs, hogs, and rats with amebas indistinguishable from *E. histolytica* have been reported.

Laboratory Diagnosis

The laboratory diagnosis rests on the identification of the parasite in the feces or tissues and on serologic studies.

Intestinal Amebiasis. Three naturally passed stools over a one-week period should be examined by the direct smear technique, using fresh saline

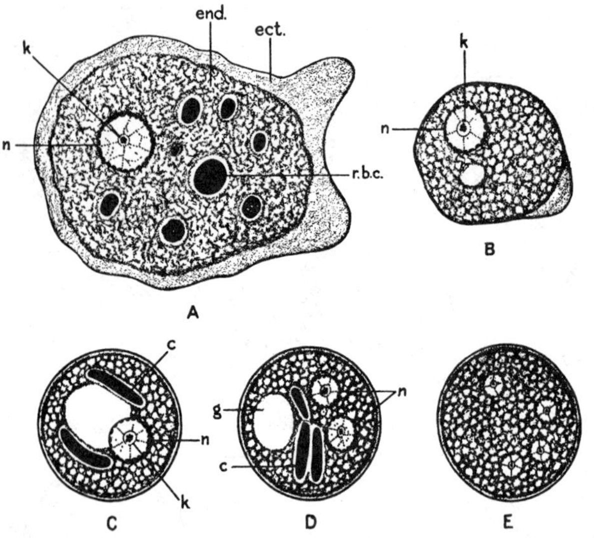

FIGURE 2. *Schematic representation of* Entamoeba histolytica. *A, trophozoite containing red blood cells; B, precystic ameba; C, immature uninucleate cyst; D, binucleate cyst; E, mature quadrinucleate cyst; c, chromatoid bodies; ect., ectoplasm; end., endoplasm; g, glycogen vacuole; k, karyosome; n, nucleus; r.b.c., red blood cells. (From Brown, H. W.: Basic Clinical Parasitology. Appleton-Century-Crofts, New York, 1975.)*

and iodine preparations. The best iodine solution is 1 per cent potassium iodide saturated with iodine crystals. Because cysts are released intermittently, three stool examinations increase the likelihood that the infection will be detected. In addition to direct examination, a permanent stain such as hematoxylin eosin and concentration technique (e.g., formalin-ether) should be done on each specimen. Finally, a saline-purged stool should also be examined and sigmoidoscopy performed to visualize lesions and to obtain aspirates for examination.

Trophozoites in aspirates and in diarrheic and purged stools may be distinguished from other intestinal amebas by their progressive motility in saline mounts (Table 1) or their characteristic nuclear morphology in iron hematoxylin or other permanent stain (Figs. 1 and 2). Formed stools are examined for typical cysts. The cysts of *E. histolytica* can be differentiated from the cysts of other intestinal amebas by the presence of typical chromatoid bodies and the type and number of nuclei (Fig. 3 and Table 2).

Several strains of *E. histolytica* have been recognized. *E. hartmanni*, a strain in which the cyst size is less than 10 μ, is considered by some to be a separate, nonpathogenic species.

Extraintestinal Amebiasis. All extraintestinal amebiasis originates from a primary intestinal infection, and evidence of such infection should be sought. Extraintestinal amebiasis without demonstrable concurrent intestinal infection is common, however. Serologic methods (complement fixation, immunodiffusion, and indirect hemagglutination) are sensitive and important diagnostic aids for confirming the clinical diagnosis of extraintestinal amebiasis. Many tests are available commercially in kit form. Microscopic examination of aspirates for motile trophozoites as well as cultural procedures are helpful.

Drug Susceptibility

Although *E. histolytica* is susceptible to many drugs, metronidazole has revolutionized the treatment of amebiasis. Metronidazole is now the treatment of choice for all forms of amebiasis, but it can be given only orally. Patients with severe intestinal amebiasis who are too ill to take oral medication may be successfully treated with emetine. The combination of emetine and chloroquine may also be needed for treatment of the occasional liver abscess caused by a metronidazole-resistant *E. histolytica*.

Immunity

Invasive amebiasis elicits the production of antibody to many different antigens of *E. histoly-*

TABLE 1. Key for the Differentiation of Amebic Trophozoites

I. Iodine-stained smear of fresh feces	
A. Presence of a ring-like nucleus	II A
B. Absence of a ring-like nucleus	II B
II. Unstained saline smear of fresh feces	
A. Trophozoites usually medium or large sized (except for small strains of *Entamoeba histolytica*)	
1. Progressive motility with cytoplasm flowing into pseudopod; motile forms have slug-like shape; may or may not contain ingested red cells; nucleus usually not visible without stain	*E. histolytica*
2. No progressive motility; cytoplasm does not flow into pseudopod; blunt pseudopodia extended and retracted; ring-like nucleus frequently visible unstained	*Entamoeba coli*
B. Trophozoites usually small or medium sized; ordinarily motility is sluggish	
1. Pseudopodia often blunt or round, resembling a yeast budding	*Endolimax nana*
2. Pseudopodia angular (triangular or tent-like), rectangular or lobulated; outline of non-motile trophozoites perfectly round or cyst-like	*Dientamoeba fragilis*
3. Pseudopodia blunt; resemble small *E. coli* trophozoites	*Iodamoeba bütschlii*

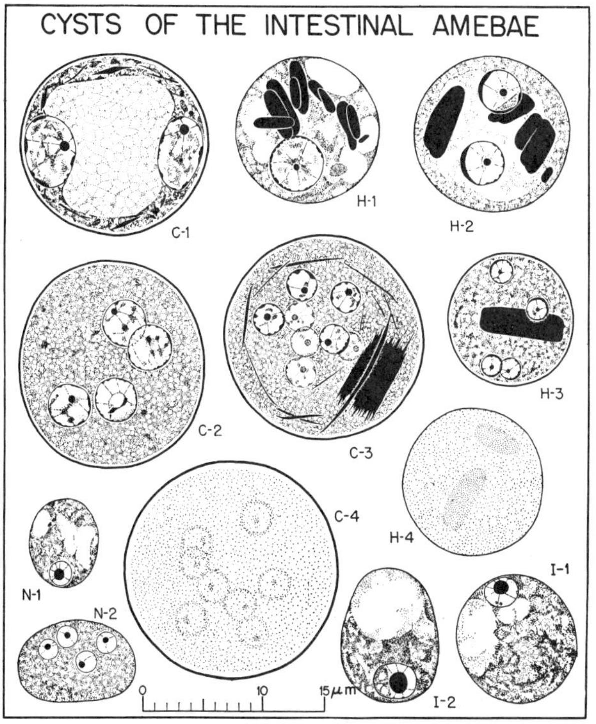

FIGURE 3. *C-1, Iron-hematoxylin-stained immature binucleate cyst of* Entamoeba coli. *C-2, Iron-hematoxylin-stained immature quadrinucleate cyst of* E. coli. *C-3, Iron-hematoxylin-stained mature cyst of* E. coli. *C-4, Unstained mature cyst of* E. coli. *H-1, Iron-hematoxylin-stained immature uninucleate cyst of* Entamoeba histolytica. *H-2, Iron-hematoxylin-stained immature binucleate cyst of* E. histolytica. *H-3, Iron-hematoxylin-stained mature cyst of* E. histolytica. *H-4, Unstained cyst of* E. histolytica *with chromatoid bodies. N-1, Iron-hematoxylin-stained uninucleate cyst of* Endolimax nana. *N-2, Iron-hematoxylin-stained mature cysts of* E. nana. *I-1, I-2, Iron-hematoxylin-stained mature cysts of* Iodamoeba bütschlii. *(From Hunter, G. W., Swartzwelder J. C., and Clyde, D. F.: Tropical Medicine. W. B. Saunders Co., Philadelphia, 1976.)*

tica. Neither these antibodies nor other immune mechanisms are known to prevent either extraintestinal invasion or reinfection.

Intestinal and Atrial Flagellates of Man

Three intestinal and two atrial flagellates are common in humans. The nonpathogenic species include *Chilomastix mesnili* and *Trichomonas hominis* of the intestine and *Trichomonas tenax* of the mouth. *Giardia lamblia* in the small intestine and *Trichomonas vaginalis* in the urogenital tract of humans are the pathogenic species. The species can be differentiated on the basis of habitat and by the morphologic comparison shown in Figure 4. *Blastocystis hominis*, a commensal yeast found in the intestine of humans, may be confused with the cysts of these flagellates.

Giardia lamblia

Morphology

Giardia lamblia is a worldwide parasite of the upper small intestine of humans. It has two stages in its life cycle: the trophozoite, which multiplies by binary fission, and the infective cyst, which is passed in the feces (Fig. 4). The trophozoite is a pear-shaped, bilaterally symmetrical flagellate, 12 to 15 μ long, with a broad rounded anterior surface and a tapering posterior extremity. The dorsal surface is convex. Most of the anterior ventral surface assumes a rigid bilobed concavity (ventral disk) in which a negative pressure is created by the beating of a pair of specialized ventral flagella. The organism is thus able to adhere to the brush border of the duodenal mucosa. Anteriorly, two large nuclei contain karyosomes. Two axostyles between the nuclei pass into the posterior extremity. Three pairs of trailing flagella are located dorsally. The ellipsoid cyst is 9 to 12 μ long and has a well-defined wall. The mature cyst contains four nuclei and the remnants of organelles found in the trophozoite. Because mitosis occurs in the cyst, two trophozoites are produced during excystation.

Life Cycle/Epidemiology

Infected persons, children more frequently than adults, pass cysts in the feces intermittently. Watery stools occasionally contain trophozoites, but they are not infective. Excystation of cysts ingested through fecal contamination of food or drink occurs in the small intestine and establishes the infection. Water-borne epidemics have

TABLE 2. Key for the Differentiation of Amebic Cysts

I. Unstained saline smear of fresh feces	
A. Presence of rod-like or sausage-shaped chromatoid bars (in some but not in all infections)	*Entamoeba histolytica*
B. Absence of chromatoid bars	II
II. Iodine-stained smear of fresh feces (and of formalin-ether concentrate)	
A. Presence of ring-like nuclei	1
1. Medium sized or small cyst with one to four nuclei	*E. histolytica*
2. Large cyst with five or more nuclei	*Entamoeba coli*
B. Absence of ring-like nuclei	3
3. Small, round, or oval cyst with nuclei appearing as vacuole-like areas	*Endolimax nana*
4. Medium sized cyst with large dark brown mass sharply delimited from cytoplasm	*Iodamoeba bütschlii*

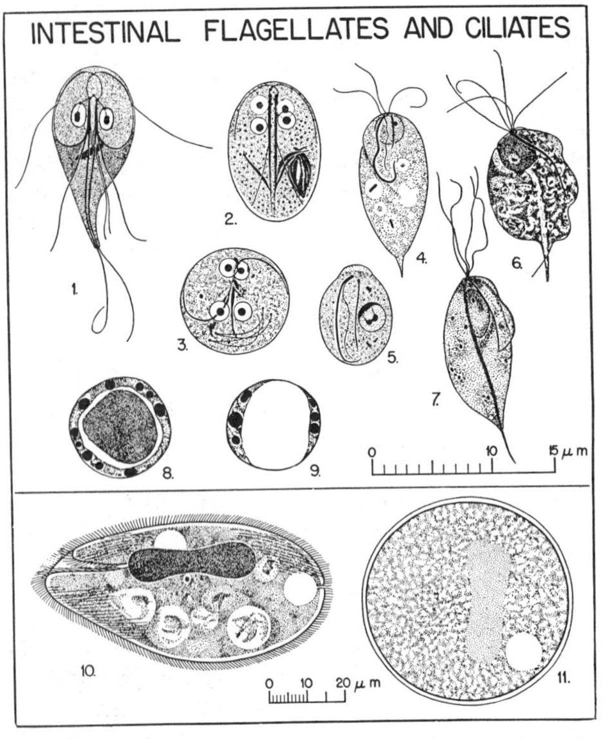

FIGURE 4. *1, Iron-hematoxylin-stained trophozoite of* Giardia lamblia. *2, Iron-hematoxylin-stained cyst of* G. lamblia. *3, Iron-hematoxylin-stained cyst of* G. lamblia, *end-view. 4, Iron-hematoxylin-stained trophozoite of* Chilomastix mesnili. *5, Iron-hematoxylin-stained cyst of* C. mesnili. *6, Iron-hematoxylin-stained trophozoite of* Trichomonas hominis. *7, Iron-hematoxylin-stained trophozoite of* Trichomonas vaginalis. *8, Iron-hematoxylin-stained* Blastocystis hominis. *9, Unstained* B. hominis. *10, Trophozoite of* Balantidium coli. *11, Unstained cyst of* B. coli. *(From Hunter, G. W., Swartzwelder, J. C., and Clyde, D. F.: Tropical Medicine. W. B. Saunders Co., Philadelphia, 1976.)*

occurred when drinking water was contaminated by feces containing cysts from human and possibly animal sources. The cysts are resistant to normal chlorination but seem to be destroyed by iodine compounds in concentrations recommended for water purification (Wolfe, 1975).

Laboratory Diagnosis

The diagnosis may be made by finding the characteristic cysts in iodine and saline mounts of fresh feces (Fig. 4). Motile trophozoites may also be distinguished by their characteristic morphology and jerky motility in watery stools. Permanently stained fecal smears are also used to distinguish *G. lamblia*. Concentration techniques should always be used. The string method (Entero-Test) of recovering trophozoites from duodenal samples is simple and useful when stool findings are negative and symptoms persist.

Immunity

Little immunologic information is available concerning *G. lamblia*. Reinfection occurs, and immunosuppressive therapy can cause an asymptomatic carrier state to change rapidly into active infection. Hypogammaglobulinemia predisposes to acute infection.

Drug Susceptibility

G. lamblia is sensitive to atabrine, metronidazole, and furazolidone.

Trichomonas vaginalis

Morphology

T. vaginalis is a pear-shaped flagellate, 10 to 30 μ long, with four flagella directed anteriorly and one directed posteriorly to form the margin of an undulating membrane (Fig. 4). The membrane extends half the length of the trophozoite. An axostyle originates anteriorly, curves about the nucleus, and terminates posteriorly as a tail-like appendage. Multiplication is by binary fission. There is no cyst stage.

Epidemiology/Life Cycle

T. vaginalis inhabits the vagina and the male urethra, epididymis, and prostate. The incidence of infection is higher in females than in males. Coitus is probably the most frequent mode of infection, but the exchange between women of toilet articles contaminated with vaginal discharge is also a common cause of infection.

Laboratory Diagnosis

The examination of fresh infected vaginal discharge in a drop of saline will disclose motile *T. vaginalis*. The erratic jerky motility of the organism is characteristic. The rippling wave-like motion of the undulating membrane can be visualized clearly when specimens begin to dry. Permanently stained specimens may also be prepared and examined for the undulating membrane and flagella.

Examination of prostatic secretions obtained by prostate massage is the best method for diagnosing infections of men. The secretion should be examined for motile trophozoites with undulating membranes. Smaller numbers of trophozoites may also be found in the urine.

For practical purposes, *Trichomonas* spp may be differentiated on the basis of the source of infected material: *T. tenax* is found in material from the oral cavity, *T. hominis* from feces, and *T. vaginalis* from vaginal discharge, prostatic fluid, or urine.

Drug Susceptibility

Oral metronidazole is highly effective in eliminating the infection in either sex. No other systemic drug is effective.

A Parasitic Ciliate of Humans: *Balantidium coli*

Morphology

B. coli is the only ciliated pathogen of man (Fig. 4). The large ovoid trophozoite, measuring 50 to 70 μ by 30 to 60 μ, is enclosed by a pellicle covered with longitudinal spiral rows of cilia. The beating cilia account for the smooth gliding motility of the organism in fresh fecal preparations. At the anterior end is a funnel-shaped peristome lined by long cilia. A large kidney-shaped macronucleus with a small rounded micronucleus lying in its concavity is usually centrally located. There are two contractile vacuoles. At the posterior end is a cytopyge for the discharge of solid waste material. The rounded trophozoites secrete a double cyst wall. Cysts measure 50 to 65 μ. The macronucleus, contractile vacuoles, and cilia are visible within the cyst.

Epidemiology/Life Cycle

B. coli trophozoites inhabit the lower ileum and large intestine, where they penetrate the mucosa and multiply by binary fission. Conjugation has been observed in this species. Infective cysts are passed in the feces. Trophozoites do not survive the extracorporeal environment. Cysts ingested in fecally contaminated food or drink produce a single trophozoite. Cysts do not multiply.

Morphologically identical forms of *B. coli* in monkeys, hogs, guinea pigs, and rats do not seem to play a major role in human infections. Instead, infection seems to result primarily from man-to-man transmission via fecal contamination.

Laboratory Diagnosis

A number of fecal specimens should be examined because the number of parasites may vary from specimen to specimen. Large, motile trophozoites with gliding motility may be found in diarrheic stools. Less frequently, typical large cysts are found in formed stools (Fig. 4). Sigmoidoscopy with aspiration of visible lesions may be helpful.

Drug Susceptibility

B. coli is most sensitive to tetracycline or Diodoquin.

BLOOD AND TISSUE PROTOZOA OF HUMANS

Parasitic Trypanosomes of Humans

Three species of hemoflagellates of the genus *Trypanosoma*, family Trypanosomatidae, are pathogenic for man: *Trypanosoma gambiense* and *T. rhodesiense* in Africa and *T. cruzi* in America. The life cycle of each is carried out partly in humans or another mammal and partly in an insect intermediate host, in which cyclic development occurs. *T. gambiense* and *T. rhodesiense* are extracellular trypomastigote parasites of the blood, lymph, or cerebrospinal fluid, whereas *T. cruzi* occurs in trypomastigote forms in the blood and amastigote forms in the tissue.

Trypanosoma gambiense

Morphology

T. gambiense trypomastigotes in the blood of humans are polymorphic. They may be either typical slender flagellates with a pointed anterior end, a blunt posterior extremity, and a long flagellum, or short stumpy forms with or without a short flagellum. The flagellum of the slender form, 8 to 30 μ in length, projects from the anterior end after passing along the edge of the undulating membrane (Fig. 5). A large, oval nucleus is centrally located. A kinetoplast is located posteriorly at the base of the flagellum. The cytoplasm contains minute refractile volutin granules. Motility is wavy and spiraling.

Life Cycle/Epidemiology

Although *T. gambiense* and *T. brucei* are morphologically identical, they are very different organisms. *T. brucei* is a parasite of wild animals that usually does not infect man, whereas *T. gambiense* is primarily a human parasite, without an important reservoir in wild or domestic animals. *T. gambiense* is transmitted from man to man by tsetse flies of the *Glossina palpalis* group. It is limited by the habitat of its vectors to forest belts, especially along rivers in tropical West and Central Africa. Trypanosomes that have been ingested by the fly during a blood meal undergo reproduction in the midgut. Long slender forms finally appear and move anteriorly through the proventriculus to the salivary glands and ducts, where epimastigote forms are produced. Eventually, short stumpy metacyclic trypanosomes are derived and pass through the channel in the hypopharynx into the bite wound when the fly feeds. During epidemics, a tsetse fly may transmit trypanosomes directly ("flying needle") from an

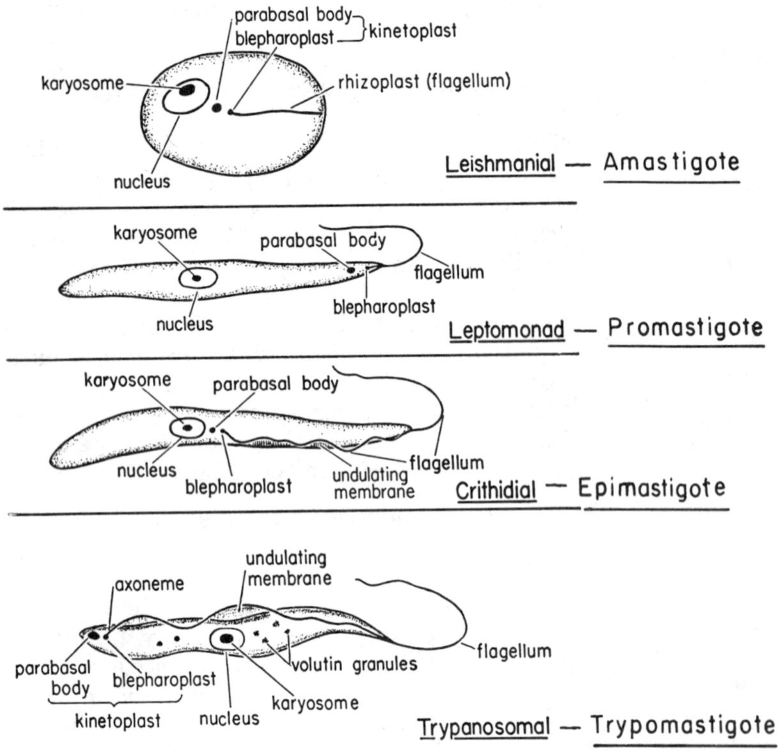

FIGURE 5. *The developmental stages in the family Trypanosomidae. (From Hunter, G. W., Swartzwelder, J. C., and Clyde, D. F., Tropical Medicine. W. B. Saunders Co., Philadelphia, 1976.)*

infected person, if the blood meal is interrupted. Congenital infections and transmission during coitus are rare.

Laboratory Diagnosis

Diagnosis is made by finding trypanosomes in the blood, lymph node aspirates, or bone marrow in early disease, and in centrifuged cerebrospinal fluid in late disease. Preparations are examined wet for motile organisms or stained with Romanowsky stains. The hematocrit tube-centrifuge technique is valuable in the diagnosis of *T. gambiense* infections (Lumsden, 1977). Small laboratory animals are less susceptible to *T. gambiense* than to the Rhodesian form. Culture of *T. gambiense* is difficult and not useful in diagnosis. Presumptive diagnosis is possible by monitoring serum IgM levels, which are greatly elevated during trypanosomiasis. Serologic techniques include the complement fixation and indirect immunofluorescent tests.

Drug Susceptibility

T. gambiense is sensitive to suramin sodium, pentamidine isethionate, and lomidine methanesulfonate. These drugs are effective in the treatment of the early stages of infection before the central nervous system is involved. In the late stages, synthetic arsenicals or antimony analogs of the melarsen group, which cross the blood-brain barrier, are used for treatment. Of these drugs, *T. gambiense* is most sensitive to tryparsamide, melarsoprol (Mel B), and MSbB. *T. gambiense* may become resistant to these compounds, which are frequently used in combination with suramin.

Immunity

T. gambiense can escape the immune response of the host by antigenic variation. The variant antigens are glycoproteins, which are the major components of the surface coat (glycocalyx) of the trypanosome. This phenomenon has delayed immunologic approaches to prevention (Doyle, 1977).

Trypanosoma rhodesiense

Morphology

The morphology of this species is identical with that of *T. gambiense* (Fig. 5).

Life Cycle/Epidemiology

T. rhodesiense is confined to East Africa in savannah areas in which cattle are raised. Its incidence and the occurrence of epidemics are lower than those for *T. gambiense*. *T. rhodesiense* has a large reservoir in wild animals, and the vectors *Glossina morsitans*, *G. pallidipes*, and *G. swynnertoni* feed primarily on ungulates and only infrequently on man. The life cycle is the same as that of *T. gambiense*.

Laboratory Diagnosis

The same procedures for *T. gambiense* infections are used for *T. rhodesiense* infections. Parasites are usually more numerous in the blood during Rhodesian trypanosomiasis, and small laboratory rodents are easily infected by intraperitoneal inoculation of whole blood that contains trypanosomes.

Drug Susceptibility

The same drugs used to treat *T. gambiense* infections are used against *T. rhodesiense*.

Immunity

Immunologic response is the same as that of *T. gambiense*.

Trypanosoma cruzi

Morphology

T. cruzi is a pleomorphic hemoflagellate that infects the blood and tissue of mammals in most countries of the Western Hemisphere. Man is less frequently infected than smaller mammals. The trypomastigote in the blood is about 20 μ long. In some cases, stumpy forms about 15 μ long may be observed. In stained blood smears the organism is usually U- or S-shaped and has a large kinetoplast near its posterior end. A flagellum arises in this area, passes through the margin of the undulating membrane, and proceeds free anteriorly for about one third of the body length. The organism does not multiply in the trypomastigote form. The amastigote form is found in various tissue cells. It is a small oval body 3 to 5 μ in diameter, containing a nucleus and a small rodlike kinetoplast (Fig. 5). In this stage the parasite divides rapidly and forms many amastigotes.

Life Cycle/Epidemiology

T. cruzi is transmitted to small mammals by various species of blood-feeding triatomids that serve as intermediate hosts. Man becomes infected when susceptible species of triatomids become domesticated. Human infections are common in South America and rare in the United States, because domestic triatomes usually live in thatched roofs or cracks in adobe walls in the former location. Trypomastigotes are ingested by triatomids during a blood meal and undergo cyclic development in the gut of the insect. Amastigotes form in the foregut, epimastigotes in the midgut, and infective trypomastigotes in the hindgut. Trypomastigote forms are deposited on the skin with the feces during feeding. The feces is rubbed either into the bite wound or onto the conjunctiva, because the insect usually feeds on the human face while a person is sleeping. Trypomastigotes invade local reticuloendothelial cells and multiply rapidly as amastigotes. Trypomastigote forms are produced and appear in the blood. They circulate to various tissues, where amastigote forms are again produced intracellularly. After an acute phase in man, the parasitemia subsides, but amastigotes may continue to multiply and may lead to serious consequences years later in the chronic stage of the infection.

Nondomestic species of triatomids live in the nests of various mammals and carry on the cycle in nature by fecal contamination or by being eaten by certain mammals. In the latter case, infective stages may be passed across the mucous membranes of the mouth. Various modifications of the life cycle are presented in Figure 6.

Significant human disease is found only in areas in which the triatomid is domesticated and defecates while feeding. Dogs and cats may be important reservoirs in endemic areas. Asymptomatic infections with *T. cruzi* occur and have been documented primarily in geographic areas in which triatomid species do not live in close association with humans. Nonpathogenic strains of *T. cruzi* may be responsible for some of these cases.

Congenital transmission has been reported.

Laboratory Diagnosis

The laboratory diagnosis depends on finding the organism in blood or tissue. Thick and thin blood films prepared during the acute stage, stained and unstained, may exhibit trypanoform stages. Parasites are never plentiful, however, because they do not multiply in peripheral blood. Concentration of large volumes of blood is helpful. Biopsies may reveal amastigote forms. Blood and bone marrow aspirates cultivated on diphasic blood agar media yield epimastigote forms in about 30 per cent of patients with positive serology. In xenodiagnosis, parasite-free triatomids are allowed to feed on the patient. Two weeks later, the intestinal contents of the triatomids are examined for hemoflagellates. This method of detecting the acute and chronic stages of the disease is not used outside of Latin America. Finally, sensitive serologic techniques (complement fixation, indirect hemagglutination, and indirect fluorescent antibody) are available for establishing a presumptive diagnosis. Many are available commercially as kits (Kagan and Norman, 1974).

T. rangeli, a nonpathogenic species of trypanosome, is found in Central and South America in humans and other mammals. It can be differentiated from *T. cruzi* by its smaller kinetoplast and

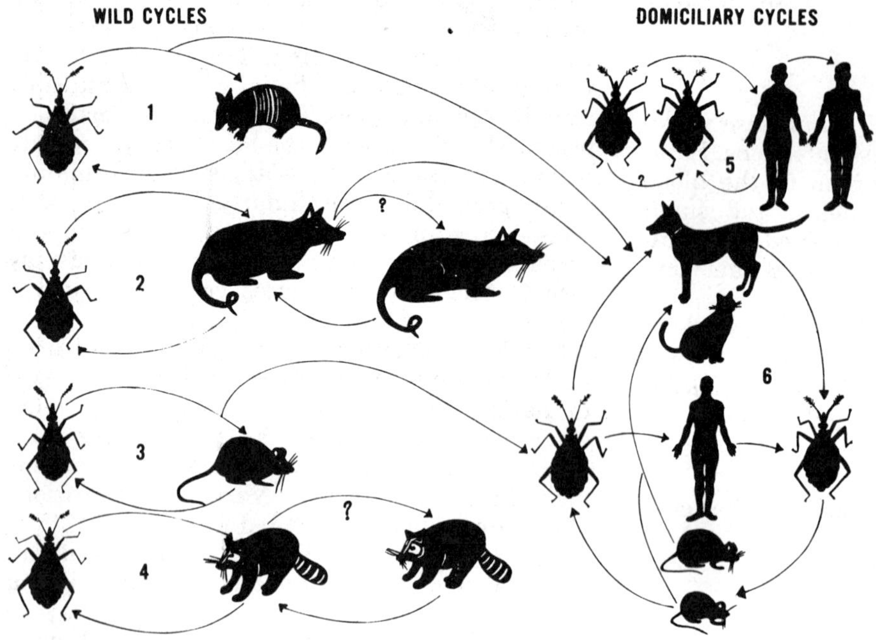

FIGURE 6. *Wild and domestic cycles of Chagas' disease – modalities and relationships. 1,* Panstrongylus geniculatus *is often associated with armadillos: adults enter houses, attracted by lights. 2, Opossums (*Didelphis *spp.) are associated with several triatomines and both marsupials and insects visit human dwellings. 3, Rats and other rodents are associated with insects that also fly to houses. 4, Raccoons may be associated with triatomines but, as with opossums, they may transmit* T. cruzi *among themselves without participation of the insect. 5, In the absence of domestic animals there is a cycle between the insect and man. Transmission from one insect to another has been suggested, and transmission from man to man is a fact, both transplacentally and transfusionally. 6, Other domestic animals might participate in the cycle and some of them may become infected by eating small rodents. (From Zeledón, R.* In *Trypanosomiasis and Leishmaniasis with special reference to Chagas' disease. Ciba Foundation Symposium 20, new series, 1974. Published by Elsevier. Excerpta Medica. North-Holland, Associated Scientific Publishers, Amsterdam.)*

by the presence of dividing forms in stained blood smears.

Drug Susceptibility

No known drug will cure *T. cruzi* infections. *T. cruzi* is probably sensitive to the delta-amine quinolines that eradicate the majority of blood forms during acute disease. No known agent is effective against the intracellular leishmanial stage.

Immunity

T. cruzi is strongly immunogenic. The highest titers of antibodies are found during the acute stage. When the parasitemia drops, the titers also decrease. Cell-mediated immunity plays an important role in resistance to *T. cruzi* but may also be responsible for the tissue damage in chronic infections.

Antigenic variation does not occur with *T. cruzi*, and there is cross-immunity between heterologous strains. Vaccination confers only partial protection.

Parasitic Leishmania of Man

The genus *Leishmania*, family Trypanosomatidae, includes four species —*Leishmania donovani*, *L. tropica*, *L. mexicana*, and *L. braziliensis* —all of which occur as intracellular amastigotes in man and promastigotes in phlebotomine sandfly intermediate hosts and in cultures (Fig. 5). All species are morphologically identical, and serologic separation has not been entirely satisfactory. They have been separated primarily by their tendency to cause visceral, cutaneous, or mucocutaneous involvement in humans.

Leishmania donovani

Morphology

In this species the organism is located in the viscera, in the reticuloendothelial cells of the spleen, liver, bone marrow, and lymph glands. *L. donovani* is a small, round, or ovoid intracellular amastigote, 2 to 5 μ in diameter. It contains a rodlike kinetoplast and a peripheral nucleus.

Life Cycle/Epidemiology

Various species of phlebotomine sandflies, depending on locality, serve as intermediate hosts. Human cells containing amastigote forms are ingested by the fly. The amastigotes are liberated in the insect's gut, transformed into promastigotes, multiply by binary fission, and move to an anterior station in the pharynx of the insect. When the insect bites, these forms pass out of the insect, enter the host's local reticuloendothelial cells, and are then distributed to the viscera.

Human infections occur in India, the Mediterranean countries, Africa, China, Middle Asia, and

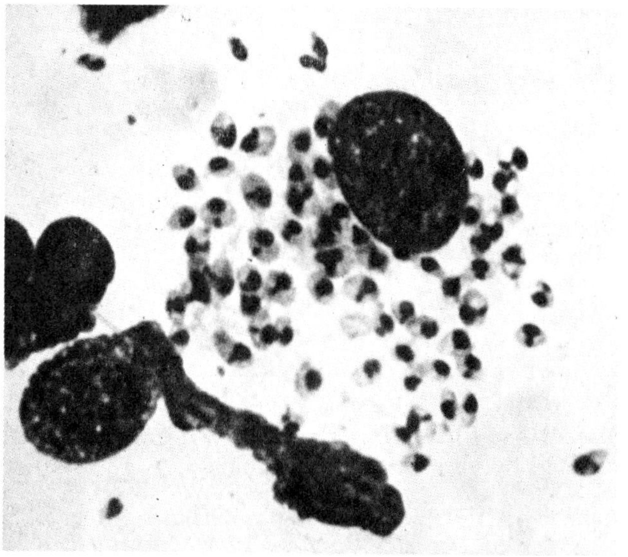

FIGURE 7. Leishmania donovani *in stained smear from spleen puncture. (From Hunter, G. W., Swartzwelder, J. C., and Clyde, D. F., Tropical Medicine. W. B. Saunders Co., Philadelphia, 1976.)*

Central and South America. In certain geographic areas man-to-man transmission is the sole method of spread of disease. In other areas, jackals, foxes, and domestic dogs may serve as reservoirs. Animals may acquire leishmaniasis by eating infected carcasses, thus perpetuating an animal reservoir.

Laboratory Diagnosis

The definitive diagnosis depends on the demonstration of *L. donovani* amastigotes in stained smears from material collected by splenic, liver, or sternal puncture (Fig. 7). Culture of the material on NNN medium is also valuable. Blood culture and examination of stained films of the buffy coat for macrophages containing amastigote forms will yield positive results in most cases. The complement fixation test is usually positive. The formol-gel test is a useful screening technique. When 1 drop of commercial formalin is added to 1 milliliter of serum from a patient with kala-azar, the serum solidifies and becomes opaque within 3 to 30 minutes.

Drug Susceptibility

L. donovani is sensitive to two groups of compounds: (1) pentavalent antimony derivatives, including sodium antimony gluconate, neostibosan, and urea stibamine; and (2) certain aromatic diamidines, including hydroxystilbamidine and pentamidine isethionate.

Immunity

Leishmania organisms are protected from antibodies by their location inside macrophages. Humans have been successfully vaccinated with promastigotes from a rodent strain.

Leishmania tropica

Morphology

L. tropica invades mononuclear and polymorphonuclear leukocytes and epithelial cells of the skin. The intracellular amastigote forms are identical to those of *L. donovani*.

Life Cycle/Epidemiology

Infection is prevalent in large areas of Asia, the countries around the Mediterranean, and the north and west coasts of Africa. Transmission is the same as for *L. donovani*. Two subspecies have been recognized: *L. tropica major*, which causes "moist" or rural leishmaniasis, and *L. tropica minor*, the cause of "dry" or urban leishmaniasis. Gerbils are a primary extrahuman reservoir of rural leishmaniasis, and dogs are a primary extrahuman reservoir of urban leishmaniasis.

Laboratory Diagnosis

Diagnosis is made by demonstrating *L. tropica* amastigotes in cells of stained smears obtained from curettage or needle aspiration of the indurated margin of the dermal lesion. Culture of aspirates on NNN medium, with bacterial growth controlled, yields promastigotes. The leishmanial skin test becomes positive early in the infection and remains positive for life.

Drug Susceptibility

Neostibosan is the most effective drug.

Immunity

Infection results in life-long immunity to homologous strains. Therefore, children are often deliberately inoculated intradermally with material from lesions.

Leishmania braziliensis

Morphology

Like *L. tropica*, *L. braziliensis* invades cells of the skin but may also spread to the mucous membranes. The amastigote stage is identical with that of *L. donovani*.

Life Cycle/Epidemiology

Infections with *L. braziliensis* are widely distributed as a zoonosis in the forests of Central and South America, with the exception of Chile and Argentina. Rodents serve as reservoirs. Man is infected by phlebotomine sandflies.

Laboratory Diagnosis

The demonstration of *T. braziliensis* amastigotes from skin lesions (see *L. tropica*) is comparatively easy in early infections because the parasites are numerous. In older infections, culture on NNN medium may be successful. The Montenegro skin test is specific and sensitive, and remains positive for life.

Drug Susceptibility

L. braziliensis is sensitive to pentavalent antimonials, cycloguanil pamoate, and amphotericin B.

Immunity

Infection produces only strain specific immunity. Experimental work suggests that immunity may be induced by vaccination with epimastigotes from culture.

Leishmania mexicana

Morphology

L. mexicana amastigotes invade cells of the skin and are identical with those of *L. donovani*.

Life Cycle/Epidemiology

Infections occur in Mexico, Belize, and northern Guatemala. However, the organism is believed to be present throughout the forests of Central and northern South America. Like *L. braziliensis*, it is also a forest zoonosis maintained in wild rodents with phlebotomine sandflies as vectors.

Laboratory Diagnosis

Amastigote forms of *L. mexicana* are easy to find in smears of early lesions. In late infections they are difficult to find and isolate. A skin test using refined antigen is helpful (Zeledón and Ponce, 1974).

Drug Susceptibility

L. mexicana is sensitive to pentavalent antimonials and cycloguanil pamoate.

Immunity

Immunity against reinfection occurs with *L. mexicana*, but little cross-immunity occurs between strains.

Parasites of Uncertain Taxonomic Status

Toxoplasma gondii

Morphology

T. gondii is an obligate intracellular sporozoan parasite of man and animals with worldwide distribution. Recent studies of its life cycle suggest a taxonomic relationship to the Eimeriidae, order Coccidia. Tachyzoites, the rapidly multiplying forms in acute infection, are cresent shaped, 6 to 7 μ long by 2 to 4 μ wide, and contain an oval nucleus (Fig. 8). Electron microscopy has revealed many organelles in common with the coccidia. Intracellular multiplication occurs by endodyogeny, with two daughter cells forming within the pellicle of the parent. Cysts formed in chronic infections are 200 to 1000 μ in diameter, oval to round, and contain numerous bradyzoites, the slowly multiplying forms. Oocysts containing eight sporozoites are passed in the feces of infected felines, primarily the domestic cat.

Life Cycle/Epidemiology

Many animals, both wild and domestic, are infected with *T. gondii*. Man becomes infected primarily by eating poorly cooked or raw meat containing cysts. Contamination of food or drink with oocysts from cat feces also occurs but probably plays only a minor role in human transmission. The ingestion of cysts or oocysts by humans gives rise to a schizogonic or proliferative phase with rapidly multiplying tachyzoites in many tissues. Eventually, as a resting stage, cysts containing many bradyzoites develop. A schizogonic and gametogenic cycle occurs in the intestinal epithelium of felines with production of oocysts. In the acute stage in humans, the organisms may be transferred transplacentally. The life cycle is presented in Figure 9.

Laboratory Diagnosis

The demonstration of cysts via biopsy specimens does not establish a causal relationship for clinical illness because cysts are found in both acute and chronic infections. Demonstration of organisms in brain impression smears of a new-

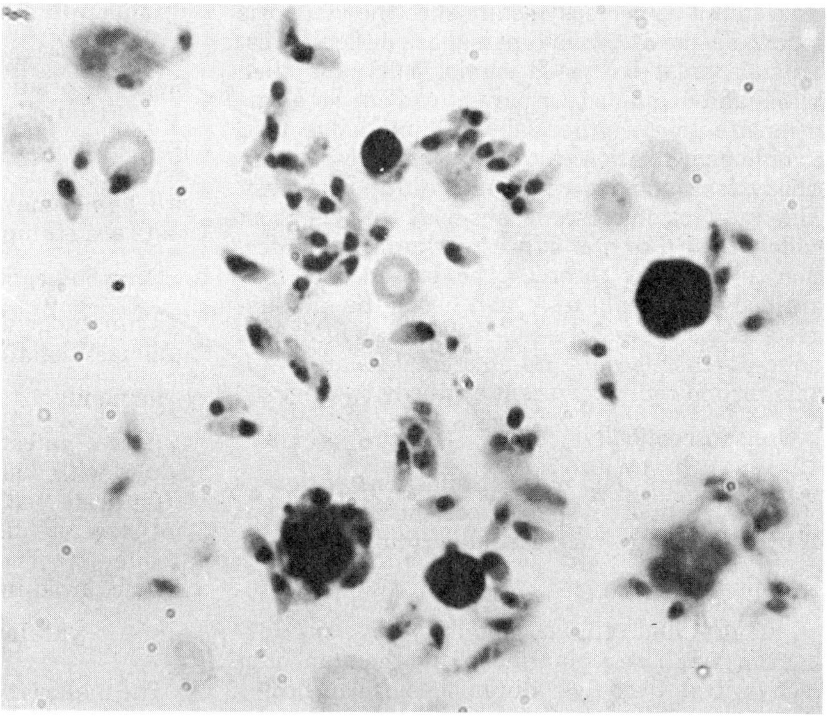

FIGURE 8. *Mouse peritoneal fluid demonstrating tachyzoites freed from cells (1000 ×). (From Hunter, G. W., Swartzwelder, J. C., and Clyde, D. F.: Tropical Medicine. W. B. Saunders Co., Philadelphia, 1976.)*

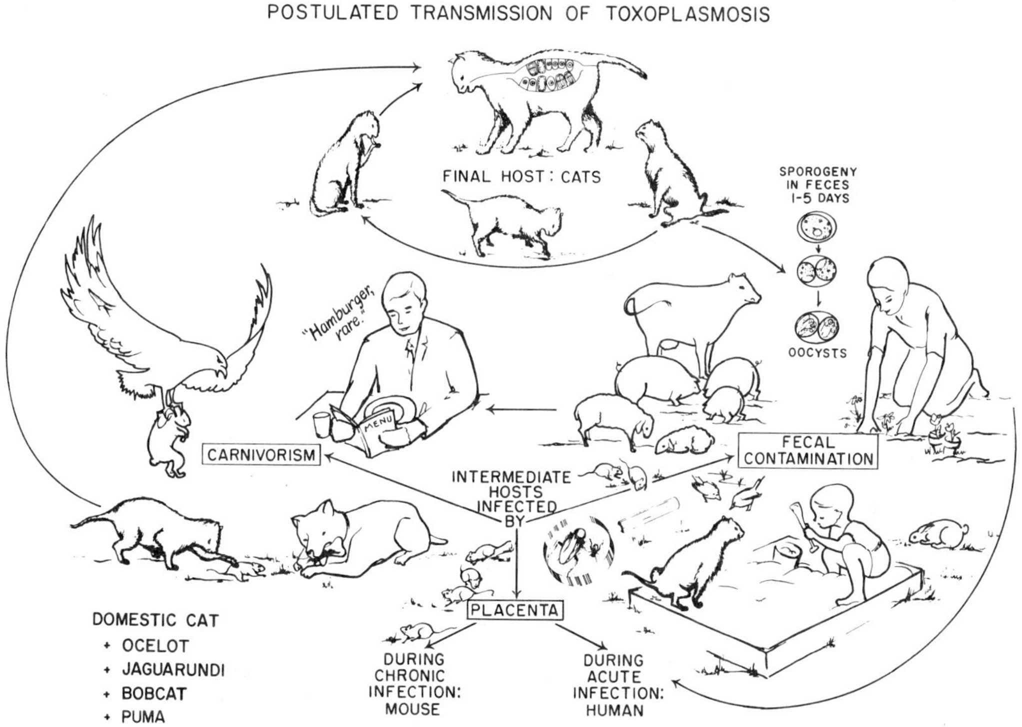

FIGURE 9. *Postulated life cycle and transmission of* Toxoplasma. *Cats and certain other felines are shown as final hosts, and other animals and humans as intermediate hosts. Flies and cockroaches can serve as transport hosts. At right, infection with oocysts is shown. At left, transmission by carnivorism is indicated. Below, the transplacental route of transmission is indicated. (Modified slightly from Frenkel, J. K.: Toxoplasmosis.* In *Marcial-Rojas, R. A. (ed.): Pathology of Protozoal and Helminthic Diseases. Williams & Wilkins, Baltimore, 1971.)*

born infant at necropsy establishes the diagnosis of toxoplasmosis. Animal passage is difficult. Test animals must be tested serologically and then killed and examined for parasites. Four serologic tests are in use: the Sabin-Feldman dye test, complement fixation test, indirect fluorescent antibody test, and indirect hemagglutination test. The indirect fluorescent antibody test is most widely used. The titer suggests whether the infection is acute or chronic. The presence of IgM antibody in a child less than 6 months old suggests acute disease because maternal IgM does not cross the placenta. Commercial kits are available for indirect fluorescent antibody testing.

Drug Susceptibility

Experimental and natural infections with toxoplasma are most responsive to the combination of pyrimethamine and trisulfapyridine.

Immunity

Humoral and cellular immunity develops during the lymphoreticular phase, and the number of tachyzoites rapidly diminishes. Antibodies against several antigens of toxoplasma can be demonstrated, and immune mononuclear cells no longer support the intracellular multiplication of *Toxoplasma* organisms. The development of immunity coincides with the transformation of toxoplasmosis into the chronic, cyst-forming stage. Cysts are isolated from the immune processes, and reactivation can occur when immunosuppression is induced by either natural or artificial means.

Pneumocystis carinii

Morphology

P. carinii is believed by some to be a coccidian. Its exact taxonomic status is unknown. The most easily recognizable stage is the cyst, which is 5 to 12 μ in diameter and contains eight ovoid or crescent-shaped parasites (sporozoites?) with eccentric nuclei. Cysts are found in a honeycombed exudate in the alveoli.

Life Cycle/Epidemiology

The life cycle of *P. carinii* is unknown. The infection was first reported in endemic form from Europe, but sporadic cases have been reported worldwide. *P. carinii* is widespread in animals, both wild and domestic, but host specificity seems high, and transmission from animals to humans seems unlikely. Aerosol dispersion of respiratory tract secretions may be involved in epidemics. Sporadic cases are thought to result from latent infections that become activated by immunosuppression.

Laboratory Diagnosis

Silver or Giemsa stains of tissue from open lung biopsy or needle aspirates have been most productive (Fig. 10). Sputum, tracheal smears, and hypopharyngeal material do not routinely show evidence of the parasite, but bronchial brushings are sometimes useful. Experimental serologic tests are of unproven value.

Drug Susceptibility

Trimethoprim-sulfamethoxazole and pentamidine isethionate are the only useful drugs.

Immunity

Active infection occurs in neonates and in adults with disturbances of either immunoglobulins or cell-mediated immunity. Normal adults do not develop disease but are probably infected frequently. The relative importance of natural and acquired immunity is unknown.

Malarial Parasites of Man

The malarial parasites of man are species of the genus *Plasmodium* in which schizogony (asexual cycle) takes place in human erythrocytes and sporogony (sexual cycle) takes place in mosquitoes of the genus *Anopheles*. All species have similar life cycles and morphology (Fig. 11). The species involved are *Plasmodium falciparum, P. vivax*, *P. malariae*, and *P. ovale*.

Morphology

Plasmodium vivax (Fig. 12). Young trophozoites appear in Giemsa-stained blood films as delicate rings of blue cytoplasm. Each contains one red nucleus and is about one third the size of the red cell. As the trophozoite grows, it becomes actively motile and its outline becomes extremely irregular. Granules of yellowish-brown pigment develop in the cytoplasm of the parasite. The erythrocyte becomes enlarged, and a bright red diffuse stippling, Schüffner's dots, may be seen in some cases. Large trophozoites undergo schizogony, producing an average of 16 merozoites.

Plasmodium malariae (Fig. 13). The ring stage is identical to that of *P. vivax*. Mature trophozoites are more compact than those of *P. vivax* and may extend as bands across the infected erythrocyte. Pigment appears earlier and in greater quantity and is darker brown and coarser than that in *P. vivax*. The erythrocyte is not enlarged and stains normally. An average of eight merozoites is produced by schizogony. The merozoites are arranged in a rosette around a centrally collected pigment mass.

Plasmodium ovale. This species resembles *P. malariae* in many ways, but the infected erythro-

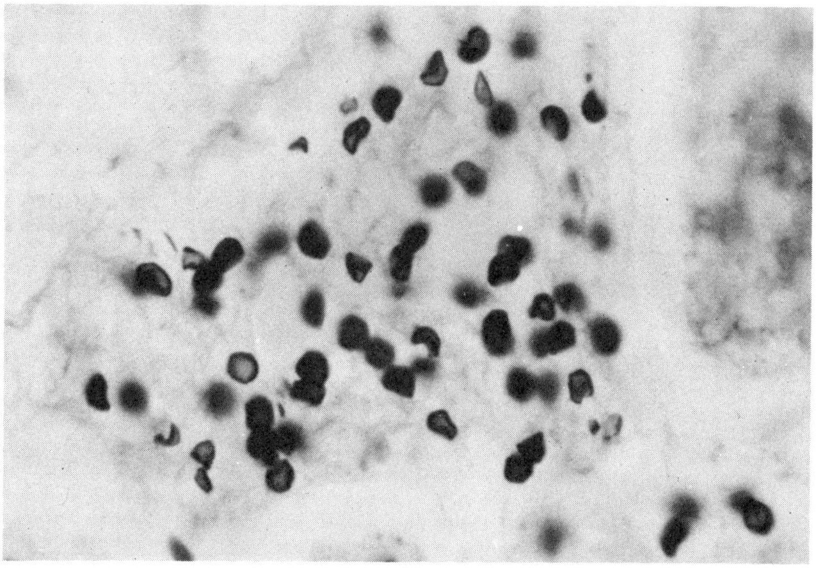

FIGURE 10. *Cysts of* Pneumocystis carinii *in alveolar exudate. Note round, crescentic, and other-shaped cysts with collapsed walls; silver methenamine-fast green stain, 1000 ×. (From Hunter, G. W., Swartzwelder, J. C., and Clyde, D. F.: Tropical Medicine. W. B. Saunders Co., Philadelphia, 1976.)*

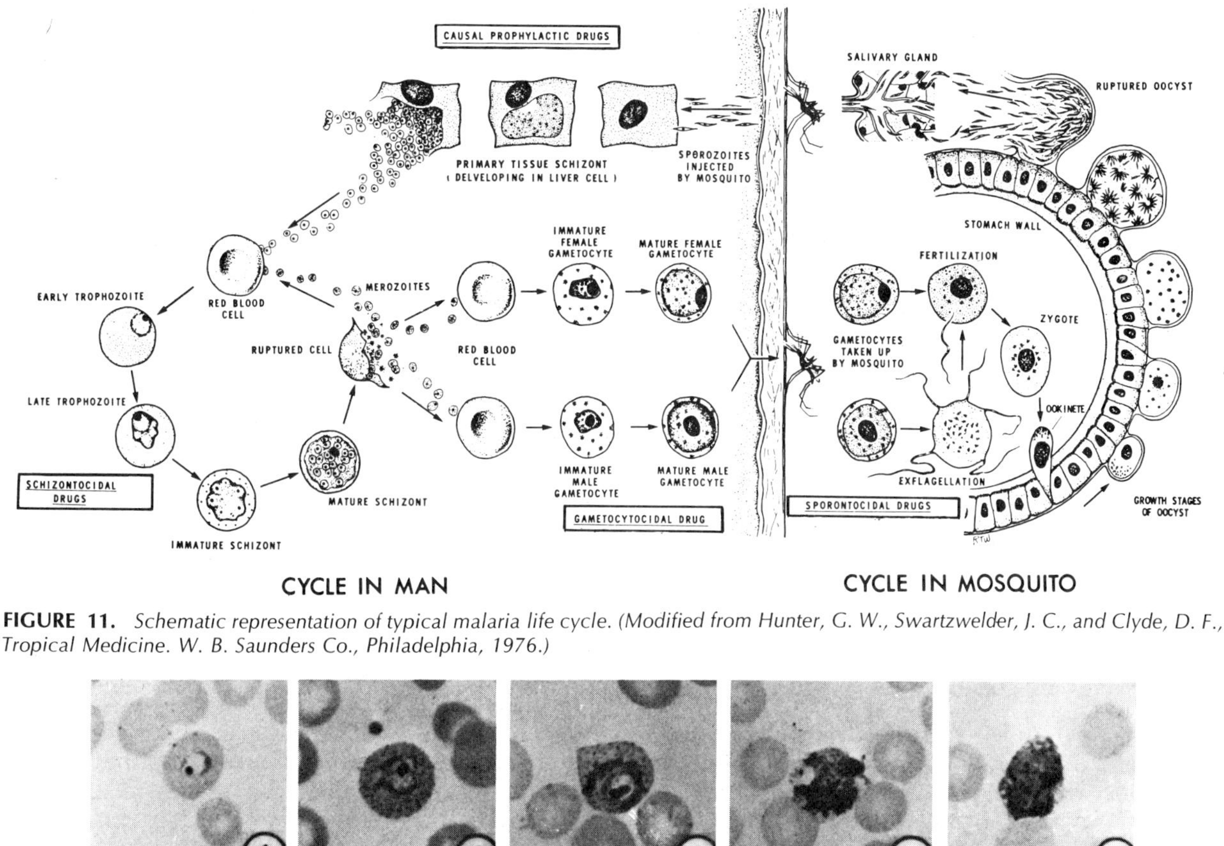

FIGURE 11. *Schematic representation of typical malaria life cycle. (Modified from Hunter, G. W., Swartzwelder, J. C., and Clyde, D. F., Tropical Medicine. W. B. Saunders Co., Philadelphia, 1976.)*

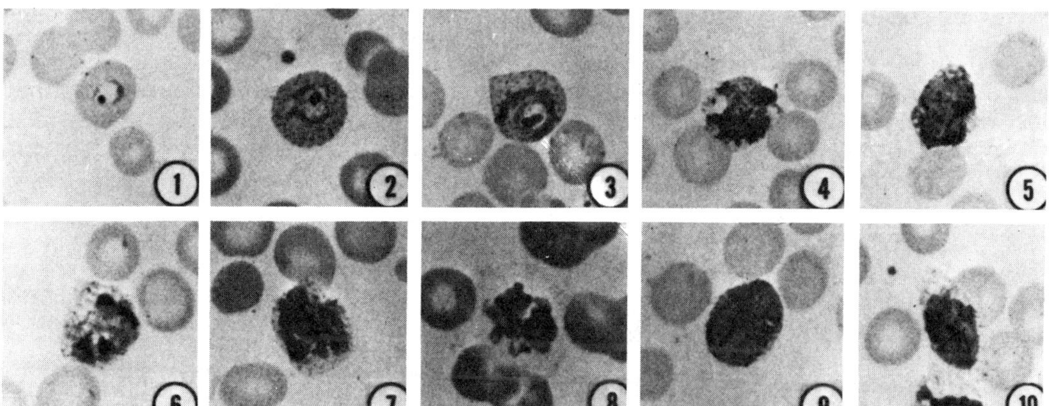

FIGURE 12. *Developmental stages of* Plasmodium vivax *in erythrocytes. 1, Young trophozoite, ring stage. 2, Developing trophozoite in enlarged erythrocyte with Schüffner's dots. 3, Older trophozoite with pseudopod. 4, Mature trophozoite with irregular margin. 5, Schizont showing initial division of nucleus. 6, Older schizont with four nuclei. 7, More mature schizont. 8, Mature schizont with 16 nuclei and clumped pigment. 9, Macrogametocyte. 10, Microgametocyte. (From Wilcox, A.: Manual for the Microscopical Diagnosis of Malaria in Man. National Institute of Health Bulletin No. 180, U.S. Government Printing Office, Washington, D.C., 1942.)*

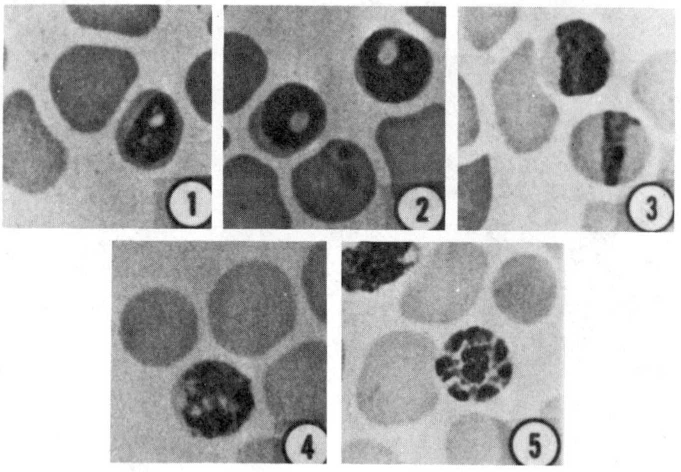

FIGURE 13. *Developmental stages of* Plasmodium malariae *in erythrocytes. 1, Maturing trophozoite. 2, One ring form with pseudopod and two mature trophozoites with heavy pigment. 3, Two trophozoites in band form. 4, A schizont with three nuclei. 5, Mature schizont in "rosette' form. (From Wilcox, A.: Manual for the Microscopical Diagnosis of Malaria in Man. National Institute of Health Bulletin No. 180, U.S. Government Printing Office, Washington, D.C., 1942.)*

cytes behave like those infected with *P. vivax*. Erythrocytes, which may show Schüffner's dots, are oval, slightly enlarged, and pale. The margin is often crenated. The trophozoite of *P. ovale* is compact and band forms may appear. An average of eight merozoites is produced by schizogony.

Plasmodium falciparum (Figs. 14 and 15). Ring forms are smaller and more delicate than those of *P. vivax* and may have double nuclei. Multiple infection of erythrocytes is common. Infected erythrocytes are not enlarged but may have comma-like red markings, called Maurer's clefts, in the cytoplasm. Infected erythrocytes are removed from the peripheral circulation, and further development and schizogony are accomplished in the capillaries of the viscera. Therefore, intermediate and mature forms of the schizogonic cycle are not seen. Crescentic gametocytes, however, may be seen in the peripheral circulation. Their shape differs from that of the rounded gametocytes of *P. vivax*, *P. malariae*, and *P. ovale*.

Life Cycle/Epidemiology

Malaria generally occurs between the latitudes 45 degrees north and 40 degrees south. *P. vivax* infections are the most widely distributed; *P. malariae* infections are comparatively rare. *P. falciparum* occurs predominantly in tropical regions, and *P. ovale* is mostly limited to Africa.

The sexual cycle in certain anopheline mosquitoes begins with the ingestion of mature gametocytes (Fig. 11). In the mosquito stomach the microgametocyte produces microgametes (flagella-like structures) that detach (exflagellation) and migrate to the macrogamete (female cell), which was formed by maturation of the macrogametocyte. Fertilization takes place, and the zygote elongates and becomes motile (ookinete). It penetrates the mosquito stomach wall and grows into an oocyst beneath the outer layer. Many filamentous sporozoites are produced by rapid multiplication, which results finally in rupture of the oocyst. Sporozoites move to the salivary glands and are injected at the next feeding.

The asexual cycle begins with the entry of sporozoites into human hosts. They rapidly enter the parenchymal cells of the liver and undergo rapid multiplication (exoerythrocytic stage). The rate of development and certain morphologic characteristics vary with the species of *Plasmodium*. Finally, infected parenchymal cells rupture, releasing myriads of merozoites that invade erythrocytes, producing ring stages. Schizogony occurs in parasitized erythrocytes, and their

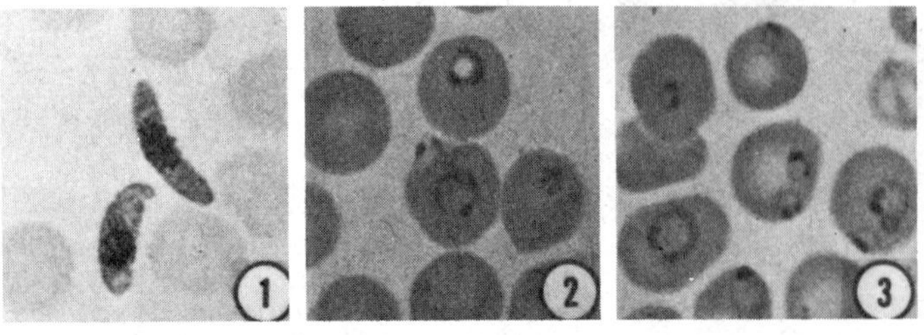

FIGURE 14. *Developmental stages of* Plasmodium falciparum *in erythrocytes. 1, Two crescent-shaped macrogametocytes. 2, Ring stages showing double nuclei and marginal forms. 3, Heavy infection showing multiple infection. (From Wilcox, A.: Manual for the Microscopical Diagnosis of Malaria in Man. National Institute of Health Bulletin No. 180, U.S. Government Printing Office, Washington, D.C., 1942.)*

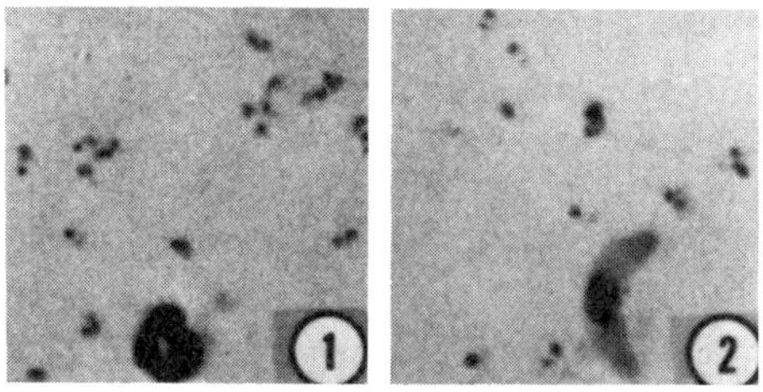

FIGURE 15. *Thick blood films of a* Plasmodium falciparum *infection. 1, Ring stage trophozoites appearing as "commas," "exclamation marks," etc. 2, Ring stages and a crescent-shaped gametocyte. (From Wilcox, A.: Manual for the Microscopical Diagnosis of Malaria in Man. National Institute of Health Bulletin No. 180, U.S. Government Printing Office, Washington, D.C., 1942.)*

eventual rupture frees merozoites to invade other erythrocytes and repeat the cycle. The schizogony is usually synchronous in all species except *P. falciparum*. Eventually gametocytes are produced to perpetuate the cycle.

Laboratory Diagnosis

Definitive diagnosis depends on identifying plasmodia in thick and thin blood smears stained with Giemsa stain (Table 3). Serologic techniques are available and are useful as an aid to diagnosis.

Drug Susceptibility

The susceptibility of plasmodia to drugs varies with their stage in the life cycle. Plasmodia in the erythrocytic stage are most sensitive to the 4-aminoquinoline drugs, chloroquine and amodiaquine. Chloroquine-resistant strains of *P. falciparum* from Southeast Asia and South America have retained their sensitivity to quinine. Neither quinine nor the 4-aminoquinolines eradicate the exoerythrocytic stage of malaria. Primaquine is the most active drug against this stage of the life cycle. The dihydrofolate reductase inhibitors, proquanil, pyrimethamine, and trimethoprim, show only weak activity against the erythrocytic stage but are sporontocidal; that is, they prevent development of sporozoites in the mosquito and in humans inoculated with sporozoites by an infected mosquito. Trimethoprim, however, acts synergistically with sulfonamides and sulfones against the erythrocytic stage. *Plasmodium* organisms become resistant to the sulfones, sulfonamides, and dihydrofolate reductase group if they are used alone.

The variation in sensitivity of different stages in the *Plasmodium* life cycle and the tendency of *Plasmodium* organisms to become resistant to a single drug has been countered by treating most forms of malaria with two or more drugs. The combination of chloroquine and primaquine results in radical cure of all forms of malaria except chloroquine-resistant *P. falciparum* malaria. Infections with chloroquine-sensitive *P. falciparum* are eradicated by chloroquine alone, since there is no persistent exoerythrocyte cycle. Chloroquine-resistant *P. falciparum* is treated with two or more schizontocides; the combination of quinine, pyrimethamine, and either a sulfonamide or sulfone is especially effective.

Immunity

P. vivax and *P. falciparum* produce partial homologous immunity that is strain-specific. There is no cross-immunity between species. Man has been immunized with irradiated sporozoites of *P. falciparum* and *P. vivax*, but immunity was

TABLE 3. Key for the Differentiation of the Three Common Malaria Species in Peripheral Blood Films

THIN FILMS	
1. No intermediate stages present (no late trophozoites or schizonts). Usually only characteristic rings (rod-like nucleus; rings located at periphery of red cell; binucleate rings; small rings; frequently multiple rings in a red cell) and/or cresent-shaped gametocytes present	*P. falciparum*
2. Intermediate stages present (late trophozoites and schizonts)	*P. vivax* or *P. malariae*
A. Parasitized red cells enlarged and pale; trophozoites with irregular outlines; pigment fine; Schüffner's dots may be present	*P. vivax*
B. Parasitized red cells normal in size and color; growing forms compact or "band" forms; pigment coarse; no stippling of the red cells	*P. malariae*
THICK FILMS	
1. No intermediate stages present. Usually only characteristic *rings* (rings, exclamation marks, swallows, comets, etc.) and/or cresent-shaped gametocytes present	*P. falciparum*
2. Intermediate stages present	*P vivax* or *P. malariae*
A. Trophozoites with irregular cytoplasmic outline	*P. vivax*
B. Trophozoites with compact cytoplasm or "band" forms	*P. malariae*
Note: Schüffner's dots are not always present in *P. vivax* infection. Crescents are not always present in *P. falciparum* infection. Band forms are not always present in *P. malariae* infection.	

short-lived. Stages in the erythrocytic cycle produce antibodies, and these forms of the parasite are currently being examined for possible use in vaccines.

Babesia

Various species of the genus *Babesia* are parasitic in the erythrocytes of many domestic and wild mammals. Ixodid ticks serve as intermediate hosts. Infected ticks may inoculate sporozoites when they feed on man. Only *Babesia bovis* was known to infect man until recent case reports appeared of human infections with *Babesia microti*, a strain found in rodents.

Morphology

Babesia spp form tiny (1.5 to 5 μ) pear-shaped, ovoid, ellipsoid, ring- or rod-shaped bodies in erythrocytes. The number, shape, and size vary with each species. No pigment is produced. Their staining characteristics are the same as those of *Plasmodium*.

Life Cycle/Epidemiology

Babesia species reproduce by binary fission in infected erythrocytes. During division, the erythrocyte ruptures and new erythrocytes are invaded. Sexual development takes place in the arthropod. When sporozoites are finally produced, they invade all the tissues of the tick, including the developing eggs and salivary glands.

Babesiosis is a rare infection in man, but it may be fatal in splenectomized patients or those who are otherwise immunologically compromised. A recent epidemic involving *Babesia microti* occurred on Nantucket Island off the northeast coast of the United States (Spielman, 1976).

Laboratory Diagnosis

Definitive diagnosis is achieved by the identification of parasites in Giemsa-stained thick and thin blood smears. Differentiation of species is difficult, and the species are easily confused with malaria organisms.

Drug Susceptibility

The antimalarial drugs have no effect on *Babesia* strains, but the organisms show some sensitivity to the antitrypanosomal drugs 4,4′-diazoaminobenzamidine and pentamidine isethionate, but parasitemia recurs after treatment. Despite the limited response, these three drugs should be used in the immunocompromised patient because infections under these circumstances are often fatal (Miller et al., 1978).

Immunity

There is serologic cross-reactivity between *B. bovis*, *Plasmodium falciparum*, and *P. vivax*, but cross-immunity between *Babesia* and *Plasmodium* has not been investigated.

References

Doyle, J. J.: Antigenic variation in the salivarian trypanosomes. In Miller, L. H., Pinto, J. A., and McKelvey, Jr., J. J. (eds.): Immunity to Blood Parasites of Animals and Man. (Advances in Experimental Medical Biology, vol. 93.) New York, Plenum Press, 1977, p. 321.

Kagan, I. G., and Norman, L.: Serodiagnosis of parasitic diseases. In Lennette, E. H., Spaulding, E. H., and Truant, J. P. (eds.): Manual of Clinical Microbiology, 2nd ed. Washington, D. C., American Society of Microbiology, 1974, p. 970.

Lumsden, W. H. R.: Field diagnosis of trypanosomiasis. Trans R Soc Trop Med Hyg 71:8, 1977.

Miller, L. H., Neva, F. A., and Gill, F.: Failure of chloroquine in human babesiosis (*Babesia microti*). Ann Intern Med 88:200, 1978.

Spielman, A.: Human babesiosis on Nantucket Island: Transmission by nymphal Ixodes ticks. Am J Trop Med 25:784, 1976.

Wolfe, M. S.: Giardiasis. JAMA 233:1362, 1975.

Zeledón, R., and Ponce, D.: Parasitological and immunological diagnosis of cutaneous leishmaniasis. Proceedings of the Third International Congress of Parasitology 3:239, 1974.

79 NEMATHELMINTHES (Roundworms)

A. O. Anya, Ph.D.

The name Nemathelminthes was first used by Gegenbaur in 1859 to describe the roundworms. In his conception, the roundworms included three zoologically distinct groups of animals: the acanthocephalans, the nematodes, and the gordiaceans (nematomorphs). Vejdowsky in 1866 used the term to refer only to the nematodes and the nematomorphs. In most zoology texts, the term includes rotifers, gastrotrichs, kinorhynchs, nematodes, and nematomorphs, which are by implication related. However, the zoologic affinities of these groups are not as clear as their categorization as nemathelminthes would suggest, and Hyman (1951) prefers to regard each as an inde-

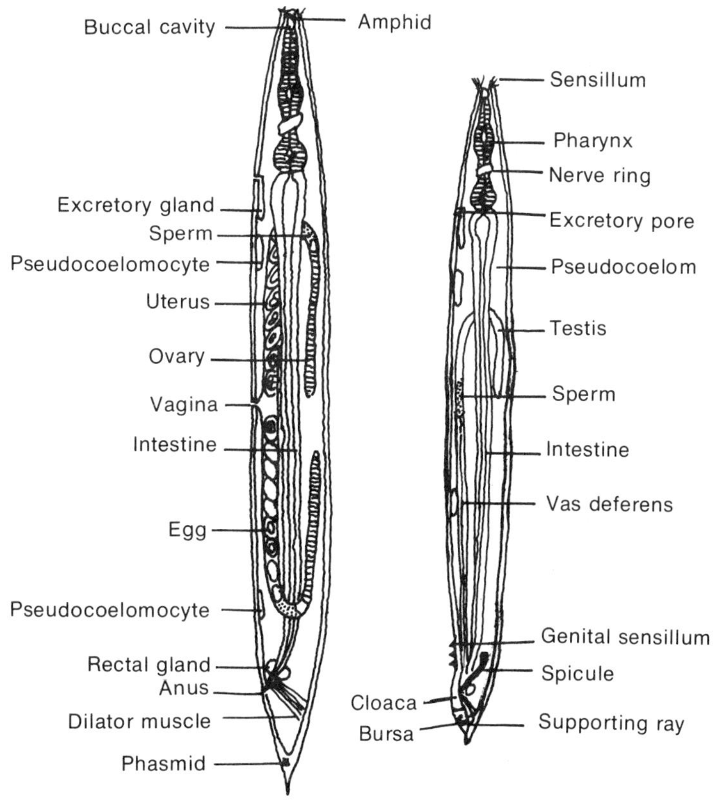

FIGURES 1 and 2. *Diagram of a hypothetical nematode: female (left) and male (right).*

pendent entity of phyletic status. Most modern authors have preferred this arrangement (Maggenti, 1976). Because the nematodes are the only Nemathelminthes of medical importance, the following discussion will concentrate on this group.

FUNCTIONAL MORPHOLOGY

Nematodes are generally small, spindle-shaped, unsegmented, bilaterally symmetric organisms. The digestive and reproductive systems lie more or less freely in the fluid-filled body cavity, which is a pseudocoelom (Figs. 1 and 2).

Within the body cavity may also be found two, four, or six ovoid or many-branched cells called pseudocoelomocytes of unknown function. Organs of respiration and circulation are absent, and the "excretory" organs are unlike those of other invertebrates. An often-quoted characteristic of nematodes is the absence of cilia or flagella. The electron microscope, which has opened up a new era in nematode morphology and anatomy, has demonstrated typical flagella in the intestine of a free-living nematode *Eudorylaimus* sp and modified flagella in the sense organs of many nematodes including parasitic species (Zmoray and Guttekova, 1972; McLaren, 1976).

The body surface is covered by a living, thin, relatively inelastic but flexible cuticle made of modified collagen (Anya, 1966). The cuticle lines part of the inner surface of the main body openings of the digestive and reproductive systems. Fine striations and annuli may be present on the cuticular surface.

The cuticle, which is a three-layered structure consisting of cortical, median, and basal layers, is often delimited by a basement lamella from the hypodermis. The latter may be either syncytial or cellular tissue. It projects into the body cavity along the middorsal, midventral, and lateral lines, giving rise to four ridges or chords (Fig. 3). When excretory canals are present, they are embedded in the tissues of the lateral chords. The main nerves run in the middorsal and midventral hypodermal chords.

The muscle layer lies beneath the hypodermis and consists of unique, spindle-shaped, and elongated cells oriented only in the longitudinal direction. There are no circular muscles. The muscle cells are characteristically divided into contractile and noncontractile portions (Fig. 4) and are obliquely striated with the filaments of actin and myosin restricted to the basal contractile portion of the cell. The noncontractile part of the muscle cell contains the main cell organelles, such as the nucleus and mitochondria, as well as stores of

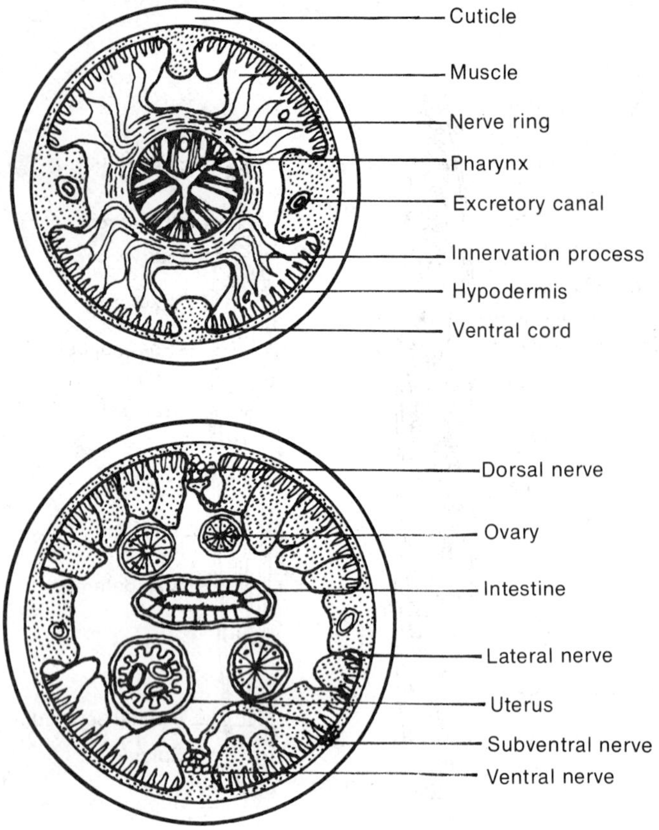

FIGURE 3. *Transverse sections through a female nematode in the pharyngeal region (above) and in the midregion (below).*

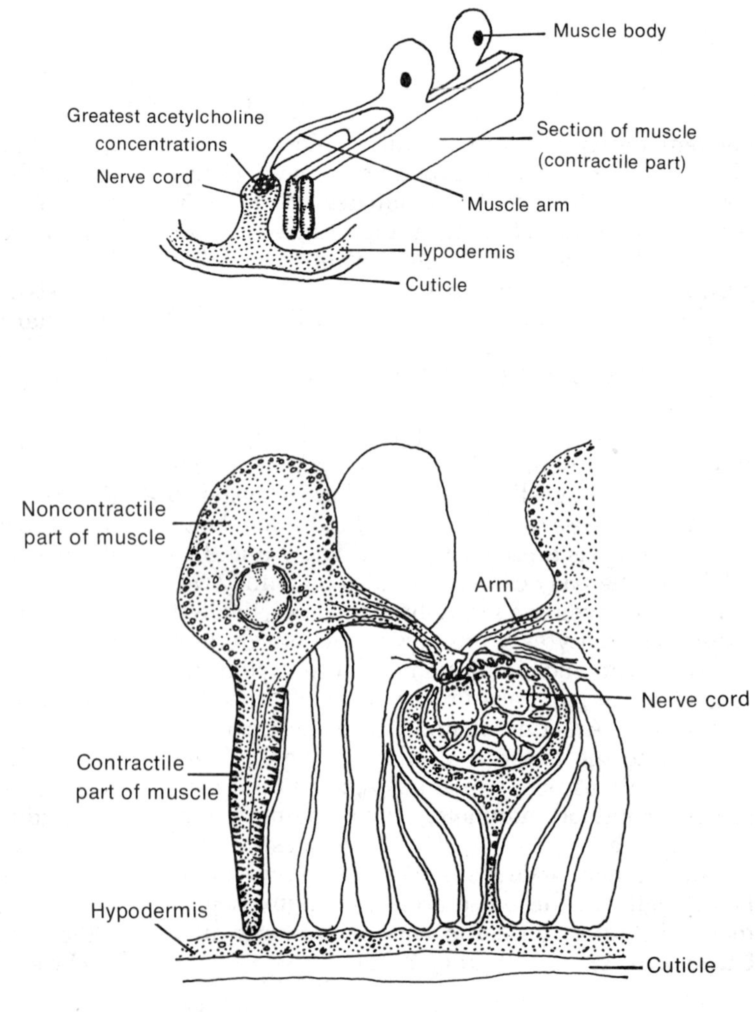

FIGURE 4. *The muscle cells in nematodes. Longitudinal aspect of two muscle cells (above) and muscle cells at myoneural junction, transverse section (below).*

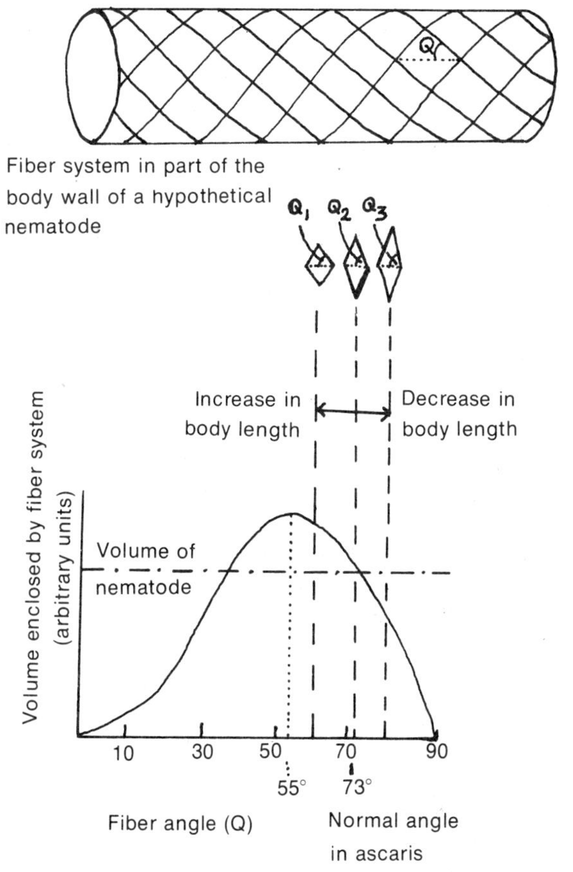

FIGURE 5. *Arrangement of fibers in cuticle of a nematode (above) and the relationship between the volume enclosed by the fiber system in the cuticle and the angle Q (after Lee and Atkinson, 1976).*

glycogen and fat. The nerve processes extend from the muscle cells to make synaptic contact with the longitudinally oriented nerve trunks in the middorsal and midventral hypodermal chords. Thus, unlike the situation in other organisms, the muscle cells transmit innervation impulses to the nerve tissue.

The fluid-filled body cavity is notable for its high but variable turgor pressure, which ensures that its incompressible fluid content acts as a hydrostatic skeleton that is antagonistic to the longitudinal muscle layer. The efficiency of this arrangement is assured by the special arrangement of the fibers in the nematode cuticle. In *Ascaris lumbricoides,* for example, the fiber layers consist of a meshwork of spiral and relatively inelastic fibers that run counter to one another in a trellis or lattice fashion. This may best be visualized as a series of parallelograms subtending an angle (Q) that will vary within defined limits (Fig. 5). Anisometric extension of the cuticle is thus possible, and the angle Q defines the critical position from which the length and diameter of the cylindrical nematode can change without altering the volume. It has been shown mathematically that as long as Q is larger than 55 degrees, an increase in length of the worm compensates for a slight decrease in diameter, and, in such an arrangement, the muscles must be oriented longitudinally. Measurements of fixed preparations of the cuticle of *Ascaris* give a value of 75 degrees 30′ for Q (Harris and Crofton, 1957). The interaction of pressure changes with the cuticle and the muscle maintain static equilibrium during variations in body volume, permitting the ingestion or expulsion of fluid despite the high pressure in the body cavity. This system may explain the relative invariability of nematode structure.

The alimentary system includes a mouth, buccal cavity, pharynx (esophagus), intestine, rectum, and anus (Fig. 1). The mouth is usually surrounded by lips arranged according to a hexamerous triradiate arrangement. In the male nematode, the distal section of the reproductive system also opens into the rectal region, converting the rectum into a cloaca. The buccal cavity, like the pharynx, is lined by cuticle, which may be modified into teeth or cutting plates. The triradiate pharynx is a highly muscular organ that may be divisible into muscular and glandular regions. Alternatively, it may contain three glands that open into the lumen at different points along its length. The intestine is lined by microvilli and may be demarcated functionally from the pharynx and rectum (or cloaca) by a valve or sphincter that controls the entry or exit of materials in the alimentary system.

Nematodes are generally bisexual and show sexual dimorphism in that males are usually smaller. The female reproductive system consists of two tubes whose diameters vary in different regions. Each tube differentiates into ovary (where the germinal cells at various stages of development are found), oviduct, seminal receptacle, and uterus. The two uteri may be joined to form a vagina uterina that leads to the exterior through the vulva (Fig. 6). The nematode egg is covered by a highly resistant three-layered eggshell in which lipoproteins, chitin, and certain glycosides form complexes that are extremely resistant to the entry of all substances except gases. The noteworthy ecologic success and evolutionary plasticity of the nematodes are probably due in large measure to the self-contained environment of the embryo provided by the eggshell, as well as to the structural stability of the adult that is provided by the unique interacting system of body fluid turgor pressure, muscle, and cuticle.

The male system consists usually of a single tube differentiated into testes, seminal vesicle, and vas deferens (Fig. 7). The vas deferens, which may consist of glandular and ejaculatory portions, opens into the cloaca. The nematode spermatozoon is generally ameboid, but recent work has indicated that the shapes, sizes, and structures of this cell vary in different groups of nematodes (Anya, 1976). However, none has flagella that are characteristic of the sperm cells of other animal groups.

Some nematodes release sex attractants, and the vas deferens secretes substances that may aid the ascent of the sperm cell prior to fertilization (Green, 1967; Anya, 1973). The availability of food in the environment is an important factor

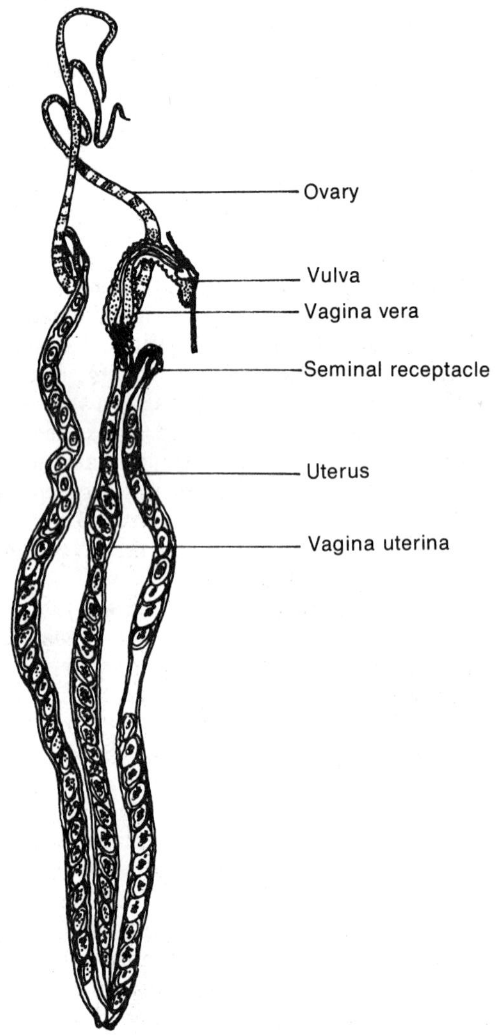

FIGURE 6. *Female reproductive system of a parasitic nematode,* Aspiculuris tetraptera.

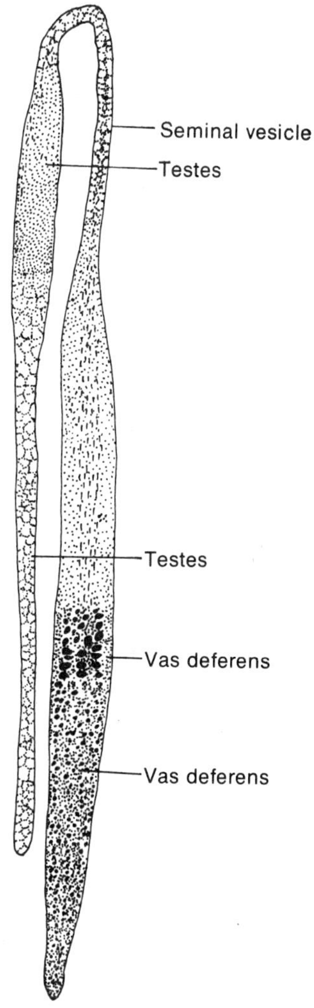

FIGURE 7. *Male reproductive system of the parasitic nematode,* Aspiculuris tetraptera.

determining the sex ratio of nematodes; more males are usually found in situations of nutritional stress (Anya, 1976).

What has often been called the "excretory" system in many nematodes has been so regarded purely on morphologic grounds. It is structurally variable, as it is absent in some nematodes and replaced by a pair of glandular cells, the renette cells, in others. The best known representative of the system, however, is the H type seen in *Rhabditis* sp, in which the main (longitudinal) trunks of the H are located in the lateral chords (Fig. 8). The cross-bar of the H is formed by transverse branches that traverse the pseudocoelom in the anterior region. These join before opening to the exterior through the ventrally located "excretory" pore. In some nematodes, the anterior arms of the H-shaped system may be lost, and the excretory system is therefore an inverted U structure. In others, such as the juvenile *Ancylostoma* sp, the inverted U type of "excretory" system may be associated with a pair of glandular renette cells.

The sense organs of nematodes may be associated with the mouth or with the genital or anal openings (and in some nematodes with the caudal region). There are, then, three main groups of sense organs in a nematode: the anteriorly located amphids, the posteriorly located phasmids, and the various papillae associated with the mouth, the genital opening, or the caudal region.

In the primitive state, according to De Conick (1942), the nematode mouth is surrounded by six lips, each of which bears a variable number of papillae (Fig. 9). Generally, 16 such papillae are supposed to be arranged in three concentric circles as follows: an inner and an outer labial papilla on each lip and four cephalic papillae situated behind the lips. The amphids are located on either side of the head region. The pair of phasmids is located posteriorly one on either side of the ventral median line, behind the anal opening in the female. The male nematode may bear (also in the posterior region) several pairs of genital or caudal papillae. The papillae are usual-

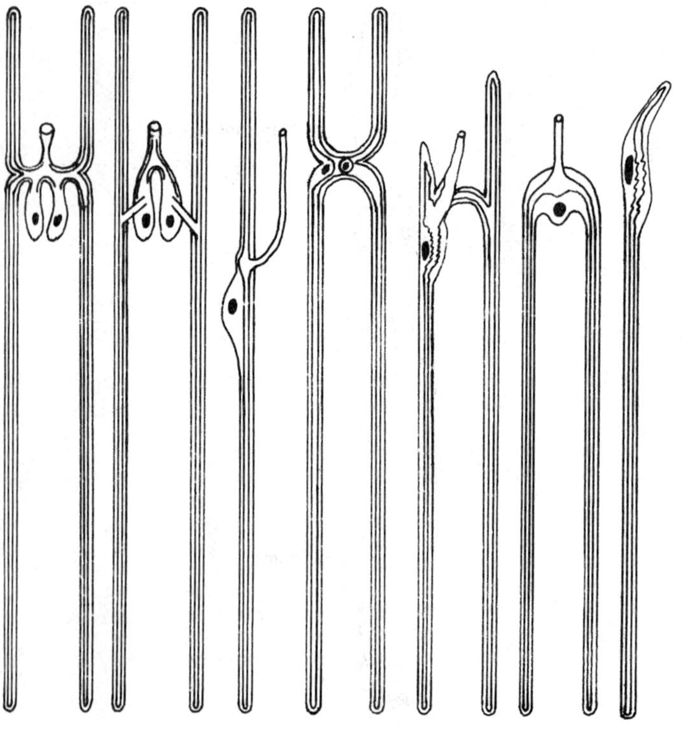

FIGURE 8. *Diagrammatic representation of the various types of excretory systems in nematodes.*

ly supported by an expansion of the caudal cuticle called alae or bursae, depending on the structure or the degree of cuticular inflation and the nematode group.

The amphids and the phasmids are cuticle-lined depressions on the body surface. Recent ultrastructural work has revealed that these depressions are associated with receptor and glandular cells with ciliated nerve axons (McLaren, 1976). Nerve processes from these cells have been traced to the circumpharyngeal nerve ring. This association suggests that amphids and phasmids may have a chemosensory and possibly a neurosecretory function, essential perhaps in monitoring the external environment.

The amphid of *Necator americanus* has been studied in greater detail, and the associated secretory cell is most active in the production of choline-esterase during the period of the life cycle when the change from the lung environment to the intestinal environment takes place. The papillae also have associated ciliated nerve axons, but do not have direct access to the external environment because they are generally covered by a rather thin cuticle. This suggests a possible role as mechanoreceptors.

LIFE CYCLES

Harris and Crofton (1957), in drawing attention to the relative invariability of nematode structure, had observed that "the elementary student may be forgiven at times for thinking that there is only one nematode but that the model comes in different sizes and with a great

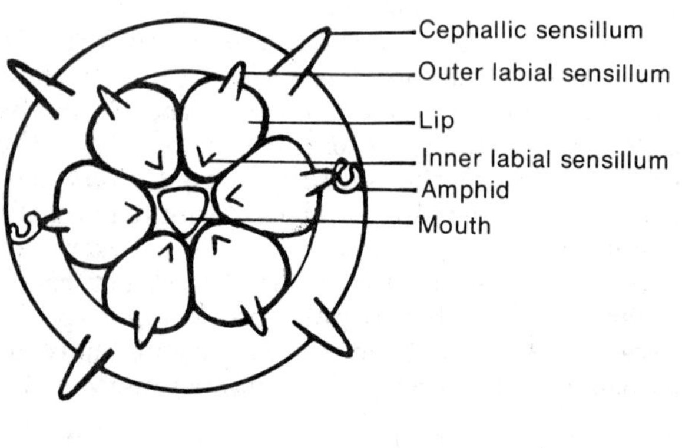

FIGURE 9. *En-face view of the hypothetical primitive nematode (after De Conick, 1942).*

variety of life histories." As we saw in the preceding section, the apparent uniformity of nematode structure is meaningful if interpreted in the context of the unique structural relations of cuticle, muscle, and body pressure changes. The apparent variety of life cycle patterns is equally logical (and reducible to basic types) if it is seen as an ecologic adaptive strategy for parasitic life and against the background of the evolutionary history of parasitism among the nematodes. For parasitism as a way of life has evolved independently several times among these organisms. The general pattern of a particular life cycle often reflects, therefore, the ecologic associations in the original habitat where the parasitic relationship first emerged; for example, the mode of entry into the final host and the migratory route in the tissues often reflect the acquisition of the first infection through food or through contact.

Moreover, it must be remembered that parasites occur as a population of organisms within an *individual* member of a host population. The regular infection of *other* members of the host population is a necessary condition for the survival of the parasite species. Successful infection of another host often dictates a period of sojourn in the environment outside the host, and an ability to survive these obviously different conditions of life outside the host is essential to continued existence as a parasitic population. A parasite must therefore be capable of surviving in the very different environments *within* and *outside* a host; alternatively, it must develop a stage or stages in its life cycle adapted to survive best in each of these environments. Nematodes have chosen the latter strategy, and the dispersive phase of the life cycle, which "runs the gauntlet of the improvident external environment," is either the resistant egg, in which a developing embryo or larva is often enclosed, or a specially adapted, hardy, and resistant infective larval stage, generally the third stage larva.

It is best to begin a consideration of the life cycle of the parasitic nematodes from the basic patterns of the nonparasitic, free-living members in the aquatic and soil environment. In these, there are an egg and five stages of growth and development: the first, second, third, and fourth larval stages and the fifth stage, in which the organism attains final maturity or adulthood with fully developed and functional reproductive organs. The six stages in this cycle are represented in Fig. 10 as E (egg), 1, 2, 3, 4 (larval), and A (adult) stages. The transition from one larval stage to the next is marked by a molt: there are four molts in the life cycle of a nematode.

The host in which a parasite attains full development and production is called the *definitive* or *final* host. Earlier stages of development take place in the *intermediate* host, which may also serve as a vector. Temporary hosts that transport the infective stage of the parasite to a suitable environment are often called the *transport* host. Life cycles that involve direct contact between the infective agent (whether egg or infective larva) and the definitive host are called *direct*. If development of any of the stages occurs in a host other than the definitive host, the life cycle is said to be *indirect*.

In the parasitic species, there are five main variations in this basic life cycle. In the first variation, exemplified by the direct life cycle of *Ascaris lumbricoides* and *Trichocephalus trichiurus (Trichuris trichiura)*, the egg is the infective agent. The mature eggs are expelled in the feces of the host. The first-stage larva develops within the *Ascaris* egg and subsequently molts. The L_2 enclosed within the eggshell is the infective stage, and it is in this stage that the parasites are commonly encountered in contaminated soils. When the infective egg with the enclosed larva is swallowed in either contaminated food or water by a potential definitive host, man, the larva hatches in the small intestine. The hatching process in the gastrointestinal tract depends upon the interplay of physicochemical factors, chiefly the presence of dissolved carbon dioxide, reducing agents, and an appropriate range of pH (Rogers and Sommerville, 1968). Under appropriate conditions, the larva hatches after digestion of the eggshell by the enzymes chitinase, lipase, and leucine aminopeptidase, which are produced within the egg. The second-stage larva now molts and penetrates the intestinal mucosa as a third-stage larva, which is carried through the lymphatics to the liver, heart, and finally to the lungs within one to three days. During these wanderings (the histotrophic phase), it molts two more times to transform itself into the young adult that breaks through the lung alveoli, ascends the trachea, migrates to the esophagus, and finally descends to the ileum, where the young adult worm reaches egg-producing maturity in about 65 days from the original infection. When evacuated in the feces, the eggs repeat the cycle.

The life cycle is similar in *Trichocephalus trichiurus (Trichuris trichiura)*. The only difference is that the first-stage larvae do not wander as extensively as *Ascaris* larvae. They merely penetrate the villi for a brief sojourn of about a week before returning to the lumen of the cecum to continue their development to full maturity, which takes about three months.

The second pattern, an indirect life cycle, is characteristic of the order Spirurida and is exemplified by the filarial worm of the subcutaneous tissues of man, *Onchocerca volvulus*. Adult females produce eggs that give rise to microfi-

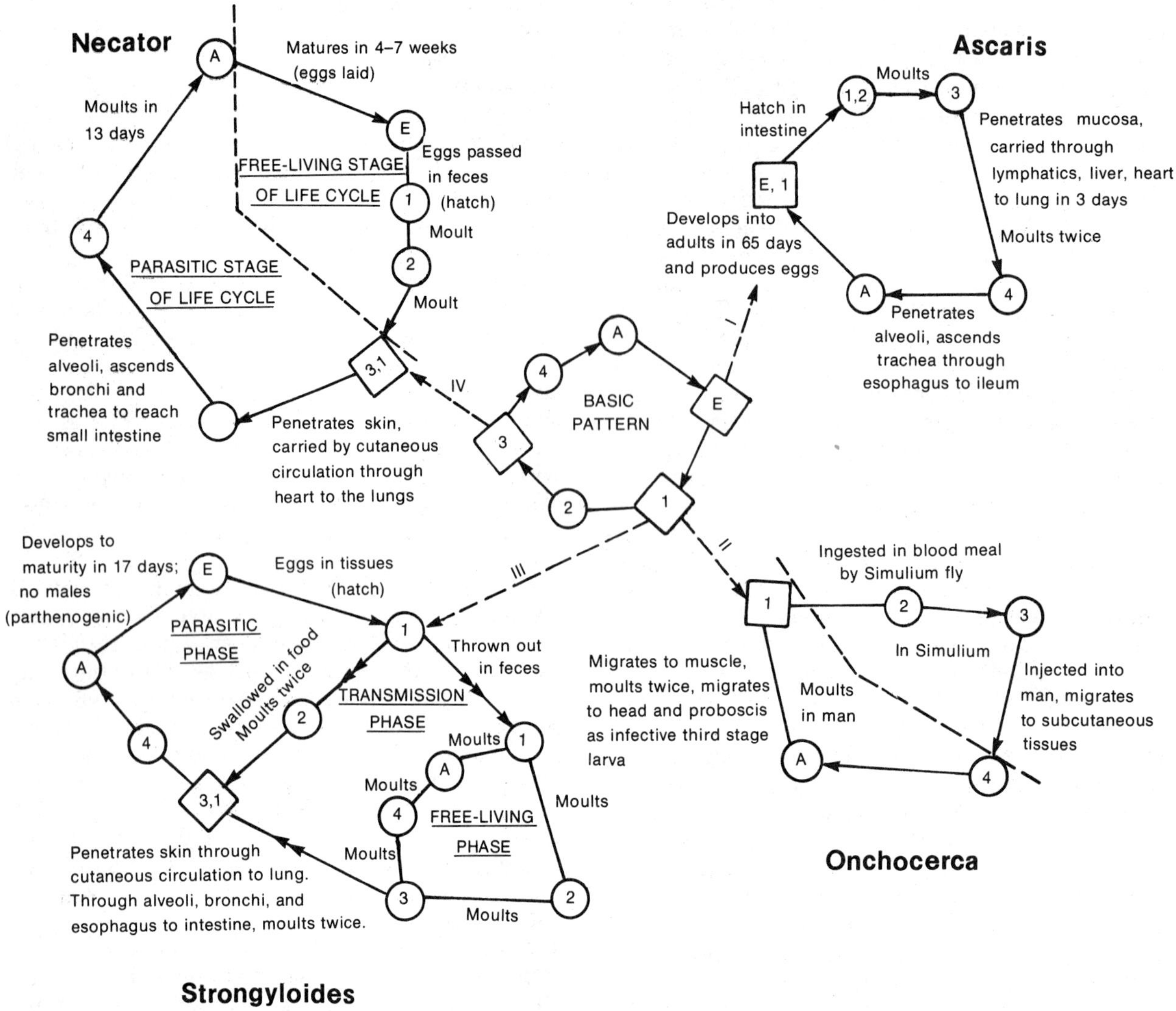

FIGURE 10. *Diagrammatic representation of the basic pattern and four of the five variations of life cycle patterns in nematodes (see text).*

lariae enclosed in the inner membrane of the egg while still within the adult females. When released into the cutaneous circulation, these microfilariae reinitiate the cycle with the next insect bite.

The sequence, summarized in Figure 10, II, is as follows. When an insect vector, usually *Simulium damnosum* in the West African region, bites an infected human, the microfilariae are ingested with the blood meal. The microfilariae migrate to the thoracic muscles of the insect, molt twice, and, as the third-stage larvae, finally come to rest in the head and proboscis region of the insect. When an infected *Simulium* fly bites the next human victim, the infective larva is reinjected into the cutaneous circulation. This larva molts, migrates to the subcutaneous tissues of the host, and molts once more before reaching maturity. Thus, the infective stage for human infection is the third-stage larva, but the first-stage larva is infective for the *Simulium* intermediate host.

The life cycle is essentially the same in all the spirurids with minor variations in the intermediate host or vector that influence the mode of entry into the human host. The intermediate hosts of *Wuchereria bancrofti* are mosquitoes of the genera *Anopheles, Aedes,* and *Culex. Anopheles* is the major vector in West Africa, while *Culex pipiens fatigans* is predominant in Southeast Asia. The vector of *Loa loa* is the day-biting tabanid fly, *Chrysops* sp. Aquatic crustaceans of the genus *Cyclops* are the intermediate hosts of *Dracunculus medinensis.* Man is infected by drinking contaminated water. *Culicoides* sp is the vector of *Dipetalonema streptocerca* in Africa and *Mansonella ozzardi* in South America, where *Simulium* sp may also transmit mansonelliasis.

The third variation that may be considered unique is found in the Rhabditida and exemplified by *Strongyloides stercoralis*. Adult female worms live in the small intestine of the definitive host. Eggs are released in the intestine, where they hatch and give rise to first-stage larvae that reach the soil in the feces. In the soil, two paths of development are open to these larvae. They may follow a free-living, nonparasitic pattern of development in which adult males and females develop from four molts, which give rise to four larval stages and finally adult males and females. The females produce eggs that repeat the cycle.

Alternatively, the first-stage larvae can undergo two successive molts and produce infective third-stage larvae that penetrate the skin of humans, reach the cutaneous circulation, migrate to the lungs, break through the alveoli, ascend the bronchi and trachea to the esophagus, and descend to the intestine. During migration, the parasitic larvae molt twice and reach maturity in about 17 days. There are no males, and egg production is through parthenogenesis. The essential elements of this cycle are summarized in Figure 10, III. The parasitic phase is an example of a direct life cycle, although a free-living phase regularly alternates with this parasitic phase. During the free-living phase, some third stage larvae may also become infective and participate in the parasitic phase by penetrating into the definitive host (Nigon, 1947).

The fourth pattern of the life cycle is exemplified by *Trichinella spiralis* (Fig. 11, I). First-stage larvae are encysted in the muscles of pigs or carnivores such as foxes and wolves. When infected meat is eaten raw or undercooked by another carnivore or human, the cyst walls are digested in the stomach, and the first-stage larvae are released. These larvae invade the mucosa of the duodenum and jejunum and undergo four molts. Some authorities insist that each larva undergoes only three molts within the mucosa, an earlier molt having taken place in the cyst. According to this view, the encysted larva is a second-stage larva.

Immature worms are found in the intestinal lumen within 24 hours of the ingestion of cysts. Adult females are ovoviviparous — i.e., first-stage larvae are released from the uterus rather than eggs. These larvae penetrate the intestinal mucosa, migrate through lymphatics and blood vessels to the skeletal muscles, penetrate the sarcolemma, and form the cyst. The life cycle is completed when the host is eaten by another carnivore (Fig. 11, I). Under normal circumstances, human infections represent a biologic dead end for the parasite. Transmission of the parasite is made possible by the zoonotic maintenance of reservoir hosts in wild carnivores. The nature of the available reservoir host in each locality introduces local variations in the life cycle in different geographic regions (Fig. 11, II to IV).

The final pattern that is also direct is characteristic of most of the Strongylidae, including *Necator americanus*. When eggs are passed out in the feces of the host, the first-stage larvae emerge, feed, and molt into second-stage larvae, which develop into infective third-stage larvae. The latter do not feed. They are enclosed in the uncast sheaths derived from the molted cuticle of the second-stage larva. The survival of third-stage larvae depends upon the reserves of lipids built up in their tissues during the feeding phase of seond-stage larvae. Recent work in our laboratories has shown that the infectivity of *Necator americanus* varies with the lipid level, activity, and age of the larva. Under the tropical conditions of Nsukka, the viability and infectivity of the larva in contaminated fields are maintained for up to 21 days (Udonsi, unpublished observations).

On contact with a human host, infective larvae penetrate the skin, are carried in the circulation through the heart to the lungs, molt within 24 hours, break through the alveoli, ascend the bronchi and trachea, and descend the esophagus to arrive at the small intestine. Here they molt again, usually within 13 days, and the female adults produce eggs in four to seven weeks. There are thus a free-living and a parasitic stage in the life cycle of each *Necator* nematode. The pattern is summarized in Figure 10, IV.

This pattern is essentially the same in other members of the Strongylidae except that most of the trichostrongylids do not infect the host by penetrating the skin. Infective larvae are swallowed with food or water. *Ancylostoma* may also develop by an identical life cycle.

IMMUNITY

The relationship between a helminth and its host is complex and influenced by physiologic, biochemical, genetic, nutritional, immunologic, and other biologic mechanisms. Host specificity, which is often observed in vertebrate-helminth interactions, may arise from natural host resistance mediated by any of these factors. Acquired immunity, on the other hand, follows the general principles elucidated in studies of viral and bacterial infections. Progress in our understanding of

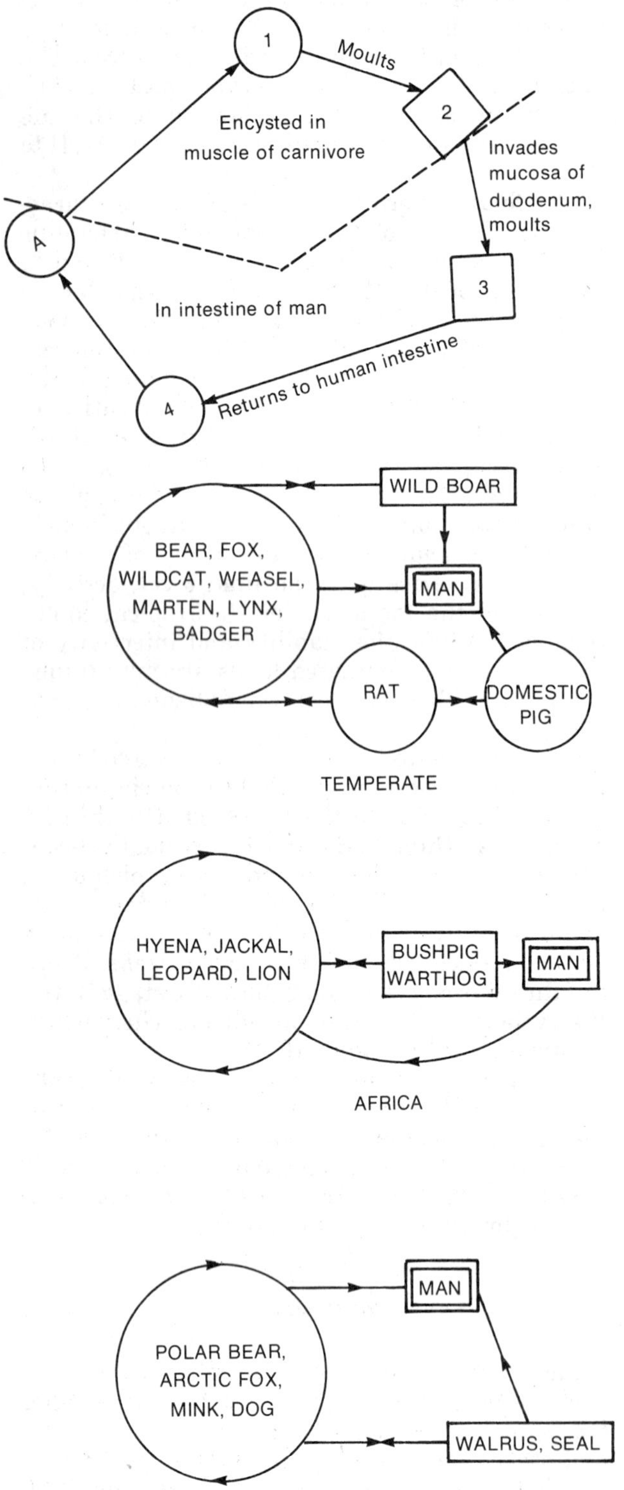

FIGURE 11. *The life cycle pattern of* Trichinella spiralis *(the fifth pattern) and the pattern of zoonotic relationship in different geographic regions: temperate, tropical (Africa), and arctic.*

the principles of immunity to helminths was relatively slow until recently.

One explanation for the complexity of helminth immunology is the fact that helminths are collections of different populations of cells with different biochemical, metabolic, and antigenic characteristics. Nevertheless, studies of various experimental systems such as *Nippostrongylus braziliensis* and *Trichuris muris* in mice (Ogilvie and Jones, 1973; Selby and Wakelin, 1973; Wakelin and Lloyd, 1976), *Ascaris suum* and *Toxocara cati* in rats and mice (Larsh and Race, 1975), *Ancylostoma caninum* in dogs (Miller, 1964), and *Onchocerca volvulus* in man (WHO, 1976) have shed considerable light on the essential features and mechanisms of helminth immunogenicity.

The immune response to helminth infections can be mediated through either circulating antibodies or delayed hypersensitivity (cell-mediated immunity) involving lymphoid cells (Larsh and Race, 1975; Muller, 1975). For example, circulating antibodies of the complement-fixing IgG and IgM classes have been detected in onchocerciasis, while *Ascaris* has stimulated the production of IgM, IgA, and IgG. *Trichinella spiralis* elicits predominantly reaginic IgE antibodies. An increase in the serum level of IgE from about 100 ng/ml to 1000 ng/ml encourages the view that IgE is an important line of immunologic defense in these infections. Indeed, a generalized eosinophilia, whose significance is at present uncertain, and these reaginic antibodies are becoming the characteristic indicators of vertebrate host-helminth interactions, especially when the association of parasite with host tissue can be regarded as close, e.g., with *Trichinella spiralis* and *Nippostrongylus braziliensis.*

A feature of helminth immunity is the inability of killed helminth material or purified extracts of helminth excretions or secretions to afford long-lasting protection. Rather, protection has been secured in some cases by the regular and persistent release of antigens. Thus, live vaccines of larvae irradiated by x-rays have provided significant protection in the veterinary field against *Dictyocaulus viviparus* bronchitis in cattle and the canine hookworm. Understandably, live vaccines have not been used in humans.

It is now generally accepted that any of the typical host reactions associated with the immune response, namely anaphylaxis, immune complex disease, and delayed hypersensitivity (cell-mediated immunity), may be manifest in nematode infections. For example, the so-called self-cure phenomenon observed in *Haemonchus contortus,* characterized by the expulsion of preexisting adult worms on the entry of new infective larvae, is explicable on the basis of anaphylaxis. It is suggested that the release of molting fluid (and thus antigens) by third-stage infective larvae results in an anaphylactic reaction brought about by the production of pharmacologically active substances such as histamine, serotonin (5-HT), and kinins, which render the intestinal mucosa unsuitable for the continued existence of the adult worms. However, recent studies in Kenya have attempted to explain the self-cure phenomenon, not on the basis of local intestinal anaphylaxis but on nutritionally induced unsuitability of the intestine for the adult worms. In *Nippostrongylus braziliensis* (infections of rats) and *Haemonchus placei* (infections of cattle), on the other hand, it has been shown that the immune reaction leads to structural changes in the intestinal cells of the nematodes, characterized by extreme vacuolation and the accumulation of lipids. These nematodes are unable to maintain their normal position in the hosts' intestine because of this interference with their normal metabolism.

After treatment for *Onchocerca volvulus* infections the histologic appearance of tissues suggests that antigens released from dying larvae form complexes with circulating antibodies in and around blood vessels. These immune complexes induce infiltration and degranulation of polymorphonuclear leukocytes. Immune complex formation in the eye or the testes may lead to blindness or infertility, respectively (WHO, 1976).

The importance of cell-mediated immunity in infections with *Trichinella, Nippostrongylus, Trichostrongylus, Trichocephalus (Trichuris),* and *Capillaria* has been demonstrated by protecting animals with immune cells from lymph nodes, spleen, and previously sensitized thymic lymphocytes.

Thus immune phenomena are associated with nematode infections, but the manifestations and modulation of the host responses are complicated by the biologic and functional complexity of the parasites. This complexity limits the effectiveness of helminth-induced antibodies and increases the helminth's chances of adaptation to immunologically induced changes in its environment (Ogilvie and Jones, 1973). Progress in the utilization of immunologic mechanisms for the control of nematode infections has been limited and may remain so, but continued research should provide additional approaches, such as the regulation of parasite reproduction.

SYSTEMATICS AND LABORATORY DIAGNOSIS

The Nematoda are broadly divided into two taxonomic classes depending upon whether or not they possess the ventrally and posteriorly located female sense organ, the phasmid. Nematodes that possess the phasmid belong to the class Phasmidia (Secernentea) and those without it belong to the class Aphasmidia (Adenophorea). Because the phasmid is not an easy structure to identify, other characteristics such as the structure of the pharynx and excretory system or the life history have often been utilized. The assignment of nematodes into the various orders varies from authority to authority and is of little practical use to the nonspecialist. For example, nematodes may be divided into 11, 16, or 18 orders depending upon the classification scheme adopted (Chabaud, 1974; Crofton, 1966; Maggenti, 1976). Nematodes parasitic in vertebrates and hence of potential medical significance are found in the following orders:

Phylum: Nematoda (Cobb, 1919)

Class: Phasmidia (Secernentea) (Dougherty, 1958)

1. Order: Rhabditida (Chitwood, 1933)—e.g., *Strongyloides*
2. Order: Ascaridida (Railliet and Henry, 1925)—e.g., *Ascaris, Enterobius, Toxocara, Anisakis, Toxoascaris*
3. Order: Strongylida (Diesing, 1815)—e.g., *Ancylostoma, Necator, Syngamus, Oesophagostomum, Trichostrongylus, Angiostrongylus*
4. Order: Spirurida (Chitwood, 1933)—e.g., *Wuchereria, Brugia, Loa, Onchocerca, Dracunculus, Dipetalonema, Mansonella, Dirofilaria, Gnathostoma, Habronema*

Class: Adenophorea (Aphasmidia) (Chitwood, 1958)

5. Order: Trichocephalida (Skryabin and Schultts, 1938)—e.g., *Trichocephalus (Trichuris), Trichinella, Dioctophyma, Capillaria*

The important and obligate parasites of man are summarized in Table 1. An outline of the life cycles with characteristics of diagnostic relevance is presented in Table 2, while an identification key generally based on the features of the life cycle, morphology, and anatomy is presented in Table 3. Table 4 and Figure 12 present aids for the identification of nematodes in tissue sections.

In addition to the adult characteristics indicated in Tables 1 through 4, three main methods are used to identify parasitic nematodes in the laboratory. These are: (1) fecal examination for eggs

TABLE 1. Summary of Obligate Parasites of Man Belonging to the Nematoda

PHYLUM	CLASS	ORDER	SUPERFAMILY	GENUS AND SPECIES
Nematoda	Phasmidia (Secernentea)	Rhabditida	Rhabditoidea	1. *Strongyloides stercoralis*
		Ascaridida	Ascaridoidea	2. *Ascaris lumbricoides*
			Oxyuroidea	3. *Enterobius vermicularis*
		Strongylida	Strongyloidea	4. *Ancylostoma duodenale*
				5. *Necator americanus*
				6. *Syngamus*[a] sp
			Trichostrongyloidea	7. *Trichostrongylus*[a] sp
				8. *Oesophagostomum*[a] sp
			Metastrongyloidea	9. *Angiostrongylus*[a] sp
		Spirurida	Spiruroidea	10. *Habronema*[a] sp
				11. *Gnathostoma spinigerum*
			Dracunculoidea	12. *Dracunculus medinensis*
			Filarioidea	13. *Brugia*[a] sp
				14. *Wuchereria bancrofti*
				15. *Dipetalonema perstans*
				16. *Onchocerca volvulus*
				17. *Dirofilaria immitis*
	Aphasmidia (Adenophorea)	Trichocephalida	Trichuroidea	18. *Trichocephalus* sp (*Trichuris*)
				19. *Capillaria phillipensis*
			Trichinelloidea	20. *Trichinella spiralis*
			Dioctophymoidea	21. *Dioctophyma*[a] renale

[a]Not obligate parasites of man but of other vertebrates; these have, however, been found often enough in man to deserve consideration as human parasites.

TABLE 2. Summary Outline of Nematode Life Cycles of Diagnostic Relevance

			DIRECT		
Vertebrate Host	Passive (per os)	Eggs contain rhabditiform larva on perianal skin	Eggs hatch in intestine when swallowed by man	Larvae (L1) develop directly to adults in intestine	*Enterobius*
		Unembryonated eggs in feces	Larvae develop in eggs but do not hatch	Larvae (L1) molt to L2 in eggs, hatch in intestine but return to intestine after sojourn in blood, lungs	*Ascaris*
				L1 develops directly to adult in intestine	*Trichocephalus* (*Trichuris*)
	Active (percutaneous)	Embryonated eggs in feces	Rhabditiform larvae develop to L3 with sheath (in soil)	L3 ingested	*Ancylostoma, Trichostrongylus, Oesophagostomum*
				L3 penetrate skin	*Necator, Ancylostoma*
		Rhabditiform larvae in feces	→Free-living generation with rhabditiform larvae in soil gives L3 →L3 develop in soil	L3 penetrate skin	Strongyloides

			INDIRECT		
Vertebrate Host	Passive (through intermediate hosts)	Rhabditiform larvae in feces	To L3 in land snail	L3 ingested	*Angiostrongylus*
		Larvae in water	L2 in *Cyclops*	L2 ingested	*Gnathostoma*
			L3 in *Cyclops*	L3 ingested	*Dracunculus*
		Larvae encysted in muscles		L2 ingested to L3 in intestine	*Trichinella*
	Active (through biting or otherwise active vectors)	Microfilariae/larvae in blood or lymph	To L3 in biting insect	L3 injected into skin	*Loa, Wuchereria, Brugia, Onchocerca, Mansonella, Dipetalonema*

Modified from Muller, 1975.

TABLE 3. Key to Nematodes Parasitic in Man

1. Filiform anterior and spindle-shaped posterior; pharynx is either a stichosome that consists of a single row of cells with a centrally located fine tube or a long and narrow structure slightly dilated posteriorly; male with or without copulatory bursa	**2**
Regular spindle-shaped, relatively stout or filiform worms with muscular pharynx that may be cylindrical or rhabditiform—i.e., having a posterior bulb that has a valvular apparatus	**3**
2. Male has muscular copulatory bursa without supporting rays; pharynx is long and narrow, not a stichosome; vulva is located anteriorly	Trichocephalida (Dioctophymoidea) *Dioctophyma* sp
Filiform anterior and spindle-shaped posterior; male without bursa; pharynx a stichosome; female vulva located at junction of anterior and posterior positions; eggs with characteristic opercular (polar) plug	Trichocephalida (Trichuroidea) *Trichocephalus (Trichuris) trichiurus*
Filiform anterior; male without bursa; pharynx a stichosome; female ovoviviparous with larvae in uterus	Trichocephalida (Trichinelloidea) *Trichinella spiralis*
3. Relatively stout round worms with muscular pharynx; cylindrical or with valvular posterior bulb	**4**
Filiform worms with cylindrical pharynx	**5**
4. Relatively stout but small worms; muscular pharynx with valvular posterior bulb; male with genital papillae; female with pointed tail; eggs embryonated and slightly flattened on one side	Ascaridida (Oxyuroidea) *Enterobius vermicularis*
Relatively stout large worms; mouth usually with three lips; muscular pharynx more or less cylindrical, not dilated posteriorly and without valvular apparatus	Ascaridida (Ascaridoidea) *Ascaris, Toxocara*
5. Filiform worm with cylindrical pharynx; no parasitic males; parasitic female parthenogenetic; small buccal capsule without teeth; larvae always found alongside adult females	Rhabditida (Rhabditoidea) *Strongyloides stercoralis*
Filiform worm with cylindrical pharynx that is muscular anteriorly and glandular posteriorly; not usually intestinal	**6**
Filiform worm with cylindrical pharynx, muscular throughout its length; copulatory bursa in males with muscular rays	**7**
6. Usually two lateral lips; cuticle of buccal capsule particularly toughened; vulva in middle of female body or posterior; characteristic and spiny head bulb; parasite of alimentary canal, respiratory system, or orbital, nasal, or oral cavity	Spirurida (Spiruroidea) *Gnathostoma* sp
Usually without lips; no buccal capsule; vulva in region of pharynx; parasites of circulatory, lymphatic, or serous cavities, not intestinal; females not more than three times longer than males	Spirurida (Filarioidea) *Loa, Wuchereria, Onchocerca, Mansonella, Dipetalonema*
Females much longer than males; atrophied vulva in gravid female	Spirurida (Dracunculoidea) *Dracunculus* sp
7. Parasites of alimentary canal, very rarely found in kidney tissue	**8**
Parasites of respiratory system	**9**
8. Buccal capsule very feebly developed or absent; thread-like worms	Strongylida (Trichostrongyloidea) *Trichostrongylus* sp
Buccal capsule well developed with ventral teeth; anterior end more or less straight	Strongylida (Ancylostomoidea) *Ancylostoma duodenale*
Buccal capsule well-developed with cutting plates; anterior end characteristically recurved	Strongylida (Ancylostomoidea) *Necator americanus*
9. Buccal capsule without teeth or cutting plates but mouth surrounded by characteristic leaf crown	Strongylida (Trichostrongyloidea) *Oesophagostomum* sp
9. Well-developed, toughened, and cuticular buccal capsule; male often attached to female	Strongylida (Strongyloidea) *Syngamus laryngeus*
Buccal capsule rudimentary or absent; filiform	Strongylida (Metastrongyloidea) *Angiostrongylus* sp

or culture of feces for the infective larval stages; (2) examination of blood, tissue fluid, or skin snips for microfilariae; and (3) immunologic techniques.

Fecal Examination

The simplest method for the detection of helminth eggs in feces is direct examination under the microscope of about 5 mg of feces diluted with tap water or saline and smeared on a slide with a coverslip. Only heavy infections are usually detected by this method.

Modifications of this simple method such as Kato's thick smear technique are also useful. In this method, about 50 mg of a fecal sample is placed on a microscope slide, and a 30 × 22 mm piece of cellophane soaked in a mixture of 100 ml of glycerine, 100 ml of water, and 1 ml of 3 per cent aqueous malachite green is placed over it. The specimen is pressed down and examined after 20 to 30 minutes at 37° C. The feces has cleared by this time, and the relatively uncleared eggs are seen in bolder relief.

Various methods for the estimation of the intensity of infection by egg counts have been developed. The best known is the Stoll dilution egg counting technique and the McMaster tech-

TABLE 4. Identification Key For Sections of Nematodes

1. A. Hypodermal chords (few to many) arranged asymmetrically. Lateral excretory canals absent. In the ovaries and testes the germinal region extends the entire length of the gonad, and gametes at various stages of development can be found at any level. Esophagus nonmuscular.	Trichurids (*Trichinella, Trichocephalus* (*Trichuris*) (*Capillaria*)
B. Hypodermal chords (two to four) arranged symmetrically. Excretory canals in lateral chords except at extremities. In the ovaries and testes the germinal region is confined to the proximal extremity of the gonad, and gametes at uniform stage of development can be found at any one level.	2
2. A. Intestine composed of a few large multinucleate cells, rarely with more than two cells in a cross-section. Anterior paired excretory gland cells present.	Strongylids (hookworms, *Trichostrongylus, Ternidens, Oesophagostomum*)
B. Intestine composed of few to many, small, uninucleate cells; excretory glands single, paired, or absent.	3
3. A. Somatic muscle cells few, large and flat. Lateral alae present or absent.	4
B. Many somatic muscle cells, U-shaped in cross-section. Lateral alae usually absent.	5
4. A. Lateral alae present. Vagina long and muscular. Esophagus has posterior bulb. Many eggs in uterus. Excretory gland cells absent	*Enterobius*
B. Lateral alae absent. Vagina very short. Ova large. Viviparous, with eggs developing into larvae. Excretory cells single or paired. Very small and often larvae (or eggs) in tissues.	*Strongyloides*
5. A. Somatic musculature divided by large lateral chords (often very wide and flattened) into two crescent-shaped folds.	6
B. Somatic musculature divided into four fields	7
6. A. Females often contain microfilariae and uterus double. Diameter less than 0.5 mm.	Filarids (*Wuchereria, Brugia, Dipetalonema, Dirofilaria, Onchocerca, Loa*)
B. Females contain larvae with long pointed tails, uterus single, diameter more than 1 mm.	*Dracunculus*
7. A. Lateral chords very large and stalked and project into body cavity; frequently unequal in size.	Spirurids (*Thelazia, Gnathostoma*)
B. Lateral chords large and project into body cavity but not stalked; usually equal in size.	Ascarids (*Ascaris, Toxocara, Toxoascaris, Anisakis*)

nique. These methods rely upon concentration techniques such as the formol-ether method or the zinc-sulfate centrifugal flotation technique (Muller, 1975). Unfortunately, efforts to relate the egg production of a female worm to the number of eggs recovered from feces have so far proved unreliable. For example, it is becoming increasingly clear that while the number of eggs laid by an individual female worm may vary from time to time and may also vary between individuals, the total egg output of a given worm population is fairly constant and is related to the nutritional status of the host (Michel, 1974). Thus, until we understand the biology of egg production in nematodes, predictions of population size from egg counts continue to be of doubtful validity.

The best known method for the differential diagnosis of nematode infections by culturing larvae in fecal matter is the Harada-Mori test tube and filter paper method (Muller, 1975). Larvae can be identified by the characteristics summarized in Table 5.

Examination of Blood, Tissue, or Skin Snips for Microfilariae

The easiest method for the diagnosis of the various filariae is to make a wet preparation from a drop of blood with a coverslip. Microfilariae can be easily detected moving under low-power magnification. Light infections are often missed by this method. However, diethylcarbamazine has improved the sensitivity of wet preparations immeasurably, especially in periodic filariasis. When this drug is administered, the microfilariae may be found with ease in the peripheral and cutaneous circulation. When microfilariae are identified, thick blood smears should be stained with hematoxylin or Giemsa stain. Differential diagnosis of the various microfilariae such as *Onchocerca, Brugia, Loa, Dirofilaria, Wuchereria, Dipetalonema,* and *Mansonella* is then possible. Finally, membrane filtration techniques using nucleopore or millipore membrane filters (pore size 5 μm) fitted to hypodermic syringes with an adaptor have considerably improved diagnostic accuracy in filariasis.

In tropical Africa, the most important filarial nematode is *Onchocerca volvulus,* whose microfilariae are easily recognized by their larger size (270 to 320 μm long compared with *Dipetalonema* or *Mansonella*) and their characteristic head and pointed tail. The most common diagnostic procedure in *Onchocerca* infections is examination of skin snip biopsies for emerging larvae. Alternatively, dermal juices obtained after scarification of the skin can be examined after staining. This procedure may pick up not only *Onchocerca* but also *Dipetalonema* and even malarial parasites.

The best site for obtaining skin snips may depend upon the geographic strain of the parasite. Skin snips can be taken with a razor blade, sharp scissors, or a scleral punch such as the Holth model. The snips are then examined for living microfilariae in water or saline, with or without teasing. The number of emerging microfilariae is usually counted after 10 to 30 minutes. Examination of emerging larvae in water is preferred, especially under dry conditions, although emergence tends to take longer.

In large-scale epidemiologic surveys, reasonably good results can be obtained by placing

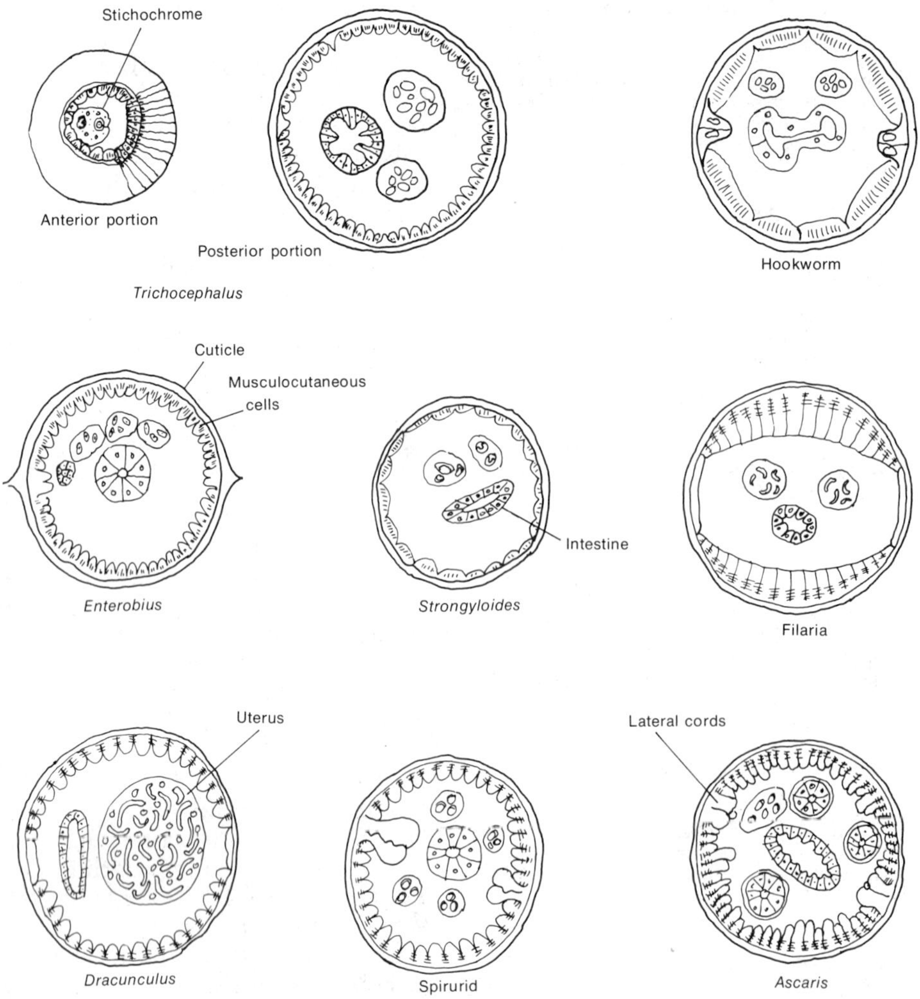

FIGURE 12. *Transverse sections of females of various nematodes (from Muller, 1975; after Chitwood and Lichtenfels, 1972).*

individual snips in a drop or two of saline in wells of an agglutination tray. The microfilariae emerge over a 24-hour period and are then preserved in the wells by the addition of a drop of 10 per cent formol-saline. The trays are sealed with a transparent plastic sealer, and the preserved material can then be examined at leisure.

Immunodiagnostic Techniques

As emphasized earlier, an individual nematode is a complex population of cells and the source of a complex assortment of antigens. This fact is the source of the two main problems with immunodiagnostic detection of nematode infections. These problems are the lack of specific, well-characterized antigens (this introduces an element of uncertainty in diagnosis) and the high level of cross-reactivity that confuses many positive results. Except in such situations as visceral larva migrans, in which the absence of adult parasites precludes easy detection of the causative agent, parasitologic diagnosis is more reliable at the present time. Nevertheless, many standard immunologic techniques have been used with good, variable, or indifferent results.

Standard serologic procedures such as precipitation tests have been used for the detection of a wide variety of nematodes and other helminths. Serologic procedures have been useful in diagnosing hookworm and *Trichinella* infections, but success often depends on a continuing challenge of the immune system with living parasites. For example, in hookworm infections, there is a lack of correlation between immunologic reactions and the presence of eggs in the feces.

Complement fixation tests (CFT) have been used for the detection of many nematode infections but particularly for visceral larva migrans. Its reproducibility and sensitivity for some nematode infections are questioned, and anticomple-

TABLE 5. The Differentiation of Infective Filariform Larvae Found in Cultures from Human Fecal Specimens

	STRONGYLOIDES[a]	TRICHOSTRONGYLUS	ANCYLOSTOMA	NECATOR	TERNIDENS
Length (microns)	500	750	660	590	680 (630–730)
Sheath	Absent	Present	Sheath 720 microns, striations not clear	Sheath 660 microns, striations clear at tail end	Present
Length of esophagus as proportion of total body length	1:2	1:4	1:3	1:3	1:3
Intestine	Straight	Intestinal lumen zigzagged	Anterior end narrower in diameter than esophageal bulb. No gap between the esophagus and the intestine.	Anterior end as wide as the esophageal bulb. Gap between the esophagus and the intestine.	Following esophagus is a pair of sphincter cells. Intestinal lumen somewhat zigzagged.
Tail	Divided into three at the tip	End of tail knob-like	Blunt	Sharply pointed	Pointed
Head	–	–	Blunt	Rounded	–
Mouth	–	–	Mouth spears not very clear and parallel	Mouth spears clear and divergent	–

[a]Free-living adults and first-stage (rhabditiform) larvae likely to be present as well.

TABLE 6. Summary of Drugs of Choice with Nematode Parasites of Man

NEMATODE SPECIES	1 THIABENDAZOLE (MINTEZOL)	2 MEBENDAZOLE (TELMIN)	3 LEVAMISOLE (KETRAX)	4 PIPERAZINE ADIPATE (ANTEPAR)	5 BEPHENIUM HYDROXYNAPHTHOATE (ALCOPAR)	6 PYRANTEL EMBONATE (COMBANTRIN)	7 PYRVINIUM EMBONATE (VANQUIN)	8 DICHLORVOS	9 BITOSCANATE	10 DIETHYLCARBAMAZINE (HETRAZAN)	11 DITHIAZANINE IODIDE
Ascaris lumbricoides		++++	++++	++++		++++					++
Trichocephalus (Trichuris trichiura)	++	+++						++			++
Ancylostoma duodenale		+++			+++	++					
Necator americanus		+++				+++					
Strongyloides stercoralis	++++	++					++				+++
Enterobius vermicularis	+++	+++		+++		++++					++
Trichostrongylus	+++				+++	+++					
Larva migrans	++	++			++				+		
Dracunculus	++									+++(mf)	
Trichinella	++	+++									+
Wuchereria (mf)										+++	
Brugia (mf)										+++	
Onchocerca (mf)										+++	
Loa										++	

Also diphetarson, suramin, metriphonate.

mentary activity in sera collected in field studies of onchocerciasis has been a problem (WHO, 1976).

The immunofluorescent antibody test (IFA) has gained wide support, especially for the diagnosis of onchocerciasis. Using frozen sections of the adult worm as antigen, up to 90 per cent of positive cases can be identified by this procedure. Titers of between 1:20 and 1:40 are considered significant. There is good correlation between this test and the results of skin biopsies in surveys of onchocerciasis.

The latest addition to the serologic diagnosis of helminth infections is the enzyme-linked immunosorbent assay (ELISA). This test utilizes the enzymes alkaline phosphatase or peroxidase, labeled with anti-immunoglobulin, in antigen-coated tubes or microtiter plates. This promising technique is as sensitive as the IFA test and is easier to perform.

Other serologic and immunologic diagnostic procedures have been attempted, but results have been so variable that no further consideration will be given to them here.

DRUG SUSCEPTIBILITY

The search for new and effective drugs against human helminthiasis has underscored the need for a greater understanding of the physiologic basis of anthelmintic action. An ideal anthelmintic should combine extreme toxicity against the parasite with complete inaction against the host tissue. Generally, action is directed against a target tissue or physiologic process whose special position in the overall physiology of the organism guarantees that if the target is rendered non-operational, the parasite will die. With nematodes, four main targets have been identified: neuromuscular function, energy metabolism, lipoproteins, and hemoproteins. Effective anthelmintics presently in use affect one or the other of these targets.

Table 6 summarizes the anthelmintics of choice for the various human nematodes. The twelve anthelmintics listed represent the best of the newer drugs, many of which have come into use in the last decade or so. In addition to these twelve, diphetarsone (an arsenical), suramin (a urea derivative), and metriphonate (an organophosphorus compound) are also used, particularly for the treatment of filariasis. Of the older drugs, such as carbon tetrachloride and tetrachloroethylene, the latter is still used occasionally for the treatment of hookworms.

Of the newer drugs, the *piperazine* derivatives, piperazine adipate or citrate and diethylcarbamazine citrate, have been particularly popular for the treatment of ascariasis and filariasis, respectively. These drugs are anticholinergic and cause hyperpolarization of the neuromuscular junction. The parasite is expelled as a result of the consequent paralysis. *Thiabendazole* is an effective broad-spectrum anthelmintic that is active against *Ascaris, Strongyloides, Enterobius, Dracunculus, Trichinella,* and larva migrans, among others. It is an inhibitor of fumarate reductase, an important enzyme of energy metabolism in nematodes. *Tetramisole* and *Levamisole* among the newer drugs also act by selective enzyme inhibition. In this case, succinic dehydrogenase, which acts as a fumarate reductase in nematode mitochondria, is also inhibited (Davis, 1973).

Bephenium, which is often given as the hydroxynaphthoate, is active against hookworms, *Trichostrongylus,* and larva migrans. The mode of drug action is not well understood, but it is assumed to have a blocking action on the neuromuscular junction. It is more effective against mucosal than luminal helminths.

EPIDEMIOLOGY

Epidemiologic techniques may be expected to provide the basis for planning and evaluating preventive health care, defining the major patterns and distribution of parasitic disease, and describing and classifying these conditions. The determinants of the distribution and maintenance of human parasitic disease, especially in the tropics, have received scant attention. Ultimate control of the major nematodes such as the hookworms and filariae, however, depends on our understanding of these factors.

The maintenance of a parasitic population in a community depends upon a complex series of interrelated factors (including host behavior) that regulate the number of parasites within individuals as well as the population size of the community. These factors may determine the level and frequency of contact between individual members of the two populations and thus the incidence and intensity of infection. The life cycle of the parasite is important in these considerations because it indicates whether one or two hosts (as distinct populations) are involved, as well as whether one or more parasitic stages should be considered in the total epidemiologic situation. It is usual to express the dynamics of such a host-parasite system in a flow chart such as that indicated in Figure 13 for *Ascaris* and *Trichuris* (direct life cycle) and the hookworms (direct life cycle with a stage consisting of free-living organisms). Although these life cycles of *Ascaris* and *Trichuris* are fairly simple, they involve three distinct populations of organisms, namely, the host, the adult

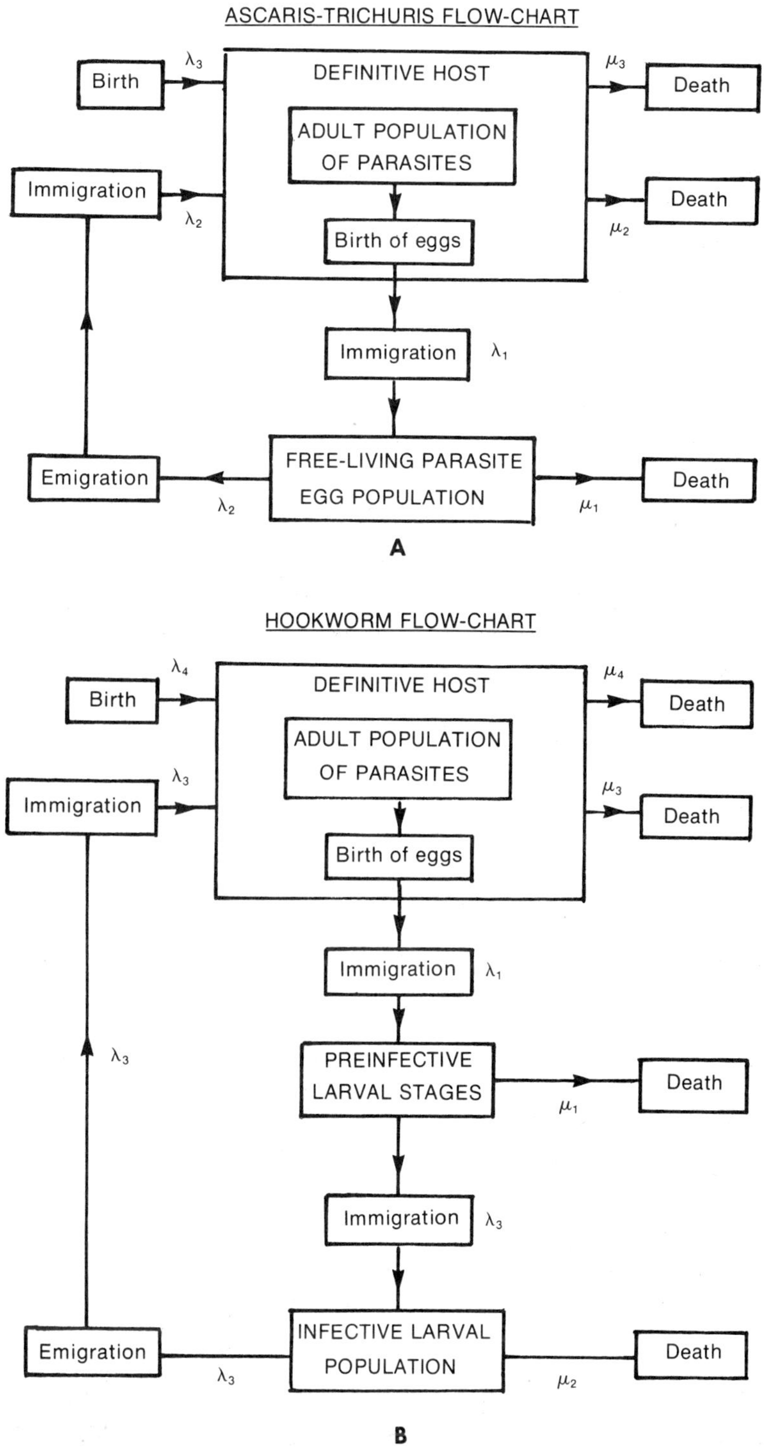

FIGURE 13. *Diagrammatic representation (as a flow-chart) of the direct life cycle of* Ascaris/Trichuris (A) *and the Hookworm* (B). *The population (rate) parameters indicated in the charts determine the flow of parasites through the system and are defined as rates per parasite per unit of time (based on Anderson, 1976).*

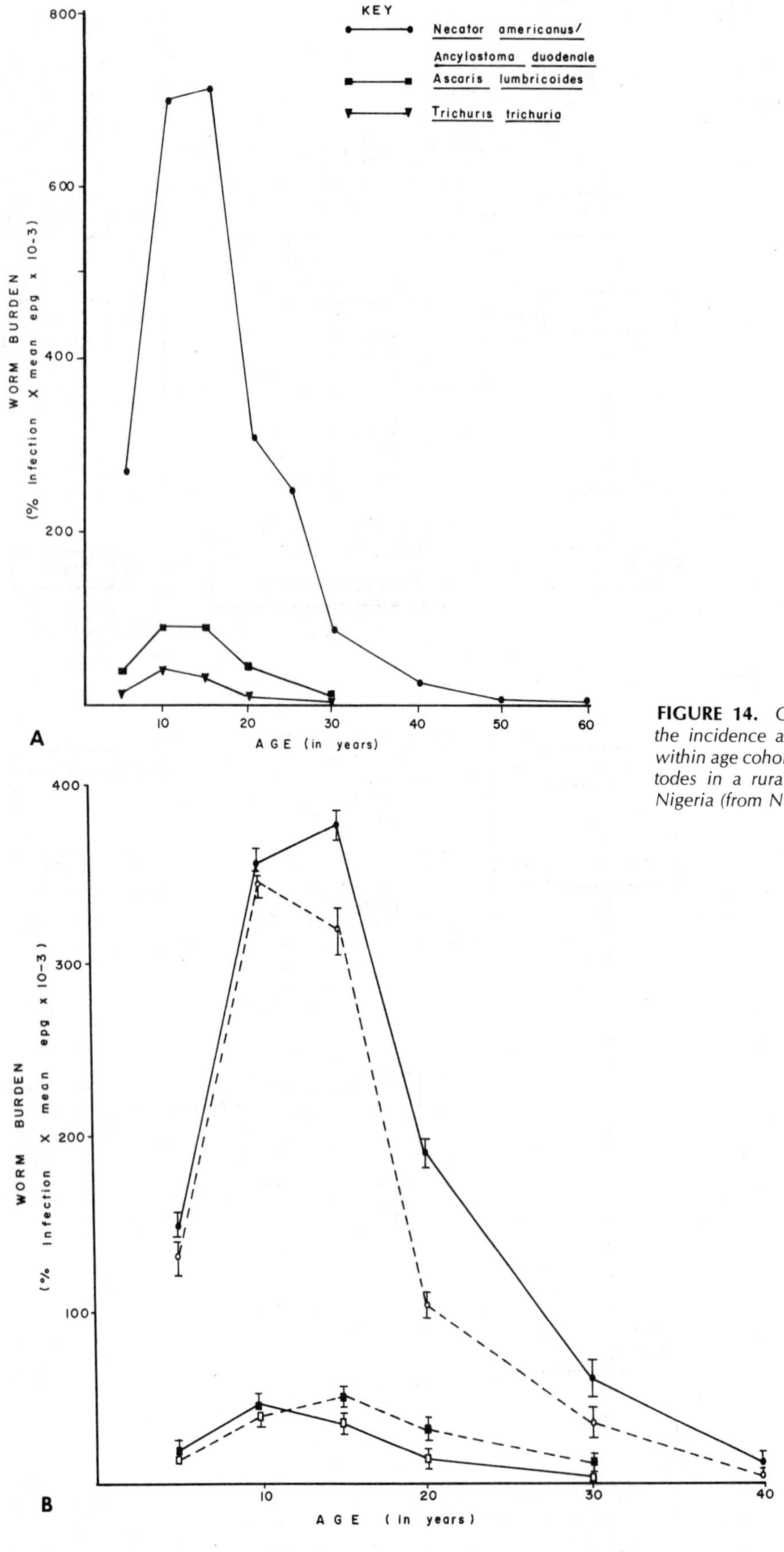

FIGURE 14. *Graphs summarizing (A) the incidence and (B) the worm burden within age cohorts of three parasitic nematodes in a rural community in Nsukka, Nigeria (from Nwosu and Anya, 1980).*

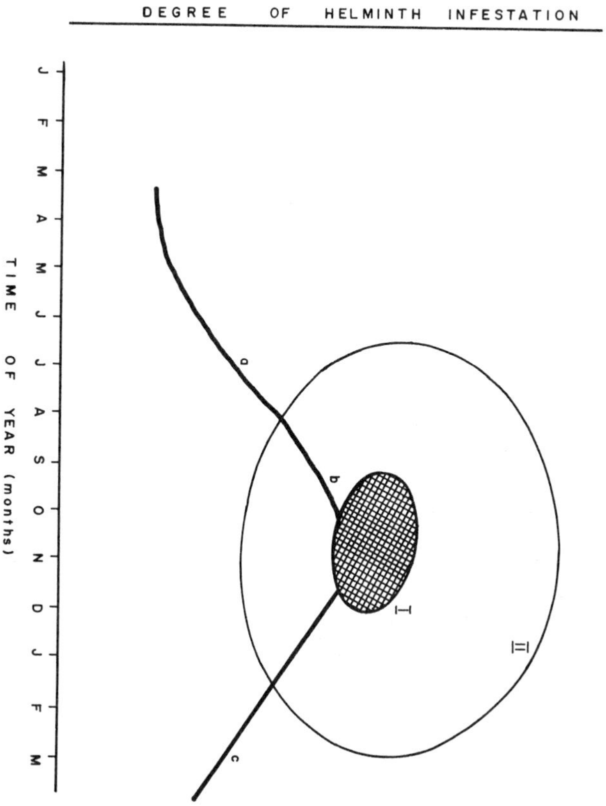

FIGURE 15. *Diagrammatic representation of the relationship between infection pressure, season, and distribution within a community of nematode infections and the exploitation of this relationship in control procedures (see text – from Nwosu and Anya, 1980).*

parasite, and the protected but free-living egg population. Any factor that changes the numbers of any of these populations will alter the infection rate and hence the epidemiologic parameters of the equilibrium between host and parasite. For the hookworm, four distinct populations are involved. In each population, such factors as age, physiology, and ecology will determine population size, contact between the populations, and, ultimately, the infection pressure (Udonsi, 1980). In recent years, the introduction of mathematical models in the analysis of host-parasite systems has increased our understanding of the epidemiologic characteristics of some parasitic diseases (Bradley, 1974; Anderson, 1976).

In our laboratories in Nsukka, this approach has been used in an epidemiologic study of parasitic infections, particularly hookworms. As can be seen from Figure 14, which represents a study of a defined rural community, a small proportion of the population (the 5- to 20-year-olds, 19.5 per cent of the population) harbors a major proportion of the community worm burden. A notable seasonality in the pattern of these hookworm infections has been observed (Nwosu and Anya, 1980). This has given rise to the concept of a target time/target population approach to the control of these helminths. The basic idea is to concentrate mass chemotherapy efforts on that segment of the population with the preponderance of the worm burden during those months that are unfavorable to transmission of the parasites. Thus, the worm population would be depressed below the threshold level for maintenance of the parasite population (within the host population) at a fraction of the normal cost. The concept is summarized in Figure 15. The circles represent the total population (I) and the segmental population (II) with the high worm burden, while a, b, and c represent the periods of normal, high, and depressed infection rates during the year. Concentration of control efforts on II and c is the strategy of choice.

Some of the original work reported in this chapter has been supported by grants from the British Medical Research Council (ODM Research Scheme R2836) and the Senate Research Grant Committee of the University of Nigeria to whom grateful acknowledgment is made. I would also like to acknowledge helpful discussions with my colleagues of the Parasitic Nematodes Research Group, especially Drs. Nwosu and Anubogu.

References

Anderson, R. M.: Dynamic aspects of parasite population ecology. In Kennedy, C. R. (ed.): Ecological Aspects of Parasitology. Amsterdam, North-Holland Publishing Company, 1976.

Anya, A. O.: Studies on the structure and chemical composition of the nematode cuticle. Observations on some oxyuroids and *Ascaris*. Parasitology 56:179, 1966.

Anya, A. O.: Serotonin (5-hydroxytryptamine) and other indolealkylamines in the male reproductive tract of a nematode. Int J Parasitol 3:573, 1973.

Anya, A. O.: Physiological aspects of reproduction in nematodes. Adv Parasitol 14:267, 1976.

Bradley, D. J.: Stability in host-parasite systems. In Usherand, M. B., and Williamson, M. H., (eds.): Ecological Stability. London, Chapman and Hall, 1974.

Chabaud, A. G.: CIH keys to the nematode parasites of vertebrates. Commonwealth Agricultural Bureau, No. 1. Slough, Buckinghamshire, England, 1974.

Crofton, H. D.: Nematodes. London, Hutchinson, 1966.

Davis, A.: Drug Treatment in Intestinal Helminthiasis. Geneva, World Health Organization, 1973.

De Conick, L.: De symmetrie — verhoudingen aan het vooreinde der vrijlevende Nematoden. Natuurwetench. Tijdschr (Ghent) 24:29, 1942.

Green, C. D.: The attraction of male cyst nematodes by their females. Nematologica 13:172, 1967.

Harris, J. E., and Crofton, H. D.: Structure and function in the nematodes: Internal pressure and cuticular structure in *Ascaris*. J Exp Biol 34:116, 1957.

Hyman, L. H.: The Invertebrates, Vol. 3. New York, McGraw-Hill Book Company, 1951.

Larsh, J. E., Jr., and Race, G. J.: Allergic inflammation as a hypothesis for the expulsion of worms from tissues: A review. Exp Parasitol 37:251, 1975.

Maggenti, A. R.: Taxonomic position of nematoda among the pseudocoelomate bilateria. In Croll, N. A. (ed.): The Organization of Nematodes. New York, Academic Press, 1976.

McLaren, D.: Nematode sense organs. Adv Parasitol 14:195, 1976.

Michel, J. F.: Arrested development of nematodes and some related phenomena. Adv Parasitol 12:279, 1974.

Miller, T. A.: Effect of x-irradiation upon the infective larvae of *Ancylostoma caninum* and the immunogenic effect in dogs of a

single infection with 40 kr-irradiated larvae. J Parasitol 50:735, 1964.
Muller, R.: Worms and Disease. London, William Heinemann Medical Books Ltd., 1975.
Nigon, V.: Le determinisme du sex et la pseudogamie chez un nematode parthenogenetique. Bull Biol Fr Belg 81:1947.
Nwosu, A. B. C., and Anya, A. O.: Seasonality in Human Hookworm infection in an endemic area of Nigeria, and its relationship to rainfall. Tropenmed Parasit 31:(1980).
Ogilvie, B. M., and Jones, V. E.: Immunity in the parasitic relationship between helminths and hosts. Prog Allergy 17:93, 1973.
Rogers, W. P., and Sommerville, R. I.: The infectious process and its relationship to the development of early parasitic stages of nematodes. Adv Parasitol 6:327, 1968.
Selby, G. R., and Wakelin, D.: Transfer of immunity against *Trichuris muris* in the mouse by serum and cells. Int J Parasitol 3:717, 1973.
Udonsi, J. K.: Studies on the ecology of the infective larvae of *Necator americanus* (Stiles, 1902) in relation to the epidemiology of human hookworm infections. Ph.D. Thesis, University of Nigeria, 1980.
Wakelin, D., and Lloyd, M.: Accelerated expulsion of adult *Trichinella spiralis* in mice given lymphoid cells and serum from infected donors. Parasitology 72:307, 1976.
World Health Organization: Epidemiology of Onchocerciasis. Technical Report Series No. 597. Geneva, 1976.
Zmoray, I., and Guttekova, A.: Ecological aspect of the study of the intestinal structure of nematodes. Z Parasitenk 39:127, 1972.

80 PLATYHELMINTHES

Zbigniew S. Pawlowski, M.D.

Some of the platyhelminthes (flat worms) are free-living animals (for example, Turbellaria), but most are exclusively parasitic and belong to the classes Trematoda (flukes) and Cestoda (tapeworms). Only a few of the Trematoda and Cestoda are parasites of man; the others are parasites of other vertebrates (Tables 1 and 7).

TREMATODA (FLUKE WORMS)

The trematodes parasitizing man can be classified according to morphologic criteria, localization in the human host, and specificity for the human host (Table 1).

Trematodes are usually classified by their anatomic location, because each family largely localizes in a certain anatomic site of the body; for example: (1) blood flukes — family Schistosomatidae; (2) liver flukes — family Opisthorchiidae and Dicrocoeliidae, Fasciolidae, except Fasciolopsis; (3) intestinal flukes — Fasciolopsis and family Heterophyidae, Echinostomatidae; and (4) lung flukes — family Troglotrematidae.

The host-specificity of trematodes correlates well with their prevalence and medical importance. For example, there are four different degrees of specificity for man among the Schistosoma: (1) man is the main definitive host of *Schistosoma haematobium, Schistosoma intercalatum,* and *Schistosoma mansoni;* (2) man shares *Schistosoma japonicum* with a wide range of animals; (3) man is only an occasional definitive host of *Schistosoma matthei* and *Schistosoma rhodhaini;* and (4) man can be invaded by other trematodes, such as Gigantobilharzia and Trichobilharzia, which cannot complete their development in man and can cause only skin lesions. The trematodes of groups 1 and 2 are of greater medical importance than the occasional or atypical parasites of groups 3 and 4 (Table 1).

MORPHOLOGY AND LIFE CYCLES

Blood Flukes — Schistosomatidae

Schistosomatidae are exceptional among trematodes in that they are dioecious (sexes separated). The female worm has a cylindric body and the male a canoe-shaped body with a gynecophoral canal in which the adult female spends most of her time. The main species of the schistosomes of man are morphologically distinct (Figs. 1 and 4).

The life cycle of *S. haematobium* is shown in Table 2. *S. mansoni* requires a shorter time of development in the snail (four to five weeks) and in man (25 to 28 days) before first egg production. *S. japonicum* has a shorter time of development in man, but a longer one in the snail (over seven weeks). It is also produces many more eggs (3500 per day per female).

Liver Flukes

Clonorchis (Opisthorchis) sinensis, Opisthorchis felineus, and *Opisthorchis viverrini* are common liver flukes (Komiya, 1966). All have a flat lanceolate body between 10 and 20 mm in length. The testes of *Opisthorchis* spp. are lobed, whereas those of *C. sinensis* are in a dendritic arrangement. The differences between *O. felineus* and *O. viverrini* are biologic rather than morphologic.

TABLE 1. Trematodes Parasitizing Man: Habitat and Hosts

SPECIES	HABITAT IN MAN	DEFINITIVE HOSTS OTHER THAN MAN	SNAILS AS FIRST INTERMEDIATE HOSTS	OTHER INTERMEDIATE HOSTS (OR PLANTS INVOLVED IN TRANSMISSION)
Schistosoma haematobium	Vesical and pelvic venous plexuses	–	*Bulinus*	–
S. intercalatum	Mesenteric veins	–	*Bulinus*	–
S. mansoni	Mesenteric veins	Baboon, rodents	*Biomphalaria*	–
S. japonicum	Mesenteric veins	Many domesticated and wild animals	*Oncomelania*	–
Clonorchis (*Opisthorchis*) *sinensis*	Bile ducts	Dog, cat, pig	*Bulinus, Paraphossalurus*	Cyprinid fishes
Opisthorchis felineus	Bile ducts	Cat, dog, pig, fox	*Bithynia*	Cyprinid fishes
Opisthorchis viverrini	Bile ducts	Civet cat	*Bithynia*	Cyprinid fishes
Fasciola hepatica, Fasciola gigantica	Bile ducts	Sheep and other herbivores	*Lymnaea*	Watercress and grass
Fasciolopsis buski	Small intestine	Pig	*Segmentina, Planorbis*	Water caltrop and water chestnut
Heterophyes heterophyes	Small intestine	Dog, cat, fish-eating wild carnivores	*Pirenella, Cerithidia*	Mullet and tilapia fishes
Metagonimus yokogawai	Small intestine	Dog, cat, pig, fish-eating birds	*Semisulcospira*	Different fresh water fishes
Gastrodiscoides hominis	Coecum, colon	Pig	*Helicorbis*	Water caltrop
Paragonimus westermani	Lungs	Cat, tiger, leopard, dog, pig, monkey	*Semisulcospira, Thiara, Oncomelania*	Fresh water crabs and crayfish

The life cycles of *C. sinensis, O. felineus*, and *O. viverrini* are similar, but the snail and fish hosts and the geographic distribution differ (Tables 1 and 3, Fig. 3).

Dicrocoelium dendriticum, an unusual parasite of man, completes its strange life cycle on the ground by using terrestrial snails and ants as hosts. *Fasciola hepatica* is not uncommon in some countries (France, Algeria, Peru, and Cuba). It has a life cycle similar to that of Fasciolopsis (see below), but encysts on watercress, lettuce, and radishes.

Intestinal Flukes

The most important intestinal fluke is *Fasciolopsis buski,* the largest fluke of man (up to 70 mm long and 20 mm wide). It has no second intermediate host (Table 4, Figs. 2 and 3).

The remaining intestinal flukes *(Heterophyes, Metagonimus*, and *Echinostoma)* are small parasites, 2 to 8 mm long. Fish are the second intermediate hosts (Table 1, Figs. 2 and 3).

Lung Flukes

Paragonimus westermani is the common representative of the family of Troglotrematidae (Yokogawa, 1965). Unlike other trematodes, it has an oval, thick body with a ventral sucker in the middle. Its life cycle includes two intermediate hosts (Table 5, Figs. 2 and 3).

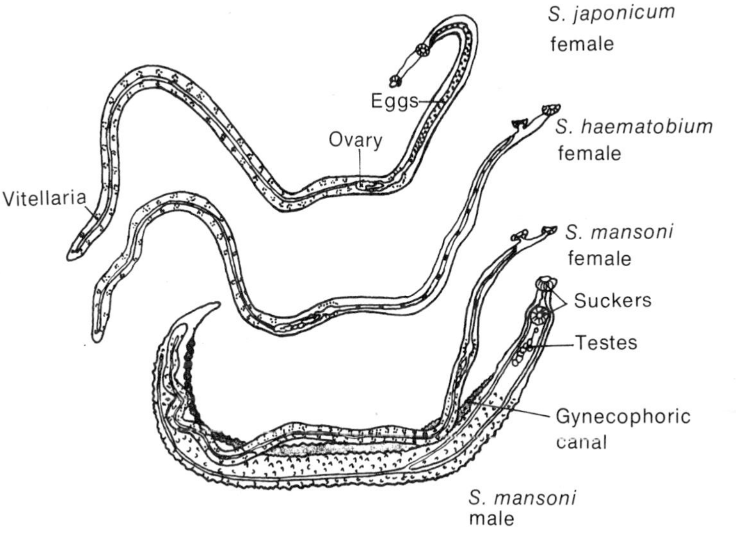

FIGURE 1. *Adult schistosomes. Diagram of morphologic characteristics.*

TABLE 2. Schistosoma haematobium: Life Cycle

HOST AND HABITAT	TIME	STAGE AND ACTIVITY	NUMBER
Man	2–25 years	EGGS fully embryonated expelled in urine (and feces)	20–260 per worm per day
Water	16–32 hrs	MIRACIDIUM hatches in water and penetrates snail	1
Snails *Bulinus* spp.	5–6 weeks	PRIMARY SPOROCYST produces SECONDARY SPOROCYSTS, which produce CERCARIAE, which leave the snail host for months	1 n n^m = ca 250,000
Water	Up to 8 hrs	cercaria swimming in water and penetrating skin	1
Man lungs, liver, venous plexuses	6–12 weeks	preadult SCHISTOSOMULA migrating and developing in lungs and liver into ADULT worm situated finally in pelvic and vesical venous plexuses and producing eggs	1 1

TABLE 3. Clonorchis (Opisthorchis) sinensis: Life Cycle

HOST AND HABITAT	TIME	STAGE AND ACTIVITY	NUMBER
Man, dog, cat, pig	Up to 20–30 years	EGGS fully embryonated in bile and feces	1000 per worm per day
Water	Up to 3 months	Eggs survive in water	
Snail *Bulimus* spp. *Parafossalurus* spp. *Semisulcospira* spp.	3–4 weeks	MIRACIDIUM hatches from egg ingested by snail SPOROCYST produces REDIAE, which produce CERCARIAE; these actively leave snail-host for months	1 1 n n^m
Water	1–2 days	cercaria, free-swimming, penetrate second intermediate host	1
Cyprinid fish	3 weeks	METACERCARIA encysts in tissue and survives for many months	1
Man, dog, cat, pig	3–4 weeks	Metacercaria excysts in duodenum and PREADULT migrates through ampulla of Vater to bile ducts and develops into an ADULT worm, producing eggs	1 1

TABLE 4. Fasciolopsis buski: Life Cycle

HOST AND HABITAT	TIME	STAGE AND ACTIVITY	NUMBER
Man, pig	6 months	EGGS undeveloped, expelled in feces	21,000–28,000 per worm per day
Water	3–7 weeks Few hours	egg embryonates in water and hatches as MIRACIDIUM which penetrates snail	1 1
Snail *Segmentina* spp. *Planorbis* spp.	Few weeks	SPOROCYST produces REDIAE of first generation, which produce REDIAE of second generation; these produce CERCARIAE, which actively leave snail	1 n n^m $(n^m)^p$
Water (Water chestnut) (Water caltrop)	Several months	cercaria shortly encysts on water vegetation and becomes METACERCARIA, which survives several months	1 1
Man, pig	3–4 months	Metacercaria excysts in duodenum and PREADULT develops into an ADULT worm	1 1

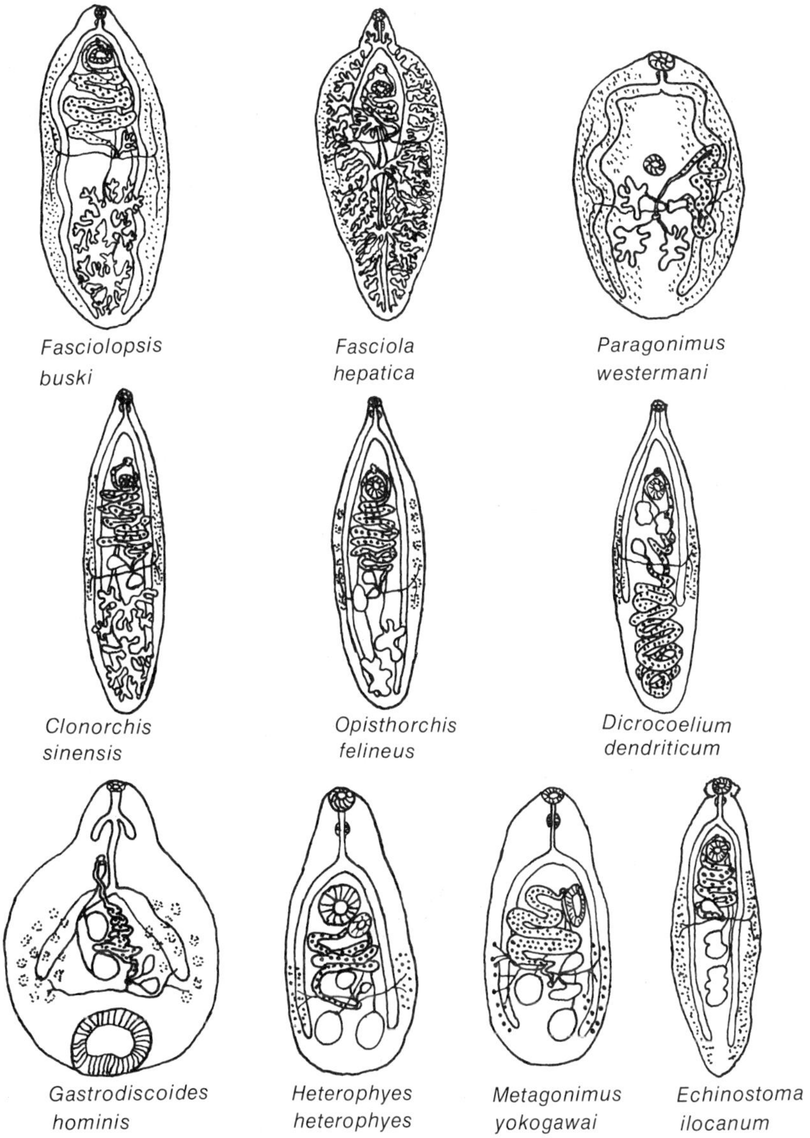

FIGURE 2. *Adult trematodes (except the schistosomes). Diagram of morphologic characteristics.*

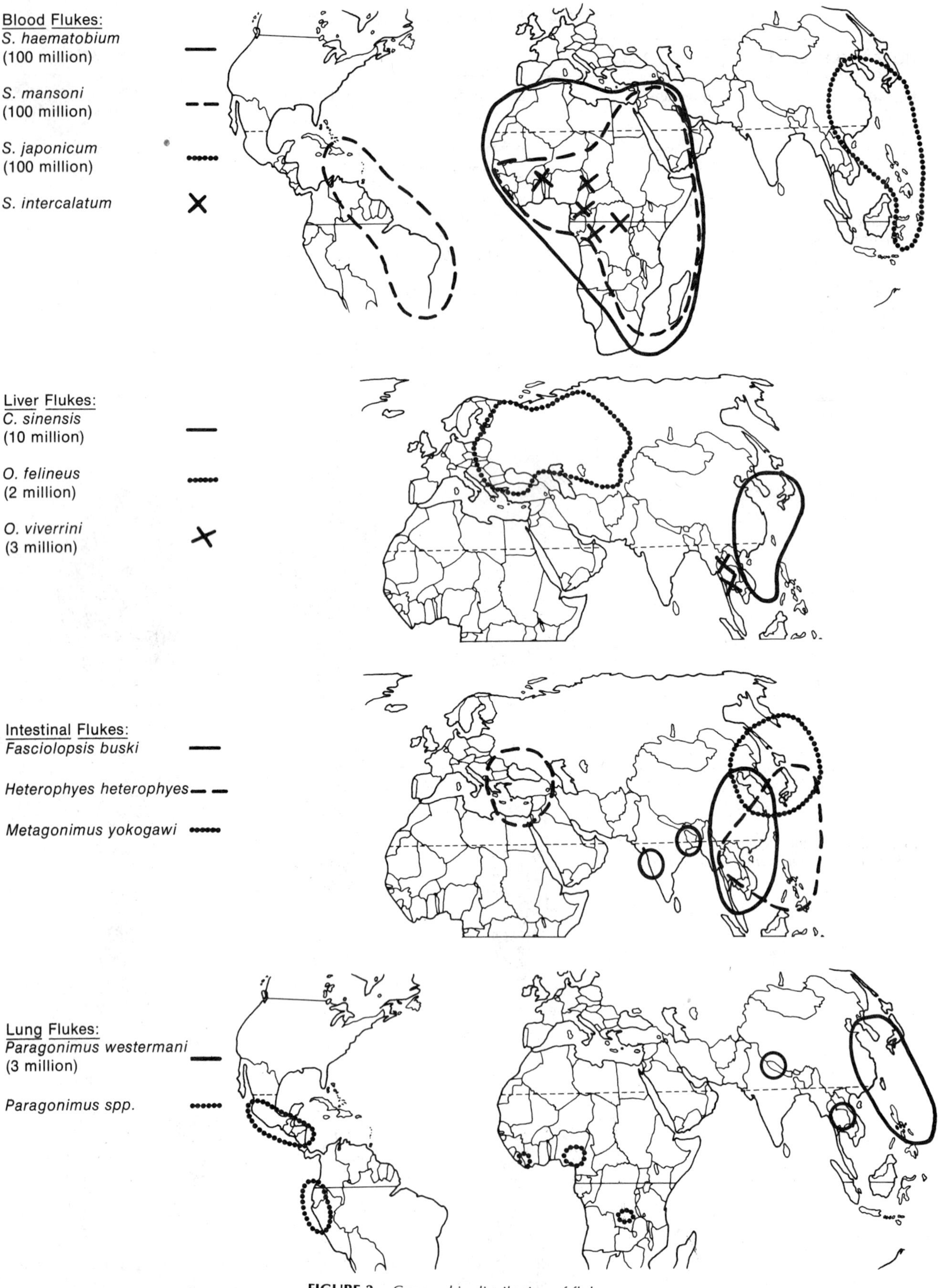

FIGURE 3. *Geographic distribution of flukes.*

TABLE 5. Paragonimus westermani: Life Cycle

HOST AND HABITAT	TIME	STAGE AND ACTIVITY	NUMBER
Man, various mammals	6–20 years	EGGS expelled in sputum or feces	?
Water	3 weeks	egg develops into MIRACIDIUM, which hatches in water and penetrates snail in 24 hrs.	1 1
Snail *Semisulcospira* spp. *Thiara* spp. *Oncomelania* spp.	3 months	SPOROCYST in snail tissue produces REDIAE I generation, which emerges as REDIAE II generation; Rediae II produce CERCARIAE, which actively leave snail	1 n n^m $(n^m)^p$
Water	24–48 hrs	cercaria, free-swimming and penetrating crabs	1
Crabs	Several months	METACERCARIA encysted in crab tissue	1
Man		Ingested by definitive host; hatches in duodenum, as	
	20 days	PREADULT migrates through gut wall, abdominal and pleural cavities, and lung tissue	1
	5–6 weeks	ADULT, produces eggs in lung cysts	1

IMMUNOLOGY

Natural resistance to trematodes depends on many biochemical, physiologic, genetic, and nutritional factors, and is low in man. Acquired immunity against trematodes rarely prevents infection but can limit its intensity. For example, new invading schistosomulae are rapidly destroyed by hosts who are already infected with adult worms. However, seven days after infection, preadult schistosomulae are no longer recognized by the host as foreign, probably owing to the incorporation of host antigens into the integument of the trematode. Living adult worms are usually not attacked by immunologic reactions but provoke a reaginic (IgE) response, releasing pharmacologic mediators of anaphylaxis. These hypersensitivity reactions are stage-specific and are intensive against eggs and cercariae. For example, schistosome eggs swept back from the mesenteric veins into the liver cause antibody-mediated reactions (Hoeppli phenomenon) first, and then a cell-mediated delayed hypersensitivity reaction producing eosinophilic granulomas. The skin reaction against penetrating cercariae is also of the cell-mediated type. Antigen-antibody complex diseases occur in schistosomiasis as glomerulonephritis (*S. mansoni*) or Katayama syndrome (*S. japonicum*), which resembles serum sickness with gastroenteritis.

There is no immunization against trematode infection. It is possible, however, that infection or immunization with some less pathogenic, but highly immunogenic, animal schistosomes might partially protect man against human schistosomes.

LABORATORY DIAGNOSIS

The symptoms or signs of human infections with trematodes are rarely pathognomonic. Therefore, the diagnosis is usually made by finding the eggs (Fig. 4) of the trematode in feces (most species), urine (*S. haematobium*), duodenal contents (*Clonorchis, Opisthorchis, Fasciola*), liver and rectal biopsies (*Schistosoma*), or sputum (*Paragonimus*). The flotation coprologic techniques are less effective than sedimentation or thick smear because the eggs are relatively heavy. Before eggs are produced it is difficult to make the diagnosis. The intradermal test, as well as the complement fixation, indirect hemagglutination, indirect fluorescent antibody, and precipitin tests, is of limited help in the early stages of infection with *Schistosoma, Clonorchis, Opisthorchis,* and *Paragonimus*.

DRUG SUSCEPTIBILITY

The intestinal flukes are susceptible to many antihelminthics. Liver, blood, and lung flukes are less susceptible, and the drugs are toxic and only partially effective. Recently some effective and less toxic drugs have been developed (praziquantel, amoscanate).

The blood flukes (Schistosomes) are susceptible to antimony compounds, which are rarely used now. The nonmetallic drugs niridazole and hycanthone are active against *S. haematobium*. Metrifonate, the organophosphorus cholinesterase inhibitor of insects, is also a potent inhibitor of nematode cholinesterase; as little as 0.1 μm

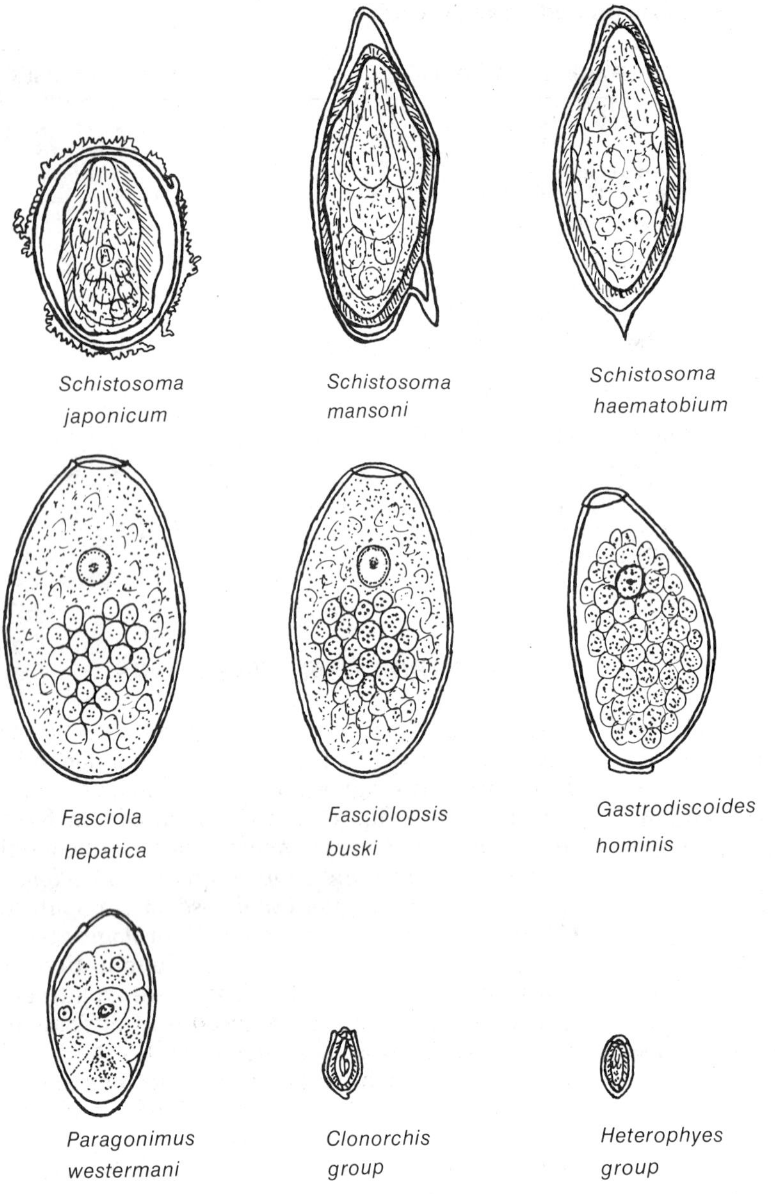

FIGURE 4. *Trematode ova. Diagram of morphologic characteristics.*

metrifonate inhibits the *S. haematobium* enzyme by 50 per cent, and lower concentrations inhibit *S. mansoni* cholinesterase. Oxamniquine is effective against *S. mansoni* only. Lung flukes and liver flukes are inhibited by bithionol or chloroquine and intestinal flukes by hexylresorcinol or tetrachlorethylene.

EPIDEMIOLOGY

Schistosomiasis is a worldwide problem second only to malaria in frequency (about 250 million people; Fig. 3). Some other trematode infections (paragonimiasis, clonorchiasis, fasciolopsiasis) are of serious, but primarily local, medical importance. Their distribution in the world is focal (Fig. 3) and depends on the presence of a suitable species of snail and certain local behavior factors, such as agricultural practices and ingestion of raw fish, crabs, or snails. In endemic areas, the transmission is usually continuous and results in frequent infection of rather low intensity, due to either infrequent exposure or some degree of concomitant immunity, or both. Epidemics of trematode infection are usually caused by changes in environmental or social factors. Thus, epidemics of *S. haematobium* occur around man-made lakes or in new irrigation systems that introduce the parasite or intermediate host into susceptible areas (Ansari, 1973).

The rise of schistosomiasis in human populations after the introduction of this fluke and its decline after the institution of control programs can be roughly predicted by using mathematical models (Fig. 5), (Macdonald, 1965).

Many of the trematodes parasitizing man are zoonotic in origin; however, the role of animal reservoirs differs and depends partly on human behavior. For example, *Paragonimus africanus* is a zoonotic infection in Nigeria and Cameroon, whereas the transmission of *P. westermani* in China, Japan, and Korea is probably independent of that in animals. In Indonesia, Malaysia, and Ceylon, the infections occur exclusively in animals.

Community control measures require education about transmission, sanitary disposal of feces and urine, and installation of safe water supplies. Mass-control programs are based primarily on improved irrigation and agricultural practices, on using molluscacides, and on mass treatment of infected people (schistosomiasis).

Personal protection is enhanced by using cercarial repellents; avoiding contaminated water for bathing, washing, and drinking (schistosomiasis); cooking fish, crabs, and crayfish (clonorchiasis, opisthorchiasis, heterophyiasis, paragonimiasis); and immersing water plants in boiling water before ingestion (fasciolopsiasis).

CESTODA (TAPEWORMS)

Tapeworms parasitizing man belong to the order Cyclophyllidea (families Taeniidae and Hymenolepididae) and Pseudophyllidea (*Diphyllobothrium* spp. and *Spirometra* spp.). All the adult tapeworms are intestinal parasites of their definitive vertebrate host; some are human tissue parasites in their larval stage (Table 6).

For some tapeworms *(T. saginata, T. solium)* man is the only definitive host. For a few species (*Diphyllobothrium* spp.) man is one of many vertebrate definitive hosts. For a larger number, man is an occasional definitive or an intermediate host (Pawlowski and Schultz, 1972) (Table 6).

MORPHOLOGY AND LIFE CYCLE

There are three large (4 to 15 m) tapeworms parasitizing man: *Taenia solium, T. saginata*, and *Diphyllobothrium latum*. Their morphologic characteristics are listed in Table 7 and Figure 6. *Hymenolepis diminuta* and *Dipylidium caninum* are of medium size (20 to 60 cm) and only *Hymenolepis nana* (dwarf tapeworm) is small (15 to 40 mm). The morphology and life cycle of *H. nana* is presented in Figure 7.

The larval stages of the tapeworms parasitizing man are as follows: (1) cysticercus (*T. solium*); (2) cysticercoid (*H. nana*); (3) coenurus (*T. multiceps*); (4) echinococcus (*Echinococcus granulosus* and *Echinococcus multilocularis*); and (5) sparganum, a plerocercoid-type larva (*Diphyllobothrium erinacei, Spirometra mansoni*). The first four are bladderworms of different sizes and different maccroscopic and microscopic structures (Table 8, Fig. 8).

Human tapeworms have three basic types of life cycles: (1) The first type has water as an environment, aquatic copepoda (crustaceans) as the first intermediate hosts, fish as the second, and many vertebrates as definitive hosts (*Diphyllobothrium*, Table 9); (2) The second type uses various herbivorous vertebrates as intermediate hosts and carnivorous vertebrates as definitive hosts (*Taenia, Echinococcus*, Table 10); (3) The third type requires only one host because the larval stages develop in the intestinal villi of the definitive host (*H. nana*) (Fig. 7). Those tape-

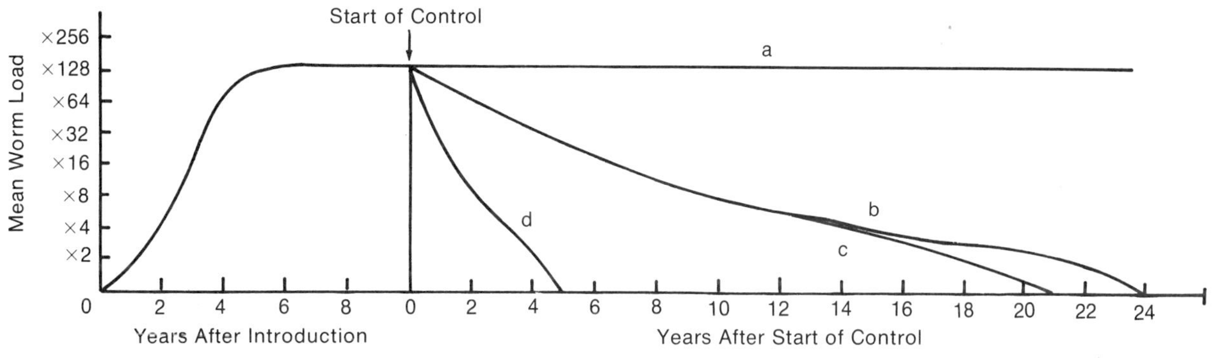

FIGURE 5. *Diagram of the rise and fall of the mean load of schistosomes in human populations as predicted by the mathematical model of Macdonald (1965). a = reduction of water contamination to 1/15 of the original by improvement of sanitation; b = reduction of snail population to 1/15 original; c = reduction of exposure to water to 1/15 by public education; d = reduction of longevity of worms to 1/15 by chemotherapy plus reducing b and c to 1/3 of their original values.*

TABLE 6. Cestodes Parasitizing Man: Hosts, Habitat, and Distribution

TAXONOMIC CLASSIFICATION	HABITAT IN MAN Adult Worm	Larval Stage	DEFINITIVE HOSTS OTHER THAN MAN	INTERMEDIATE HOSTS	WORLD DISTRIBUTION
Order Cyclophyllidea*					
Taeniidae					
Taenia saginata	Small intestine (jejunum)	No	No	Cattle and other Boviidae	Cosmopolitan
Taenia solium	Small intestine (jejunum)	CYSTICERCUS in muscle and internal organs (brain, eye)	No	Pig, man, and many other mammals	Central and South America, Mexico, South Africa, India, USSR
Taenia multiceps	–	COENURUS in brain, eye, skin	Dog and wild canidae	Sheep, other ruminants, and rabbits	Several areas in the world
Echinococcus granulosus	–	ECHINOCOCCUS in liver, lungs, brain	Dog and wild carnivora	Sheep, cattle, pig, horse, camel, goat, and many wild mammals	Cosmopolitan
E. multilocularis	–	Alveolar ECHINOCOCCUS in liver	Fox, domestic cat, and wild canidae	Wild rodents	Northern hemisphere
Hymenolepididae					
Hymenolepis nana	Ileum	Intestinal villi	Full development in man only		Cosmopolitan, warm countries
H. diminuta	Small intestine (jejunum)	No	Rats, mice, wild rodents	Fleas, flour beetles	Cosmopolitan
Order Pseudophyllidea					
Diphyllobothrium latum†	Small intestine (jejunum)	–	Cat, dog, and fish-eating carnivora	Planktonic copepoda (*Cyclops* and *Diaptomus*) and freshwater fish	Lakes, rivers, and deltas in Finland, Siberia, Canada, USA, Chile, Argentina

*Some other Cyclophyllidea occasionally invade man: e.g., the common dog tapeworm *Dipylidium caninum*.

†Several other species of Diphyllobothrium have been found in man in Alaska. Occasionally in the Far East plerocercoid larvae of *Sparganum proliferum* or *Spirometra mansoni* invade human tissues.

TABLE 7. Morphologic Differences Between Large Tapeworms in Man

	Taenia solium	*Taenia saginata*	*Diphyllobothrium latum*
ENTIRE BODY			
Length (m)	1,5–8	4–12	3–15
Maximal breadth (mm)	7–10	12–14	10–12
Proglottids (number)	700–1000	ca 2000	3000–4000
SCOLEX			
Diameter (mm)	0,6–1	1,5–2	Elongate 2–3 mm long
Suckers (number)	4	4	2 sucking grooves
Rostellum	Present	Absent	Absent
Hooks (number)	22–32	Absent	Absent
MATURE PROGLOTTIDS			
Testes (number)	375–575	800–1200	?
Ovary (number of lobes)	3	2	2
Uterus as	Blind tube	Blind tube	Coiled with a pore
Vaginal sphincter	Absent	Present	?
Genital atrium at	Lateral margin	Lateral margin	Ventral surface
GRAVID PROGLOTTIDS			
Uterus (number of branches each side)	7–12	18–32	None
Way of leaving host	In groups, passively	Single, spontaneously	

	Taenia solium	*Taenia saginata*	*Diphyllo–bothrium latum*	*Hymenolepis diminuta*	*Dipylidium caninum*	*Hymenolepis nana*
Scolex	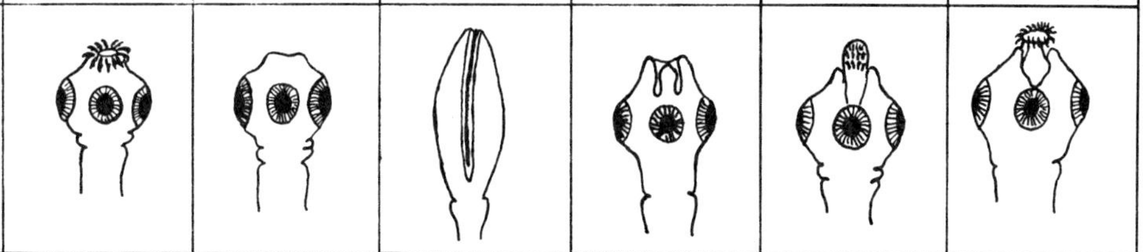					
Gravid proglottid	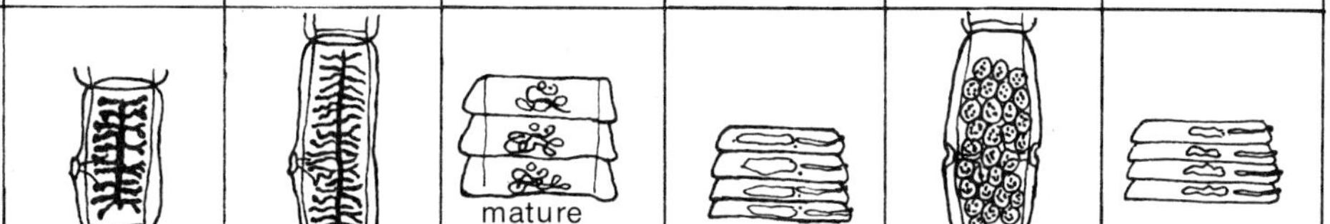		mature proglottids only			
Number of proglottids	700–1000	ca 2000	3000–4000	800–1000	60–175	ca 200
Length of strobila (cm)	150–800	400–1200	300–1500	30–60	20–40	1½–4

FIGURE 6. *Cestodes of man. Diagram of morphologic characteristics.*

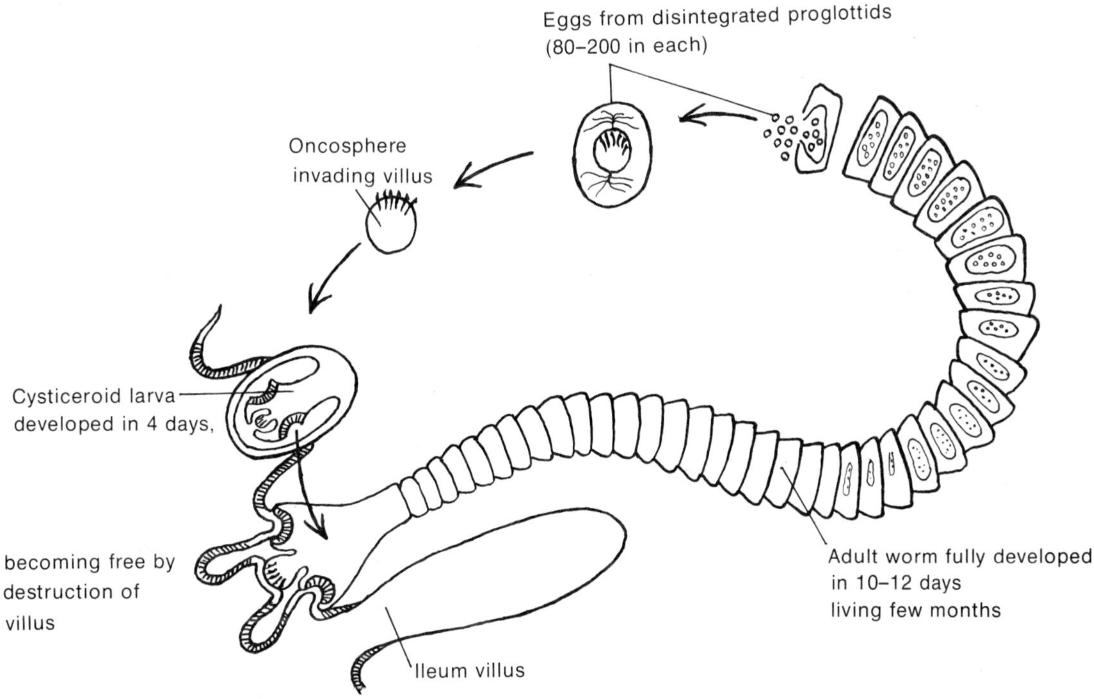

FIGURE 7. Hymenolepis nana. *Diagram of life cycle.*

Table 8. Morphologic Differences Between Teniid Larvae Parasitizing Man

	Taenia solium	*Taenia saginata*	*Taenia multiceps*	*Echinococcus granulosus*	*Echinococcus multilocularis*
TYPE OF BLADDER	Cysticercus	Cysticercus	Coenurus	Echinococcus	Alveolar Echinococcus
MACROSCOPIC STRUCTURE					
number of bladders	1	1	1	One primary cyst, many brood capsules	Many
maximum size (mm)	10	8	30	200	2
number of scolices in a bladder	1	1	100 if fertile	Many thousands if fertile	Few or none
MICROSCOPIC STRUCTURE					
hooks on scolex	Present	Absent	Present	Present	Present
surface	Cuticle	Cuticle	Cuticle	Stratified hyaline membrane	Thin membrane
make-up of wall	Wortlike processes	Rugae	Smooth and also rugae	Smooth	Smooth
superficial protuberances					
height (μm)	15–27	23–27	15–22	No	No
width at base	27–38	50–70	28–46	No	No

worms (*H. diminuta*, *Dipylidium*) with insects (fleas, beetles) as intermediate hosts rarely parasitize the intestine of man.

IMMUNOLOGY

The natural resistance of man against cestodes is low. Acquired immunity has been confirmed only in experimental *H. nana* infections. Adults are more resistant than children to human hymenolepiasis, and general immunity and nutrition are the major influences on the course of infection. Living taeniid larvae usually cause no local cellular reaction, but dead parasites invoke a strong response that produces a certain degree of resistance to reinfection. Migrating oncospheres seem to be the most immunogenic stage.

Adult *T. saginata* tapeworms frequently cause a rise in blood IgE but allergic reactions are rare in taeniasis. However, fatal anaphylactic shock may follow rupture of a cyst of *E. granulosus*.

LABORATORY DIAGNOSIS

Stool examination is useful in the diagnosis of diphyllobothriasis and hymenolepiasis, but has some limitations in taeniasis. Eggs occur in the feces irregularly and those of *T. solium* and *T. saginata* are identical (Fig. 9). The correct differential diagnosis between *T. solium* and *T. saginata* is based on the morphology of scolices (hooks), mature segments (ovarian lobes), and gravid segments (vaginal sphincter, uterine branches) (Table 7). Laboratory identification of intestinal tapeworms is not possible until they start producing eggs and/or excreting proglottids. Differentiation of larval stages usually requires macroscopic and microscopic examination (hooks, protuberances; Table 9).

Immunodiagnostic tests are useful in cysticercosis and echinococcosis. Biopsy is of diagnostic value in the identification of subcutaneously localized larvae (cysticercosis, sparganosis).

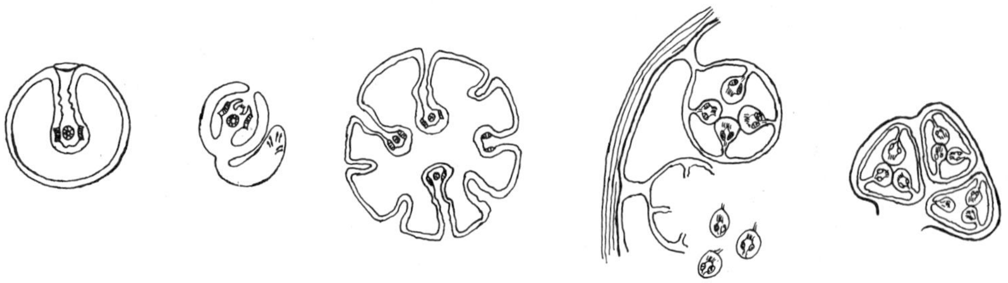

Cysticerus (*T. solium*) Cysticeroid (*H. nana*) Coenurus (*T. multiceps*) Echinococcus (*E. granulosus*) Alveolar echinococcus (*E. multilocularis*)

FIGURE 8. *Larval cestodes. Diagram of morphologic characteristics.*

TABLE 9. Diphyllobothrium latum: Life Cycle

HOST AND HABITAT	TIME	STAGE AND ACTIVITY	NUMBER
Man and many mammals	Up to 30 years	EGGS undeveloped, expelled in feces	Up to 1,000,000 per day
Water	at least 12 days 1–2 days	egg embryonates in water, and hatches free-swimming; CORACIDIUM is ingested by copepoda	1 1
Fresh water copepoda (crustaceans) *Diaptomus* spp. *Cyclops* spp.	2–3 weeks	PROCERCOID develops in body cavity; ingested by fish transforms into plerocercoid	1
Fresh-water plankton-eating fish (lota, perch)	4 weeks	PLEROCERCOID, which develops in muscle or connective tissues, invades mammals or fish when ingested	1
Fresh-water carnivorous fish (pike) as transport host	Months	plerocercoid	1
Man and many fish-eating mammals (dog, cat, bear)	3–5 weeks	PREADULT and ADULT tapeworm in small intestine, produces eggs	1 1

TABLE 10. Taenia solium: Life Cycle

HOST AND HABITAT	TIME	STAGE AND ACTIVITY	NUMBER
Man only	Up to 30 years	EGGS, fully developed, expelled in feces	500 per one proglottid
External environment	Months (years ?)	egg survives in water or soil	1
Pig; rarely in sheep, dog, cat, and man	30 minutes 60–75 days	ONCOSPHERE hatches in small intestine, migrates mainly to muscle tissue, and develops into CYSTICERCUS, a bladder larva, which lives for months	1 1
Man only	5–12 weeks	Ingested cysticercus evaginates and transforms into ADULT worm, which produces proglottids and eggs	1

T. saginata develops slower in cattle (10–12 weeks) and in man (12 weeks) and produces more proglottids (6–9 per day) and eggs (80,000) per one proglottid.

Diphyllobothrium latum — *Taenia solium*, *Taenia saginata* — *Dipylidium caninum* — *Hymenolepis nana* — *Hymenolepis diminuta*

FIGURE 9. Cestode ova.

DRUG SUSCEPTIBILITY

The intestinal tapeworms are susceptible to several taeniacides (niclosamide, paramomycin, Atabrine). Recently, some drugs (praziquantel, mebendazole) have been found that are effective against the larval stages of cestodes.

EPIDEMIOLOGY

Except for hymenolepiasis, cestode infections are zoonoses involving mainly domesticated animals. Their distribution depends chiefly on agricultural and pastoral practices (taeniasis, echinococcosis, coenurosis). When man is the only definitive host (*T. solium, T. saginata*), he is responsible for contamination of the environment with teniid eggs and proglottids of species invasive for the intermediate host (pig, cattle). Man rarely acquires infection from tapeworms circulating in feral (wild animal) cycles (*D. latum, E. multilocularis, T. multiceps, H. diminuta*).

Hands and food contaminated with cestode eggs play an essential role in transmission of infections with *E. multilocularis, T. solium* (cysticercosis), and *H. nana*. Autoinfection is rare in *T. solium* taeniasis, but is common in hymenolepiasis and is responsible for the continuation of infection for many years. National programs for the control of echinococcosis in Ireland, Cyprus, and New Zealand were based on education, control of the dog population, mass diagnosis and treatment of dogs, and sanitary practices of sheep slaughter and carcass inspection.

T. solium infections usually disappear spontaneously with improved sanitation and modern pig husbandry (Europe). Contamination of the environment by increased numbers of human *T. saginata* carriers makes the control of this taeniasis difficult, in spite of meat inspection that helps to identify and destroy many of the cysticerci.

Personal protection against cestode infections is effected by personal hygiene and avoiding the consumption of raw meat.

References

Ansari, N. (ed.): Epidemiology and Control of Schistosomiasis (bilharziasis). Basel, S. Karger, 1973.

Komiya, Y.: Clonorchis and clonorchiasis. In Ben Dawes (ed.): Advances in Parasitology. New York and London, Academic Press, 1966, vol. 4, pp. 53.

Macdonald, G.: The dynamics of helminth infections with special reference to schistosomes. Trans R Soc Trop Med Hyg 59:489, 1965.

Pawlowski, Z., and Schultz, M. G.: Taeniasis and cysticercosis (Taenia saginata). In Ben Dawes (ed.): Advances in Parasitology. New York and London, Academic Press, 1972, vol. 10, p. 269.

Yokogawa, M.: Paragonimus and paragonimiasis. In Ben Dawes (ed.): Advances in Parasitology, New York and London, Academic Press, 1965, vol. 3 p. 99, and 1969, vol. 7, p. 375.

C GENERAL RESPONSE TO INFECTION

Immunologic

81 MECHANISMS OF NATURAL RESISTANCE TO INFECTION

Abraham I. Braude, M.D., Ph.D.

Infectious agents must break through four lines of defense before they can establish themselves in the tissues, namely, the body surfaces, the circulating body fluids, the connective tissue matrix, and the cells. Resistance to infection depends on the joint activity of the processes that operate at these four levels. After recovery from an infection these processes function much more effectively, and resistance may become so great that an individual can withstand heavy exposure to the same infection with no sign of illness. This enhanced resistance is designated acquired immunity, as distinguished from the weaker native resistance or natural immunity.

AT BODY SURFACES

Mechanical Factors

The mechanical barrier offered by the epithelial surface of the skin and mucous membranes is probably the most important natural means of preventing infection.

Squamous epithelium is a more effective barrier because several layers of cells connected by desmosomes block the route of invading organisms. Not only is the single layer of columnar epithelium a weaker barrier but also the cells ingest bacteria and thus facilitate their progress to the underlying tissue. Intracytoplasmic microbes have been observed in many studies on the columnar intestinal epithelium. These consist of protozoa, *Escherichia coli, Salmonella, Shigella,* and enteroviruses. Most of these organisms penetrate the epithelial cell through the microvilli on its brush border, but *Entameba histolytica* trophozoites pass through the intracellular junctions between two adjoining cells (Takeuchi, 1975). Rupture of the dermal epithelium by trauma, ischemia, or viruses (especially varicella) is complicated by infection with *Staphylococcus aureus,* group A or B streptococci, and anaerobic bacteria. *S. aureus* and group A streptococci are carried in the nose by many healthy people and readily spread to establish infections of the face if the epidermis is injured. Group A streptococci may also colonize the extremities and trunk and produce impetigo, erysipelas, and cellulitis after the epithelium has been ruptured by insect or animal bites, scratches, accidental injury, or surgical incisions. If the epidermis breaks down after ischemia caused by pressure (bed sores) or vascular insufficiency, anaerobic streptococci and *Bacteroides fragilis* invade the tissues and produce ischemic ulcers.

The bacteria of the mouth and colon also invade broken epithelium but less readily than after injury to the skin. In the mouth, the anaerobic streptococci, *Bacteroides,* and fusobacteria frequently produce infection of necrotic gums and adjacent mucous membranes in persons suffering from poor oral hygiene. The fetid odor of these mouth infections reflects the activity of these anaerobes.

In addition to the epithelial barrier, other mechanical factors enable the body surfaces to resist infection. In the mouth, infectious particles are drawn backward by suction currents that converge at the base of the tongue and bypass the tonsils and posterior pharynx (Bloomfield, 1922). Bacteria transported through the mouth in this fashion are swallowed and exposed to the bactericidal gastric juice. Resident bacteria, however, remain firmly entrenched. These suction currents are set up by movements of the lips and tongue and by swallowing, and are mediated by saliva. Normally they remove particles in 15 to 30 minutes but weaken or disappear if saliva dries up from dehydration or if swallowing is disturbed by paralysis or weakness. Thrush and other oral infections occur in weak or paralyzed patients who lose these oral suction currents.

Cilia provide another means of sweeping mucous membranes free of pathogenic organisms. The bacteria and viruses that are carried on dust particles into the nose may be arrested by the hairs in the nares, but if the particles penetrate beyond this barrier, they are carried backward by cilia to the posterior nasopharynx and then join the oral bacteria swept to the base of the tongue. In the trachea and bronchi, currents of mucus are moved at a rate of 1 to 3 cm/hour by the ciliated epithelium and carry inspired bacteria from the small bronchioles back up to the pharynx where they can be swallowed. The cilia beat at a rate of approximately 1000 per minute and receive their energy from ATP or ADP. The cilia can be bypassed when infectious agents are carried on small particles. Thus particles carrying tubercle bacilli cause more infection experimentally if they are only 3 μ in diameter than those measuring 10 to 12 μ even though there are more bacilli on the larger particles. The larger particles are preferentially deposited in the upper airway where they are removed by cilia, whereas the smaller particles are able to pass beyond the cilia.

Particle size is also important in carrying viruses beyond the cilia. Those hygroscopic particles that are discharged in coughs shrink to 1.5 μ after they leave the nose or mouth and lose moisture, and then swell to 2 μ in the nose. These particles are small enough to pass beyond the cilia and reach the alveolar ducts, where they swell more by absorption of water from the moist air of the lung and become trapped in the narrow (0.6 μ) alveolar ducts. Larger particles will land on the ciliated respiratory epithelial cells, which have receptors for rhinoviruses, influenza viruses, and *Mycoplasma pneumoniae.* These agents may then initiate infection in the bronchial epithelium. Both influenza and *M. pneumoniae* impair mucociliary transport, so that increased susceptibility to secondary bacterial infection would be expected. Such secondary infection is prominent in influenza but does not occur in *Mycoplasma* infections, possibly because the impaired clearance is more severe in influenza.

Ciliary motion can also be reduced by cigarette smoke and alcohol. The amount of alcohol required for ciliary inhibition approaches the lethal dose and is not the important factor responsible for increased susceptibility of alcoholics to pneumonia. Formaldehyde, cyanide, and acetone are the ingredients in cigarette smoke that retard ciliary movement. When ciliary activity is damaged, the impaired ciliary clearance can be replaced by coughing. Patients with mucoviscidosis, for example, clear radioactive aerosols normally because their cough can compensate for the loss of mucociliary clearance. On the other hand, patients who suffer from congenital immotility of cilia are victims of chronic airway infections despite adequate coughs (Eliasson et al., 1977).

In the small intestine, bacteria are moved along by the action of mucosal villi, which free themselves of adherent particles by causing them to stick to the extensive lacy film of mucus that lines the bowel. The mucus is propelled along in the form of small balls by peristalsis. The absence of this protective mechanism may explain the greater tendency for *Shigella* and enteropathogenic *E. coli* to attack the colon, which has no villi. Peristalsis may be even more important than villous movement in protecting the small bowel from disease. In disorders of small bowel motility in which peristalsis is deficient, as in scleroderma, neurogenic disorders, or surgical blind loops, the colonic bacteria ascend and overgrow the small bowel. A major consequence of this bacterial overgrowth in the small bowel is deconjugation of bile salts by *B. fragilis,* the most populous member of the intestinal flora. The fall in concentration of conjugated bile salts and the corresponding rise in free bile acids (cholic, deoxycholic, and chemodeoxycholic) interfere with fat absorption and produce steatorrhea. Antibiotics that are active against *B. fragilis* will overcome the malabsorption problem. Further evidence of the importance of peristalsis is found in experiments that show that in order to infect guinea pigs with *Shigella,* it is necessary to inhibit intestinal motility with opiates. Opiates or belladonna drugs also impair resistance to human shigellosis. By restricting intestinal motility, these drugs produce more severe infection of the bowel.

Expectoration is a key process for transporting pathogenic bacteria from the respiratory tract, and the flushing action of tears, saliva, and urine removes infectious materials. Flushing the urethra by urinating after coitus will prevent cystitis in women. A sneeze or cough may contain 40,000 or more particles carrying bacteria. This reflex, as well as the epiglottis reflex, which prevents gross aspiration from reaching the pharynx, may be less efficient in adults than in children. If a contrast medium (iodized oil) is placed in the mouth of sleeping patients, chest x-rays will show the medium in the lungs of adults but not in children the following morning (Amberson, 1954). Alcohol, deep anesthesia, and exposure to cold all prevent closure of the glottis and allow aspiration pneumonia to occur.

Chemical Factors

Body surfaces also owe their freedom from infection to nonmechanical processes. Some of these depend on the action of mucin, a disaccharide of acetylneuraminil-acetylglucosamine attached to a peptide. Because these disaccharides are hydrophilic the mucin is hydrated. The negatively charged neuraminil residues repel each other and expand the hydrated molecule to form a moist lubricating blanket that moves on the surface of cilia and intestinal villi. Mucin granules can adsorb myxoviruses in mucus cells and reexcrete them as unincorporated virus into the moving mucus sheet. Susceptible cells are protected by this mechanism from infection. In a similar process, the mucin-secreting goblet cells of intestinal crypts repond to an inflammatory stimulus by rapidly discharging organisms that could infect the bowel. The *N*-acetyl neuraminic acid residue on mucin is a receptor for the hemagglutinins on myxoviruses and theoretically, at least, may prevent attachment of myxoviruses to cells by acting as a competitor for cell receptors. Should this occur, however, the viral neuraminidase could still free the virus from the mucin.

Acid kills most bacteria that reach the stomach and is the most important natural barrier to infection by intestinal pathogens (Garrod, 1937). *Brucella, Salmonella, Shigella,* and *Vibrio cholera* die quickly on exposure to gastric juice. In patients whose stomachs have been removed, the incidence of salmonellosis rises 3.3-fold, and the disease is more severe. The same applies to cholera; in a small epidemic in 1971 it was found that 25 per cent of cholera victims had had gastric resections. This association was predictable from the early experiments in 1885 of Robert Koch, who could produce infection in guinea pigs only if cholera organisms were given orally with sodium bicarbonate to neutralize the gastric juice. Bacteria are not the only agents affected by gastric acid. Although many viruses are killed by acid, the acid resistance of enteroviruses, hepatitis viruses, reoviruses, and adenoviruses is important in their ability to infect the bowel and be excreted in the feces. Among patients who get giardiasis, 42 per cent have achlorhydria. Normally the acid pH keeps the stomach and small bowel sterile or nearly so. Not until the pH approaches neutrality in the ileum does the intestinal flora assume a rich and varied character.

The low pH of normal urine may also be important in killing bacteria. Urines of pH 6 or below inhibit growth of *E. coli* and kill gonococci. This phenomenon may help prevent *E. coli* urinary infections and may also explain why the urethras of most men resist infection after exposure to gonorrhea, and why gonorrhea never spreads to the bladder or kidney (McCutchan et al., 1977).

Peroxide (H_2O_2) is another bactericidal factor found on mucous membranes. Alone or in conjunction with salivary peroxidase and thiocyanate, H_2O_2 kills various pathogenic bacteria, presumably through its powerful oxidizing ability. Peroxidase catalyzes the oxidation of thiocyanate to produce hypothiocyanite ion, which inhibits or kills bacteria (Thomas and Aune, 1978):

$$\underset{\text{Thiocyanate}}{SCN^-} + \underset{\text{Peroxide}}{H_2O_2} \xrightarrow{\text{peroxidase}} \underset{\text{Hypothiocyanite}}{OSCN^-} + H_2O_2$$

The $OSCN^-$ radical appears to kill bacteria by oxidation of their sulfhydryls to sulfenyl thiocyanate. The thiocyanate-peroxidase-H_2O_2 system is found both in the mouth and in milk.

Peroxide is elaborated by streptococci and other catalase-deficient bacteria in the mouth, bowel, and vagina. H_2O_2 from mouth bacteria correlates well with the ability of these normal resident flora to inhibit the diphtheria bacillus. These and other resident bacteria undoubtedly defend the body surfaces against infection by various mechanisms. In addition to H_2O_2, *Streptococcus mitis, Streptococcus salivarius,* and *Streptococcus mutans* elaborate antibiotics that range in molecular weight from 200 to 12 million and inhibit growth of most gram-negative and gram-positive cocci, including the pneumococcus and meningococcus (Sanders, 1975). On the other hand, pneumococci that are carried in the throat are usually not inhibited by the resident streptococcal flora, possibly because resistant pneumococci are selected for implantation or acquire resistance afterwards (Johanson et al., 1970). The normal pharyngeal flora also prevent colonization by the gram-negative bacilli that overgrow the pharnyx of persons given penicillin and other antibiotics. These antibiotics can suppress the streptococci that predominate among inhibitory strains in the pharyngeal flora.

Normal Flora

In the bowel, *E. coli* is antagonistic to enteric pathogens. *Shigella,* for example, cannot be implanted in germ-free guinea pigs if *E. coli* is introduced first, but can cause experimental shigellosis more readily in conventional animals after suppression of the normal bowel flora by antibiotics. Similarly, the infective dose of *Salmonella typhosa* was reduced from 100,000 to 1000 bacteria in volunteers who swallowed streptomycin, a drug that inhibits intestinal *E. coli* but not the anaerobic bacteria of the bowel (Hornick et al., 1970). Colicines, the peptide antibiotics elaborated by *E. coli,* can kill *Salmonella* and *Shigella* and may help prevent intestinal infection by these bacteria. There is also evidence that colicines contribute to recover from shigellosis. *E. coli* and other intestinal bacteria also produce acetic and butyric acids, which can inhibit growth of virulent enteric pathogens.

Antagonists of potential clinical significance have also been found among vaginal flora. *Candida albicans,* a common inhabitant of the vagina, inhibits most gonococci. *Staphylococcus epidermidis,* nonhemolytic streptococci, and *Pseudomonas* are also potent inhibitors of gonococci among vaginal flora. Other bacteria in the vagina can stimulate growth of the gonococcus; the stimulators are gram-positive and predominantly lactobacilli. The inhibition, which is thought to enhance natural resistance to gonorrhea in women, is mediated by acid production (killing of gonococci progressively increases as pH falls below 7.0), and by antibiotics produced by the normal flora. Stimulation of gonococci, which occurs by unknown mechanisms, might be important in promoting infection (Braude et al., 1977).

Surface Immunoglobulins

Immunoglobulins (Ig) A, M, G, and E all appear on normal mucous membranes and probably contribute to natural resistance. Secretory IgA is probably the most important of these because it is not only most abundant but also resistant to denaturation and hydrolysis by enzymes and chemicals in intestinal and other external secretions. Secretory IgA is a dimer of IgA molecules with a sedimentation coefficient of 11S. Secretory IgA is produced in submucosal plasma cells in close contact with the overlying glandular epithelium. The IgA monomers appear to be linked together by the glycopeptide J chains into stable dimers within the plasma cells. After secretion by the plasma cell, the IgA dimer forms a complex with a protein known as secretory-component. This secretory-component is manufactured by epithelial cells and from its position on the cell membrane is believed to function as a receptor protein that carries the IgA dimer into the cell and then out onto the surface of the mucous membrane (Hauptman and Tomasi, 1978). Secretory piece also appears to facilitate transport of locally synthesized IgM onto mucous membranes.

Local production of all immunoglobulins is stimulated on mucous surfaces by direct contact of B lymphocytes with microbial antigens during subclinical infection or with cross-reacting antigens present in food or in the normal microbial flora. Secretory IgA and other immunoglobulins seem to adhere to epithelial cell surfaces by sticking to the overlying mucus and provide an "antiseptic paint" that can block adherence of pathogenic agents to epithelial cells and neutralize their toxins. IgA has a limited ability, if any, to participate in reactions involving complement and, therefore, in promoting lysis or phagocytosis of bacteria. Furthermore, gonococci, meningococci, and certain streptococci produce a proteolytic enzyme that inactivates IgA1, which constitutes approximately 40 per cent of secretory IgA. Despite these limitations, the surface immunoglobulins undoubtedly contribute to natural resistance by preventing the initial attachment stage of infection by *Salmonella, Shigella,* pathogenic *E. coli,* gonococci, meningococci, *V. cholera,* group A streptococci, polio and other enteric viruses, various respiratory viruses, and certain protozoa. The importance of IgA in preventing intestinal infection is emphasized by the unique susceptibility of patients with hypogammaglobulinemia and selective IgA deficiency to giardiasis. Similarly, the presence of secretory IgA in colostrum and human milk may explain the lower incidence of gastrointestinal infections in infants, and their resistance to infection with polioviruses (Winberg and Wessner, 1971). In the respiratory tract, the IgA in mucus can inhibit the neuraminidase activity of influenza virus on the neuraminic acid moiety of mucin and thereby block penetration of the mucin barrier by the virus.

IN EXTRACELLULAR FLUIDS

After infectious organisms pass beyond the body surfaces into the subepithelial tissues, their survival depends on their ability to withstand the environment of the extracellular fluids until taken up by phagocytic cells. They may elicit inflammatory reactions that confine them to the point of invasion, or they may pass into the lymphatic channels, travel to the lymph nodes, and eventually enter the general circulation. After phagocytosis they may escape from the cells and reappear in the body fluids.

Extracellular fluids are an ideal culture medium for certain organisms, as shown by the value of human blood, serum, and ascitic fluid for the isolation and growth of some organisms in the diagnostic laboratory. On the other hand, fresh human blood may inhibit the growth of meningococci, gonococci, coliform bacteria, *Brucella, Haemophilus influenzae,* and coagulase-negative staphylococci. Survival of an infectious agent in the body fluids depends on the temperature, pH, Eh (oxidation-reduction potential), osmolality, antibodies, complement, enzymes, electrolytes, and other metabolically active substances of the given fluid.

Temperature

The oral temperature of the normal human is about 37° C; that of the internal organs is 1 to 2° C higher. Most pathogenic organisms grow well in this temperature range, but *Sporothrix schenckii* and *Mycobacterium marinum* are exceptions. These do not grow well at 37° C or higher and are the cause of infections that usually remain confined to the extremities, where the temperature is lower. As might be expected from these temperature restrictions, external heat is useful for treating sporotrichosis and *M. marinum* infections. Among viruses, rhinovirus often grows best at 33° C on primary isolation, a property that fits with the reported prevalence of rhinovirus infections in colder weather when the temperature of the nasopharynx might be lowered by cold air. Yet rhinoviruses cause widespread infection in the warm temperatures of the tropics. In any case, the failure of rhinoviruses to replicate well at body temperature probably confines them to the respiratory tract and prevents infection of the warmer internal organs.

The normal body temperature has been important in recovery from transfusion reactions with blood contaminated by psychrophilic ("cold-loving") bacteria that grow well in the refrigerator but die at 37° C. Among cryptococci, only *Cryptococcus neoformans* can grow at 37° C, and it is the only cryptococcus that is pathogenic. *C. neoformans* cannot, however, withstand temperatures above 39° C, a fact that might explain clinical improvement during malaria and other febrile disorders in patients with cryptococcosis (Kligman and Weidman, 1949). On the other hand, artificial fever therapy has been useless in the treatment of cryptococcosis. Experiments in rabbits infected with *Pasteurella multocida* suggest that fever may help control infection by lowering the plasma iron level below that needed for survival of infecting bacteria (Kluger and Rothenburg, 1979).

Acidity

Homeostatic mechanisms maintain pH of extracellular fluids within such a narrow range that even under the most extreme conditions there is not enough change to influence the reproduction of pathogenic microbes. The only exception is urine, which can become sufficiently acid (pH 4.9 to 5.8) to inhibit or kill *E. coli,* gonococci, and other bacteria (McCutchan et al., 1977).

Redox Potential

The relatively high oxidation-reduction potential (Eh) of the extracellular fluids helps prevent infection by anaerobic bacteria. Eh is a measure in millivolts of the tendency to give up electrons, or the potential that would exist between the body fluids and a hydrogen half-cell at pH O. Eh of extracellular fluids is a function not only of their inherent reducing tendency but also of their hydrogen ion concentration, and it becomes more negative as hydrogen ion concentration diminishes. The importance of Eh was first brought out in the case of tetanus by Fildes (1929), who showed that tetanus spores will not germinate readily at an Eh more oxidizing than +10 millivolts (mv) at pH 7.0 to 7.6. In living animal tissues he found the Eh to be approximately 120 mv at the prevailing pH 7.4. If the Eh is kept down by electrical methods, clostridia will grow even if a stream of air is passed through the culture (Hanke and Bailey, 1945). If the pH is kept in the acid range (6.4 to 6.8), *Clostridium perfringens* will grow at an Eh of 160 or more. In ischemic tissues, the anoxia lowers the Eh and lactic acid accumulation lowers the pH, so that anaerobic bacteria proliferate at Eh levels that are too high for their growth at the normal pH of 7.4. Since acidosis enhances proteolysis and accumulation of amino acids, the pH falls even more. Anaerobic infections do not occur unless the redox potential is lowered by one of the following disorders: (1) impairment of local blood supply by arterial occlusion; (2) pressure of foreign bodies such as sutures, clothing, dirt, and metal; (3) contamination of wounds with ionized calcium salts commonly present in fertile soil from farmlands, a powerful cause of tissue necrosis; (4) destruction of tissue by trauma, infection, or injection of quinine or epinephrine; (5) growth of aerobic or facultative bacteria in the tissues. The aerobic growth of the group A *Streptococcus* in a medium having an initial Eh of 300 mv induces a fall in redox potential to 150 to 200 mv.

Natural Antibodies and Complement

In the absence of any detectable previous antigenic stimulus, immunoglobulins appear in human sera that destroy pathogenic organisms in the presence of complement, promote their phagocytosis, or neutralize their toxins. These "natural" immunoglobulins, or antibodies, have been examined mainly for their action against meningococci, *E. coli.*, *Salmonella*, *H. influenzae,* pneumococci, staphylococci, and diphtheria bacilli. The importance of natural antibody in preventing disease was first appreciated in diphtheria. By demonstrating antitoxin with the Schick test, it has been possible to show a good correlation between resistance to diphtheria and the presence of antitoxin in people who have had no clinical history of diphtheria or active immunization. The most likely stimulus for natural antitoxin is inapparent infection.

The importance of inapparent infection in natural resistance has also been established for the meningococcus. Natural immunity against meningococci can result from antibodies that are stimulated by asymptomatic colonization of the throat with these bacteria (Goldschneider et al., 1969). The protective antibody appears to be directed against the meningococcal capsule. Natural antibodies against the capsule also prevent infection by *H. influenzae* and the pneumococcus, but these antibodies are more likely generated by stimuli from antigenically related organisms than from subclinical infection with the specific agents of influenzal and pneumococcal disease. Infants and children develop natural anticapsular antibody and immunity to *H. influenzae* type b, despite a very low colonization rate of their nasopharynx with *H. influenzae* type b. The same is true in laboratory animals that have high levels of serum antibodies to type b without having been colonized by that organism. *E. coli* K100 appears to be one of the important cross-reacting bacteria for producing *H. influenzae* type b immunity. The K100 antigen is a potent antigen for stimulating type b antibody when the whole bacteria are fed to children, and *E. coli* K100 colonizes their bowel naturally. The acidic polysaccharides in the K100 capsule appear to have the cross-reacting determinants for the type b capsule. Various other bacteria of the bowel cross-react with the *H. influenzae* type b capsule, including gram-positive bacteria with polyribitol phosphate in their cell wall teichoic acid. Ribitol is a component of the type b capsular polysaccharide antigen of *H. influenzae.*

TABLE 1. Cross-Reactions Between Pneumococci and Various Microorganisms or Plants

PNEUMOCOCCUS TYPE	SOURCE OF CROSS-REACTING ANTIGENS
1	Coliform bacilli Plant hemicelluloses (Felton et al., 1955)
2	Yeast polysaccharides Gum arabic (Heidelberger et al., 1929) Gum acacia (Marrack and Carpenter, 1938) Cherry gums *K. pneumoniae* polysaccharide B *S. typhi* *S. paratyphi* B Plant hemicelluloses
3	Gum arabic (Heidelberger et al., 1929) Gum acacia Cherry gums (Marrack and Carpenter, 1938) Meningococcus Gonococcus Plant hemicelluloses
14	Group A human red cells (Beeson and Goebel, 1939) Anthrax bacillus somatic polysaccharide
29,34,35	*H. influenzae* type b

The production of natural protective antibody against the pneumococcus is more complicated because many types of pneumococci are virulent, whereas only type b is virulent among *H. influenzae.* Nevertheless, there is enough exposure to cross-reacting antigens in bacteria and foodstuffs to provide good protection against pneumococcal infection by an early age. This is illustrated by the fact that 80 per cent of babies at the age of 1 year have antibody to type 7 pneumococcus, even though the carrier rate is only 1 per cent. The type 7 antibody response results from exposure to cross-reacting surface antigens in *S. viridans.* A partial list of other important cross-reactions with pneumococci is shown in Tables 1 and 2. These cross-reactions help explain how a broad spectrum of natural antibodies against pneumococcal infection is acquired early in life, long before subclinical infection with each of these types would be possible. Children with congenital hypogammaglobulinemia who are unable to make this natural antibody suffer from repeated pneumococcal infections of the same or different types (Bruton, 1952). The attacks of pneumococcal pneumonia and septicemia begin at 6 months after disappearance of the maternal immunoglobulins derived by transplacental passage, and affect mainly boys because the disorder is usually X-linked. The protective value of natural IgG antibody against pneumococcal infections in infancy and childhood is demonstrated by the effectiveness of commercial gamma globulin when given prophylactically to boys with hypogammaglobulinemia. Commercial gamma globulin is prepared from the blood of healthy people who have not been given pneumococcal vaccines, so that the protective IgG is naturally acquired antibody. The same susceptibility to overwhelming recurrent pneumococcal infections is seen in adults who lose their natural antibody when they develop the specific immunoglobulin disorders of multiple myeloma or acquired hypogammaglobulinemia.

Although natural antibody seems to be important against both pneumococci and *H. influenzae,* its mode of action against the two is different. Natural antibody against the pneumococcus (and other gram-positive bacteria) acts exclusively as

TABLE 2. Distribution of Immunizing Antigen Against Pneumococcal Infection in Mice Among Hemicelluloses from Commonly "Edible" and "Inedible" Plants[a]

"EDIBLE" PLANTS

Source of Hemicellulose	Immunity Against 10^3 LD		
	Pn I	Pn II	Pn III
Melon, honeydew (rind)	+++	+++	+++
Collard greens	+++	+++	++
Sunflower seed	+++	+++	+
Pumpkin pulp	+++	++	+
Squash (fruit)	+++	+++	0
Peanut meal	+++	++	0
Tomato (fruit)	++	++	+++
Watermelon rind	++	++	++
Green bean (fruit)	++	++	++
Parsnip greens	++	++	+
Pomegranate (fruit)	++	+	++
Wheat germ	++	+++	0
Cucumber	++	++	0
Kale greens	++	++	0
Soybean meal	++	++	0
Rape greens	++	+	0
Lettuce	++	0	0
Rutabaga (root)	++	0	0
Nutmeg powder	+	+++	++
Onion bulb	+	+++	0
Lima bean	+	++	0
Potato, white	+	+	+
Barley	+	0	0
Sugar cane	+	0	0
Yeast (baker's)	+	0	0
Apple, Winesap	0	+++	0
Grapefruit rind	0	+++	0
Irish moss (carrageen)	0	+++	0
Oatmeal	0	+++	0
Rhubarb, leaves	0	+++	0
Green pepper (fruit)	0	++	++
Cabbage	0	++	0
Coconut	0	+	++
Cinnamon powder	0	0	++
Carrot	0	0	+
Broccoli greens	0	0	0
Swiss chard greens	0	0	0
Eggplant (fruit)	0	0	0
Orange (rind)	0	0	0
Pear (fruit)	0	0	0
Pectin	0	0	0
Radish root	0	0	0
Rice (polished)	0	0	0

"INEDIBLE" PLANTS

Source of Hemicellulose	Immunity Against 10^3 LD		
	Pn I	Pn II	Pn III
Tomato greens (roots)	+++	++	++
Pumpkin (skin)	+++	++	+
Mosses	+++	++	0
Nightshade	+++	0	0
Grass, lawn	++	++	++
Cotton, nonabsorbent	+	+++	0
Wood, maple	+	++	0
Wood, pine	++	++	++
Squash seed	+	+++	0
Ageratum plant	+	++	0
Cottonseed meal	+	+	0
Lupine seed	+	+	0
Mullen plant	+	+	0
Iris plant	0	++	0
Spider lily	0	++	0
Gladioli (leaves and bulbs)	0	++	0
Corn husks	0	++	0
Nasturtium plant	0	0	++
Chickweed plant	0	0	0
Geranium plant	0	0	0
Hydrangea plant (leaves and small stems)	0	0	0
Ivy, English	0	0	0
Japonica plant	0	0	0
Marigold	0	0	0
Morning glory plant	0	0	0
Nicotiana plant	0	0	0
Rose	0	0	0
Spiderwort	0	0	0
Sedum plant	0	0	0
Vetch seed	0	0	0

[a]+++ indicates immunity with 0.0005 mg or 0.005 mg of hemicellulose; ++ indicates immunity with 0.05 or 0.5 mg; + indicates immunity with 1.0 or 5.0 mg.

From Felton, L., et al.: J Bacteriol 69:519, 1955.

Pn indicates pneumococcus type, I, II, or III.

an opsonin to promote phagocytosis, whereas natural antibody to *H. influenzae* can be both opsonic and bacteriolytic. Both IgM and IgG participate in opsonization, but IgM requires complement. The Fc portion of IgG attaches to the neutrophils and the Fab portion to the pneumococcal capsule. This opsonic bridge between leukocyte and pneumococcus does not require complement for phagocytosis to proceed. Complement becomes involved in a different opsonic process when Fab of either IgG or IgM reacts with capsular polysaccharide to form a complex that activates the classic complement pathway through the sequence C1, C4, C2, and C3. Activation of C3 makes it opsonic, so that it becomes the bridge between the pneumococcus and the phagocytic cell. Complement can also be opsonic for the pneumococcus when the cell wall activates C3 through the alternative pathway, which does not involve the participation of antibody. The generation of pneumococcal opsonins by the alternative pathway does not seem to be important in clinical medicine because it does not protect patients who lack gamma globulin from developing overwhelming pneumococcal infections. On the other hand, complement activation via the classic pathway is important for the opsonic activity of nonimmune human serum because most natural antibody against the pneumococcus is IgM. Because C1 and C2 are inactivated by heating at 56° C, the opsonic activity of nonimmune serum is often referred to as heat labile to distinguish it from the heat-stable, complement-independent, opsonic activity of hyperimmune antipneumococcal serum whose opsonins are IgG.

The natural antibodies against the meningococcus and *H. influenzae* are also opsonic, but in addition, they kill these two gram-negative bacteria by lysis in the absence of phagocytes. Bacteriolysis by natural antibody is a reaction between the bacterial surface antigens (polysaccharide capsules), IgM, and complement. As in opsonization, the reaction may follow either the classic or the alternative pathway of complement. After IgM attaches via the Fab to the capsular polysaccharide antigen, the first step in the classic pathway is attachment of C1 through its subunit C1q to a binding site on the Fc portion of IgM. This interaction with antibody globulin activates C1 and catalyzes the assembly of C4 and C2. Together ($C\overline{42}$) they function as the enzyme C3 convertase. This enzyme cleaves C3 into two fragments, C3a and C3b. The C3a fragment is responsible for opsonization (as described for the pneumococcus), and C3b attaches to receptors on the bacterial cell. After attachment, C3b combines with neighboring $C\overline{42}$ and apparently modifies the specificity of the latter enzyme so that it can now cleave C5 into C5a and C5b. The large fragment C5b then attaches to the cell membrane, binding in turn C6 and C7 to form the trimolecular complex $C\overline{567}$, which then becomes tetramolecular when C8 is incorporated. The terminal complex of C activation becomes complete when C9 fixes to binding sites on $C\overline{5678}$. The large complex of the five individual terminal C proteins is essential for killing susceptible gram-negative bacteria. This bactericidal activity is far more effective when it is generated via the classic pathway, but it can operate through the alternative pathway (also known as the properdin pathway). The lipopolysaccharide (LPS) in gram-negative bacteria can activate the alternative pathway by reacting directly with a noncomplement protein, C3 proactivator (properdin B) (C3PA). Upon reacting with LPS, C3PA splits into at least two fragments: One has the electrophoretic mobility of gamma globulin (mw 60,000), and a second consists of an acidic peptide (mw 20,000). The larger fragment, known as C3 activator, splits C3 into C3a and C3b (properdin A), and thereby initiates the terminal sequences of complement activation.

In addition to complement, lysozyme, calcium, and magnesium are important components of extracellular fluids that are required for killing gram-negative bacteria. Ca^{++} and Mg^{++} ions are essential for initiation of the classic complement sequence, and lysozyme is needed for disrupting the cell wall peptidoglycan. Lysozyme splits the 1–4 glycosidic linkages between *N*-acetyl muramic acid and *N*-acetylglucosamine in the peptidoglycan so that complement may have greater

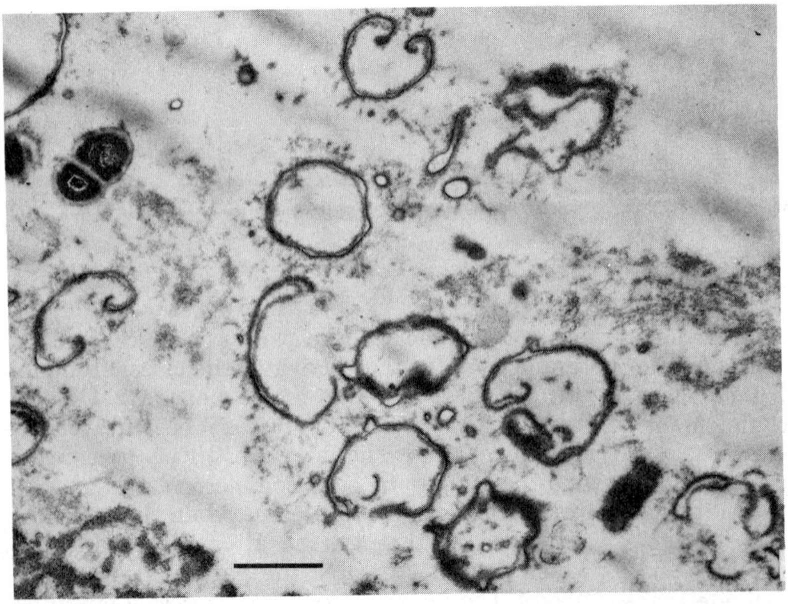

FIGURE 1. *Representative electron micrographs of markedly serum-sensitive* E. coli *bacteria incubated with undiluted fresh normal serum. Note that both cell walls and cytoplasmic membrane are ruptured by the bactericidal action of fresh serum. Compare this with Figure 2, which shows appearance of same bacteria in serum after inactivation of complement.* × *16,000 (From Olling, S.: Scand J Infect Dis (Suppl), 1977.)*

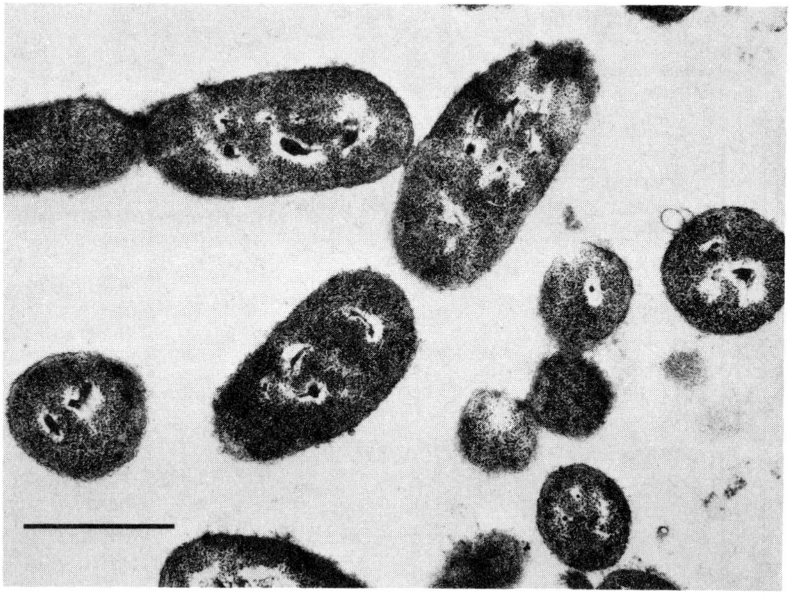

FIGURE 2. *Same as Figure 1 except that serum has been heated to inactivate complement. In the absence of complement the bactericidal system is inoperative and does not injure the bacterial cell.* × *16,000 (From Olling, S.: Scand J Infect Dis (Suppl), 1977.)*

access for its attack on the cytoplasmic membrane. The electron microscope (Figs. 1 and 2) discloses dramatic disturbances of bacterial morphology in cells that have been killed by serum. Both cell walls and cytoplasmic membranes are ruptured.

In contrast to gram-positive bacteria, which are all resistant to lysis by complement, all species of gram-negative bacteria are susceptible to lysis by the classic or alternative pathway. This difference in susceptibility to killing by complement is best explained by the concept that the cell wall of gram-positive bacteria is a barrier that hinders access of complement to the cytoplasmic membrane. This concept is reinforced by the fact that protoplasts of gram-positive bacteria whose cell walls are removed are killed by complement. The thicker peptidoglycan and the absence of LPS could explain how the cytoplasmic membrane enjoys a sanctuary in gram-positive bacteria from complement. In many gram-negative bacteria, LPS is the key intermediate factor in killing because it activates complement by either pathway. Each of the three primary structural components of LPS — O antigens, core sugars, and lipid A — can activate the classic pathway by reacting with natural antibody (Fig. 3). In the absence of antibody, lipid A can activate either the classic pathway by direct reaction with C1q or the alternate pathway. For unknown reasons, virulent enteric gram-negative bacteria often resist complement-mediated bacteriolysis, survive in the extracellular fluids, and cause bacteremia. The importance of complement in natural resistance to infection is also seen in patients who share a familial deficiency of C6, C7, and C8 and are subject to severe recurrent attacks of bacteremia due to gonococci and meningococci. Phagocytosis of these bacteria is normal because opsonins can be generated through C3, but the serum of these patients cannot kill pathogenic *Neisseria* because the terminal complex of complement is not formed.

Aside from gram-negative bacteria, certain protozoa, trematodes, and leptospiras undergo lysis in nonimmune serum in the presence of complement. Resistance to such lysis seems to be correlated with virulence. For example, the blood from nearly every person kills members of the *Trypanosoma brucei* group in vitro and protects experimental animals. The only trypanosomes of the *T. brucei* group that resist being killed by human serum are *T. gambiense* and *T. rhodesiense,* which can infect human beings and cause sleeping sickness. *Trypansoma cruzi,* the cause of Chagas' disease, also resists killing by human serum. Such a relationship also seems to exist for highly pathogenic *Schistosoma mansoni.* High dilutions of human serum will lyse cercaria of many trematodes within a few minutes but not those of *S. mansoni* unless the serum is from a patient infected with *S. mansoni* (Culbertson, 1951). Similarly, virulent leptospiras resist killing by normal human serum, which destroys nonpathogenic leptospiras and some avirulent lines of pathogenic leptospiras (Johnson and Harris, 1967).

Beta-lysin

Although not killed by antibody and complement, most gram-positive bacteria are killed by beta-lysin, a relatively heat-stable cationic pro-

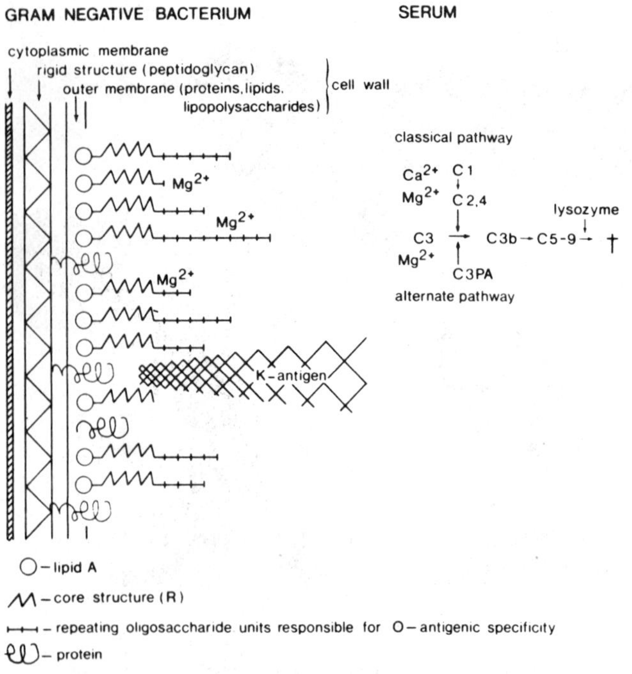

FIGURE 3. *Proposed model for the structure of the surface layers of gram-negative bacteria and the serum factors involved in bacterial killing. Each of the three primary structural components of LPS—O antigens, core sugars, and lipid A—can activate the classic pathway by reacting with natural antibody. Reactions between K antigen or other capsular antigens and antibody can probably also activate the classic pathway. In the absence of antibody lipid A can activate either the classic or the alternate pathway. Activation of complement by either pathway kills the cell. This may occur either because holes are punched in the outer membrane so that lysozyme can hydrolyze the mucopeptide and produce spheroplasts that rupture from osmotic forces, or because activated complement may impose lethal damage to the cytoplasmic membrane. (From Olling, S.: Scand J Infect Dis (Suppl), 1977.)*

tein with a molecular weight of 6,000 (Donaldson, 1975). With the exception of streptococci, most gram-positive bacteria are killed by beta-lysin, but gram-negative bacteria, molds, yeasts, viruses, and mycoplasmas are not killed. Beta-lysin does not require complement to kill and, unlike bactericidal antibody, is present in higher concentrations in serum than in plasma because it is released from platelets when blood clots. It is also present in inflammatory exudates. The bacterial cell membrane is thought to be the primary site of action of beta-lysin.

Acute Phase Substances

During a variety of unrelated acute illnesses, substances suddenly and promptly appear that react with streptococci. One of these kills group A streptococci, and the other combines with the group-specific polysaccharide C-substance of *S. pneumoniae* (C-reactive protein). Neither is an antibody. The C-reactive protein (CRP) is normally present in trace amounts in human serum but increases rapidly in infection and inflammation as a consequence of increased synthesis in hepatocytes (Kushner and Feldmann, 1978). It was first discovered in pneumococcal pneumonia but later observed in other acute infections, rheumatic fever, cancer, arthritis, serum sickness, ulcerative colitis, myocardial infarction, and various forms of tissue injury. CRP is composed of five subunits in cyclic symmetry and its chemical structure is similar to that of C1t, a serum protein that may be involved in regulation of complement (C1) function. Despite the ability to react with pneumococcal C polysaccharide and to fix complement with it, CRP has no demonstrable antimicrobial activity except for weak opsonic properties for various pathogenic bacteria. Its protective properties seem to depend instead on its ability to modulate inflammatory and immune responses by inhibiting platelet aggregation and mediator release and by suppressing lymphocyte stimulation and lymphokine production after binding to t-cells (Mortenson et al., 1977).

Both acute-phase substances were discovered by W. S. Tillett (1937), and the bactericidal factor is known as Tillett factor. Its ability to kill streptococci separates it from beta-lysin; Tillett factor is also unusual in that its activity is abolished under anaerobic conditions or in the presence of reducing agents. Both CRP and Tillett factor differ from antibody in their time of appearance and the nonspecificity of their stimuli. The two acute-phase substances are most active in the most active periods of disease, and disappear during convalescence when specific antibody usually begins to increase. For this reason, it has been suggested that acute-phase substances provide a means of body defense against bacterial invasion that is uniquely different from that of the complement-antibody apparatus and the phagocytic system.

Interferons

Various tissues can react with viruses to elaborate proteins that inhibit infection by the same or different viruses. These inhibitors, designated interferons by Isaacs and Lindenmann (1957), enhance the resistance of cells to infection by viruses. Interferon produced by one type of cell is active against a wide variety of viruses; for example, an interferon produced by treatment of cells with influenza virus will prevent infection not only by various influenza viruses but also by vaccinia viruses, togaviruses, and others. Moreover, interferon produced in cells of different organs show no tissue specificity, so that interferon produced during a pulmonary infection, for example, might inhibit viral growth in the brain. Because interferon has been found in tissues during recovery from viral infection, it has been suggested that this viral inhibitor might depress viral multiplication before specific antibody reaches effective levels. (Interferon is fully discussed in Chapter 12).

Tuftsin

This tetrapeptide of splenic origin stimulates phagocytosis by a hormonal action on the cell. Its formula is L-threonyl-L-lysyl-L-prolyl-L-arginine. It has a high positive charge that allows it to bind to negatively charged groups (e.g., sialic acid) on the neutrophil surface. An enzyme in the membrane of neutrophils splits tuftsin from a gamma globulin known as leukokinin. Deficiencies of tuftsin occur as a rare familial disorder (Constantopoulos et al., 1972) and apparently after splenectomy. The susceptibility to overwhelming pneumococcal infection in splenectomized patients is thought by some to be a consequence of tuftsin deficiency.

Enzymes, Electrolytes, and Other Metabolites

Table 3 summarizes the reaction of bacteria to various substances in normal body fluids that may affect their growth or their ability to counter the mechanisms used to withstand infection.

The peculiar susceptibility of the poorly controlled diabetic to infections by *Klebsiella pneumoniae*, tubercle bacilli, *Mucor, Rhizopus,* and *Candida albicans* suggests that chemical abnormalities of the extracellular fluids may enhance the capacity of these microorganisms to establish infection. At least 0.3 per cent glucose is necessary in culture media for the formation of well-encapsulated *K. pneumoniae* (Hoogerheide, 1939). This concentration of glucose in the tissue fluids in poorly controlled diabetes might enable these bacteria to produce larger capsules and enhance their virulence by enabling them to resist phagocystosis. High concentrations of glucose also favor growth of *Mucor, Rhizopus,* and *C. albicans.* Although there is no evidence that variations in the concentration of glucose in vivo affect the survival of tubercle bacilli, the stimulating effect of keto acids at the acid pH of inflammatory exudates may enhance the growth of these bac-

TABLE 3. Factors of Body Fluids that May Affect the Survival of Pathogenic Bacteria or Their Capacity to Injure the Host

	MICROORGANISMS	POSSIBLE INFLUENCE ON DEVELOPMENT OF INFECTION
Glucose	Friedländer's bacillus	Increased production of capsular polysaccharide at levels >0.3 per cent
Keto acids	Staphylococci, tubercle bacillus	Stimulate growth at acid pH
CO_2	Tubercle bacillus	Inhibits growth
CO_2	Pneumococci, gonococci, *Brucella abortus*	Stimulates growth
CO_2	Staphylococci	Increases production of toxin
Ca, Mg	*Clostridium perfringens*	Required for hemolytic activity
Iron	Diphtheria bacillus, *Cl. perfringens, Cl. tetani,* and *Sh. dysenteriae*	Inhibits production of toxin
Iron	Staphylococci, clostridia, *Listeria,* mycobacteria	Stimulates growth and virulence
Cholesterol	Streptococci	Inhibits streptolysin S
Nucleotides	Streptococci	Required for high concentrations of streptolysin O
Hematin	Staphylococci	Bactericidal
Hematin	*H. influenzae*	Essential for growth
Cysteine	*H. influenzae*	Can replace hematin
Cysteine or cystine	*Francisella tularensis*	Essential for growth
NAD or NADP	*H. influenzae*	Essential for growth
Oleic acid	Streptococci, pneumococci	Inhibits growth
Glutamine	Streptococci, pneumococci	Required for good growth
Spermine	Tubercle bacillus	Bactericidal after activation by spermine oxidase
Plasminogen	Streptococci, staphylococci	Converted to active fibrinolysin by streptokinase and staphylokinase
Coagulase factor	Staphylococci	Required for the production of fibrinogen by staphylococcal coagulase

teria in human tuberculosis during uncontrolled diabetes (Dubos, 1954).

In normal persons, the body fluids are well suited to meet the special growth requirements of such organisms as *H. influenzae, B. abortus,* meningococci, gonococci, *Francisella tularensis,* and *Streptobacillus moniliformis.* These bacteria grow better when carbon dioxide is present, and the physically dissolved carbon dioxide in extracellular fluid (about 3 volumes per cent) is optimum for their growth. The requirement of cysteine for growth of *F. tularensis* and *H. influenzae* is also met by body fluids during infection, and human blood can provide the requirement of *H. influenzae* for NAD, NADP, or hematin.

The same constituents of extracellular fluids may also inhibit bacterial growth or activity of toxins. Growth of the tubercle bacilli, for example, may be completely suppressed by excessive concentrations of carbon dioxide, and the beneficial effects in tuberculosis of lung collapse from therapeutic pneumothorax may be related to the high concentration of carbon dioxide in the poorly ventilated lung. Tubercle bacilli are also inhibited by low oxygen tension, and their disappearance from closed caseous areas is attributed in part to local anoxia. In sharp contrast, where an open bronchus permits good oxygenation of the cavity, tubercle bacilli flourish, and conditions are optimal for growth of drug-resistant mutants.

Iron and iron-containing compounds appear to have an important influence on virulence of bacteria. Although hematin is essential for the growth of *H. influenzae,* it may be toxic for certain gram-positive bacteria even in minute amounts. Heme is lethal for staphylococci in concentrations of 0.0003 per cent at pH 7.0 (Dubos, 1954). For the most part, however, iron or heme compounds promote virulence and growth in vivo. Virulence for mice is markedly enhanced, and disseminated lethal infections occur when gonococci are suspended in a mixture of hemoglobin and mucin, the same menstruum in which they exist during menstruation when they most commonly disseminate hematogenously in women. If elemental iron is used in place of hemoglobin the virulence of gonococci is not increased. Under other circumstances, however, iron alone will enhance and iron deprivation will lower bacterial virulence. This correlation of virulence with available iron seems to parallel the ability of iron to stimulate microbial growth. Staphylococci, clostridia, listeria, mycobacteria, various enteric bacteria, *Pseudomonas, Pasteurella, Yersinia, Vibrio, Aeromonas, Candida,* and plasmodia are among the genera of bacteria, fungi, and protozoa whose members are stimulated by excess iron to grow in body fluids (Weinberg, 1978). This growth stimulation by iron in serum and other body fluids is thought to depend on saturation of transferrin, a nonheme iron-binding protein with a molecular weight in the range of 75,000 to 80,000. It binds reversibly two ferric ions/molecule through tyrosyl and histidyl residues. Under normal conditions human transferrin is only 25 to 35 per cent saturated, so that the amount of free ionic iron in plasma is at least 100 million-fold less than that required for bacterial growth (0.4 to 4.0 μM). If circulating transferrin is saturated by injecting iron intravenously, intraperitoneally, or intramuscularly, there is an increased mortality rate of animals with experimental candidiasis, tuberculosis, staphylococcal infections, salmonellosis, and meningococcal infections. By saturating transferrin, the injection of iron makes available enough free iron to meet microbial growth requirements. Impaired resistance to infection is not limited to experimental animals treated with iron, since impaired resistance has been reported in human beings after exposure to excess iron. Thus, the incidence of serious gram-negative infections has increased in newborn infants after intramuscular injection of iron (Barry and Reeve, 1977). It should be noted that transferrin can also be saturated if its concentration falls in relation to iron, as occurs in patients with kwashiorkor whose nutritional disorder damages their ability to make iron-binding proteins. The plasma obtained from kwashiorkor victims supports growth of *S. aureus* and is bacteriostatic upon addition of transferrin. Lactoferrin, an iron-binding protein in human milk, has been given credit, along with antibodies, for the lower incidence of gastroenteritis in breast-fed compared with formula-fed infants (Weinberg, 1977). For this reason the practice of supplementing infant formulas with iron has been questioned.

Constituents of body fluids can further influence pathogenicity by affecting bacterial toxins. The virulence of staphylococci is correlated, for example, with coagulase production. Coagulase is generated from staphylococcal procoagulase by an activator in the body fluids, and coagulase converts fibrinogen to fibrin. The tendency for staphylococci to produce extensive thrombosis in the lung and other tissues and to cause severe intravascular coagulation suggests that coagulase activator has an important place in the pathogenesis of staphylococcal infection. Its importance is further implied by the greater resistance to staphylococcal infections among certain animals whose plasma lacks coagulase activator. Staphylokinase, another staphylococcal toxin, acts through plasminogen, a constituent of normal plasma. Both staphylokinase and streptokinase convert plasminogen to plasmin, which can digest fibrin clots and prevent clotting. Both enzymes are elaborated in human infections and stimulate antibodies that neutralize their activity. Staphylokinase is thought to initiate lysis of infected thrombi and the release of septic emboli to the lung during staphylococcal septicemia; streptokinase could explain the poor localization of streptococcal cellulitis by preventing fibrin deposition around an infected focus.

The chemical environment of body fluids may also disturb normal immune processes, especially in the renal medulla. Within the range found in the normal kidney and urine, high concentrations of sodium, urea, and hydrogen ion inhibit phagocytosis of human leukocytes (Chernew and Braude, 1962). These observations help explain why the kidney is uniquely susceptible to invasion by *E. coli*, *Proteus*, and other bacteria that possess little capacity to establish infection elsewhere.

IN CELLS

Transport by Lymphatics

When organisms reach the tissue fluids they may be phagocytized by neutrophils or macrophages, or carried by the lymphatics to the right or left subclavian vein. During the course of this passage to the general circulation they may be trapped in lymph nodes and phagocytized by macrophages. If they get beyond the lymph nodes and into the blood they will then be ingested by circulating neutrophils or by histiocytes in the liver, spleen, bone marrow, adrenal, or pituitary. It is unlikely that any microorganism can enter blood capillaries directly because these vessels are impermeable to substances with a molecular weight of greater than 20,000. Lymph capillaries, on the other hand, easily become permeable to particles as large as red cells, even though they appear to be closed completely by endothelium and have no direct opening into the tissues. Passage through the lymph vessels is an important process in natural immunity because infectious particles are carried to potent fixed phagocytic cells. By itself, however, the lymph has limited protective ability because it contains almost no phagocytic cells and its content of antibodies is less than that of the blood (Braude and Carlson, 1908). The impotence of lymph against gram-positive bacteria can be observed when type III pneumococci are injected intravenously. These bacteria can be cultured promptly from the lymph after intravenous injection because they migrate passively through the walls of blood capillaries into tissue fluid and then through the lymphatic capillaries. Upon intravenous injection of type III pneumococcal antiserum, the blood becomes sterile but pneumococci are still found in the lymph seven hours later (Field et al., 1937). In the absence of phagocytic cells, pneumococci can multiply in the lymph even in the presence of antibody. The meningococcus, *H. influenzae*, and other gram-negative bacteria, on the other hand, would not survive in lymph because antibody and complement can kill these bacteria without phagocytic cells. This difference accounts for the tendency of certain gram-positive bacteria to remain dormant in the lymphatics and produce recurrent infection. The best known example of this phenomenon is recurrent erysipelas, an infection in which the lymph vessels of the skin are filled with streptococci (Fehleisen, 1886).

The flow of lymph depends on body movement, massage, or the increase in tissue pressure that occurs during inflammation. The network of fibers attached to the outer wall of the lymphatic vessels is stretched by increased tissue pressure so that the lymphatics are opened and can accommodate the larger volumes of lymph that flow during inflammation (Pullinger and Florey, 1935). The importance of adequate lymph flow in the removal of organisms for the prevention of infection is obvious after the lymphatics become obstructed, when the resulting elephantiasis becomes the focus of repeated infections. The remarkable filtering capacity of lymph nodes depends for its efficiency on a network of trabeculae that fills the sinusoids. The histiocytes lining the sinuses can then ingest and destroy the trapped organisms.

Transport by Blood

Microorganisms carried by the lymphatics to the subclavian veins may be phagocytized in the blood by neutrophilic leukocytes, which are then sequestered in the capillaries of the lung and liver. Many organisms are also removed by the Küpffer cells of the liver and the macrophages in the spleen. Bacteria, fungi, protozoa, and viruses can be taken out of the circulation by both the circulating phagocytes and the macrophage (reticuloendothelial) system. The route of viruses is illustrated by mousepox (Fenner, 1949). After multiplying in the skin, mousepox virus passes within eight hours to the regional lymph node. After multiplication in the node, the virus is carried into the blood and removed by phagocytes in the liver and spleen. Multiplication in the liver may lead to another viremia with secondary spread to the skin, where the characteristic eruption appears. During viremia, pox virus is associated with lymphocytes and monocytes; in lymphocytic choriomeningitis, Rift Valley fever, and Colorado tick fever the virus is adsorbed to erythrocytes. Others, like the togaviruses and the enteroviruses, circulate free in the plasma. After removal by the macrophages in the liver, some viruses (poxviruses, yellow fever virus) multiply in these cells and then infect the hepatic cells. Some pass directly into hepatic cells without multiplication in Küpffer cells (Rift Valley fever virus), and others, such as vaccinia and influenza viruses, are rapidly destroyed within the Küpffer cells (Fenner et al., 1974).

Virulent bacteria may also multiply in the liver and spleen and reenter the circulation to cause bacteremia or fungemia. The ability of the human liver to clear some, but not all, bacteria was discovered in patients with endocarditis when it was shown that hepatic venous blood always contained many fewer colonies of streptococci than blood from arteries or other veins (Beeson et al., 1945). Bacteria may also establish a sustained bacteremia by penetrating the wall of a capillary or venule and producing an infected thrombus. This thrombus may discharge a continuous stream of organisms at a rate that exceeds the ability of phagocytic cells to remove them. Such a bacteremia, often observed with staphylococci and *Bacteroides,* is dominated by lung abscesses.

Fungi and protozoa are no exceptions to the clearance of circulating organisms by the reticuloendothelial system. Perhaps the most dramatic example of phagocytosis by tissue macrophages is the tremendous number of *Histoplasma* organisms present in these phagocytes during human infection. Similarly, malaria illustrates the prominent phagocytosis of protozoa by the reticuloendothelial cells in liver, spleen, and bone marrow. Malaria is unique, however, in that the parasitized red cells are phagocytized but not the parasite alone.

Mode of Entry

The contact between phagocytic cell and infectious particle in the blood and tissue is made either through chance collision or through a positive attraction between phagocytes and microorganisms known as chemotaxis. In chemotaxis experiments in vitro, phagocytes can be seen to crawl by ameboid motions toward the particle to be ingested. Chemotaxis in vivo has often been difficult to demonstrate, but a striking example can be found in coccidioidomycosis. In the infected tissues, the fungus *Coccidioides immitis* has the form of a thick-walled spherule containing numerous endospores. The intact spherule is surrounded by mononuclear cells, but when it begins to rupture and release its endospores, there is a chemotactic rush of neutrophilic leukocytes to the point on the spherule where the endospores will emerge (Fig. 4). The attraction by the contents of the spherule is so great that neutrophils pour into the empty shell and seem to attack the inner wall even after the endospores are gone. The mechanism of this chemotaxis is not known, but in other infections chemotaxis results from the activation of complement by microbial antigens. This activation may occur via the classic or alternative pathway and generates chemotactic properties in the small peptides C3a and C5a, and in the trimolecular complex $\overline{C567}$. Bacterial proteases and proteases released from neutrophil granules can also attack C3 and C5 to produce chemotactic derivatives. In addition to complement factors, chemotactic products of a low molecular weight

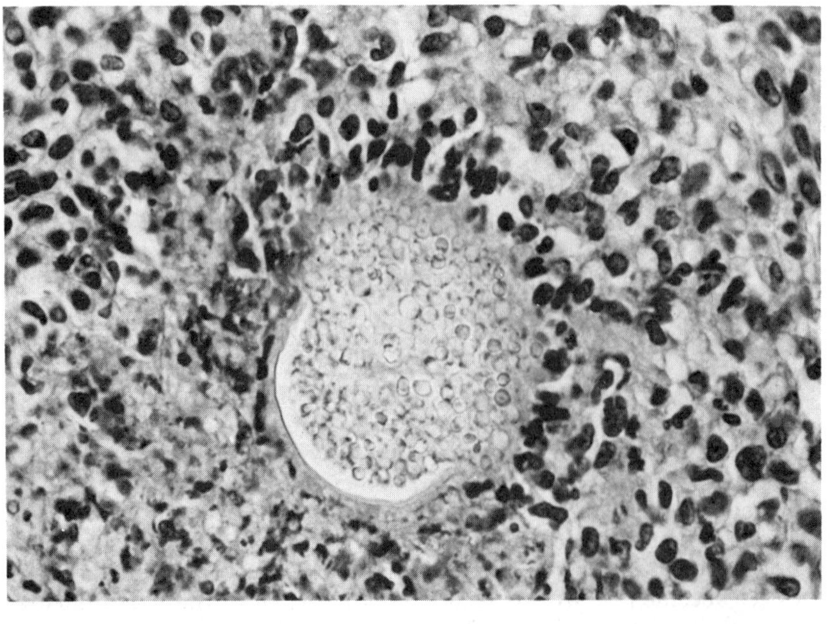

FIGURE 4. *Chemotactic rush of neutrophilic leukocytes to the point where the endospores emerge from the rupturing spherule of* Coccidioides immitis. *Notice absence of neutrophils at lower left margin of spherule where its cell wall is intact.*

are elaborated by certain bacteria. Chemotactic factors, regardless of origin, probably create a concentration gradient that directs the phagocyte to its target. It is also possible that chemotactic factors bind to specific receptors on neutrophil or monocyte membranes and activate their contractile apparatus. Anaerobic glycolysis appears to be the major energy source for chemotaxis in neutrophils because inhibitors of the glycolytic pathway depress chemotaxis but inhibitors of oxidative metabolism do not. Chemotaxis of leukocytes probably results from movements of actin and myosin, the contractile proteins of muscle, which are also present in phagocytes. Actin polymers in phagocytes form a tangled meshwork whose filaments are cross-linked by myosin and by actin-binding protein. It has been suggested that the cross-linked meshwork forms a rigid gel at the periphery of the cell ("cortical gel"), which provides the conditions for firm contractility that would be needed in pseudopods during chemotaxis and phagocytosis. This theory of phagocytic motility would take into account glycolysis as the energy source for chemotaxis, since ATP is required for movements of myosin and actin during contraction. It would also fit with the observation that cytochalasin B, which dissolves actin gels, inhibits phagocytosis by polymorphonuclear leukocytes and macrophages (Stossel, 1978).

After phagocytes reach their target and make contact, a microbe will be taken into the cell if their surfaces are properly attached. This attachment may depend partly on the hydrophobic character of the organism (Van Oss, 1978). Among the gram-negative bacteria, for example, smooth organisms with complete hydrophilic O polysaccharides are more resistant to phagocytosis than their rough hydrophilic mutants that are deficient in these surface sugars. Similarly, the hydrophobic polysaccharide capsules of pneumococci, *H. influenzae,* and *Klebsiella pneumoniae* make these bacteria more difficult for phagocytes to ingest than their unencapsulated variants. The opsonic activity of complement and specific antibody may also depend on their ability to make hydrophilic bacteria more hydrophobic. The Fc portion of IgG, which protrudes from the opsonized bacterial cell, is reported to be hydrophobic so that it can attach to the surface of the phagocytic cell. The surface protein (protein A) of *Staphylococcus aureus,* on the other hand, binds IgG by attaching to its Fc portion so that the hydrophilic Fab fragment protrudes. As a result, the opsonic property of IgG is subverted by virulent staphylococci and used instead to enhance their virulence by opposing phagocytosis.

Phagocytosis by both neutrophils and macrophages begins when proper contact is made with the microbial agent. The ingested particle is encircled by the phagocytic cell membrane through a process that combines invagination of the membrane and extrusion of pseudopods around the particle. The encirclement of the particle probably requires the participation of the contractile proteins (actin and myosin) and energy from ATP, as well as close adherence between particle and cell membrane and a sticky surface (Stossel, 1977). The sticky surface helps explain particle ingestion as well as why leukocytes stick to each other in clumps during phagocytosis and how the ends of pseudopods fuse to complete the encirclement. In neutrophils, the energy for engulfment comes primarily from anaerobic glycolysis. There are very few mitochondria in polymorphonuclear leukocytes, and Krebs cycle activity provides less than 20 per cent of ATP during glucose catabolism. Peritoneal and lung macrophages, on the other hand, have numerous mitochondria, and ATP is produced mainly from oxidative phosphorylation. Thus, inhibitors of oxidative metabolism (cyanide, dinitrophenol) inhibit ingestion by macrophages but not neutrophils, whereas inhibitors of glycolysis (iodoacetate, NaF) block phagocytosis by neutrophils but not macrophages.

Fusion of the pseudopods at their poles completes the encirclement of the particle by neutrophils and creates a phagocytic pouch, or vacuole, which separates from the cell membrane and moves centrally, accumulating cytoplasmic granules at its periphery (Fig. 5). The granules fuse to the membrane of the phagocytic vacuole and dissolve so that their contents empty into the vacuole. The azurophilic granules (purple color in Wright's stain), which have a dense appearance in the electron microscope, are actually lysosomes and release nucleases, lipases, acid phosphatase, elastase, collagenase, proteases, and other hydrolytic enzymes that digest dead organisms but are not bactericidal. The heme protein, myeloperoxidase (mw 150,000), and the mucopeptidase known as lysozyme are the important bactericidal enzymes released from azurophilic granules. Although neither kills pathogenic bacteria by itself, both help in such killing, as described elsewhere. A second group of granules of lower density is known as secondary granules because they appear second in granulocyte maturation. They contain lactoferrin, a bacteriostatic iron-binding protein that resembles transferrin in depriving bacteria of iron. Another group of antibacterial proteins, known as cationic proteins because of their basic charge, is also present in secondary granules. These basic proteins attach to the negatively charged surface of bacteria and kill them through unknown mechanisms.

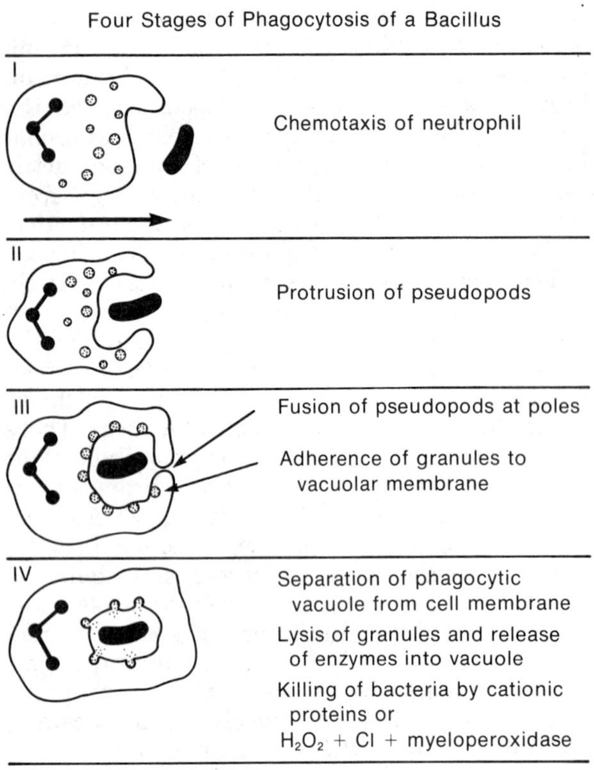

FIGURE 5. *Phagocytosis.*

Another group of bactericidal substances is produced through a series of metabolic phenomena known as the "respiratory burst" (Babior, 1978). Phagocytosis is undiminished in the absence of oxygen, but killing of the ingested organism does not occur unless there is a sharp increase in oxygen uptake, which is the first event in the respiratory burst. This rise in leukocyte oxygen uptake is stimulated by opsonized bacteria and other particles and by C5a within 30 to 60 seconds after contact with the cell membrane. The immediate consequence of this sudden utilization of oxygen is the production of superoxide (O_2^-) by the one-electron reduction of oxygen in the following reaction involving reduced nicotinamide adenine dinucleotide (NADPH):

$$2O_2 + NADPH \longrightarrow 2O_2^- + NADP^+ + H^+$$

The enzyme superoxide dismutase then converts superoxide to hydrogen peroxide (H_2O_2):

$$2O_2^- + 2H^+ \xrightarrow[\text{dismutase}]{\text{superoxide}} O_2 + H_2O_2$$

The hydrogen peroxide then acts with myeloperoxidase and a halide to kill the bacteria in the phagocytic vesicle where the myeloperoxidase has been released upon lysis of the azurophilic granules. Although hydrogen peroxide can kill bacteria, its bactericidal activity is greatly increased in the presence of myeloperoxidase (Klebanoff, 1973). Myeloperoxidase, like other peroxidases, catalyzes oxidation by hydrogen peroxide. In the bactericidal reaction, $C1^-$ appears to be the main substrate; it is oxidized to hypochlorite as follows:

$$Cl^- + H_2O_2 \xrightarrow[\text{myeloperoxidase}]{} ClO^- + H_2O$$

I^- and Br^- can also be oxidized in this reaction, which incorporates the halide in the bacterial cell. Halogenation of the bacteria is not entirely responsible for their death, however, and other unknown mechanisms must also be involved in the bactericidal effect of myeloperoxidase.

One of the key reactions in the metabolic burst is the hexose monophosphate shunt (HMP), which is the major source of NADPH, the reducing agent needed for production of superoxide. The first step in the shunt produces NADPH by the following reaction:

$$\text{Glucose-6-phosphate} + NADP^+ \xrightarrow[\text{G-6-PD}]{} \text{6 phosphogluconic acid} + NADPH$$

The reaction is catalyzed by G-6-PD (glucose-6-phosphate dehydrogenase). In the second HMP reaction more NADPH is produced as follows:

$$\text{6 phosphogluconic acid} + NADP^+ \xrightarrow[\text{6-phosphogluconate dehydrogenase}]{} NADPH + CO_2 + \text{ribulose-5-phosphate}$$

In patients whose leukocytes are deficient in G-6-PD, inadequate amounts of NADPH are generated after phagocytosis. The deficiency in NADPH results in less production of superoxide and, in turn, insufficient hydrogen peroxide for killing certain bacteria. This metabolic disturbance gives rise to increased susceptibility to pyogenic infections by *E. coli, K. pneumoniae,* and *S. aureus.*

The importance of hydrogen peroxide in killing bacteria is also evident from infections encountered in boys with chronic granulomatous disease (CGD), a disorder in which the respiratory burst and the production of hydrogen peroxide in leukocytes are selectively deficient. Their leukocytes kill streptococci but not staphylococci, so that they are prone to severe staphylococcal but not streptococcal infections. This difference is explained by the absence of catalase in streptococci and its presence in staphylococci. Catalase is needed to destroy hydrogen peroxide, so that

TABLE 4. Deficiency Diseases and Other Disorders of Natural Resistance to Infection

DISORDER	IMMUNE DEFECT	PREDOMINANT TYPE OF INFECTION
Immotile cilia	Loss of respiratory mucociliary clearance	Chronic airway infection
Achlorhydria	Loss of gastric bactericidal activity	Cholera, salmonellosis, shigellosis, giardiasis
Intestinal blind loop syndrome	Loss of small bowel motility	Overgrowth of small bowel with *B. fragilis* and deconjugation of bile salts causing malabsorption
Bruton's disease	Sex-linked agammaglobulinemia	Recurrent sinusitis, otitis, pneumonia, septicemia, and meningitis due to *Pneumococcus* and *H. influenzae;* also echovirus infections and encephalomyelitis
Acquired hypogammaglobulinema (adult men and women)	Deficient specific antibody responses	Pneumococcal and *H. influenzae* sinusitis, conjunctivitis, pneumonia, and bacteremia; giardiasis
Selective IgA deficiency	IgA less than 5 mg per cent	Recurrent infections of sinuses, bronchi, and lung (esp. middle lobe) with viruses and bacteria; giardiasis
Multiple myeloma	Absence of specific antibody globulin	Pneumococcal pneumonia and bacteremia
Wiskott-Aldrich syndrome	Low serum IgM; inability to respond to immunization with polysaccharide antigens	Meningitis, otitis media, pneumonia, and septicemia due to pneumococcus, *H. influenza,* and meningococcus
Tuftsin deficiency (congenital or after splenectomy)	Defective phagocytosis	Overwhelming pneumococcal septicemia and meningitis
Hodgkin's disease	Depressed cellular immunity	Cryptococcosis, listeriosis, herpes zoster
Chronic granulomatous disease	Inability to generate H_2O_2 in leukocytes	Fatal infections with staphylococci or gram-negative bacilli; normal resistance to pneumococci and other streptococci
Congenital thymic aplasia or hypoplasia	Absent T-cell function	Chronic candidiasis; disseminated mycobacterial, cryptococcal, aspergillus, pneumocystis, toxoplasma, strongyloides, and herpes infections
Chronic mucocutaneous candidiasis	Defective T-cell function	*Candida albicans* infection of skin and mucous membranes
Combined immunodeficiency (including Swiss-type, X-linked, or adenosine deaminase deficiency)	Complete absence of T and B cells	Candidiasis; cytomegalovirus, pneumocystis, and vaccinia infection; poliomyelitis (attenuated vaccine strains); pneumonia and chronic otitis media due to various gram-negative and gram-positive bacteria
Familial complement deficiency (C6, C8)	Loss of serum bactericidal power	Severe recurrent meningococcal and gonococcal septicemia
Cyanotic congenital heart disease	Polycythemia and cerebral thrombosis	Anaerobic brain abscess
Arteriosclerosis obliterans	Low redox potential of leg muscles	Clostridial myonecrosis

hydrogen peroxide accumulates in streptococci and replaces the deficiency in peroxide within CGD neutrophils. In other words, streptococci provide the missing reagent that is needed for their own destruction by the bactericidal effect of myeloperoxidase-Cl-H_2O_2. Staphylococci possess catalase and thus protect themselves against this reaction by destroying the peroxide component.

It must be emphasized that other systems for microbial killing exist within phagocytes and that the absence of peroxide-mediated killing does not necessarily promote infection. Alveolar macrophages, for example, have no peroxidase but kill ingested bacteria. Chickens and other birds have no peroxidase in their PMN leukocytes, but these cells kill bacteria and the birds are as resistant to infection as other animals. More important, patients with myeloperoxidase deficiency, in striking contrast to those with chronic granulomatous disease, are rarely troubled by infections (Babior, 1978).

DISEASES SECONDARY TO DISTURBANCES IN NATURAL RESISTANCE

A number of infections have been discussed in this chapter in relation to specific disturbances in natural resistance. These are summarized in Table 4. The disturbances in acquired cellular immunity described in the next chapter are also listed.

References

Amberson, J. B.: A clinical consideration of abscesses and cavities of the lung. Bull Johns Hopkins Hosp 94:227, 1954.

Babior, B.: Oxygen-dependent microbial killing by phagocytes. N Engl J Med 298:659, 1978.

Barry, D., and Reeve, A.: Increased incidence of gram-negative neonatal sepsis with intramuscular iron administration. Pediatrics 60:908, 1977.

Beeson, P., and Goebel, W.: The immunological relationship of the capsular polysaccharide of type XIV pneumococcus to the blood group A substance. J Exp Med 70:239, 1939.

Beeson, P., Brannon, E. S., and Warren, J. V.: Observations on the sites of removal of bacteria from the blood of patients with bacterial endocarditis. J Exp Med 81:9, 1945.

Bloomfield, A. L.: The dissemination of bacteria in the upper air passages. I. The circulation of foreign particles in the mouth. Am Rev Tuberc 5:903, 1922.

Braude, B., and Carlson, A. J.: The influence of lymphagogues on the relative concentration of bacterioagglutinins in serum and lymph. Am J Physiol 21:221, 1908.

Braude, A., Corbeil, L., Levine, S., Ito, J., and McCutchan, J.: Possible influence of cyclic menstrual changes on resistance to the gonococcus. In Brooks, G., Gotschlich, E., Holmes, K., Sawyer, W., and Young, F. (eds.): Immunobiology of Neisseria Gonorrhoeae. Washington, D.C., American Society of Microbiology, 1978, p. 328.

Bruton, O.: Agammaglobulinemia. Pediatrics 9:722, 1952.

Chernew, I., and Braude, A.: Depression of phagocytosis by solutes in concentrations found in the kidneys and urine. J Clin Invest 41:1945, 1962.

Constantopoulos, A., Najjar, V., and Smith, J.: Tuftsin deficiency: A new syndrome with defective phagocytosis. J Pediatr 80:564, 1972.

Culbertson, J. T.: Immunologic mechanisms in parasitic infections. In Most, H. (ed.): Parasitic Infections in Man. New York, Columbia University Press, 1951.

Donaldson, D.: Beta-lysin. In Schlesinger, D. (ed.): Microbiology, 1975. Washington, D.C., American Society Microbiology, 1975, p. 223.

Dubos, R. J.: Biochemical Determinants of Microbial Diseases. Cambridge, Mass., Harvard University Press, 1954.

Eliasson, R., Mossberg, B., Camner, P., and Afzelius, B.: A congenital ciliary abnormality as an etiologic factor in chronic airway infections and male sterility. N Engl J Med 297:1, 1977.

Fehleisen, F.: On erysipelas. In Cheyne, W. W. (ed.): Microparasites in Disease. London, New Sydenham Society, 1886, p. 261.

Felton, L., Prescott, B., Kauffmann, G., and Ottinger, B.: Antigens of vegetable origin active in pneumococcus infections. J Bacteriol 69:519, 1955.

Fenner, F.: Mouse-pox (infectious etromelia of mice): A review. J Immunol 63:341, 1949.

Fenner, F., McAuslan, B., Mims, C., Sambrook, J., and White, D.: The Biology of Animal Viruses, 2nd ed. New York, Academic Press, 1974, p. 363.

Field, M. E., Shaffer, M. F., Enders, J. F., and Drinker, C. K.: The distribution in the blood and lymph of pneumococcus type III injected intravenously into rabbits, and the effect of treatment with specific antiserum on the infection of the lymph. J Exp Med 65:469, 1937.

Fildes, P.: Tetanus: IX. The oxidation-reduction potential of the subcutaneous tissue fluid of the guinea pig: Its effect on infection. Br J Exp Pathol 10:197, 1929.

Garrod, L. P.: The susceptibility of different bacteria to destruction in the stomach. J Pathol Bacteriol 45:473, 1937.

Goldschneider, I., Gotschlich, E., and Artenstein, M.: Human immunity to the meningococcus. II. Development of natural immunity. J Exp Med 129:1385, 1969.

Hanke, M., and Bailey, J.: Oxidation-reduction potential requirements of *C. welchii* and other clostridia. Proc Soc Exp Biol Med 59:163, 1945.

Hauptman, S., and Tomasi, B.: The secretory immune system. In Fudenberg, H., Stites, D., Caldwell, J., and Wells, J. (eds.): Basic and Clinical Immunology, 2nd ed. Los Altos, Lange, 1978, p. 205.

Heidelberger, M., Avery, O., and Goebel, W.: A "soluble specific substance" derived from gum arabic. J Exp Med 49:847, 1929.

which *Klebsiella pneumoniae* (Friedländer's bacteria) forms capsules. J Bacteriol 38:367, 1939.

Hornick, R., Greisman, S., Woodward, T., DuPont, H., Dawkins, A., and Snyder, M.: Typhoid fever. I. Pathogenesis and immunologic control. N Engl J Med 283:686, 1970.

Isaacs, A., and Lindenmann, J.: Virus interference. I. The interferons. Proc R Soc B147:258, 1957.

Johanson, W., Blackstock, R., Pierce, A., and Sanford, J.: The role of bacterial antagonism in pneumococcal colonization of the human pharynx. J Lab Clin Med 75:946, 1970.

Johnson, R., and Harris, V.: Antileptospiral activity of serum. II. Leptospiral virulence factors. J Bacteriol 93:513, 1967.

Klebanoff, S.: Myeloperoxidase-halide-hydrogen peroxide antibacterial system. J Bacteriol 95:2131, 1968.

Kligman, A., and Weidman, F.: Experimental studies on treatment of human torulosis. Arch Dermatol Syph 60:726, 1949.

Kluger, M., and Rothenburg, B.: Fever and reduced iron: Their interactions as a host defense response to bacterial infections. Science 203:374, 1979.

Kushner, I., and Feldman, G.: Control of the acute phase response: Demonstration of C-reactive protein synthesis and secretion by hepatocytes during acute inflammation in the rabbit. J Exp Med 148:466, 1978.

McCutchan, J., Wunderlich, A., and Braude, A.: The role of urinary solutes in natural immunity to gonorrhea. Infect Immun 15:149, 1977.

Marrack, J., and Carpenter, B.: The cross-reactions of vegetable gums with type II antipneumococcal serum. Br J Exp Pathol 19:53, 1938.

Mortenson, R., Braun, D., and Gewurz, H.: Effects of C-reactive protein on lymphocyte functions. III. Inhibition of antigen-induced lymphocyte stimulation and lymphokine production. Cell Immunol 28:59, 1977.

Olling, S.: Sensitivity of Gram-negative bacilli to the serum bactericidal activity: A marker of the host-parasite relationship in acute and persisting infections. Scand J Infect Dis (Suppl), p. 8, 1977.

Pullinger, B. D., and Florey, H. W.: Some observations on the structure and functions of lymphatics: Their behaviour in local edema. Br J Exp Pathol 16:49, 1935.

Sanders, W. E.: Interactions between streptococci and other bacteria in the throat. In Schlesinger, D. (ed.): Microbiology, 1975. Washington, D.C., American Society of Microbiology, 1975.

Stossel, T.: How do phagocytes eat? Ann Int Med 89:398, 1978.

Stossel, T.: Phagocytosis. In Greenwalt, T., and Jamieson, G. (eds.): The Granulocyte in Function and Clinical Utilization. New York, Alan R. Liss, Inc., 1977, p. 87.

Takeuchi, A.: Electronmicroscope observations on penetration of the gut epithelial barrier by *Salmonella typhimurium*. In Schlesinger, D. (ed.): Microbiology, 1975. Washington, D.C., American Society of Microbiology, 1975, p. 174.

Thomas, E., and Aune, T.: Lactoperoxidase, peroxide, thiocyanate antimicrobial system: Correlation of sulfhydril oxidation with antimicrobial action. Infect Immun 20:456, 1978.

Tillitt, W. S.: The bactericidal action of human serum on hemolytic streptococci. I. Observations made with serum from patients with acute infections and from normal individuals. J Exp Med 65:147, 1937.

Van Oss, C.: Phagocytosis as a surface phenomenon. Ann Rev Microbiol 32:19, 1978.

Weinberg, E.: Iron and infection. Microbiol Rev 42:45, 1978.

Winberg, J., and Wessner, G.: Does breast milk protect against septicemia in the newborn? Lancet 1:1091, 1971.

82 MECHANISMS OF ACQUIRED RESISTANCE TO INFECTION

Abraham I. Braude, M.D., Ph.D.

Untreated patients do not recover from an infection unless their resistance increases enough to overcome the infecting agent or its toxins. This state of increased resistance may be manifested in the following ways:

1. The person remains well upon re-exposure to the same pathogenic organism during epidemics with high attack rates. The solid life-long immunity to measles, chickenpox, rubella, and mumps illustrates the permanence of this acquired resistance against highly contagious viral infections.
2. The patient's blood inhibits the growth of the corresponding organism in vitro. This characteristic is typical of blood from patients who have recovered from streptococcal sore throats or pneumococcal pneumonia.
3. The patient's serum will protect laboratory animals from the infectious agent. This protection is found with serum obtained after infections by bacteria (Group A streptococci, pneumococci, diphtheria bacilli, *Klebsiella pneumoniae,* leptospiral organisms, and *Treponema pallidum*), all rickettsiae, all viruses, and certain parasites (*Toxoplasma gondii*).
4. The patient's serum will prevent infection of tissue cultures by the virus that caused his disease.
5. His serum will damage the responsible microorganism (lysis of leptospira, meningococci, and *Haemophilus influenzae*; immobilization of *T. pallidum*; capsular swelling of pneumococci, *K. pneumoniae*, and *H. influenzae*; and distortion of toxoplasmas).

These signs of resistance are specific: The activity of the serum or blood and the resistance of the patient are directed only against the species or serologic type of infectious agent responsible for the illness. They are mediated by specific antibodies that are generated during the early stages of infection and increase as the patient recovers from the illness. In some infections lymphocytes and macrophages are also thought to be involved in acquired immunity to infection.

ACQUIRED HUMORAL IMMUNITY

As in any type of primary immune response, whether to living or dead antigens, the early humoral response to infection is predominantly IgM. Within ten days, however, IgG begins to appear and thereafter exceeds and outlasts IgM. IgM, which is heavier (M.W. 900,000, sedimentation coefficient 19S), tends to remain within the blood vessels and does not cross the placenta. IgG is smaller (M.W. 146,000, sedimentation coefficient 1S) and is actively transported across the placenta. Thus, IgG is responsible for the long-term immunity against many systemic infections and prevents many bacterial and viral infections in the newborn, whose immune system is undeveloped.

Opsonins

The mechanism of protection by IgG varies with the organism. As noted in the previous chapter, IgG is opsonic and protects against infections by pneumococci, *H. influenzae, K. pneumoniae,* group B *Streptococcus, Escherichia coli,* and certain other gram-negative bacilli by combining through its Fab portion with the capsular or K antigens. The Fc portion of the IgG attaches at the same time to surface receptors on neutrophilic leukocytes, so that an opsonic bridge occurs between bacterial cell and phagocyte. IgG, as well as IgM, can also promote phagocytosis after combining with bacterial antigens by activating complement via the classic pathway, so that opsonin is generated in the form of activated C3. Opsonins are also important in acquired resistance to *Streptococcus* group A, but they do not react with the streptococcal capsule. The reason for this is that the capsule of group A streptococci is composed of hyaluronic acid, a normal constituent of connective tissues, so that antibody is not produced against it. Instead, opsonins for the group A *Streptococcus* are directed against its M protein, which is carried on fimbria. Since the fimbria extend beyond the capsule, the overlying M protein antigen can react with specific anti-M IgG and thwart the antiphagocytic effect of the capsule.

Antitoxins

Most bacteria are pathogenic by virtue of their exotoxins, and antibody must neutralize these toxins to provide protection. Recovery from diphtheria is accompanied by the production of antitoxin, which appears to prevent the attachment of toxin to cells. The evidence supporting this mode of protection by antitoxin comes from observations with fragments A and B of diphtheria toxin. Fragment A has the enzymatic (or toxic) activity and fragment B is necessary for attachment and transport into the susceptible cell. When separate antisera that are specific for either the A or the B fragment are prepared, only the antibody against the B fragment will neutralize the toxin, apparently by interfering with its attachment to cells. In addition to blocking the attachment of toxin to substrate, antibody can accelerate the clearance of the toxic antigen from the circulation before it reaches susceptible cells. Recovery from diphtheria is accompanied by the production of antitoxin, which protects against toxin elaborated in the later stages of infection. Early in the infection, toxin combines with cells before antibody is produced and probably becomes inaccessible to neutralization. After recovery from diphtheria, the active production of diphtheria antitoxin gives protection for years, and its concentration in the body is probably increased periodically as a result of subclinical exposure to the diphtheria bacillus. With the Schick test, antitoxin can be demonstrated by its ability to neutralize the small amount of toxin that is injected into the skin.

Recovery from streptococcal infections is likewise accompanied by the development of antibodies against streptococcal toxins, but none of these antitoxins are known to bring about recovery. Antibodies against streptococcal hemolysin (streptolysin O), fibrinolysin (streptokinase), deoxyribonuclease (streptodornase), and hyaluronidase may all rise in titer as the streptococcal infection subsides but contribute little if anything to protection against the diseases caused by group A streptococci. Thus, sera containing high titers of antistreptolysin O cannot protect animals against *Streptococcus* group A unless antibody to M protein is also present. Antibody to streptococcal erythrogenic toxin will prevent the rash of scarlet fever but not streptococcal infections.

Aside from diphtheria there is no evidence that infection by toxin-producing bacteria produces a protective antitoxin or subsequent immunity. Second attacks of tetanus have occurred because the disease is produced by quantities of toxin that are too small to be antigenic. The lethal dose of botulinus toxin, which is even more potent than tetanus toxin, is also less than the antigenic dose. Protective antisera against the clostridial toxins can be prepared, however, by immunizing with toxoids made from the toxins of *C. botulinus, C. tetani,* and *C. perfringens.*

The endotoxins of gram-negative bacteria also produce protective antitoxins (i.e., anti-endotoxins) after immunization with the lipopolysaccharide (LPS) or the whole bacterial cell. The LPS from smooth bacteria stimulates primarily antibody to the O antigenic side chains, and LPS from rough bacteria stimulates antibody to the core (see Chapter 6). Antibody to either O or core antigens of endotoxin can prevent all toxic manifestations of endotoxin, including the local Shwartzman reaction, generalized Shwartzman reaction, and death. Antibody to endotoxin can also protect experimental animals from bacteremias due to gram-negative bacteria. The protection by O antibody is specific and is limited to homologous endotoxins or bacteria, whereas that by core antibody gives broad protection against a wide range of different bacteria and their endotoxins (Braude et al., 1977). The importance of antibody against endotoxin in human disease is still under study.

Lysins

In addition to preparing the pathogen for phagocytosis or neutralizing its toxin, antibody may destroy it by uniting with its antigens. Destruction of a microorganism, however, is not a universal feature of antibody activity; it is observed with certain gram-negative bacteria, spirochetes, leptospiras, protozoa, and viruses but not with gram-positive bacteria or fungi. Even when antibodies protect against infection by gram-positive bacteria, the antibodies are not lethal to them in vitro.

As noted in the preceding chapter, lysis of gram-negative bacteria by IgG or IgM requires the participation of complement. Perhaps the most important examples of protection by *acquired* bacteriolytic antibodies are those against the meningococcus and *H. influenzae.* In both cases, the bacteriolytic antibody acquired through infection is specifically directed against the capsules. In addition to bacteria, trypanosomes are susceptible to immune lysis. In studying this phenomenon in rats infected with *T. lewisi,* Taliaferro (1938) noted that the infected animal resorts to three mechanisms for disposing of the parasite: lysis, phagocytosis, and inhibition of reproduction. The power of reproduction was lost near the sixth day of infection, and soon afterward the parasites were killed by lysis and phagocytosis. The lysis was caused by the action of antibody plus complement, and was manifested by a precipitous fall in the number of trypanosomes in the blood after the tenth day. Despite these dramatic effects of antibodies, they soon become impotent because the surface antigens of trypanosomes are changed at each succeeding cycle of infection. Although these new antigenic variants are removed by the specific antibody they later stimulate, the next cycle of trypanosomal variants will not be susceptible to existing antibody. As many as 20 successive variants may occur, thus enabling the trypanosome to keep one step ahead of the antibody response. This phenomenon could explain why large amounts of IgM are produced in the serum and cerebrospinal fluid of patients with trypanosomiasis. Most of this immunoglobulin does not react with the trypanosome, possibly because it was stimulated by variants with different antigenic specificities.

OTHER MECHANISMS OF HUMORAL IMMUNITY

Antiparasitic

In malaria a different mechanism seems to bring recovery from infection. In this disease, recovery and protection are best correlated with antimerozoite IgG, which probably blocks the invasion of red cells by the parasite. The antibody is primarily strain-specific but also offers some protection against other strains of the same species. The antimalarial IgG persists for relatively long periods and protects newborns and infants by passing the placental barrier and by appearing in breast milk. These antibodies develop slowly, and several infections are sometimes required before immunity to malaria occurs. There are at least three reasons for this sluggish immune response: (1) Malaria suppresses immunity (Williamson and Greenwood, 1978). Humoral immunity is affected more than cellular immunity, probably because macrophages no longer process antigens normally. Immunity to both malaria and other infections is probably suppressed. (2) Immunity is probably stage-specific, so that different antigens determine immunity for each stage in the cycle of parasitic development. (3) In certain forms of malaria, variation can occur among surface antigens as in trypanosomiasis (Brown and Brown, 1965).

Immunity also develops to worms. A study by Taliaferro and Sarles (1939) of immunity to nematodes in rats demonstrated that *Nippostronglyus muris* may become exposed to lethal antibody by penetrating the intestinal wall and feeding on blood and tissue fluids. Antibody in these fluids forms a precipitate around the mouth of the parasite and seals the genital pore so that egg production is prevented. Unable to obtain nourishment or reproduce, the adult worm may be dislodged by peristalsis. Humoral antibody is probably involved in immunity to other nematodes also. Antibodies act in conjunction with eosinophils to protect against *Trichinella spiralis* by attacking adult worms in the gut and larvae en route or in the muscle. The immune response, which is elicited by products secreted by both larvae and adults, can reduce fertility of the adult, eliminate the adult before it engages in maximum production of larvae, or prevent the establishment of larvae in muscle. In other words, immunity reduces the number of larvae that can reach muscle and thus reduces the severity of the disease (Despommier, 1977).

Antibody and eosinophils also operate together in killing schistosomules, the first stage in development of *Schistosoma mansoni* infection after penetration of the skin by cercariae. In the presence of IgG antibody from infected patients, both eosinophils and macrophages kill schistosomules. This process, called *antibody-dependent cell-mediated cytotoxicity*, works differently for different cells and immunoglobulins. For example, IgG can behave like any opsonin, attaching via the Fab portion to the target and via Fc to the killer cell. IgE, on the other hand, can bind to the surface of macrophages and activate them so that they kill schistosomules. This type of macrophage activation may involve complexes composed of schistosome antigens with IgE antibody and is specific; that is, other parasites are not injured by these macrophages. Immune serum containing high levels of IgE-immune complexes can protect against experimental schistosomiasis (Capron et al., 1977).

Antiviral

Opsonization and immune-lysis are also involved in acquired resistance to virus infections after intravenous injection. Opsonic activity was demonstrated by Mims (1964), who injected complexes of mousepox virus and antibody intravenously and found that they were taken up by macrophages in the liver. Virus alone entered hepatic parenchymal cells and multiplied there, whereas virus injected with antibody disappeared after ingestion by macrophages. Immune lysis by antiviral antibody plus complement has been observed with enveloped viruses such as rubella and influenza viruses. Under electronmicroscopy, holes have been seen that appear to be punched through the envelope by complement (Almeida and Lawrence, 1969). Other structural changes have been identified in viruses after neutralization and exposure to complement. For example, studies with equine enteritis virus have shown that ribonucleases can penetrate the injured virus membrane after neutralization by antibody plus complement, and release the RNA from within the nucleocapsid (Radwan and Crawford, 1974). Although antibody on the surface of a virus particle can simply block viral attachment and penetration of susceptible cells, many exceptions to this process have been noted. There is good evidence, in fact, that the reverse may sometimes occur — namely, that antibody may help entry of virus particles into such cells. When particles of certain viruses such as vaccinia, poliovirus, and Newcastle disease virus are taken into cells in the presence of neutralizing antibody, the cells may not be infected and the virus is degraded (Doles and Kajiola, 1964; Silverstein and Marcus, 1964). Even though the target cells are not "professional" phagocytes (such as polymorphonuclears or macrophages), they confine the ingested virus to the phagocytic vesicle where nucleases and other lysosomal enzymes can degrade it. Thus, under some circumstances antibody seems to prevent the type of penetration by viruses that would enable them to appropriate the biosynthetic apparatus of the cell, and instead produces opsonization, phagocytosis, and destruction of virus particles by cells that are not customarily regarded as phagocytic.

Antibody may also prevent attachment of viruses to susceptible target cells by changing the viral surface charge and shape, by blocking critical sites for attachment of virus to cell, and by clumping virus particles so that many are sequestered within the aggregate. Autoaggregation, or clumping, of virus particles in the absence of antibody may have the reverse effect, that is, prevention of neutralization because the virus in the center of the clump is protected from antibody. Neutralization also depends on the number of immunoglobulin molecules and the fractions of complement. Many immunoglobulin molecules must attach to each virus particle to neutralize its infectivity; one antibody molecule is not enough. Complement is an important accessory factor in neutralization of viruses and can help to coat the virus surface, agglutinate the particles, or induce lysis of the virion. Studies with herpes simplex virus show that C1 is ineffective but sequential incubation with C4, C2, and C3 contributes to neutralization by coating the viral surface rather than by virolysis (Daniels, 1975). The lysis of enveloped viruses described earlier requires the terminal components (C5 through C9). Antibodies on the viral surface may not be able to neutralize infectivity if they are too few or are bound to sites that are not critical for infection. These antibody-viral complexes may circulate and remain infectious, as in experimental infection with lymphocytic choriomeningitis virus. Viral-antibody complexes can be neutralized, however, through the participation of antiglobulins such as rheumatoid factor. One of the rheumatoid factors is a large globulin (M.W. 900,000) of the IgM class that attaches to the Fc portion of IgG. Upon combining with IgG-virus complexes, rheumatoid factor itself does not reduce infectivity but allows complement to do so. Since antiglobulins appear during convalescence from cytomegalovirus, Epstein-Barr, hepatitis B, rubella, and influenza virus infection, it is possible that they contribute to recovery by participating in viral neutralization.

Acquired Secretory Immunity

The agents of most infectious diseases enter the body through mucous surfaces where secretory rather than humoral antibody may be important in recovery from and protection against infection. The importance of secretory IgA in natural resistance to infection is discussed in the preceding chapter. Its role in acquired resistance was discovered by Burrows in studies with experimental cholera (Burrows et al., 1947). He demonstrated that protection against infection after cholera vaccination was correlated with fecal antibody rather than with serum antibody. The ability of cholera vaccine to protect as many as 60 per cent of humans against cholera shows the significance of Burrow's experimental findings in clinical medicine. Furthermore, protection by vaccines that contain killed vibrios shows that antibacterial rather than antitoxic immunity can produce acquired resistance. Much evidence suggests that secretory immunity against cholera vibrios is mediated primarily by antibody that blocks their attachment to the intestinal mucosa rather than by complement-mediated reactions such as bacteriolysis or opsonization by neutrophils. IgA cannot fix C1 (Ishizaka et al., 1966), and its bactericidal activity via the alternative complement pathway has been observed only under special conditions that would not exist in the anticomplementary milieu of the gut. The idea that IgA blocks attachment of vibrios is supported by Freter's report (1969) that coproantibodies (fecal antibodies) sharply reduced the number of vibrios that adhered either to the mucosa in isolated loops of rabbit intestine or to viable slices of rabbit ileum. Coproantibody also inhibited the growth of *V. cholera* in the presence of viable mucosal cells, but the mechanism of such growth inhibition is unknown.

Other pathogenic bacteria adhere selectively to certain mucosal surfaces. Thus group A streptococci attach to human pharyngeal cells much better than *E. coli,* whereas *E. coli* adhere to bladder mucosa much better than group A streptococci (Gibbons, 1977). Similarly, gonococci and group B streptococci adhere to vaginal epithelial cells in larger numbers than *E. coli, Bacteroides fragilis,* fusobacteria, and lactobacilli (Mardh and

Westrom, 1976). Adherence to intestinal epithelium likewise appears to determine the virulence of various enteric pathogens such as *Shigella* and enteropathogenic *E. coli*. This selective adherence of bacteria to certain epithelial cells seems to parallel their relative ability to produce infection in tissues or organs covered by these cells. The surface components of bacteria that are responsible for adherence seem to vary with the organisms and the mucosal surface. There is evidence, for example, that the following surface components are involved in adherence: lipoteichoic acid and surface fibrils for group A streptococci (Beachey, 1975); pili for gonococci, *E. coli*, and *Proteus*; and flagella for *V. cholerae* (Jones and Freter, 1972). The surface components of epithelial cells responsible for adherence seem to be primarily glycoproteins (Gibbons, 1977). Similar glycoproteins are abundant in the mucus that bathes these cells and can apparently block adherence of pathogenic bacteria by competing for attachment sites on the surface of these organisms. The increased flow of mucus during respiratory, intestinal, and genital infections could thus interrupt their spread by delivering two proteins, IgA and glycoprotein, that might block further adherence of pathogenic bacteria to the diseased mucosa. Because secretory IgA is also antitoxic it may contribute to recovery from diphtheria, cholera, and diarrhea caused by the enterotoxins of *E. coli* (Stoliar et al., 1976). Thus, acquired secretory immunity might be both antitoxic and antibacterial. These attractive hypotheses, however, need more critical experimental support.

There is more convincing evidence that immunity against virus infections is acquired by the local production of secretory IgA on mucosal surfaces. This antibody, however, may be more important in providing immunity against reinfection than in recovery from infection. In influenza, for example, IgA appears in nasal washings early during the infection, but its antiviral activity is directed against strains of influenza virus that produced outbreaks in the past. This early IgA seems to be antibody stored in epithelial cells and has been found in both the respiratory and the intestinal mucosa (Kaur et al., 1972). It has no activity against the virus causing the infection and contributes nothing to recovery. Specific secretory IgA with activity against the infecting virus does not appear until two to three weeks after the onset of infection and long after the infection has ended (Alford et al., 1967). For this reason, other processes such as interferon production or cell-mediated immunity are thought to bring about recovery.

The presence of secretory antibody on mucous surfaces acquired after local infection or local inoculation of vaccines can be correlated with resistance to infection by rhinovirus, influenza virus, parainfluenza virus, and polioviruses. The IgA response to local inoculation of living viruses is better than the response to inactivated viruses, and live virus may also produce a systemic humoral immune response. The ability of local antibody to protect against mucosal viral infection was shown by studies in which local inoculation of inactivated viruses succeeded in giving protection when local but not systemic antibody was elaborated. Two striking differences have been noted between local and systemic immunization. For one thing, local immunization of mucous membranes tends to produce a very localized IgA response, which is restricted to the immunized or infected site and does not appear on other mucous membranes. Second, local immunization does not seem to evoke a secondary or booster type of secretory IgA response. Studies on the secondary IgA response are limited, but local vaccination by intranasal instillation of inactivated poliovaccine produced no difference in the quantity or duration of poliovirus antibody after primary and secondary immunization. The absence of a booster effect indicates that the secretory IgA response to poliovirus may be devoid of immunologic memory, and that long-lasting protection after primary immunization may result from continued local antibody synthesis in response to prolonged stimulation by persistent antigen (Ogra et al., 1975).

NONPROTECTIVE ANTIBODY ACQUIRED DURING INFECTION

Not all antibodies acquired after infection or vaccination can protect against infection, despite their specificity for antigens in the infectious agent. These nonprotective antibodies react with antigens that are not responsible for virulence and are often released from beneath the surface of the infecting agent. The C polysaccharide of streptococci, the structural components of virions, the teichoic acids of staphylococci, the nucleoprotein of *Brucella*, the cell-wall polysaccharide of *B. anthracis*, and the polysaccharides of mycobacteria are examples of antigens that generate specific but nonprotective antibodies during infection. Nonprotective antibodies also develop to viral enzymes that participate in the manufacture of structural proteins and nucleic acids. In fact, antibodies that develop during viral infection may be directed against antigens that are elaborated during viral synthesis but are never incorporated into the virion. The various antibodies that have no role in protection are identified by complement fixation, agglutination, or precipitation and are often useful for diagnosis.

ACQUIRED CELLULAR IMMUNITY

The increased resistance observed after certain infections, such as those caused by mycobacteria and fungi, is not always accompanied by protective antibody in the blood or body fluids. Persons who have overcome the infection are more resistant to exogenous infection than previously uninfected persons. Although precipitins and complement-fixing antibodies may be present at some stage in these infections, many healthy persons who have recovered from all clinical signs of the disease possess no detectable antibodies for the infectious agent. This type of acquired resistance without demonstrable antibody is manifested by an intense local inflammatory response to the living organisms or to their products. The local reaction is more intense and appears more rapidly than that in the nonimmune individual and prevents the extension of the organism to the regional lymph nodes. Because this reaction is marked by an intense infiltration of inflammatory cells, it is sometimes spoken of as *cellular immunity,* a term that implies a specific increase in the capacity of the cells themselves to resist infection.

Although cellular immunity has been regarded as the basis of acquired resistance to infection in various bacterial, fungal, viral, and parasitic infections, the most important original discoveries in this field were made by Robert Koch (1891) in his studies of experimental tuberculosis. Koch described his discovery of the altered response to tuberculous reinfection in the following manner:

> If a normal guinea pig is inoculated with a pure culture of tubercle bacilli, the wound, as a rule, closes and in the first few days seemingly heals. After ten to fourteen days, however, there appears a firm nodule which soon opens, forming an ulcer that persists until the animal dies. Quite different is the result if a tuberculous guinea pig is inoculated with tubercle bacilli. For this purpose it is best to use animals that have been infected four to six weeks previously. In such an animal, also, the little inoculation wound closes at first, but in this case no nodule is formed, On the next, or second day, however, a peculiar change occurs at the inoculation site. The area becomes indurated and assumes a dark color, and these changes do not remain limited to the inoculation point, but spread to involve an area 0.5 to 1.0 cm in diameter. In the succeeding days it becomes evident that the altered skin is necrotic. It finally sloughs, leaving a shallow ulcer which usually heals quickly and permanently, and the regional lymph nodes do not become infected. The action of tubercle bacilli upon the skin of a normal guinea pig is thus entirely different from their action upon the skin of a tuberculous one. This striking effect is produced not only by living tubercle bacilli but also by dead bacilli, whether killed by prolonged low temperature, by boiling, or by certain chemicals.

This remarkable observation by Koch brought out two essential features of the immune response in tuberculosis:

1. *A hypersensitivity reaction to the tubercle bacillus.* The effect of hypersensitivity is to produce an accelerated inflammatory response upon reinfection. This accelerated reaction was actually first described by Jenner with cowpox when he vaccinated people who had recovered from smallpox.

2. *Increased resistance to tuberculosis.* The bacteria are confined to the site of inoculation. In contrast to primary infection, they do not spread to the regional lymph nodes or beyond.

The hypersenstivity reaction in tuberculosis is called "delayed" hypersensitivity because it is not noticed until after 24 hours. It differs from the immediate type of hypersensitivity in anaphylaxis, which elicits tissue reactions within minutes after the antigen is injected. Delayed hypersensitivity is also different from the Arthus reaction, which begins in a few hours after challenge with antigen. These two rapid forms of hypersensitivity require a reaction between antigen and antibody, but delayed hypersensitivity does not. Delayed hypersensitivity is seen in all forms of infection but has special significance in those diseases in which immunity is not acquired through circulating antibody. This is true in all fungus infections; in such bacterial infections as leprosy, brucellosis, tularemia, lymphopathia venereum, and syphilis; in smallpox and certain other virus infections; and in various parasitic infections, including cutaneous leishmaniasis and schistosomiasis.

Because acquired immunity develops simultaneously with hypersensitivity in these diseases, because the two phenomena share the same inflammatory reaction, and because humoral antibody provides no resistance against these infections, it is generally assumed that delayed hypersensitivity can produce this form of acquired resistance. The inflammatory cells in delayed hypersensitivity are those that are important in resistance to infection. When a moderate or large dose of organisms or their antigens are injected into an infected individual, many polymorphonuclear cells appear, followed by mononuclear phagocytes (macrophages), which tend to arrange themselves in sheet-like aggregates and in tubercles. When a very small concentration of organisms is inoculated, only mononuclear phagocytes may appear. These polymorphonuclear and mononuclear cells can kill bacteria and other organisms, and their accelerated delivery to the site of infection in delayed hypersensitivity is associated with a more rapid disappearance of organisms than in controls.

In addition to this accelerated migration, reinfection stimulates macrophages to greater phagocytosis and killing. Because of these increased activities, they have been designated "activated" macrophages. Activation generally occurs after infection by organisms that reside within macrophages. Once activated by such infection, macrophages provide increased resistance against infections caused by other unrelated organisms. This increase in nonspecific resistance is accompanied by greater nonspecific killing of injected organisms. For example, macrophages from animals infected with *Listeria monocytogenes* or *Salmonella typhimurium* can kill *S. typhimurium* equally well (Blanden et al., 1966). The increased antimicrobial properties of activated macrophages are accompanied by other functional and morphologic changes. They are larger than normal macrophages, spread out more and adhere better to glass, and have more cytoplasmic granules. They also utilize more glucose through the hexose monophosphate shunt and have more activity of certain enzymes, such as the membrane enzyme adenylate cyclase and the cytoplasmic enzyme lactic dehydrogenase. They also have more lysosomes with their hydrolytic enzymes.

Lymphocytes also play an important part in cellular immunity by regulating macrophage activity. Soon after reinfection occurs with the intracellular bacteria that induce delayed hypersensitivity, specifically sensitized T lymphocytes react with the bacterial antigens and release substances, known as lymphokines, into the infected tissues. These lymphokines, which are chemotactic, attract macrophages to the infection and activate them. The specifically sensitized T lymphocytes have long lives and can provide long-lasting specific resistance to reinfection by reactivating macrophages. The macrophages, on the other hand, lose their activation within a few weeks. Generally speaking, humoral antibody is involved in these reactions only under exceptional circumstances such as trichinosis and schistosomiasis. Experimental studies indicate that acquired immunity in these parasitic infections results from a form of cellular immunity that is dependent on antibody. The role of humoral antibody in rejecting the invading larvae (schistosomules) has been established by passive transfer of immunity with immune serum. The antibody involved is IgE, which interacts with both macrophages and eosinophils, so that both cells then kill the parasite (Capron et al., 1977).

The mechanisms of cellular immunity in virus infections are still obscure, even though delayed hypersensitivity was first described with inoculation of vaccinia virus by Jenner in 1798. One reason for the difficulty in identifying the cell-mediated component of antiviral immunity is that humoral antibody usually affords prominent protection and may overshadow cellular immune processes. According to current evidence, antibody functions mainly by preventing viral infection and limiting its extracellular spread, and cell-mediated immunity functions by eliminating virus-infected cells. In patients suffering from diseases or receiving drugs that would be expected to lower cellular immunity, there is a predilection for infection with four viruses that acquire their envelope from nuclear or cytoplasmic membranes of infected cells: measles, cytomegalovirus, varicella-zoster, and herpes simplex. These viruses cause severe recurrent infections in patients given immunosuppressive therapy, and children with congenital immunodeficiencies. Since the antigens of these membrane-associated viruses are incorporated in the membrane of the infected cell, they are available on the cell surface for reactions with the inflammatory cells involved in delayed hypersensitivity. This idea gains support from the fact that infections by all membrane-associated viruses are followed by delayed hypersensitivity reactions to the specific viral antigens (Bloom and Rager-Zisman, 1975).

At least three mechanisms have been postulated to explain how cell-mediated reactions could work against virus infections. One is by a reaction of T cells against virus-infected cells. Purified T lymphocytes can lyse target cells containing specific membrane antigens (Henney, 1973). A second potential mechanism involves interferon production by activated lymphocytes. Viral antigens can cause blast cell transformation of specifically immune lymphocytes and the secretion of lymphokines, including interferon. The secretion of lymphokines with chemotactic activity could attract macrophages that can inactivate viruses, including viruses that have combined with antibody. Because macrophages have receptors for the Fc portion of IgG molecules, virus-antibody complexes can be attacked, ingested, and degraded. In the third mechanism, infected cells containing viral antigens in their membranes would be recognized as foreign and rejected by a cell-mediated reaction.

Each of these theoretical processes for mediating cellular immunity in virus infections has been demonstrated in systems examined in vitro. Until these have been found to operate in vivo as well, the role of cell-mediated immunity in human viral infections will continue to be speculative.

References

Alford, R., Rossen, R., Butler, W., and Kasel, J.: Neutralizing and hemagglutination-inhibiting activity of nasal secretions following experimental human infection with A_2 influenza virus. J Immunol 98:724, 1967.

Almeida, J., and Lawrence, G.: Heated and unheated antiserum on rubella virus. Am J Dis Child 118:101, 1969.

Beachey, E.: Binding of group A streptococci to human oral mucosal cells by lipoteichoic acid. Trans Assoc Am Phys 88:285, 1975.
Blarden, R., Mackaness, G., and Collins, F.: Mechanisms of acquired resistance in mouse typhoid. J Exp Med 124:585, 1966.
Bloom, B., and Rager-Zisman, B.: Cell-mediated immunity in viral infections. In Notkins, A. (ed.): Viral Immunology and Immunopathology. New York, Academic Press, 1975, p. 113.
Braude, A., Ziegler, E., Douglas, H., and McCutchen, J.: Protective properties of antisera to R core. In Schlesinger, D. (ed.): Microbiology-1977. Washington, D.C., American Society for Microbiology, 1977, p. 253.
Brown, K., and Brown, I.: Immunity to malaria: Antigenic variation to chronic infections of *Plasmodium knowlesi*. Nature 208:1286, 1965.
Burrows, W., Elliott, M., and Havens, I.: Studies on immunity to asiatic cholera. IV. The excretion of coproantibody in experimental enteric cholera in the guinea pig. J Infect Dis 81:261, 1947.
Capron, A., Dessaint, J., Joseph, M., Torpier, G., Capron, M., Rousseaux, R., Santoro, F., and Bazin, H.: IgE and cells in schistosomiasis. Am. J Trop Med Hyg 26:39, 1977.
Daniels, C.: Mechanism of viral neutralization. In Notkins, A. (eds.): Viral Immunology and Immunopathology. New York, Academic Press, 1975, p. 79.
Despommier, D.: Immunity to *Trichinella spiralis*. Am J Trop Med (Suppl.) 26:part 2, 68, 1977.
Dales, S., and Kajioka, R.: The cycle of multiplication of vaccinia virus in Earle's strain L cells. 1. Uptake and penetration. Virology 24:278, 1964.
Freter, R.: Studies of the mechanism of action of intestinal antibody in experimental cholera. Tex Rep Exp Biol Med 27 (Suppl 1):299, 1969.
Gibbons, R.: Adherence of bacteria to host tissue. In Schlesinger, D. (ed.): Microbiology, 1977. Washington, D.C., American Society for Microbiology, 1977, p. 395.
Henney, C.: On the mechanics of T cell mediated cytolysis. Transplant Rev 17:37, 1973.
Ishizaka, T., Ishizaka, K., Borsos, T., and Rapp, H.: C′1 fixation by human isoagglutinins: Fixation of C′1 by γG and γM but not γA antibody. J Immunol 97:711, 1966.
Jones, G., and Freter, R.: Adhesive properties of *Vibrio cholerae:* Nature of the interaction with isolated rabbit brush border membranes and human erythrocytes. Infect Immunol 6:918, 1972.
Kaur, J., McGhee, J., and Burrows, W.: Immunity to cholera: The occurrence and nature of antibody-active immunoglobulins in the lower ileum of the rabbit. J Immunol 108:387, 1972.
Koch, R.: Further communication on a remedy for tuberculosis. Dtsch Med Wochenschr 17:101, 1891
Mardh, P., and Westrom, L.: Adherence of bacteria to vaginal epithelial cells. Infect Immunol 13:661, 1976.
Mims, C.: Aspects of the pathogenesis of virus diseases. Bacteriol Rev 28:30, 1964.
Ogra, P., Morag, A., and Motti, L.: Humoral immune response to viral infections. In Notkins, A. (ed.): Viral Immunology and Immunopathology. New York, Academic Press, 1975, p. 57.
Radwan, A., and Crawford, T.: The mechanism of neutralization of sensitized equine arteritis virus by complement components. J Gen Virol 25:229, 1974.
Silverstein, S., and Marcus, P.: Early stages of Newcastle disease virus–Hela cell interaction: An electron microscopic study. Virology 23:370, 1964.
Stoliar, O., Kaniecki-Green, E., Pelley, R., Klaus, M., and Carpenter, C.: Secretory IgA against enterotoxins in breast milk. Lancet 1:1258, 1976.
Taliaferro, W. H.: Ablastic and trypanocidal antibodies against *Trypanosoma dultoni*. J Immunol 35:303, 1938.
Taliaferro, W. H., and Sarles, M.: The cellular reactions in the skin, lungs, and intestine of normal and immune rats after infection with *Nippostrongylus muris*. J Infect Dis 64:157, 1939.
Williamson, W., and Greenwood, B.: Impairment of the immune response to vaccination after acute malaria. Lancet 1:1328, 1978.

83 MECHANISMS OF IMMUNOLOGIC INJURY IN INFECTIOUS DISEASES

Abraham I. Braude, M.D., Ph.D.

Microbial antigens may injure cells or tissues during infection by the seven mechanisms listed in Table 1.

These mechanisms may operate independently or simultaneously, and some may involve both antibodies and mononuclear cells (lymphocytes and macrophages).

ANTIBODY-MEDIATED CYTOLYSIS

Lysis of cells may result from the interaction of antibody, complement, and microbial antigens on the surface of infected cells (Porter, 1971). This type of injury has been observed almost exclusively in cells infected with viruses that have envelopes such as poxviruses, herpesviruses, paramyxoviruses, arenavirus, togavirus, and rhabdovirus (Rawls and Tompkins, 1975). Viral antigens appear on the surface of infected cells after RNA viruses pass through the cellular cytoplasmic membrane during the release of the nucleocapsid from the infected cell by the process of "budding." The virus picks up its envelope from the cytoplasmic membrane and at the same time deposits viral antigen in the surface of the membrane (Rawls and Tompkins, 1975). A DNA virus such as herpes simplex picks up its envelope by budding through the nuclear membrane, but herpes antigens also appear at the cell surface. The union of IgG or IgM antibodies with viral antigen on the cell surface activates complement via the classic pathway, and cell lysis ensues through the same mechanism as that described for bacterial lysis in Chapter 81. After reacting with antibody and antigenic sites on the cell

TABLE 1. Mechanisms of Immunologic Injury in Infections

MECHANISM	EXAMPLE OF INFECTIOUS AGENTS
Antibody-mediated cytolysis	Enveloped virus
Complement-mediated chemotaxis	Pyogenic bacteria
Immune complex vasculitis	Agents of endocarditis, viral hepatitis, leprosy, and quartan malaria
Intravascular coagulation	Meningococcus and other gram-negative bacteria
Anaphylaxis	Worms and aspergilli
T cell-mediated cytolysis	Lymphocytic choriomeningitis virus
Delayed hypersensitivity	*M. tuberculosis, Brucella, Histoplasma,* vaccinia virus

membrane, C1 is activated to react with C4 and C2. This generates C3 convertase, which splits C3 into two fragments, the larger (C3b) remaining attached to the cell membrane. The protease activity of C3b binds C5, 6, and 7, thus building a complex on the cell surface composed of C1 through C7. Damage to the cell membrane is initiated when C8 is bound to the C1 through C7 complex. Cytolysis is accelerated when C9 becomes incorporated (Kolb and Müller-Eberhardt, 1975). This type of cytolysis is a theoretical mechanism not only for eliminating virus infections but also for injuring the infected organ. The effectiveness of the process in producing cytolysis depends on the type of cell and on the number of antigenic sites that appear on the cell surface (Kibler and Ter Meulen, 1975). For example, during persistent infection in tissue culture by measles virus, relatively few surface antigens appear compared with acute measles infections, and few if any cells undergo lysis. This finding correlates with the absence of necrotic foci in subacute sclerosing panencephalitis, a form of persistent measles infection of the brain.

COMPLEMENT-MEDIATED CHEMOTAXIS AND SUPPURATION

When the classic complement pathway is activated by union of antibody with microbial antigen, or the alternate pathway by bacterial lipopolysaccharides, the chemotactic factors C3a and C5a are produced. These chemotactic factors attract inflammatory cells that not only attack microbial cells and virus-infected cells but also injure the infected tissues. The chemotactic factors C3a and C5a can also stimulate the release from mast cells of histamine, which in turn increases the permeability of vessels for leukocytes. The heavy influx of granulocytes during inflammation causes the characteristic tissue damage of suppuration by release of proteolytic enzymes from neutrophil granules. These enzymes can destroy both connective tissue proteins (collagenase and elastase) and cells (Williams et al., 1977).

IMMUNE COMPLEX VASCULITIS (ARTHUS REACTION)

When microbial antigens unite with antibody, the infectivity of the organism may be inactivated so that the infection is terminated. But if the antigen is discharged from organisms protected within a "privileged sanctuary" that cannot be eliminated by immune processes, the steady supply of antigen may form soluble complexes with its antibody. Such complexes can develop from antigens released upon microbial lysis by antibody and complement or as a by-product of microbial growth. These soluble complexes can circulate in the blood until they are deposited in the tissues by phagocytosis or trapping. Phagocytosis by reticuloendothelial (RES) cells removes large complexes preferentially so that the small complexes formed in antigen excess remain in the circulation. If the large complexes contain infectious virus or toxic microbial antigens that persist within the Küpffer cells, they may impair phagocytic clearance by the RES and promote the circulation of complexes (Oldstone and Dixon, 1975). As RES clearance fails and the concentration of complexes mounts in the circulation, the complexes pass through permeable vessels and are deposited in the tissues. The permeability of these vessels is apparently increased when the antigen in the complexes reacts with IgE that is adherent to basophils or mast cells. The antigen thus causes mast cell or basophilic granules to release certain factors that can increase vascular permeability and others that cause platelets to clump and in turn release vasoactive amines. The combined effect of such basophilic and platelet permeability factors on blood vessels permits the leakage of immune complexes and their deposition in the tissue along the vessel walls (Cochrane and Koffler, 1973). In addition, glomerular capillaries have normal endothelial fenestrations that

TABLE 2. Infectious Diseases Underlying the Production of Immune Complex Glomerulonephritis

UNDERLYING DISEASE	ORGANISMS
Subacute bacterial endocarditis	*Staphylococcus aureus, Streptococcus viridans*, enterococcus, *Coxiella burnetii*
Ventriculoatrial shunt infections	*S. epidermidis*
Malaria	*Plasmodiuim malariae* and *falciparum*
Lepromatous leprosy	*Mycobacterium leprae*
Syphilis	*Treponema pallidum*
Hepatitis	Hepatitis B virus
Pneumococcal pneumonia	Type 14 pneumococcus
Streptococcal pyoderma or pharyngitis	Group A streptococci
Typhoid fever	*Salmonella typhosa*
Shistosomiasis	*Schistosoma mansoni*
Toxoplasmosis	*Toxoplasma gondii*

favor leakage of complexes and help explain the unique susceptibility of glomeruli to immune-complex deposition. Immune complexes with a density in the range of 19S or larger are most likely to be deposited along vessels and damage them (Cochrane and Koffler, 1973); complexes with a density of less than 19S not only fix complement poorly but are also cleared slowly from the blood.

One important mechanism for vessel injury requires fixation of complement by the immune complexes (Cochrane and Koffler, 1973). Activation of the complement system generates C3a and C5a, which increase permeability of vessels and attract granulocytes. The accumulation of granulocytes (neutrophilic leukocytes) is followed by the release of their lysosomal enzymes, which destroy the elastic laminae of small arteries and injure the basement membrane of glomerular capillaries. The consumption of complement by these complexes is reflected in the diminished concentration of C3 and other complement components in the serum and by the presence of complement components at the site of immune complex deposition. If complement and polymorphonuclear leukocytes are depleted in experimental animals with immune complex disease, the arterial lesions are prevented, but glomerular injury can still occur (McCluskey and Klassen, 1974) through mediators other than complement.

In bacterial endocarditis, quartan malaria, lepromatous leprosy, syphilis, and viral hepatitis B, the occurrence of immune complex disease is well established (Bayer et al., 1976; Kohler et al., 1974; Gamble and Reardon, 1975). Each of these is a subacute or chronic infection in which a bacterial, protozoan, or viral antigen is released constantly into the bloodstream at a time when high levels of antibody are already present. The immune complexes that result from the union of circulating antigen with antibody produce granular deposits along the glomerular basement membrane. These complexes, containing immunoglobulin plus antigen plus complement, produce electron-dense deposits in the glomerular basement membrane and diffuse proliferative glomerulonephritis. Those infections that have been implicated in the production of immune complex glomerulonephritis are listed in Table 2. In most of these, the specific microbial antigens have been demonstrated in the glomerular deposit. In the immune complex glomerulonephritis of syphilis, schistosomiasis, infected ventriculoatrial shunts, and quartan malaria, the clinical syndrome has usually been that of the nephrotic syndrome.

In viral hepatitis B and lepromatous leprosy there have been prominent clinical features attributed to immune complex deposition outside the kidney. In chronic carriers of hepatitis B virus, for example, hepatitis B surface antigen immune complexes were identified in vascular lesions in periarteritis nodosa. In other cases of viral hepatitis B, immune complexes do not necessarily contain viral antigens. In these immune complexes the antigen is usually IgG and the antibody is IgM (Levo et al., 1977). It appears that some of the IgG generated in various chronic infections can become antigenic itself and stimulate a form of autoimmunity to the patient's own immunoglobulin. Perhaps the antigenic groups of IgG that react with IgM antibody are normally buried within the IgG molecule and become exposed when IgG molecules unfold after union with the specific microbial antigen. Activation of the classic complement pathway by these complexes of IgG and IgM results in systemic vasculitis involving the skin, kidneys, and joints and producing vascular purpura, glomerulonephritis, and arthritis. These immune complexes composed of IgG and IgM have the distinctive property of precipitating at cold temperatures and are known as cryoglobulins.* They occur in leprosy, endocar-

*These cryoglobulins in chronic infections are composed of polyclonal IgM and IgG. They are designated Type III cryoglobulins and must be distinguished from Type I and Type II (mixed monoclonal-polyclonal) cryoglobulins that are produced in certain noninfectious diseases.

ditis, cytomegalovirus infections, and infectious mononucleosis, and can produce vascular lesions in these conditions. The most dramatic of these occur in lepromatous leprosy and take the form of painful red papules known as erythema nodosum leprosum. They may be accompanied by arthritis and glomerulonephritis.

INTRAVASCULAR COAGULATION

Tissue injury from intravascular coagulation in human infection is best illustrated by the hemorrhagic skin necrosis and renal cortical necrosis of meningococcemia. The hemorrhagic skin necrosis in meningococcemia is an example of the dermal Shwartzman reaction (Davis and Arnold, 1974), and renal cortical necrosis is typical of the generalized Shwartzman reaction. As noted in Chapter 6, either phenomenon can be produced experimentally by giving rabbits two injections of the lipopolysaccharide (LPS) antigen from any gram-negative bacteria. Two intravenous injections of LPS, given 24 hours apart, invariably produce severe disseminated intravascular coagulation (DIC) culminating in bilateral renal cortical necrosis. This dramatic form of kidney damage (Sanarelli-Schwartzman reaction) results from glomerular deposits of fibrin. The dermal Shwartzman reaction is also produced by two separate injections of LPS, the first intradermally and the second intravenously 24 hours later. Soon after the cutaneous dose, the site of the intradermal injection undergoes hemorrhagic necrosis secondary to local thrombosis and infarction. The first injection causes an Arthus-like inflammation about the skin vessels, and the intravenous dose initiates coagulation and thrombosis that are limited to the inflamed vessel.

The essential role of coagulation is illustrated by the prevention of both reactions by heparin and other anticoagulants. In the generalized Shwartzman reaction, the first dose of LPS triggers intravascular conversion of fibrinogen to fibrin by the mechanisms discussed in Chapter 6, but the reticuloendothelial system (RES) clears the fibrin before it can be deposited in the glomerular capillaries and other vessels (Lee and Stetson, 1965). The first dose of LPS seems to condition the RES so that its clearance activity is blocked by the second dose. As a result, the fibrin aggregates produced by the second dose are no longer cleared from the blood; instead, they are filtered out by the glomeruli, where they occlude the capillaries and produce infarction of the renal cortex. DIC consumes clotting factors such as fibrinogen, platelets, and prothrombin. In addition, DIC activates fibrinolysis so that fibrin degradation products are released. These fragments prevent clotting by inhibiting proteolysis of fibrinogen by thrombin and blocking polymerization of fibrin monomer to form a clot. The anticoagulant effect of these degradation products can cause serious bleeding in patients whose clotting factors are depleted by DIC, not only aggravating hemorrhages into the necrotic lesions of the skin and kidneys but also producing hemorrhages in other tissues not affected by the Shwartzman lesions.

Although meningococcemia is the most dramatic and consistent example of the two Shwartzman reactions, severe bacteremias with any gram-negative bacteria can cause DIC. One of the most devastating examples is the DIC that occurs in *E. coli* bacteremia of pregnancy. This condition is characterized by hemorrhagic lesions in the skin and by fatal renal cortical necrosis. For obscure reasons, pregnancy appears to prepare for the generalized Shwartzman reaction so that only one intravenous dose of LPS is required to evoke DIC and renal cortical necrosis.

Although gram-negative bacteria, including rickettsiae, are the primary cause of DIC because of their LPS, infections by gram-positive bacteria, viruses, fungi, and protozoa can also cause intravascular coagulation. These organisms can initiate clotting by various mechanisms such as the release of tissue factor, a lipoprotein contained in the plasma membrane of endothelial cells and monocytes. When tissue factor is liberated by damage of these cells, it can initiate clotting by complexing with Factor VII and calcium ions to activate Factor X. Bacteria can also initiate clotting through their proteolytic enzymes that break down fibrinogen. But the most relevant immunologic mechanism is the initiation of coagulation or of platelet aggregation by circulating immune complexes composed of microbial antigens and antibody. Such complexes can accelerate fibrin formation in vitro, but only in cell-rich plasma, presumably because tissue-factor is released. Similarly, soluble antigen-antibody complexes can initiate clotting in vivo by substituting for the provocative injection of LPS in the dermal Shwartzman reaction. The intravenous injection of antigen into specifically immunized rabbits also produces DIC and the generalized Shwartzman reaction (Lee, 1963).

ANAPHYLAXIS

Anaphylactic antibodies occur primarily in worm and fungus infections. These antibodies belong mainly to the IgE class and act through attachment via the Fc position to the surface of tissue mast cells and basophils. When a divalent antigen from worms or fungi combines with two

IgE molecules to form a bridge, the antibody becomes distorted. This distortion of IgE on the cell surface causes the discharge of intracellular granules, which release histamine and serotonin. In the lung, histamine and serotonin cause smooth muscle contraction and endothelial injury, resulting in bronchospasm and edema. The eosinophil chemotactic factor of anaphylaxis is released from mast cell granules and attracts eosinophils to the site of anaphylaxis. Eosinophils appear to check the allergic reaction by ingesting the antigen-antibody complexes and by inactivating histamine through the release of histaminase (Goetzl and Austen, 1977). Worms induce very high levels of IgE and eosinophils and various manifestations of anaphylaxis. *Ascaris* worms are especially potent causes of anaphylactic hypersensitivity when they invade the tissues and especially when they migrate through the lungs. They produce the syndrome of eosinophilic pneumonia, an allergic pneumonitis with a prominent asthmatic component. Eosinophilia is also prominent in strongyloidiasis, visceral larva migrans, trichinosis, certain types of filariasis, and acute schistosomiasis. *Angiostrongylus cantonensis,* a nematode that invades the central nervous system, causes eosinophilic meningitis. Systemic fatal anaphylactic shock can occur in echinococcosis upon rupture of a hydatid cyst and release of echinococcal antigen into the peritoneum, where high levels of IgE are delivered from the circulation. A more benign systemic reaction, known as Katayama fever, occurs at the onset of egg production in schistosomiasis, but the role of IgE is debatable in the induction of this reaction.

Among fungus diseases, pulmonary allergic aspergillosis is the most important example of a hypersensitivity disorder mediated by IgE (Golbert and Patterson, 1970). After colonization of the bronchial mucous secretions with aspergilli, these patients develop asthma, eosinophilia, and migratory pulmonary infiltrates suggestive of allergic pneumonitis. IgG also participates in the allergic bronchial reaction by producing an Arthus-type reaction.

CYTOLYSIS MEDIATED BY T LYMPHOCYTES

T lymphocytes can kill infected cells in the absence of antibody if three requirements are met: (1) antigen of the infecting virus must appear on the surface of the cell; (2) the T-cells must have immune-specificity for the virus; and (3) the T cells must be histocompatible with the target cell. In other words, virus-immune T lymphocytes can cause lysis of infected cells only if the T cell reacts with the viral antigen on the surface of a cell that is recognized as nonforeign. The recognition of a cell as nonforeign, or "self," is thought to involve a reaction between the lymphocyte and the histocompatibility antigen on the target cell. Thus, the attacking T lymphocyte would require two receptors to kill the infected cells: one for the viral antigen and the other for the histocompatibility antigen (Zinkernagel, 1978). T cell-mediated cytolysis may be important in infections by viruses that do not themselves destroy the infected cell. One example is lymphocytic choriomeningitis (LCM) virus. In intracerebral LCM virus infection of mice, the disease is caused by T cells rather than by the virus. The severity of this infection depends on when the T cells reach the target cells: if they reach the target cells early, they can eliminate the virus infection by destroying the cells before the virus replicates and spreads. At this point the damage to cells would be relatively insignificant. On the other hand, a cytolytic attack by T cells on advanced infections could kill the animal. In lymphocytic choriomeningitis, the choroid plexus is heavily infected, and cytolytic damage to this structure destroys the blood-brain barrier so that lethal cerebral edema occurs. A similar attack by T lymphocytes on liver tissue has been postulated as a mechanism for liver injury in viral hepatitis because the virus is assumed not to be cytotoxic. Hepatitis B surface antigen (HBsAg) can be found in the membrane of liver cells, and T cell-mediated cytotoxicity against HB_SAg has been demonstrated in both acute and chronic hepatitis B (Hirschman, 1979).

From these observations and others, it has been suggested that killer lymphocytes can be beneficial only in infections with highly cytopathic viruses, because the infected cells must be sacrificed to ward off more extensive damage by the virus itself. In infections by viruses with slight cytopathic potential, on the other hand, the onslaught of T cells may cause more tissue injury than the virus (Zinkernagel, 1978).

DELAYED HYPERSENSITIVITY

This form of hypersensitivity has been most extensively studied in tuberculosis and is also known as the "tuberculin" type of hypersensitivity. When living or dead tubercle bacilli are injected into the tissues of an animal or patient with tuberculosis, the inflammatory reaction does not begin until after a few hours; it progresses to its maximum size and intensity within 24 to 48 hours. This delay in onset and peak intensity of the reaction, compared with the anaphylactic and

Arthus types of inflammatory response, is responsible for the term *delayed hypersensitivity*. The concept that this reaction represents hypersensitivity is based on the fact that in nontuberculous animals, the injection of living bacilli produces no reaction for a week or more (until sensitization has time to develop), and dead bacilli may produce no reaction at any time. The intact bacillus is not required because antigenic tuberculous proteins can also elicit the phenomenon. The same delayed response to antigenic fractions of the infecting organism is seen in many other infections. The response is specific, occurring only when the organisms or antigens causing the infection are injected.

Because delayed hypersensitivity is characterized by necrosis at the injection site if large doses of dead organisms or antigen are used, this reaction is considered to be responsible for necrosis of tissues in natural infection after immune hypersensitivity has developed. In pulmonary tuberculosis, for example, the bacilli multiply slowly in mononuclear phagocytes that aggregate undamaged at the portal of entry until hypersensitivity develops. Then the cells are killed, the tissues undergo necrosis, and tubercle bacilli are discharged by the ulcerating focus into the lumen of the bronchioles (Canetti, 1954, 1958).

Necrosis is thought to be mediated by the lymphocytes that dominate the inflammatory reaction. A few lymphocytes and monocytes first accumulate around vessels in the area into which antigen or bacteria are injected. These specifically sensitized T lymphocytes react with the antigen to release mediators that attract and hold other mononuclear cells in the inflammatory focus. One of these mediators is a chemotactic factor, which attracts macrophages, and the other is migratory inhibitory factor (MIF), which holds them in place. In the meantime, lymphocytes themselves proliferate in situ upon stimulation by specific antigen and release a mitosis-stimulating factor that causes other lymphocytes to multiply. These reactions and others amplify the cellular response and produce the characteristic heavy infiltration of mononuclear cells that is the hallmark of delayed hypersensitivity. Polymorphonuclear leukocytes are seen very early in the inflammatory response but soon disappear before the delayed response is underway. Antibody also seems to be unimportant, because delayed hypersensitivity can be transferred by the mononuclear cells but not by serum.

At least three mechanisms have been proposed to explain cell necrosis by mononuclear cells. First, T lymphocytes attack target cells directly. They can be seen to kill cells in tissue culture, and the amount of killing can be measured by the radioactivity released from target cells labeled with ^{51}Cr. The second mechanism of killing by lymphocytes is lymphocytotoxin, a mediator released from lymphocytes after reaction with specific antigen. This cytotoxic factor is a protein-like substance with a molecular weight in the range of 85,000 and the capacity to kill a wide variety of different cells (Williams and Granger, 1973). A third process of possible importance in the production of necrosis involves macrophages. In response to lymphocyte mediators and other stimuli, macrophages release hydrolytic enzymes such as cathepsins, hyaluronidase, and collagenases, which can injure the matrix that supports the cells, the vessels that supply them, and the cells themselves.

Killing of cells by T lymphocytes can be specific or nonspecific. Specific cytolysis is carried out by T lymphocytes that have immune specificity for the antigens on the surface of the target cell, as described in the preceding section on cytolysis by T lymphocytes. The surface antigens can be derived from the viruses infecting the cells or from other microbial antigens that adhere to the cytoplasmic membrane of the target cell. Cytolysis is nonspecific when the killer lymphocytes are a product of the mitosis that follows blast transformation. This blast transformation in T cells may be induced either by specific antigen or by lymphocytic mediators released from other lymphocytes.

References

Bayer, A., Theofilopoulos, A., Eisenberg, R., Dixon, F., and Guze, L.: Circulating immune complexes in infective endocarditis. N Engl J Med 295:1500, 1976.

Canetti, G.: The Tubercle Bacillus in the Pulmonary Lesion of Man. New York, Springer Publishing Company, 1954.

Canetti, G.: Pathogenesis of tuberculosis in man. Ann NY Acad Sci 154:13, 1968.

Cochrane, C., and Koffler, D.: Immune complex disease in experimental animals and man. Adv Immunol 16:186, 1973.

Davis, C., and Arnold, K.: Role of meningococcal endotoxin in meningococcal purpura. J Exp Med 140:159, 1974.

Gamble, C., and Reardon, J.: Immunopathogenesis of syphilitic glomerulonephritis. N Engl J Med 292:449, 1975.

Goetzl, E., and Austen, K.: Cellular characteristics of the eosinophil compatible with a dual role in host defense in parasitic infections. Am J Trop Med Hyg 26:142, 1977.

Golbert, T., and Patterson, R.: Pulmonary allergic aspergillosis. Ann Int Med 72:395, 1970.

Hirschman, S.: Virologic, immunologic, and clinical correlations in viral hepatitis. Seminars Infect Dis 2:48, 1979.

Kibler, R., and Ter Meulen, V.: Antibody-mediated cytotoxicity after measles virus infection. J Immunol 114:93, 1975.

Kohler, P., Cronin, R., Hammond, W., Olen, D., and Carr, R.: Chronic membranous glomerulonephritis caused by hepatitis B antigen-antibody immune complexes. Ann Intern Med 81:448, 1974.

Kolb, W., and Müller-Eberhard, H.: The membrane attack mechanism of complement. J Exp Med 141:724, 1975.

Lee, L.: Antigen-antibody reaction in the pathogenesis of bilateral renal cortical necrosis. J Exp Med 117:365, 1963.

Lee, L., and Stetson, C.: The local and generalized Shwartzman phenomenon. In Zweifach, B., Grant, L., and McClusky, R. (eds.): The Inflammatory Process. New York, Academic Press, 1965, p. 791.

Levo, Y., Gorevic, P., Kassab, H., Zucker-Franklin, D., Gigli, I., and

Franklin, E.: Mixed cryoglobulinemia: Immune complex disease often associated with hepatitis B virus infection. Trans Assoc Am Phys 90:167, 1977.
McCluskey, R., and Klassen, J.: Immunologically mediated glomerular, tubular, and interstitial renal disease. N Engl J Med 288:564, 1973.
Odstone, M., and Dixon, F.: Immune complex disease associated with viral infections. In Notkins, A. (ed.): Viral Immunology and Immunopathology. New York, Academic Press, 1975, p. 341.
Porter, D. D.: Destruction of virus-infected cells by immunological mechanisms. Ann Rev Microbiol 25:283, 1971.
Rawls, W., and Tompkins, A.: Destruction of virus-infected cells by antibody and complement. In Notkins, A. (ed.): Viral Immunology and Immunopathology. New York, Academic Press, 1975, p. 99.
Williams, T., and Granger, G.: Lymphocyte in vitro cytotoxicity: Mechanism of human lymphotoxin-induced target cell destruction. Cell Immunol 6:171, 1973.
Williams, T., Lyons, J., and Braude, A.: In vitro lysis of target cells by rat polymorphonuclear leukocytes isolated from acute pyelonephritic exudates. J Immunol 119:671, 1977.
Zinkernagel, R. M.: Major transplantation antigens in host responses to infection. Hospital Practice 13(7):83, 1978.

84 IMMUNOPROPHYLAXIS AND IMMUNOTHERAPY

Stephen A. Spector, M.D.

In the spring of 1796, Edward Jenner took some fluid from a large cowpox vesicle on the hand of dairymaid Sarah Nelmes and inoculated it at two sites on the arms of 8-year-old James Phipps (Creighton, 1889; Parish, 1965, 1968). The boy developed typical cowpox lesions, and 2 months later, when Jenner inoculated him with smallpox, no disease developed. Less than 200 years later, on October 26, 1977, Ali Maow Maalin, a cook in Merka, Somalia, became the "last reported case" of endemic variola. The global eradication of smallpox must be considered a major accomplishment of modern science and a triumph for immunizations. The elimination of this dreaded disease, as with most immunization programs, was not without setbacks and risks to vaccinees. Prophylactic inoculation against variola (variolation) was practiced in China, India, Persia, and elsewhere for centuries before it was popularized in England by Lady Mary Wortley Montagu in the early 18th century. The practice of variolation was far from safe, and possibly 2 to 3 per cent so vaccinated died of smallpox. This fatality rate was tenfold less than that from naturally occurring smallpox, and the artificial inoculation of variola was widely practiced until Jenner developed his "improved vaccine." The concept of intentionally exposing an individual to either live or killed immunogens to prevent disease, combined with the princple of passive transfer of antibody to prevent and ameliorate infections, serves as the cornerstone of present day immunoprophylaxis.

PRINCIPLES OF PASSIVE IMMUNIZATION

Short-term immunity to many infections may be conferred by administering preformed antibody as immune serum or gamma globulin prepared from it (Stiehm, 1979). Two types of gamma globulin preparations are available: standard human immune serum globulin (ISG) for general use, and special human immune serum globulins with known antibody content for specific illnesses. Certain diseases may also be ameliorated or prevented by the use of animal sera and antitoxins.

IgG is produced primarily in plasma cells that have evolved from B-lymphocytes. Destruction of IgG-coated bacteria occurs following phagocytosis in the granulocytes of the reticuloendothelial system and in the gastrointestinal tract. IgG crosses the placenta and provides passive protection to the newborn infant for approximately six months. These passively acquired IgG antibodies may inhibit an infant's immune system and prevent an adequate response to certain vaccines.

ISG is prepared by alcohol fractionation of pooled human sera by Cohn's alcohol fractionation procedure, which removes most other serum proteins and hepatitis viruses. ISG is composed of 95 per cent IgG at 165 mg per ml (16.5 per cent solution), along with trace quantities of IgM and IgA globulins and other serum proteins. IgG has a mean half-life of 25 days. Metabolism of IgG is directly controlled by its level in serum; there is an increase in the breakdown of IgG with elevated plasma levels and a decrease in catabolism with decreased levels.

ISG is available only for intramuscular injection. In vitro, it has been shown to aggregate into large molecular weight complexes that can activate spontaneously the complement system. These aggregates are probably responsible for the occasional systemic reactions to ISG that occur in approximately 2 per 1000 injections (Ellis and Henney, 1969). The incidence of reactions is increased with repeated injection of ISG and with intravenous administration. Approximately 20 per cent of patients receiving repeated intramuscular doses of gamma globulin can be expected to experience mild reactions including dyspnea, tightness in chest, faintness, hypotension, flushing, facial swelling, and anxiety (Squire et al., 1969).

Human gamma globulin is frequently used without proven indications. Patients with certain defined immune defects or with documented exposure to one of the infections listed in Table 1 are good candidates for passive prophylaxis. There is no evidence suggesting that the child with an apparent overabundance of upper respiratory infections without documented immunologic deficits benefits from short or prolonged treatment with gamma globulin. Similarly, special human immune serum globulins are in limited supply and should be administered only when patients are likely to benefit from their use (Table 2). Animal antisera and antitoxins particularly should be used only when there are definite indications (Table 2). Refined and concentrated horse sera are still the only effective prophylaxis for diphtheria, gas gangrene, and botulism once exposure has occurred. Animal serum against tetanus and rabies should only be used when special human gamma globulin is unavailable. The use of animal sera may be associated with acute febrile reactions, serum sickness, or anaphylaxis. Febrile reactions following injections of animal serum are usually mild but may be severe, requiring vigorous antipyretic therapy. Serum sickness occurs in from 10 to 40 per cent of recipients of some animal sera and consists of rash, urticaria, arthritis, adenopathy, and fever appearing hours or several days after a second injection, or 7 to 12 days after a first injection. Anaphylaxis occurs in approximately 1 per cent of individuals receiving animal antisera. Since anaphylactic reactions are life-threatening, an intracutaneous skin test, preceded by a scratch or eye test, should be performed on every patient before any injection of animal serum. Since intradermal skin tests have resulted in fatal reactions, a skin test should never be performed unless a syringe containing 1 ml of 1:1000 epinephrine is immediately available. If the skin test is positive or a patient has a strong allergic history, desensitization of the patient should be carried out under close medical supervision if the prophylactic use of animal antiserum is strongly indicated.

A new human antiserum has been developed which attempts to protect patients from death due to gram-negative bacteremia. The antiserum is made by immunizing volunteers against a stable mutant of *Escherichia coli* that lacks the ability to form the O antigenic side chains. The antibody thus formed is against the common lipopolysaccharide core of gram-negative bacteria and in animal studies has been highly effective in protecting against lethal gram-negative bacteremic shock and disseminated intravascular coagulation (Braude et al., 1977). Preliminary human trials have been promising, but further studies are required to establish its effectiveness and value (Ziegler et al., 1978).

TABLE 1. Indications for the Use of Standard Immune Serum Globulin (Human)

DISORDER	PURPOSE	DOSE (I.M.)	COMMENT
	Proved Value		
Antibody deficiency disease (agammaglobulinemia, hypogammaglobulinemia)	Treatment	0.66 ml/kg every 3–4 wk	Maximum dose is 20–30 ml; double dose at onset of therapy
Measles	Modification, prevention	0.05 ml/kg 0.25 ml/kg	Rarely indicated; give immediately after exposure
Viral hepatitis type A (HAV)	Prevention:		
	single exposure	0.02–0.04 ml/kg	Give as soon as possible after exposure
	continuous exposure	0.06 ml/kg, repeat in 5–6 mo	
Viral hepatitis type B (HBV)	Prevention	0.12 ml/kg	Use when hepatitis B immune serum globulin is not available
	Limited Value		
Varicella	Modification	0.06–0.12 ml/kg	Give immediately after exposure; indicated only in serious underlying illness when VZIG or VZIP is not available
Rubella	Prevention	20 ml	For pregnant women in the first trimester

From American Academy of Pediatrics: Report of the Committee on Infectious Diseases, 18th ed., 1977.

TABLE 2. Passive Immunization

DISEASE	SOURCE OF ANTIBODY	INDICATION	DOSE	ADVERSE REACTIONS	COMMENT
Bacterial					
Diphtheria	Horse serum (antitoxin)	Prevention and treatment	10,000–80,000 units (I.M. and I.V.) depending on type of involvement and age of patient	Hypersensitivity, serum sickness, anaphylaxis	
Pertussis	Human pertussis immune globulin (PIG)	Treatment	<1 year old; 1.25 ml >1 year old; 2.5 ml I.M.	Pain and tenderness at injection site	Doubtful efficacy
Botulism	Horse serum against types A, B, and E	Treatment	1 vial I.V.; 1 vial I.M. Repeat in 2–4 hours if symptoms continue	Hypersensitivity, serum sickness, anaphylaxis	Has not been helpful in infantile botulism Effectiveness uncertain
Tetanus	Human tetanus immune globulin (TIG)	Prevention and treatment	Prevention: 250–500 U—half I.M., half locally Treatment: 3000–6000 U—half I.M., half locally	Pain and tenderness at injection site	If TIG unavailable, tetanus antisera from horse serum (TAT) may be used
Viral					
Hepatitis B	Human hepatitis B immune globulin (HBIG)	Prevention (see text)	0.06 ml/kg I.M.; repeat after 28 days	Pain and tenderness at injection site	See text
Varicella-zoster	Human varicella-zoster immune globulin (VZIG)	Prevention (see text); should be given within 72 hours of exposure	Newborn, 2.0 ml; others, 3–5 ml I.M.	Pain and tenderness at injection site	If VZIG unavailable, VZIP 10 ml/kg I.V. may be used Not indicated for active chickenpox or zoster
Rabies	Human rabies immune globulin (RIG)	Prevention	20 units/kg, half into wound and half I.M.	Pain and tenderness at injection site	If RIG not available, equine rabies immune serum should be used
Mumps	Human mumps immune globulin	Prevention and treatment in adult men	20 ml I.M.	Pain and tenderness at injection site	Questionable efficacy in preventing orchitis in postpubertal males
Smallpox	Human vaccinia immune globulin (VIG)	Within 24 hours after known exposure	0.3 ml/kg I.M.	Pain and tenderness at injection site	Not indicated with active infection

I.V., intravenous
I.M., intramuscular

ACTIVE IMMUNIZATION

Active immunization is the induction of immunity by inoculation of a specific organism or some fragment or toxin associated with that organism. Although passive immunizations have prevented some disease, there are clearly many advantages to active vaccination. Some active immunizations give lifelong protection and thus do not require recognition of exposure to be effective. Additionally, the immunity conferred during active immunizations frequently approaches 100 per cent, whereas passive protection is often substantially less successful. The cost of preparing immune sera is more expensive than preparing vaccines.

Most immunogens used for active immunizations are crude fractions of an organism, or killed or live whole bacteria or viruses. The most notable exceptions to this are the vaccines against tetanus and diphtheria, which contain preformed inactivated toxins as immunogens.

Inactivated "killed" vaccines may be produced by three techniques. The first of these consists of whole organisms that are killed by heat or chemicals (e.g., formaldehyde). Examples of such killed vaccines are pertussis, typhoid, cholera, inactivated poliomyelitis, and influenza. These vaccines generally do not confer lasting immunity and frequently are only partially effective. Killed measles vaccine is no longer licensed because it not only offers less protection than does the attenuated live measles vaccine but also sensitizes individuals to future challenge with live measles so that infections are more severe. The killed poliomyelitis vaccine, however, has been extremely useful. Although it has been generally replaced by the attenuated Sabin vaccine, the killed Salk vaccine was effective in substantially reducing the number of clinical cases of paralytic polio in the United States before the introduction of the live polio vaccine. The killed vaccine has virtually eradicated the disease in Scandinavia, where the live vaccine has not been used.

The second method of preparing inactivated vaccines is by using extracted cellular fractions that have been shown to induce immunity in man. Vaccines prepared from the polysaccharide contained in the cell wall of the meningococcus and the capsule of the pneumococcus appear to be effective and safe in adults and children above the age of two years (Lepow et al., 1977; Peltola et al., 1977; Klein and Mortimer, 1978; Cowan et al., 1978; Center for Disease Control, 1978; Wilkins and Wehrle, 1979).

Immunizations with toxoids are prepared by inactivating large amounts of toxin with formalin rather than acting against the organisms themselves. The antibodies induced by these vaccines are active in neutralizing the toxin produced during infection. This third method of preparing inactivated vaccines has been extremely useful in the prevention of diphtheria and tetanus.

The major disadvantages of inactivated vaccines lie in their inability to confer lasting and local immunity. Multiple doses and boosters are required for continued protection, and local IgA fails to develop from the parenteral injection. As a result, protection against respiratory and gastrointestinal infections is suboptimal.

Live attenuated vaccines are prepared by serial passage of organisms in tissue culture or embryos so that the vaccine strain organism loses virulence. Local secretory IgA production may be stimulated by administration of the vaccine at the site of entry in the gut or the respiratory tract (Rothberg et al., 1973). This method of immunization produces the strongest and most durable immunity.

Active immunization may also be achieved by inducing an infection with a nonpathogenic organism that cross-reacts with a virulent organism. For example, Bacillus Calmette-Guérin (BCG), a strain of bovine tuberculosis, is used to induce immunity to human tuberculosis (Eickhoff, 1977). Similarly, protection against smallpox has been achieved by immunization with cowpox or vaccinia (Creighton, 1889).

PROBLEMS ASSOCIATED WITH IMMUNIZATIONS

Associated with the desired benefits of vaccines are many real and potential problems. Live attenuated viral vaccines may revert to more virulent viruses and result in severe disease. In general, however, reversion has not been a problem, and serious reactions usually reflect abnormal host response rather than virus alteration. Contacts of vaccinees given live viruses may be at risk of developing an infection with the vaccine strain organism. Unfortunate cases of paralytic poliomyelitis have been well documented to occur rarely in both normal and immunosuppressed individuals following immunization or exposure to a recently immunized family member (Davis et al., 1977). Contaminating viruses or other organisms undetected in continuous cultures may lead to acute or chronic diseases and conceivably could result in malignancies in vaccine recipients (Fraumeni et al., 1963). Since many viruses cause latent or persistent infections and may produce slow viral diseases, concern has been expressed that certain vaccines may promote the development of these infections. To date, these concerns have not been substantiated. In fact, the incidence of subacute sclerosing panencephalitis (SSPE) in children following measles vaccine is nine times less common than in children with a history of naturally acquired measles (Modlin et al., 1977). The presence of unsuspected passenger viruses has been found in both live and killed vaccines. Simian viruses 40 (SV40), present in monkey cells, contaminated the polio cultures of both the Salk and Sabin vaccines and were injected with many of the original polio as well as adenovirus vaccines. Avian leukosis viruses are present in most flocks of chickens and their eggs and are subsequent contaminants of any virus grown in chick embryo cells.

Sensitization of individuals either to the organism to which immunity is desired or to the cellular antigens or antibiotics used to grow the organism has been observed. Severe reactions to vaccine prepared with chick or duck embryos are well described. In addition, repeated immunization of experimental animals with large doses of antigen has resulted in conditions similar to amyloidosis and multiple myeloma in man (White et al., 1974).

TABLE 3. Transplacental Transfer of Maternal Antibody in Newborn Infants

GOOD PASSIVE TRANSFER	POOR PASSIVE TRANSFER	NO PASSIVE TRANSFER
Diphtheria antitoxin Tetanus antitoxin Antierythrogenic toxin Antistaphylococcal antibody *Salmonella flagella* (H) antibody Antistreptolysin All the antiviral antibodies present in maternal circulation (rubeola, rubella, mumps, poliovirus) VDRL antibodies	*Haemophilus influenzae, Bordetella pertussis, Shigella flexneri, Streptococcus*	Enteric somatic (O) antibodies (*Salmonella, Shigella, E. coli*) Heterophile antibody

NATURALLY ACQUIRED PASSIVE IMMUNITY

Newborn infants receive many antibodies from the mother. Placental transfer of antibody depends on the quantity of antibodies in the maternal circulation, as well as the molecular size. Only IgG with its low molecular weight passes readily. IgA and IgM are not placentally transmitted. Viral antibodies present in high amounts are equally present in maternal and infant serum, while macroglobulins (e.g., heterophil antibody) are excluded (Table 3). Passively acquired antibody clearly protects against some diseases. Neonatal tetanus, for example, may be completely eliminated by administering at least two doses of tetanus toxoid to pregnant women. Previously immune pregnant women provide sufficient antibodies to their infants to protect them from neonatal tetanus.

Passively acquired antibody may interfere with a newborn's ability to respond adequately to immunizations. Several studies have confirmed that maternally acquired antibody to diphtheria and pertussis may actually inhibit antibody formation following active immunization. Placentally transmitted antibody may neutralize the virus in live vaccines, thus rendering the vaccine ineffective. How long passively acquired antibody may interfere with an infant's ability to respond to certain vaccines is uncertain. Low levels of maternal antibody to measles have been found to persist beyond the twelfth month of life (Albrecht et al., 1977). Several studies have shown that children vaccinated before 12 months of age have a lower seroconversion rate than those immunized after 1 year (Shelton et al., 1978; Wilkins and Wehrle, 1979). Until this question is resolved, it is recommended that children be immunized against measles at 15 months of age. Prolonged acquisition of passive antibody through breast feeding may play a role in those 5 to 10 per cent of normal children who fail to respond adequately to one dose of the live measles vaccine, and certainly may interfere with the success of live polio virus immunization.

CURRENT IMMUNIZATION SCHEDULES

At present, seven immunizations are recommended for routine use in the United States. The schedule outlined in Table 4 is the suggested sequence outlined in the Red Book, 1977, as recommended by the Committee of Infectious Diseases (American Academy of Pediatrics, 1977). The American Public Health Association and the Advisory Committee on Immunization Practices (ACIP) of the United States Public Health Service also published recommended immunization schedules. Although there is usually consistency among the three organizations, occasionally recommendations may vary slightly.

The combined vaccines recommended for routine use have been shown to be just as efficacious without increased side effects when given concomitantly as when given individually (Center for Disease Control, 1980). Single live virus vaccines should be given one month apart, since interference with the immune response to the

TABLE 4. Recommended Schedule for Active Immunization of Normal Infants and Children

2 mo	DTP	TOPV
4 mo	DTP	TOPV
6 mo	DTP	TOPV (optional)
1 yr		Tuberculin test
15 mo	Measles, rubella	Mumps
1½ yr	DTP	TOPV
4–6 yr	DTP	TOPV
14–16 yr	Td—repeat every 10 years	

DPT, diphtheria-tetanus-pertussis vaccine
Td, tetanus-diphtheria vaccine
TOPV, trivalent oral polio vaccine

From American Academy of Pediatrics: Report of the Committee on Infectious Diseases, 18th ed., 1977.

second vaccine has been demonstrated with some immunizations. Two weeks and preferably four weeks should separate administration of killed vaccines with another inactivated or live virus vaccine.

Most immunizations may be associated with various side effects. Local reactions consist of mild induration and tenderness at the injection site. Repeat dosages of tetanus toxoid or typhoid vaccine have been associated with severe local reactions including marked edema, induration, erythema, and tenderness. If there is a severe reaction, subsequent immunization with the offending antigen should be avoided unless there are extremely compelling reasons for repeating the immunization. In such cases, fractional doses should be administered under close medical supervision.

Severe febrile reactions associated with irritability, malaise, headaches, and chills may follow the use of inactivated vaccines and usually subside within 48 hours. Mild febrile reactions are much more common and are easily controlled with antipyretics. Fever and rashes associated with the measles and rubella vaccines usually appear one to several weeks after immunization and persist for one to three days. Arthralgias associated with attenuated rubella vaccines follow a similar time sequence. The persistent arthritic symptoms occasionally associated with rubella vaccines have not been shown to represent vaccine complications and at present are felt to represent coincidental disease.

IMMUNIZATIONS FOR INDIVIDUALS OUT OF STEP

Interruption of the recommended immunization schedule does not interfere with the final immunologic outcome (Phillips, 1975). It is not necessary to resume the series regardless of the time between discontinuation and reinstitution of immunizations. The schedule for primary immunization of children not immunized in infancy is shown in Table 5. Children less than 7 years old are immunized with the standard DTP triple antigens by using three dosages at intervals of four to eight weeks. Pertussis vaccine is not recommended for children older than 6 years. In addition, children older than 6 years should receive the adult-type diphtheria-tetanus toxoid. The standard DT preparation contains 7 to 25 Lf (flocculating units) and may cause severe reactions in older children and adults. The adult type of combined diphtheria-tetanus toxoid (Td) with adjuvant contains, at most, 2 Lf of diphtheria toxoid and is recommended for all adults and children older than 6 years.

TABLE 5. Primary Immunization for Children Not Immunized in Early Infancy[a]

UNDER 6 YEARS OF AGE	
First visit	DTP, TOPV, tuberculin test
Interval after first visit	
1 mo	Measles,[b] mumps, rubella
2 mo	DTP, TOPV
4 mo	DTP, TOPV[c]
10 to 16 mo or preschool	DTP, TOPV
Age 14–16 yr	Td—repeat every 10 years
6 YEARS OF AGE AND OVER	
First visit	Td, TOPV, tuberculin test
Interval after first visit	
1 mo	Measles, mumps, rubella
2 mo	Td. TOPV
8 to 14 mo	Td, TOPV
Age 14–16 yr	Td—repeat every 10 years

[a]Physicians may choose to alter the sequence of these schedules if specific infections are prevalent at the time. For example, measles vaccine might be given on the first visit if an epidemic is under way in the community.

[b]Measles vaccine is not routinely given before 15 months of age (see Table 4).

[c]Optional

DTP, diphtheria-tetanus-pertussis vaccine
TOPV, trivalent oral polio vaccine
Td, tetanus-diphtheria vaccine

From American Academy of Pediatrics: Report of the Committee on Infectious Diseases, 18th ed., 1977.

CONTRAINDICATIONS TO IMMUNIZATIONS

The decision to immunize an individual with a given vaccine should be based on the belief that the benefits accrued by that person and society outweigh the potential risks of vaccination. During the acute febrile phase of an illness, immunization should be deferred. Minor illnesses without fever, however, should not be considered contraindications. Patients with malignancies or receiving immunosuppressive therapy including chemotherapy, radiation therapy, and corticosteroids should not receive live vaccines. Similarly, individuals with immunodeficiency disorders may develop overwhelming infections when immunized with live vaccines and should not be vaccinated. Individuals allergic to eggs, chickens, or ducks should not receive vaccines grown in duck or chick embryos. Vaccines grown in fibroblast cultures derived from chicks or ducks do not contain egg albumin or yolk components and are not contraindicated in individuals with known hypersensitivity to ducks and chickens.

Individuals with neurologic disorders present considerable controversy regarding immunization. Current recommendations are that children with static neurologic disorders should generally be immunized following the usual vaccine schedules. But the child with an evolving neurologic problem should not receive immunizations likely to cause fever or to be associated with central nervous system (CNS) complications.

PREGNANCY AND IMMUNIZATIONS

In general, the use of attenuated vaccines should be avoided during pregnancy (Levine, 1974). Theoretically, immunization with live virus may be harmless to the pregnant woman and yet be hazardous to her unborn child. Rubella virus is a well-known teratogen for which live attenuated vaccines are used specifically to prevent fetal damage. There is increasing evidence that the attenuated viruses used in rubella vaccines can cause intrauterine infection (Fleet et al., 1974; Wyll and Hermann, 1973). The extent to which the attenuated virus is teratogenic is unknown. It is strongly recommended, therefore, that before any woman of childbearing potential is vaccinated, she should be shown to be serosusceptible and not pregnant. In addition, adequate contraception should be provided for three months following immunization.

It occasionally becomes necessary, because of plans to travel, to immunize a pregnant woman against anticipated high-risk exposures. Under these unavoidable circumstances immunization against yellow fever and poliomyelitis may be warranted. Since in these settings immediate prophylaxis is required for protection against polio, the live oral vaccine is recommended. Generally, all other live vaccines are discouraged during pregnancy. At present, there is no indication for any pregnant woman to be immunized against smallpox. Inactivated vaccines, however, may be used when specifically indicated. Tetanus and diphtheria toxoids (adult Td) may be given routinely to update pregnant women during their antepartum care.

A pregnancy in the family should not alter the immunization schedules of any family members or contacts of the pregnant woman. Specifically, rubella vaccination of children need not be avoided when their mother is pregnant. A woman who is seronegative for rubella during pregnancy should be given the attenuated rubella vaccine postpartum. Similarly, breastfeeding is not a contraindication to the immunization of a mother or her baby.

IMMUNIZATION FOR TRAVELERS

With increased foreign travel, it has become important that physicians be able to recommend proper immunizations to their patients to protect their health while traveling (Barrett-Connor, 1979; Medical Letter on Drugs and Therapeutics, 1979). Necessary vaccinations include not only those recommended for protection of the traveler but also those required specifically by individual nations. Detailed information is available in *Health Information For International Travel*, prepared by the Center for Disease Control and obtainable from the Superintendent of Documents, US Government Printing Office, Washington, D.C. 20402. Table 6 summarizes some of the present recommendations for foreign travelers. More detailed and up-to-date information can be obtained from the above CDC publication. Other prophylactic measures should be explored as well before the trip is begun, with particular emphasis on medications for the prevention of gastroenteritis and malaria. When children or adolescents travel, their immunization records should be reviewed and routine immunizations updated if necessary. International travelers must have their vaccinations against smallpox, yellow fever, and cholera documented on an approved version of the International Certificate of Vaccination or Revaccination. These certificates, approved by the World Health Organization, are usually available at local health departments and passport offices.

CONTROVERSIES IN IMMUNIZATIONS

Many of the vaccines currently available are widely accepted as effective and worthwhile (Table 7). Virtually no vaccine, however, has been developed and used without eliciting controversy as to its efficacy, indications, target population, and side effects (Fulginity, 1976). Some of the major controversies involving passive and active immunizations are discussed below. The reader is encouraged to review specific chapters for more detailed descriptions of immunizations available and those currently being developed for individual infectious diseases (Table 8).

Meningococcal Disease

Group A and group C meningococcal polysaccharide vaccines are currently licensed in the United States (Center for Disease Control, 1978). They are prepared as monovalent and bivalent antigens in lyophilized form requiring rehydration with diluent before use. As with all polysaccharide vaccines, they are type-specific and offer little protection against other serotypes. The recommended dosage of each polysaccharide is 50 μg administered subcutaneously. The vaccines are safe but of variable efficacy. Field trials in Egypt, the Sudan, and Finland have demonstrated that group A vaccine results in protection against group A meningococcal disease in 90 per cent of individuals, and suggest that children as young as 3 months of age may be immunized successfully. The group C vaccine has also been successful in protecting 90 per cent of adults against disease. Young children, however, do not fare as well, since the C vaccine is only 65 per cent protective in children 2 to 3 years old and provides virtually no protection to children less than 2 years of age. Persistence of antibody is also superior for the type A vaccine. Eighty per cent of children older than 2 years of age when immunized against group A have protective antibody against A disease after four years, while only 40 per cent of children immunized with group C polysaccharide will have antibody against C disease (Lepow et al., 1977; Peltola et al., 1977; Wilkins and Wehrle, 1979). Two flaws of the combined vaccine against serotypes A and C limit its usefulness. First, the highest attack rate for meningococcal disease is in infants less than 2 years old, in whom the vaccine is least effective. Second, in the United States, serogroup B presently accounts for the great majority of disease due to *Neisseria meningitidis*, and at present an adequately immunogenic preparation of the group B meningococcus is not available. In addition, in the United States meningococcal disease in civilians has tended recently to occur as single isolates or infrequently as clusters, and antibiotic prophylaxis has been successful in reducing the risk of secondary cases in close contacts.

The current meningococcal vaccines are not recommended for routine use in civilians unless an epidemic of meningococcal disease due to serogroups A or C necessitates immunizing susceptible individuals. The vaccine should be considered for travelers visiting countries having an epidemic caused by strains A or C, or as an adjunct to antibiotic prophylaxis for household contacts of cases infected with susceptible strains. A highly immunogenic meningococcal polysaccharide vaccine against group Y is currently being extensively studied and may be available shortly (Farquhar et al., 1978). In addition, a vaccine against a major outer membrane protein of meningococ-

Text continued on page 782

TABLE 6. Immunization for Foreign Travel

DISEASE	AREAS WHERE INDICATED OR REQUIRED	COMMENTS
Bacterial		
Cholera	Only where required	Risk to tourist small, vaccine of limited efficacy
Plague	Southeast Asia or frequent contact with rodents in S. America, Asia, Africa	For most tourists not necessary
Tetanus, diphtheria	All persons every 10 years regardless of travel	
Tuberculosis	Developing countries	BCG or INH prophylaxis in areas with high risk; otherwise skin test every six months
Typhoid	Countries with poor sanitation	Frequently associated with sore arm for one to two days with fever. Requires two doses four weeks apart
Rickettsial		
Typhus	Mountainous, highland, or cold areas with louse infestation in Ethiopia, Rwanda, Burundi, Mexico, Ecuador, Bolivia, Peru, and Asia	
Viral		
Hepatitis A	≥3 months in tropical areas and developing countries	ISG immunization close to departure
Poliomyelitis	Rural developing countries	If previously immunized: one dose of TOPV; if no previous immunization: primary series of TOPV or inactivated vaccine; altered immune status: inactivated vaccine
Rabies	Only where exposure to rabid animals is a constant threat	Only DEV immunization available presently. When available HDCV is immunization of choice
Smallpox	Only where required and then questionable	At present there is no indication, and a physician's letter indicating that vaccination is contraindicated on medical grounds should be seriously considered
Yellow fever	Rural S. America, tropical Africa; most countries require vaccine if traveler is coming from area with reported case of yellow fever within six days of arrival.	Should be included by traveler who might change itinerary to include areas with endemic yellow fever. Immunocompromised persons should not be vaccinated

TABLE 7. Currently Available Immunizations

DISEASE	IMMUNIZING AGENT	INDICATION	ADMINISTRATION Primary	Booster	Route	EFFICACY[a]	ADVERSE REACTIONS	COMMENTS
Bacterial								
Anthrax	Cell-free, alum-concentrated inactivated protein antigen of *Bacillus anthracis*	High-risk >6 mo old	0.5 ml for three doses given two to three weeks apart	0.5 ml annually	S.Q.[b]	++++	Local erythema, induration; fever	
Cholera	Phenol-inactivated *Vibrio cholerae*	Travel to or residence in countries with cholera	Intradermal >5 yr, 0.2 ml; S.Q. or I.M. 6 mo–4 yr, 0.2 ml; 5–10 yr, 0.3 ml; >10 yr, 0.5 ml in two doses at least one wk apart	Every 6 mo or just before cholera season	As indicated	+	Pain, erythema, and induration at injection site; fever, malaise, and headache	
Diphtheria, tetanus, pertussis (DTP)	Alum-precipitated or absorbed toxoids of diphtheria and tetanus	All persons <7 yr old	See tables 4 and 5			++++	See text	See text
	D 7 to 25 LF killed *B. pertussis*					++-+++		
Tetanus, diphtheria (Td)	T—as above d—≤2 LF	All persons >7 yr old	See tables 4 and 5; booster every 10 yr for life			++++	Local pain and induration and mild fever	
Meningococcus types A and C	Capsular polysaccharide of *Neisseria meningitidis* types A and C	>2 yr old in epidemic or high-risk setting	50 μg of each polysaccharide 0.5 ml of vaccine		S.Q.	+++	Local erythema and pain	Vaccine ineffective against other types of *N. meningitidis*
Plague	Formaldehyde-inactivated *Yersinia pestis*	High-risk	Three doses: first two doses 0.5 ml at least 4 wk apart; third dose 0.2 ml 4 to 12 wk after second dose (see comments)	Same amount as third dose at 6-month intervals to total of five doses (3 primary and two boosters) (see comments)	I.M.[c]	++++	Mild pain, erythema, and local induration with repeated doses; headache, malaise	Adjust dosage for age: <1 yr, one fifth adult dose; 1–4 yr, two fifths adult dose; 5–10 yr, three fifths adult dose
Pneumococcus	Capsular polysaccharide to types 1–4, 6, 8, 9, 12, 14, 19, 23, 25, 51, 56	Persons >2 yr old with high risk of pneumococcal infections	50 μg of each polysaccharide 0.5 ml of vaccine	Unknown	S.Q.	+++	Local erythema, induration and pain; fever	See text

Tuberculosis	BCG (an attenuated strain of *Mycobacterium bovis*)	Any age: (1) high risk for continued exposure; (2) populations where skin test conversion >1% per yr	Newborns: 0.05 ml; >1 mo old: 0.1 ml	Give full dose to newborn at 1 yr if risk still present; all others not recommended (see comments)	S.Q. or multiple percutaneous	+ − ++	Ulceration, lymphadenitis, osteomyelitis, dissemination, and death	Vaccine should be repeated at two to three months after vaccination if skin test is negative
Typhoid	Acetone-killed *Salmonella typhi*	Close contact to typhoid carrier or traveler to high-risk areas	6 mo–10 yr old: 0.25 ml; >10 yr old: 0.5 ml Two doses divided by four or more weeks	Every three years	S.Q.	++–+++	Local erythema and pain; fever, malaise, headache	See text
Rickettsiae								
Typhus	Formaldehyde-inactivated *Rickettsia prowazekii* grown in chick embryos	Travel to or residence in area with known typhus and medical personnel and laboratory workers in frequent contact with *R. prowazekii*	Two doses four or more weeks apart	Every 6–12 months	S.Q.	Unknown	Local pain, induration and erythema; fever and malaise	See text
Viruses								
Influenza	Formalin-inactivated influenza virus grown in chick embryos; usually trivalent preparations of influenza viruses expected to be prevalent	All persons ≥6 mo old at increased risk of adverse consequences from infections of lower respiratory tract	One to two doses depending on age and past exposure; doses four or more weeks apart 6–35 mo old, 0.25 ml; ≥ 3 yr old, 0.5 ml	Annually	S.Q.	++-+++	Local pain, induration and erythema; fever, malaise, allergic hypersensitivity Guillain-Barré syndrome <1/100,000 doses	Live vaccine currently being extensively studied
Measles	Live attenuated Edmonston B measles virus from chick embryos	All persons >15 mo old	One dose 0.5 ml	Unknown	S.Q.	++++	Fever, rare encephalitis and encephalopathy	See text

[a]Efficacy:
50–65% = +
65–80% = ++
80–90% = +++
>90% = ++++

[b]S.Q., subcutaneous

[c]I.M., intramuscularly

Table continued on following page

TABLE 7. Currently Available Immunizations *(Continued)*

DISEASE	IMMUNIZING AGENT	INDICATION	ADMINISTRATION			EFFICACY[a]	ADVERSE REACTIONS	COMMENTS
			Primary	Booster	Route			
Mumps	Live attenuated Jeryl-Lynn strain of mumps virus grown in chick embryo cells	All children >15 mo old (see text)	One dose 0.5 ml	Unknown	S.Q.	++++	Uncommon parotitis, allergic reactions, rash, pruritus, purpura	See text
Poliomyelitis	OPV (oral polio vaccine)—live attenuated poliovirus to types 1, 2, and 3	All children ≥6 wk to 18 yr; adults when indicated	Two doses at least 6 weeks apart followed by dose 8 to 12 months later	At 4 to 6 years old or before travel to endemic areas	Oral	++++	Rare paralysis in normal individuals	Absolutely contraindicated in any immunocompromised person (see text)
	IPV (formalin inactivated polio vaccine) of three serotypes grown in monkey cell cultures	≥6 wk old in selected instances (see text)	Three doses 4 to 8 weeks apart with fourth dose 6 to 12 months following third dose	Every 5 yr until age 18 yr	S.Q.	++++	Local pain	See text
Rabies	DEV (duck embryo vaccine) rabies virus grown in duck embryos killed by exposure to B-propiolactone.	Persons at high-risk of rabies virus exposure or following wild animal bites except rodents	Pre-exposure: either two 1 ml injections 1 month apart followed by 3rd injection 6–7 months later; or three 1 ml injections 1 wk apart followed by 4th—3 months later Post-exposure: Schedule 1: 21 daily doses; or Schedule 2: 2 daily doses first 7 days, then 7 daily doses	If no antibody response 2 booster doses 1 week apart Schedule 1: boosters on day 31 and 41; Schedule 2: boosters on day 24 and 34	S.Q.	Unknown but probably +++—++++	Local pain, pruritus, and erythema, fever, malaise, serum sickness, rare anaphylaxis; neuroparalysis 1/24,400 recipients	Inactivated nerve tissue vaccines (NTV) or inactivated suckling rodent brain vaccine (SRBV) are not recommended, since they have an incidence of neuroparalytic reactions in 1/2,000 and 1/8,000 vacinees respectively

	HDCV (human diploid cell vaccine) rabies virus grown in human diploid cell cultures and inactivated by N-tributyl phosphate	As above (preferred vaccine when available)	Pre-exposure: Two 1 ml doses 1 wk apart followed by *3rd* dose 2–3 wks later Post-exposure: Five 1 ml doses on days 0,3, 7,14 and 28 (6th dose on day 90 recommended by WHO)	If no antibody response 1 booster If no antibody response 1 booster	I.M.	Unknown but probably ++++	Local pain, erythema, pruritus, and swelling, mild headache, malaise, nausea, abdominal pain, myalgias, and dizziness	Post-exposure prophylaxis should always include 1 dose of RIG 20 IU/Kg given ½ into wound and ½ I.M. except in individuals previously vaccinated
Rubella	Live attenuated rubella virus RA 27/3 grown in human diploid fibroblast cells	See text	One dose 0.5 ml	Unknown (see text)	S.Q.	++++	Rash, lymphadenopathy, joint pain	See text
Smallpox	Live attenuated vaccinia virus in lyophilized or glycerinated vaccine	Laboratory workers studying smallpox (see text)	1 drop	1 drop every three years	Multiple pressure, multiple puncture, jet injection	++++	Fever, pain, malaise, lymphadenopathy, rare dissemination, encephalitis	Chemical agent should not be used to prepare skin
Yellow fever	Live attenuated yellow fever virus strain 17D grown in chick embryo cells	Traveling or residing in high-risk area or laboratory worker studying yellow fever	0.5 ml	Every 10 years	S.Q.	++++	Headache, myalgia, and fever	Dakar strain associated with 0.5% incidence of meningoencephalitis and is not recommended

TABLE 8. Diseases for Which Immunizations Are Currently Being Investigated or Are of Limited Availability

Adenovirus
Botulism
Cytomegalovirus
Gonorrhea
Haemophilus influenzae type B
Hepatitis A and B
Herpes simplex 1 and 2
Influenza
Mycoplasma pneumoniae
Neisseria meningitidis types B and Y
Parainfluenza 1–3
Pseudomonas
Q fever
Respiratory syncytial virus
Rotavirus
Syphilis
Trachoma
Tularemia
Varicella

cus has shown promise for protection against group B disease.

Pertussis

In the United States, immunization against pertussis is routinely begun at 2 months of age along with vaccination against diphtheria and tetanus (Center for Disease Control, 1977). Although some physicians in the United States have expressed doubts about the efficacy and overall benefits of the current pertussis vaccine, the raging debate occurring in Europe has been largely avoided (Grady and Wetterlow, 1978; Manclark, 1979; Mathias, 1978; Pittman, 1979). In England, the major reservation to the pertussis vaccine is the high incidence of severe reactions (Enrengut, 1978; Stewart, 1977; Lister, 1977). Reactions to the vaccine have been reported in the United States, but they have not been of the frequency or the severity of the British experience. Estimates of severe neurologic reaction to the pertussis vaccine have ranged from 1:2500 to 1:500,000 children. The adverse reactions described include: high fever; collapse sometimes associated with a shock-like state; prolonged periods of screaming during which the infant cannot be comforted; convulsions with or without fever; frank encephalopathy with changes in the level of consciousness, focal neurologic signs, and convulsions; and thrombocytopenic purpura. Although the incidence of these severe reactions is uncertain, mild reactions including low grade fever and pain and induration at the injection site may occur in as high as 90 per cent of vaccine recipients. Rarely, permanent neurologic sequelae have been described.

The true risk-benefit ratio remains to be established (Koplan et al., 1979). Many British physicians have argued that the current risk of contracting a serious pertussis infection is less than the risk of acquiring permanent neurologic sequelae including mental retardation from the vaccine. Proponents of the vaccine contend that the severe complications and high mortality from pertussis in infancy are major reasons for immunization early in life. They point to the dramatic decline in pertussis since routine vaccination was begun. Opponents of the vaccine, however, vehemently emphasize that the incidence of pertussis was decreasing before routine immunization was begun, and that overall cases with subsequently lower morbidity could have been predicted without the use of pertussis vaccine. The most convincing evidence for vaccine efficacy has come recently from Japan and England, where immunization rates have dropped to 10 per cent and 40 per cent, respectively. Both countries have encountered serious outbreaks of pertussis, and interest has renewed in immunizing children. In underdeveloped nations, pertussis remains a serious health problem, and immunization is the only currently viable method for control of disease.

Over the past 40 years, studies of vaccine efficacy have given results varying from 0 to 90 per cent protection. The full series of pertussis vaccinations presently confers immunity in 80 per cent of recipients. Protection is not long-lasting, however, and after 10 years only approximately 50 per cent of vaccinees have detectable protective antibody. Pertussis immunization is recommended for children from age 6 weeks up to their seventh birthday. Since most cases occur in infants and young children, and two thirds of pertussis deaths occur in infants less than 1 year of age, immunization should be instituted in infancy as part of the DPT vaccine.

Children who have experienced severe reactions following pertussis-containing vaccines should not receive further pertussis immunizations. Static neurologic conditions in infants are not reasons for deferring immunizations against pertussis. In evolving neurologic conditions, however, pertussis immunization should not be used because of the theoretical concern of exacerbating the disorder. There is no convincing evidence that pertussis immune globulin is effective in preventing or treating pertussis, and its use is not recommended.

Pneumococcus

Streptococcus pneumoniae (pneumococcus) remains the most common bacterial cause of otitis media, pneumonia, and meningitis in infants and children. Patients who have undergone splenectomy and individuals with sickle-cell disease, nephrotic syndrome, chronic liver disease, malignancy, or primary immunodeficiency are at increased risk of developing severe pneumococcal infections. A 14-valent polysaccharide vaccine is now licensed for use in the United States (Center for Disease Control, 1978). The vaccine contains purified capsular material of pneumococci extracted separately from American types 1, 2, 3, 4, 6, 8, 9, 12, 14, 19, 23, 25, 51, and 56. Each dose of vaccine contains 50 μg of each of the 14 polysaccharides that account for 80 per cent of all bacteremic pneumococcal disease in the United States. Similar vaccines have been tried successfully in Europe and South Africa. Field trials have shown that vaccination reduces by 80 per cent the incidence of bacteremic pneumococcal pneumonia caused by the bacterial types included in the vaccine. Nasopharyngeal acquisition of the pneumococcal types included in the vaccine is reduced in vaccinees, and there is no evidence that immunized individuals have any greater risk of acquiring diseases caused by other microorganisms. The duration of protection is unknown, but elevated antibody has been found to persist for at least two years.

Despite the promising findings with the presently licensed pneumococcal vaccine, several problems remain (Artenstein, 1973). A vaccine that is protective against 80 per cent of the disease-causing pneumococci and that is 80 per cent effective is capable of reducing pneumococcal disease by at most approximately 65 per cent. In addition, the current vaccine is ineffective in children less than 2 years old (Klein and Mortimer, 1978). This is particularly unfortunate for children with sickle-cell disease, who are at greatest risk of pneumococcal infection during infancy (Ammann et al., 1977; Ahonkhai et al., 1979).

Pneumococcal vaccine is not currently recommended for healthy individuals. Special populations including residents of nursing homes and institutions, and localized populations experiencing outbreaks may be considered for pneumococcal vaccination. Persons over 2 years of age with sickle-cell disease or other splenic dysfunction including splenectomy, or with chronic illnesses or conditions including diabetes mellitus, chronic cardiorespiratory disease, renal disease, hepatic dysfunction, or old age, may all benefit from pneumococcal vaccination. Caution must be advised, however, against totally relying on the present pneumococcal vaccine to prevent serious *Streptococcus pneumoniae* infections, particularly in children with sickle-cell disease (Ahonkhai, 1979; Ammann et al., 1977). Since the vaccine at best can be expected to prevent 65 per cent of pneumococcal disease, penicillin prophylaxis of children with sickle-cell disease should be continued despite vaccination until the children are 6 years old, when the risk of pneumococcal infection is markedly decreased and prophylaxis may not be necessary.

Tuberculosis

The World Health Organization has credited the BCG vaccine with playing a major role in reducing worldwide morbidity owing to tuberculosis (Eickhoff, 1977). Controlled trials, however, have found extremely variable immunity in vaccine recipients. In fact, despite the use of BCG since 1921, there is still debate about whether the vaccine is effective.

BCG is derived from a strain of *Mycobacterium bovis* that was attenuated at the Pasteur Institute in Lille, France, by Calmette and Guérin. All BCG vaccines available today were derived from the original strain. Differences in production, methods and routes of vaccination, and characteristics of the populations and environments in which the vaccine has been used have resulted in great variation in immunogenicity, efficacy, and reactogenicity among the daughter strains. At present, most laboratories producing BCG vaccine maintain production strains in a lyophilized state in an attempt to minimize genetic variation. The production strains are usually named by the city in which they are produced (e.g., BCG-London, BCG-Copenhagen). In the United States, the Bureau of Biologics, Food and Drug Administration, has specified that each freeze-dried BCG strain used for vaccination must have specified characteristics of safety and potency and be capable of inducing tuberculin sensitivity in the guinea pigs and humans (Center for Disease Control, 1979). Unfortunately, induced tuberculin sensitivity has never been proved to be related to immunity.

The rationale behind the use of the BCG vaccine to prevent tuberculosis is to confer cell-mediated immunity against an attenuated, immunologically similar strain of mycobacteria. It is known that most hosts, upon primary infection with tubercle bacilli, are able to mount an immunologic response sufficient to localize the infection and thus establish a latent or dormant infectious state. The inactive infection may continue for life or may reactivate frequently but not invariably during times of altered host cell-mediated immunity. There is evidence that the BCG vaccine prevents the establishment of a latent infection when the vaccinee is challenged with live tubercle bacilli, thus preventing not only primary tuberculosis but also the breakdown or reactivation disease.

The current lyophilized vaccines have demonstrated substantial protection in animals and are currently being tested in India. The most recent field trials, all conducted before 1955 with liquid vaccines, showed protection ranging from 0 to 80 per cent. Some argue that the potencies of the BCG strains used in these studies were sufficiently different to account for the contradictory findings. Another possibility is that populations showing little protection from BCG had a large prior exposure to atypical mycobacterial infections. These previous infections may have induced sufficient immunity to tuberculosis to mask any contribution made by BCG immunization. Interestingly, field trials performed where there is a relatively high incidence of tuberculosis in the control group have tended to show a high efficacy for the BCG vaccine, whereas areas with a low incidence of tuberculosis in the unvaccinated group have shown little benefit from BCG. Continual mycobacterial challenges may be necessary, therefore, to maintain the immunity conferred by BCG vaccination.

In the United States, it is recommended that BCG should be "seriously considered" for patients such as infants who are tuberculin skin-test-negative but who can be expected to have repeated exposure to individuals with sputum-positive pulmonary tuberculosis. In addition, BCG vaccination should be considered for groups in which the skin-test conversion rates exceed 1 per cent annually and in which the usual surveillance and treatment programs have failed or are not feasible. In developing countries where tuberculosis is epidemic and short-term INH prophylaxis or skin-test screening is not possible, BCG is indicated to attempt tuberculosis control.

In the United States, where the current annual infection rate among 6-year-olds is approximately 0.03 per cent, BCG is rarely indicated. Infants tolerate two to three months of INH prophylaxis extremely well and thus can allow time for sufficient screening and treatment of family contacts, eliminating the need for BCG vaccination. In addition, tuberculin skin tests are rendered less valuable, since it is usually impossible to distinguish between a tuberculin reaction caused by a virulent supra-infection and one resulting from persistent postvaccination sensitivity. After the immediate postvaccination period, however, caution is advised in attributing a positive skin test to BCG, and tuberculosis should always be considered a possible diagnosis.

BCG should be used only in individuals who are skin-test-negative to 5 tuberculin units of tuberculin purified protein derivative (PPD). The dosage is indicated by the manufacturer in the package insert. Infants less than 28 days old should receive one half the usual dose. If the need for immunization persists, these children should receive a full dose at 1 year of age. The World Health Organization recommends that BCG be given by intradermal injection to provide for a uniform and reliable dose. In the United States, however, both intradermal and percutaneous vaccines are licensed, and vaccination should only be by the route indicated in the package labeling (Center for Disease Control, 1979). Individuals receiving BCG should have a tuberculin skin test two to three months after immunization. If that skin test is negative and the indications for BCG remain, a second dose of vaccine should be administered.

Adverse reactions to BCG have included severe or prolonged ulceration at the vaccination site, lymphadenitis, osteomyelitis, lupoid reactions, disseminated BCG infection, and death (Passwell et al., 1976). Ulceration and lymphadenitis occur in 1 to 10 per cent of vaccinees; osteomyelitis may occur in 1 per million vaccine recipients but may be as high as 5 per 100,000 in newborns. Disseminated BCG infection and death occur in 1 to 10 per 10 million vaccinees and are seen almost exclusively in children with impaired immunity. BCG should, therefore, not be given to anyone with impaired immune status. Although no harmful effects of BCG on the fetus have been observed, immunization should be avoided during pregnancy unless there is an immediate excessive and unavoidable exposure to infectious tuberculosis.

Typhoid

Typhoid fever remains a major world health problem, particularly in developing countries, where poor sanitation and ingestion of inadequately cooked food are common. Killed typhoid vaccines have been used since the late 19th century, although their efficacy was not established until the 1960s. The Yugoslav Typhoid Commission in 1962 showed that a heat-killed, phenol-preserved vaccine gave considerable protection against typhoid fever for at least three years, whereas an alcohol-killed, alcohol-preserved vaccine was ineffective. Subsequent field trials in Guyana showed the heat-phenol and acetone-inactivated vaccines to be 65 to 90 per cent effective in preventing typhoid fever (Ashcroft et al., 1967). Although typhoid fever usually confers lifelong immunity, the nature of protection is unknown. The titers of antibodies against O, H, and Vi antigens have not been connected with protection. Naturally occurring sources of *Salmonella typhosa* usually contain 10^5 or less organisms, resulting in infection of 25 per cent of unvaccinated individuals. Following typhoid fever or immunization, most individuals are protected against a challenge of 10^5 organisms. Volunteer studies indicate, however, that vaccine-acquired immunity can be easily overcome with a challenge of 50 per cent infectious dose (10^7) of *S. typhosa* organisms (Hornick et al., 1970). The effectiveness of paratyphoid A and B vaccines has never been established, and they are not recommended for use either individually or in combination with typhoid vaccine.

Typhoid vaccine in the United States is not recommended for general use (Center for Disease Control, 1978). Persons with intimate exposure to a documented typhoid carrier or travelers to areas where there is a high risk of exposure to typhoid because of poor food or water sanitation should receive typhoid vaccine. In the United states vaccination is not indicated for use in controlling outbreaks from a common source or natural disasters such as floods.

Primary immunizations in adults and children 10 years or older consist of two doses of 0.5 ml of vaccine injected subcutaneously, three or more weeks apart, or three doses at weekly intervals. Children 6 months to 10 years old should receive 0.25 ml of vaccine following one of the same schedules. Revaccinations are recommended every three years if there is continuing exposure, with a single booster given intradermally (0.1 ml of vaccine into the flexor surface of the forearm) or subcutaneously (0.5 ml of vaccine in individuals older than 10 years or 0.25 ml in children less than 10 years old). The acetone-killed vaccine should not be given intradermally. If more than three years have elapsed since the last vaccination only one booster is still required. Reactions including local pain, malaise, headache, and fever are common.

A new live attenuated oral typhoid vaccine is currently being tested. This vaccine uses a mutant (Ty 21a) of *Salmonella typhi* that lacks the enzyme uridine 5′-diphosphate-glucose-4-epimerase. The vaccine, when grown in brain-heart infusion broth in the presence of 0.1 per cent galactose, has shown great promise in early investigations (Gilman et al., 1977).

Cytomegalovirus

Cytomegalovirus (CMV) is the most common congenital infection resulting in mental retardation and deafness. Primary CMV infection and reactivation disease are often responsible for severe illness and the death of immunocompromised patients; they may also be associated with renal allograft rejection. Pneumonia, hepatitis, encephalitis, atypical lymphocytosis, and leukopenia are all clinical manifestations of CMV infections.

It is likely that everyone develops a persistent CMV infection after primary exposure. Immunization against CMV ideally, therefore, would prevent primary as well as reactivated disease. In order to prevent transplacental transmission of CMV, a vaccine should also prevent reinfection viremia. Several theoretical problems must be faced before people are immunized against CMV. The capability of CMV to transform normal cells has stimulated concern that a vaccine might be oncogenic. Additionally, if a persistent infection is established after CMV immunization, reactivation of the vaccine virus may result in more severe disease than natural infection. Moreover, CMV strains are so heterogeneous that it is unclear if enough cross-reactivity exists for a vaccine made from one strain to establish broad protection.

Two live attenuated CMV vaccines are currently being investigated in renal transplant patients. Preliminary studies indicate that both the Towne-125 strain vaccine and the AD-169 strain vaccine can induce CMV-specific cellular immunity in previously seronegative recipients (Just et al., 1975; Plotkin et al., 1976; Glazer et al., 1979). The vaccines, however, have not been able to prevent CMV infections, but reactivation of the vaccine strain virus has not been demonstrated. The effect of immunization of seronegative women with the CMV vaccines and the subsequent incidence of infants born with congenital CMV is unknown. Additional work is being done on a specific CMV immune globulin in an attempt to ameliorate or possibly prevent primary CMV infections. Cautious investigation seems warranted to determine the risks and benefits of these approaches to immunoprophylaxis of CMV infections in selected populations.

Hepatitis

Standard ISG is effective for the prevention or modification of hepatitis A infection (Krugman et al, 1960; Center for Disease Control, 1977; Woodson and Clinton, 1969). When administered within one to two weeks after exposure to hepatitis A, it prevents illness in 80 to 90 per cent of individuals exposed. ISG should be given as soon after close contact with an infected person as possible and is not indicated if more than 2 weeks have elapsed since exposure or if clinical illness is present. Preexposure prophylaxis should be used for those at high risk of hepatitis A exposure and repeated every four to six months when risk continues. Frequently it is not necessary to continue prophylaxis in individuals with continued high contact over one year, since subclinical disease will often develop within that time.

The benefit of passive immunization against hepatitis B has been less clear. Early studies found that standard ISG was of little or no value in the prevention of post-transfusion hepatitis.

Subsequent studies, however, found that ISG was beneficial and was as useful as the newer hepatitis B immune globulin (HBIG). Initial studies of the two immune globulins suffered from the lack of standardization of the HBIG distributed by various manufacturers. At present, immune globulins with an anti-HBs titer greater than 1:100,000 by passive hemagglutination are designated HBIG.

Other explanations for disparate findings when ISG was compared with HBIG are beginning to be clarified. The original findings of the Veterans Administration Cooperative Study suggested that HBIG was superior to ISG for prophylaxis against hepatitis B following needle-stick exposure (Seeff et al., 1978). Reanalysis of their data in the light of radioimmunoassays for antibody to hepatitis core antigen (anti-HBc) and anti-HBs, however, showed no difference in serologic evidence of hepatitis B infection in exposed individuals. In fact, 12 per cent of individuals in both groups showed serologic evidence of hepatitis B infection. HBIG, however, apparently modified infection, since only 2 per cent of HBIG-treated patients had clinical disease, while 8 per cent of ISG-treated patients were symptomatic. An interesting phenomenon of passive-active immunity is postulated to be responsible for these findings. In the case of HBIG, the passive immunity transmitted is sufficient to ameliorate disease but permits infection and thus active immunity to occur also (Hoofnagle et al., 1979). Passive-active immunity has also been considered important with many special immune globulins.

Individuals with documented exposure to hepatitis B should receive HBIG. If HBIG is not available, ISG should be substituted. The indications for use of HBIG following exposure are: (1) needle-stick or mucosal contact with blood known to contain HBsAg; (2) inadvertent administration of HBsAg-positive blood; (3) sexual exposure to acute disease; and (4) infants born to mothers with acute hepatitis B (Mosley, 1979; Redeker et al., 1975; Szmuness et al., 1974). Preexposure prophylaxis may be indicated in hemodialysis units for patients and staff if adequate prevention and control are not possible and in custodial institutions under conditions of documented reported hepatitis B transmission.

Hepatitis B vaccines are currently under investation (Gerety et al., 1979; Krugman, 1975). These vaccines are prepared by plasmapheresis of suitable persons with high titers of HbsAg. Possible live virus is either heat- or formalin-inactivated during vaccine preparation. Preliminary studies are promising, but several important questions remain. The risk of inducing partial immunity in certain people and creating a pool of individuals more likely to develop chronic hepatitis must be assessed. In addition, since hepatitis B has been associated with an increased rate of hepatocellular carcinomas, there is a possibility that the vaccine might increase the incidence of hepatic cancer.

Measles (Rubeola)

Measles is a frequently severe illness associated with encephalitis in 1 of 1000 cases and commonly resulting in permanent neurologic sequelae including mental retardation. Respiratory or neurologic complications result in death in approximately 1 of 1000 cases. At present, a highly effective live vaccine is available (Center for Disease Control, 1978). The degree of measles control depends primarily on the extent to which a given population is immunized. The current further-attenuated Edmonston B strain vaccine has been so effective that the major controversies revolve around how best to immunize as many susceptible persons as possible. It is available in monovalent form (measles only) and in combination with rubella (MR) and mumps and rubella (MMR) vaccines. Follow-up studies indicate that durable immunity is achieved in 95 per cent of individuals older than 15 months of age given the live measles vaccine. Children immunized before 12 months of age have significantly lower seroconversion rates following immunization (Wilkins and Wehrle, 1978). Conflicting data exist for children between 12 and 14 months old, but it appears that these children respond adequately to measles immunization. However, because of this uncertainty, it is currently recommended that children receive their primary measles immunization at 15 months of age. Whenever there is high risk of exposure to natural measles, infants should be immunized at 6 months of age and then revaccinated when they are 15 months old.

Despite the widespread use of measles vaccine in the United States, epidemics have continued to occur, albeit to a much lesser extent than in prevaccine years (Orenstein et al., 1978; Weiner et al., 1977; Shasby et al., 1977; McCormick et al., 1977; Modlin et al., 1977). The effects of revaccination have become important with the increasing number of public schools requiring all children to have documented immunization histories before entrance (Deseda-Tous et al., 1978). Fortunately, revaccination has not been associated with any increase in complications in children who have previously received live measles vaccines or who have had natural measles. Increased reactions including local induration, pain, edema, and fever have occurred in over 50 per cent of individuals given live measles vaccine who had previously received killed vaccine (Krause et al., 1978). The risk of atypical measles in these young adults is sufficient, however, to warrant vaccination.

Exposure to measles is not a contraindication to vaccination (Ruuskanen et al., 1978). If given within 72 hours of exposure, evidence suggests that immunization may provide protection. Measles may be prevented or modified by administration of ISG within six days of exposure. Live measles vaccine should not be given until three months after administration of ISG. Measles outbreaks are best controlled by vaccination of susceptible individuals. Widespread use of ISG in these situations is not recommended.

Mumps

Clinical mumps is generally a benign disease associated with parotid swelling, tenderness, and fever. Meningoencephalitis or meningitis occurs in 0.5 to 10 per cent of all mumps cases. Although death is rare, morbidity may be severe. Bilateral orchitis occurs in approximately 2 per cent of cases in men, but sterility is uncommon. Unilateral deafness following endolymphatic labyrinthitis occurs in 2 per 10,000 cases. Arthritis, nephritis, subacute thyroiditis, pancreatitis, myocarditis, and hepatitis are all uncommon complications of mumps infections.

A highly effective, live attenuated mumps vaccine has been available in the United States since 1967 (Center for Disease Control, 1980; Hayden et al., 1978). The Jeryl-Lynn strain of mumps virus is used for immunization following attenuation of virulence in embryonated hens' eggs, and in tissue culture. Antibody appears in 95 per cent of vaccine recipients older than 12 months of age, and protection lasts for at least 9.5 years. Critics of the vaccine have argued that mumps is not severe enough to warrant widespread immunization. Others have emphasized that the vaccine was licensed before sufficient human trials were completed. The development of the combined measles-mumps-rubella (MMR) vaccine has greatly accelerated the rate of mumps immunization. Well over 30 million doses of mumps vaccine have been administered since it was licensed and have clearly shown the vaccine to be efficacious with few side effects. Uncommon adverse reactions include parotitis, low-grade fever, rash, and pruritus. Severe CNS complications following immunization are rare, occurring in 1 per million individuals immunized.

Mumps vaccine is recommended for all children over 12 months of age. As part of general well-child care, it is best given to a child as the MMR at 15 months. Mumps vaccination is of particular value to susceptible preadolescent males who have no evidence of previous infection, and they should be the target population in countries where the vaccine is not widely available.

Poliomyelitis

Poliovirus vaccines have been widely used for over 25 years. From 1955 to 1965, the inactivated vaccine was the only widely available polio immunization, and it dramatically reduced the incidence of paralytic poliomyelitis in countries where it was extensively used. Although both the IPV (formalin-inactivated polio vaccine) and the OPV (oral polio vaccine) are effective, the OPV has become the vaccine of choice in the United States (Center for Disease Control, 1979). Countries that prevent poliomyelitis with IPV are generally homogeneous populations that are able to maintain very high vaccination rates. In Sweden and Finland, where only IPV is used, intensive poliovirus surveillance has shown that not only is paralytic disease rare but circulation of the virus is virtually absent. This and animal studies suggest that blocking antibodies produced by IPV can prevent virus implantation despite failure to produce local IgA. In areas with endemic or epidemic poliovirus, however, the populations are best immunized with the live virus vaccine. OPV not only provides intestinal immunity with subsequent prevention of fecal growth of wild virus and decrease in circulating virus in the community but also establishes active infections in the nasopharynx and gastrointestinal tract of vaccinees. The virus that is shed following immunization with OPV spreads to susceptible contacts, resulting in inapparent immunizing infections. Additionally, individuals who have received IPV should be reimmunized every five years until the age of 18 years, whereas over 90 per cent of those receiving OPV have neutralizing antibodies to the three types of poliovirus after eight years. It should be noted that problems have occurred in attempting to immunize effectively some populations in tropical and semitropical areas. Viral interference of infection by other enteroviruses at the time of vaccination with OPV may play some role in reducing "take" rates to 50 per cent in these areas. There is also some evidence that 10 to 20 per cent of children in these populations have an inhibitory substance in their saliva that may prevent poliovirus replication. Probably most important in these vaccine failures is inadequate storage of the OPV, which results in inactivation of virus and decreased response rates. Repeated immunization of target populations, however, can eliminate paralytic disease, as has been demonstrated in Cuba and Puerto Rico.

In countries such as the United States where the use of OPV has made paralytic poliomyelitis uncommon, the use of IPV in certain specific situations is warranted. Current estimates indicate that less than one in three million doses of OPV distributed has been associated with paralytic disease in vaccine recipients or their close contacts. Administration of IPV should be oriented toward minimizing the chance of high-risk individuals developing paralytic poliomyelitis from OPV. IPV should be provided to persons with increased susceptibility to infections, including immunodeficient children, and other immunocompromised individuals and their household contacts. Adults undergoing initial vaccination who have the time and are committed to a full course of inoculations are also candidates for IPV since they have a slightly higher risk of developing paralysis. Adults should receive OPV if possible poliomyelitis exposure is to occur before two doses of IPV four weeks apart can be administered. If OPV is inadvertently administered to a household contact of an immunocompromised person, the patient should be isolated from the immunized person for at least two to three weeks after vaccination. Any child born to a family with a history of immunodeficiency disorders should not be given OPV until the immune status of the child is found to be normal.

Regardless of which immunization is used, the goal of polio immunization programs should be to immunize over 90 per cent of the population. When this is achieved, virtual elimination of paralytic poliomyelitis can be expected.

Rubella

Rubella is usually a mild childhood disease with rare complications. It was not until 1964, when a major epidemic in the United States resulted in more than 20,000 damaged children born of pregnancies complicated by rubella infection, that the need for a rubella vaccine was clearly established. Control of rubella by vaccination presents a unique problem, since it is the only immunization program that attempts to protect a distant fetus by preventing maternal infection during pregnancy. Two immunization strategies have been used in an attempt to decrease the number of congenitally infected infants. The "United States Program" has instituted routine immunization of all prepubertal children, boys and girls, and the selective immunization of postpubertal girls and women of childbearing age documented to be susceptible to rubella infection (Center for Disease Control, 1978; Modlin et al., 1975). The goal is to induce a high degree of herd immunity, which will prevent circulation of the agent and thus prevent susceptible pregnant women from acquiring rubella. By immunizing young children, the most likely sources of disseminating the infection are in theory eliminated. The "British Program" is directed toward the routine immunization of girls 11 to 14 years of age and selective immunization of women of childbearing potential. Thus, it attempts to prevent maternal infection without reducing circulation of the wild-type virus in the remainder of the population.

The two immunization strategies presently appear to have yielded different results. In the United States, the incidence of rubella has declined fourfold. Proponents argue that the expected epidemic in the mid-1970s was avoided by massive immunization. Along with the reduction in disease, more importantly, the number of reported cases of congenital rubella in the United States has also declined. In England, on the other hand, only approximately 70 per cent of girls have received the rubella vaccine, and the incidence of congenital rubella has not changed. It is estimated that at least 90 per cent of prepubertal girls would need to be immunized for the incidence of congenital rubella to decline using the "British Program." This approach at best would require a 10-year lag period before the immunized girls became the major portion of the women of childbearing potential, thus affecting the occurrence of congenital rubella.

Although it would appear that the "United States Program" has been more successful than the "British Program," several major problems still exist (Horstmann, 1975). It is estimated that in the United States approximately 450,000 cases of rubella occur yearly. In addition, immunization levels in women are between 85 and 90 per cent, which is essentially unchanged from prevaccine periods. The risk of fetal infection, therefore, is still very real.

Further questions have been raised concerning the duration of immunity to the rubella vaccines (Wilkins and Wehrle, 1979; Fogel et al., 1978). Of youngsters immunized with the HPV77 DE5 vaccine, between 4 and 36 per cent lack HAI antibodies three to five years after immunization. In addition, children immunized before 12 months of age have significantly lower antibody responses to rubella as well as to measles, probably secondary to persistence of passively transferred maternal antibody. Two other rubella vaccines previously licensed in the United States but now withdrawn had lower overall failure rates. The HPV77 DK12 vaccine, available from 1969 to 1973, had a failure rate of less than 1 per cent, and a 1 to 11 per cent serologic failure rate had been noted for the Cendehill vaccine, which was available from 1970 to 1976. The RA27/3 vaccine, which has been extensively used in Europe and since 1979 has been the only rubella vaccine licensed for use in the United States, is highly immunogenic (Plotkin et al., 1973) and has primary and secondary failure rates of less than 3 per cent.

Rubella reinfection is typically an inapparent

infection characterized by the absence of viremia, no or minimal transient pharyngeal virus shedding, and a prompt fourfold or greater rise in rubella HAI antibody. Subclinical reinfection has occurred more frequently after HPV77 or Cendehill immunizations than has been observed with the RA27/3 vaccine or following natural immunity. The potential risks of these subclinical reinfections are unknown but are probably small.

What to do about children, particularly girls, who were immunized with the HPV77 or Cendehill vaccines remains controversial. Revaccination of all prepubertal girls who have previously received these vaccines has been suggested by some (Balfour, 1979). Another approach is premarital screening of all women and immunization with RA27/3 of all seronegative women once adequate precautions are taken to ascertain that they are not pregnant and will not conceive for at least three months after immunization.

Smallpox

Live vaccinia virus vaccine, which is a stable hybrid of both smallpox and cowpox, is available in both glycerinated and lyophilized forms for immunization against smallpox. Both are equally protective when stored and administered properly. The glycerinated vaccine requires constant refrigeration, whereas the lyophilized vaccine is more stable until it is reconstituted. Protection lasts for approximately three years.

Smallpox vaccine may be administered either by multiple pressure, multiple puncture, or jet injection; all methods are equally effective (Center for Disease Control, 1978). The vaccine site should be examined after six to eight days to ascertain a successful result. Following primary vaccination, a typical Jennerian vesicle should be present. If a vesicle does not develop, the vaccination should be considered a failure. Following revaccination, a major reaction consisting of a vesicular or pustular lesion or an area of definite palpable induration or constant congestion surrounding a central lesion, which may be a crust or an ulcer, should be observed. All other reactions are considered equivocal and suggest vaccine failures. All primary or revaccinated persons without adequate responses should be reimmunized from another vaccine lot until a successful response is achieved.

Despite apparent global eradication of smallpox, more than 4.4 million doses of the vaccine were distributed in the United States in 1978. Severe reactions still develop in at least 1 per 100,000 vaccinees, and death occurs in 6 per 10 million doses administered (Center for Disease Control, 1979a). Any morbidity or mortality caused by smallpox vaccination seems unnecessary. Although some countries still require smallpox vaccination for arriving travelers, there seems little justification for this practice. The World Health Organization's International Regulations provide for smallpox vaccination waiver letters to be issued to travelers for whom vaccination is contraindicated for health reasons. It is the view of many physicians that such letters should be given to all travelers, unless there is a major change in the current world status of smallpox. At present, only the few individuals actively involved in doing research using the smallpox virus should be immunized. The vaccine at any time is strongly contraindicated in individuals with any impairment of immunity or persons with eczema; individuals with active skin lesions including burns, poison ivy, and impetigo; and pregnant women.

Varicella-Zoster

While varicella has a low incidence of serious complications and sequelae in normal individuals, immunocompromised patients are at high risk of developing serious and potentially fatal infections. These high-risk individuals have been the impetus for the development of means to prevent or ameliorate varicella infections. Varicella-zoster immune globulin (VZIG) prepared from outdated donor blood, or zoster immune globulin (ZIG) obtained from individuals with varicella-zoster infections has been shown to modify or prevent varicella in susceptible children when given within 72 hours of exposure (Brunell and Gershon, 1973; Center for Disease Control, 1979; Gershon et al., 1974; Judelsohn et al., 1974). At present, VZIG is available in the United States through the Division of Clinical Microbiology, Sidney Farber Cancer Institute, 44 Binney Street, Boston, Massachusetts (617-732-3121) for those meeting the criteria outlined in Table 9. Varicella-zoster immune plasma (VZIP), obtained from patients with active zoster, has also been shown to modify varicella infections in high-risk individuals (Balfour and Groth, 1979; Balfour et al., 1977; Geiser et al., 1975). There is no evidence that VZIG or VZIP has any benefit in individuals with active chickenpox or zoster and neither should be used in these situations. A recent study using dialyzable transfer factor has suggested that this method of passive immunization may be useful in preventing varicella in children with leukemia (Steele et al., 1980). The limited availability of specific varicella transfer factor at present makes this preventive approach impractical for wide use.

An experimental varicella vaccine has been extensively studied in Japan and is currently under investigation in the United States (Asano

TABLE 9. Five Criteria for Release of Varicella-Zoster Immune Globulin (VZIG) for the Prophylaxis of Varicella

I. One of the following underlying illnesses or conditions
  A. Leukemia or lymphoma
  B. Congenital or acquired immunodeficiency
  C. Under immunosuppressive medication
  D. Newly born of mother with varicella
II. One of the following types of exposure to varicella or zoster patient
  A. Household contact
  B. Playmate contact (>1 hour play indoors)
  C. Hospital contact (in same two- to four-room bedroom or adjacent beds in a large ward)
  D. Newborn contact (newborn whose mother contracted varicella less than five days before delivery or within 48 hours after delivery)
III. Negative or unknown prior disease history
IV. Age of less than 15 years
V. The request for treatment must be initiated within 72 hours of exposure

From Center for Disease Control: Varicella-zoster immune globulin. Morbid Mortal Weekly Rep 28:589, 1979.

et al., 1977; Takahashi et al., 1974). The potential of preventing varicella and its complications in childhood and the misery of herpes zoster in adult life has made some researchers believe that the varicella vaccine may be widely used. Controversy remains, however, about the eventual target population for such a vaccine. Some argue that normal children are at little risk from chickenpox and are not suitable candidates. The risk that more frequent or serious herpes zoster infections will follow vaccination, or that initial infection with varicella will be postponed from childhood when it is mild to adulthood when it is more often severe makes routine immunization against varicella potentially unattractive. High-risk, immunocompromised children are considered to be the group most likely to benefit from a varicella vaccine. This population, however, is the group in which live vaccines are usually avoided and which is most likely to respond inadequately to immunizations. Preliminary studies in Japan have indicated that children with leukemia, neuroblastoma, retinoblastoma, and others with chronic illnesses receiving steroid therapy can be successfully immunized. Clearly, further research will be necessary before benefits and risks of the live varicella vaccines are fully evaluated.

Rickettsial Diseases

The incidence of Rocky Mountain spotted fever (RMSF) continues to increase in the United States. Laboratory workers are at high risk of developing disease while working with *Rickettsia rickettsii,* the causative agent in RMSF. The first RMSF vaccine developed in 1924 was a phenol and formalin–inactivated preparation obtained from infected tick tissue. Since the 1940s, a killed vaccine produced in infected yolk sacs of embryonated eggs had been used. Although early findings in guinea pigs and human field trials suggested vaccine efficacy, after over 30 years of use studies have clearly shown the yolk sac vaccine to be ineffective in preventing infections in humans after laboratory exposure or direct challenge. Since 1970, a new formalin-inactivated vaccine, prepared from *R. rickettsii* grown in tissue culture of chick embryo fibroblasts and purified by sucrose density gradient, has been studied (Kenyon and Pedersen, 1975; Kenyon et al., 1975). The vaccine has proved to be highly immunogenic and protective in guinea pigs and rhesus monkeys and has been encouraging in preliminary human trials (Ascher et al., 1978). At present, however, no effective vaccine for RMSF is available.

Louse-borne typhus remains a major problem in rural or remote highland areas of Ethiopia, Rwanda, Burundi, Mexico, Guatemala, Ecuador, Bolivia, Peru, and the mountanous areas of Asia. A formalin-inactivated vaccine prepared from *Rickettsia prowazekii* grown in embryonated eggs is currently available (Center for Disease Center, 1978). Although no controlled studies of typhus vaccine have been carried out in humans, field and laboratory experiences suggest that the incidence and severity of typhus cases are diminished following vaccination, particularly if booster doses have been received. The vaccine protects against louse-borne (epidemic) typhus only and not against murine or scrub typhus. Typhus vaccine is recommended only for those individuals who live or visit areas where typhus cases actually occur and who will be in constant contact with the indigenous population. Medical personnel who provide care for patients in areas in which louse-borne (epidemic) typhus occurs or researchers working with *R. prowazekii* should be immunized.

At present, there is no effective vaccine against Q fever. A formalin-inactivated preparation prepared from *Coxiella burnetii* is relatively effective in eliciting an immunogenic response and providing protection in humans but is associated with an unacceptably high incidence of sterile abscesses at injection sites. A live attenuated vaccine utilizing *C. burnetii* strain M-44 initially appeared promising, but recent guinea pig studies showing microscopic evidence of myocarditis and hepatitis following vaccination suggest that the vaccine will require further attenuation before human use can be recommended (Johnson et al., 1976, 1977; Robinson and Hasty, 1974).

IMMUNIZATIONS FOR DEVELOPING COUNTRIES

Issues concerning immunizations in developed industrial countries often have little relevance for the over three billion people living in less developed nations who suffer from a multitude of infectious diseases. Preventing congenital rubella, avoiding mumps, meningoencephalitis, and orchitis, and protecting immunocompromised individuals from overwhelming varicella infections are medical luxuries shared only by inhabitants of wealthy developed nations. The collective state of ill health in these underdeveloped countries at present, unfortunately, makes comprehensive primary health care an impossibility. Selective primary health care that chooses priorities for disease control is currently the only reasonable goal (Walsh and Warren, 1979). Water and sanitation programs associated with vector control and improved nutrition must be combined with educational campaigns if infectious diseases are to be controlled. Selective immunization for children up to 3 years old and women of childbearing age should at least include vaccination against measles, diphtheria, pertussis, and tetanus. Further immunizations including those against poliomyelitis, tuberculosis, and typhoid could make a large impact on the general state of health in developing countries. The cost of these programs would be substantial and would require a worldwide commitment in order to be successful.

References

General

Artenstein, M. S.: The current status of bacterial vaccines. Hosp Prac 8:49, 1973.

Barrett-Connor, E.: Advice for young travelers. Pediatr Rev 1:25, 1979.

Center for Disease Control: Recommendation of the Immunization Practices Advisory Committee (ACIP): General recommendations on immunization. Morbid Mortal Weekly Rep 29:83, 1980.

Collins, F. M.: Vaccines and cell-mediated immunity. Bacteriol Rev 38:371, 1974.

Fraumeni, J. F., Ederer, F., and Miller, R. W.: An evaluation of the carcinogenicity of simian virus 40 in man. JAMA 185:713, 1963.

Fulginiti, V. A.: Controversies in current immunization policy and practices: One physician's viewpoint. Curr Probl Pediatr 66:3, 1976.

Harrison, H. R., and Fulginiti, V. A.: Bacterial immunizations. Am J Dis Child 134:184, 1980.

Krugman, S., and Katz, S. L.: Childhood immunization procedures. JAMA 237:2228, 1977.

Levine, M. M.: Live-virus vaccines in pregnancy risks and recommendations. Lancet 2:34, 1974.

Marks, J. S., Halpin, T. J., Irvin, J. J., Johnson, D. A., and Keller, J. R.: Risk factors associated with failure to receive vaccinations. Pediatrics 64:304, 1979.

Medical Letter on Drugs and Therapeutics: Immunizations for travelers. 21:57, 1979.

Mortimer, E. A.: Immunization against infectious disease. Science 200:902, 1978.

Phillips, C. F.: Children out of step with immunization. Pediatrics 55:877, 1975.

Rand, K. H., and Merigan, T. C.: Can immunization eradicate viral diseases?: Arch Intern Med 137:723, 1977.

Rothberg, R. M., Sumner, C. K., and Michalek, S. M.: Systemic immunity after local antigenic stimulation of the lymphoid tissue of the gastrointestinal tract. J Immunol 111:1906, 1973.

Walsh, J. A., and Warren, K. S.: Selective primary health care: An interim strategy for disease control in developing countries. N Engl J Med 301:967, 1979.

White, C. S., Adler, W. H., and McGann, V. G.: Repeated immunization: Possible adverse effects. Ann Intern Med 81:594, 1974.

Passive Immunization

Braude, A. I., Ziegler, E. J., Douglas, H., and McCutchan, J. A.: Antibody to cell wall glycolipid of gram-negative bacteria: Induction of immunity to bacteremia and endotoxemia. J Infect Dis 136:S167, 1977.

Ellis, E. F., and Henney, C. S.: Adverse reactions following administration of human gamma globulin. J Allergy 43:45, 1969.

Squire, J. R., et al.: Hypogammaglobulinaemia in the United Kingdom. Lancet 1:163, 1969.

Stiehm, E. R.: Standard and special human immune serum globulins as therapeutic agents. Pediatrics 63:301, 1979.

Stiehm, E. R., and Fudenberg, H. H.: Antibodies to gamma-globulin in infants and children exposed to isologous gamma-globulin. Pediatrics 35:229, 1965.

Ziegler, E. J., McCutchan, J. A., and Braude, A. I.: Clinical trial of core glycolipid antibody in gram-negative bacteremia. Trans Assoc Am Physicians 91:253, 1978.

Cytomegalovirus

Glazer, J. P., et al.: Live cytomegalovirus vaccination of renal transplant candidates. Ann Intern Med 91:676, 1979.

Just, M., Buergin-Wolff, A., Emoedi, G., and Hernandez, R.: Immunization trials with live attenuated cytomegalovirus, Towne 125. Infection 3:111, 1975.

Plotkin, S. A., Farquhar, J., and Hornberger, E.: Clinical trials of immunization with the Towne 125 strain of human cytomegalovirus. J Infect Dis 134:470, 1976.

Diphtheria, Pertussis, Tetanus

Barkin, R. M., and Pichichero, M. E.: Diphtheria-pertussis-tetanus vaccine: Reactogenicity of commercial products. Pediatrics 63:256, 1979.

Center for Disease Control: Diphtheria and tetanus toxoids and pertussis vaccine. Morbid Mortal Weekly Rep 26:401, 1977.

Enrengut, W.: Whooping cough vaccination. Lancet 1:370, 1978.

Grady, G. F., and Wetterlow, L. H.: Pertussis vaccine: Reasonable doubt. N Engl J Med 298:966, 1978.

Kendrick, P. L.: Can whooping cough be eradicated? J Infect Dis 132:707, 1975.

Koplan, J. P., et al.: Pertussis vaccine — An analysis of benefits, risks, and costs. N Engl J Med 301:906, 1979.

Linneman, C. C., et al.: Use of pertussis vaccine in an epidemic involving hospital staff. Lancet 2:540, 1975.

Lister, J.: The care of children — Pertussis vaccination. N Engl J Med 296:984, 1977.

Manclark, C. R.: Summary of an international symposium on pertussis. J Infect Dis 140:129, 1979.

Mathias, R. G.: Whooping cough in spite of immunization. Can J Public Health 69:130, 1978.

Peebles, T. C., et al.: Tetanus-toxoid emergency boosters. N Engl J Med 280:575, 1969.

Pittman, M.: Pertussis toxin: The cause of the harmful effects and prolonged immunity of whooping cough. Rev Infect Dis 1:401, 1979.

Stewart, G. T.: Vaccination against whooping cough. Lancet 1:234, 1977.

White, W. G., et al.: Duration of immunity after active immunization against tetanus. Lancet 2:95, 1969.

Hepatitis

Center for Disease Control: Immune globulins for protection against viral hepatitis. Morbid Mortal Weekly Rep 26:415, 1977.

Gerety, R. J., Tabor, E., Purcell, R. H., and Tyeryar, F. J.: Summary of an international workshop on hepatitis B vaccines. J Infect Dis 140:642, 1979.

Hoofnagle, J. H., et al.: Passive-active immunity from hepatitis B immune globulin. Ann Intern Med 91:813, 1979.
Krugman, S.: Viral hepatitis: Recent developments and prospects for prevention. J Pediatr 87:1067, 1975.
Krugman, S., Ward, R., Giles, J. P. and Jacobs, A. M.: Infectious hepatitis: Studies on the effect of gamma globulin and on the incidence of inapparent infection. JAMA 174:823, 1960.
Mosley, J. W.: Hepatitis B immune globulin: Some progress and some problems. Ann Intern Med 91:914, 1979.
Redeker, A. G., et al.: Hepatitis B immune globulin as a prophylactic measure for spouses exposed to acute type B hepatitis. N Engl J Med 293:1055, 1975.
Seeff, L. B., et al.: Type B hepatitis after needle-stick exposure: Prevention with hepatitis B immune globulin. Ann Intern Med 88:285, 1978.
Szmuness, W., et al.: Hepatitis B immune serum globulin in prevention of nonparenterally transmitted hepatitis B. N Engl J Med 290:701, 1974.
Woodson, R. D., and Clinton, J. J.: Hepatitis prophylaxis abroad. JAMA 209:1053, 1969.

Measles

Albrecht, P., Ennis, F. A., Saltzman, E. J., and Krugman, S.: Persistence of maternal antibody in infants beyond 12 months: Mechanism of measles vaccine failure. J Pediatr 91:715, 1977.
Center for Disease Control: Measles prevention. Morbid Mortal Weekly Rep 27:427, 1978.
Deseda-Tous, J., et al.: Measles revaccination. Am J Dis Child 132:287, 1978.
Krause, P. J., et al.: Revaccination of previous recipients of killed measles vaccine: Clinical and immunologic studies. J. Pediatr 93:565, 1978.
Landrigan, P. J., and Witte, J. J.: Neurologic disorders following live measles-virus vaccination. JAMA 223:1459, 1973.
Marks, J. S., Halpin, T. J., and Orenstein, W. A.: Measles vaccine efficacy in children previously vaccinated at 12 months of age. Pediatrics 62:955, 1978.
McCormick, J. B., Halsey, N., and Rosenberg, R.: Measles vaccine efficacy determined from secondary attack rates during a severe epidemic. J Pediatr 90:13, 1977.
Modlin, J. F., Jabbour, J. T., Witte, J. J., and Halsey, N. A.: Epidemiologic studies of measles, measles vaccine, and subacute sclerosing panencephalitis. Pediatrics 59:505, 1977.
Orenstein, W. A., et al.: Current status of measles in the United States, 1973–1977. J Infect Dis 137:847, 1978.
Ruuskanen, O., Salmi, T. T., and Halonen, P.: Measles vaccination after exposure to natural measles. J Pediatr 93:43, 1978.
Shasby, D. M., et al.: Epidemic measles in a highly vaccinated population. N Engl J Med 296:585, 1977.
Shelton, J. D., et al.: Measles vaccine efficacy: Influence of age at vaccination vs duration of time since vaccination. Pediatrics 62:961, 1978.
Weiner, L. B., et al.: A measles outbreak among adolescents. J Pediatr 90:17, 1977.
Wilkins, J., and Wehrle, P. F.: Additional evidence against measles vaccine administration to infants less than 12 months of age: Altered immune response following active/passive immunization. J Pediatr 94:865, 1979.
Wilkins, J., and Wehrle, P. F.: Evidence for reinstatement of infants 12 to 14 months of age into routine measles immunization programs. Am J Dis Child 132:162, 1978.

Meningococcus

Center for Disease Control: Meningococcal polysaccharide vaccines. Morbid Mortal Weekly Rep 27:327, 1978.
Farquhar, J. D., et al.: Clinical and serological evaluation of meningococcal polysaccharide vaccine groups A, C, and Y. Proc Soc Exp Biol Med 157:79, 1978.
Lepow, M. L., et al.: Persistence of antibody following immunization of children with groups A and C meningococcal polysaccharide vaccines. Pediatrics 60:673, 1977.
Peltola, H., et al.: Clinical efficacy of meningococcus group A capsular polysaccharide vaccine in children three months to five years of age. N Engl J Med 297:686, 1977.
Wilkins, J., and Wehrle, P. F.: Further characterization of responses of infants and children to meningococcal A polysaccharide vaccine. J Pediatr 94:828, 1979.

Mumps

Biedel, C. W.: Recurrent mumps parotitis following natural infection and immunization. Am J Dis Child 132:678, 1978.
Center for Disease Control: Mumps vaccine. Morbid Mortal Weekly Rep 29:87, 1980.
Hayden, G. F., Preblud, S. R., Orenstein, W. A., and Conrad, J. L.: Current status of mumps and mumps vaccine in the United States. Pediatrics 62:965, 1978.
Hosai, H., et al.: Studies on live attenuated mumps virus vaccine. Biken J 13:121, 1970.
Lerner, A. M.: Guide to immunization against mumps. J Infect Dis 122:116, 1970.
Yamanishi, K., et al.: Studies of live attenuated mumps virus vaccine: Biological characteristics of the strains adapted to the amniotic and chorioallantoic cavity of developing chick embryos. Biken J 13:127, 1970.
Yamauchi, T., Wilson, C., and St. Geme, J. W.: Transmission of live, attenuated mumps virus to the human placenta. N Engl J Med 290:710, 1974.

Pneumococcus

Ahonkhai, V. I., et al.: Failure of pneumococcal vaccine in children with sickle-cell disease. N Engl J Med 301:26, 1979.
Ammann, A. J., et al.: Polyvalent pneumococcal-polysaccharide immunization of patients with sickle-cell anemia and patients with splenectomy. N Engl J Med 297:897, 1977.
Center for Disease Control: Pneumococcal polysaccharide vaccine. Morbid Mortal Weekly Rep 27:25, 1978.
Cowan, M. J., et al.: Pneumococcal polysaccharide immunization in infants and children. Pediatrics 62:721, 1978.
Fikrig, S. M., et al.: Antibody response to capsular polysaccharide vaccine of *Streptococcus pneumoniae* in patients with nephrotic syndrome. J Infect Dis 137:818, 1978.
Klein, J. O., and Mortimer, E. A.: Use of pneumococcal vaccine in children. Pediatrics 61:321, 1978.
Minor, D. R., Schiffman, G., and McIntosh, L. S.: Response of patients with Hodgkin's disease to pneumococcal vaccine. Ann Intern Med 90:887, 1979.

Poliomyelitis

Center for Disease Control: Poliomyelitis — United States, Canada. Morbid Mortal Weekly Rep 28:229, 1979.
Davis, L. E., et al.: Chronic progressive poliomyelitis secondary to vaccination of an immunodeficient child. N Engl J Med 297:241, 1977.
John, T. J., et al.: Effect of breast-feeding on seroresponse of infants to oral poliovirus vaccination. Pediatrics 57:47, 1976.
Krugman, R. D., et al.: Antibody persistence after primary immunization with trivalent oral poliovirus vaccine. Pediatrics 60:80, 1977.
Melnick, J. L.: Vaccines and vaccine policy: The poliomyelitis example. Hosp Pract 13:41, 1978.
Nightingale, E. O.: Recommendations for a national policy on poliomyelitis vaccination. N Engl J Med 297:249, 1977.
Rousseau, W. E.: Persistence of poliovirus neutralizing antibodies eight years after immunization with live, attenuated-virus vaccine. N Engl J Med 289:1357, 1973.

Rabies

Center for Disease Control: Rabies prevention. Morbid Mortal Weekly Rep 29:265, 1980.
Nicholson, K. G., Turner, G. S., and Aoki, F. Y.: Immunization with a human diploid cell strain of rabies virus vaccine: Two year results. J Infect Dis 137:783, 1978.
Plotkin, S. A., and Wiktor, T.: Vaccination of children with human cell culture rabies vaccine. Pediatrics 63:219, 1979.

Rickettsial Diseases

Ascher, M. S., et al.: Initial clinical evaluation of a new Rocky Mountain spotted fever vaccine of tissue culture origin. J Infect Dis 138:217, 1978.
Center for Disease Control: Typhus vaccine. Morbid Mortal Weekly Rep 27:189, 1978.

Johnson, J. W., Eddy, G. A., and Pedersen, C. E.: Biological properties of the M-44 strain of *Coxiella burnetii*. J Infect Dis 133:334, 1976.

Johnson, J. W., et al.: Lesions in guinea pigs infected with *Coxiella burnetii* strain M-44. J Infect Dis 135:995, 1977.

Kenyon, R. H., and Pedersen, C. E.: Preparation of Rocky Mountain spotted fever vaccine suitable for human immunization. J Clin Microbiol 1:500, 1975.

Kenyon, R. E., Sammons, L. C., and Pedersen, C. E.: Comparison of three Rocky Mountain spotted fever vaccines. J Clin Microbiol 2:300, 1975.

Robinson, D. M., and Hasty, S. E.: Production of a potent vaccine from the attenuated M-44 strain of *Coxiella burnetii*. Appl Microbiol 27:777, 1974.

Rubella

Balfour, H. H.: Rubella reimmunization now. Am J Dis Child 133:1231, 1979.

Boue, A., Nicolas, A., and Montagnon, B.: Reinfection with rubella in pregnant women. Lancet 1:7712, 1971.

Center for Disease Control: Rubella vaccine. Morbid Mortal Weekly Rep 27:451, 1978.

Fleet, W. F., et al.: Fetal consequences of maternal rubella immunization. JAMA 227:621, 1974.

Fogel, A., et al.: Response to experimental challenge in persons immunized with different rubella vaccines. J Pediatr 92:26, 1978.

Horstmann, D.: Controlling rubella: Problems and perspectives. Ann Intern Med 83:412, 1975.

Medical Letter on Drugs and Therapeutics: The new rubella vaccine. 21:53, 1979.

Modlin, J. F., et al.: A review of five years' experience with rubella vaccine in the United States. Pediatrics 55:20, 1975.

Plotkin, S. A., Farquhar, J. D., and Ogra, P. L.: Immunologic properties of RA27/3 rubella virus vaccine. JAMA 225:585, 1973.

Schoenbaum, S. C., et al.: Benefit-cost analysis of rubella vaccination policy. N Engl J Med 294:306, 1976.

Wilkins, J., and Wehrle, P. F.: Further evaulation of the optimum age for rubella vaccine administration. Am J Dis Child 133:1237, 1979.

Wyll, S. A., and Herrmann, K. L.: Inadvertent rubella vaccination of pregnant women. JAMA 225:1472, 1973.

Smallpox

Center for Disease Control: Adverse reactions to smallpox vaccination, 1978. Morbid Mortal Weekly Rep 28:265, 1979a.

Center for Disease Control: Smallpox certification — East Africa. Morbid Mortal Weekly Rep 28:497, 1979b.

Center for Disease Control: Smallpox vaccine. Morbid Mortal Weekly Rep 27:156, 1978.

De Vries, R. P., Kreeftenberg, M., Loggen, H., and Van Rood, J. J.: In vitro immune responsiveness to vaccinia virus and HLA. N Engl J Med 297:692, 1977.

Tuberculosis

Center for Disease Control: BCG vaccines. Morbid Mortal Weekly Rep 28:241, 1979.

Eickhoff, T. C.: The current status of BCG immunization against tuberculosis. Annu Rev Med 28:411, 1977.

Passwell, J., et al.: Fatal disseminated BCG infection. Am J Dis Child 130:433, 1976.

Typhoid

Ashcroft, M. T., et al.: A seven year field trial of two typhoid vaccines in Guyana. Lancet 2:1056, 1967.

Center for Disease Control: Typhoid vaccine. Morbid Mortal Weekly Rep 27:231, 1978.

Gilman, R. H., et al.: Evaluation of a UDP-glucose-4-epimeraseless mutant of *Salmonella typhi* as a live oral vaccine. J Infect Dis 136:717, 1977.

Hornick, R. B., et al.: Typhoid fever: Pathogenesis and immunologic control. N Engl J Med 283:686, 739, 1970.

Varicella

Asano, Y., et al.: Protection against varicella in family contacts by immediate inoculation with live varicella vaccine. Pediatrics 59:3, 1977.

Balfour, H. H., and Groth, K. E.: Zoster immune plasma prophylaxis of varicella: A follow-up report. J Pediatr 94:743, 1979.

Balfour, H. H., Groth, K., McCullough, J., Kallis, J., Marker, S., Nesbit, M., Summons, R., and Najarian, J.: Prevention or modification of varicella using zoster immune plasma. Am J Dis Child 131:693, 1977.

Brunell, P. A., and Gershon, A. A.: Passive immunization against varicella-zoster infections and other modes of therapy. J Infect Dis 127:415, 1973.

Center for Disease Control: Varicella-zoster immune globulin. Morbid Mortal Weekly Rep 28:589, 1979.

Geiser, C. F., et al.: Prophylaxis of varicella in children with neoplastic disease: Comparative results with zoster immune plasma and gamma globulin. Cancer 35:1027, 1975.

Gershon, A. A., Steinberg, S., and Brunell, P. A.: Zoster immune globulin. N Engl J Med 290:243, 1974.

Judelsohn, R. G., et al.: Efficacy of zoster immune globulin. Pediatrics 53:476, 1974.

Steele, R. W., Myers, M. G., and Vincent, M. M.: Transfer factor for the prevention of varicella-zoster infection in childhood leukemia. N Engl J Med 303:355, 1980.

Takahashi, M. Otsuka, T., and Okuno, Y.: Live vaccine used to prevent the spread of varicella in children in hospital. Lancet 2:1288, 1974.

Uduman, S. A., Gershon, A. A., and Brunell, P. A.: Should patients with zoster receive zoster immune globulin? JAMA 234:1049, 1975.

Other Articles

Balfour, H. H., and Amren, D. P.: Rubella, measles and mumps antibodies following vaccination of children. Am J Dis Child 132:573, 1978.

Beachey, E. H., Stollerman, G. H., and Bisno, A. L.: A strep vaccine: How close? Hosp Pract 14:49, 1979.

Center for Disease Control: Plague vaccine. Morbid Mortal Weekly Rep 27:255, 1978.

Center for Disease Control: Cholera vaccine. Morbid Mortal Weekly Rep 27:173, 1978.

Center for Disease Control: Yellow fever vaccine. Morbid Mortal Weekly Rep 27:268, 1978.

Horstmann, D. M.: Viral vaccines and their ways. Rev Infect Dis 1:502, 1979.

Krugman, S.: Present status of measles and rubella immunization in the United States: A medical progress report. J Pediatr 90:1, 1977.

Melnick, J. L.: Viral vaccines: New problems and prospects. Hosp Pract 13:104, 1978.

Sato, H., et al.: Transfer of measles, mumps and rubella antibodies from mother to infant. Am J Dis Child 133:1240, 1979.

Stumpf, D. A., and Frost, M.: Immunity after rubella and measles viral vaccines. Am J Dis Child 132:748, 1978.

Weibel, R. E., et al.: Long-term follow-up for immunity after monovalent or combined live measles, mumps and rubella virus vaccines. Pediatrics 56:380, 1975.

Books

American Academy of Pediatrics: Report of the Committee on Infectious Disease, 18th ed., Evanston, Illinois, 1977.

Creighton, C.: Jenner and Vaccination: A Strange Chapter of Medical History. London, Suvan Sonnen Schein & Company, 1889.

Parish, H. J.: A History of Immunization. London, E. & S. Livingstone Ltd., 1965.

Parish, H. J.: Victory with Vaccines. London, E. & S. Livingstone Ltd., 1968.

Voller, A., and Friedman, H.: New Trends and Developments in Vaccines, Baltimore, University Park Press, 1978.

Metabolic

85 *FEVER*

Sheldon M. Wolff, M.D.

Fever has been recognized as a major manifestation of a large variety of diseases since antiquity. Although clinicians and investigators have been interested in the mechanisms of production of fever in the intact human for decades, it has been only during the past 15 years that significant progress has been made. It is not known if this commonest of all clinical symptoms is of any value in host defenses or whether it is a nonspecific response.

Normal body temperature follows a very reproducible circadian rhythm, with the maximum temperature ordinarily occurring in the late afternoon or early evening and then gradually falling to its low point in the morning. Normal temperature can be influenced by a variety of stimuli such as exercise, menstrual cycle, and environmental temperature changes, but it generally follows the same circadian pattern in a given individual. Thirty-seven degrees centigrade (98.6° F) is generally accepted as the "normal" temperature. These are mean values derived from old studies. When interpreting a temperature reading, there are a number of variables to be considered. Among these are how it was obtained (i.e., from which orifice), time of day, and whether the subject had just participated in strenuous physical activity. An isolated temperature reading in a patient is usually of little value, since so many factors can influence body temperature. Fever is a symptom of disease, and it is rare for it to be the sole sign or symptom of an illness. Normal individuals may have temperatures that occasionally rise above 38° C in the evening, and, unless multiple readings are taken, or a careful history obtained and a physical examination done, one would be incorrect in calling such an isolated reading a fever. All of the preceding facts emphasize that a temperature reading may be of no value unless it is considered merely as part of a more detailed evaluation of the patient as a whole.

Regardless of the cause of fever, most of the physiologic processes that occur when body temperature rises follow similar patterns. It is well established that the thermoregulatory center resides in the anterior hypothalamus in the floor of the third ventricle. The development of fever depends upon an intricate set of interrelated events involving heat loss, heat production, and peripheral vasoconstriction. Some of these events are mediated by neurohumoral agents such as norepinephrine and serotonin, but the exact role of such substances in the febrile response in human beings is unclear. When a patient first develops a fever, he or she may have a chill, which is accompanied by an increase in metabolic activity as measured by such things as oxygen consumption. The severity and length of the chill depend on the stimulus. Concomitant with or shortly after the increased metabolic activity, core temperature begins to rise. Peripheral vasoconstriction occurs with the chill and is associated with a decrease in heat loss, which is maximal over the more peripheral areas of the body. The increase in metabolic activity persists during the fever. With the dissipation of the fever, peripheral vasodilatation occurs, and heat loss ensues. Depending on the height of the fever, the rate of heat loss, and the rapidity of the lowering of the temperature, sweating (i.e., diaphoresis) that can be marked may occur.

Certain clinical findings accompany fever in human beings regardless of the cause. The presence or absence and the magnitude of such findings depend on the severity and duration of the febrile reaction. Headache, myalgias, arthralgias, nausea, vomiting, backache, and feelings of warmth are some of the more common complaints heard from patients with fever. Tachycardia is usual, although in certain infectious diseases a relative bradycardia may be seen. Tachypnea, widening of the pulse pressure, and flushing are additional signs that are usually noted in febrile human beings.

Fever can be a prominent symptom of all infectious and inflammatory diseases. It is also seen in a wide variety of neoplastic and metabolic illnesses. The agents that cause fever are called pyrogens, and the list of pyrogenic substances continues to increase. For example, most bacteria, viruses, and fungi are pyrogenic. Exogenous pyrogens also include the endotoxins of gram-negative bacteria, antigens in previously sensitized subjects (Root and Wolff, 1968), antigen-antibody complexes (formed either in vivo or in vitro), and some synthetic polynucleotides. Certain C-19, C-21, and C-24 steroids of endogenous human origin that share the 5-β-H configuration are potent pyrogens. The best known of these is etiocholanolone (Wolff et al., 1967). Some hormones, such as progesterone, have also been shown to be pyrogenic. What role, if any, is played by these pyrogenic steroids in human disease is not known.

It is generally accepted that the basic series of events put into motion by any exogenous pyrogen is the same. In addition, as noted above, the febrile response is similar. In other words, there is a "final common pathway" that results in fever regardless of the cause. The basic hypothesis states that pyrogens interact with host cells, which in turn release an endogenous pyrogen (or leukocytic pyrogen); this endogenously produced substance then stimulates the thermoregulatory centers in the brain, resulting in fever. Much of the information that resulted in this hypothesis has been obtained from experimental animals, but during the past few years information gained from investigations on human cells has become available. Despite the recent rapid advances, considerably more work needs to be done before we can define in exact terms the pathogenesis of fever in man (Dinarello and Wolff, 1978).

It was originally suggested that the neutrophil was the blood leukocyte that was solely responsible for endogenous pyrogen production (Wood, 1958). However, it was later demonstrated that blood leukocytes from patients with severe granulocytopenia or monocytic leukemia produced large quantities of endogenous pyrogens when stimulated in vitro. Thus, the blood monocyte was established as a potent source of endogenous pyrogen, and this finding provided an explanation for the presence of fever in patients without circulating granulocytes. The human eosinophil is also a source of endogenous pyrogen. Certain tumors, namely hypernephromas, which are often associated with febrile episodes, have been shown to release endogenous pyrogens spontaneously in vitro. Kupffer cells in the liver, splenic macrophages, alveolar macrophages, and peritoneal lining cells are sources of endogenous pyrogen in rabbits and most likely are potent producers of this mediator substance in humans. Clinically, it is important to recognize sources of endogenous pyrogen other than peripheral neutrophils or monocytes, since patients without circulating or marrow phagocytes often have marked febrile responses to infection. Kupffer and other fixed phagocytic cells are probably sources of endogenous pyrogen in such patients. It is clear that only phagocytic cells derived from bone marrow precursors produce endogenous pyrogen. Lymphocytes are not a source of endogenous pyrogen.

When blood leukocytes are incubated with inhibitors of protein synthesis such as puromycin or cycloheximide during the activation process, the subsequent production of endogenous pyrogen is prevented (Nordlund et al., 1970). Neither cycloheximide nor puromycin prevents phagocytosis. Hence, the activation process is unaffected by these drugs. Actinomycin, which prevents the transcription of DNA into messenger RNA, also prevents the production of endogenous pyrogen in vitro. It is clear that the process of activation of leukocytes to make endogenous pyrogen begins with the repression of the genome for endogenous pyrogen and that new messenger RNA must be synthesized.

Considerable progress has been made recently in the characterization of human leukocytic pyrogen. The major portion of human leukocytic pyrogen is a protein with a molecular weight of 15,000 and an isoelectric point of 6.9. A second molecule with a molecular weight of 45,000 and an isoelectric point of 5.1 is also produced. Antibodies to these molecules have been produced and a radioimmunoassay developed (Dinarello et al., 1977). These advances should lead to the measurement of this mediator protein in various disease states. Such measurements will undoubtedly improve our ability to diagnose and to treat specifically febrile states.

The preoptic region of the anterior hypothalamus appears to control body temperature. The hypothalamus also contains the monoamines that are critical for normal thermoregulation. When tumors or vascular accidents have involved this center in patients, hypothermia or, rarely, hyperthermia has occurred. Moreover, the circadian temperature rhythm may be lost in such patients.

The preoptic anterior hypothalamus contains the neurons that respond when endogenous pyrogen is injected in animals, and it seems likely that the same area contains pyrogen-sensitive neurons in man. The mechanism by which endogenous pyrogen causes thermosensitive neurons to increase their firing rate is unknown; however, it has been suggested during the past five years that prostaglandins have a critical role in the production of fever. Mounting evidence suggests that endogenous pyrogen induces synthesis of prostaglandins in the hypothalamus, in which they function as central transmitters in initiation of fever; furthermore, the well-known ability of aspirin and similar drugs to reduce fever appears to be directly related to their ability to block prostaglandin synthesis. Aspirin and other salicylate-like antipyretics do not affect either the production of endogenous pyrogen by leukocytes or the pyrogenicity of the endogenous-pyrogen molecule. The ability of an antipyretic to reduce fever is proportional to its ability to inhibit prostaglandin synthesis. In addition, antipyretics do not lower body temperature in human beings unless fever is present; thus, they do not affect the body temperature of subjects in whom the normal daily temperature is above the mean or exhibits a wide circadian range.

References

Dinarello, C. A., Renfer, L, and Wolff, S. M.: Human leukocytic pyrogen: Purification and development of a radioimmunoassay. Proc Natl Acad Sci USA 74:4624, 1977.

Dinarello, C. A., and Wolff, S. M.: Pathogenesis of fever in man. N Engl J Med 298:607, 1978.

Nordlund, J. J., Root, R. K., and Wolff, S. M.: Studies on the origin of human leukocytic pyrogen. J Exp Med 131:727, 1970.

Root, R. K., and Wolff, S. M.: Pathogenetic mechanisms in experimental immune fever. J Exp Med 128:309, 1968.

Wolff, S. M., Kimball, H. R., Perry, S., Root, R. K., and Kappas, A.: The biological properties of etiocholanolone. Ann Intern Med 67:1268, 1967.

Wood, W. B., Jr.: Studies on the cause of fever. N Engl J Med 258:1023, 1958.

86 *SHOCK IN INFECTIOUS DISEASES*

Herbert S. Heineman, M.D.

DEFINITION

Shock is a syndrome of generalized metabolic failure resulting from prolonged inadequacy of tissue perfusion. Its early clinical manifestations reflect malfunction of those organs most dependent on uninterrupted blood flow, particularly the brain, as well as compensatory adjustments designed to maintain adequate arterial pressure. As these adjustments fail, urinary output decreases and biochemical indices of distorted metabolism, specifically nonoxidative glycolysis with low yield of high energy chemical bonds, are detectable, testifying to the widespread nature of the disorder. In the end, it is the failure of energy production rather than damage to a particular organ that leads to death.

Other terms, such as "circulatory collapse," "circulatory failure," and "hypoperfusion," have been substituted for "shock" in an attempt to pinpoint the specific nature of the derangement. When it occurs as a specific complication of infection, it is referred to as "infectious shock," "septic shock," "bacteremic shock," and even "endotoxin shock." The last three terms specifically implicate bacterial infection and are therefore too restrictive. Because "infectious shock" is sufficiently broad as well as concise, this term will be used in the present chapter.

ETIOLOGY

Shock may occur in the course of almost any severe infection, but it is particularly characteristic of bacteremia due to gram-negative bacilli. In fact, although gram-negative bacteremia is a common event in seriously ill patients, shock occurs in only a small minority, and the proximate factors leading to this complication have not been identified. The importance of endotoxin, the lipopolysaccharide composing part of all gram-negative cell walls, is readily apparent because it produces a similar syndrome in experimental animals. Partly because of the extensive use of endotoxin as an investigative tool, endotoxin shock is commonly regarded as the prototype of infectious shock. Care must be taken not to use the terms interchangeably, because to do so could lead to incorrect therapy for shock of other causes.

Besides gram-negative bacilli, some of the better known etiologic agents associated with circulatory collapse are meningococci, clostridia, and staphylococci. Shock in meningococcemia is unique because of the hemorrhagic syndrome so commonly associated with it. In the case of clostridia and staphylococci, which contain no endotoxin, protein exotoxins that produce the syndrome in animals and may play an important role in human shock have been identified. The issue is complicated by the production of numerous distinct exotoxins by the same organism. For example, clostridial alpha-toxin, or phospholipase C, appears to be implicated in both hemolysis and progressive tissue destruction and has been identified in the blood during clostridial septicemia (Moore et al., 1976); however, vascular collapse does not correlate with hemolysis, and a role for other exotoxins seems likely.

Shock also occurs in infections due to fungi, rickettsiae, and viruses. Fungal endotoxins with varying effects in experimental animals have been described, but their role in humans is speculative. No shock-producing substances of microbial origin have been identified in rickettsial or viral infections.

PATHOGENESIS AND PATHOLOGY

The feature that distinguishes infectious shock from shock of other causes (cardiac, hemorrhagic, neurogenic) is the occurrence of widespread circulatory collapse without preceding loss of intravascular fluid or identifiable damage to a critical organ. In this connection, it must be emphasized that many infections do cause damage to specific organs, and this may result in vascular collapse just as if the damage had been inflicted by a noninfectious agent. The detailed mechanisms of shock in such cases will not be described here, but their various causes are listed in Table 1. Recognition of the many distinct mechanisms that produce shock is essential for proper patient evaluation because of the different therapeutic measures appropriate to each.

Understanding of the pathogenesis of true infectious shock in humans has been hampered, first, by the absence of any characteristic pathologic changes that could localize the primary insult; second, by interspecies differences among animals that make extrapolation of experimental observations to humans hazardous; and third, by overlapping but apparently distinct syndromes associated with different infections.

Common to all forms of infectious shock is a maldistribution of blood. Repeated observations and measurements have shown that hypotension and diminished tissue perfusion occur in the presence of normal blood volume. Furthermore, central venous pressure is typically low at the onset, and fluid overloading is often well tolerated, indicating that cardiac decompensation does not play a primary role. By exclusion, this places the functional deficit in the vascular system. On the basis of cardiac output measurements, two kinds of infectious shock have been described, low resistance ("hot dry") and high resistance ("cold clammy"), with features shown in Table 2. Terminally, all patients pass through the cold clammy stage, which represents compensatory sympathetic vasomotor activity.

Low resistance dynamics may be seen in the early stages of shock caused by any bacterial infection, but they are particularly characteristic of shock due to gram-positive cocci. In this stage, hot dry skin is evidence of hyperkinetic circulation, and hemodynamic measurements may reveal a markedly lowered peripheral resistance and a cardiac output well above normal. Despite this, the patient is hypotensive and obtunded, and oxygen extraction by the cells is disproportionately decreased. For a given cardiac index, variation in arteriovenous oxygen difference has been found to reflect the severity of shock and to correlate with ultimate survival (Nishijima et al., 1973). This finding suggests a specific defect in oxygen extraction at the tissue level independent of central hemodynamics. The most widely proposed mechanism is arteriovenous shunting, al-

TABLE 1. Classification of Shock Associated with Infection

MECHANISM OF SHOCK	DISEASES IN WHICH MECHANISM IS OPERATIVE
I. Shock Secondary to Identifiable Factors	
A. Loss of blood volume	
1. Fluid loss due to capillary damage	Gas gangrene; anthrax; hemorrhagic fever; allergic reaction to antibiotic
2. Fluid loss due to malfunction in fluid regulation	
a. Severe diarrhea	Cholera; staphylococcal enteritis
b. Adrenal insufficiency	Tuberculosis; histoplasmosis
c. Uncontrolled diuresis	Salt-losing pyelonephritis
d. Peritoneal exudation	Peritonitis
B. Impaired cardiac function	
1. Myocardial failure	
a. Myocarditis	Diphtheria; infective endocarditis; influenza; rickettsial infections; trichinosis; toxoplasmosis; trypanosomiasis (Chagas' disease)
b. Myocardial infarction	Coronary embolism in infective endocarditis
c. Myocardial abscess	Infective endocarditis; septicemia
d. Acute valvular insufficiency	Infective endocarditis
2. Cardiac tamponade	
a. Pericardial effusion	Pericarditis
b. Hemopericardium	Ventricular rupture, as in tuberculous myocarditis
3. Mechanical outflow obstruction	
a. Intraventricular blockage	Intraventricular rupture of echinococcus cyst
b. Pulmonary embolism	Myocarditis with mural thrombosis
C. Loss of arteriolar tone	
1. Destruction of vasomotor center	Bulbar poliomyelitis
2. Anaphylaxis	Rupture of echinococcus cyst; allergic reaction to antibiotic
II. True Infectious Shock	
Maldistribution of blood	Gram-negative bacillary infection; meningococcemia; clostridial infections; staphylococcal and other gram-positive coccal septicemia; influenza; rickettsial infection; systemic candidiasis

TABLE 2. Two Types of Infectious Shock

CHARACTERISTIC	LOW RESISTANCE	HIGH RESISTANCE
Arterial blood pressure	Low	Low
Tissue oxygenation	Decreased	Decreased
Cardiac output	High	Low
Peripheral vascular resistance	Low	High
Venous sequestration	Absent	Present
Arteriovenous shunting	Present	Absent
Arteriovenous oxygen difference	Low	High

though it is not clear whether such shunting is truly anatomic or whether normal gas exchange in the capillary bed simply does not take place, thus creating a purely physiologic shunt (Archie, 1976; Seyfer et al., 1977).

High resistance dynamics eventually supervenes in patients dying in shock. In this stage, there is a marked increase in peripheral resistance on the arterial (resistance) side of the circulation, but central venous pressure remains low until cardiac decompensation sets in, indicating pooling of blood in a dilated venous (capacitance) bed. In different animal species, particular venous beds (for example, splanchnic or pulmonary) have been implicated. In humans, specific localization has not been demonstrated, and the defect is believed to be generalized. Venous pooling leads in turn to diminished venous return to the heart, diminished cardiac output, and compensatory arteriolar constriction. Resembling hemorrhagic shock, it is manifestated by cold clammy skin, tachycardia, thready pulse, mental obtundation, and oliguria.

In addition to measurable parameters of altered circulatory dynamics, several indices of altered metabolism can be detected. One of these is disseminated intravascular coagulation, most dramatically seen in meningococcemia, in which widespread deposition of fibrin thrombi may occur in small vessels. Because coagulation factors are consumed faster than they can be replaced, the blood may become virtually incoagulable, and this results in hemorrhages into the skin (ecchymoses) and viscera. The adrenal glands are frequently the site of visceral hemorrhage, which results in the so-called Waterhouse-Friderichsen syndrome. Adrenal hemorrhage was originally thought to explain infectious shock on the basis of adrenal insufficiency. However, this explanation is rendered invalid by the inconsistency of pathologic findings in patients dying of infectious shock, the presence of a normal concentration of adrenocortical steroids in the blood, and the failure of physiologic doses of steroids to reverse the course.

The mechanism of disseminated intravascular coagulation is poorly understood. Endotoxins from gram-negative bacteria are capable of activating Factor XII (Hageman factor), which could initiate the clotting cascade. However, this leaves unexplained the occurrence of the same phenomenon in infections due to organisms that do not contain endotoxin.

Whether intravascular coagulation actually plays a role in circulatory collapse is debatable. Although the fibrin thrombi are capable of causing mechanical vascular obstruction, the extent of thrombosis seen at autopsy is highly variable, and in many cases of shock studied during life it is not possible to demonstrate significant consumption of coagulation factors. Finally, even though abnormal coagulation can be arrested by the administration of heparin, clinical trials have failed to show improvement in prognosis as a result. Rather than being part of the fundamental process of shock, intravascular coagulation may be only one of its complications.

Along similar lines, several vasoactive substances appear in the blood of endotoxin-shocked animals — for example, catecholamines, histamine, serotonin, and prostaglandins. Also, animals may become abnormally reactive to adrenalin. However, from all these observations it has not been possible to construct a comprehensive sequence of events that explains exactly how endotoxin or any other microbial product leads to circulatory collapse in infection.

Recent work suggests a possible role for endorphins in shock due to endotoxin (and other causes). In shocked animals, hypotension was dramatically reversed by naloxone, an opiate antagonist (Faden and Holaday, 1980).

Whatever the mechanism of infectious shock, the end result is metabolic failure at the cellular level. It is usually assumed that the critical substance of which cells are deprived is oxygen, and indeed, one of the universal consequences of shock is anaerobic metabolism, which can be demonstrated by increased concentrations of lactate in the blood. However, other metabolic effects not readily linked to oxygen utilization have been described — for example, failure of gluconeogenesis in the face of increased glucose utilization (Archer et al., 1976; Garcia-Barreno and Balibrea, 1978). The basic biochemical lesion has been traced to the inner mitochondrial membrane, whose transport function is believed to be inhibited by products of lipolysis (Jones, 1977).

CLINICAL MANIFESTATIONS

History and Physical Findings

In most patients suffering from infectious shock, this complication occurs in a setting in which the infection is already recognized. However, vascular collapse may be the first or only sign of severe infection, especially in elderly debilitated patients who may not demonstrate fever or local symptoms.

Gram-negative bacillary (endotoxin) shock is seen most frequently in patients with urinary infection, when trauma to the infected tissues (for example, cystoscopy) induces a sudden burst of bacteremia. The effects closely resemble those in experimental animals infused with endotoxin. Within a few hours of the traumatic event, the patient suddenly has shaking chills followed by a steep rise in temperature to 39° C or higher with corresponding tachycardia. The blood pressure may begin to fall within less than an hour or several hours later, and this is accompanied by prostration, apprehension, vomiting, and confusion. Examination at first may reveal hot dry skin, systolic pressure of less than 80 mm Hg, and diastolic pressure that is barely recordable. Hyperventilation to the point of respiratory alkalosis is typical of this stage. If the condition is untreated or unresponsive to treatment, peripheral vasoconstriction sets in with cold clammy cyanotic extremities, further diminution in pulse pressure, continued hyperventilation as metabolic acidosis ensues, and oliguria. Various therapeutic measures may prolong life for several days without reversal of shock, until death supervenes.

A similar sequence of events, likewise associated with gram-negative bacteremia, occurs following septic delivery or abortion. It has been observed that, except for pregnant women, patients under 40 infrequently develop shock from gram-negative bacillary infections (Weil, 1977). On the other hand, shock due to meningococcemia is typically a syndrome of children and young adults, because of the age distribution of meningococcal infection. The full-blown syndrome of meningococcemia is easily recognized, but this infection should be suspected even in the absence of meningitis or cutaneous manifestations whenever fever and shock occur in a previously healthy young person.

The clinical syndromes of shock in gram-positive infections differ chiefly in the settings in which they occur. The best recognized pathogens are the histotoxic clostridia — that is, those with the capacity for progressive and destructive tissue invasion — and coagulase-positive staphylococci. Shock is a typical concomitant of clostridial myonecrosis (gas gangrene), a syndrome easily recognized by the presence of large areas of edematous, often crepitant, discolored, and exquisitely tender muscle tissue. Invasion of muscle appears to be a prerequisite, as clostridial cellulitis sparing the muscle is generally not associated with this degree of systemic toxicity (MacLennan, 1962). However, shock does occur in the absence of myonecrosis when clostridia invade the blood stream from such foci as the large intestine and the uterus. In the latter cases, jaundice, hemoglobinemia, and hemoglobinuria may develop because of intravascular hemolysis, a feature that is usually absent in clostridial myonecrosis.

Shock may occur in staphylococcal septicemia with or without evidence of focal infection. The prototype of shock without focal infection was manifested in a disastrous outbreak in Bundaberg, Australia (Kellaway et al., 1928), following injection of a contaminated diphtheria toxin-antitoxin mixture. Shock and death occurred too rapidly to be explained by widespread staphylococcal infection and was probably caused by exotoxin. In more typical situations, shock occurs during the course of severe staphylococcal infection, such as pneumonia.

A wide variety of other inections may be complicated by shock (Heineman and Braude, 1961), with clinical manifestations basically reflecting the underlying cause. Finally, it must be remembered that shock may be caused by damage or malfunction of specific organs (see Table 1) — for example, myocarditis, adrenal insufficiency, pulmonary embolism, and massive diarrhea. Alertness to these possibilities is especially important because of their unique therapeutic implications.

Laboratory Tests

Hematologic findings are variable, more often reflecting the underlying infection than any characteristic of shock as such. Depending on the state of endotoxemia, for example, either leukopenia or leukocytosis may be present. Hemoglobin concentration may be normal, elevated (for example, hemoconcentration in cholera), or depressed (for example, intravascular hemolysis in clostridial septicemia). More specificially associated with shock is thrombocytopenia, which is one of the manifestations of disseminated intravascular coagulation. In such a case, other evidence of coagulopathy is also found, such as prolongation of prothrombin time and partial thromboplastin time and the presence of fibrin split products in the blood (Hardaway, 1976).

Findings specific for shock regardless of its cause relate to the deterioration of oxidative metabolism by the tissues — for example, decreased arteriovenous oxygen gradient, metabolic acidosis, and increased serum lactate concentration.

In shock due to bacterial or fungal sepsis, the causal organisms can usually be recovered from the blood. However, blood cultures are more often negative than positive in clostridial myonecrosis and may be so even in gram-negative bacillary shock, emphasizing the role of bacterial toxins apart from the presence of replicating organisms.

Arterial hypoxia has been noted in some patients with infectious shock. Radiologically visible pulmonary infiltrates may be present that resemble those of pulmonary edema. However, they occur in the absence of heart failure and are generally ascribed to "shock lung" or "adult respiratory distress syndrome." In this condition, increased capillary permeability leads to interstitial edema and eventually hyaline membrane formation, with resultant block of alveolocapillary oxygen diffusion.

COMPLICATIONS AND SEQUELAE

The mortality in infectious shock is high, its incidence generally being quoted at about 70 per cent. A lower incidence is found in series involving predominantly younger patients, specifically women with complications of pregnancy, for in this age group the prognosis is better than in elderly debilitated patients, in whom infectious shock occurs most frequently.

The road to recovery may be marked by acute renal failure due to shock or hemoglobinuria, or by infarction of organs rendered vulnerable to ischemia through occlusive arterial disease of old age, such as the myocardium or brain. Peripheral gangrene may be seen even in young subjects with shock-related coagulative disorders, as in meningococcemia.

In addition, there may be sequelae related to damage done by the infection itself or by heroic therapeutic measures, such as surgery, corticosteroids, and toxic antibiotics.

DIAGNOSIS

The diagnosis of infectious shock is easy in the typical case, and it can be made by an alert observer before serious damage is done. There are two diagnostic signs: (1) presence of infection, and (2) falling blood pressure without hemorrhage or cardiac failure. These easily recognized signs are emphasized because if a patient has already been in shock for several hours before his hypotensive state is noticed, damage to cellular metabolic functions may have already been done and such damage will be progressively harder to reverse. On the other hand, since hypotension occurs very early in the course, close monitoring of vital signs in patients with severe infections, and especially in those who have just undergone manipulation of infected pelvic organs, is the key to early diagnosis and eventual recovery.

Besides hypotension, patients in shock may exhibit hyperventilation, confusion, apprehension, prostration, and vomiting. Depending on the type of circulatory derangment, the skin may be hot and dry or cold and clammy.

Circulatory insufficiency at the tissue level is recognized most easily by oliguria (hourly urine output of less than 20 ml) and subsequent increases in concentrations of blood urea and creatinine and a decrease in bicarbonate. If laboratory facilities for its determination are available, excess lactate will be found in the serum (more than 16 mg/dl), specifically indicating inadequate tissue oxygenation.

TREATMENT

The aims of treatment in infectious shock are control of the infection and correction of the circulatory disturbance. On the basis of the relative roles of the infecting organism and the circulatory disturbance, patients may be divided into five groups.

Group 1: Specific antibiotic treatment for the infection is available; for example, shock cases due to bacteria, rickettsiae, and fungi. Infection and shock must be treated simultaneously.

Group 2: Specific antibiotics are available to combat the organism but not its effects; for example, gas gangrene and diphtheria. Surgery and administration of antitoxin may be indicated, although the value of antitoxin in the treatment of shock has not been proved in either gas gangrene or diphtheria.

Group 3: No specific antibiotic treatment for the infection is available; for example, shock cases due to viruses. The objective of therapy is to sustain life long enough to allow the host to control the infection.

Group 4: Specific antibiotic treatment is available but not critical, since invasion of tissues by the infecting organism is not a significant feature of the illness; for example, cholera.

Group 5: The potential for shock continues to exist even after the infection is under control; for example, pulmonary embolism from myocarditis, coronary embolism from endocarditis, chronic sequelae of poliomyelitis, and tuberculous adrenal insufficiency.

When it is recalled that, in addition to the foregoing possibilities, some infections can cause shock in more than one manner, the necessity for an accurate appraisal of the situation becomes obvious.

TREATMENT OF INFECTION

The urgency of treatment of patients in shock necessarily will modify the approach to selection of antibiotics. However, since the object of antibiotic therapy should always be to eliminate the infection, the guiding principle here, as in less desperate situations, should be to identify the etiologic organism as soon as possible and to prescribe an optimal antibiotic regimen to which it is sensitive. Unfortunately, there is no time to withhold treatment, so that other guides to the selection of the proper drugs must be employed. The history is often helpful in this regard. For example, a sudden temperature elevation and shock occurring in a patient with an indwelling urethral catheter is presumptive evidence of bloodstream invasion by an organism originating in the urinary tract. If the results of a recent urine culture are available, one may assume that the same organism is responsible for infection of both urine and blood. Physical examination occasionally is so typical of a specific infection that antibiotic therapy can be chosen on this basis alone. A good example is the child or young adult with an acute illness characterized by fever, nuchal rigidity, purpura, and shock. Meningococcemia should be suspected and treated with large intravenous doses of potassium penicillin G or, in the case of penicillin allergy, chloramphenicol sodium succinate.

Sometimes the smear from a lesion is sufficiently characteristic to allow positive identification, as in gas gangrene, meningococcal infection (spinal fluid or exudate from a punctured petechia), and staphylococcal pneumonia. In the infrequent cases of septicemia that occur without an obvious focus of infection, it is wise to use a combination of antibiotics that can be expected to cover most possibilities, such as a cephalosporin and an aminoglycoside. An anti-*Pseudomonas* beta-lactam antibiotic may be indicated in patients susceptible to *Pseudomonas* infection, as in leukemia, severe burns, or following long hospitalization.

When shock is due to mechanical factors, as in pulmonary embolism from a mural thrombus, myocardial infarction from coronary embolism in bacterial endocarditis, or Addisonian crisis in tuberculosis of the adrenals, the infection itself generally demands less urgent treatment, since it is not the immediate presence of the bacteria that causes circulatory collapse.

TREATMENT OF SHOCK

The prognosis in shock could probably be improved if it were possible to address the basic cellular lesions that underlie collapse of normal energy metabolism. Unfortunately, these lesions are only beginning to be understood, so that there is still no foundation for the development of a specific therapeutic strategy. Based on present knowledge, the only rational objective of therapy can be to restore the circulation. The agents employed, listed in order of importance, are (1) fluids, (2) vasoactive drugs, (3) adrenal corticosteroids, and (4) digitalis. Additional measures include assisted ventilation and, questionably, anticoagulation.

Fluids

Replenishment of the venous bed is always the first consideration and is frequently all that is needed. It is understood, of course, that the need for extra fluid is created not by dehydration (except in fluid-losing conditions such as cholera) but by an abnormal increase in the capacity of the vascular bed owing to loss of venous tone. In the early stages, simply ensuring an adequate venous return to the right atrium will result in adequate cardiac output. However, myocardial depression may limit the ability of the heart to handle the volume of fluid necessary to reverse shock, at which point further fluid loading would result in pulmonary edema. To avoid this, the pre-load of the heart should be carefully monitored. The simplest measurement is that of central venous pressure, which is done by means of a catheter opening in the superior vena cava just outside the right atrium. As long as the pressure does not exceed 10 cm saline, infusion of fluids can safely be continued; above 15 cm saline, there is danger of overloading the heart. However, since the real danger to life is pulmonary edema, many authorities feel that a more appropriate measurement is that of the left atrial filling pressure, which can be closely approximated by the pulmonary wedge pressure. The latter is obtained through a balloon-tipped catheter passed through the right cardiac chambers and pulmonary artery as far as it will go, at which point the pulmonary venous pressure is reflected back into it. Continued infusion is safe as long as the pulmonary wedge pressure is less than 15 mm Hg; above 20 mm Hg, there is danger of pulmonary edema.

The fluid most commonly used for replenishment, when there has been no extravascular fluid loss, is normal saline. If there has been exudation of fluids through capillary leakage, as in gas gangrene or peritonitis, or through inflammatory diarrhea, as in shigella dysentery, it is rational to use a colloid such as 5 per cent human serum albumin.

If the maximum tolerated amount of fluid does not reverse the shock state, the rate of infusion must be reduced to the minimum necessary to maintain patency of the access line, and therapeutic efforts are continued with vasoactive agents.

Vasoactive Drugs

In contrast to fluid loading, which simply fills the vascular bed to its exaggerated capacity, the aim of vasoactive drugs is to alter the capacity of the vessels selectively to overcome specific abnormalities of blood distribution. In theory, this should be the obvious first line of approach, but the fact is that no drug with the right selectivity has been identified. It must be remembered that the two recognized defects of distribution are dilatation of the venous bed and arteriovenous shunting. Defects in cardiac output and arteriolar tone are not primary in most cases, yet all vasoactive drugs in use affect cardiac performance and arteriolar tone predominantly. The only advance made in the past several decades has been identification of effective beta-adrenergic drugs, such as isoproterenol and dopamine, that raise the blood pressure by increasing the cardiac output while decreasing peripheral resistance, thereby increasing total blood flow. Their vasomotor effect is opposite from that of alpha-adrenergic drugs, such as noradrenalin and metaraminol, whose antihypotensive action is based primarily on increase in peripheral resistance; these drugs may actually decrease tissue perfusion, so that the metabolic insult is aggravated at the same time that central arterial pressure is increased.

Isoproterenol has been used successfully but nowadays is not favored because of its tendency to increase ventricular irritability and bring about potentially dangerous arrhythmias, to increase myocardial oxygen demand, and to shunt blood from the viscera to skeletal muscle. Dopamine lacks these adverse cardiac effects and shunts blood to the viscera, a specific distributive effect aptly termed "dopaminergic" (Reid and Thompson, 1975). This drug is administered by intravenous drip at an initial rate of 2 to 20 μg/kg body weight/minute, then titrated to the desired response (see Monitoring section below).

Adrenal Corticosteroids

Steriods are considered by many to be useful therapeutic agents in treatment of infectious shock, although their action, unlike that of fluids and vasoactive drugs, is not immediately reflected in hemodynamic measurements. It is not clear whether their mechanisms of action involve their anti-inflammatory effect (Cline and Melmon, 1966), a direct effect on the circulation (Sambhi et al., 1965), or some as yet undetected property, although the need for superpharmacologic doses suggests mechanisms distinct from those involved in other applications of these agents. Methylprednisolone and dexamethasone in respective loading doses of 30 mg/kg and 3 mg/kg, followed by fractional doses every 4 to 6 hours for up to 48 hours, illustrate the order of magnitude of the dosage used. Significant reduction in mortality was found in a prospective double-blind, controlled study with these drugs (Schumer, 1976). The possibility that steroids might impair resistance to infections is generally considered of secondary importance in these desperate situations, especially since antibiotic coverage is provided. Because their effect is delayed by several hours, the loading dose of steroids should be given as soon as it is clear that fluid loading alone is insufficient.

Digitalis

Myocardial depression can occur in sepsis, and failure of cardiac output to increase in response to increasing pulmonary wedge pressure has been shown to correlate with mortality (Weisel et al., 1977). It seems appropriate, therefore, to digitalize a patient in whom central venous or pulmonary wedge pressure rises rapidly during fluid infusion without satisfactory clinical response. Caution is necessary in patients with myocarditis or myocardial anoxia because these states heighten susceptibility to the toxic effects of digitalis, and dangerous arrhythmias may result.

Assisted Ventilation

In certain patients, arterial hypoxemia complicates shock, possibly because of intrapulmonary arteriovenous shunting or alveolocapillary block due to "shock lung." In these patients, oxygen administered by intermittent positive pressure or positive end-expiratory pressure may be helpful (Ledingham and McArdle, 1978).

Oxygen under pressure has been advocated for the treatment of severe clostridial infections because of its proposed inhibition of toxin production. Although opinion regarding its true value and potential hazards is divided, it is commonly used in centers equipped with hyperbaric chambers (Darke et al., 1977).

Anticoagulants

The abnormal measurements of coagulation factors in disseminated intravascular coagulation (prothrombin time, partial thromboplastin time, fibrin split products) can frequently be reversed by heparin. However, several clinical trials have failed to convince investigators that anticoagulation significantly affects the outcome. As a result, this measure is not recommended in infectious shock.

MONITORING THERAPY IN THE SHOCK PATIENT

Since the two most important therapeutic measures, namely, fluids and vasoactive drugs, are titrated on an hour-to-hour or minute-to-minute basis, and overtreatment may have serious adverse effects, it is important to monitor the response accurately and quantitatively. Measurement of blood pressure, which is so useful in the initial diagnosis, has only limited usefulness once vasoactive drugs are used, since central arterial pressure may correlate very poorly with tissue blood flow. A more relevant criterion of response is one that measures end-organ performance. The most readily available is urine output, increase in which is usually taken as indicative of increased renal perfusion. Less readily measured on a continuing basis but valuable as an indicator of anaerobic metabolism is serum lactate, which should decrease as the shock state is reversed.

PROPHYLAXIS

Since shock is a complication of severe infection, its prevention lies in the timely and effective treatment of infection when this is possible. One approach is carefully selected antibiotic prophylaxis for invasive procedures likely to be complicated by local infection or bacteremia, for example, surgery on the urinary tract (Sullivan et al., 1973), gastrointestinal and biliary tracts (Stone et al., 1976), and pelvic organs (Roberts and Homesley, 1978).

References

Archer, L., Benjamin B., Lane, M. D., and Hinshaw, L. B.: Renal gluconeogenesis and increased glucose utilization in shock. Am J Physiol 231:872, 1976.

Archie, J. P., Jr.: Systemic and regional arteriovenous shunting in endotoxic and septic shock in dogs. Surg Forum 27:55, 1976.

Cline, M. J., and Melmon, K. L.: A possible explanation of the anti-inflammatory action of cortisol. Science 153:1135, 1966.

Darke, S. G., King, A. M., and Slack, W. K.: Gas gangrene and related infection: Classification, clinical features and aetiology, management and mortality. A report of 88 cases. Br J Surg 64:104, 1977.

Faden, A., and Holaday, J.: Experimental endotoxin shock: The pathophysiologic function of endorphins and treatment with opiate antagonists. J Infect Dis 124:229, 1980.

Garcia-Barreno, P., and Balibrea, J. L.: Metabolic responses in shock. Surg Gynecol Obstet 146:182, 1978.

Hardaway, R. M., III: Gram-negative shock. Antibiot Chemother 21:208, 1976.

Heineman, H. S., and Braude, A. I.: Shock in infectious diseases. Disease-a-Month, October, 1961.

Jones, G. R. N.: The basic biochemical lesion in shock. Biochem Soc Trans 5:213, 1977.

Kellaway, C. H., MacCallum, P., and Tebbutt, A. H.: Report of the Royal Commission of Enquiry into the Fatalities at Bundaberg. Med J Austral 2:2, 38, 1928.

Ledingham, I., and McArdle, C. S.: Prospective study of the treatment of septic shock. Lancet 1:1194, 1978.

MacLennan, J. D.: The histotoxic clostridial infections in man. Bacteriol Rev 26:177, 1962.

Moore, A., Gottfried, E. L., Stone, P. H., and Coleman, M.: Clostridium perfringens septicemia with detection of phospholipase C activity in the serum. Am J Med Sci 271:59, 1976.

Nishijima, H., Weil, M. H., Shubin, H., and Cavanilles, J.: Hemodynamic and metabolic studies on shock associated with Gram negative bacteremia. Medicine 52:287, 1973.

Reid, P. R., and Thompson, W. L.: The clinical use of dopamine in the treatment of shock. Bull Johns Hopkins Hosp 137:276, 1975.

Roberts, J. M., and Homesley, H. D.: Low-dose carbenicillin prophylaxis for vaginal and abdominal hysterectomy. Obstet Gynecol 52:83, 1978.

Sambhi, M. P., Weil, M. H., and Udhoji, V. N.: Acute pharmacodynamic effects of glucocorticoids. Cardiac output and related hemodynamic changes in normal subjects and patients in shock. Circulation 31:523, 1965.

Schumer, W.: Steroids in the treatment of clinical septic shock. Ann Surg 184:333, 1976.

Seyfer, A. E., Zajtchuk, R., Hazlett, D. R., and Mologne, L. A.: Systemic vascular performance in endotoxic shock. Surg Gynecol Obstet 145:401, 1977.

Stone, H. H., Hooper, C. A., Kolb, L. D., Geheber, C. E., and Dawkins, E. J.: Antibiotic prophylaxis in gastric, biliary and colonic surgery. Am Surg 184:443, 1976.

Sullivan, N. M., Sutter, V. L., Mims, M. M., Marsh, V. H., and Finegold, S. M.: Clinical aspects of bacteremia after manipulation of the genitourinary tract. J Infect Dis 127:49, 1973.

Weil, M. H.: Current understanding of mechanisms and treatment of circulatory shock caused by bacterial infections. Ann Clin Rev 9:181, 1977.

Weisel, R. D., Vito, L., Dennis, R. C., Valeri, C. R., and Hechtman, H. B.: Myocardial depression during sepsis. Am J Surg 133:512, 1977.

METABOLIC EFFECTS OF INFECTION 87

William R. Beisel, M.D.

Defensive mechanisms that protect the host against invading microorganisms include the inflammatory, phagocytic, and immunologic responses. In addition, cells throughout the body undergo a variety of predictable changes in their metabolic function during a generalized infection. These metabolic responses account for some of the clinical features of infectious illness.

The predictable metabolic changes during infection can be considered physiologic or homeostatic responses, since they appear to contribute to survival. Because similar metabolic responses occur during infections caused by widely different microorganisms, the responses are said to be generalized or nonspecific. The magnitude and duration of generalized metabolic responses are governed by the severity and persistence of an infectious process rather than by its cause. Thus, generalized infectious illnesses are accompanied by fever, increased oxygen consumption, the need to generate additional carbohydrate fuels, redistribution within the body of amino acids, lipids, electrolytes, and minerals, increased utilization of vitamins, and hepatic synthesis of enzymes and "acute phase" serum proteins.

Other metabolic changes may be superimposed upon the broad group of physiologic ones. These additional changes are secondary to localization of an infectious process within a major organ. For example, pneumonia can interfere with gas exchange in the lungs, pyelonephritis can cause uremia, and diarrhea can cause a loss of intestinal fluids and electrolytes. Localized infections can thus produce both generalized illness and single organ dysfunction. In some instances, the metabolic consequences of these combinations can become life-threatening in their severity.

METABOLIC COSTS OF INFECTION

Certain daily costs must be met in order to maintain host resistance mechanisms. These costs can be expressed in metabolic terms, since they are based on the need for energy-producing substrates to manufacture cells, precursor components, and molecules important to host resistance. For example, new phagocytes, lymphocytes, and epithelial cells must be produced each day; immunoglobulins and other molecules with unique roles in host defense must also be synthesized. As discussed in the next chapter, malnourished patients become increasingly susceptible to infection if they cannot meet the daily nutritional requirements for maintaining resistance mechanisms.

Because fever causes metabolic rates to increase, additional costs are incurred by the patient with fever. These costs must be met either by using fuels already present in body stores or by increasing the intake of calories and other nutrients.

Because a loss of appetite is common during acute febrile illness, metabolic costs are generally met by mobilizing substrates from tissue stores. This mobilization process can be recognized by measurable losses from the body of nitrogen and the other principal intracellular elements (potassium, phosphorus, magnesium, sulfur, and zinc), and by a loss of body weight and muscle mass. Other metabolic costs are met by changing the patterns of cellular metabolism by using biochemical pathways already available within different tissues. Some of these pathways must be augmented by manufacturing additional enzymes. If an infection is not rapidly controlled, its metabolic costs can deplete body stores of vital nutrients.

Metabolic costs of a curable infection continue to increase until after fever has disappeared, and complete recovery from the metabolic consequences of an infection may require several additional weeks or even months. Depleted nutritional stores make the convalescing patient extremely susceptible to a secondary or superimposed infection by new microorganisms. Until stores can be replenished, it is possible for metabolic losses to initiate a vicious cycle of recurring infections and progressive malnutrition.

IMPORTANCE OF HOST METABOLIC RESPONSES DURING INFECTION

Infectious illness occurs at all ages and is managed by medical practitioners in all specialty fields. Clinicians should understand the biochemical, metabolic, and hormonal events that lead to loss of body nutrients and altered cellular functions. If the clinician can predict the most likely time of onset, magnitude, and duration of important metabolic responses, he should be able to plan optimal supportive care as an adjunct to antimicrobial drug therapy. It is important to anticipate and recognize the special metabolic complications of infection that are immediate threats to survival.

INCREASED DEMANDS FOR METABOLIZABLE ENERGY

One of the principal metabolic effects of a febrile infection is the increase in oxygen consumption. The basal metabolic rate increases about 13 per cent for each degree (Centigrade) rise in body temperature. The increased demand for cellular energy develops at the same time that anorexia causes a decline in food intake.

During simple starvation in the absence of infection, the energy needs of the tissues are met chiefly by lipids. These include free fatty acids obtained from adipose tissue and ketones synthesized from fatty acid precursors within the mitochondria of liver cells. Because the use of glucose is curtailed, less is manufactured. These metabolic adjustments to simple starvation allow the body to conserve protein and amino acid stores and to minimize nitrogen losses.

Infectious illnesses are generally accompanied by a reduction in food intake, but the usual metabolic adjustments to simple starvation are not made. Rather, glucose production is stimulated and ketone body production is inhibited. During infection, readjustments take place in metabolic pathways of individual cells and tissues, especially in muscle and liver cells. The endogenous metabolism of body fats, carbohydrates, and proteins are all involved in this process. While free fatty acids continue to be used as major sources of fuel, most of the extra requirements for cellular energy are supplied by glucose. Glucose production is speeded up within the liver by using amino acids as major additional substrates. Amino acids move more rapidly than usual from serum into liver cells. At the same time, large quantities of additional glucose-producing free amino acids, namely, alanine and glutamine, are manufactured and released by skeletal muscle for transport via plasma into the liver. Skeletal muscle contains the largest source of "labile" body protein — protein that can be degraded rapidly during an infection to yield free amino acids for use in other tissues. Some of the newly released branched-chain amino acids (leucine, isoleucine, and valine) are oxidized in situ and used for fuel in muscle cells. Nitrogen groups released by this process are used within muscle for the synthesis of alanine and glutamine. The body thus seems willing to degrade the proteins in skeletal muscle and other somatic tissues to produce amino acids that maintain visceral organ functions, generate energy, and manufacture new cells and proteins.

All the sugar-regulating hormones help stimulate or modulate the increased production and release of glucose by the liver. Plasma concentrations of glucagon, insulin, glucocorticoids, catecholamines, and growth hormone increase. Thyroid hormones are also used more rapidly.

The increased hepatic output of glucose leads to high blood glucose values and a larger glucose pool. A speedier turnover of glucose within this larger pool is caused by its increased use as a cellular fuel.

Severe hypoglycemia, accompanied by a fall of body temperature to subnormal values, can occur if the liver fails to maintain its output of glucose. Failure of hepatic glucose production is generally due to one of two basic mechanisms: (1) depletion of the supply of substrates or (2) failure of the molecular mechanisms required for producing glucose (hepatic cell failure).

Substrate depletion accounts for severe hypoglycemia during bacterial sepsis in newborns. Because infants are born with little skeletal muscle, their stores of "labile" protein are too limited to support a prolonged need for glucose production.

Failure of hepatic cells is usually caused by direct injury to the cells by viruses, such as hepatitis or yellow fever viruses, or by bacterial products such as endotoxin. Liver cells may also fail to synthesize glucose during the terminal stages of overwhelming infections.

ALTERATIONS IN LIPID METABOLISM

In addition to producing more glucose during acute infections, liver cells accelerate lipid metabolism. The hepatocellular uptake of plasma free fatty acids is increased. Production of triglycerides within liver cells is accelerated and more triglycerides move into the plasma. Some triglycerides may also accumulate as droplets in liver cells and cause fatty metamorphosis.

On the other hand, less free fatty acids are used for the synthesis of ketone bodies than would normally be expected during fasting. Increased release of insulin from the pancreas, a common secondary manifestation of generalized infection, and the lipogenic action of insulin on hepatic cells account for this curtailment in production of ketone bodies. As a result, ketone bodies are not made available to meet body energy requirements. Curtailment of ketone body production is thus a physiologic hormonal response during infection rather than a pathologic breakdown of biochemical pathways.

According to evidence obtained with isotope tracers, the liver, in addition to increasing its output of triglycerides, produces and releases more cholesterol and phospholipids during infection. Since these lipids are all released into plasma as lipoproteins, the liver must produce lipid transport proteins, but little is known about the mechanisms that regulate the rates of lipid release into plasma. However, changes in plasma lipid concentrations are controlled by the rates of lipid removal (or utilization) as well as by changes in the rates of lipid production. Plasma cholesterol, phospholipid, free fatty acid, and triglyceride values vary individually during different infections or even during different stages of a single infection. A massive accumulation of triglycerides is consistently observed during gram-negative sepsis and may give the plasma a milky appearance. Triglycerides accumulate because of increased hepatic production and decreased activity of the lipolytic enzymes that initiate cellular removal of triglycerides from plasma.

CHANGES IN PROTEIN AND AMINO ACID METABOLISM

All aspects of protein metabolism are affected by infection. Every host defense mechanism depends on the presence or function of some kind of body protein, whether it is a cellular enzyme, a structural protein, a cell membrane receptor, an immunoglobulin, a transport protein, a component of the complement, kinin, or coagulation systems, or some other plasma protein. Infection causes cells to speed up the production of some proteins, to slow the production of others and at the same time to break down "labile" body proteins into free amino acids.

During acute infections, plasma albumin values decline. The catabolism of skeletal muscle proteins exceeds their rate of production, and increased quantities of free amino acids are released into plasma. Despite the entry of more free amino acids into plasma, there is a still greater increase in uptake of amino acid by the liver. As a result, plasma concentrations of most free amino acids decline, especially the branched-chain group.

Free amino acids are used to produce phagocytic and lymphoid cells and to maintain the functional and structural integrity of other tissues. The liver also increases production of certain enzymes and the "acute phase" plasma proteins. These include C-reactive protein, haptoglobin, $alpha_1$-antitrypsin, ceruloplasmin, fibrinogen, and others.

Before amino acids are used for producing glucose, their amino nitrogen must be removed from the carbon skeleton. This nitrogen is converted into urea within the hepatic cells and excreted in the urine. Increased free tryptophan in liver cells is shunted into the serotonin pathway or degraded via the kynurenine pathway for excretion as urinary diazo-positive reactants.

Because of the increased use of amino acids to produce glucose and urea, losses of nitrogen in the urine are high during the acute stages of most febrile infections. Additional nitrogen is lost by sweating, by vomiting or diarrhea, or by nasal secretion and sputum production during respiratory tract infections. The diminished intake of nitrogen-containing foods in combination with continued losses causes negative nitrogen balance.

Nitrogen balance does not become negative during the incubation period of an infection, but begins to be negative soon after the onset of fever. Losses of body nitrogen may reach 10 to 15 g per day if fever is high; cumulative total losses may exceed 100 g during the course of an acute illness. Large daily losses of nitrogen cannot be sustained indefinitely. If an infection becomes subacute or chronic, the body again approaches a state of nitrogen equilibrium, but at a cachectic level.

After recovery, nitrogen balance should become strongly positive as degraded proteins are resynthesized. The process of rebuilding nitrogen stores can be speeded up by increasing the dietary intake of high quality protein foods, especially during early convalescence.

CHANGES IN MINERAL METABOLISM

Principal body minerals undergo a complex variety of changes during acute infection. Large amounts of some minerals are lost from the body, while others show either a physiologic redistribution or a sequestration in certain tissues. Minerals of soft tissue are lost during acute infections roughly in parallel with the losses of body nitrogen. The negative balances are caused by the combination of diminished dietary intake plus losses in the urine or feces of magnesium, phosphorus, potassium, sulfur, and zinc. Change in calcium stores is generally minimal unless long-term immobilization is a component of the illness, as in paralytic poliomyelitis. Similarly, deficits of body phosphorus appear to reflect losses from soft tissue rather than bone.

Although negative phosphorus balance generally parallels that of nitrogen, phosphorus losses during the period of early symptoms are also influenced by changes in acid-base balance. When fever occurs, respiratory rates become faster. The increased loss of carbon dioxide causes respiratory alkalosis accompanied by a transient disappearance of phosphorus through sweat and urine. During infections characterized by severe or protracted diarrhea, potassium ions escape by way of the intestine. This loss of potassium from the tissues may be large enough to produce the vacuolar changes in kidney tubules and myocardial cells characteristic of potassium deficiency.

Each of the three most widely studied essential trace minerals—iron, zinc, and copper—undergoes an abrupt redistribution within the body during infectious illnesses. These changes are under physiologic control mechanisms and are stimulated by inflammatory responses and activation of phagocytic cells. The concentrations of iron and zinc abruptly decline in the plasma as they are taken up by the liver. These two minerals become physiologically "sequestered" within cellular storage sites. Iron is held as hemosiderin or ferritin, while zinc is bound within liver cells to newly synthesized metallothionein. Copper, on the other hand, is secreted by the liver into plasma as a component of newly synthesized ceruloplasmin.

A rapid decline in plasma iron and zinc concentrations may precede the onset of fever, while the increase in plasma copper follows soon afterward. Since only the "loosely bound" fractions of plasma zinc move into the liver, plasma values of zinc rarely fall more than 50 per cent. In contrast, plasma iron may fall to undetectable values, leaving the iron-binding capacity of plasma almost entirely unsaturated. At the same time, plasma copper and ceruloplasmin values may double or triple.

Physiologic sequestration inhibits the incorporation of iron into hemoglobin and eventually causes the "anemia of infection." This process cannot be reversed by oral or parenteral iron. In addition to its temporary sequestration within the liver, appreciable quantities of zinc may be lost in the urine. The redistribution of iron, zinc, and copper are reversed after recovery from infection. Little is known about the responses of other trace minerals.

CHANGES IN ELECTROLYTE AND WATER METABOLISM

Fluid and electrolyte responses vary greatly during different infectious diseases. Imbalances can cause death, either from severe dehydration or from fluid overload.

In most acute infections, an increased adrenal output of aldosterone during fever causes renal tubular cells to reabsorb sodium and chloride. This mechanism accounts for the virtual disappearance of these electrolytes from urine and their retention in body fluids, and for a secondary increase in the extracellular fluid volume. Recovery from an acute period of fever may be followed by diuresis in early convalescence.

Retention of body water can also be caused by "inappropriate" secretion of antidiuretic hormone from the posterior pituitary. This phenomenon is common in infections of the central nervous system and may occur during severe generalized infections. Some severe infections may also be complicated by the accumulation and sequestration of sodium within cells that results in declining plasma sodium concentrations. Attempts to correct low sodium values with saline may lead to acute cardiac failure or cerebral edema. If low plasma sodium values during severe infections cannot be explained by losses from the body, the patient should be managed by careful restriction of fluid and sodium intake until the total daily urine volume consistently increases.

On the other hand, diarrhea may produce massive loss of extracellular fluids and electrolytes. In cholera or severe enterotoxic diarrhea, fluid loss may lead to hypovolemic shock and death. Therapy requires prompt resuscitation with isoosmotic saline and correction of concomitant acidosis and potassium deficiency.

ACID-BASE DISTURBANCES

Pathogenic mechanisms may produce a wide variety of acid-base abnormalities during different kinds of infections. Rapid breathing during fever causes excessive loss of dissolved carbon dioxide from the blood and a transient period of uncompensated respiratory alkalosis. In contrast, pneumonia that prevents ventilatory exchange of pulmonary gases leads to oxygen deficits and carbon dioxide retention with respiratory acidosis. Respiratory acidosis can also result from acute poliomyelitis or tetanus, both of which can cause neuromuscular abnormalities that prevent the thoracic movements required for breathing.

Increased cellular generation of organic acids may produce metabolic acidosis. Septic shock or localized capillary-bed stasis can lead to cellular hypoxia, which increases lactic acid production and the severity of metabolic acidosis. Acute intestinal loss of bicarbonate during severe diarrhea can also produce acute metabolic acidosis. On the other hand, the cumulative loss of body potassium that accompanies protracted diarrhea can lead to secondary chronic metabolic alkalosis.

ENDOCRINE RESPONSES

As alluded to above, certain hormonal actions help to modulate salt and water metabolism and energy-generating responses during infection. In contrast, several major hormones, including thyroid and parathyroid hormones, certain pituitary trophic hormones, and the gonadal steroids have no clearly defined role.

The adrenal glucocorticoid hormones serve a central but largely "permissive" role. These steroids are necessary to allow some molecular mechanisms to function in hepatic cells. The adrenal secretion of cortisol increases several fold early in the course of fever. This increase in cortisol secretion is accompanied by smaller increases in the adrenal output of ketosteroids and pregnanetriol. These increases do not persist beyond the onset of recovery. Increased adrenal output is never of sufficient magnitude or duration to produce negative nitrogen balances or the other physiologic changes known to accompany the administration of pharmacologic doses of synthetic glucocorticoid hormones.

Failure of hepatic enzyme systems during overwhelming infections may permit the plasma concentration of cortisol to reach unusally high levels. Conversely, destruction of the adrenals by infection prevents steroid production and leads to death unless this complication is recognized and treated. If an infection becomes subacute or chronic, steroid output also falls into a subnormal range.

Substances released by phagocytes and lymphocytes also have hormone-like functions that can stimulate both local and generalized physiologic responses. These biologically active products include prostaglandins, lymphokines, enzymes, and endogenous mediators such as those that initiate fever, mobilize neutrophils, or cause trace mineral redistribution.

SUMMARY

Generalized infectious illnesses are accompanied by a broad group of metabolic, biochemical, and endocrinal responses. Although caused by many different molecular mechanisms, generalized metabolic changes occur in relatively predictable patterns related to the onset, severity, and duration of fever. These changes include the catabolism of skeletal muscle proteins, acceleration of hepatic gluconeogenesis, ureagenesis and lipogenesis, inhibition of ketogenesis, retention of extracellular salt and water, loss of major intracellular elements, and redistribution within the body of certain trace minerals. Other metabolic changes develop if an infection becomes localized in a predominant organ system. Some metabolic consequences of infection, which include hypoglycemia, hypovolemia, fluid overload, and failure of various key organs, are life-threatening.

The views of the author do not purport to reflect the positions of the Department of the Army or the Department of Defense. (Para. 4-3, AR 360-5.)

References

Beisel, W. R.: Effect of infection on nutritional needs. In Rechcigl, M. (ed.): Handbook of Nutrition and Food: Nutritional Requirements. Cleveland, CRC Press, Inc., 1980.

Beisel, W. R., Blackburn, G. L., Feigin, R. D., Keusch, G. T., Long, C. L., and Nichols, B. L. (eds.): Symposium on Impact of Infection on Nutritional Status of the Host. Am J Clin Nutr 30:1203, 1439, 1977.

Suskind, R. M. (ed.): Malnutrition and the Immune Response. New York, Raven Press, 1977.

88 MALNUTRITION AND INFECTION

Gerald T. Keusch, M.D.

Sorrow may be fated, but to survive and grow is an achievement all its own.

R. COLES
Children of Crisis, 1964

Malnutrition of the affluent as well as the poor, whether because of excess, insufficiency, or inappropriate choice of foods, is a common malady throughout the world. In developing countries the most common clinical forms of malnutrition are due either to insufficiency of food per se, resulting in marasmus, or to inadequacy of specific nutrients in the diet including protein, vitamin A, or iron, resulting in kwashiorkor, xerophthalmia, and anemia, respectively. However, such "pure" inadequacies rarely exist, and it is reasonable to suggest that most malnutrition is caused by a lack of both protein and calories, for which the term protein-calorie malnutrition (PCM) will be used. Upon this base of PCM are engrafted specific nutrient inadequacies that differ in type from one country to the next.

In developing nations the youngest in society are the principal targets of PCM. In industrialized nations adults are at greatest risk, developing PCM secondary to debilitating chronic diseases, neoplasms, alcoholism, inflammatory bowel disease, or renal failure. Recently, adult PCM has also been recognized in hospitalized patients who are maintained on the routine, semistarvation regimens of parenteral fluid and electrolytes during the treatment of many acute medical and surgical illnesses. In both situations malnutrition greatly exaggerates both the susceptibility to infectious agents and the severity of the illnesses they produce. Because infection itself causes losses of nutrients by a number of mechanisms, the interaction of infection and malnutrition may lead to progressive debilitation and increased mortality. This interaction has been termed "synergistic," but it is difficult to define this synergism precisely. The minimum consequence in young children is impaired growth that may result in permanent stunting and failure to achieve optimal intellectual potential. In adult PCM, the analogous end of the spectrum may be impaired wound healing or infection that prolongs hospitalization. In both age groups, the other end of the spectrum is death. In developing countries the overall mortality in childhood from birth to age 7 often reaches 50 per cent, primarily because of the combined forces of infection and malnutrition.

Several important points have emerged from recent studies in developing countries. First, malnutrition and infectious diseases must be understood as ecologic problems that are interrelated, not only to each other but also to cultural, sociological, political, and economic factors, for which long-term corrective measures are required. The ultimate solution to these problems lies neither in a simple nutritional supplement nor in a vaccine, for example, but in the complex and difficult process of national development. Second, PCM is usually caused by a succession of insults, most often infections, over an extended period of time rather than by a single event or even multiple events at a particular time. Third, intrauterine events, both nutritional and infectious, may alter in utero development, birth weight, and postnatal growth, and postnatal morbidity and mortality from infectious agents.

The purpose of this chapter is to examine the interaction between malnutrition and infection in sufficient detail to permit the physician to make rational medical decisions for his patients in the short term. The important interactions between nutrition and infection may be broadly classified as effects of infections on nutrition and effects of nutrition on infections.

EFFECT OF INFECTIONS ON NUTRITION

It has long been known that both acute and chronic infections can cause marked debilitation; for example, acute typhoid fever can result in severe loss of muscle mass and body weight, and tuberculosis was so commonly associated with chronic wasting that it was called consumption. More recent observations, largely by Beisel and colleagues, have shown that even mild infections of well-nourished individuals cause a stereotyped metabolic response with nutritional consequences. The extent of these changes is determined in part by the severity and duration of infection, the nutritional status of the affected person at the time, and cultural responses to illness including alterations in diet and the administration of foods and drugs believed to have medicinal benefits, but which may in fact be harmful. One of the earliest physiologic responses to infection is anorexia. Neither the mechanism nor the purpose of this loss of appetite is known. Many dietary practices, including imposed starvation to "rest the bowel" or the provision of bland, poorly nutritious foods that are "easy to digest," probably originated as responses to this anorexia. However, the practices often continue long after the anorexia is past. At this time, young children will voluntarily increase food intake several-fold and achieve positive metabolic balance and "catch-up growth" that will return them to the growth curve they were following when the infection occurred. At precisely this time in early convalescence, cumulative losses of nitrogen, energy stores, trace metals, and other tissue components are at their maximum, and must be restored for the original state of health to be regained.

Fever is another early manifestation of infection. Although some have claimed that the fever response in the malnourished patient is blunted, there are no convincing data to support this anecdotal claim. On the contrary, I have observed that, except in the most severely malnourished subjects, the fever response appears to be intact. While the mechanism of fever production is largely known, its purpose is not. The clearest experiments have been performed with cold blooded lizards challenged with a natural bacterial pathogen, *Aeromonas hydrophila*. Significant increases in survival occur if the animals are allowed to raise their core body temperature (i.e., develop fever) by behavioral modifications when placed in a thermal gradient. Whatever its purpose, fever is a primitive response that would not have withstood the forces of natural selection if it did not have survival value.

Whatever the benefit, fever is not without cost. The extra heat to produce fever comes largely from involuntary work by muscles — the shivering response. It costs the febrile, septic patient an increase in caloric consumption equivalent to 25 to 40 per cent of the resting metabolic expenditure to produce fever, and the source of these extra calories is ultimately his own tissue proteins. Body stores of energy in carbohydrates (glucose and glycogen) are limited and amount to a reserve sufficient for only 12 to 24 hours of survival. Because ketones are not efficiently utilized by the septic patient, perhaps because of hyperinsulinemia, fat reserves are not useful sources of calories. Gluconeogenesis in the liver utilizing amino acids, principally alanine, from muscle is the source of the emergency energy required. Thus, infected patients literally consume themselves as muscle protein is mobilized and transported to the liver for gluconeogenesis and for new protein synthesis as well (vide infra). Alanine used for gluconeogenesis is ultimately wasted because the nitrogen removed during deamination is excreted as urea and the carbon skeleton is oxidized to carbon dioxide and re-

TABLE 1. Possible Role in Infection of Certain Acute Phase Plasma Proteins

ACUTE PHASE PROTEIN	KNOWN OR PRESUMED FUNCTION	POSSIBLE ROLE IN INFECTION
α_1-Acid glycoprotein	Increases platelet adherence to collagen and affects spacing of collagen fibrils	?Improved wound healing
α_1-Antitrypsin	Protease inhibitor; inhibition of plasmin and Hageman cofactor	Control of inflammation, induced tissue destruction, and DIC
α_2-Macroglobulin	Formation of complex with proteases to facilitate their removal	Increased granulocytopoiesis and activation of macrophages
C-reactive protein	Binds to phosphocholine groups in cell membranes	Activation of complement system
Ceruloplasmin	Copper transport	?
Haptoglobin	Complexes with free plasma hemoglobin to facilitate uptake by mononuclear phagocytes and regulate urinary losses	Reduces iron available to pathogenic bacteria for growth and production of toxins

moved by respiration. In addition, feedback control of hepatic gluconeogenesis is abrogated so that glucose infusion does not stop the process as it does in starved, noninfected volunteers. Glucose synthesis and oxidation are both increased during sepsis, the former more than the latter, accounting for an increase in the glucose pool size and fasting hyperglycemia. The glucose tolerance test curve becomes abnormal and diabetic in shape, but glucose oxidation is not diminished.

Fever is also the clinically apparent trigger for a number of diverse metabolic effects of infection including: (1) negative metabolic balances for nitrogen, magnesium, potassium, and phosphate; (2) loss of water and electrolytes through sweating and respiratory insensible losses; (3) tachypnea and respiratory alkalosis; (4) increased secretion of adrenocorticosteroids, glucagon, and insulin with numerous secondary metabolic effects. Fever itself is caused by production of a small polypeptide mediator called endogenous pyrogen (EP) that is made and released into the bloodstream by stimulated leukocytes. When EP reaches the hypothalamus it causes an upward resetting of the setpoint in the thermoregulatory center, and body temperature is raised accordingly. The apparent relationship between fever and altered metabolism, however, is not caused by EP activity but by a possibly distinct but analogous mediator (or group of mediators), also produced by white blood cells, called leukocyte endogenous mediator (LEM). Thus, stimulation of leukocytes results in both EP and LEM release, and fever and metabolic changes are therefore parallel phenomena. Administration of EP causes fever; administration of LEM causes release of amino acids from muscle, increased uptake of amino acids by the liver, synthesis of acute phase globulins in the liver, uptake and sequestration of iron, zinc, and other trace metals into hepatic macrophages, and elevations of serum insulin and glucagon. It appears that LEM plays a role in both catabolic and anabolic events during infection.

Although these changes in metabolism may serve body defense mechanisms, they are a form of nutrient wastage when there is no compensatory increase in intake. In fact, anorexia would interfere with increased intake, even if food were available. The synthesis of acute phase proteins, immunoglobulins, and complement and the production of phagocytic and lymphoid cells by the infected child deplete the available precursor materials and interfere with the synthesis of visceral protein. Hypoalbuminemia and cessation of linear growth are the result. Such wastage has been termed "functional," although many of the events, like fever itself, have no well-established beneficial function. Table 1 lists some of the postulated benefits of acute phase plasma proteins. Absolute losses, sequestration of nutrients, and diversion of nutrients to new pathways combine to produce broad disturbances in host nutriture.

EFFECTS OF MALNUTRITION ON HOST DEFENSES AND RESPONSES TO INFECTIOUS AGENTS

The key feature of the cyclical interaction of nutrition and infection is the adverse effect each has on the other. Thus, many host defense mechanisms are exquisitely sensitive to nutritional deprivation, and this nutritionally induced immunodeficiency plays an important role in increasing the susceptibility to a given inoculum of an infectious agent, as well as the severity of the disease it produces. As already noted, frequent and severe infections may preclude nutritional rehabilitation, because there is insufficient time between episodes to replace cumulative losses.

Specific host defenses can be classified into cellular (phagocytosis and cell-mediated immunity) and humoral (antibody and complement) mechanisms. Normal resistance to infection requires adequate function of all these components because they work in an integrated manner against some infectious agents, and one element of the immune response may be the critical defense against other agents. Although many functional defects can be detected in these four biologic systems in the severely malnourished host, these deficiencies may simply represent end-stage disease and anabolic failure. Developing malnutrition undoubtedly exerts selective influences on host metabolism, so that some pathways are preserved and others are curtailed. Selective changes such as these directly influence the clinical features and manifestations of malnutrition. Therefore, it is the less severely malnourished host who must be understood physiologically in order to identify critical adverse factors amenable to therapy. The limited data available at present suggest early malnutrition significantly impairs cell-mediated immunity, functioning through the T lymphocyte, and synthesis of complement. The function of phagocytes and antibody synthesis are not significantly impaired, although they may be reduced in efficiency and antigen-binding avidity, respectively (Table 2).

With increasing malnutrition, the number and function of circulating T cells diminishes, and the T-cell areas of central (i.e., thymus) and peripheral (i.e., spleen and lymph node) lymphoid organs become depleted of cells. Therefore, reactions dependent upon intact T-cell function may become impaired. This is easily demonstrated by skin test reactivity to tuberculin or other antigens that provoke delayed hypersensitivity. Malnourished individuals do not develop a positive skin test to new antigens nor do they respond normally to previously encountered ones (the recall response). Nutritional rehabilitation restores both recall and normal reactivity to new antigens. Skin responses to tuberculin are more than a diagnostic test for *Mycobacterium tuberculosis* infection. They are a cutaneous example of granuloma formation, the host defense responsible for containing tuberculosis. Thus tuberculosis is highly prevalent in the malnourished and also tends to be widespread and severe because of this failure of the host defenses.

Cell-mediated immunity (CMI) may also be responsible for immunopathology in some diseases such as schistosomiasis or even tuberculosis. Theoretically, deficiency of CMI might lessen the severity of such diseases, and malnutrition might be protective. This is difficult to document in human populations. When a clear-cut decrease in host defenses is balanced against the theoretical benefits from diminished CMI, it is the former that stands out; the adverse consequences far outweigh the possible benefits. For example, the measles rash, a manifestation of cell-mediated immunopathology, may be atypical in the child with malnutrition; however, he may die of giant cell pneumonia because of inadequate CMI in the lung.

The extensive role of complement in host defenses has become clear only in the past few years. Complement-mediated chemotaxis, opsonization, and vascular changes during the inflammatory response may be the critical early defense in pyogenic infections. The alternative pathway of activation of the complement effector sequence is now known to be markedly affected by malnutrition. Serial studies have documented a reversible deficiency in the concentration and functioning of most complement components during malnutrition and nutritional rehabilitation. This repair requires time, and in the interval death from infection may be the ultimate consequence.

Recent investigations have pointed to an ad-

TABLE 2. Effects of Malnutrition on Host Defense Mechanisms

MECHANISM	USUAL FUNCTION IN HOST DEFENSE	EFFECT OF MALNUTRITION: Severe	EFFECT OF MALNUTRITION: Moderate
Phagotosis			
Neutrophil	Ingestion and destruction of pyogenic bacteria	Impairment	None
Mononuclear phagocyte	Ingestion and destruction of facultative intracellular bacteria and fungi, virus	Impairment	None
	Processing of antigen of B-cell and T-cell responses	?	?
Cell-mediated immunity (T lymphocyte)	Activation of macrophage recruitment, scavenger and bactericidal functions, granuloma formation and delayed hypersensitivity. Killer-cell function for virus-infected cells, or helper function for antibody synthesis to certain antigens	Impairment	Impairment
Antibody synthesis (B lymphocyte)	Neutralization or killing of bacteria and other pathogens; neutralization of biologically active products and toxins	Impairment	None ?Decreased antigen-binding affinity for some antigens.
Complement	Opsonization, chemotaxis, increased vascular permeability and bacteriolysis. Participation in destruction of virus-infected cells.	Impairment	Impairment

verse effect of both iron and vitamin A deficiencies on host defense mechanisms. Additional animal and human study is needed to document the extent of this effect and its mechanism. It is likely that a combination of multiple, minor abnormalities in immune defense systems can produce a major defect in the host. Similarly, multiple nutrient deficiencies of minor to moderate degree may interact to produce a significant nutritional impairment.

THE ARENA OF THE INTERACTION

From the foregoing discussion it is evident that the problems of malnutrition and infection must be considered together because each affects the other significantly. When malnutrition is common, the cure and prevention of infectious diseases must be undertaken with this fact in mind. Traditional preventive medicine intervention such as the use of vaccines may be fraught with failure (polio) or danger (measles) in the malnourished host. A vaccination campaign should not be started without considering the impact of the nutritional state on immune responses. In addition, the problems of containment of the immunization strain of live vaccines, on the one hand, and the development of latent infections with late pathology, on the other hand, become crucial. The same reasoning process tells us that attempts to improve nutrition will be analagous to filling leaky buckets if the problem of infections is not engaged directly. Thus, whether dealing with individual patients or with large populations, whether concerned prinicipally with nutritional status or with the morbidity and mortality of infections, the physician must learn to think of nutrition and infection as interrelated processes. Concern for convalescent care and total nutritional rehabilitation becomes a part of proper therapy when this approach is taken, and the physician becomes a promoter of health as well as an alleviator of disease.

References

Beisel, W. R., Blackburn, G. L., Feigin, R. D., et al. (eds.): Symposium on Impact of Infection on Nutritional Status of the Host. Am J Clin Nutr 30:1203, 1439, 1977.

Keusch, G. T., and Katz, M.: Malnutrition and infection. In Winick, M. (ed.): Nutrition and Development. New York, Plenum Publishing Corporation, 1978.

Scrimshaw, N. S., Taylor, C. E., and Gordon, J. E.: Interactions of nutrition and infection. WHO Monograph Series No. 57, Geneva, 1968.

Suskind, R. (ed.): Malnutrition and the Immune Response. New York, Raven Press, 1977.

INDEX

Note: Page numbers in *italics* indicate figures; numbers followed by t indicate tables.